D1302801

## Timeline of Important Events in Genetics

**1856–1863** **Gregor Mendel** Conducted his famous pea experiments concerning gene segregation

**1859** **Charles Darwin** Published *On the Origin of Species*, which is identified with the modern theory of evolution

**1866** **Gregor Mendel** Published a research paper on his work establishing the basic principles of heredity

**1868** **Fredrich Miescher** Isolated nuclein from nuclei; nuclein is now known to be DNA

**1875** **O. Hertwig** Showed nucleus required for fertilization and cell division, and hence contained information for those processes

**1882–1885** **E. Strasburger, Walther Flemming** Showed that nuclei contained chromosomes

**1900** **Hugo de Vries, Carl Correns, Erich von Tschermak-Seysenegg** Independently produced results confirming Mendel's principles of heredity

**1902** **Archibald Garrod** Identified the first human genetic disease

**1902** **Walter Sutton, Theodor Boveri** Proposed the chromosome theory of heredity

**1903** **William E. Castle** First to recognize the relationship between allele and genotypic frequencies *(see 1908, Hardy and Weinberg)*

**1905** **William Bateson** Called the science of heredity "genetics"

**William Bateson, R. C. Punnett** Demonstrated linkage between genes

**1908** **Godfrey H. Hardy, Wilhelm Weinberg** Formulated the Hardy–Weinberg principle, mathematically relating the frequencies of genotypes to the frequencies of alleles in randomly mating populations

**Herman Nilsson-Ehle** Obtained experimental proof for multigene inheritance as the basis for continuous traits

**1909** **W. Johannsen** Introduced the word "gene"

**1910** **Edward M. East** Elucidated the role of sexual reproduction in evolution

**Thomas Hunt Morgan** Found the first sex-linked gene, *white*, an eye-color gene in *Drosophila melanogaster*

**1911** **Thomas Hunt Morgan** Proposed that genetic linkage was the result of the genes involved being on the same chromosome

**1913** **Alfred Sturtevant** Devised the principle for constructing a genetic linkage map

**1916** **Thomas Hunt Morgan** Proposed a theory relating mutation and selection

**1924–1932** **John B. S. Haldane** Published a series of papers on his mathematical theory of natural and artificial selection

**1927** **Herman J. Müller** Showed that X-rays can induce mutations

**1928** **Frederick Griffith** Discovered genetic transformation of a bacterium and called the agent responsible the "transforming principle"

**1930** **Ronald A. Fisher** Published his comprehensive theory of evolution, synthesizing Mendelian inheritance and Darwinian selection, as *The Genetical Theory of Natural Selection*

**1930s** **Sewall Wright** Developed his own genetical theory for natural selection, and laid the important theoretical foundation for genetic drift, the random change in gene frequency

**1931** **Harriet Creighton, Barbara McClintock** Showed that genetic recombination in maize results from a physical exchange of homologous chromosomes

**Curt Stern** Showed that genetic recombination in *Drosophila* results from a physical exchange of homologous chromosomes

**1941** **George Beadle, Edward Tatum** Proposed the one-gene–one-enzyme hypothesis

**1944** **Oswald Avery, Colin MacLeod, Maclyn McCarty** Showed that Griffith's transforming principle *(see 1928)* was DNA

**1946** **Joshua Lederberg, Edward Tatum** Discovered conjugation in bacteria

**1950** **Barbara McClintock** Reported results of maize experiments indicating movable genes, now called transposable elements

**1952** **Alfred Hershey, Martha Chase** Showed that the genetic material of bacteriophage T2 is DNA

**1953** **James Watson, Francis Crick** Proposed double helical model for DNA

**1958** **Matthew Meselson, Franklin Stahl** Proved the semiconservative model for DNA replication

**Arthur Kornberg** Isolated DNA polymerase I from *E. coli*

**1959** **Severo Ochoa** Discovered the first RNA polymerase

| | |
|---|---|
| 1961 | **Sydney Brenner, François Jacob, Matthew Meselson** Discovered messenger RNA (mRNA) |
| | **François Jacob, Jacques Monod** Put forward the operon model for the regulation of gene expression in bacteria |
| 1966 | **Marshall Nirenberg, H. Gobind Khorana** Worked out the complete genetic code |
| 1972 | **Paul Berg** Constructed the first recombinant DNA molecule *in vitro* |
| 1973 | **Herb Boyer, Stanley Cohen** First used a plasmid to clone DNA |
| 1975 | **Edward M. Southern** Developed a method for transferring DNA fragments separated in a gel to a filter, preserving the relative positioning of the fragments, which remains one of the most valuable techniques for identifying cloned genes |
| 1977 | **Walter Gilbert, Frederick Sanger** Devised methods for sequencing DNA |
| | **Phillip Sharp, and others** Discovered introns in eukaryotic genes |
| 1983 | **Thomas Cech, Sidney Altman** Discovered self-splicing of an intron RNA |
| 1986 | **Kary Mullis and others** Developed the polymerase chain reaction (PCR), a technique for amplification of selected DNA segments without cloning |
| 1989 | **L.-C Tsui and John Riordan, and Francis Collins's group** Identified and cloned the human gene responsible for cystic fibrosis |
| 1990 | **James Watson and many other scientists** Launched the Human Genome Project to map and sequence the complete genomes of a number of genetically important organisms, including humans |
| 1993 | **Huntington's Disease Collaborative Research Group** Discovered molecular basis for Huntington's disease, a human genetic trait |
| 1994 | **M. Skolnick and other scientists** Cloned the first breast cancer gene *(BRCA1)* |
| 1996 | **Many scientists in several international research groups** Published the first complete DNA sequence of a eukaryotic organism, the yeast *Saccharomyces cerevisiae* |
| | **J. Craig Venter and many other scientists in several U.S. research groups** Published the complete DNA sequence of the archaean *Methanococcus jannaschii*, confirming that the Archaea are a third major branch of life distinct from prokaryotes and eukaryotes |
| | **National Institutes of Health** Reported approval of almost 150 clinical trials for the transfer of genes into humans as part of long-term goals to treat genetic diseases by gene therapy |
| 1997 | **The Roslin Institute** Clones the first mammal, a lamb named Dolly, from an adult organism using the techniques of transgenic cloning |
| | *Escherichia coli* genome sequence completed |
| 1998 | **Celera Genomics Company** formed to sequence much of human genome in three years, using resources generated by the Human Genome Project |
| | *Caenorhabditis elegans* genome sequence completed |
| 1999 | **Human Genome Project** Announced the complete sequencing of the DNA making up human chromosome 22 |
| 1990s | **RNA interference (RNAi),** a mechanism by which a fragment of double-stranded DNA silences the expression of a gene, discovered in a number of organisms; it has subsequently become an important research tool for investigating the functions of genes |
| 2000 | **International collaborators** Published genome of fruit fly, *Drosophila melanogaster*, the largest genome sequenced to date |
| | **International research consortium** Published genome of chromosome 21, the smallest human chromosome |
| 2001 | **Human Genome Project** Announced the completion of a "working draft" DNA sequence of the entire human genome |
| 2004 | The human genome sequence is nearly finished; analysis indicates only 20,000–25,000 protein-coding genes |
| 2005 | Working draft of the chimpanzee genome sequence announced, allowing first analysis of primate sequences unique to humans |
| 2006 | **Cancer Genome Project** initiated to identify genes critical to the development of cancer |
| 2007 | **Human Microbiome Project** initiated to comprehensively characterize the microorganisms associated with humans and to analyze their roles in human health and disease |
| 2007 | Sequence of James Watson's genome completed |
| 2008 | **1,000 Genomes Project** initiated to sequence the genomes of at least a thousand people from around the world and provide a detailed map of human genetic variation to aid in studies of human diseases |

# Improve Your Grade!

## Access included with every new book.

**Get help with exam prep!**

# REGISTER NOW!

**Registration will let you:**

- Prepare for exams using multiple-choice chapter quizzes
- Grasp difficult concepts with animations and iActivities
- Learn at your pace using self-study and practice exercises
- Master key terms and vocabulary
- Access your text online 24/7 using the state-of-the-art myeBook

# www.geneticsplace.com

## TO REGISTER

1. Go to **www.geneticsplace.com**
2. Click "Register."
3. Follow the on-screen instructions to create your login name and password.

**Your Access Code is:**

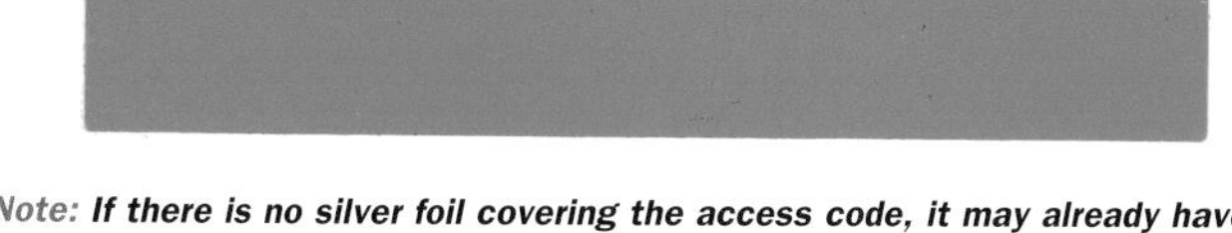

*Note: **If there is no silver foil covering the access code, it may already have been redeemed, and therefore may no longer be valid. In that case, you can purchase access online using a major credit card or PayPal account. To do so, go to** www.geneticsplace.com, **click on** "Buy Access," **and follow the on-screen instructions.***

## TO LOG IN

1. Go to **www.geneticsplace.com**
2. Click "Log In."
3. Pick your book cover.
4. Enter your login name and password.
5. Click "Log In."

**Hint:**
**Remember to bookmark the site after you log in.**

**Technical Support:**
**http://247pearsoned.custhelp.com**

# iGenetics

## *A Molecular Approach*

Third Edition

Peter J. Russell

REED COLLEGE

**Benjamin Cummings**

San Francisco Boston New York

Capetown Hong Kong London Madrid Mexico City

Montreal Munich Paris Singapore Sydney Tokyo Toronto

**Editor-in-Chief:** Beth Wilbur
**Executive Director of Development:** Deborah Gale
**Acquisitions Editor:** Gary Carlson
**Executive Marketing Manager:** Lauren Harp
**Associate Project Editor:** Rebecca Johnson
**Assistant Editor:** Kaci Smith
**Managing Editor:** Michael Early
**Production Supervisor:** Lori Newman
**Production Management:** Crystal Clifton, Progressive Publishing Alternatives
**Compositor:** Progressive Information Technologies
**Design Manager:** Marilyn Perry
**Interior and Cover Designer:** Derek Bacchus
**Illustrators:** Electronic Publishing Services
**Photo Researcher:** Eric Schrader
**Director, Image Resource Center:** Melinda Patelli
**Image Rights and Permissions Manager:** Zina Arabia
**Image Permissions Coordinator:** Silvana Attanasio
**Manufacturing Buyer:** Michael Penne
**Text printer:** Quebecor World Dubuque
**Cover printer:** Phoenix Color Corp.

**Cover Photo Credit:** Model of DNA; Will & Deni McIntyre/Science Photo Library

ISBN 0-321-61022-9 / 978-0-321-61022-5



**Benjamin Cummings**
is an imprint of

www.pearsonhighered.com

1 2 3 4 5 6 7 8 9 10—QWD—13 12 11 10 09

# Brief Contents

# Detailed Contents

## CHAPTER 15



## CHAPTER 16



## CHAPTER 17



## CHAPTER 18

# Preface

## An Approach to Teaching Genetics

The structure of DNA was first described in 1953, and since that time genetics has become one of the most exciting and ground-breaking sciences. Our understanding of gene structure and function has progressed rapidly since molecular techniques were developed to clone or amplify genes, and rapid methods for sequencing DNA became available. In recent years, the sequencing of the genomes of a large number of viruses and organisms has changed the scope of experiments performed by geneticists. For example, we can study a genome's worth of genes now in one experiment, allowing us to obtain a more complete understanding of gene expression.

I have taught genetics for over 35 years, while at the same time maintaining a molecular genetics research program involving undergraduates. Students learn genetics best if they are given a balanced approach that integrates their understanding of the abstract nature of genes (from the transmission genetics part) with the molecular nature of genes (from the molecular genetics part). My goal in this edition, as in previous editions, is to provide students with a clear and logical presentation of the material, in combination with an experimental theme that makes clear how we know what we know. The many examples of experiments used to answer questions and test hypotheses are models that show students how they might themselves develop questions and hypotheses, and design experiments. It is my hope that you will find my approach helpful to you in teaching this course successfully, as have so many colleagues who have used past editions.

The general features of *iGenetics: A Molecular Approach*, Third Edition, are as follows:

***Modern Coverage.*** The field of genetics has grown rapidly in recent years. In creating this text I have worked with experts in the field to ensure that we present these exciting developments with the highest degree of accuracy. The book covers all major areas of genetics, balancing classical and molecular aspects to give students an integrated view of genetic principles. The classical genetics material tends to be abstract and more intuitive, while the molecular genetics material is more factual and conceptual. Teaching genetics, therefore, requires teaching these two styles, as well as conveying the necessary information. The modern coverage reflects this. The molecular material, which is the material that changes most rapidly in genetics, is current and presented at a suitable level for students. Enhanced for this edition is the coverage of genomics, the analysis of the information contained within complete genomes of organisms.

***Experimental Approach.*** Research is the foundation of our present knowledge of genetics. The presentation of experiments throughout *iGenetics* allows students to learn about the formulation and study of scientific questions in a way that will be of value in their study of genetics and, more generally, in all areas of science. The amount of information that students must learn is constantly growing, making it crucial that students not simply memorize facts, but rather learn how to learn. In my classroom and in this text I emphasize basic principles, but I place them in the meaningful context of classic and modern experiments. Thus, in observing the process of science, students learn for themselves the type of critical thinking that leads to the formulation of hypotheses and experimental questions and, thence, to the generation of new knowledge.

***Classic Principles.*** Our present understanding of genes is built on the foundation of classic experiments, a number of which have led to discoveries recognized by the Nobel Prize. These classic experiments are described so that students can appreciate how ideas about genetic processes have developed to our present-day understanding. These experiments include:

- Griffith's transformation experiment
- Avery and his colleagues' transformation experiment
- Hershey and Chase's bacteriophage experiment
- Meselson and Stahl's DNA replication experiment
- Beadle and Tatum's one-gene–one-enzyme hypothesis experiments
- Mendel's experiments on gene segregation
- Thomas Hunt Morgan's experiments on gene linkage
- Seymour Benzer's experiments on the fine structure of the gene
- Jacob and Monod's experiments on the *lac* operon

***Human Applications.*** The impact of modern genetics on our daily lives cannot be understated. Gene therapy, gene mapping, genetic disorders, genetic screening, genetic engineering, and the human genome: these topics directly impact human lives. By illustrating important concepts with numerous examples of applications from human genetics, students are attracted by a natural curiosity to learn about themselves and our species. For instance, there are discussions about specific genetic diseases (in Chapter 4 on Gene Function, for example), about the sequencing of the human genome (in Chapter 8), about identifying genes in the human genome sequence and describing patterns of gene expression (in Chapter 9), and about DNA analysis approaches used to detect human gene mutations and in forensics (in Chapter 10). Human genes mentioned in the text are keyed to the OMIM (Online Mendelian Inheritance in Man) online database of human genes and genetic disorders at http://www.ncbi.nlm.nih.gov/omim, where the most up-to-date information is available about the genes.

***Using Media to Teach Genetics.*** Media for this textbook include interactive activities to allow students to self-assess their understanding of key chapter concepts, and animations to provide a dynamic representation of processes that are difficult to visualize from a static figure. I was involved in the development of most of these pieces, ensuring that their look and quality match that of the textbook.

- Twenty-four interactive activities called *iActivities* have been designed to promote interactive problem solving. Available on the *iGenetics* student website, these activities are based on case studies presented at the beginnings of the chapters. An example from Chapter 9 is the analysis of DNA microarray results for a fictional patient with breast cancer to determine gene expression differences and then determine which drugs would be useful for treating her cancer. I worked closely with the development teams for most *iActivities* to help ensure accuracy and quality. Each chapter containing an *iActivity* begins with a brief description of the *iActivity*, followed by a later reference directing students to the website at the point in the chapter at which it is appropriate to use the media.
- Fifty-six narrated animations on the *iGenetics* student website help students visualize challenging concepts or complex processes, such as DNA replication, translation, DNA cloning, analysis of gene expression using DNA microarrays, DNA molecular testing for human genetic disease mutations, meiosis, gene mapping, regulation of gene expression in bacteria and in eukaryotes, gene regulation of development, and natural selection. As with the *iActivities*, I have worked closely with the development teams for most of the animations: outlining topics, editing the storyboards, helping describe the steps for the artists, and working closely with the animators until the animations were complete. We have made a special effort to base the animations on the text figures so that students do not have to think about the processes in a different graphic format. These animations are of high quality, showing a level of detail not typical of animations that are supplements to texts. A media flag with the title of the animation appears next to the discussion of that topic in the chapter.

***Accuracy.*** An intense developmental effort, along with numerous third party reviews of both text and media, ensure the highest degree of accuracy.

## Organization

This text utilizes a molecular first presentation of materials. After the introductory chapter, a core set of nine chapters covers the molecular details of gene structure and function, and the cloning and manipulation of DNA, before the Mendelian genetics, gene segregation, and gene mapping principles are developed. However, the chapters can readily be used in any sequence to fit the needs of individual instructors.

### Changes from *iGenetics: A Molecular Approach*, Second Edition

- **All molecular material in the book was updated where necessary.**
- **Translation termination in bacteria** was expanded to provide a more complete discussion of the process (Chapter 6).
- **Discussion of ionizing radiation causing mutations was expanded to include the effects of radon (Chapter 7).**
- **Genomics coverage was reorganized and enhanced** to reflect the **increased use of genomics approaches in all areas of genetics research (Chapters 8, 9, 10)** and a **Focus on Genomics** box describing a chapter-specific example that involved a genomics study was added to each chapter (except the introductory Chapter 1). **Chapter 8** contains material derived from Chapters 8 and 9 in the Second Edition, which will be referred to here as 2e. **Chapter 9** contains material derived from 2e Chapter 10, and **Chapter 10** contains material derived from 2e Chapters 8 and 9.

In the new organization, **Genomics: The Mapping and Sequencing of Genomes (Chapter 8)** is the first of three chapters focused on **genomics and recombinant DNA technology**. Described in this chapter is DNA cloning; genomic libraries; DNA sequencing of clones and genomes; assembling and annotating genome sequences; differences in the genomes of Bacteria, Archaea, and Eukarya; and features of selected genomes of each of the three domains. Compared with material in 2e, there is a more comprehensive description of cloning vectors and their use in genome projects, a new method of DNA sequencing—pyrosequencing—is presented, analysis of DNA sequences is expanded, particularly with respect to assembling and finishing genome sequences in a genome project, annotation of variation in genome sequences, annotation of gene sequences, the analysis of cDNAs to identify gene sequences, and identifying genes in genome sequences by bioinformatics approaches. The chapter includes a discussion of the outcomes of analyses of genomes that have been sequenced, adding rice, mouse, and dog to the organisms presented in Chapter 10 of 2e.

The second chapter of the three, **Functional and Comparative Genomics (Chapter 9)** describes *functional genomics*, the analysis of the functions of genes and nongene sequences in genomes, including patterns of gene expression and their control, and *comparative genomes*, the comparison of the nucleotide sequences of entire genomes or large genome sections with the goal of understanding the functions and evolution of genes. Compared with functional genomics coverage in 2e, there is a more complete description of sequence similarity searching, the section on Assigning Gene Function Experimentally has been expanded to include the generation of gene knockouts in the mouse and in the bacterium, *Mycoplasma genitalium*, and the knock down of gene expression by RNA interference in the nematode, and the section on Describing Patterns of Gene Expression has additional examples. In 2e, the comparative genomics coverage in this part of the book was brief. In this edition, several examples of comparative genomics experiments are presented, including finding genes that make us human, identifying viruses with the Virochip microarray, and metagenomic analysis. Additional coverage of comparative genomics remains in **Chapter 23, Molecular Evolution.**

The third chapter of the three, **Recombinant DNA Technology**, contains material that was in 2e, Chapters 8, Recombinant DNA Technology, and 9, Applications of Recombinant DNA Technology. The focus is on the use of recombinant DNA technology to manipulate genes for genetic analysis, or for more practical applications such as testing for genetic disease mutations, and genetic engineering. Compared with 2e material, there is more extensive coverage of cloning vectors, and expanded coverage of PCR uses including discussion of reverse transcriptase-PCR and real-time PCR.

- **A newly created chapter on Extensions of and Deviations from Mendelian Genetic Principles (Chapter 13)** is an amalgam of the 2e chapters on Extensions of Mendelian Principles (Chapter 13) and NonMendelian Inheritance (Chapter 23). The former Chapter 13 material starts the chapter and was reorganized to deal first with examples involving single genes, and then moves to examples with two genes. The chapter then continues with the genetics-based material from the former Chapter 23 material, focused on maternal effect and non-Mendelian inheritance. The detailed description in 2e on the organization of extranuclear genomes was reduced to key concepts in this edition.
- The **Genetic Mapping in Eukaryotes** chapter **(Chapter 14)** now follows the **Extensions of and Deviations from Mendelian Genetic Principles** chapter directly. The chapter retains the content of the epinonymous chapter of 2e and adds a box to illustrate two-point mapping when one locus is a DNA marker locus, adds a section on comparing genetic and physical maps, and adds a section on constructing genetic linkage maps of the human genome (includes the lod score method for analyzing linkage, and constructing human genetic maps). The latter topic relates to the discussion of the Human Genome Project in Chapter 8, and encompasses some material presented in 2e Chapter 15.
- The chapter on Advanced Gene Mapping in Eukaryotes (Chapter 15) in 2e, which covered tetrad analysis, mitotic recombination, and mapping human genes, was deleted. The material on **tetrad analysis** (see pp. 430–435 of 2e) is now available on the companion website for the new edition, along with the corresponding *iActivity* and animation. The key material on mapping human genes is now in **Chapter 14**, as indicated above.
- The chapter on **Variations in Chromosome Structure and Number (Chapter 16)** was moved from its position between the chapters on eukaryotic gene mapping and bacterial gene mapping, to now follow the bacterial gene mapping chapter.
- The chapter on **Regulation of Gene Expression in Bacteria and Bacteriophages (Chapter 17)** was expanded to include presentation of the *ara* operon as an example of an operon that is regulated both by repression and activation.
- The chapter on **Regulation of Gene Expression in Eukaryotes (Chapter 18)** was changed to remove discussion of operons in eukaryotes (removed for space reasons), to reorganize the presentation of topics, and to include a much expanded presentation of noncoding regulatory RNAs (miRNAs and siRNAs) in RNA interference. The reorganization results in the following flow of topics: control of transcription initiation by regulatory proteins (includes a new example of combinatorial gene regulation); role of chromatin in regulating gene transcription; gene silencing and genomic imprinting; RNA processing control (includes mRNA transport control); mRNA translation control by ribosome selection; RNA interference by miRNAs and siRNAs (a completely new section to replace only a brief overview in 2e); and regulation of gene expression posttranscriptionally

by controlling mRNA degradation and protein degradation.

- The chapter on **Genetic Analysis of Development (Chapter 19)** was updated to include discussion of the roles of miRNAs in development.
- The chapter on **Genetics of Cancer (Chapter 20)** was updated to include discussion of **changes in miRNA gene expression in cancer.**
- **Chapter 21, Population Genetics,** now includes **new sections on the neutral theory and linkage disequilibrium,** as well as **discussions of large-scale sequence and SNP analysis.**
- **Quantitative Genetics** which had been located to after the core chapters on gene segregation principles, is now **Chapter 22** and follows the chapter on **Population Genetics.**

## Coverage

The four major areas of genetics—transmission genetics, molecular genetics, population genetics, and quantitative genetics—are covered in 23 chapters.

**Chapter 1** is an introductory chapter designed to summarize the main branches of genetics, describe what geneticists do and what their areas of research encompass, and introduce genetic databases and maps.

Chapters 2 through 7 are core chapters covering genes and their functions. In **Chapter 2,** we cover the structure of DNA, and the details of DNA structure and organization in prokaryotic and eukaryotic chromosomes. We cover DNA replication in prokaryotes and eukaryotes and recombination between DNA molecules in **Chapter 3.** In **Chapter 4,** we examine some aspects of gene function, such as the genetic control of the structure and function of proteins and enzymes and the role of genes in directing and controlling biochemical pathways. Examples of human genetic diseases that result from enzyme deficiencies are described to reinforce the concepts. The discussion of gene function in **Chapter 4** enables students to understand the important concept that genes specify proteins and enzymes, setting them up for the next two chapters, in which gene expression is discussed. In **Chapter 5,** we discuss transcription, and in **Chapter 6,** we describe the structure of proteins, the evidence for the nature of the genetic code, and the process of translation in both prokaryotes and eukaryotes. Then, the ways in which genetic material can change or be changed are presented in **Chapter 7.** Topics include the processes of gene mutation, some of the mechanisms that repair damage to DNA, some of the procedures used to screen for particular types of mutants, and the structures and movements of transposable genetic elements in prokaryotes and eukaryotes.

Genomics and recombinant DNA technology is described in the next three chapters. In **Chapter 8,** we present an overview of the mapping and sequencing of genomes, and an introduction to the information obtained from genome sequence analysis. Then, in **Chapter 9,** we discuss functional genomics, the comprehensive analysis of the functions of genes and of nongene sequences in genomes, and comparative genomics, the comparison of entire genomes (or of sections of genomes) from the same or different species to enhance our understanding of the functions of genomes, including evolutionary relationships. In **Chapter 10,** we discuss the applications of recombinant DNA technology in analyzing genes and other DNA, RNA and protein, including the types of DNA polymorphisms in genomes, the diagnosis of human diseases, forensics (DNA typing), gene therapy, the development of commercial products, and the genetic engineering of plants.

Chapters 11 through 18 are core chapters covering the principles of gene segregation analysis. Chapters 11 and 12 present the basic principles of genetics in relation to Mendel's laws. **Chapter 11** is focused on Mendel's contributions to our understanding of the principles of heredity, and **Chapter 12** covers mitosis and meiosis in the context of animal and plant life cycles, the experimental evidence for the relationship between genes and chromosomes, and methods of sex determination. Mendelian genetics in humans is introduced in Chapter 11 with a focus on pedigree analysis and autosomal traits. The topic is continued in Chapter 12 with respect to sex-linked genes. The exceptions to and extensions of and deviations from Mendelian principles (such as the existence of multiple alleles, the modifications of dominance relationships, essential genes and lethal alleles, gene expression and the environment, maternal effect, complementation tests, gene interactions and modified Mendelian ratios, and extranuclear inheritance) are described in **Chapter 13.** In **Chapter 14,** we discuss gene mapping in eukaryotes, describing how the order of and distance between the genes on eukaryotic chromosomes are determined in genetic experiments designed to quantify the crossovers that occur during meiosis, and outlining how human genetic maps are made. In **Chapter 15,** we discuss the ways of mapping genes in bacteria and in bacteriophages, which take advantage of the processes of conjugation, transformation, and transduction. Fine structure analysis of bacteriophage genes concludes this chapter. Chromosomal mutations—changes in normal chromosome structure or chromosome number—are discussed in **Chapter 16.** Chromosomal mutations in eukaryotes and human disease syndromes that result from chromosomal mutations, including triplet repeat mutations, are emphasized.

Gene regulation is covered in the following two chapters. **Chapter 17** focuses on the regulation of gene expression in prokaryotes. In this chapter, we discuss the operon as a unit of gene regulation, the current molecular details in the regulation of gene expression in bacterial operons, and regulation of genes in bacteriophages. **Chapter 18** focuses on the regulation of gene expression in eukaryotes, stressing molecular changes that accompany gene regulation and short-term gene regulation in simple and complex eukaryotes.

**Chapter 19** discusses genetic analysis of development. The chapter describes basic events in development, and the evidence that development results from differential gene expression, before illustrating gene regulation principles at work in case studies of well-characterized developmental processes, namely sex determination and dosage compensation, and the development of the *Drosophila* body plan. Next, **Chapter 20** discusses the relationship of the cell cycle to cancer and the various types of genes that, when mutated, play a role in the development of cancer.

In **Chapter 21**, we present the basic principles in population genetics, extending our studies of heredity from the individual organism to a population of organisms. This chapter includes an integrated discussion of the developing area of conservation genetics.

In **Chapter 22**, we discuss quantitative genetics. We consider the heredity of traits in groups of individuals that are determined by many genes simultaneously. In this chapter we also discuss heritability; the relative extent to which a characteristic is determined by genes or by the environment. Discussions of the application of molecular tools to this area of genetics is also included.

**Chapter 23** discusses evolution at the molecular level of DNA and protein sequences. The study of molecular evolution uses the theoretical foundation of population genetics to address two essentially different sets of questions: how DNA and protein molecules evolve and how genes and organisms are evolutionarily related.

## Pedagogical Features

Because the field of genetics is complex, making the study of it potentially difficult, we have incorporated a number of special pedagogical features to assist students and to enhance their understanding and appreciation of genetic principles:

- Each chapter opens with a list of **Key Questions** that prime students for the major concepts they will encounter in the chapter material.
- Throughout each chapter, strategically placed **Keynote** summaries emphasize important ideas and critical points that allow students to check their progress.
- Important **terms and concepts**—highlighted in bold—are defined where they are introduced in the text. For easy reference, they are also compiled in a glossary at the back of the book.
- Each chapter closes with a bulleted **Summary**, further reinforcing the major points that have been discussed.
- With the exception of the introductory Chapter 1, all chapters contain a section titled **Analytical Approaches to Solving Genetics Problems.** Genetics principles have always been best taught with a problem-solving approach. However, beginning students often do not acquire the necessary experience with basic concepts that would enable them to methodically resolve problems. The Analytical Approaches section, in which typical genetics problems are solved in step-by-step detail, was created to help students understand how to tackle genetics problems by applying fundamental principles.
- The **Questions and Problems** sections, which together comprise a total of approximately 750 questions and problems, including over 150 new questions, have been designed to give students further practice in solving genetics problems. The problem for each chapter represent a range of topics and difficulty levels, and have been carefully checked for accuracy. The answers to questions marked by an asterisk can be found at the back of the book, and answers to all questions are available in the separate *Study Guide and Solutions Manual* for students. The answers are also available for download on the instructor portion of the companion website for the book.
- All chapters other than the introduction include **new Focus on Genomics boxes**, written by expert genomics contributor Gregg Jongeward. These short features introduce students to genomics by connecting content in each chapter to current applications in this cutting-edge field.
- Some chapters include **boxes covering special topics** related to chapter coverage. Some of these boxed topics are *Equilibrium Density Gradient Centrifugation* (Chapter 3), *Mutants of* E. coli *DNA polymerases* (Chapter 3), *Identifying RNA–RNA interactions in pre-mRNA splicing by mutational analysis* (Chapter 5), *Labeling DNA* (Chapter 10), *Elementary Principles of Probability* (Chapter 11), *Genetic Terminology* (Chapter 11), *Investigating Genetic Relationships by mtRNA Analysis* (Chapter 13), *Determining Recombination Frequency for Linked Genes and DNA Marker Loci* (Chapter 14), and *Hardy, Weinberg, and the History of Their Contribution to Population Genetics* (Chapter 21).
- Suggested readings and selected websites for the material in each chapter are listed at the back of the book.
- Special care has been taken to provide an extensive, accurate, and well cross-referenced index.

## Supplements

### For Students

**Study Guide and Solutions Manual for *iGenetics: A Molecular Approach*, Third Edition (0-321-58101-6/978-0-321-58101-3)**
Prepared by Bruce Chase of the University of Nebraska at Omaha, the *Study Guide and Solutions Manual* contains detailed solutions for all end-of-chapter problems in the text, including a thorough explanation of the steps used to solve problems. Each chapter of the manual contains an outline of text material and a review of important terms

and concepts. The "Thinking Analytically" feature provides students with general strategies for improving their comprehension of the topic and their problem-solving skills. Finally, 1,000 additional questions for practice and review, based on chapter text as well as animations and *iActivities*, provide an extra resource for students to master chapter content.

**The Genetics Place** (www.geneticsplace.com)
This online learning environment houses the 24 *iActivities* and 59 animations developed in tandem with *iGenetics* and described above, as well as **myeBook**, an online, fully searchable version of the *iGenetics* text that allows students and instructors to add highlights, notes, bookmarks, and more. The website also contains practice quiz questions that report directly to the instructor's gradebook, RSS feeds to breaking news in genetics, links to related websites, and a glossary. The site also provides access via Pearson's **Research Navigator™** database to EBSCO, the world's leading online journal library, containing scholarly articles from over 79,000 publications. Online writing-focused **Research Navigator™ Assignments**, developed especially for students using *iGenetics,* allow students to evaluate and synthesize information from selected readings, then submit their work online directly to their instructor.

***Current Issues in Cell, Molecular Biology & Genetics***
**Volume 1: 0-8053-0568-8/978-0-8053-0568-5**
**Volume 2: 0-321-63398-9/978-0-321-63398-9**
Give your students the best of both worlds—a discussion of the most fascinating, cutting-edge topics in cell biology, genetics, and molecular biology, paired with the authority, reliability, and clarity of Benjamin Cummings' texts. This exclusive special supplement containing recent articles from *Scientific American* is available at no additional cost when packaged with select Benjamin Cummings titles. These articles have been carefully chosen to match the level of your course, and to capture some of the most exciting developments in biology today. Volume 2, the most recent edition, includes articles on the man-made PNA molecule, the genetics of mental illness, human microchimerism, and more. Each article is followed by a set of comprehension questions and class activities for both cell biology and genetics.

## For Instructors

**Instructor's Guide to Text and Media for *iGenetics: A Molecular Approach,* Third Edition (0-321-59722-2/978-0-321-59722-9)**
Written by Rebecca Ferrell of the Metropolitan State College of Denver, this guide presents sample lecture outlines, teaching tips for the text, and media tips for using and assigning the media component in class.

**Instructor's Resource CD-ROM for *iGenetics: A Molecular Approach,* Third Edition (0-321-58097-4/978-0-321-58097-9)**
This cross-platform CD-ROM features standalone files of all animations and iActivities, as well as animations pre-inserted into PowerPoint files for use in lectures. This resource also includes all illustrations, photos, and tables from the text, with each available in high-resolution JPEG and PowerPoint formats, as well as Word files of the Instructor's Guide and TestGen® software pre-loaded with test questions for each chapter of *iGenetics* (see description below).

**Computerized Test Bank for *iGenetics: A Molecular Approach,* Third Edition**
The test bank for *iGenetics*, containing over 1,100 multiple-choice questions, is available as part of the Instructor's Resource CD-ROM described above. Thoroughly revised and expanded by Indrani Bose of Western Carolina University and Heather Lorimer of Youngstown State University, and carefully checked for accuracy by Malcolm Schug of the University of North Carolina, Greensboro, it is formatted in Pearson's exclusive TestGen® software, which gives instructors the additional capability of editing questions or adding their own. In order to minimize our impact on the environment, the test bank will no longer be produced as a separate printed supplement, but will remain available for online download in Word format. However, the test bank will be available for online download in Word format.



## Acknowledgments

Publishing a textbook and all its supplements is a team effort. I have been very fortunate to have some very talented individuals working with me on this project. Thanks in particular are due to Gregg Jongeward (University of the Pacific), who contributed the extensively revised chapters on genomics and recombinant DNA technology to this edition as well as the Focus On Genomics boxes. I also would like to thank the following contributors for their talents and efforts in crafting some of the later chapters in the text: Dr. Malcolm Schug (University of North Carolina, Greensboro) for his revision of Chapter 21, "Population Genetics"; Dr. Kevin Livingstone (Trinity University) for his revision of Chapter 22, "Quantitative Genetics"; and Dr. Dan E. Krane (Wright State University) for revising Chapter 23, "Molecular Evolution." In addition, our editorial accuracy checkers Dr. Chaoyang Zeng (University of Wisconsin–Milwaukee) and Dr. Malcolm Schug (University of North Carolina, Greensboro) deserve thanks for their meticulous review of the chapter text and all end-of-chapter questions, problems, and solutions.

I would also like to thank Bruce Chase (University of Nebraska, Omaha) for his extensive and excellent work on the end-of-chapter questions, including his contribution of many new problem sets, and for his excellent work on putting together the *Study Guide and Solutions Manual*. And I am also grateful to Rebecca Ferrell (Metropolitan State College of Denver) for her careful work in revising the *Instructor's Guide*; to Indrani Bose (Western Carolina University) and Heather Lorimer (Youngstown State University) for their updating and expansion of the *Test Bank*; and to Malcolm Schug (University of North Carolina, Greensboro) for providing his advice on the *Test Bank's* clarity and accuracy.

I want to acknowledge a number of talented individuals who worked with me to develop the material found on the *iGenetics: A Molecular Approach,* Third Edition, companion website: Margy Kuntz, who did an excellent job researching this subject matter and then authoring most of the highly creative and rich *iActivities*, all of which are designed to enhance critical thinking in genetics; Dr. Todd Kelson (Ricks College; animation storyboards); Dr. Hai Kinal (Springfield College, animation storyboards); Dr. Robert Rothman (Rochester Institute of Technology; animation storyboards); Steve McEntee (*iActivity* art development, art style for the animations and text art); Kristin Mount (animations); Richard Sheppard (animations); Eric Stickney (animations); and James Costa (Western Carolina University; original website quiz questions). In addition, I thank Dr. James Caras, Principal, Jon Harmon, Content Developer, and the rest of the Science Technologies staff for developing and producing additional *iActivities* and animations for the website. I would like to thank David Kass (Eastern Michigan University) and Jocelyn Krebs (University of Alaska Anchorage) for their editorial review of the latest round of revisions to the animations, and both Jocelyn Krebs and Philip Meneely (Haverford College) for their aid in reviewing storyboards during the revision process. I would also like to thank Cheryl Ingram-Smith (Clemson University) and Robert Locy (Auburn University) for revising the website quiz questions based on the book's updated chapter content, and David Kass (Eastern Michigan University) for verifying the accuracy of the quizzes. Finally, I would like to extend my thanks to Harry Nickla for creating the new Research Navigator™ Assignments that appear on the website.

I am grateful to the literary executor of the late Sir Ronald A. Fisher, F.R.S.; to Dr. Frank Yates, F.R.S.; and to Longman Group Ltd. London, for permission to reprint Table IV from their book, *Statistical Table for Biological, Agricultural and Medical Research* (Sixth Edition, 1974).

I would like to thank Lori Newman, Production Supervisor at Benjamin Cummings, as well as Crystal Clifton and the staff at Progressive Publishing Alternatives for their handling of the production phase of the book.

Finally, I wish to thank the editorial and marketing staff at Benjamin Cummings who helped to make *iGenetics: A Molecular Approach,* Third Edition, a reality. In particular, I thank Gary Carlson, Acquisitions Editor; Beth Wilbur, Vice President and Editor-in-Chief, Biology; Deborah Gale, Director of Development; and Lauren Harp, Senior Marketing Manager. I am especially grateful to Rebecca Johnson, Project Editor, for her excellent management of the many aspects of the production of the book; her efforts have ensured that this textbook and its supplements are of the highest quality.

Finally, for all of their help in honing *iGenetics* over its several editions, I would like to thank the following reviewers: George Bajszar (University of Colorado, Colorado Springs); Ruth Ballard (California State University, Sacramento); Hank Bass (Florida State University); Tineke Berends (Houston Community College); Anna Berkovitz (Purdue University); Andrew Bohonak (San Diego State University); Paul J. Bottino (University of Maryland); Joanne Brock (Kennesaw State University); Patrick Calie (Eastern Kentucky University); Clarissa Cheney (Pomona College); Richard Cheney (Christopher Newport University); Bhanu Chowdhary (Texas A&M University); Claire Chronmiller (University of Virginia); James T. Costa (Western Carolina University); Sandra L. Davis (University of Indianapolis); Frank Doe (University of Dallas); John Doucet (Nicholls State University); David Durcia (University of Oklahoma); Larry Eckroat (Pennsylvania State University at Erie); Bert Ely (University of South Carolina); Quentin Fang (Georgia Southern University); Russ Feirer (St. Norbert College); Wayne Forrester (Indiana University); Elaine Freund (Pomona College); David Fromson (California State University, Fullerton); Gail Gasparich (Towson State University); Peter Gegenheimer (University of Kansas); Vaughn Gehle (Southwest Minnesota State); Richard C. Gethmann (University of Maryland, Baltimore County); Elliot Goldstein (Arizona State University); Mary Katherine Gonder (SUNY–University at Albany); Michael Goodisman (Georgia Tech); Pamela Gregory (Jacksonville State University); Karen Hales (Davidson College); Pamela Hanratty (Indiana University); Ernie Hannig (University of Texas, Dallas); David Haymer (University of Hawaii); Mary Healy (Springfield College); Robert Hinrichsen (Indiana University); Margaret Hollingsworth (State University of New York, Buffalo); Lynne Hunter (University of Pittsburgh); Cheryl Ingram-Smith (Clemson University); Tracie M. Jenkins (University of Georgia); Gregg Jongeward (University of the Pacific); Cheryl Jorcyk (Boise State University); Todd Kelson (Ricks College); Elliot Krause (Seton Hall University); Jocelyn Krebs (University of Alaska–Anchorage); Alexander Lai (Oklahoma State University); Sandy Latourelle (Plattsburg State University); Michael Lentz (University of North Florida); Hai Kanal (Springfield College); David Kass (Eastern Michigan University); Larry Kline (State University of New York, Brockport); Brian Kreiser (University of Southern Mississippi); Alan Leonard (Florida Institute of Technology); Robert Locy (Auburn University); Tara Macey (Washington State University–Vancouver); Mark J. M. Magbanua (University of California at Davis); Karen Malatesta (Princeton University); Russell Malmburg (University of Georgia, Athens); Patrick H. Masson (University of Wisconsin, Madison); Steven

McCommas (Southern Illinois University); David McCullough (Wartburg College); Denis McGuire (St. Cloud State University); Kim McKim (Rutgers University); Philip Meneely (Haverford College); John Merruam (University of California, Los Angeles); Stan Metzenberg (University of California, Northridge); Dwight Moore (Emporia State University); Roderick Morgan (Grand Valley State University); Muriel Nesbit (University of California, San Diego); David Nelson (University of Tennessee Health Science Center); Brent Nelson (Auburn University); Joanne Odden (Metropolitan State College of Denver); James M. Pipas (University of Pittsburgh); Jean Porterfield (St. Olaf College); Uwe Pott (University of Wisconsin–Green Bay); Diane Robbins (University of Michigan Medical School); Harry Roy (Rensselaer Polytechnic Institute); Thomas Rudge (Ohio State University); Malcolm Schug (University of North Carolina–Greensboro); Stanley Sessions (Hartwick College); Rey Antonio L. Sia (State University of New York); Randy Small (University of Tennessee, Knoxville); William Steinhart (Bowdoin College); Gary Stormo (Washington University in St. Louis); Millard Sussman (University of Wisconsin, Madison); Farshad Tamari (Kean University); Sara Tolsma (Northwestern University); Jonathan Visick (North Central College); Melina Wales (Texas A&M University); Robert West (University of Colorado); Cindy White (University of Northern Colorado); Matthew White (Ohio University); Ross Whitwam (Mississippi University for Women); Bruce Wightman (Muhlenberg College); Warren Williams (Texas Southern University); John Zamora (Middle Tennessee State University); and Chaoyang Zeng (University of Wisconsin–Milwaukee).

I would also like to thank the following media reviewers for their contributions toward ensuring the excellence of our iActivities and animations: Mary D. Healey (Springfield College); David Kass (Eastern Michigan University); Sidney R. Kushner (University of Georgia); Gayle LoPiccolo (Montgomery College); Maria Orive (University of Kansas); and Kajan Ratnakumar (Desplan Laboratory, New York University).

*Peter J. Russell*

# 1 Genetics: An Introduction

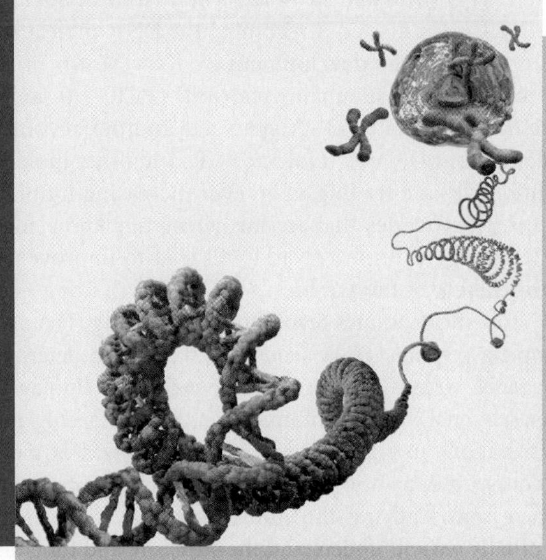

Sylized diagram of the relationship between DNA, chromosomes, and the cell.

## Key Questions

- What are the major subdivisions of genetics?
- What are geneticists, and what is genetics research?

Welcome to the study of **genetics,** the science of heredity. Genetics is concerned primarily with understanding biological properties that are transmitted from parent to offspring. The subject matter of genetics includes heredity, the molecular nature of the genetic material, the ways in which genes (which determine the characteristics of organisms) control life functions, and the distribution and behavior of genes in populations.

Genetics is central to biology because gene activity underlies all life processes, from cell structure and function to reproduction. Learning what genes are, how genes are transmitted from generation to generation, how genes are expressed, and how gene expression is regulated is the focus of this book. Genetics is expanding so rapidly that it is not possible to describe everything we know about it between these covers. The important principles and concepts are presented carefully and thoroughly; readers who want to go further are advised to look for information on the Internet, including searching for research papers using Google Scholar or the PubMed database supported by the National Library of Medicine, National Institutes of Health, at http://www.pubmed.gov.

It is assumed that your experience in your introductory biology course has given you a general understanding of genetics. This chapter provides a contextual framework for your study of genes as you read the chapters of the book.

## Classical and Modern Genetics

Humans recognized long ago that offspring tend to resemble their parents. Humans have also performed breeding experiments with animals and plants for centuries. However, the principles of heredity were not understood until the mid-nineteenth century, when Gregor Mendel analyzed quantitatively the results of crossing pea plants that varied in easily observable characteristics. He published his results, but their significance was not realized in his lifetime. Several years after his death, however, researchers realized that Mendel had discovered fundamental principles of heredity. We now consider Mendel's work to be the foundation of modern genetics.

Since the turn of the twentieth century, genetics has been an increasingly powerful tool for studying biological processes. An important approach used by many geneticists is to work with mutants of a cell or an organism affecting a particular biological process: by characterizing the differences between the mutants with normal cells or organisms, they develop an understanding of the process. Such research has gone in many directions, such as analyzing heredity in populations, analyzing evolutionary processes, identifying the genes that control the steps in a process, mapping the genes involved, determining the products of the genes, and analyzing the molecular features of the genes, including the regulation of the genes' expression.

Research in genetics underwent a revolution in 1972, when Paul Berg constructed the first recombinant DNA molecule *in vitro*, and in 1973, when Herbert Boyer and Stanley Cohen cloned a recombinant DNA molecule for the first time. The development by Kary Mullis in 1986 of the polymerase chain reaction (PCR) to amplify specific segments of DNA spawned another revolution. Recombinant DNA technology, PCR, and other molecular technologies are leading to an ever-increasing number of exciting discoveries that are furthering our knowledge of basic biological functions and will lead to improvements in the quality of human life.

Now the genomics revolution is occurring. That is, the complete genomic DNA sequences have been determined for many viruses and organisms, including humans. As scientists analyze the genomic data, we are seeing major contributions to our knowledge in many areas of biology. Of course, it is natural for us to focus on the expected outcomes from studying the human genome. For example, eventually we will understand the structure and function of every gene in the human genome. Such knowledge undoubtedly will lead to a better understanding of human genetic diseases and contribute significantly to their cures. The science-fiction scenario of each of us carrying our DNA genome sequence on a chip will become reality in the near future. However, knowledge about our genomes will raise social and ethical concerns that must be resolved carefully.

## Geneticists and Genetic Research

The material presented in this book is the result of an incredible amount of research done by geneticists working in many areas of biology. Geneticists use the standard methods of science in their studies. As researchers, geneticists typically use the **hypothetico-deductive method of investigation.** This consists of making *observations*, forming *hypotheses* to explain the observations, making experimental *predictions* based on the hypotheses, and finally *testing* the predictions. The last step provides new observations, producing a cycle that leads to a refinement of the hypotheses and perhaps, eventually, to the establishment of a theory that attempts to explain the original observations.

As in all other areas of scientific research, the exact path a research project will follow cannot be predicted precisely. In part, the unpredictability of research makes it exciting and motivates the scientists engaged in it. The discoveries that have revolutionized genetics typically were not planned; they developed out of research in which basic genetic principles were being examined. The work of Barbara McClintock on the inheritance of patches of color on corn kernels is an excellent example (see Chapter 7). After accumulating a large amount of data from genetic crosses, she hypothesized that the appearance of colored patches was the result of the movement (transposition) of a DNA segment from one place to another in the genome. Only many years later were these DNA segments—called *transposons* or *transposable elements*—isolated and characterized in detail. (A more complete discussion of this discovery and of Barbara McClintock's life is presented in Chapter 7.) We know now that transposons are ubiquitous, playing a role not only in the evolution of species but also in some human diseases.

### The Subdisciplines of Genetics

Geneticists often divide genetics into four major subdisciplines:

1. **Transmission genetics** (sometimes called classical genetics) is the subdiscipline dealing with how genes and genetic traits are transmitted from generation to generation and how genes recombine (exchange between chromosomes). Analyzing the pattern of trait transmission in a human pedigree or in crosses of experimental organisms is an example of a transmission genetics study.
2. **Molecular genetics** is the subdiscipline dealing with the molecular structure and function of genes. Analyzing the molecular events involved in the gene control of cell division, or the regulation of expression of all the genes in a genome, are examples of molecular genetics studies. Genomic analysis is part of molecular genetics.
3. **Population genetics** is the subdiscipline that studies heredity in groups of individuals for traits that are determined by one or only a few genes. Analyzing the frequency of a disease-causing gene in the human population is an example of a population genetics study.
4. **Quantitative genetics** also considers the heredity of traits in groups of individuals, but the traits of concern are determined by many genes simultaneously. Analyzing the fruit weight and crop yield in agricultural plants are examples of quantitative genetics studies.

Although these subdisciplines help us think about genes from different perspectives, there are no sharp boundaries between them. Increasingly, for example, population and quantitative geneticists analyze molecular data to determine gene frequencies in large groups. Historically, transmission genetics developed first, followed by population genetics and quantitative genetics, and then molecular genetics.

Genes influence all aspects of an organism's life. Understanding transmission genetics, population genetics, and quantitative genetics will help you understand population biology, ecology, evolution, and animal behavior. Similarly, understanding molecular genetics is useful when you study such topics as neurobiology, cell biology, developmental biology, animal physiology, plant physiology, immunology, and, of course, the structure and function of genomes.

### Basic and Applied Research

Genetics research, and scientific research in general, may be either basic or applied. In **basic research,** experiments are done to gain an understanding of fundamental phenomena, whether or not the knowledge gained leads to any immediate applications. Basic research was responsible for most of the facts we discuss in this book. For example, we know how the expression of many prokaryotic and eukaryotic genes is regulated as a result of basic research on model organisms such as the bacterium *Escherichia coli* (*E. coli*) ("esh-uh-REEK-e-uh CO-lie," shown in Figure 1.1), the yeast *Saccharomyces cerevisiae* ("sack-a-row-MY-seas serry-VEE-see-eye," shown in Figure 1.4a), and the fruit fly *Drosophila melanogaster* ("dra-SOFF-ee-la muh-LANO-gas-ter," shown in Figure 1.4b). The knowledge obtained from basic research is used largely to fuel more basic research.

In **applied research,** experiments are done with different goals in mind; namely, with an eye toward overcoming specific problems in society or exploiting discoveries. In agriculture, applied genetics has contributed significantly to improvements in animals bred for food (such as reducing the amount of fat in beef and pork) and in crop plants (such as increasing the amount of protein in soybeans). A number of diseases are caused by genetic defects, and great strides are being made in diagnosis and understanding the molecular bases of some of those diseases. For example, drawing on knowledge gained from basic research, applied genetic research involves developing rapid diagnostic tests for genetic diseases and producing new pharmaceuticals for treating diseases.

There is no sharp dividing line between basic and applied research. Indeed, in both areas, researchers use similar techniques and depend on the accumulated body of information when building hypotheses. For example, **recombinant DNA technology**—procedures that allow molecular biologists to splice a DNA fragment from one organism into DNA from another organism and to clone (make many identical copies of) the new recombinant DNA molecule—has profoundly affected both basic and applied research (see Chapters 8, 9, and 10). Many biotechnology companies owe their existence to recombinant DNA technology as they seek to clone and manipulate genes in developing their products. In the area of plant breeding, recombinant DNA technology has made it easier to introduce traits such as disease resistance from noncultivated species into cultivated species. Such crop improvement traditionally was achieved by using conventional breeding experiments. In animal breeding, recombinant DNA technology is being used in the beef, dairy, and poultry industries, for example, to increase the amount of lean meat, the amount of milk, and the number of eggs. In medicine, the results are equally impressive. Recombinant DNA technology is being used to produce a number of antibiotics, hormones, and other medically important agents such as clotting factor and human insulin (marketed under the name Humulin; Figure 1.2) and to diagnose and treat a number of human genetic diseases. In forensics, *DNA typing* (also called *DNA fingerprinting* or *DNA profiling*) is being used in paternity cases, criminal cases, and anthropological studies. In short, the science of genetics is currently in an exciting and dramatic growth phase, and there is still much to discover.

**Figure 1.1**
**Colorized scanning electron micrograph of *Escherichia coli*, a rod-shaped bacterium common in the intestines of humans and other animals.**

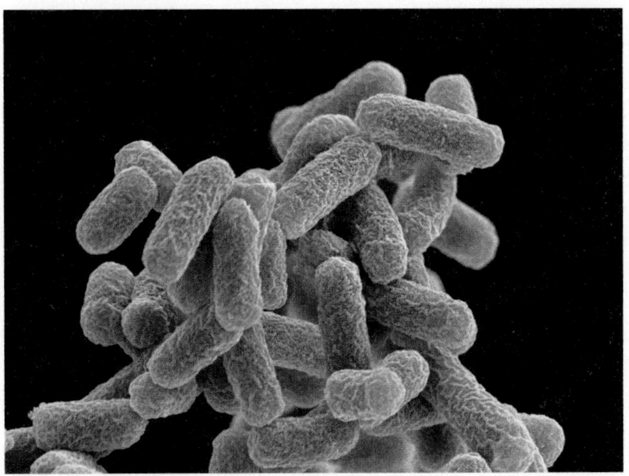

**Keynote**

Genetics can be divided into four major subdisciplines: transmission genetics, molecular genetics, population genetics, and quantitative genetics. Depending on whether the goal is to obtain a fundamental understanding of genetic phenomena or to exploit discoveries, genetic research is considered to be basic or applied, respectively.

## Genetic Databases and Maps

In this section, we talk about two important resources for genetic research: genetic databases and genetic maps. Genetic databases have become much more sophisticated and expansive as computer analysis tools have been developed and Internet access to databases has become routine. Constructing genetic maps has been part of genetic analysis for about 100 years.

**Figure 1.2**
**Example of a product developed as a result of recombinant DNA technology.** Humulin—human insulin for insulin-dependent diabetics.

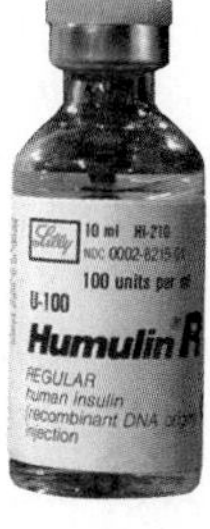

**Genetic Databases.** The amount of information about genetics has increased dramatically. No longer can we learn everything about genetics by going to a college or university library; the computer now plays a major role. For example, a useful way to look for genetic information through the Internet is by entering key terms into search engines such as Google (http://www.google.com). Typically, a vast number of hits are listed, some useful and some not.

There are many specific genetic databases on the Internet, too many to summarize all that are useful in this section. You must search for yourself and be critical about what you find. However, we can consider a set of important and extremely useful genetic databases at the National Center for Biotechnology Information (NCBI) website (http://www.ncbi.nlm.nih.gov). NCBI was created in 1988 as a national resource for molecular biology information. Its role is to "create public databases, conduct research in computational biology, develop software tools for analyzing genome data, and disseminate biomedical information—all for the better understanding of molecular processes affecting human health and disease."

Some of the search tools available at the NCBI site are as follows:

- PubMed is used to access literature citations and abstracts and provides links to sites with electronic versions of research journal articles. These articles can sometimes be viewed, or you must pay a one-time fee or obtain a free subscription. You search PubMed by entering terms, author names, or journal titles. It is highly recommended that you use PubMed to find research articles on genetic topics that interest you.
- OMIM (Online Mendelian Inheritance in Man) is a database of human genes and genetic disorders authored and edited by Dr. Victor A. McKusick and his colleagues. You search OMIM by entering terms in a textbox search window; the result is a list of linked pages, each with a specific OMIM entry number. The pages have detailed information about the gene or genetic disorder specified in the original search, including genetic, biochemical, and molecular data, along with an up-to-date list of references. Throughout the book, each time we discuss a human gene or genetic disease, we refer to OMIM entries and give the OMIM entry number.
- GenBank is the National Institutes of Health (NIH) genetic-sequence database. This database is an annotated collection of all the tens of billions of publicly available DNA sequences. You search GenBank by entering terms in the search window. For example, if you are interested in the human disease cystic fibrosis, enter the term *cystic fibrosis* into the search window, and you will find all sequences that have been entered into GenBank that include those two words in the annotations.
- BLAST (Basic Local Alignment Search Tool) is a tool used to compare a nucleotide sequence or protein sequence with all sequences in the database to find possible matches. This is useful, for example, if you have sequenced a new gene and want to find out whether anything similar has been sequenced previously. Moreover, genes with related functions may be listed in the databases, allowing you to focus your research on the function of the gene you are studying.
- Entrez is a system for searching several linked databases. The particular database is chosen from a pull-down menu. The databases include PubMed; Nucleotide, for the GenBank DNA and RNA sequences database; Protein, for amino acid sequences; Structure, for three-dimensional macromolecular structures; Genome, for complete genome assemblies; RefSeq, an annotated collection of genes, transcripts, and the proteins derived from the transcripts; OMIM, the Online Mendelian Inheritance in Man human gene database; and PopSet, population study datasets. The database can be selected from the hot links, or a pull-down menu choice on the main Entrez page will guide your search terms appropriately. For example, if you are interested in nucleotide sequences related to the human disease cystic fibrosis, you would select "Nucleotide" in the pull-down menu and enter *cystic fibrosis* in the search window. A list of relevant sequence entries will be returned.
- Books is a collection of biomedical books that can be searched directly. Included are some genetics, molecular biology, and developmental biology textbooks.

A powerful feature of the NCBI databases is that they are linked, enabling users to move smoothly between them and hence integrate the knowledge obtained in each of them. For example, a literature citation found in PubMed will have links to sequences in nucleotide and protein databases.

**Genetic Maps.** Since 1902, much effort has been made to construct **genetic maps** (Figure 1.3) for the commonly used experimental organisms in genetics. Like road maps that show the relative locations of towns along a road, genetic maps show the arrangements of genes along the chromosomes and the genetic distances between the genes. The position of a gene on the map is called a **locus** or **gene locus.** The genetic distances between genes on the same chromosome are calculated from the results of genetic crosses by counting the frequency of recombination—that is, the percentage of the time among the progeny that the genes in the two original parents exchange (i.e., recombine; see Chapter 14). The unit of genetic distance is the **map unit** (mu).

The goal of constructing genetic maps has been to obtain an understanding of the organization of genes along the chromosomes (e.g., to inform us whether genes with related functions are on the same chromosome; and if they are, whether they are close to each other). Genetic

**Figure 1.3**

**Example of a genetic map, illustrating some of the genes on chromosome 2 of the fruit fly, *Drosophila melanogaster*.** The numerical values represent the positions of the genes from the chromosome end (top) measured in map units.

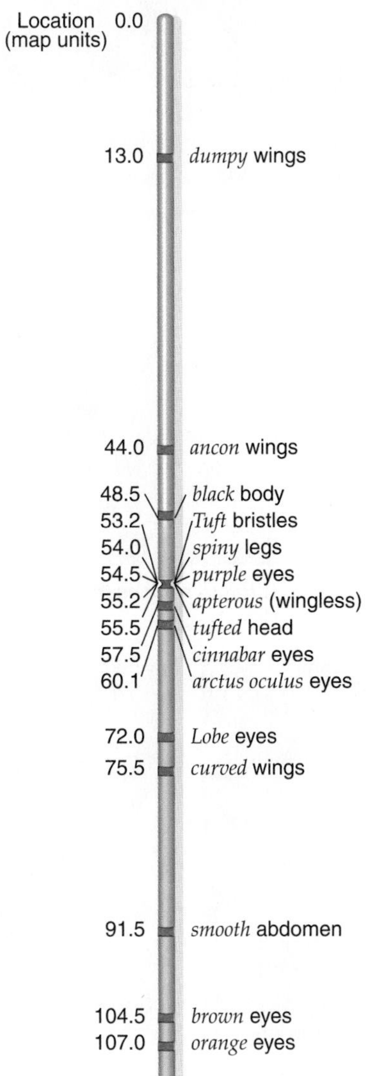


maps have also proved very useful in efforts to clone and sequence particular genes of interest—and more recently, as part of genome projects, in efforts to obtain the complete sequences of genomes.

## Keynote

Two important resources for genetic research are genetic databases and genetic maps. Databases provide the means to search for specific information about a gene, including its sequence, its function, its position in the genome, research papers written about it, and details about its product. Genetic maps show the positions of genes along a chromosome. They have proved useful in efforts to clone genes, as well as in the efforts to sequence genomes.

## Organisms for Genetics Research

The principles of heredity were first established in the nineteenth century by Gregor Mendel's experiments with the garden pea. Since Mendel's time, many organisms have been used in genetic experiments. In general, the goal of the research has been to understand gene structure and function. Because of the remarkable conservation of gene function throughout evolution, scientists have realized that results obtained from studies with a particular organism typically would apply more generally. Among the qualities that historically have made an organism a particularly good model for genetic experimentation are the following:

- The organism has a short life cycle, so that a large number of generations occur within a short time. In this way, researchers can obtain data readily over many generations. Fruit flies, for example, produce offspring in 10 to 14 days.
- A mating produces a large number of offspring.
- The organism should be easy to handle. For example, hundreds of fruit flies can be kept easily in small bottles.
- Most importantly, genetic variation must exist between the individuals in the population or be created in the population by inducing mutations so that the inheritance of traits can be studied.

Both eukaryotes and prokaryotes are used in genetics research. **Eukaryotes** (meaning "true nucleus") are organisms with cells within which the genetic material (DNA) is located in the **nucleus** (a discrete structure bounded by a nuclear envelope). Eukaryotes can be unicellular or multicellular. In genetics today, a great deal of research is done with six eukaryotes (Figure 1.4a–f): *Saccharomyces cerevisiae* (budding yeast), *Drosophila melanogaster* (fruit fly), *Caenorhabditis elegans* ("see-no-rab-DYT-us ELL-e-gans," a nematode worm), *Arabidopsis thaliana* ("a-rab-ee-DOP-sis thal-ee-AH-na," a small weed of the mustard family), *Mus musculus* ("muss MUSS-cue-lus," a mouse), and *Homo sapiens* ("homo SAY-pee-ens," human). Humans are included although they do not meet the criteria for an organism well suited for genetic experimentation, but because ultimately we want to understand as much as we can about human genes and their function. With this understanding, we will be able to combat genetic diseases and gain fundamental knowledge about our species' development and evolution.

Over the years, research with the following seven eukaryotes has also contributed significantly to our understanding of genetics (Figure 1.4g–m): *Neurospora crassa* ("new-ROSS-pore-a crass-a," orange bread mold), *Tetrahymena* ("tetra-HI-me-na," a protozoan), *Paramecium* ("para-ME-see-um," a protozoan), *Chlamydomonas reinhardtii* ("clammy-da-MOAN-as rhine-HEART-ee-eye," a green alga), *Pisum sativum* ("PEA-zum sa-TIE-vum," garden pea), *Zea mays* (corn), and *Danio rerio* (zebrafish). Of these, *Tetrahymena, Paramecium, Chlamydomonas,* and *Saccharomyces* are unicellular organisms, and the rest are multicellular.

Figure 1.4

**Eukaryotic organisms that have contributed significantly to our knowledge of genetics.**

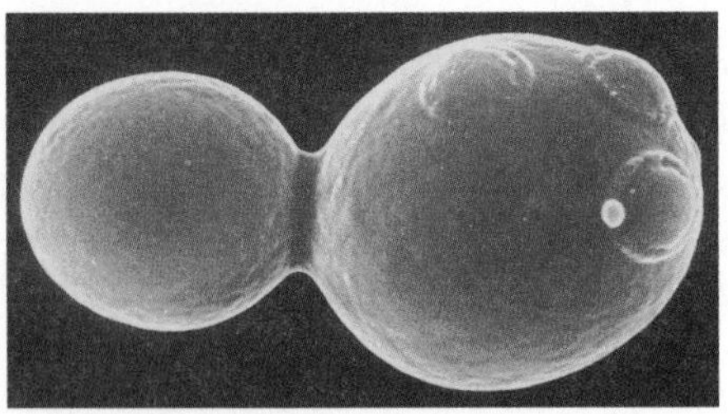

a) *Saccharomyces cerevisiae* (a budding yeast)

b) *Drosophila melanogaster* (fruit fly)

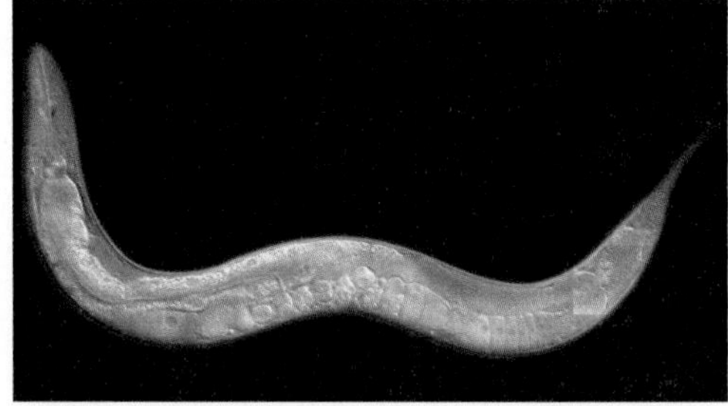

c) *Caenorhabditis elegans* (a nematode)

d) *Arabidopsis thaliana* (Thale cress, a member of the mustard family)

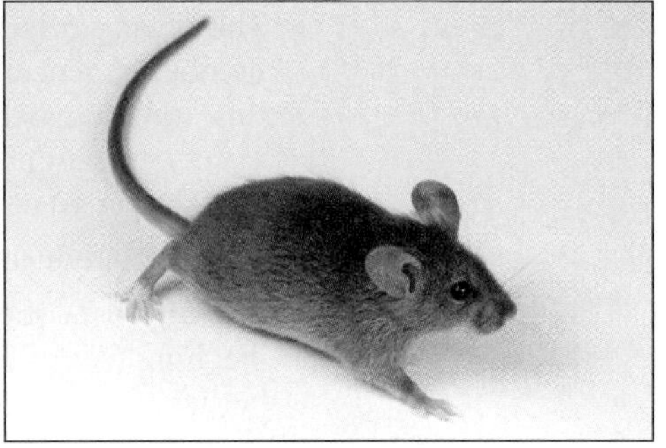

e) *Mus musculus* (mouse)

f) *Homo sapiens* (human)

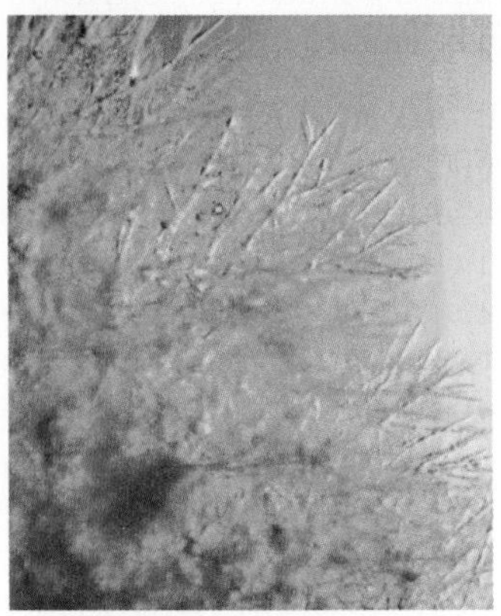

g) *Neurospora crassa* (orange bread mold)

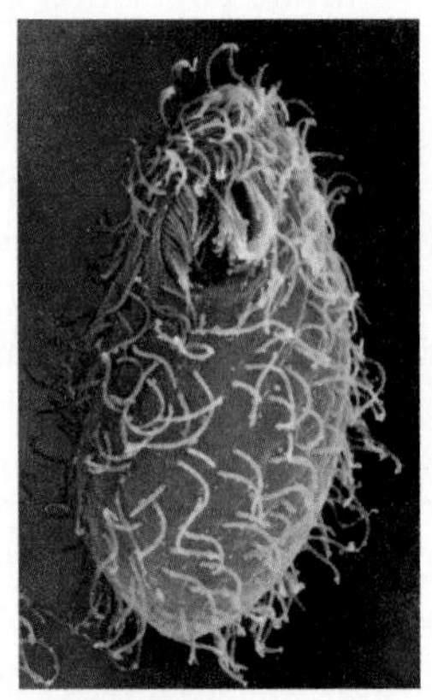

h) *Tetrahymena* (a protozoan)

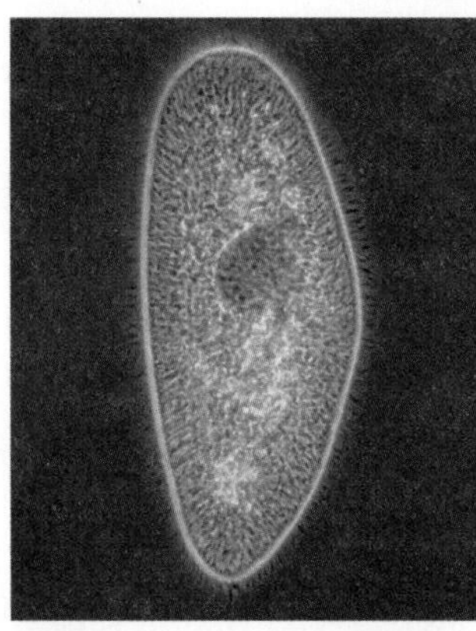

i) *Paramecium* (a protozoan)

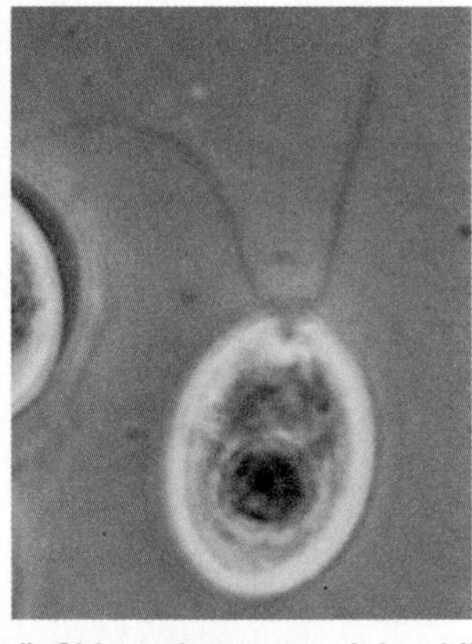

j) *Chlamydomonas reinhardtii* (a green alga)

k) *Pisum sativum* (a garden pea)

l) *Zea mays* (corn)

m) *Danio rerio* (zebrafish)

You learned about many features of eukaryotic cells in your introductory biology course. Figure 1.5 shows a generalized higher plant cell and a generalized animal cell. Surrounding the cytoplasm of both plant cells and animal cells is a lipid bilayer, the *plasma membrane*. Plant cells, but not animal cells, have a rigid cell wall outside the plasma membrane. The nucleus of eukaryotic cells contains DNA complexed with proteins and organized into a number of linear structures called chromosomes. The nucleus is separated from the rest of the cell—the cytoplasm and associated organelles—by the double membrane called the nuclear envelope. The membrane is selectively permeable and has pores about 20 to 80 nm (nm = nanometer = $10^{-9}$ meter) in diameter that allow certain materials to move between the nucleus and the cytoplasm. For example, messenger RNAs, which are translated in the cytoplasm to produce polypeptides, are synthesized in the nucleus and pass through the pores to reach the cytoplasm. In the opposite direction, enzymes for DNA replication, DNA repair, and transcription, and the proteins that associate with DNA to form the chromosomes are made in the cytoplasm and enter the nucleus via the pores.

The cytoplasm of eukaryotic cells contains many different materials and organelles. Of special interest to geneticists are the *centrioles*, the *endoplasmic reticulum* (ER), *ribosomes*, *mitochondria*, and *chloroplasts*. Centrioles (also called basal bodies) are found in the cytoplasm of nearly all animal cells (see Figure 1.5b), but not in plant cells. In animal cells, a pair of centrioles is located at the center of the centrosome, a region of undifferentiated cytoplasm that organizes the spindle fibers that are involved in chromosome segregation in mitosis and meiosis (discussed in Chapter 12).

The ER is a double-membrane structure that is part of the endomembrane system. The ER is continuous with the nuclear envelope. Rough ER has ribosomes attached to it, giving it a rough appearance, whereas smooth ER does not. Ribosomes bound to rough ER synthesize proteins to be secreted by the cell or to be localized in the plasma membrane or particular organelles within the cell. The synthesis of proteins other than those distributed via the ER is performed by ribosomes that are free in the cytoplasm.

**Mitochondria** (singular: *mitochondrion*; see Figure 1.5) are large organelles surrounded by a double membrane—the inner membrane is highly convoluted. Mitochondria play a crucial role in processing energy for the cell. They also contain DNA that encodes some of the proteins that function in the mitochondrion and some components of the mitochondrial protein synthesis machinery.

Many plant cells contain **chloroplasts**—large, triple-membraned, chlorophyll-containing organelles involved in photosynthesis (see Figure 1.5a). Chloroplasts also contain DNA that encodes some of the proteins that function in the chloroplast and some components of the chloroplast protein synthesis machinery.

In contrast to eukaryotes, **prokaryotes** (meaning "prenuclear") do not have a nuclear envelope surrounding their DNA (Figure 1.6); this is the major distinguishing



**Figure 1.5**

**Eukaryotic cells.** Cutaway diagrams of (**a**) a generalized higher plant cell and (**b**) a generalized animal cell, showing the main organizational features and the principal organelles in each.

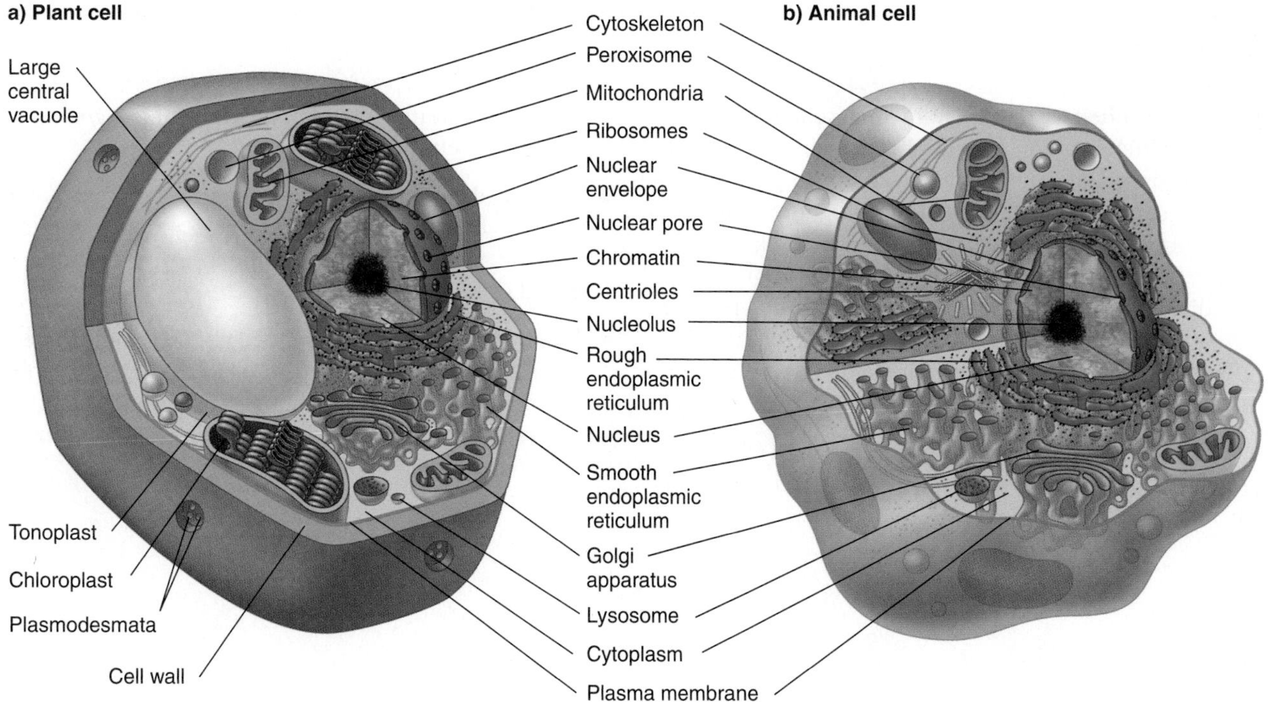

Figure 1.6

**Cutaway diagram of a generalized prokaryotic cell.**

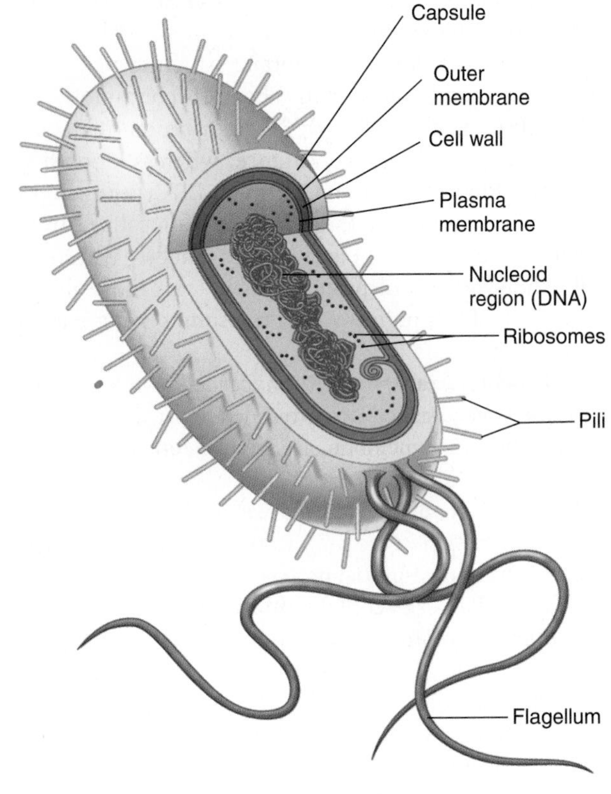


feature of prokaryotes. Included in the prokaryotes are all the bacteria, which are spherical, rod-shaped, or spiral-shaped organisms. The shape of a bacterium is maintained by a rigid cell wall located outside the cell membrane. Prokaryotes are divided into two evolutionarily distinct groups: the Bacteria and the Archaea. The Bacteria are the common varieties found in living organisms (naturally or by infection), in soil, and in water. Archaea are the prokaryotes found often in much more inhospitable conditions, such as hot springs, salt marshes, methane-rich marshes, or the ocean depths, where bacteria do not thrive. Archaea are also found under typical conditions, such as water and soil. Bacteria generally vary in size from about 100 nm in diameter to 10 μm in diameter. The largest species, the spherical *Thiomargarita namibiensis,* can reach $^3/_4$ mm in diameter, at which point it is visible to the naked eye (about the size of a *Drosophila* eye).

In most cases, the prokaryotes studied in genetics are members of the Bacteria group. The most intensely studied is *E. coli* (see Figure 1.1), a rod-shaped bacterium common in intestines of humans and other animals. Studies of *E. coli* have significantly advanced our understanding of the regulation of gene expression and the development of molecular biology. *E. coli* is also used extensively in recombinant DNA experiments.

**Keynote**

Eukaryotes are organisms that have cells in which the genetic material is located in a membrane-bound nucleus. The genetic material is distributed among several linear chromosomes. Prokaryotes, by contrast, lack a membrane-bound nucleus.

## Summary

- Genetics often is divided into four major subdisciplines: transmission genetics, which deals with the transmission of genes from generation to generation; molecular genetics, which deals with the structure and function of genes at the molecular level; population genetics, which deals with heredity in groups of individuals for traits that are determined by one or a few genes; and quantitative genetics, which deals with heredity of traits in groups of individuals wherein the traits are determined by many genes.
- Genetic research is considered to be basic when the goal is to obtain a fundamental understanding of genetic phenomena, and applied when the goal is to exploit genetics discoveries.
- Genetic databases provide the means to search for specific information about a gene and its product. Genetic maps show the positions of genes along a chromosome.
- Eukaryotes are organisms in which the genetic material is located in a membrane-bound nucleus within the cells. The genetic material is distributed among several linear chromosomes. Prokaryotes, by contrast, lack a membrane-bound nucleus.

# 2 DNA: The Genetic Material

A DNA double helix.

## Key Questions

- What is the molecular nature of the genetic material?
- What is the molecular structure of DNA and RNA?
- How is DNA organized in chromosomes?

### iActivity

IMAGINE THAT YOU ARE HANDED A SEALED black box and are told that it contains the secret of life. Determining the chemical composition, molecular structure, and function of the thing inside the box will allow you to save lives, feed the hungry, solve crimes, and even create new life-forms. What's inside the box? What tools and techniques could you use to find out?

In this chapter, you will discover how scientists identified the contents of this "black box" and, in doing so, unraveled the "secret of life." Later in the chapter, you can apply what you've learned by trying the iActivity, in which you use many of the same tools and techniques to determine the genetic nature of a virus that is ravaging rice plants in Asia.

Simple observation shows that a lot of variation exists between individuals of a given species. For example, individual humans vary in eye color, height, skin color, and hair color, even though all humans belong to the species *Homo sapiens*. The differences between individuals within and among species are mainly the result of differences in the DNA sequences that constitute the genes in their genomes. The genetic information coded in DNA is largely responsible for determining the structure, function, and development of the cell and the organism.

In the next several chapters, we explore the molecular structure and function of genetic material—both **deoxyribonucleic acid (DNA)** and **ribonucleic acid (RNA)**—and examine the molecular mechanisms by which genetic information is transmitted from generation to generation. You will see exactly what a gene is, and you will learn how genes are expressed as traits. We begin by recounting how scientists discovered the nature and structure of the genetic material. These discoveries led to an explosion of knowledge about the molecular aspects of biology.

## The Search for the Genetic Material

Long before DNA and RNA were known to carry genetic information, scientists realized that living organisms contain some substance—a genetic material—that is responsible for the characteristics that are passed on from parent to child. Geneticists knew that the material responsible for hereditary information must have three key characteristics:

1. It must contain, in a stable form, *the information* about an organism's cell structure, function, development, and reproduction.
2. It must *replicate accurately*, so that progeny cells have the same genetic information as the parental cell.
3. It must be capable of *change*. Without change, organisms would be incapable of variation and adaptation, and evolution could not occur.

The Swiss biochemist Friedrich Miescher is credited with the discovery, in 1869, of nucleic acid. He isolated a substance from white blood cells of pus in used bandages during the Crimean War. At first he believed the substance to be protein; but chemical tests indicated that it contained carbon, hydrogen, oxygen, nitrogen, and phosphorus, the last of which was not known to be a component of proteins. Searching for the same substance in other sources, Miescher found it in the nucleus of all the samples he studied—and, therefore, he called it *nuclein.* At the time, its function was unknown, and its exact location in the cell was unknown.

In the early 1900s, experiments showed that **chromosomes**—the threadlike structures found in nuclei—are carriers of hereditary information. Chemical analysis over the next 40 years revealed that chromosomes are composed of protein and **nucleic acids,** which by this time were known to include DNA and RNA. At first, many scientists believed that the protein in the chromosomes must be the genetic material. They reasoned that proteins have a great capacity for storing information because they were composed of 20 different amino acids. (Note: Twenty amino acids were known at the time. A twenty-first amino acid was identified in the 1970s, and a twenty-second was identified in 2002.) By contrast, DNA, with its four nucleotides, was thought to be too simple a molecule to account for the variation found in living organisms. However, beginning in the late 1920s, a series of experiments led to the definitive identification of DNA as genetic material.



### Griffith's Transformation Experiment

In 1928, Frederick Griffith, a British medical officer, was working with *Streptococcus pneumoniae* (also called pneumococcus), a bacterium that causes pneumonia (Figure 2.1a). Griffith used two strains of the bacterium: the *S* strain, which produces smooth, shiny colonies and is virulent (highly infectious) (Figure 2.1b); and the *R* strain, which produces rough colonies and is nonvirulent (harmless) (Figure 2.1c). Although this distinction was not known at the time, the virulence of the *S* strain is due to the presence of a polysaccharide coat—a capsule—surrounding each cell. The coat is also the reason for the smooth, shiny appearance of *S* colonies. The *R* strain is genetically identical except that it carries a *mutation* that prevents it from making the polysaccharide coat. A **mutation** is a heritable change in the genetic material (see Chapter 7). In this case, a mutation in a gene affects the ability of the bacterium to make the coat and, hence, alters the virulence state of the bacterium.

There are several types of *S* strains, each with a distinct chemical composition of the polysaccharide coat. Griffith worked with *IIS* and *IIIS* strains, which have type II and type III coats, respectively. Occasionally, *S*-type cells mutate into *R*-type cells, and *R*-type cells mutate into *S*-type cells. The mutations are type-specific—meaning that, if a *IIS* cell mutates into an *R* cell, then that *R* cell can mutate back only into a *IIS* cell, not a *IIIS* cell.

Figure 2.1
**The bacterium *Streptococcus pneumoniae*.**

**a) Electron micrograph showing individual bacteria.**

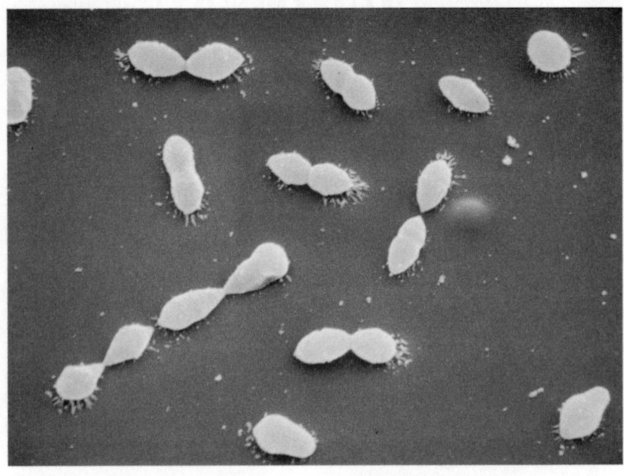

**b) Colonies of *S* (smooth) strain.**

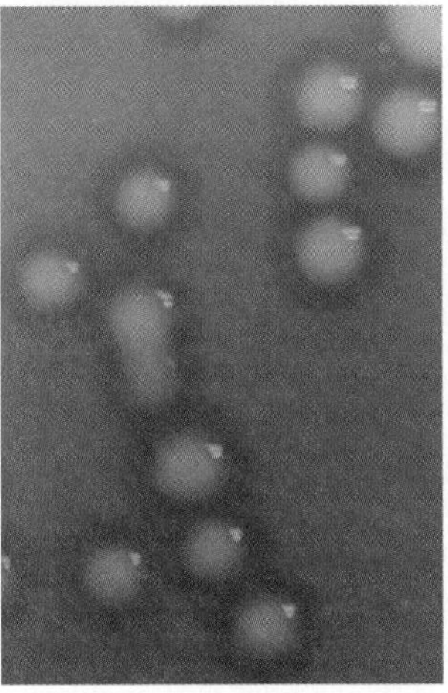

**c) Colonies of *R* (rough) strain.**

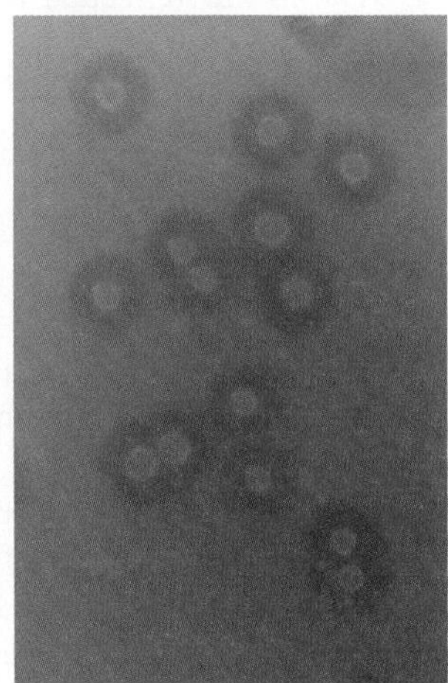

Griffith injected mice with different strains of the bacterium and observed their effects on the mice (Figure 2.2). When mice were injected with *IIR* bacteria (*R* bacteria derived by mutation from *IIS* bacteria), the mice lived. When mice were injected with living *IIIS* bacteria, the mice died, and living *IIIS* bacteria could be isolated from their blood. However, if the *IIIS* bacteria were killed by heat before injection, the mice lived. These experiments showed that the bacteria had both to be alive and to have the polysaccharide coat to be virulent and kill the mice.

In his key experiment, Griffith injected mice with a mixture of living *IIR* bacteria and heat-killed *IIIS* bacteria. The mice died, and living *IIIS* bacteria were present in the blood. These bacteria could not have arisen by mutation of the *R* bacteria, because mutation would have produced *IIS* bacteria. Griffith concluded that some *IIR* bacteria had somehow been *transformed* into smooth, virulent *IIIS* bacteria by interaction with the dead *IIIS* bacteria. Genetic

Figure 2.2

**Griffith's transformation experiment.** Mice injected with *IIIS Streptococcus pneumonia* died, whereas mice injected with either *IIR* or heat-killed *IIIS* bacteria survived. When injected with a mixture of living *IIR* and heat-killed *IIIS* bacteria, however, the mice died.

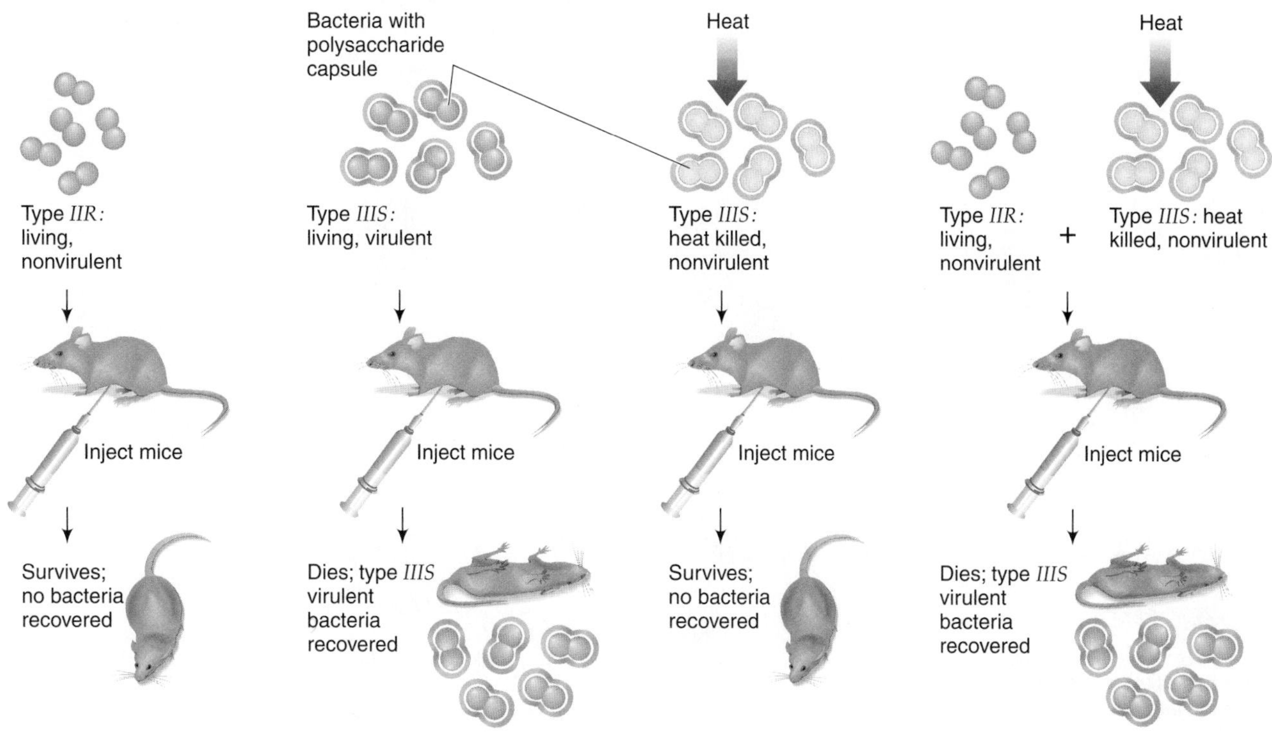




material from the dead *IIIS* bacteria had been added to the genetic material in the living *IIR* bacteria. Griffith believed that the unknown agent responsible for the change in the genetic material was a protein; but this was a hunch, and he turned out to be wrong. He had no experimental evidence one way or the other as to the material acting as the agent bringing about the genetic change. Griffith called this agent the **transforming principle.** (See Chapter 15 for a discussion of bacterial transformation. Importantly, *transformation* is an essential technique used in recombinant DNA experiments; see Chapter 8.)

## Avery's Transformation Experiment

In the 1930s and 1940s, American biologist Oswald T. Avery, along with his colleagues Colin M. MacLeod and Maclyn McCarty, tried to identify Griffith's transforming principle by studying the transformation of *R*-type bacteria to *S*-type bacteria in the test tube.

**animation**

**DNA as Genetic Material: Avery's Transformation Experiment**

They lysed (broke open) *IIIS* cells with a detergent and used a centrifuge to separate the cellular components—the cell extract—from the cellular debris. They incubated the extract with a culture of living *IIR* bacteria and then plated cells on a culture medium in a Petri dish. Colonies of *IIIS* bacteria grew on the plate, showing that the extract contained the transforming principle, the genetic material from *IIIS* bacteria capable of transforming *IIR* bacteria into *IIIS* bacteria. Avery and his colleagues knew that one of the macromolecular components in the extract—polysaccharides, proteins, RNA, or DNA—must be the transforming principle. To determine which, they treated samples of the cell extract with enzymes that could degrade one or more of the macromolecules. After an enzyme treatment, the researchers tested to see if transformation still occurred. They found that the extract failed to bring about transformation only when DNA had been degraded, despite the presence of all other remaining macromolecules in the extract. By contrast, any enzyme treatment that did not lead to digestion of the DNA did not eliminate the transforming principle. These results showed that DNA, and DNA alone, must have been the transforming principle (the genetic material). That is, removing DNA from the cell extract was the only change that could eliminate the ability of the extract to provide the *IIR* bacterium with genetic material.

Figure 2.3 shows a modern version of part of Avery's transformation experiment to illustrate the general approach. The starting point is a mixture of DNA and RNA purified from a cell extract of *IIIS* cells. Samples of the mixture are treated separately with two different kinds of **nucleases,** enzymes that degrade nucleic acids. The samples are then tested to see if they can transform *IIR* bacteria to *IIIS*. For the mixture treated with **ribonuclease**

Figure 2.3

**Experiment showing that DNA, not RNA, is the transforming principle.** When a mixture of DNA and RNA was treated with ribonuclease (RNase) and then added to living *IIR* bacteria, *IIIS* transformants resulted. However, when the DNA and RNA mixture was treated with deoxyribonuclease (DNase) and then added to living *IIR* bacteria, no *IIIS* transformants resulted. (*IIR* colonies are present on each plate in the figure but are not shown for simplicity.)

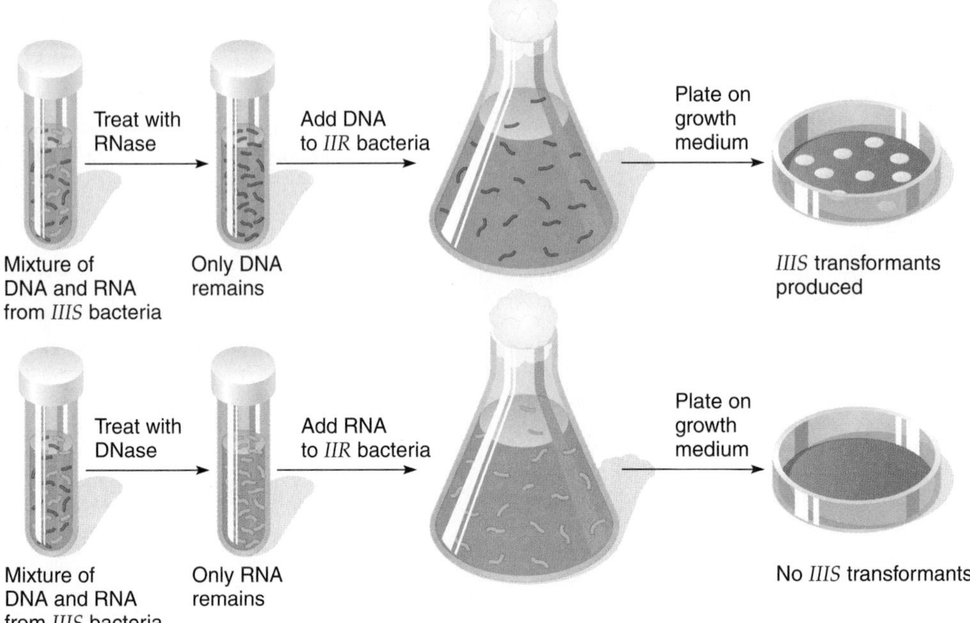


**(RNase),** which degrades RNA and not DNA, DNA is unaffected and *IIIS* transformants resulted. For the mixture treated with **deoxyribonuclease (DNase),** which degrades DNA and not RNA, RNA is unaffected but DNA is digested, and no transformants resulted. The results show that DNA is the transforming principle.

Although Avery and his colleagues' work was important, it was criticized at the time by scientists who were supporters of the hypothesis that protein was the genetic material. These scientists argued that the preparations of the various enzymes the researchers had used were only crudely purified. If proteins were the genetic material, they might have escaped digestion when protein-digesting enzymes were tested, but they might have been digested accidentally when DNases were tested.

## Hershey and Chase's Bacteriophage Experiment

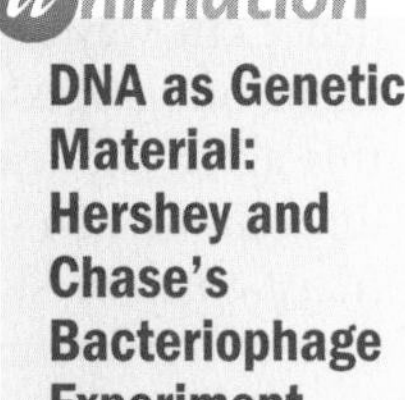


In 1953, Alfred D. Hershey and Martha Chase published a paper that provided more evidence that DNA was the genetic material. They were studying a bacteriophage called T2 (Figure 2.4). **Bacteriophages** (also called **phages**) are viruses that attack bacteria. Like all viruses, the T2

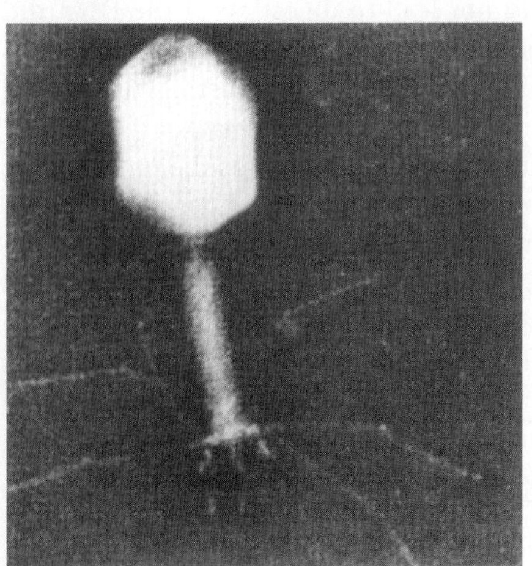

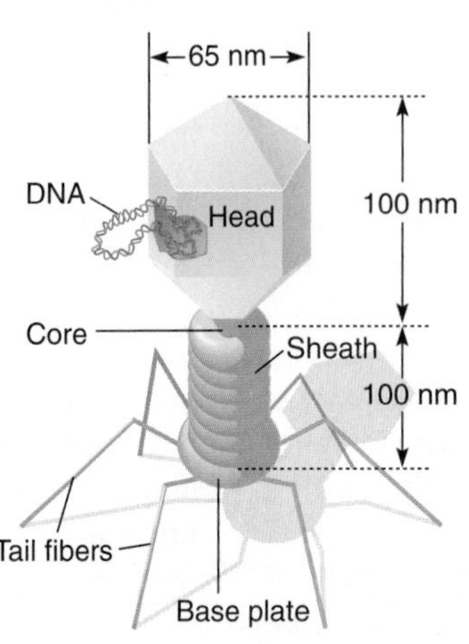


Figure 2.4

**Electron micrograph and diagram of bacteriophage T2 (1 nm = $10^{-9}$ m).**

phage must reproduce within a living cell. T2 reproduces by invading an *Escherichia coli* (*E. coli*) cell and using the bacterium's molecular machinery to make more viruses (Figure 2.5). Initially the progeny viruses are assembled inside the bacterium; but eventually the host cell ruptures, releasing 100–200 progeny phages. The suspension of released progeny phages is called a **phage lysate.** The in which a phage infects a bacterial cell and produces progeny phages that are released from the broken-open bacterium is known as the **lytic cycle.**

Hershey and Chase knew that T2 consisted of only DNA and protein, and their working hypothesis was that the DNA was the genetic material. T2 phages are very simply put together. They have an outer shell that surrounds their genetic material. When they infect a bacterium, they inject their genetic material inside the host cell but leave their outer shell on the surface of the bacterium. Once the genetic material has been injected into the host cell, the empty outer shell that is left is sometimes referred to as a *phage ghost*.

To prove that the phage genetic material was made up of DNA and not protein, Hershey and Chase grew cells of *E. coli* in media containing either a radioactive isotope of phosphorus ($^{32}P$) or a radioactive isotope of sulfur ($^{35}S$) (Figure 2.6a). They used these isotopes because DNA contains phosphorus but no sulfur, and protein contains sulfur but no phosphorus. The *E. coli* took up whichever isotope was provided and incorporated the $^{32}P$ into all the nucleic acids made inside the cell or incorporated the $^{35}S$ into all the proteins made inside the cell. Any phage inside the bacteria would use its host bacterium's nucleic acids and proteins to construct progeny phages. Hershey and Chase then infected the bacteria with T2 and collected the progeny phages. At this point, the researchers had two batches of T2, one with DNA labeled radioactively with $^{32}P$ and the other with protein labeled with $^{35}S$.

Next, they infected two cultures of *E. coli* with one or other of the two types of radioactively labeled T2 (Figure 2.6b). When the infecting phage was $^{32}P$-labeled, most of the radioactivity was found within the bacteria soon after infection. Very little was found in the phage ghosts released from the cell surface after the cells were agitated in a kitchen blender. After completion of the lytic cycle, some of the $^{32}P$ was found in the progeny phages. In contrast, after *E. coli* were infected with $^{35}S$-labeled T2, almost none of the radioactivity appeared within the cell or in the progeny phage particles, while most of the radioactivity was in the phage ghosts.

Hershey and Chase reasoned that, because *it was DNA and not protein that entered the cell*—as evidenced by the presence of $^{32}P$ and the absence of $^{35}S$ inside the bacterial cells immediately after the phage had begun the infection process by injecting their genetic material inside their host

**Figure 2.5**
**Lytic life cycle of a virulent phage, such as T2.**

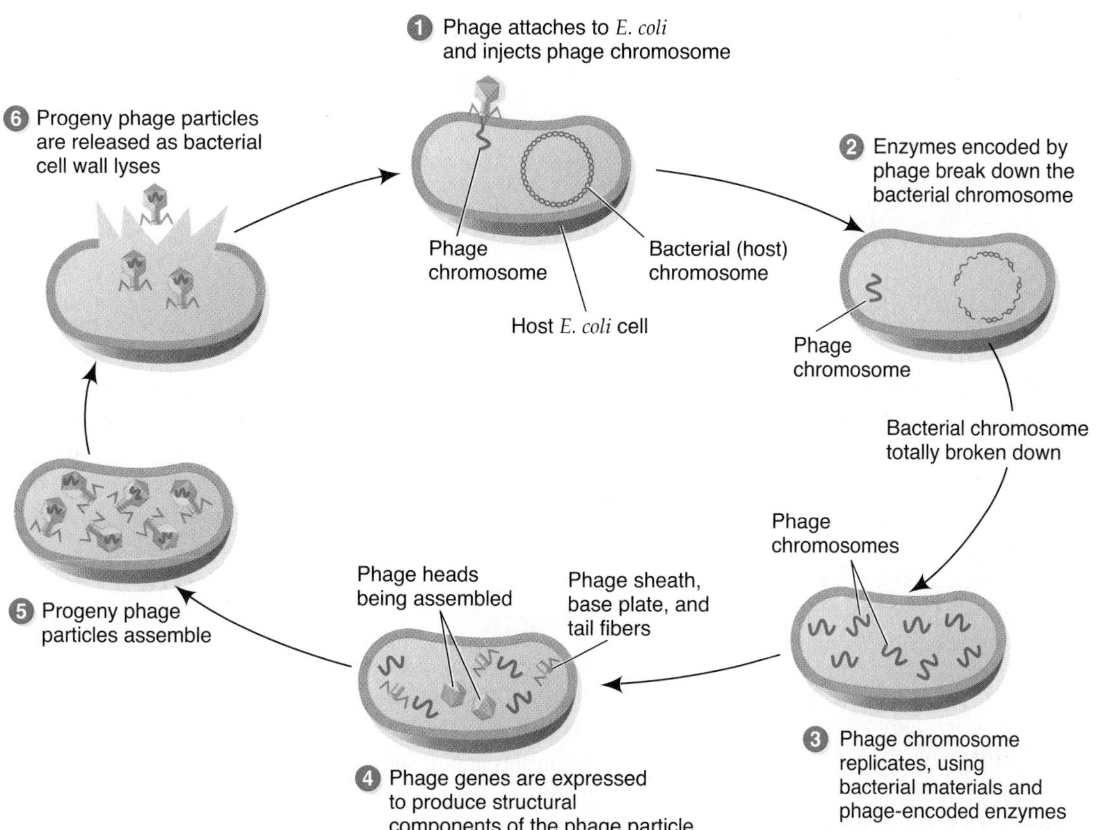
1 Phage attaches to E. coli and injects phage chromosome
Phage chromosome
Bacterial (host) chromosome
Host E. coli cell
2 Enzymes encoded by phage break down the bacterial chromosome
Phage chromosome
Bacterial chromosome totally broken down
Phage chromosomes
3 Phage chromosome replicates, using bacterial materials and phage-encoded enzymes
Phage heads being assembled
Phage sheath, base plate, and tail fibers
4 Phage genes are expressed to produce structural components of the phage particle
5 Progeny phage particles assemble
6 Progeny phage particles are released as bacterial cell wall lyses

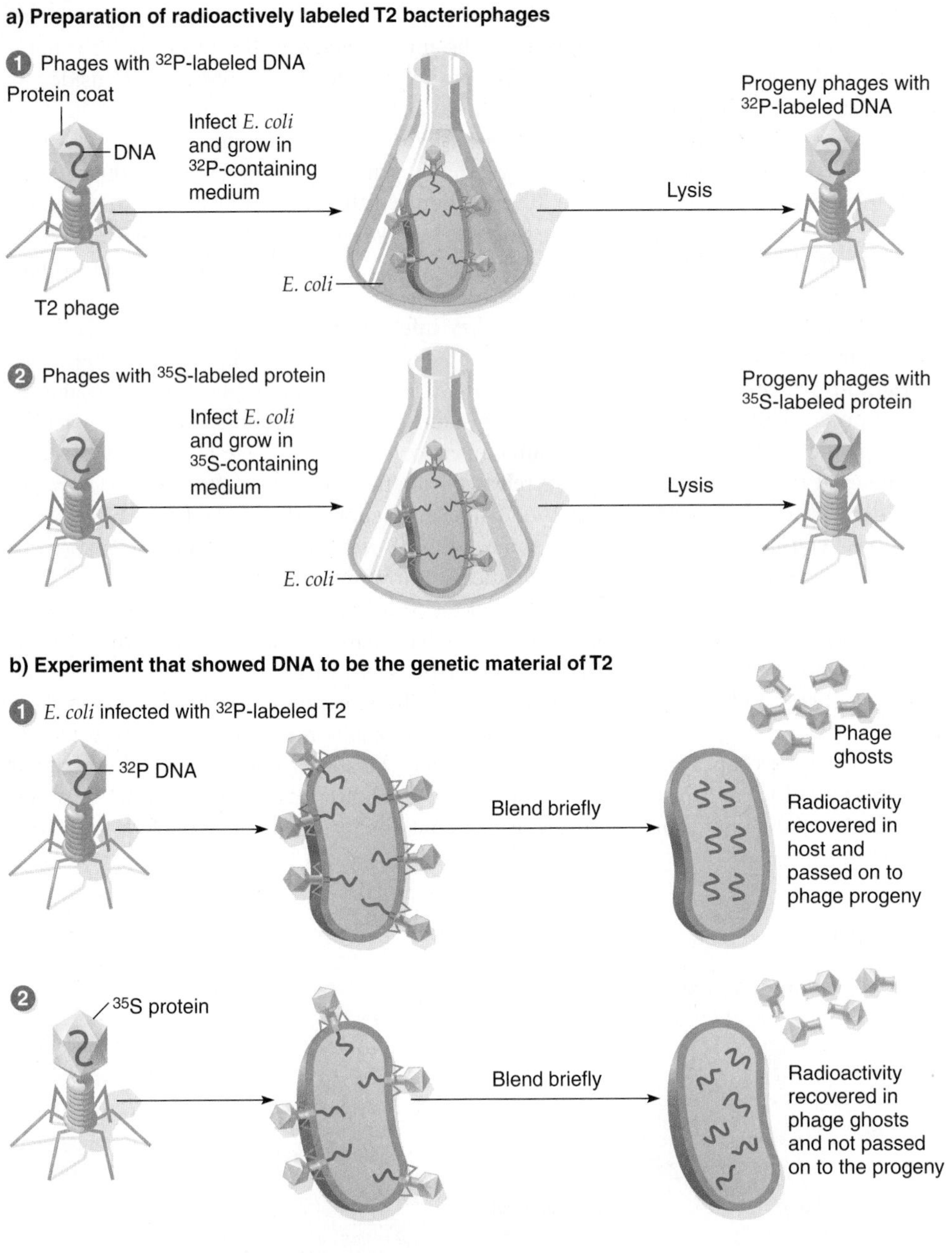

**Figure 2.6**
**The Hershey and Chase experiment.**

cells—*DNA must be the material responsible for the function and reproduction of phage T2*. That is, DNA must be the genetic material of phage T2. This was also consistent with the finding that $^{32}P$ but not $^{35}S$ was found in the progeny phages, because the phage genetic material inside the host cells would be partially repackaged in the progeny phages being assembled during the infection process. Only genetic material (DNA) is passed from parent to offspring in phage reproduction. Structural materials (the proteins) are not.

Alfred Hershey shared the 1969 Nobel Prize in Physiology or Medicine for his "discoveries concerning the genetic structure of viruses."

## RNA as Viral Genetic Material

All organisms and many viruses discussed in this book (such as a human, *Drosophila*, yeast, *E. coli*, and bacteriophage T2) have DNA as their genetic material. However, some bacteriophages (for example, MS2 and Qβ), a number of animal viruses (for instance, poliovirus and human immunodeficiency virus, HIV), and a number of plant viruses (such as tobacco mosaic virus and barley yellow dwarf virus) have RNA as their genetic material. No known prokaryotic or eukaryotic organism has RNA as its genetic material.

### Keynote

A series of experiments proved that the genetic material consists of one of two types of nucleic acids: DNA or RNA. Of the two, DNA is the genetic material of all living organisms and of some viruses, and RNA is the genetic material of the remaining viruses.

## The Composition and Structure of DNA and RNA

What is the molecular structure of DNA? DNA and RNA are *polymers*—large molecules that consist of many similar smaller molecules, called *monomers*, linked together. The monomers that make up DNA and RNA are **nucleotides.** Each nucleotide consists of a **pentose** (five-carbon) **sugar,** a **nitrogenous** (nitrogen-containing) **base** (usually just called a **base**), and a **phosphate group.**

In DNA, the pentose sugar is **deoxyribose,** and in RNA it is **ribose** (Figure 2.7). The two sugars differ by the chemical groups attached to the 2′ carbon: a hydrogen atom (H) in deoxyribose and a hydroxyl group (OH) in ribose. (The carbon atoms in the pentose sugar are numbered 1′ to 5′ to distinguish them from the numbered carbon and nitrogen atoms in the rings of the bases.)

There are two classes of nitrogenous bases: the **purines,** which are nine-membered, double-ringed structures, and the **pyrimidines,** which are six-membered, single-ringed structures. There are two purines—**adenine** (A) and **guanine** (G)—and three different pyrimidines—**thymine** (T), **cytosine** (C), and **uracil** (U) in DNA and RNA. The chemical structures of the five bases are shown in Figure 2.8 (The carbons and nitrogens of the purine rings are numbered 1 to 9, and those of the pyrimidines are numbered 1 to 6.) Both DNA and RNA contain adenine, guanine, and cytosine; however, thymine is found only in DNA, and uracil is found only in RNA.

In DNA and RNA, bases are covalently attached to the 1′ carbon of the pentose sugar. The purine bases are bonded at the 9 nitrogen, and the pyrimidines bond at the 1 nitrogen. The combination of a sugar and a base is called a **nucleoside.** Addition of a phosphate group ($PO_4^{2-}$) to a nucleoside yields a **nucleoside phosphate,** which is one kind of nucleotide. The phosphate group is attached to the 5′ carbon of the sugar in both DNA and RNA. Examples of a DNA nucleotide (a **deoxyribonucleotide**) and an RNA nucleotide (a **ribonucleotide**) are shown in Figure 2.9a. A complete list of the names of the bases, nucleosides, and nucleotides is in Table 2.1.

**Figure 2.7**

**Structures of deoxyribose and ribose, the pentose sugars of DNA and RNA, respectively.** The difference between the two sugars is highlighted.

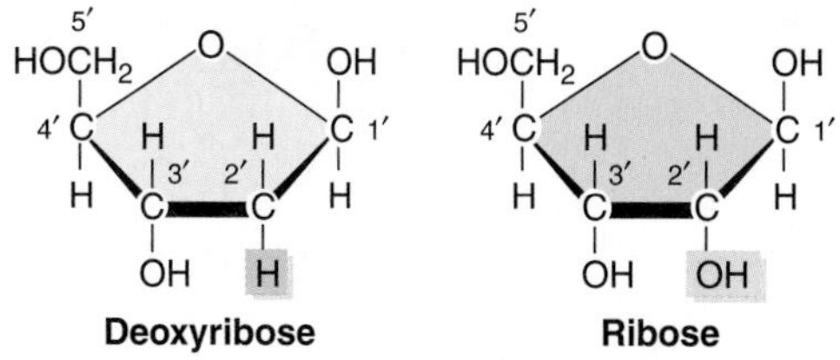

To form **polynucleotides** of either DNA or RNA, nucleotides are linked together by a covalent bond between the phosphate group of one nucleotide and the 3′ carbon of the sugar of another nucleotide. These 5′-to-3′ phosphate linkages are called **phosphodiester bonds.** The phosphodiester bonds are relatively strong, so the repeated sugar–phosphate–sugar–phosphate backbone of DNA and RNA is a stable structure. A short polynucleotide chain is diagrammed in Figure 2.9b.

Polynucleotide chains have *polarity*, meaning that the two ends are different: there is a 5′ carbon (with a phosphate group on it) at one end, and a 3′ carbon (with a hydroxyl group on it) at the other end (Figure 2.9b). The ends of a polynucleotide are routinely referred to as the 5′ end and the 3′ end.

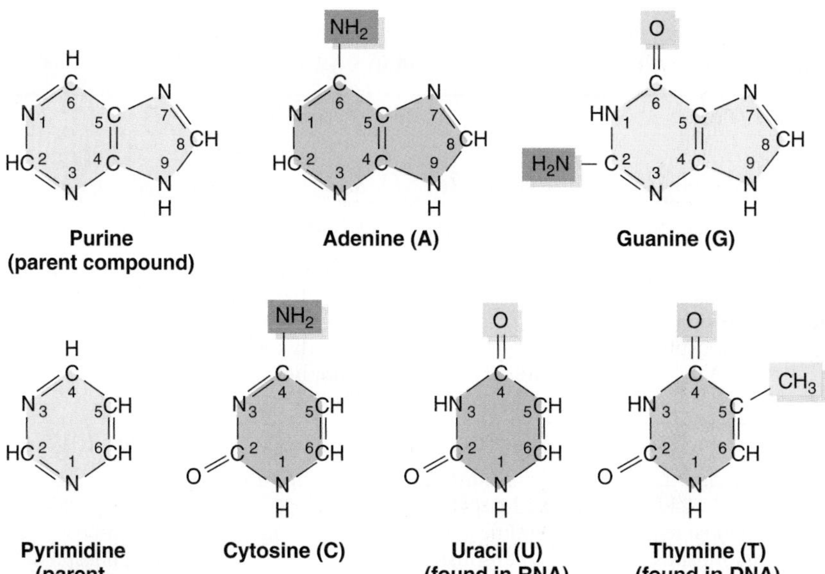


**Figure 2.8**

**Structures of the nitrogenous bases in DNA and RNA.** The parent compounds are purine (top left) and pyrimidine (bottom left). Differences between the bases are highlighted.

**a) DNA and RNA nucleotides**

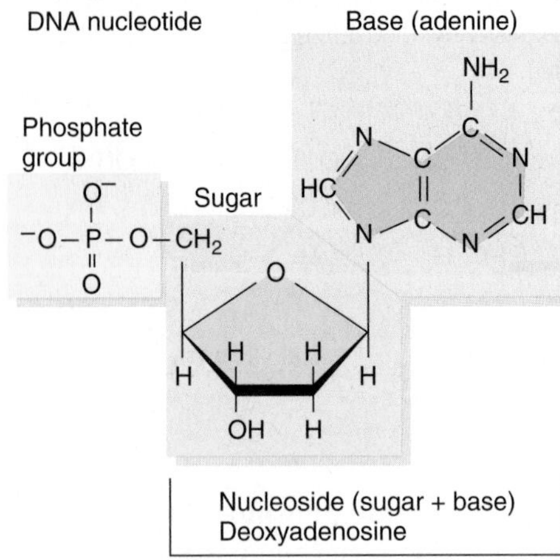




**b) DNA polynucleotide chain**

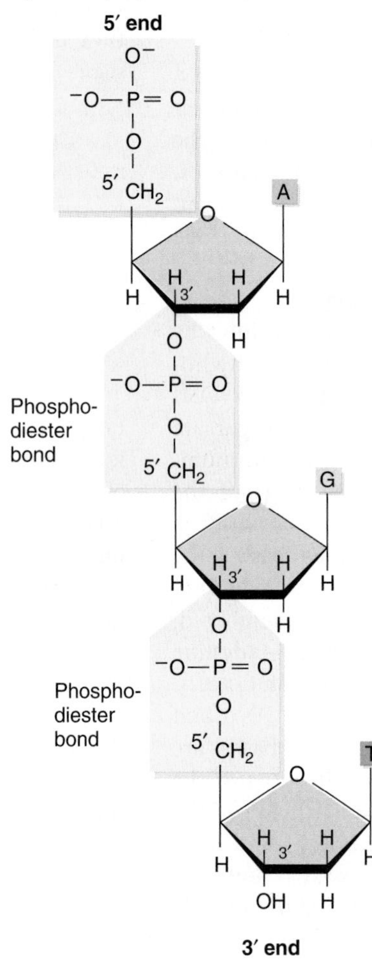


**Figure 2.9**

**Chemical structures of DNA and RNA.** **(a)** Basic structures of DNA and RNA nucleosides (sugar plus base) and nucleotides (sugar, plus base, plus phosphate group), the fundamental building blocks of DNA and RNA molecules. Here, the phosphate groups are yellow, the sugars are lavender, and the bases are peach. **(b)** A segment of a polynucleotide chain, in this case a single strand of DNA. The deoxyribose sugars are linked by phosphodiester bonds (shaded) between the 3′ carbon of one sugar and the 5′ carbon of the next sugar.

**Table 2.1 Names of the Base, Nucleoside, and Nucleotide Components Found in DNA and RNA**

| | | Base: Purines (Pu) | | Base: Pyrimidines (Py) | | |
|---|---|---|---|---|---|---|
| | | **Adenine (A)** | **Guanine (G)** | **Cytosine (C)** | **Thymine (T) (deoxyribose only)** | **Uracil (U) (ribose only)** |
| DNA | Nucleoside: deoxyribose + base | Deoxyadenosine (dA) | Deoxyguanosine (dG) | Deoxycytidine (dC) | Deoxythymidine (dT) | |
| | Nucleotide: deoxyribose + base + phosphate group | Deoxyadenylic acid or deoxyadenosine monophosphate (dAMP) | Deoxyguanylic acid or deoxyguanosine monophosphate (dGMP) | Deoxycytidylic acid or deoxycytidine monophosphate (dCMP) | Deoxythymidylic acid or Deoxythymidine monophosphate (dTMP) | |
| RNA | Nucleoside: ribose + base | Adenosine (A) | Guanosine (G) | Cytidine (C) | | Uridine (U) |
| | Nucleotide: ribose + base + phosphate group | Adenylic acid or adenosine monophosphate (AMP) | Guanylic acid or guanosine monophosphate (GMP) | Cytidylic acid or cytidine monophosphate (CMP) | | Uridylic acid or uridine monophosphate (UMP) |

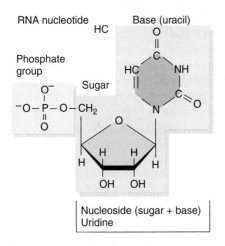

**Keynote**

DNA and RNA occur in nature as macromolecules composed of smaller building blocks called nucleotides. Each nucleotide consists of a five-carbon sugar (deoxyribose in DNA, ribose in RNA) to which is attached a phosphate group and one of four nitrogenous bases: adenine, guanine, cytosine, and thymine (in DNA) or adenine, guanine, cytosine, and uracil (in RNA).

## The DNA Double Helix

In 1953, James D. Watson and Francis H. C. Crick (Figure 2.10) proposed a model for the physical and chemical structure of the DNA molecule. The model they devised, which fit all the known data on the composition of the DNA molecule, is the now-famous double helix model for DNA. The determination of the structure of DNA was a momentous occasion in biology, leading directly to our present molecular understanding of life.

At the time of Watson and Crick's work, DNA was known to be composed of nucleotides. However, it was not known how the nucleotides formed the structure of DNA. Watson and Crick thought that understanding the structure of DNA would help determine how DNA acts as the genetic basis for living organisms. The data they used to help generate their model came primarily from base composition studies conducted by Erwin Chargaff, and X-ray diffraction studies conducted by Rosalind Franklin and Maurice H. F. Wilkins.

**Base Composition Studies.** By chemical treatment, Erwin Chargaff hydrolyzed the DNA of a number of organisms and quantified the purines and pyrimidines released. His studies showed that 50% of the bases were purines and 50% were pyrimidines. More important, the amount of adenine (A) was equal to that of thymine (T), and the amount of guanine (G) was equal to that of cytosine (C). These equivalencies have become known as Chargaff's rules. In comparisons of DNAs from different organisms, the A/T ratio is 1 and the G/C ratio is 1, but the (A + T)/(G + C) ratio (typically denoted %GC) varies. Because the amount of purines equals the amount of pyrimidines, the (A + G)/(C + T) ratio is 1 (see Table 2.2).

**Figure 2.10**

**James Watson (left) and Francis Crick (right) in 1953 with the model of DNA structure.**

**X-Ray Diffraction Studies.** Rosalind Franklin, working with Maurice H. F. Wilkins (Figure 2.11a), studied concentrated solutions of DNA pulled out into thin fibers. The analysis technique they used was X-ray diffraction, in which a beam of parallel X-rays is aimed at molecules. The beam is diffracted (broken up) by the atoms in a pattern that is characteristic of the atomic weight and the spatial arrangement of the molecules. The diffracted X-rays are recorded on a photographic plate (Figure 2.11b). By analyzing the photographs, Franklin obtained information about the molecule's atomic structure. In particular, she concluded that DNA is a helical structure with two distinctive regularities of 0.34 nm and 3.4 nm along the axis of the molecule (1 nanometer [nm] = $10^{-9}$ meter = 10 angstrom units [Å]; 1 Å = $10^{-10}$ meter).

**Watson and Crick's Model.** Watson and Crick used some of Franklin's data and some intelligent guesses of their own to build three-dimensional models of the structure of DNA. Figure 2.12a shows a three-dimensional model of the DNA molecule, and Figure 2.12b is a diagram of the same molecule, showing the arrangement of the sugar–phosphate backbone and base pairs in a stylized way. Figure 2.12c shows the chemical structure of double-stranded DNA.

**Table 2.2 Base Compositions of DNAs from Various Organisms**

| | Percentage of Base in DNA | | | | Ratios | | |
|---|---|---|---|---|---|---|---|
| **DNA origin** | A | T | G | C | A/T | G/C | (A + T)/(G + C) |
| Human (sperm) | 31.0 | 31.5 | 19.1 | 18.4 | 0.98 | 1.03 | 1.67 |
| Corn (*Zea mays*) | 25.6 | 25.3 | 24.5 | 24.6 | 1.01 | 1.00 | 1.04 |
| *Drosophila* | 27.3 | 27.6 | 22.5 | 22.5 | 0.99 | 1.00 | 1.22 |
| *Euglena* nucleus | 22.6 | 24.4 | 27.7 | 25.8 | 0.93 | 1.07 | 0.88 |
| *Escherichia coli* | 26.1 | 23.9 | 24.9 | 25.1 | 1.09 | 0.99 | 1.00 |

**Figure 2.11**

**X-ray diffraction analysis of DNA.** (a) Rosalind Franklin and Maurice H. F. Wilkins (photographed in 1962, the year he received the Nobel Prize shared with Watson and Crick). (b) The X-ray diffraction pattern of DNA that Watson and Crick used in developing their double helix model. The dark areas that form an X shape in the center of the photograph indicate the helical nature of DNA. The dark crescents at the top and bottom of the photograph indicate the 0.34-nm distance between the base pairs.

a) Rosalind Franklin

Maurice H. F. Wilkins

b) X-ray diffraction method

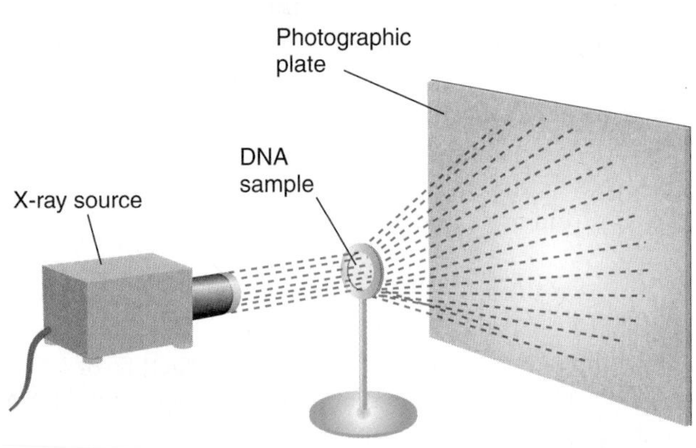


X-ray diffraction pattern

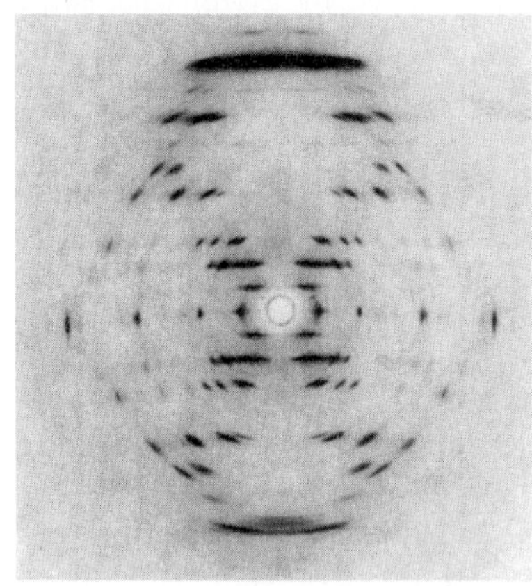

Watson and Crick's double helix model of DNA based on the X-ray crystallography data has the following main features:

1. The DNA molecule consists of two polynucleotide chains wound around each other in a right-handed double helix; that is, viewed on end (from either end), the two strands wind around each other in a clockwise (right-handed) fashion.
2. The two chains are **antiparallel** (show *opposite polarity*); that is, the two strands are oriented in opposite directions, with one strand oriented in the 5′-to-3′ way and the other strand oriented 3′ to 5′. More simply if the 5′ end is the "head" of the chain and the 3′ end is the "tail," *antiparallel* means that the head of one chain is against the tail of the other chain, and vice versa.
3. The sugar–phosphate backbones are on the outsides of the double helix, with the bases oriented toward the central axis (see Figure 2.12). The bases of both chains are flat structures oriented perpendicularly to the long axis of the DNA so that they are stacked like pennies on top of one another, following the twist of the helix.
4. The bases in each of the two polynucleotide chains are bonded together by hydrogen bonds, which are relatively weak chemical bonds. The specific pairings observed are A bonded with T (two hydrogen bonds; Figure 2.13a) and G bonded with C (three hydrogen bonds; Figure 2.13b). The hydrogen bonds make it

**Figure 2.12**
**Molecular structure of DNA.**

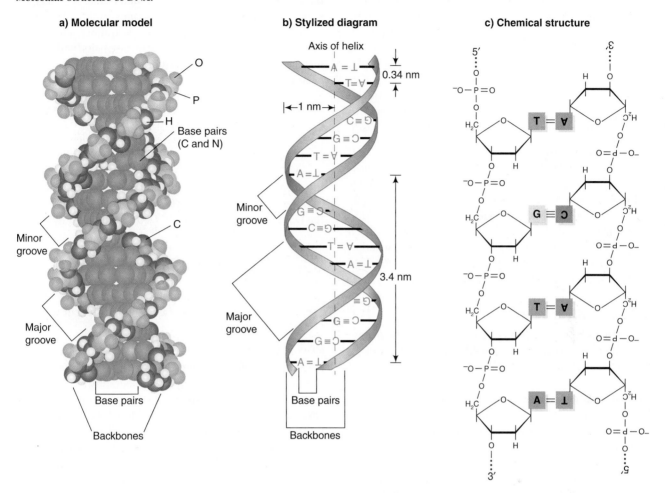

a) Molecular model

b) Stylized diagram

c) Chemical structure

relatively easy to separate the two strands of the DNA—for example, by heating. The A-T and G-C base pairs are the only ones that can fit the physical dimensions of the helical model, and their arrangement is in accord with Chargaff's rules. The specific A-T and G-C pairs are called **complementary base pairs,** so the nucleotide sequence in one strand dictates the nucleotide sequence of the other. For instance, if one chain has the sequence 5′-TATTCCGA-3′, then the opposite, antiparallel chain must bear the sequence 3′-ATAAGGCT-5′.

5. The base pairs are 0.34 nm apart in the DNA helix. A complete (360°) turn of the helix takes 3.4 nm; therefore, there are 10 base pairs (bp) per turn. The external diameter of the helix is 2 nm.

**Figure 2.13**
**Structures of the complementary base pairs found in DNA.** In both cases, a pyrimidine (left) pairs with a purine (right).

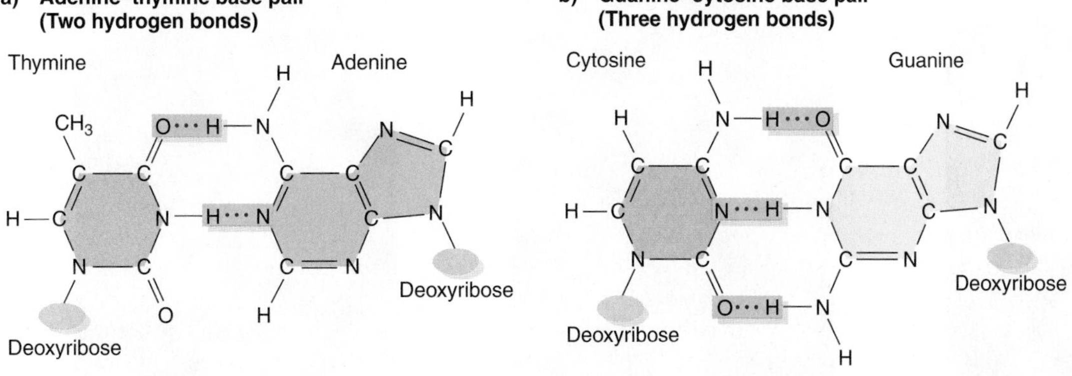

6. Because of the way the bases bond with each other, the two sugar–phosphate backbones of the double helix are not equally spaced from one another along the helical axis. This unequal spacing results in grooves of unequal size between the backbones; one groove is called the *major* (wider) *groove*, the other the *minor* (narrower) *groove* (see Figure 2.12a). The edges of the base pairs are exposed in the grooves, and both grooves are large enough to allow particular protein molecules to make contact with the bases.

For their "discoveries concerning the molecular structure of nucleic acids and its significance for information transfer in living material," the 1962 Nobel Prize in Physiology or Medicine was awarded to Francis Crick, James Watson, and Maurice Wilkins. What was Rosalind Franklin's contribution to the discovery? This has been the subject of debate, and we will never know whether she would have shared the prize. She died in 1962, and Nobel Prizes are never awarded posthumously.

## Different DNA Structures

Researchers have now shown that DNA can exist in several different forms—most notably, the A-, B-, and Z-DNA forms (Figure 2.14).

**A-DNA and B-DNA.** Early X-ray crystallography analysis of DNA fibers identified A-DNA and B-DNA, both of which are right-handed double helices with 11 and 10 bp per turn of the helix, respectively. A-DNA is seen only in conditions of low humidity. The A-DNA double helix is short and wide (diameter 2.2 nm) with a narrow, very deep major groove and a wide, shallow minor groove. (Think of these descriptions in terms of canyons: *narrow* and *wide* describe the distance from rim to rim, and *shallow* and *deep* describe the distance from the rim down to the bottom of the canyon.) B-DNA forms under conditions of high humidity and is the structure that most closely corresponds to that of DNA in the cell. The B-DNA double helix is thinner and longer than A-DNA for the same number of base pairs, with a wide major groove and a narrow minor groove; both grooves are of similar depths. B-DNA is 2 nm in diameter.

**Z-DNA.** DNA with alternating purine and pyrimidine bases can organize into left-handed as well as right-handed helices. The left-handed helix has a zigzag arrangement of the sugar–phosphate backbone, giving this helix form the name Z-DNA. Z-DNA has 12.0 bp per complete helical turn. The Z-DNA helix is thin and elongated, with a deep minor groove. The major groove is very near the surface of the helix, so it is not distinct. Z-DNA is 1.8 nm in diameter.

**i Activity**

Now, determine the molecular composition and structure of a virus infecting the rice crops of Asia. Go to the iActivity *Cracking a Viral Code* on the student website.

## DNA in the Cell

DNA in the cell is in solution, which is a different state from the DNA used in X-ray crystallography experiments. Experiments have shown that DNA in solution has 10.5 base pairs per turn, which is a little less twisted than B-DNA. Structure-wise, DNA in the cell most closely resembles B-DNA, and most of the genome is in that form. In certain DNA–protein complexes, though, the DNA assumes the A-DNA structure. Whether Z-DNA exists in cells has long been a topic of debate among scientists. In those organisms where there is some evidence for Z-DNA, its physiological significance is unknown.

**Figure 2.14**
**Space-filling models of different forms of DNA.**

a) A-DNA

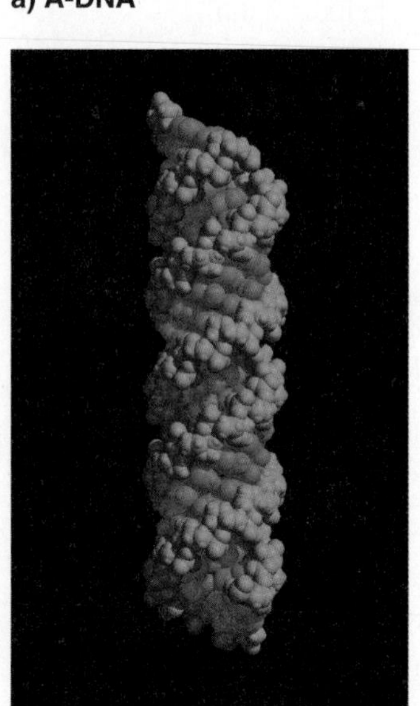

b) B-DNA

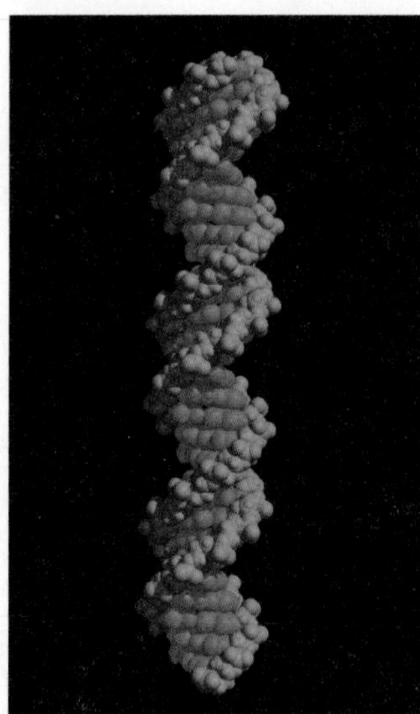

c) Z-DNA

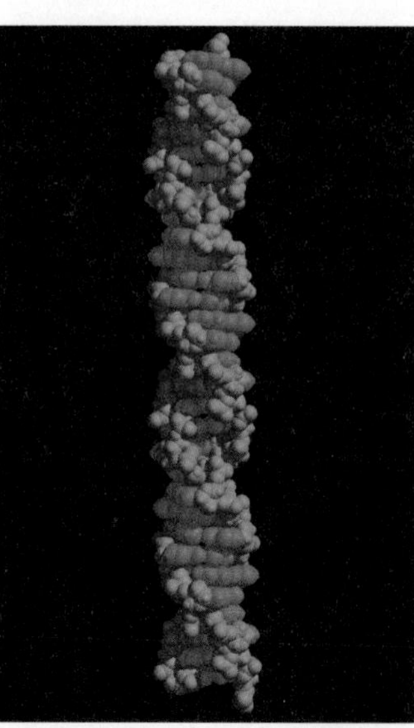

### RNA Structure

RNA is molecularly similar to DNA, differing in having ribose as the sugar rather than deoxyribose, and uracil (U) as a pyrimidine base instead of thymine.

In the cell, the functional forms of RNA such as messenger RNA (mRNA), transfer RNA (tRNA), ribosomal RNA (rRNA), small nuclear RNA (snRNA), and micro RNA (miRNA) are single-stranded molecules. However, these molecules are not stiff, linear rods. Rather, wherever bases can pair together, they will do so. This means that a single-stranded RNA molecule will fold up on itself to produce regions of antiparallel double-stranded RNA separated by segments of unpaired RNA. This configuration is called the secondary structure of the molecule.

Single-stranded RNA and double-stranded RNA molecules are the genomes of certain viruses. Double-stranded RNA has a structure similar to that of double-stranded DNA, with antiparallel strands, the sugar–phosphate backbones on the outside of the helical molecule, and complementary base pairs formed by hydrogen bonding in the middle of the helix.

#### Keynote

The DNA molecule consists of two polynucleotide chains joined by hydrogen bonds between A and T, and between G and C, in a double helix. The three major types of DNA determined by analyzing DNA fibers and crystals in vitro are the right-handed A- and B-DNAs and the left-handed Z-DNA. The common form of DNA in cells is closest in structure to B-DNA. RNA is molecularly similar that of DNA but more typically is single stranded.

## The Organization of DNA in Chromosomes

A **genome** is the full amount of genetic material found in a virus, a prokaryotic cell, a eukaryotic organelle, or in one haploid set of a haploid organism's chromosomes. In viruses, the genome may be DNA or RNA, and found in one or more pieces. In prokaryotes, the genome is usually, but not always, a single circular chromosome of DNA. In eukaryotes, the organelles—mitochondria (in all eukaryotes) and chloroplasts (in plants)—contain a single genome consisting of DNA. The main genome of eukaryotes is typically distributed among the haploid set of chromosomes in the cell nucleus. Haploid eukaryotes have one copy of the genome, whereas diploid eukaryotes have two copies of the genome. To understand the process by which the information within a gene is accessed (see Chapter 5), it is important to understand how DNA is organized in chromosomes. In the sections that follow, we discuss the organization of DNA molecules in chromosomes of viruses, prokaryotes, and eukaryotes.

### Viral Chromosomes

Depending on the virus, the genetic material may be double-stranded DNA, single-stranded DNA, double-stranded RNA, or single-stranded RNA, and it may be circular or linear. The genomes of some viruses are organized into a single chromosome, whereas other viruses have a segmented genome: The genome is distributed among a number of DNA molecules.

T2 (one of the T-even bacteriophages, which also includes T4 and T6), herpesviruses, and gemini virus are examples of viruses with double-stranded DNA genomes. Parvovirus B19, a cause of infectious redness in children; canine parvovirus, which causes a highly infectious disease in dogs that is particularly severe and often deadly in puppies; and the virulent phage ΦX174 are examples of viruses with single-stranded DNA genomes. The parvoviruses have linear genomes, while ΦX174 has a circular genome. All of these viruses, except gemini virus, have a single chromosome; the genome of the gemini virus can have either one or two DNA molecules, depending on the genus.

Reoviruses, one type of which causes mild infections of the upper respiratory tract in humans, are examples of viruses with double-stranded RNA genomes. Picornaviruses (which include poliovirus) and influenza virus are examples of viruses with single-stranded RNA genomes. The picornavirus genome consists of a single RNA molecule, while the genomes of the other RNA viruses mentioned are segmented. This leads in part to fluidity of the influenza genome and epidemiological concerns about a killer flu strain. Moreover, this viral genome organization necessitates annual flu vaccinations.

### Prokaryotic Chromosomes

Most prokaryotes contain a single, double-stranded, circular DNA chromosome. The remaining prokaryotes have genomes consisting of one or more chromosomes that may be circular or linear. In the latter cases, there is typically a main chromosome and one or more smaller chromosomes. The smaller chromosomes replicate autonomously of the main chromosome and may or may not be essential to the life of the cell. Autonomously replicating small chromosomes not essential to the life of the cell are known as **plasmids.** For example, among the bacteria, *Borrelia burgdorferi*, the causative agent of Lyme disease in humans, has a 0.91-Mb (1 Mb = 1 megabase = 1 million base pairs) linear chromosome and at least 17 small plasmids, some linear and some circular, with a combined size of 0.53 Mb. *Rhizobium radiobacter* (formerly called *Agrobacterium tumefaciens*), the causative agent of crown gall disease in some plants, has a 3.0-Mb circular chromosome and a 2.1-Mb linear chromosome. Among the archaea, chromosome organization also varies, although no linear chromosomes have yet been found. For example, *Methanococcus jannaschii* has a 1.66-Mb circular chromosome, and 58-kb and 16-kb circular plasmids, and *Archaeoglobus fulgidus* has a single 2.2-Mb circular chromosome.

In bacteria and archaea, the chromosome is arranged in a dense clump in a region of the cell known as the **nucleoid.** Unlike the case with eukaryotic nuclei, there is no membrane between the nucleoid region and the rest of the cell.

The *E. coli* genome consists of a single, circular, 4.6-Mb double-stranded DNA molecule, which is approximately 1,100 μm long (approximately 1,000 times the length of the cell). The DNA fits into the nucleoid region of the cell in part because it is **supercoiled;** that is, the double helix is twisted in space about its own axis. The twisted state of the *E. coli* chromosome can be seen if a cell is broken open gently to release its DNA (Figure 2.15).

animation
**DNA Supercoiling**

To understand supercoiling, consider a linear piece of DNA with 20 helical turns (Figure 2.16a). If we simply join the two ends, we have produced a circular DNA molecule that is *relaxed* (Figure 2.16b). If, instead, we first untwist one end of the linear DNA molecule by two turns (Figure 2.16c) and then join the two ends, the circular DNA molecule produced will have 18 helical turns and a small unwound region (Figure 2.16d). Such a structure is not energetically favored and will switch to a structure with 20 helical turns and two superhelical turns—a supercoiled form of DNA (Figure 2.16e).

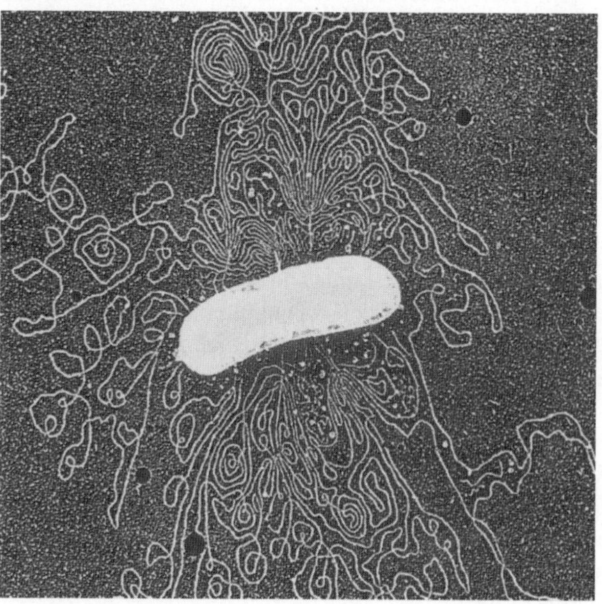

**Figure 2.15**
**Chromosome released from a lysed *E. coli* cell.**

**Figure 2.16**
**Illustration of DNA supercoiling.** (a) Linear DNA with 20 helical turns. (b) Relaxed circular DNA produced by joining the two ends of the linear molecule of (a). (c) The linear DNA molecule of (a) unwound from one end by two helical turns. (d) A possible circular DNA molecule produced by joining the two ends of the linear molecule of (c). The circular molecule has 18 helical turns and a short unwound region. (e) The more energetically favored form of (d), a supercoiled DNA with 20 helical turns and two superhelical turns.

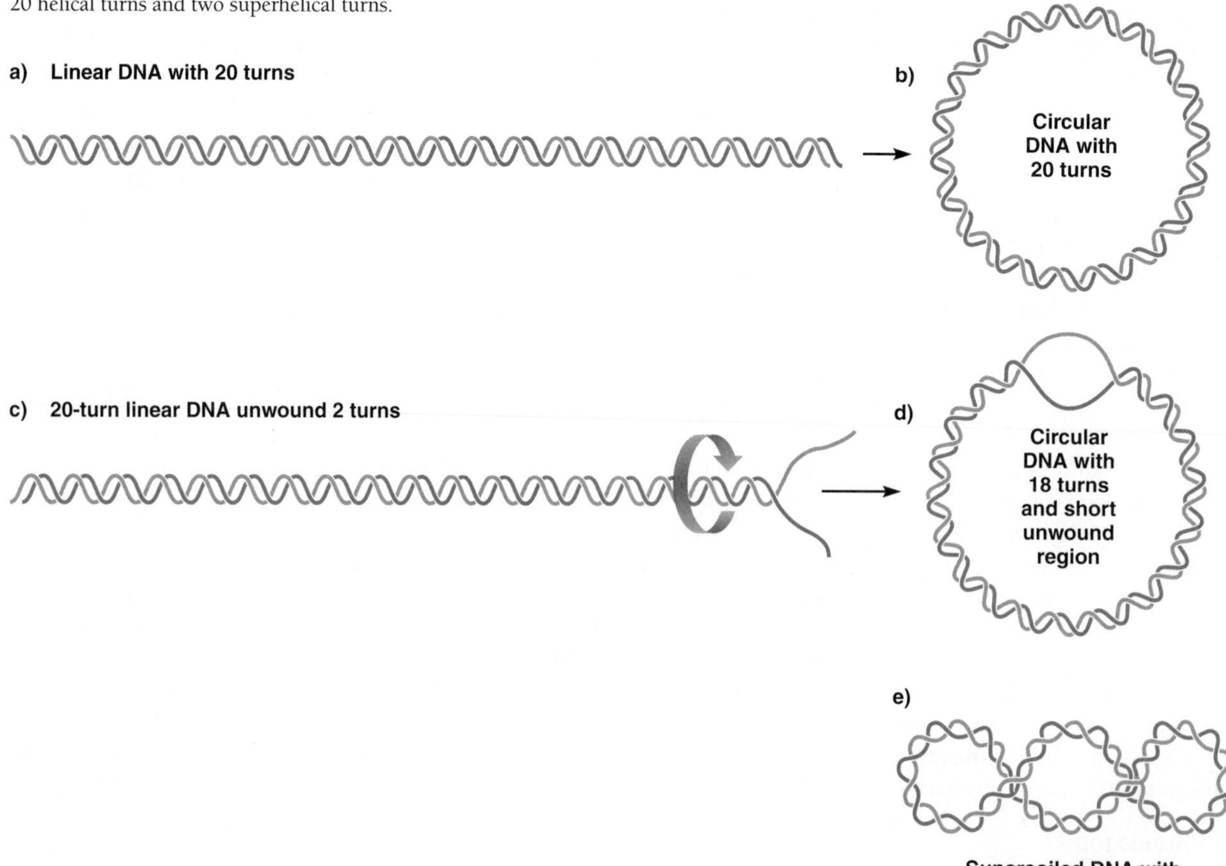

Supercoiling produces tension in the DNA molecule. Therefore, if a break is introduced into one strand of the sugar–phosphate backbone of a supercoiled circular DNA molecule—the single-stranded break is called a *nick*—the molecule spontaneously untwists and produces a relaxed DNA circle. Supercoiling can also occur in a linear DNA molecule. That is, if we twist a length of rope on one end without holding the other end, the rope just spins in the air and remains linear (relaxed). However, with a large, linear DNA molecule, supercoiling occurs in localized regions and the ends behave as if they are fixed.

Figure 2.17 shows relaxed and supercoiled circular DNA to illustrate how much more compact a supercoiled molecule is. There are two types of supercoiling: *negative supercoiling* and *positive supercoiling*. To visualize supercoiling of DNA, think of the DNA double helix as a spiral staircase that turns in a clockwise direction. If you untwist the spiral staircase by one complete turn, *you have the same number of stairs to climb, but you have one less 360° turn to make*; this is a negative supercoil. If, instead, you twist the spiral staircase by one more complete turn, *you have the same number of stairs to climb, but now there is one more 360° turn to make*; this is a positive supercoil. Either type of supercoiling causes the DNA to become more compact. The amount and type of DNA supercoiling is controlled by **topoisomerases**—enzymes that are found in all organisms.

**Figure 2.17**

**Electron micrographs of a circular DNA molecule, showing relaxed (a) and supercoiled (b) states.** Both molecules are shown at the same magnification.

**a) Relaxed circular DNA**

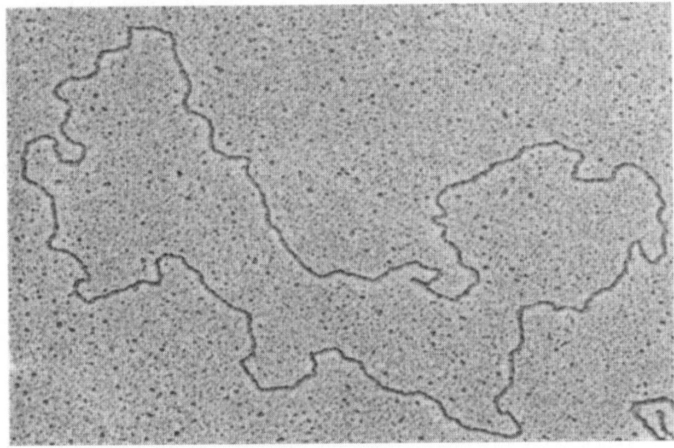

**b) Supercoiled circular DNA**

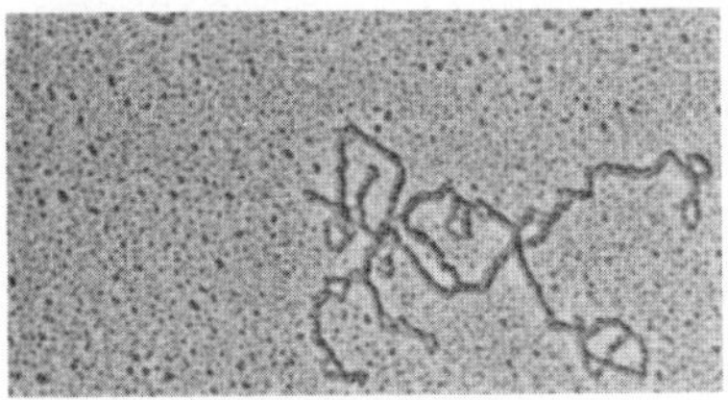

Bacterial chromosomes also become compacted because the DNA is organized into **looped domains** (Figure 2.18). In *E. coli*, there are about 400 domains of negatively supercoiled DNA per chromosome, with variable lengths for each domain. There is debate about exactly what molecules bind to the DNA to establish the domains; more than one protein type certainly is involved, along with possibly some RNA molecules. The compaction achieved by organizing into looped domains is about tenfold.

**Keynote**

Viral genomes may be either double-stranded DNA, single-stranded DNA, double-stranded RNA, or single-stranded RNA. They may be either circular or linear. The genomes of some viruses are organized into a single chromosome, whereas other viruses have a segmented genome. The genetic material of bacteria and archaea is double-stranded DNA localized into one or a few chromosomes. The *E. coli* chromosome is circular and is organized into about 400 independent looped domains of supercoiled DNA.

## Eukaryotic Chromosomes

Eukaryotic genomes typically are distributed among several linear chromosomes, with the number characteristic of each species. The complete set of metaphase chromosomes in a eukaryotic cell is called its **karyotype.** Humans, which are diploid (2N) organisms, have 46 chromosomes (two genomes), with one haploid (N) set of chromosomes (23 chromosomes: one genome) coming from the egg and another haploid set coming from the sperm.

The total amount of DNA in the haploid genome of a species is known as the species' **C-value.** (The "C" was

**Figure 2.18**

**Model for the structure of a bacterial chromosome.** The chromosome is organized into looped domains, the bases of which are anchored in an unknown way.

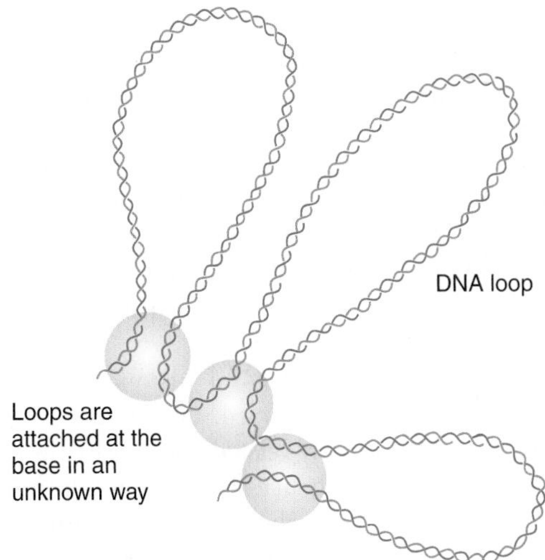
DNA loop
Loops are attached at the base in an unknown way

not defined by the coiner, but it stands for "constant.") Table 2.3 lists the C-values for some selected species. C-value data show that the amount of DNA found among organisms varies widely, and there may or may not be significant variation in the amount between related organisms. For example, mammals, birds, and reptiles show little variation, both across each other and among species within each class, whereas amphibians, insects, and plants vary over a wide range, often tenfold or more. There is also no direct relationship between the C-value and the structural or organizational complexity of the organism, a situation called the *C-value paradox*. For example, the amoeba has almost a hundred times more DNA than a human does. At least one reason for this absence of a direct link is variation in the amount of repetitive sequence DNA in the genome (see this chapter's Focus on Genomics box, as well as pp. 29–30).

As you will learn in Chapter 12 (see pp. 329–330 and Figure 12.4), eukaryotic cells reproduce in a cell cycle consisting of four phases: $G_1$, S, $G_2$, and M. During $G_1$ phase, each chromosome is a single structure. During S phase, the chromosomes duplicate to produce two sister chromatids joined by the duplicated, but not yet separated, centromeres. This state remains during $G_2$. Then, during M phase (mitosis), the centromeres separate and the sister chromatids become known as daughter chromosomes. Keep this cycle clear in your mind when you think about chromosomes. Each eukaryotic chromosome in $G_1$ consists of one linear, double-stranded DNA molecule running throughout its length and complexed with about twice as much protein by weight as DNA. Duplicated chromosomes with two sister chromatids have one linear, double-stranded DNA molecule running the length of each sister chromatid.

**The Structure of Chromatin.** **Chromatin** is the stainable material in a cell nucleus: DNA and proteins. The term is commonly used in descriptions of chromosome structure and function. The fundamental structure of chromatin is essentially identical in all eukaryotes.

**Histones** and **nonhistones** are two major types of proteins associated with DNA in chromatin. Both types of proteins play an important role in determining the physical structure of the chromosome. The histones are the most abundant proteins in chromatin. They are small basic proteins with a net positive charge that facilitates their binding to the negatively charged DNA. Five main types of histones are associated with eukaryotic nuclear DNA: H1, H2A, H2B, H3, and H4. Weight for weight, there is an equal amount of histone and DNA in chromatin.

The amino acid sequences of histones H2A, H2B, H3, and H4 are highly conserved, evolutionarily speaking, even between distantly related species. Evolutionary conservation of these sequences is a strong indicator that histones perform the same basic role in organizing the DNA in the chromosomes of all eukaryotes.

**Table 2.3 Haploid DNA Content, or C-Value, of Selected Species**

| Species | C-Value (bp) |
|---|---|
| **Viruses and Phages** | |
| λ (bacteriophage) | 48,502[a] |
| T4 (bacteriophage) | 168,904[a] |
| Feline leukemia virus (cat virus) | 8,448[a] |
| Simian virus 40 (SV40) | 5,243[a] |
| Human immunodeficiency virus-1 (HIV-1, causative agent of AIDS) | 9,750[a] |
| Measles virus (human virus) | 15,894[a] |
| **Bacteria** | |
| *Bacillus subtilis* | 4,214,814[a] |
| *Borrelia burgdorferi* (Lyme disease spirochete) | 910,724[a] |
| *Carsonella ruddii* | 159,662[a] |
| *Escherichia coli* | 4,639,221[a] |
| *Heliobacter pylori* (bacterium that causes stomach ulcers) | 1,667,867[a] |
| *Neisseria meningitis* | 2,272,351[a] |
| *Mycoplasma genitalium* | 580,076[a] |
| **Archaea** | |
| *Methanococcus jannaschii* | 1,664,970[a] |
| **Eukarya** | |
| *Saccharomyces cerevisiae* (budding yeast; brewer's yeast) | 13,105,020[a] |
| *Schizosaccharomyces pombe* (fission yeast) | 12,590,810[a] |
| *Plasmodium falciparum* (Malaria parasite) | 22,859,790[a] |
| *Lilium formosanum* (lily) | 36,000,000,000 |
| *Zea mays* (maize, corn) | 5,000,000,000 |
| *Oryza sativa* (rice) | 370.792,000[a] |
| *Amoeba proteus* (amoeba) | 290,000,000,000 |
| *Aedes aegypti* (mosquito) | 1,310,900,000[a] |
| *Drosophila melanogaster* (fruit fly) | 132,576,936[a] |
| *Caenorhabditis elegans* (nematode) | 100,269,800[a] |
| *Danio rerio* (zebrafish) | 1,527,000,581[a] |
| *Xenopus laevis* (African clawed frog) | 3,100,000,000 |
| *Mus musculus* (mouse) | 3,420,842,930[a] |
| *Rattus rattus* (rat) | 2,719,924,000[a] |
| *Loxodonta africana* (African elephant) | 3,000,000,000 |
| *Canis familiaris* (dog) | 2,443,707,000[a] |
| *Equus caballus* (horse) | 3,311,000,000 |
| *Macac mulatta* (rhesus macaque) | 3,097,179,960[a] |
| *Pan troglodytes* (chimp) | 3,350,417,645[a] |
| *Homo sapiens* (humans) | 3,253,037,807[a] |

[a]These C-values derive from the complete genome sequence; all others are estimates based on other measurements.

## Focus on Genomics

### Genome Sizes and Repetitive DNA Content

As biologists learned the sizes of haploid organismal genomes (called the C-value), they noticed that genome size tended to be smallest in viruses, larger in prokaryotes, and larger yet in eukaryotes. However, they were surprised that the genome size varied substantially within organismal groups, and it was hard to understand why particular organisms had very large or very small genomes. For instance, the largest known animal genomes are more than 6,000 times larger than the smallest animal genome, and some estimates of the variation in eukaryotic genome sizes suggested that the largest genomes were 40,000 to 200,000 times as large as the smallest eukaryotic genomes. The human genome is neither strikingly small nor large, but is solidly in the middle range of sizes. Even more surprisingly, the genomes of animals are dwarfed by those of other organisms—the largest known animal genomes are far smaller than the genomes of many protists and plants. Our initial expectations that more genes would be required for more complex lives and bodies, and that this would in turn require a larger genome, seemed to conflict with the observed genome sizes. In studying the content of the genomes, we have partially resolved this question. To a great extent, genome size is driven by repetitive DNA content—organisms with larger genomes have more repetitive DNA—while gene number has relatively less to do with genome size. Viruses and bacteria have very little repetitive DNA, but repetitive DNA content in eukaryotes can range from minimal amounts (about 15%) as found in the pufferfish, *Takifugu*, to most of the genome. As we learn more about gene content, we have seen that there is a general increase in gene number with complexity. However, plants tend to have more genes than animals do, and the number of genes in humans is quite similar to what is seen in many other animals.

Histones play a crucial role in chromatin packing. A diploid human cell, for example, has more than 1,400 times as much DNA as does *E. coli*. Without the compacting of the $6 \times 10^9$ bp of DNA in the diploid cell (two genome copies), the DNA of the chromosomes of a single human cell would be more than 2 meters long (about 6.5 feet) if the molecules were placed end to end. Several levels of packing enable chromosomes that would be several millimeters or even centimeters long to fit into a nucleus that is a few micrometers in diameter.

Nonhistones are all the proteins associated with DNA, apart from the histones. Nonhistones are far less abundant than histones. Many nonhistones are acidic proteins—proteins with a net negative charge. Nonhistones include proteins that play a role in the processes of DNA replication, DNA repair, transcription (including gene regulation), and recombination. Each eukaryotic cell has many different nonhistones in the nucleus. In contrast to the histones, the nonhistone proteins differ markedly in number and type from cell type to cell type within an organism, at different times in the same cell type, and from organism to organism.

With the electron microscope, different chromatin structures are seen. The lowest-level structures are seen while reconstituting purified DNA and histones in vitro, and the higher-level structures reflect the extra degrees of packaging necessary to compact the DNA in vivo. The least compact form seen is the **10-nm chromatin fiber,** which has a characteristic "beads-on-a-string" morphology; the beads have a diameter of about 10 nm (Figure 2.19). The beads are **nucleosomes,** the basic structural units of eukaryotic chromatin. A nucleosome is about 11 nm in diameter and consists of a core of eight histone proteins—two each of H2A, H2B, H3, and H4 (Figure 2.20a)—around which a 147-bp segment of DNA is wound about 1.65 times (Figure 2.20b). This configuration serves to compact the DNA by a factor of about six.

Individual nucleosomes are connected by strands of *linker DNA* (see Figures 2.19 and 2.20b). The length of linker DNA varies within and among organisms. Human linker DNA, for example, is 38–53 bp long.

The next level of chromatin condensation is brought about by histone H1. A single molecule of H1 binds both to the linker DNA at one end of the nucleosome and to the middle of the DNA segment wrapped around core histones. The binding of H1 causes the nucleosomal DNA to assume a more regular appearance with a zigzag arrangement (Figure 2.20c). The nucleosomes themselves then compact into a structure about 30 nm in diameter

**Figure 2.19**

**Electron micrograph of unraveled chromatin, showing the nucleosomes in a "beads-on-a-string" morphology.**

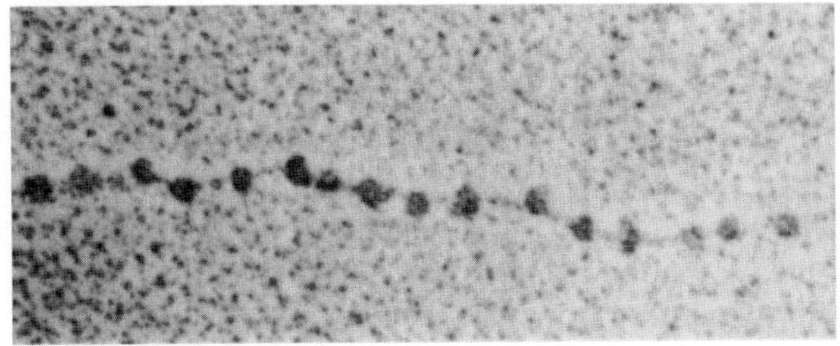

Figure 2.20

**Basic eukaryotic chromosome structure.**

a) Histone core for the nucleosome

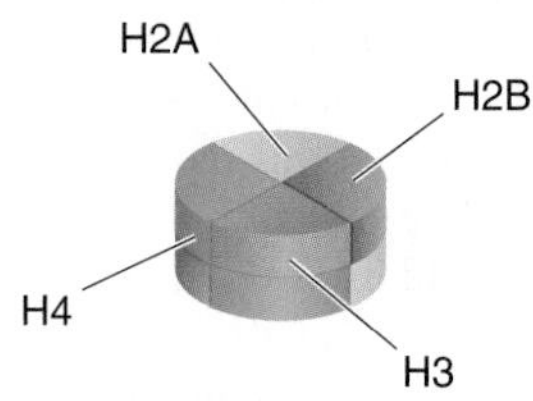


b) Basic nucleosome structure in "beads-on-a-string" chromatin

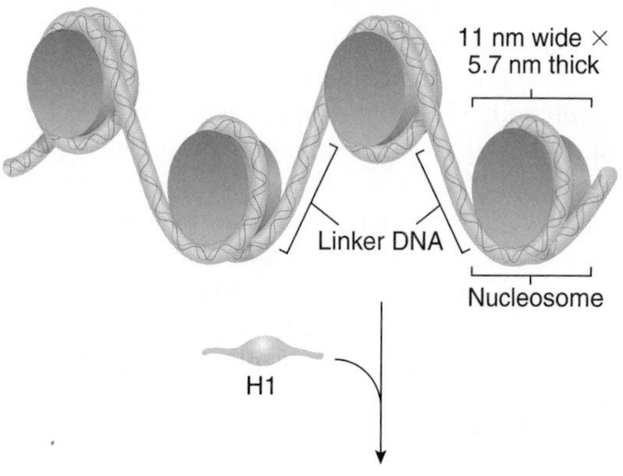


c) Chromatin condensation by H1 binding

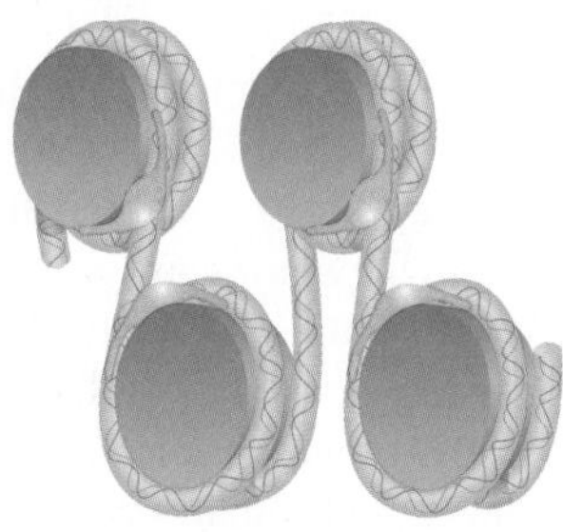

called the **30-nm chromatin fiber** (Figure 2.21a). One possible model for the 30-nm fiber—the *solenoid model*—has the nucleosomes spiraling helically (Figure 2.21b). Another, more recent, model proposes that the 30-nm fiber is an irregular zigzag of nucleosomes.

Chromatin packing beyond the 30-nm chromatin filaments is less well understood. Current models derive from 1970s-vintage electron micrographs of metaphase chromosomes depleted of histones (Figure 2.22). The photos show 30–90-kb loops of DNA attached to a protein "scaffold" with the characteristic X shape of the paired sister chromatids. If the histones are not removed, looped domains of 30-nm fibers are seen. An average human chromosome has approximately 2,000 looped domains.

Each looped domain is held together at its base by nonhistone proteins that are part of the *chromosome scaffold* (Figure 2.23a). Stretches of DNA called *scaffold-associated regions*, or SARs, bind to the nonhistone proteins to determine the loops. It is simplest to think of these loops as being arranged in a spiral fashion around the central chromosome scaffold (Figure 2.23b). In cross section, the loops would be seen to radiate out from the center like the petals of a flower. Overall, this packing produces a chromosome that is about 10,000 times shorter, and about 400 times thicker, than naked DNA.

Figure 2.21

**The 30-nm chromatin fiber.**

a) Electron micrograph of 30-nm chromatin fiber

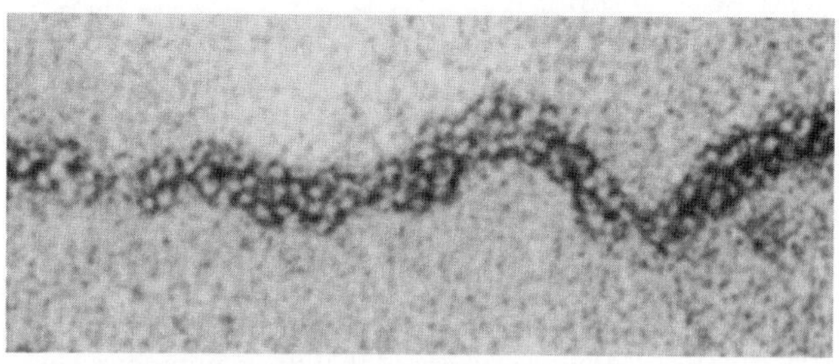

b) Solenoid model for nucleosome packaging in the 30-nm chromatin fiber (H1 is not shown)

Figure 2.22

**Electron micrograph of a metaphase chromosome depleted of histones.** Without histones, the chromosome maintains its general shape by a nonhistone protein scaffold from which loops of DNA protrude (inset).

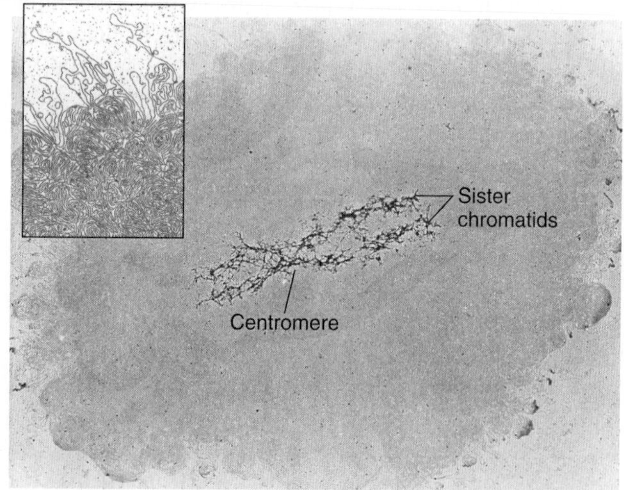

Figure 2.23

**Looped domains in metaphase chromosomes.** (a) Fiber loops 30 nm in diameter attached at scaffold-associated regions to the chromosome scaffold by nonhistone proteins. (b) Schematic of a section of the metaphase chromosome. Shown is the spiraling of looped domains. Eight looped domains are shown per turn for simplification; a more accurate estimate is 15 per turn. With that many looped domains per turn, the 700-nm diameter of the cylindrical chromatid arms of a metaphase chromosome can be accounted for.

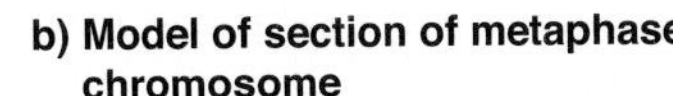

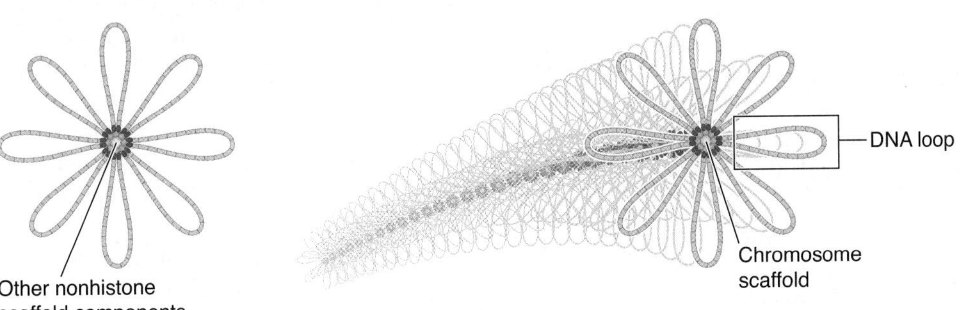


You have just learned the various levels of chromatin packing in eukaryotic chromosomes. However, the chromosomes are not organized into rigid structures. Rather, many regions of the chromosomes have dynamic structures that unpack when genes become active and pack when genes cease their activity.

**Euchromatin and Heterochromatin.** The degree of DNA packing changes throughout the cell cycle. The most dispersed state is when the chromosomes are about to duplicate (beginning of S phase of the cell cycle), and the most highly condensed is within mitosis and meiosis.

Two forms of chromatin are defined, each on the basis of chromosome-staining properties. **Euchromatin** is the chromosomes or regions of chromosomes that show the normal cycle of chromosome condensation and decondensation in the cell cycle. Visually, euchromatin undergoes a change in intensity of staining ranging from the darkest in the middle of mitosis (metaphase stage) to the lightest in the S phase. Most of the genome of an active cell is in the form of euchromatin. Typically, (1) euchromatic DNA is actively transcribed, meaning that the genes within it can be expressed; and (2) euchromatin is devoid of repetitive sequences.

**Heterochromatin,** by contrast, is the chromosomes or chromosomal regions that usually remain condensed—more darkly staining than euchromatin—throughout the cell cycle, even in interphase. Heterochromatic DNA often replicates later than the rest of the DNA in the S phase. Genes within heterochromatic DNA are usually transcriptionally inactive. There are two types of heterochromatin. **Constitutive heterochromatin** is present in all cells at identical positions on both homologous chromosomes of a pair. This form of heterochromatin consists mostly of repetitive DNA and is exemplified by centromeres and telomeres. **Facultative heterochromatin,** by contrast, varies in state in different cell types, and at different developmental stages—or sometimes, from one homologous chromosome to another. This form of heterochromatin represents condensed, and therefore inactivated, segments of euchromatin. The *Barr body*, an inactivated X chromosome in somatic cells of XX mammalian females, is an example of facultative heterochromatin (see Chapter 12, pp. 348–349).

**Keynote**

The nuclear chromosomes of eukaryotes are complexes of DNA, histone proteins, and nonhistone chromosomal proteins. Each chromosome consists of one linear, unbroken, double-stranded DNA molecule—one double helix—running throughout the length of the chromosome. Five main types of histones (H1, H2A, H2B, H3, and H4) are constant from cell to cell within an organism. Nonhistones, of which there are many, vary significantly between cell types, both within and among organisms as well as with time in the same cell type. The large amount of DNA present in the eukaryotic chromosome is compacted by its association with histones in nucleosomes and by higher levels of folding of the nucleosomes into chromatin fibers. Each chromosome contains a large number of looped domains of 30-nm chromatin fibers attached to a protein scaffold. The functional state of the chromosome is related to the extent of coiling: regions containing genes that are active are less packed than regions containing inactive genes.

**Centromeric and Telomeric DNA.** The centromere and the telomere are two areas of special function in eukaryotic chromosomes. You will learn in Chapter 12 that the behavior of chromosomes in mitosis and meiosis depends on the *kinetochores* that form on the centromeres. A **telomere,** a specific set of sequences at the end of a linear chromosome, stabilizes the chromosome and is required for replication (Chapter 3). Each chromosome has two ends and, therefore, two telomeres.

A **centromere** is the region of a chromosome containing DNA sequences to which mitotic and meiotic spindle fibers attach. Under the microscope a centromere is seen as a constriction in the chromosome. The centromere region of each chromosome is responsible for the accurate segregation of replicated chromosomes to the daughter cells during mitosis and meiosis. The centromere of a mitotic metaphase chromosome—a duplicated chromosome that is partway through the division of the cell and concomitant segregation of the chromosomes to the progeny cells—is indicated in Figure 2.22.

The DNA sequences of centromeres have been analyzed extensively in a few organisms, and notably in the yeast *Saccharomyces cerevisiae*. These sequences in yeast are called ***CEN* sequences,** after the *cen*tromere. Although each yeast centromere has the same function, the *CEN* regions are highly similar—but not identical to one another—in nucleotide sequence and organization. The common core centromere region in each yeast chromosome consists of 112–120 base pairs that can be grouped into three sequence domains (centromere DNA elements, or CDEs; Figure 2.24). CDEII, a 78–86-bp region, more than 90% of which is composed of A-T base pairs, is the largest domain. To one side is CDEI, which has an 8-bp sequence (RTCACRTG, where R is a purine—i.e., either A or G), and to the other side is CDEIII, a 26-bp sequence domain that is also AT rich. Centromere sequences have been determined for a number of other organisms and are different both from those of yeast and from each other. The centromeres of the fission yeast *Schizosaccharomyces pombe*, for example, are 40–80 kb long, with complex arrangements of several repeated sequences. Human centromeres are even longer, ranging from 240 kb to several million base pairs; the longer ones are larger than some bacterial genomes! Thus, although centromeres carry out the same function in all eukaryotes, there is no common sequence that is responsible for that function.

A telomere is required for replication and stability of a linear chromosome. In most organisms that have been examined, the telomeres are positioned just inside the nuclear envelope and often are found associated with each other as well as with the nuclear envelope.

All telomeres in a given species share a common sequence, but telomere sequences differ among species. Most telomeric sequences may be divided into two types:

1. **Simple telomeric sequences** are at the extreme ends of the chromosomal DNA molecules. Depending on the organism and its stage of life, there are on the order of 100–1,000 copies of the repeats. Simple telomeric sequences are the essential functional components of telomeric regions, in that they are sufficient to supply a chromosomal end with stability. These sequences consist of a series of simple DNA sequences repeated one after the other (called *tandemly repeated DNA sequences*). In the ciliate *Tetrahymena*, for example, reading the sequence toward the end of one DNA strand, the repeated sequence is 5′-TTGGGG-3′ (Figure 2.25a). In humans and all other vertebrates, the repeated sequence is 5′-TTAGGG-3′. Different researchers may describe the telomere repeat with other starting points, such as 5′-GGTTAG-3′ or 5′-GGGTTA-3′ for humans and other vertebrates. The telomeric DNA is not double-stranded all the way out to the end of the chromosome. In one model, the telomere DNA loops back on itself, forming a *t-loop* (Figure 2.25b). The single-stranded end invades the double-stranded telomeric sequences, causing a *displacement loop*, or *D-loop*, to form.
2. **Telomere-associated sequences** are regions internal to the simple telomeric sequences. These sequences often contain repeated, but still complex, DNA sequences extending many thousands of base pairs in from the chromosome end. The significance of such sequences is not known.

Whereas the telomeres of most eukaryotes contain short, simple, repeated sequences, the telomeres of *Drosophila* are quite different structurally. *Drosophila* telomeres consist of *transposable elements*—DNA sequences that can move to other locations in the genome (see Chapter 7, pp. 150–161).

## Unique-Sequence and Repetitive-Sequence DNA

Now that you know about the basic structure of DNA and its organization in chromosomes, we can discuss the distribution of certain sequences in the genomes of prokaryotes and eukaryotes. From molecular analyses, geneticists have found that some sequences are present

**Figure 2.24**

**Consensus sequence for centromeres of the yeast *Saccharomyces cerevisiae*.** R = a purine. Base pairs that appear in 15 to 16 of the 16 centromeres are highly conserved and are indicated by capital letters. Base pairs (bp) found in 10 to 13 of the 16 centromeres are conserved and are indicated by lowercase letters. Nonconserved positions are indicated by dashes.

```
CDE region:     I               II                         III
          RTCACRTG ←78–86 bp(>90% AT)→ tGttTttG–tTTCCGAA––––aaaaa
          ←— 8 bp —→                   ←————————— 26 bp —————————→
```

**Figure 2.25**

**Telomeres.** (a) Simple telomeric repeat sequences at the ends of human chromosomes. (b) Model of telomere structure in which the telomere DNA loops back to form a t-loop. The single-stranded end invades the double-stranded telomeric sequences to produce a displacement loop (D-loop).

**a) Human simple telomeric repeat sequences**

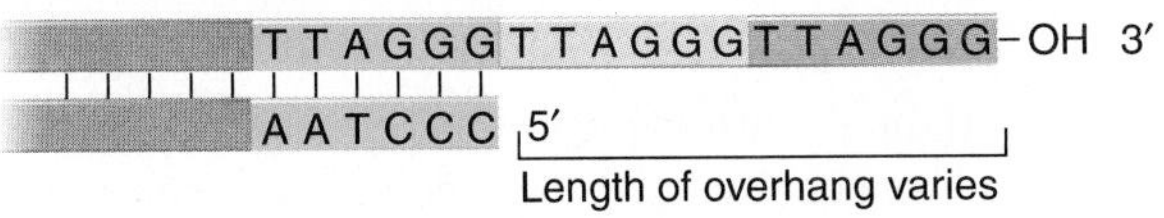


**b) t-loop model for telomeres**

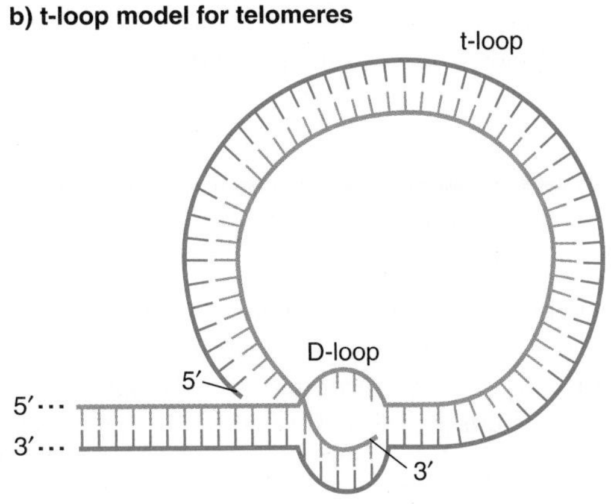


only once in the genome, whereas other sequences are repeated. For convenience, these sequences are grouped into three categories: **unique-sequence DNA** (present in one to a few copies in the genome), **moderately repetitive DNA** (present in a few to about $10^5$ copies in the genome), and **highly repetitive DNA** (present in about $10^5$ to $10^7$ copies in the genome). In prokaryotes, with the exception of the ribosomal RNA genes, transfer RNA genes, and a few other sequences, all of the genome is present as unique-sequence DNA. Eukaryotic genomes, by contrast, consist of both unique-sequence and repetitive-sequence DNA, with the latter typically being quite complex in number of types, number of copies, and distribution. To date, we have sketchy information about the distribution of the various classes of sequences in the genome. However, as the complete DNA sequences of more and more eukaryotic genomes are determined, we will develop a precise understanding of the molecular organization patterns of unique-sequence and repetitive-sequence DNA.

**Unique-Sequence DNA.** Unique sequences, sometimes called single-copy sequences, are sequences that are present as single copies in the genome. (Thus, there are two copies per diploid cell.) In current usage, the term usually applies to sequences that have one to just a few copies per genome. Most of the genes we know about—the protein-coding genes—are in the unique-sequence class of DNA. In humans, unique sequences are estimated to make up approximately 55–60% of the genome.

**Repetitive-Sequence DNA.** Both *moderately repetitive* and *highly repetitive* DNA sequences are sequences that appear many times within a genome. These sequences can be arranged within the genome in one of two ways: distributed at irregular intervals—known as **dispersed repeated DNA** or **interspersed repeated DNA**—or clustered together so that the sequence repeats many times in a row—known as **tandemly repeated DNA.**

Dispersed repeated sequences consist of families of repeated sequences interspersed through the genome with unique-sequence DNA. Each family consists of a set of related sequences characteristic of the family. Often, small numbers of families have very high copy numbers and make up most of the dispersed repeated sequences in the genome. Two types of dispersed repeated sequences are known: (1) **long interspersed elements (LINEs),** in which the sequences in the families are about 1,000–7,000 bp long; and (2) **short interspersed elements (SINEs),** in which the sequences in the families are 100–400 bp long. All eukaryotic organisms have LINEs and SINEs, with a wide variation in their relative proportions. Humans and frogs, for example, have mostly SINEs, whereas *Drosophila* and birds have mostly LINEs. LINEs and SINEs represent a significant proportion of all the moderately repetitive DNA in the genome.

Mammalian diploid genomes have about 500,000 copies of the LINE-1 (L1) family, representing about 15% of the genome. Other LINE families may be present also, but they are much less abundant than LINE-1. Full-length LINE-1 family members are 6–7 kb long, although most are truncated elements of about 1–2 kb. The full-length LINE-1 elements are *transposons*, meaning that they are DNA elements that can move from location to location in the genome. Genes they contain encode the enzymes necessary for that movement.

SINEs are found in a diverse array of eukaryotic species, including mammals, amphibians, and sea urchins. Each species with SINEs has its own characteristic array of SINE families. A well-studied SINE family is the *Alu* family of certain primates. This family is named for the cleavage site for the restriction enzyme *Alu*I ("Al-you-one"), typically found in the repeated sequence. In humans, the *Alu* family is the most abundant SINE family in the genome, consisting of 200–300-bp sequences repeated as many as a million times and making up about 9% of the total haploid DNA. One *Alu* repeat is located every 5,000 bp in the genome, on average. The SINEs are also transposons, but they do not encode the enzymes they need for movement. They can move, however, if those enzymes are supplied by an active LINE transposon.

Tandemly repeated DNA sequences are arranged one after the other in the genome in a head-to-tail organization. Tandemly repeated DNA is common in eukaryotic genomes, in some cases in short sequences 1–10 bp long and in other cases associated with genes and in much longer sequences. The tandemly repeated simple telomeric sequences shown in Figure 2.25a are not genes—whereas genes for ribosomal RNA (rRNA; see Chapter 6) are tandemly repeated genes, often organized into one or more clusters in most eukaryotes. The greatest amount of tandemly repeated DNA is associated with centromeres and telomeres. At each centromere, there are hundreds to thousands of copies of simple, short tandemly repeated sequences (highly repetitive sequences). In fact, a significant proportion of the eukaryotic genome may consist of the highly repeated sequences found at centromeres: 8% in the mouse, about 50% in the kangaroo rat, and about 5–10% in humans. See Chapter 9, pp. 229–230 for a description of what we have learned from genome sequencing about the organization of genes and repeated sequences in the human genome, and Chapter 10, pp. 272–273 for a more detailed discussion of nongenic tandemly repeated DNA.)



**Keynote**

Prokaryotic genomes consist mostly of unique-sequence DNA, with only a few sequences and genes repeated. Eukaryotes have both unique and repetitive sequences in the genome, with an extensive, complex spectrum of the repetitive sequences among species. Some of the repetitive sequences are genes, but most are not.

## Summary

- Organisms contain genetic material that governs an individual's characteristics and that is transferred from parent to progeny.
- Deoxyribonucleic acid (DNA) is the genetic material of all living organisms and some viruses. Ribonucleic acid (RNA) is the genetic material only of certain viruses. In prokaryotes and eukaryotes, the DNA is always double-stranded, whereas in viruses the genetic material may be double- or single-stranded DNA or RNA, depending on the virus.
- DNA and RNA are macromolecules composed of smaller building blocks called nucleotides. Each nucleotide consists of a five-carbon sugar (deoxyribose in DNA, ribose in RNA) to which are attached a nitrogenous base and a phosphate group. In DNA, the four possible bases are adenine, guanine, cytosine, and thymine; in RNA, the four possible bases are adenine, guanine, cytosine, and uracil.
- According to Watson and Crick's model, the DNA molecule consists of two polynucleotide (polymers of nucleotides) chains joined by hydrogen bonds between pairs of bases—adenine (A) and thymine (T); and guanine (G) and cytosine (C)—in a double helix.
- The three major types of DNA determined by analyzing DNA outside the cell are the right-handed A- and B-DNAs and the left-handed Z-DNA. The common form found in cells is closest in structure to B-DNA. A-DNA exists in cells in certain DNA–protein complexes. Z-DNA may exist in cells, but its physiological significance is unknown.
- The genetic material of viruses may be linear or circular double-stranded DNA, single-stranded DNA, double-stranded RNA, or single-stranded RNA, depending on the virus. The genomes of some viruses are organized into a single chromosome, whereas others have a segmented genome.
- The genetic material of prokaryotes is double-stranded DNA localized into one or a few chromosomes. Typically prokaryotic chromosomes are circular, but linear chromosomes are found in a number of species.
- A bacterial chromosome is compacted into the nucleoid region by the supercoiling of the DNA helix and the formation of looped domains of supercoiled DNA.
- The eukaryotic genome is distributed among several linear chromosomes. The complete set of metaphase chromosomes in a eukaryotic cell is called its karyotype.
- The nuclear chromosomes of eukaryotes are complexes of DNA and histone and nonhistone chromosomal proteins. Each unduplicated chromosome consists of one linear, unbroken, double-stranded DNA molecule running throughout its length; the DNA is variously coiled and folded. The histones are constant from cell to cell within an organism, whereas the nonhistones vary significantly between cell types.
- The large amount of DNA present in the eukaryotic chromosome is compacted by its association with histones in nucleosomes and by higher levels of folding of the nucleosomes into chromatin fibers. Highly condensed chromosomes consist of a large number of looped domains of 30-nm chromatin fibers spirally attached to a protein scaffold. The more condensed a region of a chromosome is, the less likely it is that the genes in that region will be active.
- The centromere region of each eukaryotic chromosome is responsible for the accurate segregation of the replicated chromosome to the daughter cells during mitosis and meiosis. The DNA sequences of centromeres vary a little within an organism and extensively between organisms.
- Telomeres—the ends of eukaryotic chromosomes—often are associated with each other and with the

nuclear envelope. Telomeres consist of simple, short, tandemly repeated sequences that are species-specific.

- Prokaryotic genomes consist mostly of unique DNA sequences. They have only a few repeated sequences and genes. Eukaryotes have both unique and repetitive sequences in the genome. Dispersed repetitive sequences are interspersed with unique-sequence DNA, whereas tandemly repeated DNA consists of sequences repeated one after another in the chromosome. The spectrum of complexity of repetitive DNA sequences among eukaryotes is extensive. Some repetitive sequences are transposons, meaning that they have the capability of moving to other locations in the genome.



## Analytical Approaches to Solving Genetics Problems

The most practical way to reinforce genetics principles is to solve genetics problems. In this and all subsequent chapters, we discuss how to approach genetics problems by presenting examples of such problems and discussing their answers. The problems use familiar and unfamiliar examples and pose questions designed to get you to think analytically.

**Q2.1** The linear chromosome of phage T2 is 52 μm long. The chromosome consists of double-stranded DNA, with 0.34 nm between each base pair. How many base pairs does a chromosome of T2 contain?

**A2.1** This question involves the careful conversion of different units of measurement. The first step is to put the lengths in the same units: 52 μm is 52 millionths of a meter, or $52,000 \times 10^9$ m, or 52,000 nm. One base occupies 0.34 nm in the double helix, so the number of base pairs in the chromosome of T2 is 52,000 divided by 0.34, or 152,941 base pairs.

The human genome contains $3 \times 10^9$ bp of DNA, for a total length of about 1 meter, distributed among 23 chromosomes. The average length of the double helix in a human chromosome is 3.8 cm, which is 3.8 hundredths of a meter, or 38 million nm—much longer than the T2 chromosome! There are more than 111.7 million base pairs in the average human chromosome.

**Q2.2** The following table lists the relative percentages of bases of nucleic acids isolated from different species:

| Species | Adenine | Guanine | Thymine | Cytosine | Uracil |
|---|---|---|---|---|---|
| (i) | 21 | 29 | 21 | 29 | 0 |
| (ii) | 29 | 21 | 29 | 21 | 0 |
| (iii) | 21 | 21 | 29 | 29 | 0 |
| (iv) | 21 | 29 | 0 | 29 | 21 |
| (v) | 21 | 29 | 0 | 21 | 29 |

For each species, what type of nucleic acid is involved? Is it double or single stranded? Explain your answer.

**A2.2** This question focuses on the base-pairing rules and the difference between DNA and RNA. In analyzing the data, we should determine first whether the nucleic acid is RNA or DNA and then whether it is double or single stranded. If the nucleic acid has thymine, it is DNA; if it has uracil, it is RNA. Thus, species (i), (ii), and (iii) must have DNA as their genetic material, and species (iv) and (v) must have RNA as their genetic material.

Next, we must analyze the data for strandedness. Double-stranded DNA must have equal percentages of A and T and of G and C. Similarly, double-stranded RNA must have equal percentages of A and U and of G and C. Therefore, species (i) and (ii) have double-stranded DNA, whereas species (iii) must have single-stranded DNA, because the base-pairing rules are violated, with A = G and T = C, but A ≠ T and G ≠ C. As for the RNA-containing species, (iv) contains double-stranded RNA, because A = U and G = C, and (v) must contain single-stranded RNA.

**Q2.3** Here are four characteristics of one 5′-to-3′ strand of a particular long, double-stranded DNA molecule:

i. Thirty-five percent of the adenine-containing nucleotides (As) have guanine-containing nucleotides (Gs) on their 3′ sides.
ii. Thirty percent of the As have Ts as their 3′ neighbors.
iii. Twenty-five percent of the As have Cs as their 3′ neighbors.
iv. Ten percent of the As have As as their 3′ neighbors.

Use the preceding information to answer the following questions as completely as possible, explaining your reasoning in each case:

a. In the complementary DNA strand, what will be the frequencies of the various bases on the 3′ side of A?
b. In the complementary strand, what will be the frequencies of the various bases on the 3′ side of T?
c. In the complementary strand, what will be the frequency of each kind of base on the 5′ side of T?
d. Why is the percentage of A not equal to the percentage of T (and the percentage of C not equal to the percentage of G) among the 3′ neighbors of A in the 5′-to-3′ DNA strand described?

**A2.3**

a. This question cannot be answered without more information. Although we know that the As neighbored by Ts in the original strand will correspond to As

*__2.21__ *E. coli* bacteriophage ΦX174 and parvovirus B19 (the causative agent of Fifth disease—*infectious redness*—in humans) each have a single-stranded DNA genome.

**a.** What base equalities or inequalities might we expect for these genomes?

**b.** Suppose Chargaff had analyzed *only* the genomes of ΦX174 and B19. What might he have concluded?

**c.** Suppose Chargaff had included ΦX174 and B19 in his analysis of genomes from other organisms. How might he have altered his conclusions?

**2.22** Different forms of DNA have been identified through X-ray crystallography analysis. These forms include A-DNA, B-DNA, and Z-DNA, and each has unique molecular attributes.

**a.** What are the molecular attributes of each of these forms of crystallized DNA?

**b.** Which form is closest in structure to most of the DNA found in living cells? Why isn't cellular DNA identical to this form of crystallized DNA?

**c.** When, if ever, does cellular DNA have one of the other two forms? What do you infer from this information about the potential cellular role(s) of the other DNA forms?

**2.23** If a virus particle contains 200,000 bp of double-stranded DNA, how many complete 360° turns occur in its genome? (Use the value of 10 bp per turn in your calculation.)

*__2.24__ A double-stranded DNA molecule is 100,000 bp (100 kb) long.

**a.** How many nucleotides does it contain?

**b.** How many complete turns are there in the molecule? (Use the value of 10 bp per turn in your calculation.)

**c.** How many nm long is the DNA molecule? ($1 \text{ nm} = 1 \times 10^{-9} \text{ m}$)

**2.25** The bacteriophage T4 genome is 168,900 bp long.

**a.** What are the dimensions of the genome (in nm) if the molecule remains unfolded as a linear segment of double-stranded DNA?

**b.** If the T4 protein capsid has about the same dimensions as the capsid of bacteriophage T2 (see Figure 2.4), and the thickness of the capsid is about 10 nm, about how many times must the T4 genome be folded to fit into the space available within its capsid?

**2.26** Different cellular organisms have vastly different amounts of genetic material. *E. coli* has about $4.6 \times 10^6$ bp of DNA in one circular chromosome, the haploid budding yeast (*S. cerevisiae*) has 12,067,280 bp of DNA in 16 chromosomes, and the gametes of humans have about $2.75 \times 10^9$ bp of DNA in 23 chromosomes.

**a.** For each of these organism's cells, if all of the DNA were B-DNA, what would be the average length of a chromosome in the cell?

**b.** On average, how many complete turns would be in each chromosome?

**c.** Would your answers to (a) and (b) be significantly different if the DNA were composed of, say, 20% Z-DNA and 80% B-DNA?

**d.** What implications do your answers to these questions have for the packaging of DNA in cells?

*__2.27__ If nucleotides were arranged at random in a piece of single-stranded RNA $10^6$ nucleotides long, and if the base composition of this RNA was 20% A, 25% C, 25% U, and 30% G, how many times would you expect the specific sequence 5′-GUUA-3′ to occur?

*__2.28__ Two double-stranded DNA molecules from a population of T2 phages were denatured to single strands by heat treatment. The result was the following four single-stranded DNAs:

```
1 T A G C T C C →        3 G C T C C T A →
                   and
2 ← A T C G A G G        4 ← C G A G G A T
```

These separated strands were then allowed to renature. Diagram the structures of the renatured molecules most likely to appear when (a) strand 2 renatures with strand 3 and (b) strand 3 renatures with strand 4. Label the strands, and indicate sequences and polarity.

**2.29** Define topoisomerases, and list the functions of these enzymes.

**2.30** What is the relationship between cellular DNA content and the structural or organizational complexity of the organism?

**2.31** Impressive technologies have been developed to sequence entire genomes (see Chapter 8). Some biotechnology innovators even envision low-cost ($1,000) sequencing of individual human genomes. Still, the genome of the single-celled *Ameoba proteus* might present a challenge since it has nearly 100 times the DNA content of the human genome (see Table 2.3). If we sequenced its genome, do you expect we would identify about 100-fold more genes than have been found in the human genome? Why or why not? If not, what do you expect we would learn about its genome?

**2.32** In a particular eukaryotic chromosome (choose the best answer),

**a.** heterochromatin and euchromatin are regions where genes make functional gene products (that is, where genes are active).

**b.** heterochromatin is active, but euchromatin is inactive.

**c.** heterochromatin is inactive, but euchromatin is active.

**d.** both heterochromatin and euchromatin are inactive.

*__2.33__ Compare and contrast eukaryotic chromosomes and bacterial chromosomes with respect to the following features:

**a.** centromeres

**b.** pentose sugars

c. amino acids
d. supercoiling
e. telomeres
f. nonhistone protein scaffolds
g. DNA
h. nucleosomes
i. circular chromosome
j. looping

**2.34** Discuss the components and structure of a nucleosome and the composition of a nucleosome core particle. Explain how nucleosomes are used to package DNA hierarchically.

**2.35** Histone proteins from many different eukaryotes are highly similar in their amino acid sequence, making them among the most highly conserved eukaryotic proteins. What functional properties of histone proteins might limit their diversity?

***2.36** Set up the following "rope trick": Start with a belt (representing a DNA molecule; imagine the phosphodiester backbones lying along the top and bottom edges of the belt) and a soda can. Holding the belt buckle at the bottom of the can, wrap the belt flat against the side of the can, and wind, counterclockwise three times around the can. Now remove the "core" soda can, and, holding the ends of the belt, pull the ends of the belt taut. After some reflection, answer the following questions:

a. Did you make a left- or a right-handed helix?
b. How many helical turns were present in the coiled belt before it was pulled taut?
c. How many helical turns were present in the coiled belt after it was pulled taut?
d. Why does the belt appear more twisted when pulled taut?
e. About what percentage of the length of the belt was decreased by this packaging?
f. Is the DNA of a linear chromosome that is coiled around histones supercoiled?
g. Why are topoisomerases necessary to package linear chromosomes?

***2.37** What are the main molecular features of yeast centromeres?

**2.38** Telomeres are unique repeated sequences. Where on the DNA strand are they found? Do they serve a function?

***2.39** Would you expect to find most protein-coding genes in unique-sequence DNA, in moderately repetitive DNA, or in highly repetitive DNA?

**2.40** Both histone and nonhistone proteins are essential for DNA packaging in eukaryotic cells. However, these classes of proteins are fundamentally dissimilar in a number of ways. Describe how they differ in terms of

a. their protein characteristics.
b. their presence and abundance in cells.
c. their interactions with DNA.
d. their role in DNA packaging and the formation of looped domains.

**2.41** In higher eukaryotes, what relationships exist between these elements?

a. centromeres and tandemly repeated DNA
b. constitutive heterochromatin and centromeric regions
c. euchromatin, facultative heterochromatin, constitutive heterochromatin and unique-sequence DNA

***2.42** Distinguish between LINEs and SINEs with respect to

a. their length.
b. their abundance in different higher eukaryotic genomes.
c. whether and how they are able to move within a genome.
d. their distribution within a genome.

***2.43** Chromosomal rearrangements at the end of 16p (the short arm of chromosome 16) underlie a variety of common human genetic disorders, including β-thalassemia (a defect in hemoglobin metabolism caused by mutations in the β-globin gene that lies in this region), mental retardation, and the adult form of polycystic kidney disease. Analysis of approximately 285-kb pairs of DNA sequence at the end of human chromosome 16p has allowed for a detailed understanding of the structure of this chromosome region. The first functional gene lies about 44 kb from the region of simple telomeric sequences and about 8 kb from the telomere-associated sequences. Analysis of sequences proximal (nearer the centromere) to the first gene reveals a sinusoidal variation in GC content, with GC-rich regions associated with gene-rich areas and AT-rich regions associated with *Alu*-dense areas. The β-globin gene lies about 130 kb from the telomere-associated sequences.

a. Diagram the features of the 16p telomere, and relate them to the current view of telomere structure and function as presented in the text.
b. What have the preceding data revealed about the distribution of SINEs in the terminus of 16p? (SINEs and LINEs are, respectively, short and long interspersed nuclear elements.)

# 3 DNA Replication

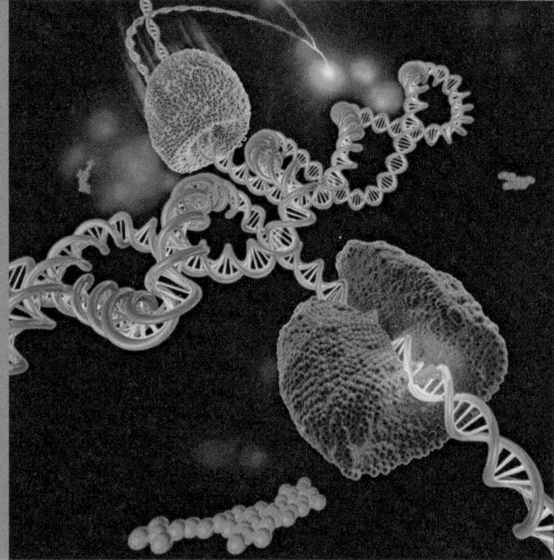

DNA polymerase (grey) replicating DNA, with topoisomerase (green) relaxing the tension in the DNA ahead of the replication fork.

## Key Questions

- How is DNA replicated?
- How does DNA polymerase synthesize a new DNA chain?
- How does DNA replication of a chromosome take place at the molecular level?
- How are circular chromosomes of prokaryotes and viruses replicated?
- How are the large genomes of eukaryotic organisms replicated in a timely fashion?
- How are the ends of eukaryotic chromosomes replicated?

### iActivity

A BASIC PROPERTY OF GENETIC MATERIAL IS ITS ability to replicate in a precise way so that the genetic information encoded in the nucleotides can be transmitted from each cell to all of its progeny. James Watson and Francis Crick recognized that the complementary relationship between DNA strands probably would be the basis for DNA replication. However, even after scientists confirmed this fact five years after Watson and Crick developed their model, many questions about the mechanics of DNA replication remained. In this chapter, you will learn about the steps and enzymes involved in the replication of prokaryotic and eukaryotic DNA molecules. Then, in the iActivity, you will have a chance to investigate the specifics of DNA replication in *E. coli.*

Replication of DNA is vital to the transmission of genomes and the genes they contain from cell generation to cell generation, and from organism generation to organism generation. Your goal in the chapter is to learn about the mechanisms of DNA replication and chromosome duplication in bacteria and eukaryotes, and about some of the enzymes and other proteins needed for replication. Some of these enzymes are also involved in the repair of damage to DNA, a topic we discuss in Chapter 7, and are used for biotechnology applications, discussed in Chapter 10.

## Semiconservative DNA Replication

When Watson and Crick proposed their double helix model for DNA in 1953, they realized that DNA replication would be straightforward if their model was correct. That is, if the DNA molecule was untwisted and the two strands separated, each strand could act as a template for the synthesis of a new, complementary strand of DNA that could then be bound to the parental strand. This DNA replication model is known as the **semiconservative model,** because each progeny molecule retains ("conserves") one of the parental strands (Figure 3.1a).

At the time, two other models for DNA replication were proposed. In the **conservative model** (Figure 3.1b), the two parental strands of DNA remain together or pair again after replication and, as a whole, serve as a template for the synthesis of new progeny DNA double helices. In this model, one of the two progeny DNA molecules is the parental double-stranded DNA molecule, and the other consists entirely of new material. In the **dispersive model** (Figure 3.1c), the parental double helix is cleaved

Figure 3.1

**Three models for DNA replication.** Parental strands are shown in red, and the newly synthesized strands are shown in blue.

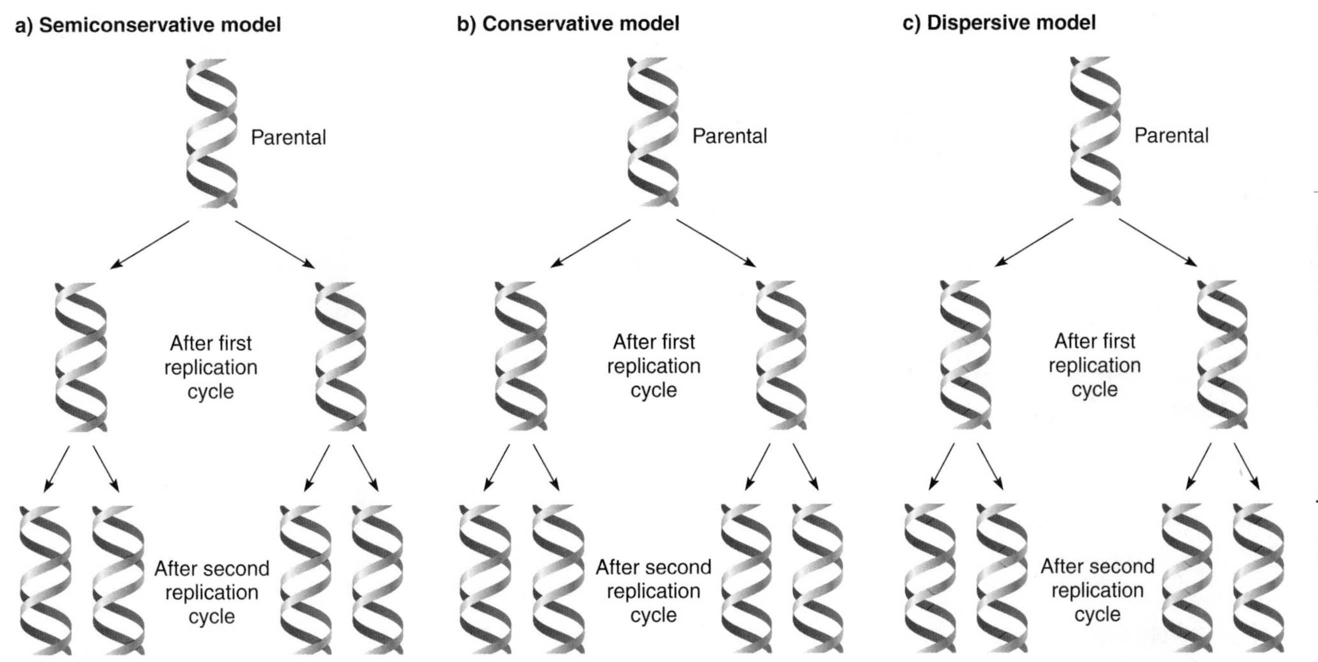




into double-stranded DNA segments that act as templates for the synthesis of new double-stranded DNA segments. Somehow, the segments reassemble into complete DNA double helices, with parental and progeny DNA segments interspersed. Although the two progeny DNAs are identical with respect to their base-pair sequence, double-stranded parental DNA has become dispersed throughout both progeny molecules. It is hard to imagine how the DNA sequences of chromosomes could be kept the same without some sophisticated regulatory mechanisms. The dispersive model is included for historical completeness.

## The Meselson–Stahl Experiment

In 1958, Matthew Meselson and Frank Stahl obtained experimental evidence that the semiconservative replication model is correct. Meselson and Stahl grew *E. coli* in a medium in which the only nitrogen source was $^{15}NH_4Cl$ (ammonium chloride; Figure 3.2). In this compound, the normal isotope of nitrogen, $^{14}N$, is replaced with $^{15}N$, the heavy isotope. (*Note*: Density is weight divided by volume, so $^{15}N$, with one extra neutron in its nucleus, is 1/14 denser than $^{14}N$.) As a result, all the bacteria's nitrogen-containing compounds, including DNA, contained $^{15}N$ instead of $^{14}N$.

**animation**
**The Meselson–Stahl Experiment**

Next, the $^{15}N$-labeled bacteria were transferred to a medium containing nitrogen in the normal $^{14}N$ form, and the bacteria were allowed to reproduce for several generations. All new DNA synthesized after the transfer was labeled, then, with $^{14}N$. As the bacteria reproduced in the $^{14}N$ medium, samples of *E. coli* were taken at various times, and the DNA was extracted and analyzed to determine its density. This was done using equilibrium density gradient centrifugation (described in Box 3.1). Briefly, in this technique, high-speed centrifugation of a solution of cesium chloride (CsCl) produces a gradient of that salt, with the least dense solution at the top of the tube and the most dense solution at the bottom. DNA that is present in the solution during centrifugation forms a band at a position where its buoyant density matches that of the surrounding cesium chloride. $^{15}N$-labeled DNA ($^{15}N$–$^{15}N$ DNA) and $^{14}N$-labeled DNA ($^{14}N$–$^{14}N$ DNA) form bands at distinct positions in a CsCl gradient, as illustrated in Box Figure 3.1.

After one replication cycle (one generation) in the $^{14}N$ medium, all of the DNA had a density that was exactly intermediate between that of $^{15}N$–$^{15}N$ DNA and that of $^{14}N$–$^{14}N$ DNA (see Figure 3.2). After two replication cycles, half the DNA was of that intermediate density and half was of the density of $^{14}N$–$^{14}N$ DNA. These observations, presented in Figure 3.2, and those obtained from subsequent replication cycles were exactly what the semiconservative model predicted.

If the conservative model for DNA replication had been correct, after one replication cycle there would have been a band of $^{15}N$–$^{15}N$ DNA (parental) and a band of $^{14}N$–$^{14}N$ DNA (newly synthesized; see Figure 3.1b). The heavy parental DNA band would have been seen at each subsequent replication cycle, in the amount found at the start of the experiment. All new DNA molecules would then have been $^{14}N$–$^{14}N$ DNA. Therefore, the relative

**Figure 3.2**

**The Meselson–Stahl experiment.** The demonstration of semiconservative replication in *E. coli.* Cells were grown in a $^{15}N$-containing medium for several replication cycles and then were transferred to a $^{14}N$-containing medium. At various times over several replication cycles, samples were taken; the DNA was extracted and analyzed by CsCl equilibrium density gradient centrifugation. Shown in the figure are a schematic interpretation of the DNA composition after various replication cycles, photographs of the DNA bands, and densitometric scans of the bands.

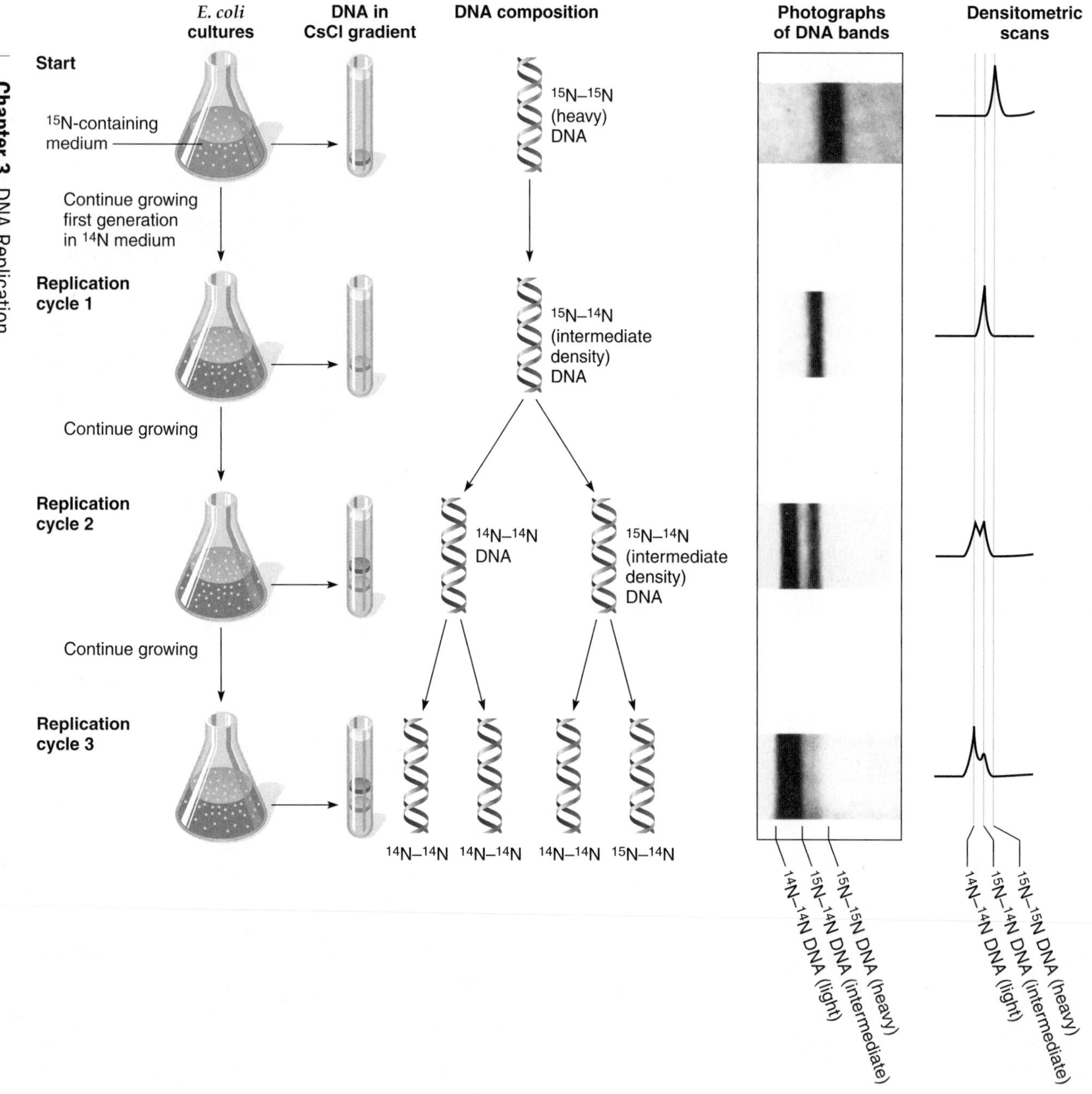




amount of DNA in the $^{14}N$–$^{14}N$ DNA position would have increased with each replication cycle. For the conservative model of DNA replication, then, the most significant prediction was that *at no time would any DNA of intermediate density be seen.* The fact that intermediate-density DNA *was* seen ruled out the conservative model.

If the dispersive model for DNA replication had been correct, then all DNA present in the $^{14}N$ medium after one replication cycle would have been of intermediate ($^{15}N$–$^{14}N$) density (see Figure 3.1c), and this was seen in the Meselson–Stahl experiment. However, the dispersive model predicted that, after a second replication cycle in the same medium, DNA segments from the first replication cycle would be dispersed throughout the progeny DNA double helices produced. Thus, the $^{15}N$–$^{15}N$ DNA segments dispersed among new $^{14}N$–$^{14}N$ DNA after one

## Box 3.1 Equilibrium Density Gradient Centrifugation

In equilibrium density gradient centrifugation, a concentrated solution of cesium chloride (CsCl) is centrifuged at high speed to produce a linear concentration gradient of the CsCl. The actual densities of CsCl at the extremes of the gradient are related to the beginning CsCl concentration that is centrifuged.

For example, to examine DNA of density 1.70 g/cm$^3$ (a typical density for DNA), a gradient is made which spans that density—for example, from 1.60 to 1.80 g/cm$^3$. If DNA is mixed with the CsCl and the mixture is centrifuged, the DNA comes to equilibrium at the point in the gradient where its buoyant density equals the density of the surrounding CsCl (see the accompanying figure). The DNA is said to have *banded* in the gradient. If DNAs that have different densities are present, as is the case with $^{15}N$–$^{15}N$ DNA and $^{14}N$–$^{14}N$ DNA, they band (come to equilibrium) in different positions. The DNA is detected in the gradient by its ultraviolet absorption.

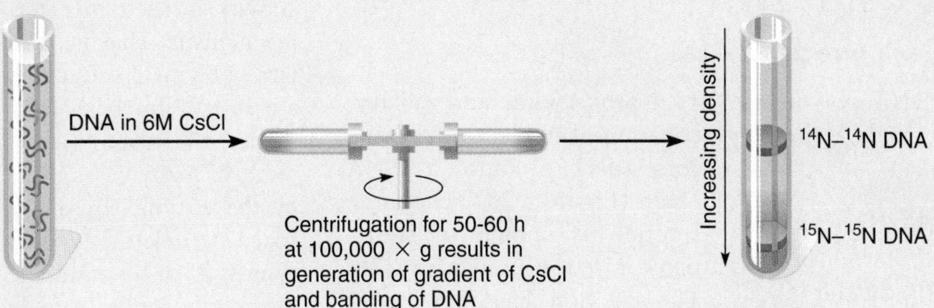
DNA in 6M CsCl
Centrifugation for 50-60 h
at 100,000 × g results in
generation of gradient of CsCl
and banding of DNA
Increasing density
14N–14N DNA
15N–15N DNA

**Box Figure 3.1**
**Schematic diagram for separating DNAs of different buoyant densities by equilibrium centrifugation in a cesium chloride density gradient.** The separation of $^{14}N$–$^{14}N$ DNA and $^{15}N$–$^{15}N$ DNA is illustrated.

replication cycle would then be distributed among twice as many DNA molecules after two replication cycles. As a result, the DNA molecules would be found in one band located halfway between the $^{15}N$–$^{14}N$ DNA and $^{14}N$–$^{14}N$ DNA positions in the gradient. With subsequent replication cycles, there would continue to be one band, and it would become lighter in density with each replication cycle. The results of the Meselson–Stahl experiment did not bear out this prediction, so the dispersive model was ruled out.

Subsequent experiments by others showed that DNA in eukaryotes replicates semiconservatively.

### Keynote

DNA replication in *E. coli* and other prokaryotes as well as in eukaryotes occurs by a semiconservative mechanism in which the strands of a DNA double helix separate and a new complementary strand of DNA is synthesized on each of the two parental template strands. Semiconservative replication results in two double-stranded DNA molecules, each having one strand from the parent molecule and one newly synthesized strand.

## DNA Polymerases, the DNA Replicating Enzymes

In 1955, Arthur Kornberg and his colleagues were the first to identify the enzymes necessary for DNA replication. Their work focused on bacteria, because the bacterial replication machinery was assumed to be less complex than that of eukaryotes. Kornberg shared the 1959 Nobel Prize in Physiology or Medicine for his "discovery of the mechanisms in the biological synthesis of deoxyribonucleic acid."

### DNA Polymerase I

Kornberg's approach was a biochemical one. He set out to identify all the ingredients needed to synthesize *E. coli* DNA in vitro. The first successful DNA synthesis was accomplished in a reaction mixture containing DNA fragments, a mixture of four deoxyribonucleoside 5′-triphosphate precursors (dATP, dGTP, dTTP, and dCTP, collectively abbreviated dNTP for *d*eoxyribo*n*ucleoside *t*ri*p*hosphate), and an *E. coli* extract (cells of the bacteria, broken open to release their contents). Kornberg used radioactively labeled dNTPs to measure the minute quantities of DNA synthesized in the reaction.

Kornberg analyzed the extract and isolated an enzyme that was capable of DNA synthesis. This enzyme was originally called the *Kornberg enzyme* but is now called **DNA polymerase I** (**DNA Pol I** for short; by definition, enzymes that catalyze DNA synthesis are called **DNA polymerases**).

With DNA Pol I isolated, researchers studied the in vitro DNA synthesis reaction in more detail. They found that five components were needed for DNA to be synthesized:

1. All four dNTPs. (If any one dNTP is missing, synthesis occurs.) These molecules are the precursors for the nucleotide (phosphate–pentose sugar–base) building blocks of DNA described in Chapter 2 (p. 16).
2. DNA Pol I.

3. *E. coli* DNA. This DNA acted as a template, that is, a molecule used to make a complementary DNA molecule in the reaction.
4. DNA to act as a primer. A primer is a short DNA chain needed to start ("prime") a DNA synthesis reaction (discussed in more detail later). For primers, Kornberg used short pieces of DNA produced by digesting *E. coli* DNA with DNase.
5. Magnesium ions ($Mg^{2+}$), needed for optimal DNA polymerase activity.

### Roles of DNA Polymerases

All DNA polymerases from prokaryotes and eukaryotes catalyze the polymerization of nucleotide precursors (dNTPs) into a DNA chain (Figure 3.3a). The same reaction is shown in shorthand notation in Figure 3.3b. The reaction has three main features:

*animation*
**DNA Biosynthesis: How a New DNA Strand Is Made**

1. At the growing end of the DNA chain, DNA polymerase catalyzes the formation of a phosphodiester bond between the 3′-OH group of the deoxyribose on the last nucleotide and the 5′-phosphate of the dNTP precursor. The energy for the formation of the phosphodiester bond comes from the release of two of three phosphates from the dNTP. The important concept here is that *the lengthening DNA chain acts as a primer in the reaction*—a preexisting polynucleotide chain to which a new nucleotide can be added at the free 3′-OH.
2. At each step in lengthening the new DNA chain, DNA polymerase finds the correct precursor dNTP that can form a complementary base pair with the nucleotide on the template strand of DNA. Nucleotides are added rapidly—850 per second in *E. coli* and 60–90 per second in human tissue culture cells. The process does not occur with 100% accuracy, but the error frequency is extremely low.
3. The direction of synthesis of the new DNA chain is only from 5′ to 3′.

One of the best understood systems of DNA replication is that of *E. coli*. For several years after the discovery of DNA polymerase I, scientists believed that it was the only DNA replication enzyme in *E. coli*. However, genetic studies disproved that hypothesis. Scientists have now identified a total of five DNA polymerases, DNA Pol I–V. Functionally, DNA Pol I and DNA Pol III are polymerases necessary for replication, and DNA Pol I, DNA Pol II, DNA Pol IV, and DNA Pol V are polymerases involved in DNA repair.

The DNA polymerases used for replication are different structurally. DNA polymerase I is encoded by a single gene (*polA*) and consists of one polypeptide. The core DNA polymerase III contains the catalytic functions of the enzyme and consists of three polypeptides: α (alpha, encoded by the *dnaE* gene), ε (epsilon, encoded by the *dnaQ* gene), and θ (theta, encoded by the *holE* gene). The complete DNA Pol III enzyme, called the DNA Pol III holoenzyme, contains an additional six different polypeptides.

Both DNA Pol I and DNA Pol III replicate DNA in the 5′-to-3′ direction. Both enzymes also have 3′-to-5′ exonuclease activity, meaning that they can remove nucleotides from the 3′ end of a DNA chain. This enzyme activity is used in error correction in a **proofreading** mechanism. That is, if an incorrect base is inserted by DNA polymerase (an event that occurs at a frequency of about $10^{-6}$ for both DNA polymerase I and DNA polymerase III, meaning that one base in a million is incorrect), in many cases the error is recognized immediately by the enzyme. By a process resembling using a backspace or delete key on a computer keyboard, the enzyme's 3′-to-5′ exonuclease activity excises the erroneous nucleotide from the new strand. Then, the DNA polymerase resumes forward movement and inserts the correct nucleotide. With this proofreading, the frequency of replication errors by DNA polymerase I or III is reduced to less than $10^{-9}$.

DNA Pol I also has 5′-to-3′ exonuclease activity and can remove either DNA or RNA nucleotides from the 5′ end of a nucleic acid strand. This activity is important in DNA replication and is examined later in this chapter.

Box 3.2 describes how early genetic studies revealed that *E. coli* cells contained DNA polymerases other than DNA polymerase I.

**Keynote**
The enzymes that catalyze the synthesis of DNA are called DNA polymerases. All known DNA polymerases synthesize DNA in the 5′-to-3′ direction. Polymerases may also have other activities, such as removing nucleotides from a strand in the 3′-to-5′ direction (also known as proofreading), or removing nucleotides from a strand in the 5′-to-3′ direction.

## Molecular Model of DNA Replication

Table 3.1 gives the functions of some of the *E. coli* DNA replication genes and key DNA sequences involved in replication. A number of the genes were identified by mutational analysis. In this section, we discuss a molecular model of DNA replication involving these genes and sequences.

### Initiation of Replication

The initiation of replication is directed by a DNA sequence called the **replicator.** The replicator usually includes the **origin of replication,** the specific region

**Figure 3.3**

**DNA chain elongation catalyzed by DNA polymerase.**

**a) Mechanism of DNA elongation**

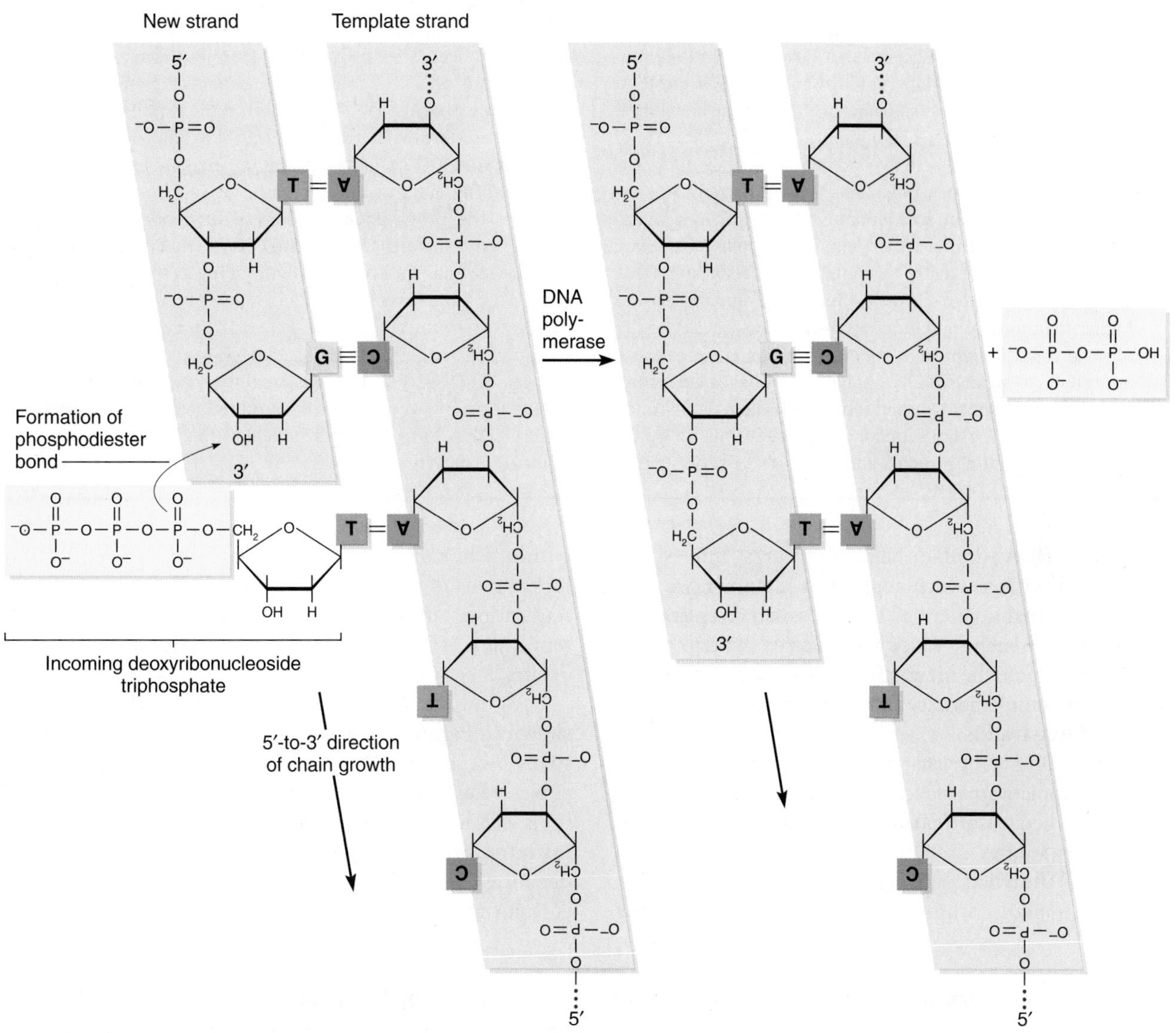


**b) DNA elongation shown using a shorthand notation for DNA**

5′
3′
Template strand
A C A T C
T G T
3′ 3′
OH OH
5′
DNA polymerase
Chain growth
3′
OH + P P
5′

## Box 3.2 Mutants of *E. coli* DNA Polymerases

One way to study the action of an enzyme *in vivo* is to induce a mutation in the gene that codes for the enzyme. In this way, the phenotypic consequences of the mutation can be compared with the wild-type phenotype. The first DNA Pol I mutant, *polA1,* was isolated in 1969 by Paula DeLucia and John Cairns. (The mutant was so named because of the alliteration of "polA" and "Paula.") This mutant shows less than 1% of normal polymerizing activity but near-normal 5′-to-3′ exonuclease activity. DNA polymerase was expected to be essential to cell function, so a mutation in the gene for that enzyme was expected to be lethal or at least crippling. However, *E. coli* cells carrying the *polA1* mutation still replicated DNA and grew and divided normally. But, *polA1* mutants have a higher than normal mutation rate when they are exposed to ultraviolet (UV) light and chemical mutagens—a property interpreted to mean that DNA polymerase I has an important function in repairing damaged (chemically altered) DNA.

To study the consequences of mutations in genes coding for essential proteins and enzymes, geneticists find it easiest to work with **temperature-sensitive mutants**—mutants that function normally until the temperature is raised past some threshold level, when some temperature-sensitive defect is manifested. At *E. coli's* normal growth temperature of 37°C, temperature-sensitive *polAex1* mutant strains produce DNA Pol I with normal polymerizing activity. Studies with the DNA Pol I from the mutant strain *in vitro* at 37°C showed that the enzyme had normal polymerizing activity but decreased 5′-to-3′ exonuclease activity (the progressive removal of nucleotides from a free 5′ end toward the 3′ end). *In vitro* at 42°C, the enzyme still shows normal polymerizing activity, but the 5′-to-3′ exonuclease activity is markedly inhibited. At 42°C, temperature-sensitive *polAex1* mutants die (the mutation is lethal), showing that 5′-to-3′ exonuclease activity of DNA Pol I is essential to DNA replication. Taken together, the results of studies of the *polA1* and *polAex1* DNA Pol I mutants indicated that there must be other DNA-polymerizing enzymes in the cell.

where the DNA double helix denatures into single strands and within which replication commences. The locally denatured segment of DNA is called a **replication bubble.** The segments of single strands in the replication bubble on which the new strands are made (in accordance with complementary base-pairing rules) are called the **template strands.**

When DNA untwists to expose the two single-stranded template strands for DNA replication, a Y-shaped structure called a **replication fork** forms. A replication fork moves in the direction of untwisting the DNA. When DNA untwists starting within a DNA molecule, as in a circular chromosome or replication starting within a linear chromosome, there are two replication forks: two Ys joined together at their tops to form a replication bubble. In many (but not all) cases, each replication fork moves, so that **bidirectional replication** occurs.

An outline of the initiation of replication in *E. coli* is shown in Figure 3.4. The *E. coli* replicator is *oriC*, which spans 245 bp of DNA and contains a cluster of three copies of a 13-bp AT-rich sequence and four copies of a 9-bp sequence. For the initiation of replication, an **initiator protein** or proteins bind to the replicator and denature the AT-rich region. The *E. coli* initiator protein is DnaA (*dnaA* gene), which binds to the 9-bp regions in

**Table 3.1 Functions of Some of the Genes and DNA Sequences Involved in DNA Replication in *E. coli***

| Gene Product or Function | Gene |
|---|---|
| DNA polymerase I | *polA* |
| DNA polymerase III | *dnaE, dnaQ, dnaX, dnaN, dnaD, holA* → E |
| Initiator protein, binds to *oriC* | *dnaA* |
| IHF protein (DNA binding protein), binds to *oriC* | *himA* |
| FIS protein (DNA binding protein); binds to *oriC* | *fis* |
| Helicase and activator of primase | *dnaB* |
| Complexes with *dnaB* protein and delivers it to DNA | *dnaC* |
| Primase; makes RNA primer for extension by DNA polymerase III | *dnaG* |
| Single-stranded binding (SSB) proteins; bind to unwound single-stranded arms of replication forks | *ssb* |
| DNA ligase; seals single-stranded gaps | *lig* |
| Gyrase (type II topoisomerase); replication swivel to avoid tangling of DNA as replication fork advances | *gyrA, gyrB* |
| Origin of chromosomal replication | *oriC* |
| Terminus of chromosomal replication | *ter* |
| TBP (*ter* binding protein), stalls replication forks | *tus* |

**Figure 3.4**

**Initiation of replication in *E. coli*.** The DnaA initiator protein binds to *oriC* (the replicator) and stimulates denaturation of the DNA. DNA helicases are recruited and begin to untwist the DNA to form two head-to-head replication forks.

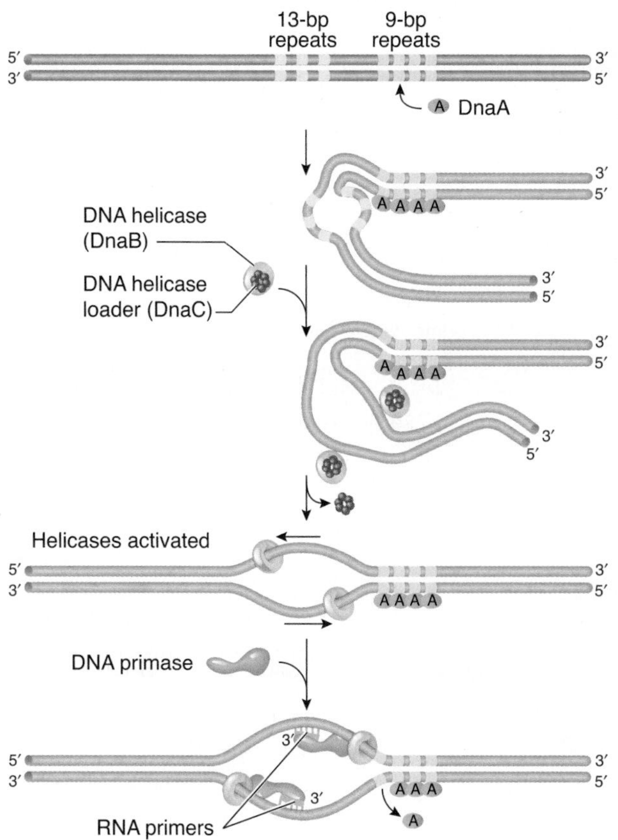


multiple copies, leading to the denaturing of the region with the 13-bp sequences. **DNA helicases** (DnaB; encoded by the *dnaB* gene) are recruited and are loaded onto the DNA by *DNA helicase loader* proteins (DnaC; encoded by the *dnaC* gene). The helicases untwist the DNA in both directions from the origin of replication by breaking the hydrogen bonds between the bases. The energy for the untwisting comes from the hydrolysis of ATP.

Next, each DNA helicase recruits the enzyme **DNA primase** (encoded by the *dnaG* gene), forming a complex called the **primosome.** DNA primase is important in DNA replication because DNA polymerases cannot initiate the synthesis of a DNA strand; they can add nucleotides only to a preexisting strand. That is, the DNA primase (which is a modified RNA polymerase) synthesizes a short **RNA primer** (about 5–10 nucleotides) to which new nucleotides are added by DNA polymerase. The RNA primer is removed later and replaced with DNA (discussed later). At this point, the bidirectional replication of DNA has begun.

You must be clear about the difference between a *template* and a *primer* with respect to DNA replication. A template strand is the one on which the new strand is synthesized according to complementary base-pairing rules. A primer is a short segment of nucleotides bound to the template strand. The primer acts as a substrate for DNA polymerase, which extends the primer and synthesizes a new DNA strand, the sequence of which is complementary to the template strand.

**Keynote**

The initiation of DNA synthesis first involves the denaturation of double-stranded DNA at an origin of replication, catalyzed by DNA helicase. Next, DNA primase binds to the helicase and the denatured DNA and synthesizes a short RNA primer. The RNA primer is extended by DNA polymerase as new DNA is made. Later, the RNA primer is removed.

## Semidiscontinuous DNA Replication

The foregoing discussion of the initiation of replication considered the production of two replication forks when DNA denatures at an origin. The replication events are identical with each replication fork, so we will now focus on the molecular events that occur at one fork (Figure 3.5). To convey clearly the concepts for this complicated series of events, our discussion simplifies the events by keeping the enzymes that synthesize the two different new DNA strands separate. In actuality, the two sets of enzymes work together in a complex; this will be described in more detail later (Figure 3.7).

**animation**
**Molecular Model of DNA Replication**

The replication fork is generated when helicase untwists the DNA to produce two single-stranded template strands. The process of separation of double-stranded DNA to two single strands is called DNA denaturation or DNA melting. **Single-strand DNA-binding (SSB) proteins** bind to each single-stranded DNA, stabilizing them (Figure 3.5) and preventing them from reforming double-stranded DNA by complementary base pairing (a process called *reannealing*). The RNA primer made by DNA primase (see Figure 3.4) is at the 5′ end of the new strand being synthesized on the bottom template strand in Figure 3.5, step 1. The DNA primase at the fork synthesizes another RNA primer, this one on the top template DNA strand (Figure 3.5, step 1). Each RNA primer is extended by the addition of DNA nucleotides by DNA polymerase III (Figure 3.5, step 1). The polymerases displace bound SSB proteins as they move along the template strands. The new DNAs synthesized are complementary to the template strands.

Recall that DNA polymerases can synthesize DNA only in the 5′-to-3′ direction, yet the two DNA strands are of opposite polarity. To maintain the 5′-to-3′ polarity of DNA synthesis on each template, and to maintain one overall direction of replication fork movement, DNA is made in opposite directions on the two template strands (see Figure 3.5, step 1). The new strand being made in

**Figure 3.5**

**Model for the events occurring around a single replication fork of the *E. coli* chromosome.** RNA is green, parental DNA is blue, and new DNA is red.

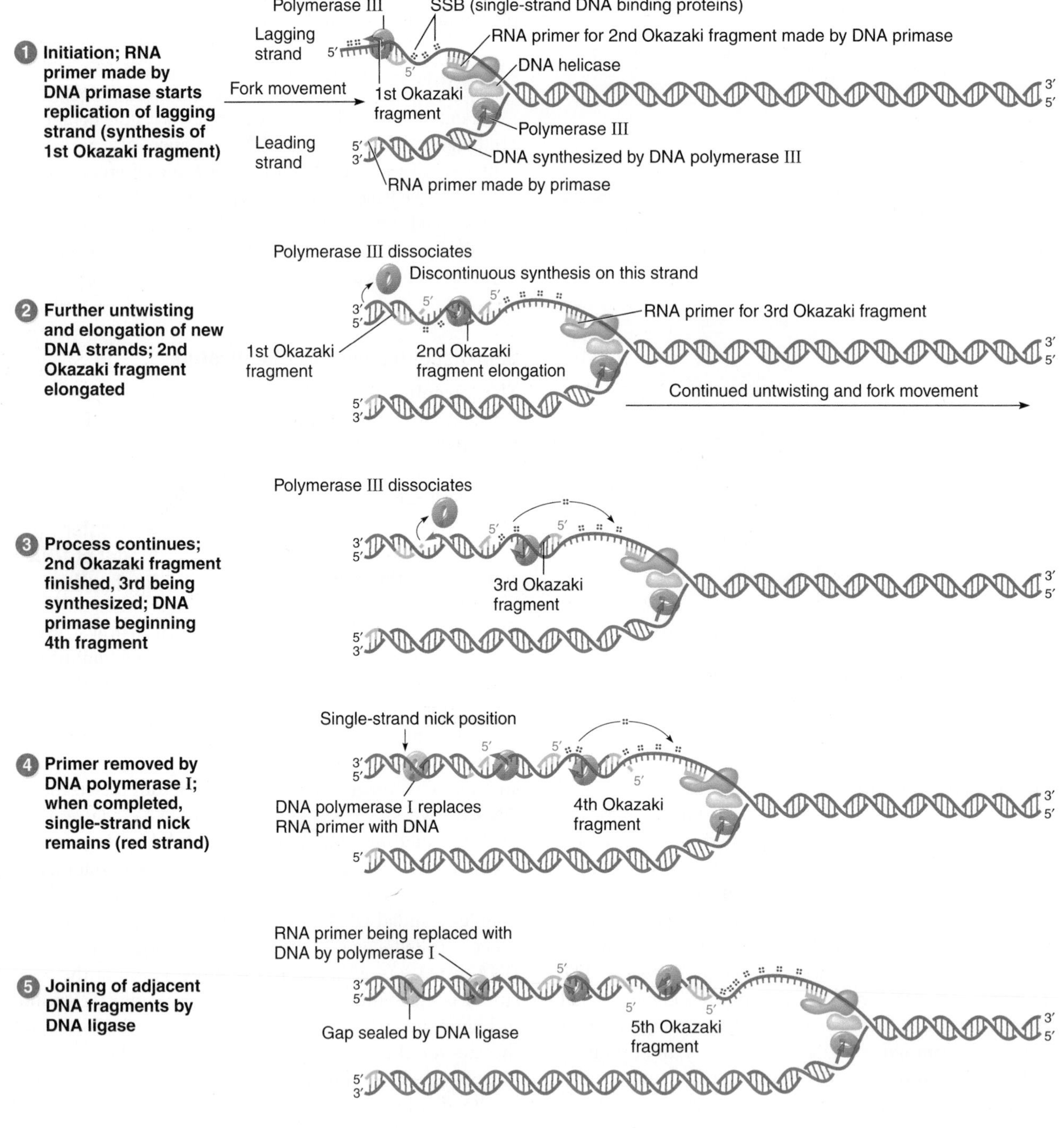


the *same* direction as the movement of the replication fork is the **leading strand** (its template strand—the bottom strand in Figure 3.5—is the *leading-strand template*), and the new strand being made in the direction *opposite* that of the movement of the replication fork is the **lagging strand** (its template strand—the top strand in Figure 3.5—is the *lagging-strand template*). The leading strand needs a single RNA primer for its synthesis, whereas the lagging strand needs a series of primers, as we will see.

Helicase untwists more DNA, causing the replication fork to move along the chromosome (Figure 3.5, step 2). DNA gyrase (a form of topoisomerase) relaxes the tension produced in the DNA ahead of the replication fork. This tension is considerable because the replication fork rotates at about 3,000 rpm. On the leading-strand template (the bottom strand in Figure 3.5), DNA polymerase III synthesizes the leading strand continuously toward the replication fork. Because of the 5′-to-3′ direction of DNA synthesis, however, synthesis of the lagging strand has gone as far as it can. For DNA replication to continue on the lagging-strand template, a new initiation of DNA synthesis occurs: an RNA primer is synthesized by the DNA primase at the replication fork (see Figure 3.5, step 2). DNA polymerase III adds DNA to the RNA primer to make another DNA fragment. Because the leading strand is synthesized *continuously*, whereas the lagging strand is synthesized in pieces, or *discontinuously*, DNA replication as a whole occurs in a **semidiscontinuous** manner.

The fragments of lagging-strand DNA made in semidiscontinuous replication are called **Okazaki fragments** after their discoverers, Reiji and Tuneko Okazaki and colleagues. Experimentally, the Okazakis added a radioactive DNA precursor ($^{3}$H-thymidine) to cultures of *E. coli* for 0.5% of a generation time. They then added a large amount of nonradioactive thymidine to prevent the incorporation of any more of the radioactive precursor into the DNA. At various times (up to 10% of a generation time), they extracted the DNA and determined the size of the newly labeled molecules. At times very soon after the labeling period, most of the radioactivity was present in DNA about 100 to 1,000 nucleotides long. As time increased, a greater and greater proportion of the labeled molecules was found in DNA of much larger size. These results indicated that DNA replication normally involves the synthesis of short DNA segments—the Okazaki fragments—that are subsequently joined together.

The replication process continues in the same way (Figure 3.5, step 3): Helicase continues to untwist the DNA, DNA is synthesized continuously on the leading-strand template, and DNA is synthesized discontinuously on the lagging-strand template, each lagging-strand Okazaki fragment starting with a new RNA primer. Eventually, the Okazaki fragments are joined into a continuous DNA strand. Joining them requires the activities of two enzymes, DNA polymerase I and **DNA ligase.** Consider two adjacent Okazaki fragments: The 3′ end of the newer fragment is adjacent to the primer at the 5′ end of the previously made fragment. DNA polymerase III leaves the newer DNA fragment, and DNA polymerase I binds. The DNA polymerase I simultaneously digests the RNA primer strand ahead of it and extends the DNA strand behind it (Figure 3.5, step 4, and shown in enlarged form in Figure 3.6). Digesting the RNA strand ahead of it involves using the enzyme's 5′-to-3′ exonuclease activity to



**Figure 3.6**

**Joining of Okazaki fragments.** Detail of the replacement of the RNA primer with DNA.

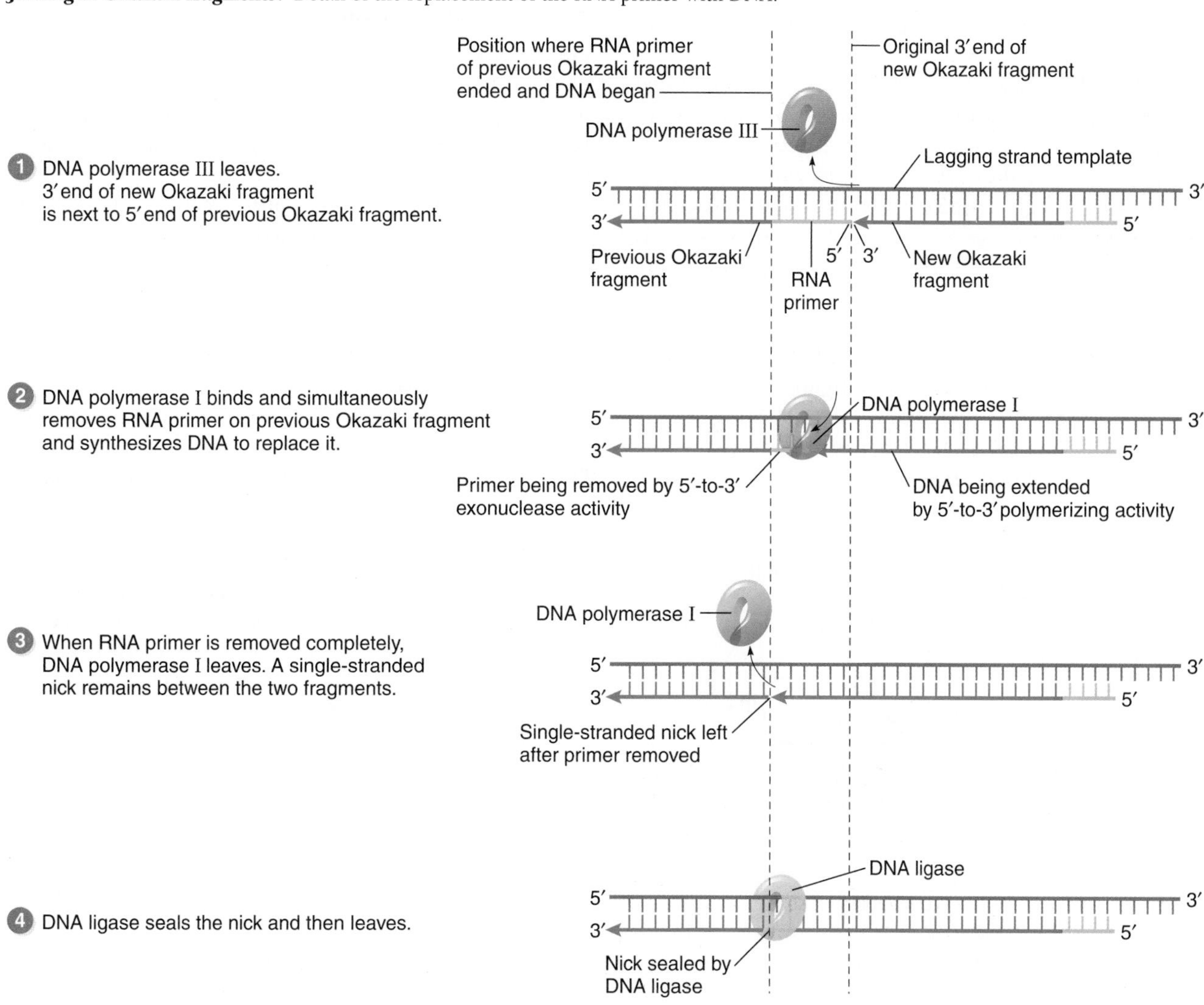

**Figure 3.7**

**Action of DNA ligase in sealing the nick between adjacent DNA fragments (e.g., Okazaki fragments) to form a longer, covalently continuous chain.** The DNA ligase catalyzes the formation of a phosphodiester bond between the 3′-OH and the 5′-phosphate groups on either side of a nick, sealing the nick.

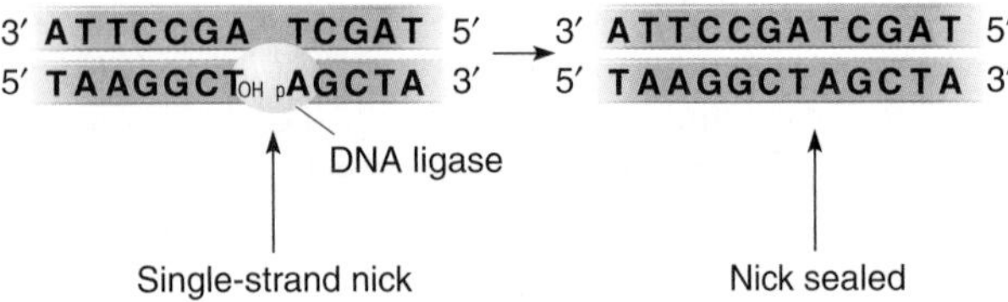


remove nucleotides from the primer's 5′ end, which also exposes template nucleotides. Extending the DNA strand behind it involves the enzyme's 5′-to-3′ polymerase activity to add nucleotides to the DNA strand's 3′ end, whose sequence is directed by the newly exposed template nucleotides. When DNA polymerase I has replaced all the RNA primer nucleotides with DNA nucleotides, a single-stranded nick (a point at which the sugar–phosphate backbone between two adjacent nucleotides is unconnected) is left between the two DNA fragments. DNA ligase joins the two fragments, producing a longer DNA strand (Figure 3.5, step 5). The reaction DNA ligase catalyzes is diagrammed in Figure 3.7. The steps are repeated until all the DNA is replicated.

Figure 3.5 shows DNA replication in a simplified way. In fact, the key replication proteins are closely associated to form a replication machine called a **replisome,** which is bound to the replicating DNA where it is being unwound into single strands. Figure 3.8 shows the lagging-strand DNA, looped so that its DNA polymerase III is complexed with the DNA polymerase III on the leading strand. These are two copies of the core enzyme described earlier (see p. 40), held together by the six other polypeptides to form the DNA Pol III holoenzyme. Only the core enzymes are shown in the figure, for simplicity. The looping of the lagging-strand template brings the 3′ end of each completed Okazaki fragment near the site where the next Okazaki fragment will start. The primase stays near the replication fork, synthesizing new RNA primers intermittently on the leading-strand template. Similarly, because the lagging-strand polymerase is complexed with the other replication proteins at the fork, that polymerase can be reused over and over at the same replication fork, synthesizing a string of Okazaki fragments as it moves with the rest of the replisome. That is, the complex of replication proteins that forms at the replication fork moves as a unit along the DNA and synthesizes new DNA simultaneously on both the leading-strand and lagging-strand templates.

The discussion has focused on a single replication fork, while in reality two replication forks are involved in a replication bubble. Figure 3.9 shows how the leading strands and lagging strands are synthesized in the early stages of bidirectional replication. Figure 3.10 shows bidirectional replication of a circular chromosome, such as that of *E. coli*.

**i Activity**

Identify some of the specific elements and processes needed for DNA replication in the iActivity *Unraveling DNA Replication* on the student website.

## Rolling Circle Replication

For some virus chromosomes, such as that of bacteriophage λ, a circular, double-stranded DNA replicates to produce linear DNA; the process is called **rolling circle replication** (Figure 3.11). The first step in rolling circle replication is the generation of a specific nick in one of the two strands at the origin of replication (Figure 3.11, step 1). The 5′ end of the nicked strand is then displaced from the circular molecule to create a replication fork (Figure 3.11, step 2). The free 3′ end of the nicked strand acts a primer for DNA polymerase to synthesize new DNA, using the single-stranded segment of the circular DNA as a template (Figure 3.11, step 3).

The displaced single strand of DNA rolls out as a free "tongue" of increasing length as replication proceeds. New DNA is synthesized by DNA polymerase on the displaced

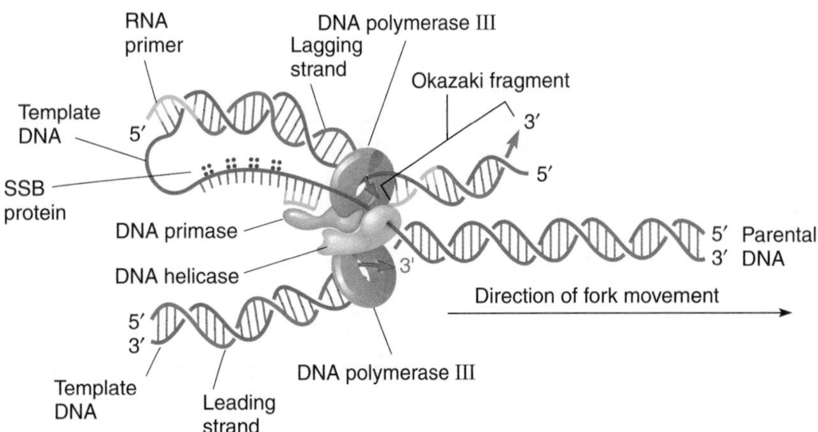


**Figure 3.8**

**Model for the replisome, the complex of key replication proteins, with the DNA at the replication fork.** The DNA polymerase III on the lagging-strand template (top of figure) is just finishing the synthesis of an Okazaki fragment.

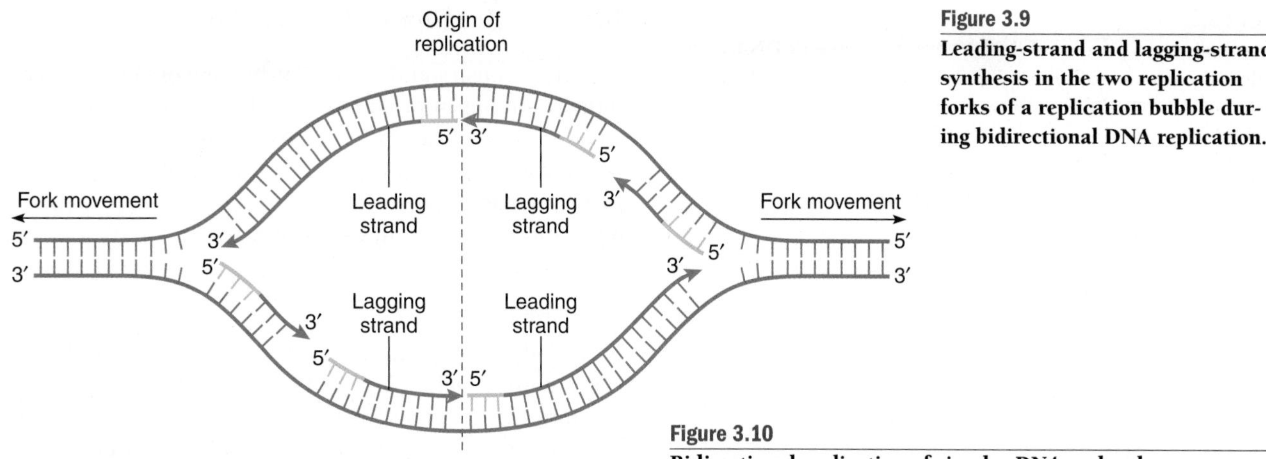


**Figure 3.9**
**Leading-strand and lagging-strand synthesis in the two replication forks of a replication bubble during bidirectional DNA replication.**

DNA in the 5′-to-3′ direction, meaning from the circle out toward the 5′ end of the displaced DNA. With further displacement, new DNA is synthesized again, beginning at the circle and moving outward along the displaced DNA strand (Figure 3.11, step 4). Thus, synthesis on this strand is discontinuous because the displaced strand is the lagging-strand template (Figure 3.5). As the single-stranded DNA tongue rolls out, new DNA synthesis proceeds continuously on the circular DNA template. Because the parental DNA circle can continue to "roll," a linear double-stranded DNA molecule can be produced that is longer than the circumference of the circle.

Let us consider the rolling circle mechanism of DNA replication for phage λ. (A full description of the life cycle of phage λ is in Chapter 15, pp. 440–445, and is diagrammed in Figure 15.12, p. 441.) Phage λ has a linear, mostly double-stranded DNA chromosome with 12-nucleotide-long, single-stranded ends (Figure 3.12). The two ends have complementary sequences—they are referred to as "sticky" ends because they can pair with one another. When phage λ infects *E. coli*, the linear chromosome is injected into the cell and the complementary ends pair. To produce copies of the chromosome to package in progeny phages, the now-circular phage chromosome replicates by the rolling circle mechanism. The result is a multi-genome-length "tongue" of head-to-tail copies of the λ chromosome. A DNA molecule like this, made up of repeated chromosome copies, is called a *concatamer.* From this concatameric molecule, unit-length progeny phage λ chromosomes are generated as follows: The phage λ chromosome has a gene called *ter* (for *ter*minus-generating activity, Figure 3.12b), which codes for a DNA endonuclease (an enzyme that digests a nucleic acid chain by cutting somewhere along its length rather than at the termini). The endonuclease binds to the *cos* sequence (see Figure 3.12b) and makes a staggered cut such that linear λ chromosomes with the correct complementary, 12-base-long, single-stranded ends are produced. The chromosomes are then packaged into the progeny λ phages.

**Figure 3.10**
**Bidirectional replication of circular DNA molecules.**

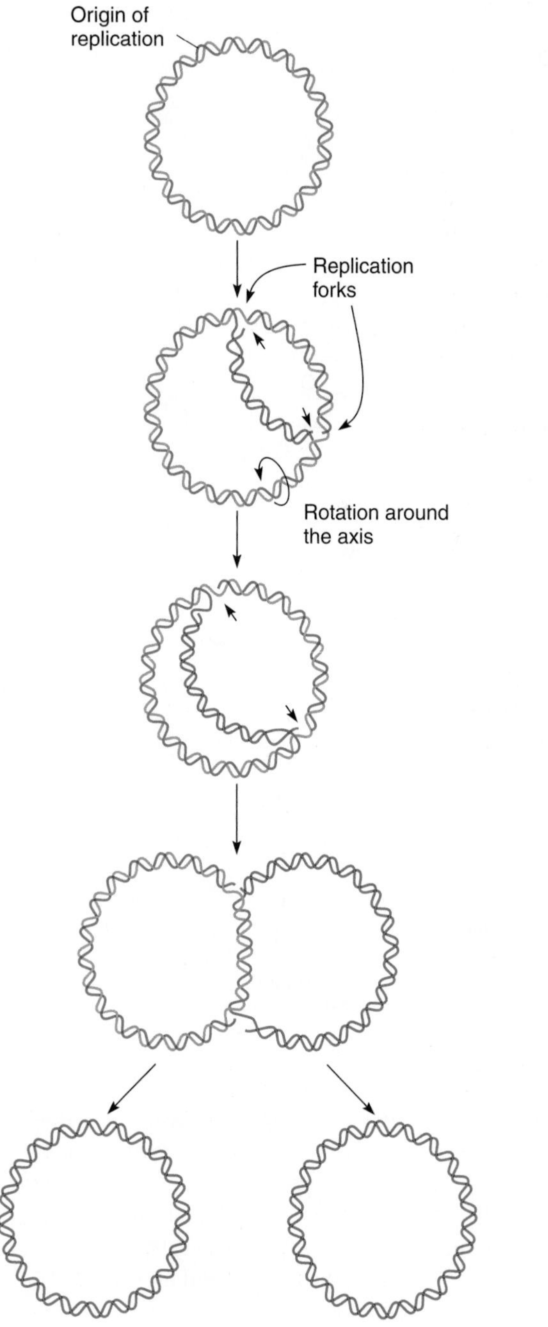

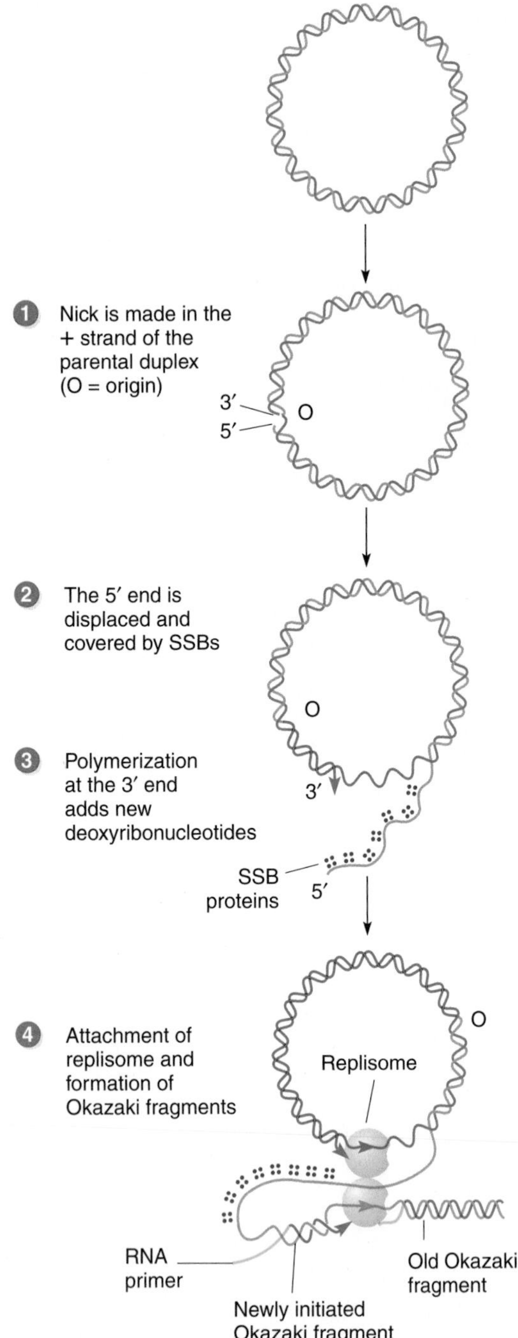
1 Nick is made in the + strand of the parental duplex (O = origin)
3′
5′
O
2 The 5′ end is displaced and covered by SSBs
O
3 Polymerization at the 3′ end adds new deoxyribonucleotides
3′
SSB proteins
5′
O
4 Attachment of replisome and formation of Okazaki fragments
Replisome
RNA primer
Newly initiated Okazaki fragment
Old Okazaki fragment

**Figure 3.11**

**The replication process of double-stranded circular DNA molecules through the rolling circle mechanism.** The active force that unwinds the 5′ tail is the movement of the replisome propelled by its helicase components.

### Keynote

During DNA replication, new DNA is made in the 5′-to-3′ direction, so chain growth is continuous on one strand and discontinuous (i.e., in segments that are later joined) on the other strand. This semidiscontinuous model is applicable to many other prokaryotic replication systems, each of which differs in the number and properties of the enzymes and proteins needed.

## DNA Replication in Eukaryotes

The biochemistry and molecular biology of DNA replication are similar in prokaryotes and eukaryotes. However, an added complication in eukaryotes is that DNA is distributed among many chromosomes rather than just one. In this section, some of the important aspects of DNA replication in eukaryotes are summarized.

### Replicons

Each eukaryotic chromosome consists of one linear DNA double helix. For example, the haploid human genome (24 chromosomes) consists of about 3 billion base pairs of DNA, meaning that the average chromosome is roughly $10^8$ base pairs long, about 25 times longer than the *E. coli* chromosome. Replication fork movement is much slower in eukaryotes than in *E. coli*; so, if there was only one origin of replication per chromosome, replicating each chromosome would take many days. In fact, eukaryotic chromosomes replicate efficiently and relatively quickly because DNA replication is initiated at many origins of replication throughout the genome. At each origin of replication, as in *E. coli*, the DNA unwinds to single strands, and replication proceeds bidirectionally. Eventually, each replication fork runs into an adjacent replication fork, initiated at an adjacent origin of replication.

The stretch of DNA from the origin of replication to the two termini of replication (where adjacent replication forks fuse) on each side of the origin is called a **replicon** or **replication unit** (Figure 3.13). The *E. coli* genome consists of one replicon, of size 4.6 Mb (million base pairs, the entire genome size), with a rate of movement of each replication fork of about 1,000 bp per second. Replicating the entire chromosome takes 42 minutes. By contrast, eukaryotic replicons are smaller. For example, there are an estimated 10,000–100,000 replicons in humans, for an average of 30–300 kb; the rate of fork movement is about 100 bp per second. Replicating the entire genome takes 8 hours, but each replicon is replicating for only part of that time.

There is a cell-specific timing of initiation of replication at the various origins of replication. Figure 3.14 shows a (theoretical) segment of one chromosome in which three replicons begin replicating at distinct times. When the replication forks fuse at the margins of adjacent replicons, the chromosome has replicated into two sister chromatids.

### Initiation of Replication

Replicators (recall from earlier discussions that they are DNA sequences that direct the initiation of replication) are less well defined in eukaryotes than in prokaryotes. In the yeast *Saccharomyces cerevisiae,* replicators are approximately 100-bp sequences called **autonomously replicating sequences (ARSs).** Replicators of more complex, multicellular organisms are less well characterized. The Focus

**Figure 3.12**

**λ chromosome structure at different stages of the phage's life cycle in *E. coli*.** (a) Parts of the λ chromosome, showing the nucleotide sequence of the two single-stranded, complementary ("sticky") ends and the chromosome circularizing after infection by pairing of the ends, with the single-stranded nicks filled in to produce a covalently closed circular chromosome. (b) Generation of the "sticky" ends of the λ DNA during replication. Replication produces a giant concatameric DNA molecule containing many tandem repeats of the λ genome. The diagram shows the joining of two adjacent λ chromosomes and the extent of the *cos* sequence. The *cos* sequence is recognized by the *ter* gene product, an endonuclease that makes two cuts at the sites shown by the arrows. These cuts produce a complete λ chromosome from the concatamer.

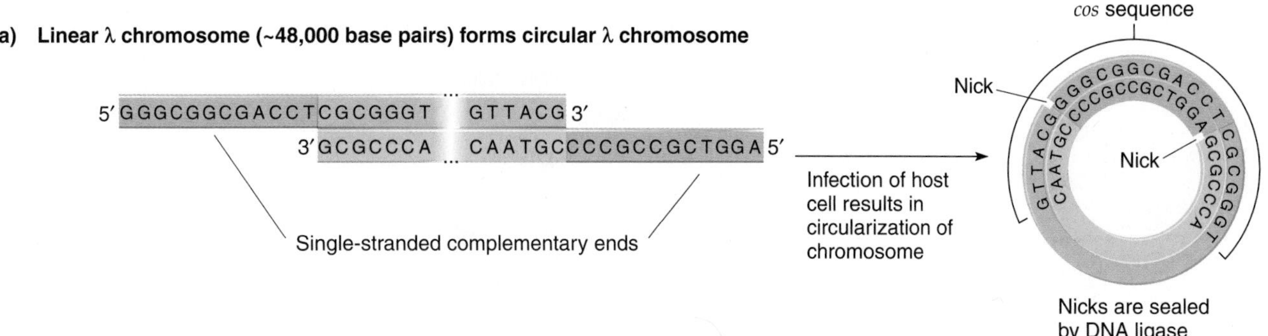


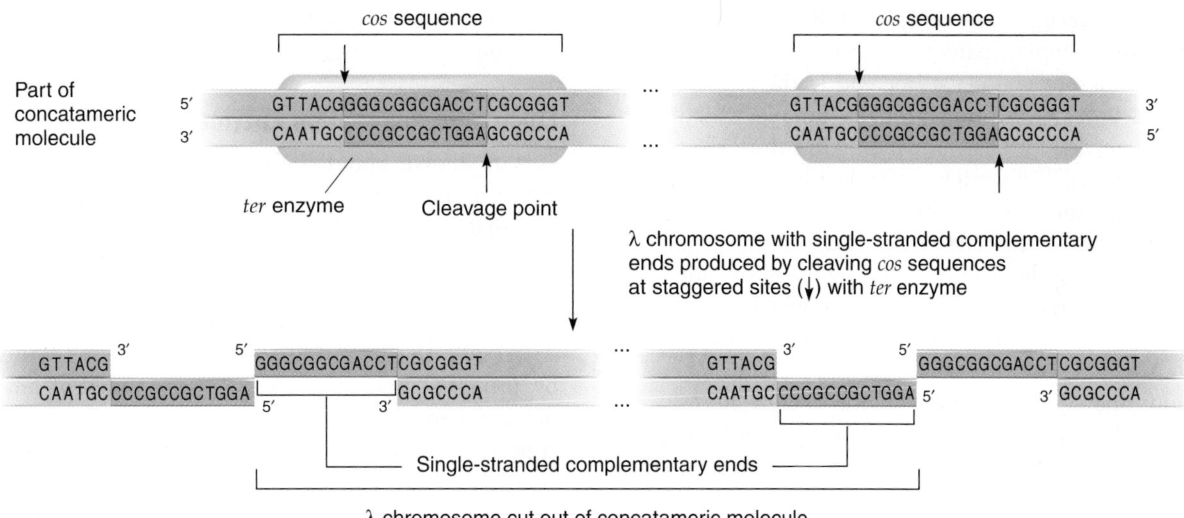


**Figure 3.13**

**Replicating DNA of *Drosophila melanogaster*.**

a) Electron micrograph of replicons

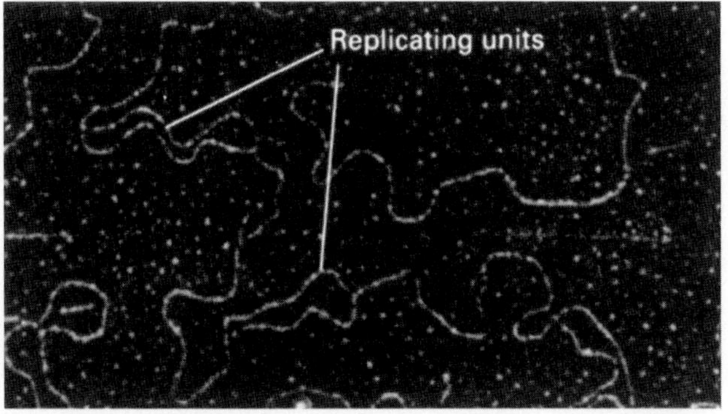


b) Schematic drawing of replicons

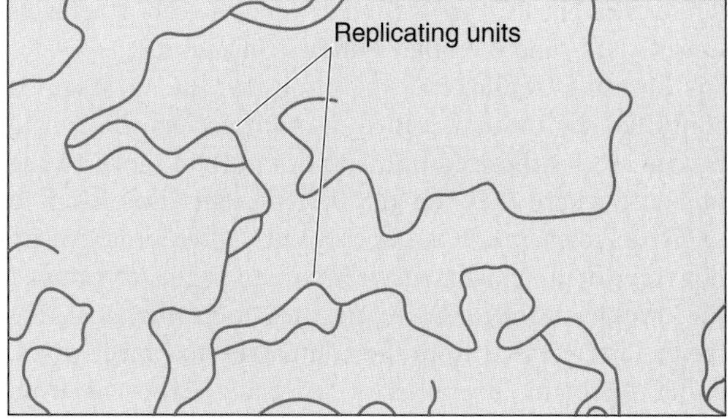

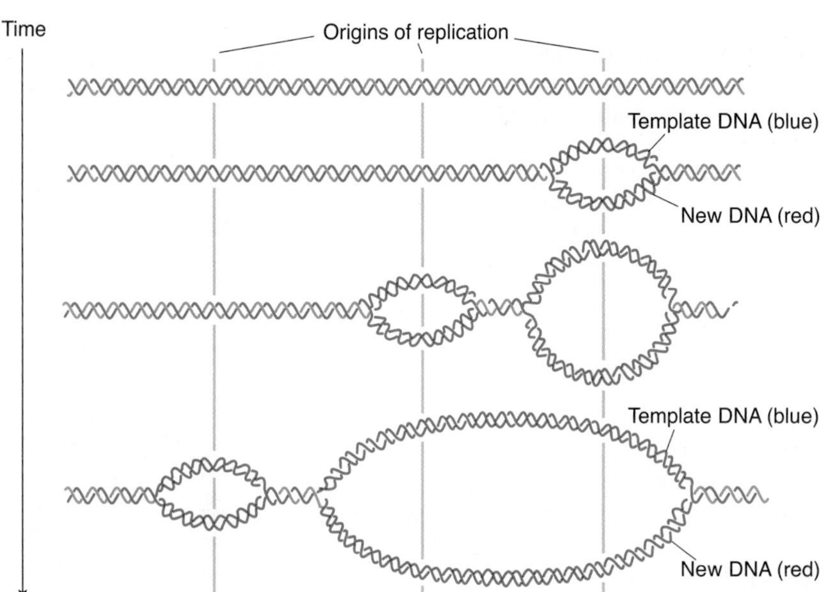


**Figure 3.14**
**Temporal ordering of DNA replication initiation events in replication units of eukaryotic chromosomes.**

on Genomics box on p. 54 describes a genomics approach to identifying replication origins in yeast.

The initiator protein in eukaryotes is the multisubunit **origin recognition complex (ORC).** The yeast replicator, for example, spans about 100 bp. The ORC binds to two different regions at one end of the replicator and recruits other replication proteins, among which is the protein needed for DNA unwinding in a third region near the other end. The origin of replication is between the first two regions and the third region.

DNA replication takes place in a specific stage of the cell division cycle. The cell cycle consists of four stages (see Figure 12.4, p. 329): $G_1$, during which the cell prepares for DNA replication; S, during which DNA replication occurs; $G_2$, during which the cell prepares for cell division; and M, the division of the cell by mitosis. For correct duplication of the chromosomes, each origin of replication must be used only once in the cell cycle. This is accomplished by a complicated series of events. In outline, the initiation of replication involves two temporally separate steps. The first step is *replicator selection*, in which ORC binds to each replicator in the $G_1$ stage and recruits other proteins to form *prereplicative complexes (pre-RCs)*. Unwinding of the DNA does not occur yet, in contrast to the case in bacteria when an initiator binds to a replicator. Rather, the pre-RCs are activated when the cell progresses from $G_1$ to S, and then they initiate replication.

Limiting replication initiation to the S stage is controlled by proteins called *licensing factors*. Licensing factors are synthesized only in $G_1$ and then move to the nucleus, where they are the first proteins that bind to ORCs to form pre-RCs (see above). Other proteins are now recruited, and the entire complex begins to untwist the double-stranded DNA. At this point the licensing factors are released from the complexes and inactivated, either by being degraded or by being exported from the nucleus, depending on the organism. Overall, the combination of the synthesis of licensing factors only in $G_1$, the way in which they function within the pre-RCs, and their directed inactivation serves to limit replication initiation at each origin to once per cell cycle.

## Eukaryotic Replication Enzymes

Less is known about the detailed functions of the enzymes and proteins involved in eukaryotic DNA replication than is the case for prokaryotic DNA replication. Eukaryotic cells may have 15 or more DNA polymerases. Typically, replication of nuclear DNA requires three of these: Pol $\alpha$ (alpha)/primase, Pol $\delta$ (delta), and Pol $\varepsilon$ (epsilon). Pol $\alpha$/primase initiates new strands in replication by primase, making about 10 nucleotides of an RNA primer; then Pol $\alpha$ adds 10–20 nucleotides of DNA. Pol $\varepsilon$ appears to synthesize the leading strand, whereas Pol $\delta$ synthesizes the lagging strand. Other eukaryotic DNA polymerases are involved in specific DNA repair processes, and yet others replicate mitochondrial and chloroplast DNA.

As in prokaryotes, joining of Okazaki fragments on the lagging-strand template involves removing the primer on the older Okazaki fragment and replacing it with DNA by extension of the newer Okazaki. Primer removal does not involve the progressive removal of nucleotides, as is the case in prokaryotes. Rather, Pol $\delta$ continues extension of the newer Okazaki fragment; this activity displaces the RNA/DNA ahead of the enzyme, producing a flap. Nucleases remove the flap. The two Okazaki fragments are then joined by the eukaryotic DNA ligase.

## Replicating the Ends of Chromosomes

Because DNA polymerases can synthesize new DNA only by extending a primer, there are special problems in

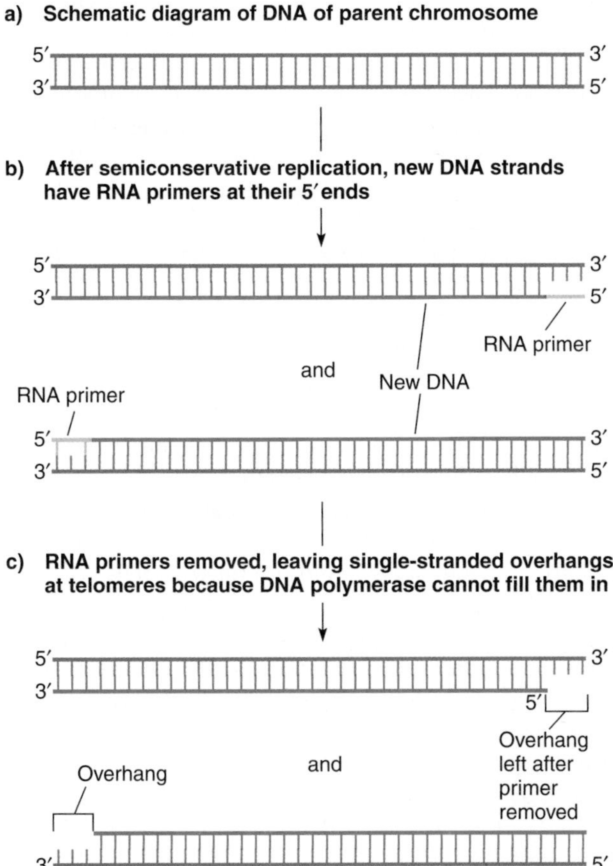
a) Schematic diagram of DNA of parent chromosome
5′
3′
3′
5′
b) After semiconservative replication, new DNA strands have RNA primers at their 5′ ends
5′
3′
3′
5′
RNA primer
and
New DNA
RNA primer
5′
3′
3′
5′
c) RNA primers removed, leaving single-stranded overhangs at telomeres because DNA polymerase cannot fill them in
5′
3′
3′
5′
Overhang left after primer removed
Overhang
and
3′
5′

**Figure 3.15**

**The problem of replicating completely a linear chromosome in eukaryotes.**

replicating the ends—the telomeres—of eukaryotic chromosomes (Figure 3.15). Replication of a parental chromosome (Figure 3.15a) produces two new DNA molecules, each of which has an RNA primer at the 5′ end of the newly synthesized strand in the telomere region (Figure 3.15b). By contrast, the numerous RNA primers in each lagging strand have been replaced by DNA during the normal DNA replication steps (Figure 3.6). Notice that the Okazaki fragment 5′ to the RNA primer is extended in 5′ to 3′ direction to replace the RNA primer. Since there is no Okazaki fragment 5′ to the primers at the 5′ ends, the same mechanism would not work at the 5′ ends. Removal of the RNA primers at the 5′ ends of the new DNA strands leaves a single-stranded stretch of parental DNA—an overhang—extending beyond the 5′ end of each new strand. DNA polymerase cannot fill in the overhang. If nothing were done about these overhangs, the chromosomes would get shorter and shorter with each replication cycle.

A special mechanism is used for replicating the ends of chromosomes. Most eukaryotic chromosomes have species-specific, tandemly repeated, simple sequences at their telomeres (see Chapter 2, p. 28). Elizabeth Blackburn and Carol W. Greider have shown that an enzyme called **telomerase** maintains chromosome lengths by adding telomere repeats to one strand (the one with the 3′ end), which serves as template on previous DNA replication at each end of a linear chromosome. The complementary strand to the one synthesized by telomerase must be added by the regular replication machinery.

Figure 3.16 is a simplified diagram of the mechanism for the addition of telomere repeats to the end of a human chromosome. The repeated sequence in humans and all other vertebrates is `5′-TTAGGG-3′`, reading toward the end of the overhanging DNA (the top strand in the figure). The actual 3′ end varies from chromosome to chromosome; shown here is the most common end sequence. Telomerase acts at the stage shown in Figure 3.15c—that is, where a chromosome end has been produced after primer removal with an overhang extending beyond the 5′ end of the new DNA (Figure 3.16a). Telomerase is an enzyme made up of both protein and RNA. The RNA component (451 bases long in humans) includes an 11-base template RNA sequence that is used for the synthesis of new telomere repeat DNA. The telomerase binds specifically to the overhanging telomere repeat on the strand of the chromosome with the 3′ end (Figure 3.16b). The 3′ end of the RNA template sequence in the telomerase—here, `3′-CAAUC-5′`—base-pairs with the `5′-GTTAG-3′` sequence at the end of the overhanging DNA strand. Next, the telomerase catalyzes the addition of new nucleotides to the 3′ end of the DNA—here, `5′-GGGTTAG-3′`—using the telomerase RNA as a template (Figure 3.16c). The telomerase then slides to the new end of the chromosome, so that the 3′ end of the RNA template sequence—`3′-CAAUC-5′`, as before—now pairs with some of the newly synthesized DNA (Figure 3.16d). Then, as before, telomerase synthesizes telomere DNA, extending the overhang (Figure 3.16e). If the telomerase leaves the DNA now, the chromosome will have been lengthened by two telomere repeats (Figure 3.16f). But, the process can recur to add more telomere repeats. In this way, the chromosome can be lengthened by the addition of a number of telomere repeats. Then, when the chromosome is replicated using the elongated strand as a template, and the primer of the new DNA strand is removed, there will still be an overhang—but any net shortening of the chromosome will have been more than compensated for due to the action of telomerase (Figure 3.16g). In most cells, the telomere DNA then loops back on itself to form a t-loop, with the single-stranded end invading the double-stranded telomeric repeat sequences to form a D-loop (see Chapter 2, p. 28, and Figure 2.25, p. 29).

The synthesis of DNA from an RNA template is called *reverse transcription*, so telomerase is an example of a **reverse transcriptase** enzyme. (The *t*elomerase *r*everse *t*ranscriptase is abbreviated TERT. Other reverse

**Figure 3.16**

**Synthesis of telomeric DNA by telomerase.** The example is of human telomeres, and the overall process is shown in a simplified way.

**a) Chromosome end after primer removal**

Overhang left after primer removal

5′ TTAGGGTTAGGGTTAG 3′

3′ AATCCC 5′

**b) Binding of telomerase to the overhanging 3′ end of the chromosome**

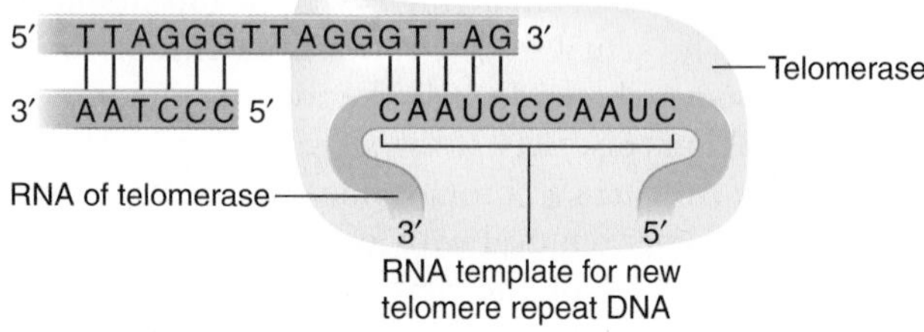


**c) Synthesis of new telomere DNA using telomerase RNA as template**

New DNA

5′ TTAGGGTTAGGGTTAGGGTTAG 3′

3′ AATCCC 5′ CAAUCCCAAUC

3′ 5′

**d) Telomerase movement to 3′ end of newly synthesized telomere DNA**

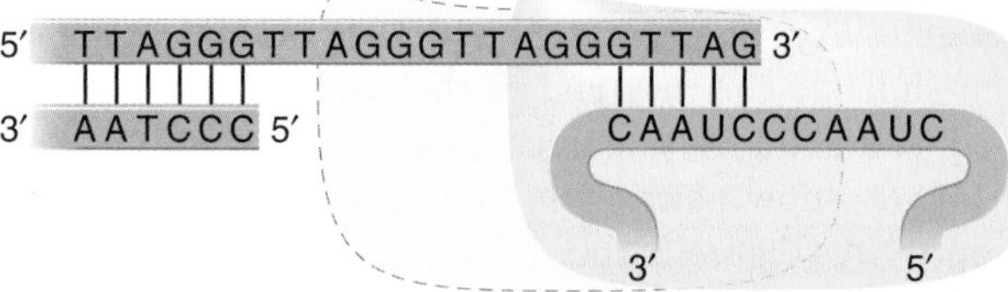


**e) Synthesis of new telomere DNA**

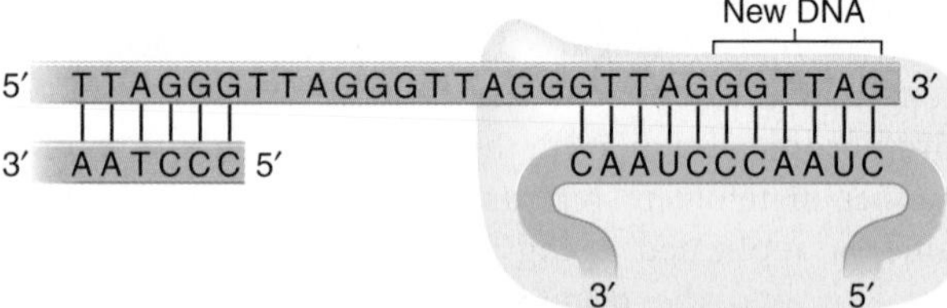


**f) Chromosome end after telomerase leaves**

**g) New end of the chromosome after replication and primer removal**

Overhang left after primer removal

5′ TTAGGGTTAGGGTTAGGGTTAGGGTTAG 3′

3′ AATCCCAATCCCAATCCC 5′

Longer 5′ end of chromosome due to telomerase activity

transcriptase enzymes are used in biotechnology applications such as reverse transcription-polymerase chain reaction—RT-PCR—described in Chapter 10, p. 264.)

Telomere length, while not identical from chromosome end to chromosome end, nonetheless is regulated to an average length for the organism and cell type. In wild-type yeast, for example, the simple telomeric sequences ($TG_{1-3}$, a repeating sequence of one T followed by one to three Gs) occupy an average of about 300 bp. Mutants are known that affect telomere length. For example, deletion of the *TLC1* gene (telomerase component 1: encodes the telomerase RNA) or mutation of the *EST1* or *EST3* (ever shorter telomeres) genes causes telomeres to shorten continuously until the cells die. This phenotype provides evidence that telomerase activity is necessary for long-term cell viability. Mutations of the *TEL1* and *TEL2* genes cause cells to maintain their telomeres at a new, shorter-than-wild-type length, making it clear that telomere length is regulated genetically.

There are many levels of regulation of telomerase activity and telomere length. For example, telomerase activity in mammals is found in immortal cells (such as tumor cells) and in some proliferative cells (such as some stem cells and sperm). The absence of telomerase activity in other cells not only results in progressive shortening of chromosome ends during successive divisions, because of the failure to replicate those ends, but also results in a limited number of cell divisions before the cell dies.

**Keynote**

Special enzymes—telomerases—replicate the ends of chromosomes in eukaryotes. A telomerase is a complex of proteins and RNA. The RNA acts as a template for synthesizing the complementary telomere repeat of the chromosome, so telomerase is a type of reverse transcriptase enzyme.

## Assembling Newly Replicated DNA into Nucleosomes

Eukaryotic DNA is complexed with histones in nucleosomes, which are the basic units of chromosomes (see Chapter 2, p. 25). Recall that there are eight histones in the histone core of the nucleosome—two each of H2A, H2B, H3, and H4. Therefore, when the DNA is replicated, the histone complement must be doubled so that all nucleosomes are duplicated. Doubling involves two processes: the synthesis of new histone proteins and the assembly of new nucleosomes. Most histone synthesis occurs during the S stage of the cell cycle, so as to be coordinated with DNA replication.

For replication to proceed, nucleosomes must disassemble during the short time when a replication fork passes; the newly replicated DNA assembles into nucleosomes almost immediately. The new nucleosomes are

**Figure 3.17**

**Assembly of new nucleosomes at a replication fork.** New nucleosomes are assembled first with the use of either a parental or a new H3–H4 tetramer and then by completing the structure with a pair of H2A–H2B dimers.

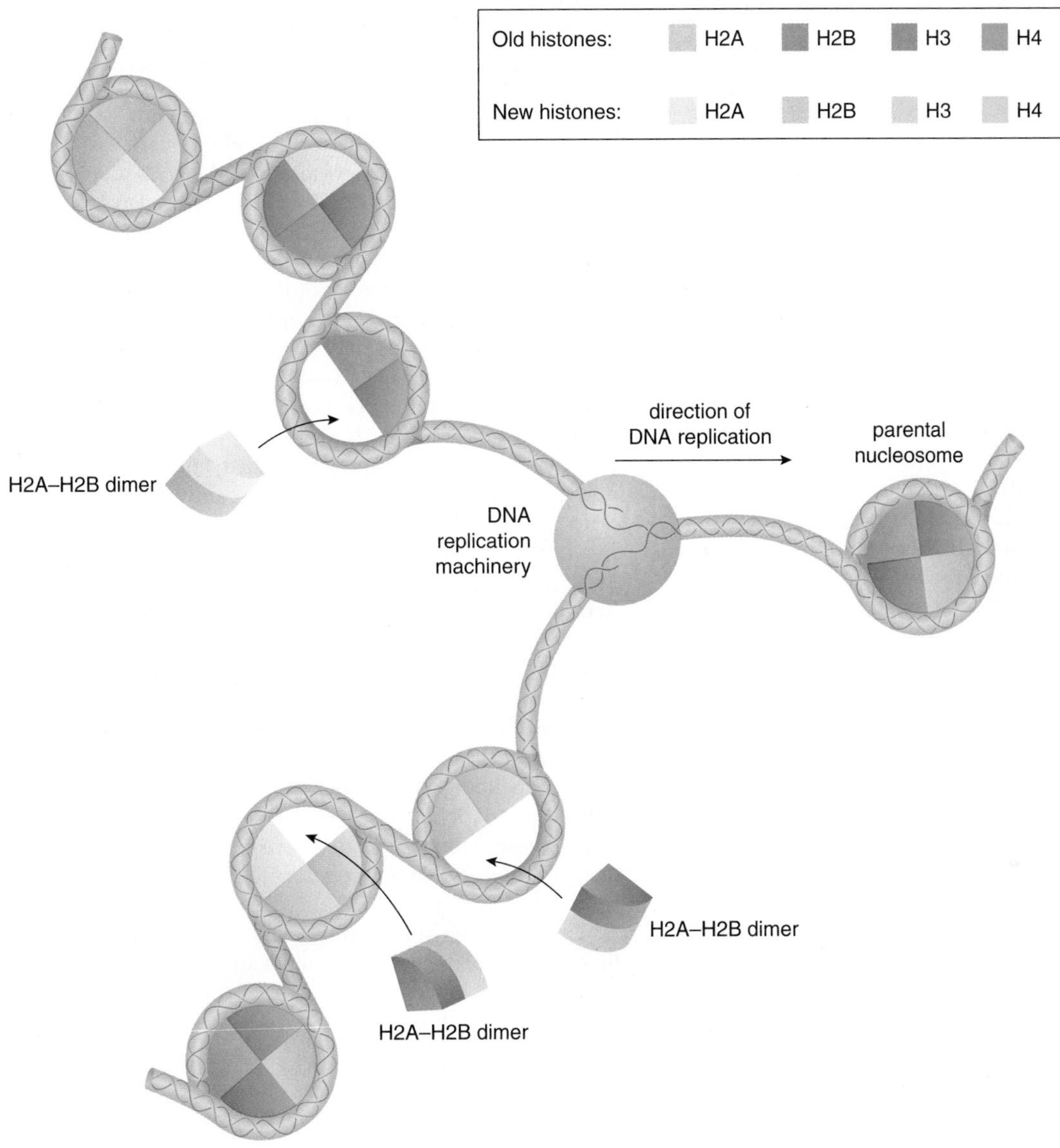


assembled as follows (Figure 3.17): Each parental histone core of a nucleosome separates into an H3–H4 tetramer (two copies each of H3 and H4) and two copies of an H2A–H2B dimer. The H3–H4 tetramer is transferred directly to one of the two replicated DNA double helices past the fork, where it begins nucleosome assembly. The H2A–H2B dimers are released, adding to the pool of newly synthesized H2A–H2B dimers. A pool of new H3–H4 tetramers is also present, and one of these tetramers initiates nucleosome assembly on the other DNA double helix past the fork. The rest of the new nucleosomes are assembled from H2A–H2B dimers, which may be parental or new. Thus, a new nucleosome will have either a parental or new H3–H4 tetramer, and a pair of H2A–H2B dimers that may be parental–parental, parental–new, or new–new. *Histone chaperone* proteins in the nucleus direct the process of nucleosome assembly.

## Focus on Genomics
### Replication Origins in Yeast

Scientists first found replication origins in brewer's yeast (*Saccharomyces cerevisiae*) by looking for pieces of DNA that triggered replication of yeast plasmids. Origins contain a 200-bp ACS (*a*utonomously repli-cating sequence *c*onsensus *s*equence) region, where a group of polypeptides (the origin recognition complex, or ORC) binds as replication begins. Using traditional molecular approaches, scientists found only about 10 percent of the origins (30 of about 400) predicted to function in the yeast genome.

Genomics made it possible to exhaustively catalog origins in yeast. When the yeast genome was sequenced, about 12,000 possible ACS regions were found, far more than the expected 400. Clearly, it takes more than an ACS to be an origin. Several experimenters used DNA microarrays (Chapter 8, pp. 192–193) to analyze many DNA sequences simultaneously. To create a DNA microarray, millions of identical, single-stranded copies of a particular DNA sequence are attached to a unique, known position on a glass slide (creating a "spot" of many copies of that one sequence). Thousands of different DNA sequences, representing genes and non-gene regions, can be placed as unique "spots" on a single glass slide (creating a large array of tiny, individual spots that we call a microarray).

The investigators "spotted" random sequences from the yeast genome onto the glass slide. Some of these spots contained origins or sequences near origins, but most did not, and the investigators needed to identify the sequences on the microarray that were origins or were near origins. Here is how they found those sequences. First, they needed a supply of DNA from cells that had just begun to replicate. They then grew yeast cells in the presence of heavy isotopes to produce denser DNA. They transferred the cells to a medium with normal, light isotopes and allowed the cells to start DNA replication. After a few minutes, they collected DNA from these cells. The newly made DNA contained one strand with light isotopes and one strand with heavy isotopes, but the unreplicated DNA contained only heavy isotopes (this is similar to part of the Meselson–Stahl experiment). They cut the DNA into small pieces and collected the less dense (replicated) DNA—because it had already replicated, it must be near an origin. The investigators labeled this DNA with a fluorescent tag, denatured it to make it single-stranded, and added it to the DNA microarray. The fluorescently labeled DNA could anneal to DNA bound to the microarray if the two DNA sequences were complementary. Pairing two DNA strands experimentally is called hybridization or probing. Fluorescent probe DNA bound to some sequences on the DNA microarray and ignored other sequences. The investigators used a laser to detect the locations of the fluorescent tags. Because they knew the exact DNA sequence at that location on the microarray, the researchers knew what sequences in the genome hybridized to the fluorescently labeled (replicated) DNA. These genome sequences are near an origin or replication. These investigators identified 332 candidate origin regions in this way.

This and other studies ultimately allowed scientists to clone 228 *S. cerevisiae* replication origins. Each of these cloned replication origins was shown to be functional in yeast cells.

## Summary

- DNA replication in prokaryotes and eukaryotes occurs by a semiconservative mechanism in which the two strands of a DNA double helix are separated and a new complementary strand of DNA is synthesized in the 5′-to-3′ direction on each of the two parental template strands. This mechanism ensures that genetic information will be copied faithfully at each cell division.
- The enzymes called DNA polymerases catalyze the synthesis of DNA. Using deoxyribonucleoside 5′-triphosphate (dNTP) precursors, all DNA polymerases make new strands in the 5′-to-3′ direction.
- DNA polymerases cannot initiate the synthesis of a new DNA strand. Most newly synthesized DNA uses RNA, the synthesis of which is catalyzed by the enzyme DNA primase.
- DNA replication in *E. coli* requires two DNA polymerases and several other enzymes and proteins. In both prokaryotes and eukaryotes, the synthesis of

DNA is continuous on one template strand and discontinuous on the other template strand—a process called semidiscontinuous replication.

- In eukaryotes, DNA replication occurs in the S phase of the cell cycle and is biochemically and molecularly similar to replication in prokaryotes.
- In prokaryotes, DNA replication begins at a single replication origin and proceeds bidirectionally. In eukaryotes, DNA replication is initiated at many replication origins along each chromosome and proceeds bidirectionally from each origin.
- Special enzymes—telomerases—replicate the ends of chromosomes in many eukaryotic cells. A telomerase is a complex of proteins and RNA. The RNA acts as a template for the synthesis of the complementary telomere repeat of the chromosome. In mammals, telomerase activity is limited to immortal cells (such as stem cells, germline cells, or tumor cells). The absence of telomerase activity in a cell results in a progressive shortening of chromosome ends as the cell divides, thereby limiting the number of somatic cell divisions.
- The nucleosome organization of eukaryotic chromosomes must be duplicated as replication forks move. Nucleosomes are disassembled to allow the replication fork to pass, and then new nucleosomes are assembled soon after a replication fork passes. Nucleosome assembly is an orderly process directed with the aid of histone chaperones.



## Analytical Approaches to Solving Genetics Problems

**Q3.1**

**a.** Meselson and Stahl used $^{15}$N-labeled DNA to prove that DNA replicates semiconservatively. The method of analysis was cesium chloride equilibrium density gradient centrifugation, in which bacterial DNA labeled in both strands with $^{15}$N (the heavy isotope of nitrogen) bands to a different position in the gradient than DNA labeled in both strands with $^{14}$N (the normal isotope of nitrogen). Starting with a mixture of $^{15}$N-containing and $^{14}$N-containing DNAs, then, two bands result after CsCl density gradient centrifugation.

When double-stranded DNA is heated to 100°C, the two strands separate because the hydrogen bonds between the strands break—a process called denaturation. When the solution is cooled slowly, any two complementary single strands will find each other and reform the double helix—a process called renaturation or reannealing. If the mixture of $^{15}$N-containing and $^{14}$N-containing DNAs is first heated to 100°C and then cooled slowly before centrifuging, the result is different. In this case, two bands are seen in exactly the same positions as before, and a new third band is seen at a position halfway between the other two. From its position relative to the other two bands, the new band is interpreted to be intermediate in density between the other two bands. Explain the existence of the three bands in the gradient.

**b.** DNA from *E. coli* containing $^{15}$N in both strands is mixed with DNA from another bacterial species, *Bacillus subtilis,* containing $^{14}$N in both strands. Two bands are seen after CsCl density gradient centrifugation. If the two DNAs are mixed, heated to 100°C, slowly cooled, and then centrifuged, two bands again result. The bands are in the same positions as in the unheated DNA experiment. Explain these results.

**A3.1**

**a.** When DNA is heated to 100°C, it is denatured to single strands. If denatured DNA is allowed to cool slowly, complementary strands renature to produce double-stranded DNA again. Thus, when mixed, denatured $^{15}$N–$^{15}$N DNA and $^{14}$N–$^{14}$N DNA from the same species is cooled slowly, the single strands pair randomly during renaturation so that $^{15}$N–$^{15}$N, $^{14}$N–$^{14}$N, and $^{15}$N–$^{14}$N double-stranded DNA are produced. The latter type of DNA has a density intermediate between those of the two other types, accounting for the third band. Theoretically, if all DNA strands pair randomly, there should be a 1:2:1 distribution of $^{15}$N–$^{15}$N, $^{15}$N–$^{14}$N, and $^{14}$N–$^{14}$N DNAs, and this ratio should be reflected in the relative intensities of the bands.

**b.** DNA molecules from different bacterial species have different sequences. In other words, DNA from one species typically is not complementary to DNA from another species. Therefore, only two bands are seen because only the two *E. coli* DNA strands can renature to form $^{15}$N–$^{15}$N DNA, and only the two *B. subtilis* DNA strands can renature to form $^{14}$N–$^{14}$N DNA. No $^{15}$N–$^{14}$N hybrid DNA can form, so in this case there is no third band of intermediate density.

**Q3.2** What would be the effect on chromosome replication in *E. coli* strains carrying deletions of the following genes?

**a.** *dnaE*
**b.** *polA*
**c.** *dnaG*
**d.** *lig*
**e.** *ssb*
**f.** *oriC*

**A3.2** When genes are deleted, the function encoded by those genes is lost. All the genes listed in the question are involved in DNA replication in *E. coli,* and their functions are briefly described in Table 3.1 and discussed further in the text.

**a.** *dnaE* encodes a subunit of DNA polymerase III, the principal DNA polymerase in *E. coli* that is responsible for elongating DNA chains. A deletion of the *dnaE* gene undoubtedly would lead to a nonfunctional DNA polymerase III. In the absence of DNA polymerase III activity, DNA strands could not be synthesized from RNA primers; therefore, new DNA strands could not be synthesized, and there would be no chromosome replication.

**b.** *polA* encodes DNA polymerase I, which is used in DNA synthesis to extend DNA chains made by DNA polymerase III while simultaneously excising the RNA primer by 5′-to-3′ exonuclease activity. As discussed in the text, in mutant strains lacking the originally studied DNA polymerase—DNA polymerase I—chromosome replication still occurred. Thus, chromosomes would replicate normally in an *E. coli* strain carrying a deletion of *polA*.

**c.** *dnaG* encodes DNA primase, the enzyme that synthesizes the RNA primer on the DNA template. Without the synthesis of the short RNA primer, DNA polymerase III cannot initiate DNA synthesis, so chromosome replication will not take place.

**d.** *lig* encodes DNA ligase, the enzyme that catalyzes the ligation of Okazaki fragments. In a strain carrying a deletion of *lig*, DNA would be synthesized. However, stable progeny chromosomes would not result, because the Okazaki fragments could not be ligated together, so the lagging strand synthesized discontinuously on the lagging-strand template would be in fragments.

**e.** *ssb* encodes the single-strand binding proteins that bind to and stabilize the single-stranded DNA regions produced as the DNA is unwound at the replication fork. In the absence of single-strand binding proteins, DNA replication would be impeded or absent, because the replication bubble could not be kept open.

**f.** *oriC* is the origin-of-replication region in *E. coli*—that is, the location at which chromosome replication is initiated. Without the origin, the initiator protein cannot bind, and no replication bubble can form, so chromosome replication cannot take place.



# Questions and Problems

**3.1** Describe the Meselson–Stahl experiment, and explain how it showed that DNA replication is semiconservative.

***3.2** In the Meselson–Stahl experiment, $^{15}$N-labeled cells were shifted to a $^{14}$N medium at what we can designate as generation 0.

**a.** For the semiconservative model of replication, what proportion of $^{15}$N–$^{15}$N, $^{15}$N–$^{14}$N, and $^{14}$N–$^{14}$N DNA would you expect to find after one, two, three, four, six, and eight replication cycles?

**b.** Answer (a) in terms of the conservative model of DNA replication.

**3.3** A spaceship lands on Earth, bringing with it a sample of extraterrestrial bacteria. You are assigned the task of determining the mechanism of DNA replication in this organism.

You grow the bacteria in an unlabeled medium for several generations and then grow it in the presence of $^{15}$N for exactly one generation. You extract the DNA and subject it to CsCl centrifugation. The banding pattern you find is as follows:

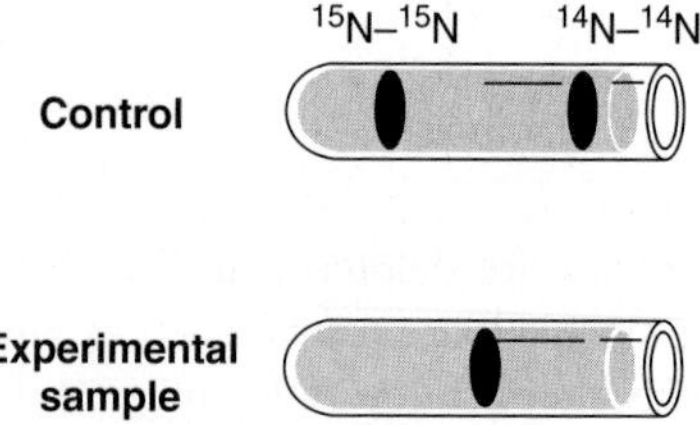


It appears to you that this pattern is evidence that DNA replicates in the semiconservative manner, but you are wrong. Why? What other experiment could you perform (using the same sample and technique of CsCl centrifugation) that would further distinguish between semiconservative and dispersive modes of replication?

***3.4** The elegant Meselson–Stahl experiment was among the first experiments to contribute to what is now a highly detailed understanding of DNA replication. Consider this experiment again in light of current molecular models by answering the following questions:

**a.** Does the fact that DNA replication is semiconservative mean that it must be semidiscontinuous?

**b.** Does the fact that DNA replication is semidiscontinuous ensure that it is also semiconservative?

**c.** Do any properties of known DNA polymerases ensure that DNA is synthesized semiconservatively?

***3.5** List the components necessary to make DNA in vitro, using the enzyme system isolated by Kornberg.

***3.6** Each of the following templates is added to an in vitro DNA synthesis reaction using the enzyme system isolated by Kornberg with 5′-ATG-3′ as a primer.

3′-TACCCCCCCCCCCCC-5′

3′-TACGCATGCATGCAT-5′

3′-TACTTTTTTTTTTTT-5′

In what ways besides their sequence will the synthesized molecules differ if a trace amount of each of the following nucleotides is added to the reaction?

**a.** $\alpha$-$^{32}$P-dATP (dATP where the phosphorus closest to the 5′-carbon is radioactive)

**b.** $^{32}$P-dAMP (dAMP where the phosphorus is radioactive)

**c.** $\gamma$-$^{32}$P-dATP (dATP where the phosphorus furthest from the 5′-carbon is radioactive)

***3.7** How do we know that the Kornberg enzyme is not the main enzyme involved in DNA synthesis for chromosome duplication in the growth of *E. coli*?

**3.8** Kornberg isolated DNA polymerase I from *E. coli*. What is the function of the enzyme in DNA replication?

**3.9** Suppose you have a DNA molecule with the base sequence `TATCA`, going from the 5′ to the 3′ end of one of the polynucleotide chains. The building blocks of the DNA are drawn as in the following figure:

A G C T

PPP OH PPP OH PPP OH PPP OH

Use this shorthand system to diagram the completed double-stranded DNA molecule, as proposed by Watson and Crick.

**3.10** Use the shorthand notation of Question 3.9 to diagram how a strand with the sequence `3′-GGTCTAA-5′` would anneal to a primer having the sequence `5′-AGA-3′`. Then answer the following questions.

**a.** What chemical groups do you expect to find at the 5′ and 3′ ends of each DNA strand?

**b.** What nucleotides would be used to extend the primer if the annealed DNA molecules are added to an in vitro DNA synthesis reaction using the system established by Kornberg?

**c.** What is the source of the energy used to catalyze the formation of phosphodiester bonds in the synthesis reaction in part (b)?

**d.** On a distant planet, cellular life is found to have a novel DNA polymerase that synthesizes a complementary DNA strand from a primed, single-stranded template, but does so only in the 3′-to-5′ direction. What nucleotides would be added to the primer if the annealed DNAs were present in a cell with this polymerase?

**e.** Reflect on your answer to part (c). Do you think the novel DNA polymerase catalyzes the formation of phosphodiester bonds in the same way as Earth DNA polymerases? If not, how might it catalyze the formation of phosphodiester bonds?

**f.** It would be faster if DNA polymerases could synthesize DNA in both the 3′-to-5′ and 5′-to-3′ directions. Speculate on why no known Earth DNA polymerase can synthesize DNA in both directions even though this seems to be a desirable trait.

**3.11** Listed below are three enzymatic properties of DNA polymerases.

1. All DNA polymerases replicate DNA only 5′ to 3′.
2. During DNA replication, DNA polymerases synthesize DNA from an RNA primer.
3. Only some DNA polymerases have 5′-to-3′ exonuclease activity.

Explain whether each of these properties constrains DNA replication to be

**a.** semiconservative.

**b.** semidiscontinuous.

***3.12** Base analogs are compounds that resemble the natural bases found in DNA and RNA but are not normally found in those macromolecules. Base analogs can replace their normal counterparts in DNA during in vitro DNA synthesis. Researchers studied four base analogs for their effects on in vitro DNA synthesis using *E. coli* DNA polymerase. The results were as follows, with the amounts of DNA synthesized expressed as percentages of the DNA synthesized from normal bases only:

| | Normal Bases Substituted by the Analog | | | |
|---|---|---|---|---|
| **Analog** | A | T | C | G |
| A | 0 | 0 | 0 | 25 |
| B | 0 | 54 | 0 | 0 |
| C | 0 | 0 | 100 | 0 |
| D | 0 | 97 | 0 | 0 |

Which bases are analogs of adenine? of thymine? of cytosine? of guanine?

**3.13** Concerning DNA replication:

**a.** Describe (draw) models of continuous, semidiscontinuous, and discontinuous DNA replication.

**b.** What was the contribution of Reiji and Tuneko Okazaki and colleagues with regard to these replication models?

**3.14** The following events, steps, or reactions occur during *E. coli* DNA replication. For each entry in column A, select its match(es) from column B. Each entry in A may have more than one match, and each entry in B can be used more than once.

| | Column A | Column B |
|---|---|---|
| _____ | **a.** Unwinds the double helix | **A.** Polymerase I |
| _____ | **b.** Prevents reassociation of complementary bases | **B.** Polymerase III |
| _____ | **c.** Is an RNA polymerase | **C.** Helicase |
| _____ | **d.** Is a DNA polymerase | **D.** Primase |
| _____ | **e.** Is the "repair" enzyme | **E.** Ligase |
| _____ | **f.** Is the major elongation enzyme | **F.** SSB protein |
| _____ | **g.** Is a 5′-to-3′ polymerase | **G.** Gyrase |
| _____ | **h.** Is a 3′-to-5′ polymerase | **H.** None of these |
| _____ | **i.** Has 5′-to-3′ exonuclease function | |
| _____ | **j.** Has 3′-to-5′ exonuclease function | |
| _____ | **k.** Bonds the free 3′-OH end of a polynucleotide to a free 5′-monophosphate end of polynucleotide | |
| _____ | **l.** Bonds the 3′-OH end of a polynucleotide to a free 5′ nucleotide triphosphate | |
| _____ | **m.** Separates daughter molecules and causes supercoiling | |

***3.15** Distinguish between the actions of helicase and topoisomerase on double-stranded DNA and their roles during DNA replication.

**3.16** How long would it take *E. coli* to replicate its entire genome ($4.2 \times 10^6$ bp), assuming a replication rate of 1,000 nucleotides per second at each fork with no pauses?

***3.17** A diploid organism has $4.5 \times 10^8$ bp in its DNA. The DNA is replicated in 3 minutes. Assuming that all replication forks move at a rate of $10^4$ bp per minute, how many replicons (replication units) are present in the organism's genome?

***3.18** Describe the molecular action of the enzyme DNA ligase. What properties would you expect an *E. coli* cell to have if it had a temperature-sensitive mutation in the gene for DNA ligase?

***3.19** Chromosome replication in *E. coli* commences from a constant point, called the origin of replication. It is known that DNA replication is bidirectional. Devise a biochemical experiment to prove that the *E. coli* chromosome replicates bidirectionally. (Hint: Assume that the amount of gene product is directly proportional to the number of genes.)

**3.20** Reiji Okazaki concluded that both DNA strands could not replicate continuously. What evidence led him to this conclusion?

***3.21** A space probe returns from Jupiter and brings with it a new microorganism for study. It has double-stranded DNA as its genetic material. However, studies of replication of the alien DNA reveal that, although the process is semiconservative, DNA synthesis is continuous on both the leading-strand and the lagging-strand templates. What conclusions can you draw from this result?

**3.22** A space probe returning from Europa, one of Jupiter's moons, carries back an organism having linear chromosomes composed of double-stranded DNA. Like Earth organisms, its DNA replication is semiconservative. However, it has just one DNA polymerase, and this polymerase initiates DNA replication only at one, centrally located site using a DNA-primed template strand.

**a.** What enzymatic properties must its DNA polymerase have?

**b.** How is DNA replication in this organism different from DNA replication in *E. coli,* which is also initiated at just one site?

**3.23** Some phages, such as λ, are packaged from concatamers.

**a.** What is a concatamer, and what type of DNA replication is responsible for producing a concatamer?

**b.** In what ways does this type of DNA replication differ from that used by *E. coli?*

***3.24** Although λ is replicated into a concatamer, linear unit-length molecules are packaged into phage heads.

**a.** What enzymatic activity is required to produce linear unit-length molecules, how does it produce molecules that contain a single complete λ genome, and what gene encodes the enzyme involved?

**b.** What types of ends are produced when this enzyme acts on DNA, and how are these ends important in the λ life cycle?

***3.25** M13 is an *E. coli* bacteriophage whose capsid holds a closed circular DNA molecule with 2,221 T, 1,296 C, 1,315 G, and 1,575 A nucleotides. M13 lacks a gene for DNA polymerase and so must use bacterial DNA polymerases for replication. Unlike λ this phage does not form concatamers during replication and packaging.

**a.** Suppose the M13 chromosome were replicated in a manner similar to the way the *E. coli* chromosome is replicated, using semidiscontinuous replication from a double-stranded circular DNA template. How would the semidiscontinuous DNA replication mechanism discussed in the text need to be modified?

**b.** Suppose the M13 chromosome were replicated in a manner similar to the way the λ chromosome is replicated, using rolling circle replication. How would the rolling circle replication mechanism discussed in the text need to be modified?

***3.26** Compare and contrast eukaryotic and prokaryotic DNA polymerases.

**3.27** What mechanism do eukaryotic cells employ to keep their chromosomes from replicating more than once per cell cycle?

**3.28** A mutation occurs that results in the failure of licensing factors to be inactivated after they are released from prereplicative complexes. What molecular consequences do you predict for this mutation?

***3.29** Autoradiography is a technique that allows radioactive areas of chromosomes to be observed under the microscope. The slide is covered with a photographic emulsion, which is exposed by radioactive decay. In regions of exposure, the emulsion forms silver grains on being developed. The tiny silver grains can be seen on top of the (much larger) chromosomes. Devise a method to find out which regions in the human karyotype replicate during the last 30 minutes of the S phase. (Assume a cell cycle in which the cell spends 10 hours in $G_1$, 9 hours in S, 4 hours in $G_2$, and 1 hour in M.)

**3.30** In typical human fibroblasts in culture, the $G_1$ period of the cell cycle lasts about 10 hours, S lasts about 9 hours, $G_2$ takes 4 hours, and M takes 1 hour. Suppose you added radioactive ($^3H$) thymidine to the medium, left it there for 5 minutes, and then washed it out and replaced it with an ordinary medium.

**a.** What percentage of cells would you expect to become labeled by incorporating the $^3H$-thymidine into their DNA?

**b.** How long would you have to wait after removing the $^3H$ medium before you would see labeled metaphase chromosomes?

**c.** Would one or both chromatids be labeled?

**d.** How long would you have to wait if you wanted to see metaphase chromosomes containing $^3H$ in the regions of the chromosomes that replicated at the beginning of the S period?

**3.31** Suppose you performed the experiment in Question 3.30, but left the radioactive medium on the cells for 16 hours instead of 5 minutes. How would your answers change?

**3.32** How is chromosomal organization related to the chromosome's temporal pattern of replication?

***3.33** A trace amount of a radioactively labeled nucleotide is added to a rapidly dividing population of *E. coli*. After a minute, and again after 30 minutes, nucleic acid is isolated and analyzed for the presence of radioactivity. Explain whether you expect to find radioactivity in small (<1,000 nucleotide) or large (>10,000 nucleotide) DNA fragments, or neither, at each time point if the radioactively labeled nucleotide is

**a.** UTP uniformly labeled with $^3H$ (tritium)

**b.** dATP uniformly labeled with $^3H$ (tritium)

**c.** $\alpha$-$^{32}P$-dATP (dATP where the phosphorus closest to the 5′-carbon is radioactive)

**d.** $\alpha$-$^{32}P$-UTP (UTP where the phosphorus closest to the 5′-carbon is radioactive)

**e.** $\gamma$-$^{32}P$-dATP (dATP where the phosphorus furthest from the 5′-carbon is radioactive)

**3.34** When the eukaryotic chromosome duplicates, the nucleosome structures must duplicate.

**a.** How is the synthesis of histones related to the cell cycle?

**b.** One possibility for the assembly of new nucleosomes on replicated DNA is that it is semiconservative. That is, parental nucleosomes are assembled on one daughter double helix and newly synthesized nucleosomes are synthesized on the other daughter double helix. Is this what happens? If not, what does occur?

***3.35** A mutant *Tetrahymena* has an altered repeated sequence in its telomeric DNA. What change in the telomerase enzyme would produce this phenotype?

**3.36** What is the evidence that telomere length is regulated in cells, and what are the consequences of the misregulation of telomere length?

# 4 Gene Function

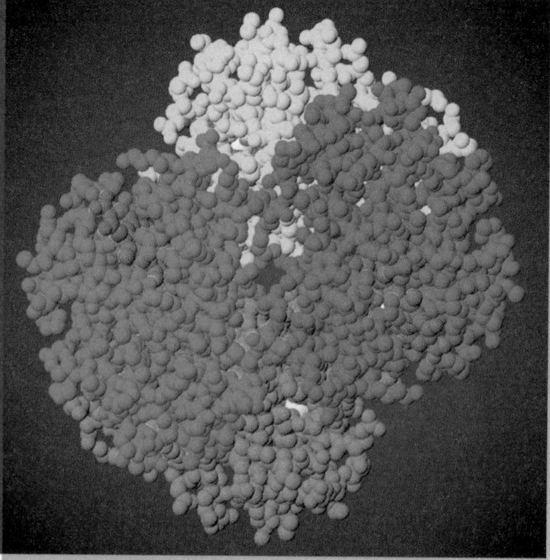
The protein hemoglobin.

## Key Questions

- What is the relationship between genes and enzymes?
- How do genes control biochemical pathways?
- What is the relationship between genes and nonenzymatic proteins?
- How can people be tested for mutations causing genetic diseases?

WITHIN THE FIRST FEW MINUTES OF LIFE, MOST newborns in the United States are subjected to a battery of tests: Reflexes are tested, respiration and skin color assessed, and blood samples collected and rushed to a lab. Assays of the blood samples help health practitioners determine whether the child has a debilitating or even lethal genetic disease. What are genetic diseases? What is the relationship between genes, enzymes, and genetic disease? How can understanding gene function help prevent or minimize the risk of such diseases?

What do bread mold and certain human genetic disorders have in common? In the iActivity for this chapter, you will use Beadle and Tatum's experimental procedure to learn the answer to that question.

In this chapter, we examine gene function. We present some of the classic evidence that genes code for enzymes and for nonenzymatic proteins. Through examining the genetic control of biochemical pathways, you will see that genes do not function in isolation, but in cooperation with other genes for cells to function properly. Understanding the functions of genes and how genes are regulated are fundamental goals for geneticists.

The experiments discussed in this chapter represent the beginnings of molecular genetics, historically speaking, in that their goal was to understand better a gene at the molecular level. In following chapters, we develop our modern understanding of gene structure and expression.

## Gene Control of Enzyme Structure

### Garrod's Hypothesis of Inborn Errors of Metabolism

In 1902, Archibald Garrod, an English physician, and geneticist William Bateson studied *alkaptonuria* (Online Mendelian Inheritance in Man [OMIM], http://www.ncbi.nlm.nih.gov/omim, entry 203500), a human disease characterized by urine that turns black upon exposure to the air and by a tendency to develop arthritis later in life. Because of the urine phenotype, the disease is easily detected soon after birth. The researchers' results suggested that alkaptonuria is a genetically controlled trait caused by homozygosity for a recessive allele. In 1908 Garrod reported the results of studying a larger number of families and provided proof that alkaptonuria is a recessive genetic disease. Many human genetic diseases are recessive—meaning that, to develop the disease, an individual must inherit one recessive mutant allele for the gene responsible for the disease from each parent, making that individual homozygous for the allele.

Garrod found that people with alkaptonuria excrete homogentisic acid (HA) in their urine, whereas people without the disease do not; it is the HA in urine that turns it black in air. This result indicated to Garrod that normal people can metabolize HA, but that people with alkaptonuria cannot. In Garrod's terms, the disease is an example of an **inborn error of metabolism;** that is, alkaptonuria is a genetic disease caused by the absence of a particular enzyme necessary for HA metabolism. Figure 4.1 shows part of the phenylalanine–tyrosine metabolic pathway: the HA-to-maleylacetoacetic acid step cannot be carried out in people with alkaptonuria. The mutation responsible for alkaptonuria is recessive, so only people homozygous for the mutant gene express the defect. Later analysis has pinpointed the location of this gene on chromosome 3. Garrod's work provided the first evidence of a specific relationship between genes and enzymes.

An important aspect of Garrod's analysis of alkaptonuria and of three other human genetic diseases that affected biochemical processes was his understanding that the position of a block in a metabolic pathway can be determined by the accumulation of the chemical compound (HA in the case of alkaptonuria) that precedes the blocked step. However, the significance of Garrod's work was not appreciated by his contemporaries.

## The One-Gene–One-Enzyme Hypothesis

In 1942, George Beadle and Edward Tatum heralded the beginnings of biochemical genetics, a branch of genetics that combines genetics and biochemistry to explain the nature of metabolic pathways. Results of their studies involving the haploid fungus *Neurospora crassa* (orange bread mold) showed a direct relationship between genes and enzymes and led to the *one-gene–one-enzyme hypothesis*, a landmark in the history of genetics. Beadle and Tatum shared one-half of the 1958 Nobel Prize in Physiology or Medicine for their "discovery that genes act by regulating definite chemical events."

**animation**
**The One-Gene–One-Enzyme Hypothesis**

**Isolation of Nutritional Mutants of *Neurospora*.** To understand Beadle and Tatum's experiment, we must understand the life cycle of *Neurospora crassa*, the orange bread mold (Figure 4.2). *Neurospora crassa* is a mycelial-form fungus, meaning that it spreads over its growth medium in a weblike pattern (Figure 1.04g, p. 6). The mycelium produces asexual spores called conidia; their orange color gives the fungus its common name. *Neurospora* has important properties that make it useful for genetic and biochemical studies including the fact that it is a haploid organism, so the effects of mutations may be seen directly, and that it has a short life cycle, enabling rapid study of the segregation of genetic defects.

*Neurospora* can be propagated vegetatively (asexually) by inoculating either pieces of the mycelial growth or the asexual spores (conidia) on a suitable growth medium to give rise to a new mycelium. *Neurospora crassa* can also reproduce by sexual means. There are two mating types ("sexes," in a loose sense), called *A* and *a*. The two mating types look identical and can be distinguished only because strains of the *A* mating type do not mate with other *A* strains, and *a* strains do not mate with other *a* strains. The sexual cycle is initiated by mixing *A* and *a* mating-type strains on nitrogen-limiting medium. Under these conditions, cells of the two mating types fuse, followed by fusion of two haploid nuclei to produce

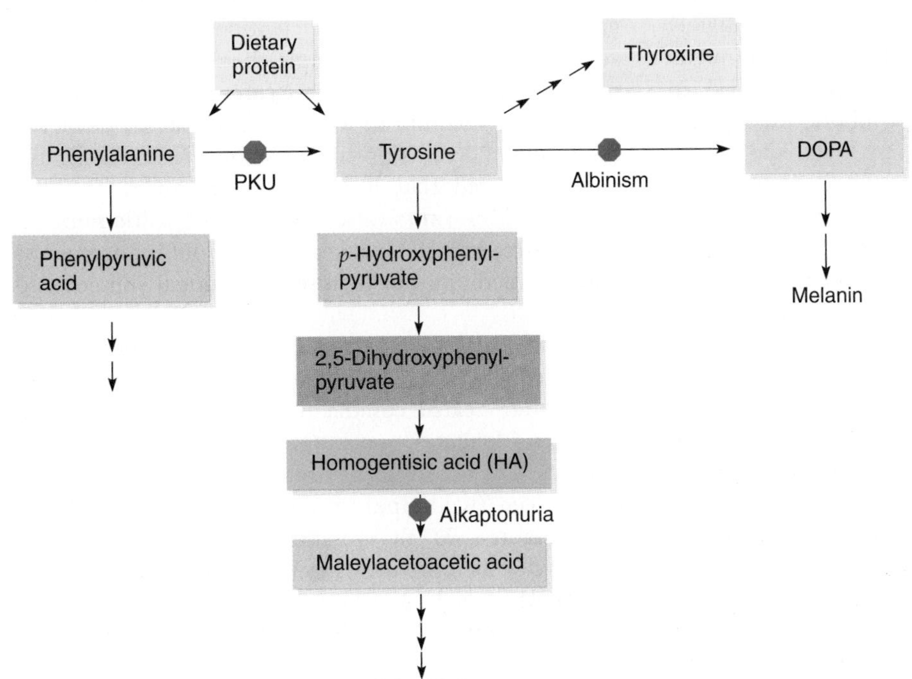


**Figure 4.1**
**Phenylalanine–tyrosine metabolic pathways.** People with alkaptonuria cannot metabolize homogentisic acid (HA) to maleylacetoacetic acid, causing HA to accumulate. People with PKU cannot metabolize phenylalanine to tyrosine, causing phenylpyruvic acid to accumulate. People with albinism cannot synthesize much melanin from tyrosine.

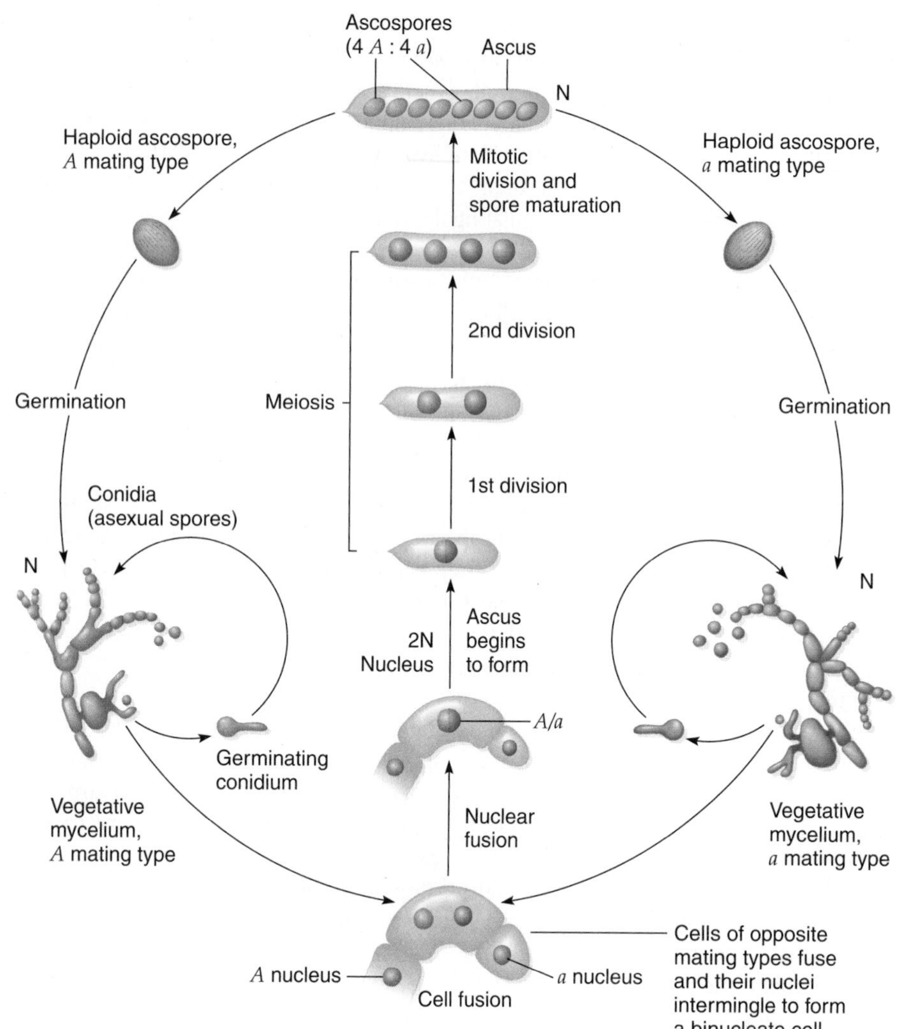


**Figure 4.2**
**Life cycle of the haploid, mycelial-form fungus *Neurospora crassa*.** (Parts not to scale.)

a transient *A/a* diploid nucleus, which is the only diploid stage of the life cycle. The diploid nucleus immediately undergoes meiosis and produces four haploid nuclei (two *A* and two *a*) within an elongating sac called an *ascus* (plural = *asci*). A subsequent mitotic division results in a linear arrangement of eight haploid nuclei around which spore walls form to produce eight sexual ascospores (four *A* and four *a*). Each ascus, then, contains all the products of the initial, single meiosis. Several asci develop within a fruiting body. When an ascus is ripe, the ascospores (sexual spores) are shot out of it and out of the fruiting body to be dispersed by wind currents. Germination of an ascospore begins the formation of a new haploid mycelium.

The simple growth requirements of *Neurospora* were important for Beadle and Tatum's experiments. Wild-type *Neurospora* grows on a *minimal medium*, that is, on the simplest set of chemicals needed for the organism to grow and survive. The minimal medium for *Neurospora* contains only inorganic salts (including a source of nitrogen), an organic carbon source (such as glucose or sucrose), and the vitamin biotin. A strain that can grow on the minimal medium is called a **prototrophic strain** or a **prototroph.** Beadle and Tatum reasoned that *Neurospora* synthesized the other materials it needed for growth (e.g., amino acids, nucleotides, vitamins, nucleic acids, proteins) from the simple chemicals present in the minimal medium. Wild-type *Neurospora* can also grow on minimal medium to which nutritional supplements, such as amino acids or vitamins, are added. Beadle and Tatum realized that it should be possible to isolate **nutritional mutants** (also called **auxotrophic mutants** or **auxotrophs**) of *Neurospora* that would not grow on minimal medium, but required nutritional supplements to grow.

Beadle and Tatum isolated and characterized auxotrophic mutants. To isolate auxotrophic mutants, Beadle and Tatum treated conidia with X-rays. An X-ray is a **mutagen** ("*mutation generator*"), an agent that induces mutants. They crossed the mutants they obtained with a *prototrophic* (wild-type) *strain* of the opposite mating type (Figure 4.3). By crossing the mutagenized spores with the wild type, they ensured that any auxotrophic mutant they isolated was heritable and therefore had a genetic basis, rather than a nongenetic reason, for requiring the nutrient.

The researchers allowed one progeny per ascus from the crosses to germinate in a nutritionally *complete*

**Figure 4.3**

**Method devised by Beadle and Tatum to isolate auxotrophic mutations in *Neurospora*.** Here, the mutant strain isolated is a tryptophan auxotroph.

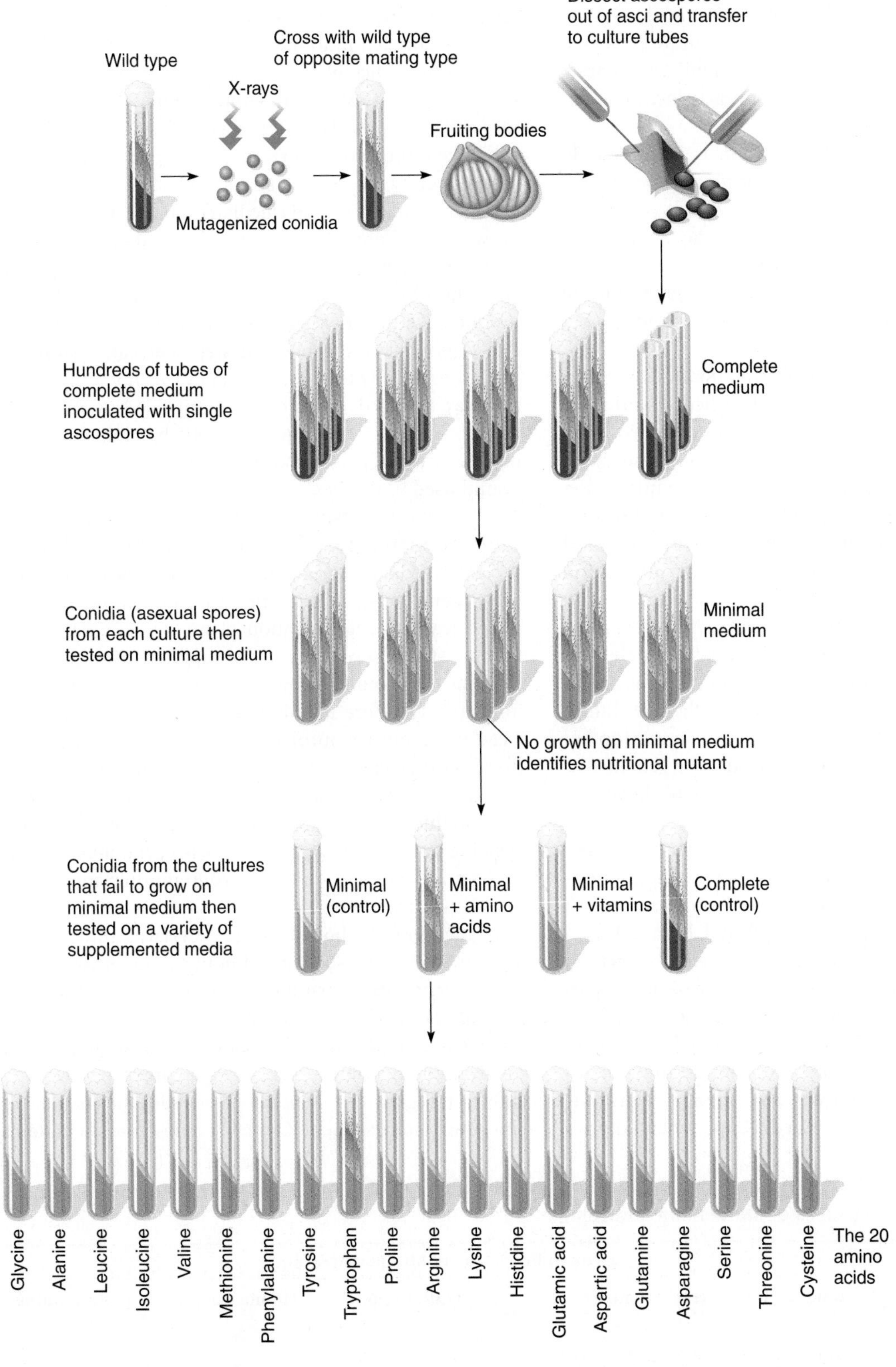


*medium*—that is, a medium containing all the amino acids, purines, pyrimidines, and vitamins—in addition to the sucrose, salts, and biotin found in minimal medium. In complete medium, any strain that could not make any amino acid, purine, pyrimidine, or vitamin from the basic ingredients in minimal medium could still grow by using the compounds supplied in the growth medium. Each culture grown on the complete medium was then tested for growth on minimal medium. The strains that did not grow were the auxotrophs. Those mutants, in turn, were tested individually for their ability to grow on minimal medium plus amino acids and on minimal medium plus vitamins. Theoretically, an amino acid auxotroph—a mutant strain that has lost the ability to synthesize a particular amino acid—would grow on minimal medium plus amino acids, but not on minimal medium plus vitamins or on minimal medium alone. Similarly, vitamin auxotrophs would grow only on minimal medium plus vitamins.

Suppose an amino acid auxotroph is identified. To determine which of the 20 amino acids is required by the mutant, the strain is inoculated into 20 tubes, each containing minimal medium plus one of the 20 different amino acids. In the example shown in Figure 4.3, a tryptophan auxotroph is identified because it grew only in the tube containing minimal medium plus tryptophan.

**Genetic Dissection of a Biochemical Pathway.** Once Beadle and Tatum had isolated and identified auxotrophic mutants, they investigated the biochemical pathways affected by the mutations. They assumed that *Neurospora* cells, like all other cells, function through the interaction of the products of a very large number of genes. Furthermore, they reasoned that wild-type *Neurospora* converted the simple constituents of minimal medium into amino acids and other required compounds by a series of reactions that were organized into pathways. In this way, the synthesis of cellular components occurred through a series of small steps, each catalyzed by an enzyme. As an example of the analytical approach Beadle and Tatum used that led to an understanding of the relationship between genes and enzymes, let us consider the genetic dissection of the pathway for the biosynthesis of the amino acid methionine in *Neurospora crassa*.

Starting with a set of methionine auxotrophs—mutants that require the addition of methionine to minimal medium to grow—genetic analysis (complementation tests; see Chapter 13, pp. 377–378 and Figure 13.12, p. 377) identifies four separate genes: *met-2*$^+$, *met-3*$^+$, *met-5*$^+$, and *met-8*$^+$. A mutation in any one of them gives rise to auxotrophy for methionine. Note that the number associated with each gene is no reflection of where the product encoded by each gene is found in its metabolic pathway. Next, the growth pattern of the four mutant strains is determined on media supplemented with chemicals thought to be intermediates involved in the methionine biosynthetic pathway—*O*-acetyl homoserine, cystathionine, and homocysteine—with the results shown in Table 4.1. By definition, all four mutant strains can grow on methionine, and none can grow on unsupplemented minimal medium.

The sequence of steps in a pathway can be deduced from the pattern of growth supplementation. The principles are as follows: The later in a pathway a mutant strain is blocked, the fewer intermediate compounds permit the strain to grow. If a mutant strain is blocked at early steps, a larger number of intermediates enable the strain to grow, because any of the intermediates after the blocked step can be processed by the enzymes in the pathway after the block, resulting in the production of the final product. That is, the earlier the block, the more intermediates exist after the blocked step that can restore the final product. Thus, in these analyses, not only is the pathway deduced, but the steps controlled by each gene are determined. In addition, a genetic block in a pathway may lead to an accumulation of the intermediate compound used in the step that is blocked.

The *met-8* mutant strain grows when supplemented with methionine, but not when supplemented with any of the intermediates (see Table 4.1). This means that the *met-8* gene must control the last step in the pathway, which leads to the formation of methionine. The *met-2* mutant strain grows on media supplemented with methionine or homocysteine, so homocysteine must be immediately before methionine in the pathway, and the *met-2* gene must control the synthesis of homocysteine from another chemical. The *met-3* mutant strain grows on media supplemented with methionine, homocysteine, or cystathionine, so cystathionine must precede homocysteine in the pathway, and the *met-3* gene must control the synthesis of cystathionine from another compound. The *met-5* strain grows on media supplemented with either methionine, homocysteine, cystathionine, or *O*-acetyl homoserine, so *O*-acetyl homoserine must precede cystathionine in the pathway, and the *met-5* gene must control the synthesis of *O*-acetyl homoserine from another compound. The methionine biosynthetic pathway involved here (which is part of a larger pathway) is shown in Figure 4.4. Gene *met-5*$^+$ encodes the enzyme for converting homoserine to *O*-acetyl homoserine, so mutants

**Table 4.1 Growth Responses of Methionine Auxotrophs**

| | Growth Response on Minimal Medium + | | | | |
|---|---|---|---|---|---|
| **Mutant Strains** | **Nothing** | ***O*-Acetyl Homoserine** | **Cystathionine** | **Homocysteine** | **Methionine** |
| Wild type | + | + | + | + | + |
| *met-5* | – | + | + | + | + |
| *met-3* | – | – | + | + | + |
| *met-2* | – | – | – | + | + |
| *met-8* | – | – | – | – | + |

Figure 4.4

**Methionine biosynthetic pathway showing four genes in *Neurospora crassa* that code for the enzymes that catalyze each reaction.** (The *met-5* and *met-2* genes are on the same chromosome; *met-3* and *met-8* are on two other chromosomes.)

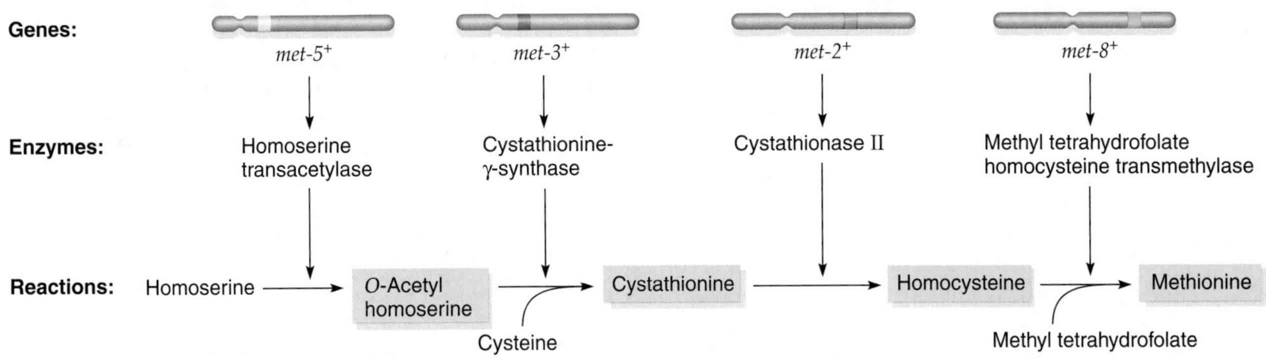


for this gene can grow on a minimal medium plus either *O*-acetyl homoserine, cystathionine, homocysteine, or methionine. Gene *met-3*$^+$ codes for the enzyme that converts *O*-acetyl homoserine to cystathionine, so a *met-3* mutant strain can grow on a minimal medium plus either cystathionine, homocysteine, or methionine, and so on.

Based on results of experiments of this kind, Beadle and Tatum proposed that a specific gene encodes each enzyme. This hypothetical relationship between an organism's genes and the enzymes that catalyze the steps in a biochemical pathway was called the **one-gene–one-enzyme hypothesis.** Gene mutations that result in the loss of enzyme activity lead to the accumulation of precursors in the pathway (and to possible side reactions) and to the absence of the end product of the pathway. With the approach described, then, a biochemical pathway can be dissected genetically; through the study of mutants and their effects, the sequence of steps in the pathway can be determined and each step related to a specific gene or genes.

However, researchers subsequently learned that more than one gene may control each step in a pathway. That is, an enzyme[1] may have two or more different polypeptide chains, each of them coded for by a specific gene. An example is the *E. coli* enzyme, DNA polymerase III, which has several subunits (see Table 3.1, p. 42). In such a case, more than one gene specifies that enzyme and thus that step in the pathway. Therefore, the one-gene–one-enzyme hypothesis was updated to the **one-gene–one-polypeptide hypothesis**. That hypothesis is not completely supported based on our present knowledge. That is, some genes do not encode proteins. And, expression of particular protein-coding genes in eukaryotes can result in more than one polypeptide. Examples of these will be seen later in the book.

Biochemical pathways are key to cell function and metabolism in all organisms. Some pathways synthesize compounds needed by the cell—such as amino acids, purines, pyridimines, fats, lipids, and vitamins—while other pathways break down compounds into simpler molecules, such as for recycling DNA, RNA, or protein, or for digesting food. Insofar as biochemical pathways are run by enzymes, they are under gene control. But, because of gene differences between organisms, biochemical pathways are not the same in all organisms.

The sum of all of the small chemicals that are intermediates or products of metabolic pathways is the metabolome, and the study of the metabolome is called metabolomics. The Focus on Genomics box in this chapter presents the results of a metabolomics investigation involving prokaryotes in the mammalian gut.

**Keynote**

A specific relationship between genes and enzymes is embodied in Beadle and Tatum's one-gene–one-enzyme hypothesis, which stated that each gene controls the synthesis or activity of a single enzyme. Some enzymes may consist of more than one polypeptide each coded by a different gene. Because of this, historically the hypothesis was changed to the one-gene–one-polypeptide hypothesis. Present-day knowledge indicates exceptions to that hypothesis also.

**i Activity**

Use the Beadle and Tatum experimental procedure to identify a nutritional mutant in the iActivity *Pathways to Inherited Enzyme Deficiencies* on the student website.

## Genetically Based Enzyme Deficiencies in Humans

Many human genetic diseases result when a single gene mutation alters the function of an enzyme that, typically, functions in a metabolic pathway (Table 4.2). In general, an enzyme deficiency caused by a mutation may have either simple effects or **pleiotropic** (multiple distinct) effects. Studies of these diseases have offered further evidence that many genes code

[1]We will see later in the book that some enzymes are RNA molecules, not proteins (see Chapter 5, pp. 95–96).

# Focus on Genomics

## Metabolomics in the Gut

Many species of Bacteria, and a few Archaea, live in the mammalian gut. The only abundant gut archaean is *Methanobrevibacter smithii*, and it plays a key metabolic role. Mammals cannot digest complex dietary carbohydrates (fibers), but members of the gut bacterial community can (by fermentation). As an end product of this fermentation, the bacteria release a number of short-chain fatty acids (SCFAs), which the mammalian host absorbs and metabolizes. These SCFAs comprise up to 10% of the calories taken in by the host. By consuming several of the end products of bacterial fermentation, including hydrogen gas and formate, *M. smithii* makes the bacterial community function more efficiently and increases the rate of production of SCFAs.

Genomic analyses—*transcriptomics* and *metabolomics*—have shown that *M. smithii* and the bacteria *Bacteriodes thetaiotomicron* change their transcriptional and metabolic states when both are present in the gut, and that these changes improve the digestion of fiber and provide more calories to the host. **Transcriptomics** is the study of gene expression at the level of the entire genome. The transcriptome is all of the RNAs expressed under a particular set of conditions and is thus a measure of which genes are transcribed and which proteins are likely to be produced. **Metabolomics** is the study of all of the small chemicals that are intermediates or products of metabolic pathways. Collectively, these cellular or extracellular chemicals constitute the *metabolome*. Metabolomics studies use chemical techniques to determine the identity of the small organic molecules present in or around the cell. The goal is to understand the functions of cellular enzymes and their pathways, as well as the effects that drugs and environmental conditions have on these processes.

To study the interaction of these organisms and their hosts, investigators delivered cultures of prokaryotes to colons of mice with germ-free guts. Some mice were given both *B. thetaiotomicron* and *M. smithii* (Bt/Ms), while other mice got control cultures lacking *M. smithii*. The investigators gave the cells several days to colonize the colon, and they fed the mice a diet high in fructans, a specific class of indigestible fiber. The Bt/Ms gut community degraded the fructans more efficiently than the control gut communities did. Transcriptome analysis showed that *B. thetaiotomicron* in the Bt/Ms community had increased the transcription of genes involved in degradation of fructans and decreased transcription of genes for degradation of other complex carbohydrates compared to the control. *B. thetaiotomicron* also increased production of acetate (an SCFA). Models based on transcription suggested that more formate should be produced as well, but that was not observed. One reason the formate levels did not increase was found when the transcriptome of *M. smithii* was characterized. When *M. smithii* is in a Bt/Ms mouse, *M. smithii* increases transcription of genes encoding enzymes in the formate metabolism pathway. Presumably, excess formate production by *B. thetaiotomicron* is balanced by increased formate consumption by *M. smithii*. On the whole, Bt/Ms guts were more effective metabolizers of fructans, because both species underwent changes in gene expression and metabolism to work together to break down these carbohydrates.

Did the mouse benefit from all of this activity? The answer is yes—the host recovered more calories from the food because it absorbed the SCFAs released by *B. thetaiotomicron*. Further, the investigators found increased acetate levels in the blood of mice with a Bt/Ms gut (acetate is one of the SCFAs released by *B. thetaiotomicron*). These Bt/Ms mice also had more fats in their livers and in their fat pads. Other studies have suggested that the presence of a large colony of *M. smithii* in the gut may predispose mice (and, presumably, humans) to obesity. Therefore, scientists are studying the genome of *M. smithii* in the hopes of finding genes that could be targeted by drugs. Someday we may be able to use drugs that interfere with *M. smithii* to help overweight people lose weight!

for enzymes. Some genetic diseases are discussed in the sections that follow.

### Phenylketonuria

*Phenylketonuria* (PKU, OMIM 261600) occurs in about 1 in 12,000 Caucasian births; it is most commonly caused by a recessive mutation of a gene on the long arm of chromosome 12 (an **autosome**—that is, a chromosome other than a sex chromosome) at position 12q24.1. To exhibit the condition, people must therefore be homozygous for the mutation. (The terminology for positions along chromosomes is described in the discussion of *karyotypes* in Chapter 12, pp. 327–329.) In brief, the first number is the chromosome number; each chromosome has a short arm, p, and a long arm, q. Each arm is subdivided into numbered regions and subregions based on particular staining patterns; here 24 is a region, and the 1 after the period is the subregion. The mutation is in the gene for phenylalanine hydroxylase. The absence of that enzyme activity

**Table 4.2 Selected Human Genetic Disorders with Demonstrated Enzyme Deficiencies**

| Genetic Defect | Locus | Enzyme Deficiency | OMIM Entry |
|---|---|---|---|
| Alkaptonuria | 3q21–q23 | Homogentisic acid oxidase | 203500 |
| Cystic fibrosis | 7q31.2 | Cystic fibrosis transmembrane conductance regulator (CFTR) | 602421 |
| Cataract | 17q24 | Galactokinase | 230200 |
| Citrullinemia[a] | 9q34 | Argininosuccinate synthetase | 215700 |
| Disaccharide intolerance I | 3q25–q26 | Invertase | 222900 |
| Fructose intolerance | 9q22.3 | Fructose-1-phosphate aldolase | 229600 |
| Galactosemia[a] | 9p13 | Galactose-1-phosphate uridyl transferase | 230400 |
| Gaucher disease[a] | 1q21 | Glucocerebrosidase | 230800 |
| G6PD deficiency (favism)[a] | Xq28 | Glucose-6-phosphate dehydrogenase | 305900 |
| Glycogen storage disease I | 17q21 | Glucose-6-phosphatase | 232200 |
| Glycogen storage disease II[a] | 17q25.2–q25.3 | α-1,4-Glucosidase | 232300 |
| Glycogen storage disease III[a] | 1p21 | Amylo-1, β-glucosidase | 232400 |
| Glycogen storage disease IV[a] | 3p12 | Glycogen branching enzyme | 232500 |
| Hemolytic anemia[a] | 3p21.1, 8p21.1, 20q11.2, 1q21 | Glutathione peroxidase, glutathione reductase, glutathione synthetase, hexokinase, or pyruvate kinase | 138320, 138300, 231900, 266200 |
| Intestinal lactase deficiency (adult) | | Lactase | 223000 |
| Ketoacidosis[a] | 5p13 | Succinyl CoA:3-Ketoacid CoA-transferase | 245050 |
| Lesch–Nyhan syndrome[a] | Xq26–q27.2 | Hypoxanthine guanine phosphoribosyltransferase | 308000 |
| Maple sugar urine disease, type IA[a] | 19q13.1–q13.2 | Keto acid decarboxylase | 248600 |
| Muscular dystrophy, Duchenne and Becker types | Xp21.2 | Dystrophin absent or defective; serum acetylcholinesterase, acetylcholine trans-ferase, or creatine phosphokinase elevated | 310200 |
| Phenylketonuria[a] | 12q24.1 | Phenylalanine hydroxylase | 261600 |
| Porphyria, congenital erythropoietic[a] | 10q25.2–q26.3 | Uroporphyrinogen III synthase | 263700 |
| Pulmonary emphysema | 14q32.1 | α-I-Antitrypsim | 107400 |
| Ricketts, vitamin D-dependent | | 25-Hydroxycholecalciferol 1-hydroxylase | 277420 |
| Tay–Sachs disease[a] | 15q23–q24 | Hexosaminidase A | 272800 |
| Tyrosinemia, type III | 12q24–qter | *p*-Hydroxyphenylpyruvate oxidase | 276710 |

[a]Prenatal diagnosis possible.

prevents the amino acid phenylalanine from being converted to the amino acid tyrosine (see Figure 4.1). Phenylalanine is one of the *essential amino acids*, meaning it is an amino acid that must be included in the diet because humans are unable to synthesize it. Phenylalanine is needed to make proteins, but excess amounts are harmful and are converted to tyrosine for further metabolism. Children born with PKU accumulate the phenylalanine they ingest because they are unable to metabolize it. The accumulated phenylalanine is converted to phenylpyruvic acid, which drastically affects the cells of the central nervous system and produces serious symptoms including severe mental retardation, a slow growth rate, and early death. (Children with PKU whose mothers do not have PKU are unaffected before or during birth, because any excess phenylalanine that accumulates is metabolized by maternal enzymes.)

PKU has pleiotropic effects. People with PKU cannot make tyrosine, an amino acid needed for protein synthesis, production of the hormones thyroxine and adrenaline, and production of the skin pigment melanin. This aspect of the phenotype is not very serious, because tyrosine can be obtained from food. Yet food does not normally contain a lot of tyrosine. As a result, people with PKU make little melanin and therefore tend to have very fair skin and blue eyes (even if their genes specify brown eye color). In addition, people with PKU have low levels of epinephrine (adrenaline), a hormone produced in a biochemical pathway starting with tyrosine.

The adverse symptoms of PKU depend on the amount of phenylpyruvic acid that is generated when phenylalanine accumulates, so the disease can be managed by controlling the dietary intake of phenylalanine. A mixture of individual amino acids with a controlled amount of phenylalanine is used as a protein substitute in the PKU diet. The diet must maintain a level of phenylalanine in the blood that is high enough to facilitate normal development of the nervous system, yet low enough to prevent mental retardation. Treatment must begin in the first month or two after birth, or the brain will be damaged and treatment will be ineffective. The diet is expensive, costing more than $5,000 per year. A difference of opinion exists as to whether the diet must be continued for life or whether it can be discontinued by about 10 years of age without subsequent defects developing in mental capacity or behavior. In addition, women with PKU are advised either to maintain the restricted diet for life or to return to the diet before becoming pregnant and maintain the diet through pregnancy. The reason is that children born to women with PKU living on normal diets are mentally retarded because high levels of phenylalanine in the maternal blood pass to the developing fetus across the placenta and adversely affect nervous system development independently of the genotype of the fetus.

Given the serious consequences of allowing PKU to go untreated, all U.S. states require that newborns be screened for the condition. The screen—the Guthrie test—is conducted by placing a drop of blood on a filter paper disc and situating the disc on a solid culture medium containing the bacterium *Bacillus subtilis* and the chemical β-2-thienylalanine, which inhibits the growth of the bacterium. If phenylalanine is present, the inhibition is prevented; therefore, continued growth of the bacterium is evidence of the presence of high levels of phenylalanine in the blood and indicates the need for further tests to determine whether the infant has PKU.

Some foods and drinks containing the artificial sweetener aspartame (trade name NutraSweet®) carry a warning that people with PKU should not use them. Aspartame is a dipeptide consisting of aspartic acid and phenylalanine. This combination signals to your taste receptors that the substance is sweet (yet it is not sugar and does not have the calories of sugar). Once ingested, aspartame is broken down to aspartic acid and phenylalanine, so it can have serious effects on people with PKU.

The gene for phenylalanine hydroxylase has been characterized at the molecular level. A variety of mutations in the gene result in loss of enzyme activity in individuals with PKU, including mutants that alter an amino acid in the protein, mutants that result in a truncated protein, and mutants that affect splicing of the pre-mRNA transcribed from the gene.

## Albinism

The classic form of *albinism* (see Figure 11.18b, p. 316; OMIM 203100) is caused by an autosomal recessive mutation. About 1 in 33,000 Caucasians and 1 in 28,000 African Americans in the United States have albinism. A gene for tyrosinase is mutated in individuals with albinism. Tyrosinase is an enzyme used in the conversion of tyrosine to DOPA, from which the brown pigment melanin derives (see Figure 4.1). Melanin absorbs light in the ultraviolet (UV) range and protects the skin against harmful UV radiation from the sun. People with albinism produce no melanin, so they have white skin and white hair, as well as eyes whose irises appear red (due to a lack of pigment) and are highly sensitive to light.

There are at least two other kinds of albinism (see OMIM 203200 and OMIM 203290) because a number of biochemical steps occur during biosynthesis of melanin from tyrosine. Thus, two parents with albinism who are each homozygous for a mutation in a different gene in the pathway can produce normal children.

## Kartagener Syndrome

As in albinism, several genes can be mutated to cause a rare disease called either *Kartagener syndrome* (OMIM 244400) or Kartagener's triad. This autosomal recessive disease affects about 1 in 32,000 live births. It is characterized by sinus and lung abnormalities, sterility, and in some cases, *dextrocardia*—a condition where the heart is shifted to the right rather than to the left of center. On the surface, without a molecular understanding of the genes involved, these pleiotropic symptoms seem to have very little to do with each other. The genes known to be mutated in these individuals all encode parts of the dynein motors of flagella and cilia. Dynein motor proteins slide microtubules of flagella and cilia over each other to produce movements of those structures. Without functional dynein, neither flagella nor cilia can move properly. As a result, sinus and lung infections are common in individuals with Kartagener's syndrome because they have a defective cilia lining of their respiratory passages and, therefore, they cannot remove bacteria and spores from their respiratory systems efficiently. Sterility in males occurs because the sperm cannot swim; sterility in females occurs because the cilia that should help draw the oocyte into the reproductive tract are unable to do so.

The causes of dextrocardia were less obvious until mouse models with defects in the gene were developed. Mice carrying certain mutations of the gene developed a similar set of defects, and studies on the early embryos of these mice illuminated the cause of dextrocardia. In the developing embryo, researchers saw that cilia on a structure called the node rotate in a clockwise direction and generate a "leftward" flow of extraembryonic fluids. This flow can be detected by the surrounding cells, which respond by moving either left or right, a response that determines their future developments. In Kartagener syndrome, the flow of fluids cannot be generated, and the tissues move "left" or "right" at random.

## Tay–Sachs Disease

*Tay–Sachs disease* (Figure 4.5; OMIM 272800), also called *infantile amaurotic idiocy*, is caused by homozygosity for a rare recessive mutation of a gene on chromosome 15 at 15q23–q24. Although Tay–Sachs disease is rare in the population as a whole, it has a higher incidence in Ashkenazi Jews of central European origin—among whom about 1 in 3,600 children have the disease.

**Figure 4.5**
**Child with Tay–Sachs disease.**

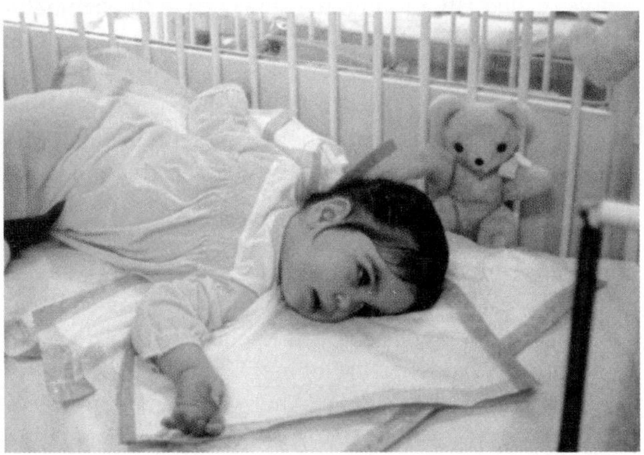

The gene that is defective in individuals with Tay–Sachs disease codes for an enzyme in the lysosome. *Lysosomes* are membrane-bound organelles in the cell; they contain 40 or more different digestive enzymes that catalyze the breakdown of nucleic acids, proteins, polysaccharides, and lipids. When a lysosomal enzyme is nonfunctional or partially functional, normal breakdown of the substrate for the enzyme cannot occur. The gene that is mutated in individuals with Tay–Sachs disease is *HEXA*, which codes for the enzyme *N*-acetylhexosaminidase A (Hex-A). This enzyme cleaves a terminal *N*-acetylgalactosamine group from a brain ganglioside (Figure 4.6). (A *ganglioside* is one of a group of complex glycolipids found mainly in nerve membranes.) In infants with Tay–Sachs disease, the enzyme is nonfunctional; the unprocessed ganglioside accumulates in brain neurons, causing them to swell and thereby producing several different clinical symptoms. Typically, the symptom first recognized is an unusually enhanced reaction to sharp sounds. A cherry-colored spot on the retina, surrounded by a white halo, also aids early diagnosis of the disease. About a year after birth, a rapid neurological degeneration occurs as the unprocessed ganglioside accumulates and the brain begins to lose control over normal function and activities. This degeneration produces generalized paralysis, blindness, a progressive loss of hearing, and serious feeding problems. By 2 years of age the children are essentially immobile, and death occurs at about 3 to 4 years of age, often from respiratory infections. There is no known cure for Tay–Sachs disease; but because carriers (*heterozygotes*, who have one normal and one mutant allele of the gene) can be detected, the incidence of this disease can be controlled.

**Keynote**

Many human genetic diseases are caused by deficiencies in enzyme activities. Most of these diseases are inherited as recessive traits.

## Gene Control of Protein Structure

While most enzymes are proteins, not all proteins are enzymes. To understand completely how genes function, we next look at the experimental evidence that genes also are responsible for the structure of nonenzymatic proteins such as hemoglobin. Nonenzymatic proteins often

**Figure 4.6**
**Diagram of the biochemical step for the conversion of the brain ganglioside $G_{M2}$ to the ganglioside $G_{M3}$, catalyzed by the enzyme N-acetylhexosaminidase A (Hex-A).**

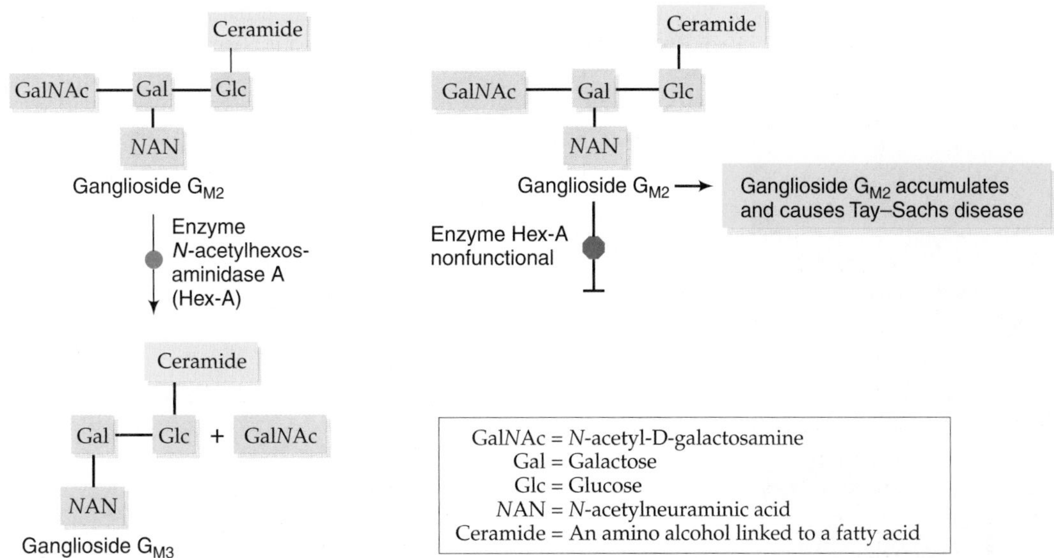

are easier to study than enzymes. This is because enzymes usually are present in small amounts, whereas nonenzymatic proteins can occur in large quantities in the cell so they are easier to isolate and purify.

## Sickle-Cell Anemia

*Sickle-cell anemia* (SCA; OMIM 603903) is a genetic disease affecting hemoglobin, the oxygen-transporting protein in red blood cells. Sickle-cell anemia was first described in 1910 by J. Herrick, who found that in conditions of low oxygen tension, red blood cells from people with the disease lose their characteristic disc shape and assume the shape of a sickle (Figure 4.7). The sickled red blood cells are fragile and break easily, resulting in the anemia. Sickled cells also are not as flexible as normal cells and therefore tend to clog capillaries rather than squeeze through them. As a result, blood circulation is impaired and tissues become deprived of oxygen. Although oxygen deprivation occurs particularly at the extremities, the heart, lungs, brain, kidneys, gastrointestinal tract, muscles, and joints can also become oxygen deprived and be damaged. A person with sickle-cell anemia therefore may suffer from a variety of health problems, including heart failure, pneumonia, paralysis, kidney failure, abdominal pain, and rheumatism. Some people have a milder form of the disease called *sickle-cell trait*.

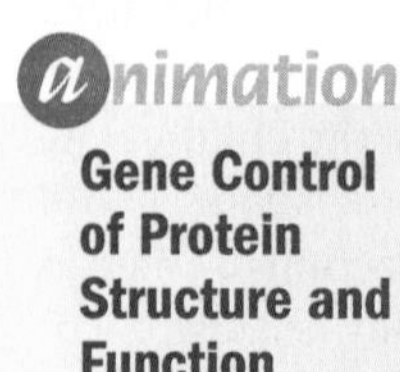


**Gene Control of Protein Structure and Function**

In 1949, E. A. Beet and J. V. Neel independently hypothesized that sickling was caused by a single mutant allele that was homozygous in sickle-cell anemia and heterozygous in sickle-cell trait. In the same year, Linus Pauling and coworkers showed that the hemoglobins in normal, sickle-cell anemia, and sickle-cell trait blood differ when they are subjected to electrophoresis—a technique for separating molecules based on their electrical charges and/or masses. Under the electrophoresis conditions they used, both forms of hemoglobin acted as cations (positively charged molecules) and migrated toward the negative pole. The hemoglobin from normal people (called Hb-A) migrated slower than the hemoglobin from people with sickle-cell anemia (called Hb-S; Figure 4.8). Hemoglobin from people with sickle-cell trait had a 1:1 mixture of Hb-A and Hb-S, indicating that heterozygous people make both types of hemoglobin. Pauling concluded that sickle-cell anemia results from a mutation that alters the chemical structure of the hemoglobin molecule. This experiment was one of the first rigorous proofs that protein structure is controlled by genes.

Hemoglobin, the molecule affected in sickle-cell anemia, consists of four polypeptide chains—two α-globin polypeptides and two β-globin polypeptides—each of which is associated with a heme group (a nonprotein chemical group involved in oxygen binding and added to each polypeptide after the polypeptide is synthesized; Figure 4.9). In 1956, V. M. Ingram analyzed some amino acid sequences of the polypeptides of Hb-A and Hb-S and found that the molecular defect in the Hb-S hemoglobin is a change from the acidic amino acid glutamic acid (Glu: hydrophilic [water loving], with a negative electric charge) at the sixth position from the N-terminal end of the β polypeptide to the neutral amino acid valine (Val: hydrophobic [water hating], with no electrical charge; Figure 4.10). This particular substitution causes the β polypeptide to fold up in a different way. (You will learn in Chapter 6 that the three-dimensional shape of a polypeptide is determined by its amino acid sequence.) Red blood cells are packed full of hemoglobin protein. Hemoglobin with this mutant version of the β polypeptide aggregates readily, falling out of solution and leading to extreme sickling of the red blood cells in people with sickle-cell anemia and mild sickling of the red blood cells in people with sickle-cell trait.

**Figure 4.7**
**Scanning electron micrograph of three normal red blood cells next to a sickled cell.**

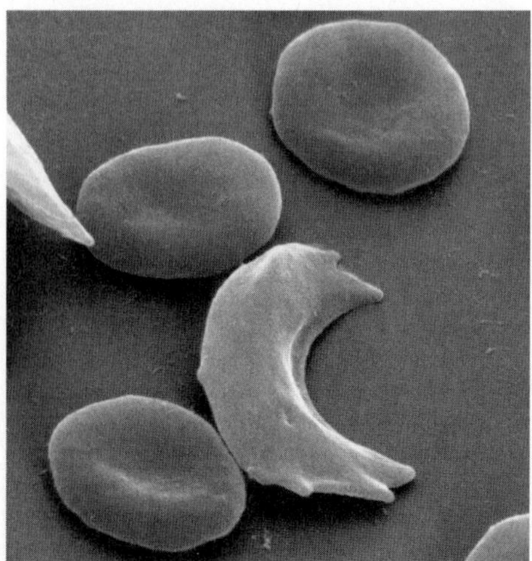

**Figure 4.8**
**Electrophoresis of hemoglobin variants.** Hemoglobin found (left) in normal $\beta^A\beta^A$ individuals, (center) in $\beta^A\beta^S$ individuals who have sickle-cell trait, and (right) in $\beta^S\beta^S$ individuals who have sickle-cell anemia. The two hemoglobins migrate to different positions in an electric field and therefore must differ in electric charge.

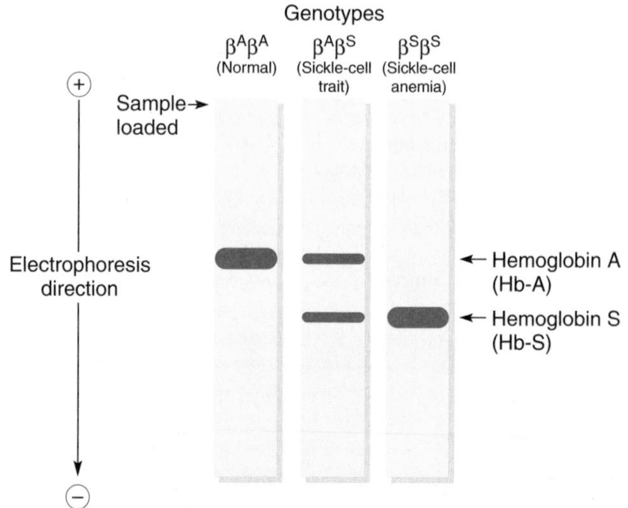

**Figure 4.9**

**The hemoglobin molecule.** The diagram shows the two α polypeptides and two β polypeptides, each associated with a heme group. Each α polypeptide contacts both β polypeptides, but there is little contact between the two α polypeptides or between the two β polypeptides.

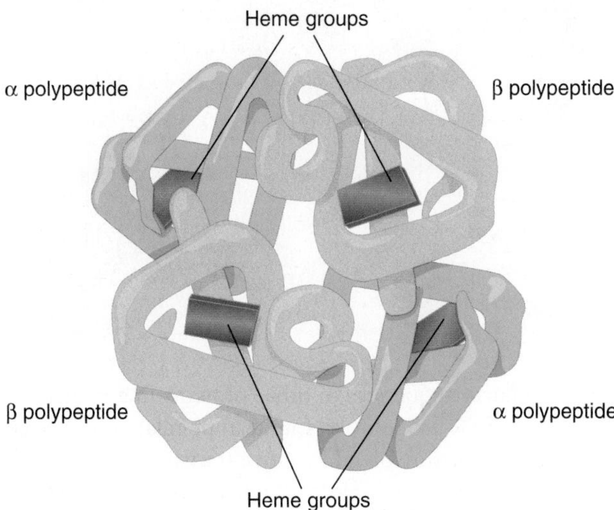


The genetics and the products of the genes involved are as follows. The β polypeptide sickle-cell mutant allele is $\beta^S$, and the normal allele is $\beta^A$. Homozygous $\beta^A\beta^A$ people make normal Hb-A with two normal α chains encoded by the wild-type α-globin gene and two normal β chains encoded by the normal β-globin $\beta^A$ allele. Homozygous $\beta^S\beta^S$ people make Hb-S, the defective hemoglobin, with two normal α chains specified by wild-type α-globin genes and two abnormal β chains specified by the mutant β-globin $\beta^S$ allele: these people have sickle-cell anemia. Heterozygous $\beta^A\beta^S$ people make *both* Hb-A and Hb-S and have sickle-cell trait. Because only one type of β chain is found in any one hemoglobin molecule, only two types of hemoglobin molecules are possible—one with two normal β chains, the other with two mutant β chains. Under normal conditions, people with sickle-cell trait usually show few symptoms of the disease. However, after a sharp drop in oxygen tension (as in an unpressurized aircraft climbing into the atmosphere, in high mountains, or after intense exercise), sickling of red blood cells may occur, giving rise to some symptoms similar to those found in people with severe anemia.

The one-gene–one-polypeptide hypothesis is consistent with the hemoglobin example just described because proteins, like enzymes, can be made up of more than one polypeptide chain. However, in eukaryotes a process known as *alternative splicing* (see Chapter 18, pp, 534-536) can result in one gene producing more than one polypeptide, rendering the one-gene–one-polypeptide hypothesis a simplification.

## Other Hemoglobin Mutants

More than 200 hemoglobin mutants have been detected in general screening programs in which hemoglobin is isolated from red blood cells and analyzed for different migration compared with normal hemoglobin in electrophoresis. Figure 4.11 lists some of these mutants, along with the amino acid substitutions that have been identified. Some mutations affect the α chain and others the β chain, and there is wide variation in the types of amino acid substitutions that occur. From the changes in DNA that are assumed to be responsible for the substitutions, a single base-pair change is involved in each case.

The identified hemoglobin mutants have various effects, depending on the amino acid substitution involved and its position in the polypeptide chains. Most have effects that are not as drastic as those of the sickle-cell anemia mutant. For example, in the Hb-C hemoglobin molecule, the same β-polypeptide glutamic acid that is altered in sickle-cell anemia is changed to a lysine. Compared with the Hb-S change, however, this change is not as serious a defect—because both amino acids are hydrophilic, the conformation of the hemoglobin molecule is not as drastically altered. People homozygous for the $\beta^C$ mutation experience only a mild form of anemia.

## Cystic Fibrosis

Cystic fibrosis (CF; OMIM 219700 and 602421) is a human disease that causes pancreatic, pulmonary, and digestive dysfunction in children and young adults. Typical of the disease is an abnormally high viscosity of secreted mucus. In some male patients, the vas deferens (part of the male reproductive system) does not form properly, resulting in sterility. Cystic fibrosis is managed by pounding the chest and back of a patient to help shake mucus free in different parts of the lungs (Figure 4.12) and by giving antibiotics to treat any infections that develop. Cystic fibrosis is a lethal disease; with present management procedures, life expectancy is about 40 years.

Cystic fibrosis is caused by homozygosity for an autosomal recessive mutation located on the long arm of chromosome 7 at position 7q31.2–q31.3. Cystic fibrosis is the most common lethal autosomal recessive disease among Caucasians—among whom about 1 in 2,000 newborns have the disease. Approximately 1 in 23 Caucasians is estimated to be a heterozygous carrier. In the African American population, about 1 in 17,000 newborns have cystic fibrosis; in Asian-Americans, the cystic fibrosis frequency is about 1 in 31,000 newborns.

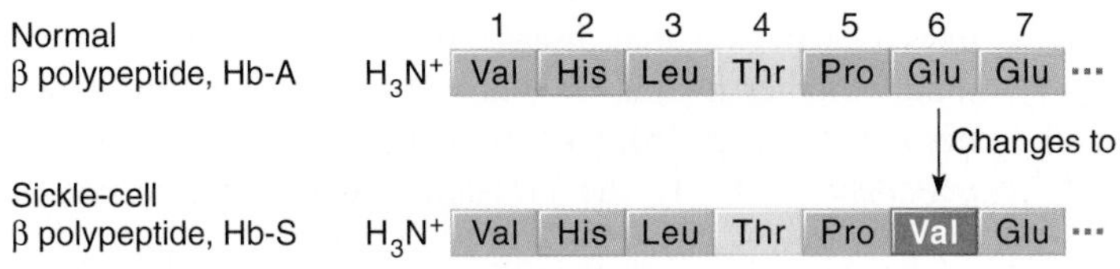


**Figure 4.10**

**The first seven N-terminal amino acids in normal and sickled hemoglobin β polypeptides.** There is a single amino acid change from glutamic acid to valine at the sixth position in the sickled hemoglobin polypeptide.

Figure 4.11

**Examples of amino acid substitutions found in (a) the 141-amino acid long α-globin polypeptide and (b) the 146-amino acid β-globin polypeptide of various human hemoglobin variants.**

**a) α-chain**

| | Amino acid position | | | | | | |
|---|---|---|---|---|---|---|---|
| | 1 | 2 | 16 | 30 | 57 | 68 | 141 |
| Normal | Val | Leu | Lys | Glu | Gly | Asn | Arg |
| Hb variants: | | | | | | | |
| HbI | Val | Leu | **Asp** | Glu | Gly | Asn | Arg |
| Hb-G Honolulu | Val | Leu | Lys | **Gln** | Gly | Asn | Arg |
| Hb Norfolk | Val | Leu | Lys | Glu | **Asp** | Asn | Arg |
| Hb-G Philadelphia | Val | Leu | Lys | Glu | Gly | **Lys** | Arg |

**b) β-chain**

| | Amino acid position | | | | | | |
|---|---|---|---|---|---|---|---|
| | 1 | 2 | 6 | 26 | 63 | 121 | 146 |
| Normal | Val | His | Glu | Glu | His | Glu | His |
| Hb variants: | | | | | | | |
| Hb-S | Val | His | **Val** | Glu | His | Glu | His |
| Hb-C | Val | His | **Lys** | Glu | His | Glu | His |
| Hb-E | Val | His | Glu | **Lys** | His | Glu | His |
| Hb-M Saskatoon | Val | His | Glu | Glu | **Tyr** | Glu | His |
| Hb Zurich | Val | His | Glu | Glu | **Arg** | Glu | His |
| Hb-D β Punjab | Val | His | Glu | Glu | His | **Gln** | His |

The defective gene product in patients with cystic fibrosis was identified not by biochemical analysis, as was the case for PKU and many other diseases, but by a combination of genetic and modern molecular biology techniques. The gene was localized to chromosome 7, and then it was molecularly cloned from a normal subject and from patients with cystic fibrosis. In patients with a serious form of cystic fibrosis, the most common mutation—ΔF508 (Δ = delta, for a deletion)—is the deletion of three consecutive base pairs in the gene. Since each amino acid in a protein is specified by three base pairs in the DNA, this means that one amino acid is missing, in this case phenylalanine at position 508. But what does the cystic fibrosis protein do? From the DNA sequence of the gene, researchers deduced the amino acid sequence of the protein and then made some predictions about the type and three-dimensional structure of that protein. Their analysis indicated that the 1,480-amino acid cystic fibrosis protein is associated with cell membranes. The proposed structure for the cystic fibrosis protein—called cystic fibrosis transmembrane conductance regulator (CFTR)—is shown in Figure 4.13. The ΔF508 mutation affects the adenosine triphosphate (ATP)-binding, nucleotide-binding fold (NBF) region of the protein near the left membrane-spanning region. Through a comparison of the amino acid sequence of the cystic fibrosis protein with the amino acid sequences of other proteins in a computer database, CFTR protein was found to be related to a large family of proteins involved in active transport of materials across cell membranes. We now know that this protein is a chloride channel in certain cell membranes. In people with cystic fibrosis, the abnormal CFTR protein results in impaired ion transport across membranes. The symptoms of cystic fibrosis ensue, starting with abnormal mucus secretion and accumulation.

Figure 4.12

**Child with cystic fibrosis having the back pounded to dislodge accumulated mucus in the lungs.**

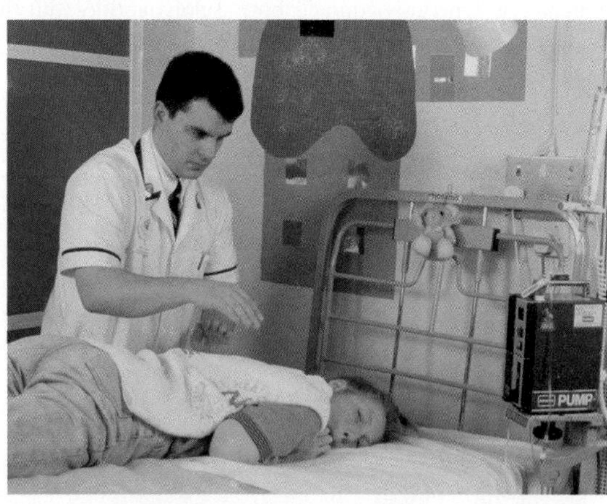


Cystic fibrosis is being studied in mice genetically engineered to have the same defect in their CFTR gene. The hope is that, through work with the mice modeling the disease, researchers will obtain a better understanding of the disease and be able to develop effective treatment, perhaps even an effective gene therapy cure.

### Keynote

From the study of alterations in proteins other than enzymes—such as those in hemoglobin, which are responsible for sickle-cell anemia—convincing evidence was obtained that genes control the structures of all polypeptides, one or more of which are used to make all proteins.

## Genetic Counseling

You have learned that many human genetic diseases are caused by enzyme or protein defects that ultimately result from mutations at the DNA level. Several other genetic diseases arise from chromosome defects that, in some way, affect gene expression. Scientists can now test for many enzyme or protein deficiencies, as well as for many of the DNA changes associated with genetic diseases, and thereby determine whether a person has a genetic disease or is a carrier for that disease. It is also possible to determine whether people have any chromosomal abnormalities (see Chapter 16). **Genetic counseling** is advice based on analysis of: (1) the probability that patients have a

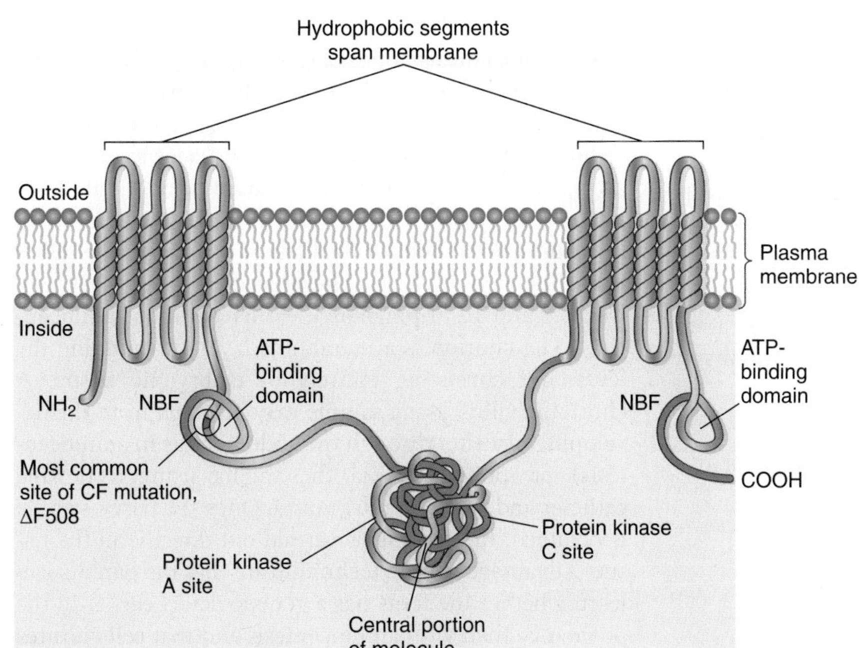


**Figure 4.13**

**Proposed structure for cystic fibrosis transmembrane conductance regulator (CFTR).** The protein has two hydrophobic segments that span the plasma membrane, and after each segment is a nucleotide-binding fold (NBF) region that binds ATP. The site of the amino acid deletion resulting from the three-nucleotide-pair deletion in the cystic fibrosis gene most commonly seen in patients with severe cystic fibrosis is in the first (toward the amino end) NBF; this is the ΔF508 mutation. The central portion of the molecule contains sites that can be phosphorylated by the enzymes protein kinase A and protein kinase C.



genetic defect; or (2) of the risk that prospective parents may produce a child with a genetic defect. In the latter case, genetic counseling involves presenting the available options for avoiding or minimizing those risks. If a serious genetic defect is identified in a fetus, one option is abortion. Genetic counseling gives people an understanding of the genetic problems that are or may be in their families or prospective families. The health professional who offers genetic counseling is a *genetic counselor*. Typically a genetic counselor has specialized degrees and experience in medical genetics and counseling.

Genetic counseling includes a wide range of information on human heredity. In many instances the risk of having a child with a genetic condition may be stated as precise probabilities; in others, where the role of heredity is not completely clear, the risk is estimated only generally. It is the responsibility of genetic counselors to give their clients clear, unemotional, and nonprescriptive statements based on the family history and on their knowledge of all relevant scientific information and the probable risks of giving birth to a child with a genetic defect.

Genetic counseling often starts with **pedigree analysis**—the study of a family tree and the careful compilation of phenotypic records of both families over several generations. (Pedigree analysis is described in more detail in Chapters 12 and 13.) Pedigree analysis is used to determine the likelihood that a particular allele is present in the family of either parent. A genetic condition is detected in one (or both) of two ways: by detection of **carriers** (individuals heterozygous for recessive mutations) or by fetal analysis. Assays for enzyme activities or protein amounts are limited to genetic diseases in which the biochemical condition is expressed in the parents or the developing fetus. Tests that measure disease-associated alleles in the DNA do not depend on expression of the gene in the parents or the fetus.

Although carriers of many mutant alleles may be identified, and fetuses can be analyzed to see if they have a genetic condition, in most cases there is no way to correct the genetic condition. Carrier detection and fetal analysis serve mainly to inform parents of the risks and probabilities of having a child with the mutation.

## Carrier Detection

*Carrier detection* identifies people who are heterozygous for a recessive gene mutation. The heterozygous carrier of a mutant gene typically is normal in phenotype. If homozygosity for the mutation results in serious deleterious effects, there is great value in determining whether two people who are contemplating having a child are both carriers—because in that situation they have a one in four chance of having a child with that genetic disease. Carrier detection can be used in cases in which a gene product (a protein or an enzyme) can be assayed. In those cases, the heterozygote (carrier) is expected to have approximately half the enzyme activity or protein amount as do homozygous normal individuals, although this is not observed for all mutations. In Chapter 10, we see how carriers can be identified by DNA tests.

## Fetal Analysis

Another important aspect of genetic counseling is finding out whether a fetus is normal. **Amniocentesis** is one way this can be done (Figure 4.14). As a fetus develops in the amniotic sac, amniotic fluid surrounds it, serving as a cushion against shock. In amniocentesis, a syringe needle is inserted carefully through the mother's uterine wall and into the amniotic sac, and a sample of amniotic fluid is taken. The fluid contains cells that the fetus's skin has sloughed off; these cells can be cultured in the laboratory and then examined for protein or enzyme alterations or deficiencies, DNA changes, and chromosomal abnormalities. Amniocentesis is possible at any stage of pregnancy, but the small quantity of amniotic fluid available and the risk to the fetus makes it impractical to perform the procedure before week 12 of gestation. Because amniocentesis is complicated and costly, it is used primarily in high-risk cases.

Another method for fetal analysis is **chorionic villus sampling** (Figure 4.15). The procedure is done between weeks 8 and 12 of pregnancy, earlier than for amniocentesis. The chorion is a membrane layer surrounding the fetus and consisting entirely of embryonic tissue. A chorionic villus tissue sample may be taken from the developing placenta through the abdomen (as in amniocentesis) or, preferably, via the vagina using a flexible catheter and aided by ultrasound. Once the tissue sample is obtained, the analysis is carried out directly on the tissue. Advantages of the technique are that the parents can learn whether the fetus has a genetic defect earlier in the pregnancy than with amniocentesis and that cell cultures are not required to do the biochemical assays. Fetal death and inaccurate diagnoses caused by the presence of maternal cells are more common in chorionic villus sampling than in amniocentesis, however.

**Figure 4.14**
**Amniocentesis, a procedure used for prenatal diagnosis of genetic defects.**

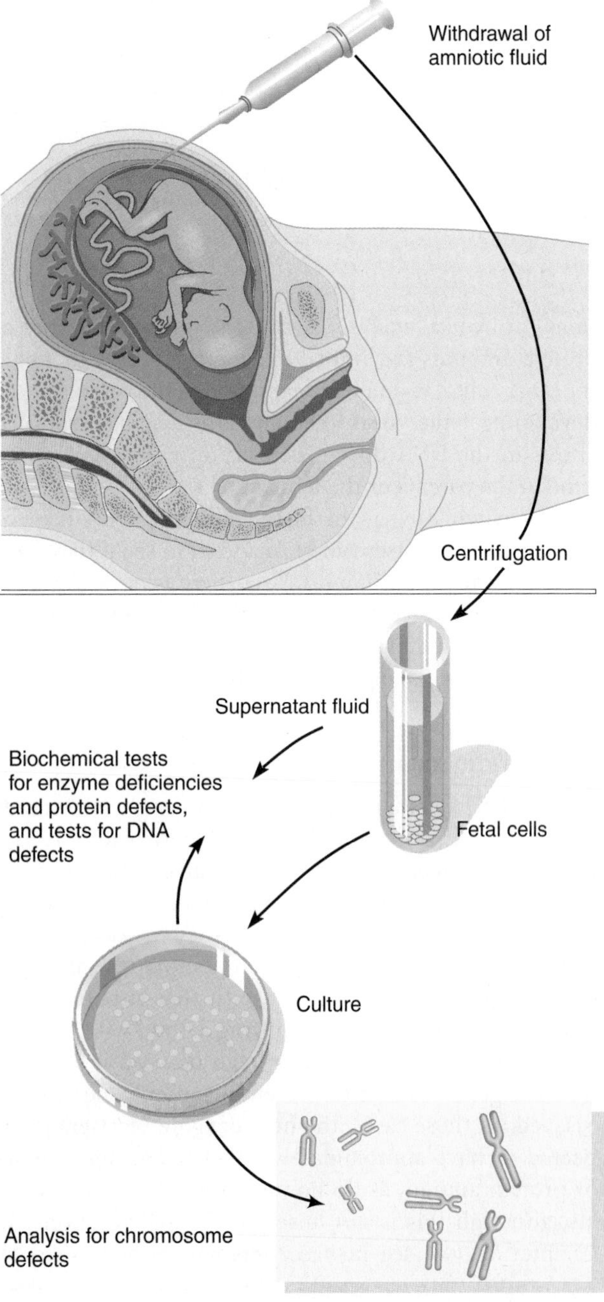


### Keynote

Genetic counseling is advice based on analyzing the probability that patients have a genetic defect or calculating the risk that prospective parents may produce a child with a genetic defect. Carrier detection and fetal analysis result in early detection of a genetic disease.

**Figure 4.15**
**Chorionic villus sampling, a procedure used for early prenatal diagnosis of genetic defects.**

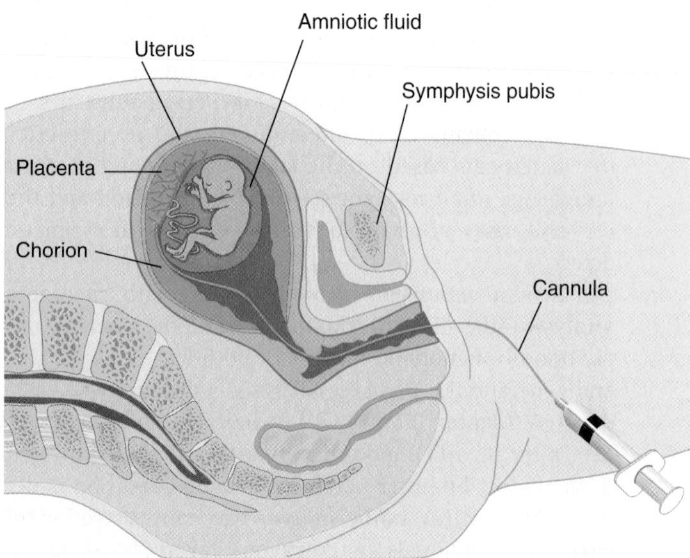

# Summary


- There is a specific relationship between genes and enzymes, initially embodied in the one-gene–one-enzyme hypothesis stating that each gene controls the synthesis or activity of a single enzyme. Since some enzymes consist of more than one polypeptide, and genes code for individual polypeptide chains, this relationship historically was updated to the one-gene–one-polypeptide hypothesis. We know now that some genes do not code proteins, and that some eukaryotic protein-coding genes are expressed to produce more than one polypeptide.
- Many human genetic diseases are caused by deficiencies in enzyme activities. Although some of these diseases are inherited as dominant traits, most are inherited as recessive traits.
- From the study of alterations in proteins other than enzymes, convincing evidence was obtained that genes control the structures of all proteins, not just those that are enzymes.
- Genetic counseling consists of an analysis of the risk that prospective parents may produce a child with a genetic defect, together with a presentation to appropriate family members of the available options for avoiding or minimizing those risks. Carrier detection and fetal analysis allow for early detection of a genetic disease.




# Analytical Approaches to Solving Genetics Problems

**Q4.1** A number of auxotrophic mutant strains were isolated from wild-type, haploid yeast. These strains responded to the addition of certain nutritional supplements to minimal culture medium with either growth (+) or no growth (0). The following table gives the growth patterns for single-gene mutant strains:

| | Supplements Added to Minimal Culture Medium | | | | |
|---|---|---|---|---|---|
| **Mutant Strains** | **B** | **A** | **R** | **T** | **S** |
| 1 | + | 0 | + | 0 | 0 |
| 2 | + | + | + | + | 0 |
| 3 | + | 0 | + | + | 0 |
| 4 | 0 | 0 | + | 0 | 0 |

Diagram a biochemical pathway that is consistent with the data, indicating where in the pathway each mutant strain is blocked.

**A4.1** The data to be analyzed are similar to those discussed in the text for Beadle and Tatum's analysis of *Neurospora* auxotrophic mutants, from which they proposed the one-gene–one-enzyme hypothesis. Recall that the later in the pathway a mutant is blocked, the fewer nutritional supplements must be added to allow growth. In the data given, we must assume that the nutritional supplements are not necessarily listed in the order in which they appear in the pathway.

Analysis of the data indicates that all four strains will grow if given R and that none will grow if given S. From this, we can conclude that R is likely to be the end product of the pathway (all mutants should grow if given the end product) and that S is likely to be the first compound in the pathway (none of the mutants should grow if given the first compound in the pathway). Thus, the pathway, as deduced so far, is

$$\text{S} \longrightarrow [\text{B,A,T}] \longrightarrow \text{R}$$

where the order of B, A, and T is as yet undetermined.

Now let us consider each of the mutant strains and see how their growth phenotypes can help define the biochemical pathway.

Strain 1 will grow only if given B or R. Therefore, the defective enzyme in strain 1 must act somewhere before the formation of B and R and after the substances A, T, and S. Since we have deduced that R is the end product of the pathway, we can propose that B is the immediate precursor to R and that strain 1 cannot make B. The pathway so far is

$$\text{S} \longrightarrow [\text{A,T}] \xrightarrow{1} \text{B} \longrightarrow \text{R}$$

Strain 2 will grow on all compounds except S, the first compound in the pathway. Thus, the defective enzyme in strain 2 must act to convert S to the next compound in the pathway, which is either A or T. We do not know yet whether A or T follows S in the pathway, but the growth data at least allow us to conclude where strain 2 is blocked in the pathway—that is,

$$\text{S} \xrightarrow{2} [\text{A,T}] \xrightarrow{1} \text{B} \longrightarrow \text{R}$$

Strain 3 will grow on B, R, and T, but not on A or S. We know that R is the end product and S is the first compound in the pathway. This mutant strain allows us to determine the order of A and T in the pathway. That is, because strain 3 grows on T, but not on A, T must be later in the pathway than A, and the defective enzyme in

3 must be blocked in the yeast's ability to convert A to T. The pathway now is

$$S \xrightarrow{2} A \xrightarrow{3} T \xrightarrow{1} B \longrightarrow R$$

Strain 4 will grow only if given the deduced end product R. Therefore, the defective enzyme produced by the mutant gene in strain 4 must act before the formation of R and after the formation of A, T, and B from the first compound S. The mutation in 4 must be blocked in the last step of the biochemical pathway in the conversion of B to R. The final deduced pathway, and the positions of the mutant blocks, are as follows:

$$S \xrightarrow{2} A \xrightarrow{3} T \xrightarrow{1} B \xrightarrow{4} R$$



## Questions and Problems

**4.1** Most enzymes are proteins, but not all proteins are enzymes. What are the functions of enzymes, and why are they essential for living organisms to carry out their biological functions?

**4.2** What was the significance of Archibald Garrod's work, and why do you expect that it was not appreciated by his contemporaries?

**4.3** Phenylketonuria (PKU) is an inherited human metabolic disorder whose effects include severe mental retardation and death. This phenotypic effect results from

**a.** the accumulation of phenylketones in the blood.
**b.** the absence of phenylalanine hydroxylase.
**c.** a deficiency of phenylketones in the blood.
**d.** a deficiency of phenylketones in the diet.

*__4.4__ If a person were homozygous for both PKU and alkaptonuria (AKU), would you expect him or her to exhibit the symptoms of PKU, AKU, or both? Refer to the following pathway:

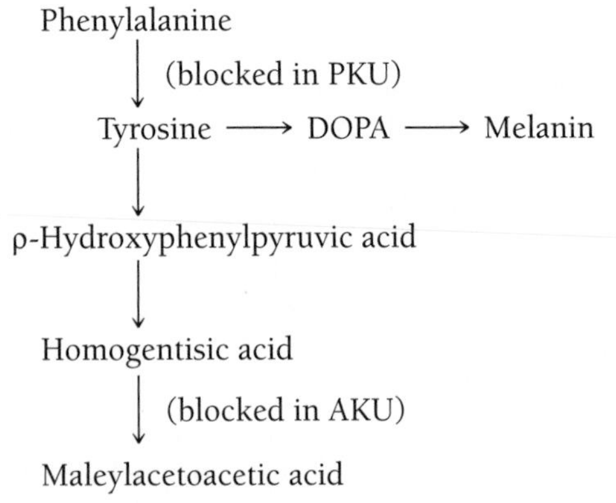


**4.5** Refer to the pathway shown in Question 4.4. What effect, if any, would you expect PKU or AKU to have on pigment formation? Explain your answer.

*__4.6__ Define the term *autosomal recessive mutation*, and give some examples of diseases that are caused by autosomal recessive mutations. Explain how two parents who display no symptoms of a given disease (albinism or any of the diseases you have named) can have two or even three children who have the disease. How can these same parents have no children with the disease?

*__4.7__ Consider sickle-cell anemia as an example of a devastating disease that is the result of an autosomal recessive genetic mutation on a specific chromosome. Explain what a molecular or genetic disease is. Compare and contrast this disease with a disease caused by an invading microorganism such as a bacterium or virus.

**4.8** A breeder of Irish setters has a particularly valuable show dog that he knows is descended from the famous bitch Rheona Didona, who carried a recessive gene for atrophy of the retina. Before he puts the dog to stud, he must ensure that it is not a carrier of this allele. How should he proceed?

**4.9** As geneticists, what problems might we encounter if we accept the one-gene–one-enzyme hypothesis as completely accurate? What further information have we discovered about this hypothesis since its formulation? What work led to that discovery?

*__4.10__ Upon infection of *E. coli* with bacteriophage T4, a series of biochemical pathways result in the formation of mature progeny phages. The phages are released after lysis of the bacterial host cells. Suppose that the following pathway exists:

enzyme enzyme
↓ ↓
A ⟶ B ⟶ mature phage

Suppose also that we have two temperature-sensitive mutants that involve the two enzymes catalyzing these sequential steps. One of the mutations is cold sensitive (*cs*), in that no mature phages are produced at 17°C. The other is heat sensitive (*hs*), in that no mature phages are produced at 42°C. Normal progeny phages are produced when phages carrying either of the mutations infect bacteria at 30°C. However, let us assume that we do not know the sequence of the two mutations. Two models are therefore possible:

$$(1)\ A \xrightarrow{hs} B \xrightarrow{cs} \text{phage}$$

$$(2)\ A \xrightarrow{cs} B \xrightarrow{hs} \text{phage}$$

Outline how you would determine experimentally which model is the correct model without artificially lysing phage-infected bacteria.

***4.11** Four mutant strains of *E. coli* (*a*, *b*, *c*, and *d*) all require substance X to grow. Four plates were prepared, as shown in the following figure:

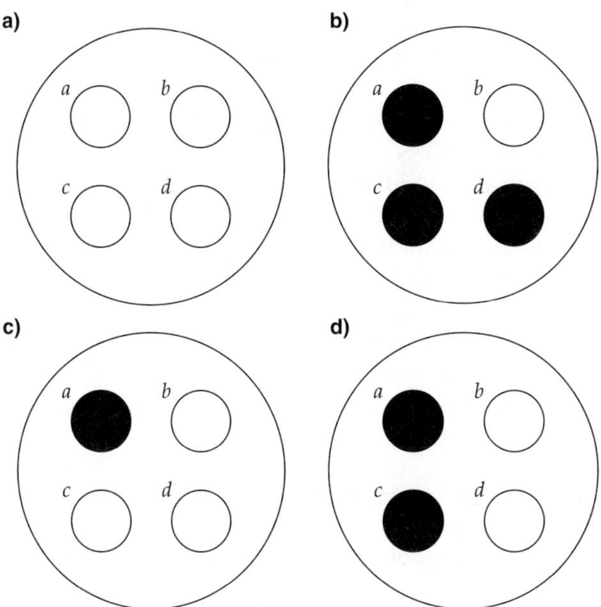


In each case the medium was minimal, with just a trace of substance X, to allow a small amount of growth of the mutant cells. On plate *a*, cells of mutant strain *a* were spread over the entire surface of the agar and grew to form a thin lawn (continuous bacterial growth over the plate). On plate *b*, the lawn was composed of mutant *b* cells, and so on. On each plate, cells of the four mutant types were inoculated over the lawn, as indicated by the circles. Dark circles indicate luxuriant growth. This experiment tests whether the bacterial strain spread on the plate can feed the four strains inoculated on the plate, allowing them to grow. What do these results show about the relationship of the four mutants to the metabolic pathway leading to substance X?

***4.12** Wax moths can be cultured by allowing adult females to lay their eggs onto an artificial medium. The eggs hatch into larvae and, as they eat the medium, the larvae grow and molt through several larval stages. After the larval period, the animals enter a pupal stage during which they metamorphose into an adult moth. Two independently isolated moth mutants, *rose-1* and *rose-2*, have eyes that are rose colored instead of the normal dark-red color. When *rose-1* adults are ground up, mixed with artificial medium, and fed to *rose-2* larvae, moths with dark-red eyes are produced. However, when *rose-2* adults are ground up, mixed with artificial medium, and fed to *rose-1* larvae, the resulting moths have rose-colored eyes. Propose a hypothesis to explain these results.

***4.13** The following growth responses (where + = growth and 0 = no growth) of mutants *1–4* were seen on the related biosynthetic intermediates A, B, C, D, and E:

| | **Growth on** | | | | |
|---|---|---|---|---|---|
| **Mutant** | **A** | **B** | **C** | **D** | **E** |
| *1* | + | 0 | 0 | 0 | 0 |
| *2* | 0 | 0 | 0 | + | 0 |
| *3* | 0 | 0 | + | 0 | 0 |
| *4* | 0 | 0 | 0 | + | + |

Assume that all intermediates are able to enter the cell, that each mutant carries only one mutation, and that all mutants affect steps after B in the pathway. Which of the following schemes best fits the data with regard to the biosynthetic pathway?

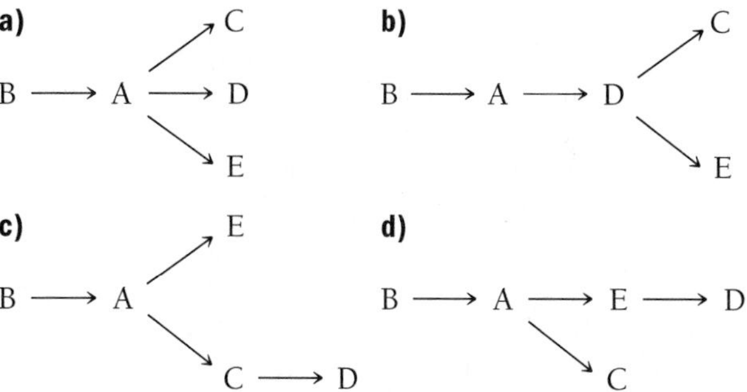


***4.14** A *Neurospora* mutant has been isolated in the laboratory where you are working. This mutant cannot make an amino acid we will call Y. Wild-type *Neurospora* cells make Y from a cellular product X through a biochemical pathway involving three intermediates called c, d, and e. How would you demonstrate that your mutant contains a defective gene for the enzyme that catalyzes the d → e reaction?

**4.15** In *Neurospora crassa*, the amino acid lysine can be synthesized using either of two completely independent pathways. One pathway uses aspartate as an initial precursor, while the other uses α-ketoglutarate. Four biochemical intermediates in the α-ketoglutarate-initiated pathway are α-aminoadipate, homocitrate, α-aminoadipate semialdehyde, and saccharopine. Precisely describe the experiments you would carry out to answer each of the following questions.

**a.** How would you obtain lysine auxotrophs in *Neurospora crassa*?

**b.** Can a lysine auxotrophic strain be blocked in just one of the two biosynthetic pathways for lysine?

**c.** How would you determine the order of the four intermediates used in the α-ketoglutarate-initiated pathway?

***4.16** Upon learning that the diseases listed in the following table are caused by a missing enzyme activity, a medical student proposes the therapies shown in the rightmost column:

| Disease | Missing Enzyme Activity | Proposed Therapy |
|---|---|---|
| Tay–Sachs disease | *N*-acetylhexosaminidase A, which catalyzes the formation of ganglioside $G_{M3}$ from ganglioside $G_{M2}$ | Administer ganglioside $G_{M3}$ (by feeding or injection) |
| Phenylketonuria | Phenylalanine hydroxylase, which catalyzes the formation of tyrosine from phenylalanine | Administer tyrosine |

**a.** Explain why each of the proposed therapies will be ineffective in treating the associated disease. For which disease would symptoms worsen if the proposed therapy were followed?

**b.** Vitamin D–dependent ricketts results in muscle and bone loss and is caused by a deficiency of 25-hydroxycholecalciferal 1 hydroxylase, an enzyme that catalyzes the formation of 1,25-dihydroxycholecalciferol (vitamin D) from 25-hydroxycholecalciferol. Unlike any of the situations in part (a), for this condition patients can be effectively treated by daily administration of the product of the enzymatic reaction, 1,25-dihydroxycholecalciferol (vitamin D). If you assayed for levels of serum 25-hydroxycholecalciferol in patients, what would you expect to find? Why is treatment with the product of the enzymatic reaction effective here, but not in the situations described in part (a)?

**4.17** Two couples in which both partners have albinism each have three children. All of the first couple's children likewise have albinism, while all of the second couple's children have normal pigmentation. How can you explain these findings?

***4.18** Glutathione (GSH) is important for a number of biological functions, including the prevention of oxidative damage in red blood cells, the synthesis of deoxyribonucleotides from ribonucleotides, the transport of some amino acids into cells, and the maintenance of protein conformation. Mutations that have lowered levels of glutathione synthetase (GSS), a key enzyme in the synthesis of glutathione, result in one of two clinically distinguishable disorders. The severe form is characterized by massive urinary excretion of 5-oxoproline (a chemical derived from a synthetic precursor to glutathione), metabolic acidosis (an inability to regulate physiological pH appropriately), anemia, and central nervous system damage. The mild form is characterized solely by anemia. The characterization of GSS activity and the GSS protein in two affected patients, each with normal parents, is given in the following table:

| Patient | Disease Form | GSS Activity in Fibroblasts (percentage of normal) | Effect of Mutation on GSS Protein |
|---|---|---|---|
| 1 | Severe | 9% | Arginine at position 267 replaced by tryptophan |
| 2 | Mild | 50% | Aspartate at position 219 replaced by glycine |

**a.** What pattern of inheritance do you expect these disorders to exhibit?

**b.** Explain the relationship of the form of the disease to the level of GSS activity.

**c.** How can two different amino acid substitutions lead to dramatically different phenotypes?

**d.** Why is 5-oxoproline produced in significant amounts only in the severe form of the disorder?

**e.** Is there evidence that the mutations causing the severe and mild forms of the disease are allelic (in the same gene)?

**f.** How might you design a test to aid in prenatal diagnosis of this disease?

**4.19** You have been introduced to the functions and levels of proteins and their organization. List as many protein functions as you can, and give an example of each.

**4.20** We know that the function of any protein is tied to its structure. Give an example of how a disruption of a protein's structure by mutation can lead to a distinctive phenotypic effect.

**4.21** The human β-globin gene provides an excellent example of how the sequence of nucleotides in a gene is eventually expressed as a functional protein. Explain how mutations in the β-globin gene can cause an altered phenotype. How can two different mutations in the same gene cause very different disease phenotypes?

***4.22** Consider the human hemoglobin variants shown in Figure 4.11. What would you expect the phenotype to be in people heterozygous for the following two hemoglobin mutations?

**a.** Hb Norfolk and Hb-S

**b.** Hb-C and Hb-S

**4.23** α-Tubulin and β-tubulin are structural (nonenzymatic) proteins that polymerize together to form microtubules. In the nematode *Canenorhabditis elegans*, mutations in either of these proteins can result in recessive male sterility.

**a.** Generate a hypothesis to explain why the tubulin mutants are male-sterile.

**b.** What would you do to gather evidence to support your hypothesis?

**4.24** Devise a rapid screen to detect new mutations in hemoglobin, and critically evaluate which types of mutations your screen can and cannot detect.

*__4.25__ Glutamate oxaloacetic transaminase-2 (GOT-2) is a mitochondrial enzyme that synthesizes glutamate from aspartate and α-ketoglutarate. GOT-2 is a homodimer—a protein made of paired identical polypeptides. The *Got-2^M^* mutation introduces a single amino-acid change that alters the charge of the polypeptide produced by the normal *Got-2^+^* allele. When enzymes from $Got\text{-}2^+\ Got\text{-}2^+$ homozygotes, $Got\text{-}2^+\ Got\text{-}2^M$ heterozygotes, and $Got\text{-}2^M\ Got\text{-}2^M$ homozygotes are separated by charge using gel electrophoresis, the gel shows the following pattern of bands (thicker bands indicate more protein):

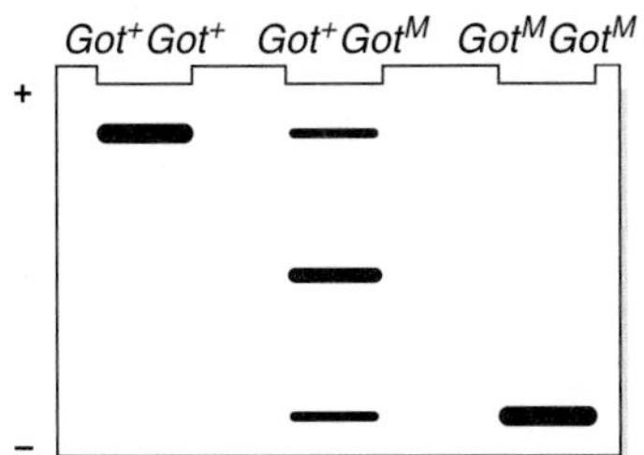


**a.** Compared to the normal GOT-2 polypeptide, is the polypeptide produced in $Got\text{-}2^M\ Got\text{-}2^M$ mutants more basic or more acidic?

**b.** Explain the pattern and relative intensities of the bands seen in each *Got-2* genotype.

**c.** Figure 4.8 shows the pattern of bands seen when hemoglobin of individuals with sickle-cell trait is separated by charge using gel electrophoresis. Why do $\beta^A\beta^S$ heterozygotes have only two types of hemoglobin, while $Got\text{-}2^+Got\text{-}2^M$ heterozygotes have three types of GOT-2 protein?

**4.26**

**a.** What is a mouse model for a human disease, and what is its utility?

**b.** What genetic and phenotypic properties would you require in a mouse model for Tay–Sachs disease?

**c.** How might a mouse model for Tay–Sachs disease be helpful in evaluating alternative therapeutic strategies.

**4.27** What can prospective parents do to reduce the risk of bearing offspring who have genetically based enzyme deficiencies?

**4.28** Some methods used to gather fetal material for prenatal diagnosis are invasive and therefore pose a small, but very real, risk to the fetus.

**a.** What specific risks and problems are associated with chorionic villus sampling and amniocentesis?

**b.** How are these risks balanced with the benefits of each procedure?

**c.** Fetal cells are reportedly present in the maternal bloodstream after about 8 weeks of pregnancy. However, the number of cells is very low, perhaps no more than one in several million maternal cells. To date, it has not been possible to isolate fetal cells from maternal blood in sufficient quantities for routine genetic analysis. If the problems associated with isolating fetal cells from maternal blood were overcome, and sufficiently sensitive methods were developed to perform genetic tests on a small number of cells, what would be the benefits of performing such tests on these fetal cells?

*__4.29__ Many autosomal recessive mutations that cause disease in newborns can be diagnosed and treated. However, only a few inherited diseases are routinely tested for in newborns. Explore the basis upon which tests are performed by answering the following questions concerning testing for PKU, which is required on newborns throughout the United States, and testing for CF, which is done only if a newborn or infant shows symptoms consistent with a diagnosis of CF.

**a.** What are the relative frequencies of PKU and CF in newborns, and how—if at all—are these frequencies related to mandated testing?

**b.** What is the basis of the Guthrie test used for detecting PKU, and what features of the test make it useful for screening large numbers of newborns efficiently?

**c.** Multiple diagnostic tests have been developed for CF. Some are DNA-based while others indirectly assess CFTR protein function. An example of the latter is a test that measures salt levels in sweat. In CF patients, salt levels are elevated due to diminished CFTR protein function. Although the ΔF508 mutation discussed in the text is common in patients with a severe form of CF, other CF mutations are associated with less severe phenotypes. Tests assessing CFTR protein function may not reliably distinguish normal newborns from newborns with mild forms of CF. What challenges do the types of available tests and the range of disease phenotypes present in a population pose for implementing diagnostic testing?

**d.** Discuss the importance of testing for PKU and CF at birth relative to the time that therapeutic intervention is required. Under what circumstances is testing newborns for CF warranted?

**4.30** Reflecting on your answers to Question 4.29, state why newborns are *not* routinely tested for recessive mutations that cause uncurable diseases such as Tay–Sachs disease.

*__4.31__ Mr. and Mrs. Chávez have a son who was found to have PKU at birth. Mr. and Mrs. Lieberman have a son who developed Tay–Sachs disease at about 7 months of age. Each couple is now expecting a second child, is concerned that their second child might develop the disease seen in their son, and so discusses their situation with a genetic counselor. After taking their family histories, the counselor describes a set of tests that can provide information about whether the second child will develop disease.

**a.** What different types of tests can be done to aid in carrier detection and fetal analysis, and what are their advantages and disadvantages?

**b.** How would you determine whether the disease seen in each couple's son results from a new mutation or has been transmitted from one or both of the parents?

**c.** Place yourself in each couple's predicament. Would you ask that fetal analysis be performed in each situation? Explain your reasoning.

*__4.32__ Neuronal development has essentially ceased by the time humans reach their early twenties. Why then are all women with PKU who become pregnant, including women over 25, advised to return to a phenylalanine-restricted diet throughout their pregnancy?

**4.33** In evaluating my teacher, my sincere opinion is that

**a.** he or she is a swell person whom I would be glad to have as a brother-in-law or sister-in-law.

**b.** he or she is an excellent example of how tough it is when you do not have either genetics or environment going for you.

**c.** he or she must be missing a critical enzyme and is accumulating some behavior-altering intermediate.

**d.** he or she ought to be preserved in tissue culture for the benefit of other generations.

# 5 Gene Expression: Transcription

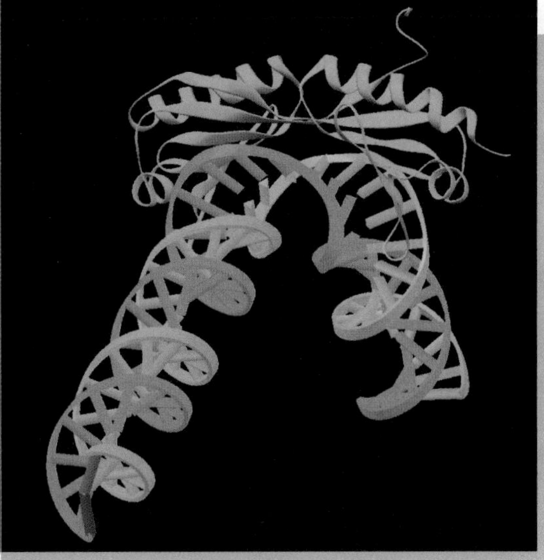

Yeast TBP (TATA-binding protein) binding to a promoter region in DNA.

## Key Questions

- What is the central dogma?
- What are the four main types of RNA molecules in cells?
- How is an RNA chain synthesized?
- How is transcription initiated, elongated, and terminated in bacteria?
- How does transcription occur in eukaryotes?
- How is functional mRNA produced from the initial transcript of a protein-coding gene in eukaryotes?

### iActivity

DO YOU WANT TO MAKE A CLONE? MIX GENES to create a new organism? Treat genetic disease with DNA? Investigate a murder? These biotechnology techniques, and many others, are made possible by an understanding of gene expression, the first step of which is transcription, during which information is transferred from the DNA molecule to a single-stranded RNA molecule. In this chapter, you will learn about how DNA is transcribed into RNA and about the structure and properties of different forms of RNA. Then, in the iActivity, you can investigate how mutations that affect the process of transcription can lead to an inherited disease.

The structure, function, development, and reproduction of an organism depend on the properties of the proteins present in each cell and tissue. A protein consists of one or more chains of amino acids. Each chain is a polypeptide, and the sequence of amino acids in a polypeptide is coded for by a gene. When a protein is needed in the cell, the genetic code for the amino acid sequence of that protein must be read from the DNA and the protein made. Two major steps occur during protein synthesis: transcription and translation. **Transcription** is the synthesis of a single-stranded RNA copy of a segment of DNA. In the case of protein synthesis, a protein-coding gene is transcribed to give a messenger RNA. **Translation** (protein synthesis) is the conversion of the messenger RNA base-sequence information into the amino acid sequence of a polypeptide. In this chapter, you will learn about the transcription process.

## Gene Expression—The Central Dogma: An Overview

In 1956, three years after Watson and Crick proposed their double helix model of DNA, Crick gave the name *central dogma* to the two-step process denoted DNA → RNA → protein (transcription followed by translation). Transcription is the synthesis of an RNA copy of a segment of DNA; only one of the two DNA strands is transcribed into an RNA. This is logical because the RNA has to function in the cell, and its function depends on its base sequence. A transcript of the other DNA strand would have a complementary RNA sequence that would not be the correct sequence for function.

The production of an RNA by transcription of a gene is one step of **gene expression.** There are four main types of RNA molecules, each encoded by its own type of gene, but only one of them is translated:

1. **mRNA (messenger RNA)** encodes the amino acid sequence of a polypeptide. mRNAs are the transcripts of *protein-coding genes*. Translation of an mRNA produces a polypeptide.
2. **rRNA (ribosomal RNA),** with ribosomal proteins, makes up the ribosomes—the structures on which mRNA is translated.
3. **tRNA (transfer RNA)** brings amino acids to ribosomes during translation.
4. **snRNA (small nuclear RNA),** with proteins, forms complexes that are used in eukaryotic RNA processing to produce functional mRNAs.

A number of other small RNA molecules occur in the cell and will be introduced in later chapters. In the remainder of this chapter, you will learn about transcription in both bacteria and eukaryotes, with a focus on protein-coding genes.

## The Transcription Process

How is an RNA chain synthesized? Associated with each gene are sequences called **gene regulatory elements,** which are involved in regulating transcription. The enzyme **RNA polymerase** catalyzes the process of transcription (Figure 5.1). (More rigorously, the enzyme is known as **DNA-dependent RNA polymerase** because it uses a DNA template for the synthesis of an RNA chain.) The DNA double helix unwinds for a short region next to the gene before transcription begins. In bacteria, RNA polymerase is responsible for unwinding; in eukaryotes, unwinding is done by other proteins that bind to the DNA near the start point for transcription.

*animation*
**RNA Biosynthesis**

In transcription, RNA is synthesized in the 5′-to-3′ direction. The 3′-to-5′ DNA strand that is read to make the RNA strand is called the **template strand.** The 5′-to-3′ DNA strand complementary to the template strand, and having the same polarity as the resulting RNA strand, is called the *nontemplate strand.* By convention, in the literature and databases of gene sequences, the sequence presented is of the nontemplate DNA strand. From this strand, the sequence of the RNA transcript can be directly derived and, if it is an mRNA, the encoded amino acids can be directly read from the genetic code dictionary.

The RNA precursors for transcription are the ribonucleoside triphosphates ATP, GTP, CTP, and UTP, collectively called NTPs (*n*ucleoside *t*ri*p*hosphates). RNA synthesis occurs by polymerization reactions similar to those involved in DNA synthesis (Figure 5.2; DNA polymerization is shown in Figure 3.3, p. 41). RNA polymerase selects the next nucleotide to be added to the chain by its ability to pair with the exposed base on the DNA template strand. Unlike DNA polymerases, RNA polymerases can initiate new RNA chains; in other words, no primer is needed.

Recall that RNA chains contain nucleotides with the base uracil instead of thymine and that uracil pairs with adenine. Therefore, where an A nucleotide occurs on the DNA template chain, a U nucleotide is placed in the RNA chain instead of a T. For example, if the template DNA strand reads

3′-ATACTGGAC-5′

then the RNA chain will be synthesized in the 5′-to-3′ direction and will have the sequence

5′-UAUGACCUG-3′

**Keynote**

Transcription is the process of transferring the genetic information in DNA into RNA base sequences. The DNA unwinds in a short region next to a gene, and an RNA polymerase catalyzes the synthesis of an RNA molecule in the 5′-to-3′ direction along the 3′-to-5′ template strand of the DNA. Only one strand of the double-stranded DNA is transcribed into an RNA molecule.

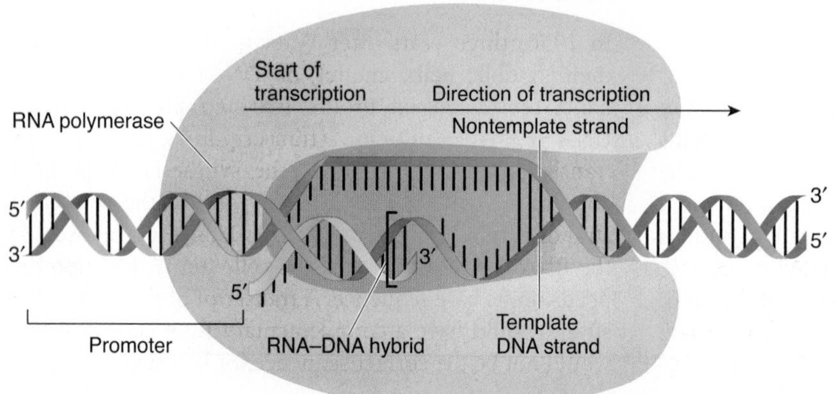


**Figure 5.1**
**Transcription process.** The DNA double helix is denatured by RNA polymerase in prokaryotes and by other proteins in eukaryotes. RNA polymerase then catalyzes the synthesis of a single-stranded RNA chain, beginning at the "start of transcription" point. The RNA chain is made in the 5′-to-3′ direction, with only one strand of the DNA used as a template to determine the base sequence.

**Figure 5.2**
**Chemical reaction involved in the RNA-polymerase-catalyzed synthesis of RNA on a DNA template strand.**

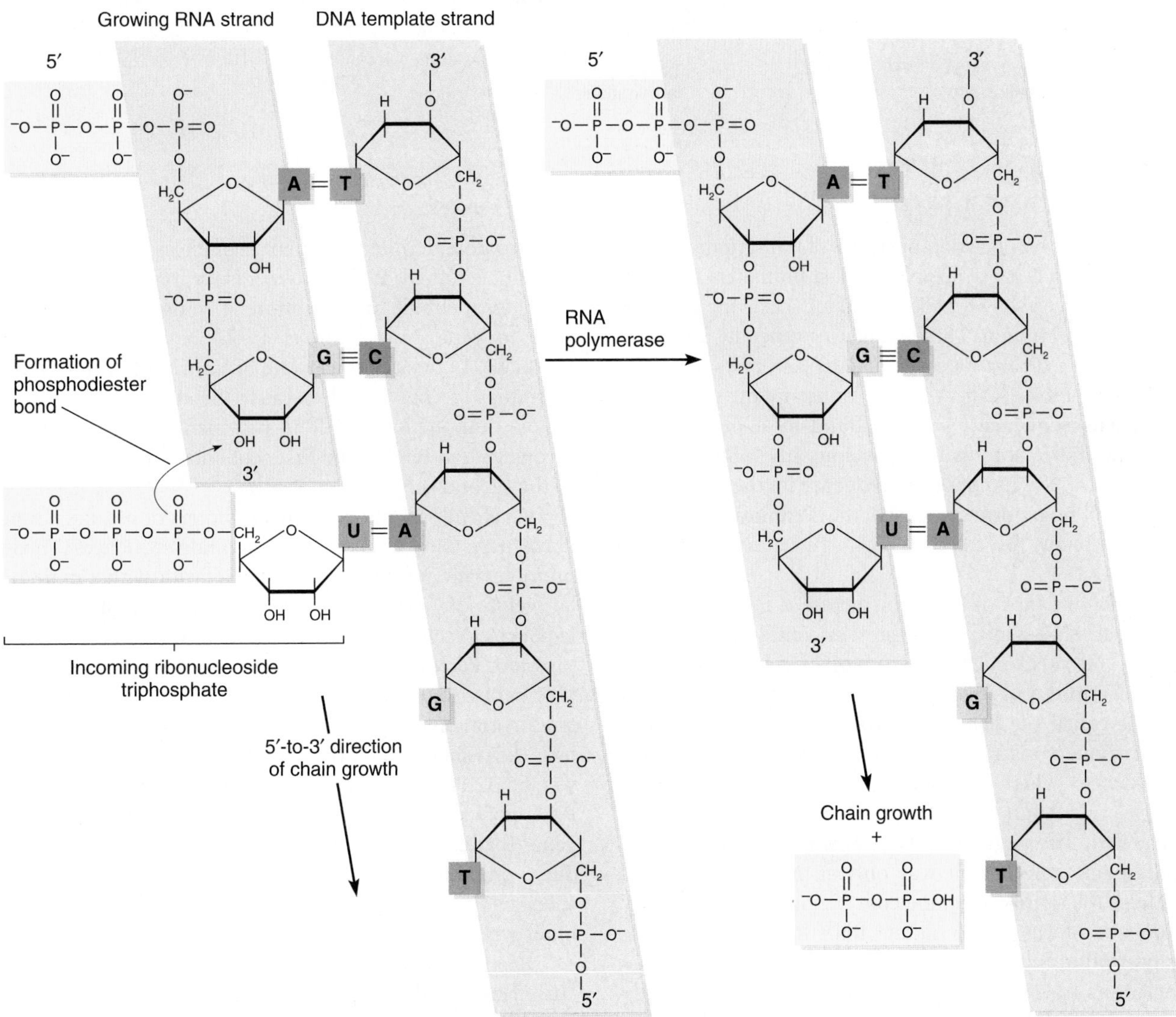




## Transcription in Bacteria

The process of transcription occurs in three stages: initiation, elongation, and termination. In this section we focus on transcription in the model bacterium, *E. coli*.

### Initiation of Transcription at Promoters

What is the mechanism of transcription in initiation in *E. coli*? A bacterial gene may be divided into three sequences with respect to its transcription (Figure 5.3):

1. A **promoter,** a sequence upstream of the start of the gene that encodes the RNA. The RNA polymerase interacts with the promoter. The way the RNA polymerase interacts, spatially speaking, defines the direction for transcription and, thus, dictates to the enzyme which DNA strand is the template strand and where transcription is to begin. That is, the promoter sequence serves to orient the RNA polymerase to start transcribing at the beginning of the gene and ensures that the initiation of synthesis of every RNA occurs at the same site.

   A gene with its promoter is an independent unit. This means that the strand of the double helix that is the template strand is gene specific. In other words, some genes use one strand of the DNA as the template strand, while other genes use the other strand. The present organization of genes in this regard is the result of the evolution of present-day genomes.
2. The RNA-coding sequence itself—that is, the DNA sequence transcribed by RNA polymerase into the RNA transcript.
3. A **terminator,** specifying where transcription stops.

From comparisons of sequences upstream of coding sequences and from studies of the effects of specific base-pair

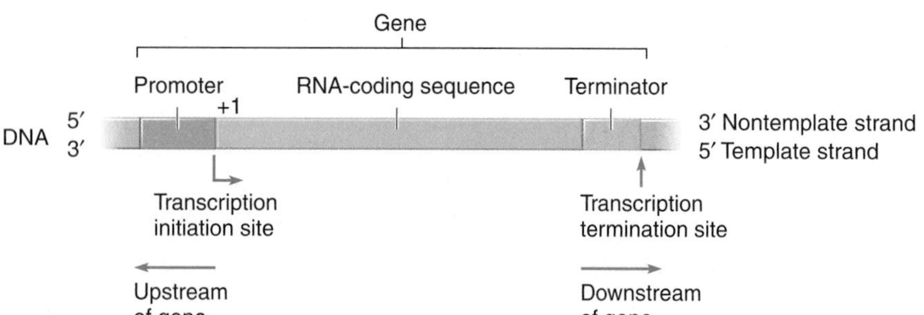


**Figure 5.3**
**Promoter, RNA-coding sequence, and terminator regions of a gene.** The promoter is upstream of the coding sequence, the terminator downstream. The coding sequence begins at nucleotide +1.



mutations at every position upstream of transcription initiation sites, two DNA sequences in most promoters of *E. coli* genes have been shown to be critical for specifying the initiation of transcription. These sequences generally are found at −35 and −10, that is, at 35 and 10 base pairs upstream from the +1 base pair at which transcription starts. The **consensus sequence** (the base found most frequently at each position) for the −35 region (the −35 box) is 5′-TTGACA-3′. The consensus sequence for the −10 region (the **−10 box,** formerly called the **Pribnow box,** after David Pribnow, the researcher who first discovered it) is 5′-TATAAT-3′.

Only one type of RNA polymerase is found in bacteria, so all classes of genes—protein-coding genes, tRNA genes, and rRNA genes—are transcribed by it. Initiation of transcription of a gene requires a form of RNA polymerase called the *holoenzyme* (or *complete enzyme*). The holoenzyme consists of the **core enzyme** form of RNA polymerase, which consists of two $\alpha$, one $\beta$, and one $\beta'$ polypeptide, bound to another polypeptide called a *sigma factor* ($\sigma$). The sigma factor ensures that the RNA polymerase binds in a stable way only at promoters. That is, without the sigma factor, the core enzyme can bind to any sequence of DNA and initiate RNA synthesis, but this transcription initiation is not at the correct sites. The association of the sigma factor with the core enzyme greatly reduces the ability of the enzyme to bind to DNA nonspecifically and establishes the promoter-specific binding properties of the holoenzyme. A sigma factor is not required for the elongation and termination stages of transcription.

The RNA polymerase holoenzyme binds to the promoters of most genes as shown in Figure 5.4. First, the holoenzyme contacts the −35 sequence and then binds to the full promoter while the DNA is still in standard double helix form, a state called the *closed promoter complex* (Figure 5.4a). Then the holoenzyme untwists the DNA in the −10 region (Figure 5.4b). The untwisted form of the promoter is called the *open promoter complex.* The sigma factor of the holoenzyme plays a key role in these steps by contacting the promoter directly at the −35 and −10 sequences. Once the RNA polymerase is bound at the −10 box, it is oriented properly to begin transcription at the correct nucleotide of the gene. At this point the RNA polymerase is contacting about 75 bp of the DNA from −55 to +20.

Promoters differ in their sequences, so the binding efficiency of RNA polymerase varies. As a result, the rate at which transcription is initiated varies from gene to gene. For example, a −10 region sequence of 5′-GATACT-3′ has a lower rate of transcription initiation than does 5′-TATAAT-3′ because the ability of the sigma factor component of the RNA polymerase holoenzyme to recognize and bind to the first sequence is lower than it is to the second sequence.

As already mentioned, the promoters of most genes in *E. coli* have the −35 and −10 recognition sequences. Those promoters are recognized by a sigma factor with a molecular weight of 70,000 Da, called $\sigma^{70}$. There are other sigma factors in *E. coli* with important roles in regulating gene expression. Each type of sigma factor binds to the core RNA polymerase and permits the holoenzyme to recognize different promoters. For example, under conditions of high heat (heat shock) and other forms of stress, $\sigma^{32}$ (molecular weight 32,000 Da) increases in amount, directing some RNA polymerase molecules to bind to the promoters of genes that encode proteins needed to cope with the stress. Such promoters have consensus recognition sequences specific to the $\sigma^{32}$ factor at −39 and −15. There are several other types of sigma factors with various roles.

Regulation of expression of bacterial genes will be discussed in Chapter 17. In brief, the transcription of many bacterial genes is controlled by the interaction of regulatory proteins with regulatory sequences upstream of the RNA-coding sequence in the vicinity of the promoter. There are two classes of regulatory proteins: *activators* stimulate transcription by making it easier for RNA polymerase to bind or elongate an RNA strand, while *repressors* inhibit transcription by making it more difficult for RNA polymerase to bind or elongate an RNA strand.

## Elongation of an RNA Chain

RNA synthesis takes place in a region of DNA that has separated into single strands to form a transcription bubble. Once initiation succeeds and the elongation stage is established, the RNA polymerase begins to move along the DNA and the sigma factor is released (Figure 5.4c). The core enzyme alone is able to complete the transcription of the gene. In *E. coli* growing at 37°C, transcription occurs at about 40 nucleotides/sec. During the transition from initiation to elongation, the RNA polymerase

**Figure 5.4**

**Action of *E. coli* RNA polymerase in the initiation and elongation stages of transcription.**

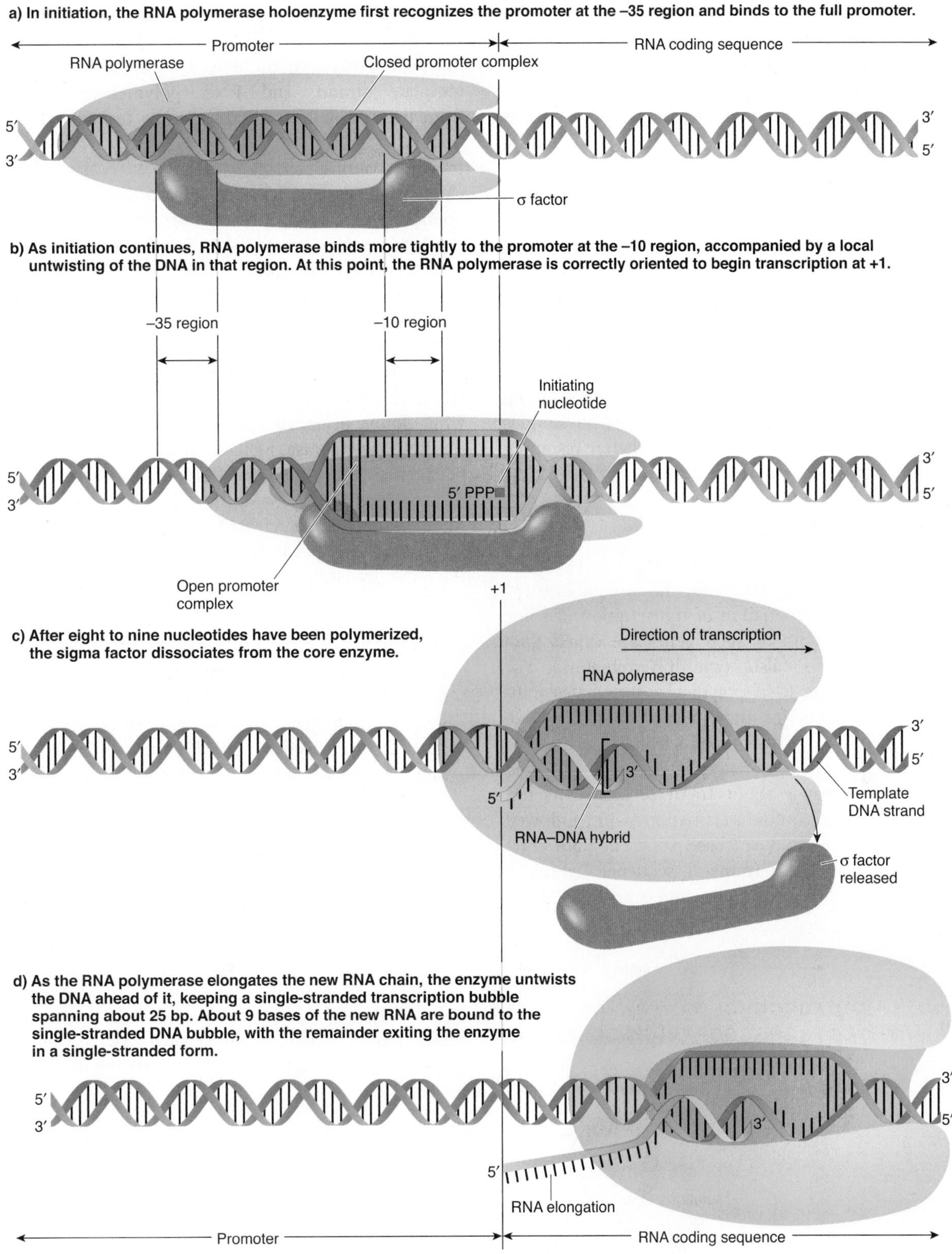
a) In initiation, the RNA polymerase holoenzyme first recognizes the promoter at the –35 region and binds to the full promoter.
Promoter
RNA coding sequence
RNA polymerase
Closed promoter complex
5′
3′
3′
5′
σ factor
b) As initiation continues, RNA polymerase binds more tightly to the promoter at the –10 region, accompanied by a local untwisting of the DNA in that region. At this point, the RNA polymerase is correctly oriented to begin transcription at +1.
–35 region
–10 region
Initiating nucleotide
5′ PPP
Open promoter complex
+1
c) After eight to nine nucleotides have been polymerized, the sigma factor dissociates from the core enzyme.
Direction of transcription
RNA polymerase
RNA–DNA hybrid
Template DNA strand
σ factor released
d) As the RNA polymerase elongates the new RNA chain, the enzyme untwists the DNA ahead of it, keeping a single-stranded transcription bubble spanning about 25 bp. About 9 bases of the new RNA are bound to the single-stranded DNA bubble, with the remainder exiting the enzyme in a single-stranded form.
RNA elongation
Promoter
RNA coding sequence

becomes more compact, contacting less of the DNA. Once the elongation stage is established, the RNA polymerase contacts about 40 bp of the DNA with approximately 25 bp in the transcription bubble.

During the elongation stage, the core RNA polymerase moves along, untwisting the DNA double helix ahead of itself to expose a new segment of single-stranded template DNA. Behind the untwisted region, the two DNA strands reform into double-stranded DNA (Figure 5.4d). Within the untwisted region, about 9 RNA nucleotides are base-paired to the DNA in a temporary RNA–DNA hybrid; the rest of the newly synthesized RNA exits the enzyme as a single strand (see Figure 5.4d).

RNA polymerase has two proofreading activities. One of these is similar to the proofreading by DNA polymerase, in which the incorrectly inserted nucleotide is removed by the enzyme reversing its synthesis reaction, backing up one step, and then replacing the incorrect nucleotide with the correct one in a forward step. In the other proofreading process, the enzyme moves back one or more nucleotides and then cleaves the RNA at that position before resuming RNA synthesis in the forward direction.

## Termination of an RNA Chain

The termination of bacterial gene transcription is signaled by *terminator sequences.* In *E. coli*, the protein Rho (ρ) plays a role in the termination of transcription of some genes. The terminators of such genes are called *Rho-dependent terminators* (also, type II terminators). For other genes, the core RNA polymerase terminates transcription; terminators for those genes are called *Rho-independent terminators* (also, type I terminators).

Rho-independent terminators consist of an inverted repeat sequence that is about 16 to 20 base pairs upstream of the transcription termination point, followed by a string of about 4 to 8 A-T base pairs. The RNA polymerase transcribes the terminator sequence, which is part of the initial RNA-coding sequence of the gene. Because of the inverted repeat arrangement, the RNA folds into a hairpin loop structure (Figure 5.5). The hairpin structure causes the RNA polymerase to slow and then pause in its catalysis of RNA synthesis. The string of U nucleotides downstream of the hairpin destabilizes the pairing between the new RNA chain and the DNA template strand, and RNA polymerase dissociates from the template; transcription has terminated. Mutations that disrupt the hairpin partially or completely prevent termination.

Rho-dependent terminators are C-rich, G-poor sequences that have no hairpin structures like those of rho-independent terminators. Termination at these terminators is as follows: Rho binds to the C-rich terminator sequence in the transcript upstream of the transcription termination site. Rho then moves along the transcript until it reaches the RNA polymerase, where the most recently synthesized RNA is base paired with the template DNA. Rho is a helicase enzyme, meaning that it can unwind double-stranded nucleic acids. When Rho reaches the RNA polymerase, helicase unwinds the helix formed between the RNA and the DNA template strand, using ATP hydrolysis to provide the needed energy. The new RNA strand is then released, the DNA double helix reforms, and the RNA polymerase and Rho dissociate from the DNA; transcription has terminated.

### Keynote

In *E. coli*, the initiation and termination of transcription are signaled by specific sequences that flank the RNA-coding sequence of the gene. The promoter is recognized by the sigma factor component of the RNA polymerase–sigma factor complex. Two types of termination sequences are found, and a particular gene has one or the other. One type of terminator is recognized by the RNA polymerase alone, and the other type is recognized by the enzyme in association with the Rho factor.

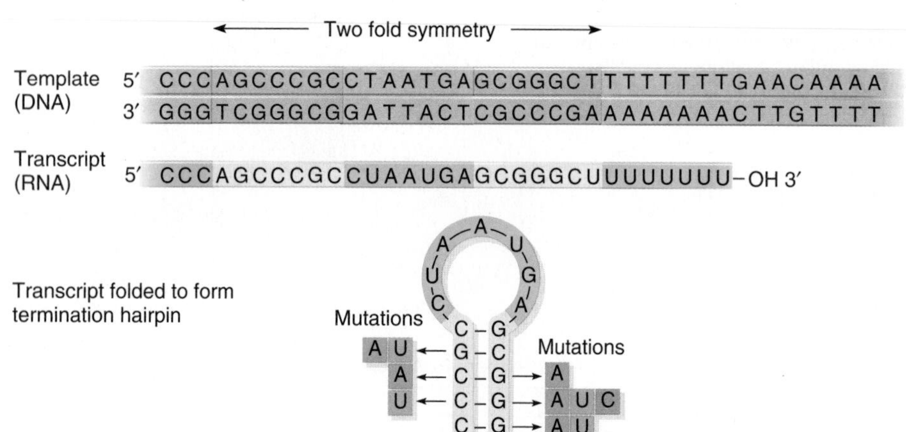


**Figure 5.5**

**Sequence of a Rho-independent terminator and structure of the terminated RNA.** The mutations in the stem (yellow section) partially or completely prevent termination.

## Transcription in Eukaryotes

Transcription is more complicated in eukaryotes than in bacteria. This is because eukaryotes possess three different classes of RNA polymerases and because of the way in which transcripts are processed to their functional forms. The focus in this section is on the transcription of protein-coding genes.

### Eukaryotic RNA Polymerases

In eukaryotes, three different RNA polymerases transcribe the genes for four main types of RNAs. **RNA polymerase I,** located in the nucleolus, catalyzes the synthesis of three of the RNAs found in ribosomes: the 28S, 18S, and 5.8S rRNA molecules. (The S values derive from the rate at which the rRNA molecules sediment during centrifugation and give a very rough indication of molecular sizes.) **RNA polymerase II,** located in the nucleoplasm, synthesizes messenger RNAs (mRNAs) and some small nuclear RNAs (snRNAs). **RNA polymerase III,** located in the nucleoplasm, synthesizes: (1) transfer RNAs (tRNAs); (2) 5S rRNA, a small rRNA molecule found in each ribosome; and (3) the snRNAs not made by RNA polymerase II.

All eukaryotic RNA polymerases consist of multiple subunits. For example, yeast RNA polymerase II consists of 12 subunits and has a U-shaped structure; the open end of the U leads the polymerase as it moves along the DNA (Figure 5.6). A similar type of structure is seen for eukaryotic RNA polymerase II enzymes of other species. Bacterial RNA polymerases are smaller but have a relatively similar structure to eukaryotic RNA polymerases.

**Keynote**

In *E. coli*, a single RNA polymerase synthesizes mRNA, tRNA, and rRNA. Eukaryotes have three distinct nuclear RNA polymerases, each of which transcribes different gene types: RNA polymerase I transcribes the genes for the 28S, 18S, and 5.8S ribosomal RNAs; RNA polymerase II transcribes mRNA genes and some snRNA genes; and RNA polymerase III transcribes genes for the 5S rRNAs, the tRNAs, and the remaining snRNAs.

### Transcription of Protein-Coding Genes by RNA Polymerase II

In this section, we discuss the sequences and molecular events involved in transcribing a protein-coding gene by RNA polymerase II. Eukaryotic genes transcribed by RNA polymerase II have specific promoter sequences but, in contrast to bacterial genes, they do not have specific terminator sequences. The product of transcription is a **precursor mRNA (pre-mRNA)** molecule—a transcript that must be modified, processed, or both to produce the mature, functional mRNA molecule that can be translated to generate a polypeptide.

Figure 5.6

**Three-dimensional structure of RNA polymerase II from yeast.** Each color represents a different polypeptide.

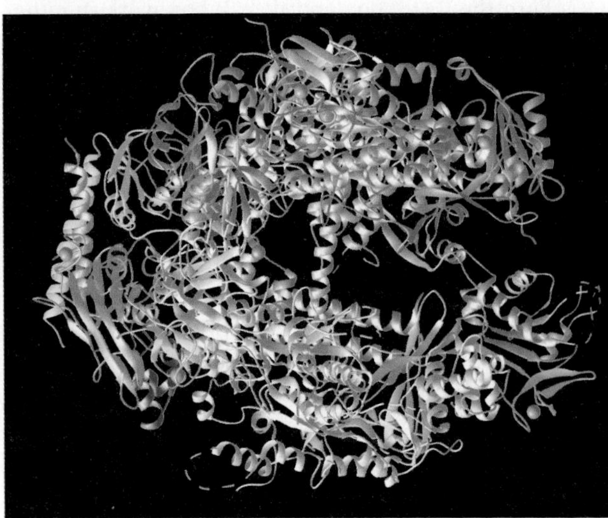

**Promoters and Enhancers.** Promoters of protein-coding genes are analyzed in two principal ways. One way is to examine the effect of mutations that delete or alter base pairs upstream from the starting point of transcription and to see whether those mutants affect transcription. Mutations that significantly affect transcription define important promoter elements. The second way is to compare the DNA sequences upstream of a number of protein-coding genes to see whether any regions have similar sequences. The results of these experiments show that the promoters of protein-coding genes encompass about 200 base pairs upstream of the transcription initiation site and contain various sequence elements. Two general regions of the promoter are: (1) the core promoter; and (2) promoter-proximal elements.

The **core promoter** is the set of *cis*-acting sequence elements needed for the transcription machinery to start RNA synthesis at the correct site. ('*Cis*' means "on the same side." A *cis*-acting sequence element affects the activity only of a gene on the same molecule of DNA.) These elements are typically within no more than 50 bp upstream of that site. The best-characterized core promoter elements are: (1) a short sequence element called *Inr* (*initiator*), which spans the transcription initiation start site (defined as +1); and (2) the TATA **box,** or TATA **element** (also called the **Goldberg-Hogness box,** after its discoverers), located at about position −30. The TATA box has the seven-nucleotide consensus sequence 5′-TATAAAA-3′. The Inr and TATA elements specify where the transcription machinery assembles and determine where transcription will begin. However, in the absence of other elements, transcription will occur only at a very low level.

**Promoter-proximal elements** are upstream from the TATA box, in the area from −50 to −200 nucleotides from the start site of transcription. Examples of these elements are the CAAT ("*cat*") box, named for its consensus sequence and located at about −75; and the GC box, with consensus sequence 5′-GGGCGG-3′, located at about −90. Both the CAAT box and the GC box work in either orientation (meaning with the sequence element oriented either toward or away from the direction of transcription). Mutations in either of these elements (or other promoter-proximal elements not mentioned) markedly decrease transcription initiation from the promoter, indicating that they play a role in determining the efficiency of the promoter.

Promoters contain various combinations of core promoter elements and promoter-proximal elements that together determine promoter function. The promoter-proximal elements are important in determining how and when a gene is expressed. Key to this regulation are transcription regulatory proteins called **activators,** which determine the efficiency of transcription initiation. For example, genes that are expressed in all cell types for basic cellular functions—"housekeeping genes"—have promoter-proximal elements that are recognized by activators found in all cell types. Examples of housekeeping genes are the actin gene and the gene for the enzyme glucose 6-phosphate dehydrogenase. By contrast, genes that are expressed only in particular cell types or at particular times have promoter-proximal elements recognized by activators in those cell types or at those particular times.

Other sequences—**enhancers**—are required for the maximal transcription of a gene. Enhancers are another type of cis-acting element. By definition, enhancers function either upstream or downstream from the transcription initiation site—although, commonly, they are upstream of the gene they control, sometimes thousands of base pairs away. In other words, enhancers modulate transcription from a distance. Enhancers contain a variety of short sequence elements, some of them the same as those found in the promoter. Activators also bind to these elements and with other protein complexes. The DNA containing the enhancer is brought close to the promoter DNA to which the transcription machinery is bound, stimulating transcription to the maximal level for the particular gene.

We will discuss activators, promoters, and enhancers and how eukaryotic protein-coding genes are regulated in more detail in Chapter 18. This chapter's Focus on Genomics box describes how researchers identify promoters in genomic DNA sequences.

**Transcription Initiation.** Accurate initiation of transcription of a protein-coding gene involves the assembly of RNA polymerase II and a number of other proteins called **general transcription factors (GTFs)** on the core promoter. In contrast to bacterial RNA polymerase enzymes, none of the three eukaryotic RNA polymerases can bind directly to DNA. Instead, particular GTFs bind first and recruit the RNA polymerase to form a complex. Other GTFs then bind, and transcription can begin. The GTFs are numbered for the RNA polymerase with which they work and are lettered to reflect their order of discovery. For example, TFIID is the fourth general transcription factor (D) discovered that works with RNA polymerase II.

For protein-coding genes, the GTFs and RNA polymerase II bind to promoter elements in a particular order in vitro to produce the *complete transcription initiation complex,* also called the *preinitiation complex* (PIC) because it is ready to begin transcription (Figure 5.7). As mentioned earlier, the binding of activators to promoter-proximal elements and to enhancer elements determines the overall efficiency of transcription initiation at a particular promoter.

While *in vitro* experiments indicate a sequential order of loading of GTFs and RNA polymerase II onto the promoter, the situation is less clear *in vivo.* Some data indicate that the initiation complex comes to the promoter in a single complex. Whether or not that is the case, transcription initiation *in vivo* is clearly more

## Focus on Genomics
### Finding Promoters

Promoters are obviously important for gene function. Earlier in the chapter, we defined consensus sequences for promoters and other upstream regulatory regions, for instance the TATA and CAAT boxes described in the chapter. The sequence of these elements as well as their spacing relative to each other and the transcriptional start site are functionally important. Not all genes have great matches to these sequences in their promoters, either because they bind more poorly to the transcription machinery or because other proteins assist RNA polymerase to bind. One early application of genomics was to scan a sequence for candidate promoter sequences and then to look for a gene associated with those sequences. This can be helpful, especially in conjunction with the scans for the open reading frames (amino acid-coding regions) described in Chapter 6, as well as other scans for regions such as termination signals.

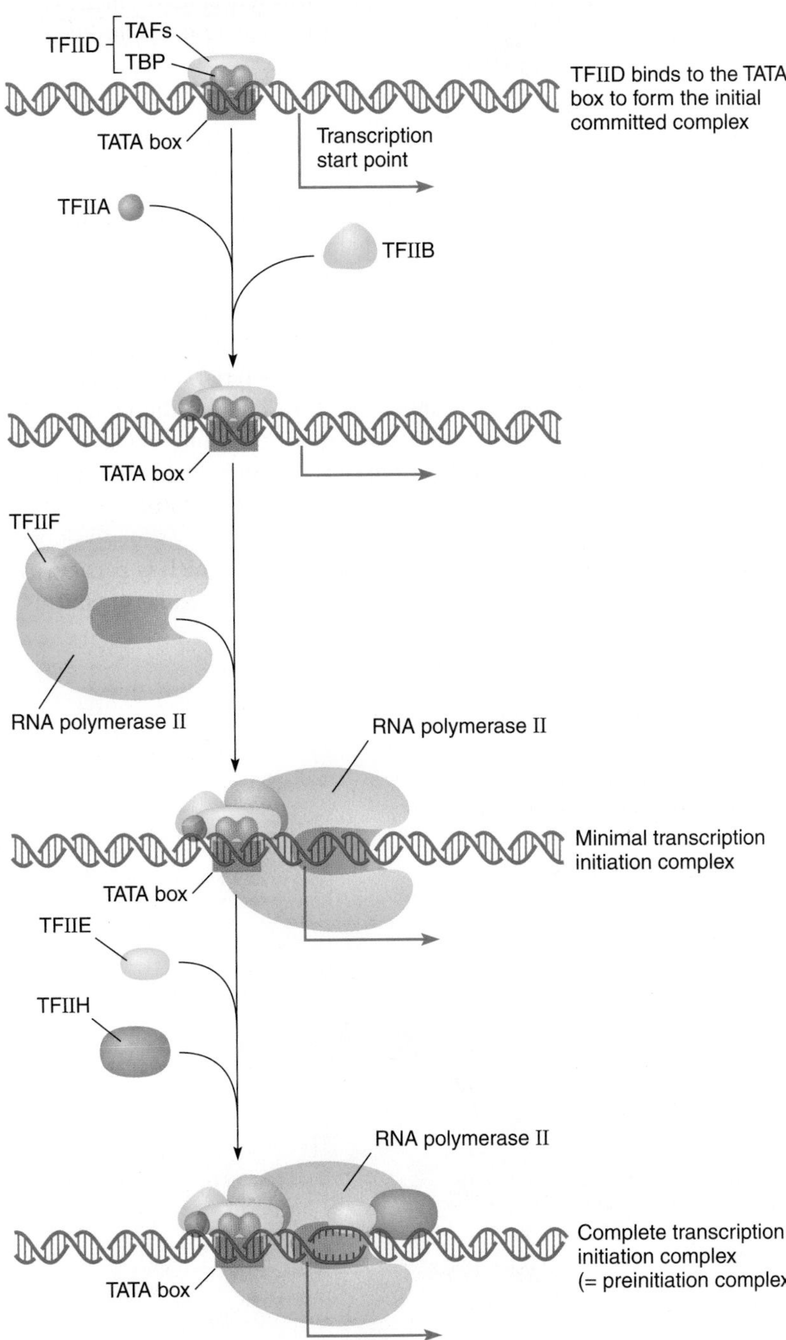


**Figure 5.7**

**Assembly of the transcription initiation machinery.** First, TFIID binds to the TATA box to form the *initial committed complex.* The multi-subunit TFIID has one subunit called the TATA-binding protein (TBP), which recognizes the TATA box sequence and several other proteins called TBP-associated factors (TAFs). *In vitro*, the TFIID–TATA box complex acts as a binding site for the sequential addition of other transcription factors. Initially, TFIIA and then TFIIB bind, followed by RNA polymerase II and TFIIF, to produce the *minimal transcription initiation complex.* (RNA polymerase II, like all eukaryotic RNA polymerases, cannot directly recognize and bind to promoter elements.) Next, TFIIE and TFIIH bind to produce the *complete transcription initiation complex,* also called the *preinitiation complex* (PIC). TFIIH's helicase-like activity now unwinds the promoter DNA, and transcription is ready to begin.



complicated because of the nucleosome organization of chromosomes (this complication is addressed in Chapter 18).

**iActivity**

Investigate how mutations at different regions in the β-globin gene affect mRNA transcription and the production of β-globin in the iActivity *Investigating Transcription in Beta-Thalassemia Patients* on the student website.

## The Structure and Production of Eukaryotic mRNAs

The mature, biologically active mRNA in both prokaryotic and eukaryotic cells has three main parts (Figure 5.8): (1) A **5′ untranslated region (5′ UTR;** also called a **leader sequence)** at the 5′ end; (2) the **protein-coding sequence**, which specifies the amino acid sequence of a protein during translation; and (3) a **3′ untranslated region (3′ UTR;** also called a **trailer sequence)**.

**mRNA Production in Eukaryotes**

The 3′ UTR sequence may contain sequence information

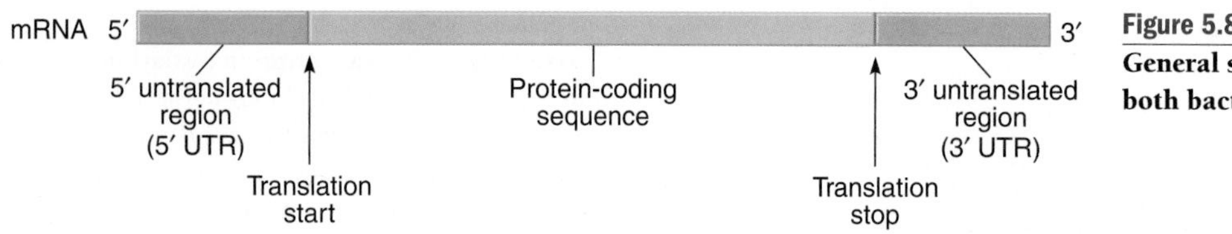


**Figure 5.8**

**General structure of mRNA found in both bacterial and eukaryotic cells.**



that signals the stability of the particular mRNA (see Chapter 18).

mRNA production is different in bacteria and eukaryotes. In bacteria (Figure 5.9a), the RNA transcript functions directly as the mRNA molecule; that is, the base pairs of a bacterial gene are colinear with the bases of the translated mRNA. In addition, because bacteria lack a nucleus, an mRNA begins to be translated on ribosomes before it has been transcribed completely; this process is called *coupled transcription and translation*. In eukaryotes (Figure 5.9b), the RNA transcript (the pre-mRNA) is modified in the nucleus by *RNA processing* to produce the mature mRNA. Also, the mRNA must migrate from the nucleus to the cytoplasm (where the ribosomes are located) before it can be translated. Thus, a eukaryotic mRNA is always transcribed completely and then processed before it is translated.

Another fundamental difference between bacterial and eukaryotic mRNAs is that bacterial mRNAs often are *polycistronic,* meaning that they contain the amino acid-coding information from more than one gene, whereas eukaryotic mRNAs typically are *monocistronic,* meaning that they contain the amino acid-coding information from just one gene. The eukaryotic system allows for additional levels of control of gene expression, which is particularly important in the more complex, multicellular organisms.

**Figure 5.9**

**Processes for the synthesis of functional mRNA in bacteria and eukaryotes.** (a) In bacteria, the mRNA synthesized by RNA polymerase does not have to be processed before it can be translated by ribosomes. Also, because there is no nuclear membrane, mRNA translation can begin while transcription continues, resulting in a coupling of transcription and translation. (b) In eukaryotes, the primary RNA transcript is a precursor-mRNA (pre-mRNA) molecule, which is processed in the nucleus by the addition of a 5′ cap and a 3′ poly(A) tail and the removal of introns. Only when that mRNA is transported to the cytoplasm can translation occur.

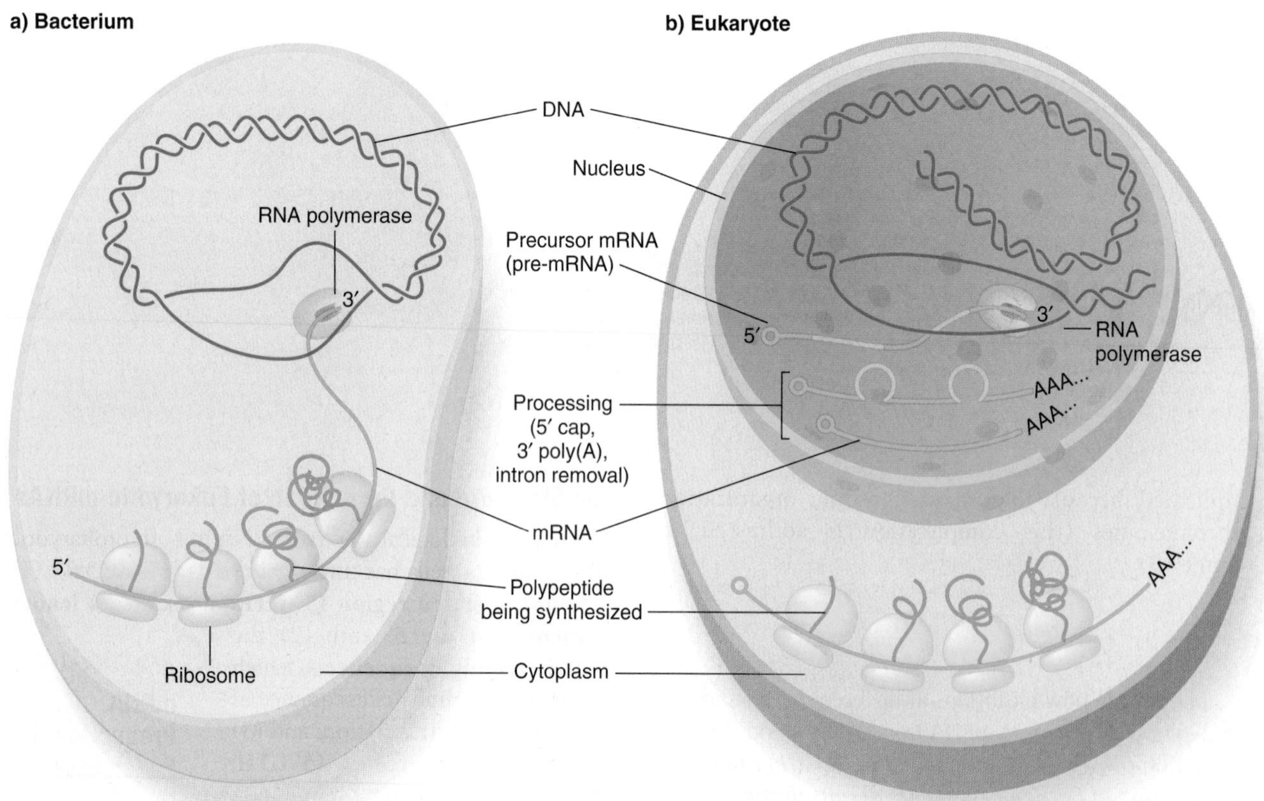

**Production of Mature mRNA in Eukaryotes.** Unlike bacterial mRNAs, eukaryotic mRNAs are modified at both the 5′ and 3′ ends. In addition, an exciting discovery in the history of molecular genetics took place in 1977 when Richard Roberts, Tom Broker, and Louie Chow—and, separately, Philip Sharp and Susan Berger—showed that the genes of certain animal viruses contain internal sequences that are not expressed in the amino acid sequences of the proteins they encode. Subsequently, the same phenomenon was seen in eukaryotes. We now know that, in eukaryotes in general, protein-coding genes typically have non-amino acid–coding sequences called **introns** between the other sequences that are present in mRNA, the **exons.** The term *intron* is derived from *intervening sequence*—a sequence that is not translated into an amino acid sequence—and the term *exon* is derived from *expressed sequence.* Exons include the 5′ and 3′ UTRs, as well as the amino acid-coding portions. In the processing of pre-mRNA to the mature mRNA molecule, the introns are removed. The 1993 Nobel Prize in Physiology or Medicine was awarded to Roberts and Sharp for their independent discoveries of genes with introns.

***5′ Modification.*** Once RNA polymerase II has made about 20 to 30 nucleotides of pre-mRNA, a *capping enzyme* adds a guanine nucleotide—most commonly, 7-methyl guanosine ($m^7G$)—to the 5′ end. The addition involves an unusual 5′-to-5′ linkage, rather than a 5′-to-3′ linkage (Figure 5.10). The process is called **5′ capping.** The sugars of the next two nucleotides are also modified by methylation. The 5′ cap remains throughout processing and is present in the mature mRNA, protecting it against degradation by exonucleases because of the unusual 5′-to-5′ linkage. The 5′ cap is also important for the binding of the ribosome as an initial step of translation.

***3′ Modification.*** Most eukaryotic pre-mRNAs become modified at their 3′ ends by the addition of a sequence of about 50 to 250 adenine nucleotides called a **poly(A) tail.** There is no DNA template for the poly(A) tail. The poly(A) tail remains when the pre-mRNA is processed to mature mRNA. mRNA molecules with 3′ poly(A) tails are called **poly(A)+ mRNAs.** The poly(A) tail is required for efficient export of the mRNA from the nucleus to the cytoplasm. Once in the cytoplasm, the poly(A) tail protects the 3′ end of the mRNA by buffering coding sequences against early degradation by exonucleases. The poly(A) tail also plays important roles in the initiation of translation by ribosomes and in processes that regulate the stability of mRNA.

Addition of the poly(A) tail defines the 3′ end of an mRNA strand and is associated with the termination of transcription of protein-coding genes. Addition of the poly(A) tail is signaled when mRNA transcription proceeds past the **poly(A) site,** a site in the RNA transcript that is about 10 to 30 nucleotides downstream of the poly(A) consensus sequence 5′-AAUAAA-3′. A number of proteins, including CPSF (cleavage and polyadenylation specificity factor) protein, CstF (cleavage stimulation factor) protein, and two cleavage factor proteins (CFI and CFII), then bind to and cleave the RNA at the poly(A) site (Figure 5.11a). Then, the enzyme **poly(A) polymerase (PAP),** which is bound to CPSF, adds A nucleotides to the 3′ end of the RNA using ATP as the substrate to produce the poly(A) tail. Poly(A) binding protein II (PABII) molecules bind to the poly(A) tail as it is synthesized.

Meanwhile, RNA polymerase II is still synthesizing RNA although, of course, that RNA is not part of the mRNA. Protein-coding genes do not have specific terminator sequences, as is the case in bacteria. (In contrast, eukaryotic genes transcribed by RNA polymerases I and III do have specific terminators.) So, how does the post-poly(A) site transcription terminate? A number of models have been proposed. In one model, a 5′-to-3′ exonuclease binds to the post-poly(A) site RNA and starts to degrade it. When it catches up to the RNA polymerase II,

**Figure 5.10**

**Cap structure at the 5′ end of a eukaryotic mRNA.** The cap results from the addition of a guanine nucleotide and two methyl groups.

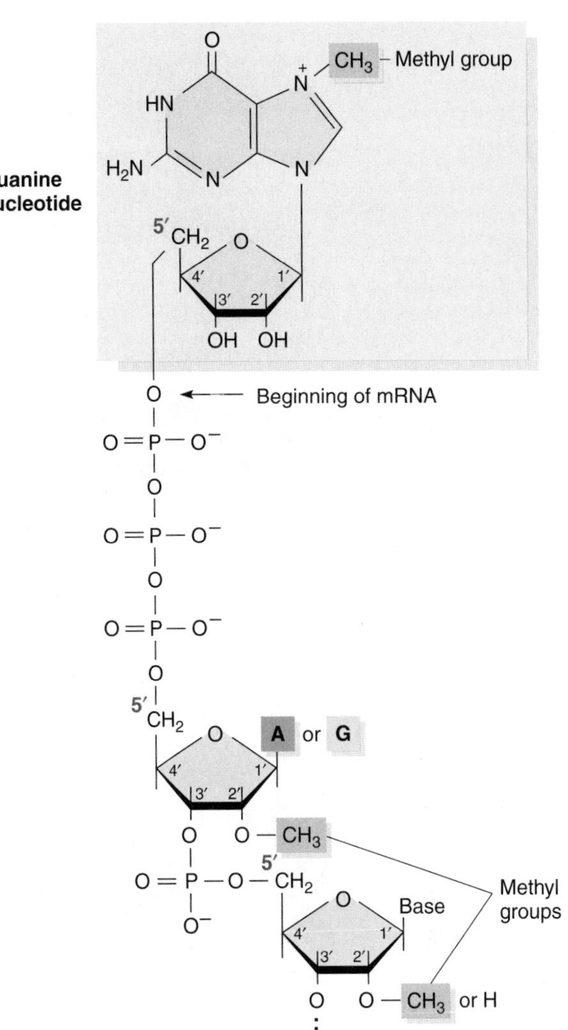

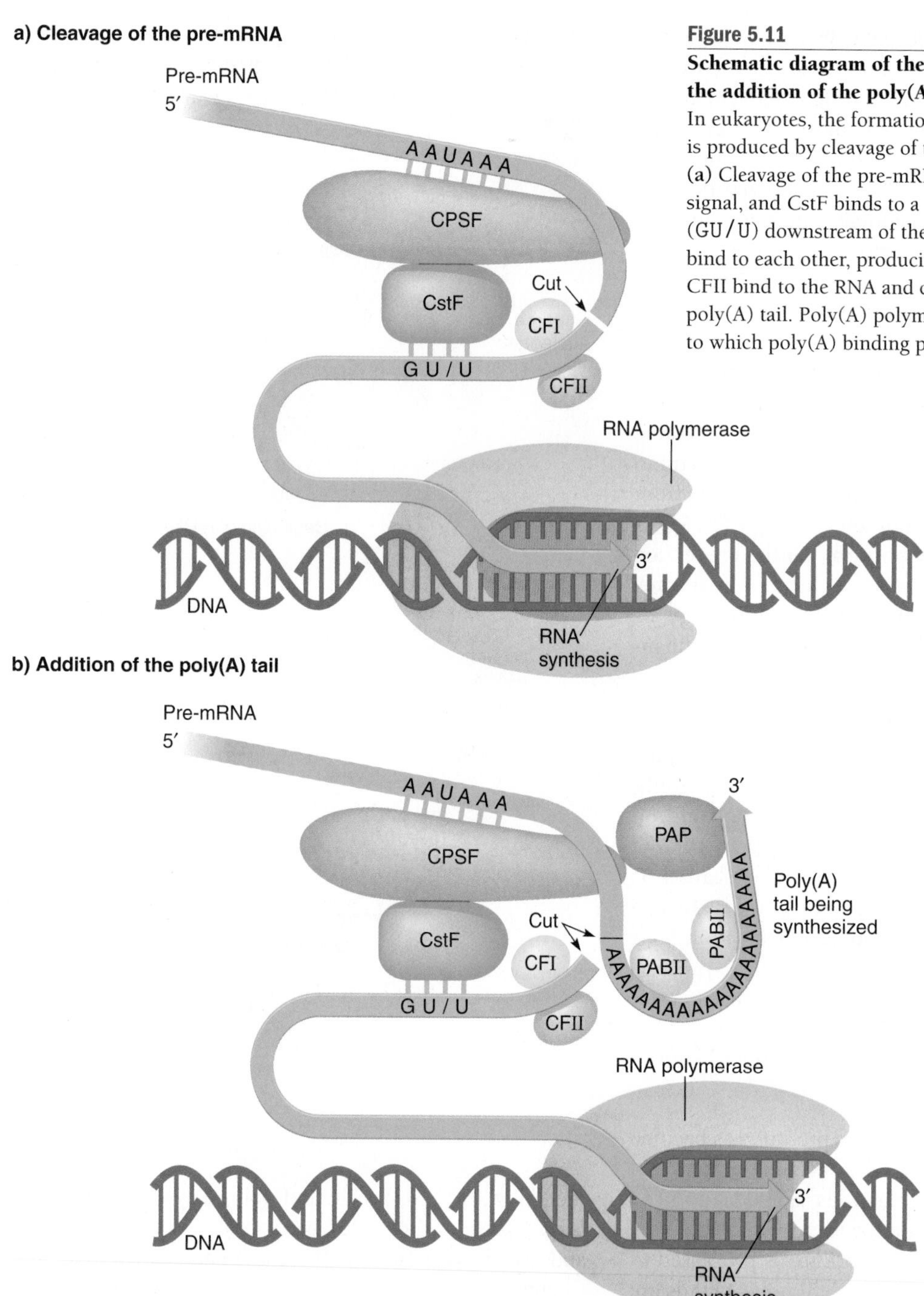


**Figure 5.11**

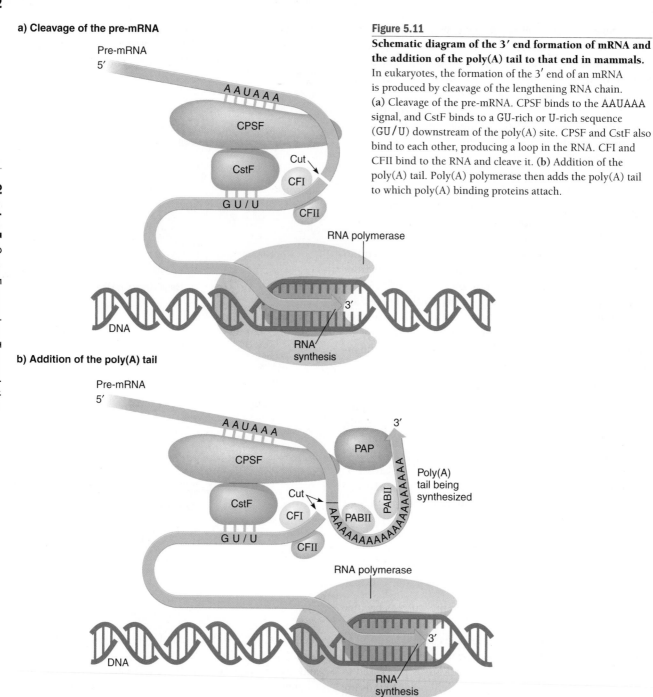

**Schematic diagram of the 3′ end formation of mRNA and the addition of the poly(A) tail to that end in mammals.** In eukaryotes, the formation of the 3′ end of an mRNA is produced by cleavage of the lengthening RNA chain. (a) Cleavage of the pre-mRNA. CPSF binds to the AAUAAA signal, and CstF binds to a GU-rich or U-rich sequence (GU/U) downstream of the poly(A) site. CPSF and CstF also bind to each other, producing a loop in the RNA. CFI and CFII bind to the RNA and cleave it. (b) Addition of the poly(A) tail. Poly(A) polymerase then adds the poly(A) tail to which poly(A) binding proteins attach.

the degradation somehow stimulates termination of transcription, probably by destabilizing the enzyme–transcription factor–DNA complex.

***Introns.*** Pre-mRNAs often contain a number of introns. Introns must be excised from each pre-mRNA to produce a mature mRNA that can be translated into the encoded polypeptide. The mature mRNA, then, contains RNA copies of the exons in the gene, now contiguously arranged in the RNA molecule without being separated by intron sequences.

At the time introns were discovered, researchers knew that the nucleus contains a large population of RNA molecules of various sizes, known as **heterogeneous nuclear RNAs (hnRNAs).** They correctly assumed that hnRNAs include pre-mRNA molecules. In 1978, Philip Leder's group was studying the β-globin gene in cultured mouse cells. This gene encodes the 146-amino-acid β-globin polypeptide that is part of a hemoglobin protein molecule. Leder's group isolated a 1.5-kb RNA molecule of nuclear hnRNA that was the β-globin pre-mRNA. Like the 0.7-kb mature mRNA, the pre-mRNA has a 5′ cap and a 3′ poly(A) tail. Leder's group demonstrated that the 1.5-kb pre-mRNA is colinear with the gene that encoded it, whereas the 0.7-kb β-globin mRNA is not. The scientists interpreted their results to mean that the β-globin gene has an intron of about 800 bp. Transcription of the gene results in a 1.5-kb pre-mRNA contain-ing both exon and intron sequences. This RNA is found only in the nucleus. The intron sequence is excised by processing events, and the flanking exon sequences are spliced together to produce a mature mRNA. (Subsequent research showed that the β-globin gene contains two introns; the second, smaller intron was not detected in the early research.)

At the time of this discovery, scientists had accepted that the gene sequence was completely colinear with the amino acid sequence of the encoded protein. Thus, finding that genes could be "in pieces" was most surprising. It was one of those highly significant discoveries that changed our thinking about genes. In the years since the discovery of introns, we have learned that many eukaryotic genes contain introns. Introns are rare in prokaryotes, though; they occur only in some tRNA and rRNA genes.

## Keynote

The transcripts of protein-coding genes are messenger RNAs or their precursors. These molecules are linear and vary widely in length with the size of the polypeptides they specify and whether they contain introns. Prokaryotic mRNAs are not modified once they are transcribed, whereas most eukaryotic mRNAs are modified by the addition of a cap at the 5′ end and a poly(A) tail at the 3′ end. Many eukaryotic pre-mRNAs contain introns, which must be excised from the mRNA transcript to make a mature, functional mRNA molecule. The segments separated by introns are called exons.

***Processing of Pre-mRNA to Mature mRNA.*** Messenger RNA production from genes with introns involves transcription of the gene by RNA polymerase II, addition of the 5′ cap and poly(A) tail to produce the pre-mRNA molecule, and processing of the pre-mRNA in the nucleus to remove the introns and splice the exons together to produce the mature mRNA (Figure 5.12).

Introns typically begin with 5′-GU and end with AG-3′, although more than just those nucleotides are needed to specify a junction between an intron and an exon. Introns in pre-mRNAs are removed and exons joined in the nucleus by **mRNA splicing.** The splicing events occur in a **spliceosome,** a complex of the pre-mRNA bound to **small nuclear ribonucleoprotein particles (snRNPs;** pronounced *snurps*). snRNPs are small nuclear RNAs (snRNAs) associated with proteins. The five principal snRNAs are U1, U2, U4, U5, and U6; each is associated with a number of proteins to form the snRNPs. U4 and U6 snRNAs are found within the same snRNP (U4/U6 snRNP), and the others are found within their own special snRNPs. Each snRNP type is abundant in the nucleus, with at least $10^5$ copies per cell.

Figure 5.13 shows a simplified stepwise model of splicing for two exons separated by an intron:

1. U1 snRNP binds to the 5′ splice junction of the intron. This binding is primarily the result of base pairing of U1 snRNA in the snRNP to the 5′ splice junction.
2. U2 snRNP binds to a sequence called the **branch-point sequence,** which is located upstream of the 3′ splice junction. This binding occurs as a result of the

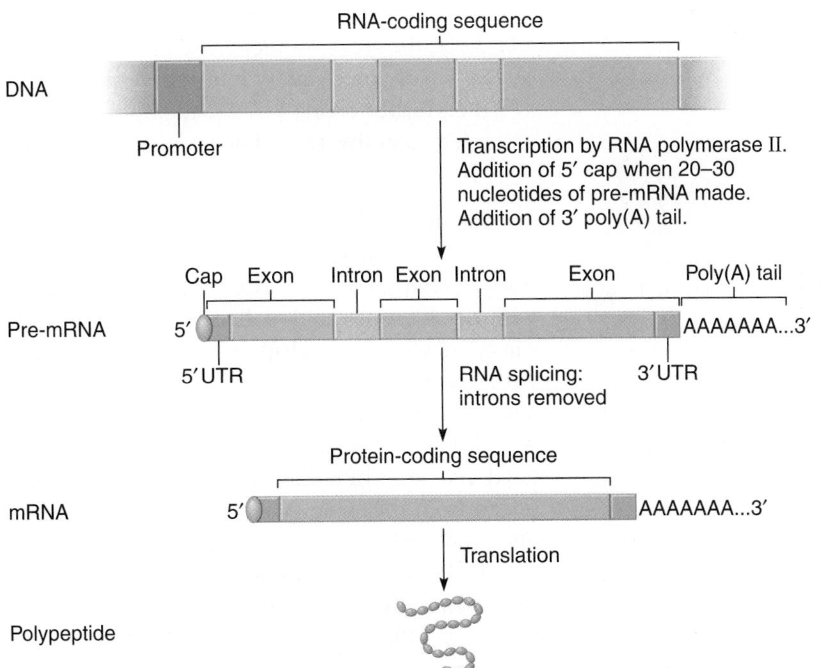


**Figure 5.12**
**General sequence of steps in the formation of eukaryotic mRNA.** Not all steps are necessary for all mRNAs.

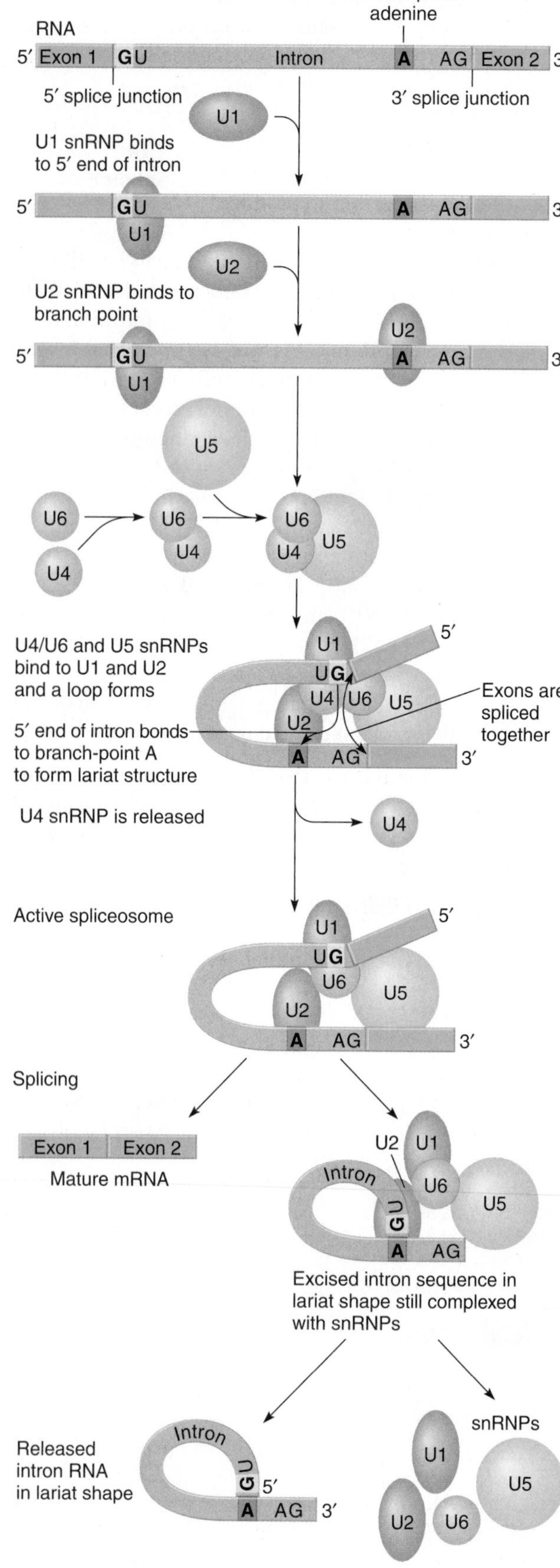


**Figure 5.13**

**Model for intron removal by the spliceosome.** At the 5′ end of an intron is the sequence GU and at the 3′ end is the sequence AG. Near the 3′ end of the intron is an A nucleotide located within the branch-point sequence, which in mammals is YNCURAY, where Y = pyrimidine, N = any base, R = purine, and A = adenine, and in yeast is UACUAAC (the italic A is where the 5′ end of the intron bonds). With the aid of snRNPs, intron removal begins with a cleavage at the first exon–intron junction. The G at the released 5′ of the intron folds back and forms an unusual 2′–5′ bond with the A of the branch-point sequence. This reaction produces a lariat-shaped intermediate. Cleavage at the 3′ intron–exon junction and ligation of the two exons completes the removal of the intron.

base pairing of U2 snRNA in the snRNP to the branch-point sequence.

3. A U4/U6 snRNP and a U5 snRNP interact, and the combination binds to the U1 and U2 snRNPs, causing the intron to loop and thereby bringing its two junctions close together.
4. U4 snRNP dissociates, resulting in the formation of the *active spliceosome.*
5. The snRNPs in the spliceosome cleave the intron from exon 1 at the 5′ splice junction, and the now-free 5′ end of the intron bonds to a particular A nucleotide in the branch-point sequence. Because of its resemblance to the rope cowboys use, the looped-back structure is called an *RNA lariat structure.* The branch point in the RNA that produces the lariat structure involves an unusual 2′–5′ phosphodiester bond formed between the 2′ OH of the adenine nucleotide in the branch-point sequence and the 5′ phosphate of the guanine nucleotide at the end of the intron. The A itself remains in normal 3′–5′ linkage with its adjacent nucleotides of the intron.
6. Next, the spliceosome excises the intron (still in lariat shape) by cleaving it at the 3′ splice junction and then ligates exons 1 and 2 together. The snRNPs are released at this time. The process is repeated for each intron.

In the splicing steps, the snRNPs function through RNA–RNA, RNA–protein, and protein–protein interactions. Examples of RNA–RNA interactions are U1 snRNA with the RNA at the 5′ splice junction, U2 snRNA with the RNA of the branch-point sequence, and U6 snRNA with U2 snRNA. Box 5.1 summarizes some mutational studies that revealed the RNA–RNA interactions.

In Chapter 18 you will learn that splicing is regulated and that, in some cases, different mRNAs are produced from the same gene as a result of a process called *alternative splicing.* A consequence of alternative splicing is that different polypeptides can be produced from the same gene. These polypeptides have regions of similarity but are not identical; that is, they have variant functions. For example, muscle proteins produced by alternative

**Box 5.1 Identifying RNA–RNA Interactions in pre-mRNA Splicing by Mutational Analysis**

Conceptually, showing that RNA–RNA interactions were important in RNA splicing was straightforward. Gene mutants were isolated that were defective in pre-mRNA splicing. Many of those mutants had alterations of the key intron sequences for pre-mRNA splicing, namely in the 5′ splice junction region, in the branch-point sequence, and in the 3′ splice junction sequence. (Indeed, such mutants help define the roles of those sequences in pre-mRNA splicing.) Researchers hypothesized that the snRNAs of snRNPs were important in recognizing the three sequences. This hypothesis is supported by models indicating that the mutants with alterations in the splicing sequences theoretically would bond more weakly with segments of snRNA molecules than would normal sequences. Experimental support for snRNA–intron sequence RNA interactions came from making mutants of snRNAs that restored strong binding. That is, the mutant splicing sequence was used to design specific compensatory mutations in snRNAs such that the binding of mutant snRNA with mutant splicing sequence was now as good as that of normal snRNA with normal splicing sequence. The compensatory mutants restored splicing activity of the mutant gene, providing functional evidence that specific RNA–RNA interactions are important for pre-mRNA splicing.

splicing might have optimal functions in different tissues, such as heart muscle, smooth muscle, and so on.

***Coupling of Pre-mRNA Processing to Transcription and to mRNA Export from the Nucleus.*** Evidence from research of the past few years has shown that expression of a eukaryotic protein-coding gene—transcription through the production of the functional protein—is a continuous process rather than a series of independent events. Key results supporting this view include the fact that proteins responsible for steps in the process are functionally, and sometimes structurally, connected; and that regulation of the process occurs at several stages. And, importantly, the machinery involved is conserved evolutionarily from yeast to humans. In short, for expression of a eukaryotic protein-coding gene, transcription is coupled to pre-mRNA processing, which is coupled to mRNA export from the nucleus through the nuclear pores.

**Keynote**

Introns are removed from pre-mRNAs in a series of well-defined steps. Intron removal begins with the cleavage of the pre-mRNA at the 5′ splice junction. The free 5′ end of the intron loops back and bonds to a site upstream of the 3′ splice junction. Cleavage at that junction releases the intron, which is shaped like a lariat. Once the intron is excised, the exons that flanked it are spliced together. The removal of introns from eukaryotic pre-mRNA occurs in the nucleus in complexes called spliceosomes, which consist of several snRNPs bound specifically to each intron. Pre-mRNA processing is coupled both to transcription and to mRNA export from the nucleus as part of a continuous, rather than discontinuous, process of expression of a protein-coding gene in eukaryotes.

### Self-Splicing Introns

In some species of the ciliated, free-living protozoan *Tetrahymena*, the genes for the 28S rRNA found in the large ribosomal subunit (see Chapter 6, p. 113–114) are interrupted by a 413-bp intron. Transcription of this gene produces a pre-rRNA molecular analogous to a pre-mRNA molecule in the sense that the intron must be removed to produce a functional rRNA. The excision of this intron—now called a *group I intron*—was shown to occur by a *protein-independent reaction* in which the RNA intron folds into a secondary structure that promotes its own excision. The process, called **self-splicing,** was discovered in 1982 by Tom Cech and his research group. In 1989, Cech shared the Nobel Prize in Chemistry for his discovery.

Figure 5.14 diagrams the self-splicing reaction for the group I intron in *Tetrahymena* pre-rRNA. The steps are as follows:

1. The pre-rRNA is cleaved at the 5′ splice junction as guanosine is added to the 5′ end of the intron.
2. The intron is cleaved at the 3′ splice junction.
3. The two exons are spliced.
4. The excised intron circularizes to produce a lariat molecule, which is cleaved to produce a circular RNA and a short, linear piece of RNA.

The self-splicing activity of the intron RNA sequence does not meet the definition of an enzyme activity. That is, although the RNA carries out the reaction, it is not regenerated in its original form at the end of the reaction, as is the case with protein enzymes. Modified forms of the *Tetrahymena* intron RNA and of other self-cleaving RNAs that function catalytically have been produced in the lab. These **RNA enzymes** are called **ribozymes;** they are useful experimentally for cleaving RNA molecules at specific sequences.

The self-splicing of the *Tetrahymena* pre-rRNA intron was the first example of what is now called *group I intron self-splicing*. Group I introns are rare. Other self-splicing group I introns have been found in nuclear rRNA genes, in some mitochondrial protein-coding genes, and in some protein-coding and tRNA genes of certain bacteriophages. Another class of self-splicing introns are the *group II introns*. These introns, which use a different

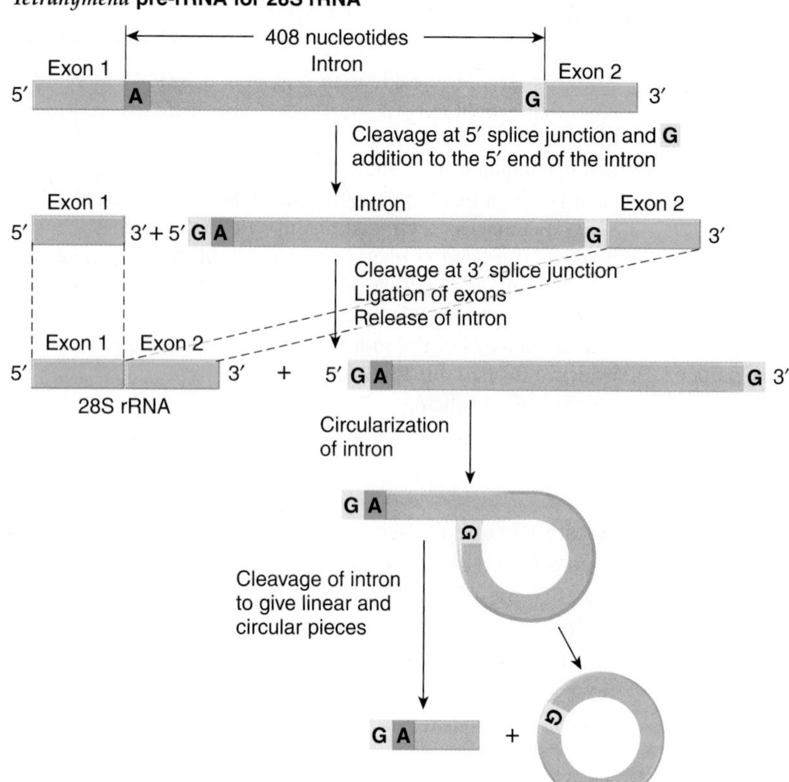


**Figure 5.14**

**Self-splicing reaction for the group I intron in *Tetrahymena* pre-rRNA.**



molecular mechanism for self-splicing than do group I introns, are found in some genes of bacteria and of organelles in protists, fungi, algae, and plants.

The discovery that RNA can act like a protein was an extremely important landmark in biology and has revolutionized theories about the origin of life. Previous theories proposed that proteins were required for replication of the first nucleic acid molecules. The **RNA world hypothesis** now proposes that RNA-based life predates the present-day DNA-based life, with the RNA carrying out the necessary catalytic reactions required for life in the presumably primitive cells of the time and store the genetic information at the same time.

**Keynote**

In some precursor RNAs, there are introns whose RNA sequences fold into a secondary structure that excises itself in a process called self-splicing. The self-splicing reaction does not involve any proteins.

## RNA Editing

**RNA editing** involves the posttranscriptional insertion or deletion of nucleotides or the conversion of one base to another. As a result, the functional RNA molecule has a base sequence that does not match the base-pair sequence of its DNA coding sequence.

RNA editing was discovered in the mid-1980s in some mitochondrial mRNAs of trypanosomes, the parasitic protozoa that cause sleeping sickness. For example, the sequences of the *COIII* gene for subunit III of cytochrome oxidase and its mRNA transcripts for the protozoans *Trypanosome brucei (Tb)*, *Crithridia fasiculata (Cf)*, and *Leishmania tarentolae (Lt)* are shown in Figure 5.15. Although the mRNA sequences are highly similar among the three organisms, only the *Cf* and *Lt* gene sequences are colinear with the mRNAs. Strikingly, the *Tb* gene has a sequence that cannot produce the mRNA it apparently encodes. The differences between the two are U nucleotides in the mRNA that are not encoded in the DNA and T nucleotides in the DNA that are not found in the transcript. Once it is made, the transcript of the *Tb COIII* gene is edited to add U nucleotides in the appropriate places and remove the U nucleotides encoded by the T nucleotides in the DNA. As the figure shows, there are extensive insertions of U nucleotides. The magnitude of the changes is even more apparent when the whole sequence is examined: More than 50% of the mature mRNA consists of U nucleotides added posttranscriptionally. This RNA editing must be accurate in order to reconstitute the appropriate sequence for translation into the correct protein. A special RNA molecule, called a *guide RNA* (gRNA), is involved in the process. The gRNA pairs with the mRNA transcript and cleaves the transcript, templating the missing U nucleotides, and ligating the transcript back together again.

RNA editing is not confined to trypanosomes. In the slime mold *Physarum polycephalum*, single C nucleotides are added posttranscriptionally at many positions of several mitochondrial mRNA transcripts. In higher plants,

**Figure 5.15**

**Comparison of the DNA sequences of the cytochrome oxidase subunit III gene *(COIII)* in the protozoans *Trypanosome brucei (Tb)*, *Crithridia fasiculata (Cf)*, and *Leishmania tarentolae (Lt)*, aligned with the conserved mRNA for *Tb*.** The lowercase u's are the U nucleotides added to the transcript by RNA editing. The template T's in *Tb* DNA that are not in the RNA transcript are yellow.

Region of *COIII* gene transcript

| | | |
|---|---|---|
| *Tb* DNA | G GTTTTTGG AGG G | GTTTTG G G A A GA GAG |
| *Tb* RNA | uuGuGUUUUUGGuuuAGGuuuuuuuGuuG | UUGuuGuuuuGuAuuAuGAuuGAGu |
| *Cf* DNA | TTTTTATTTTGATTTCGTTTTTTTTTATG | TGTATTATTTGTGCTTTGATCCGCT |
| *LT* DNA | TTTTTATTTTGATTTCGTTTTTTTTTATG | TGTTTTATTTATGTTATGAGTAGGA |
| *Tb* Protein | Leu Cys Phe Trp Phe Arg Phe Phe Cys Cys | Cys Cys Phe Val Leu Trp Leu Ser |

the sequences of many mitochondrial and chloroplast mRNAs are edited by C-to-U changes. C-to-U editing is also involved in producing an AUG initiation codon from an ACG codon in some chloroplast mRNAs in a number of higher plants. In mammals, C-to-U editing occurs in the nuclear gene-encoded mRNA for apolipoprotein B and results in tissue-specific generation of a stop codon. Also in mammals, A-to-G editing has been shown to occur in the glutamate receptor mRNA, and pyrimidine editing occurs in a number of tRNAs.

## Summary



- Transcription is the process of copying genetic information in DNA into RNA base sequences. The DNA unwinds in a short region next to a gene, and an RNA polymerase catalyzes the synthesis of an RNA molecule in the 5′-to-3′ direction. Only one strand of the double-stranded DNA is transcribed into an RNA molecule.
- Transcription of four main classes of genes produces messenger RNA (mRNA), transfer RNA (tRNA), ribosomal RNA (rRNA), and small nuclear RNA (snRNA). snRNA is found only in eukaryotes, and the other three classes are found in both prokaryotes and eukaryotes. Only mRNA is translated to produce a protein molecule.
- In *E. coli*, the initiation of transcription of protein-coding genes requires a complex of RNA polymerase and the sigma factor protein binding to the promoter. Once transcription has begun, the sigma factor dissociates and RNA synthesis is completed by the RNA polymerase core enzyme. Termination of transcription is signaled by specific sequences in the DNA.
- In bacteria, a single RNA polymerase synthesizes mRNA, tRNA, and rRNA. Eukaryotes have three distinct nucleus-located RNA polymerases, each of which transcribes different gene types: RNA polymerase I transcribes the genes for the 18S, 5.8S, and 28S ribosomal RNAs; RNA polymerase II transcribes mRNA genes and some snRNA genes; and RNA polymerase III transcribes genes for the 5S rRNAs, the tRNAs, and the other snRNAs.
- Eukaryotic RNA polymerases are unable to bind to promoters directly. For transcription to be initiated, then, general transcription factors first bind and then recruit the RNA polymerase to form a complex. Other transcription factors then bind and transcription can commence.
- mRNAs have three main parts: a 5′ untranslated region (UTR), the amino acid coding sequence, and the 3′ untranslated region.
- In prokaryotes the gene transcript functions directly as the mRNA molecule, whereas in eukaryotes the RNA transcript must be modified in the nucleus to produce mature mRNA. Modifications include the addition of a 5′ cap and a 3′ poly(A) tail and the removal of any introns. Spliceosomes perform intron removal and exon splicing through specific interactions of snRNPs with the pre-mRNA. Only when all processing events have been completed can the mRNA function; at that point, once it is exported from the nucleus, it can be translated.
- In some organisms with introns, the precursor-RNA sequences fold into a secondary structure that excises itself, a process called self-splicing. This process does not involve protein enzymes.

Chapter 5 Gene Expression: Transcription

the mRNA is annealed to ovalbumin-gene DNA, RNA–DNA hybrids are formed. The following figure shows an interpretive diagram of these hybrids as visualized by electron microscopy:

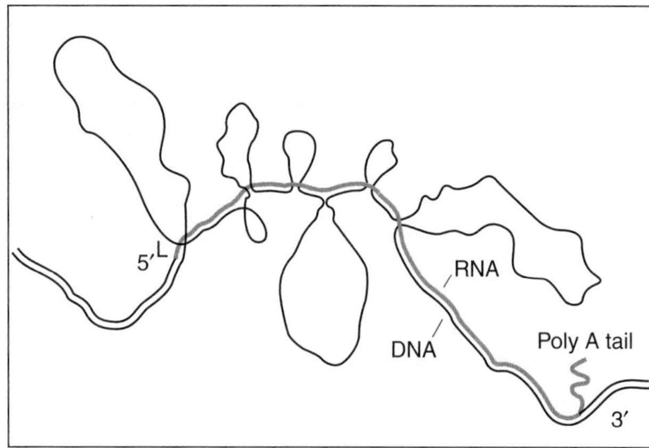


**a.** For what does the image provide evidence?
**b.** Based on the figure, how many introns and exons does the gene for ovalbumin have?
**c.** Was the mRNA for this experiment purified from the nucleus or from the cytoplasm? Explain your reasoning.

***5.16** A pre-mRNA for a yeast gene contains two exons separated an intron. Figure 5.C shows the lengths of its exons and intron, its sequence in the regions near the 5′ splice site and branch point, and the alignment of its sequence with the sequence of U1 snRNA. Capital letters denote exonic mRNA sequence, and the branch-point nucleotide is underlined.
**a.** If there is a poly(A) site near at the end of exon 2 and a poly(A) tail of 200 nucleotides is added, about what size mRNA will be produced from this gene in a normal yeast cell?
**b.** What size transcript will be produced if the U1 snRNA has an A-to-G base substitution at the position marked with an asterisk? Explain your reasoning.
**c.** What mutation in the gene would result in a normal-sized transcript in a cell with the U1 snRNA described in part (**b**)?

**5.17** For the pre-mRNA of the yeast gene diagrammed in Question 5.16, diagram the shape and dimensions of the RNAs that will be produced in
**a.** a normal yeast strain.
**b.** a strain carrying a mutated gene where the 5′-GU-3′ at the 5′ end of its intron is changed to a 5′-AC-3′.
**c.** a strain carrying a mutated gene where its branch point sequence is changed from 5′-UACUAAC-3′ to 5′-UACUCTC-3′.
**d.** a strain carrying a mutated gene where the 5′-AG-3′ at the 3′ end of its intron is changed to 5′-UU-3′.

**5.18** How is the mechanism of group I intron removal different from the mechanism used to remove the introns in most eukaryotic mRNAs? Speculate as to why these different mechanisms for intron removal might have evolved and how each might be advantageous to a eukaryotic cell.

**5.19** What is the RNA world hypothesis, and what led to its formulation?

**5.20** Small RNA molecules such as snRNAs and gRNAs play essential roles in eukaryotic transcript processing.
**a.** Where are these molecules found in the cell, and what roles do they have in transcript processing?
**b.** How is the abundance of snRNAs related to their role in transcript processing?

***5.21** Which of the mutations that follow are likely to be recessive lethal mutations (i.e., mutations causing lethality when they are the only alleles present in a homozygous individual) in humans? Explain your reasoning.
**a.** deletion of the U1 genes
**b.** a single base-substitution mutation in the U1 gene that prevented U1 snRNP from binding to the 5′-GU-3′ sequence found at the 5′ splice junctions of introns
**c.** deletion within intron 2 of β-globin
**d.** deletion of four bases at the end of intron 2 and three bases at the beginning of exon 3 in β-globin

**Figure 5.C**

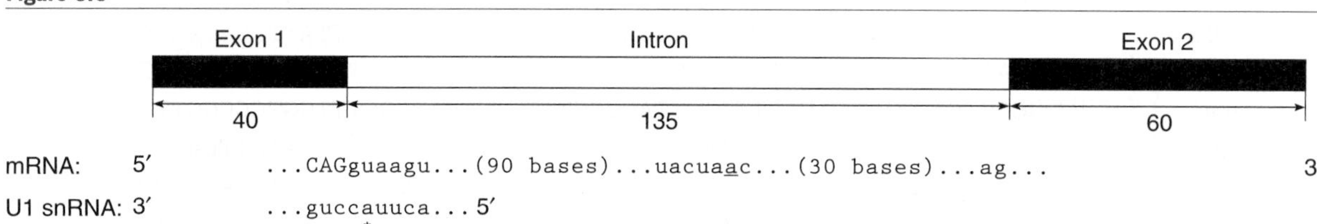

**Figure 5.D**

```
RNA: 5′-GUGGAGAAGU GGUCCAUGGA GCGGCUGCAG GCAGCUCCCC GGUCCGAGUC-3′

DNA: 5′-GTGGAGAAGT GGTCCATGGA GCTGCTGCAG GCAGCTCCCC GGTCCGAGTC-3′
     3′-CACCTCTTCA CCAGGTACCT CGACGACGTC CGTCGAGGGG CCAGGCTCAG-5′
```

***5.22** In Figure 5.D above, part of the sequence of an exon from the human *GRIK3* gene, which codes for a subunit of one type of glutamate receptor, is aligned with the mRNA used for translation.

**a.** Which is the coding strand and which is the template strand?

**b.** Propose an explanation for why the mRNA sequence is not identical to the coding strand (after allowing for T in DNA to be replaced by U in RNA).

***5.23** The following figure shows the transcribed region of a typical eukaryotic protein-coding gene:

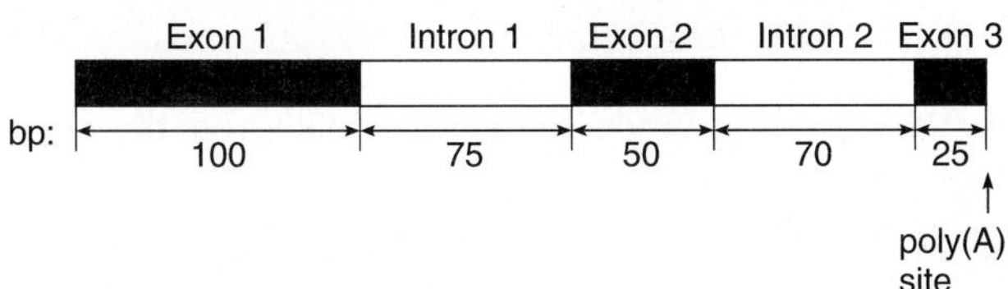


What is the size (in bases) of the fully processed, mature mRNA? Assume a poly(A) tail of 200 As in your calculations.

***5.24** Most human obesity does not follow Mendelian inheritance patterns, because body fat content is determined by a number of interacting genetic and environmental variables. Insights into how specific genes function to regulate body fat content have come from studies of mutant, obese mice. In one mutant strain, *tubby (tub)*, obesity is inherited as a recessive trait. A comparison of the DNA sequence of the $tub^+$ and *tub* alleles has revealed a single base-pair change: within the transcribed region, a 5′ G-C base pair has been mutated to a T-A base pair. The mutation causes an alteration of the initial 5′ base of the first intron. Therefore, in the homozygous *tub/tub* mutant, a longer transcript is found. Propose a molecularly based explanation for how a single base change causes a nonfunctional gene product to be produced, why a longer transcript is found in *tub/tub* mutants, and why the *tub* mutant is recessive.

# 6 Gene Expression: Translation

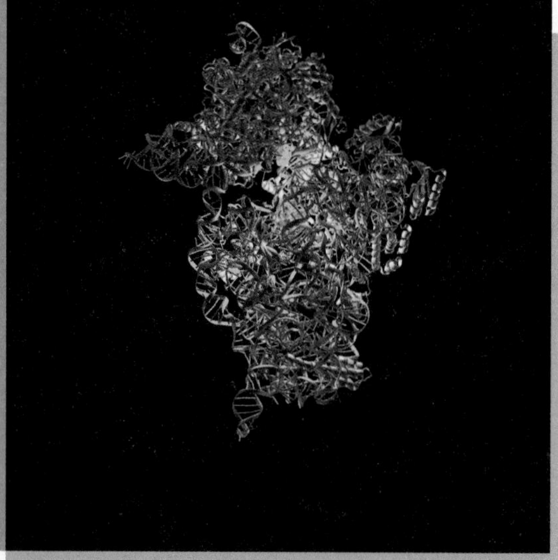

Three-dimensional structure of the 30S ribosomal subunit.

## Key Questions

- What is the chemical composition of a protein?
- What is the structure of a protein?
- What is the nature of the genetic code?
- What is the structure and function of transfer RNA (tRNA)?
- What is the structure and function of ribosomal RNA (rRNA)?
- How is polypeptide synthesis initiated on the ribosome?
- How is a polypeptide elongated on the ribosome?
- How is a polypeptide terminated in translation of messenger RNA (mRNA)?
- How are proteins sorted in the cell?

### iActivity

CHANGING A SINGLE LETTER IN A WORD CAN completely change the meaning of the word. This, in turn, can change the meaning of the sentence containing that word. In living organisms, a sequence of three nucleotide "letters" produces an amino acid "word." The amino acids are strung together to form polypeptide "sentences." In this chapter, you will study the process by which nucleotide "letters" are translated into polypeptide "sentences."

One of the most important applications of human genome research is the use of sequence information to track down the causes of genetic diseases. In the iActivity for this chapter, you will investigate part of the gene responsible for cystic fibrosis, the most common fatal genetic disease in the United States, and try to identify possible causes of the disease.

The information for the proteins found in a cell is encoded in genes of the genome of the cell. A protein-coding gene is expressed by transcription of the gene to produce an mRNA (discussed in Chapter 5), followed by translation of the mRNA. Translation involves the conversion of the base sequence of the mRNA into the amino acid sequence of a polypeptide. The base sequence information that specifies the amino acid sequence of a polypeptide is called the genetic code. In this chapter, you will learn about the structure of proteins, and about how the nucleotide sequence of mRNA is translated into the amino acid sequence of a polypeptide.

## Proteins

### Chemical Structure of Proteins

A **protein** is a high-molecular-weight, nitrogen-containing organic compound of complex shape and composition. A protein consists of one or more macromolecular subunits called **polypeptides,** which are composed of smaller building blocks: the **amino acids.** Each cell type has a characteristic set of proteins that gives it its functional properties.

With the exception of proline, the amino acids have a common structure, shown in Figure 6.1. The structure consists of a central carbon atom (α-carbon) to which is bonded an amino group ($NH_2$), a carboxyl group

**Figure 6.1**
**General structural formula for an amino acid.**

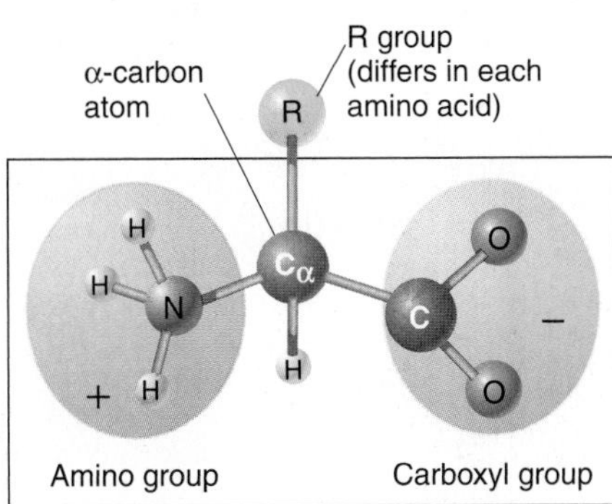


(COOH), and a hydrogen atom. At the pH commonly found within cells, the $NH_2$ and COOH groups of free amino acids are in a charged state, $-NH_3^+$ and $-COO^-$ respectively (as drawn in Figure 6.1). Also bound to the α-carbon is the *R group*, which is specific for each amino acid, giving that amino acid its distinctive properties. Different polypeptides have different sequences and proportions of amino acids; the sequence of amino acids, and thus the sequence of R groups, determines the chemical properties of each polypeptide.

Twenty amino acids are used to make proteins in all living cells—their names, three-letter and one-letter abbreviations, and chemical structures are shown in Figure 6.2. The 20 amino acids are divided into subgroups on the basis of whether the R group is acidic, basic, neutral and polar, or neutral and nonpolar.

Amino acids of a polypeptide are joined by a **peptide bond**—a covalent bond formed between the carboxyl group of one amino acid and the amino group of an adjacent amino acid (Figure 6.3). Every polypeptide has a free amino group at one end (called the N terminus, or the N-terminal end) and a free carboxyl group at the other end (called the C terminus, or the C-terminal end). The N-terminal end is defined as the beginning of a polypeptide chain because it is the end first made by translation of an mRNA molecule in the cell.

## Molecular Structure of Proteins

Proteins can have four levels of structural organization (Figure 6.4).

1. The *primary structure* of a polypeptide chain is the amino acid sequence (Figure 6.4a). The amino acid sequence is directly determined by the base-pair sequence of the gene that encodes the polypeptide.
2. The *secondary structure* of a protein is the regular folding and twisting of a portion of polypeptide chain into a variety of shapes (Figure 6.4b). A polypeptide's secondary structure is the result of weak bonds, such as electrostatic or hydrogen bonds, between NH and CO groups of amino acids that are near each other on the chain. The particular type of secondary structure seen for a polypeptide, or part of a polypeptide, is primarily the result of the amino acid sequence of the polypeptide or the region of the polypeptide.

   One type of secondary structure found in regions of many polypeptides is the *α-helix* (see Figure 6.4b), a structure discovered by Linus Pauling and Robert Corey in 1951. The R groups in a segment of a polypeptide determine whether an α-helix can form. Note the hydrogen bonding between the NH group of one amino acid (i.e., an NH group that is part of a peptide bond) and the CO group (also part of a peptide bond) of an amino acid that is four amino acids away in the chain. The repeated formation of this bonding results in the helical coiling of the chain. As will all secondary structure types, the α-helix content of proteins varies.

   Another type of secondary structure is the β-pleated sheet. The β-pleated sheet involves a polypeptide chain or chains folded in a zigzag way, with parallel regions or chains linked by hydrogen bonds. Many proteins contain a mixture of α-helical and β-pleated sheet regions.
3. A protein's *tertiary structure* (Figure 6.4c) is the three-dimensional structure of a single polypeptide chain. The three-dimensional shape of a polypeptide often is called its *conformation*. The tertiary structure of a polypeptide is directly determined by the distribution of the R groups along the chain. That is, the tertiary structure forms as a result of interactions between the R groups. Those interactions include hydrogen bonds, ionic interactions, sulfur bridges, and van der Waals forces. In an aqueous environment, the tertiary structure typically forms with polar and charged groups on the outside and nonpolar groups on the inside. Figure 6.4c shows the tertiary structure of the β polypeptide of hemoglobin. (The 1962 Nobel Prize in Chemistry was awarded to Max Perutz and Sir John Kendrew for their studies of the structures of proteins, and the 1972 Nobel Prize in Chemistry was awarded to Christian Anfinsen for his work on the RNA-degrading enzyme, ribonuclease, especially concerning the connection between the amino acid sequence and the biologically active conformation.)
4. The *quaternary structure* is the complex of polypeptide chains in a multisubunit protein, so quaternary structure is found only in proteins having more than one polypeptide chain (Figure 6.4d). Interactions between R groups and between NH and CO groups of peptide bonds on different polypeptides leads to the folding into a quaternary structure. Shown in Figure 6.4d is the quaternary structure of a heteromultimeric (*hetero*, "different"; *multimeric*, "many-subunit") protein, the oxygen-carrying protein hemoglobin. Hemoglobin consists of four polypeptide chains (two 141-amino acid α polypeptides and

**Figure 6.2**

**Structures of the 20 naturally occurring amino acids, organized according to chemical type.**
Below each amino acid name are its three-letter and one-letter abbreviations.

**Acidic**

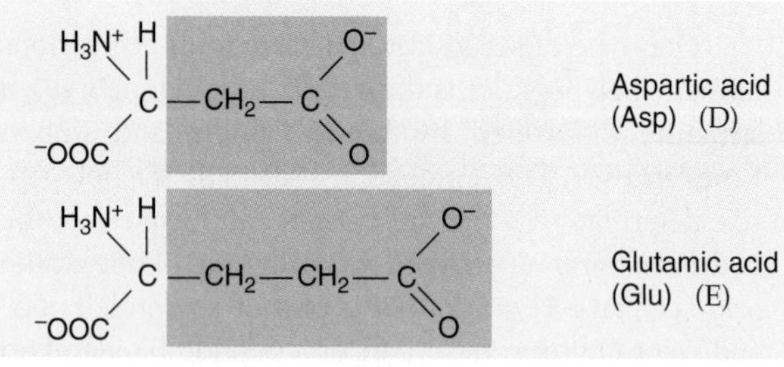


**Neutral, nonpolar**

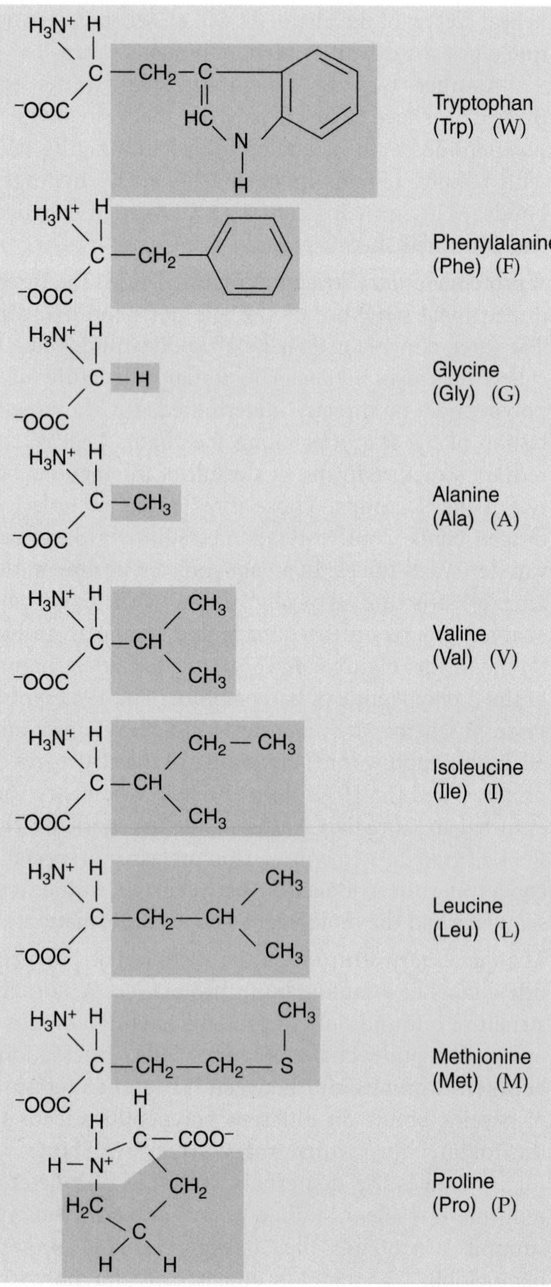


**Basic**

Lysine
(Lys) (K)

Arginine
(Arg) (R)

Histidine
(His) (H)

**Neutral, polar**

Tyrosine
(Tyr) (Y)

Serine
(Ser) (S)

Threonine
(Thr) (T)

Asparagine
(Asn) (N)

Glutamine
(Gln) (Q)

Cysteine
(Cys) (C)

**Figure 6.3**
**Peptide bond formation.**

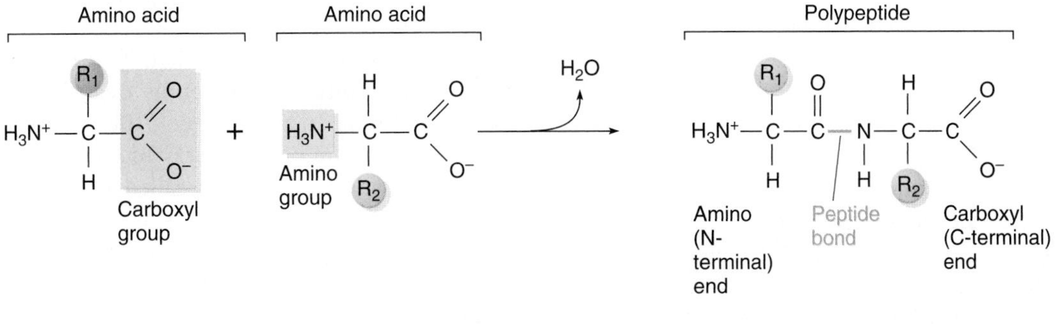

Amino acid + Amino acid → Polypeptide + $H_2O$

$H_3N^+—C(R_1)(H)—COO^-$ (Carboxyl group) + $H_3N^+$ (Amino group) $—C(H)(R_2)—COO^-$ → $H_3N^+—C(R_1)(H)—C(=O)—N(H)—C(H)(R_2)—COO^-$

Amino (N-terminal) end | Peptide bond | Carboxyl (C-terminal) end

two 146-amino acid β polypeptides), each of them associated with a heme group that is involved in the binding of oxygen. In the quaternary structure of hemoglobin, each α chain is in contact with each β chain, but there is little interaction between the two α chains or between the two β chains.

For many years, it was thought that the amino acid sequence alone was sufficient to specify how a protein folds into its functional state. We know that polypeptides fold cotranslationally; that is, they fold during the translation process rather than after they are released from the ribosome. Clearly, the amino acid sequence determines what structures can form. But, for many proteins, folding into their functional states depends on one or more of a family of proteins called *chaperones* (also called *molecular chaperones*). Chaperones act analogously to enzymes in

**Figure 6.4**
**Four levels of protein structure.**

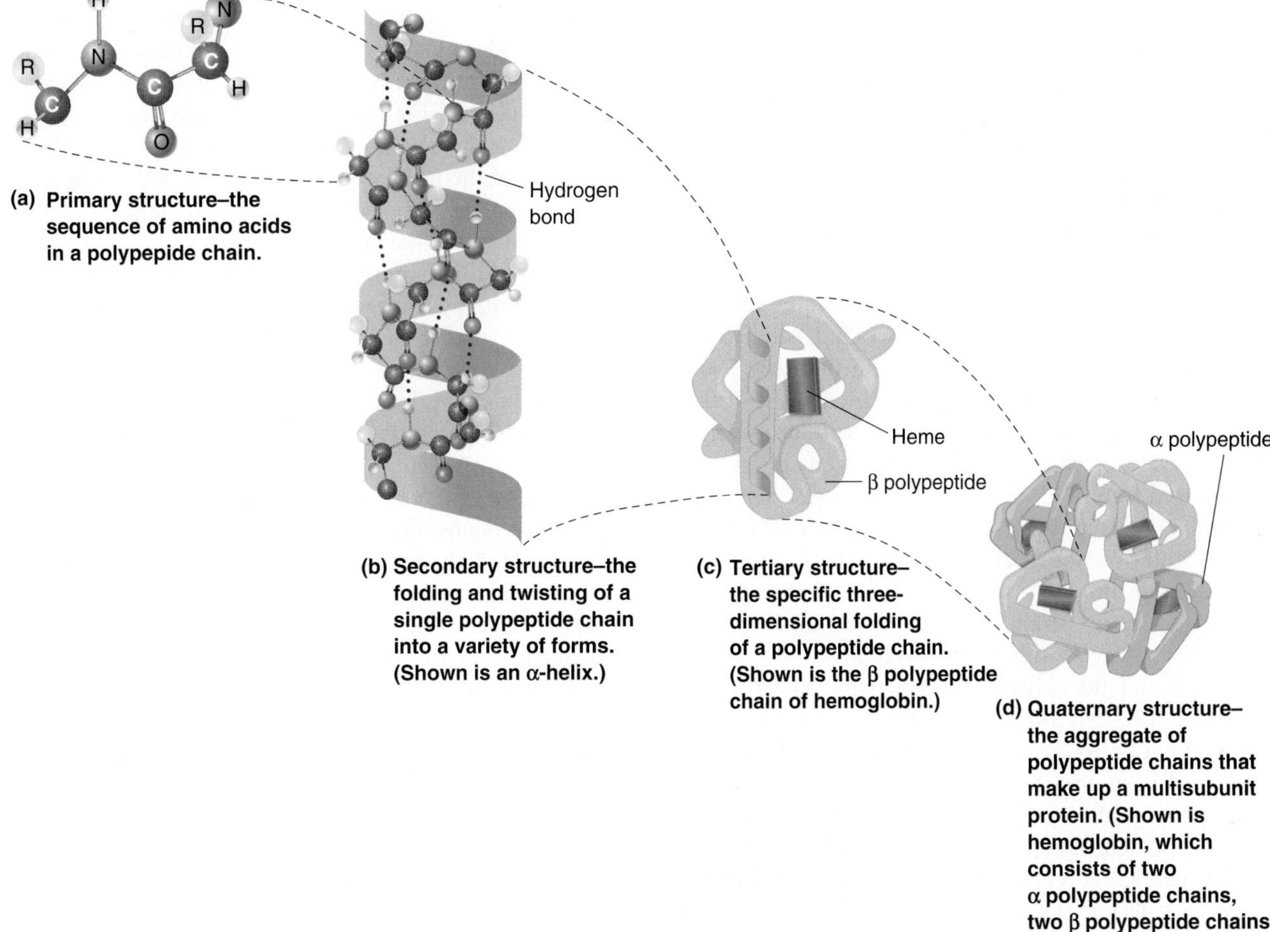
H
R
N
R
N
C
C
C
H
H
O
Hydrogen bond
Heme
β polypeptide
α polypeptide
(a) Primary structure–the sequence of amino acids in a polypepide chain.
(b) Secondary structure–the folding and twisting of a single polypeptide chain into a variety of forms. (Shown is an α-helix.)
(c) Tertiary structure–the specific three-dimensional folding of a polypeptide chain. (Shown is the β polypeptide chain of hemoglobin.)
(d) Quaternary structure–the aggregate of polypeptide chains that make up a multisubunit protein. (Shown is hemoglobin, which consists of two α polypeptide chains, two β polypeptide chains, and four heme groups.)

that they interact with the proteins they help fold—the amino acid sequence of the protein determines the interaction—but do not become part of the functional protein produced. A detailed discussion of chaperones is beyond the scope of this book.

**Keynote**

A protein consists of one or more molecular subunits called polypeptides, which are themselves composed of smaller building blocks, the amino acids, linked together by peptide bonds to form long chains. The primary amino acid sequence of a protein determines its secondary, tertiary, and quaternary structure and hence its functional state.

## The Nature of the Genetic Code

How do nucleotides in the mRNA molecule specify the amino acid sequence in proteins? With four different nucleotides (A, C, G, U), a three-letter code generates 64 possible codons. If it were a one-letter code, only four amino acids could be encoded. If it were a two-letter code, then only 16 ($4 \times 4$) amino acids could be encoded. A three-letter code, however, generates 64 ($4 \times 4 \times 4$) possible codes, more than enough to code for the 20 amino acids found in living cells. Since there are only 20 different amino acids, the assumption of a three-letter code suggests that some amino acids may be specified by more than one codon, which is in fact the case.

### The Genetic Code Is a Triplet Code

The evidence that the **genetic code** is a triplet code—that a set of three nucleotides (a **codon**) in mRNA code form one amino acid in a polypeptide chain—came from genetic experiments done by Francis Crick, Leslie Barnett, Sydney Brenner, and R. Watts-Tobin in the early 1960s. The experiments used bacteriophage T4. T4 is a virulent phage, meaning that, when it infects *E. coli*, it undergoes the lytic cycle, producing 100 to 200 progeny phages that are released from the cell when the cell lyses. Some mutants of T4 affect the lytic cycle: *rII* mutants produce clear plaques on the strain *E. coli B*, whereas the wild-type $r^+$ strain produces turbid plaques. Furthermore, in contrast to the $r^+$ strain, *rII* mutants are unable to undergo the lytic cycle in strain *E. coli K12*(λ).

Crick and his colleagues began with an *rII* mutant strain that had been produced by treating the $r^+$ strain with the mutagen proflavin, a chemical that induces mutations (discussed in more detail in Chapter 7, p. 143). Proflavin causes the addition or deletion of a base pair in the DNA. When such mutations occur in the amino acid-coding part of a gene, the mutations are **frameshift mutations.** That is, if a series of three-nucleotide "words" is read by the translation machinery to assemble the correct polypeptide chain, then if a single base pair is deleted or added in this region, the words after the deletion or addition are now different—they are in another frame—and a different set of amino acids will be specified.

Crick and his colleagues reasoned that, if an *rII* mutant resulted from an addition or a deletion, treatment of the *rII* mutant with proflavin could reverse the mutation to the wild-type—$r^+$—state. The process of changing a mutant back to the wild-type state is called **reversion,** and the wild type produced in this way is called a *revertant*. If the original mutation was an addition, it could be corrected by a deletion; and if the original mutation was a deletion, it could be corrected by an addition. The researchers isolated a number of $r^+$ revertant strains by plating a population of *rII* mutant phages that had been treated with proflavin onto a lawn of *E. coli K12*(λ), in which only $r^+$ phages can undergo the lytic cycle and produce plaques. This approach made it easy to select for and isolate the low number of $r^+$ revertants produced by the proflavin treatment.

One type of revertant resulted from an exact correction of the original mutation; that is, an addition corrected the deletion, or a deletion corrected the addition. A second type of revertant was much more useful for determining the nature of the genetic code in that it resulted from a second mutation within the *rII* gene very close to, but distinct from, the original mutation site. For example, if the first mutation was a deletion of a single base pair, the reversion of that mutation involved an addition of a base pair nearby. Figure 6.5a shows a hypothetical segment of DNA. For the purposes of discussion, we will assume that the code is a triplet code. Thus, the mRNA transcript of the DNA would be read ACG ACG ACG, etc., giving a polypeptide with a string of identical amino acids—threonine—each specified by ACG. This is our starting **reading frame**—the codons (words) that are read sequentially to specify the amino acids. If proflavin treatment causes a deletion of the second A-T base pair, the mRNA will now read ACG CGA CGA CGA, and so on, giving a polypeptide starting with the amino acid specified by ACG (threonine), followed by a string of amino acids that are specified by the repeating CGA (arginine; Figure 6.5b). This mutation is a frameshift mutation because the codons after the deletion are changed. That is, after the ACG, the reading frame of the message is now a string of CGA codons. In that repeated CGA codon sequence, the repeated ACG sequence is still present, with the A as the last letter of the CGA codon and the CG as the first two letters of the CGA. This deletion mutation can revert by the addition of a base pair nearby. For example, the insertion of a G-C base pair after the GC in the third triplet results in an mRNA that is read as ACG CGA CGG ACG ACG, and so on (Figure 6.5c). This gives a polypeptide consisting mostly of the amino acid specified by ACG (threonine), but with two wrong amino acids: those specified by CGA and CGG (both arginine). Thus, the second mutation has restored the reading frame, and a nearly

**Figure 6.5**

**Reversion of a deletion frameshift mutation by a nearby addition mutation. (a)** Hypothetical segment of normal DNA, mRNA transcript, and polypeptide in the wild type. **(b)** Effect of a deletion mutation on the amino acid sequence of a polypeptide. The reading frame is disrupted. **(c)** Reversion of the deletion mutation by an addition mutation. The reading frame is restored, leaving a short segment of incorrect amino acids.

a) Wild type

DNA 5′ ACG ACG ACG ACG ACG 3′
3′ TGC TGC TGC TGC TGC 5′

mRNA 5′ ACG ACG ACG ACG ACG 3′

Polypeptide ··· Thr Thr Thr Thr Thr ···

b) Frameshift mutation by deletion

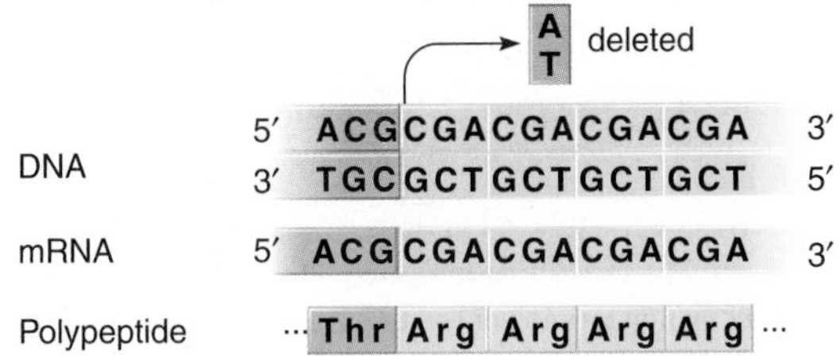


c) Reversion of deletion mutation by addition

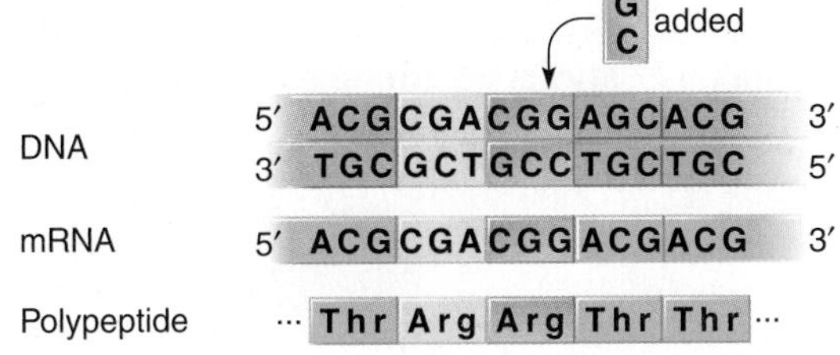


wild-type polypeptide is produced. As long as the incorrect amino acids in the short segment between the mutations do not significantly affect the function of the polypeptide, the double mutant will have a normal or near-normal phenotype.

Addition mutations are symbolized as + mutations and deletion mutations as – mutations. The next step Crick and his colleagues took was to combine genetically distinct *rII* mutations of the same type (either all + or all – mutations)[1] in various numbers to see whether any combinations reverted the *rII* phenotypes. Figure 6.6 is a hypothetical presentation of the type of results they obtained, showing the effects of the mutations just on the mRNA. The figure shows a 30-nucleotide segment of mRNA that codes for 10 different amino acids in the polypeptide. If we add three base pairs at nearby locations in the DNA coding for this mRNA segment, the result will be a 33-nucleotide segment that codes for 11 amino acids, one more than the original. However, the amino acids between the first and third insertions are not the same as the wild-type mRNA. In essence, the reading frame is correct before the first insertion and again after the third insertion. The incorrect amino acids between those points may result in a not-quite wild-type phenotype for the revertant.

Crick and his colleagues found that the combination of three nearby + mutations or three nearby – mutations gave $r^+$ revertants. No multiple combinations worked, except multiples of three. Therefore, they concluded that the simplest explanation was that the genetic code is a triplet code.

## Deciphering the Genetic Code

The exact relationship of the 64 codons to the 20 amino acids was determined by experiments done mostly in the laboratories of Marshall Nirenberg and H. Gobind Khorana, who shared the 1968 Nobel Prize in Physiology or Medicine with Robert Holley. Essential to these experiments was the use of *cell-free, protein-synthesizing systems* with components isolated and purified from *E. coli*. These systems contain ribosomes, tRNAs with amino acids attached, and all the necessary protein factors for polypeptide synthesis. Radioactively labeled amino acids were used to measure the incorporation of amino acids into new proteins.

In one approach to establishing which codons specify which amino acids, synthetic mRNAs containing one,

[1] Crick and his colleagues did not know whether an *rII* mutant resulted from a + or a – mutation. But they did know which of their single-mutant *rII* strains were of one sign and which were of the other sign. That is, all mutants of one sign (e.g., +) could be reverted by nearby mutants of the other sign (i.e., –) and vice versa.

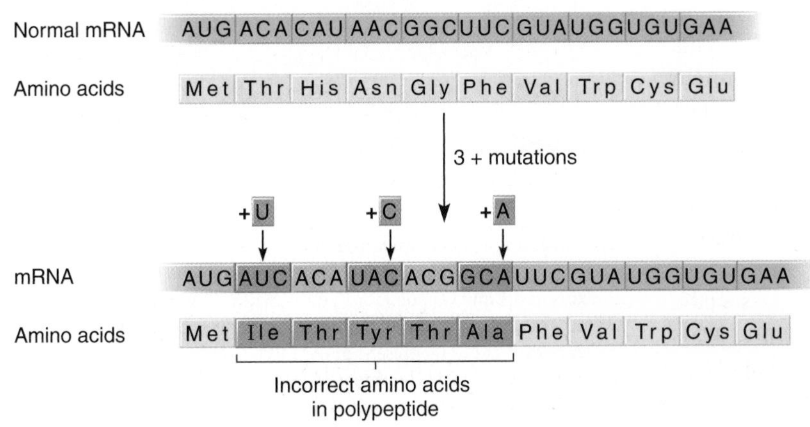


**Figure 6.6**

**Hypothetical example showing how three nearby + (addition) mutations restore the reading frame, giving normal or near-normal function.** The mutations are shown here at the level of the mRNA.

two, or three different types of bases were made and added to the cell-free protein-synthesizing systems. The polypeptides produced in these systems were then analyzed. When the synthetic mRNA contained only one type of base, the results were unambiguous. Synthetic poly(U) mRNA, for example, directed the synthesis of a polypeptide consisting of a chain of phenylalanines. Since the genetic code is a triplet code, this result indicated that UUU is a codon for phenylalanine. Similarly, a synthetic poly(A) mRNA directed the synthesis of a lysine chain, and poly(C) directed the synthesis of a proline chain, indicating that AAA is a codon for lysine and CCC is a codon for proline. The results from poly(G) were inconclusive because the poly(G) folds up upon itself, so it cannot be translated in vitro.

Researchers also analyzed synthetic mRNAs made by the random incorporation of two different bases (called *random copolymers*). For example, poly(AC) molecules contain the eight different codons CCC, CCA, CAC, ACC, CAA, ACA, AAC, and AAA. In the cell-free protein-synthesizing system, poly(AC) synthetic mRNAs resulted in the incorporation of asparagine, glutamine, histidine, and threonine into polypeptides, in addition to the lysine expected from AAA codons and the proline expected from CCC codons. The proportions of asparagine, glutamine, histidine, and threonine incorporated into the polypeptides that were produced depended on the A : C ratio used to make the mRNA and were used to deduce information about the codons that specify the amino acids. For example, because an AC random copolymer containing much more A than C resulted in the incorporation of many more asparagines than histidines, researchers concluded that asparagine is coded by two A's and one C and histidine by two C's and one A. With experiments of this kind, the base composition (but *not* the base sequence) of the codons for a number of amino acids was determined.

Another experimental approach also used synthetic copolymers of known sequences. For example, when a 5′-UCUCUCUCUCUC-3′ copolymer was used in a cell-free protein-synthesizing system, the resulting polypeptide had a repeating amino acid pattern of leucine–serine–leucine–serine. Therefore, UCU and CUC specify leucine and serine, although which coded for which cannot be determined from the result.

Yet another approach used a *ribosome-binding assay*, developed in 1964 by Nirenberg and Philip Leder. This assay depends on the fact that, in the absence of protein synthesis, specific tRNA molecules bind to ribosome–mRNA complexes. For example, when a synthetic mRNA codon, UUU, is mixed with ribosomes, it forms a UUU–ribosome complex, and only a phenylalanine tRNA (the tRNA with an AAA anticodon that brings phenylalanine to an mRNA) binds to the UUU codon. This codon-binding property made it possible to determine the specific relationships between many codons and the amino acids for which they code. Note that in this particular approach, *the specific nucleotide sequence of the codon is determined.* Using the ribosome-binding assay, Nirenberg and Leder resolved many ambiguities that had arisen from other approaches. For example, UCU was found to be a codon for serine, and CUC was found to be a codon for leucine. All in all, about 50 codons were identified with this approach.

In sum, no single approach produced an unambiguous set of codon assignments. But information obtained through all of the approaches enabled 61 codons to be assigned to the 20 amino acids found in all living cells; the other 3 codons do not specify amino acids (Figure 6.7)[2]. Each codon is written as it appears in mRNA and reads in a 5′-to-3′ direction.

## Characteristics of the Genetic Code

The genetic code has these characteristics:

1. *The code is a triplet code.* Each mRNA codon that specifies an amino acid in a polypeptide chain consists of three nucleotides.

**Figure 6.7**

**The genetic code.** Of the 64 codons, 61 specify one of the 20 amino acids. The other 3 codons are chain-terminating codons and do not specify any amino acid. AUG, one of the 61 codons that specify an amino acid, is used in the initiation of protein synthesis.

| First letter | Second letter: U | C | A | G | Third letter |
|---|---|---|---|---|---|
| U | UUU Phe (F) | UCU Ser (S) | UAU Tyr (Y) | UGU Cys (C) | U |
| U | UUC | UCC | UAC | UGC | C |
| U | UUA Leu (L) | UCA | UAA Stop | UGA Stop | A |
| U | UUG | UCG | UAG Stop | UGG Trp (W) | G |
| C | CUU Leu (L) | CCU Pro (P) | CAU His (H) | CGU Arg (R) | U |
| C | CUC | CCC | CAC | CGC | C |
| C | CUA | CCA | CAA Gln (Q) | CGA | A |
| C | CUG | CCG | CAG | CGG | G |
| A | AUU Ile (I) | ACU Thr (T) | AAU Asn (N) | AGU Ser (S) | U |
| A | AUC | ACC | AAC | AGC | C |
| A | AUA | ACA | AAA Lys (K) | AGA Arg (R) | A |
| A | AUG Met (M) | ACG | AAG | AGG | G |
| G | GUU Val (V) | GCU Ala (A) | GAU Asp (D) | GGU Gly (G) | U |
| G | GUC | GCC | GAC | GGC | C |
| G | GUA | GCA | GAA Glu (E) | GGA | A |
| G | GUG | GCG | GAG | GGG | G |

= Chain termination codon (stop)

= Initiation codon

[2]Two other amino acids are found rarely in proteins and are specified by the genetic code. The amino acid selenocysteine is found in all three domains of life and is coded for by UGA, which is normally a stop codon. This coding is not direct, however. Rather, it requires a specific sequence element to be present in the mRNA to direct the UGA to encode selenocysteine. The amino acid pyrrolysine is found in enzymes for methane production in some archaeans. In these organisms, pyrrolysine is encoded by UAG, which is normally a stop codon.

2. *The code is comma free; that is, it is continuous.* The mRNA is read continuously, three nucleotides at a time, without skipping any nucleotides of the message.
3. *The code is nonoverlapping.* The mRNA is read in successive groups of three nucleotides.
4. *The code is almost universal.* Almost all organisms share the same genetic language. It is arbitrary in the sense that many other codes are possible, but the vast majority of organisms share this one (this is a major piece of evidence that all living organisms share a common ancestor). Therefore, we can isolate an mRNA from one organism, translate it by using the machinery from another organism, and produce the protein as if it had been translated in the original organism. The code is not completely universal, however. For example, the mitochondria of some organisms, such as mammals, have minor changes in the code, as does the nuclear genome of the protozoan *Tetrahymena*.
5. *The code is "degenerate."* With two exceptions, more than one codon occurs for each amino acid; the exceptions are AUG, which alone codes for methionine, and UGG, which alone codes for tryptophan. This multiple coding is called the **degeneracy** or *redundancy* of the code. There are particular patterns in this degeneracy (see Figure 6.7). When the first two nucleotides in a codon are identical and the third letter is U or C, the codon always codes for the same amino acid. For example, UUU and UUC specify phenylalanine, and CAU and CAC specify histidine. Also, when the first two nucleotides in a codon are identical and the third letter is A or G, the same amino acid often is specified. For example, UUA and UUG specify leucine, and AAA and AAG specify lysine. In a few cases, when the first two nucleotides in a codon are identical and the base in the third position is U, C, A, or G, the same amino acid often is specified. For example, CUU, CUC, CUA, and CUG all code for leucine.
6. *The code has start and stop signals.* Specific start and stop signals for protein synthesis are contained in the code. In both eukaryotes and prokaryotes, AUG (which codes for methionine) is almost always the start codon for protein synthesis.

   Only 61 of the 64 codons specify amino acids; these codons are called *sense codons* (see Figure 6.7). The other three codons—UAG (amber), UAA (ochre), and UGA (opal)—do not specify an amino acid, and no tRNAs in normal cells carry the appropriate anticodons. (The three-nucleotide anticodon pairs with the codon in the mRNA by complementary base pairing during translation.) These three codons are the **stop codons,** also called **nonsense codons** or **chain-terminating codons.** They are used to specify the end of translation of a polypeptide chain. Thus, when we read a particular mRNA sequence, we look for a stop codon located at a multiple of three nucleotides—*in the same reading frame*—from the AUG start codon to determine where the amino acid-coding sequence for the polypeptide ends. This is called an **open reading frame (ORF).**
7. *Wobble occurs in the anticodon.* Since 61 sense codons specify amino acids in mRNA, a total of 61 tRNA molecules could have the appropriate anticodons. According to the **wobble hypothesis** proposed by Francis Crick, the complete set of 61 sense codons can be read by fewer than 61 distinct tRNAs, because of pairing properties of the bases in the anticodon (Table 6.1). Specifically, the base at the 5′ end of the anticodon complementary to the base at the 3′ end of the codon—the third letter—is not as constrained three dimensionally as the other two bases. As a result, less exact base pairing can occur: the base at the 5′ end of the anticodon can pair with more than one type of base at the 3′ end of the codon—in other words, the 5′-base of the anticodon can *wobble.* As the table shows, a single tRNA molecule can recognize at most three different codons. Figure 6.8 gives an example of how a single leucine tRNA can read two different leucine codons by base-pairing wobble.

**Table 6.1 Wobble in the Genetic Code**

| Nucleotide at 5′ End of Anticodon | | Nucleotide at 3′ End of Codon |
|---|---|---|
| G | can pair with | U or C |
| C | can pair with | G |
| A | can pair with | U |
| U | can pair with | A or G |
| I (inosine) | can pair with | A, U, or C |

One characteristic of the genetic code just mentioned is that it is almost universal. This chapter's Focus on Genomics box expands on this point and describes the variations in the code that have been identified in genomes.

**Activity**

Learn how to use sequencing information to track down part of the gene responsible for cystic fibrosis in the iActivity *Determining Causes of Cystic Fibrosis* on the student website.

**Figure 6.8**

**Example of base-pairing wobble.** Two different leucine codons (CUC, CUU) can be read by the same leucine tRNA molecule, contrary to regular base-pairing rules.

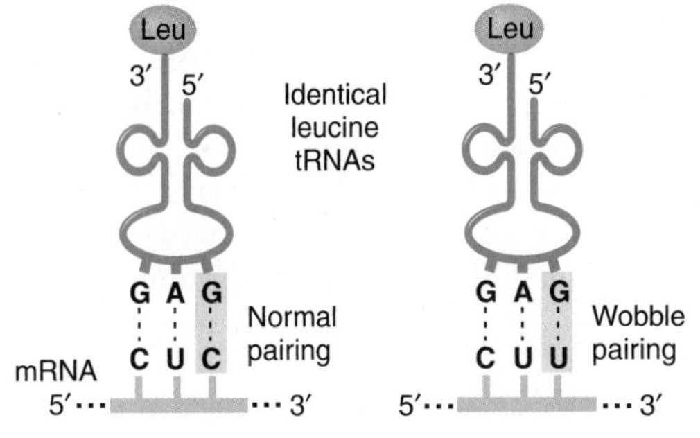

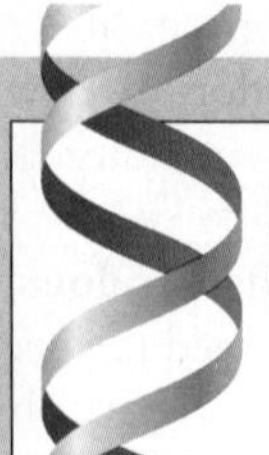

## Focus on Genomics
### Other Genetic Codes

The genetic code is almost universal. How much do other codes vary, and where are they found? The greatest divergence is seen in organelle genomes. That is, in the known organelle (mitochondria and chloroplast) genetic codes (12, as of early 2008), 53 of the 64 codons are invariant in all 12 codes. Variations have been found at only 11 codons, and a total of only 28 variations have been found. Fourteen of the 28 known variations concern stop codons, where either a codon that normally codes for an amino acid now codes for a stop, or one of the standard three stop codons now codes for an amino acid. The others reassign one or more codons from one amino acid to another. The greatest variation known is in the genome of yeast mitochondria, where UGA codes for tryptophan, rather than stop, and CTN codes for threonine, rather than for leucine. Nuclear genomes have far less variation. Only six total changes are known, and these affect only three codons. All are found at codons that serve as stop codons in the standard code, and all are changes consistent with mutations in a tRNA gene that alter the anticodon of tRNA in one position.

There is a surprising amount of variation in start codons. It is true that most genes start translation on an AUG codon, but in both mitochondrial and nuclear genomes, at least seven other codons have been seen to serve as start codons for certain proteins. All but one of these is similar to AUG at two of the three bases.

**Keynote**

The genetic code is a triplet code in which each codon (a set of three contiguous bases) in an mRNA specifies one amino acid. The code is degenerate: some amino acids are specified by more than one codon. The genetic code is nonoverlapping and almost universal. Specific codons are used to signify the start and end of protein synthesis.

## Translation: The Process of Protein Synthesis

Polypeptide synthesis takes place on ribosomes, where the genetic message encoded in mRNA is translated. The mRNA is translated in the 5′-to-3′ direction, and the polypeptide is made in the N-terminal–to–C-terminal direction. Amino acids are brought to the ribosome bound to tRNA molecules.

### Transfer RNA

During translation of mRNA, each transfer RNA (tRNA) brings a specific amino acid to the ribosome to be added to a growing polypeptide chain. The correct amino acid sequence of a polypeptide is achieved as a result of: (1) the binding of each amino acid to a specific tRNA; and (2) the binding between the codon of the mRNA and the complementary anticodon in the tRNA.

**Structure of tRNA.** tRNAs are 75 to 90 nucleotides long, each type having a different sequence. The differences in nucleotide sequences explain the ability of a particular tRNA molecule to bind a specific amino acid. The nucleotide sequences of all tRNAs can be arranged into what is called a cloverleaf (Figure 6.9a). The cloverleaf results from complementary base pairing between different sections of the molecule, producing four base-paired "stems" separated by four loops: I, II, III, and IV. Loop II contains the three-nucleotide **anticodon** sequence, which pairs with a three-nucleotide codon sequence in mRNA by complementary base pairing during translation. This codon–anticodon pairing is crucial to the addition of the amino acid specified by the mRNA to the growing polypeptide chain. Figures 6.9b and 6.9c show the tertiary structure of phenylalanine tRNA from yeast; the latter space-filling depiction is the three-dimensional form that functions in cells. All other tRNAs that have been examined exhibit similar upside-down L-shaped structures in which the 3′ end of the tRNA—the end to which the amino acid attaches—is at the end of the L that is opposite from the anticodon loop.

All tRNA molecules have the sequence 5′-CCA-3′ at their 3′ ends. All tRNA molecules also have a number of bases modified chemically by enzyme reactions, with different arrays of modifications on each tRNA type (examples of modified bases are given in Figure 6.9a).

**Transfer RNA Genes.** Bacterial tRNA genes are found in one or at most a few copies in the genome, whereas

**Figure 6.9**

**Transfer RNA.** Py = pyrimidine. Modified bases: I = inosine, T = ribothymidine, ψ = pseudouridine, D = dihydrouridine, GMe = methylguanosine, $GMe_2$ = dimethylguanosine, IMe = methylinosine.

**a) Cloverleaf model of tRNA**

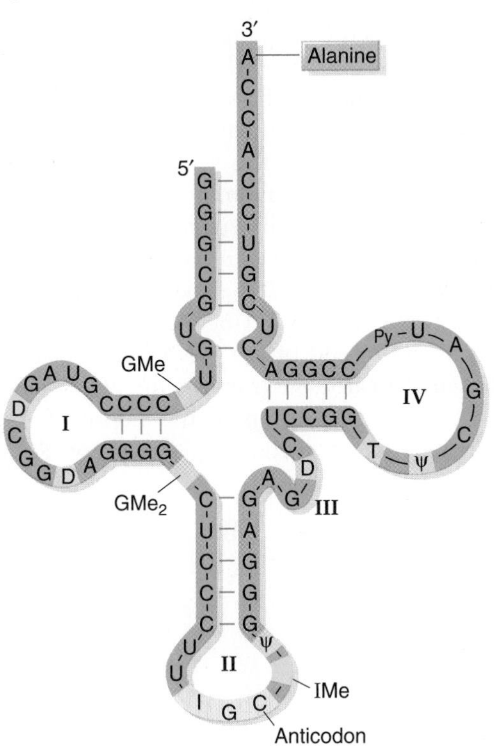


**b) Schematic of the three-dimensional L-shaped structure of a tRNA, here yeast phenylalanine tRNA**

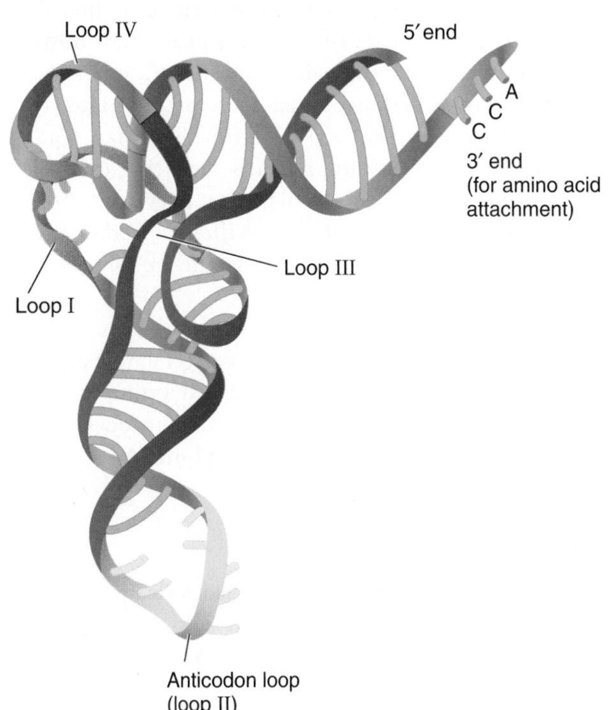


eukaryotic tRNA genes are repeated many times in the genome. In the South African clawed toad *Xenopus laevis*, for example, there are about 200 copies of each tRNA gene. Bacterial tRNA genes are transcribed by the only RNA polymerase found in bacteria; eukaryotic tRNA genes are transcribed by RNA polymerase III. Transcription of tRNA genes in both bacteria and eukaryotes produces **precursor tRNA (pre-tRNA)** molecules, each of which has extra sequences at each end that are removed posttranscriptionally. 5′-CCA-3′ addition at the 3′ end, and modification of bases throughout the molecule, then take place.

Some tRNA genes in certain eukaryotes contain introns. The intron is almost always located between the first and second nucleotides 3′ to the anticodon. Removal of the introns occurs by a mechanism different from that of pre-mRNA splicing.

**Recognition of the tRNA Anticodon by the mRNA Codon.** That the mRNA codon recognizes the tRNA anticodon, and not the amino acid carried by the tRNA, was proved by G. von Ehrenstein, B. Weisblum, and S. Benzer. These researchers attached cysteine in vitro to tRNA.Cys (this terminology indicates the amino acid specified by the anticodon of the tRNA—in this case,

**c) Space-filling molecular model of yeast phenylalanine tRNA**

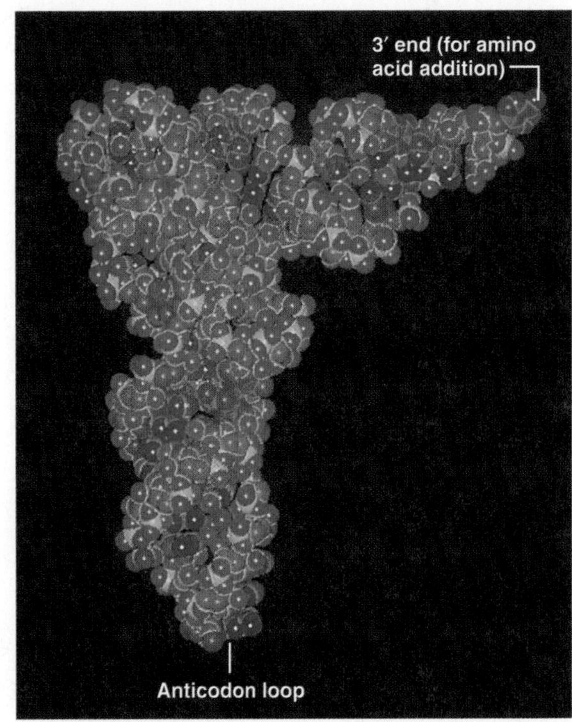

cysteine); then they chemically converted the attached cysteine to alanine. The resulting Ala–tRNA.Cys (the amino acid alanine attached to the tRNA with an anticodon for a codon specifying cysteine) was used in the in vitro synthesis of hemoglobin. *In vivo*, the α and β chains of hemoglobin each contain one cysteine. When the hemoglobin made in vitro was examined, however, the amino acid alanine was found in both chains at the positions normally occupied by cysteine. This result could only mean that the Ala–tRNA.Cys had read the codon for cysteine and had inserted the amino acid it carried—in this case, alanine. Therefore, the researchers concluded that *the specificity of codon recognition lies in the tRNA molecule, not in the amino acid it carries.*

**Adding an Amino Acid to tRNA.** The correct amino acid is attached to the tRNA by an enzyme called **aminoacyl–tRNA synthetase.** The process is called aminoacylation, or **charging,** and produces an **aminoacyl–tRNA** (or **charged tRNA**). Aminoacylation uses energy from ATP hydrolysis. There are 20 different aminoacyl–tRNA synthetases, one for each of the 20 different amino acids. Each enzyme recognizes particular structural features of the tRNA or tRNAs it aminoacylates.

Figure 6.10 shows the charging of a tRNA molecule to produce valine–tRNA (Val–tRNA). First, the amino acid and ATP bind to the specific aminoacyl–tRNA synthetase enzyme. The enzyme then catalyzes a reaction in which the ATP is hydrolyzed to AMP, which joins to the amino acid as AMP to form aminoacyl–AMP. Next, the tRNA molecule binds to the enzyme, which transfers the amino acid from the aminoacyl–AMP to the tRNA and displaces the AMP. The enzyme then releases the aminoacyl–tRNA molecule. Chemically, the amino acid attaches at the 3′ end of the tRNA by a covalent linkage between the carboxyl group of the amino acid and the 3′-OH or 2′-OH group of the ribose of the adenine nucleotide found at the 3′ end of every tRNA (Figure 6.11).



**Figure 6.10**
**Aminoacylation (charging) of a tRNA molecule by aminoacyl–tRNA synthetase to produce an aminoacyl–tRNA (charged tRNA).**

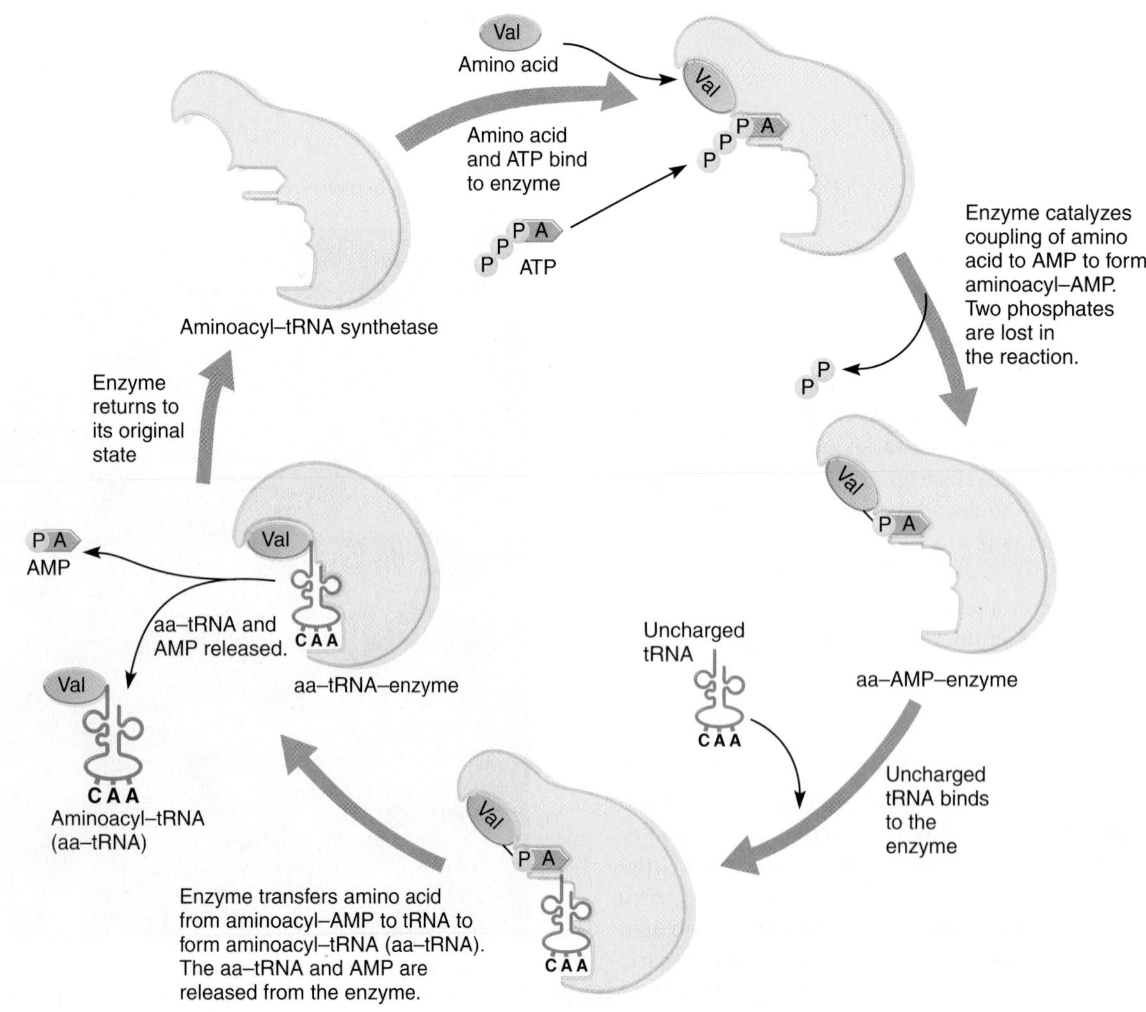

Figure 6.11

**Attachment of an amino acid to a tRNA molecule.** In an aminoacyl–tRNA molecule (charged tRNA), the carboxyl group of the amino acid is attached to the 3′-OH or 2′-OH group of the 3′ terminal adenine nucleotide of the tRNA.

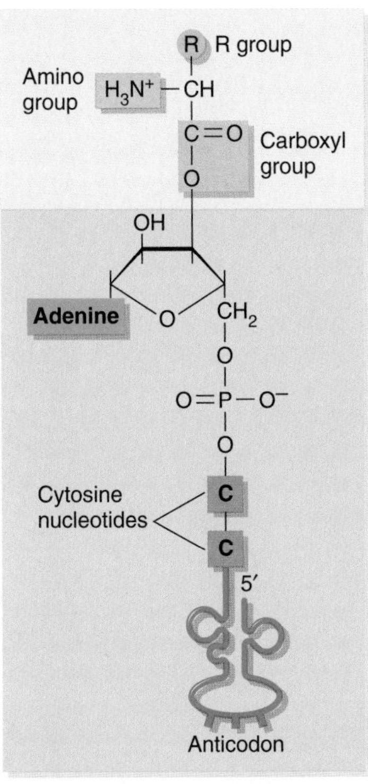


## Keynote

Each tRNA molecule brings a specific amino acid to the ribosome to be added to the growing polypeptide chain. The amino acid is added to a tRNA by an amino acid-specific aminoacyl–tRNA synthetase enzyme. All tRNAs are similar in length (75 to 90 nucleotides), have a 5′-CCA-3′ sequence at their 3′ ends, have a number of tRNA-specific modifications of the bases, and have a similar tertiary structure. The anticodon of a tRNA is keyed to the amino acid it carries, and it pairs with a complementary codon in an mRNA molecule. Functional tRNA molecules are produced by processing of pre-tRNA transcripts of tRNA genes to remove extra sequences at each end, the addition of the CCA sequence to the 3′ end, and enzyme-catalyzed modification of some bases. For some tRNA genes in certain eukaryotes, introns are present and are removed during processing of the pre-tRNA molecule.

## Ribosomes

Polypeptide synthesis takes place on ribosomes, many thousands of which occur in each cell. Ribosomes bind to mRNA and facilitate the binding of the tRNA to the mRNA so that a polypeptide chain can be synthesized.

**Ribosomal RNA and Ribosomes.** In both prokaryotes and eukaryotes, **ribosomes** consist of two unequally sized subunits—the large and small ribosomal subunits—each of which consists of a complex between RNA molecules and proteins. Each subunit contains one or more rRNA molecules and a large number of **ribosomal proteins.**

The bacterial ribosome has a size of 70S and consists of two subunits of sizes 50S (large subunit) and 30S (small subunit)[3] (Figure 6.12). Eukaryotic ribosomes are larger and more complex than their prokaryotic counterparts, and they vary in size and composition among eukaryotic organisms. Mammalian ribosomes, for example, have a size of 80S and consist of a large 60S subunit and a small 40S subunit.

Each ribosomal subunit contains one or more specific rRNA molecules and a number of ribosomal proteins (Figure 6.13; also shown in the molecular model in Figure 6.12). Bacterial ribosomes contain three rRNA molecules—the 23S rRNA and 5S rRNA in the large subunit, and the 16S rRNA in the small subunit. Eukaryotic ribosomes contain four rRNA molecules—the 28S rRNA, 5.8S rRNA, and 5S rRNA in the large subunit, and the 18S rRNA in the small subunit. The rRNA molecules play a structural role in ribosome and have a functional role in several steps of translation.

Figure 6.12

**Molecular model of the complete (70S) bacterial ribosome.** The ribosome is from *Thermus thermophilus*. Visible are the rRNAs and proteins of the two subunits, as well as a tRNA in its binding site.

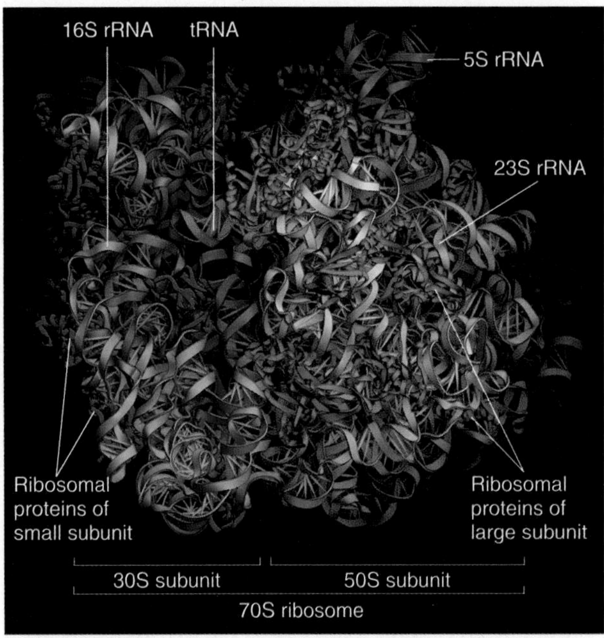


[3]The S value is a measure of sedimentation rate in a centrifuge. Sedimentation rate depends not only on mass, but on the three-dimensional shape of the object. Hence, given two objects with the same mass but different shapes, the more compact one will sediment faster and therefore have a higher S value than the less compact one. For ribosomes, 50S + 30S ≠ 70S because, when the two subunits come together to form the whole ribosome, the shape changes to a less compact one and sedimentation is slower than expected from the sum of the two subunits.

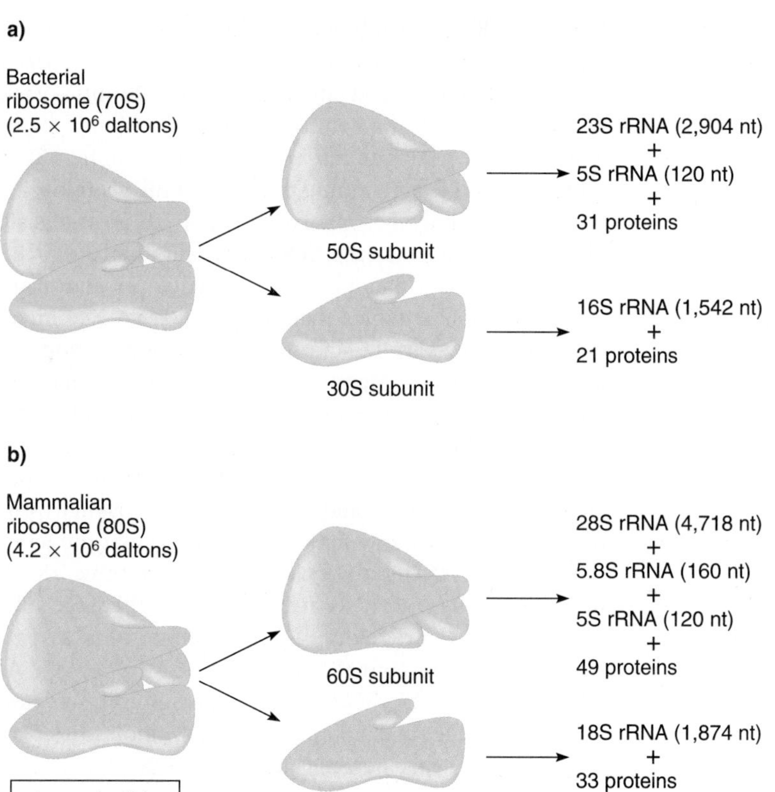


**Figure 6.13**

**Composition of whole ribosomes and of ribosomal subunits in (a) bacterial and in (b) mammalian cells.**

During translation, the mRNA passes through the small subunit of the ribosome (Figure 6.14). Specific sites of the ribosome bind tRNAs at different stages of polypeptide synthesis: the A (aminoacyl) site is where an incoming aminoacyl–tRNA binds, the P (peptidyl) site is where the tRNA carrying the growing polypeptide chain is located, and the E (exit) site is where a tRNA binds on its path from the P site to leaving the ribosome. The P and A sites consist of regions of both the large and small subunits, whereas the E site is a region of the large subunit. We will learn more about these sites in the discussion of the steps of translation in the next three sections.

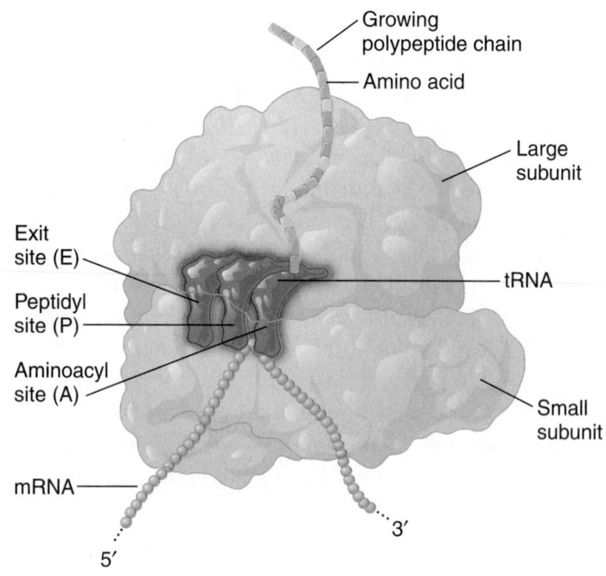


**Figure 6.14**

**Structure of the ribosomes showing the path of mRNA through the small subunit, and the three sites to which tRNAs bind at different stages of polypeptide synthesis and the exit path for the polypeptide chain.**

**Ribosomal RNA Genes.** In prokaryotes and eukaryotes, the regions of DNA that contain the genes for rRNA are called **ribosomal DNA (rDNA)** or **rRNA transcription units.** *E. coli* has seven rRNA transcription units scattered in the *E. coli* chromosome. Each rRNA transcription unit contains one copy each of the 16S, 23S, and 5S rRNA coding sequences, arranged in the order 16S–23S–5S. There is a single promoter for each rRNA transcription unit, and transcription by RNA polymerase produces a **precursor rRNA (pre-rRNA)** molecule with the organization 5′-16S–23S–5S-3′, with non-rRNA sequences called *spacer sequences* between each rRNA sequence and at the 5′ and 3′ ends. Processing by specific ribonucleases removes the spacers, releasing the three rRNAs. Ribosomal proteins associate with the pre-rRNA molecule as it is being transcribed to form a large ribonucleoprotein complex. The transcript-processing events take place in that complex and specific associations of the rRNAs with ribosomal proteins generate the functional ribosomal subunits.

Most eukaryotes have many copies of the genes for each of the four rRNA species 18S, 5.8S, 28S, and 5S. The genes for 18S, 5.8S, and 28S rRNAs are found adjacent to one another in the order 18S–5.8S–28S, with each set of three genes typically tandemly repeated 100 to 1,000 times (depending on the organism), to form one or more clusters of **rDNA repeat units.** Due to active transcription of the repeat units, a nucleolus forms around each cluster. Typically, the multiple nucleoli so formed fuse to form one nucleolus.

Each eukaryotic rDNA repeat unit is transcribed by RNA polymerase I to produce a pre-rRNA molecule with the organization 5′-18S–5.8S–28S-3′, which has spacer sequences between each rRNA and at the 5′ and 3′ ends. Processing by specific ribonucleases generates the three rRNAs by removing the spacers. The pre-rRNA-processing events take place in complexes formed between the pre-rRNA, 5S rRNA, and ribosomal proteins. The 5S rRNA is produced by transcription of the 5S rRNA genes (typically located elsewhere in the genome) by RNA polymerase III. As pre-rRNA processing proceeds, the complexes undergo changes in shape, resulting in formation of the functional 60S and 40S ribosomal subunits, which are then transported to the cytosol.

It is important to be clear about the distinction between an intron and a spacer. The removal of a spacer releases the flanking RNAs, and they remain separate. Intron removal, by contrast, results in the splicing together of the RNA sequences that flanked the intron.

**Keynote**

Ribosomes consist of two unequally sized subunits, each containing one or more ribosomal RNA molecules and ribosomal proteins. The three prokaryotic rRNAs and three of the four eukaryotic rRNAs are encoded in rRNA transcription units. The fourth eukaryotic rRNA is encoded by separate genes. The transcription of rRNA transcription units by RNA polymerase produces pre-rRNA molecules that are processed to mature rRNAs by the removal of spacer sequences. The processing events occur in complexes of the pre-rRNAs with ribosomal proteins and other proteins and are part of the formation of the mature ribosomal subunits.

### Initiation of Translation

The three basic stages of protein synthesis—*initiation, elongation*, and *termination*—are similar in bacteria and eukaryotes. In this section and the two sections that follow, we discuss each of these stages in turn, concentrating on the processes in *E. coli*. In the discussions, significant differences in translation between bacteria and eukaryotes are noted.

**Initiation of Translation**

Initiation encompasses all of the steps preceding the formation of the peptide bond between the first two amino acids in the polypeptide chain. Initiation involves an mRNA molecule, a ribosome, a specific initiator tRNA, protein **initiation factors (IF),** and GTP (guanosine triphosphate).

**Initiation in Bacteria.** In bacteria, the first step in the initiation of translation is the interaction of the 30S (small) ribosomal subunit to which IF-1 and IF-3 are bound with the region of the mRNA containing the AUG initiation codon (Figure 6.15). IF-3 aids in the binding of the subunit to mRNA and prevents binding of the 50S ribosomal subunit to the 30S subunit.

The AUG initiation codon alone is not sufficient to indicate where the 30S subunit should bind to the mRNA; a sequence upstream (to the 5′ side in the leader of the mRNA) of the AUG called the **ribosome-binding site (RBS)** is also needed. In the 1970s, John Shine and Lynn Dalgarno hypothesized that the purine-rich RBS sequence (5′-AGGAG-3′ or some similar sequence) and sometimes other nucleotides in this region could pair with a complementary pyrimidine-rich region (always containing the sequence 5′-UCCUCC-3′) at the 3′ end of 16S rRNA (Figure 6.16). Joan Steitz was the first to demonstrate this pairing experimentally. The mRNA RBS region is now commonly known as the **Shine–Dalgarno sequence.**

Most of the RBSs are 8 to 12 nucleotides upstream from the initiation codon. The model is that the formation of complementary base pairs between the mRNA and 16S rRNA allows the small ribosomal subunit to locate the true sequence in the mRNA for the initiation of protein synthesis. Genetic evidence supports this model. If the Shine–Dalgarno sequence of an mRNA is mutated so that its possible pairing with the 16S rRNA sequence is significantly diminished or prevented, the mutated mRNA cannot be translated. Likewise, if the rRNA sequence complementary to the Shine–Dalgarno sequence is mutated, mRNA translation cannot occur. Since it can be argued that the loss of translatability as a result of mutations in one or the other RNA partner could be caused by effects unrelated to the loss of pairing of the two RNA segments, a more elegant experiment was done. That is, mutations were made in the Shine–Dalgarno sequence to abolish pairing with the wild-type rRNA sequence, and compensating mutations were made in the rRNA sequence so that the two mutated sequences could pair. In this case, mRNA translation occurred normally, indicating the importance of the pairing of the two RNA segments. (This type of experiment, in which compensating mutations are made in two sequences that are hypothesized to interact, has been used in a number of other systems to explore the roles of specific interactions in biological functions.)

The next step in the initiation of translation is the binding of a special initiator tRNA to the AUG start codon to which the 30S subunit is bound. In both prokaryotes and eukaryotes, the AUG initiator codon specifies methionine. As a result, newly made proteins in both types

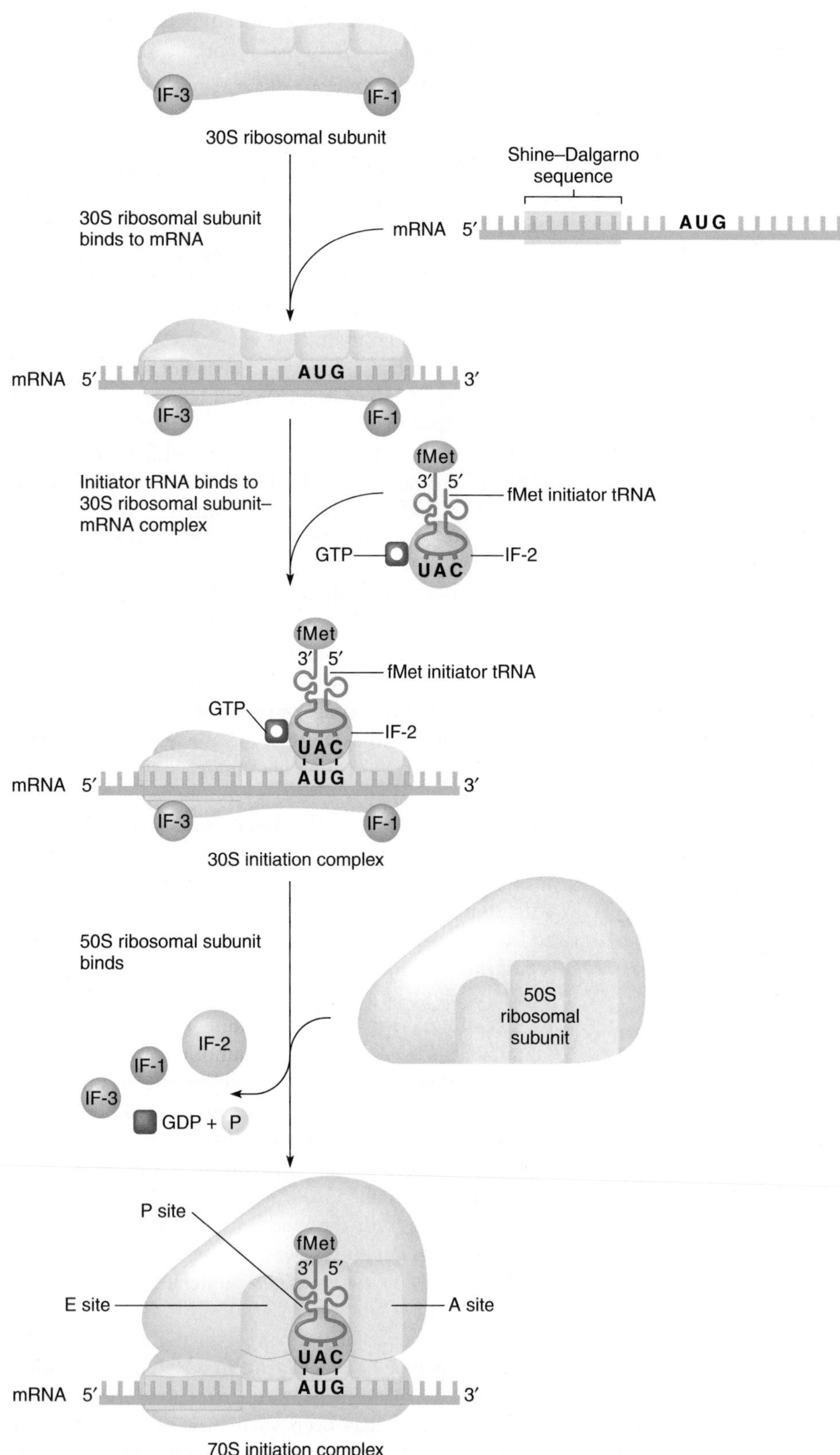


**Figure 6.15**

**Initiation of protein synthesis in bacteria.** A 30S ribosomal subunit, mRNA, initiator fMet–tRNA, and initiation factors form a 30S initiation complex. Next, the 50S ribosomal subunit binds, forming a 70S initiation complex. During this event, the initiation factors are released and GTP is hydrolyzed.

of organisms begin with methionine. In many cases, the methionine is removed later.

In bacteria, the initiator tRNA is tRNA.fMet, which has the anticodon 5′-CAU-3′ to bind to the AUG start codon. This tRNA carries a modified form of methionine, **formylmethionine (fMet),** in which a formyl group has been added to the amino group of methionine. That is, first, methionyl–tRNA synthetase catalyzes the addition of methionine to the tRNA. Then the enzyme *transformylase* adds the formyl group to the methionine.

Figure 6.16

**Sequences involved in the binding of ribosomes to the mRNA in the initiation of protein synthesis in prokaryotes.**

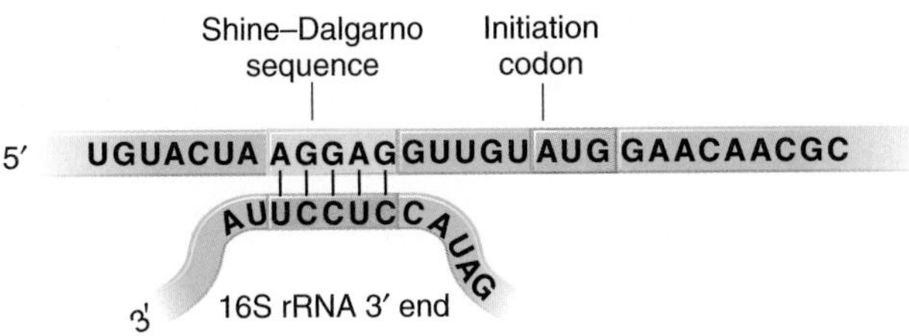


The resulting molecule is designated fMet–tRNA.fMet. (This nomenclature indicates that the tRNA is specific for the attachment of fMet and that fMet is attached to it.)

Note that, when an AUG codon in an mRNA molecule is encountered at a position other than the start of the amino acid-coding sequence, a different tRNA, called tRNA.Met, is used to insert methionine at that point in the polypeptide chain. This tRNA is charged by the same aminoacyl–tRNA synthetase as is tRNA.fMet to produce Met–tRNA.Met. However, tRNA.Met and tRNA.fMet molecules are coded for by different genes and have different sequences. We will see later in the chapter how the two tRNAs are used differently.

The initiator tRNA, fMet–tRNA.fMet, is brought to the 30S subunit–mRNA complex by IF-2, which also carries a molecule of GTP. The initiator tRNA binds to the subunit in the P site. We will see later that, subsequently, all aminoacyl–tRNAs that come to the ribosome bind to the A site. However, IF-1 bound to the 30S subunit is blocking the A site so that only the P site is available for the initiator tRNA to bind to. Formed at this point is the *30S initiation complex*, consisting of the mRNA, 30S subunit, initiator tRNA, and the initiation factors (see Figure 6.15). Next, the 50S ribosomal subunit binds, leading to GTP hydrolysis and the release of the three initiation factors. The final complex is called the *70S initiation complex* (see Figure 6.15).

**Initiation in Eukaryotes.** The initiation of translation is similar in eukaryotes, although the process is more complex and involves many more initiation factors, called *eukaryotic initiation factors (eIF)*, than is the case in bacteria. The main differences are that: (1) the initiator methionine is unmodified, although a special initiator tRNA still brings it to the ribosome; and (2) Shine–Dalgarno sequences are not found in eukaryotic mRNAs. Instead, the eukaryotic ribosome uses another way to find the AUG initiation codon. First, a eukaryotic initiator factor eIF-4F—a multimer of several proteins, including eIF-4E, the *cap-binding protein* (CBP)—binds to the cap at the 5′ end of the mRNA (see Chapter 5). Then, a complex of the 40S ribosomal subunit with the initiator Met–tRNA, several eIF proteins, and GTP binds, together with other eIFs, and moves along the mRNA, scanning for the initiator AUG codon. The AUG codon is embedded in a short sequence—called the Kozak sequence, after Marilyn Kozak—which indicates that it is the initiator codon. This process is called the *scanning model* for initiation. The AUG codon is almost always the first AUG codon from the 5′ end of the mRNA; but, to be an initiator codon, it must be in an appropriate sequence context. Once the 40S subunit finds this AUG, it binds to it, and then the 60S ribosomal subunit binds, displacing the eIFs (except for eIF-4F, which is needed for the subsequent initiation of translation), producing the 80S initiation complex with the initiator Met–tRNA bound to the mRNA in the P site of the ribosome.

The poly(A) tail of the eukaryotic mRNA also plays a role in translation. Poly(A) binding protein II (PABPII; see Figure 5.11b, p. 92) bound to the poly(A) tail also binds to eIF-4G, one of the proteins of eIF-4F at the cap, thereby looping the 3′ end of the mRNA close to the 5′ end. In this way, the poly(A) tail stimulates the initiation of translation.

## Elongation of the Polypeptide Chain

After initiation is complete, the next stage is elongation. Figure 6.17 depicts the elongation events—the addition of amino acids to the growing polypeptide chain one by one—as they take place in bacteria. This phase has three steps:

1. Aminoacyl–tRNA (charged tRNA) binds to the ribosome in the A site.
2. A peptide bond forms.
3. The ribosome moves (translocates) along the mRNA one codon.

As with initiation, elongation requires accessory protein factors, here called **elongation factors (EF),** and GTP. Elongation is similar in eukaryotes.

**Binding of Aminoacyl–tRNA.** At the start of elongation, the anticodon of fMet–tRNA is hydrogen bonded to the AUG initiation codon in the P site of the ribosome (Figure 6.17, step 1). The next codon in the mRNA is in the A site; in Figure 6.17, this codon (UCC) specifies the amino acid serine (Ser).

Next, the appropriate aminoacyl–tRNA (here, Ser–tRNA.Ser) binds to the codon in the A site (Figure 6.17, step 2). This aminoacyl–tRNA is brought to the ribosome bound to EF-Tu–GTP, a complex of the protein elongation

**Figure 6.17**

**Elongation stage of translation in bacteria.** For the EF-Tu and EF-Ts proteins, the "u" stands for unstable, while the "s" stands for stable.

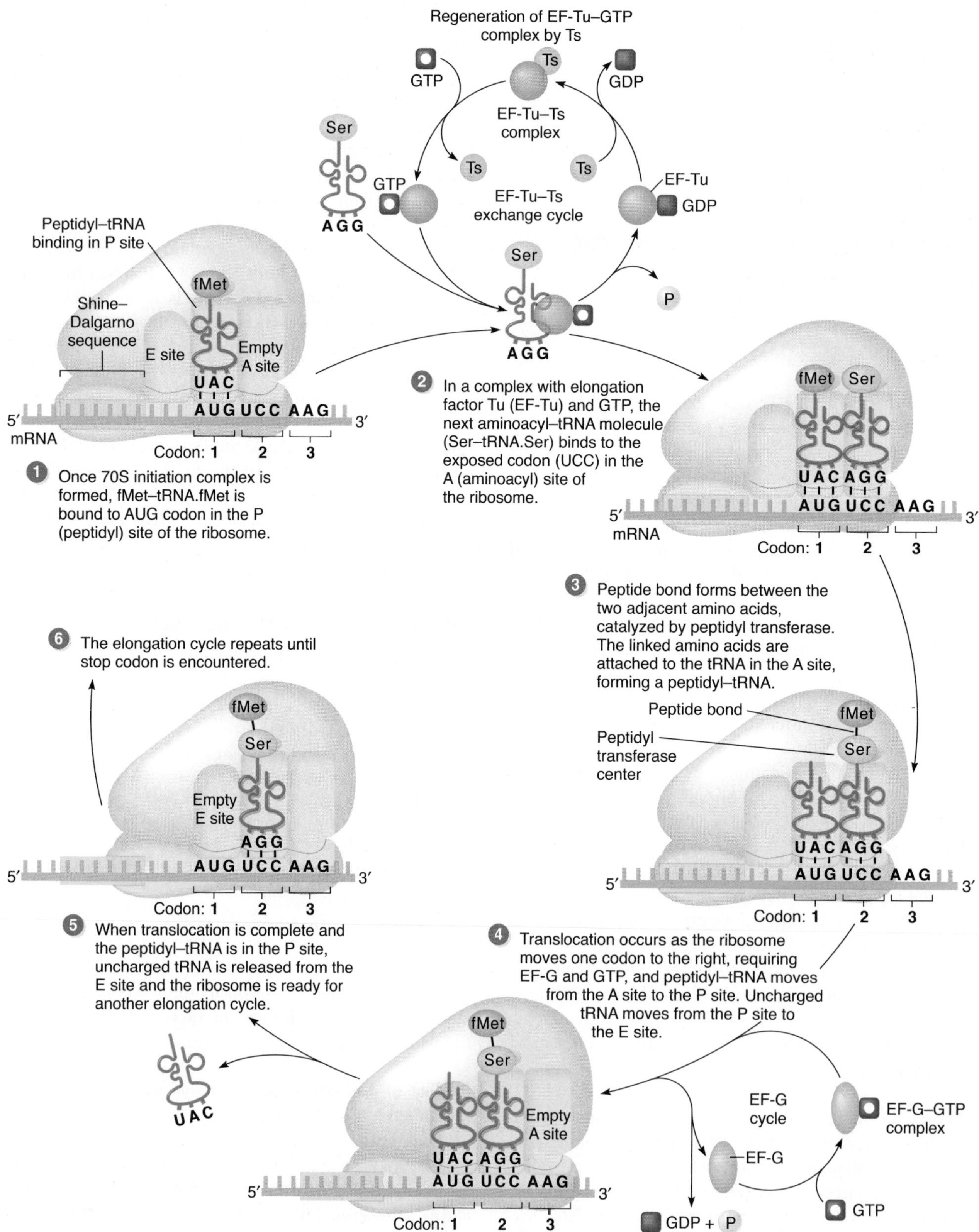

Regeneration of EF-Tu–GTP complex by Ts
GTP
Ts
GDP
EF-Tu–Ts complex
Ser
Ts
Ts
GTP
EF-Tu–Ts exchange cycle
EF-Tu
GDP
AGG
Peptidyl–tRNA binding in P site
Ser
fMet
P
Shine–Dalgarno sequence
E site
Empty A site
UAC
AGG
AUG UCC AAG
5′
3′
mRNA
Codon: 1 2 3
1 Once 70S initiation complex is formed, fMet–tRNA.fMet is bound to AUG codon in the P (peptidyl) site of the ribosome.
2 In a complex with elongation factor Tu (EF-Tu) and GTP, the next aminoacyl–tRNA molecule (Ser–tRNA.Ser) binds to the exposed codon (UCC) in the A (aminoacyl) site of the ribosome.
fMet Ser
UAC AGG
AUG UCC AAG
3 Peptide bond forms between the two adjacent amino acids, catalyzed by peptidyl transferase. The linked amino acids are attached to the tRNA in the A site, forming a peptidyl–tRNA.
Peptide bond
Peptidyl transferase center
6 The elongation cycle repeats until stop codon is encountered.
Empty E site
5 When translocation is complete and the peptidyl–tRNA is in the P site, uncharged tRNA is released from the E site and the ribosome is ready for another elongation cycle.
4 Translocation occurs as the ribosome moves one codon to the right, requiring EF-G and GTP, and peptidyl–tRNA moves from the A site to the P site. Uncharged tRNA moves from the P site to the E site.
UAC
EF-G cycle
EF-G–GTP complex
EF-G
GTP
GDP + P

factor EF-Tu and a molecule of GTP. When the aminoacyl-tRNA binds to the codon in the A site, GTP hydrolysis releases EF-Tu–GDP. As shown in Figure 6.17, step 2, EF-Tu is recycled. First, a second elongation factor, EF-Ts, binds to EF-Tu and displaces the GDP. Next, GTP binds to the EF-Tu–EF-Ts complex to produce an EF-Tu–GTP complex simultaneously with the release of EF-Ts. An aminoacyl-tRNA binds to the EF-Tu–GTP, and that complex can bind to the A site in a ribosome when the complementary codon is exposed. The process is highly similar in eukaryotes, with eEF-1A playing the role of EF-Tu, and eEF-1B playing the role of EF-Ts.

**Peptide Bond Formation.** The ribosome maintains the two aminoacyl–tRNAs in the P and A sites in the correct positions, so that a peptide bond can form between the two amino acids (Figure 6.17, step 3). Two steps are involved in the formation of this peptide bond (Figure 6.18). First, the bond between the amino acid and the tRNA in the P site is cleaved. In this case, the breakage is between the fMet and its tRNA. Second, the peptide bond is formed between the now-freed fMet and the Ser attached to the tRNA in the A site in a reaction catalyzed by **peptidyl transferase.**

For many years, this enzyme activity was thought to be a result of the interaction of a few ribosomal proteins of the 50S ribosomal subunit. However, in 1992, Harry Noller and his colleagues found that when most of the proteins of the 50S ribosomal subunit were removed, leaving only the ribosomal RNA, peptidyl transferase activity could still be measured. In addition, this activity was inhibited by the antibiotics chloramphenicol and carbomycin, both of which are known to inhibit peptidyl transferase activity specifically. Furthermore, when the rRNA was treated with ribonuclease T1, which degrades RNA but not protein, the peptidyl transferase activity was lost. These results suggested that the 23S rRNA molecule of the large ribosomal subunit is intimately involved with the peptidyl transferase activity and may in fact be that enzyme. In this case, the rRNA would be acting as a ribozyme (catalytic RNA; see Chapter 5, p. 95). From the structure of the large ribosomal subunit determined at high resolution, it has been deduced that the peptidyl transferase consists entirely of RNA. Ribosomal RNA also plays key roles in interacting with the tRNAs as they bind and release from the ribosome. Thus, in a reversal of what was once thought, the ribosomal proteins are the structural units that help organize the rRNA into key functional elements in the ribosomes.

Once the peptide bond has formed (see Figure 6.17, step 3), a tRNA without an attached amino acid (an uncharged tRNA) is left in the P site. The tRNA in the A site, now called peptidyl–tRNA, has the first two amino acids of the polypeptide chain attached to it—in this case, fMet–Ser.

**Translocation.** In the last step in the elongation cycle, **translocation** (Figure 6.17, step 4), the ribosome moves one codon along the mRNA toward the 3′ end. In bacteria, translocation requires the activity of another protein



**Figure 6.18**

**The formation of a peptide bond between the first two amino acids (fMet and Ser) of a polypeptide chain is catalyzed on the ribosome by peptidyl transferase.**

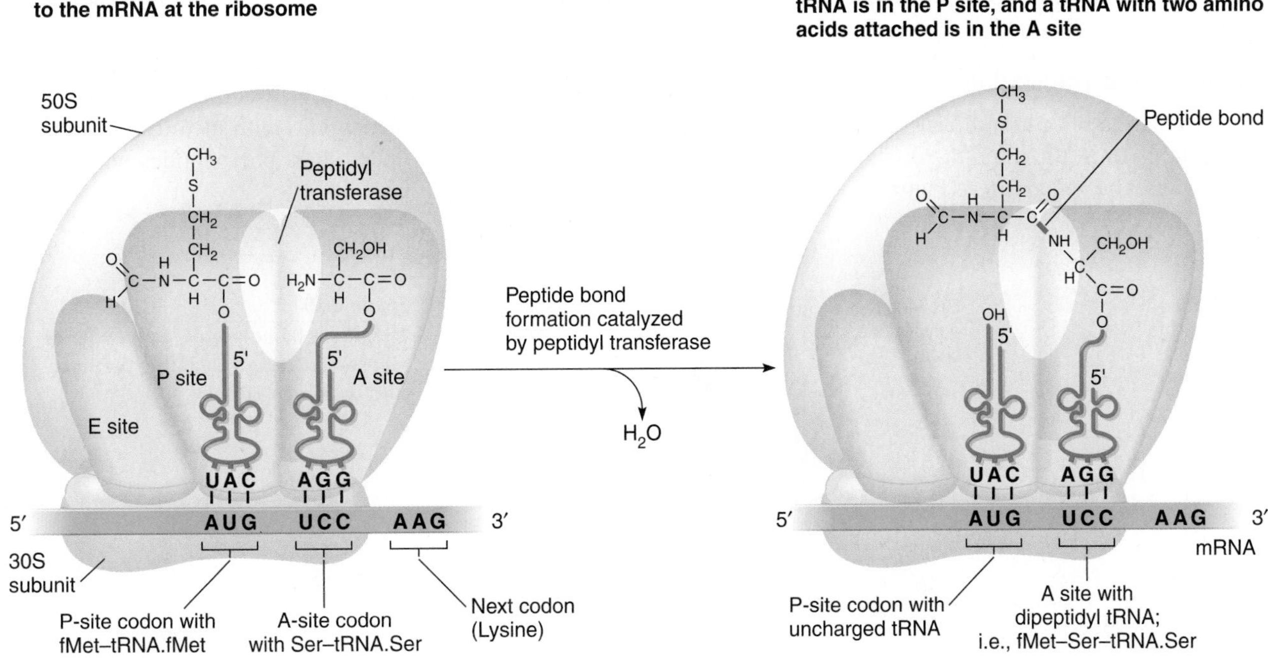

elongation factor, EF-G. An EF-G–GTP complex binds to the ribosome, GTP is hydrolyzed, and translocation of the ribosome occurs along with displacement of the uncharged tRNA away from the P site. It is possible that GTP hydrolysis changes the structure of EF-G, which facilitates the translocation event. Translocation is similar in eukaryotes; the elongation factor in this case is eEF-2, which functions like bacterial EF-G.

The uncharged tRNA moves from the P site and then binds transiently to the E site in the 50S ribosomal subunit, blocking the next aminoacyl–tRNA from binding to the A site until translocation is complete and the peptidyl–tRNA is bound correctly in the P site. Once that has occurred, the uncharged tRNA is then released from the ribosome. After translocation, EF-G is released and then reused, as shown in Figure 6.17, step 4. During the translocation step, the peptidyl–tRNA remains attached to its codon on the mRNA; and because the ribosome has moved, the peptidyl–tRNA is now located in the P site (hence the name *peptidyl site*).

After the completion of translocation, the A site is vacant. An aminoacyl–tRNA with the correct anticodon binds to the newly exposed codon in the A site, reiterating the process already described. The whole process is repeated until translation terminates at a stop codon (Figure 6.17, step 5).

In both bacteria and eukaryotes, once the ribosome moves away from the initiation site on the mRNA, another initiation event occurs. The process is repeated until, typically, several ribosomes are translating each mRNA simultaneously. The complex between an mRNA molecule and all the ribosomes that are translating it simultaneously is called a **polyribosome,** or **polysome** (Figure 6.19). Each ribosome in a polysome translates the entire mRNA and produces a single, complete polypeptide. Polyribosomes enable a large number of polypeptides to be produced quickly and efficiently from a single mRNA.

### Termination of Translation

The termination of translation is signaled by one of three stop codons (UAG, UAA, and UGA), which are the same in prokaryotes and eukaryotes (Figure 6.20, step 1). The stop codons do not code for any amino acid, so no tRNAs in the cell have anticodons for them. The ribosome recognizes a stop codon with the help of proteins called **release factors (RF),** which have shapes mimicking that of a tRNA including regions that read the codons (Figure 6.20, step 2) and then initiate a series of specific termination events. In *E. coli*, there are three RFs, two of which read the stop codons: RF1 recognizes UAA and UAG, and RF2 recognizes UAA and UGA—RF1 is shown binding to UAG in the figure. The binding of RF1 or RF2 to a stop codon triggers peptidyl transferase to cleave the polypeptide from the tRNA in the P site (Figure 6.20, step 3). The polypeptide then leaves the ribosome.

**animation**
**Termination of Translation**

Next, RF3–GDP binds to the ribosome, stimulating the release of the RF from the stop codon and the ribosome (Figure 6.20, step 4). GTP now replaces the GDP on RF3, and RF3 hydrolyses the GTP, which allows RF3 to be released from the ribosome.

An additional important step is the deconstruction of the remaining complex of ribosomal subunits, mRNA, and uncharged tRNA so that the ribosome and tRNA may be recycled. In *E. coli*, **ribosome recycling factor (RRF)**—the shape of which mimics that of a tRNA—binds to the A site (Figure 6.20, step 5). Then EF-G binds, causing translocation of the ribosome and thereby moving RRF to the P site and the uncharged tRNA to the E site (Figure 6.20, step 6). The RRF releases the uncharged tRNA, and EF-G releases RRF, causing the two ribosomal subunits to dissociate from the mRNA (Figure 6.20, step 7).

In eukaryotes, the termination process is similar to that in bacteria. In this case, a single release factor—eukaryotic release factor 1 (eRF1)—recognizes all three stop codons, and eRF3 stimulates the termination events. Ribosome recycling occurs in eukaryotes, but there is no equivalent of RRF.

As mentioned earlier, a polypeptide folds during the translation process. Box 6.1 discusses recent research showing that two polypeptides with identical amino acid sequences can fold to produce polypeptides with different structures and functions.

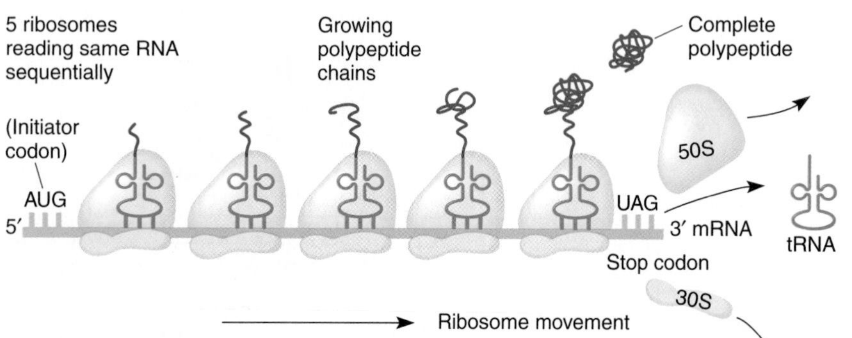


**Figure 6.19**
**Diagram of a polysome—a number of ribosomes, each translating the same mRNA sequentially.**

**Figure 6.20**

**Termination of translation.** The ribosome recognizes a chain termination codon (UAG) with the aid of release factors. A release factor reads the stop codon, initiating a series of specific termination events leading to the release of the completed polypeptide. Subsequently, the ribosomal subunits, mRNA, and uncharged tRNA separate. In bacteria, this event is stimulated by ribosome recycling factor (RRF) and EF-G.

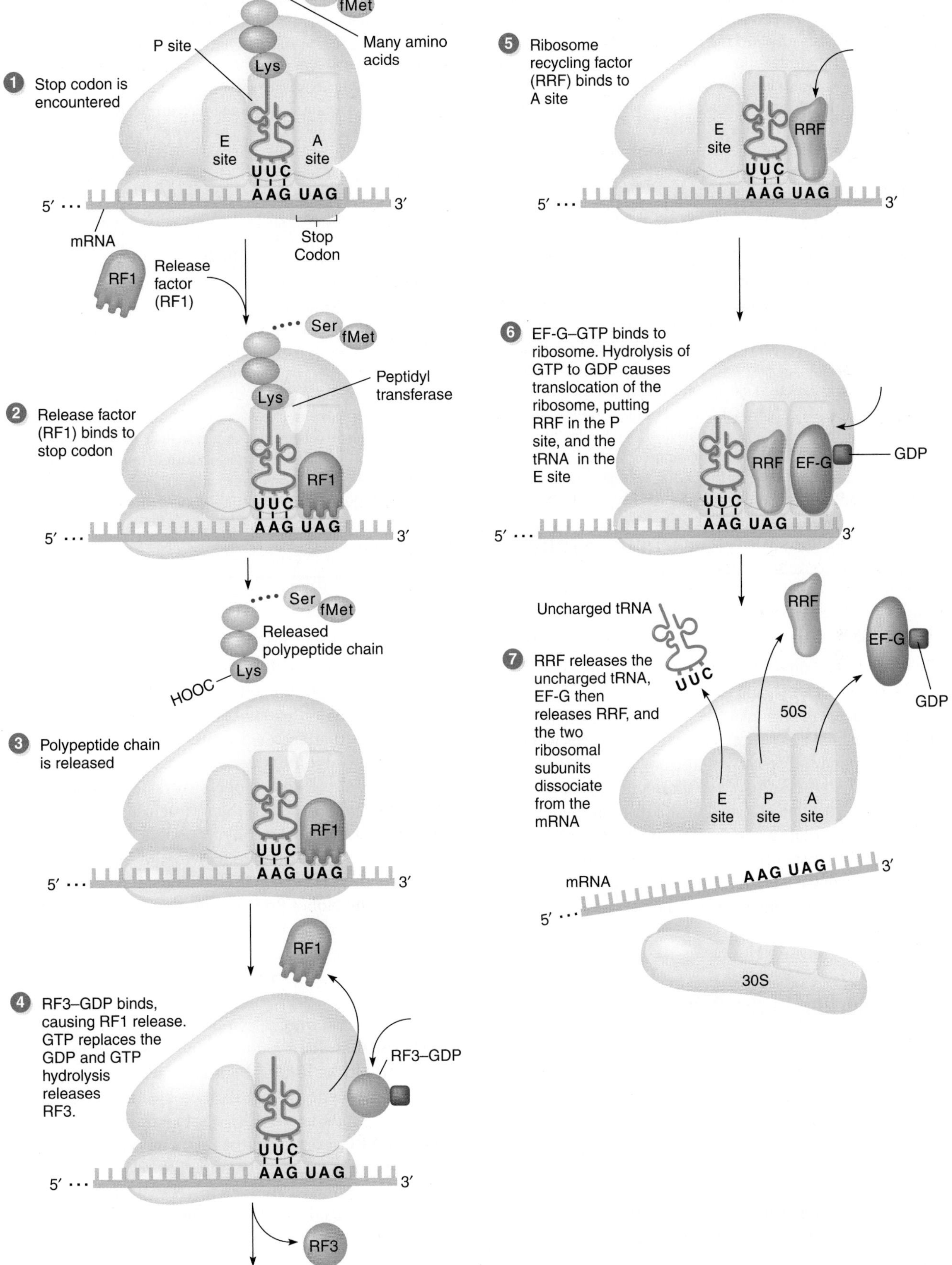

## Box 6.1 Same Amino Acid Sequence, Different Structures and Functions

We have learned in this chapter that the amino acid sequence of a polypeptide is determined by the sequence of codons in the mRNA which, in turn, is specified by the base-pair sequence of the protein-coding region of the gene. We also learned that the amino acid sequence of a polypeptide governs how the polypeptide folds and, hence, determines the three-dimensional, functional form of the polypeptide. Scientists have believed this to be true for decades. However, new research has shown that it is possible for two polypeptides with identical amino acids sequences to fold into different conformations and, therefore, to have different functions. How can that occur? One of the features we discussed for the genetic code (Figure 6.7) is degeneracy, in which, for most amino acids, more than one codon specifies the same amino acid. Thus, a base-pair change in the protein-coding region of a gene could change a codon in the mRNA to one that specifies the same amino acid. Such a base-pair mutation is called a *silent mutation*, and the new codon in this case is said to be *synonymous* to the wild-type codon. While the two codons specify the same amino acid, they could have different effects on translation. That is, aminoacyl–tRNA molecules are not all equally abundant. If the synonymous codon is read by a relatively rare aminoacyl–tRNA while the wild-type codon is read by a common aminoacyl–tRNA, then the rate of translation through the codon will be slower for the mutant mRNA compared with the wild-type mRNA. Why should that matter? We learned in the chapter that polypeptide folding is not solely a property of the polypeptide itself. Rather, accessory proteins such as chaperones often are involved. And, the folding process occurs cotranslationally—that is, during translation, rather than after the polypeptide is completed. About 20 years ago, some researchers hypothesized that the rates at which regions of some polypeptides are translated in the cell affect the ways in which those polypeptides fold. Certainly it is known that the rate of ribosome movement along a particular mRNA is not constant. Now, some recent research has produced results supporting the hypothesis. The researchers studied two different silent mutations in the human *MDR1* (multidrug resistance 1) gene. This gene encodes a membrane transporter protein called P-glycoprotein. This protein acts as a pump to transport various drugs out of cells. The extent to which it functions therefore can alter the efficiency of particular drug treatments, including certain chemotherapy treatments. Each of the silent mutations changed a codon from one read rapidly during translation to one read slowly. The P glycoproteins produced in the mutant cells were shown to have different structures compared with the wild-type protein, in particular showing alterations in binding sites for drugs and inhibitors. Thus, indeed, polypeptides with the same amino acid sequence can fold differently during their translation, producing polypeptides with different structures and functions. This means that silent mutations could affect the progression of diseases, and they could also affect how patients respond to drug treatments.

### Keynote

Translation is a complicated process requiring many RNAs, protein factors, and energy. The AUG (methionine) initiator codon signals the start of translation in prokaryotes and eukaryotes. Elongation proceeds when a peptide bond forms between the amino acid attached to the tRNA in the A site of the ribosome and the growing polypeptide attached to the tRNA in the P site. Translocation occurs when the now-uncharged tRNA in the P site is released from the ribosome and the ribosome moves one codon down the mRNA. Termination occurs as a result of the interaction of a protein release factor with a stop codon.

## Protein Sorting in the Cell

In bacteria and eukaryotes, some proteins may be secreted; and in eukaryotes, some other proteins must be placed in different cell compartments, such as the nucleus, a mitochondrion, a chloroplast, and a lysosome. The sorting of proteins to their appropriate compartments is under genetic control, in that specific "signal" or "leader" sequences on the proteins direct them to the correct organelles. Similarly, in bacteria, certain proteins become localized in the membrane and others are secreted.

Let us consider briefly how proteins are secreted from a eukaryotic cell. Such proteins are passaged through the endoplasmic reticulum (ER) and Golgi apparatus. In 1975, Günther Blobel, B. Dobberstein, and colleagues found that secreted proteins and other proteins sorted by the Golgi initially contain extra amino acids at the amino terminal end. Blobel's work led to the **signal hypothesis,** which states that proteins sorted by the Golgi bind to the ER by a hydrophobic amino terminal extension (the **signal sequence**) to the membrane that is subsequently removed and degraded (Figure 6.21). Blobel won the Nobel Prize in Physiology or Medicine in 1999 for this work.

The signal sequence of a protein destined for the ER consists of about 15 to 30 N-terminal amino acids. When the signal sequence is produced by translation and exposed on the ribosome surface, a cytoplasmic **signal recognition particle** (**SRP,** an RNA–protein complex) binds to the sequence and blocks further translation of the mRNA until the growing polypeptide–SRP–ribosome–mRNA complex reaches and binds to the ER (see Figure 6.21). The SRP binds to an **SRP receptor** in the ER membrane, causing the firm binding of the ribosome to the ER, release of the SRP, and the resumption of translation. The growing polypeptide extends through the ER membrane into the cisternal space of the ER.

Figure 6.21
**Model for the translocation of proteins into the endoplasmic reticulum in eukaryotes.**

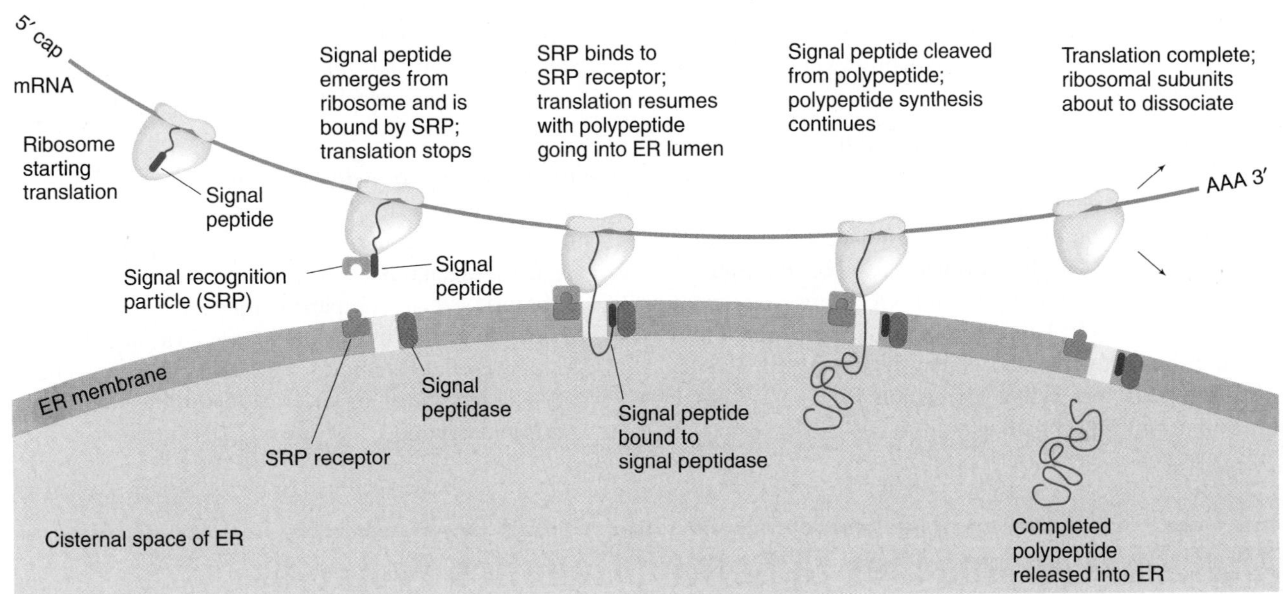


Once the signal sequence is fully into the cisternal space of the ER, it is removed from the polypeptide by the enzyme **signal peptidase.** When the complete polypeptide is entirely within the ER cisternal space, it is typically modified by the addition of specific carbohydrate groups to produce *glycoproteins*. The glycoproteins are then transferred in vesicles to the Golgi apparatus, where most of the sorting occurs. Proteins destined to be secreted, for example, are packaged into secretory storage vesicles, which migrate to the cell surface, where they fuse with the plasma membrane and release their packaged proteins to the outside of the cell.

**Keynote**

Eukaryotic proteins that enter the endoplasmic reticulum, have signal sequences at their N-terminal ends, which target them to that organelle. The signal sequence first binds to a signal recognition particle (SRP), arresting translation. The complex then binds to an SRP receptor in the outer ER membrane, translation resumes, and the polypeptide is translocated into the cisternal space of the ER. Once in the ER, the signal sequence is removed by signal peptidase. The proteins are then sorted to their final destinations by the Golgi complex.

## Summary

- A protein consists of one or more subunits called polypeptides, which are composed of smaller building blocks called amino acids. The amino acids are linked together in the polypeptide by peptide bonds.
- The amino acid sequence of a protein (its primary structure) determines its secondary, tertiary, and quaternary structures and, in most cases, its functional state.
- The genetic code is a triplet code in which each three-nucleotide codon in an mRNA specifies one amino acid or translation termination. Some amino acids are represented by more than one codon. Three codons are used for termination of polypeptide synthesis during translation. The code is almost universal, and it is read without gaps in successive, nonoverlapping codons.
- An mRNA is translated into a polypeptide chain on ribosomes. Amino acids for polypeptide synthesis come to the ribosome on tRNA molecules. The correct amino acid sequence is achieved by specific binding of each amino acid to its specific tRNA and by specific binding between the codon of the mRNA and the complementary anticodon of the tRNA.
- In bacteria and eukaryotes, AUG (methionine) is the initiator codon for the start of translation. In bacteria, the initiation of protein synthesis requires a sequence upstream of the AUG codon, to which the small ribosomal subunit binds. This sequence is the Shine–Dalgarno sequence, which binds specifically to the 3′ end of the 16S rRNA of the small ribosomal subunit, thereby associating the small subunit with the mRNA. No functionally equivalent sequence occurs in eukaryotic mRNAs; instead, the ribosomes load onto the mRNA at its 5′ end and scan toward the 3′ end, initiating translation at the first AUG codon.

- In both bacteria and eukaryotes, the initiation of polypeptide synthesis requires protein factors called initiation factors (IF). Bound to the ribosome–mRNA complex during the initiation phase, IFs dissociate once the polypeptide chain has been started.
- Elongation of the protein chain involves peptide bond formation between the amino acid on the tRNA in the A site of the ribosome and the growing polypeptide on the tRNA in the adjacent P site. Once the peptide bond has formed, the ribosome translocates one codon along the mRNA in preparation for the next tRNA. The incoming tRNA with its amino acid binds to the next codon occupying the A site. Protein factors called elongation factors (EF) play important roles in elongation.
- Translation continues until a stop codon (UAG, UAA, or UGA) is reached in the mRNA. These codons are read by release factor proteins and then the polypeptide is released from the ribosome. Subsequently, the other components of the protein synthesis machinery dissociate and are recycled in other translation events.
- In eukaryotes, proteins are found free in the cytoplasm and in various cell compartments, such as the nucleus, mitochondria, chloroplasts, and secretory vesicles. Mechanisms exist that sort proteins to their appropriate cell compartments. For example, proteins that are to be secreted have N-terminal signal sequences that facilitate their entry into the endoplasmic reticulum for later sorting in the Golgi apparatus and beyond.

# Analytical Approaches to Solving Genetics Problems

**Q6.1**

**a.** How many of the 64 codons can be made from the three nucleotides A, U, and G?

**b.** How many of the 64 codons can be made from the four nucleotides A, U, G, and C with one or more Cs in each codon?

**A6.1**

**a.** This question involves probability. There are four bases, so the probability of a cytosine at the first position in a codon is $^1/_4$. Conversely, the probability of a base other than cytosine in the first position is $(1 - {}^1/_4) = {}^3/_4$. These same probabilities apply to the other two positions in the codon. Therefore, the probability of a codon without a cytosine is $({}^3/_4)^3 = {}^{27}/_{64}$.

**b.** This question involves the relative frequency of codons that have one or more cytosines. We have already calculated the probability of a codon not having a cytosine, so all the remaining codons have one or more cytosines. The answer to this question, therefore, is $(1 - {}^{27}/_{64}) = {}^{37}/_{64}$.

**Q6.2** Random copolymers were used in some of the experiments directed toward deciphering the genetic code. For each of the following ribonucleotide mixtures, give the expected codons and their frequencies, and give the expected proportions of the amino acids that would be found in a polypeptide directed by the copolymer in a cell-free protein-synthesizing system:

**a.** 2 U : 1 C

**b.** 1 U : 1 C : 2 G

**A6.2**

**a.** The probability of a U at any position in a codon is $^2/_3$, and the probability of a C at any position in a codon is $^1/_3$. Thus, the codons, their relative frequencies, and the amino acids for which they code are as follows:

UUU = (2/3)(2/3)(2/3) = 8/27 = 0.296 = 29.6% Phe
UUC = (2/3)(2/3)(1/3) = 4/27 = 0.148 = 14.8% Phe
UCC = (2/3)(1/3)(1/3) = 2/27 = 0.0741 = 7.41% Ser
UCU = (2/3)(1/3)(2/3) = 4/27 = 0.148 = 14.8% Ser
CUU = (1/3)(2/3)(2/3) = 4/27 = 0.148 = 14.8% Leu
CUC = (1/3)(2/3)(1/3) = 2/27 = 0.0741 = 7.41% Leu
CCU = (1/3)(1/3)(2/3) = 2/27 = 0.0741 = 7.41% Pro
CCC = (1/3)(1/3)(1/3) = 1/27 = 0.037 = 3.7% Pro

In sum, we have 44.4% Phe, 22.21% Ser, 22.21% Leu, and 11.11% Pro. (The total does not quite add up to 100%, because of rounding.)

**b.** The probability of a U at any position in a codon is $^1/_4$, the probability of a C at any position in a codon is $^1/_4$, and the probability of a G at any position in a codon is $^1/_2$. Thus, the codons, their relative frequencies, and the amino acids for which they code are as follows:

UUU = (1/4)(1/4)(1/4) = 1/64 = 1.56% Phe
UUC = (1/4)(1/4)(1/4) = 1/64 = 1.56% Phe
UCU = (1/4)(1/4)(1/4) = 1/64 = 1.56% Ser
UCC = (1/4)(1/4)(1/4) = 1/64 = 1.56% Ser
CUU = (1/4)(1/4)(1/4) = 1/64 = 1.56% Leu
CUC = (1/4)(1/4)(1/4) = 1/64 = 1.56% Leu
CCU = (1/4)(1/4)(1/4) = 1/64 = 1.56% Pro
CCC = (1/4)(1/4)(1/4) = 1/64 = 1.56% Pro
UUG = (1/4)(1/4)(1/2) = 2/64 = 3.13% Leu
UGU = (1/4)(1/2)(1/4) = 2/64 = 3.13% Cys
UGG = (1/4)(1/2)(1/2) = 4/64 = 6.25% Trp
GUU = (1/2)(1/4)(1/4) = 2/64 = 3.13% Val
GUG = (1/2)(1/4)(1/2) = 4/64 = 6.25% Val
GGU = (1/2)(1/2)(1/4) = 4/64 = 6.25% Gly

GGG = (1/2)(1/2)(1/2) = 8/64 = 12.5% Gly
CCG = (1/4)(1/4)(1/2) = 2/64 = 3.13% Pro
CGC = (1/4)(1/2)(1/4) = 2/64 = 3.13% Arg
CGG = (1/4)(1/2)(1/2) = 4/64 = 6.25% Arg
GCC = (1/2)(1/4)(1/4) = 2/64 = 3.13% Ala
GCG = (1/2)(1/4)(1/2) = 4/64 = 6.25% Ala
GGC = (1/2)(1/2)(1/4) = 4/64 = 6.25% Gly
UCG = (1/4)(1/4)(1/2) = 2/64 = 3.13% Ser
UGC = (1/4)(1/2)(1/4) = 2/64 = 3.13% Cys
CUG = (1/4)(1/4)(1/2) = 2/64 = 3.13% Leu
CGU = (1/4)(1/2)(1/4) = 2/64 = 3.13% Arg
GUC = (1/2)(1/4)(1/4) = 2/64 = 3.13% Val
GCU = (1/2)(1/4)(1/4) = 2/64 = 3.13% Ala

In sum, 3.12% Phe, 6.25% Ser, 9.38% Leu, 6.25% Pro, 6.26% Cys, 6.25% Trp, 12.51% Val, 25% Gly, 12.51% Arg, 12.51% Ala.



## Questions and Problems

**6.1** Most genes encode proteins. What exactly is a protein, structurally speaking? List some of the functions of proteins.

***6.2** In each of the following cases stating how a certain protein is treated, indicate what level(s) of protein structure would change as the result of the treatment:

**a.** Hemoglobin is stored in a hot incubator at 80°C.
**b.** Egg white (albumin) is boiled.
**c.** RNase (a single-polypeptide enzyme) is heated to 100°C.
**d.** Meat in your stomach is digested (gastric juices contain proteolytic enzymes).
**e.** In the β-polypeptide chain of hemoglobin, the amino acid valine replaces glutamic acid at the number-six position.

***6.3** Bovine spongiform encephalopathy (BSE; mad cow disease) and the human version, Creutzfeldt–Jakob disease (CJD), are characterized by the deposition of amyloid—insoluble, nonfunctional protein deposits—in the brain. In these diseases, amyloid deposits contain an abnormally folded version of the prion protein. Whereas the normal prion protein has lots of α-helical regions and is soluble, the abnormally folded version has α-helical regions converted into β-pleated sheets and is insoluble. Curiously, small amounts of the abnormally folded version can trigger the conversion of an α-helix to a β-pleated sheet in the normal protein, making the abnormally folded version infectious.

**a.** Some cases of CJD may have arisen from ingesting beef having tiny amounts of the abnormally folded protein. What would you expect to find if you examined the primary structure of the prion protein in the affected tissues? What levels of protein structural organization are affected in this form of prion disease?
**b.** Answer the questions posed in part (a) for cases of CJD in which susceptibility to CJD is inherited due to a rare mutation in the gene for the prion protein.

***6.4** The form of genetic information used directly in protein synthesis is (choose the correct answer)

**a.** DNA.
**b.** mRNA.
**c.** rRNA.
**d.** tRNA.

**6.5** If codons were four bases long, how many codons would exist in a genetic code?

***6.6** What would the minimum word (codon) size need to be if, instead of four, the number of different bases in mRNA were

**a.** two?
**b.** three?
**c.** five?

**6.7** Suppose that, at stage A in the evolution of the genetic code, only the first two nucleotides in the coding triplets led to unique differences and that any nucleotide could occupy the third position. Then, suppose there was a stage B in which differences in meaning arose, depending on whether a purine (A or G) or pyrimidine (C or U) was present at the third position. Without reference to the number of amino acids or the multiplicity of tRNA molecules, how many triplets of different meaning can be constructed out of the code at stage A? at stage B?

***6.8** Key experiments indicating that the genetic code was a triplet code came from the work of Crick and his colleagues with proflavin-induced *rII* mutants in T4 phage. Answer the following questions to explore the reasoning behind Crick's experiments.

**a.** What types of DNA changes does proflavin induce? What are the effects of these mutations if they occur within a gene?
**b.** Suppose you expose $r^+$ T4 phage to proflavin, and infect the phage into *E. coli*. What type of *E. coli* would you infect the phage into to select for *rII* mutants? How would you know if you had recovered an *rII* mutant?
**c.** Suppose you isolate two proflavin-induced *rII* mutations at exactly the same site in the *rII* gene. Mutation $rII^X$ is caused by the insertion of one base pair (a + mutation), while mutation $rII^Y$ is caused by the deletion of one base pair (a − mutation). How would you select for revertants of these mutations?

**d.** Suppose you isolate five revertants of $rII^X$. Using a diagram, explain whether all of them are likely to affect the same DNA base pair.

**e.** A colleague in your lab analyzes your revertants, and tells you that none of them result from the deletion of the base pair that was inserted in the $rII^X$ mutation. Does this mean that all of the revertants are double mutants? If so, explain how a double mutant can have a $r^+$ phenotype.

**f.** Your colleague uses recombination (see Chapter 14) to separate the nucleotide changes induced in your revertants from the chromosome with the original $rII^X$ mutation, and gives you five phage, each of which has only the DNA change introduced by the reversion event. Will these phage show an *rII* phenotype, that is, are these phage *rII* mutants? If they are, what type of mutations are present in them, how would you select for revertants, and what type of additional mutation in a revertant would lead to an $r^+$ phenotype?

**g.** Your colleague uses recombination to combine the $rII^Y$ mutation with each of the five mutations that led to reversion of the $rII^X$ mutation. Explain whether the five double mutants she gives you will have an $r^+$ phenotype. If not, and you treat the double mutants with proflavin and select for revertants, what type of mutation would lead to an $r^+$ phenotype? Use diagrams in your answers.

**h.** Use diagrams to explain which of your answers in part (g) require the genetic code to be a triplet code. For example, could you recover proflavin-induced revertants in part (g) if the genetic code were not a triplet code?

***6.9** Random copolymers were used in some of the experiments that revealed the characteristics of the genetic code. For each of the following ribonucleotide mixtures, give the expected codons and their frequencies, and give the expected proportions of the amino acids that would be found in a polypeptide directed by the copolymer in a cell-free protein-synthesizing system:

**a.** 4 A : 6 C
**b.** 4 G : 1 C
**c.** 1 A : 3 U : 1 C
**d.** 1 A : 1 U : 1 G : 1 C

***6.10** Two populations of RNAs are made by the random combination of nucleotides. In population 1 the RNAs contain only A and G nucleotides (3 A : 1 G), whereas in population 2 the RNAs contain only A and U nucleotides (3 A : 1 U). In what ways other than amino acid content will the proteins produced by translating the population 1 RNAs differ from those produced by translating the population 2 RNAs?

**6.11** The term *genetic code* refers to the set of three-base code words (codons) in mRNA that stand for the 20 amino acids in proteins. What are the characteristics of the code?

**6.12** How do the structures of mRNA, rRNA, and tRNA differ? Hypothesize a reason for the difference.

***6.13** Match each term (1–4) with its corresponding description(s) in a–g, noting both that each term may have more than one description and each description may apply to more than one term.

1. Eukaryotic mRNAs
2. Prokaryotic mRNAs
3. Transfer RNAs
4. Ribosomal RNAs

**a.** _____ have a cloverleaf structure
**b.** _____ are synthesized by RNA polymerases
**c.** _____ display one anticodon each
**d.** _____ are the template of genetic information during protein synthesis
**e.** _____ contain exons and introns
**f.** _____ are of four types in eukaryotes and only three types in *E. coli*
**g.** _____ are capped on their 5′ end and polyadenylated on their 3′ end

**6.14** The structure and function of the rRNA and protein components of ribosomes have been investigated by separating those components from intact ribosomes and then using reconstitution experiments to determine which of the components are required for specific ribosomal activities.

**a.** Contrast the components of prokaryotic ribosomes with those of eukaryotic ribosomes.
**b.** What is the function of ribosomes, what steps are used by ribosomes to carry out that function, and which components of ribosomes are active in each step?

***6.15** A gene encodes a polypeptide 30 amino acids long containing an alternating sequence of phenylalanine and tyrosine. What are the sequences of nucleotides corresponding to this sequence in each of the following?

**a.** the DNA strand that is read to produce the mRNA, assuming that Phe = UUU and Tyr = UAU in mRNA
**b.** the DNA strand that is not read
**c.** tRNAs

***6.16** Base-pairing wobble occurs in the interaction between the anticodon of the tRNAs and the codons. On a theoretical level, determine the minimum number of tRNAs needed to read the 61 sense codons.

**6.17** A segment of a polypeptide chain is Arg-Gly-Ser-Phe-Val-Asp-Arg. It is encoded by the following segment of DNA:

```
—G G C T A G C T G C T T C C T T G G G G A—
 | | | | | | | | | | | | | | | | | | | | |
—C C G A T C G A C G A A G G A A C C C C T—
```

Which strand is the template strand? Label each strand with its correct polarity (5′ and 3′).

*6.18 Antibiotics have been highly useful in elucidating the steps of protein synthesis. If you have an artificial messenger RNA with the sequence AUGUUUUUUUUUUUUU. . ., it will produce the following polypeptide in a cell-free protein-synthesizing system: fMet–Phe–Phe–Phe . . . Suppose that, in your search for new antibiotics, you find one called putyermycin, which blocks protein synthesis. When you try it with your artificial mRNA in a cell-free system, the product is fMet–Phe. What step in protein synthesis does putyermycin affect? Why?

*6.19 One feature of the genetic code is that it is degenerate.

**a.** What do we mean when we say that the genetic code is degenerate?

**b.** Which amino acids have codons where a mutation in the first nucleotide can result in a synonymous codon? Which, and how many, codons show this property?

**c.** Which amino acids have codons where a mutation in the second nucleotide can result in a synonymous codon? Which, and how many, codons show this property?

**d.** Which amino acids have codons where a mutation in the third nucleotide never generates a synonymous codon? Which, and how many, codons show this property?

**e.** Calculate the fraction of sense codons that can be changed by a single nucleotide mutation to a synonymous codon. What does this tell you about the degree to which the genetic code is degenerate? What implications does this have?

**f.** Since silent mutations do not alter the amino acid inserted into a polypeptide chain, how might they alter gene function?

**6.20** As discussed in Box 6.1, organisms often show a preference for using one of the several codons that encode the same amino acid. By obtaining and analyzing the sequence of an entire genome (see Chapters 8 and 9), the amino acid composition of all of its proteins can be compared to the codons used in their synthesis, so that this *codon usage bias* can be tabulated. The following table gives the number of times particular codons for alanine and arginine are used in 1,611,503 codons found in a one strain of *E. coli*.

| Amino Acid | Codon | Usage |
|---|---|---|
| Alanine | GCU | 24,855 |
| | GCC | 40,571 |
| | GCA | 33,343 |
| | GCG | 52,091 |
| Arginine | CGU | 32,590 |
| | CGC | 33,547 |
| | CGA | 6,166 |
| | CGG | 9,955 |
| | AGA | 4,656 |
| | AGG | 2,915 |

The *E. coli* gene *ECs4312* makes a protein that functions during cell division. A researcher has hypothesized that the rate of synthesis of its protein affects the rate of cell division. He wants to test this hypothesis by replacing the wild-type gene with a modified version whose mRNA is translated more slowly and then measuring the rate of cell division. Part of the protein's amino acid sequence and the wild-type and two variant coding-strand nucleotide sequences, given 5′ to 3′, are shown below.

Amino acid sequence:
Arg Arg Arg Val Ser Ala Ala Leu

Wild-type nucleotide sequence:
CGC CGC CGG GUG UCG GCG GCA AUC

Nucleotide sequence variant 1:
AGG AGA AGG GUG UCG GCU GCA AUC

Nucleotide sequence variant 2:
CGA CGC CGG GUG UCG GCC GCC AUC

Using the data about codon usage bias, which nucleotide sequence variant should the researcher use in trying to diminish the rate of translation of the *ECs4312* mRNA? Explain your reasoning.

**6.21** In *E. coli*, a particular tRNA normally has the anticodon 5′-GGG-3′, but because of a mutation in the tRNA gene, the tRNA has the anticodon 5′-GGA-3′.

**a.** What codon would the normal tRNA recognize?

**b.** What codon would the mutant tRNA recognize?

*6.22 A protein found in *E. coli* normally has the N-terminal amino acid sequence Met–Val–Ser–Ser–Pro–Met–Gly–Ala–Ala–Met–Ser. . . A mutation alters the anticodon of a tRNA from 5′-GAU-3′ to 5′-CAU-3′. What would be the N-terminal amino acid sequence of this protein in the mutant cell? Explain your reasoning.

**6.23** The gene encoding an *E. coli* tRNA containing the anticodon 5′-GUA-3′ mutates so that the anticodon is now 5′-UUA-3′. What will be the effect of this mutation? Explain your reasoning.

**6.24** Describe the reactions involved in the aminoacylation (charging) of a tRNA molecule.

**6.25** If the initiating codon of an mRNA were altered by mutation, what might be the effect on the transcript?

**6.26** What differences are found in the initiation of protein synthesis between prokaryotes and eukaryotes? What differences are found in the termination of protein synthesis between prokaryotes and eukaryotes?

**6.27** Small protein factors that are not intrinsic parts of the ribosome are essential for each of the initiation, elongation, and termination stages of translation.

**a.** What protein factors are used in each of these stages in bacteria, and what functions do they serve?

**b.** In which stages of translation in eukaryotes are similar protein factors used? What are these factors?

**c.** In the stages of translation in eukaryotes where similar protein factors are not used, what protein factors are used and what functions do they serve?

***6.28** What is the evidence that the rRNA component of the ribosome serves more than a structural role?

***6.29** In Chapter 5, we saw that eukaryotic mRNAs are posttranscriptionally modified at their 5′ and 3′ ends. What role does each of these modifications play in translation?

**6.30** Translation is usually initiated at an AUG codon near the 5′ end of an mRNA, but mRNAs often have multiple AUG triplets near their 5′ ends. How is the initiation AUG codon correctly identified in prokaryotes? How is it correctly identified in eukaryotes?

***6.31** The following diagram shows the normal sequence of the coding region of an mRNA, along with six mutant versions of the same mRNA:

| | |
|---|---|
| Normal | AUGUUCUCUAAUUAC(...)AUGGGGUGGGUGUAG |
| Mutant *a* | AUGUUCUCUAAUUAG(...)AUGGGGUGGGUGUAG |
| Mutant *b* | AGGUUCUCUAAUUAC(...)AUGGCGUGGGUGUAG |
| Mutant *c* | AUGUUCUCGAAUUAC(...)AUGGCGUGCGUGUAG |
| Mutant *d* | AUGUUCUCUAAAUAC(...)AUGGGGUGGGUGUAG |
| Mutant *e* | AUGUUCUCUAAUUC(...)AUCGGGUGGGUGUAG |
| Mutant *f* | AUGUUCUCUAAUUAC(...)AUGGGGUGGGUGUCG |

Indicate what protein would be formed in each case, where (...) denotes a multiple of three unspecified bases.

**6.32** The following diagram shows the normal sequence of a particular protein, along with several mutant versions of it:

Normal: Met-Gly-Glu-Thr-Lys-Val-Val-...-Pro
Mutant 1: Met-Gly
Mutant 2: Met-Gly-Glu-Asp
Mutant 3: Met-Gly-Arg-Leu-Lys
Mutant 4: Met-Arg-Glu-Thr-Lys-Val-Val-...-Pro

For each mutant, explain what mutation occurred in the coding sequence of the gene, where (...) denotes a multiple of three unspecified bases.

**6.33** The N-terminus of a protein has the sequence Met-His-Arg-Arg-Lys-Val-His-Gly-Gly. A molecular biologist wants to synthesize a DNA chain that can encode this portion of the protein. How many DNA sequences can encode this polypeptide?

**6.34** In the recessive condition in humans known as sickle-cell anemia, the b-globin polypeptide of hemoglobin is found to be abnormal. The only difference between it and the normal b-globin is that the sixth amino acid from the N-terminal end is valine, whereas the normal b-globin has glutamic acid at this position. Explain how this amino acid substitution occurred in terms of differences in the DNA and the mRNA.

***6.35** Cystic fibrosis is an autosomal recessive disease in which the cystic fibrosis transmembrane conductance regulator (CFTR) protein is abnormal. The transcribed portion of the cystic fibrosis gene spans about 250,000 base pairs of DNA. The CFTR protein, with 1,480 amino acids, is translated from an mRNA of about 6,500 bases. The most common mutation in this gene results in a protein that is missing a phenylalanine at position 508 (DF508).

**a.** Why is the RNA coding sequence of this gene so much larger than the mRNA from which the CFTR protein is translated?

**b.** About what percentage of the mRNA together makes up 5′ untranslated leader, and 3′ untranslated trailer, sequences?

**c.** At the DNA level, what alteration would you expect to find in the DF508 mutation?

**d.** What consequences might you expect if the DNA alteration you describe in (c) occurred at random in the protein-coding region of the cystic fibrosis gene?

***6.36** The human *ADAM12* gene encodes a membrane-bound protein that functions in muscle and bone cell development. The N-terminal sequence of the protein encoded by the *ADAM12* mRNA is not identical to the N-terminal sequence of the polypeptide found in the cell membrane: the polypeptide found in the cell membrane is missing the first 28 amino acids of the polypeptide encoded by the mRNA. The following alignment is obtained when the two sequences are compared using the single-letter code for amino acids (see Figure 6.2).

mRNA-encoded sequence:

MAARPLPVSPARALLLALAGALLAPCEARGVSLWNQGRADEVVSAS...

polypeptide in membrane:

----------------------------RGVSLWNQGRADEVVSAS...

**a.** Explain why the N-terminal sequence of the polypeptide that is present within the cell membrane is not identical to the polypeptide encoded by its mRNA.

**b.** Suppose a small deletion occurred within the gene and, when an mRNA was synthesized, resulted in the elimination of codons for the amino acids PLPVSPARALLLALAGALL from the 5′ end of the mRNA. What effect would you expect this mutation to have on the subcellular distribution of the ADAM12 protein?

**6.37** All of the following steps are part of the process of gene expression in eukaryotes. Number them to reflect the approximate order in which each occurs during this process.

____ A complex of the 40S ribosomal subunit, an initiator Met–tRNA, several eIF proteins, and GTP scan for an AUG codon embedded within a Kozak sequence.

____ An intron is removed from the Val–pre-tRNA.

____ Poly(A) polymerase adds 200 A nucleotides onto the 3′ end of the mRNA.

____ Introns are removed from the mRNA by a spliceosome.

____ A specific aminoacyl–tRNA synthetase charges initiator Met–tRNA.

____ A specific aminoacyl–tRNA synthetase charges Val–tRNA.

____ An activator protein binds an enhancer.

____ eRF1 recognizes a nonsense codon.

____ Peptidyl transferase catalyzes the formation of a peptide bond.

____ The mRNA is transported out of the nucleus into the cytoplasm.

____ Cap-binding protein binds the 7-$^{m}$G cap at the 5′ end of the mRNA.

____ RNA Pol II initiates mRNA synthesis.

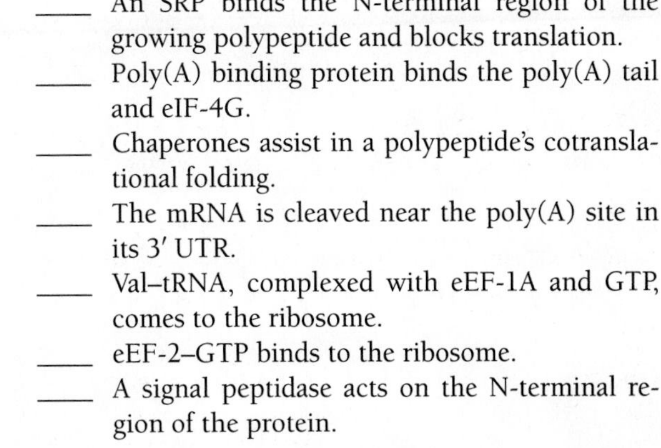

____ An SRP binds the N-terminal region of the growing polypeptide and blocks translation.

____ Poly(A) binding protein binds the poly(A) tail and eIF-4G.

____ Chaperones assist in a polypeptide's cotranslational folding.

____ The mRNA is cleaved near the poly(A) site in its 3′ UTR.

____ Val–tRNA, complexed with eEF-1A and GTP, comes to the ribosome.

____ eEF-2–GTP binds to the ribosome.

____ A signal peptidase acts on the N-terminal region of the protein.

*__6.38__ Antibiotics have been useful in determining whether cellular events depend on transcription or translation. For example, actinomycin D is used to block transcription, and cycloheximide is used (in eukaryotes) to block translation. In some cases, though, surprising results are obtained after antibiotics are administered. Adding actinomycin D, for example, may result in an increase, not a decrease, in the activity of a particular enzyme. Discuss how this result might come about.

# 7 DNA Mutation, DNA Repair, and Transposable Elements

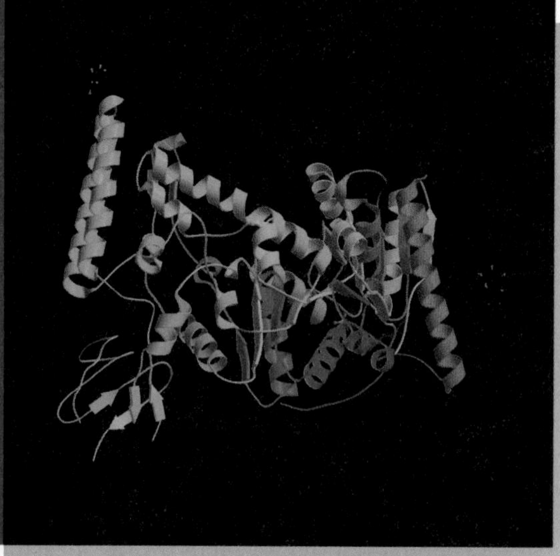

UvrB protein, a nucleotide excision repair enzyme.

## Key Questions

- Does genetic variation occur by adaptation or mutation?
- How do mutations affect polypeptide structure and function?
- How can mutations be reversed?
- How can mutations be induced in DNA?
- How can potential mutagens that are carcinogens be detected?
- How can mutants be detected?
- How is DNA damage repaired?
- What are transposable elements?
- How do transposable elements move between genome locations?
- What transposable elements are found in bacteria?
- What transposable elements are found in eukaryotes?

### iActivity

A MUTATION IN A GENE CAN LEAD TO A CHANGE in a phenotype. What types of mutations can occur in our DNA? And what effect do DNA mutations have on our health? In the first iActivity in this chapter, you will investigate the possible health hazards, including mutations, associated with contaminated ground water. In a second iActivity, you will examine another way that DNA can change. In the 1940s Barbara McClintock found that "jumping genes," or transposable elements, can create gene mutations, affect gene expression, and produce various types of chromosome mutations. In this iActivity, you will have the opportunity to explore further how a transposable element in *E. coli* moves from one location to another.

DNA can be changed in a number of ways, including through spontaneous changes, errors in the replication process, or the action of radiation or particular chemicals. We consider **chromosomal mutations**—changes involving whole chromosomes or sections of them—in Chapter 16. Another broad type of change in the genetic material is the **point mutation**, a change of one or a few base pairs. A point mutation may change the phenotype of the organism if it occurs within the coding region of a gene or in the sequences regulating the gene. Thus, the point mutations that have been of particular interest to geneticists are **gene mutations**, mutations which affect the function of genes. A gene mutation can alter the phenotype by changing the function of a protein, as illustrated in Figure 7.1. In this chapter, you will learn about some of the mechanisms that cause point mutations, some of the repair systems that can fix genetic damage, and some of the methods used to detect genetic mutants. As you learn about the specifics of point mutations, be aware that mutations are a major source of genetic variation in a species and therefore are important elements of the evolutionary process.

Genetic change also can occur when certain genetic elements in the chromosomes of prokaryotes and eukaryotes move from one location to another in the genome. These mobile genetic elements are known as **transposable elements**, because the term reflects the **transposition** (change in position) events associated

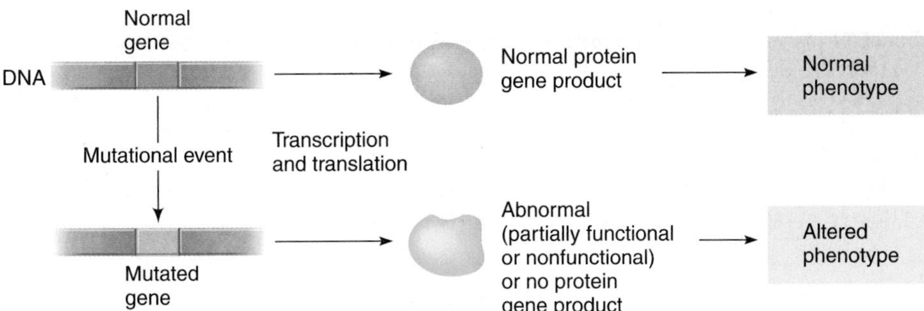


**Figure 7.1**
**Concept of a mutation in the protein-coding region of a gene.** (Note that not all mutations lead to altered proteins and that not all mutations are in protein-coding regions.)



with them. The discovery of transposable elements was a great surprise that altered our classic picture of genes and genomes and brought to our attention a new phenomenon to consider in developing theories about the evolution of genomes. In this chapter, you will learn about the nature of transposable elements and about how they move.

## DNA Mutation

### Adaptation versus Mutation

In the early part of the twentieth century, there were two opposing schools of thought concerning the variation in heritable traits. Some geneticists thought that variation among organisms resulted from random mutations that sometimes happened to be adaptive. Others thought that variations resulted from *adaptation*; that is, the environment induced an adaptive inheritable change. The adaptation theory was based on Lamarckism, the doctrine of the inheritance of acquired characteristics. Some observations made in experiments with bacteria fueled the debate. For instance, if a culture of wild-type *E. coli* started from a single cell is plated in the presence of an excess of the virulent bacteriophage T1, most of the bacteria are killed. However, a few survive and produce colonies because they are resistant to infection by T1. The resistance trait is heritable. Supporters of the adaptation theory argued that the resistance trait arose as a result of the presence of the T1 phage in the environment. Supporters of the mutation theory argued that mutations occur randomly such that, at any time in a large enough population of cells, some cells have mutated to make them resistant to T1 (in the example at hand), even though they have never been exposed to the bacteriophage. When T1 is subsequently added to the culture, the T1-resistant bacteria are selected for.

In 1943 Salvador Luria and Max Delbrück used the acquisition of resistance to T1 to determine whether the mutation mechanism or the adaptation mechanism was correct. They used the *fluctuation test*: Consider a dividing population of wild-type *E. coli* that started with a single cell (Figure 7.2). Assume that phage T1 is added at generation 4, when there are 16 cells. (This number is for illustration; in the actual experiment, the number of cells was much higher.) If the adaptation theory is correct, a certain proportion of the generation-4 cells will be induced *at that time* to become resistant to T1 (Figure 7.2a). Most importantly, *that proportion will be the same for all identical cultures, because adaptation would not commence until T1 was added.* However, if the mutation theory is correct, then the number of generation-4 cells that are resistant to T1 depends on when in the culturing process the random mutational event occurred that confers resistance to T1. If the mutational event occurs in generation 3 in our example, then 2 of the 16 cells in generation 4 will be T1 resistant (Figure 7.2b). However, if the mutational event occurs instead at generation 1, then 8 of the 16 generation-4 cells will be T1 resistant (Figure 7.2b). That is, if the mutation theory is correct, there should be a *fluctuation in the number of T1-resistant cells in generation 4 because the mutation to T1 resistance occurred randomly in the population and did not require the presence of T1.*

Luria and Delbrück observed a large fluctuation in the number of resistant colonies among identical cultures. Those results supported the mutation mechanism.

**Keynote**

Heritable adaptive traits result from random mutation, rather than by adaptation as a result of induction by environmental influences.

### Mutations Defined

**Mutation** is the process by which the sequence of base pairs in a DNA molecule is altered. A mutation may result in a change to either a DNA base pair or a chromosome.

A cell with a mutation is a mutant cell. If a mutation happens to occur in a somatic cell (in multicellular organisms), it is a **somatic mutation**—the mutant characteristic affects only the individual in which the mutation occurs and is not passed on to the succeeding generation. In contrast, a mutation in the germ line of sexually reproducing organisms—a **germ-line mutation**—may be transmitted by the gametes to the next generation, producing an individual with the mutation in both its somatic and its germ-line cells.

Figure 7.2

**Representation of a dividing population of T1 phage-sensitive wild-type *E. coli*.** At generation 4, T1 phage is added. (a) If the adaptation theory is correct, cells mutate only when T1 phage is added, so the proportions of resistant cells in duplicate cultures are the same. (b) If the mutation theory is correct, cells mutate independently of when T1 phage is added, so the proportions of resistant cells in duplicate cultures are different. *Left:* If one cell mutates to become resistant to T1 phage infection at generation 3, then 2 of the 16 cells at generation 4 are resistant to T1. *Right:* If one cell mutates to become resistant to T1 phage infection at generation 1, then 8 of the 16 cells at generation 4 are resistant to T1.



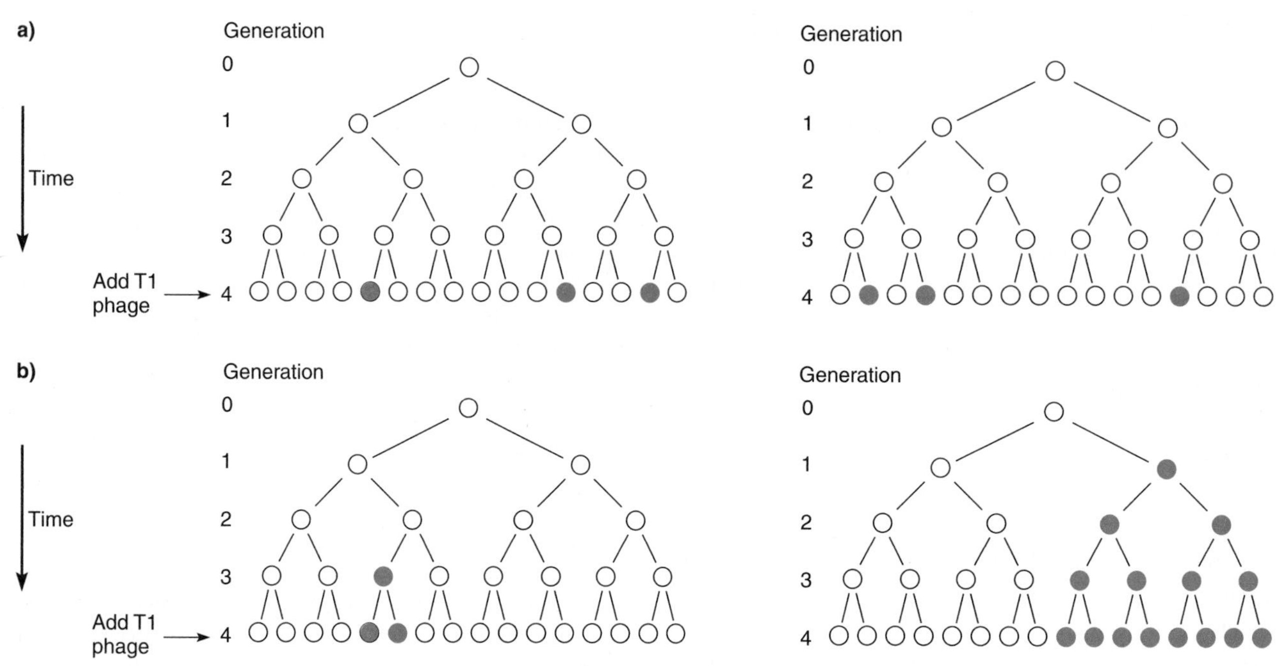


Two terms are used to give a quantitative measure of the occurrence of mutations. The **mutation rate** is the probability of a particular kind of mutation as a function of time, such as the number of mutations per nucleotide pair per generation, or the number per gene per generation. The **mutation frequency** is the number of occurrences of a particular kind of mutation, expressed as the proportion of cells or individuals in a population, such as the number of mutations per 100,000 organisms or the number per 1 million gametes.

**Types of Point Mutations.** Point mutations fall into two general categories: base-pair substitutions and base-pair insertions or deletions. A **base-pair substitution mutation** is a change from one base pair to another in DNA, and there are two general types. A **transition mutation** (Figure 7.3a) is a mutation from one purine–pyrimidine base pair to the other purine–pyrimidine base pair, such as A-T to G-C. Specifically, this means that the purine on one strand of the DNA (A in the example) is changed to the other purine, while the pyrimidine on the complementary strand (T, the base paired to the A) is changed to the other pyrimidine. A **transversion mutation** (Figure 7.3b) is a mutation from a purine–pyrimidine base pair to a pyrimidine–purine base pair, such as G-C to C-G, or A-T to C-G. Specifically, this means that the purine on one strand of the DNA (A in the second example) is changed to a pyrimidine (C in this example), while the pyrimidine on the complementary strand (T, the base paired to the A) is changed to the purine that base pairs with the altered pyrimidine (G in this example).

Base-pair substitutions in protein-coding genes also are defined according to their effects on amino acid sequences in proteins. Depending on how a base-pair substitution is translated via the genetic code, the mutations can result in no change to the protein, an insignificant change, or a noticeable change.

A **missense mutation** (Figure 7.3c) is a gene mutation in which a base-pair change causes a change in an mRNA codon so that a different amino acid is inserted into the polypeptide. A phenotypic change may or may not result, depending on the amino acid change involved. In Figure 7.3c, an AT-to-GC transition mutation changes the DNA from $\frac{5'\text{-AAA-}3'}{3'\text{-TTT-}5'}$ to $\frac{5'\text{-GAA-}3'}{3'\text{-CTT-}5'}$ by changing a base in the mRNA codon from one purine to the other purine. In this case the mRNA codon is changed from 5′-AAA-3′ (lysine) to 5′-GAA-3′ (glutamic acid).

A **nonsense mutation** (Figure 7.3d) is a gene mutation in which a base-pair change alters an mRNA codon for an amino acid to a stop (nonsense) codon

**Figure 7.3**

**Types of base-pair substitution mutations.** Transcription of the segment shown produces an mRNA with the sequence 5′. . . UCUCAAAAAUUUACG. . . 3′, which encodes . . . -Ser-Gln-Lys-Phe-Thr- . . .

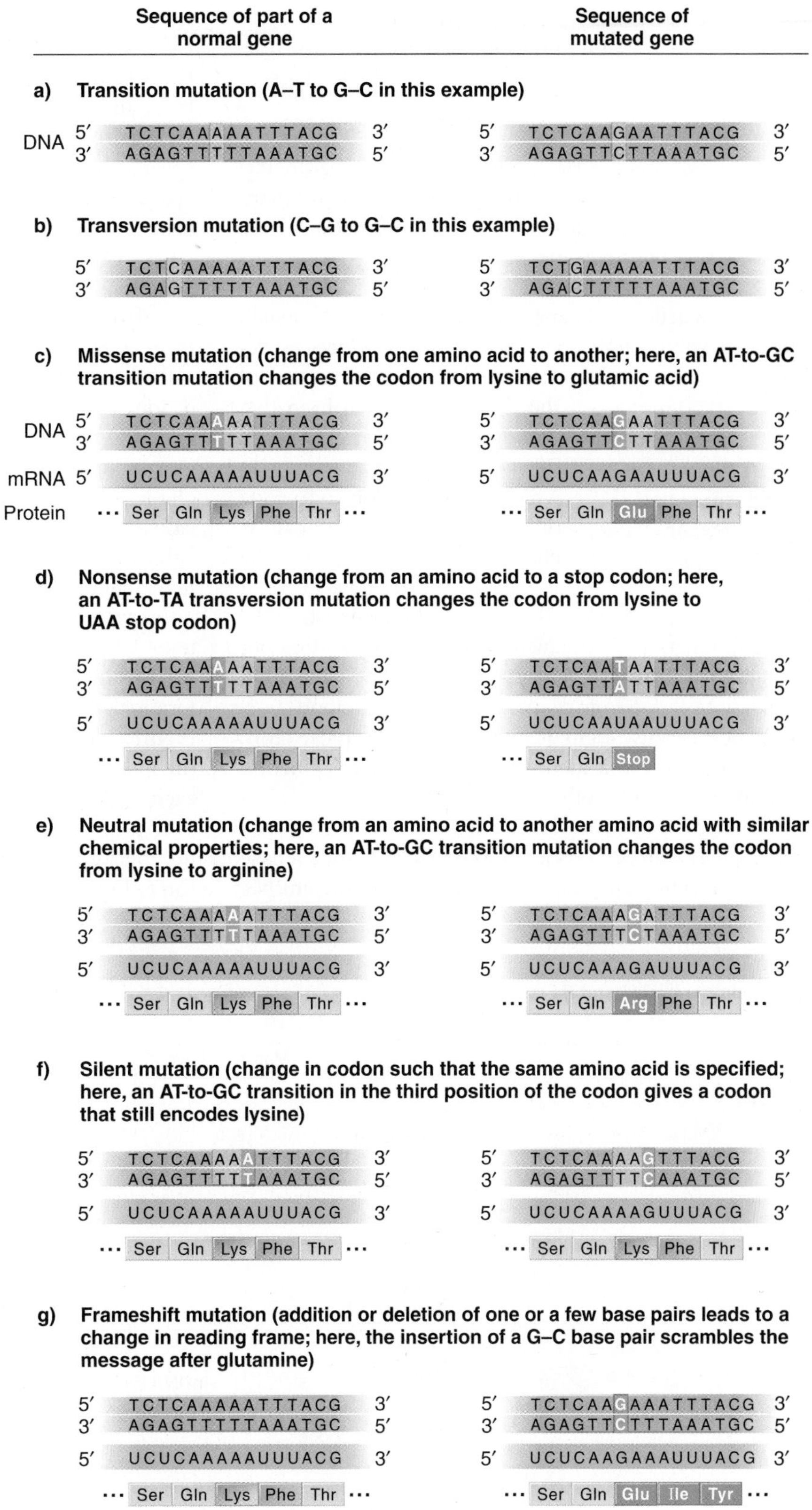

(UAG, UAA, or UGA). For example, in Figure 7.3d, an AT-to-TA transversion mutation changes the DNA from $\begin{matrix}\text{5}'\text{-AAA-3}' \\ \text{3}'\text{-TTT-5}'\end{matrix}$ to $\begin{matrix}\text{5}'\text{-TAA-3}' \\ \text{3}'\text{-ATT-5}'\end{matrix}$, and this changes the mRNA codon from 5′-AAA-3′ (lysine) to 5′-UAA-3′, which is a stop codon. A nonsense mutation causes premature termination of polypeptide chain synthesis, so shorter-than-normal polypeptide fragments (often nonfunctional) are released from the ribosomes (Figure 7.4).

A **neutral mutation** (Figure 7.3e) is a base-pair change in a gene that changes a codon in the mRNA such that the resulting amino acid substitution produces no detectable change in the *function* of the protein translated from that message. A neutral mutation is a subset of missense mutations in which the new codon codes for a different amino acid that is chemically equivalent to the original or the amino acid is not functionally important and therefore does not affect the protein's function. Consequently, the phenotype does not change. In Figure 7.3e, an AT-to-GC transition mutation changes the codon from 5′-AAA-3′ (lysine) to 5′-AGA-3′ (arginine). Because arginine and lysine have similar properties —both are basic amino acids—the protein's function may not alter significantly.

> **animation**
> **Nonsense Mutations and Nonsense Suppressor Mutations**

A **silent mutation** (Figure 7.3f)—also known as a *synonymous mutation*—is a mutation that changes a base pair in a gene, but the altered codon in the mRNA specifies the *same* amino acid in the protein. In this case, the protein obviously has a wild-type function. For example, in Figure 7.3f, a silent mutation results from an AT-to-GC transition mutation that changes the codon from 5′-AAA-3′ to 5′-AAG-3′, both of which specify lysine. Silent mutations most often occur by changes such as this at the third—wobble—position of a codon. This makes sense from the degeneracy patterns of the genetic code (see Figure 6.7 and Chapter 6, p. 109).

If one or more base pairs are added to or deleted from a protein-coding gene, the reading frame of an mRNA can change downstream of the mutation. An addition or deletion of one base pair, for example, shifts the mRNA's downstream reading frame by one base so that incorrect amino acids are added to the polypeptide chain after the mutation site. This type of mutation, called a **frameshift mutation** (Figure 7.3g), usually results in a nonfunctional protein. Frameshift mutations may generate new stop codons, resulting in a shortened polypeptide; they may result in longer-than-normal proteins because the normal stop codon is now in a different reading frame; or they may result in a significant alteration of the amino acid sequence of a polypeptide. In Figure 7.3g, an insertion of a G-C base pair scrambles the message after the codon specifying glutamine. Since each codon consists of three bases, a frameshift mutation is produced by the insertion or deletion of any number of base pairs in the DNA that is not divisible by three. Frameshift mutations were instrumental in scien-

**Figure 7.4**
**A nonsense mutation and its effect on translation.**

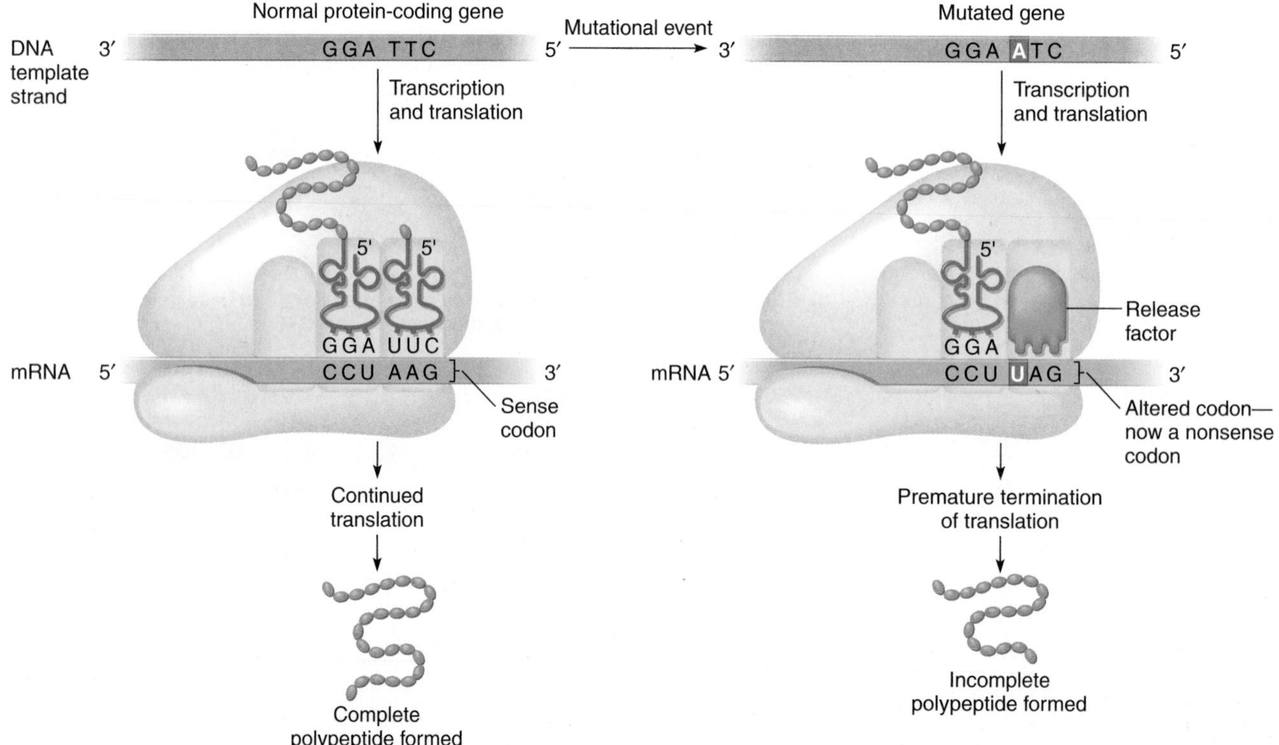

tists' determining that the genetic code is a triplet code (see Chapter 6, pp. 106–107).

In sum, mutations can be classified according to different criteria. That is, mutations are classified by their cause (spontaneous vs. induced), effect on DNA (point vs. chromosomal, substitution vs. insertion/deletion, transition vs. transversion) or by their effect on an encoded protein (nonsense, missense, neutral, silent, and frameshift).

**Keynote**

Mutation is the process by which the sequence of base pairs in a DNA molecule is altered. Mutations that affect a single base pair of DNA are called base-pair substitution mutations. Base-pair substitutions and single base-pair insertions or deletions are called point mutations. Mutations in the sequences of genes are called gene mutations.

**Reverse Mutations and Suppressor Mutations.** Point mutations are divided into two classes, based on how they affect the phenotype: (1) A **forward mutation** changes a wild-type gene to a mutant gene; and (2) a **reverse mutation** (also known as a **reversion** or **back mutation**) changes a mutant gene at the same site so that it functions in a completely wild-type or nearly wild-type way. Reversion of a nonsense mutation, for instance, occurs when a base-pair change results in a change of the mRNA nonsense codon to a codon for an amino acid. If this reversion is back to the wild-type amino acid, the mutation is a **true reversion**. If the reversion is to some other amino acid, the mutation is a **partial reversion,** and complete or partial function may be restored, depending on the change. Reversion of missense mutations occurs in the same way.

The effects of a mutation may be diminished or abolished by a **suppressor mutation**—a mutation at a different site from that of the original mutation. A suppressor mutation masks or compensates for the effects of the initial mutation, but it does not reverse the original mutation.

Suppressor mutations may occur within the same gene where the original mutations occurred, but at a different site (in which case they are known as **intragenic** [*intra* = within] **suppressors**), or they may occur in a different gene (where they are called **intergenic** [*inter* = between] **suppressors**). Both intragenic and intergenic suppressors operate to decrease or eliminate the deleterious effects of the original mutation. However, the mechanisms of the two suppressors are completely different.

Intragenic suppressors act by altering a different nucleotide in the same codon where the original mutation occurred or by altering a nucleotide in a different codon. An example of the latter is the suppression of a base-pair addition frameshift mutation by a nearby base-pair deletion (see Figure 6.5, p. 107).

Intergenic suppression is the result of a second mutation in another gene. Genes that cause the suppression of mutations in other genes are called **suppressor genes.** For example, in the case of nonsense suppressors, particular tRNA genes mutate so that their anticodons recognize a chain-terminating codon and put an amino acid into the chain. Thus, instead of polypeptide chain synthesis being stopped prematurely because of a nonsense mutation, the altered (suppressor) tRNA inserts an amino acid at that position, and full or partial function of the polypeptide is restored. This suppression process is not very efficient, but sufficient functional polypeptides are produced to reverse or partially reverse the phenotype.

There are three classes of nonsense suppressors, one for each of the stop codons UAG, UAA, and UGA. For example, if a gene for a tyrosine tRNA (which has the anticodon 3′-AUG-5′) is mutated so that the tRNA has the anticodon 3′-AUC-5′, the mutated suppressor tRNA (which still carries tyrosine) reads the nonsense codon 5′-UAG-3′. So, instead of chain termination occurring, tyrosine is inserted at that point in the polypeptide (Figure 7.5).

But there is a dilemma: If the suppressor tRNA.Tyr gene has mutated so that the encoded tRNA's anticodon can read a nonsense codon, it can no longer read the original codon that specifies the amino acid it carries. This turns out not to be a problem, because nonsense suppressor tRNA genes typically are produced by mutations of tRNA genes that are present in two or more copies in the genome. If there is a mutation in one of the genes to produce a suppressor tRNA, then the other gene(s) produce(s) a tRNA molecule that reads the normal Tyr codon.

**Keynote**

Reverse mutations occur at the same site as the original mutation and cause the genotype to change from mutant to wild type. A suppressor mutation is one that occurs at a second site and completely or partially restores a function that was lost or altered because of a primary mutation. Intragenic suppressors are suppressor mutations that occur within the same gene where the original mutation occurred, but at a different site. Intergenic suppressors are suppressor mutations that occur in a suppressor gene—a gene different from the one with the original mutation.

## Spontaneous and Induced Mutations

**Mutagenesis,** the creation of mutations, can occur spontaneously or can be induced. **Spontaneous mutations** are naturally occurring mutations. **Induced mutations** occur when an organism is exposed either deliberately or accidentally to a physical or chemical agent, known as a **mutagen,** that interacts with DNA to cause a mutation. Induced mutations typically occur at a much higher frequency than do spontaneous mutations and hence have been useful in genetic studies.

**Spontaneous Mutations.** All types of point mutations occur spontaneously. Spontaneous mutations can occur during DNA replication, as well as during other stages of cell growth and division. Spontaneous mutations also can

**Figure 7.5**

**Mechanism of action of an intergenic nonsense-suppressor mutation that results from the mutation of a tRNA gene.** In this example, a tRNA.Tyr gene has mutated so that the anticodon of the tRNA is changed from 3′-AUG-5′ to 3′-AUC-5′, which can read a UAG nonsense codon, inserting tyrosine in the polypeptide chain at that codon.

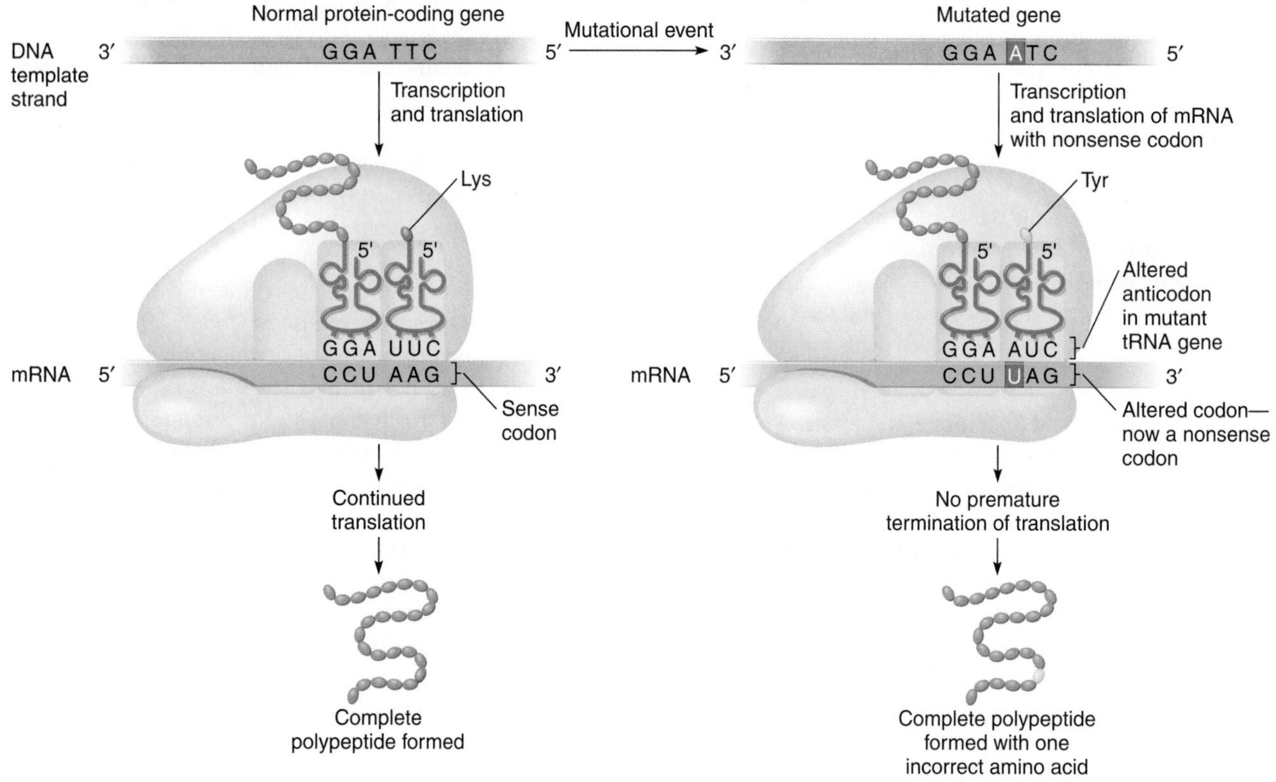




result from the movement of transposable genetic elements, a process you will learn about later in the chapter.

In humans, the spontaneous mutation rate for individual genes varies between $10^{-4}$ and $4 \times 10^{-6}$ per gene per generation. For eukaryotes in general, the spontaneous mutation rate is $10^{-4}$ to $10^{-6}$ per gene per generation, and for bacteria and phages the rate is $10^{-5}$ to $10^{-7}$ per gene per generation. (The spontaneous mutation frequencies at specific loci for various organisms are presented in Table 21.6, p. 623.) *These rates and frequency values represent the mutations that become fixed—heritable—in DNA.* Most spontaneous errors are corrected by cellular repair systems, which you will learn about later in this chapter; only some errors remain uncorrected as permanent changes.

**DNA Replication Errors.** *Base-pair substitution mutations*—point mutations involving a change from one base pair to another—can occur if mismatched base pairs form during DNA replication. Chemically, each base can exist in alternative states, called **tautomers.** When a base changes state, it has undergone a *tautomeric shift*. In DNA, the *keto* form of each base is usually found and is responsible for the normal Watson-Crick base pairing of T with A and C with G (Figure 7.6a). However, non–Watson-Crick base pairing can result if a base is in a rare tautomeric state, the *enol* form. Figure 7.6b and Figure 7.6c respectively show mismatched base pairs that can occur if purines are in their rare tautomeric states or if pyrimidines are in their rare tautomeric states.

Figure 7.7 illustrates how a mismatch caused by a base shifting to a rare tautomeric state can result in a mutation. Here, the rare form of T forms a mismatched base pair with G in the template strand of the DNA. If this mismatch is not repaired, a GC-to-AT transition mutation is produced after replication.

Small additions and deletions also can occur spontaneously during replication (Figure 7.8). They occur because of displacement—looping out—of bases from either the template or the growing DNA strand, generally in regions where a run of the same base or of a repetitive sequence is present. If DNA loops out from the template strand, DNA polymerase skips the looped-out base or bases, producing a deletion mutation; if DNA polymerase synthesizes an untemplated base or bases, the new DNA loops out from the template, producing an addition. An addition or deletion mutation in the coding region of a structural gene is a frameshift mutation if it involves other than 3 bp or a multiple of 3 bp.

DNA replication errors may be repairable by mismatch repair systems (see later in this chapter).

**Spontaneous Chemical Changes.** Depurination and deamination of particular bases are two common chemical events

**Figure 7.6**

**Normal Watson-Crick and non–Watson-Crick base pairing in DNA.**

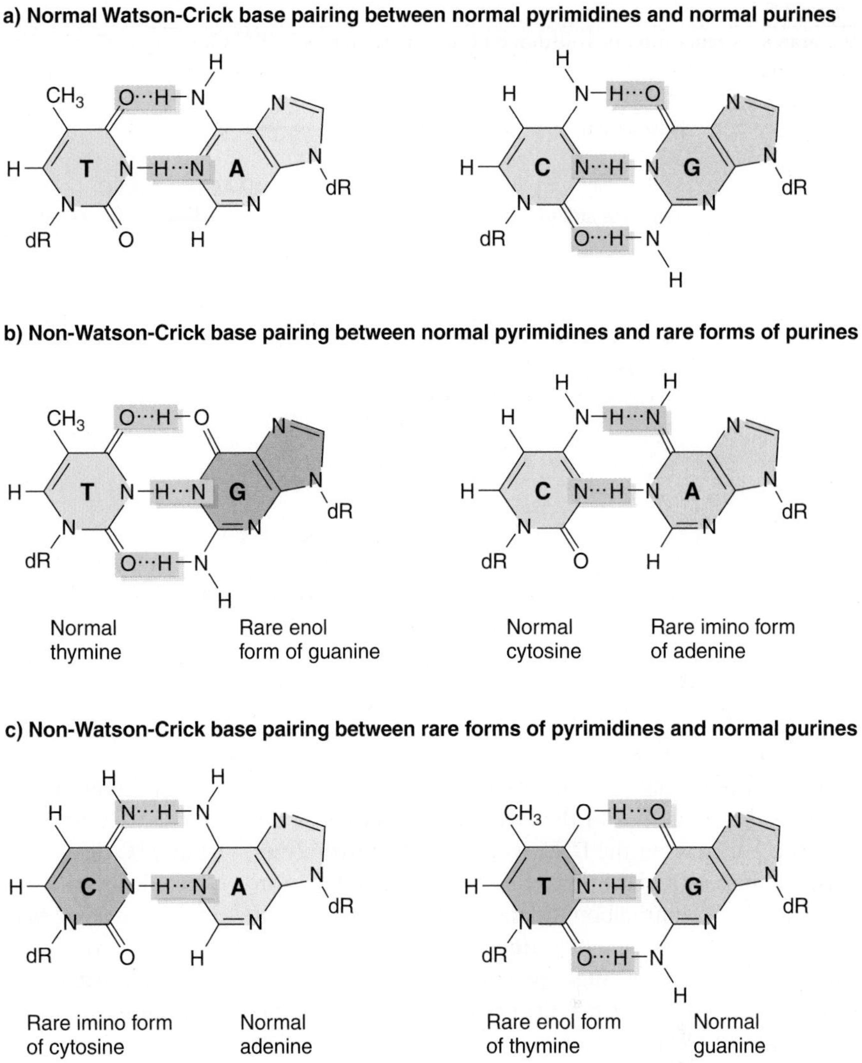

**a) Normal Watson-Crick base pairing between normal pyrimidines and normal purines**

**b) Non-Watson-Crick base pairing between normal pyrimidines and rare forms of purines**

Normal thymine — Rare enol form of guanine

Normal cytosine — Rare imino form of adenine

**c) Non-Watson-Crick base pairing between rare forms of pyrimidines and normal purines**

Rare imino form of cytosine — Normal adenine

Rare enol form of thymine — Normal guanine

**Figure 7.7**

**Production of a mutation as a result of a mismatch caused by non–Watson-Crick base pairing.** The details are explained in the text.

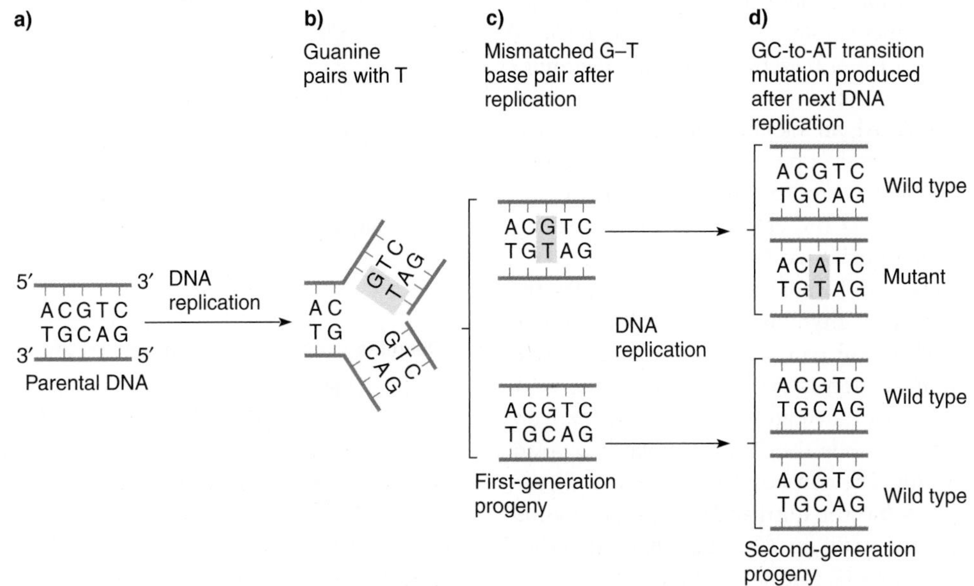

**Figure 7.8**

**Spontaneous generation of addition and deletion mutants by DNA looping-out errors during replication.**

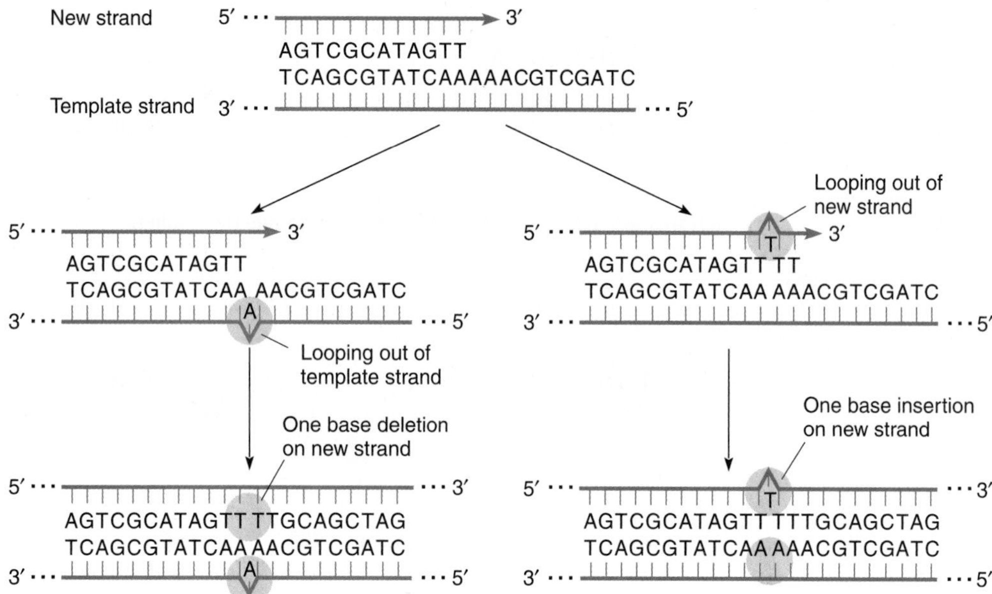


that produce spontaneous mutations. These events create lesions—damaged sites in the DNA. **Depurination** is the loss of a purine from the DNA when the bond hydrolyzes between the base and the deoxyribose sugar, resulting in an *apurinic site*. Depurination occurs because the covalent bond between the sugar and purine is much less stable than the bond between the sugar and pyrimidine and is very prone to breakage. A mammalian cell typically loses thousands of purines in an average cell generation period. If such lesions are not repaired, there is no base to specify a complementary base during DNA replication, and the DNA polymerase may stall or dissociate from the DNA.

**Deamination** is the removal of an amino group from a base. For example, the deamination of cytosine produces uracil (Figure 7.9a), which is not a normal base in DNA, although it is a normal base in RNA. A repair system replaces most of the uracils in DNA, thereby minimizing the mutational consequences of cytosine deamination. However, if the uracil is not replaced, an adenine will be incorporated into the new DNA strand opposite it during replication, eventually resulting in a CG-to-TA transition mutation.

DNA of both bacteria and eukaryotes contains small amounts of the modified base 5-methylcytosine ($5^{m}C$) (Figure 7.9b) in place of the normal base cytosine. Deamination of $5^{m}C$ produces thymine (Figure 7.9b), thereby changing the G-$5^{m}C$ base pair to the mismatched base pair, G-T. If the mismatch is not corrected, at the next replication cycle the G of the pair is the template for C on the new DNA strand, while the T is a template for A on the new DNA strand. The consequence is that one of the new DNA molecules has the normal G-C base pair, while the other is mutant, with an A-T base pair. In other words, deamination of $5^{m}C$ can result in a GC-to-AT transition mutation. Because significant proportions of other kinds of mutations are corrected by repair mechanisms, but $5^{m}C$ deamination mutations are less likely to be corrected, locations of $5^{m}C$ in the genome often appear as *mutational hot spots*—that is, nucleotides where a higher-than-average frequency of mutation occurs.

Depurination and deamination mutations may be repairable by base excision repair systems (see later in the chapter).

**Figure 7.9**

**Changes of DNA bases as a result of deamination.**

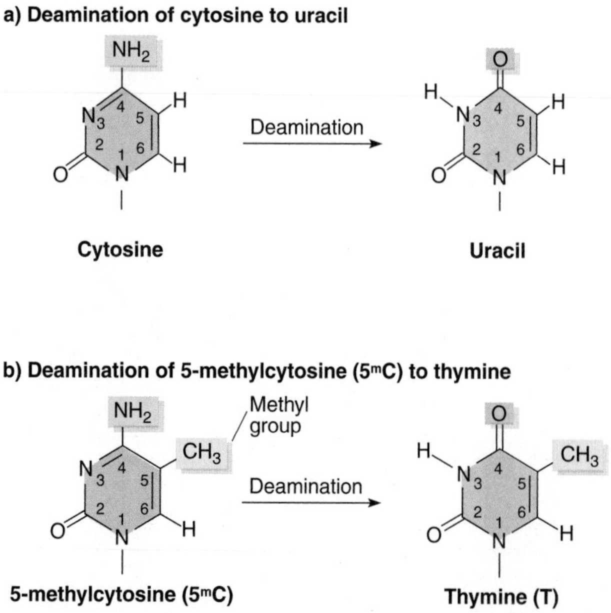

**a) Deamination of cytosine to uracil**

$NH_2$ — Deamination → — O

**Cytosine** **Uracil**

**b) Deamination of 5-methylcytosine ($5^{m}C$) to thymine**

Methyl group — $CH_3$ — Deamination

**5-methylcytosine ($5^{m}C$)** **Thymine (T)**

**Induced Mutations.** Mutations can be induced by exposing organisms to physical mutagens, such as radiation, or to chemical mutagens. Deliberately induced mutations have played, and continue to play, an important role in the study of mutations. Since the rate of spontaneous mutation is so low, geneticists use mutagens to increase the frequency of mutation so that a significant number of organisms have mutations in the gene being studied.

**Radiation.** All forms of life are exposed continuously to radiation. We are exposed to various sources of radiation. Among the natural sources are cosmic rays from space, radon, and radioactivity from decay of natural radioisotopes in rocks and soil. Among the man-made sources are X-rays (e.g., for medical uses), cathode ray tube displays (present in older-style computer monitors and television sets), and watches and other devices that glow in the dark.

Radiation occurs in nonionizing or ionizing forms. Ionization occurs when energy is sufficient to knock an electron out of an atomic shell and hence break covalent bonds. Except for ultraviolet light (UV), nonionizing radiation does not induce mutations; but all forms of ionizing radiation, such as X-rays, cosmic rays, and radon, can induce mutations.

UV light causes mutations by increasing the chemical energy of certain molecules, such as pyrimidines, in DNA. One effect of UV radiation on DNA is the formation of abnormal chemical bonds between adjacent pyrimidine molecules in the same strand of the double helix. This bonding is induced mostly between adjacent thymines, forming what are called **thymine dimers** (Figure 7.10), usually designated T^T. (C^C, C^T, and T^C pyrimidine dimers are also produced by UV radiation but in much lower amounts.) This unusual pairing produces a bulge in the DNA strand and disrupts the normal pairing of T bases with corresponding A bases on the opposite strand. Replication cannot proceed past the lesion, so the cell will die if enough pyrimidine dimers remain unrepaired.

Ionizing radiation penetrates tissues, colliding with molecules and knocking electrons out of orbits, thereby creating ions. The ions can result in the breakage of covalent bonds, including those in the sugar–phosphate backbone of DNA. In fact, ionizing radiation is the leading cause of gross chromosomal mutations in humans. High dosages of ionizing radiation kill cells—hence their use in treating some forms of cancer. At certain low levels of ionizing radiation, point mutations are commonly produced; at these levels, there is a linear relationship between the rate of point mutations and the radiation dosage. Importantly, for many organisms, including humans, the effects of ionizing radiation doses are cumulative. That is, if a particular dose of radiation results in a certain number of point mutations, the same number of point mutations will be induced whether the radiation dose is received over a short or over a long period of time. Interestingly some organisms are highly resistant to radiation damage. The genetics of this phenotype in one such organism, an archaean, is described in this chapter's Focus on Genomics box.

The X-ray is a form of ionizing radiation that has been used to induce mutations in laboratory experiments. For his pioneering work in this area in the 1930s, Hermann Joseph Müller received the 1946 Nobel Prize in Physiology or Medicine for "the discovery of the production of mutations by means of X-ray irradiation."

Radon is an invisible, inert radioactive gas with no smell or taste. The decay of radon produces ionizing radiation, which can induce mutations. In the United States, radon is the second most frequent cause of lung cancer after cigarette smoking. Radon-induced lung cancer, with more than 20,000 deaths per year, is thought to be the sixth leading cause of death among all forms of cancer.

The ultimate source of radon is uranium. All rocks, and hence nearly all soil, contain some uranium. As a result, we can be exposed to radon essentially anywhere in the world. Radon exposure can occur in homes and dwellings when surrounding or underlying soil, or materials used in construction, contain uranium. Decay of the uranium leads to the accumulation of radon within the home. The danger of radon exposure was discovered in 1984 when a nuclear power plant worker in the United States set off radiation alarms at the plant. However, the worker had not been exposed to radiation at the plant, but to radon in the basement of his house. Because of this incident, national radon safety standards are now in place, and radon detection systems and ventilation devices are available for homeowners. In January 2005 the U.S. Surgeon General issued a National Health Advisory on Radon, notifying the public of the risks of breathing indoor radon and advising them to take action to be sure they are not being exposed.

**Figure 7.10**

**Production of thymine dimers by ultraviolet light irradiation.** The two components of the dimer are covalently linked in such a way that the DNA double helix is distorted at that position.

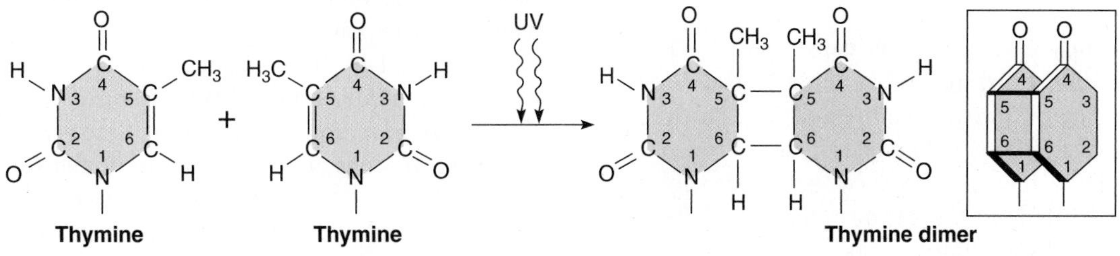

## Focus on Genomics

### Radiation Resistance in the Archaea: Conan the Bacterium

The Archaean *Deinococcus radiodurans* is highly resistant to radiation damage. This resistance is common to most members of the *Deinococcus-Thermus* group to which it belongs. This group includes *Thermus aquaticus*, which you will learn more about when studying the polymerase chain reaction (PCR), a technique to amplify DNA in vitro, in Chapter 8. Members of this group can survive acute doses of ionizing radiation in excess of 10,000 grays, where a gray (Gy) is defined as the absorption of one joule (J) of radiation energy by one kilogram of matter. They also can survive chronic ionizing radiation exposure of 60 Gy/hour, and ultraviolet light doses of 1 kJ/m$^2$. By comparison, doses of 10 Gy can kill a human, and the common bacterium *E. coli* is killed by a dose of 60 Gy. Members of the *Deinococcus-Thermus* group all live at high to very high temperatures, growing best at temperatures in excess of 50°C. They also can survive long periods of desiccation.

Classical genetics identified a number of genes that were required for radiation resistance in *D. radiodurans*. That is, mutants were isolated that had decreased radiation resistance. The wild-type genes corresponding to the mutants were molecularly cloned and sequenced, and most were found to be similar to DNA repair genes from other organisms, including repair genes in *E. coli*. Surprisingly, orthologs (genes in a different species that evolved from a common ancestor) from *E. coli* could be used to replace the mutated genes in *D. radiodurans*. In other words, orthologous genes from *E. coli* introduced into mutants of *D. radiodurans* were able to restore the radiation resistance to a level similar to that of the wild-type strain. This result suggested that these genes were necessary for the radiation resistance, but not sufficient. In other words, the result explained how *D. radiodurans* resisted, but not why. To study further the why, *D. radiodurans* was chosen as one of the first genomes for sequencing.

The genomic sequence revealed that the genome is relatively small, at about 3.28 million base pairs (Mb). The genome of *E. coli* is about 1.5 times larger than this, and the human genome is 1,000 times larger. There is one large, circular chromosome and three minichromosomes, or plasmids—two of the three are much larger than most plasmids (nearly the size of the chromosome itself) and are called megaplasmids. Scientists studying these organisms used transcriptomics to identify genes that were transcribed at high rates after exposure to ionizing radiation. Transcriptomics is a genomics-based approach using computers and molecular techniques to profile when, to what extent, and why genes are expressed. The researchers also used proteomics to identify proteins that became more abundant after radiation. **Proteomics** is another genomics-based approach used to characterize the abundance, identity, and function of all of the proteins in a cell or an organism. However, mutations in most of these genes did not slow or stop recovery from radiation.

Recently, other members of the *Deinococcus-Thermus* group have been sequenced, including *Deinococcus geothermalis* and two strains of *Thermus thermophilus*. This work has allowed scientists to use comparative genomics as well. In comparative genomics, two or more genomes are compared, under the assumption that genes found in both organisms probably play similar roles and that genes unique to one of the organisms are probably for functions found only in that organism. In this case, since all four genomes are from closely related, highly radiation-resistant organisms, it stands to reason that all would have a similar radiation-resistance mechanism. Several genes have been identified that are found in members of this group but are absent in genomes of nonresistant prokaryotes, and scientists are now determining whether these genes can explain why *Deinococcus radiodurans* can survive such massive doses of radiation.

The carcinogenic (cancer-causing) effects of certain types of radiation, including UV light and ionizing radiation, are discussed in Chapter 20, pp. 597–598.

### Keynote

Radiation may cause genetic damage by producing chemicals that affect the DNA (as in the case of X-rays) or by causing the formation of unusual bonds between DNA bases, such as thymine dimers (as in the case of ultraviolet light). If radiation-induced genetic damage is not repaired, mutations or cell death may result. Radiation may also break chromosomes.

**Chemical Mutagens.** Chemical mutagens include both naturally occurring chemicals and synthetic substances. These mutagens can be grouped into different classes based on their mechanism of action. Here we discuss base analogs, base-modifying agents, and intercalating agents and explain how they induce mutations. Mutations induced by base analogs and intercalating agents depend on replication, whereas base-modifying agents can induce mutations at any point of the cell cycle.

**Base analogs** are bases that are similar to those normally found in DNA. Like normal bases, base analogs exist in normal and rare tautomeric states. In each of the two states, the base analog pairs with a different normal base in DNA. Because base analogs are so similar to the normal nitrogen bases, they may be incorporated into DNA in place of the normal bases.

One base analog mutagen is 5-bromouracil (5BU), which has a bromine residue instead of the methyl group of thymine. In its normal state, 5BU resembles thymine and pairs with adenine in DNA (Figure 7.11a). In its rare state, it pairs with guanine (Figure 7.11b). 5BU induces mutations by switching between its two chemical states once the base analog has been incorporated into the DNA (Figure 7.11c).

*animation* **Mutagenic Effects of 5BU**

If 5BU is incorporated in its normal state, it pairs with adenine. If it then changes into its rare state during replication, it pairs with guanine instead. In the next round of replication, the 5BU-G base pair is resolved into a C-G base pair instead of the T-A base pair. By this process, a TA-to-CG transition mutation is produced. 5BU can also induce a CG-to-TA transition mutation if it is first incorporated into DNA in its rare state and then switches to the normal state during replication (Figure 7.11c.) Thus, 5BU-induced mutations can be reverted by a second treatment of 5BU.

Not all base analogs are mutagens. For example, AZT (azidothymidine), an approved drug given to patients with AIDS, is an analog of thymidine—but it is not a mutagen, because it does not cause base-pair changes.

**Base-modifying agents** are chemicals that act as mutagens by modifying the chemical structure and properties of bases. Figure 7.12 shows the action of three types of mutagens that work in this way: a deaminating agent, a hydroxylating agent, and an alkylating agent.

Nitrous acid, $HNO_2$ (Figure 7.12a), is a deaminating agent that removes amino groups ($-NH_2$) from the bases guanine, cytosine, and adenine. Treatment of guanine



**Figure 7.11**
**Mutagenic effects of the base analog 5-bromouracil (5BU).**

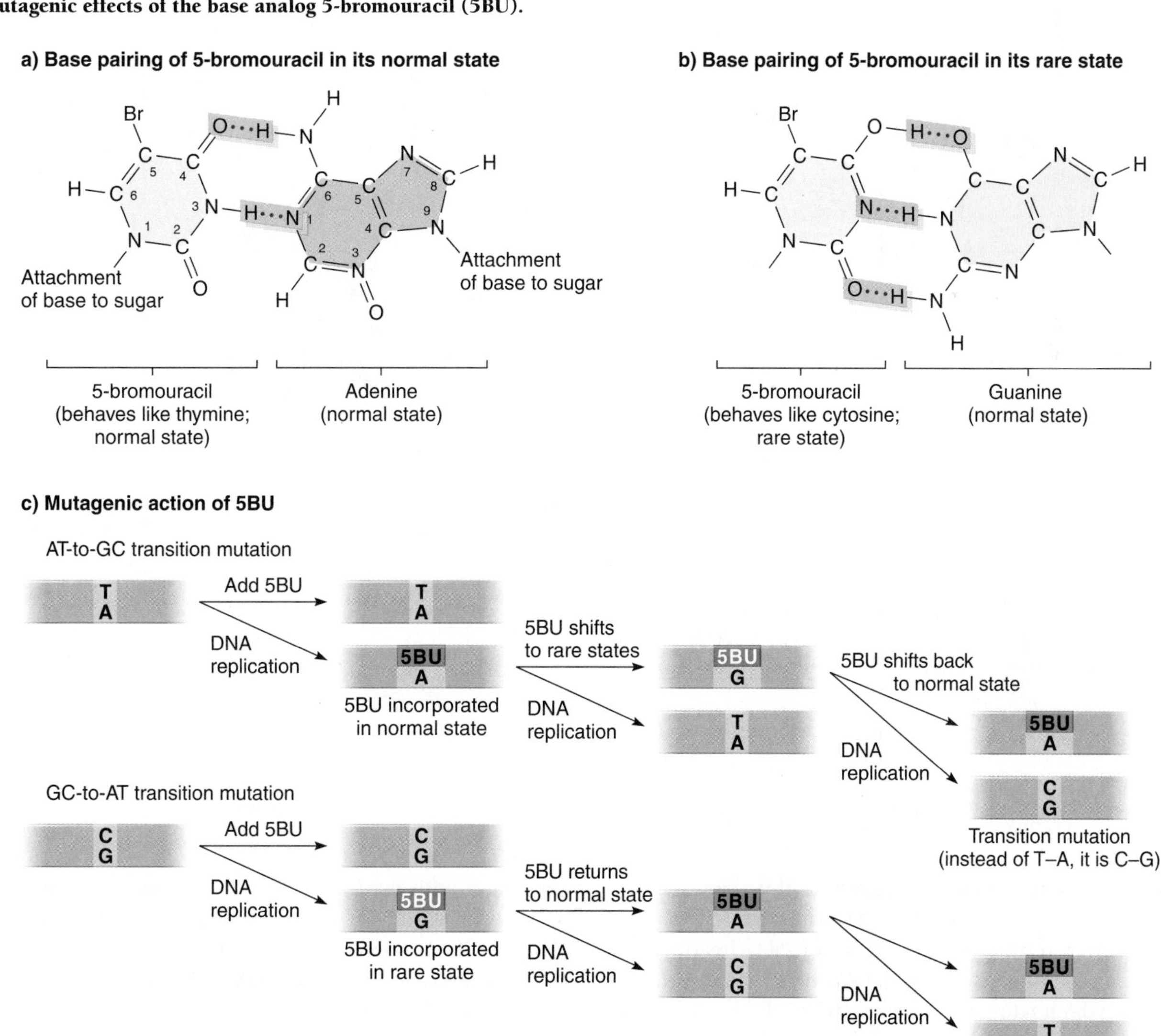

**Figure 7.12**

**Action of three base-modifying agents: (a) nitrous acid, (b) hydroxylamine, and (c) methylmethane sulfonate.**

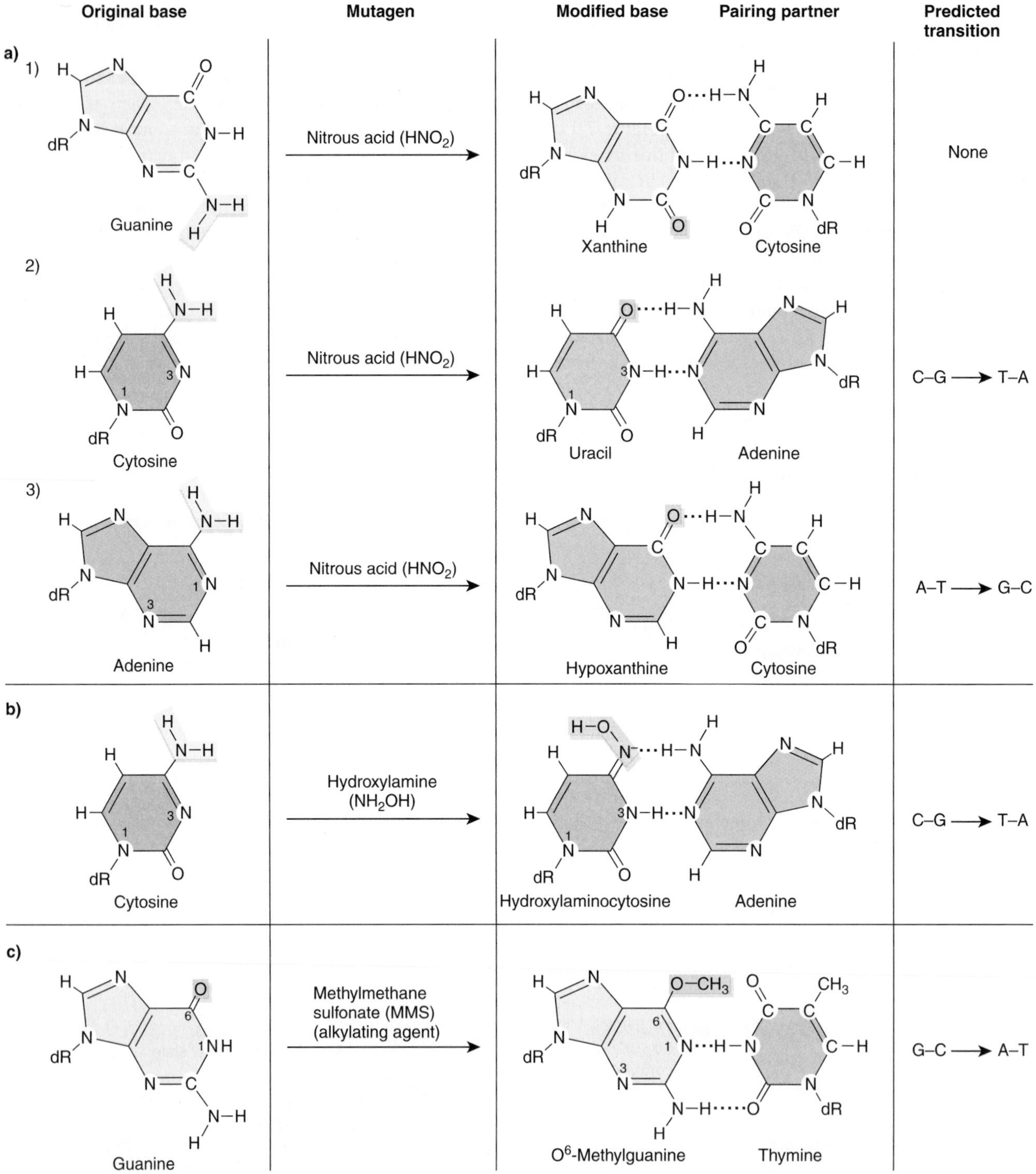

| | Original base | Mutagen | Modified base | Pairing partner | Predicted transition |
|---|---|---|---|---|---|
| a) 1) | Guanine | Nitrous acid ($HNO_2$) | Xanthine | Cytosine | None |
| 2) | Cytosine | Nitrous acid ($HNO_2$) | Uracil | Adenine | C–G ⟶ T–A |
| 3) | Adenine | Nitrous acid ($HNO_2$) | Hypoxanthine | Cytosine | A–T ⟶ G–C |
| b) | Cytosine | Hydroxylamine ($NH_2OH$) | Hydroxylaminocytosine | Adenine | C–G ⟶ T–A |
| c) | Guanine | Methylmethane sulfonate (MMS) (alkylating agent) | $O^6$-Methylguanine | Thymine | G–C ⟶ A–T |

with nitrous acid produces xanthine, but because this purine base has the same pairing properties as guanine, no mutation results (Figure 7.12a, part 1). Treatment of cytosine with nitrous acid produces uracil (Figure 7.12a, part 2), which pairs with adenine to produce a CG-to-TA transition mutation during replication. Likewise, nitrous acid modifies adenine to produce hypoxanthine, a base that pairs with cytosine rather than thymine, which results in an AT-to-GC transition mutation (Figure 7.12a, part 3). Therefore, a nitrous acid-induced mutation can be reverted by a second treatment with nitrous acid.

Hydroxylamine ($NH_2OH$) is a hydroxylating mutagen that reacts specifically with cytosine, modifying it by adding a hydroxyl group (OH) so that it pairs with adenine instead of guanine (Figure 7.12b). Mutations induced by hydroxylamine can only be CG-to-TA transitions, so hydroxylamine-induced mutations cannot be reverted by a second treatment with this chemical. However, they can be reverted by treatment with other mutagens (such as 5BU and nitrous acid) that cause TA-to-CG transition mutations.

Methylmethane sulfonate (MMS) is one of a diverse group of alkylating agents that introduce alkyl groups (e.g., $-CH_3$, $-CH_2CH_3$) onto the bases at a number of locations (Figure 7.12c). Most mutations caused by alkylating agents result from the addition of an alkyl group to the 6-oxygen of guanine to produce $O^6$-alkylguanine. For example, after treatment with MMS, some guanines are methylated to produce $O^6$-methylguanine. The methylated guanine pairs with thymine rather than cytosine, giving GC-to-AT transitions (Figure 7.12c).

**Intercalating agents**—such as proflavin, acridine, and ethidium bromide (commonly used to stain DNA in gel electrophoresis experiments)—insert (*intercalate*) themselves between adjacent bases in one or both strands of the DNA double helix, causing the helix to relax (Figure 7.13). If the intercalating agent inserts itself between adjacent base pairs of the DNA strand that is the template for new DNA synthesis (Figure 7.13a), an extra base (chosen at random; G in the figure) is inserted into the new DNA strand opposite the intercalating agent. After one more round of replication, during which the intercalating agent is lost, the overall result is a base-pair addition mutation. (C-G is added in Figure 7.13a.) If the intercalating agent inserts itself into the new DNA strand in place of a base (Figure 7.13b), then when that DNA double helix replicates after the intercalating agent is lost, the result is a base-pair deletion mutation. (T-A is lost in Figure 7.13b.)

If a base-pair addition or base-pair deletion point mutation occurs in a protein-coding gene, the result is a frameshift mutation. Since intercalating agents can cause either additions or deletions, frameshift mutations induced by intercalating agents can be reverted by a second treatment with those same agents.

**Figure 7.13**

**Intercalating mutations.** (**a**) Frameshift mutation by addition, when agent inserts itself into template strand. (**b**) Frameshift mutation by deletion, when agent inserts itself into newly synthesizing strand.

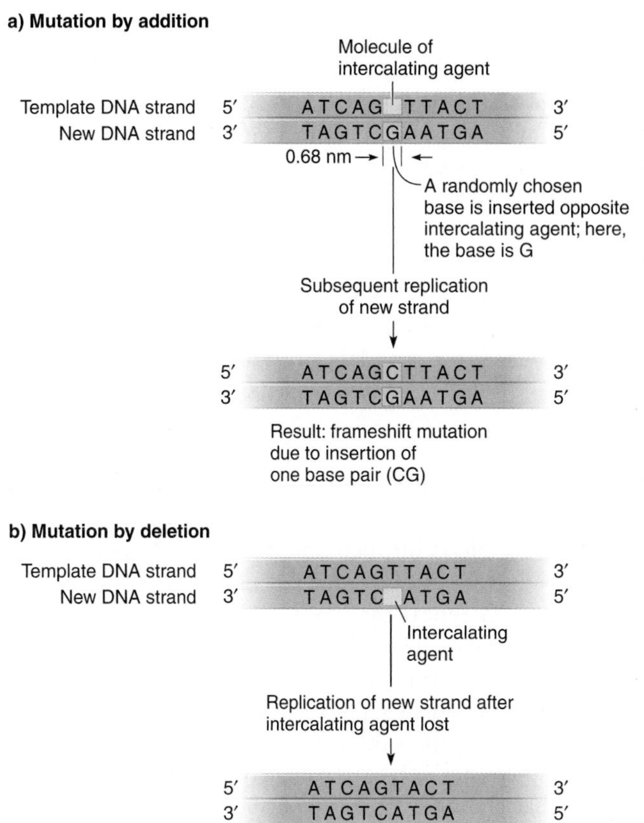


## Keynote

Mutations can be produced by exposure to chemical mutagens. If the genetic damage caused by the mutagen is not repaired, mutations result. Chemical mutagens act in a variety of ways, such as by substituting for normal bases during DNA replication, modifying the bases chemically, and intercalating themselves between adjacent bases during replication.

**Site-Specific *In Vitro* Mutagenesis of DNA.** Spontaneous and induced mutations occur not only in specific genes, but are scattered randomly throughout the genome. However, most geneticists want to study the effects of mutations in particular genes. With recombinant DNA technology, we can clone genes and produce large amounts of DNA for analysis and manipulation. This means that it is now possible to mutate a gene at specific positions in the base-pair sequence by **site-specific mutagenesis** in the test tube and then introduce the mutated gene back into the cell and investigate the phenotypic changes produced by the mutation *in vivo*. Such techniques enable geneticists to study, for example, genes with unknown function and specific sequences involved in regulating the expression of a gene.

**Environmental Mutagens.** Every day, we are heavily exposed to a wide variety of chemicals in our environment. The chemicals may be natural ones, such as those synthesized by plants and animals that we eat as food, or man-made ones, such as drugs, cosmetics, food additives, pesticides, and industrial compounds. Our exposure to chemicals occurs primarily through eating food, absorption through the skin, and inhalation. Many of these chemicals are, or can be, mutagenic. For a mutagenic chemical to cause DNA changes, it must enter cells and penetrate to the nucleus, which many chemicals cannot do.

Some chemicals are converted from nonmutagenic to mutagenic by our metabolism. That is, when these chemicals are directly tested for mutagenic activity on, say, a bacterial species, no mutations result. But, after they are processed in the body, they become mutagens. For example, benzpyrene, a polycyclic aromatic hydrocarbon found in cigarette smoke, coal tar, automobile exhaust fumes, and charbroiled food, is nonmutagenic. But its metabolite, benzpyrene diol epoxide, which is both a mutagen and a *carcinogen*, can induce cancer. Many other polycyclic aromatic hydrocarbons similarly become mutagenic when activated by metabolism.

**The Ames Test: A Screen for Potential Mutagens.** Some chemicals induce mutations that result in tumorous or cancerous growth. These chemical agents are a subclass of mutagens called chemical **carcinogens.** The mutations typically are base-pair substitutions that produce missense or nonsense mutations, or base-pair additions or deletions that produce frameshift mutations. Directly testing chemicals for their ability to cause tumors in animals is time-consuming and expensive. However, the fact that most chemical carcinogens are mutagens led Bruce Ames to develop a simple, inexpensive, indirect assay for mutagens. The **Ames test** assays the ability of chemicals to revert mutant strains of the bacterium *Salmonella typhimurium* to the wild type.

@nimation
**Ames Test Protocol**

In the Ames test, approximately $10^8$ cells of tester bacteria that are auxotrophic for histidine (*his* mutants) are spread with or without a mixture of rat, mouse, or hamster liver enzymes on a culture plate lacking histidine (Figure 7.14). Histidine (*his*) auxotrophs require histidine in the growth medium in order to grow; normal ($his^+$) individuals do not. An array of tester bacterial strains are available that allow detection of base-pair substitution mutations and frameshift mutations in the test. The liver enzymes, called the S9 extract, are used because, as just described, many chemicals are not mutagenic themselves but are metabolized to mutagens (and carcinogens) in the body, often in the liver and other tissues. A filter disk impregnated with the test chemical is then placed on the plate, which is incubated overnight and then examined for colony formation. Control plates lack the chemical being tested. After the incubation period, the control plates have a few colonies due to spontaneous reversion of the *his* strain to wild type. A similar result is seen with chemicals that are not mutagenic in the Ames test. A positive result in the Ames test is a significantly higher number of revertants near the test chemical disk than is seen on the control plate.

The Ames test is so straightforward that it is used routinely in many laboratories around the world. The test has identified a large number of mutagens, including many industrial and agricultural chemicals. In general the Ames test is an excellent indicator of whether a chemical is a carcinogen, but some carcinogenic chemicals assay negative in the test. For example, Ziram, which is used as an agricultural fungicide, gives a positive Ames test for both base substitution and frameshift reversion when S9 extract is present, but a negative test when S9 extract is absent. Thus this chemical presumably is turned into a mutagen by metabolism. In contrast, nitrobenzene is negative in the Ames test with or without the S9 extract. Most nitrobenzene is used to manufacture aniline, which is used in the manufacture of polyurethane. Styrene, used in producing polystyrene polymers and resins, similarly tests negative with or without the S9 extract, yet animal tests

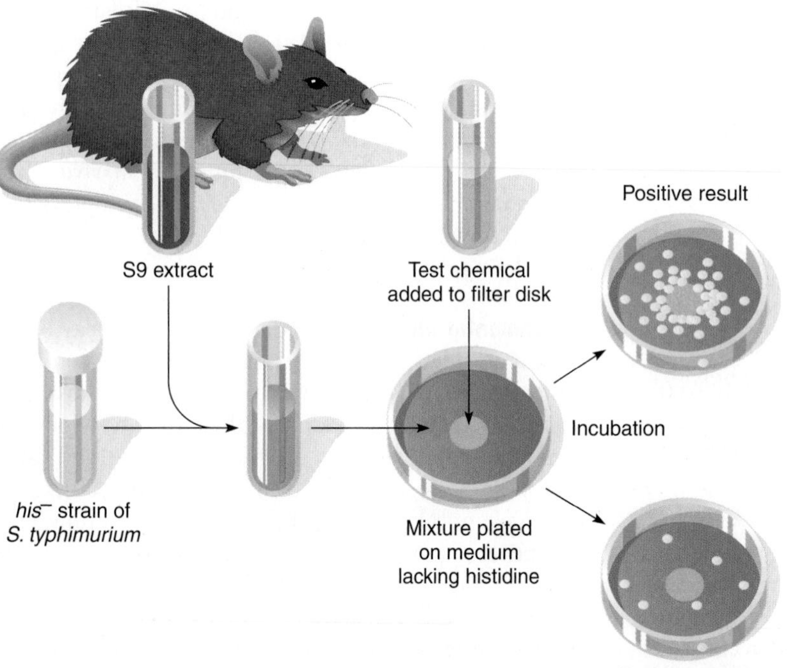


**Figure 7.14**
**The Ames test for assaying the potential mutagenicity of chemicals.**

indicate that it is a carcinogen. Because of results like this, the Ames test is not the sole test relied upon in determining whether a compound is mutagenic.

Finally, the Ames test can be quantified by using different amounts of chemicals to produce a dose–response curve. With this approach, the relative mutagenicity of different chemicals can be compared.

**iActivity**

Now it is your turn to investigate the health problems plaguing the inhabitants of Russellville. Conduct your own Ames test in the iActivity *A Toxic Town* on the student website.

### Detecting Mutations

Geneticists have made great progress over the years in understanding how normal processes take place, primarily by studying mutants that have defects in those processes. Researchers have used mutagens to induce mutations at a greater rate than the one at which spontaneous mutations occur. However, mutagens change base pairs at random, without regard to the positions of the base pairs in the genetic material. Once mutations have been induced, they must be detected if they are to be studied. Mutations of haploid organisms are readily detectable because there is only one copy of the genome. In a diploid experimental organism such as *Drosophila*, dominant mutations are also readily detectable, and X-linked recessive mutations can be detected because they are expressed in half of the sons of a mutated, heterozygous female. However, autosomal recessive mutations can be detected only if the mutation is homozygous.

Detecting mutations in humans is much more difficult than in *Drosophila*, because geneticists cannot make controlled crosses. Dominant mutations can be readily detected, of course, but other types of mutations may be revealed only by pedigree analysis or by direct biochemical or molecular probing.

Fortunately, for some organisms of genetic interest—particularly microorganisms—selection and screening procedures historically helped geneticists isolate mutants of interest from a heterogeneous mixture in a mutagenized population. Brief descriptions of some of these procedures follow.

**Visible Mutants.** **Visible mutants** affect the morphology or physical appearance of an organism. Examples of visible mutants are eye-color and wing-shape mutants of *Drosophila*, coat-color mutants of animals (such as albino organisms), colony-size mutants of yeast, and plaque morphology mutants of bacteriophages. Since visible mutants, by definition, are readily apparent, screening is done by inspection.

**Nutritional Mutants.** An **auxotrophic (nutritional) mutant** is unable to make a particular molecule essential for growth (see Chapter 4, p. 62). Auxotrophic mutants are most readily detected in microorganisms such as *E. coli* and yeast that grow on simple and defined growth media from which they synthesize the molecules essential to their growth. A number of selection and screening procedures are available to isolate auxotrophic mutants.

One simple procedure called **replica plating** can be used to screen for auxotrophic mutants of any microorganism that grows in discrete colonies on a solid medium (Figure 7.15). In replica plating, samples from a culture of a mutagenized or an unmutagenized colony-forming organism or cell type are plated onto a medium containing the nutrients appropriate for the mutants desired. For example, to isolate arginine auxotrophs, we would plate the culture on a master plate of minimal medium plus

**Figure 7.15**

**Replica-plating technique to screen for mutant strains of a colony-forming microorganism.**

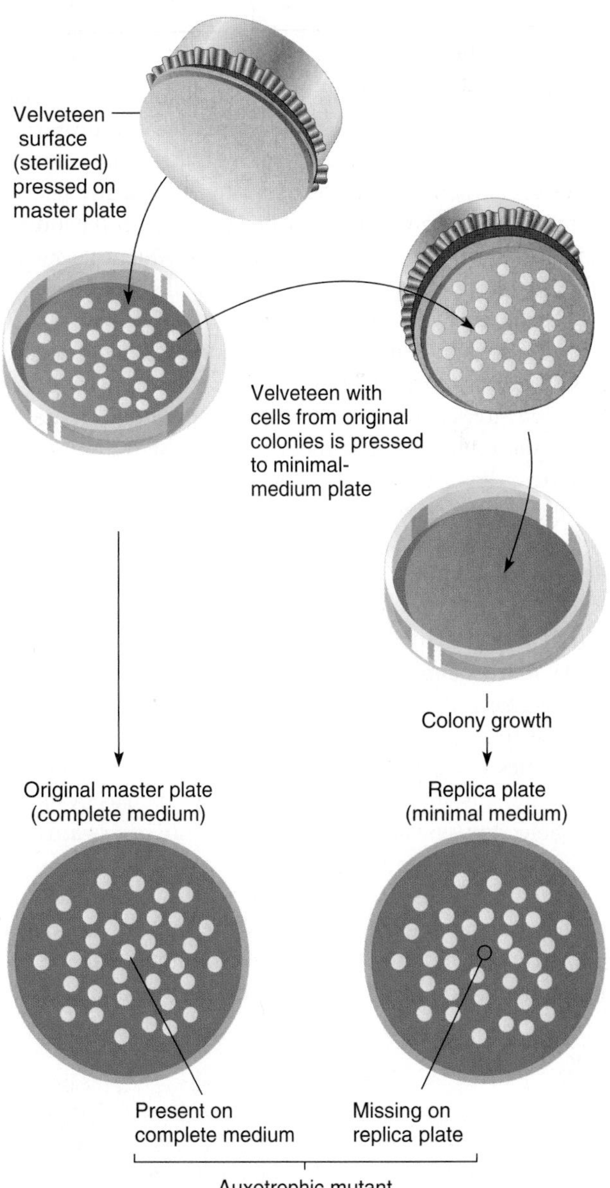

arginine (see Figure 7.15). On this medium, wild-type and arginine auxotrophs grow, but no other auxotrophs grow. The pattern of the colonies that grow is transferred onto sterile velveteen cloth, and replicas of the colony pattern on the cloth are then made by gently pressing new plates onto the velveteen. If the new plate contains minimal medium, the wild-type colonies can grow but the arginine auxotrophs cannot. By comparing the patterns on the original minimal medium plus arginine master plate with those on the minimal medium replica plate, researchers can readily identify the potential arginine auxotrophs. They can then be picked from the original master plate and cultured for further study.

**Conditional Mutants.** The products of many genes—DNA polymerases and RNA polymerases, for example—are important for the growth and division of cells, and knocking out the functions of such genes by introducing mutations typically is lethal. The structure and function of such genes can be studied by inducing **conditional mutants,** which reduce the activity of gene products only under certain conditions. A common type of conditional mutation is a temperature-sensitive mutation. In yeast, for instance, many **temperature-sensitive mutants** that grow normally at 23°C but grow very slowly or not at all at 36°C can be isolated. Heat sensitivity typically results from a missense mutation causing a change in the amino acid sequence of a protein so that, at the higher temperature, the protein assumes a nonfunctional shape.

Essentially the same procedures are used to screen for heat-sensitive mutations of microorganisms as for auxotrophic mutations. For example, replica plating can be used to screen for temperature-sensitive mutants when the replica plate is incubated at a higher temperature than the master plate. That is, such mutants grow on the master plate, but not on the replica plate.

**Resistance Mutants.** In microorganisms such as *E. coli*, yeast, and cells in tissue culture, mutations can be induced for resistance to particular viruses, chemicals, or drugs. For example, in *E. coli*, mutants resistant to phage T1 have been induced (recall the discussion at the beginning of this chapter), and some mutants are resistant to antibiotics such as streptomycin. In yeast, for example, some mutants are resistant to antifungals such as nystatin.

Selecting resistance mutants is straightforward. To isolate azide-resistant mutants of *E. coli*, for example, mutagenized cells are plated on a medium containing azide, and the colonies that grow are resistant to azide. Similarly, antibiotic-resistant *E. coli* mutants can be selected by plating on antibiotic-containing medium.

**Keynote**

A number of screening procedures have been developed to isolate mutants of interest from a heterogeneous mixture of cells in a mutagenized population of cells.

## Repair of DNA Damage

Mutagenesis involves damage to DNA. Especially with high doses of mutagens, the mutational damage can be considerable. What we see as mutations are DNA alterations that are not corrected by various DNA damage repair systems; that is, "mutations = DNA damage − DNA repair." Both prokaryotic and eukaryotic cells have a number of enzyme-based systems that repair DNA damage. If the repair systems cannot correct all the lesions, the result is a mutant cell (or organism) or, if too many mutations remain, death of the cell (or organism).

There are two general categories of repair systems, based on the way they function. *Direct reversal repair systems* correct damaged areas by reversing the damage, whereas *excision repair systems* cut out a damaged area and then repair the gap by new DNA synthesis. Selected repair systems are described in this section.

### Direct Reversal Repair of DNA Damage

**Mismatch Repair by DNA Polymerase Proofreading.** The frequency of base-pair substitution mutations in bacterial genes ranges from $10^{-7}$ to $10^{-11}$ errors per generation. However, DNA polymerase inserts incorrect nucleotides at a frequency of $10^{-5}$. Most of the difference between the two values is accounted for by the 3′-to-5′ exonuclease proofreading activity of the DNA polymerase in both bacteria and eukaryotes (see Chapter 3, p. 40). When an incorrect nucleotide is inserted, the polymerase often detects the mismatched base pair and corrects the area by "backspacing" to remove the wrong nucleotide and then resuming synthesis in the forward direction.

The *mutator* mutations in *E. coli* illustrate the importance of the 3′-to-5′ exonuclease activity of DNA polymerase for maintaining a low mutation rate. Mutator mutants have a much higher than normal mutation frequency for all genes. These mutants have mutations in genes for proteins whose normal functions are required for accurate DNA replication. For example, the *mutD* mutator gene of *E. coli* encodes the ε (epsilon) subunit of DNA polymerase III, the primary replication enzyme of *E. coli*. The *mutD* mutants are defective in 3′-to-5′ proofreading activity, so that many incorrectly inserted nucleotides are left unrepaired.

**Repair of UV-Induced Pyrimidine Dimers.** Through **photoreactivation,** or **light repair,** UV light-induced thymine (or other pyrimidine) dimers (see Figure 7.10) are reverted directly to the original form by exposure to near-UV light in the wavelength range from 320 to 370 nm. Photoreactivation occurs when an enzyme called *photolyase* (encoded by the *phr* gene) is activated by a photon of light and splits the dimers apart. Strains with mutations in the *phr* gene are defective in light repair. Photolyase has been found in bacteria and in simple eukaryotes, but not in humans.

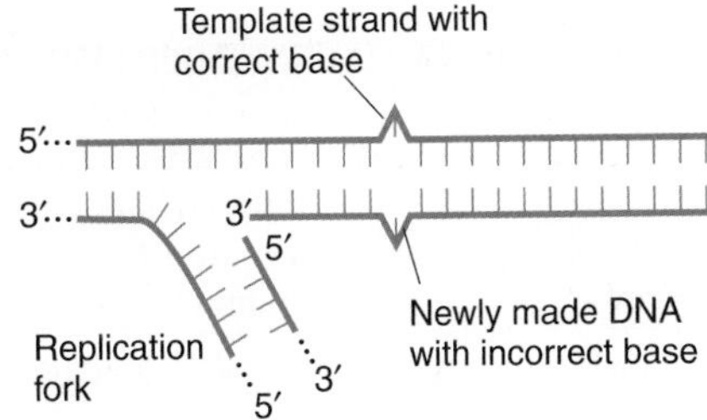


1 MutS binds to the mismatch.

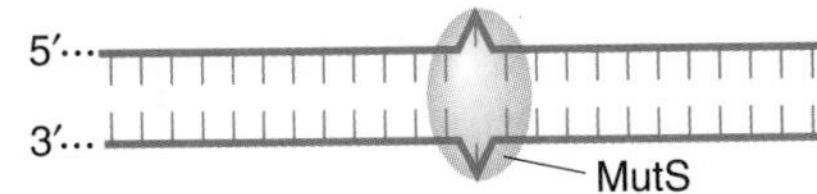


2 MutS bound to mismatch forms a complex with MutL and MutH to bring the unmethylated GATC close to the mismatch.

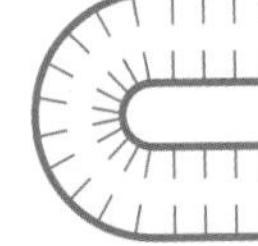

3 MutH nicks the unmethylated DNA strand, and an exonuclease excises a section of the new DNA strand, including the mismatch.

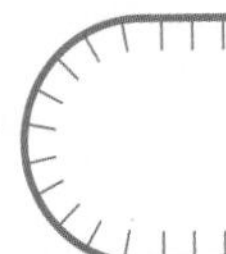

4 DNA polymerase III and ligase repair the gap, producing the correct base pair.

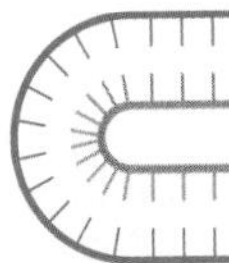

and dealing with various kinds of DNA da
sufficient damage to DNA, the *recA*-enco
RecA, is activated. Activated RecA stimula
protein to cleave itself, which in turn reliev
sion of the DNA repair genes. As a result, th
genes are expressed, and DNA repair proce
DNA damage is dealt with, RecA is ina
newly synthesized LexA protein again repre
repair genes.

Among the gene products made durin
sponse is the DNA polymerase for translesi
thesis. This polymerase continues replicat
past the lesion, although it does so by inco
or more nucleotides that are not specified by
strand into the new DNA across from the
nucleotides may not match the wild-type
quence; therefore, the SOS response itself is
system because mutations will be introdu
DNA as a result of its activation. Such muta
harmful than the potentially lethal alternat
incompletely replicated DNA.





**Repair of Alkylation Damage.** Alkylating agents transfer alkyl groups (usually methyl or ethyl groups) onto the bases. The mutagen MMS methylates the oxygen of carbon-6 in guanine, for example (see Figure 7.12c). In *E. coli*, this alkylation damage is repaired by an enzyme called $O^6$-methylguanine methyltransferase, encoded by the *ada* gene. The enzyme removes the methyl group from the guanine, thereby changing the base back to its original form. A similar specific system exists to repair alkylated thymine. Mutations of the genes encoding these repair enzymes result in a much higher rate of spontaneous mutations.

## Excision Repair of DNA Damage

Many mutations affect only one of the two strands. In such cases, the DNA damage can be excised and the normal strand used as a template for producing a corrected strand. Depending on the damage, excision may involve a single base or nucleotide, or two or more nucleotides. Each excision repair system involves a mechanism to recognize the specific DNA damage it repairs.

**Base Excision Repair.** Damaged single bases or nucleotides are most commonly repaired by removing the base or the nucleotide involved and then inserting the correct base or nucleotide. In **base excision repair,** a repair glycosylase enzyme removes the damaged base from the DNA by cleaving the bond between the base and the deoxyribose sugar. Other enzymes then cleave the sugar–phosphate backbone before and after the now baseless sugar, releasing the sugar and leaving a gap in the DNA chain. The gap is filled with the correct nucleotide by a repair DNA polymerase and DNA ligase, with the opposite DNA strand used as the template. Mutations caused by depurination or deamination are examples of damage that may be repaired by base excision repair.

**Nucleotide Excision Repair.** In 1964, two groups of scientists—R. P. Boyce and P. Howard-Flanders, and R. Setlow and W. Carrier—isolated mutants of *E. coli* that, after UV irradiation, showed a higher than normal rate of induced mutation in the dark. These UV-sensitive mutants were called *uvrA* mutants (*uvr* for "*UV* repair"). The *uvrA* mutants can repair thymine dimers only with the input of light, meaning they have a normal photoreactivation repair system. However, *uvrA*$^+$ (wild-type) *E. coli* can repair thymine dimers in the dark. Because the normal photoreactive repair system cannot operate in the dark, the investigators hypothesized that there must be another light-independent repair system. They called this system the **dark repair** or **excision repair system,** now typically referred to as the **nucleotide excision repair (NER)** system. The NER system in *E. coli* also corrects other serious damage-induced distortions of the DNA helix.

The NER system involves four proteins—UvrA, UvrB, UvrC, and UvrD—encoded by the genes *uvrA, uvrB, uvrC,* and *uvrD* (Figure 7.16). A complex of two UvrA proteins and one UvrB protein slides along the DNA (Figure 7.16, step 1). When the complex recognizes a pyrimidine dimer or another serious distortion in the DNA, the UvrA subunits dissociate and a UvrC protein binds to the UvrB protein at the lesion (Figure 7.16, step 2). The resulting UvrBC protein bound to the lesion makes one cut about four nucleotides to the 3′ side in the damaged DNA strand (done by UvrB) and about seven nucleotides to the 5′ side of the lesion (done by UvrC) (Figure 7.16, step 3). UvrB is then released, and UvrD binds to the 5′ cut (Figure 7.16, step 4). UvrD is a helicase that unwinds the region between the cuts, releasing the short single-stranded segment. DNA polymerase I fills in the gap in the 5′-to-3′ direction (Figure 7.16, step 5), and DNA ligase seals the final gap (Figure 7.16, step 6).

Nucleotide excision repair systems have been found in most organisms that have been studied. In yeast and mammalian systems, about 12 genes encode proteins involved in nucleotide excision repair.

**Methyl-Directed Mismatch Repair.** Despite proofreading by DNA polymerase, a number of mismatched base pairs remain uncorrected after replication has been completed. In the next round of replication, these errors will become fixed as mutations if they are not repaired.

Many mismatched base pairs left after DNA replication can be corrected by **methyl-directed mismatch repair**. This system recognizes mismatched base pairs, excises the incorrect bases, and then carries out repair synthesis. In *E. coli*, the products of three genes—*mutS*, *mutL*, and *mutH*—are involved in the initial stages of mismatch repair (Figure 7.17, p. 149). First, the *mutS*-encoded protein, MutS, binds to the mismatch (Figure 7.17, step 1). Then the repair system determines which base is the correct one (the base on the parental DNA strand) and which is the erroneous one (the base on the new DNA strand). In *E. coli*, the two strands are distinguished by methylation of the A nucleotide in the sequence GATC. This sequence has an axis of symmetry; that is, the same sequence is present 5′-to-3′ on both DNA strands to give

5′-GATC-3′
3′-CTAG-5′. Both A nucleotides in the sequence usually are methylated. However, after replication, the parental DNA strand has a methylated A in the GATC sequence, whereas the A in the GATC of the *newly replicated DNA strand* is not methylated until a short time after its synthesis. Therefore, the MutS protein bound to the mismatch forms a complex with the *mutL*- and *mutH*-encoded proteins, MutL and MutH, to bring the unmethylated GATC sequence close to the mismatch (Figure 7.17, step 2). The MutH protein then nicks the unmethylated DNA strand at the GATC site, the mismatch is removed by an exonuclease (Figure 7.17, step 3), and the gap is repaired by DNA polymerase III and ligase (Figure 7.17, step 4).

Mismatch repair also takes place in eukaryotes. However, it is unclear how the new DNA strand is distinguished from the parental DNA strand (no methylation is involved). In humans, four genes, respectively named

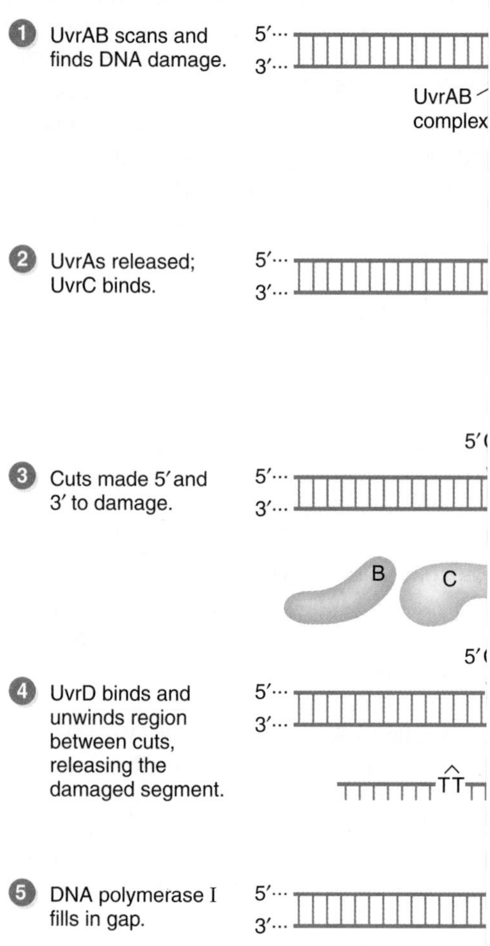

1 UvrAB scans and finds DNA damage.
5′…
3′…
UvrAB complex
2 UvrAs released; UvrC binds.
5′…
3′…
3 Cuts made 5′ and 3′ to damage.
5′…
3′…
B
C
4 UvrD binds and unwinds region between cuts, releasing the damaged segment.
5′…
3′…
TT
5 DNA polymerase I fills in gap.
5′…
3′…
6 DNA ligase joins the DNA segments; repair is complete.
5′…
3′…

*hMSH2*, *hMLH1*, *hPMS1*, and *hPMS2*, have bee
*hMSH2* is homologous to *E. coli mutS*, an
three genes have homologies to *E. coli mutL*
are known as *mutator genes*, because loss of
such a gene results in an increased accumul
tations in the genome. Mutations in any one
human mismatch repair genes confer a p
hereditary predisposition to a form of c
called hereditary nonpolyposis colon cance
OMIM 120435). The role of mutator genes
described in Chapter 20, p. 594.

**Translesion DNA Synthesis and the SOS Respo** that block the replication machinery from
past that point can be lethal if unrepaired. F

**Table 7.1 Some Examples of Naturally Occurring Human Cell Mutants That Are Defective in DNA Replication or Repair**

| Disease and Mode of Inheritance | Symptoms | Functions Affected | Chromosome Location[a] and OMIM number |
|---|---|---|---|
| Xeroderma pigmentosum (XP)—autosomal recessive | Sensitivity to sunlight, with skin freckling and cancerous growths on skin; lethal at early age as a result of the malignancies | Repair of DNA damaged by UV irradiation or chemicals | 9q34.1—278700 |
| Ataxia-telangiectasia (AT)—autosomal recessive | Muscle coordination defect; propensity for respiratory infection; progressive spinal muscular atrophy in significant proportion of patients in second or third decade of life; marked hypersensitivity to ionizing radiation, cancer prone, high frequency of chromosome breaks leading to translocations and inversions | Repair replication of DNA | 11q22.3—208900 |
| Fanconi anemia (FA)—autosomal recessive | Aplastic anemia;[b] pigmentary changes in skin; malformations of heart, kidney, and limbs; leukemia is a fatal complication, genital abnormalities common in males; spontaneous chromosome breakage | Repair replication of DNA, UV-induced pyrimidine dimers, and chemical adducts not excised from DNA; a repair exonuclease, DNA ligase, and transport of DNA repair enzymes have been hypothesized to be defective in patients with FA | 16q24.3—227650 |
| Bloom syndrome (BS)—autosomal recessive | Pre- and postnatal growth deficiency; sun-sensitive skin disorder, predisposition to malignancies; chromosome instability; diabetes mellitus often develops in second or third decade of life | Elongation of DNA chains intermediate in replication: candidate gene is homologous to *E. coli* helicase Q | 15q26.1—210900 |
| Cockayne syndrome (CS)—autosomal recessive | Dwarfism; precociously senile appearance; optic atrophy; deafness; sensitivity to sunlight; mental retardation; disproportionately long limbs; knee contractures produce bowlegged appearance, early death | Precise molecular defect is unknown, but may involve transcription-coupled repair | 5—216400 |
| Hereditary nonpolyposis colon cancer (HNPCC)—autosomal dominant | Inherited predisposition to non-polyp-forming colorectal cancer | Defect in mismatch repair develops when the remaining wild-type allele of the inherited mutant allele becomes mutated; homozygosity for mutations in any one of four genes (*hMSH2*, *hMLH1*, *hPMS1*, and *hPMS2*, known as mutator genes) has been shown to give rise to HNPCC | 2p22-p21—114500 |

[a]If multiple complementation groups exist, the location of the most common defect is given.
[b]Individuals with aplastic anemia make no or very few red blood cells.

skin that have been exposed to light show intense pigmentation, freckling, and warty growths that can become malignant. Those afflicted are deficient in excision repair of damage caused by ultraviolet light or chemical treatment. Thus individuals with xeroderma pigmentosum are unable to repair radiation damage to DNA and often die as a result of malignancies arising from the damage.

## Transposable Elements

In this section, we learn about the nature of transposable elements and about the genetic changes they cause.

### General Features of Transposable Elements

Transposable elements are normal, ubiquitous components of the genomes of prokaryotes and eukaryotes.

Figure 7.18

**An individual with xeroderma pigmentosum.**

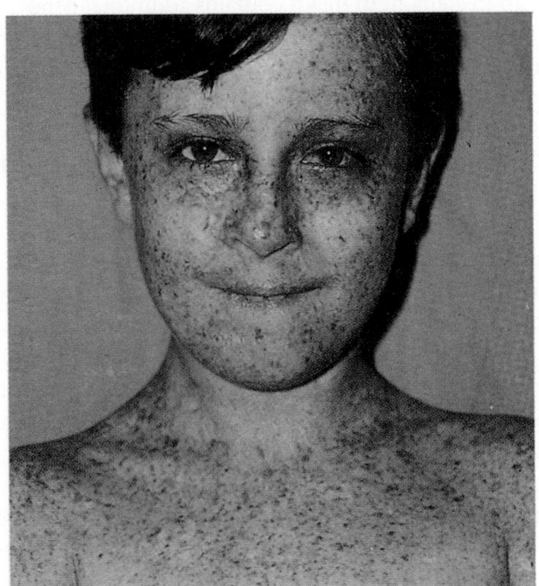

Transposable elements fall into two general classes based on how they move from location to location in the genome. One class—found in both prokaryotes and eukaryotes—moves as a DNA segment. Members of the other class—found only in eukaryotes—are related to retroviruses and move via an RNA. First an RNA copy of the element is synthesized; then a DNA copy of that RNA is made, and it integrates at a new site in the genome.

In bacteria, transposable elements can move to new positions on the same chromosome (because there is only one chromosome) or onto plasmids or phage chromosomes; in eukaryotes, transposable elements may move to new positions within the same chromosome or to a different chromosome. In both bacteria and eukaryotes, transposable elements insert into new chromosome locations with which they have no sequence homology; therefore, transposition is a process different from homologous recombination (recombination between matching DNA sequences) and is called *nonhomologous recombination*. Transposable elements are important due to the genetic changes they cause. For example, they can produce mutations by inserting into genes (a process called *insertional mutagenesis*), they can increase or decrease gene expression by inserting into gene regulatory sequences (such as by disrupting promoter function or stimulating a gene's expression through the activity of promoters on the element), and they can produce various kinds of chromosomal mutations through the mechanics of transposition. In fact, transposable elements have made important contributions to the evolution of the genomes of both bacteria and eukaryotes through the chromosome rearrangements they have caused.

The frequency of transposition, though typically low, varies with the particular element. If the frequency were high, the genetic changes caused by the transpositions would likely kill the cell.

## Transposable Elements in Bacteria

Two examples of transposable elements in bacteria are insertion sequence (IS) elements and transposons (Tn).

**Insertion Sequences.** An **insertion sequence (IS),** or **IS element**, is the simplest transposable element found in bacteria. An IS element contains only genes required to mobilize the element and insert it into a new location in the genome. IS elements are normal constituents of bacterial chromosomes and plasmids.

IS elements were first identified in *E. coli* as a result of their effects on the expression of three genes that control the metabolism of the sugar galactose. Some mutations affecting the expression of these genes did not have properties typical of point mutations or deletions, but rather had an insertion of an approximately 800-bp DNA segment into a gene. This particular DNA segment is now called *insertion sequence 1*, or IS*1* (Figure 7.19), and the insertion of IS*1* into the genome is an example of a *transposition event*.

*E. coli* contains a number of IS elements (e.g., IS*1*, IS2, and IS*10R*), each present in up to 30 copies per genome and each with a characteristic length and unique nucleotide sequence. IS*1* (see Figure 7.19), for instance, is 768 bp long and is present in 4 to 19 copies on the *E. coli* chromosome. Among bacteria as a whole, the IS elements range in size from 768 bp to more than 5,000 bp and are found in most cells.

All IS elements end with perfect or nearly perfect terminal inverted repeats (IRs) of 9 to 41 bp. This means that essentially the same sequence is found at each end of an IS, but in opposite orientations. The inverted repeats of IS*1* are 23 bp long (see Figure 7.19).

When IS elements integrate at random points along the chromosome, they often cause mutations by disrupting either the coding sequence of a gene or a gene's regulatory region. Promoters within the IS elements themselves may also have effects by altering the expression of nearby genes. In addition, the presence of an IS element in the chromosome can cause mutations such as deletions and inversions in the adjacent DNA. Finally, deletion and

Figure 7.19

**The insertion sequence (IS) transposable element IS*1*.** The 768-bp IS element has inverted repeat (IR) sequences at the ends. Shown below the element are the sequences for the 23-bp terminal inverted repeats (IR).

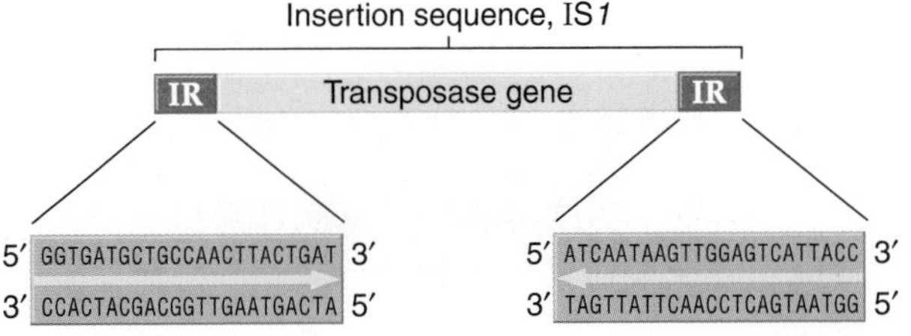

insertion events can also occur as a result of crossing-over between duplicated IS elements in the genome.

The transposition of an IS element requires an enzyme encoded by the IS element called **transposase.** The transposase recognizes the IR sequences of the element to initiate transposition. The frequency of transposition is characteristic of each IS element and ranges from $10^{-5}$ to $10^{-7}$ per generation.

Figure 7.20 shows how an IS element inserts into a new location in a chromosome. Insertion takes place at a *target site* with which the element has no sequence homology. First, a staggered cut is made in the target site and the IS element is then inserted, becoming joined to the single-stranded ends. DNA polymerase and DNA ligase fill in the gaps, producing an integrated IS element with two direct repeats of the target-site sequence flanking the IS element. In this case, *direct* means that the two sequences are repeated in the same orientation (see Figure 7.20). The direct repeats are called *target-site duplications*. Their size is specific to the IS element, but they tend to be small (4 to 13 bp).

**Transposons.** Like an IS element, a **transposon (Tn)** contains genes for the insertion of the DNA segment into the chromosome and mobilization of the element to other locations on the chromosome. A transposon is more complex than an IS element in that it contains additional genes.

There are two types of bacterial transposons: composite transposons and noncomposite transposons (Figure 7.21). *Composite transposons* (Figure 7.21a), exemplified by Tn*10*, are complex transposons with a central region containing genes (for example, genes that confer resistance to antibiotics), flanked on both sides by IS elements (also called *IS modules*). Composite transposons may be thousands of base pairs long. The IS elements are both of the same type and are called IS*L* (for "left") and IS*R* (for "right"). Depending on the transposon, IS*L* and IS*R* may be in the same or inverted orientation relative to each other. Because the ISs themselves have terminal inverted repeats, the composite transposons also have terminal inverted repeats.

Transposition of composite transposons occurs because one or both IS elements supply the transposase, which recognizes the inverted repeats of the IS elements at the two ends of the transposon and initiates transposition (as with the transposition of IS elements). Transposition of Tn*10* is rare, occurring once in $10^7$ cell generations. Like IS elements, composite transposons produce target-site

**Figure 7.20**

**Process of integration of an IS element into chromosomal DNA.** As a result of the integration event, the target site becomes duplicated, producing direct target repeats. Thus, the integrated IS element is characterized by its inverted repeat (IR) sequences, flanked by direct target-site duplications. Integration involves making staggered cuts in the host target site. After insertion of the IS, the gaps that result are filled in with DNA polymerase and DNA ligase. (*Note*: The base sequences given for the IR are for illustration only and are neither the actual sequences found nor their actual length.)

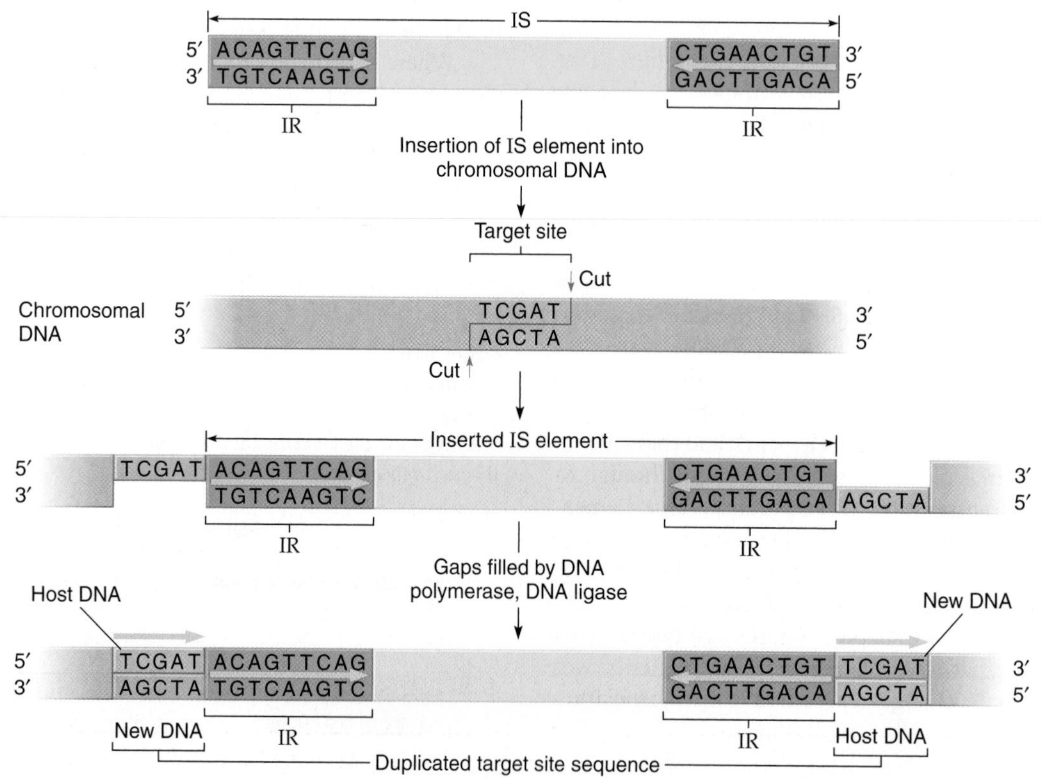

**Figure 7.21**

**Structures of bacterial transposons.** (a) The composite transposon Tn*10*. The general features of composite transposons are a central region carrying a gene or genes, such as a gene for drug resistance, flanked by either direct or inverted IS elements. The Tn*10* transposon is 9,300 bp long and consists of 6,500 bp of central, nonrepeating DNA containing the tetracycline resistance gene, flanked at each end with 1,400-bp IS elements IS*10L* and IS*10R* arranged in an inverted orientation. The IS elements themselves have terminal inverted repeats. (b) The noncomposite transposon Tn*3*. The 4,957-bp Tn*3* has genes for three enzymes in its central region: *bla* encodes β-lactamase (destroys antibiotics such as penicillin and ampicillin), *tnpA* encodes transposase, and *tnpB* encodes resolvase. Transposase and resolvase are involved in the transposition process. Tn*3* has 38-bp terminal inverted repeats that are unrelated to IS elements.

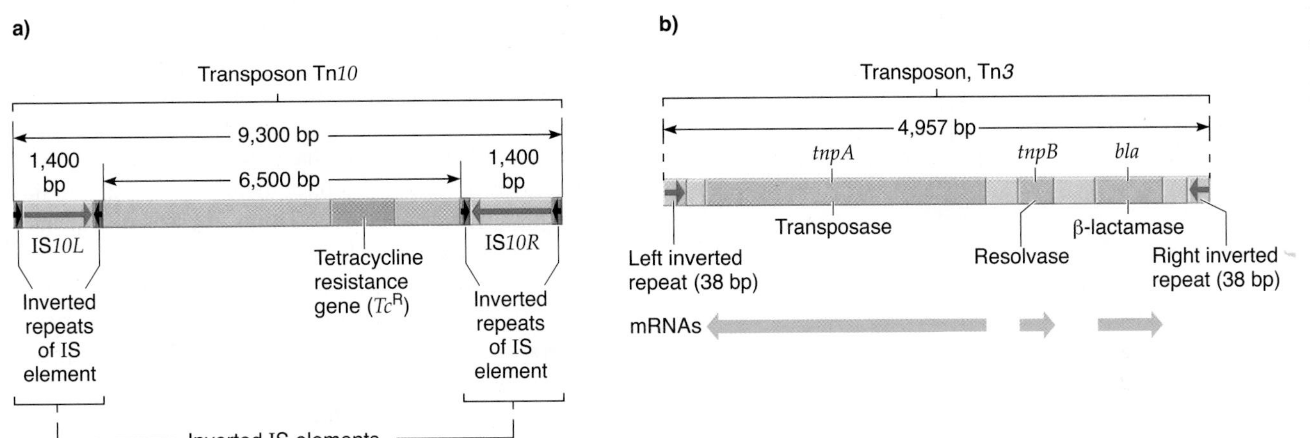


duplications after transposition. In the case of Tn*10*, the target-site duplications are 9bp long.

*Noncomposite transposons* (Figure 7.21b), exemplified by Tn*3*, also contain genes such as those conferring resistance to antibiotics, but they do not terminate with IS elements. However, at their ends they have inverted repeated sequences that are required for transposition. Enzymes for transposition are encoded by genes in the central region of noncomposite transposons. Transposase catalyzes the insertion of a transposon into new sites, and resolvase is an enzyme involved in the particular recombinational events associated with transposition. Like composite transposons, noncomposite transposons cause target-site duplications when they move. For example, Tn*3* produces a 5-bp target-site duplication when it inserts into the genome.

Figure 7.22 shows a *cointegration* mechanism for the transposition of a transposon from one DNA to another (e.g., from a plasmid to a bacterial chromosome, or vice versa). Similar events can occur between two locations on the same chromosome. First, the donor DNA containing the transposable element fuses with the recipient DNA to form a *cointegrate*. Because of the way this occurs, the transposable element is duplicated and one copy is located at each junction between donor and recipient DNA. Next, recombination between the duplicated transposable elements resolves the cointegrate into two genomes, each with one copy of the element. Because the transposable element becomes duplicated, the process is called *replicative transposition* (also called *copy-and-paste transposition*). Tn*3* and related noncomposite transposons move by replicative transposition.

A second type of transposition mechanism involves the movement of a transposable element from one location to another on the same or different DNA *without* replication of the element. This mechanism is called *conservative (nonreplicative) transposition* (also called *cut-and-paste transposition*). In other words, the element is lost from the original position when it transposes. Tn*10* transposes by conservative transposition.

As with the movement of IS elements, the transposition of transposons can cause mutations. The insertion of a transposon into the reading frame of a gene disrupts it, causing a loss-of-function mutation of that gene. Insertion into a gene's controlling region can cause changes in the level of expression of the gene, depending on the promoter elements in the transposon and how they are oriented with respect to the gene. Deletion and insertion events also result from the activities of the transposons and from crossing-over between duplicated transposons in the genome.

**iActivity**

Go to the iActivity *The Genetics Shuffle* on the student website, where you will assume the role of a researcher in a genetics lab investigating how the Tn *10* transposon is transposed.

## Transposable Elements in Eukaryotes

Transposable elements have been identified in many eukaryotes. They have been studied extensively, with most research being done with yeast, *Drosophila*, corn, and

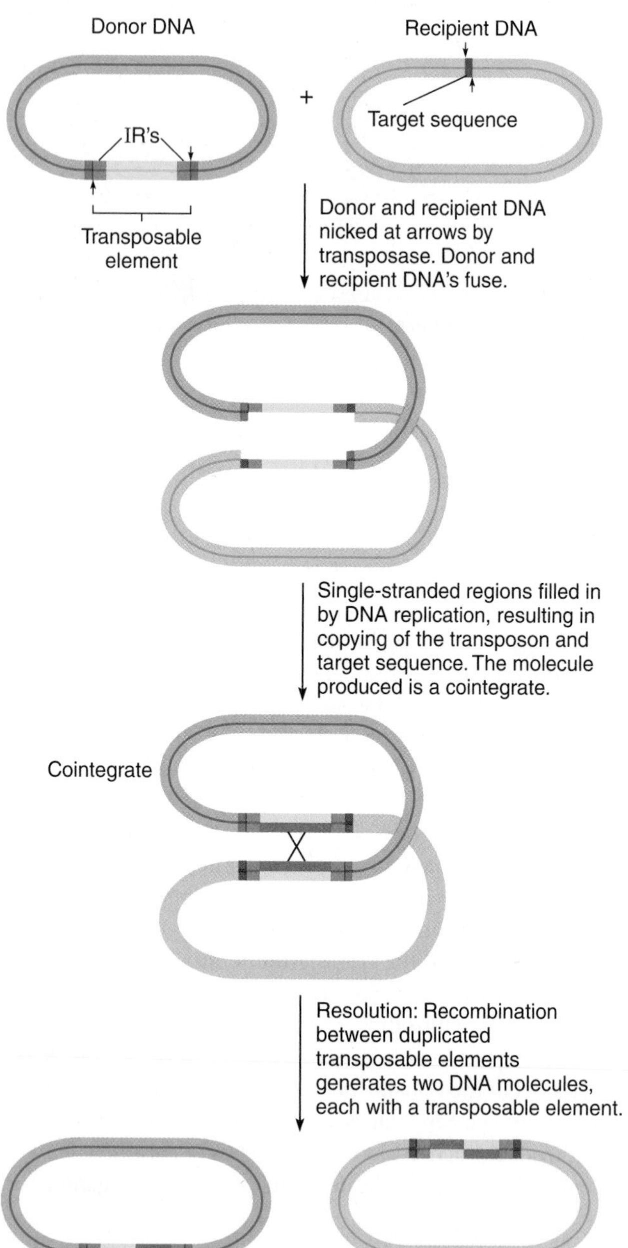
Donor DNA
Recipient DNA
+
IR's
Target sequence
Transposable element
Donor and recipient DNA nicked at arrows by transposase. Donor and recipient DNA's fuse.
Single-stranded regions filled in by DNA replication, resulting in copying of the transposon and target sequence. The molecule produced is a cointegrate.
Cointegrate
Resolution: Recombination between duplicated transposable elements generates two DNA molecules, each with a transposable element.

**Figure 7.22**

**Cointegration model for the replicative transposition of a transposable element.** A donor DNA with a transposable element fuses with a recipient DNA. During the fusion, the transposable element is duplicated, so that the product is a cointegrate molecule with one transposable element at each junction between donor and recipient DNA. The cointegrate is resolved by recombination into two molecules, each with one copy of the transposable element.

humans. In general, their structure and function are similar to those of prokaryotic transposable elements. Functional eukaryotic transposable elements have genes that encode enzymes required for transposition, and they can integrate into chromosomes at a number of sites. Thus, such elements may affect the function of any gene. Typically, the effects range from activation or repression of adjacent genes to chromosome mutations such as duplications, deletions, inversions, translocations, or breakage. That is, as with bacterial IS elements and transposons, the transposition of transposable element into genes generally causes mutations. Disruption of the amino acid-coding region of a gene typically results in a *null mutation*, which is a mutation that reduces the expression of the gene to zero. If a transposable element moves into the promoter of a gene, the efficiency of that promoter can be decreased or obliterated. Alternatively, the transposable element may provide promoter function itself and lead to an *increase* in gene expression.

**General Properties of Plant Transposable Elements.** Like some of the transposable elements discussed earlier, plant transposable elements have inverted repeated (IR) sequences at their ends and generate short, direct repeats of the target-site DNA when they integrate.

Transposable elements have been particularly well studied in corn. Geneticists have identified several families of transposable elements. Each family consists of a characteristic array of transposable elements with respect to numbers, types, and locations. Each family has two forms of transposable elements: *autonomous elements*, which can transpose by themselves, and *nonautonomous elements*, which cannot transpose by themselves because they lack the gene for transposition. The nonautonomous elements require an autonomous element to supply the missing functions. Often, the nonautonomous element is a defective derivative of the autonomous element in the family. When an autonomous element is inserted into a host gene, the resulting mutant allele is *unstable*, because the element can excise and transpose to a new location. This transposition event results in restoration of function of the gene. The frequency of transposition out of a gene is higher than the spontaneous reversion frequency for a regular point mutation; therefore, the allele produced by an autonomous element is called a *mutable allele*.

animation

**Transposable Elements in Plants**

By contrast, mutant alleles resulting from the insertion of a nonautonomous element in a gene are *stable*, because the element is unable to transpose out of the locus by itself. However, if the autonomous element of its family is also either already present in, or introduced into, the same genome, the autonomous element can provide the enzymes needed for transposition, and the nonautonomous element can then transpose.

**McClintock's Study of Transposable Elements in Corn.** In the 1940s and 1950s, Barbara McClintock did a series of elegant genetic experiments with *Zea mays* (corn) that led her to hypothesize the existence of what she called "controlling elements," which modify or suppress gene activity in corn and are mobile in the genome. Decades later, the controlling elements she studied were shown to be transposable elements. McClintock was awarded the 1983 Nobel Prize in Physiology or Medicine for her "discovery of mobile genetic elements." A fascinating and moving biographical sketch of Barbara McClintock is given in Box 7.1.

## Box 7.1 Barbara McClintock (1902–1992)

Barbara McClintock's remarkable life spanned the history of genetics in the twentieth century. She was born in Hartford, Connecticut, to Sara Handy McClintock, an accomplished pianist, poet, and painter, and Thomas Henry McClintock, a physician. Both parents were quite unconventional in their attitudes toward rearing children: They were interested in what their children would and could be rather than what they should be.

During her high school years, Barbara discovered science, and she loved to learn and figure things out. After high school, Barbara attended Cornell University, where she flourished both socially and intellectually. She enjoyed her social life, but her comfort with solitude and the tremendous joy she experienced in knowing, learning, and understanding things were to be the defining themes of her life. The decisions she made during her university years were consistent with her adamant individuality and self-containment. In Barbara's junior year, after a particularly exciting undergraduate course in genetics, her professor invited her to take a graduate course in genetics. After that, she was treated much like a graduate student. By the time she had finished her undergraduate course work, there was no question in her mind: She had to continue her studies of genetics.

At Cornell, genetics was taught in the plant-breeding department, which at the time did not take female graduate students. To circumvent this obstacle, McClintock registered in the botany department with a major in cytology and a minor in genetics and zoology. She began to work as a paid assistant to Lowell Randolph, a cytologist. McClintock and Randolph did not get along well and soon dissolved their working relationship, but as McClintock's colleague and lifelong friend Marcus Rhoades later wrote, "Their brief association was momentous because it led to the birth of maize cytogenetics." McClintock discovered that metaphase or late-prophase chromosomes in the first microspore mitosis were far better for cytological discrimination than were root-tip chromosomes. In a few weeks, she prepared detailed drawings of the maize chromosomes, which she published in *Science*.

This was McClintock's first major contribution to maize genetics, and it laid the groundwork for a veritable explosion of discoveries that connected the behavior of chromosomes to the genetic properties of an organism, defining the new field of cytogenetics. McClintock was awarded a Ph.D. in 1927 and appointed an instructor at Cornell, where she continued to work with maize. The Cornell maize genetics group was small. It included Professor R. A. Emerson, the founder of maize genetics, as well as McClintock, George Beadle, C. R. Burnham, Marcus Rhoades, and Lowell Randolph, together with a few graduate students. By all accounts, McClintock was the intellectual driving force of this talented group.

In 1929, a new graduate student, Harriet Creighton, joined the group and was guided by McClintock. Their work showed, for the first time, that genetic recombination is a reflection of the physical exchange of chromosome segments. A paper on their work, published in 1931, was perhaps McClintock's first seminal contribution to the science of genetics.

**Barbara McClintock in 1947.**

Although McClintock's fame was growing, she had no permanent position. Cornell had no female professors in fields other than home economics, so her prospects were dismal. She had already attained international recognition, but as a woman, she had little hope of securing a permanent academic position at a major research university. R. A. Emerson obtained a grant from the Rockefeller Foundation to support her work for two years, allowing her to continue to work independently. McClintock was discouraged and resentful of the disparity between her prospects and those of her male counterparts. Her extraordinary talents and accomplishments were widely appreciated, but she was also seen as difficult by many of her colleagues, in large part because of her quick mind and intolerance of second-rate work and thinking.

In 1936, Lewis Stadler convinced the University of Missouri to offer McClintock an assistant professorship. She accepted the position and began to follow the behavior of maize chromosomes that had been broken by X irradiation. However, soon after her arrival at Missouri, she understood that hers was a special appointment. She found herself excluded from regular academic activities, including faculty meetings. In 1941, she took a leave of absence from Missouri and departed with no intention of returning. She wrote to her friend Marcus Rhoades, who was planning to go to Cold Spring Harbor, New York, for the summer to grow his corn. An invitation for McClintock was arranged through Milislav Demerec (a member, and later the director, of the genetics department at the Carnegie Institution of Washington, then the dominant research laboratory at Cold Spring Harbor), who offered her a year's research appointment. Though hesitant to commit herself, McClintock accepted. When Demerec later offered

**Box 7.1 continued**

her an appointment as a permanent member of the research staff, McClintock accepted, still unsure whether she would stay. Her dislike of making commitments was a given; she insisted that she would never have become a scientist in today's world of grants, because she could not have committed herself to a written research plan. It was the unexpected that fascinated her, and she was always ready to pursue an observation that didn't fit. Nevertheless, McClintock did stay at Carnegie until 1967.

At Carnegie, McClintock continued her studies on the behavior of broken chromosomes. She was elected to the National Academy of Sciences in 1944 and to the presidency of the Genetics Society of America in 1945. In those same two years, McClintock reported observing "an interesting type of chromosomal behavior" involving the repeated loss of one of the broken chromosomes from cells during development. What struck her as odd was that, in this particular stock, it was always chromosome 9 that broke, and it always broke at the same place. McClintock called the unstable chromosome site Dissociation (*Ds*), because "the most readily recognizable consequence of its actions is this dissociation." She quickly established that the *Ds* locus would "undergo dissociation mutations only when a particular dominant factor is present." She named the factor Activator (*Ac*), because it activated chromosome breakage at *Ds*. She also reached the extraordinary conclusion that *Ac* not only was required for *Ds*-mediated chromosome breakage but also could destabilize previously stable mutations. But more than that, and unprecedentedly, the chromosome-breaking *Ds* locus could "change its position in the chromosome," a phenomenon she called *transposition*. Moreover, she had evidence that the *Ac* locus was required for the transposition of *Ds* and that, like the *Ds* locus, the *Ac* locus was mobile.

Within several years, McClintock had established beyond a doubt that both the *Ac* and *Ds* loci were capable not only of changing their positions on the genetic map, but also of inserting into loci to cause unstable mutations. She presented a paper on her work at the Cold Spring Harbor Symposium of 1951. Reactions to her presentation ranged from perplexed to hostile. Later she published several papers in refereed journals, but from the paucity of requests for reprints, she inferred an equally cool reaction on the part of the larger biological community to the astonishing news that genes could move.

McClintock's work had taken her far outside the scientific mainstream, and in a profound sense she had lost her ability to communicate with her colleagues. By her own admission, McClintock had neither a gift for written exposition nor a talent for explaining complex phenomena in simple terms. But more important factors underlay her isolation: The very notion that genes can move contradicted the assumption of the regular relationships between genes that serves as a foundation for the construction of linkage maps and the physical mapping of genes onto chromosomes. The concept that genetic elements can move would undoubtedly have met with resistance regardless of its author and presentation.

McClintock was deeply frustrated by her failure to communicate, but her fascination with the unfolding story of transposition was sufficient to keep her working at the highest level of physical and mental intensity she could sustain. By the time of her formal retirement, she had accumulated a rich store of knowledge about the genetic behavior of two markedly different transposable-element families—and beginning about the time her active fieldwork ended, transposable genetic elements began to surface in one experimental organism after another.

These later discoveries came in an altogether different age. In the two decades between McClintock's original genetic discovery of transposition and its rediscovery, genetics had undergone as profound a change as the cytogenetic revolution that had occurred in the second and third decades of the century. The genetic material had been identified as DNA, the manner in which information is encoded in the genes had been deciphered, and methods had been devised to isolate and study individual genes. Genes were no longer abstract entities known only by the consequences of their alteration or loss; they were real bits of nucleic acids that could be isolated, visualized, subtly altered, and reintroduced into living organisms.

By the time the maize transposable elements were cloned and their molecular analysis initiated, the importance of McClintock's discovery of transposition was widely recognized, and her public recognition was growing. For example, she received the National Medal of Science in 1970, she was named Prize Fellow Laureate of the MacArthur Foundation and received the Lasker Basic Medical Research Award in 1981, and in 1982 she shared the Horwitz Prize. Finally, in 1983, 35 years after the publication of the first evidence for transposition, McClintock was awarded the Nobel Prize in Physiology or Medicine.

McClintock was sure she would die at 90, and a few months after her ninetieth birthday she was gone, drifting away from life gently, like a leaf from an autumn tree. What Barbara McClintock was and what she left behind are eloquently expressed in a few short lines written many years earlier by her friend and champion, Marcus Rhoades, whose death preceded hers by a few months:

> *One of the remarkable things about Barbara McClintock's surpassingly beautiful investigations is that they came solely from her own labors. Without technical help of any kind she has by virtue of her boundless energy, her complete devotion to science, her originality and ingenuity, and her quick and high intelligence made a series of significant discoveries unparalleled in the history of cytogenetics. A skilled experimentalist, a master at interpreting cytological detail, a brilliant theoretician, she has had an illuminating and pervasive role in the development of cytology and genetics.*

*Adapted by permission of Nina Fedoroff and by courtesy of the National Academy of Sciences, Washington, DC.*

McClintock studied the genetics of corn kernel pigmentation. A number of different genes must function together to synthesize of red anthocyanin pigment, which gives the corn kernel a purple color. Mutation of any one of these genes causes a kernel to be unpigmented. McClintock studied kernels that, rather than being either of a solid color or colorless, had spots of purple pigment on an otherwise colorless kernel (Figure 7.23). She knew that the phenotype was the result of an unstable mutation. From her careful genetic and cytological studies, she concluded that the spotted phenotype was not the result of any conventional kind of mutation (such as a point mutation), but rather the result of a controlling element, which we now know is a transposon.

The explanation for the spotted kernels McClintock studied is as follows: If the corn plant carries a wild-type *C* gene, the kernel is purple; *c* (colorless) mutations are defective in purple pigment production, so the kernel is colorless. During kernel development, revertants of the mutation occur, leading to a spot of purple pigment. The earlier in development the reversion occurs, the larger is the purple spot. McClintock determined that the original *c* (colorless) mutation resulted from a "mobile controlling element" (in modern terms, a transposable element), called *Ds* for "dissociation," being inserted into the *C* gene (Figures 7.24a and 7.24b). We now know *Ds* is a nonautonomous element. Another mobile controlling element, an autonomous element called *Ac* for "activator," is required for transposition of *Ds* into the gene. *Ac* can also result in *Ds* transposing (excising perfectly in this case)

**Figure 7.23**

**Corn kernels, some of which show spots of pigment produced by cells in which a transposable element had transposed out of a pigment-producing gene, thereby allowing the gene's function to be restored.** The cells in the white areas of the kernel lack pigment because a pigment-producing gene continues to be inactivated by the presence of a transposable element within that gene.



**a) Purple kernels**

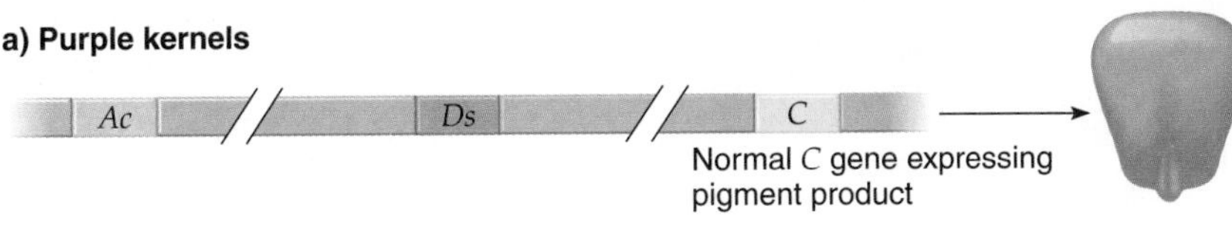


**b) Colorless kernels**

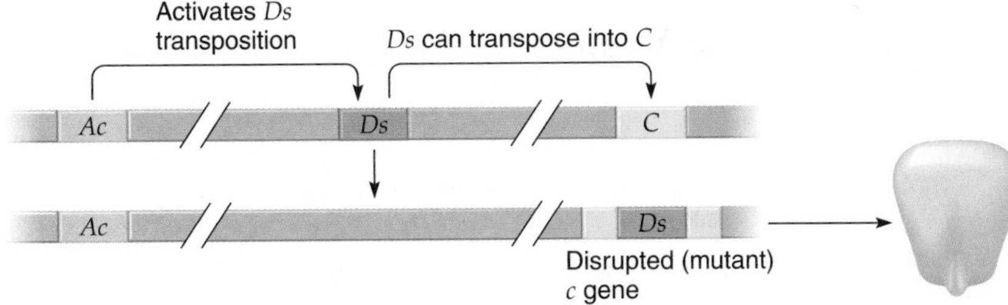


**c) Spotted kernels**

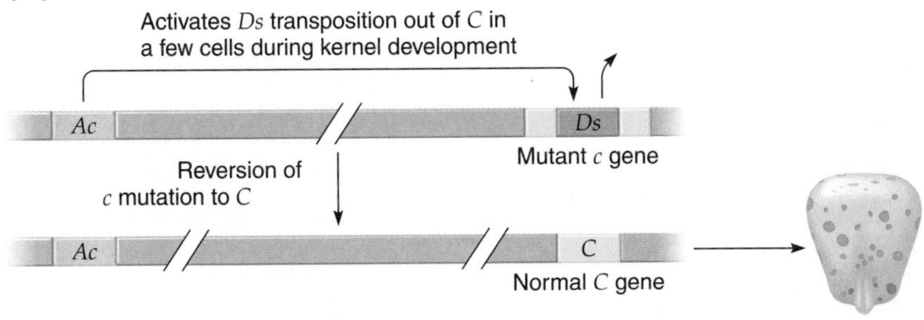


**Figure 7.24**

**Kernel color and transposable element effects in corn.** (**a**) Purple kernels result from the active *C* gene. (**b**) Colorless kernels can result when the *Ac* transposable element activates *Ds* transposition and *Ds* inserts into *C*, producing a mutation. (**c**) Spotted kernels result from reversion of the *c* mutation during kernel development when *Ac* activates *Ds* transposition out of the *C* gene.

out of the *c* gene, giving a wild-type revertant with a purple spot (Figure 7.24c).

The remarkable fact of McClintock's conclusion was that, at the time, there was no precedent for the existence of transposable genetic elements. Rather, the genome was thought to be static with regard to gene locations. Only much more recently have transposable genetic elements been widely identified and studied, and only in 1983 was direct evidence obtained for the movable genetic elements proposed by McClintock.

**The *Ac-Ds* Transposable Elements in Corn.** The *Ac-Ds* family of controlling elements has been studied in detail. The autonomous *Ac* element is 4,563 bp long, with short terminal inverted repeats and a single gene encoding the transposase. Upon insertion into the genome, it generates an 8-bp direct duplication of the target site. *Ds* elements are heterogeneous in length and sequence, but all have the same terminal IRs as *Ac* elements, because most have been generated from *Ac* by the deletion of segments or by more complex sequence rearrangements. As a result, *Ds* elements have no complete transposase gene; hence, these elements cannot transpose on their own.

Transposition of the *Ac* element occurs only during chromosome replication and is a result of the cut-and-paste (conservative) transposition mechanism (Figure 7.25). Consider a chromosome with one copy of *Ac* at a site called the *donor site*. When the chromosome region containing *Ac* replicates, two copies of *Ac* result, one on each progeny chromatid. There are two possible results of *Ac* transposition, depending on whether it occurs to a replicated or an unreplicated chromosome site.

If one of the two *Ac* elements transposes to a replicated chromosome site (Figure 7.25a), an empty donor site is left on one chromatid, and an *Ac* element remains in the homologous donor site on the other chromatid. The transposing *Ac* element inserts into a new, already replicated recipient site, which is often on the same chromosome. In Figure 7.25a, the site is shown on the same chromatid as the parental *Ac* element. Thus, in the case of transposition to an already replicated site, there is no net increase in the number of *Ac* elements.

Figure 7.25b shows the transposition of one *Ac* element to an unreplicated chromosome site. As in the first case, one of the two *Ac* elements transposes, leaving an empty donor site on one chromatid and an *Ac* element in



**Figure 7.25**

**The Ac transposition mechanism.** (**a**) Transposition to an already replicated recipient site results in no net increase in the number of *Ac* elements in the genome. (**b**) Transposition to an unreplicated recipient site results in a net increase in the number of *Ac* elements when the region of the chromosome containing the transposed element is replicated.

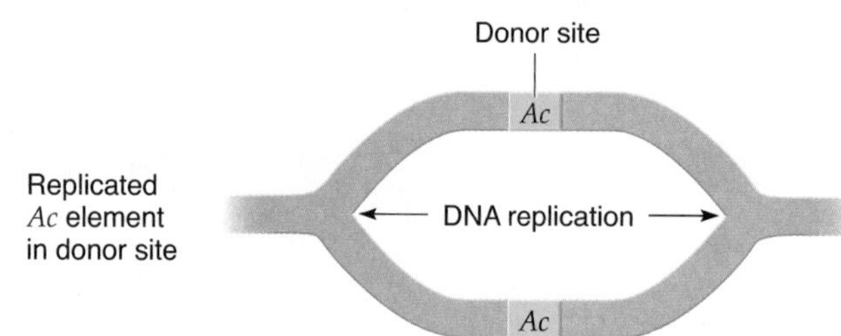


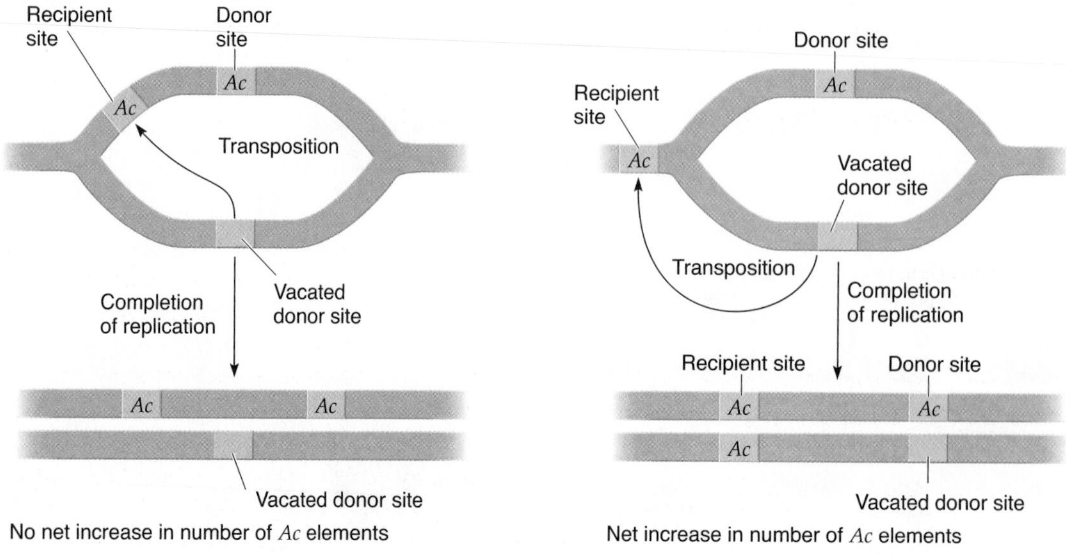

the homologous donor site on the other chromatid. But now the transposing element inserts into a nearby recipient site that has yet to be replicated. When that region of the chromosome replicates, the result will be a copy of the transposed *Ac* element on both chromatids, in addition to the one original copy of the *Ac* element at the donor site on one chromatid. Thus, in the case of transposition to an unreplicated recipient site, there is a net increase in the number of *Ac* elements.

The transposition of most *Ds* elements occurs in the same way as *Ac* transposition, using transposase supplied by an *Ac* element in the genome.

**Keynote**

The transposition mechanism of plant transposable elements is similar to that of bacterial IS elements or transposons. Transposable elements integrate at a target site by a precise mechanism, so that the integrated elements are flanked at the insertion site by a short duplication of target-site DNA of a characteristic length. Many plant transposable elements occur in families, the autonomous elements of which are able to direct their own transposition and the nonautonomous elements of which are able to transpose only when activated by an autonomous element in the same genome. Most nonautonomous elements are derived from autonomous elements by internal deletions or complex sequence rearrangements.

***Ty* Transposable Elements in Yeast.** A *Ty* transposable element is about 5.9 kb long and includes two directly repeated terminal sequences called *long terminal repeats* (LTR) or deltas (δ) (Figure 7.26). Each delta contains a promoter and sequences recognized by transposing enzymes. The *Ty* elements encode a single, 5,700-nucleotide mRNA that begins at the promoter in the delta at the left end of the element (see Figure 7.26). The mRNA transcript contains two open reading frames (ORFs), designated *TyA* and *TyB*, that encode two different proteins required for transposition. On average, a strain contains about 35 *Ty* elements.

*Ty* elements are similar to **retroviruses**—single-stranded RNA viruses that replicate via double-stranded DNA intermediates. That is, when a retrovirus infects a cell, its RNA genome is copied by *reverse transcriptase*, an enzyme that enters the cell as part of the virus particle. **Reverse transcriptase** is an RNA-dependent DNA polymerase, meaning that the enzyme uses an RNA template to produce a DNA copy. The enzyme then catalyzes the synthesis of a complementary DNA strand, in the end producing a double-stranded DNA copy of the RNA genome. The DNA integrates into the host's chromosome, where it can be transcribed to produce progeny RNA viral genomes and mRNAs for viral proteins. HIV, the virus responsible for AIDS in humans, is a retrovirus. As a result of their similarity to retroviruses, *Ty* elements were hypothesized to transpose not by a DNA-to-DNA mechanism, but by making an RNA copy of the integrated DNA sequence and then creating a new *Ty* element by reverse transcription. The new element would then integrate at a new chromosome location.

Evidence substantiating the hypothesis was obtained through experiments with *Ty* elements modified by DNA manipulation techniques to have special features enabling their transposition to be monitored easily. One compelling piece of evidence came from experiments in which an intron was placed into the *Ty* element (there are no introns in normal *Ty* elements) and the element was monitored from its initial placement through the transposition event. At the new location, the *Ty* element no longer had the intron sequence. This result could only be interpreted to mean that transposition occurred via an RNA intermediate.

Subsequently, it was shown that *Ty* elements encode a reverse transcriptase. Moreover, *Ty* viruslike particles containing *Ty* RNA and reverse transcriptase activity have been identified in yeast cells. Because of their similarity to retroviruses in this regard, *Ty* elements are called **retrotransposons**, and the transposition process is called **retrotransposition**.

**Figure 7.26**
**The *Ty* transposable element of yeast.**

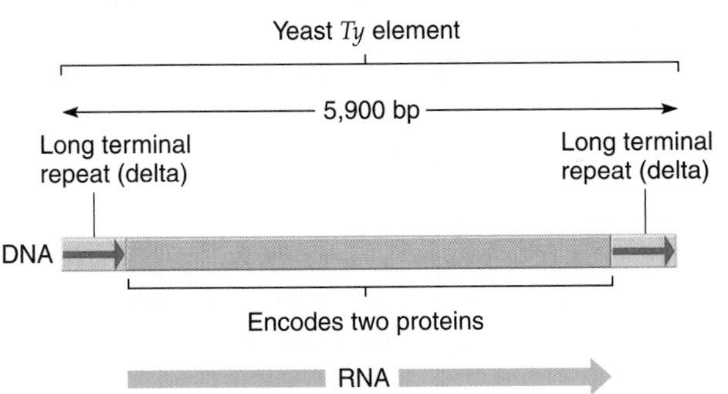


***Drosophila* Transposable Elements.** A number of classes of transposable elements have been identified in *Drosophila*. In this organism, it is estimated that about 15% of the genome is mobile—a remarkable percentage.

The *P element* is an example of a family of transposable elements in *Drosophila*. *P* elements vary in length from 500 to 2,900 bp, and each has terminal inverted repeats. The shorter *P* elements are nonautonomous elements, while the longest *P* elements are autonomous elements that encode a transposase needed for transposition of all the *P* elements (Figure 7.27). Insertion of a *P* element into a new site results in a direct repeat of the target site.

*P* elements are important vectors for transferring genes into the germ line of *Drosophila* embryos, allowing genetic manipulation of the organism. Figure 7.28 illustrates an experiment by Gerald M. Rubin and Allan C. Spradling in which the wild-type $rosy^+$ gene was introduced into a strain homozygous for a mutant *rosy* allele (which has a red-brown eye color). The $rosy^+$ gene was

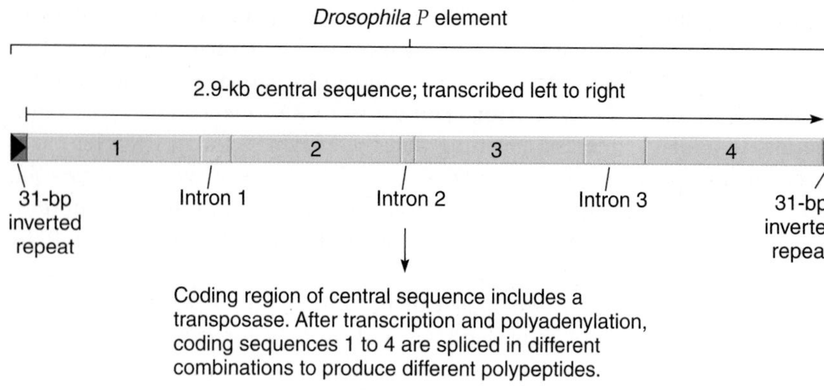
Drosophila P element
2.9-kb central sequence; transcribed left to right
1
2
3
4
31-bp inverted repeat
Intron 1
Intron 2
Intron 3
31-bp inverted repeat
Coding region of central sequence includes a transposase. After transcription and polyadenylation, coding sequences 1 to 4 are spliced in different combinations to produce different polypeptides.

**Figure 7.27**
**Structure of the autonomous *P* transposable element found in *Drosophila melanogaster*.**

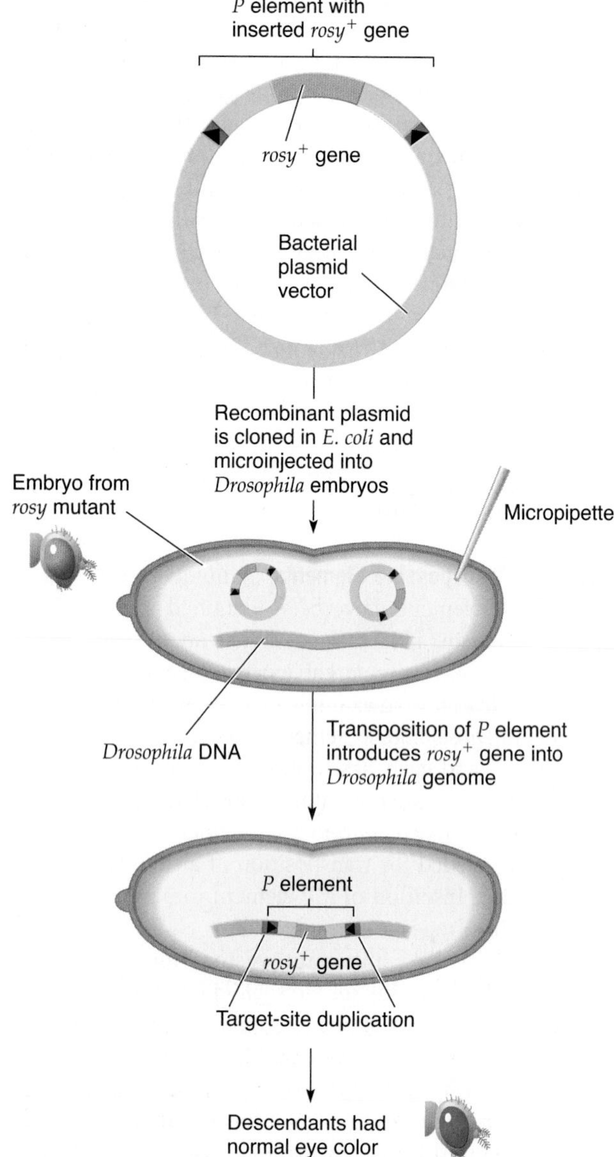
P element with inserted rosy+ gene
rosy+ gene
Bacterial plasmid vector
Recombinant plasmid is cloned in E. coli and microinjected into Drosophila embryos
Embryo from rosy mutant
Micropipette
Drosophila DNA
Transposition of P element introduces rosy+ gene into Drosophila genome
P element
rosy+ gene
Target-site duplication
Descendants had normal eye color

**Figure 7.28**
**Illustration of the use of *P* elements to introduce genes into the *Drosophila* genome.**

introduced into the middle of a *P* element by recombinant DNA techniques and cloned in a plasmid vector (see Chapter 8, pp. 175–176.) The plasmids were then microinjected into *rosy* embryos in the regions that would become the germ-line cells. *P* element-encoded transposase then catalyzed the movement of the *P* element, along with the *rosy*$^+$ gene it contained, to the *Drosophila* genome in some of the germ-line cells. When the flies that developed from these embryos produced gametes, they contained the *rosy*$^+$ gene, so descendants of those flies had normal eye color. In principle, any gene can be transferred into the genome of the fly in this way.

## Keynote

Transposable elements in eukaryotes can transpose to new sites while leaving a copy behind in the original site, or they can excise themselves from the chromosome. When the excision is imperfect, deletions can occur; and by various recombination events, other chromosomal rearrangements such as inversions and duplications can occur. Whereas most transposable elements move by using a DNA-to-DNA mechanism, some eukaryotic transposable elements, such as yeast *Ty* elements, transpose via an RNA intermediate (using a transposable elements-encoded reverse transcriptase) and so resemble retroviruses.

**Human Retrotransposons.** In Chapter 2, pp. 28–30, we discussed the different repetitive classes of DNA sequences found in the genome. Of relevance here are the LINEs (long interspersed sequences) and SINEs (short interspersed sequences) found in the moderately repetitive class of sequences. **LINEs** are repeated sequences 1,000–7,000 bp long, interspersed with unique-sequence DNA. **SINEs** are 100–400-bp repeated sequences interspersed with unique-sequence DNA. Both LINEs and SINEs occur in DNA families whose members are related by sequence.

Like the yeast *Ty* elements, LINEs and SINEs are retrotransposons. Full-length LINEs are autonomous elements that encode the enzymes for their own retrotransposition and for that of LINEs with internal deletions—nonautonomous derivatives. Those enzymes are also required for the transposition of SINEs, which are nonautonomous elements.

About 20% of the human genome consists of LINEs, with one-quarter of them being L1, the best-studied LINE. The maximum length of L1 elements is 6,500 bp, although only about 3,500 of them in the genome are of that full length, the rest having internal deletions of various length (much as corn *Ds* elements have). The full-length L1 elements contain a large open reading frame that is homologous to known reverse transcriptases. When the yeast *Ty* element reverse transcriptase gene was replaced with the putative reverse transcriptase gene from L1, the *Ty* element was able to transpose. Point mutations introduced into the sequence abolished the enzyme activity, indicating that the L1 sequence can indeed make a functional reverse transcriptase. Thus, like corn *Ac* elements, full-length L1 elements (and full-length LINEs of other families) are autonomous elements. L1 and other LINEs do not have LTRs, so they are not closely related to the retrotransposons we have already discussed. Therefore, while transposition is via an RNA intermediate, the mechanism is different. Interestingly, in 1991, two unrelated cases of hemophilia (OMIM 306700) in children were shown to result from insertions of an L1 element into the factor VIII gene, the product of which is required for normal blood clotting. Molecular analysis showed that the insertion was not present in either set of parents, leading to the conclusion that the L1 element had newly transposed. More generally, these results show that L1 elements in humans can transpose and that they can cause disease by insertional mutagenesis (that is, by inserting into genes).

SINEs are also retrotransposons, but none of them encodes enzymes needed for transposition. These nonautonomous elements depend upon the enzymes encoded by LINEs for their transposition. In humans, a very abundant SINE family is the *Alu family*. The repeated sequence in this family is about 300 bp long and is repeated 300,000 to 500,000 times in the genome, amounting to up to 3% of the total genomic DNA. The name for the family refers from the fact that the sequence contains a restriction site for the enzyme *Alu*I ("Al-you-one").

Evidence that Alu sequences can transpose has come from the study of a young male patient with neurofibromatosis (OMIM 162200), a genetic disease caused by an autosomal dominant mutation. Individuals with neurofibromatosis develop tumorlike growths (neurofibromas) over the body (see Chapter 13, p. 372). DNA analysis showed that an Alu sequence was present in one of the introns of the neurofibromatosis gene of the patient. RNA transcripts from this gene are longer than those from normal individuals. The presence of the Alu sequence in the intron disrupts the processing of the transcript, causing one exon to be lost completely from the mature mRNA. As a result, the protein encoded is 800 amino acids shorter than normal and is nonfunctional. Neither parent of the patient has neurofibromatosis, and neither has an Alu sequence in the neurofibromatosis gene. Individual members of the Alu family are not identical in sequence, having diverged over evolutionary time. This divergence made it possible to track down the same Alu sequence in the patient's parents. The analysis showed that an Alu sequence probably inserted into the neurofibromatosis gene by retrotransposition in the germ line of the father from a different chromosomal location.



# Summary

- Mutations can result in changes in heritable traits.
- Mutation is the process that alters the sequence of base pairs in a DNA molecule. The alteration can be as simple as a single base-pair substitution, insertion, or deletion or as complex as rearrangement, duplication, or deletion of whole sections of a chromosome. Mutations may occur spontaneously, such as through the effects of natural radiation or errors in replication, or they may be induced experimentally by the application of mutagens.
- Mutations at the level of the chromosome are called chromosomal mutations (see Chapter 12). Mutations in the sequences of genes and in other DNA sequences at the level of the base pair are called point mutations.
- The consequences to an organism of a mutation in a gene depend on a number of factors, especially the extent to which the amino acid-coding information for a protein is changed.
- By studying mutants that have defects in certain cellular processes, geneticists have made great progress in understanding how those processes take place. Various screening procedures have been developed to help find mutants of interest after mutagenizing cells or organisms.
- The effects of a gene mutation can be reversed either by reversion of the mutated base-pair sequence or by a mutation at a site distinct from that of the original mutation. The latter is called a suppressor mutation.
- High-energy radiation may damage genetic material by producing chemicals that interact with DNA or by causing unusual bonds between DNA bases. Mutations result if the genetic damage is not repaired. Ionizing radiation may also break chromosomes.

- Gene mutations may be caused by exposure to a variety of chemicals called chemical mutagens, a number of which exist in the environment and can cause genetic diseases in humans and other organisms.
- The Ames test can indicate whether chemicals (such as environmental or commercial chemicals) have the potential to cause mutations in humans. A large number of potential human carcinogens have been found in this way.
- In bacteria and eukaryotes, a number of enzymes repair different kinds of DNA damage. Not all DNA damage is repaired; therefore, mutations do appear, but at low frequencies. At high dosages of mutagens, repair systems cannot correct all of the damage, and mutations occur at high frequencies.
- Transposable elements are DNA segments that can insert themselves at one or more sites in a genome, and can move to other sites in that genome. Transposable elements in a cell usually are detected by the changes they bring about in the expression and activities of the genes at or near the chromosomal sites into which they integrate.
- In bacteria, two important types of transposable elements are insertion sequence (IS) elements and transposons (Tn). Each of these elements has inverted repeated sequences at its ends and encodes proteins, such as transposases, that are responsible for its transposition. Transposons also carry genes that encode other functions, such as drug resistance.
- Many transposable elements in eukaryotes resemble bacterial transposons in both general structure and transposition properties. Eukaryotic transposable elements may transpose either while leaving a copy behind in the original site or by excision from the chromosome. They integrate at a target site by a precise mechanism, so that the integrated elements are flanked at the insertion site by a short duplication of target-site DNA. Some transposable elements are autonomous elements that can direct their own transposition, and some are nonautonomous elements that can transpose only when activated by an autonomous element in the same genome.
- Although most transposons move by means of a DNA-to-DNA mechanism, some eukaryotic transposable elements move via an RNA intermediate (using a transposable element-encoded reverse transcriptase). Such transposable elements resemble retroviruses in genome organization and other properties and are called retrotransposons.

# Analytical Approaches to Solving Genetics Problems

**Q7.1** Five strains of *E. coli* containing base-substitution mutations that affect the tryptophan synthetase A polypeptide have been isolated. Figure 7.A shows the changes produced in the protein itself in the indicated mutant strains. In addition, *A23* can be further mutated to insert Ile, Thr, Ser, or the wild-type Gly into position 210.

In the following questions, assume that only a single base change can occur at each step:

**a.** Using the genetic code (see Figure 6.7, p. 108), explain how the two mutations *A23* and *A46* can result in two different amino acids being inserted at position 210. Give the nucleotide sequence of the wild-type gene at that position and of the two mutants.

**b.** Can mutants *A23* and *A46* recombine? Why or why not? If recombination can occur, what would be the result?

**c.** From what you can infer of the nucleotide sequence in the wild-type gene, indicate, for the codons specifying amino acids 48, 210, 233, and 234, whether a nonsense mutant could be generated by a single nucleotide substitution in the gene.

**A7.1**

**a.** There are no simple ways to answer questions like this one. The best approach is to scrutinize the genetic-code dictionary and use a pencil and paper to try to define the codon changes that are compatible with all the data. The number of amino acid changes in position 210 of the polypeptide is helpful in this case. The wild-type amino acid is Gly, and the codons for Gly are GGU, GGC, GGA, and GGG. The *A23* mutant has Arg at position 210, and the arginine codons are AGA, AGG, GGU, GGC, GGA, and GGG. Any Arg

**Figure 7.A**

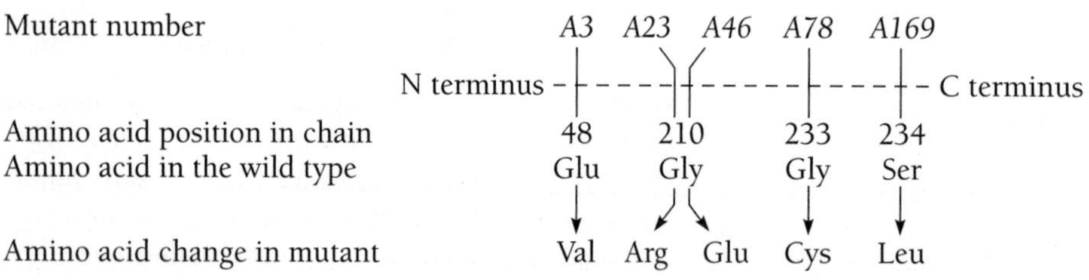

| Mutant number | *A3* | *A23* | *A46* | *A78* | *A169* |
|---|---|---|---|---|---|
| N terminus – – – C terminus | | | | | |
| Amino acid position in chain | 48 | 210 | | 233 | 234 |
| Amino acid in the wild type | Glu | Gly | | Gly | Ser |
| | ↓ | ↙ | ↘ | ↓ | ↓ |
| Amino acid change in mutant | Val | Arg | Glu | Cys | Leu |

codon could be generated by a single base change. We have to look at the amino acids at 210 generated by further mutations of *A23*. In the case of Ile, the codons are AUU, AUC, and AUA. The *only* way to get from Gly to Arg in one base change and then to Ile in a subsequent single base change is GGA (Gly) → AGA (Arg) → AUA (Ile). Is this change compatible with the other mutational changes from *A23*? There are four possible Thr codons—ACU, ACC, ACA, and ACG—so a mutation from AGA (Arg) to ACA (Thr) would fit. There are six possible Ser codons—UCU, UCC, UCA, UCG, AGU, and AGC—so a mutation from AGA to either AGU or AGC would fit.

As regards the *A46* mutant, the possible codons for Glu are GGA and GAG. Given that the wild-type codon is GGA (Glu), the only possible single base change that gives Glu is if the Glu codon in the mutant is GAA. So the answer to the question is that the wild-type sequence at position 210 is GGA, the sequence in the *A23* mutant is AGA, and the sequence in the *A46* mutant is GAA. In other words, the *A23* and *A46* mutations are in different bases of the codon.

**b.** The answer to this question follows from the answer deduced in part (**a**). Mutants *A23* and *A46* can recombine because the mutations in the two mutant strains are in different base pairs. The results of a single recombination event (at the DNA level) between the first and second base of the codon in AGA × GAA are a wild-type GGA codon (Gly) and a double mutant AAA codon (Lys). Recombination can also occur between the second and third bases of the codon, but the products are AGA and GAA—that is, identical to the parents.

**c.** Amino acid 48 had a Glu-to-Val change. This change must have involved GAA to GUA or GAG to GUG. In either case, the Glu codon can mutate with a single base-pair change to a nonsense codon, UAA or UAG, respectively.

Amino acid 210 in the wild type has a GGA codon, as we have already discussed. This gene could mutate to the UGA nonsense codon with a single base-pair change.

Amino acid 233 had a Gly-to-Cys change. This change must have involved either GGU to UGU or GGC to UGC. In either case, the Gly codon cannot mutate to a nonsense codon with one base change.

Amino acid 234 had a Ser-to-Leu change. This change was either UCA to UUA or UCG to UUG. If the Ser codon was UCA, it could be changed to AGA in one step, but if the Ser codon was UCG, it cannot change to a nonsense codon in one step.

**Q7.2** The chemically induced mutations *a*, *b*, and *c* show specific reversion patterns when subjected to treatment by the following mutagens: 2-aminopurine (AP), 5-bromouracil (BU), proflavin (PRO), and hydroxylamine (HA). AP is a base-analog mutagen that induces mainly AT-to-GC changes and can cause GC-to-AT changes also. BU is a base-analog mutagen that induces mainly GC-to-AT changes and can cause AT-to-GC changes. PRO is an intercalating agent that can cause a single base-pair addition or deletion with no specificity. HA is a base-modifying agent that modifies cytosine, causing one-way GC-to-AT transitions. The reversion patterns are shown in the following table:

| | Mutagens Tested in Reversion Studies | | | |
|---|---|---|---|---|
| **Mutation** | **AP** | **BU** | **PRO** | **HA** |
| *a* | – | – | + | – |
| *b* | + | + | + | + |
| *c* | + | + | + | – |

(Note: + indicates that many reversions to wild type were found; – indicates that no reversions or very few reversions to wild type were found.)

For each original mutation ($a^+$ to *a*, $b^+$ to *b*, etc.), indicate the probable base-pair change (A-T to G-C, deletion of G-C, etc.) and the mutagen that was probably used to induce the original change.

**A7.2** This question tests your knowledge of the base-pair changes that can be induced by the various mutagens used.

Mutagen AP induces mainly AT-to-GC changes and can cause GC-to-AT changes. Thus, AP-induced mutations can be reverted by AP.

Base-analog mutagen BU induces mainly GC-to-AT changes and can cause AT-to-GC changes, so BU-induced mutations can be reverted by BU.

Proflavin causes single base-pair deletions or additions, so proflavin-induced changes can be reverted by a second treatment with proflavin.

Mutagen HA causes one-way GC-to-AT transitions from, so HA-induced mutations cannot be reverted by HA.

With these mutagen specificities in mind, we can answer the questions about each mutation in turn.

Mutation $a^+$ to *a*: The *a* mutation was reverted only by proflavin, indicating that it was a deletion or an addition (a frameshift mutation). Therefore, the original mutation was induced by an intercalating agent such as proflavin, because it is the only class of mutagen that can cause an addition or a deletion.

Mutation $b^+$ to *b*: The *b* mutation was reverted by AP, BU, or HA. A key here is that HA causes only GC-to-AT changes. Therefore, *b* must be GC, and the original $b^+$ must have been AT. Thus, the mutational change of $b^+$ to *b* must have been caused by treatment with AP or BU, because these are the only two mutagens in the list able to induce that change.

Mutation $c^+$ to *c*: The *c* mutation was reverted only by AP and BU. Since it could not be reverted by HA, *c* must be AT and $c^+$ must be GC. The mutational change from $c^+$ to *c* therefore involved a GC-to-AT transition and could have resulted from treatment with AP, BU, or HA.

**Q7.3** Imagine that you are a corn geneticist. You are interested in a gene you call *zma*, which is involved in the formation of the tiny hairlike structures on the upper surfaces of leaves. You have a cDNA clone of this gene. In a particular strain of corn that contains many copies of *Ac* and *Ds*, but no other transposable elements, you observe a mutation of the *zma* gene. You want to figure out whether this mutation involves the insertion of a transposable element into the *zma* gene. How would you proceed? Suggest at least two approaches, and state how your expectations for an inserted transposable element would differ from your expectations for an ordinary gene mutation.

**A7.3.** One approach would be to make a detailed examination of leaf surfaces in mutant plants. Since there are many copies of *Ac* in the strain, if a transposable element has inserted into *zma*, it should be able to leave again, so the mutation of *zma* would be unstable. The leaf surfaces should then show a patchy distribution of regions with, and regions without, the hairlike structures. A simple point mutation would be expected to be more stable.

A second approach would be to digest the DNA from mutant plants and the DNA from normal plants with a particular restriction endonuclease, run the digested DNA on a gel, prepare a Southern blot, and probe the blot using the cDNA. If a transposable element has inserted into the *zma* gene in the mutant plants, then the probe should bind to fragments of different molecular weight in mutant, compared with normal, DNA. This would not be the case if a simple point mutation had occurred.

# Questions and Problems

***7.1** Mutations are (choose the correct answer)

**a.** caused by genetic recombination.
**b.** heritable changes in genetic information.
**c.** caused by faulty transcription of the genetic code.
**d.** usually, but not always, beneficial to the development of the individuals in which they occur.

***7.2** Answer true or false: Mutations occur more frequently if there is a need for them.

**7.3** Which of the following is *not* a class of mutation?

**a.** frameshift
**b.** missense
**c.** transition
**d.** transversion
**e.** none of the above; all are classes of mutation

***7.4** Ultraviolet light usually causes mutations by a mechanism involving (choose the correct answer)

**a.** one-strand breakage in DNA.
**b.** light-induced change of thymine to alkylated guanine.
**c.** induction of thymine dimers and their persistence or imperfect repair.
**d.** inversion of DNA segments.
**e.** deletion of DNA segments.
**f.** all of the above.

**7.5** The amino acid sequence shown in the following table was obtained from the central region of a particular polypeptide chain in the wild-type and several mutant bacterial strains:

| | Codon | | | | | | | | |
|---|---|---|---|---|---|---|---|---|---|
| | 1 | 2 | 3 | 4 | 5 | 6 | 7 | 8 | 9 |
| a. Wild type: . . . | Phe | Leu | Pro | Thr | Val | Thr | Thr | Arg | Trp |
| b. Mutant 1: . . . | Phe | Leu | His | His | Gly | Asp | Asp | Thr | Val |
| c. Mutant 2: . . . | Phe | Leu | Pro | Thr | Met | Thr | Thr | Arg | Trp |
| d. Mutant 3: . . . | Phe | Leu | Pro | Thr | Val | Thr | Thr | Arg | |
| e. Mutant 4: . . . | Phe | Pro | Pro | Arg | | | | | |
| f. Mutant 5: . . . | Phe | Leu | Pro | Ser | Val | Thr | Thr | Arg | Trp |

For each mutant, say what change has occurred at the DNA level, whether the change is a base-pair substitution mutation (transversion or transition, missense or nonsense) or a frameshift mutation, and in which codon the mutation occurred. (Refer to the codon dictionary in Figure 6.7, p. 108.)

***7.6** In mutant strain X of *E. coli*, a leucine tRNA that recognizes the codon 5′-CUG-3′ in normal cells has been altered so that it now recognizes the codon 5′-GUG-3′. A missense mutation that affects amino acid 10 of a particular protein is suppressed in mutant X cells.

**a.** What are the anticodons of the two Leu tRNAs, and what mutational event has occurred in mutant X cells?
**b.** What amino acid would normally be present at position 10 of the protein (without the missense mutation)?
**c.** What amino acid is put in at position 10 if the missense mutation is not suppressed (i.e., in normal cells)?
**d.** What amino acid is inserted at position 10 if the missense mutation is suppressed (i.e., in mutant X cells)?

**7.7** A researcher using a model eukaryotic experimental system has identified a temperature-sensitive mutation, *rpIIA*$^{ts}$, in a gene that encodes a protein subunit of RNA polymerase II. This mutation is a missense mutation. Mutants have a recessive lethal phenotype at the higher, restrictive temperature, but grow at the lower, permissive (normal) temperature. To identify genes whose products interact with the subunit of RNA polymerase II, the researcher designs a screen to isolate mutations that will act as dominant suppressors of the temperature-sensitive recessive lethal mutation.

**a.** Explain how a new mutation in an interacting protein could suppress the lethality of the temperature-sensitive original mutation.
**b.** In addition to mutations in interacting proteins, what other type of suppressor mutations might be found?

c. Outline how the researcher might select for the new suppressor mutations.
d. Do you expect the frequency of suppressor mutations to be similar to, much greater than, or much less than the frequency of new mutations at a typical eukaryotic gene?
e. How might this approach be used generally to identify genes whose products interact to control transcription?

*7.8 The mutant *lacZ-1* was induced by treating *E. coli* cells with acridine, whereas *lacZ-2* was induced with 5BU. What kinds of mutants are these likely to be? Explain. How could you confirm your predictions by studying the structure of the β-galactosidase in these cells?

*7.9

a. The sequence of nucleotides in an mRNA is

5′-AUGACCCAUUGGUCUCGUUAG-3′

Assuming that ribosomes could translate this mRNA, how many amino acids long would you expect the resulting polypeptide chain to be?

b. Hydroxylamine is a mutagen that results in the replacement of an A-T base pair for a G-C base pair in the DNA; that is, it induces a transition mutation. When hydroxylamine was applied to the organism that made the mRNA molecule shown in part (a), a strain was isolated in which a mutation occurred at the 11th position of the DNA that coded for the mRNA. How many amino acids long would you expect the polypeptide made by this mutant to be? Why?

7.10 In a series of 94,075 babies born in a particular hospital in Copenhagen, 10 were achondroplastic dwarfs (an autosomal dominant condition). Two of these 10 had an achondroplastic parent. The other 8 achondroplastic babies each had two normal parents. What is the apparent mutation rate at the achondroplasia locus?

*7.11 Three of the codons in the genetic code are chain-terminating codons for which no naturally occurring tRNAs exist. Just like any other codons in the DNA, though, these codons can change as a result of base-pair changes in the DNA. Confining yourself to single base-pair changes at a time, and referring to the genetic code listed in Figure 6.7, p. 108, determine which amino acids could be inserted into a polypeptide by mutation of these chain-terminating codons:

a. UAG
b. UAA
c. UGA

7.12 Nonsense mutations change sense codons into chain-terminating (nonsense) codons. Another class of mutation alters the sequence of a tRNA's anticodon so that the mutant tRNA now recognizes a nonsense codon and inserts an amino acid into an elongating polypeptide chain. When the mutant tRNA is able to suppress a nonsense mutation, it is called a tRNA nonsense suppressor.

a. Which sense codons can be changed by a single nucleotide mutation to nonsense codons? Which amino acids are encoded by these codons? (Compare this question, and your answer, to those of Question 7.11.)
b. Ignoring the effects of wobble, which amino acids have tRNAs with anticodons that can be changed by a single nucleotide mutation to a tRNA nonsense suppressor?
c. Will tRNA nonsense suppressors always insert the correct (wild-type) amino acid into the elongating polypeptide chain?

7.13 The amino acid substitutions in the following figure occur in the α and β chains of human hemoglobin:

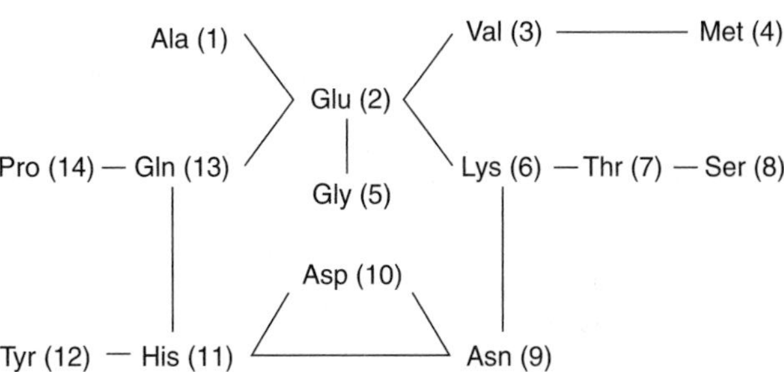


Those amino acids connected by lines are related by single-nucleotide changes. Propose the most likely codon or codons for each of the numbered amino acids. (Refer to the genetic code in Figure 6.7, p. 108.)

*7.14 Charles Yanofsky studied the tryptophan synthetase of *E. coli* in an attempt to identify the base sequence specifying this protein. The wild type gave a protein with a glycine in position 38. Yanofsky isolated two *trp* mutants: *A23* and *A46*. Mutant *A23* had Arg instead of Gly at position 38, and mutant *A46* had Glu at position 38. Mutant *A23* was plated on minimal medium, and four spontaneous revertants to prototrophy were obtained. The tryptophan synthetase from each of the four revertants was isolated, and the amino acids at position 38 were identified. Revertant 1 had Ile, revertant 2 had Thr, revertant 3 had Ser, and revertant 4 had Gly. In a similar fashion, three revertants from *A46* were recovered, and the tryptophan synthetase from each was isolated and studied. At position 38, revertant 1 had Gly, revertant 2 had Ala, and revertant 3 had Val. A summary of these data is given in the following figure:

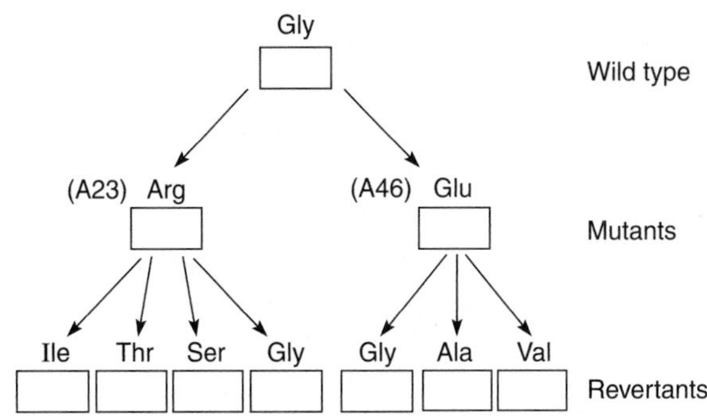

Using the genetic code in Figure 6.7, p. 108, deduce the codons for the wild type, for the mutants *A23* and *A46*, and for the revertants, and place each designation in the space provided in the figure.

**7.15** Consider an enzyme chewase from a theoretical microorganism. In the wild-type cell, chewase has the following sequence of amino acids at positions 39 to 47 (reading from the amino end) in the polypeptide chain:

```
-Met-Phe-Ala-Asn-His-Lys-Ser-Val-Gly-
 39  40  41  42  43  44  45  46  47
```

A mutant organism that lacks chewase activity was obtained. The mutant was induced by a mutagen known to cause single base-pair insertions or deletions. Instead of making the complete chewase chain, the mutant made a short polypeptide chain only 45 amino acids long. The first 38 amino acids were in the same sequence as the first 38 of the normal chewase, but the last seven amino acids were as follows:

```
-Met-Leu-Leu-Thr-Ile-Arg-Val
 39  40  41  42  43  44  45
```

A partial revertant of the mutant was induced by treating it with the same mutagen. The revertant that made a partly active chewase has the following sequence of amino acids at positions 39 to 47 in its amino acid chain:

```
-Met-Leu-Leu-Thr-Ile-Arg-Gly-Val-Gly-
 39  40  41  42  43  44  45  46  47
```

Using the genetic code given in Figure 6.7, p. 108, deduce the nucleotide sequences for the mRNA molecules that specify this region of the protein in each of the three strains.

***7.16** The Ames test can effectively evaluate whether compounds or their metabolites are mutagenic.

**a.** What type of genetic selection is used by the Ames test? Explain why this type of selection allows for a highly sensitive test.

**b.** Describe how you would use the Ames test to assess whether a widely used herbicide or its animal metabolites are mutagenic.

**c.** In a crop field, the herbicide decays to compounds that are not identical to its animal metabolites. How does this information affect your interpretation of any Ames test results from part (**b**)? If it poses additional concerns, how might you address them?

**7.17** DNA polymerases from different organisms differ in the fidelity of their nucleotide insertion; however, even the best DNA polymerases make mistakes, usually mismatches. If such mismatches are not corrected, they can become fixed as mutations after the next round of replication.

**a.** How does DNA polymerase attempt to correct mismatches during DNA replication?

**b.** What mechanism is used to repair such mismatches if they escape detection by DNA polymerase?

**c.** How is the mismatched base in the newly synthesized strand distinguished from the correct base in the template strand?

**7.18** Two mechanisms in *E. coli* were described for the repair of thymine dimer formation after exposure to ultraviolet light: photoreactivation and excision (dark) repair. Compare these mechanisms, indicating how each achieves repair.

***7.19** DNA damage by mutagens has serious consequences for DNA replication. Without specific base pairing, the replication enzymes cannot specify a complementary strand, and gaps are left after the passing of a replication fork.

**a.** What response has *E. coli* developed to large amounts of DNA damage by mutagens? How is this response coordinately controlled?

**b.** Why is the response itself a mutagenic system?

**c.** What effects would loss-of-function mutations in *recA* or *lexA* have on *E. coli*'s response?

***7.20** After a culture of *E. coli* cells was treated with the chemical 5-bromouracil, it was noted that the frequency of mutants was much higher than normal. Mutant colonies were then isolated, grown, and treated with nitrous acid; some of the mutant strains reverted to wild type.

**a.** In terms of the Watson-Crick model, diagram a series of steps by which 5BU may have produced the mutants.

**b.** Assuming that the revertants were not caused by suppressor mutations, indicate the steps by which nitrous acid may have produced the back mutations.

***7.21** The mutagen 5-bromouracil (5BU) was added to a rapidly dividing culture of wild-type *E. coli* cells growing in a liquid medium containing a rich variety of nutrients, including arginine. After one cell division, the cells were washed free of the mutagen, resuspended in sterile water, and plated onto master plates containing minimal medium supplemented only with arginine. Plates were obtained having well-separated colonies, so that each colony derived from just one progenitor cell. The colonies were then replica-plated from the master plates onto plates containing minimal medium. One colony that grew in the presence of arginine but failed to grow on minimal medium was selected from the master plate. The cells of this colony were suspended in sterile water, and each of 20 tubes containing minimal medium supplemented with arginine was inoculated with a few cells from this suspension. After the 20 cultures grew to a density of $10^8$ cells/mL, 0.1 mL from each was plated on plates containing minimal medium. The following table shows the number of bacterial colonies that grew on each plate.

| Plate | Number of Colonies |
|---|---|
| 1 | 1 |
| 2 | 0 |
| 3 | 4 |
| 4 | 0 |
| 5 | 15 |
| 6 | 116 |
| 7 | 1 |
| 8 | 45 |
| 9 | 160 |
| 10 | 0 |
| 11 | 3 |
| 12 | 1 |
| 13 | 130 |
| 14 | 1 |
| 15 | 0 |
| 16 | 0 |
| 17 | 7 |
| 18 | 9 |
| 19 | 320 |
| 20 | 0 |

**a.** In which stage(s) of this process did mutations occur? What is the evidence that a mutational event occurred?

**b.** At each stage where mutations occurred, were the mutations induced or spontaneous? Were they forward or reverse mutations?

**c.** At each stage where mutations were recovered, how were they selected for?

**d.** Though all of the 20 cultures started from a single colony that failed to grow on minimal medium were treated identically, they produced different numbers of bacterial colonies when they were plated. Why did this occur?

**e.** Suppose that 5BU had been added to the medium in the 20 tubes. Would plating the 20 cultures have given the same results? If not, how would they have differed?

**f.** Supposing that methylmethane sulfonate (MMS) rather than 5BU had been added to the medium in the 20 tubes, answer the questions given above in part **(e)**.

**7.22** A single, hypothetical strand of DNA is composed of the following base sequence, where A indicates adenine, T indicates thymine, G indicates guanine, C denotes cytosine, U denotes uracil, BU is 5-bromouracil, 2AP is 2-amino-purine, BU-enol is a tautomer of 5BU, 2AP-imino is a rare tautomer of 2AP, HX is hypoxanthine, X is xanthine, and 5′ and 3′ are the numbers of the free, OH-containing carbons on the deoxyribose part of the terminal nucleotides:

5′-T-HX-U-A-G-BU-enol-2AP-C-BU-X-2AP-imino-3′

**a.** Opposite the bases of the hypothetical strand, and using the shorthand of the base sequence, indicate the sequence of bases on a complementary strand of DNA.

**b.** Indicate the direction of replication of the new strand by drawing an arrow next to the new strand of DNA from part (**a**).

**c.** When postmeiotic germ cells of a higher organism are exposed to a chemical mutagen before fertilization, the resulting offspring expressing an induced mutation are almost always mosaics for wild-type and mutant tissue. Give at least one reason that these mosaics, and not so-called complete or whole-body mutants, are found in the progeny of treated individuals.

The following information applies to Problems 7.23 through 7.27: A solution of single-stranded DNA is used as the template in a series of reaction mixtures and has the base sequence

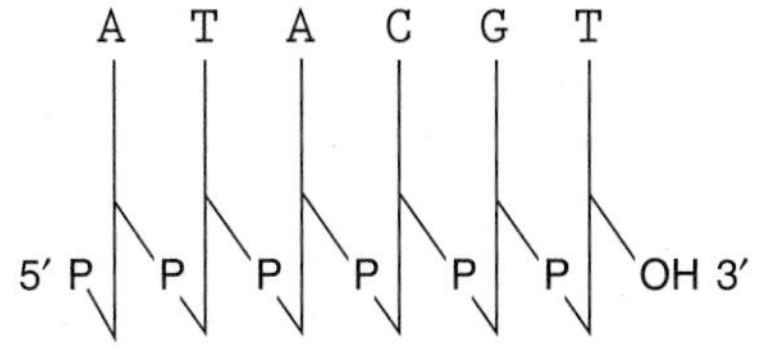

A T A C G T

5′ P P P P P P OH 3′

where A = adenine, G = guanine, C = cytosine, T = thymine, H = hypoxanthine, and $HNO_2$ = nitrous acid. Use the shorthand system shown in the sequence, and draw the products expected from the reaction mixtures. Assume that a primer is available in each case.

**7.23** The DNA template + DNA polymerase + dATP + dGTP + dCTP + dTTP + $Mg^{2+}$.

*__7.24__ The DNA template + DNA polymerase + dATP + dGMP + dCTP + dTTP + $Mg^{2+}$.

**7.25** The DNA template + DNA polymerase + dATP + dHTP + dGMP + dTTP + $Mg^{2+}$.

*__7.26__ The DNA template is pretreated with $HNO_2$ + DNA polymerase + dATP + dGTP + dCTP + dTTP + $Mg^{2+}$.

**7.27** The DNA template + DNA polymerase + dATP + dGMP + dHTP + dCTP + dTTP + $Mg^{2+}$.

**7.28** A strong experimental approach to determining the mode of action of mutagens is to examine the revertibility of the products of one mutagen by other mutagens. The following table presents data on the revertibility of *rII* mutations in phage T2 by various mutagens ("+" indicates majority of mutants reverted, " –" indicates almost no reversion; BU = 5-bromouracil, AP = 2-aminopurine, NA = nitrous acid, and HA = hydroxylamine):

| Mutation Induced by | Proportion of Mutations Reverted by BU | AP | NA | HA | Base-pair Substitution Inferred |
|---|---|---|---|---|---|
| BU | + | ___ | ___ | − | ___ |
| AP | ___ | − | + | ___ | ___ |
| NA | + | + | ___ | + | ___ |
| HA | ___ | ___ | + | − | GC → AT |

Fill in the empty spaces.

**7.29**

**a.** Nitrous acid deaminates adenine to form hypoxanthine, which forms two hydrogen bonds with cytosine during DNA replication. After a wild-type strain of bacteria is treated with nitrous acid, a mutant is recovered that is caused by an amino acid substitution in a protein: wild-type methionine (Met) has been replaced with valine (Val) in the mutant. What is the simplest explanation for this observation?

**b.** Hydroxylamine adds a hydroxyl (OH) group to cytosine, causing it to pair with adenine. Could mutant organisms like those in part **(a)** be back-mutated (returned to normal) using hydroxylamine? Explain.

***7.30** A wild-type strain of bacteria produces a protein with the amino acid proline (Pro) at one site. Treatment of the strain with nitrous acid, which deaminates C to make it U, produces two different mutants. At the site, one mutant has a substitution of serine (Ser), and the other has a substitution of leucine (Leu).

Treatment of the two mutants with nitrous acid now produces new mutant strains, each with phenylalanine (Phe) at the site. Treatment of these new Phe-carrying mutants with nitrous acid then produces no change. The results are summarized in the following figure:

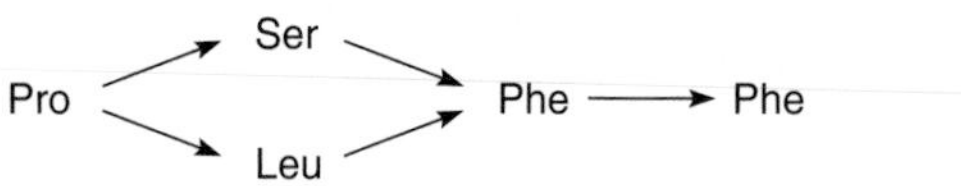


Using the appropriate codons, show how it is possible for nitrous acid to produce these changes and why further treatment has no influence. (Assume that only single-nucleotide changes occur at each step.)

***7.31** Three *ara* mutants of *E. coli* were induced by mutagen X. The ability of other mutagens to cause the reverse change (*ara* to *ara*$^+$) was tested, with the results shown in Table 7.A.

Assume that all *ara*$^+$ cells are true revertants. What base changes were probably involved in forming the three original mutations? What kinds of mutations are caused by mutagen X?

**7.32** As genes have been cloned for a number of human diseases caused by defects in DNA repair and replication, striking evolutionary parallels have been found between human and bacterial DNA repair systems. Discuss the features of DNA repair systems that appear to be shared in these two types of organism.

***7.33** MacConkey-lactose medium contains a dye indicator that detects the fermentation of the sugar lactose. When *E. coli* cells able to metabolize lactose are plated on this medium, they produce red-colored colonies. Cells unable to metabolize lactose (due to a point mutation) mostly produce completely white colonies. However, occasionally they produce a white colony having a red sector whose size varies.

**a.** How can you explain the appearance of red sectors within the otherwise white colonies? Why does the size of the red sectors vary?

**b.** What kinds of colonies would be seen in a doubly mutant *E. coli* strain having a point mutation preventing it from metabolizing lactose and a *mutator* mutation?

**c.** Explain what functions are affected by *mutator* mutations and how the absence of one of these functions would lead to the colony phenotype you described for part **(b)**.

**7.34** Distinguish between prokaryotic insertion elements and transposons. How do composite transposons differ from noncomposite transposons?

**7.35** What properties do bacterial and eukaryotic transposable elements have in common?

**7.36** An IS element became inserted into the *lacZ* gene of *E. coli*. Later, a small deletion occurred that removed 40 base pairs near the left border of the IS element. The deletion removed 10 *lacZ* base pairs, including the left copy of the target site, and the 30 leftmost base pairs of the IS element. What will be the consequence of this deletion?

**Table 7.A**

| Mutant | Frequency of *ara*$^+$ Cells among Total Cells after Treatment: None | Mutagen: BU | AP | HA | Frameshift |
|---|---|---|---|---|---|
| *ara-1* | $1.5 \times 10^{-8}$ | $5 \times 10^{-5}$ | $1.3 \times 10^{-4}$ | $1.3 \times 10^{-8}$ | $1.6 \times 10^{-8}$ |
| *ara-2* | $2 \times 10^{-7}$ | $2 \times 10^{-4}$ | $6 \times 10^{-5}$ | $3 \times 10^{-5}$ | $1.6 \times 10^{-7}$ |
| *ara-3* | $6 \times 10^{-7}$ | $10^{-5}$ | $9 \times 10^{-6}$ | $5 \times 10^{-6}$ | $6.5 \times 10^{-7}$ |

**7.37** Although the detailed mechanisms by which transposable elements transpose differ widely, some features underlying transposition are shared. Examine the shared and different features by answering the following questions:

**a.** Use an example to illustrate different transposition mechanisms that require
  i. DNA replication of the element.
  ii. no DNA replication of the element.
  iii. an RNA intermediate.

**b.** What evidence is there that the inverted or direct terminal repeat sequences found in transposable elements are essential for transposition?

**c.** Do all transposable elements generate a target-site duplication after insertion?

**7.38** In addition to single gene mutations caused by the insertion of transposable elements, the frequency of chromosomal aberrations such as deletions or inversions can be increased when transposable elements are present. How?

***7.39** A geneticist was studying glucose metabolism in yeast and deduced both the normal structure of the enzyme glucose-6-phosphatase (G6Pase) and the DNA sequence of its coding region. She was using a wild-type strain called A to study another enzyme for many generations when she noticed that a morphologically peculiar mutant had arisen from one of the strain A cultures. She grew the mutant up into a large stock and found that the defect in this mutant involved a markedly reduced G6Pase activity. She isolated the G6Pase protein from these mutant cells and found that it was present in normal amounts but had an abnormal structure. The N-terminal 70% of the protein was normal. The C-terminal 30% was present, but altered in sequence by a frameshift reflecting the insertion of 1 base pair, and the N-terminal 70% and the C-terminal 30% were separated by 111 new amino acids unrelated to normal G6Pase. These amino acids represented predominantly the AT-rich codons (Phe, Leu, Asn, Lys, Ile, and Tyr). There were also two extra amino acids added at the C-terminal end. Explain these results.

***7.40** Consider two theoretical yeast transposable elements, A and B. Each contains an intron, and each transposes to a new location in the yeast genome. Suppose you then examine the transposable elements for the presence of the intron. In the new locations, you find that A has no intron, but B does. From these facts, what can you conclude about the mechanisms of transposable element movement for A and B?

**7.41** After the discovery that *P* elements could be used to develop transformation vectors in *Drosophila melanogaster*, attempts were made to use them for the development of germ-line transformation in several different insect species. Charalambos Savakis and his colleagues successfully used a different transposable element found in *Drosophila*—the *Minos* element—to develop germ-line transformation in that organism and in the medfly, *Ceratitus capitata*, a major agricultural pest present in Mediterranean climates.

**a.** What is the value of developing a transformation vector for an insect pest?

**b.** What basic information about the *Minos* element would need to be gathered before it could be used for germ-line transformation?

# 8 Genomics: The Mapping and Sequencing of Genomes

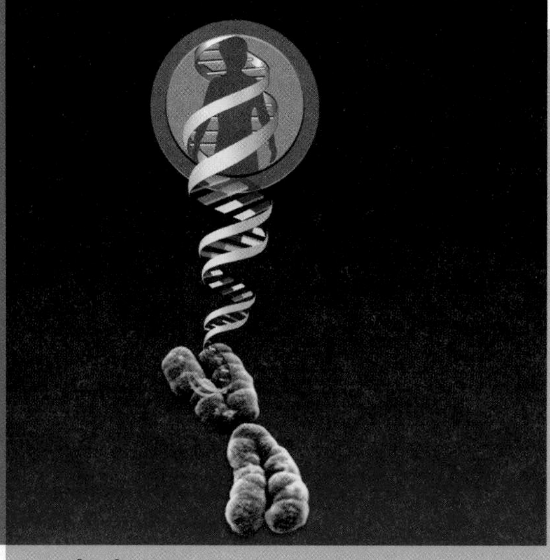
Logo for the Human Genome Project.

## Key Questions

- What was the Human Genome Project?
- What are the steps for determining the sequence of a genome?
- How is DNA cloned?
- What are genomic libraries and chromosome libraries?
- How is sequencing of DNA done?
- How is the complete sequence of a genome or a chromosome determined?
- How are genes and other important regions in genome sequences identified and described?
- How is genome organization similar and different in Bacteria, Archaea, and Eukarya?
- What are the future directions for genomics studies?
- What are the ethical, legal, and social implications of sequencing the human genome?

### iActivity

GENOMICS IS THE SCIENCE OF OBTAINING AND analyzing the sequences of complete genomes. At the core of genomics is recombinant DNA technology, the ability to construct and clone individual fragments of a genome, and to manipulate the cloned DNA in various ways, including sequencing it or expressing it in a foreign cell. In this chapter, you will learn about the cloning of genomic DNA fragments as it applies to obtaining the sequences of whole genomes. Then you can apply what you have learned by trying the iActivity, in which you can use recombinant DNA techniques to create a genetically modified brewing yeast for beer.

The development of molecular techniques for analyzing genes and gene expression has revolutionized experimental biology. Once DNA sequencing techniques were developed, scientists realized that determining the sequences of whole genomes was possible, although not necessarily easy. Why sequence a genome? The answer is that you then have the complete genetic blueprint for the organism in your hands—well, in the computer. The sequence of nucleotides in the genome, and their distribution among the chromosomes, is information that can be analyzed to determine how genes and functional nongenic regions of the genome control the development and function of an organism.

The first complete nonviral genome sequenced was the 16,159-bp circular genome of the human mitochondrion in 1981. But the human nuclear genome is 200,000 times larger, making the determination of its sequence daunting. However, major advances in automating DNA sequencing and developing computer programs to analyze large amounts of sequence data made the sequencing of large genomes a real possibility by the mid-1980s. The field of **genomics**—obtaining and analyzing the sequences of complete genomes—was born! This and the next chapter describe aspects of genomics and techniques used for genomic analysis. In this chapter you will learn about the branch of genomics that involves the cloning and sequencing of entire genomes, and genomic annotation, the identification and description of putative genes and other important sequences in these genomes.

In Chapter 9, you will learn about *functional genomics* and *comparative genomics*. In **functional genomics,** biologists attempt to understand how and when each gene in the genome is used, while in **comparative genomics,** biologists compare entire genomes to understand evolution and fundamental biological differences between species. Several of the organisms that geneticists understand best were among the first whose genomes were sequenced: *E. coli* (representing prokaryotes), the yeast *Saccharomyces cerevisiae* (representing single-celled eukaryotes), *Drosophila melanogaster* and *Caenorhabditis elegans* (fruit fly and nematode worm, respectively, representing multicellular animals of moderate genome complexity), and *Mus musculus* (the mouse). The genome of *Homo sapiens* (humans) was also included in the initial set of genomes for sequencing, for obvious reasons.

This chapter is an overview of the mapping and sequencing of genomes, and an introduction to the information obtained from genome sequence analysis. Your goal in this chapter is to understand how **cloning**—the production of many identical copies of a DNA molecule by replication in a suitable host—is done, with specific emphasis on how cloning is used in a genome project, how the DNA sequence of these clones is determined, how these DNA sequences are assembled into a full genomic sequence, and how genes and gene regulators are identified in the assembled genomic sequence.

As you read through this chapter, recognize that sequencing the genome of an organism is *descriptive science* rather than *hypothesis-driven science*. Clearly there can be no hypotheses in collecting the primary data of an organism's genome. But hypothesis-driven experiments are a major part of researchers' efforts to understand the genome data being generated, especially what genes are present and how they direct the structure and function of the organism.

## The Human Genome Project

In the mid 1980s, a number of scientists came to the conclusion that sequencing the human genome might be a reachable goal. Significant roadblocks existed, with cost and technology being the most significant. When the project started in 1990, the cost was estimated to be $3 billion over 15 years. These scientists ultimately assembled a massive international collaboration—called HUGO, the Human Genome Organization—and sought funding from various sources, including the Department of Energy and the National Institutes of Health in the United States, and the governments of a number of other countries, including Great Britain, France, and Japan. As a part of the **Human Genome Project** (HGP), the genomes of several well-studied organisms (*E. coli*, budding yeast, the nematode *Caenorhabditis elegans,* the fruit fly, and the mouse) were also sequenced, in part as trial runs, since most of these organisms have genomes that are simpler than the human genome, and also as genomes for comparison with the human genome. Ultimately, scientists published a draft version of the human genome in 2000, and a final version was released in 2003, well ahead of schedule. By the time this group completed their genomic sequence, scientists at a private company, Celera Genomics, also had produced a similar sequence for the human genome.

**Keynote**

The ambitious and expensive plan to sequence the human genome was proposed less than 25 years ago. When the project started, researchers were not certain that it was either affordable or possible. Despite that, the human genome was sequenced ahead of schedule, along with the genomes of several other organisms of genetic interest.

## Converting Genomes into Clones, and Clones into Genomes

Even the smallest cellular genomes are far too large and complex to work with in an intact form. For instance, the human genome is nearly 3 billion base pairs in length, and human chromosome 1 is over 250 million base pairs long (fully stretched out, this would be several centimeters long). To study a genome, we must first break it into much smaller fragments that can be worked with in the lab, and we need to use an easily cultured host cell, such as the easy-to-handle and manipulate microorganisms, *E. coli* or yeast, to take up and maintain these small fragments so that we can isolate many thousands of identical copies of each fragment. Most frequently, we need to make a **physical map** of the genome; that is, a map of the chromosomes showing the positions of important landmarks like genes and promoters, as well as specific DNA base pairs, sequences, and regions that vary between individuals. In a physical map, distances are measured in base pairs. To make a physical map, we must determine where these landmarks come from in the intact genome. This means taking the small fragments and then reassembling a "virtual chromosome" from them. The first step is to construct a **genomic library,** a collection of clones that contains at least one copy of every DNA sequence in the genome of an organism. Since most genomes contain millions or billions of base pairs, and a clone contains a relatively small piece of DNA, genomic libraries must have a great many clones (thousands to millions), with each clone containing a random small fragment of genomic DNA carried by a **cloning vector,** an artificially constructed DNA molecule capable of replication in a host organism such as a bacterium. A cloning vector allows us to make a great many copies of the small fragment of genomic DNA.

In this section, we examine how genomic libraries are made and then how the smaller clones are sequenced. In the following sections, we then discuss how the sequence data generated are used to reconstruct the sequence of the entire genome, how genes are found in the sequence, and how comparing different genomes informs us about genes, proteins, organisms, and evolutionary relationships.

## DNA Cloning

In brief, DNA is cloned molecularly typically by the following steps:

1. Isolate DNA from an organism.
2. Cut the DNA into pieces with a *restriction enzyme*—an enzyme that recognizes and cuts within a specific DNA sequence—and insert (*ligate*) each piece individually into a cloning vector cut with the same restriction enzyme to make a **recombinant DNA molecule,** a DNA molecule constructed *in vitro* containing sequences from two or more distinct DNA molecules.
3. Introduce (transform) the recombinant DNA molecules into a host such as *E. coli*. Replication of the recombinant DNA molecule—the process of **molecular cloning**—occurs in the host cell, producing many identical copies called *clones*. As the host organism reproduces, the recombinant DNA molecules are passed on to all the progeny, giving rise to a population of cells carrying the cloned sequences.

There are many reasons for cloning DNA beyond studying genomes. You will see cloning being used as an important technique in several chapters, and you will notice that different experiments use different cloning strategies and different types of cloning vectors.

**Restriction Enzymes.** To analyze genomic DNA, we must first cut it into smaller, more manageable pieces. The tools for this are restriction enzymes. A **restriction enzyme** (or **restriction endonuclease**) recognizes a specific nucleotide-pair sequence in DNA called a **restriction site** and cleaves the DNA (hydrolyzes the phosphodiester backbones) within or near that sequence. All restriction enzymes cut DNA between the 3′ carbon and the phosphate moiety of the phosphodiester bond so that fragments produced by restriction enzyme digestion have 5′ phosphates and 3′ hydroxyls. Most restriction enzymes function optimally at 37°C.

Restriction enzymes are used to produce a pool of DNA fragments to be cloned. Restriction enzymes are also used to analyze the positions of restriction sites in a piece of cloned DNA or in a segment of DNA in the genome (see Chapter 10, pp. 262–263). In most laboratory uses of restriction enzyme digestions (usually shortened to *restriction digests*), we attempt to "cut to completion," meaning that the enzyme is allowed to cut at every one of its restriction sites in the DNA. Such a digest will cut each genome copy of the same organism into the same large set of pieces. As we will see, in certain genomics applications it is desirable, instead, to do a "partial digest" in which the enzyme does not have enough time to complete its job. As a result, only some of the restriction sites are cut, and many are left uncut. Because we are cutting millions of identical DNA molecules, in a partial digest each will be cut at a unique subset of the available restriction sites.

***General Properties of Restriction Enzymes.*** Most restriction enzymes are found naturally in bacteria, although a handful have been found in eukaryotes. In bacteria, restriction enzymes protect the host organism against viruses by cutting up—restricting—invading viral DNA. The bacterium modifies its own restriction sites (by methylation) so that its own DNA is protected from the action of the restriction enzyme(s) it makes. Werner Arber, Daniel Nathans, and Hamilton O. Smith received the 1978 Nobel Prize in Physiology or Medicine "for their discovery of restriction enzymes and their application in problems of molecular genetics."

More than 400 different restriction enzymes have been isolated, and at least 2,000 more have been characterized partially. They are named for the organisms from which they are isolated. Conventionally, a three-letter system is used. Commonly the first letter is that of the genus, and the second and third letters are from the species name. The letters are italicized or underlined, followed by roman numerals that signify a specific restriction enzyme from that organism. Additional letters sometimes are added just before the number to signify a particular bacterial strain from which the enzymes were obtained. For example, *Eco*RI and *Eco*RV are both from <u>*E*</u>*scherichia* <u>*co*</u>*li* strain *RY*13, but recognize different restriction sites; *Hin*dIII is from <u>*H*</u>*aemophilus* <u>*in*</u>*fluenzae* strain Rd. The Roman numerals indicate the order in which the restriction enzymes from that strain were identified. Hence, *Eco*RI and *Eco*RV are the first and fifth restriction enzymes identified for *E. coli* strain RY13. The names are pronounced in ways that follow no set pattern. For example, *Bam*HI is "bam-H-one," *Bgl*II is "bagel-two," *Eco*RI is "echo-R-one" or "eeko-R-one," *Hin*dIII is "hin-D-three," *Hha*I is "ha-ha-one," and *Hpa*II is "hepa-two."

Many restriction sites have an axis of symmetry through the midpoint. Figure 8.1 shows this symmetry for the *Eco*RI restriction site: the nucleotide sequence from 5′ to 3′ on one DNA strand is the same as the nucleotide sequence from 5′ to 3′ on the complementary DNA strand. Thus, the sequences are said to have *twofold rotational symmetry.* A number of restriction sites are shown in Table 8.1. The most commonly used restriction enzymes recognize four nucleotide pairs (for example, *Hha*I) or six nucleotide pairs (for example, *Bam*HI, *Eco*RI). Some enzymes recognize eight-nucleotide pair sequences (for example, *Not*I ["not-one"]). Other classes of enzymes do not fit our model because the restriction site is not symmetrical about the center. *Hin*fI ("hin-f-one"), for example, recognizes a five-nucleotide pair sequence in which there is symmetry in the two nucleotide pairs on either side of the central nucleotide pair, but the

Figure 8.1

**Restriction site in DNA, showing the twofold rotational symmetry of the sequence.** The sequence reads the same from left to right (5′ to 3′) on the top strand (GAATTC, here) as it does from right to left (5′ to 3′) on the bottom strand. Shown is the restriction site for *Eco*RI.

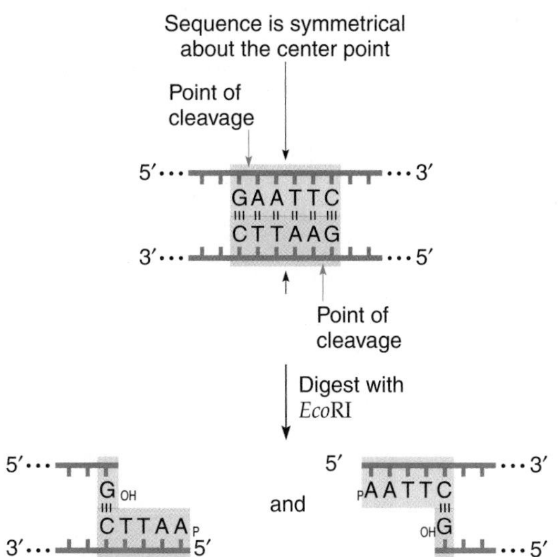


central nucleotide pair is obviously asymmetrical within the sequence. *Bst*XI ("b-s-t-x-one") is representative of a number of restriction enzymes with a nonspecific spacer region between symmetrical sequences (see Table 8.1).

***Frequency of Occurrence of Restriction Sites in DNA.*** Since each restriction enzyme cuts DNA at an enzyme-specific sequence, the number of cuts the enzyme makes in a particular DNA molecule depends on the number of times that particular restriction site occurs. When we cut a number of copies of the same genome with a particular restriction enzyme, the DNA is cleaved at the specific restriction sites for the enzyme, which are distributed throughout the genome. Although this produces millions of fragments of different sizes from one genome copy, all copies of the same genome will be cut at identical places.

Based on probability principles, the frequency of a short nucleotide pair sequence in the genome theoretically will be greater than the frequency of a long nucleotide pair sequence, so an enzyme that recognizes a four-nucleotide pair sequence will cut a DNA molecule more frequently than one that recognizes a six-nucleotide pair sequence, and both enzymes will cut more frequently than one that recognizes an eight-nucleotide pair sequence.

Consider DNA with a 50% GC content (meaning that 50% of the nucleotides in the DNA carry a G or C base) and that nucleotide pairs are distributed uniformly. For that DNA, there is an equal chance of finding one of the four possible nucleotide pairs $\frac{G}{C}$, $\frac{C}{G}$, $\frac{A}{T}$, and $\frac{T}{A}$ at any one position. The restriction enzyme *Hpa*II recognizes the sequence $\begin{matrix}5'\text{-GGCC-}3'\\3'\text{-CCGG-}5'\end{matrix}$. The probability of this sequence occurring in DNA is computed as follows:

1st nucleotide pair: $\frac{G}{C}$, probability = 1/4

2nd nucleotide pair: $\frac{G}{C}$, probability = 1/4

3rd nucleotide pair: $\frac{C}{G}$, probability = 1/4

4th nucleotide pair: $\frac{C}{G}$, probability = 1/4

The probability of finding any one of the nucleotide pairs at a particular position is independent of the probability of finding a particular nucleotide pair at another position. Therefore, the probability of finding the *Hpa*II restriction site in DNA with a uniform distribution of nucleotide pairs is $1/4 \times 1/4 \times 1/4 \times 1/4 = 1/256$. In short, the recognition sequence for *Hpa*II occurs on average once every 256 base pairs in such a piece of DNA, and the average DNA fragment produced by digestion with *Hpa*II (a "*Hpa*II fragment") would be 256 base pairs (bp).

In general, the probability of occurrence of a restriction site in uniformly distributed nucleotide pairs with 50% GC content is given by the formula $(1/4)^n$, where $n$ is the number of nucleotide pairs in the recognition sequence. These values are given in Table 8.2. In practice, however, genomes usually do not have exactly 50% GC content, nor are the base pairs uniformly distributed. Thus, a range of sizes of fragments result when genomic DNA is cut with a restriction enzyme, so the theoretical predictions typically are not seen.

***Restriction Sites and Creation of Recombinant DNA Molecules.*** One major class of restriction enzymes recognizes a specific DNA sequence and then cuts within that sequence. Another class of restriction enzymes recognize a specific nucleotide-pair sequence, and then cut the two strands of DNA outside of that sequence. This latter class of restriction enzymes is not useful for creating recombinant DNA molecules and will not be considered further.

Restriction enzymes in the first class cut DNA in different general ways. As Table 8.1 indicates, some enzymes, such as *Sma*I ("sma-one"), cut both strands of DNA between the same two nucleotide pairs to produce DNA fragments with blunt ends (Figure 8.2a). Other enzymes, such as *Bam*HI, make staggered cuts in the symmetrical nucleotide-pair sequence to produce DNA fragments with *sticky* or *staggered ends*, either 5′ overhanging ends, as in the case of cleavage with *Bam*HI (Figure 8.2b) or *Eco*RI, or 3′ overhanging ends, as in the case of cleavage with *Pst*I ("P-S-T-one"; Figure 8.2c).

Restriction enzymes that produce sticky ends are of particular value in cloning DNA because every DNA fragment generated by cutting a piece of DNA with the same restriction enzyme has the same single-stranded nucleotide sequence at the two overhanging ends. If the ends of two pieces of DNA produced by the action of

**Table 8.1 Characteristics of Some Restriction Enzymes**

| | Enzyme Name | Pronunciation | Organism in Which Enzyme Is Found | Recognition Sequence and Position of Cut[a] |
|---|---|---|---|---|
| Enzymes with 6-bp Recognition Sequences | *Bam*HI | "bam-H-one" | *Bacillus amyloliquefaciens H* | 5′-G↓G A T C C-3′<br>3′-C C T A G↑G-5′ |
| | *Bgl*II | "bagel-two" | *Bacillus globigi* | 5′-A↓G A T C T-3′<br>3′-T C T A G↑A-5′ |
| | *Eco*RI | "echo-R-one" | *Escherichia coli RY13* | 5′-G↓A A T T C-3′<br>3′-C T T A A↑G-5′ |
| | *Hae*II | "hay-two" | *Haemophilus aegypticus* | 5′-R G C G C↓Y-3′<br>3′-Y↑C G C G R-5′ |
| | *Hind*III | "hin-D-three" | *Haemophilus influenzae* $R_d$ | 5′-A↓A G C T T-3′<br>3′-T T C G A↑A-5′ |
| | *Pst*I | "P-S-T-one" | *Providencia stuartii* | 5′-C T G C A↓G-3′<br>3′-G↑A C G T C-5′ |
| | *Sal*I | "sal-one" | *Streptomyces albus* | 5′-G↓T C G A C-3′<br>3′-C A G C T↑G-5′ |
| | *Sma*I | "sma-one" | *Serratia marcescens* | 5′-C C C↓G G G-3′<br>3′-G G G↑C C C-5′ |
| Enzymes with 4-bp Recognition Sequences | *Hae*III | "hay-three" | *Haemophilus aesypticus* | 5′-G G↓C C-3′<br>3′-C C↑G G-5′ |
| | *Hha*I | "ha-ha-one" | *Haemophilus haemolyticus* | 5′-G C G↓C-3′<br>3′-C↑G C G-5′ |
| | *Hpa*II | "hepa-two" | *Haemophilus parainfluenzae* | 5′-C↓G G C-3′<br>3′-G G C↑C-5′ |
| | *Sau*3A | "sow-three-A" | *Staphylococcus aureus 3A* | 5′-↓G A T C -3′<br>3′- C T A G↑-5′ |
| Enzyme with 8-bp Recognition Sequences | *Not*I | "not-one" | *Nocardia otitidis-caviarum* | 5′-G C↓G G C C G C-3′<br>3′-C G C C G G↑C G-5′ |
| Enzyme with Recognition Sequence Containing a Nonspecific Spacer Sequence | *Bst*XI | "b-s-t-x-one" | *Bacillus stearothermophilus* | 5′-C C A N N N N N↓N T G G-3′<br>3′-G G T N↑N N N N N A C C-5′ |

[a]In this column the two strands of DNA are shown with the sites of cleavage indicated by arrows. Since there is an axis of twofold rotational symmetry in each recognition sequence, the DNA molecules resulting from the cleavage are symmetrical. Key: R = purine; Y = pyrimidine; N = any base.

the same restriction enzyme (such as *Eco*RI)—a cloning vector and a chromosomal DNA fragment, for example—come together in solution, base pairing occurs between the overhanging ends; the two single-stranded DNA ends are said to *anneal* (Figure 8.3). Using DNA ligase, the two DNAs can be covalently linked (ligated) to produce a longer DNA molecule with the restriction sites reconstituted at the junction of the two fragments. (Recall from our discussion of DNA replication that DNA ligase seals nicks in a DNA strand by forming a phosphodiester bond when the two nucleotides have a free 5′ phosphate and a free 3′ hydroxyl group, respectively (see Figure 3.7, p. 46). Even DNA fragments with blunt ends can be ligated together by DNA ligase at high concentrations of the enzyme. The ligation of two DNA fragments is the principle behind the formation of recombinant DNA molecules. Paul Berg received part of the 1980 Nobel Prize in Chemistry "for his fundamental studies of the biochemistry of nucleic acids, with particular regard to recombinant-DNA."

**Table 8.2 Occurrence of Restriction Sites for Restriction Enzymes in DNA with Randomly Distributed Nucleotide Pairs**

| Nucleotide Pairs in Restriction Site | Probability of Occurrence |
|---|---|
| 4 | $(1/4)^4$ = 1 in 256 bp |
| 5 | $(1/4)^5$ = 1 in 1,024 bp |
| 6 | $(1/4)^6$ = 1 in 4,096 bp |
| 8 | $(1/4)^8$ = 1 in 65,476 bp |
| $n$ | $(1/4)^n$ |

**Keynote**

Genomics is the study of the complete DNA sequence of an organism or virus. First, genomic DNA is fragmented, each fragment is cloned and then the sequence of each clone is determined. DNA is cloned by inserting fragmented DNA from an organism into a cloning vector to make a recombinant DNA molecule and then introducing that molecule into a host cell in which it will replicate. Essential to cloning are restriction enzymes. Restriction enzymes that are useful for cloning recognize specific nucleotide-pair sequences in DNA (restriction sites) and cleave at a specific point within the sequence. If the DNA to be cloned and the vector are cleaved by the same restriction enzyme, the two different molecules can base-pair together and be ligated to produce a recombinant DNA molecule. A blunt-ended DNA fragment can also be cloned by ligating it to a blunt-ended vector.

## Cloning Vectors and DNA Cloning

To determine the sequence of a genome, we need to break the genome into fragments and clone each fragment to produce multiple copies to use for DNA sequencing. Several types of vectors have been constructed specially for cloning DNA. They include plasmids, bacteriophages (e.g., λ and certain single-stranded DNA species), cosmids (vectors with features of both plasmid and bacteriophage vectors), and artificial chromosomes. The vector types differ in their molecular properties and in the maximum amount of inserted DNA they can hold. Each type of vector has been specially constructed in the laboratory. We focus on plasmid and artificial chromosome vectors in this section, as they have been workhorses in genomics.

**Plasmid Cloning Vectors.** Bacterial **plasmids** are extrachromosomal elements that replicate autonomously within cells (see Chapter 15). Plasmid DNA is double-stranded and (often) circular, and contains an origin sequence (*ori*) required for plasmid replication and genes for the other functions of the plasmid. Plasmid cloning vectors are derivatives of circular

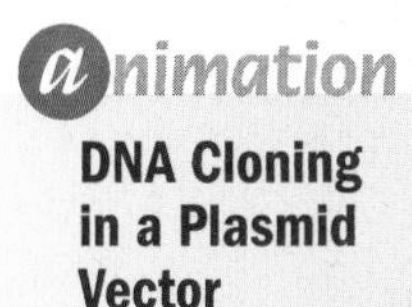


**DNA Cloning in a Plasmid Vector**

**Figure 8.2**

**Examples of how restriction enzymes cleave DNA.** (a) *Sma*I results in blunt ends. (b) *Bam*HI results in 5′ overhanging ("sticky") ends. (c) *Pst*I results in 3′ overhanging ("sticky") ends.

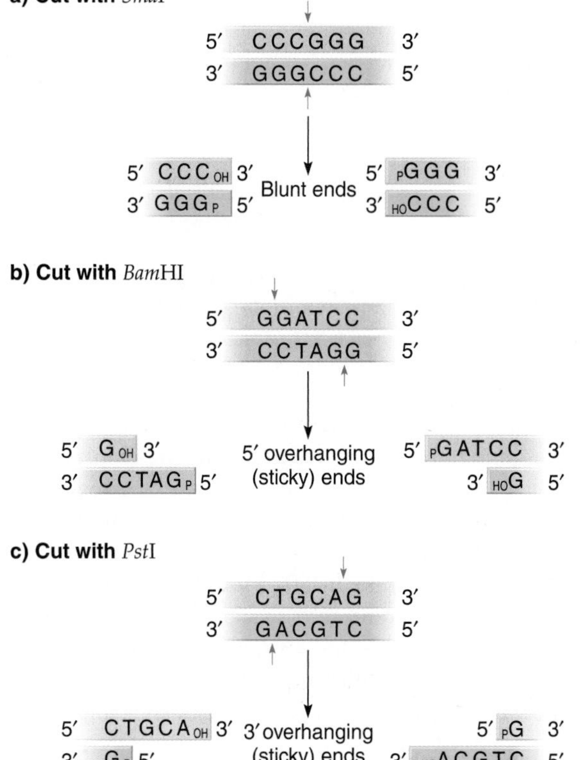


natural plasmids "engineered" to have features useful for cloning DNA. We focus here on features of *E. coli* plasmid cloning vectors.

An *E. coli* plasmid cloning vector must have three features:

1. An *ori* (origin of DNA replication) sequence, needed for the plasmid to replicate in *E. coli*.
2. A **selectable marker,** so that *E. coli* cells with the plasmid can be distinguished easily from cells that lack the plasmid. A selectable marker is a gene that allows us to determine easily if a cell does or does not contain the cloning vector. For bacterial plasmid cloning vectors, typically the selectable marker is a gene for resistance to an antibiotic, such as the $amp^R$ gene for ampicillin resistance or the $tet^R$ gene for tetracycline resistance. When plasmids carrying antibiotic-resistance genes are added to a population of plasmid-free and therefore antibiotic-sensitive *E. coli,* the cells that take up the plasmid can be selected for by culturing the cells on a solid medium containing the appropriate antibiotic; only bacteria with the plasmid will grow on the medium.
3. *One or more unique restriction enzyme cleavage sites*—sites present just once in the vector—for the insertion of the DNA fragments to be cloned. Typically, a

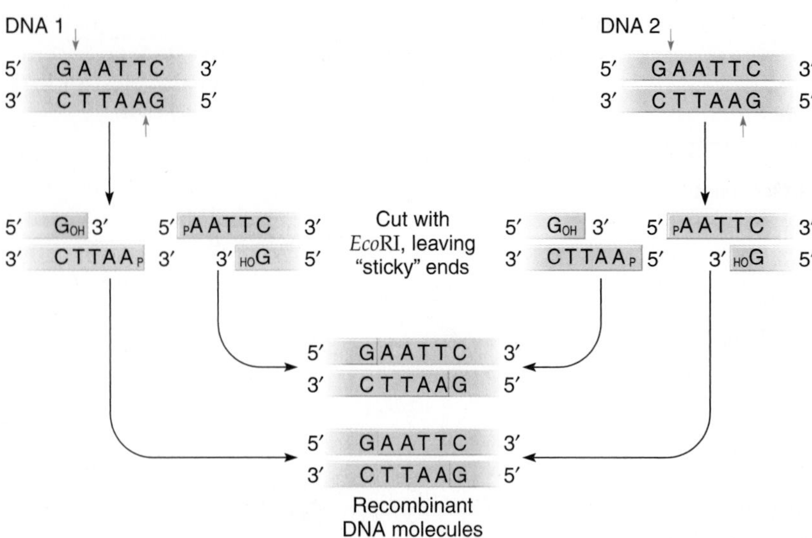


**Figure 8.3**

**Cleavage of DNA by the restriction enzyme *Eco*RI.** *Eco*RI makes staggered, symmetrical cuts in DNA, leaving "sticky" ends. A DNA fragment with a sticky end produced by *Eco*RI digestion can bind by complementary base pairing (anneal) to any other DNA fragment with a sticky end produced by *Eco*RI cleavage. The nicks can then be sealed by DNA ligase.



number of sites are present in the vector, and these sites tend to be engineered as a **multiple cloning site** or *polylinker*. A multiple cloning site is a region of DNA containing several unique restriction sites where a fragment of foreign DNA (not originally part of the vector) can be inserted into the vector. With a number of different sites available in the multiple cloning site of a vector, an investigator can use the same vector in different cloning experiments by choosing different restriction sites for the cloning.

As an example, Figure 8.4 diagrams the plasmid cloning vector pBluescript II. This 2,961-bp vector has the following features that make it useful for cloning DNA in *E. coli*:

1. It has a high copy number, approaching 100 copies per cell because it has a very active *ori*. As a result, many copies of a cloned piece of DNA can be generated readily in a small number of host cells.
2. It has the $amp^R$ selectable marker for ampicillin resistance.
3. It has a multiple cloning site containing 18 restriction sites.
4. The multiple cloning site is embedded in part of the *E. coli* β-galactosidase ($lacZ^+$) gene (see Figure 8.4). pBluescript II, like other plasmids similarly constructed with such a *lacZ* gene fragment, is usually introduced into an *E. coli* strain with a mutated *lacZ* gene. When the (unmodified) plasmid is present in the cell, functional β-galactosidase is produced. However, when a piece of DNA is cloned into the multiple cloning site, the *lacZ* fragment on the plasmid is disrupted and no functional β-galactosidase can be produced. Therefore, the presence or absence of β-galactosidase activity indicates whether the plasmid introduced into *E. coli* is the empty pBluescript II vector (no inserted DNA fragment: functional enzyme present) or pBluescript II with an inserted DNA fragment (functional enzyme absent). The chemical X-gal—a colorless artificial substrate for β-galactosidase—is included in the medium on which the cells containing plasmids are plated as an indicator for β-galactosidase activity in cells of a colony. Cleavage of X-gal by β-galactosidase leads to the production of a blue dye. Thus, if functional enzyme is present (vector with no insert), the colony turns blue, whereas if nonfunctional β-galactosidase is made (vector with inserted DNA), the colony is white. This protocol is called *blue–white colony screening*.

**Figure 8.4**

**The plasmid cloning vector pBluescript II.** This plasmid cloning vector has an origin of replication (*ori*), an $amp^R$ selectable marker, and a multiple cloning site located within part of the β-galactosidase gene $lacZ^+$.

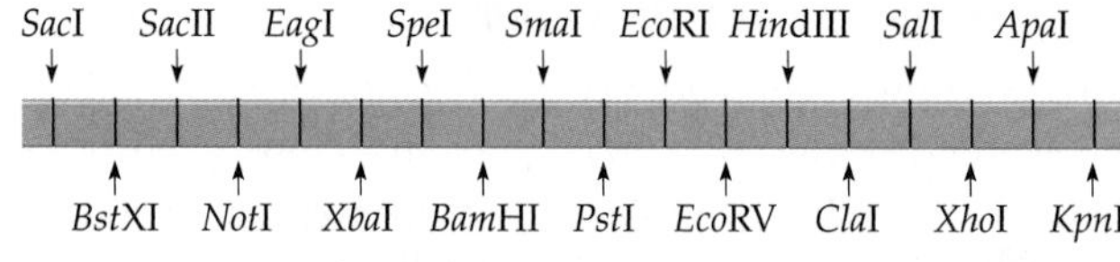


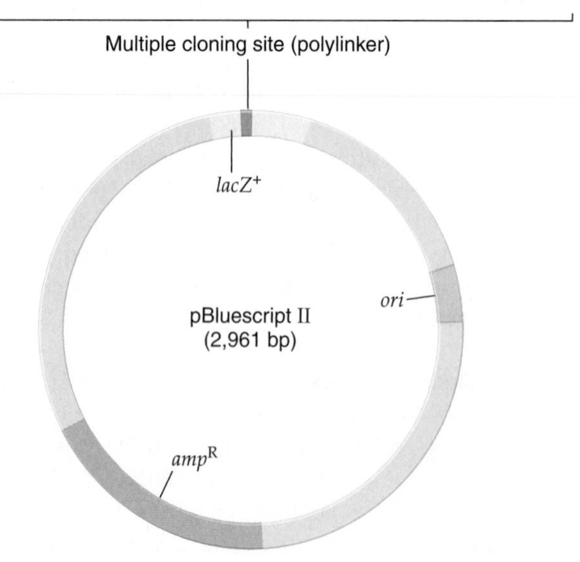

Figure 8.5 illustrates how a piece of DNA can be inserted into a plasmid cloning vector such as pBluescript II. In the first step, pBluescript II is cut with a restriction enzyme that has a site in the multiple cloning site. Next, the piece of DNA to be cloned is generated by cutting high-molecular-weight DNA with the same restriction enzyme. Since restriction sites are nonuniformly arranged in DNA, fragments of various sizes are produced. The DNA fragments are mixed with the cut vector in the presence of DNA ligase; in some cases, the DNA fragment becomes inserted between the two cut ends of the plasmid and DNA ligase joins the two molecules covalently. The resulting recombinant DNA plasmid is introduced into an *E. coli* host by transformation. (By definition, **transformation** is a process in which new genetic information is introduced into a cell via extracellular pieces of DNA: see Chapter 15, pp. 437–440.) Transformation is done either by incubating the recombinant DNA plasmids with *E. coli* cells treated chemically (such as with $CaCl_2$) to take up DNA, or by *electroporation*, a method in which an electric shock is delivered to the cells, causing temporary disruptions of the cell membrane to let the DNA enter. Transformed cells are plated onto media containing ampicillin and X-gal. Cells that can grow and divide on this medium, forming a colony, must have been transformed by a plasmid. Colonies containing plasmids with an insert can be identified by the blue–white colony screening method.

In a ligation reaction, the restriction enzyme-digested vector alone can recircularize. Such recircularization is quite common because it is a reaction involving only one DNA molecule, and thus more likely than ligation of two DNA molecules, such as vector and insert. This can make it more difficult to find the desired recombinant plasmids from amongst all the plasmids. Fortunately, vector recircularization can be minimized by treating the digested vector with the enzyme alkaline phosphatase to remove the 5′ phosphates, leaving a 5′–OH group at the two ends of the DNA. DNA ligase can only join a 3′–OH to a 5′–phosphate, so if we remove both 5′ phosphates from a vector, it cannot recircularize. DNA to be inserted into the vector—*insert DNA*—is not treated with phosphatase, so the insert DNA retains 5′ phosphate groups and the 5′ ends of the insert DNA can be ligated to the 3′ ends of the vector DNA. This ligation reaction creates a circular molecule with two nicks where the phosphodiester backbone is broken but, since these nicks are far apart, the complex holds together as a single molecule. If the digested vector is treated with alkaline phosphatase before the ligation reaction, then, the proportion of blue colonies among transformants is reduced drastically. (Why not completely? No enzymatic reaction is 100% effective, so some vectors are are not affected and are still able to recircularize.) In other words, the alkaline phosphatase treatment makes the identification of the desired clones more efficient.

DNA fragments of up to 15 kb may be cloned efficiently in *E. coli* plasmid cloning vectors. Plasmids carrying larger DNA fragments often are unstable *in vivo* and tend to lose most of the insert DNA. This size limitation means that plasmid vectors are of limited use in genomic analysis, since millions of clones would be needed to contain a single genome of a complex multicellular organism such as a human. To clone larger DNA inserts, different vectors are used such as *cosmids* and *artificial chromosomes* (see the next section). A cosmid can accommodate DNA inserts in the range of 40–45 kb for genomics uses. A cosmid cloning vector is similar to a plasmid cloning vector, with an origin, a drug resistance marker, and a multiple cloning site, but it is introduced into host cells differently. Cosmids are frequently used as vectors when libraries are made, because they are able to hold larger inserts.

**Artificial Chromosomes.** Artificial chromosomes are cloning vectors that can accommodate very large pieces



**Figure 8.5**

**Insertion of a piece of DNA into the plasmid cloning vector pBluescript II to produce a recombinant DNA molecule.** The vector pBluescript II contains several unique restriction enzyme sites localized in a multiple cloning site that are convenient for constructing recombinant DNA molecules. The insertion of a DNA fragment into the multiple cloning site disrupts part of the β-galactosidase (*lacZ*⁺) gene, leading to nonfunctional β-galactosidase in *E. coli*. The blue-white colony screening method described in the text can be used to identify vectors with or without inserts.

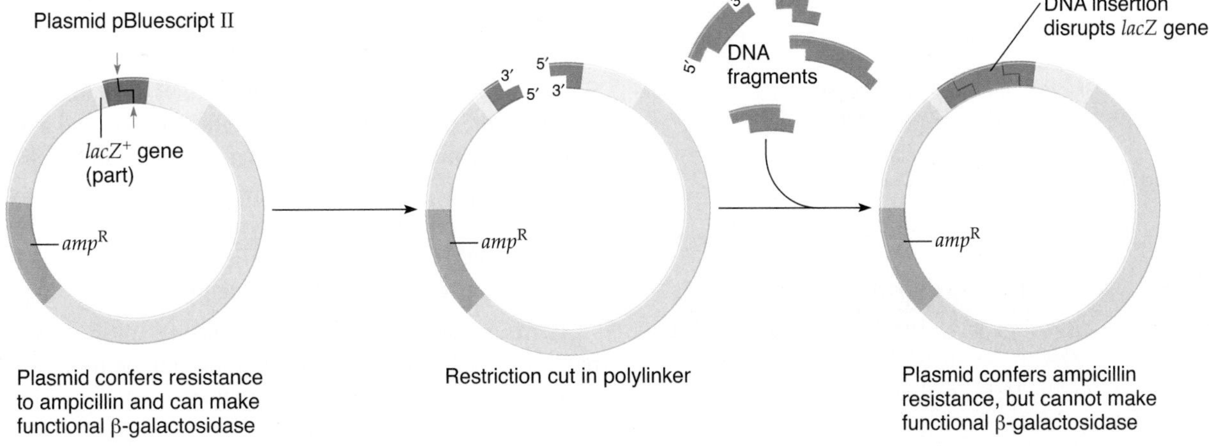

of DNA, producing recombinant DNA molecules resembling small chromosomes. Artificial chromosomes are useful in genomics applications because we can use them to study large segments of chromosomes, and they can contain an entire genome in a manageable number of clones. We consider two examples here, bacterial artificial chromosomes and yeast artificial chromosomes.

***Bacterial Artificial Chromosomes.*** **Bacterial artificial chromosomes** (BACs, "backs") are cloning vectors containing the origin of replication from a natural plasmid found in *E. coli* called the *F* factor (see Chapter 15, p. 432), a multiple cloning site, and one or more selectable markers. One BAC vector, pBeloBAC11, is shown in Figure 8.6a. This particular vector can be used with the blue–white colony screening method, just like a plasmid. The selectable marker for this BAC is $cam^R$. This gene encodes an enzyme that degrades the antibiotic chloamphenicol, and thus, cells carrying this vector (with or without an insert) can grow in the presence of chloramphenicol while cells lacking this vector are unable to grow if chloramphenicol is present. BACs accept inserts up to 300 kb and have the advantage that they can be manipulated like giant bacterial plasmids. One major difference between BACs and the plasmids you have already learned about is that once transformed into *E. coli,* the *F* factor origin of replication keeps the copy number of the BAC at one per cell, while the origins of typical plasmid cloning vectors drive multiple rounds of DNA replication to generate many copies of the plasmid in each cell. Unlike yeast artificial chromosomes that will be described next, BACs do not undergo rearrangements in the host. Therefore, they have become the preferred vector for making large clones in physical mapping studies of genomes. Two disadvantages of BACs (and with other cloning vectors for *E. coli*) are that AT-rich DNA fragments (DNA fragments with a high proportion of A and T nucleotides) typically do not clone well, and some DNA sequences are toxic to *E. coli* and, hence, are unclonable in that organism.

**Figure 8.6**

**Examples of artificial chromosome cloning vectors.** **(a)** A BAC (bacterial artificial chromosome) vector, such as pBeloBAC11, is similar to a plasmid vector, with one or more selectable markers (here, $cam^R$ for chloramphenicol resistance), a multiple cloning site in part of the $lacZ^+$ gene, but uses an origin derived from the *F* factor, which limits the copy number of the BAC to one per *E. coli* cell. **(b)** A YAC (yeast artificial chromosome) vector contains a yeast telomere (*TEL*) at each end, a yeast centromere sequence (*CEN*), a yeast selectable marker for each arm (here, *TRP1* and *URA3*), a sequence that allows autonomous replication in yeast (*ARS*), and restriction sites for cloning.

**a) A bacterial artificial chromosome (BAC) vector**

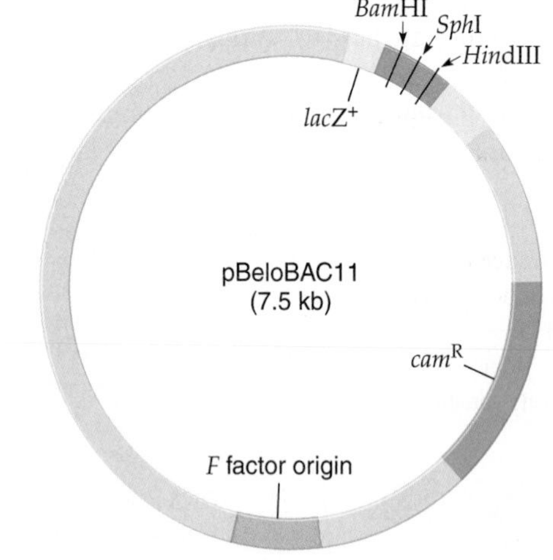


**b) A yeast artificial chromosome (YAC) vector**

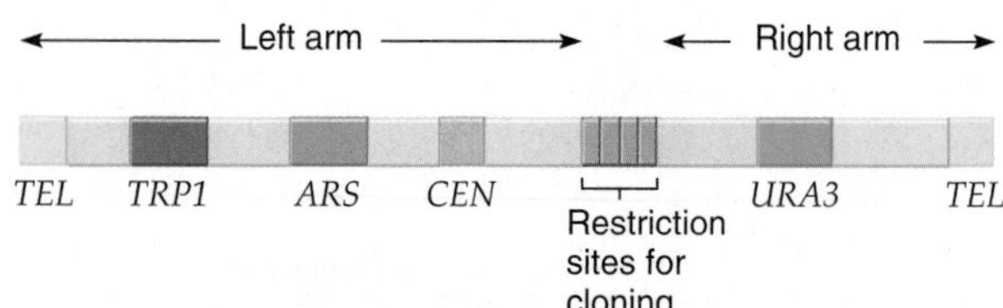


***Yeast Artificial Chromosomes.*** **Yeast artificial chromosomes** (YACs; "yaks") are cloning vectors that enable artificial chromosomes to be made and replicated in yeast cells. YAC vectors can accommodate DNA fragments that are several hundred kilobase pairs long, much longer than the fragments that can be cloned in the plasmid, cosmid, or BAC vectors we have discussed. Therefore, YAC vectors have been used to clone very large DNA fragments (between 0.2 and 2.0 Mb [Mb = megabase = 1,000,000 bp = 1,000 kb]), for example, in creating physical maps of large genomes such as the human genome. A YAC (shown in its linear form) has the following features (Figure 8.6b):

1. A yeast telomere (*TEL*) at each end. (Recall that all eukaryotic chromosomes need a telomere at each end.)
2. A yeast centromere sequence (*CEN*) allowing regulated segregation during mitosis.
3. A selectable marker on each arm for detecting and maintaining the YAC in yeast (for example, *TRP1* and *URA3* to enable transformed *trp1* [tryptophan requiring] *ura3* [uracil requiring] mutant yeast to grow on a medium lacking tryptophan and uracil).
4. An origin of replication sequence—*ARS* (autonomously replicating sequence)—that allows the vector to replicate in a yeast cell.
5. An origin of replication (*ori*) that allows a circular version of the empty vector to replicate in *E. coli*, and a selectable marker such as $amp^R$ that functions in *E. coli*.
6. A cloning region that contains one or more restriction sites; the restriction enzymes cutting in this region should not have any other sites in the YAC. This region is used for inserting foreign DNA.

There are two disadvantages associated with these very large YAC-based clones. First, during the cloning process, a fraction of the YAC vectors accept two or more inserts, rather than one, creating a chimeric YAC. A second problem is that portions of the insert DNA are frequently deleted or otherwise modified by the host cell, or undergo recombination with other DNA in the host cell. The altered inserts in chimeric and rearranged YACs will confound the assembly of the genome (described on pp. 189–191), because assembly requires that we compare how different inserts in our library overlap. The alterations in these inserts will cause us to misinterpret how they overlap with other clones, because a chimeric clone might contain, for instance, DNA from chromosome 5 ligated to DNA from chromosome 18. Determining which YACs are modified is often a very slow and labor-intensive process, making the assembly of a genome sequence more difficult.

Empty YAC vectors—ones that have yet to contain a DNA insert—are propagated in *E. coli* as circular plasmids; in this form the two telomeres are end-to-end. This propagation step makes use of the bacterial origin of replication and the bacterial selectable marker. Recall that bacterial and eukaryotic origins of replication are not functionally similar, which means that the yeast *ARS* sequence will not work in a bacterial cell, just as the bacterial *ori* sequence will not function in a yeast cell. In addition, bacterial and eukaryotic promoters are different, meaning that the bacterial RNA polymerase cannot transcribe the yeast *TRP1* and *URA3* genes, so those selectable markers will function only in yeast, not in bacteria. Likewise, yeast RNA polymerase II is unable to transcribe the $amp^R$ gene.

For cloning experiments, a circular YAC is cut with one restriction enzyme that cuts in the multiple cloning site and with another restriction enzyme that cuts between the two *TELs*. In this way, the left and right arms are produced. High-molecular-weight DNA, cut with the same restriction enzyme used to cut the YAC multiple cloning site, is ligated to the two arms and the recombinant molecules are transformed into yeast. By selecting for both *TRP1* and *URA3*, it can be ensured that the transformants have both the left and right arms.

**iActivity**

Better beer through science? Go to the iActivity *Building a Better Beer* on the student website and discover how genetically modified yeasts can improve your brew.

### Keynote

Many different kinds of vectors have been developed to construct and clone recombinant DNA molecules. These vectors differ in several key ways–most importantly, the size of insert that they will accept and the types of host cells that can propagate the clone. Cloning vectors also have unique restriction sites for inserting foreign DNA fragments, as well as one or more dominant selectable markers. The choice of the vector to use depends on the sizes of the fragments to clone which, in turn, depends on the experimental goals.

## Genomic Libraries

A genomic library is a collection of clones that, when successfully made, theoretically contains at least one copy of every DNA sequence in the genome. (The word "theoretically" is used because practically speaking, not all of the sequences in the genome can be cloned, but our goal is always to get as complete a library as is reasonably possible.) Genomic libraries have many uses in molecular biology and in genomics. Remember that a key step in analysis of a genome is breaking the genomic DNA into smaller, more easily manipulated fragments. A genomic library will contain these smaller fragments, which are used in many types of genetic analysis. You will see in Chapter 10 (pp. 258–260) that a genomic library can also be used to isolate and study a particular clone, such as that for a gene of interest. In this section we focus on the construction of genomic libraries of eukaryotic DNA.

Genomic libraries are made using the basic cloning procedures already described. A restriction enzyme is used to cut up the genomic DNA, and a vector is chosen so that the entire genome is represented in a manageable number of clones. You might assume that it is as simple as digesting the genomic DNA completely with a restriction enzyme and cloning the resulting DNA fragments in a cloning vector. This will create a genomic library, but this library will have serious functional limitations for four important reasons: (1) If the specific gene the researcher wants to study contains one or more restriction sites for the enzyme used to create the library, the gene will be split into two or more fragments when genomic DNA is digested completely by the restriction enzyme. As a result, the gene would then be cloned in two or more pieces. (2) The average size of the fragment produced by digestion of eukaryotic DNA with restriction enzymes is small (about 4 kb for restriction enzymes that have 6-bp recognition sequences; see Table 8.2). Not only are many genes larger than 4 kb (especially those in mammals), but also an entire genomic library would have to contain a very large number of recombinant DNA molecules, and screening for a specific gene would be very laborious. (3) The number of base pairs between adjacent restriction sites can vary significantly; so, for instance, cutting a 10-kb fragment of DNA with *Bam*HI might yield fragments of 500, 2,500, and 7,000 base pairs. When genomic DNA is digested, the resultant fragments will fall in a range of sizes. Some of these fragments will be too large to clone. As a result, part of the genome would be unclonable in this type of library. (4) The most troublesome aspect of this sort of library is the loss of information. If we have a library made, say, of the *Bam*HI-generated fragments of the 10-kb fragment described above, it would contain three clones. We would have no idea how the individual fragments were positioned in the original fragment, and we could never determine that order from the library itself. Extrapolating this issue to the thousands of clones in a genomic library made using complete digestion of genomic DNA, we would not be able to reassemble the cloned fragments into their arrangement in the genome.

To deal with these functional limitations, we need to break the genomic DNA differently. Specifically, we need to break the genomic DNA into fragments that are of the correct size for our cloning vector and that overlap each other. (Remember that we are breaking millions of copies of the genome in question, so each genome will be broken in a unique pattern, and the fragments we make from one copy of the genome will not be the same as the fragments that we make from another copy of the genome). To generate these overlapping fragments, we can either mechanically break (shear) the genomic DNA, or we can use a restriction enzyme under conditions such that the genomic DNA is digested partially.

DNA is sheared by passing it through a syringe needle to produce a population of overlapping DNA fragments of a particular size. However, because the ends of the resulting DNA fragments have been generated by physical means and not by cutting with restriction enzymes, additional enzymatic manipulations are necessary to add appropriate ends to the molecules for their insertion into a restriction site of a cloning vector.

Large, overlapping DNA fragments of appropriate size for constructing a genomic library can also be generated by using a partial digestion of the genomic DNA with a restriction enzyme that recognizes a frequently occurring 6- or 4-bp recognition sequence (Figure 8.7a). *Partial digestion* means that only a random portion of the available restriction sites is cut by the enzyme. This is achieved by limiting the amount of the enzyme used and/or the time of incubation with the DNA.

DNA fragments generated by partial digestion with a restriction enzyme can be cloned directly. For example, if the DNA is digested with the enzyme *Sau*3A, which has the recognition sequence $\frac{\text{5'-GATC-3'}}{\text{3'-CTAG-5'}}$, the ends are complementary to the ends produced by digestion of a cloning vector with *Bam*HI, which has the recognition sequence $\frac{\text{5'-GGATCC-3'}}{\text{3'-CCTAGG-5'}}$ (Figure 8.7b).

That is, in

```
 ↓
5'-GATC-3'
3'-CTAG-5'
       ↑
```

*Sau*3A cuts to the left of the upper G and to the right of the lower G to give a 5′ overhang with the sequence 5′-GATC. . . 3′, as follows:

```
5'-            and  5'GATC-3'
3'-CTAG5'               -5'
```

In the sequence

```
   ↓
5'-GGATCC-3'
3'-CCTAGG-5'
        ↑
```

*Bam*HI cuts between the two G nucleotides also to give a 5′ overhang with the sequence 5′-GATC. . . 3′, as follows:

```
5'-G           and  5'GATCC-3'
3'-CCTAG5'              G-5'
```

**Figure 8.7**

**Partial digestion with a restriction enzyme to produce overlapping DNA fragments of appropriate size for constructing a genomic library.**

**a) Partial digestion of DNA by a restriction enzyme (for example *Sau*3A) generates a series of overlapping fragments, each with identical 5′ GATC sticky ends**

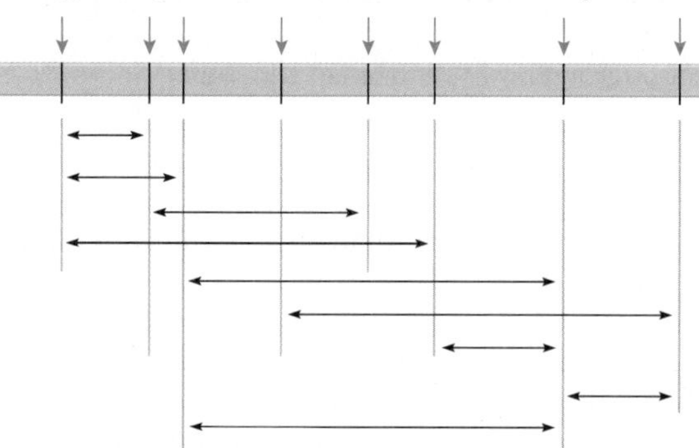

**b) Resulting fragments may be inserted into *Bam*HI site of cloning vector**

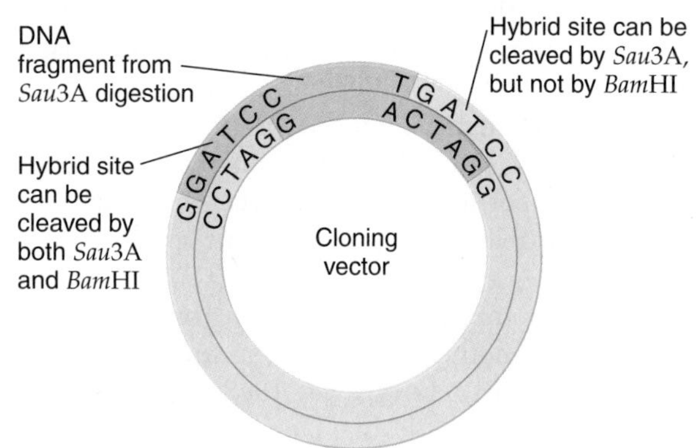


The *Sau*3A and *Bam*HI "sticky" ends can pair to produce a hybrid recognition site.[1]

The recombinant DNA molecules produced by ligating the *Sau*3A-cut fragments and the *Bam*HI-cut vectors together are then introduced into *E. coli,* where the molecules are cloned (see earlier discussion in "Cloning Vectors and DNA Cloning").

Regardless of how we broke the DNA into overlapping fragments, there will be a broad distribution of fragment sizes. Now it is necessary to select the fragments that are the right size for cloning in the vector being used, and to eliminate those that are either too small or too large. Consider a population of overlapping fragments generated by

---

[1]Since the hybrid site contains a 5′-GATC-3′ sequence, it can be cleaved by *Sau*3A. However, whether it can be cleaved by *Bam*HI depends on the base pair "inside" the cloned *Sau*3A-digested fragment. If it is a C-G nucleotide pair, then the hybrid site is

```
5'-GGATCC-3'
3'-CCTAGG-5'
```

which is the recognition site for *Bam*HI. This is the case with the left-hand hybrid site in Figure 8.7b. If any other nucleotide pair is next along the *Sau*3A fragment, the hybrid site is not a *Bam*HI cleavage site (e.g., the right-hand hybrid site in Figure 8.7b).

partial digestion with a restriction enzyme (Figure 8.8a). One common way to sort fragments of the desired size for cloning is to use **agarose gel electrophoresis** (see Figure 8.8a). In agarose gel electrophoresis, an electric field is used to move the negatively charged DNA fragments through a gel matrix of agarose from the negative pole to the positive pole. The gel, a horizontal slab of agarose and a liquid buffer, is made by pouring a hot, liquid agarose/buffer mix into a mold. A toothed comb is added, which creates "wells" in the gel. As the agarose mixture cools, the agarose itself forms a "sieve" through which the DNA transits. The DNA fragments (produced by shearing or restriction digestion) are placed in a well in the gel. Other wells may contain a **DNA ladder** (also called *DNA size markers*), a set of DNA molecules of known size. For example, a complete digestion of the phage lambda chromosome with *Hin*dIII, which yields fragments of 23.1 kb, 9.4 kb, 6.6 kb, 4.4 kb, 2.3 kb, 2.0 kb, and 0.56 kb, is frequently used as a DNA ladder and is often called a lambda ladder). An electric field is then applied to the gel and the DNA migrates toward the positive pole. Smaller molecules are able to move through the gel more rapidly, and larger molecules move more slowly (see Figure 8.8a).

The separated DNA fragments are invisible to the eye. They are made visible by adding either ethidium bromide or SYBR® Green to stain the DNA. Both chemicals bind tightly to DNA and emit visible light when excited with the correct wavelength of light. Ethidium bromide emits visible light after being excited with ultraviolet light, and SYBR® Green, when bound to DNA, emits green light after being excited with blue light. The emission of visible light makes the position of the DNA in the gel obvious. Since the wells are rectangular, the DNA fragments form "bands" on the gel. Figure 8.8b shows an actual agarose gel electrophoresis analysis that shows partial digestion of genomic DNA. The vertical "lanes" of the gel show how the DNA fragments in the samples loaded into the wells at the top separated during the electrophoresis. Lane 1 contains the DNA ladder, in this case the lambda ladder. Note the discrete set of bands of known sizes in the lane. Lane 2 shows a sample of genomic DNA not treated with a restriction enzyme. There is not a highly discrete band, but a concentrated mass of DNA in a region of the lane corresponding to the large DNA fragments of the lambda ladder, and a smear of DNA going down the lane from that point. The mass of DNA is the large DNA fragments of genomic DNA that came out of the cell. It is unavoidable to break the genomic DNA mechanically during isolation, so the size of the large DNA is much smaller than the sizes of chromosomes. The mechanical shearing during isolation is also responsible for the many bands of various sizes of DNA fragments that are seen as a smear down the lane. Lane 3 shows genomic DNA digested completely with a restriction enzyme. There are no discrete bands of DNA fragments here either. Instead, a smear of fragments is seen, most of which are smaller than the smallest visible lambda ladder fragment at 2.0 kb. Lanes 4 and 5 show

**Figure 8.8**

**Separation of DNA fragments by agarose gel electrophoresis.** (a) Partial digestion of genomic DNA with a restriction enzyme, and separation of the DNA fragments by agarose gel electrophoresis. (b) Agarose gel electrophoresis analysis of genomic DNA partially digested with a restriction enzyme. Lane 1: Lambda ladder (a type of DNA ladder). The sizes for the DNA bands of the ladder are indicated on the left side of the gel. Lane 2: Genomic DNA undigested by a restriction enzyme. Lane 3: Genomic DNA digested completely with a restriction enzyme. Lanes 4 and 5: Genomic DNA digested partially with a restriction enzyme. Enzyme reaction conditions allowed for less DNA digestion for the DNA in lane 5 than the DNA in lane 4.

**a) Partial restriction digestion of genomic DNA.**

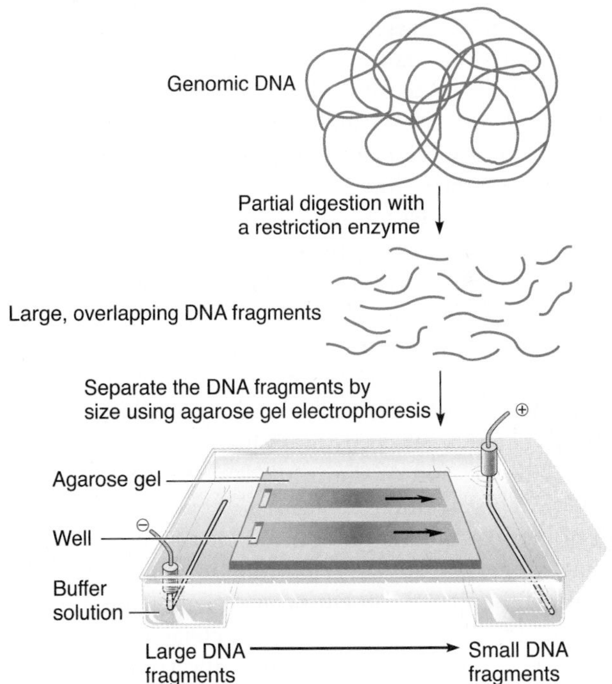


**b) Agarose gel electrophoresis analysis of genomic DNA partially digested with restriction enzyme.**

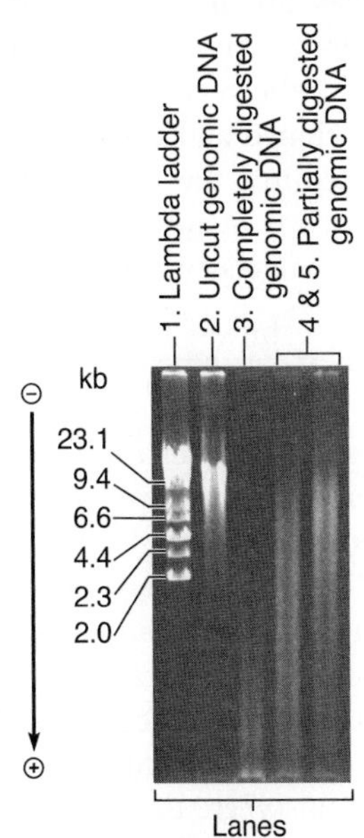

the results of digesting the genomic DNA partially using the same restriction enzyme. In both cases the DNA is of much larger size than that seen in the complete digest lane, this being the expected outcome of partial digestion. The partial digestion conditions were different for the samples loaded in the two lanes, with more digestion carried out for the DNA in lane 4 than for the DNA in lane 5. The difference in partial digestion conditions is reflected in the range of DNA fragment sizes on the gel; that is, larger DNA fragments are seen in lane 5 than in lane 4. As for the complete digestion of genomic DNA, partial digestion does not result in discrete bands when the digested DNA is analyzed by agarose gel electrophoresis. Rather, there is a smear of DNA fragments of different sizes. Since there is a DNA ladder in the gel showing where DNA fragments of particular sizes migrated, researchers can use that information and isolate DNA fragments of the desired size for cloning from the partial digest lanes. The isolation is done simply by cutting out a block of agarose containing the DNA fragments of the desired size and then extracting the DNA from the gel piece.

Agarose gel electrophoresis is an important technique used commonly in the lab to separate and visualize DNA fragments. It is useful for analyzing partial digests of genomic DNA as we have discussed here as well as for analyzing complete restriction digests of a variety of DNA molecules, including specific clones, virus genomes, and organelle genomes. You will see further examples of the use of agarose gel electrophoresis in other chapters.

While the aim of the methods just described is to produce a library of recombinant molecules that contains all of the sequences in the genome, that is not possible. Some sequences are very difficult to clone and, as a result, will either be absent or underrepresented in our library. For example, some regions of eukaryotic chromosomes may contain sequences that affect the ability of vectors containing them to replicate in *E. coli*; these sequences are lost from the library.

How many clones are needed to contain all sequences in the genome? The number of clones needed to include all sequences in the genome depends on the size of the genome being cloned and the average size of the DNA fragments inserted into the vector. The probability of having at least one copy of any DNA sequence in the genomic library can be calculated from the formula

$$N = \frac{\ln(1 - P)}{\ln(1 - f)}$$

where $N$ is the necessary number of recombinant DNA molecules, $P$ is the probability desired, $f$ is the fractional proportion of the genome in a single recombinant DNA molecule (that is, $f$ is the average size, in kilobase pairs, of the fragments used to make the library divided by the size of the genome, in kilobase pairs), and ln is the natural logarithm. For example, for a 99% chance that a particular yeast DNA fragment is represented in a genomic library of 10-kb fragments, where the yeast genome size is about 12,000 kb, 5,524 recombinant DNA molecules would be needed. For the approximately 3,000,000-kb human genome, more than 1,380,000 plasmid clones would be needed, while an artificial chromosome library, with an average insert size of 250 kb, would require only 56,000 clones, hence the use of YAC or BAC vectors for making libraries of large genomes. This formula can also be used to calculate the fraction of the genome likely to be present in a newly constructed library, since the number of clones, $N$, and average insert size are all easily determined after a library is made, and the size of the genome is probably a known value. In this case, we would know $N$ and $f$, and we would solve for $P$. Whatever the genome or vector, to have confidence that all genomic sequences are represented, one must make a library with several times more than the calculated minimum number of clones.

## Chromosome Libraries

As seen above, a genomic library must contain a very large number of clones to achieve nearly complete representation of the genome. This is a particularly major problem for larger genomes, like the human genome. One solution to this problem is to simplify the library by making several smaller libraries, each from an individual chromosome. A library consisting of a collection of cloned DNA fragments derived from one chromosome is called a **chromosome library.** In humans, this means 24 different libraries, one each for the 22 autosomes, the X, and the Y. Since each chromosome is far smaller than the total genome, the resulting libraries can also be smaller. Using these chromosomal libraries can simplify later organizational steps, as the genomic sequence is assembled, because all of the clones in a given chromosome library are, by definition, from the same chromosome and thus from the same large piece of DNA. These libraries proved to be quite useful in certain aspects of the Human Genome Project, as several research teams had been assigned specific chromosomes to sequence, and they turned to these smaller, less complex libraries to make their analysis simpler. Both genomic libraries and chromosome libraries have other uses, as you will see in later chapters. If you wish to clone a specific gene but do not have genomic sequences, libraries (either genomic or chromosome) will be important tools for finding and cloning that gene.

Individual chromosomes can be separated if their morphologies and sizes are distinct enough, as is the case for human chromosomes. In one separation procedure, *flow cytometry*, chromosomes from cells in mitosis are stained with a fluorescent dye and passed through a laser beam connected to a light detector. This system sorts the chromosomes based on the differences in dye intensity that result from subtle differences in the abilities of the various chromosomes to bind the dye. Once the chromosomes have been sorted and collected from a number of cells, a library of each chromosome type can be made in the manner just described.

No matter how the library was made, or whether it was a chromosome or genomic library, at least some of the DNA sequence of the inserts ultimately must be determined. For genomic analysis, we generally start with a genomic library and sequence many clones to determine the sequence of the entire genome.

**Keynote**

A genomic library is a collection of clones that contains at least one copy of every DNA sequence in an organism's genome. Like regular book libraries, genomic libraries are great resources of information; in this case, the information is about the genome. Library size is highly dependent on insert size and genome size, and so more clones are required for libraries that contain smaller inserts, especially for larger genomes. A chromosome library is similar conceptually to a genomic library, except that the collection of clones is made of just one chromosome of the genome.

## DNA Sequencing and Analysis of DNA Sequences

A clone from a genomic library, or any other clone, can be analyzed to determine the nucleotide sequence of the DNA insert, as well as to determine the distribution and location of restriction sites. Its nucleotide sequence is the most detailed information one can obtain about a DNA fragment. The information is useful, for example, in computer database analyses for comparing sequences from different genomes, which can tell us how closely related two organisms are, or for identifying gene sequences and the regulatory sequences—like promoters, silencers, and enhancers—that control gene expression. Furthermore, the DNA sequence of a protein-coding gene can be translated by computer to provide information about the properties of the protein for which it codes. Such information can be helpful for an investigator who wants to isolate and study a protein product of a gene for which a clone is available. Walter Gilbert and Frederick Sanger shared one half of the 1980 Nobel Prize in Chemistry for their "contributions concerning the determination of base sequences in nucleic acids." The DNA sequence of protein-coding genes is also useful for comparing the sequences of homologous genes from different organisms. These analyses can compare either the DNA sequences from the organisms, or the predicted protein sequences. Comparative genomics is a field that is growing as more and more genomic sequences become available.

### Dideoxy Sequencing

The most commonly used method of DNA sequencing, called **dideoxy sequencing** (developed by Fred Sanger in the 1970s), is based on DNA replication. Using a sequence of interest already cloned into a vector as a template, DNA polymerase adds nucleotides to a short primer, until extension of the new DNA strand is stopped by inclusion of a modified nucleotide. This generates an array of short fragments, which can be interpreted by gel electrophoresis either in an automated DNA sequencer or in a standard gel apparatus. Both linear DNA and circular DNA can be sequenced using the dideoxy DNA sequencing method. Linear DNA fragments can be generated, for example, by cutting plasmid DNA with a restriction enzyme or enzymes, or by using the polymerase chain reaction (PCR: see Chapter 9, pp. 221–223).

**Sequencing Primers.** In dideoxy DNA sequencing, the template DNA first is denatured to single strands by heat treatment. Next, an **oligonucleotide** (short DNA strand) primer is annealed to one of the two DNA strands (Figure 8.9a). Typically the primer is 10–20 nucleotides long. For simplicity, the primers shown in the DNA sequencing figure are 3 nucleotides long. The oligonucleotide primer is designed so that its 3′ end is next to the DNA sequence the investigator wishes to determine. The oligonucleotide acts as a *primer* for DNA synthesis catalyzed by a DNA polymerase enzyme (recall from Chapter 3, p. 43, that DNA polymerase requires a primer to begin DNA synthesis), and its 5′-to-3′ orientation ensures that the DNA made is a complementary copy of the DNA sequence of interest (see Figure 8.9a).

Commonly, the DNA sequence a researcher wishes to determine is that of the insert in a cloning vector. This is the case for the inserts in a genomic library when a complete genome sequence is the goal. Consider as an example a DNA fragment cloned into the plasmid cloning vector pBluescript II (see Figure 8.4). For this discussion, the fragment cloned had a *Kpn*I sticky end at one end and a *Sac*I sticky end at the other and was cloned into pBluescript II that had been cut in the multiple cloning site with both *Kpn*I and *Sac*I (Figure 8.9b). With an oligonucleotide primer complementary to a DNA sequence adjacent to the multiple cloning site, we can sequence into the DNA insert. In fact, most plasmid cloning vectors have the same sequences flanking their multiple cloning sites, so that with only two *universal sequencing primers* we can sequence into any cloned insert in those vectors. Two such primers are the SP6 and T7 universal sequencing primers (several other universal primers are also used) and sites to which they anneal are at the ends of the multiple cloning site in pBluescript II (see Figure 8.9b). Both universal sequencing primers are ultimately useful in sequencing. For instance, after a pBluescript II-based clone is denatured with heat, the SP6 universal sequencing primer will anneal to one of the two strands, in this case to a DNA region at the left end of the multiple cloning site (see Figure 8.9b). Using this primer, we can sequence into the DNA insert from this side. With a second reaction that uses using the T7 universal sequencing primer, which is complementary to a short segment of DNA on the other side of the multiple cloning site, we can sequence into the DNA insert from that side. If the DNA insert is small, the two sequencing

**Figure 8.12**

**Pyrosequencing.** (a) In a pyrosequencing reaction, a single-stranded DNA template is attached to a bead. A sequencing primer and several enzymes, including DNA polymerase, are added. dNTPs are added to this mix one at a time. In this example, dCTP has just been added to the reaction. DNA polymerase can add a deoxy C nucleotide to the 3′ end of the growing strand. This reaction releases pyrophosphate (PPi), which is converted to ATP by a second enzyme in the mixture and then a third enzyme in the mix breaks this ATP to release light. The pyrosequencer quantifies the amount of light released. Excess dCTP is consumed by yet another enzyme in the mixture and then another dNTP is added. If the next dNTP is dTTP or dATP, no reaction occurs, since neither can be added to the growing strand. Only when dGTP is added can the new DNA strand be extended. In this case two units of light will be created since the template has two adjacent C nucleotides, so two deoxy G nucleotides can be added. (b) The pyrogram shows how much light was made. It is used to determine the sequence of the new DNA strand that was synthesized.

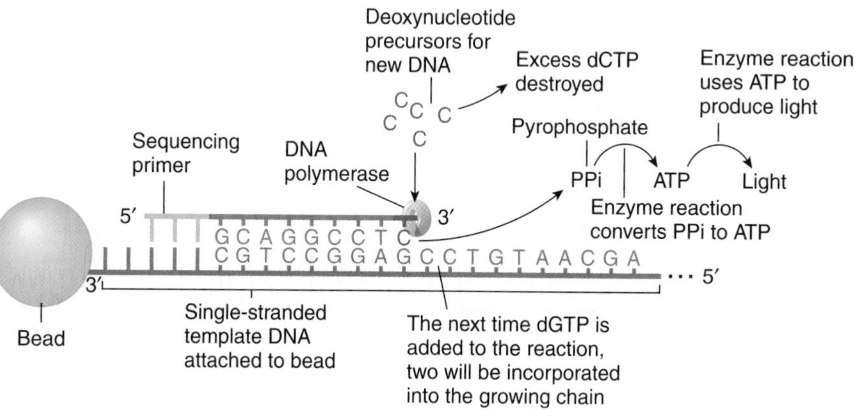
a) A pyrosequencing reaction
Deoxynucleotide precursors for new DNA
Excess dCTP destroyed
Enzyme reaction uses ATP to produce light
Pyrophosphate
Sequencing primer
DNA polymerase
PPi
ATP
Light
5′
3′
Enzyme reaction converts PPi to ATP
GCAGGCCTC
CGTCCGGAGCCTGTAACGA
3′
... 5′
Single-stranded template DNA attached to bead
Bead
The next time dGTP is added to the reaction, two will be incorporated into the growing chain

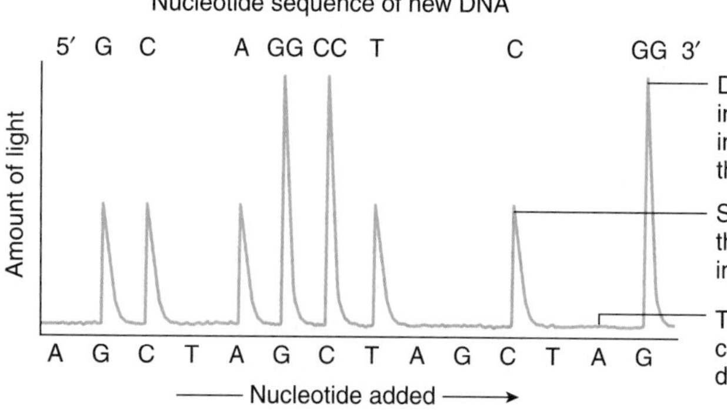
b) Pyrogram result of pyrosequencing
Nucleotide sequence of new DNA
5′ G C A GG CC T C GG 3′
Amount of light
A G C T A G C T A G C T A G
Nucleotide added
Double-height peak indicates that two nucleotides were incorporated into the new DNA when the precursor was added. In this case a G was incorporated meaning that there were two adjacent C nucleotides on the template.
Single-height peak indicates that one nucleotide was incorporated into the new DNA when the precursor was added. In this case a T was incorporated meaning that the template had an A.
The absence of peaks when nucleotides are added means that they could not be incorporated into new DNA, meaning that the template did not have complementary bases.

to the bead (Figure 8.12a). Since the first unpaired base in the template strand is a G, the dCTP can be added to the 3′ end of the primer by DNA polymerase, and a molecule of pyrophosphate ($PP_i$) is released. Another enzyme in the mix uses this pyrophosphate in a reaction that produces ATP, and a third enzyme uses the energy stored in the newly produced ATP to produce light. The pyrosequencer detects and quantifies the amount of light released and correlates it to which dNTP was present in the reaction. Thus, for this example, since light was emitted when dCTP was present, we know that C was incorporated into the growing strand. Excess dCTP is destroyed by another enzyme in the reaction. Now another dNTP is added, for example, dTTP. In our example, no light is emitted when dTTP is added, because a dTTP will not base-pair with the C on the template. The excess dTTP is degraded enzymatically, and the pyrosequencer will next add dATP. Once again, this cannot be added to the growing strand, so the dATP is destroyed without powering the creation of light. The next addition is dGTP. Since the next two bases on the template strand are both C, DNA polymerase adds two molecules of dGTP to the growing strand after the C. This means that new DNA with the sequence 5′-CGG-3′ has been synthesized. We can tell that two G residues were incorporated, since adding two G residues to the growing strand releases two molecules of pyrophosphate, which are in turn used to create two molecules of ATP, and twice as much light is produced as is the case when one nucleotide is added to the strand. The pyrosequencer measures exactly how much light is made as a particular dNTP is added, and, based on the output of light, we can determine the exact sequence of the DNA that has been synthesized based on the pyrogram (Figure 8.12b). The pyrosequencer continues this cyclical process, adding dCTP, then returning to dTTP, dATP, dGTP, and so on. As for dideoxy sequencing, the DNA sequence obtained is the complement of the sequence of the DNA template.

We have described the pyrosequencing reaction with one bead. The pyrosequencer has about 200,000 microscopic wells, in each of which a different pyrosequencing reaction with a different single-stranded template DNA attached to a bead is carried out. Thus, the sequencing of many DNA templates is done simultaneously, making it possible to obtain about 20 million nucleotides of genome sequence in about 6 hours. The pyrosequencing technique is still quite new and expensive, but it should become an important technique as the equipment becomes refined and more affordable.

### Analysis of DNA Sequences

Since the best sequencing reaction will generate only a few hundred base pairs of sequence, it is generally necessary to assemble the results of many reactions, each starting with a different primer, to determine the sequence of a larger piece of DNA and, further, to assemble the sequences of many individual small cloned fragments into an entire chromosome or a genome. It is relatively simple to compare by computer two (or more) sequences that have been generated by DNA sequencing. If these sequences overlap, then a series of bases will be found in both sequences. If the overlap is long enough, it can be tentatively assumed that the two fragments sequenced partially overlap. For instance, if sequencing clone 1 tells us that the insert has a sequence of 5′-AGCTTACGCCGATATTATGCGTTTA-3′, and sequencing clone 2 tells us that it has an insert with the sequence 5′-ATGCGTTTAGGGCGCAATAATTAGCGCAAT-3′, then these sequences overlap (overlapping sequences are in bold), and the true sequence of the DNA as it would be found in the genome would be 5′-AGCTTACGCCGATATTATGCGTTTAGGGCGCAATAATTAGCGCAAT-3′ (overlapping region is highlighted in bold). Additional overlaps can be discovered as more clones are sequenced, allowing assembly of long sequences. This is a critical step in nearly all DNA sequence analysis, not just genomics. If a gene of interest is cloned from a library (Chapter 10, pp. 258–261), we will need to sequence the insert to understand the gene we have just cloned. Only a few genes are small enough to be sequenced completely in a single reaction, so this assembly typically is needed even when we are working with a single clone.

**Keynote**

Methods have been developed for determining the sequence of a cloned piece of DNA. A commonly used method, the dideoxy procedure, uses enzymatic synthesis of a new DNA chain on a cloned template DNA strand. With this procedure, synthesis of new strands is stopped by the incorporation of a dideoxy analog of the normal deoxyribonucleotide. Using four different dideoxy analogs, the new strands stop at all possible nucleotide positions, thereby allowing the complete DNA sequence to be determined. A newer DNA sequencing technique, pyrosequencing, also is based on DNA synthesis. In this technique a single-stranded template DNA is attached to a microscopic bead and a reaction mix containing primer, DNA polymerase, and other enzymes is added. dNTPs are added sequentially one at a time and, if a particular dNTP can extend the new DNA strand, pyrophosphate is released and, by the action of the other enzymes in the reaction, this release is detected by light emission. The pattern of light emission correlated with the particular dNTP present gives the DNA sequence complementary to the template DNA. Whichever DNA sequencing technique is used, the DNA sequence obtained from a reaction is relatively limited in length. To obtain the sequence of long stretches of DNA, it is necessary to assemble the results of many reactions by using computer algorithms to identify overlap between adjacent DNA sequences.

## Assembling and Annotating Genome Sequences

Now that we have discussed the techniques for cloning and sequencing DNA, we turn to considering them in the context of obtaining the sequences of complete genomes. The current approach to sequencing genomes is called the whole-genome shotgun approach. We also discuss in this section the annotation of genome sequences, meaning the analysis of the sequences to identify putative genes and other important sequences.

### Genome Sequencing Using a Whole-Genome Shotgun Approach

In the **whole-genome shotgun approach for genome sequencing,** the whole genome is broken into partially overlapping fragments, each fragment is cloned and sequenced, and the genome sequence is assembled using a computer. This approach to sequencing genomes has become the most common because it has proven to be both fast and efficient, and it can be used even if very little is known about the genome.

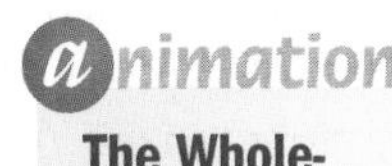


**The Whole-Genome Shotgun Approach to Sequencing**

Figure 8.13 outlines the whole-genome shotgun approach for genome sequencing. First, random, partially overlapping fragments of genomic DNA are generated by mechanical shearing and the fragments are cloned to form a library. In contrast to the libraries described earlier, the insert size for each clone is small—about 2 kb—enabling the clones to be made using simple plasmid vectors. This does mean that a huge library, with thousands or millions of clones, is required. A few hundred nucleotides are sequenced from each end of each insert, and the sequence data are entered into the computer. For the sake of discussion, let us consider that 500 nucleotides are sequenced in each reaction. This would mean that, because the clones partially overlap, the sequence of the central approximately 1 kb of DNA is obtained only when an overlapping clone is sequenced. For example, if a second clone overlapped the first clone by 500 bp, then sequencing the second clone would generate 500 bp of sequence from the middle unsequenced section of the first clone. The computer compiles a genomic sequence from these short sequences by assembling them based on the overlaps. The result of sequencing this library is a relatively small number of assembled sequences covering most of the genome. There are gaps between the assembled sequences because some sequences are missing in the library.

**Figure 8.13**
**The whole-genome shotgun approach to obtaining the genomic DNA sequence of an organism.**

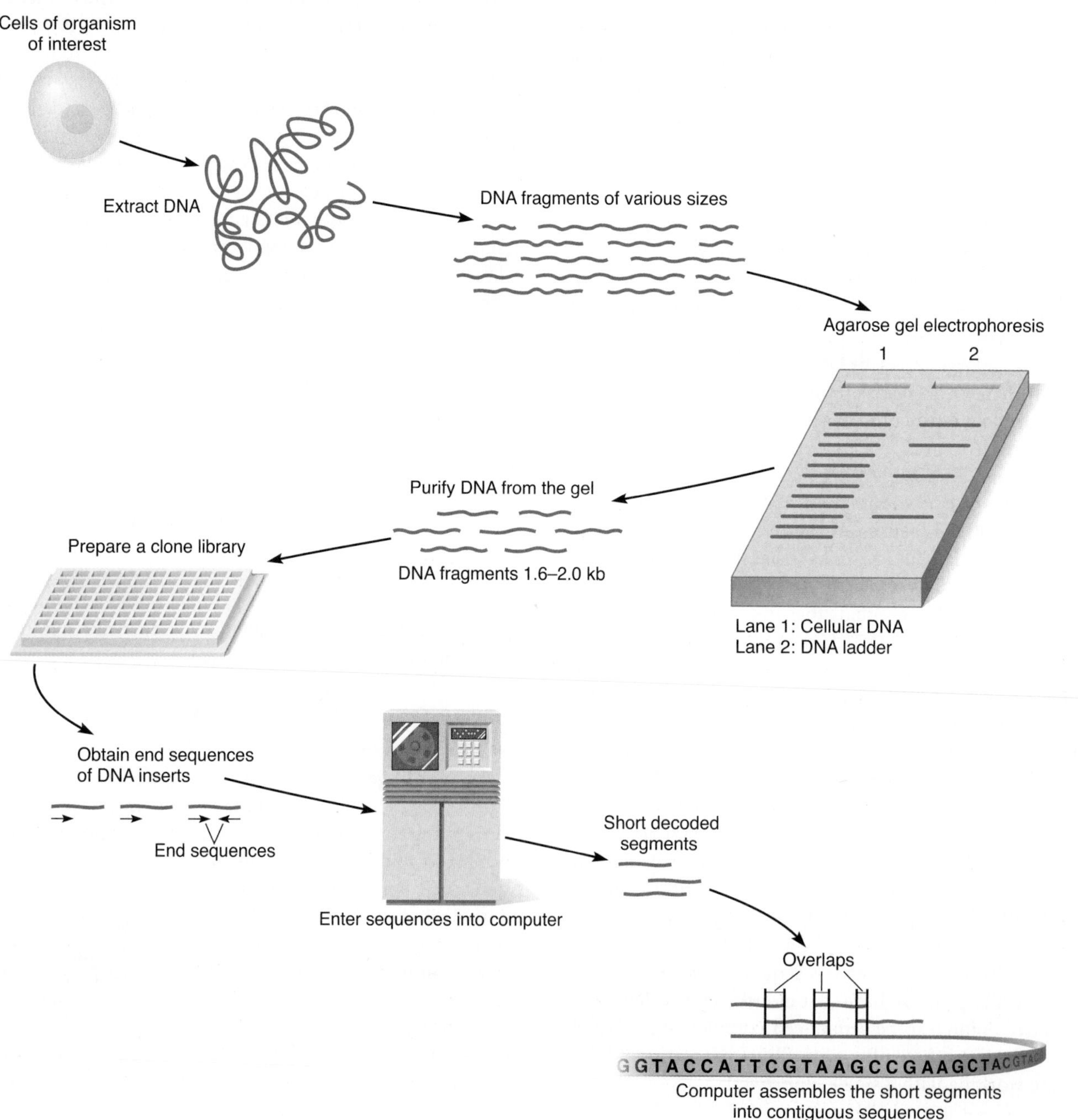

A second library is used in the shotgun approach consisting of a random, partially overlapping library of genomic DNA fragments of about 10 kb in size in a simple plasmid vector. One important purpose for this library is to sequence regions of the genome containing repeated sequences. Many repeated sequences are around 5 kb in size, so a 10-kb clone can contain one of these units and non-repetitive flanking DNA both before and after the repeat, which cannot both be found in a single 2-kb clone. Here is the dilemma with the 2-kb clone library. In assembling a genome sequence from the 2-kb clones, a clone with an insert consisting of some unique sequence DNA followed by part of a copy of a repeated sequence causes a dead stop in sequence assembly. This is because many clones in the library contain parts of the repeated sequence family, and they come from all over the genome. The computer algorithms will be unable to define the correct overlapping partner for this clone, as many clones will look like possible matches. Each of these possible matches will have flanking unique sequence DNA, but we cannot determine which clone is the true overlapping one from the genome. The 10-kb clone library allows us to get around this problem because some clones have unique sequence DNA flanking a repeated DNA sequence. When we sequence one of these clones, we will be able to connect the smaller clones, essentially jumping over the repetitive region—the large clone acts as a bridge to connect the gap. This allows us to proceed with the genome sequence assembly, provided that the 10-kb library clone contains only a single insert and is not contaminated with clones that have multiple inserts, as discussed earlier for YAC clones. Another purpose of the library is to obtain sequence information to provide independent confirmation of assembled sequence structure.

Computer assembly of a genome sequence from sequencing data is similar to that described earlier, but on a much larger scale. The quality of the assembled sequence is closely related to the *coverage* of the genome, the average number of times a given sequence will appear in the sequencing reads, with higher coverage meaning a higher-quality assembled sequence. For example, for a 7-fold coverage of a 100-Mb genome, 700 Mb of DNA sequence is collected. The quality of the genomic sequence is closely related to the coverage, because the clones that are sequenced are selected at random, so higher coverage means there is a smaller chance a given region will never be selected. Thus, a higher coverage value indicates that a greater percentage of the genome has been sequenced (and that most of the genome has been sequenced more than once, which allows us to have more confidence in the quality of the sequence), while a lower coverage value indicates that there will be many more gaps in the sequence and that much of the genome has been sequenced only once. Many of the high-quality genome sequences were generated from 7-to-8–fold coverage, while some genome sequences have only 2-to-3–fold coverage, and, as a result, the data are less complete for these genomes.

Initially, the whole-genome shotgun approach for genome sequencing was thought to be of limited usefulness for sequencing whole genomes greater than 100 kb. This was due to two concerns: (1) that the labor involved to reach high coverage was overwhelming for nonautomated sequencing; and (2) because the computer analysis becomes very complex as the number of sequences increases. In recent years, robotic procedures for preparing DNA for sequencing, and powerful automated sequencers and sophisticated computer algorithms for assembling sequences from hundreds to millions of 300–500-bp sequences, opened the door for sequencing large genomes using this shotgun approach. The final proof that this approach would work for large genomes was when a draft sequence of the human genome was released by Celera Genomics. This sequence, built using the whole-genome shotgun approach, had 5-fold coverage (each nucleotide had been sequenced, on average, five times). The draft sequence covered about 97% of the genome, but gaps were present in the compiled sequence. Why were these gaps present? Even at 5-fold coverage, a few regions will not be sequenced. This accounts for some, but not all of the gaps. As you have learned, our genome contains repetitive sequences. In many cases, we have long stretches containing many copies of a single type of repetitive sequence, and assembly across these regions is very difficult as a result. Furthermore, cloned DNA sometimes undergoes recombination or deletion in its bacterial host, and certain sequences, especially highly repetitive sequences, undergo these processes frequently. While some of these gaps have been resolved recently, they are not viewed as a high priority since they tend to contain very few genes.

Advances continue to be made in DNA sequencing automation and in computer algorithms for analyzing sequences obtained. The whole-genome shotgun approach is now used almost exclusively in genome sequencing projects, even for large genomes.

## Assembling and Finishing Genome Sequences

The raw sequences obtained from genome sequencing projects must be *assembled* into larger sequences; that is, the bases must be pieced together in their correct order as they are found in the genome. Once assembly is complete, that is often the point when "working drafts" of genome sequences are announced. The work is not completed at that point, because there are still many gaps in the sequences to fill in as well as errors from the sequencing. *Finishing* the genome sequence is the next step, producing a highly accurate sequence with fewer than one error per 10,000 bases, and as many gaps as possible filled in.

### Keynote

Sequencing a genome by the whole-genome shotgun approach involves constructing a partially overlapping library of genomic DNA fragments, and sequencing each clone. The DNA sequences obtained are assembled into larger sequences by computer based on the sequence overlaps. Gaps remaining at this point are filled in by subsequent sequencing in a process known as finishing.

## Annotation of Variation in Genome Sequences

The next step after obtaining the complete sequence of a genome in a genome project is *annotation*, the identification and description of putative genes and other important sequences. Annotation begins the process of assigning functions to all the genes of an organism. Once an entire genome has been sequenced, scientists can also begin to study all the differences found between individuals of a species. This can help scientists understand where natural variation in populations comes from, and helps us identify which DNA sequences are responsible for particular traits in a population, Though sequencing technology is improving daily, for many eukaryotic species it is still prohibitive to sequence the entire genome of many individuals. One way around this is to analyze many small regions of DNA scattered throughout the genome to build up maps of genetic differences between individuals that can be studied, such as haplotype maps.

**SNPs and Haplotypes.** The most detailed maps use **single nucleotide polymorphisms (SNPs).** A SNP is a type of *DNA marker* with a simple, single base-pair alteration in some individuals at a site; that site is the SNP locus. **DNA markers** are sequence variations among individuals in a specific region of DNA that are detected by molecular analysis of the DNA and can be used in genetic analysis. SNP loci are abundant in the human genome and can be found, on average, about once every 1,000 bp (and are even more abundant in some regions). Thus, each polymorphic SNP locus will have other polymorphic SNP loci nearby.

The abundance of SNP loci has allowed researchers to develop highly detailed maps showing the location of the SNPs on the chromosome. For SNP loci that are close to each other, genetic recombination rarely scrambles the pattern of SNP alleles present on a particular chromosome. This means that if your father gave you allele one of *SNP-A* (*SNP-A1*) and allele one of *SNP-B* (*SNP-B1*), and your mother gave you allele two of each SNP (*SNP-A2* and *SNP-B2*), your children most likely will either *inherit SNP-A1* and *SNP-B1*, or *SNP-A2* and *SNP-B2* (so it is very unlikely that you will pass a new mixture of these SNPs to your offspring). If another SNP, *SNP-C*, is far from either *SNP-A* or *SNP-B*, then you will not be able to make a similar prediction about the inheritance of versions of *SNP-C* relative to *SNP-A* or *SNP-B*. A **haplotype** is a set of specific SNP alleles at particular SNP loci that are close together in one small region of a chromosome, so in any particular family, these haplotypes are rarely scrambled by genetic recombination. In the example above, *SNP-A1* and *SNP-B1* would form a small haplotype. Genetic recombination tends to happen in regions called *recombination hot spots*, and it is far rarer in *recombination cold spots*. In general, all of the SNP loci in a haplotype will reside in a single recombination cold spot. As a result, the inheritance of one SNP allele in the haplotype predicts the inheritance of other haplotype SNP alleles. Since each recombination cold spot is a small region of a chromosome, all of the SNP loci in a haplotype are close to each other on the same chromosome. This is, in essence, a small group of genetically linked SNPs.

If we know that a group of several SNPs tend to be inherited together, we can test for only a diagnostic subset of them—called *tag SNPs*—rather than all of them. By definition, a **tag SNP** is one (or more) SNP locus used to test for and represent an entire haplotype. If all members of one haplotype are inherited together, then testing only a couple members of the group will tell us what happened with the untested members. For example, assume that SNP loci *A*, *F*, *L*, *M*, *X*, and *Z* are all in the same recombination cold spot and form a haplotype. Your father inherited SNP alleles *A1*, *F2*, *L2*, *M2*, *X1*, and *Z2* from his mother (this would be one haplotype) and SNP alleles *A2*, *F1*, *L2*, *M1*, *X2*, and *Z1* from his father (this would be another haplotype). We wish to determine which haplotype you inherited from your father, so instead of looking at every SNP locus (*A*, *F*, *L*, *M*, *X*, and *Z*), we test the inheritance of just SNP *A* and *Z* alleles. We determine that your father gave you *A1* and *Z2*, so we may tentatively assume that *F2*, *L2*, *M2*, and *X1* were inherited as part of that haplotype. If your sister inherited *A2* and *Z1* from your father, we would assume that she inherited the other haplotype. Furthermore, if SNP loci *A*, *F*, *L*, *M*, *X*, and *Z* are inherited together, any clones from a genomic library containing one or more of these SNPs must be close to each other in the physical map.

We have identified more than 13 million human SNPs. Many of these SNPs fall into known haplotypes with defined tag SNPs, so we can test the tag SNPs only (there are only about 500,000 of these) and predict the inheritance of all the SNPs from each haplotype based on the inheritance of just the tag SNPs that define the haplotypes. Testing half a million SNPs may seem impossibly labor-intensive, but *DNA microarrays* (see Chapter 9, pp. 230–232) allow us to test thousands at once. **DNA microarrays** (also called *DNA chips*) are glass slides spotted with thousands of different DNA probes. (A DNA probe is a molecule in an experiment used to determine if a complementary DNA or RNA target molecule is present. Pairing of probe with target is detected using the properties of the label.) A SNP DNA microarray (often called a *SNP chip*) is a specific type of DNA microarray that has single-stranded, unlabeled tag SNP allele oligonucleotide probes affixed to the slide. Fluorescently labeled, single-stranded target DNA from an individual to be tested is mixed with the tag SNP probe on the SNP DNA microarray. If probe and target DNA sequences are complementary, then they will form base pairs with each other in a process called **hybridization** (since we are forming a hybrid double helix with two different single-stranded pieces of DNA). Hybridization always involves a probe that can form base pairs with target DNA, and in typical experiments the probe DNA molecules are labeled in some way while the target DNA is unlabeled. For a DNA microarray experiment, however, the probes are unlabeled and are each affixed to a specific, known location on the slide while the target DNA is labeled.

For a SNP DNA microarray, the labeled target DNA, which is fluorescently labeled genomic DNA from a single individual, is added to the microarray, and if some of the target DNA can form base pairs with one or more probes on the slide, the labeled DNA will be present at the site of that probe. For SNP DNA microarrays, the hybridization conditions are set to be very demanding, so that just a single mismatch between the probe and the target prevents the formation of base pairs between the probe and the target. That is, the fluorescently labeled target DNA of an individual will stick to tag SNP allele probes that match perfectly the SNP alleles present in his or her DNA, but will not stick to tag SNP allele probes that test for SNP alleles that are imperfect matches for his or her DNA (Figure 8.14a). In a SNP DNA microarray experiment, a laser quantifies the intensity of the fluorescent signal at each of the thousands of locations on the slide, and the resulting profile is cross referenced by computer with the locations of the individual tag SNP probes on the slide (Figure 8.14b). The result of this experiment is the identification of all the specific tag SNP alleles in this person's genome, which tells us ultimately which haplotypes are present in that individual.

What is the value of knowing all of a specific individual's haplotypes? Well, this analysis can help scientists isolate the particular gene or genes associated with specific human genetic diseases, since this technique allows for the rapid analysis of human pedigrees for the study of disease inheritance. We might observe that the inheritance of five linked sets of tag SNPs correlates with the inheritance of a particular genetic disease in a family, while unaffected individuals in the family never inherit these tag SNPs. This would suggest that the gene that causes the disease was near the tag SNPs on that chromosome. Since we know the physical location of each of these tag SNPs, we can analyze these regions of the genome for nearby genes that may be altered in people with this disease.

**The Haplotype Map.** Experiments like the tag SNP DNA microarray just described can help identify all the haplotypes a particular individual has inherited. Scientists can then begin to look at all the combinations of haplotypes present in many human populations and build a **haplotype map (hapmap).** The haplotype map is a complete description of all of the haplotypes known in all human populations tested, as well as the chromosomal location of each of these haplotypes. If two haplotypes are neighbors on a chromosome, separated by a recombination cold spot, then these haplotypes will generally be inherited together. If two haplotypes are neighbors on a chromosome and are separated by one or more recombination hot spots, then these haplotypes will tend to be inherited together. However, the correlation will not be as strong as the correlation seen for SNP loci within the same haplotype, since there will be some recombination at the hot spot that separates them. Haplotypes that are very far apart from each other will be passed from one generation to the next independently of each other. Thus, a haplotype map is a very fine structure physical and genetic map of a chromosome. Haplotype maps can be used to study the inheritance of complex traits such as heart disease and obesity in humans, which may be caused by the additive effects of multiple genes that would be hard to find using classical genetic analyses. They can also be used to study evolutionary relationships (see the Focus on Genomics box for this chapter).

**Keynote**

SNPs, or single nucleotide polymorphisms, are small regions of DNA that vary between individuals. These SNPs can be studied individually or as haplotypes, which are sets of SNP alleles that tend to be inherited as a group. DNA microarrays allow us to determine the SNP genotype for thousands of SNP loci at once. This allows us to develop haplotype maps. Studying haplotype maps can tell us about the differences between individuals and can teach us about variation found in both non-protein-coding regions as well as the sequences that encode functional proteins.

### Identification and Annotation of Gene Sequences

The regions of particular interest to scientists are the protein-coding genes since they are the functional units of an organism. We now focus our attention on several methods used to find these protein-coding regions specifically. We can look for protein-coding genes by analyzing cDNAs or by searching for likely coding regions in the genomic DNA. Each of these approaches has its strengths and weaknesses, but the combination has proven to be quite reliable.

**Analysis of cDNAs to Identify Gene Sequences.** Theoretically, the simplest way to find genes is to look at messenger RNAs (mRNAs), since every messenger RNA, by definition, comes from a gene. One problem with this direct method is the nature of transcription itself—a given

**Figure 8.14**

**Tag SNP (single nucleotide polymorphism) testing.** (a) Principle of typing a tag SNP by hybridization. Hybridization conditions are used so that a single mismatch destabilizes the hybrid, thereby preventing the two strands from base-pairing. (b) A microarray test of tag SNPs. Hybridization of tag SNPs using the labeled target DNA and the unlabeled tag SNP allele probes on the microarray can be detected because of the fluorescent label (in this case, a red dye) on the individual's DNA.

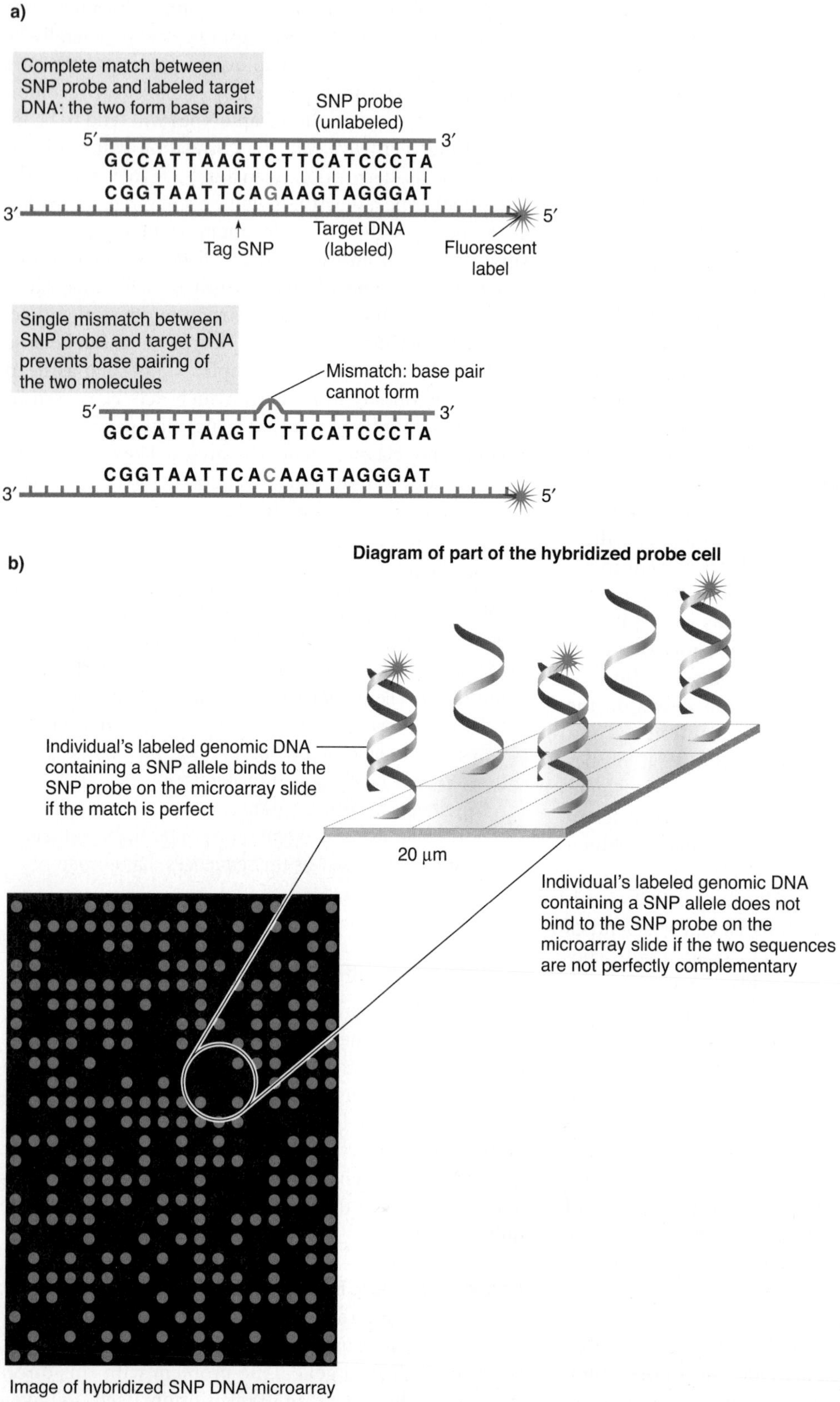

a)

Complete match between SNP probe and labeled target DNA: the two form base pairs

SNP probe (unlabeled)

5′ GCCATTAAGTCTTCATCCCTA 3′

3′ CGGTAATTCAGAAGTAGGGAT 5′

Tag SNP

Target DNA (labeled)

Fluorescent label

Single mismatch between SNP probe and target DNA prevents base pairing of the two molecules

Mismatch: base pair cannot form

5′ GCCATTAAGT C TTCATCCCTA 3′

3′ CGGTAATTCACAAGTAGGGAT 5′

b)

Diagram of part of the hybridized probe cell

Individual's labeled genomic DNA containing a SNP allele binds to the SNP probe on the microarray slide if the match is perfect

20 μm

Individual's labeled genomic DNA containing a SNP allele does not bind to the SNP probe on the microarray slide if the two sequences are not perfectly complementary

Image of hybridized SNP DNA microarray

## Focus on Genomics
### The Real Old Blue Eyes

One use of haplotype maps is to study the inheritance of traits in humans. Blue eyes are found in many human populations, and, while rare in many regions, blue-eyed people make up a large fraction of the population in many parts of Europe. For example, up to 95% of some Scandinavian populations have blue eyes. Since blue-eyed people are found in many populations that have historically been partially isolated from their neighbors by geography, language, religion, or culture, it was assumed that the gene that controls eye color had been mutated a number of times, at least once in each population containing blue-eyed individuals, giving rise to small, unrelated blue-eyed subgroups in different, isolated ethnic groups. This "multiple mutation" model seems to explain the origins of red hair. Under this model, blue-eyed Danes and blue-eyed Turks would not share a blue-eyed common ancestor. Using haplotype maps, scientists analyzed the DNA of more than 800 blue-eyed individuals. The surprising result was that all blue-eyed people shared the same haplotype for a region of chromosome 15, where the genes OCA2 and HERC2 are found. This suggests that all of the tested blue-eyed individuals share a common ancestor. This ancestor probably lived between 6,000 and 10,000 years ago. She or he carried the same haplotype and has passed it on, generation after generation, to his or her descendants. How did it become so common in such a short period of time? There are two possible explanations. The mutation that leads to blue eyes also decreases skin and hair pigmentation. In Europe, the sunlight is less intense than in the tropical parts of Africa where we evolved. When the sunlight is intense, skin pigments are of critical importance to protect us from damaging rays of the sun. These pigments interfere with a crucial, light-requiring step in the production of vitamin D. Under this intense light, synthesizing vitamin D is easy, despite the protective pigments. In Europe, and other regions far from the tropics, the sunlight is far less intense. The protective role of the pigments is, therefore, less critical because the light is less damaging. However, the pigments continue to interfere with vitamin D production. Thus, it is possible that this mutation increased the availability of vitamin D for people living out of the tropics. Sexual selection also may have played a role in the process. Sexual selection can occur when one sex, generally females, prefers a particular set of appearances in a partner. Partners matching that appearance have more children and pass on their haplotypes to their offspring. The tail of the peacock is a classic example of sexual selection. Males derive only one benefit from the tail—females (peahens) prefer males with flashy tails, so bigger tails lead to more mating success. So European women may have preferred blue-eyed men, and sexual selection did the rest. It may have been a combination of both types of selection; females simply might have picked healthier males in all populations. This would lead to blue-eyed people far from the tropics, where the lighter pigmentation allows production of vitamin D; and in tropical regions, where vitamin D synthesis is possible even with darker skin, and the extra pigment served as protection from the damaging solar radiation. No matter how it happened, if you have blue eyes, you can count Reese Witherspoon, Brad Pitt, Paul Newman, Cameron Diaz, Cate Blanchett, and Steve McQueen as (very, very) distant cousins!

cell will transcribe only a small fraction of the genes in its DNA, and some genes are transcribed far less frequently than others, so some mRNAs will be very rare in a sample. A second problem is that mRNAs are chemically unstable, and cloning and sequencing techniques do not work with mRNAs. This problem can be surmounted by working with **cDNA libraries.** Like any DNA library, a cDNA library is a large collection of cloned sequences. In this case, the inserts are **complementary DNAs (cDNAs),** which are double-stranded DNA molecules: one of the strands is a DNA molecule complementary to an mRNA, and the other strand is the partner to this DNA molecule. This second strand is almost identical in sequence to the mRNA, differing only where a T replaces a U in the sequence.

***Synthesis of cDNAs.*** cDNA molecules are made in a two-step process. In the first step, mRNA molecules are used as a template for the production of a DNA partner strand. This step uses **reverse transcriptase (RT),** an enzyme that synthesizes a DNA molecule using RNA as a template. The enzyme was named because it "reversed" the transcription described in central dogma. That is, in classical transcription, DNA is used as a template for RNA production, whereas reverse transcriptase reverses roles for the molecules by using RNA as the template for DNA production.

To make cDNA, we start with an mRNA template. cDNA libraries are most often made from eukaryotic mRNAs (which, as you will recall, differ from the genes that encode them by the removal of intron sequences). This is partly because eukaryotes tend to have larger genomes with more noncoding regions and more genes, so a cDNA library offers a way to sort through only the transcribed regions. Most prokaryotic genomes contain very little DNA that is not part of a gene, so making a cDNA library is often extra work with very little reward because most of the genome will be transcribed and would therefore be represented in the cDNA library. It is generally easier, faster, and less expensive to sequence prokaryotic genomes directly and find the genes by examining the genomic DNA sequences. Luckily, mRNAs are the only RNA molecules in a eukaryotic cell that contain a poly(A) tail (see Chapter 5, pp. 91–92). Other eukaryotic RNAs (rRNA, tRNA, snRNA) and all prokaryotic RNAs lack these tails. The poly(A)+ (shorthand for "molecules with a poly(A) tail") mRNAs can be purified from a mixture of cellular RNAs by passing the RNA molecules over a column to which short chains of deoxythymidylic acid, called *oligo(dT) chains*, have been attached. As the RNA molecules pass through the column, the poly(A) tails on the mRNA molecules base-pair to the oligo(dT) chains. As a result, the mRNAs are captured on the column while the other RNAs pass through. The captured mRNAs are released and collected, for example, by decreasing the ionic strength of the buffer passing through the column so that the hydrogen bonds are disrupted. This method results in significant enrichment of poly(A)+ mRNAs in the mixed RNA population to about 50% versus about 3% in the cell.

Figure 8.15 shows how a cDNA molecule can be made from the mRNA molecules. Key to this synthesis is the presence of the 3′ poly(A) tails on the mRNAs. After the mRNA has been isolated, the first step in cDNA synthesis is annealing a short oligo(dT) primer to the poly(A) tail. The primer is extended by reverse transcriptase to make a DNA copy of the mRNA strand. The result is a DNA–mRNA double-stranded molecule. Next, RNase H ("R-N-aze H," a type of ribonuclease), DNA polymerase I, and DNA ligase are used to synthesize the second DNA strand. RNase H partially degrades the RNA strand in the hybrid DNA–mRNA, DNA polymerase I makes new DNA fragments using the partially degraded RNA fragments on the single-stranded DNA as primers, and finally DNA ligase ligates the new DNA fragments to make a complete chain. The result is a double-stranded

**Figure 8.15**

**The synthesis of double-stranded complementary DNA (cDNA) from a polyadenylated mRNA, using reverse transcriptase, RNase H, DNA polymerase I, and DNA ligase.**

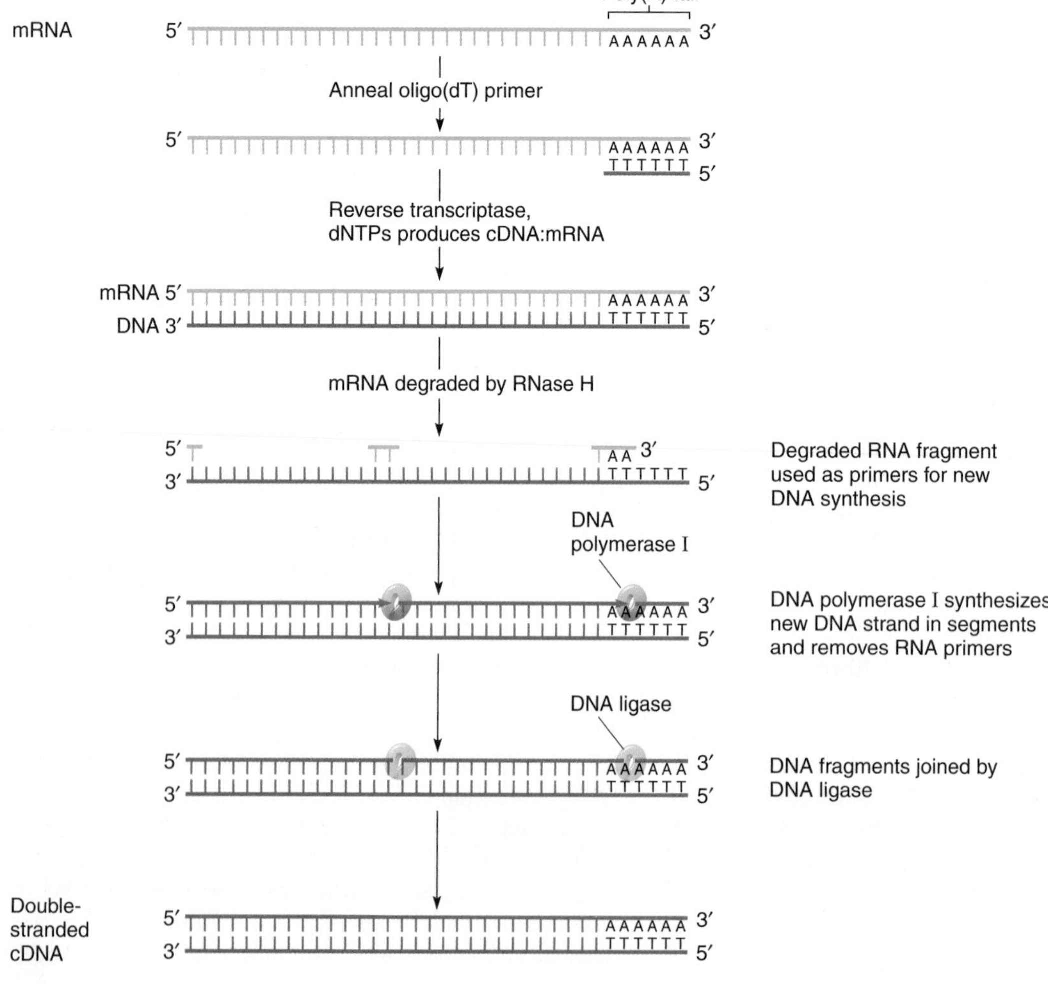

cDNA molecule that is a faithful DNA copy of the starting mRNA.

***Building cDNA Libraries.*** Once double-stranded cDNAs are made, as described above, we must first select only the most complete cDNAs and then clone them into a vector so they can be propagated in a host cell. Because reverse transcriptase has the frustrating tendency to finish only part of its job (thus creating a shortened cDNA that contains only the 3′ end of the gene), we first need to eliminate any truncated cDNAs. We do this by size selection. The cDNAs are separated by gel electrophoresis, visualized, and the part of the gel containing large cDNAs (for instance, everything larger than 1 kb) is excised. The cDNAs are then recovered from this gel slice.

How can we clone cDNA molecules? We cannot clone in the ways described for genomic DNA. That is, cutting these cDNAs to get sticky ends would be both counterproductive and pointless—counterproductive because we want to recover cDNAs as similar to their template mRNAs as possible, and cutting them would break the molecule into pieces. Furthermore, these molecules are small, and we would not be certain that any restriction enzyme would cut all of them to give sticky ends. It would also be pointless to cut them. Recall that we cut genomic DNA to make small, easily manipulated fragments. The cDNAs are much smaller than genomic DNAs, in most cases averaging only 1–5 kb in length. We need to make these intact, uncut fragments clonable. Figure 8.16 illustrates the cloning of cDNA using a **restriction site linker,** or **linker,** which is a short, double-stranded piece of DNA (oligodeoxyribonucleotide) about 8-to-12 nucleotide pairs long that includes a restriction site, in this case the site for *Bam*HI. Both the cDNA molecules and the linkers have blunt ends, and they can be ligated at high concentrations of T4 DNA ligase. Sticky ends are produced in the cDNA molecule by cleaving the cDNA (with linkers now at each end) with *Bam*HI. The resulting DNA is inserted into a cloning vector that has also been cleaved with *Bam*HI, and the recombinant DNA molecule produced is transformed into an *E. coli* host cell for cloning.

A problem with using linkers for cloning cDNAs is that there may be a restriction site within the cDNA for the enzyme used to cleave the linkers. This would mean the cDNA would also be cut when the linkers are cut, resulting in cloning the cDNA in pieces. This problem can be avoided by using one or more methylated nucleotides, in place of their normal analogs, during the synthesis of the cDNA. Some restriction enzymes are unable to cut at restriction sites that contain methylated bases. The linker, which is unmethylated, can be cut. Thus, internal sites will be protected while linker sites will be cut, leaving the cDNA complete and placing sticky ends on both ends of the molecule. Another way to get around this potential problem is to use an *adapter* instead of a linker. An adapter already has one sticky end on it suitable for

**Figure 8.16**

**The use of linkers in cDNA cloning.**

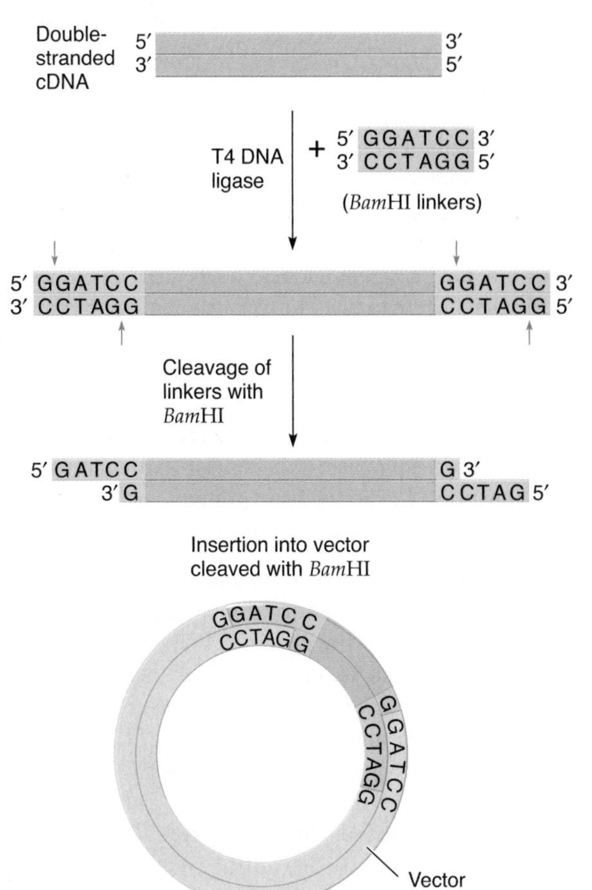


cloning, so the cDNA is never digested with a restriction enzyme. The adapter cannot use this sticky end to connect to the cDNA, because the cDNA has blunt ends. For example, if we make the following adapter, formed by annealing `5′-GATCCAGAC-3′` with `5′-GTCTG-3′`,

```
5′-GATCCAGAC-3′
        GTCTG-5′
```

and ligate it to a cDNA, the blunt end of the adapter will covalently attach to the blunt end of the cDNA, leaving the 5′ overhang `GATC` at each end. You might wonder why two adapters do not ligate using their sticky ends. The 5′ end of the longer strand is modified during synthesis. The phosphate is intentionally left off. As a result, it cannot ligate to a 3′ end. This is exactly what you learned earlier when phosphatase was used to limit certain types of ligations. The overhang will base-pair with a vector digested with *Bam*HI (see Figure 8.16), which has phosphate groups at the 5′ ends of its overhangs, and the cDNA will be cloned in one piece.

You may wonder why cDNA molecules are not cloned directly into the vector by blunt end cloning. That is, the cDNA molecules have blunt ends, so they can be inserted into a vector that has been cut with a restriction enzyme such as *Sma*I (see Table 8.1) that generates blunt ends. On the surface, this seems easier, but linkers and adapters are inexpensive and easy to use under conditions that favor blunt ligations, while properly cut vectors are expensive and much more difficult to work with at conditions that favor blunt ligations.

Regardless of how the ligation is completed, the clones in the cDNA library represent the mature mRNAs found in the cell. In eukaryotes, mature mRNAs are processed molecules, so the sequences obtained are not equivalent to gene clones. In particular, intron sequences are present in gene clones but not in cDNA clones; hence, cDNA clones are typically smaller than the equivalent gene clone. For any mRNA, cDNA clones can be useful for subsequently isolating the gene that codes for that mRNA. The gene clone can provide more information than can the cDNA clone, for example, on the presence and arrangement of introns and on the regulatory sequences that control expression of the gene. However, predicting the protein encoded by the cDNA is far easier when the introns are absent.

***Using a cDNA Library to Annotate Genes.*** Obviously, the clones in the cDNA library can be sequenced to identify expressed genes in the genome. A single cDNA library will not be sufficient to identify all of the genes in the genome, since the starting tissue (from which the mRNA was isolated) will transcribe only a subset of the genes in the genome. Most of these clones are not full length, as conversion of the 5′ end of the mRNA into cDNA tends to be very difficult, but they do identify regions on the chromosome that are transcribed. Furthermore, since these libraries contain neither introns nor non-transcribed sequences, this is the most reliable way to define the exact boundaries of exons. Sequences derived from these cDNAs can be compared to genomic sequences to identify regions of the genomic sequences that are transcribed. Even if the cDNA is incomplete, the region can be annotated as containing a gene, and computer algorithms can take advantage of this and predict the rest of the coding region.

**Keynote**

DNA copies, called complementary DNA or cDNA, can be made of the population of mRNAs purified from a cell. First, a primer and the enzyme reverse transcriptase are used to make a single-stranded DNA copy of the mRNA; then RNase H, DNA polymerase I, and DNA ligase are used to make a double-stranded DNA copy called cDNA. This cDNA can be inserted into cloning vectors and cloned. These cDNAs can be sequenced and then compared to the sequenced genome of the organism as one way of annotating gene sequences in the genome.

**Identifying Genes in Genome Sequences by Computation.** Procedurally, annotation involves using computer algorithms to search both DNA strands of the sequence for protein-coding genes. Putative protein-coding genes are found by searching for open reading frames (ORFs), that is, start codons (AUG) in frame (separated by a multiple of three nucleotides) with a stop codon (UAG, UAA, or UGA). ORFs are searched for particularly in regions that have more G-C and C-G base pairs than the rest of the genome, because noncoding regions tend to be AT-rich. The searching process is straightforward with prokaryotic genomes because there are no introns. However, the presence of introns in many eukaryotic protein-coding genes necessitates the use of more sophisticated algorithms designed to include the identification of junctions between exons and introns in scanning for ORFs, as well as algorithms designed to find exons that are only part of the coding region of a gene. For instance, a gene might have three exons and two introns and code for a polypeptide containing 102 amino acids. Assume that the first exon contains the 5′ untranslated region, then the start codon and 15 more codons, that the second exon contains codons 16 to 95 (and no untranslated regions), and the third exon contains codons 95 to 102, the stop codon, and the 3′ untranslated region. A simple algorithm would not detect this gene in the genomic sequence, since the ORF after the start codon is quite short, and the algorithm will be fooled by any stop codons that might be present in the intron after the first exon. The second exon will probably lack an in-frame AUG (start) codon and stop codon, so it will also be ignored by a simple algorithm. However, if the algorithm is told to search for long stretches without an in-frame stop codon, it would find this second exon. Once one candidate exon is found, that region can be scanned carefully for intron–exon boundaries and other possible exons.

ORFs of all sizes are found in the computer scan, so a size must be set below which it is deemed unlikely that the ORF encodes a protein *in vivo* and it is not analyzed further. For the yeast genome, for instance, the lower limit was set to 100 codons. However, a few genes may be below this limit, and not all ORFs above 100 codons encode proteins. The plasma membrane proteolipid gene *PMP1*, for instance, encodes a protein of only 40 amino acids. It is estimated that of the 6,607 ORFs in the yeast genome, 6–7% do not correspond to actual genes, leaving approximately 5,700 actual protein-coding genes. One way of testing these candidate genes further is by comparison. If another organism has an ORF that encodes a similar predicted protein, or if the ORF encodes a predicted protein similar to a known protein in the databases, it suggests that this ORF is more likely to be part of a real gene, rather than a random sequence that happens to resemble a real gene. Analysis of the human genome initially identified more than 1,000 genes not seen in other genomes. Reanalysis suggested that most of these (nearly 1,000), were ORFs that probably did not correspond to a true gene. This uncertainty makes it difficult to determine the exact number of genes in the genome. This problem of annotation is made even more complex by the genes encoding microRNAs and other small, non-translated RNA molecules. These small RNAs are critical regulators of transcription and RNA stability in many eukaryotes (see Chapter 18, pp. 537–540). Hundreds of genes for small RNAs have been identified in the human genome, and there may be many, many more. The genes encoding these RNAs cannot be identified by ORF scans, however, because they do not code for proteins (so no ORF). Furthermore, generally speaking we will not be able to find cDNAs corresponding to any of these RNAs in cDNA libraries because most of them do not have a poly(A) tail and we select larger cDNAs for cloning, so their genes are difficult to identify in that way, too. It is clear that our gene tallies will be revised extensively as we annotate the genome to include the genes encoding these small RNAs, and genes that encode small proteins, and to eliminate the ORFs that do not correspond to genes.

**Keynote**

Computer analysis of genomic DNA allows us to identify possible genes. These computer programs look for open reading frames (ORFs) or other hallmarks of genes, like intron–exon boundaries. These programs are quite accurate with prokaryotic genomes, but they are less accurate with eukaryotes because the genomes tend to be more complex and because the introns confound the simplest types of analysis. As a result, they generate both false positives (an identified candidate gene region that probably does not function as a gene) and false negatives (true genes that the program fails to find).

## Insights from Genome Analysis: Genome Sizes and Gene Densities

In Chapter 2 (pp. 23–24), we discussed the C-value paradox, where there is no direct relationship between the C-value—the amount of DNA in the haploid genome—and the structural or organizational complexity of the organism. This is an old concept based on measuring the amount of DNA in the nuclei of haploid cells. Having a number of genomes sequenced makes it possible to make comparisons about genome organizations, particularly with respect to the arrangement of genes and intergenic regions. Such comparisons have revealed some differences in genome organizations that are responsible for the C-value paradox, including the gene density (the number of genes for a given length of DNA). The genome sizes, estimated number of genes, and gene densities for selected Bacteria, Archaea, and Eukarya are shown in Table 8.3. An overview of the organizations of the genomes of each of these kingdoms is presented in this section.

### Genomes of Bacteria

Organisms of the Bacteria evolutionary group have genomes that vary in size over quite a large range. Of the completely sequenced bacterial genomes, *Carsonella ruddii* (a symbiotic bacterium living in the guts of certain insects) has the smallest genome, with a size of only 160,000 base pairs (0.16 Mb) and fewer than 200 genes. This is the smallest known cellular genome. *Sorangium cellulosum* has the largest sequenced bacterial genome, with a size of 13 Mb (see Table 8.3), or more than 80 times as large as the genome of *Carsonella*.

Bacterial genomes have similar gene densities of one gene per 1–2 kb. For example, *Mycoplasma genitalium*'s 0.58-Mb genome has 523 genes, for a density of one gene per 1.15 kb, and the 4.6-Mb genome of *E. coli* has 4,397 genes for a density of one gene per 1.05 kb. The combination of high gene density and a relatively small number of genes required for a cell to survive in the lab has brought up a fascinating new challenge—it seems possible that we could soon create custom cells by synthesizing a novel genome.

*Carsonella ruddii* has 182 genes spread across 160,000 base pairs, for a density of one gene every 880 base pairs. Gene number and genome size tend to correlate, at least roughly, so that bacteria with larger genomes have more genes, and those with smaller genomes have fewer genes. The *Carsonella ruddii* genome forced scientists to reconsider the minimum number of genes required for life, as all previous estimates had suggested that about 400 genes were needed. This bacterium seems to lack genes that we have always thought to be needed for life, so it is possible that this organism is becoming an organelle before our eyes.

The spaces between genes are relatively small (110–125 bp for *Mycoplasma genitalium*), meaning that the genes are very densely packed in the genome. In fact, it is typical of Bacteria and of Archaea that approximately 85–90% of their genomes consist of coding DNA. *Carsonella* DNA is 97% coding, an almost impossible number given the sizes required for promoters and terminators. Bacterial genomes tend to have very little repetitive DNA, and introns are almost completely absent in prokaryotes in general. Both repetitive DNA and introns contribute to the amount of noncoding DNA, so gene density can obviously be higher if noncoding DNA content is minimized.

### Genomes of Archaea

The Archaea are a group of prokaryotes that share significant similarities with both eubacteria and eukaryotes. Current models suggest that eukaryotes (the Eukarya) are more closely related to the Archaea than to the Bacteria. The Archaea are best known for the extremophiles, those cells that "love" extreme environments, such as very high temperature, high pressure, extreme pH, high metal ion concentration, and high salt. Members of the Archaea resemble Bacteria morphologically, occurring with shapes such as spheres, rods, and spirals. However, physiological and molecular studies showed that they resemble Eukarya in a number of respects. Indeed, genes for DNA replication, RNA transcription, and protein synthesis machinery more closely resemble those of Eukarya than those of Bacteria. There are no introns in protein-coding genes as

**Table 8.3 Genome Sizes, Estimated Number of Genes, and Gene Densities for Selected Bacteria, Archaea, and Eukarya**

| Organism | Genome Size (Mb) | Number of Protein-Coding Genes | Gene Density (kb per gene) |
|---|---|---|---|
| **Bacteria** | | | |
| *Carsonella ruddii* | 0.16 | 182 | 0.87 |
| *Nanoarcheum equitans* | 0.49 | 552 | 0.88 |
| *Mycoplasma genitalium* | 0.58 | 523 | 1.11 |
| *Escherichia coli K12* | 4.6 | 4,200 | 1.03 |
| *Agrobacterium tumefaciens* | 5.7 | 5,482 | 1.04 |
| *Bradyrhizobium japonicum* | 9.1 | 8,322 | 1.10 |
| *Sorangium cellulosum* | 13 | 9,367 | 1.39 |
| **Archaea** | | | |
| *Thermoplasma acidophilum* | 1.56 | 1,509 | 1.03 |
| *Methanosarcina acetivorans* | 5.75 | 4,662 | 1.23 |
| **Eukarya** | | | |
| Fungi | | | |
| *Saccharomyces cerevisae* (yeast) | 12 | ~6,000 | 2.0 |
| *Neurospora crassa* (orange bread mold) | 40 | ~10,100 | 3.8 |
| Protozoa | | | |
| *Tetrahymena thermophila* | 220 | >20,000 | 11 |
| Invertebrates | | | |
| *Caenorhabditis elegans* (nematode) | 100 | 20,443 | 5 |
| *Drosophila melanogaster* (fruit fly) | 180 | 14,015 | 13 |
| Vertebrates | | | |
| *Takifugu rubripes* (pufferfish) | 393 | >31,000 | 13 |
| *Mus musculus* (mouse) | 2,700 | ~22,000 | 90 |
| *Rattus norvegicus* (rat) | 2,750 | ~30,200 | 91 |
| *Homo sapiens* (human) | 2,900 | ~20,067 | 107 |
| Plants | | | |
| *Arabidopsis thaliana* | 125 | 25,900 | 4.9 |
| *Oryza sativa* (rice) | 430 | ~56,000 | 9.6 |

there are in eukaryotic genes, but there are introns in tRNA genes as has been found in Eukarya.

Considering the genomes as a whole, archaean genomes also show a wide range of sizes, from 0.49 Mb for *Nanoarchaeum equitans* to 5.75 Mb for *Methanosarcina acetivorans* (see Table 8.3). As for Bacteria, genes are densely packed in the genome; the two examples just given have one gene per 880 bp and 1.23 kb, respectively. As in bacteria, larger genomes tend to reflect increased gene number rather than significant alterations in gene density.

## Genomes of Eukarya

The Eukarya vary enormously in form and complexity, from single-celled organisms such as yeast to multicellular organisms such as humans. There is a weak trend of increasing genomic DNA content with increasing complexity, although as already mentioned, there is by no means a direct relationship. For example, the two insects *Drosophila melanogaster* (fruit fly) and *Locusta migratoria* (locust) have similar complexity, yet the 5,000-Mb locust genome is 50 times larger than that of the fruit fly, and twice that of the mouse (see Table 8.3). Extreme differences in gene density are observed in eukaryotes. In this particular example there is one gene every 13 kb in the fruit fly genome and, assuming there are a similar number of genes in the locust genome (the number is not known at present), there is one gene every 365 kb in the locust, a substantial difference in gene density. Similar variation is seen in other groups, with a 50-fold or more variation in genome size in the genus *Allium*, which contains onions and their relatives. Some genomes, like those of some amphibians and some ferns, are about 200 times that of the human or mouse genome. Other eukaryotes, like yeast, have comparatively tiny genomes—the yeast genome is only 0.4% (1/250) the size of the human genome. For genomes that have been annotated, variation in gene number cannot account for variation in genome size. Again, we assume that these differences are due to variations in gene density. Most of the variation in gene density seems to be due to differences in amount of repetitive DNA in the genome.

In general, gene density in the Eukarya is lower and shows more variability than in Bacteria and Archaea (see Table 8.3). The Eukarya show a great range in gene density, although with a general trend of decreasing gene density with increasing complexity. Figure 8.17 illustrates the gene density differences in yeast, the fruit fly, and humans and compares them with *E. coli*. Yeast has a gene density closest to that of prokaryotes, one gene per 2 kb versus one gene per 1.03 kb for *E. coli*. Compared with yeast, the fruit fly has a 7-fold and humans have a 56-fold lower gene density. Organisms with genomes larger than that of humans are assumed to have lower gene densities than humans.

Of course, the gene density values given are averages. In any particular organism there will be stretches of chromosomes with significantly more genes than average—*gene-rich regions*—and stretches with significantly fewer genes than average—*gene deserts*. Eukaryotes seem to have these deserts, but deserts appear to be uncommon in prokaryotes. In humans, for example, the most gene-rich region of the genome has about 25 genes per megabase, and gene deserts (regions with no identified genes) of more than 1 Mb are common. Defining a gene desert as a region of 1 Mb or more without any genes, there are about 80 gene deserts in the human genome. This means that more than 25% of the human genome is desert.

In short, humans and other complex organisms have a minority of their genomes dedicated to exons, the remainder being introns and intergenic regions. In humans at least, most of the intergenic sequences consist of repetitive DNA (see Chapter 2, p. 25 and pp. 28–30). With a gene-sparse genome such as this, it is difficult and sometimes impossible to find genes of interest. Potentially, another vertebrate with high gene density may help with this problem. The vertebrate is *Takifugu rubripes*, the pufferfish (Figure 8.18), the genome of which has been sequenced completely. *Takifugu* is a spotted fish that puffs up into a ball when threatened. Particularly in Japan, this fish is a delicacy. It has a tangy taste but brings with it risk; if not prepared properly, it can paralyze and kill. As Table 8.3 shows, *Takifugu* has a genome size of 393 Mb, about 8-fold smaller than that of humans, but with an estimated gene number higher than that of humans. In other words, the gene density of *Takifugu* is at least 8-fold higher than in humans. In part, this density results from smaller and fewer introns in genes, so homologous genes in humans tend to take up more space on the chromosome. In addition, high gene density occurs because there is very little repetitive DNA, and much less intergenic DNA is present. The

**Figure 8.18**

**The pufferfish, *Takifugu rubripes*.**

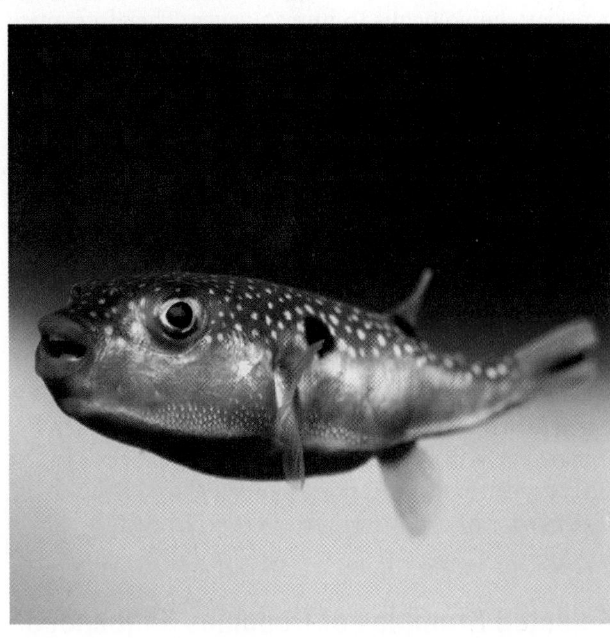

**Figure 8.17**

**Regions of the chromosomes of *E. coli*, yeast, fruit fly, and human showing the differences in gene density.**

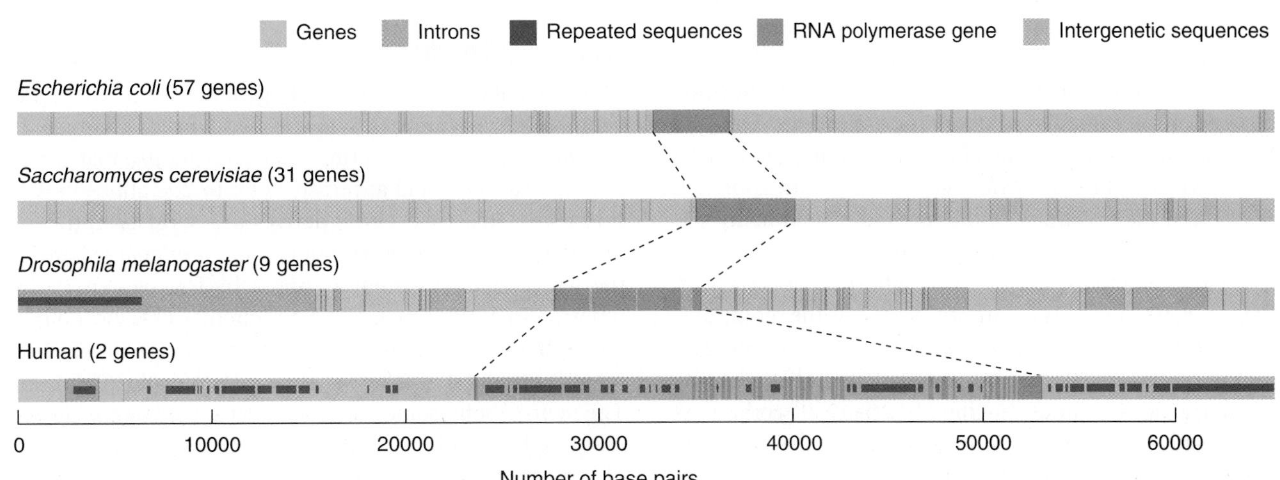

higher gene density makes *Takifugu* DNA much easier to study than human DNA. Happily, many of the *Takifugu* genes are homologous to human genes. Therefore, once genes are identified in *Takifugu*, the homologous genes in humans can be identified and studied. Scientists are hopeful that decoding the functions of pufferfish genes will aid in understanding the functions of human genes.

**Keynote**

Genome sequences are resources that inform us about the number of genes and the organization of genes in different organisms. Genomes show a trend of increasing DNA amount with increasing complexity of the organism, although the relationship is not perfect. In Bacteria and Archaea, genes make up most of their genomes; that is, gene density is very high. In Eukarya there is a wide range of gene densities, showing a trend of decreasing gene density with increasing complexity.

## Selected Examples of Genomes Sequenced

We now discuss some of the genomes that have been sequenced as well as why the particular organisms were chosen or what the sequences are likely to contribute to our knowledge about those organisms. Genome sequences are becoming available at an increasing rate, with hundreds of genomic sequences available as of early 2008. For sequencing information about your favorite organism, check the Internet sites for the Genome News Network (http://genomenewsnetwork.net), the Genome Online Database (GOLD, http://www.genomesonline.org/), the National Center for Biotechnology Information (http://www.ncbi.nlm.nih.gov/Genomes/index.html), and the Institute for Genomic Research (http://www.tigr.org/).

### Genomes of Bacteria

***Haemophilus influenzae.*** The first cellular organism to have its genome sequenced was the eubacterium *H. influenzae*. This organism was chosen because its genome size is typical among bacteria, and the GC content of the genome is close to that of humans. This task was completed by the Institute for Genomic Research in 1995. The only natural host for *H. influenzae* is the human; in some cases, it causes ear and respiratory tract infections. The 1.83 Mb (1,830,137 bp) genome of this bacterium was the first to be sequenced by the whole-genome shotgun approach as a test of the feasibility of the method, which many scientists considered was unlikely to succeed.

The annotated genome of *H. influenzae* is shown in Figure 8.19. With the current state of the computer searching algorithms and the amount of defined information in sequence databases, a complete microbial genome sequence can be annotated for essentially all coding regions and other elements, such as repeated sequences, operons, and transposable elements.

For *H. influenzae*, genome analysis predicted 1,737 protein-coding genes comprising 87% of the genome. Of these predicted genes, 469 either did not match any protein in the databases or matched only proteins designated hypothetical. The remaining 1,268 predicted ORFs matched genes in the databases that have known functions. This sort of result is typical of genome projects. Many genes have predicted functions, while a significant fraction has unknown functions, requiring much hypothesis-driven science to determine those functions.

***Escherichia coli.*** *E. coli* (see Figure 1.1, p. 3) is an extremely important organism. It is found in the lower intestines of animals, including humans, and survives well when introduced into the environment. Pathogenic *E. coli* strains make the news all too frequently as humans develop sometimes deadly enteric and other infections after contacting the bacterium at restaurants (e.g., in tainted meat or on vegetables exposed to raw sewage) or in the environment (e.g., in lakes with contamination). In the laboratory, nonpathogenic *E. coli* has been an extremely important model system for molecular biology, genetics, and biotechnology. Thus, the complete genome sequence of this bacterium was awaited eagerly.

In 1997, the annotated genome sequence of lab strain *E. coli K12* was reported by researchers at the *E. coli* Genome Center at the University of Wisconsin, Madison. It was the first genomic sequence of a cellular organism that had undergone extensive genetic analysis. An unannotated sequence of the *E. coli* genome made up of sequence segments from more than one strain was reported at the same time by Takashi Horiuchi of Japan. Subsequently, several other *E. coli* strains have been sequenced. One of the strains sequenced by Horiuchi was O157:H7, the strain that is responsible for approximately 70,000 cases of foodborne illness, and about 60 deaths, per year in the United States.

The circular strain *K12* genome was sequenced using the whole-genome shotgun approach. The genome of *E. coli* is 4.64 Mb (4,639,221 bp). The 4,288 ORFs make up 87.8% of the genome. Thirty-eight percent of the ORFs had unknown functions.

### Genomes of Archaea

The *Methanococcus jannaschii* genome was the first genome of an archaean to be sequenced completely. *M. jannaschii* is a hyperthermophilic methanogen that grows optimally at 85°C and at pressures up to 200 atmospheres. It is a strict anaerobe, and it derives its energy from the reduction of carbon dioxide to methane. Sequencing was by the whole-genome shotgun approach. The sequence was reported in 1996. The large, main circular chromosome is 1,664,976 bp; in addition, there is a circular plasmid of 58,407 bp and a smaller, circular plasmid of 16,550 bp. The main chromosome has 1,682 ORFs, the larger plasmid has 44 ORFs, and the smaller plasmid has 12 ORFs. Most of the genes involved in energy production, cell division,

**Figure 8.19**

**The annotated genome of *H. influenzae*.** The figure shows the location of each predicted ORF containing a database match as well as selected global features of the genome. Outer perimeter: Key restriction sites. Outer concentric circle: Coding regions for which a gene identification was made. Each coding region location is color coded with respect to its function. Second concentric circle: Regions of high GC content are shown in red (> 42%) and blue (> 40%), and regions of high AT content are shown in black (> 66%) and green (> 64%). Third concentric circle: The locations of the six ribosomal RNA gene clusters (green), the tRNAs (black) and the cryptic mu-like prophage (blue). Fourth concentric circle: Simple tandem repeats. The origin of replication is illustrated by the outward-pointing arrows (green) originating near base 603,000. Two possible replication termination sequences are shown near the opposite midpoint of the circle (red).

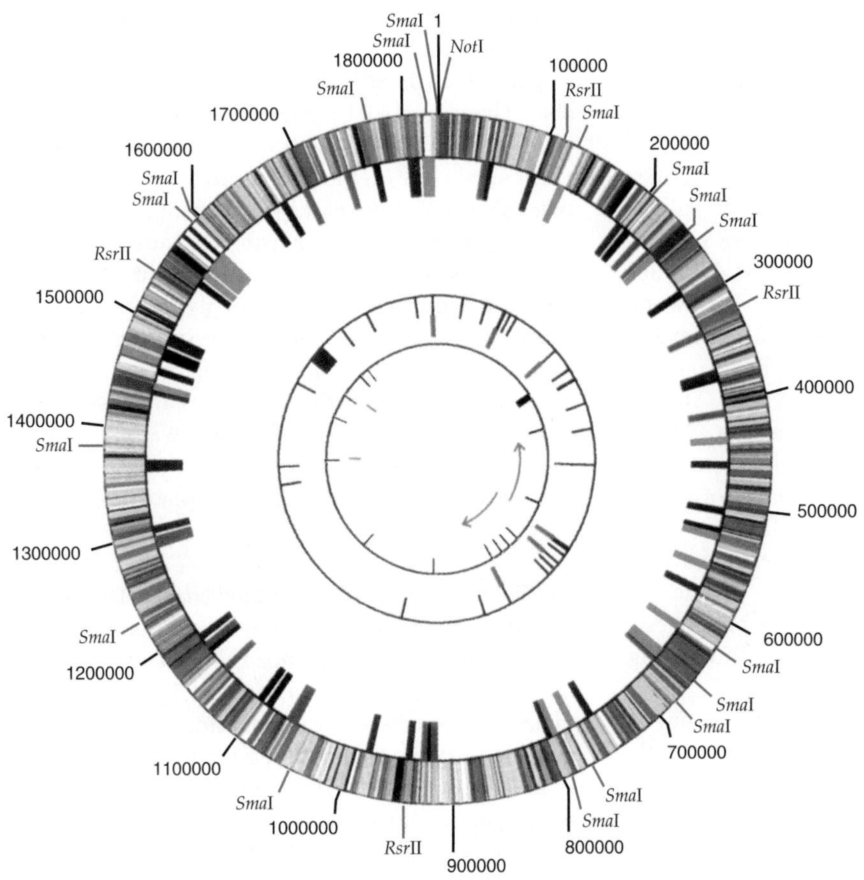


and metabolism are similar to their counterparts in the Bacteria, whereas most of the genes involved in DNA replication, transcription, and translation are similar to their counterparts in the Eukarya. Clearly this organism was neither a bacterium nor a eukaryote. The genome sequence of this organism therefore affirmed the existence of a third major branch of life on Earth.

## Genomes of Eukarya

**The Yeast, *Saccharomyces cerevisiae*.** For decades, the budding yeast *Saccharomyces cerevisiae* (Figure 8.20) has been a model eukaryote for many kinds of research. Some reasons for its usefulness are that it can be cultured on simpler media, it is highly amenable to genetic analysis, and it is highly tractable for sophisticated molecular manipulations. Moreover, functionally it resembles

**Figure 8.20**

**Scanning electron micrograph of the yeast *Saccharomyces cerevisiae*.**

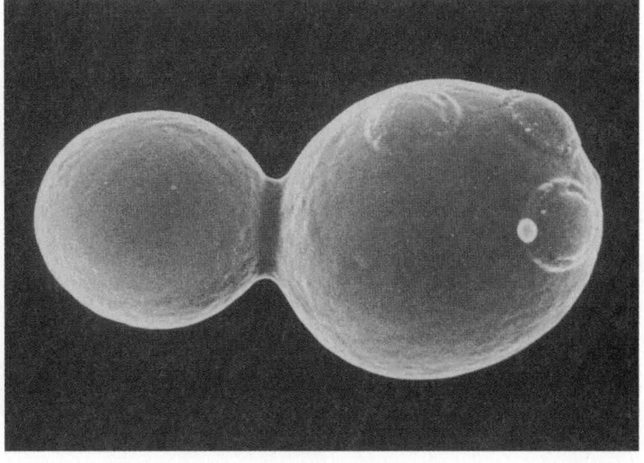

mammals in many ways. Therefore, its genome was a logical target for early genome sequencing efforts. In fact, the *S. cerevisiae* genome was the first eukaryotic genome to be sequenced completely; the sequence was reported in 1996. The 16-chromosome genome was reported to be 12,067,280 bp. Approximately 969,000 bp of repeated sequences were estimated not to be included in the published sequence. The sequence revealed 6,607 ORFs; only 233 of the ORFs have introns. Best estimates suggest that about 5,700 of these ORFs truly code for proteins, and the rest are not true protein-coding genes. At the outset of the yeast genome project, only about 1,000 genes had been defined by genetic analysis. About a third of the protein-coding genes have no known function.

**The Nematode Worm, *Caenorhabditis elegans*.** The genome of the nematode *C. elegans* (Figure 8.21), also called the "worm," was the first multicellular eukaryotic genome to be sequenced. Nematodes are smooth, nonsegmented worms with long, cylindrical bodies. *C. elegans* is about 1 mm long; it lives in the soil, where it feeds on microbes. There are two sexes: a self-fertilizing XX hermaphrodite and an XO male. The former has 959 somatic cells and the latter has 1,031 cells. The lineage of each adult cell through development is well understood. The worm has a simple nervous system, exhibits simple behaviors, and is even capable of simple learning tasks. Sydney Brenner was the first geneticist to study *C. elegans*, and this worm has become an important model organism for studying the genetic and molecular aspects of embryogenesis, morphogenesis, development, nerve development and function, aging, and behavior.

The *C. elegans* genome project was carried out by labs at Washington University in St. Louis and at the Sanger Center in England. The genome is 100.3 Mb, with 20,443 genes, 1,270 of which are not protein coding. Several major projects have built on these data, including a genome-wide knockout project that is attempting to generate distinct mutations in every identified gene. These projects are discussed further in Chapter 9.

**Figure 8.21**
**The nematode worm *Caenorhabditis elegans*.**

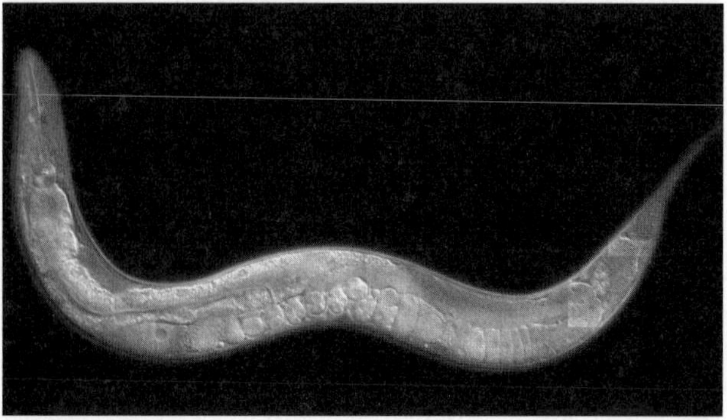

**The Fruit Fly, *Drosophila melanogaster*.** The genome sequence of an organism of particular historical importance in genetics, the fruit fly *D. melanogaster* (see Figure 1.4b, p. 6), was reported in March 2000. The fruit fly has been the subject of much genetics research and has contributed to our understanding of the molecular genetics of development. This genome sequence was as eagerly awaited as that of yeast. The genome of this organism was sequenced using the whole-genome shotgun approach.

The sequence of the euchromatic part of the *Drosophila* genome is 118.4 Mb in size. Another ~60 Mb of the genome consists of highly repetitive DNA that is essentially unclonable, making the sequences unobtainable. There are 14,015 genes, fewer than the number of genes in the worm but with similar diversity of functions. Surprisingly, the number of fruit fly genes is just over twice that found in yeast, yet the fruit fly seems to be a much more complex organism. We must conclude that higher complexity in animals such as flies and humans does not require a correspondingly larger repertoire of gene products, or that alternative splicing allows additional complexity without adding new genes to the genome. The value of the fruit fly as a model system for studying human biology and disease was affirmed by the finding that *D. melanogaster* has homologs for well over half of the genes currently known to be involved in human disease, including cancer.

**The Flowering Plant, *Arabidopsis thaliana*.** The genome of *A. thaliana* (see Figure 1.4d, p. 6) was the first flowering plant genome to be sequenced. *Arabidopsis* has been an important model organism for studying the genetic and molecular aspects of plant development. The 120-Mb genome contains about 25,900 genes. This gene number is almost twice that found in the fruit fly *Drosophila melanogaster* and exceeds the lower estimates for the number of genes in the human genome. Interestingly, about 100 *Arabidopsis* genes are similar to disease-causing genes in humans, including the genes for breast cancer and cystic fibrosis. The next step is to fill in the gaps in the sequence and explore the structure and function of the genome in detail. Toward this end, an initiative called the "Arabidopsis 2010 Project" has been set up. It has an ambitious set of goals, including defining the function of every gene, determining where and when every gene is expressed, showing where the encoded protein ends up in the plant, and defining all protein–protein interactions.

**Rice, *Oryza sativa*.** The 389-Mb genome of rice was reported in 2005 and is one of several crop plants subjected to genomic sequencing. The genome of rice is much smaller than that of humans, at only about one seventh the size, but its estimated gene number, currently 56,000 (of which 15,000 are from transposable elements), suggests that rice has about twice as many genes as humans.

The goal here is to identify genes that relate to disease, pest, and herbicide resistance as well as genes that influence yield and nutritive qualities.

**The Human, *Homo sapiens*.** As mentioned earlier, the genomics era began with the ambitious plan to sequence the 3 billion base pair (3,000-Mb) genome of *Homo sapiens*. Whose DNA was sequenced? The researchers collected samples from a large number of donors but used only some of the samples to extract DNA for sequencing. The human genome sequence generated is a mixture of sequences that is not an exact match for any one person's genome in the human population.

The draft genome sequences and initial interpretations of assembled sequences were published in 2001, several years ahead of schedule. Within two years, the human genome sequence was finished and announced to the public in 2003. How many genes make a human? Current estimates are for about 20,067 protein-coding genes, far fewer than the 50,000 to 100,000 protein-coding genes often predicted before sequencing began. An additional 4,800 genes code for RNAs that are not translated, including rRNAs, tRNAs, snRNAs, and microRNAs. Interestingly, this means that we have about as many protein-coding genes as *C. elegans*. This low number is drastically changing the way scientists think about organism complexity and development. All in all, the human genome sequence is proving a great resource for scientists to learn about our species. *Data mining*, searching through genome sequences for information, will continue for many years. Undoubtedly there will be a strong focus on human disease genes, with an eye toward treatment and therapy.

**The Mouse, *Mus musculus*.** Another early target of genomics researchers was the genome of the mouse (see Figure 1.4e, p. 6), as it is the genetically best understood nonhuman mammal. The mouse genome, at 2.7 billion base pairs (2,700 Mb), is slightly smaller than that of the human and has over 22,000 protein-coding genes and nearly 3,200 genes coding for RNAs. Most of the genes in the mouse are also found in humans, and vice versa. This result is not unexpected, as mice are used as models of human disease and can suffer from many of the same disorders found in humans. Many genetic manipulations are possible in mice that are either impossible or unethical in humans, so the mouse serves as the model organism for many of the analyses of genes identified in these processes.

**The Dog, *Canis familiaris*.** The dog genome is a bit smaller than ours, at 2.5 billion base pairs (2,500 Mb); it seems to contain less repetitive DNA. Annotation of this genome is not yet complete, but scientists working on the dog genome project estimate that there are at least 15,000 protein-coding genes and 2,500 genes coding for RNAs. Dogs were selected for a variety of reasons. Like mice, dogs have most of the same genes that we have. Dogs are one of the few mammals to have undergone fairly extensive genetic analysis due to extensive artificial selection and inbreeding for many generations, resulting in the breeds that we all know, like dachshunds and German shepherds. These breeds have both behavioral differences and genetic predispositions to disease. For instance, some breeds tend to develop muscular dystrophy, while several others are at elevated risk for Ehler-Danlos syndrome, a disease that alters skin elasticity and strength, and Doberman pinschers are at higher risk to develop narcolepsy, a disturbing neurological disorder characterized by sudden uncontrollable sleep attacks. In fact, at least 220 human diseases have natural models in one or more dog breeds. DNA from particular breeds can be compared to the genomic sequence, and regions that differ in the two can be studied to see if the genes in these regions are responsible for the disease correlations.

## Future Directions in Genomics

Current plans by the National Human Genome Research Institute (NHGRI) are for high-coverage, high-quality sequences of at least seven mammalian genomes (cow, dog, chimpanzee, human, macaque, mouse, and rat), and these projects are all complete or nearly complete. More than 40 other mammalian genomes are in progress, including the tammar wallaby (a kangaroo), the cat, the horse, two species of bats, dolphins, elephants, and rabbits. NHGRI is also supporting the sequencing of many bacteria that inhabit our bodies, as well as the sequencing of a number of pathogenic bacteria and fungi that cause human disease. Many other genomes are to be sequenced by other organizations. Some organisms have been selected for their economic importance, while others were chosen for their position in our family tree. Some conclusions can be made, such as (1) the genome size of most mammals is not too different from the size of the human genome; and (2) for the mammals that have completed genomic sequences and annotated genes, the number of genes is fairly similar as well. Importantly, both the mouse and the rat have been model organisms for studies of mammalian physiology, including those involved in diseases. The mouse, in particular, has been a model for mammalian genetics due to its genetic tractability, including the ability to use molecular techniques to create a specific mutation in any selected mouse gene (this is done in mouse cells grown in the laboratory), and then to use these modified culture cells to create new, mutant mice (see Chapter 9, pp. 225–227). Sequence analysis reveals that approximately 99% of the genes of the mouse and the rat have direct counterparts in the human, including genes associated with disease. Studies of the mouse and rat genomes will undoubtedly provide valuable knowledge about human diseases and other areas of human biology. Many of the other organisms will also offer valuable insights into human and animal disease,

sequences the insert of just one clone from each library? Based on this information, which library should he use if he wants to sequence the fewest number of clones?

**b.** When he tries to sequence the insert of the first clone he picks from the library by a calleague suggested by a colleague in (a), he realizes that he does not enjoy this type of lab work. So, he hires a technician with experience in genomics, assigns the project to her, and goes to Antarctica to catch more fish. He tells her to sequence the inserts of enough clones to be 95% certain of obtaining at least one insert containing the antifreeze gene and says he will analyze all of the sequence data for the presence of the antifreeze gene after he returns. How many clones should she sequence to satisfy this requirement if he constructed the genomic library in a plasmid vector? a BAC vector? a YAC vector?

**c.** What advantages and disadvantages does each of the different vectors have for constructing libraries with cloned genome DNA?

**d.** Suppose the Antarctic fish has a very AT-rich genome and the biochemist propagated the genomic library using *E. coli*. Will the library be representative of all the sequences in the genome of the fish?

***8.18** When Celera Genomics sequenced the human genome, they obtained 13,543,099 reads of plasmids having an average insert size of 1,951 bp, and 10,894,467 reads of plasmids having an average insert size of 10,800 bp.

**a.** Dideoxy sequencing provides only about 500–550 nucleotides of sequence. About how many nucleotides of sequence did cetera obtain from sequencing these two plasmid libraries? To what fold coverage does this amount of sequence information correspond?

**b.** Why did they sequence plasmids from two libraries with different-sized inserts?

**c.** They sequenced only the ends of each insert. How did they determine the sequence lying between the sequenced ends?

***8.19**

**a.** What features of pBluescript II facilitate obtaining the sequence at the ends of an insert?

**b.** Devise a strategy to obtain the entire sequence of a 7-kb insert in pBluescript II.

**c.** Devise a strategy to obtain the entire sequence of a 200-kb insert in pBeloBAC11.

**8.20** Explain how the whole-genome shotgun approach to sequencing a genome differs from the biochemist's approach described in Question 8(c). What information does it provide that the biochemist's approach does not? What does it mean to obtain 7-fold coverage, and why did his colleague advise him to do this?

***8.21** In a sequencing reaction using dideoxynucleotides that are labeled with different fluorescent dyes, the DNA chains produced by the reaction are separated by size using capillary gel electrophoresis and then detected by a laser eye as they exit the capillary. A computer then converts the differently colored fluorescent peaks into a pseudocolored trace. Suppose green is used for A, black for G, red for T, and blue for C. What pattern of peaks do you expect to see on a sequencing trace if you carry out a dideoxy sequencing reaction after the primer 5′-CTAGG-3′ is annealed to the following single-stranded DNA fragment?

3′-GATCCAAGTCTACGTATAGGCC-5′

**8.22** How does pyrosequencing differ from dideoxy chain-termination sequencing? What advantages does it have for large-scale sequencing projects?

**8.23** Do all SNPs lead to an alteration in phenotype? Explain why or why not.

**8.24** Researchers at Perlegen Sciences sought to identify tag SNPs on human chromosome 21. After determining the genotypes at 24,047 common SNPs in 20 hybrid cell lines containing a single, different human chromosome 21, they used computerized algorithms to identify haplotypes containing between 2 and 114 SNPs that cover the entire chromosome. A total of 2,783 tag SNPS were selected from SNPs within these blocks.

**a.** What is a SNP marker?

**b.** How do haplotypes arise in members of a population?

**c.** What is a hapmap?

**d.** What is a tag SNP?

**e.** What advantages were there for the researchers to use hybrid cell lines instead of genomic DNA from 20 different individuals?

**f.** The 20 individuals whose chromosome 21 was used in this analysis were unrelated and had different ethnic origins. Do you expect the haplotypes and number of tag SNPs to differ if

**i.** the cell lines were established from blood samples drawn at a large family reunion.

**ii.** the cell lines were established from unrelated individuals, but their ancestors originated in the same geographical region.

***8.25** A set of hybrid cell lines containing a single copy of the same human chromosome from 10 different individuals was genotyped for 26 SNPs, A through Z. The SNPs are present on the chromosome in the order A, B, C, . . . Z. Table 8.C lists the SNP alleles present in each cell line. State which SNPs can serve as tag SNPs, and which haplotypes they identify. What is the minimum number of tag SNPs needed to differentiate between the haplotypes present on this chromosome?

**8.26** Some features that we commonly associate with racial identity, such as skin pigmentation, hair shape, and facial morphology, have a complex genetic basis. However, it turns out that these features are not representative of the

Table 8.C

| Cell Line | | | | | | | | | |
|---|---|---|---|---|---|---|---|---|---|
| **1** | **2** | **3** | **4** | **5** | **6** | **7** | **8** | **9** | **10** |
| A1 | A1 | A2 | A3 | A1 | A3 | A2 | A3 | A1 | A2 |
| B1 | B1 | B2 | B3 | B2 | B3 | B2 | B3 | B1 | B2 |
| C3 | C3 | C1 | C2 | C1 | C2 | C1 | C2 | C3 | C1 |
| D4 | D4 | D3 | D2 | D1 | D2 | D3 | D2 | D4 | D3 |
| E1 | E1 | E2 | E2 | E3 | E2 | E2 | E2 | E1 | E2 |
| F2 | F1 | F2 | F2 | F2 | F1 | F2 | F2 | F2 | F2 |
| G3 | G2 | G3 | G3 | G1 | G2 | G1 | G3 | G1 | G3 |
| H1 | H1 | H1 | H1 | H2 | H1 | H2 | H1 | H2 | H1 |
| I3 | I1 | I3 | I3 | I2 | I1 | I2 | I3 | I2 | I3 |
| J2 | J1 | J2 | J2 | J2 | J1 | J2 | J2 | J2 | J2 |
| K1 | K1 | K1 | K1 | K2 | K1 | K2 | K1 | K1 | K1 |
| L2 | L1 | L2 | L2 | L1 | L1 | L1 | L2 | L2 | L2 |
| M1 | M1 | M2 | M1 | M1 | M2 | M2 | M1 | M2 | M1 |
| N2 | N2 | N1 | N2 | N2 | N1 | N1 | N2 | N1 | N2 |
| O1 | O1 | O1 | O1 | O1 | O2 | O1 | O1 | O1 | O2 |
| P2 | P1 | P2 | P1 | P2 | P1 | P1 | P1 | P2 | P1 |
| Q2 | Q2 | Q2 | Q2 | Q2 | Q1 | Q2 | Q2 | Q2 | Q1 |
| R3 | R1 | R3 | R1 | R3 | R2 | R1 | R1 | R3 | R2 |
| S1 | S2 | S1 | S2 | S1 | S1 | S2 | S2 | S1 | S1 |
| T1 | T1 | T1 | T1 | T1 | T1 | T1 | T1 | T1 | T1 |
| U2 | U1 | U2 | U1 | U2 | U2 | U1 | U1 | U2 | U2 |
| V2 | V2 | V2 | V2 | V2 | V2 | V2 | V2 | V2 | V2 |
| W2 | W3 | W1 | W2 | W1 | W3 | W1 | W1 | W3 | W1 |
| X1 | X2 | X1 | X1 | X3 | X2 | X3 | X1 | X2 | X3 |
| Y2 | Y1 | Y4 | Y2 | Y3 | Y1 | Y3 | Y4 | Y1 | Y3 |
| Z1 | Z1 | Z2 | Z1 | Z2 | Z1 | Z2 | Z2 | Z1 | Z2 |

genetic differences between racial groups—individuals assigned to different racial categories share many more DNA polymorphisms than not—supporting the contention that race is a social and not a biological construct. How could you use DNA chips to quantify the percentage of SNPs that are shared between individuals assigned to different racial groups?

***8.27** Mutations in the dystrophin gene can lead to Duchenne muscular dystrophy. The dystrophin gene is among the largest known: it has a primary transcript that spans 2.5 Mb, and it produces a mature mRNA that is about 14 kb. Many different mutations in the dystrophin gene have been identified. What steps would you take if you wanted to use a DNA microarray to identify the specific dystrophin gene mutation present in a patient with Duchenne muscular dystrophy?

**8.28** Three of the steps in the analysis of a genome's sequence are assembly, finishing, and annotation. What is involved in each step, and how do they differ from each other?

**8.29** What is a cDNA library, and from what cellular material is it derived? How is a cDNA synthesized, and how do the steps used to clone a cDNA differ from the steps used to clone genomic DNA? How are cDNA sequences used to help annotation of a sequenced genome?

***8.30** Eukaryotic genomes differ in their repetitive DNA content. For example, consider the typical euchromatic 50-kb segment of human DNA that contains the human β T-cell receptor. About 40% of it is composed of various genome-wide repeats, about 10% encodes three genes (with introns), and about 8% is taken up by a pseudogene. Compare this to the typical 50-kb segment of yeast DNA containing the *HIS4* gene. There, only about 12% is composed of a genome-wide repeat, and about 70% encodes genes (without introns). The remaining sequences in each case are untranscribed and either contain regulatory signals or have no discernible information. Whereas some repetitive sequences can be interspersed throughout gene-containing euchromatic regions, others are abundant near centromeres. What problems do these repetitive sequences pose for sequencing eukaryotic genomes? When can these problems be overcome, and how?

**8.31** What is the difference between a gene and an ORF? Explain whether all ORFs correspond to a true gene, and if they do not, what challenges this poses for genome annotation.

***8.32** Once a genomic region is sequenced, computerized algorithms can be used to scan the sequence to identify potential ORFs.

**a.** Devise a strategy to identify potential prokaryotic ORFs by listing features accessible by an algorithm checking for ORFs.

**b.** Why does the presence of introns within transcribed eukaryotic sequences preclude direct application of this strategy to eukaryotic sequences?

**c.** The average length of exons in humans is about 100–200 bp, while the length of introns can range from about 100 to many thousands of base pairs. What challenges do these findings pose for identifying exons in uncharacterized regions of the human genome?

**d.** How might you modify your strategy to overcome some of the problems posed by the presence of introns in transcribed eukaryotic sequences?

**8.33** Annotation of genomic sequences makes them much more useful to researchers. What features should be included in an annotation, and in what different ways can they be depicted? For some examples of current annotations in databases, see the following websites:

http://www.yeastgenome.org/

http://flybase.org (*Drosophila*)

http://www.tigr.org/tdb/e2k1/ath1/ (*Arabidopsis*)

http://www.ncbi.nlm.nih.gov/genome/guide/human/ (humans)

http://genome.ucsc.edu/cgi-bin/hgGateway (humans)

http://www.h-invitational.jp/

***8.34** One powerful approach to annotating genes is to compare the structures of cDNA copies of mRNAs to the genomic sequences that encode them. Indeed, a large collaboration involving 68 research teams analyzed 41,118 full-length cDNAs to annotate the structure of 21,037 human genes (see http://www.h-invitational.jp/).

**a.** What types of information can be obtained by comparing the structures of cDNAs with genomic DNA?

**b.** During the synthesis of cDNA (see Figure 8.15), reverse transcriptase may not always copy the entire length of the mRNA and so a cDNA that is not full-length can be generated. Why is it desirable, when possible, to use full-length cDNAs in these analyses?

**c.** The research teams characterized the number of loci per Mb of DNA for each chromosome. Among the autosomes, chromosome 19 had the highest ratio of 19 loci per Mb while chromosome 13 had the lowest ratio of 3.5 loci per Mb. Among the sex chromosomes, the X had 4.2 loci per Mb while the Y had only 0.6 loci per Mb. What does this tell you about the distribution of genes within the human genome? How can these data be reconciled with the idea that chromosomes have gene-rich regions as well as gene deserts?

**d.** When the research teams completed their initial analysis, they were able to map 40,140 cDNAs to the available human genome sequence. Another 978 cDNAs could not be mapped. Of these 978 cDNAs, 907 cDNAs could be roughly mapped to the mouse genome. Why might some (human) cDNAs be unable to be mapped to the human genome sequence that was available at the time although they could be mapped to the mouse genome sequence? (Hint: Consider where errors and limited information might exist.)

***8.35** How has genomic analysis provided evidence that Archaea is a branch of life distinct from Bacteria and Eukarya?

**8.36** The genomes of many different organisms, including bacteria, rice, and dogs, have been sequenced. Choose three phylogenetically diverse organisms. Compare the rationales for sequencing their genomes, and describe what we have learned from sequencing each genome.

**8.37** In which type of organisms does gene number appear to be related to genome size? Explain why this is not the case in all organisms.

**8.38** The C-value paradox (see Chapter 2, pp. 23–24) states that there is no obvious relationship between an organism's haploid DNA content and its organizational and structural complexity. Discuss, citing data from the genome sequencing, whether there is also a gene-number paradox or a gene-density paradox.

**8.39** In the United States, 3–5% of public funds used to support the Human Genome Project were devoted to research to address its ethical, legal, social, and policy implications. Some of the results are described in the website http://www.ornl.gov/sci/techresources/Human_Genome/elsi/elsi.shtml. After exploring this website, answer the following questions.

**a.** Summarize the main ethical, legal, social, and policy issues associated with the human genome project.

**b.** Why is legislation necessary to protect an individual's genetic privacy? What such legislation currently exists?

**c.** What are the pros and cons of gene testing?

**d.** Both presymptomatic and symptomatic individuals are subject to gene testing for an inherited disease. How are gene tests used in each situation, and how do the concerns about using gene testing differ in these situations?

**e.** Are laboratories that conduct genetic testing regulated by law?

# 9 Functional and Comparative Genomics

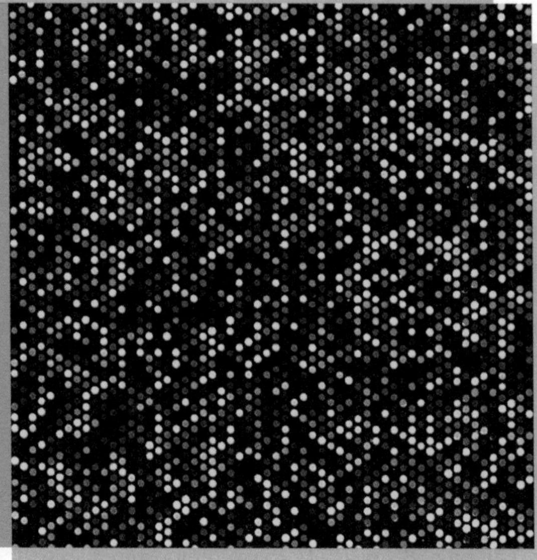

A DNA microarray.

## Key Questions

- How are the functions of genes in a genome determined from sequence data?
- How are newly identified genes compared to those studied previously?
- How can the functions of newly identified genes be determined experimentally?
- Are genes and other sequences organized in the genome in a particular way?
- How do the transcripts and protein products of all genes in the genome vary in different cell types, or in different conditions?
- How can genomics studies make drug therapies more effective?
- How can the comparison of the genome sequences of different organisms provide information about evolutionary relationships?
- How can the comparison of genome sequences indicate gene changes in cancer, and the nature of infectious agents in disease?
- How can we use genomics to understand complex communities in microbes in environmental samples?

## iActivity

IF YOU ARE LIKE MOST PEOPLE IN THE UNITED States, at some point in your life you have taken a prescription drug. Although your doctor may have considered your medical history when selecting the drug, it is very unlikely that he or she could predict fully how you would react to the medication before you took it. In fact, because of inherited variations in your genes, your ability to metabolize any given drug and the side effects you may experience from that drug differ greatly from those of other people. But in the near future, doctors may be able to prescribe medications, adjust dosage, and select treatments based on the patient's genetic information. The DNA microarrays that you learned about in Chapter 8 make this possible. In this chapter, you will learn more about DNA microarrays and other tools and techniques used to analyze the entire genomes of organisms. Then, in the iActivity, you will discover how DNA microarrays can be used to create a personalized drug therapy regimen for a patient with cancer.

The sequencing of complete genomes has opened new doors to our understanding of gene and cellular function, organismal evolution, and many other aspects of biology In this chapter, you will learn about applications of genomics, specifically **functional genomics,** the comprehensive analysis of the functions of genes and of nongene sequences in entire genomes; and **comparative genomics,** the comparison of entire genomes (or parts of genomes) from different species, strains, or individuals, with the goal of enhancing our understanding of the functions of each genome (or parts of each genome), including evolutionary relationships. Comparative genomics approaches are used also to determine which organisms or viruses are present in a sample. In the functional genomics section, you will learn how we look at functional genomics and assign functions to genes in a genome by either computer modeling or gene knockout analysis, how we analyze global transcription in cells, and how we can use functional genomics to regulate drug therapies. Then, in the comparative genomics section, you will learn how we compare genomes, and how these comparisons have helped us to understand gene function and evolution. You will also learn how comparative genomics can be used in a clinical setting to help us understand how infections have spread. Much of what you will read about is at the cutting edge of biology, where new techniques and approaches are developed almost daily.



## Functional Genomics

The successes of the HGP (Human Genome Project; see Chapter 8, p. 171) have empowered researchers working with a wide range of organisms, providing them with the techniques to obtain genome sequences for those organisms quickly. Research questions about gene expression, physiology, development, and so on can now be asked at the genomic level. In other words, the ability to sequence genomes efficiently and quickly has changed how research in biology, and in genetics in particular, is being done.

Of course, the complete genome sequence for an organism is just a very long string of the letters A, T, G, and C. The sequence must be analyzed in detail. One important research direction is to describe the functions of all the genes in the genomes, including studying gene expression and its control, and this defines the field of functional genomics. The difficulty in assigning gene function is that going from gene sequence to function is the reverse direction of that classically taken in genetic analysis, in which researchers start with a phenotype and set out to identify and study the genes responsible. In fact, many of the techniques you will learn about in this chapter were developed for **reverse genetics.** In reverse genetics, investigators attempted to find what phenotype, if any, would be associated with a gene. Generally, the investigators attempted first to create mutations in cloned genes, and then tried to introduce those mutations into the organism. Present-day functional genomics relies on laboratory experiments by molecular biologists as well as sophisticated computer analysis by researchers in the rapidly growing field of **bioinformatics.** Bioinformatics fuses biology with mathematics and computer science. It is used for many things, including finding genes within a genomic sequence, aligning sequences in databases to determine how similar they are (or their degree of similarity), predicting the structure and function of gene products, describing the interactions between genes and gene products at a global level within the cell, between cells, and between organisms, and postulating phylogenetic relationships for sequences.

**Keynote**

Functional genomics has the goal of describing the functions of all genes in a genome, including their expression and control of that expression. Functional genomics involves both molecular analysis in the laboratory and computer analysis of sequences (also called bioinformatics).

### Sequence Similarity Searches to Assign Gene Function

Once candidate genes have been annotated in a fully sequenced genome (see Chapter 8), it is important to assign probable functions to the proteins encoded by these genes. Most organisms that undergo genomic sequence have not undergone extensive "classical" genetic analysis, so generally there will not be extensive banks of mutant strains with well-characterized phenotypes. In such a case, our knowledge may be limited to the genomic sequence only. If we do not understand what the protein encoded by a gene does, we cannot make any sense of when and where the gene is expressed. In contrast, if we can assign some likely function to the protein encoded by the gene, we can begin to predict how, and why, the gene is used by the organism. The function of an ORF, or open reading frame, identified in genome scans may be assigned by searching databases for a sequence match with a gene whose function has been defined. (As introduced in Chapter 6, p. 109, an ORF is a segment of DNA that is a potential polypeptide-coding sequence identified by a start codon in frame with a stop codon. We make the assumption that most large ORFs are part of a gene that is transcribed at some time.) An ORF in genomic DNA analysis typically is defined as a segment of DNA that could encode a polypeptide of 100 amino acids or more. As you learned in Chapter 8, ORFs in eukaryotes can be much more difficult to find, because introns in the genomic sequence confound this simple definition. As a result, we often turn to cDNAs (see Chapter 8, pp. 193–197) as a way of finding these genes.

Searches for sequence matches are called *sequence similarity searches* and involve computer-based comparisons of an input sequence with all sequences in the database. The searches can be done using an Internet browser to access the computer programs. For example, the BLAST (Basic Local Alignment Search Tool) program at the National Center for Biotechnology Information (http://blast.ncbi.nlm.nih.gov/blast.cgi) enables a user to paste the identified ORF sequence to be studied into a window. BLAST will accept either the DNA sequence of the ORF or the sequence of the protein encoded by the ORF. BLAST comparisons based on the protein sequences tend to be somewhat easier to interpret, because many DNA mismatches may not alter the encoded protein due to the degeneracy of the genetic code. Furthermore, sequence similarity searching with an amino acid sequence tends to be preferred because, with 20 different amino acids and only four different nucleotides, a similar sequence of 10 or 12 amino acids is far less likely to be a random match than a DNA match of similar length. The BLAST program searches the databases of known sequences and returns the best matches, indicating the degree to which the sequence of interest is similar to sequences in the database. BLAST even aligns the entered sequence with some of the matching sequences it has found. The search does not simply look for a perfect match, since a perfect match across tens of hundreds of amino acids in two different species would be very rare. Instead, the analysis software searches for partial matches, and calculates the chance that this match would happen at random. The candidate matches are then listed in order, starting with the match least likely to occur at random (this is also the best match for our query). Obviously, if two polypeptides are highly similar, they likely function in a similar way, while if they are similar over only a small region, they may not fulfill the same function in the cell.

Figure 9.1 shows a small part of one alignment generated by using BLAST to compare protein sequences. In this case, the program searched for protein sequences in the database that match the amino acid sequence of human fibronectin, an important protein in the extracellular matrix that surrounds many cells. The entered sequence is called the *query sequence*. The BLAST program has found a match and has returned a subject (Sbjct) sequence for bovine fibronectin. The BLAST program also shows how the two sequences align. In between the two sequences, the BLAST program lists matching amino acids (this case is noted by placing the one-letter code in the middle when the amino acid in query and subject are exactly alike), or when very similar amino acids are used (this case is denoted by a "+" between the query and subject—for example, this code might be used when one protein uses leucine and the other uses isoleucine, since both amino acids have moderately bulky, hydrophobic side chains). BLAST can even adjust if one of the proteins is longer than the other—this is shown in Figure 9.1 in regions where either the query or subject sequence has the code "–", which means that a particular sequence is shorter in a small region than its partner.

Similarity searching is an effective way to assign gene function because homology—descent from a common ancestor—is a reflection of evolutionary relationships. That is, if a pair of homologous genes in different organisms has a common evolutionary ancestor, then the nucleotide sequences of the two genes will be similar. Any differences between the gene sequences have resulted from mutational changes that have occurred over evolutionary time. Thus, if a newly sequenced gene (e.g., from a genome sequence project) is similar to a previously sequenced gene, the two genes are related in an evolutionary sense, so the function of the new gene probably is the same as, or at least similar to, the function of the previously sequenced gene.

Given the information in current databases, most new genes are similar, but not identical, to at least one predicted gene in another organism. In many cases, this gene does not have a known function. For example, in 2005 the genome of the nematode *C. elegans* was analyzed. Most of the predicted *C. elegans* genes (56%) were similar to genes with known or predicted protein function from other organisms. As indicated above, this sequence similarity suggests that the pairs of genes have similar functions. Similarity searches with the remaining predicted genes were less informative. Those predicted genes were similar either to other nematode genes with no known or predicted functions (23%), or to nothing in the database (21%). Since that time, many more sequences have been added to the databases, so the fraction with no match has decreased significantly. When a predicted protein sequence matches a region of genomic sequence from another organism in the database, but neither of these predicted proteins have a clearly defined function, it is difficult, if not impossible, to predict what the protein might do in the cell.

A sequence similarity search can indicate a match for either the whole protein sequence or for parts of it (see Figure 9.1). In the figure, the first part of the entered query protein sequence does not match the subject



**Figure 9.1**

**The outcome of a sequence similarity search.** In this example, the program BLASTp, which compares protein sequences, was used to compare human fibronectin (the Query sequence) and bovine fibronectin (the Subject, or Sbjct sequence). Numbers indicate the position of the amino acids in the protein sequence. Letters entered on the middle line indicate that the two sequences match perfectly at that amino acid, while the "+" indicates that the proteins have chemically similar amino acids at that position. If nothing is entered on the middle line, the amino acids in the query and subject are not similar. Dashes in either the query or subject sequence indicate that one of the sequences (the one with the dashes) is missing one or more amino acids. [Sequences from NCBI Database, http://blast.ncbi.nlm.nih.gov/ (retrieved June 1, 2008). See Figure 6.2, p. 104 for the one-letter abbreviations for amino acids.]

```
Query 2072 RPRPY--PPNVGQEALSQTTISWAPFQDT 2098
           +  P       GQEALSQTTISW PFQ++
Sbjct 1982 KSEPLIGRKKTGQEALSQTTISWTPFQES 2010
```

sequence very well, but the second part of the query sequence matches the subject sequence very well in another region. In the latter case, this might mean that a domain of the new gene product matches a domain of a previously identified gene product. A domain is a part of a polypeptide sequence that tends to fold and function independent of the rest of the polypeptide. Many domains have a well-understood function. For example, a number of domains are known to be involved in DNA binding, while other domains are used to bind calcium. This means that at least part of the new protein's function can be inferred, as long as the match between the two proteins spans a domain of known function. Evolutionarily speaking, such a result means that the domains have a common ancestor, but the genes as a whole may not.

Sequence similarity searching plays an important part in assigning gene function. When the budding yeast genome was first sequenced and annotated, about 30% of the genes were already known as the result of standard genetic analysis, including direct assays for function. The remaining 70% of genes needed to have a function assigned, if possible, using sequence similarity searches. From such searches, 30% of the genes in the yeast genome encode a protein that matched a protein in the database with a known function, and it is tentatively assumed that the function of the yeast gene product is similar to that of the homolog. Ten percent of the yeast genes encode proteins that have homologs in databases, but the functions of those homologs are unknown. Such yeast ORFs are called *FUN* (*f*unction *un*known) genes, and those genes and their homologs are called *orphan families*. The remaining 30% of candidate yeast genes have no homologs in the databases. Within this class are the 6–7% of candidate yeast genes that are questionable in terms of being real genes; that is, some of these ORFs are probably not transcribed. The remainder of the unknown function ORFs are probably real genes, but at present are unique to yeast. These genes are called *single orphans*. In the years since this analysis was first done, functions have been assigned to many of the orphan families and single orphans, but there are still a large number of yeast genes (about 14%) that encode proteins for which a function cannot be predicted. This is not to say that these genes encode proteins with no function; rather, these genes encode a protein that we do not yet understand.

If we consider the genes that encode proteins with a predicted function, we can ask what percentage of the genes in the yeast genome are used for a particular function. Figure 9.2 shows this sort of analysis for the annotated genes in the yeast genome. We can ask how many genes encode proteins involved in particular molecular functions (Figure 9.2a). For instance, about 10% of the genes in the yeast genome encode proteins that bind RNAs, and about 6% encode transporter proteins that are involved in moving small molecules across membranes. We can also ask how many genes encode proteins involved in particular biological processes in the cell (Figure 9.2b). For example, about 10% of the yeast genes encode proteins that are involved in translation, and about 5% of the genes encode proteins involved in meiosis or sporulation.

The problem of "function unknown" genes applies to the genomes of other organisms, both prokaryotic and eukaryotic ones. However, as more and more genes with defined functions are added to the databases, the percentage of ORFs with no matches to database sequences is decreasing. A surprisingly large number of human genes (nearly a thousand) were placed in the single orphan class and were not found in the genomes of other mammals as those genomic sequences became available. While we may have a number of genes not found in either the mouse or the dog, at least some of the single orphan candidate genes should have been found in our closest relative, the chimpanzee (*Pan troglodytes*), since some of these potential new genes should have evolved in the millions of years between the time primate ancestors diverged from other mammals and the time when humans and chimps diverged. An extensive analysis of these single orphans suggested that most of them are probably not true genes, but regions that resembled a gene enough that they were detected as candidate genes by the computer programs.

**Keynote**

To assign gene function by computer analysis, the sequence of an unknown gene from one organism is compared to sequences of genes with known function in databases. For the unknown gene, the sequence compared may be the DNA sequence of the gene itself or the amino acid sequence of the polypeptide encoded by the gene. A sequence similarity search such as this may return a match for the whole sequence or part of it, the latter indicating that a domain of a gene's product has a known function.

## Assigning Gene Function Experimentally

One key approach to assigning gene function experimentally is to knock out the function of a gene and determine what phenotypic changes occur. Major projects have been undertaken to eliminate systematically the function of each gene identified in several organisms, including yeast, mouse, the fruit fly, *Mycoplasma genitalium*, and the nematode worm *Caenorhabditis elegans*.

There are several ways to knock out the functions of protein-coding genes. Two of the most common techniques are *gene knockouts* and *RNA interference* (RNAi). A *gene knockout* is made by disrupting the gene on the chromosome. We will look at strategies for knocking out chromosomal genes in yeast, mouse, and *M. genitalium*. **RNA interference (RNAi),** also called RNA silencing, is a technique where small regulatory RNAs are used to

Figure 9.2

**The predicted functions of proteins encoded in the yeast genome.** (a) Predicted yeast proteins grouped by probable enzymatic function. (b) Predicted yeast proteins grouped by the cellular process in which the protein acts. [Data for (a) and (b) from "*Saccharomyces* Genome Database Genome Overview," http://www.yeastgenome.org/ (retrieved June 1, 2008).]

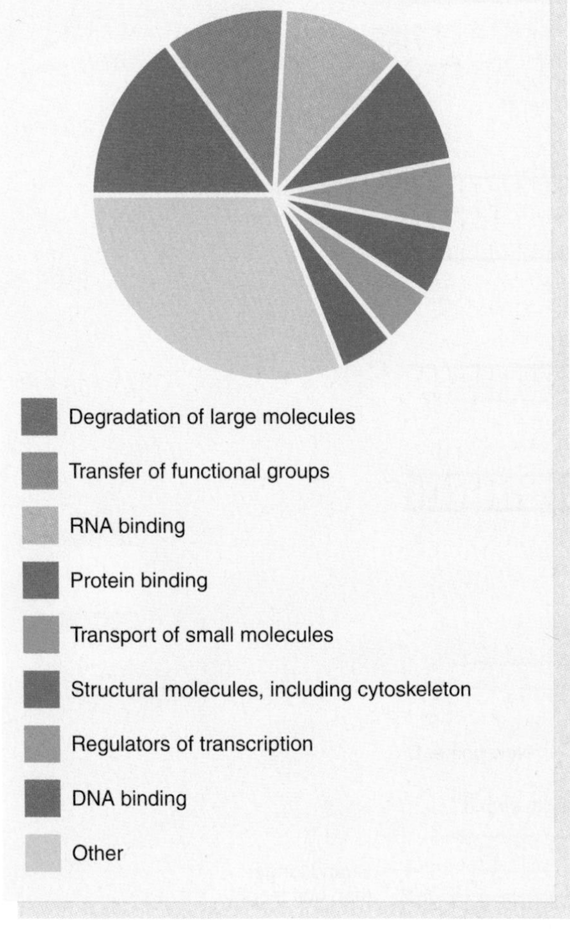


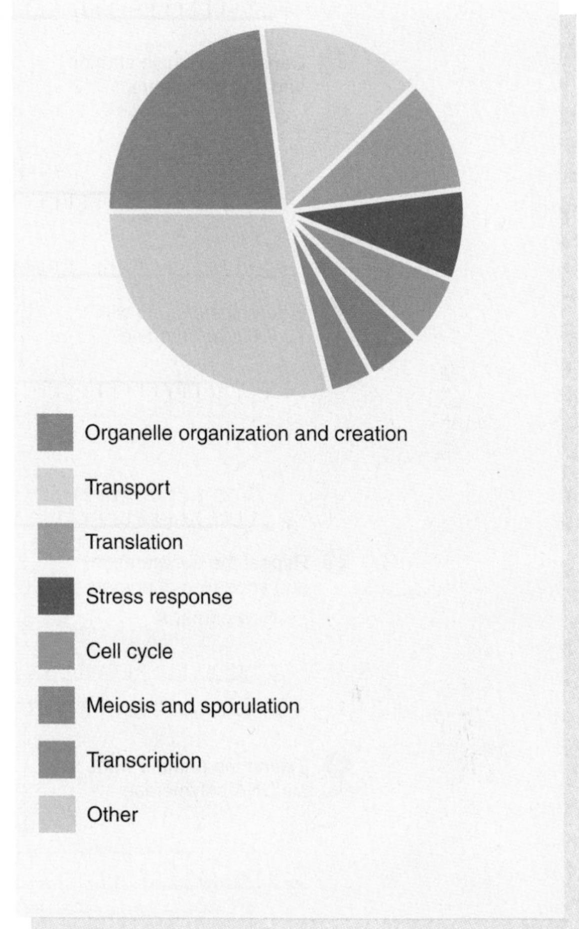


silence gene expression in eukaryotes (see also Chapter 18, pp. 537–540). This technique does not create a permanent chromosomal change, but does prevent a targeted gene from functioning correctly for as long as the small regulatory RNA is present in the cell. We will see how this technique is used in the study of genes from the worm. In both techniques, the goal is to see what happens if the protein encoded by the gene of interest is not made.

**Gene Knockouts in Yeast.** Gene function can be knocked out in yeast using a PCR-based strategy. The **polymerase chain reaction,** or **PCR,** is one of the most frequently used genetics techniques. PCR is a way of amplifying a small (generally less than 10 kb) region of DNA—the *target DNA sequence*—allowing us to make an essentially unlimited number of copies of that DNA without cloning the region. Once generated, these copies could be cloned, separated using gel electrophoresis, or quantified, depending on the needs of the investigator. PCR is, at its heart, a modification of DNA replication. PCR is carried out using a PCR machine, or thermal cycler, which takes samples through a series of carefully controlled temperature changes for very specific periods of time. Kary Mullis received part of the 1993 Nobel Prize in Chemistry "for his invention of the polymerase chain reaction (PCR) method."

**Animation**
**Polymerase Chain Reaction (PCR)**

Figure 9.3 illustrates the polymerase chain reaction. To amplify a specific target DNA sequence using the polymerase chain reaction, we start with a template, which is generally double-stranded. This template can be large and complex—it can even be an entire genome. It really does not matter that the target DNA sequence is a tiny, tiny fraction of the entire template. Two primers are designed and synthesized to make the desired polymerase chain reaction possible. These primers must be

**Figure 9.3**

**The polymerase chain reaction (PCR) for selective amplification of DNA sequences.**

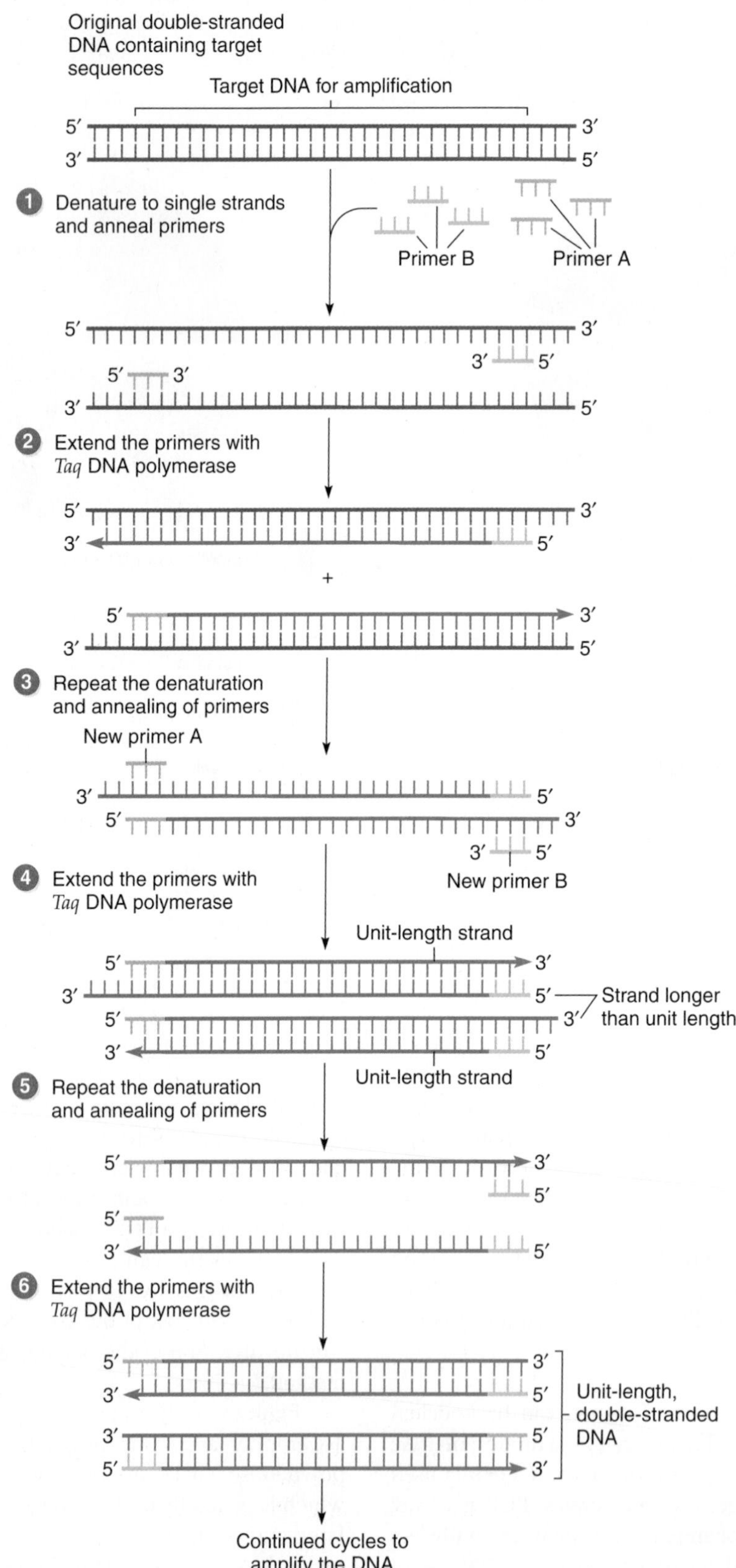

complementary to the two ends of the target DNA sequence to be amplified. The primers are added to the template DNA along with dNTP precursors (dATP, dCTP, dGTP, and dTTP) and a buffer, and the reaction mixture is heated to 95°C. The heat denatures the DNA to single strands. The reaction mixture is allowed to cool to a temperature at which the primers will anneal to the template (Figure 9.3, step 1). That temperature will vary with the primers and template used, but typically will be in the range 55–65°C. The orientation of the primers on the templates is crucial for the amplification of target DNA. That is, the two primers are designed so that they anneal to the opposite strands of the template DNA at the two ends of the target DNA sequence. That is, the 3′ end of each primer must be oriented to "point" at the 3′ end of the other primer.

Next, a heat-stable DNA polymerase is added. Such enzymes have been isolated from bacteria or archaea that have evolved to survive in very hot environments, so their enzymes must therefore function and retain proper structure at high temperatures. One example is *Taq* ("tack") polymerase, an enzyme isolated from *Thermus aquaticus*. In the PCR, the DNA polymerase extends each of the primers from their 3′ ends at 72°C (the optimal temperature for the enzyme) (Figure 9.3, step 2). After a specified amount of time for the DNA synthesis step (determined by the size of the target DNA to be amplified, as the enzyme can add about 1,000 bases per minute), the denaturation step is repeated at 95°C (the reason for the heat-stable enzyme, which is still in the reaction mixture) and the mixture is cooled to allow the primers to anneal (Figure 9.3, step 3). (Further amplification of the original strands is omitted in the remainder of the figure.) Here is the beauty of PCR—extension from primer A created a DNA fragment that can now bind to primer B, and extension from primer B created a DNA fragment that can bind primer A. Thus, in this second round of amplification, twice as many primers and enzymes can be involved. Now extension of the primers with DNA polymerase is done (Figure 9.3, step 4). Note that, in each of the two double-stranded molecules produced in the figure, one strand is of unit length; it is the length of DNA between the 5′ end of primer A and the 5′ end of primer B, which is the length of the target DNA. The other strand in both molecules is longer than unit length. The denaturation step and primer annealing is again repeated (Figure 9.3, step 5). (For simplification, the further amplification of those strands that are longer than unit length is omitted in the rest of the figure.) The primers then are extended with DNA polymerase (Figure 9.3, step 6). This amplification step produces unit-length, double-stranded DNA. Note that it took three cycles to produce the two molecules of amplified unit-length DNA. Repeated denaturation, annealing, and extension cycles result in the exponential increase in the amount of unit-length DNA. Typically the PCR amplification cycle is repeated for 30–35 rounds.

PCR is a useful technique, as you will see in several of the later genomic analyses. PCR is also used diagnostically and is a key step in quantification of transcriptional activity, as you will learn in Chapter 10.

Figure 9.4a shows the use of a PCR-based gene knockout strategy in yeast. We start by designing PCR primers based on the known genome sequence and then construct and amplify an artificial *linear DNA deletion module*, also called a *target vector*. This module consists of part of the sequence of the gene of interest upstream of and including the start codon and part of the gene sequence downstream of and including the stop codon, flanking a selectable marker. In this example, the selectable marker is a DNA fragment containing the $kan^R$ (kanamycin) selectable marker that confers resistance to the inhibitory chemical G418. In essence, the $kan^R$ marker replaces most of the coding region in the middle part of the gene of interest. As you might expect, this altered gene can no longer code for its protein. This linear DNA is transformed into yeast, and G418-resistant colonies are selected. Unlike the plasmids we have discussed previously, this linear piece of DNA will not replicate in the host cell, because it lacks an origin of replication. If that is the case, how can we recover colonies that clearly carry sequences from our plasmid? The linear plasmid integrates into the yeast chromosome by a process called **homologous recombination.** Homologous recombination is the recombination between similar sequences, and it is most common during meiosis. It can occur (but is generally very rare) in nonmeiotic cells. In this circumstance, we are looking for homologous recombination between the copy of the gene of interest on the chromosome and the fragments of the gene of interest on the linear plasmid. Luckily, yeast has a high rate of homologous recombination between plasmids and chromosomes. The small linear deletion construct will also be changed by the recombination event. It will carry a functional copy of the gene of interest that it picks up from the chromosome but will lack the $kan^R$ selectable marker. Since this linear construct lacks the proper sequences for replication and segregation, it will be lost by most of the cells generated as the recombinant yeast divides.

The homologous recombination event completely inactivates—knocks out—the chromosomal copy of the gene of interest because most of the coding region is replaced by the $kan^R$ selectable marker. In genetic terms, a *null allele* (an allele unable to code for any functional polypeptide) is produced when the $kan^R$ gene replaces most of the gene of interest. Recall that yeast is generally haploid, so these cells will not have a second copy of the gene. This means that if the gene is required for a specific function in the cell, the new mutant cell will have a defect in that function as a result of the knockout mutation. Furthermore, if this gene is essential for viability, the cell carrying the knockout mutation will die. Since these mutant cells would die before they were able to replicate, it would seem as if the experiment failed completely, since no G418 resistant colonies would be recovered.

Figure 9.4

**Creating and verifying a gene knockout in yeast.** (a) Schematic of a PCR-based gene deletion strategy involving a DNA fragment constructed by PCR from gene sequences flanking the *kan*$^R$ selectable marker that is transformed into yeast and replaces a large segment of the chromosomal gene by homologous recombination. (b) Verification of gene deletion. PCR-based screening method to confirm (1) unsuccessful deletion (gene still present) and (2) successful deletion (gene replaced with *kan*$^R$ DNA segment).

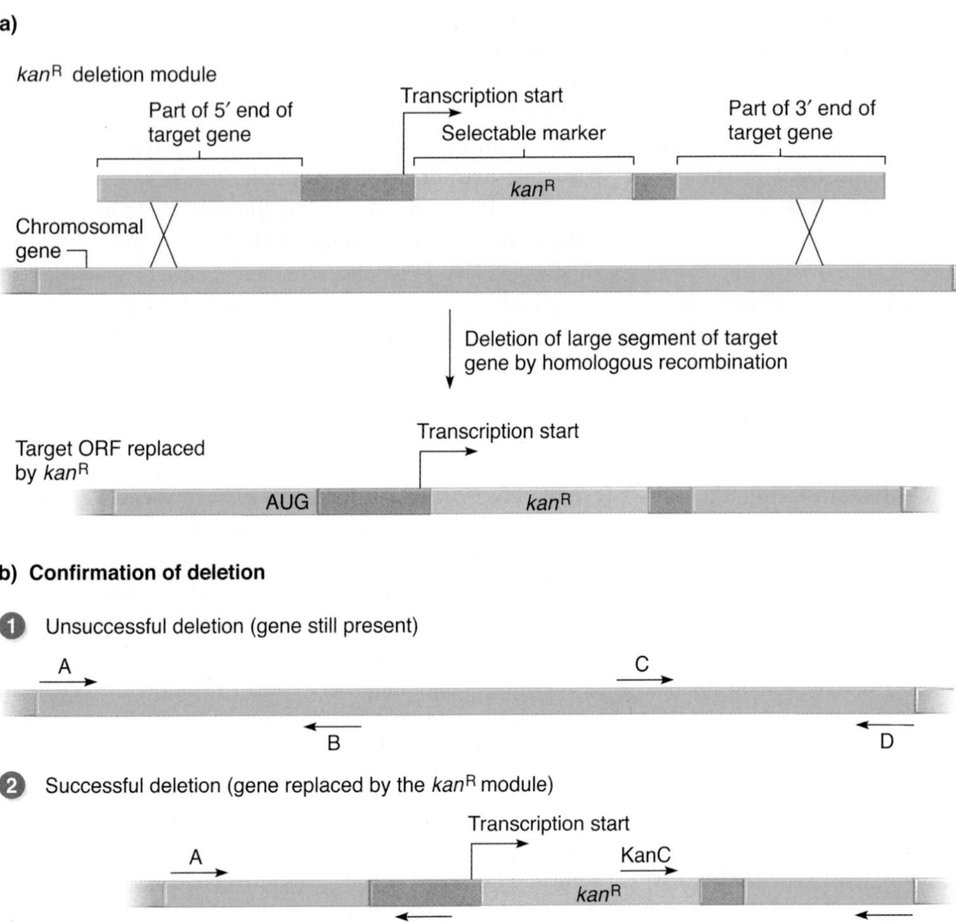


This gene knockout approach is efficient in yeast because homologous recombination is quite common in this organism. However, homologous recombination is rare in most other organisms. Thus this type of approach must be modified when working with other organisms, otherwise it would be nearly impossible to generate enough transformants to find the rare homologous recombinants.

A molecular screen must be used to confirm that the yeast transformant has resulted in deletion of the gene of interest. That is, the target vector may integrate elsewhere in the genome. This occurs by **nonhomologous recombination**, a process involving crossing-over between sequences that are not similar. Such integration also produces a G418-resistant transformant. PCR is used for the screen, as illustrated in Figure 9.4b. First, let us consider the condition of an unsuccessful deletion in which the gene is still present (Figure 9.4b.1). Four different PCR primers, A–D, are used. Primers A and D are 200 to 400 bases upstream or downstream, respectively, of the gene. Primers B and C are from within the gene itself. DNA is isolated from transformants, and separate PCRs are done with primers A and B and with primers C and D. If the gene is still present, these reactions produce DNA fragments of predictable sizes. If the gene is deleted, no PCR products are seen. However, it is still necessary to show definitively that the deletion has been made, and the scheme is shown in Figure 9.4b.2. Primers A and D are as in Figure 9.4b.1, and there are two other primers—KanB and KanC—that are specific for the *kan*$^R$ DNA fragment. If deletion has been successful, the *kan*$^R$ module has replaced the gene and PCR using primers A and KanB, and primers KanC and D generate fragments of predictable sizes.

Using the gene deletion approach, a yeast knockout (YKO) project has been completed in which each yeast gene has been deleted one at a time. Because some genes have essential functions, deleting them gave a lethal phenotype. However, about 4,200 of the approximately 6,600 ORFs are nonessential, since knocking each of them out individually results in a viable phenotype. This set of 4,200 strains in the yeast deletion collection is a genomic resource for investigating the functions of nonessential genes in this organism. For example, to assign function to the knocked-out genes, the deletion strains are being studied under various conditions and examined for changes in their phenotype. The work involved in this task is substantial due to the many areas of cell function that must be screened for a change in phenotype, including cell cycle events, meiosis, DNA synthesis, RNA synthesis and processing, protein synthesis, DNA repair, energy metabolism, and molecular transport mechanisms. From such work, it has been shown that approximately one-half of the deletion strains show no significant changes in phenotype for the functions that have been analyzed, and the other one-half do.

**Gene Knockouts in the Mouse.** The mouse is an important organism for genetic study. This is because a mouse is quite similar to humans, and because it is one of the few mammals that can be kept in the lab in large numbers and studied using genetic techniques. Gene knockouts in mice, for example, are being used as models to identify the functions of mouse homologs of unknown human genes (because it is unethical to knock out human genes), and to address more basic questions about how mammals function.

Figure 9.5 shows how gene knockouts can be made in the mouse. The procedure is somewhat similar to that described for yeast, although the experiments are more involved. First a cloned copy of the target gene to be knocked out is modified to replace a central portion with a selectable marker. In our example the marker is *neo*$^{R}$, which is a gene that confers upon mouse cells the ability to grow on the drug neomycin. To the modified gene is added a segment of DNA containing a second selectable marker, in this case *tk*, a viral gene that encodes the enzyme thymidine kinase. If the chemical ganciclovir is added to mouse cells in culture expressing the *tk* gene, growth of those cells becomes inhibited. That is, the thymidine kinase phosphorylates ganciclovir, modifying it to become an inhibitory chemical for DNA replication. The complete DNA segment with the disrupted target gene and the two selectable markers is the target vector (Figure 9.5a).

The deletion module is transformed into mouse **embryonic stem (ES) cells** in culture.[1] An ES cell is a cell derived from a very early embryo that retains the ability to differentiate into a cell type characteristic of any part of the organism. ES cells can be grown as single cells in culture in the lab without differentiating and, importantly, they can be moved back into a very young mouse embryo, where they can make any part of the embryo. The transformed ES cells are grown in medium containing neomycin, which selects for cells containing the integrated target vector. Two different paths may be followed to produce stable transformants. In one path (Figure 9.5a, left side), homologous recombination between the target vector and the target gene in the chromosome leads to the replacement of the complete, normal copy of the target gene in the chromosome with the disrupted target gene from the target vector. The now knocked-out target gene in the chromosome is nonfunctional, while the target vector recombinant with the complete gene does not replicate and is lost as the cells divide. The other path that can be followed is transformation by random integration. In random integration, the target vector integrates into a chromosome by nonhomologous recombination. As shown in Figure 9.5a, right side, random integration can involve the insertion of most of the target vector into a chromosome, including the disrupted target gene and the *tk* gene. Of the two paths, random integration is far more common. Fortunately the transformants desired—those from homologous recombination—can be selected for by exploiting the selectable markers. That is, the transformed ES cells are grown on culture medium containing both neomycin and ganciclovir. The homologous recombination replaces one copy of the target gene with *neo*$^{R}$ but the *tk* marker gene is lost, because *tk* is outside of the homologous recombination region. As a result, these homologous recombinants are able to grow because they contain the target gene containing the *neo*$^{R}$ gene, but no *tk* gene. The random integration transformants cannot grow because, even though they contain the *neo*$^{R}$ gene and thus are resistant to neomycin, they also contain the *tk* gene, which inhibits their growth on ganciclovir.

The transformants that grow on neomycin plus ganciclovir are then tested to make sure that the target gene has been knocked out as expected (see Figure 9.5a). Typically this is done using a PCR approach, which is conceptually similar to that described and illustrated for the yeast knockout system.

The correct, transformed ES cells are injected into blastocysts (an early embryonic stage of development) derived from a mouse strain with a different coat color than that from which the ES cells came. In our example, the ES cells came from an agouti mouse and the blastocysts came from a black mouse. Agouti is the greyish color of wild rodents (see Chapter 13, p. 380); agouti is genetically dominant to black. The introduced cells become part of the developing embryo, including at times producing some of the germ-line cells. The embryo is introduced into a surrogate mother, where it continues to develop. The resulting mouse pup will be a *chimera*, meaning that it has a mixture of two distinct tissue types.

[1]In 2007, Mario Capecchi, Oliver Smithies, and Sir Martin Evans were awarded the Nobel Prize in Medicine and Physiology "for their discoveries of principles for introducing specific gene modifications in mice by the use of embryonic stem cells."

Figure 9.5

**Creating a gene knockout in the mouse.**

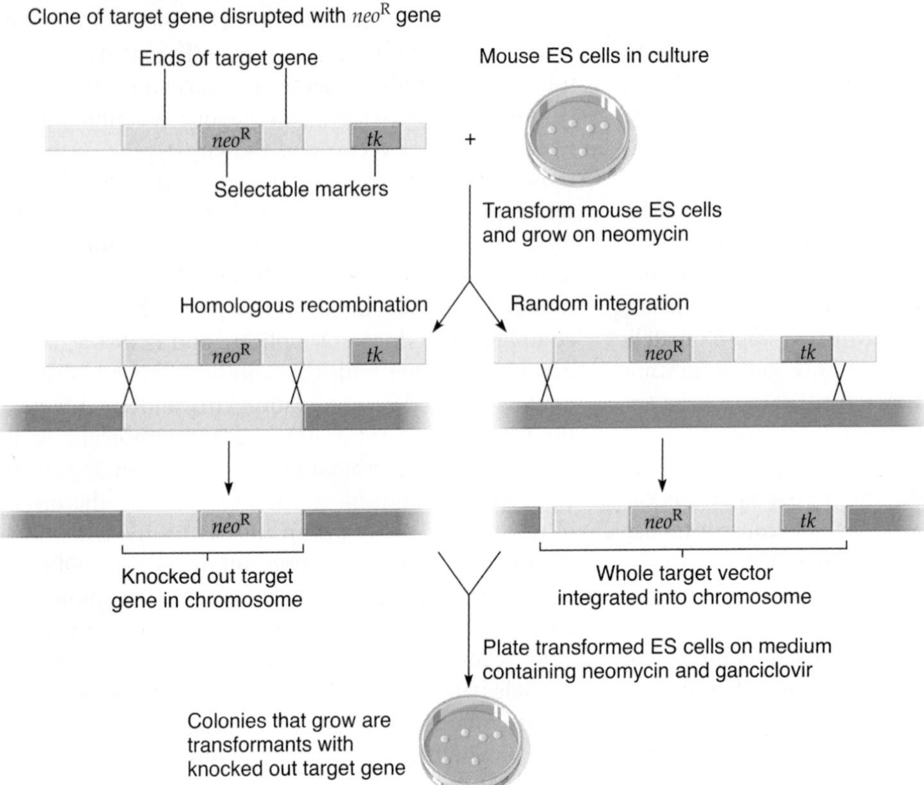


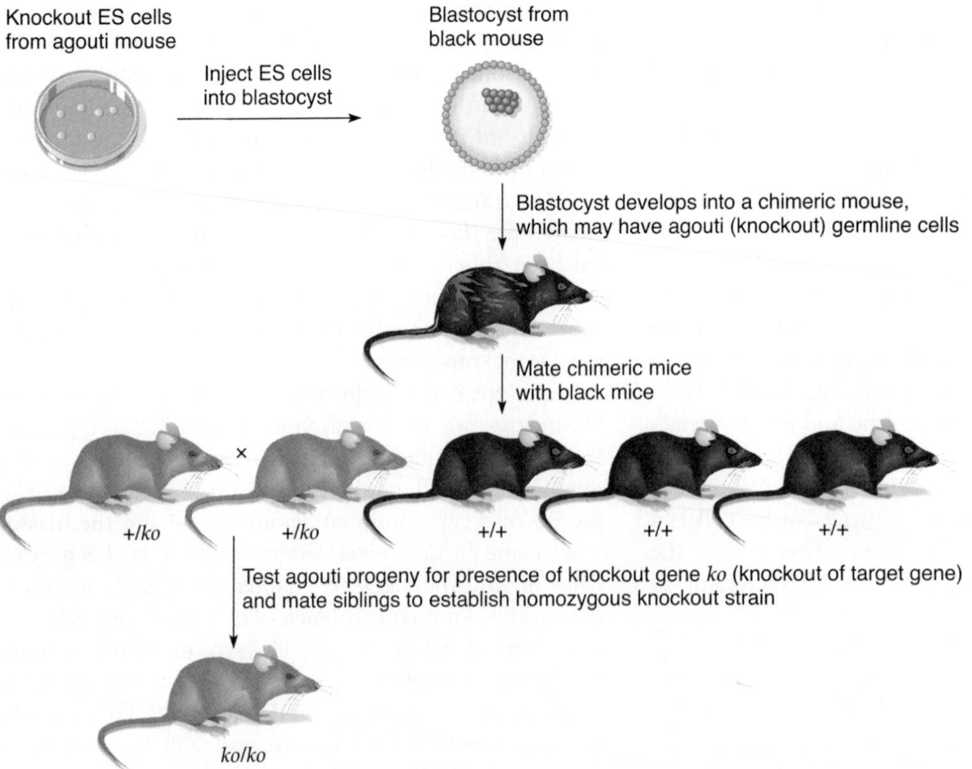

One type is derived from the knockout ES cells, and the other type is derived from the blastocyst cells. Since coat color differences were used for the origins of the two cell types, chimeric pups are readily identified by the presence of patches of agouti and black hair.

When the chimeric mice mate with normal black mice, they will pass the gene knockout to some of their progeny provided that some of their germ line consists of the transformed cells (see Figure 9.5b). These progeny will have one copy of the agouti gene and one copy of the black gene, but they will be agouti due to the dominance of agouti over black. These mice can be tested by PCR to confirm that they carry the *neo*$^R$ gene in their DNA. Those carrying *neo*$^R$ have one copy of the knocked-out target gene (*+/ko* in Figure 9.5b). Breeding these *+/ko* mice to each other produces offspring, 25% of which are homozygous *ko/ko*; that is, they have knockouts of both copies of the target gene (see Figure 9.5b). This is the knockout mouse strain that was wanted. Since we often do not know what the phenotype of our new mutation will be, or even if there will be an obvious phenotype, PCR is often used to determine which mice are homozygous for the knockout. That is, primer pairs are used to prove that the disrupted gene containing *neo*$^R$ is present and that no chromosomal copy of the normal target gene is present.

The knockout technique produces a loss-of-function or null allele of the target gene. The homozygous knockouts can be studied to determine what happens when the animal is unable to make the protein encoded by our target gene. As you might expect, animals unable to make a protein encoded by our gene of interest may be unable to survive. If homozygous pups cannot be found, investigators have to carefully monitor the embryos that form when two heterozygotes mate, and, with careful observation, can determine when and why these embryos die. This is frequently the case, and characterization of when and how these embryos die can often tell us much about what the gene product does in normal development.

**Gene Knockouts in the Bacterium *Mycoplasma genitalium*.** One of the smallest characterized genomes, that of *Mycoplasma genitalium*, contains about 500 protein-coding genes. Scientists used transposons to identify which of these genes were required for the bacteria to survive in lab culture. As you learned in Chapter 7, transposons are mobile DNA elements, and the insertion of a transposon into a gene tends to disrupt the function of that gene, much as insertion of a DNA fragment into a multiple cloning site of a plasmid cloning vector disrupts function of the *lacZ* gene (see Chapter 8, p. 176). Over 2,000 new transposon insertions were generated and characterized, and these insertion sites were mapped to the annotated genome. It was assumed that if a transposon integrated into the coding region of an essential gene, the new mutation would be lethal and would disappear from the population before it could be characterized. In essence, only viable transposon insertions could be identified. Insertions in at least 100 genes were viable, suggesting that most of the protein-coding genes (estimates ranged from 265 to 340) are required for the organism to survive in the lab. In this case, the goal was to identify the minimal gene set for a project to create an artificial cell. This organism was selected because it had the smallest genome of any organism that was known to be able to survive without a host.

**Knocking Down Expression of a Gene by RNA Interference.** In this section we learn how gene knockouts or gene knockdowns may be made using RNA interference. RNA interference (RNAi) is a normal cellular process in which small regulatory RNA molecules silence gene expression in eukaryotes. The key features of the RNAi process are shown in Figure 9.6a; the process is described in detail in Chapter 18, pp. 537–540 and in Figure 18.15. First, a double-stranded RNA (dsRNA) molecule forms in the cell. Recall from Chapter 6 that RNA typically is single-stranded; the unusual double-stranded form of RNA triggers the RNAi process. Cellular proteins bind to the dsRNA and cut it into lengths of about 21–23 bp. A protein known as Slicer binds to the short dsRNA molecule and unwinds one of the two strands, which is then discarded. The remaining short single-stranded RNA in the complex with Slicer (the small regulatory RNA molecule mentioned earlier) will pair with any single-stranded RNA molecule in the cell with which it is complementary; that molecule is the target RNA for RNAi. When pairing occurs, either translation of the mRNA is repressed, or Slicer cleaves the target RNA and the pieces are degraded. In either case, the target RNA, which typically is a mRNA molecule, is rendered nonfunctional. That is, the protein encoded by the mRNA no longer can be made from that mRNA and, effectively, the expression of the gene that encoded that mRNA has been silenced (interfered with) at the translation step.

There are different sources for the dsRNA from which the small single-stranded regulatory RNAs are made in the RNAi process. For instance, some dsRNA molecules are encoded by genes. Expression of those genes results in single-stranded RNAs that fold up into a hairpin structure by complementary base pairing involving different parts of the molecule. The paired RNA segments in the hairpin is the dsRNA that starts the RNAi process. The role of the small regulatory RNAs made from gene-encoded dsRNAs is to regulate the expression of other genes by silencing expression of the mRNAs of those genes.

Silencing gene expression by RNAi is highly specific because it depends upon the complementary base pairing of the small regulatory RNA with the target mRNA. Because of this specificity, RNAi has been adapted for use as a laboratory technique to knock out or knock down the expression of genes in a variety of eukaryotes, including

**Figure 9.6**

**Knocking out gene expression by RNA interference (RNAi).** (**a**) Outline of the mechanism for silencing gene expression at the mRNA level by RNAi. (**b**) Use of an engineered gene to produce a hairpin RNA transcript with a double-stranded RNA section that can initiate the RNAi process to silence a specific target gene.

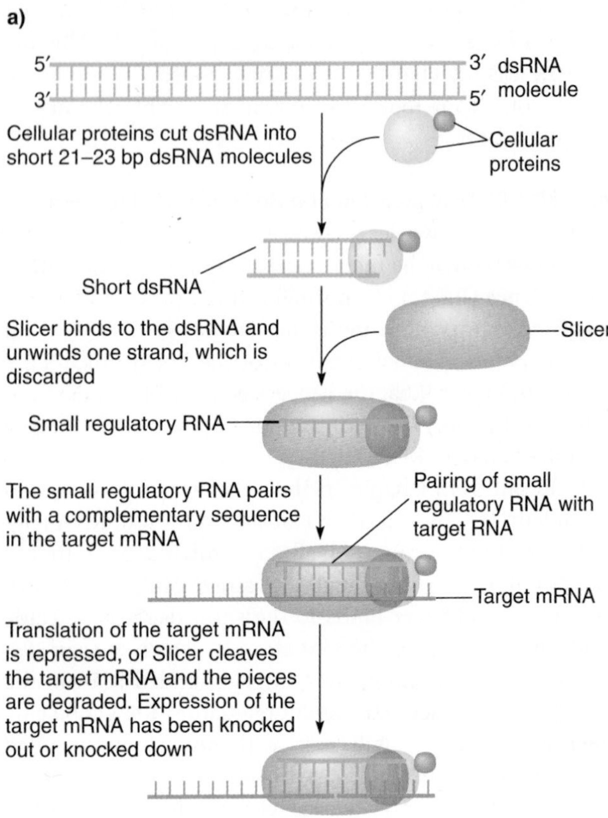
a)
5′
3′
3′
5′
dsRNA molecule
Cellular proteins cut dsRNA into short 21–23 bp dsRNA molecules
Cellular proteins
Short dsRNA
Slicer binds to the dsRNA and unwinds one strand, which is discarded
Slicer
Small regulatory RNA
The small regulatory RNA pairs with a complementary sequence in the target mRNA
Pairing of small regulatory RNA with target RNA
Target mRNA
Translation of the target mRNA is repressed, or Slicer cleaves the target mRNA and the pieces are degraded. Expression of the target mRNA has been knocked out or knocked down

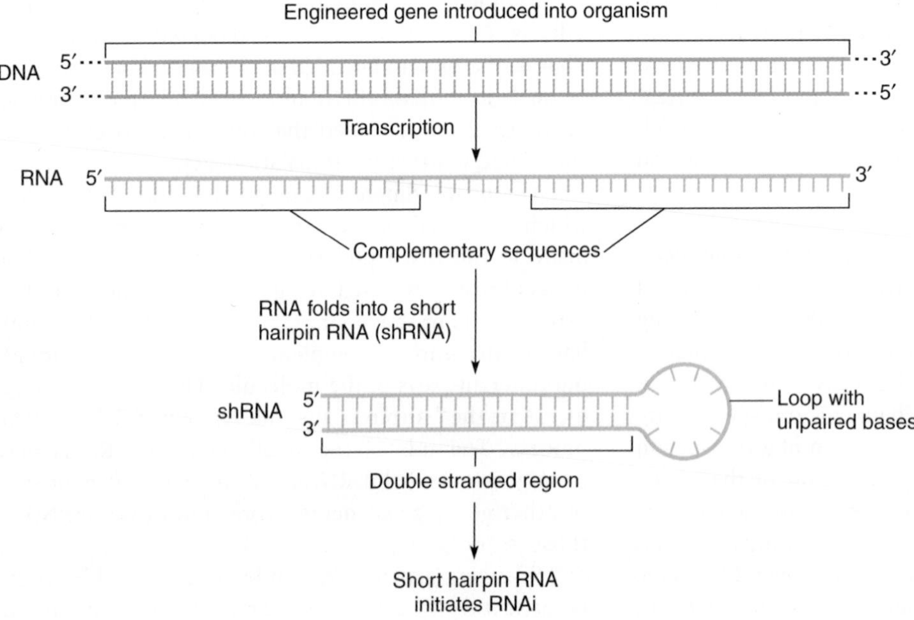
b)
Engineered gene introduced into organism
DNA
5′...
3′...
...3′
...5′
Transcription
RNA 5′
3′
Complementary sequences
RNA folds into a short hairpin RNA (shRNA)
shRNA
5′
3′
Loop with unpaired bases
Double stranded region
Short hairpin RNA initiates RNAi

the nematode worm *Caenorhabditis elegans*, *Drosophila*, mouse, and plants. "Knock out" here means blocking the expression of a gene's mRNA completely, while "knock down" means inhibiting the expression of a gene's mRNA incompletely so that some functional protein product results.

To knock out or knock down the expression of a specific target gene, a small, single-stranded regulatory RNA molecule complementary to the mRNA encoded by that gene must be introduced into the cell or organism. For example, several ways are used to deliver dsRNA to cells of *C. elegans*. In one way, an engineered gene transformed into the organism is transcribed to produce an RNA molecule that begins the RNAi process. (A gene introduced by artificial means into a cell or organism is called a **transgene.** The cell or organism, having received a transgene, is a transgenic cell or organism.) The sequence of the engineered gene is designed so that the RNA transcript base-pairs with itself to form a hairpin structure, called a *short hairpin RNA* or shRNA (Figure 9.6b). As described earlier, the double-stranded part of the hairpin initiates RNAi. Alternatively, the RNA transcript that can fold into a hairpin can be microinjected into the gonads of a hermaphrodite, where it will be incorporated into its offspring, or young animals can be soaked in a solution of the small RNA (in that case, they will absorb it into their cells). The RNA can even be delivered to the cells by letting the worms eat bacteria that produce the double-stranded hairpin RNA. For most organisms, the first two methods of dsRNA delivery are possible, but the latter two are not. Regardless of how the RNA is delivered, cells containing the interfering RNA are generally unable to make the protein encoded by the target, even though the gene itself is unchanged in the genome. Analysis of these animals can tell us what happens when the gene is nonfunctional, even though this technique does not create a permanent, chromosomal mutation.

Using this RNAi approach, screens have been set up in *C. elegans* to systematically knock out, or at least knock down, each of the approximately 20,000 protein-coding genes and to characterize the resulting phenotypes. A similar screen using RNAi systematically on every known gene in *Drosophila* has been completed. Obviously, it can be very difficult to examine each of 20,000 individual experimental samples and determine which, if any, aspects of normal life are disrupted for the animals that have lost the function of one gene. In several screens, anywhere from 10–25% of the RNAi gene knockouts or knockdowns resulted in a detectable phenotype. More specific genome-wide tests, where RNAi was used on all 20,000 genes but very specific phenotypes were selected (for instance, by screening specifically for genes involved in fat metabolism, and regulation of transposon activity), have been successful at suggesting functions for some of the genes that did not seem to have a clear defect in initial genome-wide screens.

### Keynote

Gene function can be assigned experimentally by knocking out a gene or knocking down its expression and investigating the phenotypes that result. Different methods are used to knock out a gene, including replacing the normal chromosomal copy of the gene with a disrupted copy (used in many organisms) and inactivating the gene by inserting a transposon into it (typically used with bacteria). The outcome in either case is a gene with no, or markedly reduced, function. Gene knockouts made in this way are permanent changes in the chromosome. Alternatively, gene expression can be silenced (knocked down) in many eukaryotes at the translation level by RNA interference in which a specific small regulatory RNA targets a specific mRNA for degradation. This method does not cause permanent change, but prevents the translation of the mRNA of a targeted gene for as long as the small regulatory RNA molecule is present.

## Organization of the Genome

As the human genome has been sequenced and annotated, one interesting question that can now be addressed systematically is whether the organization of the genome is somewhat random or whether the genes and other sequences present are organized in a specific way. Recent analysis suggests that the genome is highly organized both at the chromosomal level and in how it is arranged in the nucleus, at least when the cell is in interphase. When we look at the arrangement of genes and repetitive sequences in the human genome, we can note several interesting organizational aspects. Many of the abundantly transcribed genes are grouped together in small clusters where the gene density tends to be high and the introns tend to be small. In contrast, genes that are less frequently transcribed tend also to be clustered together, but the gene density in these areas is low and these genes tend to have larger introns. Several other trends can be seen in these groupings. Certain repetitive sequences, called SINEs (short interspersed elements; see Chapter 2, p. 29), are more common in areas with frequently transcribed genes. This includes the Alu family of repeat sequences (see Chapter 2, p. 29, and Chapter 7, p. 161). In contrast, regions containing less frequently transcribed genes are enriched in sequences called LINEs, or long interspersed elements (see Chapter 2, p. 29). In the interphase nucleus, the clusters with lower gene density tend to be found near the nuclear membrane, while the high-density regions tend to be more central in the nucleus. These studies have shown that the genome is more organized than once thought, both at the chromosomal level and in the nucleus. Gene-dense parts of the chromosome contain more of the highly transcribed genes and are held in the center of the nucleus, while less dense chromosomal regions tend to contain less frequently transcribed genes and are pushed out of the central parts of the nucleus.

### Describing Patterns of Gene Expression

In classical genetic analysis, research begins with a phenotype and leads to the gene or genes responsible. Once the gene is found and isolated, experiments can be done to study the expression of the gene in normal and mutant organisms as a way to understand the role of the gene in determining the phenotype. When the complete genome sequence is obtained for an organism, exciting new lines of research are possible, such as the analysis of expression of *all* genes in a cell at the transcriptional and translational levels as well as the analysis of all protein–protein interactions. Measuring the levels of RNA transcripts (usually focusing on mRNA transcripts), for example, gives us insight into the global gene expression state of the cell. To go along with this new research, a new term has been coined for the set of mRNA transcripts in a cell: the **transcriptome.** The study of the transcriptome is **transcriptomics.** Because the mRNAs specify the proteins responsible for cellular function, the transcriptome is a major indicator of cellular phenotype and function. By extension, the complete set of proteins in a cell is called the **proteome.** The study of the proteome is **proteomics.** Studies of the transcriptome and the proteome are described in this section.

**The Transcriptome.** The transcriptome is not the same in all the cells in an organism. A muscle cell and a liver cell will transcribe overlapping but very different subsets of the genes in the genome. Furthermore, while a given cell type typically will have a fairly constant transcriptome, the transcriptome of a particular cell may change if the cell changes. For instance, a yeast cell that undergoes a change in its growth conditions, or a human stem cell that differentiates into a muscle cell, will change which genes are transcribed. By defining exactly which genes are expressed, when they are expressed, and their levels of expression, we can begin to understand cellular function at a global level. In this case, understanding the global level would mean that we understood the entirety of the cellular response to a particular condition, at least at the level of transcription—we would know how all transcription changes, rather than just how the transcription of one gene changes. Studies of the transcriptome have allowed us to begin asking questions about these global responses and have thus added to our understanding of basic cellular and organismal processes as well as helping us understand the effects of disease and environmental hazards. Most commonly, these studies use DNA microarrays (also called gene chips or DNA chips) to ask about global gene expression. Here we will discuss two examples of the use of transcriptomics to understand changes in gene expression.

One example of the use of transcriptomics to understand changes in gene expression is a collaborative study by Pat Brown and Ira Herskowitz of yeast sporulation, the process of producing haploid spores from a diploid cell by meiosis (Figure 9.7a). Yeast sporulation involves four major stages: DNA replication and recombination, meiosis I, meiosis II, and spore maturation. (Meiosis is described in Chapter 12, pp. 333–336.) The sequential transcription of at least four classes of genes—early, middle, mid-late, and late—correlates with these stages.

*animation*
**Analysis of Gene Expression Using DNA Microarrays**

When these DNA microarray experiments began, about 150 genes had been identified that are differentially expressed during sporulation. In the new study, the researchers induced diploid yeast cells to sporulate. At seven timed intervals, they took cell samples and used DNA microarrays containing 97% of the known or predicted yeast genes to analyze the timing of gene expression during meiosis and spore formation. Light and electron microscopy were used to correlate the sampling time with the exact stage of sporulation.

For quantifying gene expression, the researchers isolated mRNAs from the cell samples and synthesized fluorescently labeled cDNAs from them by reverse transcription in the presence of Cy5-labeled dUTP (Figure 9.7b). (The synthesis of cDNA from mRNA by reverse transcription is described in Chapter 8, pp. 195–197, and shown in Figure 8.15.) Cy5 is a fluorescent dye that can be added to a nucleotide—in this case a precursor for RNA synthesis—without changing the base-pairing properties. It emits a specific wavelength of red light when excited by ultraviolet light. For a nonsporulating cell control, they isolated mRNAs from cells at a time point immediately before inducing sporulation and synthesized fluorescently labeled cDNAs, this time using Cy3-labeled dUTP. Cy3, like Cy5, is a fluorescent dye, but Cy3 emits light at a slightly different wavelength than does Cy5.

For each time point, the researchers hybridized a mixture of experimental Cy5-labeled cDNAs and reference Cy3-labeled cDNAs to DNA microarrays. The DNA microarrays were made by using PCR to amplify each ORF (using primers based on the genome sequence) and printing the denatured PCR products onto DNA microarrays. After completing the hybridization, the researchers scanned the microarrays with a laser detector device to detect the Cy5 and Cy3 fluorescence locations and to quantify their intensities (see Figure 9.7b). Because only a small amount of light is emitted, and because each ORF is printed in a tiny spot, the results are greatly magnified and presented on a computer screen. The software converts the Cy5 signal to red on the screen (the same color it really emits), and converts the Cy3 to green, rather than its actual color—a different red than that of Cy5. The relative abundance of transcripts from each gene in sporulating versus nonsporulating yeast cells is seen by the ratio of red to green fluorescence on the microarray.

**Figure 9.7**

**Global gene expression analysis of yeast sporulation using a DNA microarray.** (a) The stages of sporulation in yeast, correlated with the sequential transcription of at least four classes of genes. (b) Outline of the DNA microarray experiment. (c) Example of results of a global gene expression analysis of yeast sporulation, obtained using a DNA microarray. The entire yeast genome is represented on the DNA chip, and the colored dots represent levels of gene expression, as described in the text.

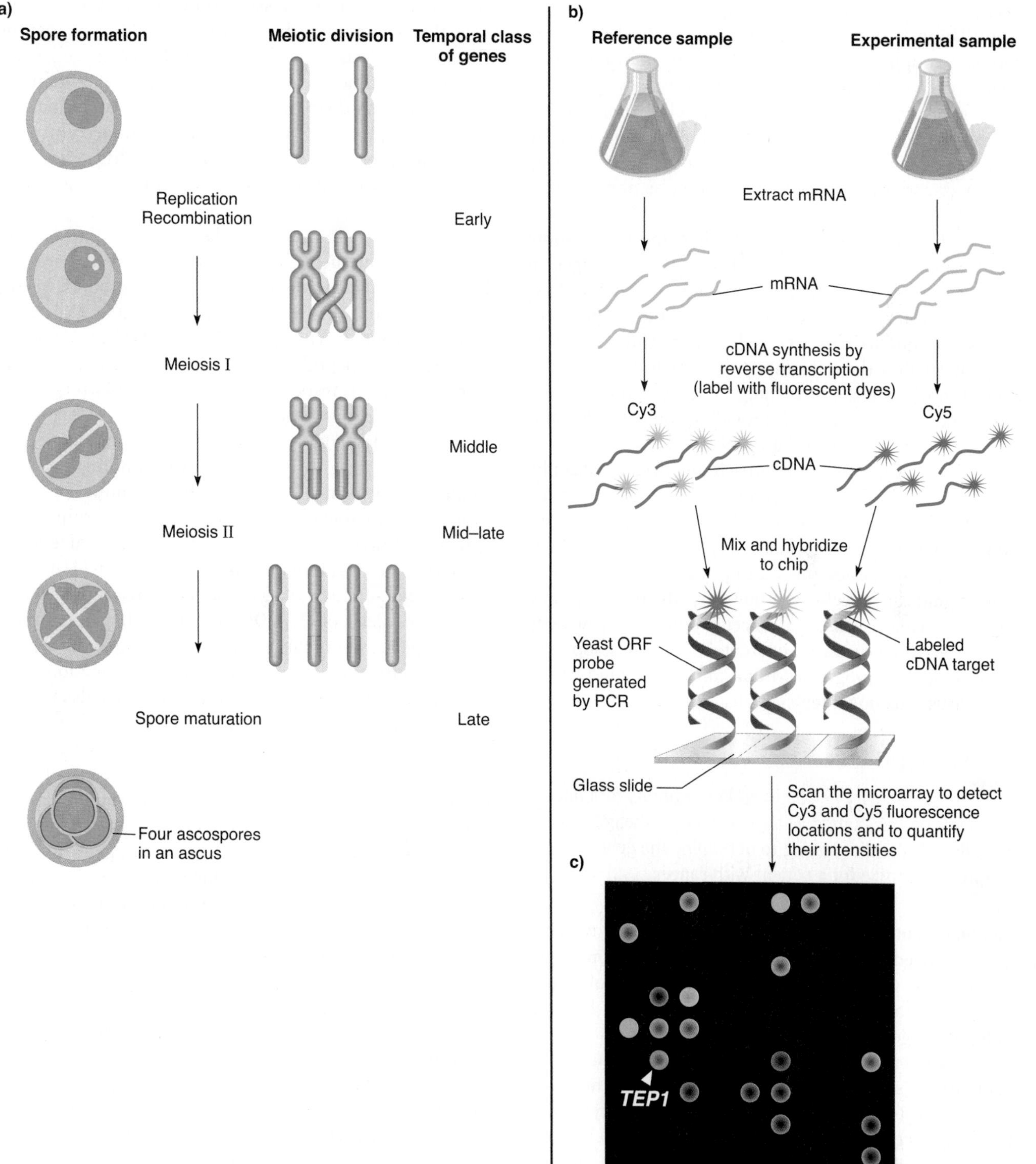

If an mRNA is more abundant in sporulating cells than in nonsporulating cells, as is the case for the *TEP1* gene (Figure 9.7c), this results in a higher ratio of red-labeled to green-labeled cDNAs prepared from the two types of cells and, therefore, in the same higher ratio of red to green fluorescence detected on the array. In general, a gene whose expression is induced by sporulation is seen as a red spot, and a gene whose expression is repressed by sporulation is seen as a green spot. Genes that are expressed at approximately equal levels in nonsporulating cells and during sporulation are seen as yellow spots. Orange spots might indicate that the level of transcription changed during the experiment, and black spots indicate that the gene represented in that spot of the microarray is not transcribed in either sporulating or nonsporulating cells.

With this approach, the researchers found that more than 1,000 yeast genes showed significant changes in mRNA levels during sporulation. About one-half of these genes are transcribed less during sporulation than other times, and one-half are transcribed more (induced) during sporulation than at other times. At least seven distinct timing patterns of turning on gene expression are seen, and this observation is providing some insights into the functions of many orphan genes.

The DNA microarray approach just described can be used to analyze the transcriptome to answer a wide variety of questions. For instance, how does the transcriptome vary in different normal cell types in a multicellular organism? How does the transcriptome differ between normal and cancer cells, and how does the transcriptome change in cancer cells as a cancer progresses? How does the transcriptome vary at different stages of development as an organism progresses from embryo to adult? How does virus infection alter the transcriptome?

**iActivity**

In the iActivity *Personalized Prescriptions for Cancer Patients* on the student website, you are a researcher at the Russellville clinic trying to determine the gene expression profile for a patient with cancer.

**Pharmacogenomics.** One very promising area involving genome-based gene expression research is **pharmacogenomics.** The word *pharmacogenomics* is a blend of "pharmacology" and "genomics"; it is the study of how an individual's genome affects the body's response. That is, medicine operates mostly on the assumption that all humans are the same, and pharmaceuticals are administered to treat diseases based on that assumption. However, a variety of factors affect a person's response to medicines, notably the genome (including the expression of that genome), as well as nongenetic factors such as age, state of health, diet, and the environment. The promise of pharmacogenomics is that drugs may be customized for individuals—that is, adapted to each person's genome.

Research in pharmacogenomics is based in biochemistry (a major component of pharmaceutical sciences) enhanced with information about genes, proteins, and DNA polymorphisms. The goal is to develop drugs based on the RNA molecules and proteins that are associated with genes and diseases. If successful, the drugs used to treat an individual would be much more specifically tailored to the misexpression observed in the diseased cells than is presently the case. This would mean that the therapeutic effects of the drugs would be maximized, while at the same time the side effects would be minimized. Moreover, drug dosages would be tailored to an individual's genetic makeup; that is, taking into account how and at what rate a person metabolizes a drug. Presently, dosages are decided upon largely on the basis of weight and age.

Pharmacogenomics is a relatively young area of research at the moment, so mostly there is a lot of promise but very few demonstrated successes. One productive area of research concerns the cytochrome p450 (CYP) family of liver enzymes. The gene *CYP2D6* (OMIM +124030) encodes a polypeptide called debrisoquine hydroxylase, an enzyme that is responsible for the metabolic removal of a great many drugs introduced into humans. Drugs used to treat a wide variety of disorders, including depression and other mental disorders, nausea, vomiting, motion sickness, and heart disorders are broken down by these proteins, as are opiate family members like morphine and codeine. However, variations in the genes that encode these enzymes result in enzymes with different abilities to metabolize particular drugs. That is, there are more than 70 known alleles of *CYP2D6*, and, depending on an individual's genotype at this locus, he or she may be a poor metabolizer (such as people who make no functional debrisoquine hydroxylase), an intermediate metabolizer (such as people who carry one null allele and one allele that encodes a crippled version of debrisoquine hydroxylase), an extensive metabolizer (such as people who carry at least one fully functional allele), or even an ultra-rapid metabolizer (such as people who carry more than the normal number of copies of the gene as a result of gene duplication events). The metabolic profile of a patient is of critical importance in determining appropriate dosage. That is, a poor metabolizer is likely to be at greater risk of harmful side effects or overdose because the body clears the drug poorly, while an ultra-rapid metabolizer will probably need a higher dose to benefit from a drug due to their increased ability to modify and remove the drug.

Another exciting pharmacogenomics development concerns chemotherapeutic drugs, the drugs used to kill cancer cells. One study involved diffuse large B-cell lymphoma patients. Diffuse large B-cell lymphoma is one of many cancers of the lymphatic system collectively classified as non-Hodgkin's lymphoma, a common class of lymphoma (Figure 9.8). Diffuse large B-cell lymphomas make up a significant number of all diagnosed non-Hodgkin's lymphomas, and as such, this is a fairly common cancer.

Figure 9.8
**A diffuse large B-cell lymphoma of the epididymis (*) and testis (arrow).**

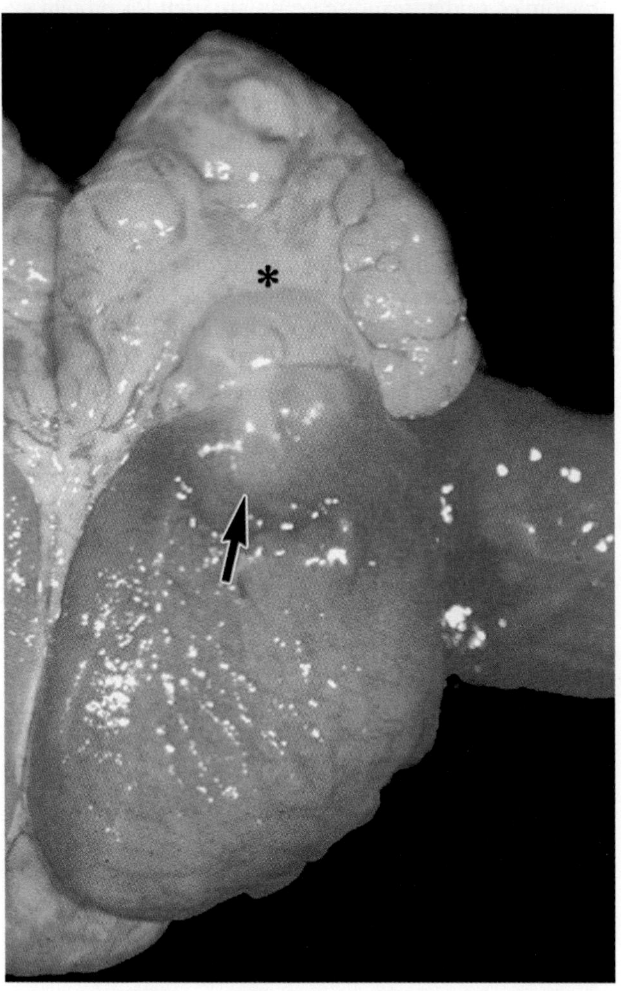


Left untreated, it is a rapidly fatal disease. Investigators studied the transcriptomes of diffuse large B-cell lymphomas in a group of patients and related the transcriptomes to the effectiveness of chemotherapy treatment. For these patients, the cancer type had been identified by histological analysis. Tumor samples were collected, frozen, and stored before any chemotherapy was started. All of the patients underwent similar treatment with the same chemotherapy drugs. Some of the patients responded well to chemotherapy and their cancers went into remission. Other patients had tumors that were less affected by the initial chemotherapeutic treatment, and most of those patients died. When the stored tumor samples were studied, it was determined that all of the tumors that responded to treatment had a similar transcriptome. Likewise, all of the nonresponsive tumors had a similar transcriptome but, importantly, the two transcriptomes were different. This means that the responsive tumors expressed a different set of genes than did the nonresponsive tumors. Thus, even though all of the cancers looked the same histologically, the DNA microarray results showed that there were two very different tumor types at the molecular level, and only one of them responds to the current chemotherapeutic treatment. Most significantly, the results show that the DNA microarray is a more sensitive diagnostic tool than is the classic histological analysis. Therefore, if the transcriptome of a newly diagnosed diffuse large B-cell lymphoma tumor can be determined quickly, the appropriate treatment path can be followed. That is, if the transcriptome is that of the responsive tumors, it can be treated with the standard chemotherapy drugs, since tumors of this type tend to respond well to this regimen. However, if the transcriptome is that of the nonresponsive tumors, patients can be subjected to other, more aggressive treatments. Similar tests are under development for other cancer types, and, in the near future, it seems likely that a DNA microarray will be one of the first tests performed on a newly diagnosed cancer.



**Keynote**

The transcriptome is the complete set of mRNA transcripts in a cell. Transcriptomics, the study of the transcriptome, involves characterizing the transcriptome in cell types and organisms, and determining how it changes quantitatively as the cell changes. An example is understanding how gene expression changes in cancer cells. An overarching goal of transcriptomics is to understand gene expression at a global level. The technique for analyzing the transcriptome is the DNA microarray.

**The Proteome.** The proteome is the complete set of expressed proteins in a cell at a particular time. Proteomics is the cataloging and analysis of those proteins to determine when a protein is expressed, how much is made, and what other proteins it can interact with. The approaches in proteomics are mostly biochemical and molecular. The goals of proteomics are: (1) to identify every protein in the proteome; (2) to determine the sequences of each protein and to enter the data into databases; (3) to analyze globally protein levels in different cell types and at different stages in development; and (4) to understand the biochemical functions of all of the proteins in the proteome. Of course, we can use what we learn about the proteome to help us annotate the genome (see Chapter 8, pp. 192–199), and our annotation of the genome will help us understand the proteome.

Identifying and sequencing all of the proteins from a cell is much more complex than mapping and sequencing a genome. Craig Venter's Celera Genomics company is also playing a large role in this area, as it did in genome sequencing, working hard to speed up dramatically the identification and sequencing of proteins and the computer analysis of the data. In addition, coinciding with the publication of the human genome sequences, a global Human Proteome Organisation (HUPO) was launched. HUPO is intended to be the postgenomic analog of the Human Genome Organisation (HUGO), with a mission to increase awareness of and support for proteomics research at scientific, political, and financial levels.

Proteomics is an extremely important field because it focuses on the functional products of genes, which play important roles in determining the phenotypes of a cell. Of particular human interest are diseases, and proteins and peptides are more intimately related to the actual disease process than are the genes that encode them since, at some level, disease can be viewed as a disruption of normal cellular processes, which means that the cellular proteins are somehow misbehaving. However, the challenges for proteomics are much greater than those for genomics. This may seem counterintuitive since the genome must be larger than the proteome, but recall that many genes encode mRNAs that can undergo alternative splicing and that many proteins can undergo posttranslational modification, so a single gene could theoretically code for many related, but subtly different, proteins. Thus, although there are an estimated 20,000 genes in the human genome, there may be about 500,000 different proteins.

Conventional proteome analysis is by two-dimensional acrylamide gel electrophoresis and mass spectrometry. These procedures are not well suited for analyzing large numbers of proteins at once, and they are not sensitive enough to detect proteins expressed at low levels. Fortunately there is a new, sensitive tool for analyzing large numbers of proteins at once—**protein arrays.** Protein arrays—also called *protein microarrays* and *protein chips*—are similar in concept to DNA microarrays. They are rapidly becoming the best way to detect proteins, measure their levels in cells, and characterize their functions and interactions, all on a very large scale. As such, they are a central proteomics technology, valuable both for basic research and for biotechnology applications. As with DNA microarrays, the use of protein arrays is becoming highly automated, making it possible to do large numbers of measurements in parallel.

Protein arrays involve proteins immobilized on solid substrates, such as glass, membranes, or microtiter wells. At the moment the density of proteins on the arrays is much lower than for DNA on DNA microarrays. However, with technological advances, we can expect the density of proteins in the arrays to increase. As with DNA microarrays, target proteins are labeled fluorescently (e.g., with Cy5 and Cy3 as used for DNA), and binding to spots on the arrays is measured by automated laser detection. The resulting complex data are analyzed by computer. Because of the similarities with DNA microarray technology, the same instrumentation used for analyzing DNA microarrays can be used for analyzing protein arrays.

One type of protein array is the *capture array*, in which a set of antibodies (usually) bound to the array surface is used to detect target molecules, for example in cell or tissue extracts. The antibodies are made either by conventional immunization procedures or using recombinant DNA techniques to make clones from which antibody fragments are made. A capture array can be used as a diagnostic device, for example, to screen for the presence of tumors (detecting tumor-specific markers in extracts of biopsied material). In proteomics studies, capture arrays are used for protein expression profiling, that is, defining the proteome qualitatively and quantitatively. For example, one can quantify proteins in different cell types and different tissues as well as compare proteins in different conditions, such as during differentiation, with and without a drug treatment, and with and without a disease.

In sum, protein arrays are a promising new technology. There are still technological hurdles to be overcome before protein arrays are as useful as DNA microarrays, but in the future we can expect protein arrays to "take off" and become routine for high-throughput analysis of proteins in proteomic studies, and their use will further our understanding of the proteome greatly.

### Keynote

The complete set of proteins in a cell is the proteome, and the study of the proteome is proteomics. The goals of proteomics are to identify every protein in the proteome, to understand each of their functions, to develop a database of protein sequences, and to analyze proteomes in different cell types and in different stages of development.

## Comparative Genomics

**Comparative genomics** involves comparing entire genomes (or parts of genomes) of different species, strains, or individuals with the goal of enhancing our understanding of the functions and evolutionary relationships of each genome. Comparative genomics approaches are also used for determining which organisms or viruses are present in a sample. Comparative genomics is rooted in the tenet that all present-day genomes have evolved from common ancestral genomes. Therefore, studying a gene in one organism can provide meaningful information about the homologous gene in another organism. More broadly, comparing the overall arrangements of genes and nongene sequences of different organisms can tell us about the evolution of genomes. Since direct experimentation with humans is unethical, comparative genomics provides a valuable way to determine the functions of human genes by studying homologous genes in nonhuman organisms. Identifying and studying homologs to human disease genes in another organism is potentially valuable for developing an understanding of the biochemical function and malfunction of the human gene.

In comparative genomics studies, genomes from two or more species, strains, or individuals are analyzed with the goal of defining the extent and specifics of similarities and differences between sequences, either gene sequences or nongene sequences. An obvious question that comparative genomics can address is the evolutionary relationships between two or more genomes. For example, as we discussed earlier, complete genome sequence analysis affirmed the evolutionary relationships and distinctions among the Bacteria, Archaea, and Eukarya. (You will learn about the use of comparative genomics to understand evolutionary relationships in Chapter 23, "Molecular Evolution.")

## Examples of Comparative Genomics Studies and Uses

Comparative genomics approaches have become incredibly powerful as multiple genome sequences are completed. Some recent studies that show some of the power of this approach are discussed in this section.

**Finding the Genes That Make Us Human.** The chimpanzee genome was compared to the genomes of the mouse and rat to find regions where at least 96 of 100 bases were perfect matches. Over 30,000 such regions were found. Presumably, natural selection has acted on these regions to select against most changes, since chimpanzees and mice do not share a recent common ancestor. Researchers then compared these regions to the human genome, looking for the small set of regions that were similar in mouse, rat, and chimp, but significantly more dissimilar in humans. If the DNA region is strikingly similar in the other mammals, it can be assumed that this region plays an important role and that most changes are harmful. However, if it has changed in humans, this change presumably occurred in the 6-million-year period since humans last shared a common ancestor with the chimpanzee, our nearest relative, and this small set of genes might help explain the changes that occurred as modern humans evolved. One of the genes identified in this analysis was named *HAR-1,* for *h*uman-*a*ccelerated *r*egion *1*. The chimpanzee *HAR-1* gene is nearly identical to the chicken *HAR-1* gene, with exact matches at 116 out of 118 bases. This means that only two bases changed in about 310 million years since chimpanzees and chickens shared a common ancestor. However, only 100 of 118 bases match in the human and chimpanzee *HAR-1* genes. This region of the human genome has clearly changed a great deal in the last 6 million years.

The *HAR-1* gene encodes a small, noncoding RNA, but it does not seem to be a small regulatory RNA that regulates gene expression, so the precise function of the gene is as yet unknown. When the investigators looked for the RNA encoded by *HAR-1* in sections of developing brains, they found that it is expressed in a region of the brain that undergoes a unique developmental process in humans, unlike the developmental processes seen in other primate brains. The same cells that express *HAR-1* also express the protein reelin. This protein is known to regulate proper development of the cortex of the brain. Ongoing studies are being done to define the function of the RNA encoded by *HAR-1*, and to determine if there is any functional significance to the coexpression of the *HAR-1* gene and the gene for reelin in the same cells.

Several other key human genes have been identified in other comparative genomics screens, including the genes encoding the proteins FOXP2 and ASPM. The FOXP2 protein seems to play a critical role in speech production, while the ASPM protein regulates brain size. Presumably other genes changed as we evolved, and identification and study of these genes will help us understand how we differ from our nearest relatives. The Focus on Genomics box for this chapter describes an experiment with similar goals—the sequencing of the Neanderthal genome to find how their genome differs from ours.

**Recent Changes in the Human Genome.** An analysis of the human haplotype map (discussed in Chapter 8, p. 193) looked for regions that had undergone rapid changes after human populations split. In this case, the investigators studied *linkage disequilibrium.* **Linkage disequilibrium** describes the condition when specific alleles of two or more different genes tend to appear together more frequently than random chance predicts.[2] If a new mutation occurs in a population and creates a new allele of a specific gene, it will be associated with a specific set of haplotypes because the new alleles will be flanked by specific SNP (single nucleotide polymorphism) alleles. (See Chapter 8, pp. 192–193, for information on SNPs and haplotypes.) This set of haplotypes is also called a *haplotype block*. Recall that each haplotype is a set of SNP alleles that are rarely rearranged by recombination, so a haplotype block is a series of neighboring haplotypes. Genetic recombination within the small region defined by this haplotype block can occur but is very rare. When the haplotype block carrying the new allele is passed from parent to child, the new allele will also segregate with the haplotype block. This is linkage disequilibrium, and it will tend to persist for many generations until very rare recombination events scramble the association of the haplotypes in the haplotype block and the new allele.

Researchers looked for large haplotype blocks that were very common in one or more populations. A large haplotype block is almost certainly of recent origin because genetic recombination will remove some of the haplotypes from the haplotype block at only a very slow rate. These large haplotype blocks almost always correspond to regions that have undergone positive selection in the recent past. In other words, some mutation in the region conferred a selective benefit, and carriers of this mutation (and the haplotype block associated with it) tended to have more offspring. These offspring also carried the mutation and the associated haplotype. First, the researchers compiled

[2]The discussion of linkage disequilibrium here focuses on the case where a high degree of linkage disequilibrium indicates genetic linkage. In Chapter 21, you will learn that high linkage disequilibrium can result in other ways.

## Focus on Genomics

### The Neanderthal Genome Project

Our closest relative was Neanderthal man, now extinct for about 28,000 years. Fossil evidence suggests that modern man and Neanderthals coexisted for quite some time. About 50,000 years ago, Neanderthals appeared to make a cultural advance. The archaeological record shows that they made more use of symbols and that their culture became more complex in Europe, Africa, and Australasia. It would be fascinating to know how similar their genome was to ours, and to know if the two groups intermixed in their history. This task is becoming possible, as genomics techniques become more and more sensitive. A small fraction of Neanderthal remains still contain DNA, although the DNA is highly degraded. One sample, found in the Vindija Cave in Croatia, is about 38,000 years old but still contains enough DNA that scientists were able to sequence over 1 million base pairs of Neanderthal DNA. The techniques used were very sensitive. The researchers had to look over all of their data carefully to remove contaminating human DNA that came from the archaeologists and the investigators themselves, and to account for degradation of the DNA. As you learned in Chapter 7 (p. 138), DNA tends to undergo deamination reactions. In living cells, these are mostly repaired because these reactions create bases that do not belong in DNA. Once the cell is dead, deamination of cytosine, which creates uracil, is unrepairable, and will cause errors in the sequencing reaction, where the deaminated `C` is interpreted to be a `T`. Most of the fragments isolated were very short, with the average piece being only about 60–200 base pairs long. Nonetheless, the scientists were able to compare the Neanderthal sequences to those of the human and the chimpanzee. Like the human genome, the chimp genome is sequenced fully, and chimps are the closest living relative of humans. Based on the success of the sequencing, these scientists have decided to pursue sequencing the entire Neanderthal genome, a task they believe is possible with about 20 grams of bone (they were able to get 1 million base pairs from 0.1 grams of bone). What did they learn from the preliminary data? Comparisons of the three genomes allowed them to estimate how long ago we diverged from our Neanderthal relatives. Most of their models suggested a divergence about 0.5 million years ago, with Neanderthals being much more similar to us than either group is to chimps. Another group was able to clone and study the gene *FOXP2*, a gene known to play a role in the ability to speak. Chimps and humans differ at only two amino acids in the FOXP2 proteins, but this is an important difference between humans and chimps—defects in this gene tend to result in profound difficulties with speech and language. One group of scientists sequenced the *FOXP2* gene from Neanderthal DNA and found that it was identical to ours, and unlike that of chimps, so it is possible that Neanderthals spoke a more complex language than we had imagined. Other scientists have analyzed both human and Neanderthal DNA to estimate how "clean" the split was between the two species, and the results have been mixed. Some studies suggest that very little mixing occurred, while other studies have suggested that at least some **introgression** (transfer of genes across species barriers) has occurred.

haplotype information from individuals from different, isolated human populations. They collected data from 89 members of an Asian population (a mix of Japanese and Han Chinese individuals), 60 Africans (all Yoruba from Nigeria), and 60 individuals of northern and central European ancestry. They then looked for, and found, specific haplotypes that spanned a much larger region of DNA than most of the other haplotypes, and that were relatively common in at least one of the populations. The thinking was as follows. If a rare haplotype conferred no benefit, it would spread very slowly, if at all, in a population and either would never become common or would become common only after a very long time. On the other hand, if a haplotype contains an allele that confers a benefit, both the haplotype and the allele will tend to become more common in the population because of positive selection. A region that confers a selective benefit can become common very rapidly in a population. Linkage disequilibrium tends to disappear over time as recombination trims away haplotypes from the haplotype block, so a large, common haplotype block probably contains an internal mutation of recent origin that is undergoing positive selection in the population. The large, common haplotype blocks that the investigators found had presumably undergone recent positive selection in one or more of the tested populations.

The investigators then set out to see what gene or genes were present in this region of DNA in the large haplotype blocks they had identified. Since each haplotype region contains, on average, about a million base pairs, typically a number of protein-coding genes will be present in each of the blocks. The investigators attempted to identify the gene or genes in the region that might have been the target of the positive selection. In some cases, this was relatively simple. For instance, in the European population, one candidate region contained the gene encoding the enzyme lactase. This enzyme breaks down milk sugars in the gut and is normally neither transcribed nor translated in adult mammals. Several human populations, including most European populations, have relied on milk from domesticated cattle as a major food source and consume dairy products well into adulthood. A person without active lactase is lactose-intolerant, and will feel quite ill after consuming milk. In a population where dairy is not consumed, there is no benefit associated with a mutation that allows lactase to be expressed throughout life, while in a culture with domesticated cattle, this mutation would allow the carrier access to a new food source. Thus, this region has probably undergone recent selection associated with expression changes in the lactase gene.

Several other selected haplotype blocks were identified, and many contained genes thought to play a role in olfaction, sperm function, gamete development, and fertilization. All of these gene classes have been seen to be targets of selection according to other studies comparing human and chimpanzee genomes. The study also identified other large haplotype blocks that are common in Europeans. These contained genes that regulate skin and eye color. One of the haplotype blocks they found is associated with the allele for blue eyes discussed in the Focus on Genomics box in Chapter 8, p. 195. This presumably relates to the selective loss of normal pigmentation as humans spread to Europe. They also found that a haplotype block containing a gene for the metabolism of the sugar mannose has undergone recent selection in Yoruba populations, while other haplotype blocks that have undergone positive selection in Asian populations contain genes that encode proteins for the metabolism of sucrose. Haplotypes containing cytochrome genes, which encode proteins involved in detoxification of various chemicals, have also undergone recent selection in particular populations. Presumably, these changes reflect selective pressures, probably imposed by dietary differences, in the different groups.

This analysis, and others like it, will help us find the genetic changes that have been critical in human (or any other organism with a haplotype map) adaptation. We could look for mutations that conferred resistance to an epidemic disease, like the plague or typhoid, in the past. We could also look for the mutations that allowed us to domesticate and modify animals and crop plants. For instance, we could use the bovine (cow) haplotype map to find the mutations that increased milk production, or we could use the rice or wheat haplotype maps to find the mutations that increased grain production as we domesticated these crops.

**Characterization of Gene Amplifications and Deletions in Cancer Using DNA Microarrays.** The genome tends to become unstable in cancer cells, accumulating a number of mutations. These mutations can affect a single base pair, creating a point mutation, or can change the copy number of a gene, a part of a gene, or a larger fragment of the chromosome. Deletions and duplications are common among copy number changes, both in random regions of the genome and in areas with genes. (Deletions and duplications are discussed in more detail in Chapter 16, pp. 464–468.) Particular genes regulate cell growth and division, and altering their copy number can stimulate a cell to follow a path to unregulated growth and division, a characteristic of cancer. For instance, if a gene encodes a polypeptide that functions to slow cell division, deletion of this gene might confer a growth advantage to a tumor cell. In contrast, if a gene encodes a polypeptide that promotes cell division, then duplication or higher amplification of that gene, with a corresponding increase in the amount of protein made by the gene, could result in the tumor cell growing more rapidly than its neighbors.

Michael Wigler and Robert Lucito have developed a method to identify genomic copy number variation in cancer and in other diseases in which a change in gene copy number is characteristic. The method called *representational oligonucleotide microarray analysis*, or ROMA, is a comparative genomics approach in that whole genomes are compared. Figure 9.9 illustrates the use of ROMA to identify genes with altered copy number in cancer cells. First, clinicians biopsy a tumor. Genomic DNA isolated from the tumor cells is digested with a restriction enzyme such as *Bgl*II that leaves a single-stranded overhang (see Chapter 8, p. 174). A single-stranded adapter molecule is ligated to each end of all the restriction fragments (see Figure 9.9, inset). The adapter is designed with a sequence at one end that is complementary to the overhang sequence of the restriction fragments. The rest of the adapter is a sequence that is complementary to a primer designed to amplify the restriction fragment using PCR. That is, adding the same adapter sequence at the two ends of each restriction fragment enables all the restriction fragments in the mixture to be amplified by PCR by using the same PCR primer. During the PCR amplification step, the restriction fragments are labeled with Cy5 (red) to create the labeled target DNA for microarray analysis. For a control, a sample of normal (noncancerous) tissue from the same individual is obtained and taken through the same steps, except that in this case the amplified restriction fragments are labeled with Cy3 (green).

The two target DNAs are now mixed and added to a DNA microarray containing oligonucleotide probes (~70 nucleotides in length) representing thousands of individual human genes (see Figure 9.9). As we have described before for DNA microarray analysis, the labeled target DNAs pair with the unlabeled oligonucleotide probes with which they are complementary. The DNA microarrays are then scanned with a laser and the Cy5 and Cy3 labels are quantified. The results indicate if changes in gene copy number have occurred in the tumor (see Figure 9.9). That is, if a spot on the microarray is yellow, Cy5- and Cy3-labeled target DNAs have bound equally, meaning that the copy number of the particular gene represented by the oligonucleotide probe is not changed in the tumor. If a spot is red, more Cy5-labeled (tumor)

Figure 9.9

**Characterizing genes amplified and deleted in cancer cells using representational oligonucleotide microarray analysis (ROMA).**

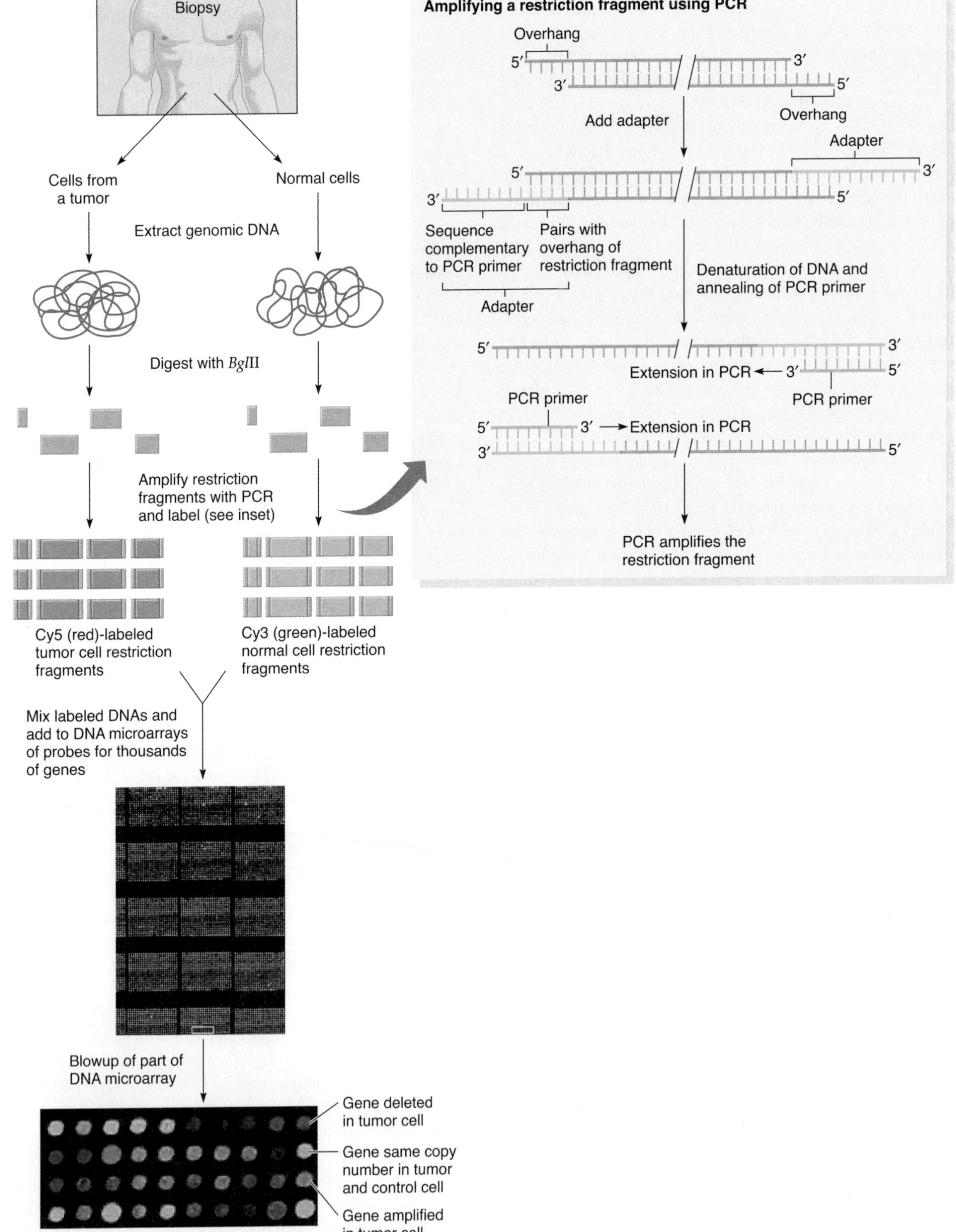

DNA has bound than Cy3-labeled (control) DNA, meaning that the copy number of the gene represented by the probe is increased in the tumor. If a spot is green, more Cy3-labeled (control) DNA has bound than Cy5-labeled (tumor) DNA, meaning that the copy number of the gene represented by the probe is decreased in the tumor.

In sum, the ROMA technique can show if a particular gene or genes is duplicated and/or amplified to a higher-than-normal copy number in a cancer type. The genes so identified can then be studied in more detail to obtain a more complete understanding of the cancer. Moreover, the identified genes with altered copy number are potential targets for the development of new diagnostic procedures and therapeutic strategies for the cancer.

**Identifying a Virus in a Viral Infection Using DNA Microarray Analysis.** A wide variety of viruses cause infections in humans and in animals of veterinary importance. In some cases, the type of virus causing the infection is easy to identify based on symptoms, but for many viruses such identification is challenging. Recently a comparative genomics approach using DNA microarrays has been developed by Joseph DeRisi to make virus identification simple and effective.

Key to the virus identification process is a DNA microarray called a Virochip, which has oligonucleotide probes on it for about 20,000 genes representing the very large number of viruses with sequenced viral genomes. That includes the viruses with which you are likely familiar, such as those that cause herpes, chicken pox, smallpox, warts, and many, many more. When a patient has a virus disease that cannot be diagnosed easily, phlegm, or another body fluid likely to carry cells containing the virus, is collected from an infected tissue. Messenger RNA is isolated from the sample, and reverse transcriptase is used to make cDNA copies of the mRNA. Some of the RNA will be of viral origin, and some will be of host cell origin. By using a dNTP precursor tagged with Cy5 in the reverse transcription step, the DNA copies become fluorescently labeled. (Cy3 could be used instead of Cy5.) These labeled target DNAs are then added to the Virochip. If the virus causing the infection is a known virus, the target DNA will base-pair with one or more probes on the Virochip; this hybridization is revealed by laser scanning as described before for DNA microarray analysis. Which spot or spots are fluorescent indicates which virus or viruses are involved in the viral infection.

Soon after its development, the Virochip was used to characterize a new infection. In 2003, the World Health Organization declared a travel alert for a new disease, SARS (sudden acute respiratory syndrome), a potentially fatal human infection. Just 7 days later, samples from infected patients were tested using the Virochip, and the next day, the investigators determined that SARS patients all had a novel coronavirus. Diagnostic sequences from this particular virus were not present on the Virochip, but the labeled target DNA made from the infected cells hybridized to probes on the Virochip from known coronaviruses. When the investigators compared the sequences to which the SARS sequences hybridized, they were able to reconstruct a section of SARS sequence by determining the sequence similarities of the spots on the virochip. This shows the amazing diagnostic power of the Virochip—identifying an infection by a known virus or determining the identity of a new virus (as long as it is related to known viruses) both quickly and acurately.

**Keynote**

Comparative genomics is the comparison of complete genomes of different species with the goal of increasing our understanding of the gene and nongene sequences of each genome and their evolutionary relationships. Comparative genomic analysis can define genes that are evolving rapidly or genes that have undergone changes as a disease progresses.

**Metagenomic Analysis.** **Metagenomics** (also called **environmental genomics**) is a branch of comparative genomics involving the analysis of genomes in entire communities of microbes isolated from the environment. Essentially it is an extension of genomic analysis of the individual to mixed populations of microbes, bypassing the need to isolate and culture individual microbial species to analyze them. Indeed, we do not know the conditions under which many microbial species will grow in the lab and, therefore, one outcome of metagenomic analysis is the identification and characterization of new species.

In sequence-based metagenomics analysis, an environmental sample is collected and DNA is isolated directly from the sample. This DNA will have derived from all the microbes in the sample, including bacteria, viruses, protists, and fungi. The DNA is then cloned and subjected to whole-genome shotgun sequencing (see Chapter 8, pp. 189–191). Sequences are reassembled (see Chapter 8, p. 191) and, after extensive sequencing and aligning, each microbial organism or virus should be represented by one or more reassembled sequences. How is this possible starting from a mixed population? Recall that the whole-genome shotgun technique described in Chapter 8 uses complex computer algorithms to reassemble chromosomal sequences from small sequence fragments of a single genome. These same algorithms are able to sort out the different genomes, since the DNA from organism A is unlikely ever to have a long enough stretch of bases that perfectly matches the bases in the DNA of organism B. There may be short stretches of nearly perfect matches, but the algorithm can demand longer stretches of perfect alignment than are normally found across species.

Each of the reassembled sequences can then be compared to the DNA sequences in databases. The goal here is to find the closest match(es) in the database. This can help us identify the organisms or the closest relatives of the organisms in our sample.

Another type of metagenomic analysis is function based. In this case, researchers screen the DNA extracted from the environmental sample for genes with specific biological functions, such as antibiotic production. New antibiotics have already been discovered using function-based metagenomic analysis.

One area of metagenomic analysis is focused on the human gut *microbiome*. A **microbiome** is the community of microorganisms in a particular environment. In this case, the environment is the human gut. A human microbiome project has recently been established, with the aim of characterizing the human microbiome, understanding how it changes with the health of the human host, and determining how much variation exists between individual humans and human populations. In one case study focused on bacteria, DNA was collected from the gut microbiomes of two healthy volunteers. The DNA was collected from fecal material, since most of the bacteria in the large intestine will also be present in feces. The analysis of the bacteria did not involve culturing them in the lab, because we know that many of the bacteria in our guts will not survive in lab conditions. Instead, the DNA was sequenced directly. (Typically pyrosequencing [see Chapter 8, pp. 187–189] is the sequencing method of choice for studies like this because it does not require culturing the bacteria as do other methods.) Over 100 million bases of DNA sequence was generated using the gut microbiome DNA as a template, and the sequences were analyzed using the algorithms developed for whole-genome shotgun sequencing. Assembled sequences (these were generally only parts of genomes, not entire genomes) were compared to the databases, and the investigators were able to infer that about two-thirds of the assembled seqeunces contained DNA from members of Domain Bacteria, while about 3% of assembled sequences contained DNA from Domain Archaea, and the remainder could not be clearly identified. Two well-characterized human gut inhabitants, the bacterium *Bifidobacterium longum* and the archaean *Methanobrevibacter smithii*, were both abundant in their samples.

To understand how the gut microbes are related to other, known organisms, the genes encoding the 16S rRNA (the ribosomal RNA found in the small ribosomal subunit; see Chapter 6, pp. 113–114) were amplified from the gut DNA using the PCR, and these DNA fragments were sequenced. The genes encoding rRNAs are used frequently for studying evolutionary relationships because ribosomes are made by all organisms, and some regions of the rRNAs undergo genetic change over time (allowing us to compare them), while other regions are essentially identical in all organisms. (The latter property making the genes easy to amplify by PCR.) If we compare the sequences for the 16S rRNA gene from two species, and these two sequences are highly similar, the organisms probably had a recent common ancestor. In contrast, if the two regions have significant internal differences, they probably diverged from each other long ago, and their common ancestor was farther in the past. The analysis of the 16S rRNA genes in the gut genomic DNA samples identified 72 distinct bacterial sequence types and a single archaean sequence type. The archaean matched the 16S rRNA sequece of *Methanobrevibacter smithii*. Presumably, it came from either *M. smithii* or a close relative. Only 12 of the 72 bacterial sequences corresponded to organisms that had been cultured in the lab, and 16 were unique enough that they must represent previously uncharacterized species. The PCR analysis identified an additional 79 sequence types. Statistical analysis suggested that a minimum of 300 species of bacteria were present in the analyzed stool samples. The analysis of the two samples was not extensive enough to determine exactly how similar the two gut microbiomes were, but significant overlap was noted in the sequences.

The DNA sequences obtained from the intestinal microbiomes were analyzed further to identify ORFs, that is, potential protein-coding genes. Recall from Chapter 8 that the human genome had fewer genes than most scientists had predicted. One partial explanation for this is that many human gut bacteria are beneficial partners, rather than harmful pests. These bacteria synthesize certain chemicals that we then absorb and use, including some vitamins. The interactions between us and our bacterial partners are more complex than we currently understand. We do know that people lacking the normal intestinal microbes have some defects in immune system function and in wound healing. The exact causes for this are not yet fully understood. However, based on other interactions between us and our gut microbiomes, the most likely explanation seems to be that these people lack certain chemicals normally provided by the gut bacteria. When ORFs from the microbiome sequences were compared to genes with known functions, investigators found that the gut microbiome had a significant enrichment of genes coding for enzymes involved in transport and metabolism of carbohydrates, amino acids, nucleotides, and coenzymes compared to the abundance of these genes in the databases. Furthermore, the gut microbiome was enriched in genes coding for enzymes with these activities compared to the human genome. Presumably, our microbiome is enriched in these enzymatic activities because bacteria with these enzymatic abilities are beneficial to their hosts, and furthermore, that we have probably lost some genes that code for these enzymatic activities as we have come to rely on our microbiome to complete certain enzymatic tasks for us.

### Keynote

Metagenomics, a branch of comparative genomics, is the analysis of the genomes of entire communities. At the core of metagenomics analysis is whole-genome shotgun sequencing. Metagenomics can be used to understand complex relationships between organisms in the environment, such as cataloging the microbes and viruses in a particular place or identifying a disease-causing agent.

# Summary


- Describing one (or more) function for each gene found in an organism's genome, including the expression pattern of each gene and how it is controlled, is the goal of functional genomics. Functional genomics involves molecular analysis in the laboratory as well as computer analysis (also called bioinformatics).
- To assign a gene's function by computer analysis, the sequence of an unknown gene from one organism is compared to the sequences of well-characterized genes found in a wide variety of well-studied organisms to identify a similarity between the unknown gene and one of known function.
- A key approach to assigning gene function experimentally is to knock out or knock down the function of a gene and then to determine what phenotypic change or changes occur. Gene knockouts are permanent changes in chromosomal copies of the targeted gene made typically by replacing the normal gene with a disrupted copy (used in many organisms) or by introducing a transposon into the gene (typically used in bacteria). Gene knockdowns do not involve a permanent change in the targeted gene. Instead, RNA interference is used to reduce the level of the mRNA encoded by the target gene.
- The transcriptome is the complete set of mRNA transcripts in a cell and the study of the transcriptome is transcriptomics. The transcriptome changes as the state of the cell changes, so by defining the transcriptome quantitatively, an understanding of cellular function at a global level can be obtained. Typically the transcriptome is studied using a DNA microarray. This technique allows scientists to analyze the expression pattern of thousands of genes at once.
- The proteome is the complete set of proteins in a cell, and the study of the proteome is proteomics. Proteomics seeks not only to identify and catalog all of the proteins in the proteome but also to understand the functions of each protein and to characterize how the proteome varies in different cell types and in different stages of development. Since proteins govern the phenotypes of a cell, a study of the proteome provides much more information about cellular function at a global level than does a study of the transcriptome.
- Comparative genomics involves the comparison of entire genomes (or parts of genomes) from different individuals or species. The goal of comparative genomics is to enhance our understanding of all parts of the genome, including the various functions of the noncoding sequences as well as the RNA- or protein-coding regions of the genome, and this information can be used in many ways. In all cases, two or more genomes are compared to detect subtle, or not so subtle, differences. Comparative genomics can also help scientists develop a better understanding of evolutionary relationships, since all present-day genomes have evolved from common ancestral genomes. This type of comparison can be used to identify genes that are unique to one species, to identify genes that have changed since two populations diverged, to determine the changes in the transcriptome of mutated cells or to detect mutations in specific genes, and even to identify an infectious agent when normal diagnostics fail. Comparative genomics is important for studies of the human genome because direct human experimentation is unethical, so obtaining information about a gene in closely related organisms can inform researchers about the function of the equivalent gene in humans.
- Metagenomic analysis is a branch of comparative genomics that does not just compare two organisms with each other, but instead analyzes entire communities of microbes or viruses. In this approach, all the different types of microbes or viruses in a particular community are identified by the presence of particular gene sequences in a sample of DNA isolated from the community.




# Analytical Approaches to Solving Genetics Problems

**Q9.1** A spontaneous A to G mutation at a previously nonpolymorphic site ($x$) produces two SNP alleles, $x^A$ and $x^G$. Figure 9.A depicts the haplotype block into which $x^G$ was introduced. In it, the extent of haplotypes in the population is represented by white and black segments, SNPs are represented by letters and the SNP superscripts identify nucleotides present in the haplotype block in which the $x^G$ mutation occurred.

**Figure 9.A**

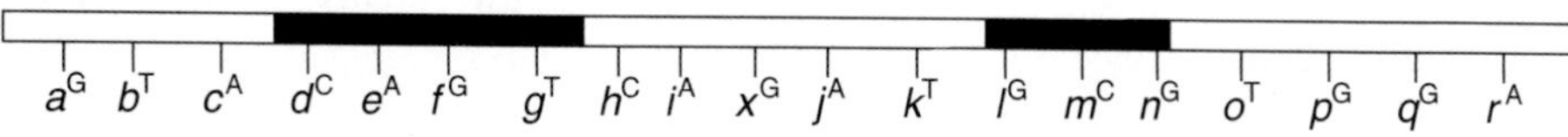

**a.** Explain whether you expect to find a $h^C\, i^A\, x^A\, j^A\, k^T$ haplotype in the population near the time that the $x^G$ mutation occurred. If you do, would you expect it always to be associated with the $d^C\, e^A\, f^G\, g^T$ haplotype?

**b.** The first individual having a $x^G$ allele transmits it to four of his children. If none of his meioses had recombination within the haplotype block shown in the figure, what haplotype would those children receive from him?

**c.** Assume that, over time, there is a low, constant rate of recombination near $x$. After a small number of generations (say, 10 to 20 generations), a set of random events leads to an increase in the frequency of the $x^G$ allele so that it is now found in about 2% of the population. Which SNPs are expected to show the highest amount of linkage disequilibrium with $x^G$?

**d.** Explain how the size of the region that shows linkage disequilibrium with $x^G$ will change under each of the following scenarios.

 **i** Over many generations, the frequency of $x^G$ increases to 40% due to random chance.

 **ii** Over a relatively small number of generations, the frequency of $x^G$ increases to 40% due to positive selection.

**A9.1.** This problem probes your understanding of how recent changes in the human genome are identified. It requires you to understand the difference between a *haplotype*—a set of specific SNP alleles at particular SNP loci that are close together in one small region of a chromosome—and a *haplotype block*—a set of neighboring haplotypes that are seen in an individual. Haplotypes are seen because, within a small chromosomal region, recombination is rare. Not all possible combinations of different alleles at nearby loci are seen in a population, so two (or more) individuals can have a segment of a chromosome with the same set of SNP alleles. Those two individuals share a haplotype in that segment. Now consider the set of SNP alleles present in each individual in a neighboring segment. If these differ, the two individuals have different haplotypes in the neighboring segment—they have different haplotype blocks. However, each of the two haplotypes in the neighboring segment may be found in other individuals in the population.

This problem asks you to reflect on what happens when a new mutation occurs within an existing haplotype block. Since there is a low rate of recombination within a haplotype block, the new mutation will tend to be transmitted along with the haplotype block it originated on—recombination will only infrequently separate it from that haplotype block. At the population level, this results in linkage disequilibrium—the condition when specific alleles at two or more different genes tend to appear together more frequently than random chance would predict. Linkage disequilibrium will be strongest in the region nearest to the new mutation and persist for many generations. However, over time, linkage disequilibrium will decay as recombination gradually separates the mutant allele from specific alleles at nearby loci.

**a.** The problem statement indicates that polymorphism at site $x$ results from the A to G change introduced by the mutation. Therefore, we can infer that only the $x^A$ allele was present in the population before the mutation occurred. The figure shows that the $h^C$, $i^A$, $j^A$, and $k^T$ alleles belong to a haplotype found in the population, so before the mutation introduced a polymorphism at site $x$, the $x^A$ allele must have been part of that haplotype. Therefore, the $h^C\, i^A\, x^A\, j^A\, k^T$ haplotype is part of the haplotype block on which the $x^G$ mutation occurred, and it would be found in the population.

The figure indicates that the haplotype block in this chromosomal region consists of a set of four neighboring haplotypes. Therefore, we can infer that when alleles at the loci in this haplotype block were determined in a population of individuals, a specific set of alleles at the loci within one haplotype were associated with each other, but that set of alleles was not always associated with a specific set of alleles at loci found in neighboring haplotypes. Therefore, though the $h^C\, i^A\, x^A\, j^A\, k^T$ and $d^C\, e^A\, f^G\, g^T$ alleles were associated in one individual, the $h^C\, i^A\, x^A\, j^A\, k^T$ alleles may be associated with a different set of alleles at the $d$, $e$, $f$, and $g$ loci in a different individual.

**b.** The figure indicates that the man with the original $x^G$ mutation has the extended haplotype $a^G\, b^T\, c^A\, d^C\, e^A\, f^G\, g^T\, h^C\, i^A\, x^G\, j^A\, k^T\, l^G\, m^C\, n^G\, o^T\, p^G\, q^G\, r^A$. Since recombination did not occur within the region of the haplotype block, the four children receiving the $x^G$ allele also received this haplotype.

**c.** During a time span that encompasses a relatively small number of generations, recombination will only rarely separate $x^G$ from the alleles at the loci close to it. Consequently, the closer a locus is to $x^G$, the greater the level of linkage disequilibrium between a specific allele at that locus and $x^G$. Therefore, the $h^C$, $i^A$, $j^A$, and $k^T$ alleles should show the highest levels of linkage disequilibrium with $x^G$. Though linkage disequilibrium is expected to decay as the distance of a locus from $x^G$ increases, it may exist throughout a haplotype block if sufficient time has not yet passed for recombination to separate $x^G$ from neighboring loci. This is why a large haplotype block is almost certainly of recent origin.

**d.** **i.** As the frequency of $x^G$ increases in the population over many generations, the chromosome containing $x^G$ will have repeated opportunities to recombine with a variety of chromosomes having different haplotypes. As recombination separates $x^G$ from the alleles at loci closest to it, the size of the region that shows significant linkage disequilibrium with $x^G$ will diminish. Only alleles at the loci closest to $x^G$ will show linkage disequilibrium, and this may not be large.

ii. If $x^G$ confers an advantage that leads to positive selection for it in the population, it might increase in frequency faster than local recombination can reduce the range of linkage disequilibrium between it and specific alleles at nearby loci. In this case, a large haplotype block will remain associated with the new mutation. This is why searching for large haplotype blocks can identify regions that have undergone positive selection in the recent past.

## Questions and Problems



**9.1** What is bioinformatics, and what is its role in functional and comparative genomics?

*__9.2__ What is the difference between a gene and an ORF? How might you identify the functions of ORFs whose functions are not yet known?

**9.3** A dot plot provides a straightforward way to identify similar regions in pairs of sequences. In a dot plot, one sequence is written along the X-axis on a sheet of graph paper, and the second sequence is written along the Y-axis. A dot is placed in the plot whenever the nucleotide in a column on the X-axis matches the nucleotide in a row on the Y-axis.

**a.** Construct a dot plot for each of the following pairs of sequences, and then state where the plot reveals regions of similarity between each pair of sequences.

**i.** GCATTTAGAGCCCTAGTCGTGACAG
ATTCAGTTAGAGCCCTAGCTGATTGC

**ii.** AGCGATTGGTCCTGTACGAGCTAA
GATGCACCTGTACGAGCCTTA

**b.** Consider the results of your dot plots. What are some of the issues that the BLAST program, which performs sequence similarity searches between a query sequence and sequences in a database, must address?

**9.4** The BLAST program can use either a DNA or an amino acid sequence as a query sequence, and can search either a database containing all known sequences or a database with sequences from just one organism. Suppose you are taking a reverse genetics approach to identify the function of your favorite human gene (*YFG*). You have sequenced a *YFG* cDNA, identified its ORF, and translated the ORF to obtain the amino acid sequence of the protein it encodes. For each of the following goals, state: (1) whether you would use the cDNA sequence or an amino acid sequence as a query in a BLAST search; (2) the type of database you would search; and (3) the kind of information you would hope to obtain from the search.

**a.** Identify the sequence coordinates and chromosomal location of *YFG* within the human genome, so that you can determine whether any disease mutations are in that region.

**b.** Identify the approximate locations of the intron and exon boundaries of *YFG*.

**c.** Predict the function of *YFG*.

**9.5** What is meant by a conserved domain? Give an example to illustrate how identifying conserved domains within a protein can provide clues about its function.

*__9.6__ When a DNA fragment from a newly identified bacterium was sequenced and the DNA sequence was used in a BLAST search, the best match was to the *HprK* gene in *Streptococcus pneumoniae*. The *HprK* gene encodes a kinase that regulates carbohydrate metabolism. Can you conclude that the DNA fragment contains a gene encoding a kinase? Can you conclude that the DNA contains a gene homologous to *HprK*? Can you conclude that the DNA contains a gene that functions to regulate carbohydrate metabolism? For any one of these inferences that you cannot make, state why you cannot make it, and what would you do to investigate the issue further.

**9.7**

**a.** What is a *single orphan gene*? What is an *orphan family*?

**b.** In humans, the full name of the *RORC* gene is *RAR-related orphan receptor C*. A BLAST analysis of the RORC amino acid sequence reveals a protein with two domains, a zf-C4 domain (a DNA-binding domain that contains a protein motif known as a zinc finger) and a HOLI domain (a ligand-binding domain found in hormone receptors). The *RORC* gene is similar to a gene in mice that is essential for the formation of lymphoid tissue. Given all of this information, why do you think the *RORC* gene might still be considered to be an orphan gene?

*__9.8__ What information and materials are needed to amplify a segment of DNA using PCR?

**9.9** In the polymerase chain reaction (PCR), a DNA polymerase that can withstand short periods at very high (near boiling) temperatures is used. Why?

*__9.10__ Both PCR and cloning allow for the production of many copies of a DNA sequence. What are the advantages of using PCR instead of cloning to amplify a DNA template?

*__9.11__ If you assume that each step of the PCR process is 100% efficient, how many copies of a template would be amplified after 30 cycles of a PCR reaction if the number of starting template molecules were

**a.** 10?
**b.** 1,000?
**c.** 10,000?

**9.12** Describe the steps you would take to obtain a null allele in your favorite yeast gene (*YFG*) using homologous recombination if you have available a *YFG*$^+$ yeast strain that is sensitive to the antibiotic kanamycin, pBluescript II plasmids (see Chapter 8, p. 176) with the DNA inserts diagrammed in Figure 9.B, and are able to transform yeast with a targeting vector, once you construct it. In Figure 9.B, *Eco*RI, *Hae*II, *Hin*dIII, and *Pst*I are restriction enzymes (see Chapter 8, p. 174) that cleave these DNAs at the sites shown, and the distances between the sites are given in kb.

As part of your answer, diagram the targeting vector you would construct and the structure of the chromosomal region once *YFG* is knocked out using this targeting vector. Also, describe how you would use PCR to confirm that you had obtained a null allele at the gene, and indicate on your diagrams the regions you would use for designing PCR primers. Remember that the absence of a PCR product does not provide strong evidence for a specific DNA arrangement, as a PCR could fail for any number of reasons.

***9.13**
**a.** What are ES cells, and how are they used in generating targeted gene knockouts in mice?
**b.** What is a chimera? How do chimeras arise during the generation of a knockout mouse?
**c.** How can you confirm that an offspring of a chimeric mouse is heterozygous for the knocked-out target gene?

**9.14** After the gene for an autosomal dominant human disease was identified, sequence analysis of the mutant allele revealed it to be a missense mutation. Two alternate hypotheses are proposed for how the mutant allele could cause disease. In one hypothesis, the missense mutation alters a critical amino acid in the protein so that the protein is no longer able to function: heterozygotes with just one copy of the normal allele develop the disease because they have half of the normal dose of this protein's function. In the second hypothesis, the missense mutation alters the protein so that it interferes with a normal process: heterozygotes develop the disease because the mutant allele actively disrupts a required function. How could you gather evidence to support one of these alternate hypotheses using knockout mice?

**9.15**
**a.** In yeast, a gene can be targeted using a target vector with just one selectable marker. In contrast, target vectors used to knockout mouse genes typically have two selectable markers. Why is the second selectable marker necessary?
**b.** Using the target gene diagrammed in Figure 9.5a, describe how you would use PCR to confirm that an ES cell able to grow in the presence of both neomycin and ganciclovir is a transformant resulting from homologous recombination. Specifically indicate the regions you choose for designing PCR primers.

***9.16** Generating gene knockouts using gene-targeting vectors requires the development of experimental approaches that are tailored to an organism—the approach described in this chapter for generating gene knockouts in yeast cannot be used to generate gene knockouts in mice. Describe two experimental approaches for knocking out or knocking down gene function that do not require gene-targeting vectors. Do either of these approaches have the potential to be used in a number of organisms without extensive modification?

**9.17** Systematic screens have been undertaken in some organisms to individually knockout or knock down the function of each of the organism's genes. Summarize the results of these screens, and critically evaluate what we have learned from them.

***9.18** Comparative genomics offers insights into the relationship between homologous genes and the organization of genomes. When the genome of *C. elegans* was

**Figure 9.B**

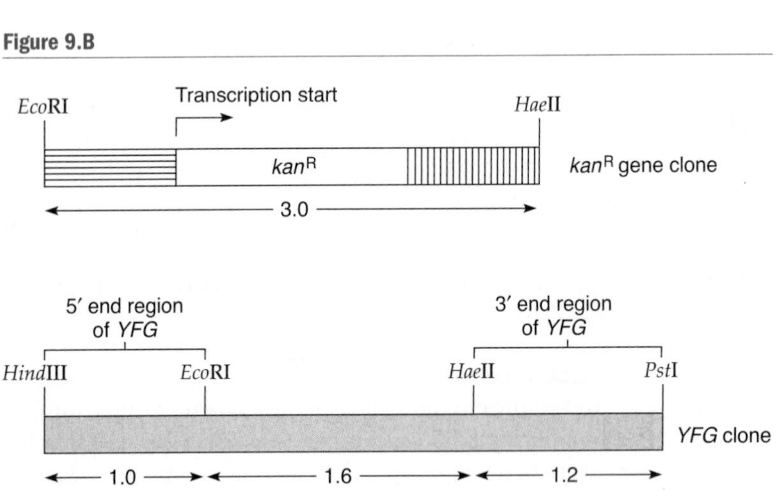

sequenced, it was striking that some types of sequences were distributed nonrandomly. Consider the data obtained for chromosome V and the X chromosome shown below. The following figure shows the distribution of genes, the distribution of inverted and tandem repeat sequences, and conserved genes (the location of transcribed sequences in *C. elegans* that are highly similar to yeast genes).

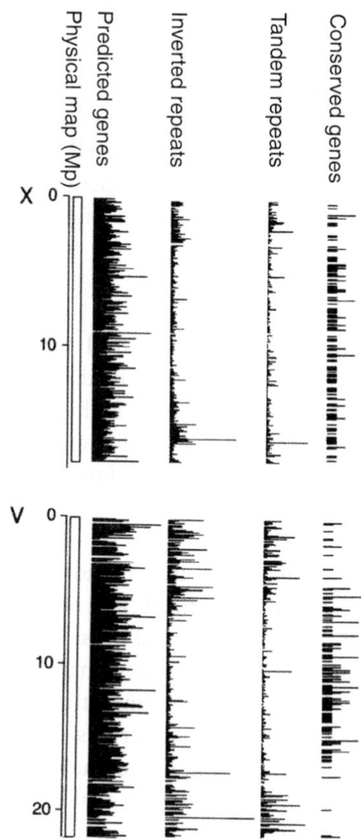


**a.** How do the distributions of genes, inverted and tandem repeat sequences, and conserved genes compare?

**b.** Based on your analysis in (a), what might you hypothesize about the different rates of DNA evolution (change) on the arms and central regions of autosomes in *C. elegans*?

**c.** Curiously, meiotic recombination (crossing-over, discussed in Chapter 12, p. 333) is higher on the arms of autosomes, with demarcations between regions of high and low crossing-over at the boundaries between conserved and nonconserved genes seen in the physical map. Does this information support your hypothesis in (b)?

*__9.19__ How does a cell's transcriptome compare with its proteome?

**a.** For a specific eukaryotic cell, can you predict which has more total members? Can you predict which has more unique members?

**b.** Suppose you are interested in characterizing changes in the pattern of gene expression in the mouse nervous system during development. Describe how you would efficiently assess changes in the transcriptome from the time the nervous system forms during embryogenesis to its maturation in the adult.

**c.** How would your analyses differ if you were studying the proteome?

**9.20** When cells are exposed to short periods of heat (heat shock), they alter the set of genes they transcribe as part of a protective response.

**a.** What steps would you take to characterize alterations to the yeast transcriptome following a heat shock?

**b.** Suppose the transcriptome analyses identify a set of genes whose transcript levels increase following heat shock. How might you experimentally determine which of these genes are required for a protective response following heat shock?

*__9.21__ Pathologists categorize different types of leukemia, a cancer that affects cells of the blood, using a set of laboratory tests that assess the different types and numbers of cells present in blood. Patients classified into one category using this method had very different responses to the same therapy: some showed dramatic improvement while others showed no change or worsened. This finding raised the hypothesis that two (or more) different types of leukemia were present in this set of patients, but that these types were indistinguishable using existing laboratory tests. How would you test this hypothesis using DNA microarrays and mRNA isolated from blood cells of these leukemia patients?

*__9.22__ A central theme in genetics is that an organism's phenotype results from an interaction between its genotype and the environment. Because some diseases have strong environmental components, researchers have begun to assess how disease phenotypes arise from the interactions of genes with their environments, including the genetic background in which the genes are expressed. (See http://pga.tigr.org/desc.shtml for additional discussion.) How might DNA microarrays be useful in a functional genomic approach to understanding human diseases that have environmental components, such as some cancers?

*__9.23__ What is the difference between a DNA chip and a protein chip? How is a protein chip used to analyze the proteome?

**9.24**

**a.** What is a haplotype block, and why do researchers believe that large haplotype blocks have more recent origins?

**b.** In Yoruba individuals from Ibadan, Nigeria, a large haplotype block near the β-globin gene (*HBB*) shows positive selection. While any individual homozygous

**Figure 10.1**
**Cloning in an expression vector.**

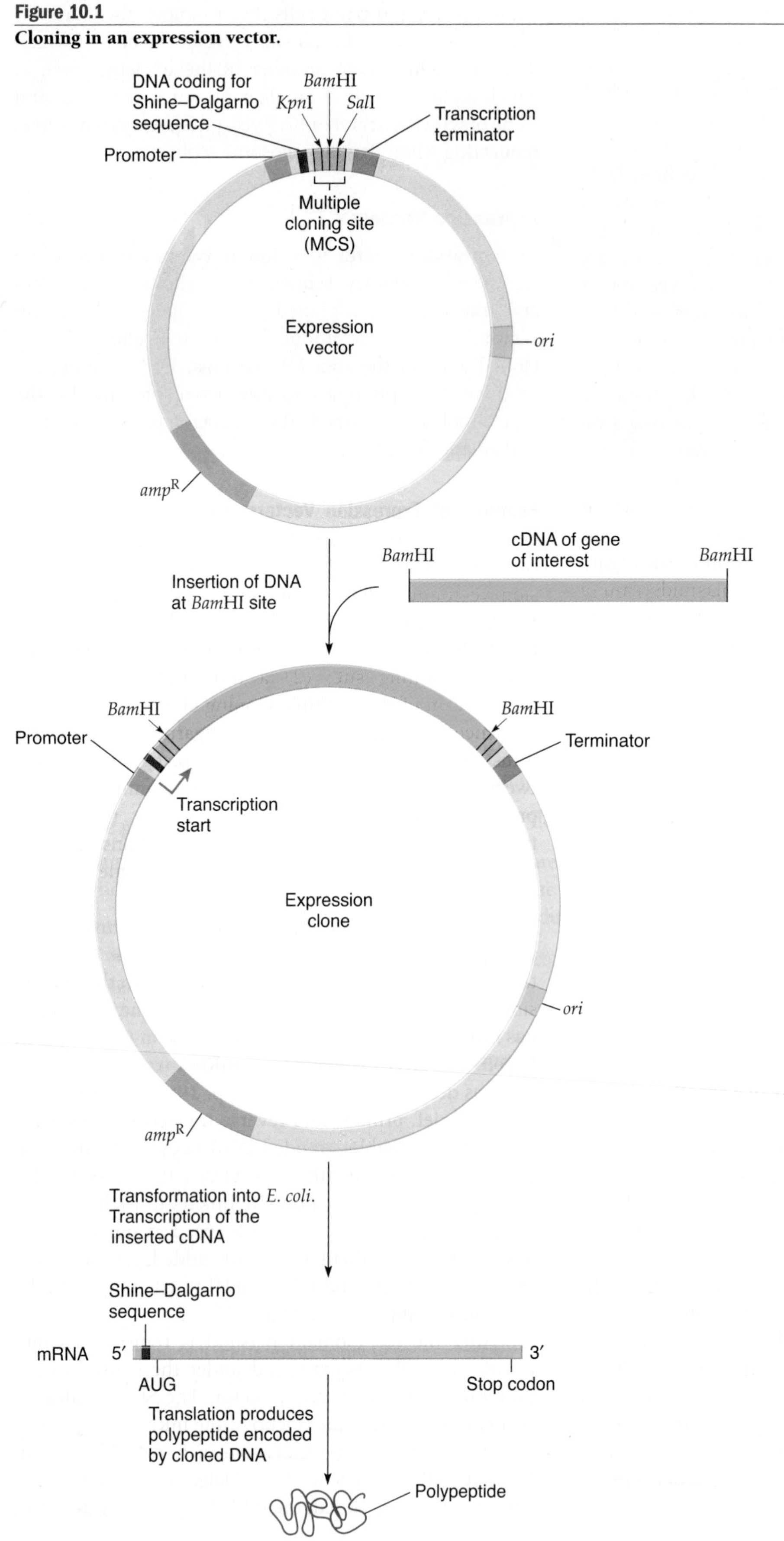

Translation generates the polypeptide encoded by the cloned cDNA.

**Practical Issues for Constructing Clones Using an Expression Vector.** Regardless of the host, the key issue for expressing a gene is inserting the gene into the expression vector so that transcription of the gene produces the mRNA for the desired protein. In the strategy shown in Figure 10.1, the cDNA was inserted into the expression vector by cutting the restriction sites at each end of the cDNA with *Bam*HI, and inserting the digested cDNA into the vector cut at the *Bam*HI site in the multiple cloning site. Practically speaking, the *Bam*HI-digested cDNA can become inserted into the vector in two possible orientations. In the orientation shown in Figure 10.1, the cDNA is in the correct orientation so that transcription produces an mRNA that encodes the desired polypeptide. However, if the *Bam*HI-digested cDNA inserts into the vector in the opposite orientation, the mRNA transcribed from the promoter will be complementary to the correct mRNA. This mRNA does not encode the desired polypeptide.

How can the correct clone be distinguished from the incorrect clone? One way to do this is by DNA sequencing. Typically, expression vectors have binding sites for universal sequencing primers flanking the multiple cloning site. This enables researchers to sequence into the insert DNA of a clone and thereby determine the orientation of the insert. An alternative approach is to use **restriction mapping,** the determination of the number and positions of restriction sites for a restriction enzyme or enzymes in a DNA fragment or clone. The outcome of restriction mapping is a *restriction map* showing the locations and positions of the mapped restriction sites. Restriction mapping uses the tools described in Chapter 8. That is, DNA is digested by restriction enzymes, the fragments are separated by agarose gel electrophoresis, and their patterns and sizes are used to construct the map.

Figure 10.2 illustrates in a theoretical way the use of restriction mapping to distinguish between correct and incorrect clones. The example is based on Figure 10.1. Suppose the cDNA with the *Bam*HI linkers is 2,000 bp long, and that we know from sequencing experiments that there is a restriction site for *Aat*II ("a-a-t-two") 1,800 bp from the beginning of the cDNA. Suppose we clone this cDNA into the *Bam*HI site in the MCS of a 3,500-bp expression vector that has an *Aat*II site 500 bp counterclockwise from that *Bam*HI site. If the cDNA inserts in the correct orientation so that its encoded polypeptide can be expressed, we get the 5,500-bp clone at the bottom left of Figure 10.2. The clone with the opposite, incorrect orientation is at the bottom right of Figure 10.2. *Aat*II digestion of the clones created would enable us to screen the clones to determine which are the correct ones for polypeptide expression. That is, for a correct clone, *Aat*II digestion produces two fragments of 3,200 and 2,300 bp, whereas for an incorrect clone, *Aat*II digestion produces two fragments of 4,800 and 700 bp (see Figure 10.2). These alternative results can be distinguished readily by agarose gel electrophoresis.

All in all, it would be preferable to avoid having to deal with clones with inserts in the wrong orientation if at all possible. As we have just seen, if a single restriction enzyme is used to prepare the insert and the vector, then any given clone can have the insert in either of the two possible orientations. However, if we use two restriction enzymes, we can insert a DNA fragment into a vector in a directional way; that is, a clone with an opposite, incorrect orientation of the insert cannot be created with a two-enzyme approach.

Let us look again at the expression vector shown in Figure 10.1. There is a *Kpn*I ("k-p-n-one") site near the promoter end of the multiple cloning site and a *Sal*I ("sall-one") site in the multiple cloning site near the terminator end. Therefore, if we created a cDNA with a *Kpn*I site added at the start codon end and a *Sal*I site added at the stop codon end, that cDNA can be inserted into the vector digested with *Kpn*I and *Sal*I *only* in the correct orientation for polypeptide expression. That is, the two *Kpn*I sticky ends can pair and the two *Sal*I sticky ends can pair, but a *Kpn*I sticky end cannot pair with a *Sal*I sticky end. This cloning approach is often called *forced cloning*, because we "force" the fragments to connect in only one orientation. It is also called *directional cloning*.

How can we make such a cDNA? Recall that, when we use PCR (the polymerase chain reaction; see Chapter 9, pp. 221–223 and Figure 9.3), we design the ends of our amplified region when we design the primers. Therefore, through the design of the PCR primers, the restriction sites can be added to the ends of the cDNA during DNA amplification (Figure 10.3). The starting point is a cDNA made from a mRNA in a reverse transcriptase reaction (see Figure 8.15). That cDNA is a double-stranded DNA copy of the single-stranded mRNA. By cloning the cDNA as already described, its sequence can be determined. The sequence can then be used to design the two primers for PCR. The left primer is designed to have two regions. A region of approximately twenty nucleotides at the 3′ end can base-pair with the left end of the cDNA (identical to the PCR primers discussed in Chapter 9, pp. 221–223), while the 5′ end of the primer contains the sequence of the *Kpn*I restriction site, which cannot base-pair with the template cDNA (see Figure 10.3). Similarly, the right primer also has two regions. The 20 nucleotides at the 3′ end can base-pair with the right end of the cDNA, while the 5′ end contains the sequence of the *Sal*I restriction, which cannot base-pair with the template cDNA (see Figure 10.3). When these primers anneal to the template (see Figure 10.3) the enzyme will be able to extend using the 3′ end of the primers. However, when the fragment produced by this extension is in turn used

Figure 10.2

**Theoretical example of restriction mapping to confirm that a correct plasmid clone has been constructed.**

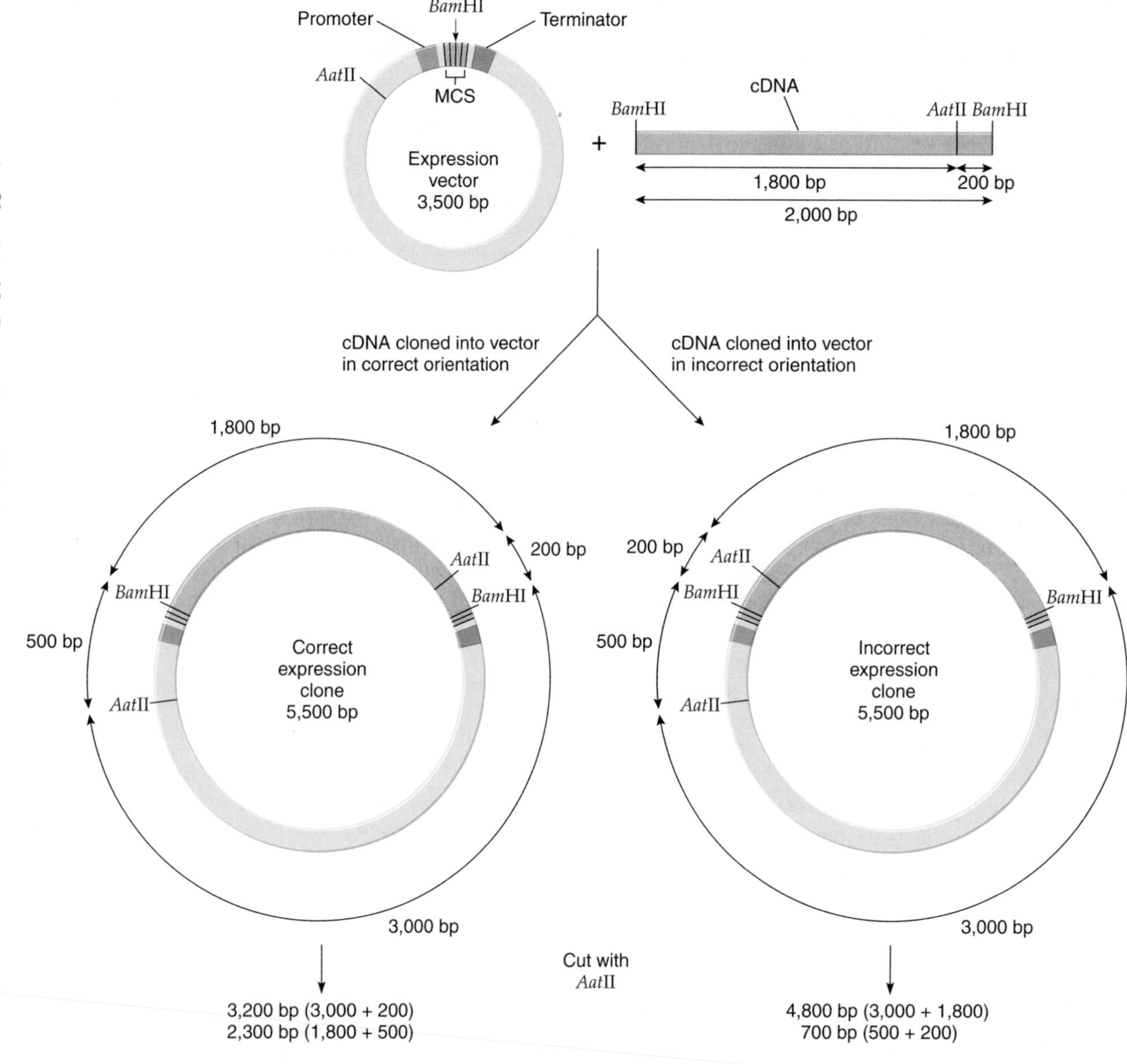


as a template for the next round of PCR, extension will make a sequence complementary to all of the primer, even the parts that did not initially anneal to the template. The amplified PCR products can be cut with *Kpn*I and *Sal*I, creating one large fragment (the cDNA) with two different sticky ends and two very small fragments. The large fragment is purified and inserted into the expression vector cut with *Kpn*I and *Sal*I. That digestion also produces a large fragment and a small fragment (the part of the multiple cloning site between the *Kpn*I and *Sal*I restriction sites). The large vector fragment is purified and then ligated with the amplified cDNA fragment to produce the correct clone for polypeptide expression in *E. coli*.

## PCR Cloning Vectors

It may seem simple to clone a fragment produced by the polymerase chain reaction because you would assume that these fragments are blunt ended. But, in fact, some of the thermostable DNA polymerases commonly used in PCR create an overhang. In most of these cases, the enzyme adds an unpaired A nucleotide at the 3′ ends of the DNA made during PCR that is not specified by the template DNA. In essence, PCR fragments generated with these enzymes have what can be thought of as a tiny sticky end. Unfortunately, no known restriction enzyme creates a sticky end that works with this single A overhang. Some commercially available vectors are designed to work with these sticky ends. These vectors are delivered in linear form with a

**Figure 10.3**

**Use of specially designed primers in the polymerase chain reaction (PCR) to create restriction sites at the ends of a cDNA to be cloned into an expression vector.**

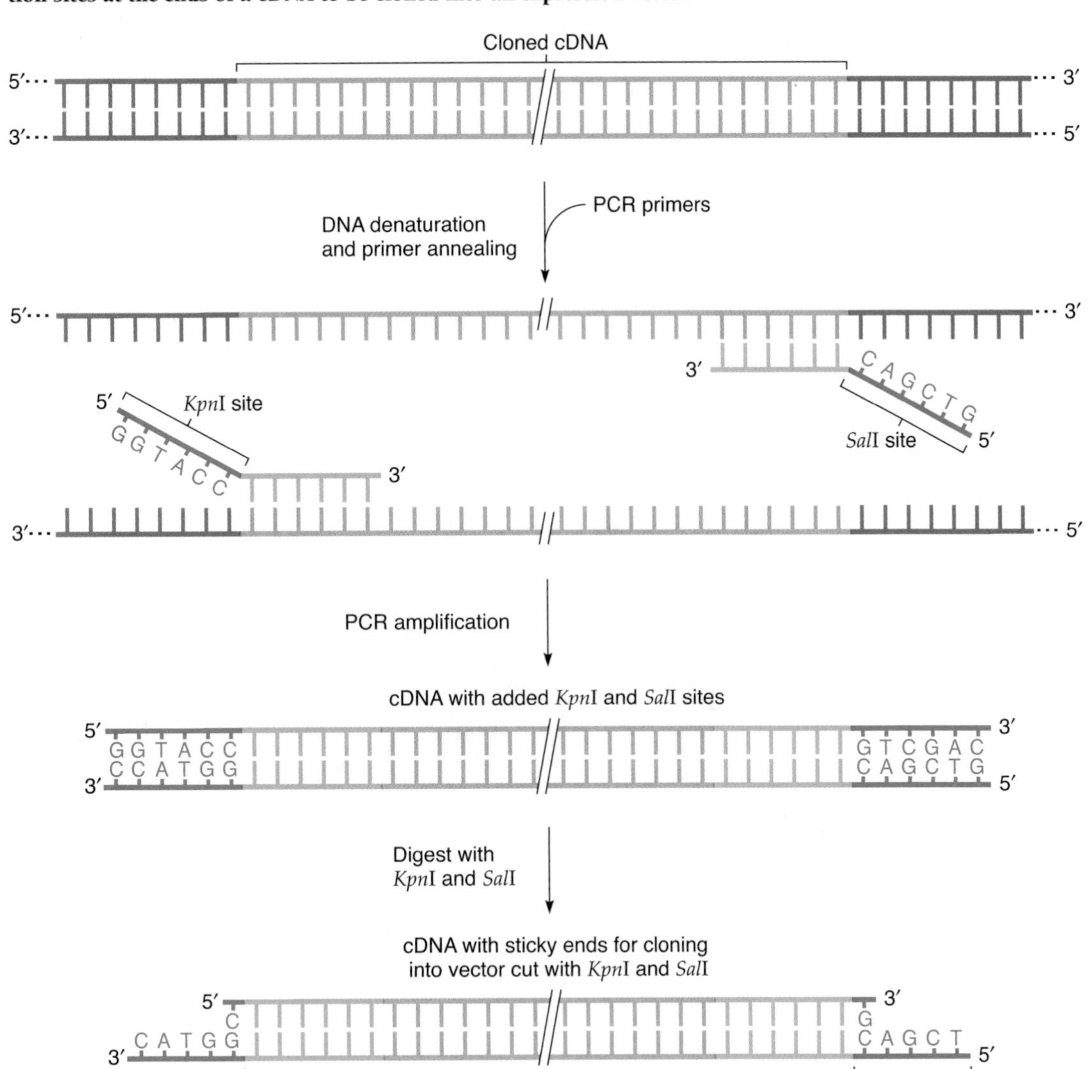


single T nucleotide overhang at each 5′ end. The vectors cannot circularize in a ligation reaction, but a PCR fragment with a single A nucleotide overhang can be inserted into the vector to make a circular recombinant DNA plasmid.

## Transcribable Vectors

A **transcribable vector** is a plasmid vector that has a promoter for an RNA polymerase just upstream of the multiple cloning site (Figure 10.4). The other features of plasmid cloning vectors (such as the one shown in Figure 8.4, p. 176) are generally also present if the vector will be carried by a host cell. Transcribable vectors are designed for the transcription of the insert *in vitro* and, for some systems, also *in vivo*. By contrast, expression vectors are designed only for *in vivo* expression of a cloned gene. The promoters of transcribable vectors, therefore, are chosen for efficient transcription of a cloned gene *in vitro*. Typically the promoter is from one of three bacteriophages, T7, T3, or SP6. The promoter shown in Figure 10.4 is for T7 RNA polymerase. Why use a bacteriophage RNA polymerase? The answer is that these enzymes are highly active and can synthesize a lot of RNA in a relatively short time.

To transcribe a cloned gene from a transcribable vector *in vitro*, a purified sample of the plasmid is digested with a restriction enzyme at a site in what remains of the multiple cloning site downstream of the gene (see Figure 10.4). This is done because the phage RNA polymerase works more efficiently *in vitro* if the plasmid is not supercoiled, as is the case for an undigested plasmid. T7 RNA polymerase, NTPs, and a buffer are added, and the reaction is incubated at 37°C. The mRNA transcripts are synthesized beginning

**Figure 10.4**

**A transcribable vector containing a cDNA insert.** The transcribable vector shown has a T7 promoter immediately adjacent to the multiple cloning site (MCS). Transcribable vectors can be used either *in vitro* by linearizing them and adding the appropriate RNA polymerase (T7 RNA polymerase in this case) and NTPs, or *in vivo* by transforming them into a host cell (*E. coli* here) that expresses the appropriate RNA polymerase.

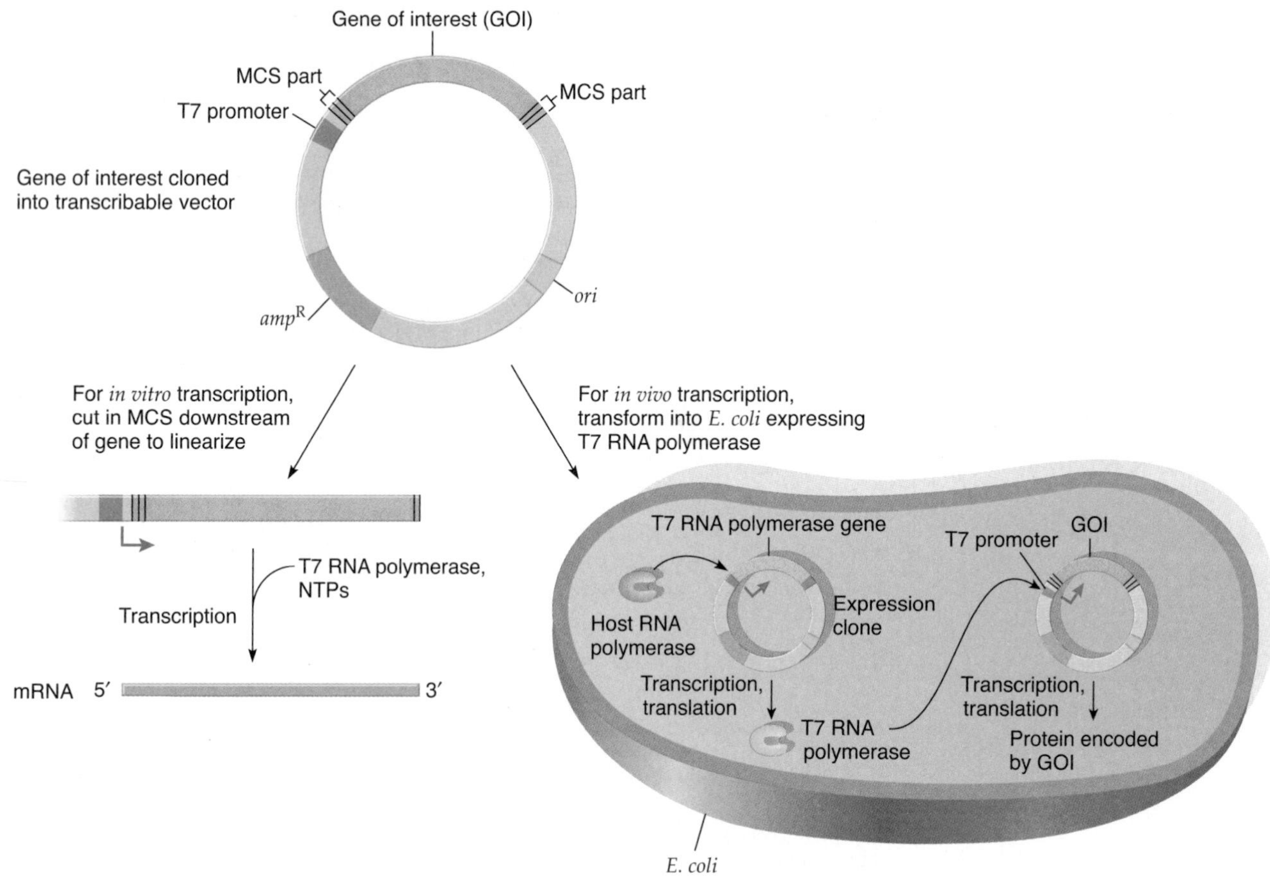


at the nucleotide just downstream of the promoter and ending at the end of the linearized plasmid; that is, just downstream of the end of the cloned gene.

The mRNA molecules made in this way are used for different purposes. In one use, they are added to a cell-free, *in vitro*, translation system to synthesize the polypeptide encoded by the cloned gene or cDNA. A cell-free translation system is a purified mix of the amino acids, proteins, tRNAs, and ribosomes needed for translation, but lacking any mRNAs. Adding mRNAs sets the translation system in operation. In another use, the RNA made is used as a probe—called a *riboprobe*—in various analytical techniques (some of these techniques are presented later in this chapter). In this case, the RNA transcribed from the vector must be labeled, either radioactively or nonradioactively. This is achieved by including in the transcription reaction radioactive or modified NTPs to add the label. For example, for radioactive labeling, $^{32}$P-NTPs commonly are used.

The gene or cDNA cloned in a transcribable vector can be expressed *in vivo* if the clone is transformed into a cell that expresses the RNA polymerase specific to the promoter in the vector (see Figure 10.4). For instance, by transforming the clone into *E. coli* that contains, in addition, an expression vector with the gene for T7 RNA polymerase, the gene in the transcribable vector can be transcribed specifically. That is, the T7 promoter is specific for the T7 RNA polymerase, making transcription of the cloned gene dependent on the synthesis of the T7 RNA polymerase. In this case, transcription occurs from the circular, supercoiled plasmid. As mentioned earlier, the high activity of the T7 RNA polymerase leads to high levels of transcripts. Moreover, since only the transcribable vectors transformed into the cell have the T7 promoter, there is great specificity for transcription since all the T7 RNA polymerases made will transcribe the cloned gene. The mRNA transcripts made in this way are then translated by the cellular translation machinery to produce large amounts of the encoded polypeptide. *In vivo* transcription of a gene is possible in this way in any cell type in which the T7 RNA polymerase can be introduced and expressed.

### Non-Plasmid Vectors

A great many other vectors are available for specific purposes. Many expression vectors are based on phage lambda rather than plasmids. In many ways, plasmid vectors are easier to work with than phage vectors—phage vectors lack extensive multiple cloning sites, and blue-white selection and ampicillin resistance are not used in phage cloning, for instance. However, there is one major advantage to phage that merits their use in genomic, cDNA, and expression library applications. Phage clones are propagated in a different manner than plasmid clones. Plasmid-containing bacteria form colonies when grown on agar plates. The phage vector regions contain all the genes required for the clone to lyse (or kill) the host cell, but does not have the selectable markers we have discussed. This means that a cell carrying a phage clone will be killed by the phage, and will release about 300 copies of the clone, which then infect neighboring cells.

It may seem counterintuitive to build a clone that kills the host cell, but this can be advantageous. Since host cells are constantly killed by the clone, we start with a lawn of bacterial cells. A *lawn* is the term used to describe the appearance of a plate with bacterial cells covering the entire available surface of the plate. Unlike the previous plates discussed, the cells in the lawn are mixed with some warm, molten agar, and then this mix is poured onto an agar plate. As a result, the bacterial cells are embedded in the top few millimeters of agar, rather than sitting on top of the agar. A tiny fraction of these embedded cells are infected with phages. Cells infected with a phage clone undergo lysis (see Chapter 2, pp. 12–13), and this lysis releases phages that infect and kill neighboring cells in the lawn. This repeated process leads to the killing of all of the cells in a small region, leaving a clear hole in the opaque lawn. This hole is called a **plaque.** It is a region in the lawn on a plate where there are no living cells. The clear area contains large numbers of the released phages which can be collected to continue work with the cloned DNA they contain. (Figure 15.11, p. 440 shows a photograph of a plate with plaques.)

There are two major advantages that make phage vectors useful. First, phage vectors accept larger inserts than plasmids. Second, many more plaques can "fit" on a plate than can colonies, so we can work with a much larger set of clones than is the case if a plasmid vector is used.

Some of the other vectors you have learned about have other uses. For instance, because of their ability to accommodate large DNA inserts, BACs (Chapter 8, p. 178) form the basis of vectors for studying gene regulation in vertebrates such as mouse and zebrafish. That is, the promoter and regulatory sequences of many vertebrate genes are known often to span a large section of DNA. Therefore, a gene and a large segment of DNA upstream of the gene can be cloned in a BAC, and the clone transformed into the organism. Hopefully the clone has all the sequences present for normal regulation of the gene, making the study of that regulation feasible.

#### Keynote

Many different kinds of vectors have been developed to manipulate cloned DNA sequences. Shuttle vectors can be moved from one host species to another. Expression vectors carry sequences that allow the protein encoded by the insert to be expressed by the host cell. PCR cloning vectors have specialized ends to facilitate cloning of DNA amplified by PCR in which the DNA polymerase adds a single-nucleotide sticky end to each strand. Transcribable vectors allow either *in vitro* transcription or *in vivo* transcription and translation. Phage vectors offer certain advantages, specifically larger inserts and the ability to place more clones on a plate, making it easier to grow large numbers of clones in a small space. Most plasmid vectors replicate within their host organism. Those that do not replicate extrachromosomally or integrate into the genome and are replicated when the genome replicates. Phage vectors kill their bacterial host while replicating the DNA insert. The choice of the vector to use depends on the experimental goal and the organisms involved.

## Cloning a Specific Gene

Often, researchers want to study a particular gene or DNA fragment. Many researchers work with an organism with a sequenced genome. When they want to clone a gene of interest, it is often as simple as looking up the genomic sequence in a database, designing PCR primers that will amplify their gene of interest, and then using either genomic DNA or cDNA as the template for a polymerase chain reaction. The PCR fragment can be cloned directly or, if restriction sites were designed in the primers (see Figure 10.3), the PCR product can be cut and cloned as described earlier. If the genome is sequenced, genes associated with a specific disease can even be found without any phenotypic information. One such example is described in the Focus on Genomics box for this chapter.

What happens if a researcher wants to clone a gene from an organism without a sequenced genome? With no sequence information, the gene cannot be cloned by the simple types of PCR that we have discussed. There are several approaches that can be used—most are different ways of looking for our gene of interest in a pool of cDNAs (see Chapter 8, pp. 193–198). Each way of finding the cDNA requires very specific molecular tools, and the availability of these tools will be important in deciding which strategy will be most successful. Furthermore, the strategy used may influence the nature of the clone recovered, and this will also inform our choices.

### Finding a Specific Clone Using a DNA Library

If we have, or make, a cDNA library (see Chapter 8, pp. 197–198) from our organism, we can then start to look

## Focus on Genomics

### Finding a New Gene Linked to Type 1 Diabetes

The human genome and the haplotype map can be used to find new genes associated with well-known diseases. In one study, investigators set out to find additional genes associated with type 1 diabetes. *Type 1 diabetes*, also called *juvenile diabetes*, is characterized by an attack on the β cells of the pancreas by the immune system. Several genes involved in certain aspects of immune system function have been implicated in the development of this disorder. The investigators wondered if any additional genes were involved in the development of this disease. The β cells make insulin, and release it when the blood sugar is high (generally after a meal). Insulin can instruct the liver and muscles to increase their rates of glucose uptake, and ultimately, the rate of glycogen production is increased in both tissues as the glucose is converted into glycogen, a more easily stored polymer of glucose. The liver glycogen will be degraded as the blood sugar concentration decreases. Insulin, then, plays an essential role in regulating blood sugar. It is the signal to decrease blood sugar and is also responsible for the production of the stored glycogen that will be used to increase low blood sugar. In type 1 diabetes, the death of the β cells prevents the normal release of, and response to, insulin, leading to high blood sugar after meals and limited glycogen production. Since very little glycogen is made, it is not possible for a person with type 1 diabetes to use stored glycogen to raise the blood sugar if a meal is skipped. People with this disorder are generally treated with insulin, either isolated from animals or produced in the lab.

The investigators found one region, about 230 kb on chromosome 16, that was significantly associated with an increased risk of type 1 diabetes. Only a single gene (named *KIAA0350*) is in this region, and it encodes a lectin specific type of sugar-binding protein called *lectin*. Proteins of this type are often involved in immune-system function, and it seems possible that the mutations in the lectin that predispose to type 1 diabetes may act by making an inapropriate attack on the β cells more likely.



in this library for the cDNA corresponding to the gene of interest. Unlike libraries of books, clone libraries have no catalog, so they must be searched through (screened) to find the desired clone. Fortunately, a number of screening procedures have been developed, and some are discussed in this section. We will assume that we have antibodies that recognize (bind to) a protein of interest, in addition to a cDNA library (in an expression vector) and a genomic library. Our goal is to find a cDNA clone, and a genomic clone containing the entire gene.

**Screening a cDNA Library.** We can screen a cDNA library in a number of ways to identify a cDNA clone we are interested in studying. Later we will screen this library twice in our theoretical cloning experiment. Our first screening of the cDNA expression library will be a search for a cDNA clone that encodes a specific protein (Figure 10.5). This approach entails using antibodies that can bind to the protein encoded by our gene of interest. Recall that the cDNAs are cloned in an expression vector (Figure 10.5, step 1; and see p. 249). This means that the cDNA is inserted between a promoter and a transcription termination signal, both of which are part of the vector. In the bacterial host cell, an mRNA is transcribed corresponding to the cDNA, and the mRNA is translated to produce the encoded protein.

For screening, first *E. coli* is transformed with cDNA clones made in an expression vector (Figure 10.5, step 2), and then the cells are plated so that each bacterium gives rise to a colony (Figure 10.5, step 3). These clones are preserved, for example, by picking each colony off the plate and placing it into the medium in a well of a microtiter dish (Figure 10.5, step 4; 16 wells are shown in the example). Replicas of the set of clones are placed (printed) onto a membrane filter that has been placed on a culture plate of selective medium appropriate for the recombinant molecules—for example, ampicillin for plasmids carrying the ampicillin resistance gene (Figure 10.5, step 5). Colonies grow on the filter in the same pattern as the clones in the microtiter dish. The filter is peeled from the dish, and the cells are lysed *in situ* (Figure 10.5, step 6). The proteins that were within the cell, including those expressed from the cDNA, become stuck to the filter. The filter is then incubated with an antibody to the protein of interest (Figure 10.5, step 7).

If the antibody is labeled radioactively, any clones that expressed the protein of interest can be identified by placing the dried filter against X-ray film, leaving it in the dark for a period of time (from 1 hour to overnight) to produce an *autoradiogram* (Figure 10.5, step 8). The process is called *autoradiography*. When the film is developed, dark spots are seen wherever the radioactive probe is bound to the filter in the antibody reaction. (The dark spots result from the decay of the radioactive atoms, which changes silver grains in the film.) These spots correspond to the cDNA clones expressing the protein of interest. You might assume that this clone must contain the entire cDNA, corresponding to the complete mRNA transcribed in your organism of interest. Unfortunately,

Figure 10.5

**Screening for specific cDNA plasmids in a cDNA library by using an antibody probe.**

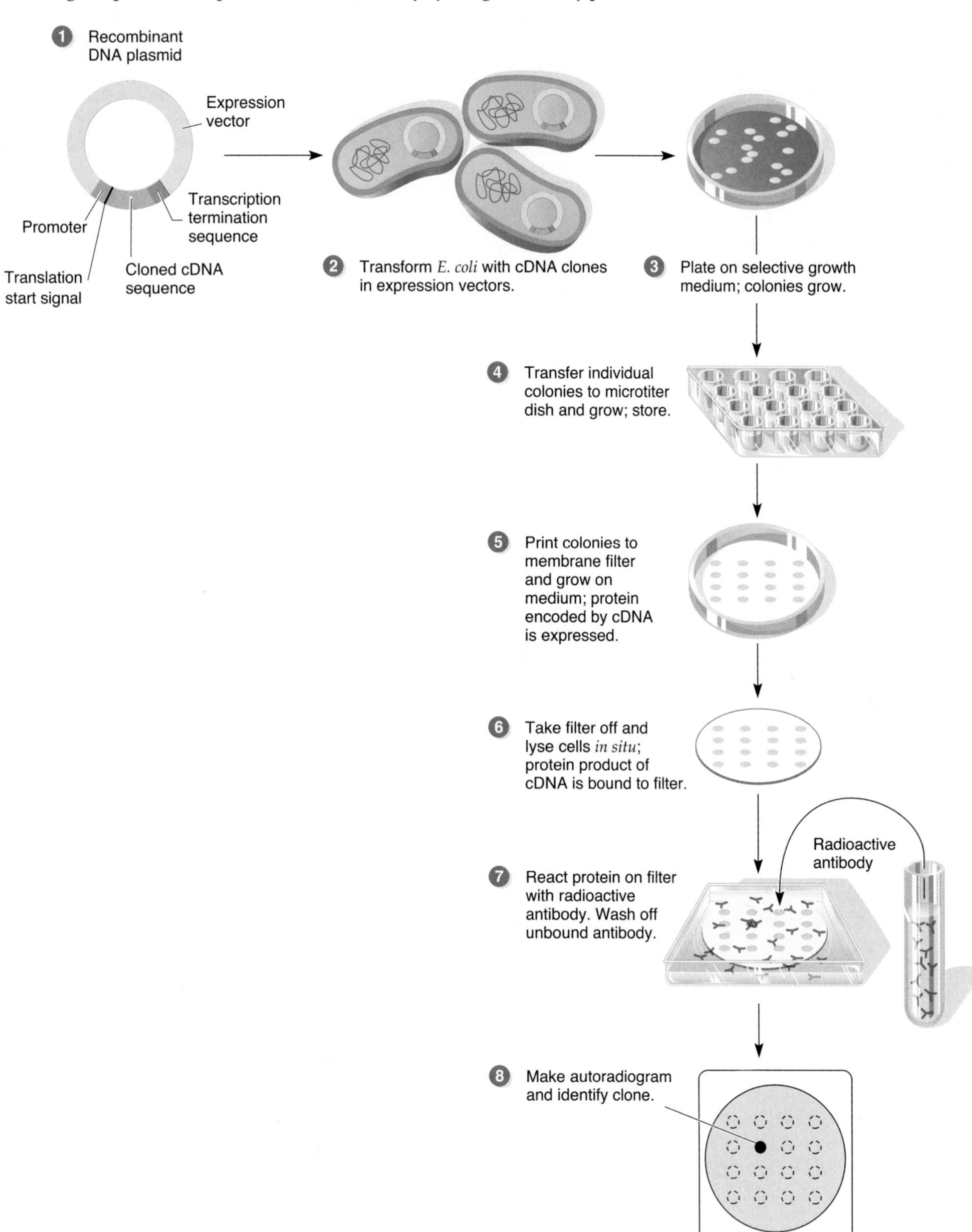
1 Recombinant DNA plasmid
Expression vector
Promoter
Transcription termination sequence
Translation start signal
Cloned cDNA sequence
2 Transform *E. coli* with cDNA clones in expression vectors.
3 Plate on selective growth medium; colonies grow.
4 Transfer individual colonies to microtiter dish and grow; store.
5 Print colonies to membrane filter and grow on medium; protein encoded by cDNA is expressed.
6 Take filter off and lyse cells *in situ*; protein product of cDNA is bound to filter.
7 React protein on filter with radioactive antibody. Wash off unbound antibody.
Radioactive antibody
8 Make autoradiogram and identify clone.

an antibody recognizes only a small **epitope.** An epitope is the specific short region of a protein (or other molecule recognized by an antibody) that is bound specifically by the antibody. Epitopes are often less than ten amino acids in length, so our selected clone definitely contains the part of the cDNA encoding the epitope, but may or may not contain the entire cDNA. Once a cDNA clone for a protein of interest has been identified (we will assume that our selected clone contains the entire cDNA for encoding our protein of interest), it can be used for other applications, for example, to analyze the genome of the same or other organisms for homologous sequences, to isolate the nuclear gene for the mRNA from a genomic library, or to quantify mRNA synthesized from the gene.

**Screening a Genomic Library.** Given the existence of a probe, such as a cloned cDNA, it is now possible to identify the genomic DNA, including the promoter region and introns, that corresponds to the gene of interest by screening a genomic library. Once the correct genomic clone has been identified, we can isolate the DNA insert in the clone and ask further questions about how the gene functions. For instance, we could compare the genomic sequence to the cDNA sequences to study how the mRNA for the gene is spliced, or we could study the promoter and regulatory sequences to see how transcription of the gene of interest is controlled. Here we discuss the screening of genomic libraries made in a phage cloning vector.

Screening a genomic library made using a vector derived from phage lambda is similar to that just described for screening a cDNA library. First, *E. coli* cells are infected with the genomic library (Figure 10.6, step 1, and the cells are plated as a lawn, where plaques are produced (Figure 10.6, step 2). Then a membrane filter is placed on the plate. Phage particles, which are present in the plaques, stick to the membrane filter. The filter is then processed to lyse any bacterial cells, to remove the proteins protecting the phage DNA, to denature the DNA to single strands, and then bind that DNA firmly to the filter (Figure 10.6, step 3).

Next, the filter is placed in a heat-sealable plastic bag and incubated with the cDNA probe (Figure 10.6, step 4), which has been labeled radioactively or nonradioactively. Box 10.1 describes one method for the creation of radioactive DNA probes. Riboprobes (which are made of RNA) can also be used as radioactive probes. Since these are created by *in vitro* transcription, they can be labeled very

**Figure 10.6**

**Using a DNA probe to screen a phage genomic library for specific DNA sequences.**

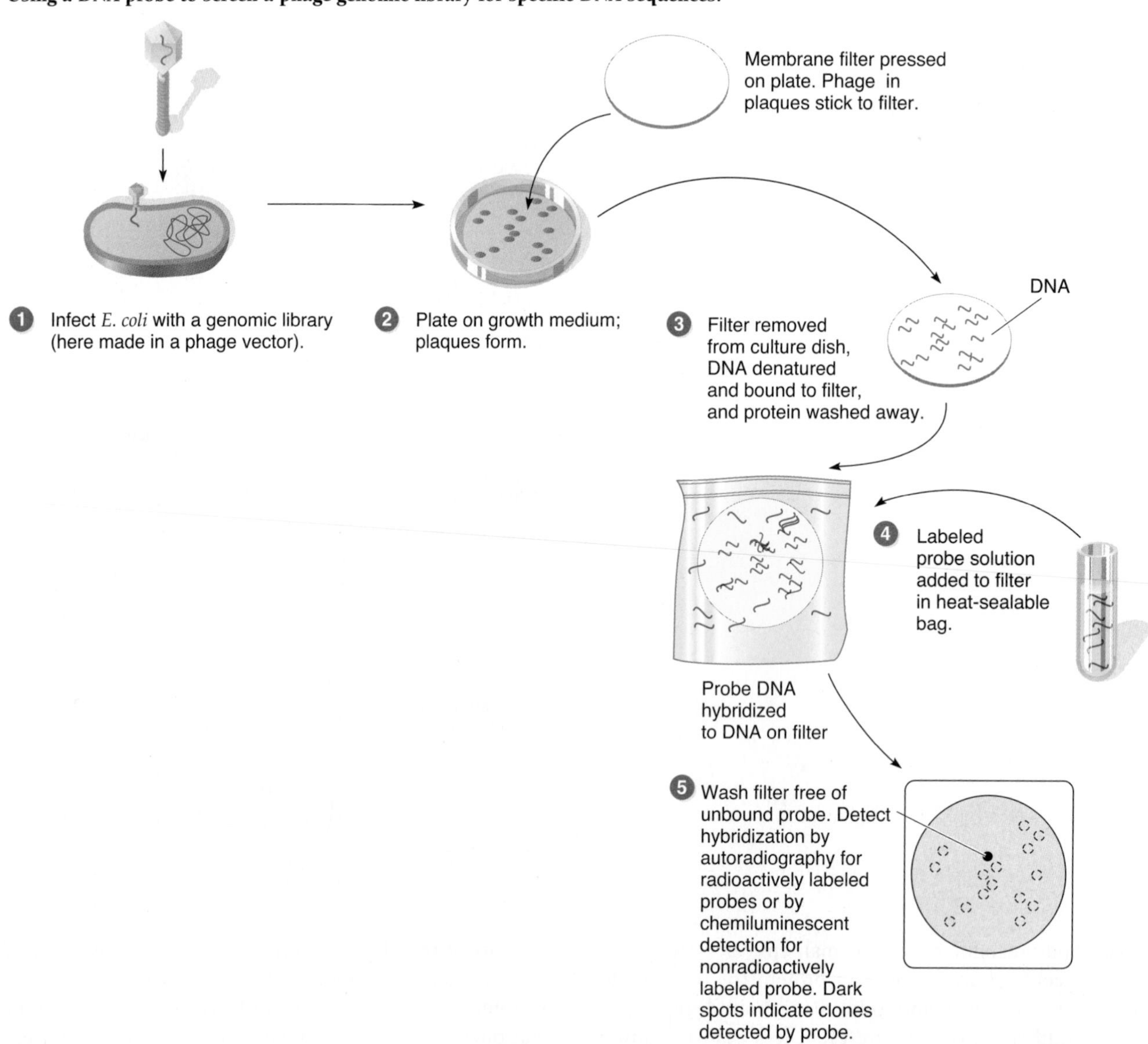
Membrane filter pressed on plate. Phage in plaques stick to filter.
1 Infect E. coli with a genomic library (here made in a phage vector).
2 Plate on growth medium; plaques form.
3 Filter removed from culture dish, DNA denatured and bound to filter, and protein washed away.
DNA
4 Labeled probe solution added to filter in heat-sealable bag.
Probe DNA hybridized to DNA on filter
5 Wash filter free of unbound probe. Detect hybridization by autoradiography for radioactively labeled probes or by chemiluminescent detection for nonradioactively labeled probe. Dark spots indicate clones detected by probe.

## Box 10.1 Labeling DNA

DNA can be labeled either radioactively or nonradioactively. Typically, it has been more common to label DNA radioactively, but with increasing regulations pertaining to the disposal of radioactive material and the health risks of exposure to radioactive compounds, great strides have been made in developing nonradioactive DNA labeling methods which produce probes that are as sensitive as radioactive probes in seeking out the target DNA. Thus, it is now possible to detect as little as 0.1 picogram ($0.1 \times 10^{-12}$ g) of DNA with either radioactive or nonradioactive probes. We now discuss briefly some methods for preparing radioactively labeled and nonradioactively labeled DNA probes.

### Radioactive Labeling of DNA

A DNA probe can be labeled radioactively by the *random-primer method* (Box Figure 10.1). In this approach, the DNA is denatured to single strands by boiling and quick cooling on ice. DNA primers six nucleotides long (hexanucleotides), synthetically made by the random incorporation of nucleotides, are annealed to the DNA. The *hexanucleotide random primers* pair with complementary sequences in the DNA, and such pairing occurs at many locations because all possible hexanucleotide sequences are present. The primers are elongated by the Klenow fragment of DNA polymerase I, which uses radioactively labeled precursors (dNTPs). (The Klenow fragment, named for the person who discovered it, lacks 5′-to-3′ exonuclease activity, which would otherwise remove the short primers, but still has the 3′-to-5′ proofreading activity.) Typically, the label is $^{32}$P, located in the phosphate group that is attached to the 5′ carbon of the deoxyribose sugar. This phosphate group is called the *α-phosphate*, because it is the first in the chain of three; the α-phosphate is used in forming the phosphodiester bonds of the sugar–phosphate backbone.

**Box Figure 10.1**
**Random primer method of radioactively labeling DNA.**

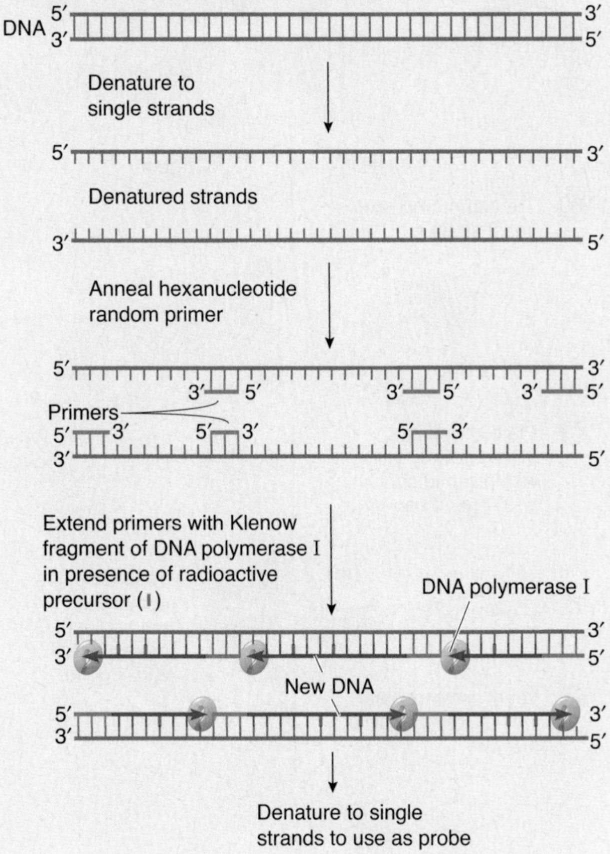

After the radioactive DNA probe is applied in an experiment, detection depends on the properties of the radioactive isotope. For example, if a $^{32}$P-labeled probe has hybridized with a target DNA sequence on a membrane filter, the filter is placed against a piece of X-ray film and the sandwich is placed in the dark. Every location on the filter where there is $^{32}$P (a spot, band, etc.) is detected as a black region on the X-ray film after it is developed. This process is called *autoradiography*, and the resulting picture of radioactive signals is called an *autoradiogram*.

### Nonradioactive Labeling of DNA

Random-primer labeling also can be used to prepare nonradioactively labeled DNA probes. The difference from preparing radioactively labeled DNA is that a special DNA precursor molecule, rather than a $^{32}$P-labeled precursor, is used. For example, in one of many labeling systems, digoxigenin-dUTP (DIG-dUTP) is added to the dATP, dCTP, dGTP, and dTTP precursor mixture. Digoxigenin is a steroid, and it is linked to dUTP (deoxyuridine 5′-triphosphate). During DNA synthesis, DIG-dUTP can be incorporated opposite to A nucleotides on the template DNA strand.

The nonradioactively labeled DNA can be used in experiments in the same way as is radioactively labeled DNA. Detection is different, however. Once the DIGdUTP-labeled probe has bound to target DNA on a filter, for example, an anti-DIG-AP conjugate is added. The anti-DIG part of the conjugate is an antibody that reacts specifically with DIG, and the AP part of the conjugate is the enzyme alkaline phosphatase. Wherever the DIG-labeled DNA is hybridized to target DNA on the filter, the anti-DIG-AP conjugate binds to form a DNA-DIG–anti-DIG-AP complex. The location of the probe-target hybrid is then visualized by substrates that react with the alkaline phosphatase. To achieve sensitivity that matches radioactively labeled probes, a chemiluminescent substrate is used. Such a substrate produces light in a reaction catalyzed by alkaline phosphatase, and detection involves exposing X-ray film much like making an autoradiogram. If great sensitivity is *not* necessary, colorimetric substrates for the enzyme are used. In this case, spots or bands develop directly on the filter as purple or blue regions as the enzyme reaction proceeds.

simply—we just have to add radioactive NTPs to the *in vitro* transcription reaction to make a radioactive riboprobe. Nonradioactive probes take advantage of enzyme-based detection systems in which the labeled probe undergoes a reaction with a chemical substrate to create either light or a colored precipitate. One type of nonradioactive probe labeling and detection system is described in Box 10.1. To use labeled DNA as a probe, the DNA is denatured by boiling and then cooled quickly on ice to produce single-stranded DNA molecules. These labeled molecules are added to the membrane filters to which the denatured (single-stranded) DNA from each colony has been bound. The labeled molecules diffuse over the filter and, with time, they will find the DNA bound to the filter with which they can pair by complementary base pairing. By this hydrogen bonding, probe–target DNA hybrids form. For example, if the cDNA probe is derived from the mRNA for β-globin, that probe will hybridize with DNA bound to the filter that encodes the β-globin mRNA, that is, the genomic β-globin gene. After the hybridization step, the filters are washed to remove unbound probe and subjected to the detection procedure appropriate for whether the probe was radioactive or nonradioactive: autoradiography for a radioactive probe, or chemiluminescent or colorimetric detection for a nonradioactive probe (Figure 10.6, step 5). From the positions of the spots on the film or filter, the locations of the phage plaque or plaques on the original plate can be determined and the clones of interest isolated for further characterization.

**Comparing the cDNA Clone and Genomic Clones.** After we recover a cDNA clone and a genomic clone, we can sequence both (Chapter 8, pp. 183–187) and compare the sequences. Obviously, both cDNA and genomic clones will have exon sequences, but only the genomic sequence will contain introns and upstream regulatory sequences. We can use these comparisons, then, to identify candidate promoter sequences, and to understand how the exons and introns are arranged in the genome.

## Identifying Genes in Libraries by Complementation of Mutations

For microorganisms in which genetic systems of analysis have been well-developed and for which there are well-defined mutations, it is possible to clone genes by complementation of those mutations. In brief, this approach relies on the expression of the wild-type gene introduced into the cell by transformation overcoming the defect of a mutant form of the gene in the genome. (Complementation is discussed in more detail in Chapter 13, pp. 377–378.) This can be done with the yeast *Saccharomyces cerevisiae*, for example, which is easy to manipulate genetically and for which efficient integrative and replicative transformation systems using yeast–*E. coli* shuttle vectors are available.

To clone a yeast gene by complementation, first a genomic library is made of DNA fragments from the wild-type yeast strain in a yeast–*E. coli* shuttle vector. The library is used to transform a host yeast strain carrying two mutations: a mutation to allow transformants to be selected (*ura3*, for example, which gives a uracil growth requirement) and a mutation in the gene for which the wild-type gene clone is sought. Consider the cloning of the *ARG1* gene, the wild-type gene for an enzyme needed for arginine biosynthesis (Figure 10.7), by complementation of an *arg1* mutation. A yeast strain carrying the *arg1* mutation has an inactive enzyme for arginine biosynthesis

**Figure 10.7**

**Example of cloning a gene by complementation of mutations: cloning of the yeast *ARG1* gene.**

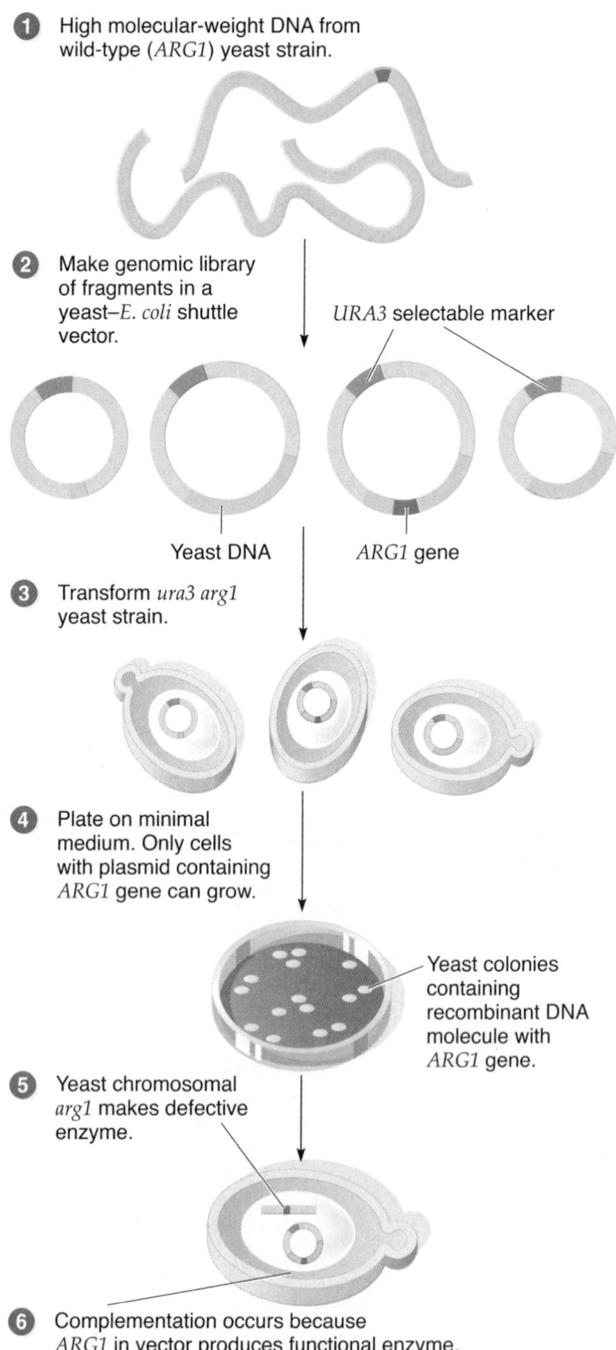

1 High molecular-weight DNA from wild-type (*ARG1*) yeast strain.
2 Make genomic library of fragments in a yeast–*E. coli* shuttle vector.
*URA3* selectable marker
Yeast DNA
*ARG1* gene
3 Transform *ura3 arg1* yeast strain.
4 Plate on minimal medium. Only cells with plasmid containing *ARG1* gene can grow.
Yeast colonies containing recombinant DNA molecule with *ARG1* gene.
5 Yeast chromosomal *arg1* makes defective enzyme.
6 Complementation occurs because *ARG1* in vector produces functional enzyme.

and therefore needs arginine to grow. A genomic library is made using DNA from a wild-type (*ARG1*) yeast strain (Figure 10.7, steps 1 and 2). When a population of *ura3 arg1* yeast cells is transformed with the genomic library prepared in the shuttle vector (Figure 10.7, step 3), some cells receive plasmids containing the normal (*ARG1*) gene for the arginine biosynthesis enzyme. The plasmid's *ARG1* gene is expressed, enabling the cell to grow on minimal medium—that is, in the absence of arginine—despite the presence of a defective *arg1* gene in the cell's genome (Figure 10.7, step 4). The *ARG1* gene is said to overcome the functional defect of the *arg1* mutation by *complementation* of that mutation (Figure 10.7, steps 5 and 6). The plasmid is then isolated from the cells, and the cloned gene is characterized.

### Identifying Specific DNA Sequences in Libraries Using Heterologous Probes

cDNA probes can be used to identify and isolate specific genes, and a large number of genes have been cloned from both prokaryotes and eukaryotes in this way. It is also possible to identify specific genes in a genomic library by using clones of similar genes from other organisms as probes. For example, a mouse probe could be used to probe a human genomic library. Such probes are called *heterologous probes*, and their effectiveness depends on a good degree of homology between the probes and the genes. For that reason, the greatest success with this approach has come with highly conserved genes or with probes from a species closely related to the organism from which a particular gene is to be isolated.

### Identifying Genes or cDNAs in Libraries Using Oligonucleotide Probes

A number of genes have been isolated from libraries by using synthetically made oligonucleotide probes. In this method, at least some of the amino acid sequence must be known for the protein encoded by the gene. In that case, it may be possible that a *consensus sequence* (the most common nucleotide at each position) can be determined from previously cloned versions of the gene that are available in *GenBank* (http://www.ncbi.nlm.nih.gov/Genbank/index.html), a computer database where sequences are deposited and made available to researchers worldwide. Then, because the genetic code is universal, oligonucleotides about 20 nucleotides long can be designed that, if translated, would give the known amino acid sequence. Because of the degeneracy of the genetic code—up to six different codons can specify a given amino acid—a number of different oligonucleotides are made, all of which could encode the targeted amino acid sequence. These probes are known as *guessmers*. These mixed oligonucleotides are labeled and used as probes to search the libraries with the hope that at least one of the oligonucleotides will detect the gene or cDNA of interest. If the probe is labeled radioactively, detection is by autoradiography, whereas if the probe is labeled nonradioactively, detection is done colorimetrically or with chemiluminescence (see Box 8.1). Though not successful all of the time, oligonucleotide-based library screening has been extremely fruitful and has allowed many genes that were missing molecular information to be cloned.

**Keynote**

Specific sequences in cDNA libraries and genomic libraries can be identified using a number of approaches, including the use of specific antibodies, cDNA probes, complementation of mutations, heterologous probes, and oligonucleotide probes.

## Molecular Analysis of Cloned DNA

Cloned DNA sequences are resources for experiments designed to answer many kinds of biological questions. Two examples are given in this section: *Southern blotting* and *northern blotting*.

### Southern Blot Analysis of Sequences in the Genome

As part of the analysis of genes, it can be helpful to determine the arrangement and specific locations of restriction sites in the genome. This information is useful, for example, for comparing homologous genes in different species, analyzing intron organization, planning experiments to clone parts of a gene (such as its promoter or controlling sequences) into a vector, or screening individuals for restriction endonuclease site differences associated with disease genes. The arrangement of restriction sites in a genome can be analyzed directly by using a gene probe, a cDNA probe, or by using as a probe the same gene cloned from a closely related organism. The process of analysis is as follows:

1. Samples of genomic DNA are cut with different restriction enzymes (Figure 10.8, steps 1 and 2), each of which produces DNA fragments of different lengths depending on the locations of the restriction sites.
2. The DNA restriction fragments are separated by size using agarose gel electrophoresis (Figure 10.8, step 3). After electrophoresis, the DNA is stained with ethidium bromide so that it can be seen under ultraviolet light. When genomic DNA is digested with a restriction enzyme, the result is a continuous smear of fluorescence down most of the length of the gel lane because the enzyme produces many fragments ranging in size from large to small.
3. The DNA fragments are transferred to a membrane filter (Figure 10.8, step 4). In brief, the gel is soaked in an alkaline solution to denature the double-stranded DNA into single strands. The gel is neutralized and placed on a piece of blotting paper on a glass plate.

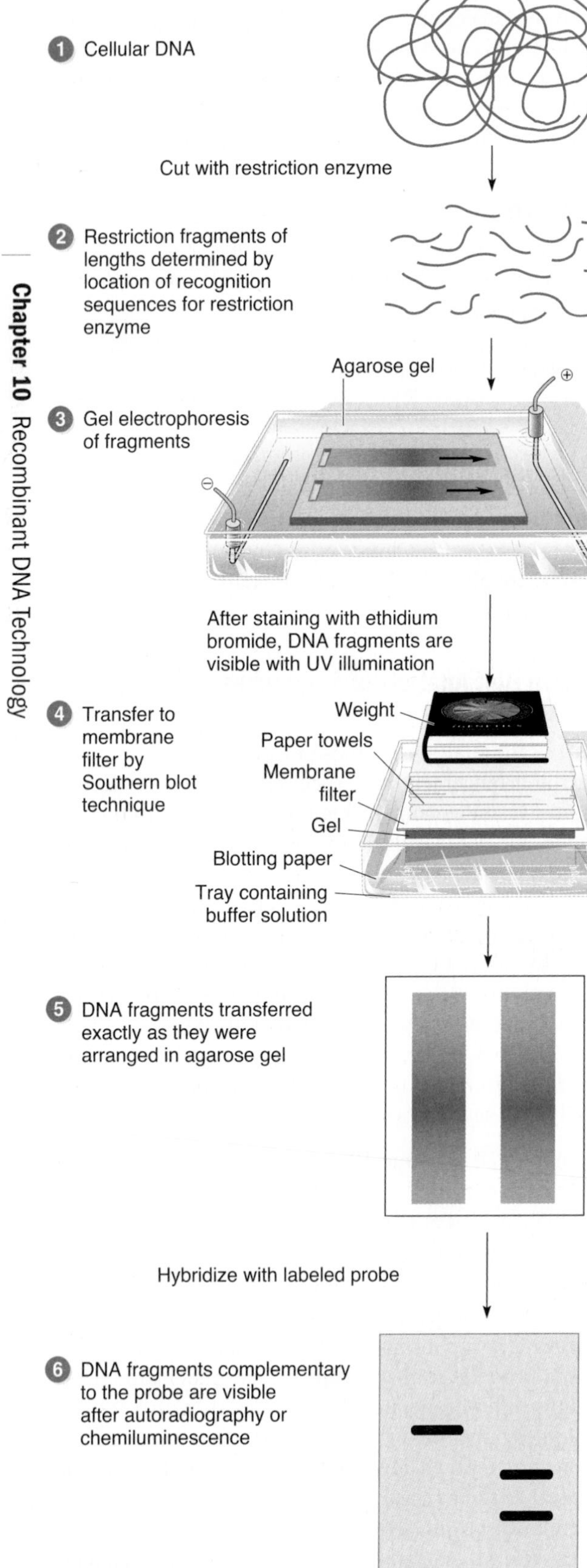
1 Cellular DNA
Cut with restriction enzyme
2 Restriction fragments of lengths determined by location of recognition sequences for restriction enzyme
Agarose gel
3 Gel electrophoresis of fragments
After staining with ethidium bromide, DNA fragments are visible with UV illumination
4 Transfer to membrane filter by Southern blot technique
Weight
Paper towels
Membrane filter
Gel
Blotting paper
Tray containing buffer solution
5 DNA fragments transferred exactly as they were arranged in agarose gel
Hybridize with labeled probe
6 DNA fragments complementary to the probe are visible after autoradiography or chemiluminescence

**Figure 10.8**

**Southern blot procedure for analyzing cellular DNA for the presence of sequences complementary to a labeled probe, such as a cDNA molecule made from an isolated mRNA template.** The hybrids, shown as three bands in this theoretical example, are visualized by autoradiography or chemiluminescence.

The ends of the paper are in a container of buffer and act as wicks. A piece of membrane filter is laid down so that it covers the gel. Sheets of blotting paper (or paper towels) and a weight are stacked on top of the membrane filter. The buffer solution in the bottom tray is wicked up by the blotting paper, passing through the gel and the membrane filter and finally into the stack of blotting paper. During this process, the DNA fragments are picked up by the buffer and transferred from the gel to the membrane filter, to which they bind because of the membrane filter's chemical properties. The fragments on the filter are arranged in exactly the same way as they were in the gel (Figure 10.8, step 5).

4. A labeled probe is added to the membrane filter; it hybridizes to any complementary DNA fragment(s) (Figure 10.8, step 6). Detection of the probe is carried out in a way appropriate for whether the probe is radioactive or nonradioactive to determine the positions of the hybrids (Figure 10.8, step 6). If a sample of DNA size markers is separated in a different lane in the agarose gel electrophoresis process, the sizes of the genomic restriction fragments that hybridized with the probe can be calculated. From the fragment sizes obtained, a restriction map can be generated to show the relative positions of the restriction sites. Suppose, for example, that using only *Bam*HI produces a DNA fragment of 3-kb that hybridizes with the labeled probe. If a combination of *Bam*HI and *Pst*I is then used and produces two DNA fragments, one of 1 kb and the other of 2 kb, we would deduce that the 3-kb *Bam*HI fragment contains a *Pst*I restriction site 1 kb from one end and 2 kb from the other end. Further analysis with other enzymes, individually and combined, enables the researcher to construct a map of all the enzyme sites relative to all other sites.

The whole process of separating DNA fragments by agarose gel electrophoresis, transferring (blotting) the fragments onto a filter, and hybridizing them with a labeled complementary probe is called **Southern blot analysis** or **Southern blotting** (named after its inventor, Edward Southern). Applications of Southern blot analysis will be described later in the chapter.

## Northern Blot Analysis of RNA

A technique that is very similar to Southern blot analysis—called **northern blot analysis** or **northern blotting**—is for the study of RNA rather than DNA. (The name is not derived from a person but indicates, that the technique is related to Southern blot analysis.) In northern blot analysis, RNA extracted from cells or a tissue is separated by size using denaturing gel electrophoresis (a type of electrophoresis in which the buffer used disrupts the secondary structure of RNA, that is, regions that have formed double-stranded sections). The RNA molecules are then transferred and bound to a filter in a procedure that is essentially identical to that used in Southern blot analysis. After hybridization with a labeled probe and use of the appropriate detection system, bands show the locations of RNA fragments that were complementary to the probe. Given appropriate RNA size markers, the sizes of the RNA fragments identified with the probe can be determined.

Northern blot analysis is useful for revealing the size or sizes of the mRNA encoded by a gene. In some cases, a number of different mRNA species encoded by the same gene have been identified in this way, suggesting that different promoter sites or different terminator sites are used or that alternative mRNA processing can occur. Northern blot analysis can also be used to investigate whether an mRNA is present in a cell type or tissue and how much of it is present. This type of experiment is useful for determining levels of gene activity, such as during development, in different cell types of an organism, in cancer cells vs noncancerous cells, or in cells before and after they are subjected to various physiological stimuli.

**Keynote**

Cloned genes and other DNA sequences often are analyzed to determine the arrangement and specific locations of restriction sites. The analytical process involves cleavage of the DNA with restriction enzymes, followed by separation of the resulting DNA fragments by agarose gel electrophoresis. The sizes of the DNA fragments are calculated, enabling restriction maps to be constructed. The many DNA fragments produced by cleavage of genomic DNA show a wide range of sizes, resulting in a continuous smear of DNA fragments in the gel. In this case, specific gene fragments can be visualized only by transferring the DNA fragments to a membrane filter by Southern blotting, hybridizing a specific labeled probe with the DNA fragments, and detecting the hybrids. A similar procedure—northern blotting—is used to analyze the sizes and quantities of RNAs isolated from a cell.

## The Wide Range of Uses of the Polymerase Chain Reaction (PCR)

PCR is one of the most commonly used techniques in the modern genetics research lab, if not the most commonly used technique. This is because PCR allows us to make unlimited copies of a fragment of interest, even if we do not know all that much about the region, or if we do not have much starting template. With a few modifications, we can even use PCR to quantify the abundance of a specific mRNA in a sample. PCR is also a rapid procedure with most reactions completed in less than a few hours. PCR was introduced in Chapter 9, pp. 221–223 and shown in Figure 9.3.

### Advantages and Limitations of PCR

PCR is a powerful technique for amplifying segments of DNA. Such amplification is very similar to cloning DNA by using vectors. However, PCR is a much more sensitive and quicker technique than cloning. Starting with just one molecule, PCR can produce millions of copies of a DNA segment in just a few hours. By contrast, cloning requires a significant amount of starting DNA for restriction digestion, and then at least a week is needed to go through all the cloning steps. There are two major limitations of PCR, however. First, PCR requires the use of specific primers and this means that there must be sequence information available for the DNA to be amplified in order for primers to be designed. Second, the length of DNA that can be amplified by PCR is limited by the enzyme and conditions to about 40 kb. In fact, amplifications of this size are technically demanding, and, in most cases, investigators use PCR to amplify much smaller fragments, if possible.

A further issue with PCR is that the *Taq* polymerase used by many researchers has no proofreading activity. This means that base pair mismatches that occur during DNA synthesis go uncorrected in this *in vitro* procedure so any clone made using *Taq* polymerase in PCR must be analyzed carefully to ensure that there are no mutations introduced by the enzyme. Alternative thermostable DNA polymerases that have proofreading activity are available for PCR and such enzymes significantly decrease the error frequency. One such enzyme is Vent polymerase, which was extracted from an archaean growing around high-temperature deep-sea oceanic vents.

Finally, the tremendous sensitivity of PCR is its liability in some applications. Because PCR can produce many copies from a single DNA molecule, great care has to be taken that it is the right DNA molecules that are amplified. In forensic applications, for example, it is crucial that DNA used for evidence has no chance of being contaminated by DNA from the investigators or researchers handling the DNA.

### Applications of PCR

There are many applications for PCR, including, as discussed earlier (Chapter 9, pp. 221–225), amplifying DNA for cloning, amplifying DNA for subcloning (moving part of a cloned sequence to a new vector), amplifying DNA from genomic DNA preparations for sequencing without cloning, mapping DNA segments, disease diagnosis, sex determination of embryos, forensics, and studies of molecular evolution. In disease diagnosis, for example, PCR can be used to detect bacterial pathogens or viral pathogens such as HIV (human immunodeficiency virus, the causative agent of AIDS) and hepatitis B virus. PCR can also be used in genetic disease diagnosis, which is discussed later in the chapter.

PCR is useful for subcloning a segment of cloned DNA. This example follows from the discussion of cloning a yeast gene by complementation earlier in the chapter (see pp. 260–261 and Figure 10.7). The concept presented was that a yeast genomic library can be used to identify a particular wild-type gene by complementation of a mutation. Experimentally, a clone in the library is identified because it confers a wild-type phenotype to the mutant cell it transforms. In the specific example, the wild-type *ARG1* yeast gene was identified. Let us now refine the analysis. The plasmid clone that complements the *arg1* mutant must contain the *ARG1* gene. The plasmid is extracted from the yeast, and the sequence of the cloned fragment is determined. If there is only one gene in the fragment, then of course it must be the *ARG1* gene. However, if there is more than one gene, further steps are needed to identify the *ARG1* gene. Since we have determined the sequence of the cloned fragment, we can design PCR primers and amplify each gene individually. These genes can then be cloned separately into a vector just as for the genomic library construction in Figure 10.3. Now each cloned gene can be tested separately for its ability to complement the *arg1* mutant gene and, in this way, the *ARG1* gene is found.

In forensics, PCR can be used, for example, to amplify trace amounts of DNA in samples such as hair, blood, or semen collected from a crime scene. The amplified DNA can be analyzed and compared with DNA from a victim and a suspect, and the results can be used to implicate or exonerate suspects in the crime. This analysis, called *DNA typing*, *DNA fingerprinting*, or *DNA profiling*, is discussed in more detail later in the chapter.

## RT-PCR and mRNA Quantification

PCR is used for some experiments in which the starting material is RNA. The two examples described are *reverse transcription-PCR*, and *real-time PCR*.

**Reverse Transcription-PCR. Reverse transcription-PCR (RT-PCR)** is a method in which RNA first is converted to cDNA and then the cDNA is amplified by PCR. RT-PCR is a very sensitive technique for detecting and quantifying RNA, often mRNA. There are three steps to RT-PCR. First, cDNA is synthesized from RNA using a primer (oligo(dT), for example, for mRNA) and the enzyme reverse transcriptase (RT). The synthesis of cDNA from RNA was described in Chapter 8, pp. 195–197 and Figure 8.15. Then the specific cDNA made is amplified by PCR (see Figure 9.3, p. 222) using primers complementary to the two ends of the molecule, and the PCR products are analyzed using gel electrophoresis. The RT-PCR technique, like regular PCR, is a very sensitive technique, in that it will be able to tell us that our mRNA is present, even if our mRNA makes up only a tiny fraction of the starting RNA pool.

RT-PCR is used either for testing for the presence of a particular RNA, or for roughly quantifying the amount of an RNA. For example, some viruses have RNA genomes and theoretically RT-PCR could be used to detect whether an individual has been infected by the virus. Such tests have been developed for HIV, measles virus, and mumps virus. If we wish to determine the abundance of the mRNA for our gene of interest, we can get some idea of relative abundance—whether the mRNA for our gene is common, rare, or very rare, for instance. The primary limitation is that it is difficult to figure out exactly how much template was present in the starting mixture by looking at a band on a gel after 30 or more cycles. Thus, it is generally impossible to determine exact abundance of the mRNA for a gene of interest.

**Real-time PCR. Real-time PCR** (also called *real-time quantitative PCR*) is a PCR method for measuring the increase in the amount of DNA as it is amplified (which gives the technique its "real-time" name). An important application of real-time PCR is the accurate quantification of mRNAs levels in a sample (Figure 10.9). As with RT-PCR, this application of real-time PCR involves using RNA as a template for the reverse transcriptase-catalyzed synthesis of cDNA (Figure 10.9, step 1), and then using this cDNA as template for PCR. For the PCR steps, the cDNA is denatured (Figure 10.9, step 2), primers are annealed (Figure 10.9, step 3), and the primers are extended by a thermostable DNA polymerase such as *Taq* polymerase (Figure 10.9, step 4). It is during the extension phase that real-time PCR differs from RT-PCR. In the version shown in the figure (there are several versions), the DNA synthesis reaction mixture contains SYBR® Green, a very sensitive DNA dye. SYBR® Green fluoresces very strongly when bound to DNA, but emits very little fluorescence when not bound to DNA. When the DNA is single stranded in the PCR, then, there is essentially no fluorescence detectable. But, as the primers are extended and double-stranded DNA is being made, the SYBR® Green dye binds to the double-stranded regions (Figure 10.9, step 4). Thus, as extension continues, more and more SYBR® Green molecules are bound to the DNA molecules, meaning that fluorescence increases. By quantifying that fluorescence, a researcher can measure, in real time as new DNA is being synthesized, the amount of double-stranded DNA in the reaction. This measurement requires the use of a special thermal cycler that uses laser detection of the fluorescence produced after each PCR cycle. The rate of production of SYBR® Green-labeled amplified DNA is compared with controls that

**Figure 10.9**

**The use of real-time PCR (and SYBR® Green) to determine the abundanc be of the mRNA for a gene of interest.**

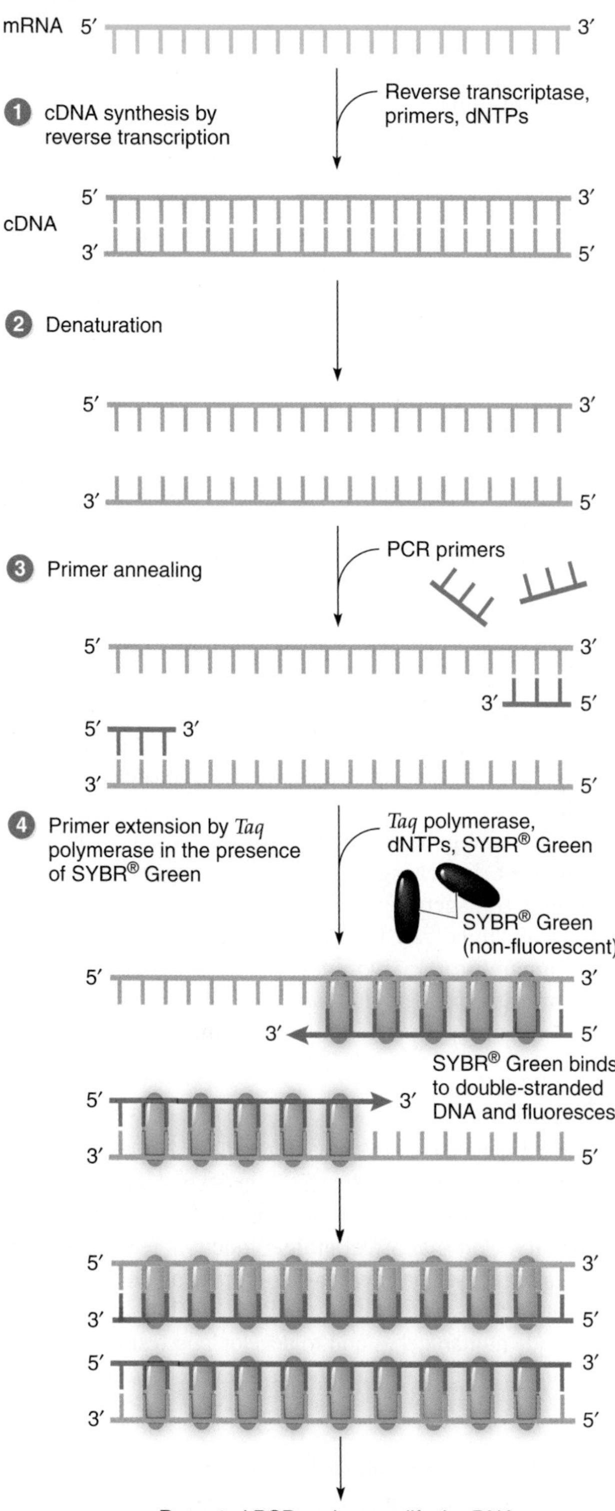


contain known amounts of a control mRNA. This allows the amount of mRNA in the experimental sample to be quantified.

Real-time PCR is used extensively to quantify mRNA levels for many genes in a wide range of cells and tissues in many organisms. For example, real-time PCR is used diagnostically to detect HIV (the virus that causes AIDS) and hepatitis C virus (this virus attacks the liver, causing inflammation and scarring of the liver, and damage caused by hepatitis C infections is the most common reason for liver transplants).

**Keynote**

The polymerase chain reaction (PCR) uses specific oligonucleotide primers to amplify a specific segment of DNA many thousandfold in an automated procedure. PCR has many applications in research and in the commercial arena, including generating specific DNA segments for cloning or sequencing, amplifying DNA to detect specific genetic defects, and amplifying DNA for DNA fingerprinting for crime scene investigations. If cDNA is used as a template for PCR (RT-PCR and real-time PCR), mRNAs can be detected and quantified.

## Applications of Molecular Techniques

In this section, we will discuss some basic applications of recombinant DNA technologies, going from DNA manipulation and analysis, to gene expression, then to protein analysis, and on to more specialized applications such as gene therapy. The applications are so broad-ranging that we can only scratch the surface. The examples have been chosen to describe some of the applications as case studies so that you can learn about the specific example while looking beyond it to see more generally the types of questions and hypotheses that can be investigated.

### Site-Specific Mutagenesis of DNA

The study of mutants is a cornerstone of genetics research. We learned in Chapter 7 that mutations can be induced in experimental organisms by treatment with mutagens. In making mutations this way, the whole genome is the target for the mutagen. Thus, each survivor of the mutagenesis likely has many mutations and the challenge is to find the mutants of interest by an appropriate screen or selection. Further, while mutations of a particular gene might well produce an altered phenotype that can be used in a screen or selection, the precise mutation in the gene is undirected because mutagenesis is random. However, if a researcher is studying the function of a particular gene, for example, and that gene has been cloned, then specific mutations can be targeted to any part of the gene *in vitro*. This is **site-specific mutagenesis.**

There are many procedures for site-specific mutagenesis, a number of them using PCR. Figure 10.10 shows one way in which a point mutation or small addition or deletion can be made in cloned DNA (such as a cloned gene) using a PCR-based mutagenesis approach. Four primers are used. Primer 1 is at the left end of the sequence to be amplified, and primer 2 is at the right end. Two other primers, 1M and 2M, match the target DNA sequence within its length, except where the mutation (M) is desired; 1M and 2M are complementary to each other. The mutation is symbolized in the figure as a "blip" in the primers. First, a PCR is done with primers 1 and 1M, and a second PCR is done with primers 2 and 2M. Then the primers are removed, and the two products A and B are mixed and the DNAs are denatured and allowed to reanneal. In some cases this results in pairing of a molecule of single-stranded A with a molecule of single-stranded B. DNA polymerase can then extend the 3′ ends of the strands in the central paired region, giving a full-length double-stranded DNA. This full-length molecule with the introduced mutation in the central region is then amplified using primers 1 and 2 and transformed into a cell to replace the wild-type sequence.

**Figure 10.10**

**An example of site-specific mutagenesis using PCR.**

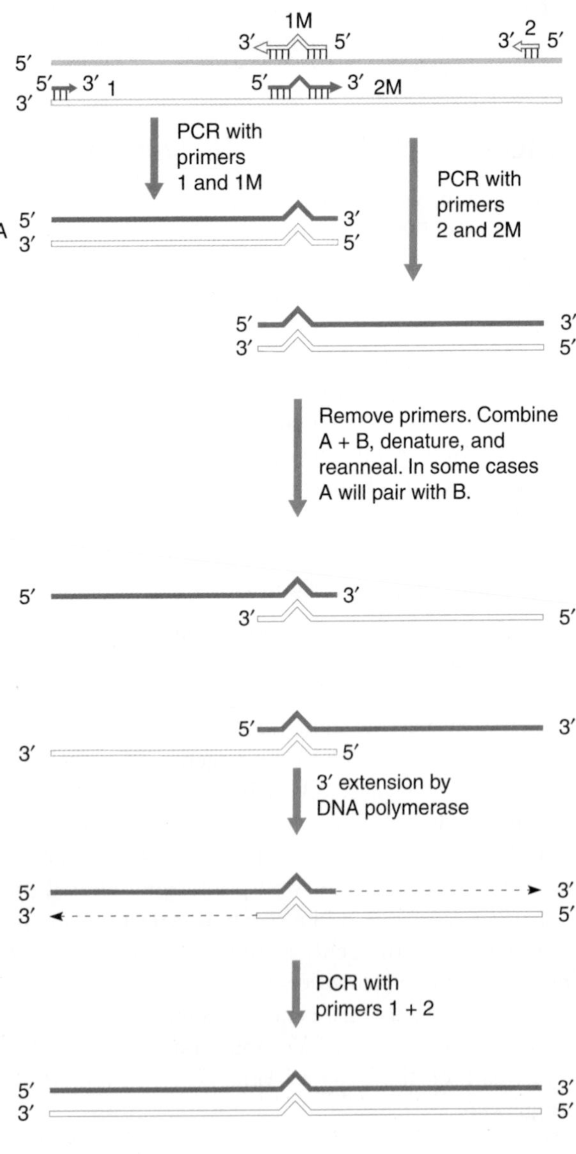


One application of site-specific mutagenesis is the creation of mutant mice. Since we cannot perform mutational studies with humans, researchers often attempt to mimic human mutations in mice. Such mouse models of human mutations are valuable for furthering our understanding of the gene involved and, in the case of disease genes, may move us toward diagnosis and a cure.

How could we study the function of a human gene in a mouse? Let us assume that we have cloned a human gene, and want to study the function of a similar gene in mice. We can easily clone or locate on the sequenced genomes the equivalent mouse gene because the two genes likely have a high degree of similarity. The cloned mouse gene can then be knocked out as described earlier (see Chapter 9, pp. 225–227). We can characterize the phenotypic defects apparent in these knockout mice, or, if we wish to study the human homologs of the gene in a model organism (this is done most frequently in the mouse), we can replace the mouse gene with a transgenic copy of the human gene. This process is called *humanization*, and is done either by modifying the mouse gene (using site-directed mutagenesis to change a cloned copy into something more similar to the human gene, and then using knockout techniques to replace the genomic copy with the mutated version) or by first knocking out the mouse gene and then adding a transgene expressing the human gene. These transgenic mice can be used, for instance, to test how the human protein would react to a candidate drug, without the concerns that would normally be associated with exposing people to a drug that might harm them.

**Keynote**

When a gene has been cloned, specific mutations can be made in that gene *in vitro*, and then studied *in vivo*. The mutations may be site-specific changes in the protein-coding region that affect protein function. Techniques can also be used to alter a gene in the genome of a model organism—making it similar to the human gene—to study human genes and to develop and test therapeutic treatments for genetic diseases.

## Analysis of Expression of Individual Genes

In this section, two illustrative examples of the use of recombinant DNA and PCR techniques to study gene expression are presented.

**Regulation of Transcription: Glucose Repression of the Yeast *GAL1* Gene.** In Chapter 18 we will discuss in detail the regulation of gene expression in eukaryotes. This example illustrates how recombinant DNA technology can be used to study gene transcription.

In the yeast *Saccharomyces cerevisiae*, the expression of the *GAL* (galactose) genes is induced by the carbon source, galactose, in the growth medium. The products of the *GAL* genes are enzymes that catalyze the breakdown of galactose. However, when yeast is grown on the preferred carbon source, glucose, the *GAL* genes are not transcribed. (The genetics of transcriptional regulation of the *GAL* genes is described in Chapter 18, pp. 522–523.) What happens if glucose is added to a culture of yeast already growing in medium containing galactose? The *GAL* genes are turned off. Not only is transcription of the *GAL* genes stopped, but the *GAL* mRNAs in the cell are rapidly degraded. The latter was shown in the following experiment, the results of which are illustrated in Figure 10.11. Yeast cells were grown so that their *GAL* genes were turned on. Then, at time zero, glucose was added and samples were taken at various times thereafter. RNA was extracted from each of the samples, and the RNAs were separated by agarose gel electrophoresis. Northern blotting was then performed and the blot was probed for the mRNA of one of the *GAL* genes, *GAL1*, using a radioactive probe, such as a riboprobe made as described earlier. Visually, it is easy to see in Figure 10.11 that the amount of hybridization decreased rapidly in the 45-minute span of the sampling period. When these results were quantified and plotted on a graph, it was seen that there is a very rapid loss of mRNA in the first 10 minutes and a more gradual loss thereafter.

**Alternative pre-mRNA Splicing: *P* Element Transposition in *Drosophila*.** In Chapter 18 we will discuss in detail alternative splicing—the removal of different amounts of a pre-mRNA molecule as a result of the use of different splice sites—as one of the levels of regulation of gene expression in eukaryotes. The result is different mRNA molecules which encode proteins with different functions. Here we discuss the expression of a gene that is alternatively spliced. The gene encodes an enzyme responsible for transposition of *P* elements (a type of transposable element) in *Drosophila melanogaster* (see Chapter 7, pp. 159–160, and Figure 7.27).

*P* elements are a common transposable element in many strains of *Drosophila melanogaster.* The *P* elements are generally stable in the fly—in most circumstances, their rate of transposition is very low. *P* elements are almost never able to transpose in body tissues, but for a fly with a father who carried *P* elements (and passed them to his offspring) and a mother who did not carry *P* elements, the *P* elements are able to transpose only in the germline (reproductive) tissues. This activation of *P* elements is called *hybrid dysgenesis*.

The *P* element itself carries a single gene, and this gene encodes *P* transposase, the enzyme required for transposition of *P* elements. The gene encoding *P* transposase has been cloned molecularly; it is quite small, spanning less than 3 kb and contains 4 exons and 3 introns (Figure 10.12). Two transcripts have been identified using northern blot analysis of poly(A)+ RNAs isolated from flies undergoing hybrid dysgenesis and the cloned *P* element as a probe. The smaller is about 200 bases smaller than the larger transcript. Normal flies have only the larger transcript. DNA sequencing of cDNAs prepared from the mRNAs using reverse transcriptase (see Chapter 8, pp. 195–197 and Figure 8.15) indicates that the transcripts are produced by alternative splicing. Specifically, in the bodies of all flies, the third intron is ignored by the splicing machinery and left in the final mRNA (see Figure 10.12, left side). This results in a larger mRNA, but it codes for a smaller protein, because there is an in-frame stop codon in the retained intron. This protein is unable to act as a transposase. In the germline of a fly undergoing hybrid dysgenesis, all of the introns are spliced out, and the resulting mRNA encodes an active *P* element transposase (see Figure 10.12, right side).

**Figure 10.11**
**Regulation of transcription of the yeast *GAL1* gene by glucose.** Glucose was added at time zero, and the amount of *GAL1* transcribed was analyzed at various times thereafter by blotting and probing, as described in the text.

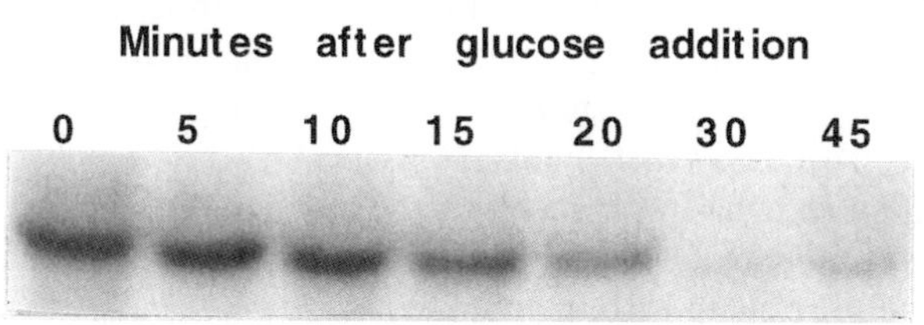


## Analysis of Protein–Protein Interactions

We study genes and their products because we want to understand the structure and function of cells and organisms. As we have been learning about proteins and their roles in the cell, we have discovered that many cellular functions are carried out by proteins that contact one another. We have already seen some examples of this, such as the α-globin and β-globin polypeptides in hemoglobin and the transcription factors interacting with one another and with RNA polymerase to form a complex that initiates transcription (see Chapter 5, pp. 88–89).

**animation**
**The Yeast Two-Hybrid System**

One experimental procedure to find genes which encode proteins that interact with a known protein is the **yeast two-hybrid system** (also called the *interaction trap assay*) developed by Stanley Fields and his coworkers

**Figure 10.12**

**Alternative tissue-specific splicing in the *P* transposase gene of *Drosophila melanogaster*.** In the body, the third intron is not spliced out; as a result, the transcript does not encode a functional transposase enzyme, while in the germline, all introns are spliced out, and the mRNA encodes the functional transposase enzyme.

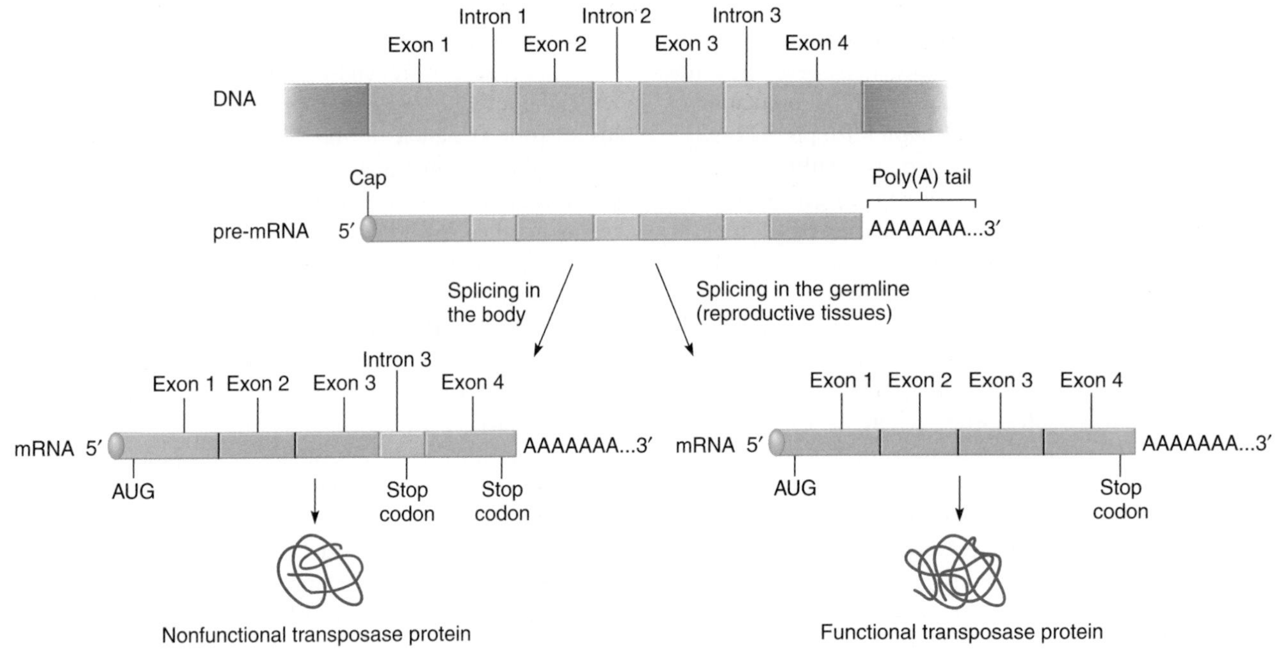




(Figure 10.13). Here is how it works. For the yeast galactose metabolizing gene *GAL1* to be transcribed, a regulatory protein called Gal4p (encoded by the *GAL4* gene) binds to a promoter element called the upstream activator sequence G, or $UAS_G$ (see Figure 10.13). Gal4p has two domains: a DNA-binding domain (BD) that binds directly to $UAS_G$ and an activation domain (AD) that facilitates the binding of RNA polymerase to the promoter and the initiation of transcription.

In the two-hybrid system, two types of yeast expression plasmids are used. One type contains the sequence for the Gal4p BD fused to the sequence for the known protein (X). The other type contains the Gal4p AD sequence fused to protein-coding sequences encoded by a library of cDNAs (Y). A yeast strain is cotransformed with the BD plasmid and the AD plasmid library so that each transformant has the BD plasmid and one of the plasmids from the AD library. In the chromosome of the yeast strain into which the plasmids are transformed is a *reporter gene*—a gene that encodes a readily assayable product—with a $UAS_G$. In Figure 10.13, the reporter gene is the *lacZ* gene from *E. coli* that encodes β-galactosidase. Yeast colonies expressing this enzyme turn blue in the presence of the colorless substrate X-gal (see Chapter 8, p. 176). The reporter gene is expressed only if the unknown protein (Y) of the AD fusion protein interacts with the known protein (X) of the BD fusion protein, thereby bringing the AD and BD domains close together and activating transcription of the reporter gene. If X and Y do not interact, the AD and BD parts of Gal4p stay separate, and transcription of the reporter gene is not activated. In other words, the BD fusion protein acts as a *bait* for the interacting protein or proteins. When an interaction is seen, as evidenced by expression of the reporter gene, the AD fusion plasmid from that yeast transformant can be isolated and the cDNA sequence used to find the genomic gene for study.

One example of the use of the two-hybrid system involves studies of interactions between human proteins called peroxins—encoded by *PEX* genes—that are required for peroxisome biogenesis. (The peroxisome is a single-membrane organelle present in nearly all eukaryotic cells; one of the most important metabolic processes of the peroxisome is the β-oxidation of long-chain fatty acids.) The two-hybrid system has shown that the PEX1 and PEX6 proteins interact in normal individuals, but disruption of that interaction is the most common cause of a variety of neurological disorders, such as Zellweger syndrome (OMIM 214100). Individuals with Zellweger syndrome have lost many peroxisome enzyme functions, they have severe neurological, liver, and renal abnormalities and mental retardation, and they die in early infancy.

Figure 10.13

**Detecting protein–protein interactions using the yeast two-hybrid system.**

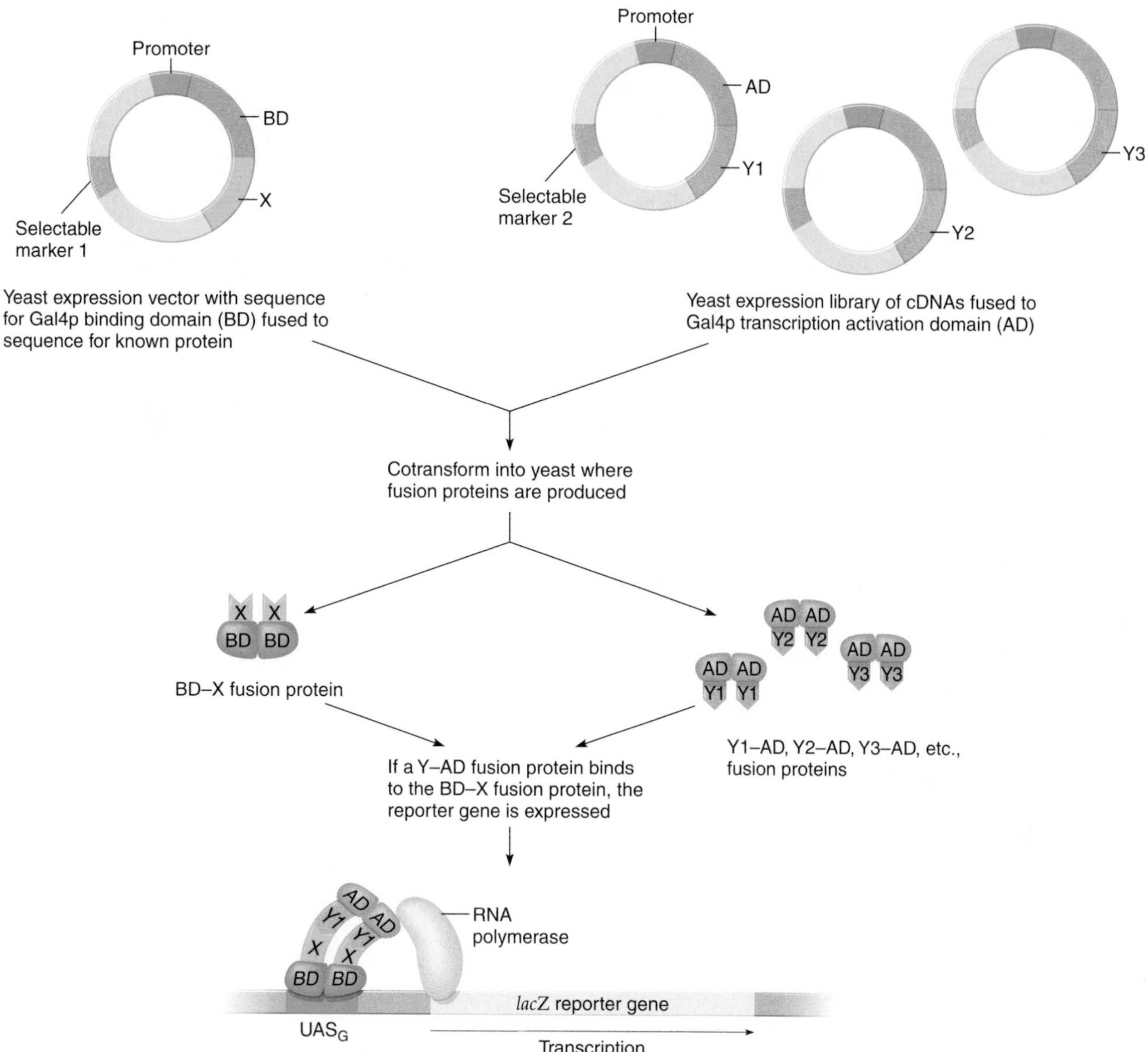




**Keynote**

Recombinant DNA techniques and PCR are widely used in the analysis of basic biological processes. For example, DNA can be analyzed in detail (such as in the construction of restriction maps), RNA transcripts can be sized and quantified, RNA processing events can be monitored, and protein–protein interactions can be studied.

## Uses of DNA Polymorphisms in Genetic Analysis

To this point in the book we have mostly focused on **genes** as markers for genetic analysis. Genes have different **alleles** that produce different phenotypes that can be followed in crosses. For example, to build a picture of the arrangement of genes in the genome—a genetic map—crosses are made between parents differing in alleles of two or more genes and the fraction of recombinant phenotypes among the progeny is determined. Mapping genes will be discussed in detail in Chapter 14. The results from genetic mapping crosses indicate the location on the chromosome—the **locus**—for each gene that is mapped.

A **DNA polymorphism** is one of two or more alternate forms (alleles) of a chromosomal locus that differ either in nucleotide sequence (like the SNPs you learned about in Chapter 8, pp. 192–193) or have variable numbers of tandemly repeated nucleotide units or *indels.* (**Indel** is a word created from the words “insertion” and “deletion” and refers to short stretches of insertions or deletions in the genome.) This definition introduces the concept of an allele that sometimes is something other than a form of a gene, because a DNA polymorphism can be anywhere in the genome, not necessarily as part of a gene. In addition, in order to include both the location of genes and of DNA polymorphisms in our definition of locus, we must broaden the concept of a locus to include any chromosomal location. Many DNA polymorphisms are useful for genetic mapping studies (and other uses), and those are called **DNA markers.** Since there are no products that interact to give a phenotype, the alleles of DNA markers are codominant; that is, they do not show dominance or recessiveness, as is seen for the alleles of most genes. DNA markers are detected using molecular tools (generally hybridization on Southern blots or DNA microarrays, or by PCR tests) that focus on the DNA itself rather than on the gene product or associated phenotype. With genes and DNA markers, map distances can be calculated between genes, between DNA markers, or between a gene and a DNA marker. DNA polymorphisms have a number of other useful applications apart from mapping, as we shall see later in this section.

## Classes of DNA Polymorphisms

We will consider three major classes of DNA polymorphisms—*single nucleotide polymorphisms (SNPs)*, *short tandem repeats (STRs)*, and *variable number of tandem repeats (VNTRs)*—and describe ways in which they may be analyzed. Our focus is on the human genome, but these polymorphisms occur also in the genomes of other organisms.

### Single Nucleotide Polymorphisms (SNPs, "Snips").

As described in Chapter 8, pp. 192–193), SNPs can be used for genomic characterization and mapping. Here we will discuss the use of individual SNPs in more detail.

***Detection of SNPs That Alter Restriction Sites.*** A small fraction of SNPs affect restriction sites, either creating them or eliminating them. Such SNPs can be detected using the restriction enzyme for the site and either Southern blot analysis or, more typically these days, by PCR. The different patterns of restriction sites in different genomes result in **restriction fragment length polymorphisms** (RFLPs, "riff-lips"), which are restriction enzyme-generated fragments of different lengths. The usefulness of RFLPs will be apparent in the following examples.

Figure 10.14 illustrates the Southern blot analysis approach to study SNPs that affect restriction sites. Figure 10.14 shows a theoretical 7-kb segment in the genome with a pair of SNP alleles, one of which (SNP allele 1) is a T-A base pair in a *Bam*HI restriction site, and the other of which (SNP allele 2) is a C-G base pair that eliminates that site. The site is 2 kb from the left-hand *Bam*HI site. Determining which SNP alleles are present involves the Southern blot analysis steps shown in Figure 10.8. That is, genomic DNA is isolated, digested with *Bam*HI, and the fragments are separated by agarose gel electrophoresis. After transferring the fragments to a membrane filter, DNA fragments of interest

**Figure 10.14**

**Southern blot analysis method for studying SNPs that affect restriction sites.** A 7-kb section of the chromosome has *Bam*HI sites at each end. SNP allele 1 (top) has a *Bam*HI site 2 kb from the left end, whereas SNP allele 2 (bottom) has a C-G base pair in place of a T-A base pair, so that the *Bam*HI site has been lost. *Bam*HI digestion with DNA samples from individuals with different SNP genotypes, followed by Southern blot analysis using the probe shown, gives the DNA banding patterns at the bottom.

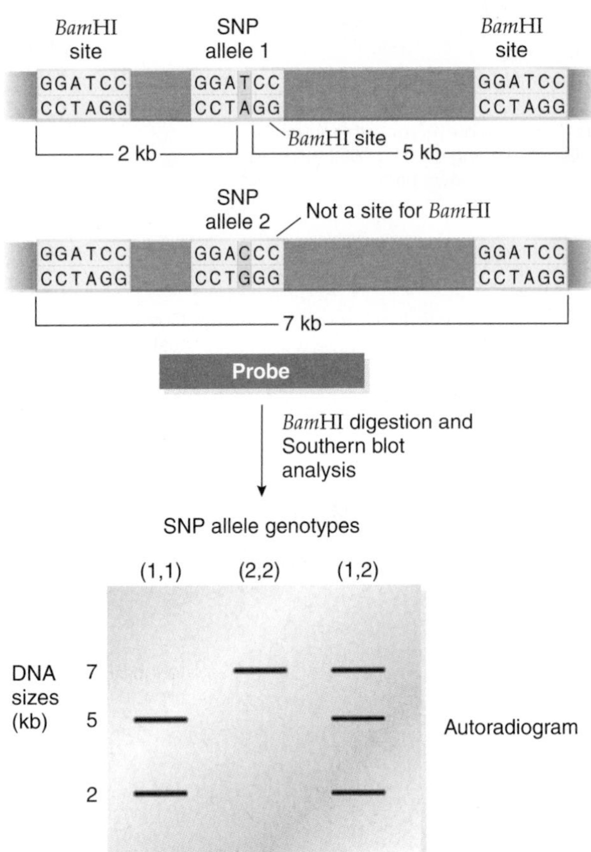


are visualized by hybridization with a labeled probe (which, here, spans a large part of the DNA shown in Figure 10.14), followed by autoradiography. The results are shown for possible genotypes in the bottom of the figure. When we probe a Southern blot, the probe will anneal to any fragment that can form base pairs with the probe, and so the probe can bind to more than one band, as shown in Figure 10.14. A homozygote for SNP allele 1 (1,1), which has the intact *Bam*HI site, will show two bands, one of 5 kb and one of 2 kb. A homozygote for SNP allele 2 (2,2), which has lost the *Bam*HI site, will show one band of 7 kb. A heterozygote for the two SNP alleles (1,2) will show three bands of 7 kb (from the homolog with allele 2), 5 kb, and 2 kb (the latter two from the homolog with allele 1).

Figure 10.15 illustrates the *PCR-RFLP analysis method*. We consider a 2,000-bp segment of the genome with a similar pair of SNP alleles as above affecting a *Bam*HI site that is 500 bp from the left end of the segment. Primers for PCR are available that recognize the DNA at

**Figure 10.15**

**PCR method for studying SNPs that affect restriction sites.** A 2,000-bp section of the chromosome has SNP alleles 500 bp from the left end. The TA-to-CG change from SNP allele 1 (top) to SNP allele 2 (bottom) alters a *Bam*HI site to a sequence that is not recognized by a restriction enzyme. PCR of DNA samples from individuals with different SNP allele genotypes using the left and right primers shown, followed by *Bam*HI digestion, gives the DNA-banding pattern at the bottom.

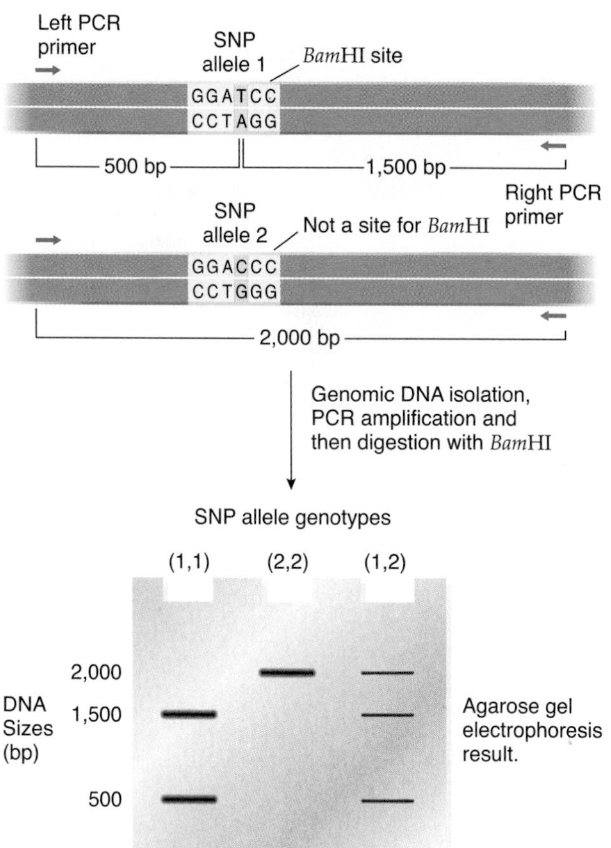


the left and right ends. PCR analysis of SNP alleles affecting restriction sites involves isolating genomic DNA, amplifying the DNA segment of interest using the left and right primers, digesting the amplified fragment with the restriction enzyme (*Bam*HI, here), and using agarose gel electrophoresis to examine the sizes of the fragments produced. For our example, the results for possible genotypes are shown in the bottom of the figure. A homozygote for the SNP allele 1 (1,1) will give an amplified DNA fragment that can be digested with *Bam*HI to produce 1,500-and 500-bp fragments. A homozygote for the SNP allele 2 (2,2) will give a 2,000-bp fragment, and a heterozygote for the two alleles (1,2) will give 2,000-, 1,500-, and 500-bp fragments.

***Detection of All SNPs.*** Since most SNPs do not affect restriction sites, other methods of analysis were needed to analyze SNPs generally. You can imagine that analyzing one particular SNP locus is a challenge because, in humans, this is one base pair difference in the 3 billion base pairs genome.

Individual SNPs can be analyzed by **allele-specific oligonucleotide (ASO) hybridization analysis** (Figure 10.16). In this procedure, short oligonucleotide probes are synthesized that are complementary to each SNP allele, and each oligonucleotide is spotted onto (and then chemically linked to) a membrane filter. We can then take DNA from the individual whose genotype we want to determine, and use this DNA as a template for PCR. The primers for this PCR are designed to amplify the region containing the SNP. Some of the nucleotides used for the PCR are radioactively or chemically labeled, giving a labeled PCR product (the target DNA). The labeled target DNA molecules are then separated to single strands and added to the filter with the unlabeled SNP allele probes. A target DNA strand can hybridize with an SNP probe if their sequences are complementary. This hybridization step is performed under *high stringency*, meaning that the conditions favor *only* a perfect match between probe and target DNAs. If hybridization occurs,



**Figure 10.16**

**Typing of an SNP by allele-specific oligonucleotide (ASO) hybridization analysis.** SNP allele oligonucleotide probes are bound to a filter. PCR is used to amplify the target DNA region containing the SNP locus. During the amplification, the DNA is labeled radioactively or chemically. The labeled target DNA is hybridized to the unlabeled SNP probes on the filter under conditions in which the target DNA can base pair with the probe only if the sequences are completely complementary (top of figure). The hybridization is visualized by detecting the label of the target DNA now bound to the probe on the filter. Under the hybridization conditions used, even a single base-pair mismatch—an SNP polymorphism—is enough to prevent hybridization of target DNA and probe (bottom of figure). No label is detected in this case.

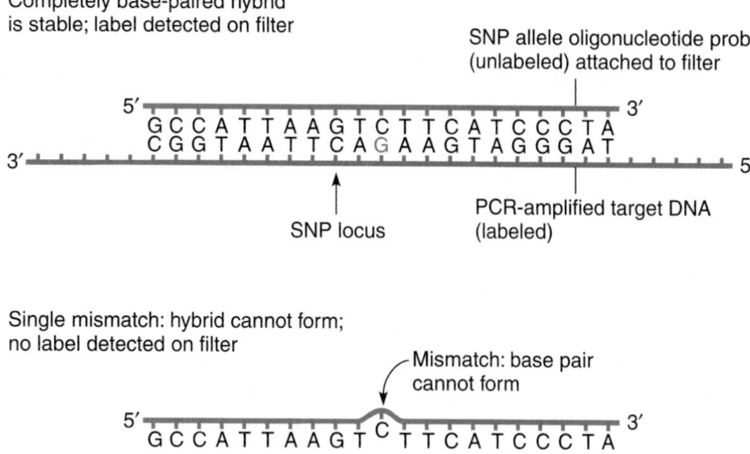

the target DNA matches the SNP allele probe on that particular filter. That hybridization is visualized by detecting the presence of the label on the target DNA at a particular, known SNP allele probe spot on the filter. Under these same high-stringency conditions, an SNP allele probe will not hybridize with a target DNA that has even just a single base-pair mismatch. That is, an SNP allele probe will not hybridize with target DNA containing any other SNP allele for the locus.

**Short Tandem Repeats (STRs). Short tandem repeats (STRs)**—also called **microsatellites** or *simple sequence repeats* (SSRs)—are 2–6-bp DNA sequences that are tandemly repeated. At each STR locus, an STR sequence can be repeated anywhere from just a few times up to about 100 times. Examples are the dinucleotide repeat, $(GT)_n$, and the trinucleotide repeat, $(CAG)_n$. A recent count for STRs in the human genome is 128,000 two-nucleotide sites, 8,740 three-nucleotide, 23,680 four-nucleotide, 4,300 five-nucleotide, and 230 six-nucleotide repeat sites. The six-nucleotide repeats include the repeated sequences found at the telomeres.

Many STRs are polymorphic in a population, so they have become valuable in several types of study, including genetic mapping and forensics. Because the overall length of an STR is relatively short, PCR is the preferred method for analyzing STR polymorphic loci (Figure 10.17). Two alleles of an STR locus are shown, one with 6 copies of the GATA repeat, and the other with 10 copies. In a population, there will be many different length alleles at an STR locus. One particular human STR locus with the GATA repeat has alleles from 6 to 15 copies, for example. The analysis uses primers that flank the locus. PCR will produce DNA fragments of different lengths, consisting of the STR span plus the DNA from the STR to the left and right primers. For these two alleles, the DNA fragments will differ by 16 bp due to the four-repeat difference in the repeat length. In analyzing genomic DNA from different individuals, this PCR approach can distinguish homozygotes and heterozygotes, as well as defining the actual copy number of each repeat, both from the lengths of the DNAs amplified.

**Figure 10.17**

**Using PCR to determine which STR (microsatellite) alleles are present.** Genomic DNA is isolated, and PCR primers flanking an STR locus are used to amplify the repeats. The sizes of the DNA fragments produced are determined by agarose gel electrophoresis. In the figure, STR allele 1 has 6 repeats of GATA, and STR allele 2 has 10 repeats of GATA. The gel shows the three possible genotypes for these two alleles: (6,6) [i.e., both homologs have the six-repeat allele], (10,10), and (6,10). In reality, there typically is a lot of variation in repeat numbers at an STR locus.

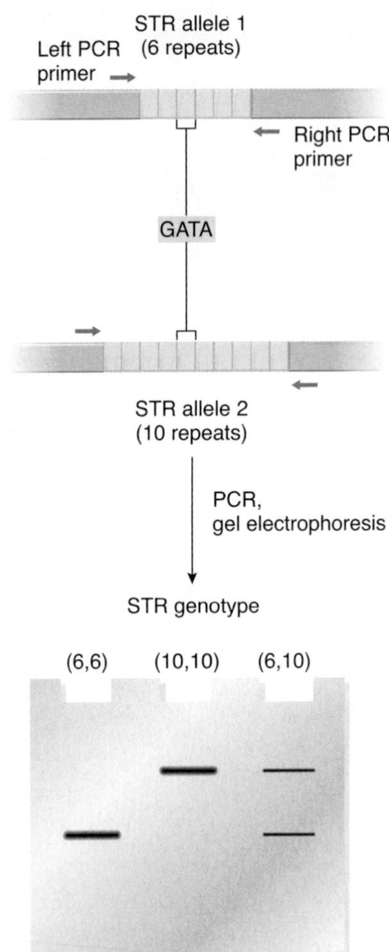


**Variable Number Tandem Repeats (VNTRs). Variable number tandem repeats (VNTRs)**—also called **minisatellites**—are similar to STRs, but the repeating unit is larger than that for STRs, ranging from 7 to a few tens of base pairs in length per repeat. VNTRs were first discovered by Alec J. Jeffreys in 1985. This was the first demonstration of DNA sequence polymorphism in the human genome. There are far fewer VNTR loci in the human genome than STR loci.

VNTR loci also show polymorphisms. VNTR repeat lengths are longer than those in STRs, so PCR is usually not a convenient way to analyze VNTRs because of the overall length of DNA that would have to be amplified for the VNTR locus. Instead, restriction digestion and Southern blot analysis is more typically used to study VNTRs. That is, genomic DNA is isolated and cut with a restriction enzyme that cuts on either side of the VNTR locus. The restriction fragments are separated by gel electrophoresis and transferred to a membrane filter by Southern blotting. The length of the VNTR allele is then determined by using a probe for the particular repeat sequence of the VNTR locus. As in the STR analysis described above, the results indicate the allele(s) present in the genome being studied. For example, an individual could be homozygous or heterozygous for alleles at a particular VNTR locus. In a population study, the range of alleles for a locus can be determined.

There are two types of VNTR loci: unique loci and multicopy loci. In other words, there may be only one copy of a VNTR locus in an organism's genome (with its own unique repeat sequence), or there may be a number of copies of this repeat scattered around the genome. If a probe detects only one VNTR locus, it is called a *monolocus*, or *single-locus, probe.* Probes that detect VNTR loci at a number of sites in the genome are known as *multilocus probes.*

**Keynote**

A DNA polymorphism is one of two or more alternate forms of a locus that either differ in nucleotide sequence or have variable numbers of tandemly repeated sequences. Polymorphic loci are DNA markers that, analogous to genes, can be used in mapping experiments, as well as for other applications. The phenotypes of polymorphic loci are the DNA variations that are analyzed molecularly. Examples of DNA polymorphisms are single nucleotide polymorphisms (SNPs), short tandem repeats (STRs), and variable number of tandem repeats (VNTRs).

## DNA Molecular Testing for Human Genetic Disease Mutations

DNA polymorphisms of all types can be used in diagnostic testing for human diseases. DNA polymorphisms are both abundant and easily tested, so if we know the general chromosomal location of a gene that causes a specific genetic disease, it is generally possible to find polymorphic DNA markers near the gene, or even polymorphic markers that are contained within the gene. We can then test for the inheritance of these polymorphic markers, and attempt to predict whether the individual did, or did not, inherit the disease allele. Obviously, this is easier when the DNA polymorphism is part of the gene itself, rather than near the gene. Throughout this text there are many examples of human genetic diseases. These diseases are caused by enzyme or other protein defects that, in turn, are the result of mutations at the DNA level. For an increasingly large number of genetic diseases, including Huntington disease (OMIM 143100), hemophilia (OMIM 306700), cystic fibrosis (OMIM 219700), Tay-Sachs disease (OMIM 272800), and sickle-cell anemia (SCA; OMIM 141900), we can perform DNA molecular tests for the presence of mutations associated with the disease. In this section, practical issues of DNA molecular testing are discussed along with some examples of the testing approaches used. The mutations involved fall into the classes of DNA polymorphisms we have just discussed, so we will be able to see some practical applications of the methods that use those polymorphisms.

**Concept of DNA Molecular Testing.** **Genetic testing** determines whether an individual who has symptoms, or is at a high risk of developing a genetic disease because of family history, actually has a particular gene mutation. **DNA molecular testing** is a type of genetic testing that focuses on the molecular nature of mutations associated with disease. Designing DNA molecular tests, then, depends on having knowledge about the types of mutations that occur in the specific gene that causes the disease of interest. This information comes from sequencing the gene involved (once it has been identified).

A complication of genetic testing is that many different mutations of a gene can cause loss of function and therefore lead to the development of the disease. Often, no single molecular test can detect all mutations of the disease gene in question. For example, two genes, *breast cancer one* (*BRCA1* [OMIM 113705]) and *breast cancer two* (*BRCA2* [OMIM 600185]) are implicated in the development of breast and ovarian cancer. When functioning normally, the *BRCA1* and *BRCA2* gene products help control the cell growth in breast and ovarian tissue. However, mutations that cause loss of or abnormal function of the genes' products can increase the chance of the development of cancer. (See Chapter 20, p. 593 for more information on *BRCA* genes and cancer.) Hundreds of mutations have been identified in *BRCA1* and *BRCA2*, but the risk of developing breast cancer varies widely among individuals depending on the mutations they carry. Obviously this makes it impossible to develop a single DNA molecular test for *BRCA* gene mutations. Later in this section we discuss a microarray test to test for *BRCA* gene mutations.

It is important to recognize that a genetic test primarily tells an investigator whether an individual has a mutation known to be associated with a genetic disease. However, genetic testing is distinct from screening for a disease. That is, screening usually is done on people without symptoms or a family history for the disease, whereas genetic testing is done on a targeted population of people with symptoms of, or a significant family history of, the disease. For example, mammograms are clinical screening tests that detect breast lesions that might lead to cancer before there are clinical symptoms. Genetic testing for breast cancer, by contrast, reveals the presence or absence of mutations potentially associated with the development of breast cancer, although it cannot predict whether or when breast cancer will develop.

In the same vein, genetic tests are different from diagnostic tests for a disease. Diagnostic tests reveal whether a disease is present and to what extent the disease has developed. For example, a biopsy of a lump in the breast is a diagnostic test to determine whether the lesion is benign or cancerous.

**Purposes of Human Genetic Testing.** Genetic testing is done for three main purposes: *prenatal diagnosis*, *newborn screening*, and *carrier (heterozygote) detection*.

Prenatal diagnosis is done to assess whether a fetus is at risk for a genetic disorder. Amniocentesis or chorionic villus samples can be taken and analyzed for a specific gene mutation or for biochemical or chromosomal abnormalities (see Chapter 4, p. 74). If both parents are asymptomatic carriers (heterozygotes) for a genetic disease gene, for example, there is a $^{1}/_{4}$ chance of the fetus being homozygous for the mutant allele, and the risk of developing the disease is likely to be very high. More recently, techniques have been developed to test embryos produced by *in vitro* fertilization for genetic disorders before implanting them in the mother. Embryos containing mutated genes that could lead to serious genetic disease can then be removed before implantation is performed.

Testing individuals to see whether they are carriers (heterozygotes) for a recessive genetic disease is done to identify those who may pass on a deleterious gene to their offspring. Carriers can be detected now for a large number of genetic diseases, including Huntington disease (a disease that causes progressive neural degeneration; OMIM 143100), Duchenne muscular dystrophy (a progressive disease resulting in muscle atrophy and muscle dysfunction; OMIM 310200), and cystic fibrosis (a disease that interferes with normal mucus formation in the lungs, and results in respiratory difficulties; OMIM 602421). Newborns can also be tested for specific mutations. For example, we mentioned in Chapter 4 (pp. 67–68) that all newborns in the United States are tested for PKU (phenylketonuria: OMIM 261600) using the Guthrie test with blood taken from the newborn. Other tests are available for groups at high risk for other genetic disorders, such as sickle-cell anemia (OMIM 141900) in African Americans and Tay-Sachs disease (OMIM 272800) in Ashkenazi Jews. These genetic tests, including the DNA molecular tests described below, typically are done using blood samples or cheek swabs.

**Examples of DNA Molecular Testing.** For DNA molecular testing, DNA samples typically are analyzed by restriction enzyme digestion and Southern blotting, by procedures involving PCR, or by DNA microarray analysis. In this section we discuss some examples of these testing approaches.

***Testing by Restriction Fragment Length Polymorphism Analysis.*** A mutation associated with a genetic disease may cause the loss or addition of a restriction site either within the gene or in a flanking region. As we learned in the previous section, the chromosomal site where the mutation occurs is a SNP locus, and the different patterns of restriction sites result in restriction fragment length polymorphisms (RFLPs). Remember that DNA markers are codominant, so we can determine the exact genotype of an tested individual, even when the disease itself is seen only in individuals homozygous for a recessive allele.

*animation*
**DNA Molecular Testing for Human Disease Gene Mutations**

A small number of RFLPs are associated with genes known to cause diseases, as the following example about sickle-cell anemia illustrates. In sickle-cell anemia (SCA), a single base-pair change in the gene for the β-globin polypeptide of hemoglobin results in an abnormal form of hemoglobin, Hb-S, instead of the normal Hb-A (see Chapter 4, pp. 70–71). Hb-S molecules associate abnormally, leading to sickling of the red blood cells, tissue damage, and possibly death.

**Figure 10.18**
**The beginning of the β-globin gene, mRNA, and polypeptide showing the normal Hb-A sequences and the mutant Hb-S sequences.** The sequence differences between Hb-A and Hb-S are shown in red. The mutation alters a *Dde*I site (boxed in the Hb-A DNA).

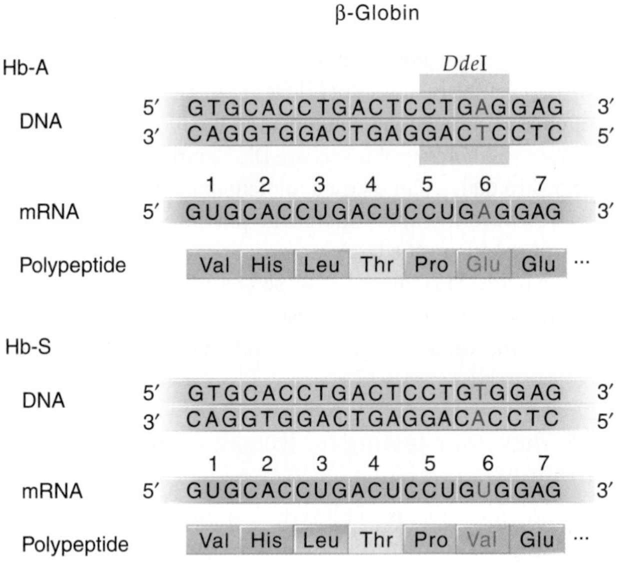

β-Globin

Hb-A
*Dde*I
DNA 5′ GTGCACCTGACTCCTGAGGAG 3′
3′ CAGGTGGACTGAGGACTCCTC 5′
1 2 3 4 5 6 7
mRNA 5′ GUGCACCUGACUCCUGAGGAG 3′
Polypeptide Val His Leu Thr Pro Glu Glu ...

Hb-S
DNA 5′ GTGCACCTGACTCCTGTGGAG 3′
3′ CAGGTGGACTGAGGACACCTC 5′
1 2 3 4 5 6 7
mRNA 5′ GUGCACCUGACUCCUGUGGAG 3′
Polypeptide Val His Leu Thr Pro Val Glu ...

The sickle-cell mutation changes an A-T base pair to a T-A base pair so that the sixth codon for β-globin is changed from GAG to GUG. As a result of this SNP allele, valine is inserted into the polypeptide instead of glutamic acid (Figure 10.18). The mutational change also generates an RFLP for the restriction enzyme *Dde*I ("D-D-E-one"). The *Dde*I restriction site is

5′-CTNAG-3′
3′-GANTC-5′

where the central base pair can be any of the four possible base pairs. The A-T to T-A mutation changes the fourth base pair in the restriction site. Thus, in the normal β-globin gene, $\beta^A$, there are three *Dde*I sites, one upstream of the start of the gene and the other two within the coding sequence (Figure 10.19a). In the sickle-cell mutant β-globin gene, $\beta^S$, the mutation has removed the middle *Dde*I site (see Figure 10.18), leaving only two *Dde*I sites (see Figure 10.19a). When DNA from normal individuals is cut with *Dde*I and the fragments separated by gel electrophoresis are transferred to a membrane filter by the Southern blot technique and then probed with the 5′ end of a cloned β-globin gene, two fragments of 175 bp and 201 bp are seen (Figure 10.19b). DNA from individuals with SCA analyzed in the same way gives one fragment of 376 bp because of the loss of the *Dde*I site. Heterozygotes are detected by the presence of three bands of 376 bp, 201 bp, and 175 bp.

Not all RFLPs result from changes in restriction sites directly related to the gene mutations. Many result from changes to the DNA flanking the gene, sometimes a fair distance away. This is the case for a RFLP that is related to

**Figure 10.19**

**Detection of sickle-cell gene by the *Dde*I restriction fragment length polymorphism.** (a) DNA segments showing the *Dde*I restriction sites. (b) Results of analysis of DNA cut with *Dde*I, subjected to gel electrophoresis, blotted, and probed with a β-globin probe.

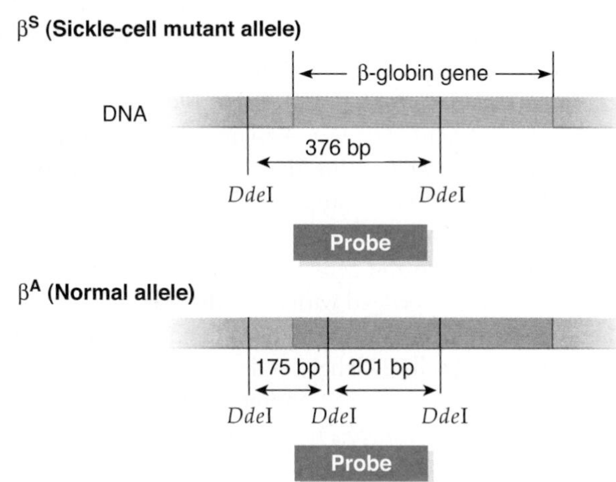


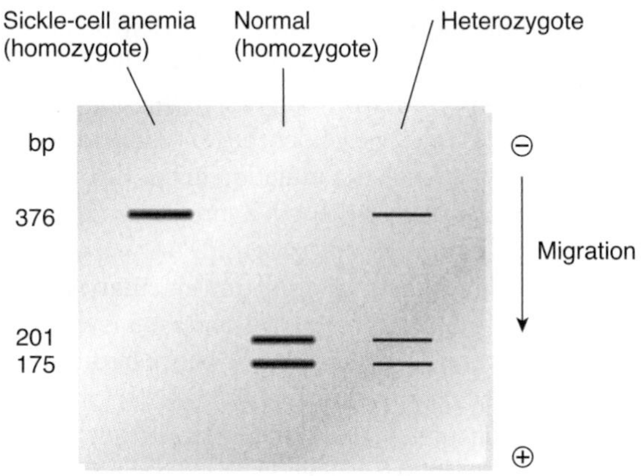


**Figure 10.20**

**DNA molecular testing for mutations of the open-angle glaucoma gene *GLC1A*, using PCR and allele-specific oligonucleotide (ASO) hybridization.** (a) Sequence of part of the *GLC1A* gene from a heterozygote showing a mutation from C to T, causing a Pro-to-Leu change in the polypeptide at amino acid 370. (b) Sequences of the two allele-specific oligonucleotides (ASOs), one for the wild-type allele and one for the mutant allele. (c) Results (theoretical) of hybridization with radioactive PCR copies of the *GLC1A* gene used as a probe on dots containing either the wild-type or mutant ASOs for homozygous normal, homozygous mutant, and heterozygous individuals.

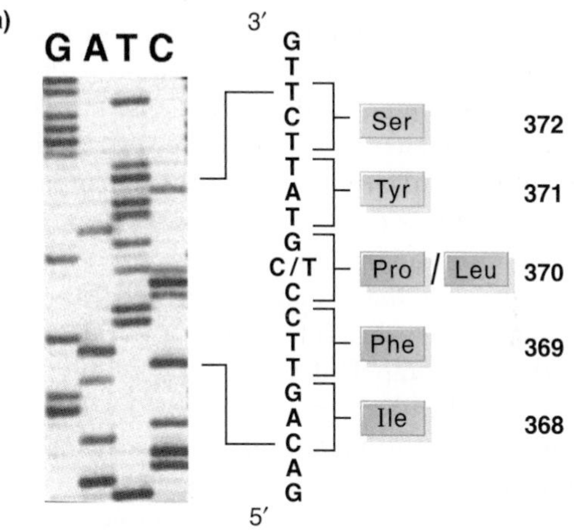


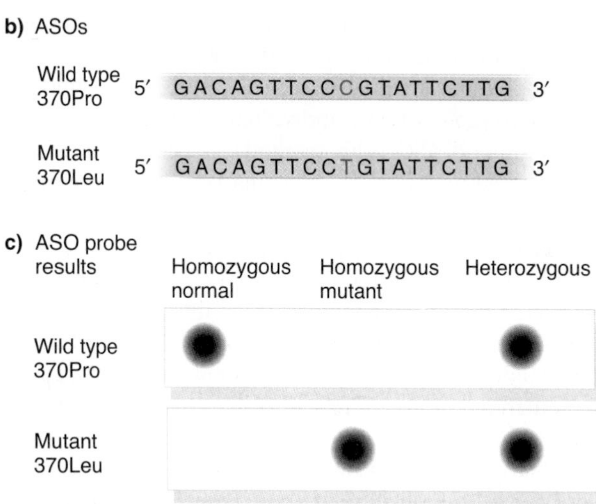


the genetic disease PKU (OMIM 261600; see Chapter 4, pp. 66–68). Recall that PKU results from a deficiency in the activity of the enzyme phenylalanine hydroxylase. After digestion of genomic DNA with *Hpa*I ("hepa-one"), Southern blotting, and probing with a cDNA probe derived from phenylalanine hydroxylase mRNA, different-sized restriction fragments are produced from DNA isolated from individuals with PKU and from DNA isolated from homozygous normal individuals. This RFLP results from a difference outside the coding region of the gene, in this case to the 3′ side of the gene. The RFLP can be used to test for the PKU mutant gene in fetuses after amniocentesis or chorionic villus sampling. In these cases, detection of the mutation relies on the flanking RFLP segregating most of the time with the gene mutation. In rare cases, recombination occurs between the RFLP and the gene of interest, and is one potential difficulty in interpreting the results.

***Testing Using PCR Approaches.*** DNA molecular tests using PCR can be developed only when sequence information is available because the PCR primers cannot be designed without such information. One common test using PCR is allele-specific oligonucleotide (ASO) hybridization (see Figure 10.16). The principles are illustrated in this example of testing for mutations of the *GLC1A* gene (OMIM 137750), one of several genes that, when mutated, cause open-angle glaucoma (Figure 10.20). Glaucoma generally is caused by increasing pressure in the eye. Open-angle glaucoma is by far the most common form of glaucoma. This form of glaucoma has no symptoms initially, but as the pressure in the eye builds peripheral vision is lost and, if it is not diagnosed and treated, total blindness can occur.

The *GLC1A* gene has been sequenced, and a number of glaucoma-causing mutations have been identified. One of the mutations involves a change from C-G to T-A in the DNA, resulting in a codon change from CCG (Pro) to CUG (Leu). (These two alleles define a SNP locus.) Figure 10.20a presents the sequence of part of the *GLC1A* gene to show the mutation; the DNA from a heterozygote was sequenced so both the wild-type C and the mutant T are seen at the mutation location.

Based on the *GLC1A* gene sequence, primers were designed for PCR amplification of the region of the gene containing the mutation. The PCR product was dotted onto two membrane filters under conditions that denatured the DNA to single strands. Two ASOs were made, one for the wild-type allele and one for the mutant allele (Figure 10.20b). In this case, each ASO was 19 nt long, with the mutation positioned approximately in the middle. And, in contrast to the example in Figure 10.16, here the ASO probes are labeled (radioactively in the example) rather than the amplified DNA. Each labeled ASO was then hybridized with the unlabeled *GLC1A* DNA immobilized on one of the filters. When compared, the resulting autoradiograms indicated whether the individual from whom the DNA was taken was homozygous normal, heterozygous, or homozygous mutant. As Figure 10.20c shows, for a homozygous normal individual a signal is seen only for the wild-type ASO, for the heterozygous individual a signal is seen for both ASOs, and for the homozygous mutant individual a signal is seen only for the mutant ASO. This method has been used to analyze affected members of glaucoma families for the presence of particular mutations.

As used here, ASO hybridization used one radioactively labeled ASO as a probe for hybridization with a PCR product immobilized on a membrane filter. This approach allows one allele to be probed for on each filter and therefore is used to test individuals for the presence of a single particular mutation. A related procedure, called *reverse ASO hybridization*, by contrast uses labeled PCR product as a probe for hybridization with many different unlabeled ASOs bound to a membrane filter (this matches the approach in Figure 10.16). This approach is useful for testing DNA samples for the presence of any one of several mutations simultaneously. For example, there are hundreds of mutations in the gene for cystic fibrosis. *Multiplex PCR* can be used to amplify several regions of the gene in DNA samples from patients. The resulting PCR products are labeled and hybridized with wild-type or mutant oligonucleotides bound to membrane filters. On the autoradiograms, the dot to which the PCR product binds indicates the allele that the individual has. This method, then, tells us whether an individual has any of the mutant alleles used in the test and, if so, whether the individual is homozygous for that allele or heterozygous. It cannot rule out an individual having a mutation in the gene that is not covered by the array of ASOs being used.

***DNA Microarrays in Disease Diagnosis.*** In addition to the applications we have discussed previously in Chapters 8 and 9, DNA microarrays are also useful for screening for genetic diseases, including cancers. Of particular interest are genetic diseases that are characterized by a large number of possible mutations, making simple DNA typing methods inefficient. For example, mutations in the genes *BRCA1* (OMIM 113705) and *BRCA2* (OMIM 600185) are responsible for approximately 60% of all cases of hereditary breast and ovarian cancers. However, at least 500 different mutations have been discovered in *BRCA1* that can lead to the development of cancer. Assaying for that many different mutations is well within the scope of DNA microarray technology, and such microarrays are used to test women with a strong family history of breast cancer to see if they have a mutation in the gene. Similar tests are being developed for alleles associated with other diseases, including pediatric acute lymphoblastic leukemia (a childhood cancer of the white blood cells), Alzheimer disease, and cystic fibrosis.

In the *BRCA* test, the genome of the patient is compared with the genome of a normal individual following the general principles of microarray analysis we have discussed previously in Chapter 8. In this application of the technique, blood is taken from a patient, and Cy3(green)-labeled DNA, corresponding to the *BRCA1* and *BRCA2* genes, is produced by PCR and mixed with Cy5(red)-labeled DNA from a normal individual. The DNA microarray in this case consists of a number of small oligonucleotide probes that collectively represent the entirety of the *BRCA1* and *BRCA2* genes. Under the hybridization conditions used, if the patient has a mutation in one or other of the genes, the red (normal) DNA will hybridize to the DNA on the microarray, but the green (patient) DNA will not hybridize to oligonucleotides that are complementary to the region where the mutation is located. The reason is that the mutation prevents complete base pairing between the DNA being tested and the oligonucleotide probe on the microarray. Equal hybridization, when both samples match the oligonucleotide, is seen as a yellow (red/green) spot, and a mutation is seen as a red spot. Because the position of the spot on the array is known and because the oligonucleotide for each spot is known, the results localize the mutation to within a very narrow region of the *BRCA1* or *BRCA2* gene and can be analyzed in more detail.

**Availability of DNA Molecular Testing.** Genetic testing is not always available for a disease. There may be one or more reasons, such as the following:

1. The disease-causing gene may not yet have been identified, or the gene has been cloned but not yet sequenced, so that molecular testing tools have not been developed. For obvious reasons the more common genetic diseases tend to be the first for which genes have been cloned and molecular tests have been developed.
2. The gene has been cloned and sequenced, but there are many different mutations within the gene, making a

single molecular test impossible to develop. In this case there may be tests for a subset of the known mutations so that a positive test result confirms the presence of a disease gene mutation but a negative test result does not rule out the presence of such a mutation. You have just learned about one test for mutations in the *BRCA1* and *BRCA2* genes, designed to succeed despite these conditions. However, many genes implicated in human disease have many known mutations, and tests have not been developed for all of these genes.

3. For some diseases, mutations in the gene involved do not necessarily cause the disease to develop in every individual. A prime example concerns gene mutations that predispose individuals to the development of cancer (discussed in more detail in Chapter 20, pp. 582–595 and pp. 595–596). In such cases, testing might be limited to high-risk families.
4. Many diseases are caused by multiple gene interactions.

**Keynote**

Recombinant DNA techniques, PCR techniques, and microarray approaches are used in DNA molecular testing for human genetic disease mutations. These tests have become possible as knowledge about the molecular nature of mutations associated with human genetic diseases has increased. In general, human genetic testing is done for prenatal diagnosis, newborn screening, or carrier detection. Many DNA molecular tests are based on restriction fragment length polymorphisms (RFLPs), or on PCR amplification followed by allele-specific oligonucleotide (ASO) hybridization.

## DNA Typing

No two human individuals have exactly the same genome, base pair for base pair (not even identical twins—see Focus on Genomics, Chapter 11, p. 315—although the testing you will learn about below would probably not detect those tiny differences), and this has led to the development of **DNA typing** (also called **DNA fingerprinting,** or **DNA profiling**) techniques for use in forensic science, in paternity and maternity testing, and elsewhere. DNA typing relies on DNA analysis of DNA polymorphisms (molecular markers) described earlier in the chapter.

**DNA Typing in a Paternity Case.** Let us consider an example of using DNA typing in a paternity case. In this fictional scenario, a mother of a new baby has accused a particular man of being the father of her child, and the man denies it. The court will decide the case based on evidence from DNA typing. The DNA typing proceeds as follows (Figure 10.21): DNA samples are obtained from all three individuals involved (Figure 10.21, step 1). In a paternity case, the usual source of DNA is from a blood sample or a cheek swab. The DNA is cut with the restriction enzyme for the marker to be analyzed, and the resulting fragments

**Figure 10.21**
**DNA typing to determine paternity.**

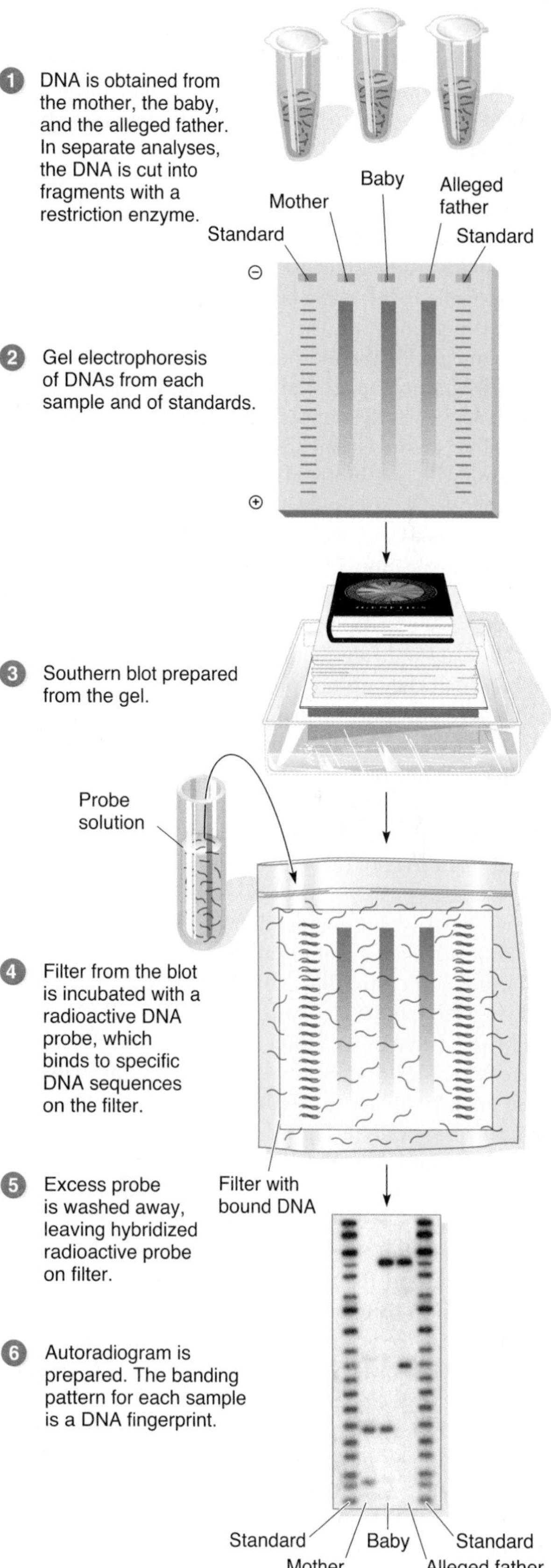

are separated by electrophoresis (Figure 10.21, step 2), transferred to a membrane filter by Southern blotting (Figure 10.21, step 3), and probed with a labeled monolocus STR or VNTR probe (Figure 10.21, steps 4 and 5). After autoradiography or chemiluminescent detection, the DNA-banding pattern—the DNA fingerprint, or DNA profile—is then analyzed to compare the samples (Figure 10.21, step 6).

The data can be interpreted as follows: Two DNA fragments are detected for the mother, so she is heterozygous for one particular pair of alleles at the STR or VNTR locus under study. Likewise, two DNA fragments are detected for the baby, so the baby is also heterozygous. One of the fragments for the baby matches the larger of the fragments for the mother, and the other fragment for the baby is much larger, indicating many more repeats in that allele. Each allele in the child must come from either the mother or the father, so an allele present in the child (but absent in the mother) *must* have been provided by the father. Keep in mind that both the mother and father will have alleles that were not passed to the child. In our example, inspection of that lane in the autoradiogram leads us to conclude that the allele that must have come from the father is also present in the alleged father.

The data indicate that the man shares an allele with the baby, but they do not prove that he is the father—he might have contributed that allele to the genome of the baby, but many other men also carry this allele, and it is possible that one of these other men could be the father. If the man lacked the allele that must have come from the baby's father, then the DNA typing data would have proved that he is not the father; this is the *exclusion* result. To establish positive identity—the *inclusion* result—through DNA typing is more difficult. It requires calculating the relative odds that the allele came from the accused or from another person. This calculation depends on knowing the frequencies of STR or VNTR alleles identified by the probe in the ethnic population from which the man comes. Most legal arguments focus on this matter because good estimates of STR or VNTR allele frequencies are known for only a limited array of ethnic groups, so that calculations of probability of paternity give numbers of questionable accuracy in many cases. To minimize possible inaccuracy, investigators use a number of different probes (often five or more) so that the combined probabilities calculated for the set of STRs or VNTRs can be high enough to convince a court that the accused is actually the parent (or is guilty in a criminal case), even allowing for problems with knowing true STR or VNTR allele frequencies for the population in question.

It is these combined probabilities that you hear or read about in the media with respect to DNA typing in court cases (see next section). In court, usually the scientific basis for the method is not in question; rather, DNA evidence is most commonly rejected for reasons such as possible errors in evidence collection or processing, or weak population statistics. In our paternity case, we would probably be more persuaded that the accused was the father of the child if the data for each of five different monolocus probes indicated that he contributed a particular allele to the child.

Recently, PCR testing for paternity determination has become the chosen method of commercial laboratories performing such tests.

**Crime Scene Investigation: DNA Forensics.** The reason that DNA typing can be used in paternity testing is that, with the exception of identical twins, no two individuals in the human population start life with identical genomes (and we have recently learned that mitotic mutations create subtle differences even between identical twins; see Focus on Genomics, Chapter 11, p. 315). The very large number of loci we have that contain DNA polymorphisms make each of our genomes almost unique. On these principles, it is possible to compare two DNA samples to determine the likelihood that they are from the same individual. In crime investigations these days, it is routine to seek out and analyze DNA samples as a means of building a case against, or of exonerating, a suspect. If DNA samples match, probability calculations are made as described in the previous section. Of course, in court cases, DNA evidence is only one type of evidence that is considered.

**iActivity**

You are the forensic scientist using STR analysis to solve a murder case in the iActivity, *Combing Through "Fur"ensic Evidence*, on the student website.

The methods used in DNA forensics are the ones we have already discussed. The usefulness of DNA typing in forensics is illustrated in the following selected case studies. The examples include cases in which the DNA evidence helped establish the guilt of a suspect, and cases in which it proved a suspected, or already convicted, individual to be innocent.

***The Narborough Murders: The First Murder Exoneration and Conviction Due to DNA Evidence.*** In 1983 and 1986, two schoolgirls were murdered in the small town of Narborough, Leicestershire ("less-ter-shear"), England. Both girls had been sexually assaulted, and semen samples recovered from the bodies indicated the murderer or murderers had the same blood type. The prime suspect in the second murder had that blood type and eventually confessed to the killing but denied involvement in the first murder. The police were convinced that he had done both murders and so they contacted Alec Jeffreys (Figure 10.22) at nearby Leicester University to perform DNA typing on samples they had taken. As was mentioned earlier, Alec Jeffreys—now Sir Alec Jeffreys—had discovered VNTRs. He had also just demonstrated that DNA could be extracted from stains at crime scenes and typed for particular VNTR loci. Using

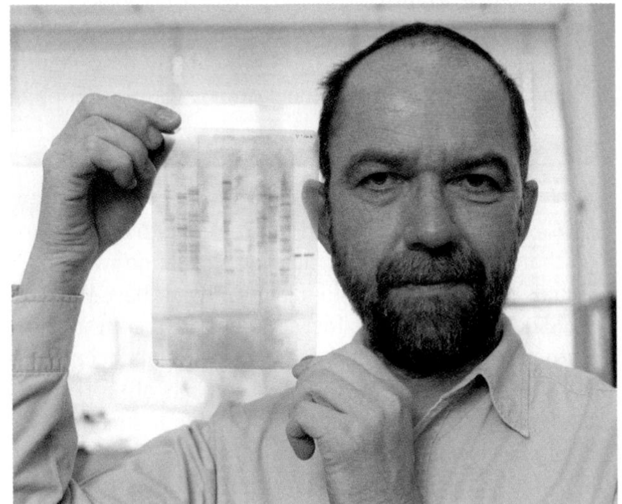

**Figure 10.22**
**Sir Alec Jeffreys, the discoverer of VNTRs.** He is holding examples of DNA fingerprints.

Southern blot analysis with multilocus VNTR probes, Dr. Jeffreys showed that the DNA in the semen samples from both murders did not match the police's suspect. He was released, the first person in the world to be exonerated of murder through the use of DNA fingerprinting. In the absence of the DNA evidence, it was almost certain that a court would have convicted him.

What of the real murderer? The Chief Superintendent of Police overseeing the case decided to embark on the world's first mass screening of DNA in a population. A total of 5,000 adult males in nearby towns were asked to provide blood or saliva samples for forensic analysis. About ten percent of the samples showed the blood type as the killer, and those were followed up using DNA typing. No DNA profiles matched the crime scene profiles, a frustrating result for the police. In a strange twist, though, a woman overheard her work colleague saying that he had given his sample in the name of his friend, Colin Pitchfork. The police arrested Pitchfork. His DNA profile matched the semen samples' profile and in 1988 he was convicted of the murders and sentenced to life in prison.

***The Green River Murders: Conviction.*** On July 8th, 1982, Wendy Lee Coffield, age 16, disappeared in Tacoma, Washington. Her body was found in the Green River, in King County, Washington on July 15th, 1982. She had been strangled. Over the next few years many other young women, usually prostitutes, also disappeared and were found strangled, a number of them in the Green River. A serial killer was loose. Interviews of many prostitutes in the Seattle area revealed that some had been raped or had been threatened with being killed by a man driving a blue and white truck. The evidence made Gary Ridgway a suspect. When King County Sheriffs searched his home in 1987, they had him chew a piece of gauze. At the time, DNA forensics was in its infancy, but increasingly crime investigators collected samples in anticipation of future applications of DNA fingerprinting in forensics. Fortunately, the sample was handled and stored properly, so the DNA in it did not degrade. Ridgway was the prime suspect, but material evidence was not sufficient to arrest him. However, in September 2001, PCR-based STR analysis was able to be used with the collected evidence, with the result that Ridgway's DNA profile matched that in sperm samples taken from Carol Christensen, one of the Green River victims. In November 2003, Ridgway admitted in court to killing 48 women, pleading guilty to 48 counts of murder in the first degree. Apparently he "hated prostitutes" and said that "strangling young women was his career."

***The Central Park Jogger Case: Exoneration.*** In April, 1989, a 28-year-old female investment banker was violently raped and beaten while jogging in Central Park in New York City. She was left tied up, bleeding, and unconscious with severe injuries. Eventually she regained consciousness and began a slow recovery. The public was outraged by the savagery of the crime. Police investigators discovered that, at the time of the crime, a group of teenage men had been "wilding"—attacking people at random. Five suspects were arrested in connection with the woman's rape and beating. Four of them confessed, and all five were convicted and imprisoned in 1990. However, supporters of the men argued that the confessions had been coerced and, in addition, there was no other physical evidence to connect any of the men to the crime scene. Then, in 2002, Matias Reyes, a convict serving time for another rape and murder, confessed to the Central Park Jogger rape. His DNA, and not that of any of the convicted five, was shown to match that of the semen sample taken from the victim. Based on Reyes' confession, the convictions of the five men were overturned.

Clearly, DNA typing is a powerful tool in criminal investigation. Used properly, it can convict or exonerate an individual of a crime, or free a wrongly convicted person. To the latter end, Barry Scheck and Peter Neufeld in 1992 set up The Innocence Project, a non-profit legal organization that takes cases where post-conviction DNA typing of evidence can result in proof of innocence. Through March 2008, 215 convicted people have been exonerated by the efforts of The Innocence Project (http://www.innocenceproject.org/).

**Other Applications of DNA Typing.** There are many uses of DNA typing with present-day samples. The following is a list of a few examples to illustrate the scope of usefulness of DNA typing for human testing and for tests involving other organisms.

1. Population genetics studies to establish variability in populations or ethnic groups.
2. Proving pedigree status in certain breeds of horses for breed registration purposes.
3. Conservation biology studies of endangered species to determine genetic variability.

4. Forensic analysis in wildlife crimes. Wild animals sometimes are killed illegally, and DNA typing is increasingly helping solve the crimes. For example, a set of six STR markers was used in a poaching investigation in Wyoming involving pronghorn antelope. Six headless pronghorn antelope carcasses were discovered and reported to authorities. An investigation turned up a suspect who had a skull with horns. DNA samples were taken from the skull and compared to DNA samples from carcass samples and a match was found. At the trial, the suspect was convicted on six counts of wanton destruction of big or trophy game. He received 30 days in jail, was fined $1,300 and ordered to pay $12,000 in restitution, and had his hunting license suspended for 36 years.
5. PCR using strain-specific primers to test for the presence of pathogenic *E. coli* strains in food sources such as hamburger meat.
6. Detecting genetically modified organisms (GMOs). GMOs have been introduced widely into agriculture in the United States. Genetically modified crops typically contain genes that were introduced in the development of the new crop. Often these genes are expressed using a particular promoter and a particular transcription terminator, enabling PCR primers designed based on these sequences to be used to test for their presence. We can do these tests with plants themselves or with processed foods. A positive PCR result indicates that the plant is genetically modified or that the food contains one or more GMOs. However, a negative result does not rule out the presence of a GMO. That is, the plant may be genetically modified using genes that have a different promoter or terminator, and the food may have been made from such organisms, or the DNA may have been destroyed in processing. Between 50 and 75 percent of produce and processed foods in a supermarket may be genetically modified or contain GMOs.

There are also an increasing number of interesting applications of DNA typing with non-present-day samples:

1. Analyzing the DNA extracted from ancient organisms, such as a 40 million-year-old insect in amber, a 17 million-year-old fossil leaf, and a 40,000-year-old mammoth to compare them molecularly with present-day descendants.
2. Some historical controversies and mysteries have been resolved by DNA typing. For example, in 1795 a 10-year-old boy died of tuberculosis in the tower of the Temple Prison in France. The great mystery was whether the boy was the dauphin, the sole surviving son of Louis XVI and Marie Antoinette, who were executed by republicans on the guillotine, or whether he was a stand-in while the true heir to the throne escaped. The dead boy's heart was saved after the autopsy and, despite some very rough handling and storage conditions since his death, in December 1999 two small tissue samples were taken from the heart; remarkably, DNA could be extracted from them. This DNA was typed against DNA extracted from locks of the dauphin's hair kept by Marie Antoinette, from two of the queen's sisters, and from present-day descendants. The results showed that the dead boy was the dauphin.

**Keynote**

DNA typing, or DNA fingerprinting, is done to distinguish individuals based on the concept that no two individuals of a species, save for identical twins, have the same genome sequence. The variations are manifested in restriction fragment length polymorphisms and length variations resulting from different numbers of short tandemly repeated sequences. DNA typing has many applications, including basic biological studies, forensics, detecting infectious species of bacteria, and analysis of old or ancient DNA.

## Gene Therapy

Is it possible to modify the genome to treat genetic diseases? Theoretically, two types of gene therapy are possible: *somatic cell therapy*, in which somatic cells are modified genetically to prevent a genetic defect in the individual receiving the therapy; and *germ-line cell therapy*, in which germ-line cells are modified to correct a genetic defect. Somatic cell therapy results in a treatment for the genetic disease in the individual, but progeny could still inherit the mutant gene. Germ-line cell therapy, however, could prevent the disease because the mutant gene can be replaced by the normal gene and that normal gene would be inherited by the offspring. Both somatic cell therapy and germ-line cell therapy have been used successfully in nonhuman organisms, including mice, but only somatic cell therapy has been used in humans because of ethical issues raised by germ-line cell therapy.

The most promising candidates for somatic cell therapy are genetic disorders that result from a simple defect of a single gene and for which the cloned normal gene is available. Gene therapy involving somatic cells proceeds as follows. A sample of the individual's cells carrying the defective gene is taken. Then normal, wild-type copies of the mutant gene are introduced into the cells, and the cells are reintroduced into the individual. There, it is hoped, the cells will produce a normal gene product and the symptoms of the genetic disease will be completely or partially reversed.

The source of the cells varies with the genetic disease. For example, blood disorders, such as thalassemia or sickle-cell anemia, require modification of bone marrow cells that produce blood cells. For genetic diseases affecting circulating proteins, a promising approach is the gene therapy of skin fibroblasts, cells that are constituents of the dermis (the lower layer of the skin). Modified fibroblasts can easily be implanted back into the dermis, where blood vessels invade the tissue, allowing gene products to be distributed.

A cell that has had a gene introduced into it by artificial means is said to be **transgenic,** and the gene involved is called a **transgene.** The introduction of normal genes into a mutant cell poses several problems. First, procedures to introduce DNA into cells (transformation, although actually called *transfection* for eukaryotic cells) typically are inefficient; perhaps only one in 1,000 or 100,000 cells will receive the gene of interest. Thus, a large population of cells is needed to attempt gene therapy. Present procedures use special virus-related vectors to introduce the transgene. Second, in cells that take up the cloned gene, the fate of the foreign DNA cannot be predicted. In some cases the mutant gene is replaced by the normal gene, and in others the normal gene integrates into the genome elsewhere. In the first case, the gene therapy is successful provided that the gene is expressed. In the second case, successful treatment of the disease results only if the introduced gene is expressed and the resident mutant gene is recessive, so that it does not interfere with the normal gene.

Successful somatic gene therapy has been demonstrated repeatedly in experimental animals such as mice, rats, and rabbits. However, in humans, there have been more failures than successes. In addition, a recent concern is the development of leukemias in therapy patients as a result of the viral vectors used for introducing the transgene.

One successful human somatic gene therapy treatment was done in 1990 with a 4-year-old girl suffering from severe combined immunodeficiency (SCID; OMIM 102700) caused by a deficiency in adenosine deaminase (ADA), an enzyme needed for normal function of the immune system. T cells (cells involved in the immune response) were isolated from the girl and grown in the laboratory, and the normal ADA gene was introduced using a viral vector. The "engineered" cells were then reintroduced into the patient. Since T cells have a finite life in the body, continued infusions of engineered cells have been necessary. The introduced ADA gene is expressed, probably throughout the life of the T cell. As a result, the patient's immune system is functioning more normally, and she now gets no more than the average number of infections. The gene therapy treatment has enabled her to live a more normal life. Recently some patients who received gene therapy for ADA have developed leukemia for reasons unknown.

With time, many other genetic diseases are expected to be treatable with somatic gene therapy, including thalassemias, phenylketonuria, cancer, Duchenne muscular dystrophy, and cystic fibrosis. For example, after successful experiments with rats, human clinical trials are under way for transferring the normal CF gene to patients with cystic fibrosis. As methods are learned for targeting genes to replace their mutant counterparts and regulating the expression of the introduced genes, increasing success in treating genetic diseases is expected. However, many scientific, ethical, and legal questions must be addressed before the routine implementation of gene therapy.

### Keynote

Gene therapy is the curing of a genetic disorder by introducing into the individual a normal gene to replace or overcome the effects of a mutant gene. For ethical reasons, only somatic gene therapy is being developed for humans. There are few examples of successful somatic gene therapy in humans, but there is great hope for treating many genetic diseases in this way in the future.

## Biotechnology: Commercial Products

The development of cloning and other DNA manipulation techniques has spawned the formation of many *biotechnology* companies, some of which focus on using DNA manipulations for making a wide array of commercial products. Although the details vary, the general approach to making a product is to express a cloned gene or cDNA in an organism that will transcribe the cloned sequence and translate the mRNA. The gene or cDNA is placed into an expression vector (see pp. 249–251) appropriate for the organism into which it will be transformed. Many different organisms are used, from *E. coli* to mammals, so the expression vectors differ in the promoters used for transcription, in the translation start signals, and in the selectable markers. Recall from earlier in the chapter that, for expression in *E. coli,* for example, the promoter must be recognized by that bacterium's RNA polymerase, and there must be a Shine–Dalgarno sequence so that ribosomes will read the mRNA from the correct AUG. In mammals such as goats or sheep, the simplest way to isolate the product is to have it secreted into the milk. The milk is easy to collect, of course, and the protein product can then be extracted. The production of recombinant protein products in transgenic mammals (in this case sheep) is illustrated in Figure 10.23. Here the gene of interest (GOI) has been manipulated so that it is adjacent to a promoter that is active only in mammary tissue, such as the β-lactoglobulin promoter. The recombinant DNA molecules are microinjected into sheep ova, and each ovum is then implanted into a foster mother. Transgenic offspring are identified using PCR to detect the recombinant DNA sequences. When these transgenic animals mature, the β-lactoglobulin promoter begins to express the associated gene in the mammary tissue, the milk is collected, and the protein of interest is obtained by biochemical separation techniques.

A few examples of the many products produced by biotechnology companies are as follows:

1. Tissue plasminogen activator (TPA), used to prevent or dissolve blood clots, therefore preventing strokes, heart attacks, or pulmonary embolisms
2. Human growth hormone, used to treat pituitary dwarfism
3. Tissue growth factor-beta (TGF-β), which promotes new blood vessel and epidermal growth and thus is potentially useful for wound and burn healing

**Figure 10.23**

**Production of a recombinant protein product (here, the protein encoded by the gene of interest, *GOI*) in a transgenic mammal—in this case, a sheep.**

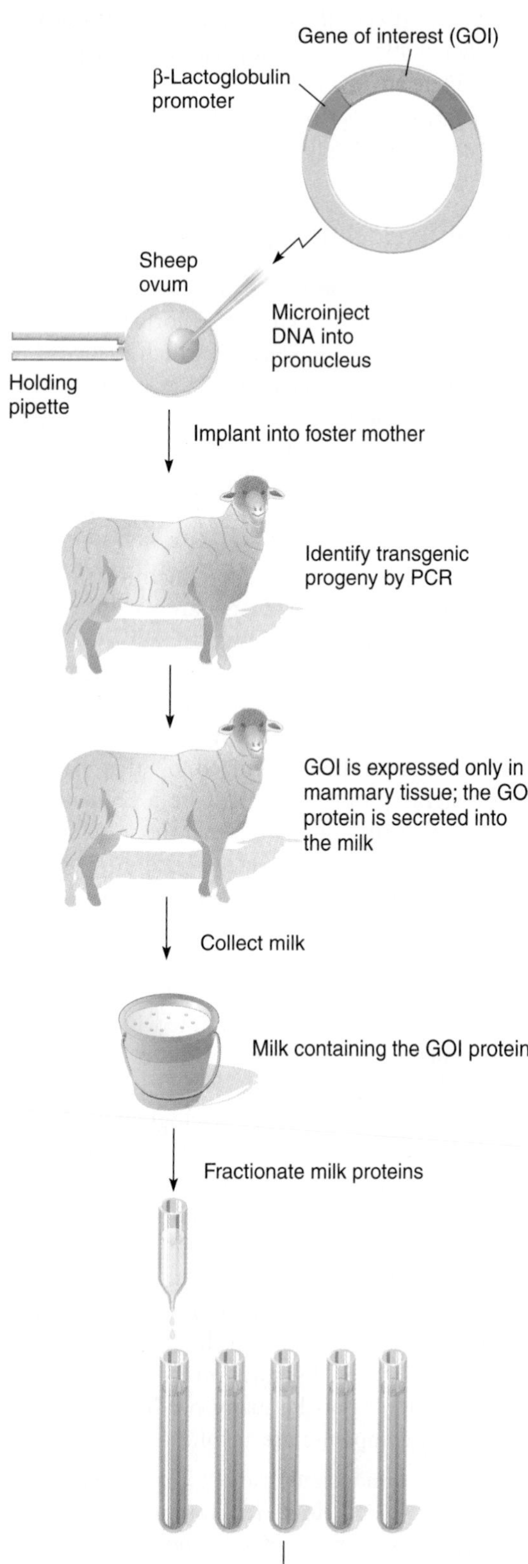

Gene of interest (GOI)
β-Lactoglobulin promoter
Sheep ovum
Microinject DNA into pronucleus
Holding pipette
Implant into foster mother
Identify transgenic progeny by PCR
GOI is expressed only in mammary tissue; the GOI protein is secreted into the milk
Collect milk
Milk containing the GOI protein
Fractionate milk proteins
GOI protein

4. Human blood clotting factor VIII, used to treat hemophilia
5. Human insulin ("humulin"), used to treat insulin-dependent diabetes
6. DNase, used to treat cystic fibrosis
7. Recombinant vaccines, used to prevent human and animal viral diseases (such as hepatitis B in humans)
8. Bovine growth hormone, used to increase cattle and dairy yields
9. Platelet-derived growth factor (PDGF), used to treat chronic skin ulcers in patients with diabetes
10. Genetically engineered bacteria and other microorganisms used to improve production of, for example, industrial enzymes (such as amylases to break down starch to glucose), citric acid (flavoring), and ethanol
11. Genetically engineered bacteria that can accelerate the degradation of oil pollutants or certain chemicals in toxic wastes (such as dioxin)

**Keynote**

With the same kinds of recombinant DNA and PCR techniques used in basic biological analysis, DNA molecular testing, gene cloning, DNA typing, and gene therapy, biotechnology and pharmaceutical companies develop useful products. Many types of products are now available or are in development, including pharmaceuticals and vaccines for humans and for animals and genetically engineered organisms for improved production of important compounds in the food industry or for cleaning up toxic wastes.

## Genetic Engineering of Plants

For many centuries the traditional genetic engineering of plants involved selective breeding experiments in which plants with desirable traits were selectively allowed to produce offspring. As a result, humans have produced hardy varieties of plants (for example, corn, wheat, and oats) and increased yields, all using long-established plant breeding techniques. (Similar techniques have also been used with animals, such as dogs, cattle, and horses, to produce desired breeds.) Now, vectors developed by recombinant DNA technology are available for transforming cells of crop plants; this has made possible the genetic engineering of plants for agricultural use.

animation

**Plant Genetic Engineering**

### Transformation of Plant Cells

Introducing genes into plant cells is more difficult in some respects than introducing genes into bacteria, yeast, and animals, and this has slowed plant genetic engineering's rate of progress. Typical plant transformation approaches exploit features of a soil bacterium, *Agrobacterium tumefaciens*, which infects many kinds of plants. Specifically, they take advantage of a natural mechanism in the bacterium for transferring a defined segment of DNA into the chromosome of the plant.

*Agrobacterium tumefaciens* causes crown gall disease, characterized by tumors (the gall) at wounding sites. Most dicotyledonous plants (called *dicots*) are susceptible to crown gall disease, but monocotyledonous plants are not. *Agrobacterium tumefaciens* transforms plant cells at the wound site, causing the cells to grow and divide autonomously and therefore to produce the tumor.

The transformation of plant cells is mediated by a natural plasmid in the *Agrobacterium* called the *Ti plasmid* (the *Ti* stands for *tumor-inducing;* Figure 10.24). Ti plasmids are circular DNA plasmids somewhat analogous to pUC19, but, in comparison, Ti plasmids are huge (about 200 kb versus 2.96 kb for pBluescript II).

The interaction between the infecting bacterium and the plant cell of the host stimulates the bacterium to excise a 30-kb region of the Ti plasmid called T-DNA (so called because it is *transforming DNA*). T-DNA is flanked by two repeated 25-bp sequences called *borders* that are involved in T-DNA excision. Excision is initiated by a nick in one strand of the right-hand border sequence. A second nick in the left-hand border sequence releases a single-stranded T-DNA molecule, which is then transferred from the bacterium to the nucleus of the plant cell by a process analogous to bacterial conjugation. Once in the plant cell nucleus, the T-DNA integrates into the nuclear genome. As a result, the plant cell acquires the genes found on the T-DNA, including the genes for plant cell transformation. However, the genes needed for the excision, transfer, and integration of the T-DNA into the host plant cell are not part of the T-DNA. Instead, they are found elsewhere on the Ti plasmid, in a region called the *vir* (for *virulence*) region.

Using recombinant DNA approaches, researchers have found that excision, transfer, and integration of the T-DNA require only the 25-bp terminal repeat sequences. As a result, the Ti plasmid and the T-DNA it contains is a useful vector for introducing new DNA sequences into the nuclear genome of somatic cells from susceptible plant species. Since any genes placed between the 25-bp borders will integrate into the host genome, a variety of transformation vectors have been derived from the Ti plasmid and T-DNA.

Although the T-DNA-based transformation system is very effective for dicotyledonous plants, it is not effective for monocotyledonous plants because they are not part of the normal host range of *Agrobacterium tumefaciens.* This is a serious limitation because most crop plants are monocotyledonous. Fortunately, alternative transformation procedures have been developed in which the DNA is delivered into the cell physically rather than by a plasmid vector. In the *electroporation* method, DNA is added to plant cell protoplasts and the mixture is "shocked" with high voltage to introduce the DNA into the cell. After the cells are grown in tissue culture to allow them to regenerate their cell walls and begin growing again, appropriate procedures can be applied to select for the cells that were successfully transformed. Another method involves the *gene gun* (made by Bio-Listics). In this method, DNA is coated onto the surface of tiny tungsten beads, which are placed on the end of a plastic bullet. The bullet is fired by a special particle gun. The bullet hits a plate, and the tungsten beads are propelled through a small hole in the plate into a chamber in which target cells have been placed. The force of the "shot" is sufficient to introduce the DNA-carrying beads into the cells. Selection techniques can then be applied to isolate successfully transformed cells, and these can be used to regenerate whole plants.



**Figure 10.24**

**Formation of tumors (crown galls) in plants by infection with certain species of *Agrobacterium*.** Tumors are induced by the Ti plasmid, which is carried by the bacterium and integrates some of its DNA (the T, or transforming, DNA) into the plant cell's chromosome.

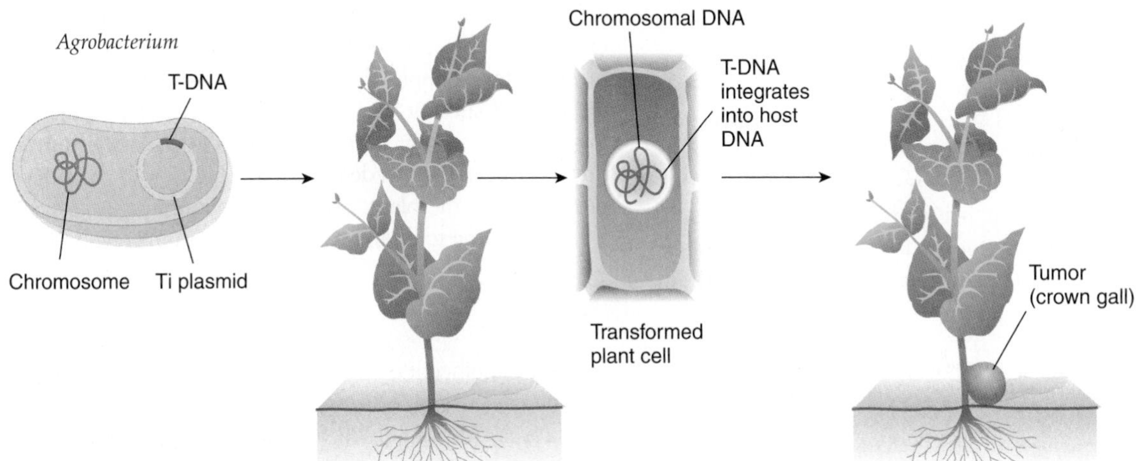

## Applications for Plant Genetic Engineering

We mentioned in the section on DNA typing that a very large number of genetically modified crops already have been developed, and a lot of the processed food we buy contains them. Let us briefly consider approaches to generating transgenic plants that are tolerant to the broad-spectrum herbicide Roundup™ to illustrate the types of approaches that are possible. Roundup contains the active ingredient glyphosate, which kills plants by inhibiting EPSPS, a chloroplast enzyme required for the biosynthesis of essential aromatic amino acids. Roundup is used widely to kill weeds because it is active in low doses and is degraded rapidly in the environment by microbes in the soil. If a crop plant is resistant to Roundup, a field can be sprayed with the herbicide to kill weeds without affecting the crop plant. Approaches for making transgenic, Roundup-tolerant plants include: (1) introducing a modified bacterial form of EPSPS that is resistant to the herbicide, so that the aromatic amino acids can still be synthesized even when the chloroplast enzyme is inhibited (Figure 10.25); and (2) introducing genes that encode enzymes for converting the herbicide to an inactive form. Monsanto brought Roundup Ready soybeans to market in 1996, although their use has been controversial because of opposition by groups questioning the safety of genetically engineered plants for human consumption.

With more sophisticated approaches it will be possible to make transgenic plants that control the expression of genes in different tissues. Examples include controlling the rate at which cut flowers die or the time at which fruit ripens. Approved for market in 1994 was the Flavr Savr tomato, genetically engineered by Calgene Inc. in collaboration with the Campbell Soup Company. Commercially produced, genetically unaltered tomatoes are picked while unripe so they can be shipped without bruising. Prior to shipping, they are exposed to ethylene gas, which initiates the ripening process so that they arrive in the ripened state at the store. Such prematurely picked, artificially ripened tomatoes do not have the flavor of tomatoes picked when they are ripe. Calgene scientists devised a way to block the tomato from making the normal amount of polygalacturonase (PG), a fruit-softening enzyme. They introduced into the plant a copy of the PG gene that was backward in its orientation with respect to the promoter. When this gene is transcribed, the mRNA is complementary to the mRNA produced by the normal gene; it is called an **antisense mRNA.** In the cell, the antisense mRNA binds to the normal, "sense" mRNA, with the result that much of that mRNA is prevented from being translated.[1] As a result, much less PG enzyme is produced, and tomato ripening is slowed, allowing it to remain longer on the vine without getting too soft for handling and shipping. Once picked, the Flavr Savr tomato was also less susceptible to bruising in shipping or to overripening in the store. The Flavr Savr tomato was advertised as tasting better than store-ripened tomatoes and more like home-grown tomatoes. However, it was expensive and did not achieve commercial success. For economic reasons, it is no longer on the market.

In the past few years, more and more genetically modified crop plants have been brought to market. In addition to herbicide resistance, other crops have been modified to increase insect-resistance. Many of these crop plants express a protein called Bt. Bt is normally made by certain bacteria. When a susceptible insect ingests Bt protein (either as a protein outside of a cell, or as part of a bacterial cell or a plant cell), the Bt protein kills or injures the insect. Purified Bt proteins and bacteria that naturally express Bt have been used as insecticides in organic farming for years. In theory, these modified crop plants could allow farmers to decrease their reliance on pesticides, without decreasing yield. Other genetically modified crop plants have been altered to increase the production of amino acids or vitamins, with the goal of making the crop more nutritious. Such plants potentially could help in alleviating world hunger. However, there is significant public resistance to genetically modified plants in many countries, including a growing resistance in the United States. As a result, most of the genetically modified plants grown are not used for human food, but are instead for either animal feed or for nonfood products.

Transgenic plants may also be useful for delivering vaccines. The cost of an injected vaccine is relatively high, making it a significant issue in inoculating people in developing countries. Furthermore, vaccines require refrigeration and sterile needles, both of which can be either very expensive or impossible to find in parts of the world. However, potentially it could cost just pennies to deliver vaccines in a plant. Such vaccines have been termed *edible vaccines*, and the area of biotechnology dealing with pharmaceuticals in plants or animals has been whimsically termed *pharming*. Basically, transgenic plants are made that express antigens for infections or diseases of interest so that, when the plant is eaten, the individual potentially will develop antibodies. Indeed, after successful trials with animals, human early stage clinical trials have shown that eating raw potatoes can elicit the expected immune responses when those potatoes are expressing, for example, the hepatitis B virus surface antigen, the toxin B subunit of enterotoxigenic *E. coli* (responsible for diarrhea), or the capsid protein of the Norwalk virus. Further research is needed to obtain high levels of antigen production in the plants so that sufficient antigen is available after eating to mount a protective immune response.

### Keynote

Genetic engineering of plants is also possible using recombinant DNA. It is expected that many more types of improved crops will result from future applications of this new technology.

[1]With our present-day knowledge, the mechanism of knocking down translation was most likely RNA interference (RNAi). That is, the double-stranded RNA formed between sense and antisense mRNAs would be processed to produce a single-stranded, small regulatory RNA that binds to the mRNA, leading to knocking down or knocking out expression of that mRNA (see Chapter 9, pp. 227–229, and Ch 18, pp. 537–540).

Figure 10.25

**Making a transgenic, Roundup™-tolerant tobacco plant by introducing a modified form of the bacterial gene for the enzyme EPSPS that is resistant to the herbicide.** The gene encoding the bacterial EPSPS was spliced to a petunia sequence encoding a transit peptide for directing polypeptides into the chloroplast, and the modified gene was inserted into a T-DNA vector and introduced into tobacco by *Agrobacterium*-based transformation. Both the native and the modified bacterial EPSPS are transported into the chloroplast. When plants are sprayed with Roundup, wild-type plants die because only the native chloroplast EPSPS is present, and it is sensitive to the herbicide, but the transgenic plants live because they contain the bacterial EPSPS that is resistant to the herbicide.

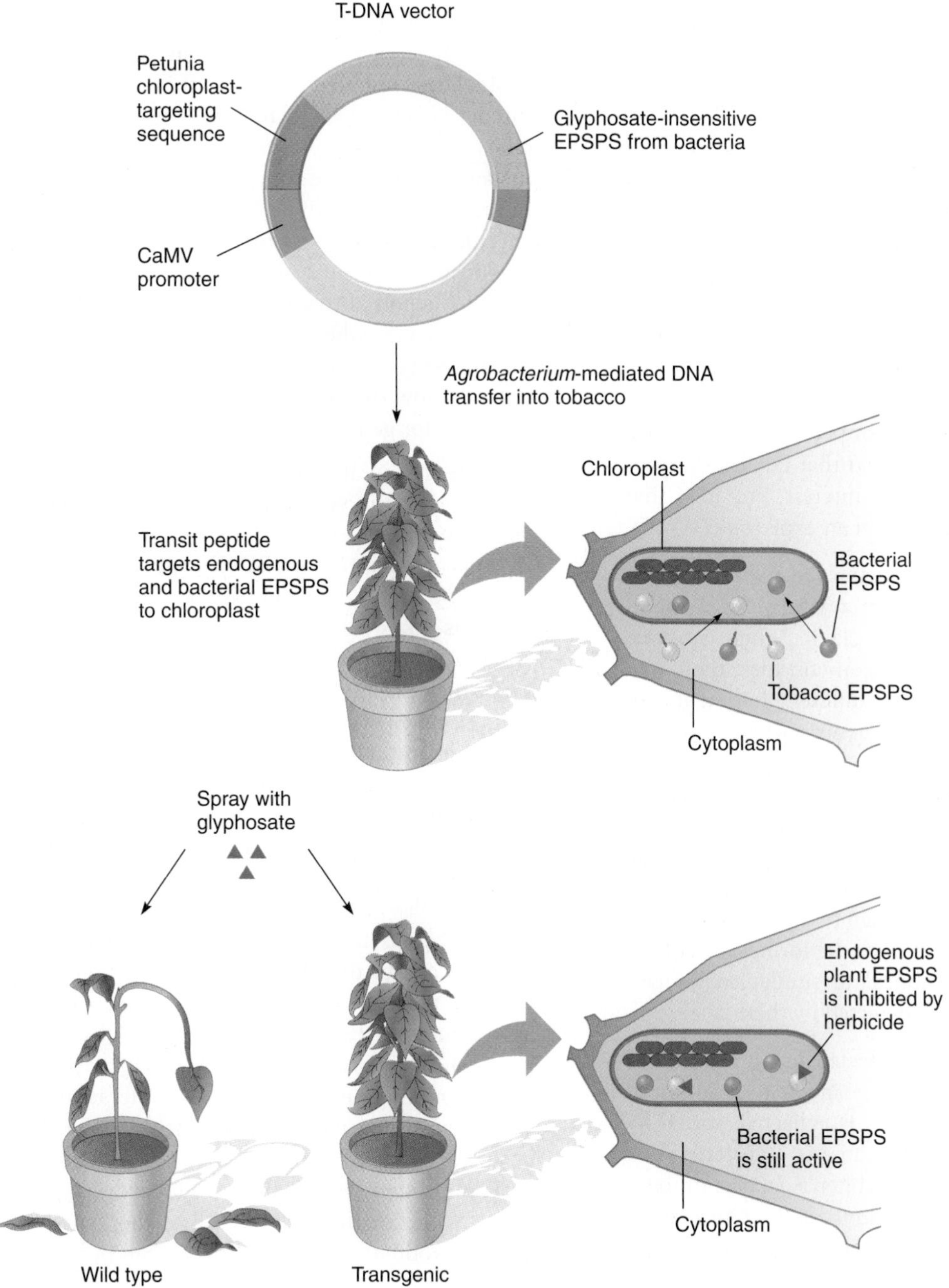

## Summary

- Many specific vectors have been developed for the manipulation of cloned DNA. Some are shuttle vectors that allow a cloned sequence to be moved from one host organism to another. Other vectors, called expression vectors, are designed so that the inserted gene will be expressed in the host cell. Many vectors are designed so that the inserted gene can be transcribed *in vitro*. Not all vectors are based on plasmids. Phage vectors accept larger inserts and can be propagated at higher densities. Some vectors integrate into the host chromosome, while others are maintained extrachromosomally. Vectors are chosen based on the needs of the experimenter.
- To find a specific gene in a library, a DNA or RNA probe is used that will detect either all or part of the gene. An entire gene, a fragment of a gene, all or part of a cloned gene from a related species, or an oligonucleotide designed to be similar to a part of the gene can be used as the probe depending on the experiment. A gene can be found even if the match between probe and gene is not perfect. Alternatively, an antibody probe can be used that detects the protein encoded by the gene of interest, provided that the library being screened is in an expression vector.
- Southern blotting is used to analyze a specific piece of DNA in the genome, or in any large DNA molecule. Since the genome is so large, when genomic DNA is digested with restriction enzymes, there will be thousands to millions of different fragments. In order to see only those fragments corresponding to the gene of interest, agarose gel electrophoresis is used to sort the fragments by size, and then transfer the sorted fragments to a membrane filter. Using chemical treatment, the DNA is converted to single strands which then bind tightly to the filter. A labeled single-stranded probe can be added to the filter and the conditions set to favor the formation of base pairs. The probe will anneal to similar sequences, and these hybrids can be detected by their label.
- A specific mRNA can be detected using a northern blot, which is technically very similar to a Southern blot. In a northern blot, RNA is collected, sorted by size using agarose gel electrophoresis, transferred to a filter, and bound tightly to the filter. A labeled probe is added and allowed to anneal to the RNA. Once again, detecting the label indicates where the probe found a similar sequence. This indicates whether or not a specific mRNA is present in the starting RNA pool.
- The polymerase chain reaction (PCR) has many uses in the research lab. PCR can be used as a step in cloning and/or sequencing a particular gene in an individual. PCR can also be used in the analysis of the genome of an individual, either to determine the genotype of that individual or to determine if two DNA samples match. Two very powerful PCR-based techniques are reverse-transcription PCR (RT-PCR) and real-time PCR. Both techniques allow very sensitive detection of whether a specific mRNA is present in a pool of mRNA. Real-time PCR can be used to quantify accurately the amount of the mRNA in question.
- PCR can be used to create specific mutations in a cloned gene, in a process called site-specific mutagenesis. These mutated genes can then be reintroduced into a host cell. This technique is used to make specific changes in the protein encoded by the gene, for instance to study the function of the altered protein in the cell. A gene can be "humanized" by using site-specific mutagenesis to make a gene from a model organism, such as a mouse, more similar to the human version of that gene. A transgenic mouse can then be made, where the humanized gene replaces the mouse gene, and these humanized mice are used to study how the gene functions and to test possible therapies for genetic diseases.
- Protein–protein interactions in the cell can be revealed using the yeast two-hybrid system. This test uses two expression plasmids in a single yeast cell. One plasmid expresses a BD–X fusion protein, where BD is the binding domain for a regulatory protein and X is a known protein being used as the bait to identify proteins with which it interacts in the cell. The other plasmid expresses an AD–Y fusion protein, where AD is the activation domain for the same regulatory protein, and Y is the protein encoded by one cDNA in a cDNA library. The AD is needed to activate transcription but does not itself bind to a promoter element. The nature of Y is different in each transformed yeast cell because it is encoded by the particular cDNA clone in the library that cell receives. If proteins X and Y normally interact in the cell, this causes the BD–X and AD–Y fusion proteins to bind together. When that happens, the BD of BD–X binds to the promoter element of a reporter gene (such as *lacZ*), and the AD (which is now in close proximity because of the X–Y interaction) activates transcription of the reporter gene. Reporter gene expression, then, is the positive signal of protein–protein interaction and analysis of the cDNA clone in the cells identifies the gene whose protein product interacted with the bait protein.
- Several types of DNA polymorphisms are present in a genome. DNA polymorphisms are regions of DNA where two or more allelic versions can be found in a population. These polymorphisms can be the result of either variations in base-pair sequence, such as SNPs (single nucleotide polymorphisms), or differences in the number of tandemly repeated sequences,

such as STRs (short tandem repeats) and VNTRs (variable number tandem repeats).

- DNA polymorphisms can be used in disease diagnosis and in the analysis of the DNA of an individual; for example, DNA polymorphisms that are present can be tested to determine whether a fetus or a newborn infant is likely to develop a specific genetic disease. These polymorphisms also can be used to determine if an individual is a carrier of a genetic disease.
- DNA typing, or DNA fingerprinting, compares polymorphic regions in two or more individuals. DNA typing can be used to determine whether a DNA sample could have come from a given person, such as to determine if a suspected rapist matches a semen sample. Such tests can unambiguously prove innocence, but can never offer absolute proof of guilt, since it can never be proven that no other person in the world has the same pattern of polymorphisms. DNA typing also can be used to assess whether a particular person might be the parent of a child, since all polymorphisms in a child must come from either the mother or the father. Once again, it cannot prove absolutely that a man is the father of a child but can prove definitively that he is not.
- Gene therapy is the treatment of a genetic disease by direct alteration of the DNA. In humans, this has been limited to somatic gene therapy, where the somatic tissues are modified, but the reproductive tissue has not been altered. Gene therapy has been tried for a handful of human genetic diseases, but successes have been limited, and many roadblocks remain before gene therapy becomes a common medical treatment.
- Recombinant DNA is used in the biotechnology and pharmaceutical industries. This has led to the development of many products, including vaccines and pharmaceuticals, as well as modified organisms that can be used in the food industry or that can be helpful in destroying dangerous chemicals.
- Genetic engineering of plants using recombinant DNA techniques is agriculturally important. Genetic modifications of plants have included changes that alter the timing of fruit ripening and that alter the resistance of the plants to herbicides. Future applications of these techniques should radically alter the productivity of crop plants.



## Analytical Approaches to Solving Genetics Problems

**Q10.1** ROC is a hypothetical polymorphic STR (microsatellite) locus in humans with a repeating unit of CAGA. The locus is shown in Figure 10.A as a box with 25 bp of flanking DNA sequences.

**a.** You plan to use PCR to type individuals for the ROC locus. If PCR primers must be 18 nucleotides long, what are the sequences of the pair of primers required to amplify the ROC locus?

**b.** Consider ROC alleles with 10 and 7 copies of the repeating unit. Using the primers you have designed, what will be the sizes of the amplified PCR products for each allele?

**c.** There are four known alleles of ROC with 15, 12, 10, and 7 copies of the repeating unit. How many possible human genotypes are there for these alleles, and what are they?

**d.** If one parent is heterozygous for the 15 and 10 alleles of the ROC locus and the other parent is heterozygous for the 10 and 7 alleles, what are the possible genotypes of their offspring for this locus, and in what proportion will they be found?

**e.** Growing up in the house with the two parents mentioned in (d) are three children. When you type them for the ROC locus, you find that their genotypes are (10,10), (15,10), and (12,7). What can you conclude?

**A10.1** This problem requires that you to understand multiple properties of STR (short tandem repeat, or microsatellite) loci. First, it requires you to understand that, in a population of individuals, chromosomes can be polymorphic at a particular STR locus. That is, the length of the repeated sequence at the STR locus varies among different chromosomes. The number of times the sequence is repeated defines which STR allele is present on a particular chromosome. Second, this problem requires you to understand that STR alleles are inherited in the same manner as any other nuclear gene—offspring receive one allele from each of their parents. It is important to realize that, even though members of a population have different alleles at an STR locus, the repeat length usually does not change when it is inherited. Third, this problem requires you to understand that the sequences that flank the repeat are identical on different chromosomes, and that this allows for PCR to be used to detect the repeat length. PCR primers can be designed based on the sequences that flank the repeat. After they are used to amplify the repeat, the PCR products are sized by gel electrophoresis. The alleles present in one individual are determined by the sizes of the PCR products that are produced. If the PCR amplification produces a single band, the individual has two identical alleles and, therefore, is homozygous, while if

**Figure 10.A**

```
5'-CTGATTCTTGATCTCCTTTAGCTTC[     ]GTATAATTCATTATGTGATAATGCC-3'
                            [ ROC ]
3'-GACTAAGAACTAGAGGAAATCGAAG[     ]CATATTAAGTAATACACTATTACGG-5'
```

| A Fragment | Fragments Produced by Digestion with B | B Fragment | Fragment Produced by Digestion with A |
|---|---|---|---|
| 2,100 bp → | 1,900, 200 bp | 2,500 bp → | 1,900, 600 bp |
| 1,400 bp → | 800, 600 bp | 1,300 bp → | 800, 500 bp |
| 1,000 bp → | 1,000 bp | 1,200 bp → | 1,000, 200 bp |
| 500 bp → | 500 bp | | |

Construct a restriction map of the 5,000-bp DNA fragment.

*__10.12__ A colleague has sent you a 4,500-bp DNA fragment excised from a plasmid cloning vector with the enzymes *Pst*I and *Bgl*II (see Table 8.1, p. 174, for a description of these enzymes and the sites they recognize). Your colleague tells you that within the fragment there is an *Eco*RI site that lies 490 bp from the *Pst*I site.

**a.** List the steps you would take to clone the *Pst*I-*Bgl*II DNA fragment into the plasmid vector pBluescript II (described in Figure 8.4, p. 176).

**b.** How would you verify that you have cloned the correct fragment and how would you determine its orientation within the pBluescript II cloning vector?

**10.13** A researcher has a cDNA for a human gene.

**a.** How should she proceed if she wants to clone the genomic sequence for that gene?

**b.** What kinds of information can be obtained from the analysis of genomic DNA clones that cannot be obtained from the analysis of cDNA clones?

**10.14** A molecular genetics research laboratory is working to develop a mouse model for bovine spongiform encephalopathy (BSE) ("mad cow") disease, which is caused by misfolding of the prion protein. As part of their investigation, they want to investigate the structure of the gene for the prion protein in mice. They have a mouse genomic DNA library made in a BAC vector and a 2.1-kb long cDNA for the gene. List the steps they should take to screen the BAC library with the cDNA probe.

**10.15** A scientist has carried out extensive studies on the mouse enzyme phosphofructokinase. He has purified the enzyme and studied its biochemical and physical properties. As part of these studies, he raised antibodies against the purified enzyme. What steps should he take to clone a cDNA for this enzyme?

*__10.16__ A researcher interested in the control of the cell cycle identifies three different yeast mutants whose rate of cell division is temperature-sensitive. At low, permissive temperatures, the mutant strains grow normally and produce yeast colonies having a normal size. However, at elevated, restrictive temperatures, the mutant strains are unable to divide and produce no colonies. She has a yeast genomic library made in a plasmid *E. coli*–yeast shuttle vector, and wants to clone the genes affected by the mutants. What steps should she take to accomplish this objective?

**10.17** The amino acid sequence of the actin protein is conserved among eukaryotes. Outline how you would use a genomic library of yeast prepared in a bacterial plasmid vector and a cloned cDNA for human actin to identify the yeast actin gene.

*__10.18__ It is 3 a.m. Your best friend has awakened you with yet another grandiose scheme. He has spent the last two years purifying a tiny amount of a potent modulator of the immune response. He believes that this protein, by stimulating the immune system, could be the ultimate cure for the common cold. Tonight, he has finally been able to obtain the sequence of the first seven amino acids at the N-terminus of the protein: Met–Phe–Tyr–Trp–Met–Ile–Gly–Tyr. He wants your help in cloning a cDNA for the gene so that he can express large amounts of the protein and undertake further testing of its properties. After you drag yourself out of bed and ponder the sequence for a while, what steps do you propose to take to obtain a cDNA for this gene?

*__10.19__ A 10-kb genomic DNA *Eco*RI fragment from a newly discovered insect is ligated into the *Eco*RI site of the pBluescript II plasmid vector (described in Figure 8.4, p. 176) and transformed into *E. coli*. Plasmid DNA and genomic DNA from the insect are prepared and each DNA sample is digested completely with the restriction enzyme *Eco*RI. The two digests are loaded into separate wells of an agarose gel, and electrophoresis is used to separate the products by size.

**a.** What will be seen in the lanes of the gel after it is stained to visualize the size-separated DNA molecules?

**b.** What will be seen if the gel is transferred to a membrane to make a Southern blot, and the blot is probed with the 10-kb *Eco*RI fragment? (Assume the fragment does not contain any repetitive DNA sequence.)

*__10.20__ During Southern blot analysis, DNA is separated by size using gel electrophoresis, and then transferred to a membrane filter. Before it is transferred, the gel is soaked in an alkaline solution to denature the double-stranded DNA, and then neutralized. Why is it important to denature the double-stranded DNA? (Hint: Consider how the membrane will be probed.)

*__10.21__ A researcher digests genomic DNA with the restriction enzyme *Eco*RI, separates it by size on an agarose gel, and transfers the DNA fragments in the gel to a membrane filter using the Southern blot procedure. What result would she expect to see if the source of the DNA and the probe for the blot is described as follows?

**a.** The genomic DNA is from a normal human. The probe is a 2.0-kb DNA fragment excised by the enzyme *Eco*RI from a plasmid containing single-copy genomic DNA.

**b.** The genomic DNA is from a normal human. The probe is a 5.0-kb DNA fragment that is a copy of a LINE ("long interspersed element", a type of repetitive sequence; see

Chapter 2, p. 29 and Chapter 7, pp. 160–161) that has an internal *Eco*RI site.

**c.** The genomic DNA is from a normal human. The probe is a 5.0-kb DNA fragment that is a copy of a LINE that lacks an internal *Eco*RI site.

**d.** The genomic DNA is from a human heterozygous for a translocation (exchange of chromosome parts) between chromosomes 14 and 21. The probe is a 3.0-kb DNA fragment that is obtained by excision with the enzyme *Eco*RI from a plasmid containing single-copy genomic DNA from a normal chromosome 14. The translocation breakpoint on chromosome 14 lies within the 3.0-kb genomic DNA fragment.

**e.** The genomic DNA is from a normal female. The probe is a 5.0-kb DNA fragment containing part of the *testis determining factor TDF* gene, a gene located on the Y chromosome.

*__10.22__ The investigators described in Question 10.14 were successful in purifying a BAC clone containing the gene for the mouse prion protein. To narrow down which region of the BAC DNA contains the gene for the prion protein gene, they purified the BAC DNA, digested it with the restriction enzyme *Not*I, and separated the products of the enzymatic digestion by size using gel electrophoresis. Then they purified each of the relatively large *Not*I DNA fragments from the gel, digested each individually with the restriction enzyme *Bam*HI, and separated the products of each enzymatic digestion by size using gel electrophoresis (see Table 8.1, p. 174, for a description of the sites recognized by *Not*I and *Bam*HI). Finally, they transferred the size-separated DNA fragments from the agarose gel onto a membrane filter using the Southern blot technique, and allowed the DNA fragments on the filter to hybridize with a labeled cDNA probe. Figure 10.C shows the results that were obtained: The pattern of DNA bands seen after the BAC DNA is digested with *Not*I is shown in Panel A, the pattern of DNA bands seen after each *Not*I fragment is digested with *Bam*HI is shown in Panel B, and the pattern of hybridizing DNA fragments visible after probing the Southern blot is shown in Panel C.

**a.** Note the scales (in kb) on the left of each figure. Why are relatively larger DNA fragments obtained with *Not*I than with *Bam*HI?

**b.** An alternative approach to identify the *Bam*HI fragments containing the prion-protein gene would be to digest the BAC DNA directly with *Bam*HI, separate the products by size using gel electrophoresis, make a Southern blot, and probe it with the labeled cDNA clone. Why might the researchers have added the additional step of first purifying individual large *Not*I fragments, and then separately digesting each with *Bam*HI before making the Southern blot?

**c.** Which *Not*I DNA fragment contains the gene for the mouse prion protein?

**d.** Which *Bam*HI fragments contain the gene for the mouse prion protein?

**e.** About what size is the RNA-coding region of the gene for the mouse prion protein? Why is it so much larger than the cDNA?

**10.23** Sara is an undergraduate student who is doing an internship in the research laboratory described in Questions 10.14 and 10.22. Just before Sara started working in the lab, the restriction map in Figure 10.D was made of the 47-kb *Not*I restriction fragment containing the prion-protein gene (distances between restriction sites are in kb). Since smaller DNA fragments cloned into plasmids are more easily analyzed than large DNA fragments cloned

**Figure 10.C**

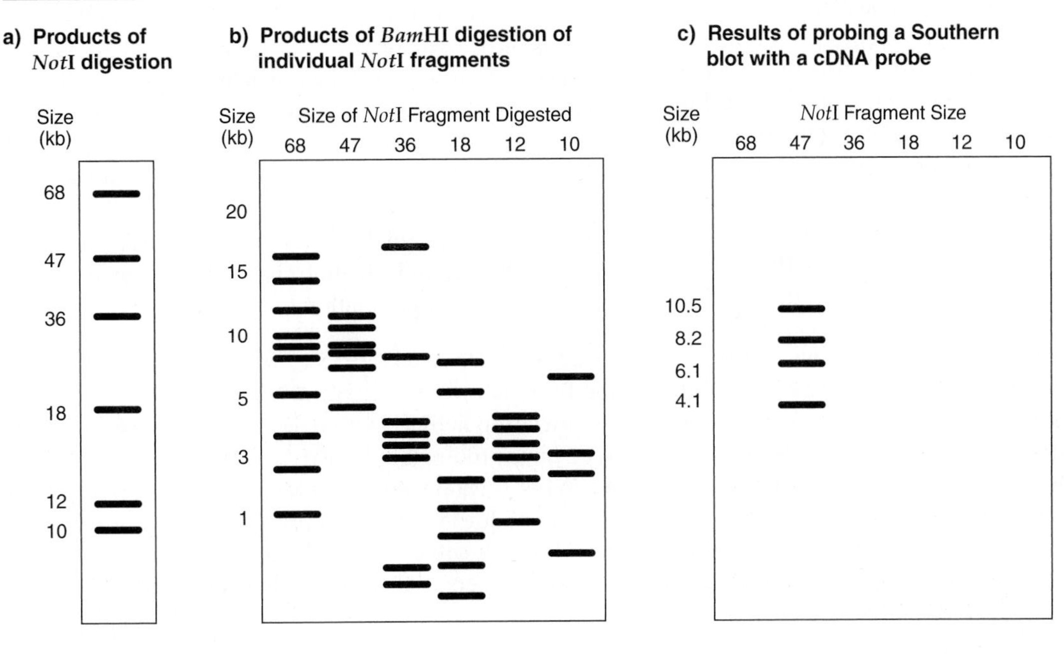

Figure 10.D

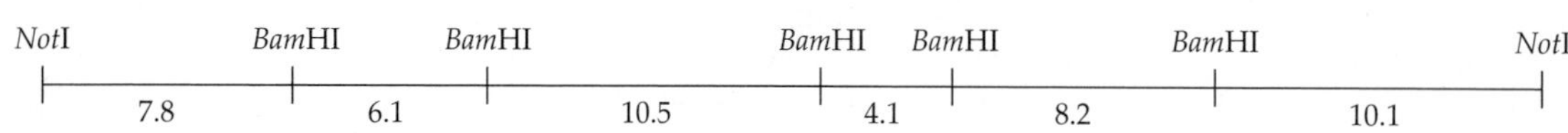




into BACs, Sara has been asked to "subclone" the 6.1-, 10.5-, 4.1- and 8.2-kb *Bam*HI DNA fragments containing the prion-protein gene into the pBluescript II plasmid vector (see Figure 8.4, p. 174, for a description of pBluescript II). Her mentor gives her some intact pBluescript II plasmid DNA, some of the purified 47-kb *Not*I fragment, and shows her where the stocks of DNA ligase, *Bam*HI, and reagents for PCR are stored in the lab.

**a.** Describe the steps Sara should take to complete her task if she has no information about the sequence of the 47-kb *Not*I fragment. In your answer, address how she will identify plasmids that contain genomic DNA inserts, and how she will verify that she has identified clones containing each of the desired genomic *Bam*HI fragments.

**b.** Describe an alternative approach that Sara could take to complete her task if she first performs a bioinformatic analysis utilizing DNA sequence information available from the mouse genome project, and identifies the sequence of the 47-kb *Not*I fragment.

***10.24** Imagine that you have cloned the structural gene for an enzyme that functions in the biosynthesis of catecholamines in the adrenal gland of rats. How could you use this cloned DNA as a probe to determine whether this same gene functions in the rat brain?

**10.25** A cDNA library is made with mRNA isolated from liver tissue and the vector shown in Figure 10.4. When a cloned cDNA insert from that library is digested with the enzymes *Eco*RI (E), *Hin*dIII (H), and *Bam*HI (B) (described in Table 8.1, p. 174), the restriction map shown in the following figure, part (a), is obtained. When this cDNA is used to screen a cDNA library made with mRNA from brain tissue and the vector shown in Figure 10.4, three identical cDNAs with the restriction map shown in the following figure, part (b), are obtained. When a uniformly labeled, $^{32}$P-labeled riboprobe made using T7 RNA polymerase is prepared using either cDNA and the probe is allowed to hybridize to a Southern blot prepared from genomic DNA digested singly with the enzymes *Eco*RI, *Hin*dIII, and *Bam*HI, an autoradiograph shows the pattern of bands in the following figure, part (c). When any of the $^{32}$P-labeled riboprobes are used to probe a northern blot prepared with poly(A)+ mRNA isolated from liver and brain tissues, no signal is seen. However, when the same northern blot is probed with a uniformly labeled, $^{32}$P-labeled probe is prepared using the random primer method (described in Box 10.1), the pattern of bands in part (d) of the figure is seen. Fully analyze these data and then answer the following questions.

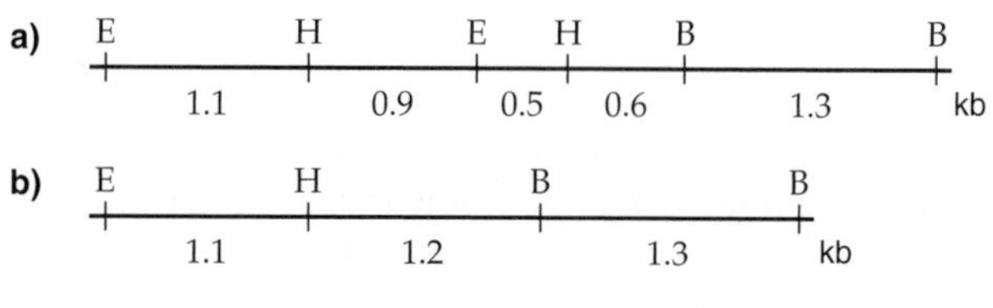


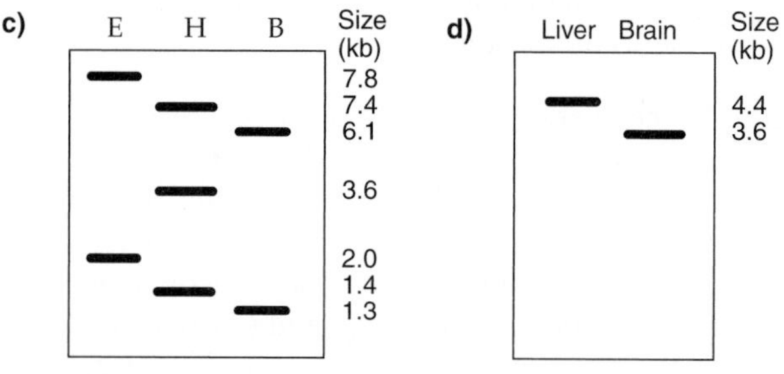


**a.** Do these cDNAs derive from the same gene?

**b.** Why are different-sized bands seen on the northern blot?

**c.** Why could hybridization signal be detected on the Southern blot but not on the northern blot when riboprobes were used? Why could hybridization signal be detected on the northern blot when a random-primed probe was used? (Hint: consider how these probes are made and which nucleic acid strands become labeled.)

**d.** Why do the cDNAs have different restriction maps?

**e.** Why are some of the bands seen on the whole-genome Southern blot different sizes than some of the restriction fragments in the cDNAs?

***10.26** A scientist is interested in understanding the physiological basis of alcoholism. She hypothesizes that the levels of the enzyme alcohol dehydrogenase, which is involved in the degradation of ethanol, are increased in individuals who routinely consume alcohol. She develops a rat model system to test this hypothesis. What steps should she take to determine if the transcription of the gene for alcohol dehydrogenase is increased in the livers of rats who are fed alcohol chronically compared to a control, abstinent population?

***10.27** *Taq* DNA polymerase, which is commonly used for PCR, is a thermostable DNA polymerase that lacks proofreading activity. Other DNA polymerases, such as *Vent*, have proofreading activity.

**a.** What advantages are there to using a DNA polymerase for PCR that has proofreading activity?

**b.** Although some DNA polymerases are more accurate than others, all DNA polymerases used in PCR introduce errors at a low rate. Why are errors introduced in the first few cycles of a PCR amplification more problematic than errors introduced in the last few cycles of PCR amplification?

***10.28** Katrina purified a clone from a plasmid library made using genomic DNA and sequenced a 500-bp long segment using the dideoxy sequencing method. Her twin-sister Marina used PCR with *Taq* DNA polymerase to amplify the same 500-bp fragment from genomic DNA. Marina sequenced the fragment using the dideoxy sequencing method, and obtained the same sequence as Katrina did. She then cloned the fragment into a plasmid vector and, following ligation and transformation into *E. coli*, sequenced several, independently isolated plasmids to verify that she had cloned the correct sequence. Most of them have the same sequence as Katrina's clone, but Marina finds that about $^1/_3$ of them have a sequence that differs in one or two base-pairs. None of the clones that differ from Katrina's clone are identical. Fearing she has done something wrong, Marina repeats her work, only to obtain the same results: about $^1/_3$ of the fragments cloned from the PCR product have single base-pair differences. Explain this discrepancy.

**10.29** What modifications are made to the polymerase chain reaction (PCR) to use this method for site-specific mutagenesis?

***10.30** Chapter 9 presented a description of how DNA microarray analysis was used to characterize changes to the transcriptome during yeast sporulation. That analysis found that more than 1,000 yeast genes showed significant changes in mRNA levels during sporulation, identified at least seven distinct temporal patterns of gene induction, and provided insights into the functions of many orphan genes. It is important to confirm findings from microarray analyses using independent methods. How would you confirm independently that three orphan genes display altered expression during yeast sporulation?

**10.31** Metalloproteases are enzymes that require a metal ion as a cofactor when they cleave peptide bonds. Members of one family of metalloproteases share the following consensus amino acid sequence in their catalytic site: His–Glu–X–Gly–His–Asp–X–Gly–X–X–His–Asp (X is any amino acid). Structural models of the catalytic site developed from X-ray crystallographic data suggest that the second amino acid, glutamate, is essential for proteolytic activity. Outline the experimental steps you would take to test this hypothesis. Assume you possess a cDNA encoding a metalloprotease having the consensus sequence, and can measure metalloprotease activity in a biochemical assay.

**10.32** What is meant by humanization, and how is it used to evaluate candidate drugs for treating a disease?

***10.33** DNA was prepared from small samples of white blood cells from a large number of people. These DNAs were individually digested with *Eco*RI, subjected to electrophoresis and Southern blotting, and the blot was probed with a radioactively labeled cloned human sequence. Ten different patterns were seen among all of the samples. The following figure shows the results seen in ten individuals, each of whom is representative of a different pattern.

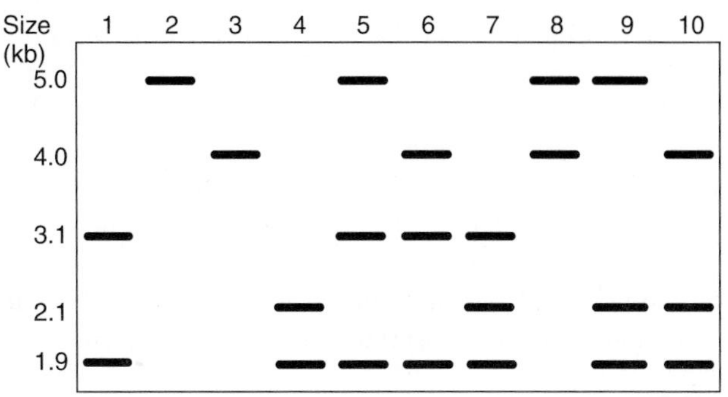


**a.** Explain the hybridization patterns seen in the 10 representative individuals in terms of variation in *Eco*RI sites.

**b.** If the individuals whose DNA samples are in lanes 1 and 6 on the blot were to produce offspring together, what bands would you expect to see in DNA samples from these offspring?

***10.34** The maps of the sites for restriction enzyme R in the wild type and the mutated cystic fibrosis genes are shown schematically in the following figure:

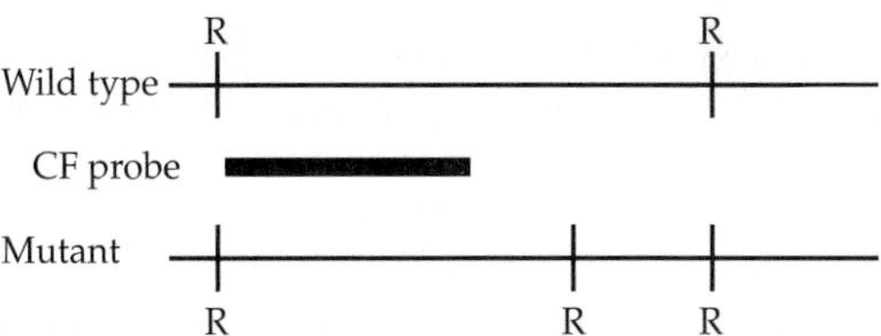


Samples of DNA obtained from a fetus (F) and her parents (M and P) were analyzed by gel electrophoresis followed by the Southern blot technique and hybridization with the radioactively labeled probe designated "CF probe" in the previous figure. The autoradiographic results are shown in the following figure:

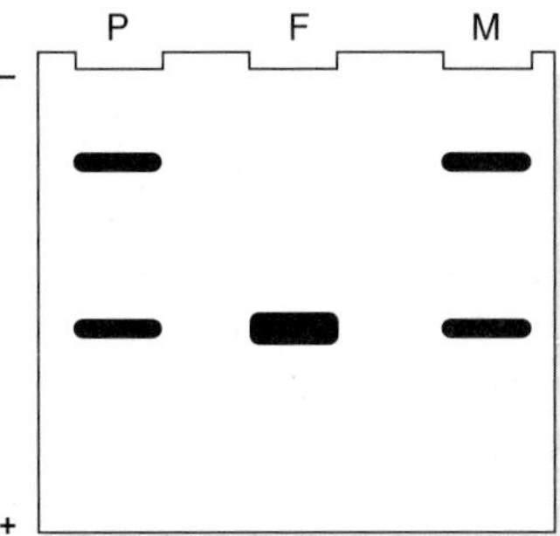


Given that cystic fibrosis results from a recessive trait and affected individuals always have two mutant alleles, will the fetus be affected? Explain.

***10.35** The enzyme *Tsp*45I recognizes the 5-bp site 5′-G-T-(either C or G)-A-C-3′. This site appears in exon 4 of the human gene for α-synuclein, where, in a rare form of Parkinson disease, it is altered by a single G-to-A mutation. (Note: Not all forms of Parkinson disease are caused by genetic mutations.)

**a.** Suppose you have primers that can be used in PCR to amplify a 200-bp segment of exon 4 containing the *Tsp*45I site, and that the *Tsp*45I site is 80 bp from the right primer. Describe the steps you would take to determine if a patient with Parkinson disease has this α-synuclein mutation.

**b.** What different results would you see in homozygotes for the normal allele, homozygotes for the mutant allele, and in heterozygotes?

**c.** How would you determine, in heterozygotes, if the mutant allele is transcribed in a particular tissue?

***10.36** For rare genetic disorders that have only one mutant allele, genetic tests can be tailored to detect the mutant and normal alleles specifically. However, for more prevalent genetic disorders, such as anemia caused by mutations in α- and β-globin, Duchenne muscular dystrophy caused by mutations in the dystrophin gene, and cystic fibrosis caused by mutations in CFTR, there are many different alleles at one gene that can lead to different disease phenotypes. These diseases present a challenge to genetic testing because, for these diseases, a genetic test that identifies only a single type of DNA change is inadequate. How can this challenge be overcome?

**10.37** What different types of DNA polymorphisms exist and what different methods can be used to detect them?

***10.38** Abbreviations used in genomics typically facilitate the quick and easy representation of longer tongue-twisting terms. Explore the nuances associated with some abbreviations by stating whether an RFLP, VNTR, or STR could be identified as an SNP? Explain your answers.

**10.39** A research team interested in social behavior has been studying different populations of laboratory rats. By using a selective breeding strategy, they have developed two populations of rats that differ markedly in their behavior: one population is abnormally calm and placid, while the second population is hyperactive, nervous, and easily startled. Biochemical analyses of brains from each population reveal different levels of a catecholamine neurotransmitter, a molecule used by neurons to communicate with each other. Relative to normal rats, the hyperactive population has increased levels while the calm population has decreased levels. Based on these results, the researchers have hypothesized that the behavioral and biochemical differences in the two populations are caused by variations in a gene that encodes an enzyme used in the synthesis of the catecholamine. Suppose you have a set of SNPs that are distributed throughout the rat genomic region containing this gene, including its promoter, coding region, enhancers, and silencers. How could you use these SNPs to test this hypothesis?

***10.40** The frequency of individuals in a population with two different alleles at a DNA marker is called the marker's heterozygosity. Why would an STR DNA marker with nine known alleles and a heterozygosity of 0.79 be more useful for mapping and DNA fingerprinting studies than a nearby STR having three alleles and a heterozygosity of 0.20?

**10.41** What is DNA fingerprinting and what different types of DNA markers are used in DNA fingerprinting? How could this method be used to establish parentage? How is it used in forensic science laboratories?

***10.42** One application of DNA fingerprinting technology has been to identify stolen children and return them to their parents. Bobby Larson was taken from a supermarket parking lot in New Jersey in 1978, when he was 4 years old. In 1990, a 16-year-old boy called Ronald Scott was found in California, living with a couple named Susan and James Scott, who claimed to be his parents. Authorities suspected that Susan and James might be the kidnappers and that Ronald Scott might be Bobby Larson. DNA samples were obtained from Mr. and Mrs. Larson and from Ronald, Susan, and James Scott. Then DNA fingerprinting was done, using a multilocus probe for a particular VNTR family, with the results shown in the following figure. From the information in the figure, what can you say about the parentage of Ronald Scott? Explain.

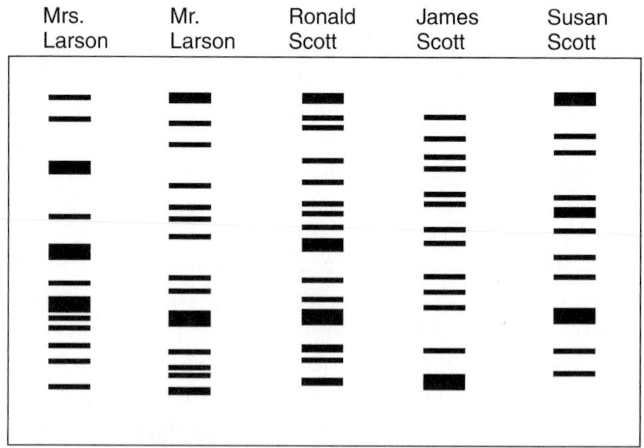


***10.43** As described in the text and demonstrated in Question 10.42, VNTRs can robustly distinguish between different individuals. Five well-chosen, single-locus VNTR probes used together can almost uniquely identify one individual because, statistically, they are able to discriminate 1 in $10^9$ individuals. However, the use of VNTR markers has largely been supplanted by the use of STR markers. For example, the FBI uses a set of 13 STR markers in forensic analyses. Different fluorescently labeled primers and reaction conditions have been developed so that this marker set can be multiplexed—all of the markers can be amplified in one PCR reaction. The marker set used by the FBI, the number of alleles at each marker, and the probability of obtaining a random match of a marker in Caucasians is listed in the following table:

| STR Marker | Number of Alleles | Probability of a Random Match (Based on an Analysis of Caucasians) |
|---|---|---|
| CSF1PO | 11 | 0.112 |
| FGA | 19 | 0.036 |
| TH01 | 7 | 0.081 |
| TPOX | 7 | 0.195 |
| VWA | 10 | 0.062 |
| D3S1358 | 10 | 0.075 |
| D5S818 | 10 | 0.158 |
| D7S820 | 11 | 0.065 |
| D8S1179 | 10 | 0.067 |
| D13S317 | 8 | 0.085 |
| D16S539 | 8 | 0.089 |
| D18S51 | 15 | 0.028 |
| D21S11 | 20 | 0.039 |

**a.** Consider the types of DNA samples that the FBI analyzes and the requirements concerning DNA samples in the methods used to analyze STR and VNTR markers. Why is the use of STR markers preferable to the use of VNTR markers?

**b.** Why is it advantageous to be able to multiplex the PCR reactions used in forensic STR analyses?

**c.** Suppose the first four STR markers listed in the table are used to characterize the genotype of an individual, and the genotype is an exact match with results obtained from a hair sample found at a crime scene. What is the probability that the individual has been misidentified, that is, what is the chance of a random match when just these four markers are used? About how often do you expect an individual to be misidentified if only these four markers are used?

**d.** Answer the questions posed in (c) if all 13 STR markers are used.

**10.44** About midnight on Saturday, the strangled body of a regular patron of the Seedy Lounge is found in an alleyway near the bar. The police interview the workers and patrons A–R remaining in the bar. A few of the patrons indicate that several individuals, including the bartender, owed money to the victim. The police notice that the bartender and patrons A, C, D, F, K, L, O, and R all have recent cuts and scratches on their faces and backs of their necks, but are told that these happened during mud-wrestling matches earlier in the evening. DNA samples are obtained from the bartender and the bar's patrons, from tissues of the victim, and from scrapings of her fingernails. STR analyses are performed on the DNA samples using three of the markers described in Question 10.43: THO1, D18S51, and D21S11. The sizes of the PCR products obtained in each DNA sample for each marker are shown in Table 10.A.

**Table 10.A**

| DNA Sample | STR | | |
|---|---|---|---|
| | THO1 | D21S11 | D18S51 |
| Victim | 162, 170 | 221, 239 | 292, 304 |
| Victim's fingernail scraping | 162, 170, 174 | 221, 225, 233, 239 | 280, 292, 300, 304 |
| Patron A | 159, 174 | 221, 225 | 292, 316 |
| Patron B | 162 | 221, 235 | 296, 304 |
| Patron C | 162, 174 | 225, 233 | 280, 300 |
| Patron D | 170, 174 | 229, 231 | 300, 304 |
| Patron E | 170, 174 | 225, 233 | 288, 292 |
| Patron F | 162, 166 | 229, 243 | 284, 288 |
| Patron G | 174 | 225, 235 | 292, 308 |
| Patron H | 159, 174 | 221, 233 | 296 |
| Patron I | 159, 174 | 233, 235 | 300, 308 |
| Patron J | 170, 174 | 225 | 284, 296 |
| Patron K | 170, 174 | 231, 235 | 288, 292 |
| Patron L | 159 | 237, 239 | 276, 304 |
| Patron M | 159, 170 | 221, 229 | 304, 308 |
| Patron N | 166, 174 | 229, 239 | 292, 304 |
| Patron 0 | 170 | 221, 225 | 288, 308 |
| Patron P | 162, 170 | 221 | 296, 300 |
| Patron Q | 159, 174 | 235, 239 | 284, 304 |
| Patron R | 170, 174 | 225, 233 | 288, 292 |
| Bartender | 170, 174 | 221, 231 | 300, 308 |

**a.** How many different alleles are present at each marker in these samples and how does this compare to the total number of alleles that exist? How do you explain the appearance of only one marker allele in some individuals? How do you explain the appearance of three and four marker alleles in the DNA sample obtained from the victim's fingernails?
**b.** Who should the police investigate further if they consider the results obtained using only the D21S11 marker? Explain your reasoning.
**c.** Who should the police investigate further if the consider the results obtained using all three STR markers? Explain your reasoning.

***10.45** Male sexual behavior in *Drosophila* (fruit fly) is under the control of several regulatory genes, including a gene called *fruitless*. This gene has been cloned, and both genomic and cDNA clones are available. It encodes proteins that appear to function as male-specific transcriptional regulators. One means to understand more fully the function of *fruitless* in male sexual behavior is to identify genes for proteins that interact with its protein product. Describe the steps you would take to accomplish this goal.

**10.46** Genetic variability is important for maintaining the ability of a species to adapt to different environments. Therefore, it is important to understand how much genetic variation there is in an endangered species, as this type of information can be used to design better strategies to help the species from becoming extinct. Listed below are four strategies that have been proposed for detecting a SNP in a known DNA sequence in several hundred individuals from an endangered species. Evaluate them critically, and explain why each is, or is not, a good strategy for this purpose.
**a.** Sacrifice each of the animals or plants in the name of science. Isolate their genomic DNA, prepare libraries from each, and screen for clones containing the sequence. Sequence each clone individually. Then compare the sequences of the different clones.
**b.** Isolate a few cells (e.g., by using a cheek scraping or leaf sampling) from each of the individuals. Prepare DNA from the samples and use the ASO hybridization method.
**c.** Isolate a few cells from each of the individuals. Prepare DNA from each of the samples and then use the yeast two-hybrid system.
**d.** Search the literature to find a restriction enzyme that cleaves the sequence containing the SNP and that cleaves the site when only one SNP allele is present. Use the restriction enzyme to measure the site as an RFLP marker. After isolating a few cells from the individuals, prepare DNA from the cells, digest it with the restriction enzyme, separate it by size using electrophoresis, make a Southern blot, and perform an RFLP analysis.

**10.47** Just as VNTRs and STRs can be used in forensic analyses to determine human genetic identity, they can be used to determine the genetic identity of members of an endangered species. This can be helpful to track animals poached from protected reserves and associate parts of endangered animals that are sold illegally with their source. How would you identify a set of polymorphic STR loci containing a CAG repeat in an endangered species?

**10.48** In 1990, the first human gene therapy experiment on a patient with adenosine deaminase deficiency was done. Patients who are homozygous for a mutant gene for this enzyme have defective immune systems and risk death from diseases as simple as a common cold. Which cells were involved, and how were they engineered?

**10.49** What methods are used to introduce genes into plant cells, and how are these methods different than those used to introduce genes into animal cells?

**10.50** The ability to place cloned genes into plants raises the possibility of engineering new, better strains of crops such as wheat, maize, and squash. It is possible to identify useful genes, isolate them by cloning, and insert them directly into a plant host. Usually these genes bring out desired traits that allow the crops in question to flourish. Why then is there such concern by consumers about this process? Do you feel that the concern is justified? Defend your answer.

# 11 Mendelian Genetics

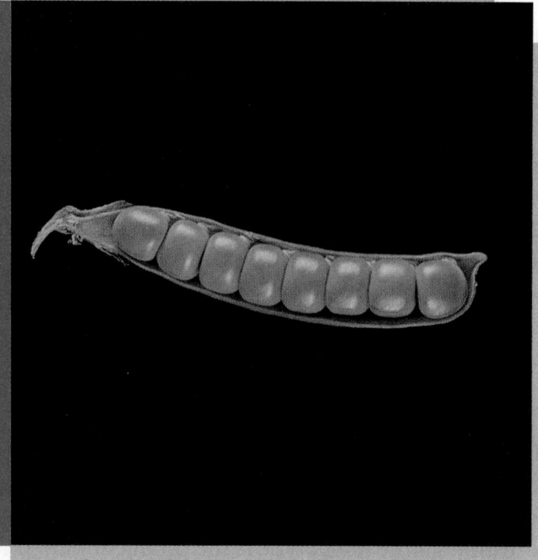

Smooth seeds of the garden pea, *Pisum sativum*.

## Key Questions

- How do single genes segregate in genetic crosses?
- How do two genes segregate in genetic crosses?
- How is the inheritance of a gene analyzed in humans?

### iActivity

PEOPLE HAVE BRED ANIMALS AND PLANTS FOR specific traits for many centuries. Through breeding pea plants, Gregor Mendel developed his theory to explain the transmission of hereditary characteristics from generation to generation. What were Mendel's experiments? What is the relationship between genes and traits? How can knowing the way in which characteristics are inherited allow people to breed for specific traits?

Later on, you can try the iActivity for this chapter, which allows you to apply the knowledge you've gained in the effort to breed a very special pet.

Genetics is the study of the structure and function of genes. Historically, scientists were limited in the type of genetic analysis they could do. They focused on basic questions about heredity, notably: Is a trait inherited? How is a genetic trait inherited? How are genes transmitted from generation to generation? How do genes recombine? What are the specific locations of genes in the genome? These questions are about the subdiscipline of genetics known as *transmission genetics*. This chapter is the first of a series of chapters about transmission genetics. Once biochemical and molecular methodologies were developed, genetics researchers were then able to ask new questions such as: What is the structure of a gene at the molecular level? What are the processes for expressing a gene? What mechanisms cause mutations of genes? Studies of the structure and function of genes at the molecular level fall within the subdiscipline of *molecular genetics*. The molecular structure of the gene, and the molecular aspects of DNA replication, gene expression, and DNA mutation are discussed in Chapters 2–7.

The understanding of how genes are transmitted from parent to offspring began with the work of Gregor Johann Mendel (1822–1884), an Augustinian monk. The goal of this chapter is for you to learn the basic principles of the transmission of genes by examining Mendel's work. Be aware that, even though Mendel analyzed the segregation of hereditary traits, he did not know about the nature of genes, that genes are located in chromosomes, or even that chromosomes existed.

## Genotype and Phenotype

The characteristics of an individual are called **traits** (also called **characters**). Some traits are heritable—they are transmitted from generation to generation—while others are not heritable. Traits are under the control of **genes** (Mendel called them *factors*). The genetic constitution of an organism is called its **genotype,** and the **phenotype** is an observable trait or set of traits (structural and functional) of an organism produced by the interaction between its genotype and the environment. A phenotype may be visible, for example, an eye color; or not readily visible but measurable, for example, a molecular characteristic such as blood type, or an altered protein or enzyme.

Genes provide only the potential for developing a particular phenotype. The extent to which that potential is realized, in many cases, depends on environmental influences and random developmental events (Figure 11.1). A person's height, for example, is controlled by many genes, the expression of which can be significantly affected by environmental influences such as the effects of hormones during puberty (an internal environmental influence) and nutrition (an external environmental influence). In other words, genotypes set the range of possible phenotypes, while the environment determines where in that range the phenotype ends up.

Although the phenotype is the product of interaction between genes and environment, the contribution of the environment varies. In some cases, the environmental influence is great, but in others, the environmental contribution is nonexistent. We will develop the relationship between genotype and phenotype in more detail as the text proceeds.

### Keynote

The genotype is the genetic constitution of an organism. The phenotype is the observable characteristics of the organism. The genes give the potential for the development of traits; this potential often is affected by interactions with other genes and with the environment. Thus, individuals with the same genotype can have different phenotypes, and individuals with the same phenotypes may have different genotypes.

**Figure 11.1**
**Relationship between genotype and phenotype.**

Genotype (genetic constitution)

Environmental influences and random developmental events

Phenotype (expression of physical trait)

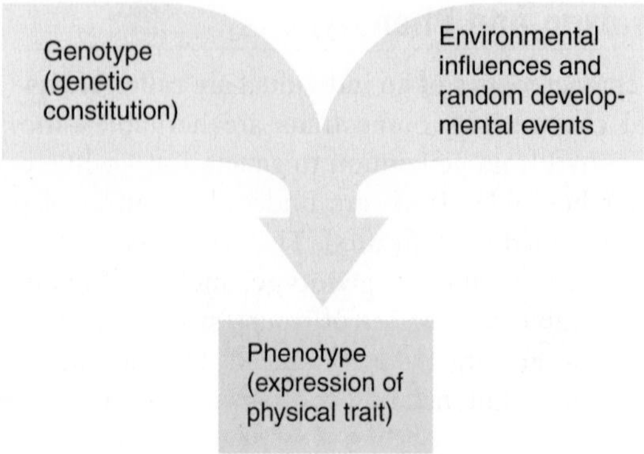

## Mendel's Experimental Design

The work of Gregor Johann Mendel (Figure 11.2) is considered the foundation of modern genetics. In 1843, he was admitted to the Augustinian Monastery in Brno (now Brünn, Czech Republic). In 1854, he began a series of breeding experiments with the garden pea *Pisum sativum* to learn something about the mechanisms of heredity. As a result of his creativity, Mendel discovered some fundamental principles of genetics.

From the results of crossbreeding pea plants with different traits involving seed shape, seed color, and flower color, Mendel developed a simple theory to explain the transmission of hereditary traits from generation to generation. (Mendel had no knowledge of mitosis and meiosis, so he did not know that genes segregate according to chromosome behavior.) Mendel reported his conclusions in 1865, but their significance was not fully realized until the late 1800s and early 1900s.

Mendel's experimental approach was effective because he made simple interpretations of the ratios of the types of progeny he obtained from his crosses and because he then carried out direct and convincing experiments to test his hypotheses. In his initial breeding experiments, he took the simplest approach of studying the inheritance of one trait at a time. (This is how you should work genetics problems.) He made carefully controlled matings (crosses) between pea strains that had obvious differences in heritable traits and, most importantly, he kept very careful records of the outcomes of the crosses. The numerical data he obtained enabled him to do a rigorous analysis of the hereditary transmission of traits.

**Figure 11.2**
**Gregor Johann Mendel, founder of the science of genetics.**

Generally, genetic crosses with eukaryotes are done as follows: two diploid individuals are allowed to produce haploid gametes by meiosis. Fusion of male and female gametes produces zygotes from which the diploid progeny individuals are generated. The phenotypes of the parents and offspring are analyzed to provide clues to the heredity of those phenotypes.

Mendel did all his significant genetic experiments with the garden pea (see Figure 1.4k, p. 6). The garden pea was a good choice because it fits many of the criteria that make an organism suitable for use in genetic experiments: it is easy to grow, bears flowers and fruit in the same year a seed is planted, and produces a large number of seeds.

Figure 11.3, which presents the procedure for crossing pea plants, begins with a cross section of a flower, showing the stamens (male reproductive organs) and the pistils (female reproductive organs). The pea normally reproduces by **self-fertilization** (also called **selfing**); that is, the anthers at the ends of the stamen produce pollen (microspore of a flowering plant that germinates to form the male [♂] gametophyte), which lands on the pistil (containing the female [♀] gametophyte) within the same flower and fertilizes the plant. Fortunately for the success of his experiments, Mendel was able to prevent self-fertilization of the pea by removing the stamens from a developing flower bud before their anthers produced any mature pollen. Next, he took pollen from the stamens of another flower and dusted them onto the pistil of the emasculated one to pollinate it.

**Cross-fertilization**, or simply **cross**, is the fusion of male gametes (in this case, pollen) from one individual and female gametes (eggs) from another. Once cross-fertilization has occurred, the zygote develops in the seeds (peas). Certain phenotypes are analyzed by inspecting the seeds themselves; others are analyzed by examining the plants that grow from the seeds.

Mendel obtained 34 strains of pea plants that differed in a number of traits. He allowed each strain to self-fertilize for many generations to ensure that the traits he wanted to study were inherited. This preliminary work ensured that Mendel worked only with pea strains in which the trait under investigation remained unchanged from parent to offspring for many generations. Such strains are called **true-breeding** or **pure-breeding strains.**

Next, Mendel selected seven pairs of traits to study in breeding experiments. Each pair affected one characteristic of the plant, with each member of a pair being clearly distinguishable (Figure 11.4):

1. Flower and seed coat color: grey versus white seed coats, and purple versus white flowers (a single gene controls both these color properties of seed coats and flowers)
2. Seed color: yellow versus green
3. Seed shape: smooth versus wrinkled
4. Pod color: green versus yellow
5. Pod shape: inflated versus pinched
6. Stem height: tall versus short
7. Flower position: axial versus terminal

**Figure 11.3**

**Procedure for crossing pea plants.**

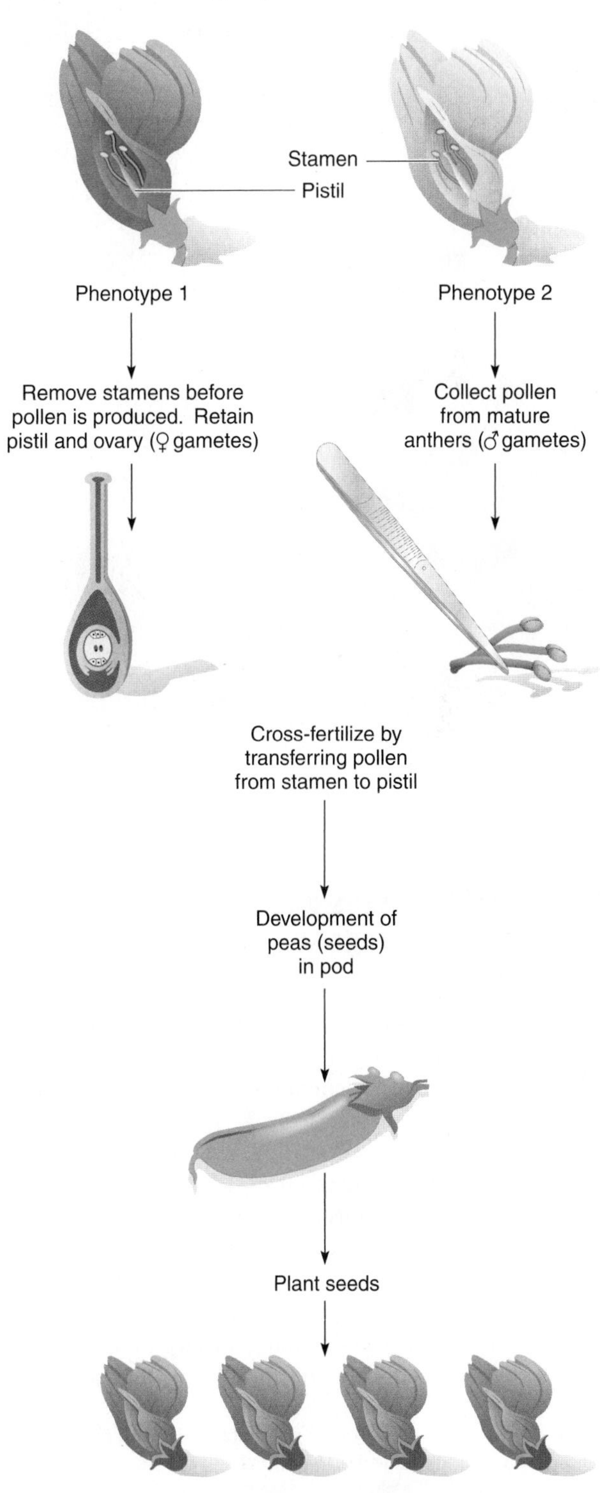
Stamen
Pistil
Phenotype 1
Phenotype 2
Remove stamens before pollen is produced. Retain pistil and ovary (♀ gametes)
Collect pollen from mature anthers (♂ gametes)
Cross-fertilize by transferring pollen from stamen to pistil
Development of peas (seeds) in pod
Plant seeds
Observe phenotypes of offspring

**Figure 11.4**

**Seven character pairs in the garden pea that Mendel studied in his breeding experiments.**

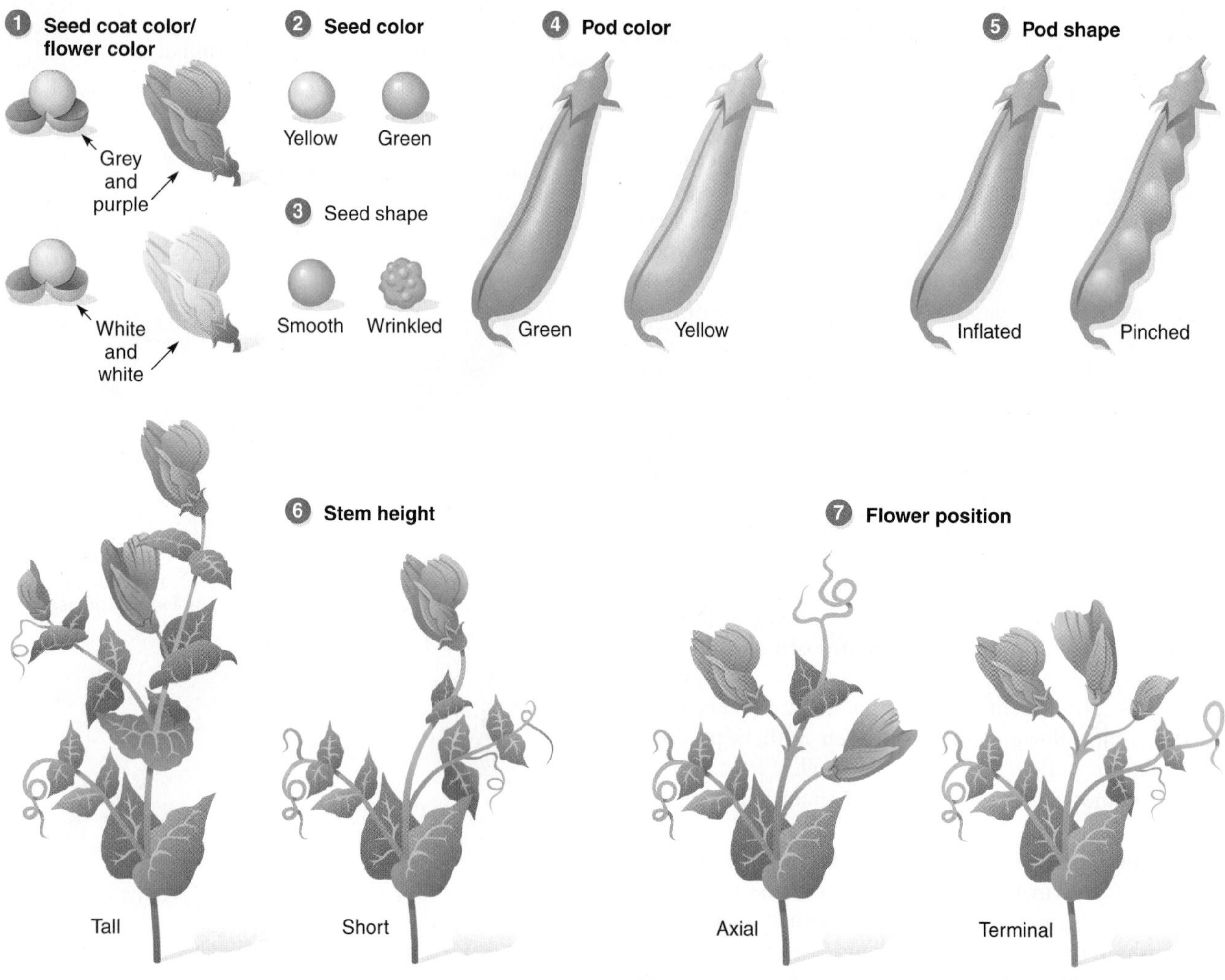


## Monohybrid Crosses and Mendel's Principle of Segregation

Let us be clear on the terminology used in breeding experiments. The parental generation is the **P generation.** The progeny of the P mating is the **first filial generation,** or $\mathbf{F_1}$. The subsequent generation produced by breeding together the $F_1$ offspring is the $\mathbf{F_2}$ **generation (second filial generation).** Interbreeding the offspring of each generation results in generations $F_3$, $F_4$, $F_5$, and so on.

Mendel first performed **monohybrid crosses**—crosses between true-breeding strains of peas that had alternative forms of a single trait. For example, when he pollinated pea plants that gave rise only to smooth seeds[1] with pollen from a true-breeding variety that produced only wrinkled seeds, the result was all smooth seeds (Figure 11.5). When the parental types were reversed—that is, when the pollen from a smooth-seeded plant was used to pollinate a pea plant that gave wrinkled seeds—the result was the same: all smooth seeds. Matings that are done both ways—here, smooth female [♀] × wrinkled male [♂] and wrinkled female [♀] × smooth male [♂]—are called **reciprocal crosses.** Conventionally, the female is given first in crosses of plants. If the results of reciprocal crosses are the same, it means that the inheritance of the trait does not depend on sex.

The significant point of this cross is that all the $F_1$ progeny seeds of the smooth × wrinkled reciprocal crosses were smooth: they exactly resembled only one of the parents in this cross rather than being a blend of both parental phenotypes. The finding that all offspring of true-breeding parents are alike is sometimes referred to as the *principle of uniformity in* $F_1$.

Next, Mendel planted the seeds and allowed the $F_1$ plants to self-fertilize to produce the $F_2$ seed. Both smooth and wrinkled seeds appeared in the $F_2$ generation, and both types could be found within the same pod. Typical of

[1]Seeds are the diploid progeny of sexual reproduction. If a phenotype concerns the seed itself, the results of the cross can be seen directly by looking at the seeds. If a phenotype concerns a part of the mature plant, such as flower color, then the seeds must be germinated and grown to maturity before that phenotype can be seen.

**Figure 11.5**

**Results of one of Mendel's breeding crosses.** In the parental generation, he crossed a true-breeding pea strain that produced smooth seeds with one that produced wrinkled seeds. All the $F_1$ progeny seeds were smooth.

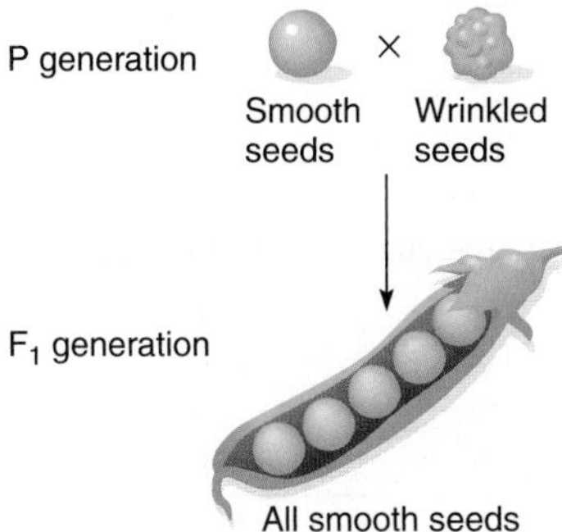


his analytical approach to the experiments, Mendel counted the number of seeds of each type. He found that 5,474 were smooth and 1,850 were wrinkled (Figure 11.6). The calculated ratio of smooth seeds to wrinkled seeds was 2.96:1, which is very close to a 3:1 ratio.

Mendel observed that, although the $F_1$ resembled only one of the parents in their phenotype, they did not breed true—a fact that distinguished the $F_1$ from the parent they resembled. Moreover, the $F_1$ could produce some $F_2$ progeny with the parental phenotype that had disappeared in the $F_1$. But how can a trait present in the P generation disappear in the $F_1$ and then reappear in the $F_2$? Mendel concluded that the alternative traits in the cross—smoothness or wrinkledness of the seeds—were determined by **particulate factors.** He reasoned that these factors, which were transmitted from parents to progeny through the gametes, carried hereditary information. Importantly, the two factors remain distinct in crosses, rather than blending together. We know factors now by another name: *genes.*

Since Mendel was examining a pair of traits (wrinkled and smooth seeds), each factor was considered to exist in alternative forms (which we now call **alleles**), each of which specified one of the traits. For the gene that controls the pea seed shape traits, there is one allele that results in a smooth seed and another allele that results in a wrinkled seed.

Mendel reasoned further that a true-breeding strain of peas must contain a pair of identical factors. In modern terms, this is the case because peas are diploid so that there are two copies of each gene on a pair of homologous chromosomes. Because the $F_2$ exhibited both traits and the $F_1$ exhibited only one of those traits, each $F_1$ individual must have contained both factors, one for each of the alternative traits. In other words, crossing two different true-breeding strains brings together in the $F_1$ one factor from each strain: the eggs (which are haploid, meaning having one set of chromosomes) contain one factor from one strain, and the pollen grains (which also are haploid, meaning having one set of chromosomes) contain one factor from the other

**Figure 11.6**

**The $F_2$ progeny of the cross shown in Figure 11.5.** When the plants grown from the $F_1$ seeds were self-pollinated, both smooth and wrinkled $F_2$ progeny seeds were produced. Commonly, both seed types were found in the same pod. In his experiments, Mendel counted 5,474 smooth and 1,850 wrinkled $F_2$ progeny seeds for a ratio of 2.96:1.

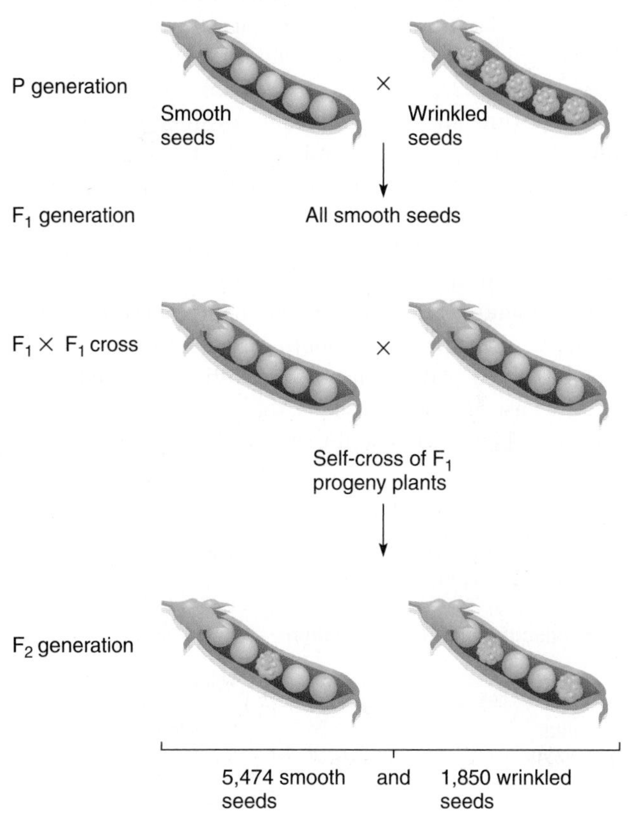


strain. Furthermore, because only one of the traits was seen in the $F_1$ generation, the expression of the missing trait must somehow have been masked by the visible trait; this masking is called *dominance.* For the smooth × wrinkled cross, the $F_1$ seeds were all smooth. Thus, the allele for smoothness is masking or **dominant** to the allele for wrinkledness, and the smooth seed trait is considered to be the *dominant trait*, and the allele associated with it is called the *dominant allele.* Conversely, wrinkled is **recessive** to smooth because the factor for wrinkled is masked, and the wrinkled seed trait is considered to be the *recessive trait*, and the allele associated with it is called the *recessive allele.* Note that the terms *dominant* and *recessive* as applied to alleles have no meaning in isolation; in other words, they only have meaning with respect to another allele.

Crosses are visualized by using symbols for the alleles, as Mendel did. For the smooth × wrinkled cross, we can give the symbol *S* to the allele for smoothness and the symbol *s* to the allele for wrinkledness. The letter used is based on the dominant phenotype, and the convention in this case is that the dominant allele is given the uppercase letter and the recessive allele the lowercase letter. (This convention was used for many years, particularly in plant genetics. Now it is more conventional to base the letter assignment on the recessive phenotype. We will use the newer convention later.)

Using these symbols, we denote the genotype of the parental plant grown from the smooth seeds by *SS* and that of the wrinkled parent by *ss*. Individuals that contain two copies of the same specific allele of a particular gene are said to be **homozygous** for that gene (Figure 11.7). When diploid plants produce haploid gametes by meiosis (see Chapter 12), each gamete contains only one copy of the gene (one allele); the plants from smooth seeds produce *S*-bearing gametes, and the plants from wrinkled seeds produce *s*-bearing gametes. When the gametes fuse during fertilization, the resulting diploid zygote has one *S* allele and one *s* allele, a genotype of *Ss*. Plants that have two different alleles of a particular gene are said to be **heterozygous.** Because of the dominance of the smooth *S* allele, *Ss* plants produce smooth seeds (see Figure 11.7).

Figure 11.8 diagrams the smooth × wrinkled cross with the use of genetic symbols; the production of the $F_1$ is shown in Figure 11.8a and that of the $F_2$ in Figure 11.8b.



**Figure 11.7**

**Dominant and recessive alleles of a gene for seed shape in peas.**

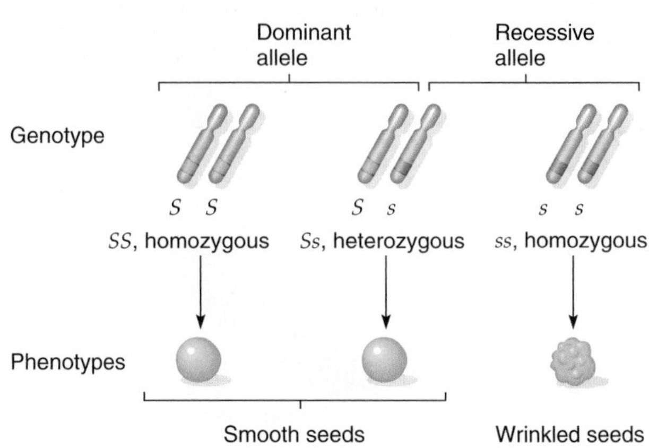


**Figure 11.8**

**The same cross as in Figures 11.5 and 11.6, using genetic symbols to illustrate the principle of segregation of Mendelian factors.**

**a) Production of the $F_1$ generation**

| P generation | Parent 1 ♀ | Parent 2 ♂ |
|---|---|---|
| Parental phenotype | Smooth seeds | Wrinkled seeds |
| Diploid parental genotype | *SS* | *ss* |
| Haploid gametes | *S* *S* | × *s* *s* |

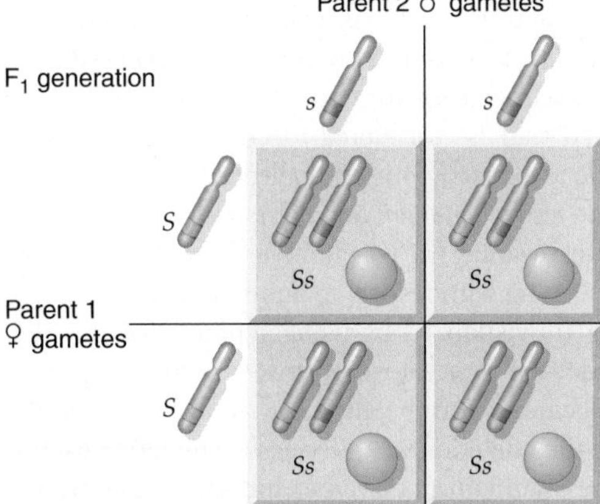


$F_1$ genotypes: all *Ss*

$F_1$ phenotypes: all smooth (smooth is dominant to wrinkled)

**b) Production of the $F_2$ generation**

| $F_1$ generation | ♀ | ♂ |
|---|---|---|
| $F_1$ phenotypes | Smooth seeds | Smooth seeds |
| Diploid $F_1$ genotypes | *Ss* | *Ss* |
| Haploid $F_1$ gametes | *S* *s* | × *S* *s* |

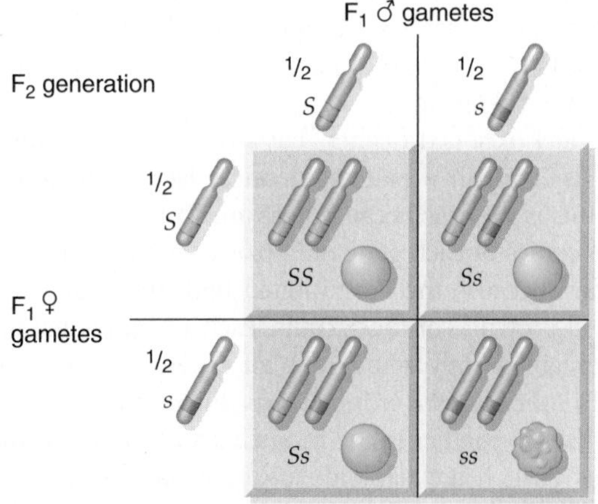


$F_2$ genotypes: 1/4 *SS*, 1/2 *Ss*, 1/4 *ss*

$F_2$ phenotypes: 3/4 smooth seeds, 1/4 wrinkled seeds

(In Figures 11.7 and 11.8, the genes are shown on chromosomes because the segregation of genes from generation to generation follows the behavior of chromosomes in meiosis and fertilization.) The true-breeding, smooth-seeded parent has the genotype *SS*, and the true-breeding, wrinkle-seeded parent has the genotype *ss*. Because each parent is true breeding and diploid (that is, has two sets of chromosomes), each must contain two copies of the same allele. All the $F_1$ plants produce smooth seeds, and all are *Ss* heterozygotes.

The plants grown from the $F_1$ seeds differ from the smooth parent in that they produce equal numbers of two types of gametes: *S*-bearing gametes and *s*-bearing gametes. All possible fusions of $F_1$ gametes are shown in the matrix in Figure 11.8b, called a **Punnett square** after its originator, Reginald Punnett. These fusions give rise to the zygotes that produce the $F_2$ generation.

In the $F_2$ generation, three types of genotypes are produced: *SS*, *Ss*, and *ss*. As a result of the random fusing of gametes, the relative proportion of these zygotes is 1:2:1, respectively. However, because the *S* factor is dominant to the *s* factor, both the *SS* and *Ss* seeds are smooth, and the $F_2$ generation seeds show a phenotypic ratio of 3 smooth : 1 wrinkled.

Mendel also analyzed the behavior of the six other pairs of traits. Qualitatively and quantitatively, the same results were obtained (Table 11.1). From the seven sets of crosses, he made the following general conclusions about his data:

1. The results of reciprocal crosses were always the same.
2. All $F_1$ progeny resembled one of the parental strains, indicating the dominance of one allele over the other.
3. In the $F_2$ generation, the parental trait that had disappeared in the $F_1$ generation reappeared. Furthermore, the trait seen in the $F_1$ (the dominant trait) was always found in the $F_2$ at about three times the frequency of the other trait (the recessive trait).

## The Principle of Segregation

From the sort of data just discussed, Mendel proposed what has become known as his first law, the **principle of segregation**: *Recessive traits, which are masked in the $F_1$ from a cross between two true-breeding strains, reappear in a specific proportion in the $F_2$.* In modern terms this means that *the two members of a gene pair (alleles) segregate (separate) from each other during the formation of gametes in meiosis.* As a result, half the gametes carry one allele, and the other half carry the other allele. In other words, each gamete carries only a single allele of each gene. The progeny are produced by the random combination of gametes from the two parents.

In proposing the principle of segregation, Mendel had differentiated between the factors (genes) that determined the traits (the genotype) and the traits themselves (the phenotype). We know now, of course, that genes are on chromosomes. The specific location of a gene on a chromosome is called its **locus** (or **gene locus;** plural *loci*). Furthermore, Mendel's first law means that, at the gene level, the members of a pair of alleles segregate during meiosis and that each offspring receives only one allele from each parent. Thus, **gene segregation** parallels the separation of homologous pairs of chromosomes at anaphase I in meiosis (see Chapter 12, pp. 334–335).

Box 11.1 presents a summary of the genetics concepts and terms we have discussed so far in this chapter. A thorough familiarity with these terms is essential to your study of genetics.

**Table 11.1 Mendel's Results in Crosses between Plants Differing in One of Seven Characters**

| Character[a] | $F_1$ | Dominant | $F_2$ (Number) Recessive | Total | $F_2$ (Ratio) Dominant : Recessive |
|---|---|---|---|---|---|
| Seeds: smooth versus wrinkled | All smooth | 5,474 | 1,850 | 7,324 | 2.96:1 |
| Seeds: yellow versus green | All yellow | 6,022 | 2,001 | 8,023 | 3.01:1 |
| Seed coats: grey versus white[b] / Flowers: purple versus white | All grey / All purple | 705 | 224 | 929 | 3.15:1 |
| Flowers: axial versus terminal | All axial | 651 | 207 | 858 | 3.14:1 |
| Pods: inflated versus pinched | All inflated | 882 | 299 | 1,181 | 2.95:1 |
| Pods: green versus yellow | All green | 428 | 152 | 580 | 2.82:1 |
| Stem: tall versus short | All tall | 787 | 277 | 1,064 | 2.84:1 |
| Total or average | | 14,949 | 5,010 | 19,959 | 2.98:1 |

[a]The dominant trait is always written first.
[b]A single gene controls both the seed coat and the flower color trait.

## Box 11.1 Genetic Terminology

**Alleles:** Different forms of a gene. For example, *S* and *s* alleles represent the smoothness and wrinkledness of the pea seed. (Like gene symbols, allele symbols are italicized.)

**Character:** A characteristic of an individual that is transmitted from generation to generation. Synonym of **trait.**

**Cross:** A mating between two individuals, leading to the fusion of gametes.

**Diploid:** A eukaryotic cell or organism with two homologous sets of chromosomes.

**$F_1$ generation (the first filial generation):** The progeny of mating of individuals of the P generation.

**$F_2$ generation (the second filial generation):** The progeny resulting from interbreeding $F_1$ generation individuals.

**Gamete:** A mature reproductive cell that is specialized for sexual fusion. Each gamete is haploid and fuses with a cell of similar origin, but of opposite sex, to produce a diploid zygote.

**Gene (Mendelian factor):** The determinant of a characteristic of an organism. (Gene symbols are italicized.) A gene's nucleotide sequence specifies a polypeptide or an RNA.

**Genotype:** The genetic constitution of an organism. A diploid organism in which both alleles are the same at a given gene locus is said to be **homozygous** for that allele. Homozygotes produce only one gametic type with respect to that locus. For example, true-breeding, smooth-seeded peas have the genotype *SS*, and true-breeding wrinkle-seeded peas have the genotype *ss*; both are homozygous. The smooth parent is **homozygous dominant;** the wrinkled parent is **homozygous recessive.**

Diploid organisms that have two different alleles at a specific gene locus are said to be **heterozygous.** Thus, $F_1$ hybrid plants from the cross of *SS* and *ss* parents have one *S* allele and one *s* allele. Individuals heterozygous for two allelic forms of a gene produce two kinds of gametes (*S* and *s*).

**Haploid:** A cell or an individual with set of chromosomes.

**Locus (gene locus; plural *loci*):** The specific place on a chromosome where a gene is located.

**P generation:** Parental generation in breeding experiments.

**Phenotype:** The physical manifestation of a genetic trait that results from a specific genotype and its interaction with the environment. In our example, the *S* allele was dominant to the *s* allele, so in the heterozygous condition the seed is smooth. Therefore, both the homozygous dominant *SS* and the heterozygous *Ss* seeds have the same phenotype (smooth), even though they differ in genotype.

**Trait:** A characteristic of an individual. A heritable trait is transmitted from generation to generation. Synonym of **character.**

**True-breeding:** When a trait being studied remains unchanged from parent to offspring for many generations. Typically this means that there is homozygosity for the allele responsible for the trait.

**Zygote:** The cell produced by the fusion of male and female gametes.

### Keynote

Mendel's first law, the principle of segregation, states that the two members of a gene pair (alleles) segregate (separate) from each other in the formation of gametes; half the gametes carry one allele, and the other half carry the other allele.

## Representing Crosses with a Branch Diagram

The use of a Punnett square to represent the pairing of all possible gamete types from two parents in a cross (see Figure 11.8) is a simple way to predict the relative frequencies of genotypes and phenotypes in the next generation. There is an alternative method, one you are encouraged to master: the branch diagram. (Box 11.2 discusses some elementary principles of probability that will help you understand this approach.) To use the branch diagram approach, it is necessary to know the dominance–recessiveness relationship of the allele pair so that the progeny phenotypic classes can be determined. Figure 11.9 illustrates the application of the branch diagram to analysis of the $F_1$ selfing of the smooth × wrinkled cross diagrammed in Figure 11.8.

**Figure 11.9**

**Using the branch diagram approach to calculate the ratios of phenotypes in the $F_2$ generation of the cross in Figure 11.8.**

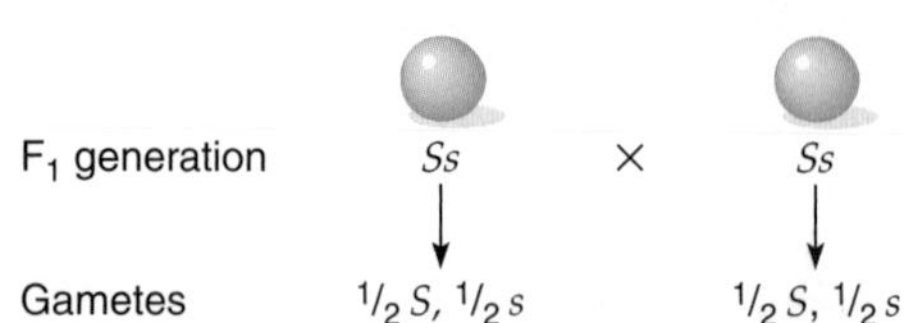


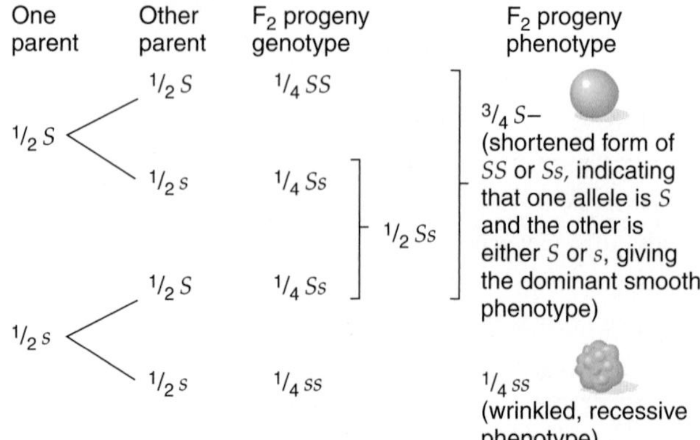

**Box 11.2 Elementary Principles of Probability**

A **probability** is the ratio of the number of times a particular event is expected to occur to the number of trials during which the event could have happened. For example, the probability of picking a heart from a deck of 52 cards, 13 of which are hearts, is *P*(heart) = $^{13}/_{52}$ = $^{1}/_{4}$. That is, we would expect, on the average, to pick a heart from a deck of cards once in every four trials.

Probabilities and the *laws of chance* are involved in the transmission of genes. As a simple example, consider a couple and the chance that their child will be a boy or a girl. Assume that an exactly equal number of boys and girls are born (which is not precisely true, but we can assume it to be so for the sake of discussion). The probability that the child will be a boy is $^{1}/_{2}$ or 0.5. Similarly, the probability that the child will be a girl is also $^{1}/_{2}$.

Now a rule of probability can be introduced: the **product rule.** The product rule states that the probability of two independent events occurring simultaneously is the product of each of their individual probabilities. Thus, the probability that both children in a family with two children will be girls is $^{1}/_{4}$. That is, the probability of the first child being a girl is $^{1}/_{2}$, the probability of the second being a girl is also $^{1}/_{2}$, and, by the product rule, the probability of the first and second being girls is $^{1}/_{2} \times {}^{1}/_{2} = {}^{1}/_{4}$. Similarly, the probability of having three boys in a row is $^{1}/_{2} \times {}^{1}/_{2} \times {}^{1}/_{2} = {}^{1}/_{8}$.

Another rule of probability, the **sum rule,** states that the probability of occurrence of any of several mutually exclusive events is the sum of the probabilities of the individual events. For example, if one die is thrown, what is the probability of getting a one or a six? The individual probabilities are calculated as follows: The probability of rolling a one, *P*(one), is $^{1}/_{6}$, because there are six faces to a die. For the same reason, the probability of rolling a six, *P*(six), is also $^{1}/_{6}$. To roll a one or a six with a single throw of the die involves two mutually exclusive events, so the sum rule is used. The sum of the individual probabilities is $^{1}/_{6} + {}^{1}/_{6} = {}^{2}/_{6} = {}^{1}/_{3}$. To return to our family example, the probability of having two boys or two girls is $^{1}/_{4} + {}^{1}/_{4} = {}^{1}/_{2}$.

The $F_1$ seeds from the cross in Figure 11.8 have the genotype *Ss*. In meiosis, we expect half of the gametes to be *S* and half to be *s* (see Figure 11.9). Thus, $^{1}/_{2}$ is the predicted frequency of each of these two types.

From the rules of probability, we can predict the expected frequencies of the three possible genotypes in the $F_2$ generation using a branch diagram. From one parent, the frequency of an *S* gamete is $^{1}/_{2}$, and the frequency of an *s* gamete is $^{1}/_{2}$. The *S* gamete from that parent fuses with a gamete from the other parent. From that other parent, the frequency of an *S* gamete is also $^{1}/_{2}$, and the frequency of an *s* gamete is also $^{1}/_{2}$. To produce an $F_2$ *SS* plant requires fusion of an *S* gamete from one parent and an *S* gamete from the other parent. The frequency of this occurring is $^{1}/_{2} \times {}^{1}/_{2} = {}^{1}/_{4}$. Similarly, to produce an $F_2$ *ss* plant requires fusion of an *s* gamete from one parent and an *s* gamete from the other parent. The frequency of this occurring is $^{1}/_{2} \times {}^{1}/_{2} = {}^{1}/_{4}$.

What about the *Ss* progeny? Again, the frequency of *S* in a gamete from one parent is $^{1}/_{2}$. and the frequency of *s* in a gamete from the other parent is also $^{1}/_{2}$. However, there are two ways in which *Ss* progeny can be obtained. The first involves the fusion of an *S* egg with *s* pollen, and the second is a fusion of an *s* egg with *S* pollen. Using the product rule (see Box 11.2), we find that the probability of *each* of these events occurring is $^{1}/_{2} \times {}^{1}/_{2} = {}^{1}/_{4}$. Using the sum rule (see Box 11.2), we see that the probability of *one or the other* occurring is the sum of the individual probabilities, or $^{1}/_{4} \times {}^{1}/_{4} = {}^{1}/_{2}$.

The overall prediction, then, is that one-fourth of the $F_2$ progeny will be *SS*, half will be *Ss*, and one-fourth will be *ss*, exactly as we found with the Punnett square method shown in Figure 11.8. Either method—the Punnett square or the branch diagram—may be used with any cross, but as crosses become more complicated, the Punnett square method becomes cumbersome.

## Confirming the Principle of Segregation: The Use of Testcrosses

When formulating his principle of segregation, Mendel did a number of genetic tests to ensure the correctness of his results. He continued the self-fertilizations at each generation up to the $F_6$ and found that, in every generation, both the dominant and recessive trait were found. He concluded that the principle of segregation was valid no matter how many generations were involved.

Another important test concerned the $F_2$ plants. As shown in Figure 11.9, a ratio of 1:2:1 occurs for the genotypes *SS*, *Ss*, and *ss* for the smooth × wrinkled example. Phenotypically, the ratio of smooth to wrinkled is 3:1. At the time of Mendel's experiments, the presence of segregating factors that were responsible for the smooth and wrinkled phenotypes was only a hypothesis. To test his factor hypothesis, Mendel allowed the $F_2$ plants to self-pollinate. As he expected, the plants produced from wrinkled seeds bred true, supporting his conclusion that they were pure (homozygous) for the *s* factor (gene).

Selfing the plants derived from the $F_2$ smooth seeds produced two different types of progeny: one-third of the smooth $F_2$ seeds produced all smooth-seeded progeny, whereas the other two-thirds produced both smooth and wrinkled seeds in each pod in a ratio of 3 smooth : 1 wrinkled, the same ratio as seen for the $F_2$ progeny (Figure 11.10). These results support the principle of gene segregation. The random combination of gametes that form the

Figure 11.10

**Determining the genotypes of the $F_2$ smooth progeny of Figure 11.8 by selfing the plants grown from the smooth seeds.**

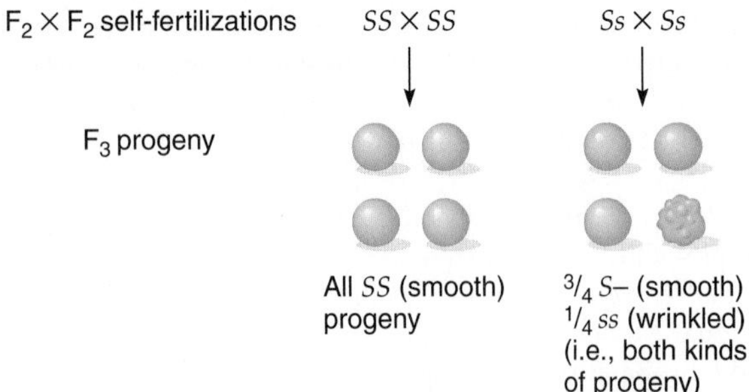


zygotes of the original $F_2$ produces two genotypes that give rise to the smooth phenotype (see Figures 11.8 and 11.9); the relative proportion of the two genotypes *SS* and *Ss* is 1:2. The *SS* seeds give rise to true-breeding plants, whereas the *Ss* seeds give rise to plants that behave exactly like the $F_1$ plants when they are self-pollinated in that they produce a 3:1 ratio of smooth : wrinkled progeny. *Mendel explained these results by proposing that each plant had two factors, whereas each gamete had only one. He also proposed that the random combination of the gametes generated the progeny in the proportions he found. Mendel obtained the same results in all seven sets of crosses.*

The *SS* and *Ss* plants have different genotypes but the same dominant phenotype. The self-fertilization test of the $F_2$ progeny proved a useful way to determine whether a plant with the dominant phenotype was homozygous or heterozygous. A more common test to do this is to perform a **testcross,** a cross of an individual expressing the dominant phenotype with a homozygous recessive individual to determine its genotype.

Consider again the cross shown in Figure 11.8. We can predict the outcome of a testcross of the $F_2$ progeny showing the dominant, smooth-seed phenotype. If the $F_2$ individuals are smooth because they are homozygous *SS*, then the result of a testcross with an *ss* plant will be all smooth seeds. As Figure 11.11a shows, the Parent 1 smooth *SS* plants produce only *S* gametes. Parent 2 is homozygous recessive wrinkled, *ss*, so it produces only *s* gametes. Therefore, all zygotes are *Ss*, and all the resulting seeds have the smooth phenotype. In actual practice, then, if a plant with a dominant trait is testcrossed and only the dominant phenotype is seen among the progeny, then the plant must have been homozygous for the dominant allele. In contrast, if the $F_2$ plants are smooth because they are heterozygous *Ss* $F_2$, then the result of a testcross with a homozygous *ss* plant will be a 1:1 ratio of dominant : recessive phenotypes. As Figure 11.11b shows, the Parent 1 smooth *Ss* produces both *S* and *s* gametes in equal proportion, and the homozygous *ss* Parent 2 produces only *s* gametes. As a result, half the progeny of the testcross are *Ss* heterozygotes and have a smooth phenotype because of the dominance of the *S* allele, and the other half are *ss* homozygotes and have a wrinkled phenotype. In actual practice, then, if a plant with a dominant trait is testcrossed and the progeny exhibit a 1:1 ratio of dominant : recessive phenotypes, then the plant must have been heterozygous. Considered another way, if the outcome of a testcross is a mixture of dominant and recessive phenotypes, then the parent with the dominant phenotype must have been heterozygous since that is the only way progeny with a recessive phenotype can be generated.

In sum, testcrosses of the $F_2$ progeny from Mendel's crosses that showed the dominant phenotype resulted in a 1:2 ratio of homozygous dominant : heterozygous genotypes in the $F_2$ progeny. That is, when crossed with the homozygous recessive, one-third of the $F_2$ progeny with the dominant phenotype gave rise only to progeny with the dominant phenotype and were therefore homozygous for the dominant allele. The other two-thirds of the $F_2$ progeny with the dominant phenotype produced progeny with a 1:1 ratio of dominant phenotype : recessive phenotype and therefore were heterozygous.

## The Wrinkled-Pea Phenotype

Why is the wrinkled phenotype recessive? To answer this question, we must think about genes at the molecular level. The functional allele of a gene that predominates (is present in the highest frequency) in the population of an organism found in the "wild" is called the **wild-type allele.** Wild-type alleles typically encode a product for a particular biological function. Therefore, if a mutation in the gene causes the protein product of a gene to be absent, partially functional, or nonfunctional, then the associated biological function is likely to be lost or decreased significantly. Such mutations are called **loss-of-function mutations** and are usually recessive because the function of a single copy of a wild-type allele in a heterozygote is usually sufficient to produce enough protein to allow the normal phenotype. Loss-of-function mutations may be caused in various ways but, most commonly, the base-pair sequence of the gene is altered, resulting in either a protein with impaired function due to an altered amino acid sequence, a truncated protein with little or no function, or no protein at all. A mutation that results in no protein or a protein with no function is known as a **null mutation.**

Mendel's wrinkled peas result from a loss-of-function mutation. In *SS* and *Ss* (smooth or wild-type) peas, enough functional protein is produced to result in large starch grains, while in *ss* (wrinkled) peas the starch grains are small and deeply fissured. *SS* and *Ss* seeds contain larger amounts of starch and lower levels of sucrose than do *ss* seeds. The sucrose difference leads to a higher water content and larger size of developing *ss* seeds. When the seeds mature, the *ss* seeds lose a larger proportion of their volume, leading to the wrinkled phenotype. At the molecular level, the seed-shape gene encodes one form of starch-branching enzyme (SBEI) in developing embryos. SBEI is important in determining the starch content of embryos so that, in *ss* plants, starch content is reduced. The wrinkled peas in Mendel's experiments did

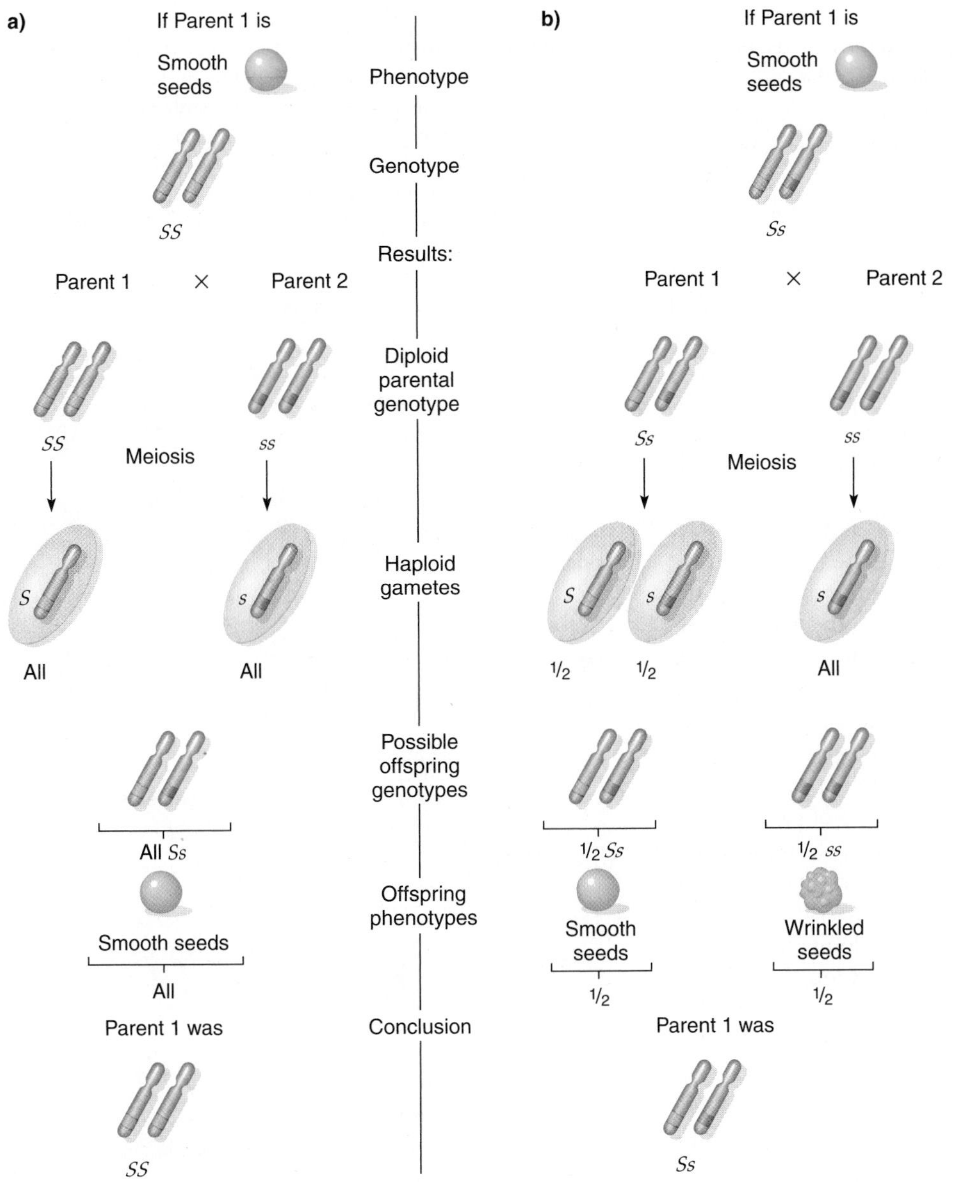

a) If Parent 1 is Smooth seeds — Phenotype — Genotype: *SS* — Results: Parent 1 × Parent 2 — Diploid parental genotype: *SS*, *ss* — Meiosis — Haploid gametes: *S* All; *s* All — Possible offspring genotypes: All *Ss* — Offspring phenotypes: Smooth seeds, All — Conclusion: Parent 1 was *SS*

b) If Parent 1 is Smooth seeds — Genotype: *Ss* — Results: Parent 1 × Parent 2 — Diploid parental genotype: *Ss*, *ss* — Meiosis — Haploid gametes: *S* 1/2, *s* 1/2; *s* All — Possible offspring genotypes: 1/2 *Ss*, 1/2 *ss* — Offspring phenotypes: Smooth seeds 1/2, Wrinkled seeds 1/2 — Conclusion: Parent 1 was *Ss*

**Figure 11.11**
**Determining the genotypes of the $F_2$ generation smooth seeds (Parent 1) of Figure 11.8 by testcrossing plants grown from the seed with a homozygous recessive wrinkled (*ss*) strain (Parent 2).**



not have a simple base-pair change in the seed-shape gene that inactivated SBEI, however. Rather, molecular analysis of *ss* plant lines directly descended from those that Mendel used in his experiments shows that the *s* allele has an 800-bp extra piece of DNA inserted into the *S* gene, disrupting the gene and its function. This inserted piece of DNA is a *transposable element* (see Chapter 7), a piece of DNA that can move ("transpose") to different locations in the genome.

## Keynote

A testcross is a cross of an individual of unknown genotype, usually expressing the dominant phenotype, with a known homozygous recessive individual to determine the genotype of the unknown individual. The phenotypes of the progeny of the testcross indicate the genotype of the individual tested.

# Dihybrid Crosses and Mendel's Principle of Independent Assortment

## The Principle of Independent Assortment

Mendel also analyzed a number of crosses in which two pairs of alternative traits were involved simultaneously. In each case, he obtained the same results. From these experiments, he proposed his **second law**, the **principle of independent assortment**, which states that *the factors for different pairs of traits assort independently of one another*. In modern terms, this means that *pairs of alleles for genes on different chromosomes segregate independently in the formation of gametes*.

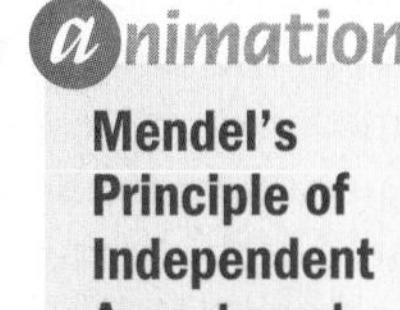


Consider an example involving the pair of traits for seed shape, smooth (*S*) and wrinkled (*s*), and the pair of traits for seed color, yellow ($Y$) and green ($y$). (Yellow is dominant to green.) When Mendel made crosses between true-breeding smooth, yellow plants ($SS\ YY$) and wrinkled, green plants ($ss\ yy$), he got the results shown in Figure 11.12. All the $F_1$ seeds from this cross were smooth and yellow, as the results of the monohybrid crosses predicted. As Figure 11.12a shows, the smooth, yellow parent produces only $S\ Y$ gametes, which give rise to $Ss\ Yy$ zygotes upon fusion with the $s\ y$ gametes from the wrinkled, green parent. Because of the dominance of the smooth and yellow alleles, all $F_1$ seeds are smooth and yellow.

The $F_1$ are heterozygous for two pairs of alleles at two different loci. Such individuals are called dihybrids, and a cross between two of these dihybrids of the same type is called a **dihybrid cross.**

When Mendel self-pollinated the dihybrid $F_1$ plants to give rise to the $F_2$ generation (Figure 11.12b), there were two possible outcomes. One was that the alleles determining seed shape and seed color in the original parents would be transmitted together to the progeny. In this case, a phenotypic ratio of 3:1 smooth, yellow : wrinkled, green would be predicted. The other possibility was that the alleles determining seed shape and seed color would be inherited independently of one another. In this case, the dihybrid $F_1$ would produce four types of gametes: $S\ Y$, $S\ y$, $s\ Y$, and $s\ y$. Given the independence of the two pairs of alleles, each gametic type is predicted to occur with equal frequency. In $F_1 \times F_1$ crosses, the four types of gametes would be expected to fuse randomly in all possible combinations to give rise to the zygotes and hence the progeny seeds. All the possible gametic fusions are represented in the Punnett square in Figure 11.12b. In a dihybrid cross, there are 16 possible gametic fusions. The result is nine different genotypes but, because of dominance, only four phenotypes are predicted:

1 $SS\ YY$, 2 $Ss\ YY$, 2 $SS\ Yy$, 4 $Ss\ Yy$ = 9 smooth, yellow
1 $SS\ yy$, 2 $Ss\ yy$ = 3 smooth, green
1 $ss\ YY$, 2 $ss\ Yy$ = 3 wrinkled, yellow
1 $ss\ yy$ = 1 wrinkled, green

According to the rules of probability, if the alleles for two pairs of traits are inherited independently in a dihybrid cross, then the $F_2$ from an $F_1 \times F_1$ cross will give a 9:3:3:1 ratio of the four possible phenotypic classes. Such a ratio is the result of the independent assortment of the two pairs of alleles for the two genes into the gametes and of the random fusion of those gametes. The 9:3:3:1 ratio may be considered as two separate 3:1 ratios multiplied together—the multiplication being done because of the product rule for independent events. That is, $(3{:}1) \times (3{:}1)$ involves multiplying the two terms within one set of brackets in turn with the two terms within the other set of brackets: $3 \times 3$, $3 \times 1$, $1 \times 3$, and $1 \times 1$. The result is 9:3:3:1. Further, independent assortment in our example means that while both pairs of traits involve the seeds, seed shape and seed color are independent of one another in terms of the genes involved and how those genes function in the generation of the phenotypes.

**Figure 11.12a**

**The principle of independent assortment in a dihybrid cross.** This cross, actually done by Mendel, involves the smooth, wrinkled and yellow, green character pairs of the garden pea. (Note that, compared with previous figures of this kind, only one box is shown in the $F_1$ instead of four. This is because only one class of gametes exists for Parent 2 and only one class for Parent 1. Previously, we showed two gametes from each parent, even though those gametes were identical.)

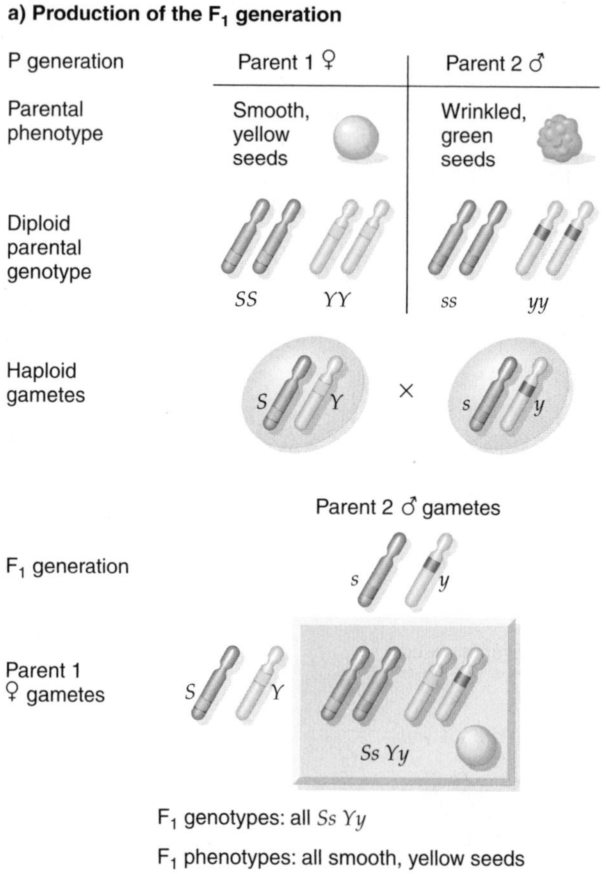

**a) Production of the $F_1$ generation**

P generation: Parent 1 ♀ | Parent 2 ♂

Parental phenotype: Smooth, yellow seeds | Wrinkled, green seeds

Diploid parental genotype: $SS$ $YY$ | $ss$ $yy$

Haploid gametes: $S$ $Y$ × $s$ $y$

Parent 2 ♂ gametes: $s$ $y$

$F_1$ generation

Parent 1 ♀ gametes: $S$ $Y$

$Ss\ Yy$

$F_1$ genotypes: all $Ss\ Yy$

$F_1$ phenotypes: all smooth, yellow seeds

This prediction was met in all the dihybrid crosses Mendel performed. In every case, the $F_2$ ratio was close to 9:3:3:1. For our example, he counted 315 smooth, yellow; 108 smooth, green; 101 wrinkled, yellow; and 32 wrinkled, green seeds—very close to the predicted ratio. To Mendel, this result meant that the factors (genes) determining the two pairs of traits he was analyzing were transmitted independently. Thus, in effect, Mendel rejected the possibility that the factors for the two pairs of traits were inherited together.

**Keynote**

Mendel's second law, the principle of independent assortment, states that pairs of alleles for genes on different chromosomes segregate independently in the formation of gametes.

Figure 11.12b

b) $F_1 \times F_1$ cross producing the $F_2$ generation

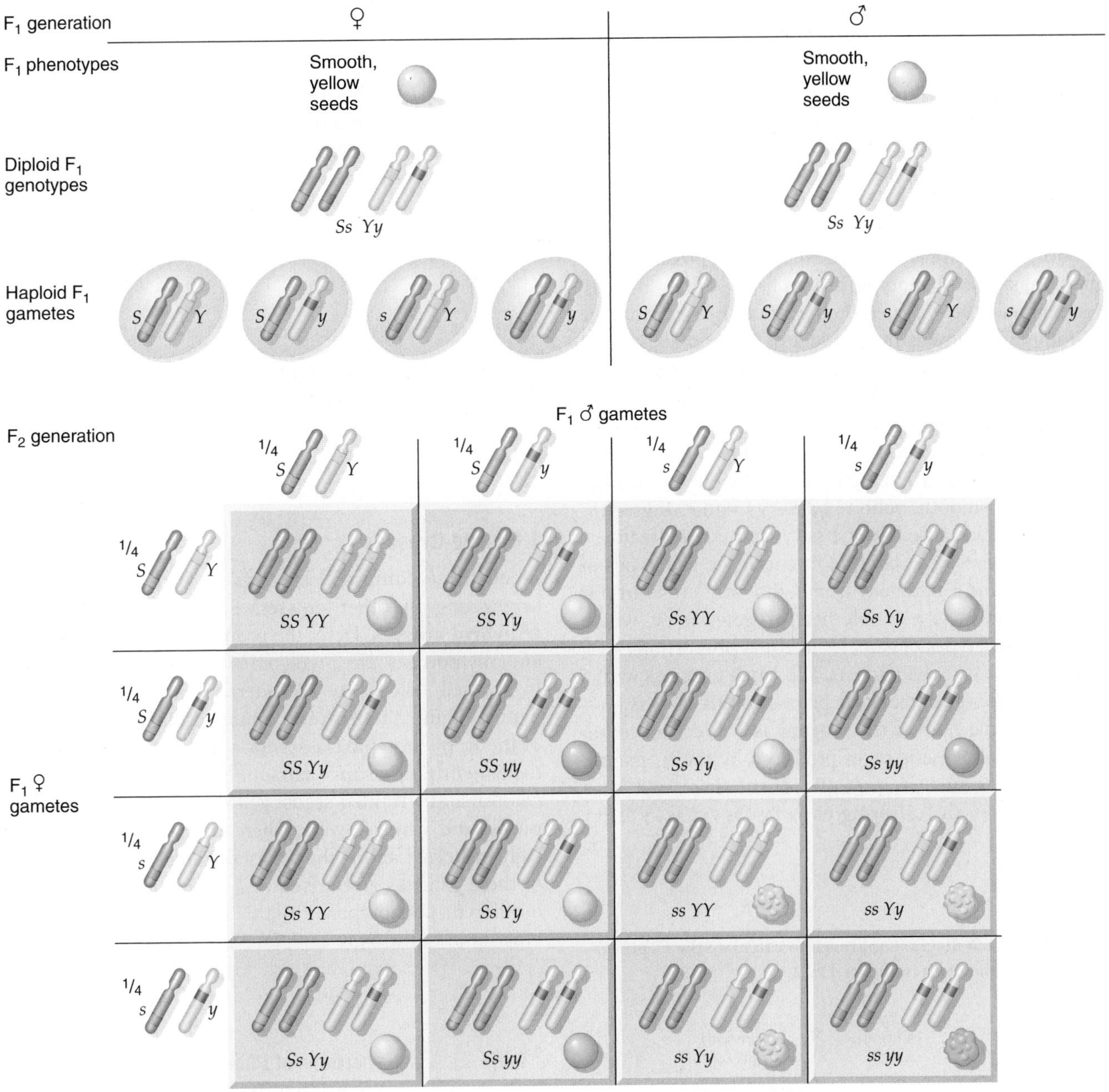

| | ♀ | ♂ |
|---|---|---|
| $F_1$ generation | | |
| $F_1$ phenotypes | Smooth, yellow seeds | Smooth, yellow seeds |
| Diploid $F_1$ genotypes | *Ss Yy* | *Ss Yy* |
| Haploid $F_1$ gametes | *S Y*, *S y*, *s Y*, *s y* | *S Y*, *S y*, *s Y*, *s y* |

$F_2$ generation

| $F_1$ ♀ gametes \ $F_1$ ♂ gametes | 1/4 *S Y* | 1/4 *S y* | 1/4 *s Y* | 1/4 *s y* |
|---|---|---|---|---|
| 1/4 *S Y* | *SS YY* | *SS Yy* | *Ss YY* | *Ss Yy* |
| 1/4 *S y* | *SS Yy* | *SS yy* | *Ss Yy* | *Ss yy* |
| 1/4 *s Y* | *Ss YY* | *Ss Yy* | *ss YY* | *ss Yy* |
| 1/4 *s y* | *Ss Yy* | *Ss yy* | *ss Yy* | *ss yy* |

$F_2$ genotypes: $F_2$ phenotypes:

1/16 (*SS YY*) + 2/16 (*Ss YY*) + 2/16 (*Ss Yy*) + 4/16 (*Ss Yy*) = 9/16 smooth, yellow seeds

1/16 (*SS yy*) + 2/16 (*Ss yy*) = 3/16 smooth, green seeds

1/16 (*ss YY*) + 2/16 (*ss Yy*) = 3/16 wrinkled, yellow seeds

1/16 (*ss yy*) = 1/16 wrinkled, green seeds



## Branch Diagram of Dihybrid Crosses

As for monohybrid crosses, a branch diagram can be used with dihybrid crosses to calculate the expected ratios of phenotypic or genotypic classes. In this approach, we apply the laws of probability to each pair of alleles in turn. With practice you should be able to calculate the probabilities of outcomes of various crosses just by using the laws of probability without drawing out the branch diagram. Diligently working problems helps to hone this skill.

Using the same example, in which the two pairs of alleles assort independently into the gametes, we consider each pair of alleles in turn. Earlier, we saw that an

$F_1$ self of an *Ss* heterozygote gave rise to progeny of which three-fourths were smooth and one-fourth were wrinkled. Genotypically, the former class had at least one dominant *S* allele; that is, they were *SS* or *Ss*. A convenient way to signify this situation is to use a dash to indicate an allele that has no effect on the phenotype. Thus, *S*– means that, phenotypically, the seeds are smooth and, genotypically, they are either *SS* or *Ss*.

Now consider the $F_2$ produced from a selfing of *Yy* heterozygotes: a 3:1 ratio is seen, with $^3/_4$ of the seeds being yellow and $^1/_4$ being green. Because this segregation occurs independently of the segregation of the smooth, wrinkled pair, we can consider all possible combinations of the phenotypic classes in the dihybrid cross. For example, the expected proportion of $F_2$ seeds that are smooth and yellow is the product of the probability that an $F_2$ seed will be smooth and the probability that it will be yellow, or $^3/_4 \times {}^3/_4 = {}^9/_{16}$. Similarly, the expected proportion of $F_2$ progeny that are wrinkled and yellow is $^3/_4 \times {}^1/_4 = {}^3/_{16}$. Extending the calculation to all possible phenotypes, as shown in Figure 11.13, we obtain the ratio of 9 *S*– *Y*– (smooth, yellow) : 3 *S*– *yy* (smooth,green) : 3 *ss Y*– (wrinkled, yellow) : 1 *ss yy* (wrinkled, green).

The testcross can be used to check the genotypes of $F_1$ progeny and $F_2$ progeny from a dihybrid cross. In our example, the $F_1$ is a double heterozygote, *Ss Yy*, which produces four types of gametes in equal proportions: *S Y*, *S y*, *s Y*, and *s y*. (See Figure 11.12b.) In a testcross with a doubly homozygous recessive plant—in this case, *ss yy*—the phenotypic ratio of the progeny is a direct reflection of the ratio of gametic types produced by the $F_1$ parent. In a testcross such as this one, then, there will be a 1:1:1:1 ratio in the offspring of *Ss Yy* : *Ss yy* : *ss Yy* : *ss yy* genotypes, which means a ratio of 1 smooth, yellow : 1 smooth, green : 1 wrinkled, yellow : 1 wrinkled, green phenotypes. The 1:1:1:1 phenotypic ratio is diagnostic of testcrosses in which the "unknown" parent is a double heterozygote.

In the $F_2$ of a dihybrid cross, there are nine different genotypic classes but only four phenotypic classes. The genotypes can be ascertained by testcrossing, as we have shown. Table 11.2 lists the expected ratios of progeny phenotypes from such testcrosses. No two patterns are the same, so here the testcross is truly a diagnostic approach to confirm genotypes.

**Figure 11.13**

**Using the branch diagram approach to calculate the $F_2$ phenotypic ratio of the cross in Figure 11.12.**

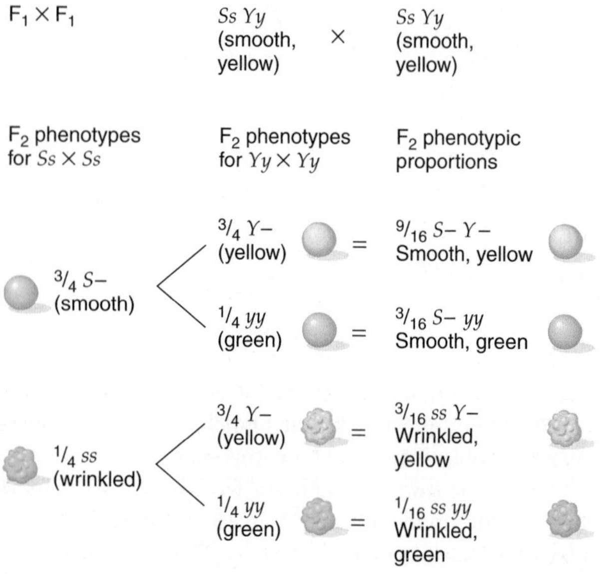

$F_1 \times F_1$ — *Ss Yy* (smooth, yellow) × *Ss Yy* (smooth, yellow)

| $F_2$ phenotypes for *Ss* × *Ss* | $F_2$ phenotypes for *Yy* × *Yy* | $F_2$ phenotypic proportions |
|---|---|---|
| $^3/_4$ *S*– (smooth) | $^3/_4$ *Y*– (yellow) | = $^9/_{16}$ *S*– *Y*– Smooth, yellow |
| | $^1/_4$ *yy* (green) | = $^3/_{16}$ *S*– *yy* Smooth, green |
| $^1/_4$ *ss* (wrinkled) | $^3/_4$ *Y*– (yellow) | = $^3/_{16}$ *ss Y*– Wrinkled, yellow |
| | $^1/_4$ *yy* (green) | = $^1/_{16}$ *ss yy* Wrinkled, green |

**iActivity**

Go to the iActivity *Tribble Traits* on the student website to discover how, as a Tribble breeder, you can choose the right combination of traits to produce the cuddliest creature.

## Trihybrid Crosses

Mendel also confirmed his laws for three pairs of traits segregating in other crosses. Such crosses are called **trihybrid crosses.** Here, the proportions of $F_2$ genotypes and phenotypes are predicted with precisely the same logic used before: by considering each trait independently. Figure 11.14 shows a branch diagram derivation of the $F_2$ phenotypic classes for a trihybrid cross. The independently assorting pairs of traits in the cross are smooth and wrinkled seeds, yellow and green seeds, and purple and white flowers. There are 64 combinations of eight maternal and eight paternal gametes. Combination of these gametes gives rise to 27 different genotypes and 8 different phenotypes in the $F_2$ generation. The phenotypic ratio in the $F_2$ is 27:9:9:9:3:3:3:1.

**Table 11.2 Proportions of Phenotypic Classes Expected from Testcrosses of Strains with Various Genotypes for Two Gene Pairs**

| | Proportion of Phenotypic Classes | | | |
|---|---|---|---|---|
| **Testcrosses** | ***A*– *B*–** | ***A*– *bb*** | ***aa B*–** | ***aa bb*** |
| *AA BB* × *aa bb* | 1 | 0 | 0 | 0 |
| *Aa BB* × *aa bb* | $^1/_2$ | 0 | $^1/_2$ | 0 |
| *AA Bb* × *aa bb* | $^1/_2$ | $^1/_2$ | 0 | 0 |
| *Aa Bb* × *aa bb* | $^1/_4$ | $^1/_2$ | $^1/_4$ | $^1/_4$ |
| *aa bb* × *aa bb* | 0 | 1 | 0 | 0 |
| *Aa bb* × *aa bb* | 0 | $^1/_2$ | 0 | $^1/_2$ |
| *aa BB* × *aa bb* | 0 | 0 | 1 | 0 |
| *aa Bb* × *aa bb* | 0 | 0 | $^1/_2$ | $^1/_2$ |
| *aa bb* × *aa bb* | 0 | 0 | 0 | 1 |

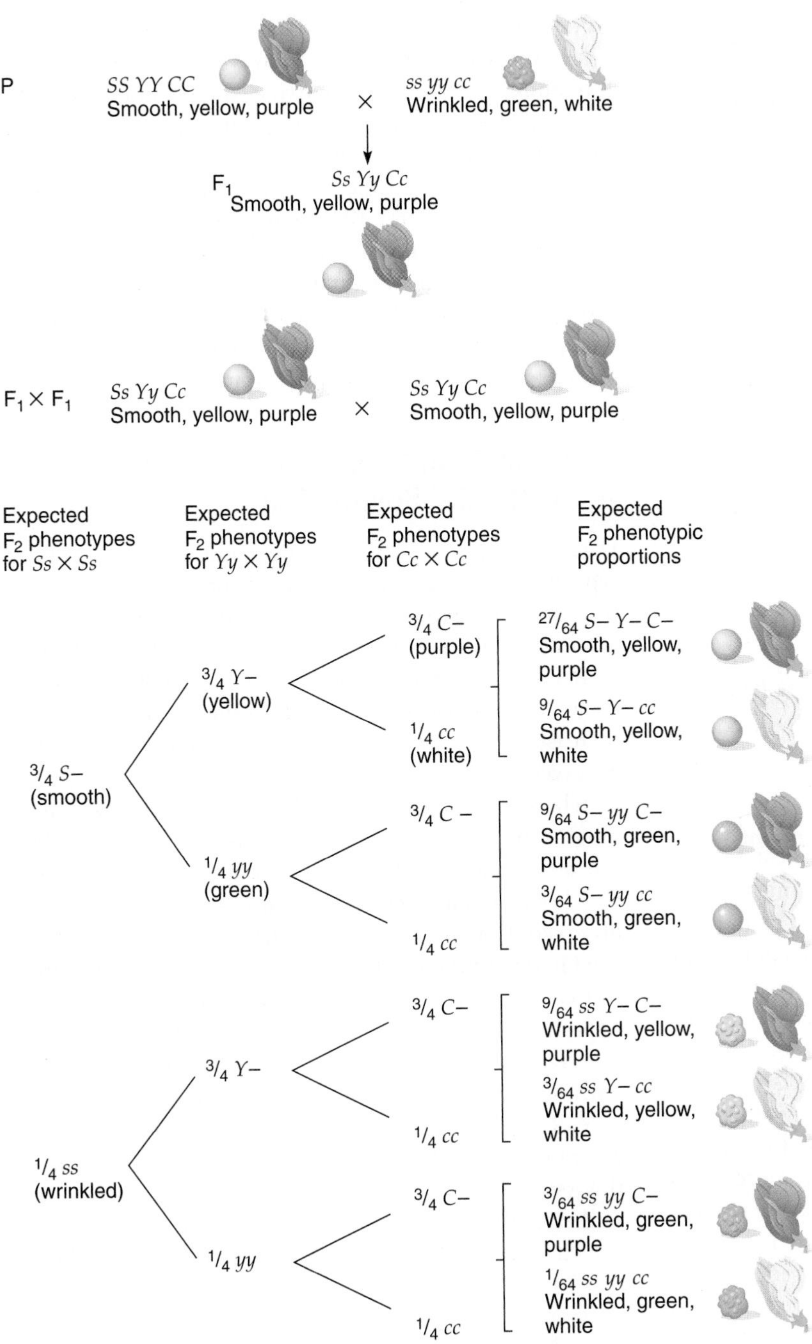


**Figure 11.14**

**Branch diagram derivation of the relative frequencies of the eight phenotypic classes in the $F_2$ of a trihybrid cross.**



Now that we have considered enough examples, we can make some generalizations about phenotypic and genotypic classes. In each example discussed, the $F_1$ is heterozygous for each gene involved in the cross, and the $F_2$ is generated by selfing (when possible) or by allowing the $F_1$ progeny to interbreed. In monohybrid crosses, there are two phenotypic classes in the $F_2$; in dihybrid crosses, there are four; and in trihybrid crosses, there are eight. The general rule is that there are $2^n$ phenotypic classes in the $F_2$ where $n$ is the number of independently assorting, heterozygous gene pairs (Table 11.3). (This rule holds *only* when a true dominant–recessive relationship holds for each of the gene pairs.) Furthermore, we saw that there are 3 genotypic classes in the $F_2$ of monohybrid crosses, 9 in dihybrid crosses, and 27 in trihybrid crosses. A simple rule is that the number of genotypic classes is $3^n$, where $n$ is the number of independently assorting, heterozygous gene pairs (see Table 11.3).

Incidentally, the phenotypic rule ($2^n$) can also be used to predict the number of classes that will come from a multiple heterozygous $F_1$ used in a testcross. Here, the number of genotypes in the next generation will be the same as the number of phenotypes. For example, from *Aa Bb* × *aa bb* there are four progeny genotypes ($2^n$, where $n$ is 2)—*Aa Bb*, *Aa bb*, *aa Bb*, and *aa bb*—and four phenotypes:

**Table 11.3 Number of Genotypic Classes Expected from Self-Crosses of Heterozygotes and Number of Phenotypic Classes If All Genes Show Complete Dominance**

| Number of Segregating Gene Pairs | Number of Phenotypic Classes | Number of Genotypic Classes |
|---|---|---|
| 1[a] | 2 | 3 |
| 2 | 4 | 9 |
| 3 | 8 | 27 |
| 4 | 16 | 81 |
| $n$ | $2^n$ | $3^n$ |

[a]For example from *Aa* × *Aa*, two phenotypic classes are expected, with genotypic classes of *AA*, *Aa*, and *aa*.

1. Both dominant phenotypes, *A* and *B*.
2. The *A* dominant phenotype and *b* recessive phenotype.
3. The *a* recessive phenotype and *B* dominant phenotype.
4. Both recessive phenotypes, *a* and *b*.

## The "Rediscovery" of Mendel's Principles

Mendel published his treatise on heredity in 1866 in *Verhandlungen des Naturforschenden Vereines* in Brünn, but it received little attention from the scientific community at the time. In 1985, Iris and Laurence Sandler proposed one possible reason. They contend that it may have been impossible for the scientific community from 1865 to 1900 to understand the significance of Mendel's work because it did not fit into that community's conception of the relationship of heredity to other sciences. To Mendel's contemporaries, heredity included not only those ideas that are today considered as genetic but also those that are considered developmental. In other words, their concept of heredity included what we now know as genetics and embryology. More pertinently, they also viewed heredity as simply a particular moment in development and not as a distinct process requiring special analysis. By 1900, conceptions had changed enough that the significance of Mendel's work was more apparent.

In 1900, three botanists—Carl Correns, Hugo de Vries, and Erich von Tschermark—independently came to the same conclusions as Mendel. Each was working with different plant hybrids: Correns with maize (corn) and peas, de Vries with several different plant species, and von Tschermark with peas. From their experiments, each botanist deduced the basic laws of genetic inheritance, thinking he was the first to do so. However, in preparing their conclusions for publication, they discovered that those laws had already been published by Mendel several decades earlier. Nonetheless, their work was important in that their rediscovery of Mendelian principles brought to the now more mature scientific world an awareness of the laws of genetic inheritance. They set in motion the research on gene structure and function that was so productive in the twentieth century.

That Mendelism applied to animals came in 1902 from the work of William Bateson, who experimented with fowl. Bateson also coined the terms *character, genetics, zygote*, $F_1$, $F_2$, and **allelomorph** (literally, "alternative form," meaning one of an array of different forms of a gene), which other researchers shortened to *allele*. The term *gene* as a replacement for Mendelian *factor* was introduced by W. L. Johannsen in 1909. *Gene* derives from the Greek word *genos*, meaning "birth."

## Statistical Analysis of Genetic Data: The Chi-Square Test

Data from genetic crosses are quantitative. A geneticist typically uses statistical analysis to interpret a set of data from crossing experiments to understand the significance of any deviation of observed results from the results predicted by the hypothesis being tested. (As with all statistical analyses, large sets of data are valuable so that we can increase our confidence in the results of the analyses.) The observed phenotypic ratios among progeny rarely exactly match expected ratios due to chance factors inherent in biological phenomena. A hypothesis is developed based on the observations and is presented as a **null hypothesis,** which states that there is no real difference between the observed data and the predicted data. Statistical analysis is used to determine whether the difference is due to chance. If it is not, then the null hypothesis is rejected, and a new hypothesis must be developed to explain the data.

A simple statistical analysis used to test null hypotheses is called the **chi-square ($\chi^2$)** test, which is a type of *goodness-of-fit test*. In the genetic crosses we have examined so far, the progeny seemed to fit particular ratios (such as 1:1, 3:1, and 9:3:3:1), and this is where a null hypothesis can be posed and where the chi-square test can tell us whether the data support that hypothesis.

To illustrate the use of the chi-square test, we will analyze theoretical progeny data from a testcross of a smooth, yellow double heterozygote (*Ss Yy*) with a wrinkled, green homozygote (*ss yy*); see pp. 309–310 and Table 11.2. (Additional applications of the chi-square test are given in Chapter 14.) The progeny data are as follows:

| | |
|---|---|
| | 154 smooth, yellow |
| | 124 smooth, green |
| | 144 wrinkled, yellow |
| | 146 wrinkled, green |
| Total | 568 |

The hypothesis is that a testcross should give a 1:1:1:1 ratio of the four phenotypic classes if the two genes assort independently. The chi-square test is then used to test the hypothesis, as shown in Table 11.4.

First, in column 1, the four classes expected in the progeny of the cross are listed. Then the observed (*o*)

**Table 11.4 Chi-Square Test Example**

| (1) | (2) | (3) | (4) | (5) | (6) |
|---|---|---|---|---|---|
| Phenotypes | Observed Number (*o*) | Expected Number (*e*) | *d* (= *o* − *e*) | $d^2$ | $d^2/e$ |
| Smooth, yellow | 154 | 142 | +12 | 144 | 1.01 |
| Smooth, green | 124 | 142 | −18 | 324 | 2.28 |
| Wrinkled, yellow | 144 | 142 | +2 | 4 | 0.03 |
| Wrinkled, green | 146 | 142 | +4 | 16 | 0.11 |
| Total | 568 | 568 | 0 | | 3.43 |

(7) $\chi^2 = 3.43$ (8) Degrees of freedom (df) = 3

numbers for each phenotype are listed, using actual numbers, not percentages or proportions (column 2). Next, we calculate the expected number (*e*) for each phenotypic class, given the total number of progeny (568) and the hypothesis under evaluation (in this case, a ratio of 1:1:1:1). Thus, in column 3 we list $1/4 \times 568 = 142$. Now we subtract the expected number (*e*) from the observed number (*o*) for each class to find differences, called the deviation value (*d*).

In column 5, the deviation squared ($d^2$) is computed by multiplying each deviation value in column 4 by itself. In column 6, the deviation squared is then divided by the expected number (*e*). The chi-square value, $\chi^2$ (item 7 in the table), is the total of all the values in column 6. The more the observed data deviate from the data expected on the basis of the hypothesis being tested, the higher chi-square is. In our example, $\chi^2 = 3.43$. The general formula is

$$\chi^2 = \Sigma\frac{d^2}{e}, \text{where } \Sigma \text{ means "sum" and } d^2 = (o - e)^2$$

The last value in the table, item 8, is the degrees of freedom (df) for the set of data. The degrees of freedom in a test involving *n* classes are usually equal to $n - 1$. There are four phenotypic classes here, so in this case, df = 3.

The chi-square value and the degrees of freedom are next used to determine the probability (*P*) that the deviation of the observed values from the expected values is due to chance. For example, in tossing coins, a deviation from a 1 head : 1 tail ratio can occur because of chance. However, if a coin was weighted on one side, then an observed deviation from 1 head : 1 tail would *not* be the result of chance, but due to the asymmetry of weight distribution in the coin. The *P* value for a set of data is obtained from tables of chi-square values for various degrees of freedom. Table 11.5 is part of a table of chi-square probabilities. For

**Table 11.5 Chi-Square Probabilities**

| | Probabilities | | | | | | | | | |
|---|---|---|---|---|---|---|---|---|---|---|
| df | 0.95 | 0.90 | 0.70 | 0.50 | 0.30 | 0.20 | 0.10 | 0.05 | 0.01 | 0.001 |
| 1 | 0.004 | 0.016 | 0.15 | 0.46 | 1.07 | 1.64 | 2.71 | 3.84 | 6.64 | 10.83 |
| 2 | 0.10 | 0.21 | 0.71 | 1.39 | 2.41 | 3.22 | 4.61 | 5.99 | 9.21 | 13.82 |
| 3 | 0.35 | 0.58 | 1.42 | 2.37 | 3.67 | 4.64 | 6.25 | 7.82 | 11.35 | 16.27 |
| 4 | 0.71 | 1.06 | 2.20 | 3.36 | 4.88 | 5.99 | 7.78 | 9.49 | 13.28 | 18.47 |
| 5 | 1.15 | 1.61 | 3.00 | 4.35 | 6.06 | 7.29 | 9.24 | 11.07 | 15.09 | 20.52 |
| 6 | 1.64 | 2.20 | 3.83 | 5.35 | 7.23 | 8.56 | 10.65 | 12.59 | 16.81 | 22.46 |
| 7 | 2.17 | 2.83 | 4.67 | 6.35 | 8.38 | 9.80 | 12.02 | 14.07 | 18.48 | 24.32 |
| 8 | 2.73 | 3.49 | 5.53 | 7.34 | 9.52 | 11.03 | 13.36 | 15.51 | 20.09 | 26.13 |
| 9 | 3.33 | 4.17 | 6.39 | 8.34 | 10.66 | 12.24 | 14.68 | 16.92 | 21.67 | 27.88 |
| 10 | 3.94 | 4.87 | 7.27 | 9.34 | 11.78 | 13.44 | 15.99 | 18.31 | 23.21 | 29.59 |
| 11 | 4.58 | 5.58 | 8.15 | 10.34 | 12.90 | 14.63 | 17.28 | 19.68 | 24.73 | 31.26 |
| 12 | 5.23 | 6.30 | 9.03 | 11.34 | 14.01 | 15.81 | 18.55 | 21.03 | 26.22 | 32.91 |
| 13 | 5.89 | 7.04 | 9.93 | 12.34 | 15.12 | 16.99 | 19.81 | 22.36 | 27.69 | 34.53 |
| 14 | 6.57 | 7.79 | 10.82 | 13.34 | 16.22 | 18.15 | 21.06 | 23.69 | 29.14 | 36.12 |
| 15 | 7.26 | 8.55 | 11.72 | 14.34 | 17.32 | 19.31 | 22.31 | 25.00 | 30.58 | 37.70 |
| 20 | 10.85 | 12.44 | 16.27 | 19.34 | 22.78 | 25.04 | 28.41 | 31.41 | 37.57 | 45.32 |
| 25 | 14.61 | 16.47 | 20.87 | 24.34 | 28.17 | 30.68 | 34.38 | 37.65 | 44.31 | 52.62 |
| 30 | 18.49 | 20.60 | 25.51 | 29.34 | 33.53 | 36.25 | 40.26 | 43.77 | 50.89 | 59.70 |
| 50 | 34.76 | 37.69 | 44.31 | 49.34 | 54.72 | 58.16 | 63.17 | 67.51 | 76.15 | 86.66 |

← Fail to reject | Reject →
at 0.05 level

our example—$\chi^2 = 3.43$, with 3 degrees of freedom—the $P$ value is between 0.30 and 0.50. This is interpreted to mean that, with the hypothesis being tested, in 30 to 50 out of 100 trials (that is, 30–50% of the time) we could expect chi-square values of such magnitude or greater due to chance. We can reasonably regard this deviation as simply due to chance. We must be cautious how we use the result obtained, however, because a result like this does not tell us that the hypothesis is *correct*: it indicates only that the experimental data provide no statistically compelling argument against the hypothesis.

As a general rule, if the probability of obtaining the observed chi-square values is greater than 5 in 100 (5% of the time, $P > 0.05$), then the deviation of expected from observed is not considered statistically significant, and the data do not indicate that the hypothesis should be rejected.

Suppose that, in another chi-square analysis of a different set of data, we obtained $\chi^2 = 15.85$, with 3 degrees of freedom. By looking up the value in Table 11.5, we see that the $P$ value is less than 0.01 and greater than 0.001 ($0.001 < P < 0.01$), which means that, from 0.1 to 1 times out of 100 (0.1–1% of the time), we could expect chi-square values of this magnitude or greater due to chance with the hypothesis being true. That this $P$ value is less than 0.05 indicates that, because of the poor fit, the results are not statistically consistent with the 1:1:1:1 hypothesis being tested.

## Mendelian Genetics in Humans

After the rediscovery of Mendel's laws, geneticists found that the inheritance of genes follows the same principles in all sexually reproducing eukaryotes, including humans. W. Farabee, in 1905, was the first to document a genetic trait in humans, *brachydactyly* (OMIM 112500 at http://www.ncbi.nlm.nih.gov/omim), which results in abnormally broad and short fingers (Figure 11.15). By analyzing the trait in human families, Farabee learned that brachydactyly is inherited as a simple dominant trait. In this section, we explore some of the methods used to determine the mechanism of hereditary transmission in humans and learn about some inherited human traits. The Focus on Genomics box for this chapter gives some genomics perspectives on human genetics traits in twins.

**Figure 11.15**
**Hands of an individual with brachydactyly.**

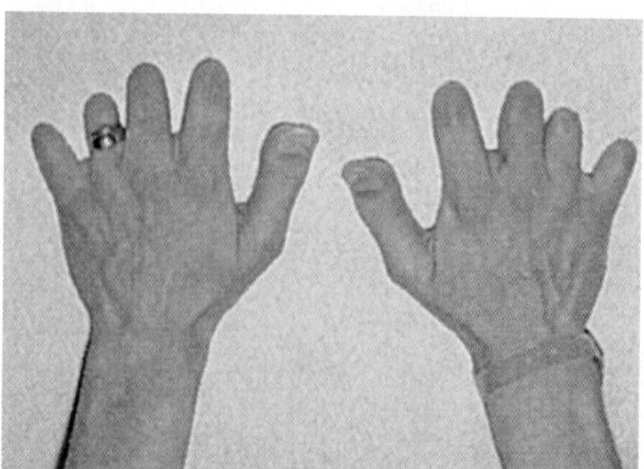

### Pedigree Analysis

The study of human genetics is complicated because controlled matings of humans are not possible for ethical reasons. The inheritance patterns of human traits usually are identified by examining the way the trait occurs in the family trees of individuals who clearly exhibit the trait. Such a study of a family tree, called **pedigree analysis,** involves carefully assembling phenotypic records of the family over several generations. The affected individual through whom the pedigree is discovered is called the **proband** (**propositus** if a male, **proposita** if a female).

Figure 11.16 summarizes the basic symbols used in pedigree analysis. (The terms *autosomal* and *sex-linked*

**Figure 11.16**
**Symbols used in human pedigree analysis.**

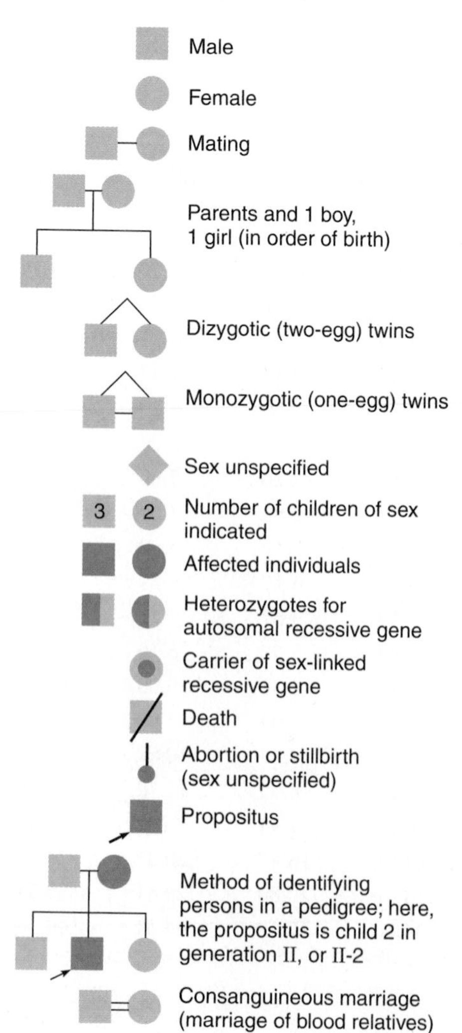

Hi manly man

I love you!

LEE

Capital Women's Health Care

2141 K st.
suite #808

202-331-9293

1 2 3 4 5 6 7 8 9 10 11 12

1 2 3 4 5 6 7 8 9 10 11 12

1 2 3 4 5 6 7 8 9 10 11 12

1 2 3 4 5 6 7 8 9 10 11 12

## Focus on Genomics

### Sometimes Identical Just Isn't That Similar

Identical twins are products of the same fertilization event, and they start life with exactly the same DNA. Genomics investigators compared the entire genomes of pairs of identical twins to determine just how much the DNA changes as cells divide by mitosis. These investigators used DNA microarrays to test copy number variation (CNV) in white blood cell DNA from pairs of twins. This test is also called a CNV test and is similar to ROMA, which was described in Chapter 9, pp. 237–239. A CNV test is a very sensitive way of detecting small duplications and deletions in the genome. If DNA replication were perfect, two twins should have exactly identical DNA and should, presumably, have the same genetic diseases. In this CNV test, the investigators compared the DNA from nineteen pairs of identical twins. Nine pairs were selected because one twin differed phenotypically from the other. In each twin pair, one of the twins had a neurological disease, and the other twin either did not have the disease or had very mild symptoms. The other ten pairs were controls and did not have such a phenotypic mismatch.

The investigators found copy number variation between twins in both the experimental group and the control group. Most regions were the same in both twins, while other regions had undergone some sort of duplication or deletion, with one twin having more copies than the other. The twins that had phenotypic mismatches had more chromosomal variation, but investigators found small duplications and deletions in the control twins as well. Furthermore, the investigators showed that copy number variation was present within individual samples. When they looked more closely at the DNA from the blood cells of just one individual, they were able to detect copy number variations that were present in some cells but absent in other cells—some white blood cells had deletions or duplications, and other white blood cells lacked them. The investigators pursued this further by comparing DNA from disparate parts of a single person. Once again, they saw that copy number variations were present, even within a single person. Changes present in the liver, for instance, are not necessarily the same as the changes in the skin. This cell-to-cell variability meant that they could not conclude that the observed copy number variations were the cause of the phenotypic neural variation between the twins, since, for obvious reasons, the investigators were not able to analyze DNA from brain tissue. It seems reasonable to conclude that variation occurs in all cells, not just white blood cells. In any case, it is clear that deletions and duplications occur quite frequently as DNA replicates.

are explained in Chapter 12; they are included here for completeness.) Figure 11.17 presents a hypothetical pedigree to show how the symbols are assigned to the family tree.

The trait presented in Figure 11.17 is determined by a recessive mutant allele *a*. (Note that recessive mutant alleles may be rare or common in a population.) Generations are numbered with Roman numerals, and individuals are numbered with Arabic numerals, which makes it easy to refer to particular people in the pedigree. The trait in the pedigree presented in Figure 11.17 results from homozygosity for the allele, in this case resulting from cousins mating. Since cousins share a fair proportion of their genes, a number of alleles are homozygous in their offspring. Here, one mutant recessive allele became homozygous and resulted in an identifiable genetic trait.

Gene symbols are included in this pedigree to show the deductive reasoning possible with such analysis. The trait appears first in generation IV. Since neither parent (the two cousins) had the trait, but they produced two children with the trait (IV-2 and IV-4), the simplest hypothesis is that the trait is caused by a recessive allele. Thus, IV-2 and IV-4 would both have the genotype *aa*, and their parents (III-5 and III-6) must have the genotype *Aa*. All other individuals who did not have the trait must have at least one *A* allele—that is, they must be *A*–

**Figure 11.17**
**A human pedigree, illustrating the use of pedigree symbols.**

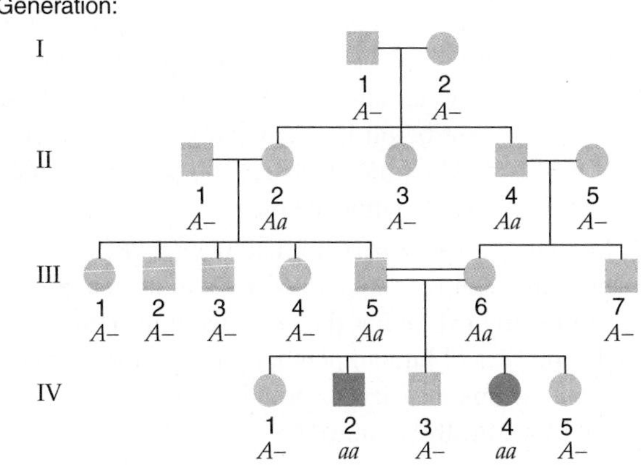

(either *AA* or *Aa*). Because III-5 and III-6 are both heterozygotes, at least one of each of their parents must have carried an *a* allele. Furthermore, because the trait appeared only after cousins had children, the simplest assumption is that the *a* allele was inherited from individuals with bloodlines shared by III-5 and III-6. This means that II-2 and II-4 probably are both *Aa* and that one of I-1 and I-2 is *Aa*.

## Examples of Human Genetic Traits

**Recessive Traits.** Many human traits are known to be caused by homozygosity for mutant alleles that are recessive to the normal allele. Such recessive mutant alleles produce mutant phenotypes because of a *loss of function* or a modified function of the gene product, either of them resulting from the mutation involved.

Many serious abnormalities or diseases result from homozygosity for recessive mutant alleles. Two individuals expressing the recessive trait of *albinism* (deficient pigmentation; OMIM 203100) are shown in Figure 11.18a, and a pedigree for this trait is shown in Figure 11.18b. Individuals with albinism do not produce the pigment melanin, which protects the skin from harmful ultraviolet radiation. Consequently, their skin and eyes are very sensitive to sunlight. Frequencies of harmful recessive mutant alleles usually are higher than frequencies of harmful dominant mutant alleles because heterozygotes for the recessive mutant allele are not at a significant selective disadvantage (see Chapter 21). Nonetheless, individuals homozygous for harmful recessive mutant alleles usually are rare. In the United States, approximately 1 in 17,000 of the white population and 1 in 28,000 of the African American population have albinism.

The following are some general characteristics of recessive inheritance for a rare trait:

1. Most affected individuals have two normal parents, both of whom are heterozygous. The trait appears in the $F_1$ because a quarter of the progeny are expected to be homozygous for the recessive allele. If the trait is rare, an individual expressing the trait is likely to mate with a homozygous normal individual. The next generation from such a mating would be heterozygotes who do not express the trait. In other words, recessive traits often skip generations. In the pedigree in Figure 11.18b, for example, II-6 and II-7 must both be *aa,* and this means both parents (I-3 and I-4) must be *Aa* heterozygotes. I-1 is also *aa,* so II-4 must be *Aa*. Since II-4 and II-5 produce some *aa* children, II-5 also must be *Aa*.
2. Matings between two normal heterozygotes should produce an approximately 3:1 ratio of normal progeny to progeny exhibiting the recessive trait. However, in the analysis of human populations (families), it is difficult to obtain a large enough sample to make the data statistically significant.

**Figure 11.18**
**Albinism, an autosomal recessive trait.**

a) Individuals with albinism: musicians Johnny (left) and Edgar Winter (right)

b) Pedigree for the autosomal recessive trait of albinism

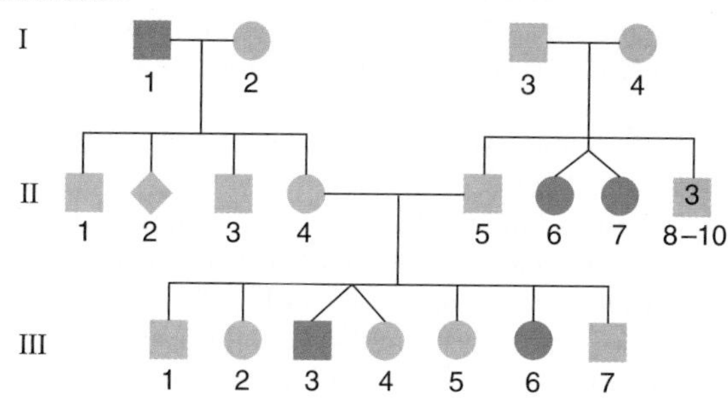


3. When both parents are affected, they are homozygous for the recessive trait, and all their progeny usually exhibit the trait.

**Dominant Traits.** There are many known dominant human traits. Dominant mutant alleles may produce mutant phenotypes because of **gain-of-function mutations** that result in gene products with new functions. In other words, the dominant mutant phenotype is a new or increased property of the mutant gene rather than a decrease in its normal activity. Figure 11.19a illustrates one such trait, *achondroplasia* (OMIM 100800; dwarfism resulting from defects in long-bone growth), the most common form of short-limb dwarfism. Individuals with achondroplasia have short stature, disproportionately short arms and legs; short fingers and toes; a large head with prominent forehead; and, often, bowleg or knock-knee deformities. Life span typically is normal. A pedigree of a family with achondroplasia is shown in Figure 11.19b.

Achondroplasia results from heterozygosity for a dominant mutation in the *FGFR3* (fibroblast growth factor receptor 3) gene on chromosome 4. Homozygosity for the mutation is lethal. The normal product of the *FGFR3* gene is a membrane-embedded receptor for particular growth factors that control growth and development. When a growth factor binds to the receptor, the receptor is acti-

Figure 11.19

**Achondroplasia, an autosomal dominant trait.**

a) Individual with achondroplasia

b) Pedigree for the autosomal dominant trait of achondroplasia

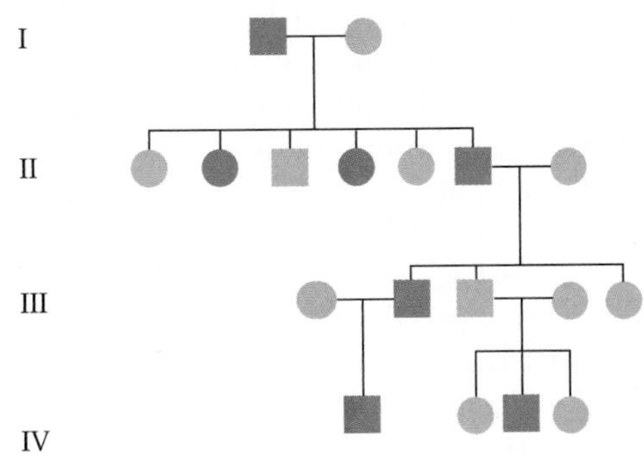


vated, which triggers a cascade of molecular reactions in the cell that leads to specific cellular responses. The FGFR3 protein is involved in the development and maintenance of bone and brain tissue. The normal form of the protein is thought to regulate bone growth by acting in a negative pathway to limit ossification, the formation of bone from cartilage. The regulatory effects are strongest on the long bones. In these effects, the function of the FGFR3 protein is controlled carefully. The dominant gain-of-function mutations in the *FGFR3* gene in people with achondroplasia cause the FGFR3 protein to be continuously active, which leads to the significantly shortened long bones. There is no treatment for achondroplasia; but greater than 99% of the individuals with the disease have one of two mutations in *FGFG3*, which has lead to the development of an effective molecular genetic test for the mutations associated with the disease.

The mutant phenotype is seen in heterozygotes for a dominant mutant allele and a wild-type allele. Because many dominant mutant alleles that give rise to recognizable traits are rare or lethal, it is highly unusual to find individuals homozygous for the dominant allele. An affected person in a pedigree is likely to be a heterozygote, and most pairings that involve the mutant allele are between a heterozygote and a homozygous recessive (wild type). Most dominant mutant genes that are clinically significant (that is, cause medical problems) fall into this category.

The following are some general characteristics of dominant inheritance for a rare trait (refer to Figure 11.19b):

1. Every affected person in the pedigree must have at least one affected parent.
2. The trait usually does not skip generations.
3. On average, an affected heterozygous individual will transmit the mutant gene to half of his or her progeny. If the dominant mutant allele is designated *A* and its wild-type allele is *a*, then most crosses will be *Aa* × *aa*. From basic Mendelian principles, half the progeny will be *aa* (wild type), and the other half will be *Aa* (and show the trait).

Other examples of human dominant traits are *autosomal dominant polycystic kidney disease* (ADPKD, OMIM 173900; formation of fluid-filled cysts in the kidneys potentially causing death by kidney failure—ADPKD is one of the most common, life-threatening diaseases), *brachydactyly* (malformed hands with short fingers), and Marfan syndrome (OMIM 154700; connective tissue defects, potentially causing death by aortic rupture).

**Keynote**

Mendelian principles apply to humans and all other eukaryotes. Study of the human inheritance of genetic traits is more difficult because no controlled crosses can be done. Instead, human geneticists analyze genetic traits by pedigree analysis.



# Summary

- The genotype is the genetic makeup of an organism, whereas the phenotype is the observable trait or set of traits (structural and functional) of an organism produced by the interaction between its genotype and the environment.
- The genotype provides the potential for the phenotype of an individual; this potential can be affected by the environment.
- Mendel's first law, the principle of segregation, states that the two members of a single gene pair (alleles)

segregate from each other in the formation of gametes. For each gene with two alleles, half the gametes carry one allele, and the other half carry the other allele. For the principle of segregation, in a monohybrid cross between two true-breeding parents, one exhibiting a dominant phenotype and the other a recessive phenotype, the $F_2$ phenotypic ratio is 3:1 for the dominant : recessive phenotypes.

- To determine an unknown genotype (usually in an individual expressing the dominant phenotype), a cross is made between that individual and a homozygous recessive individual. This cross is called a testcross.
- Mendel's second law, the principle of independent assortment, states that pairs of alleles for genes on different chromosomes segregate independently in the formation of gametes. For the principle of independent assortment, in a dihybrid cross, the $F_2$ phenotypic ratio is 9:3:3:1 for the four phenotypic classes.
- Mendelian principles apply to all eukaryotes. Study of the inheritance of genetic traits in humans is more difficult because controlled crosses cannot be done within ethical bounds. Instead, human geneticists examine genetic traits by pedigree analysis—that is, by following the occurrence of a trait in family trees in which the trait is segregating.

## Analytical Approaches to Solving Genetics Problems

The most practical way to reinforce genetics principles is to solve genetics problems. In this and all following chapters, we discuss how to approach genetics problems by presenting examples of such problems and by discussing the answers to those problems. The following problems use familiar and unfamiliar examples and pose questions designed to get you to think analytically.

**Q11.1** A purple-flowered pea plant is crossed with a white-flowered pea plant. All the $F_1$ plants produce purple flowers. When the $F_1$ plants are allowed to self-pollinate, 401 of the $F_2$ plants have purple flowers and 131 have white flowers. What are the genotypes of the parental and $F_1$ generation plants?

**A11.1** The ratio of plant phenotypes in the $F_2$ is 3.06:1, which is very close to the 3:1 ratio expected of a monohybrid cross. More specifically, this ratio is expected to result from an $F_1 \times F_1$ cross in which both are heterozygous for a specific gene pair. In addition, because the two parents differed in phenotype and only one phenotypic class appeared in the $F_1$, it is likely that both parental plants were true breeding. Furthermore, because the $F_1$ phenotype exactly resembled one of the parental phenotypes, we can say that purple is dominant to white flowers. Assigning the symbol *P* to the gene that determines purpleness of flowers and the symbol *p* to the alternative form of the gene that determines whiteness, we can write the genotypes:

| | |
|---|---|
| P generation: | *PP*, for the purple-flowered plant; *pp*, for the white-flowered plant |
| $F_1$ generation: | *Pp*, which, because of dominance, is purple flowered |

We could further deduce that the $F_2$ plants have an approximately 1:2:1 ratio of *PP* : *Pp* : *pp* by performing testcrosses.

**Q11.2** Consider three gene pairs *Aa*, *Bb*, and *Cc*, each of which affects a different character. In each case, the uppercase letter signifies the dominant allele and the lowercase letter the recessive allele. These three gene pairs assort independently of each other. Calculate the probability of obtaining the following:

**a.** an *Aa BB Cc* zygote from a cross of individuals that are *Aa Bb Cc* × *Aa Bb Cc*

**b.** an *Aa BB cc* zygote from a cross of individuals that are *aa BB cc* × *AA bb CC*

**c.** an *A B C* phenotype (that is, having the dominant phenotypes for each of the three genes) from a cross of individuals that are *Aa Bb CC* × *Aa Bb cc*

**d.** an *a b c* phenotype (that is, having the recessive phenotypes for each of the three genes) from a cross of individuals that are *Aa Bb Cc* × *aa Bb cc*

**A11.2** We must break down the question into simple parts in order to apply basic Mendelian principles. The key is that the genes assort independently, so we must multiply the probabilities of the individual occurrences to obtain the answers.

**a.** First, we must consider the *Aa* gene pair. The cross is *Aa* × *Aa*, so the probability of the zygote being *Aa* is $^2/_4 = {}^1/_2$, because the expected distribution of genotypes is 1 *AA* : 2 *Aa* : 1 *aa*. Then, following the same logic, the probability of *BB* from *Bb* × *Bb* is $^1/_4$, and that of *Cc* from *Cc* × *Cc* is $^2/_4 = {}^1/_4$. Using the product rule (see Box 2.2), we find that the probability of an *Aa BB Cc* zygote is $^1/_2 \times {}^1/_4 \times {}^1/_2 = {}^1/_{16}$.

**b.** Similar logic is needed here; although, because they differ from one gene pair to another, we must be sure of the genotypes of the parental types. For the *Aa* pair, the probability of getting *Aa* from *AA* × *aa* has to be 1. Next, the probability of getting *BB* from *BB* × *bb* is 0, so on these grounds alone, we cannot get the zygote asked for from the cross given.

**c.** This question and the next ask for the probability of getting a particular phenotype, so we must start thinking about dominance. Again, we consider each character pair in turn. From basic Mendelian principles, the probability of an $A$ phenotype from $Aa \times Aa$ is $^3/_4$. Similarly, the probability of a $B$ phenotype from $Bb \times Bb$ is $^3/_4$. Lastly, the probability of a $C$ phenotype from $CC \times cc$ is 1. Overall, the probability of an $A\ B\ C$ phenotype is $^3/_4 \times {}^3/_4 \times 1 = {}^9/_{16}$.

**d.** The probability of an $a\ b\ c$ phenotype from $Aa\ Bb\ Cc \times aa\ Bb\ cc$ is $^1/_2 \times {}^1/_4 \times {}^1/_2 = {}^1/_{16}$.

**Q11.3** In chickens, the white plumage of the leghorn breed is dominant over colored plumage, feathered shanks are dominant over clean shanks, and pea comb is dominant over single comb. Each of the gene pairs segregates independently. If a homozygous white, feathered, pea-combed chicken is crossed with a homozygous colored, clean, single-combed chicken and the $F_1$ birds are allowed to interbreed, what proportion of the birds in the $F_2$ will produce only white, feathered, pea-combed progeny if mated to colored, clean-shanked, single-combed birds?

**A11.3** This example is typical of a question that presents the unfamiliar in an attempt to get at the familiar. The best approach to such questions is to reduce them to their simplest parts and, whenever possible, to assign gene symbols for each character. We are told which character is dominant for each of the three gene pairs, so we can use $W$ for white and $w$ for colored, $F$ for feathered and $f$ for clean shanks, and $P$ for pea comb and $p$ for single comb. The cross involves true-breeding strains and can be written as follows:

P generation: $WW\ FF\ PP \times ww\ ff\ pp$

$F_1$ generation: $Ww\ Ff\ Pp$

Now, the question asks for the proportion of the birds in the $F_2$ that will produce only white, feathered, pea-combed progeny if mated to colored, clean-shanked, single-combed birds. The latter are homozygous recessive for all three genes—that is, $ww\ ff\ pp$, as in the parental generation. For the requested result, the $F_2$ birds must be white, feathered, and pea-combed, and they must be homozygous for the dominant alleles of the respective genes to produce only progeny with the dominant phenotype. What we are seeking, then, is the proportion of the $F_2$ chickens that are $WW\ FF\ PP$ in genotype. We know that each gene pair segregates independently; thus, the answer can be calculated by using simple probability rules. We consider each gene pair in turn. For the white versus colored case, the $F_1 \times F_1$ is $Ww \times Ww$, and we know from Mendelian principles that the relative proportion of $F_2$ genotypes is $1\ WW : 2\ Ww : 1\ ww$. Therefore, the proportion of the $F_2$ birds that will be $WW$ is $^1/_4$. The same relationship holds for the other two pairs of genes. Because the segregation of the three gene pairs is independent, we must multiply the probabilities of each occurrence to calculate the probability for $WW\ FF\ PP$ individuals. The answer is $^1/_4 \times {}^1/_4 \times {}^1/_4 = {}^1/_{64}$.



## Questions and Problems

***11.1** In tomatoes, red fruit color is dominant to yellow. Suppose a tomato plant homozygous for red is crossed with one homozygous for yellow. Determine the appearance of

**a.** the $F_1$ tomatoes,

**b.** the $F_2$ tomatoes,

**c.** the offspring of a cross of the $F_1$ tomato plants back to the red parent,

**d.** the offspring of a cross of the $F_1$ tomato plants back to the yellow parent.

**11.2** In maize, a dominant allele $A$ is necessary for seed color, as opposed to colorless ($a$). Another gene has a recessive allele $wx$ that results in waxy starch, as opposed to normal starch ($Wx$). The two genes segregate independently. An $Aa\ WxWx$ plant is testcrossed. What are the phenotypes and relative frequencies of offspring?

***11.3** $F_2$ plants segregate $^3/_4$ colored : $^1/_4$ colorless. If a colored plant is picked at random and selfed, what is the probability that both colored and colorless plants will be seen among a large number of its progeny?

***11.4** In guinea pigs, rough coat ($R$) is dominant over smooth coat ($r$). A rough-coated guinea pig is bred to a smooth one, giving 8 rough and 7 smooth progeny in the $F_1$ generation.

**a.** What are the genotypes of the parents and their offspring?

**b.** If one of the rough $F_1$ animals is mated to its rough parent, what progeny would you expect?

**11.5** In cattle, the polled (hornless) condition ($P$) is dominant over the horned ($p$) phenotype. A particular polled bull is bred to three cows. Cow A, which is horned, produces a horned calf; polled cow B produces a horned calf; and horned cow C produces a polled calf. What are the genotypes of the bull and the three cows, and what phenotypic ratios do you expect in the offspring of these three matings?

***11.6** In jimsonweed, purple flowers are dominant to white. Self-fertilization of a particular purple-flowered jimsonweed produces 28 purple-flowered and 10 white-flowered progeny. What proportion of the purple-flowered progeny will breed true?

***11.7** Two black female mice are crossed with the same brown male. In a number of litters, female X produced 9 blacks and 7 browns, and female Y produced 14 blacks.

What is the mechanism of inheritance of black and brown coat color in mice? What are the genotypes of the parents?

**11.8** Bean plants may have different symptoms when infected with a virus. Some show local lesions that do not seriously harm the plant; others show general systemic infection. The following genetic analysis was made:

P local lessions × systemic infection
$F_1$ all local lesions
$F_2$ 785 local lesions : 269 systemic infection

What is the likely genetic basis of this difference in beans? Evaluate your hypothesis using a chi-square test. Assign gene symbols to the genotypes occurring in the genetic analysis. Design a testcross to verify your assumptions.

**11.9** A normal *Drosophila* (fruit fly) has both brown and scarlet pigment granules in its eyes, which appear red as a result. Brown (*bw*) is a recessive allele on chromosome 2 that, when homozygous, results in brown eyes due to the absence of scarlet pigment granules. Scarlet (*st*) is a recessive allele on chromosome 3 that, when homozygous, results in scarlet eyes due to the absence of brown pigment granules. Any fly homozygous for recessive alleles at both genes produces no eye pigment and has white eyes. The following results were obtained from crosses:

P brown-eyed fly × scarlet-eyed fly
$F_1$ red eyes (both brown and scarlet pigment present)
$F_2$ $^9/_{16}$ red : $^3/_{16}$ scarlet : $^3/_{16}$ brown : $^1/_{16}$ white

**a.** Assign genotypes to the P and $F_1$ generations.
**b.** Design a testcross to verify the $F_1$ genotype, and predict the results.

***11.10** Grey seed color (*G*) in garden peas is dominant to white seed color (*g*). In the following crosses, the indicated parents with known phenotypes, but unknown genotypes, produced the progeny listed:

| Parents | Progeny | | Female Parent |
|---|---|---|---|
| Female × Male | Grey | White | Genotype |
| grey × white | 81 | 82 | ? |
| grey × grey | 118 | 39 | ? |
| grey × white | 74 | 0 | ? |
| grey × grey | 90 | 0 | ? |

Based on the segregation data, give the possible genotypes of each female parent.

***11.11** Fur color in the babbit, a furry little animal and popular pet, is determined by a pair of alleles, *B* and *b*. *BB* and *Bb* babbits are black, and *bb* babbits are white. A farmer wants to breed babbits for sale. True-breeding white (*bb*) female babbits breed poorly. The farmer purchases a pair of black babbits, and these mate and produce 6 black and 2 white offspring. The farmer immediately sells his white babbits, and then he comes to consult you for a breeding strategy to produce more white babbits.

**a.** If he performed random crosses between pairs of $F_1$ black babbits, what proportion of the $F_2$ progeny would be white?
**b.** If he crossed an $F_1$ male to the parental female, what is the probability that this cross would produce white progeny?
**c.** What would be the farmer's best strategy to maximize the production of white babbits?

**11.12** Explain the difference between loss-of-function and gain-of-function mutations, giving a specific example of each.

***11.13** The following diagram illustrates the structure of a gene for an essential liver enzyme in humans: In the DNA, E1, E2, and E3 are enhancers that are required for the proper transcription of the gene in the liver; P indicates the promoter region; and the numbers 1–8 indicate mutation sites. The transcribed region is aligned below the DNA, and the 5′-to-3′ polarity of the mRNA is shown. In the mRNA, filled rectangles represent exons, with grey rectangles representing 5′ and 3′ UTRs and black rectangles representing protein-coding regions, and white rectangles represent introns.

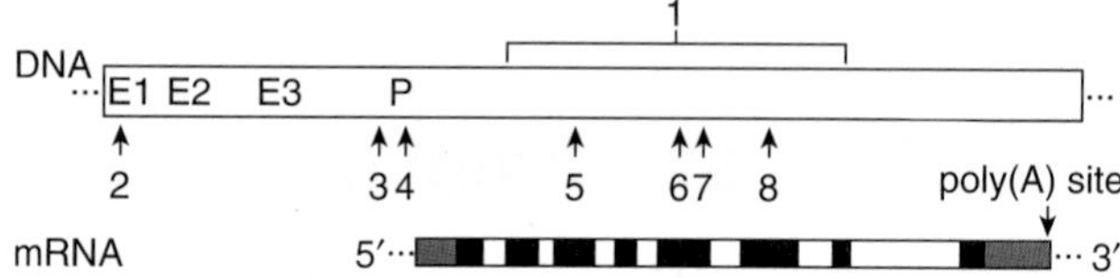


The mutations have the following characteristics:

1. a deletion of exons 2–7
2. a point mutation in enhancer E1 causing the transcription in the heart as well as in the liver
3. a point mutation decreasing the efficiency of transcription initiation by RNA polymerase
4. a point mutation increasing the efficiency of transcription initiation by RNA polymerase
5. a nonsense mutation in exon 3
6. a missense mutation in exon 5 resulting in a protein with 200% of normal enzymatic activity
7. a missense mutation in exon 5 resulting in a protein with 10% of normal enzymatic activity
8. a missense mutation in exon 6 resulting in an enzyme that acts on additional substrates

**a.** Predict which of the mutations will be gain-of-function mutations and which will be loss-of-function mutations.
**b.** When enzyme activity is assayed in phenotypically normal individuals, enzyme levels range between 50 and 150% of a standard, reference level. Individuals with less than 50% or more than 150% of the

reference level are sick, as are individuals who have enzyme activity in non-liver tissues. Assume that the total amount of enzyme activity is the sum of that produced by each allele of the gene. Which mutations do you expect to show recessive inheritance, which mutations do you expect to show dominant inheritance, and for which mutations are you unable to predict a clear inheritance pattern?

**11.14** Analyze the information given for each of the following *Drosophila* mutations, state whether each is most likely a loss-of-function or a gain-of-function mutation, and explain why.

**a.** Flies homozygous for the recessive allele *vermillion* (*v*) have bright red eyes instead of the darker deep-red color seen in animals with the wild-type (*V*) allele. Homozygous *vv* flies lack tryptophan 2,3-dioxygenase, an enzyme used in eye-pigment biosynthesis. A distinct mutation *del*(*v*) deletes the *v* gene. When the phenotypes of three types of heterozygotes, *Vv*, *del*(*v*) *V*, and *del*(*v*) *v*, are compared, *Vv* and *del*(*v*) *V* animals have normal-colored eyes just like *VV* animals, but *del*(*v*) *v* have bright-red eyes just like *vv* animals.

**b.** The dominant allele *Lobe* (*L*) results in small, kidney-bean-shaped eyes. Wild-type animals (*ll*) have oval-shaped eyes. Homozygotes for the *L* allele die. A distinct mutation *del*(*L*) deletes the *L* gene. When the phenotypes of two types of heterozygotes, *Ll* and *del*(*L*) *l*, are compared, *Ll* animals have the *Lobe* phenotype, but *del*(*L*) *l* have normal eyes.

**c.** The dominant allele *Notch* (*N*) results in wings that have notches along their edges instead of the normal smooth shape (*n*). Homozygotes for the *N* allele die. A distinct mutation *del*(*N*) deletes the *N* gene. When the phenotypes of two types of heterozygotes, *Nn* and *del*(*N*) *n*, are compared, both have identical, mutant phenotypes.

**11.15** In jimsonweed, purple flower (*P*) is dominant to white (*p*), and spiny pods (*S*) are dominant to smooth (*s*). A true-breeding plant with white flowers and spiny pods is crossed to a true-breeding plant with purple flowers and smooth pods. Determine the phenotype of

**a.** the $F_1$ generation;

**b.** the $F_2$ generation;

**c.** the progeny of a cross of the $F_1$ plants back to the white, spiny parent; and

**d.** the progeny of a cross of the $F_1$ back to the purple, smooth parent.

**11.16** Use the information in Problem 11.15 to determine what progeny you would expect from the following jimsonweed crosses (you are encouraged to use the branch diagram approach):

**a.** *PP ss* × *pp SS*

**b.** *Pp ss* × *pp ss*

**c.** *Pp Ss* × *Pp Ss*

**d.** *Pp Ss* × *Pp Ss*

**e.** *Pp Ss* × *Pp ss*

**f.** *Pp Ss* × *pp ss*

***11.17** Cleopatra normally is a very refined cat. When she finds even a small amount of catnip, however, she purrs madly, rolls around in the catnip, becomes exceedingly playful, and appears intoxicated. Cleopatra and Antony, who walks past catnip with an air of indifference, have produced five kittens who respond to catnip just as Cleopatra does. When the kittens mature, two of them mate and produce four kittens that respond to catnip and one that does not. When another of Cleopatra's daughters mates with Augustus (a nonrelative), who behaves just like Antony, three catnip-sensitive and two catnip-insensitive kittens are produced. Propose a hypothesis for the inheritance of catnip sensitivity that explains these data.

***11.18** In summer squash, white fruit (*W*) is dominant over yellow (*w*), and disk-shaped fruit (*D*) is dominant over sphere-shaped fruit (*d*). Determine the genotypes of the parents in each of the following crosses:

**a.** White, disk × yellow, sphere gives 1/2 white, disk and 1/2 white, sphere.

**b.** White, sphere × white, sphere gives 3/4 white, sphere and 1/4 yellow, sphere.

**c.** Yellow, disk × white, sphere gives all white, disk progeny.

**d.** White, disk × yellow, sphere gives 1/4 white, disk; 1/4 white, sphere; 1/4 yellow, disk; and 1/4 yellow, sphere.

**e.** White, disk × white, sphere gives 3/8 white, disk; 3/8 white, sphere; 1/8 yellow, disk; and 1/8 yellow, sphere.

***11.19** Genes *a*, *b*, and *c* assort independently and are recessive to their respective alleles *A*, *B*, and *C*. Two triply heterozygous (*Aa Bb Cc*) individuals are crossed.

**a.** What is the probability that a given offspring will be phenotypically *A B C*—that is, will exhibit all three dominant traits?

**b.** What is the probability that a given offspring will be homozygous for all three dominant alleles?

**11.20** In garden peas, tall stem (*T*) is dominant over short stem (*t*), green pods (*G*) are dominant over yellow pods (*g*), and smooth seeds (*S*) are dominant over wrinkled seeds (*s*). Suppose a homozygous short, green, wrinkled pea plant is crossed with a homozygous tall, yellow, smooth one.

**a.** What will be the appearance of the $F_1$ generation?

**b.** If the $F_1$ plants are interbred, what will be the appearance of the $F_2$ generation?

**c.** What will be the appearance of the offspring of a cross of the $F_1$ back to the short, green, wrinkled parent?

**d.** What will be the appearance of the offspring of a cross of the $F_1$ back to the tall, yellow, smooth parent?

**11.21** *C* and *c*, *O* and *o*, and *I* and *i* are three independently segregating pairs of alleles in chickens. *C* and *O* are dominant alleles, both of which are necessary for pigmentation. *I* is a dominant inhibitor of pigmentation. Individuals of genotype *cc*, *oo*, *Ii*, or *II* are white, regardless of what other genes they possess.

Assume that white leghorns are *CC OO II*, white wyandottes are *cc OO ii*, and white silkies are *CC oo ii*. What types of offspring (white or pigmented) are possible, and what is the probability of each, from the following crosses?

**a.** white silkie × white wyandotte
**b.** white leghorn × white wyandotte
**c.** (wyandotte–silkie $F_1$) × white silkie

*__11.22__ Mendel found that in peas, yellow seeds are dominant to green seeds, purple flowers are dominant to white flowers, axially positioned flowers are dominant to terminally positioned flowers, and inflated pods are dominant to pinched pods. A single yellow pea, when sown, produces a plant with axially positioned purple flowers. When self-fertilized, this plant produces inflated pods. All of the 140 peas collected from this plant's pods are yellow. When two of these peas are sown, one produces a plant with terminally positioned purple flowers while the other produces a plant with axially positioned white flowers. Each plant, when self-fertilized, produces pinched pods containing only yellow seeds.

**a.** Invent symbols for the traits, and determine the genotypes of the three plants.
**b.** Suppose that the remaining 138 peas were sown, self-fertilized, and produced pods containing peas. What phenotypes and frequencies would you expect to see in the plants, their pods, and the peas they produce?

**11.23** Two homozygous strains of corn are hybridized. They are distinguished by six different pairs of genes, all of which assort independently and produce an independent phenotypic effect. The $F_1$ hybrid is selfed to give an $F_2$ generation.

**a.** What is the number of possible genotypes in the $F_2$ plants?
**b.** How many of these genotypes will be homozygous at all six gene loci?
**c.** If all gene pairs act in a dominant recessive fashion, what proportion of the $F_2$ plants will be homozygous for all dominants?
**d.** What proportion of the $F_2$ will show all dominant phenotypes?

*__11.24__ The coat color of mice is controlled by several genes. The agouti pattern, characterized by a yellow band of pigment near the tip of the hairs, is produced by the dominant allele *A*; homozygous *aa* mice do not have the band and are nonagouti. The dominant allele *B* determines black hairs, and the recessive allele *b* determines brown. Homozygous $c^hc^h$ individuals allow pigments to be deposited only at the extremities (e.g., feet, nose, and ears) in a pattern called Himalayan. The genotype *C*– allows pigment to be distributed over the entire body.

**a.** If a true-breeding black mouse is crossed with a true-breeding brown, agouti, Himalayan mouse, what will be the phenotypes of the $F_1$ and $F_2$ generation?
**b.** What proportion of the non-Himalayan black agouti $F_2$ animals will be $Aa\ BB\ Cc^h$?
**c.** What proportion of the Himalayan mice in the $F_2$ generation is expected to show brown pigment?
**d.** What proportion of all agoutis in the $F_2$ generation is expected to show black pigment?

**11.25** In cocker spaniels, solid coat color is dominant over spotted coat. Suppose a true-breeding, solid-colored dog is crossed with a spotted dog, and the $F_1$ dogs are interbred.

**a.** What is the probability that the first puppy born will have a spotted coat?
**b.** What is the probability that, if four puppies are born, all of them will have solid coats?

**11.26** In cats, alleles at one gene determine whether a cat has pigmented fur and dark eyes or has white fur and blue eyes: White cats with blue eyes have the dominant *W* allele, while dark-eyed pigmented cats are *ww*. Alleles at another gene determine hair length: Short-haired cats have the dominant *L* allele, while long-haired cats are *ll*. A stray long-haired white female cat with blue eyes delivers a litter of four kittens under your neighbor's porch. All of the kittens look different from each other: one is just like its mother, one is a dark-eyed grey cat with short hair, one is a dark-eyed grey cat with long hair, and one is a blue-eyed white cat with short hair. Your neighbor learns that litters of stray urban female cats often show multiple paternities, and is certain that this is why the kittens look different from each other. Is this the only possible explanation for the phenotypic differences among the kittens? Is there evidence for multiple paternities in this litter?

**11.27** In the $F_2$ of his cross of red-flowered × white-flowered *Pisum* (pea plant), Mendel obtained 705 plants with red flowers and 224 with white.

**a.** Is this result consistent with his hypothesis of factor segregation, which predicts a 3:1 ratio?
**b.** In how many similar experiments would a deviation as great as or greater than this one be expected? (Calculate $\chi^2$ and obtain the approximate value of *P* from Table 11.5.)

**11.28** In tomatoes, cut leaf and potato leaf are alternative characters, with cut (*C*) dominant to potato (*c*). Purple stem and green stem are another pair of alternative characters, with purple (*P*) dominant to green (*p*). A true-breeding cut, green tomato plant is crossed with a true-breeding potato, purple plant, and the $F_1$ plants are allowed to interbreed. The 320 $F_2$ plants were phenotypically 189 cut, purple; 67 cut, green; 50 potato, purple; and 14 potato, green. Propose a hypothesis to explain the data, and use the chi-square test to test the hypothesis.

**11.29** A true-breeding tall pea plant with axially positioned purple flowers was crossed to a true-breeding dwarf pea plant with terminally positioned white flowers. When the $F_1$ plants, which were tall with axially positioned purple flowers, were crossed to true-breeding plants that were short with terminally positioned white flowers, the following offspring were obtained:

164 tall, terminal, white
144 tall, axial, white
156 tall, terminal, purple
176 tall, axial, purple
138 short, terminal, white
149 short, axial, white
182 short, terminal, purple
166 short, axial, purple

Propose a hypothesis to explain the data and use the chi-square test to test your hypothesis.

***11.30** The simple case of two mating types (male and female) is by no means the only sexual system known. The ciliated protozoan *Paramecium bursaria* has a system of four mating types, controlled by two genes (*A* and *B*). Each gene has a dominant and a recessive allele. The four mating types are expressed according to the following scheme:

| Genotype | Mating Type |
|---|---|
| *AA BB* | A |
| *Aa BB* | A |
| *AA Bb* | A |
| *Aa Bb* | A |
| *AA bb* | D |
| *Aa bb* | D |
| *aa BB* | B |
| *aa Bb* | B |
| *aa bb* | C |

It is clear, therefore, that some of the mating types result from more than one possible genotype. We have four strains of known mating type—"A," "B," "C," and "D"—but unknown genotype. The following crosses were made, with the indicated results:

| | Mating Type of Progeny | | | |
|---|---|---|---|---|
| Cross | A | B | C | D |
| "A" × "B" | 24 | 21 | 14 | 18 |
| "A" × "C" | 56 | 76 | 55 | 41 |
| "A" × "D" | 44 | 11 | 19 | 33 |
| "B" × "C" | 0 | 40 | 38 | 0 |
| "B" × "D" | 6 | 8 | 14 | 10 |
| "C" × "D" | 0 | 0 | 45 | 45 |

Assign genotypes to "A," "B," "C," and "D".

***11.31** In bees, males (drones) develop from unfertilized eggs and are haploid. Females (workers and queens) are diploid and come from fertilized eggs. *W* (black eyes) is dominant over *w* (white eyes). Workers of genotype *RR* or *Rr* use wax to seal crevices in the hive; *rr* workers use resin instead. A *Ww Rr* queen founds a colony after being fertilized by a black-eyed drone bearing the *r* allele.

**a.** What will be the appearance and behavior of workers in the new hive, and what are their relative frequencies?

**b.** Give the genotypes of male offspring, with relative frequencies.

**c.** Fertilization normally takes place in the air during a nuptial flight, and any bee unable to fly would effectively be rendered sterile. Suppose a recessive mutation, *c*, occurs spontaneously in a sperm that fertilizes a normal egg, and suppose also that the effect of the mutant gene is to cripple the wings of any adult not bearing the normal allele *C*. The fertilized egg develops into a normal queen named Madonna. What is the probability that wingless males will be *found in a hive* founded two generations later by one of Madonna's granddaughters?

**d.** By one of Madonna's great-great-granddaughters?

**11.32** For the pedigrees A and B, indicate whether the trait involved in each case could be recessive or dominant, and explain your answers.

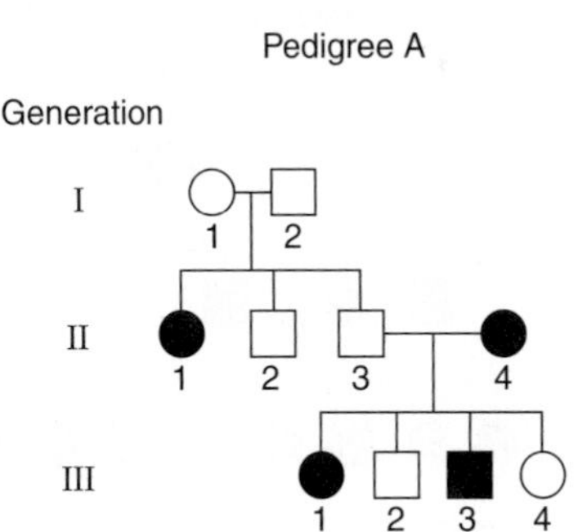


Pedigree B

Generation

I 1 2

II 1 2 3 4 5 6 7

III 1 2 3

***11.33** Consider the following pedigree, in which the allele responsible for the trait (*a*) is recessive to the normal allele (*A*):

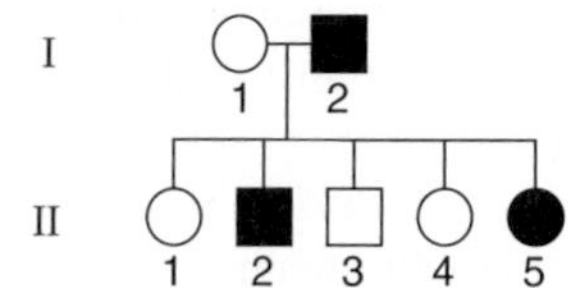


**a.** What is the genotype of the mother?

**b.** What is the genotype of the father?

**c.** What are the genotypes of the children?

**d.** Given the mechanism of inheritance involved, does the ratio of children with the trait to children without the trait match what would be expected?

**11.34**

**a.** What possible mode(s) of inheritance can explain the pattern of affected individuals in each of Pedigrees A and B?

**b.** How would your answers to part (a) change, if at all, if you also knew the following? Affected individuals in each pedigree have the same disease; homozygotes for the disease allele never have offspring; and Pedigree A is from a relatively small, isolated community where the disease is common while Pedigree B is from a large, diverse community where the disease is quite rare.

**c.** If in each pedigree, III-5 and III-6 have another child, what is the probability that their offspring will exhibit the trait?

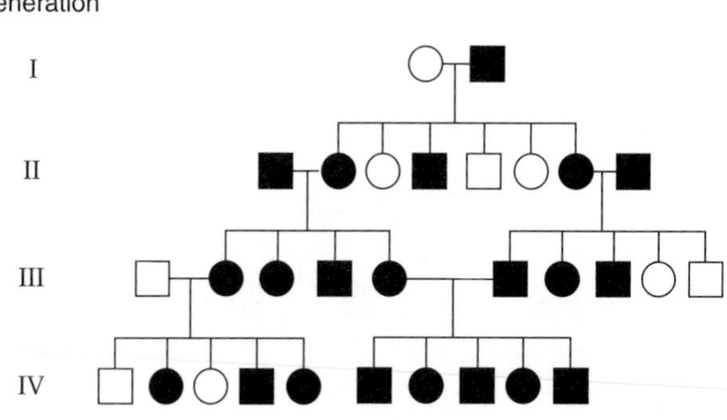


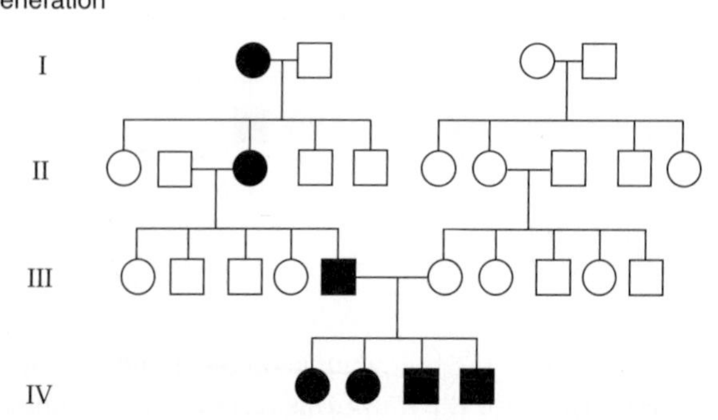


**11.35** After a few years of marriage, a woman comes to believe that, among all of the reasonable relatives in her and her husband's families, her husband, her mother-in-law, and her father have so many similarities in their unreasonableness that they must share a mutation. A friend taking a course in genetics assures her it is unlikely that this trait has a genetic basis and that, even if it did, all of her children would be reasonable. Diagram and analyze the relevant pedigree to evaluate whether the friend's advice is accurate.

***11.36** Gaucher disease is caused by a chronic enzyme deficiency that is more common among Ashkenazi Jews than in the general population. A Jewish man has a sister afflicted with the disease. His parents, grandparents, and three siblings are not affected. Discussions with relatives in his wife's family reveal that the disease is not likely to be present in her family, although some relatives recall that the brother of his wife's paternal grandmother suffered from a similar disease. Diagram and analyze the relevant pedigree to determine (a) the genetic basis for inheriting the trait for Gaucher disease and (b) the highest probability that, if this couple has a child, the child will be affected (i.e., what is the chance of the worst-case scenario occurring?).

The remaining questions use a different way of symbolizing dominant and recessive alleles, one that will be introduced formally in Chapter 12. In this system, the recessive mutant allele is in lower case, while the dominant wild-type (normal) allele uses the same lower case letter(s) with a superscript plus. Thus, $a^+$ and $a$ in this system are comparable to $A$ and $a$ used earlier in this chapter.

***11.37** $a^+$, $b^+$, $c^+$, and $d^+$ are independently assorting Mendelian genes controlling the production of a black pigment. The alternate alleles that give abnormal functioning of these genes are $a$, $b$, $c$, and $d$. A black individual of genotype $a^+/a^+\ b^+/b^+\ c^+/c^+\ d^+/d^+$ is crossed with a colorless individual of genotype $a/a\ b/b\ c/c\ d/d$ to produce a black $F_1$. $F_1 \times F_1$ crosses are then done. Assume that $a^+$, $b^+$, $c^+$, and $d^+$ act in a pathway as follows:

$$\text{colorless} \xrightarrow{a^+} \text{colorless} \xrightarrow{b^+} \text{colorless} \xrightarrow{c^+} \text{brown} \xrightarrow{d^+} \text{black}$$

**a.** What proportion of the $F_1$ progeny is colorless?

**b.** What proportion of the $F_2$ progeny is brown?

**11.38** Using the genetic information given in Problem 11.37, now assume that $a^+$, $b^+$, and $c^+$, act in a pathway as follows:

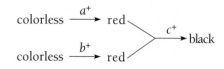

$$\text{colorless} \xrightarrow{a^+} \text{red}$$
$$\text{colorless} \xrightarrow{b^+} \text{red}$$
$$\text{red} + \text{red} \xrightarrow{c^+} \text{black}$$

Black can be produced only if both red pigments are present; that is, $c^+$ converts the two red pigments together into a black pigment.

**a.** What proportion of the $F_2$ progeny is colorless?
**b.** What proportion of the $F_1$ progeny is red?
**c.** What proportion of the $F_2$ progeny is black?

***11.39** Three genes on different chromosomes are responsible for three enzymes that catalyze the same reaction in corn:

$$\text{colorless compound} \xrightarrow{a^+,\, b^+,\, c^+} \text{red compound}$$

The normal functioning of any one of these genes is sufficient to convert the colorless compound to the red compound. The abnormal functioning of these genes is designated by *a*, *b*, and *c*, respectively.

**a.** A red $a^+/a^+\ b^+/b^+\ c^+/c^+$ is crossed with a colorless $a/a\ b/b\ c/c$ to give a red $F_1$ $a^+/a\ b^+/b\ c^+/c$. The $F_1$ is selfed. What proportion of the $F_2$ progeny is colorless?
**b.** It turns out that another step is involved in the pathway—one that is controlled by gene $d^+$, which assorts independently of $a^+$, $b^+$, and $c^+$.

$$\text{colorless compound 1} \xrightarrow{d^+} \text{colorless compound 2} \xrightarrow{a^+,\, b^+,\, c^+} \text{red compound}$$

The inability to convert colorless 1 to colorless 2 is designated *d*. A red $a^+/a^+\ b^+/b^+\ c^+/c^+\ d^+/d^+$ is crossed with a colorless $a/a\ b/b\ c/c\ d/d$. The $F_1$ corn are all red. The red $F_1$ corn are now selfed. What proportion of the $F_2$ corn is colorless?

***11.40** In J. R. R. Tolkien's *The Lord of the Rings*, the Black Riders of Mordor ride steeds with eyes of fire. As a geneticist, you are very interested in the inheritance of the fire-red eye color. You discover that the eyes contain two types of pigments—brown and red—that are usually bound to core granules in the eye. In wild-type steeds, precursors are converted by these granules to the aforesaid pigments; but in steeds homozygous for the recessive X-linked gene *w* (white eye), the granules remain unconverted and a white eye results. The metabolic pathways for the synthesis of the two pigments are shown in Figure 11.A.

Each step of the pathway is controlled by a gene: Mutation *v* results in vermilion eyes, *cn* results in cinnabar eyes, *st* results in scarlet eyes, *bw* results in brown eyes, and *se* results in black eyes. All these mutations are recessive to their wild-type alleles, and all are unlinked. For the following genotypes, show the proportions of steed eye phenotypes that would be obtained in the $F_1$ of the given matings:

**a.** $w/w\ bw^+/bw^+\ st/st \times w^+/Y\ bw/bw\ st^+/st^+$
**b.** $w^+/w^+\ se/se\ bw/bw \times w/Y\ se^+/se^+\ bw^+/bw^+$
**c.** $w^+/w^+\ v^+/v^+\ bw/bw \times w/Y\ v/v\ bw/bw$
**d.** $w^+/w^+\ bw^+/bw\ st^+/st \times w/Y\ bw/bw\ st/st$

**Figure 11.A**

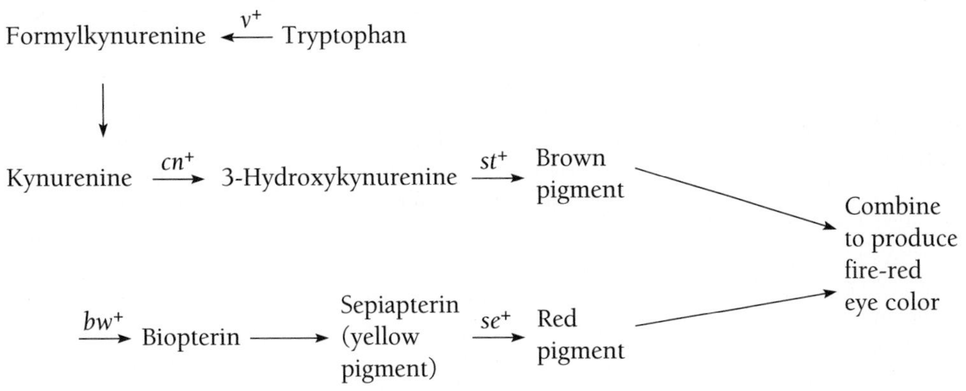

Formylkynurenine $\xleftarrow{v^+}$ Tryptophan

↓

Kynurenine $\xrightarrow{cn^+}$ 3-Hydroxykynurenine $\xrightarrow{st^+}$ Brown pigment → Combine to produce fire-red eye color

$\xrightarrow{bw^+}$ Biopterin ⟶ Sepiapterin (yellow pigment) $\xrightarrow{se^+}$ Red pigment → Combine to produce fire-red eye color

# 12 Chromosomal Basis of Inheritance

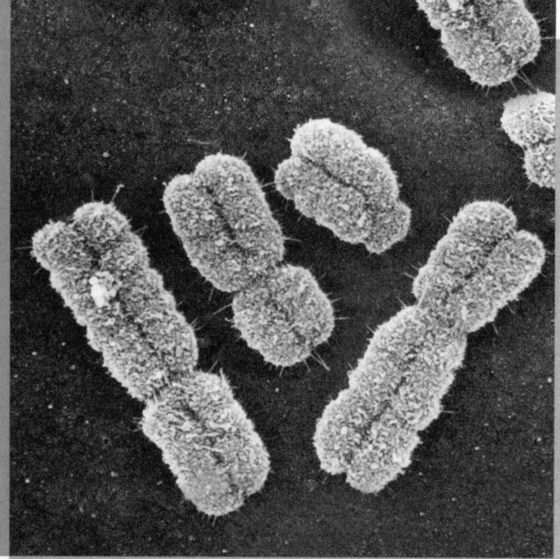

Human X and Y chromosomes.

## Key Questions

- How is the genome organized into chromosomes in eukaryotes?
- How is the chromosome complement of a eukaryote analyzed?
- How are eukaryotic chromosomes transmitted from generation to generation in mitosis?
- How are eukaryotic chromosomes transmitted from a diploid cell into haploid gametes in meiosis?
- How do genes segregate in meiosis?
- What is the role of meiosis in animals and plants?
- How does chromosome segregation explain gene segregation in meiosis?
- How do sex chromosomes affect gene segregation patterns?
- How do sex chromosomes relate to the sex of an organism?
- How are sex-linked traits analyzed in humans?

## iActivity

WHEN A CHILD IS BORN, THE FIRST QUESTION most people ask is "Is it a boy or a girl?" The answer, at the chromosomal level, depends on the sex chromosomes: Two X chromosomes produce a girl, while an X and a Y chromosome produce a boy. But the genes contained on these chromosomes determine more than just the sex of an individual; they are responsible for the inheritance of a number of other traits.

In this chapter, you will learn about the behavior of chromosomes during nuclear division in eukaryotes, the ways in which sex is determined in humans and other organisms, and sex-linked traits in humans. After you have read and studied this chapter, you can apply what you've learned by trying the iActivity, in which you will investigate the inheritance of deafness within a family.

On Mendel's foundation, early geneticists began to build genetic hypotheses that could be tested by appropriate crosses, and they began to investigate the nature of Mendelian factors. We now know that Mendelian factors are genes and that genes are located on chromosomes. In this chapter, we focus on the behavior of genes and chromosomes. We start by learning about the transmission of chromosomes from cell division to cell division and from generation to generation by the processes of mitosis and meiosis, respectively. We then consider the evidence for the association of genes and chromosomes. In so doing, we will learn about the segregation of genes located on the sex chromosomes. Next, we learn about various mechanisms of sex determination, and finally, we discuss sex-linked traits in humans. The goal of the chapter is for you to learn how to think about gene segregation in terms of chromosome inheritance patterns.

## Chromosomes and Cellular Reproduction

The association between chromosomes and genes was determined as a result of the efforts of cytologists, who examined the behavior of chromosomes, and geneticists, who examined the behavior of genes. In this section, we discuss the general structure of eukaryotic chromosomes and the transmission of chromosomes from cell division to cell division and from generation to generation by the processes of mitosis and meiosis, respectively.

## Eukaryotic Chromosomes

The genome of a eukaryote is distributed among multiple, linear chromosomes; the number of chromosomes typically is characteristic of the species. Many eukaryotes have two copies of each type of chromosome in their nuclei, so their chromosome complement is said to be **diploid,** or 2N. Diploid eukaryotes are produced by the fusion of two haploid gametes (mature reproductive cells that are specialized for sexual fusion), one from the female parent and one from the male parent. The fusion produces a diploid **zygote,** which then undergoes embryological development. Each gamete has only one set of chromosomes and is said to be **haploid** (N). The complete compendium of genetic information in a haploid chromosome set is the *genome*. Two examples of diploid organisms are humans, with 46 chromosomes (23 pairs), and *Drosophila melanogaster*, with 8 chromosomes (4 pairs). By contrast, laboratory strains of the yeast *S. cerevisiae*, with 16 chromosomes, are haploid.

Figure 12.1 illustrates the chromosomal organization of haploid and diploid organisms. In diploid organisms, the members of a chromosome pair that contain the same genes and that pair during meiosis are called **homologous chromosomes;** each member of a pair is called a **homolog,** and one homolog is inherited from each parent. Chromosomes that contain different genes and that do not pair during meiosis are called **nonhomologous chromosomes.**

In animals and in some plants, male and female cells are distinct with respect to their complement of **sex chromosomes**—the chromosomes that are represented differently in the two sexes in many eukaryotic organisms. One sex has a matched pair of sex chromosomes, and the other sex has an unmatched pair of sex chromosomes or a single sex chromosome. For example, human females have two X chromosomes (XX), whereas human males have one X and one Y (XY). Chromosomes other than sex chromosomes are called **autosomes.** We discuss X chromosomes in more detail later in the chapter.

Under the microscope, chromosomes are seen to differ in size and morphology (appearance) within and between species. Each chromosome has a constriction along its length called a **centromere,** which is important for the behavior of the chromosomes during cellular division. The location of the centromere in one of four general positions in the chromosome is useful in classifying eukaryotic chromosomes (Figure 12.2). A **metacentric chromosome** has the centromere at about the center, so the chromosome appears to have two approximately equal arms. **Submetacentric chromosomes** have one arm longer than the other, **acrocentric chromosomes** have one arm with a stalk and often with a "bulb" (called a *satellite*) on it, and **telocentric chromosomes** have only one arm, because the centromere is at the end. Chromosomes also vary in relative size. Chromosomes of mice, for example, are all similar in length, whereas those of humans have a wide range of relative lengths. Chromosome length and centromere position are constant for each chromosome and help in identifying individual chromosomes.

A complete set of all the metaphase chromosomes in a cell is called the cell's **karyotype** ("carry-o-type"; literally, "nucleus type"). Chromosomes are typically identified during metaphase (see p. 332), the stage of mitosis at which they are most condensed, which makes them easier to see under the microscope following staining. The karyotype is species specific, so a wide range of numbers, sizes, and shapes of chromosomes are seen among eukaryotic organisms. Even closely related organisms may have quite different karyotypes.

Figure 12.3a shows the karyotype for the cell of a normal human male. It is customary in karyotypes to arrange chromosomes in order according to their sizes and positions of their centromere. This karyotype shows 46 chromosomes: two pairs of each of the 22 autosomes and one of each of the X and Y sex chromosomes (which differ greatly in size). In a human karyotype, the chromosomes

**Figure 12.1**
**Chromosomal organization of haploid and diploid organisms.**

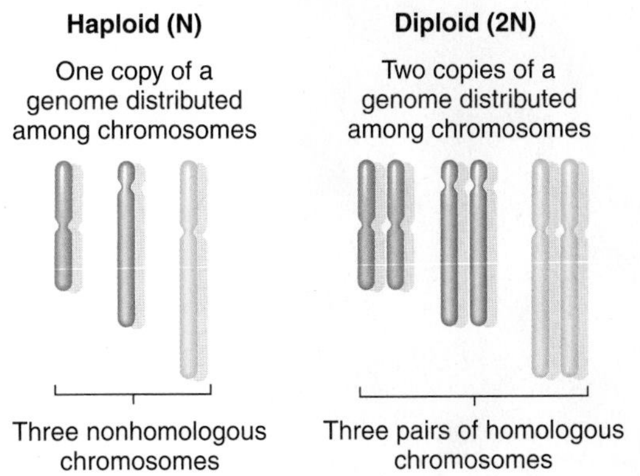


**Figure 12.2**
**General classification of eukaryotic chromosomes as metacentric, submetacentric, acrocentric, and telocentric types, based on the position of the centromere.**

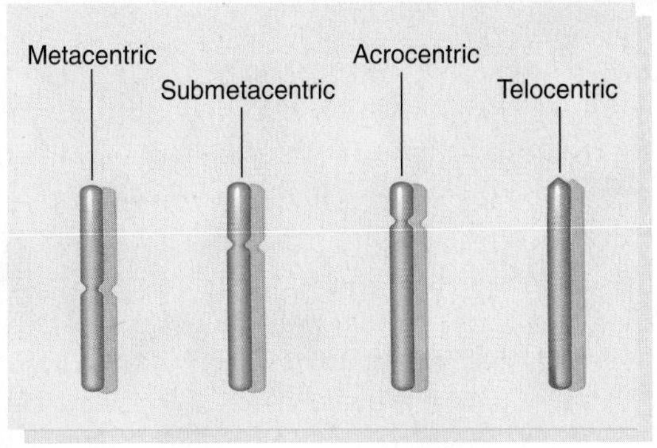

**Figure 12.3**
**Human karyotypes.**

a) G banding in a karyotype of a male

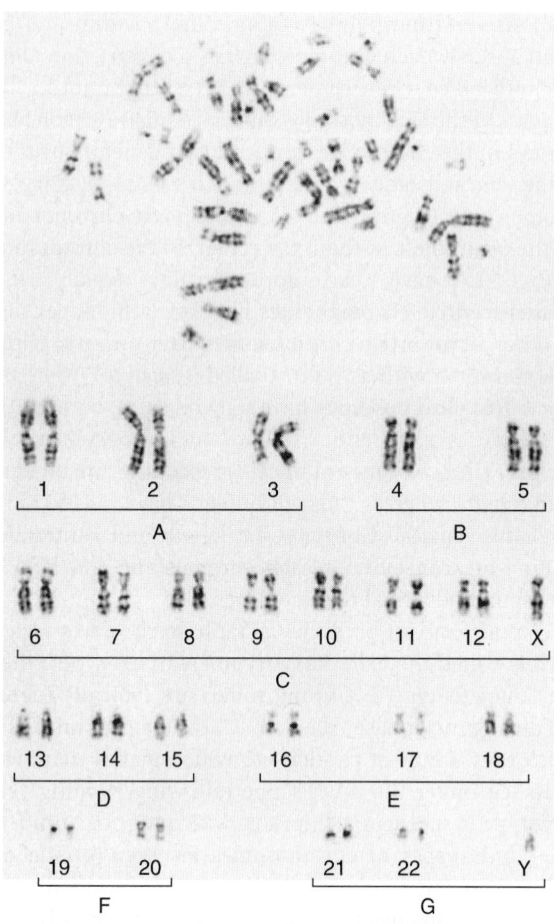


b) Chromosome painting in a male karyotype

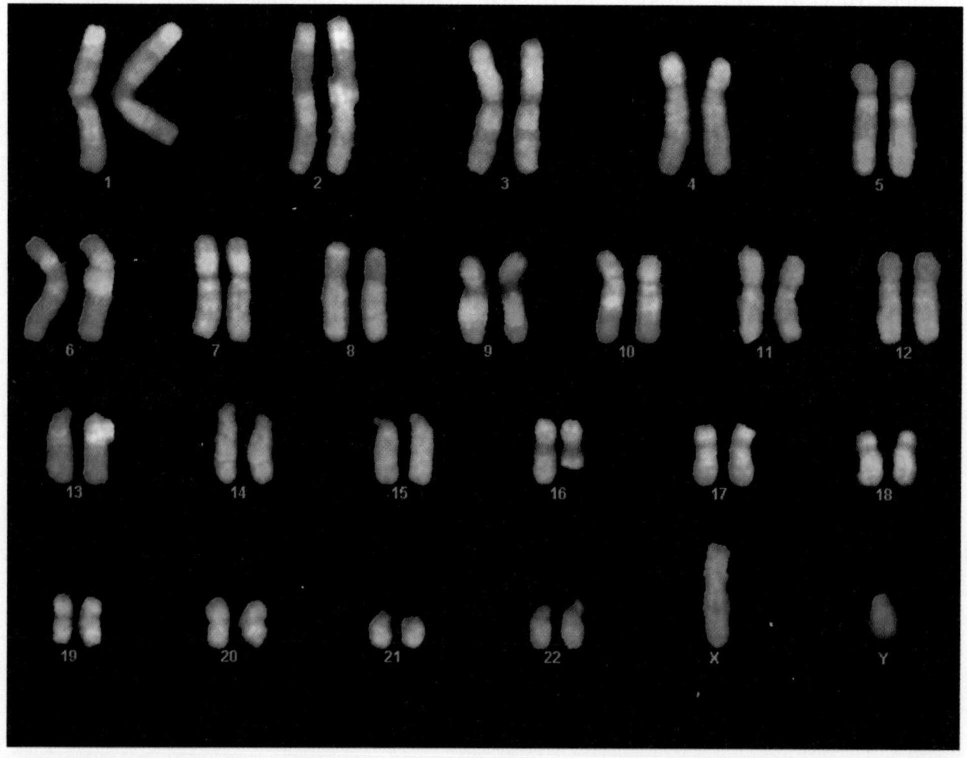


are numbered for easy identification. Conventionally, the largest pair of homologous chromosomes is designated 1, the next largest 2, and so on. Although chromosome 21 is smaller than chromosome 22, it is called 21 for historical reasons. As shown in Figure 12.3a, chromosomes with similar morphologies may be arranged under the letter designations A through G.

Based on size and morphology alone, the different chromosomes are hard to distinguish unambiguously when they are stained evenly. Fortunately, a number of procedures stain certain regions or *bands* of the chromosomes more intensely than other regions. Banding patterns are specific to each chromosome, enabling us to distinguish each chromosome clearly in the karyotype. One of these staining techniques is called G banding. This technique is used commonly in the generation of karyotypes in clinical analysis of human chromosomes. In G banding, chromosomes are treated with mild heat or proteolytic enzymes (enzymes that digest proteins) to digest the chromosomal proteins partially and then are stained with Giemsa stain to produce dark bands called G bands (see Figure 12.3a). In humans, approximately 300 G bands can be distinguished in metaphase chromosomes, and approximately 2,000 G bands can be distinguished in chromosomes from the prophase stage of mitosis. Conventionally, drawings (*ideograms*) of human chromosomes show the G banding pattern. Furthermore, a standard nomenclature based on the banding patterns has been established for the chromosomes so that scientists can talk about gene and marker locations with reference to specific regions and subregions. Each chromosome has two arms separated by the centromere. The smaller arm is designated p, and the larger arm is designated q. Numbered regions and numbered subregions are then assigned from the centromere outward; that is, region 1 is closest to the centromere. For example, the breast cancer susceptibility gene *BRCA1* is at location 17q21, meaning that it is on the long arm of chromosome 17 in region 21. Subregions are indicated by decimal numerals after the region number. For instance, the cystic fibrosis gene spans subregions 7q31.2–q31.3; that is, it spans both subregions 2 and 3 of region 31 of the long arm of chromosome 7.

Human chromosomes in a karyotype can also be distinguished using a more recent method in which DNA probes specific to regions of particular chromosomes are hybridized to chromosomes spread on a microscope slide. The DNA probes are labeled with fluorescent molecules that have various wavelengths of fluorescence emissions. The fluorescence emissions are processed by computer to generate a colorized image of each chromosome. In the karyotype shown in Figure 12.3b, the combination of probes that hybridize to each chromosome is responsible for the unique pattern of colors. The method, known as *chromosome painting*, has several variations that enable users to paint each chromosome either with several colors (as in the example in the figure), or with one distinct color. Chromosome painting is used only rarely in karyotyping performed in clinical analysis of human chromosomes.

**Keynote**

Diploid eukaryotic cells have two haploid sets of chromosomes, one set from each parent. The members of a pair of chromosomes, one from each parent, are called homologous chromosomes. Haploid eukaryotic cells have only one set of chromosomes. The complete set of chromosomes in a cell is called its karyotype. The karyotype is species specific. Staining with particular dyes results in characteristic banding patterns, thereby defining chromosome regions and subregions by number.

## Mitosis

In both unicellular and multicellular eukaryotes, cellular reproduction is a cyclical process of growth, **mitosis** (*nuclear division* or *karyokinesis*), and (usually, but not always) **cell division** (*cytokinesis*). The cycle of growth, mitosis, and cell division is called the **cell cycle.** In proliferating somatic cells, the cell cycle consists of two phases: the mitotic (or division) phase (M) and an interphase between divisions (Figure 12.4). Interphase consists of three stages: $G_1$ (gap 1), S, and $G_2$ (gap 2). During $G_1$ (the presynthesis stage), the cell prepares for DNA and chromosome replication, which take place in the S stage. In $G_2$ (the postsynthesis stage), the cell prepares for cell division, or the M stage.

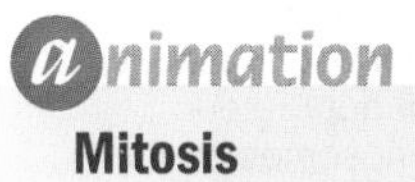


**Mitosis**

**Figure 12.4**

**Eukaryotic cell cycle.** This cycle assumes a period of 24 hours, although great variation exists between cell types and organisms.

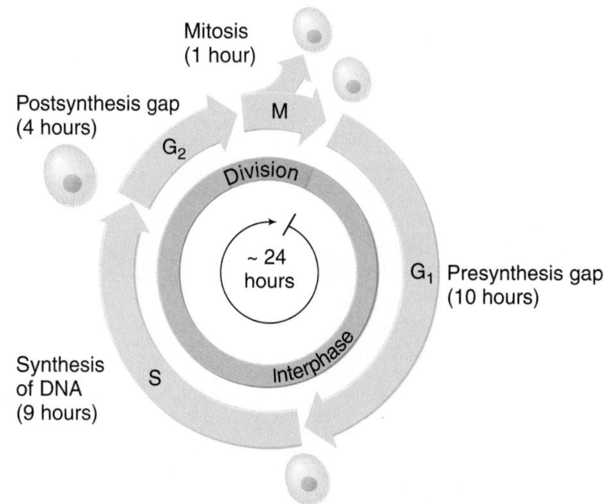


Put another way, chromosome replication takes place in interphase and then mitosis occurs, resulting in the distribution of a complete chromosome set to each of two progeny nuclei.

The relative time spent in each of the four stages of mitosis varies greatly among cell types. In a given organism, variation in the length of the cell cycle depends primarily on the duration of $G_1$; the duration of S plus $G_2$ plus M is approximately the same in all cell types. For example, some cancer cells and early fetal cells of humans spend minutes in $G_1$, whereas some differentiated adult cells (such as nerve cells) spend years in $G_1$. Some cells exit the cell cycle from $G_1$ and enter a quiescent, nondividing state called $G_0$.

During interphase, the individual chromosomes are elongated and difficult to see under the light microscope. The DNA of each chromosome is replicated in the S phase, giving two exact copies, called **sister chromatids,** that are held together by the replicated but unseparated centromeres. (Because the centromeres have not separated, only one centromere is visible under the light microscope.) More precisely, a **chromatid** is one of the two distinct longitudinal subunits of all replicated chromosomes that becomes visible between early prophase and metaphase of mitosis. Later, when the centromeres separate, the sister chromatids become known as **daughter chromosomes.** So, in a diploid cell there are pairs of homologous chromosomes. When the DNA of a pair of homologous chromosomes replicates in interphase, the result is two pairs of sister chromatids. After the centromeres separate, two pairs of daughter chromosomes are produced.

Mitosis occurs in both haploid and diploid cells. It is a continuous process, but for purposes of discussion, it is commonly divided into five stages called *prophase, prometaphase, metaphase, anaphase*, and *telophase*. Figure 12.5 shows the five stages in simplified diagrams. The photographs in Figure 12.6 show the typical chromosome morphology in interphase and in the five stages of mitosis in animal cells.

**Figure 12.5**

**Interphase and mitosis in an animal cell.**

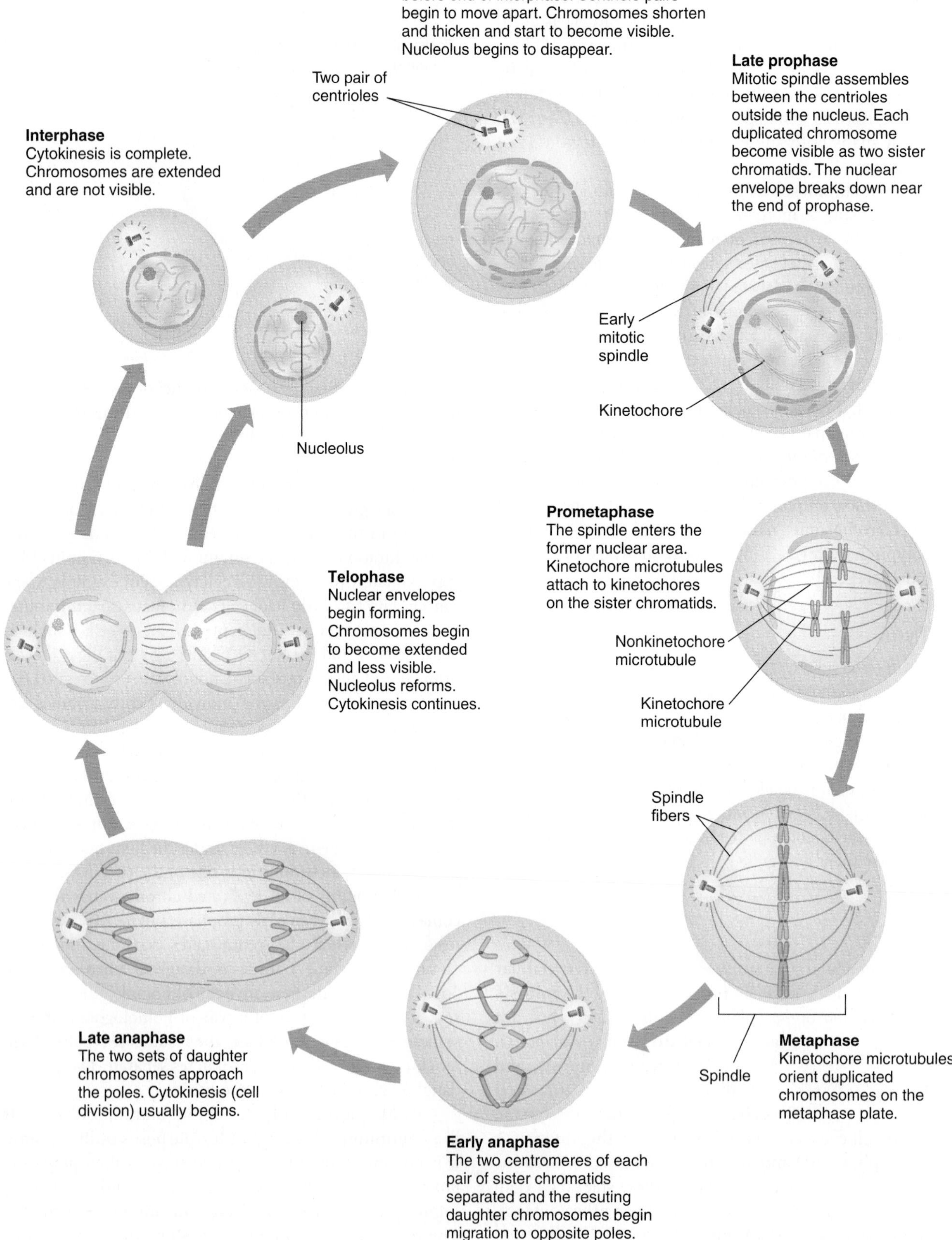
Early prophase
Chromosomes and centrioles replicate before end of interphase. Centriole pairs begin to move apart. Chromosomes shorten and thicken and start to become visible. Nucleolus begins to disappear.
Two pair of centrioles
Late prophase
Mitotic spindle assembles between the centrioles outside the nucleus. Each duplicated chromosome become visible as two sister chromatids. The nuclear envelope breaks down near the end of prophase.
Interphase
Cytokinesis is complete. Chromosomes are extended and are not visible.
Early mitotic spindle
Kinetochore
Nucleolus
Prometaphase
The spindle enters the former nuclear area. Kinetochore microtubules attach to kinetochores on the sister chromatids.
Telophase
Nuclear envelopes begin forming. Chromosomes begin to become extended and less visible. Nucleolus reforms. Cytokinesis continues.
Nonkinetochore microtubule
Kinetochore microtubule
Spindle fibers
Late anaphase
The two sets of daughter chromosomes approach the poles. Cytokinesis (cell division) usually begins.
Spindle
Metaphase
Kinetochore microtubules orient duplicated chromosomes on the metaphase plate.
Early anaphase
The two centromeres of each pair of sister chromatids separated and the resuting daughter chromosomes begin migration to opposite poles.

**Prophase.** In the $G_2$ stage of the cell cycle, just prior to the start of M, each chromosome consists of two sister chromatids, and the centrioles have duplicated to produce two pairs (see Figures 12.5 and 12.6a). In **prophase** (see Figures 12.5 and 12.6b) the chromatids condense, so they gradually appear shorter and fatter under the microscope. By late prophase, each chromosome, which was duplicated during the preceding S phase of interphase, can be seen to consist of two sister chromatids. While condensation is occurring, the nucleolus shrinks and eventually disappears in most species.

Many mitotic events depend on the *mitotic spindle* (spindle apparatus), a structure consisting of fibers composed of microtubules made of special proteins called *tubulins*. The mitotic spindle assembles outside the nucleus during prophase. In most animal cells, the centrioles (see Figure 1.5b, p. 7) are the focal points for spindle assembly; higher plant cells usually lack centrioles, but they do have a mitotic spindle. Centrioles are arranged in pairs and, before the S phase, the cell's centriole pair is duplicated. Then, during mitosis, each new centriole pair becomes the focus of a radial array of microtubules called the *aster*. Early in prophase, the two asters are next to one another and close to the nuclear envelope; by late prophase, the two asters have moved far apart along the outside of the nucleus and are spanned by the microtubular spindle fibers.

**Prometaphase.** The nuclear envelope breaks down at the end of prophase, denoting the beginning of **prometaphase** (see Figures 12.5 and 12.6c). The developing spindle now enters the former nuclear area. A specialized multiprotein complex called a **kinetochore** binds to each centromere. The kinetochores are the sites for the attachment of the chromosomes to spindle microtubules known as *kinetochore microtubules*. For a pair of sister chromatids, one to many kinetochore microtubules from one pole

**Figure 12.6**

**Interphase and the stages of mitosis in whitefish early embryo cells.**

a) Interphase

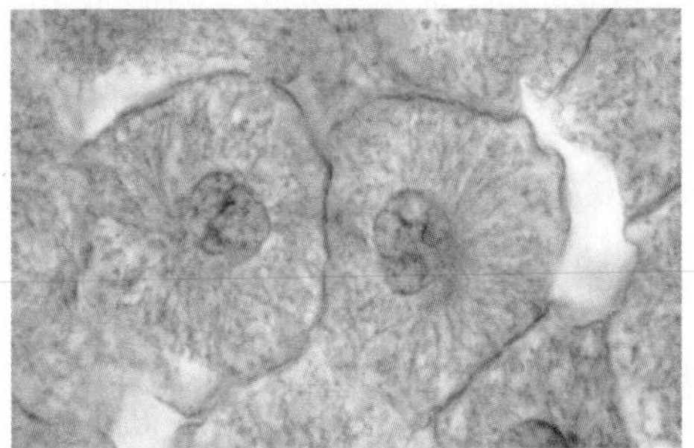

b) Prophase

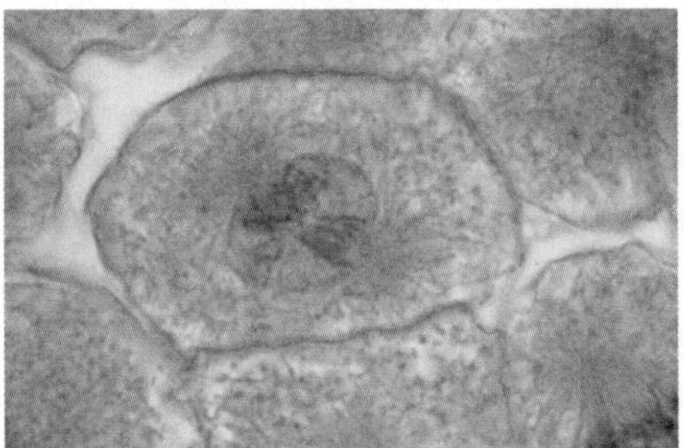

c) Prometaphase

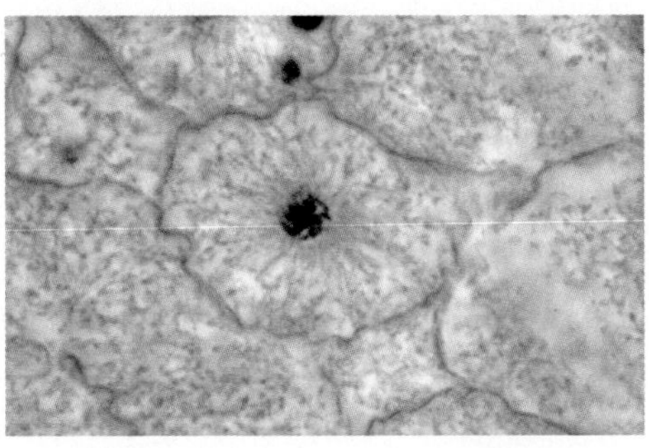

d) Metaphase

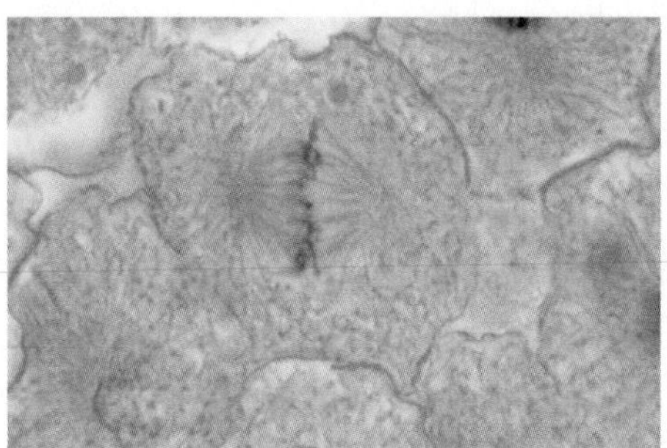

e) Anaphase

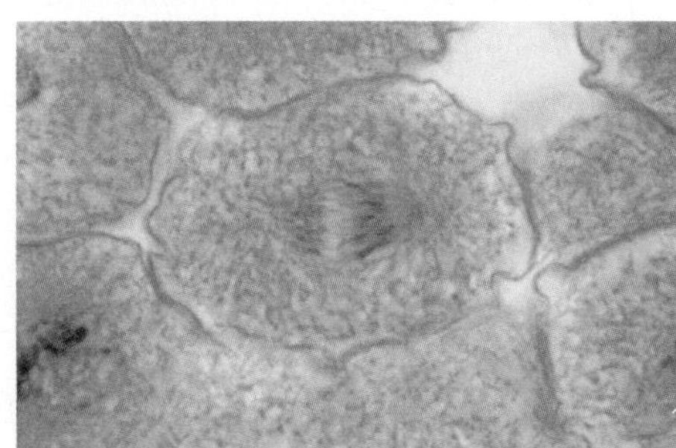

f) Telophase

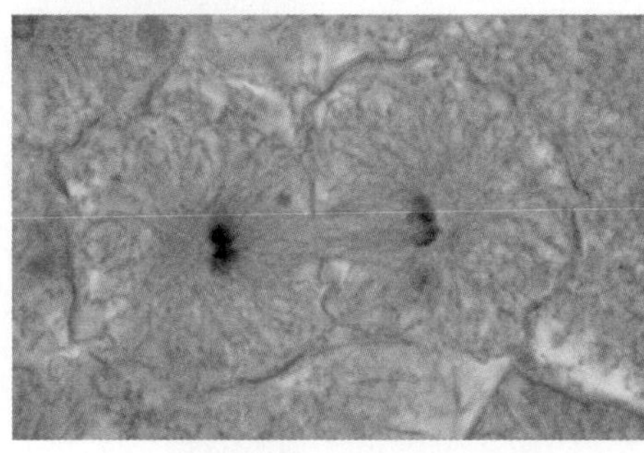

attach to the kinetochore of one chromatid, and an equivalent number of kinetochore microtubules from the other pole attach to the kinetochore of the other chromatid. *Nonkinetochore microtubules*—spindle microtubules that do not bind to kinetochores—also originate from each spindle pole and overlap in the middle of the spindle.

**Metaphase.** During **metaphase** (see Figures 12.5 and 12.6d), the kinetochore microtubules orient the chromosomes so that their centromeres become aligned at the **metaphase plate**, a plane halfway between the two spindle poles, with the long axes of the chromosomes oriented at 90 degrees to the spindle axis. It had been accepted for many years that metaphase chromosomes (Figure 12.7) are the most condensed form of chromosomes in mitosis (and meiosis). Recently, examination of 3D reconstructions of microscope images of living mammalian cells taken with high-power microscopes has revealed that further chromosome condensation occurs just after the chromosomes have finished separating in the subsequent anaphase stage. This late condensation serves to minimize the potential problem of chromosome arms extending over the plane of division, which could result in mechanical damage to the chromosomes.

**Anaphase.** During **anaphase** (see Figures 12.5 and 12.6e), the joined centromeres of sister chromatids separate, giving rise to two daughter chromosomes. Once the paired kinetochores on each chromosome separate, the sister chromatid pairs undergo disjunction (separation), and the daughter chromosomes move toward the opposite poles. In anaphase, the daughter chromosomes are pulled toward the opposite poles of the cell by the shortening microtubules attached to the kinetochores. As they are pulled, the chromosomes assume characteristic shapes related to the location of the centromere along the chromosome's length. For example, a metacentric chromosome is V shaped as the two roughly equal-length chromosome arms trail the centromere in its migration toward the pole, and a submetacentric chromosome has a J shape with a long and short arm. The movement continues until the separated daughter chromosomes have reached the two poles, at which point chromosome segregation has been completed. Cytokinesis usually begins in the latter stages of anaphase.

**Telophase.** At the start of **telophase** (see Figures 12.5 and 12.6f), the two sets of daughter chromosomes are assembled into two groups at opposite ends of the cell. The chromosomes begin to uncoil and assume the elongated state characteristic of interphase. A nuclear envelope forms around each group of chromosomes, the spindle microtubules disappear, and the nucleolus or nucleoli reform. At this point, nuclear division is complete and the cell now has two nuclei.

**Cytokinesis. Cytokinesis** is division of the cytoplasm; usually, it follows the nuclear division stage of mitosis and is completed by the end of telophase. Cytokinesis compartmentalizes the two new nuclei into separate daughter cells, completing mitosis and cell division (Figure 12.8). In cytokinesis in animal cells, a constriction forms in the middle of the cell; the constriction continues until two daughter cells are produced (Figure 12.8a). In cytokinesis in plant cells, a new cell membrane and cell wall are assembled between the two new nuclei to form a *cell plate* (Figure 12.8b). Cell wall material coats each side of the plate, and the result is two progeny cells.

**Gene Segregation in Mitosis.** In mitosis, one copy of each duplicated chromosome segregates into both daughter cells. Thus, for a haploid (N) cell, chromosome duplication produces a cell in which each chromosome has doubled its content. Mitosis then results in two progeny haploid cells, each with one complete set of chromosomes (a genome).

**Figure 12.7**
**Human metaphase chromosome.**

a) Metaphase chromosome (transmission electron micrograph)

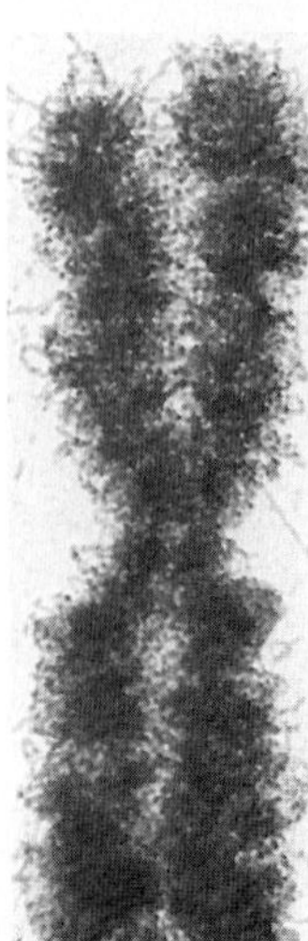

b) Metaphase chromosome (scanning electron micrograph)

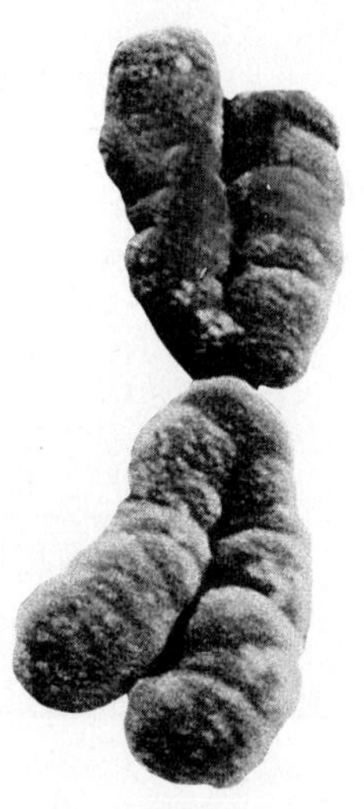

**Figure 12.8**
**Cytokinesis (cell division).**

a) Cytokinesis in an animal cell b) Cytokinesis in a plant cell

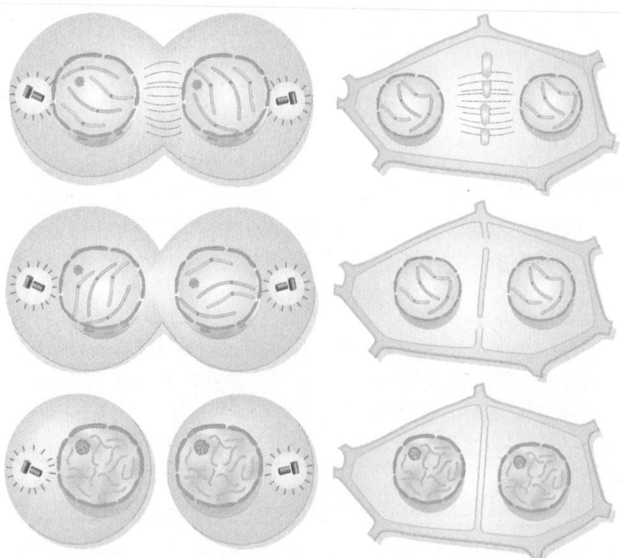

For a diploid (2N) cell, which has two sets of chromosomes (two genomes), chromosome duplication produces a cell in which each chromosome set has doubled its content. Mitosis thus produces two genetically identical progeny diploid cells, each with two sets of chromosomes (two genomes). As a result, an equal distribution of genetic material occurs, and no genetic material is lost.

**Keynote**

Mitosis is the process of nuclear division in eukaryotes. It is one part of the cell cycle ($G_1$, S, $G_2$, and M), and it results in the production of daughter nuclei that contain identical chromosome numbers and that are genetically identical to one another and to the parent nucleus from which they arose. Before mitosis, the chromosomes duplicate. Mitosis usually is followed by cytokinesis. Both haploid and diploid cells proliferate by mitosis.

### Meiosis

**Meiosis** is the two successive divisions of a *diploid* nucleus after only one DNA replication (chromosome duplication) cycle. The original diploid nucleus contains one haploid set of chromosomes from the mother and one set from the father (with the exception of self-fertilizing organisms—for example, many plants—in which both sets of chromosomes come from the same parent). Meiosis occurs only at a special point in an organism's life cycle. In animals, it results in the formation of haploid gametes (eggs and sperm by **gametogenesis**); in plants, it results in the formation of haploid meiospores (by *sporogenesis*). (A meiospore undergoes mitosis to produce a gamete-bearing, multicellular stage called the *gametophyte*.) Before meiosis, the DNA that makes up homologous chromosomes replicates, and during meiosis these chromosomes pair and then undergo two divisions—meiosis I and meiosis II—each consisting of a series of stages (Figure 12.9). Meiosis I results in a reduction in the number of chromosomes in each cell from diploid to haploid (reductional division—each resulting pair of attached sister chromatids counts as a single chromosome), and meiosis II results in the separation of the sister chromatids. As a consequence, each of the four nuclei that come about from the two meiotic divisions receives one chromosome of each chromosome set (that is, one complete haploid genome). In most cases, the divisions are accompanied by cytokinesis, so the meiosis of a single diploid cell produces four haploid cells.

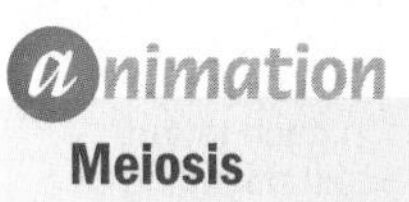


**Meiosis I: The First Meiotic Division.** **Meiosis I**, in which the chromosome number is reduced from diploid to haploid, consists of five stages: *prophase I, prometaphase I, metaphase I, anaphase I,* and *telophase I* (see Figure 12.9).

***Prophase I.*** When **prophase I** begins, the chromosomes have already duplicated, with each consisting of two sister chromatids attached at a centromere (see Figure 12.9). Prophase I is divided into a number of substages. Prophase I of meiosis is similar to prophase of mitosis. The important difference between prophase I of meiosis and prophase of mitosis is that homologous chromosomes pair with each other in meiosis, and crossing-over occurs only in meiosis.

In **leptonema** (early prophase I, the leptotene stage), the extended chromosomes begin to condense and become visible as long, thin threads. Once a cell enters leptonema, it is committed to the meiotic process.

In **zygonema** (early to middle prophase I, the zygotene stage), the chromosomes continue to condense. The homologous pairs of chromosomes actively find each other and align roughly along their lengths. Each pair of homologs then undergoes **synapsis**—the formation along the length of the chromatids of a zipperlike structure called the **synaptonemal complex,** which aligns the two homologs precisely, base pair for base pair.

The telomeres of chromosomes play a key role in the initiation of synapsis. That is, during meiosis I, telomeres are clustered on the nuclear envelope to produce an arrangement called a *bouquet* because of its resemblance to the stems from a bouquet of cut flowers. In some way, the telomeres move the chromosomes around so that homologous chromosomes align and undergo synapsis.

**Pachynema** (middle prophase I, the pachytene stage) starts when synapsis is completed. Because of the replication that occurred earlier, each synapsed set of homologous chromosomes consists of four chromatids and is called a **bivalent** or a **tetrad.** During pachynema, a most significant event for genetics occurs: **crossing-over**—the reciprocal physical exchange of chromosome segments at corresponding positions along pairs of homologous chromosomes (see Figure 12.9). The positions at which crossing-over occur along the chromosomes are largely random, and vary from one meiosis to another. The physical exchange that occurs in crossing-over is facilitated by the alignment of the homologous chromosomes brought about by the synaptonemal complex. If there are genetic differences between the homologs, crossing-over can produce new gene combinations in a chromatid. There is usually no loss or addition of genetic material to either chromosome, since crossing-over involves reciprocal exchanges. A chromosome that emerges from meiosis with a combination of alleles that differs from the combination with which it started is called a **recombinant chromosome.** Therefore, crossing-over is a mechanism that can give rise to **genetic recombination.** At the end of pachynema, the synaptonemal complex is disassembled, and the chromosomes have started to elongate.

In **diplonema** (middle to late prophase I, the diplotene stage), the synaptonemal complex disassembles and the homologous chromosomes begin to move apart. The result of crossing-over becomes visible during diplonema as a cross-shaped structure called a **chiasma** (plural, *chiasmata*; see Figures 12.9 and 12.10). At each chiasma, the homologous chromosomes are very tightly associated. Because all four chromatids may be involved in crossing-over events along the length of the homologs, the chiasma pattern at this stage may be quite complex.

Figure 12.9

**The stages of meiosis in an animal cell.**

**Early prophase I**

Chromosomes, already duplicated, become visible. Centriole pairs begin separation and a spindle forms between them.

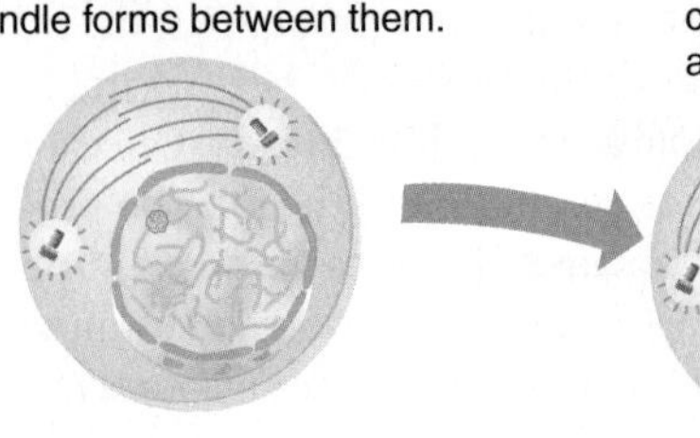

**Middle prophase I**

Homologous chromosomes shorten and thicken. The chromosomes synapse and crossing-over occurs.

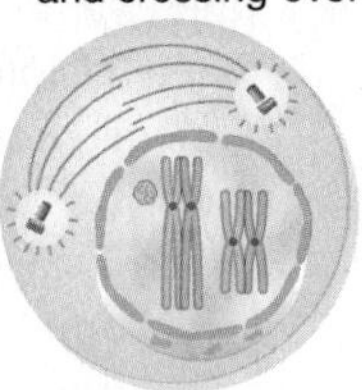

**Late prophase I/Prometaphase I**

Results of crossing-over become visible as chiasmata. Nuclear envelope breaks down. Meiotic spindle enters the former nuclear area. Kinetochore microtubules attach to the chromosomes.

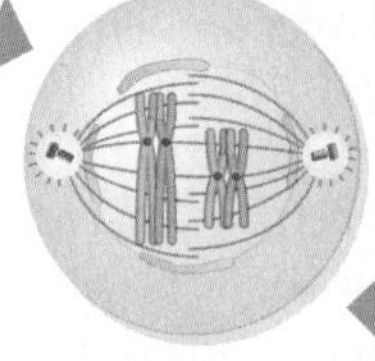

**Metaphase I**

Kinetochore microtubules align each chromosome pair (the tetrads) on the metaphase plate.

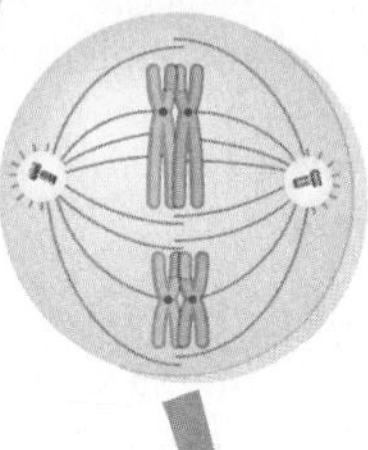

**Anaphase I**

Chromosomes in each tetrad separate and begin migrating toward opposite poles.

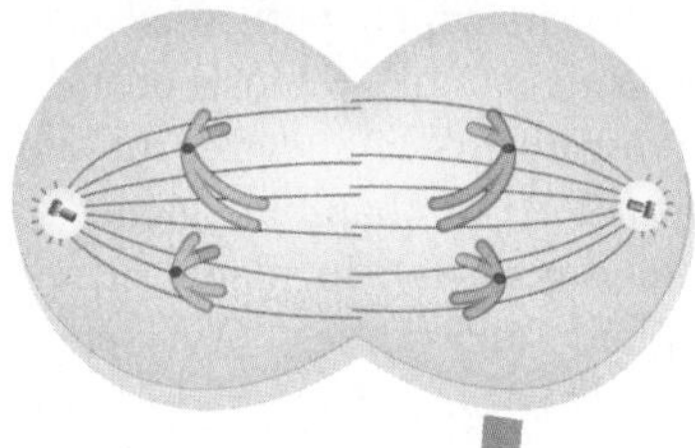

**Telophase I**

Chromosomes (each with two sister chromatids) complete migration to the poles and new nuclear envelopes may form. (Other sorting patterns are possible.)

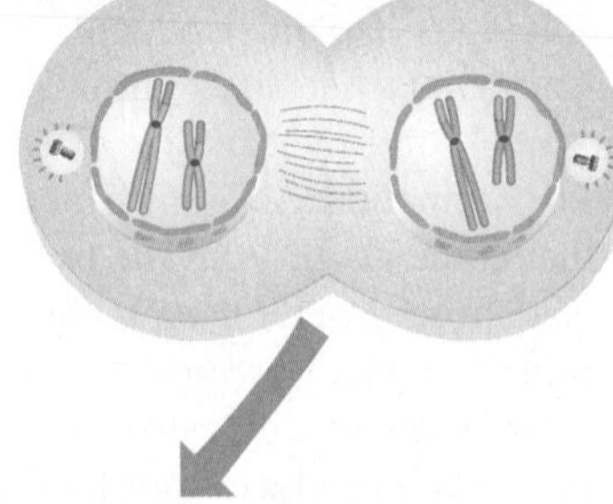

**Cytokinesis**

In most species, cytokinesis occurs to produce two cells. Chromosomes do not replicate before meiosis II.

**Prophase II**

**Metaphase II**

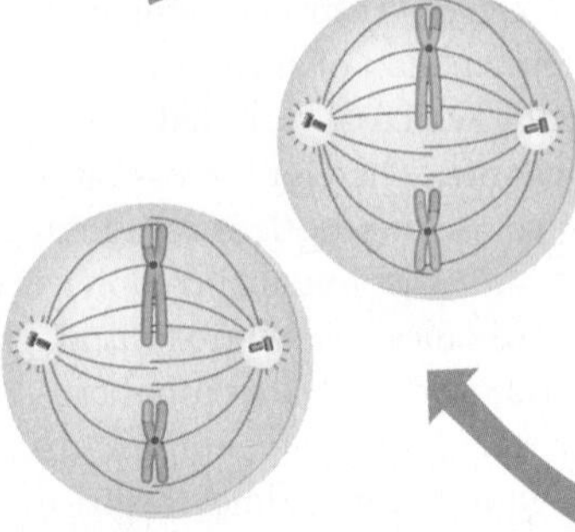

**Anaphase II**

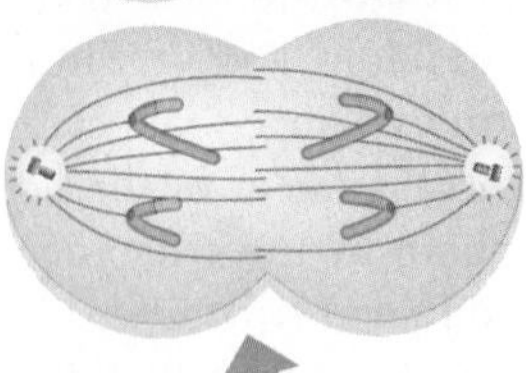

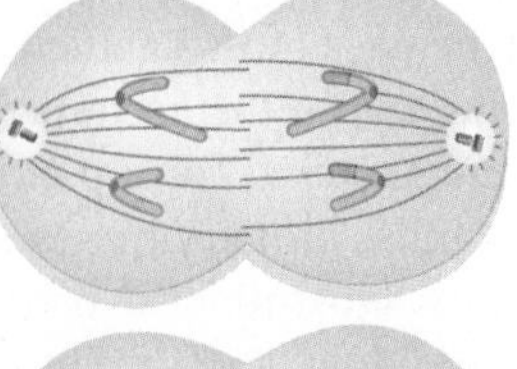

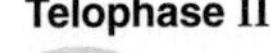

**Telophase II**

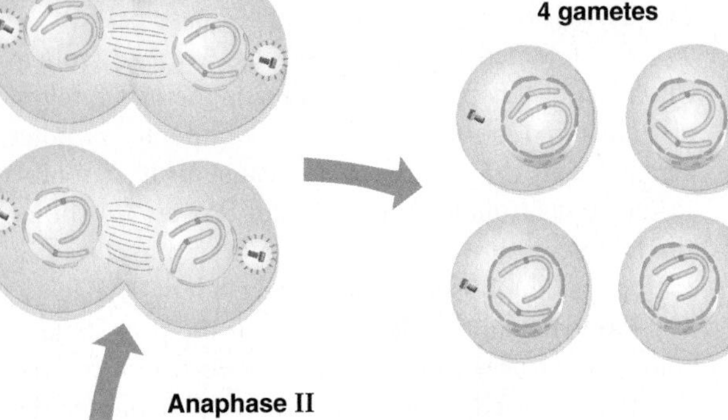

**4 gametes**

Figure 12.10

**Appearance of chiasmata, the visible evidence of crossing-over, in diplonema.**

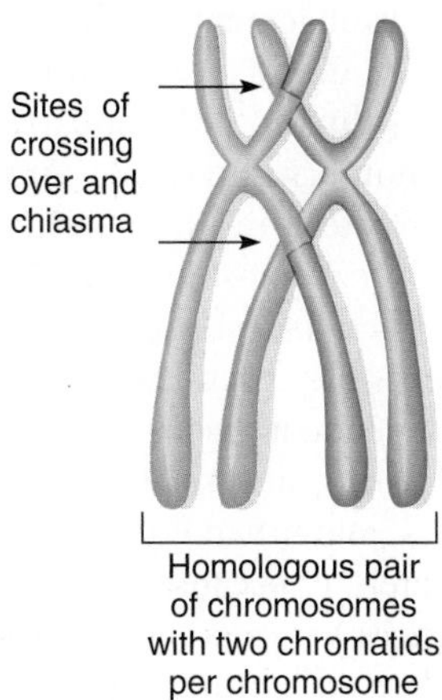


In most organisms, diplonema is followed rapidly by the remaining stages of meiosis. However, in many animals, the oocytes (egg cells) can remain in diplonema for very long periods. In human females, for example, oocytes go through meiosis I up to diplonema by the seventh month of fetal development and then remain arrested in this stage for many years. At the onset of puberty and until menopause, one oocyte per menstrual cycle completes meiosis I and is ovulated. If the oocyte is fertilized by a sperm as it passes down the fallopian tube, it quickly completes meiosis II, and, by fusion with a haploid sperm, a functional zygote is produced.

In **diakinesis** (late prophase I), the chromosomes condense even more, making it now possible to see the four members of the tetrads. The chiasmata are clearly visible at this stage.

The synapsis and crossing-over phenomena that take place in prophase I apply to homologous chromosomes—namely, the autosomes. Even though the sex chromosomes are not homologous, the Y chromosome of eutherian (placental) mammals has small regions at each end that are homologous to regions on the X chromosome. These *pseudoautosomal regions* (PARs) pair in male meiosis, and crossing-over occurs between them. When the PAR is deleted from the short arm of the Y chromosome, pairing between the X and Y chromosomes does not occur, and the male is sterile. Thus, pairing and crossing-over of the PARs have been considered necessary for the correct segregation of X and Y chromosomes as meiosis proceeds. Interestingly, the genes found in the PARs are variable, even among primates. Even the mouse and human PARs are completely different. PARs are not found in all mammals, however: PARs are absent from some rodents and from all marsupial chromosomes, and the X and Y chromosomes of these animals do not pair or show crossing-over in meiosis. Still, the X and Y chromosomes segregate normally in marsupial meiosis, indicating that a PAR is not essential for sex chromosome pairing and male fertility in these mammals.

***Prometaphase I.*** In **prometaphase I**, the nucleoli disappear, the nuclear envelope breaks down, and the meiotic spindle that has been forming between the separating centriole pairs enters the former nuclear area (see Figure 12.9). As in mitosis, kinetochore microtubules attach to the chromosomes; that is, kinetochore microtubules from one pole attach to both sister kinetochores of one duplicated chromosome, and kineteochore microtubules from the other pole attach to both sister kinetochores of the other duplicated chromosome in a tetrad. Nonkinetochore microtubules from each pole overlap in the center of the cell.

***Metaphase I.*** In **metaphase I** (see Figure 12.9), the kinetochore microtubules align the tetrads on the metaphase plate. Importantly, the *pairs* of homologs (the tetrads) are found at the metaphase plate. In contrast, in mitosis, replicated homologous chromosomes (sister chromatid pairs) align *independently* of one another at the metaphase plate (compare Figures 12.5 and 12.9).

***Anaphase I.*** In **anaphase I** (see Figure 12.9), the chromosomes in each tetrad separate, so the chromosomes of each homologous pair disjoin and migrate toward opposite poles, the areas in which new nuclei will form. (At this stage, each of the separated chromosomes is called a *dyad*.) This migration assumes that maternally derived and paternally derived centromeres segregate randomly to each pole (of course, parts of chromosomes may have exchanged during the crossing-over process) and that, at each pole, there is a haploid complement of replicated centromeres with associated sister chromatids. *At this time, homologous chromosomes have segregated from each other, but sister chromatids remain attached at their respective centromeres. In other words, a key difference between meiosis I and mitosis is that sister chromatids remain joined after metaphase in meiosis I, whereas they separate in mitosis.*

***Telophase I.*** In **telophase I** (see Figure 12.9), the dyads complete their migration to opposite poles of the cell and the spindle disassembles. In some species, but not all, new nuclear envelopes form around each haploid grouping. In most species, cytokinesis follows telophase I, producing two haploid cells.

Thus, meiosis I, which begins with a diploid cell that contains one maternally derived and one paternally derived set of chromosomes, ends with two nuclei, each of which is haploid and contains one mixed-parental set of dyads. After cytokinesis, each of the two progeny cells has a nucleus with a haploid set of dyads.

**Meiosis II: The Second Meiotic Division.** No DNA replication occurs between meiosis I and meiosis II. **Meiosis II** is similar to a mitotic division (see Figure 12.9).

In **prophase II** (see Figure 12.9), the chromosomes condense and a spindle forms.

In **prometaphase II** (not shown), the nuclear envelopes (if formed in telophase I) break down, and the spindle organizes across the cell. Kinetochore microtubules from the opposite poles attach to the kinetochores of each chromosome.

In **metaphase II** (see Figure 12.9), the movement of the kinetochore microtubules aligns the chromosomes on the metaphase plate.

During **anaphase II** (see Figure 12.9), the centromeres separate, and the now-daughter chromosomes are pulled to the opposite poles of the spindle. One sister chromatid of each pair goes to one pole, and the other goes to the opposite pole. The separated chromatids are now considered chromosomes in their own right.

In the last stage, **telophase II** (see Figure 12.9), the chromosomes begin decondensing, a nuclear envelope forms around each set of chromosomes, and cytokinesis takes place. After telophase II, the chromosomes continue decondensing, eventually becoming invisible under the light microscope.

The end products of the two meiotic divisions are four haploid cells (gametes in animals) from one original diploid cell. Each of the four progeny cells has one chromosome from each homologous pair of chromosomes. Because of crossing-over, these chromosomes are not exact copies of the original chromosomes.

**Gene Segregation in Meiosis.** Meiosis has three significant results:

1. Meiosis generates haploid nuclei with half the number of chromosomes found in the diploid cell that entered the process. This occurs because two division cycles follow only one cycle of DNA replication (the S period). The fusion of haploid nuclei in fertilization restores the diploid number. Therefore, through a cycle of meiosis and fusion, the chromosome number is maintained in sexually reproducing organisms.
2. In metaphase I, each maternally derived chromosome and each paternally derived chromosome has an equal chance of aligning on one or the other side of the equatorial metaphase plate. (The random alignment of maternal and paternal chromosomes is the basis of the independent assortment of genes—Mendel's second law—described in Chapter 11.) As a result, each nucleus generated by meiosis usually has some combination of maternal and paternal chromosomes. (Due to the earlier crossing-over, the chromosomes are a mixture of paternal and maternal sequences. For simplicity, we call the chromosome that contains a maternal centromere a *maternal chromosome*, and the chromosome that contains a paternal centromere a *paternal chromosome*.)

   The number of possible chromosome combinations in the haploid nuclei resulting from the independent assortment of chromosomes in meiosis is large, especially when the number of chromosomes in an organism is large. Consider a hypothetical organism with two pairs of chromosomes in a diploid cell entering meiosis. Figure 12.11 shows the two combinations of maternal and paternal chromosomes that can occur at the metaphase plate. Understanding this concept is useful in considering gene segregation (see Chapter 11).

   The general formula for the number of possible chromosome arrangements at the metaphase plate in meiosis is $2^{n-1}$, where $n$ is the number of chromosome pairs in a diploid cell. Similarly, the general formula for the number of possible chromosome combinations in the nuclei resulting from the independent assortment of chromosomes in meiosis is $2^n$. In *Drosophila,* which has four pairs of chromosomes, the number of possible combinations in nuclei resulting from the independent assortment of chromosomes in meiosis is $2^3$, or 8; in humans, which have 23 chromosome pairs, more than 4 million combinations are possible. Therefore, because there are many possible allele differences between the maternally derived and paternally derived chromosomes, the nuclei produced by meiosis are genetically quite different from the parental cell and from one another.
3. The crossing-over between maternal and paternal chromatid pairs during meiosis I generates still more variation in the final combinations. Crossing-over occurs during every meiosis,[1] and because the sites of crossing-over vary from one meiosis to another, the number of different kinds of progeny nuclei produced by the process is extremely large. That is, with the exception of identical twins, the genome sequence of a human individual almost certainly has never existed before and will not exist again in the future by natural means.

Given its genetic features, an understanding of meiosis is of critical importance for understanding the behavior of genes. Indeed, the events that occur in meiosis are the bases for the segregation and independent assortment of genes according to Mendel's laws, discussed in Chapter 11. This chapter's Focus on Genomics box discusses how genomics revealed some of the genes whose functions are involved in meiotic chromosome segregation.

**Keynote**

Meiosis occurs in all sexually reproducing eukaryotes. It is a process by which a specialized diploid (2N) cell or cell nucleus with two sets of chromosomes is transformed, through one round of chromosome replication and two rounds of nuclear division, into four haploid (N) cells or nuclei, each with one set of chromosomes. In the first of two divisions, pairing, synapsis, and the crossing-over of homologous chromosomes occur. The meiotic process, in combination with fertilization, conserves the number of chromosomes from generation to generation. It also generates genetic variability through the various ways in which maternal and paternal chromosomes are combined in the progeny nuclei and by crossing-over (the physical exchange of chromosome segments at corresponding positions along pairs of homologous chromosomes).

[1]There are some exceptions. For instance, there is no crossing-over in meiosis in *Drosophila* males.

## Focus on Genomics

### Genes Involved in Meiotic Chromosome Segregation

In Chapter 9. p. 220, we discussed the problem of the *FUN* (*f*unction *un*known) genes of the yeast, *Saccharomyces cerevisiae. FUN* genes are those identified from analysis of the yeast genome that are homologous to genes of other organisms in sequence databases but whose functions are unknown. The related species, *Schizosacchromyces pombe* (fission yeast), has many *FUN* genes as well. The genomics techniques described below were employed to identify the meiotic functions of the products of several of these genes.

DNA microarray analysis (see Chapter 8, pp. 192–194, and Chapter 9, pp. 230–233) was used to find genes that were expressed more in the middle of sporulation (as meiosis occurs) than at other times. Nearly 200 genes were found that had this expression profile as well as no known function. The investigators then knocked out (see Chapter 9, pp. 221–224) each of these genes individually—that is, deleted the reading frame of the gene so its function was lost completely—and analyzed the result of the knockout mutation. Two of the knockouts totally disrupted meiosis. One of the genes so identified, *sgo1*$^+$, encodes a protein of the shugoshin family. Members of this family prevent separation of the sister chromatids in both mitosis and meiosis. The protein appears to protect another protein, cohesin, that holds the sister chromatids together. Meiotically, both the cohesin and shugoshin proteins are required for the sister chromatids to stay attached at the centromere after anaphase I. Disruption of this gene in *S. pombe* results in major failures in meiotic chromosome segregation. A second gene was also found to be required for proper chromosome segregation. This gene, *mde2*, encodes a protein that is required for the formation of double-strand breaks in the DNA. These double-strand breaks are crucial for crossing-over and the formation of the resulting chiasmata, and chiasmata are required for proper meiotic chromosome segregation in nearly all organisms. This functional genomic approach identified one protein that held sister chromatids together and another protein that was required for the formation of chiasmata. Both processes are needed for normal chromosome segregation in meiosis.

**Meiosis in Animals and Plants.** Lastly, we discuss briefly the role of meiosis in animals and plants.

***Meiosis in Animals.*** Most multicellular animals are diploid through most of their life cycles. In such animals, meiosis produces haploid gametes, the fusion of two haploid gametes produces a diploid zygote when their nuclei fuse in fertilization, and the zygote then divides by mitosis to produce the new diploid organism; this series of events, involving an alternation of diploid and haploid phases, is **sexual reproduction.** Thus the gametes are the only haploid stages of the life cycle. Gametes are formed only in specialized cells. In males, the gamete is the sperm, produced through a process called **spermatogenesis;** in females, the gamete is the egg, produced by **oogenesis** (Figure 12.12).

In male animals, the **sperm cells (spermatozoa)** are produced within the testes, which contain the primordial germ cells (*primary spermatogonia*). Via mitosis, the primordial germ cells produce *secondary spermatogonia*, which then transform into *primary spermatocytes* (*meiocytes*), each of which undergoes meiosis I and gives rise to two *secondary spermatocytes*. Each secondary spermatocyte in turn undergoes meiosis II. The results of these two divisions are four haploid *spermatids* that eventually differentiate into the male gametes: the spermatozoa.

In female animals, the ovary contains the primordial germ cells (*primary oogonia*) that, by mitosis, give rise to *secondary oogonia*. These cells transform into **primary oocytes,** which grow until the end of oogenesis. The diploid primary oocyte goes through meiosis I and unequal cytokinesis to give two cells: a large one called the **secondary oocyte** and a very small one called the *first*

**Figure 12.11**

**The two possible arrangements of two pairs of homologous chromosomes on the metaphase plate of the first meiotic division.** Paternal chromosomes are shown in yellow and green, maternal chromosomes in blue and red.

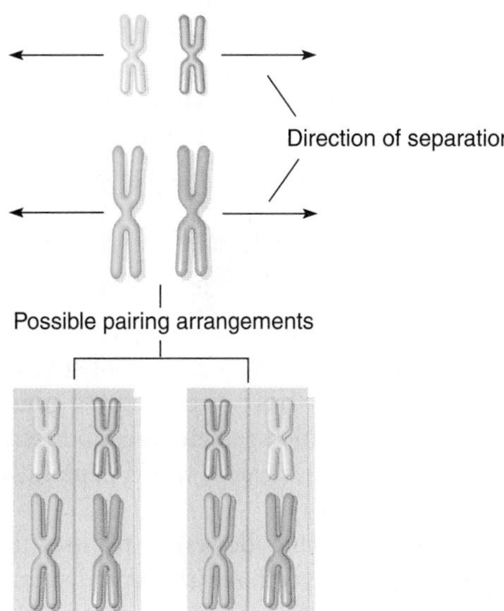

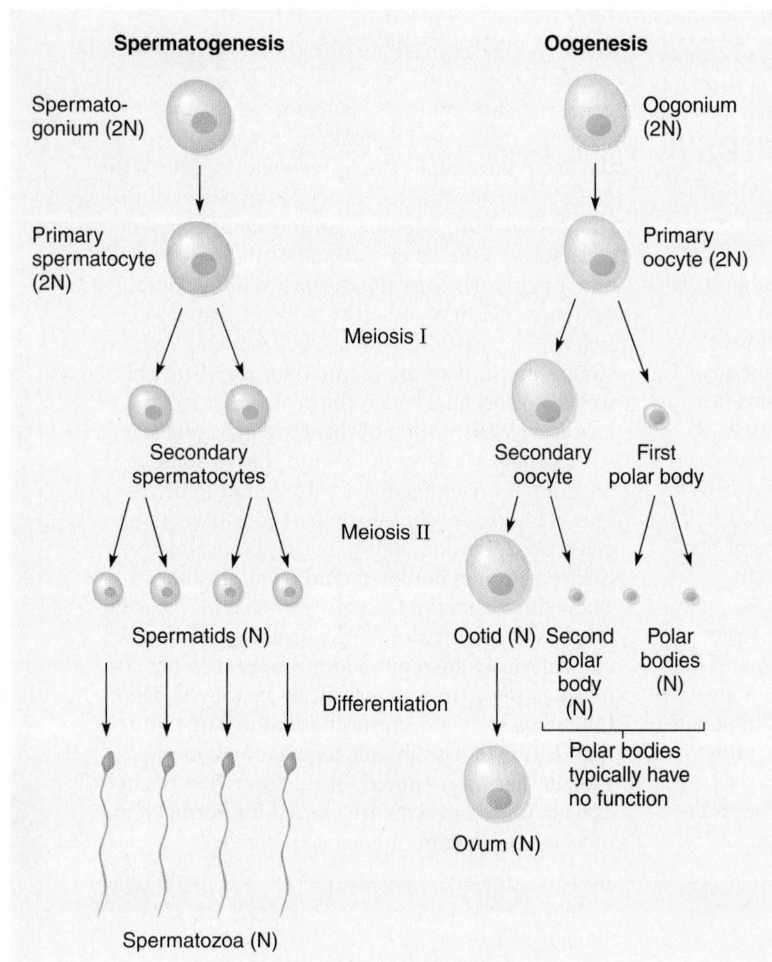


**Figure 12.12**
**Spermatogenesis and oogenesis in an animal cell.**

*polar body*. In meiosis II, the secondary oocyte produces two haploid cells. One is a very small cell called a *second polar body;* the other is a large cell that rapidly matures into the mature egg cell, or **ovum.** The first polar body may or may not divide during meiosis II. The polar bodies have no function in most species and degenerate; only the ovum is a viable gamete. (In many animals, including humans, the cell that is actually fertilized is the secondary oocyte; however, nuclear fusion must await completion of meiosis by that oocyte.) Thus, in the female animal, only one mature gamete (the ovum) is produced by meiosis of a diploid cell. In humans, all oocytes are formed in the fetus, and one oocyte completes meiosis I each month in the adult female, but does not progress further unless stimulated to do so through fertilization by a sperm.

***Meiosis in Plants.*** The life cycle of sexually reproducing plants typically has two phases: the **gametophyte** or haploid stage, in which gametes are produced, and the **sporophyte** or diploid stage, in which haploid spores are produced by meiosis.

In angiosperms (the flowering plants), the flower is the structure in which sexual reproduction occurs. Figure 12.13 shows a generalized flower containing both male and female reproductive organs—the **stamens** and **pistils,** respectively. Each stamen consists of a single stalk—the filament—on top of which is an *anther*. Pollen grains, which are immature male gametophytes (formed in the gamete-producing phase), are released from the anther. The pistil, which contains the female gametophytes,

**Figure 12.13**
**Generalized structure of a flower.**

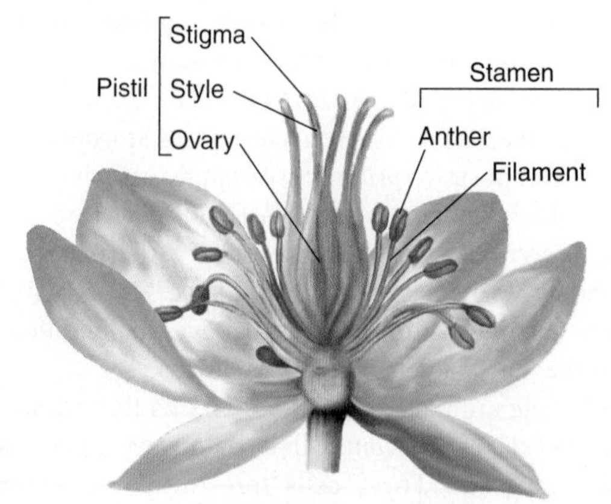

typically consists of the *stigma*, a sticky surface specialized to receive the pollen; the *style*, a thin stalk down which a pollen tube grows from a pollen grain that adheres to the stigma; and, at the base of the structure, the *ovary*, within which are the ovules. Each ovule encloses a female gametophyte (the embryo sac) containing a single egg cell. When the egg cell is fertilized, the ovule develops into a seed.

Among living organisms, only plants produce gametes from special bodies called gametophytes. Thus plant life cycles have two distinct reproductive phases, called the **alternation of generations** (Figure 12.14). Meiosis and fertilization are the transitions between these stages. The haploid *gametophyte generation* begins with spores that are produced by meiosis. In flowering plants, the spores are the cells that ultimately become pollen and the embryo sac. Fertilization initiates the diploid *sporophyte generation,* which produces the specialized haploid cells called spores, completing the cycle.

## Chromosome Theory of Inheritance

Around the turn of the twentieth century, cytologists had established that, within a given species, the total number of chromosomes is constant in all cells, whereas the chromosome number varies widely among species (Table 12.1). In 1902, Walter Sutton and Theodor Boveri independently recognized that the transmission of chromosomes from one generation to the next closely paralleled the pattern of inheritance of Mendelian factors from one generation to the next. This correlation has become known as the **chromosome theory of inheritance.** The theory states that genes are located on chromosomes. In this section, we consider some of the evidence cytologists and geneticists obtained to support that theory.

**Figure 12.14**
**Alternation of gametophyte and sporophyte generations in flowering plants.**

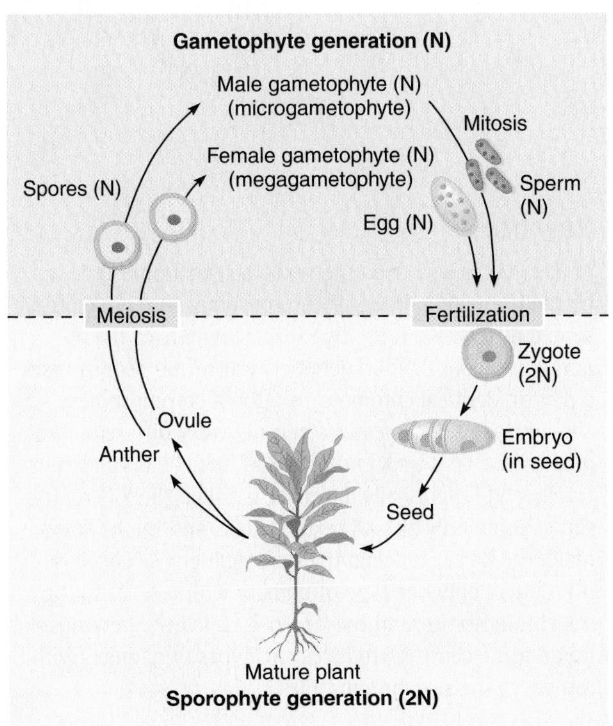


**Table 12.1 Chromosome Number in Various Organisms**[a]

| Organism | Total Chromosome Number |
|---|---|
| Human | 46 |
| Chimpanzee | 48 |
| Dog | 78 |
| Cat | 72 |
| Mouse | 40 |
| Horse | 64 |
| Chicken | 78 |
| Toad | 36 |
| Goldfish | 94 |
| Starfish | 36 |
| Fruit fly (*Drosophila melanogaster*) | 8 |
| Mosquito | 6 |
| Australian ant (*Myrecia pilosula*) | ♂1, ♀2 |
| Nematode | ♂11, ♀12 |
| *Neurospora* (haploid) | 7 |
| Sphagnum moss (haploid) | 23 |
| Field horsetail | 216 |
| Giant sequoia | 22 |
| Tobacco | 48 |
| Cotton | 52 |
| Potato | 48 |
| Tomato | 24 |
| Bread wheat | 42 |
| Yeast (*Saccharomyces cerevisiae*) (haploid) | 16 |

[a]Except as noted, all chromosome numbers are for diploid cells.

### Sex Chromosomes

Support for the chromosome theory of inheritance came from experiments that related the hereditary behavior of particular genes to the transmission of the *sex chromosomes*, which, as you will recall, are the chromosomes in eukaryotes that are represented differently in the two sexes. The other chromosomes in eukaryotes, which are not represented differently in the two sexes, are the *autosomes*. In eukaryotes with sex chromosomes, the sex chromosomes and autosomes are found in all cells. It is a misconception to think that sex chromosomes are found only in gametes.

Sex chromosomes were discovered in the early 1900s when Clarence E. McClung, Nettie Stevens, and Edmund B. Wilson, all experimenting with insects, independently obtained evidence that particular chromosomes determined the sex of an organism. In one of those studies, performed in 1905, Stevens found that, in grasshoppers, the female has an even number of chromosomes while the male has an odd number. There are two copies of one of the chromosomes in the female but only one copy in the male. Stevens called the extra chromosome an **X chromosome.** Since this chromosome is directly related to the sex of the organism, the X chromosome is an example of a sex chromosome. The sex of progeny grasshoppers, then, is determined by whether a sperm contains an X chromosome. All eggs have one X chromosome. If the sperm carries an X chromosome, then the resulting fertilized egg will have a pair of X chromosomes and will give rise to a female. If the sperm does not have an X, the fertilized egg will have an unpaired X and will give rise to a male.

Unlike grasshoppers, some insects have two different types of sex chromosomes. For example, Stevens found that in the common mealworm, *Tenebrio molitor,* the male has a partner chromosome for the X chromosome. That partner is much smaller than and clearly distinguishable from the X chromosome. Stevens called the partner chromosome the **Y chromosome,** and like the X chromosome, it is a sex chromosome. The sperm cells of the mealworm contain either an X or a Y chromosome, and the sex of the offspring is determined by the type of sperm that fertilizes the X-chromosome-bearing egg: XX mealworms are female, and XY mealworms are male.

Similar X–Y sex chromosome complements are found in other organisms, including humans and the fruit fly, *Drosophila melanogaster.* In most cases, the female has two X chromosomes (she is XX with respect to the sex chromosomes), and the male has one X chromosome and one Y chromosome (he is XY). Figure 12.15a shows male and female *Drosophila,* and Figure 12.15b shows the chromosome sets of the two sexes. Because the male produces two kinds of gametes with respect to sex chromosomes (X or Y), the male is called the **heterogametic sex,** and because the female produces only one type of gamete (X), the female is called the **homogametic sex.** In *Drosophila*, the X and Y chromosomes are similar in size, but their shapes are different. (Note that in some organisms the male is homogametic and the female is heterogametic.)

The pattern of transmission of X and Y chromosomes from generation to generation is straightforward (Figure 12.16). In this figure, the X is represented by a straight structure much like a forward slash mark, and the Y is represented by a similar structure topped by a hook to the right. The female produces only X-bearing gametes, and the male produces both X-bearing and Y-bearing gametes. Random fusion of male and female gametes produces progeny with 1/2 XX (female) and 1/2 XY (male) flies.

**Figure 12.15**
***Drosophila melanogaster* (fruit fly), an organism used extensively in genetics experiments.**

**a) Female (left) and male (right) adult *Drosophila* (top), and ventral views of their abdomens (bottom).**

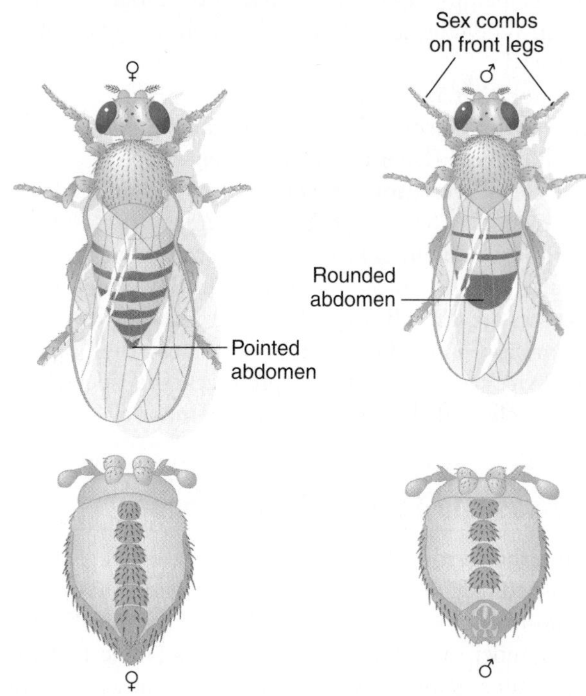


**b) Chromosomes of female (left) and male (right) *Drosophila*.**

## Keynote

In eukaryotes with separate sexes, a sex chromosome is a chromosome or a group of chromosomes that are represented differently in the two sexes. In many of the organisms encountered in genetic studies, one sex possesses a pair of identical chromosomes (the X chromosomes). The opposite sex possesses a pair of visibly different chromosomes: One is an X chromosome, and the other, structurally and functionally different, is called the Y chromosome. Commonly, the XX sex is female, and the XY sex is male. The XX sex is called the homogametic sex because it produces only one type of gamete with respect to the sex chromosomes, and the XY sex is called the heterogametic sex because it produces two types of gametes with respect to the sex chromosomes.

**Figure 12.16**

**Inheritance pattern of X and Y chromosomes in organisms where the female is XX and the male is XY.**

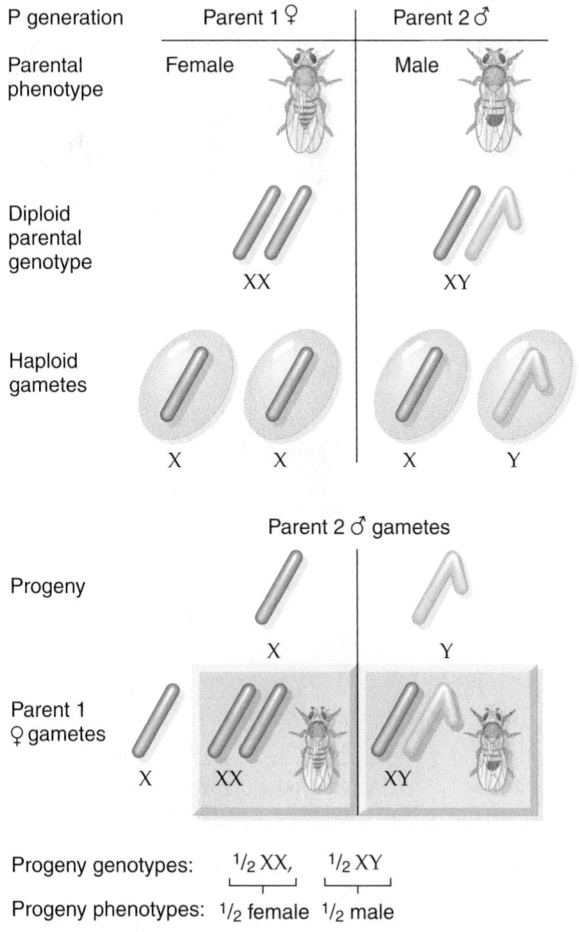

## Sex Linkage

Evidence to support the chromosome theory of heredity came in 1910 when Thomas Hunt Morgan of Columbia University reported the results of genetics experiments with *Drosophila*. Morgan received the 1933 Nobel Prize in Physiology or Medicine for "his discoveries concerning the role played by the chromosome in heredity."

In one of his true-breeding strains, Morgan found a male fly that had white eyes instead of the brick red eyes characteristic of the **wild type.** The term *wild type* refers to a strain, an organism, or an allele that is most prevalent in the "wild" population of the organism with respect to genotype and phenotype. For example, a *Drosophila* strain with all wild-type alleles of genes that determine eye color has brick red eyes. Variants of a wild-type strain arise from mutational changes of the wild-type alleles that produce **mutant alleles;** the result is strains with mutant characteristics. Mutant alleles may be recessive or dominant to the wild-type allele; for example, the mutant allele that causes white eyes in *Drosophila* is recessive to the wild-type (red eye) allele.

Morgan crossed the white-eyed male with a red-eyed female from the same strain and found that all the $F_1$ flies were red-eyed. He concluded that the white-eyed trait was recessive. Next, he allowed the $F_1$ progeny to interbreed and counted 3,470 red-eyed and 782 white-eyed flies in the $F_2$ generation. (The number of individuals with the recessive phenotype was too small to fit the Mendelian 3:1 ratio. Later, Morgan determined that the lower-than-expected number of flies with the recessive phenotype was the result of the lower viability of white-eyed flies). In addition, *Morgan noticed that all the white-eyed flies were male.* This was a novel result; that is, in all other genetic crosses performed with other mutants up until then, the mutant phenotype had never been confined just to one sex.

Figure 12.17a diagrams the crosses. The *Drosophila* gene symbolism used here is different from the symbolism we adopted for Mendel's crosses and is described in Box 12.1 The *Drosophila* symbolism is more typical of what is used in most genetic systems, and you should understand it before proceeding with this discussion. As we continue, note that the mother–son inheritance pattern presented in Figure 12.17 is the result of the segregation of genes located on a sex chromosome.

Morgan proposed that the gene for the eye color variant is located on the X chromosome. The condition of X-linked genes in males is said to be **hemizygous,** because the gene is present only once in the organism and there is no homologous gene on the Y. For example, the white-eyed *Drosophila* males have an X chromosome with a white allele and no other allele of that gene in their genomes; these males are hemizygous for the white allele. Because the white allele of the gene is recessive, the original white-eyed male must have had the recessive allele for white eyes (designated *w;* see Box 12.1) on his X chromosome. The red-eyed female came from a true-breeding strain, so both of her X chromosomes must have carried the dominant allele for red eyes, $w^+$ ("w plus").

The $F_1$ flies are produced in the following way (see Figure 12.17a): The males receive their only X chromosome from their mother and hence have the $w^+$ allele and are red-eyed. The $F_1$ females receive a dominant $w^+$ allele from their mother and a recessive *w* allele from their father, making them also red-eyed.

In the $F_2$ produced by interbreeding the $F_1$ flies, the males that received an X chromosome with the *w* allele from their mother are white-eyed; those that received an X chromosome with the $w^+$ allele are red-eyed (see Figure 12.17a). The gene transmission shown in this cross—from a male parent to a female offspring ("child") to a male "grandchild"—is called **crisscross inheritance.**

Morgan also crossed a true-breeding white-eyed female (homozygous for the *w* allele) with a red-eyed male (hemizygous for the $w^+$ allele; Figure 12.17b). This cross is the *reciprocal cross* of Morgan's first cross—white male × red female—shown in Figure 12.17a. All the $F_1$ females receive a $w^+$-bearing X from their father and a $w^+$-bearing X from their mother (see Figure 12.17b). Consequently,

**Figure 12.17**

**The X-linked inheritance of red eyes and white eyes in *Drosophila melanogaster*.** The symbols *w* and $w^+$ indicate the white- and red-eyed alleles, respectively. The figure shows the difference in outcomes for reciprocal crosses.

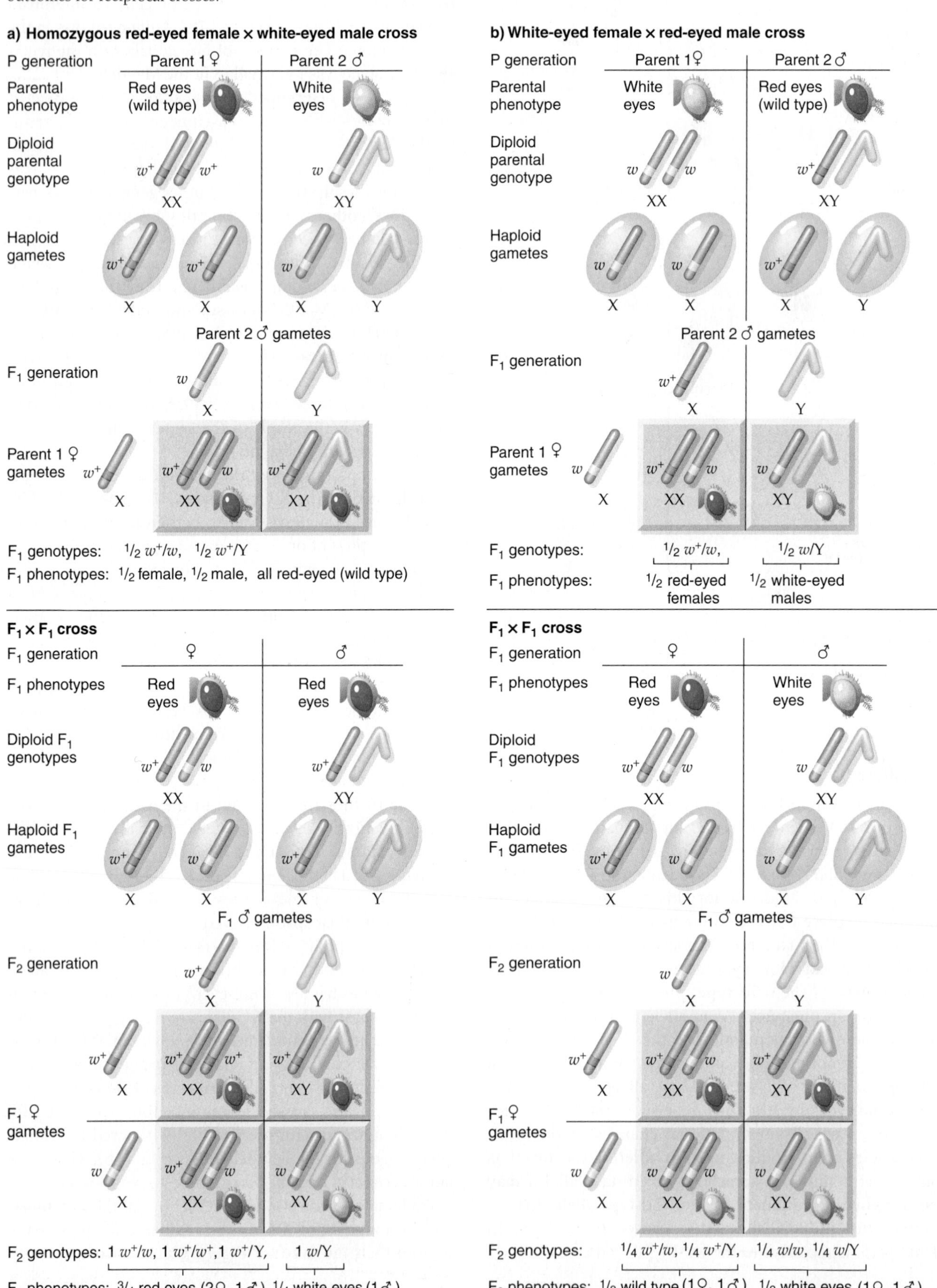

**Box 12.1 Genetic Symbols Revisited**

Unfortunately, no single system of gene symbols has been adopted by geneticists; the gene symbols used for *Drosophila* are different from those used for peas in Chapter 11. The *Drosophila* symbolism is commonly, but not exclusively, employed in genetics today. In this system, the symbol [+] indicates a wild-type allele of a gene. A mutant allele that is recessive to the wild-type allele of a particular gene is designated by a lowercase letter or letters. A mutant allele that is dominant to the wild-type allele of a particular gene is designated by an uppercase letter, or an initial uppercase letter followed by a lowercase letter or letters. *The letters are chosen based on the phenotype of the organism expressing the mutant allele.* For example, a variant strain of *Drosophila* has bright orange eyes instead of the usual brick red. The mutant allele involved is recessive to the wild-type brick red allele, and because the bright orange eye color is close to vermilion in tint, the allele is designated *v* and is called the vermilion allele. The wild-type allele of *v* is $v^+$, but when there is no chance of confusing it with other genes in the cross, it is often shortened to +. In the "Mendelian" terminology used up to now, the recessive mutant allele would be *v*, and its wild-type allele would be *V*.

A conventional way to represent the chromosomes (instead of the way we have been doing so in the figures) is to use a slash (/). Thus $v^+/v$ or +/*v* indicates that there are two homologous chromosomes, one with the wild-type allele ($v^+$ or +) and the other with the recessive allele (*v*). The Y chromosome usually is symbolized as a Y or a bent slash /̸. Morgan's cross of a true-breeding red-eyed female fly with a white-eyed male could be written $w^+/w^+ \times w/Y$ or +/+ × *w*/̸.

The same rules apply when the alleles involved are dominant to the wild-type allele. For instance, some *Drosophila* mutants, called *Curly*, have wings that curl up at the end rather than the normal straight wings. The symbol for this mutant allele is *Cy*, and the wild-type allele is $Cy^+$ or + in the shorthand version. Thus, a heterozygote would be $Cy^+/Cy$ or +/*Cy*.

In the rest of the book, the *A*/*a* ("Mendelian"), $a^+/a$ (*Drosophila*), and other symbols will be used. Because it is easier to verbalize the "Mendelian" symbols (e.g., big *A*, little *a*), many of our examples follow that symbolism, even though the *Drosophila* symbolism in many ways is more informative and is more typical of the symbolism used in many genetic systems. That is, with the *Drosophila* system, the wild-type and mutant alleles are readily apparent because the wild-type allele is indicated by $a^+$. The "Mendelian" system is still used in animal and plant breeding. A good reason for this is that, after many years (sometimes centuries) of breeding, it is no longer apparent what the "normal" (wild-type) allele is.

---

they are heterozygous $w^+/w$ and have red eyes. All the $F_1$ males receive a *w*-bearing X from their mother and a Y from their father, so they have white eyes (see Figure 12.17b). This result is distinct from that of the cross in Figure 12.17a. Furthermore, all the results obtained are different from the normal results of a reciprocal cross because of the inheritance pattern of the X chromosome.

Interbreeding of the $F_1$ flies (see Figure 12.17b) involves a *w*/Y male and a $w^+/w$ female, giving approximately equal numbers of male and female red- and white-eyed flies in the $F_2$. This ratio differs from the approximately 3:1 ratio of red-eyed : white-eyed flies obtained in the first cross, in which none of the females and approximately half the males exhibited the white-eyed phenotype. The difference in phenotypic ratios in the two sets of crosses reflects the transmission patterns of sex chromosomes and the genes they contain.

Morgan's crosses of *Drosophila* involved eye-color characteristics that we now know are coded for by a gene found on the X chromosome. These characteristics and the genes that give rise to them are called **sex-linked**—or, more correctly, **X-linked**—because the gene locus is part of the X chromosome. *X-linked inheritance* is the term used for the pattern of hereditary transmission of X-linked genes. When the results of reciprocal crosses are not the same, and different ratios are seen for the two sexes of the offspring, a sex-linked trait may well be involved. By comparison, the results of reciprocal crosses are always the same when they involve genes located on the autosomes, with the same distribution of dominant and recessive phenotypes in males and females. Most significantly, Morgan's results that the inheritance pattern of the *w* gene paralleled the inheritance pattern of the X chromosome strongly supported the hypothesis that genes were located on chromosomes. Morgan found many other examples of genes on the X chromosome in *Drosophila* and in other organisms, thereby showing that his observations were not confined to a single species. Later in this chapter, we discuss the analysis of X-linked traits in humans.

*a*nimation

**X-Linked Inheritance**

**Keynote**

Sex linkage is the linkage of genes with the sex chromosomes of eukaryotes. Such genes, as well as the phenotypic characteristics these genes control, are called sex-linked. Genes that are on the X chromosome are called X-linked or sex-linked. Morgan's pioneering work with the inheritance of X-linked genes of *Drosophila* strongly supported, but did not prove, the chromosome theory of inheritance.

### Nondisjunction of X Chromosomes

Proof of the chromosome theory of inheritance came from the work of Morgan's student Calvin Bridges. Morgan's work showed that, from a cross of a white-eyed female (*w*/*w*) with a red-eyed male ($w^+$/Y), all the $F_1$ males

**Figure 12.22**
**Turner syndrome (XO).**

a) Individual with Turner syndrome

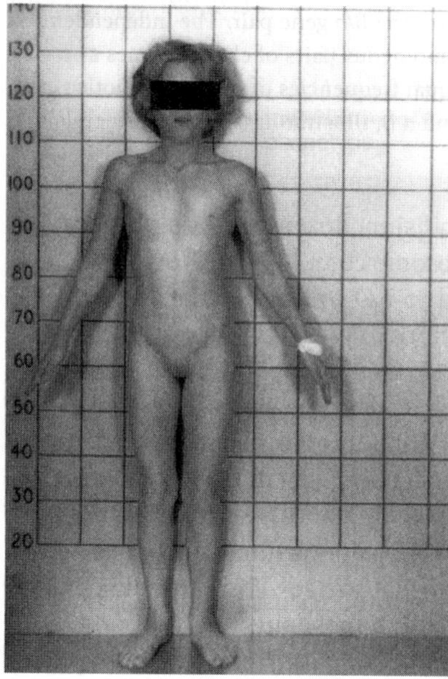

b) Karyotype for Turner syndrome

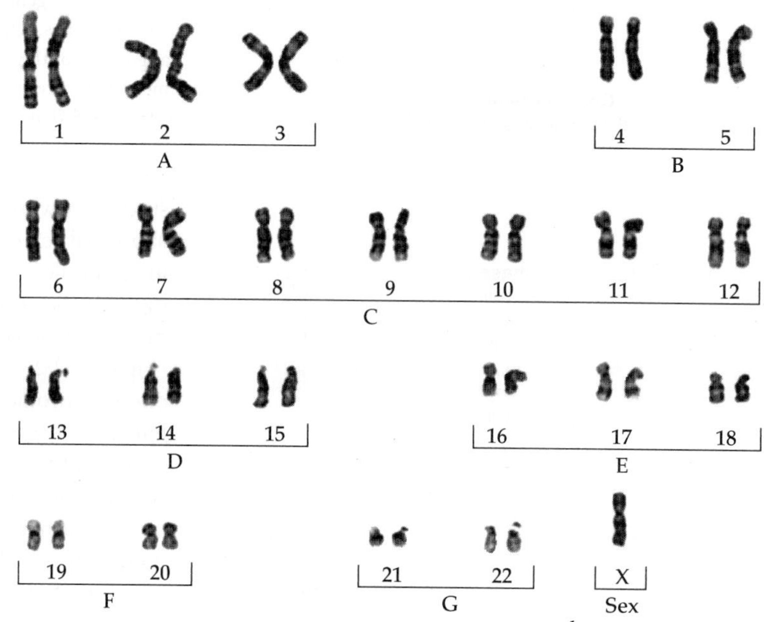


**Figure 12.23**
**Klinefelter syndrome (XXY).**

a) Individual with Klinefelter syndrome

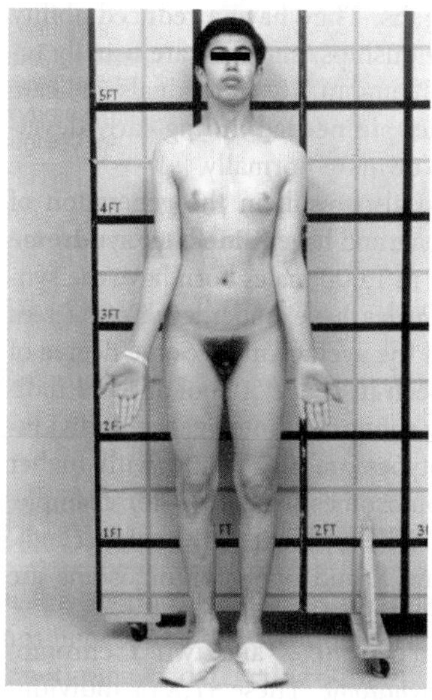

b) Karyotype for Klinefelter syndrome

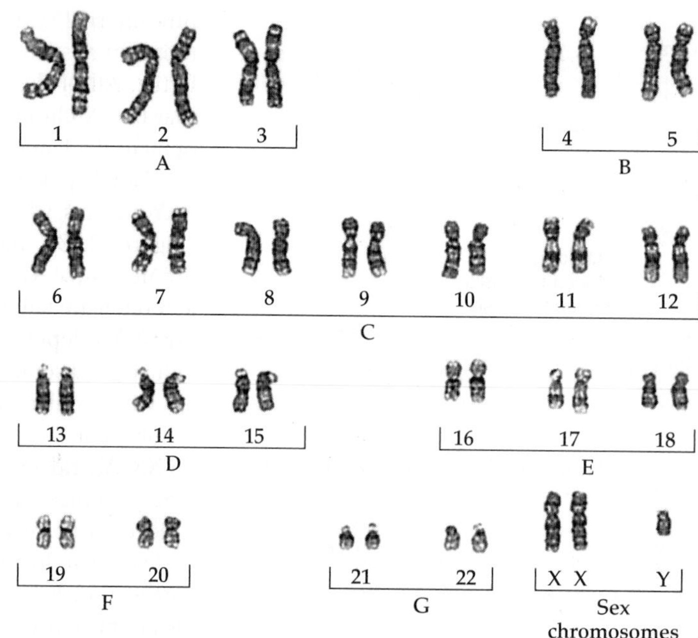


Table 12.2 summarizes the consequences of exceptional X and Y chromosomes in humans. In every case, the normal two sets of autosomes are associated with the sex chromosomes. The Barr bodies mentioned in the table are discussed next.

***Dosage Compensation Mechanism for X-Linked Genes in Mammals.*** Organisms with sex chromosomes have an inequality in gene dosage (the number of gene copies) between the sexes; that is, there are two copies of X-linked genes in females and one copy in males. In many such organisms, if gene expression on the X chromosome is not equalized, the condition is lethal early in development. Several different systems for **dosage compensation** have evolved. In mammals, the somatic cell nuclei of normal XX females contain a highly condensed mass of chromatin—

**Table 12.2 Consequences of Various Numbers of X- and Y-Chromosome Abnormalities in Humans, Showing Role of the Y in Sex Determination**

| Chromosome Constitution[a] | Designation of Individual | Expected Number of Barr Bodies |
|---|---|---|
| 46,XX | Normal ♀ | 1 |
| 46,XY | Normal ♂ | 0 |
| 45,X | Turner syndrome ♀ | 0 |
| 47,XXX | Triplo-X ♀ | 2 |
| 47,XXY | Klinefelter syndrome ♂ | 1 |
| 48,XXXY | Klinefelter syndrome ♂ | 2 |
| 48,XXYY | Klinefelter syndrome ♂ | 1 |
| 47,XYY | XYY syndrome ♂ | 0 |

[a]The first number indicates the total number of chromosomes in the nucleus, and the Xs and Ys indicate the sex chromosome complement.

named the **Barr body** after its discoverer, Murray Barr—not found in the nuclei of normal XY male cells. That is, somatic cells of XX individuals have one Barr body, and somatic cells of XY individuals have no Barr bodies (Figure 12.24 and Table 12.2). In 1961, Mary Lyon and Lillian Russell expanded this concept into what is now called the **Lyon hypothesis,** which proposed the following:

1. The Barr body is a highly condensed and (mostly) genetically inactive X chromosome. (It has become "lyonized" in a process called **lyonization.**) This leaves a single X chromosome that is transcriptionally equivalent to the single X chromosome of the male.
2. The X chromosome that is inactivated is randomly chosen from the maternally derived and paternally derived X chromosomes in a process that is independent from cell to cell. (But, once a maternal or paternal X chromosome is inactivated in a cell, all descendants of that cell inherit the inactivation pattern.)

X inactivation is an example of an **epigenetic** phenomenon—a heritable change in gene expression that occurs without a change in DNA sequence. In other words, X inactivation is an epigenetic silencing of one X chromosome. X inactivation occurs at about the 16th day after fertilization in humans (when the developing embryo is composed of about 500 to 1,000 cells), and between 3.5 and 6.5 days after fertilization in mice. Due to X inactivation, mammalian females that are heterozygous for X-linked traits are effectively *genetic mosaics*; that is, some cells show the phenotypes of one X chromosome, and the other cells show the phenotypes of the other X chromosome. This mosaicism is readily visible in, for example, the orange and black patches on calico cats (Figure 12.25). A calico cat is a female cat with the genotype *Oo B–*. That is, a calico is homozygous or heterozygous for the dominant *B* allele of an autosomal gene for black hair, and heterozygous for an X-linked gene for orange hair. If the dominant *O* allele of the X-linked gene is expressed, orange hair results no matter what other genes for coat color the cat has. So orange and black patches are produced because of random X inactivation as the female develops. The orange patches are where the chromosome with the *O* allele was *not* inactivated, so that the active *O* allele masks the *B* alleles, and the black patches are where the chromosome with the *O* allele *was* inactivated, allowing the *B* alleles to be expressed. (The white areas on calico cats are the result of the activity of yet another coat-color gene that, when expressed, masks the expression of any other color gene, leaving white hairs. Very rarely is a calico cat male; it is an XXY cat with the appropriate coat color gene genotype.)

A similar, but less visible, phenotype is seen in human females who are heterozygous for an X-linked mutation that causes the absence of sweat glands (anhidrotic ectodermal dysplasia; OMIM 305100). In this condition, there is a mosaic of skin patches lacking sweat glands.

The X inactivation process explains how mammals tolerate abnormalities in the number of sex chromosomes quite well, whereas, with rare exceptions, mammals with

a) Nuclei of XX female cells–one Barr body

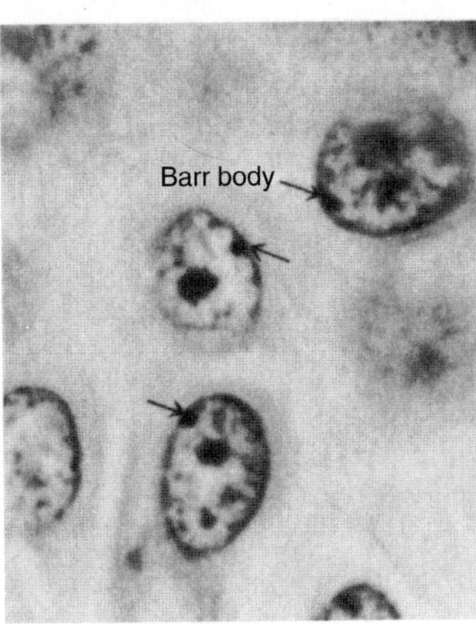


b) Nuclei of XY male cells–no Barr bodies

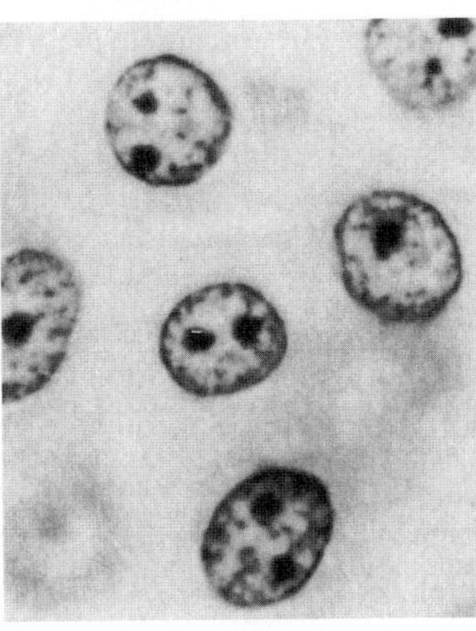

**Figure 12.24**
Barr bodies.

**Figure 12.25**

**A calico kitten.**

**Table 12.3 Sex Balance Theory of Sex Determination in *Drosophila melanogaster***

| Sex Chromosome Complement | Autosome Complement (A) | X:A Ratio[a] | Sex of Flies |
|---|---|---|---|
| XX | AA | 1.00 | ♀ |
| XY | AA | 0.50 | ♂ |
| XXX | AA | 1.50 | Metafemale (sterile) |
| XXY | AA | 1.00 | ♀ |
| XXX | AAAA | 0.75 | Intersex (sterile) |
| XX | AAA | 0.67 | Intersex (sterile) |
| X | AA | 0.50 | ♂ (sterile) |

[a]If the X chromosome autosome ratio is greater than, or equal to, 1.00 (X:A ≥ 1.00), the fly is a female. If the X chromosome : autosome ratio is less than, or equal to, 0.50(X:A ≤ 0.50), the fly is male. Between these two ratios, the fly is an intersex.

an unusual number of autosomes usually die. When lyonization operates in cells with extra X chromosomes, all but one of the X chromosomes typically become inactivated to produce Barr bodies; no such mechanism exists for extra autosomes. A general formula for the number of Barr bodies is the number of X chromosomes minus one. Table 12.2 lists the number of Barr bodies associated with abnormal human X chromosome numbers we have discussed. The molecular events involved in X inactivation are discussed in Chapter 19.

**Sex Determination in *Drosophila* and *Caenorhabditis*.** In the fruit fly *Drosophila melanogaster* and the nematode *Caenorhabditis elegans* (*C. elegans*), sex is determined by the *ratio of the number of X chromosomes to the number of sets of autosomes*. In this **X chromosome–autosome balance system** of sex determination, the Y chromosome (if present) has no effect on sex determination. When a Y chromosome is present, it may be required for male fertility.

In *Drosophila*, the homogametic sex is the female (XX) and the heterogametic sex is the male (XY). That the Y chromosome is not sex determining is seen by the fact that an XXY fly is female and an XO fly is male. Table 12.3 presents some chromosome complements and the sex of the resulting flies, to illustrate the relationship between sex and the ratio of X chromosomes to sets of autosomes. A normal female has two X chromosomes and two sets of autosomes; the X:A ratio is 1.00. A normal male has a ratio of 0.50. If the X:A ratio is greater than or equal to 1.00, the fly is female; if the X:A ratio is less than or equal to 0.50, the fly is male. If the ratio is between 0.50 and 1.00, the fly is neither male nor female; it is an intersex. Intersex flies are variable in appearance, generally having complex mixtures of male and female attributes for the internal sex organs and external genitalia. Such flies are sterile. Some molecular details of the complicated regulatory cascade for sex determination in *Drosophila* are presented in Chapter 19, pp. 559–564.

Dosage compensation of X-linked genes also occurs in *Drosophila*, but in a different way than in mammals. That is, the transcription of X-linked genes in males is higher than in females, to equal the sum of the expression levels of the two X chromosomes in females.

In *C. elegans*, there are two sexual types: hermaphrodites and males. Genetically, hermaphrodites are XX and males are XO with respect to sex chromosomes; both have five pairs of autosomes. That is, an X chromosome : autosome ratio of 1.00 results in hermaphrodites, and a ratio of 0.50 results in males. Most individuals are **hermaphroditic**—they have both sex organs: an ovary and two testes. They make sperm when they are larvae and store those sperm as development continues. In adults, the ovary produces eggs that are fertilized by the stored sperm as the eggs migrate to the uterus. Self-fertilization in this way almost always produces more hermaphrodites. However, 0.2% of the time, XO males are produced from self-fertilization as a result of nondisjunction. These males can fertilize hermaphrodites if the two mate, and such matings result in about equal numbers of hermaphrodite and male progeny, because the sperm from males has a competitive advantage over the sperm stored in the hermaphrodite.

Dosage compensation of X-linked genes in *C. elegans* occurs by yet another mechanism. In this case, genes on both X chromosomes in an XX hermaphrodite are transcribed at half the rate of the same gene on the single X chromosome in the XO male.

**Sex Chromosomes in Other Organisms.** In birds, butterflies, moths, and some fish, the sex chromosome composition is the opposite of that in mammals. The male is the homogametic sex, and the female is the heterogametic sex. To prevent confusion with the X and Y chromosome convention, the sex chromosomes in these organisms are designated as Z and W: Males are ZZ, and females are ZW. Genes on the Z chromosome behave just like X-linked genes, except that hemizygosity is found only in females. All the daughters of a male homozygous for a Z-linked recessive gene express the recessive trait. Interestingly, examination of the locations of genes on the sex chromosomes has revealed that the W and Z chromosomes of birds are quite different from the X and Y chromosomes of mammals. That is, mammalian X- and Y-chromosome genes typically are on bird chromosomes 1 and 4, while bird W- and Z-chromosome genes are on chromosomes 5 and 9 of mammals. The interpretation is that mammalian and bird sex chromosomes have evolved from different autosomal pairs.

Plants exhibit a variety of arrangements of sex organs. Some species (the ginkgo, for example) have plants of separate sexes, with male plants producing flowers that contain only stamens and female plants producing flowers that contain only pistils. These species are called **dioecious** ("two houses"). Other species have both male and female sex organs on the same plant; such plants are said to be **monoecious** ("one house"). If both sex organs are in the same flower, as in the rose and the buttercup, the flower is said to be a *perfect flower*. If the male and female sex organs are in different flowers on the same plant, as in corn, the flower is said to be an *imperfect flower*.

Some dioecious plants have sex chromosomes that differ between the sexes, and a large proportion of these plants have an X–Y system. Such plants typically have an X chromosome–autosome balance system of sex determination like that in *Drosophila*. However, we see many other sex determination systems in dioecious plants.

### Genic Sex Determination

Many other eukaryotic species, particularly eukaryotic microorganisms, do not have sex chromosomes but instead rely on a *genic system* for sex determination. In this system, the sexes are specified by simple allelic differences at one or a small number of gene loci. For example, the yeast *Saccharomyces cerevisiae* is a haploid eukaryote that has two "sexes"—a and α—called **mating types.** The mating types have the same morphologies, but crosses can occur only between individuals of opposite type. These mating types are determined by the *MAT*a and *MAT*α alleles, respectively, of a single gene.

**Keynote**

Many eukaryotic organisms have sex chromosomes that are represented differentially in the two sexes; in humans and many other mammals, the male is XY and the female is XX. In other eukaryotes with sex chromosomes, the male is ZZ and the female is ZW. In many cases, sex determination is related to the sex chromosomes. For humans and many other mammals, for instance, the presence of the Y chromosome confers maleness, and its absence results in femaleness. *Drosophila* and *Caenorhabditis* have an X chromosome–autosome balance system of sex determination: The sex of the individual is related to the ratio of the number of X chromosomes to the number of sets of autosomes. Several other sex-determining systems are known in the eukaryotes, including genic systems, found particularly in the lower eukaryotes.

## Analysis of Sex-Linked Traits in Humans

In Chapter 11, we introduced the analysis of recessive and dominant traits in humans; those traits were not sex-linked, but instead were the result of alleles carried on autosomes. In this section, we discuss examples of the analysis of X-linked and Y-linked traits in humans.

For the analysis of all pedigrees, whether the trait is autosomal or sex-linked, collecting reliable human pedigree data is a difficult task. For example, researchers often must rely on a family's recollections. Also, there may not be enough affected people to enable a clear determination of the mechanism of inheritance involved, especially when the trait is rare and the family is small. Furthermore, the expression of a trait may vary, resulting in some individuals erroneously being classified as normal. Finally, because the same mutant phenotype could result from mutations in more than one gene, it is possible that different pedigrees will indicate, correctly, that different mechanisms of inheritance are involved in the "same" trait.

**iActivity**

Go to the iActivity *It Runs in the Family* on the student website, where you will assume the role of a genetic counselor helping a couple determine whether deafness could be passed on to their children.

### X-Linked Recessive Inheritance

A trait resulting from a recessive mutant allele carried on the X chromosome is called an **X-linked recessive trait.** At least 100 human traits are known for which the gene has been traced to the X chromosome. Most of the traits involve X-linked recessive alleles. The best known X-linked recessive trait is hemophilia A (OMIM 306700), which occurred most famously in the family of Britain's Queen Victoria (Figure 12.26). Hemophilia is a serious ailment in which the blood lacks a clotting factor, thus, a cut or even a bruise can be fatal to a hemophiliac. In Queen Victoria's pedigree, the first instance of hemophilia was in one of her sons. Since she passed the mutant allele on to some of her other children (carrier daughters), she must have been a carrier (heterozygous) herself. Scientists think the mutation occurred on an X chromosome in the germ cells of one of her parents.

In X-linked recessive traits, females usually must be homozygous for the recessive allele in order to express the mutant trait. The trait is expressed in males who possess only one copy of the mutant allele on the X chromosome. Therefore, affected males normally transmit the mutant gene to all their daughters but to none of their sons. The instance of father-to-son inheritance of a rare trait in a pedigree tends to rule out X-linked recessive inheritance.

Other characteristics of X-linked recessive inheritance are the following (see Figure 12.26):

1. For X-linked recessive mutant alleles, many more males than females should exhibit the trait due to the different number of X chromosomes in the two sexes.
2. All sons of an affected (homozygous mutant) mother should show the trait because males receive their only X chromosome from their mothers.
3. The sons of heterozygous (carrier) mothers should show an approximately 1:1 ratio of normal individuals to individuals expressing the trait; that is, $a^+/a \times a^+/Y$ gives $\frac{1}{2}$ $a^+/Y$ and $\frac{1}{2}$ $a/Y$ sons.
4. From a mating of a carrier female with a normal male, all daughters will be normal phenotypically, but $\frac{1}{2}$ will be carriers; that is, $a^+/a \times a^+/Y$ gives $\frac{1}{2}$ $a^+/a^+$ and $\frac{1}{2}$ $a^+/a$ females. In turn, $\frac{1}{2}$ the sons of these carrier females will exhibit the trait.

**Figure 12.26**

**X-linked recessive inheritance.** Painting of Queen Victoria as a young woman. **(b)** Pedigree of Queen Victoria (III-2) and her descendants, showing the inheritance of hemophilia. (See Figure 11.16, p. 314, for an explanation of symbols used in pedigrees. In the pedigree shown here, marriage partners who were normal with respect to the trait may have been omitted to save space.) Since Queen Victoria was heterozygous for the sex-linked recessive hemophilia allele, but no cases occurred in her ancestors, the trait may have arisen as a mutation in one of her parents' germ cells (the cells that give rise to the gametes).

**a) Queen Victoria**

**b) Pedigree of Queen Victoria**

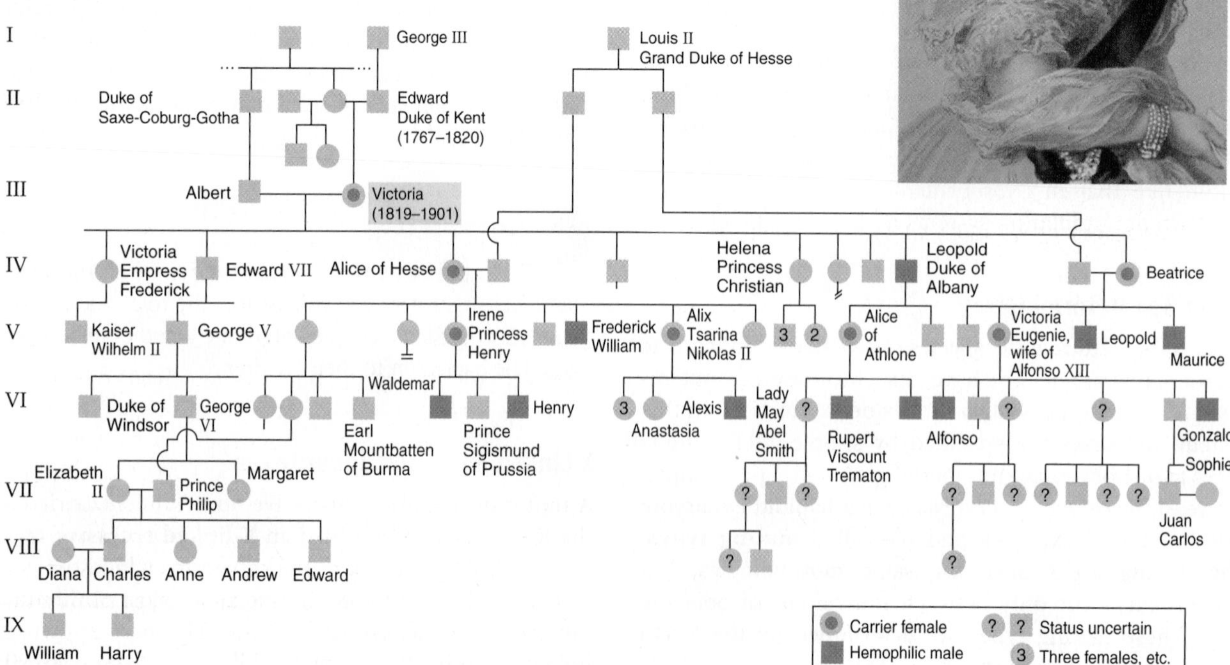

5. A male expressing the trait, when mated with a homozygous normal female, will produce all normal children. But all the female progeny will be carriers; that is, $a^+/a^+ \times a/Y$ gives $a^+/a$ females and $a^+/Y$ (normal) males.

Other examples of human X-linked recessive traits are Duchenne muscular dystrophy (progressive muscle degeneration that shortens the person's life) and two forms of color blindness.

## X-Linked Dominant Inheritance

A trait resulting from a dominant mutant allele carried on the X chromosome is called an **X-linked dominant trait.** Only a few X-linked dominant traits have been identified.

One example of an X-linked dominant trait is faulty tooth enamel and dental discoloration (hereditary enamel hypoplasia, OMIM 130900; Figure 12.27a). In the pedigree (Figure 12.27b), all the daughters and none of the sons of an affected father (III-1) are affected, and heterozygous mothers (IV-3) transmit the trait to half of their sons and half of their daughters. Other X-linked dominant mutant traits include web-tipped toes found in a specific South Dakota family that was studied in the 1930s and a severe bleeding anomaly called constitutional thrombopathy. In the latter (also studied in the 1930s), bleeding is caused not by the absence of a clotting factor (as in hemophilia) but instead by interference with the formation of blood platelets, which are needed for clotting.

X-linked dominant traits follow the same sort of inheritance rules as do X-linked recessives, except that heterozygous females express the trait. In general, X-linked dominant traits tend to be milder in females than in males. Also, because females have twice the number of X chromosomes as males, X-linked dominant traits are more frequent in females than in males. If the trait is rare, females with the trait are likely to be heterozygous. These females pass on the trait to 1/2 of their male progeny and 1/2 of their female progeny. Males with an X-linked dominant trait pass on the trait to all of their daughters and none of their sons.

**Figure 12.27**

**X-linked dominant inheritance.** The pedigree in part **(b)** illustrates a shorthand convention that omits parents who do not exhibit the trait. Thus, it is a given that the female in generation I paired with a male who did not exhibit the trait.

**a) X-linked dominant trait of faulty enamel**

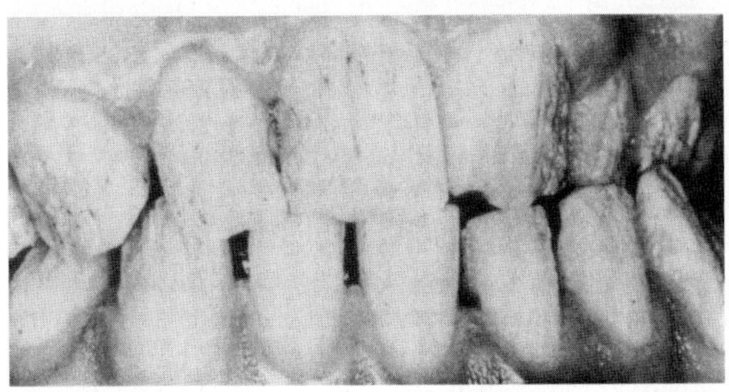

**b) Pedigree of a family with faulty enamel**

Generation:

I 1

II 1

III 1 ? 2 ? 3

IV 1 2 3 4 5 6 7 8 9 10 11

V 1 2 3 4 5 6 7 8

## Y-Linked Inheritance

A trait resulting from a mutant gene that is carried on the Y chromosome but has no counterpart on the X is called a **Y-linked,** or **holandric** ("wholly male"), **trait.** Such traits should be easily recognizable because every son of an affected male should have the trait, and no females should ever express it. Several traits with Y-linked inheritance have been suggested. In most cases, the genetic evidence for such inheritance is poor or nonexistent. A number of genes on the Y chromosome have been identified, however, including the gene for testis-determining factor mentioned earlier and other testis-specific genes present in multiple copies.

A possible example of Y-linked inheritance is the hairy ears trait (OMIM 425500), in which bristly hairs of atypical length grow from the ears. This trait is common in parts of India, and some other populations also exhibit it. Although the trait shows father-to-son inheritance, there is no doubt that it is a complex phenotype. However, many of the collected pedigrees can be interpreted in other ways, such as autosomal inheritance. The trait could also be the result of the interaction of a gene with the male hormone testosterone, which is known to cause the appearance of hair on the face and chest.

**Keynote**

Analysis of the inheritance of genes in humans typically relies on pedigree analysis: the careful study of the phenotypic records of the family extending over several generations. Data obtained from pedigree analysis enable geneticists to make judgments, with varying degrees of confidence, about whether a mutant gene is inherited as an autosomal recessive, an autosomal dominant, an X-linked recessive, an X-linked dominant, or a Y-linked allele.

## Summary

- Diploid eukaryotic cells have two haploid sets of chromosomes, one set coming from each parent. The members of a pair of chromosomes, one from each parent, are called homologous chromosomes. The complete set of chromosomes in a eukaryotic cell is called its karyotype.
- Mitosis is the process of nuclear division in eukaryotic cells represented by M in the cell cycle (that is, $G_1$, S, $G_2$, and M). Mitosis involves one round of DNA replication followed by one round of nuclear division (often accompanied by cell division). Mitosis results in the production of daughter nuclei that contain identical chromosome numbers and that are genetically identical to one another and to the parent nucleus from which they arose.
- Meiosis occurs in all sexually reproducing eukaryotes. A specialized diploid cell (or cell nucleus) with two haploid sets of chromosomes is transformed through one round of DNA replication and two rounds of nuclear division into four haploid nuclei (often in four cells), each with one set of chromosomes.
- Meiosis generates genetic variability through the processes by which maternal and paternal chromosomes are reassorted in progeny nuclei and through crossing-over between members of a homologous pair of chromosomes.
- The chromosome theory of inheritance states that genes are located on chromosomes. Support for the chromosome theory of inheritance came from experiments that related the hereditary behavior of particular genes to the transmission of a sex chromosome from generation to generation.
- In eukaryotes with separate sexes, a sex chromosome is a chromosome or a group of chromosomes that is represented differently in the two sexes. In organisms with sex chromosomes, one sex is homogametic and the other is heterogametic.
- Sex linkage is the physical association of genes with the sex chromosomes of eukaryotes. Such genes are called sex-linked genes. Genes on the X sex chromosome are called X-linked genes, and genes on the Y sex chromosome are called Y-linked genes.
- In many eukaryotic organisms, sex determination is related to the sex chromosomes. In humans and other mammals, for example, the presence of a Y chromosome specifies maleness, and its absence results in femaleness. Several other sex determination mechanisms are known in eukaryotes.
- In humans, the allele responsible for a trait can be inherited in one of five main ways: autosomal recessive, autosomal dominant, X-linked recessive, X-linked dominant, or Y-linked. As with autosomally inherited traits, sex-linked traits are studied in humans by using pedigree analysis.

## Analytical Approaches to Solving Genetics Problems

The concepts introduced in this chapter may be reinforced by solving genetics problems similar to those introduced in Chapter 11. Remember that, when sex linkage is involved, one sex has two kinds of sex chromosomes, whereas the other sex has only one; this feature alters the inheritance patterns.

**Q12.1** A female from a true-breeding strain of *Drosophila* with vermilion-colored eyes is crossed with a male from a true-breeding, wild-type, red-eyed strain. All the $F_1$ males have vermilion-colored eyes, and all the females have wild-type red eyes. What conclusions can you draw about the mechanism of inheritance of the vermilion trait, and how could you test them?

**A12.1** The observation is the classic one which suggests that a sex-linked trait is involved. That is, because none of the $F_1$ daughters have the trait and all the $F_1$ males do, the trait is presumably X-linked recessive. The results fit this hypothesis because the $F_1$ males receive the X chromosome with the *v* gene from their homozygous *v/v* mother. Furthermore, the $F_1$ females are $v^+/v$, because they receive a $v^+$-bearing X chromosome from the wild-type male parent and a *v*-bearing X chromosome from the female parent. If the trait were autosomal recessive, all the $F_1$ flies would have had wild-type eyes. If it were autosomal dominant, both the $F_1$ males and females would have had vermilion-colored eyes. If the trait were X-linked dominant, all the $F_1$ flies would have had vermilion eyes.

The easiest way to verify the hypothesis is to let the $F_1$ flies interbreed. This cross is $v^+/v$ ♀ × $v$/Y ♂, and the expectation is that there will be a 1:1 ratio of wild-type : vermilion eyes in both sexes in the $F_2$ flies. That is, half the females are $v^+/v$ and half are *v/v*; half the males are $v^+$/Y and half are *v*/Y. This ratio is certainly not the 3:1 ratio that would result from an $F_1 \times F_1$ cross for an autosomal gene.

**Q12.2** In humans, hemophilia is caused by an X-linked recessive gene. A woman who is a nonbleeder—that is, she does not display the blood-clotting irregularities associated with hemophilia—had a father who was a hemophiliac.

She marries a nonbleeder, and they plan to have children. Calculate the probability of hemophilia in the female and male offspring.

**A12.2** Because hemophilia is an X-linked trait, and because her father was a hemophiliac, the woman must be heterozygous for this recessive gene. If we assign the symbol *h* to this recessive mutation and $h^+$ to the wild-type (nonbleeder) allele, she must be $h^+/h$. The man she marries is normal with regard to blood clotting and hence must be hemizygous for $h^+$—that is, $h^+/Y$. All their daughters receive an X chromosome from the father, so each must have an $h^+$ gene. In fact, half the daughters are $h^+/h^+$ and the other half are $h^+/h$. Because the wild-type allele is dominant, none of the daughters are hemophiliacs. However, all the sons of the marriage receive their X chromosome from their mother. Therefore, the probability is 1/2 that they will receive the chromosome carrying the *h* allele, in which case they will be hemophiliacs. Thus, the probability of hemophilia among daughters of this marriage is 0; among sons, it is 1/2.

**Q12.3** Tribbles are hypothetical animals that have an X–Y sex determination mechanism like that of humans. The trait blotchy (*b*), with pigment in spots, is X-linked and recessive to solid color ($b^+$), and the trait light color (*l*) is autosomal and recessive to dark color ($l^+$). If you make reciprocal crosses between true-breeding blotchy, light-colored tribbles and true-breeding solid, dark-colored tribbles, do you expect a 9:3:3:1 ratio in the $F_2$ of either or both of these crosses? Explain your answer.

**A12.3.** This question focuses on the fundamentals of X-chromosome and autosome segregation during a genetic cross, and it tests whether you have grasped the principles involved in gene segregation. Figure 12.A diagrams the two crosses involved, and we can discuss the answer by referring to the figure.

First, consider the cross of a wild-type female tribble ($b^+/b^+\ l^+/l^+$) with a male double-mutant tribble ($b/Y\ l/l$). Part (a) of the figure diagrams this cross. These F1 tribbles are all normal—that is, they are solid and light colored—because, for the autosomal character, both sexes are heterozygous, and for the X-linked character, the female is heterozygous and the male is hemizygous for the $b^+$ allele donated by the normal mother. For the production of the $F_2$ progeny, the best approach is to treat the X-linked and autosomal traits separately. For the X-linked

**Figure 12.A**

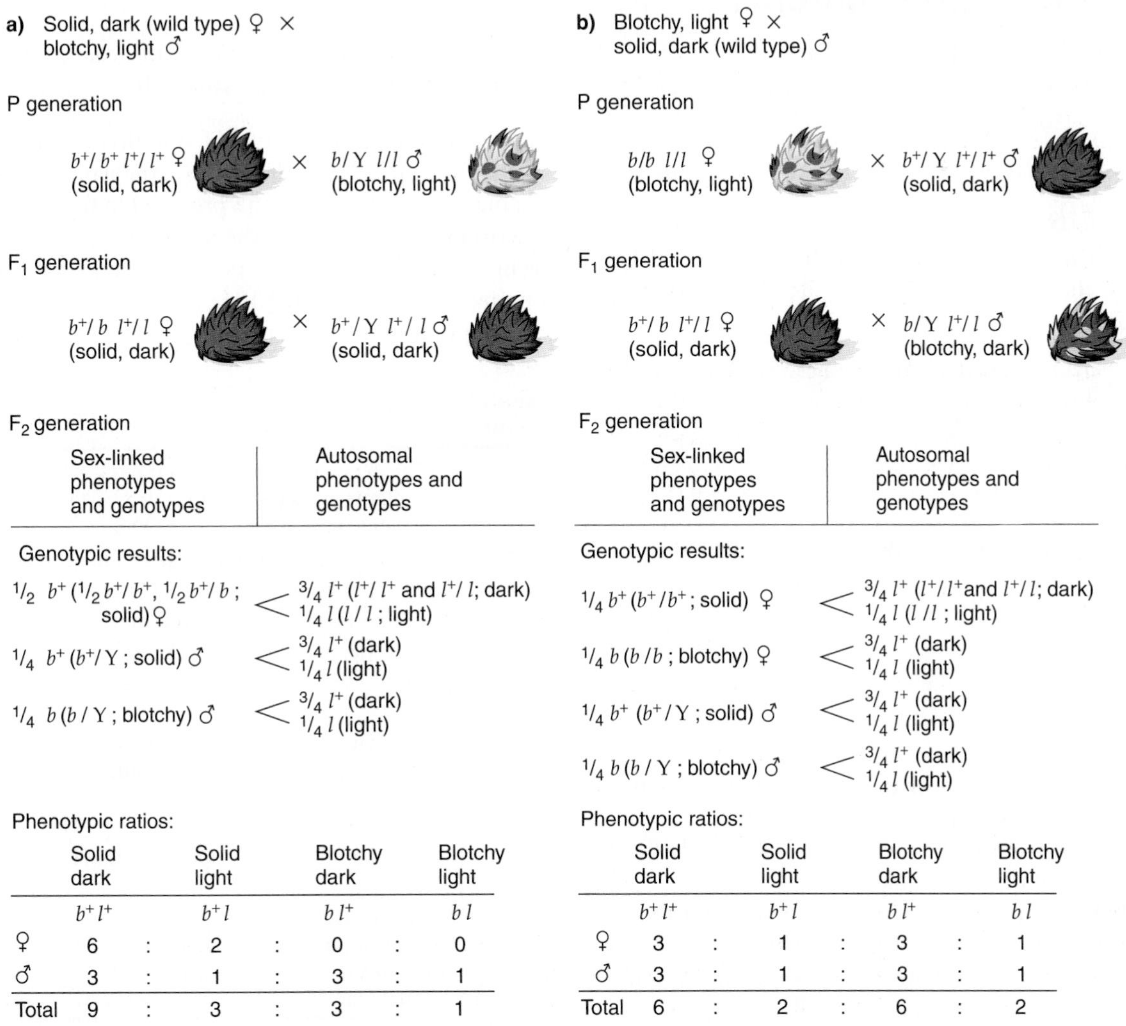

**a)** Solid, dark (wild type) ♀ × blotchy, light ♂

P generation

$b^+/b^+\ l^+/l^+$ ♀ (solid, dark) × $b/Y\ l/l$ ♂ (blotchy, light)

$F_1$ generation

$b^+/b\ l^+/l$ ♀ (solid, dark) × $b^+/Y\ l^+/l$ ♂ (solid, dark)

$F_2$ generation

| Sex-linked phenotypes and genotypes | Autosomal phenotypes and genotypes |
|---|---|
| Genotypic results: | |
| 1/2 $b^+$ (1/2 $b^+/b^+$, 1/2 $b^+/b$; solid) ♀ | 3/4 $l^+$ ($l^+/l^+$ and $l^+/l$; dark)<br>1/4 $l$ ($l/l$; light) |
| 1/4 $b^+$ ($b^+/Y$; solid) ♂ | 3/4 $l^+$ (dark)<br>1/4 $l$ (light) |
| 1/4 $b$ ($b/Y$; blotchy) ♂ | 3/4 $l^+$ (dark)<br>1/4 $l$ (light) |

Phenotypic ratios:

| | Solid dark | | Solid light | | Blotchy dark | | Blotchy light |
|---|---|---|---|---|---|---|---|
| | $b^+\ l^+$ | | $b^+\ l$ | | $b\ l^+$ | | $b\ l$ |
| ♀ | 6 | : | 2 | : | 0 | : | 0 |
| ♂ | 3 | : | 1 | : | 3 | : | 1 |
| Total | 9 | : | 3 | : | 3 | : | 1 |

**b)** Blotchy, light ♀ × solid, dark (wild type) ♂

P generation

$b/b\ l/l$ ♀ (blotchy, light) × $b^+/Y\ l^+/l^+$ ♂ (solid, dark)

$F_1$ generation

$b^+/b\ l^+/l$ ♀ (solid, dark) × $b/Y\ l^+/l$ ♂ (blotchy, dark)

$F_2$ generation

| Sex-linked phenotypes and genotypes | Autosomal phenotypes and genotypes |
|---|---|
| Genotypic results: | |
| 1/4 $b^+$ ($b^+/b^+$; solid) ♀ | 3/4 $l^+$ ($l^+/l^+$and $l^+/l$; dark)<br>1/4 $l$ ($l/l$; light) |
| 1/4 $b$ ($b/b$; blotchy) ♀ | 3/4 $l^+$ (dark)<br>1/4 $l$ (light) |
| 1/4 $b^+$ ($b^+/Y$; solid) ♂ | 3/4 $l^+$ (dark)<br>1/4 $l$ (light) |
| 1/4 $b$ ($b/Y$; blotchy) ♂ | 3/4 $l^+$ (dark)<br>1/4 $l$ (light) |

Phenotypic ratios:

| | Solid dark | | Solid light | | Blotchy dark | | Blotchy light |
|---|---|---|---|---|---|---|---|
| | $b^+\ l^+$ | | $b^+\ l$ | | $b\ l^+$ | | $b\ l$ |
| ♀ | 3 | : | 1 | : | 3 | : | 1 |
| ♂ | 3 | : | 1 | : | 3 | : | 1 |
| Total | 6 | : | 2 | : | 6 | : | 2 |

trait, a random combination of the gametes produced gives a genotypic ratio of 1 $b^+/b^+$ (solid female) : 1 $b^+/b$ (solid female) : 1 $b^+$/Y (solid male) : 1 $b$/Y (blotchy male) progeny. Categorizing by phenotypes, 1/2 of the progeny are solid females, 1/4 are solid males, and 1/4 are blotchy males. For the autosomal leg trait, the $F_1 \times F_1$ is a cross of two heterozygotes, so we expect a 3:1 phenotypic ratio of dark : light tribbles in the $F_2$ progeny. Since autosome segregation is independent of the inheritance of the X chromosome, we can multiply the probabilities of occurrence of the X-linked and autosomal traits to calculate their relative frequencies. The calculations are presented at the bottom of part (a) of the figure.

The first cross, then, has a 9:3:3:1 ratio of the four possible phenotypes in the $F_2$. However, due to the inheritance pattern of the X chromosome, the ratio in each sex is not 9:3:3:1. This result contrasts markedly with the pattern of two autosomal genes segregating independently, in which the 9:3:3:1 ratio is found for both sexes.

The second cross (a reciprocal cross) is diagrammed in part (b) of the figure. Because the parental female in this cross is homozygous for the sex-linked trait, all the $F_1$ males are blotchy. Genotypically, the $F_1$ males and females differ from those in the first cross with respect to the sex chromosome, but they are just the same with respect to the autosome. Again, considering the X chromosome first as we go to the $F_2$ progeny, we find a genotypic ratio of 1 solid females : 1 blotchy females : 1 solid males : 1 blotchy males. In this case, then, half of both males and females are solid, and half are blotchy, in contrast to the results of the first cross, in which no blotchy females were produced in the $F_2$ progeny. For the autosomal trait, we expect a 3:1 ratio of dark : light in the $F_2$ progeny, as before. Putting the two traits together, we get the calculations presented in part (b) of the figure. (*Note*: We use the total 6:2:6:2 here rather than 3:1:3:1 because the numbers add to 16, as does 9:3:3:1.) Hence, in this case, we do not get a 9:3:3:1 ratio; moreover, the ratio is the same in both sexes.

This question has forced us to think through the segregation of two types of chromosomes and has shown that we must be careful about predicting the outcomes of crosses in which sex chromosomes are involved. Nonetheless, the basic principles for the analysis are the same as those used before: Reduce the questions to their basic parts, and then put the puzzle together step-by-step.

# Questions and Problems

***12.1** Interphase is a period corresponding to the cell cycle phases of

**a.** mitosis.
**b.** S.
**c.** $G_1 + S + G_2$.
**d.** $G_1 + S + G_2 + M$.

**12.2** What are the differences between the $G_0$, $G_1$, and $G_2$ phases of the cell cycle? Do all cells proceed through each of these phases?

**12.3** Chromatids joined together by a centromere are called

**a.** sister chromatids.
**b.** homologs.
**c.** alleles.
**d.** bivalents (tetrads).

***12.4** Mitosis and meiosis always differ in regard to the presence of

**a.** chromatids.
**b.** homologs.
**c.** bivalents.
**d.** centromeres.
**e.** spindles.

**12.5** State whether each of the following statements is true or false, and explain your choice:

**a.** The chromosomes in a somatic cell of any organism are all morphologically alike.
**b.** During mitosis, the chromosomes divide and the resulting sister chromatids separate at anaphase, ending up in two nuclei, each of which has the same number of chromosomes as the parental cell.
**c.** At zygonema, a chromosome can synapse with any other chromosome in the same cell.

**12.6** Descriptions of a series of mitotic events are provided in the table below. First, write the name of the corresponding event in the blank provided to the left of each description. Then, number the blanks to the right of each event according to the sequence in which they occur, beginning with a 1 for interphase and ending with a 7 for the last event in the sequence.

| Name of Event | | Order of Event |
|---|---|---|
| ______ | The cytoplasm divides and the cell contents are separated into two separate cells. | ______ |
| ______ | Chromosomes become aligned along the equatorial plane of the cell. | ______ |
| ______ | Chromosome replication occurs. | ______ |
| ______ | The migration of the daughter chromosomes to the two poles is complete. | ______ |
| ______ | Replicated chromosomes begin to condense and become visible under the microscope. | ______ |
| ______ | Sister chromatids begin to separate and migrate toward opposite poles of the cell. | ______ |
| ______ | The nuclear envelope breaks down, a developing mitotic spindle enters the former nuclear area, and kinetochores bind to centromeres. | ______ |

*12.7 Answer yes or no to the following questions, and then explain the reasons for your answer:

**a.** Can meiosis occur in haploid species?
**b.** Can meiosis occur in a haploid individual?

**12.8** Which of the following sequences describes the general life cycle of a eukaryotic organism?

**a.** 1N → meiosis → 2N → fertilization → 1N
**b.** 2N → meiosis → 1N → fertilization → 2N
**c.** 1N → mitosis → 2N → fertilization → 1N
**d.** 2N → mitosis → 1N → fertilization → 2N

*12.9 Which statement is true?

**a.** Gametes are 2N; zygotes are 1N.
**b.** Gametes and zygotes are 2N.
**c.** The number of chromosomes can be the same in gamete cells and in somatic cells.
**d.** The zygotic and the somatic chromosome numbers cannot be the same.
**e.** Haploid organisms have haploid zygotes.

**12.10** Which of the following does **not** occur in prophase I of meiosis?

**a.** chromosome condensation
**b.** pairing of homologs
**c.** chiasma formation
**d.** formation of a telomere bouquet
**e.** segregation

*12.11 Give the name of each stage of mitosis and meiosis at which each of the following events occurs:

**a.** Chromosomes are located in a plane at the center of the spindle.
**b.** The chromosomes move away from the spindle equator to the poles.

**12.12** Consider the diploid, meiotic mother cell shown below. Diagram the chromosomes as they would appear

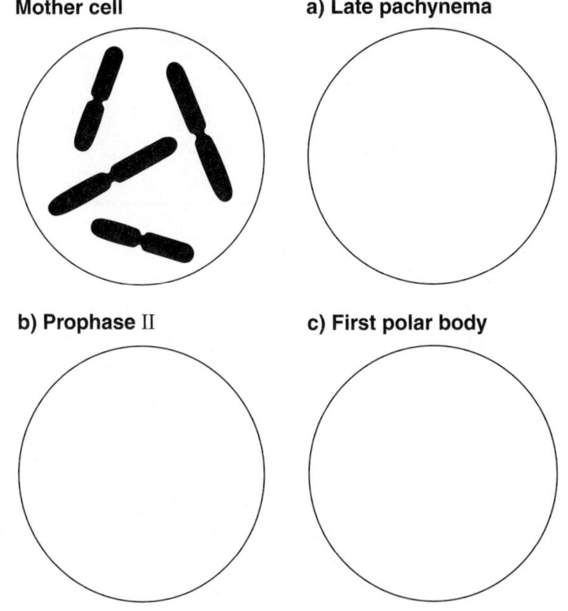


**a.** in late pachynema.
**b.** in a nucleus at prophase of the second meiotic division.
**c.** in the first polar body resulting from oogenesis in an animal.

**12.13** The cells in the following figure were all taken from the same individual (a mammal):

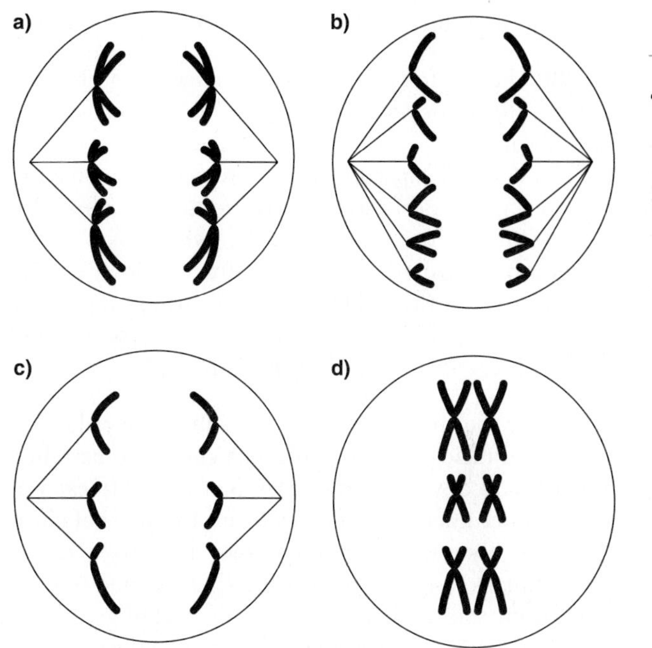


Identify the cell division events occurring in each cell, and explain your reasoning. What is the sex of the individual? What is the diploid chromosome number?

**12.14** Does mitosis or meiosis have greater significance in the study of heredity? Explain your answer.

*12.15 Consider a diploid organism that has three pairs of chromosomes. Assume that the organism receives chromosomes A, B, and C from the female parent and A′, B′, and C′; from the male parent. Answer the following questions, assuming that crossing-over does not occur:

**a.** What proportion of the gametes of this organism would be expected to contain all the chromosomes of maternal origin?
**b.** What proportion of the gametes would be expected to contain some chromosomes of both maternal and paternal origin?

*12.16 Normal diploid cells of a theoretical mammal are examined cytologically at the mitotic metaphase stage for their chromosome complement. One short chromosome, two medium-length chromosomes, and three long chromosomes are present. Explain how the cells might have such a set of chromosomes.

**12.17** Explain whether the following statement is true or false: "Meiotic chromosomes can be seen after appropriate staining in nuclei from rapidly dividing skin cells."

*12.18 Explain whether the following statement is true or false: "All the sperm from one human male are genetically identical."

12.19 The horse has a diploid set of 64 chromosomes, and the donkey has a diploid set of 62 chromosomes. Mules are the viable, but usually sterile, progeny of a mating between a male donkey and a female horse. How many chromosomes will a mule cell contain?

*12.20 The red fox has 17 pairs of large, long chromosomes. The arctic fox has 26 pairs of shorter, smaller chromosomes.

**a.** What do you expect to be the chromosome number in somatic tissues of a hybrid between these two foxes?

**b.** The first meiotic division in the hybrid fox shows a mixture of paired and single chromosomes. Why do you suppose this occurs? Can you suggest a possible relationship between the mixed chromosomes and the observed sterility of the hybrid?

*12.21 At the time of synapsis preceding the reduction division in meiosis, the homologous chromosomes align in pairs, and one member of each pair passes to each of the daughter nuclei. In an animal with five pairs of chromosomes, assume that chromosomes 1, 2, 3, 4, and 5 have come from the father, and 1′, 2′, 3′, 4′, and 5′ have come from the mother. Assuming no crossing over, in what proportion of the gametes of this animal will all the paternal chromosomes be present together?

12.22 Depict each of the crosses that follow, first using Mendelian and then using *Drosophila* notation (see Box 12.1). Give the genotype and phenotype of the $F_1$ progeny that can be produced

**a.** in humans, from a mating between two individuals, each heterozygous for the recessive trait phenylketonuria, whose locus is on chromosome 12.

**b.** in humans, from a mating between a female heterozygous for both phenylketonuria and X-linked color blindness and a male with normal color vision and heterozygous for phenylketonuria.

**c.** in *Drosophila*, from a mating between a female with white eyes, curled wings, and normal long bristles and a male that has normal red eyes, normal straight wings, and short, stubbly bristles. In these individuals, curled wings result from a heterozygous condition at a gene whose locus is on chromosome 2, whereas the short, stubbly bristles result from a heterozygous condition at a gene whose locus is on chromosome 3.

**d.** in *Drosophila*, from a mating between a female from a true-breeding line that has eyes of normal size that are white, black bodies (a recessive trait on chromosome 2), and tiny bristles (a recessive trait called *spineless* on chromosome 3) and a male from a true-breeding line that has normal red eyes, normal grey bodies, normal long bristles, and a reduced eye size (a dominant trait called *eyeless* on chromosome 4).

12.23 In *Drosophila*, white eyes are an X-linked character. The mutant allele for white eyes ($w$) is recessive to the wild-type allele for brick red eye color ($w^+$). A white-eyed female is crossed with a red-eyed male. An $F_1$ female from this cross is mated with her father, and an $F_1$ male is mated with his mother. What will be the eye color of the offspring of these last two crosses?

*12.24 One form of color blindness ($c$) in humans is caused by an X-linked recessive mutant gene. A woman with normal vision ($c^+$) whose father was color blind marries a man with normal vision whose father was also color blind. What proportion of their offspring will be color blind? (Give your answer separately for males and females.)

12.25 In humans, red-green color blindness is recessive and X linked, whereas albinism is recessive and autosomal. What will be the genotypes and phenotypes of the children resulting from a marriage between a woman with albinism and normal vision, and a man with normal skin pigmentation and color blindness, each of whom is homozygous for the traits displayed?

*12.26 In *Drosophila*, vestigial (partially formed) wings ($vg$) are recessive to normal long wings ($vg^+$), and the gene for this trait is autosomal. The gene for the white-eye trait is on the X chromosome. Suppose a homozygous white-eyed, long-winged female fly is crossed with a homozygous red-eyed, vestigial-winged male.

**a.** What will be the genotypes and phenotypes of the $F_1$ flies?

**b.** What will be the genotypes and phenotypes of the $F_2$ flies?

**c.** What will be the genotypes and phenotypes of the offspring of a cross of the $F_1$ flies back to each parent?

12.27 In *Drosophila*, two red-eyed, long-winged flies are bred together and produce offspring with the following proportion of characters:

| | Females | Males |
|---|---|---|
| red eyed, long winged | 3/4 | 3/8 |
| red eyed, vestigial winged | 1/4 | 1/2 |
| white eyed, long winged | — | 3/8 |
| white eyed, vestigial winged | — | 1/8 |

What are the genotypes of the parents?

*12.28 In cats, the dominant allele $B$ at an autosomal locus results in a black coat while homozygosity for the recessive allele $b$ results in a dark brown, chocolate coat. The dominant $O$ allele at an X-linked locus results in an orange coat, no matter what alleles are present at the $B/b$ locus, while the recessive $o$ allele allows the alleles at the $B/b$ locus to be expressed. A calico cat having black and orange patches had a father with a chocolate coat. She mates with a chocolate male whose parents were both solid black.

a. What are the genotypes of the animals in this mating, and what phenotypes and frequencies are expected from it?
b. What phenotypes would be produced if there is paternal sex-chromosome nondisjunction?

**12.29** In chickens, a dominant sex-linked gene (*B*) produces barred feathers, and the recessive allele (*b*), when homozygous, produces nonbarred (solid-color) feathers. Suppose a nonbarred cock is crossed with a barred hen.
a. What will be the phenotype of the $F_1$ birds?
b. If an $F_1$ female is mated with her father, what will be the phenotype of their offspring?
c. If an $F_1$ male is mated with his mother, what will be the phenotype of their offspring?

***12.30** A man (A) suffering from defective tooth enamel, which results in brown-colored teeth, marries a normal woman. All their daughters have brown teeth, but the sons are normal. The sons of man A marry normal women, and all their children are normal. The daughters of man A marry normal men, and 50% of their children have brown teeth. Explain these facts genetically.

***12.31** In humans, differences in the ability to taste phenylthiourea result from a pair of autosomal alleles. Inability to taste is recessive to ability to taste. A child who is a nontaster is born to a couple who can both taste the substance. What is the probability that their next child will be a taster?

***12.32** Cystic fibrosis is inherited as an autosomal recessive. Two parents without cystic fibrosis have two children with cystic fibrosis and three children without. The parents come to you for genetic counseling.
a. What is the probability that their next child will have cystic fibrosis?
b. Their unaffected children are concerned about being heterozygous. What is the probability that a given unaffected child in the family is heterozygous?

**12.33** Huntington disease is a human disease inherited as a Mendelian autosomal dominant. The disease results in choreic (uncontrolled) movements, progressive mental deterioration, and eventually death. In carriers of the trait, the disease appears at between 15 and 65 years of age. The American folk singer Woody Guthrie died of Huntington disease, as did just one of his parents. Marjorie Mazia, Woody's wife, had no history of this disease in her family. The Guthries had three children. What is the probability that a particular Guthrie child will die of Huntington disease?

**12.34** Suppose gene *A* is on the X chromosome, and genes *B*, *C*, and *D* are on three different autosomes. Thus, *A*– signifies the dominant phenotype in the male or female. An equivalent situation holds for *B*–, *C*–, and *D*–. The cross *AA BB CC DD* ♀ × *a*Y *bb cc dd* ♂ is made.
a. What is the probability of obtaining an *A*– individual in the $F_1$ progeny?
b. What is the probability of obtaining an *a* male in the $F_1$ progeny?
c. What is the probability of obtaining an *A*– *B*– *C*– *D*– female in the $F_1$ progeny?
d. How many different $F_1$ genotypes will there be?
e. What proportion of $F_2$ individuals will be heterozygous for the four genes?
f. Determine the probabilities of obtaining each of the following types in the $F_2$ individuals (1) *A*– *bb CC dd* (female), (2) *a*Y *BB Cc Dd* (male), (3) *A*Y *bb CC dd* (male), (4) and *aa bb Cc Dd* (female).

***12.35** In humans and the fly *Drosophila melanogaster*, XX animals are female and XY animals are male; in the nematode *Caenorhabditis elegans*, XX animals are hermaphroditic and XO animals are male.
a. In what different ways is sex determined in each of these organisms?
b. How is the expression of X-linked genes equalized in the sexes of each of these organisms?

**12.36** In the nematode *Caenorhabditis elegans*, mutations at different genes can lead to an *uncoordinated* phenotype where animals exhibit abnormal locomotion. Two recessive mutations with this phenotype are *unc-115*, on the X chromosome, and *unc-26*, on chromosome IV. What genotypic and phenotypic frequencies do you expect to see in each of the following crosses?
a. A hermaphrodite heterozygous for *unc-115* mates with an *unc-115* male.
b. A hermaphrodite heterozygous for *unc-115* produces progeny by self-fertilization.
c. A hermaphrodite heterozygous for both *unc-115* and *unc-26* mates with a doubly mutant *unc-115; unc-26 unc-26* male.

***12.37** As a famous mad scientist, you have cleverly devised a method to isolate *Drosophila* ova that have undergone primary nondisjunction of the sex chromosomes. In one experiment, you used females homozygous for the X-linked recessive mutation causing white eyes (*w*) as your source of nondisjunction ova. The ova were collected and fertilized with sperm from red-eyed males. The progeny of this "engineered" cross were then backcrossed separately to the two parental strains. What classes of progeny (genotype and phenotype) would you expect to result from these backcrosses? (The genotype of the original parents may be denoted as *ww* for the females and $w^+$/Y for the males.)

**12.38** In *Drosophila*, the bobbed gene ($bb^+$) is located on the X chromosome: *bb* mutants have shorter, thicker bristles than do wild-type flies. Unlike most X-linked genes, however, a bobbed gene is also present on the Y chromosome. The mutant allele *bb* is recessive to $bb^+$. If a wild-type $F_1$ female that resulted from primary nondisjunction in oogenesis in a cross of bobbed female with a wild-type male is mated to a bobbed male, what will be the phenotypes and their frequencies in the offspring? List males and females separately in your answer. (*Hint*: Refer to the chapter for information about the frequency of nondisjunction in *Drosophila*; see p. 345.)

*__12.39__ An individual with Turner syndrome would be expected to have how many Barr bodies in the majority of cells?

**12.40** An XXY individual with Klinefelter syndrome would be expected to have how many Barr bodies in the majority of cells?

*__12.41__ State whether each of the following observations is most likely the result of a genetic or an epigenetic phenomenon, and explain why.

**a.** Female calico cats have orange and black patches: cells in the orange patches exhibit the *O*-allele phenotype and cells in black patches exhibit the *o*-allele phenotype.

**b.** In *Drosophila*, an XY male with just a single copy of the X-linked $w^+$ gene produces the same amount of $w^+$ gene transcripts as an XX female with two copies of the $w^+$ gene.

**c.** Among the 122 offspring of two straight-winged *Drosophila* with normal phenotypes, 121 have straight wings and one male has curly wings. When the curly-winged male is crossed with a wild-type female, the progeny are 42 straight-winged males, 46 straight-winged females, 44 curly-winged males, and 37 curly-winged females.

**d.** There is an increased incidence of tumors in animals exposed to the compound diethylstilbestrol, which is not positive in the Ames test.

**e.** When a cow bears twins in which one calf is chromosomally male and the other is chromosomally female, the female calf is usually what is called a *freemartin*—it displays masculine characteristics as a result of exposure to hormones from its male sibling while in utero.

**f.** Over a period of several years, the matings of two "wild-type" parakeets produce offspring colored just like the parents: They develop solid green bodies and dark stripes on otherwise yellow heads and wings. One chick, however, develops cinnamon-colored stripes. When she matures and mates with a wild-type male, all of her offspring have wild-type coloration. However, when any of her sons mate with wild-type parakeets, half of their daughters have cinnamon-colored stripes.

*__12.42__ In human genetics, pedigrees are used to analyze inheritance patterns. Females are represented by a circle, males by a square. The figure that follows presents three 2-generation family pedigrees for a trait in humans. Normal individuals are represented by unshaded symbols, people with the trait by shaded symbols. For each pedigree (A, B, and C), state (by answering "yes" or "no" in the appropriate blank space) whether transmission of the trait can be accounted for based on each of the listed simple modes of inheritance:

Generation: Pedigree A Pedigree B Pedigree C

I

II

| | Pedigree A | Pedigree B | Pedigree C |
|---|---|---|---|
| Autosomal recessive | ______ | ______ | ______ |
| Autosomal dominant | ______ | ______ | ______ |
| X-linked recessive | ______ | ______ | ______ |
| X-linked dominant | ______ | ______ | ______ |

**12.43** Shaded symbols in the following pedigree represent a trait:

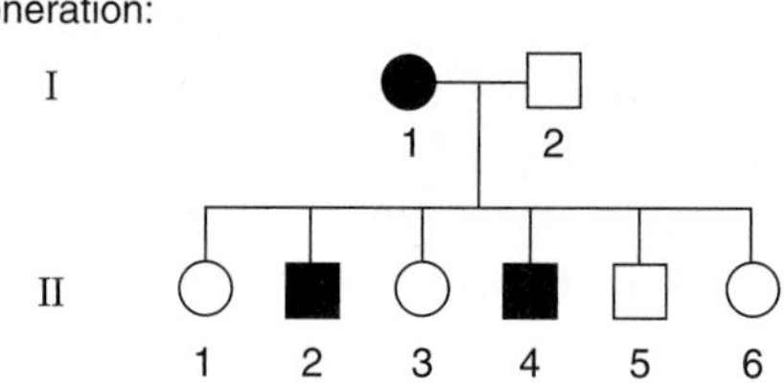


Which of the progeny eliminate X-linked recessiveness as a mode of inheritance for the trait?

**a.** I-1 and I-2

**b.** II-4

**c.** II-5

**d.** II-2 and II-4

*__12.44__ In the following pedigree, individuals with Duchenne muscular dystrophy are shaded.

Generation:

I 1 2

II 1 2 3 4 5 6 7 8

III 1 2 3 4 5 6 7 8 9 10

IV 1 2 3 4

**a.** What inheritance pattern is shown by Duchenne muscular dystrophy?

**b.** Which members of the pedigree must be heterozygous for the Duchenne mutation?

**c.** If IV-1 and IV-2 have another child, what is the chance the child will develop the disease?

**d.** If IV-3 and IV-4 have another child, what is the chance the child will develop the disease?

*__12.45__ When constructing human pedigrees, geneticists often refer to particular individuals by a number. The generations are labeled with Roman numerals, the individuals in each generation with Arabic numerals. For example, in the pedigree in the following figure, the female with the asterisk is I-2:

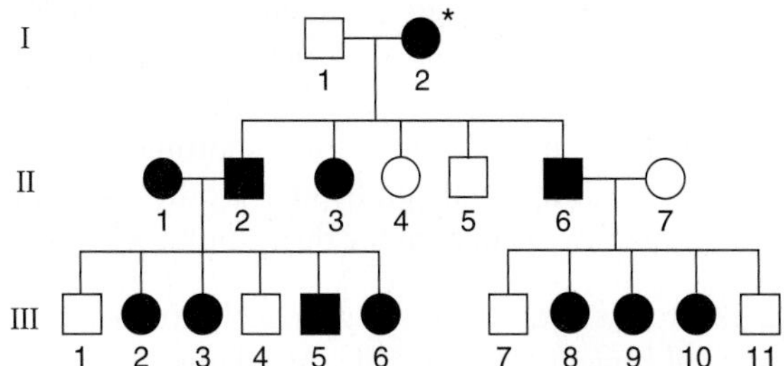


Use this method to designate specific individuals in the pedigree. Determine the probable inheritance mode for the trait shown in the affected individuals (the shaded symbols) by answering the following questions (assume that the condition is caused by a single gene):

**a.** Y-linked inheritance can be excluded at a glance. What two other mechanisms of inheritance can be definitely excluded? Why can these be excluded?
**b.** Of the remaining mechanisms of inheritance, which is the most likely? Why?

**12.46** A three-generation pedigree for a particular human trait is shown in the following figure:

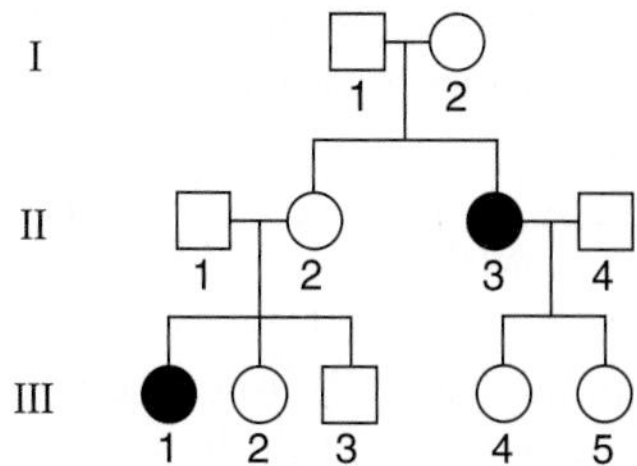


**a.** What is the mechanism of inheritance for the trait?
**b.** Which persons in the pedigree are known to be heterozygous for the trait?
**c.** What is the probability that III-2 is a carrier (heterozygous)?
**d.** If III-3 and III-4 marry, what is the probability that their first child will have the trait?

**12.47** For each of the more complex pedigrees shown in Figure 12.B, determine the probable mechanism of inheritance: autosomal recessive, autosomal dominant, X-linked recessive, X-linked dominant, or Y linked.

**Figure 12.B**

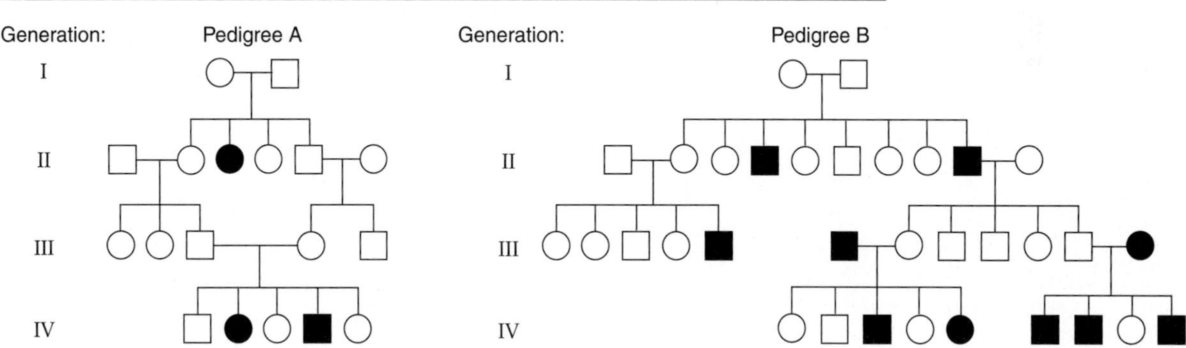


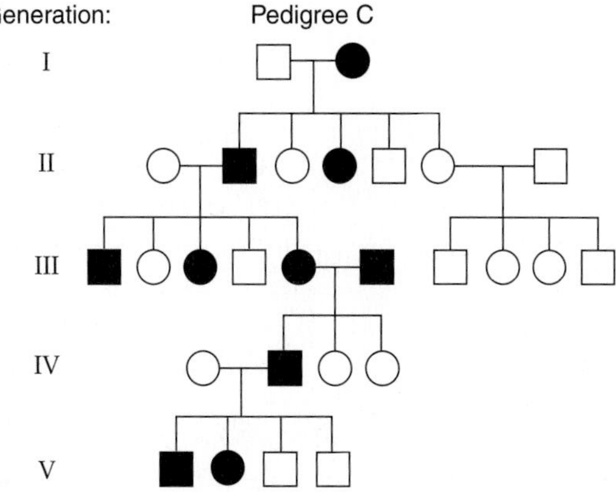

***12.48** A genetic disease is inherited on the basis of an autosomal dominant gene. State whether each of the statements that follows is true or false with regard to this disease, explaining your reasoning in each case.

**a.** Affected fathers have only affected children.

**b.** Affected mothers never have affected sons.

**c.** If both parents are affected, all of their offspring have the disease.

**d.** If a child has the disease, one of his or her grandparents also had the disease.

***12.49** A genetic disease is inherited as an autosomal recessive. State whether each of the statements that follows is true, false, or neither with regard to this disease, explaining your reasoning in each case.

**a.** Two affected individuals never have an unaffected child.

**b.** Two affected individuals have affected male offspring, but no affected female children.

**c.** If a child has the disease, one of his or her grandparents had it.

**d.** In a marriage between an affected individual and an unaffected one, all the children are unaffected.

**12.50** State whether each of the following statements is true or false with regard to a disease that is inherited as a rare X-linked dominant trait, explaining your reasoning in each case.

**a.** All daughters of an affected male will inherit the disease.

**b.** Sons will inherit the disease only if their mothers have it.

**c.** Both affected males and affected females will pass the trait to half the children.

**d.** Daughters will inherit the disease only if their fathers have it.

***12.51** Women who were known to be carriers of the X-linked, recessive hemophilia gene were studied to determine the amount of time required for blood to clot. It was found that the time required for clotting was extremely variable from individual to individual. The values obtained ranged from normal clotting time at one extreme to clinical hemophilia at the other. What is the most probable explanation for these findings?

**12.52** Hurler syndrome is a genetically transmitted disorder of mucopolysaccharide metabolism resulting in short stature, mental retardation, and various bony malformations. Two specific types are described with extensive pedigrees in the medical genetics literature:

Type I: recessive autosomal
Type II: recessive X linked

You are a consultant in a hospital ward with several patients with Hurler syndrome. They have asked you for advice about their relatives' offspring. Being aware that both types are extremely rare and that afflicted individuals almost never reproduce, what counsel would you give to a woman with type I Hurler syndrome (whose normal brother's daughter is planning marriage) about the offspring of the proposed marriage? In your answer, state the probabilities that the offspring will be affected and whether male and female offspring have an equal probability of being affected.

# 13 Extensions of and Deviations from Mendelian Genetic Principles

Palomino horse. The coat color is an intermediate phenotype resulting from incomplete dominance.

## Key Questions

- How many alleles can a gene have?
- How do incomplete dominance and codominance affect phenotypic ratios?
- What is the effect of a mutation in a gene that is essential for cell or organism function?
- How does the internal and external environment affect gene expression?
- What is maternal effect?
- How do you determine the number of genes involved in a set of mutants that all have the same phenotype?
- How do interactions between two genes affect Mendelian ratios in crosses?
- How does a modifier gene affect the phenotypic expressions associated with alleles of another gene?
- What is the inheritance pattern of extranuclear genes (genes found in the genomes of mitochondria and chloroplasts)?

### iActivity

A HALF-CENTURY BEFORE WATSON AND CRICK determined the structure of DNA, Karl Landsteiner discovered that different individuals have different blood types and that these blood types are inherited. However, the inheritance of blood type does not always follow the inheritance patterns predicted by Mendel's principles. As it turns out, blood types are one example of a trait that has an inheritance pattern more complex than Mendel described. In this chapter, you will learn about the inheritance of blood types and other traits that represent exceptions to, and extensions of, Mendel's principles. Then, in one iActivity, you can use your understanding of these inheritance patterns to help solve a paternity suit involving the actor Charlie Chaplin. You will also learn in this chapter about the non-Mendelian inheritance patterns of extranuclear genes, those genes located on chromosomes in mitochondria and chloroplasts. Then, in the second iActivity, you can use your understanding of both Mendelian and non-Mendelian inheritance patterns to study an unusual human disorder.

Mendel's principles apply to all diploid eukaryotic organisms and form the foundation for predicting the outcome of crosses in which segregation and independent assortment occur. As more and more geneticists did experiments, though, they found that Mendel's principles did not apply exactly. Sex linkage, discussed in the previous chapter, is one example of an extension of Mendelian principles. In sex linkage, genes segregate because of chromosome segregation—as is the case with the genes Mendel studied—but the pattern of segregation differs from the patterns Mendel saw in his experiments. Several cases of extensions of Mendelian principles are discussed in this chapter with the goal of gaining a broader knowledge of genetic analysis, particularly in terms of how genes relate to the phenotypes of an organism. Examples are also presented of segregation patterns that deviate from Mendelian principles because the genes for the traits involved are in organellar genomes rather than in the nuclear genome. This chapter first discusses extensions of Mendelian principles involving single genes (multiple alleles, modifications of dominance relationships, essential genes and lethal alleles, gene expression and the environment, and maternal effect). The chapter then discusses extensions of Mendelian principles involving two genes (determining the number of genes for mutations with the same phenotype, gene interactions and modified Mendelian ratios, and gene interactions involving modifier genes) and concludes by discussing deviations from Mendelian principles involving genes that are outside of the nucleus.



## Multiple Alleles

So far in our genetic analyses, we have talked about genes as if they have only two alleles—a normal allele and a mutant allele. An example is smooth versus wrinkled seeds in peas. In a population of individuals, however, a given gene may have several alleles (often one wild type and the rest mutant), not just two. Such genes are said to have **multiple alleles,** and the alleles are said to constitute a *multiple allelic series* (Figure 13.1). At the molecular level, multiple alleles represent different forms of the DNA sequence of the gene; the figure illustrates this with a short theoretical DNA segment. Although a gene may have multiple alleles in a given population of individuals, *a single diploid individual can have only a maximum of two of these alleles, one on each of the two homologous chromosomes carrying the gene locus.*

The number of possible genotypes in a multiple allelic series depends on the number of alleles involved (Table 13.1). With one allele, only one genotype is possible (such as *A*). With two alleles, three genotypes are possible, namely the two homozygotes and the heterozygote (e.g., *AA*, *aa*, and *Aa*). The general formula for $n$ alleles is $n(n + 1)/2$ possible genotypes, of which $n$ are homozygotes and $n(n - 1)/2$ are heterozygotes. For the four alleles (one wild-type and three mutant) in Figure 13.1, there are 10 possible genotypes.

**Figure 13.1**

**Conceptual illustration of multiple alleles of a gene.** Shown is a theoretical short segment of a gene's DNA for the wild-type allele, *A*, and three mutant alleles in a population, $a_1$, $a_2$, and $a_3$, each of which has a base-pair change compared with the wild-type allele.

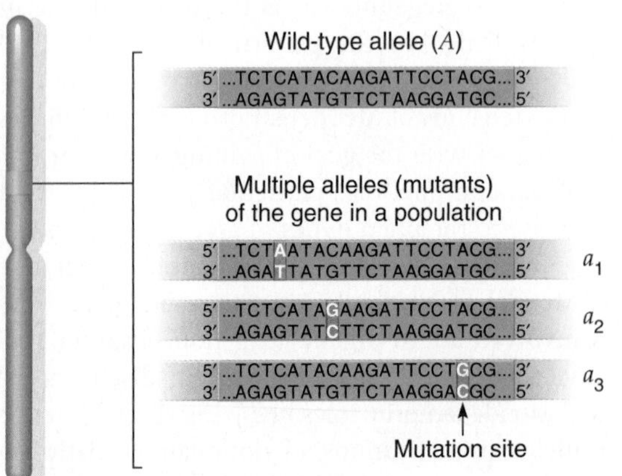


**Table 13.1 Genotype Number of Multiple Alleles**

| Number of Alleles | Kinds of Genotypes | Kinds of Homozygotes | Kinds of Heterozygotes |
|---|---|---|---|
| 1 | 1 | 1 | 0 |
| 2 | 3 | 2 | 1 |
| 3 | 6 | 3 | 3 |
| 4 | 10 | 4 | 6 |
| 5 | 15 | 5 | 10 |
| $n$ | $n(n + 1)/2$ | $n$ | $n(n - 1)/2$ |

### ABO Blood Groups

An example of multiple alleles of a gene is found in the human ABO blood group series, which was discovered by Karl Landsteiner in the early 1900s. He received the 1930 Nobel Prize in Physiology or Medicine for his discovery of human blood groups. Since certain ABO blood groups are incompatible, these alleles are particularly important when blood transfusions are done. (There are many blood group series other than ABO, and they may also cause problems in blood transfusions.)

O, A, B, and AB are the four blood group phenotypes in the ABO system. Different combinations of three alleles of the ABO blood group gene—$I^A$, $I^B$, and *i*—give rise to the four phenotypes (Table 13.2). People of blood group O are homozygous for the recessive *i* allele. Both $I^A$ and $I^B$ are dominant to *i*. Individuals of blood group A are either $I^A/I^A$ or $I^A/i$, and those of blood group B are either $I^B/I^B$ or $I^B/i$. Heterozygous $I^A/I^B$ individuals are of blood group AB—that is, essentially of both blood groups A and B (see the discussion of codominance in this chapter, pp. 368–369).

The genetics of this system follows Mendelian principles. An individual who expresses blood group O, for example, must be *i/i* in genotype. Therefore, each parent must be either homozygous *i* or heterozygous, with *i* as one of the two alleles. That is, both parents could be O ($i/i \times i/i$), both could be A ($I^A/i \times I^A/i$ to produce one-fourth *i/i* progeny), both could be B ($I^B/i \times I^B/i$), or one could be A and one could be B ($I^A/i \times I^B/i$).

Blood typing (determining an individual's blood group) and analyzing blood group inheritance sometimes are used in cases of disputed paternity or maternity. In such cases, genetic analysis based on blood group can be used to show that an individual is *not* the parent of a

**Table 13.2 ABO Blood Groups In Humans, Determined by the Alleles $I^A$, $I^B$, and $i$**

| Phenotype (Blood Group) | Genotype |
|---|---|
| O | $i/i$ |
| A | $I^A/I^A$ or $I^A/i$ |
| B | $I^B/I^B$ or $I^B/i$ |
| AB | $I^A/I^B$ |

particular child, but not to prove the individual *is* the parent. For example, a child of phenotype AB (genotype $I^A/I^B$) could not be the child of a parent of phenotype O (genotype $i/i$). (In most states, blood type data alone usually are not sufficient for a legal decision about paternity or maternity. The more precise results available from DNA fingerprinting, discussed in Chapter 10, pp. 277–280, are typically required for this.)

With blood transfusions, the blood types of donors and recipients must be carefully matched, because the blood group alleles specify molecular groups, called *cellular antigens*, that are attached to the outsides of the red blood cells. An **antigen** (*anti*body-*gen*erating substance) is any molecule that is recognized as foreign by an organism and that therefore stimulates the production of specific protein molecules called antibodies, which bind to the antigen. An **antibody** is a protein molecule that recognizes and binds to the foreign substance (antigen) introduced into the organism as part of the immune response to remove the foreign antigen from the body. Any given individual has a large number of antigens on cells and tissues, many of which will be foreign to any other given individual—hence the concern over blood type in blood transfusions and tissue type in organ transplants. Except in autoimmune diseases, antigens are not recognized as foreign by the organism expressing them.

The $I^A$ allele of the ABO blood group gene encodes a product that is needed for the biosynthesis of the A antigen but that is not involved in the biosynthesis of the B antigen. People of blood type A (genotype $I^A/I^A$ or $I^A/i$) have only the A antigen on their red blood cells and therefore the B antigen is foreign to them. Blood serum prepared from them contains naturally occurring antibodies against the B antigen (called anti-B antibodies), but none against the A antigen. Antibodies against the B antigen agglutinate, or clump, any red blood cells that have the B antigen on them. Since clumped cells cannot move through the fine capillaries, agglutination may lead to organ failure and, possibly, death.

The $I^B$ allele of the ABO blood group gene encodes a product needed for biosynthesis of the B antigen, but it is not involved in biosynthesis of the A antigen. Therefore, people of blood type B (genotype $I^B/I^B$ or $I^B/i$) have the B antigen on their red blood cells, and their blood serum contains naturally occurring anti-A antibodies but no anti-B antibodies. People of AB blood type (genotype $I^A/I^B$) have both A and B antigens on the blood cells and neither anti-A nor anti-B antibodies in their blood serum.

Lastly, the $i$ allele encodes no functional products involved in the biosynthesis of either the A or the B antigen. Therefore, in people with blood type O ($i/i$), the red blood cells have neither A nor B antigen, and their blood serum contains both anti-A and anti-B antibodies. The antigen–antibody relationships are summarized in Figure 13.2. Agglutination (clumping) of the red blood cells is seen in each case where an antibody interacts with the antigen for which it is specific.

What transfusions are safe between people with different blood groups in the ABO system?

1. People with blood type A produce the A antigen, so their blood can be transfused only into recipients who do not have the anti-A antibody—that is, people of blood type A or AB.
2. People with blood type B produce the B antigen, so their blood can be transfused only into recipients who do not have the anti-B antibody—that is, people of blood type B or AB.
3. People with blood type AB produce both the A and B antigens, so their blood can be transfused only into recipients who do not have either the anti-A antibody or the anti-B antibody—that is, people of blood type AB.

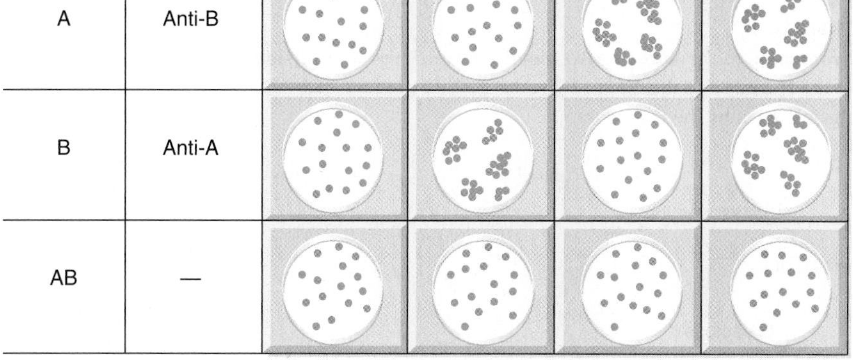

**Figure 13.2**

**Antigenic reactions that characterize the human ABO blood types.** Blood serum from each of the four blood types was mixed with blood cells from the four types in all possible combinations. In some cases, such as a mix of B serum with A cells, the cells become clumped.

| Serum from blood type | Antibodies present in serum | Cells from blood type | | | |
|---|---|---|---|---|---|
| | | O | A | B | AB |
| O | Anti-A Anti-B | | | | |
| A | Anti-B | | | | |
| B | Anti-A | | | | |
| AB | — | | | | |

4. People with blood type O produce neither A nor B antigens, so their blood can be transfused into any recipient—that is, people of blood type A, B, AB, or O.

The preceding discussion indicates that people with blood type AB are *universal recipients* because they can receive blood from people of any of the four blood types, and that people with blood type O are *universal donors* because their blood elicits no reaction in people of any of the four blood types.

The relationship between the ABO multiple alleles and the antigens on the red blood cells is as follows: The ABO blood group gene encodes *glycosyltransferases,* enzymes that add sugar groups to a preexisting polysaccharide that is combined with a lipid to form a glycolipid (Figure 13.3). The glycolipids then associate with red blood cell membranes to form the blood group antigens. Most people produce a glycolipid called the H antigen. The $I^A$ allele encodes a glycosyltransferase enzyme that adds a particular type of sugar to the H antigen to produce the A antigen. The $I^B$ allele encodes a different glycosyltransferase, which adds a different sugar to the H antigen to produce the B antigen. An important point to understand here is that it is the differences in the DNA sequences of the $I^A$ and $I^B$ alleles that result in two functionally different, but highly related, glycosyltransferases. The small difference in the structure of the A and B antigens produced by these enzymes is recognized by the immune system.

In an $I^A/I^B$ heterozygote, both enzymes are produced and, therefore, some H antigen is converted to the A antigen and some is converted to the B antigen. The red blood cell has both antigens on the surface, so the person is of blood group AB.

People who are homozygous for the *i* allele produce no enzymes to convert the H antigen glycolipid. Therefore, their red blood cells carry only the H antigen. This antigen does not elicit an antibody response in people of other blood groups, because its polysaccharide component is also the basic component of the A and the B antigens and, therefore, it is not detected as a foreign substance. People who are heterozygous for the *i* allele have the blood type of the other allele. For example, in $I^B/i$ people, the $I^B$ allele results in the conversion of some of the H antigen to the B antigen, determining the person's blood type.

The H antigen is encoded by the dominant *H* allele of locus distinct from the ABO blood group gene. People who are homozygous for the recessive mutant allele, *h*, do not make the H antigen; therefore, regardless of the presence of $I^A$ or $I^B$ alleles at the ABO blood group gene, no A or B antigens can be produced. These very rare *h/h* people are like blood group O people in the sense that they lack A and B antigens; they are said to have the Bombay blood type. However, people in the Bombay blood group produce anti-O antibodies (antibodies against the H antigen), whereas people in blood group O do not.

**Figure 13.3**

**Production of the human ABO blood type antigens.** Conversion of the H antigen to the A antigen by the $I^A$ allele product, and to the B antigen by the $I^B$ product.

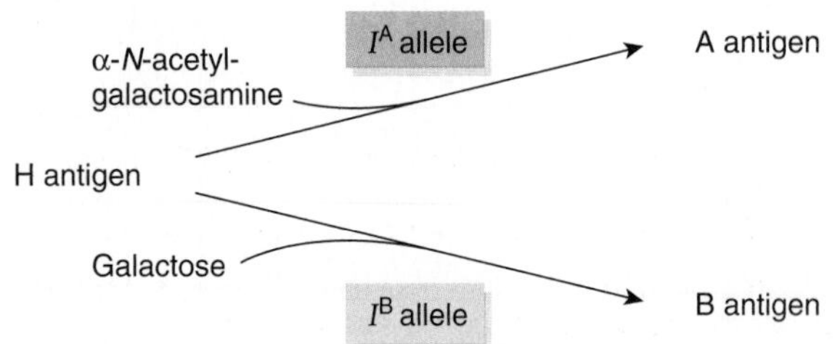


**iActivity**

Go to the iActivity *Was She Charlie Chaplin's Child?* on the student website, where you will use your expertise to interpret the results of blood group tests that may prove whether silent-movie great Charlie Chaplin is the father of Carol Ann Berry.

### *Drosophila* Eye Color

Another example of multiple alleles concerns the white (*w*) locus of *Drosophila*. Recall from Chapter 12 (pp. 341–343) that the $w^+$ allele results in wild-type brick-red eyes and that the recessive *w* allele, when homozygous or hemizygous, results in white eyes. There are more than 100 recessive mutant alleles at the white locus. Each allele, when homozygous, has a distinctive color on the spectrum between white and red. The specific eye color of each mutant depends on how much function the encoded protein of the *w* allele involved has lost, which determines how much pigment becomes deposited in the eye.

One allele of the white locus is *eosin* which, when homozygous, gives a reddish-orange eye color. The symbol for this allele is $w^e$. Genetic crosses done by Alfred Sturtevant in 1913 showed that: (1) red (wild-type) eye color is dominant to *eosin* and to *white*; and (2) *eosin* is recessive to the wild type, but dominant to *white*. Figure 13.4 illustrates these properties. In Figure 13.4a, a homozygous eosin-eyed female is crossed with a white-eyed male. The $F_1$ females are $w^e/w$ and have eosin eyes, because $w^e$ is dominant over *w*. When these $F_1$ females are crossed with red-eyed males, who are $w^+/Y$ (Figure 13.4b), all the female progeny are heterozygous and red-eyed, because they contain the $w^+$ allele; they are either $w^+/w^e$ or $w^+/w$. Half the male progeny are eosin-eyed ($w^e/Y$), and the other half are white-eyed ($w/Y$).

### Relating Multiple Alleles to Molecular Genetics

The base-pair sequence of a gene specifies the amino acid sequence of a protein, and the function of a protein depends on its amino acid sequence. From this modern perspective, we should not be surprised to find multiple alleles of a gene. For example, an amino acid change at one of many places in the protein could affect its function adversely, and the position and type of the change would determine the extent of loss of function of the protein.

**Figure 13.4**

**Results of crosses of *Drosophila melanogaster* involving two mutant alleles of the same locus: white (*w*) and white-eosin ($w^e$). (a)** White-eosin-eyed ($w^e/w^e$) ♀ × white-eyed ($w$/Y) ♂. **(b)** $F_1$ ($w^e/w$) ♀ × red-eyed (wild type) ($w$/Y) ♂.

**a) White-eosin-eyed female × white-eyed male**

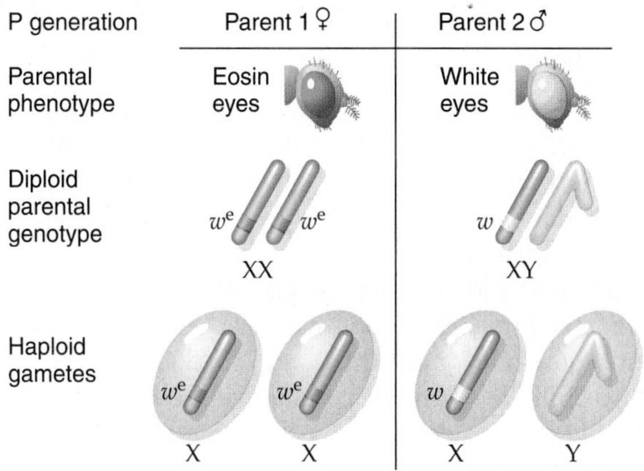


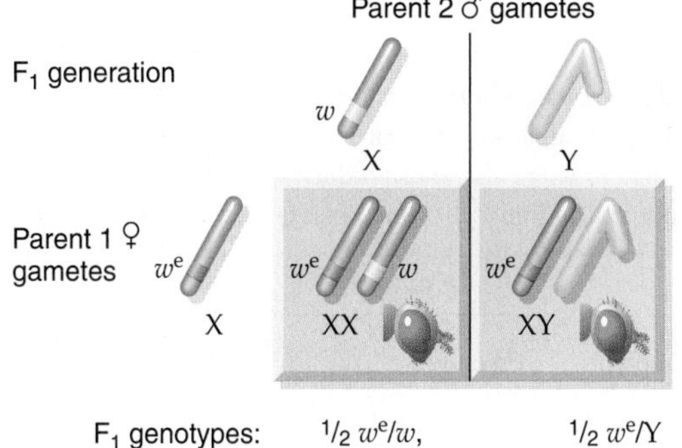


**b) $F_1$ female × wild-type male**

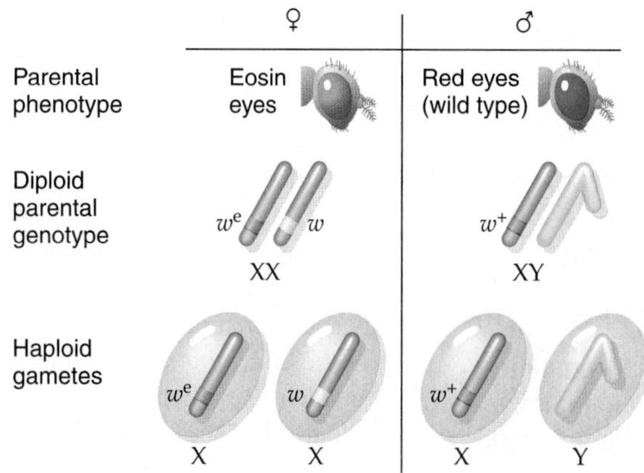


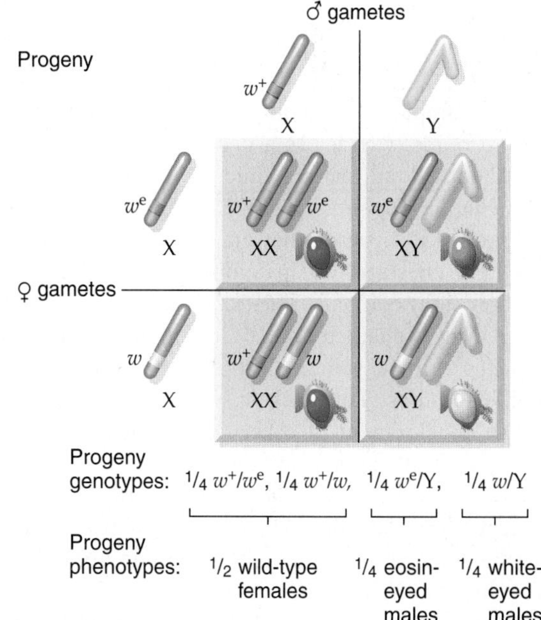


The *Drosophila white* gene alleles illustrates this concept. If you look at the entries for many human genetic diseases in OMIM (http://www.ncbi.nlm.nih.gov/omim), such as the breast cancer susceptibility gene *BRCA1* (OMIM 113705); the *APC* (adenomatous polyposis of the colon) gene (OMIM 175100), which is mutated in familial adenomatous polyposis, an autosomal dominant disease typically first detected as colorectal cancer; and the phenylalanine hydroxylase gene (OMIM 261600), which is mutated in individuals with phenylketonuria, you will often find that several alleles (listed under "Allelic Variants" in the entries) have been identified and associated with the diseases. Two practical consequences of multiple alleles in the case of human genetic diseases are that the symptoms of a disease may vary with the allele and therefore it is important to determine the exact alleles of the patients in such cases.

**Keynote**

Many allelic forms of a gene can exist in a population. When they do, the gene is said to have multiple alleles, and the alleles involved constitute a multiple allelic series. However, any given diploid individual can possess a maximum of two different alleles of a given gene. Multiple alleles obey the same rule of transmission as do alleles of which there are only two types, although the dominance relationships among multiple alleles vary from one group to another.

## Modifications of Dominance Relationships

**Complete dominance** is the phenomenon in which one allele is dominant to another, so that the phenotype of the heterozygote is the same as that of the homozygous dominant. With **complete recessiveness,** the recessive allele is phenotypically expressed only when it is homozygous. Complete dominance and complete recessiveness are the two extremes of a range of dominance relationships for two alleles. Whereas all the allelic pairs Mendel studied showed complete dominance–complete recessiveness relationships, many allelic pairs do not.

### Incomplete Dominance

When one allele of a gene is not completely dominant to another allele of the same gene, it is said to show **incomplete dominance,** also referred to as *semidominance* or *partial dominance*. With incomplete dominance, the phenotype of the heterozygote lies in the range between the phenotypes of individuals that are homozygous for either allele involved. The phenotype of the heterozygote is typically referred to as an *intermediate* phenotype, even though it may not be exactly in the middle between the phenotypes of the two homozygotes.

**Animation**
**Incomplete Dominance and Codominance**

An example of incomplete dominance is the palomino horse, which has a golden-yellow body color and a mane and tail that are almost white (see the photo on this chapter's opening page). Palominos do not breed true (Figure 13.5). When they are interbred, the progeny are 1/4 light chestnuts, 1/2 palominos, and 1/4 cremellos (extremely light colored). This 1:2:1 ratio resulting from interbreeding is characteristic of incomplete dominance. Two alleles of the *C* gene, *C* and $C^{cr}$, are involved in producing palominos. The *C* allele, when homozygous, allows full development of other coat color genes in the horse. The $C^{cr}$ allele is a modifier allele that dilutes the expression of other coat color genes in the horse in a dose-dependent fashion. That is, heterozygous $C/C^{cr}$ horses have less diluted coat colors than do homozygous $C^{cr}/C^{cr}$ horses. When the combination of alleles of other genes for coat color in the horse would specify light chestnut, the $C/C^{cr}$ genotype dilutes that potential color to give a palomino. Palomino intercrosses then are $C/C^{cr} \times C/C^{cr}$, which results in a progeny ratio of 1 *C/C*, light chestnuts (full coat color) : 2 $C/C^{cr}$, palominos (partly diluted light chestnut color) : 1 $C^{cr}/C^{cr}$, cremellos (fully diluted light chestnut color; see Figure 13.5).

There are many examples of incomplete dominance in plants, such as flower color in the snapdragon, which involves two alleles of the *C* gene of the plant, $C^R$ and $C^W$. $C^R/C^R$ snapdragons have red flowers, $C^R/C^W$ plants have pink flowers, and $C^W/C^W$ plants have white flowers. The allelic symbols in this example are useful when diagramming crosses involving incompletely dominant traits because they are designed to give equal weight to the two alleles, an indication that neither dominates the phenotype. In this particular example, the *C* signifies color, while "R" and "W" indicate red and white, respectively.

Some human diseases show incomplete dominance. One example we have already discussed is sickle-cell anemia (OMIM 603903; see Chapter 4, pp. 70–71) in which homozygotes for the sickle-cell mutant alleles have sickle-cell anemia, while heterozygotes for that allele have the milder sickle-cell trait.

### Codominance

Another modification of the dominance relationship is **codominance**. In codominance, the heterozygote exhibits the phenotypes of *both* homozygotes. By contrast, in incomplete dominance, the heterozygote exhibits a phenotype intermediate between the two homozygotes.

The ABO blood group series discussed earlier in this chapter provides a good example of codominance. Heterozygous $I^A/I^B$ individuals are of blood group AB because both the A antigen (product of the $I^A$ allele) and the B antigen (product of the $I^B$ allele) are produced. Thus, the $I^A$ and $I^B$ alleles are codominant.

**Figure 13.5**

**Incomplete dominance in horses.** Palomino horses are heterozygotes for a coat color dilution gene that shows incomplete dominance. Intercrossing palominos produces progeny in the ratio 1 light chestnut : 2 Palomino : 1 cremello (extremely light coat color).

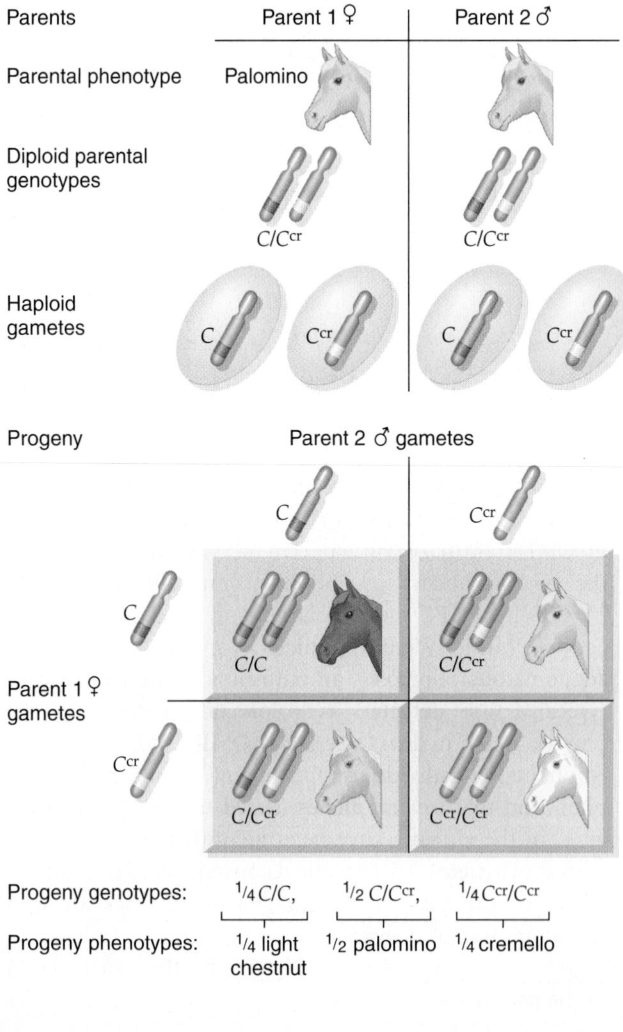

The human M–N blood group system is another example of codominance. For transfusion compatibility, this system is of less clinical importance than the ABO system. In the M–N system, alleles $L^{M}$ and $L^{N}$ of one gene determine the blood types. Three blood types occur: M, MN, and N, respectively specified by the genotypes $L^{M}/L^{M}$, $L^{M}/L^{N}$, and $L^{N}/L^{N}$. As in the ABO system, the M–N alleles result in the formation of antigens on the surface of the red blood cell. The heterozygote in this case has both the M and the N antigens and shows the phenotypes of both homozygotes.

### Molecular Explanations of Incomplete Dominance and Codominance

What explains incomplete dominance and codominance at the molecular level? A general interpretation is that, in codominance, products result from both alleles in a heterozygote, and the nature of the phenotype in question allows the distinct phenotypes seen in the two homozygotes to be observed simultaneously. For example, $L^{M}/L^{N}$ individuals express both M and N blood group antigens. There are many cases of complete dominance in which products result from both alleles, but the phenotype is not of a kind that enables the two homozygote phenotypes to be observed simultaneously. Rather, it is just one phenotype seen in the heterozygote; that is, the one we designate as dominant.

In incomplete dominance, in cases involving a loss-of-function allele, only one allele in a heterozygote is expressed to produce a product and the quantity of the product is important for the phenotype. A homozygote for the expressed allele, then, has two doses of the gene product, and full phenotypic expression results (for example, light chestnut horses or red snapdragons). In a homozygote for the allele that is not expressed, a phenotype characteristic of no gene expression results (cremello horses or white snapdragons). In a heterozygote, the single allele expressed results in only enough product for an intermediate phenotype (palomino horses or pink snapdragons). By contrast, in a heterozygote for an allele showing normal dominance, half the amount of protein produced by the homozygote is sufficient for normal cell function. Alternatively, the expression of the one normal allele in the heterozygote may be increased to produce protein levels that give normal cell function. In both of these cases the gene is said to be **haplosufficient**, defined as the condition where one copy of a gene in a diploid organism is sufficient to give a normal phenotype.

Incomplete dominance involving a gain-of-function allele also occurs. In this case, homozygosity for the wild-type allele gives the wild-type phenotype, two doses of the gain-of-function allele in a homozygote results in the fully mutant phenotype, and one dose of the gain-of-function allele in a heterozygote results in a mutant phenotype intermediate between the wild-type and fully mutant phenotypes. A number of dominant human genetic disorders follow this pattern, with individuals homozygous for the dominant mutant allele exhibiting much more severe (sometimes lethal) symptoms than do individuals who are heterozygous.

#### Keynote

With complete dominance, the same phenotype results whether the dominant allele is heterozygous or homozygous. With complete recessiveness, the allele is phenotypically expressed only when the genotype is homozygous recessive; the recessive allele has no effect on the phenotype of the heterozygote. Complete dominance and complete recessiveness are two extremes between which all transitional degrees of dominance are possible. In incomplete dominance, the phenotype of the heterozygote is intermediate between those of the two homozygotes, whereas in codominance the heterozygote exhibits the phenotypes of both homozygotes.

## Essential Genes and Lethal Alleles

For a few years after the rediscovery of Mendel's principles, geneticists believed that mutations only changed the appearance of a living organism. But then they discovered that a mutant allele could cause death. In a sense, this mutation is still a change in phenotype, with the new phenotype being lethality. An allele that results in the death of an organism is called a **lethal allele,** and the gene involved is called an essential gene. **Essential genes** are genes that, when mutated, can result in a lethal phenotype. If the mutation is caused by a **dominant lethal allele,** both homozygotes and heterozygotes for that allele show the lethal phenotype. If the mutation is caused by a **recessive lethal allele,** only homozygotes for that allele have the lethal phenotype.

An example of a recessive lethal gene is the allele for yellow body color in mice. No true-breeding yellow mice exist. From a yellow × yellow cross, progeny are produced with a phenotypic ratio of 2 yellow : 1 nonyellow (the nonyellow color depends on which other coat color genes are present). The living yellow mice are all heterozygotes for a yellow allele; homozygotes for the yellow allele die at the embryo stage. In other words, the yellow allele has a *dominant* effect with regard to coat color, but acts as a *recessive lethal* allele—that is, individuals homozygous for it die.

The recessive lethal yellow allele is an allele of the agouti locus (*a;* see discussion of recessive epistasis later in the chapter) and has been given the symbol $A^{Y}$. The yellow × yellow cross genotypically is $A^{Y}/A \times A^{Y}/A$ (Figure 13.6). The expected genotypic ratio in the progeny is $^{1}/_{4}\ A^{Y}/A^{Y} : {}^{2}/_{4}\ A^{Y}/A : {}^{1}/_{4}\ A/A$. The $^{1}/_{4}\ A^{Y}/A^{Y}$ mice die before birth, giving a birth ratio of $^{2}/_{3}\ A^{Y}/A$ (yellow) : $^{1}/_{3}\ A/A$ (nonyellow). Characteristically, recessive lethal alleles are recognized by a 2:1 ratio of progeny types from crosses of two heterozygotes.

The agouti gene of mouse has been molecularly cloned, permitting analysis of the lethal yellow allele. In wild-type agouti mice, the agouti gene is expressed in skin samples taken a few days after birth, when the yellow band

**Figure 13.6**

**Inheritance of a lethal gene $A^Y$ in mice.** A mating of two yellow mice gives 1/4 nonyellow (black) mice, 1/2 yellow mice, and 1/4 dead embryos. The viable yellow mice are heterozygous $A^Y/A$, and the dead individuals are homozygous $A^Y/A^Y$.

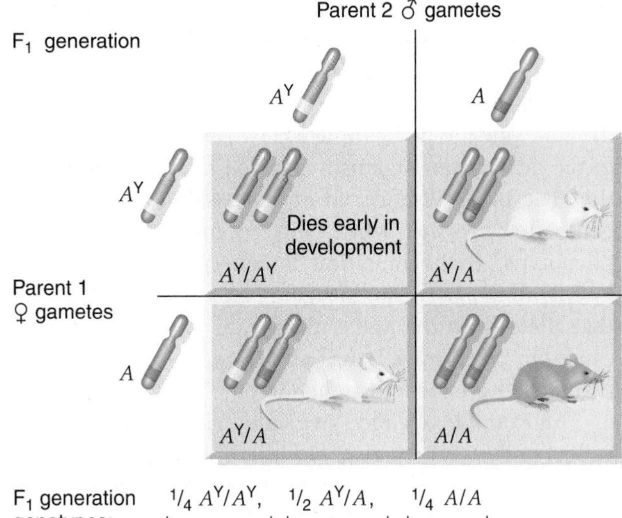


in the hair is being produced, in skin during regeneration of hair after plucking, and in no other tissues and at no other time. In heterozygous $A^Y/A$ mice, the yellow allele is expressed at high levels in all tissues and at all developmental stages, indicating that tissue-specific regulation of expression has been lost. The explanation is that the $A^Y$ allele has resulted from the deletion of a large DNA segment between the agouti gene and an upstream gene called *Raly*, such that the *Raly* promoter and the first part of that gene are now fused to the agouti gene. The *Raly* promoter thus controls the expression of the attached agouti gene. The expression in all tissues is caused by regulatory signals in the *Raly* promoter. The embryonic lethality of yellow homozygotes probably results from the absence of *Raly* gene activity rather than a defective agouti gene.

Essential genes are found in all organisms. You would expect, for instance, that at least some of the genes involved in key organismal functions, such as DNA replication, transcription, and translation, are essential. In a methodical study in which each protein-coding gene of yeast was knocked out one-by-one, researchers found that about 1,800 of the approximately 5,700 genes were essential.

There are many known recessive lethal alleles in humans. One example is Tay–Sachs disease (OMIM 272800; see Chapter 4, pp. 68–69). The gene involved, *HEXA*, encodes the enzyme hexosaminidase A. Homozygotes appear normal at birth, but before about 1 year of age they show symptoms of central nervous system deterioration. Progressive mental retardation, blindness, and loss of neuromuscular control follow. Afflicted children usually die at 3 to 4 years of age. The genetic defect in Tay–Sachs results in an enzyme deficiency that prevents proper nerve function. Most disease-causing mutations of the *HEXA* gene are single base-pair substitutions, either causing amino acid changes in the protein or altering the splicing of the gene's pre-mRNA.

There are X-linked lethal mutations as well as autosomal lethal mutations. There are also dominant lethal mutations as well as recessive lethals. In humans, for example, the genetic disease hemophilia (OMIM 306700) is caused by an X-linked recessive allele. Untreated hemophilia is lethal. Dominant lethals exert their effect in heterozygotes, resulting in the death of the organism. Dominant lethals cannot be studied over multiple generations in a family unless death occurs after the organism has reached reproductive age. For example, the symptoms of the autosomal dominant trait Huntington disease (OMIM 143100)—involuntary movements and progressive central nervous system degeneration—may not begin until affected individuals reach their early thirties; as a result, parents may unknowingly pass on the gene to their offspring. Death usually occurs when the afflicted persons are in their forties or fifties. The American folk singer Woody Guthrie died from Huntington disease in 1967.

**Keynote**

A lethal allele is fatal to the individual. There are recessive lethal and dominant lethal alleles, and they can be X-linked or autosomal. The existence of lethal alleles of a gene indicates that the gene's normal product is essential to the functioning of the organism; therefore, the gene is an essential gene.

## Gene Expression and the Environment

The *development* of a multicellular organism from a zygote is a process of *regulated growth and differentiation* that results from the interaction of the organism's genome with both the internal cellular environment and the external environment. Development is a tightly controlled, programmed series of phenotypic changes that, under normal environmental conditions, are essentially irreversible. Four major processes interact to constitute the complex process of development: replication of the genetic material, growth, differentiation of the various cell types, and the arrangement of differentiated cells into defined tissues and organs.

Think of development as a series of complex, intertwined biochemical pathways. The internal or external environment may influence any of these pathways by affecting the products of the genes controlling the pathways. This phenomenon is most readily studied in experimental organisms whose genotypes are unequivocally known. The extent to which the gene manifests its effects under varying environmental conditions can then be seen. We consider some examples in the next section. They reinforce the important concept presented at the beginning of Chapter 11 that *genes provide only the potential for developing a particular phenotype, and that the extent to which that potential is realized depends on interactions with other genes and their products—and, in many cases, on environmental influences and random developmental events* (see Figure 11.1, p. 298).

## Penetrance and Expressivity

In some cases, not all individuals with a particular genotype show the expected phenotype due to influences of the internal and/or external environment. The percentage of individuals with a given genotype who exhibit the phenotype associated with that genotype is called the **penetrance** of the genotype (Figure 13.7a). Penetrance depends on both the genotype (for example, the presence of epistatic or other genes) and the environment. Penetrance is *complete* (100%) when all the homozygous recessives show one phenotype, when all the homozygous dominants show another phenotype, and when all the heterozygotes are alike. For example, if all individuals carrying a dominant mutant allele show the mutant phenotype, the allele is completely penetrant. Many genes show complete penetrance; examples include the seven gene pairs in Mendel's experiments and the alleles in the human ABO blood group system.

If less than 100% of the individuals with a particular genotype exhibit the phenotype expected, penetrance is *incomplete*. If, for instance, 80% of the individuals carrying a particular gene show the corresponding phenotype, there

**Figure 13.7**

**Illustrations of the concepts of penetrance and expressivity in the phenotypic expression of a genotype.**

**a) Complete penetrance compared with incomplete penetrance**

<u>Complete penetrance</u>

Identical known genotypes yield 100% expected phenotype

<u>Incomplete penetrance</u>

Identical known genotypes yield <100% expected phenotype

**b) Constant expressivity compared with variable expressivity**

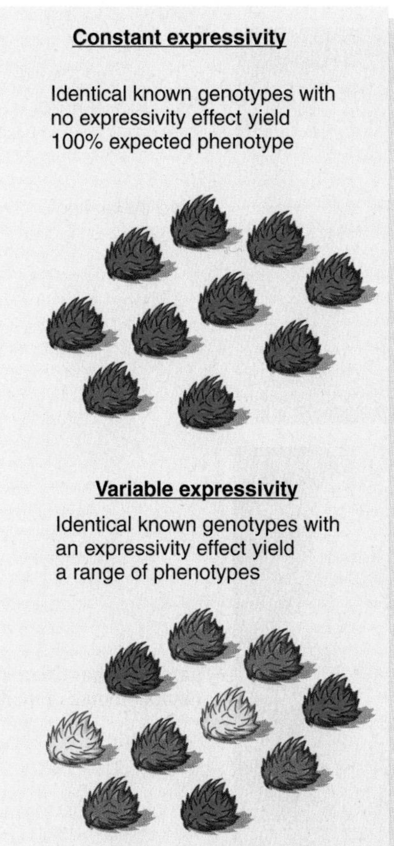
Constant expressivity
Identical known genotypes with no expressivity effect yield 100% expected phenotype
Variable expressivity
Identical known genotypes with an expressivity effect yield a range of phenotypes

**c) Incomplete penetrance with variable expressivity**

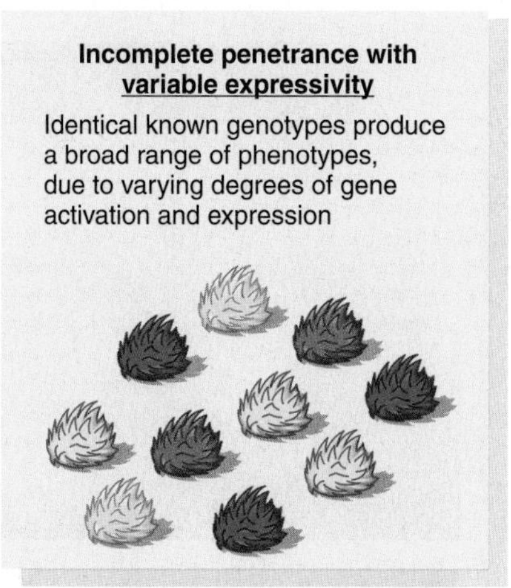
Incomplete penetrance with variable expressivity
Identical known genotypes produce a broad range of phenotypes, due to varying degrees of gene activation and expression

is 80% penetrance. In humans, many genes show reduced penetrance. For example, brachydactyly (OMIM 112500), an autosomal dominant trait that causes shortened and malformed fingers, shows 50–80% penetrance. A number of genes that confer a predisposition to cancer also exhibit low to moderate penetrance, adding to the difficulty of identifying and characterizing those genes.

Genes may influence a phenotype to different degrees. **Expressivity** is the degree to which a penetrant gene or genotype is phenotypically expressed in an individual (Figure 13.7b). Like penetrance, expressivity depends on both the genotype and the environment and thus may be *constant* or *variable*. Molecularly, we can think of expressivity at a simple level as the result of different degrees of function of the protein encoded by the gene.

An example of variation in expressivity is found in the human condition called osteogenesis imperfecta (OMIM 166200). The three main features of this disease are blueness of the sclerae (the whites of the eyes), very fragile bones, and deafness. Osteogenesis imperfecta is inherited as an autosomal dominant with almost 100% penetrance. However, the trait shows variable expressivity: a person with the mutation may have any one or any combination of the three traits. Moreover, the fragility of the bones for those who exhibit the condition is also highly variable.

Lastly, some genotypes exhibit both incomplete penetrance *and* variable expressivity. Figure 13.7c illustrates this concept. For example, neurofibromatosis (OMIM 162200) is an autosomal dominant disease that shows 50–80% penetrance as well as exhibiting variable expressivity (Figure 13.8). In its mildest form, the disease causes individuals to have only a few pigmented areas on the skin (called *café-au-lait spots* because they are the color of coffee with milk). In more severe cases, one or more other symptoms may be seen, including neurofibromas (tumorlike growths) of various sizes over the body; high blood pressure; speech impediments; headaches; a large head; short stature; tumors of the eye, brain, or spinal cord; and curvature of the spine. Because of these issues, in medical genetics it is important to recognize that an allele may vary widely in its expression, a qualification making the task of genetic counseling that much more difficult.

### Keynote

Penetrance is the frequency with which a genotype manifests itself in individuals in a given population. Expressivity is the type or degree of phenotypic manifestation of a penetrant allele or genotype in a particular individual.

## Effects of the Environment

In this section, we consider some examples of environmental influences on phenotype. In each case, the environment typically is not the only influence; other genes may also affect the phenotype.

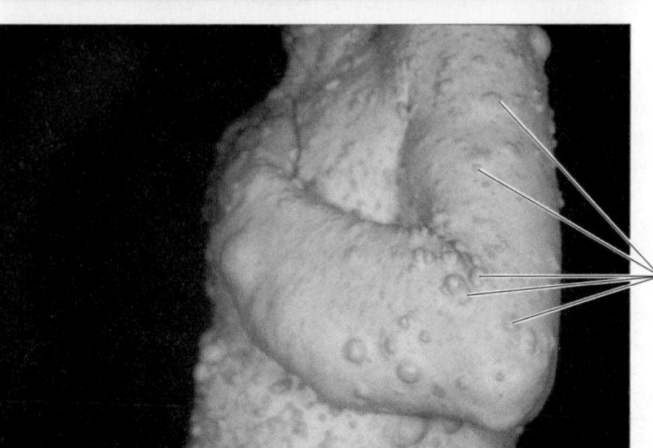


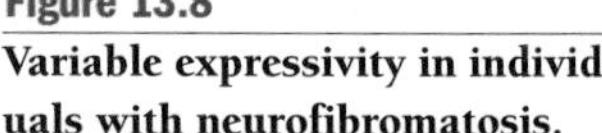

**Figure 13.8**
**Variable expressivity in individuals with neurofibromatosis.**

**Age of Onset.** The age of the organism creates internal environmental changes that can affect the phenotypic expression associated with an allele. Not all genes are active all the time. Rather, over time, programmed activation and deactivation of genes occurs as the organism develops and functions. Many human genetic traits are not exhibited at birth, despite the presence of the genotypes for those traits, but instead are exhibited in an age-dependent manner, For instance, pattern baldness (OMIM 109200) typically appears in males between 20 and 30 years of age, and symptoms of Duchenne muscular dystrophy (DMD; OMIM 310200) develop in children between 2 and 5 years of age. In most cases, the nature of the age dependency is not understood.

**Sex.** Phenotypic expression associated with an allele may be influenced by the sex of the individual. In the case of sex-linked genes, as mentioned earlier, differences in the phenotypes of the two sexes are related to different complements of genes on the sex chromosomes. However, in some cases, genes that are on autosomes affect a particular phenotype that appears in one sex, but not the other. Traits of this kind are called **sex-limited traits.**

Examples of sex-limited traits in animals are milk production in dairy cattle (the genes involved obviously operate in females but not in males), the appearance of horns in certain species of sheep (males with genes for horns have horns, and females with genes for horns do not have horns), and the ability to produce eggs or sperm. An example in humans is the distribution of facial hair.

A slightly different situation is found in **sex-influenced traits,** which, like sex-limited traits, often are controlled by autosomal genes. Such traits appear in both sexes, but either the frequency of occurrence in the two sexes is different or the relationship between genotype and phenotype is different.

Pattern baldness is an example of a sex-influenced trait in humans (Figure 13.9). Pattern baldness is controlled by an *autosomal* gene with an allele for baldness, *b*, that acts as a dominant in males and as a recessive (or at least it is expressed at lower levels) in females. That is, the *b/b* genotype specifies pattern baldness in both males and females, and the $b^+/b^+$ genotype gives a nonbald phenotype in both sexes. The difference lies in the heterozygote: In males $b^+/b$ leads to the bald phenotype, and in females it leads to the nonbald phenotype. In other words, the *b* allele acts as a dominant in males but as a recessive in females. The expression of the *b* allele is influenced by the sex hormones of the individual—the male hormone testosterone is responsible for expression of the pattern baldness allele *b* when it is present in one dose. The sex-influenced pattern of inheritance and gene expression explains why pattern baldness is far more common among men than among women. From a large sample of the progeny of matings between two heterozygotes, $^3/_4$ of the daughters are nonbald and $^1/_4$ are bald, and $^3/_4$ of the sons are bald and $^1/_4$ are nonbald. Finally, baldness is not a straightforward trait to study. For one thing, there is variable expressivity in the baldness phenotype: as is apparent in the adult population, baldness may occur early in life or late, it may appear first on the crown (for example, Prince Charles) or on the forehead, and the degree of baldness varies from minimal to total. In fact, several genes can affect the presence of hair on the head, including the pattern baldness gene. The final phenotype, then, is mostly the result of interaction between the environment and the particular set of those genes present. For example, although *b/b* females show pattern baldness, the onset of baldness in these women occurs much later in life than it does in men, due to the influence of hormones in the female internal environment.

Other human examples of sex-influenced traits are cleft lip and palate (incomplete fusion of the upper lip and palate; OMIM 119530), in which there is a 2:1 ratio of the trait in males : females; clubfoot (OMIM 119800; 2:1 ratio); gout (OMIM 138900; 8:1 ratio); rheumatoid arthritis (OMIM 180300; 1:3 ratio); osteoporosis (OMIM 166710; 1:3 ratio); and systemic lupus erythematosus (OMIM 152700, an autoimmune disease; 1:9 ratio).

**Temperature.** Biochemical reactions in the cell are catalyzed by enzymes. Normally, enzymes are unaffected by temperature changes within a reasonable range. However, some alleles of an enzyme-coding gene may give rise to an enzyme that is temperature sensitive; that is, it may function normally at one temperature but be nonfunctional at another temperature. An example of a temperature effect on gene expression is fur color in Siamese cats (Figure 13.10). Siamese cats are homozygous for the recessive $c^s$, siamese, allele of the *C* (albino) locus, which encodes tyrosinase, an enzyme involved in melanin biosynthesis. This genotype blocks melanin synthesis in the fur over most of the body due to the heat-sensitive nature of the tyrosinase encoded by the allele. At birth, Siamese kittens are cream or white in color due to the uniformly warm temperature of their bodies. However, as they grow, their extremities or "points" (ears, nose, paws, and tail) become relatively cooler as they increase in distance from the body core. The lower local surface temperature enables tyrosinase activity, so that melanin is synthesized, and the points become darker in color. The rest of the body retains a much lighter (though not white) color, due to a much lower level of tyrosinase activity. (A similar situation applies to Himalayan rabbits, which are homozygous for the $c^h$, Himalayan, allele of the *C* locus.) The color of the points depends on other coat color genes in the breed. For example, a sealpoint Siamese has extremely dark brown—almost black—points due to the presence of genes for black coat color.

**Chemicals.** Certain chemicals can affect an organism significantly. For example, the human disease phenylketonuria (PKU; OMIM 261600) is an autosomal recessive trait, with a defect in the biochemical pathway for the metabolism of the amino acid phenylalanine (see Chapter 4,

**Figure 13.9**

**Sex-influenced inheritance of pattern baldness in humans.** The $b$ allele is recessive in one sex and dominant in the other.

**a) Nonbald female × $b/b$ bald male**

| | Parent 1 ♀ | Parent 2 ♂ |
|---|---|---|
| Parental phenotype | Nonbald | Bald |
| Diploid parental genotype | $b^+/b^+$ | $b/b$ |
| Haploid gametes | $b^+$, $b^+$ | $b$, $b$ |

Progeny

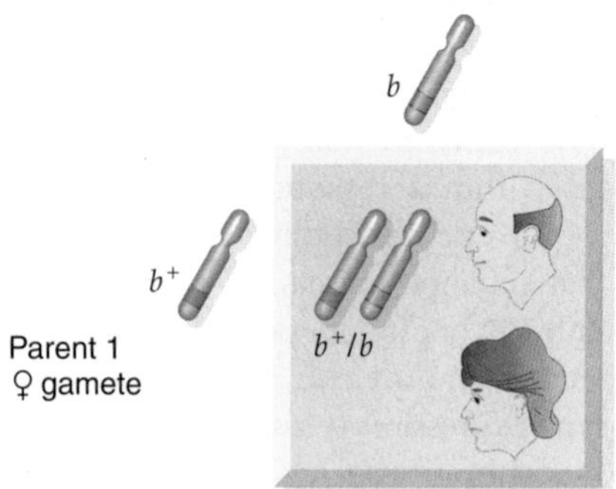


Progeny genotypes: All $b^+/b$

Progeny phenotypes: Nonbald ♀; bald ♂ because the trait is sex influenced

**b) $F_1 \times F_1$ cross**

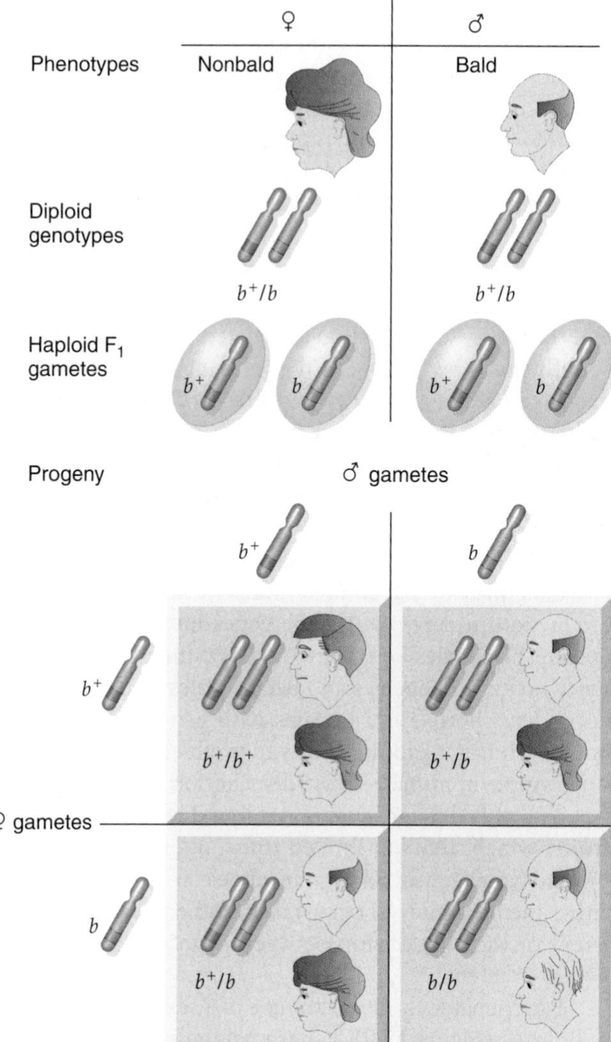


$F_2$ genotypes: $^1/_4$ $b^+/b^+$, $^1/_2$ $b^+/b$, $^1/_4$ $b/b$

$F_2$ phenotypes: Nonbald whether male or female; Nonbald if female, bald if male; Bald whether female or male

3 bald : 1 nonbald in males
1 bald : 3 nonbald in females

p. 61). In individuals who are homozygous for the recessive allele, various symptoms appear, most notably mental retardation at an early age. However, their diet determines how severe the symptoms of PKU will be. Problem foods include protein containing phenylalanine, such as the protein of mother's milk. PKU can be treated by restricting the amount of phenylalanine in the diet. (PKU is discussed further in Chapter 4, pp. 66–68.)

### Keynote

The phenotypic expression of a gene depends on several factors, including its dominance relationships, the genetic constitution of the rest of the genome, and the influences of the internal and external environments.

## Nature versus Nurture

We are left with the nature–nurture question: What are the relative contributions of genes and the environment to the phenotype? (The nature–nurture question is discussed more fully in Chapter 22.) The variation in most of the traits we have examined up to this point has been determined largely by differences in genotype; that is, phenotypic differences have reflected genetic differences. However, we have already seen that the phenotypes of many traits are influenced by both genes and the environment. Let us consider the nature–nurture issue in the context of some human examples.

Human height, or stature, is definitely influenced by genes. On the average, tall parents tend to have tall offspring, and short parents tend to have short offspring. There are also a number of genetic forms of dwarfism in humans. Achondroplasia is a type of dwarfism in which the bones of the arms and legs are shortened but the trunk and head are of normal size; achondroplasia results from a single dominant allele. But the environment also plays a role in determining height. For example, human height has increased about 1 inch per generation over the past 100 years as a result of better diets and improved health care. Genes and the environment have interacted in determining human height.

For a trait such as height, genes set certain limits (or specify a potential) for the phenotype. Within these limits, the phenotype an individual develops depends on the environment. The range of potential phenotypes that a single genotype could develop if exposed to a range of environmental conditions is called the **norm of reaction.** For some genotypes, the norm of reaction is small; that is, the phenotype produced by a genotype is nearly the same in different environments. For other genotypes, the norm of reaction is large, and the phenotype produced by the genotype varies greatly in different environments.

Many human *behavioral* traits are the result of interaction between genes and the external environment. One example is alcoholism, which is a major health problem in the United States: about 14 million Americans have alcohol use disorders. Numerous studies have shown that alcoholism is influenced by genes. For instance, sons of alcoholic fathers who are separated from their biological parents at birth and adopted into a family with nonalcoholic parents are four times more likely to become alcoholic than are sons adopted at birth whose biological fathers were not alcoholic. However, no gene *forces* a person to drink alcohol. That is, people cannot become alcoholic unless they are exposed to an environment in which alcohol is available and drinking is encouraged. What genes do is make certain people more or less susceptible to alcohol abuse; they increase or decrease the risk of developing alcoholism. How genes influence our susceptibility to alcohol abuse is not yet clear. They may affect the way we metabolize alcohol, which in turn might affect how much we drink. Or genes may influence certain of our personality traits, making us more or less likely to drink heavily. The important point is that a behavioral trait such as alcoholism may be influenced by genes, but the genes alone do not produce the phenotype.

**Figure 13.10**

**Effect of temperature on gene expression.** A sealpoint Siamese cat showing light color over most of the body and dark color at the points (ears, nose, paws, and tail) where temperature is lower.

Nowhere has the role of genes and environment been more controversial than in the study of human intelligence. In the past, people tended to think of human intelligence as either genetically preprogrammed or produced entirely by the environment. The clash of these opposing views was called the nature–nurture controversy. Today, geneticists recognize that neither of these extreme views is correct; human intelligence is the product of both genes and the environment.

That genes influence human intelligence is clearly evidenced by genetic conditions that produce mental retardation, such as PKU (OMIM 261600; see Chapter 4, pp. 66–68) and Down syndrome (OMIM 190685; see Chapter 16, pp. 478–480). Many studies also indicate that genes influence differences in IQ among nonretarded people. (IQ, or intelligence quotient, is a standardized measure of mental age compared with chronological age; it is fairly stable over time. However, it is important to note that what we generally consider as intelligence is far more complex than what is measured in an IQ test.) For example, adoption studies show that the IQs of adopted children are closer to those of their biological parents than to those of their adoptive parents.

However, IQ is also influenced by environment. Identical twins frequently differ in IQ, a fact that can be explained most likely by environmental differences. Family size, diet, and culture are environmental factors known to affect IQ. Thus, IQ results from the interaction of genes and the environment. Consequently, if two people (other than identical twins) differ in IQ, it is impossible to attribute that difference either solely to genes or the environment, because both interact in determining the phenotype. So, although we cannot change our genes, we can alter the environment and thus affect a phenotypic trait such as intelligence.

**Keynote**

Variation in most of the genetic traits considered in the discussion of Mendelian principles is determined predominantly by differences in genotype; that is, phenotypic differences result from genotypic differences. For many traits, however, the phenotypes are influenced by both genes and the environment. The debate over the relative contribution of genes and the environment to the phenotype has been called the nature–nurture controversy.

## Maternal Effect

**Maternal effect** is the phenomenon in which a phenotype of the offspring is determined not by the genotype of the offspring but by the nuclear genotype of the mother, with no influence by the paternal nuclear genome. Maternal effect occurs as the result of mRNA or proteins deposited in the oocyte before fertilization that direct early development of the embryo. The genes that encode those products are known as *maternal effect genes*.

Maternal effect is seen in the inheritance of the coiling direction in the shell of the snail *Limnaea peregra*. The shell coiling trait is determined by a single pair of nuclear alleles: the dominant *D* allele for coiling to the right (dextral coiling) and the recessive *d* allele for coiling to the left (sinistral coiling). The shell coiling phenotype is always determined by the genotype of the mother. The latter is shown by the results of reciprocal crosses between a true-breeding, dextral-coiling and a sinistral-coiling snail (Figure 13.11). All the $F_1$ snails have the same genotype because a nuclear gene is involved, yet the phenotype is different for the reciprocal crosses.

In the cross of a dextral (*D/D*) female with a sinistral (*d/d*) male (Figure 13.11a), the $F_1$ snails are all *D/d* in genotype and dextral in phenotype. Selfing the $F_1$ produces $F_2$ snails with a 1:2:1 ratio of *D/D*, *D/d*, and *d/d* genotypes. All the $F_2$ snails are dextral, even the *d/d* snails whose genotype seems to indicate sinistral phenotype. Here is our first encounter with maternal effect; the *d/d* snails have a coiling phenotype specified not by their own genotype, but by the genotype of their mother (*D/d*).

**Figure 13.11**
**Inheritance of the direction of shell coiling in the snail *Limnaea peregra* is an example of maternal effect.**

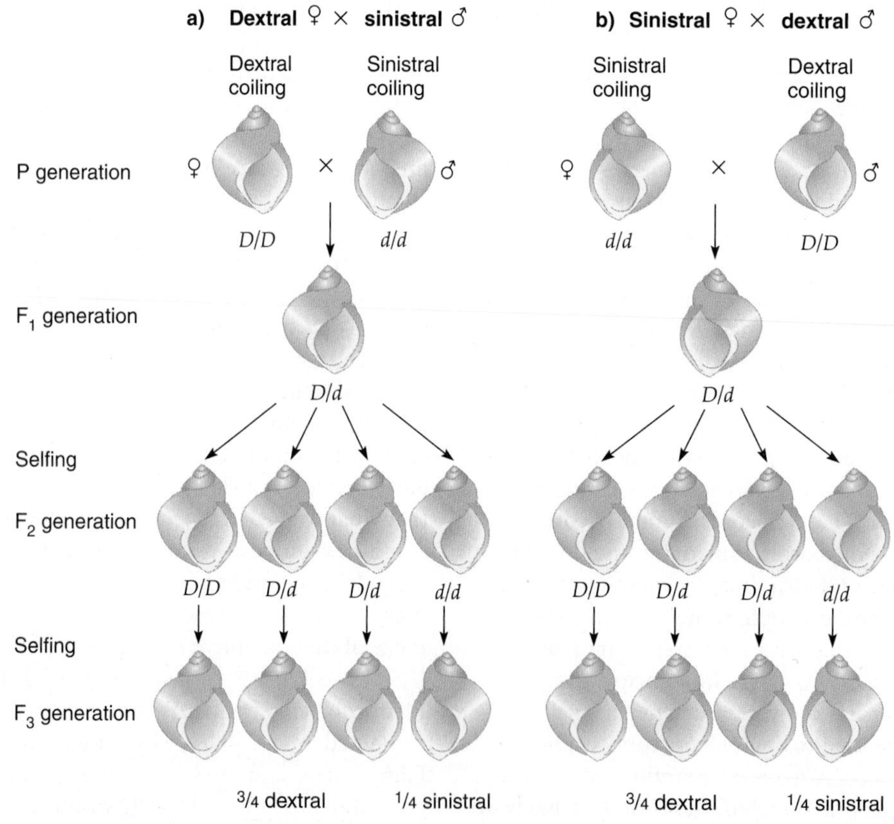

Selfing the $F_2$ snails gives $F_3$ progeny, 3/4 of which are dextral and 1/4 of which are sinistral. The latter are the *d/d* progeny of the $F_2$ *d/d* female snails; these $F_3$ snails are sinistral because their phenotype reflects the $F_2$ genotype of their mother.

The reciprocal cross is of a sinistral (*d/d*) female with a dextral (*D/D*) male (Figure 13.11b). The $F_1$ snails are all *D/d* in genotype, yet they are sinistral in phenotype because the mother is genotypically *d/d*. Selfing the $F_1$ produces $F_2$ snails, all of which are dextral for the same reason as the reciprocal cross just described. The genotypes and phenotypes of the $F_2$ and $F_3$ generations are the same as for the reciprocal cross, again for the same reasons.

What is the basis for the coiling? The orientation of the mitotic spindle in the first mitotic division after fertilization controls the direction of coiling. The mother encodes products, deposited in the oocyte, that direct the orientation of the mitotic spindle and therefore the direction of cell cleavage. Thus, a mother of genotype *D/–* deposits a gene product that specify a dextral coiling. A mother of genotype *d/d* either does not produce a gene product, or that product is nonfunctional, and this results in default sinistral coiling.

Maternal effect is also seen for certain genes involved in axis formation during embryo development in *Drosophila melanogaster*. Those genes are discussed in Chapter 19.

**Keynote**

In the maternal effect, an inherited trait is controlled by the maternal nuclear genotype before the egg is fertilized and is not influenced by the paternal genotype.

## Determining the Number of Genes Involved in a Set of Mutations with the Same Phenotype

Up to this point in the book, each mutation we have discussed has affected a different gene. We will now begin to encounter cases where that is not so. To help us analyze and understand those cases, we need to understand the relationship between the phenotype and the gene in more detail.

We have learned that the general genetic approach to studying a biological process is to isolate mutants which affect that process. Those mutants are identified by their phenotype—the mutant phenotype—which is distinct from the wild-type phenotype. Consider a genetic study in which a large number of mutants are isolated, with each mutant having the same altered phenotype. Our aim is to understand the structures and functions of the genes controlling the biological process involved. Does each mutant define a different gene, or not? We can answer that question with the **complementation test**, also called the **cis-trans test**, which determines whether two independently isolated mutants with the same phenotype have mutations in the same or different genes. The complementation test was developed by Edward Lewis to study genes in *Drosophila*.

In a complementation test, two mutants resulting in the same phenotype are crossed, and the phenotype of the progeny is observed. If the two mutations involved are in different genes, then the progeny will be wild-type/mutant heterozygotes for each of the two genes involved. Because there is a wild-type copy of each gene, the phenotype will be wild type, not mutant (Figure 13.12a). We say that the two mutants complement each other. However, if the two mutations are in the same gene, then the progeny will have a different mutant version of the gene on each of the two homologues, and the phenotype will be mutant

**Figure 13.12**

**Complementation test to determine whether two mutations resulting in the same phenotype are in the same or different genes.**

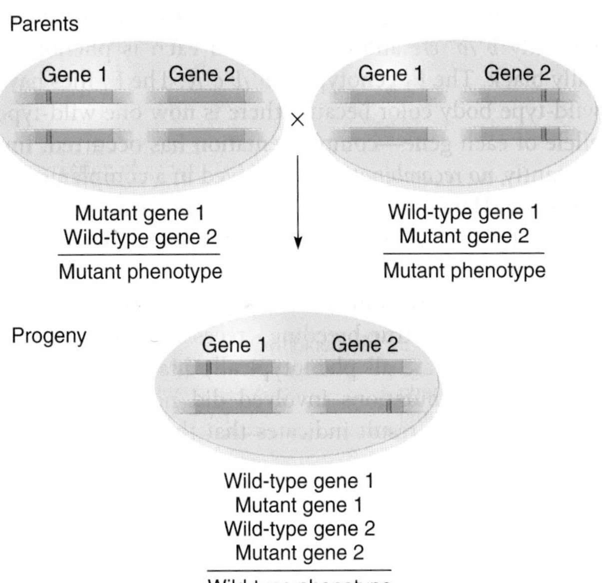


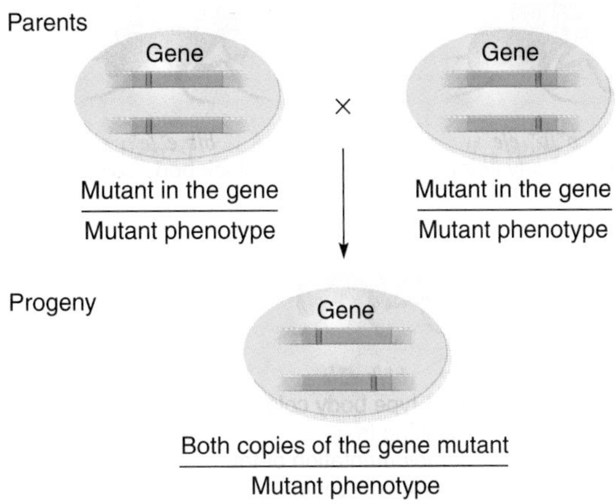

*bw/bw st*$^+$*/–* = brown; and *bw/bw st/st* = white. Many other pairs of eye-color genes in *Drosophila* have conceptually similar genotype–phenotype relationships in which the doubly homozygous recessive gives a new eye color.

The molecular explanation for the white eye phenotype in the above example is as follows. The wild-type eye color is the result of a combination of a brown pigment and a red pigment to produce a brick red color. The brown and red pigments are the end products of two distinct biosynthetic pathways, each of which is controlled by a number of genes. Flies of genotype *bw*$^+$*/– st/st* do not make the red pigment but do make the brown pigment and, therefore, have brown eyes. Flies of genotype *bw/bw st*$^+$*/–* do not make the brown pigment but do make the red pigment and, therefore, have scarlet eyes. Flies of genotype *bw/bw st/st* make neither brown nor red pigment and, therefore, have white, unpigmented eyes.[1]

Another example of this type of gene interaction involves comb shape in chickens, which involves two independently assorting genes, *R* and *P*. Chickens with the genotype *R/– P/–* have a walnut comb, so named because it resembles half a walnut meat. An *R/r P/p* × *R/r P/p* cross gives these progeny: 9 *R/– P/–*, walnut comb : 3 *R/– p/p*, rose comb (broad, nearly flat on top and fleshy, with a tapering spike) : 3 *r/r P/–*, pea comb (medium length, low, with three lengthwise ridges) : 1 *r/r p/p*, single comb (thinnish, fleshy, extending relatively high up from head and beak with multiple serrations).

## Epistasis

**Epistasis** (literally, "to stand upon") is the interaction between alleles of two or more genes to control a single phenotype. The interaction involves one gene masking or modifying the phenotypic expression of another gene. No new phenotypes are produced by this type of gene interaction. A gene that masks the expression of another gene is said to be *epistatic,* and a gene whose expression is masked is said to be *hypostatic*. If we think about the $F_2$ genotypes *A/– B/–*, *A/– b/b*, *a/a B/–*, and *a/a b/b*, epistasis may be caused by the presence of homozygous recessives of one gene pair, so that *a/a* masks the effect of the *B* allele. Or epistasis may result from the presence of one dominant allele in a gene pair. For example, the *A* allele might mask the effect of the *B* allele. Epistasis can also occur in both directions between two pairs of alleles of two genes. All these possibilities produce a number of modifications of the 9:3:3:1 ratio in a dihybrid cross. Some examples of epistasis follow.

**Recessive Epistasis.** In *recessive epistasis*, *A/– b/b* and *a/a b/b* individuals have the same phenotype, which results in a phenotypic ratio in the $F_2$ of 9:3:4 rather than 9:3:3:1. An example is coat color in rodents. Wild mice have a greyish color, because the hairs in the fur have a band of yellow between two bands of black. This coloration—the agouti pattern—aids in camouflage and is found in many other wild rodents, including guinea pigs and grey squirrels. Several other coat colors are seen in domesticated rodents. Albinos, for example, have no pigment in the fur or in the irises of the eyes, giving them a white coat and pink eyes. Albinos are true breeding, and this variation acts as a complete recessive to any other color. Another variant has black coat color as the result of the absence of the yellow pigment found in the agouti pattern. Black is recessive to agouti.

When true-breeding agouti mice are crossed with albinos, the $F_1$ progeny are all agouti, and when these $F_1$ agoutis are interbred, the $F_2$ progeny consist of $^9/_{16}$ agouti animals, $^3/_{16}$ black, and $^4/_{16}$ albino (Figure 13.15). This pattern occurs because the parents differ in whether they have a dominant allele, *C*, of a gene for the development of any color (black mice are *C/–* and albinos are *c/c*), and in whether they have a dominant allele *A* of a gene for the agouti pattern, which is a yellow banding of the black hairs (*A/–* are agouti, and *a/a* are nonagouti). (Note: The symbols here are the actual ones used for rodent coat color genes. Do not confuse this *a* locus with the theoretical *a* locus referred to in our continuing discussion of modified ratios that began on p. 378.) Phenotypically, *A/– C/–* are agouti, *a/a C/–* are black, and *A/– c/c* and *a/a c/c* are albino, giving a 9:3:4 phenotypic ratio of agouti : black : albino. In other words, this is recessive epistasis of *c/c* over *A/–* and *a/a*. That is, white hairs are produced in *c/c* mice, regardless of the genotype at the other locus.

In fact, three genes are involved in the phenotypes of rodent coat color described here. At one gene, the dominant *C* allele encodes the enzyme tyrosinase, a key enzyme in the pathway for the synthesis of the black pigment, eumelanin. The recessive *c* allele, when homozygous, results in no eumelanin pigment formation, regardless of the genotypes of other coat color genes. (In humans, loss of function of the tyrosinase gene results in one form of oculocutaneous albinism [OMIM 203100], a type of albinism characterized by little or no pigment in skin, hair, and eyes.)

At a second gene, the agouti locus, the dominant allele *A* determines agouti; and its recessive allele *a* in the homozygous state produces nonagouti mice. The product of the agouti locus is agouti signal protein. This protein regulates the pigments produced by the melanin-producing cells of the hair follicles. In rodents with genotype *A/–*, the black-and-yellow banding pattern results because the *A* allele is not always expressed during hair growth. In the early part of hair growth, agouti signal protein is not

[1]Earlier in this chapter, we discussed multiple alleles of the X-linked *w* (*white*) locus in *Drosophila*. We learned that the *w* allele, when homozygous in females or hemizygous in males, results in white eyes. In that example, a mutation of a *single* gene results in white eyes. The mechanism for the white eye phenotype in this case is as follows: The wild-type allele of the *white* locus encodes a product needed for the deposition of the red and brown pigments into the eye. In a *w/w* or *w/Y* fly, red and brown pigments *are* synthesized as a result of the activity of genes controlling the brown pigment and red pigment biosynthesis pathways, but they are not deposited in the eye.

**Figure 13.15**

**Recessive epistasis: generation of an $F_2$ 9 agouti : 3 black : 4 white ratio for coat color in rodents.**

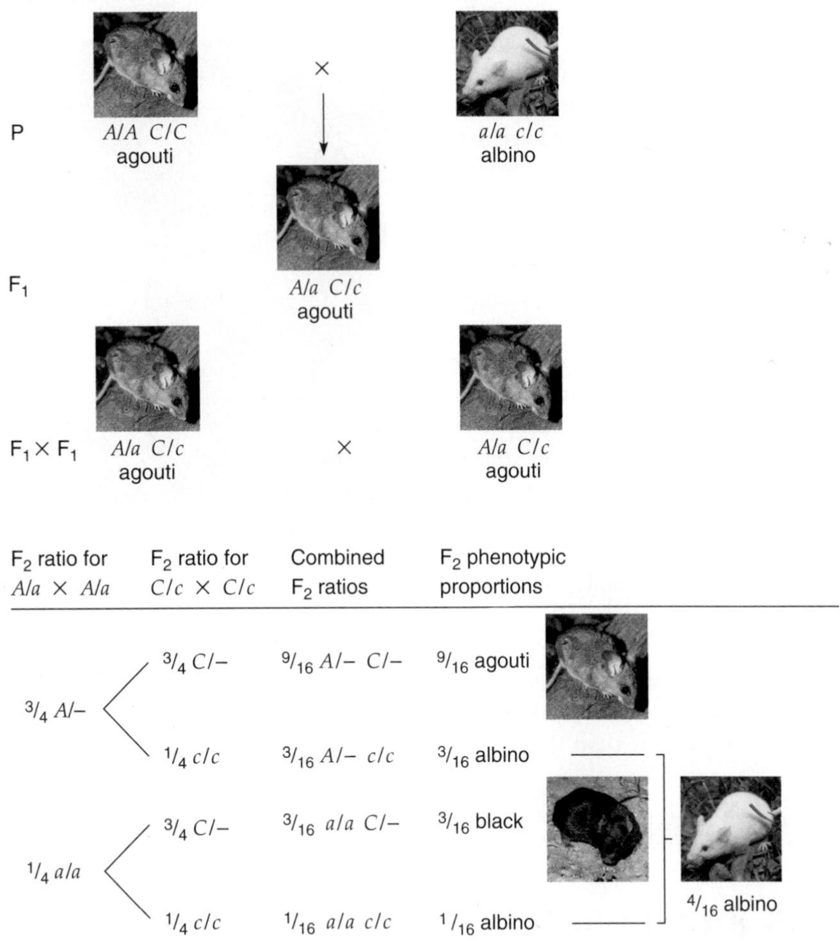


expressed from the *A* allele. This means the black pigment, eumelanin, is synthesized. In the middle part of hair growth, the *A* allele becomes active. The resulting agouti signal protein switches the synthesis of melanin to the pheomelanin form, which in rodents is yellow in color. Finally, the *A* allele again becomes inactive, and black pigment is deposited in the hair. Overall, the action of a dominant agouti allele produces a hair that has a yellow band between two black bands. Homozygous nonagouti *a/a* rodents have solid black hair because no active agouti signal peptide is produced and, therefore, pigment production never switches from melanin to pheomelanin. Note that this discussion has assumed that the rodent is otherwise wild type for coat color genes, meaning that eumelanin synthesis gives black hair color. There are over 60 coat color genes in rodents, so many variations in hair color are seen in these animals. The third gene for our example of epistasis—discussed in the next paragraph—illustrates one of these other genes.

The human homolog of the mouse agouti gene is the *agouti signaling protein* gene (OMIM 600201). This gene is expressed in the testis, ovary, and heart, and at lower levels in the liver, kidney, and foreskin. The agouti signaling protein is secreted and is involved in the regulation of melanin production. In contrast to the action of the agouti product in rodents, the human agouti signaling protein may affect the quality of hair pigmentation rather than the pattern of melanin pigment deposition.

Lastly, at a third gene, the dominant allele *B* encodes a product involved in the production of the black pigment, eumelanin. Homozygosity for the recessive allele, *b*, results in a brown pigment; that is, a visibly lower amount of eumelanin. All the mice in Figure 13.15 must have had at least one *B* allele; otherwise some brown mice would have been seen—for example, *A/– C/– b/b* are brown. The *B* gene encodes tyrosinase related protein 1. The exact function of this protein in coat color pathways is not clear; one model is that the protein acts to stabilize the key eumelanin synthesizing enzyme, tyrosinase. (In humans a significant decrease in function, or loss of function, of the tyrosinase-related protein results in decreased or absent coloration of skin, hair, and irises of the eyes like that seen in oculocutaneous albinism.)

Another example of recessive epistasis involves coat color in labrador retrievers (labs). At one gene, *B/–* specifies the formation of a black pigment, while *b/b* results in a brown pigment. At an independent gene, *E/–* allows the expression of the *B* gene, while *e/e* does not allow the expression of the *B* gene; this latter situation gives yellow. Therefore, genotype *B/– E/–* produces a black lab, *b/b E/–* produces a chocolate (brown) lab, and *–/– e/e* (*–/–* means either *B* or *b* on each chromosome) produces a yellow lab (Figure 13.16). An interesting slight variation among the yellow labs is that dogs with the *B/– e/e* genotype have dark noses and lips, whereas dogs with the *b/b e/e* genotype have pale noses and lips. From a cross of a true-breeding black lab of genotype *B/B E/E* with a true-breeding yellow lab of genotype *b/b e/e*, the $F_1$ progeny are black labs of genotype *B/b E/e*. Intercrossing dogs with this genotype produces 9 *B/– E/–* black : 3 *B/– e/e* yellow : 3 *b/b E/–* chocolate : 1 *b/b e/e* yellow, for a phenotypic ratio of 9 black : 3 chocolate : 4 yellow. The *E* gene encodes a protein called the melanocortin 1 receptor (MC1R). MC1R is a key regulator of hair and skin color. Dogs with the *E/–* genotype produce normal MC1R and, hence, have dark fur, whereas dogs with the *e/e* genotype produce nonfunctional MC1R have yellow fur. This chapter's Focus on Genomics box describes studies of the gene encoding MC1R in woolly mammoths and Neanderthals.

**Figure 13.16**

**Recessive epistasis in labrador retrievers.**
Left: black lab, genotype *B/– E/–*. Middle: yellow lab, genotype *–/– e/e*. Right: chocolate lab, genotype *b/b E/–*.

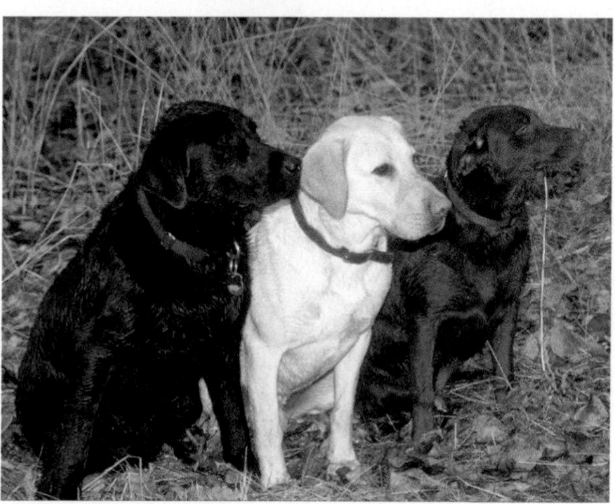

**Dominant Epistasis.** In dominant epistasis, *A/– B/–* and *A/– b/b* individuals have the same phenotype, so the phenotypic ratio in the $F_2$ is 12:3:1 rather than 9:3:3:1. In other words, in dominant epistasis, one gene, when dominant—*A* here—*is* epistatic to the other gene.

An example of dominant epistasis may be seen in the fruit color of summer squash, which has three common fruit colors: white, yellow, and green. In crosses of white and yellow and of white and green, white is always expressed.

## Focus on Genomics
### Redheads of the Past

The gene *MC1R* encodes a membrane protein that acts as a switch between production of pheomelanin, which is reddish yellow, and blackish brown eumelanin. At least some human redheads can attribute their fair skin and red hair to a *MC1R* allele that encodes a less functional version of this enzyme. Scientists using genomics techniques including SNP analysis, and modified standard molecular techniques like the polymerase chain reaction, on ancient DNA have studied samples from woolly mammoths and Neanderthals and found similar mutations. A 43,000-year-old woolly mammoth bone yielded enough DNA to sequence the *MC1R* gene, and researchers found that the mammoth carried two distinct versions of the gene. One version coded for a very active version of the *MC1R* protein, and the other coded for a protein with very limited ability to respond to signals instructing the cell to produce eumelanin. Many mummified mammoths have been found, and some have had lighter hair than others, so it seems likely that this genetic variation may have resulted in color variation in the mammoths. The alterations in the gene are very similar to changes observed in the *MC1R* gene in a population of beach mouse, where both dark-and light-haired individuals are common.

At least two Neanderthals, one from Spain and one from Italy, carried similarly altered versions of *MC1R*. When they tested the activity of the Neanderthal *MC1R* gene product in human cells in the lab, it behaved much like known human alleles that cause fair skin and red hair. These two individuals were not close relatives, as dating of the remains suggested that the Italian bones were about 7,000 years older than those from Spain. This led the investigators to conclude that the allele was at least fairly common in the population, since it was found in two individuals widely separated in space and time. Researchers then asked if this specific allele was common in modern humans. They sequenced the *MC1R* gene from DNA obtained from the cells in the human genome diversity cell line panel, a collection of cultured cells from more than 1,000 modern humans of diverse genetic background, and looked at the sequence of this gene from 2,800 additional modern humans. They found many alleles for the *MC1R* gene, but they never found an allele identical to the one they found in the Neanderthal DNA. They concluded that if this allele is present in modern humans, it is far rarer than it was in Neanderthals.

**Figure 13.17**

**Dominant epistasis: generation of an $F_2$ 12 white : 3 yellow : 1 green ratio for fruit color in summer squash.**

$F_1 \times F_1$ — $W/w\ Y/y$ white fruit $\times$ $W/w\ Y/y$ white fruit

| $F_2$ ratio for $W/w \times W/w$ | $F_2$ ratio for $Y/y \times Y/y$ | Combined $F_2$ ratios | $F_2$ phenotypic proportions | |
|---|---|---|---|---|
| $^3/_4\ W/-$ | $^3/_4\ Y/-$ | $^9/_{16}\ W/-\ Y/-$ | $^9/_{16}$ white | $^{12}/_{16}$ white |
| | $^1/_4\ y/y$ | $^3/_{16}\ W/-\ y/y$ | $^3/_{16}$ white | |
| $^1/_4\ w/w$ | $^3/_4\ Y/-$ | $^3/_{16}\ w/w\ Y/-$ | $^3/_{16}$ yellow | |
| | $^1/_4\ y/y$ | $^1/_{16}\ w/w\ y/y$ | $^1/_{16}$ green | |

In crosses of yellow and green, yellow is expressed. Yellow thus is recessive to white, but dominant to green.

Consider two genes, each with a pair of alleles: *W/w* and *Y/y*. In squashes that are *W/–* in genotype, the fruit is white no matter what genotype is at the other locus. In *w/w* plants, the fruit is yellow if a dominant allele of the other locus is present, but green if it is absent. In other words, *W/– Y/–* and *W/– y/y* plants have white fruits, *w/w Y/–* plants have yellow fruits, and *w/w y/y* plants have green fruits. The $F_2$ progeny of an $F_1$ self of doubly heterozygous individuals shows a ratio of 12 white : 3 yellow : 1 green fruits in the plants (Figure 13.17).

A theoretical biochemical pathway to explain the 12:3:1 ratio of squash color is shown in Figure 13.18. The hypothesis is that a green substance is converted to a yellow substance in a reaction requiring the product of the dominant *Y* allele, and the product of the dominant *W* allele can convert either the green or yellow substance to a white substance. Thus, all plants that have at least one *W* allele will have white fruit, no matter which alleles are present at the *Y* locus, because green and/or yellow is converted to white. The nonwhite, *w/w*, fruits are either yellow, if they are *Y/–*, or green, if they are *y/y*.

Another example of dominant epistasis causes greying in horses. If a horse is genotypically *G/–*, it will show a progressive silvering of the coat color with which it is born until the coat becomes grey (really, essentially white) as a mature animal. Horses with the genotype *gg* do not go grey as they mature; instead, they remain the color they were at birth.

**Epistasis Involving Duplicate Genes.** A gene or genotype at one locus may produce an identical phenotype to that produced by a gene or genotype at a second locus. In that case we conclude that *duplicate genes* are involved. When recessive or dominant epistasis is involved with both of these genes, modification of the 9:3:3:1 ratio occurs.

***Duplicate Recessive Epistasis.*** In *duplicate recessive epistasis* (also called *complementary gene action*), *a/a* is epistatic to *B* and *b*, and *b/b* is epistatic to *A* and *a*. An example concerns flower color in the sweet pea. Purple flower color is dominant to white. A recessive mutation in either gene *C* or *P*, when homozygous, results in white flowers. Purple flowers are produced only in genotypes with at least one normal *C* allele and one normal *P* allele. Therefore, a cross of a true-breeding white variety of genotype *C/C p/p* with a true-breeding white variety of genotype *c/c P/P* produces $F_1$ plants that are purple-flowered, with genotype *C/c P/p* (Figure 13.19). When the $F_1$ hybrids are self-fertilized, they produce an $F_2$ generation with the following genotype–phenotype relationships: 9 *C/– P/–* (purple flowers) : 3 *C/– p/p* (white flowers) : 3 *c/c P/–* (white flowers) : 1 *c/c p/p* (white flowers), for an overall phenotypic ratio of 9 purple flowers : 7 white flowers. That is, the recessive genotype *c/c* is epistatic to *P* and *p*, and the recessive genotype *p/p* is epistatic to *C* and *c*.

The following purely theoretical pathway can be envisioned for the production of the purple pigment:

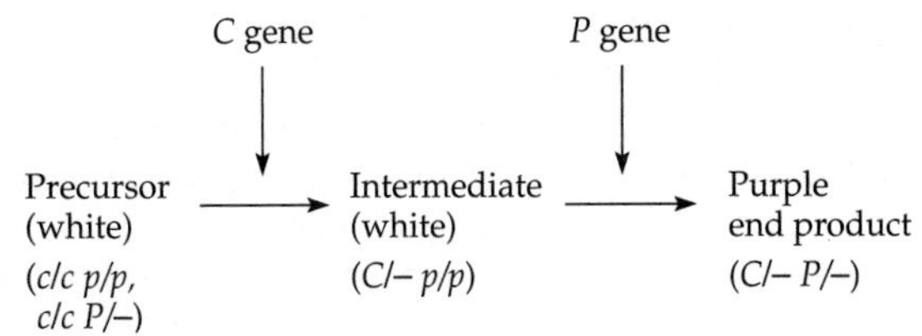


**Figure 13.18**

**Dominant epistasis: hypothetical pathway to explain the $F_2$ ratio of 12 white : 3 yellow : 1 green color in summer squash.**

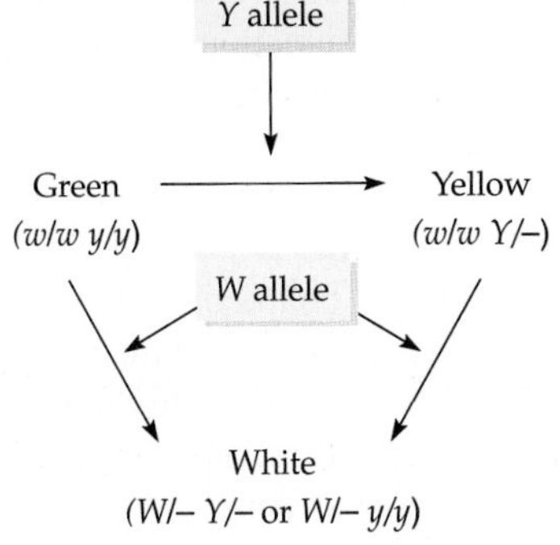

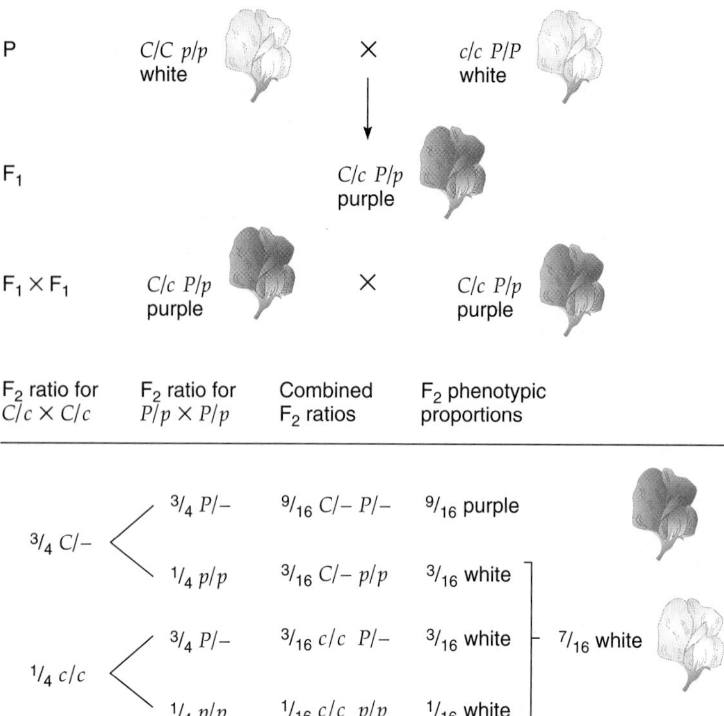


**Figure 13.19**

**Duplicate recessive epistasis: generation of an $F_2$ 9 purple : 7 white ratio for flower color in sweet peas.**

Genes *C* and *P* encode products that control different steps in the pathway. Only in *C/– P/–* plants can the purple end product be produced, whereas all other plants fail to make purple and have white flowers, being blocked either in the conversion of the white intermediate to purple—genotype *C/– p/p*—or in the conversion of the white precursor to the white intermediate—genotypes *c/c P/–* and *c/c p/p*.

***Duplicate Dominant Epistasis.*** In *duplicate dominant epistasis*, *A* is epistatic to *B* and *b*, and *B* is epistatic to *A* and *a*. An example concerns fruit shape in the shepherd's purse plant. When a true-breeding plant that produces heart-shaped fruit is crossed with a narrow fruit plant that produces narrow fruit, the $F_1$ plants all produce heart-shaped fruit. When the heart-shaped $F_1$ plants are crossed, the $F_2$ show a ratio of 15 heart-shaped fruit plants : 1 narrow fruit plant. This is a modification of the 9:3:3:1 ratio, with the genotypes *A/– B/–*, *A/– b/b*, and *a/a B/–* all producing the phenotype—heart-shaped fruit—and the genotype *a/a b/b* producing the other phenotype of narrow fruit. In other words, there are duplicate genes involved in the fruit shape phenotype, and the dominant allele of either gene is epistatic to either allele of the other gene.

In sum, many types of phenotypic modifications are possible as a result of interactions between the products of different allele pairs that contribute to the same trait. Geneticists detect such interactions when they observe deviations from the expected phenotypic ratios in crosses. Table 13.3 summarizes the epistatic interactions we have discussed in the chapter.

Epistasis plays a role in many human genetic diseases, further complicating their analysis. In these cases, complex interrelationships exist. For example, most cases of bipolar disorder (also called manic–depressive illness), a complex human genetic disorder involving pathological mood disturbances, involve epistasis between multiple genes and may include other, more complex genetic mechanisms.

**Keynote**

In many instances, alleles of different genes interact to determine a phenotypic characteristics. Sometimes the interaction between genes results in new phenotypes without modification of typical Mendelian ratios. In epistasis, interaction between genes causes modifications of Mendelian ratios because one genotype interferes with the phenotypic expression of another genotype (or genotypes). The phenotype is controlled largely by the former genotype, and not the latter, when both genotypes occur together. The analysis of epistasis is complicated further when one or both allele pairs involve incomplete dominance or codominance, or when allele pairs do not assort independently.

## Gene Interactions Involving Modifier Genes

In our discussions of epistasis, we learned about alleles that have major effects in altering the phenotype by masking the phenotype associated with a particular genotype. A **modifier gene** also interacts with another nonallelic gene but, instead of masking it, it affects—modifies—the phenotype associated with expression of the alleles of that gene in a milder way. Based on the type of modification caused, modifier genes are subdivided into two groups: *enhancers* intensify the phenotype controlled by the other gene, and *reducers* decrease the phenotypic expression of the other gene. (Note that the term enhancer is used in Chapter 18 to

**Table 13.3 Summary of Epistatic $F_2$ Phenotypic Ratios from an $F_1$ *A/a B/b* × $F_1$ *A/a B/b* Cross in Which Complete Dominance Is Shown for Each Allele Pair**

| | $F_2$ Phenotypic Ratio from an *A/a B/b* × *A/a B/b* Cross | | | |
|---|---|---|---|---|
| **Gene Interaction** | ***A/– B/–*** | ***A/– b/b*** | ***a/a B/–*** | ***a/a b/b*** |
| None | 9 | 3 | 3 | 1 |
| **Recessive epistasis**<br>*a/a* epistatic to *B* and *b* | 9 | 3 | 4 | |
| **Dominant epistasis**<br>*A* epistatic to *B* and *b* | 12 | | 3 | 1 |
| **Duplicate recessive epistasis (complementary gene action)**<br>*a/a* epistatic to *B* and *b*; and *b/b* epistatic to *A* and *a* | 9 | 7 | | |
| **Duplicate dominant epistasis**<br>*A* epistatic to *B* and *b*; and *B* epistatic to *A* and *a* | 15 | | | 1 |

mean DNA elements in eukaryotes that mediate the activation of genes. Those enhancers are unrelated to the enhancer genes discussed here.) When a modifier gene shifts the phenotype associated with a mutant allele of another gene toward the phenotype associated with the wild-type allele of that gene, it is called a **suppressor gene.**

A simple illustration of the effect of a modifier gene involves coat color in cats (and rodents). The dominant allele, *D*, of the *dense pigment* gene encodes a protein needed for transporting and depositing pigment in growing hairs. Cats of genotype *D/–* have full color, that color depending on the other coat colors present in the cat. For example, cats of genotype *B/– D/–* are black since the dominant *B* allele produces a full amount of black pigment and the dominant *D* allele leads to full transport and deposition of that pigment in the hairs. However, cats of genotype *d/d* have reduced ability to transport and deposit pigment in the growing hairs. The decreased pigment density in the hairs produces a lightened color compared with the color in a *D/–* cat that shares the same set of other coat color genes. For example, cats of genotype *B/– d/d* are grey, rather than black. Similarly, the *d/d* genotype modifies brown coat color to light brown, cinnamon to light tan, and orange to cream.

Modifier genes are known that affect the phenotypic expression of mutant alleles responsible for a number of human genetic diseases. It is likely that many symptoms of human genetic disease are affected by modifier genes. Conversely, the variability in symptoms among individuals with the same genetic disease could potentially be due to the action of modifier genes. For example, the phenotypes seen in cystic fibrosis (CF) are highly variable. They are more extensive than could be explained even by the fact that there are more than 1,000 disease-causing mutations in the gene responsible. The simplest hypothesis is that modifier genes influence the phenotypic expression of mutant cystic fibrosis alleles. Modifier genes also are known to affect the severity of one genetic form of hearing loss. That is, mutations in the *cadherin 3* gene cause hearing loss in humans (as well as in other mammals). Researchers have shown that alterations of another gene, *V586M*, increase the severity of hearing loss caused by mutations in the *cadherin 3* gene. In this example, *V586M* is a modifier gene that affects the phenotypic expression of the *cadherin 3* gene.

### Keynote

A modifier gene interacts with another nonallelic gene, affecting the phenotype associated with the expression of the alleles of that gene rather than masking their expression, as is seen in the gene interactions of epistasis.

## Extranuclear Inheritance

Outside the nucleus, DNA is found in the mitochondrion and the chloroplast. The genes in these mitochondrial and chloroplast genomes are known as extranuclear genes, extrachromosomal genes, cytoplasmic genes, non-Mendelian genes, or organellar genes. The term *non-Mendelian* is informative because extranuclear genes do *not* follow the rules of Mendelian inheritance, as do nuclear genes. The mitochondrial and chloroplast genes encode the rRNA components of the ribosomes of these organelles for many (if not all) of the tRNAs used in organellar protein synthesis, and for a few proteins that remain in the organelles and perform functions specific to the organelles. All other proteins required by these organelles are nuclear encoded.

Cytoplasm is inherited from the mother in many organisms, so the inheritance of extranuclear genes in these organisms is strictly maternal. Extranuclear inheritance differs from maternal effect in two related respects: (1) the phenotype in extranuclear inheritance is determined by an individual's organellar gene, whereas the phenotype in maternal effect is determined by a nuclear gene in the mother of the individual; and (2) an individual's phenotype in extranuclear inheritance matches its genotype, whereas an individual's phenotype in maternal effect does not match its own genotype, instead matching that of its mother.

## Extranuclear Genomes

Mitochondria, organelles found in the cytoplasm of all aerobic eukaryotic cells, are involved in cellular respiration. They oxidize pyruvate—the product of glycolysis in the cytosol—to carbon dioxide and water, with the concomitant production of ATP. The genomes of mitochondria of many organisms are circular, double-stranded, supercoiled DNA molecules. Linear mitochondrial genomes are found in some protozoa and some fungi. In general, mitochondrial (mt)DNA contains information for a number of mitochondrial components such as tRNAs, rRNAs, and *some* of the polypeptide subunits of the proteins cytochrome oxidase, NADH-dehydrogenase, and ATPase. The other components found in the mitochondria—most of the proteins in the organelles—are encoded by nuclear genes and are imported into the mitochondria. These components include the DNA polymerase and other proteins for mtDNA replication, RNA polymerase and other proteins for transcription, ribosomal proteins for ribosome assembly, protein factors for translation, the aminoacyl–tRNA synthetases, and the other polypeptide subunits for cytochrome oxidase, NADH-dehydrogenase, and ATPase.

Chloroplasts are cellular organelles found only in green plants and photosynthetic protists; they are the site of *photosynthesis* in the cells containing them. Photosynthesis is carried out in *light reactions* and *dark reactions*. In the light reactions, *chlorophyll* is used to convert light energy into chemical energy, specifically ATP and NADPH. In the dark reactions, carbon dioxide and water are converted into carbohydrate using chemical energy in the form of ATP and NADPH. In all cases, chloroplast (cp)DNA is double-stranded, circular, and supercoiled. The chloroplast genome contains genes for the rRNAs of chloroplast ribosomes, for tRNAs, and for *some* of the proteins required for transcription and translation of the cp-encoded genes (such as ribosomal proteins, RNA polymerase subunits, and translation factors) and for photosynthesis. Most of the proteins found in the chloroplast are encoded by nuclear genes.

## Rules of Extranuclear Inheritance

The pattern of inheritance shown by extranuclear genes is known as **extranuclear inheritance** or **non-Mendelian inheritance,** and it differs strikingly from the pattern shown by nuclear genes. In fact, if the results obtained from genetic crosses do not conform to predictions based on the inheritance of nuclear genes, there is a good reason to suspect extranuclear inheritance.

Here are the four main characteristics of extranuclear inheritance:

1. Ratios typical of Mendelian segregation are not found, because meiosis-based Mendelian segregation is not involved.
2. In multicellular eukaryotes, the results of reciprocal crosses involving extranuclear genes are not the same as reciprocal crosses involving nuclear genes, because meiosis-based Mendelian segregation is not involved. (Recall from Chapters 11 and 12 that in a reciprocal cross, the sexes of the parents are reversed in each case. For example, if A and B represent contrasting genotypes, A ♀ × B ♂ and B ♀ × A ♂ would be a pair of reciprocal crosses.)

   Mitochondrial and chloroplast genes usually show **uniparental inheritance** from generation to generation. In uniparental inheritance, all progeny (both males and females) have the phenotype of only one parent. Usually for multicellular eukaryotes, the phenotype of the mother is inherited exclusively, a phenomenon called **maternal inheritance.** Maternal inheritance occurs because the amount of cytoplasm in the female gamete usually greatly exceeds that in the male gamete. Therefore, the zygote receives most of its cytoplasm (containing the extranuclear genomes of the mitochondria and, where applicable, of the chloroplasts) from the female parent and a negligible amount from the male parent. (Note that maternal inheritance is distinct from maternal effect because in maternal inheritance, the progeny always have the maternal phenotype. By contrast, in maternal effect, the progeny always have the phenotype specified by the maternal nuclear genotype.) Box 13.1 discusses how the the maternal inheritance of mitochondrial genomes can be used to study genetic relationships in human populations.

   In contrast, the results of reciprocal crosses between a wild-type and a mutant strain are identical if the genes are located on nuclear chromosomes. One exception occurs when X-linked genes are involved (see Chapter 12), but even then the results follow nuclear chromosome segregation patterns and are distinct from those for extranuclear inheritance.
3. Extranuclear genes cannot be mapped to the chromosomes in the nucleus.
4. Extranuclear inheritance is not affected by substituting a nucleus with a different genotype.

### Keynote

The inheritance of extranuclear genes follows rules different from those for nuclear genes. In particular, no meiotic segregation is involved, uniparental (and often maternal) inheritance is common, extranuclear genes are not mappable to chromosomes in the nucleus, and the phenotype persists even after nuclear substitution.

## Examples of Extranuclear Inheritance

In this section, we discuss the properties of a selected number of mutations in extranuclear chromosomes to illustrate the principles of extranuclear inheritance.

**The [*poky*] Mutant of *Neurospora*—Maternal Inheritance.** The fungus *Neurospora crassa* (see Chapter 4, pp. 61–62) is an

## Box 13.1 Investigating Genetic Relationships by mtDNA Analysis

In Chapter 10, we discussed some types of DNA analysis involving polymorphisms in nuclear DNA sequences. Briefly, a DNA polymorphism is one of two or more alternate forms (alleles) of a chromosomal locus that differ either in nucleotide sequence or have variable numbers of tandemly repeated DNA sequences. This definition introduces the concept of an allele being something other than a form of a gene, because a DNA polymorphism can be anywhere in the genome, not necessarily in a gene. DNA polymorphisms occur also in extranuclear genomes, such as those in the mitochondria and the chloroplasts. In human mtDNA, one 400-bp region is highly polymorphic. This polymorphism, along with the fact that most mitochondria are maternally inherited, means that maternal lineages are practically unique. Thus maternal line relationships between individuals can be investigated by using PCR to analyze mtDNA for polymorphisms.

An example of using mtDNA analysis involves the last czar and czarina of Russia and their children. During the Bolshevik Revolution of 1917, Czar Nicholas Romanov II was overthrown and exiled, and in 1918 the czar and his family were executed by Bolshevik guards. Rumors persisted that one of the czar's daughters, Princess Anastasia, escaped the execution. In 1922, a woman came forward in Berlin claiming to be Anastasia. In 1928, using the name Anna Anderson, she came to the United States, where she lived until her death in 1984. Although she claimed until she died that she was Anastasia, there was insufficient information available to prove or disprove her claim during her lifetime. In 1993, mtDNA analysis was done on bones believed to be those of the czar's family, found two years earlier in a shallow grave in a Russian town. The DNA samples were compared with a blood sample provided by Prince Philip, Duke of Edinburgh, who is the grand nephew of the Czarina Alexandra. (Prince Philip's grandmother was Princess Victoria, Alexandra's sister.) The mtDNA patterns of the bones matched perfectly the mtDNA of Prince Philip, indicating that they all belonged to the same maternal lineage. Further investigation showed unequivocally that the bones were the remains of the czarina and three of her five children. The bones of the czar were identified in a similar way by matching mtDNA patterns with those of two living relatives. Soon afterward, mtDNA analysis proved that Anna Anderson was not Anastasia, because her mtDNA pattern did not match that of Prince Philip. It is not clear whether any of the three children was Anastasia, although a Russian government commission has stated that there is "definite proof" that one of the skeletons is that of Anastasia.

The case of the Romanovs is an example in which mtDNA analysis has been a powerful tool for analyzing maternal lineages in humans. Mitochondrial DNA analysis is being used to study genetic relationships in many other organisms as well (see Chapter 23, p. 699). Mitochondrial DNA analysis is also being used in conservation biology studies to assess the extent of genetic variability in natural populations. One such study is analyzing the threatened grizzly bear in Yellowstone National Park as a model population for many endangered predator species.

obligate aerobe; it requires oxygen to grow and survive, so mitochondrial functions are essential for its growth. The [*poky*] mutant grows much more slowly than the wild type. The mutant results from a change in the mtDNA, and the convention in such cases is to place square brackets around the mutant symbol. The [*poky*] mutant is defective in aerobic respiration as a result of changes in the cytochrome complement of the mitochondria. The change in the cytochrome spectrum affects the ability of the mitochondria to generate sufficient ATP to support rapid growth, which explains the slow growth of the mutant.

The *Neurospora* life cycle was presented in Figure 4.2, p. 62. The sexual phase of the life cycle is initiated after a fusion of nuclei from mating type *A* and *a* parents. A sexual cross can be made in one of two ways: by putting both parents on the crossing medium simultaneously or by inoculating the medium with one strain and, after three or four days at 25°C, adding the other parent. In the latter case, the first parent on the medium produces all the protoperithecia ("proto-perry-THEECE-e-ah"), the bodies that will give rise to the true-fruiting bodies containing the asci with the sexual ascospores.

Compared with the conidia (the asexual spores), the protoperithecia have a large amount of cytoplasm. Thus they can be considered the female parent, in much the same way that an egg of a plant or an animal is the product of the female parent. Therefore, by using a strain to produce the protoperithecia as the female parent and conidia of another strain as a male parent, geneticists can make reciprocal crosses to determine whether any trait shows extranuclear inheritance.

Reciprocal crosses between [*poky*] and the wild type produce the following results:

[*poky*] ♀ × wild type ♂ → all [*poky*] progeny
wild type ♀ × [*poky*] ♂ → all wild-type progeny

In other words, all progeny show the same phenotype as the maternal parent, indicating maternal inheritance as a characteristic for the [*poky*] mutation.

The cytochrome deficiency phenotype in [*poky*] results from a defect in mitochondrial protein synthesis. The [*poky*] mutant has been shown to be a small deletion in the promoter for the gene for the 19S rRNA of the small mitochondrial ribosomal subunit. The mutation results in a decreased amount of small ribosomal subunits in the organelle and hence to a greatly diminished protein synthesis capability. As a result, synthesis of all mitochondrial proteins, including cytochromes, is reduced and the fungus grows slowly.

**Keynote**

The slow-growing [*poky*] mutant of *Neurospora crassa* shows maternal inheritance and deficiencies for some mitochondrial cytochromes. The molecular defect in [*poky*] is a deletion in the promoter for the rRNA gene of the small mitochondrial ribosomal subunit, which leads to deficiencies for some mitochondrial cytochromes and the slow-growth phenotype.

**Human Genetic Diseases and Mitochondrial DNA Defects.** A number of human genetic diseases result from mtDNA gene mutations. These diseases show maternal inheritance. The following are some brief examples.

- *Leber's hereditary optic neuropathy* (LHON; OMIM 535000, http://www.ncbi.nlm.nih.gov/omim). This disease affects midlife adults and results in complete or partial blindness from optic nerve degeneration. Mutations in the mitochondrial genes for eight electron transport chain proteins, and ATPase 6, all lead to LHON. The electron transport chain drives cellular ATP production by oxidative phosphorylation. It appears that death of the optic nerve in LHON is a common result of oxidative phosphorylation defects, here brought about by inhibition of the electron transport chain.
- *Kearns-Sayre syndrome* (OMIM 530000). People with this syndrome have three major types of neuromuscular defects: progressive paralysis of certain eye muscles, abnormal accumulation of pigmented material on the retina leading to chronic inflammation and degeneration of the retina, and heart disease. The syndrome is caused by large deletions at various positions in the mtDNA. One model is that each deletion removes one or more tRNA genes, so mitochondrial protein synthesis is disrupted. In some unknown way, this leads to development of the syndrome.
- *Myoclonic epilepsy and ragged-red fiber (MERRF) disease* (OMIM 545000). Individuals with this disease exhibit "ragged-red fibers," an abnormality of tissue when seen under the microscope. The most characteristic symptom of MERRF disease is myoclonic seizures (sudden, short-lived, jerking spasms of limbs or the whole body). Other principal symptoms are ataxia (defect in movement coordination) and the accumulation of lactic acid in the blood. The disease is caused by a single nucleotide substitution in the gene for a lysine tRNA. The mutated tRNA adversely affects mitochondrial protein synthesis, and somehow this gives rise to the various phenotypes of the disease.

In most diseases resulting from mtDNA defects, the cells of affected individuals are **heteroplasmons** (also called *cytohets*); that is, they have a mixture of mutant and normal mitochondria. Characteristically the proportions of the two mitochondrial types vary from tissue to tissue and from individual to individual within a pedigree. The severity of the disease symptoms correlates approximately with the relative amount of mutant mitochondria.

**iActivity**

Go to the iActivity *Mitochondrial DNA and Human Disease* on the student website, where you will construct a pedigree to help determine whether a neurological disease has been inherited.

**Cytoplasmic Male Sterility and Hybrid Seed Production.** Hybrid crops are very important to commercial agriculture. A hybrid is produced by crossing two varieties of the crop plant that are not closely related. The hybrids typically grow more vigorously and produce more seeds than does either parent. This phenomenon, called *heterosis* or *heterozygote superiority*, is described further in Chapter 21, pp. 636–637. Farmers are sold the hybrid seed—the seed that germinates to produce the hybrids—which means that plant breeders need to make controlled crosses between two parental varieties on a commercial scale.

Corn was the first crop plant used to generate hybrid seed. Corn plants can self-fertilize, but the male (tassel) and female (ear) parts are separate. Therefore, the manual way to make a controlled cross is to detassel (emasculate) the plant to be used as the female parent and fertilize it with pollen from another plant that will be the male parent. (Recall that emasculation was involved in setting up Mendel's controlled crosses.) Emasculation is relatively easy to do in corn, but laborious in many other crop plants. Fortunately, plant breeders can exploit mutations that cause male sterility. Male sterility may result from mutations of nuclear genes or of extranuclear genes, producing *genic male sterility* and *cytoplasmic male sterility* (CMS), respectively.

The mutation in CMS is in the mitochondrial genome. Like the chloroplast, the mitochondrion is inherited in plants in a maternal fashion. That is, all the mitochondria in the zygote come from the egg and not the pollen. The CMS mutation results in a defect in pollen formation, so the plant is male sterile. However, when this CMS plant is used as the female parent in a controlled cross, the hybrid seed germinates and produces progeny plants. Those plants are male sterile because they have inherited the CMS mutation and cannot produce seeds by self-fertilization. Understandably, the farmer would be unhappy with the hybrid plants because male-fertile plants would also have to be planted so that the hybrids can be fertilized. The solution to this problem involves a nuclear *restorer of fertility* (*Rf*) gene. The dominant *Rf* allele overrides the CMS mutation, whereas the recessive *rf* allele cannot.

Figure 13.20 shows how hybrid seed can be generated using CMS and the *Rf* gene. The male-sterile female parent is [CMS] *rf*/*rf*, and the fertile male is [*CMS*] *Rf*/*rf* (where [*CMS*] indicates the cytoplasm for cytoplasmic

**Figure 13.20**

**Production of hybrid seed using cytoplasmic male sterility (CMS) and a nuclear *restorer of fertility* gene.**

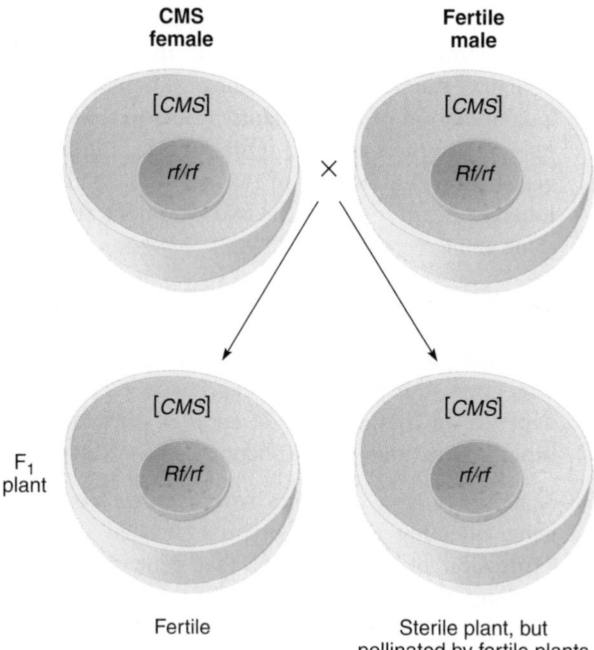


male sterility). The $F_1$ progeny exhibits the desired hybrid vigor, and all have the [*CMS*] cytoplasm. The hybrids will segregate 1 *Rf/rf* : 1 *rf/rf*. The *Rf/rf* plants are male fertile because the restorer gene has overriden the CMS mutation. The *rf/rf* plants are male sterile, though, because there is no effect on the [*CMS*] cytoplasm. However, these latter plants are fertilized readily by pollen from the *Rf/rf* plants in the field.

There is now a genetic engineering approach for making male-sterile plants. This approach does not involve extranuclear genes; it involves making transgenic plants using standard plant transformation methods. Two genes are needed, both from the soil bacterium *Bacillus amyloliquefaciens*. One gene encodes *barnase*, an RNase, that is secreted from the bacterium as a defense mechanism against other organisms. The other gene encodes *barnstar*, a protein that binds to and inhibits barnase, thereby protecting the bacterium from its own enzyme. A wild-type plant is transformed with the barnase gene fused to the TA29 promoter. The TA29 promoter is from a gene that is expressed only in the tapetum, a tissue that surrounds the pollen sac that is essential for pollen production. In transgenic plants with the TA29-*barnase* gene, barnase is made in the tapetum and destroys the RNA molecules in that tissue, making the plant male sterile. This plant is used as the female parent in a cross with a transgenic plant containing a TA29-*barnstar* gene. The seeds from the female produce hybrid plants in which both the barnase and barnstar proteins are produced in the tapetum. The barnstar binds to the barnase, inhibiting the RNase activity and thereby preventing male sterility. The hybrids are male fertile.

**Exceptions to Maternal Inheritance.** Strict maternal inheritance has been considered to be the case for extranuclear mutations in animals and plants where the female gamete contributes most of the cytoplasm to the zygote. However, exceptions are coming to light. Here are some examples.

1. By exploiting DNA sequence differences between mtDNAs of two inbred lines of mice, researchers have used PCR (polymerase chain reaction; see Chapter 9, pp. 221–222) to demonstrate that paternally inherited mtDNA molecules are present at a frequency of $10^{-4}$ relative to maternal mtDNA molecules. This heteroplasmy of paternal and maternal mitochondria has potentially significant evolutionary implications. That is, it has been generally considered that maternal and paternal mtDNA remain distinct due to the strict maternal inheritance of mitochondria. However, if heteroplasmy can occur, then there is a likelihood of genetic recombination between maternally derived and paternally derived mtDNA molecules. Such recombination will lead to significant diversity of mtDNA in an individual. The extent to which this phenomenon occurs is unknown, but the fact that it exists at all makes it necessary to be cautious about conclusions based on a purely maternal inheritance of mtDNA.
2. In most angiosperms (flowering plants), the chloroplasts are inherited only from the maternal parent. In some angiosperms, however, chloroplasts are inherited at high frequency from both parents, or mostly from the paternal parent. For example, biparental inheritance of chloroplasts is seen in the evening primrose, *Oenothera*. Paternal inheritance of chloroplasts is the rule in conifers, which are gymnosperms.



# Summary


- A gene may have many allelic forms in a population, and these alleles are called multiple alleles. However, any given diploid individual can possess only two different alleles of a given gene.
- With complete dominance, the same phenotype results whether the dominant allele is heterozygous or homozygous. In incomplete dominance, the phenotype of the heterozygote is intermediate between those of the two homozygotes. In codominance, the heterozygote exhibits the phenotypes of both homozygotes.
- Alleles of certain genes result in the failure to produce a necessary functional gene product, and this

deficiency gives rise to a lethal phenotype. Such lethal alleles may be recessive or dominant. The existence of lethal alleles of a gene indicates that the normal product of the gene is essential to the viability the organism.

- Penetrance is the frequency (in percent) at which an allele manifests itself phenotypically within a population. Expressivity is the kind or degree of phenotypic manifestation of a gene or genotype in a particular individual. Both penetrance and expressivity can occur for the same trait. Both penetrance and expressivity depend on the genotype and the external environment.
- The zygote's genetic constitution specifies only the organism's potential to develop and function. As the organism develops and cells differentiate, many things can influence phenotypic expression associated with an allele. One such influence is the organism's environment, both internal and external. Examples of the internal environment include age and sex; examples of the external environment include nutrition, light, chemicals, temperature, and infectious agents.
- Variation in most of the genetic traits considered in the earlier discussion of Mendelian principles is determined predominantly by differences in genotype; that is, phenotypic differences result from genotypic differences. For many traits, however, the phenotypes are influenced by both genes and the environment.
- Maternal effect is the phenomenon in which the nuclear genotype of the mother is expressed in the phenotype of the offspring, with no influence by the paternal nuclear genome. Maternal effect occurs as the result of gene products that are deposited in the oocyte before fertilization and that direct early development of the embryo. The genes that encode those products are maternal effect genes.
- A complementation test determines whether two independently isolated mutants with the same phenotype have mutations in the same or different genes. If a combination of two mutants results in a wild-type phenotype, then the two mutations are in different genes. If a combination of two mutants results in a mutant phenotype, then the two mutations are in the same gene.
- In many cases, different genes interact to determine phenotypic characteristics. In epistasis, for example, modified Mendelian ratios occur because of gene interactions: the phenotypic expression of one gene depends on the genotype of another gene locus. In other interactions, a new phenotype is produced.
- A modifier gene interacts with another nonallelic gene, affecting the phenotype associated with the expression of the alleles of that gene rather than masking their expression, as is seen in the gene interactions of epistasis.
- Both mitochondria and chloroplasts contain their own DNA genomes, which contain genes for some of the molecular components in those organelles. The inheritance of mitochondrial and chloroplast genes follows rules different from those for nuclear genes: meiosis-based Mendelian segregation is not seen, uniparental (usually maternal) inheritance is typically exhibited, extranuclear genes cannot be mapped to the known nuclear linkage groups, and a phenotype resulting from an extranuclear mutation persists after nuclear substitution.



## Analytical Approaches to Solving Genetics Problems

**Q13.1** In snapdragons, red flower color ($C^R$) is incompletely dominant to white flower color ($C^W$); the heterozygote has pink flowers. Also, normal broad leaves ($L^B$) are incompletely dominant to narrow, grasslike leaves ($L^N$); the heterozygote has an intermediate leaf breadth. If a red-flowered, narrow-leaved snapdragon is crossed with a white-flowered, broad-leaved one, what will be the phenotypes of the $F_1$ and $F_2$ generations, and what will be the frequencies of the different classes?

**A13.1.** This basic question about gene segregation involves the issue of incomplete dominance. In the case of incomplete dominance, remember that the genotype can be determined directly from the phenotype. Therefore, we do not need to ask whether a strain is true breeding, because all phenotypes have a different (and therefore known) genotype.

The best approach here is to assign genotypes to the parental snapdragons. Let $C^R/C^R\ L^N/L^N$ represent the red, narrow plant and $C^W/C^W\ L^B/L^B$ represent the white, broad plant. The $F_1$ plants from this cross will all be double heterozygotes, $C^R/C^W\ L^B/L^N$. Because of the incomplete dominance, these plants are pink flowered and have leaves of intermediate breadth. Interbreeding the $F_1$ plants gives the $F_2$ generation, but it does not have the usual 9:3:3:1 ratio. Instead, there is a different phenotype for each genotype. These genotypes and phenotypes and their relative frequencies are shown in Figure 13.A.

Figure 13.A

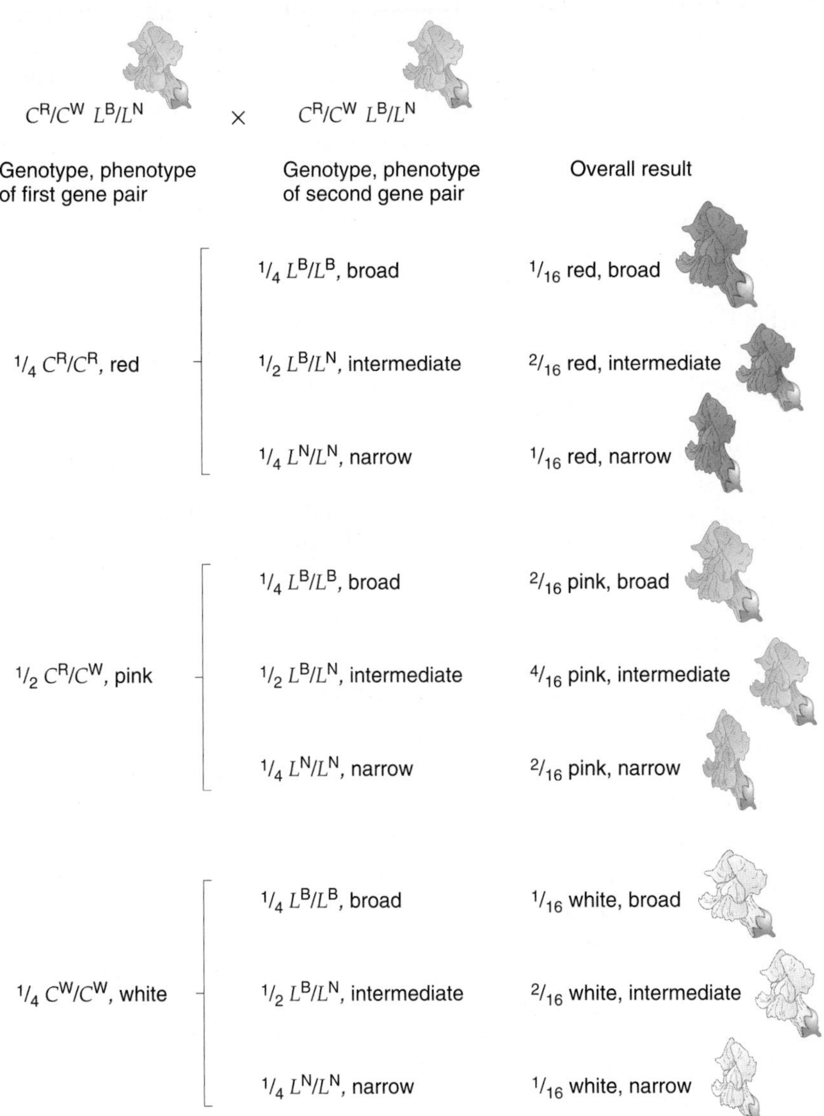


**Q13.2** In snapdragons, red flower color is incompletely dominant to white, with the heterozygote being pink; normal flowers are completely dominant to peloric-shaped ones; and tallness is completely dominant to dwarfness. The three gene pairs segregate independently. If a homozygous red, tall, normal-flowered plant is crossed with a homozygous white, dwarf, peloric-flowered one, what proportion of the $F_2$ plants will resemble the $F_1$ plants in appearance?

**A13.2.** Let us assign symbols: $C^R$ = red and $C^W$ = white; $N$ = normal flowers and $n$ = peloric; $T$ = tall and $t$ = dwarf. The initial cross, then, becomes $C^R/C^R$ $T/T$ $N/N$ × $C^W/C^W$ $t/t$ $n/n$. From this cross, we see that all the $F_1$ plants are triple heterozygotes with the genotype $C^R/C^W$ $T/t$ $N/n$ and with the phenotype pink, tall, normal flowered. Interbreeding the $F_1$ generation will produce 27 different genotypes among the $F_2$ plants; this answer follows from the rule that the number of genotypes is $3^n$, where $n$ is the number of heterozygous gene pairs involved in the $F_1 \times F_1$ cross (see Chapter 11).

Here, we are asked specifically for the proportion of $F_2$ progeny that resemble the $F_1$ plants in appearance. We can calculate this proportion directly without needing to display all the possible genotypes and then grouping the progeny in classes according to phenotype. First, we calculate the frequency of pink-flowered plants in the $F_2$; then we determine the proportion of these plants that have the other two attributes. From a $C^R/C^W \times C^R/C^W$ cross, we calculate that half of the progeny will be heterozygous $C^R/C^W$ and therefore pink. Next, we determine the proportion of $F_2$ plants that are phenotypically like the $F_1$ with respect to height (tall). Either $T/T$ or $T/t$ plants will be tall, so $3/4$ of the $F_2$ will be tall. Similarly, $3/4$ of the $F_2$ plants will be normal flowered like the $F_1$ plants. To obtain the probability of all three of these phenotypes occurring together (pink, tall, normal), we must multiply the individual probabilities, because the gene pairs segregate independently. The answer is $1/2 \times 3/4 \times 3/4 = 9/32$.

**Q13.3**

**a.** An $F_1 \times F_1$ self gives a 9:7 phenotypic ratio in the $F_2$. What phenotypic ratio would you expect if you testcrossed the $F_1$ individuals?

**b.** Answer the same question for an $F_1 \times F_1$ cross that gives a 9:3:4 ratio.

**c.** Answer the same question for a 15:1 ratio.

**A13.3.** This question deals with epistatic effects. In answering the question, we must consider the interaction between the different genotypes in order to proceed with the testcross. Let us set up the general genotypes that we will deal with throughout. The simplest are allelic pairs $a^+$ and $a$ and $b^+$ and $b$, where the wild-type alleles are completely dominant to the other member of the pair.

**(a)** A 9:7 ratio in the $F_2$ implies that both members of the $F_1$s are double heterozygotes and that epistasis is involved. Essentially, any genotype with a homozygous recessive condition has the same phenotype, so the 3, 3, and 1 parts of a 9:3:3:1 ratio are phenotypically combined into one class. In terms of genotype, $^9/_{16}$ are $a^+/– \; b^+/–$ types, and the other $^7/_{16}$ are $a^+/– \; b/b$, $a/a \; b^+/–$, and $a/a \; b/b$. (Recall that the dash after a wild-type allele signifies that the same phenotype results, whether the missing allele is a wild type or a mutant.) The testcross asked for is $a^+/a \; b^+/b \times a/a \; b/b$, and we can predict a 1:1:1:1 ratio of $a^+/a \; b^+/b$ : $a^+/a \; b/b$ : $a/a \; b^+/b$ : $a/a \; b/b$. The first genotype will have the same phenotype as the $^9/_{16}$ class of the $F_2$; but because of epistasis, the other three genotypes will have the same phenotype as the $^7/_{16}$ class of the $F_2$. In sum, the answer is a phenotypic ratio of 1:3 in the progeny of a testcross of the $F_1$ individuals.

**(b)** We are asked to answer the same question for a 9:3:4 ratio in the $F_2$. Again, this question involves a modified dihybrid ratio where two classes of the 9:3:3:1 have the same phenotype. Complete dominance for each of the two gene pairs occurs here also, so the $F_1$ individuals are are $a^+/a \; b^+/b$. Perhaps both the $a^+/– \; b^+/b$ and $a/a \; b/b$ classes in the $F_2$ will have the same phenotype, whereas the $a^+/– \; b^+/–$ and $a/a \; b^+/–$ classes will have phenotypes distinct from each other and from the interaction class. The genotypic ratio of a testcross of the $F_1$ individuals is the same as in part (a) of this question. Considering them in the same order as we did there, we find that the second and fourth classes would have the same phenotype because of epistasis. So there are only three possible phenotypic classes, instead of the four found in the testcross of a dihybrid $F_1$, where there is complete dominance and no interaction. The phenotypic ratio here is 1:1:2.

**(c)** This question is yet another example of epistasis. Since 15 + 1 = 16, this number gives the outcome of an $F_1$ self of a dihybrid where there is complete dominance for each gene pair and interaction between the dominant alleles. In this case, the $a^+/– \; b^+/–$, $a^+/– \; b/b$, and $a/a \; b^+/–$ classes have one phenotype and include $^{15}/_{16}$ of the $F_2$ progeny, and the $a/a \; b/b$ class has the other phenotype and $^1/_{16}$ of the $F_2$. The genotypic results of a testcross of the $F_1$ individuals are the same as in parts (a) and (b) of this question; that is, the $F_2$ progeny exhibit a 1:1:1:1 ratio of $a^+/a \; b^+/b$ : $a^+/a \; b/b$ : $a/a \; b^+/b$ : $a/a \; b/b$. The first three classes have the same phenotype, which is the same as that of the $^{15}/_{16}$ of the $F_2$, and the last class has the other phenotype. The answer, then, is a 3:1 phenotypic ratio.

**Q13.4** Four slow-growing mutant strains of *Neurospora crassa*, coded *a*, *b*, *c*, and *d*, have been isolated. All have an abnormal system of respiratory mitochondrial enzymes. The inheritance patterns of these mutants were tested in controlled crosses with the wild type, with the following results:

| Female Parent | | Male Parent | Progeny (Ascospores) Wild Type | Slow Growing |
|---|---|---|---|---|
| Wild type | × | *a* | 847 | 0 |
| *a* | × | Wild type | 0 | 659 |
| Wild type | × | *b* | 1,113 | 0 |
| *b* | × | Wild type | 0 | 2,071 |
| Wild type | × | *c* | 596 | 590 |
| Wild type | × | *d* | 1,050 | 1,035 |

Give a genetic interpretation of these results.

**A13.4.** This question asks us to consider the expected transmission patterns for nuclear and extranuclear genes. The nuclear genes will have a 1:1 segregation in the offspring because this organism is a haploid organism and therefore should exhibit no differences in the segregation patterns, whichever strain is the maternal parent. On the other hand, a distinguishing characteristic of extranuclear genes is a difference in the results of reciprocal crosses. In *Neurospora*, this characteristic usually is manifested by all progeny having the phenotype of the maternal parent. With these ideas in mind, we can analyze each mutant in turn.

Mutant *a* shows a clear difference in its segregation in reciprocal crosses and is a classic case of maternal inheritance. The interpretation here is that the gene is extranuclear; therefore, the gene must be in the mitochondrion. The [*poky*] mutant described in this chapter shows this type of inheritance pattern.

By the same reasoning, the mutation in strain *b* must also be extranuclear.

Mutants *c* and *d* segregate 1:1, indicating that the mutations involved are in the nuclear genome. In these cases we need not consider the reciprocal cross, because there is no evidence of maternal inheritance. In fact, the actual mutations that are the basis for this question cause sterility, so the reciprocal cross cannot be done. We can confirm that the mutations are in the nuclear genome by doing mapping experiments using known nuclear markers. Evidence of linkage to such markers would confirm that the mutations are not extranuclear.

## Questions and Problems

**13.1** In rabbits, $C$ = agouti coat color, $c^{ch}$ = chincilla, $c^h$ = Himalayan, and $c$ = albino. The four alleles constitute a multiple allelic series. The agouti $C$ is dominant to the other three alleles, $c$ is recessive to the other three alleles, and chinchilla is dominant to Himalayan. Determine the phenotypes of progeny from the following crosses:

**a.** $C/C \times c/c$
**b.** $C/c^{ch} \times C/c$
**c.** $C/c \times C/c$
**d.** $C/c^h \times c^h/c$
**e.** $C/c^h \times c/c$

***13.2** If a given population of diploid organisms contains only three alleles of a particular gene (say, $w^1$, $w^2$, and $w^3$), how many different diploid genotypes are possible in the populations? List all possible genotypes of diploids. (Consider only the three given alleles.)

**13.3** In humans, the three alleles $I^A$, $I^B$, and $i$ constitute a multiple allelic series that determines the ABO blood group system, as described in this chapter. In each of the following instances, state whether a child of the given blood type could be produced by the parents described, and explain your answer.

**a.** An O child from two A parents
**b.** An O child from an A parent and a B parent
**c.** An AB child from an A parent and an O parent
**d.** An O child from an AB parent and an A parent
**e.** An A child from an AB parent and a B parent

**13.4** A man is blood type O, M. A woman is blood type A, M, and her child is type A, MN. The man cannot be the father of this child because

**a.** O men cannot have type-A children.
**b.** O men cannot have MN children.
**c.** An O man and an A woman cannot have an A child.
**d.** An M man and an M woman cannot have an MN child.

***13.5** A woman of blood group AB marries a man of blood group A whose father was of group O. What is the probability that

**a.** their two children will both be of group A?
**b.** one child will be of group B, the other of group O?
**c.** the first child will be a son of group AB and the second child a son of group B?

**13.6** In snapdragons, red flower color ($C^R$) is incompletely dominant to white ($C^W$); the $C^R/C^W$ heterozygotes are pink. A red-flowered snapdragon is crossed with a white-flowered one. Determine the flower color of

**a.** the $F_1$ snapdragons.
**b.** the $F_2$ snapdragons.
**c.** the progeny of a cross of the $F_1$ snapdragons to the red parent.
**d.** the progeny of a cross of the $F_1$ snapdragons to the white parent.

***13.7** In shorthorn cattle, the heterozygous condition of the alleles for red coat color ($C^R$) and white coat color ($C^W$) is roan coat color. If two roan cattle are mated, what proportion of the progeny will resemble their parents in coat color?

**13.8** In guinea pigs, short hair ($L$) is dominant to long hair ($l$), and the heterozygous conditions of yellow coat ($C^Y$) and white coat ($C^W$) give cream coat. A short-haired, cream guinea pig is bred to a long-haired, white guinea pig, and a long-haired, cream baby guinea pig is produced. When the baby grows up, it is bred back to the short-haired, cream parent. What phenotypic classes, and in what proportions, are expected among the offspring?

**13.9** The shape of radishes may be long ($S^L/S^L$), oval ($S^L/S^S$), or round ($S^S/S^S$), and the color of radishes may be red ($C^R/C^R$), purple ($C^R/C^W$), or white ($C^W/C^W$). If a long, red radish plant is crossed with a round, white plant, what will be the appearance of the $F_1$ and the $F_2$ plants?

***13.10** In four-o'clock plants, two genes, $Y$ and $R$, affect flower color. Neither is completely dominant, and the two interact with each other to produce seven different flower colors:

$Y/Y\ R/R$ = crimson  $Y/y\ R/R$ = magenta
$Y/Y\ R/r$ = orange-red  $Y/y\ R/r$ = magenta-rose
$Y/Y\ r/r$ = yellow  $Y/y\ r/r$ = pale yellow
$y/y\ R/R$, $y/y\ R/r$, *and* $y/y\ r/r$ = white

**a.** In a cross of a crimson-flowered plant with a white one ($y/y\ r/r$), what will be the appearances of the $F_1$ plants, the $F_2$ plants, and the offspring of the $F_1$ plants backcrossed to their crimson parent?
**b.** What will be the flower colors in the offspring of a cross of orange-red × pale yellow?
**c.** What will be the flower colors in the offspring of a cross of a yellow with a $y/y\ R/r$ white?

**13.11** Two four-o'clock plants were crossed and gave the following offspring: 1/8 crimson, 1/8 orange-red, 1/4 magenta, 1/4 magenta-rose, and 1/4 white. Unfortunately, the person who made the crosses was color blind and could not record the flower colors of the parents. From the results of the cross, deduce the genotypes and flower colors of the two parents.

**13.12** The allele $l$ in *Drosophila* is recessive, sex-linked, and lethal when homozygous or hemizygous (the condition in the male). If a female of genotype $L/l$ is crossed with a normal male, what is the probability that the first two surviving progeny will be males?

***13.13** A locus in mice is involved in pigment production; when parents heterozygous at this locus are mated, 3/4 of the progeny are colored and 1/4 are albino. Another

Figure 13.B

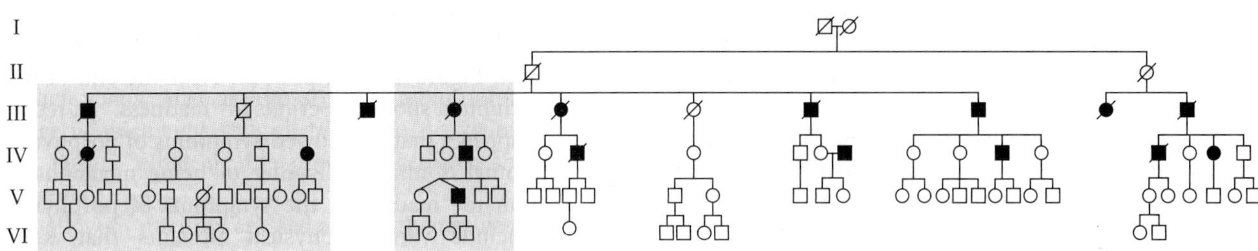




of parkinsonism in a family of European descent. The shaded portions of the pedigree indicate family members who reside in the United States. The remaining portions of the pedigree reside in Europe. Members of the U.S. branches of the family have not visited Europe for any extensive period since the initial emigration from Europe.

**a.** If the disease in this family has a genetic basis, what is its basis? Explain your answer.

**b.** Why might this pedigree be particularly helpful in distinguishing between an environmental and a genetic cause of Parkinson disease?

**c.** What reservations, if any, do you have about concluding that the disease has a genetic basis in some individuals?

***13.26** The inheritance of shell-coiling direction in the snail *Limnaea peregra* has been studied extensively. A snail produced by a cross between two individuals has a shell with a dextral (right-handed) coil. This snail produces only sinistral (left-handed) progeny on selfing. What are the genotypes of the $F_1$ snail and its parents?

**13.27** In *Drosophila melanogaster*, a recessive autosomal allele, black (*b*), produces a black body color when homozygous. An independently assorting autosomal allele, ebony (*e*), also produces a black body color when homozygous. Flies with genotypes $b^+/–\ e/e$, $b/b\ e^+/–$, and $b/b\ e/e$ are phenotypically identical with respect to body color. Flies with genotype $b^+/–\ e^+/–$ have a grey body color. True-breeding $b^+/b^+\ e/e$ ebony flies are crossed with true-breeding $b/b\ e^+/e^+$ black flies.

**a.** What will be the phenotype of the $F_1$ files?

**b.** What phenotypes and what proportions would occur in the $F_2$ generation?

**c.** What phenotypic ratios would you expect to find in the progeny of these backcrosses?

**i.** $F_1 \times$ true-breeding ebony

**ii.** $F_1 \times$ true-breeding black

***13.28** In *Drosophila*, recessive mutants *a*, *b*, *c*, *d*, *e*, *f*, and *g* all have the same phenotype; namely, the absence of red pigment in the eyes. In pairwise combinations in complementation tests, the following results were produced, where + = complementation and − = no complementation.

| | *a* | *b* | *c* | *d* | *e* | *f* | *g* |
|---|---|---|---|---|---|---|---|
| *g* | + | − | + | + | + | + | − |
| *f* | − | + | + | − | + | − | |
| *e* | + | + | − | + | − | | |
| *d* | − | + | + | − | | | |
| *c* | + | + | − | | | | |
| *b* | + | − | | | | | |
| *a* | − | | | | | | |

**a.** What genotypes were crossed to show complementation between mutants g and *a*?

**b.** What genotypes were crossed to show no complementation between mutants g and *b*?

**c.** How many genes are present?

**d.** Which mutants have defects in the same gene?

***13.29** In Question 13.28, all of the mutant alleles evaluated in complementation tests were recessive. Can a complementation test be used to determine whether two dominant alleles affect the same gene? Can a complementation test be used to determine whether a dominant allele affects the same gene as a recessive allele? Explain your answers.

**13.30** In poultry, the genotype–phenotype relationships for comb shape are *R/– P/–*, walnut; *R/– p/p*, rose, *r/r P/–*, pea; and *r/r p/p*, single. What will be the comb characters of the offspring of the following crosses?

**a.** *R/R P/p* × *r/r P/p*

**b.** *r/r P/P* × *R/r P/p*

**c.** *R/r p/p* × *r/r P/p*

**13.31** For the following crosses involving the comb character in poultry (see previous question), determine the genotypes of the two parents:

**a.** A walnut crossed with a single produces offspring that are 1/4 walnut, 1/4 rose, 1/4 pea, and 1/4 single.

**b.** A rose crossed with a walnut produces offspring that are 3/8 walnut, 3/8 rose, 1/8 pea, and 1/8 single.

**b.** A rose crossed with a pea produces five walnut and six rose offspring.

**d.** A walnut crossed with a walnut produces one rose, two walnut, and one single offspring.

**13.32** In cats, two alleles (*B*, *O*) at an X-linked gene control whether black or orange pigment is deposited. A dominant allele at an autosomal gene *I/i* partially inhibits the deposition of pigment, lightening the coat color from black to grey or from orange to pale orange. A dominant allele at the autosomal gene *T/t* determines whether a tabby, or vertically striped, pattern is present. The tabby pattern depends on a dominant agouti (*A*) allele for its expression, with nonagouti (*a*) epistatic to tabby. The agouti allele also causes a speckled, rather than solid, coat color. Judy, a stray cat, gives birth to four kittens. Judy has a speckled coat with small grey and pale orange spots that are distributed in a pattern like that of a tortoiseshell. She shows no trace of the tabby pattern. Of the three female offspring, two are solid grey and the third is speckled grey and light orange like her mother, but also shows traces of a tabby pattern. The single male offspring is solid grey.

**a.** Explain how the tortoiseshell pattern arises in cats. That is, how can a cat have distinct patches of fur with different deposits of pigment?

**b.** Cats with a tortoiseshell pattern usually are female. Explain why this is the case. Also explain why, when an unusual male tortoiseshell male cat is found, he is atypically large and typically not very swift.

**c.** What genotype(s) might Judy and her kittens have?

**d.** Assuming the kittens all have the same father, what phenotype(s) should be considered in assessing the neighborhood males for paternity?

*__13.33__ White cats are produced by three distinct genotypes. First, the dominant *W* allele blocks the production of melanin-producing cells called melanocytes. This allele does not block the formation of pigment per se, so *W*– cats are white with blue eyes. *W*– cats develop with a hearing deficit, as the cochlea of the ear contains a band of melanocytes that regulate ion balance needed for hearing. Second, the dominant piebald-spotting allele *S* hampers the migration of melanocytes during development, and areas that lack melanocytes appear as white spots. The homozygote phenotype is more severe than that of the heterozygote, as a result of which some *SS* cats appear completely white. Except for a few pigmented hairs somewhere, they develop with one big white spot. Third, the recessive allele *c* at the albino locus completely blocks the expression of pigment, so that *cc* cats are white with red eyes. They have normal hearing.

**a.** A very regal cat is white except for a small, pigmented patch of hair surrounding her left eye. This eye is brown, while her right eye is blue. What is her genotype with respect to the *W/w*, *S/s*, and *C/c* genes, why does she have eyes of two different colors, and why does she acknowledge her human servant only when he is kneeling directly in front of her?

**b.** Explain whether any of these alleles are pleiotropic.

**c.** Explain whether it is ever possible to obtain offspring with pigmented coats from

  **i.** a mating between two completely white cats, one with blue eyes and one with red eyes.

  **ii.** a mating between two blue-eyed cats, one completely white and one white with a few grey hairs behind one ear.

**d.** What types of epistatic interactions do you predict for *W* and *S*, *W* and *cc*, and *S* and *cc*?

*__13.34__ $F_2$ plants segregate $^9/_{16}$ colored : $^7/_{16}$ colorless. If just one colored plant from the $F_2$ generation is chosen at random and selfed, what is the probability that there will be *no* segregation of the two phenotypes among its progeny?

*__13.35__ In peanuts, a plant may be either "bunch" or "runner." Two different strains of peanut, V4 and G2, in which "bunch" occurred were crossed, with the following results:

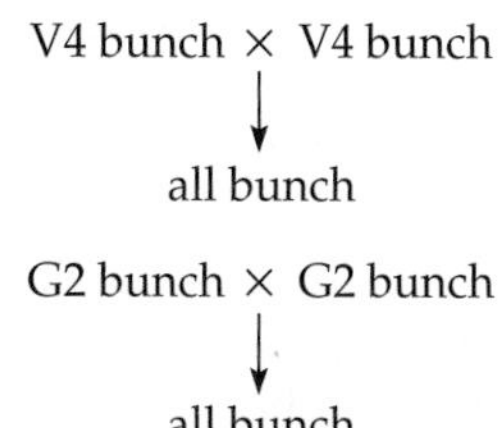


The two true-breeding strains of bunch were crossed in the following way:

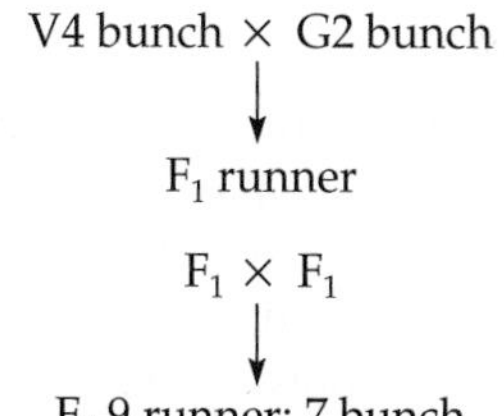


What is the genetic basis of the inheritance pattern of runner and bunch in the $F_2$ peanuts?

*__13.36__ In rabbits, one enzyme (the product of a functional gene *A*) is needed to produce a substance required for hearing. Another enzyme (the product of a functional gene *B*) is needed to produce another substance required for hearing. The genes responsible for the two enzymes are not linked. Individuals homozygous for either one or both of the nonfunctional recessive alleles, *a* or *b*, are deaf.

**a.** If a large number of matings were made between two double heterozygotes, what phenotypic ratio would be expected in the progeny?

**b.** The phenotypic ratio found in part (a) is a result of what well-known phenomenon?

**c.** What phenotypic ratio would be expected if rabbits homozygous recessive for trait A and heterozygous for trait B were mated to rabbits heterozygous for both traits?

*__13.37__ Genes *A*, *B*, and *C* are independently assorting and control the production of a black pigment.

**a.** Suppose that *A*, *B*, and *C* act in the following pathway:

$$\text{colorless} \xrightarrow{A} \xrightarrow{B} \xrightarrow{C} \text{black}$$

The alternative alleles that give abnormal functioning of these genes are designated *a*, *b*, and *c*, respectively. A black *A/A B/B C/C* is crossed with a colorless *a/a b/b c/c* to give a black $F_1$. The $F_1$ is selfed. What proportion of the $F_2$ individuals is colorless? (Assume that the products of each step except the last are colorless, so only colorless and black phenotypes are observed.)

**b.** Suppose instead that a different pathway is utilized. In it, the *C* allele produces an inhibitor that prevents the formation of black by destroying the ability of *B* to carry out its function. The mechanism is as follows:

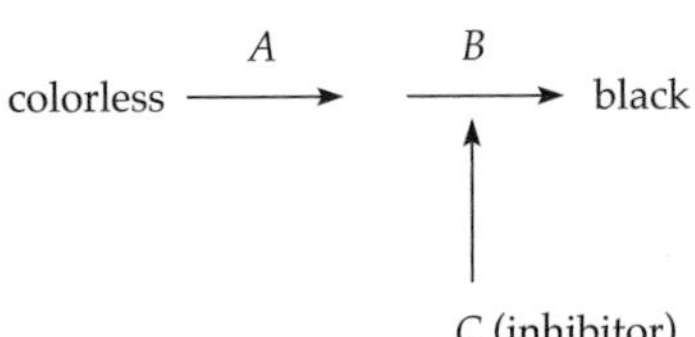


A colorless *A/A B/B C/C* individual is crossed with a colorless *a/a b/b c/c*, giving a colorless $F_1$. The $F_1$ is selfed to give an $F_2$. What is the ratio of colorless to black in the $F_2$ individuals? (Only colorless and black phenotypes are observed, as in part (**a**).)

**c.** How would you evaluate which of the biochemical pathways hypothesized in parts (**a**) and (**b**) is more likely?

**13.38** In doodlewags (hypothetical creatures), the dominant allele *S* causes solid coat color; the recessive allele *s* results in white spots on a colored background. The black coat color allele *B* is dominant to the brown allele *b*, but these genes are expressed only in the genotype *a/a*. Individuals that are *A/–* are yellow regardless of *B* alleles. Six pups are produced in a mating between a solid yellow male and a solid brown female. Their phenotypes are 2 solid black, 1 spotted yellow, 1 spotted black, and 2 solid brown.

**a.** What are the genotypes of the male and female parents?

**b.** What is the probability that the next pup will be spotted brown?

***13.39** A substantial body of evidence indicates that defects in mitochondrial energy production may contribute to the neuronal cell death seen in a number of late-onset neurodegenerative diseases, including Alzheimer disease, Parkinson disease, Huntington disease, and amyotrophic lateral sclerosis (ALS, or Lou Gehrig disease). Some of these diseases have been associated with mutations in the nuclear genome. One experimental system that has been developed to evaluate the contributions of the mitochondrial genome to these diseases uses a cytoplasmic hybrid known as a cybid. Cybids are made by repopulating a tissue culture cell line that has become deficient in mitochondria with mitochondria from the cytoplasm of a human platelet cell. The cybids thus have nuclear DNA from the tissue culture cell and mitochondrial DNA from the human platelet cell.

The mitochondrial protein cytochrome oxidase has subunits encoded by both nuclear and mitochondrial genes. Patients with Alzheimer disease have been reported to have lower levels of cytochrome oxidase than do age-matched controls.

**a.** Given the means to assay cytochrome oxidase activity, how would you investigate whether the decreased levels of cytochrome oxidase activity in patients with Alzheimer disease could be ascribed to nuclear or mitochondrial genetic defects? What controls would you create?

**b.** If you demonstrated that the mitochondrial contribution to cytochrome oxidase is responsible for lowered cytochrome oxidase activity, could you conclude that each mitochondrion of an affected individual has an identical defect?

**13.40** Reciprocal crosses between two types of green-leaved evening primroses, *Oenothera hookeri* and *Oenothera muricata*, produce the following effects on the chloroplasts:

*O. hookeri* female × *O. muricata* male → Yellow chloroplasts

*O. muricata* female × *O. hookeri* male → Green chloroplasts

Explain the difference between these results, noting that the chromosome constitution is the same in both types.

***13.41** A series of crosses are performed with a recessive mutation in *Drosophila* called *tudor*. Homozygous *tudor* animals appear normal and can be generated from the cross of two heterozygotes, but a true-breeding *tudor* strain cannot be maintained. When homozygous *tudor* males are crossed to homozygous *tudor* females, both of which appear to be phenotypically normal, a normal-appearing $F_1$ is produced. However, when $F_1$ males are crossed to wild-type females, or when $F_1$ females are crossed to wild-type males, no progeny are produced. The same results are seen in the $F_1$ progeny of homozygous *tudor* females crossed to wild-type males. The $F_1$ progeny of homozygous *tudor* males crossed to wild-type females appear normal, and they are capable of producing progeny when mated either with each other or with wild-type animals.

**a.** How would you classify the *tudor* mutation? Why?

**b.** What might cause the *tudor* phenotype?

**13.42** Several investigators have demonstrated that chemical and environmental treatments of plants and animals can lead to abnormalities that persist for several generations before disappearing. For example, Hoffman found that treating the bean *Phaseolus vulgaris* with chloral hydrate led to abnormalities in leaf shape that persisted in the female (but not in the male) line for almost six generations before disappearing.

**a.** In what different ways could you explain the origin of these abnormalities and their disappearance after several generations?

**b.** What broader implications might these findings have?

**13.43** *Drosophila melanogaster* has a sex-linked, recessive, mutant gene called *maroon-like* (*mal*). Homozygous *mal* females or hemizygous *mal* males have light-colored eyes due to the absence of the active enzyme xanthine dehydrogenase, which is involved in the synthesis of eye pigments. When heterozygous $mal^+/mal$ females are crossed with *mal* males, all the offspring are phenotypically wild type. However, half the female offspring from this cross, when crossed back to *mal* males, give all *mal* progeny. The other half of the females, when crossed to *mal* males, give all phenotypically wild-type progeny. What is the explanation for these results?

***13.44** When females of a particular mutant strain of *Drosophila melanogaster* are crossed to wild-type males, all the viable progeny flies are females. Hypothetically, this result could be the consequence of either a sex-linked, male-specific lethal mutation or a maternally inherited factor that is lethal to males. What crosses would you perform to distinguish between these alternatives?

**13.45** Reciprocal crosses between two *Drosophila* species, *D. melanogaster* and *D. simulans*, produce the following results:

*melanogaster* ♀ × *simulans* ♂ → Females only

*simulans* ♀ × *melanogaster* ♂ → Males, with few or no females

Propose a possible explanation for these results.

***13.46** Some *Drosophila* are very sensitive to carbon dioxide; administering it to them anesthetizes them. The sensitive flies have a cytoplasmic particle called sigma that has many properties of a virus. Resistant flies lack sigma. The sensitivity to carbon dioxide shows strictly maternal inheritance. What would be the outcome of the following two crosses: (**a**) sensitive ♀ × resistant ♂ and (**b**) sensitive ♂ × resistant ♀?

**13.47** The pedigree in the following figure shows a family in which a rare inherited disease called Leber hereditary optic atrophy is segregating. This condition causes blindness in adulthood. Studies have recently shown that the mutant gene causing Leber hereditary optic atrophy is located in the mitochondrial genome.

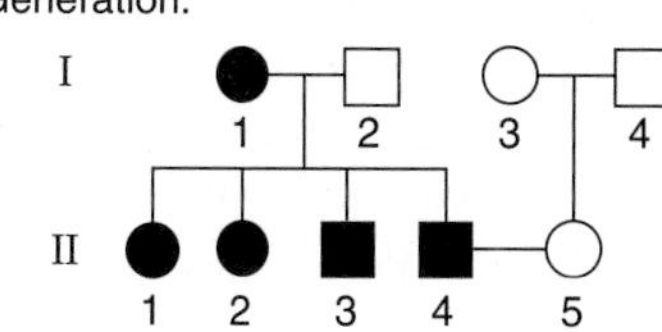


**a.** What other modes of inheritance (e.g., autosomal dominant, X-linked recessive) are consistent with the inheritance of this rare disease? How could you provide evidence that this disease is not inherited using these modes?

**b.** Assuming II-5 is normal, what proportion of the offspring of II-4 and II-5 are expected to inherit Leber's hereditary optic atrophy?

**c.** Assuming that II-2 marries a normal male, what proportion of their sons should be affected? What proportion of their daughters should be affected?

***13.48** Figure 13.C shows a second pedigree segregating Leber hereditary optic atrophy.

**a.** Which individuals must have transmitted the mutation?

**b.** Which individuals in generation IV have the potential to transmit the mutation?

**c.** How can you explain the decreased penetrance of the disease in members of this pedigree?

***13.49** A form of male sterility in corn is maternally inherited. Plants of a male-sterile line crossed with normal pollen give male-sterile plants. Some lines of corn carry a dominant, so-called restorer (*Rf*) nuclear gene that restores pollen fertility in male-sterile lines.

**a.** If a male-sterile plant is crossed with pollen from a plant homozygous for gene *Rf*, what will be the genotype and phenotype of the $F_1$?

**b.** If the $F_1$ plants of (**a**) are used as females in a testcross with pollen from a normal plant (*rf*/*rf*), what would be the result? Give genotypes and phenotypes, and designate the type of cytoplasm.

**Figure 13.C**

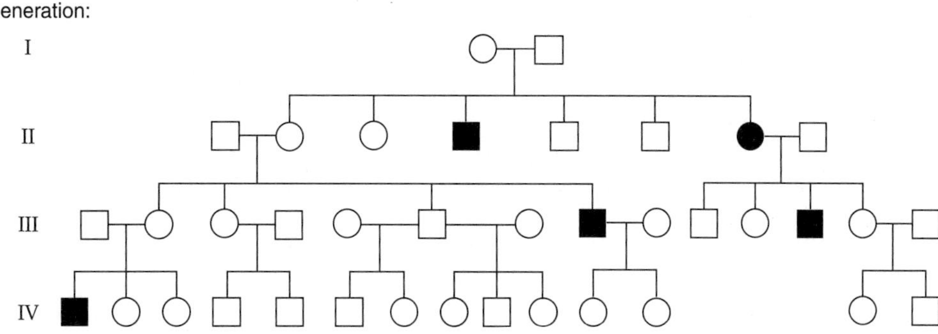

***13.50** A few years ago, Chile allowed its government agents to kidnap, torture, and kill many young adults who opposed the regime in control of the country. The children of abducted women were often taken and given to government supporters to raise as their own. Now that the political situation has changed, grandparents of these stolen children are trying to locate and reclaim them as their legitimate grandchildren. Imagine that you are the judge in a trial centering on the custody of a child. Mr. and Mrs. Escobar believe Carlos Mendoza is the son of their abducted, murdered daughter. If this is true, then Mr. and Mrs. Sanchez are the paternal grandparents of the child because their son (also abducted and murdered) was the husband of the Escobars' daughter. Mr. and Mrs. Mendoza claim that Carlos is their natural child. The attorney for the Escobar and Sanchez couples informs you that scientists have discovered a series of restriction fragment length polymorphims (RFLPs) in human mitochondrial DNA. He tells you that his clients are eager to be tested and asks you to order that Mr. and Mrs. Mendoza and Carlos also be tested.

**a.** Can mitochondrial RFLP data be helpful in this case? In what way?

**b.** Does the mitochondrial DNA of all seven parties need to be tested to resolve the case? If not, whose mitochondrial DNA actually needs to be tested in this case? Explain your choices.

**c.** Assume that the mitochondrial DNA of critical people has been tested, and you have received the results. How would the results resolve the question of Carlos's parentage?

**13.51** The analysis of mitochondrial DNA has been very useful in assessing the history of specific human populations. For example, a 9-bp deletion in a small intergenic region between the genes for cytochrome oxidase subunit II and tRNA.Lys has been an informative marker for tracing the origins of Polynesians. The deletion is widely distributed across Southeast Asia and the Pacific and is present in 80 to 100% of individuals in the different populations within Polynesia. One of the most polymorphic regions of the mitochondrial genome is found in another, approxmately 450 bp-long intergenic region between the genes for tRNA.Phe and tRNA.Thr. Unlike other regions of the mitochondrial genome that are densely packed with genes, this region lacks gene-coding sequences. Some of its sequences are important for DNA replication and transcription. In Asians with the 9-bp deletion, a specific set of DNA sequence polymorphisms in this region is found. Using the 9-bp deletion and the DNA sequence polymorphisms as markers, comparative analysis of Asian populations has found a genetic trail of mitochondrial DNA variation. The trail begins in Taiwan, winds through the Philippines and Indonesia, proceeds along the coast of New Guinea, and then moves into Polynesia. Based on an estimated rate of mutation in the tRNA.Phe to tRNA.Thr region, this expansion of mitochondrial DNA variants is thought to be about 6,000 years old. This is consistent with linguistic and archeological evidence that associates Polynesian origins with the spread of the Austronesian language family out of Taiwan between 6,000 and 8,000 years ago.

**a.** Why are these types of mitochondrial DNA polymorphisms such good markers for tracing human migration patterns?

**b.** Why is it important to correlate findings from mitochondrial DNA polymorphisms with other (non-DNA) assessment methods?

**c.** Why might sequences in the tRNA.Phe to tRNA.Thr region be more polymorphic than other sequences in the mitochondrial genome?

**d.** The 9-bp deletion has also been found in human populations in Africa. What different explanations are possible for this, and how might these explanations be evaluated?

# 14 Genetic Mapping in Eukaryotes

The fruit fly, *Drosophila melanogaster*, with the *vestigial* wing mutation.

## Key Questions

- How is linkage between genes determined?
- How are genetic maps of genes and DNA markers created for experimental organisms?
- How are genetic maps of genes and DNA markers created for the human genome?

### iActivity

THE HUMAN GENOME PROJECT MAY BE THE best-known gene-mapping project in the world. Its goal is to determine the locations of all the genes in the human genome and the exact nucleotide sequence of the 3 billion nucleotide pairs that make up the genome. But long before developing the recombinant technologies that allowed the Human Genome Project to come into being, scientists were creating genetic maps of eukaryotic organisms. The first map of genes in a eukaryotic organism—the fruit fly—was made in the second decade of the twentieth century. Since then, genetic maps have been created for many eukaryotes. What do these maps tell us? How are they constructed? How can they be used? After you have read and studied this chapter, you can explore the answers to these and other questions further in the iActivity, as you map tomato genes.

Genes on nonhomologous chromosomes assort independently during meiosis. In many instances, however, certain genes (and hence the phenotypes they control) are inherited together because they are located on the same chromosome. Genes that are on the same chromosome are said to be *syntenic*. Genes that do not appear to assort independently because they are located on the same chromosome exhibit **linkage** and are called **linked genes.** These genes belong to a *linkage group*.

Genetic analysis is the dissection of the structure and function of the genetic material. In classic genetic analysis, progeny from crosses between parents with different genetic characters are analyzed to determine the frequency with which differing parental alleles are associated in new combinations. Progeny showing the parental combinations of alleles are called *parentals*, and progeny showing nonparental combinations of alleles are called *recombinants*. The process by which the recombinants are produced is called **genetic recombination.** Through testcrosses, we can determine which genes are linked to each other and can then construct a *linkage map*, or *genetic map*, of each chromosome.

Classic genetic mapping has provided information that is useful in many aspects of genetic analysis. For example, knowing the locations of genes on chromosomes has been useful in recombinant DNA research and in experiments directed toward understanding the DNA sequences in and around genes. Genetic maps are constructed using both gene markers and DNA markers. A *marker*, or **genetic marker,** is another name for a mutation or variant that gives a distinguishable phenotype. In other words, it is an allele that marks a chromosome or a gene. **Gene markers** are alleles of genes, whereas **DNA markers** are molecular markers—that is, DNA regions in the genome that are *polymorphic* (differ among individuals) and thus can be detected by the molecular analysis of DNA.

Genes, of course, are DNA sequences that are part of the overall sequence of a chromosome. Having the complete sequence of a chromosome or a genome makes it possible to determine exactly the positions and spacings of genes. Typically, that is how maps of genes are drawn in the current era of genomics; such maps are examples of *physical maps* because they involve measurements made with molecular tools, and they do not depend on the exchange of homologous chromosome parts during meiosis by crossing-over.

The goal for this chapter is to learn about genetic linkage and how genes have been mapped classically, and how genetic mapping of genes and DNA markers is done today. Determining the sequence of a genome and analyzing that sequence for genes are processes described in Chapter 8.

## Early Studies of Genetic Linkage: Morgan's Experiments with *Drosophila*

By 1911, Thomas Hunt Morgan had identified a number of X-linked genes, including *w* (*white* eye) and *m* (*miniature* wing; the wing is smaller than normal). Morgan crossed a female fly ($w\ m/w\ m$) with white eyes and miniature wings with a wild-type male ($w^+\ m^+/Y$; Figure 14.1). For the former genotype, the slash signifies the pair of homologous chromosomes and indicates that the genes on either side of the slash are linked. For the latter genotype, because the genes are X linked, a slash indicates the X chromosome and Y indicates a Y chromosome. We will also use another special genetic symbol for genes on the same chromosome: $\frac{a\,b}{a\,b}$, which signifies that genes *a* and *b* are on the same chromosome, with the chromosome represented by the horizontal line. In this system, X-linked genes in a female are indicated by allele symbols separated by one or two continuous lines to denote the homologous chromosomes, as in

$$\frac{w\ m}{w\ m} \text{ or } \frac{w\ m}{\overline{w\ m}}$$

and X-linked genes in a male are shown as, for example,

$$\frac{w\ m}{\rightharpoondown}$$

where the straight line designates the X chromosome and the bent line the Y chromosome. This genotype represen-

**Figure 14.1**

**Morgan's experimental crosses of white eye and miniature wing variants of *Drosophila melanogaster*, showing evidence of linkage and recombination in the X chromosome.** (Figure from *Genetics*, 2/e by Ursala W. Goodenough, copyright © 1978 Brooks/Cole, a part of Cengage Learning, Inc. Reproduced by permission.)

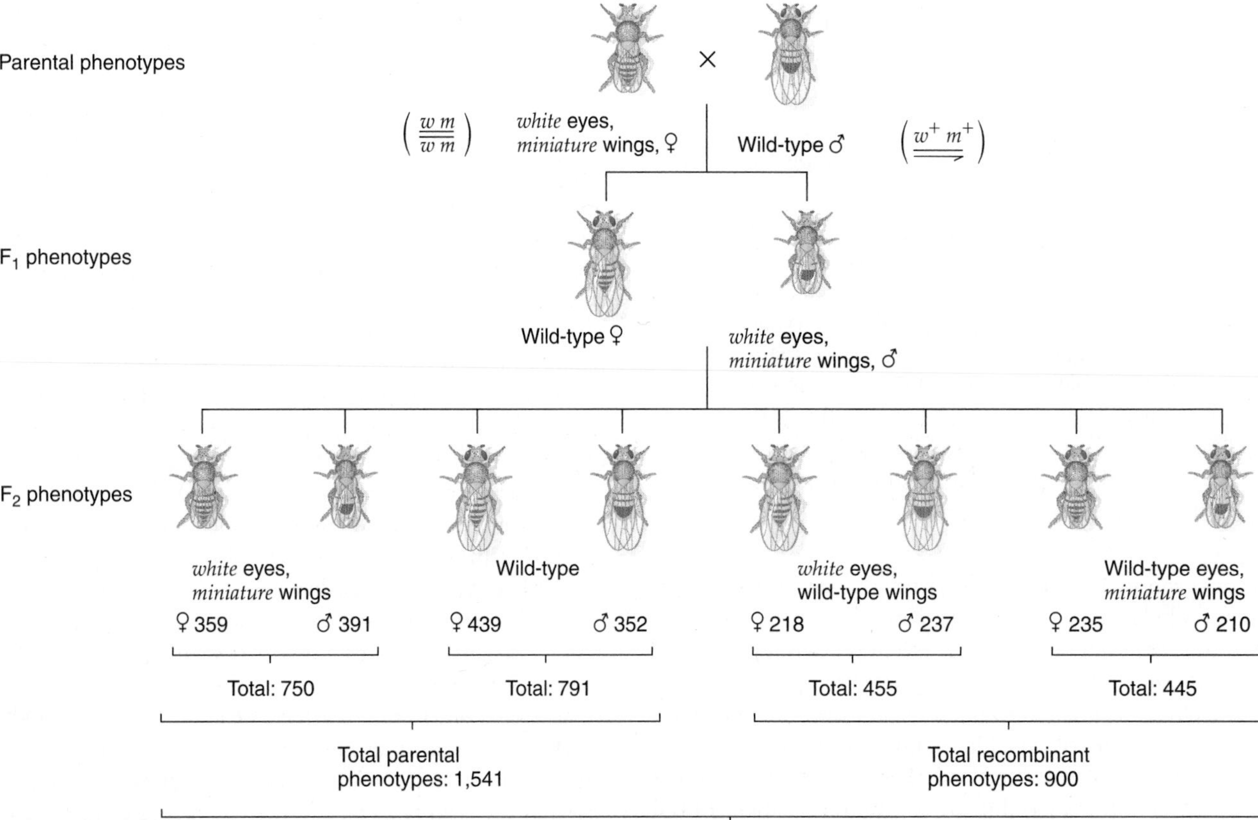

tation is the same as *w m*//^ (where the bent slash is the Y chromosome) or *w m*/Y. (*Note*: If a discontinuous line is used between a series of allele pairs, the extent of each segment signifies a different chromosome.) In the cross, the $F_1$ males had white eyes and miniature wings (genotype *w m*/Y), whereas all $F_1$ females were heterozygous and wild type for both eye color and wing size (genotype $w^+ m^+/w\ m$). The $F_1$ flies were interbred, and 2,441 $F_2$ flies flies were analyzed. In crosses of X-linked genes set up as in Figure 14.1, the $F_1 \times F_1$ is equivalent to doing a testcross, because the $F_1$ males produce X-bearing gametes with recessive alleles of both genes and Y-bearing gametes that have no alleles for the genes being studied. In the $F_2$, the most common phenotypic classes in both sexes were the *grandparental phenotypes* of white eyes plus miniature wings, and wild-type red eyes plus normal wings. Conventionally, we call the original genotypes of the two chromosomes **parental genotypes, parental classes,** or, more simply, **parentals.** The term is also used to describe phenotypes, so the original white, miniature females and wild-type males in these particular crosses are defined as the parentals. Morgan observed that 900 of the 2,441 $F_2$ flies, or 36.9%, had nonparental phenotypic combinations of white eyes plus normal wings, and red eyes plus miniature wings. Nonparental combinations of linked alleles are called **recombinants.** A total of 50% recombinant phenotypes is expected in the case of independent assortment; thus, the lower percentage observed is evidence of linkage of the two genes. To explain the recombinants, Morgan proposed that, in meiosis, exchanges of genes had occurred between the two X chromosomes of the $F_1$ females.

Morgan's group analyzed a large number of other crosses of this type. *In each case, the parental phenotypic classes were the most frequent, and the recombinant classes occurred less frequently*. Approximately equal numbers of each of the two parental classes, and approximately equal numbers of each of the two recombinant classes, were obtained. Morgan's general conclusion was that, *during meiosis, alleles of some genes assort together because they lie near each other on the same chromosome*. To turn this statement around, the closer two genes are on the chromosome, the more likely they are to remain together during meiosis; hence, they will not assort independently. The reason is that the recombinants are produced as a result of crossing-over between homologous chromosomes during meiosis, and the closer two genes are together, the less likely there will be a recombination event between them.

The terminology related to the physical exchange of homologous chromosome parts can be confusing. To clarify,

1. A chiasma (plural, *chiasmata;* see Figure 12.10, p. 333) is the place on a homologous pair of chromosomes at which a physical exchange is occurring; it is the site of crossing-over.
2. Crossing-over is the reciprocal exchange of chromatid segments at corresponding positions along homologous chromosomes; the process involves breakage and rejoining of two chromatids.
3. Crossing-over is also defined as the events leading to genetic recombination between linked genes in both prokaryotes and eukaryotes.

Crossing-over occurs at the four-chromatid stage in prophase I of meiosis. Each crossover involves two of the four chromatids. Along the length of a chromosome, all chromatids can be involved in crossing-over.

**Keynote**

The production of genetic recombinants results from physical exchanges between homologous chromosomes during meiotic prophase I. A chiasma is the site of crossing-over. Crossing-over is the reciprocal exchange of chromosome parts at corresponding positions along homologous chromosomes by the breaking and rejoining of two chromatids. The term *crossing-over* is also used to describe the events leading to genetic recombination between linked genes. Crossing-over in eukaryotes takes place at the four-chromatid stage in prophase I of meiosis.

## Gene Recombination and the Role of Chromosomal Exchange

Two key experiments in the 1930s, one using corn and the other using *Drosophila*, established that the appearance of genetic recombinants is associated with crossing-over and the consequent exchange of homologous chromosome parts. In both experiments, the researchers used genetic markers and *physical markers* (also called *cytological markers*) to analyze genetic recombination in meiosis. **Physical markers** are cytologically detectable visible changes in the chromosomes that make it possible to distinguish the chromosomes and, hence, the results of crossing-over under the microscope.

The corn (*Zea mays*) experiment was done by Harriet B. Creighton and Barbara McClintock. The *Drosophila melanogaster* experiment was done by Curt Stern (Figure 14.2). He studied two X-linked genes, *car* (*carnation*) and *B* (*bar*-eye). Homozygous *car* mutants have carnation-colored eyes instead of the wild-type red. *B* mutants are incompletely dominant; homozygotes or heterozygotes result in a narrow, bar-shaped eye instead of the round eye of the wild type. In this cross, Stern used a *car* $B^+$/Y male, which had carnation-colored and nonbar (round) eyes. The female parent had two physical markers: One X chromosome, genotype *car B*, was shorter than normal X, because part of it had broken off and was attached to the small chromosome 4. The other X chromosome, genotype $car^+$ $B^+$, had a portion of the Y chromosome attached to it. In

**animation**
**Genetic Recombination and the Role of Chromosomal Exchange**

**Figure 14.2**

**Stern's experiment to demonstrate the relationship between genetic recombination and chromosomal exchange in *Drosophila melanogaster*.**

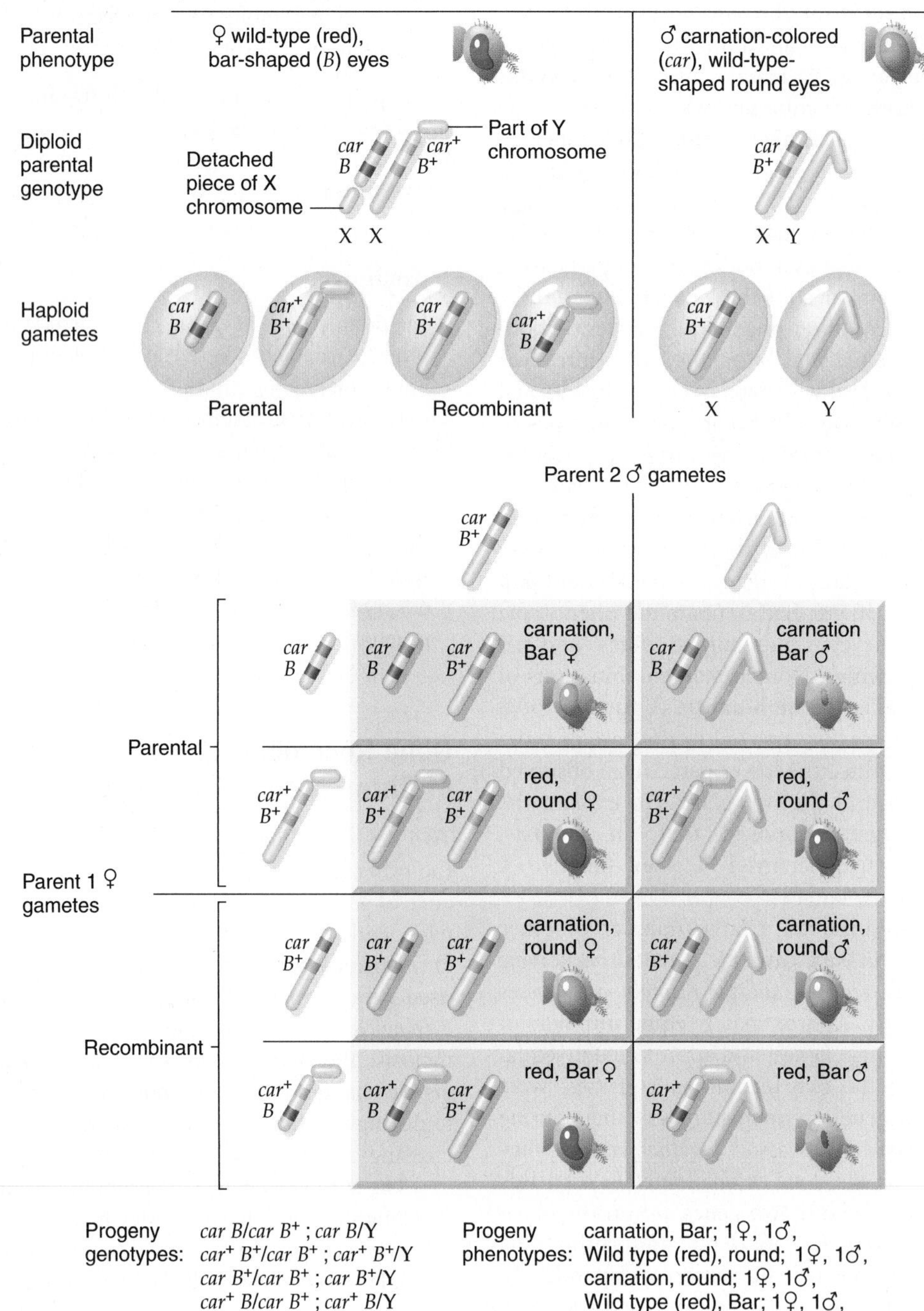


Progeny genotypes: *car B*/*car B*$^+$ ; *car B*/Y
*car*$^+$ *B*$^+$/*car B*$^+$ ; *car*$^+$ *B*$^+$/Y
*car B*$^+$/*car B*$^+$ ; *car B*$^+$/Y
*car*$^+$ *B*/*car B*$^+$ ; *car*$^+$ *B*/Y

Progeny phenotypes: carnation, Bar; 1♀, 1♂,
Wild type (red), round; 1♀, 1♂,
carnation, round; 1♀, 1♂,
Wild type (red), Bar; 1♀, 1♂,

females, the shape of the eye depends on the number of copies of the mutant *B* allele: the eye is a narrow bar in *B*/*B* homozygotes, whereas it is a kidney-shaped bar in *B*/*B*$^+$ heterozygotes. Thus, the *car*$^+$ *B*$^+$/*car B* parental females had red, kidney-shaped bar eyes. (In the figure, all variants of the bar phenotype are referred to as Bar for simplicity.)

In the progeny, every case in which genetic recombination occurred was accompanied by an exchange of identifiable chromosome segments. That is, if no recombination occurred, the two phenotypic classes of progeny were: (1) carnation, bar eyes (kidney-shaped bar in females and narrow bar shape in males), genotypically *car B*/*car B*$^+$ females and *car B*/Y males; and (2) red, round eyes (i.e., wild type for both genes), genotypically *car*$^+$ B$^+$/*car B*$^+$ females and *car*$^+$ *B*$^+$/Y males. No exchanges of chromosome parts were seen among these nonrecombinants. The two classes of recombinants were (1) carnation, round eyes, genotypically *car* $B^+$/*car* $B^+$ females and *car* $B^+$/Y males; and (2) red, bar eyes, genotypically *car*$^+$ *B*/*car* $B^+$ females and *car*$^+$ *B*/Y males. The flies with carnation eyes had a complete X chromosome, and the flies with bar eyes had a shorter-than-normal X chromosome to which a piece of the Y chromosome was attached, whereas the rest of the X chromosome was attached to chromosome 4. This chromosomal makeup could have resulted only from physical exchanges of homologous chromosome parts.

There is no doubt, therefore, that genetic recombination results from crossing-over causing physical exchanges between chromosomes.

**Keynote**

The proof that genetic recombination occurs when crossing-over takes place during meiosis came from breeding experiments in which the parental chromosomes differed with respect to both genetic and cytological markers. These experiments showed that whenever recombinant phenotypes occurred, the cytological markers indicated that crossing-over had also occurred.

## Constructing Genetic Maps

We have learned that the number of genetic recombinants produced is characteristic of the two linked genes involved. We now examine how genetic experiments can be used in genetic mapping—the process of constructing a **genetic map** (also called a **linkage map**) of the relative positions of genes on a chromosome.

### Detecting Linkage through Testcrosses

Unlinked genes assort independently. Therefore, a way to test for linkage is to analyze the results of crosses to see whether the genetic data deviate significantly from those expected by independent assortment.

The best cross to use to test for linkage is the testcross, a cross of an individual with another individual homozygous recessive for all genes involved. A testcross between $a^+/a\ b^+/b$ and $a/a\ b/b$, where genes $a$ and $b$ are unlinked, gives a progeny phenotypic ratio of 1 $a^+\ b^+$ : 1 $a^+\ b$ : 1 $a\ b^+$ : 1 $a\ b$ (see Chapter 11, p. 310). A significant deviation from this ratio in the direction of too many parental types and too few recombinant types therefore indicates that the two genes do not assort independently. The simplest alternative hypothesis is that the two genes are linked. How large a deviation is considered significant? The *chi-square test* can be used to find the significance (see Chapter 11, p. 312).

The Chi-Square Test

Consider data from a testcross involving fruit flies. In *Drosophila*, *b* is a recessive autosomal mutation that results in black body color, and *vg* is a recessive autosomal mutation that results in vestigial (short, crumpled) wings. Wild-type flies have grey bodies and long, uncrumpled (normal) wings. Crossing true-breeding black, normal-winged ($b/b\ vg^+/vg^+$) flies with true-breeding grey, vestigial ($b^+/b^+\ vg/vg$) flies produces $F_1$ flies that are phenotypically grey with normal wings ($b^+/b\ vg^+/vg$). Testcrossing $F_1$ female flies with black, vestigial ($b/b\ vg/vg$) male flies produced the following progeny. (Note: The female is the heterozygote in this testcross because, in *Drosophila*, no crossing-over occurs between any homologous pair of chromosomes in males.)

| | |
|---|---|
| | 283 grey, normal |
| | 1,294 grey, vestigial |
| | 1,418 black, normal |
| | 241 black, vestigial |
| Total | 3,236 files |

We hypothesize that the two genes are unlinked (the null hypothesis) and use the chi-square test to test the hypothesis, as shown in Table 14.1. We use this particular null hypothesis because the hypothesis must be testable; that is, we must be able to make meaningful predictions. A hypothesis that "two genes are linked" is not testable, because we cannot predict what the progeny ratios would be.

**Table 14.1 Chi-Square Test Used with Testcross Data to Test the Hypothesis That Two Genes Are Unlinked**

| (1) Phenotypes | (2) Observed Number (*o*) | (3) Expected Number (*e*) | (4) *d* (= *o* − *e*) | (5) $d^2$ | (6) $d^2/e$ |
|---|---|---|---|---|---|
| Parentals: (black, normal and grey, vestigial) | 2,712 | 1,618 | 1,094 | 1,196,836 | 739.7 |
| Recombinants (black, vestigial and grey, normal) | 524 | 1,618 | −1,094 | 1,196,836 | 739.7 |
| Total | 3,236 | 3,236 | | | 1,479.4 |
| | | (7) $\chi^2$ = 1,479.4 | (8) df 1 | | |

If the two genes are unlinked, then a testcross should result in a 1:1 ratio of parentals : recombinants. Column 1 lists the parental and recombinant phenotypes expected in the progeny of the cross, column 2 lists the observed ($o$) numbers, and column 3 lists the expected ($e$) numbers for the parentals and recombinants, given the total number of progeny (3,236) and the hypothesis being tested (1:1 in this case). Column 4 lists the deviation ($d$), calculated by subtracting the expected number ($e$) from the observed number ($o$) for each class. The sum of the $d$ values is always zero.

Column 5 lists the deviation squared ($d^2$), and column 6 lists the deviation squared divided by the expected number ($d^2/e$). The chi-square value, $\chi^2$ (item 7 in the table), is given by the formula

$$\chi^2 = \Sigma\frac{d^2}{e}, \quad \text{where} \quad d^2 = (o - e)^2$$

and where $\Sigma$ means "sum of."

In the table, chi-square is the sum of the two values in column 6. In our example, $\chi^2 = 1{,}479.4$. The last value in the table, item 8, is the degrees of freedom (df) for the set of data; there is $n - 1 = 1$ degree of freedom in this case.

The chi-square value and the degrees of freedom are used with a table of chi-square probabilities (see Table 11.5, p. 313) to determine the probability ($P$) that the deviation of the observed values from the expected values is due to chance. For $\chi^2 = 1{,}479.4$ with 1 degree of freedom, the $P$ value is much lower than 0.001; in fact, it is not in the table. This means that independent repetitions of the experiment would produce chance deviations from what was expected as large as those observed in many fewer than 1 out of 1,000 trials. As a reminder, if the probability of obtaining the observed chi-square values is greater than 5 in 100 ($P > 0.05$), the deviation is considered not statistically significant and could have occurred by chance alone. If $P \leq 0.05$, the deviation from the expected values is statistically significant and not due to chance alone; then the hypothesis may well be invalid. If $P \leq 0.01$, the deviation is highly statistically significant, and the data are not consistent with the null hypothesis. In that case, we would reject the independent assortment hypothesis, and, genetically, the only alternative hypothesis that could logically apply is that the genes are linked.

**The Concept of a Genetic Map.** In an individual that is doubly heterozygous for the $w$ and $m$ alleles, for example, the alleles can be arranged in two ways:

$$\frac{w^+\,m^+}{w\;\;m} \quad \text{or} \quad \frac{w^+\,m}{w\;\;m^+}$$

In the arrangement on the left, the two wild-type alleles are on one homolog and the two recessive mutant alleles are on the other homolog, an arrangement called **coupling** (or the *cis* configuration). Crossing-over between the two loci produces $w^+\,m$ and $w\,m^+$ recombinants. In the arrangement on the right, each homolog carries the wild-type allele of one gene and the mutant allele of the other gene, an arrangement called **repulsion** (or the *trans* configuration). Crossing-over between the two genes produces $w^+\,m^+$ and $w\,m$ recombinants.

The data obtained by Morgan from *Drosophila* crosses indicated that the frequency of crossing-over (and hence of recombinants) for linked genes is characteristic of the gene pairs involved: For the X-linked genes *white* ($w$) and *miniature* ($m$), he calculated a recombination frequency of 36.9%. The recombination frequency for two linked genes is the same, regardless of whether the alleles of the two genes involved are in coupling or in repulsion. *Although the actual phenotypes of the recombinant classes are different for the two arrangements, the percentage of recombinants among the total progeny will be the same in each case (within experimental error).*

In 1913, a student of Morgan's, Alfred Sturtevant, determined that recombination frequencies could be used as a quantitative measure of the genetic distance between two genes on a genetic map. The genetic distance between genes is measured in **map units (mu),** where 1 map unit is defined as the interval in which 1 percent crossing-over takes place. The map unit is also called a **centimorgan (cM),** a term named by Sturtevant in honor of Morgan.

It is important to understand that, for a pair of linked genes, the crossover frequency is *not* the same as the recombination frequency. The former refers to the frequency of physical exchanges between chromosomes in meiosis for the region between the genes, and the latter refers to the frequency of recombination of genetic markers in a cross, as determined by analyzing the phenotypes of the progeny. Geneticists follow genetic markers in crosses, so the data obtained are in the form of recombination frequencies. In our discussions, we will use recombination frequencies as geneticists often do: as working estimates of map distances between genes, where a map unit is equivalent to a recombination frequency of 1%. Later we will discuss how such data relate to crossover frequencies and therefore to true map units.

The genes on a chromosome, then, can be represented by a one-dimensional genetic map that shows, in linear order, the genes belonging to the chromosome. Crossover and recombination frequencies give the linear order of the genes on a chromosome and provide information about the genetic distance between any two genes. The farther apart two genes are, the greater is the *crossover frequency*.

The first genetic map ever constructed was based on *recombination frequencies* from *Drosophila* crosses involving the X-linked genes $w$ (*white* eyes), $m$ (*miniature* wings), and $y$ (*yellow* body). From these mapping experiments, the recombination frequencies for the $w \times m$, $w \times y$, and $m \times y$ crosses were established as 32.7, 1.0, and 33.7%, respectively.

(In this independent experiment, the recombination frequency for *w* and *m* is a little lower than in the experiment discussed previously on pp. 402–403.) The percentages are quantitative measures of the distances between the genes involved.

We can construct a genetic map based on the recombination frequency data. The recombination frequencies show that genes *w* and *y* are closely linked and that gene *m* is quite far from the other two genes. Since the *w–m* genetic distance is less than the *y–m* distance (as shown by the smaller recombination frequency in the $w \times m$ cross), the order of genes must be *y w m* (or *m w y*); thus, the three genes are ordered and spaced with 1.0 mu between *y* and *w* and 32.7 mu between *w* and *m*. We can draw a map for these three genes as follows:

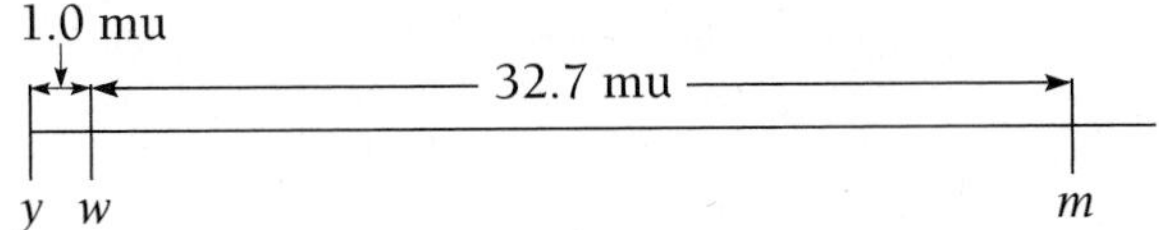


## Gene Mapping with Two-Point Testcrosses

We have seen that the recombination frequency may be used to obtain an estimate of the genetic distance between two linked genes. By carrying out two-point testcrosses such as those shown in Figure 14.3, we can determine the relative numbers of parental and recombinant classes in the progeny. For linked autosomal genes with recessive mutant alleles (as in the figure), a double heterozygote is crossed with a doubly homozygous recessive mutant strain. When the double heterozygous $a^+ b^+/a\ b$ $F_1$ progeny from a cross of $a^+ b^+/a^+ b^+$ with $a\ b/a\ b$ are testcrossed with $a\ b/a\ b$, four phenotypic classes are found among the $F_2$ progeny. Two of these classes have the parental phenotypes $a^+ b^+$ and $a\ b$, and the other two have the recombinant phenotypes $a^+ b$ and $a\ b^+$.

Testcrosses are used for mapping because the homozygous recessive parent produces only one type of gamete, with alleles that are recessive to the alleles in gametes produced by the heterozygous parent. Thus, the phenotypes of the progeny directly result from the genotypes of the parental and recombinant gametes generated by meiosis in that heterozygous parent. Using two doubly heterozygous parents to map genes is not done, because both parents produce parental and recombinant gametes, making analysis of the progeny complex.

The principle, then, in constructing a testcross is to use one parent that is heterozygous for the genes being mapped and another parent that has the recessive alleles for those genes. Thus, two-point testcrosses for mapping X-linked genes would be a doubly heterozygous female crossed with a hemizygous male carrying the recessive alleles:

$$\frac{a^+\ b^+}{a\ \ b} \times \frac{a\ \ b}{\longrightarrow}$$

In all cases, a two-point testcross should yield a pair of parental types that occur with about equal frequency

**Figure 14.3**

**Testcross to show that two genes are linked.** Genes *a* and *b* are linked on the same autosome. A homozygous $a^+ b^+/a^+ b^+$ individual is crossed with a homozygous recessive $a\ b/a\ b$ individual, and the doubly heterozygous $F_1$ progeny ($a^+ b^+/a\ b$) are testcrossed with homozygous $a\ b/a\ b$ individuals.

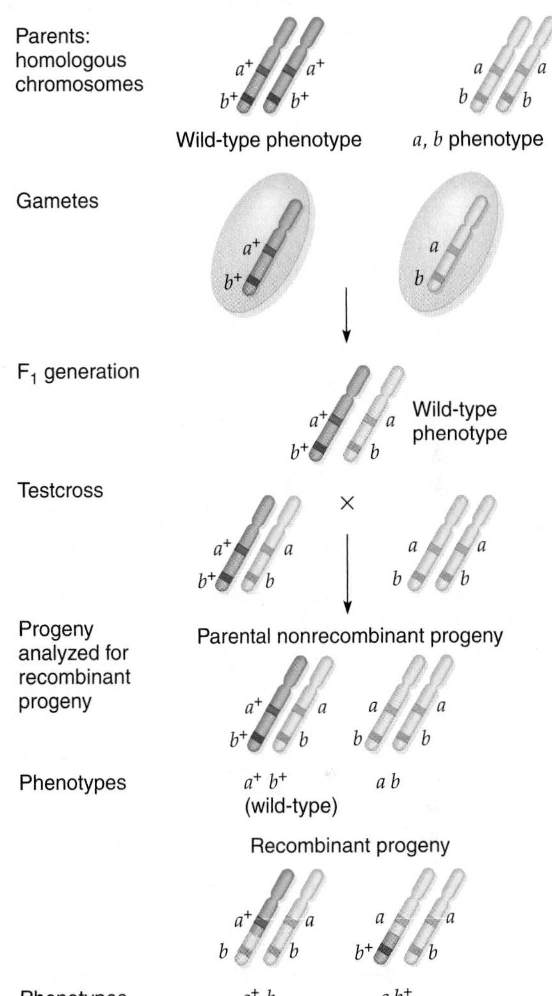


and a pair of recombinant types that also occur with about equal frequency. The actual phenotypes depend whether the two allelic pairs in the homologous chromosomes are in coupling (*cis*) or in repulsion (*trans*). The following formula is used to calculate the recombination frequency:

$$\frac{\text{number of recombinants}}{\text{total number of testcross progeny}} \times 100 = \begin{array}{l}\text{recombination frequency}\\ = \text{map units}\end{array}$$

The recombination frequency is used directly as an estimate of map units.

The same principles of mapping are used whether gene markers and/or DNA markers are used in two-point mapping. Box 14.1 outlines how recombination frequency is determined between a gene locus and a DNA marker locus.

The two-point method of mapping is most accurate when the two genes examined are close together; when genes are far apart, there are inaccuracies, as we will see later. Large numbers of progeny must also be counted (scored) to ensure a high degree of accuracy. From mapping experiments carried out in all types of organisms, we know that genes are linearly arranged in linkage groups. There is a one-to-one correspondence between linkage groups and chromosomes, so the sequence of genes on the linkage group reflects the sequence of genes on the chromosome.

### Generating a Genetic Map

We can now discover how a genetic map is generated from an estimate of the number of times crossing-over occurred in a particular segment of the chromosome out of all meioses examined. In many cases, the probability of a crossing-over event is not uniform along a chromosome,

## Box 14.1 Determining Recombination Frequency for Linked Gene and DNA Marker Loci

The principles for determining the recombination frequency for linked gene and DNA marker loci are the same as those used for the two-gene loci example in the text. That is, geneticists set up a testcross and determine the percentage of recombinants in the progeny. We will see in this box how the DNA marker alleles are handled in such a genetic analysis.

Consider a theoretical diploid organism (a MendAlien). Normal eye color is black, while an autosomal recessive allele, *o*, when homozygous, results in orange eyes (Box Figure 14.1a). Known to be linked to the orange eye gene locus is a DNA marker locus that is polymorphic for *short tandem repeats*, or STRs (Box Figure 14.1b, and see the discussion in Chapter 10, pp. 271–272).

*Polymorphic* means there is a variation in the DNA at the locus. In this case, the variation concerns short DNA sequences 2–6 bp long that are tandemly repeated from a few times up to about 100 times in that location in the genome. That is, different alleles of the locus have different numbers of the STRs, resulting in length differences for the alleles. In a population, there may be many different alleles of an STR DNA marker locus like this; but any one individual can be only homozygous for one allele, or heterozygous for two alleles. The length of an STR allele is the phenotype of the allele, in the same way as eye color is the phenotype of the gene marker. The length of an STR is measured by using the polymerase chain reaction (PCR; discussed in Chapter 9, pp. 221–227). The PCR amplifies a specific section of DNA in the genome, using DNA primers that define the ends of that section. The amplified DNA is analyzed for size using agarose gel electrophoresis, in which shorter DNA fragments migrate toward the positive pole faster than do longer DNA fragments. With DNA fragments of known length separated in the gel at the same time as experimental DNA fragments, the lengths of the experimental DNA fragments can be determined. (Agarose gel electrophoresis analysis of DNA is discussed in Chapter 8, pp. 181–182.) In our example, two STR alleles are involved, one with 6 copies of the repeat (the 6 allele), and the other with 10 alleles (the 10 allele). DNA markers behave like codominant alleles in the sense that the phenotype of both alleles is seen in heterozygotes. Practically speaking, homozygotes and heterozygotes are detected directly by the DNA analysis. In our example, the three possible genotypes and phenotypes are (6,6), (6,10), and (10,10), where the two numbers separated by the comma

**Box Figure 14.1**

**Phenotypes of alleles of linked gene and DNA marker loci.** **(a)** Gene locus: Normal black eyes, and recessive orange eyes, scored visually. **(b)** DNA marker locus with STR alleles: (10) and (6) repeat alleles scored by PCR amplification of the locus and analysis of DNA fragments by agarose gel electrophoresis.

**a) Gene locus**

MendAliens (2N)

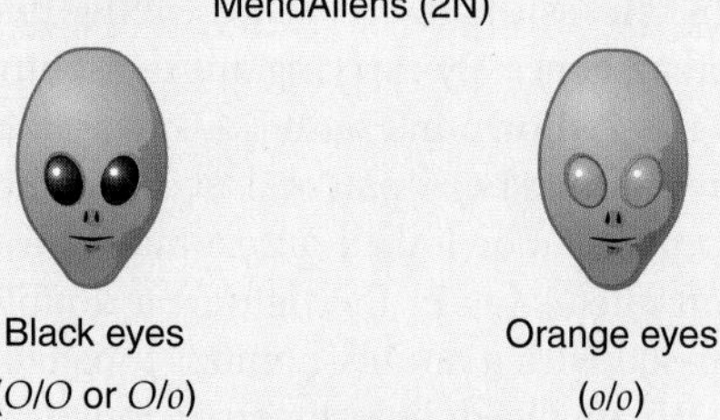

Black eyes
(*O/O* or *O/o*)

Orange eyes
(*o/o*)

**b) DNA marker locus with STR alleles**

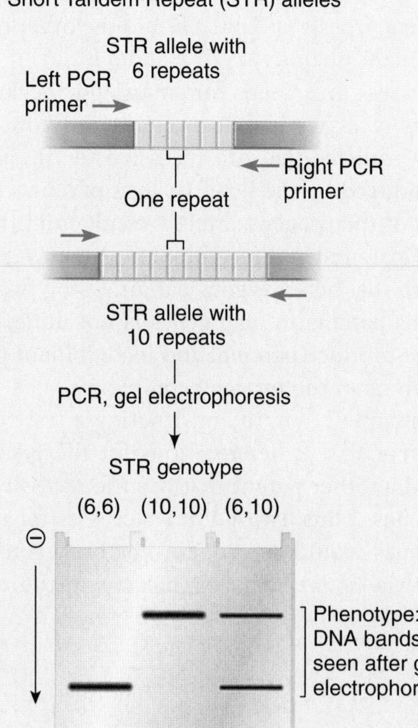

**Box 14.1 (*continued*)**

represent the number of repeats in the STR on the two homologous chromosomes.

Box Figure 14.2 shows how to calculate the map distance between the eye-color and STR loci. First, a true-breeding black-eyed, parent with 10 repeats, genotype *O* (10)/*O* (10), is crossed with a true-breeding orange-eyed, parent with 6 repeats, genotype *o* (6)/*o* (6). The $F_1$ progeny are black-eyed, and exhibit both 10 and 6 repeats, genotype *O* (10)/*o* (6). The $F_1$ double heterozygote is then testcrossed. Because the STR alleles are codominant, either of two homozygous "recessive" parents can be used here: *o* (10)/*o* (10), or *o* (6)/*o* (6), that is orange-eyed and homozygous for either the (10) or the (6) repeat allele. Let us use the *o* (6)/*o* (6) as the parent in the testcross. The parental progeny of this testcross are black-eyed, with 10 and 6 repeats (genotype *O* (10)/*o* (6)), and orange-eyed, with 6 repeats (genotype *o* (6)/*o* (6)). The recombinant progeny are black-eyed, with 6 repeats (genotype *O* (6)/*o* (6)), and orange-eyed, with 10 and 6 repeats (genotype, *o* (10)/*o* (6)). The eye phenotype is scored visually, while the STR repeat alleles are scored by PCR. Recombination frequency is then calculated using the standard formula.

**Box Figure 14.2**

**Crosses used to calculate map distance between the eye color gene locus and the STR DNA marker locus.**

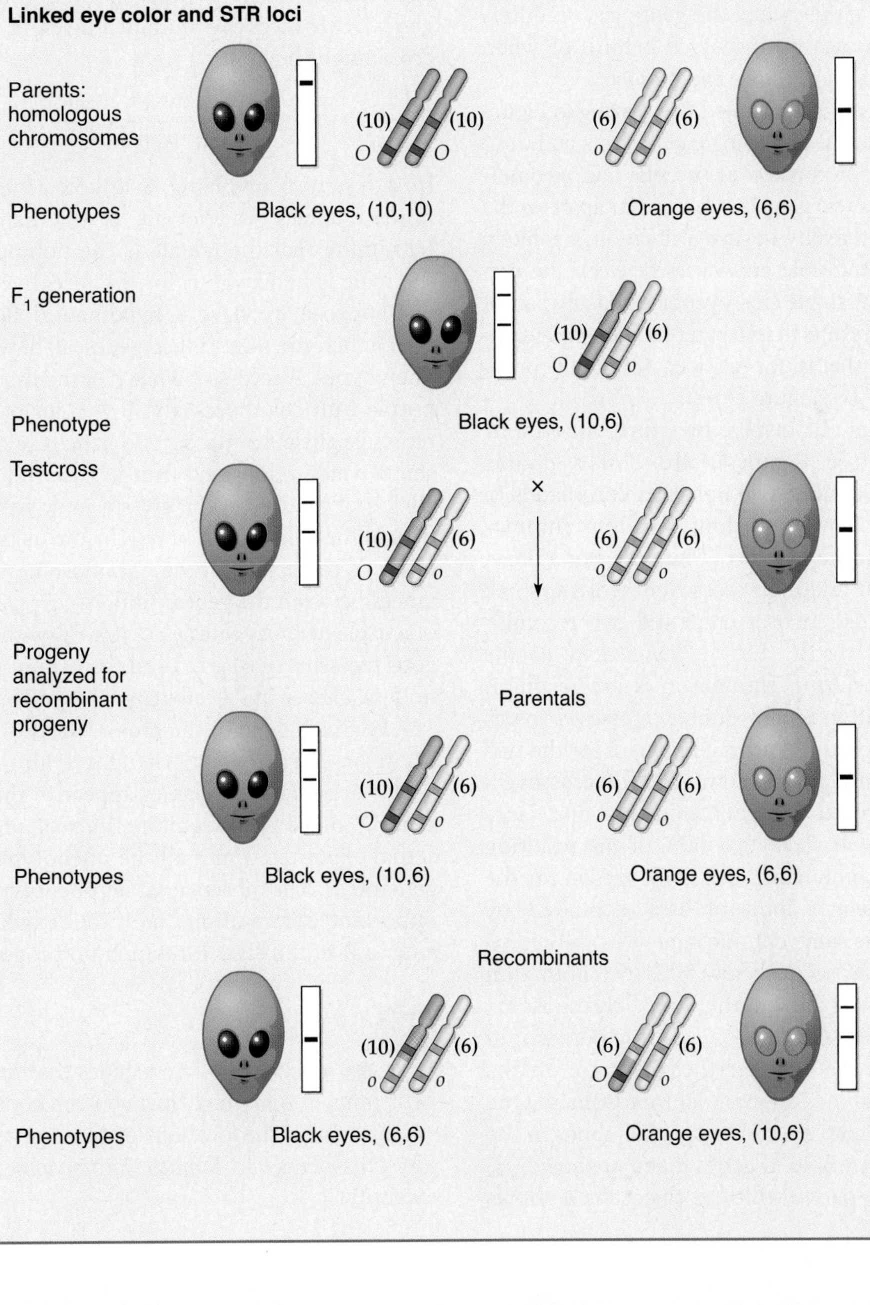

so we must be cautious about how far we extrapolate the genetic map (derived from data produced by genetic crosses) to the physical map of the chromosome (derived from determinations of the exact locations of genes along the chromosome itself from sequencing the DNA).

The recombination frequencies observed between genes may also be used to predict the outcome of genetic crosses. For example, a recombination frequency of 20% between genes indicates that, for a doubly heterozygous genotype (such as $a^+ \; b^+/a \; b$), 20% of the gametes produced, on average, will be recombinants ($a^+ \; b$ and $a \; b^+$ in the example, with 10% of each expected).

For any testcross, the recombination frequency in the progeny cannot exceed 50%. That is, if the genes assort independently, an equal number of recombinants and parentals are *expected* in the progeny, so the recombination frequency is 50%. If we get a recombination frequency of 50% from a cross, then we state that the two genes are unlinked. Genes may be unlinked (that is, show 50% recombination) either when the genes are on different chromosomes (a case we discussed before) or when *the genes are far apart on the same chromosome.*

The second case can be illustrated by referring to Figure 14.4, which shows the effects of single crossovers and double crossovers on the production of parental and recombinant chromosomes for two gene loci that are far apart on the same chromosome. (In reality, in such a situation, it is likely that there would be multiple crossovers between the two loci in each meiosis.) A single crossover between any pair of nonsister chromatids results in two parental and two recombinant chromosomes; that is, for two loci, 50% of the products are recombinant (see Figure 14.4a.)

Double crossovers can involve two, three, or all four of the chromatids (see Figure 14.4b.) For a double crossover involving the same two nonsister chromatids (a *two-strand double crossover*), all four resulting chromosomes are parental for the two loci of interest. For a *three-strand double crossover* (a double crossover involving three of the four chromatids), two parental and two recombinant chromosomes result. For a *four-strand double crossover*, all four resulting chromosomes are recombinant. Considering all possible double crossover types together, 50% of the products are recombinant for the two loci. Similarly, for any multiple number of crossovers between loci that are far apart, examination of a large number of meioses will show that 50% of the resulting chromosomes are recombinant. This is the reason for the recombination frequency limit of 50% exhibited by unlinked genes on the same chromosome.

The point is, if two genes show 50% recombination in a cross, they could be on the same chromosome rather than on different chromosomes. More data would be needed to determine whether the genes are on the same chromosome or on different chromosomes. One way to find out is to map a number of other genes in the linkage group. For example, if genes $a$ and $m$ show 50% recombination, perhaps we will find that gene $a$ shows 27% recombination with gene $e$ and that gene $e$ shows 36% recombination with gene $m$. This result would indicate that genes $a$ and $m$ are in the same linkage group approximately 63 mu apart, as shown here:

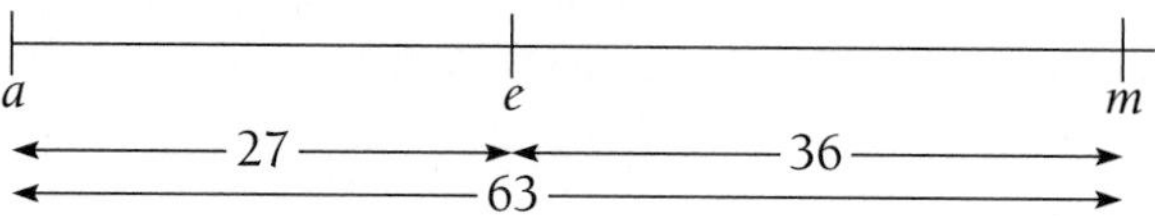


## Gene Mapping with Three-Point Testcrosses

Although genetic maps can be built by using a series of two-point testcrosses, geneticists typically have mapped several linked genes at a time in single testcrosses. Illustrated here is the more complex type of mapping analysis for three linked genes using a **three-point testcross.** In diploid organisms, the three-point testcross is a cross of a triple heterozygote with a triply homozygous recessive. For genes with recessive mutant alleles, a three-point testcross might be

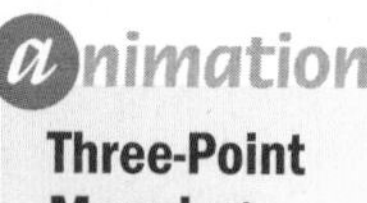


$$\frac{a^+ \; b^+ \; c^+}{a \;\; b \;\; c} \times \frac{a \, b \, c}{a \, b \, c}$$

In a testcross involving X-linked genes with recessive mutant alleles, the female is the heterozygous strain (assuming that the female is the homogametic sex) and the male is hemizygous for the recessive alleles.

Suppose we have a hypothetical flowering plant in which there are three linked genes, all of which control fruit phenotypes. A recessive allele $p$ of the first gene determines purple fruit color, versus yellow color of the wild type. A recessive allele $r$ of the second gene results in a round fruit shape, versus elongated fruit in the wild type. A recessive allele $j$ of the third gene gives a juicy fruit, versus the dry fruit of the wild type. The task before us is to determine the order of the genes on the chromosome and the map distances between the genes. To do so, we perform a testcross of a triple heterozygote ($p^+ \; r^+ \; j^+/p \; r \; j$) with a triply homozygous recessive ($p \; r \; j/p \; r \; j$) and then count the different phenotypic classes in the progeny (Figure 14.5).

For each gene in the cross, two different phenotypes occur in the progeny; therefore, for the three genes, $(2)^3 = 8$ phenotypic classes appear in the progeny, representing all possible combinations of phenotypes. In an actual experiment, not all the phenotypic classes may be generated. The absence of a phenotypic class is also important information, and the experimenter should enter a 0 in the class for which no progeny are found.

**iActivity**

You've discovered new genes that determine different traits in tomatoes. Now you can construct a genetic map showing the locations of those genes in the iActivity *Crossovers and Tomato Chromosomes* on the student website.

**Figure 14.4**

**Demonstration that the recombination frequency between two genes located far apart on the same chromosome cannot exceed 50%.** (a) Single crossovers produce one-half parental and one-half recombinant chromatids. (b) Two-strand, three-strand, and four-strand double crossovers collectively produce one-half parental and one-half recombinant chromatids.

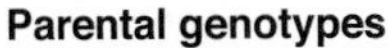


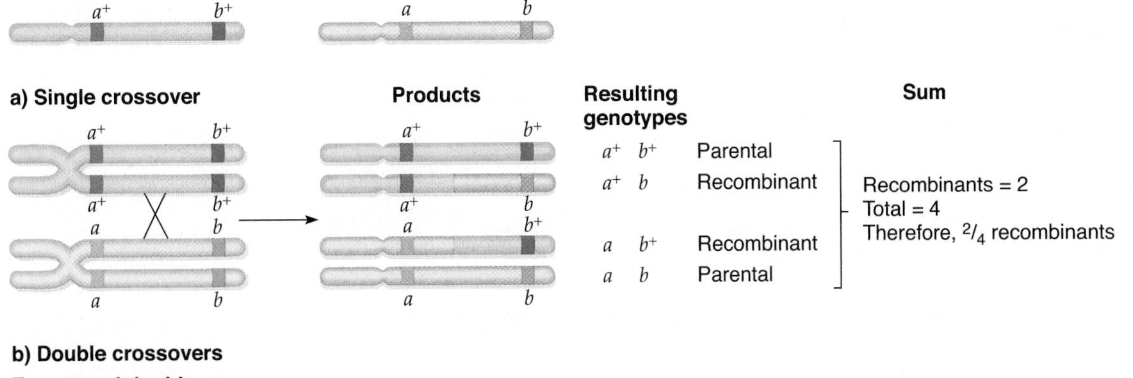


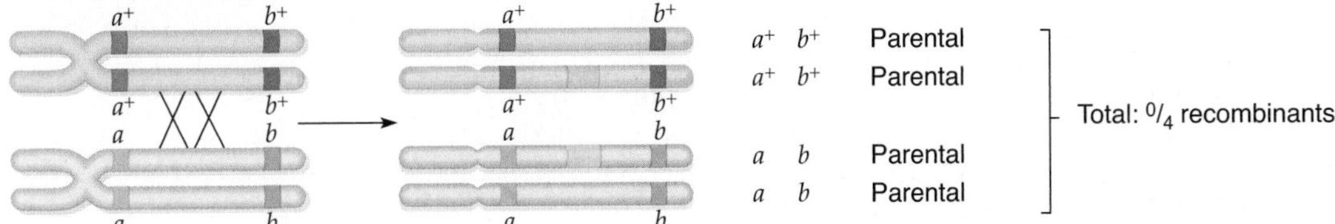

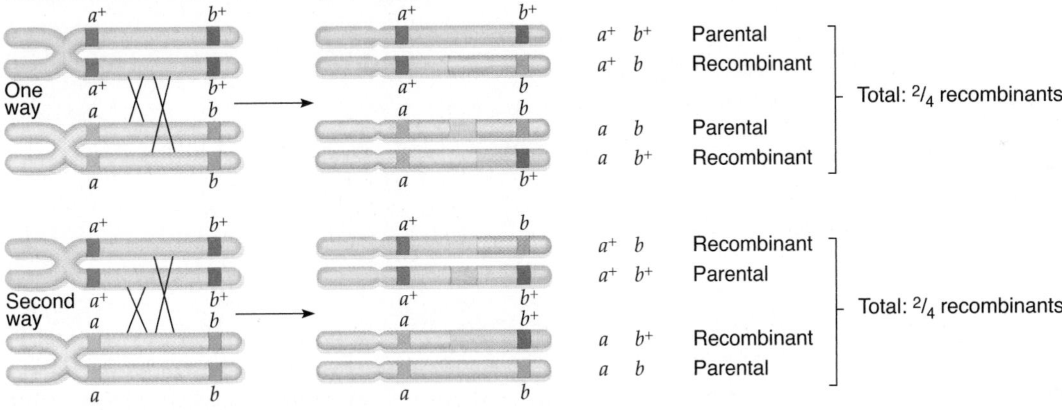


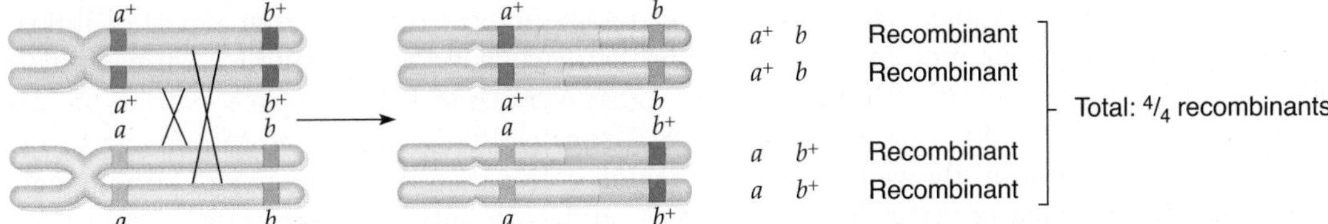


Sum: Recombinants = 0 + 2 + 2 + 4 = 8
Total = 4 + 4 + 4 + 4 = 16
Therefore, recombinants = 50%

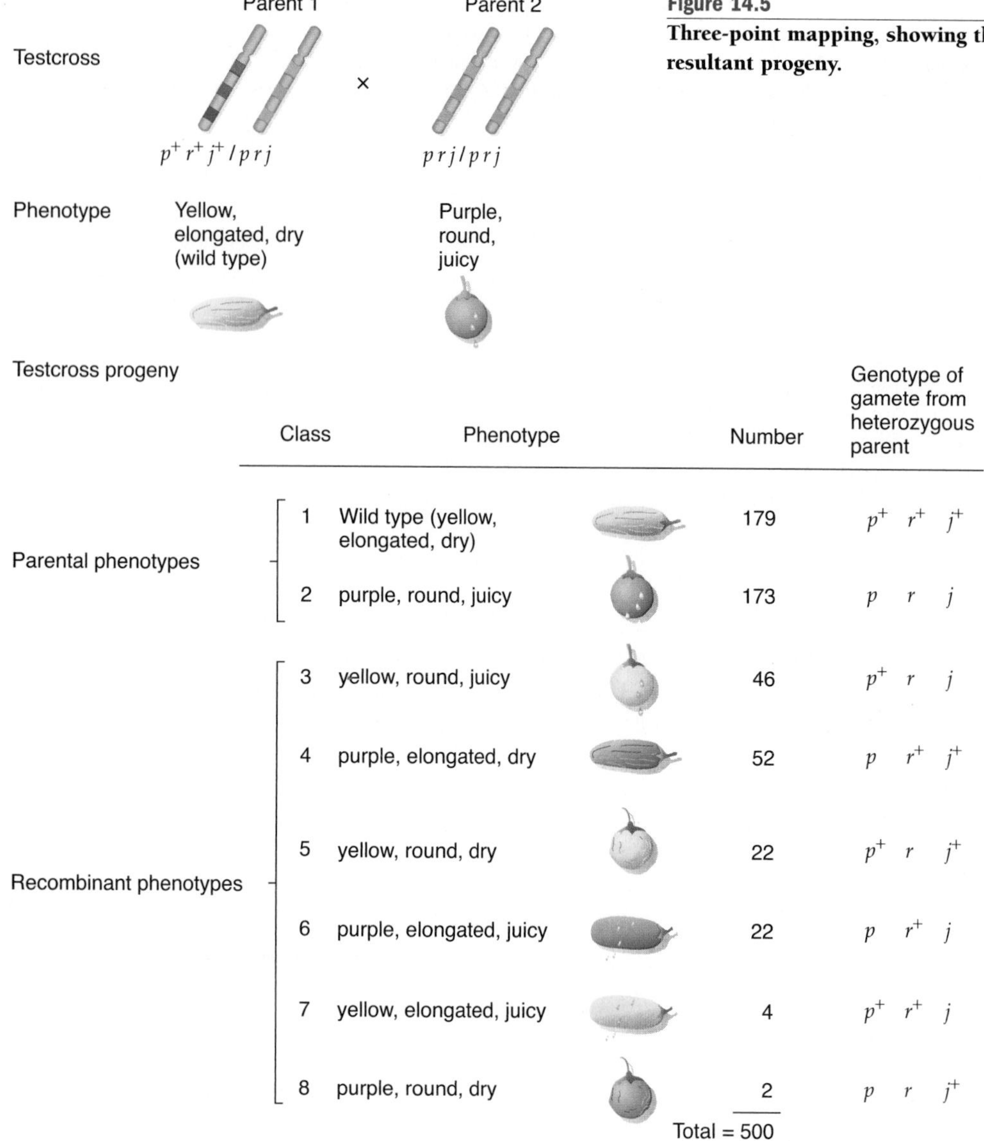


**Figure 14.5**
**Three-point mapping, showing the testcross used and the resultant progeny.**

**Establishing the Order of Genes.** The first step in mapping the three genes is to determine the order of the genes on the chromosome. One parent carries the recessive alleles for all three genes; the other is heterozygous for all three. Therefore, the phenotype of each of the progeny is determined by the alleles in the gamete from the triply heterozygous parent; the gamete from the other parent carries only recessive alleles. We know from the genotypes of the original parents that all three genes are in coupling. Since the heterozygous parent in the testcross was $p^+ r^+ j^+/p\,r\,j$, classes 1 and 2 in Figure 14.5 are parental progeny: class 1 is produced by the fusion of a $p^+ r^+ j^+$ gamete with a $p\,r\,j$ gamete from the triply homozygous recessive parent, and class 2 is produced by the fusion of a $p\,r\,j$ gamete from the heterozygous parent and a $p\,r\,j$ gamete. These classes are generated from meioses in which no crossing-over occurs in the region of the chromosome where the three genes are located.

The other six progeny classes result from crossovers within the region spanned by the three genes that gave rise to recombinant gametes. There may have been a single crossover between a pair of linked genes, or there may have been a double crossover—that is, two crossovers, one between *each* pair of linked genes. Statistically, the frequency of double crossovers in the region is less than the frequency of either single crossover, so *double-crossover gametes are the least frequent pair found.* Therefore, to identify the double-crossover progeny, we examine the progeny to find the pair of classes that has the lowest number of representatives. In Figure 14.5, classes 7 and 8 are such a pair. The genotypes of the gametes from the heterozygous parent that give rise to these phenotypes are $p^+ r^+ j$ and $p\,r\,j^+$.

Figure 14.6 illustrates the consequences of a double crossover in a triple heterozygote for three linked

**Figure 14.6**

**Consequences of a double crossover in a triple heterozygote for three linked genes.** In a double crossover, the middle allelic pair changes its orientation relative to the outside allelic pairs.

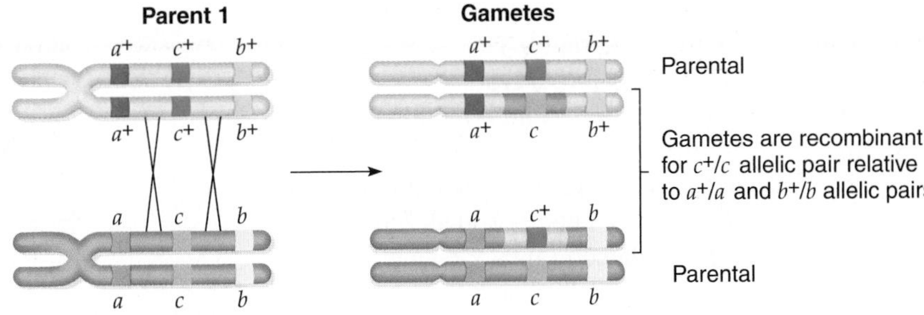


genes $a$, $b$, and $c$, where the alleles are in coupling and the $c$ gene is in the middle. A double crossover changes the orientation of the allelic pair in the middle of the three genes (here, $c^+/c$) with respect to the two flanking allelic pairs. That is, after the double crossover, the $c$ allele is now on the chromatid with the $a^+$ and $b^+$ alleles, and the $c^+$ allele is on the chromosome with the $a$ and $b$ alleles. Therefore, genes $p$, $r$, and $j$ must be arranged in such a way that the center gene switches from the parental arrangement to give classes 7 and 8. To determine the arrangement, we first check the relative organization of the genes in the parental heterozygote to be sure which alleles are in coupling and which are in repulsion. In this example, the parental (noncrossover) gametes are $p^+\ r^+\ j^+$ and $p\ r\ j$, so all are in coupling. The double-crossover gametes are $p^+\ r^+\ j$ and $p\ r\ j^+$, so the only possible gene order compatible with the data is $p\ j\ r$, with the genotype of the heterozygous parent being $p^+\ j^+\ r^+/p\ j\ r$.

**Calculating the Recombination Frequencies for Genes.** Figure 14.7 shows the cross data rewritten to reflect the newly determined gene order. For convenience in the analysis, the region between genes $p$ and $j$ is called region I, and that between genes $j$ and $r$ is called region II.

The recombination frequency can now be calculated for two genes at a time. For the $p$–$j$ distance, all the crossovers that occurred in region I are added together. Thus, we must add the recombinant progeny resulting from a single crossover in that region (classes 3 and 4) and the recombinant progeny produced by a double crossover in which one crossover is between $p$ and $j$ and the other is between $j$ and $r$ (classes 7 and 8). The double crossovers must be included because each double crossover includes a single crossover in region I and therefore involves recombination between genes $p$ and $j$. From Figure 14.7, there are 98 recombinant progeny in classes 3 and 4, and 6 in classes 7 and 8, giving a total of 104 progeny that result from recombination in region I. There are 500 progeny in all, so the percentage of progeny generated by crossing-over in region I is 20.8%, determined as follows (sco = single crossovers; dco = double crossovers):

$$\frac{\text{sco in region I } (p - j) + \text{dco}}{\text{total progeny}} \times 100$$

$$= \frac{(52 + 46) + (4 + 2)}{500} \times 100$$

$$= \frac{98 + 6}{500} \times 100$$

$$= \frac{104}{500} \times 100$$

$$= 20.8\%$$

In other words, the recombination frequency for genes $p$ and $j$ is 20.8, which gives us an estimated map distance of 20.8 mu. This map distance, which is quite large, is chosen mainly for illustration. We will see later that a recombination frequency of 20.8 in an actual cross would underestimate the true map distance.

**Figure 14.7**

**Rewritten form of the testcross and testcross progeny in Figure 14.6, based on the determined gene order $p\ j\ r$.**

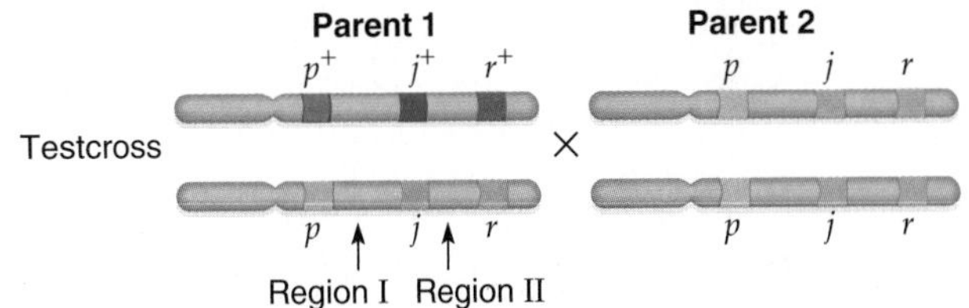


**Testcross progeny**

| Class | Genotype of gamete from heterozygous parent | Number | Origin |
|---|---|---|---|
| 1 | $p^+\ j^+\ r^+$ | 179 | Parentals, no crossover |
| 2 | $p\ j\ r$ | 173 | |
| 3 | $p^+\ j\ r$ | 52 | Recombinants, single crossover region I |
| 4 | $p\ j^+\ r^+$ | 46 | |
| 5 | $p^+\ j^+\ r$ | 22 | Recombinants, single crossover region II |
| 6 | $p\ j\ r^+$ | 22 | |
| 7 | $p^+\ j\ r^+$ | 4 | Recombinants, double crossover |
| 8 | $p\ j^+\ r$ | 2 | |
| | | Total = 500 | |

The same method is used to calculate the recombination frequency for the $j$–$r$ distance. That is, we calculate the frequency of crossovers in the cross that gave rise to progeny recombinant for genes $j$ and $r$ and directly relate that frequency to map distance. In this case, all the crossovers that occurred in region II (see Figure 14.7), that is, classes 5, 6, 7, and 8, must be added. The percentage of crossovers is calculated in the following manner:

$$\begin{aligned}&\frac{\text{sco in region II } (j - r) + \text{dco}}{\text{total progeny}} \times 100\\ &= \frac{(22 + 22) + (4 + 2)}{500} \times 100\\ &= \frac{44 + 6}{500} \times 100\\ &= \frac{50}{500} \times 100\\ &= 10.0\%\end{aligned}$$

Thus, the recombination frequency for genes $j$ and $r$ is 10.0, which gives us an estimated map distance of 10.0 map units.

To summarize, we have generated a genetic map of the three genes in the example (Figure 14.8). The example has illustrated that the three-point testcross is an effective way to establish the order of genes and calculate map distances.

To compute the map distance between the two outside genes, we simply add the two map distances. In the example, the $p - r$ distance is $20.8 + 10.0 = 30.8$ mu. This map distance also can be computed directly from the data by combining the two formulas discussed previously:

$$\begin{aligned}\text{distance} &= \frac{(\text{sco in region I}) + (\text{dco}) + (\text{sco in region II}) + (\text{dco})}{\text{total progeny}} \times 100\\ &= \frac{(\text{sco in region I}) + (\text{sco in region II}) + (2 \times \text{dco})}{\text{total progeny}} \times 100\\ &= \frac{52 + 46 + 22 + 22 + 2(4 + 2)}{500} \times 100\\ &= \frac{98 + 44 + 2(6)}{500} \times 100\\ &= 30.8 \text{ map units}\end{aligned}$$

**Keynote**

The map distance between genes can be calculated from the results of testcrosses between strains carrying appropriate genetic markers. The unit of genetic distance is the map unit (mu), where 1 mu is defined as the interval in which 1% crossing-over takes place. Gene mapping crosses produce data in the form of recombination frequencies, which are used to estimate map distance, where 1 mu is equivalent to a recombination frequency of 1%. Recombination frequencies are not identical to crossover frequencies and typically underestimate the true map distance.

**Interference and Coincidence.** The recombination frequencies determined by three-point mapping are useful in elaborating the overall organization of genes on a chromosome

**Figure 14.8**

**Genetic map of the $p$–$j$–$r$ region of the chromosome computed from the recombination data in Figure 14.7.**

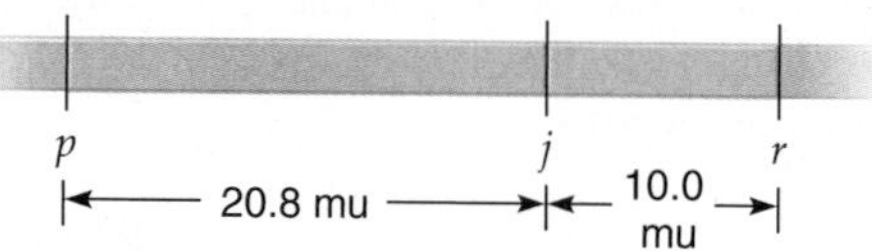


and in telling us a little about the recombination mechanisms themselves. For example, we computed a recombination frequency of 20.8 between genes $p$ and $j$ and a recombination frequency of 10.0 between genes $j$ and $r$ in the previous three-point testcross example. If crossing-over in region I is independent of crossing-over in region II, then the probability of a double crossover in the two regions is equal to the product of the probabilities of the two events occurring separately; that is,

$$\frac{\text{recombination frequency, region I}}{100} \times \frac{\text{recombination frequency, region II}}{100}$$
$$= 0.208 \times 0.100 = 0.208$$

or 2.08% double crossovers are expected to occur. However, only $^6/_{500} = 1.2\%$ double crossovers occurred in this cross (classes 7 and 8).

It is characteristic of mapping crosses that double-crossover progeny typically do not appear as often as the map distances between the genes lead us to expect. That is, in some way, the presence of one crossover interferes with the formation of another crossover nearby, in a phenomenon called **interference.** The extent of interference is expressed as a **coefficient of coincidence;** that is,

$$\text{coefficient of coincidence} = \frac{\text{observed double crossover frequency}}{\text{expected double crossover frequency}}$$

and

$$\text{interference} = 1 - \text{coefficient of coincidence}$$

For the portion of the map in our example, the coefficient of coincidence is

$$0.012/0.0208 = 0.577$$

A coefficient of coincidence of 1 means that, in a given region, all double crossovers occurred that were expected on the basis of two independent events; there is no interference, so the interference value is zero. If the coefficient of coincidence is zero, none of the expected double crossovers occurred. Here, there is total interference, with one crossover completely preventing a second crossover in the region under examination; the interference value is 1. These examples show that coincidence values and interference values are inversely related. In our mapping example, the coefficient of coincidence of 0.577 means that the interference value is 0.423. Only 57.7% of the expected double crossovers took place in the cross.

## Keynote

The occurrence of a crossover may interfere with the occurrence of a second crossover nearby. The extent of interference is expressed by the coefficient of coincidence, which is calculated by dividing the number of observed double crossovers by the number of expected double crossovers. The coefficient of coincidence ranges from zero to one, and the extent of interference is measured as one minus the coefficient of coincidence.

## Calculating Accurate Map Distances

Map units between linked genes, strictly speaking, are defined in terms of the crossover frequency, whereas, operationally, geneticists quantify the frequency of recombinants in genetic crosses. The crossover frequency and the recombination frequency are not identical, so the latter often leads to an underestimation of the true map distance. How, then, do we obtain accurate map distances for linked genes?

To answer this question, we need to focus on the consequences of crossovers between linked genes. Consider a hypothetical case of two allelic pairs ($a^+/a$ and $b^+/b$) linked in coupling and separated by quite a distance on the same chromosome. Figure 14.9a shows that a single crossover results in recombination of the two allelic pairs, producing two parental and two recombinant gametes. The same result will occur for any odd number of crossovers in the region between the genes. Figure 14.9b shows that a two-strand double crossover (one that involves two of the four chromatids) does not result in recombination of the allelic pairs, so only parental gametes result. Parental gametes also result for any even number of crossovers between the two linked genes. However, the crossover frequency between genes is a measure of the distance between them.

**Figure 14.9**

**Progeny of single and double crossovers.**

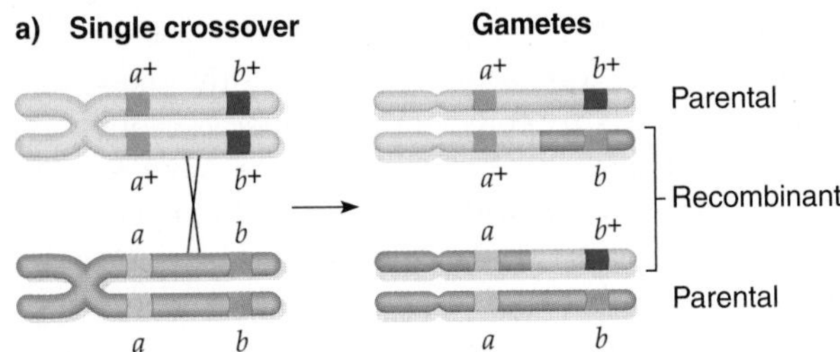


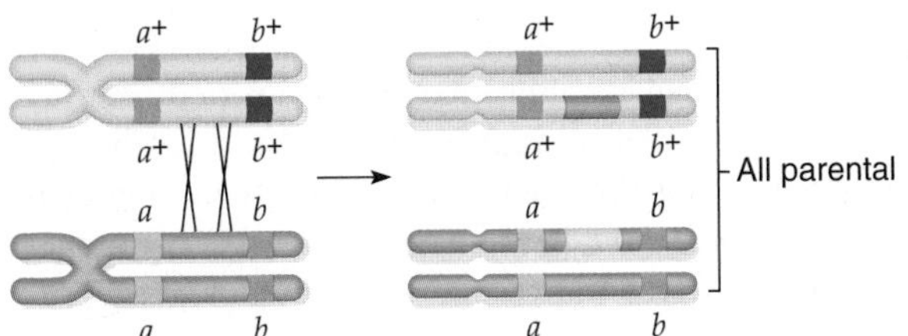


Therefore, because the double crossover in Figure 14.9b did not generate recombinant gametes, two crossover events will go uncounted, and the map distance based on recombination frequency between genes $a$ and $b$ will be underestimated.

In gene mapping, if no more than a single crossover occurs between linked genes, there is a direct linear relationship between the genetic map distance and the observed recombination frequency, because the recombination frequency then equals the crossover frequency. In practice, we see this relationship only when genetic map distances are small—when genes are between 0 and approximately 7 mu apart. To turn this statement around, map distances based on recombination frequencies of 7% or less are highly accurate. As the distance between genes increases beyond this point, the chance of multiple crossovers increases, and there is no longer an exact linear relationship between map distance and recombination frequency, because some crossovers go uncounted. As a result, it is difficult to obtain an accurate measure of map distance when multiple crossovers are involved.

Fortunately, mathematical formulas, called **mapping functions,** have been derived and define the relationship between map distance and recombination frequency. A particular mapping function that assumes no interference between crossovers is shown in Figure 14.10. You can see the direct relationship between map distance and recombination frequency at 7 mu or less, and the curve slowly approaches the limit recombination frequency of 50%, Just to pick a couple of points, when the recombination frequency is 20%, the true map distance is almost 30 mu, and when the recombination frequency is 30%, the true map distance is almost 50 mu. In general, mapping functions require some basic assumptions about the frequency of crossovers compared with distance between genes. Therefore, the usefulness of applying the mapping functions depends on the validity of the assumptions.

**Figure 14.10**

**A mapping function for relating map distance and recombination frequency.** This particular mapping function was developed by J. B. S. Haldane and assumes no interference between crossovers. The variable $d$ is the crossover frequency, and $e$ is the base of natural logarithms.

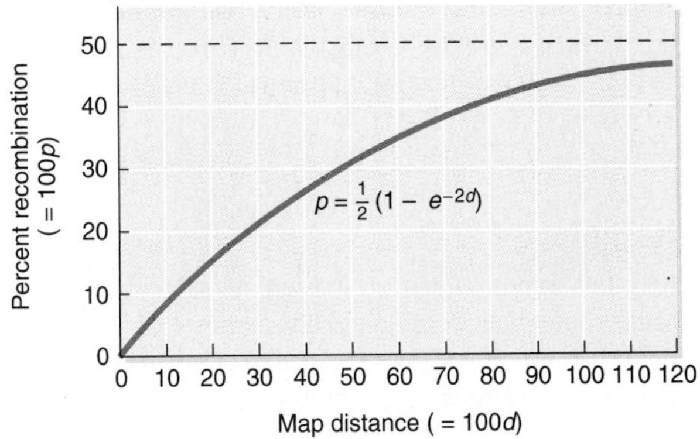

**Keynote**

At genetic distances greater than about 7 mu, the incidence of multiple crossovers causes the recombination frequency to be an underestimate of the crossover frequency and hence of the true map distance. Mapping functions can be used to correct for the effects of multiple crossovers and thereby give a more accurate map distance.

### Genetic Maps and Physical Maps Compared

We have just described the construction of genetic maps. The map distance between two markers (gene or DNA) depends on the frequency of crossing-over that occurs between them in meioses. The simplifying assumption that Sturtevant made when he constructed the first genetic map and that typically is made today in genetic mapping experiments is that crossovers occur at random along chromosomes. This would mean there is an equal likelihood that a crossover will occur at any point along the chromosome. However, the assumption is not completely correct, because we know there are hot spots and cold spots for crossing-over throughout the genome. That is, for some chromosome regions (the hot spots), crossing-over occurs at a higher-than-average frequency, while for some other chromosome regions (the cold spots), crossing-over occurs at a lower-than-average frequency. Because crossing-over frequency is directly used to determine map distance in genetic mapping, the nonrandom distribution of crossovers throughout genomes leads to genetic maps that have limited accuracy.

Physical maps are maps of chromosomes generated using molecular approaches, rather than by analyzing the results of genetic crosses. Thus, crossing-over is not an issue. The ultimate physical map is the DNA sequence of a chromosome, and of a genome. How does a genomic sequence map compare with a genetic map? In most cases, the order of gene and DNA markers is the same, although in some cases where gene order was seen to be different in the two maps. However, typically, the relative positioning of markers in a genome is not the same in the two maps because of variation in crossing-over frequency. For instance, in the human genome, one study shows that recombination frequency varies greatly along each chromosome, with values from 0 to at least 9 map units per megabase ($10^6$ base pairs).

**Keynote**

Recombination frequency varies greatly along chromosomes, potentially resulting in a wide range of map units (measured by genetic mapping) per megabase (measured by physical mapping).

## Constructing Genetic Linkage Maps of the Human Genome

This section is an overview of some methods used in the construction of genetic linkage maps of the human genome. Nowadays, researchers rely on the human genome sequence to locate genes and determine distances between linked genes.

### The lod Score Method for Analyzing Linkage of Human Genes

For practical and ethical reasons, with humans it is not possible to do genetic mapping experiments of the kind performed with experimental organisms. Nonetheless, historically geneticists have been very interested in mapping genes in human chromosomes, since so many known diseases and traits that have a genetic basis. In Chapters 11 and 12, we saw that pedigree analysis could be used to determine the mode by which a particular genetic trait is inherited. In this way, many genes were localized to the X chromosome, and map distances between some could be counted. However, pedigree analysis cannot show on which chromosome a particular autosomal gene is located.

Because suitable pedigrees for conventional genetic mapping are rare, a statistical test known as the **lod (logarithm of odds) score method** is instead used to analyze pedigrees for possible linkage between two loci. The lod score method was invented by mathematical geneticist Newton Morton in 1955. It is usually done by computer programs that use pooled data from a number of pedigrees. A full discussion of the method is beyond the scope of this text, so only a brief presentation is given here.

The lod score method compares: (1) the probability of obtaining the pedigree results if two genetic markers (gene or DNA) are linked with a certain amount of recombination between them to; (2) the probability that the results would have been obtained if there was no linkage (i.e., 50% recombination) between the markers. The results are expressed as the $\log_{10}$ of the ratio of the two probabilities. By convention, a hypothesis of linkage between two genes is accepted if the lod score at a particular recombination frequency is +3 or more because a score of +3 means that the odds are $10^3$ to 1 (1,000:1) in favor of linkage between two genes or markers (the $\log_{10}$ of 1,000 is +3). Similarly, a hypothesis of linkage between two genes is rejected when the lod score reaches –2 or more because a score of –2 means that the odds are $10^{-2}$ to 1 (100:1) against the two genes or markers being linked.

Once linkage is established between genetic markers, the map distance is computed from the recombination frequency giving the highest lod score. (The higher the lod score, the closer to each other are the two genes.) This is done by solving lod scores for a range of proposed map units.

## Human Genetic Maps

The Human Genome Project (HGP) for sequencing the human genome was described in Chapter 8 (p. 171). Two different sequencing approaches were used in the HGP. One, the mapping approach, involved three steps: (1) making detailed genetic maps to provide a relatively sparse set of landmarks; (2) making high-resolution physical maps to provide a detailed set of landmarks; and (3) constructing the sequencing map of the genome. The last step was made possible because each segment of the genome that was sequenced came from a known place on the maps that had been developed. The mapping approach is analogous to ordering the headings in a book first, and then finding the words between the headings. The other approach is the whole-genome shotgun approach in which the entire genome is broken up into random, overlapping fragments that are then sequenced (see Chapter 8, pp. 189–191). The genome sequence is then assembled on the basis of the sequence overlaps between fragments. This approach is analogous to taking 10 books that have been torn randomly into smaller leaflets of a few pages each and, by matching overlapping pages of the leaflets, assembling a complete copy of the book with the pages in the correct order. Nowadays, the whole-genome shotgun approach is the technique used routinely for genome sequencing.

While the mapping approach is no longer used for genome sequencing, the principles of constructing genetic maps of the human genome are useful to learn in the context of the focus in this chapter on mapping genes in eukaryotes. As we have learned in this chapter, certain model experimental organisms were used with great success for constructing detailed linkage maps of genes by using designed genetic crosses involving two or more genes. Designed genetic crosses are not possible, of course, with humans; besides the frequency of matings between individuals with allelic differences for two genes is small. Hence, human geneticists considered constructing a human genetic map based on genes an impossible task. We know now from analysis of sequence data that only about 2% of the human genome consists of genes, which means that genes are generally scattered widely in the genome.

Fortunately, the discovery of DNA markers elevated genetic mapping in humans to a new level. Each DNA marker corresponds to a particular sequence at a site in the genome. If more than one type of sequence is found in the population at the site, then the DNA marker is polymorphic; in essence, we have alleles of a locus that differ in a *molecular* phenotype, rather than a visible or biochemical phenotype such as eye color, plant height, or enzyme activity. And polymorphic DNA markers are far more frequent in the human genome than are genes. (Polymorphic DNA markers and the methods used to analyze them are discussed in detail in Chapter 10, pp. 270–280.) The alleles of polymorphic DNA marker loci are analyzed—typed—in DNA samples isolated from individuals in a large number of pedigrees. The DNA samples are collectively referred to as a *panel of DNAs*. Linkage between DNA marker loci is typically determined by the lod score method using computer algorithms. The use of polymorphisms to identify alleles that predispose an individual to develop multiple sclerosis is described in this chapter's Focus on Genomics box.

The first detailed genetic map of human DNA markers was published in 1987. The map consists of 403 polymorphic loci in a panel of DNAs from 21 three-generation families. Of the loci, 393 were *restriction fragment length polymorphism (RFLP)* loci, meaning that single base-pair differences created or abolished cleavage sites for particular restriction enzymes. The spacing between markers on the map averaged about 10 mu.

Higher-resolution human genetic maps subsequently were generated by the analysis of *short tandem repeats (STRs)*, also called *microsatellites*. STRs are 2–6-bp DNA sequences repeated tandemly; polymorphic STR loci vary in the number of repeats. PCR is used to type them. STR loci are far more frequent in the genome than are RFLP loci. Making genetic maps of a large number of DNA markers is too much for a single research lab, due to the great numbers of manipulations needed. For example, to type 5,000 DNA marker loci in 500 individuals would require performing 2,500,000 typing tests, each involving molecular techniques, and then entering 2,500,000 results into a database. Therefore, geneticists set up a collaboration among many laboratories to do the work and, most importantly, had *the consortium work on the same set of DNA samples from the same set of individuals*. To generate a high-resolution map of STRs, a panel of DNAs was used from a human DNA collection held at the Centre d'Étude du Polymorphisme Humain (CEPH), a research center in Paris, France. This collection is from 517 individuals representing 40 three-generation families. From this work, a genetic map of 814 STR loci with an average resolution of about 5 mu was published in 1992. Subsequently a comprehensive human genetic map of 5,840 loci—which included 3,617 STR loci and 427 genes—with resolution of about 0.7 mu was published in 1994. The comprehensive human genetic map was a highly valuable resource for the subsequent development of high-resolution physical maps that were the foundations for human genome sequencing by the mapping approach.

### Keynote

Genetic maps of genomes are constructed by using recombination data from genetic crosses in the case of experimental organisms or from pedigree analysis in the case of humans. Both gene markers and DNA markers are used in genetic-mapping analysis. In one approach used for the Human Genome Project, high-resolution genetic maps were stepping-stones to the construction of physical maps and then to the actual sequencing of the human genome.

and *cho* (chocolate-brown eyes, map position 13.0), with *w* epistatic to *cho*.

**a.** A white-eyed female is crossed to a chocolate-eyed male, and the normal, red-eyed $F_1$ females are crossed to either wild-type or white-eyed males. Determine the frequency of the progeny types produced in each cross.

**b.** The recessive mutation *st* causes scarlet (bright red) eyes and maps to the third chromosome at position 44. Mutant flies with only *st* and *cho* alleles have white eyes, and *w* is epistatic to *st*. Suppose a true-breeding *w* male is crossed to a true-breeding *cho, st* female. Determine the frequency of the progeny types you would expect if the $F_1$ females are crossed to true-breeding scarlet-eyed males.

***14.33** Breeders of thoroughbred horses used for racing keep extensive information on pedigrees. Such information can be useful in determining simple inheritance patterns (e.g., chestnut coat color has been determined to be recessive to bay coat color) and in speculating whether racehorses that win competitive races ("classy" horses) share genetic traits. Sharpen Up was a chestnut stallion that was only somewhat successful as a racehorse. At age 4, he was retired from horse racing and put out to stud. His progeny were very successful: of 367 foals fathered in the United States, 43 were prizewinners in highly competitive races, and of 200 foals fathered in England, 40 were prizewinners in highly competitive races. A commentator who analyzed Sharpen Up's progeny (and that of other chestnut prizewinners) has suggested that whatever gene combinations produced class (winning horses) were tied to the horses' chestnut coat. Indeed, of the 83 progeny that have shown class (won highly competitive races), about 45 were also chestnut in color. Use a chi-square test to assess whether there is any reason to believe that if there is a gene (or genes) for class, it is linked to the gene for chestnut coat color. Examine this issue using two different assumptions: (1) Sharpen Up was mated equally frequently to homozygous bay, heterozygous bay/chestnut, and homozygous chestnut mares; and (2) Sharpen Up was mated equally frequently to heterozygous bay/chestnut and homozygous chestnut horses. Be careful to state any additional assumptions before giving your hypothesis.

***14.34** In an $a^+ \, b^+/a \, b$ individual, a physical exchange between the *a* and *b* loci occurs in 14% of meioses. What percentage of gametes will be $a^+ \, b$ *or* $a \, b^+$? Explain your answer.

***14.35** Some dogs love water, while others avoid it. A dog that loved water was mated to a dog that avoided it, and their $F_1$ progeny were interbred to give an $F_2$. The parental, $F_1$, and $F_2$ generations were evaluated by DNA typing, and the lod score method was used to assess linkage between DNA markers and genes for water affection (*waf* genes). Suppose that the following data were obtained for one marker, where $\theta$ gives the value of the recombination frequency between the marker and a *waf* gene used in calculating the lod score:

| $\theta$ | lod Score |
|---|---|
| 0 | $-\infty$ |
| 0.05 | −12.51 |
| 0.10 | −2.34 |
| 0.15 | −1.32 |
| 0.20 | 2.66 |
| 0.25 | 4.01 |
| 0.30 | 3.21 |
| 0.35 | 2.14 |
| 0.40 | 1.56 |
| 0.50 | 0 |

Graph these lod scores, and evaluate whether the marker is linked to a *waf* gene. If it is, estimate the physical distance between the marker and the gene, assuming that 1 mu corresponds to 1 megabase pair.

***14.36** In humans, the three STR loci *A*, *B*, and *C* are linked on the long arm of the X chromosome in the order *centromere–A–B–C–telomere*. *A* is 0.36 megabases (Mb) from *B*, and *B* is 0.64 Mb from *C*. In the pedigree shown in Figure 14.A, the STR alleles present in each individual are shown below their symbol.

**Figure 14.A**

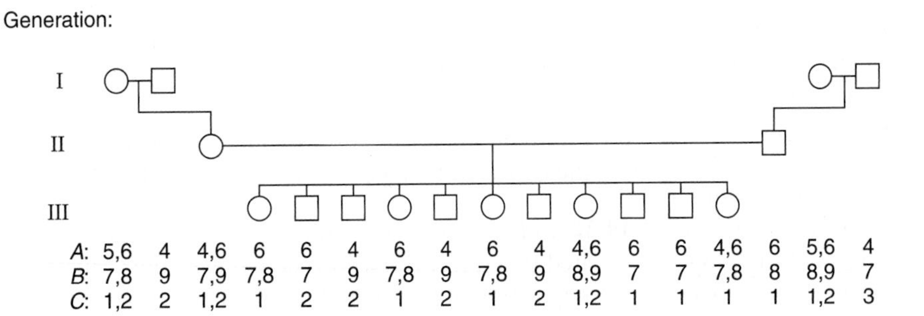

Generation:

I

II

III

| | | | | | | | | | | | | | | | | | |
|---|---|---|---|---|---|---|---|---|---|---|---|---|---|---|---|---|---|
| *A*: | 5,6 | 4 | 4,6 | 6 | 6 | 4 | 6 | 4 | 6 | 4 | 4,6 | 6 | 6 | 4,6 | 6 | 5,6 | 4 |
| *B*: | 7,8 | 9 | 7,9 | 7,8 | 7 | 9 | 7,8 | 9 | 7,8 | 9 | 8,9 | 7 | 7 | 7,8 | 8 | 8,9 | 7 |
| *C*: | 1,2 | 2 | 1,2 | 1 | 2 | 2 | 1 | 2 | 1 | 2 | 1,2 | 1 | 1 | 1 | 1 | 1,2 | 3 |

**a.** Analyze the inheritance of the STR alleles, and answer the following questions:

**i.** Why do males always have just one STR allele at each of the three loci? Why do females sometimes have just one STR allele at a locus, but at other times have two STR alleles at the same locus?

**ii.** Which STR alleles are present on each X chromosome of individual II-1?

**iii.** Can you tell whether the X chromosomes of II-1 and II-2 are recombinant chromosomes? Explain why or why not.

**iv.** Which individuals in generation III have recombinant X chromosomes that arose from meiotic crossovers between the STR loci in II-1?

**v.** For each recombinant individual in generation III, did a crossover occur between *A* and *B* or between *B* and *C*? Did any double crossovers occur?

**b.** Draw a map showing the genetic distances between the three STR loci.

**c.** How confident are you of the accuracy of your map, and how might you improve its accuracy?

**d.** What is the average rate of recombination (in mu/Mb) in the 1-Mb interval between STRs *A* and *C*?

**14.37** In Question 14.36, you analyzed one pedigree to estimate the map distances between three linked STR markers and rate of recombination rate in the interval defined by these markers. The company deCODE Genetics analyzed the inheritance of 5,136 STR markers through 1,257 meiotic events in 146 multigenerational families to gain a more comprehensive understanding of how the rate of recombination varies along the length of human chromosomes. In their calculations of recombination rate, they assumed a linear genetic distance across immediately flanking STR markers. Figure 14.B graphs their data on the average rate of recombination in each 1 mB window (in mu per Mb of DNA) in females along the approximately 93 Mb of the long arm of the X chromosome (Xq).

**a.** deCODE Genetics followed the inheritance of 5,136 STR loci in 1,257 meioses. What did they gain by tracking the inheritance of these STR loci through so many meioses? If these STR loci were evenly distributed over the approximately 2.85 billion base pairs of euchromatic DNA, how many base pairs, on average, lie between neighboring STR loci?

**b.** As a rule of thumb, a map distance of 1 mu in humans is said to correspond to about 1 Mb of DNA. Based on your inspection of Figure 14.B, how reasonable is this statement? How much variation is there in the frequency of recombination along Xq?

**c.** In the interval on Xq corresponding to sequence coordinates 87,000,000 to 100,000,000, the recombination rate is about 1 mu/Mb; in the interval corresponding to sequence coordinates 149,000,000 to 150,000,000, the recombination rate is over 8 mu/Mb. Which of these intervals has a greater average number of crossovers per Mb of DNA?

**d.** Suppose two genes lie 500,000 bp apart. Would these genes be more likely to be separated by crossing-over in a chromosomal region exhibiting a recombination rate of 1 mu/Mb or in a chromosomal region with a recombination rate of 8 mu/Mb?

**e.** Here, as in Question 14.36, the recombination rate in X chromosome intervals is based on how often X-linked STR loci recombine in females. On autosomes, crossing-over occurs in males as well as females. Suppose you wanted to ascertain whether the recombination rate in an autosomal region differs between males and females. How would you proceed?

**f.** Use the information from your answers to summarize the key differences between a genomic sequence map and a genetic map.

***14.38** Charcot-Marie-Tooth disease is an inherited neurological disease where patients have progressive muscular and sensory loss in their legs and arms. lod score analyses in different families exhibiting inherited disease have led

**Figure 14.B**

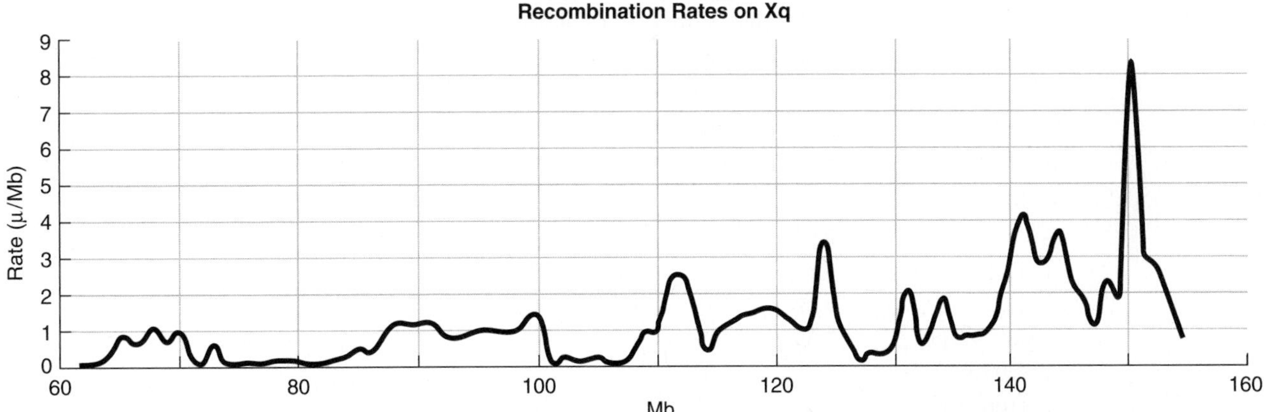

to the identification of more than 30 loci and 19 genes where mutations can lead to this disease. Mutations at different loci show different inheritance patterns: autosomal dominant, autosomal recessive, and X linked. When lod score analysis was done in two Middle Eastern families exhibiting autosomal recessive inheritance of this disease, several STR markers on chromosome 12 showed evidence of linkage. Table 14.B describes these STR markers and their cytological, genetic, and sequence map positions. Table 14.C shows the results of a lod score analysis using these markers. For each STR marker, Table 14.C gives the lod scores between the disease locus and the marker at the different recombination frequencies ($\theta$) used to calculate the lod score. Analyze the data in Tables 14.B and 14.C, and then answer the following questions.

**a.** Why are lod scores calculated at different recombination frequencies?

**b.** Consider the results obtained with STR markers AFM296YG5 and AFMB283XH5. How do you explain the gradual but steady increase in lod score values for these markers as the recombination frequency $\theta$ is decreased from 0.40 to 0.001?

**c.** Consider the results obtained with the other STR markers. For these markers, lod score values initially increase as the recombination frequency $\theta$ is decreased from 0.40, but before $\theta$ reaches 0.001, the lod score values decrease, sometimes dramatically. How do you interpret the initial increase and then the decrease in lod score values? Why, for different markers, do the lod score values decrease at different values of $\theta$? (Hint: Plotting the lod score values for each STR marker as a function of $\theta$ may help you visualize these results.)

**d.** Which STR markers show evidence of linkage to the disease locus?

**e.** What is the maximum lod score seen in this analysis, and at what value of $\theta$ is it seen? How do you interpret this result?

**f.** What cytological interval has the disease locus been localized to by this analysis? Does this interval include the centromere of chromosome 12? How large is this interval in mu? How large is this interval in megabase pairs of DNA?

**Table 14.B Features of STR Markers Used in the lod Score Analysis**

| STR | Cytological Map Position (band coordinates) | Genetic Map Location (sex-averaged, in mu) | Sequence Map Position (in megabase pairs[Mb]) |
|---|---|---|---|
| AFMB013YB1 | 12p11.22 | 51.90 | 29.2385 |
| AFMA288WD5 | 12p11.22 | 51.90 | 29.4208 |
| AFM337TF5 | 12p11.21 | 51.99 | 30.7557 |
| AFMB041XB9 | 12p11.21 | 52.54 | 30.9756 |
| AFM296YG5 | 12p11.21 | 53.09 | 32.1160 |
| AFMB283XH5 | 12q12 | 56.38 | 40.9934 |
| AFM122XF6 | 12q13.11 | 61.34 | 45.5229 |
| AFMB314YH5 | 12q13.11 | 63.89 | 46.7925 |
| AFM294WC5 | 12q13.12 | 64.43 | 47.3824 |
| AFM196XA3 | 12q13.13 | 64.96 | 49.2233 |
| AFMB347VB9 | 12q13.13 | 65.49 | 50.4991 |

**Table 14.C lod Scores between the Disease Locus and STR Markers**

| | lod Score at Recombination Frequency $\theta$ = | | | | | | |
|---|---|---|---|---|---|---|---|
| STR | 0.001 | 0.01 | 0.05 | 0.1 | 0.2 | 0.3 | 0.4 |
| AFMB013YB1 | 0.71 | 2.01 | 2.80 | 2.77 | 2.11 | 1.24 | 0.45 |
| AFMA288WD5 | –0.54 | 0.93 | 1.86 | 1.99 | 1.60 | 0.94 | 0.30 |
| AFM337TF5 | –0.51 | 0.89 | 1.76 | 1.79 | 1.24 | 0.56 | 0.09 |
| AFMB041XB9 | 5.64 | 6.28 | 6.27 | 5.71 | 4.20 | 2.52 | 0.96 |
| AFM296YG5 | 6.97 | 6.82 | 6.20 | 5.39 | 3.72 | 2.07 | 0.70 |
| AFMB283XH5 | 5.81 | 5.68 | 5.12 | 4.42 | 2.99 | 1.63 | 0.56 |
| AFM122XF6 | 5.09 | 4.97 | 4.51 | 3.92 | 2.70 | 1.50 | 0.54 |
| AFMB314YH5 | 3.35 | 3.92 | 4.02 | 3.63 | 2.51 | 1.31 | 0.36 |
| AFM294WC5 | $-\infty$ | 1.93 | 2.85 | 2.88 | 2.26 | 1.37 | 0.52 |
| AFM196XA3 | $-\infty$ | 1.12 | 2.49 | 2.85 | 2.55 | 1.75 | 0.80 |
| AFMB347VB9 | $-\infty$ | 1.01 | 1.98 | 2.14 | 1.79 | 1.15 | 0.46 |

# 15 Genetics of Bacteria and Bacteriophages

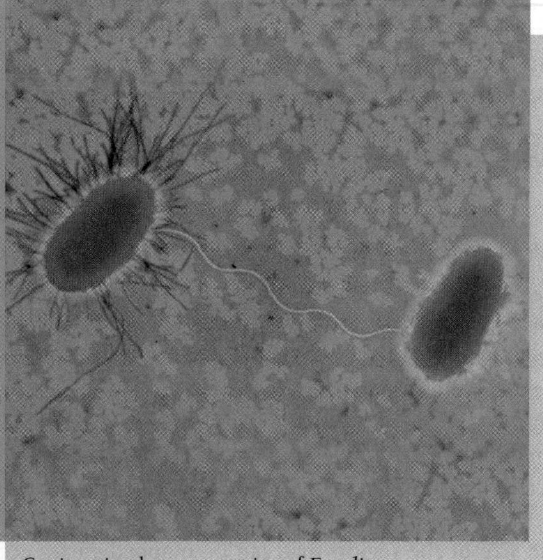

Conjugation between strains of *E. coli.*

## Key Questions

- How are genes mapped in bacteria?
- How are genes mapped in bacteriophages?
- How are mutations with the same phenotypes sorted into different genes?

### Activity

BACTERIA AND BACTERIOPHAGES HAVE LONG played a key role in genetics. Griffith's and Avery's experiments with *Streptococcus* were key in the discovery of DNA as the genetic material. Herbert Boyer and Stanley Cohen used bacteria in the first recombinant DNA molecule. Alfred Hershey and Martha Chase used bacteriophage T2 in their experiments showing that DNA is the genetic material.

In this chapter, you will learn about the genetics of bacteria and bacteriophages: how these organisms and viruses reproduce, how new strains are produced, and the experimental techniques that geneticists use to map bacterial and viral genes. After you have read and studied the chapter, you can apply what you have learned by trying the iActivity, in which you will create a genetic map of the *E. coli* chromosome.

In Chapter 14, we considered the principles of gene mapping in eukaryotic organisms. To map genes in bacteria and bacteriophages, geneticists use essentially the same experimental strategies. Crosses are made between strains that differ in genetic markers, and recombinants—the products of the exchange of genetic material—are detected and counted. The major difference lies in the experimental techniques involved.

Recently, emphasis in genetic research has shifted from localizing individual genes on chromosomes by making crosses to determining the complete sequence of nucleotides in the genome. With such DNA sequence information, scientists can identify genes directly, and place those genes with precise coordinates on the genome sequence, producing the most accurate genetic map of a species that can be achieved. When all the genes of an organism are identified, at least at the nucleotide level, we can study the function of each gene. In the case of pathogenic microorganisms, the genome sequence information is an extremely valuable resource in efforts to identify and understand the genes responsible for pathogenesis. Complete genome sequences have been determined for many bacteriophages and for many species of Bacteria and Archaea (see Chapter 8). Bacterial genomes sequenced include the 4.6-million-base-pair (4.6-megabase-pair (Mb)) genome of *E. coli*, the 1.44-Mb genome of the Lyme disease causative agent *Borrelia burgdorferi*, the 1.66-Mb genome of the stomach-ulcer-causing *Helicobacter pylori*, and the 1.14-Mb genome of the syphilis bacterium *Treponema pallidum*. Among the archaean genomes sequenced is the 1.66-Mb genome of *Methanococcus jannaschii*, a hyperthermophilic methanogen that grows optimally at 85°C and at pressures up to 200 atmospheres.

In this chapter, you will learn about the classic genetic studies of bacteria and bacteriophages that gave researchers the first insights into the positions of genes on their chromosomes. You will also learn about a series of classic genetic experiments that investigated the fine structure of the gene—that is, the detailed molecular organization of the gene as it relates to the mutational, recombinational, and functional events in which the gene is involved. A bacteriophage gene was the subject of these experiments.

## Genetic Analysis of Bacteria

Genetic material can be transferred between bacteria by three main processes: conjugation, transformation, and transduction. Bacterial genes may be mapped using classical methods involving any one of these processes. In each case, (1) the transfer is unidirectional, and (2) no complete diploid stage is formed (in contrast with what occurs in eukaryotes). However, not all methods of genetic analysis can be used for all bacterial species, and the size of the region that can be mapped varies according to the method.

Among bacteria, *E. coli* has been used extensively for genetic analysis. This bacterium is a good subject for study because it can be grown on a simple, defined medium and can be handled with simple microbiological techniques. *E. coli* is a cylindrical organism about 1–3 μm long and 0.5 μm in diameter (see chapter opening photo). *E. coli* has a single circular DNA chromosome.

Like other bacteria, *E. coli* can be grown both in a liquid culture medium and on the surface of a growth medium solidified with agar. Genetic analysis of bacteria typically is done by spreading (plating) cells on the surface of an agar medium. Wherever a single bacterium lands on the agar surface, it will grow and divide repeatedly, ultimately forming a visible cluster of genetically identical cells called a *colony* (Figure 15.1). Each colony consists of a clone of cells, each of which is genetically identical to the parental cell that initiated the colony. The concentration of bacterial cells in a liquid culture—the *titer*—can be determined by spreading known volumes of the culture or of a known dilution of the culture on the agar surface, incubating the plates at a temperature of 37°C, and then counting the number of resulting colonies. The number obtained is converted to colony-forming units (cfu) per milliliter (mL). For example, if 100 microliters (μL) of a thousandfold dilution of a culture is spread on a plate and 165 colonies are produced, then there were 165 bacteria in 100 μL of the thousandfold dilution. Thus, in the original culture, there were 165 (colonies) × 1,000 (dilution factor) × 10 (because 0.1 mL was plated) = 1,650,000 cfu/mL = $1.65 \times 10^6$ cfu/mL.

**Figure 15.1**

**Bacterial colonies growing on a nutrient medium in a Petri dish.**

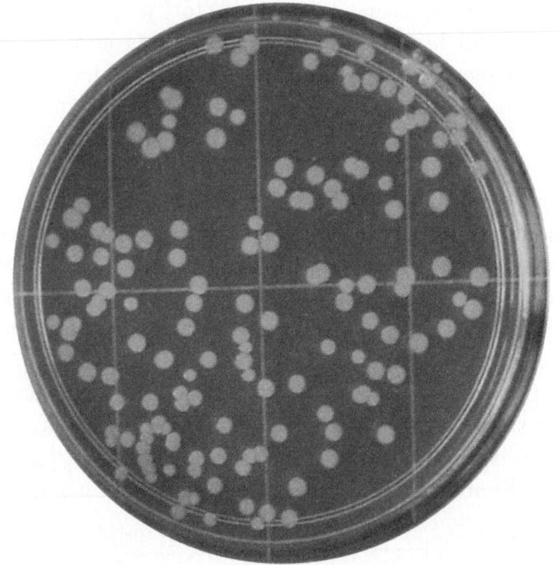

The composition of the culture medium used depends on the experiment and the genotypes of the strains being examined. Each bacterial species (or any other microorganism, such as yeast) has a characteristic **minimal medium** on which it will grow. A minimal medium is the simplest set of chemicals needed for the organism to grow and survive. The minimal medium for wild-type *E. coli*, for example, consists of a sugar (a carbon source) and some salts and trace elements. From the minimal medium, the organism can synthesize all the other components it needs for growth and reproduction, including amino acids, vitamins, and the precursors for DNA and RNA. By contrast, the **complete medium** for a microorganism supplies vitamins and amino acids and all kinds of substances that might be expected to be essential metabolites and whose biosynthesis might be interfered with by mutation.

Historically, the genetic analysis of bacteria (and other microorganisms) typically involved studying mutants that are defective in their abilities to make one or more molecules essential for growth and that are perhaps also defective in genes affecting other metabolic processes. Strains unable to synthesize essential nutrients are called **auxotrophs** (also called **auxotrophic mutants** or **nutritional mutants**). A strain that is wild type and thus that can synthesize all essential nutrients is a **prototroph** (also called a **prototrophic strain**). Prototrophs need no nutritional supplements in the growth medium. By definition, the wild type or prototroph grows on the minimal medium for that organism, whereas an auxotroph grows on a complete medium or on a minimal medium plus the appropriate nutritional supplement or supplements.

For example, an *E. coli* strain with the genotype *trp ade thi*$^+$ will not grow on a minimal medium, because it has mutations for tryptophan and adenine biosynthesis. It will grow either on a complete medium or on a minimal medium supplemented with the amino acid, tryptophan (because of the *trp* mutation), and the purine, adenine (because of the *ade* mutation). It does not need the vitamin thiamine to grow, because it carries the wild-type *thi* allele, as signified by the superscript +.

Some genes are involved not in biosynthetic pathways, but in utilization pathways. For example, a number of different genes use various carbon sources, such as lactose, arabinose, and maltose. In this case, the superscript + next to the gene symbol means that the gene is wild type and therefore that the bacterium can metabolize the substance. For instance, a *lac*$^+$ strain can metabolize lactose, whereas a *lac* mutant strain cannot.

In genetic experiments with microorganisms such as *E. coli*, crosses are made between strains differing in genotype (and, therefore, phenotype), and progeny are analyzed for parental and recombinant phenotypes. When auxotrophic mutations are involved, the determination of parental and progeny phenotypes (and, therefore, genotypes, because bacteria are haploid) involves testing colonies for their growth requirements. One convenient procedure for such testing is *replica plating*, invented by Joshua and Esther Lederberg (see Figure 7.15, p. 145). In replica plating, some of the bacteria of colonies on a plate of complete medium (the master plate) are transferred onto a sterile velveteen cloth mounted on a replica plater. Replicas of the original colony pattern on the cloth are then made by gently pressing new plates onto the velveteen. If the new plate contains minimal medium, only prototrophic colonies can grow. Then researchers can readily identify auxotrophic colonies because they will be on the master plate, but not on the minimal medium plate. Using other plates containing minimal medium plus combinations of nutritional supplements appropriate for the strain or strains involved, the phenotypes and genotypes of all auxotrophic colonies can be determined.

## Gene Mapping in Bacteria by Conjugation

### Discovery of Conjugation in *E. coli*

**Conjugation** is a process in which there is a unidirectional transfer of genetic information through direct cellular contact between a donor bacterial cell and a recipient bacterial cell. The contact is followed by the formation of a physical bridge joining the cells. Then a segment (rarely all) of the donor's chromosome may be transferred into the recipient cell and may undergo genetic recombination with a homologous chromosome segment of that cell. Recipients that have incorporated a piece of donor DNA into their chromosomes are called **transconjugants.**

**animation**
**Mapping Bacterial Genes by Conjugation**

Conjugation was discovered in 1946 by Joshua Lederberg and Edward Tatum. They studied two *E. coli* strains that differed in their nutritional requirements. Strain A had the genotype *met bio thr*$^+$ *leu*$^+$ *thi*$^+$—it could grow only on a medium supplemented with the amino acid methionine (*met*) and the vitamin biotin (*bio*), but it did not need the amino acids threonine (*thr*) or leucine (*leu*) or the vitamin thiamine (*thi*). Strain B had the genotype *met*$^+$ *bio*$^+$ *thr leu thi*—it could grow only on a medium supplemented with threonine, leucine, and thiamine, but it did not require methionine or biotin.

Lederberg and Tatum mixed *E. coli* strains A and B together and plated them onto minimal medium (Figure 15.2). The mixed culture gave rise to some prototrophic colonies (*met*$^+$ *bio*$^+$ *thr*$^+$ *leu*$^+$ *thi*$^+$) at a frequency of about 1 in 10 million cells. Mutation was ruled out as the cause of the prototrophic colonies because no colonies appeared when each strain was plated separately on minimal medium. The mixing, then, was a genetic cross that produced recombinants.

In a separate experiment, Bernard Davis placed strains A and B in a liquid medium on either side of a U-tube apparatus (Figure 15.3) separated by a filter with pores too small to allow bacteria to move through. The medium was moved between compartments by alternating suction and pressure, and then the cells were plated on minimal medium to check for the appearance of prototrophic colonies. No prototrophic colonies appeared, leading to the

**Figure 15.2**

**Lederberg and Tatum experiment showing that sexual recombination occurs between cells of *E. coli*.** After the cells from strain A and strain B were mixed and the mixture plated, a few colonies grew on the minimal medium, indicating that they could now make the essential onstituents. These colonies are recombinants produced by an exchange of genetic material between the strains.

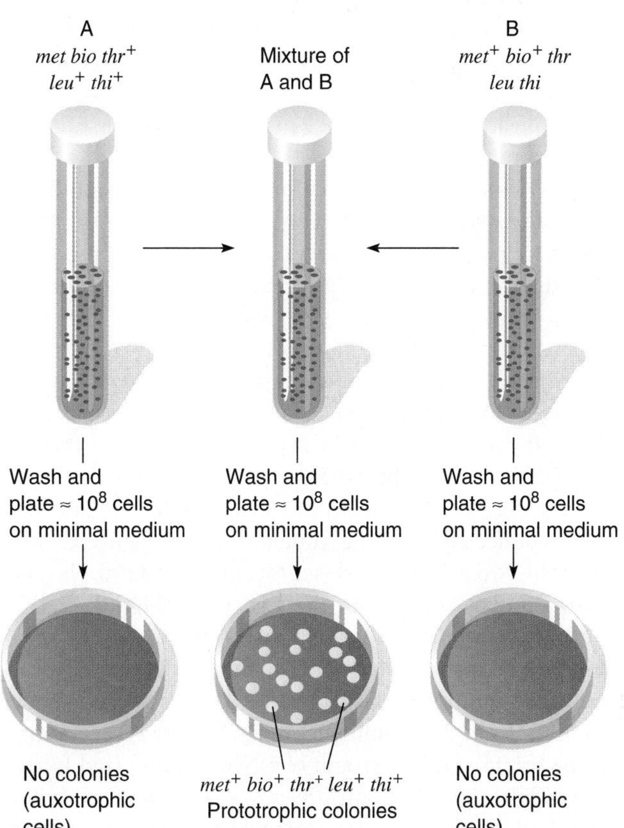

**Figure 15.3**

**U-tube experiment showing that physical contact between the two bacterial strains of the Lederberg and Tatum experiment was needed for genetic exchange to occur.**

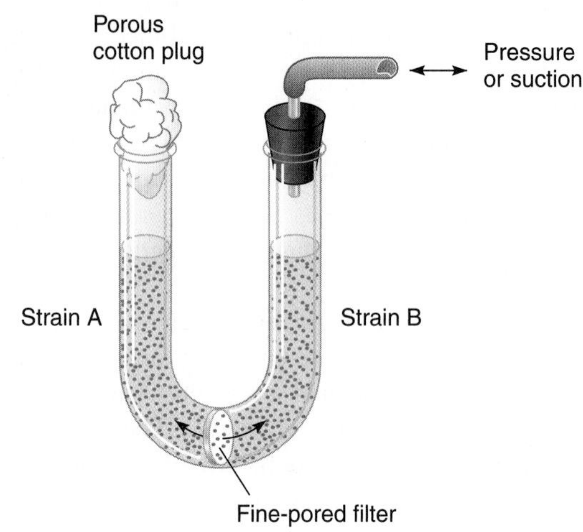


conclusion that cell-to-cell contact was required for the genetic exchange to occur. These experiments indicated that *E. coli* has the type of mating system called *conjugation*.

## The Sex Factor *F*

In 1953, William Hayes showed that genetic exchange in *E. coli* occurs in only one direction, with one cell acting as a *donor* and the other cell acting as a *recipient*. Hayes proposed that the transfer of genetic material between the strains is mediated by a *sex factor* named the **F factor** that the donor cell possesses ($F^+$) and the recipient cell lacks ($F^-$). The *F* factor found in *E. coli* is a *plasmid*—a self-replicating, circular DNA distinct from the main bacterial chromosome. About 1/40 of the size of the host chromosome, the *F* factor contains a region of DNA called the **origin** (or O)—the point where DNA transfer to the recipient begins—as well as a number of genes, including those which specify hairlike host cell surface components called **F-pili** (singular *F*-pilus), or *sex-pili*, which permit the physical union of $F^+$ and $F^-$ cells.

When $F^+$ and $F^-$ cells are mixed, they may conjugate ("mate") (Figure 15.4 and Figure 15.5a, part 1). No conjugation can occur between two cells of the same mating type (that is, two $F^+$ bacteria or two $F^-$ bacteria). In conjugation, an $F^+$ cell contacts an $F^-$ cell, and they initially become connected by a long, tubular *F*-pilus (Figure 15.4a). Next the two cells move closer together and a cytoplasmic bridge forms, connecting the two cells (Figure 15.4b). During conjugation, genetic material is transferred from donor to recipient when one DNA strand of the *F* factor is nicked at the origin; rolling circle DNA replication then proceeds from that point (Figure 15.5a, part 2). Beginning at the origin, a single strand of DNA is transferred to the $F^-$ as replication maintains the remaining circular *F* factor in a double-stranded form (Figure 15.5a, part 3; see also Figure 3.11, p. 48, and Chapter 3, pp. 46–47). Think of the process loosely like a roll of paper towels unraveling. Once the *F* factor single-stranded DNA enters the $F^-$ recipient, DNA polymerase in the recipient synthesizes the complementary strand (Figure 15.5a, part 4). If the complete *F* factor is transferred and circularizes, which commonly happens in $F^+ \times F^-$ crosses, the $F^-$ cell becomes an $F^+$ cell (Figure 15.5a, part 5). In $F^+ \times F^-$ crosses, none of the bacterial chromosome is transferred; only the *F* factor is.

**Figure 15.4**

**Conjugation between an $F^+$ donor and $F^-$ recipient *E. coli* bacterium.**

a) Two bacterial cells connected by a long, tubular *F*-pilus

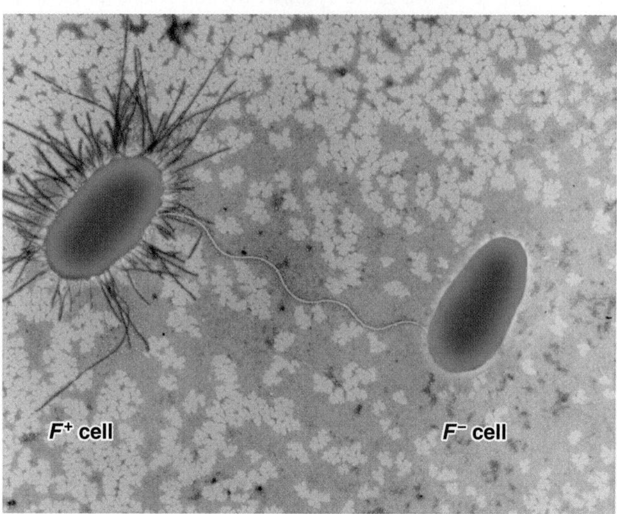


b) Conjugating *E. coli* cells showing the cytoplasmic bridge

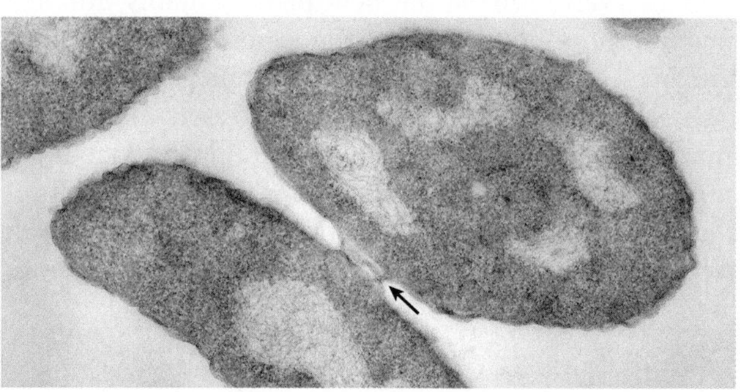

### Keynote

Some *E. coli* bacteria possess a plasmid, the *F* factor, that is required for mating. *E. coli* cells containing the *F* factor are designated $F^+$ and those without it are $F^-$. The $F^+$ cells (donors) can mate with $F^-$ cells (recipients) in a process called conjugation, which leads to the one-way transfer of a copy of the *F* factor from donor to recipient during replication of the *F* factor. As a result, both donor and recipient are $F^+$. None of the bacterial chromosome is transferred during $F^+ \times F^-$ conjugation.

**Figure 15.5**

**Transfer of genetic material during conjugation in *E. coli*.** (a) Transfer of the *F* factor from donor to recipient cell during $F^{+} \times F^{-}$ matings. (b) Production of *Hfr* strain by integration of *F* factor and transfer of bacterial genes from donor to recipient cell during $Hfr \times F^{-}$ matings.

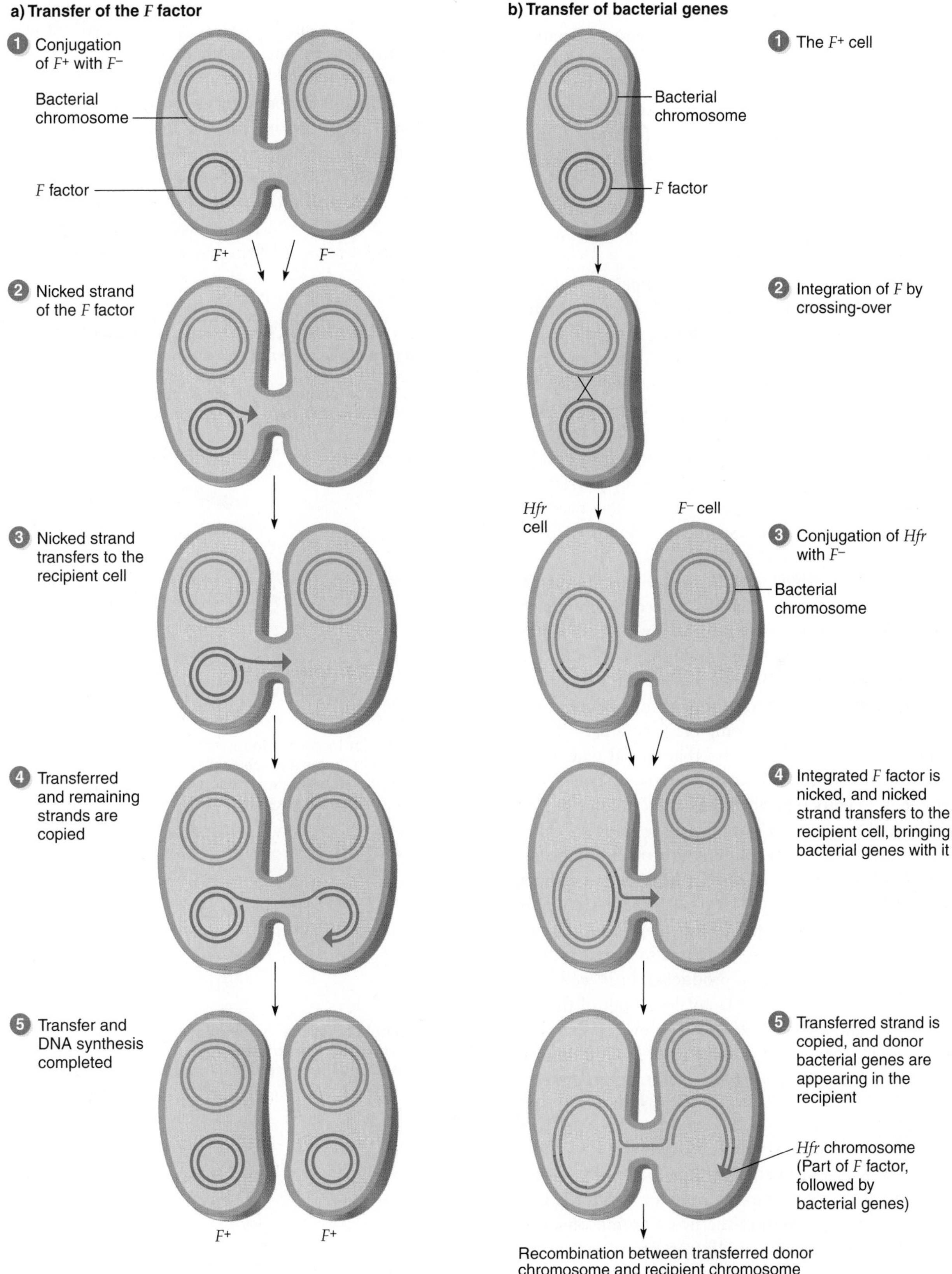
a) Transfer of the F factor
1 Conjugation of F+ with F−
Bacterial chromosome
F factor
F+
F−
2 Nicked strand of the F factor
3 Nicked strand transfers to the recipient cell
4 Transferred and remaining strands are copied
5 Transfer and DNA synthesis completed
F+
F+
b) Transfer of bacterial genes
1 The F+ cell
Bacterial chromosome
F factor
2 Integration of F by crossing-over
Hfr cell
F− cell
3 Conjugation of Hfr with F−
Bacterial chromosome
4 Integrated F factor is nicked, and nicked strand transfers to the recipient cell, bringing bacterial genes with it
5 Transferred strand is copied, and donor bacterial genes are appearing in the recipient
Hfr chromosome (Part of F factor, followed by bacterial genes)
Recombination between transferred donor chromosome and recipient chromosome

### High-Frequency Recombination Strains of *E. coli*

Producing recombinants for chromosomal genes by conjugation involves special derivatives of $F^+$ strains, called ***Hfr* (high-frequency recombination)** strains. Discovered separately by William Hayes and Luca Cavalli-Sforza, *Hfr* strains originate by a rare crossover event in which the F factor integrates into the bacterial chromosome (Figure 15.5b, parts 1–2). Plasmids such as F that are also capable of integrating into the bacterial chromosome are called **episomes.** When the F factor is integrated, it no longer replicates independently but is replicated as part of the host chromosome.

Because of the F factor genes, *Hfr* cells can conjugate with $F^-$ cells (Figure 15.5b, part 3). When such mating happens, events similar to those in the $F^+ \times F^-$ mating occur. The integrated F factor becomes nicked at the origin, and replication begins (Figure 15.5b, part 4). During replication, part of the F factor starting with the origin moves into the recipient cell, where the transferred strand is copied. In a short time, the donor bacterial chromosome begins to transfer into the recipient. If there are allelic differences between donor genes and recipient genes, recombinants can be isolated (Figure 15.5b, part 5). The recombinants are produced by double crossovers between the linear donor DNA and the circular recipient chromosome, which switch a segment of donor DNA for the homologous segment of recipient DNA.

In *Hfr* $\times$ $F^-$ matings, the $F^-$ cell almost never acquires the *Hfr* phenotype. That is, to become *Hfr*, the recipient cell must receive a complete copy of the F factor. However, only part of the F factor is transferred at the beginning of conjugation; the rest lies at the end of the donor chromosome. All of the donor chromosome would have to be transferred for a complete functional F factor to be found in the recipient, and that would take about 100 minutes at 37°C. This is an extremely rare event, because mating pairs typically break apart long before the second part of the F factor is transferred.

The low-frequency recombination of chromosomal gene markers in $F^+ \times F^-$ crosses can be understood when we consider that only about 1 in 10,000 $F^+$ cells in a population become *Hfr* cells by F factor integration. The reverse process, excision of the F factor, also occurs spontaneously and at low frequency, producing an $F^+$ cell from an *Hfr* cell. In excision, the F factor loops out of the *Hfr* chromosome, and by a single crossing-over event (just like the integration event), a circular host chromosome and a circular extrachromosomal F factor are generated.

### *F′* Factors

Occasionally, excision of the F factor from the chromosome of an *Hfr* cell is not precise, and an F factor is produced with a small section of the host chromosome that was adjacent to the integrated F factor. Because the F factor integrates at one of many sites on the chromosome, many different host chromosome segments can be picked up in this way. Consider an *E. coli* strain in which the F factor has integrated next to the $lac^+$ region, a set of genes required for the breakdown of lactose (Figure 15.6a). If the looping out is not precise, then the adjacent $lac^+$ host chromosomal genes may be included in the loop (Figure 15.6b). Then, by a single crossover, the looped-out DNA is separated from the host chromosome (Figure 15.6c) to produce an F factor carrying the $lac^+$ genes of the host. F factors containing bacterial genes are called *F′* (F prime) factors, and they are named for the genes they have picked up. An *F′* with the *lac* genes is called *F′* (*lac*).

Cells with *F′* factors can conjugate with $F^-$ cells. As in $F^+ \times F^-$ conjugation, a copy of the *F′* factor is transferred to the $F^-$ cell, which then becomes *F′*. The recipient also receives a copy of the bacterial gene(s) on the *F′*

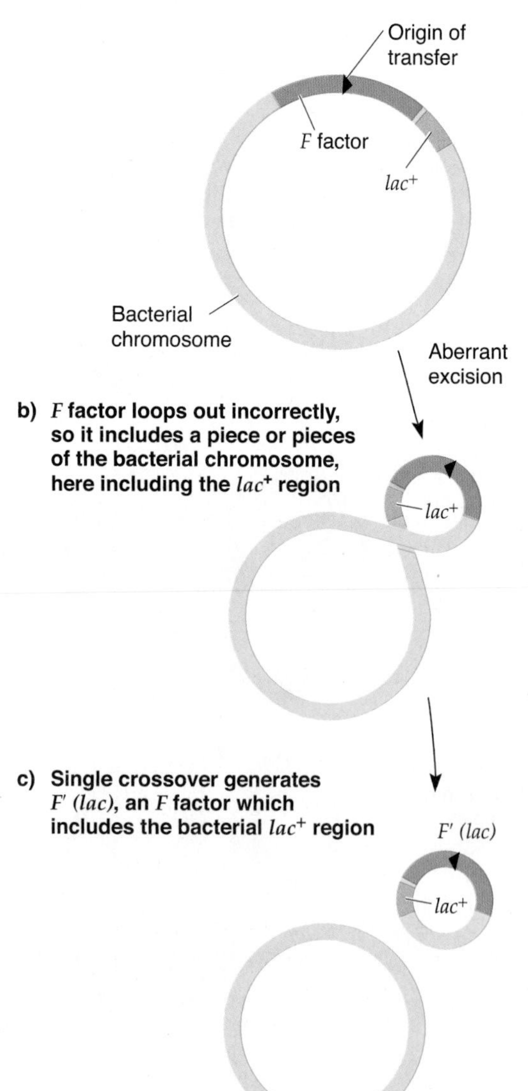
a) F factor integrated into the bacterial chromosome, here adjacent to the lac+ region
Origin of transfer
F factor
lac+
Bacterial chromosome
Aberrant excision
b) F factor loops out incorrectly, so it includes a piece or pieces of the bacterial chromosome, here including the lac+ region
lac+
c) Single crossover generates F′ (lac), an F factor which includes the bacterial lac+ region
F′ (lac)
lac+

**Figure 15.6**

**Production of an *F′* factor, here *F′* (*lac*), an F factor including the $lac^+$ region of the bacterial chromosome.**

factor (*lac* in our example). Since the recipient has its own copy of that DNA, the resulting cell line is partially diploid (*merodiploid*), having two copies of one or a few genes and only one copy of all the others in the genome. This particular type of conjugation is called **F-duction,** or *sexduction,* and it provides a way to study particular genes in a diploid state in *E. coli*.

## Using Conjugation to Map Bacterial Genes

In the late 1950s, François Jacob and Elie Wollman studied the transfer of chromosomal genes from *Hfr* strains to $F^-$ cells that had allelic differences for a number of genes. Their experimental design involved making an $Hfr \times F^-$ mating and, at various times after conjugation began, using a kitchen blender to break apart the conjugating pairs and then analyzing the transconjugants for which donor genes they had received. This approach is called an *interrupted-mating experiment.*

The use of interrupted mating to map bacterial genes is illustrated by the following cross (Figure 15.7a):

Donor:

*HfrH* $thr^+$ $leu^+$ $azi^R$ $ton^R$ $lac^+$ $gal^+$ $str^S$

Recipient:

$F^-$ *thr leu* $azi^S$ $ton^S$ *lac gal* $str^R$

(The superscript S means "sensitive," and the superscript R means "resistant.")

The *HfrH* strain is prototrophic and is resistant to growth inhibition by the chemical sodium azide ($azi^R$), resistant to infection by the bacteriophage T1 ($ton^R$), and sensitive to the antibiotic streptomycin ($str^S$). The $F^-$ strain is auxotrophic for threonine (*thr*) and leucine (*leu*), sensitive to growth inhibition by the chemical sodium azide ($azi^S$), sensitive to infection by the bacteriophage T1 ($ton^S$), unable to ferment lactose (*lac*) or galactose (*gal*), and resistant to growth inhibition by the antibiotic streptomycin ($str^R$).

In a conjugation experiment, the two cell types are mixed together in a liquid medium at 37°C. Samples are removed from the mating mixture at various times and are then agitated to break the pairs apart. Through plating on selective agar media, recombinant recipients (the transconjugants) are then searched for and analyzed with respect to the time at which the first donor genes entered the recipient and produced recombinants.

For this particular cross, the medium contains streptomycin to kill the *HfrH* parent and lacks threonine and leucine, so that the $F^-$ parent cannot grow. The threonine ($thr^+$) and leucine ($leu^+$) genes are the first donor genes to be transferred to the $F^-$ to produce a merodiploid, so recombinants formed by the exchange of those genes with the *thr leu* genes of the $F^-$ recipient grow on the selective medium. Appropriate media can be used to test for the appearance of other donor genes ($azi^R$, $ton^R$, $lac^+$, and $gal^+$) among the selected $thr^+$ $leu^+$ $str^R$ transconjugants. For example, a medium with sodium azide added can test for the presence of $azi^R$ from the donor.

Figure 15.7b shows the results. The threonine ($thr^+$) and leucine ($leu^+$) genes are the first donor genes to be transferred to the $F^-$, at 8 minutes. (The two genes are inseparable timewise in a conjugation experiment because they are physically very close to one another.) Recombinants for the next gene to be transferred, $azi^R$, are seen at 9 minutes after the start of conjugation—that is, 1 minute after the $thr^+$ and $leu^+$ genes entered. Then $ton^R$ recombinants are seen at 10 minutes, followed by $lac^+$ recombinants at about 16 minutes and $gal^+$ recombinants at about 25 minutes. The maximum frequency of recombinants becomes smaller the later the gene enters the recipient because, with time, there is an increasing chance that mating pairs will break apart.

In this experiment, each gene from the *Hfr* bacterium appears in recombinants at a different, but reproducible, time after mating begins. Thus, from the time intervals for the experiment described, the map in Figure 15.7c may be constructed, with map units in minutes; the entire *E. coli* chromosome takes about 100 minutes to transfer.

Note that we analyzed the conjugation results while knowing the order of the genes, so that the principles of analysis could be made clear. For an actual experiment, the order of genes may not be known at the outset. In that case, the order is determined by screening for all possible recombinants and seeing the time at which each enters. The progressive times of entry indicate the order and the distance between genes in time units.

## Circularity of the *E. coli* Map

Only one F factor is integrated into each *Hfr* strain. Different *Hfr* strains have the F factor integrated into the chromosome at different locations and in different orientations. Therefore, *Hfr* strains differ with respect to where the transfer of donor genes begins and in what order donor genes transfer. Figure 15.8a shows the order of chromosomal gene transfer for four different *Hfr* strains: *H, 1, 2,* and *3*. In each case, only one *Hfr* strain was used to cross with the recipient, and the order of gene transfer and the time between the appearance of each gene in the recipient were determined. The genetic distance in time units between a particular pair of genes is constant, no matter which *Hfr* strain is used as donor; for example, the genetic distance between *thr* and *pro* is the same in *H, 1, 2,* and *3*. This sameness validates the use of time units as a measure of distance between genes in *E. coli*.

From the preceding sort of data, a map of the chromosome is constructed by aligning the genes transferred by each *Hfr* as shown in Figure 15.8b. As a result of the overlap of the genes, the simplest map that can be drawn from these data is a circular one, as shown in Figure 15.8c. The map is a composite of the results of the individual matings. The circularity of the map was itself a significant finding,

**Figure 15.7**

**Interrupted-mating experiment involving the cross *HfrH thr*⁺ *leu*⁺ *azi*ᴿ *ton*ᴿ *lac*⁺ *gal*⁺ *str*ˢ × F⁻ *thr leu azi*ˢ *ton*ˢ *lac gal str*ᴿ.** The progressive transfer of donor genes with time is illustrated. Recombinants are generated by an exchange of a donor fragment with the homologous recipient fragment resulting from a double crossover event. (a) At various times after mating commences, the conjugating pairs are broken apart and the transconjugant cells are plated on selective agar media to determine which genes have been transferred from the *Hfr* to the F⁻. (b) The graph shows the frequency (percentage) of *Hfr* genetic markers among *thr*⁺, *leu*⁺ recombinants and their time of appearance in the recipient. (c) Map of the genes based on the time of entry of the donor genes into the recipient during the experiment.

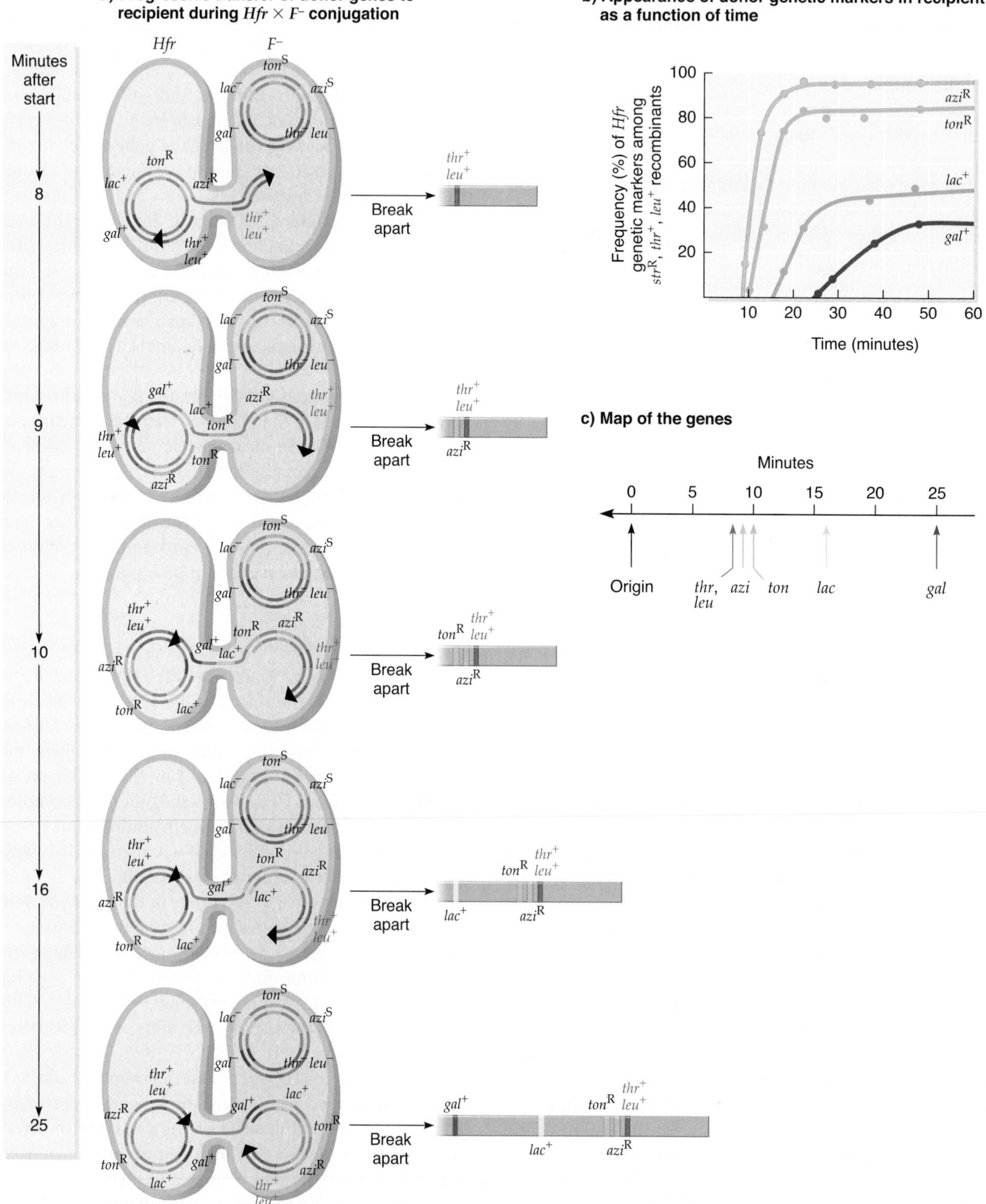

Figure 15.8

**Interrupted-mating experiments with a variety of *Hfr* strains, showing that the *E. coli* linkage map is circular.** (a) Orders of gene transfer for the *Hfr* strains *H*, *1*, *2*, and *3*; (b) Alignment of gene transfer for the *Hfr* strains. (c) Circular *E. coli* chromosome map derived from the *Hfr* gene transfer data. The map is a composite showing various locations of integrated F factors. A given *Hfr* strain has only one integrated F factor.

**a) Orders of gene transfer**

*Hfr* strains:

*H* *origin–thr–pro–lac–pur–gal*
*1* *origin–thr–thi–gly–his*
*2* *origin–his–gly–thi–thr–pro–lac*
*3* *origin–gly–his–gal–pur–lac–pro*

**b) Alignment of gene transfer for the *Hfr* strains**

*H* *thr–pro–lac–pur–gal*
*1* *his–gly–thi–thr*
*2* *his–gly–thi–thr–pro–lac*
*3* *pro–lac–pur–gal–his–gly*

**c) Circular *E. coli* chromosome map derived from *Hfr* gene transfer data**

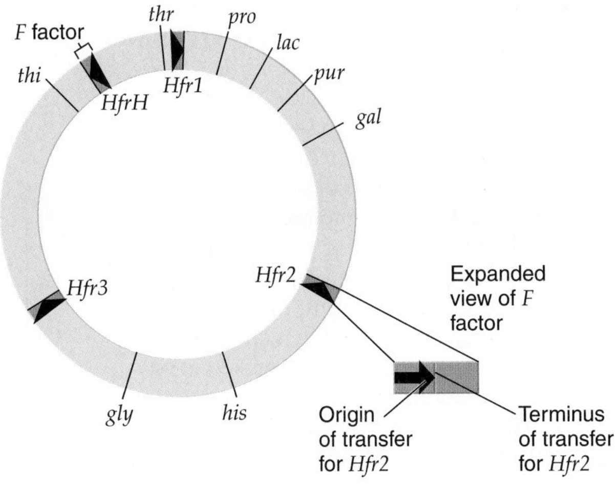


because all previous genetic maps of eukaryotic chromosomes were linear.

A complete genetic map of the *E. coli* chromosome eventually was constructed from conjugation experiments; the map is 100 minutes long. Like genetic maps of other organisms, this one provides information about the relative locations of *E. coli* genes on the circular chromosome. In 1997, the ultimate genetic map of *E. coli* was completed—that of the $4.6 \times 10^6$ base-pair (4.6-Mb) sequence of the bacterium's genome.

**Activity**

You are assisting Elie Wollman and François Jacob as they construct a genetic map of *E. coli*, using the newly discovered interrupted-mating procedure, in the iActivity *Conjugation in E. coli* on the student website.

**Keynote**

The circular *F* factor can integrate into the circular bacterial chromosome by a single crossover event. Strains in which this integration has happened can conjugate with $F^-$ strains, and transfer of the bacterial chromosome occurs. The strains containing the integrated *F* factor are called *Hfr* (high-frequency recombination) strains. In *Hfr* × $F^-$ matings, the chromosome is transferred in a one-way fashion from the *Hfr* cell to the $F^-$ cell, beginning at a specific site called the origin (O). The farther a gene is from O, the later it is transferred to the $F^-$, and this temporal difference is the basis for mapping genes by their times of entry into the $F^-$ cell. Conjugation and interrupted mating allow mapping of the chromosome.

## Genetic Mapping in Bacteria by Transformation

**Transformation** is the unidirectional transfer of extracellular DNA into cells, resulting in a phenotypic change in the recipient. Bacterial transformation was first observed by Frederick Griffith in 1928, and in 1944 Oswald Avery and his colleagues showed that DNA was responsible for the genetic change that was observed (see Chapter 2). Bacterial transformation has been used to map the genes of certain bacterial species in which mapping by other methods (conjugation or transduction) was not possible. In mapping experiments using transformation, DNA from a donor bacterial strain is extracted, purified, and broken into small fragments. This DNA is then added to recipient bacteria with a different genotype. If the donor DNA is taken up by a recipient cell and recombines with the homologous parts of the recipient's chromosome, a recombinant chromosome is produced. Recipients whose phenotypes are changed by transformation are called **transformants.**

Bacterial species vary in their ability to take up DNA. To enhance the efficiency of transformation, cells typically are treated chemically or are exposed to a strong electric field in a process called *electroporation,* making the cell membrane more permeable to DNA. Cells prepared to take up DNA by transformation are called *competent cells.*

There are two types of bacterial transformation. In *natural transformation,* bacteria are naturally able to take up DNA and be transformed genetically by it. In *engineered transformation,* bacteria are altered to enable them to take up and be transformed genetically by added DNA. *Bacillus subtilis,* a cylindrical, spore-forming bacterium about about 3–8 μm long and 1–1.5 μm wide, exemplifies bacteria amenable to natural transformation. *E. coli* exemplifies bacteria responsive to engineered transformation. This chapter's Focus on Genomics box describes the transformation of a complete genome into a cell, as well as the synthesis of an entire chromosome.

Only a small proportion of the cells involved in transformation actually take up DNA. Consider an example of

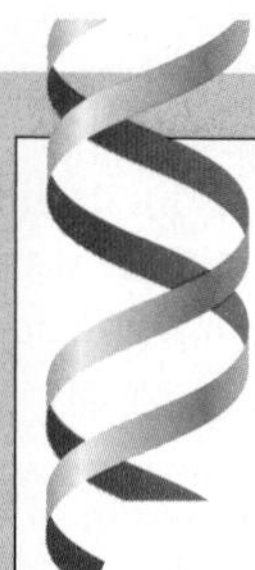

## Focus on Genomics

### Artificial Life: Artificial Genomes and Genome Transfer

It seems likely that within our lifetimes, scientists will engineer cells with custom genomes. These genomes will enable the cells to carry out useful new biochemical pathways, for instance, the synthesis of methane or ethanol from cellulose. Scientists have successfully crossed two major technical hurdles to the creation of artificial life. One group has successfully synthesized an entire genome, and another has transplanted the bacterial genome of one species into a cell from another.

Chemical synthesis of small pieces of DNA is simple, but the synthesis of an entire genome as a single piece of DNA is currently impossible. However, an experiment with *Mycoplasma genitalium*, which has the smallest known genome among organisms that can be grown in the lab without a host, may anticipate future advances. Investigators first synthesized 101 small cassettes that contained 5,000–7,000 bp each. Most of these cassettes were identical to the corresponding genomic regions of *M. genitalium*, but a few carried some changes. The investigators deleted a gene coding for a protein required for parasitic interactions and inserted a few unique sequences in non-protein-coding regions to "tag" the artificial genome. The cassettes overlapped their neighbors by 80 to 360 base pairs. The investigators connected the cassettes using these overlap regions to create four overlapping "quarter-genome" assemblies, each part of a yeast artificial chromosome. All four quarter-genomes were introduced into a single yeast host cell. The recombination machinery of the yeast host cell recombined the quarter-genomes at the overlap regions, and assembled them into a novel circular *M. genitalium* genome. The investigators tested the genome and found that it was both complete and carried their modifications. Although they had hoped to transplant this genome into a bacterial host cell, as described below, the vector interrupted a gene thought to be required for viability. As a result, this particular reassembled genome might not be suitable for transplantation.

Other scientists transplanted a genome from one species to another. The donor organism was *Mycoplasma mycoides*, a bacterium that forms large colonies on plates and carries the *tetM* gene. *tetM* codes for a protein that confers tetracycline resistance on the cell. The tetracycline-sensitive host strain, *Mycoplasma capricolum,* forms small colonies and lacks the *tetM* gene. The scientists isolated intact, protein-free genomes from *M. mycoides* and transformed this genomic DNA into *M. capricolum* cells. After the transformation, some cells had both genomes, and others had only the host genome. The investigators grew the transformed cells in the presence of tetracycline. This antibiotic treatment killed all of the non-transformed *M. capricolum* cells. Under these selective conditions, the fastest-growing cells will have a single genome (two genomes take longer to replicate) and must have the *tetM* gene in this genome—perfect conditions to select for the loss of the host genome and the retention of the donor genome. The investigators observed that some cells grew very rapidly, forming tetracycline-resistant colonies. To prove that these cells contained only the *M. mycoides* genome and that the *M. capricolum* genome was no longer present, they cloned and sequenced several genes that differed between the two species. In each case, gene matched *M. mycoides*. Visualizations of the cells' proteins on gels established that the proteins in the cells were identical to the proteins of *M. mycoides* and unlike the proteins of *M. capricolum*. While the investigators could not completely rule out that some small pieces of the host genome might still be present, they were able to show that the donor genome was present and functional, and that the cells had become phenotypically similar to the donor species.

the transformation of *B. subtilis* (Figure 15.9). (Other systems may differ in the details of the process.) The donor double-stranded DNA fragment is wild type ($a^+$) for a mutant allele $a$ in the recipient cell (Figure 15.9a). The two DNA strands are shown in the figure and, because subsequent stages of transformation involve unusual strand pairing, *each DNA strand* is labeled with an allele.

During DNA uptake, one of the two DNA strands is degraded, so only one intact linear DNA strand is left inside the cell (Figure 15.9b). This single linear strand pairs with the homologous DNA of the recipient cell's circular chromosome to form a triple-stranded region (Figure 15.9c). Recombination then occurs by a double crossover event involving the single-stranded DNA strand of the donor and the double-stranded DNA of the recipient (Figure 15.9d). The result is a recombinant recipient chromosome: In the region between the two crossovers, one DNA strand has the donor $a^+$ DNA segment, and the other strand has the recipient $a$ DNA segment. In other words, in that region, *the two DNA strands*

Figure 15.9

**Natural transformation in *Bacillus subtilis*.**

a) Linear donor double-stranded DNA fragment carries the $a^+$ allele, and the recipient bacterium carries the $a$ allele.

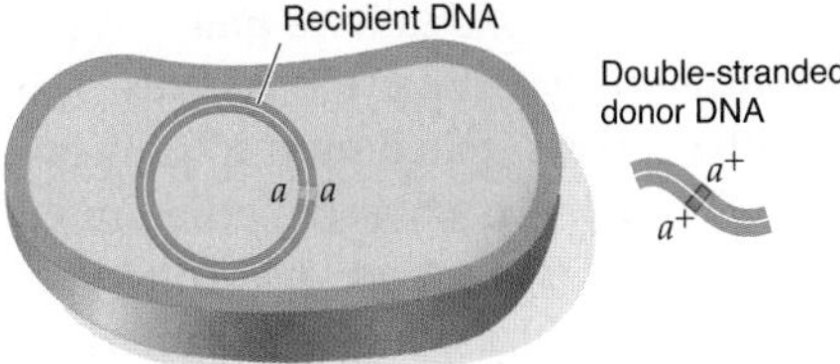


b) One strand of donor DNA enters cell as the other is degraded.

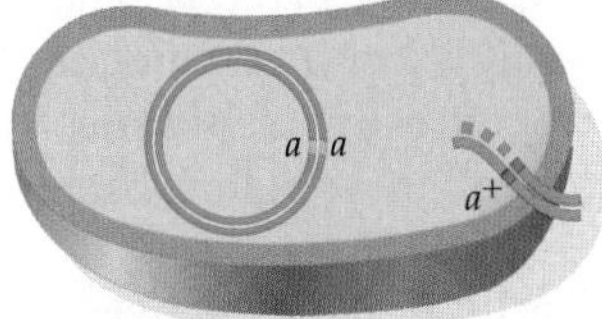


c) The single linear DNA strand pairs with the homologous region of the recipient's chromosome, forming a triple-stranded structure.

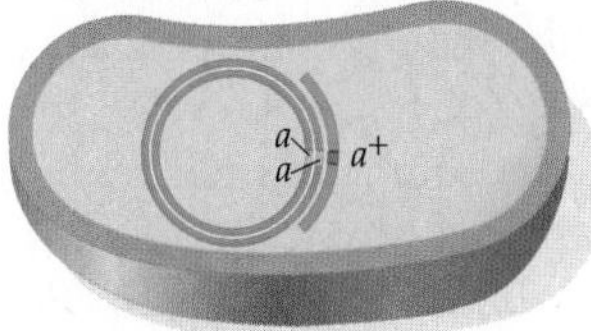


d) A double crossover produces a recombinant recipient chromosome with an $a^+$ allele on one strand and an $a$ allele on the other strand. The other product of recombination is a linear DNA strand with an $a$ allele.

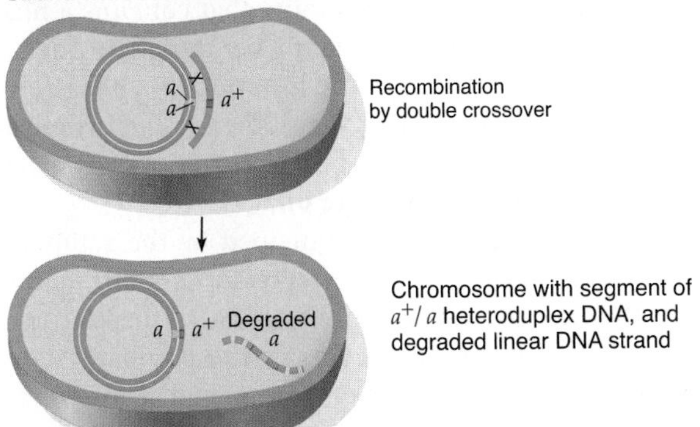


e) Replication of recipient chromosome.

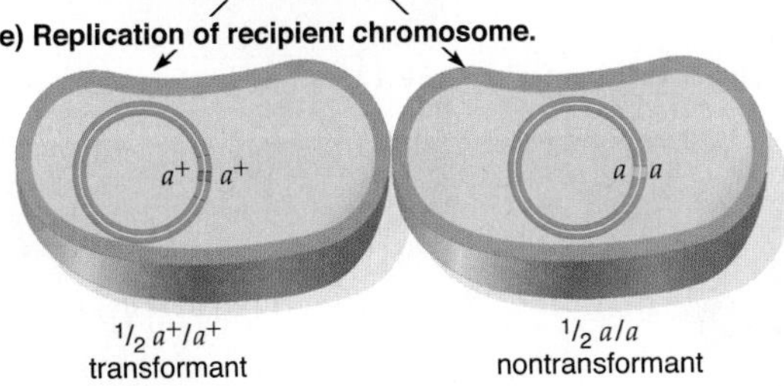


*are part donor, part recipient, for the genetic information.* A region of DNA with different sequence information on the two strands is called **heteroduplex DNA.** The other product of the double crossover event, a single-stranded piece of DNA carrying an $a$ DNA segment, is degraded.

After replication of the recipient chromosome, one progeny chromosome has donor genetic information on both DNA strands and is an $a^+$ transformant (Figure 15.9e). The other progeny chromosome has recipient genetic information on both DNA strands and is an $a$ nontransformant. Equal numbers of $a^+$ transformants and $a$ nontransformants are produced. Given highly competent recipient cells, the transformation of most genes occurs at a frequency of about 1 cell in every $10^3$ cells.

Transformation can be used to determine whether genes are linked (in this case, meaning physically close to one another on the single bacterial chromosome), to determine the order of genes on the genetic map, and to determine map distance between genes. The principles of determining whether two genes are linked are as follows: The efficient transformation of DNA involves fragments with a size sufficient to include only a few genes. If two genes, $x^+$ and $y^+$, are far apart on the donor chromosome, they will always be found on different DNA fragments. Thus, given an $x^+ \ y^+$ donor and an $x \ y$ recipient, the probability of simultaneous transformation (cotransformation) of the recipient to $x^+ \ y^+$ (from the product rule) is the product of the probability of transformation with each gene alone. If transformation occurred at a frequency of 1 in $10^3$ cells per gene, $x^+ \ y^+$ transformants would be expected to appear at a frequency of 1 in $10^6$ recipient cells ($10^{-3} \times 10^{-3}$). So if two genes are close enough that they often are carried on the same DNA fragment, the cotransformation frequency would be close to the frequency of transformation of a single gene. As determined experimentally, if the frequency of cotransformation of two genes is substantially higher than the products of the two individual transformation frequencies, the two genes must be close together.

Gene order can be determined from cotransformation data (Figure 15.10). If genes $p$ and $q$ are often transmitted to the recipient together, then these two genes

Figure 15.10

**Demonstration of the determination of gene order by cotransformation.**

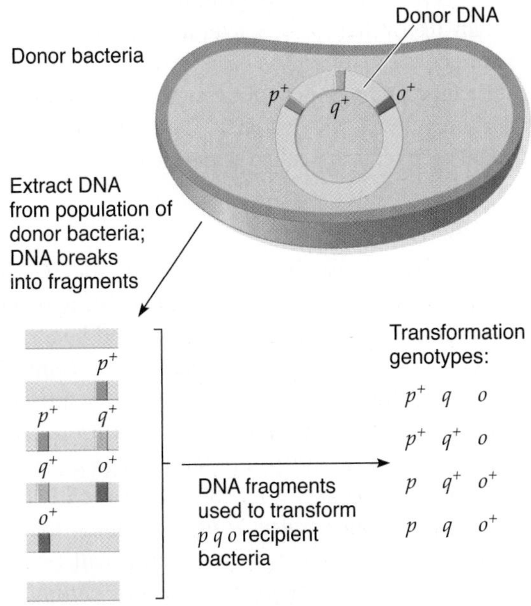

must be closely linked. Similarly, if genes *q* and *o* are often transmitted together, those two genes must be close to one another. To determine gene order, we now need information about genes *p* and *o*. Theoretically, there are two possible orders: *p-o-q* and *p-q-o*. If the order is *p-o-q*, then *p* and *o* should be cotransformed because they are more closely linked than *p* and *q*, whereas if the order is *p-q-o*, then *p* and *o* should be cotransformed rarely or not at all, because they are far apart. The data show no cotransformants for *p* and *o*, indicating that the gene order must be *p-q-o*.



**Keynote**

Transformation is the transfer of small, extracellular pieces of DNA between organisms. In transformation, DNA is extracted from a donor strain and added to recipient cells. A DNA fragment taken up by the recipient cell may associate with the homologous region of the recipient's chromosome. Part of the transforming DNA molecule can exchange with part of the recipient's chromosomal DNA. Frequent cotransformation of donor genes indicates close physical linkage of those genes. Cotransformants can be analyzed to determine gene order. Transformation has been used to construct genetic maps of bacterial species for which conjugation or transduction analyses are not possible.

## Genetic Mapping in Bacteria by Transduction

**Transduction** (literally, "leading across") is a process by which bacteriophages (bacterial viruses; phages, for short) transfer genes from one bacterium (the donor) to another (the recipient); such phages are called **phage vectors.** Since the amount of DNA a phage can carry is limited, the amount of genetic material that can be transferred usually is less than 1% of that in the bacterial chromosome. Once the donor genetic material has been introduced into the recipient, it may undergo genetic recombination with a homologous region of the recipient chromosome. The recombinant recipients are called **transductants.**

### Bacteriophages

Most bacterial strains can be infected by specific phages. A phage has a relatively simple structure consisting of a single chromosome of DNA or RNA surrounded by a coat of protein molecules. Variation in the number and organization of the proteins gives the phages their characteristic appearances. Phages T2 and T4 were introduced in Chapter 2 (pp. 12–13 and p. 21), and phage λ (lambda) was introduced in Chapter 3 (pp. 46–47). Phages T2 (Figure 2.4, p. 12) and T4 are *virulent phages*, meaning that they follow the *lytic cycle* when they infect *E. coli* (see Figure 2.5, p. 13). That is, the phage injects its chromosome into the cell, and phage genes take over the function of the cell. The activities of the phage gene products lead to the assembly of progeny phages that are released from the bacterium when the cell is broken open (lysed). The suspension of released progeny phages is called a **phage lysate.**

We can follow the phage lytic cycle visually. A mixture of phages and bacteria is plated on a solid medium. The concentration of bacteria is chosen so that an entire "lawn" of bacteria grows. Phages are present in much lower concentrations. Each phage infects a bacterium on the plate surface. Progeny phages released from the first infected bacterium infect neighboring bacteria, and the lytic cycle is repeated. The result is a cleared patch in the lawn of bacteria. The clearing is called a **plaque,** and each plaque derives from one of the original bacteriophages that was plated (Figure 15.11).

The λ life cycle (Figure 15.12) is more complex than that of a T2 phage. When phage λ DNA is injected into *E. coli*, the phage follows one of two alternative paths. One is a lytic cycle, exactly like that of the T phages. The other is the **lysogenic pathway** (or *lysogenic cycle*). In the lysogenic pathway, the λ chromosome does not replicate; instead, it inserts (integrates) itself physically into a specific region of the host cell's chromosome, much like *F* factor integration. In this integrated state, the phage chromosome is called a **prophage.** Every time the host cell chromosome replicates, the integrated λ chromosome replicates as part of it. A bacterium that contains a phage in the prophage state is said to be **lysogenic** for that phage; the phenomenon of the insertion of a phage chromosome into a bacterial chromosome is called **lysogeny.** Phages that have a choice between lytic and lysogenic pathways are called **temperate phages.**

The prophage state is maintained by the action of a specific phage gene product (a repressor protein) that prevents the expression of λ genes essential to the lytic cycle. When the repressor that maintains the prophage state is destroyed—for example, by environmental factors such as

**Figure 15.11**
**Plaques of the *E. coli* bacteriophage T2.**

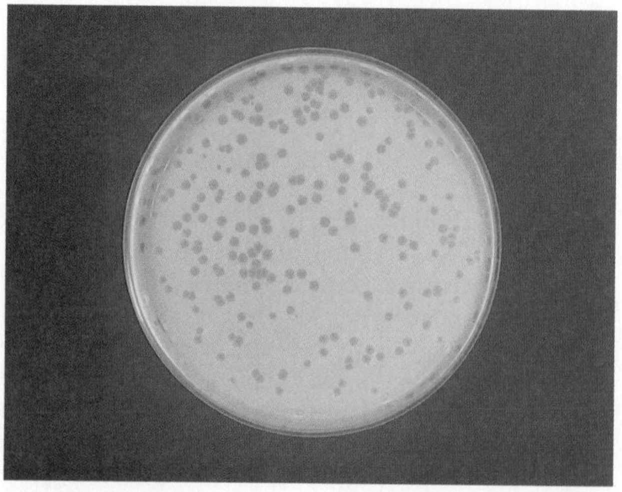

**Figure 15.12**
**Life cycle of the temperate phage λ.** When a temperate phage infects a cell, the phage may go through the lytic or lysogenic cycle.

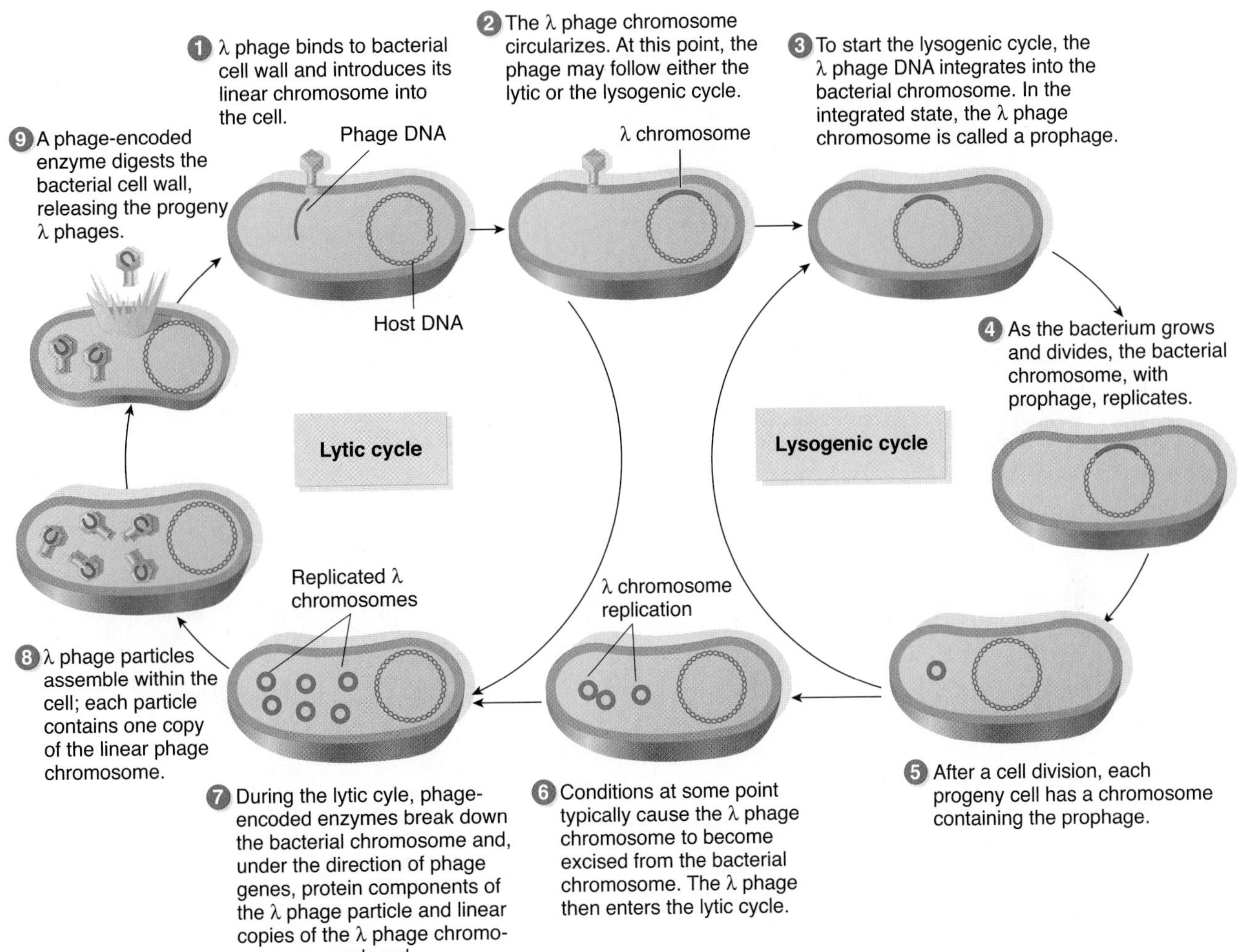
1 λ phage binds to bacterial cell wall and introduces its linear chromosome into the cell.
Phage DNA
Host DNA
2 The λ phage chromosome circularizes. At this point, the phage may follow either the lytic or the lysogenic cycle.
λ chromosome
3 To start the lysogenic cycle, the λ phage DNA integrates into the bacterial chromosome. In the integrated state, the λ phage chromosome is called a prophage.
4 As the bacterium grows and divides, the bacterial chromosome, with prophage, replicates.
5 After a cell division, each progeny cell has a chromosome containing the prophage.
6 Conditions at some point typically cause the λ phage chromosome to become excised from the bacterial chromosome. The λ phage then enters the lytic cycle.
λ chromosome replication
7 During the lytic cyle, phage-encoded enzymes break down the bacterial chromosome and, under the direction of phage genes, protein components of the λ phage particle and linear copies of the λ phage chromosome are produced.
Replicated λ chromosomes
8 λ phage particles assemble within the cell; each particle contains one copy of the linear phage chromosome.
9 A phage-encoded enzyme digests the bacterial cell wall, releasing the progeny λ phages.
Lytic cycle
Lysogenic cycle

ultraviolet light irradiation—the lytic cycle is induced. Upon induction, the integrated λ chromosome excises from the bacterial chromosome and the lytic cycle begins, resulting in the production and release of progeny λ phages from the cell.

## Transduction Mapping of Bacterial Chromosomes

Transduction may be used to map bacterial genes. Two types of transduction occur: in **generalized transduction,** any gene can be transferred between bacteria; in **specialized transduction,** only specific genes are transferred.

**Generalized Transduction.** Joshua Lederberg and Norton Zinder discovered transduction in 1952. These researchers tested whether conjugation occurred in the bacterial species *Salmonella typhimurium*. Their experiment was similar to the one showing that conjugation existed in *E. coli*. They mixed together two multiple auxotrophic strains, *phe*$^+$ *trp*$^+$ *tyr*$^+$ *met his* (required methionine and histidine) and *phe trp tyr met*$^+$ *his*$^+$ (required phenylalanine, tryptophan, and tyrosine) and found prototrophic recombinants—*phe*$^+$ *trp*$^+$ *tyr*$^+$ *met*$^+$ *his*$^+$—at a low frequency. However, unlike what happened in the conjugation experiment, when they used the U-tube apparatus (see Figure 15.3), they still found prototrophs. This result indicated that recombinants were being produced by a mechanism that did not require cell-to-cell contact. The interpretation was that the agent responsible for the formation of recombinants was a *filterable agent*, because it could pass through a filter with pores small enough to block bacteria. In this particular case, the filterable agent was identified as the temperate phage P22.

As an example, Figure 15.13, shows the mechanism for the generalized transduction of *E. coli* by temperate phage P1. Normally, the P1 phage enters the lysogenic state when it infects *E. coli* (Figure 15.13, part 1). If the lysogenic state is not maintained, the phage goes through the lytic cycle and produces progeny phages (Figure 15.13, part 2). During the lytic cycle, the bacterial DNA is degraded and, rarely, a piece of bacterial DNA is packaged into a phage head instead of phage DNA (Figure 15.13, part 3).

**Figure 15.13**
**Generalized transduction between strains of *E. coli*.**

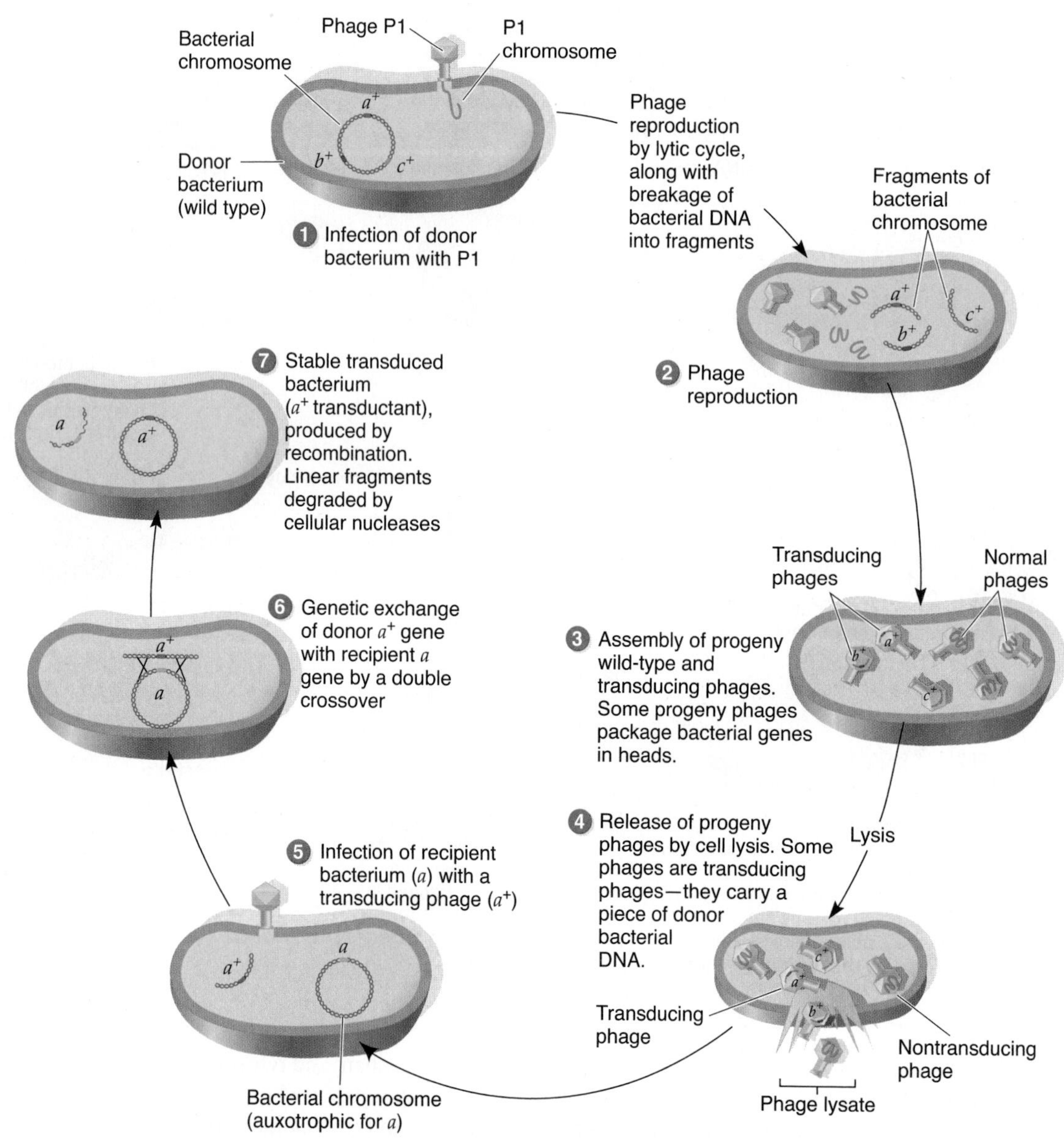


These phages are called **transducing phages** because they are the vehicles by which genetic material is carried between bacteria. In the example shown in the figure, the transducing phages are those carrying the donor bacterial genes genes $a^+$, $b^+$, or $c^+$. The population of phages in the phage lysate (Figure 15.13, part 4), consisting mostly of normal phages, but with about 1 in $10^5$ transducing phages present, can now be used to infect a new population of bacteria (Figure 15.13, part 5). The recipient bacteria are *a* in genotype. If a transducing phage carrying the $a^+$ gene infects the recipient, genetic exchange of the donor $a^+$ gene with the recipient *a* gene can occur by a double crossover (Figure 15.13, part 6). The result is a stable transduced bacterium called a *transductant*—in this case, an $a^+$ transductant.

Typically, a transduction experiment is designed so that the donor cell type and the recipient cell type have different genetic markers; then the transduction events can be followed. For instance, if the donor cell is $thr^+$ and the recipient cell is *thr*, prototrophic transductants can be detected because the cell no longer requires threonine to grow. In this way, researchers can pick out the extremely low number of transductants from among all the cells present by selecting for those cells that are able to do something the nontransduced cells cannot, namely, grow on a minimal medium. In such case, $thr^+$ is called a *selected marker*. Other markers in the experiment are termed *unselected markers*.

The process just described is called *generalized transduction*; the piece of bacterial DNA that the phage erroneously picks up is a *random* piece of the fragmented bacterial chromosome. Thus, any genes can be transduced; only an appropriate phage and bacterial strains carrying different genetic markers are needed.

Gene order and map distances between cotransduced genes can be determined by generalized transduction, and it is also by this procedure that fine-structure (i.e., detailed) linkage maps of bacterial chromosomes have been constructed. The logic is identical to that for mapping genes by transformation. For example, consider the mapping of some *E. coli* genes by using transduction with the temperate phage P1. The donor *E. coli* strain is $leu^+$ $thr^+$ $azi^R$ (able to grow on a minimal medium and resistant to the metabolic poison sodium azide). The recipient cell is *leu thr* $azi^S$ (requires leucine and threonine as supplements in the medium and is sensitive to sodium azide). The P1 phages are grown on the bacterial donor cells, and the phage lysate is used to infect the recipient bacterial cells. Transductants are selected for any one of the donor markers and are then analyzed for the presence of the other unselected markers (in this case, two). Typical data from such an experiment are shown in Table 15.1.

Consider the $leu^+$ selected transductants. We look to see if other donor markers are also present—that is, whether they have been *cotransduced* with the selected marker. Such *cotransductants* can occur in one of two possible ways: (1) if two genes are close enough so that they can be packaged physically into a phage head and be injected into a cell by a single phage; or (2) if two genes are not closely linked and are introduced into the same bacterium by simultaneous infection with two different phages. The transduction of two genes into a single bacterium by two phages is rare. Therefore, if two genes are close enough that they often are packaged into the phage head on the same DNA fragment, the **cotransduction** frequency would be close to the frequency of transduction of a single gene; cotransduction of two or more genes is a good indication that the genes are closely linked. Of the $leu^+$ transductants, 50% were $azi^R$ and 2% were $thr^+$. This means that the *leu* and *azi* genes often are cotransduced on the same DNA molecule; much less frequently, the *thr* gene is on the same transducing DNA with the *leu* gene. For the $thr^+$ transductants, 3% are $leu^+$ and 0 percent are $azi^R$. This distribution confirms that the *thr* and *leu* genes can be on the same transducing DNA and also indicates that the *azi* gene is distant enough never to be included on the same DNA. Taken together, these two sets of results tell us that the *leu* gene is closer to the *thr* gene than is the *azi* gene and that the *leu* and *azi* genes are closer together than are the *leu* and *thr* genes. The order of genes and rough map must then be as follows:

```
thr                          leu   azi
 |                            |     |
______________________________________
```

**Table 15.1 Transduction Data for Deducing Gene Order**

| Selected Marker | Unselected Markers |
|---|---|
| $leu^+$ | 50% = $azi^R$ |
| | 2% = $thr^+$ |
| $thr^+$ | 3% = $leu^+$ |
| | 0% = $azi^R$ |

The transductants are produced by crossing-over between the piece of donor bacterial chromosome brought in by the infecting phage and the homologous region on the recipient bacterial chromosome. The infected donor DNA finds the region of the recipient chromosome to which it is homologous, and the exchange of parts is accomplished by double (or some other even-numbered) crossovers (see Figure 15.13).

Map distance can be obtained from transduction experiments involving two or more genes. As before, transductants for one of two or more donor markers are selected, and these transductants are then analyzed for the presence or absence of other donor markers. For example, transduction from an $a^+$ $b^+$ donor to an $a$ $b$ recipient produces various transductants for $a^+$ and $b^+$, namely $a^+ b$, $a\,b^+$, and $a^+ b^+$. If we select for one or other donor markers, we can determine linkage information for the two genes. If we select for $a^+$ transductants, map distance between genes $a$ and $b$ is given by

$$\frac{\text{number of single-gene transductants}}{\text{number of total transductants}} \times 100\%$$

$$= \frac{(a^+ b)}{(a^+ b) + (a^+ b^+)} \times 100\%$$

If we select for $b^+$ transductants, map distance between $a$ and $b$ is given by

$$\frac{(a\, b^+)}{(a\, b^+) + (a^+ b^+)} \times 100\%$$

This method of gene mapping produces map distances only if the genes involved are close enough on the chromosome so that they can be cotransduced. This is because, to be cotransduced, they must both be on a piece of DNA that is about the same size as the phage's genome.

**Specialized Transduction.** Some temperate bacteriophages can transduce only certain sections of the bacterial chromosome, in contrast to generalized transducing phages, which can carry any part of the bacterial chromosome. An example of such a **specialized transducing phage** is λ, which infects *E. coli*.

The life cycle of λ was described earlier (see Figure 15.12). In the lysogenic cycle cell, the λ genome integrates into the bacterial chromosome at a specific site between the *gal* region and the *bio* region, producing a *lysogen* (Figure 15.14a). That site on the *E. coli* chromosome is

**Figure 15.14**

**Specialized transduction by bacteriophage λ.**

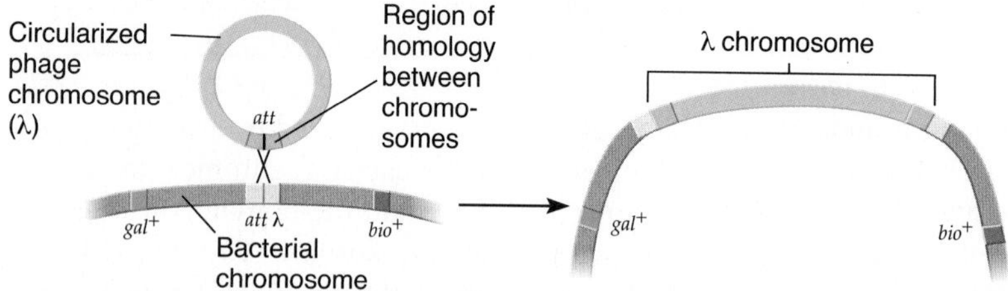
a) Production of lysogen by crossing-over between circular bacterial chromosome and circularized phage chromosome
Circularized phage chromosome (λ)
Region of homology between chromo-somes
att
gal+
att λ
bio+
Bacterial chromosome
λ chromosome
gal+
bio+

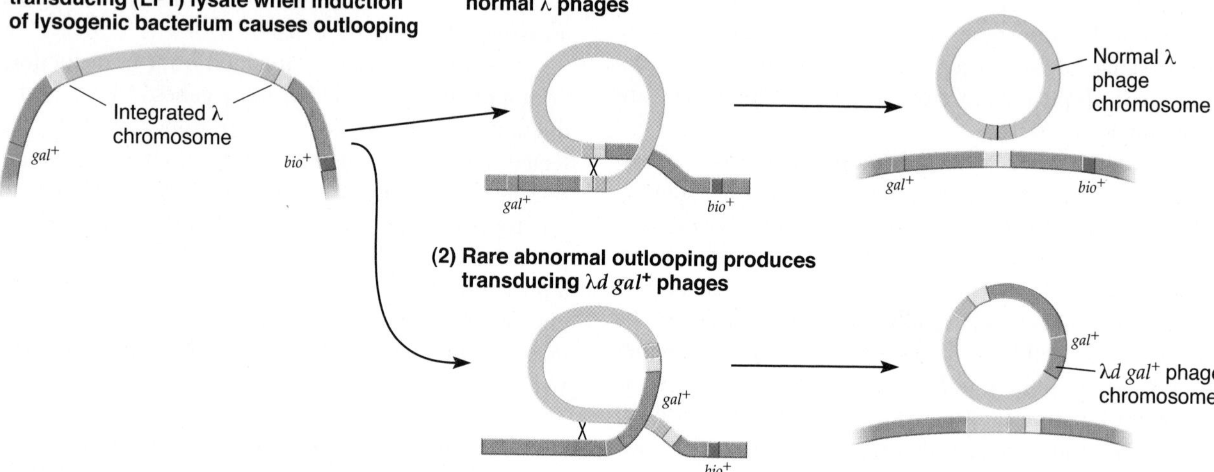
b) Production of initial low-frequency transducing (LFT) lysate when induction of lysogenic bacterium causes outlooping
Integrated λ chromosome
gal+
bio+
(1) Normal outlooping produces normal λ phages
gal+
bio+
Normal λ phage chromosome
gal+
bio+
(2) Rare abnormal outlooping produces transducing λd gal+ phages
gal+
bio+
gal+
λd gal+ phage chromosome

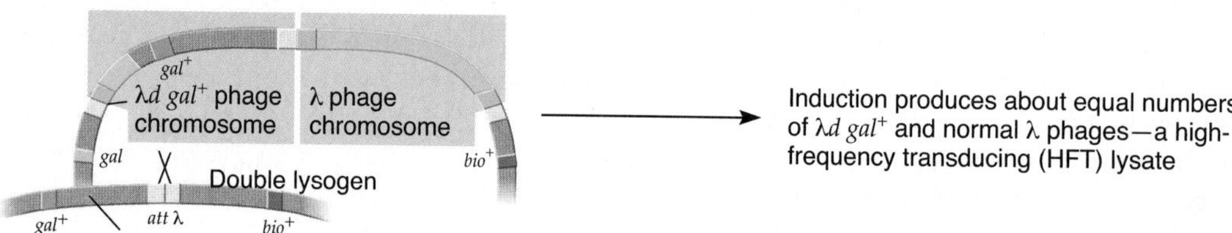
c) Transduction of gal bacteria by initial lysate, consisting of λ and λd gal+ phage
1) If both λ and λd gal+ integrate, an unstable transductant—a double lysogen—results
gal+
λd gal+ phage chromosome
λ phage chromosome
gal
bio+
Double lysogen
gal+
att λ
bio+
Induction produces about equal numbers of λd gal+ and normal λ phages—a high-frequency transducing (HFT) lysate

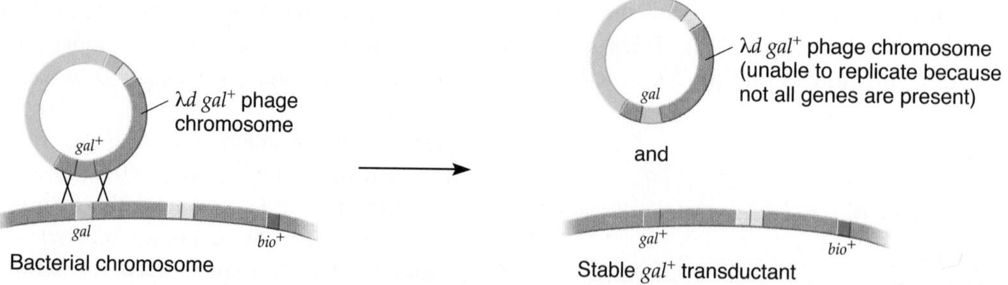
2) Stable transductant produced by recombination
λd gal+ phage chromosome
gal+
gal
bio+
Bacterial chromosome
λd gal+ phage chromosome (unable to replicate because not all genes are present)
gal
and
gal+
bio+
Stable gal+ transductant

called *att* λ (attachment site for lambda) and is homologous with a site called *att* in the λ DNA. By a single crossover, the λ chromosome integrates. In the integrated state, the phage, now called a *prophage*, is maintained by the action of a phage-encoded repressor protein.

The particular *E. coli* strain that λ lysogenizes is *E. coli K12*, and when it contains the λ prophage, it is called *E. coli K12*(λ). Let us focus just on the *gal* gene and assume that the particular *K12* strain that lysogenized is $gal^+$; that is, it can ferment galactose as a carbon source. This phenotype is readily detectable by plating the cells on a solid medium containing galactose as a carbon source, together with a dye that changes color in response to the products of galactose fermentation. On this medium, the $gal^+$ colonies are pink and the *gal* colonies are white. If we induce the prophage (see p. 441)—that is, reverse the inhibition of phage functions—the lytic cycle is initiated.

When the lytic cycle initiates, the phage chromosome loops out, generating a separate circular λ chromosome by a single crossover at the *att* λ/*att* sites (Figure 15.14b). In most cases, the phage chromosome excises precisely, producing the complete λ chromosome (Figure 15.14b, part 1). In rare cases, crossing-over occurs at sites other than the homologous recognition sites, giving rise to an abnormal circular DNA product (Figure 15.14b, part 2). In the case diagrammed, a piece of λ chromosome has been left in the bacterial chromosome, and a piece of bacterial chromosome which includes the $gal^+$ gene, has been added to the rest of the λ chromosome. Because a bacterial gene (or genes) is included in a progeny phage, we have a transducing phage—here, λ*d* $gal^+$. The *d* stands for "defective," because not all phage genes are present, and the *gal* indicates that the bacterial host cell *gal* gene has been acquired. This outcome of transduction is similar to *F*′ production by defective excision of the *F* factor. The λ*d* $gal^+$ can replicate and lyse the host cell in which it is produced, however, because all λ genes are still present: some are on the phage chromosome, and the others are in the bacterial chromosome.

The abnormal looping-out phenomenon is a rare event, so the phage lysate produced from the initial infection diagrammed contains mostly normal phages and a few $gal^+$ transducing phages ($1/10^5$). Due to the small proportion of transducing phages, the lysate is called a *low-frequency transducing* (*LFT*) lysate. Infection of *gal* bacterial cells with the LFT lysate produces two types of transductants (Figure 15.14c). In one type, the wild-type λ integrates at its normal *att* λ site and then the λ*d* $gal^+$ phage integrates by crossing-over within the common λ sequences to produce a double lysogen (Figure 15.14c1). In this case, both types of phages are integrated into the bacterial chromosome. As a result, the bacterium is heterozygous $gal^+$/*gal* and therefore can ferment galactose.

This type of transductant is unstable, because the lytic cycle can be initiated by induction. The wild-type λ has a complete set of genes for virus replication, so it controls the outlooping and replication of itself and the λ*d* $gal^+$. In this capacity, the wild-type λ phage acts as a *helper phage*. Since as many as one-half of the progeny phages could be λ*d* $gal^+$, this new lysate is called a *high-frequency transducing* (*HFT*) lysate.

The second type of transductant produced by the initial lysate is stable: It is produced when only a λ*d* $gal^+$ phage infects a cell (Figure 15.14c2). The $gal^+$ gene carried by the phage may be exchanged for the bacterial *gal* gene by a double crossover. Such a transductant is stable because the bacterial chromosome contains only one type of *gal* gene, and no phage genes are integrated.

Because of the mechanisms involved, specialized transduction can transduce only small segments of the bacterial chromosome that are on either side of the prophage. Specialized transduction is used to move *specific* genes between bacteria—for example, for constructing strains with particular genotypes. (A discussion of the process is beyond the scope of this text.)

**Keynote**

Transduction is the process by which bacteriophages mediate the transfer of genetic information from one bacterium (the donor) to another (the recipient). The capacity of the phage particle is limited, so the amount of DNA transferred usually is less than 1% of that in the bacterial chromosome. In generalized transduction, any bacterial gene can be incorporated accidentally into the transducing phage during the life cycle of the phage and subsequently transferred to a recipient bacterium. Specialized transduction is mediated by temperate phages (such as λ) in which prophages associate with only one site of the bacterial chromosome. In this case, the transducing phage is generated by abnormal excision of the prophage from the host chromosome, so the prophage includes both bacterial and phage genes. Transduction allows the fine-structure mapping of small chromosome segments.



## Mapping Bacteriophage Genes

The same principles used to map eukaryotic genes are used to map phage genes. Crosses are made between phage strains that differ in genetic markers, and the proportion of recombinants among the progeny is determined. The basic procedure for mapping phage genes in two-, three-, or four-gene crosses involves the mixed infection of bacteria with phages of different genotypes and the analysis of progeny—that is, the plaques.

To perform a genetic analysis of bacteriophages, we must have phage phenotypes to study. Several mutations affect the phage life cycle, giving rise to differences in the appearance of plaques on a bacterial lawn. For example, there are strains of T2 that differ in either plaque morphology (the size and shape of the edge of the plaque) or host range (which bacterial strain the phage can lyse). Consider two phage strains. One has the genotype $h^+$ *r*, meaning that it is wild type for the host range gene ($h^+$—able to lyse the *B* strain, but not the *B*/2 strain, of *E. coli*; that is, strain *B* is the *permissive host* and strain *B*/2 is the *nonpermissive host* for $h^+$ phages) and mutant for the plaque morphology gene (*r*—producing large plaques with distinct borders). The other phage strain has the genotype *h* $r^+$, meaning that it is mutant for the host range gene (able to lyse both the *B* and *B*/2 strains of *E. coli*) and wild type for the plaque morphology gene ($r^+$—producing small plaques with fuzzy borders). When plated on a lawn containing both the *B* and *B*/2 strains, any phage carrying the mutant host range allele *h* (infects both *B* and *B/2*) produces clear plaques, whereas phages carrying the wild-type $h^+$ allele produce cloudy plaques. The latter characteristic arises because phages bearing the $h^+$ allele can infect only the *B* bacteria, leaving a background cloudiness of uninfected *B/2* bacteria.

To map these two genes, we make a genetic cross by infecting *E. coli* strain *B* with the two (parental) phages $h^+\ r$ and $h\ r^+$ (Figure 15.15a). Once the two genomes are within the bacterial cell, each of them replicates (Figure 15.15b). If an $h^+\ r$ and an $h\ r^+$ chromosome come together, a crossover can occur between the two gene loci to produce $h^+\ r^+$ and $h\ r$ recombinant chromosomes (Figure 15.15c), which are assembled into progeny phages. When the bacterium lyses, the recombinant progeny are released into the medium, along with nonrecombinant (parental) phages (Figure 15.15d).

After the life cycle is completed, the progeny phages are plated onto a bacterial lawn containing a mixture of *E. coli* strains *B* and *B/2*. Four plaque phenotypes—two parental types and two recombinant types—are found from the experiment. The parental type $h\ r^+$ gives a small, clear plaque with a fuzzy border; the other parental $h^+\ r$ gives a large, cloudy plaque with a distinct border (Figure 15.16). The reciprocal recombinant types give recombined phenotypes: The $h^+\ r^+$ plaques are cloudy and small with a fuzzy border, and the $h\ r$ plaques are clear and large with a distinct border (see Figure 15.16.).

Once the progeny plaques are counted, we can find the recombination frequency between *h* and *r* from the formula

$$\frac{(h^+\ r^+) + (h\ r)\ \text{plaques}}{\text{total plaques}} \times 100$$

As with eukaryotes, this recombination frequency reflects the relative genetic distance between the phage genes. When the genes are close enough together so that multiple crossovers are not likely to occur, the recombination frequency equals the crossover frequency. In that case, the recombination frequency can be converted directly to map units.

**Figure 15.15**

**The principles of performing a genetic cross with bacteriophages.**

a) Coinfect bacteria with the two parental phages, $h^+\ r$ and $h\ r^+$

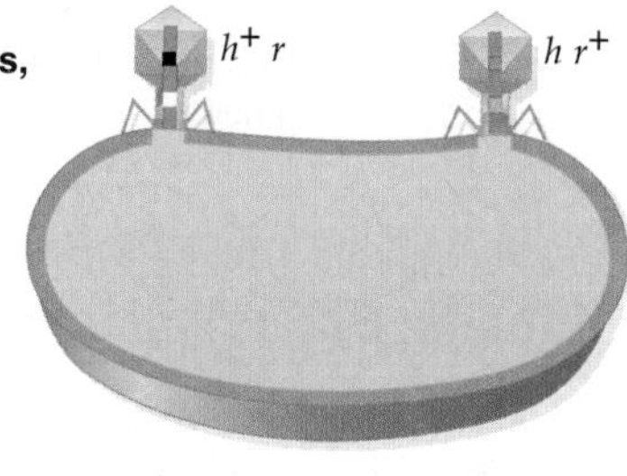


b) Replication of phage chromosomes in cell

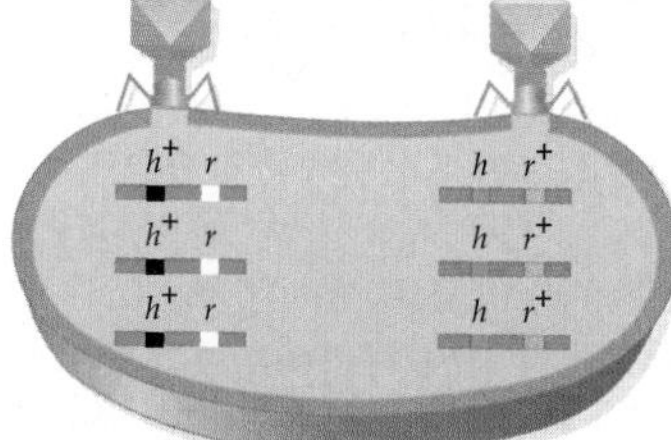


c) Recombination between some parental chromosomes resulting in $h^+\ r^+$ and $h\ r$ recombinants

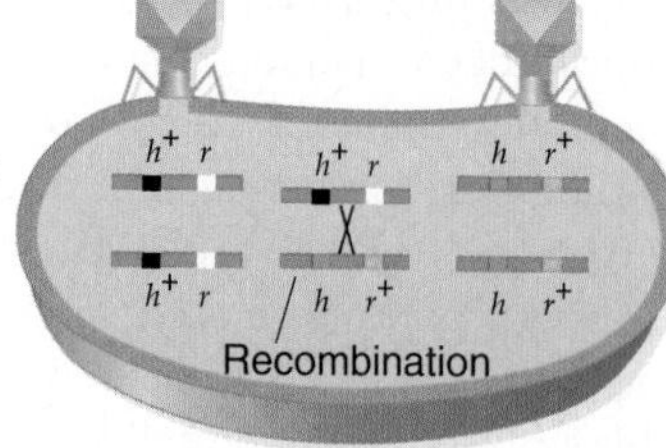


d) Phage assembly, bacterial lysis, and release of parental and recombinant progeny phages

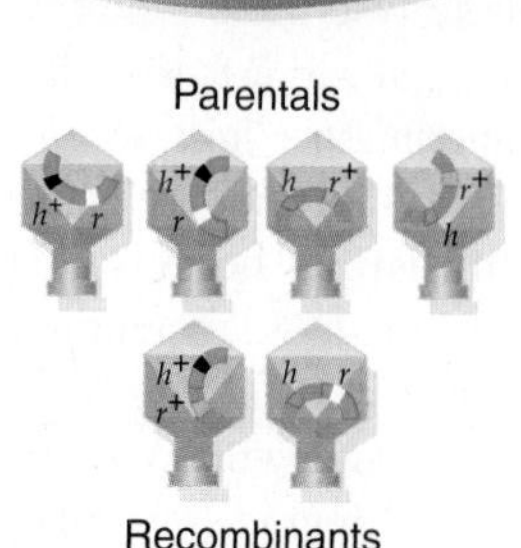


**Figure 15.16**

**Plaques produced by progeny of a cross of T2 strains $h\ r^+ \times h^+\ r$.** Four plaque phenotypes, representing both parental types and the two recombinants, can be discerned. The parental $h\ r^+$ phage produces a small, clear plaque with a fuzzy border; the other parental $h^+\ r$ phage produces a large, cloudy plaque with a distinct border. The recombinant $h^+\ r^+$ phage produces a small, cloudy plaque with a fuzzy border. The recombinant $h\ r$ phage produces a large, clear plaque with a distinct border.

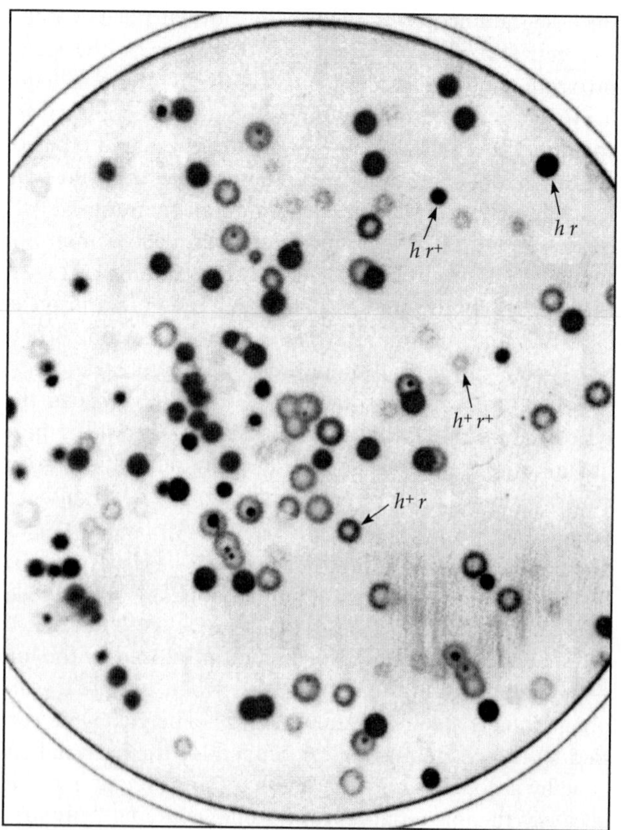

**Keynote**

The same principles used to map eukaryotic genes are used to map phage genes. That is, genetic material is exchanged between strains differing in genetic markers, and recombinants are detected and counted.

## Fine-Structure Analysis of a Bacteriophage Gene

The recombinational mapping of the distance between genes, called *intergenic mapping*, can be used to construct chromosome maps for both eukaryotic and prokaryotic organisms. Historically, the early picture of a gene was that it was like a bead on a string with mutation changing the bead from wild type to mutant or vice versa and with recombination occurring between the beads. We now know, of course, that the gene is subdivisible by mutation and recombination and that the same general principles of recombinational mapping can be applied to mapping the distance between mutational sites within the same gene. Recombinational mapping used to map distances between alleles within the same gene is called *intragenic mapping*.

The first evidence that the gene was subdivisible by mutation and recombination came from the work of C. P. Oliver in 1940. Oliver studied two mutations that were considered to be alleles of the X-linked *lozenge* (lz) locus of *Drosophila;* that is, females heterozygous for the two mutations showed the mutant lozenge-shaped eye phenotype. When female flies heterozygous for these two alleles were crossed with male flies hemizygous for either allele, progeny flies with wild-type eyes were seen with a frequency of about 0.2%. Oliver showed that these wild-type offspring had resulted from recombination between the alleles. In other words, he had shown that the gene was divisible by recombination, rather than being an indivisible "bead on a string." Using genetic symbols, we can represent the last cross as

$$\frac{lz^A \; +}{+ \; lz^B} \times \frac{lz^A \; +}{\longrightarrow}$$

where $lz^A$ and $lz^B$ are the two lozenge alleles. Recombination in the female between the two alleles produces ++ gametes and hence wild-type progeny.

Oliver's discovery spawned investigations of the detailed organization of alleles within a gene. As we now know, such intragenic mapping is possible because each gene consists of many nucleotide pairs of DNA, linearly arranged along the chromosome. The impetus to analyze the fine details of gene structure came largely from the elegantly detailed work by Seymour Benzer in the 1950s and 1960s with bacteriophage T4. His genetic experiments revealed much about the relationship between mapping and gene structure. His initial experiments involved **fine-structure mapping:** the detailed genetic mapping of sites within a gene.

Benzer used strains of phage T4 carrying mutations of the *rII* region. *rII* mutants have both a distinct *plaque morphology* phenotype and distinct *host range properties*. Specifically, when cells of *E. coli* growing on a solid medium are infected with wild-type ($r^+$) T4, small, turbid plaques with fuzzy edges are produced; in contrast, plaques produced by *rII* mutants are large and clear (Figure 15.17). Regarding host range properties, wild-type T4 can grow in and lyse cells of either *E. coli* strain *B* or *K12*(λ), whereas *rII* mutants can grow in *B* but not in *K12*(λ). That is, strain *B* is the permissive host for *rII* mutants, and strain *K12*(λ) is the nonpermissive host.

### Recombination Analysis of *rII* Mutants

Benzer realized that the growth defect of *rII* mutants on *E. coli K12*(λ) could serve as a powerful selective tool for detecting the presence of a very small proportion of $r^+$ phages within a large population of *rII* mutants. Initially, he set out to construct a fine-structure genetic map of the *rII* region. Using *E. coli B* as the permissive host, he crossed 60 independently isolated *rII* mutants in all possible combinations and then collected the progeny phages once the cells had lysed. For each cross *rIIx* × *rIIy*, where *x* and *y* are different mutations, there can be four types of progeny: two parental classes, *rIIx* and *rIIy;* and two recombinant classes, the double mutant *rIIx,y* and the $r^+$ wild type. Roughly equal numbers of the two parental classes will be produced, and roughly equal numbers of the two recombinant classes will be produced. The relative frequencies of the parentals and recombinants will depend on how far apart the two alleles are. For his analysis, Benzer plated a sample of the phage progeny on *E. coli B*, the permissive host. Then, from the number of plaques produced, the total number of progeny phage per milliliter was calculated. He plated another sample on *E. coli K12*(λ), the nonpermissive host, to find the frequency of $r^+$ recombinants. In this way, Benzer calculated the percentage of very rare $r^+$ recombinants produced by crossing-over

**Figure 15.17**

**The $r^+$ and mutant *rII* plaques on a lawn of *E. coli B*.** The $r^+$ plaque is turbid, with a fuzzy edge; the *rII* plaque is larger and clear and has a distinct boundary.

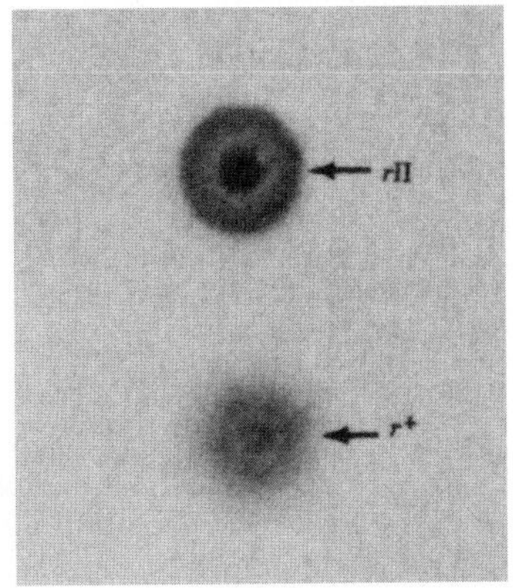

between closely linked alleles. The recombination frequency for the two alleles is given by the formula

$$\frac{2 \times \text{number of } r^+ \text{ recombinants}}{\text{total number of progeny}} \times 100\%$$

The number of $r^+$ recombinants is multiplied by 2 to account for the other class of recombinants—the double mutants—because they have the same phenotype as single mutants.

A control was set up for each cross. Each *rII* parent alone was used to infect the permissive *E. coli B* host, and the progenies were tested on plates of *B* and of *K12*(λ). That is, just as a mutation can generate an *rII* mutant from the $r^+$, a mutation can occur whereby an *rII* mutant changes back (reverts) to the $r^+$. Thus, it is extremely important to calculate the reversion frequencies for the two *rII* mutations in a cross and subtract the combined value from the computed recombination frequency. Fortunately, the reversion frequency for an *rII* mutation is at least an order of magnitude lower than the smallest recombination frequency that was found.

Benzer constructed a linear genetic map from the recombination data obtained from all possible pairwise crosses of the 60 *rII* mutants (Figure 15.18). Some pairs produced no $r^+$ recombinants when they were crossed, meaning that those pairs carried mutations at exactly the same site. Mutations that change the same nucleotide pair within a gene are called *homoallelic*. However, most pairs of *rII* mutants did produce $r^+$ recombinants when crossed, indicating that they carried different altered nucleotide pairs in the DNA. Mutations that change different nucleotide pairs within a gene are called *heteroallelic*. The map showed that the lowest frequency with which $r^+$ recombinants were formed in any pairwise crosses of *rII* mutants carrying heteroallelic mutations was 0.01%.

The minimum map distance of 0.01% can be used to make a rough calculation of the molecular distance—the distance in nucleotide pairs—between mutant markers. The genetic map of phage T4 is about 1,500 map units. If two *rII* mutants produce 0.01% $r^+$ recombinants, then the mutations are separated by 0.02 map unit, or by about $0.02/1{,}500 = 1.3 \times 10^{-5}$ of the total T4 genome. Since the total T4 genome contains about $2 \times 10^5$ nucleotide pairs (base pairs), the smallest recombination distance that was observed was $(1.3 \times 10^{-5}) \times (2 \times 10^5)$, or about 3 base pairs. This means that Benzer's data showed that genetic recombination can occur within distances on the order of 3 base pairs. Later experiments by others demonstrated that recombination can occur between mutations that affect adjacent base pairs in the DNA. That is,

**Figure 15.18**

**Preliminary fine-structure genetic map of the *rII* region of phage T4; map derived by Benzer from crosses of an initial set of 60 *rII* mutants.** Lower levels in the figure show finer detail of the map. In the lowest level, the numbered vertical lines indicate individual *rII* point mutants; the blue rectangles indicate the individual *rII* deletion mutants 47, 312, 295, 164, 196, 187, and 102; and the decimals indicate the percentage of $r^+$ recombinants found in crosses between the two *rII* mutants connected by an arrow.

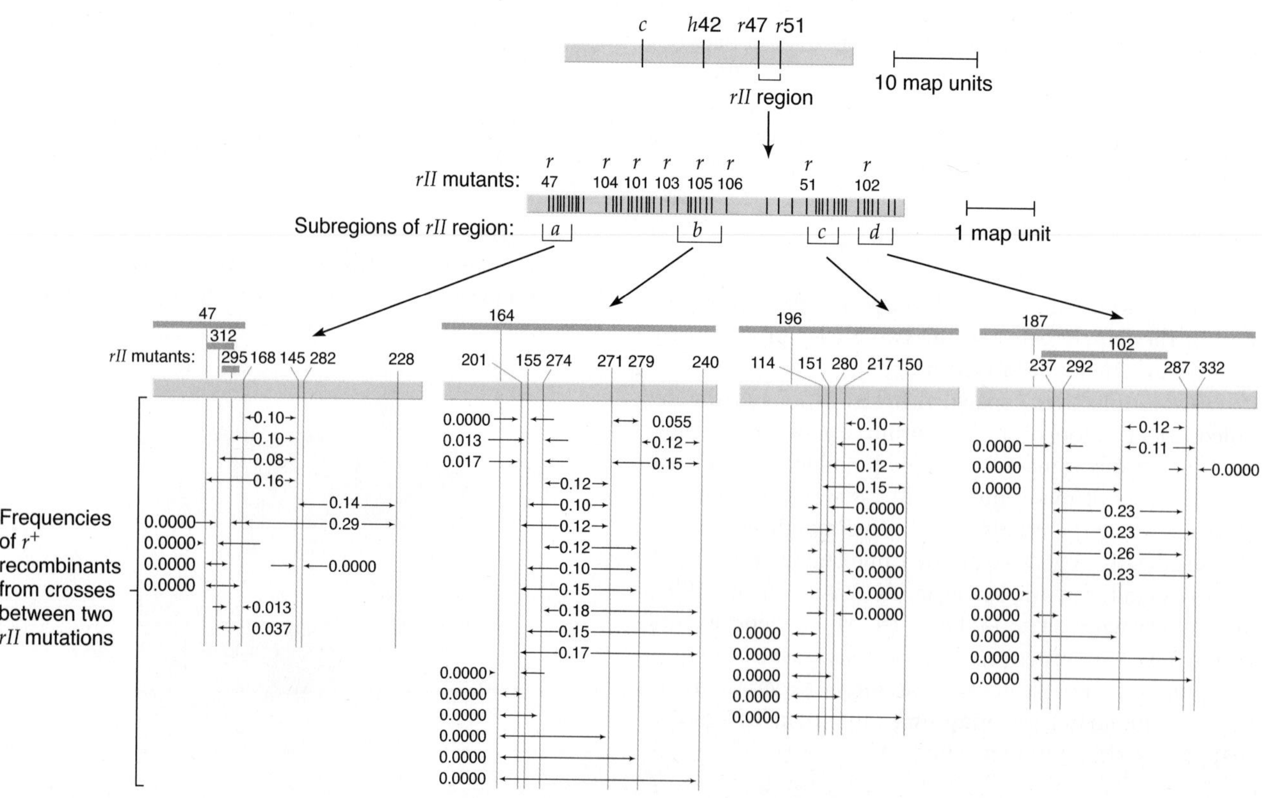

genetic experiments have shown that the *base pair* is both the *unit of mutation* and the *unit of recombination*. These definitions replaced the classic definitions that the *gene* was the unit of mutation and the unit of recombination. In that definition, the gene was considered to be indivisible by the processes of mutation and recombination.

> **Keynote**
>
> The same general principles of recombinational mapping can be applied to mapping the distance between mutational sites in different genes (intergenic mapping) and to mapping mutational sites within the same gene (intragenic mapping). Through fine-structure analysis of the *rII* region of bacteriophage T4 and other experiments, it was determined that the unit of mutation and the unit of recombination is the base pair in DNA.

### Deletion Mapping

After his initial series of crossing experiments, Benzer continued to map more than 3,000 *rII* mutants to complete his fine-structure map. Mapping that number of mutants would have required approximately 5 million crosses, an overwhelming task even with phages, with which up to 50 crosses can be done per day. Therefore, Benzer developed some elegant genetic procedures to simplify his mapping studies. These procedures involved using *deletion mapping* to localize unknown mutations, as we will now see.

Most of the *rII* mutants Benzer isolated were **point mutants;** their phenotype resulted from an alteration of a single nucleotide pair. A point mutant can revert to the wild-type state spontaneously or after treatment with an appropriate mutagen. However, some of Benzer's *rII* mutants did not revert, nor did they produce $r^+$ recombinants in crosses with a number of *rII* point mutants that were known to be located at different places on the *rII* map. These mutants were interpreted to be *deletion mutants*—mutants that had lost a segment of DNA. Benzer found a wide range in the extent and location of deleted genetic material among the *rII* deletion mutants he studied. Some deletion mutants are shown in Figure 15.19.

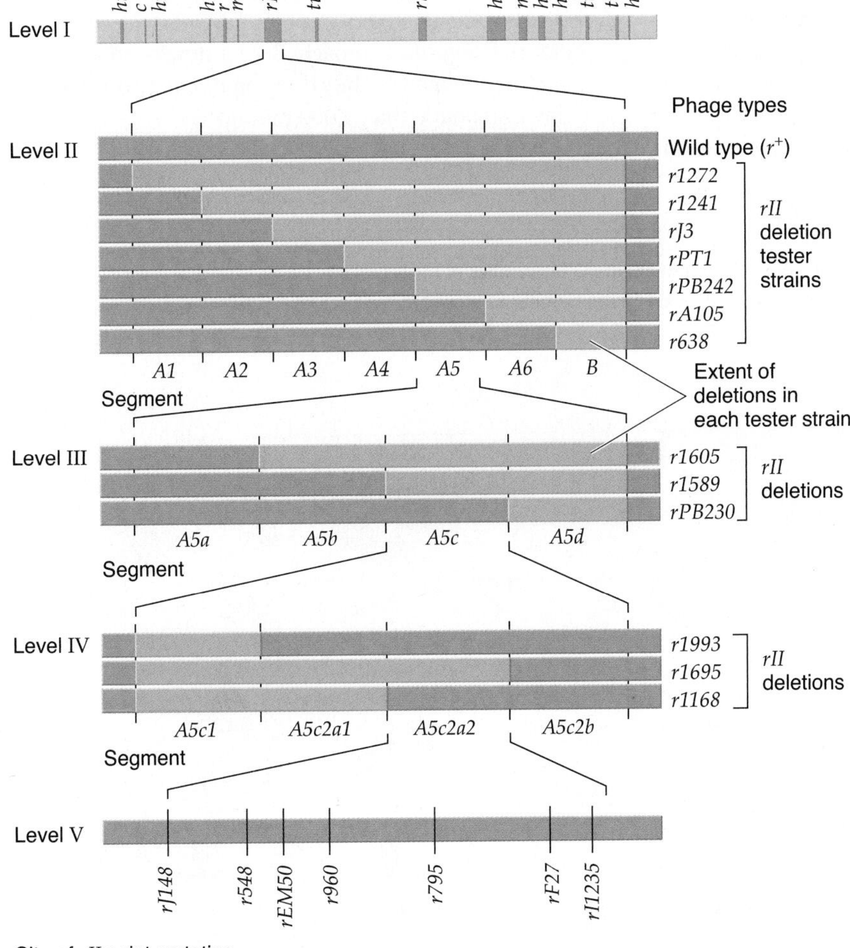


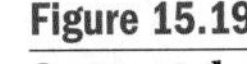

**Figure 15.19**

**Segmental subdivision of the *rII* region of phage T4 by means of deletion.** Level I shows the whole T4 genetic map. In Level II, seven deletions define seven segments of the *rII* region. In Level III, three deletions define four subsegments of the *A5* segments. In Level IV, three deletions define four subsegments of the *A5c* subsegment. Level V shows the order and spacing of the sites of the *rII* mutations in the *A5c2a2* subsegment, as established by pairwise crosses of seven point mutants. Level VI is a model of the DNA double helix, indicating the approximate scale of the level V map.

In actual practice, an unknown *rII* point mutant was first crossed with each of the seven standard deletion mutants that defined seven main segments of the *rII* region (segments *A1–A6* and *B* in Figure 15.19). For example, if an *rII* point mutant produced $r^+$ recombinants when crossed with deletion mutants *rA105* (deficient in *A6* and *B*) and *r638* (deficient in *B*), but did not produce $r^+$ recombinants when crossed with deletion mutants *r1272* (deficient in all segments), *r1241* (deficient in *A2–A6* and *B*), *rJ3* (deficient in *A3–A6* and *B*), *rPT1* (deficient in *A4–A6* and *B*), and *rPB242* (deficient in *A5–A6* and *B*), then the point mutation had to be in the segment of DNA that the five nonrecombinant deletion mutants lacked. $r^+$ recombinants cannot be produced in crosses with deletion mutants if the deleted segment contains the site of the point mutation. In Benzer's experiment, all five nonrecombinant deletion mutants lacked the segment *A5*, and both recombinant deletion mutants contained that segment, so the point mutation had to be in the *A5* region.

Once the main segment in which the mutation occurred was known, the point mutant was crossed with each of the relevant secondary set of reference deletions: *r1605*, *r1589*, and *rPB230* (see Figure 15.19). With segment *A5*, for example, three deletions divide *A5* into the four subsegments *A5a* through *A5d*. The presence or absence of $r^+$ recombinants in the progeny of the crosses of the *A5 rII* mutant with the secondary set of deletions enabled Benzer to localize the mutation more precisely to a smaller region of the DNA. For example, if the mutation was in segment *A5c*, then $r^+$ recombinants were produced with deletion *rPB230*, but not with either of the other two deletions. Other deletion mutants defined even smaller regions of each of the four subsegments *A5a* through *A5d*; for instance, *A5c* was divided into *A5c1*, *A5c2a1*, *A5C2a2*, and *A5c2b* by deletions *r1993*, *r1695*, and *r1168*.

In all, Benzer's deletions divided the *rII* region into 47 segments. Any given *rII* point mutant can be localized to one of these segments in three sequential sets of crosses of point mutants with deletion mutants. Then, all those point mutants within a given segment can be crossed in all possible pairwise crosses to construct a detailed genetic map. In this way, Benzer used the more than 3,000 *rII* mutants to prove that the *rII* region is subdivisible into more than 300 mutable sites that were separable by recombination (Figure 15.20). The distribution of mutants is not random: certain sites, called *hot spots*, are represented by a large number of independently isolated point mutants.

**Keynote**

Using deletions with defined ends, Benzer devised a genetic analysis scheme that enabled any given *rII* point mutant to be localized to one of 47 segments in three sequential sets of crosses of point mutants with deletion mutants. Using this approach, Benzer developed a fine-structure map of the *rII* region of bacteriophage T4 involving more than 3,000 mutants.

**Figure 15.20**

**Fine-structure map of the *rII* region derived from Benzer's experiments.** The number of independently isolated mutations that mapped to a given site is indicated by the number of blocks at the site. Hot spots are represented by a large number of blocks.

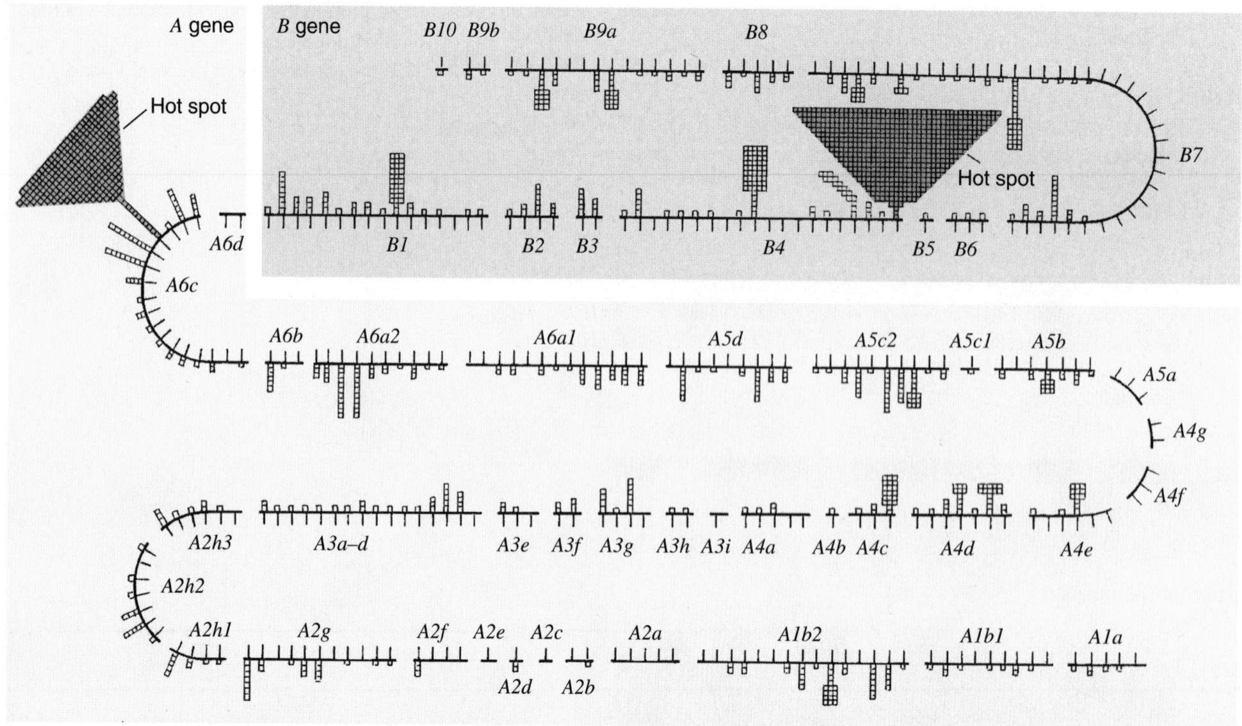

## Defining Genes by Complementation (*Cis-Trans*) Tests

From the classic point of view, the gene is a unit of function; that is, each gene specifies one function. Benzer designed genetic experiments to determine whether this classic view was true of the *rII* region. To find out whether two different *rII* mutants belonged to the same gene (unit of function), Benzer adapted the ***cis-trans* test**, or **complementation test**, developed by Edward Lewis to study the nature of the functional unit of the gene in *Drosophila*. We introduced the complementation test in Chapter 13, pp. 377–378. As you read the discussion that follows, it will help you to know that the complementation tests showed that the *rII* region actually consists of two genes (units of function): *rIIA* and *rIIB* (hence, the *A* and *B* regions in Figures 15.19 and 15.20). A mutation anywhere in either gene produces the *rII* plaque morphology phenotype and host range property. In other words, *rIIA* and *rIIB* each specify a different product needed for growth in *E. coli K12*(λ).

**@nimation**
**Defining Genes by Complementation Tests**

The complementation test is used to establish how many units of function (genes) are defined by a given set of mutations that express the same mutant phenotypes. In Benzer's work with the *rII* mutants, the nonpermissive strain *K12*(λ) was infected with a pair of *rII* mutant phages to see whether the two mutants, each unable by itself to grow in strain *K12*(λ), could work together to produce progeny phages. If the phages do produce progeny, the two mutants are said to complement each other, meaning that the two mutations must be in different genes (units of function) that encode different products. That is, those two products work together to allow progeny to be produced. If no progeny phages are produced, the mutants are not complementary, indicating that the mutations are in the same functional unit. In this case, both mutants produce the same defective product, so the phage life cycle cannot proceed and no progeny phages result. (Note that genetic recombination is not necessary for complementation to occur; if genetic recombination does take place, a few plaques may occur on the lawn; but if complementation occurs, the entire lawn of bacteria will be lysed.)

The two situations are diagrammed in Figure 15.21. In the first case, the bacterium is infected with two phage genomes, one with a mutation in the *rIIA* gene and the other with a mutation in the *rIIB* gene (Figure 15.21a). The *rIIA* mutant makes a nonfunctional *A* product and a functional *B* product, whereas the *rIIB* mutant makes a functional *A* product and a nonfunctional *B* product. Complementation occurs because the *rIIA* mutant still makes a functional *B* product and the *rIIB* mutant makes a functional *A* product, so phage propagation in *E. coli K12*(λ) can occur. In the second case, the bacterium is infected with two phage genomes, each with a different mutation in the same gene, *rIIA* (Figure 15.21b). No



**Figure 15.21**
**Complementation tests for determining the units of function in the *rII* region of phage T4; the nonpermissive host *E. coli K12*(λ) is infected with two different *rII* mutants.**

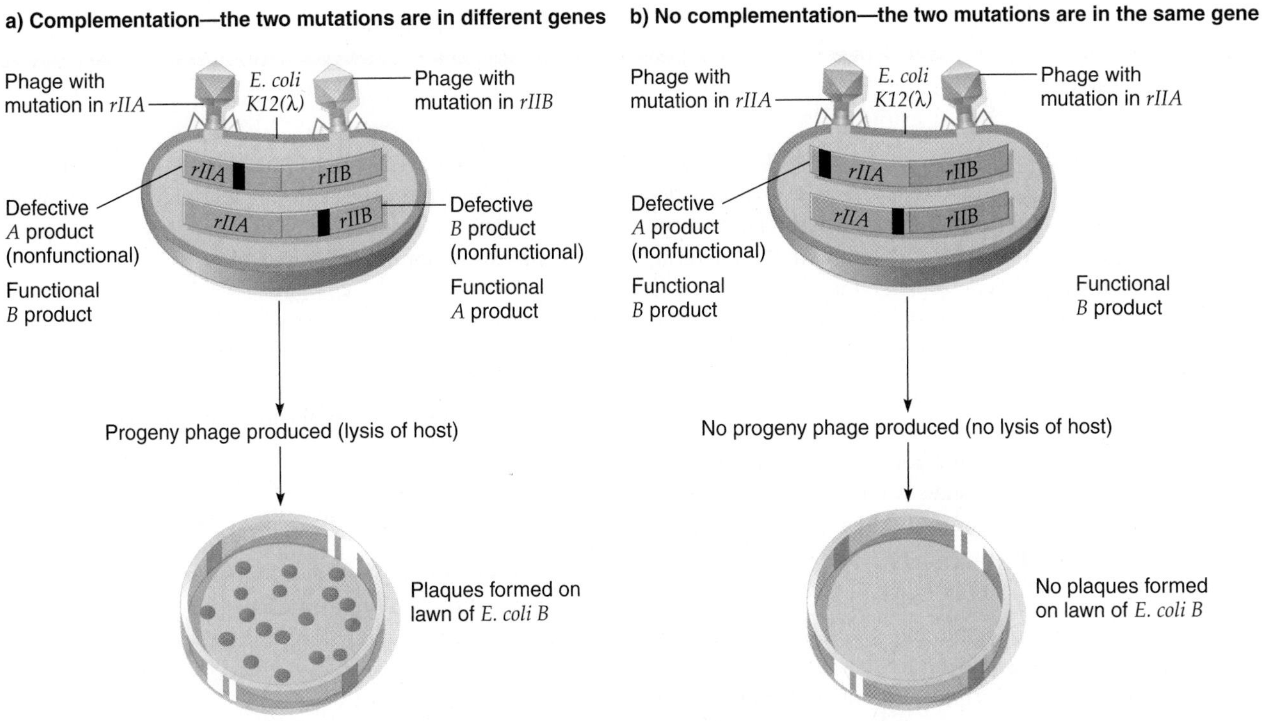

complementation occurs in this case because, although both mutants produce a functional *B* product, neither makes a functional *A* product. Thus the *A* function cannot take place, and phage propagation in *E. coli K12*(λ) cannot occur.

Based on the results of such complementation tests, Benzer found that each *rII* mutant falls into one of two units of function: *rIIA* and *rIIB* (also called *complementation groups*, which directly correspond to genes). That is, all *rIIA* mutants complement all *rIIB* mutants, but *rIIA* mutants fail to complement other *rIIA* mutants, and *rIIB* mutants fail to complement other *rIIB* mutants. The dividing line between the *rIIA* and *rIIB* units of function is indicated in the fine-structure map in Figure 15.20. Point mutants and deletion mutants in the *rII* region obey the same rules in the complementation tests. The only exceptions are deletions that span parts of both the *A* and the *B* functional units. Such deletion mutants do not complement either *A* or *B* mutants.

In the complementation test examples shown in Figure 15.21, each phage that coinfects the nonpermissive *E. coli* strain *K12*(λ) carries one *rII* mutation, which is actually a configuration of mutations called the *trans* configuration. In this configuration, the two mutations are carried by different phages. As a control, it is usual to coinfect *E. coli K12*(λ) with an $r^+$ (wild-type) phage and an *rII* mutant phage carrying both mutations to see whether the expected wild-type function results. When both mutations under investigation are carried on the same chromosome, the configuration is called the *cis* configuration of mutations. (It is because of the *cis* and *trans* configurations of mutations used in the complementation test that it is also called the *cis-trans* test.) In the *cis* test, the $r^+$ is expected to be dominant over the two mutations carried by the *rII* mutant phage, so progeny phages are produced. Therefore, the failure to produce progeny would not prove that the mutations are in different functional genes.

Benzer called the genetic unit of function revealed by the *cis-trans* test the *cistron*. A cistron is the smallest segment of DNA that encodes a piece of RNA. At present, *gene* is commonly used and *cistron* is used far less often. Genetically, the *rIIA* cistron is about 6 mu and 800 bp long, and the *rIIB* cistron is about 4 mu and 500 bp long. Presumably, their two products act in common processes necessary for T4 propagation in strain *K12*(λ).

The principles underlying a complementation test are the same in other organisms; only the practical details of performing the test are organism specific. For example, in yeast, one could select two haploid cells that are of different mating types (*MAT***a** and *MAT*α) and that carry different mutations conferring the same mutant phenotype. Mating these two types would produce a diploid, which would then be analyzed for complementation of the two mutations. In animal cells, two cells, each exhibiting the same mutant phenotype, can be fused and analyzed; a wild-type phenotype indicates that complementation has occurred. Again, in neither of these cases is recombination necessary for complementation to occur.

**Keynote**

The complementation, or *cis-trans*, test is used to determine how many units of function (genes) define a given set of mutations expressing the same mutant phenotypes. If two mutants, each carrying a mutation in a different gene, are combined, the mutations complement and a wild-type function results. If two mutants, each carrying a mutation in the same gene, are combined, the mutations do not complement and the mutant phenotype is exhibited.

# Summary


- The same experimental strategy is used for all gene mapping; that is, genetic material is exchanged between strains differing in genetic markers, and recombinants are detected and counted. In bacteria, the mechanism of gene transfer may be transformation, conjugation, or transduction. In each process, there is a donor strain and a recipient strain.
- Conjugation is a process in which there is a unidirectional transfer of genetic information through direct cellular contact between a donor and a recipient bacterial cell. The donor state is conferred on that cell by the presence of a plasmid called an F factor. Conjugation results in the unidirectional transfer of a copy of the F factor from donor to recipient.
- The F factor can integrate into the bacterial chromosome. Strains in which this has occurred—*Hfr* strains—can conjugate with recipient strains and transfer part of the bacterial chromosome. The sequence and distances between genes can be determined by the order and time of acquisition of the genes by the recipient from the donor during conjugation.
- Transformation is the transfer of genetic material between organisms by small extracellular pieces of DNA. Through genetic recombination, part of the transforming DNA molecule can exchange with a portion of the recipient's chromosomal DNA. Transformation can be used experimentally to determine gene order and map distances between genes.
- Transduction is a process whereby bacteriophages (phages) mediate the transfer of bacterial DNA from one bacterium (the donor) to another (the recipient). Transduction can be used experimentally to map bacterial genes.

- The same principles used to map eukaryotic genes are used to map phage genes. That is, a bacterial host is simultaneously infected with two strains of phages differing from each other in one or more gene loci. The percentages of recombinants are determined, and the sequence and distances between genes are then inferred.
- The same principles of recombinational mapping in eukaryotes can be applied to mapping the distance between mutational sites in different genes (intergenic mapping) and to mapping mutational sites within the same gene (intragenic mapping)
- From fine-structure analysis of the *rII* region of bacteriophage T4, it was determined that the unit of mutation and recombination is the DNA base pair.
- The number of genes that cause a particular mutant phenotype is determined by the complementation, or *cis-trans*, test. If two viral mutants, each carrying a mutation in a different gene, are combined in a single host cell, the mutations make up for each other's defect (that is, they complement) and a wild-type phenotype results. If two mutants, each carrying a mutation in the same gene, are combined, the mutations do not complement and the mutant phenotype is still expressed.



## Analytical Approaches to Solving Genetics Problems

**Q15.1** In *E. coli,* the following *Hfr* strains donate the genes shown in the order given:

| *Hfr* Strain | Order of Gene Transfer |
|---|---|
| 1 | *G E B D N A* |
| 2 | *P Y L G E B* |
| 3 | *X T J F P Y* |
| 4 | *B E G L Y P* |

All the *Hfr* strains were derived from the same $F^+$ strain. What is the order of genes in the original $F^+$ chromosome?

**A15.1** This question is an exercise in piecing together various segments of the circumference of a circle. The best approach is to draw a circle and label it with the genes transferred from one *Hfr* and then see which of the other *Hfr* strains transfers an overlapping set. For example, *Hfr 1* transfers E, then B, then D, and so on, and *Hfr 4* transfers B, then E, and so forth. Now we can juxtapose the two sets of genes transferred by the two *Hfrs* and deduce that the polarities of transfer are opposite:

*Hfr* 1 *G E B D N A*

*Hfr* 4 *P Y L G E B*

Extending this reasoning to the other *Hfr* strains, we can draw an unambiguous map (see the following figure), with the arrowheads indicating the order of transfer:

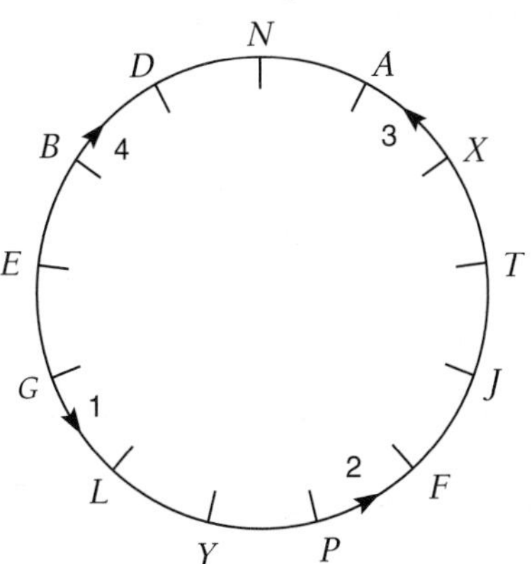


The same logic would apply if the question gave the relative time units of entry of each of the genes. In that case, we would expect that the time distance between any two genes would be approximately the same, regardless of the order of transfer or how far the genes were from the origin.

**Q15.2** In a transformation experiment, donor DNA from an $a^+ \ b^+$ strain was used to transform a recipient strain of genotype *a b*. The transformed classes were isolated and their frequencies determined to be

| | |
|---|---|
| $a^+ \ b^+$ | 307 |
| $a^+ \ b$ | 215 |
| $a \ b^+$ | 278 |

The total number of transformants was 800. What is the frequency with which the *b* locus is cotransformed with the *a* locus?

**A15.2** The frequency with which $b^+$ is cotransformed with the $a^+$ gene is calculated by using of values for the total number of $a^+$ transformants and the number of transformants for both $a^+$ and $b^+$. The formula is

$$\frac{\text{number of } a^+ \ b^+ \text{ cotransformants}}{\text{total number of } a^+ \text{ transformants}} \times 100\%$$

The $a^+ \ b^+$ cotransformants number 307. The $a^+$ transformants are represented by two classes: $a^+ \ b^+$ (307) and $a^+ \ b$ (215), for a total of 522. The $a \ b^+$ class is irrelevant to the question because its members are not transformants for $a^+$. Thus, the cotransformation frequency for $a^+$ and $b^+$ is 307/522 × 100 = 58.8%.

**Q15.3** In a transduction experiment, the donor was $c^+ \ d^+ \ e^+$ and the recipient was *c d e*. Selection was for $c^+$. The four classes of transductants from this experiment are shown in the following table:

| Class | Genetic Composition | Number of Individuals |
|---|---|---|
| 1 | $c^+ d^+ e^+$ | 57 |
| 2 | $c^+ d^+ e$ | 76 |
| 3 | $c^+ d\ e$ | 365 |
| 4 | $c^+ d\ e^+$ | 2 |
| | | Total 500 |

**a.** Determine the cotransduction frequency for $c^+$ and $d^+$.
**b.** Determine the cotransduction frequency for $c^+$ and $e^+$.
**c.** Which of the cotransduction frequencies calculated in (a) and (b) represents the greater actual distance between genes? Why?

**A15.3**

**a.** The analysis is similar to the cotransformation frequency analysis described in Analytical Question 15.2. The formula for the cotransduction frequency for $c^+$ and $d^+$ is

$$\frac{\text{number of } c^+\ d^+ \text{ contransductants}}{\text{total number of } c^+ \text{ transductants}} \times 100\%$$

From the data presented, classes 1 and 2 are the $c^+\ d^+$ cotransductants, and the total number of $c^+$ transductants is the sum of classes 1 through 4. Thus the number of $c^+$ and $d^+$ transductants is $57 + 76 = 133$, and the cotransductant frequency is $133/500 \times 100 = 26.6\%$.

**b.** The analysis is identical in approach to (a). The formula for the cotransduction frequency for $c^+$ and $e^+$ is

$$\frac{\text{number of } c^+\ e^+ \text{ contransductants}}{\text{total number of } c^+ \text{ transductants}} \times 100\%$$

From the data presented, classes 1 and 4 are the $c^+\ e^+$ cotransductants, and the total number of $c^+$ transductants is the sum of classes 1 through 4. Thus the number of $c^+$ and $e^+$ transductants is $57 + 2 = 59$, and the cotransductant frequency is $59/500 \times 100 = 11.8\%$.

**c.** The greater actual distance is between the $c^+$ and $e^+$ genes. The principle involved is that the closer two genes are on the chromosome, the greater is the likelihood that they will be cotransduced. Thus, as the distance between genes increases, the cotransduction frequency decreases. Since the $c^+\ e^+$ cotransduction frequency is 11.8% and the $c^+\ d^+$ cotransduction frequency is 26.6%, genes $c^+$ and $e^+$ are farther apart than genes $c^+$ and $d^+$.

**Q15.4** Five different *rII* deletion strains of phage T4 were tested for recombination by pairwise crossing in *E. coli B*. The following results were obtained, where + = $r^+$ recombinants produced, and 0 = no $r^+$ recombinants produced:

| | A | B | C | D | E |
|---|---|---|---|---|---|
| E | 0 | + | 0 | + | 0 |
| D | 0 | 0 | 0 | 0 | |
| C | 0 | 0 | 0 | | |
| B | + | 0 | | | |
| A | 0 | | | | |

Draw a deletion map compatible with these data.

**A15.4** The principle here is that if two deletion mutations overlap, then no $r^+$ recombinants can be produced. Conversely, if two deletion mutations do not overlap, then $r^+$ recombinants can be produced. To approach a question of this kind, we must draw overlapping and nonoverlapping lines from the given data.

Starting with *A* and *B*, these two deletions do not overlap, because $r^+$ recombinants are produced. Therefore, these two mutations can be represented as follows:

A B

The next deletion, *C*, does not produce $r^+$ recombinants with any of the other four deletions. We must conclude, therefore, that *C* is an extensive deletion that overlaps the other four, with endpoints that cannot be determined from the data given. One possibility is as follows:

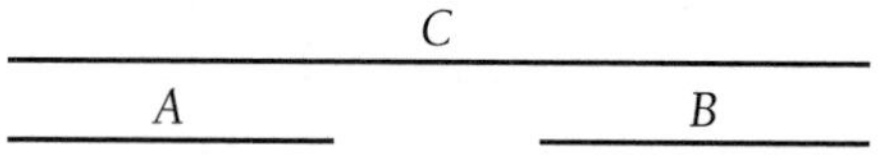

C

A B

Deletion *D* does not produce $r^+$ recombinants with *A*, *B*, or *C*, but it does with *E*. In turn, *E* produces $r^+$ recombinants with *B* and *D* but not with *A* or *C*. Thus *D* must overlap both *A* and *B* but not *E*, and *E* must overlap *A* and *C* but not *B*. A compatible map for this situation is as follows:

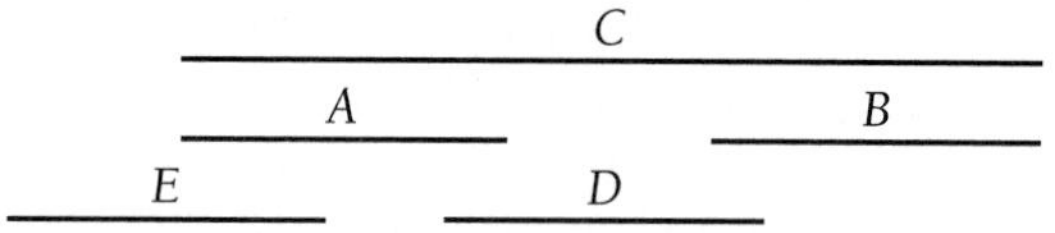

C

A B

E D

Other maps can be drawn in terms of the endpoints of the deletions.

**Q15.5** Seven different *rII* point mutants (*1* to *7*) of phage T4 were tested for recombination crosses in *E. coli B* with the five deletion strains described in question 15.4. The following results were obtained, where + = $r^+$ recombinants produced and 0 = no $r^+$ recombinants produced:

| | A | B | C | D | E |
|---|---|---|---|---|---|
| *1* | 0 | + | 0 | + | + |
| 2 | + | 0 | 0 | + | + |
| 3 | 0 | + | 0 | + | 0 |
| 4 | + | + | 0 | + | 0 |
| 5 | + | 0 | 0 | 0 | + |
| 6 | 0 | + | 0 | 0 | + |
| 7 | + | + | 0 | 0 | + |

In which regions of the map can you place the seven point mutations?

**A15.5** If an $r^+$ recombinant is produced, the *rII* point mutation cannot overlap the region missing in the deletion mutation with which it was crossed. Thus, the table of results enables us to localize the point mutations to the regions defined by the deletion mutants. Potentially, the results define the relative extent of deletion overlap. For example, point mutation *7* produces $r^+$ recombinants with *A*, *B*, and *E*, but not with *D*. Logically, then, *7* is located in the region defined by the part of deletion *D* that is not involved in the overlap with *A* and *B*. Similarly, point mutation *4* gives $r^+$ recombinants with *A*, *D*, and *B*, but not with *E*. Hence, *4* must be in a region defined by a segment of deletion *E* that does not overlap deletion *A*. Furthermore, because *4* does not produce $r^+$ recombinants with *C* either, deletion *C* must overlap the site defined by point mutation *4*. This result, then, refines the deletion map with regard to the *E*, *C*, and *A* endpoints. The map we can draw from the matrix of results is as follows:

C
A B
E D
4 3 1 6 7 5 2

# Questions and Problems

*__15.1__ In $F^+ \times F^-$ crosses, the $F^-$ recipient is converted to a donor with very high frequency. However, it is rare for a recipient to become a donor in *Hfr* $\times F^-$ crosses. Explain why.

**15.2** The growing resistance of bacteria to antibiotics is a significant health concern. As early as the 1950s, physicians identified hospital patients afflicted with severe diarrhea, resulting from bacterial dysentery, who did not respond to previously effective antibiotics. Some strains of *Shigella*, the pathogen that causes bacterial dysentery, had developed resistance to antibiotics. In the 1970s the basis for this resistance was discovered: plasmids containing multiple antibiotic resistance genes were isolated from *Shigella*. Researchers then found that the same genes conferring antibiotic resistance in *Shigella* were also present in other species of pathogenic bacteria.

**a.** In each of the following crosses, where would genes conferring antibiotic resistance need to be located in order for them to be transferred reliably from one bacterial cell to another cell?
 **i.** *Hfr* (resistant) $\times F^-$ (sensitive)
 **ii.** $F^+$ (resistant) $\times F^-$ (sensitive)

**b.** Which cross would be more efficient at spreading antibiotic resistance between cells? Why?

**c.** That the same genes conferring resistance in *Shigella* were found in other bacterial species suggests that these genes were transferred across species. Generate hypotheses to explain how this might have occurred.

*__15.3__ With the technique of interrupted mating, four *Hfr* strains were tested for the sequence in which they transmitted a number of different genes to an $F^-$ strain. Each *Hfr* strain was found to transmit its genes in a unique order, as shown in the accompanying table. (Only the first six genes transmitted were scored for each strain.)

| Order of Transmission | *Hfr* Strain 1 | 2 | 3 | 4 |
|---|---|---|---|---|
| First | O | R | E | O |
| | F | H | M | G |
| | B | M | H | X |
| | A | E | R | C |
| | E | A | C | R |
| Last | M | B | X | H |

What is the gene sequence in the original strain from which these *Hfr* strains derive? On a diagram, indicate the origin and polarity of each of the four *Hfr* strains.

**15.4** At time zero, an *Hfr* strain (*Hfr* 1) was mixed with an $F^-$ strain, and at various times after mixing, samples were removed and agitated to separate conjugating cells. The cross may be written as

$$Hfr\ 1:\ a^+\ b^+\ c^+\ d^+\ e^+\ f^+\ g^+\ h^+\ str^S$$
$$F^-:\ a\ b\ c\ d\ e\ f\ g\ h\ str^R$$

(No order is implied in listing the markers.) The samples were then plated onto selective media to measure the frequency of $h^+$ $str^R$ recombinants that had received certain genes from the *Hfr* cell. A graph of the number of recombinants against time is shown in the accompanying figure.

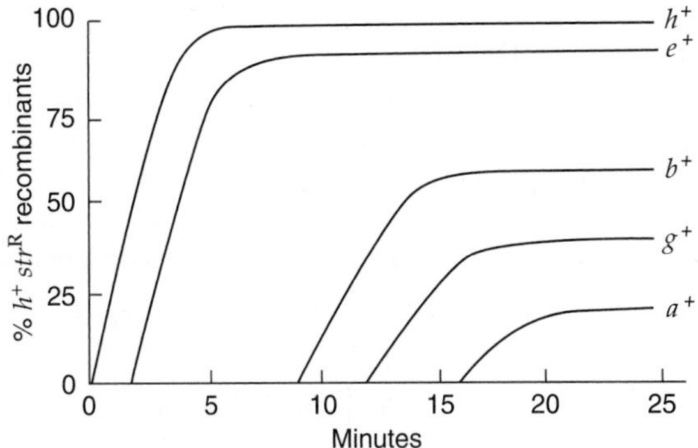

**a.** Indicate whether each of the following statements is true or false:

**i.** All $F^+$ cells that received $a^+$ from the *Hfr* in the chromosome transfer process must also have received $b^+$.

**ii.** The order of gene transfer from *Hfr* to $F^-$ was $a^+$ (first), then $g^+$, then $b^+$, then $e^+$, and, finally, $h^+$.

**iii.** Most $e^+$ $str^R$ recombinants are likely to be *Hfr* cells.

**iv.** None of the $b^+$ $str^R$ recombinants plated at 15 minutes are also $a^+$.

**b.** Draw a linear map of the *Hfr* chromosome, indicating

**i.** the point of nicking (the origin) and the direction of DNA transfer.

**ii.** the order of the genes $a^+$, $b^+$, $e^+$, $g^+$, and $h^+$.

**iii.** the shortest distance between consecutive genes on the chromosomes.

**15.5** What steps would you take to selectively grow each of the bacterial cell types found in the following mixtures A through D?

| Mixture | Genotypes Present | Phenotypes |
|---|---|---|
| A | *his*, *his*$^+$ | *his* cells require supplemental histidine; *his*$^+$ cells are able to grow without supplemental histidine. |
| B | *azi*$^R$, *azi*$^S$ | *azi*$^R$ cells are able to grow even in the presence of the poison sodium azide; *azi*$^S$ cells die in the presence of sodium azide. |
| C | *lac*, *lac*$^+$ | *lac*$^+$ cells can grow even if lactose is the only sugar present. *lac*$^-$ cells cannot utilize lactose for growth; they require a sugar other than lactose for growth whether or not lactose is present. |
| D | *pcsA*$^+$, *pcsA* | *pcsA* cells are cold sensitive and grow at 37°C, but not at 30°C; *pcsA*$^+$ cells can grow at both 37°C and 30°C. |

***15.6** Three different prototrophic strains (1, 2, and 3) that are all sensitive to the antibiotic streptomycin are isolated. Each is individually mixed with an auxotrophic $F^-$ strain that is *a b c d e f g h* (and therefore requires compounds A, B, C, D, E, F, G, and H to grow) and that is also resistant to the antibiotic streptomycin. At 1-minute intervals after the initial mixing, a sample of the mixture is removed, shaken violently, and plated on media to select for $c^+$ $str^R$ recombinants. Recombinants are then tested for the presence of other genes. The following results are obtained:

Strain 1 × $F^-$: No $c^+$ recombinants are ever obtained, even after 25 minutes.

Strain 2 × $F^-$: $c^+$ recombinants are obtained at 6 minutes, $g^+$ at 8 minutes, $h^+$ at 11 minutes, $a^+$ at 14 minutes, and $b^+$ at 16 minutes. No $d^+$, $e^+$, or $f^+$ recombinants are obtained.

Strain 3 × $F^-$: $c^+$ recombinants are obtained at the 1-minute time point, and $c^+$ $g^+$ recombinants are obtained on or after the 3-minute time point. No $a^+$, $b^+$, $d^+$, $e^+$, $f^+$, or $h^+$ recombinants are obtained.

If $c^+$ recombinants obtained at the 16-minute time point from the cross involving strain 2 are mixed with an $amp^R$ (ampicillin-resistant) $F^-$ strain, no $c^+$ $amp^R$ recombinants are ever recovered. However, if $c^+$ recombinants obtained at the 16-minute time point from the cross involving strain 3 are mixed with an $amp^R$ $F^-$ strain, $amp^R$ $c^+$ recombinants can be recovered after 1 minute of mating.

**a.** How was the initial selection for $c^+$ $str^R$ recombinants done? How were the subsequent selections done?

**b.** Use these data to ascertain, as best you can, whether each strain is $F^-$, *Hfr*, $F^+$, or $F'$. If these data do not allow you to make an unambiguous determination, indicate the possibilities.

**c.** To the extent you can, draw a map of the chromosomes that might be present in each of strains 1, 2, and 3. Indicate the location and distance between genes $a^+$, $b^+$, $d^+$, $e^+$, $f^+$, and $h^+$ as best you can.

***15.7** If an *E. coli* auxotroph *A* could grow only on a medium containing thymine, and an auxotroph *B* could grow only on a medium containing leucine, how would you test whether DNA from *A* could transform *B*?

**15.8** You are given a prototrophic $str^R$ (streptomycin-resistant) *Hfr* strain and an $amp^R$ (ampicillin-resistant) $F^-$ auxotrophic strain that requires leucine (*leu*), arginine (*arg*), lysine (*lys*), purine (*pur*), and biotin (*bio*).

**a.** Devise a strategy to determine quickly which gene ($leu^+$, $arg^+$, $lys^+$, $pur^+$, or $bio^+$) lies closest to the F factor origin of replication.

**b.** Even when the prototrophic *Hfr* strain is mixed with the $F^-$ strain for very long periods of time, streptomycin resistance is not transferred. State two hypotheses that explain this finding.

***15.9** $F^-$ *leu arg* $str^R$ (see Question 15.8 for a description of these marker mutations) cells were treated with the mutagen MMS and plated at low density on plates containing medium with leucine, arginine, and the antibiotic streptomycin. Colonies were picked individually with a sterile toothpick and gently stabbed into a grid-like pattern on plates with the same type of medium to produce 50 plates, each with 100 colonies in a 10 × 10 array. After new colonies grew, cells from each gridded colony were picked with a fresh, sterile toothpick and mixed with *Hfr* $leu^+$ $arg^+$ $str^S$ cells. The mating mixtures were gently stabbed in a similar 10 × 10 gridded array onto plates containing minimal medium and streptomycin. Following incubation, growth was seen in nearly all of the squares of the 10 × 10 grids on the 50 plates. However, a few squares had no growth.

**a.** What can you infer about the location and orientation of the F factor origin of replication relative to the *leu*⁺ and *arg*⁺ genes in the *Hfr* strain?

**b.** What sort of functions could have been mutated by treatment of the $F^-$ *leu arg str*$^R$ cells with MMS to prevent growth under these conditions? (Hint: Consider what happens to the DNA that is donated by the *Hfr* strain.)

**15.10** Four independently isolated auxotrophic mutations, *met1*, *met2*, *met3*, and *met4*, each require supplemental methionine to grow.

**a.** Each *met* mutant was crossed into an $F^-$ *str*$^R$ genetic background. The four $F^-$ strains bearing different *met* mutants were then crossed to $F'$ strains bearing either the *metD*⁺ or *metC*⁺ genes. Three minutes after the $F^-$ and $F'$ strains were mixed, the cultures were violently shaken, and exconjugants were plated on medium containing streptomycin but no methionine. The following table indicates whether or not colonies were recovered from each cross.

| Cross | Colonies Recovered? |
|---|---|
| $F'$ *metC*$^+$ *str*$^S$ × $F^-$ *met1 str*$^R$ | No |
| $F'$ *metC*$^+$ *str*$^S$ × $F^-$ *met2 str*$^R$ | Yes |
| $F'$ *metC*$^+$ *str*$^S$ × $F^-$ *met3 str*$^R$ | No |
| $F'$ *metC*$^+$ *str*$^S$ × $F^-$ *met4 str*$^R$ | No |
| $F'$ *metD*$^+$ *str*$^S$ × $F^-$ *met1 str*$^R$ | Yes |
| $F'$ *metD*$^+$ *str*$^S$ × $F^-$ *met2 str*$^R$ | No |
| $F'$ *metD*$^+$ *str*$^S$ × $F^-$ *met3 str*$^R$ | Yes |
| $F'$ *metD*$^+$ *str*$^S$ × $F^-$ *met4 str*$^R$ | Yes |

What phenotype was selected for when cells were plated on this medium? What phenotype was selected against? Which *met* mutants affect function(s) at *metD*, and which *met* mutants affect function(s) at *metC*?

**b.** Each of the four *met* mutants were crossed into two different genetic backgrounds, an *Hfr lac*⁺ *str*$^S$ background and an $F^-$ *lac str*$^R$ background. All possible pairwise crosses were then performed between the different *Hfr* and $F^-$ strains. For each cross, the culture was violently shaken six minutes after *Hfr* and $F^-$ cells were mixed, and exconjugants were plated onto medium A, which had lactose as the sole carbon source and was supplemented with streptomycin and methionine. Colonies that grew on medium A were then replica-plated onto medium B, which had lactose as the sole carbon source and was supplemented only with streptomycin. The following table gives the fraction of colonies that grew on medium A that were able to grow on medium B.

| Cross | Fraction of Colonies Grown on Medium A Also Able to Grow on Medium B |
|---|---|
| *Hfr met1 lac*$^+$ *str*$^S$ × $F^-$ *met2 lac str*$^R$ | 0.000 |
| *Hfr met2 lac*$^+$ *str*$^S$ × $F^-$ *met1 lac str*$^R$ | 0.180 |
| *Hfr met1 lac*$^+$ *str*$^S$ × $F^-$ *met3 lac str*$^R$ | 0.004 |
| *Hfr met3 lac*$^+$ *str*$^S$ × $F^-$ *met1 lac str*$^R$ | 0.020 |
| *Hfr met1 lac*$^+$ *str*$^S$ × $F^-$ *met4 lac str*$^R$ | 0.015 |
| *Hfr met4 lac*$^+$ *str*$^S$ × $F^-$ *met1 lac str*$^R$ | 0.003 |
| *Hfr met2 lac*$^+$ *str*$^S$ × $F^-$ *met3 lac str*$^R$ | 0.160 |
| *Hfr met3 lac*$^+$ *str*$^S$ × $F^-$ *met2 lac str*$^R$ | 0.000 |
| *Hfr met2 lac*$^+$ *str*$^S$ × $F^-$ *met4 lac str*$^R$ | 0.185 |
| *Hfr met4 lac*$^+$ *str*$^S$ × $F^-$ *met2 lac str*$^R$ | 0.000 |
| *Hfr met3 lac*$^+$ *str*$^S$ × $F^-$ *met4 lac str*$^R$ | 0.035 |
| *Hfr met4 lac*$^+$ *str*$^S$ × $F^-$ *met3 lac str*$^R$ | 0.006 |

What genotype is selected for by each medium? Construct a genetic map showing as much information as you can about the order of the *met* mutations relative to the *lac* locus.

**15.11** In a cross between a donor strain that is *Hfr leu*⁺ *arg*⁺ *str*$^S$ and a recipient that is $F^-$ *leu arg str*$^R$ (see Question 15.8 for a description of these marker mutations), an occasional recipient cell is $F^+$ *leu*⁺ *arg str*$^R$.

**a.** Explain what events must have occurred to generate this type of recipient cell. Did the events occur in the *Hfr* or $F^-$ parent, or both?

**b.** What properties would an $F^+$ *leu*⁺ *arg str*$^R$ cell have, and what steps would you take to select for a colony produced by this type of cell?

***15.12** Label each of the characteristics or processes below according to the following codes:

GT: Occurs in or is a characteristic of generalized transduction

ST: Occurs in or is a characteristic of specialized transduction

B: Occurs in both

N: Occurs in neither

**a.** Phage carries DNA of bacterial or viral DNA origin, never both.

**b.** Phage carries viral DNA covalently linked to bacterial DNA.

**c.** Phage integrates into a specific site on the host chromosome.

**d.** Phage integrates at a random site on the host chromosome.

**e.** "Headful" of bacterial DNA is packaged into phage.

**f.** Host is lysogenized.

**g.** Prophage state exists.

**h.** Temperate phage is involved.

**i.** Virulent phage is involved.

**15.13** Consider the following transduction data:

| Donor | Recipient | Selected Marker | Unselected Marker | % |
|---|---|---|---|---|
| *aceF*$^+$ *dhl* | *aceF dhl*$^+$ | *aceF*$^+$ | *dhl* | 88 |
| *aceF*$^+$ *leu* | *aceF leu*$^+$ | *aceF*$^+$ | *leu* | 34 |

Is *dhl* or *leu* closer to *aceF*?

***15.14** Consider the following data pertaining to P1 transduction:

| Donor | Recipient | Selected Marker | Unselected Marker | % |
|---|---|---|---|---|
| *aroA pyrD*$^+$ | *aroA*$^+$ *pyrD* | *pyrD*$^+$ | *aroA* | 5 |
| *aroA*$^+$ *cmlB* | *aroA cmlB*$^+$ | *aroA*$^+$ | *cmlB* | 26 |
| *cmlB pyrD*$^+$ | *cmlB*$^+$ *pyrD* | *pyrD*$^+$ | *cmlB* | 54 |

Choose the correct gene order:

**a.** *aroA cmlB pyrD*
**b.** *aroA pyrD cmlB*
**c.** *cmlB aroA pyrD*

**15.15** Order the mutants *trp*, *pyrF*, and *qts* based on the following three-factor transduction cross:

Donor *trp*$^+$ *pyr*$^+$ *qts*
Recipient *trp pyr qts*$^+$
Selected Marker *trp*$^+$

| Unselected Markers | Number |
|---|---|
| *pyr*$^+$ *qts*$^+$ | 22 |
| *pyr*$^+$ *qts* | 10 |
| *pyr qts*$^+$ | 68 |
| *pyr qts* | 0 |

***15.16** Order *cheA*, *cheB*, *eda*, and *supD* from the following data:

| Markers | % Cotransduction |
|---|---|
| *cheA–eda* | 15 |
| *cheA–supD* | 5 |
| *cheB–eda* | 28 |
| *cheB–supD* | 2.7 |
| *eda–supD* | 0 |

**15.17** Cells from a F$^-$ *gal* strain of *E. coli* (a strain unable to grow on medium having galactose as the sole carbon source) were treated with the mutagen 5BU and plated onto a rich medium at 37°C. The colonies were then replica plated onto a rich medium and grown at 41°C. Twenty colonies that grew at 37°C but not 41°C were individually picked and resuspended in tubes numbered 1 through 20. Cells from each suspension were mixed with a lysate produced by growing λ phage on wild-type *E. coli*, plated on medium having galactose as the sole carbon source, and incubated at 41°C. Sample 12, but none of the other samples, produced colonies under these conditions.

**a.** Why did the other 19 samples not produce the same result seen with sample 12?
**b.** Use diagrams to explain the events that occurred in sample 12 and allowed colonies to grow at 41°C on plates having galactose as the sole carbon source.
**c.** What can you infer about the location of the temperature-sensitive mutation in cells of sample 12? Can you infer anything about the nature of this mutation?
**d.** Assume that each of the 20 colonies that were initially picked harbor different temperature-sensitive loss-of-function mutations. Describe an experimental approach to determine the location of their mutations.

***15.18** Wild-type phage T4 grows on both *E. coli B* and *E. coli K12*(λ), producing turbid plaques. The *rII* mutants of T4 grow on *E. coli B*, producing clear plaques, but do not grow on *E. coli K12*(λ). This host range property permits the detection of a very low number of $r^+$ phages among a large number of *rII* phages. With such a sensitive system, it is possible to determine the genetic distance between two mutations within the same gene—in this case, the *rII* locus. Suppose *E. coli B* is mixedly infected with *rIIx* and *rIIy*, two separate mutants in the *rII* locus. Suitable dilutions of progeny phages are plated on *E. coli B* and *E. coli K12*(λ). A 0.1-mL sample of a thousandfold dilution plated on *E. coli B* produces 672 plaques. A 0.2-mL sample of undiluted phage plated on *E. coli K12*(λ) produces 470 turbid plaques. What is the genetic distance between the two *rII* mutations?

**15.19** Construct a map from the following two-factor phage cross data (show the map distance):

| Cross | % Recombination |
|---|---|
| r1 × r2 | 0.10 |
| r1 × r3 | 0.05 |
| r1 × r4 | 0.19 |
| r2 × r3 | 0.15 |
| r2 × r4 | 0.10 |
| r3 × r4 | 0.23 |

***15.20** The following two-factor crosses were made to analyze the genetic linkage between four genes in phage λ: *c*, *mi*, *s*, and *co*.

| Parents | Progeny |
|---|---|
| *c* + × + *mi* | 1,213 *c* +, 1,205 + *mi*, 84 + +, 75 *c mi* |
| *c* + × + *s* | 566 *c* +, 808 + *s*, 19 + +, 20 *c s* |
| *co* + × + *mi* | 5,162 *co* +, 6,510 + *mi*, 311 + +, 341 *co mi* |
| *mi* + × + *s* | 502 *mi* +, 647 + *s*, 65 + +, 56 *mi s* |

Construct a genetic map of the four genes.

**15.21** Wild-type ($r^+$) strains of T4 produce turbid plaques, whereas *rII* mutant strains produce larger, clearer plaques. Five *rII* mutations (*a–e*) in the A cistron of the *rII* region of T4 give the following percentages of wild-type recombinants in two-point crosses:

| Cross | % of Wild-Type Recombinants | Cross | % of Wild-Type Recombinants |
|---|---|---|---|
| $a \times b$ | 0.2 | $e \times d$ | 0.7 |
| $a \times c$ | 0.9 | $e \times c$ | 1.2 |
| $a \times d$ | 0.4 | $e \times b$ | 0.5 |
| $b \times c$ | 0.7 | $b \times d$ | 0.2 |
| $e \times a$ | 0.3 | $d \times c$ | 0.5 |

What is the order of the mutational sites, and what are the map distances between the sites?

***15.22** Given the following map with point mutants and given the data in the following table, draw a topological representation of deletion mutants *r*21, *r*22, *r*23, *r*24, and *r*25. (Be sure to show the endpoints of the deletions. + = $r^+$ recombinants are obtained, 0 = $r^+$ recombinants are not obtained.)

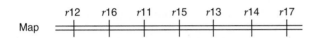

Map: *r*12 — *r*16 — *r*11 — *r*15 — *r*13 — *r*14 — *r*17

| Deletion Mutants | r11 | r12 | r13 | r14 | r15 | r16 | r17 |
|---|---|---|---|---|---|---|---|
| *r*21 | 0 | + | 0 | + | 0 | + | + |
| *r*22 | + | + | 0 | 0 | + | + | 0 |
| *r*23 | 0 | 0 | 0 | + | 0 | 0 | + |
| *r*24 | + | + | 0 | 0 | + | + | + |
| *r*25 | + | + | 0 | 0 | 0 | + | + |

***15.23** Given the following deletion map with deletions *r*31, *r*32, *r*33, *r*34, *r*35, and *r*36, place the point mutants *r*41, *r*42, etc., on the map. (Be sure to show where they lie with respect to end points of the deletions.)

*r*31
*r*32
*r*33
*r*34
*r*35 *r*36

| Point Mutants | r31 | r32 | r33 | r34 | r35 | r36 |
|---|---|---|---|---|---|---|
| | Deletion Mutants (+ = recombinants produced; 0 = no $r^+$ recombinants produced) | | | | | |
| *r*41 | 0 | 0 | 0 | 0 | + | 0 |
| *r*42 | 0 | 0 | 0 | + | 0 | + |
| *r*43 | 0 | 0 | + | + | + | 0 |
| *r*44 | 0 | 0 | 0 | 0 | + | + |
| *r*45 | 0 | + | 0 | + | + | + |
| *r*46 | 0 | 0 | + | 0 | + | 0 |

Show the dividing line between the *A* cistron and the *B* cistron on your map from the following data [+ = growth on strain *K12*(λ), 0 = no growth on strain *K12*(λ)]:

| Mutant | Complementation with *rIIA* | *rIIB* |
|---|---|---|
| *r*31 | 0 | 0 |
| *r*32 | 0 | 0 |
| *r*33 | 0 | + |
| *r*34 | 0 | 0 |
| *r*35 | 0 | + |
| *r*36 | 0 | 0 |
| *r*41 | 0 | + |
| *r*42 | 0 | + |
| *r*43 | + | 0 |
| *r*44 | 0 | + |
| *r*45 | 0 | + |
| *r*46 | 0 | + |

**15.24** Some adenine-requiring mutants of yeast are pink because of the intracellular accumulation of a red pigment. Diploid strains were made by mating haploid mutant strains. The diploids exhibited the following phenotypes:

| Cross | Diploid Phenotypes |
|---|---|
| 1 × 2 | pink, adenine requiring |
| 1 × 3 | white, prototrophic |
| 1 × 4 | white, prototrophic |
| 3 × 4 | pink, adenine requiring |

How many genes are defined by the four different mutants? Explain.

**15.25** Specialized transduction can be used to develop fine-structure maps. Five different λ*d gal*$^+$ phage were isolated (*1, 2, 3, 4, 5*). Each was infected into five different *gal* point mutants (*a, b, c, d, e*), and *gal*$^+$ recombinants were selected by plating the cells on media containing galactose as the sole carbon source. The results are shown in the following table (+ indicates that wild-type *gal*$^+$ recombinants were obtained; – indicates that no *gal*$^+$ recombinants were obtained):

| *E. coli gal* mutant | λ*d gal*$^+$ phage 1 | 2 | 3 | 4 | 5 |
|---|---|---|---|---|---|
| *a* | – | + | – | – | – |
| *b* | – | + | – | + | – |
| *c* | + | + | + | + | + |
| *d* | + | + | + | + | – |
| *e* | + | + | – | + | – |

Each λ*d gal*$^+$ phage was then coinfected into *E. coli* with each of five lambda point mutants (*j, k, l, m, n*), and a selection was performed for wild-type lambda progeny. The following table shows the results (+ indicates that wild-type λ recombinants were obtained; – indicates that no wild-type λ recombinants were obtained):

| | λd $gal^+$ phage | | | | |
|---|---|---|---|---|---|
| **λ mutant** | **1** | **2** | **3** | **4** | **5** |
| *j* | + | – | + | + | + |
| *k* | + | – | – | – | + |
| *l* | + | – | + | – | + |
| *m* | + | – | – | – | – |
| *n* | – | – | – | – | – |

Draw the *gal–bio* region of an *E. coli* λ lysogen, and label the location of the *att* site, the *gal* and *bio* genes, and λ. Below your drawing, indicate the relative map positions of the five *gal* mutations, the relative map positions of the five *gal* mutations, and the regions of the *gal* gene and λ genome that are present in each λ*d* $gal^+$ phage.

***15.26** In *E. coli*, eight spontaneously and independently arising *leu* mutants were isolated from a parental $F^-$ $str^R$ strain. Each mutant requires supplemental leucine to grow, but is resistant to streptomycin. Interrupted-mating experiments were performed with each of the eight *leu* mutants and a prototrophic *E. coli Hfr* strain sensitive to streptomycin. In each cross, $str^R$ $leu^+$ recombinants were recovered just after 4 minutes of mating. The fine structure of the *leu* region was then evaluated with the use of generalized transduction. Each of the eight mutants was individually infected with a generalized transducing phage. The resulting lysate was used to infect the other mutants, and $leu^+$ recombinants were selected. The following table shows the results (+ indicates that $leu^+$ recombinants were recovered, – indicates that $leu^+$ recombinants were not recovered):

| ***leu* mutant** | *1* | *2* | *3* | *4* | *5* | *6* | *7* | *8* |
|---|---|---|---|---|---|---|---|---|
| *1* | – | – | + | – | + | + | + | + |
| *2* | | – | – | – | + | – | – | + |
| *3* | | | – | + | + | + | – | + |
| *4* | | | | – | – | + | + | + |
| *5* | | | | | – | + | + | + |
| *6* | | | | | | – | + | + |
| *7* | | | | | | | – | – |
| *8* | | | | | | | | – |

**a.** Draw a map showing the relative order and locations of the mutant sites in this region. (*Hint*: First identify the deletions.)

**b.** Can you infer whether any of these mutations are point mutants? If not, how would you address this issue?

**c.** Explain whether you can infer how many cistrons involved in leucine biosynthesis are present in the region of interest.

**15.27** A homozygous white-eyed Martian fly ($w_1/w_1$) is crossed with a homozygous white-eyed fly from a different stock ($w_2/w_2$). It is well known that wild-type Martian flies have red eyes. The cross produces all white-eyed progeny. State whether each of the following is true or false, and explain your answer:

**a.** $w_1$ and $w_2$ are allelic genes.

**b.** $w_1$ and $w_2$ are nonallelic.

**c.** $w_1$ and $w_2$ affect the same function.

**d.** The cross was a complementation test.

**e.** The cross was a *cis-trans* test.

**f.** $w_1$ and $w_2$ are allelic by the terms of the functional test.

The $F_1$ white-eyed flies are allowed to interbreed, and when you classify the $F_1$, you find 20,000 white-eyed flies and 10 red-eyed progeny. Concerned about contamination, you repeat the experiment and get exactly the same results. How can you best account for the presence of the red-eyed progeny? As part of your explanation, give the genotypes of the $F_1$ and $F_2$ generation flies.

**15.28** Propose a genetic explanation for the ugly-duckling phenomenon: two white parents have a rare black offspring amid a prolific number of white offspring.

***15.29** Both *trpA* and *trpB* mutants of *E. coli* lack tryptophan synthetase activity. All *trpA* mutants complement all *trpB* mutants. Explain how two different complementing mutants (*trpA* and *trpB*) can affect the activity of the same enzyme.

**15.30** Four strains of *Neurospora*, all of which require arginine but have an unknown genetic constitution, have the following nutrition and accumulation characteristics:

| | Growth on | | | | |
|---|---|---|---|---|---|
| **Strain** | **Minimal Medium** | **Ornithine** | **Citrulline** | **Arginine** | **Accumulates** |
| 1 | – | – | + | + | ornithine |
| 2 | – | – | – | + | citrulline |
| 3 | – | – | – | + | citrulline |
| 4 | – | – | – | + | ornithine |

Pairwise complementation tests of the four strains gave the following results (+ = growth on minimal medium and – = no growth on minimal medium):

| | 4 | 3 | 2 | 1 |
|---|---|---|---|---|
| 1 | 0 | + | + | 0 |
| 2 | 0 | 0 | 0 | |
| 3 | 0 | 0 | | |
| 4 | 0 | | | |

Crosses among mutants yielded prototrophs in the following percentages:

1 × 2: 25%
1 × 3: 25%
1 × 4: none detected among 1 million ascospores
2 × 3: 0.002%
2 × 4: 0.001%
3 × 4: none detected among 1 million ascospores

Analyze the data and answer the following questions:

**a.** How many distinct mutational sites are represented among these four strains?

**b.** In this collection of strains, how many types of polypeptide chains (normally found in the wild type) are affected by mutations?

**c.** Write the genotypes of the four strains, using a consistent and informative set of symbols.

**d.** Determine the map distances between all pairs of linked mutations.

**e.** Determine the percentage of prototrophs expected among ascospores of the following types: (1) strain 1 × wild type; (2) strain 2 × wild type; (3) strain 3 × wild type; (4) strain 4 × wild type.

***15.31** Herpes simplex virus type 1 (HSV-1) is a large eukaryotic virus whose growth proceeds sequentially. Progression from one stage to the next requires completion of the earlier stage. Understanding how different viral genes are used at each stage should aid in the development of therapies for viral infection. Nine different mutations (*B2, B21, B27, B28, B32, 901, LB2, D, c75*) block viral growth at a very early stage. Each mutant grows at the permissive temperature of 34°C and fails to grow at the restrictive temperature of 39°C. All of the mutants except for *c75* spontaneously revert to wild type at about the same low frequency; *c75* reverts to wild type, but much less frequently than the others. Complementation analysis of these mutants and a tenth temperature-sensitive mutation that blocks growth at a later stage, *J12g*, was performed by coinfecting pairs of mutants into cells at 39°C, collecting the cell culture media, and assaying their virus content by infecting cells at 34°C. Virus production was quantified with an index *I*. For two mutants *A* and *B*, *I* = [yield (coinfection of *A* and *B*)]/[yield(infection of *A*) + yield (infection of *B*)]. *I* must be over 2 to be considered positive. The following table gives the values of *I* for pairs of coinfected mutants:

| Virus | *B21* | *B27* | *B28* | *B32* | *901* | *LB2* | *D* | *c75* | *J12g* |
|---|---|---|---|---|---|---|---|---|---|
| *B2* | 1.5 | 0.70 | 0.81 | 0.37 | 0.40 | 0.55 | 0.48 | 0.70 | 170 |
| *B21* | | 0.31 | 0.27 | 0.86 | 0.76 | 0.88 | 0.33 | 0.32 | 4.9 |
| *B27* | | | 0.28 | 0.18 | 1.9 | 1.5 | 0.61 | 0.68 | 19 |
| *B28* | | | | 0.20 | 0.42 | 0.68 | 0.13 | 0.50 | 72 |
| *B32* | | | | | 1.4 | 0.84 | 0.28 | 0.38 | 580 |
| *901* | | | | | | 0.45 | 0.10 | 0.20 | 570 |
| *LB2* | | | | | | | 0.44 | 0.91 | 22 |
| *D* | | | | | | | | 0.35 | 444 |
| *c75* | | | | | | | | | 30 |

Pairwise recombination frequencies of the eight mutants that block very early growth were determined. The recombination frequency for two mutants *A* and *B* was calculated as RF = [yield (coinfection of *A* and *B*) at 39°C]/[yield (coinfection of *A* and *B*) at 34°C] × 2 × 100. The following table gives RF values for pairs of mutants:

| Mutant | *B21* | *B27* | *B28* | *B32* | *901* | *LB2* | *D* | *c75* |
|---|---|---|---|---|---|---|---|---|
| *B2* | 0.36 | 0.51 | 2.6 | 2.5 | 6.0 | 1.7 | 4.0 | 0.87 |
| *B21* | | 0.91 | 3.0 | 2.8 | 6.4 | 2.3 | 4.5 | 1.6 |
| *B27* | | | 2.1 | 2.2 | 5.5 | 1.4 | 3.6 | 1.8 |
| *B28* | | | | 0.11 | 3.8 | 0.71 | 1.55 | 0.31 |
| *B32* | | | | | 3.4 | 0.73 | 1.45 | 0.55 |
| *901* | | | | | | 4.1 | 1.9 | 0.89 |
| *LB2* | | | | | | | 2.2 | 0.40 |
| *D* | | | | | | | | 0.0 |

Analyze these results and answer the following questions about the nine mutants that block HSV-1 growth at a very early stage:

**a.** Are the mutants point mutations or deletions?

**b.** How many functions are affected by the mutants?

**c.** Do any of the mutants affect the same site?

**d.** Do any of the mutants affect multiple sites?

**e.** What is the rationale behind the calculations of *I* and RF?

**f.** What are the relative map positions of *B2, B21, B27, B28, B32, 901, LB2*, and *D*?

**g.** RF values for the *c75* mutant are inconsistent with those of the other mutants. Assuming that there are no technical errors, what might explain this inconsistency?

***15.32** A large number of mutations in *Drosophila* alter the wild-type brick red eye color. The phenotypes, map locations, and genetic characteristics of eye color and many other *Drosophila* mutations are described at the Flybase website (http://www.flybase.org). As we discussed in Chapter 13, even alleles at one gene (*white, w*) can display a variety of phenotypes.

You are given a wild-type, red-eyed strain and six independently isolated, true-breeding mutant strains that have varying shades of brown eyes, with the assurance that each mutant strain has only a single mutation. How would you determine

**a.** whether the mutation in each strain is dominant or recessive?

**b.** how many different genes are affected in the six mutant strains?

**c.** which mutants, if any, are allelic?

**d.** whether any of these mutants are alleles of genes already known to affect eye color?

**15.33** In *Drosophila*, the *kar, ry*, and *l(3)26* loci are located on chromosome 3 at map positions 51.7, 52.0, and 52.2, respectively. Mutants at *kar* have karmoisin (bright red) eyes. Mutants at *l(3)26* are recessive lethal. Mutants at *ry* lack the enzyme xanthine dehydrogenase.

They survive and have rosy eyes if their dietary purine is limited, but they die if it is not. Wild-type $ry^+$ flies have brick red eyes and survive if fed a diet rich in purine. You want to test whether Benzer's findings at the *rII* locus in the T4 phage can be replicated in eukaryotes, so embark on a fine-structure analysis of the *ry* locus. Over the years, hundreds of mutants with rosy eyes have been identified by different researchers, and you have obtained many of them. Describe your experimental design, and address each of the following concerns:

**a.** Many loci affect eye color in *Drosophila*. What methods will you use to ensure that a rosy-eyed mutant is caused by a mutation at the *ry* locus?

**b.** What sets of crosses would you perform? In general terms, what progeny and frequencies do you expect to see in each cross?

**c.** How will you efficiently select for intragenic recombinants at the *ry* locus?

**d.** If you undertake both fine-structure recombination and a complementation analyses, what results do you expect to see if Benzer's findings are replicated?

# 16 Variations in Chromosome Structure and Number

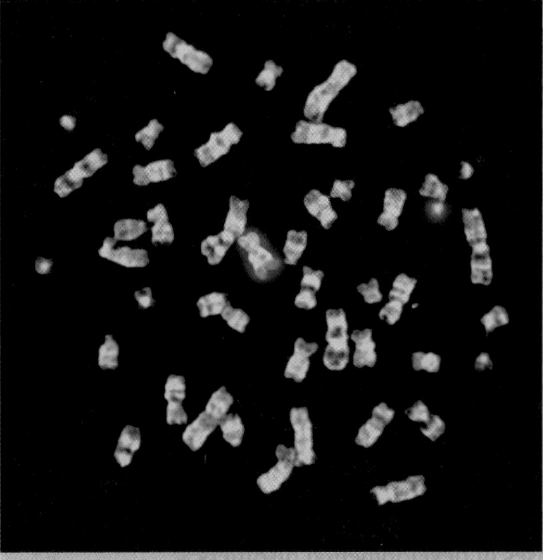
Human karyotype showing abnormal chromosomes (blue) resulting from a reciprocal exchange of parts of two nonhomologous chromosomes.

## Key Questions

- What changes occur in chromosome structure in eukaryotes?
- What changes occur in chromosome number in eukaryotes?
- What are the consequences of changes in chromosome structure and number on phenotypes?

### iActivity

YOU LIE ON A TABLE IN A SOFTLY LIT ROOM watching a black-and-white monitor. You see the image of a long needle being inserted into your uterus as you simultaneously feel the pressure of the needle against your abdomen. The doctor collects some of the amniotic fluid that surrounds your 16-week-old fetus. When the procedure is done, you get up, get dressed, and go home. Six weeks later, you go back to the clinic, where a counselor gently informs you that your unborn child has an extra chromosome 21, which causes Down syndrome.

Down syndrome is just one example of a number of human disorders that are the result of variations in the normal set of chromosomes. In this chapter, you will learn about the causes and effects of different chromosomal mutations. After you have read and studied the chapter, you can try the iActivity, in which you use your understanding of chromosomal mutations to help a couple who are trying to conceive a child.

In previous chapters, we learned many fundamental principles of transmission genetics, as applied to both eukaryotes and bacteria. With our understanding of the relationship between genes and chromosomes, we now consider chromosomal mutations—changes in normal chromosome structure or chromosome number. Changes in normal chromosome structure involve losses, additions, rearrangements of genes in the genome.

Chromosomal mutations affect both prokaryotes and eukaryotes, as well as viruses. The association of genetic defects with changes in chromosome structure or chromosome number indicates that not all genetic defects result from simple mutations of single genes. The study of normal and mutated chromosomes and their behavior is called *cytogenetics*. Your goal in this chapter is to learn about the various types of chromosomal mutations in eukaryotes and about some of the human disease syndromes that result from chromosomal mutations.

## Types of Chromosomal Mutations

**Chromosomal mutations** (or **chromosomal aberrations**) are *variations from the normal (wild-type) condition in chromosome structure or chromosome number.* In Bacteria, Archaea, and Eukarya, chromosomal mutations arise spontaneously or can be induced experimentally by certain chemicals or radiation. Chromosomal mutations traditionally were detected by genetic analysis—that is, by observing changes in the linkage arrangements of genes. A more precise description of chromosomal mutations is now possible in cases where genome sequences can be compared. In many eukaryotes, significant chromosomal

**Figure 16.6**

**Chromosome constitutions of *Drosophila* strains, showing the relationship between duplications of region 16A of the X chromosome and the production of reduced-eye-size phenotypes.**

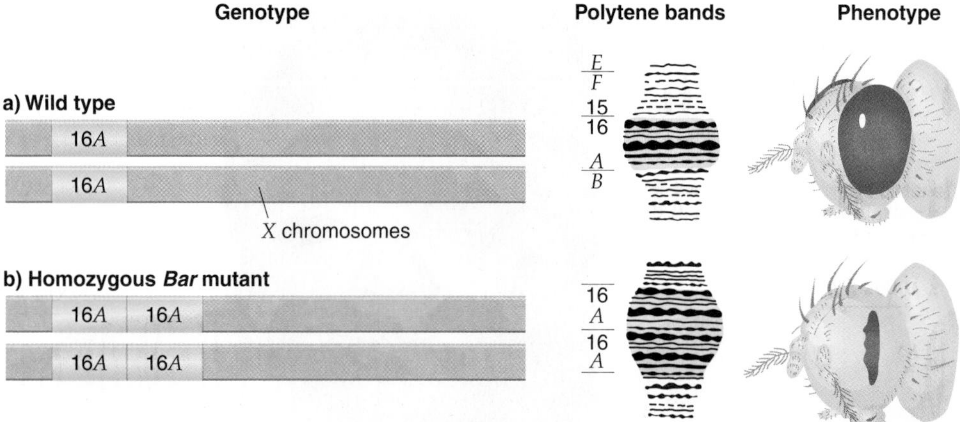


The sequences of the α-globin genes are all similar, as are the sequences of the β-globin genes. It is thought that each assembly of genes evolved from a different ancestral gene by duplication and subsequent divergence in the sequences of the duplicated genes (see Chapter 23, pp. 700–701). This chapter's Focus on Genomics box describes the duplications of the genes in the androgen-binding protein family that have occurred during evolution of some mammalian lineages.

## Inversion

An **inversion** is a chromosomal mutation that results when a segment of a chromosome is excised and then reintegrated at an orientation 180 degrees from the original orientation (Figure 16.7). There are two types of inversions: A **paracentric inversion** does not include the centromere (Figure 16.7a), and a **pericentric inversion** includes the centromere (Figure 16.7b).

**Figure 16.7**

**Inversions.**

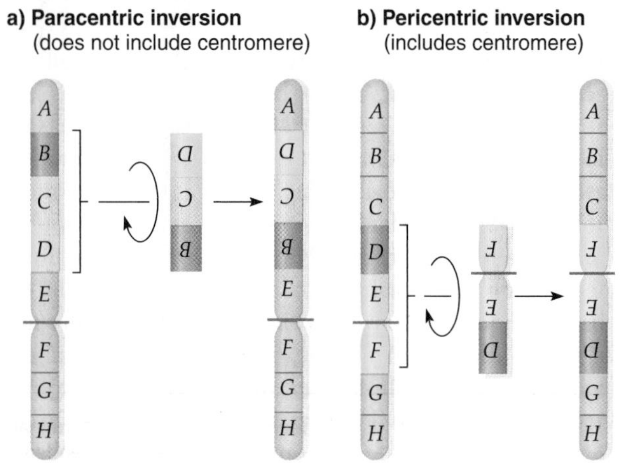

a) Paracentric inversion (does not include centromere)

b) Pericentric inversion (includes centromere)

Typically, genetic material is not lost when an inversion takes place, although there can be phenotypic consequences when the breakpoints (inversion ends) occur within genes or within regions that control gene expression. Homozygous inversions can be identified through the non-wild-type linkage relationships that result between the genes within the inverted segment and the genes that flank the inverted segment. For example, if the order of genes on the normal chromosome is *ABCDEFGH* and the *BCD* segment is inverted (shown next in bold), the gene order will be ***ADCBEFGH***, with *D* now more closely linked to *A* than to *E* and *B* now more closely linked to *E* than to *A* (see Figure 16.7a).

The meiotic consequences of a chromosome inversion depend on whether the inversion occurs in a homozygote or a heterozygote. If the inversion is homozygous, then meiosis is normal and there are no problems related to gene duplications or deletions. For an inversion heterozygote, there are no meiotic problems if crossing-over is absent in the inversion, but serious genetic consequences ensue if crossing-over occurs in the inversion, as we will now see.

Let us consider a paracentric inversion heterozygote, genotype ○*A**BCD**EFGH*/○*A**DCB**EFGH*, with the centromere (○) to the left of gene *A*. In meiosis, the homologous chromosomes attempt to pair such that the best possible base pairing occurs. Because of the inverted segment on one homolog, pairing of homologous chromosomes requires the formation of loops containing the inverted segments, called *inversion loops*. Inversion heterozygotes, then, may be identified by looking for those loops. If no crossovers occur in the inversion loop of a paracentric inversion heterozygote, then all resulting gametes receive a complete set of genes (two gametes with a normal gene order, ○*A**BCD**EFGH*, and two gametes with the inverted segment, ○*A**DCB**EFGH*), and they are all viable. Figure 16.8 shows the effects of a single crossover in the inversion loop, here between genes *B* and *C*. During the first meiotic

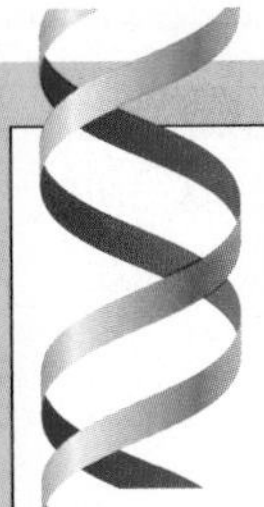

## Focus on Genomics

### Gene Duplications and Deletions in the Androgen-Binding Protein Family

In mice, a secreted protein complex called ABP (androgen-binding protein) is involved in mate choice. Females prefer males that make an ABP (present in his saliva) similar to the ABP made by the female, and reject males that make different ABP variants. ABP contains three subunits: α, β, and γ. The mouse (*Mus musculus*) genome has a large collection of ABP-like genes, which encode similar proteins, and this ABP family consists of 14 protein-coding α-like genes, 16 nonfunctional α-like *pseudogenes* (collectively, the *Abpa* genes), 13 protein-coding β- or γ-like genes and 21 nonfunctional β- or γ-like genes (collectively, the *Abpbg* genes). A **pseudogene** is very similar to a functional gene when you look at the DNA, but one or more mutations have changed the gene so that it cannot encode a functional product. Most pseudogenes are derived from protein-coding genes. Some of these pseudogenes lack regulatory sequences such as promoters, while others have frameshift mutations or nonsense mutations that prevent the gene from functioning.

The biochemical functions of the ABP proteins are not yet known. Most mammals have one *Abpa* gene and one *Abpbg* gene. Rats and mice shared a common ancestor about 12 million years ago, and comparison of the rat genome and the mouse genome suggests that this common ancestor had one *Abpa* gene and one *Abpbg* gene, so these genes have undergone extensive duplication in the mouse lineage since mice and rats diverged. In fact, some species of mice, including *Mus pahari,* still have only one *Abpa* gene and one *Abpbg* gene. *Mus pahari* and *Mus musculus* diverged about 7 million years ago, so the duplications in the *M. musculus* lineage occurred very rapidly. The investigators studying this gene family in mice then turned their attention to the other mammalian genomes and made a striking observation. Duplications in the *Abpa* and *Abpbg* gene are not limited to mice. Large-scale duplications of these genes have occurred in at least three different evolutionary lineages. Specifically, in addition to mice, the genes have undergone duplications in the lineage leading to rabbits and in the lineage leading to cattle. Comparison of the genes in rabbits, mice, and cattle suggests that the duplications occurred independently in each lineage. Other mammalian genomes—including bats, cats, dogs, squirrels, and tree shrews—show no signs of duplication of the *Abpa* and *Abpbg* genes. Some mammals, including humans and chimps, have only a single *Abpa* pseudogene and a single *Abpbg* pseudogene. Other animals—including hedgehogs, elephants, and armadillos—completely lack all versions of these genes. This is clearly a gene family undergoing very rapid duplication and deletion, although the reason for this rapid change is as yet unknown. Organisms that use ABP family members for mate selection may have more copies of these genes, but this correlation has not been tested in rabbits or cattle.

anaphase, the two centromeres migrate to opposite poles of the cell. Because of the crossover, one recombinant chromatid becomes stretched across the cell as the two centromeres begin to migrate in anaphase, forming a **dicentric bridge**—that is, a chromosome with two centromeres (a **dicentric chromosome**). With continued migration, the dicentric bridge breaks due to tension. The other recombinant product of the crossover event is a chromosome without a centromere (an *acentric fragment*). This acentric fragment is unable to continue through meiosis and is usually lost. (It is not found in the gametes.)

In the second meiotic division, each daughter cell receives a copy of each chromosome. Two of the gametes—the gamete with the normal order of genes (○***ABCD****EFGH*) and the gamete with the inverted segment of genes (○***ADCB****EFGH*)—have complete sets of genes and are viable. The other two gametes are inviable because they are unbalanced: many genes are deleted. Thus, *the only gametes that can give rise to viable progeny are those containing the chromosomes that did not involve crossing-over*. However, in many cases in female animals, the dicentric chromosomes or acentric fragments arising as a result of inversion are shunted to the polar bodies, so the reduction in fertility may not be so great. In short, for paracentric inversion heterozygotes, viable recombinants are reduced significantly or suppressed altogether. That is, the frequency of crossing-over is not lower in the heterozygotes than in normal cells, but gametes or zygotes derived from recombined chromatids are inviable.

The consequences of a single crossover in the inversion loop of an individual heterozygous for a pericentric inversion are shown in Figure 16.9. The normal chromosome is ***ABC***○***D****EFGH* and the inversion chromosome is ***AD***○***CB****EFGH*; the centromere is between *C* and *D*. The crossover event and the ensuing meiotic divisions result in two viable gametes with the nonrecombinant chromosomes ***ABC***○***D****EFGH* (normal) and ***AD***○***CB****EFGH* (inversion) and in two recombinant

**Crossing-over in an Inversion Heterozygote**

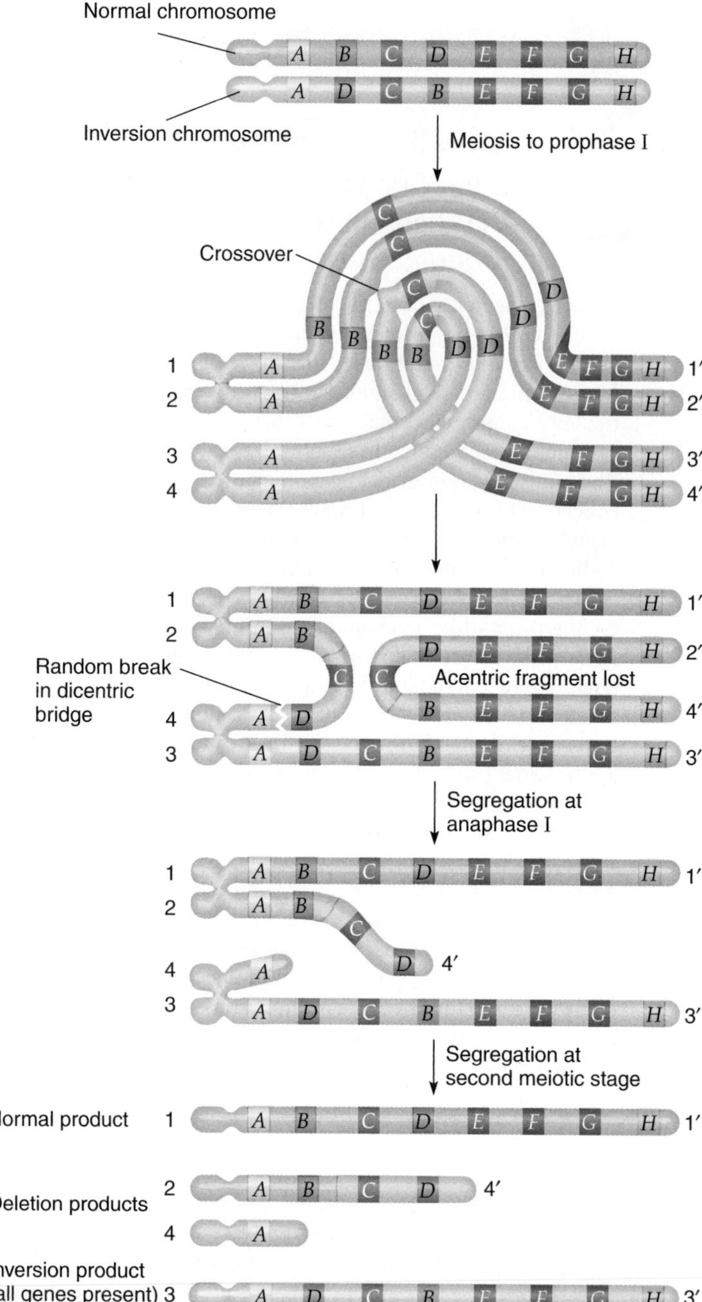


**Figure 16.8**

**Consequences of a paracentric inversion.** Meiotic products resulting from a single crossover within a heterozygous, paracentric inversion loop. Crossing-over occurs at the four-strand stage involving two nonsister homologous chromatids.

gametes that are inviable, each as a result of the deletion of some genes and the duplication of other genes.

Some crossover events within an inversion loop do not affect gamete viability. For example, a double crossover close together and involving the same two chromatids (a two-strand double crossover; see Chapter 14) produces four viable gametes. A second exception occurs when the duplicated and deleted segments of the recombinant chromatids do not affect essential genes, and hence viability, to a significant degree, as when the chromosome segments involved are very small. Also, recent studies with mammals show that inverted segments may remain unpaired. Since crossing-over cannot occur between unpaired segments, no inviable gametes are generated.

## Translocation

A **translocation** is a chromosomal mutation in which there is a change in position of chromosome segments and the gene sequences they contain to a different location in the genome (Figure 16.10). No gain or loss of genetic material is involved in a translocation. If a chromosome segment changes position within the same chromosome, the translocation is a *nonreciprocal intrachromosomal* (within a chromosome) *translocation* (Figure 16.10a). If a chromosome segment is transferred from one chromosome to another, the translocation is a *nonreciprocal interchromosomal* (between chromosomes) *translocation* if a one-way transfer is involved (Figure 16.10b) and a *reciprocal interchromosomal translocation* if

Figure 16.9

**Meiotic products resulting from a single crossover within a heterozygous, pericentric inversion loop.** Crossing-over occurs at the four-strand stage involving two nonsister homologous chromatids.

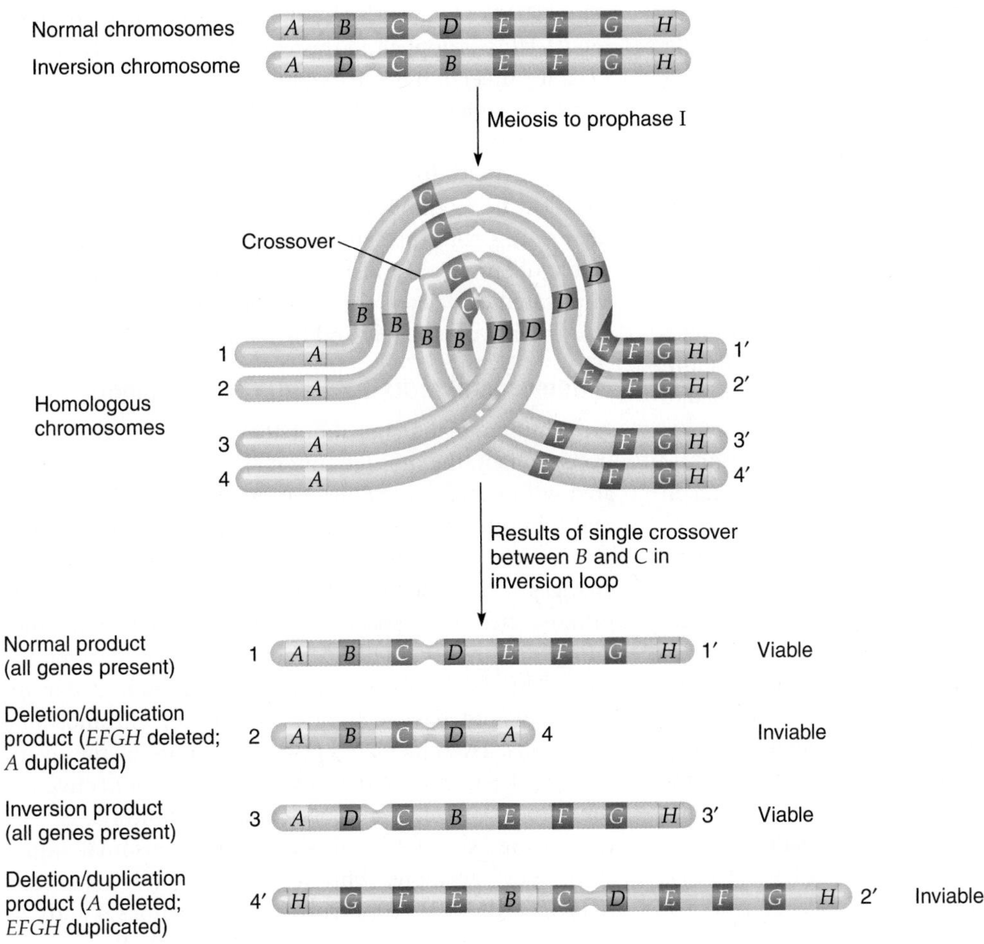


Figure 16.10

**Translocations.**

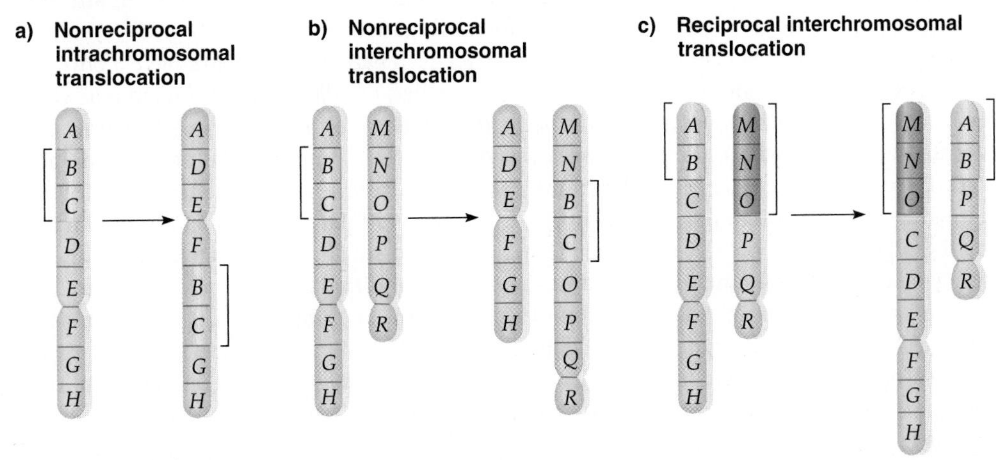

an exchange of segments between the two chromosomes is involved (Figure 16.10c).

In organisms homozygous for the translocations (i.e., when both copies of the genome in the diploid have the translocation), the genetic consequence is an alteration in the linkage relationships of genes. For example, in the nonreciprocal intrachromosomal translocation shown in Figure 16.10a, the *BC* segment has moved to the other chromosome arm and has become inserted between the *F* and *G* segments. As a result, genes in the *F* and *G* segments are now farther apart than they are in the normal strain, and genes in the *A* and *D* segments are now more closely linked. Similarly, in reciprocal translocations, new linkage relationships are produced.

Translocations typically affect the products of meiosis. In many cases, some of the gametes produced are unbalanced, in that they have duplications or deletions, and consequently are inviable. In other cases, such as familial Down syndrome resulting from a duplication stemming from a translocation, the gametes are viable (see later in the chapter). We focus here on reciprocal translocations.

In strains *homozygous* for a reciprocal translocation, meiosis takes place normally because all chromosome pairs can pair properly, and crossing-over does not produce any abnormal chromatids. In strains heterozygous for a reciprocal translocation, however, all homologous chromosome parts pair as best they can. Since one set of normal chromosomes (N) and one set of translocated chromosomes (T) are involved, the result is a crosslike configuration in meiotic prophase I (Figure 16.11). These crosslike figures consist of four associated chromosomes, each partially homologous to two other chromosomes in the group.

**animation**
**Meiosis in a Translocation Heterozygote**

Segregation at anaphase I may occur in three different ways. (We ignore the complication of crossing-over in this discussion.) In one way, called *alternate segregation*, alternate centromeres migrate to the same pole (Figure 16.11, left: $N_1$ and $N_2$ migrate to one pole, $T_1$ and $T_2$ to the other pole). This produces two gametes, each of which is viable because it contains a complete set of genes—no more, no less. One of these gametes has two normal chromosomes, and the other has two translocated chromosomes. In the second way, called *adjacent-1 segregation*, adjacent *nonhomologous* centromeres migrate to the same pole (Figure 16.11, middle: $N_1$ and $T_2$ migrate to one pole, $N_2$ and $T_1$ to the other pole). Both gametes produced contain gene duplications and deletions and are often inviable. Adjacent-1 segregation occurs about as frequently as alternate segregation. In the third way, called *adjacent-2 segregation*, different pairs of adjacent *homologous* centromeres migrate to the same pole (Figure 16.11, right: $N_1$ and $T_1$ migrate to one pole, $N_2$ and $T_2$ to the other pole). Both products have gene duplications and deletions and are always inviable. Adjacent-2 segregation seldom occurs.

In sum, of the six theoretically possible gametes, the two from alternate segregation are functional, the two from adjacent-1 segregation usually are inviable (because of gene duplications and deficiencies), and the two from adjacent-2 seldom occur and are inviable if they do. Moreover, because alternate segregation and adjacent-1 segregation occur with about equal frequency, the term *semisterility* is applied to this condition. (The term is also used for inversion heterozygotes.)

In practice, animal gametes that have large duplicated or deleted chromosome segments may function, but the zygotes formed by such gametes typically die. In contrast, if the duplicated and deleted chromosome segments are small, the gametes may function normally and viable offspring may result. In plants, pollen grains with duplicated or deleted chromosome segments typically do not develop completely and hence are nonfunctional.

## Chromosomal Mutations and Human Tumors

Most human malignant tumors have chromosomal mutations. In fact, the most common class of mutation associated with cancer is a translocation. The exact chromosomal abnormality actually varies quite a bit among tumors, ranging from simple rearrangements to complex changes in chromosome structure and number. In many tumors, there is no specific associated chromosomal mutation. Rather, a variety of chromosomal mutations are seen. This is the case with most solid tumors, for instance, which have complex patterns of chromosomal mutations. Examples are epithelial tumors of the ovary, lung, and pancreas and many sarcomas (connective-tissue tumors), such as osteosarcoma. By contrast, certain tumors are associated with specific chromosomal anomalies. For example, chronic myelogenous leukemia (CML; OMIM 151410; involves chromosomes 9 and 22) and Burkitt lymphoma (BL; OMIM 113970; involves chromosomes 8 and 14) are associated with reciprocal translocations. Untreated, CML is an invariably fatal cancer involving the uncontrolled replication of myeloblasts (stem cells of white blood cells). A new targeted drug developed recently, Gleevec®, is now the standard first treatment for CML patients.

Ninety percent of patients with CML have a chromosomal mutation called the *Philadelphia chromosome* ($Ph^1$) in the leukemic cells. The mutation was so named because it was discovered in Philadelphia. The Philadelphia chromosome results from a reciprocal translocation involving the movement of part of the long arm of chromosome 22 (the second-smallest human chromosome) to chromosome 9 and the movement of a small part from the tip of the long arm of chromosome 9 to chromosome 22 (Figure 16.12). This reciprocal translocation apparently converts a **proto-oncogene**—a gene that, in normal cells, controls the normal proliferation of cells—to an **oncogene** (see Chapter 20)—a gene that encodes a protein which plays a role in the transition from a differentiated cell to a tumor cell with an uncontrolled pattern of growth. Specifically, the *ABL* ("able"; named for *Abelson*) proto-oncogene, normally located on chromosome 9, is translocated to chromosome 22 in patients with CML

**Figure 16.11**

**Meiosis in a translocation heterozygote in which no crossover occurs.**

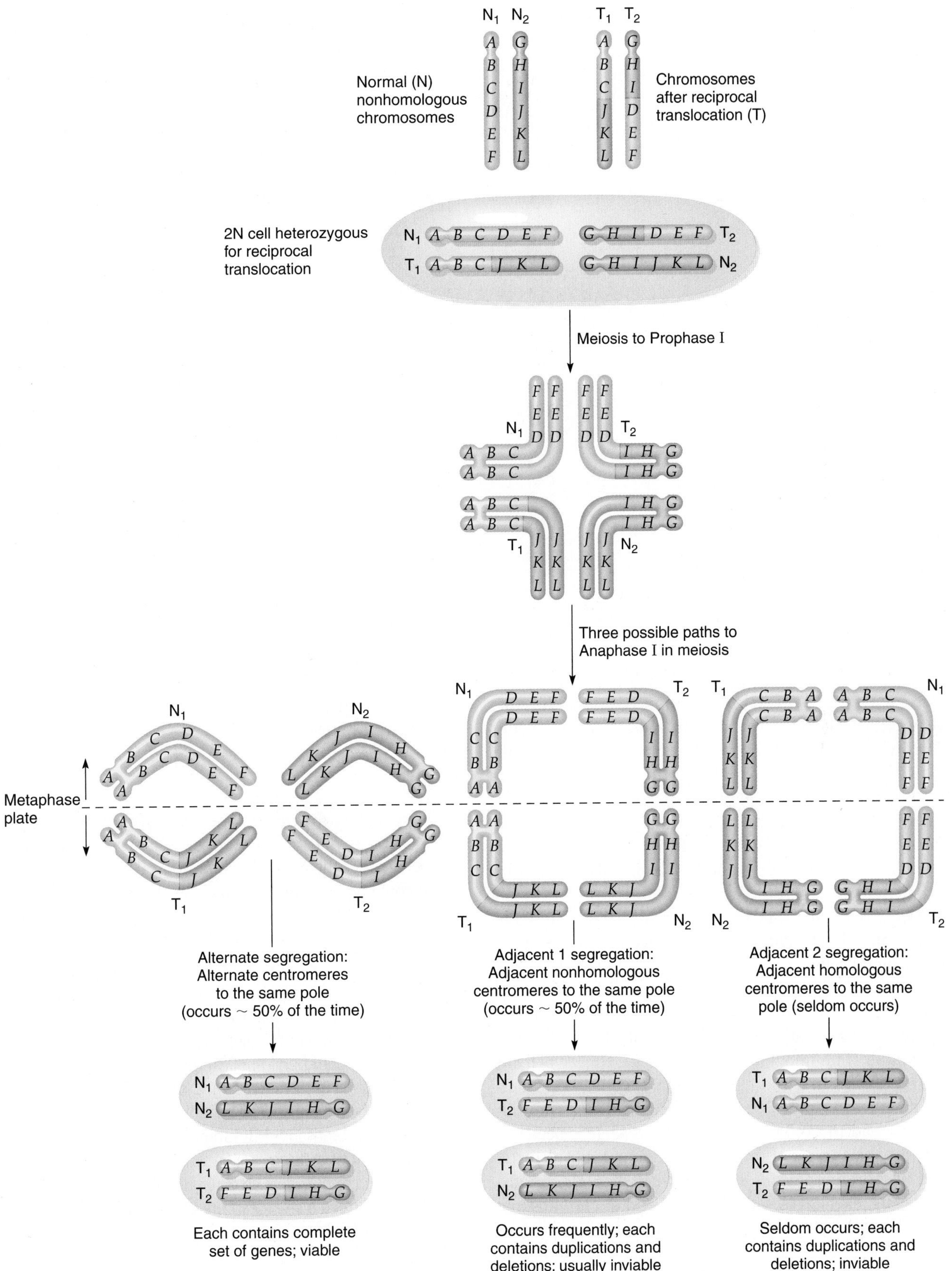

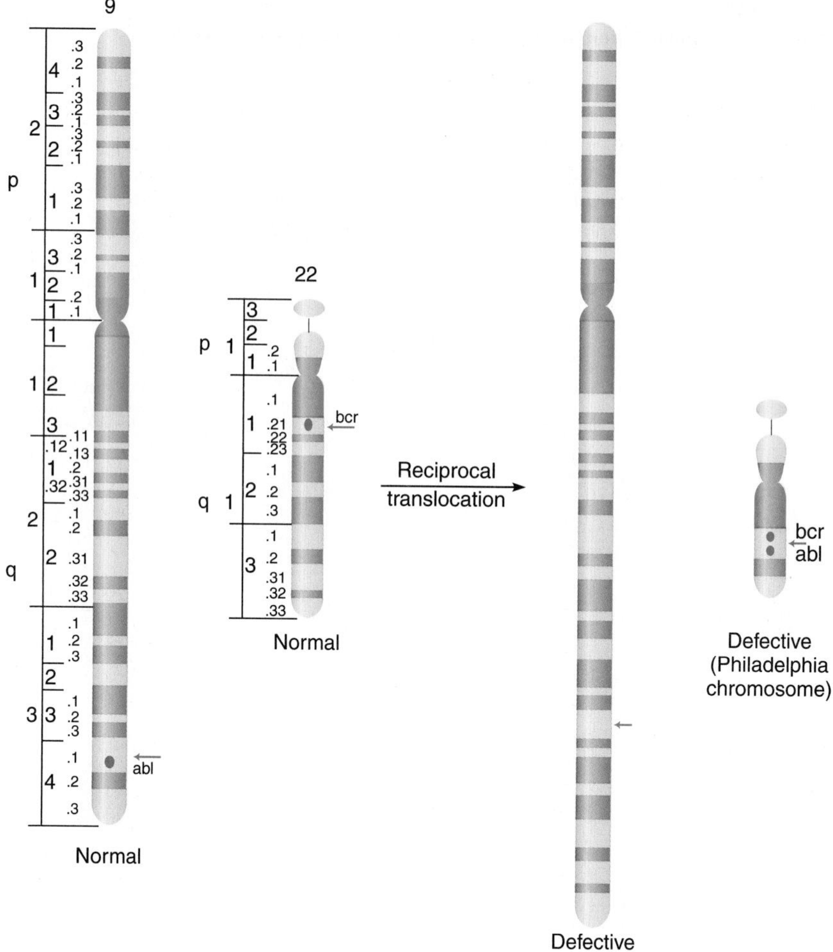


**Figure 16.12**
**Origin of the Philadelphia chromosome in chronic myelogenous leukemia (CML) by a reciprocal translocation involving chromosomes 9 and 22.** The arrows show the sites of the breakage points.

(see Figure 16.12). The translocation positions the *ABL* gene within the *BCR* (breakpoint cluster region) gene. The hybrid *BCR–ABL* gene is the oncogene responsible for CML; that is, the hybrid gene expresses a constitutively activated tyrosine kinase. ("Constitutively activated" means activated all the time.) Activated tyrosine kinase is one kind of gene product that contributes to stimulating a cell to grow and divide, in this case resulting in too many white blood cells. The drug Gleevec® works by blocking the tyrosine kinase so that the body stops (or at least reduces) the manufacture of too many white blood cells.

Burkitt lymphoma, a particularly common disease in Africa, is a virus-induced tumor that affects cells of the immune system called B cells. Normally, B cells secrete antibodies (immunoglobulin molecules). Ninety percent of the tumors in Burkitt lymphoma patients are associated with a reciprocal translocation involving chromosomes 8 and 14. As with CML, a proto-oncogene becomes activated as a result of the translocation event: The distal end of chromosome 8, starting with the *MYC* proto-oncogene, exchanges with the distal end of chromosome 14. The *MYC* gene becomes positioned next to a transcriptionally active immunoglobulin gene, resulting in overexpression of the *MYC* gene. The overexpressed *MYC* gene is the oncogene responsible for the uncontrolled cell growth and division that leads to the development of Burkitt lymphoma.

### *i* Activity

In the iActivity *Deciphering Karyotypes* on your student website, you are a genetic counselor who must determine whether there are any chromosomal abnormalities that could be affecting a couple's ability to have children.

### Keynote

Chromosomal mutations may involve parts of individual chromosomes, rather than whole chromosomes or sets of chromosomes. The four major types of structural alterations are deletions and duplications (both of which involve a change in the amount of DNA on a chromosome), inversions (which involve no change in the amount of DNA on a chromosome, but rather a change in the arrangement of a chromosomal segment), and translocations (which also involve no change in the amount of DNA, but instead a change in the chromosomal location of one or more DNA segments). Problems associated with inversions and translocations are often manifested only as a result of crossing-over in heterozygotes during meiosis.

## Position Effect

Unless inversions or translocations involve breaks within a gene, those chromosomal mutations do not produce mutant phenotypes. Rather, as we have seen, they have significant consequences in meiosis when they are heterozygous with normal sequences. In some cases, however, phenotypic effects of inversions or translocations occur because of a different phenomenon called **position effect**—a change in the phenotypic expression of one or more genes as a result of a change in position in the genome. This is an example of an *epigenetic* phenomenon, where **epigenetics** refers to a meiotically or mitotically heritable change in gene expression that does not involve a change in the DNA sequence of the affected gene(s).

For example, position effect may be exhibited if a gene that is normally located in euchromatin (chromosomal regions, representing most of the genome, that are condensed during division, but become uncoiled during interphase) is brought near heterochromatin (chromosomal regions that remain condensed throughout the cell cycle and are gene poor and transcriptionally inactive) by a chromosomal rearrangement. (Euchromatin and heterochromatin are described more fully in Chapter 2, p. 27). Note that gene transcription typically occurs in euchromatin, but not in heterochromatin; the difference in chromosome condensation is responsible for the distinction (see Chapter 18). An example of this kind of position effect involves the X-linked white-eye ($w$) locus in *Drosophila*. One inversion moves the $w^+$ gene from a euchromatic region near the end of the X chromosome to a position next to the heterochromatin at the centromere of the X. In a $w^+$ male, or in a $w^+/w$ female in which the $w^+$ is involved in the inversion, the eye exhibits a mottled pattern of red and white rather than being completely red as expected. The explanation is that, in flies with the inversion, some eye cell clones have the $w^+$ allele inactivated because of the position effect of $w^+$ near heterochromatin—the chromatin with the moved gene becomes more condensed, inhibiting transcription of that gene. Those clones produce white spots in the eyes. Clones of cells in which the $w^+$ allele is not inactivated produce red spots in the eye. Since the inactivation event is variable, the eye exhibits a mottled pattern of red and white spots.

Some human genetic diseases are associated with position effects. An example is aniridia (literally, "without iris"; OMIM 106210), a congenital eye condition characterized by severe hypoplasia (underdevelopment) of the iris, typically associated with cataracts and clouding of the cornea. Aniridia is caused by loss of function of the *PAX6* gene, which is involved in eye development. In individuals with a nonfunctional *PAX6* gene, eye development stops too early, and, at the time of birth, most of the eye is underdeveloped. The loss of function may be due to a deletion of the gene or simple mutations within the gene. In addition, some affected individuals have translocations with chromosomal breakpoints somewhat distant from the *PAX6* gene. It appears that, in this case, the expression of *PAX6* is suppressed by a position effect brought about by the new chromosomal environment surrounding the gene generated by the translocation.

## Fragile Sites and Fragile X Syndrome

When human cells are grown in culture, some of the chromosomes develop narrowings or unstained areas (gaps) called *fragile sites*. The chromosome may break spontaneously at a fragile site, resulting in deletion of the chromosome distal to the site. More than 40 fragile sites have been identified since the first one was discovered in 1965. One particular fragile site on the long arm of the X chromosome at position Xq27.3 (Figure 16.13) is associated with *fragile X syndrome* (also called *fragile site mental retardation*). After Down syndrome, fragile X syndrome is the leading genetic cause of mental retardation in the United States, with an incidence of about 1 in 4,000 males and 1 in 6,000 females (heterozygotes) of all races and ethnic groups. As with all recessive X-linked traits, the majority of those exhibiting this type of mental retardation are males. An individual with fragile X syndrome is shown in Figure 16.14.

The fragile X chromosome is inherited in a typical Mendelian fashion. Male offspring of carrier females have a 50% chance of receiving a fragile X chromosome. However, only 80% of males with a fragile X chromosome are mentally retarded; the rest are normal. These phenotypically normal males are called *normal transmitting males* and carry a *premutation* because they can pass on the fragile X chromosome to their daughters. (A premutation

**Figure 16.13**
**Fragile site on the X chromosome.**

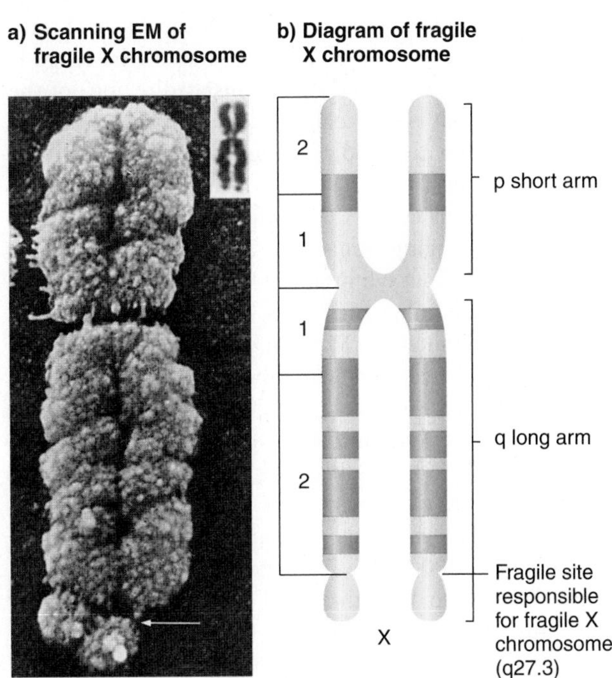

Figure 16.14
**Individual with fragile X syndrome.**

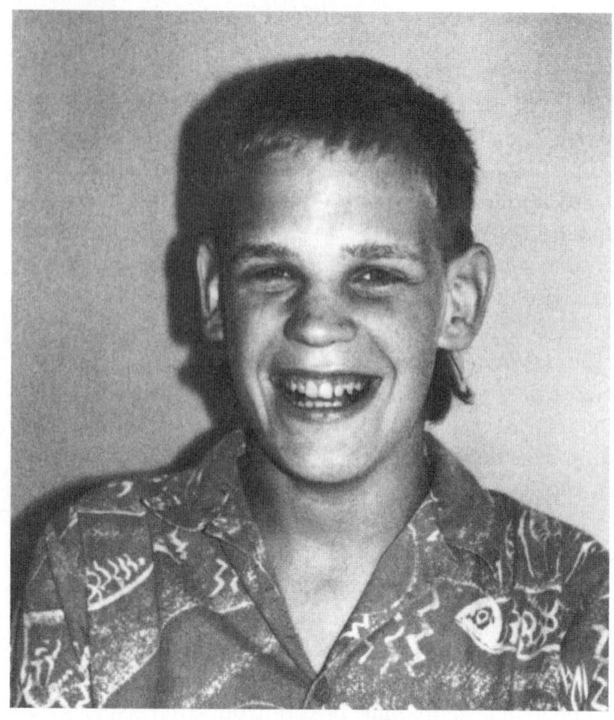



could be considered a silent mutation.) The sons of those daughters frequently show symptoms of mental retardation. Female offspring of carrier (heterozygous) females also have a 50% chance of inheriting a fragile X chromosome. Up to 33% of the carrier females show mild mental retardation.

Modern molecular techniques brought to bear on this disease have resulted in an understanding of the disease at the DNA level. There is a repeated 3 base-pair sequence, CGG, in a gene called *FMR-1* (*f*ragile X *m*ental *r*etardation-1; OMIM 309550) located at the fragile X site. Normal individuals have an average of 29 CGG repeats (the range is from 6 to 54) in the 5′ UTR (untranslated region) of the *FMR-1* gene; that is, the region which, in the mRNA, precedes the amino acid-coding region. Phenotypically normal transmitting males and their daughters, as well as some carrier females, have a significantly larger number of CGG repeats, ranging from 55 to 200 copies. These individuals do not show symptoms of fragile X syndrome, and the increased number of repeats they have is the aforementioned premutation. Males and females with fragile X syndrome have even larger numbers of the CGG repeats, ranging from 200 to 1,300 copies; these are considered to be the full mutations. In other words, the triplet repeat, CGG, in the *FMR-1* gene becomes duplicated (amplified) in a tandem manner. The process has been termed *triplet repeat amplification*. Below a certain threshold number of copies (about 200 or fewer) there are no clinical symptoms, and above that threshold number (greater than 200) clinical symptoms are seen. Interestingly, amplification of the CGG repeats does not occur in males, but only in females. Therefore, a phenotypically normal transmitting male (who has the premutation) transmits his X chromosome to his daughter. In a slipped mispairing process during DNA replication in his daughter, perhaps, the triplets may amplify, and she can transmit the amplified X to her offspring. Thus, affected males inherit the mutation from their grandfather.

The *FMR-1* gene encodes FMRP, an RNA-binding protein that binds to target mRNAs in the cell. The current model is that FMRP regulates the translation of certain target mRNAs in the cell by binding to those RNAs and blocking protein synthesis. Recent studies have shown that FMRP is active at synapses in the brain, participating in processes that control synaptic plasticity, the ability of the strength of the synaptic signal between two neurons to change. The triplet repeat expansion in *FMR-1* upstream of the protein-coding region affects the expression of *FMR-1*. In individuals with the full mutation, the C nucleotides of the CGG repeats become extensively methylated, leading to silencing of the *FMR-1* gene. The loss of gene activity affects the control of synaptic plasticity and, in ways that are incompletely understood, mental retardation results.

Triplet repeat amplification has also been shown to cause other human diseases, such as myotonic dystrophy (MD; OMIM 160900), spinobulbar muscular atrophy (also called Kennedy disease; OMIM 313200), and Huntington disease (HD; OMIM 143100; see Chapter 13). In each of these cases, no fragility of the associated chromosome is seen. Also, they differ from fragile X syndrome in that the amplification can occur in both sexes at each generation. For each, there is a threshold number of triplet repeat copies above which symptoms of the disease are produced.

## Variations in Chromosome Number

When an organism or a cell has one complete set of chromosomes or an exact multiple of complete sets, that organism or cell is said to be **euploid.** Thus, eukaryotic organisms that are normally diploid (such as humans and fruit flies) and eukaryotic organisms that are normally haploid (such as yeast) are euploids. Chromosome mutations that result in variations in the number of chromosome sets occur in nature, and the resulting organism or cells are also euploid. Chromosome mutations resulting in variations in the number of individual chromosomes are examples of **aneuploidy.** An **aneuploid** organism or cell has a chromosome number that is not an exact multiple of the haploid set of chromosomes. Both euploid and aneuploid variations affecting whole chromosomes are discussed in this section.

### Changes in One or a Few Chromosomes

**Generation of Aneuploidy.** Changes in chromosome number can occur in both diploid and haploid organisms. The nondisjunction of one or more chromosomes during meiosis I or meiosis II typically is responsible for generating gametes with abnormal numbers of chromosomes. Nondisjunction was discussed in Chapter 12 in the context of unusual complements of X chromosomes, with Figure 12.18 (p. 344) illustrating the consequences of nondisjunction at the first and second meiotic divisions. Referring to that figure and considering just one particular chromosome, one can see that nondisjunction at meiosis I produces four abnormal gametes: two with a chromosome duplicated and two with the corresponding chromosome missing. In a male, nondisjunction at meiosis I can produce a gamete with both the X and the Y chromosome; in a female, it produces a gamete with both sets of homologs (and thus possible heterozygotes). Fusion of the former gamete type with a normal gamete produces a zygote with three copies of the particular chromosome instead of the normal two and, unless nondisjunction also involves other chromosomes, there will be two copies of all other chromosomes. The latter gamete type may well be inviable. If it is viable, fusion with a normal gamete produces a zygote with only one copy of the particular chromosome instead of the normal two, and two copies of all other chromosomes. Nondisjunction in meiosis II (see Figure 12.18) is different from nondisjunction in meiosis I in that some normal gametes are produced. As Figure 12.19 shows, nondisjunction in meiosis II results in two normal gametes and two abnormal gametes—that is, a single gamete with two daughter chromosomes and one gamete with that same chromosome missing. Fusion of these with normal gametes gives the zygote types just discussed. Nondisjunction can occur in mitosis, giving rise to somatic cells with unusual chromosome complements.

**Figure 16.15**

**Normal (theoretical) set of metaphase chromosomes in a diploid (2N) organism *(top)* and examples of aneuploidy *(bottom)*.**

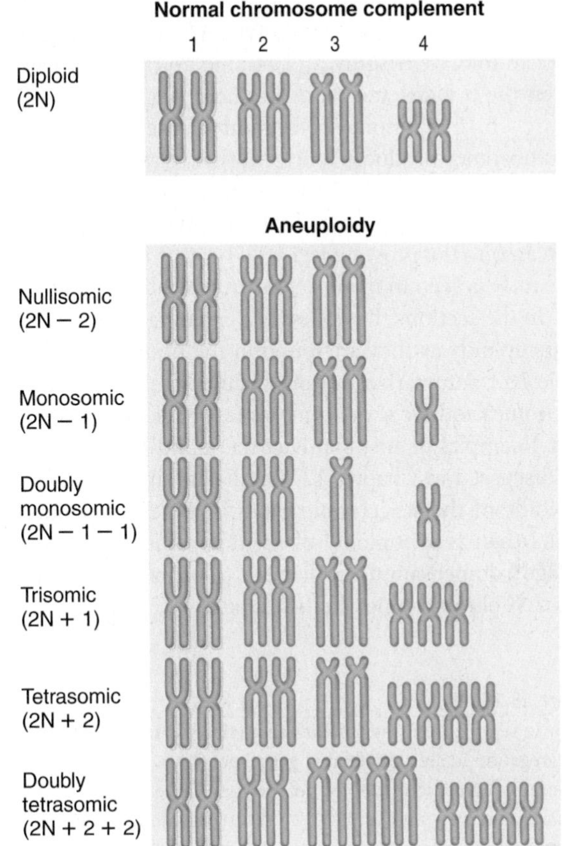


**Types of Aneuploidy.** In aneuploidy, one or more chromosomes are lost from or added to the normal set of chromosomes (Figure 16.15). Aneuploidy can occur, for example, from the loss of individual chromosomes in meiosis or (rarely) in mitosis by nondisjunction. In animals, autosomal aneuploidy is almost always lethal, so in mammals it is detected mainly in aborted fetuses. Aneuploidy is tolerated more by plants, especially in species that are considered polyploid (having more sets of chromosomes than the usual two).

In diploid organisms, there are four main types of aneuploidy (see Figure 16.15):

1. **Nullisomy** (a nullisomic cell) involves a loss of one homologous chromosome pair—the cell is 2N – 2. (Nullisomy can arise, for example, if nondisjunction occurs for the same chromosome in meiosis in both parents, producing gametes with no copies of that chromosome and one copy of all other chromosomes in the set.)
2. **Monosomy** (a monosomic cell) involves a loss of a single chromosome—the cell is 2N – 1. (Monosomy can arise, for example, if nondisjunction in meiosis in a parent produces a gamete with no copies of a particular chromosome and one copy of all other chromosomes in the set.)
3. **Trisomy** (a trisomic cell) involves a single extra chromosome—the cell has three copies of a particular chromosome and two copies of all other chromosomes. A trisomic cell is 2N + 1. (Trisomy can arise, for example, if nondisjunction in meiosis in a parent produces a gamete with two copies of a particular chromosome and one copy of all other chromosomes in the set.)
4. **Tetrasomy** (a tetrasomic cell) involves an extra chromosome pair; that is, there are four copies of one particular chromosome and two copies of all other chromosomes—the cell is 2N + 2. (Tetrasomy can arise, for example, if nondisjunction occurs for the same chromosome in meiosis in both parents, producing gametes with two copies of that chromosome and one copy of all other chromosomes in the set.)

Aneuploidy may involve the loss or the addition of more than one specific chromosome or chromosome pair. For example, a *double monosomic* has two separate chromosomes present in only one copy each; that is, it is 2N – 1 – 1. A *double tetrasomic* has two chromosomes present in four copies each; that is, it is 2N + 2 + 2. In both cases, meiotic nondisjunction involved two different chromosomes in one parent's gamete production.

Most forms of aneuploidy have serious consequences in meiosis. Monosomics, for example, produce two kinds of haploid gametes: N and N − 1. Alternatively, the odd, unpaired chromosome in the 2N − 1 cell may be lost during meiotic anaphase and not be included in either daughter nucleus, thereby producing two N − 1 gametes. For trisomics, there are more segregation possibilities in meiosis. Consider a trisomic of genotype +/+/*a* in an organism that can tolerate trisomy, and assume no crossing-over between the *a* locus and its centromere. Then, as shown in Figure 16.16, random segregation of the three types of chromosomes produces four genotypic classes of gametes: 2 (+ *a*) : 2 (+) : 1 (+ +) : 1 (*a*). In a cross of a +/+/*a* trisomic with an *a*/*a* individual, the predicted phenotypic ratio among the progeny is 5 wild type (+) : 1 mutant (*a*). This ratio is seen in many actual crosses of this kind.

In the sections that follow, we examine some examples of aneuploidy as they are found in the human population. Table 16.1 summarizes various aneuploid abnormalities for autosomes and for sex chromosomes in the human population. Examples of aneuploidy of the X and Y chromosomes are discussed in Chapter 12. Recall that, in mammals, aneuploidy of the sex chromosomes is more often found in adults than is aneuploidy of the autosomes, because of a dosage compensation mechanism (lyonization) by which excess X chromosomes are inactivated.

**Figure 16.16**

**Meiotic segregation possibilities in a trisomic individual.** Shown is segregation in an individual of genotype +/+/*a* when two chromosomes migrate to one pole and one goes to the other pole, and assuming no crossing-over between the *a* locus and its centromere. The two + alleles are labeled $+_1$ and $+_2$ to distinguish them.

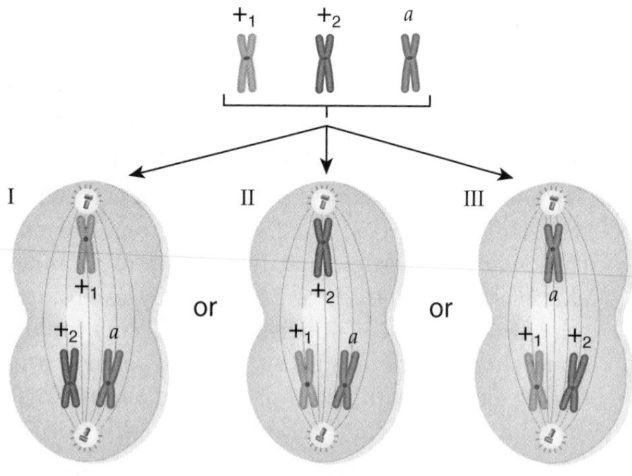


Gametes produced after 2nd meiotic division

| | haploid | disomic |
|---|---|---|
| I | $+_1$ | $+_2/a$ |
| II | $+_2$ | $+_1/a$ |
| III | *a* | $+_1/+_2$ |

In sum: 2 +/*a* : 2 + : 1 +/+ : 1 *a*

**Table 16.1 Aneuploid Abnormalities in the Human Population**

| Chromosomes | Syndrome | Frequency at Birth |
|---|---|---|
| **Autosomes** | | |
| Trisomic 21 | Down | 14.3/10,000 |
| Trisomic 13 | Patau | 2/10,000 |
| Trisomic 18 | Edwards | 2.5/10,000 |
| **Sex chromosomes, females** | | |
| XO, monosomic | Turner | 4/10,000 females |
| XXX, trisomic | Viable; most | |
| XXXX, tetrasomic | are fertile | 14.3/10,000 females |
| XXXXX, pentasomic | | |
| **Sex chromosomes, males** | | |
| XYY, trisomic | Normal | 25/10,000 males |
| XXY, trisomic | | |
| XXYY tetrasomic | Klinefelter | 40/10,000 |
| XXXY, tetrasomic | | |

In humans, autosomal monosomy is rare. Presumably, monosomic embryos do not develop significantly and are lost early in pregnancy. In contrast, autosomal trisomy accounts for about one-half of chromosomal abnormalities producing fetal deaths. In fact, only a few autosomal trisomies are seen in live births. Most of these (trisomy-8, -13, and -18) result in early death. Only in trisomy-21 (Down syndrome) does survival to adulthood occur.

***Trisomy-21.*** **Trisomy-21** (OMIM 190685) occurs when there are three copies of chromosome 21 (Figure 16.17a) and with a frequency of about 3,510 per 1 million conceptions and about 1,430 per 1 million live births. Individuals with trisomy-21 have Down syndrome (Figure 16.17b), characterized by such abnormalities as low IQ, epicanthal folds (in which the skin of the upper eyelid forms a layer that covers the inner corner of the eye), short and broad hands, and below-average height. Down syndrome is named for the late-nineteenth-century English physician John Langdon Down, who, in 1866, became the first to publish an accurate description of a person with the condition.

A direct relationship exists between maternal age and probability of giving birth to an individual with trisomy-21. (Table 16.2). (For many years, it was thought that there was no correlation with age of the father. Recent evidence, however, indicates that paternal age has an effect on Down syndrome if the mother is 35 years old or older; in younger women, there is no paternal effect.) During the development of a female fetus before birth, the primary oocytes in the ovary undergo meiosis, but stop at prophase I. In a fertile female, each month at ovulation the nucleus of a secondary oocyte (see Chapter 12) begins the second meiotic division, but progresses only to metaphase, when division

Figure 16.17
**Trisomy-21 (Down syndrome).**

a) Karyotype (G banding)

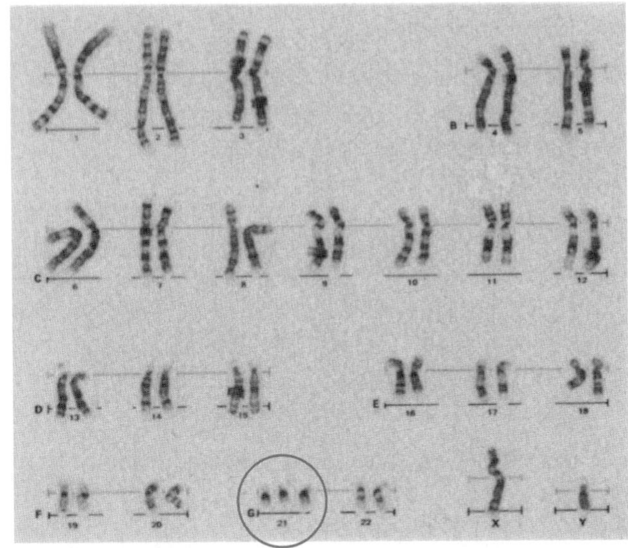

b) Individual with trisomy-21 (Down syndrome)

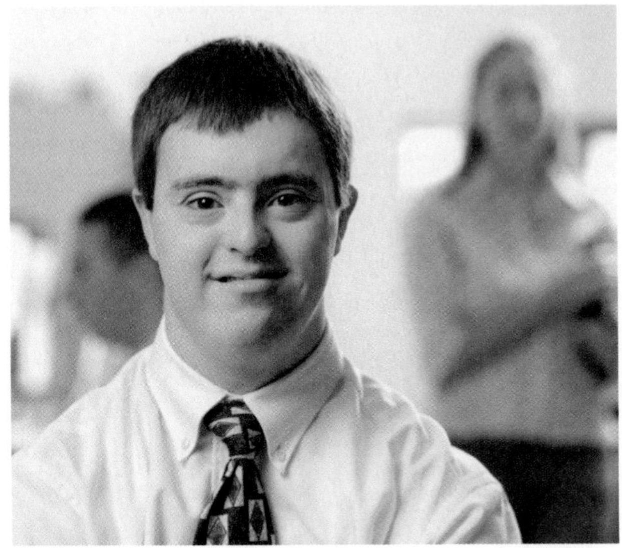

**Table 16.2 Relationship Between Age of Mother and Risk of Trisomy-21**

| Age of Mother | Risk of Trisomy-21 in Child |
|---|---|
| 16–26 | 7.7/10,000 |
| 27–34 | 4/10,000 |
| 35–39 | 29/10,000 |
| 40–44 | 100/10,000 |
| 45–47 | 333/10,000 |
| All mothers combined | 14.3/10,000 |

again stops. If a sperm penetrates the secondary oocyte, the second meiotic division is completed. The probability of nondisjunction increases with the length of time the primary oocyte is in the ovary. It is important, then, that older mothers-to-be consider testing—for example, by undergoing amniocentesis or chorionic villus sampling (see Chapter 4, p. 74)—to determine whether the fetus has a normal complement of chromosomes.

Are there other risk factors for having a Down syndrome baby? Where a person lives, social class, and race have no influence on the chance of having a baby with Down syndrome. However, mothers under 35 years of age who smoke are at an increased risk of having children with the syndrome. If mothers with these characteristics use cigarettes and oral contraceptives, the risk is increased over using cigarettes alone. Oral contraceptive use alone for this class of mothers has no effect on the incidence of Down syndrome.

Down syndrome can also result from a different sort of chromosomal mutation called centric fusion or **Robertsonian translocation,** which produces three copies of the long arm of chromosome 21. (The translocation is named for W. R. B. Robertson, an insect geneticist who first described this type of chromosomal mutation.) This form of Down syndrome, called familial Down syndrome, is responsible for 2–3% of Down syndrome cases. A Robertsonian translocation is a type of reciprocal translocation in which two nonhomologous acrocentric chromosomes (chromosomes with centromeres near their ends) break at their centromeres and then the long arms become attached to a single centromere (Figure 16.18). The short arms also join to form the reciprocal product, which typically contains nonessential genes and usually is lost within a few cell divisions. In humans, when a Robertsonian translocation joins the long arm of chromosome 21 with the long arm of chromosome 14 (or 15), the heterozygous carrier is phenotypically normal, because there are two copies of all major chromosome arms and hence two copies of all essential genes.

**Down Syndrome Caused by a Robertsonian Translocation**

There is a high risk of Down syndrome among the offspring of pairings between heterozygous carriers and normal individuals (Figure 16.19). The normal parent produces gametes with one copy each of chromosomes 14 and 21. The heterozygous carrier parent produces three reciprocal pairs of gametes, each as a result of different segregation of the three chromosomes involved: (1) 14/21 (translocated 14 and 21) + 21, and 14; (2) 14/21 + 14, and 21; and (3) 14/21, and 14 + 21 (The three gamete pairs do not occur with equal frequency.) The zygotes are produced by pairing these gametes with gametes of normal chromosomal constitution: 14 and 21. Figure 16.19 shows the result of the gamete fusions. In only one case is

**Figure 16.18**

**Robertsonian translocation.** Production of a Robertsonian translocation (centric fusion) by breakage of two acrocentric chromosomes at their centromeres (indicated by arrows) and fusion of the two large chromosome arms and of the two small chromosome arms.

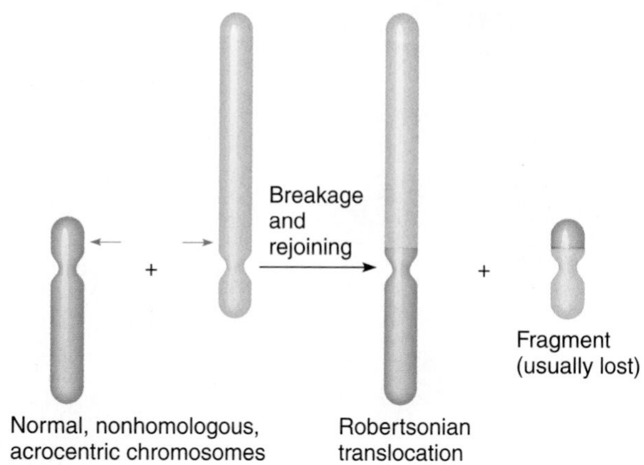


a normal zygote produced with chromosomes 14, 14, 21, and 21. One other zygote that leads to a normal phenotype is a carrier zygote with chromosomes 14, 21, and 14/21. A viable trisomy-21 zygote is produced with chromosomes 14, 14/21, 21, and 21. Three inviable zygotes are produced, one with monosomy-21, one with trisomy-14, and one with monosomy-14.

***Trisomy-13.*** **Trisomy-13** produces Patau syndrome (Figure 16.20). About 2 in 10,000 live births produce individuals with trisomy-13. Characteristics of individuals with trisomy-13 include cleft lip and palate, small eyes, polydactyly (extra fingers and toes), mental and developmental retardation, and cardiac anomalies, among many other abnormalities. Most infants die before the age of 3 months.

***Trisomy-18.*** **Trisomy-18** produces Edwards syndrome (Figure 16.21), which occurs in about 2.5 in 10,000 live births. For reasons that are not known, about 80 percent of infants with Edwards syndrome are female. Individuals with trisomy-18 are small at birth and have multiple congenital malformations affecting almost every organ in the body. Clenched fists, an elongated skull, low-set malformed ears, mental and developmental retardation, and many other abnormalities are associated with the syndrome. Ninety percent of infants with trisomy-18 die within 6 months, often from cardiac problems.

## Changes in Complete Sets of Chromosomes

**Monoploidy** and **polyploidy** involve variations from the normal state in the number of complete sets of chromosomes. Because the number of complete sets of chromosomes is involved in each case, monoploids and polyploids are euploids. Monoploidy and polyploidy are lethal in most animal species, but are less consequential in plants.

**Figure 16.19**

**The three segregation patterns of a heterozygous Robertsonian translocation involving human chromosomes 14 and 21.** Fusion of the resulting gametes with gametes from a normal parent produces zygotes with various combinations of normal and translocated chromosomes.

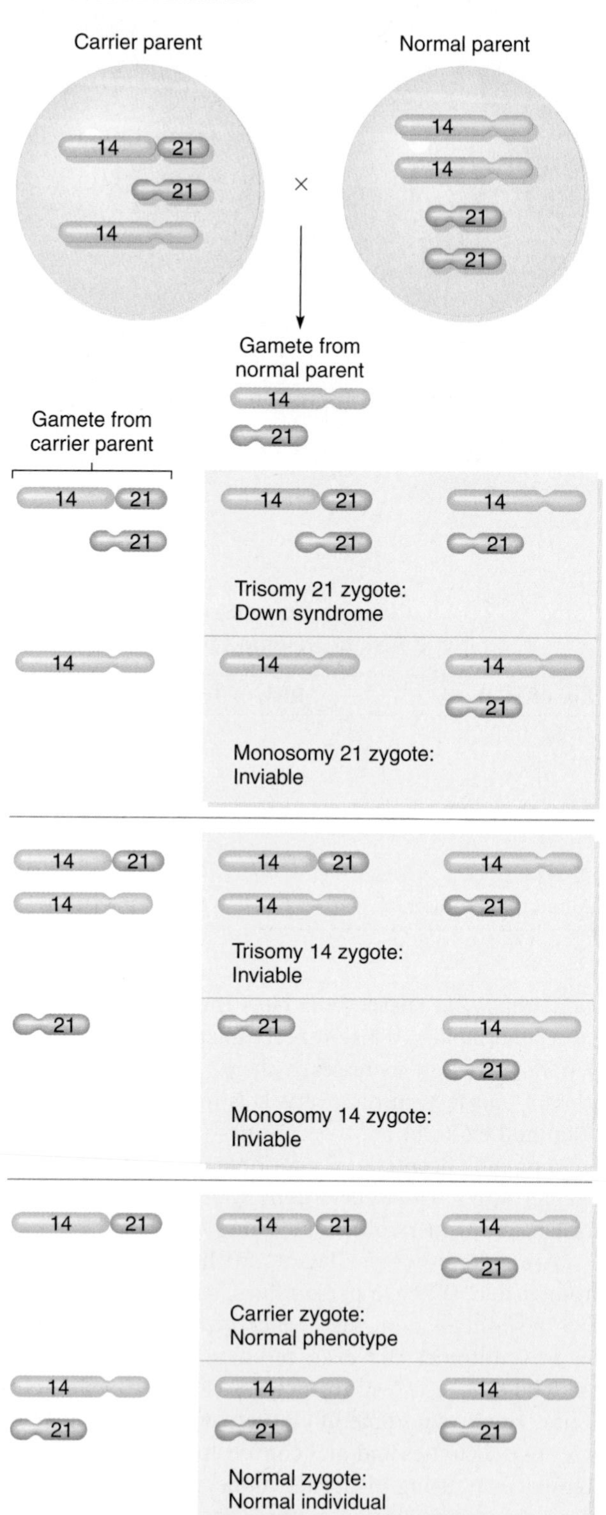

**Figure 16.20**
**Trisomy-13 (Patau syndrome).**

a) Karyotype (G banding)

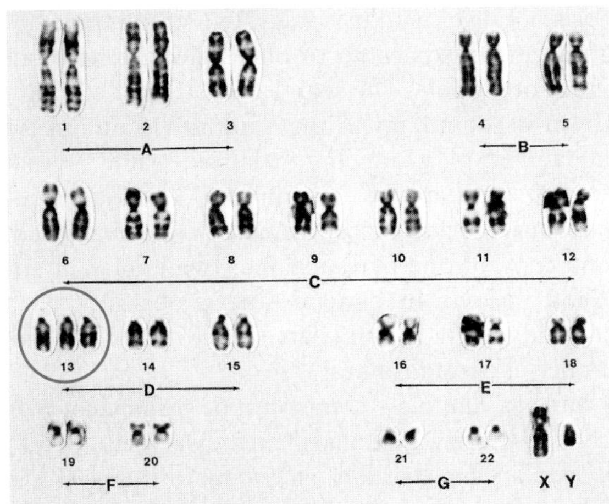


b) Individual with trisomy-13 (Patau syndrome)

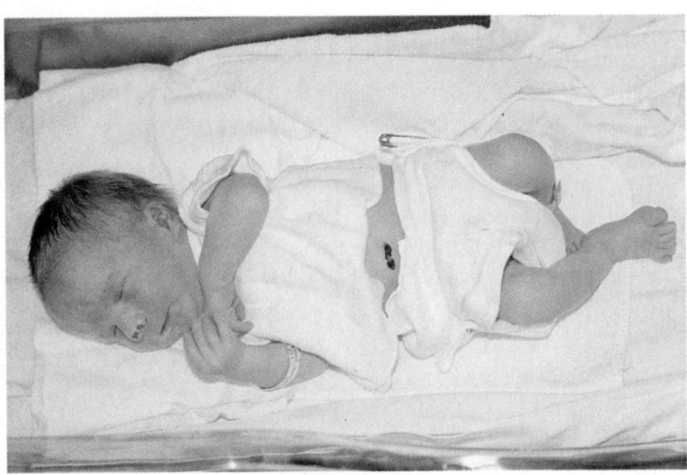

Both have played significant roles in plant speciation and diversification.

Changes in complete sets of chromosomes result when the first or second meiotic division is abortive (resulting in a lack of cytokinesis) or when meiotic nondisjunction occurs for all chromosomes, for example. If such nondisjunction occurs at meiosis I, half of the gametes have no chromosome sets, and half have two chromosome sets (see Figure 12.18b, p. 344). If such nondisjunction occurs at meiosis II, half of the gametes have the normal one set of chromosomes, one-quarter have two sets of chromosomes, and one-quarter have no chromosome sets (see Figure 12.18c). Fusion of a gamete with two chromosome sets with a normal gamete produces a polyploid zygote—in this case, one with three sets of chromosomes, which is a *triploid* (3N). Similarly, fusion of two gametes, each with two chromosome sets, produces a *tetraploid* (4N) zygote. Polyploidy of somatic cells can also occur following the mitotic nondisjunction of complete chromosome sets. Monoploid (haploid) individuals, by contrast, typically develop from unfertilized eggs.

**Monoploidy.** A monoploid individual has only one set of chromosomes instead of the usual two sets (Figure 16.22a). Monoploidy is sometimes called haploidy, although the term *haploidy* typically is used to describe the chromosome complement of gametes. Some fungi and males of haploid/diploid species (ants, bees, wasps) are haploid, for example.

Monoploidy is seen only rarely among adults in normally diploid organisms. As a result of the presence of recessive lethal mutations (which are usually counteracted by dominant wild-type alleles in heterozygous individuals) in the chromosomes of many diploid eukaryotic organisms, many monoploids probably do not survive. Certain species produce monoploid organisms as a normal part of their life cycle. Some male wasps, ants, and bees, for example, are monoploid because they develop from unfertilized eggs.

**Figure 16.21**
**Trisomy-18 (Edwards syndrome).**

a) Karyotype (G banding)

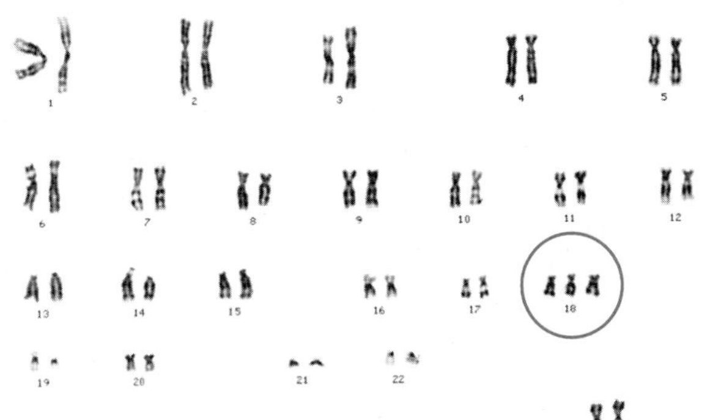


b) Individual with trisomy-18 (Edwards syndrome)

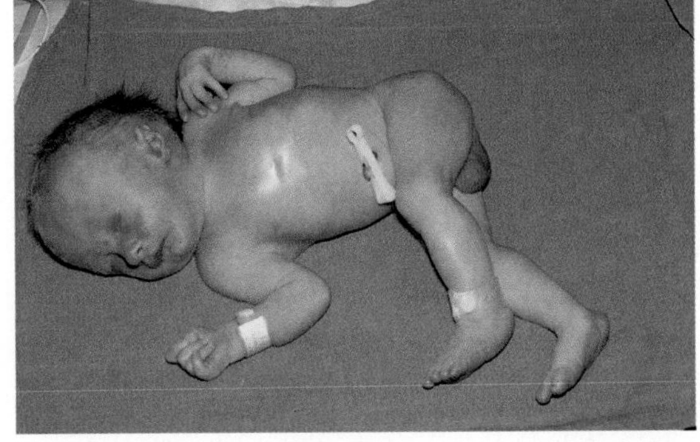

**Figure 16.22**

**Variations in number of complete chromosome sets.**

**Normal chromosome complement**

1 2 3

Diploid (2N)

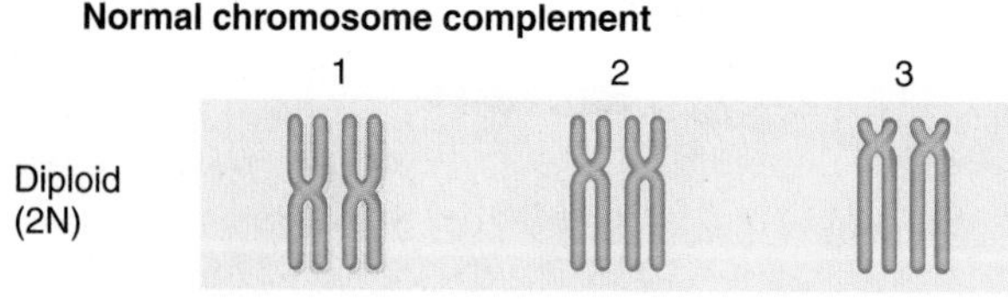

**a) Monoploidy**
(only one set of chromosomes)

Monoploid (N)

**b) Polyploidy**
(more than the normal number of sets of chromosomes)

Triploid (3N)

Tetraploid (4N)

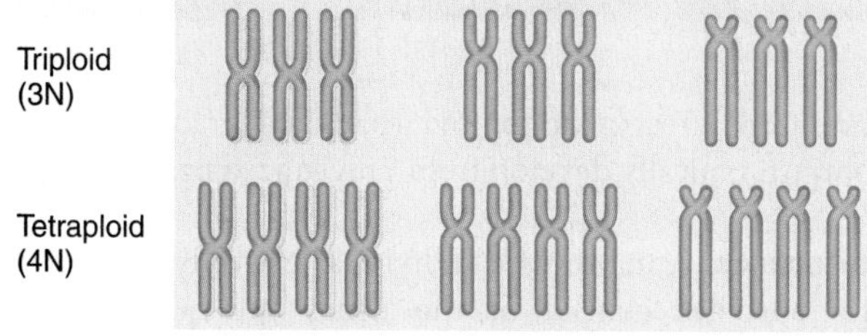

Cells of a monoploid individual are very useful for producing mutants, because there is only one dose of each of the genes. Thus, mutants can be isolated directly without dominance/recessiveness complications.

**Polyploidy.** Polyploidy is the chromosomal constitution of a cell or an organism that has more than the normal two sets of homologous chromosomes (Figure 16.22b). Polyploids may arise spontaneously or be induced experimentally. They often result from a breakdown of the spindle apparatus in one or more meiotic divisions or in mitotic divisions. Almost all plants and animals probably have some polyploid tissues. For example, the endosperm of plants is triploid, the liver of mammals and perhaps other vertebrates is polyploid, and the giant abdominal neuron of the sea hare *Aplysia* has about 75,000 copies of the genome. Plants that are completely polyploid include wheat, which is hexaploid (6N), and the strawberry, which is octaploid (8N). Some animal species, such as the North American sucker (a freshwater fish), salmon, and some salamanders, are polyploid.

Polyploids fall into two general classes: those with an *even* number of chromosome sets and those with an *odd* number of sets. Polyploids with an even number of chromosome sets have a better chance of being at least partially fertile, because there is the potential for homologs to be segregated equally during meiosis. Polyploids with an odd number of chromosome sets always have an unpaired chromosome for each chromosome type, so the probability of producing a balanced euploid gamete is extremely low; such organisms usually are sterile or have an increased incidence of abortion of zygotes.

In triploids, the nucleus of a cell has three sets of chromosomes. As a result, triploids are highly unstable in meiosis because, as in trisomics, two of the three homologous chromosomes go to one pole and the other goes to the other pole. The segregation of each chromosome from its homologs in the triploid is random, so the probability of producing balanced gametes that contain either a haploid or a diploid set of chromosomes is small; many of the gametes are unbalanced, with one copy of one chromosome, two copies of another, and so on. In general, the probability of a triploid producing a haploid gamete is $(^1/_2)^n$, where $n$ is the number of chromosomes.

In humans, the most common type of polyploidy is triploidy, and it is always lethal. Triploidy is seen in 15 to 20% of spontaneous abortions and about 1 in 10,000 live births, but most affected infants die within 1 month. Triploid infants have many abnormalities, including a characteristically enlarged head. Tetraploidy in humans is also always lethal, usually before birth. It is seen in about 5% of spontaneous abortions. Very rarely is a tetraploid human born, but such an individual does not survive long.

Polyploidy is less consequential to plants. One reason is that many plants undergo self-fertilization, so if a plant is produced with an even polyploid number of chromosome sets (for example, 4N) it can still produce fertile gametes and reproduce.

Two types of polyploidy are encountered in plants. In **autopolyploidy,** all the sets of chromosomes originate in the same species. The condition probably results from a defect in meiosis that leads to diploid or triploid gametes. If a diploid gamete fuses with a normal haploid gamete, the zygote and the organism that develops from it will have three sets of chromosomes; in other words, it will be triploid. The cultivated banana is an example of a triploid autopolyploid plant. Because it has an odd number of chromosome sets, the gametes have a variable number of chromosomes, and few fertile seeds are set, thereby making most bananas seedless and highly palatable. Because of the triploid state, cultivated bananas are propagated vegetatively (by cuttings). In general, the development of "seedless" fruits such as grapes and watermelons relies on odd-number polyploidy. Triploidy has also been found in grasses, garden flowers, crop plants, and forest trees.

In **allopolyploidy,** the sets of chromosomes involved come from different, though usually related, species. This situation can arise if two different species interbreed to produce an organism with one haploid set of each parent's chromosomes (one set from each species) and then both chromosome sets double. For example, the fusion of haploid gametes of two diploid plants that can cross may produce an $N_1 + N_2$ hybrid plant that has a haploid set of chromosomes from plant species 1 and a haploid set from plant species 2. However, because of the differences between the two chromosome sets, no chromosomes pair at meiosis, and no viable gametes are produced. As a result, the hybrid plants are sterile. Rarely, through a division error, the two sets of chromosomes double, producing tissues of $2N_1 + 2N_2$ genotype. (That is, the cells in the tissue have a diploid set of chromosomes from plant species 1 and a diploid set from plant species 2.) Each diploid set can function normally in meiosis, so that gametes produced from the $2N_1 + 2N_2$ plant are $N_1 + N_2$. Such fusion of two gametes can produce fully fertile, allotetraploid, $2N_1 + 2N_2$ plants.

A classic example of allopolyploidy resulted from crosses made between cabbages (*Brassica oleracea*) and radishes (*Raphanus sativus*) by Karpechenko in 1928. Both parents have a chromosome number of 18, and the $F_1$ hybrids also have 18 chromosomes, 9 from each parent. The hybrids produced are morphologically intermediate between cabbages and radishes. The $F_1$ plants are mostly sterile as a result of the failure of chromosomes to pair at meiosis. However, a few seeds are produced through meiotic errors, and some of those seeds are fertile. The somatic cells of the plants produced from those seeds have 36 chromosomes—that is, full diploid sets of chromosomes from both the cabbage and the radish. These plants are completely fertile and belong to a breeding species named *Raphanobrassica*, a fusion of the two parental genus names. Morphologically, the plants look a lot like the $F_1$ hybrids.

Finally, many commercial grains, most crops, and many common commercial flowers are polyploid. In fact, polyploidy is the rule rather than the exception in agriculture and horticulture. For example, the cultivated bread wheat, *Triticum aestivum*, is an allohexaploid with 42 chromosomes. This plant species is descended from three distinct species, each with a diploid set of 14 chromosomes. Meiosis is normal because only homologous chromosomes pair, so the plant is fertile.

**Keynote**

Variations in the chromosome number of a cell or an organism give rise to aneuploidy, monoploidy, and polyploidy. In aneuploidy, a cell or organism has one, two, or a few whole chromosomes more or less than the basic number of the species under study. In monoploidy, an organism that is usually diploid has only one set of chromosomes. In polyploidy, an organism has more than the normal number of complete sets of chromosomes. Any or all of these abnormal conditions may have serious consequences for the organism.



## Summary

- Chromosomal mutations are variations from the normal condition in chromosome number or chromosome structure. Chromosomal mutations can occur spontaneously, or they can be induced by chemicals or radiation.
- Deletion is the loss of a DNA segment, duplication is the addition of one or more extra copies of a DNA segment, inversion is a reversal of orientation of a DNA segment in a chromosome, and translocation is the movement of a DNA segment to another chromosomal location in the genome.
- Variations in the chromosome number of a cell or an organism include aneuploidy, monoploidy, and polyploidy. In aneuploidy there are one, two, or more whole chromosomes greater or fewer than the diploid number. In monoploidy, each body cell of the organism has only one set of chromosomes, and in polyploidy more than two sets of chromosomes are present.
- A change in chromosome number or chromosome structure can have serious, and even lethal, consequences for the organism. In eukaryotes, abnormal phenotypes typically result from abnormal chromosome segregation during meiosis, from gene disruptions where chromosomes break, or from altered gene expression levels when the number of copies of a gene or genes (gene dosage) is altered or when rearrangement separates a gene from its regulatory sequence. Some human tumors, for example, have chromosomal mutations associated with them—either a change in the number of chromosomes or a change in chromosome structure.

## Analytical Approaches to Solving Genetics Problems

**Q16.1** Diagram the meiotic pairing behavior of the four chromatids in an inversion heterozygote *a b c d e f g*/*a′ b′ f′ e′ d′ c′ g′*. Assume that the centromere is to the left of gene *a*. Next, diagram the early anaphase configuration if a crossover occurred between genes *d* and *e*.

**A16.1.** Answering this question requires a knowledge of meiosis (see Figure 12.9, p. 334) and the ability to draw and manipulate an appropriate inversion loop. Part (**a**) of the following figure shows the diagram for the meiotic pairing:

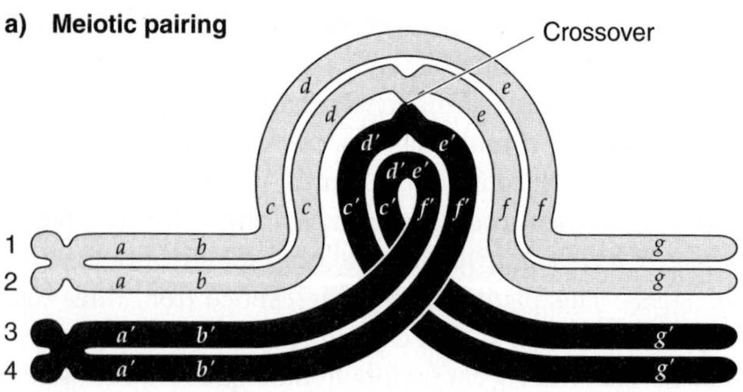


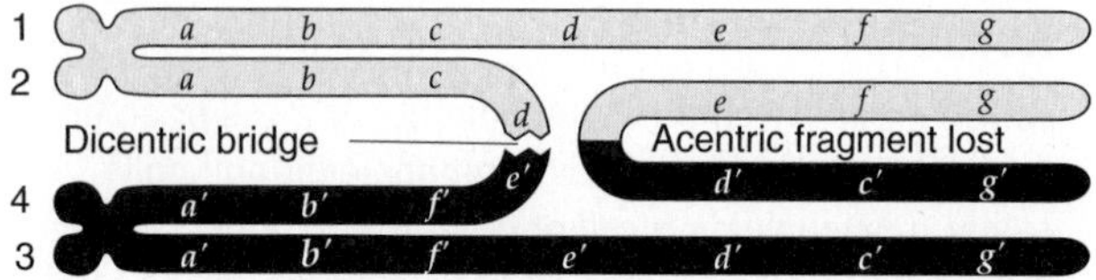


Note that the lower pair of chromatids ($a'$, $b'$, etc.) must loop over in order for all the genes to align; this looping is characteristic of the pairing behavior expected for an inversion heterozygote.

Once the first diagram has been constructed, answering the second part of the question is straightforward. We diagram the crossover and then trace each chromatid from the centromere end to the other end. It is convenient to distinguish maternal and paternal genes, perhaps by $a'$ versus $a$, and so on, as we did in part **(a)** of the figure. The result of the crossover between $d$ and $e$ is shown in part **(b)**.

In anaphase I of meiosis, the two centromeres, each with two chromatids attached, migrate toward the opposite poles of the cell. At anaphase, the noncrossover chromatids (top and bottom chromatids in the figure) segregate to the poles normally. As a result of the single crossover between the other two chromatids, however, unusual chromatid configurations are produced, and these configurations are found by tracing the chromatids from left to right. If we begin by tracing the second chromatid from the top, we get

$$\underset{\circ\!-\!-\!-\!-\!-\!-\!-\!-\!-\!-\!\circ}{a\;b\;c\;d\;e\;f'\;b'\;a'}$$

which is a dicentric chromatid (where ○ is a centromere); in other words, we have a single chromatid attached to two centromeres. This chromatid also has duplications and deletions for some of the genes. Thus, during anaphase, this so-called dicentric chromosome becomes stretched between the two poles of the cells as the centromeres separate, and the chromosome eventually breaks at a random location. The other product of the single crossover is an acentric fragment (a fragment without a centromere) that can be traced starting from the right with the second chromatid from the top. This chromatid,

$$\underline{g\;f\;e\;d'\;c'\;g'}$$

contains neither a complete set of genes nor a centromere; it is an acentric fragment that will be lost as meiosis continues.

Thus, the consequence of a crossover within the inversion in an inversion heterozygote is the production of gametes with duplicated or deleted genes. These gametes often are inviable. However, viable gametes are produced from the noncrossover chromatids: one of these chromatids (1 in part [**b**] of the figure) has the normal gene sequence, and the other (3 in part [**b**] of the figure) has the inverted gene sequence.

**Q16.2** *Eyeless* is a recessive gene ($ey$) on chromosome 4 of *Drosophila melanogaster*. Flies homozygous for $ey$ have tiny eyes or no eyes at all. A male fly trisomic for chromosome 4 with the genotype $+/+/ey$ is crossed with a normal diploid, *eyeless* female of genotype $ey/ey$. What expected genotypic and phenotypic ratios would result from random assortment of the chromosomes to the gametes?

**A16.2** To answer this question, we must apply our understanding of meiosis to the unusual situation of a trisomic cell. Regarding the $ey/ey$ female, only one gamete class can be produced, namely, eggs of genotype $ey$. Gamete production with respect to the trisomy for chromosome 4 occurs by a random segregation pattern in which, during meiosis I, two chromosomes migrate to one pole and the other chromosome migrates to the other pole. (This pattern is similar to the meiotic segregation pattern shown in the secondary nondisjunction of XXY cells; see Chapter 12.) Three types of segregation are possible in the formation of gametes in the trisomy, as shown in part **(a)** of the following figure: The random union of these sperm with eggs of genotype $ey$ occurs as shown in part **(b)**, and the resulting genotypic and phenotypic ratios are listed in part **(c)**.

**a) Segregation**

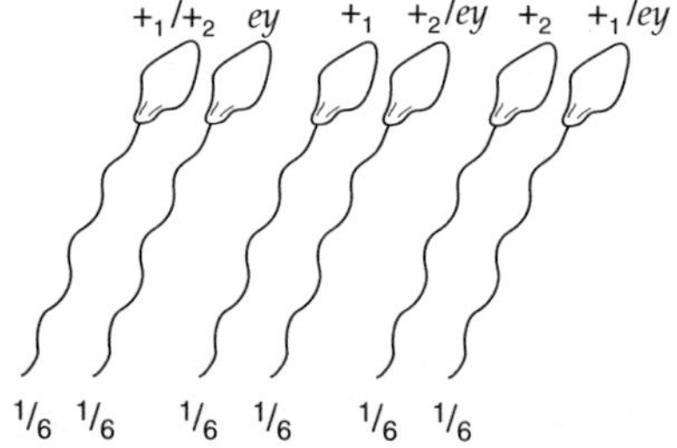


**b) Union**

| | | Eggs $ey$ | Phenotype |
|---|---|---|---|
| Sperm | $+/+$ | $+/+/ey$ | + |
| | $ey$ | $ey/ey$ | $ey$ |
| | + | $+/ey$ | + |
| | $+/ey$ | $+/ey/ey$ | + |
| | + | $+/ey$ | + |
| | $+/ey$ | $+/ey/ey$ | + |

**c) Summary of genotypes and phenotypes**

| Ratios: | Genotypes | Phenotypes |
|---|---|---|
| | $1/6$ $+/+/ey$ | $5/6$ wild type |
| | $1/3$ $+/ey/ey$ | $1/6$ eyeless |
| | $1/3$ $+/ey$ | |
| | $1/6$ $ey/ey$ | |

# Questions and Problems

*__16.1__ A normal chromosome has the following gene sequence:

A B C D ○ E F G H

Determine the chromosomal mutation illustrated by each of the following chromosomes:

**a.** A B C F E ○ D G H
**b.** A D ○ E F B C G H
**c.** A B C D ○ E F E F G H
**d.** A B C D ○ E F F E G H
**e.** A B D ○ E F G H

*__16.2__ Distinguish between pericentric and paracentric inversions.

**16.3** In some instances, very small deletions behave like recessive mutations. Why are some recessive mutations known not to be deletions?

*__16.4__ Inversions are known to affect crossing-over. The following homologs have the indicated gene order (the filled and open circles are homologous centromeres):

● A B C D E

○ A D C B E

**a.** Considering the position of the centromere, what is this sort of inversion called?
**b.** Diagram the alignment of these chromosomes during meiosis.
**c.** Diagram the results of a single crossover between homologous genes *B* and *C* in the inversion.

**16.5** Single crossovers within the inversion loop of inversion heterozygotes give rise to chromatids with duplications and deletions. What happens if, within the loop, there is a two-strand double crossover in such an inversion heterozygote when the centromere is outside the loop?

**16.6** An inversion heterozygote possesses one chromosome with genes in the normal order:

○ a b c d e f g h

It also contains one chromosome with genes in the inverted order:

○ a b f e d c g h

A four-strand double crossover occurs in the areas *e–f* and *c–d*. Diagram and label the four strands at synapsis (showing the crossovers) and at the first meiotic anaphase.

*__16.7__ The following gene arrangements in a particular chromosome are found in *Drosophila* populations in different geographic regions:

**a.** A B C D E F G H I
**b.** H E F B A G C D I
**c.** A B F E D C G H I
**d.** A B F C G H E D I
**e.** A B F E H G C D I

Assuming that the arrangement in part (a) is the original arrangement, in what sequence did the various inversion types probably arise?

**16.8** The following figure shows map distances observed for genes in one chromosomal region (the open circle represents a centromere):

| A | ○ | B | C | D | E | F | G | H | I |
|---|---|---|---|---|---|---|---|---|---|
| 2 | 3 | 3 | 6 | 2 | 7 | 5 | 6 | 3 | |

A new paracentric inversion bears a recessive mutation *d* at the *D* locus. Its proximal breakpoint lies between *B* and *C*, 1 mu from *B*, and its distal breakpoint lies between *G* and *H*, 1 mu from G. A heterozygote for this inversion and the wild-type arrangement mates with a *dd* individual homozygous for the inversion.

**a.** In the absence of multiple crossovers, what is the chance that a *dd* offspring will have a homolog with a wild-type arrangement?
**b.** What type of event is required to produce a *dd* offspring having a homolog with a wild-type arrangement? What is the likelihood of such an event?
**c.** Based on your answers to (a) and (b), how might spontaneously arising inversions contribute to the maintenance of genetic differences between subpopulations of a species?

*__16.9__ The brick red eye color of normal *Drosophila* results from pigment deposition controlled by the *white* gene, which lies on the X chromosome at map position 1.5, far from centromeric heterochromatin (which starts at about map position 66). Hermann Müller screened for new *white* mutants by irradiating wild-type *Drosophila* males ($w^+$/Y) and mating them to white-eyed (*w*/*w*) females. He isolated several mutant females bearing mottled red eyes—red eyes with varying amounts of white spotting. One mutant, $w^{M5}$, was associated with a reciprocal translocation with breakpoints near the *white* locus and centromeric heterochromatin of chromosome 4. A different mutant, $w^{M4}$, was associated with an X-chromosome inversion with breakpoints near the *white* locus and centromeric heterochromatin. Kenneth Tartof screened for revertants of the mottled eye phenotype of $w^{M4}$ by crossing irradiated $w^{M4}$/Y males with *w*/*w* females. He recovered three different normal-eyed female revertants, each associated with a new X-chromosome inversion. In addition to having the original $w^{M4}$ breakpoints near the *white* locus and in centromeric heterochromatin, each had a third euchromatic breakpoint.

**a.** Based on these results, is the mottled eye phenotype in Müller's mutants due to a mutation within the *white* gene? If not, what is its most likely cause?

**b.** How can the $w^{M4}$ mutation be reverted by an additional inversion with a euchromatic breakpoint?

***16.10** Human abnormalities associated with chromosomal mutations often exhibit a range of symptoms, of which only some subsets appear in a particular individual. Recombinant 8 [Rec(8)] syndrome is an inherited chromosomal abnormality found primarily in individuals of Hispanic origin. Phenotypic characteristics associated with the syndrome include congenital heart disease, urinary system abnormalities, eye abnormalities, hearing loss, and abnormal muscle tone. Most reported cases of Rec(8) have been found in the offspring of phenotypically normal parents who are heterozygous for an inversion of chromosome 8 with breakpoints at p23.1 and q22.1. Individuals with Rec(8) syndrome typically have a duplication of part of 8q (from q22.1 to the terminus of the q arm) and a deletion of 8p (from p23.1 to the terminus of the p arm).

**a.** Using diagrams, explain why individuals with Rec(8) syndrome typically have a duplication and a deletion for part of chromosome 8.

**b.** An individual is heterozygous for an inversion on chromosome 8 with breakpoints at p23.1 and q22.1. If a crossover occurs within the inverted region during a particular meiosis, what is the chance that the resulting offspring will have Rec(8) syndrome?

**c.** Why might the phenotypes of Rec(8) individuals vary?

**d.** A child with some of the symptoms of Rec(8) syndrome is referred to a human geneticist. The karyotype of the child reveals heterozygosity for a large pericentric inversion in chromosome 8 with breakpoints at p23.1 and q22.1. Cytogenetic analysis of her phenotypically normal mother and phenotypically normal maternal grandmother reveals a similar karyotype. According to the child's mother, the father has a normal phenotype, but he is unavailable for examination. Propose at least two explanations for why the child, but not her mother or maternal grandmother, is affected with some of the symptoms of Rec(8) syndrome. (Hint: Consider the limitations of karyotype analysis using G-banding methods [see Chapter 12, pp. 328–329], and also consider what is unknown about the father.)

***16.11** A particular plant species that had been subjected to radiation for a long time in order to produce chromosomal mutations was then inbred for many generations until it was homozygous for all of these mutations. It was then crossed to the original unirradiated plant, and the meiotic process of the $F_1$ hybrids was examined. It was noticed that a cell with a dicentric chromosome (bridge) and a fragment occurred at low frequency in anaphase I of the hybrid.

**a.** What kind of chromosomal mutation occurred in the irradiated plant? In your answer, indicate where the centromeres are.

**b.** Explain, in words and with a clear diagram, where crossover(s) occurred and how the bridge chromosome of the cell arose.

**16.12** On a normal-ordered chromosome, two loci, *a* and *b*, lie 15 map units apart on the left arm of a metacentric chromosome. A third locus, *c*, lies 10 map units to the right of *b* on the right arm of the chromosome. What frequency of progeny phenotypes do you expect to see in a testcross of an $a\ b\ c/a^+\ b^+\ c^+$ individual if the $a^+\ b^+\ c^+$ chromosome

**a.** has a normal order?

**b.** has an inversion with breakpoints just proximal (toward the centromere) to *a* and just distal (away from the centromere) to *b*?

**c.** has an inversion with breakpoints just proximal to *a* and just proximal to *c*?

**d.** has an inversion with breakpoints just distal to *a* and just distal to *c*?

***16.13** Mr. and Mrs. Lambert have not yet been able to produce a viable child. They have had two miscarriages and one severely defective child who died soon after birth. Studies of banded chromosomes of father, mother, and child showed that all chromosomes were normal except for pair number 6. The number 6 chromosomes of mother, father, and child are shown in the following figure:

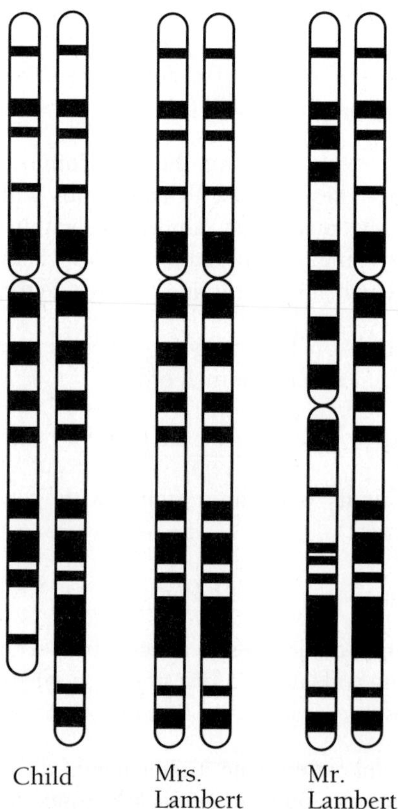

**a.** Does either parent have an abnormal chromosome? If so, what is the abnormality?
**b.** How did the chromosomes of the child arise? Be specific as to what events in the parents gave rise to those chromosomes.
**c.** Why is the child not phenotypically normal?
**d.** What can be predicted about future conceptions by this couple?

**16.14** Mr. and Mrs. Simpson have been trying for years to have a child, but have been unable to conceive. They consulted a physician, and tests revealed that Mr. Simpson had a markedly low sperm count. His chromosomes were studied, and a testicular biopsy was done. His chromosomes proved to be normal, except for pair 12. The following figure shows Mrs. Simpson's normal pair of number 12 chromosomes and Mr. Simpson's number 12 chromosomes.

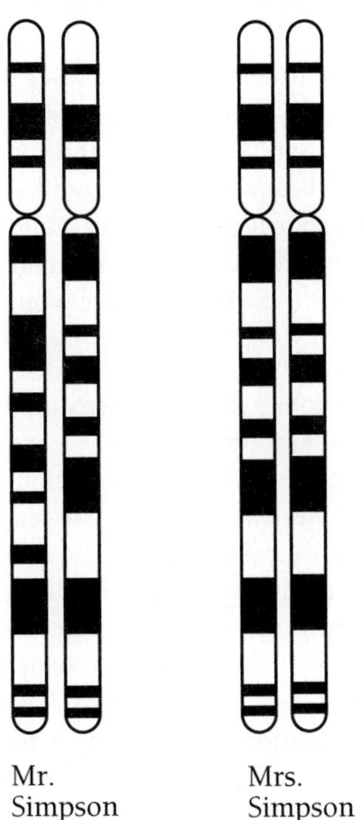


**a.** What is the nature of the abnormality of pair number 12 in Mr. Simpson's chromosomes?
**b.** What abnormal feature would you expect to see in the testicular biopsy? (Cells in various stages of meiosis can be seen.)
**c.** Why is Mr. Simpson's sperm count low?
**d.** What can be done about Mr. Simpson's low sperm count?

***16.15** Chromosome I in maize has the gene sequence *ABCDEF*, whereas chromosome II has the sequence *MNOPQR*. A reciprocal translocation resulted in *ABCPQR* and *MNODEF*. Diagram the expected pachytene (see Chapter 12, p. 333) configuration in the $F_1$ of a cross of homozygotes of these two arrangements.

**16.16** Diagram the pairing behavior at prophase of meiosis I (see Chapter 12, p. 333) of a translocation heterozygote that has normal chromosomes of gene order *abcdefg* and *tuvwxyz* and has the translocated chromosomes *abcdvwxyz* and *tuefg*. Assume that the centromere is at the left end of all chromosomes. What types of chromosome segregation can occur in this individual? Ignoring the complication of crossing-over, what gametes will each type of segregation produce?

***16.17** Mr. and Mrs. Denton have been trying for several years to have a child. They have experienced a series of miscarriages, and last year they had a child with multiple congenital defects. The child died within days of birth. The birth of this child prompted the Dentons' physician to order a chromosome study of parents and child. The results of the study are shown in the accompanying figure. Chromosome banding was done, and all chromosomes were normal in these individuals, except some copies of number 6 and number 12. The number-6 and number-12 chromosomes of mother, father, and child are shown in the figure (the number 6 chromosomes are the larger pair):

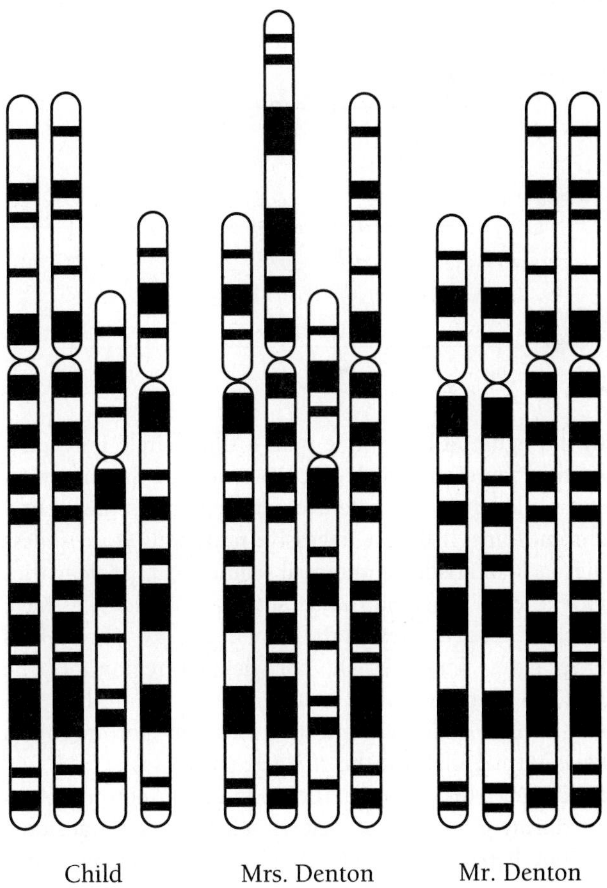


**a.** Does either parent have an abnormal karyotype? If so, which parent has it, and what is the nature of the abnormality?
**b.** How did the child's karyotype arise? (What pairing and segregation events took place in the parents?)
**c.** Why is the child phenotypically defective?
**d.** What can this couple expect to happen in subsequent conceptions?
**e.** What medical help, if any, can be offered to the couple?

**16.18** Irradiation of *Drosophila* sperm produces translocations between the X chromosome and autosomes, between the Y chromosome and autosomes, and between different autosomes. Translocations between the X and Y chromosomes are not produced. Explain the absence of X–Y translocations.

**16.19** In *Drosophila*, the Y chromosome is small, heterochromatic, and acrocentric while chromosome 2 is large and metacentric. *T(Y;2)A* is a reciprocal interchromosomal translocation with breakpoints in the long arm of the Y chromosome and the very distal region of the left arm of the second chromosome. Another chromosome, *Del(2)al*, has a tiny deletion within the left arm of chromosome 2 that removes just the *aristaless (al)* locus. The following figure illustrates the Y chromosome, a cytologically normal second chromosome with four recessive mutations, and the *T(Y;2)A* and *Del(2)al* chromosomes.

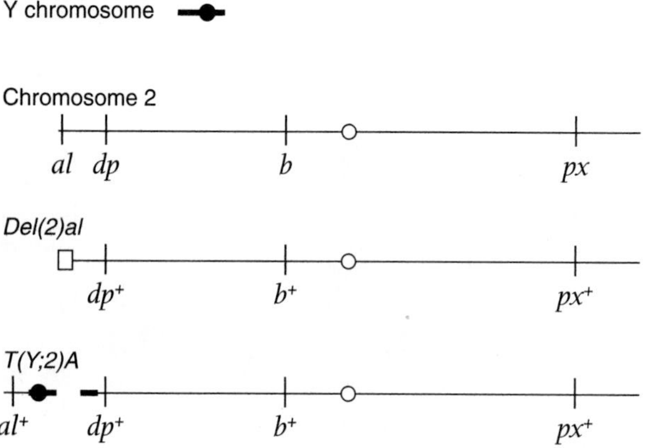


In the figure, a heavy line with a filled circle for a centromere represents the Y chromosome. The thin line with an unfilled circle for a centromere represents the second chromosome. It has the recessive mutations *al (aristaless* flies lack an arista, an antennal segment), *dp* (*dumpy* flies have altered wing and thorax shapes), *b* (*black* flies have black instead of grey body color), and *px* (*plexus* flies have extra wing veins). The deleted region on the *Del(2)al* chromosome is represented by an unfilled rectangle. The deletion and translocation chromosomes have normal alleles at the *al*, *dp*, *b*, and *px* loci.

**a.** Diagram the pairing of the second and Y chromosomes during prophase I of meiosis in a *T(Y;2)A/+* translocation heterozygote (+ is a normal ordered second chromosome bearing $al^+$, $dp^+$, $b^+$, and $px^+$).

**b.** What types of chromosomal segregation will occur in meiosis a *T(Y;2)A/+* male, and what gametes will each produce?

**c.** What progeny phenotypes will be produced in each of the following crosses? (Flies missing one copy of the distal portion of chromosome 2 containing the *al* locus, flies homozygous for *Del(2)al*, and flies lacking some or all of the Y chromosome are viable. XO flies, and single-X-bearing flies with a single X who are missing a large part of the Y chromosome are sterile males. XX flies with part or all of a Y chromosome are fertile females.)

**i.** *Del(2)al/+* ♂ × *al dp b px/+* ♀
**ii.** *T(Y;2)A/+* ♂ × *al dp b px/+* ♀
**iii.** *T(Y;2)A/+* ♂ × *Del(2)al/+* ♀

**16.20** Although humans have 22 autosomes, only three autosomal trisomies are viable, and even these are not always viable, as they are also seen in miscarried fetuses. Which chromosomes are associated with viable trisomies and how often is each type of trisomy seen? What phenotypes are associated with each type of trisomy?

**16.21** Down syndrome is associated with trisomy-21.

**a.** What are the known risk factors for having a child with Down syndrome? What factors are known not to increase a couple's risk of having a child with Down syndrome?

**b.** What are two different genetic causes of trisomy-21?

**c.** A phenotypically normal couple with no family history of Down syndrome has a child with Down syndrome. Should they have their own karyotypes examined before conceiving another child? Under what circumstances would you recommend that their siblings also have their karyotypes examined?

***16.22**

**a.** How does a Robertsonian translocation differ from a reciprocal translocation?

**b.** Suppose a Robertsonian translocation occurs in the germline of a human male, and a Y-bearing sperm carrying this translocation fertilizes a genetically normal oocyte. How many chromosomes will this zygote have?

**c.** Closely examine the normal human karyotype shown in Figure 12.3 (p. 328) and state which human chromosomes theoretically could be involved in a Robertsonian translocation if the zygote is to be phenotypically (but not necessarily chromosomally) normal. Explain your choices.

**d.** Suppose the son bearing the Robertsonian translocation matures and fathers a child. Diagram the events that will lead either to inviable or viable, but phenotypically abnormal offspring. Is it possible for him to father a phenotypically normal child? Is it possible for him to father a genetically normal child?

***16.23** In the pedigree shown in Figure 16.A, mental retardation is indicated by shaded symbols, with individuals shaded black being more severely affected than individuals shaded grey.

**a.** What features of this pedigree suggest that the mental retardation phenotype may be associated with fragile X syndrome?

**b.** What molecular and cytological means would you employ to evaluate if mental retardation in this pedigree is due to fragile X syndrome, and what would you expect to find if it is?

Figure 16.A

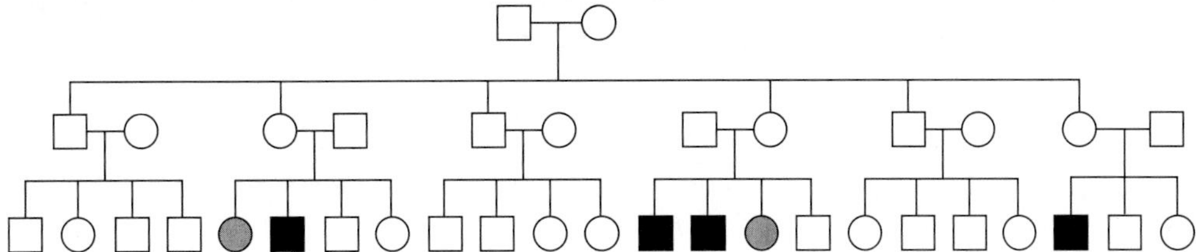

c. If the mental retardation in this pedigree is due to fragile X syndrome, which individual(s)
   i. must carry a premutation?
   ii. must be a normal transmitting male?
   iii. must carry a fragile X chromosome?
   iv. in generation III are phenotypically normal but may still carry a fragile X chromosome?
d. Why might females in this pedigree be less severely affected?

**16.24** Define the terms *aneuploidy*, *monoploidy*, and *polyploidy*.

**16.25** If a normal diploid cell is 2N, what is the chromosome content of the following?

a. a nullisomic
b. a monosomic
c. a double monosomic
d. a tetrasomic
e. a double trisomic
f. a tetraploid
g. a hexaploid

*__16.26__ In humans, how many chromosomes would be typical of nuclei of cells that are

a. monosomic?
b. trisomic?
c. monoploid?
d. triploid?
e. tetrasomic?

*__16.27__ An individual with 47 chromosomes, including an additional chromosome 15, is said to be

a. triplet.
b. trisomic.
c. triploid.
d. tricycle.

*__16.28__ A color-blind man marries a homozygous normal woman, and after four joyful years of marriage they have two children. Unfortunately, both children have Turner syndrome, although one has normal vision and one is color blind. The type of color blindness involved is a sex-linked recessive trait.

a. For the color-blind child, did nondisjunction occur in the mother or the father? Explain your answer.
b. For the child with normal vision, in which parent did nondisjunction occur? Explain your answer.

*__16.29__ The frequency of chromosome loss in *Drosophila* can be increased by a recessive chromosome 2 mutation called *pal*. The mutation causes the preferential loss of chromosomes contributed to a zygote by *pal/pal* fathers. The paternally contributed chromosomes are lost during the first few mitotic divisions after fertilization. What phenotypic consequences do you expect in offspring of the following crosses? (Keep in mind how sex is determined in *Drosophila*, that the loss of an entire chromosome 2 or chromosome 3 is lethal, and that the loss of one copy of the small chromosome 4 is tolerated.)

a. X chromosome loss at the first mitotic division in a cross between a true-breeding *yellow* (recessive, X-linked mutation causing yellow body color) female and a *pal/pal* father
b. Chromosome 4 loss at the first mitotic division in a cross between a true-breeding *eyeless* (recessive mutation on chromosome 4 causing reduced eye size) female and a *pal/pal* father
c. Chromosome 3 loss at the first mitotic division in a cross between a true-breeding *ebony* (recessive mutation on chromosome 3 causing black body color) female and a *pal/pal* father

**16.30** Assume that *x* is a new mutant gene in corn. A female *x/x* plant is crossed with a triplo-10 individual (trisomic for chromosome 10) carrying only dominant alleles at the *x* locus. Trisomic progeny are recovered and crossed back to the *x/x* female plant.

a. What ratio of dominant to recessive phenotypes is expected if the *x* locus is not on chromosome 10?
b. What ratio of dominant to recessive phenotypes is expected if the *x* locus is on chromosome 10?

**16.31** Why are polyploids with even multiples of the chromosome set generally more fertile than polyploids with odd multiples of the chromosome set?

*__16.32__ In her novel *The Cleft*, Nobel Prize-winning writer Doris Lessing explores the consequences of the increasingly frequent birth of male "monsters" within an isolated human-like species, called clefts, whose population previously consisted only of females. Based on our knowledge of how sex is determined in humans (for review, see pp. 346–347 in Chapter 12), it may seem far-fetched that a population composed entirely of females could reproduce. However, female-only species of animals do exist. For example, a species of whiptail lizards living in the southwestern U.S. desert is a haploid female species that reproduces by parthenogenesis. Use your knowledge of the genetic basis for sex determination in humans, chromosomal rearrangements, and polyploidy to hypothesize at least one detailed *genetic mechanism* that could explain the birth of male "monsters" in the cleft species. In your hypothetical mechanism, would heterosexual matings lead to viable offspring with a male : female ratio of 1:1?

**16.33** One plant species (N = 11) and another (N = 19) produced an allotetraploid. Which of the lettered options below correctly describes both of the following statements?

**I.** The chromosome number of this allotetraploid is 30.

**II.** The number of linkage groups of this allotetraploid is 30.

**a.** Statement I is true and Statement II is true.

**b.** Statement I is true but Statement II is false.

**c.** Statement I is false but Statement II is true.

**d.** Statement I is false and Statement II is false.

*__16.34__ According to Mendel's first law, genes *A* and *a* segregate from each other and appear in equal numbers among the gametes. But Mendel did not know that his plants were diploid. In fact, because plants are frequently tetraploid, he could have been unlucky enough to have started with peas that were 4N rather than 2N. Let us assume that Mendel's peas were tetraploid, that every gamete contains two alleles, and that the distribution of alleles to the gamete is random. Suppose we have a cross of *AAAA* × *aaaa* where *A* is dominant, regardless of the number of *a* alleles present in an individual.

**a.** What will be the genotype of the $F_1$ peas?

**b.** If the $F_1$ peas are selfed, what will be the phenotypic ratios in the $F_2$ peas?

**16.35** What phenotypic ratio of *A* to *a* is expected if *AAaa* plants are testcrossed with *aaaa* individuals? (Assume that the dominant phenotype is expressed whenever at least one *A* is present, that no crossing-over occurs, and that each gamete receives two chromosomes.)

**16.36** The root-tip cells of an autotetraploid plant contain 48 chromosomes. How many chromosomes were contained by the gametes of the diploid from which this plant was derived?

**16.37** A number of species of the birch genus have a somatic chromosome number of 28. The paper birch is reported as occurring with several different chromosome numbers; *fertile* individuals with the somatic numbers 56, 70, and 84 are known. How should the 28-chromosome individuals be designated with regard to chromosome number?

*__16.38__ How many chromosomes would be found in somatic cells of an allotetraploid derived from two plants, one with N = 7 and the other with N = 10?

**16.39** Plant species A has a haploid complement of four chromosomes. A related species, B, has five. In a geographic region where A and B are both present, C plants are found that have some characters of both species and somatic cells with 18 chromosomes. What is the chromosome constitution of the C plants likely to be? With what plants would they have to be crossed to produce fertile seed?

# 17 Regulation of Gene Expression in Bacteria and Bacteriophages

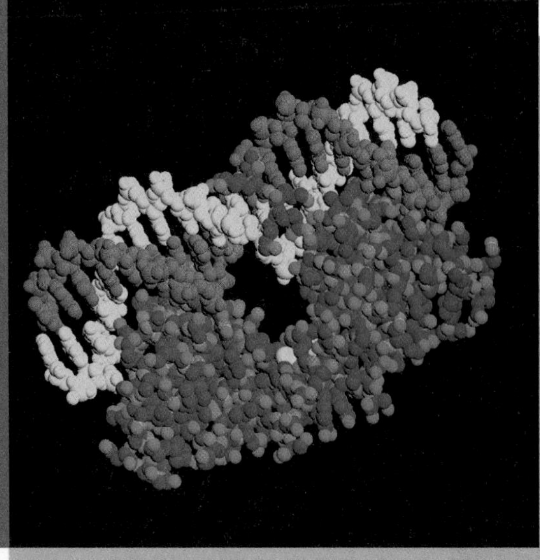

*lac* operon repressor protein binding to DNA.

## Key Questions

- How is gene expression regulated in bacteria?
- How is gene expression regulated in bacteriophages?

### iActivity

ONE OF THE BEST STRATEGIES FOR AN organism's survival is to be able to adapt quickly to changes in its environment. In fact, a fundamental property of living cells is their ability to turn their genes on and off in response to extracellular signals. This control of gene expression makes it possible for cells to produce specific kinds of proteins when and where they are needed. In this chapter, you will learn some of the ways in which gene expression in microorganisms is regulated. Then, in the iActivity, you can investigate how mutations affect the process of regulation in *E. coli.*

Most bacteria are free-living organisms that grow by increasing in mass and then divide by binary fission. Growth and division are controlled by genes, the expression of which must be regulated appropriately. Genes whose activity is controlled in response to the needs of a cell or organism are called **regulated genes.** All organisms also have a large number of genes whose products are essential to the normal functioning of a growing and dividing cell, no matter what the conditions are. These genes are always active in growing cells and are known as **constitutive genes** or *housekeeping genes*; examples include genes that code for the enzymes needed for protein synthesis and glucose metabolism. Note that all genes are regulated on some level. If normal cell function is impaired for some reason, the expression of all genes, including constitutive genes, is reduced by regulatory mechanisms. Thus, the distinction between regulated and constitutive genes is somewhat arbitrary.

The goal of this chapter is to learn about some of the mechanisms by which gene expression is regulated in bacteria and bacteriophages. Significantly, genes which encode proteins that work together in the cell typically are organized into operons; that is, the genes are adjacent to each other and are transcribed together onto a *polycistronic mRNA*, so called because it contains the information from more than one gene. (Here, the word *cistron* is used synonymously with *gene.*) Regulation of the synthesis of this mRNA depends on interactions between regulatory proteins and regulatory sequences that are next to the gene array. Such studies of bacterial and bacteriophage gene regulation also have provided important insights into how genes are regulated in higher organisms, including humans. Of course, much remains to be done to understand completely the regulation of gene expression in bacteria. The 4.6-megabase ($4.6 \times 10^6$ bp) genome of *E. coli,* for example, has 4,288 protein-coding genes according to the genomic sequence. Genomics researchers are able to say something about the function of approximately 80% of those genes, but much remains to be determined about their complete functions and how they are regulated. And, the functions of the remaining approximately 20% of the genes remain unknown. This chapter's Focus on Genomics box describes a computer model for regulation of gene expression in one prokaryote, *Halobacterium salinum*.

## Focus on Genomics

### Models of Gene Expression

The combination of genomics, transcriptomics, proteomics, and DNA microarrays has allowed scientists to begin building models that predict how a cell will respond to environmental change. These models can predict which genes will be expressed under certain conditions, and which DNA-binding proteins will regulate the expression of these genes. In one such study, bioinformatics was used to predict the likely function, where possible, for each of the 2,400 proteins encoded in the complete genomic sequence of the archaean *Halobacterium salinum*. The investigators then performed an extensive series of DNA microarray experiments examining the expression of all 2,400 genes under specific environmental conditions. For instance, they compared gene expression in cells grown in the presence of high levels of nickel with that of cells grown in environments without nickel. They also used DNA microarrays to test the functions of specific genes. For example, they compared gene expression in wild-type cells with cells containing a mutation that prevented the production of the transcription factor TFBf. Using this kind of data, the researchers were able to build a computer model that accurately predicted the transcriptional response of the cell to new environmental or genetic challenges. When they tested their model against real-life experiments using new combinations of environmental or genetic changes, they found that the model accurately predicted the response of 80% of the expressed genes in the cell. As a result, they were able to organize the genes into *biclusters*, or sets of genes that respond in the same manner to a series of environmental and/or genetic changes, whose expression is likely regulated by the same cellular proteins. For instance, transcription of the 34 genes in bicluster 66 is regulated by the environmental factors oxygen and light and the transcription factors TFBf and Cspd1.

What can we learn from these models? We can extrapolate possible functions for genes that have no bioinformatics clues. For instance, bioinformatics told the investigators nothing about the function of the protein encoded by gene VNG1459H. This gene was placed in a bicluster with genes known to encode proteins that help the cell respond to light. This suggested a role for VNG1459H in light response, and the researchers were then able to find this protein localized to a region of the cell involved in sensing light.

We can also use this sort of analysis to modify microorganisms to help us with certain chemical or environmental processes. For example, if this approach is used to define biclusters in bacteria that degrade environmental toxins, we could then engineer bacteria to use these biclusters more effectively.

## The *lac* Operon of *E. coli*

When gene expression is turned on in a bacterium by adding a substance (such as lactose) to the medium, the genes involved are said to be *inducible*. The regulatory substance that brings about this gene induction is called an **inducer,** and the phenomenon of producing a gene product in response to an inducer is called **induction.** The inducer is an example of a class of small molecules, called **effectors** or **effector molecules,** that help control the expression of many regulated genes. An inducible gene is transcribed in response to a regulatory event occurring at a specific regulatory DNA sequence adjacent to or near the protein-coding sequence (Figure 17.1). The regulatory event typically involves an inducer and a regulatory protein; and when it occurs, RNA polymerase initiates transcription at the promoter (usually upstream of the regulatory sequence). The gene is turned on, mRNA is made, and the protein encoded by the gene is produced. The regulatory sequence itself does not code for any product. As an example of such gene regulation, let us examine regulation of the genes of the *E. coli lac* operon, an **inducible operon.**

### Lactose as a Carbon Source for *E. coli*

*E. coli* can grow in a simple medium containing salts (including a nitrogen source) and a carbon source such as glucose. The energy for biochemical reactions in the cell comes from glucose metabolism. The enzymes required for glucose metabolism are coded for by constitutive genes. If lactose is provided to *E. coli* as a carbon source instead of glucose, a number of enzymes that are required to metabolize lactose are rapidly synthesized. (A similar series of events, each involving a sugar-specific set of enzymes, is triggered by other sugars as well.) The enzymes are synthesized because the genes that code for them become actively transcribed in the presence of the sugar; the same genes are inactive if the sugar is absent. In other words, the genes are regulated genes whose products are needed only under certain conditions.

Lactose is a disaccharide consisting of the monosaccharides D-galactose and D-glucose. When lactose is present as the sole carbon source in the growth medium, three proteins are synthesized:

1. *β-Galactosidase*. This enzyme breaks down lactose into galactose and glucose, as well as catalyzing the

**Figure 17.1**
**General organization of an inducible gene.**

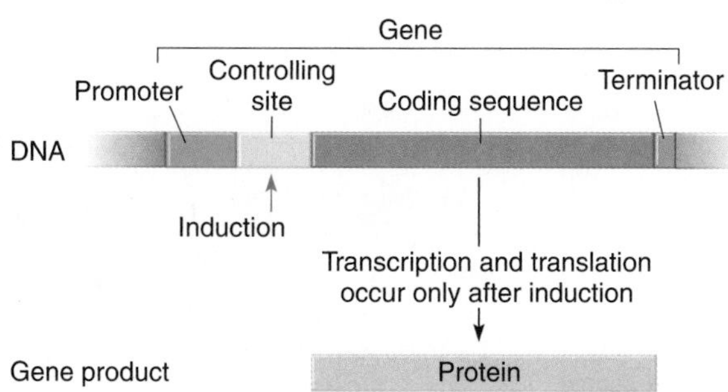


Inducible genes are expressed only in the absence of a repressor and/or presence of an effector/inducer molecule.

isomerization (conversion to a different form) of lactose to *allolactose*, a compound that is important in regulating expression of the *lac* operon (Figure 17.2). (In the cell, the galactose is converted to glucose by enzymes encoded by a gene system specific to galactose catabolism. The glucose is then utilized by constitutively produced enzymes.)

2. *Lactose permease* (also called *M protein*). This protein, found in the *E. coli* cytoplasmic membrane, actively transports lactose into the cell.
3. *β-Galactoside transacetylase*. This enzyme transfers an acetyl group from acetyl-CoA to β-galactosides. The function of this enzyme in the *lac* operon is not understood.

In wild-type *E. coli* growing in a medium containing glucose, only a low concentration of each of these three proteins is produced. For example, only an average of three molecules of β-galactosidase is present in the cell under these conditions. In the presence of lactose, and the absence of glucose, the amount of each enzyme increases coordinately (simultaneously) about a thousandfold (e.g., to about 3,000 molecules of β-galactosidase), because the three essentially inactive genes are now actively transcribed. The process is called **coordinate induction.** Allolactose, not lactose, is the inducer molecule directly responsible for the increased production of the three enzymes (see Figure 17.2). Furthermore, the mRNAs for the enzymes have a short half-life, so the transcripts must be made continually in order for the enzymes to be produced. When lactose is no longer present, transcription of the three genes is stopped and any mRNAs already present are broken down, so no more of these proteins are made. Existing proteins are degraded and diluted out by cell growth and division.



**Figure 17.2**
**Reactions catalyzed by the enzyme β-galactosidase.** Lactose brought into the cell by the permease is converted to glucose and galactose *(top)* or to allolactose *(bottom)*, the true inducer for the lactose operon of *E. coli*.

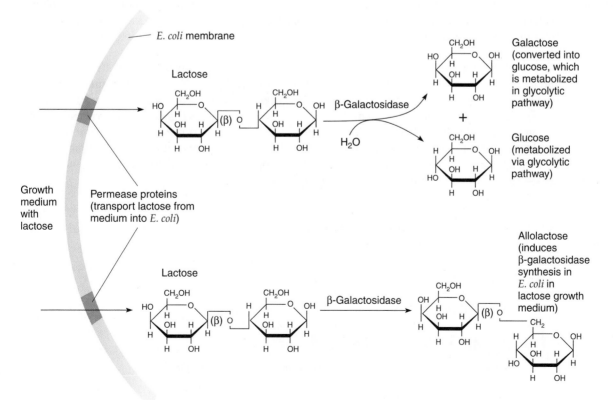

*E. coli* membrane

Lactose

β-Galactosidase

$H_2O$

Galactose (converted into glucose, which is metabolized in glycolytic pathway)

\+

Glucose (metabolized via glycolytic pathway)

Growth medium with lactose

Permease proteins (transport lactose from medium into *E. coli*)

Lactose

β-Galactosidase

Allolactose (induces β-galactosidase synthesis in *E. coli* in lactose growth medium)

**Figure 17.3**

**Structure of the polycistronic mRNA encoded by the three clustered *lac* utilization genes in *E. coli* and its translation to produce β-galactosidase, permease, and transacetylase.**

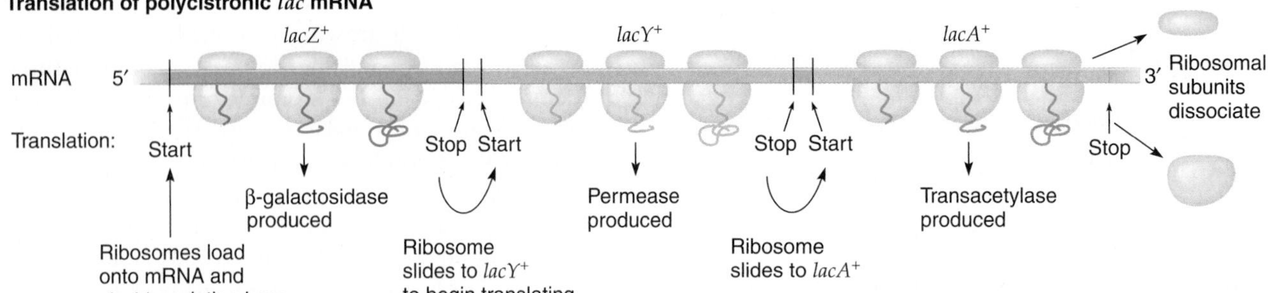


## Experimental Evidence for the Regulation of *lac* Genes

Our basic understanding of the organization of the genes, the regulatory sequences involved in lactose utilization, and the control of expression of the *lac* genes of *E. coli* came largely from the genetic experiments of François Jacob and Jacques Monod, for which they shared (along with Andre Lwoff, for his work on the genetic control of virus synthesis) the 1965 Nobel Prize in Physiology or Medicine. Let us summarize their experiments.

**animation**

**Regulation of Expression of the *lac* Operon Genes**

**Mutations in the Protein-Coding Genes.** Mutations in the protein-coding (structural) genes were obtained after treatment of cells with chemical mutagens. The mutations were characterized to determine which enzyme activity was affected. The β-galactosidase gene was named *lacZ*, *the permease gene lacY*, and the transacetylase gene *lacA*. The *lacZ*⁻, *lacY*⁻, and *lacA*⁻ mutations obtained were used to map the locations of the three genes via standard mapping experiments (see Chapter 15). The experiments showed that the three genes are tightly linked in the order *lacZ–lacY–lacA*. The three clustered genes were shown to be transcribed onto a single mRNA molecule—called a **polycistronic mRNA**—rather than onto three separate mRNAs (Figure 17.3). That is, RNA polymerase initiates transcription at a single promoter, and a polycistronic mRNA is synthesized with the gene transcripts in the order 5′-*lacZ*⁺–*lacY*⁺–*lacA*⁺–3′. In translation, a ribosome loads onto the polycistronic mRNA at the 5′ end and synthesizes β-galactosidase; it then reinitiates translation at the permease sequence and synthesizes permease, reinitiates at the transacetylase sequence and synthesizes transacetylase, and finally dissociates from the mRNA.

**Mutations Affecting the Regulation of Gene Expression.** In wild-type *E. coli*, the three gene products are induced coordinately when lactose is present. Jacob and Monod isolated mutants in which all gene products of the operon were synthesized *constitutively*; that is, they were synthesized regardless of whether the inducer was present. The researchers hypothesized that the mutations were regulatory mutations that affected the normal mechanisms controlling the expression of the structural genes for the enzymes. They identified two classes of constitutive mutations: one class mapped to a small region just upstream of the *lacZ* gene they called the **operator** (*lacO*), and the other class mapped to a gene upstream of the operator they called the *lacI* gene or Lac **repressor gene.** Figure 17.4 depicts the organization of the *lac* structural gene cluster and the associated regulatory sequences. The promoter, operator, and the three structural genes constitute the *lac* operon.

**Figure 17.4**

**Organization of the *lac* genes of *E. coli* and the associated regulatory elements: the operator, promoter, and regulatory gene.** The promoter, operator, and three adjacent *lac* genes together constitute the *lac* operon.

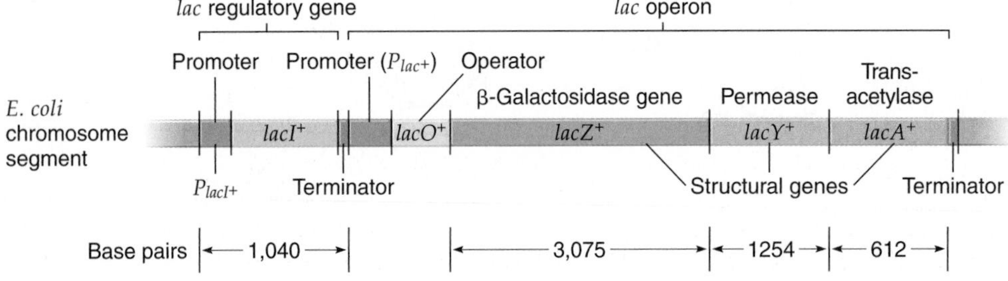

***Operator Mutations.*** The mutations of the operator were called operator-constitutive, or *lacO*$^c$, mutations. Through the use of partial diploid strains ($F'$ strains in which a few chromosomal genes on an extrachromosomal genetic element called the F factor are introduced into a bacterial cell; see Figure 15.6, p. 434), Jacob and Monod were able to define better the role of the operator in regulating expression of the *lac* genes. One such partial diploid was $\frac{F'\ lacO^+\ lacZ^-\ lacY^+}{lacO^c\ lacZ^+\ lacY^-}$. (Both gene sets have a normal promoter, and the *lacA* gene is omitted because it is not important to our discussion.)

One *lac* region in the partial diploid has a normal operator (*lacO*$^+$), a mutated β-galactosidase gene (*lacZ*$^-$), and a normal permease gene (*lacY*$^+$). The other *lac* region has a constitutive operator mutation (*lacO*$^c$), a normal β-galactosidase gene (*lacZ*$^+$), and a mutated permease gene (*lacY*$^-$). This partial diploid was tested for production of β-galactosidase (from the *lacZ*$^+$ gene) and of permease (from the *lacY*$^+$ gene), both in the presence and the absence of the inducer.

Jacob and Monod found that active β-galactosidase is synthesized in the *absence* of the inducer and that permease is synthesized, but is inactive because of the mutation. Only when lactose is added to the culture and the allolactose inducer is produced is active permease synthesized. That is, the *lacZ*$^+$ gene (which is on the same DNA molecule as *lacO*$^c$) is *constitutively expressed* (meaning that the gene is active in the presence or absence of inducer), whereas the *lacY*$^+$ gene is under normal inducible control: it is inactive in the absence of inducer and active in the presence of inducer. In other words, a *lacO*$^c$ *mutation affects only the genes downstream from it on the same DNA molecule.* Similarly, the *lacO*$^+$ region controls only *lac* structural genes adjacent to it and has no effect on the genes on the other DNA molecule. A gene or DNA sequence that controls only genes located on the same, contiguous piece of DNA is said to be ***cis*-dominant.** The *lacO*$^c$ mutation is *cis*-dominant because the defect affects the adjacent genes only and cannot be overcome by a normal *lacO*$^+$ region elsewhere in the cell. In other words, the operator must not encode a diffusible product. If it did, then in the *lacO*$^+$/*lacO*$^c$ diploid state, one would have controlled all the lactose utilization genes regardless of their location.

***lacI Gene Regulatory Mutations.*** The second class of *lac* constitutive mutants defined the *lacI* gene. That is, *lacI*$^-$ mutants in a haploid cell have a constitutive phenotype. Again, the use of partial diploid strains illuminated the normal function of the gene.

The partial diploid here is $\frac{lacI^+\ lacO^+\ lacZ^-\ lacY^+}{lacI^-\ lacO^+\ lacZ^+\ lacY^-}$; both gene sets have normal operators and normal promoters. In the absence of the inducer, no β-galactosidase or permease was produced; both were synthesized in the presence of the inducer. In other words, the expression of both operons was inducible. This means that the *lacI*$^+$ gene in the cell can overcome the defect of the *lacI*$^-$ mutation. Because the two *lacI* genes are located on different DNA molecules (that is, they are in a *trans* configuration), the *lacI*$^+$ gene is said to be ***trans*-dominant** to the *lacI*$^-$ gene.

Because the *lacI*$^+$ gene controls the genes on the other DNA molecule, Jacob and Monod proposed that the *lacI*$^+$ gene is a repressor gene that encodes a **repressor** molecule, the Lac repressor. No functional repressor molecules are produced in *lacI*$^-$ mutants. Thus, in a haploid *lacI*$^-$ bacterial strain, the *lac* operon is constitutive. In a partial diploid with both a *lacI*$^+$ and a *lacI*$^-$, however, the functional Lac repressor molecules produced by the *lacI*$^+$ gene control the expression of both *lac* operons in the cell, making both operons inducible.

***Promoter Mutations.*** The promoter for the structural genes (located at the *lacZ* end of the cluster of *lac* genes; see Figure 17.4) is also affected by mutations. Promoter mutants ($P_{lac}$–) affect all three structural genes. Even in the presence of inducer, the lactose utilization enzymes are not made or are made only at very low rates. Since the promoter is the recognition sequence for RNA polymerase and does not code for any product, the effect of a *P* mutation is confined to the genes that it controls on the same DNA strand. The $P_{lac}$– mutations are another example of *cis*-dominant mutations.

## Jacob and Monod's Operon Model for the Regulation of *lac* Genes

On the basis of their results, Jacob and Monod proposed their now-classic *operon model.* By definition, an **operon** is a *cluster of genes, the expressions of which are regulated together by operator–repressor protein interactions, plus the operator region itself and the promoter.* The promoter was not part of Jacob and Monod's original model; its existence was demonstrated in later studies. The order of the controlling elements and genes in the *lac* operon is promoter–operator–*lacZ*–*lacY*–*lacA*, and the regulatory gene *lacI* is located close to the structural genes, just upstream of the promoter (see Figure 17.4). The *lacI* gene has its own promoter and terminator and encodes the Lac repressor.

The description that follows of the Jacob–Monod model for regulation of the *lac* operon has been embellished with up-to-date molecular information. Figure 17.5 depicts the state of the *lac* operon in wild-type *E. coli* growing in the absence of lactose. The repressor gene (*lacI*$^+$) is transcribed constitutively, and the translation of its mRNA produces a 360-amino acid polypeptide. Four of these polypeptides associate together to form a tetramer, the functional Lac repressor protein (Figure 17.6). The promoter for the *lacI* gene is weak, so few repressor molecules are found in the cell.

The Lac repressor binds to the operator (*lacO*$^+$). The DNA sequence covered by the repressor protein overlaps

**Figure 17.5**

**Functional state of the *lac* operon in wild-type *E. coli* growing in the absence of lactose.**

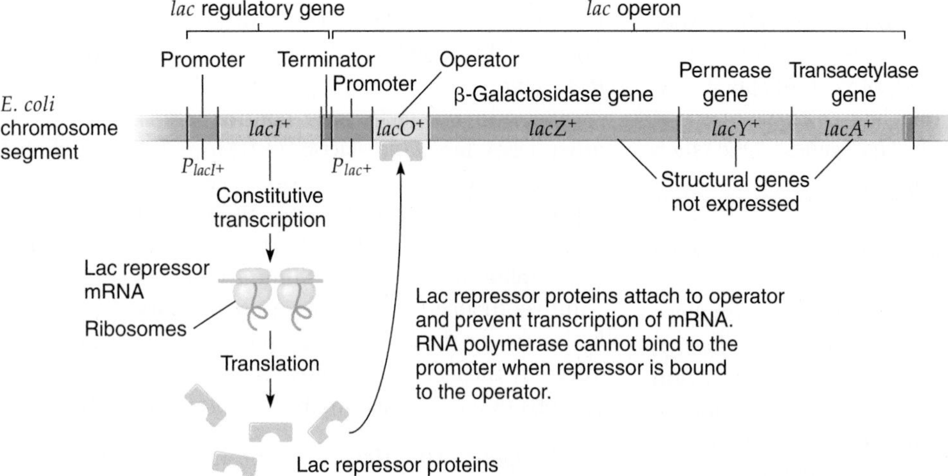


the DNA sequence recognized by RNA polymerase. Therefore, when the repressor is bound to the operator, RNA polymerase cannot bind to the promoter and transcription cannot occur—because of this, the *lac* operon is said to be under *negative control.* The low level of gene transcription that produces a few molecules of the enzymes, even in the absence of the inducer, occurs because repressors do not just bind and stay; they bind and dissociate. In the split second after one repressor unbinds and before another binds, an RNA polymerase could initiate transcription of the operon, even in the absence of the inducer. This "leaky" expression generates a few molecules of the three enzymes encoded by the *lac* operon, which is necessary to allow the initial transportation of lactose into the cell, and the initial conversion of lactose to allolactose when lactose is first added.

When wild-type *E. coli* grows in the presence of lactose as the sole carbon source (Figure 17.7), some lactose is converted by β-galactosidase into allolactose (see Figure 17.2). Allolactose binds to the Lac repressor and changes its shape; the shape change is called an *allosteric shift*. As a result, the repressor loses its affinity for the *lac* operator and dissociates from the site. Free repressor proteins are also altered by binding to allolactose so that they cannot bind to the operator. In this way, allolactose induces production of the *lac* operon-encoded enzymes.

With no Lac repressor bound to the operator, RNA polymerase initiates synthesis of a single polycistronic mRNA molecule for the $lacZ^+$, $lacY^+$, and $lacA^+$ genes. The polycistronic mRNA for the *lac* operon is translated by a string of ribosomes to produce the three enzymes specified by the operon. This efficient mechanism ensures the coordinate production of proteins of related function.

**Figure 17.6**

**Molecular model of the *lac* repressor tetramer.** The four monomers are colored green, violet, red, and yellow.

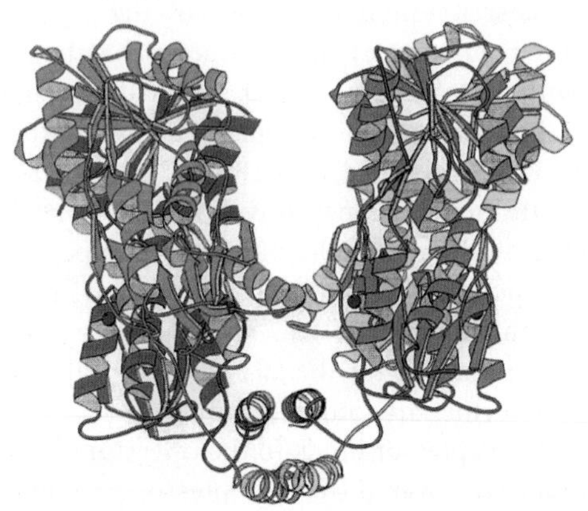

**Effect of *lacO*^c Mutations.** The $lacO^c$ mutations lead to constitutive expression of the *lac* operon genes and are *cis*-dominant to $lacO^+$ (Figure 17.8). This is because base-pair alterations of the operator DNA sequence make it unrecognizable to the repressor protein. Because the repressor cannot bind, the structural genes physically linked to the $lacO^c$ mutation become constitutively expressed.

**Effects of *lacI* Gene Mutations.** The *lacI* mutations map within the coding region of the repressor gene and result in changes to amino acids in the repressor polypeptide. The Lac repressor's shape consequently is changed or the translation in prematurely terminated, and it can neither recognize nor bind to the operator. As a consequence, in a haploid strain, RNA polymerase is not blocked from binding to the promoter and transcription cannot be prevented, even in the absence of lactose. As a result, constitutive expression of the *lac* operon occurs (Figure 17.9a).

The dominance of the $lacI^+$ (wild-type) gene over $lacI^-$ mutants is illustrated for the partial diploid $\frac{lacI^+\ lacO^+\ lacZ^-\ lacY^+}{lacI^-\ lacO^+\ lacZ^+\ lacY^-}$ described earlier. In the absence

**Figure 17.7**

**Functional state of the *lac* operon in wild-type *E. coli* growing in the presence of lactose as the sole carbon source.**

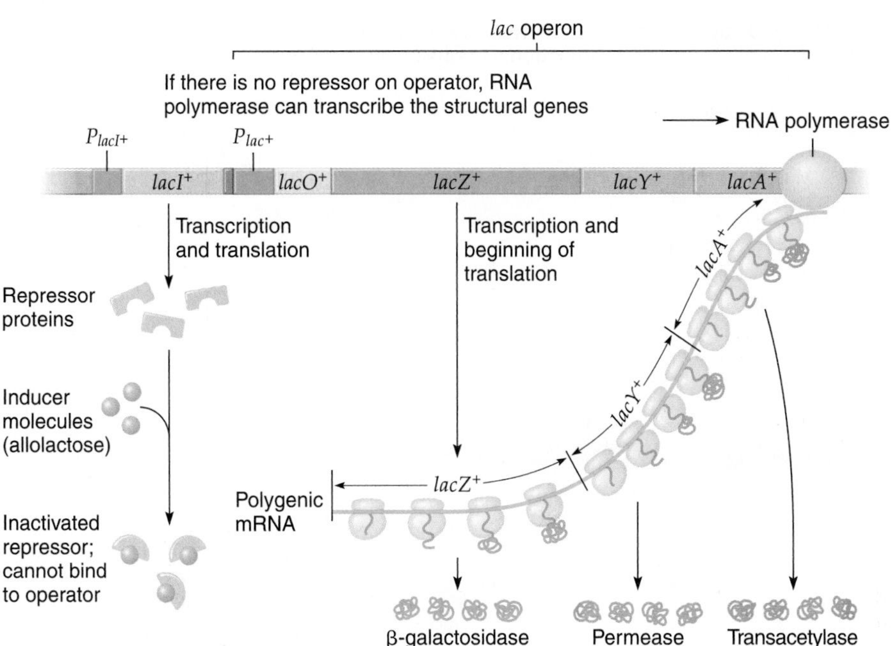


of the inducer (Figure 17.9b), the defective *lacI*$^-$ repressor cannot bind to either normal operator (*lacO*$^+$) in the cell. But sufficient normal Lac repressors, produced from the *lacI*$^+$ gene, are present, and they bind to the two operators and block transcription of both operons. When the inducer is present (Figure 17.9c), the wild-type repressors are inactivated, so both operons are transcribed. One produces a defective β-galactosidase and a normal permease, and the other produces a normal β-galactosidase and a defective permease; between them, active β-galactosidase and permease are produced. Thus, in *lacI*$^+$/*lacI*$^-$ partial diploids, both operons present in the cell are under inducible control.

Other classes of *lacI* gene mutants have been identified since Jacob and Monod studied the *lacI*$^-$ class of mutants. One of these classes, the *lacI*$^S$ (*superrepressor*) mutants, shows no production of *lac* enzymes in the presence or absence of lactose. In partial diploids with a *lacI*$^+$/*lacI*$^S$ genotype, the *lacI*$^S$ allele is *trans*-dominant, affecting both operon copies (Figure 17.10). In this situation, the mutant repressor gene produces a superrepressor protein that can bind to the operator, but cannot recognize the inducer allolactose. Therefore, the mutant superrepressors bind to the operators even in the presence of the inducer, and the operons can never be transcribed. The presence of normal repressors in the cell has no effect, because, once a *lacI*$^S$ repressor is on the operator, the repressor cannot be induced to fall off. Cells with a *lacI*$^S$ mutation cannot use lactose as a carbon source.

A third type of repressor gene mutation is the *lacI*$^{-d}$ (*dominance*) class. In haploid cells, the *lacI*$^{-d}$ mutants have a constitutive phenotype like the other *lacI*$^-$ mutants; the *lac* enzymes are made in the presence or absence of lactose. Unlike the *lacI*$^-$ mutations, the *lacI*$^{-d}$ mutations are *trans*-dominant to *lacI*$^+$ in *lacI*$^{-d}$/*lacI*$^+$ partial diploids, so *lac* enzymes are produced constitutively even in the presence of the normal repressor.

The dominance of *lacI*$^{-d}$ mutants is explained as follows: The Lac repressor protein is a tetramer consisting of four identical polypeptides. In *lacI*$^{-d}$ mutants, the repressor subunits do not combine normally, so no functional repressor tetramer is formed and no operator-specific binding is possible. The *lacI*$^{-d}$/*lacI*$^+$ diploids have a mixture of normal and mutant polypeptides, which combine randomly to form repressor tetramers. There are only about a dozen repressor molecules in the cell. The presence of one or more defective polypeptide subunits in the repressor tetramer is enough to block normal binding to the operator, so there is a good chance that no normal repressor proteins will be produced, because there are so few molecules per cell. As a result of the absence or near absence of complete, functional repressors, a constitutive enzyme phenotype results.

Finally, some mutations in the repressor gene promoter affect the expression of the repressor gene. Earlier, we indicated that the extent of transcription of a gene is a function of the affinity of that gene's promoter for RNA polymerase molecules. Since few repressor molecules are synthesized in wild-type *E. coli* cells, the repressor gene promoter must be of low affinity (i.e., it is a weak promoter). Base-pair mutations have been found that decrease and that increase transcription rates. For example,

Figure 17.8

***Cis*-dominant effect of *lacO*$^c$ mutation in a *lacI*$^+$ *lacO*$^+$ *lacZ*$^-$ *lacY*$^+$/*lacI*$^+$ *lacO*$^c$ *lacZ*$^+$ *lacY*$^-$ partial diploid strain of *E. coli*.** (The *lacZ*$^-$ asqd *lacY*$^-$ mutations are missense mutations.)

**a) Partial diploid in the absence of inducer. The *lacO*$^+$ operon is turned off, whereas the *lacO*$^c$ operon produces functional β-galactosidase from the *lacZ*$^+$ gene and nonfunctional permease molecules from the *lacY*$^-$ gene with a missense mutation.**

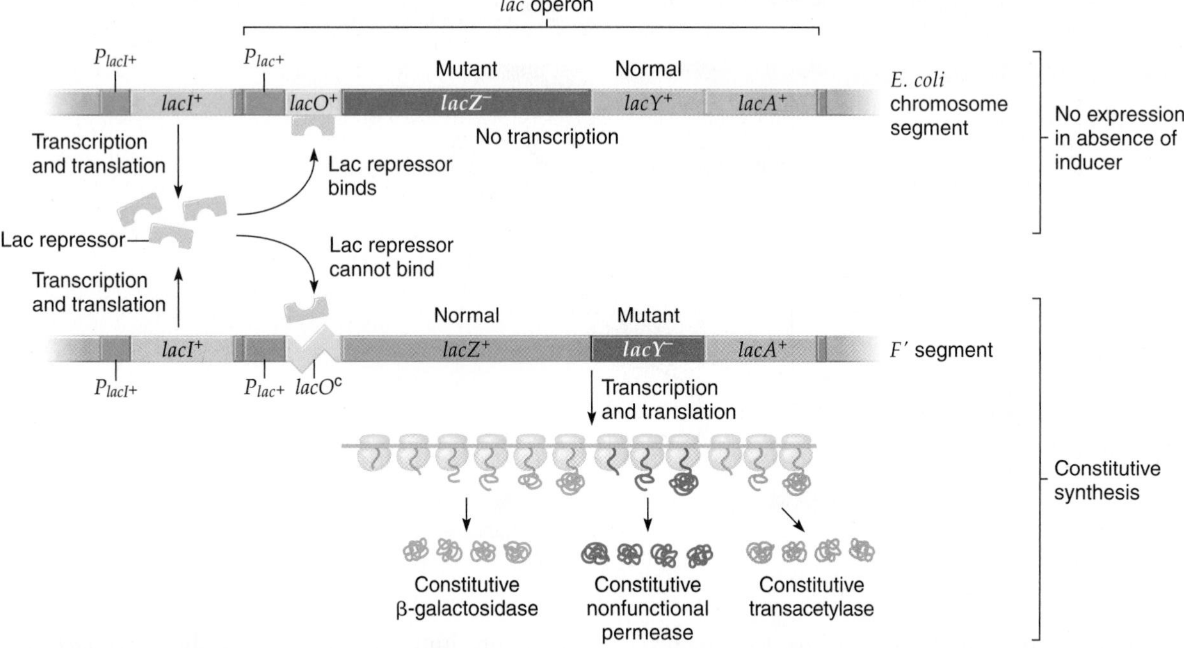


**b) Partial diploid in the presence of inducer. The *lacO*$^+$ operon is turned on and produces nonfunctional β-galactosidase from the *lacZ*$^-$ gene and functional permease from the *lacY*$^+$ gene. The constitutive *lacO*$^c$ operon produces functional β-galactosidase from the *lacZ*$^+$ gene and nonfunctional permease from the *lacY*$^-$ gene. Between the two operons, functional β-galactosidase and permease are produced.**

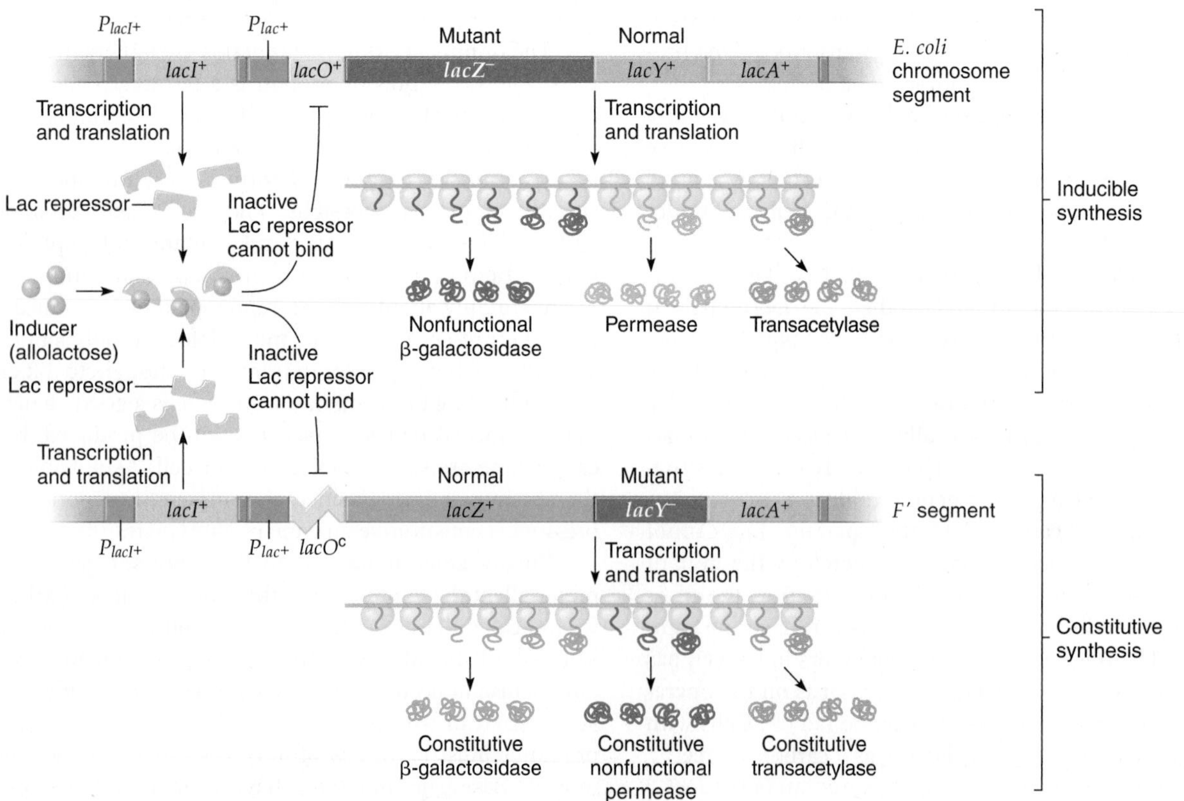

$lacI^Q$ and $lacI^{SQ}$ mutants (where Q stands for "quantity" and SQ stands for "super quantity") result in an increase in the rate of transcription of the repressor gene, with the $lacI^{SQ}$ giving the greater increase. These mutants were useful historically because they produce large numbers of repressor molecules, which facilitated their isolation and purification and, consequently, the determination of the amino acid sequence of the repressor polypeptide. Because $lacI^Q$ and $lacI^{SQ}$ mutants produce more Lac repressor molecules than the wild type produces, these mutants reduce the efficiency of induction of the *lac* operon. Note that $lacI^Q$ and $lacI^{SQ}$ can be induced at high lactose concentrations.

The mutants of the *lacI* gene point out the three different recognition interactions of the Lac repressor: (1) binding of the repressor to the operator region; (2) binding of the inducer to the repressor; and (3) binding of individual repressor polypeptides to each other to form the active repressor tetramer.

## Positive Control of the *lac* Operon

The Lac repressor protein exerts a negative effect on the expression of the *lac* operon by blocking RNA polymerase's binding to the promoter if the inducer is absent. Several years after Jacob and Monod proposed their operon model, researchers found a positive control system that also regulates the *lac* operon—a system that functions to turn on the expression of

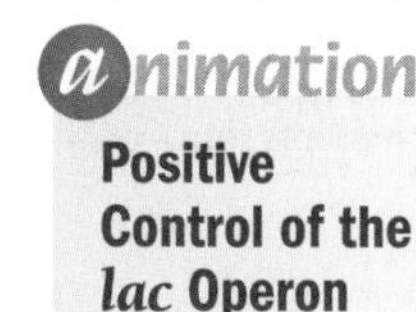




**Figure 17.9**

**Effects of a $lacI^-$ mutation in a $lacI^-$ $lacO^+$ $lacZ^+$ $lacY^+$ haploid cell, and in a $lacI^+$ $lacO^+$ $lacZ^-$ $lacY^+$/$lacI^-$ $lacO^+$ $lacZ^+$ $lacY^-$ partial diploid strain of *E. coli*.** (The $lacZ^-$ and $lacY^-$ mutations are missense mutations.)

**a) Haploid strain (in presence or absence of inducer). The mutant Lac repressor cannot bind to the $lacO^+$ operator, resulting in constitutive synthesis of *lac* operon enzymes.**

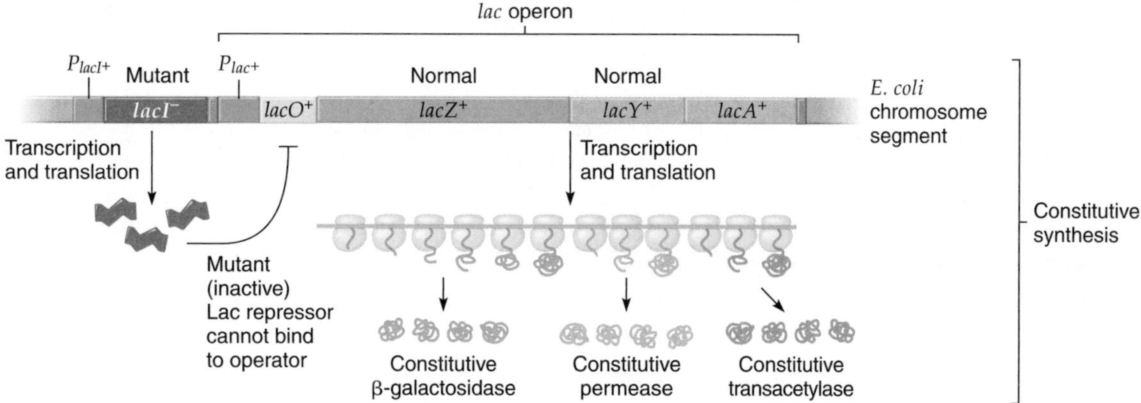


**b) Partial diploid in the absence of inducer. The $lacI^+$ operon produces wild-type Lac repressors, whereas the $lacI^-$ operon produces inactive Lac repressors. The mutant Lac repressor cannot bind to the $lacO^+$ operators, but the wild-type Lac repressors can, so no transcription occurs from either operon.**

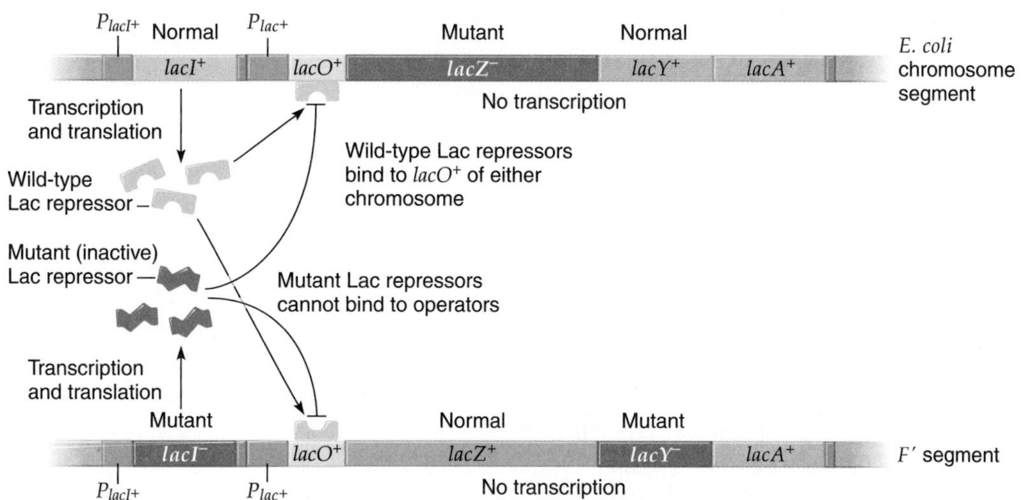

**c) Partial diploid in the presence of inducer. The inducer inactivates the wild-type Lac repressor, preventing it from binding to the *lacO*⁺ operators. The mutant Lac repressor is unable to bind to those operators. The result is transcription of both operons: nonfunctional β-galactosidase and functional permease are produced from the *lacI*⁺ operon, and functional β-galactosidase and nonfunctional permease are produced from the *lacI*⁻ operon.**

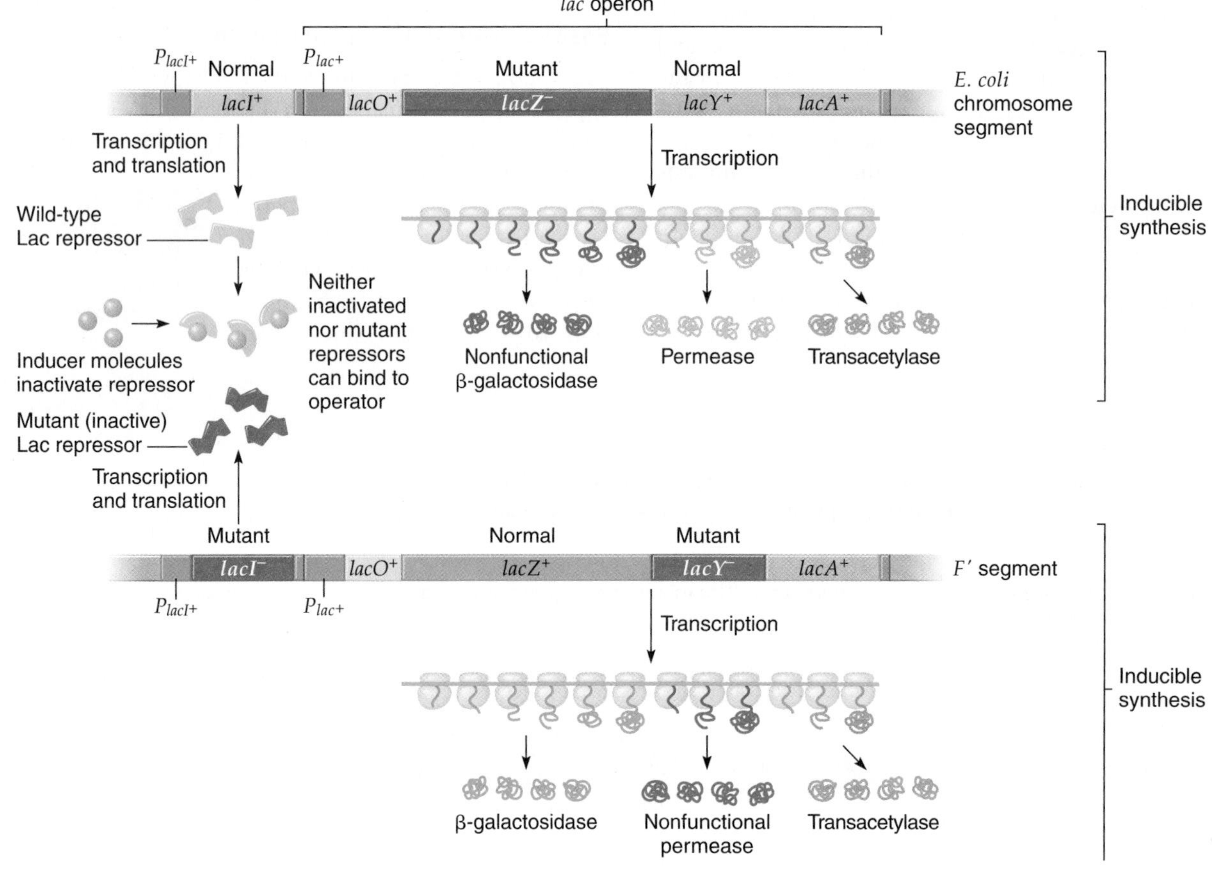




**Figure 17.10**

**Dominant effect of *lacI*ˢ mutation over wild-type *lacI*⁺ in a *lacI*⁺ *lacO*⁺ *lacZ*⁺ *lacY*⁺ *lacA*⁺/*lacI*ˢ *lacO*⁺ *lacZ*⁺ *lacY*⁺ *lacA*⁺ partial diploid cell growing in the presence of lactose.**

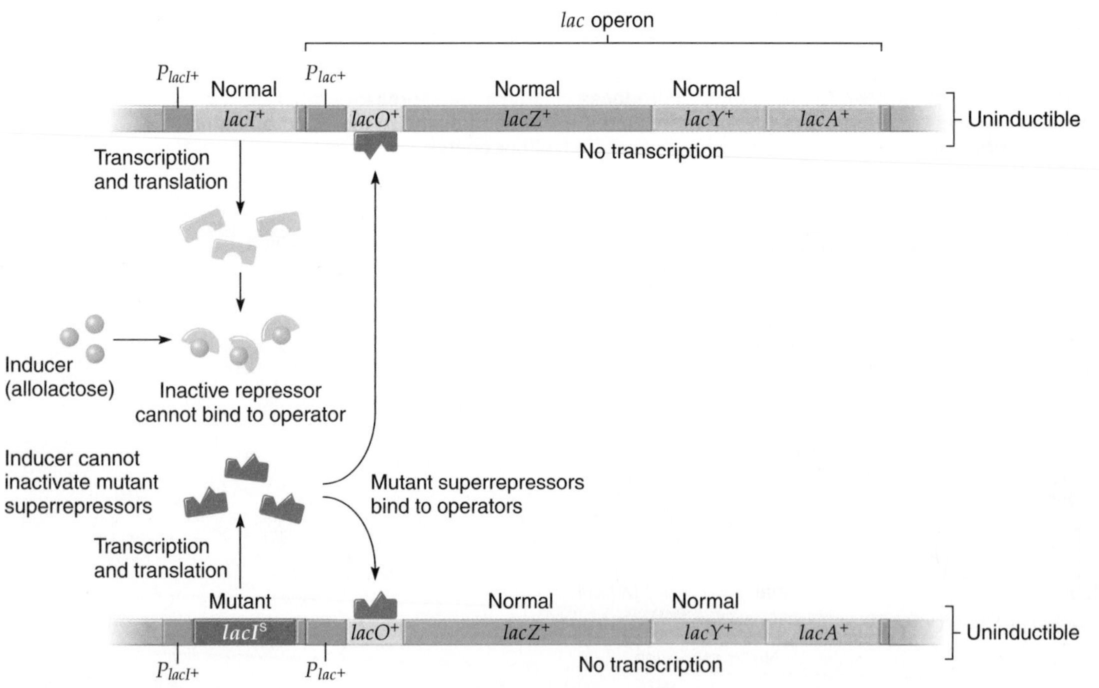

the operon. This system ensures that the *lac* operon will be expressed at high levels only if lactose is the sole carbon source *and not if glucose is present as well.* Glucose is a preferred carbon source because it can be used directly by the glycolytic pathway to produce "energy" for the cell. Lactose and other sugars converted to glucose consume energy. Therefore, more energy can be obtained for the cell from glucose than from other sugar sources.

Figure 17.11 shows the positive regulation of the *lac* operon if lactose is present and glucose is absent. First, a protein called **catabolite activator protein (CAP)** binds with **cAMP (cyclic AMP,** or cyclic adenosine 3′,5′-monophosphate; see Figure 17.12) to form a CAP–cAMP complex. This complex is the positive-regulator molecule. The CAP protein is a dimer of two identical polypeptides. Next, the CAP–cAMP complex binds to the *CAP site,* which is upstream of the site where RNA polymerase binds to the promoter. CAP then recruits RNA polymerase to the promoter, and transcription is initiated.

When glucose is in the medium along with lactose, the glucose is used preferentially because **catabolite repression** (also called the **glucose effect**) occurs. In catabolite repression, the *lac* operon is expressed at only very low levels even though lactose is present in the medium. This occurs because glucose causes the amount of cAMP in the cell to be reduced greatly. As a result, insufficient CAP–cAMP complex is available to recruit RNA polymerase to the *lac* promoter, and transcription is lowered significantly, even though repressors are removed from the operator by the presence of allolactose. In other words, RNA polymerase cannot bind efficiently to the promoter without the aid of the CAP–cAMP complex. That cAMP plays a crucial role in catabolite repression was shown by a number of experiments, including one in which transcription of the *lac* operon was restored by the addition of cAMP to the cell, even though glucose was still present.

The model is that catabolite repression acts on adenylate cyclase, the enzyme that makes cAMP (see Figure 17.12). In *E. coli*, adenylate cyclase is activated by the active form of an enzyme called $III^{Glc}$. When glucose is transported across the cell membrane into the cell, it triggers a series of events that inactivates $III^{Glc}$. Without active $III^{Glc}$, adenylate cyclase is inactivated and no new cAMP is produced. This, along with the breakdown of cAMP by phosphodiesterase, reduces the level of cAMP in the cell. Thus, cAMP is an indicator of glucose levels: when glucose levels are high, cAMP concentration is low, and when glucose levels are low, cAMP concentration is high.

Catabolite repression occurs in the same way in a number of other bacterial operons related to the catabolism of sugars other than glucose. These operons all have in common a CAP site in their promoters to which a specific CAP–cAMP complex binds to facilitate RNA polymerase binding.



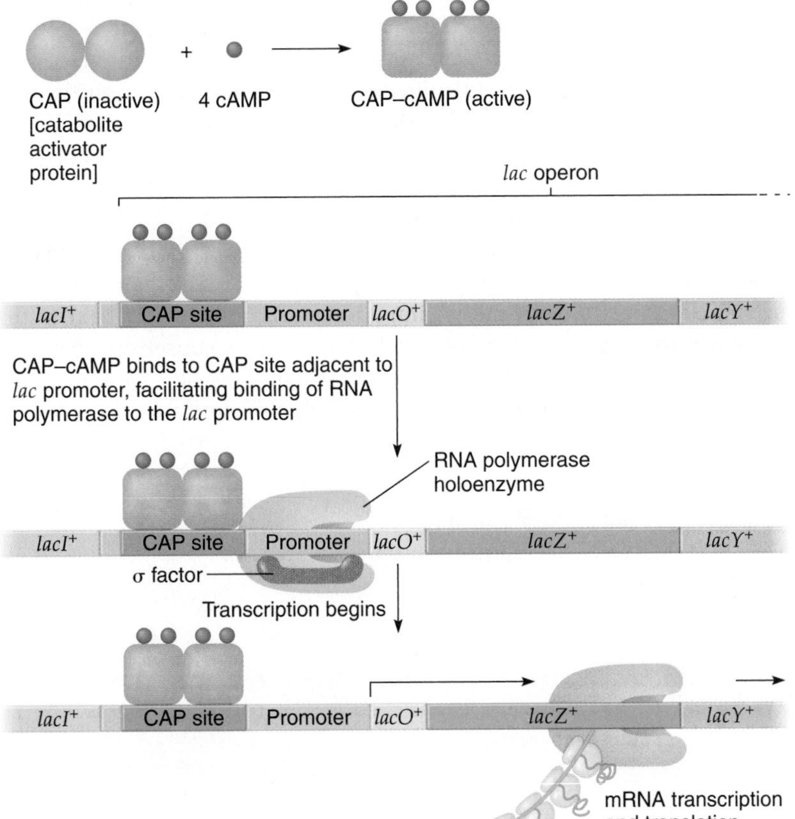


**Figure 17.11**

**Role of cyclic AMP (cAMP) in the functioning of glucose-sensitive operons such as the *lac* operon of *E. coli*.** Shown is the condition in which lactose is present and glucose is absent.

**Figure 17.12**

**Structure, synthesis, and breakdown of cyclic AMP (cAMP, or cyclic adenosine 3′, 5′ monophosphate).**

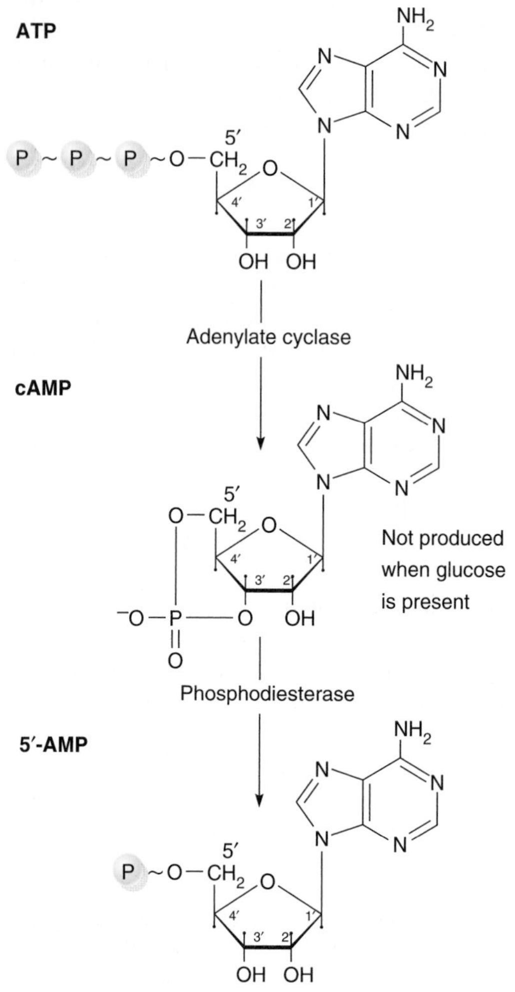

ATP

Adenylate cyclase

cAMP

Not produced when glucose is present

Phosphodiesterase

5′-AMP

## Molecular Details of *lac* Operon Regulation

From DNA- and RNA-sequencing experiments, we know the nucleotide sequences of the significant *lac* operon regulatory sequences. One general approach to obtaining this information has been to purify the protein known to bind to a regulatory sequence and to let it bind to isolated *lac* operon DNA *in vitro*. For example, if the repressor is bound to the *lac* operator, it will protect that region of the operon from deoxyribonuclease digestion. If DNase is allowed to digest the rest of the DNA, the intact operator sequence, cloned by using recombinant DNA technology, and sequenced.

**Promoter Region of the *lac* Repressor Gene (*lacI*).** Figure 17.13 shows the nucleotide pair sequence for the *lacI* gene promoter region, the sequence for the 5′ end of the repressor mRNA, and the first few amino acids of the repressor protein itself. The nucleotide sequence of the repressor mRNA can be aligned with this promoter sequence, with its start approximately in the middle. As with all gene transcripts, translation does not start right at the end of the mRNA molecule. The ribosome binding site is a Shine–Dalgarno sequence (AGGG) at 12 to 9 bases upstream from the start codon (Chapter 6, p. 115). In this unusual case, the start codon is GUG rather than AUG, at nucleotides 27 to 29 from the 5′ end of the messenger. The figure also shows the single base-pair change found for a particular *lacI*$^Q$ mutant; this change, from C-G to T-A, brings about a tenfold increase in repressor production.

***lac* Operon Regulatory Sequences.** Figure 17.14 shows the nucleotide pair sequence of the *lac* operon regulatory sequences. The orientation of this sequence was put together from several different pieces of information. First, the amino acid sequences of the repressor protein and of β-galactosidase were completely known, and that information made it possible to identify the coding regions of the *lacI* gene and of the *lacZ*$^+$ gene. Then the other regions were identified on the basis of "protection" experiments of the kind described previously. Here, CAP–cAMP complex, RNA polymerase, and repressor protein were used separately to bind to the DNA, and DNase-resistant regions were then sequenced.

The beginning of the promoter region is defined as position −84 in the figure (i.e., 84 base pairs upstream from the mRNA initiation site), immediately next to the stop codon for the *lacI* gene. The consensus sequence matches for the CAP–cAMP binding site are nucleotide pairs −54 to −58 and −65 to −69, and the DNA covered

**Figure 17.13**

**Base-pair sequences of the *lac* operon *lacI*$^+$ gene promoter ($P_{lac}{}^+$) and of the 5′ end of the repressor mRNA.** Also shown is the amino acid sequence of the first part of the repressor protein itself. Note that GUG is the initiation codon for methionine in this case.

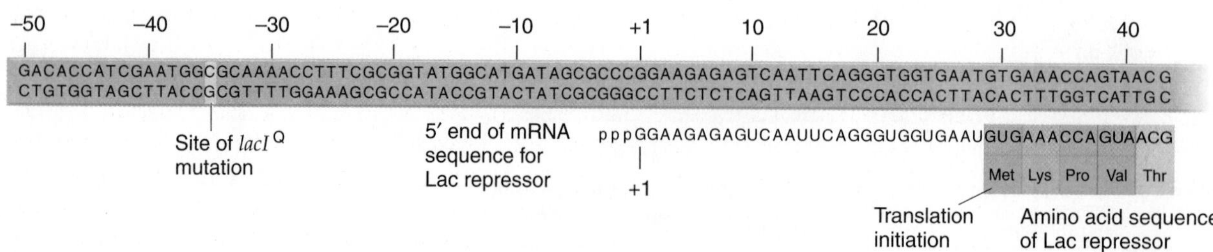

**Figure 17.14**
**Base-pair sequence of the promoter and operator for the *lac* operon of *E. coli*.**

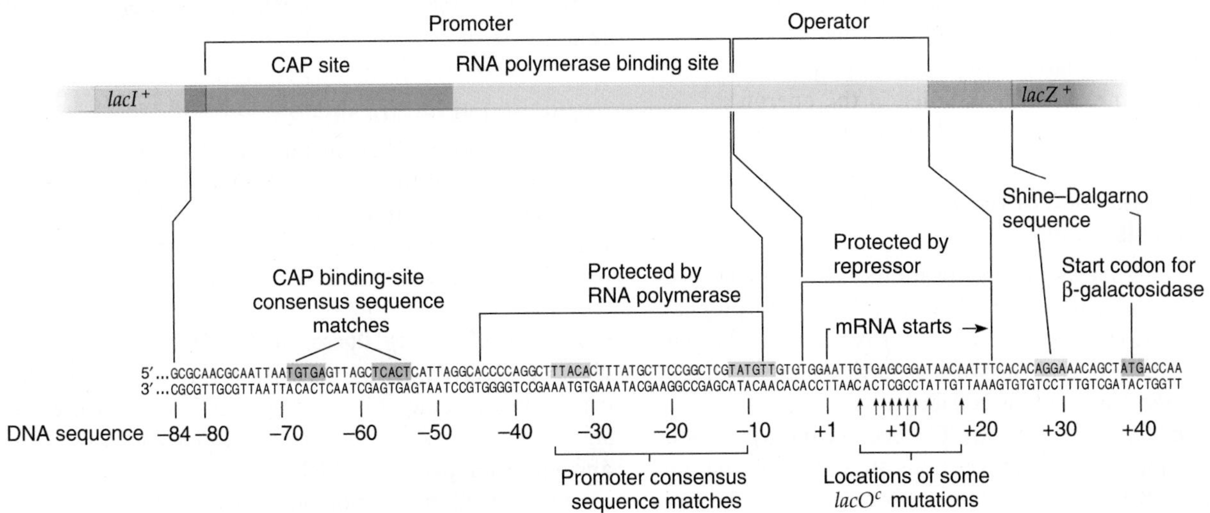


by RNA polymerase spans nucleotide pairs −44 to −8, including −10 and −35 consensus sequence matches (see Figure 17.11). Together, the region from −84 to −8, which includes the CAP protein and the RNA polymerase interaction sites (including a Pribnow box), essentially defines the *lac* operon promoter region.

Adjacent to the promoter region is the operator. The region protected by the Lac repressor protein is the area containing nucleotide pairs −3 to +21. When the Lac repressor is bound to the operator, RNA polymerase cannot bind to the promoter.

The β-galactosidase mRNA has a *leader region* before the start codon is encountered. The actual start of the mRNA here is nucleotide pair +1 in Figure 17.14, which is very close to the beginning of the repressor binding site. Transcription of the *lac* operon includes a large proportion of the operator region, in addition to the protein-coding genes themselves. The AUG start codon for β-galactosidase, which defines the beginning of the *lacZ* coding sequence, is at nucleotide pairs +39 to +41. Thus, the first 38 bases of the *lac* mRNA are not translated.

Figure 17.14 also shows the sites of base-pair substitutions that have been identified for some of the $lacO^c$ mutations studied. In each case, a single base-pair change is responsible for the altered regulatory control of the *lac* operon.

In conclusion, the *lac* operon has proved to be a model system for understanding gene regulation in prokaryotic organisms. Jacob and Monod's original work on this system had a great impact on further studies. As the first molecular model for the regulation of gene expression in any organism, it sparked numerous studies in both prokaryotes and eukaryotes to see whether operons were generally present. We now know that operons are prevalent in bacteria and bacteriophages, but they are very rarely encountered in eukaryotes.

**iActivity**

Your job is to determine the location and effect of a mutation in strains of *E. coli* in the iActivity *Mutations and Lactose Metabolism* on the student website.

**Keynote**

Studies of the synthesis of the lactose-utilizing enzymes of *E. coli* generated a model that is the basis for the regulation of gene expression in a large number of bacterial and bacteriophage systems. In the lactose system, the addition of lactose to the cells brings about a rapid synthesis of three enzymes. The genes for these enzymes are contiguous on the *E. coli* chromosome and are adjacent to two regulatory sequences: a promoter and an operator. The promoter, operator, and genes constitute an operon, which is transcribed as a single unit. In the absence of lactose, the operon is turned off by a repressor.

A positive control system also regulates the *lac* operon. That is, CAP–cAMP binds to the promoter, and this facilitates the binding of RNA polymerase to the promoter. If glucose is present, however, no CAP–cAMP is produced, so RNA polymerase cannot bind efficiently and the *lac* genes are not transcribed.

## The *trp* Operon of *E. coli*

*E. coli* has certain operons and other gene systems that enable it to manufacture any amino acid that is lacking in the medium in which it is placed, so that it can grow and reproduce. When an amino acid is present in the growth medium, though, the genes encoding the enzymes for biosynthetic pathway for that amino acid are turned off. Unlike the *lac* operon, wherein gene activity is induced when a chemical (lactose) is added to the medium, in this case gene activity is repressed when a chemical (an amino acid) is added. We call amino acid biosynthesis operons controlled in this way **repressible operons.** In general, operons for anabolic (biosynthetic) pathways are repressed (turned off) when the end product is readily available. One repressible operon in *E. coli* that has been extensively studied is the operon for the biosynthesis of the amino acid tryptophan (Trp).

## Gene Organization of the Tryptophan Biosynthesis Genes

Figure 17.15 shows the organization of the regulatory sequences and of the genes that code for the tryptophan biosynthetic enzymes and how they relate to the biosynthetic steps. Much of the work we will discuss is that of Charles Yanofsky and his collaborators.

Five structural genes (*A–E*) occur in the *trp* operon. The promoter and operator regions are upstream from the *trpE* gene. Between the promoter–operator region and *trpE* is a short region called *trpL*, the leader region. Within *trpL*, close to *trpE*, is an *attenuator site* (*att*) that plays an important role in the regulation of the *trp* operon.

The entire *trp* operon is approximately 7,000 base pairs long. Transcription of the operon results in the production of a polycistronic mRNA for the five structural genes.

## Regulation of the *trp* Operon

Two regulatory mechanisms are involved in controlling the expression of the *trp* operon. One mechanism uses a repressor–operator interaction, and the other determines whether initiated transcripts include the structural genes or are terminated before those genes are reached.

**Expression of the *trp* Operon in the Presence of Tryptophan.** The regulatory gene for the *trp* operon is *trpR*, located some distance from the operon (and therefore not shown in Figure 17.15). The product of *trpR* is an **aporepressor protein,** which is basically an inactive repressor that alone cannot bind to the operator. When tryptophan is abundant within the cell, it interacts with the aporepressor and converts it to an active Trp repressor. (Tryptophan is an example of an effector molecule,

**Figure 17.15**

**The regulatory sequences and structural genes of the *E. coli trp* operon, and the functions of the structural gene products.**

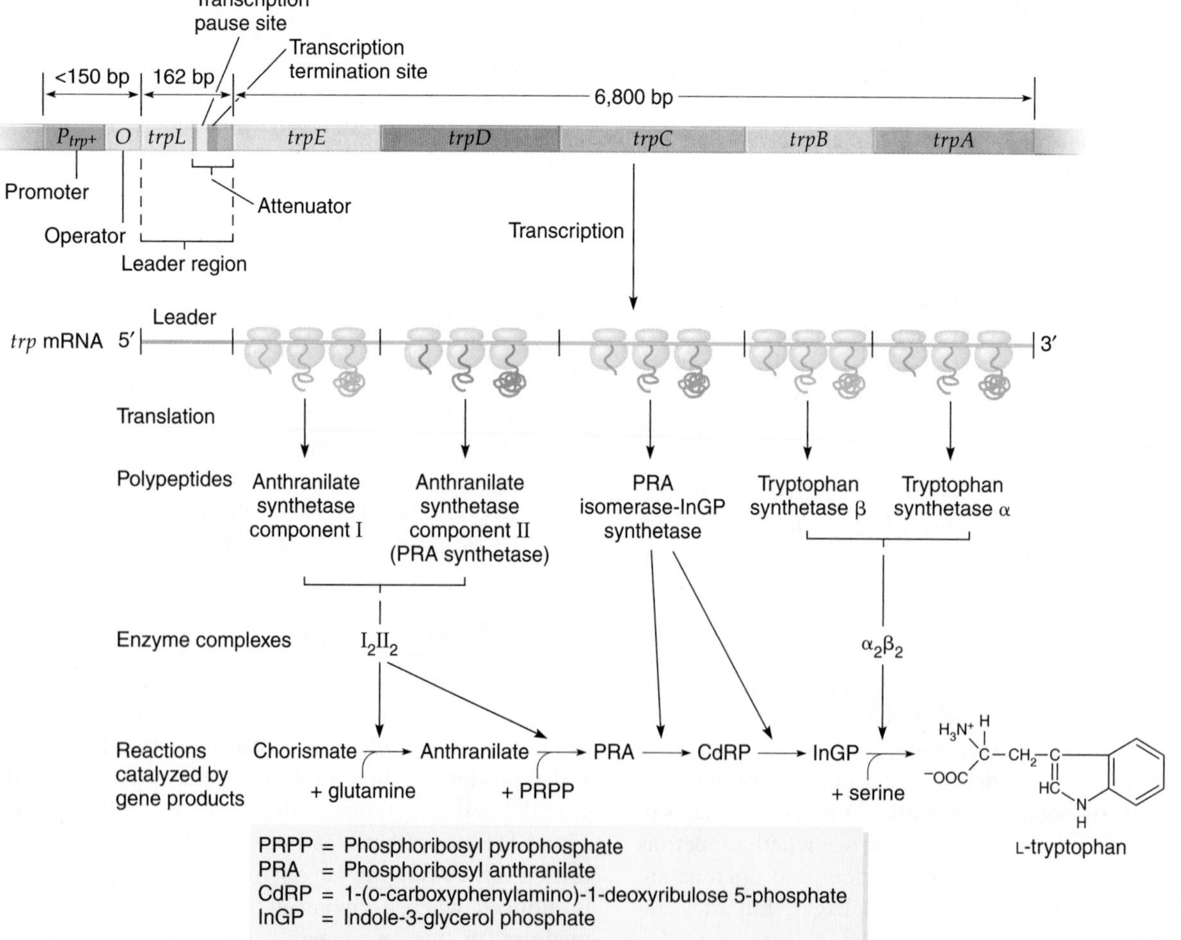

just as allolactose is the effector molecule for the *lac* repressor.) The active Trp repressor binds to the operator and prevents the initiation of transcription of the *trp* operon protein-coding genes by RNA polymerase. As a result, the tryptophan biosynthesis enzymes are not produced. By repression, transcription of the *trp* operon can be reduced about seventy-fold.

**Expression of the *trp* Operon in the Presence of Low Concentrations of Tryptophan.** The second regulatory mechanism is involved in the expression of the *trp* operon under conditions of tryptophan starvation or tryptophan limitation. Under severe tryptophan starvation, the *trp* genes are expressed maximally; under less severe starvation conditions, the *trp* genes are expressed at less than maximal levels. This is accomplished by a mechanism that controls the ratio of full-length transcripts that include the five *trp* structural genes to short, 140-bp transcripts that have terminated at the attenuator site within the *trpL* region (see Figure 17.15). The short transcripts are terminated by a process called **attenuation.** The proportion of the transcripts that include the structural genes is inversely related to the amount of tryptophan in the cell; the more tryptophan there is, the greater is the proportion of short transcripts. Attenuation can reduce transcription of the *trp* operon by a factor of 8 to 10. Thus, repression and attenuation together can regulate the transcription of the *trp* operon by a factor of about 560 to 700.

**Molecular Model for Attenuation.** The mRNA transcript of the leader region includes a sequence that can be translated to produce a short polypeptide. Just before the stop codon in the transcript are two adjacent codons for tryptophan that play an important role in attenuation.

There are four regions of the leader peptide mRNA that can fold and form secondary structures by complementary base pairing (Figure 17.16). The pairing of regions 1 and 2 results in a transcription *pause signal*, that of 3 and 4 is a *termination of transcription signal*, and the pairing of 2 and 3 is an *antitermination signal* for transcription to continue.

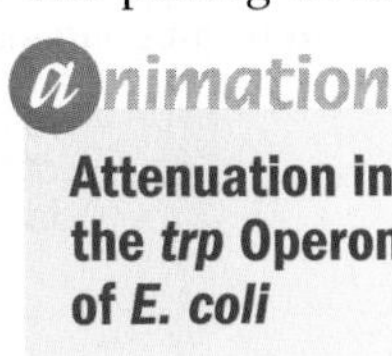


Crucial to the attenuation model is the fact that transcription and translation are tightly coupled in prokaryotes, made possible by the absence of a nuclear envelope and the lack of processing of mRNA transcripts. In the *trp* regulatory system, a pause of the RNA polymerase is caused by the pairing of RNA regions 1 and 2 just after they have been synthesized (see Figure 17.16). The pause enables the ribosome to load onto the mRNA and to begin translating the leader peptide so that translation of the leader mRNA transcript occurs just behind transcription by RNA polymerase.

As coupled transcription and translation continues, the position of the ribosome on the leader transcript plays an important role in the regulation of transcription termination at the attenuator. If the cells are starved for tryptophan, the amount of Trp–tRNA molecules (charged tryptophanyl–tRNA) drops dramatically, since very few tryptophan molecules are available for the aminoacylation of the tRNA. A ribosome translating the leader transcript stalls at the tandem Trp codons in region 1 because the next specified amino acid in the peptide is in short supply; the leader peptide cannot be completed (Figure 17.17a). Since the ribosome now "covers" region 1 of the attenuator region, the 1:2 pairing cannot happen, as region 1 is no longer available. However, RNA region 2 will pair with RNA region 3



**Figure 17.16**

**Four regions of the *trp* operon leader mRNA and the alternative secondary structures they can form by complementary base pairing.**

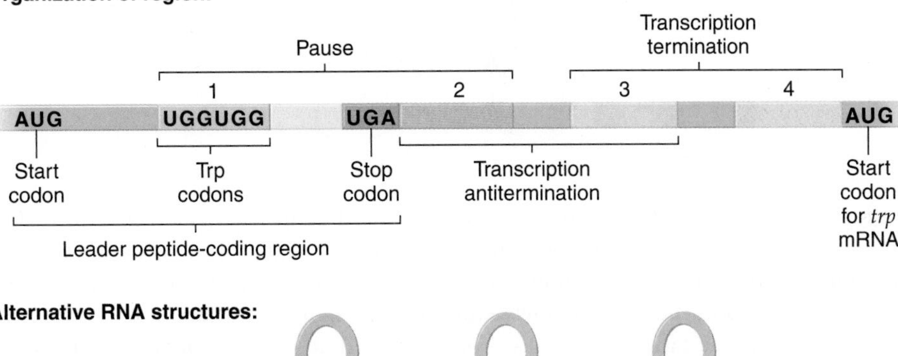


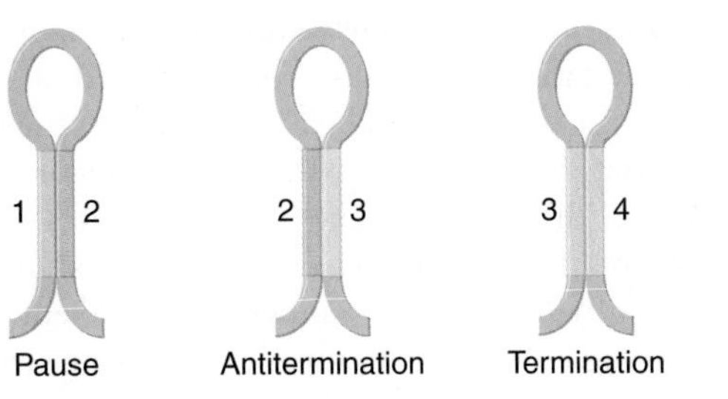

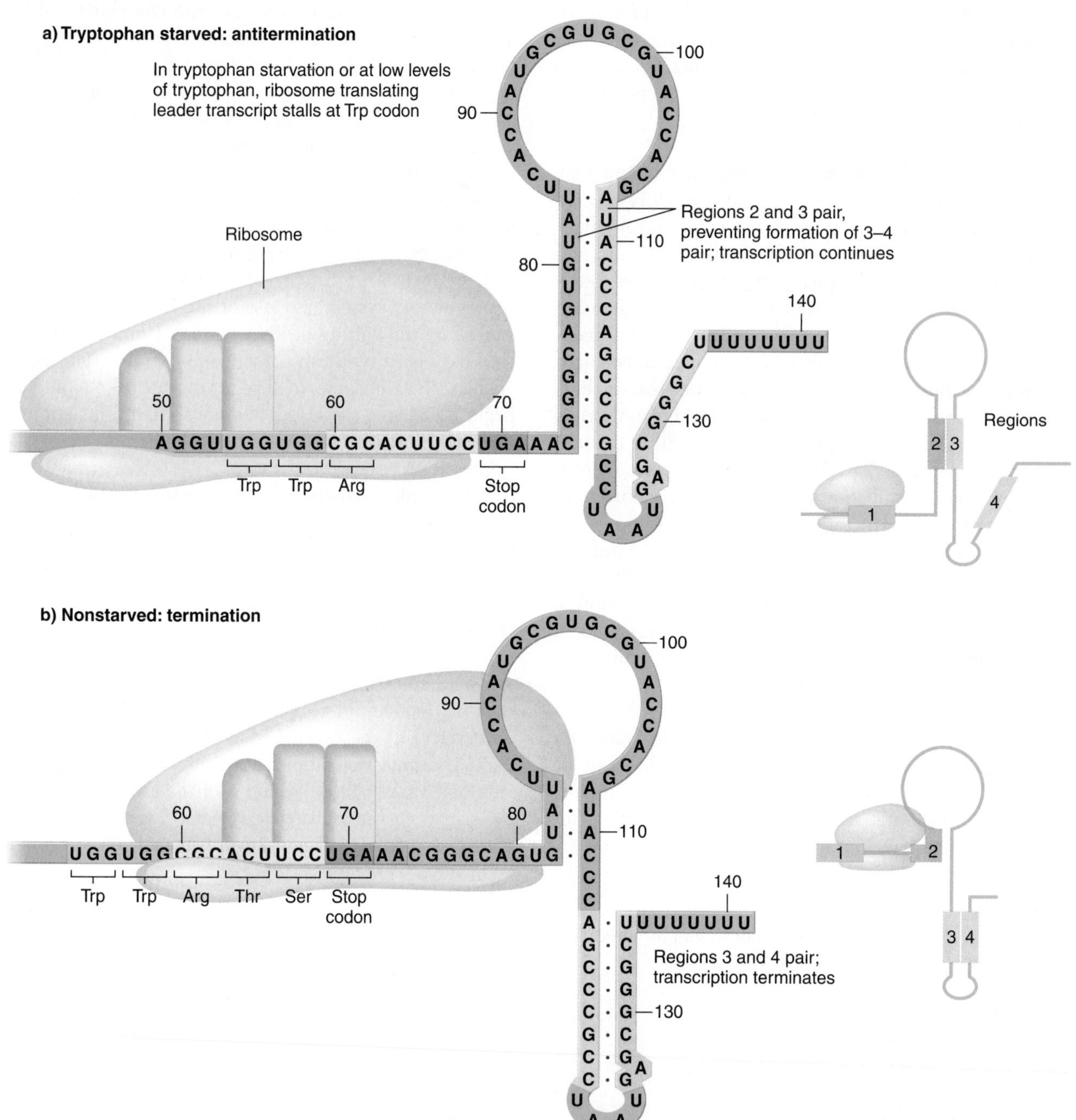
a) Tryptophan starved: antitermination
In tryptophan starvation or at low levels of tryptophan, ribosome translating leader transcript stalls at Trp codon
Ribosome
Regions 2 and 3 pair, preventing formation of 3–4 pair; transcription continues
AGGUUGGUGGCGCACUUCCUGAAAC
Trp
Trp
Arg
Stop codon
UUUUUUUU
Regions
b) Nonstarved: termination
UGGUGGCGCACUUCCUGAAACGGGCAGUG
Trp
Trp
Arg
Thr
Ser
Stop codon
Regions 3 and 4 pair; transcription terminates

**Figure 17.17**
**Model for attenuation in the *trp* operon of *E. coli*.**

once region 3 is synthesized. Because region 3 is paired with region 2, region 3 cannot pair with region 4 when it is synthesized. The 2:3 pairing is an antitermination signal, since the termination signal of 3 paired with 4 does not form, thereby allowing RNA polymerase to continue past the attenuator and transcribe the structural genes.

If, instead, enough tryptophan is present so that the ribosome can translate the Trp codons (Figure 17.17b), then the ribosome continues to the stop codon for the leader peptide. Since the ribosome is then covering part of RNA region 2, that region is unable to pair with region 3, and region 3 is then able to pair with region 4 when it is transcribed. The bonding of region 3 with region 4 is a transcription termination signal (see rho-independent terminator, Figure 5.5, p. 86). The 3:4 structure is called the *attenuator.* The key signal for attenuation is the concentration of Trp–tRNA in the cell because that determines how far the ribosome gets on the leader transcript, either to the Trp codons or to the stop codon.

Genetic evidence for the attenuation model has been obtained through the study of mutants. One type of mutant shows less efficient transcription termination at the attenuator, increasing structural gene expression. The mutations involved are single base-pair changes leading

**Figure 17.18**

**In the *trpL* region, mutation sites that show less efficient transcription at the attenuator site.** The mutations map to DNA regions that correspond to regions 3 and 4 in the RNA.

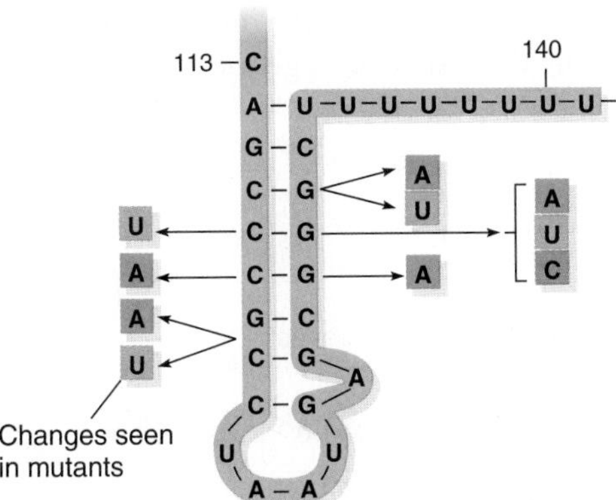


to the base changes in the leader transcript shown in Figure 17.18. In each case, the change is in the regions of 3:4 pairing; each causes a disruption of the pairing so that the structure is less stable. In the less stable state, the structure is less able to prevent transcription from proceeding into the structural genes.

Further direct evidence for the attenuation model came from DNA manipulations in which the DNA sequences for the two Trp codons were changed to encode another amino acid. In those mutant strains, attenuation was not seen in response to changing levels of tryptophan, but it was seen in response to changing levels of the amino acid now specified by the codons.

Attenuation is involved in the genetic regulation of a number of other amino acid biosynthetic operons of *E. coli* and *Salmonella typhimurium*. In every case, there is a leader sequence with two or more codons for the particular amino acid, the synthesis of which is controlled by the enzymes encoded by the operon (Figure 17.19). For example, the *his* operon of *E. coli* has a string of seven histidines in the leader peptide, and 7 of the 15 amino acids in the leader for the *pheA* operon of *E. coli* are phenylalanine. Attenuation has also been shown to regulate a number of genes not involved with amino acid biosynthesis, such as the *ampC* gene of *E. coli* (for resistance to ampicillin).

**Keynote**

Regulation of the tryptophan (*trp*) operon of *E. coli* is at the level of initiating completing a transcript of the operon. This is accomplished through a repressor–operator system, which responds to free tryptophan levels, and through attenuation at a second regulatory sequence called an attenuator, which responds to Trp–tRNA levels. The attenuator is located in the leader region between the operator region and the first *trp* structural gene. The attenuator acts to terminate transcription, depending on the concentration of tryptophan. In the presence of large amounts of tryptophan, attenuation is highly effective; that is, enough Trp–tRNA is present so that the ribosome can move past the attenuator and allow the leader transcript to form a secondary structure that causes transcription to be blocked. In the absence of tryptophan or at low amounts of the amino acid, the ribosomes stall at the attenuator and the leader transcript forms a secondary structure that permits transcription to continue.

## The *ara* Operon of *E. coli*: Positive and Negative Control

Jacob and Monod's operon model for the regulation of gene expression convinced most researchers of the time that negative control was involved in the regulation of gene expression in all systems. In the *lac* and *trp* operons discussed earlier, the negative control is the result of the action of a repressor blocking transcription. In the case of the *lac* operon, the repressor produced by translation of the mRNA transcribed from *lacI* is active, blocking transcription in that form. When the inducer is present, the repressor is inactivated, allowing transcription of the structural genes to

**Figure 17.19**

**Predicted amino acid sequences of the leader peptides of a number of attenuator-controlled bacterial operons.** Shown are leader peptides for the *pheA, his, leu, thr,* and *ilv* (isoleucine and valine) operons of *E. coli* or *Salmonella typhimurium*. The amino acids that regulate the respective operons are highlighted in orange.

| Operon | Leader peptide sequence |
|---|---|
| *pheA:* | Met – Lys – His – Ile – Pro – Phe – Phe – Phe – Ala – Phe – Phe – Phe – Thr – Phe – Pro – – |
| *his:* | Met – Thr – Arg – Val – Gln – Phe – Lys – His – His – His – His – His – His – His – Pro – Asp – – |
| *leu:* | Met – Ser – His – Ile – Val – Arg – Phe – Thr – Gly – Leu – Leu – Leu – Leu – Asn – Ala – Phe – Ile – Val – Arg – Gly – Arg – Pro – Val – Gly – Ile – Gln – His – – |
| *thr:* | Met – Lys – Arg – Ile – Ser – Thr – Thr – Ile – Thr – Thr – Thr – Ile – Thr – Ile – Thr – Thr – Gly – Asn – Gly – Ala – Gly – – |
| *ilv:* | Met – Thr – Ala – Leu – Leu – Arg – Val – Ile – Ser – Leu – Val – Val – Ile – Ser – Val – Val – Val – Ile – Ile – Ile – Pro – Pro – Cys – Gly – Ala – Ala – Leu – Gly – Arg – Gly – Lys – Ala – – |

occur. In the case of the *trp* operon, the repressor is inactive on its own, becoming active in blocking transcription only when bound to tryptophan. Also in the 1960s, at about the same time Jacob and Monod were doing their experiments, Ellis Englesberg and his coworkers were using genetic analysis to study the regulation of the arabinose (*ara*) operon of *E. coli*. Their results indicated that the *ara* operon was under positive control; that is, an *activator* is needed for transcription to occur. While that conclusion was not immediately accepted at the time, subsequent biochemical and molecular analysis supported it. Positive regulation involving activators is now known to occur in a variety of prokaryotic systems (including the *lac* operon, as already detailed) and in all eukaryotes.

Figure 17.20a shows the organization of the *ara* operon. The three structural genes, *araB*, *araA*, and *araD*,

**Figure 17.20**
**Regulation of the *ara* operon of *E. coli*.**

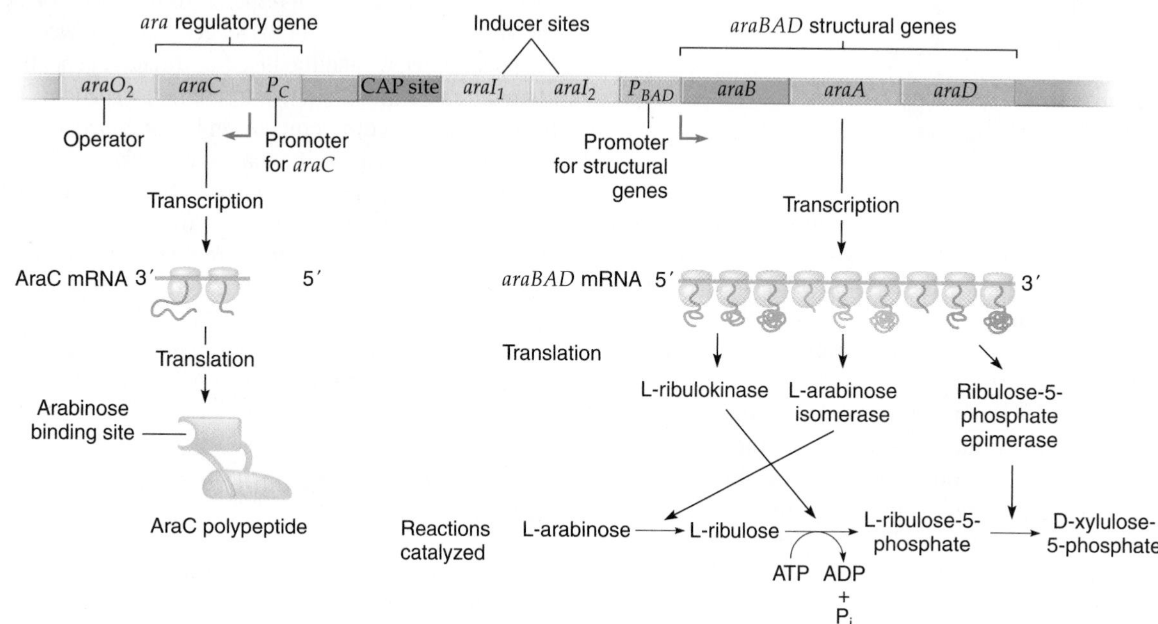


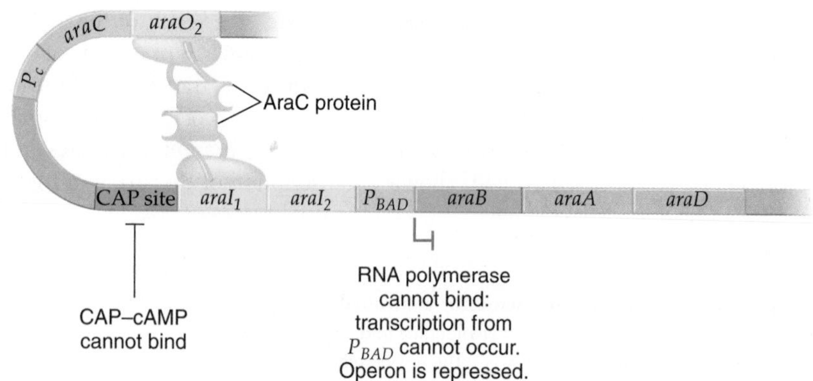


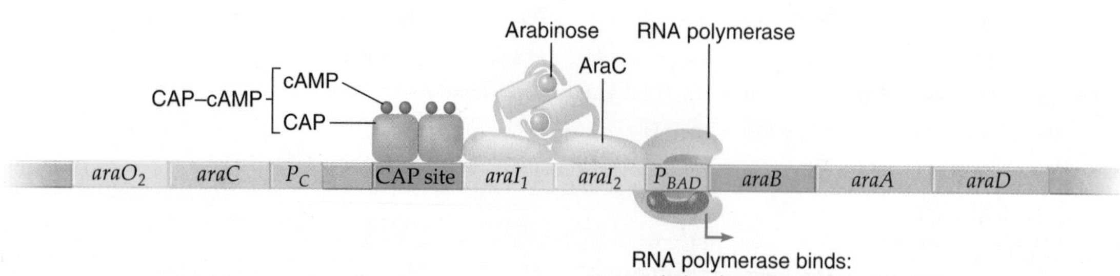

encode three enzymes that convert the pentose sugar L-arabinose to D-xylulose-5-phosphate, which is then metabolized by other biochemical pathways. Like lactose, arabinose can be used as the sole source of carbon and energy for *E. coli*. In this case, arabinose is the inducer of the operon. And, also like the *lac* operon and other sugar utilization operons, this operon can be transcribed only if glucose is absent, because cAMP–CAP must bind to the CAP site in order for RNA polymerase to bind (see "Positive Control of the *lac* Operon," pp. 499–501). Upstream of the structural genes is the regulatory gene, *araC*, and several controlling sites for transcription of that gene and the structural genes. The polypeptide product of the regulatory gene forms a dimer of two identical subunits, named AraC; it is the key regulator of gene expression. $P_{BAD}$ is the promoter where RNA polymerase binds to transcribe the structural genes (from left to right), and $P_C$ is the promoter where RNA polymerase binds to transcribe *araC* (from right to left). The other sites will be explained as the regulatory events are described.

When arabinose and glucose are absent from the growth medium, the *araBAD* structural genes are not transcribed (Figure 17.20b). This is accomplished by one subunit of the AraC protein binding to the inducer site, $araI_1$, and the other subunit binding to the operator, $araO_2$. Binding of AraC in this way causes the DNA to form a loop. The loop blocks CAP–cAMP from binding to the CAP site and RNA polymerase from binding to $P_{BAD}$; transcription of the structural genes therefore does not occur. This part of the regulation of the *ara* operon involves negative control in which AraC acts as a repressor.

When arabinose is present in the growth medium (and glucose is absent), the *ara* operon is induced (Figure 17.20c). The switch is thrown to transcribing the *araBAD* genes when arabinose, acting as an inducer, binds to each subunit of AraC, causing an allosteric shift in the protein. In the changed conformation, one subunit of AraC remains bound to the $araI_1$ site, but the other subunit releases from $araO_2$ and binds instead to $araI_2$. As a result, the DNA no longer forms a loop. Therefore, CAP–cAMP binds to the CAP site, and RNA polymerase binds to $P_{BAD}$ and transcribes the structural genes. This part of the regulation of the *ara* operon involves positive control in which AraC acts as an activator.

If glucose is present as well as arabinose, the *araBAD* genes are not transcribed due to catabolite repression. That is, as for the *lac* operon, the decrease in cAMP levels that occurs when glucose is present means CAP–cAMP complexes are not formed. In the absence of CAP–cAMP binding to the CAP site, initiation of transcription by RNA polymerase at $P_{BAD}$ is highly inefficient.

## Regulation of Gene Expression in Phage Lambda

Bacteriophages exist by invading and manipulating bacterial cells. Many or all of the essential components for phage reproduction are provided by the bacterial host cell, and the use of those components is controlled by the products of phage genes. Most genes of a phage, then, code for products that control the life cycle and the production of progeny phage particles. Much is known about gene regulation in a number of bacteriophages. In this section, we discuss the regulation of gene expression as it relates to the lytic cycle and lysogeny in bacteriophage lambda (λ). (Recall from Chapter 15, p. 440–442, that in the lytic cycle, the phage takes over the bacterium and directs its growth and reproductive abilities so that it expresses the phage's genes and produces progeny phages. Lysogeny involves the insertion of a temperate phage chromosome into a bacterial chromosome, with the former replicating whenever the latter does. In this state, the phage genome is repressed and is said to be in the *prophage* state.)

### Early Transcription Events

Figure 17.21 shows the genetic map of λ. The mature λ chromosome is linear and has complementary "sticky" ends. Once free in the host cell, the λ chromosome circularizes, so we show the genetic map in a circular form. Recall that λ is a temperate phage (see Figure 15.12, p. 441), so when it infects a bacterial cell, the phage has a choice of whether to enter the lytic pathway (when progeny phages are assembled and released from the cell) or the lysogenic pathway (when the λ chromosome integrates into the chromosome and no progeny phages are produced). The regulatory system involved in this choice is an excellent model for a genetic switch and, as such, has contributed to our thinking about how genetic switches might operate in eukaryotic systems.

The choice between the lytic and lysogenic pathways occurs soon after λ infects the cell and its genome circularizes. The choice involves a sophisticated *genetic switch*. First, transcription begins at promoters $P_L$ and $P_R$ (Figure 17.22, part 1). Promoter $P_L$ is for leftward transcription of the left early operon, and promoter $P_R$ is for rightward transcription of the right early operon.

The first gene to be transcribed from $P_R$ is *cro* (control of repressor and other), the product of which is the Cro protein. This protein plays an important role in setting the genetic switch to the lytic pathway. The first gene to be transcribed from $P_L$ is *N*. The resulting N protein is a transcription *antiterminator* that allows RNA synthesis to proceed past certain transcription terminators, in this case leftward of *N* and rightward of *cro*, thereby including all the early genes (Figure 17.22, part 2). Genes transcribed due to the action of the N protein are *cII*, *O*, *P*, and *Q*. Gene *cII* encodes protein cII, which can turn on gene *cI* (which encodes the λ repressor) and gene *int* (which encodes the integrase required for integrating the lambda chromosome into the host chromosome during the lysogenic pathway). However, cII protein performs this function only when the phage follows the lysogenic pathway. Genes *O* and *P* encode two DNA replication proteins, and gene *Q* encodes a protein needed to turn on late genes for

**Figure 17.21**

**A map of phage λ, showing the major genes.** (Promoters discussed in text: $P_L$ = promoter for leftward transcription of the left early operon, $P_R$ = promoter for rightward transcription of the right early operon, $P_{RE}$ = promoter for repressor establishment, and $P_{RM}$ = promoter for repressor maintenance.)

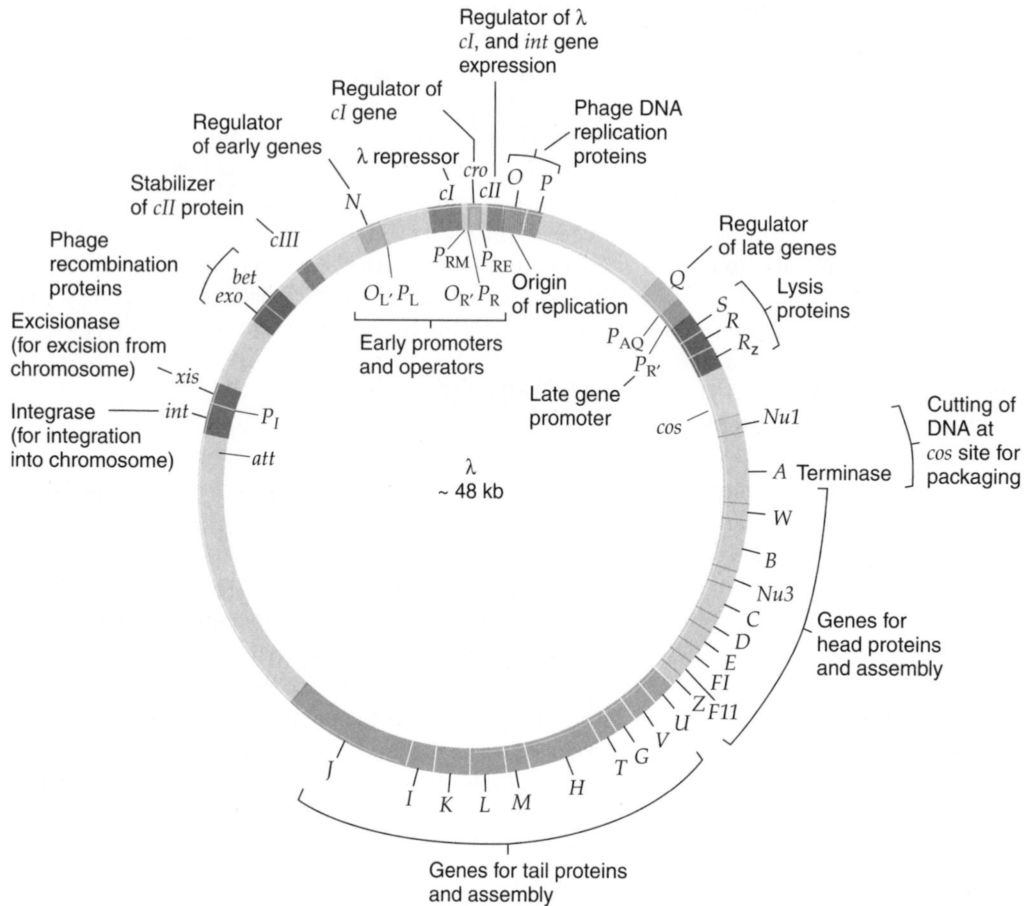


lysis and phage particle proteins. The Q protein is another antiterminator, permitting transcription to continue into the late genes involved in the lytic pathway. However, only when the switch is set to the lytic pathway and transcription continues from $P_R$ for a sufficient time does enough Q protein accumulate to function effectively.

## The Lysogenic Pathway

After the early transcription events, either the lysogenic or lytic pathway is followed (Figure 17.22, part 3). The switch is set for the lysogenic pathway as follows.

The establishment of lysogeny requires the protein products of the *cII* (right early operon) and *cIII* (left early operon; see Figure 17.21) genes. The cII protein (stabilized by cIII protein) activates transcription of the *cI* gene (located between the $P_L$ and $P_R$ promoters; see Figure 17.22, part 4a) leftward from a promoter called $P_{RE}$ (promoter for repressor establishment). The *cro* gene is not transcribed during this event, however, because transcription of *cro* occurs in the rightward direction under the control of a promoter on its left. The product of the *cI* gene, the λ repressor, binds to two operator regions, $O_L$ and $O_R$ (see Figure 17.21 and Figure 17.22, part 5a), whose sequences overlap the $P_L$ and $P_R$ promoters, respectively. The binding of the λ repressor prevents the further transcription by RNA polymerase of the early operons controlled by $P_L$ and $P_R$. As a result, transcription of the *N* and *cro* genes is blocked, and because the two proteins specified by these genes are unstable, the concentrations of those two proteins in the cell drop dramatically. Furthermore, a repressor bound to $O_R$ stimulates the synthesis of more repressor mRNA from a different promoter, $P_{RM}$ (promoter for repressor maintenance), thereby maintaining repressor concentrations in the cell (see Figure 17.22, part 5a). Thus, if enough λ repressors are present, lysogeny is established by the binding of the repressor to operators $O_L$ and $O_R$, followed by the integration of λ DNA catalyzed by integrase, which is the product of the cII-regulated promoter $P_I$. As the concentration of cII drops, $P_I$ transcription shuts off, leaving $P_{RM}$ as the only active promoter.

In sum, the lysogenic pathway is favored when enough λ repressor is made so that early promoters are turned off, thereby repressing all the genes needed for the

**Figure 17.22**

**Expression of λ genes after infection of *E. coli*, and the transcriptional events that occur when either the lysogenic or lytic pathway is followed.** In the figure, stimulation of transcription is indicated by green arrows, and repression of transcription by red arrows.

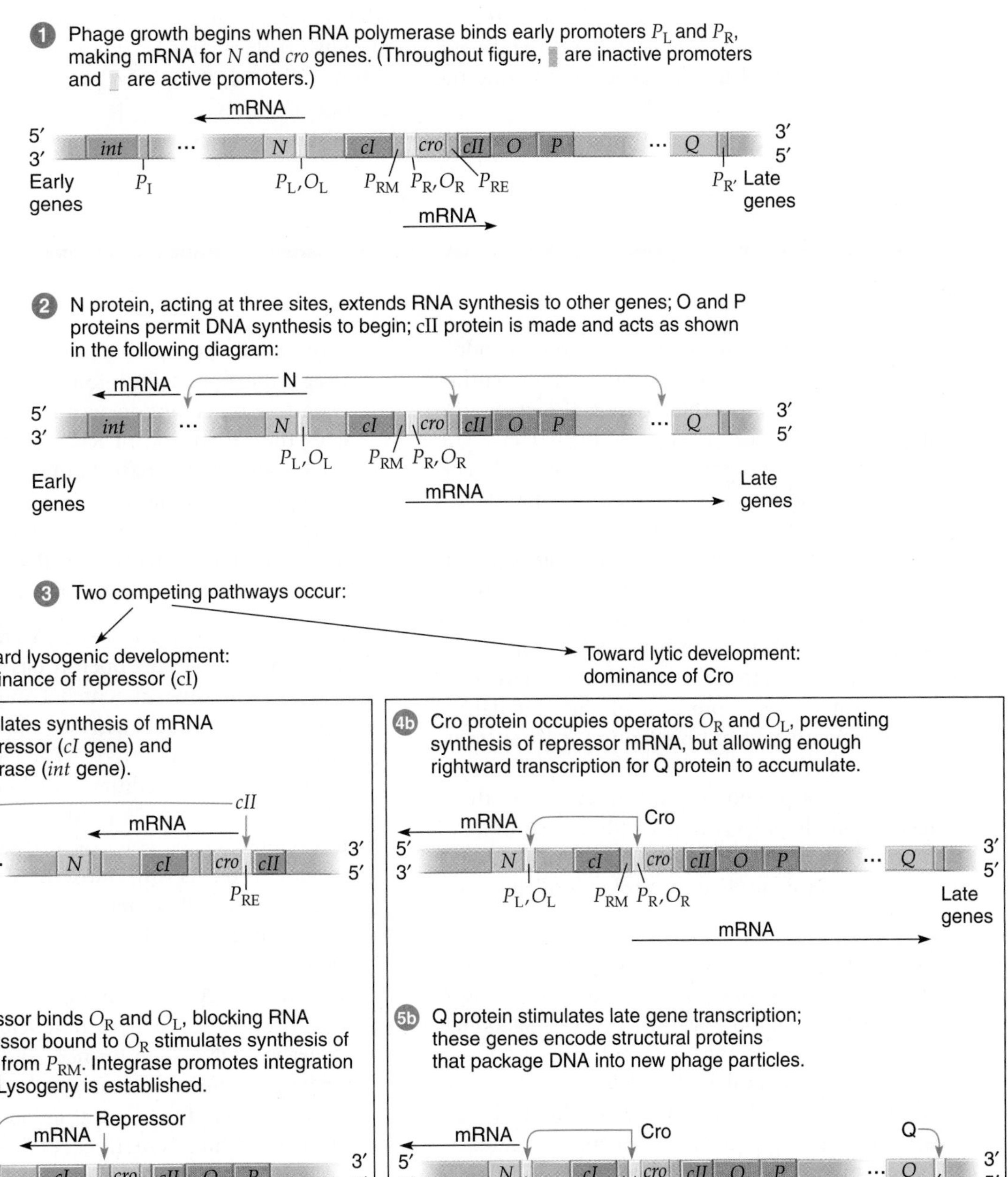
1 Phage growth begins when RNA polymerase binds early promoters $P_L$ and $P_R$, making mRNA for N and cro genes. (Throughout figure, are inactive promoters and are active promoters.)
mRNA
5′ 3′
int
N
cI
cro
cII
O
P
Q
3′ 5′
Early genes
$P_I$
$P_L, O_L$
$P_{RM}$ $P_R, O_R$ $P_{RE}$
$P_{R'}$ Late genes
mRNA
2 N protein, acting at three sites, extends RNA synthesis to other genes; O and P proteins permit DNA synthesis to begin; cII protein is made and acts as shown in the following diagram:
mRNA
N
Early genes
Late genes
mRNA
3 Two competing pathways occur:
Toward lysogenic development: dominance of repressor (cI)
Toward lytic development: dominance of Cro
4a cII protein stimulates synthesis of mRNA from $P_{RE}$ for repressor (cI gene) and from $P_I$ for integrase (int gene).
cII
mRNA
mRNA
$P_I$
$P_{RE}$
Early genes
4b Cro protein occupies operators $O_R$ and $O_L$, preventing synthesis of repressor mRNA, but allowing enough rightward transcription for Q protein to accumulate.
mRNA
Cro
Late genes
mRNA
5a Sufficient repressor binds $O_R$ and $O_L$, blocking RNA synthesis; repressor bound to $O_R$ stimulates synthesis of more repressor from $P_{RM}$. Integrase promotes integration of phage DNA. Lysogeny is established.
Repressor
mRNA
$P_{RM}$ $P_R, O_R$
5b Q protein stimulates late gene transcription; these genes encode structural proteins that package DNA into new phage particles.
mRNA
Cro
Q
$P_{R'}$ Late genes
mRNA
mRNA

lytic pathway. One important lytic pathway gene that is repressed is *Q*; the Q protein is a positive regulatory protein required for the production of phage coat proteins and lysis proteins (see the next section).

## The Lytic Pathway

Let us consider the induction of the lytic pathway caused by ultraviolet light irradiation. Inducers such as ultraviolet light typically damage DNA, and this somehow causes a change in the function of the bacterial protein RecA (the product of the *recA* gene). Normally RecA functions in DNA recombination, but when DNA is damaged, RecA stimulates the λ repressor polypeptides to cleave themselves in two and therefore become inactivated. The resulting absence of repressor at $O_R$ allows RNA polymerase to bind at $P_R$, and the *cro* gene is then further transcribed. The Cro protein that is produced then acts to decrease

RNA synthesis from $P_L$ and $P_R$, and this reduces the synthesis of the cII protein, the regulator of λ repressor synthesis, and blocks the synthesis of λ repressor mRNA from $P_{RM}$ (Figure 17.22, part 4b). At the same time, transcription of the right early operon genes from $P_R$ is decreased, but enough Q proteins are accumulated to set the genetic switch for transcription of the late genes for starting the lytic pathway (Figure 17.22, part 5b).

In sum, lambda uses complex regulatory systems to choose either the lytic or the lysogenic pathway. The decision depends on a sophisticated genetic switch that involves competition between the products of the *cI* gene (the repressor) and the *cro* gene (the gene for the Cro protein). If the repressor dominates, the lysogenic pathway is followed; if the Cro protein dominates, the lytic pathway is followed.



# Summary

- In the lactose utilization system of *E. coli,* the addition of lactose to cells brings about a rapid synthesis of three enzymes. In the absence of lactose, the synthesis of the three enzymes is turned off. The genes for the enzymes are contiguous on the *E. coli* chromosome and are adjacent to two regulatory sequences: a promoter and an operator. The promoter, operator, and genes constitute an operon. Transcription of the genes results in a single polycistronic mRNA. A regulatory gene is associated with an operon. To turn on gene expression in the lactose system, a lactose metabolite binds with a repressor protein (the product of the regulatory gene), inactivating it and preventing it from binding to the operator. As a result, RNA polymerase can bind to the promoter and transcribe the three genes as a single polycistronic mRNA. Operons are commonly involved in the regulation of gene expression in a large number of prokaryotic and bacteriophage systems.

- If both glucose and lactose are present in the medium, the lactose operon is not induced, because glucose (which requires less energy to metabolize than does lactose) is the preferred energy source. This phenomenon is called catabolite repression and involves cellular levels of cyclic AMP. That is, in the presence of lactose and in the absence of glucose, cAMP complexes with CAP to form a positive regulator needed for RNA polymerase to bind to the promoter efficiently. The addition of glucose results in a lowering of cAMP concentration, so no CAP–cAMP complex is produced, and therefore RNA polymerase cannot bind efficiently to the promoter transcribe the *lac* genes.

- The expression of a number of bacterial amino acid synthesis operons is controlled by a repressor–operator system and through attenuation at a second regulatory sequence, called an attenuator. The repressor–operator system functions essentially like that for the *lac* operon, except that the addition of amino acid to the cell activates the repressor, thereby turning off the operon. An attenuator located between the operator region and the first structural gene is a transcription termination site that modulates the proportion of RNA polymerases that continue transcription past that point based on the level of tryptophan in the cell. Attenuation requires a tight coupling between transcription and translation, and the formation of particular RNA secondary structures that signal whether transcription can continue.

- The *ara* operon of *E. coli* encodes the genes for arabinose utilization. This operon has a regulatory gene, *araC*, the product of which (AraC) functions as a repressor and as an activator of transcription depending on the conditions. When arabinose is absent from the growth medium, AraC binds to regulatory sites, causing the DNA to loop and thereby blocking access of RNA polymerase to the structural gene's promoter. In this way, transcription of the structural genes is blocked by repression. When arabinose is present in the growth medium and glucose is absent, the operon is induced. Acting as an inducer, arabinose binds to AraC, changing its conformation. In its new form, AraC changes how it binds to DNA. No longer is a DNA loop formed, and RNA polymerase binding to the structural gene's promoter is facilitated by AraC functioning as an activator. In this sugar utilization system, catabolite repression operates as it does for the *lac* operon so the operon cannot be induced if glucose is present.

- Bacteriophages such as lambda are especially adapted for undergoing reproduction within a bacterial host. Many genes related to the production of progeny phages or to the establishment or reversal of lysogeny in temperate phages are organized into operons. Like bacterial operons, these operons are controlled through the interaction of regulatory proteins with operators and promoters that are adjacent to clusters of structural genes. Phage lambda has been an excellent model for studying the genetic switch that controls the choice between lytic and lysogenic pathways in a temperate phage.

# Analytical Approaches to Solving Genetics Problems

**Q17.1** In the laboratory, you are given 10 strains of *E. coli* with the following *lac* operon genotypes, where $I$ = *lacI* (the Lac repressor gene), $P = P_{lac}$ (the promoter), $O$ = *lacO* operator), and $Z$ = *lacZ* (the β-galactosidase gene):

1. $I^+\ P^+\ O^+\ Z^+$
2. $I^-\ P^+\ O^+\ Z^+$
3. $I^+\ P^+\ O^C\ Z^+$
4. $I^-\ P^+\ O^C\ Z^+$
5. $I^+\ P^+\ O^c\ Z^-$
6. $\dfrac{F'\ I^+\ P^+\ O^C\ Z^-}{I^+\ P^+\ O^+\ Z^+}$
7. $\dfrac{F'\ I^+\ P^+\ O^+\ Z^-}{I^+\ P^+\ O^C\ Z^+}$
8. $\dfrac{F'\ I^+\ P^+\ O^+\ Z^+}{I^-\ P^+\ O^+\ Z^-}$
9. $\dfrac{F'\ I^+\ P^+\ O^C\ Z^-}{I^-\ P^+\ O^+\ Z^+}$
10. $\dfrac{F'\ I^-\ P^+\ O^+\ Z^-}{I^-\ P^+\ O^C\ Z^+}$

For each strain, predict whether β-galactosidase will be produced **(a)** if lactose is absent from the growth medium and **(b)** if lactose is present in the growth medium. Glucose is absent from the medium in every case. (*Note*: In the partial diploid strains (6–10), one copy of the *lac* operon is in the host chromosome and the other copy is in the extrachromosomal F factor.)

**A17.1** The answers are as follows, where "+" = β-galactosidase is produced and "–" = β-galactosidase is not produced:

| Genotype | Noninduced: Lactose Absent | Induced: Lactose Present |
|---|---|---|
| (1) | – | + |
| (2) | + | + |
| (3) | + | + |
| (4) | + | + |
| (5) | – | – |
| (6) | – | + |
| (7) | + | + |
| (8) | – | + |
| (9) | – | + |
| (10) | + | + |

To answer this question completely requires a good understanding of how the *lac* operon is regulated in the wild type and of the consequences of particular mutations on the regulation of the operon.

Strain (1) is the standard wild-type operon. No enzyme is produced in the absence of lactose because the Lac repressor produced by the $I^+$ gene binds to the operator ($O^+$) and blocks the initiation of transcription. When lactose is added, it binds to the repressor, changing its conformation so that it no longer can bind to the $O^+$ region, thereby facilitating transcription of the structural genes for RNA polymerase.

Strain (2) is a haploid strain with a mutation in the *lacI* gene ($I^-$). The consequence is that the Lac repressor protein cannot bind to the (normal) operator region, so there is no inhibition of transcription, even in the absence of lactose. This strain, then, is constitutive, meaning that β-galactosidase is produced by the *lacZ*$^+$ gene in the presence or absence of lactose.

Strain (3) is another constitutive mutant. In this case, the repressor gene is a wild type and the β-galactosidase gene *lacZ*$^+$ is a wild type, but there is a mutation in the operator region ($O^c$). Therefore, the Lac repressor protein cannot bind to the operator, and transcription occurs in the presence or absence of lactose.

Strain (4) carries both regulatory mutations of the previous two strains. Functional Lac repressor is not produced, but even if it were, the operator is changed so that it cannot bind. The consequence is the same: constitutive enzyme production.

Strain (5) produces functional Lac repressor, but the operator ($O^c$) is mutated. Therefore, transcription cannot be blocked, and *lac* polycistronic mRNA is produced in the presence or absence of lactose. However, because there is also a mutation in the β-galactosidase gene ($Z^-$), no functional enzyme is generated.

In the partial diploid strain (6), one *lac* operon is completely wild type and the other carries a constitutive operator mutation and a mutant β-galactosidase gene. In the absence of lactose, no functional enzyme is produced. For the wild-type operon, the Lac repressor binds to the operator and blocks transcription. For the operon with the two mutations, the operator region is mutated and cannot bind repressor, so the mRNA for the mutated operon is produced; however, the *lacZ* gene is also mutated, so that functional enzyme cannot be produced. In the presence of lactose, functional enzyme *is* produced, because repression of the wild-type operon is relieved, so that the $Z^+$ gene can be transcribed. This type of strain provided one of the pieces of evidence that the operator region does not produce a diffusible substance.

In partial diploid (7), functional enzyme is produced in the presence or absence of lactose because one of the operons has an $O^C$ mutation that does not respond to a repressor and that is linked to a wild-type $Z^+$ gene. That operon is transcribed constitutively. The other operon is inducible, but because there is a $Z^-$ mutation, no functional enzyme is produced.

Partial diploid (8) has a wild-type operon and an operon with an $I^-$ regulatory mutation and a $Z^-$ mutation. The $I^+$ gene product is diffusible, so that it can bind to the $O^+$ region of both operons, thereby putting both operons under inducer control. This strain demonstrates that the $I^+$ gene is *trans*-dominant to an $I^-$ mutation. In this case, the particular location of the one $Z^-$ mutation is irrelevant: The same result would have been obtained had the $Z^+$ and $Z^-$ been switched between the two operons. In this partial diploid, β-galactosidase is not produced unless lactose is present.

In strain (9), β-galactosidase is produced only when lactose is present, because the $O^c$ region controls only the genes that are adjacent to it on the same chromosome (*cis* dominance) and in this case one of the adjacent genes is $Z^-$, which codes for a nonfunctional enzyme. The partial diploid is heterozygous $I^+/I^-$, but $I^+$ is *trans*-dominant, as discussed for strain (8). Thus, the only normal $Z^+$ gene is under inducer control.

Partial diploid (10) has a defective repressor protein as well as an $O^c$ mutation adjacent to a $Z^+$ gene. On the latter ground alone, this partial diploid is constitutive. The other operon is also constitutively transcribed, but because there is a $Z^-$ mutation, no functional enzyme is generated from it.

# Questions And Problems

**17.1** What is meant by constitutive gene expression? How is constitutive gene expression unlike regulated gene expression? What can you infer about the site of a mutation that causes constitutive expression of a protein that normally shows regulated expression?

***17.2** Give two examples of effector molecules, and discuss how effector molecules function to regulate gene expression.

**17.3** Operons produce polycistronic mRNA when they are active. What is a polycistronic mRNA? What advantages, if any, does it confer in terms of function of the cell?

**17.4** How does lactose trigger the coordinate induction of the synthesis of β-galactosidase, permease, and transacetylase? Why does the synthesis of these enzymes not occur when glucose is also in the medium?

***17.5** An *E. coli* mutant strain synthesizes β-galactosidase whether or not the inducer is present. What genetic defect(s) might be responsible for this phenotype?

***17.6** Some complete loss-of-function mutations result from a deletion of just one gene or a missense mutation that results in a completely nonfunctional protein.

**a.** If such mutations were obtained in each of the *lacA*, *lacI*, *lacY*, and *lacZ* genes, which would show a *lac* phenotype (be unable to grow using lactose as a sole carbon source)? What properties would the other(s) show?

**b.** If such mutations were obtained in each of the *trpA*, *trpB*, *trpC*, *trpD*, and *trpR* genes, which would show a *trp* phenotype (require supplemental tryptophan for growth)? What properties would the other(s) show?

**c.** If such mutations were obtained in each of the *araA*, *araB*, *araC*, and *araD* genes, which would show an *ara* phenotype (be unable to grow using arabinose as the sole carbon source)? What properties would the other(s) show?

**17.7** Elucidation of the regulatory mechanisms associated with the enzymes of lactose utilization in *E. coli* was a landmark in our understanding of regulatory processes in microorganisms. In formulating the operon hypothesis as applied to the lactose system, Jacob and Monod found that results from particular partial diploid strains were invaluable. In terms of their operon hypothesis, what specific information did analyses of partial diploids provide that analyses of haploids could not?

***17.8** For the *E. coli lac* operon, write the partial diploid genotype for a strain that will produce β-galactosidase constitutively and permease by induction.

**17.9** Mutants were instrumental in elaborating the model for regulation of the *lac* operon.

**a.** Discuss why $P_{lac^-}$ and *lacO*$^c$ mutants are *cis*-dominant but not *trans*-dominant.

**b.** Explain why *lacI*$^S$ and *lacI*$^{-d}$ mutants are *trans*-dominant to the wild-type *lacI*$^+$ allele but *lacI*$^-$ mutants are recessive.

**c.** Discuss the consequences of mutations in the repressor gene promoter as compared with mutations in the structural gene promoter.

***17.10** This question involves the *lac* operon of *E. coli*, where $I$ = *lacI* (the repressor gene), $P = P_{lac}$ (the promoter), $O$ = *lacO* (the operator), $Z$ = *lacZ* (the β-galactosidase gene), and $Y$ = *lacY* (the permease gene). Complete Table 17.A, using + to indicate that the enzyme in question will be synthesized and – to indicate that the enzyme will not be synthesized.

***17.11** What consequences would a loss-of-function mutation in the catabolite activator protein (*CAP*) gene of *E. coli* have for the expression of wild-type *lac* and *ara* operons? Would a constitutive mutation in the *CAP* gene have any effect on the expression of these operons?

**17.12** The presence of glucose in the medium along with lactose leads to catabolite repression. Explain why

Table 17.A

| Genotype | Inducer Absent | | Inducer Present | |
|---|---|---|---|---|
| | β-Galactosidase | Permease | β-Galactosidase | Permease |
| a. $I^+\ P^+\ O^+\ Z^+\ Y^+$ | | | | |
| b. $I^+\ P^+\ O^+\ Z^-\ Y^+$ | | | | |
| c. $I^+\ P^+\ O^+\ Z^+\ Y^-$ | | | | |
| d. $I^-\ P^+\ O^+\ Z^+\ Y^+$ | | | | |
| e. $I^S\ P^+\ O^+\ Z^+\ Y^+$ | | | | |
| f. $I^+\ P^+\ O^c\ Z^+\ Y^+$ | | | | |
| g. $I^S\ P^+\ O^c\ Z^+\ Y^+$ | | | | |
| h. $I^+\ P^+\ O^c\ Z^+\ Y^-$ | | | | |
| i. $I^{-d}\ P^+\ O^+\ Z^+\ Y^+$ | | | | |
| j. $\frac{I^-\ P^+\ O^+\ Z^+\ Y^+}{I^+\ P^+\ O^+\ Z^-\ Y^-}$ | | | | |
| k. $\frac{I^-\ P^+\ O^+\ Z^+\ Y^-}{I^+\ P^+\ O^+\ Z^-\ Y^+}$ | | | | |
| l. $\frac{I^S\ P^+\ O^+\ Z^+\ Y^-}{I^+\ P^+\ O^+\ Z^-\ Y^+}$ | | | | |
| m. $\frac{I^+\ P^+\ O^c\ Z^-\ Y^+}{I^+\ P^+\ O^+\ Z^+\ Y^-}$ | | | | |
| n. $\frac{I^-\ P^+\ O^c\ Z^+\ Y^-}{I^+\ P^+\ O^+\ Z^-\ Y^+}$ | | | | |
| o. $\frac{I^S\ P^+\ O^+\ Z^+\ Y^+}{I^+\ P^+\ O^c\ Z^+\ Y^+}$ | | | | |
| p. $\frac{I^{-d}\ P^+\ O^+\ Z^+\ Y^-}{I^+\ P^+\ O^+\ Z^-\ Y^+}$ | | | | |
| q. $\frac{I^+\ P^-\ O^c\ Z^+\ Y^-}{I^+\ P^+\ O^+\ Z^-\ Y^+}$ | | | | |
| r. $\frac{I^+\ P^-\ O^+\ Z^+\ Y^-}{I^+\ P^+\ O^c\ Z^-\ Y^+}$ | | | | |
| s. $\frac{I^-\ P^-\ O^+\ Z^+\ Y^+}{I^+\ P^+\ O^+\ Z^-\ Y^-}$ | | | | |
| t. $\frac{I^-\ P^+\ O^+\ Z^+\ Y^-}{I^+\ P^-\ O^+\ Z^-\ Y^+}$ | | | | |

catabolite repression is considered to be a form of positive control, while repression by the *lac* repressor is considered to be a form of negative control.

***17.13** DNase protection experiments were helpful to elucidate the functions of different DNA sequences in the *lac* promoter.

**a.** What is a DNase protection experiment, and how does it provide this information?

**b.** How are the binding sites for the *lac* repressor, RNA polymerase, and CAP–cAMP arranged at the 5′ end of the *lac* operon?

**c.** What effects would you expect each of the following mutations to have on the coordinate induction of the *lac* operon by lactose (in the absence of glucose)? Explain your reasoning. [The base-pair coordinates used here are those specified in Figure 17.14.]

**i.** a deletion of base pairs from +3 to +18

**ii.** a TA-to-GC transversion at −12

**iii.** a TA-to-GC transversion at −69

**iv.** a GC-to-AT transition at +28

**v.** a GC-to-AT transition at +9

**d.** Would any of the mutations listed in (c) affect catabolite repression of the *lac* operon?

**17.14** A new sugar, sugarose, induces synthesis of two enzymes from the *sug* operon of *E. coli*. Some properties of deletion mutations affecting the appearance of these enzymes are as follows (here, + = enzyme induced normally, i.e., synthesized only in the presence of the inducer; C = enzyme synthesized constitutively; 0 = enzyme cannot be detected):

| Mutation of | Enzyme 1 | Enzyme 2 |
|---|---|---|
| Gene *A* | + | 0 |
| Gene *B* | 0 | + |
| Gene *C* | 0 | 0 |
| Gene *D* | C | C |

**a.** The genes are adjacent, in the order *A B C D*. Which gene is most likely to be the structural gene for enzyme 1?

**b.** Complementation studies using partial diploid ($F'$) strains were made. The extrachromosomal element ($F'$) and chromosome each carried one set of *sug* genes. The results were as follows:

| Genotype of $F'$ | Chromosome | Enzyme 1 | Enzyme 2 |
|---|---|---|---|
| $A^+ B^- C^+ D^+$ | $A^- B^+ C^+ D^+$ | + | + |
| $A^+ B^- C^- D^+$ | $A^- B^+ C^+ D^+$ | + | 0 |
| $A^- B^+ C^- D^+$ | $A^+ B^- C^+ D^+$ | 0 | + |
| $A^- B^+ C^+ D^+$ | $A^+ B^- C^+ D^-$ | + | + |

From all the evidence given, determine whether the following statements are true or false:

**i.** It is possible that gene *D* is a structural gene for one of the two enzymes.

**ii.** It is possible that gene *D* produces a repressor.

**iii.** It is possible that gene *D* produces a cytoplasmic product required to induce genes *A* and *B*.

**iv.** It is possible that gene *D* is an operator locus for the *sug* operon.

**v.** The evidence is also consistent with the possibility that gene *C* could be a gene that produces a cytoplasmic product required to induce genes *A* and *B*.

**vi.** The evidence is also consistent with the possibility that gene *C* could be the controlling end of the *sug* operon (the end from which mRNA synthesis presumably commences).

**17.15** The *lac* operon is an inducible operon, whereas the *trp* operon is a repressible operon. Discuss the differences between these two types of operons.

**17.16** Transcription of the *trp* operon can be reduced through a combination of repression using an aporepressor and attenuation.

**a.** How much of a reduction in transcription can be achieved using the aporepressor, and how much of a reduction in transcription can be achieved using attenuation? Speculate why the aporepressor might be unable to silence expression of the *trp* operon completely, and why this might be advantageous to *E. coli*.

**b.** Explain how the mechanism of attenuation is dependent on translation of transcripts at the *trp* operon.

***17.17** In the presence of high intracellular concentrations of tryptophan, only short transcripts of the *trp* operon are synthesized because of attenuation of transcription 5′ to the structural genes. This is mediated by the recognition of two Trp codons in the leader sequence. What effect would mutating these two codons to UAG stop codons have on the regulation of the operon in the presence or absence of tryptophan? Explain.

***17.18** The mutant *E. coli* strains described in the following table are individually inoculated into two different media, one with supplemental tryptophan and one without:

| Mutant | Phenotype |
|---|---|
| *1* | Aporepressor is unable to bind to tryptophan. |
| *2* | A point mutation in the *trp* operator prevents binding by an active Trp repressor. |
| *3* | The *trpE* gene has a nonsense mutation. |
| *4* | The levels of Trp–tRNA.Trp are decreased due to a mutation in a gene for Trp-aminoacyl synthetase. |
| *5* | Three adjacent G-C base pairs in region 4 (see Figures 17.16 and 17.17) are mutated to A-T base pairs. |

For each mutant and medium, state whether the level of tryptophan synthetase will be increased or decreased relative to the level found in a wild-type strain and, where possible, by how much, and why.

***17.19** In *E. coli*, the *ilvGMEDA* operon is regulated by attenuation. The mRNA transcript of its leader region has a stretch of 17 codons that includes four for leucine, five for isoleucine, and six for valine. Generate a hypothesis to explain why attenuation can be relieved equally well by low levels of Leu–tRNA, Ile–tRNA, or Val–tRNA. How would you test your hypothesis experimentally?

**17.20** In the bacterium *Salmonella typhimurium*, seven of the genes coding for histidine biosynthetic enzymes are located adjacent to one another in the chromosome. If excess histidine is present in the medium, the synthesis in all seven enzymes is coordinately repressed, whereas in the absence of histidine all seven genes are coordinately expressed. Most mutations in this region of the chromosome result in the loss of activity of only one of the enzymes. However, mutations mapping to one end of the gene cluster result in the loss of all seven enzymes, even though none of the structural genes have been lost. What is the counterpart of these mutations in the *lac* operon system?

**17.21** On infecting an *E. coli* cell, bacteriophage λ has a choice between the lytic and lysogenic pathways. Discuss the molecular events that determine which pathway is taken.

**17.22** How do the lambda repressor protein and the Cro protein regulate their own synthesis?

*17.23 A mutation in the phage lambda *cI* gene results in a nonfunctional *cI* gene product. What phenotype would you expect the phage to exhibit?

*17.24 Which of the protein products of the *CAP, araC, lacI, cI,* and *trpR* genes are positive regulators, which are negative regulators, and which can be both? How can one protein serve both positive and negative regulatory roles? Which of these proteins interact directly with RNA polymerase, and how does this contribute to their regulatory function?

*17.25 As a last-resort emergency response to severe DNA damage, *Bacillus subtilis,* like *E. coli,* activates an SOS response (see Chapter 7, pp. 148–149). The DNA damage activates the RecA protein, and this results in the LexA protein cleaving itself. Intact LexA binds to a consensus sequence, 5'-CGAACNNNNGTTCG-3' (N is any nucleotide), found in the promoter region of about 17 genes used to repair DNA damage. Cleaved LexA is unable to bind this sequence and so cleavage of LexA leads to the transcription of these genes.

**a.** Is LexA a positive or negative regulatory protein?

**b.** How would you classify the 5'-CGAACNNNNGTTCG-3' sequence?

**c.** Two allelic mutations, designated *A* and *B*, alter the 5'-CGAACNNNNGTTCG-3' sequence at one gene bound by LexA. LexA cannot bind to the *A* sequence, while it binds to the *B* sequence much more tightly than it does to the wild-type sequence. Suppose an SOS response is triggered in each mutant. What phenotype do you expect each to exhibit?

**d.** A plasmid bearing a mutant *lexA* gene is introduced into an otherwise normal strain of *B. subtilis*. The mutant gene makes a LexA protein that binds tightly to the 5'-CGAACNNNNGTTCG-3' sequence but is unable to undergo self-cleavage. What phenotype do you expect this strain to have?

17.26 Bacteriophage λ can form a stable association with the bacterial chromosome because the virus manufactures a repressor. This repressor prevents the virus from replicating its DNA and making lysozyme and all the other tools used to destroy the bacterium. When you induce the virus with ultraviolet (UV) light, you destroy the repressor, and the virus undergoes its normal lytic cycle. The repressor is the product of a gene called the *cI* gene and is a part of the wild-type viral genome. A bacterium that is lysogenic for $\lambda^+$ is full of repressor protein, which confers immunity against any λ virus added to these bacteria. Added viruses can inject their DNA, but the repressor from the resident virus prevents replication, presumably by binding to an operator on the incoming virus. Thus, this system has many elements analogous to the *lac* operon. We could diagram a virus as shown in the following figure. Several mutations of the *cI* gene are known. The $c_i$ mutation results in an inactive repressor.

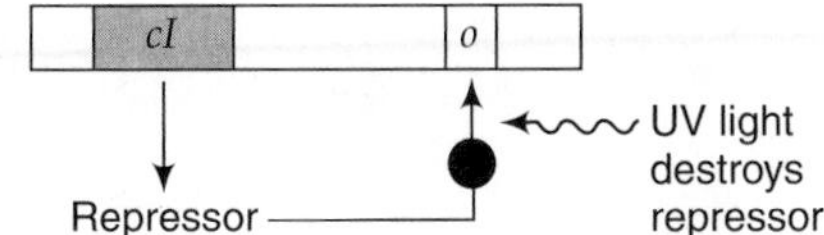


**a.** If you infect *E. coli* with λ containing a $c_i$ mutation, can it lysogenize (form a stable association with the bacterial chromosome)? Why or why not?

**b.** If you infect a bacterium simultaneously with a wild-type $c^+$ and a $c_i$ mutant of λ, can you obtain stable lysogeny? Why or why not?

**c.** Another class of mutants called $c^{IN}$ makes a repressor that is insensitive to UV destruction. Will you be able to induce a bacterium lysogenic for $c^{IN}$ with UV light? Why or why not?

**17.27** In phage λ, the *N* and *Q* genes encode antiterminator proteins.

**a.** What are antiterminator proteins, and how are they used during the λ life cycle?

**b.** $N^{ts}$ and $Q^{ts}$ are temperature-sensitive missense mutations that lead to nonfunctional N and Q proteins, respectively, at the restrictive temperature. In each of the following situations, state which steps of the λ life cycle will be affected.

   **i.** *E. coli* is infected with $N^{ts}$ λ, $Q^{ts}$ λ, or $N^{ts}$ $Q^{ts}$ λ (a double mutant) and immediately shifted to the restrictive temperature.

   **ii.** An *E. coli* strain lysogenic for $N^{ts}$ λ, $Q^{ts}$ λ, or $N^{ts}$ $Q^{ts}$ λ growing at the permissive temperature is irradiated with UV lights and then shifted to the restrictive temperature.

*17.28 Five λ mutants have the molecular phenotypes shown in the left column of the following table:

| Mutant | Molecular Phenotype | Lytic Growth | Lysogenic Growth | Inducible by UV Light |
|---|---|---|---|---|
| 1 | The Cro protein is unable to bind DNA. | | | |
| 2 | The N protein does not function. | | | |
| 3 | The cII protein does not function. | | | |
| 4 | The Q protein does not function. | | | |
| 5 | $P_{RM}$ is unable to bind RNA polymerase. | | | |

Fill in the table to indicate whether each mutant will be able to undergo lytic or lysogenic growth. For mutants able to follow a lysogenic pathway, state whether they can be induced by UV light.

**Figure 18.5**
**Mechanisms of action of polypeptide hormones and steroid hormones.**

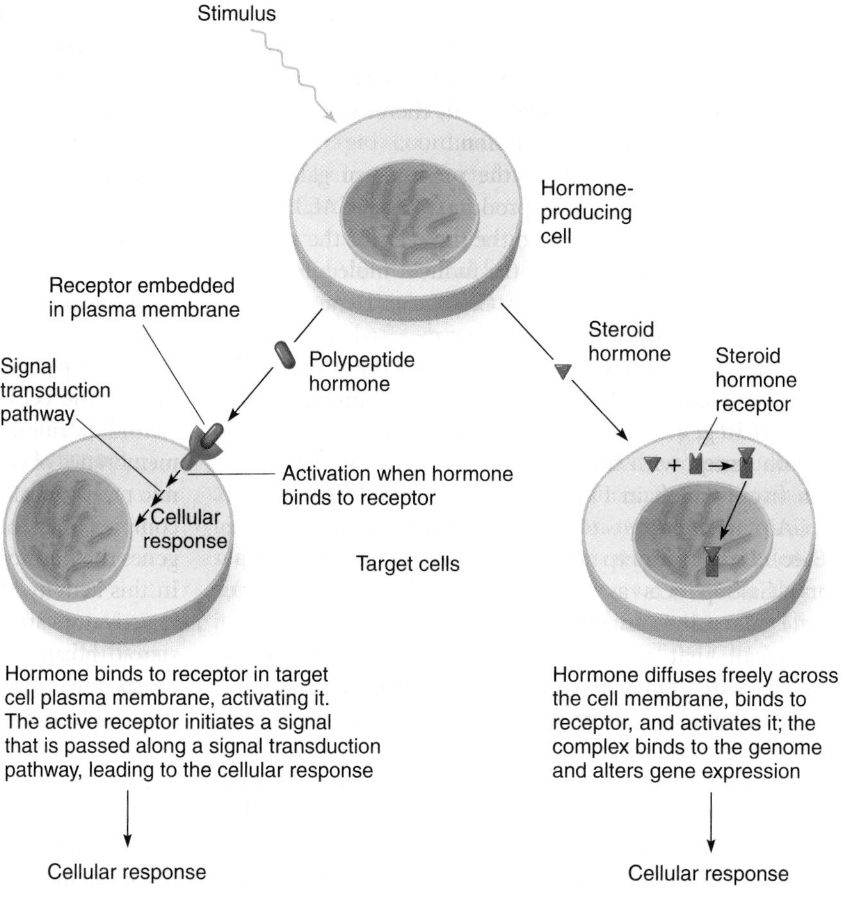


domain, depending on the particular SHR. In addition, an SHR has a third domain: the binding domain for the steroid hormone for which it is specific.

Most steroid hormones exert their effect in the same general way. For example, in the absence of a particular steroid hormone, the appropriate SHR is found in an inactive state in the cytoplasm associated with a large complex of proteins called *chaperones*, one of which is *Hsp90*. Data from studies of mutant yeast strains suggest that chaperone proteins have an active role in keeping the SHRs functional. When a steroid hormone such as glucocorticoid passes through the plasma membrane and

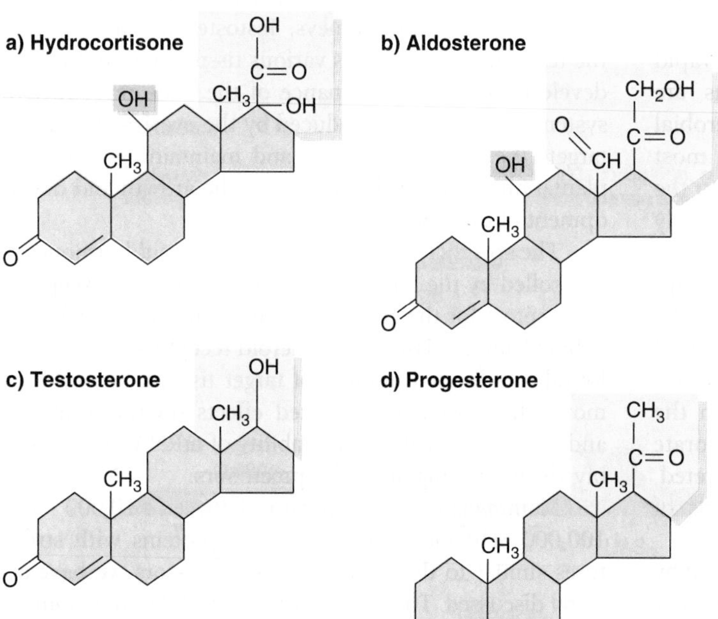


**Figure 18.6**
**Structures of some mammalian steroid hormones.** (a) Hydrocortisone, which helps regulate carbohydrate and protein metabolism. (b) Aldosterone, which regulates salt and water balance. (c) Testosterone, which is used for the production and maintenance of male sexual characteristics. (d) Progesterone, which, with estrogen, prepares and maintains the uterine lining for implantation of an embryo.

enters a cell, it binds to the hormone-binding domain (HBD) of its specific SHR molecule, displacing Hsp90 bound there (Figure 18.7). Hormone binding brings about a conformational change in the receptor, activating it. The resulting active glucocorticoid–receptor complex now is able to enter the nucleus where it binds to specific DNA regulatory sequences, activating or repressing the transcription of the specific genes controlled by the hormone. For genes turned on by the hormone, the new mRNAs appear within minutes after a steroid hormone encounters its target cell, enabling new proteins to be produced rapidly. The DNA-binding domains of many steroid hormone receptor proteins are zinc fingers.

All genes regulated by a specific steroid hormone have in common a DNA sequence to which the steroid–receptor complex binds. The binding regions are called **steroid hormone response elements (HREs).** The "H" in the abbreviation is replaced with another letter to indicate the specific steroid involved. Thus, GRE is the glucocorticoid response element and ERE is the estrogen response element. The HREs are located, often in multiple copies, in the promoters of genes they control, typically within 1 kb of the transcription start site. The GRE, for example, is located about 250 bp upstream from the transcription start point. The consensus sequence for GRE is AGAACANNNTGTTCT, where N is any nucleotide. The ERE consensus sequence is AGGTCANNNTGACCT. Note that, for both of these HREs, the sequences on each side of the N's are complementary; that is, the sequences show twofold symmetry.

How the hormone–receptor complexes, once bound to the correct HREs, regulate transcriptional levels is not completely known. Potentially, functional interactions arise among the hormone–receptor complexes and with coactivators and general transcription factors in the transcription initiation complex. To this end, recall that multiple HREs are present for many genes, so multiple hormone–receptor complexes can bind to each gene. These interactions may facilitate the initiation of transcription by RNA polymerase II. Each steroid hormone is presumed to regulate its specific transcriptional activation by the same general mechanism. The unique action of each type of steroid results from the different receptor proteins and HREs involved.

Finally, it is of particular interest that, in different types of cells, the same steroid hormone may activate different sets of genes, even though the various cells have the same SHR. This is because many regulatory proteins bind to both promoter elements and enhancers to regulate gene expression. (See the discussions earlier in this chapter and in Chapter 5.) Thus, a steroid–receptor complex can activate a gene only if the correct array of other



**Figure 18.7**

**Model for the action of the steroid hormone glucocorticoid in mammalian cells.** For the receptor, AD = activation domain, BD = DNA-binding domain, and HBD = hormone-binding domain.

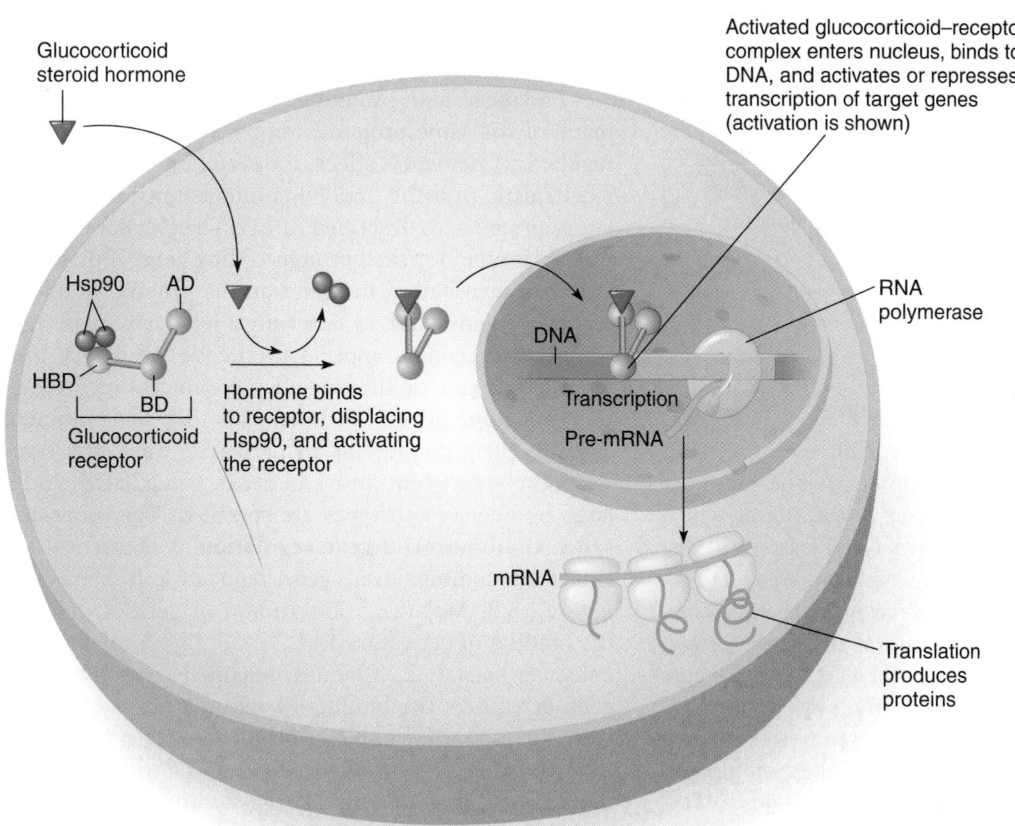

regulatory proteins is present. Since the other regulatory proteins are specific to the cell type, different patterns of gene expression can result.

In sum, steroid hormones act as effector molecules and SHRs act as regulatory molecules. When the two combine, the resulting complex binds to DNA and regulates gene transcription, producing a large and specific increase or decrease in cellular mRNA levels of target genes. The specific responses characteristic of each steroid hormone result from the fact that receptors are found only in certain cell types, and each of those cell types contains different arrays of other cell-type-specific regulatory proteins that interact with the steroid–receptor complex to activate specific genes.



**i Activity**

Go to the iActivity *Sorting the Signals of Gene Regulation* on the student website and assume the role of a researcher tracking down some methods used by a eukaryotic cell for the synchronized regulation of genes.

**Keynote**

In multicellular eukaryotes, one well-studied system of short-term gene regulation is the control of protein synthesis by hormones. A polypeptide hormone binds to a specific cell surface receptor, activating it, and thereby triggering a signal transduction pathway that produces a cellular response. A steroid hormone exerts its action by forming a complex with a specific receptor protein in the cytoplasm, thereby activating the receptor; the complex then enters the nucleus and binds directly to particular sequences on the cell's genome to regulate the expression of specific target genes. The specificity of hormone action is caused by the presence of hormone receptors only in certain cell types and by interactions of steroid–receptor complexes with cell-type-specific regulatory proteins.

### Combinatorial Gene Regulation: The Control of Transcription by Combinations of Activators and Repressors

Protein-coding eukaryotic genes contain both promoter elements and enhancers (see Chapter 5). The promoter elements are located just upstream of the site at which transcription begins. The enhancers usually are some distance away, either upstream or downstream. We can think of the different promoter elements as modules that function in the regulation of expression of the gene. Certain promoter elements, such as the TATA element in the core promoter, are required to specify where transcription is to begin. Other promoter elements, in the promoter-proximal region, control whether transcription of the gene occurs; specific regulatory proteins bind to these elements.

A regulatory promoter element is specialized with respect to the gene (or genes) it controls because it binds a signaling molecule—activator or repressor—that is involved in the regulation of that gene's expression. Depending on the particular gene, there can be one, a few, or many regulatory promoter elements because, under various conditions, there may be one, a few, or many regulatory proteins that control the gene's expression. The remarkable specificity of regulatory proteins in binding to their specific regulatory element in the DNA and to no others ensures careful control of which genes are turned on and which are turned off.

Whereas promoter elements are crucial for determining whether transcription can occur, enhancer elements ensure maximal transcription of the gene. At an enhancer element, depending on its sequence, an activator or a repressor will bind.

Both promoters and enhancers are important in regulating the transcription of a gene. Each regulatory promoter element and enhancer element binds special regulatory proteins. Some regulatory proteins are found in most or all cell types, whereas others are found in only a limited number of cell types. Because some of the regulatory proteins activate transcription when they bind to the enhancer or promoter element, whereas others repress transcription, the net effect of a regulatory element on transcription depends on the combination of different proteins bound. If activators are bound at both the enhancer and promoter elements, the result is activation of transcription. However, if a repressor binds to the enhancer and an activator binds to the promoter element, the result depends on the interaction between the two regulatory proteins. If the repressor has a strong effect, the gene is repressed. In this case, the enhancer is called a **silencer element.**

Enhancer and promoter elements appear to bind many of the same proteins, implying that both types of regulatory elements affect transcription by a similar mechanism, probably involving interactions of the regulatory proteins, as described earlier. There is not one regulatory protein for each protein-coding gene. If that were the case, then half of the genes in the genome would be regulatory genes and, in fact, only a relatively small fraction of the genome appears to encode regulatory proteins. How, then, do those regulatory proteins regulate all protein-coding genes? The answer is that, by combining a few regulatory proteins in particular ways, the transcription of different arrays of genes is regulated, and a large number of cell types are specified. The process is called **combinatorial gene regulation.** A theoretical example of combinatorial gene regulation is shown in Figure 18.8. Maximal transcription of gene *A* involves the binding of activators 1, 2, 3, and 4 to corresponding enhancer sites 1, 2, 3, and 4. Maximal transcription of gene *B* requires the binding of activators 2, 4, and 6 to their corresponding enhancer sequences. That is, each of the two genes requires activators 2 and 4 for full activation, in combination with different additional activators.

**Figure 18.8**

**Combinatorial gene regulation.** A theoretical example in which (**a**) the transcription of gene *A* is controlled by activators 1, 2, 3, and 4 and (**b**) the transcription of gene *B* is controlled by activators 2, 4, and 6.

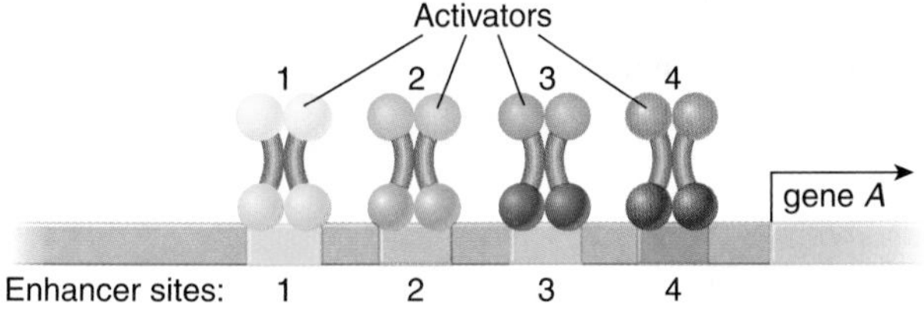


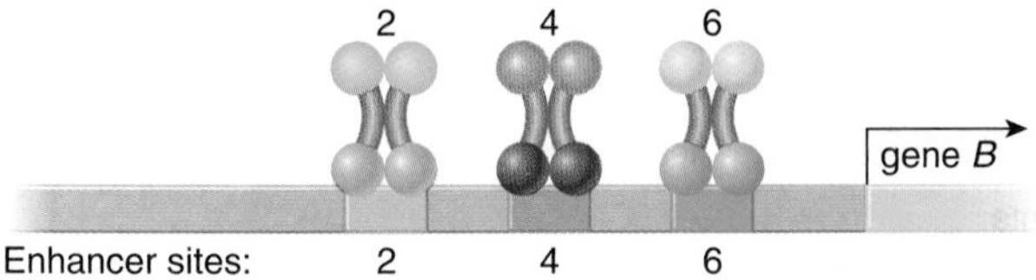


Combinatorial controls are seen for regulation of transcription of the *even-skipped* (*eve*) gene in *Drosophila*. Expression of *eve* is an important event in the development of the *Drosophila* embryo. An *eve* mutant does not develop a number of parts of the embryo, and it dies early in development. The following is a brief overview of *Drosophila* development to enable the significance of regulation of *eve* to be understood. *Drosophila* embryonic development, which is unusual, is described in more detail in Chapter 19, pp. 564–571.

A number of regulatory genes control the establishment of the embryo's body plan. These genes regulate the expression of other genes. Two classes of genes, the maternal-effect genes and the segmentation genes, work sequentially. First, a set of *maternal-effect genes* expressed by the mother during oogenesis determine the polarity of the egg by generating gradients of the regulatory proteins (transcription factors) that they encode along the anterior-posterior and dorsal-ventral axes of the egg. Two important maternal-effect genes for our discussion are *bicoid* and *hunchback*. Once the axis of the embryo is established, the expression of at least 24 *segmentation genes* progressively subdivides the embryo into regions, determining the segments of the embryo and the adult. These genes begin to be expressed at an early stage of embryo development when the embryo is a single large cell—a *syncytial blastoderm*—with many nuclei at the periphery, all in a common cytoplasm (Figure 18.9a). Which segmentation gene is expressed in the embryo and where it expressed is determined by the gradients of regulatory proteins in the embryo. That is, the nuclei respond to the particular sets and concentrations of regulatory proteins they are encountering. If a set of activators at a high enough concentration are present, then the genes controlled by those activators will be expressed, and so on.

Three sets of segmentation genes are regulated in a cascade of gene activations. The first set to be expressed is the *gap genes*, which are responding to the gradients of regulatory proteins encoded by maternal-effect genes. The function of gap genes, through their regulation of expression of the next genes in the regulatory cascade, is to subdivide the embryo along the anterior-posterior axis into several broad regions. Two important gap genes for our discussion are *giant* and *Krüppel*. The regulatory protein products of gap genes activate *pair-rule genes*, the products of which are also regulatory proteins. The function of pair-rule genes is to generate 14 *parasegments* in the early embryo (Figure 18.9b). Through the action of the segment polarity genes, the 14 segments of the late embryo and, hence, of the larval stages and the adult are then defined. Each segment (except C1) derives from the posterior part of one parasegment (labeled "p" in Figure 18.9b) and the anterior part of the next parasegment (labeled "a" in Figure 18.9b).

Starting at the syncytial blastoderm stage, two pair-rule genes, *eve* and *fushi-tarazu*, are each expressed in seven stripes that alternate with one another to produce a repeating series of stripes. The *eve* gene specifies odd parasegments (1, 3, and so on), while the *fushi-tarazu* gene specifies even parasegments. The seven *eve* stripes are controlled by five distinct enhancers of the *eve* gene, each about 500 bp long. Each enhancer contains binding sites for transcriptional activators and repressors and controls the expression of one or two stripes.

The regulation of transcription of *eve* for the expression of stripe 2 (the second stripe of expression, which corresponds to parasegment 3) has been well studied. The enhancer for stripe 2 expression contains binding sites for the regulatory proteins, Bicoid, Hunchback, Giant, and Krüppel, which are encoded by the *bicoid*, *hunchback*, *giant*, and *Krüppel* genes, respectively (Figure 18.9c). Bicoid and Hunchback are transcription activators, while Giant and Krüppel are transcription repressors. Some of the binding sites are unique for each activator or repressor. Other binding sites overlap for an activator and a repressor so that binding of one protein prevents binding of the other. Due to previous regulatory gene activity, there are specific gradients of each of these four regulatory proteins along the anterior-posterior axis of the syncytial blastoderm (Figure 18.9d). The *eve* gene is transcribed to a high level to produce stripe 2 because the two activators are present, while the Giant repressor is absent and Krüppel is at a low level. That is, the combination of the activator and repressor proteins leads to activation of *eve* transcription, producing a stripe of the *eve* regulatory protein.

Expression of *eve* in the other six stripes is controlled independently. Conceptually, the same principles are involved; namely, the expression of the gene is determined by regulatory events involving particular combinations of gene regulatory proteins binding to the enhancer controlling expression of each stripe. And, in the even parasegments, *eve* is not expressed, which means

peptide), with its amino acid sequence encoded by part of exon 5. (The remainder of exon 5 is the 3′ trailer part of the mRNA.) The outcome of alternative polyadenylation and alternative splicing in this case is two different polypeptides encoded by the same gene and synthesized in two different tissues. The thyroid hormone calcitonin is a circulating calcium ion homeostatic hormone that aids the kidney in retaining calcium. CGRP is found in the hypothalamus and appears to have neuromodulatory and trophic (growth-promoting) activities.

Many genes exhibit alternative splicing. In humans, for example, researchers consider that perhaps three-quarters of genes are alternatively spliced. Often, alternative splicing produces relatively few variants. The "record" amongst studied cases of alternative slicing occurs for pre-mRNA transcribed from the *Drosophila Dscam* gene, which encodes proteins required for the formation of neuronal connections. By alternative splicing, the *Dscam* gene has the potential to produce 38,016 different protein isoforms. How many actually are produced is not known. Alternative splicing also plays a key role in sex determination in *Drosophila*. In outline, the number of X chromosomes are counted and, as a result, a molecular switch is set that governs male or female sexual differentiation. The setting of the molecular switch sets in motion a cascade of regulated alternative splicing events that are different in XX and XY flies. Those events ultimately lead to differentiation into female-specific or male-specific cells. Sex determination in *Drosophila*, including the regulated splicing events, is described in detail in Chapter 19, pp. 559–564.

After mRNAs are generated by RNA processing, they are exported from the nucleus through the nuclear pore complex to the cytoplasm. The mRNA is exported in a complex with a number of proteins. Some of those proteins are recruited to an mRNA during the transcription and splicing processes, illustrating the tight linkage between transcription, splicing, and nuclear export. The mRNA export process is perhaps the most intricate of the export systems using the nuclear pore complexes, involving many quality control steps. Once in the cytoplasm, the mRNA may be translated immediately, or stored for later translation (see next section).

**Keynote**

Gene expression in eukaryotes can be regulated at the level of RNA processing. This type of regulation operates to direct the production of mature RNA molecules from precursor-RNA molecules. Two regulatory events that exemplify this level of control are the choice of poly(A) site and the choice of splice site. In both cases, different types of mRNAs are produced, depending on the choices made. Once a mature mRNA is produced, in a controlled process it is exported from the nucleus through the nuclear pore complex to the cytoplasm in a complex with several proteins.

## mRNA Translation Control

Messenger RNA molecules are subject to *translational control*. Differential translation can greatly affect gene expression. For example, mRNAs are stored in many unfertilized vertebrate and invertebrate eggs. In the unfertilized egg, the rate of protein synthesis is very slow; however, protein synthesis increases significantly after fertilization. This increase occurs without new mRNA synthesis because of translational control. That is, mRNAs are altered so that they may be stored in a state in which they are not translated. Reversal of the alterations renders the mRNAs translatable. Typically, for storage, proteins bind to the mRNAs that both protect them and inhibit their translation. The length of the poly(A) tail of the mRNA also affects its translatibility. That is, the poly(A) tail is known to promote the initiation of translation. In general, stored, inactive mRNAs have shorter poly(A) tails (15–90 As) than active mRNAs have (100–300 As). mRNAs synthesized in growing oocytes that are destined for storage and later translation have short poly(A) tails.

In principle, an mRNA molecule can have a short poly(A) tail either because only a short string of A nucleotides was added at the time of polyadenylation or because a normal-length poly(A) tail was added that was subsequently trimmed. At least for some messages stored in growing oocytes of mouse and frog, the latter mechanism is involved. In one example, the examination of one particular mRNA in this class has shown that the pre-mRNA still in the process of intron removal has a long poly(A) tail (300–400 As), whereas the mature, stored message has a short poly(A) tail (40–60 As). The decrease in length of the poly(A) tail for this message class occurs rapidly in the cytoplasm by a deadenylation enzyme. What pinpoints a particular mRNA for rapid deadenylation, rather than a default, slow decrease in poly(A) length, is a sequence in the 3′ untranslated region (3′ UTR) of the mRNA upstream of the AAUAAA polyadenylation sequence. This signal for deadenylation is called the *adenylate/uridylate (AU)-rich element (ARE)* and has the consensus sequence UUUUUAU. To activate a stored mRNA in this class, a cytoplasmic polyadenylation enzyme recognizes the ARE and adds 150 A nucleotides or so. Thus, the same sequence element is used to control the poly(A) tail length, and therefore mRNA translatability, at different times and in opposite ways.

**Keynote**

mRNAs entering the cytoplasm may be translated immediately or stored for later translation. Stored mRNAs characteristically are complexed with proteins that inhibit translation and have a short poly(A) tail compared with the same active mRNA. Signals in the 3′ UTR control the shortening and lengthening of poly(A) tails.

## RNA Interference: Silencing of Gene Expression at the Posttranscriptional Level by Small Regulatory RNAs

The paradigm of protein control of expression of the *E. coli lac* operon led to the commonly accepted view that regulation of gene expression in prokaryotes and eukaryotes involved protein-based mechanisms. However, experiments performed in the past 20 years have shown that small regulatory RNA molecules can silence gene expression in eukaryotes in a process called **RNA interference (RNAi)** or **RNA silencing.** RNAi was first discovered in plants, where it was called *posttranscriptional gene silencing* (PTGS), although the involvement of RNA in silencing was not demonstrated at the time. In 1998 Andrew Fire and Craig Mello published the results of their studies on gene regulation in the nematode worm, *Caenorhabditis elegans*. They injected mRNA molecules encoding a muscle protein into the worm and saw no difference in behavior. However, when they injected both sense and antisense RNA together ("sense RNA" is the mRNA with the coding information for the gene-encoded polypeptide, while "antisense RNA" is the complementary RNA to the mRNA), the worms began moving in a way similar to worms that contained a defective gene for the same muscle protein. Fire and Mello hypothesized that double-stranded RNA formed by pairing of the sense and antisense RNAs was responsible for silencing expression of the gene encoding the same mRNA. They obtained support for this hypothesis by injecting double-stranded RNA (dsRNA) for a number of genes and observing that, in every case, the expression of the corresponding gene was silenced. They deduced that dsRNA could silence genes and named the phenomenon RNA interference; they received the 2006 Nobel Prize in Physiology or Medicine "for their discovery of RNA interference—gene silencing by double-stranded RNA." Their discovery revolutionized scientists' thinking about the regulation of gene expression and we are now seeing that RNA-based regulation of gene expression is widespread among eukaryotes. We know now that dsRNAs are not the direct regulators of silencing, but are precursors to single-stranded RNA (ssRNA) molecules that are the actual regulators of silencing. This section discusses the key principles of RNAi for silencing transcription, and for silencing gene expression posttranscriptionally. While not discussed here, recent research indicates that an RNAi pathway is involved also with chromatin remodeling associated with at least some gene silencing at the transcriptional level, including genomic imprinting, as well as with a number of other aspects of genome structure and maintenance, such as heterochromatin formation.

### The Roles of Small Regulatory RNAs in Posttranscriptional Gene Silencing

Small regulatory RNAs in eukaryotes fall into two main groups, **microRNAs (miRNAs)** and **short interfering RNAs (siRNAs).** Both of these RNAs are noncoding, meaning that they are untranslated and, therefore, do not specify a polypeptide product.

**MicroRNAs.** MicroRNAs are ssRNA regulatory molecules about 21–23 nucleotides (nt) long that derive from RNA transcripts. MicroRNAs are encoded by genes in the genomes of all multicellular eukaryotes, as well as some unicellular ones (the budding yeast, *Saccharomyces cerevisiae*, is a notable exception). As of November 2007, more than 5,000 miRNA genes have been identified among eukaryotes. The several hundred miRNA genes in humans are scattered throughout all the chromosomes with the exception of the Y. About 30% of mammalian miRNA genes are located in intergenic regions (*intergenic* here meaning "between protein-coding genes" of the genome; they are transcribed by RNA polymerase II resulting in capped, polyadenylated transcripts. Some of these genes are located in transposons. The remainder of the mammalian miRNA genes are located within other genes—many are in introns of protein-coding genes, while some are in introns and exons of non-protein-coding genes. In all of these cases, the miRNA sequence is transcribed by an RNA polymerase as part of the transcript of the "host" gene. In a few cases, an intron-located miRNA gene is transcribed independently by RNA polymerase II.

Figure 18.15a illustrates the production and functions of miRNAs in posttranscriptional gene silencing in animals. The production of miRNAs in plants occurs using a similar pathway to that of animals, although there are differences in the details of the steps. The transcript containing an miRNA is called the *primary miRNA transcript*, or *pri-miRNA*. As just mentioned, in many cases, this is a pre-mRNA molecule or a precursor RNA for a noncoding RNA in the cell. The pri-miRNA molecule contains a hairpin structure about 70 nt long, within which is the eventual miRNA. The hairpin is cut out of the pri-miRNA in the nucleus by the dsRNA-specific endonuclease Drosha complexed to an accessory protein (Pasha in *Drosophila*). Drosha makes staggered cuts, resulting in a ~2 nt 3′ single-stranded overhang. The excised hairpin—*pre-miRNA*—is exported rapidly to the cytoplasm.

In the cytoplasm, another dsRNA-specific endonuclease, Dicer, complexed to an accessory protein (Loq in *Drosophila*), makes staggered cuts in the pre-miRNA, releasing a short miRNA:miRNA* dsRNA consisting of some of the former paired sides of the hairpin. The two RNA strands are imperfectly paired: "miRNA" is the mature miRNA strand that subsequently functions in the cell for RNA silencing (see later), while "miRNA*" is its partial complement and does not function in RNA silencing. Because the miRNA directs RNA silencing, it is termed the *guide strand*, while the miRNA* is termed the *passenger strand.* Next the dsRNA, Dicer, and accessory protein bind to Ago1, a member of the Argonaute family

**Figure 18.15**

**RNA interference (RNAi) by small regulatory RNAs.**

a) Production and functions of microRNAs (miRNAs) in posttranscriptional silencing in animals
miRNA gene—standalone, or within another gene
Cap
5′
Transcription by RNA polymerase II
Pri-miRNA
Poly(A) tail
AAA
3′
Processing by Drosha and accessory protein
Nucleus
5′
3′
Pre-miRNA
Cytoplasm
Export to cytoplasm
5′
3′
Pre-miRNA
Dicer, with accessory protein, binds and makes staggered cuts
miRNA (guide strand)
Accessory protein
miRNA*
(passenger strand)
Dicer
Ago1 (Slicer), with accessory proteins (not shown) binds
Ago1
Pre-miRISC (pre-microRNA-induced silencing complex)
Ago1 cleaves miRNA*, a helicase unwinds the two pieces, and they dissociate from the complex
miRNA
miRISC
miRISC binds to target mRNAs
Imperfect pairing of miRNA to target site
5′
AAAA 3′
Target mRNA
Target site is 3′UTR of mRNA
Due to imperfect pairing of miRNA and target site, binding of miRISC represses translation of mRNA; mRNA moves to P body where it is either degraded or stored
P body
mRNA degradation
or
mRNA storage
b) Production and functions of short interfering RNAs (siRNAs) in posttranscriptional silencing in animals
Nucleus
Cytoplasm
5′
3′
3′
5′
Long dsRNA molecule
Dicer, with accessory protein, binds and makes staggered cuts. Long dsRNA molecule is cut into many ~22 nt duplexes (one is shown)
Accessory protein
Dicer
~22 nt RNA duplex
Ago2 (Slicer), with accessory proteins (not shown), binds
Ago2
Pre-siRISC (pre-siRNA-induced silencing complex)
Ago2 cleaves one of the two RNA strands, a helicase unwinds the two pieces, and they dissociate from the complex
siRNA
siRISC
siRISC binds to target RNA, i.e., one strand of dsRNA molecule from which the siRNA derived
Perfect pairing of siRNA to target RNA
Target mRNA
Region of target RNA complementary to siRNA
Due to perfect pairing of siRNA and target RNA, the target RNA is cleaved by Ago2
Target RNA pieces are moved to P body for degradation
P body
Target RNA fragment degradation

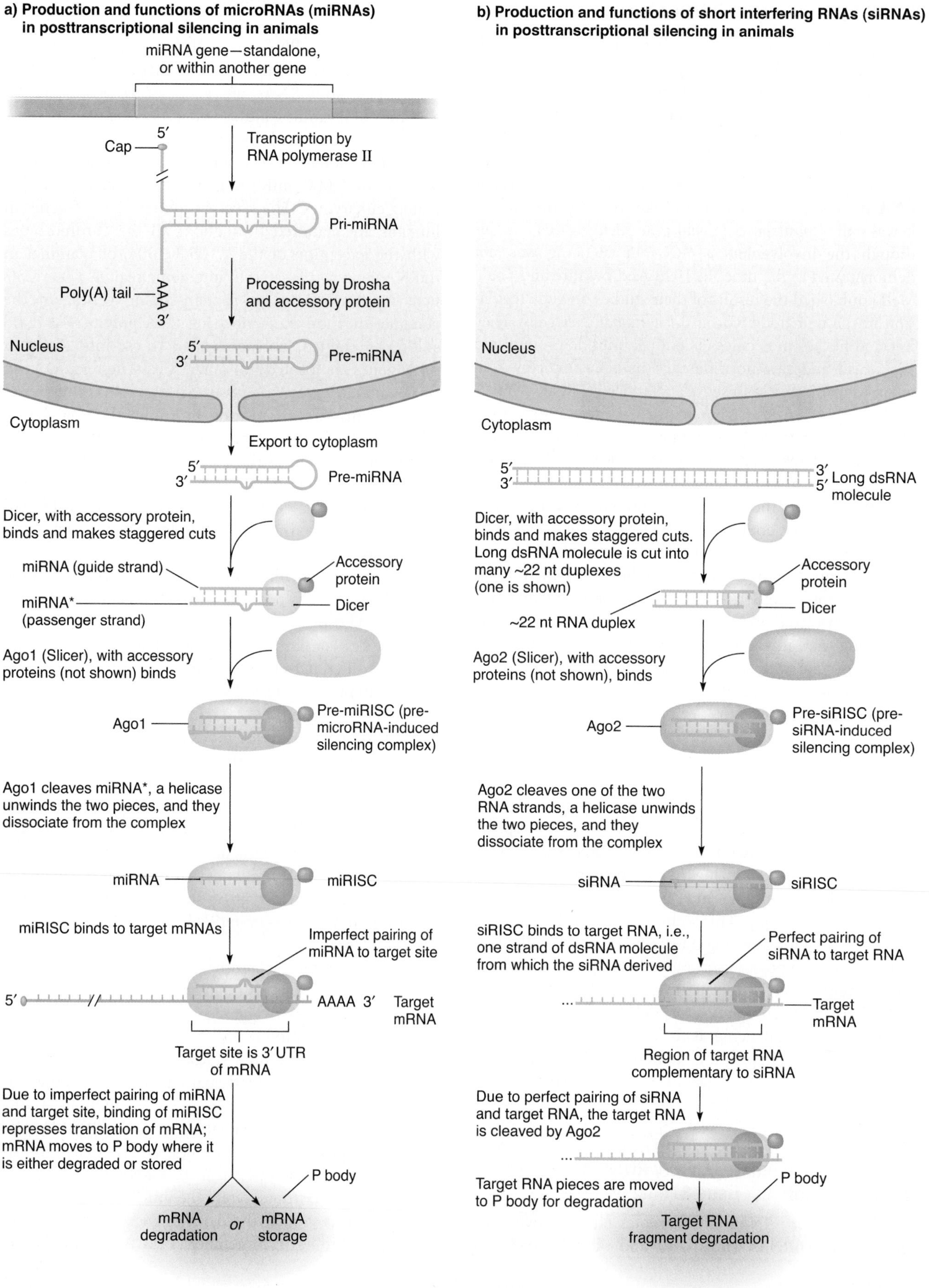

of protein, and other proteins to form the *pre-microRNA-induced silencing complex*, or *pre-miRISC*. Ago1 is another RNA endonuclease; more generally called Slicer, it makes a single cut within the miRNA* passenger strand. A helicase that is part of the pre-miRISC then unwinds to two pieces from the miRNA guide strand, and they dissociate from the complex. The result is the mature miRISC, the ribonucleoprotein complex that can silence gene expression.

How does an miRISC function in posttranscriptional gene silencing? The miRNA in the miRISC is a *trans*-acting RNA regulatory molecule, meaning that it targets mRNAs that are not the same as the RNA molecules from which the miRNA is derived. This is one distinguishing feature of miRNAs compared with siRNAs. An miRISC binds to a target mRNA through complementary base pairing involving the miRNA. Usually, the sequences to which the miRNA binds are short sequences in the 3′ UTR of the mRNA. An mRNA molecule may have one or more sequences in its 3′ UTR to which the same miRNA can bind, and/or it may have several sequences in its 3′ UTR to which several different miRNAs can bind. The latter raises the possibility of regulating the expression of the same gene (through its mRNA) by various combinations of miRNA regulator molecules. In Figure 18.15a, one miRISC is shown binding to a 3′ UTR sequence for simplicity. In animals, binding of most miRISCs to their target mRNAs involves imperfect pairing between the miRNA and the 3′ UTR region of the mRNA. Such pairing triggers translational repression—translation of that mRNA becomes inhibited. The translationally repressed mRNA with its associated miRISC(s) is then sequestered from the translation machinery by becoming, or moving into a *P body*. A P body is a cytoplasmically located aggregate of translationally repressed mRNAs complexed with proteins, and proteins for mRNA decapping, and mRNA degradation. The mRNAs in P bodies may be degraded using the contained mRNA degradation machinery or stored in ribonucleoprotein complexes. Stored mRNAs can be returned to translation at a later time. Whether degraded or stored, the effect of miRNA action is to reduce the expression of the gene encoding the targeted mRNA at the translation level.

In plants, binding of most miRISCs to their target mRNAs involves perfect or near-perfect pairing between much of the miRNA and the 3′ UTR region of the mRNA. Perfect pairing triggers mRNA degradation rather than translational repression. Here, the Ago1 Slicer protein cuts the target mRNA into two and the mRNA-miRISC complex forms, or it is moved to a P body where degradation of the mRNA is completed.

MicroRNAs play central roles in controlling gene expression in a variety of cellular, physiological, and developmental processes. For example, miRNAs are involved in a variety of developmental events in animals and plants. In animals, for instance, miRNAs participate in the regulation of specific developmental timing events, of cell proliferation, of gene expression in neurons, of brain morphogenesis, and of stem cell division.

**Short Interfering RNAs.** Short interfering RNAs (siRNAs) are ssRNA regulatory molecules about 22 nt long. They are found in eukaryotes spanning the phylogenetic spectrum (again with the notable exception of *Saccharomyces cerevisae*). Short interfering RNAs are produced by processing of long dsRNA molecules.

Figure 18.15b illustrates the production and functions of siRNAs in posttranscriptional gene silencing. The pathway producing siRNAs begins with cytoplasmically located dsRNA molecules that are hundreds to thousands of base pairs long. Sources of these RNA molecules include intermediates in the replication of viruses with RNA genomes, naturally generated molecules from complementary or partially complementary sense and antisense transcripts from regions of the genome, and transcripts that fold into long, extended hairpins. The long dsRNA is processed using a pathway highly similar to that for processing pre-miRNA in the cytoplasm. First, the molecule is processed by a Dicer–protein complex into many ~22 nt duplexes, each with 2 nt 3′ overhangs. One strand of each duplex is the siRNA guide strand that will carry out RNA silencing, while the complementary strand (perfectly complementary for these molecules) is the passenger strand that will be discarded. Then the dsRNA–Dicer–protein complex binds to Ago2, another member of the Argonaute family, and other proteins to form the *pre-siRNA-induced silencing complex*, or *pre-siRISC*. The next steps then parallel those for the miRNA pathway, resulting in the mature siRISC.

The siRISC functions in posttranscriptional gene silencing by recognizing single-stranded RNAs that are complementary to one strand or the other of the dsRNAs from which the siRNAs were produced. The siRNA in the siRISC pairs with the target RNA with perfect base pairing, and Ago2 cleaves the target RNA into two. Degradation of the two RNA pieces occurs in a P body. If we think about the target RNA as a viral genome or a viral transcript, the consequence would be that the viral life cycle is inhibited. This example of siRNA-based RNA interference, then, is an RNA-directed immune system.

**Keynote**

In RNA interference (RNAi), particular small, noncoding, regulatory RNA molecules silence the expression of individual genes posttranscriptionally. The two main groups of small RNAs are microRNAs (miRNAs) and short interfering RNAs (siRNAs). miRNAs are single-stranded molecules that derive from RNA transcripts encoded by specific nuclear genes. siRNAs are single-stranded molecules derived by processing of long double-stranded RNAs, such as viral replication intermediates. Both miRNA and siRNA are produced by similar pathways with some shared components in which a double-stranded RNA, or a double-stranded region of an RNA, is processed by enzymes to produce the short RNA. The RNA is associated with proteins to form an RNA-induced silencing complex, or RISC—miRISC and siRISC for the two types of interfering RNAs. miRISC binds specifically to target mRNAs—the gene transcripts regulated by the miRNA—by base pairing of the miRNA to a sequence in the 3′ UTR. If the pairing is imperfect, translation is inhibited, and the mRNA may be stored or degraded. If the pairing is perfect, or near perfect, an enzyme in the miRISC cleaves the mRNA within the region paired with the miRNA, and the mRNA fragments are subsequently degraded. siRISC recognizes single-stranded RNAs that are one strand or the other of the double-stranded RNA from which the siRNA in the complex is derived. The pairing consequently is always perfect and, hence, the single-stranded RNA is cleaved and subsequently degraded. miRNA targets specific cellular mRNAs. siRNA targets a single-stranded RNA, such as a viral RNA, related to the dsRNA from which the siRNA was made.

## Regulation of Gene Expression Posttranscriptionally by Controlling mRNA Degradation and Protein Degradation

Gene expression is regulated posttranscriptionally both by control of mRNA degradation and by control of protein degradation.

### Control of mRNA Degradation

All RNA species are subjected to **degradation control,** in which the rate of RNA breakdown (also called RNA turnover) is regulated. Usually, both rRNA (in ribosomes) and tRNA are highly stable species. By contrast, mRNA molecules exhibit a diverse range of stability, with some mRNA types known to be stable for many months whereas others degrade within minutes. The stability of particular mRNA molecules may change in response to regulatory signals. For example, the addition of a regulatory molecule to a cell type can lead to an increase in the synthesis of a particular protein or proteins, accomplished by an increase in the rate of transcription of the genes involved or an increase in the stability of the mRNAs produced. Table 18.1 presents examples of systems in which changes in mRNA stability for a number of cell types occur in the presence and absence of specific effector molecules.

mRNA degradation is an important control point in the regulation of gene expression in eukaryotes. Various sequences or structures have been shown to affect the half-lives of mRNAs, including the AU-rich elements (ARE) discussed earlier and various secondary structures. Two major mRNA decay pathways are the *deadenylation-dependent decay pathway* and the *deadenylation-independent decay pathway*. In the deadenylation-dependent decay pathway, the poly(A) tails are deadenylated until the tails are too short (5–15 As) to bind PAB (poly(A) binding protein). In yeast, the product of the *PAN1* gene, PAB-dependent poly(A) nuclease, catalyzes the deadenylation. Once the tail is almost removed, the 5′ cap structure is removed in a step called *decapping*, an enzyme-catalyzed process. In yeast, the decapping enzyme, or at least an essential part of it, is encoded by the *DCP1* gene. After an mRNA molecule is decapped, it is degraded from the 5′ end by a 5′-to-3′ exonuclease. In yeast, this enzyme—encoded by the *XRN1* gene—degrades decapped mRNAs extremely rapidly, attesting to the importance of the 5′ cap in protecting active mRNAs in the cell.

Yeast strains with a mutant *DCP1* gene are viable, and mRNA degradation still occurs, providing evidence for the existence of mRNA degradation pathways other

**Table 18.1 Examples of Tissues or Cells in Which Regulation of mRNA Stability Occurs in Response to Specific Effector Molecules[a]**

| mRNA | Tissue or Cell | Regulatory Y Single (= Effector Molecule) | Half-Life of mRNA With Effector | Half-Life of mRNA Without Effector |
|---|---|---|---|---|
| Vitellogenin | Liver (frog) | Estrogen | 500 h | 16 h |
| Vitellogenin | Liver (hen) | Estrogen | −24 h | <3 h |
| Apo-very low density-lipoprotein | Liver (hen) | Estrogen | ~20–24 h | <3 h |
| Ovalbumin, conalbumin | Oviduct (hen) | Estrogen, progesterone | >24 h | 2–5 h |
| Casein | Mammary gland (rat) | Prolactin | 92 h | 5 h |
| Prostatic steroid-binding protein | Prostate (rat) | Androgen | Increases 30 × | |

[a]Note that the effector molecule in each case results in an increase in transcription, as well as stabilization, of the mRNA.

than the pathway just described. In these deadenylation-independent decay pathways, mRNAs may be decapped without being deadenylated, thereby exposing them to rapid degradation by 5′-to-3′ exonucleases, or they may be cleaved internally by endonucleases without being deadenylated and then may be broken down further.

Note that, although our understanding of mRNA degradation in yeast is becoming clearer, the details of mRNA degradation in mammalian cells are not as well known. Both deadenylation-dependent and deadenylation-independent decay pathways exist in mammals, and decapping is an important step in at least the former pathway.

### Control of Protein Degradation

Regulatory mechanisms also exist at the posttranslational level. These mechanisms determine the lifetime of a protein. This topic is peripheral to our discussion of gene expression, so it is discussed only very briefly here.

A wide variety of possibilities exist to regulate the amount of a particular protein in a cell. A constitutively produced mRNA may be translated continuously, with the level of protein product controlled by the degradation rate of that protein, or a short-lived mRNA may encode a protein that is highly stable so that it persists for very long periods in the cell. Proteins in the lens of higher vertebrate eyes, for example, are long lived. Their mRNAs have long since been degraded, but the protein itself persists, usually for the lifetime of the individual. By contrast, steroid receptors and heat-shock proteins have short half-lives.

The degradation of proteins (*proteolysis*) in eukaryotes requires the addition of *ubiquitin* proteins (a protein consisting of 76 amino acids found ubiquitously in essentially all eukaryotes) to the proteins. The ubiquitinated proteins are then transported to the *proteasome*, a large multisubunit complex containing proteases, in which the tagged proteins are degraded into short peptides. The peptides subequently are broken down into amino acids by proteolytic enzymes in the cytoplasm. Ubiquitin is released intact during degradation in the proteasome, enabling it to be used to tag other proteins for degradation. Aaron Ciechanover, Avram Hershko, and Irwin Rose received a Nobel Prize in Chemistry in 2004 "for the discovery of ubiquitin-mediated protein degradation." Monoubiquitination—the addition of a single ubiquitin protein to a protein—plays a very different role in the cell, as described in this chapter's Focus on Genomics box.

The amino acid at the N-terminus of a protein is the key to how the protein initially is targeted for ubiquitin binding. In what has become known as the *N-end rule*, the particular N-terminal amino acid relates directly to the half-life of the protein. In a yeast test system in which the lifetime of the same protein was measured with different N-terminal amino acids, arginine, lysine, phenylalanine, leucine, and tryptophan each specified a half-life of 3 minutes or less, whereas cysteine, alanine, serine, threonine, glycine, valine, proline, and methionine all specified a half-life of more than 20 hours. The same general hierarchy is seen in an *E. coli* system. The N-terminal amino acid directs the rate at which ubiquitin molecules are added to the protein, which, in turn, determines the half-life of the protein.

In sum, in prokaryotes, gene expression is controlled mainly at the transcriptional level, in association with the rapid turnover of mRNA molecules. In eukaryotes, gene expression is regulated at transcriptional, posttranscriptional, and posttranslational levels. The intertwining of the regulatory events at the different levels leads to the fine-tuning of the amount of protein in the eukaryotic cell.

**Keynote**

Gene expression is also regulated by the control of mRNA translation and degradation. mRNA degradation is believed to be a major control point in regulating gene expression. Structural features of individual mRNAs have been shown to be responsible for the range of mRNA degradation rates, although the precise roles of cellular factors and enzymes have yet to be determined. Protein degradation is also regulated. Which amino acid is at the protein's terminus correlates with the stability of the protein and directs the rate at which ubiquitin binds to the protein. In turn, that rate determines the rate of protein breakdown.



## Summary

- In eukaryotes, gene expression is regulated at a number of levels. That is, there are regulatory systems for the control of transcription, precursor-RNA processing, transport of the mature RNA out of the nucleus, translation of the mRNAs, degradation of the mature RNAs, and the processing and degradation of protein.
- Activation of transcription initiation for a eukaryotic protein-coding gene requires three classes of proteins: general transcription factors, activators, and coactivators. The general transcription factors bind to the core promoter and are required for basal transcription. Activators bind to enhancers and promoter-proximal elements and stimulate transcription initiation by recruiting a coactivator—a large multiprotein complex that does not by itself bind to DNA, but bridges between activator proteins and general transcription factors. Once bound to activators, coactivators recruit RNA polymerase II, which, through interaction with the general transcription factors, is oriented correctly for transcription initiation.

- While much of the regulation of gene expression occurs by a positive regulatory system—activation of transcription—for some genes negative regulation occurs using repressors. These repressors bind to the DNA and act in various ways to block or limit transcription initiation.

- Each regulatory promoter element and enhancer element binds specific regulatory proteins. However, there are fewer regulatory proteins than there are genes. In combinatorial gene regulation, particular combinations of a few regulatory proteins—activators and/or repressors—control the transcription of different arrays of genes.

- Chromatin configuration poses a barrier to transcription. Transcriptionally active regions have looser chromatin structure than do transcriptionally inactive regions. It is the chromatin structure at the core promoter of a nonexpressed gene that is repressive to transcription. Remodeling of this chromatin region is necessary for activation of transcription. Chromatin remodeling occurs when activators bind to enhancers and recruit large, multiprotein remodeling complexes that either acetylate nucleosomes, loosening their association with the DNA, or move or restructure nucleosomes, allowing the transcription machinery to access the promoter.

- Gene silencing is the phenomenon of turning off the transcription of a gene as a result of its position in the chromosome. Gene silencing involves changes in chromatin structure to produce heterochromatin, which usually affects a zone of genes. A gene may also be silenced through the methylation of cytosines in the promoter upstream of the gene. Sometimes the methylation pattern is associated with genomic imprinting, an epigenetic phenomenon in which the expression of a gene is determined by whether the gene is inherited from the female or male parent.

- Gene expression is also regulated at the RNA processing. This type of regulation operates to determine the production of mature RNA molecules from precursor-RNA molecules. Two regulatory events that exemplify this level of control are the choice of poly(A) site and the choice of splice site. In both cases, different types of mRNAs are produced, depending on the choice made. As a result, proteins can be produced that are encoded by the same gene, but that differ structurally and functionally. Once a mature mRNA is produced, it is exported from the nucleus to the cytoplasm.

- Mature mRNA in the cytoplasm is subject to translational control. That is, an mRNA may be translated directly by ribosomes, or it may be stored in a complex with proteins in a form that is not translatable. Typically stored mRNAs have short poly(A) tails. Activation of stored mRNAs involves removing the associated proteins and lengthening the poly(A) tail. Signals in the 3′ UTR of an mRNA control the shortening and lengthening of the poly(A) tail.

- RNA interference (RNAi) involves the silencing of the expression of individual genes posttranscriptionally with small, noncoding, regulatory RNA molecules. The two main groups of small RNAs are microRNAs (miRNAs) and short interfering RNAs (siRNAs). miRNAs derive from transcripts of nuclear genes and, in a complex with specific proteins, the miRNAs bind to specific sequences in the 3′ UTRs of target mRNAs. When the match is imperfect, the interaction blocks translation of the target mRNA. If the match is perfect, the interaction leads to cleavage of the target mRNA. siRNAs are derived from long, double-stranded RNAs, such as intermediates of viral replication. In a complex with specific proteins, the siRNA binds to a single strand of the RNA from which it was derived and, because the match is perfect, cleaves that RNA.

- Gene expression is also regulated by mRNA degradation control. The latter is believed to be a major control point in the regulation of gene expression, as evidenced by the wide range of mRNA stabilities found within organisms. It is clear that nucleases are ultimately responsible for the degradation of the RNAs, and the signals for the differential mRNA stabilities are a property of the structural features of individual mRNAs. Regulation at the level of proteins involves a mechanism that specifies the lifetime and rate of degradation of a protein.

# Analytical Approaches to Solving Genetics Problems

**Q18.1** A region of one yeast chromosome specifies three enzyme activities in the histidine biosynthesis pathway; these activities are synthesized coordinately. How would you distinguish between the following three models? (Hint: Recall that the complete sequence of the yeast genome has been determined.)

**a.** Three genes are not organized into an operon. They code for three discrete mRNAs that are translated into three different enzymes.

**b.** Three genes are arranged in an operon. The operon is transcribed to produce a single polycistronic mRNA whose translation produces three distinct enzymes.

c. One gene (a supergene) is transcribed to produce a single mRNA whose translation produces a single polypeptide with three different enzyme activities.

**A18.1** Examination of the region of the yeast genome that encodes the three enzymes activities could distinguish between the three-independent-gene model (a) and the other two models. That is, if three different genes specify the three enzyme activities, then each gene will have its own promoter. If putative promoter sequences can be identified adjacent to each protein-coding sequence, then likely model (a) is correct. If only one putative promoter sequence can be identified at one end or other of the cluster of three protein-coding sequences, then either model (b) or (c) is correct.

Northern blotting can also help distinguish between the models. That is, isolate mRNAs from yeast under conditions when the histidine biosynthesis enzymes are made. Separate the mRNAs by gel electrophoresis and transfer the mRNA fragments to a membrane filter by the northern blotting technique. Next, probe that filter separately with labeled single-stranded DNA derived from each region of the chromosome that encodes the three enzymes. If model (a) is correct, the probing will reveal one band for each of the three labeled single-stranded DNAs. The size of that band will vary depending upon the length of the mRNA, that length being related to the size of the enzyme it encodes. If model (b) or model (c) are correct, the probe will also reveal one band. That band will be the same size for each model, and its size will be predicted to be much larger than that for any of the three bands predicted if model (a) is correct.

Characterizing the enzyme activities coded for by the three genes would enable us to distinguish between models (b) and (c). That is, in the operon model (b), three distinct polypeptides would be produced. These polypeptides could be isolated and purified individually by using standard techniques. Thus, if the operon model (b) is correct, we could show that there are three distinct polypeptides, each exhibiting only one of the enzyme activities—that is, three polypeptides and three enzyme activities. By contrast, if the supergene model (c) is correct, it should be possible only to isolate a large polypeptide with all three enzyme activities; no polypeptides with only one of the enzyme activities should exist.



# Questions and Problems

**18.1** Critically evaluate the following contention: Prokaryotes and eukaryotes use fundamentally different mechanisms to control gene expression.

**18.2** Promoters, enhancers, general transcription factors, activators, coactivators, and repressors that regulate the expression of one gene often have structural features that are similar to those regulating the expression of other genes. Nonetheless, the transcriptional control of a gene can be exquisitely specific: it will be specifically transcribed in some tissues at very defined times. Explore how this specificity arises by addressing the following questions:

a. Distinguish between the functions of promoters and enhancers in transcriptional regulation.
b. Distinguish between the functions of general transcription factors, activators, coactivators, and repressors in transcriptional regulation.
c. What structural features are found in activators, and what role do these play in transcriptional activation?
d. How is the mechanism by which eukaryotic repressors function different from that by which prokaryotic repressors function?
e. How can the same enhancer stimulate as well as quench transcription?
f. Given that several different genes may contain the same types of promoter and enhancer elements, and a number of the proteins that bind these elements contain the same structural features, how is transcriptional specificity generated?

***18.3** A temperature-sensitive mutation in yeast results in the production of a Gal80p protein in cells grown at the permissive temperature of 18°C but not at the restrictive temperature of 29°C. How will transcription levels of each of the *MIG1*, *GAL1*, *GAL4*, *GAL7*, and *GAL10* genes change when yeast growing at the permissive temperature are shifted to the restrictive temperature when

a. both glucose and galactose are present
b. glucose, but not galactose, is present
c. galactose, but not glucose, is present
d. neither glucose nor galactose is present

**18.4** Both peptide and steroid hormones can affect gene regulation of a targeted population of cells.

a. What is a hormone?
b. Distinguish between the mechanisms by which a peptide and a steroid hormone affect gene expression.
c. What role does each of the following have in a physiological response to a peptide or a steroid hormone?
   i. steroid hormone receptor (SHR)
   ii. chaperone
   iii. steroid hormone response element (HRE)
   iv. second messenger
   v. cAMP and adenylate cyclase
d. How can the same steroid hormone simultaneously activate distinct patterns of gene expression in two different cell types and have no effect on a third cell type?

***18.5** In *Drosophila*, pulses of the steroid hormone, ecdysone, trigger molting between the larval stages and then, at the end of the larval stages, trigger the formation of a pupa, where the larva will metamorphose into an adult fly. Immediately after the ecdysone pulse at the end of the larval stages, transcription of several genes, including *Eip93F*, is dramatically increased. To investigate how ecdysone regulates *Eip93F*, chromatin is isolated from staged wild-type animals just before and just after the ecdysone pulse at the end of the larval stages. The chromatin is distributed to separate test tubes, where it is digested for 2 minutes with different concentrations of DNase I. DNA is then purified from each sample and digested with *Eco*RI. The resulting DNA fragments are then resolved by size using gel electrophoresis, and a Southern blot is made. The Southern blot is probed with two *Eco*RI fragments from the *Eip93F* gene: a 4.0-kb fragment from its promoter and a 3.0-kb fragment from its protein-coding region. The following figure shows the results, where the thickness of the band corresponds to the intensity of hybridization signal:

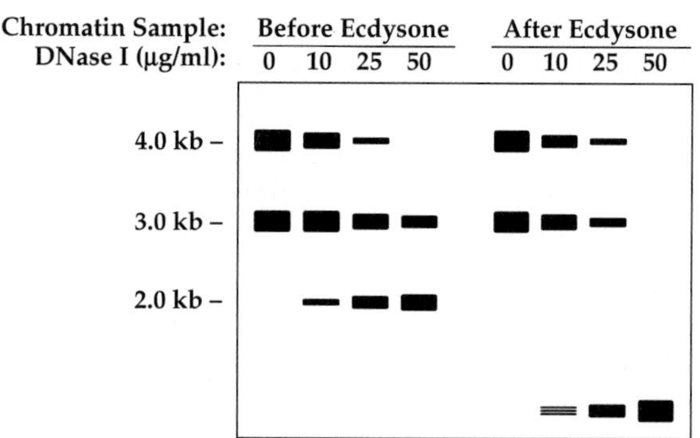


**a.** Explain why the 4-kb band, but not the 3-kb band, diminishes in intensity when chromatin that was isolated before the pulse of ecdysone is treated with increasing concentrations of DNase I. How do you explain the increasing amounts of the 2-kb band in these samples?

**b.** Explain why both the 4-kb and 3-kb bands diminish in intensity when chromatin isolated after the pulse of ecdysone is treated with increasing amounts of DNase I. How do you interpret the increasing amounts of low molecular weight digestion products in these samples?

***18.6** DNA, histones, promoter-binding proteins, and enhancer-binding proteins are mixed together in the following orders:

**a.** first histones and DNA, then promoter-binding proteins

**b.** first histones and promoter-binding proteins, then DNA

**c.** first DNA and promoter-binding proteins, then histones

**d.** first histones, promoter-binding proteins, and enhancer-binding proteins, then DNA

For each case, state whether transcription can occur. Explain your answers.

**18.7** Chromatin remodeling is essential for gene activation and can be achieved using different mechanisms.

**a.** What different types of enzymes are used to modify histones, and how do these enzymatic modifications lead to chromatin remodeling?

**b.** In what other ways can chromatin be remodeled?

**c.** What phenotype(s) would you expect to see in a mutant where a protein involved in chromatin remodeling failed to function?

***18.8** The following figure shows the effect of the hormone estrogen on ovalbumin synthesis in the oviduct of 4-day-old chicks. Chicks were given daily injections of estrogen ("Primary stimulation") and then after 10 days the injections were stopped. Two weeks after withdrawal (25 days), the injections were resumed ("Secondary stimulation").

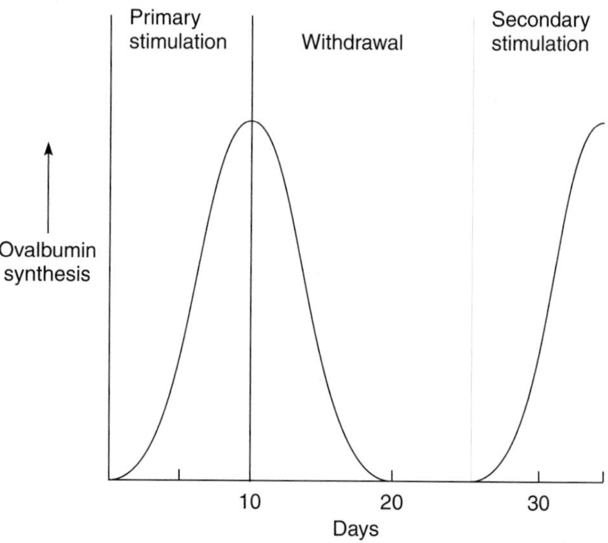


Provide possible explanations of these data.

**18.9** In what different ways can DNA methylation affect gene expression?

***18.10** Genetic mechanisms underlying gene silencing involve DNA sequence alterations, while epigenetic mechanisms do not. What are three epigenetic mechanisms that can lead to gene silencing? What features do they share in common, and how are they different?

**18.11** When male mice heterozygous for a small deletion on chromosome 2 are mated to normal females, deletion-bearing offspring have thin bodies and are slow moving, while non-deletion-bearing offspring are normal. However, when females heterozygous for the same deletion are mated to normal males, all offspring are normal.

**a.** How can these findings be explained in terms of imprinting?

**b.** When, and in what cell types, does imprinting occur?

**c.** Do you expect imprinted genes will show a dominant or recessive pattern of inheritance? Why?

**d.** *Neuronatin* is a gene that lies within the deleted region. A DNA polymorphism exists in the 3′ UTR of the *Neuronatin* gene that can be distinguished using PCR-RFLP (see Chapter 10, pp. 270–271 for a discussion of PCR-RFLP). How would you determine if the *Neuronatin* gene is expressed in a manner consistent with its being imprinted in embryos produced by the cross? Explain what results you would expect if it is imprinted, and what results you would expect if it is not imprinted.

**18.12** The Human Epigenome Project (HEP) seeks to identify, catalog, and interpret genome-wide DNA methylation patterns in all major human tissues. Investigate the rationale for these goals and how these data are collected by viewing the information at http://www.epigenome.org and the links it presents, and then answer the following questions:

**a.** What are the anticipated benefits of meeting the goals of the HEP?

**b.** Why is it important to assess methylation patterns of the same gene in different tissues?

**c.** In general terms, what methods are used by the HEP to assess methylation patterns?

*__18.13__ Both fragile X syndrome and Huntington disease are caused by trinucleotide repeat expansion. Individuals with fragile X syndrome have at least 200 CGG repeats at the 5′ end of the *FMR-1* gene. In contrast, individuals with Huntington disease have 36 or more in-frame CAG repeats within the protein-coding region of the Huntington gene.

**a.** Do you expect gene expression at the two genes to be affected in the same way by these repeat expansions? Explain your answer.

**b.** Based on your answer to (a), why might fragile X syndrome be recessive, whereas Huntington disease is dominant?

**c.** Generate a hypothesis to explain why the number of trinucleotide repeats needed to cause a disease phenotype is different at each gene.

**18.14** Although the primary transcript of a gene may be identical in two different cell types, the translated mRNAs can be quite different. Consequently, in different tissues, distinct protein products can be produced from the same gene. Discuss two different mechanisms by which the production of mature mRNAs can be regulated to this end; give a specific example for each mechanism.

*__18.15__ Four different cDNAs were identified when a cDNA library was screened with a probe from one gene. The locations of introns and exons in the gene were determined by comparing the cDNA and genomic DNA sequences. The results are summarized in Figure 18.A: Exons are represented by filled rectangles with protein-coding regions shaded black and 5′ UTR and 3′ UTR regions shaded grey; introns are represented by thin lines.

**a.** How many different protein isoforms are encoded by this gene?

**b.** Carefully inspect these data and generate a specific hypothesis about the type(s) of posttranscriptional control that could generate these different protein isoforms.

**c.** How might you experimentally investigate whether the different protein isoforms produced by this gene have distinct functions? (*Hint:* Consider how RNA interference was discovered.)

**18.16** Discuss the similarities and differences between miRNAs and siRNAs in terms of

**a.** the location and structure of the genes that encode them,

**b.** the mechanisms leading to their production in cells,

**c.** how they function in posttranscriptional gene silencing, and

**d.** the type of RNAs they regulate with respect to their roles in gene regulation and viral infection.

**18.17** Although many mRNAs are present in the cytoplasm of unfertilized vertebrate and invertebrate embryos, the rate of protein synthesis is very low. After fertilization, the rate of protein synthesis increases dramatically without new mRNA transcription.

**a.** What differences are seen in the length of poly(A) tails between inactive, stored mRNAs and actively translated mRNAs?

**b.** What role does cytoplasmic polyadenylation have in this process?

**c.** What signals are present in mRNAs that control polyadenylation and deadenylation?

**d.** In what way is deadenylation also important for controlling mRNA degradation?

*__18.18__ During *Drosophila* oogenesis, mRNAs from the *bicoid*, *nanos*, and *toll* genes are deposited into the developing oocyte and stored regionally in the cytoplasm for future translation after the oocyte is fertilized. Each of these genes produces a protein that helps establish the embryo's body plan. Mutations in these genes show maternal effects and exhibit phenotypes related to where each gene's mRNA is localized within the embryo: mutant *bicoid* mothers produce embryos with defects in anterior (head, thorax) structures; mutant *toll* mothers produce embryos with defects in dorsoventral structures; and mutant *nanos* mothers produce embryos with defects in posterior structures. Two other mutations, *cortex* and *grauzone*, show defective translation of multiple maternal

**Figure 18.A**

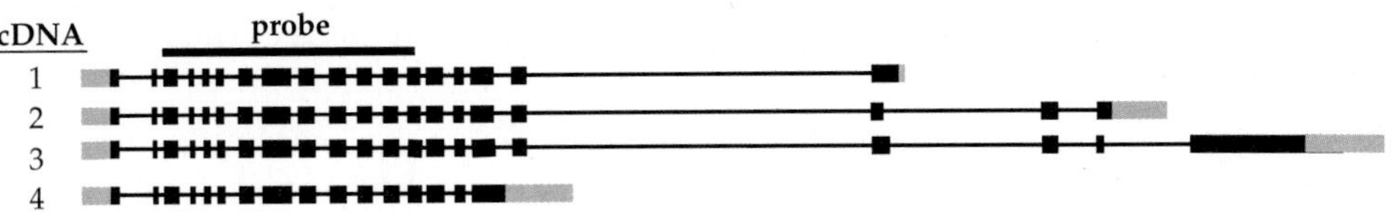


Questions and Problems

mRNAs. Even though *cortex*, *grauzone*, and wild-type embryos contain *bicoid* and *toll* mRNAs that are identical in amount, structure, and localization, dramatically less Bicoid and Toll protein accumulates in *cortex* and *grauzone* mutant embryos than in wild-type embryos. In contrast, the *cortex* and *grauzone* mutants contain normal amounts of the Nanos protein.

**a.** Generate a hypothesis about the regulatory process that the *cortex* and *grauzone* genes are used in, and explain how deficits in this process would lead to decreased production of the Bicoid and Toll proteins.

**b.** Under your hypothesis, why is the production of the Nanos protein unaffected by the *cortex* and *grauzone* mutants?

**c.** Under your hypothesis, what would you expect to find if you compared the length of the poly(A) tails in each of the *bicoid*, *toll*, and *nanos* mRNAs in wild-type, *cortex*, and *grauzone* embryos?

**d.** Suppose you injected structurally normal *bicoid* mRNA into embryos and then assessed the amount of Bicoid protein produced from the injected mRNA. Under your hypothesis, how would the levels of Bicoid compare in the following situations?

**i.** The mRNA is injected into wild-type embryos and its poly(A) tail is the same length as that in wild-type embryos.

**ii.** The mRNA is injected into wild-type embryos and its poly(A) tail is considerably shorter than that in wild-type embryos.

**iii.** The mRNA is injected into mutant *cortex* embryos and its poly(A) tail is the same length as that in wild-type embryos.

**iv.** The mRNA is injected into mutant *cortex* embryos and its poly(A) tail is considerably shorter than that in wild-type embryos.

**18.19** As shown in Figure 18.9, the *even-skipped* (*eve*) gene is transcribed in a pattern of seven stripes during *Drosophila* embryonic development. Address how combinatorial gene regulation leads to this striped pattern by answering each of the following questions.

**a.** How is the spatial expression pattern of two transcriptional repressors, Giant and Krüppel, related to the pattern of *eve* stripe-2 transcription?

**b.** How is the spatial expression pattern of two transcriptional activators, Bicoid and Hunchback, related to the pattern of *eve* stripe-2 transcription?

**c.** The *eve* stripe-2 enhancer has multiple binding sites for each of the Giant, Krüppel, Bicoid, and Hunchback proteins. Will all of these sites be bound simultaneously?

**d.** How would you expect the filling of the binding sites in part **(c)** and the pattern of *eve* stripe-2 transcription to be altered in mutants having these traits?

**i.** a broader distribution of the Giant protein

**ii.** a broader distribution of the Krüppel protein

**iii.** Hunchback protein expression restricted to parasegments 1 and 2

**iv.** dramatically elevated Bicoid protein expression

**e.** Is the expression of *eve* in one stripe, say stripe 2, dependent upon or independent from its expression in another stripe, say stripe 6? How is this achieved?

**f.** How is the spatial pattern of *eve* expression simultaneously dependent upon both transcriptional activators and transcriptional repressors?

***18.20** Although most eukaryotes lack operons such as those found in prokaryotes, the exceptional conserved organization of the *ChAT/VAChT* locus in *Drosophila* is reminiscent of a prokaryotic operon. *ChAT* is the gene for the enzyme choline acetyltransferase, which synthesizes acetylcholine, a neurotransmitter released by one neuron to signal another neuron. *VAChT* is the gene for the vesicular acetylcholine transporter protein, which packages acetylcholine into vesicles before its release from a neuron. Both *ChAT* and *VAChT* are expressed in the same neuron.

Part of the *VAChT* gene is nested within the first intron of the *ChAT* gene, and the two genes share a common regulatory region and a first exon. The structure of a primary mRNA and two processed mRNA transcripts produced by this locus are diagrammed in the following figure. The common regulatory region important for transcription of the locus in neurons is shown in the DNA, black rectangles in RNA represent exons, lines connecting the exons represent spliced intronic regions, and AUG indicates the translation start codons within the *ChAT* and *VAChT* mRNAs. Polyadenylation sites are not shown.

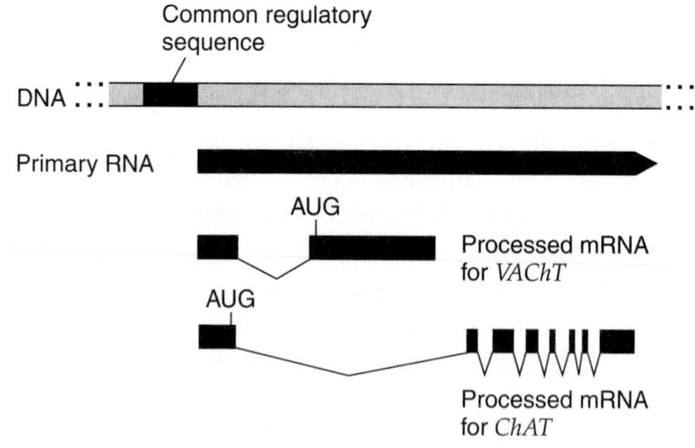


**a.** In what ways is the organization of the *VAChT/ChAT* locus reminiscent of a bacterial operon?

**b.** Why is the organization of this locus not structurally equivalent to a bacterial operon?

**c.** Based on the transcript structures shown, what modes of regulation might be used to obtain two different protein products from the single primary mRNA?

# 19 Genetic Analysis of Development

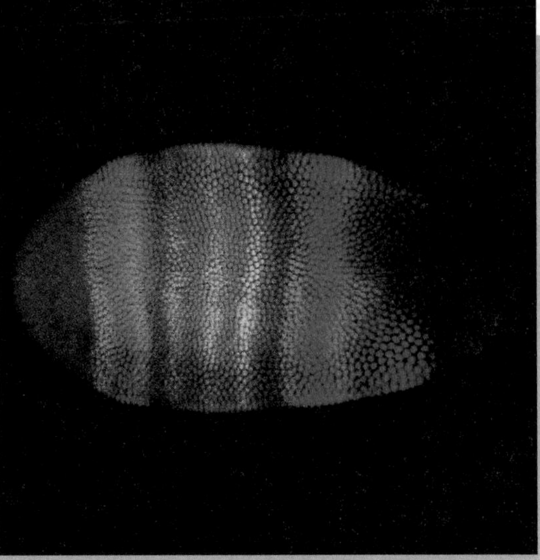

Differential expression of three genes—shown by a red, blue, and yellow immunofluorescence reaction to the proteins they produce—in a developing *Drosophila* embryo.

## Key Questions

- What are development and differentiation?
- What model organisms are used for the genetic analysis of development?
- Does the genome remain constant during development, or is there a loss of DNA?
- How do gene rearrangements generate antibody diversity?
- How is sex determined in mammals?
- How is the inequality of gene dosage on the X chromosomes in male and female mammals compensated for?
- How is sex determined in *Drosophila?*
- How is the inequality of gene dosage on the X chromosomes in male and female *Drosophila* compensated for?
- How is the development of the *Drosophila* body plan regulated genetically?
- How do microRNAs (miRNAs) regulate development?

### Activity

HOW DOES THE SINGLE CELL CREATED BY THE fusion of sperm and egg transform itself into a complex organism? How do the cells of a developing human "know" how to arrange themselves in the shape of a human? What makes dividing cells form a leg rather than an eye, a heart, or a hand? As you will learn in this chapter, the development of humans and other eukaryotic organisms requires the precise regulation of groups of genes. After you have read and studied this chapter, you can apply what you've learned by trying the iActivity, in which you will attempt to identify some of the genes responsible for changing embryonic stem cells into different forms of tissue.

It is natural to follow the chapter on regulation of gene expression in eukaryotes with a chapter on developmental genetics because genes program development, and an understanding of how genes are regulated therefore helps researchers in their genetic analysis of development. Developmental genetics is a subfield of developmental biology, and the amount of important information known is far too much to cover in one chapter for an introductory genetics course. Therefore, we will focus on some key principles and discuss a few examples to illustrate aspects of the genetic analysis of development.

## Basic Events of Development

**Development** is the process of regulated growth that results from the interaction of the genome with cytoplasm and the cellular external environment and that involves a programmed sequence of cellular-level phenotypic events that are typically irreversible. The total of the phenotypic changes constitutes the life cycle of an organism.

For a multicellular organism, development starts when a zygote is formed by fusion of sperm and egg. The zygote is **totipotent,** meaning that that cell has the potential to develop into any cell type of the complete organism. Of course, it must be able to do that. Cells later in development may also be totipotent; this is common in plants, but uncommon in animals past the four-cell embryo stage. The ability of a cell to become different cell types during development is called its *developmental* *potential*. As development progresses, the developmental potential of most individual cells decreases.

By following a cell through development, researchers can discover what that cell will become. This is called the *fate* of the cell. More specifically, the fates of all the cells in an embryo can be followed, resulting in the construction of a **fate map,** which is a diagram of the fate of each cell of an embryo. For instance, in 1983 John Sulston and his coworkers painstakingly observed the development of embryonic cells of *C. elegans* under the microscope and produced a fate map showing the complete lineage of each adult cell.

When the genetic program sets the fate of a cell, the cell is said to be *determined*, and the process is called **determination.** This is still a relatively early stage of development, so although a determined cell is molecularly different, it is not morphologically distinct from its neighbors. The cellular changes that occur during determination are directed, and lead to a stable state. That is, once the fate of the cell is determined, it does not change. Of course, the corollary is that a determined cell now has zero developmental potential: there is no longer a range of cell types that the cell can become.

There are two principal mechanisms for cell determination. In most cases, determination occurs by **induction;** that is, an inductive signal produced by one cell or group of cells affects the development of another cell or group of cells. For example, the signal can move by diffusion through the space between cells and be detected by a surface receptor on the target cell. Or, cells in contact can lead to interaction of transmembrane proteins in the plasma membranes, resulting in the production of the signal in one cell type. In some cases, there is an asymmetric distribution of cell-determining molecules when a cell divides. As a result, the two daughter cells differ in the signals they have for future differentiation.

After determination, the most spectacular aspect of development takes place: **differentiation.** Differentiation is the process by which determined cells undergo cell-specific developmental programs to produce cell types with specific identities, such as nerve cells, antibody-producing cells, skin cells, and so on, in animals; and phloem cells, leaf guard cells, meristematic cells, and so on, in plants. Differentiation in most cases results from differential gene expression, rather than from a differential loss of DNA that leaves different sets of genes in different cell types. That is, expression of different sets of genes in different kinds of determined cells leads to different proteins in the cells, and the proteins guide the progression to the various differentiated states.

Related to differentiation is **morphogenesis,** literally the "generation of form" and, by definition, the developmental process by which anatomical structures or cell shape and size are generated and organized. In both animals and plants, morphogenesis involves regulated patterns of cell division and changes in cell shapes. In animal morphogenesis, cell movement is an important component.

**Keynote**

Development is regulated growth resulting from the interaction of the genome with the cytoplasm and with the extracellular environment. Development begins when a zygote is formed. The zygote, and cells in the subsequent few generations, are totipotent, meaning they can develop into any cell type of the organism. At some point, the genetic program sets the fate of a cell in a process called determination. After determination, differentiation occurs, in which determined cells undergo developmental programs to produce their specific cell types. A related process to differentiation is morphogenesis, in which anatomical structures or cell shape and size are produced by a regulated pattern of cell division and changes in cell shape.

## Model Organisms for the Genetic Analysis of Development

For genetic analysis of development, researchers need mutants that affect development. These mutants may be naturally occurring or they may be induced but, either way, it must be possible to study the mutants genetically and molecularly. Thus, while a wide range of organisms have been the subjects for descriptive studies of development, relatively few organisms qualify as models for the genetic analysis of development. Several of the organisms that have contributed most to our understanding of the genetics of development are introduced here. For many of these organisms, the genome has been completely sequenced.

*Saccharomyces cerevisiae.* The single-celled organism yeast (see Figure 1.4a, p. 6) has a limited developmental repertoire, but notably yeast cells signal each other through secreted extracellular pheromones as an essential part of mating. In this way, a *MAT***a** cell and a *MAT*α cell can identify each other; only mating between these two mating types can produce a zygote. Moreover, the actual differentiation of yeast cells into the two mating types has similarities to developmental processes found in multicellular organisms.

*Drosophila melanogaster.* The fruit fly (see Figure 1.4b, p. 6) has been a model organism for genetics since Morgan's work prior to and following 1910. Among the thousands of mutants isolated are many that affect development; Figure 19.1 shows a mutant with four eyes instead of the normal two. The study of *Drosophila* developmental mutants is providing a rich array of data about the molecular aspects of development. Later in this chapter, we will discuss how mutants have helped us understand sex determination and pattern formation in embryogenesis, and how genes program segments in the adult organism.

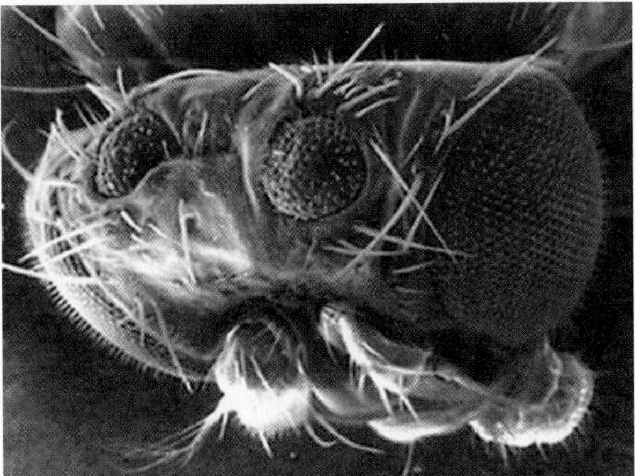

**Figure 19.1**
***Drosophila* developmental mutant with four eyes instead of the normal two.**

*Caenorhabditis elegans.* This nematode worm is transparent, permitting developmental processes to be followed easily under the microscope. As already mentioned, the fate map for every adult cell is known, so it is easy to see where mutations affect developmental processes. Figure 19.2 shows three stages of *C. elegans* development from the two-cell stage to the adult.

*Arabidopsis thaliana.* This small plant (see Figure 1.4d, p. 6) has become popular for genetic and molecular analysis. Its genome has been sequenced completely, and genetic analysis is relatively straightforward. As a model for the genetic analysis of the development of plants, *Arabidopsis* has been valuable in particular for a genetic dissection of floral development. Figure 19.3 shows a wild-type flower next to the developmental mutant *agamous* (*ag*), in which petals have formed instead of stamens, and sepals have formed instead of carpels.

*Danio rerio.* The zebrafish (Figure 19.4, left) is a model vertebrate for developmental genetics. The embryos are transparent, facilitating observation of developmental stages (Figure 19.4, right). Genetic crosses can be made, large numbers of fish can be bred in the laboratory, and genetic screens have been developed to search for genes that affect embryogenesis and other biological processes. The genome of the zebrafish currently is being sequenced.

*Mus musculus.* The mouse (see Figure 1.4e, p. 6) is a mammal, of course, which makes it a model organism particularly close to humans. The mouse has been used for many years as a subject for genetic analysis, including the genetic analysis of development. The genome sequence of the mouse is known, and making gene knockouts (see Chapter 9, pp. 225–227) is straightforward technically. While many developmental mutants are known, their study *in vivo* is difficult because embryogenesis takes place *in utero.*

**Figure 19.2**
**Three stages of *Caenorhabditis elegans* development from the two-cell stage to the adult.**

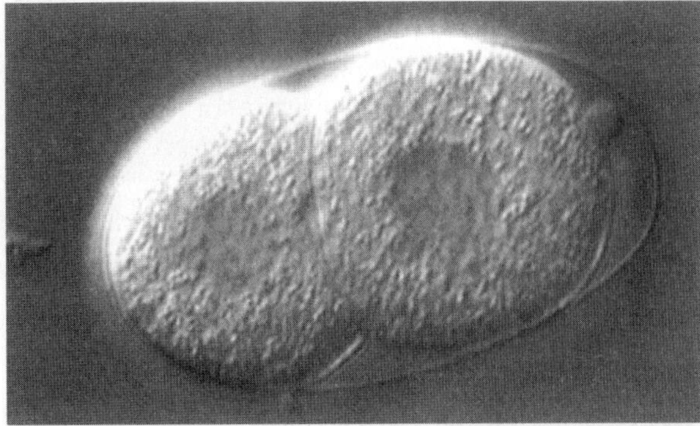

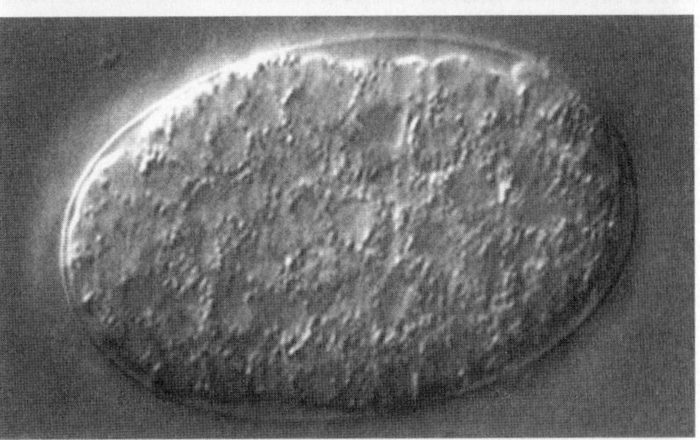

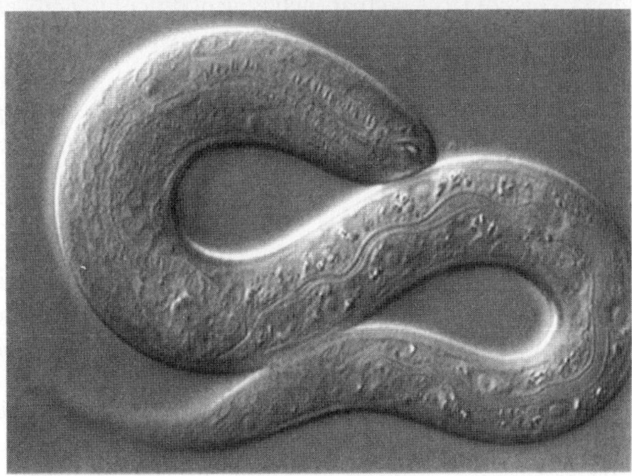

**Figure 19.3**
**Wild-type *Arabidopsis* flower (*left*) and flower of the developmental mutant, *agamous* (*ag*) (*right*).** Flowers with a mutation in *ag* have petals replacing stamens in one whorl, and sepals replacing carpels in another whorl.

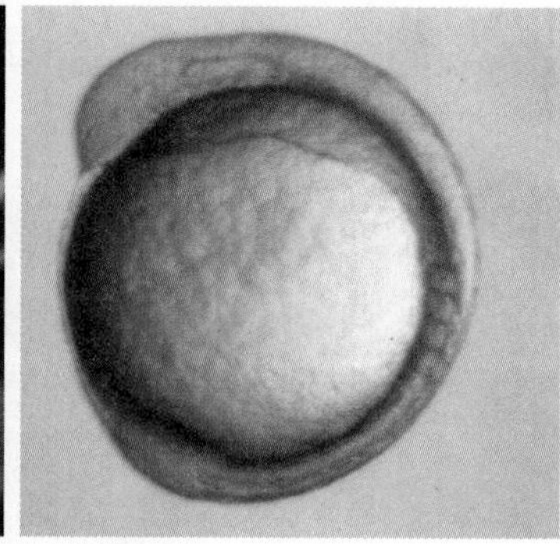

**Figure 19.4**
**Adult (*left*) and embryo (*right*) of the zebrafish, *Danio rerio*.**

## Development Results from Differential Gene Expression

In this section, we discuss selected experiments showing that, in most cases, development is the result of differential gene expression.

### Constancy of DNA in the Genome during Development

In early studies of the genetic control of development, an important question was whether development involves differential gene expression of a genome that is the same in all adult cells as it is in the zygote, or whether it involves a *loss* of DNA so that each type of differentiated cell retained only those genes required for that cell type, the remaining DNA having been discarded as part of the differentiation process. Experiments involving the **cloning** of plants and animals, that is, generating individuals genetically identical to the starting individual, indicate that the DNA remains constant during development.

**Regeneration of Carrot Plants from Mature Single Cells.** In the 1950s, Frederick Steward dissociated phloem tissue of a carrot into single cells and then attempted to culture new carrot plants from those cells by using plant tissue culture techniques. Mature plants with edible carrots were successfully produced (cloned) from the phloem cells. That the mature cells had the potential to act as zygotes and develop into complete plants indicated that mature cells had all the DNA found in zygotes. Steward's findings supported the hypothesis that the DNA content of a cell remains constant during development.

**Cloning Animals.** In 1975, John Gurdon and his colleagues showed that a nucleus from a skin cell of an adult frog injected into an enucleated egg could direct development to the tadpole stage. In those experiments, very few adults were produced, and all of them were sterile.

Gurdon's results left unresolved the question of whether a nucleus from adult differentiated tissue is genetically capable of directing development from the egg cell stage to fertile adulthood. That question was answered in the affirmative in 1997, when Ian Wilmut (now Sir Ian Wilmut, as of January 2008) and his colleagues reported the cloning of a mammal (a sheep), starting with an adult cell.

Wilmut's group tested the ability of nuclei from embryonic, fetal, and adult cells to direct the development of sheep. Their experimental approach was as follows (Figure 19.5):

1. Embryonic cells, fetal fibroblast (muscle-forming) cells, and mammary epithelial cells from donor ewes (poll Dorset, black Welsh, and Finn Dorset breeds, respectively) were grown in tissue culture and then induced to enter a quiescent state (the $G_0$ phase of the cell cycle) by reducing the concentration of the growth serum.
2. The cells were fused with enucleated oocytes (egg cells), and the fusion cells were allowed to grow and divide for 6 days to produce embryos.
3. The embryos were implanted into recipient ewes, and the establishment and progression of pregnancy were monitored.

The results were as follows: Four of 385 embryo-derived cells, two of 172 fetal fibroblast-derived cells, and one of 277 adult mammary epithelium-derived cells gave rise to live lambs. The most significant of these results is the last because it demonstrates that the adult nucleus contains all the genetic information required to specify a new organism. That lamb, designated 6LL3 and named Dolly, progressed normally to sexual maturity and became pregnant with offspring Bonnie, born in 1998. In 1999, Dolly delivered a set of triplets. Dolly was euthanized at the age of 6 after being diagnosed with a fatal, virus-induced lung disease that commonly affects sheep of her age. Wilmut's group believes that cloning was not a factor in Dolly becoming infected.

Evidence that Dolly was truly the result of the cell fusion experiment is of two kinds. First, the fusion cell contained the nucleus from a (whiteface) Finn Dorset ewe and was implanted into a Scottish blackface recipient ewe; Dolly is morphologically Finn Dorset. Second, and more definitively, analysis of polymorphic STR (short tandem repeat [microsatellite]) DNA markers (see Chapter 10, p. 272) at four loci showed that the DNA of Dolly matched that of the donor mammary epithelial cells perfectly, but did not match that of the recipient ewe.

Figure 19.5

**Representation of Wilmut's sheep cloning experiment, which showed the totipotency of the nucleus of a differentiated, adult cell.**

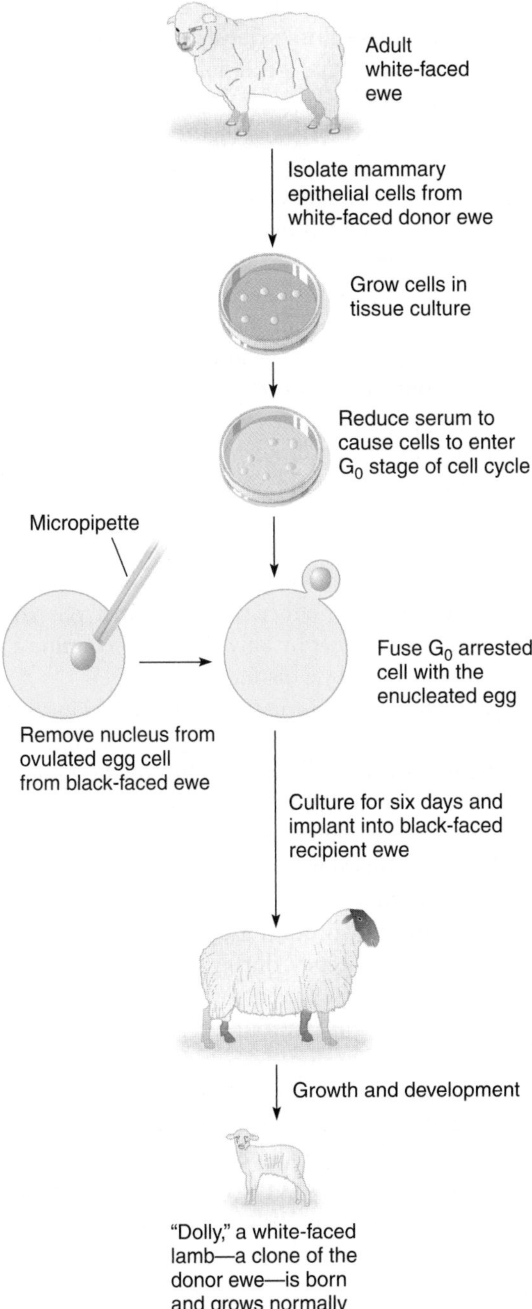
Adult
white-faced
ewe
Isolate mammary
epithelial cells from
white-faced donor ewe
Grow cells in
tissue culture
Reduce serum to
cause cells to enter
$G_0$ stage of cell cycle
Micropipette
Fuse $G_0$ arrested
cell with the
enucleated egg
Remove nucleus from
ovulated egg cell
from black-faced ewe
Culture for six days and
implant into black-faced
recipient ewe
Growth and development
"Dolly," a white-faced
lamb—a clone of the
donor ewe—is born
and grows normally

In sum, although the success rate for the experiment was not high (for technical reasons), the highly significant accomplishment here was the development of a live lamb directed by an adult nucleus. When the result of the experiment was published, concerns were raised internationally about the possible cloning of humans. Cloning technology is undoubtedly applicable to humans, and the ethical issues it raises will continue to be debated.

**Mammal Cloning Problems.** Since the cloning of Dolly, the cloning of a number of other mammals has been achieved, including cats, cattle, deer, dogs, ferrets, goats, gaurs (a type of ox), horses, cows, mice, mouflons (a species of wild sheep), mules, pigs, rabbits, rats, rhesus monkeys, sheep, water buffalo, and wolves. Biotechnology companies have a particular interest in cloning certain mammals because, once they have invested large sums of monies in making transgenic mammals (e.g., for producing pharmaceuticals, or modeling human genetic diseases), those animals can be mass produced.

The cloning of mammals has not been as straightforward as was hoped, however. Not only is the process itself very inefficient, but some problems have arisen with the clones produced. Consider, for instance, the cat Cc (carbon copy) that was cloned at Texas A&M University (Figure 19.6a). This calico female has a coat pattern that is not identical to that of Rainbow (Figure 19.6b), the parent that donated the nucleus from which Cc developed. Rainbow has the typical calico pattern of patches of black tabby and orange on white, while her clone Cc has black tabby patches on white with no orange patches. Recall from Chapter 12, p. 349, that a calico results because of the process of X chromosome inactivation in a female that is heterozygous for an X-linked gene for orange pigment production (*O*/*o*) and homozygous or heterozygous for an autosomal gene for black pigment production (*B*/–). (The molecular basis of X chromosome inactivation is explained later in this chapter.) When the dominant *O* allele is expressed, orange pigment is produced regardless of other color genes present in the cat. The coat color pattern differences can be explained by the fact that X chromosome inactivation is a random process in different cells, and that the movements of pigment-producing cells in the skin are mostly environmentally determined rather than genetically determined. However, Cc also differs from Rainbow in body shape and personality, both of which have genetic components. Thus, while Cc and Rainbow are genetically identical, they are not phenotypically identical. This fact argues that the genetic program is not alone in specifying the adult organism. Notably, environmental factors play an important role.

Figure 19.6

**Problems with cloning mammals.** The cloned cat, Cc (**a**), has a different calico pattern from her mother, Rainbow (**b**).

**a) Cc, the cloned cat**

**b) Rainbow, mother of Cc**

More serious problems than variations in coat color and personality have turned up in cloned mammals. As we have already mentioned, mammal cloning is extremely inefficient; usually, most clones die before or soon after birth. The few survivors seem to exhibit varying degrees of developmental abnormalities, suggesting problems at the gene expression level. Rudolph Jaenisch and his colleagues at the Whitehead Institute for Biomedical Research used DNA microarray analysis to study the expression of more than 10,000 genes in the livers and placentas of cloned mice. They found hundreds of genes in the set regulated abnormally; those genes represent about 4% of the protein-coding genes in the mouse's genome. Notably, the same genes showed abnormal expression whether taken from the cloned mice or from cultured cells containing the donor nuclei (prior to implanting in a surrogate mother). The interpretation is that the transfer of a donor nucleus into an enucleated cell is the cause of many gene expression changes. Practically speaking, it means that the clones that survive are unlikely to be normal. The problem is that the donor nucleus is taken from a differentiated cell, and it must become reprogrammed to start the determination/differentiation process anew. This is a major issue, and researchers have no tools to apply to this problem currently. We can expect the production of cloned mammals to continue to be inefficient, and the living clones potentially to show various problems resulting from abnormal gene expression. In view of this, any serious attempts to clone a human should be put out of mind at present.

**Activity**

Assume the role of a researcher at the Institute of Animal Development to investigate how gene expression patterns change as mouse stem cells differentiate into specific tissues in the iActivity *The Great Divide* on the student website.

## Examples of Differential Gene Activity During Development

The following classic examples illustrate differential gene activity during development.

**Hemoglobin Types and Human Development.** Human adult hemoglobin, Hb-A, is examined in this book in many contexts. Hb-A is a tetrameric protein made up of two α and two β polypeptides. Each type of polypeptide is coded by a separate gene, the α-globin and β-globin genes. The two genes appear to have arisen during evolution by duplication of a single ancestral gene, followed by alteration of the base sequences in each gene. Each gene contains two introns (intron 1 and intron 2; Figure 19.7).

Hb-A is only one type of hemoglobin found in humans. Genetic studies have shown that several distinct genes code for α- and β-like globin polypeptides, which form different types of hemoglobin at different times during human development (Figure 19.8). In the human embryo, the hemoglobin initially made in the yolk sac consists of two ζ (zeta) polypeptides and two ε (epsilon) polypeptides. From comparisons of the amino acid sequences, ζ is an α-like polypeptide, and ε is a β-like polypeptide. After about 3 months of development, synthesis of embryonic hemoglobin ceases (i.e., the ζ and ε genes are no longer transcribed), and the site of hemoglobin synthesis shifts to the fetal liver and spleen. Here, *fetal hemoglobin* (Hb-F) is made. Hb-F contains two α polypeptides and two β-like γ (gamma) polypeptides, either two γA or two γG. γA and γG differ from each other by only 1 out of 146 amino acids, and each is coded for by a distinct gene.

Fetal hemoglobin is made until just before birth, when synthesis of the two types of γ chains stops, and the site of hemoglobin synthesis switches to the bone marrow. In that tissue, α and β polypeptides are made, along with some β-like δ (delta) polypeptides. In the newborn through adult human, most of the hemoglobin is our familiar $\alpha_2\beta_2$ tetramer (Hb-A), with about one in 40 molecules having the constitution $\alpha_2\delta_2$ (Hb-A2). Thus, globin gene expression switches during human development, and this switching involves a sophisticated gene regulatory system that turns appropriate globin genes on and off over a long time period.

In the genome, the α-like genes (two α genes and one ζ gene) are all on chromosome 16, and the β-like genes (ε, γA, γG, δ, and β) are all on chromosome 11. Significantly, the α-like genes and the β-like genes are arranged in the chromosome in an order that exactly parallels the

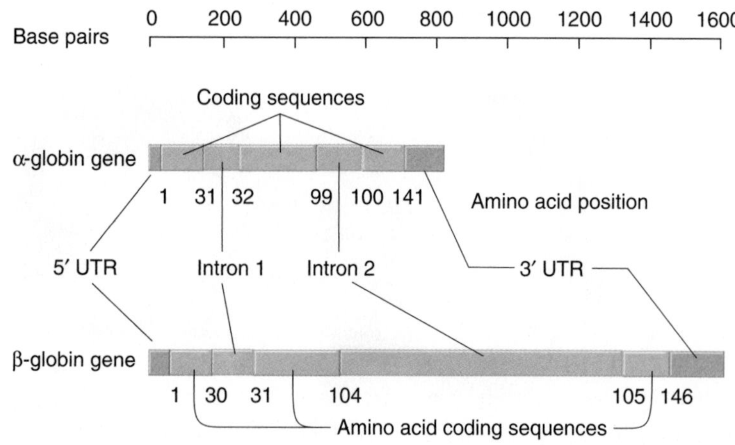


**Figure 19.7**
**Molecular organization of the human α-globin and β-globin genes.**

Figure 19.8

**Comparison of synthesis of different globin chains at given stages of embryonic, fetal, and postnatal development.**

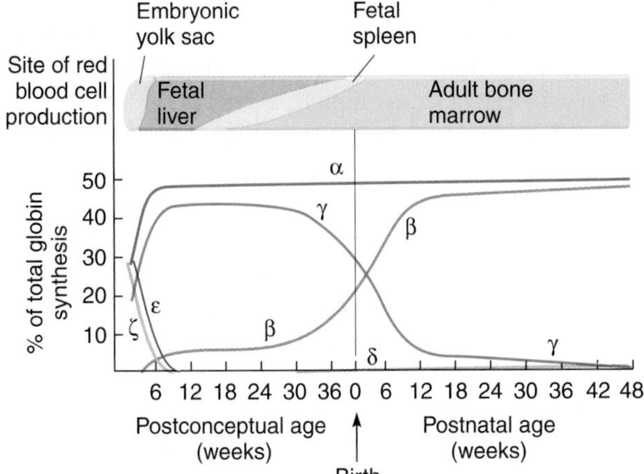


Figure 19.9

**Light micrograph of a polytene chromosome from *Chironomus* showing two puffs that result from localized uncoiling of the chromosome structure and indicate transcription of those regions.** DNA is shown in blue, RNA in red/violet.

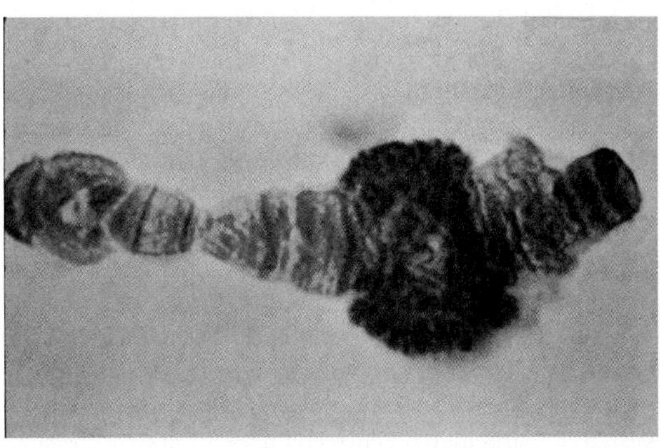

timing in which the genes are transcribed during human development. Recall that embryonic hemoglobin consists of ζ and ε polypeptides; these genes are the first functional genes at the left of the clusters. Next, the α and γ genes are transcribed to produce Hb-F, and these genes are the next functional genes that can be transcribed from the clusters. Finally, the δ and β polypeptides are produced, and these genes are last in line in the β-like globin gene cluster.

**Polytene Chromosome Puffs during Dipteran (Two-Winged Fly) Development.** Recall (Chapter 16, pp. 464–465, and Figure 16.1) that **polytene chromosomes** are a special type of chromosome that consists of a bundle of chromatids produced by repeated cycles of chromosome duplication without nuclear division, and that they are readily visible after staining under the light microscope. After staining, distinct and characteristic bands are visible along the chromosomes. Genes are located both in the bands and in the interband regions. Polytene chromosomes are found, for instance, in *Diptera* in the salivary glands in the larval stages or in the nuclei of other somatic cells.

At characteristic times during development, specific bands unwind locally to form *puffs* (Figure 19.9). The fact that puffs appear and disappear in specific patterns at certain chromosomal loci as development proceeds indicates they are developmentally controlled.

The puffing occurs as a result of very high levels of gene transcription. (Most genes are expressed at low levels and do not puff.) That is, when a polytene chromosome gene is being transcribed during development, the chromosome structure loosens to permit efficient transcription of that region of the DNA. When transcription is completed, the puff disappears and the chromosome resumes its compact configuration.

Puffing is under hormonal control in many cases, the key hormone being the steroid ecdysone. (The regulation of gene expression by steroid hormones was described in Chapter 18, pp. 523–526.) A model for the control of sequential gene activation by ecdysone is as follows. Ecdysone binds to a receptor protein, and this complex binds to both early (the early-puffing genes) and late genes (the expression of which is seen later in development). The complex turns on the early genes and represses the late genes. One or more early genes encode a protein, which accumulates during development. When the level of this protein reaches a certain threshold, it displaces the ecdysone–receptor complex from both early and late genes. This turns off the early genes and removes the repression of (i.e., turns on) the late genes. In support of the model, some of the early genes have been shown to encode DNA binding proteins, products expected of regulatory genes. Furthermore, an ecdysone receptor gene has been cloned and shown to encode a steroidlike receptor protein.

**Keynote**

Development results from differential gene activity of a genome that contains a constant amount of DNA from the zygote stage to the mature organism stage. Nonetheless, the genes are only part of the equation for development; environmental factors can affect the phenotype of an adult organism, as evidenced by phenotypic differences in cloned mammals from the parent that donated the nucleus for cloning.

## Exception to the Constancy of Genomic DNA during Development: DNA Loss in Antibody-Producing Cells

There are a few exceptions to the rule that no DNA is lost during development. One such example involves the loss of genetic information during the development of cells that produce antibodies.

**Antibody Molecules.** The cells responsible for immune specificity are *lymphocytes*, specifically *T cells* and *B cells*. We focus our discussion on B cells. B lymphocytes develop in the adult bone marrow. When activated by an **antigen,** B cells develop into plasma cells that make proteins called **antibodies.** Antibody molecules are inserted in the plasma membrane of the plasma cells, and they are also released into the blood and lymph, where they are responsible for the humoral (*humor*, meaning "fluid") immune responses. The antibodies bind specifically to the antigens that stimulated their production.

The establishment of immunity against a particular antigen results from **clonal selection.** This is a process whereby cells that have antibodies displayed on their surfaces that are specific for the antigen are stimulated to proliferate and secrete that antibody. During development, each lymphocyte becomes committed to react with a particular antigen, even though the cell has *never been exposed* to the antigen. For the humoral response system, there is a population of B cells, *each of which can recognize a single antigen*. A particular B cell recognizes an antigen because the B cell has made antibody molecules, which are attached to the outer membrane of the cell and act as receptor molecules. When an antigen encounters a B cell that has the appropriate antibody receptor capable of binding to the antigen, that B cell is stimulated selectively to proliferate. This produces a clonal population of plasma cells, each of which makes and secretes the identical antibody. It is important to note that *any given cell makes only one specific kind of antibody toward one specific antigen*. However, the actual immune response may involve the binding of many different antibodies to an array of antigens on invaders such as an infecting virus. This binding mediates a variety of other mechanisms that inactivate the invading antigen.

All antibody molecules made by a given plasma cell are identical—they have the same protein chains and bind the same antigen. There are millions of B cells in the whole organism, and millions of different antibody types can be produced, each with a different amino acid sequence and a different antigen-binding specificity. As a group, antibodies are proteins called **immunoglobulins (Igs).** A stylized antibody (immunoglobulin) molecule of the type IgG is shown in Figure 19.10a, and a model of an antibody molecule based on X-ray crystallography is shown in Figure 19.10b. Both figures show the molecule's two short polypeptide chains, called *light (L) chains*, and two long polypeptide chains, called *heavy (H) chains*. (All antibody molecules also have carbohydrates attached to the regions of H chains not involved in binding with L chains.) The two H chains are held together by disulfide (–S–S–) bonds, and an L chain is bonded to each H chain by disulfide bonds. Other disulfide bonds within each L and H chain cause the chains to fold up into their characteristic shapes.

The overall structure resembles a Y, with the two arms containing the two antigen-binding sites. The two L chains in each Ig molecule are identical, as are the two H chains, so the two antigen-binding sites are identical. The hinge region (see Figure 19.10a) allows the two arms to move in space, making it easier for the antibody to bind an antigen. Also, one arm can bind an antigen on, say, one virus, while the other arm binds the same antigen on a different virus of the same type. Such cross-linking of antibody molecules helps inactivate infecting agents.

Five major classes of antibodies are found in mammals: IgA, IgD, IgE, IgG, and IgM. Each class has a different type of H chain: $\alpha$ (alpha), $\delta$ (delta), $\varepsilon$ (epsilon), $\gamma$ (gamma), and $\mu$ (mu), respectively. Two types of L chains are found: $\kappa$ (kappa) and $\lambda$ (lambda). Both L chain types are found in all Ig classes, but a given antibody molecule has either two identical $\kappa$ chains or two identical $\lambda$ chains. A complete discussion of the functions of the five Ig classes is beyond the scope of this text. For our purposes, we need to be aware that the most abundant class of immunoglobulin in the blood is IgG, and that IgM plays an important role in the early stages of an antibody response to a previously unrecognized antigen. We will focus on these two antibody classes from now on.

Each polypeptide chain in an antibody is organized into domains of about 110 amino acids (see Figure 19.10a). Each L chain ($\kappa$ or $\lambda$) has two domains, and the H chain of IgG (the $\gamma$ chain) has four domains, whereas the IgM's H chain (the $\mu$ chain) has five domains. The N-terminal domains of the H and L chains have highly variable amino acid sequences that constitute the antigen-binding sites. These domains, representing in IgG the N-terminal half of the L chain and the N-terminal quarter of the H chain, are called the *variable*, or V, regions. The V regions are symbolized generically as $V_L$ (for the light chain) and $V_H$ (for the heavy chain). The $V_L$ and $V_H$ regions comprise the antigen-binding sites (see Figure 19.10a). The amino acid sequence of the rest of the L chain is constant for antibodies with the same L chain type (i.e., $\kappa$ or $\lambda$) and is called $C_L$. Similarly, the amino acid sequence of the rest of the H chain is constant and is called $C_H$. For IgG, there are three approximately equal domains of $C_H$ called $C_H1$, $C_H2$, and $C_H3$ (see Figure 19.10a). For IgM, there are four approximately equal domains of $C_H$ called $C_H1$, $C_H2$, $C_H3$, and $C_H4$. Thus, the production of antibody molecules involves synthesizing polypeptide chains, one part of which varies in molecules from different cells and the other part of which is constant. How this occurs is discussed in the following section.

**Assembly of Antibody Genes from Gene Segments during B Cell Development.** A mammal may produce $10^6$ to $10^8$ different antibodies. Since each antibody molecule consists of one kind of L chain and one kind of H chain, these antibodies theoretically would require $10^3$ to $10^4$ different L chains and $10^3$ to $10^4$ different H chains, if L and H chains paired randomly. However, there are not nearly enough genes in the human genome to specify that many different molecules. Instead, variability in L and H chains results from particular DNA rearrangements that occur during B cell development. These rearrangements involve the joining of different gene segments to form a gene that is transcribed to produce an Ig chain; the process is called

**Figure 19.10**
**IgG antibody molecule.**

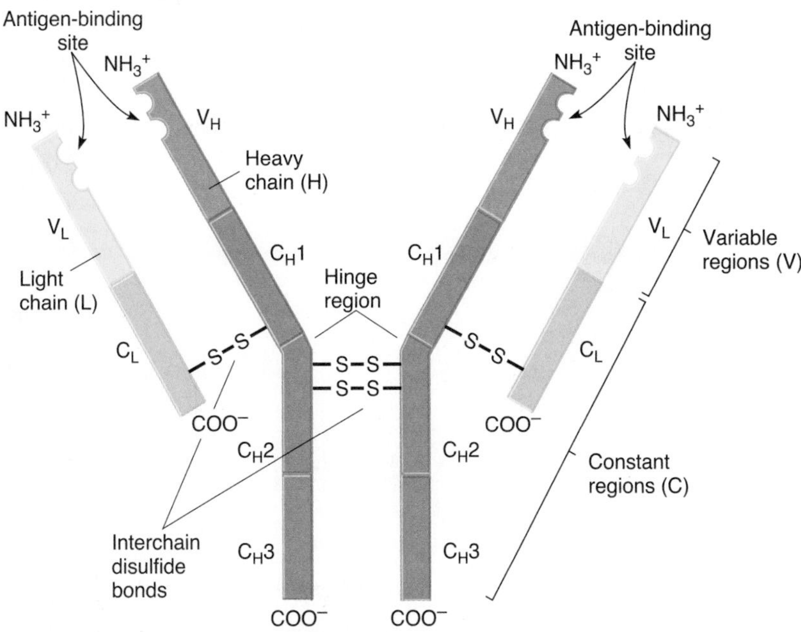


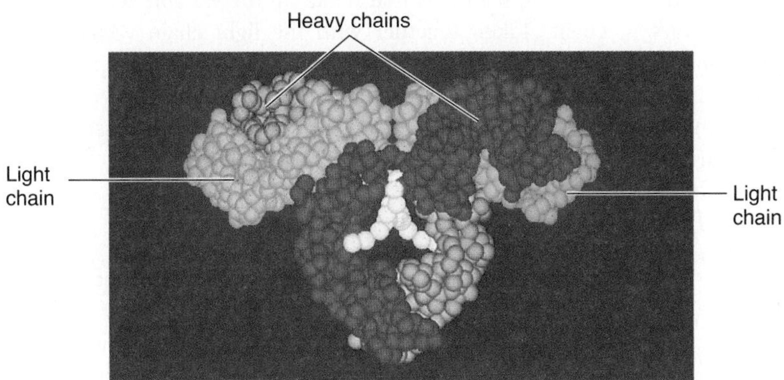


*somatic recombination*. The process is now illustrated for mouse immunoglobulin chains.

***Light Chain Gene Somatic Recombination.*** In mouse germ-line DNA, there are three types of gene segments, and one of each type is needed to make a complete, functional κ light chain gene (Figure 19.11):

1. Each of the 350 or so L–$V_\kappa$ segments consists of a leader sequence (L) and a $V_\kappa$ segment, which varies from segment to segment. Each $V_\kappa$ segment encodes most of the amino acids of the light chain variable domain. The leader sequence also encodes a special sequence called a signal sequence (see Chapter 6) that is required for secretion of the Ig molecule; this signal sequence is subsequently removed and is not part of the functional antibody molecule.
2. A $C_\kappa$ segment specifies the constant domain of the κ light chain.
3. Four $J_\kappa$ segments (joining segments) are used to join $V_\kappa$ and $C_\kappa$ segments to produce a functional κ light chain gene.

In the pre-B cell, the L–$V_\kappa$, $J_\kappa$, and $C_\kappa$ segments, in that order, are widely separated on the chromosome. As the B cell develops, a particular L–$V_\kappa$ segment becomes associated with one of the $J_\kappa$ segments and with the $C_\kappa$ segment. In the example in Figure 19.11, L–$V_{\kappa 2}$ has recombined next to $J_{\kappa 3}$. Transcription of this new DNA arrangement produces the primary RNA transcript, which includes a poly(A) tail. Removal of the intron from the primary RNA transcript produces the mature mRNA, which has the organization L–$V_{\kappa 2}J_{\kappa 3}C_\kappa$; translation and leader removal produces the κ light chain that the B cell is committed to make.

In the mouse, there are about 350 L–$V_\kappa$ gene segments, four functional $J_\kappa$ segments, and one $C_\kappa$ gene segment. Thus, the number of possible κ chain variable regions that can be produced by this mechanism is

**Figure 19.11**

**Production of the kappa (κ) light chain gene in mouse by recombination of V, J, and C gene segments during development.** The rearrangement shown is only one of many possible recombinations.

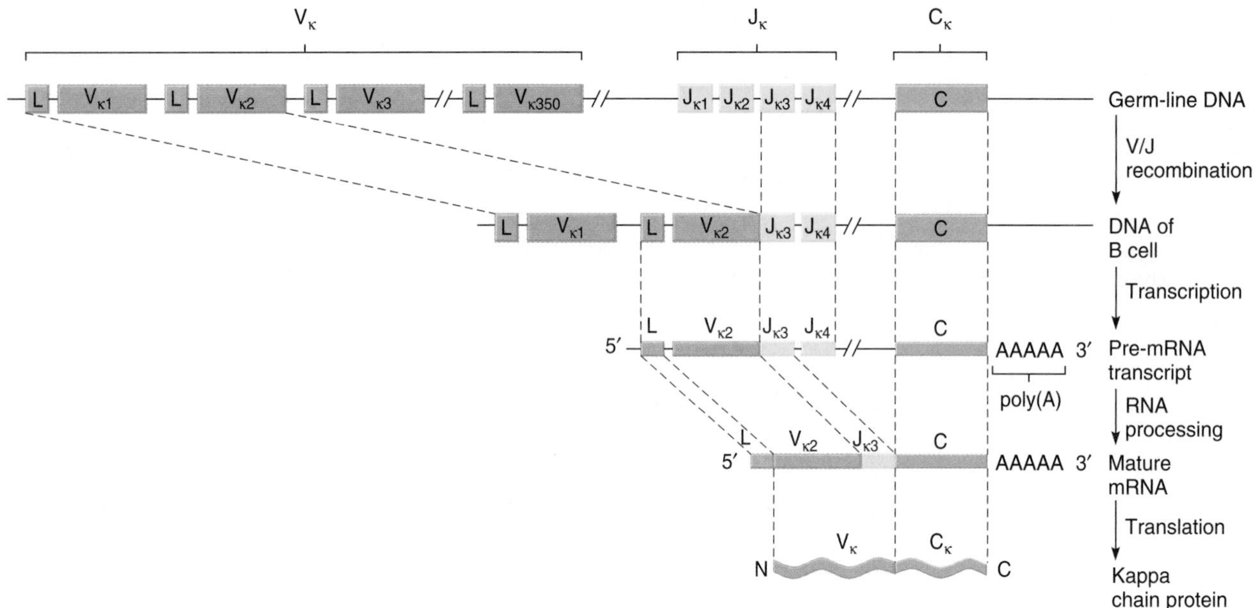




350 × 4 = 1,400. Further diversity results from imprecise joining of the $V_\kappa$ and $J_\kappa$ gene segments. That is, during the joining process, a few nucleotide pairs from $V_\kappa$ and a few nucleotide pairs from $J_\kappa$ are lost from the DNA at the $V_\kappa J_\kappa$ joint, generating significant diversity in sequence at that point. Thus, diversity of κ light chains results from: (1) variability in the sequences of the multiple $V_\kappa$ gene segment; (2) variability in the sequences of the four $J_\kappa$ gene segments; and (3) variability in the number of nucleotide pairs deleted at $V_\kappa J_\kappa$ joints.

A similar mechanism exists for mouse λ light chain gene assembly. In this case, there are only two L–$V_\lambda$ gene segments and four $C_\lambda$ gene segments, each with its own $J_\lambda$ gene segment. Thus, fewer λ variable regions can be produced than is the case for κ chains.

***Heavy Chain Gene Somatic Recombination.*** The immunoglobulin heavy chain gene is also encoded by $V_H$, $J_H$, and $C_H$ segments. In this case, additional diversity is provided by another gene segment, D (diversity), which is located between the $V_H$ segments and the $J_H$ segments (Figure 19.12). For an IgG heavy chain, in the germ line of mouse DNA there is a tandem array of about 500 L–$V_H$ segments, then a spacer, then 12 D segments, then a spacer, and then 4 $J_H$ segments. After another spacer, the constant region gene segments are arranged in a cluster that, in mouse, has the order μ, δ, γ (four different sequences for four different, but similar, IgG H chain constant domains), ε, and α for the H chain constant domains of IgM, IgD, IgG, IgE, and IgA, respectively. Thus, for the assembly of a heavy chain, there are 500 × 12 × 4 = 24,000 possibilities for each heavy chain type. As in L chain gene rearrangements, further antibody diversity results from imprecise joining of the gene segments that make up the variable region of the chain. Taken together with the light chain variation, an enormous variety of antibody molecules can be produced. Just for antibodies with one of the heavy chain types and a κ light chain, 24,000 × 1,400 = 33,600,000 possible antibody molecules can be produced.

**Keynote**

Antibodies are specialized proteins called immunoglobulins, which bind specifically to antigens. Immunity against a particular antigen results from clonal selection, in which cells already making the required antibody are stimulated to proliferate by the specific antigen. Antibody molecules consist of two light (L) chains and two heavy (H) chains. The amino acid sequence of one domain of each type of chain is variable; this variation is responsible for the different antigen-binding sites on different antibody molecules. The other domains of each chain are constant in amino acid sequence. In germ-line DNA, the coding regions for immunoglobulin chains are scattered in tandem arrays of gene segments. Thus, for light chains, there are many variable region (V) gene segments, a few joining (J) gene segments, and one constant region (C) gene segment. During development, somatic recombination occurs to bring particular gene segments together into a functional L chain gene. A large number of different L chain genes result from the many possible ways in which the gene segments can recombine. Similar rearrangements occur for H chain genes but with the addition of several D (diversity) segments that are between V and J, which increase the possible diversity of H chain genes.

**Figure 19.12**

**Production of heavy chain genes in mouse by recombination of V, D, J, and C gene segments during development.** Depending on the $C_H$ segment used, the resulting antibody molecule is IgM, IgD, IgE, or IgA. Shown here is the assembly of an IgG heavy chain. This rearrangement is only one of the many thousands possible.

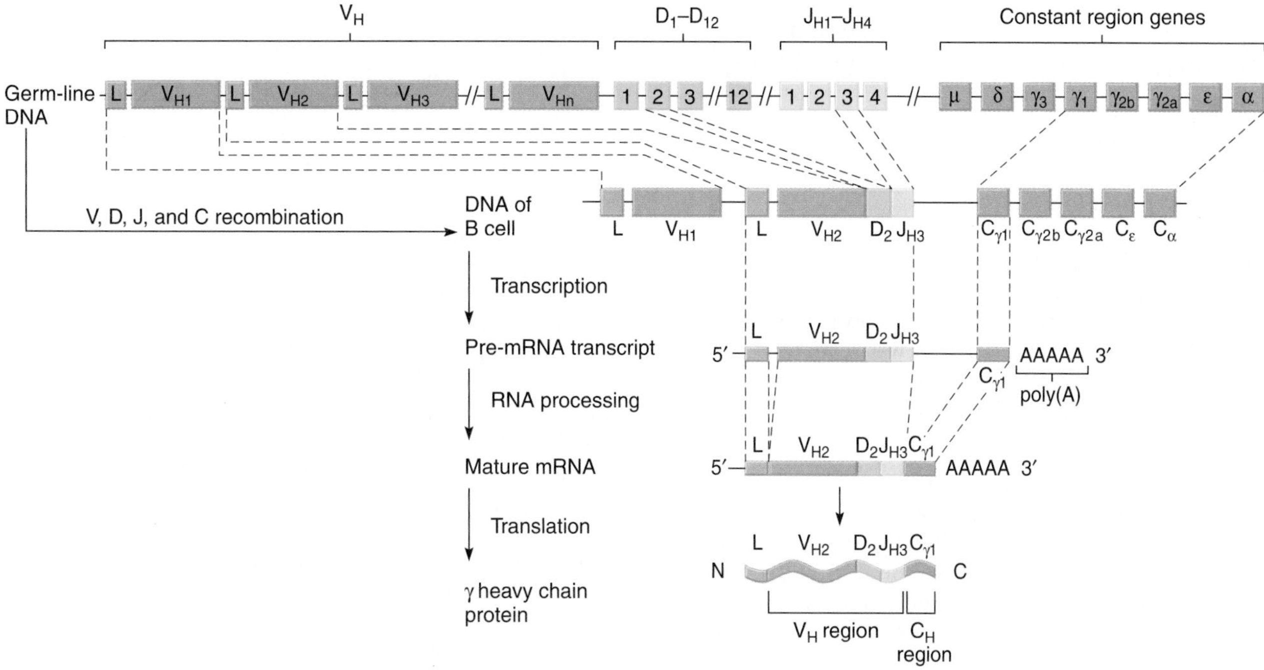


## Case Study: Sex Determination and Dosage Compensation in Mammals and *Drosophila*

For the remainder of this chapter, we discuss specific examples of the genetic regulation of developmental processes. In this section, we discuss sex determination and dosage compensation in mammals and *Drosophila*. These two topics are related because the sex chromosomes play a role in sex determination in both organisms, yet the different copy number of X chromosomes in the two sexes necessitates regulation of X-linked genes to equalize their expression in males and females.

### Sex Determination in Mammals

In humans and other placental mammals, sex is determined by the Y chromosome mechanism of sex determination. That is, the presence of the Y chromosome specifies maleness in that gonads develop as testes, while in the absence of a Y chromosome, the gonads develop as ovaries. The Focus on Genomics box in this chapter explores sex determination in the duck-billed platypus, an egg-laying mammal in which males have five X and five Y chromosomes and females have five pairs of X chromosomes.

The Y chromosome mechanism of sex determination means that the Y uniquely carries an important gene (or perhaps genes) that sets the switch toward male sexual differentiation. The product of this gene is called **testis-determining factor,** and the corresponding hypothesized gene is the *testis-determining factor* gene, or *TDF*. Testis-determining factor causes the tissue that will become gonads to differentiate into testes instead of ovaries. This is the central event in sex determination of many mammals; all other differences between the sexes are secondary effects resulting from hormone action or from the action of factors produced by the gonads. Therefore, sex determination is equivalent to testis determination.

The testis-determining factor gene was found by studying rare so-called *sex reversal* individuals; that is, males who are XX (instead of XY) and females who are XY (instead of XX). In the XX males, a small fragment from near the tip of the small arm of the Y chromosome had broken off during the production of gametes and become attached to one of the X chromosomes. The XY females had deletions of the same region of the Y chromosome. These findings suggested that the testis-determining factor gene is in that small segment of the Y chromosome. The gene is the *SRY* (sex-determining region of the Y) gene. An equivalent gene, *Sry*, has been cloned from mice.

Several lines of evidence indicate that the *SRY* gene is the testis-determining gene. First, the mouse *Sry* gene is expressed only at the time and place expected for the testis-determining factor, that is, in the undifferentiated genital ridges of the embryo just before the formation of testes. Second, when a 14-kb DNA fragment with the mouse *Sry* gene is introduced into XX mouse embryos by microinjection, the transgenic mice produced are males

## Focus on Genomics

### The Platypus: An Odd Mammal with a Very Odd Genome

Australia's duck-billed platypus (*Ornithorhynchus anatinus*) is one of the oddest mammals. Not only is it one of the monotremes, or egg-laying mammals, but it is also poisonous (the males make a poison that coats spurs on their hind legs). Even sex determination in the platypus is odd, and involves multiple X and Y chromosomes. Recently, the genome of the platypus was sequenced, and several of these oddities are now somewhat easier to understand.

**Eggs**

Several genes related to egg laying and egg development are present in the platypus genome. Some of the genes are common to all mammals, while others are found only in egg-laying animals. For example, the platypus can produce four proteins that are very similar to the four proteins that make the human zona pellucida, a critical structure that forms shortly after fertilization in most amniotes. The platypus also has two *ZPAX* genes, which encode egg envelope proteins. These *ZPAX* genes have not been found in other mammals, but are present in birds, amphibians, and fish. The platypus also has a vitellogenin gene. In egg-laying animals, these genes encode egg yolk precursors.

**Venom**

Platypus venom is a cocktail of different peptides. Several genes in the platypus genome have duplicated and diverged so that the "new" copy of the gene encodes a likely venom peptide. This includes the genes encoding β-defensin, C-type natriuretic factor, and nerve growth factor. Intriguingly, similar duplication and divergence events have occurred in the same genes in poisonous reptiles. The DNA sequences suggest that the duplications in the platypus lineage were distinct from the reptile duplications, so this is an example of convergent evolution. Similar mutations have occurred in each group, but these mutations were not present in the common ancestor of the platypus and reptiles.

**Sex Chromosomes**

The female platypus has five different pairs of X chromosomes, and the male has 5 X chromosomes (one from each pair) and 5 Y chromosomes. As you might predict, this makes meiosis very complex in males. The sex chromosomes form a linked chain during meiosis, and, as a result, sperm either inherit all 5 Y chromosomes or all 5 X chromosomes. When the investigators looked at the genomic sequences of the X and Y chromosomes, they made two very surprising observations. First, the X chromosome(s) from the platypus are not at all similar to the X chromosome from other mammals. Most of the genes found on the mouse X chromosome are on a platypus autosome. The platypus X chromosomes are far more similar to the Z chromosome of birds. (Birds, and some other organisms, have Z and W sex chromosomes rather than X and Y chromosomes [see Chapter 12, p. 351]. Males are ZZ and females are ZW in sex chromosome constitution.) Furthermore, *SRY*, the Y chromosome gene used in most mammals to trigger male development, is absent from the platypus genome. The platypus probably uses a gene called *DMRT1* instead. This gene is thought to function in sex determination in birds. It appears that the *SRY* gene, as well as the more familiar mammalian X and Y chromosomes, evolved after monotremes split from the other mammals (marsupials and placentals).

with normal testis differentiation and subsequent normal male secondary sexual development. In other words, *Sry* alone is sufficient to cause a full phenotypic sex reversal in an XX chromosomally female mouse. Third, there are rare XY human females who, instead of having lost a section of the Y chromosome as described earlier, have a simple mutation in the *SRY* gene.

*SRY* encodes a transcription factor that specifies development of the gonad into a testis. The testes produce the masculinizing steroid hormone testosterone. If the *SRY* gene is absent, the gonad develops into an ovary by default. Ovaries produce the feminizing steroid hormone estrogen.

## Dosage Compensation Mechanism for X-Linked Genes in Mammals

Organisms with sex chromosomes have an inequality in gene dosage (number of gene copies) between the sexes: there are two copies of X-linked genes in females and one copy in males. In many such organisms, if gene expression on the X chromosome is not equalized, lethality results early in development. Fortunately, there is a **dosage compensation** mechanism for dealing with this problem. In female mammals, this involves inactivating one of the two X chromosomes in somatic cells at an early stage in development (see Chapter 12, pp. 348–350). The X chromosome inactivated is randomly chosen from the maternally derived and paternally derived X chromosomes in a process that is independent from cell to cell. Once a maternal or paternal X chromosome is inactivated in a cell, all descendants of that cell inherit the inactivation pattern. This is another example of an epigenetic phenomenon.

Three steps are involved in X inactivation: chromosome counting (determining the number of X chromosomes in the cell), selection of an X for inactivation, and X inactivation itself. A key region on each X chromosome called the *X inactivation center* (*XIC* in humans, *Xic* in mice) is involved in the chromosome counting mechanism. Two or more *XIC*s must be present for X inactivation to occur. Some evidence for this has come from an experiment in which a 450-kb (450,000 base pairs) piece of the mouse X chromosome containing *Xic* was introduced into autosomes of male mouse cells in tissue culture. In normal male cells with one X chromosome, X inactivation does not occur. However, in the transgenic mouse cells with a *Xic* added to an autosome, chromosome inactivation was turned on, with either the X or the autosome becoming inactivated in a random fashion. This means that the genetically modified male cells showed properties typical of X inactivation in normal female cells. The 450 kb of DNA with *Xic* must contain the sequences for chromosome counting and for the initiation of X inactivation.

Female somatic cells have a choice mechanism that determines which X chromosome is inactivated and which X chromosome remains active. The choice is made at the *X-controlling element* (*Xce*), which is in the *XIC/Xic* region. A gene called *XIST*/(humans)/*Xist* (mice), for *X inactivation-specific transcripts*, is also located in the *XIC/Xic* region. *XIST* is expressed from the *inactive* X (Xi) rather than from the active X (Xa), which is the opposite of the expression pattern of other X-linked genes. The *Xist* gene has been shown to be essential for X inactivation in cultured cells and in mice. *XIST/Xist* is transcribed into an unusually large (17 kb) noncoding RNA that is not translated. Similar to the noncoding miRNAs and siRNAs introduced in Chapter 18 (pp. 537–540), this RNA has a negative effect on gene expression, in this case bringing about heterochromatinization of the Xi chromosome and, therefore, silencing the genes on that chromosome at the transcriptional level. Exactly how *XIST* RNA brings about X chromosome inactivation is not fully understood. It is known that *XIST* RNA coats the X from which it is transcribed, spreading out in both directions from the *XIC/Xic*. *XIST* RNA recruits a number of enzymes responsible for chromatin remodeling, notably enzymes that deacetylate histones, and that methylate or demethylate specific lysine residues on histones H3 and H4. The chromatin remodeling along the length of the Xi silences the genes on that chromosome—the Xi becoming heterochromatic, namely, a Barr body (see Chapter 12, pp. 348–349, and Figure 12.24). Maintenance of gene silencing on Xi does not depend on *XIST* RNA, but involves epigenetic modifications.

**Keynote**

In mammals, the sex of the individual is determined by the presence or absence of the Y chromosome. If the Y chromosome is present, the *SRY* gene on that chromosome is transcribed to produce a transcription factor that regulates genes required for directing the gonad to form a testis. In the absence of the *SRY* gene, as in an XX female, the gonad develops into an ovary by default.

Genes on the X chromosome are unusual in that their dosage is different between males and females. In mammals, a dosage compensation mechanism operates to equate expression of X-linked genes in males and females. This mechanism involves the inactivation of one of the two X chromosomes in a female early in development, leaving only one X chromosome transcriptionally active, as is the case in males. The inactivation process is complex, involving the transcription of the *XIST* gene on the chromosome to be inactivated; the noncoding, regulatory RNA made then coats that chromosome in both directions from *XIC*, triggering chromatin remodeling that silences the genes at the transcriptional level.

## Sex Determination in *Drosophila*

The number of X chromosomes : sets of autosomes (X:A) ratio determines sex in *Drosophila* (see Chapter 12, p. 350). Our understanding of sex determination in this organism has come from studies of a number of mutations that disrupt normal sex determination. These studies have led to a *regulation cascade model* for sex determination in *Drosophila*, summarized in Figure 19.13. First, the X:A ratio is read during development. For wild-type *Drosophila*, the ratio that sets the initial switch for development into females (XX) is 2X : 2 sets of autosomes = 1.0, and the ratio that sets the initial switch for development into males (XY) is 1X : 2 sets of autosomes = 0.5. This information is transmitted to the sex determination genes, which make the choice between the alternative female and male developmental pathways, starting with the master regulatory gene *Sex-lethal* (*Sxl*). Loss-of-function mutants of *Sxl* are lethal for female embryo development (meaning that *Sxl* needs to be active in females), but they have no effect on male embryo development (meaning that *Sxl* expression is not necessary for male development). However, gain-of-function mutants are lethal for male embryo development, which means that *Sxl* needs to be inactive in males. Alternative splicing of the *Sxl* pre-mRNA in embryos destined to become females or males sets in motion the two different pathways. Steps in each pathway are regulated by alternative splicing of pre-mRNAs, as we will see.

**Sex Determination and Dosage Compensation in *Drosophila***

How is the X:A ratio detected? On the X chromosome are the *sisterless* numerator genes *sis-a*, *sis-b*, and *sis-c*, and

**Figure 19.15**

**Expression of *Sex-lethal* (*Sxl*) during embryogenesis.** (a) In early embryogenesis in females, the numerator-numerator dimer transcription factors activate transcription of *Sxl* from $P_E$. Splicing of the pre-mRNA skips exons 2 and 3 (the angled lines above the pre-mRNA indicate the segments that are removed during splicing); the resulting mRNA is translated to produce SXL early protein. (b) Later in embryogenesis, the *Sxl* gene is transcribed constitutively from $P_L$ in both female and male embryos. The pre-mRNA is spliced in a regulated fashion in female embryos owing to the presence of SXL early protein, and in a default fashion in male embryos owing to the absence of SXL early protein. As a result, SXL late protein is produced in female embryos, but not in male embryos.

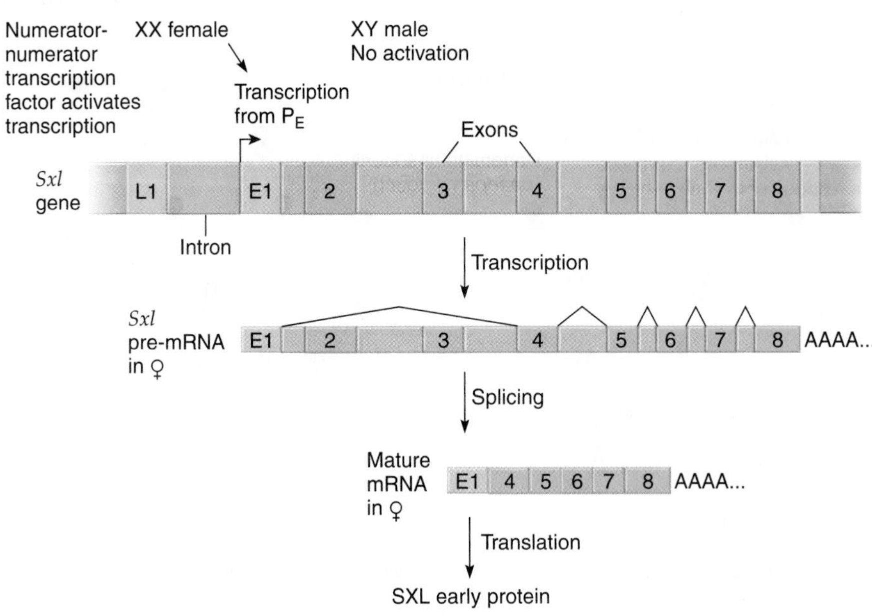


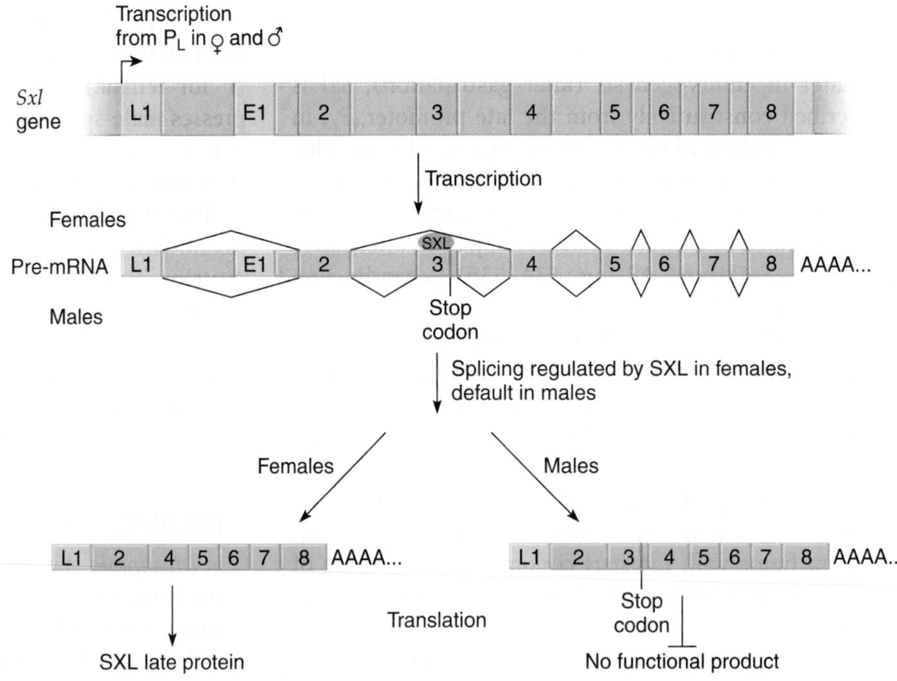


## Dosage Compensation in *Drosophila*

In mammals, dosage compensation occurs by decreasing transcriptional activity of X chromosome genes in females to match that in males. In *Drosophila*, the opposite occurs: Transcriptional activity is increased twofold in males to match that in females, who have twice the number of X chromosomes. In both cases, chromatin remodeling is involved in regulating transcriptional activity.

An understanding of dosage compensation in *Drosophila* has been illuminated by studies of mutants of genes that are essential for male viability. Key male-specific lethal genes are *mle (maleness)*, *msl-1 (male-specific lethal-1)*, *msl-2*, *msl-3*, and *mof (males absent on the first)*. Males with mutations in these genes die at the late larval stage, while females with the same mutations develop normally. The products of these genes are collectively called the male-specific lethal (MSL) proteins.

The SXL late protein (see previous section) plays a key role in dosage compensation. In females, the SXL

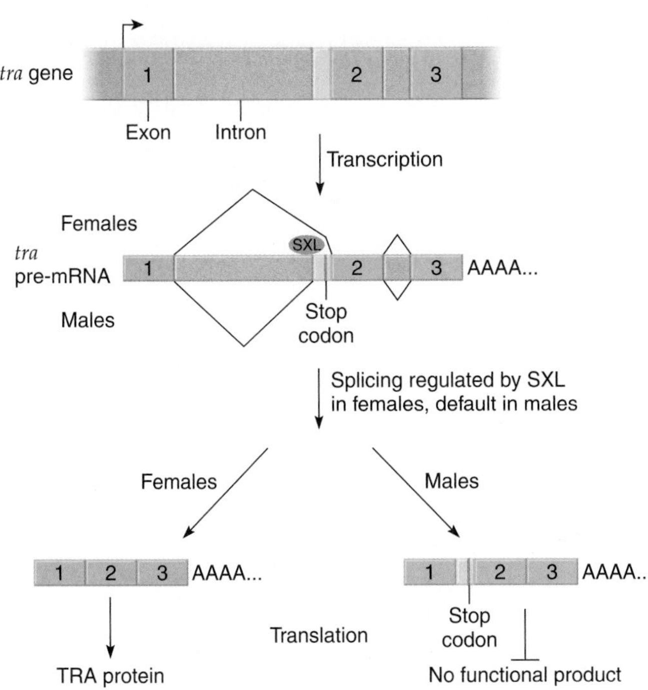


**Figure 19.16**

**Expression of *transformer* (*tra*) during embryogenesis.** SXL late protein in female embryos regulates splicing of *tra* pre-mRNA; the resulting mRNA is translated to generate the TRA protein that regulates splicing of the *doublesex* transcript (see Figure 19.17). In male embryos, *tra* pre-mRNA is spliced in a default fashion because SXL late protein is absent. The resulting mRNA has a stop codon prior to exon 2 and so no TRA protein is made.

late protein binds to the transcript of *msl-2*, blocking its translation; no MSL2 protein is produced. In males, the *msl-2* transcript can be translated because SXL late protein is absent. MSL2 forms a complex with the other MSL proteins, MLE, MSL1, MSL3, and MOF. This MSL complex binds to about 35 chromatin entry sites (CES) on the *Drosophila* male X chromosome and then MSL complexes spread from those sites in both directions into the flanking chromatin. The MOF protein of the MSL complex is a histone acetyltransferase (HAT), and its chromatin remodeling activity (see Chapter 18, pp. 529–531) spreads along the X chromosome and is responsible for the twofold higher level of transcription of X chromosome genes in males than in females.

In females, the MSL proteins other than MSL2 are produced. However, because MSL2 is essential for the binding of the MSL complex to the X chromosome, no chromatin remodeling can occur in XX females.

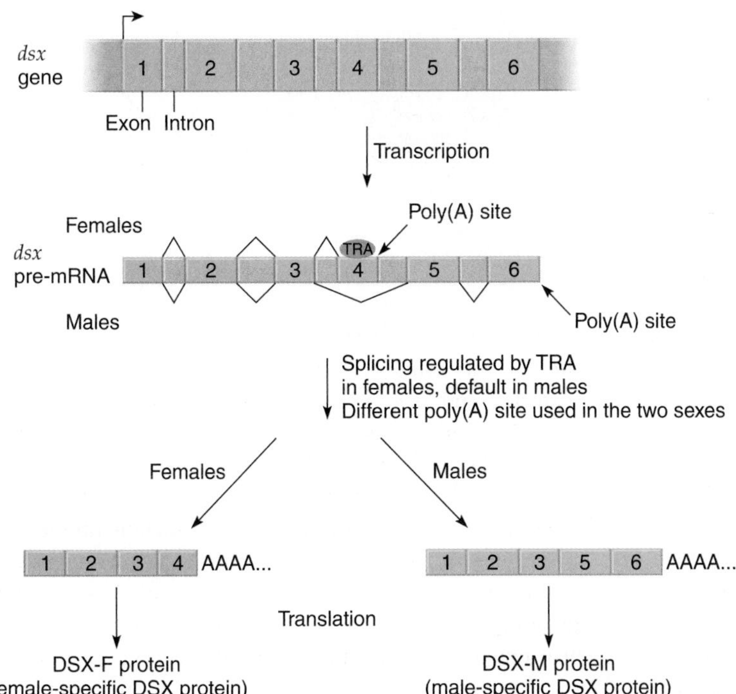


**Figure 19.17**

**Expression of *doublesex* (*dsx*) during embryogenesis.** TRA protein in female embryos regulates splicing, so exon 4 is included and cleavage and polyadenylation occurs at the poly(A) site following exon 4. In male embryos, default splicing occurs in the absence of TRA protein, leading to the exclusion of exon 4, but to the inclusion of exons 5 and 6 because of cleavage and polyadenylation at the poly(A) site following exon 6. Translation of the two different mRNAs produces the female-specific DSX protein, DSX-F, in females, and the male-specific DSX protein, DSX-M, in males.

**Keynote**

As with mammals, there is a different dosage of genes on the X chromosomes in female and male *Drosophila*. Dosage compensation occurs in fruit flies also, but in this case the transcriptional level of genes on the male's X chromosome is increased twofold to match the gene expression of X-linked genes in the female. The mechanism for dosage compensation relates to the molecular steps for sex determination. That is, the absence of the *Sxl*-encoded protein in males enables a key protein to be translated from its mRNA. That protein associates with other proteins to form a complex that binds to many sites on the X chromosome. The complex triggers chromatin remodeling events spreading in each direction from the binding sites until the whole chromosome is affected. The chromatin remodeling is responsible for the twofold increase in transcriptional activity.

## Case Study: Genetic Regulation of the Development of the *Drosophila* Body Plan

Significant progress has been made in understanding the genetic regulation of development in *Drosophila*. Many developmental mutants have been isolated after extensive genetic screens, so that almost all areas of *Drosophila* development can be studied in detail at genetic and molecular levels. The discoveries made from such studies have become even more important as discoveries in other systems (e.g., nematode, mouse) indicate that many genes discovered in *Drosophila* have counterparts in all higher organisms, including humans. This implies that the same mechanisms that control development in *Drosophila* could be used in higher organisms as well. In this section, we provide a brief overview of what is known.

**animation**
**Gene Regulation of the Development of the *Drosophila* Body Plan**

### *Drosophila* Developmental Stages

The production of an adult *Drosophila* from a fertilized egg involves a well-ordered sequence of developmentally programmed events under strict genetic control (Figure 19.18). About 24 hours after fertilization, a *Drosophila* egg hatches into a larva, which undergoes three molts, after which it is called a pupa. The pupa metamorphoses into an adult fly. The whole process from egg to adult fly takes about 10 to 12 days at 25°C.

**Figure 19.18**
**Development of an adult *Drosophila* from a fertilized egg.**

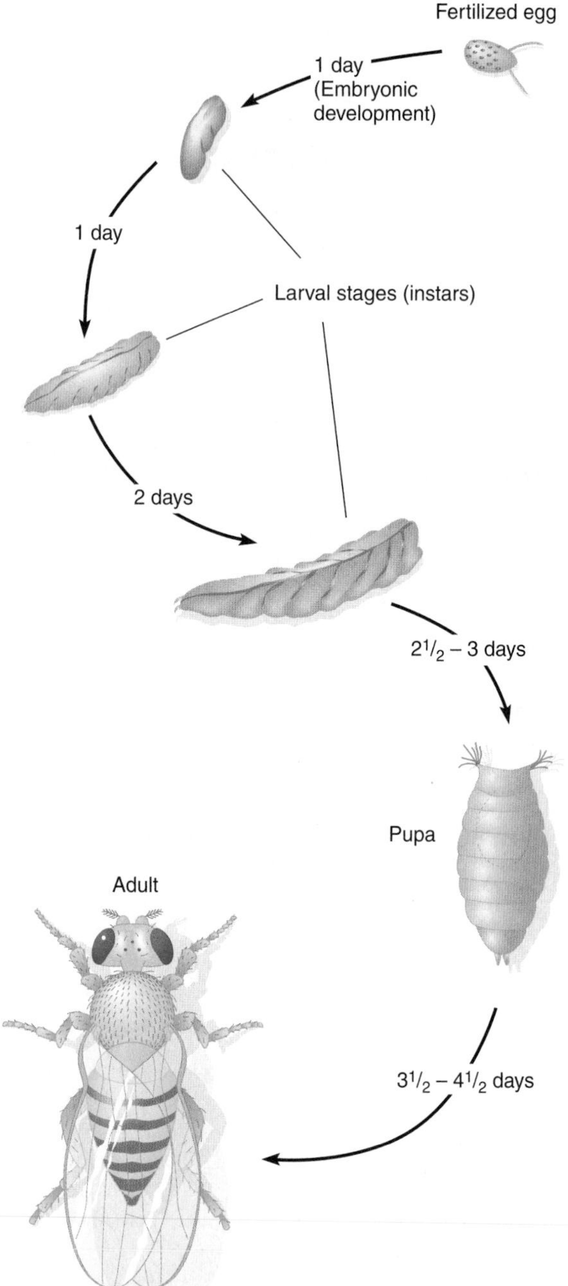


### Embryonic Development

Development commences with a fertilized egg, which gives rise to cells that have different developmental fates. What follows is a brief discussion of the information that has been obtained about the relationship between the developmental events in the egg and the determination of adult body parts.

Before a mature egg is fertilized, particular molecular gradients are established within it, as we will discuss. The posterior end is indicated by the presence of a region called the *polar cytoplasm* (Figure 19.19a). At fertilization, the two parental nuclei are roughly centrally located in the egg. The two nuclei fuse to produce a 2N zygote nucleus (Figure 19.19b). The zygote nucleus divides mitotically eight times in a common cytoplasm—cytokinesis does not occur—to produce a *multinucleate syncytium*, meaning nuclei all in the same cytoplasm with no nuclear envelopes around them (Figure 19.19c). These 256 nuclei migrate to the periphery of the egg, producing the *syncytial blastoderm* (Figure 19.19d), where mitotic

**Figure 19.19**

**Embryonic development in *Drosophila*.**

**a) Fertilized egg with two parental nuclei**

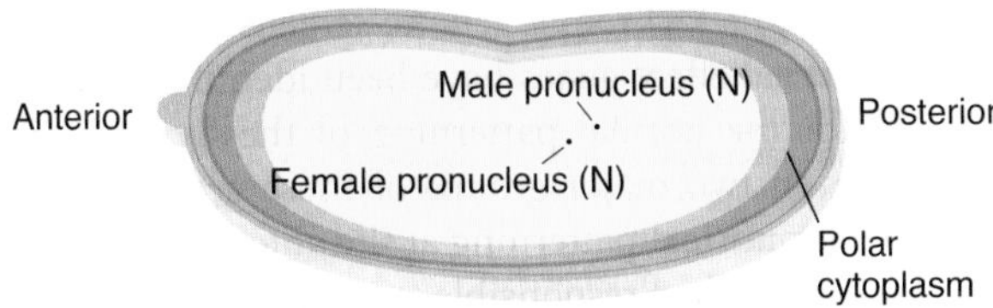


**b) Parental nuclei fuse and produce a diploid zygote nucleus**

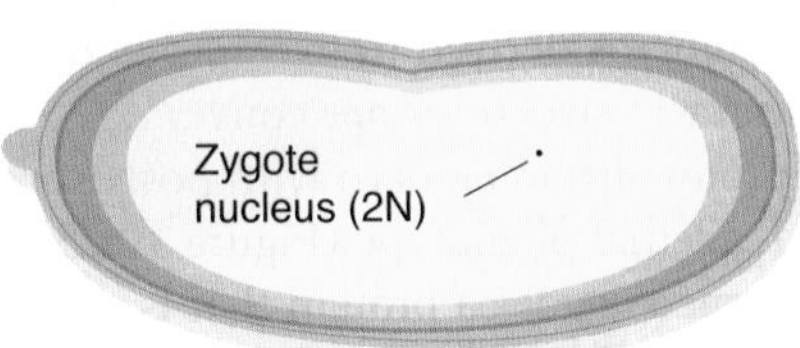


**c) The nucleus divides for eight divisions in a common cytoplasm to give a multinucleate syncytium**

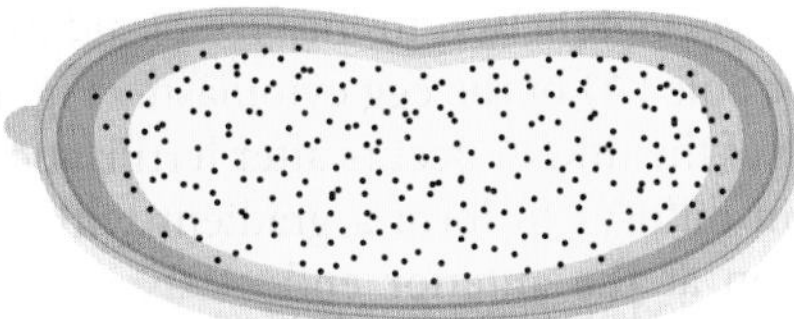

**d) The nuclei migrate to the periphery of the egg to produce a syncytial blastoderm. Mitotic divisions continue.**

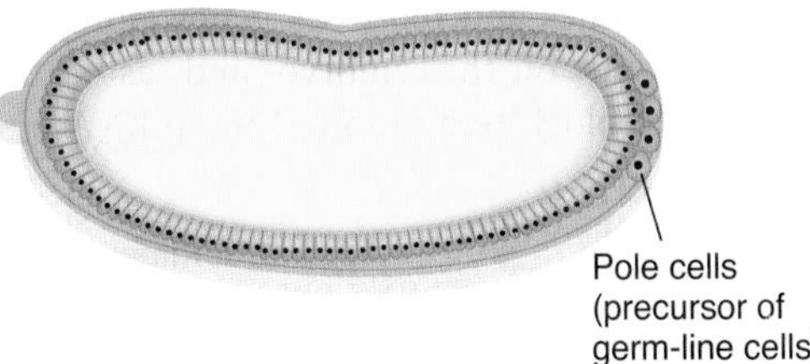


**e) After the thirteenth division, membranes form around the nuclei producing the somatic cells of the cellular blastoderm**

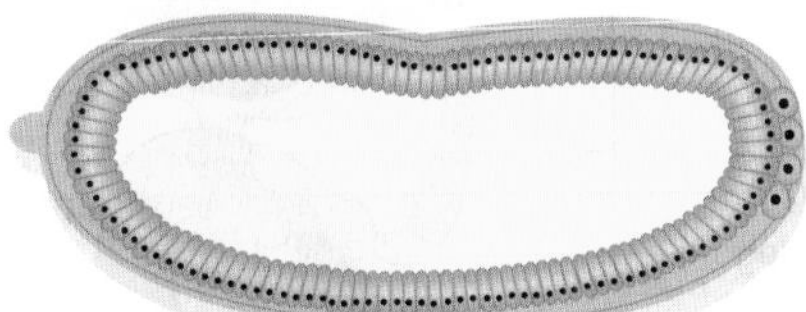

divisions continue. After the ninth division, about five nuclei reach the *polar cytoplasm* at the posterior pole of the embryo. Plasma membranes form around these nuclei to produce the *pole cells*, which are precursors to germ-line cells. After the thirteenth division, membranes form around the nuclei to produce somatic cells, creating the approximately 6,000-cell *cellular blastoderm* by about 4 hours after fertilization (Figure 19.19e).

Development of body structures depends on two processes (Figure 19.20):

**Figure 19.20**

***Drosophila* development results from gradients in the egg that define parasegments in the cellular blastoderm and segments in the embryo and adult.** The adult segment organization directly reflects the segment pattern of the embryo. A and P are the anterior and posterior compartments of the segments.

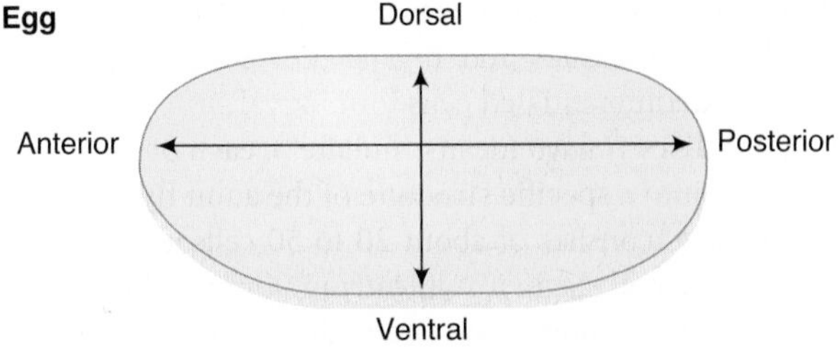


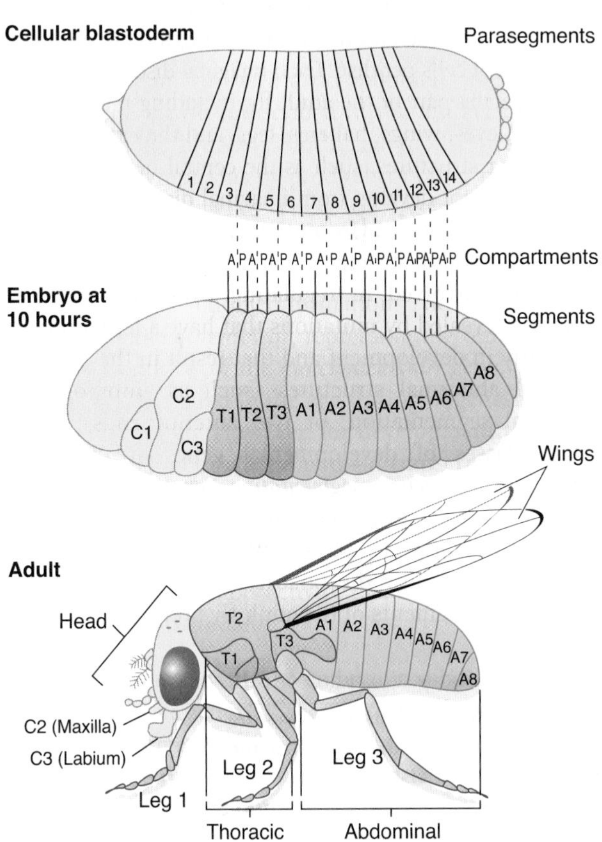


1. Gradients of proteins are produced along the anterior-posterior and the dorsal-ventral axes of the egg. Gene expression is affected by the position of a nucleus in the concentration of proteins in the two intersecting gradients.
2. Regions are determined in the embryo that correspond to adult body segments. In the cellular blastoderm stage *parasegments* form. Each parasegment is an indistinct region that includes what will become the posterior compartment of one segment and the anterior compartment of the next segment in the embryo. The *segments* of the embryo are visible, forming a striped pattern along the anterior-posterior axis.

## The Roles of miRNAs in Development

In Chapter 18, pp. 537–539, you learned about microRNAs (miRNAs) and their role in RNA interference (RNAi), the silencing of gene expresson at the posttranscriptional level. Recall that miRNAs are short, single-stranded RNA regulatory molecules encoded by genes. A complex of a miRNA and several proteins, including Argonaute protein Ago1, silences gene expression by binding to the 3′ untranslated region (UTR) of one or more target mRNAs. Base pairing between the miRNA and the mRNA causes either inhibition of translation or degradation of the mRNA. In animals, most miRNAs act by inhibiting the initiation of mRNA translation.

Experimental evidence of various kinds indicates that miRNAs are essential for development. The first such evidence came from a study by Rosalind Lee, Rhonda Feinbaum, and Victor Ambros in 1993 of loss-of-function mutants in the *lin-4* gene of *C. elegans*. The researchers proposed that a 22-nucleotide (nt) transcript of *lin-4* regulated *lin-14* mRNA expression by pairing between the *lin-4* RNA and a region in the 3′ UTR of the *lin-14* RNA. The *lin-4* gene is now known to be an miRNA gene; the 22-nt transcript is the mature, single-stranded miRNA. It regulates the target gene *lin-14* by silencing its expression at the translation level. The *lin-14* is a so-called *heterochronic gene*, meaning that it is involved in developmental timing. In normal worms, particular stem cells undergo a regulated cell division pattern synchronized with the four larval molts of the animal. Then, in the adult, those cells differentiate. In *lin-4* mutants, the stem cells keep repeating the cell division pattern characteristic of the first larval stage. In other words, the cells remain stuck in their larval stage 1 form. Loss-of-function mutations in the *lin-14* gene result in a phenotype in which the stem cells differentiate earlier than normal. Another miRNA that is involved in developmental timing in *C. elegans* is encoded by the *let-7* gene.

Studies of miRNA loss-of-function mutants, and of organisms with defective miRNA biogenesis, have revealed many other important roles of miRNAs in development and differentiation, including embryogenesis, organogenesis, and germ-line development. For example, in *C. elegans*, miRNAs regulate developmentally important cell division (described above), germ-line development, and vulval development. In *Drosophila*, miRNAs are required for development of both somatic tissues and the germ line, and in the maintenance of stem cells in the germ line. In zebrafish, miRNAs are essential for development; for example, a knockout of Dicer (see Chapter 18, p. 537) results in a developmental arrest at 7 to 10 days postfertilization. miRNAs are also involved in gastrulation, brain formation, somitogenesis (generation of somites, the tissue blocks that give rise to vertebrate muscle), and heart development. In mice (and, by extrapolation, other mammals), miRNAs are essential for development; that is, a knockout of Dicer dies at 7.5 days of gestation. Dicer is also required for embryonic stem cell differentiation *in vitro* and, therefore, probably *in vivo* also.

In sum, miRNAs play vital roles in development. No longer can we remain protein-centric in our thinking about the genetic control of development. Indeed, proteins do play critical roles in development and differentiation; but they are one player, not the only player.

# Summary

- Development is regulated growth resulting from the interaction of the genome with the cytoplasm and with the extracellular environment. Development begins when a zygote is formed. The zygote, and the cells in the subsequent few generations, are totipotent, meaning they can develop into any cell type of the organism. At some point, the genetic program sets the fate of a cell in a process called determination. After determination, differentiation occurs, in which determined cells undergo developmental programs to produce their specific cell types. A process related to differentiation is morphogenesis, in which anatomical structures or cell shape and size are produced by a regulated pattern of cell division, migration, programmed cell death, and changes in cell shape.

- Development results from differential gene activity of a genome that, in most cases, contains a constant amount of DNA from the zygote stage to the mature organism stage. Nonetheless, the genes are only part of the equation for development; environmental factors can affect the phenotype of an adult organism, as evidenced by phenotypic differences in cloned mammals from the parent that donated the nucleus for cloning.

- There are some exceptions to differentiated cells containing the same genome as the zygote from which they are derived. One example is B cells, a type of cell involved in antibody production. Antibodies are specialized proteins called immunoglobulins, which bind specifically to antigens (*anti*body *gen*erators: chemicals, recognized as foreign by an organism, that induce an immune response). Antibody molecules consist of two light chains and two heavy chains. In germ-line DNA, the coding regions for immunoglobulin chains are scattered in tandem arrays of gene segments. During development, somatic recombination occurs to bring particular gene segments together to form functional antibody chain genes. A large number of different antibody chain genes result from the many possible ways in which the gene segments can recombine.

- In mammals, the sex of the individual is determined by the presence or absence of the Y chromosome. If the Y chromosome is present, the *SRY* gene on that chromosome is transcribed to produce the testis-determining factor, a transcription factor that regulates genes required for directing the gonad to form a testis. In the absence of the *SRY* gene, as in an XX female, the gonad develops into an ovary by default.
- There is a different dosage of genes on the X chromosome in males and females. In mammals, a dosage compensation mechanism equates expression of X-linked genes in males and females by inactivating one of the two X chromosomes in a female early in development, leaving only one X chromosome transcriptionally active, as is the case in males. The inactivation process is complex, involving the transcription of the *XIST* gene on the chromosome to be inactivated; the noncoding RNA made then coats that chromosome, triggering chromatin changes that silence the genes.
- In *Drosophila*, the sex of the individual is determined by the ratio of the number of X chromosomes to the number of sets of autosomes. A ratio of 1 results in a female, and a ratio of 0.5 results in a male. The different ratios of the two types of chromosomes result in different levels of proteins encoded by genes on the chromosomes. These proteins influence a master regulatory gene, *Sxl*, for sex determination that controls a cascade of regulated alternative RNA splicing events, which ultimately leads to differentiation into female-specific or male-specific cells.
- As with mammals, there is a different dosage of genes on the X chromosomes in female and male *Drosophila*. In *Drosophila*, dosage compensation for X-linked genes occurs by increasing the transcriptional level of male X-linked genes twofold to match the gene expression of X-linked genes in the female. The mechanism for dosage compensation relates to the molecular steps for sex determination. That is, the absence of the *Sxl*-encoded protein in males enables a key protein to be translated from its mRNA. That protein associates with other proteins to form a complex that binds to many sites on the X chromosome. The complex triggers chromatin remodeling events spreading in each direction from the binding sites until the whole chromosome is affected. The key chromatin remodeling event involved is acetylation of histone H4; this modification is responsible for the twofold increase in transcriptional activity.
- *Drosophila* has become an important model system in which to study the genetic control of development. *Drosophila* body structures result from specific gradients in the egg and the subsequent determination of embryo segments that directly correspond to adult body segments. Both processes are under genetic control, as shown by mutations that disrupt the development events. Studies of the mutations indicate that *Drosophila* development is directed by a temporal regulatory cascade.
- Once the basic segmentation pattern has been laid down in *Drosophila*, homeotic genes determine the developmental identity of the segments. Homeotic genes share common DNA sequences called homeoboxes. Homeoboxes have been found in developmental genes in other organisms, and homeodomains—the regions of the proteins the homeoboxes encode—play a role in regulating transcription by binding to specific DNA sequences.
- Proteins play key roles in regulating the processes of development and differentiation; and, of course, proteins are key components of the structures that are the outcome of developmental and differentiation processes. In addition, microRNAs (miRNAs) play essential roles in development and differentation, including embryogenesis, organogenesis, and germ-line development.



# Analytical Approaches to Solving Genetics Problems

**Q19.1** We learned in this chapter that in humans, there are several distinct genes that code for α- and β-like globin polypeptides. These α-, β-, γ-, δ-, ε-, and ζ-globin genes are transcriptionally active at specific stages of development, resulting in the synthesis of polypeptides that are assembled in specific combinations to form different types of hemoglobin (see pp. 552–553 and Figure 19.8). Fill in the following table, indicating whether the globin gene in question is sensitive (S) or resistant (R) to DNase I digestion at each of the developmental stages listed.

| | Tissue | | |
|---|---|---|---|
| **Globin Gene** | **Embryonic Yolk Sac** | **Fetal Spleen** | **Adult Bone Marrow** |
| α | | | |
| β | | | |
| γ | | | |
| δ | | | |
| ζ | | | |
| ε | | | |

**A19.1.** The correctly filled-in table is as follows:

| | Tissue | | |
|---|---|---|---|
| Globin Gene | Embryonic Yolk Sac | Fetal Spleen | Adult Bone Marrow |
| $\alpha$ | R | S | S |
| $\beta$ | R | R | S |
| $\gamma$ | R | S | R |
| $\delta$ | R | R | S |
| $\zeta$ | S | R | R |
| $\varepsilon$ | S | R | R |

The explanation for the answers is as follows: DNase I typically digests regions of DNA that are transcriptionally active, while not digesting regions of DNA that are transcriptionally inactive. This is because transcriptionally inactive DNA is more highly coiled than transcriptionally active DNA. R means, then, that the gene was transcriptionally inactive, while S means that the gene was transcriptionally active.

To consider each globin gene in turn, the $\alpha$ gene is transcriptionally inactive in the embryonic yolk sac, but active in fetal spleen and adult bone marrow. That is, in the spleen fetal hemoglobin (Hb-F) is made, and Hb-F contains two $\alpha$ polypeptides and two $\gamma$ polypeptides. In the bone marrow, Hb-A is made, which contains two $\alpha$ and two $\beta$ polypeptides.

The $\beta$-globin gene is inactive in yolk sac and spleen and is active in bone marrow, making one of the two polypeptides found in Hb-A, the main adult form of hemoglobin. The $\beta$-like $\gamma$ polypeptide is found in Hb-F, which is made only in the liver and the spleen; thus, the $\gamma$ gene is active in spleen and inactive in yolk sac and bone marrow. The $\beta$-like $\delta$ polypeptide is found in $\alpha_2\delta_2$ hemoglobin, which is a minor class of hemoglobin found in adults; thus, the $\delta$ gene is active only in adult bone marrow.

The $\zeta$ gene makes an $\alpha$-like polypeptide found only in the hemoglobin of the embryo, so the $\zeta$ gene is active in the yolk sac but inactive in spleen and bone marrow. Finally, the $\varepsilon$ gene encodes the $\beta$-like polypeptide of the embryo's hemoglobin, so this gene is also active in the yolk sac but inactive in spleen and bone marrow.

# Questions and Problems

**19.1** Distinguish between the terms *development*, *determination*, and *differentiation*.

**19.2** What is totipotency? Give an example of the evidence for the existence of this phenomenon. What two mechanisms are used to restrict a cell's totipotency during development?

***19.3** It is possible to excise small pieces of early embryos of the frog, transplant them to older embryos, and follow the course of development of the transplanted material as the older embryo develops. A piece of tissue is excised from a region of the late blastula or early gastrula that would later develop into an eye and is transplanted to three different regions of an older embryo host (see part **[a]** of Figure 19.A). If the tissue is transplanted to the head region of the host, it will form eye, brain, and other material characteristic of the head region. If the tissue is transplanted to other regions of the host, it will form organs and tissues characteristic of those regions in normal development (e.g., ear, kidney). In contrast, if tissue destined to be an eye is excised from a neurula and transplanted into an older embryo host to exactly the same places as used for the blastula or gastrula transplants, in every case the transplanted tissue differentiates into an eye (see part **[b]** of Figure 19.A). Explain these results.

**19.4** With respect to the genetic analysis of development, what is meant by a *model* organism? Describe the features that model organisms possess to make them attractive for the genetic analysis of development, using specific examples.

**Figure 19.A**

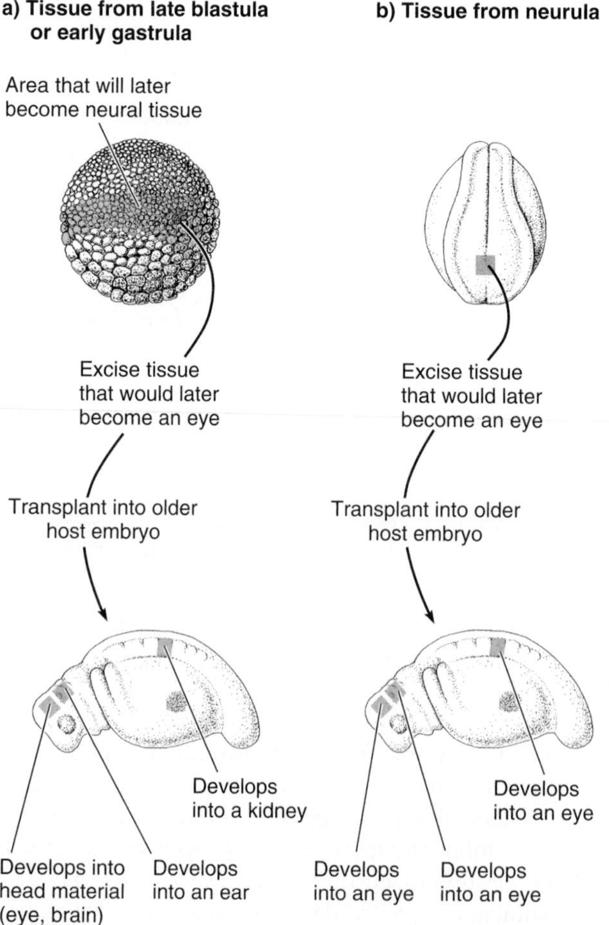
a) Tissue from late blastula or early gastrula
b) Tissue from neurula
Area that will later become neural tissue
Excise tissue that would later become an eye
Excise tissue that would later become an eye
Transplant into older host embryo
Transplant into older host embryo
Develops into a kidney
Develops into an eye
Develops into head material (eye, brain)
Develops into an ear
Develops into an eye
Develops into an eye

**19.5** In the set of experiments used to clone Dolly, six additional live lambs were obtained. Why is the production of Dolly more significant than the production of the other lambs? What is the evidence that Dolly resulted from the fusion of a nucleus from one cell with the cytoplasm of another?

***19.6** In Woody Allen's 1973 film *Sleeper*, the aging leader of a futuristic totalitarian society has been dismembered in a bomb attack. The government wants to clone the leader from his only remaining intact body part, a nose. The characters Miles and Luna thwart the cloning by abducting the nose and flattening it under a steamroller.

**a.** In light of the 1996 cloning of the sheep Dolly, how should the cloning have proceeded if Miles and Luna had not intervened?

**b.** If methods like those used for Dolly had been successful, in what genetic ways would the cloned leader be unlike the original?

**c.** Suppose that instead of a nose, only mature B cells (B lymphocytes of the immune system) were available. What genetic deficits would you expect in the "new leader"?

**d.** Based on what has been discovered about cloned cats and mice, in what nongenetic ways might the cloned leader differ from the original? What (constructive) advice would you give the totalitarian government based on these findings?

**e.** If the cloning of the leader had succeeded, can you make any prediction about whether the "cloned leader" would be interested in perpetuating the totalitarian state?

***19.7** The first cloned horse was a Haflinger—a chestnut horse with a flaxen mane and tail—female named Prometea. Prometea resulted from fusing the nucleus of a skin cell of a Haflinger mare with an enucleated donor egg, culturing the reconstructed embryo until it developed to the blastocyst stage, and then reimplanting the embryo into the same mare who donated the nucleus. These cloning experiments were part of a larger set of experiments where skin cell nuclei from one male Arabian thoroughbred horse or one Halfinger mare were fused with eggs that were enucleated after being taken from slaughtered abattoir horses. Of the 841 reconstructed male and female embryos that were grown in culture, just 8 male and 14 female embryos survived seven days and reached the blastocyst stage. Of these, 17 were implanted into 9 recipient mares. Prometea, who was the sole survivor, developed from one of the two embryos implanted into the same mare from which the donor nucleus was taken.

**a.** Explain whether Prometea's mother gave birth to her identical twin.

**b.** To demonstrate that Dolly the sheep was indeed a clone, researchers took particular care to distinguish Dolly from her surrogate mother using molecular and phenotypic markers. Do you have any concerns about this issue for Prometea given the source of the donor nucleus and the recipient mare? How would you experimentally demonstrate that Prometea was indeed a clone of her mother?

**c.** Propose a molecular genetic hypothesis to explain why only one of the 841 cloned embryos survived. What data would you gather to test your hypothesis?

**19.8** Discuss some of the evidence for differential gene activity during development. How have microarray analyses enhanced our understanding of this process?

**19.9** Discuss the expression of human hemoglobin genes during development.

**19.10** How are the hemoglobin genes organized in the human genome, and how is this organization related to their temporal expression during development?

***19.11** In humans, β-thalassemia is a disease caused by failure to produce sufficient β-globin chains. In many cases, the mutation causing the disease is a deletion of all or part of the β-globin structural gene. Individuals homozygous for certain of the β-thalassemia mutations are able to survive because their bone marrow cells produce γ-globin chains. The γ-globin chains combine with α-globin chains to produce fetal hemoglobin. In these people, fetal hemoglobin is produced by the bone marrow cells throughout life, whereas normally it is produced in the fetal liver. Use your knowledge about gene regulation during development to suggest a mechanism by which this expression of γ globin might occur in β-thalassemia.

***19.12** What are polytene chromosomes? Discuss the molecular nature of the puffs that occur in polytene chromosomes during development.

**19.13** Puffs of regions of the polytene chromosomes in salivary glands of *Drosophila* are surrounded by RNA molecules. How would you show that this RNA is single-stranded and not double-stranded?

***19.14** In experiment A, $^3$H-thymidine (a radioactive precursor of DNA) is injected into larvae of *Chironomus*, and the polytene chromosomes of the salivary glands are later examined by autoradiography. The radioactivity is seen to be distributed evenly throughout the polytene chromosomes. In experiment B, $^3$H-uridine (a radioactive precursor of RNA) is injected into the larvae, and the polytene chromosomes are examined. The radioactivity is first found only around puffs; later, radioactivity is also found in the cytoplasm. In experiment C, actinomycin D (an inhibitor of transcription) is injected into larvae and then $^3$H-uridine is injected. No radioactivity is found associated with the polytene chromosomes, and few puffs are seen. The puffs that are present are much smaller than the puffs found in experiments A and B. Interpret these results.

# 20 Genetics of Cancer

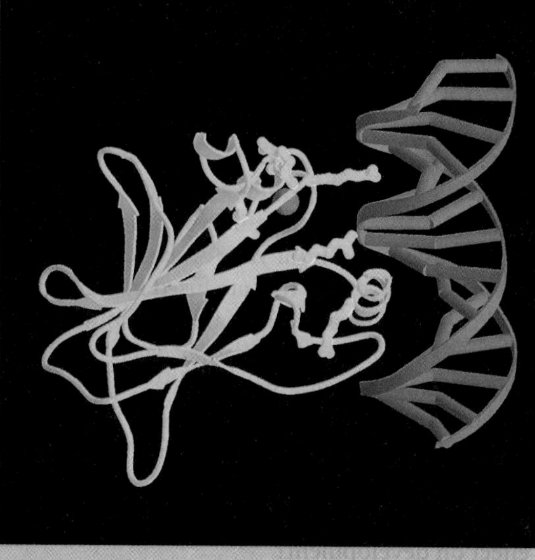

p53 protein binding to DNA.

## Key Questions

- How does a normal cell transform into a cancer cell?
- What is the difference between a tumor and a cancer?
- How does cancer relate to the cell cycle?
- What is the evidence that cancers are genetic diseases?
- What are oncogenes, and how are they involved in the development of cancer?
- What are tumor suppressor genes, and how are they involved in the development of cancer?
- How are miRNA genes involved in the development of cancer?
- What are mutator genes, and how are they involved in the development of cancer?
- What are carcinogens, and how do they contribute to the development of cancer?

### iActivity

AT CURRENT RATES, OVER A THIRD OF THE people who read this will die of cancer. Cancers are diseases characterized by the uncontrolled and abnormal division of eukaryotic cells. When cells divide unchecked within the body, they can give rise to tissue masses known as tumors. Some of these are not life threatening, but others can invade and disrupt surrounding tissues. What are the mechanisms that regulate cell growth and division? What causes uncontrolled growth in a cell? What genes are involved in the development of cancer? Is cancer inherited? In this chapter, you will learn the answers to these, and other, questions. Then, in the iActivity, you can apply what you learned as you investigate the origins of a form of bladder cancer.

In Chapter 19 some of the genetically controlled processes involved in development and differentiation were described. During development, specific tissues and organs arise by genetically programmed cell division and differentiation. In an adult, the many different types of cells of the body proliferate only in a controlled way. For example, for many tissues, programmed cell division occurs only to replace cells lost normally or through injury. Other cells, such as those of the intestinal lining and those that give rise to blood cells, must divide routinely to replace cells that have died.

Occasionally, dividing and differentiating cells deviate from their normal genetic program and give rise to tissue masses called **tumors,** or *neoplasms* ("new growth"). Figure 20.1 shows a mammogram indicating the presence of a tumor. The process by which a cell loses its ability to remain constrained in its growth properties is called **transformation** (not to be confused with transformation of a cell by uptake of exogenous DNA). If the transformed cells stay together in a single mass, the tumor is said to be *benign*. Benign tumors usually are not life threatening, and their surgical removal generally results in a complete cure. Exceptions include many brain tumors, which are life threatening because they impinge on essential cells.

The unregulated cell division of transformed cells can be seen easily in the culture dish. Normal fibroblast cells

Figure 20.1

**A mammogram showing a tumor.**

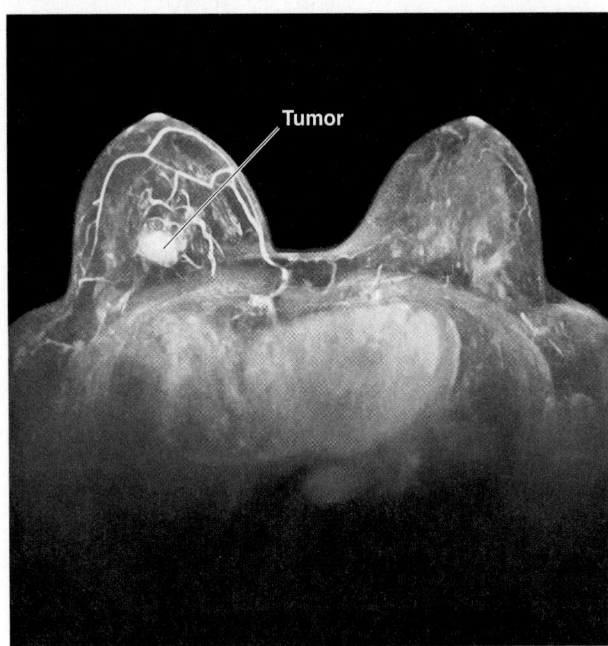


(cells that make the structural fibers and ground substance of connective tissue) grown in culture attach to the dish surface and divide until they contact each other. The result is a *monolayer*—a single layer of cells—covering the dish. This phenomenon is brought about by *contact inhibition*, a process whereby cells in contact communicate with one another and cell division is stopped. Contact inhibition is reflective of the regulated nature of cell division shown by normal cells. By contrast, transformed cells do not show contact inhibition, instead continuing to grow and divide after contacting neighbors and piling up in multiple layers.

If the cells of a tumor can invade and disrupt surrounding tissues, the tumor is said to be *malignant* and is identified as a **cancer.** Sometimes, cells from malignant tumors can also break off and move through the blood system or lymphatic system, forming new tumors at other locations in the body. The spreading of malignant tumor cells throughout the body is called **metastasis** ("change" of "state"). Malignancy can result in death because of damage to critical organs, secondary infection, metabolic problems, second malignancies, or hemorrhage.

The initiation of tumors in an organism is called **oncogenesis** (*onkos*, "mass" or "bulk"; *genesis*, "birth"). In this chapter, we focus on the genetic basis of oncogenesis.

## Relationship of the Cell Cycle to Cancer

During development, a tissue is produced by cell proliferation. During a series of divisions, progeny cells begin to express genes that are specific for the tissue, a process called cell differentiation. In some tissues, cell differentiation is also associated with the progressive loss of the ability of cells to proliferate: the most highly differentiated cell, the one that is fully functional in the tissue, can no longer divide. Such cells are known as *terminally differentiated cells*. They have a finite life span in the tissue and are replaced with younger cells produced by division of unspecialized cells, called *stem cells*, which are a small fraction of cells in the tissue that are capable of *self-renewal*. To understand both benign and malignant neoplastic diseases, we must realize that the linkage of growth with differentiation of any tissue is not necessary. That is, cells *can* divide without undergoing terminal differentiation. For example, in malignant neoplasms most daughter cells from any replication event fail to express fully the genetic programs that regulate terminal differentiation.

### Molecular Control of the Cell Cycle

In every cell cycle, all chromosomes must be duplicated faithfully and a copy of each distributed to both progeny cells. The cell cycle in most somatic cells of higher eukaryotes is divided into four stages: gap 1 ($G_1$), synthesis (S), gap 2 ($G_2$), and mitosis (M) (see Figure 12.4, p. 329).

Progression through the cell cycle is tightly controlled by the activities of many genes in an elaborate system of checks and balances (Figure 20.2). **Checkpoints** at different points in the cell cycle are control points at which the cell cycle is arrested if there is damage to the genome or cell cycle machinery. This allows the damage to be repaired or, if it is not, the cell is destroyed. These processes are necessary to prevent the possibility of damaged cells dividing in an unprogrammed way, that is, from becoming cancerous.

As a cell proceeds through $G_1$, it prepares for DNA replication and chromosome duplication in the S phase. A major checkpoint in $G_1$—called *START* in yeast and *$G_1$-to-S checkpoint* in mammalian cells—determines whether the cell is able to or should continue into S. The cell stays in $G_1$ unless it grows large enough and the environment is favorable. Another major checkpoint, the *$G_2$-to-M checkpoint*, occurs at the junction between $G_2$ and M. Unless all the DNA has replicated, the cell is big enough, and the environment is favorable, the cell cannot enter the mitotic phase of the cell cycle. A third checkpoint occurs during M: the chromosomes must be attached properly to the mitotic spindle to trigger the separation of chromatids and the completion of mitosis.

Proteins known as **cyclins** (named because their concentration increases and decreases in a regular pattern through the cell cycle) and enzymes known as **cyclin-dependent kinases** (Cdks) are the key components in the regulatory events that occur at checkpoints. At the $G_1$-to-S checkpoint, two different $G_1$ cyclin–Cdk complexes form, resulting in activation of the kinases. The kinases catalyze a series of phosphorylations (addition of phosphate groups) of cell cycle control proteins, affecting the functions of those proteins and leading, therefore, to transition into the S phase.

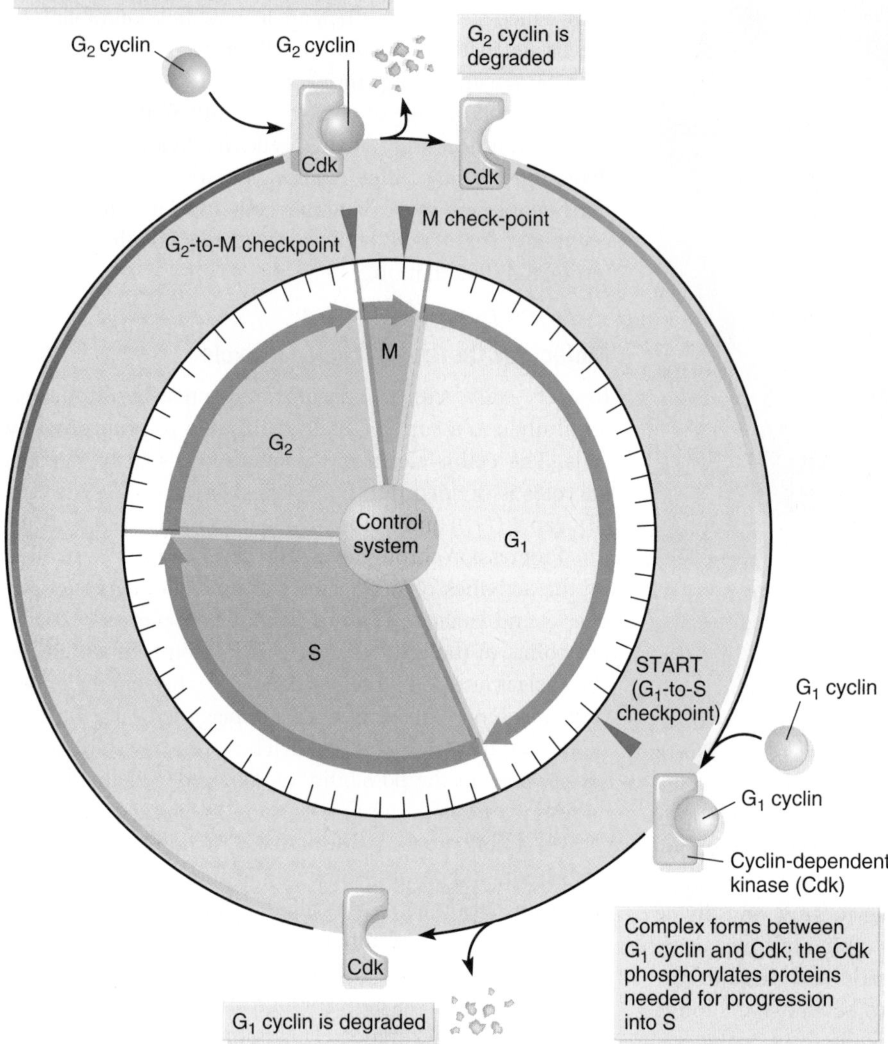


**Figure 20.2**
**Some of the molecular events that control the cell cycle.**



A similar process occurs at the $G_2$-to-M checkpoint. A $G_2$ cyclin binds to a Cdk to form a complex. Until the cell is ready to enter mitosis, phosphorylation of the Cdk by another kinase keeps the Cdk inactive. At that time, a phosphatase removes the key phosphate from the Cdk, activating the enzyme. Phosphorylations of proteins by the Cdk move the cell into mitosis.

## Regulation of Cell Division in Normal Cells

Control of cell division of a normal cell is handled by both extracellular and cellular molecules that operate in a complicated signaling system. The extracellular molecules are steroids and polypeptide hormones made in other tissues that influence the growth and division of cells in other tissues. For example, a *growth factor* is a molecule that stimulates cell division of a target cell (Figure 20.3a). Growth factors have specific effects because they bind to specific receptors on their target cells. The receptors are proteins that span the plasma membrane. The growth factor binds to the part of the receptor that is outside of the cell. The signal is then transmitted through the membrane-embedded part of the receptor to an intracellular part, and the receptor becomes activated. The signal then is relayed through a series of proteins, eventually activating nuclear genes that encode proteins for stimulating cell growth and division. The last step is brought about by transcription factors. There are similar pathways for *growth-inhibitory factors*, which lead to inhibition of cell growth and division (Figure 20.3b). The process of relaying a growth-stimulatory or growth-inhibitory signal after an extracellular factor binds to a cell is called **signal transduction**, and the proteins involved are known as *signal transducers*. Normal, healthy cells give rise to progeny cells only when the balance of stimulatory and inhibitory signals from outside the cell favors cell division. A neoplastic cell, on the other hand, has lost control of cell

*animation*
**Regulation of Cell Division in Normal Cells**

Figure 20.3

**General events for regulation of cell division in normal cells.** (**a**) When a growth factor binds to its cell membrane receptor, it acts as a signal to stimulate cell growth. To do that, the signal is transduced into the cell and relayed to the nucleus, activating the expression of a gene or genes that encode a protein or proteins required for the stimulation of cell division. (**b**) When a growth-inhibiting factor binds to its cell membrane receptor, it acts as a signal to inhibit cell growth. In the case illustrated, the signal is transduced into the cell and relayed to the nucleus, activating the expression of a gene or genes that encode a protein or proteins required for the inhibition of cell division. (Other events may occur instead. For instance, the end product of the signal transduction pathway may repress the expression of cell division-stimulating genes directly.)

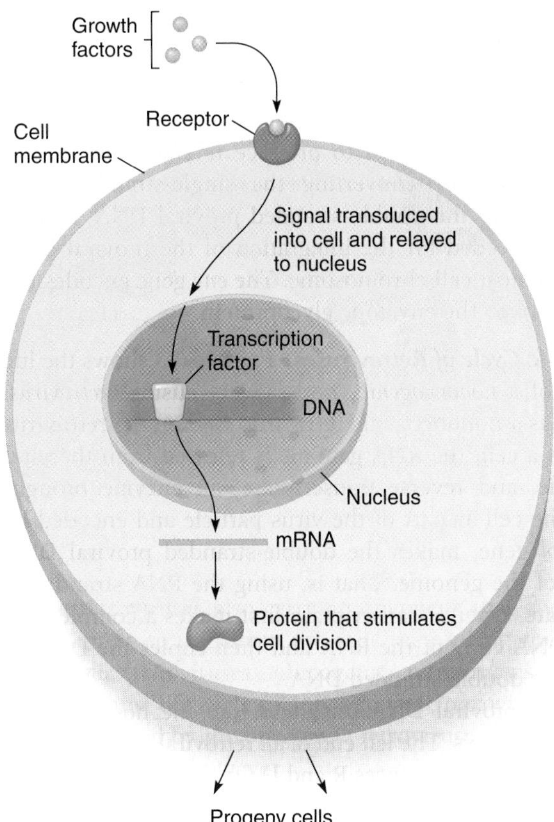


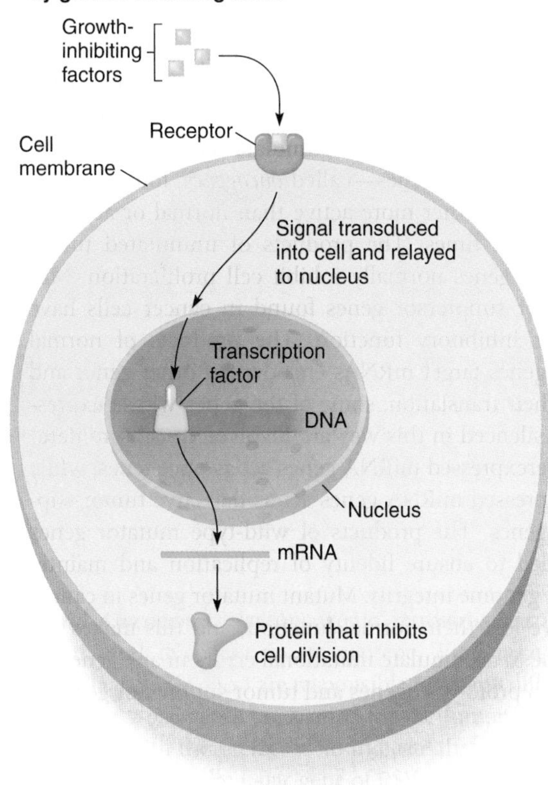


division and reproduces without constraints. This can occur when genes either for stimulatory factors, inhibitory factors, or signal transducers mutate, or when genes encoding cell surface receptors or signal transducers involved in cell cycle control are mutated.

## Keynote

Progression through the cell cycle is tightly controlled by the activities of many genes. Checkpoints at key points determine whether a cell has DNA damage or has problems with its cell cycle machinery and permits only normal cells to continue. The key molecules used at these checkpoints are cyclins and cyclin-dependent kinases (Cdks). In addition, healthy cells grow and divide only when the balance of stimulatory and inhibitory signals received from outside the cell favor cell proliferation. A cancerous cell does not respond to the usual signals and reproduces without constraints.

## Cancers Are Genetic Diseases

Several lines of evidence indicate that cancers are genetic disorders:

- There is a high incidence of particular cancers in some human families. Cancers that run in families are known as *familial (hereditary) cancers*; cancers that do not appear to be inherited are known as *sporadic* (or *nonhereditary*) *cancers*. Sporadic cancers are more frequent than are familial cancers.
- Some viruses can induce cancer. This means that expression of viral genes introduced into the host is able to disrupt normal cell cycle controls.
- Descendants of cancerous cells are all cancerous. In fact, it is the clonal descendants of a cell that become cancerous, forming a tumor.
- The incidence of cancers increases upon exposure to mutagenic agents. This is the case with experimental organisms treated under controlled laboratory

**Table 20.2 Classes of Proto-Oncogene Products**

| | |
|---|---|
| **Growth factors** | |
| *sis* | PDGF B-chain growth factor |
| *int-2* | FGF-related growth factor |
| **Receptor and nonreceptor protein-tyrosine and protein-serine/threonine kinases** | |
| *src* | Membrane-associated nonreceptor protein-tyrosine kinase |
| *fgr* | Membrane-associated nonreceptor protein-tyrosine kinase |
| *fps/fes* | Nonreceptor protein-tyrosine kinase |
| *kit* | Truncated stem cell receptor protein-tyrosine kinase |
| *pim-1* | Cytoplasmic protein-serine kinase |
| *mos* | Cytoplasmic protein-serine kinase (cytostatic factor) |
| **Receptors lacking protein kinase activity** | |
| *mas* | Angiotensin receptor |
| **Membrane-associated G proteins activated by surface receptor** | |
| H-*ras* | Membrane-associated GTP-binding/GTPase |
| K-*ras* | Membrane-associated GTP-binding/GTPase |
| *gsp* | Mutant-activated form of G$\alpha$ |
| **Cytoplasmic regulators** | |
| *crk* | SH-2/3 protein that binds to (and regulates?) phosphotyrosine-containing proteins |
| **Nuclear transcription factors (gene regulators)** | |
| *myc* | Sequence-specific DNA-binding protein |
| *fos* | Combines with c-*jun* product to form AP-1 transcription factor |
| *jun* | Sequence-specific DNA-binding protein; part of AP-1 |
| *erbA* | Dominant negative mutant thyroxine (T3) receptor |
| *ski* | Transcription factor? |

proto-oncogenes might be regulatory genes involved with the control of cell multiplication during differentiation. Evidence supporting this hypothesis came when the product of the viral oncogene v-*sis* was shown to be identical to part of platelet-derived growth factor (PDGF, a factor found in blood platelets in mammals), which is released after tissue damage. PDGF affects only one type of cell, fibroblasts, causing them to grow and divide. The fibroblasts are part of the wound-healing system. The causal link between PDGF and tumor induction was demonstrated in an experiment in which the cloned PDGF gene was introduced into a cell that normally does not make PDGF (i.e., a fibroblast); that cell was transformed into a cancer cell.

We can generalize and say that some cancer cells can result from the excessive or untimely synthesis of growth factors in cells that do not normally produce those factors. An altered growth factor gene such as a v-*onc* or mutation of a cellular proto-oncogene are examples of changes that can alter growth factor levels.

***Protein Kinases.*** Many proto-oncogenes encode protein kinases, enzymes that add phosphate groups to target proteins, thereby modifying the function of the proteins. Protein kinases are integral members of signal transduction pathways. The *src* gene product, for example, is a nonreceptor protein kinase called pp60*src*. The viral protein, pp60v-*src*, and the protein encoded by the cellular oncogene, pp60c-*src*, differ in only a few amino acids, and both proteins bind to the inner surface of the plasma membrane.

What is particularly interesting about the *src* protein kinases is that both versions add a phosphate group to the amino acid tyrosine; that is, they are *tyrosine protein kinases*. Before this discovery, the protein kinases that had been characterized had all been shown to add phosphates only to the amino acids serine or threonine. Because protein phosphorylation was known to be important in effecting a multitude of metabolic changes in cells, the *src* discovery was exciting. It suggested a possible explanation of how the *src* and other tyrosine protein kinase-coding oncogenes might transform a normal cell into a metabolically different cancer cell. For example, a large class of proteins, including the receptors for growth factors, uses protein phosphorylation to transmit signals through the membrane. Thus, the action of protein kinases is also linked to growth factors and their activities.

***Membrane-Associated G Proteins Activated by Surface Receptors.*** Earlier we discussed how a signal produced by binding of a growth factor to its membrane-embedded receptor is transduced through the cell, activating key nuclear genes that control the cell cycle. The steps from the growth factor receptor to the nucleus are many, forming what is known as a *signaling cascade*. Membrane-associated G proteins are involved in this cascade; they are activated by the binding of the growth factor to the cell surface receptor.

An example of such a G protein is the Ras protein, encoded by the *ras* gene. This gene is mutated in many cancers. Figure 20.6 shows part of the signaling cascade involving Ras. Binding of the growth factor EGF to the EGF receptor stimulates autophosphorylation of the receptor. Protein Grb2 can then bind, and the complex recruits SOS protein to the plasma membrane. SOS displaces GDP from Ras (which is anchored to the inner side of the plasma membrane), and allows Ras now to bind GTP. Ras–GTP recruits Raf-1 and activates it. This initiates a cascade of cytoplasm-based phosphorylations of proteins (the MAP kinase cascade), eventually producing phosphorylated ERK. ERK moves from the cytoplasm into the nucleus, where it phosphorylates several transcription factors, including Elk-1, activating them. The

**Figure 20.6**

**Role of the membrane-associated G protein, Ras, in the activation of transcription of cell cycle-specific target genes.**

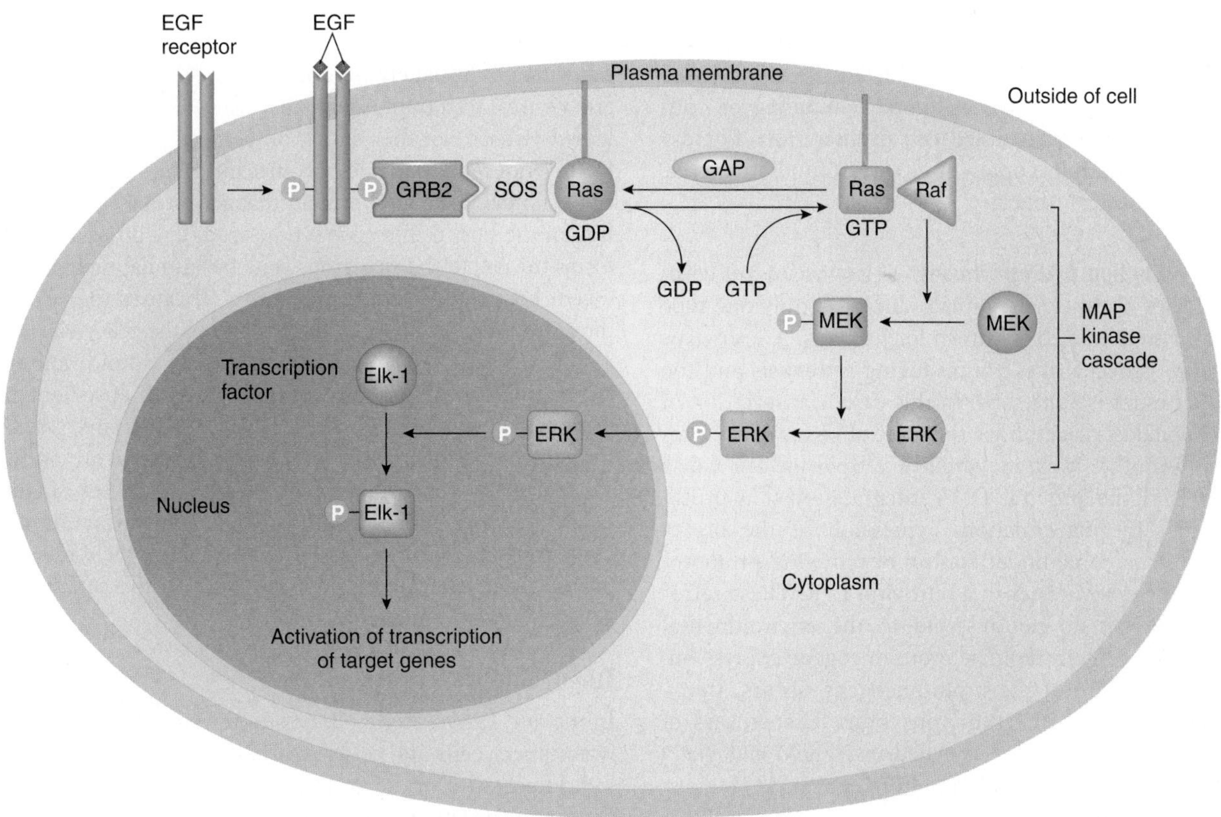




activated transcription factors then turn on the transcription of specific sets of cell division-stimulating genes.

Turning the Ras signal off in normal cells involves GAP (GTPase activating protein), which makes Ras hydrolyze the GTP bound to it back to GDP. This inactivates Ras and cancels the cell cycle stimulatory signal.

How does *ras* become an oncogene? One way is by a mutation that abolishes its ability to hydrolyze GTP to GDP. Therefore, even with stimulation from GAP, the Ras–GTP complex remains and the signal is continuously on.

**Changing Cellular Proto-Oncogenes into Oncogenes.** In normal cells, expression of proto-oncogenes is tightly controlled so that cell growth and division occur only as appropriate for the cell type involved. However, when proto-oncogenes are changed into oncogenes, the tight control can be lost, and unregulated cell proliferation can take place.

Three examples of the types of changes that have been found are:

1. *Point mutations* (base-pair substitutions). Point mutations in the coding region or in the controlling sequences (promoter, regulatory elements, enhancers) can change a proto-oncogene into an oncogene by causing an increase in either the activity of the gene product or the expression of the gene, the latter leading in turn to an increase in the amount of gene product. For example, the *ras* mutations described in the previous section typically are point mutations.
2. *Deletions*. Deletions of part of the coding region, or part of the controlling sequences of a proto-oncogene, have been found frequently in oncogenes. The deletions cause changes in the amount or activity of the encoded growth stimulatory protein, causing unprogrammed activation of some cell proliferation genes.

For example, the *myc* oncogene can arise from its proto-oncogene by deletion. The normal proto-oncogene consists of three exons and two introns; in some commonly found *myc* oncogenes, the first exon and most of the first intron are deleted. Transcription is then controlled from sequences in exon 2, which can function as a promoter. The *myc* proto-oncogene encodes a nuclear transcription factor that positively regulates genes involved in cell proliferation. Thus, the deletions in the oncogene forms have brought about a change in the amount or activity of the remaining *myc* protein chain that activates those genes.

3. *Gene amplification* (increased number of copies of the gene). Some tumors have multiple (sometimes hundreds of) copies of proto-oncogenes. These probably result from a random overreplication of small segments of the genomic DNA. Extra copies of the proto-oncogene in the cell result in an increased amount of gene product, thereby inducing or contributing to unscheduled cell proliferation. For example, multiple copies of *ras* are found in mouse adrenocortical tumors.



**Cancer Induction by Retroviruses.** Retroviruses are common causes of cancer in animals, although only one type of cancer thus caused is known for humans. A retrovirus can cause cancer if it is a transducing retrovirus and the v-*onc* it carries is expressed. In this case, transcription of the v-*onc* takes place under the control of retroviral promoters. Another way in which a retrovirus can cause cancer is if the proviral DNA integrates near a proto-oncogene. In this situation, expression of the proto-oncogene can come under control of retroviral promoter and enhancer sequences in a retroviral LTR. These retroviral sequences do not respond to the environmental signals that normally regulate proto-oncogene expression, so overexpression of the proto-oncogene occurs, transforming the cell to the tumorous state. The process of proto-oncogene activation is called *insertional mutagenesis*. It is rare in both animals and humans.

### Keynote

After retrovirus infection, tumor induction can occur as a result of the activity of a viral oncogene (v-*onc*) in the retroviral genome. Retroviruses carrying an oncogene are known as transducing retroviruses. Normal cellular genes, called proto-oncogenes, have DNA sequences that are similar to those of the viral oncogenes. Proto-oncogenes encode proteins that stimulate cell growth and division. In their mutated state, proto-oncogenes are called cellular oncogenes (c-*oncs*), and they may induce tumors. Retroviral oncogenes are modified copies of the cellular proto-oncogenes that have been picked up by the retrovirus.

**DNA Tumor Viruses.** DNA tumor viruses are oncogenic—they induce cell proliferation—but, as mentioned previously (p. 582), their mechanism for transforming cells is completely different. DNA tumor viruses transform cells to the cancerous state through the action of an oncogene or oncogenes in the viral genome. DNA tumor virus oncogenes are essential viral genes that have no relationship to cellular genes. In this respect, these oncogenes are distinct from those of RNA tumor viruses. Examples of DNA tumor viruses are found among five of six major families of DNA viruses: papovaviruses, hepatitis B viruses, herpes viruses, adenoviruses, and pox viruses.

DNA tumor viruses normally progress through their life cycles without transforming the cell to a cancerous state. Typically, the virus produces a viral protein that activates DNA replication in the host cell. Then, through the use of host proteins, the viral genome is replicated and transcribed, ultimately producing a large number of progeny viruses, resulting in lysis and death of the cell. The released viruses can then infect other cells. Rarely, the viral DNA is not replicated and becomes integrated into the host cell genome. If the viral protein that activates DNA replication of the host cell is now synthesized, this protein transforms the cell to the cancerous state by stimulating the quiescent host cell to proliferate; that is, it causes the cell to move from the $G_0$ phase to the S phase of the cell cycle.

Examples of DNA tumor viruses are found among the papillomaviruses. Some of these viruses cause benign tumors such as skin and venereal warts in humans. Other human papillomaviruses (*HPV-16*, *HPV-18*, or both) cause cervical cancer, which is a leading cause of cancer deaths among women worldwide. Transformation is caused by the key viral genes, *E6* and *E7*, which encode proteins that activate progression through the cell cycle. A vaccine is now available for HPV.

## Tumor Suppressor Genes

In the late 1960s, Henry Harris fused normal rodent cells with cancer cells and observed that some of the resultant hybrid cells did not form tumors but instead established a normal growth pattern. Harris hypothesized that the normal cells contained gene products that had the ability to suppress the uncontrolled cell proliferation that is characteristic of cancer cells. The genes involved were called **tumor suppressor genes.**

Further evidence for the existence of tumor suppressor genes came from data indicating that, in certain cancers, specific chromosome regions were deleted from both homologues. Logically, if the loss of function of particular genes is correlated with tumor development, then the normal alleles of those genes must suppress tumor formation. In other words, the normal products of tumor suppressor genes have an inhibitory role in cell growth and division. Thus, when tumor suppressor genes are inactivated, the inhibitory activity is lost, and unprogrammed cell proliferation can begin. Inactivation of tumor suppressor genes has been linked to the development of a wide variety of human cancers, including breast, colon, and lung cancers.

**Finding Tumor Suppressor Genes.** In contrast to mutations that change proto-oncogenes to oncogenes, mutations of tumor suppressor genes are recessive—cell proliferation can be affected only if both alleles are inactivated. Thus, oncogenes can be identified in the laboratory because they can stimulate the growth of cells in culture, but that is not the case for tumor suppressor genes. Introducing tumor suppressor genes into cells in culture either results in no change or kills the cell. A number of tumor suppressor genes have been cloned molecularly. Table 20.3 lists some of the known tumor suppressor genes in humans. The products of tumor suppressor genes are found throughout the cell.

### The Retinoblastoma Tumor Suppressor Gene, *RB*

***Retinoblastoma and Knudson's Two-Hit Mutation Model for Cancer.*** Retinoblastoma (OMIM, http://www.ncbi.nlm.nih.gov/OMIM) is a childhood cancer of the eye (Figure 20.7). Retinoblastoma develops during the period from birth to age 4 years and is the most common eye tumor in children. If discovered early enough, more than 90% of the eye tumors can be permanently destroyed, for example by laser therapy or by radiation therapy.

There are two forms of retinoblastoma. In *sporadic retinoblastoma* (60% of cases), an eye tumor develops spontaneously in a patient from a family with no history of the disease. In these cases, a *unilateral tumor* develops in one eye only. In *hereditary retinoblastoma* (40% of cases), the susceptibility to develop the eye tumors is inherited. Patients with this form of retinoblastoma typically develop multiple eye tumors involving both eyes (*bilateral tumors*), usually at an earlier age than is the case for unilateral tumor formation in patients with sporadic retinoblastoma.

In 1971, Alfred Knudson proposed the following to explain the occurrence of hereditary and sporadic forms of retinoblastoma: "Retinoblastoma is a cancer caused by two mutational events. . . . One mutation is inherited via the germinal cells and the second occurs in somatic cells. In the nonhereditary form, both mutations occur in somatic cells."[2] This model relates in a general way to all forms of familial cancer.

Knudson's two-hit mutational model, as exemplified by retinoblastoma, is illustrated in Figure 20.8. In sporadic retinoblastoma (Figure 20.8a), a child is born with two wild-type copies of the retinoblastoma gene (genotype *RB/RB*), and mutation of each to a mutant *rb* allele must then occur in the same eye cell.[3] Since the chance of having two independent mutational events in the same cell is very low, sporadic retinoblastoma patients would be expected to develop mostly unilateral tumors, as is the case. Furthermore, the rarity of the mutation event means that the two gene copies are mutated at different times, the first mutation producing an *RB/rb* cell, and the second mutation in that cell giving rise to the *rb/rb* genotype that results in eye tumor development. In hereditary

[2]Knudson, A. G., Jr. 1971. Mutation and cancer: statistical study of retinoblastoma. *Proc. Natl. Acad. Sci. USA* 68:820–823.

[3]Retinoblastoma is among a very few cancers for which only one gene is critical for its development, in this case a mutation in a gene for a growth inhibitory factor, that is, a tumor suppressor gene. In most cases, cancer develops as a multistep process involving mutations in several different key genes related to cell growth and division.



**Table 20.3 Examples of Tumor Suppresser Genes**

| Gene | Cancer Type | Protein Function | Hereditary Syndrome | Chromosome Location |
|---|---|---|---|---|
| *APC* | Colon carcinoma | Cell adhesion | Familial adenomatous polyposis (FAP) | 5q21–q22 |
| *BRCA1* | Breast cancer | Has possible transcription activation domain; interacts with DNA damage repair machinery | Breast cancer and ovarian cancer | 17q21 |
| *BRCA2* | Breast cancer | Has possible transcription activation domain; interacts with DNA damage repair machinery | Breast cancer | 13q12–q13 |
| *DCC* | Colon carcinoma | Cell adhesion | Involved in colorectal cancer | 18q21.3 |
| *NF1* | Neurofibromas | GTPase activating protection | Neurofibromatosis type I | 17q11.2 |
| *NF2* | Schwannomas and meningiomas | Links cell surface glycoprotein to the actine cytoskeleton? | Neurofibromatosis type II | 22q12.2 |
| *p16* | Melanoma and others | Cdk inhibitor | Familial melanoma | 9p21 |
| *RB* | Retinoblastoma | Cell cycle and transcription regulation | Retinoblastoma | 13q14.1–q14.2 |
| *TP53* | Wide variety | p53 is a transcription factor | Li–Fraumeni syndrome | 17p13.1 |
| *VHL* | Kidney carcinoma, pancreatic tumors, and others | Transcription elongation | von Hippel–Lindau syndrome | 3p26–p25 |
| *WTI* | Nephroblastoma | Transcription activator or repressor depending on cell | Wilms tumor 1 | 11p13 |

Figure 20.7

**An eye tumor in a patient with retinoblastoma.**

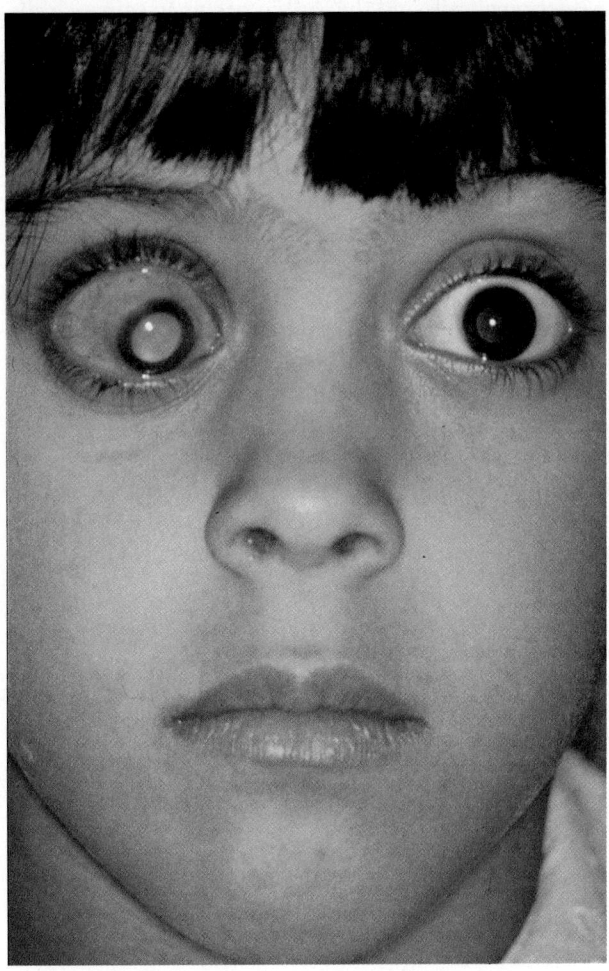



retinoblastoma, patients inherit one copy of the mutated retinoblastoma gene through the germ line; that is, they are *RB/rb* heterozygotes (Figure 20.8b). Only a single additional mutation of the retinoblastoma gene in an eye cell is needed to produce an *rb/rb* homozygote that would result in tumor formation. Given the number of cells in a developing retina and the rate of mutation per cell, *loss of heterozygosity* (LOH; here, a mutation in the *RB* allele) is very likely for at least a few cells. Furthermore, because only a single mutation is needed to produce homozygosity for *rb*, hereditary retinoblastoma is characterized on the average by earlier onset than in sporadic retinoblastoma and by multiple bilateral tumors.

According to Knudson's model, the *retinoblastoma mutation is recessive* because cancer develops only if both alleles are mutant. However, if one mutation is inherited through the germ line, tumor formation requires only a mutational event in the remaining wild-type allele in any one of the cells in that particular tissue. Due to the high likelihood of such an event, the *disease appears dominant* in pedigrees. Thus, for hereditary retinoblastoma and in hereditary neoplasms in general, inheritance of just one gene mutation predisposes a person to cancer but does not cause it directly; a second mutation is required for loss of heterozygosity. Commonly, therefore, it is considered that there is a hereditary disposition for cancer in such families.

Support for Knudson's hypothesis came in the 1980s from the analysis of the chromosomes of tumor cells and normal tissues in patients with retinoblastoma. Many patients carried deletions of a region of chromosome 13, and through genetic analysis the *RB* gene was mapped to chromosome location 13q14.1–q14.2.

**Keynote**

The two-hit mutation model for cancer explains the difference between familial (hereditary) cancers and sporadic (nonhereditary) cancers. In familial cancers, one mutation is inherited, thereby predisposing the person to cancer. When the second mutation occurs later in somatic cells, cancer may then develop. In sporadic cancers both mutations occur in the somatic cells, so such cancers typically occur later in life than do familial cancers because the probability of two mutations is lower than the probability of one mutation.

***Function of the* RB *Tumor Suppressor Gene.*** The human *RB* tumor suppressor gene (OMIM 180200) has been mapped to 13q14.1–q14.2. The *RB* gene was cloned in 1986. It spans 180 kb of DNA and encodes 4.7-kb mRNA that is translated to produce the 928-amino acid nuclear phosphoprotein (a protein that can be phosphorylated), pRB.

pRB is involved in regulating the cell cycle at the $G_1$-to-S checkpoint (see Figure 20.2) as follows. Two Cdk–cyclin complexes are formed during $G_1$: Cdk4–cyclin D and Cdk2–cyclin E (Figure 20.9). These complexes cause the progression to the S phase by catalyzing a series of phosphorylations of cell cycle controlling proteins, including pRB. pRB is found in a complex with transcription factor E2F and, when pRB is unphosphorylated, the activity of E2F is inhibited. Then, when pRB becomes phosphorylated, the inhibition of E2F activity is removed and the now-active transcription factor turns on the transcription of genes for DNA synthesis. As the cell begins DNA synthesis, cyclins D and E are degraded, and cyclin A is made. A Cdk2–cyclin A complex forms and activates DNA replication. After the S phase, pRB is dephosphorylated again, rendering E2F inactive.

In a cell with two mutant *rb* alleles, pRB often is truncated and unstable and does not bind to E2F, which is then free to activate DNA synthesis genes. As a result, unprogrammed cell division takes place. Interestingly, several different DNA tumor viruses (e.g., adenovirus, SV40) exert their tumorigenic effects in part by a process in which proteins encoded by their oncogenes form complexes with pRB in the cell, thereby blocking its ability to bind to E2F and inactivating the suppressive function of the protein. In other words, these tumor viruses transform cells to the neoplastic state by inactivating a mechanism that inhibits cell cycle progression.

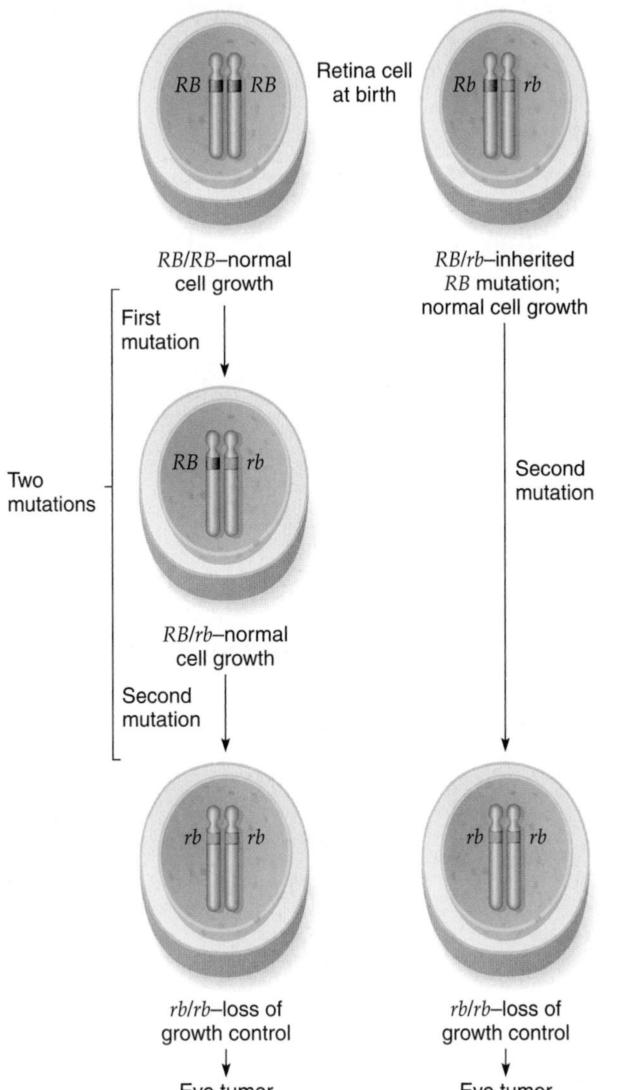


**Figure 20.8**

**Knudson's two-hit mutation model for familial cancers.**

**Figure 20.9**

**Role of pRB in regulating the cell cycle at the $G_1$-to-S checkpoint.**

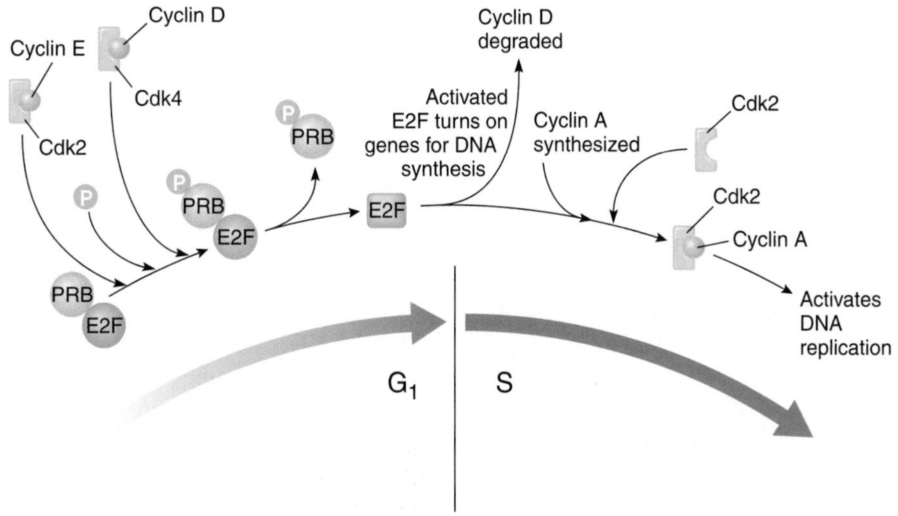

Cyclin D
Cyclin E
Cdk4
Cdk2
P
PRB
E2F
P
PRB
E2F
P
PRB
E2F
Cyclin D degraded
Activated E2F turns on genes for DNA synthesis
Cyclin A synthesized
Cdk2
Cdk2
Cyclin A
Activates DNA replication
$G_1$
S

Both point mutations and deletions in the gene have been shown to lead to loss of function of pRB in patients with retinoblastoma. In approximately 5% of patients with retinoblastoma, the genetic abnormality can be detected by karyotype analysis. The remainder are more difficult to detect, even with molecular analysis.

**The *TP53* Tumor Suppressor Gene.** The tumor suppressor gene *TP53* (*Tumor protein 53:* OMIM 191170) encodes a protein of molecular weight 53 kDa called p53. When both alleles are mutated, *TP53* may be involved in the development of perhaps 50% of all human cancers, including breast, brain, liver, lung, colorectal, bladder, and blood cancers. This does not mean that *TP53* causes 50% of human cancers, but rather that mutations in *TP53* are among the several genetic changes usually found in those cancers.

**animation** The Tumor Suppressor Gene, *TP53*

***Genetics of the* TP53 *Tumor Suppressor Gene.*** The human *TP53* gene is at chromosome location 17p13.1. Individuals who inherit one mutant copy of *TP53* develop Li–Fraumeni syndrome, a rare form of cancer that is an autosomal dominant trait because the cancer develops when the second copy of *TP53* becomes mutated. Individuals with this syndrome develop cancers in a number of tissues, including breast and blood.

***Function of p53.*** The 393-amino acid p53 tumor suppressor protein is a transcription factor that is regulated by phosphorylation and by its interaction with another phosphoprotein, the negative regulator Mdm2 (Figure 20.10). In a normal cell, both proteins are unphosphorylated, which allows them to bind together. Mdm2 stimulates degradation of p53, and as a result, the amount of p53 in the cell is low. When DNA damage occurs, p53 initiates a cascade of events leading to arrest in $G_1$. DNA damage results in phosphorylation of both p53 and Mdm2 on the domains where they normally interact. Therefore, a p53–Mdm2 complex cannot form and p53 degradation is not promoted, so p53 accumulates. Functioning as a transcription factor, p53 turns on transcription of DNA repair genes and of *WAF1*, which encodes a 21-kDa protein called p21. The p21 protein binds to the $G_1$-to-S checkpoint Cdk4–cyclin D complexes and inhibits their activity.[4] As a result, pRB in the pRB–E2F complex does not become phosphorylated, thereby keeping E2F inhibited. Entry into S is blocked (see Figure 20.10), and the cell arrests in $G_1$.

p53 provides some protection against oncogenes. Expression of viral or cellular oncogenes such as *ras* induces expression of the *ARF* gene. The *ARF* gene product, p14 protein, binds to Mdm2 in the p53–Mdm2 complex and blocks the stimulation of p53 degradation by $Mdm_2$. Somehow, the requirement of phosphorylation of p53 for activation of gene transcription is bypassed.

p53 also plays a role in **programmed cell death (apoptosis)**, a process by which a cell with a high level of DNA damage commits suicide. During apoptosis, DNA

**Figure 20.10**
**Function of p53 in cell cycle control.**

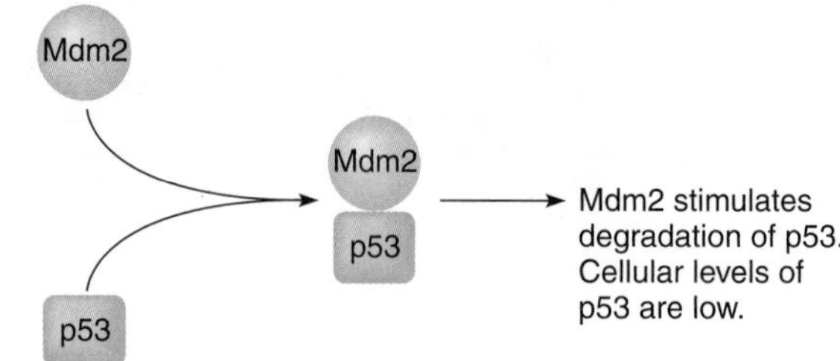


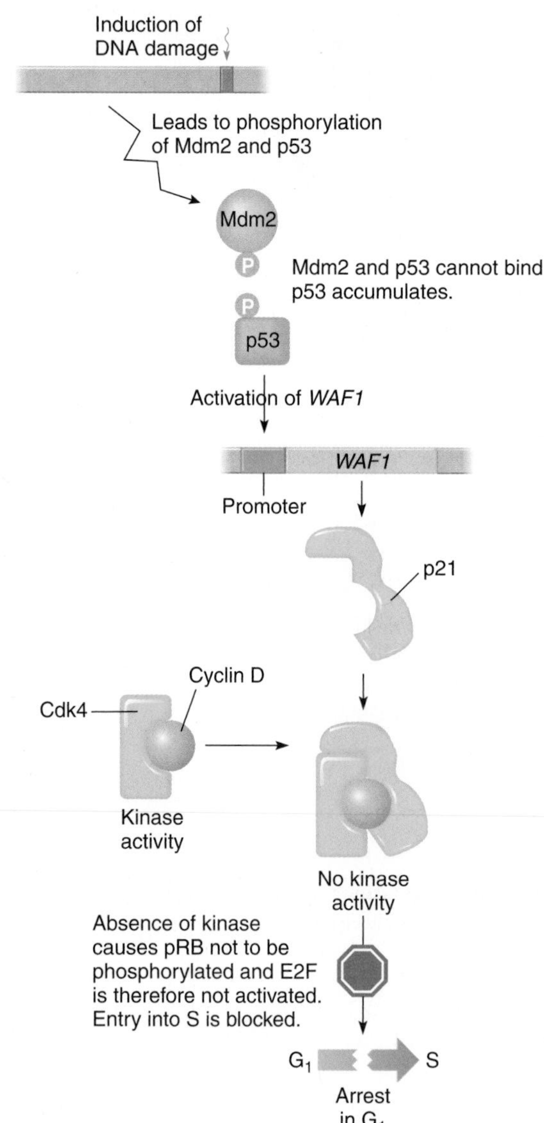


[4]The p21 protein can also bind to other checkpoint Cdk–cyclin complexes and inhibit their activity, thereby blocking the cell cycle at any stage. The example here focuses on the retinoblastoma protein and the $G_1$-to-S checkpoint.

is degraded and the nucleus condenses, and the cell may be devoured by phagocytes. In this process, p53 does not induce DNA repair genes or *WAF1*, but activates the *BAX* gene for the apoptosis pathway. The BAX protein blocks the function of the BCL-2 protein, which is a repressor of the apoptosis pathway. Without an active BCL-2 repressor, the apoptotic pathway is activated and the cell commits suicide.

If both alleles of *TP53* carry loss-of-function mutations, no active p53 can be produced. Thus, *WAF1* cannot be activated, and no p21 is available to block Cdk activity, so the cell is unable to arrest in $G_1$. Therefore, the cell may proceed to S, which is undesirable. Similarly, a cell with a high level of DNA damage will not be able to undergo programmed cell death. Most loss-of-function mutations of *TP53* occur in the part of the gene that encodes the DNA-binding domain of the p53 transcription factor. The mutant p53 molecules produced are unable to activate transcription of its target genes.

Transgenic mice have been produced with deletions of both *TP53* alleles. These *TP53*$^-$/*TP53*$^-$ knockout mice actually developed and were fully viable, indicating that the *TP53* gene is not essential for the processes of cell growth, cell division, or cell differentiation, at least in the mouse. The knockout mice showed only one major phenotype, that of a very high frequency of cancers from the sixth month (in 75% of the mice) to the tenth month (in 100% of the mice). These results support the roles of p53 in tumor suppression and in maintaining the genetic stability of cells.

**Breast Cancer Tumor Suppressor Genes.** In the United States, more than 185,000 new cases of breast cancer are diagnosed each year, representing more than 31% of all new cancers in women, and more than 46,000 women die each year from this cancer. In developed countries, there is a 1 in 10 chance that a woman will be diagnosed with breast cancer during her lifetime. The average age of onset is 55. Approximately 5% of breast cancers are hereditary. As with hereditary retinoblastoma, this form of the cancer has an earlier age of onset than the sporadic form, and the cancer often is bilateral.

Among the several genes that play a role in familial breast cancer, two genes—*BRCA1* (OMIM 113705) and *BRCA2* (OMIM 600185)—have been hypothesized to be tumor suppressor genes. (Some studies have led to the alternative hypothesis that these two genes are mutator genes; see next section.) It is believed that most hereditary breast cancer in the United States results from mutations in *BRCA1* or *BRCA2*, with most of the mutations occurring in *BRCA1*.

The *br*east *ca*ncer susceptibility gene *BRCA1* is at chromosome location 17q21. Mutations of the *BRCA1* gene also lead to susceptibility to ovarian cancer. The *BRCA1* gene encompasses more than 100 kb of DNA; it is transcribed in numerous tissues, including breast and ovary, to produce a 7.8-kb mRNA that is translated to produce a 190-kDa protein of 1,863 amino acids. The BRCA1 protein is involved in a number of functions, including cellular responses to DNA damage (the protein is essential for DNA damage repair), transcription regulation, and the addition of ubiquitin to proteins (ubiquitinated proteins are targeted for degradation).

*BRCA2* is at chromosome location 13q12–q13. Unlike *BRCA1*, *BRCA2* does not have an associated high risk of ovarian cancer. The *BRCA2* encompasses approximately 70 kb of DNA and encodes a 3,418-amino acid protein. The various functions proposed for BRCA2 are similar to that of BRCA1.

**Keynote**

Tumor suppressor genes, like proto-oncogenes, are involved in the regulation of cell growth and division. Whereas the normal products of proto-oncogenes have a stimulatory role in those processes, the normal products of tumor suppressor genes have inhibitory roles. Therefore, when both alleles of a tumor suppressor gene are inactivated or lost, the inhibitory activity is lost, and unprogrammed cell proliferation can occur. Inactivation of tumor suppressor genes is involved in the development of a wide variety of human cancers, including breast, colon, and lung cancers.

### MicroRNA Genes

MicroRNAs (miRNAs) are short, single-stranded, noncoding RNA regulatory molecules that are derived from RNA transcripts of eukaryotic nuclear genes (see Chapter 18, pp. 537–539). MicroRNAs are one type of short, RNA regulatory molecule that silences genes posttranscriptionally in eukaryotes by RNA interference (RNAi). Mechanistically, the silencing process involves binding of an miRNA to complementary or near-complementary sequences in the 3′ untranslated regions (UTRs) of target mRNAs, thereby inhibiting translation of those mRNAs and targeting them for storage or degradation.

Researchers are learning that miRNAs have important roles in regulating gene expression related to many biological processes. For example, miRNAs are involved in the control of many fundamental cellular and physiological processes, including cell proliferation and differentiation, apoptosis, and tissue and organ development. Relevant to this chapter, evidence is mounting that miRNAs also play a significant role in cell transformation and carcinogenesis in humans, and is leading to exciting new research in the field of cancer biology. For example, for human cancers that have been studied, numerous miRNA genes show altered expression patterns. More specifically, each form of cancer has a distinct miRNA expression pattern that differs from that of the equivalent normal tissue, and from that of other types of cancers. In the various cancers in which a given miRNA shows an altered pattern of expression, the change in expression often is the same—either toward increased or decreased expression. For example, relative to their expression in normal

cells. Explain whether telomerase activity alone can lead to cancer. If it cannot, why is it present in most cancerous cells, and what would be the biological consequences of eliminating it from cancerous cells?

**20.30** Explain whether the following statement is correct, and why: "Even though metastatic cancer cells are clonal descendants of a somatic cell that became cancerous, they are genetically distinct from that cell."

***20.31** Material that has been biopsied from tumors is useful for discerning both the type of tumor and the stage to which a tumor has progressed. It has been known for a long time that biopsied tissues with more differentiated cellular phenotypes are associated with less advanced tumors. Explain this finding in view of the multistep nature of cancer.

***20.32** Some tumor types have been very frequently associated with specific chromosomal translocations. In some cases, these translocations are found as the only cytogenetic abnormality. In each case examined to date, the chromosome breaks that occur result in a chimeric gene (pieces of two genes fused together) that encodes a fusion protein (a protein consisting of parts of two proteins fused together, corresponding to the coding sequences of the chimeric gene). A partial list of tumor-specific chromosomal translocations in bone and soft tissue tumors is given here.

| Tumor Type | Translocation | Characteristic | Genes |
|---|---|---|---|
| Ewing sarcoma | t(11;22) (q24;q12) | Malignant | *FLI1, EWS,* |
| | t(21;22) (q22;q12) | | *ERD, EWS,* |
| | t(7;21) (p22;q12) | | *ATV1, EWS* |
| Soft tissue clear cell carcinoma | t(12;22) (q13;q12) | Malignant | *ATF1, EWS* |
| Myxoid chondrosarcoma | t(9;22)(q22–31;q12) | Malignant | *CHN, EWS* |
| Synocial sarcoma | t(11;22) (p13;q12) | Malignant | *SSX1, SSX2, SYT* |
| Lipoma | t(var;12) (var;q13–15) | Benign | *HMGI-C* |
| Leiomyoma | t(12;14) (q13–15;q23–24) | Benign | *HMGI-C* |

**a.** What conclusions might you draw from the fact that in some cases these translocations are found as the only cytogenetic abnormality?

**b.** How might the formation of a chimeric protein result in tumor formation?

**c.** Based on the data presented here, can you infer whether the genes near the breakpoints of these translocations are tumor suppressor genes or proto-oncogenes? If so, which?

**d.** Can you speculate on how multiple translocations involving the *EWS* gene result in different sarcomas?

**e.** It is often difficult to diagnose individual sarcoma types based solely on tissue biopsy and clinical symptoms. How might the cloning of the genes involved in translocation breakpoints associated with specific tumors have a practical value in improving tumor diagnosis and management?

***20.33** What mechanisms ensure that cells with heavily damaged DNA are unable to replicate?

***20.34** Distinguish between direct-acting carcinogens, procarcinogens, and ultimate carcinogens in the induction of cancer.

**20.35** What sources of radiation exist, and how does radiation induce cancer?

# 21 Population Genetics

Extensive phenotypic variation in the color patterns of the Cuban tree snail.

## Key Questions

- How do we extend basic Mendelian principles to estimate the frequency of genotypes and alleles in a population?
- What is the Hardy–Weinberg law, and what assumptions does it make to predict how allele and genotype frequencies will change in populations?
- How do geneticists use the Hardy–Weinberg law to infer the evolutionary forces that may cause populations to change over time?
- How do we test whether genetic data we collect from a natural population is in agreement with the predictions of the Hardy–Weinberg law?
- How can we use the Hardy–Weinberg law to estimate the percentage of individuals in a population who are carriers of a trait?
- How do we measure genetic variation at the protein and DNA level in organisms from natural populations?
- What is the role of new mutations in changing allele frequencies in natural populations?
- What is the role of chance in changing allele frequencies in large and small populations?
- How does migration change allele frequencies between populations?
- How does natural selection affect changes in allele frequencies in natural populations?
- How does crossing-over influence the segregation of alleles at adjacent loci?

### iActivity

SOON AFTER MENDEL'S PRINCIPLES WERE rediscovered, geneticists began to look not only at the genetic makeup of individuals but also the genetic makeup of populations. Population genetics is one way that scientists determine whether evolution is occurring in groups of individuals and identify the forces that cause populations to evolve. In this chapter, you will learn about changes in the genetic makeup of populations, how such changes are measured, and the factors that cause these changes. Then, in the iActivity, you will explore the genetics of a type of mussel that is rapidly spreading through North American waterways.

The science of genetics can be broadly divided into four major subdisciplines: transmission genetics, molecular genetics, population genetics, and quantitative genetics. Each of these four areas focuses on a different aspect of heredity. **Transmission genetics** is concerned primarily with genetic processes that occur within individuals and how genes are passed from one individual to another. Thus, the unit of study for transmission genetics is the *individual*. In **molecular genetics,** we are interested largely in the molecular nature of heredity: how genetic information is encoded within the DNA and how biochemical processes of the cell translate the genetic information into influencing the phenotype. Consequently, in molecular genetics we focus on the *cell.* **Population genetics,** the subject of this chapter, applies the principles of transmission genetics to large groups of individuals, focusing on the transmission processes at one or a few genetic loci. **Quantitative genetics,** the subject of Chapter 22, also considers the transmission of traits simultaneously determined by many genes. Both population and quantitative genetics apply Mendelian principles, and they are amenable to mathematical treatment. In fact, these areas provide some of the oldest and richest examples of the success of mathematical theory in biology. The impetus for the development of these areas came after the rediscovery of Mendel's work and its great implications for Darwinian theory. In fact, the fusion of Mendelian theory with Darwinian theory is called the neo-Darwinian synthesis and was championed by Sir Ronald Fisher, Sewall Wright, and J. B. S. Haldane (Figure 21.1). The neo-Darwinian synthesis is now the foundation of a large part of modern biology.

Population geneticists investigate the patterns of genetic variation found among individuals within groups (the **genetic structure of populations**) and how these patterns vary geographically and change over time. In this discipline, our perspective shifts away from the individual and the cell and focuses instead on a large group of individuals, a Mendelian population. A **Mendelian population** is a group of *interbreeding* individuals who share a common set of genes. The genes shared by the individuals of a Mendelian population are called the **gene pool.** The principal aim of population genetics is to understand the genetics of **evolution,** a change in a population or species over time. The methods used focus on the gene pool of a Mendelian population rather than the genotypes of its individual members. An understanding of the genetic structure of a population is also a key to our understanding of the importance of genetic resources and the importance of genes for the conservation of species and biodiversity.

Questions frequently studied by population geneticists include the following:

1. How much genetic variation is found in natural populations, and what processes control the amount of variation observed?
2. What processes are responsible for producing genetic divergence among populations?
3. How do biological characteristics of a population, such as mating system, fecundity, and age structure, influence the genetic structure of the population?

To answer these questions, population geneticists sometimes make direct measurements of genetic variability within and among populations. Often they also develop mathematical models to describe how the gene pool of a population will change under various conditions. An example is the set of equations that describes the influence of random mating on the allele and genotypic frequencies of an infinitely large population, a model called the **Hardy–Weinberg law,** which we discuss later in this chapter. It is important to note that, while the models are simple and require numerous assumptions, many of which seem unrealistic, such models are useful because they strip a process to its essence and allow scientists to test particular attributes of a system in isolation. With such models we can examine what happens to the genetic structure of a population when we deliberately violate one assumption after another and then in combination. Once we understand the results of the simple models, we can

**Figure 21.1**

**Sir Ronald Fisher, Sewall Wright, and J. B. S. Haldane, considered the major architects of neo-Darwinian theory.**

Sir Ronald Fisher

Sewall Wright

J. B. S. Haldane

incorporate more realistic conditions into the equations, and we can use these more realistic models to help us understand historical evolutionary mechanisms that drove the changes that resulted in the patterns of differences we see among present-day natural populations. In the end, we will see that many attributes of genetic variation in populations can be explained by surprisingly simple models and that more complex models are required to explain other patterns of genetic variation.

**Keynote**

Population genetics seeks to understand the underlying causes of the observed patterns of genetic variation within populations (or gene pools) and divergence among populations. It uses both empirical tools, measuring variation in natural populations, and theoretical tools, which attempt to explain the observed variation with quantitative modeling.

## Genetic Structure of Populations

### Genotype Frequencies

To study the genetic structure of a Mendelian population, population geneticists must first describe it quantitatively. They do this by calculating genotype frequencies and allele frequencies within the population. A frequency is a proportion, and it always ranges between 0 and 1. If 43% of the people in a group have red hair, the frequency of red hair in the group is 0.43, and the frequency of people who *do not* have red hair is

$$1 - 0.43 = 0.57 \text{ or } 57\%$$

To calculate the **genotype frequencies** at a specific locus, we count the number of individuals with one particular genotype and divide this number by the total number of individuals in the population. We do this for each of the genotypes at the locus. The sum of the genotype frequencies should be 1. Consider a locus that determines the pattern of spots in the scarlet tiger moth, *Panaxia dominula* (Figure 21.2). Three genotypes are present in most populations, and each genotype produces a different phenotype. E. B. Ford collected moths at one locality in England and found the following numbers of genotypes: 452 *BB*, 43 *Bb*, and 2 *bb*, for a total of 497 moths. The genotype frequencies are therefore

$$f(BB) = 452/497 = 0.909$$
$$f(Bb) = 43/497 = 0.087$$
$$f(bb) = 2/497 = 0.004$$

where $f$ = *frequency*.

### Allele Frequencies

Although genotype frequencies at a single locus are useful for examining the effects of certain evolutionary processes on a population, in most cases population geneticists use frequencies of alleles to describe how the gene pool changes over time. The use of allele frequencies offers several advantages over the genotypic frequencies. First, in sexually reproducing organisms, genotypes break down to alleles when gametes are formed, and alleles, not genotypes, are passed from one generation to the next. Consequently, only alleles have continuity over time, and the gene pool evolves when allele frequencies change.

**Allele frequencies** can be calculated in two equivalent ways: from the observed numbers of different genotypes at a particular locus or from the genotype frequencies. First, we can calculate the allele frequencies directly from the *numbers* of genotypes. In this method, we count the number of alleles of one type at a particular locus and divide it by the total number of alleles at that locus in the population. This method is called *gene counting* and works for a wide variety of cases, including X-linked genes and mitochondrial genes. Expressing the gene counting method as a formula for a nuclear gene with two alleles, we get

$$\text{Allele frequency} = \frac{\text{Number of copies of a given allele}}{\text{Sum of counts of all alleles in the population}}$$

**Figure 21.2**

***Panaxia dominula*, the scarlet tiger moth.** The top two moths are normal homozygotes (*BB*), those in the middle two rows are heterozygotes (*Bb*), and the bottom moth is the rare homozygote (*bb*).

For example, imagine a population of 1,000 diploid individuals with 353 *AA*, 494 *Aa*, and 153 *aa* individuals. Each *AA* individual has two *A* alleles, whereas each *Aa* heterozygote possesses only a single *A* allele. Therefore, the number of *A* alleles in the population is (2 × the number of *AA* homozygotes) + (the number of *Aa* heterozygotes), or (2 × 353) + 494 = 1,200. Since every diploid individual has two alleles, the total number of alleles in the population is twice the number of individuals, or 2 × 1,000 for autosomal genes. Using the equation just given, the allele frequency is 1,200/2,000 = 0.60. When two alleles are present at a locus, we can use the following formula to calculate allele frequencies:

$$p = f(A) = \frac{(2 \times \text{count of } AA) + (\text{count of } Aa)}{2 \times \text{total number of individuals}}$$

The second method of calculating allele frequencies goes through the step of first calculating genotype frequencies as demonstrated previously. In this example $f(AA) = 0.353$, $f(Aa) = 0.494$, and $f(aa) = 0.153$. From these genotype frequencies we calculate the allele frequencies as follows:

$$p = f(A) = (\text{frequency of the } AA \text{ homozygote}) + (1/2 \times \text{frequency of the } Aa \text{ heterozygote})$$

$$q = f(a) = (\text{frequency of the } aa \text{ homozygote}) + (1/2 \times \text{frequency of the } Aa \text{ heterozygote})$$

The frequencies of two alleles, $f(A)$ and $f(a)$, are commonly symbolized as $p$ and $q$. The allele frequencies for a locus, like the genotype frequencies, should always add up to 1. This is because in a one-locus model that has only two alleles, 100% (i.e., the frequency = 1) of the alleles are accounted for by the sum of the percentages of the two alleles ($p + q = 1$). Therefore, once $p$ is calculated, $q$ can be easily obtained by subtraction: $1 - p = q$.

**Allele Frequencies with Multiple Alleles.** Suppose we have three alleles—$A^1$, $A^2$, and $A^3$—at a locus, and we want to determine the allele frequencies. Here, we use the same rule that we used with two alleles: we add up the number of alleles of each type and divide by the total number of alleles in the population:

$$p = f(A^1) = \frac{(2 \times \text{count of } A^1A^1) + (A^1A^2) + (A^1A^3)}{(2 \times \text{total number of individuals})}$$

$$q = f(A^2) = \frac{(2 \times \text{count of } A^2A^2) + (A^1A^2) + (A^2A^3)}{(2 \times \text{total number of individuals})}$$

$$r = f(A^3) = \frac{(2 \times \text{count of } A^3A^3) + (A^1A^3) + (A^2A^3)}{(2 \times \text{total number of individuals})}$$

To illustrate the calculation of allele frequencies when more than two alleles are present, we will use data from a study on genetic variation in milkweed beetles. Walter Eanes and his coworkers examined allele frequencies at a locus that codes for the enzyme phosphoglucomutase (PGM). Three alleles were found at this locus; each allele codes for a different molecular variant of the enzyme. In one population sample, the following numbers of genotypes were collected:

| | |
|---|---|
| $A^1A^1$ = | 4 |
| $A^1A^2$ = | 41 |
| $A^2A^2$ = | 84 |
| $A^1A^3$ = | 25 |
| $A^2A^3$ = | 88 |
| $A^3A^3$ = | 32 |
| Total = | 274 |

The frequencies of the alleles are calculated as follows:

$$p = f(A^1) = \frac{(2 \times 4) + (41) + (25)}{(2 \times 274)} = 0.135$$

$$q = f(A^2) = \frac{(2 \times 84) + (41) + (88)}{(2 \times 274)} = 0.542$$

$$r = f(A^3) = \frac{(2 \times 32) + (88) + (25)}{(2 \times 274)} = 0.323$$

As seen in these calculations, we add twice the number of homozygotes that possess the allele and one times the count of each of the heterozygotes that have the allele. We then divide by twice the number of individuals in the population, which represents the total number of alleles present. In the top part of the equation, notice that for each allele frequency, we do not add all the heterozygotes because some of the heterozygotes do not have the allele; for example, in calculating the allele frequency of $A^1$, we do not add the number of $A^2A^3$ heterozygotes in the top part of the equation. $A^2A^3$ individuals do not have an $A^1$ allele. We can use the same procedure for calculating allele frequencies when four or more alleles are present.

The second method for calculating allele frequencies from genotypic frequencies can also be used here. This calculation may be quicker if we have already determined the frequencies of the genotypes. The frequency of the homozygote is added to half of the heterozygote frequency because half of the heterozygote's alleles are *A* and half are *a*. If three alleles ($A^1$, $A^2$, and $A^3$) are present in the population, the allele frequencies are

$$p = f(A^1) = f(A^1A^1) + \frac{f(A^1A^2)}{2} + \frac{f(A^1A^3)}{2}$$

$$q = f(A^2) = f(A^2A^2) + \frac{f(A^1A^2)}{2} + \frac{f(A^2A^3)}{2}$$

$$r = f(A^3) = f(A^3A^3) + \frac{f(A^1A^3)}{2} + \frac{f(A^2A^3)}{2}$$

Although calculating allele frequencies from genotypic frequencies may be quicker than calculating them directly from the numbers of genotypes, more rounding error will occur. As a result, calculations from direct counts usually are preferred. Calculating genotype frequencies and allele frequencies is illustrated for a one-locus, three-allele example in Box 21.1.

**Box 21.1 Sample Calculation of Genotype and Allele Frequencies for Hemoglobin Variants Among Nigerians Where Multiple Alleles Are Present**

**Hemoglobin Genotypes**

| *AA* | *AS* | *SS* | *AC* | *SC* | *CC* | *Total* |
|---|---|---|---|---|---|---|
| 2,017 | 783 | 4 | 173 | 14 | 11 | 3,002 |

**Calculation of Genotype Frequencies**

$$\text{Genotype frequency} = \frac{\text{Number of individuals with the genotype}}{\text{Total number of individuals}}$$

$$f(SS) = \frac{4}{3{,}002} = 0.0013 \qquad f(AA) = \frac{2{,}017}{3{,}002} = 0.672 \qquad f(AC) = \frac{173}{3{,}002} = 0.058$$

$$f(AS) = \frac{783}{3{,}002} = 0.261 \qquad f(SC) = \frac{14}{3{,}002} = 0.0047 \qquad f(CC) = \frac{11}{3{,}002} = 0.0037$$

**Calculation of Allele Frequencies from the Number of Individuals with a Particular Genotype**

$$\text{Allele frequency} = \frac{\text{Number of copies of a given allele in the population}}{\text{Sum of all alleles in the population}}$$

$$f(S) = \frac{(2 \times \text{number of } SS \text{ individuals}) + (\text{number of } AS \text{ individuals}) + (\text{number of } SC \text{ individuals})}{2 \times \text{total number of individuals}}$$

$$f(S) = \frac{(2 \times 4) + 783 + 14}{(2 \times 3{,}002)} = \frac{805}{6{,}004} = 0.134$$

$$f(A) = \frac{(2 \times \text{number of } AA \text{ individuals}) + (\text{number of } AS \text{ individuals}) + (\text{number of } AC \text{ individuals})}{2 \times \text{total number of individuals}}$$

$$f(A) = \frac{(2 \times 2{,}017) + 783 + 173}{(2 \times 3{,}002)} = \frac{4{,}990}{6{,}004} = 0.831$$

$$f(C) = \frac{(2 \times \text{number of } CC \text{ individuals}) + (\text{number of } AC \text{ individuals}) + (\text{number of } SC \text{ individuals})}{2 \times \text{total number of individuals}}$$

$$f(C) = \frac{(2 \times 11) + 173 + 14}{(2 \times 3{,}002)} = \frac{209}{6{,}004} = 0.035.$$

**Calculation of Allele Frequencies from the Frequencies of Particular Genotypes**

$$f(S) = f(SS) + {}^1\!/\!_2 f(AS) + {}^1\!/\!_2 f(SC)$$

$$f(S) = 0.0013 + ({}^1\!/\!_2 \times 0.261) + ({}^1\!/\!_2 \times 0.0047) = 0.134$$

$$f(A) = f(AA) + {}^1\!/\!_2 f(AS) + {}^1\!/\!_2 f(AC)$$

$$f(A) = 0.672 + ({}^1\!/\!_2 \times 0.261) + ({}^1\!/\!_2 \times 0.058) = 0.831$$

$$f(C) = f(CC) + {}^1\!/\!_2 f(SC) + {}^1\!/\!_2 f(AC)$$

$$f(C) = 0.0037 + ({}^1\!/\!_2 \times 0.0047) + ({}^1\!/\!_2 \times 0.058) = 0.035$$

**Allele Frequencies at an X-Linked Locus.** Calculating allele frequencies at an X-linked locus (a locus found on the X chromosome) is slightly more complicated because males have only a single X-linked allele (in mammals and flies, for example). However, we can apply the same principles we used for autosomal loci. Remember that each homozygous female carries two X-linked alleles; heterozygous females have only one of that particular allele, and all males have only a single X-linked allele. To determine the number of alleles at an X-linked locus, we multiply the number of homozygous females by 2, then add the number of heterozygous females and the number of hemizygous males. We next divide by the total number of alleles in the population. When determining the total number of alleles, we add twice the number of females (because each female has two X-linked alleles) to the number of males (who have a single allele at X-linked loci). Using this reasoning, the frequencies of two alleles at an X-linked locus ($X^A$ and $X^a$) are determined with the following equations:

$$p = f(X^A) = \frac{(2 \times X^AX^A \text{ females}) + (X^AX^a \text{ females}) + (X^AY \text{ males})}{(2 \times \text{number of females}) + (\text{number of males})}$$

$$q = f(X^a) = \frac{(2 \times X^aX^a \text{ females}) + (X^AX^a \text{ females}) + (X^aY \text{ males})}{(2 \times \text{number of females}) + (\text{number of males})}$$

If the population has the same number of males and females, then the allele frequencies at an X-linked locus (averaged across sexes) can be determined from the genotypic frequencies as follows:

$$p = f(X^A) = {}^2/_3[f(X^AX^A) + {}^1/_2 f(X^AX^a)] + {}^1/_3 f(X^AY)$$

$$q = f(X^a) = {}^2/_3[f(X^aX^a) + {}^1/_2 f(X^AX^a)] + {}^1/_3 f(X^aY)$$

This formula assumes that the genotypic frequencies were calculated separately for each sex, so that $f(X^AY) + f(X^aY) = 1$. Be sure that you understand the logic behind the gene counting method; do not just memorize the formulas. If you understand fully the basis of the calculations, you will not need to remember the exact equations and will be able to determine allele frequencies for any situation.

### Keynote

The genetic structure of a population is determined by the total of all alleles (the gene pool). In the case of diploid, sexually interbreeding individuals, the structure is also characterized by the distribution of alleles into genotypes. The genetic structure can be described in terms of allele and genotypic frequencies. Except for rare mutations, individuals are born and die with the same set of alleles; what changes genetically over time (evolves) is the hereditary makeup of a group of individuals, reproductively connected from generation to generation as a Mendelian population.

## The Hardy–Weinberg Law

The Hardy–Weinberg law serves as a foundation for population genetics because it offers a simple explanation for how the Mendelian principle of segregation influences allele and genotype frequencies in a population. The Hardy–Weinberg law is named after the two individuals who independently discovered it in the early 1900s (Box 21.2). We begin our discussion of the Hardy–Weinberg law by simply stating what it tells us about the gene pool of a population. We then explore the implications of this principle and briefly discuss how the Hardy–Weinberg law is derived. Finally, we present some applications of the Hardy–Weinberg law and test a population to determine whether the genotypes are in Hardy–Weinberg proportions.

The Hardy–Weinberg law is divided into three parts: a set of assumptions and two major results. A simple statement of the law follows:

*Part 1 (Assumptions):* In an infinitely large, randomly mating population, free from mutation, migration, and natural selection (note that there are five assumptions);

*Part 2 (Result 1):* the frequencies of the alleles do not change over time where $p$ is the allele frequency of $A$ and $q$ is the allele frequency of $a$; and

*Part 3 (Result 2):* the genotypic frequencies remain in the proportions $p^2$ (frequency of $AA$), $2pq$ (frequency of $Aa$), and $q^2$ (frequency of $aa$). The sum of the genotype frequencies equals 1 (that is, $p^2 + 2pq + q^2 = 1$).

In short, the Hardy–Weinberg law explains what happens to the allele and genotype frequencies of a population as the alleles are passed from generation to generation in the absence of evolutionary forces. In other words, if the assumptions listed in part 1 are met, alleles are expected to combine into genotypes following the simple laws of probability described by the Hardy–Weinberg law. Under these

### Box 21.2 Hardy, Weinberg, and the History of Their Contribution to Population Genetics

Godfrey H. Hardy (1877–1947), a mathematician at Cambridge University, often met R. C. Punnett, the Mendelian geneticist, at the faculty club. One day in 1908 Punnett told Hardy of a problem in genetics that he attributed to a strong critic of Mendelism, G. U. Yule (Yule later denied having raised the problem). Supposedly, Yule said that if the allele for short fingers (brachydactyly) was dominant (which it is) and its allele for normal-length fingers was recessive, then short fingers ought to become more common with each generation. In time almost everyone in Britain should have short fingers. Punnett believed the argument was incorrect, but he could not prove it.

Hardy was able to write a few equations showing that, given any particular frequency of alleles for short fingers and alleles for normal fingers in a population, the relative number of people with short fingers and people with normal fingers will stay the same generation after generation if no natural selection is involved that favors one phenotype or the other in producing offspring. Hardy published a short paper describing the relationship between genotypes and phenotypes in populations, and within a few weeks a paper was published by Wilhelm Weinberg (1862–1937), a German physician of Stuttgart, that clearly stated the same relationship. The Hardy–Weinberg law signaled the beginning of modern population genetics.

To be complete, we should note that in 1903 American geneticist W. E. Castle of Harvard University was the first to recognize the relationship between allele and genotype frequencies, but it was Hardy and Weinberg who clearly described the relationship in mathematical terms. Therefore, the law is sometimes called the Castle–Hardy–Weinberg law.

circumstances a population is in Hardy–Weinberg equilibrium, and genotype frequencies can be predicted from allele frequencies.

## Assumptions of the Hardy–Weinberg Law

Part 1 of the Hardy–Weinberg law presents certain conditions, or assumptions that must be present for the law to apply. First, the law indicates that the population must be infinitely large. If a population is limited in size, chance deviations from expected ratios can result in changes in allele frequency, a phenomenon called **genetic drift.** It is true that the assumption of infinite size in part 1 is unrealistic: no population has an infinite number of individuals. But statistically, large populations look very similar to populations that are infinitely large. (We discuss this phenomenon later, when we examine genetic drift in more detail.) At this point, it is important to understand that populations need not be infinitely large for the Hardy–Weinberg law to provide a good approximation of genotypic frequencies. In fact, small departures from the assumptions of the law lead only to small departures from the Hardy–Weinberg proportions. Later we will see that in the end one must do a statistical test for goodness of fit of the data to determine whether the observed proportions are really different from those that are expected. We will see that finite populations with rare mutations, rare migrants, and weak selection depart from Hardy–Weinberg proportions only a little and that only when the deviations from the assumptions become large or when our sample size is very large can we detect departures from the Hardy–Weinberg proportions.

A second condition of the Hardy–Weinberg law is that mating must be random. **Random mating** is mating between genotypes occurring in proportion to the frequencies of the genotypes in the population. More specifically, the probability of a mating between two genotypes is equal to the product of the two genotypic frequencies.

To illustrate random mating, consider the *M-N* blood types in humans discussed in Chapter 13. The *M-N* blood type results from an antigen on the surface of a red blood cell, similar to the ABO antigens, except that incompatibility in the *M-N* system does not cause problems during blood transfusion. The *M-N* blood type is determined by one locus with two codominant alleles, $L^M$ and $L^N$. In a population of Eskimos, the frequencies of the three *M-N* genotypes are $L^M/L^M = 0.835$, $L^M/L^N = 0.156$, and $L^N/L^N = 0.009$. If Eskimos interbreed randomly, the probability of a mating between an $L^M/L^M$ male and an $L^M/L^M$ female is equal to the frequency of $L^M/L^M$ times the frequency of $L^M/L^M = 0.835 \times 0.853 = 0.697$. Similarly, the probabilities of other possible matings are equal to the products of the genotypic frequencies when mating is random.

The requirement of random matings for the Hardy–Weinberg law often is misinterpreted. Many students assume, incorrectly, that the population must be interbreeding randomly for all traits for the Hardy–Weinberg law to be valid. If this were true, human populations would never obey the Hardy–Weinberg law, because humans do not mate randomly. Humans mate preferentially for height, IQ, skin color, socioeconomic status, and other traits. However, although mating is nonrandom for some traits, most humans still choose mates at random for the *M-N* blood types; few of us even know what our *M-N* blood type is. The principles of the Hardy–Weinberg law apply to any trait (locus) for which random mating occurs, even if mating is nonrandom for other traits.

Finally, for the Hardy–Weinberg law to be a valid description of a trait, the population must be free from mutation, migration, and natural selection (described in detail later). In other words, the gene pool must be closed to the addition or subtraction of alleles, and we are interested in how allele frequencies are related to genotypic frequencies solely on the basis of meiosis and sexual reproduction. Remember though, that the assumptions of the Hardy–Weinberg law (evolutionary processes act on the population) apply only to the locus in question: A population may be subject to evolutionary processes acting on some traits while still meeting the Hardy–Weinberg assumptions at other traits.

## Predictions of the Hardy–Weinberg Law

If the conditions of the Hardy–Weinberg law are met, the population will be in genetic equilibrium, and two results are expected. First, the frequencies of the alleles will not change from one generation to the next. Second, the genotypic frequencies will be in the proportions $p^2$, $2pq$, and $q^2$ after one generation of random mating, and they will remain in those proportions in every generation that follows as all the conditions of the Hardy–Weinberg law continue to be met. When the genotypes are in these proportions, the population is said to be in Hardy–Weinberg equilibrium. An important use of the Hardy–Weinberg law is that it allows us to calculate the genotypic frequencies from the allele frequencies when the population is in equilibrium. If the observed genotype proportions are different from what we expect, we know that one or more of the Hardy–Weinberg assumptions has been violated.

To summarize, the Hardy–Weinberg law makes several predictions about the allele frequencies and the genotypic frequencies of a population when certain conditions are satisfied. The necessary conditions are that the population is large, randomly mating, and free from mutation, migration, and natural selection. When these conditions are met, the Hardy–Weinberg law indicates that allele frequencies will not change from generation to generation, and genotypic frequencies will be determined by the allele frequencies, occurring in the proportions $p^2$, $2pq$, and $q^2$.

## Derivation of the Hardy–Weinberg Law

The Hardy–Weinberg law states that when a population is in equilibrium, the genotypic frequencies will be in the proportions $p^2$, $2pq$, and $q^2$. To understand why, consider a hypothetical population in which the frequency of allele $A$ is $p$ and the frequency of allele $a$ is $q$. In producing gametes, each genotype passes on both alleles that it possesses with equal frequency; therefore, the frequencies of $A$ and $a$ in the gametes are also $p$ and $q$. If one thinks of the gametes as being in a pool, the random formation of zygotes involves reaching into the pool and drawing two gametes at random. The genotypes that then form after repeatedly drawing two gametes at a time will be in frequencies that are predicted by the probabilities of drawing the particular allele bearing gametes (see product rule, Chapter 11, p. 305). Table 21.1 shows the combinations of gametes when mating is random. This table illustrates the relationship between the allele frequencies and the genotypic frequencies, which forms the basis of the Hardy–Weinberg law. We see that when gametes pair randomly, the genotypes will occur in the proportions $p^2(AA)$, $2pq(Aa)$, and $q^2(aa)$ and are thus random mating frequencies. These genotypic proportions result from the expansion of the square of the allele frequencies $(p + q)^2 = p^2 + 2pq + q^2$, and the genotypes reach these proportions after one generation of random mating.

The Hardy–Weinberg law also states that allele and genotype frequencies remain constant generation after generation if the population remains large, randomly mating, and free from mutation, migration, and natural selection (i.e., evolutionary forces). This result can also be understood by considering a hypothetical, randomly mating population, as illustrated in Table 21.2. In Table 21.2, all possible matings are given. By definition, random mating means that the frequency of mating between two genotypes is equal to the product of the genotype frequencies. For example, the frequency of an $AA \times AA$ mating is equal to $p^2$ (the frequency of $AA$) × $p^2$ (the frequency of $AA$) = $p^4$. The frequencies of the offspring produced from each mating are also presented in Table 21.2.

We see that the sum of the probabilities of $AA \times Aa$ (or $2p^3q$) and $Aa \times AA$ (or $2p^3q$) matings is $4p^3q$, and we know from Mendelian principles that these crosses produce 1/2 $AA$ and 1/2 $Aa$ offspring. Therefore, the probability of obtaining $AA$ offspring from these matings is $4p^3q \times 1/2 = 2p^3q$. The frequencies of offspring produced by each type of mating are presented in the body Table 21.2. At the bottom of the table, the total frequency for each genotype is obtained by addition. As we can see,

**Table 21.1 Possible Combinations of *A* and *a* Gametes from Gametic Pools for a Population**

| Female gametes | Male gametes: $A(p)$ | Male gametes: $a(q)$ |
|---|---|---|
| $A(p)$ | $AA$ $(p^2)$ | $Aa$ $(pq)$ |
| $a(q)$ | $Aa$ $(pq)$ | $aa$ $(q^2)$ |

In sum, $p^2\,AA + 2pq\,Aa + q^2\,aa = 1.00$

**Table 21.2 Algebraic Proof of Genetic Equilibrium in a Randomly Mating Population for One Gene Locus with Two Alleles**

| Type of Mating ♀ ♂ | Mating Frequency | Offspring Frequencies Contributed to the Next Generation by a Particular Mating: $AA$ | $Aa$ | $aa$ |
|---|---|---|---|---|
| $p^2\,AA \times p^2\,AA$ | $p^4$ | $p^4$ | — | — |
| $p^2\,AA \times 2pq\,Aa$ [a]<br>$2pq\,Aa \times p^2\,AA$ | $4p^3q$ | $2p^3q$ | $2p^3q$ | — |
| $p^2\,AA \times q^2\,aa$<br>$q^2\,aa \times p^2\,AA$ | $2p^2q^2$ | — | $2p^2q^2$ | — |
| $2pq\,Aa \times 2pq\,Aa$ | $4p^2q^2$ | $p^2q^2$ | $2p^2q^2$ | $p^2q^2$ |
| $2\,pq\,Aa \times q^2\,aa$<br>$q^2\,aa \times 2\,pq\,Aa$ | $4pq^3$ | — | $2pq^3$ | $2pq^3$ |
| $q^2aa \times q^2aa$ | $q^4$ | — | — | $q^4$ |
| Totals | $(p^2 + 2pq + q^2)^2 = 1$ | $p^2(p^2 + 2pq + q^2) = p^2$ | $2pq(p^2 + 2pq + q^2) = 2pq$ | $q^2(p^2 + 2pq + q^2) = q^2$ |

Genotype frequencies = $(p + q)^2 = p^2 + 2pq + q^2 = 1$ in each generation afterward.

Gene (allele) frequencies = $p(A) + q(a) = 1$ in each generation afterward.

[a]For example, matings between $AA$ and $Aa$ will occur at $p^2 \times 2pq = 2p^3q$ for $AA \times Aa$ and at $2pq \times p^2 = 2p^3q$ for $Aa \times AA$ for a total of $4p^3q$. Two progeny types, $AA$ and $Aa$, result in equal proportions from these matings. Therefore, offspring frequencies are $2p^3q$ (i.e., $1/2 \times 4p^3q$) for $AA$ and for $Aa$.

after random mating the genotype frequencies are still $p^2$, $2pq$, and $q^2$ and the allele frequencies remain at $p$ and $q$. The frequencies of the population can thus be represented in the zygotic and gametic stages as follows:

| Zygotes | Gametes |
|---|---|
| $p^2(AA) + 2pq(Aa) + q^2(aa)$ | $p(A) + q(a)$ |

Each generation of zygotes produces $A$ and $a$ gametes in proportions $p$ and $q$. The gametes unite to form $AA$, $Aa$, and $aa$ zygotes in the proportions $p^2$, $2pq$, and $q^2$, and the cycle is repeated indefinitely as long as the assumptions of the Hardy–Weinberg law hold. This short proof gives the theoretical basis for the Hardy–Weinberg law.

The Hardy–Weinberg law indicates that at equilibrium, the genotype frequencies depend only on the frequencies of the alleles. This relationship between allele frequencies and genotype frequencies for a locus with two alleles is represented in Figure 21.3. Several aspects of this relationship should be noted: (1) The maximum frequency of the heterozygote is 0.5, and this maximum value occurs only when the frequencies of $A$ and $a$ are both 0.5; (2) if allele frequencies are between 0.33 and 0.66, the heterozygote is the most numerous genotype; and (3) when the frequency of one allele is less than 0.33, the homozygote for that allele is the rarest of the three genotypes.

**Figure 21.3**

**Relationship of the frequencies of the genotypes *AA*, *Aa*, and *aa* to the frequencies of alleles *A* and *a* (in values of *p* [top abscissa] and *q* [bottom abscissa], respectively) in populations that meet the assumptions of the Hardy–Weinberg law.** Any single population is defined by a single vertical line such as $p = 0.3$ and $q = 0.7$.

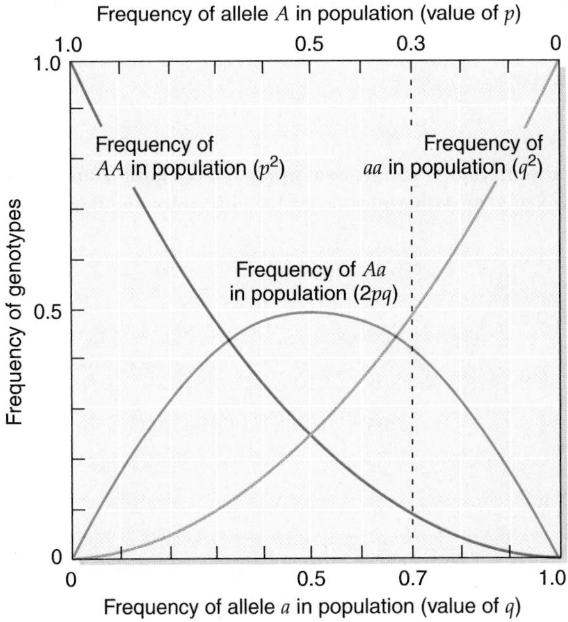


This point is also illustrated by the distribution of genetic diseases in humans, which are frequently rare and recessive. For a rare recessive trait, the frequency of the allele causing the trait is much higher than the frequency of the trait itself because most of the rare alleles are in unaffected heterozygotes (i.e., carriers). Albinism, for example, is a rare recessive condition in humans. In *tyrosinase-negative albinism*, affected individuals have no tyrosinase activity, which is required for normal production of melanin pigment. Among North American whites, the frequency of tyrosinase-negative albinism is roughly 1 in 40,000, or 0.000025. Since albinism is a recessive condition, the genotype of affected individuals is $aa$. If we assume that the population meets the assumptions of the Hardy–Weinberg law, then the frequency of the $aa$ genotypes equals $q^2$. If $q^2 = 0.000025$, then $q = 0.005$ and $p = 1 - q = 0.995$. The heterozygote frequency is therefore $2pq = 2(0.995)0.005 = 0.00995$ (almost 1%). Thus, although the frequency of albinism is low (1 in 40,000), individuals heterozygous for albinism are much more common (almost 1 in 100). When an allele is rare, almost all copies of that allele are in heterozygotes, and in this case recessive phenotypes often are very rare.

**Keynote**

The Hardy–Weinberg law describes what happens to allele and genotype frequencies of a large population, when gametes fuse randomly and there is no mutation, migration, or natural selection. If these conditions are met, allele frequencies do not change from generation to generation, and the genotype frequencies stabilize after one generation in the proportions $p^2$, $2pq$, and $q^2$, where $p$ and $q$ equal the frequencies of the alleles in the population. Under these conditions, the population is in Hardy–Weinberg equilibrium.

## Extensions of the Hardy–Weinberg Law to Loci with More than Two Alleles

When two alleles are present at a locus, the Hardy–Weinberg law tells us that at equilibrium the frequencies of the genotypes are $p^2$, $2pq$, and $q^2$, which is the square of the allele frequencies $(p + q)^2$. This is a simple binomial expansion, and this principle of probability theory can be extended to any number of alleles that are sampled two at a time into a diploid zygote. For example, if three alleles are present (e.g., alleles $A$, $B$, and $C$) with frequencies equal to $p$, $q$, and $r$, the frequencies of the genotypes at equilibrium are also given by the square of the allele frequencies:

$$(p + q + r)^2 = p^2(AA) + 2pq(AB) + q^2(BB) + 2pr(AC) + 2qr(BC) + r^2(CC)$$

In the blue mussel found along the Atlantic coast of North America, three alleles are common at a locus coding for the enzyme leucine aminopeptidase (LAP).

For a population of mussels inhabiting Long Island Sound (discussed later; see Figure 21.6), Richard K. Koehn and colleagues determined that the frequencies of the three alleles were as follows:

| Allele | Frequency |
|---|---|
| $LAP^{98}$ | $p = 0.52$ |
| $LAP^{96}$ | $q = 0.31$ |
| $LAP^{94}$ | $r = 0.17$ |

If the population were in Hardy–Weinberg equilibrium, the expected genotype frequencies would be

| Genotype | Expected Frequency | |
|---|---|---|
| $LAP^{98}/LAP^{98}$ | $p^2 = (0.52)^2$ | $= 0.27$ |
| $LAP^{98}/LAP^{96}$ | $2pq = 2(0.52)(0.31)$ | $= 0.32$ |
| $LAP^{96}/LAP^{96}$ | $q^2 = (0.31)^2$ | $= 0.10$ |
| $LAP^{96}/LAP^{94}$ | $2qr = 2(0.31)(0.17)$ | $= 0.11$ |
| $LAP^{94}/LAP^{98}$ | $2pr = 2(0.52)(0.17)$ | $= 0.18$ |
| $LAP^{94}/LAP^{94}$ | $r^2 = (0.17)^2$ | $= 0.03$ |

The square of the allele frequencies can be used in the same way to estimate the expected frequencies of the genotypes when four or more alleles are present at a locus.

## Extensions of the Hardy–Weinberg Law to X-Linked Alleles

In species like humans or *Drosophila*, in which females are XX and males are XY, the Hardy–Weinberg law must be modified. If alleles are X-linked, females may be homozygous or heterozygous, but males carry only a single allele for each X-linked locus. For X-linked alleles in females, the Hardy–Weinberg frequencies are the same as those for autosomal loci: $p^2(X^AX^A)$, $2pq(X^AX^B)$, and $q^2(X^BX^B)$. In males, however, the genotype frequencies are $p(X^AY)$ and $q(X^BY)$, the same as the allele frequencies in the population. For this reason, recessive X-linked traits are more frequent among males than among females. To illustrate this concept, consider red-green color blindness, which is an X-linked recessive trait. We actually know that many different defective alleles cause red-green color blindness, but for now we well lump them together. The frequency of the color-blindness allele varies among human ethnic groups; the frequency among African-Americans is 0.039. At equilibrium, the expected frequency of color-blind males in this group is $q = 0.039$, but the frequency of color-blind females is only $q^2 = (0.039)^2 = 0.0015$.

When random mating occurs within a population, the equilibrium genotype frequencies are reached in one generation. However, if the alleles are X-linked and the sexes differ in allele frequency, the equilibrium frequencies are approached over several generations. This is because males receive their X chromosome from their mother only, whereas females receive an X chromosome from both the mother and the father. Consequently, the frequency of an X-linked allele in males is the same as the frequency of that allele in their mothers, whereas the frequency in females is the average of that in mothers and fathers. With random mating, the allele frequencies in the two sexes alternate each generation, and the difference in allele frequency between the sexes is reduced by half each generation, as shown in Figure 21.4. Once the allele frequencies of the males and females are equal, the genotype frequencies are in Hardy–Weinberg proportions after one more generation of random mating.

### Testing for Hardy–Weinberg Proportions

When we take a sample from a population and calculate genotype frequencies, the match to Hardy–Weinberg proportions rarely is exact. To test whether the fit to Hardy–Weinberg is acceptable, we ask, "What is the chance that we would get this big a departure by chance alone?" If the observed genetic structure does not match the expected structure based on the law, we can begin to ask about which of the assumptions are being violated. To determine whether the genotypes of a population are in Hardy–Weinberg proportions, we first compute the allele frequencies from the observed genotype frequencies. (Note: It is important not to take the square roots of the homozygote frequencies to obtain allele frequencies, because to do so already assumes that the population is in equilibrium. Thus, allele frequencies should be calculated by the gene counting method. After obtaining the allele frequencies, we can calculate the expected genotype frequencies ($p^2$, $2pq$, and $q^2$) and compare these frequencies with the actual observed genotype frequencies using a chi-square test (see Chapter 2). The chi-square test gives us the probability that the difference between what we observed and what we expect under the Hardy–Weinberg law is due to chance. To illustrate this procedure, consider a locus that codes for transferrin (a blood protein) in the red-backed vole, *Clethrionomys gapperi*.

**Figure 21.4**

**Representation of the gradual approach to equilibrium of an X-linked gene with an initial of 1.0 in females and 0 in males.**

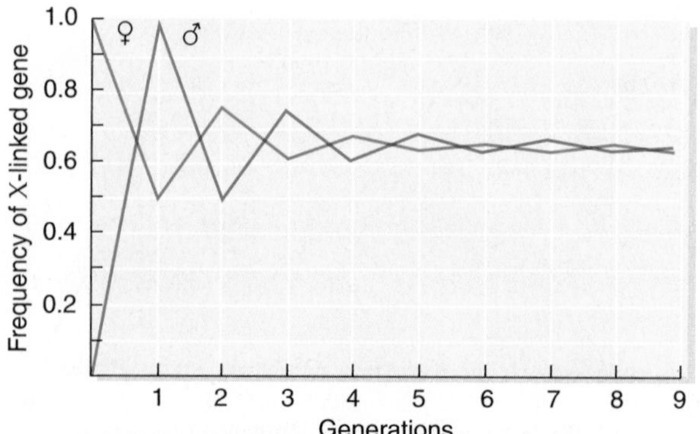

Three genotypes are found at the transferrin locus: *MM*, *MJ*, and *JJ*. In a population of voles trapped during 1976 in the Northwest Territories of Canada, 12 *MM* individuals, 53 *MJ* individuals, and 12 *JJ* individuals were found. To determine whether the genotypes are in Hardy–Weinberg proportions, we first calculate the allele frequencies for the population using our familiar formula:

$$p = \frac{(2 \times \text{number of homozygotes}) + (\text{number of heterozygotes})}{2 \times \text{total number of individuals}}$$

Therefore,

$$p = f(M) = \frac{(2 \times 12) + (53)}{(2 \times 77)} = 0.50$$

$$q = 1 - p = 0.50$$

Using $p$ and $q$ calculated from the observed genotypes, we can now compute the expected Hardy–Weinberg proportions for the genotypes: $f(MM) = p^2 = (0.50)^2 = 0.25$, $f(MJ) = 2pq = 2(0.50)(0.50) = 0.50$, and $f(JJ) = q^2 = (0.50)^2 = 0.25$. However, for the chi-square test, actual numbers of individuals are needed, not the proportions. To obtain the expected numbers, we simply multiply each expected proportion times the total number of individuals counted ($N$), as follows:

| | Expected | Observed |
|---|---|---|
| $f(MM) = p^2 \times N$ | | |
| $= 0.25 \times 77 =$ | 19.3 | 12 |
| $f(MJ) = 2pq \times N$ | | |
| $= 0.50 \times 77 =$ | 38.5 | 53 |
| $f(JJ) = q^2 \times N$ | | |
| $= 0.25 \times 77 =$ | 19.3 | 12 |

With observed and expected numbers, we can compute a chi-square value to determine the probability that the differences between observed and expected numbers could be the result of chance. The chi-square ($\chi^2$) is computed using the same formula we used for analyzing genetic crosses. That is, $d$, the deviation, is calculated for each class as (observed − expected); $d^2$, the deviation squared, is divided by the expected number $e$ for each class; and chi-square ($\chi^2$) is computed as the sum of all $d^2/e$ values. For this example, $\chi^2 = 10.98$. We now need to find this value in the chi-square probability table (see Table 11.5, p. 313) under the appropriate degrees of freedom. In contrast to the previous chi-square analyses, two degrees of freedom are lost—one for every parameter ($p$ in this case) that must be calculated from the data and another for the fixed number of individuals. Therefore, with three genotypic classes (*MM*, *MJ*, and *JJ*), two degrees of freedom are lost, leaving one degree of freedom (df $= n - 2$).

In the chi-square table under the column for one degree of freedom, the $\chi^2$ value of 10.98 indicates a $P$ value less than 0.01. Thus the probability that the differences between the observed and expected values is due to chance is very low, less than 1%. We can thus conclude that the observed numbers of genotypes are not similar to the expected numbers under Hardy–Weinberg law, the trait is not consistent with the expectations of random mating in a large population, and it deviates from the expectations of Hardy–Weinberg equilibrium.

## Using the Hardy–Weinberg Law to Estimate Allele Frequencies

An important application of the Hardy–Weinberg law is the calculation of allele frequencies when one or more alleles is recessive. For example, we have seen that albinism in humans results from an autosomal recessive gene. Normally this trait is rare, but among the Hopi Indians of Arizona, albinism is remarkably common. Charles M. Woolf and Frank C. Dukepoo visited the Hopi villages in 1969 and observed 26 cases of albinism in a total population of about 6,000 Hopis (Figure 21.5). This gave a frequency for the trait of 26/6,000, or 0.0043, which is much higher than the frequency of albinism in most populations. Although we have calculated the frequency of the trait, we cannot directly determine the frequency of the gene for albinism because we cannot distinguish between heterozygous individuals and those homozygous for the normal allele. Recall that our computation of the allele frequency involves counting the number of alleles:

$$p = \frac{(2 \times \text{number of homozygotes}) + (\text{number of heterozygotes})}{2 \times \text{total number of individuals}}$$

But because heterozygotes for a recessive trait such as albinism cannot be visually identified, counting the

**Figure 21.5**

**Three Hopi girls, photographed about 1900.** The middle child has albinism, an autosomal recessive disorder that occurs with high frequency among the Hopi Indians of Arizona.

number of alleles is not possible. Nevertheless, we can determine the allele frequency from the Hardy–Weinberg law if we assume that genotypes in the population are found in Hardy–Weinberg proportions. When genotypes are in Hardy–Weinberg proportions, the frequency of the homozygous recessive genotype is $q^2$. For albinism among the Hopi, $q^2 = 0.0043$, and $q$ can be obtained by taking the square root of the frequency of the affected genotype. Therefore, $q = \sqrt{0.0043} = 0.065$, and $p = 1 - q = 0.935$. Following the Hardy–Weinberg law, the frequency of heterozygotes in the population is $2pq = 2 \times 0.935 \times 0.065 = 0.122$, or about $1/8$. Thus, one out of eight Hopis, on average, carries an allele for albinism.

We should not forget that this method of calculating allele frequency rests on the assumption that genotypes are in Hardy–Weinberg proportions. In this case, there are reasons to expect departures from Hardy–Weinberg because strong social factors affect albino Hopi individuals which cause a violation of the random mating assumption. If the conditions of the Hardy–Weinberg law do not apply, then our estimate of allele frequency is inaccurate. Also, once we calculate allele frequencies with these assumptions, we cannot then test the population to determine whether the genotype frequencies are in the Hardy–Weinberg expected proportions. To do so would involve circular reasoning, for we assumed Hardy–Weinberg proportions in the first place to calculate the allele frequencies. Before we explore how the model can be used to discern the causes of deviations of observed populations that may not be in Hardy–Weinberg equilibrium from theoretically expected populations that are in Hardy–Weinberg equilibrium, we will look at the genetic structure of some real populations.

## Genetic Variation in Space and Time

The genetic structure of populations can vary in space and time. This means that the frequencies and distribution of alleles can vary in samples of the same species in different geographic areas or samples from the same geographic area collected at different generation times. Figure 21.6 shows how the frequencies of three alleles at the locus for the enzyme leucine aminopeptidase gradually change in a geographic series of samples of blue mussels that inhabit the East Coast of the United States. Many populations of plants and animals that have widespread geographic distributions show differences of this sort in the allele frequencies of populations in different geographic regions. In some cases, the spatial variation shows clear patterns or trends across geographic space. When allele frequencies change in a systematic way across a geographic transect, we call this an allele frequency **cline.** Often clines are associated with changes in a physical attribute in the environment, such as temperature or water availability. In the case of Figure 21.6, there is a clear thermal cline. Atlhough such clines suggest that the geographic pattern is caused by differential selection for the alternate alleles, additional studies are required to exclude other possibilities and to provide evidence for selectively maintained clines. Just as the genetic composition of a species may vary geographically, the genetic structure of individual populations can change over time, as shown in Figure 21.7.

As a result of the importance of geographic variation in allele frequencies, population geneticists have devised many statistical tools for quantifying the spatial patterns of genetic variation. The simplest of these quantifies the partitioning of total genetic variance into component parts. At the simplest level we can think of a fraction of the genetic variance that exists within each local population and another fraction of the genetic variance that results from differences between distinct local populations. By this kind of measure, for example, generally we find that only about 12–13% of the genetic variance in humans is found between different populations, whereas 87–88% of the total genetic variance in humans is found within populations. Geographic patterns of genetic variation may also be immensely important for conservation. To evaluate the potential for evolution of a species and conservation of its current genetic resources, the spatial component to genetic variation among current populations must be accounted for carefully. Conservation of genetic diversity thus demands a quantitative understanding of the geographic patterns of variability.

**Keynote**

The genetic structure of a species can vary both geographically and temporally.

## Genetic Variation in Natural Populations

One of the most significant questions addressed in population genetics is how much genetic variation exists within natural populations. Genetic variation within populations is important for several reasons. First, it determines the potential for evolutionary change and adaptation. The amount of variation also gives us important clues about the relative importance of various evolutionary processes because some processes increase variation while others decrease it. The manner in which new species arise and contemporary populations become extinct may depend on the amount of genetic variation harbored within populations. In addition, the ability of a population to persist over time can be influenced by how much genetic variation it has to draw on should environments change. For all these reasons, population geneticists are interested in measuring genetic variation, attempting to understand the evolutionary processes that affect it, and understanding the

Figure 21.6

**Geographic variation in frequencies of the alleles *LAP*$^{94}$, *LAP*$^{96}$, and *LAP*$^{98}$ of the locus coding for the enzyme leucine amino peptidase (LAP) in the blue mussel.**

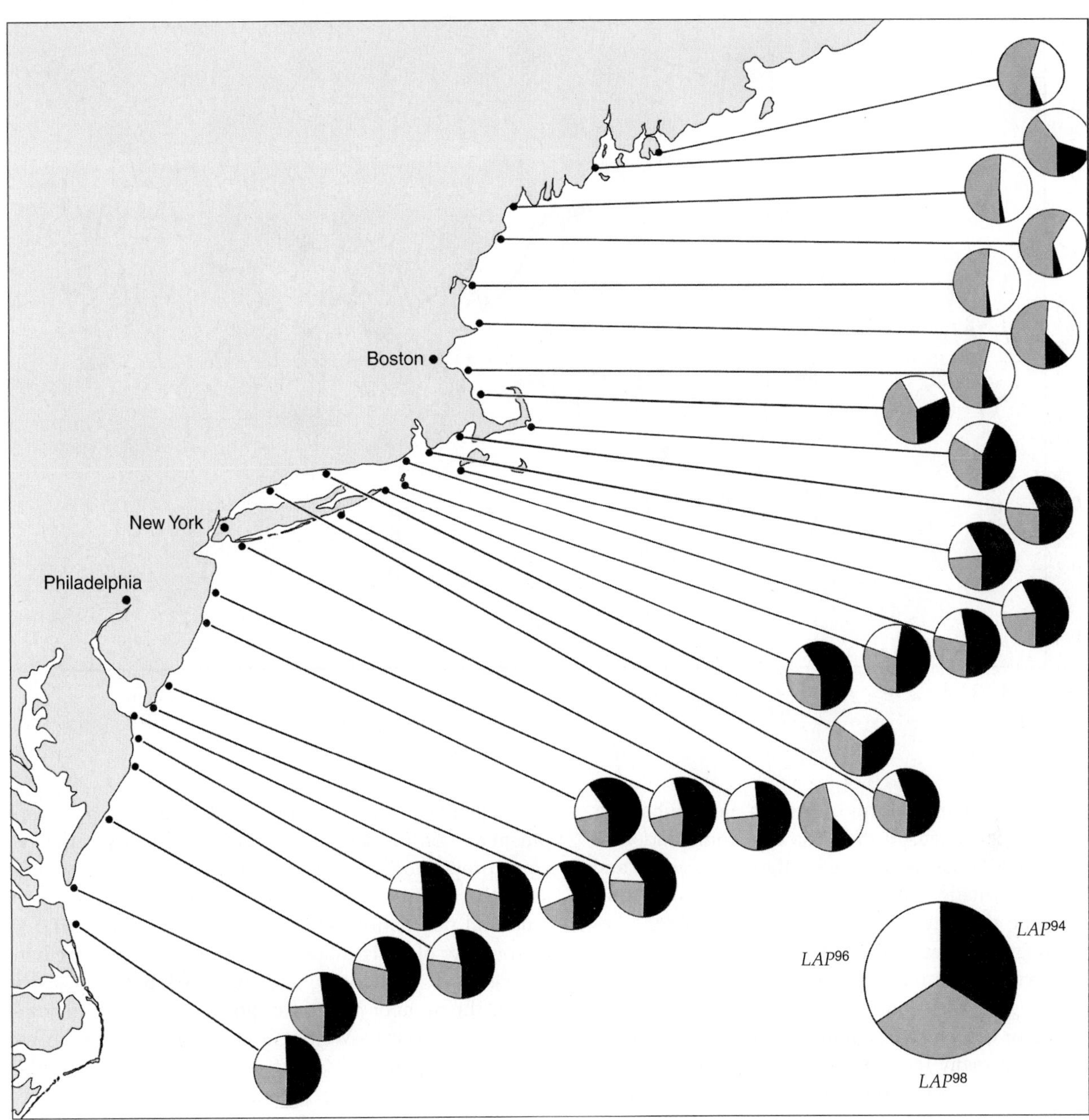




effects of human environmental disturbance that may alter it.

## Measuring Genetic Variation at the Protein Level

For many years, population geneticists were constrained in quantifying how much variation existed within natural populations. Naturalists recognized that plants and animals in nature frequently differ in phenotype, but the genetic basis of most phenotypic traits is too complex to assign specific genotypes to individuals. A few traits and alleles that behaved in a Mendelian fashion, such as spot patterns in butterflies and shell color in snails, provided observable genetic variation, but these isolated cases were too few to provide any general estimate of genetic variation. In 1966, Richard Lewontin and John Hubby published a study that applied protein electrophoresis to the study of polymorphism in natural populations. As in the electrophoretic separation of DNA molecules, protein electrophoresis works by separating proteins as they move through a gel matrix. Once they are separated, a specific stain is added to the gel to visualize the protein

Figure 21.7

**Temporal variation in the locus coding for the enzyme esterase 4F in the prairie vole, *Microtus ochrogaster*.** The four populations are close to each other and near Lawrence, Kansas.

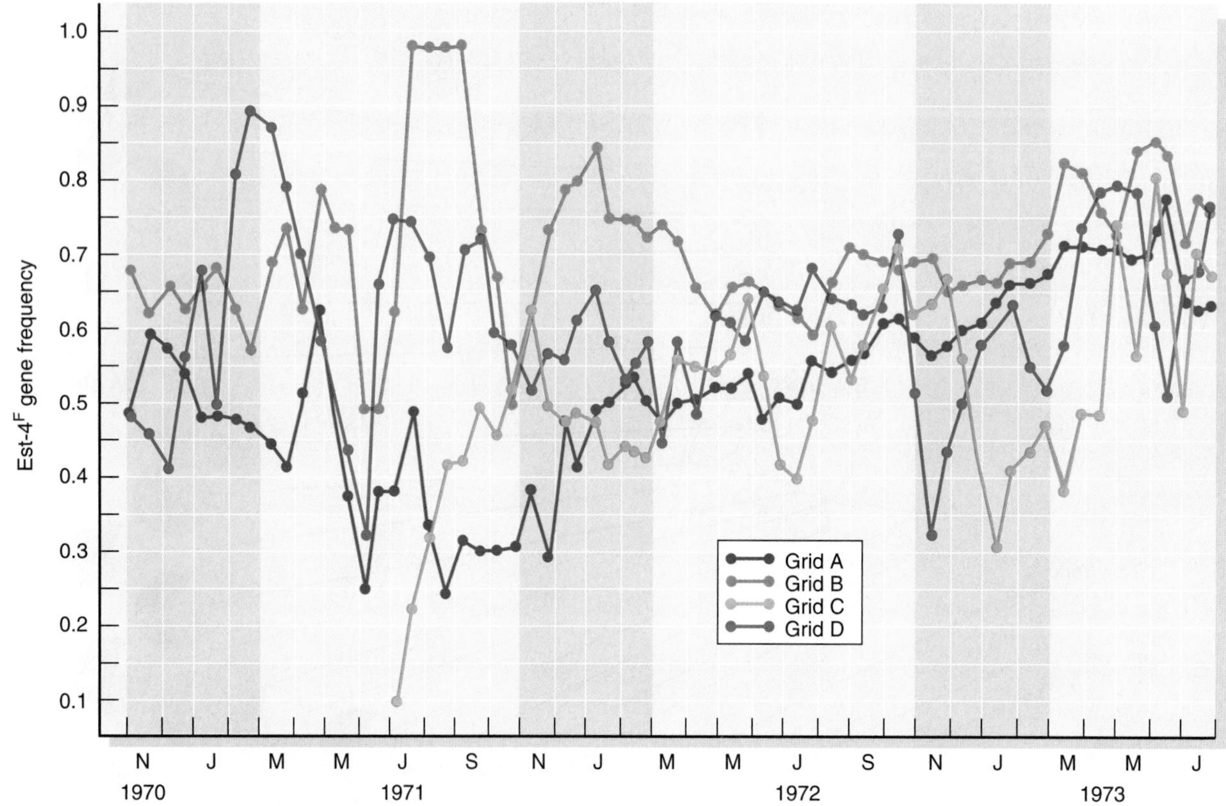


bands. In one common form of protein electrophoresis, proteins are separated on the basis of a combination of charge, which varies depending on the amino acids, and folding conformation of the protein. The fact that proteins can be separated based on these features implies that as long as alleles of the same gene produce protein products with different charge, they can be separated. For population geneticists, protein electrophoresis provides a technique for quickly determining the genotypes of many individuals at many loci. This procedure was soon used to examine genetic variation in hundreds of plant and animal species, and it paved the way for modern applications of DNA sequencing to answer questions about the forces underlying this genetic variation. The amount of genetic variation within a population was commonly measured with two parameters, the proportion of polymorphic loci and heterozygosity.

A polymorphic locus is any locus that has more than one allele present within a population. The **proportion of polymorphic loci (*P*)** is calculated by dividing the number of polymorphic loci by the total number of loci examined. For example, suppose we found that of 33 loci in a population of green frogs, 18 were polymorphic. The proportion of polymorphic loci would be $^{18}/_{33} = 0.55$. It is important to realize that the proportion of polymorphic loci depends on the technique used to identify polymorphism and on the sample size. **Heterozygosity** is the average proportion of heterozygous individuals in a population for many loci. When considering a single locus in a natural population, **Observed heterozygosity ($H_o$)** is the number of individuals in the population that are heterozygous at that locus, and **expected heterozygosity ($H_e$)** is the number of heterozygotes expected if the population is in Hardy–Weinberg equilibrium. Suppose we analyzed the genotypes of green frogs from one population at a locus coding for esterase and found that the frequency of heterozygotes ($2pq$) was 0.09. $H_o$ for this locus is 0.09. We would average this heterozygosity with those for other loci and obtain an estimate of mean observed heterozygosity for the population. Note that protein electrophoresis misses much of the variation that is detected when the DNA sequence of the same gene is determined, because silent or synonymous nucleotides may vary at the DNA level but leave no trace of variation at the protein level. This means that estimates of heterozygosity and proportion of polymorphic loci are typically different from estimates based on nucleotide sequence variation for the same gene.

Table 21.3 presents estimates from protein electrophoresis of the proportion of polymorphic loci and

**Table 21.3 Genic Variation in Some Major Groups of Animals and Plants**

| Group | Number of Species or Forms | Mean Number of Loci Examined per Species | Mean Proportion of Loci: Polymorphic per Population | Mean Proportion of Loci: Heterozygous per Individual |
|---|---|---|---|---|
| Insects | | | | |
| *Drosophila* | 28 | 24 | 0.529 ± 0.030[a] | 0.150 ± 0.010 |
| Others | 4 | 28 | 0.531 | 0.151 |
| Haplodiploid wasps | 6 | 15 | 0.243 ± 0.039 | 0.062 ± 0.007 |
| Marine invertebrates | 9 | 26 | 0.587 ± 0.084 | 0.147 ± 0.019 |
| Snails | | | | |
| Land | 5 | 18 | 0.437 | 0.150 |
| Marine | 5 | 17 | 0.175 | 0.083 |
| Fish | 14 | 21 | 0.306 ± 0.047 | 0.078 ± 0.012 |
| Amphibians | 11 | 22 | 0.336 ± 0.034 | 0.082 ± 0.008 |
| Reptiles | 9 | 21 | 0.231 ± 0.032 | 0.047 ± 0.008 |
| Birds | 4 | 19 | 0.145 | 0.042 |
| Rodents | 26 | 26 | 0.202 ± 0.015 | 0.054 ± 0.005 |
| Large Mammals[b] | 4 | 40 | 0.233 | 0.037 |
| Plants[c] | 8 | 8 | 0.464 ± 0.064 | 0.170 ± 0.031 |

[a]Values are mean ± standard error.
[b]Human, chimpanzee, pigtailed macaque, and southern elephant seal.
[c]Predominantly outcrossing species; mean gene diversity is 0.233 ± 0.029.

heterozygosity for many species that have been surveyed with electrophoresis. The results of these studies are unambiguous: most species have large amounts of genetic variation in their proteins. When these estimates first emerged during the 1960s and 1970s, they were quite surprising because they ruled out the **classical model** of genetic variation that had emerged primarily from the work of laboratory geneticists. According to this model, most natural populations have a "wild-type" allele with only rare mutant variants existing in the population. The large amounts of genetic variation observed in natural populations begs the question that if the classical model is wrong, then what maintains so much genetic variation within populations?

Initially it was assumed that the large amount of genetic variation was maintained by some form of natural selection; that is, it was advantageous for populations to possess genetic variation. However, in 1968, Motoo Kimura proposed that much of the pattern of evolutionary changes in protein molecules could be explained by the opposing forces of mutation and random genetic drift. This model was also applied to patterns of variation within populations, and initially many predictions from the model seemed to provide good statistical fits to the data. Because of the explanatory power of this model, Kimura called it the **neutral theory.** This theory acknowledges the presence of extensive genetic variation in proteins but proposes that most of the variation is neutral with regard to natural selection. This does not mean that the proteins detected by electrophoresis have no function, but rather that the different forms of the proteins that are seen as genotypes on gels are physiologically equivalent. Therefore, natural selection does not act on the neutral alleles, and random processes such as mutation and genetic drift shape the vast majority of genetic variation that we see in natural populations. According to this theory only a small proportion of new mutations have some effect on fitness, and because natural selection causes rapid fixation or loss of alternative alleles at such loci, most of the genetic variation present at any given time is effectively neutral and not acted on by natural selection.

The neutral theory was one of the most important advances in our understanding of why there is so much genetic variation in natural populations. It led to a great debate in evolutionary biology between those who believed that most genetic variation is maintained by natural selection and those who believed that most variation was selectively neutral. With the advances in DNA sequencing and more refined statistical methods for testing the influence of mutation, genetic drift, and natural selection, we now understand that a complex interaction of these forces determines levels of genetic variation in natural populations. In fact, the major areas of current research in empirical population genetics are aimed at determining how the relative impact of mutation, genetic drift, natural selection, and migration interact in complex ways to determine levels of genetic variation we observe in natural populations.

## Keynote

In population genetics one often encounters competing models that seek to explain the amount of variation in natural populations. In the 1920s and 1930s, the classical model predicted that there was an archetypal "wild type" and that most alleles were of this sort, with a few mutant variants in the population. In the 1960s and 1970s, protein electrophoresis revealed abundant genetic variation in natural populations, disproving the classical model and raising the question of what maintains so much genetic variation in natural populations. The neutral theory proposes that the genetic variation in proteins is neutral with regard to natural selection. We now know that there are complex interactions between mutation, random genetic drift, natural selection, and migration that determine the levels of genetic variation in natural populations.

## iActivity

Now, use protein electrophoresis to measure genetic variation in mussel populations in the iActivity, *Measuring Genetic Variation*, on the student website.

### Measuring Genetic Variation at the DNA Level

The development of the polymerase chain reaction (PCR; see Chapter 9, pp. 221–223) made it easy for population geneticists to obtain large numbers of copies of gene fragments from many individuals. These fragments can then be separated directly on gels to determine size differences, they can be cut with restriction enzymes to reveal differences, or the fragments can be directly sequenced.

To illustrate the quantification of genetic variation at the DNA level to estimate heterozygosity, we use an example where we assay nucleotide polymophism using restriction enzymes (this is a historically important technique). Today, it is more common to use automated DNA sequencing or single nucleotide polymorphism (SNP) detection methods as discussed later in the chapter. Suppose that two individuals differ in one or more nucleotides at a particular DNA sequence and that the differences occur at a site recognized by a restriction enzyme. One individual has a DNA molecule with the restriction site, but the other individual does not because the sequences of DNA nucleotides differ. If the DNA from these two individuals is mixed with the restriction enzyme and the resulting fragments are separated on a gel, the two individuals produce different patterns of fragments. The different patterns on the gel are called **restriction fragment length polymorphisms,** or **RFLPs** ("RIFF-lips"; see Chapter 10, pp. 270–271). They indicate that the DNA sequences of the two individuals differ. RFLPs are inherited in the same way that alleles coding for other traits are inherited, except that the RFLPs do not produce any outward phenotypes; their phenotypes are the fragment patterns produced on a gel when the DNA is cut by the restriction enzyme.

To illustrate the use of RFLPs for estimating genetic variation, suppose we isolate DNA from five wild mice and amplify a polymorphic DNA region we want to test by PCR using oligonucleotide primers complementary to each end of the region. Next, we cut the amplified DNA fragments with the restriction enzyme *Bam*HI and separate the fragments using agarose gel electrophoresis. A typical set of restriction patterns that might be obtained is shown in Figure 21.8. Thus, a mouse could be 1/1 (the polymorphic restriction site is present on both chromosomes), 1/2 (the restriction site is present on one chromosome and is absent on the other), or 2/2 (the restriction site is absent on both chromosomes). For the 10 chromosomes present among these particular five mice, four have the restriction site and six do not. Heterozygosity at the nucleotide level can be estimated from restriction site patterns.

Nucleotide heterozygosity has been studied for a number of different organisms. A few examples are shown in Table 21.4. Nucleotide heterozygosity typically varies from 0.001 to 0.02 in eukaryotic organisms. Recent estimates of nucleotide heterozygosity across the entire human genome average around 0.0008, meaning that an individual is heterozygous (contains different nucleotides on the two homologous chromosomes) at about one in every 1,000 nucleotides.

**DNA Sequence Variation.** We saw earlier that protein electrophoresis misses genetic differences that do not change the protein charge or conformation. The best method for identifying all genetic variation is to obtain the DNA sequence of the gene from each individual. In the first study to apply this approach, Martin Kreitman sequenced 11 copies (obtained from different fruit flies) of the alcohol dehydrogenase (*Adh*) gene in *Drosophila melanogaster*. Among the 11 copies, he found different nucleotides at 43 positions within the 2,659 base-pair segment. Furthermore, only 3 of the 11 copies were identical at all nucleotides examined; thus, there were 8 different alleles (at the nucleotide level) among the 11 copies of this gene. This suggests that populations harbor a tremendous amount of genetic variation in their DNA sequences.

Different regions of a gene apparently are subject to different evolutionary processes, and this is reflected in their different levels of nucleotide diversity. Table 21.5 shows nucleotide diversity estimates for different parts of the *Drosophila Adh* gene. As is observed within most functional genes, the highest diversity occurs at sites that do not change the amino acid sequence of the resulting protein, and these are known as **synonymous** changes. The level of synonymous nucleotide diversity is greater than the observed diversity at nucleotides that change the resulting amino acid sequence of the protein (known as **nonsynonymous** positions). Thus the large amount of nucleotide diversity seen in synonymous sites is not unexpected, because these changes do not affect the

**Figure 21.8**

**Restriction patterns from five mice.** The patterns differ in the presence (1) or absence (2) of a particular restriction site (middle one of the three shown). Each mouse has two homologous chromosomes, each of which potentially carries the restriction site. Thus, a mouse may be 1/1 (has the restriction site on both chromosomes), 1/2 (has the restriction site on one chromosome), or 2/2 (has the restriction site on neither chromosome). When the restriction site is present, the DNA is broken into two fragments after digestion with the restriction enzyme and separation with electrophoresis.

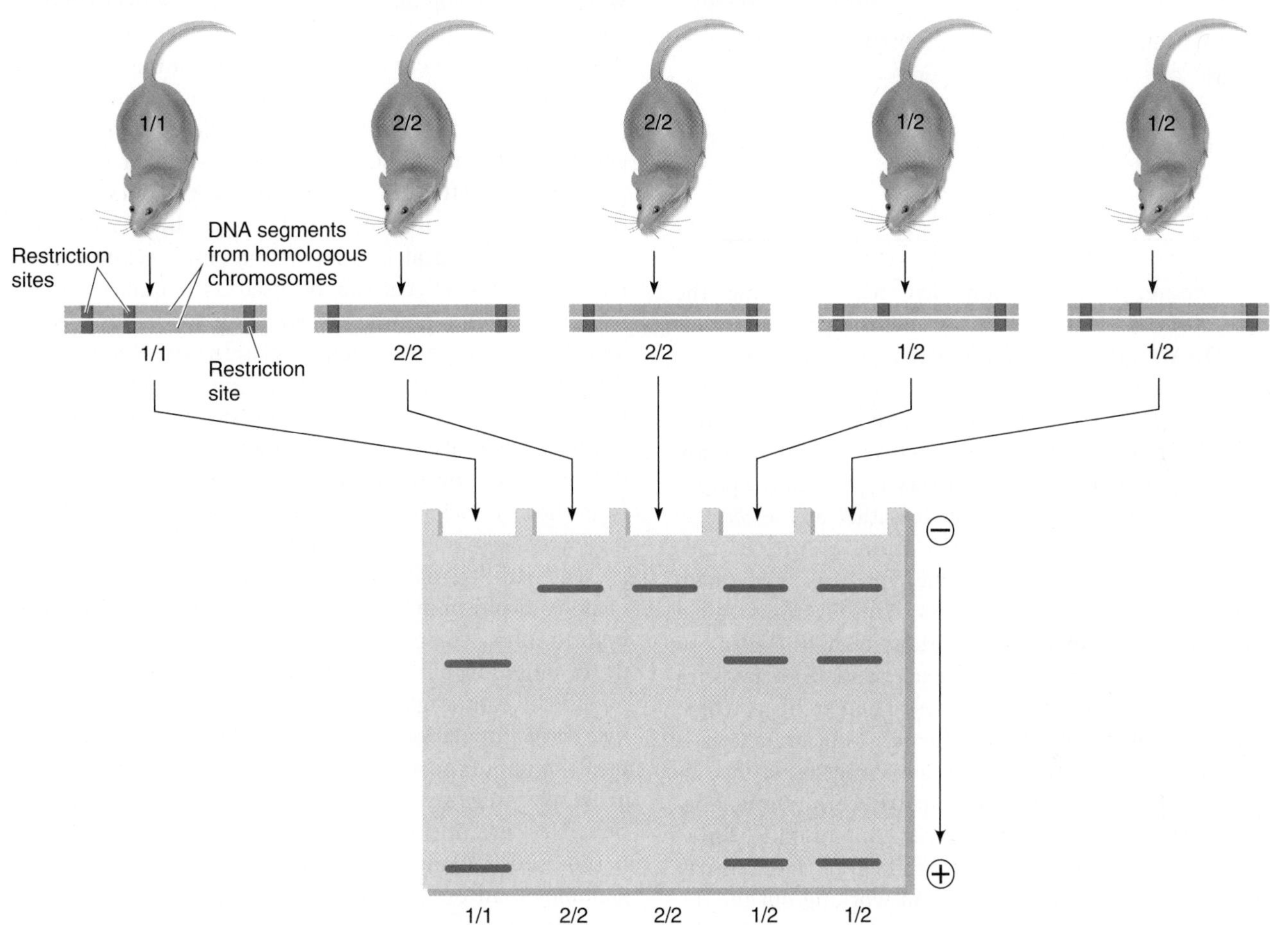


functioning of a protein. Roughly 3/4 of the positions in a gene are nonsynonymous, so 3/4 of all random mutations ought to be nonsynonymous, but in fact the variation one sees is much more likely to be synonymous. For example, in Kreitman's study, there were 13 synonymous sites that varied. If nonsynonymous mutations were equally likely to be seen, then we ought to see three times this number of nonsynonymous differences, or 39 positions. Instead, Kreitman found only a single nonsynonymous polymorphism. The conclusion is inescapable: most nonsynonymous mutations are visible to natural selection and have been eliminated from the population, leaving the currently observed excess of synonymous sites that vary.

**Table 21.4 Estimates of Nucleotide Heterozygosity for DNA Sequences**

| DNA Sequences | Organism | $H_{nuc}$ |
|---|---|---|
| β-Globin genes | Humans | 0.002 |
| Growth hormone gene | Humans | 0.002 |
| Alcohol dehydrogenase gene | Fruit fly | 0.006 |
| Mitochondrial DNA | Humans | 0.004 |
| H4 gene region | Sea urchin | 0.019 |

As was the case for the neutral theory, the development of more refined statistical methods and the advancements in technology allowing us to sequence large regions of genomes rapidly have revealed that the mechanisms determining levels of DNA sequence variation are more complex than originally thought. For example, we now know there is a bias in the codon usage for amino acids that subjects what appear to be synonomous sites to weak natural selection, which may also determine how much variation can exist at the third positions of codons.

**Table 21.5 Number of Varying Nucleotides and Diversity in Five Different Regions of the Alcohol Dehydrogenase (*Adh*) Gene of *Drosophila melanogaster***

| Gene Region | Variable Sites | Total Sites | Diversity |
|---|---|---|---|
| 5′ + 3′ flank | 3 | 335 | 0.002 |
| Exons | | | |
| Synonymous | 13 | 192 | 0.013 |
| Nonsynonymous | 1 | 576 | 0.001 |
| Introns | 18 | 789 | 0.004 |
| 3′ nontranscribed | 5 | 767 | 0.003 |

Furthermore, recombination that occurs during the process of crossing-over during meiosis profoundly affects DNA sequence variation across chromosomes. Because new mutations can occur anywhere along the length of a DNA sequence, an allele that is advantageous or detrimental and is thus a target of natural selection may sweep to fixation or be lost very rapidly in the population. During this process, variants that are selectively neutral, or nearly so, and lie in positions on the chromosome nearby the new mutation may hitchhike along with the mutation to fixation or loss. This event is called **genetic hitchhiking** and has a strong mediating effect on the distribution of DNA sequence variation across a genome and within specific genes. The extent to which selection on new mutations influences adjacent alleles is determined by how often crossing over occurs in that region, which effectively breaks up the chromosomal region, making it less subject to genetic hitchhiking. Theoretical models and empirical studies in modern-day population genetics are focusing on working out the details of how all these forces interact in natural populations to determine the level and pattern of variation we observe in the genomes of organisms in nature.

**Single Nucleotide Polymorphisms (SNPs).** Rapid advances in technology developed to assay DNA sequences has led to an ability to measure the frequency of alleles at many polymorphic nucleotides across broad expanses of the genome very quickly and inexpensively. Single nucleotide polymorphisms, called SNPs, can be detected using a variety of high-throughput methods including single base extension (SBE) sequencing reactions, special dyes that detect alternative bases at a specific nucleotide position (Taqman), and various other techniques that make use of intrumentation to assay thousands of SNPs in a very short time. In humans, these techniques have been used to catalogue literally millions of SNPs across the entire genome and estimate their allele frequencies in populations with different ethnic origins. These data have been made available to the public for research purposes as part of the efforts by a consortium of scientists to build haplotype maps of the human genome (see Chapter 8, p. 193). The primary motivation for cataloguing SNPs across the human genome is for linkage and association studies designed to identify genes that underlie human genetic traits, such as diseases. However, the data also provided an unprecedented opportunity to explore the genetic structure and evolutionary history of human populations using population genetic models. Scientists are currently working to catalog all of the SNPs in the human genome and extend the haplotype map even further, an endeavor described in the Focus on Genomics box for this chapter.

One goal of population geneticists is to sort out the effects of natural selection from the effects of migration, genetic drift, and mutation that have left their traces on patterns of genetic variation we observe in natural populations today. Since migration and genetic drift have effects on polymorphism at loci across the entire genome, estimates of these effects can be used as a framework on which to overlay patterns of variation we see at specific loci that may have been targets of selection. Similarity in the patterns of genetic variation at such a target locus with loci across the genome would suggest that the locus may not have been influenced by natural selection. In contrast, differences in the patterns of variation at a target locus from those at loci across the genome would suggest that it may have been a target of natural selection in the past or present. The SNP database provides us with measurements of polymorphism at such a large scale that we can use it both to infer demographic history (migration and genetic drift), mutation, and selection at specific loci. Recent studies use high-throughput SNP detection methods to confirm serial population bottlenecks as humans migrated out of Africa approximately 200,000 years ago into most other geographic regions of the world. A similar high-throughput SNP detection approach has shown that Europeans harbor substantially more deleterious mutations in their genomes than do the ancestral African individuals. This condition is most likely due to the population genetic bottlenecks after the migration out of Africa that provide opportunities for deleterious mutations to rise in frequency.

**DNA Length Polymorphisms and Short Tandem Repeats.** In addition to evolution of nucleotide sequences through nucleotide substitution, variation frequently occurs in the number of nucleotides found within a gene. These variations are called DNA length polymorphisms, and they arise through deletions and insertions of short stretches of nucleotides. For example, DNA length polymorphisms have been observed in the alcohol dehydrogenase gene of *Drosophila melanogaster*. In addition to extensive variation in nucleotide sequence, Kreitman found 6 insertions and deletions in the 11 copies of the gene he examined. All of these were confined to introns and flanking regions of the DNA; none was found within exons. Insertions and deletions within exons usually alter the reading frame, so they are selected against. As a result, insertions and deletions are most common in noncoding regions of the DNA. However, some insertions and deletions have been found

## Focus on Genomics
### The 1,000 Genome Project

After the human genome sequence was completed, scientists set another very ambitious goal. They realized that since genomic sequencing had become both faster and less expensive, they could now use genomic sequencing to understand far more about the genetic variation present in human populations, so they set out to sequence 1,000 human genomes. As of late 2008, this project had only just begun, but expectations for the project are high. As sequencing has become both faster and far less expensive in the last 20 years, this project is expected to cost somewhere between 50 and 500 million dollars, substantially less than it cost to sequence the first human genome, and estimates suggest that the project will be completed in only a few years. When the project is in full swing, the investigators anticipate that they will generate more than 8 billion base pairs of sequence per day, or nearly three full human genomes each day. The genomes to be sequenced will be selected from diverse human populations, including African, Asian, and European groups, as well as Native Americans, to ensure that we can measure the extent of diversity in human populations. Once these sequences are completed, we should have a much deeper understanding of how human populations differ from each other. The investigators think that they will be able to catalog all common alleles present in humans. In this case, common means alleles present at a frequency greater than 0.5 to 1 percent. Rarer alleles might, or might not, be missed in the sample set. Although this will not be enough information to understand the allele distributions in every population fully, it should allow us to begin to understand not only how much variation is present in some of the larger human populations, but also how the populations relate to each other. This should allow us, for instance, to confirm earlier and less extensive observations that there is far more variation in African populations than in either European or Asian populations, which is consistent with the theory that modern humans evolved in Africa and then spread across the world. This is a very exciting project, and should, in the next few years, allow us to learn much more about our own variation.

in the coding regions of certain genes. Another class of DNA length polymorphisms involves variation in the number of copies of a particular gene. For example, among individual fruit flies, the number of copies of ribosomal genes varies widely.

A particularly useful kind of polymorphism that occurs in nearly all organisms is seen in **short tandem repeats (STRs),** also called microsatellites. STRs are 2–6-bp DNA sequences tandemly repeated a few times up to about 100 times (see Chapter 10, p. 272). More than 8,000 STRs have been mapped in the human genome, and their use in genetic mapping in humans has been critical to the discovery of many genes associated with genetic disorders. In conservation genetics, we are often faced with a need to obtain information on genetic variation in organisms for which there is almost no prior knowledge about gene sequences. In these cases STRs are often used to quantify patterns of genetic variability.

**Keynote**

The Hardy–Weinberg principle applies to alleles at loci defined in a number of ways, including single nucleotide polymorphisms (SNPs), protein variants, or any other factor that segregates as a Mendelian gene.

## Forces That Change Gene Frequencies in Populations

We have looked at the many ways in which genetic variation can be quantified in natural populations and have concluded that populations harbor enormous amounts of variation. We have also touched on some of the evolutionary forces that may influence why so much genetic variation exists. Now it is time to return to the question of what maintains this variation and introduce some of the formal models that population geneticists use to infer what forces are responsible. If a population is large, randomly mating, and free from mutation, migration, and natural selection, then whatever variation there is in the population will remain at the same frequency forever. This is the Hardy–Weinberg law. For many populations, however, the conditions required by the Hardy–Weinberg law do not hold. Populations frequently are small, mating may be nonrandom, and mutation, migration, and natural selection may occur. In these circumstances, allele frequencies do change, and the gene pool of the population evolves in response to the interplay of these processes. In a sense then, the Hardy–Weinberg law is a null model that we can use to dissect which forces may be causing deviations from its predictions, and thus may underlie the process of evolution in natural populations.

In the following sections we discuss the role of four evolutionary processes—mutation, genetic drift, migration, and natural selection—in changing the allele frequencies of a population. We also discuss the effects of nonrandom mating on genotype frequencies. We first consider violations of the Hardy–Weinberg equilibrium assumptions one by one. We then consider several cases in which two assumptions are violated simultaneously. We also briefly discuss how Hardy–Weinberg equilibrium is affected by two or more loci that do not segregate independently. Be aware that in real populations it is possible for all of the Hardy–Weinberg assumptions to be violated simultaneously. However, our understanding of the effects of the violations is enhanced by considering one or two violations at a time.



**Keynote**

Mutation, random genetic drift, migration, non-random mating, and natural selection are processes that can alter allele frequencies in a population.

### Mutation

One process that can alter the frequencies of alleles within a population is mutation. As we discussed in Chapter 7, **gene mutations** consist of heritable changes in the DNA that occur within a locus. Usually a mutation converts one allele form of a gene to another. The rate at which mutations arise is generally low but varies between loci and between species (Table 21.6). Certain genes modify overall mutation rates, and many environmental factors, such as chemicals, radiation, and infectious agents, may increase the number of mutations.

Ultimately, mutation is the source of all new genetic variation; new combinations of alleles may arise through recombination, but new alleles occur only as a result of mutation. Thus mutation provides the raw genetic material on which evolution acts. Some mutations are entirely neutral and do not affect the reproductive fitness of the organisms. Other mutations are detrimental and are eliminated from the population. However, a few mutations convey some advantage to the individuals that possess them and spread through the population. Whether a mutation is neutral, detrimental, or advantageous depends on the specific environment, and if the environment changes, previously harmful or neutral mutations may become beneficial. For example, after the widespread use of the insecticide DDT, insects with mutations that conferred resistance to DDT were capable of surviving and reproducing; because of this advantage, the mutations spread, and many insect populations quickly evolved resistance to DDT. Mutations are fundamentally important to the process of evolution because they provide the genetic variation on which other evolutionary processes act.

In general, rates of mutation are so low that once a mutant allele enters a population, the fate of the mutation is determined almost entirely by forces. To see why this is so, we can consider a model in which mutation is the only evolutionary force. The mutation of $A$ to $a$ is called a **forward mutation.** Mutations also occur in the reverse direction; $a$ may mutate to $A$. These mutations are called **reverse mutations,** and they typically occur at a lower rate than forward mutations. The forward mutation rate—the rate at which $A$ mutates to $a$ ($A \rightarrow a$)—is symbolized with $u$. The reverse mutation rate—the rate at which $a$ mutates to $A$ ($A \leftarrow a$)—is symbolized with $v$. Consider a hypothetical population in which the frequency of $A$ is $p$ and the frequency of $a$ is $q$. We assume that the population is large and that no selection occurs on the alleles. In each generation, a proportion $u$ of all $A$ alleles mutates to $a$. The actual number mutating depends on both $u$ and the frequency of $A$ alleles. For example, suppose the population consists of 100,000 alleles. If $u$ equals $10^{-4}$, one out of every 10,000 $A$ alleles mutates to $a$. When $p = 1.00$, all 100,000 alleles in the population are $A$ and free to mutate to $a$, so $10^{-4} \times 100{,}000 = 10$ $A$ alleles should mutate to $a$. However, if $p = 0.10$, only 10,000 alleles are $A$ and free to mutate to $a$. Therefore, with a mutation rate of $10^{-4}$ only 1 of the $A$ alleles will undergo mutation. The decrease in the frequency of $A$ resulting from mutation of $A \rightarrow a$ is equal to $up$; the increase in frequency resulting from $A \leftarrow a$ is equal to $vq$. As a result of mutation, the amount by which $A$ decreases in one generation is equal to the increase in $A$ alleles caused by reverse mutations minus the decrease in $A$ alleles caused by forward mutations. Since we have a forward mutation rate increasing the frequency of $a$ and a reverse mutation rate decreasing the frequency of $a$, it is intuitively easy to see that eventually the population achieves equilibrium, in which the number of alleles undergoing forward mutation is exactly equal to the number of alleles undergoing reverse mutation. At this point, no further change in allele frequency occurs, even though forward and reverse mutations continue to take place. With some simple algebra, population genetics theorists have shown that the equilibrium frequency for $a$ is

$$\hat{q} = \frac{u}{u + v}$$

and, therefore

$$\hat{p} = 1 - \hat{q}$$

Now consider how slowly this process of pure mutation changes allele frequencies. In a population with the initial allele frequencies $p = 0.9$ and $q = 0.1$ and mutation rates $u = 5 \times 10^{-5}$ and $v = 2 \times 10^{-5}$, we can calculate the change in allele frequency in the first generation:

$$\Delta p = vq - up$$

$$= (2 \times 10^{-5} \times 0.1) - (5 \times 10^{-5} \times 0.9)$$

$$\Delta p = -0.000043$$

**Table 21.6 Spontaneous Mutation Frequencies at Specific Loci for Various Organisms[a]**

| Organism | Trait | Mutation per 100,000 Gametes[b] |
|---|---|---|
| T2 Bacteriophage (virus) | To rapid lysis ($r^+ \rightarrow r$) | 7 |
| | To new host range ($h^+ \rightarrow h$) | 0.001 |
| *E. coli K12* (bacterium) | To streptomycin resistance | 0.00004 |
| | To phage T1 resistance | 0.003 |
| | To leucine independence | 0.00007 |
| | To arginine independence | 0.0004 |
| | To tryptophan independence | 0.006 |
| | To arabinose dependence | 0.2 |
| *Salmonella typhimurium* (bacterium) | To threonine resistance | 0.41 |
| | To histidine dependence | 0.2 |
| | To tryptophan independence | 0.005 |
| *Diplococcus pneumoniae* (bacterium) | To penicillin resistance | 0.01 |
| *Neurospora crassa* | To adenine independence | 0.0008–0.029 |
| | To inositol independence | 0.001–0.010 |
| | (One *inos* allele, JH5202) | 1.5 |
| *Drosophila melanogaster* males | $y^+$ to *yellow* | 12 |
| | $bw^+$ to *brown* | 3 |
| | $e^+$ to *ebony* | 2 |
| | $ey^+$ to *eyeless* | 6 |
| Corn | *Wx* to *waxy* | 0.00 |
| | *Sh* to *shrunken* | 0.12 |
| | *C* to *colorless* | 0.23 |
| | *Su* to *sugary* | 0.24 |
| | *Pr* to *purple* | 1.10 |
| | *I* to *i* | 10.60 |
| | $R^r$ to $r^r$ | 49.20 |
| Mouse | $a^+$ to *nonagouti* | 2.97 |
| | $b^+$ to *brown* | 0.39 |
| | $c^+$ to *albino* | 1.02 |
| | $d^+$ to *dilute* | 1.25 |
| | $ln^+$ to *leaden* | 0.80 |
| | Reverse mutations for above genes | 0.27 |
| Chinese hamster somatic cell tissue culture | To azaguanine resistance | 0.0015 |
| | To glutamine independence | 0.014 |
| Humans | Achondroplasia | 0.6–1.3 |
| | Aniridia | 0.3–0.5 |
| | Dystrophia myotonica | 0.8–1.1 |
| | Epiloia | 0.4–1 |
| | Huntington disease | 0.5 |
| | Intestinal polyposis | 1.3 |
| | Neurofibromatosis | 5–10 |
| | Osteogenesis imperfecta | 0.7–1.3 |
| | Pelger anomaly | 1.7–2.7 |
| | Retinoblastoma | 0.5–1.2 |

[a]Mutations to independence for nutritional substances are from the auxotrophic condition (e.g., *leu*) to the prototrophic condition [e.g., $leu^+$).
[b]Mutation frequency estimates of viruses, bacteria, *Neurospora*, and Chinese hamster somatic cells are based on particle or cell counts rather than gametes.

The frequency of *A* decreases by only four-thousandths of 1%. Because mutation rates are so low, the change in allele frequency due to mutation pressure is exceedingly slow. To change the frequency from 0.50 to 0.49 would require 2,000 generations, and to change it from 0.1 to 0.09 would require 10,000 generations. If some reverse mutation occurs, the rate of change is even slower. In practice, mutation by itself changes the allele frequencies at such a slow rate that populations are rarely in mutational equilibrium. Other processes affect allele frequencies more profoundly, and mutation alone rarely determines the allele frequencies of a population. For example, achondroplastic dwarfism is an autosomal dominant trait in humans that arises through recurrent mutation. However, the frequency of this disorder in human populations is determined by an interaction of mutation pressure and natural selection.

### Keynote

When we study what happens when we violate the assumption of the Hardy–Weinberg equilibrium of the absence of mutation, we see that mutation is the only way in which novel genetic material can come to exist within a species. The larger a population, the more potential there is for a novel mutation to arise, but mutation by itself changes allele frequencies only a little.

### Random Genetic Drift

Another major assumption of the Hardy–Weinberg law is that the population is infinitely large. Real populations are not infinite in size, but frequently they are large enough that chance factors have small effects on allele frequencies. Some populations are small, however, and in this case chance factors may produce large changes in allele frequencies. Random change in allele frequency due to chance is called random genetic drift, or simply genetic drift. Ronald Fisher and Sewall Wright (see Figures 21.1a and Figure 21.1b, respectively), brilliant population geneticists who laid much of the theoretical foundation of the discipline, first described how genetic drift affects the evolution of populations.

Changes in allele frequency resulting from random events can have important evolutionary implications in small populations. In addition, such changes can have important consequences for the conservation of a rare or endangered species. To see how chance can play a big role in altering the genetic structure of a population, imagine a small group of humans inhabiting a South Pacific island. Suppose this population consists of only 10 individuals, 5 of whom have green eyes and 5 of whom have brown eyes. For this example, we assume that eye color is determined by a single locus (although actually a number of genes control eye color) and that the allele for green eyes is recessive to brown (*BB* and *Bb* specify brown eyes, and *bb* specifies green eyes). The frequency of the allele for green eyes is 0.6 in the island population. A typhoon strikes the island, killing 50% of the population, all of whom have brown eyes. Eye color in no way affects the probability of surviving; the fact that only those with green eyes survive is strictly the result of chance. After the typhoon, the allele frequency for green eyes is 1.0. Evolution has occurred in this population: the frequency of the green eye allele has changed from 0.6 to 1.0, simply as a result of chance.

Now imagine the same scenario, but this time with a population of 1,000 individuals. As before, 50% of the population has green eyes and 50% has brown eyes. A typhoon strikes the island and kills half the population. How likely is it that, just by chance, all 500 people who perish will have brown eyes? In a population of 1,000 individuals, the probability of this occurring by chance is extremely remote. This example illustrates an important characteristic of genetic drift: chance factors are likely to produce rapid changes in allele frequencies only in small populations.

Random factors producing mortality in natural populations, such as the typhoon in the preceding example, is only one of several ways in which genetic drift arises. Chance deviations from expected ratios of gametes and zygotes also produce genetic drift. We have seen the importance of chance deviations from expected ratios in the genetic crosses we studied in earlier chapters. For example, when we cross a heterozygote with a homozygote ($Aa \times aa$), we expect 50% of the progeny to be heterozygous and 50% to be homozygous. We do not expect to get exactly 50% every time, however, and if the number of progeny is small, the observed ratio may differ greatly from the expected. Recall that the Hardy–Weinberg law is based on random mating and expected ratios of progeny resulting from each type of mating (see Table 21.2). If the actual number of progeny differs from the expected ratio due to chance, genotypes may not be in Hardy–Weinberg proportions. Simply put, random genetic drift may also result in changes in allele frequencies.

Chance deviations from expected proportions arise from a general phenomenon called **sampling error.** Imagine that a population produces an infinitely large pool of gametes, with alleles in the proportions $p$ and $q$. If random mating occurs and all the gametes unite to form viable zygotes, the proportions of the genotypes will be equal to $p^2$, $2pq$, and $q^2$, and the frequencies of the alleles in these zygotes will remain $p$ and $q$. If the number of progeny is limited, however, the gametes that unite to form the progeny constitute a sample from the infinite pool of potential gametes. Just by chance, or by "error," this sample may deviate from the larger pool; the smaller the sample, the larger the potential deviation.

Flipping a coin is analogous to the situation in which sampling error occurs. When we flip a coin, we expect 50% heads and 50% tails. If we flip the coin 1,000 times, we will get very close to that expected 50:50 ratio. But if we flip the coin only four times, we would not be surprised if by chance we obtain 3 heads and 1 tail, or even all tails. When the sample—in this case, the number of flips—is small, the sampling error can be large. All genetic drift arises from such sampling error.

Mathematicians have worked out the exact probability of getting, for example, 499 heads and 501 tails. This probability has what is called a binomial distribution. For our purposes, it is important to notice that a population with frequency $p$ of allele $A$ in one generation samples alleles for the next generation, and the expected frequency of that next generation is still $p$, but there is some variation around $p$ due to drift. In fact, the sampling variance for the allele frequency is simply $pq/2N$, where $N$ is the number of individuals in the population. This is another way to see that small populations have a larger sampling variance, and larger populations have a small sampling variance, so that drift works more slowly in large populations.

**Effective Population Size.** Genetic drift is random, so we cannot predict what the allele frequencies will be after drift has occurred. However, since sampling error is related to the size of the population, we can make predictions about the magnitude of genetic drift. Ecologists often measure population size by counting the number of individuals, but not all individuals contribute gametes to the next generation. To determine the magnitude of genetic drift, we must know the **effective population size,** which equals the equivalent number of adults contributing gametes to the next generation. If the sexes are equal in number and all individuals have an equal probability of producing offspring, the effective population size equals the number of breeding adults in the population. However, when males and females are not present in equal numbers, the effective population size ($N_e$) is

$$N_e = \frac{4 \times N_f \times N_m}{N_f + N_m}$$

where $N_f$ equals the number of breeding females and $N_m$ equals the number of breeding males.

It is not difficult to see why the effective population size is not simply the number of breeding adults. The reason is that males, as a group, contribute half of all genes to the next generation and females, as a group, contribute the other half. Therefore, in a population of 70 females and 2 males, the two males are not genetically equivalent to two females; each male contributes $1/2 \times 1/2 = 0.25$ of the genes to the next generation, whereas each female contributes $1/2 \times 1/70 = 0.007$ of all genes. The small number of males disproportionately influences what alleles are present in the next generation. Using the preceding equation, the effective population size is $N_e = (4 \times 70 \times 2)/(70 + 2) = 7.8$, or approximately 8 breeding adults. This means that in a population of 70 females and 2 males, genetic drift occurs as if the population had only 4 breeding males and 4 breeding females. Therefore, genetic drift has a much greater effect in this population than in one with 72 breeding adults equally divided between males and females.

Other factors, such as differential production of offspring, fluctuating population size, and overlapping generations, can further reduce the effective population size. Most of these factors result in an effective size that is smaller than the count of individual adults in the population.

**Bottlenecks and Founder Effects.** All genetic drift arises from sampling error, but there are several ways in which sampling error occurs in natural populations. First, as already discussed, genetic drift arises when population size remains continuously small over many generations. Undoubtedly this situation is common, particularly where populations occupy marginal habitats or when competition for resources limits population growth. In such populations, genetic drift plays an important role in the evolution of allele frequencies. Many species are spread out over a large geographic range. This can result in a species consisting of many populations of small size, each undergoing drift independently. In addition, human intervention, such as the clear-cutting of forests, can result in the fragmentation of previously large continuous populations.

The effect of genetic drift arising from small population size is seen in a classic laboratory experiment conducted by P. Buri with *Drosophila melanogaster*. Buri examined the frequency of two alleles, $bw^{75}$ and $bw$, at a locus that determines eye color in the fruit flies. He set up 107 experimental populations, and the initial frequency of $bw^{75}$ was 0.5 in each. The flies in each population interbred randomly, and in each generation Buri randomly selected 8 males and 8 females to be the parents for the next generation. Thus, population size was always 16 individuals. The distribution of allele frequencies in these 107 populations is presented in Figure 21.9. Notice that the allele frequencies in the early generations were clumped around 0.5, but genetic drift caused the frequencies in the populations to spread out or diverge over time. By generation 19, the frequency of $bw^{75}$ was 0 or 1 in most populations. What is most elegant about this experiment is that it closely matched the theoretically expected effects of drift demonstrated by Wright and Fisher. The Wright–Fisher model considers the changes in allele frequency each generation as a binomial sampling, but at each successive generation the allele frequency can change, so the binomial parameter (in this case, the allele frequency) changes each generation. The fit of this elegant model to Buri's data was very good, provided that one accounts for the effective population size (which was smaller than 16).

Another way in which genetic drift arises is through a **founder effect.** A founder effect occurs when a population is initially established by a small number of breeding individuals. Although the population may subsequently grow in size and later consist of a large number of individuals, the gene pool of the population is derived from the genes present in the original founders. Chance may

Figure 21.9

**Results of Buri's study of genetic drift in 107 populations of *Drosophila melanogaster*.** Shown are the distributions of the frequency of the $bw^{75}$ allele among the populations in 19 consecutive generations. Each population consisted of 16 individuals.

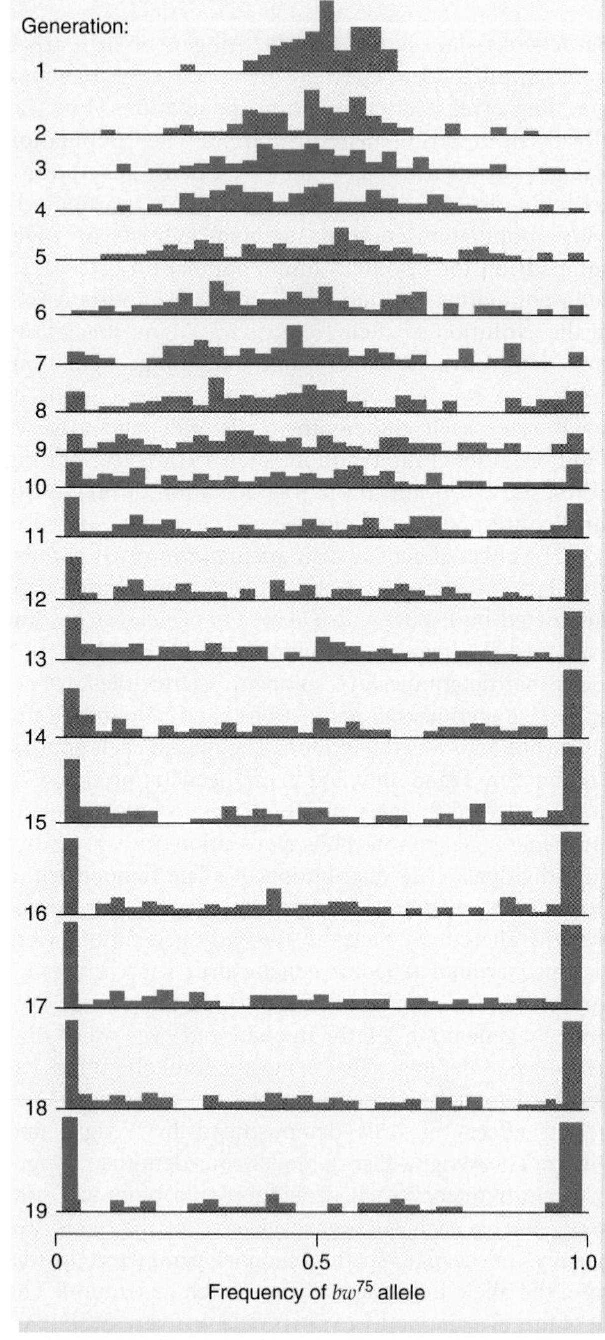


play a significant role in determining which genes were present among the founders, and chance has a profound effect on the gene pool of subsequent generations. Founder effects have frequently been used to explain the subsequent evolution of new species, but their importance to the process of species formation is currently under intense study and debate.

Many examples of founder effect come from the study of human populations. Consider the inhabitants of Tristan da Cunha, a small, isolated island in the South Atlantic. This island was first permanently settled by William Glass, a Scotsman, and his family in 1817. (Several earlier attempts at settlement failed.) They were joined by a few additional settlers, some shipwrecked sailors, and a few women from the distant island of St. Helena, but for the most part the island remained a genetic isolate. In 1961, a volcano on Tristan da Cunha erupted, and the population of almost 300 inhabitants was evacuated to England. During the two years that the islanders were in England, geneticists studied the islanders and reconstructed the genetic history of the population. These studies revealed that the gene pool of Tristan da Cunha was strongly influenced by genetic drift.

Three forms of genetic drift occurred in the evolution of the island's population. First, founder effect took place at the initial settlement. By 1855, the population of Tristan da Cunha consisted of about 100 individuals, but 26% of the genes of the population in 1855 were contributed by William Glass and his wife. Even in 1961, these original two settlers contributed 14% of all the genes in the 300 individuals of the population. The particular genes that Glass and other original founders carried heavily influenced the subsequent gene pool of the population. Second, population size remained small throughout the history of the settlement, and sampling error continually occurred.

A third form of sampling error, called **bottleneck effect,** also played an important role in the population of Tristan da Cunha. Bottleneck effect is a form of genetic drift that occurs when a population is drastically reduced in size. During such a population reduction, some genes may be lost from the gene pool as a result of chance. Recall our earlier example of the population consisting of 10 individuals inhabiting a South Pacific island. When a typhoon struck the island, the population size was reduced to 5, and by chance all individuals with brown eyes perished in the storm, changing the frequency of green eyes from 0.6 to 1.0. This is an example of bottleneck effect. Bottleneck effect can be viewed as a type of founder effect because the population is refounded by the few individuals that survive the reduction.

Two severe bottlenecks occurred in the history of Tristan da Cunha. The first took place around 1856 and was precipitated by two events: the death of William Glass and the arrival of a missionary who encouraged the inhabitants to leave the island. At that time many islanders emigrated to America and South Africa, and the population dropped from 103 individuals at the end of 1855 to 33 in 1857. A second bottleneck occurred in 1885. The island of Tristan da Cunha has no natural harbor, and the islanders intercepted passing ships for trade by rowing out in small boats. On November 28, 1885, 15 of the adult males on the island put out in a small boat to make contact with a passing ship. In full view of the entire island community, the boat capsized and all 15 men drowned. Following this disaster, only four adult males were left on the island, one of whom was insane and two of whom were old. Many of the widows and their families left the island during the next few years, and the population size dropped from 106 to 59. Both bottlenecks had a major effect on the gene pool of the population. All the genes contributed by several settlers were lost, and the relative contributions of others were altered by these events. Thus, the gene pool of Tristan da Cunha has been influenced by genetic drift in the form of founder effect, small population size, and bottleneck effect.

As we shall see later, when we discuss migration, gene flow in populations reduces the effects of genetic drift. Small breeding units that lack gene flow are genetically isolated from other groups and often experience considerable genetic drift, even though they are surrounded by much larger populations. A good example is a religious sect known as the Dunkers, found in eastern Pennsylvania. Between 1719 and 1729, 50 Dunker families emigrated from Germany and settled in the United States. Since that time, the Dunkers have remained an isolated group, rarely marrying outside of the sect, and the number of individuals in their communities has always been small.

During the 1950s geneticists studied one of the original Dunker communities in Franklin County, Pennsylvania. At the time of the study, this population had about 300 members, and the population size had remained nearly constant for many generations. The investigators found that some of the allele frequencies in the Dunkers were very different from the frequencies found among the general population of the United States. The Pennsylvania frequencies were also different from the frequencies of the West German population from which the Dunkers descended. Table 21.7 presents some of the allele frequencies at the ABO blood group locus. The ABO allele frequencies among the Dunkers are not the same as those in the U.S. population or the West German population. Nor are the Dunker frequencies intermediate between those of the United States and Germany. (Intermediate frequencies might be expected if intermixing of Dunkers and Americans had occurred.) The most likely explanation for the unique Dunker allele frequencies observed is that genetic drift has produced random change in the gene pool. Founder effect probably occurred when the original 50 families emigrated from Germany, and genetic drift probably has continued to influence allele frequencies in each generation since 1729 because the population size has remained small.

**Effects of Genetic Drift.** Genetic drift produces changes in allele frequencies, and these changes have several effects on the genetic structure of populations. First, genetic drift causes the allele frequencies of a population to change over time. This is illustrated in Figure 21.10. The different lines represent allele frequencies in several populations over a number of generations. Although all populations begin with an allele frequency equal to 0.50, the frequencies in each population change over time as a result of sampling error. In each generation, the allele frequency may increase or decrease, and over time the frequencies wander randomly or drift (hence the name *genetic drift*). Sometimes, within 30 generations, just by chance, the allele frequency reaches a value of 0.0 or 1.0. At this point, one allele is lost from the population and the population is said to be *fixed* for the remaining allele in a one locus, two allele example. Once an allele has reached fixation, no further change in allele frequency can occur unless the other allele is reintroduced through mutation or migration. The probability of fixation in a population increases with time, as shown theoretically by Motoo Kimura in Figure 21.11. If the initial allele frequencies are equal, which allele becomes fixed is strictly random. On the other hand, if initial allele frequencies are not equal, the rare allele is more likely to be lost. During this process of genetic drift and fixation, the number of heterozygotes in the population also decreases; and after fixation, the population heterozygosity is zero. As heterozygosity decreases and alleles become fixed, populations lose genetic variation; thus, the second effect of genetic drift is a reduction in genetic variation within populations. The probability of fixation by genetic drift is equal to the current frequency of the allele.

Since genetic drift causes random change in allele frequency, the allele frequencies in separate, individual populations do not change in the same direction. Therefore, populations diverge in their allele frequencies through genetic drift. This is illustrated in Figure 21.9 and Figure 21.10; all the populations begin with $p$ and $q$ equal to 0.5. After a few generations, the allele frequencies of the populations diverge, and this divergence

**Table 21.7 Frequencies of Alleles Controlling the ABO Blood Group System in Three Human Populations**

| | Allele Frequencies | | | Phenotype (Blood Group) Frequencies | | | |
|---|---|---|---|---|---|---|---|
| **Population** | $I^A$ | $I^B$ | $i$ | A | B | AB | O |
| Dunker | 0.38 | 0.03 | 0.59 | 0.593 | 0.036 | 0.023 | 0.348 |
| United States | 0.26 | 0.04 | 0.70 | 0.431 | 0.058 | 0.021 | 0.490 |
| West Germany | 0.29 | 0.07 | 0.64 | 0.455 | 0.095 | 0.041 | 0.410 |

**Figure 21.10**

**The effect of genetic drift on the frequency ($q$) of allele $A^2$ in four populations.** Each population begins with $q$ equal to 0.5, and the effective population size for each is 20. The mean frequency of allele $A^2$ for the four replicates is indicated by the red line. These results were obtained by a computer simulation.

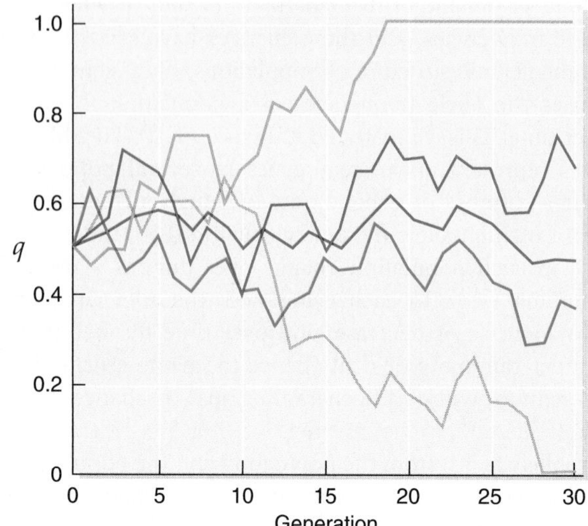


**Figure 21.11**

**The average time to fixation or loss of an allele from a population as a function of population size and initial allele frequency as predicted by Kimura.** For example, if the initial allele frequency is 0.3 and the population size is 10, it would take just under 2.5 × 10 generations on average for the allele to be lost or fixed in the population.

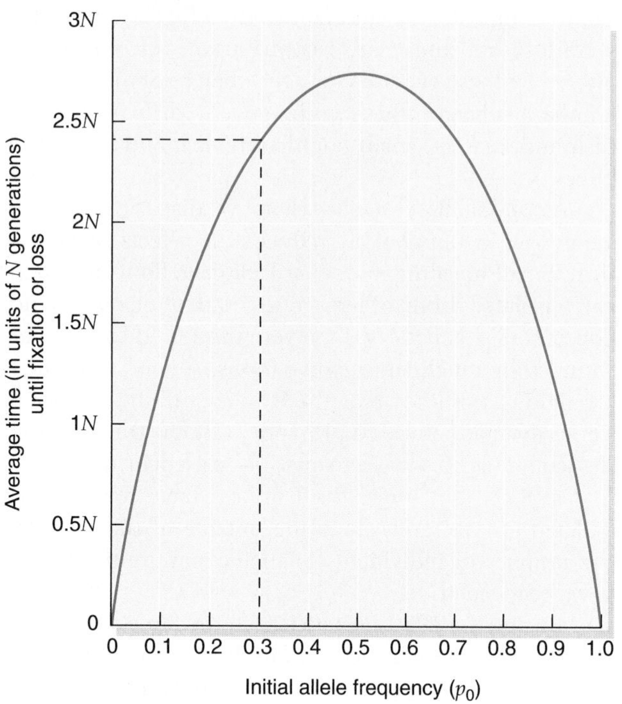


increases over generations. The maximum divergence in allele frequencies is reached when all populations are fixed for one or the other allele. If allele frequencies are initially 0.5, approximately half of the populations will be fixed for one allele, and half will be fixed for the other.

Since genetic drift is greater in small populations and leads to genetic divergence, we expect more variance in allele frequency among small populations than among large populations. Such a relationship has been observed in studies of natural populations. Robert K. Selander, for example, studied genetic variation in populations of the house mouse inhabiting barns in Texas. Through systematic trapping, he was able to estimate population size; and using electrophoresis, he examined the variance in allele frequency at two loci, a locus coding for the enzyme esterase (*Est-3*) and a locus coding for hemoglobin (*Hbb*). Selander found that the variance in allele frequency between small populations was several times larger than that between large populations (Table 21.8), an observation consistent with our understanding of how genetic drift leads to population divergence.

## Keynote

Genetic drift, or chance changes in allele frequency caused by sampling error, can have important evolutionary and survival implications for small populations. Genetic drift leads to loss of genetic variation within populations, genetic divergence among populations, and random fluctuation in the allele frequencies of a population over time. Genetic drift can also explain how molecules in different species accumulate differences on a seemingly regular basis, forming the basis for the neutral theory of molecular evolution.

**Table 21.8 Variance in Allele Frequency Among Populations as a Function of Population Size**

| Type of Population | Number of Populations | Mean Allele Frequency | | Variance in Allele Frequency | |
|---|---|---|---|---|---|
| | | *Est-3*$^b$ | *Hbb*$^8$ | *Est-3*$^b$ | *Hbb*$^8$ |
| Small ($N < 50$) | 29 | 0.418 | 0.849 | 0.051 | 0.188 |
| Large ($N > 50$) | 13 | 0.372 | 0.843 | 0.013 | 0.008 |

**Balance Between Mutation and Random Genetic Drift.** We have seen that mutation continues to introduce new variants into a population, and the net effect of random genetic drift is to remove that variation from the population. We also know that many organisms currently or historically exist in small populations. Thus, both mutation and genetic drift have likely been important mechanisms determining the allele frequencies at many of the genes in their genome. What happens if we create models to combine these two forces, mutation and drift? Will there be a balance achieved such that the rate of gain of variability by mutation is precisely matched by the rate of loss due to chance fixations by drift? Several models have been devoted to this problem, and one of the simplest is the *infinite alleles model*. In this model, each mutation that occurs in a gene is assumed to generate a novel allele. If you imagine a very large gene, consisting of perhaps 10,000 bp, then the chance that two mutations will generate the same allele is very small. So to a first approximation, this assumption seems reasonable. The model also assumes that random genetic drift occurs by the repeated sampling described earlier. In this situation, the forces of mutation and drift balance each other, and we end up with a curious kind of steady state in which new mutations keep on being generated, but alleles are also continuously lost in the population generation after generation. In this steady state, the number of alleles changes slightly each generation but tends not to stray too far from an equilibrium value. The heterozygosity of the locus, which you can think of as the frequency of heterozygotes in the population, or the chance of drawing two alleles and having them be different, is

$$H = \frac{4N_e\mu}{4N_e\mu + 1}$$

This equation establishes an important point about many models in population genetics: the roles of neutral mutations and population size often are combined into a single term, $4N_e\mu$. (Recall that $N_e$ is the effective population size.) This means that if one population is twice as large but has a mutation rate half that of another, the two populations will have the same level of heterozygosity. The reciprocal role is fairly easy to understand: the rate at which a population loses heterozygosity as a result of drift is inversely proportional to its size, while the number of new mutations introduced in each generation is directly proportional to population size. Figure 21.12 shows a plot of the heterozygosity for a range of values of $4N_e\mu$ and shows that plausible population sizes and mutation rates give plausible levels of heterozygosity. One problem with this model was pointed out by John Gillespie, who noted that whereas population sizes vary by several orders of magnitude among organisms, the mutation rates vary only a little. One would expect organisms with higher population sizes to have higher heterozygosity. There are striking exceptions to this prediction, suggesting that a simple mutation–drift balance does not explain everything.

**Figure 21.12**

**Relationship between the neutral parameter $\theta = 4N_e\mu$ and the expected heterozygosity under the infinite alleles models, which gives a balance between mutation and random drift.**

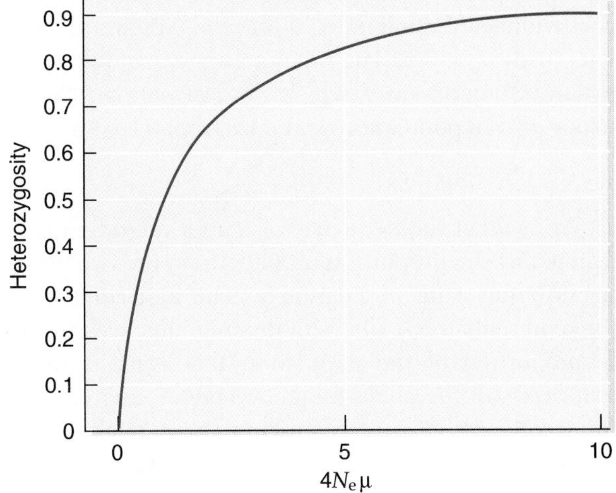


## Migration

One assumption of the Hardy–Weinberg law is that the population is isolated and not influenced by other populations. Many populations are not completely isolated, however, and exchange genes with other populations of the same species. Individuals migrating into a population may introduce new alleles to the gene pool and alter the frequencies of existing alleles. Thus migration has the potential to disrupt Hardy–Weinberg equilibrium and may influence the evolution of allele frequencies within populations.

The term *migration* usually implies movement of organisms. In population genetics, however, we are interested in the movement of genes, which may or may not occur when organisms move. Movement of genes takes place only when organisms or gametes migrate and contribute their genes to the gene pool of the recipient population. This process is also called **gene flow.**

Gene flow has two major effects on a population. First, it introduces new alleles to the population. Since mutation generally is a rare event, a specific mutant allele may arise in one population and not in another. Gene flow spreads these new alleles to other populations and, like mutation, is a source of genetic variation for the recipient population. Second, when the allele frequencies of migrants and the recipient population differ, gene flow changes the allele frequencies within the recipient population. Through exchange of genes, different populations remain similar, and thus migration is a homogenizing force that tends to prevent populations from accumulating genetic differences among them.

To illustrate the effect of migration on allele frequencies, we consider a simple model in which gene flow occurs in only one direction, from population $x$ to population $y$. Suppose the frequency of allele $A$ in population $x$

($p_x$) is 0.8 and the frequency of $A$ in population $y$ ($p_y$) is 0.5. In each generation, some individuals migrate from population $x$ to population $y$, and these migrants are a random sample of the genotypes in population $x$. After migration, population $y$ actually consists of two groups of individuals: the migrants, with $p_x = 0.8$, and the residents, with $p_y = 0.5$. The migrants now make up a proportion of population $y$, which we designate $m$. The frequency of $A$ in population $y$ after migration ($p'_y$) is

$$p'_y = mp_x + (1 - m)p_y$$

We see that the frequency of $A$ after migration is determined by the proportion of $A$ alleles in the two groups that now make up population $y$. The first component, $mp_x$, represents the $A$ alleles in the migrants; we multiply the proportion of the population that consists of migrants ($m$) by the allele frequency of the migrants ($p$). The second component represents the $A$ alleles in the residents and equals the proportion of the population consisting of residents ($1 - m$) multiplied by the allele frequency in the residents ($p_y$). Adding these two components together gives us the allele frequency of $A$ in population $y$ after migration. This model of gene flow is diagrammed in Figure 21.13.

The change in allele frequency in population $y$ as a result of migration ($\Delta p$) equals the original frequency of $A$ subtracted from the frequency of $A$ after migration:

$$\Delta p = p'_y - p_y$$

In the previous equation, we found that $p'_y$ equaled $mp_x + (1 - m)p_y$, so the change in allele frequency can be written as

$$\Delta p = mp_x + (1 - m)p_y - p_y$$

**Figure 21.13**

**Theoretical model illustrating the effect of migration on the gene pool of a population.** After migration, population $y$ consists of two groups: the migrants with allele frequency of $p_x$ and the original residents with allele frequency $p_y$.

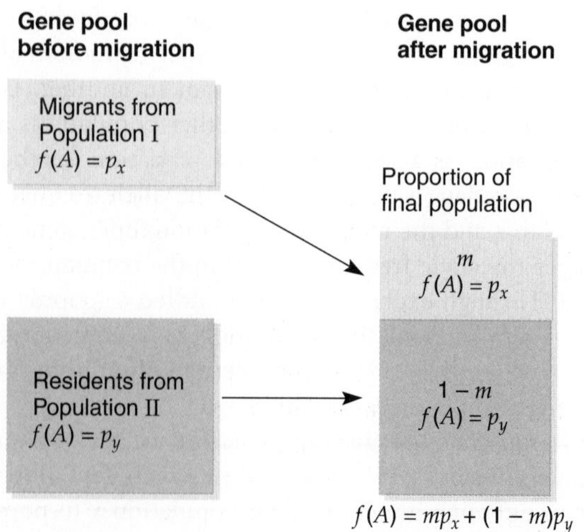


Multiplying $(1 - m)$ by $p_y$ in this equation, we obtain

$$\Delta p = mp_x + p_y - mp_y - p_y$$

$$\Delta p = mp_x - mp_y$$

$$\Delta p = m(p_x - p_y)$$

This final equation indicates that the change in allele frequency from migration depends on two factors: the proportion of the migrants in the final population and the difference in allele frequency between the migrants and the residents. If no differences exist in the allele frequency of migrants and residents ($p_x - p_y = 0$), then we can see that the change in allele frequency is zero. Populations must differ in their allele frequencies for migration to affect the makeup of the gene pool. With continued migration, $p_x$ and $p_y$ become increasingly similar; and, as a result, the change in allele frequency due to migration decreases. Eventually, allele frequencies in the two populations will be equal, and no further change will occur. However, this is true only when other factors besides migration do not influence allele frequencies.

The effects of gene flow have important ramifications, not only for the evolution of species but also for the conservation of species. As discussed earlier, many species that have wide geographic ranges show variation in genetic structure over the species range. Part of the natural genetic structure of a species could include population subdivision in which populations are loosely connected to each other by gene flow. Since gene flow is important in maintaining genetic diversity, this feature of population genetic structure must be taken into account by those interested in conserving the genetic identity of a species.

**Keynote**

Migration of individuals into a population may alter the makeup of the population gene pool if the allele frequency of the migrants differs from that of the resident population. Migration, also called gene flow, tends to reduce genetic divergence among populations and increases the effective size of the population. The amount of migration among populations of the same species determines how much genetic substructuring exists and whether different populations of the same species become very different from each other genetically.

## Natural Selection

We have now examined three major evolutionary processes capable of changing allele frequencies and contributing to evolution: mutation, genetic drift, and migration. Together, these processes introduce new variation (mutation) and alter the allele frequencies (genetic drift and migration) in the gene pool of populations. However, mutation, migration, and genetic drift do not result in

adaptation. *Adaptation* refers to the tendency for plants and animals to be well suited to the environments in which they are found. **Natural selection** is the process by which traits evolve that make organisms more suited to their immediate environment; these traits increase the organism's chances of surviving and reproducing. Natural selection is responsible for the many extraordinary traits seen in nature: wings that enable a hummingbird to fly backward, leaves of the pitcher plant that capture and devour insects, and brains that allow humans to speak, read, and love. These biological features and countless other exquisite traits are the product of adaptation (Figure 21.14). Genetic drift, mutation, and migration all influence the pattern and process of adaptation, but adaptation arises from natural selection. Natural selection is the force responsible for adaptation in all living organisms, and it has shaped much of the phenotypic variation observed in nature.

Charles Darwin and Alfred Russel Wallace (Figure 21.15) independently developed the concept of natural selection in the mid-nineteenth century, although some earlier naturalists had similar ideas. In 1858, Darwin and Wallace's theory was presented to the Linnaean Society of London and was enthusiastically received by other scientists. Darwin pursued the theory of evolution further than Wallace did, amassing hundreds of observations to support it and publishing his ideas in the book *On the Origin of Species* in 1859. For his innumerable contributions to our understanding of natural selection, Darwin often is regarded as the father of evolutionary theory. What is amazing about this theory is that Darwin had no clue about how genetic transmission worked. All that was necessary for his theory was that somehow offspring resemble their parents. Knowing the details of genetic transmission of traits makes the theory much deeper and richer and provides much more satisfying tests of how evolutionary change works at the genetic level.

Natural selection can be defined as differential reproduction of genotypes. It simply means that individuals with certain genotypes produce more offspring than do others; therefore, those genotypes increase in frequency in the next generation, as discussed in Chapter 22. Through natural selection, traits that contribute to survival and reproduction increase over time. In this way, organisms adapt to their environment.

**Figure 21.14**

**Spiny lizard (genus *Sceloporus*) whose cryptic coloration allows it to blend in with its environment and avoid predators, a product of natural selection.**

**Selection in Natural Populations.** A classic example of selection in natural populations is the evolution of melanic (dark) forms of moths in association with industrial pollution, by a phenomenon known as industrial melanism. Melanic phenotypes have appeared in a number of different species of moths found in the industrial regions of continental Europe, North America, and England. One of the best-studied cases involves the peppered moth, *Biston betularia*. The common phenotype of this species, called the *typical* form, is a greyish white color with black mottling over the body and wings.

Before 1848, all peppered moths collected in England possessed this *typical* phenotype; but in 1848, a single black moth was collected near Manchester, England. This new phenotype, called *carbonaria*, presumably arose by mutation and rapidly increased in frequency around Manchester and in other industrial regions. By 1900, the *carbonaria* phenotype had reached a frequency of more than 90% in several populations. High frequencies of *carbonaria* appeared to be associated with industrial regions, whereas the *typical* phenotype remained common in more rural districts. Laboratory studies by a number of investigators, including E. B. Ford and R. Goldschmidt, demonstrated that the *carbonaria* phenotype was dominant to the *typical* phenotype. A third phenotype was also discovered, which was somewhat intermediate to *typical* and *carbonaria*; this phenotype, *insularia*, was produced by a dominant allele at a different locus.

H. B. D. Kettlewell investigated color polymorphism in the peppered moth, demonstrating that the increase in the *carbonaria* phenotype occurred as a result of strong selection against the *typical* form in polluted woods. Peppered moths are nocturnal; during the day they rest on the trunks of lichen-covered trees. Birds often prey on the moths during the day, but because the lichens that cover the trees are naturally grey in color, the *typical* form of the peppered moth is well camouflaged against this background (Figure 21.16a). In industrial areas, however, extensive pollution beginning with the industrial revolution in the mid-nineteenth century had killed most of the lichens and covered the tree trunks with black soot. Against this black background, the *typical* phenotype was conspicuous and was readily consumed by birds. In contrast, the *carbonaria* form was well camouflaged against the blackened trees and had a higher rate of survival than did the *typical* phenotype in polluted areas (Figure 21.16b). Because *carbonaria* survived better in polluted woods, more *carbonaria* genes were transmitted to the next generation; thus, the *carbonaria* phenotype

**Figure 21.15**
**Charles Darwin and Alfred Russel Wallace, who independently developed the theory of evolution through natural selection.**

increased in frequency in industrial areas. In rural areas, where pollution was absent, the *carbonaria* phenotype was conspicuous and the *typical* form was camouflaged; in these regions the frequency of the *typical* form remained high.

Kettlewell demonstrated that selection affected the frequencies of the two phenotypes by conducting a series of mark-and-recapture experiments involving dark and light moths in smoky, industrial Birmingham, England, and in nonindustrialized Dorset. As predicted, the *typical* phenotype was favored in Dorset, and *carbonaria* was favored in Birmingham.

**Fitness and Coefficient of Selection.** Darwin described natural selection primarily in terms of survival. Even today, many nonbiologists think of natural selection in terms of a struggle for existence. However, what is most important in the process of natural selection is the relative number of genes that are contributed to future generations. Certainly the ability to survive is important, but survival alone does not ensure that genes are passed on; reproduction must also occur. Therefore, we measure natural selection by assessing reproduction. Natural selection is measured in terms of **Darwinian fitness,** which is defined as the relative reproductive ability of a genotype.

Darwinian fitness is often symbolized as $w$. Since it is a measure of the relative reproductive ability, population geneticists usually assign a fitness of $w = 1$ to a genotype that produces the most offspring. The fitnesses of the other genotypes are assigned relative to this. For example, suppose that the genotype $G^1G^1$ on the average produces eight offspring, $G^1G^2$ produces an average of four offspring, and $G^2G^2$ produces an average of two offspring. The $G^1G^1$

**Figure 21.16**
***Biston betularia*, the peppered moth, and its dark form *carbonaria* on the trunk of a lichened tree in the unpolluted countryside and on the trunk of a tree with dark bark.** On the lichened tree, the dark form of the moth is readily seen, whereas the light form is well camouflaged. On the dark tree, the dark form of the moth is well camouflaged.

a) Peppered moths on a lichened tree trunk

b) Peppered moths on a dark tree trunk

genotype has the highest reproductive output, so its fitness is 1($w_{11}$ = 1.0). Genotype $G^1G^2$ produces on the average four offspring for the eight produced by the most fit genotype, so the fitness of $G^1G^2(w_{12})$ is 4/8 = 0.5. Similarly, $G^2G^2$ produces two offspring for the eight produced by $G^1G^1$, so the fitness of $G^2G^2(w_{22})$ is 2/8 = 0.25. Table 21.9 illustrates the calculation of relative fitness values.

Fitness values for genotypes must be estimated with great care. For example, equating fitness to the number of offspring is an oversimplification, because we need to know about the survival probability of those offspring as well. For example, David Lack found that starlings had an optimal number of eggs that they could rear to mature offspring successfully. If they laid too many eggs, they produced fewer chicks that survived to maturity. Assigning higher fitness values to birds that laid more eggs would have been incorrect. The fitness associated with a genotype is difficult to pin down by looking at a snapshot of a part of the organism's life. Genotypes that have a high survival probability can have on average fewer offspring. Single genes can have effects on different aspects of the life cycle that affect fitness. Such **pleiotropic** effects make a big difference in the overall genotypic fitnesses. Despite these challenges, Darwinian fitness tells us how well a genotype is doing in terms of natural selection. A related measure is the **selection coefficient,** which is a measure of the relative intensity of selection against a genotype. The selection coefficient is symbolized by s and equals 1 − w. In our example, the selection coefficients for $G^1G^1$ are s = 0; for $G^1G^2$, s = 0.5; for $G^2G^2$, s = 0.75.

**Effect of Selection on Allele Frequencies.** Natural selection produces a number of different effects. At times, natural selection eliminates genetic variation; at other times, it maintains variation. It can change allele frequencies or prevent allele frequencies from changing; it can produce genetic divergence between populations or maintain genetic uniformity. Which of these effects occurs depends primarily on the relative fitness of the genotypes and on the frequencies of the alleles in the population.

The change in allele frequency that results from natural selection can be calculated by constructing a table such as Table 21.10. This table method can be used for any type of single-locus trait, whether the trait is dominant, codominant, recessive, or overdominant. To use the table method, we begin by listing the genotypes ($A^1A^1$, $A^1A^2$, and $A^2A^2$) and their initial frequencies. If random mating has just taken place, the genotypes are in Hardy–Weinberg proportions, and the initial frequencies are $p^2$, $2pq$, and $q^2$. We then list the fitnesses for each of the genotypes, $w_{11}$, $w_{12}$, and $w_{22}$. Now, suppose that selection occurs and only some of the genotypes survive. The contribution of each genotype to the next generation is equal to the initial frequency of the genotype multiplied by its fitness. For $A^1A^1$ this is $p^2 \times w_{11}$. Notice that the contributions of the three genotypes do not add up to 1. We calculate the relative contributions of each genotype by dividing each by the mean fitness of the population. The *mean fitness of the population* equals $p^2w_{11} + 2pqw_{12} + q^2w_{22} = \overline{w}$. The mean fitness is the average fitness of individuals in the population. After dividing the relative contributions of each genotype by the mean fitness, we have the frequencies of the genotypes after selection, where $P'$ is the frequency of $A^1A^1$ genotypes, $H'$ is the frequency of $A^1A^2$ genotypes, and $Q'$ is the frequency of $A^2A^2$ genotypes. We then calculate the new allele frequency ($p'$) from the genotypes after selection, using our familiar formula, $p'$ = (frequency of $A^1A^1$) + (1/2 × frequency of $A^1A^2$. Finally, the change in allele frequency resulting from selection equals $p' - p$. A sample calculation using some actual allele frequencies and fitness values is presented in Table 21.11.

The wide range of generally unappreciated effects of natural selection discussed earlier can now be understood in terms of the figures in Table 21.10. Again remembering that we will arbitrarily set the genotype with highest fitness to 1.0, we get a variety of classes of natural selection, each with its own effects, by permuting all possible relationships among fitness values for the genotypes. These include the following:

1. $w_{11} = w_{12} = w_{22} = 1.0$. All fitnesses are equal, and there is no selection.
2. $w_{11} = w_{12} < 1.0$ and $w_{22} = 1.0$. The heterozygote has a fitness equal to a homozygote but less than the best fitness of the other homozygote. Natural selection is operating against a dominant allele $A^1$.

**Table 21.9 Computation of Fitness Values and Selection Coefficients of Three Genotypes**

| | Genotypes | | |
|---|---|---|---|
| | $G^1G^1$ | $G^1G^2$ | $G^2G^2$ |
| Number of breeding adults in one generation | 16 | 10 | 20 |
| Number of offspring produced by all aduNlts of the genotype in the next generation | 128 | 40 | 40 |
| Average number of offspring produced per breeding adult | 128/16 = 8 | 40/10 = 4 | 40/20 = 2 |
| Fitness w (relative number of offspring produced) | 8/8 = 1 | 4/8 = 0.5 | 2/8 = 0.25 |
| Selection coefficient (s = 1 − w) | 1 − 1 = 0 | 1 − 0.5 = 0.5 | 1 − 0.25 = 0.75 |

**Table 21.10 General Method of Determining Change in Allele Frequency Caused by Natural Selection**

| | Genotypes | | |
|---|---|---|---|
| | $A^1A^1$ | $A^1A^2$ | $A^2A^2$ |
| Initial genotype frequencies | $p^2$ | $2pq$ | $q^2$ |
| Fitness[a] | $w_{11}$ | $w_{12}$ | $w_{22}$ |
| Frequency after selection | $p^2w_{11}$ | $2pqw_{12}$ | $q^2w_{22}$ |
| Relative genotype frequency after selection[b] | $P' = \frac{p^2w_{11}}{\overline{w}}$ | $H' = \frac{2pqw_{12}}{\overline{w}}$ | $Q' = \frac{q^2w_{22}}{\overline{w}}$ |
| Allele frequency after selection $= p' = P' + 1/2(H')$ | | | |
| $q' = 1 - p'$ | | | |
| Change in allele frequency caused by selection $= \Delta p = p' - p$ | | | |

[a]For simplicity, fitness in this example is considered to be the probability of survival. Change in allele frequency caused by differences in the number of offspring produced by the genotypes is calculated in the same manner.
[b]$\overline{w} = p^2w_{11} + 2pqw_{12} + q^2w_{22}$

**3.** $w_{11} = w_{12} = 1.0$ and $w_{22} < 1.0$. The heterozygote along with a homozygote has the highest fitness, which is greater than that of the other homozygote. Natural selection is operating against a recessive allele $A^2$.

**4.** $w_{11} < w_{12} < w_{22}$ and $w_{22} = 1.0$. The heterozygote has an intermediate fitness. Natural selection is operating against $A^2$ without effects of dominance.

**5.** $w_{11}$ and $w_{22} < 1.0$ and $w_{12} = 1.0$. The heterozygote has the highest fitness, and the two homozygotes have a lower fitness that may or may not be the same. Natural selection is favoring the heterozygote.

**6.** $w_{12} < w_{11}$ and $w_{22} = 1.0$. The heterozygote has lower fitness than both homozygotes. Only one of the homozygotes must have a fitness equal to 1.0. Natural selection is favoring the homozygotes.

Each of the five cases of natural selection results in a characteristic pattern of change in the genetic structure of a population. Cases 2, 3, and 4 are all a type of natural selection called *directional selection* and result in the

**Table 21.11 General Method of Determining Change In Allele Frequency Caused by Natural Selection When Initial Allele Frequencies Are $p = 0.6$ and $q = 0.4$**

| | Genotypes | | |
|---|---|---|---|
| | $A^1A^1$ | $A^1A^2$ | $A^2A^2$ |
| Initial genotype frequencies | $p^2$<br>$(0.6)^2 = 0.36$ | $2pq$<br>$2(0.6)(0.4) = 0.48$ | $q^2$<br>$(0.4)^2 = 0.16$ |
| Fitness | $w_{11} = 0$ | $w_{12} = 0.4$ | $w_{22} = 1$ |
| Frequency after selection | $p^2w_{11} =$<br>$(0.36)(0) = 0$ | $2pqw_{12} =$<br>$(0.48)(0.4) = 0.19$ | $q^2w_{22} =$<br>$(0.16)(1) = 0.16$ |
| Relative genotype frequency after selection[a] | $P' = \frac{p^2w_{11}}{\overline{w}}$<br>$P' = 0/0.35 = 0$ | $H' = \frac{2pqw_{12}}{\overline{w}}$<br>$H' = 0.19/0.35$<br>$= 0.54$ | $Q' = \frac{q^2w_{22}}{\overline{w}}$<br>$Q' = 0.16/0.35$<br>$= 0.46$ |
| Allele frequency after selection $p' = P' + 1/2(H')$<br>$p' = 0 + 1/2(0.54) = 0.27$<br>$q' = 1 - p' = 1 - 0.27 = 0.73$ | | | |
| Change in allele frequency caused by selection $= \Delta p = p' - p$<br>$\Delta p = 0.27 - 0.6 = -0.33$ | | | |

[a]$\overline{w} = p^2w_1 + 2pqw_{12} = q^2w_{22}$
$\overline{w} = 0 + 0.19 + 0.16$
$\overline{w} = 0.35$

elimination or reduction in frequency of one of the alleles. Case 5 is very different and results in no evolutionary change once a stable equilibrium has been reached. Case 6 results in what looks like a directional change in allele frequency, but the allele that is selected against depends on the initial allele frequency. We now consider how several of these cases affect the genetic structure of a population.

**Selection Against a Recessive Trait.** Cases 2, 3, and 4 are similar in that they result in a directed change in the allele frequency of a population. This directional effect is the one that is most often associated with natural selection. The case of the peppered moth discussed earlier falls into the category of selection against a recessive allele. We will discuss one of these cases in more detail because most new mutations are recessive and have reduced fitness. When a trait is completely recessive (case 3), both the heterozygote and the dominant homozygote have a fitness of 1, whereas the recessive homozygote has reduced fitness, where s represents the strength of selection, as shown here.

| Genotype | Fitness |
|---|---|
| *AA* | 1 |
| *Aa* | 1 |
| *aa* | $1 - s$ |

If the genotypes are initially in Hardy–Weinberg proportions, the contribution of each genotype to the next generation is the frequency times the fitness.

| | |
|---|---|
| *AA* | $p^2 \times 1 = p^2$ |
| *Aa* | $2pq \times 1 = 2pq$ |
| *aa* | $q^2 \times (1 - s) = q^2 - sq^2$ |

The mean fitness of the population is $p^2 + 2pq + q^2 - sq^2$. Since $p^2 + 2pq + q^2 = 1$, the mean fitness becomes $1 - sq^2$, and the normalized genotypic frequencies *after* selection are

| | |
|---|---|
| *AA* | $\dfrac{p^2}{1 - sq^2}$ |
| *Aa* | $\dfrac{2pq}{1 - sq^2}$ |
| *aa* | $\dfrac{q^2 - sp^2}{1 - sq^2}$ |

To obtain $q'$, the frequency after selection, we add the frequency of the *aa* homozygote and half the frequency of the heterozygote.

$$q' = \frac{q^2 - sq^2}{1 - sq^2} + \frac{1}{2} \times \frac{2pq}{1 - sq^2}$$

With a bit of algebra and recalling that $(q + p) = 1$, we find that

$$q' = \frac{q - sq^2}{1 - sq^2}$$

Therefore, the change in the frequency of *a* after one generation of selection is

$$\Delta q = q' - q$$

and with a bit of algebraic simplification, this reduces to

$$\Delta q = -spq^2/(1 - sq^2)$$

When $\Delta q = 0$, no further change occurs in allele frequencies and an equilibrium has been reached. Notice that there is a negative sign in the equation to the left of $spq^2$, because the values of $s$, $p$, and $q$ are always positive or zero, $\Delta q$ is negative or zero. Thus, the value of $q$ decreases with selection.

Selection also depends on the actual frequencies of the allele in the population. This is because the relative proportions of *Aa* and *aa* individuals at various frequencies of allele *a* influence how effectively selection can reduce a detrimental recessive trait. When the frequency of a recessive detrimental allele is high, many homozygous recessive individuals are present in the population and have low fitness, causing a large change in the allele frequency. When the allele frequency is low, however, the homozygous recessive genotype is rare, and little change in allele frequency occurs. In fact, when a detrimental recessive allele is not lethal, it may segregate in a population for many generations, which explains why some genetic diseases continue to be present in human populations at very low frequencies. It may also segregate as a result of a balance between mutation and natural selection.

Figure 21.17 shows the magnitude of change in allele frequency for each generation in three populations with different initial allele frequencies. Population 1 begins with allele frequency $q$ equal to 0.9, population 2

**Figure 21.17**
**Effectiveness of selection against a recessive lethal genotype at different initial allele frequencies.** The three populations had initial frequencies of 0.9, 0.5, and 0.1.

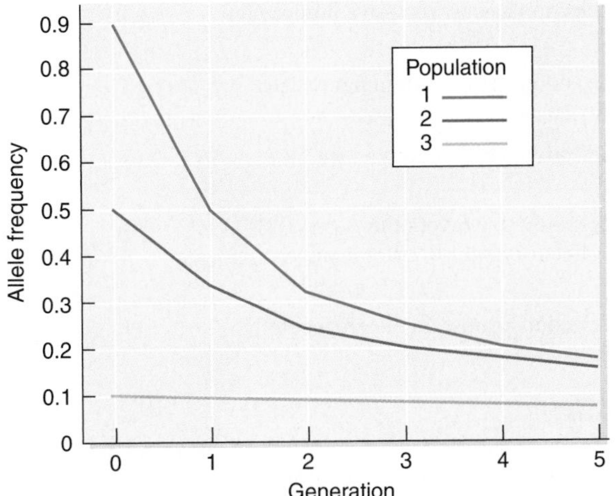

begins with $q$ equal to 0.5, and population 3 begins with $q$ equal to 0.1. In this example, the homozygous recessive genotype (*aa*) has a fitness of 0, and the other two genotypes (*AA* and *Aa*) have a fitness of 1 (recessive lethal condition). When the frequency of $q$ is high, as in population 1, the change in allele frequency is large; in the first generation, $q$ drops from 0.9 to 0.47. However, when $q$ is small, as in population 3, the change in $q$ is much less; here $q$ drops from 0.1 to 0.091. Therefore, as $q$ becomes smaller, the change in $q$ becomes less. Because of this diminishing change in frequency, it is almost impossible to eliminate a recessive trait from the population entirely. This is easily understood if one realizes that the final recessive alleles in a population will almost always find themselves in the heterozygote condition. However, this result applies only to completely recessive traits; if the fitness of the heterozygote is also reduced (case 4), the change in allele frequency will be more rapid because now selection also acts against the heterozygote in addition to the homozygote. The effect of dominance on changes in allele frequency as a result of selection is illustrated in Figure 21.18.

We have discussed at length the effects of selection on a recessive trait to illustrate how the formula for change in allele frequency can be derived from our general table method of allele frequency change under selection. Similar derivations can be carried out for dominant traits and codominant traits (cases 2 and 4). We will not present those derivations here, but the appropriate formula for calculating changes in allele frequency under different types of dominance are presented in Table 21.12. However, by using the table method it is possible to calculate changes in allele frequency for any type of trait.

**Figure 21.18**

**Fitnesses of the genotypes *AA*, *Aa*, and *aa* are 1, 0.5, and 0.5 for the dominant case, 1, 0.75, and 0.5 for the additive case, and 1, 1, and 0.5 for the recessive case.** Frequency of the *a* allele is plotted.

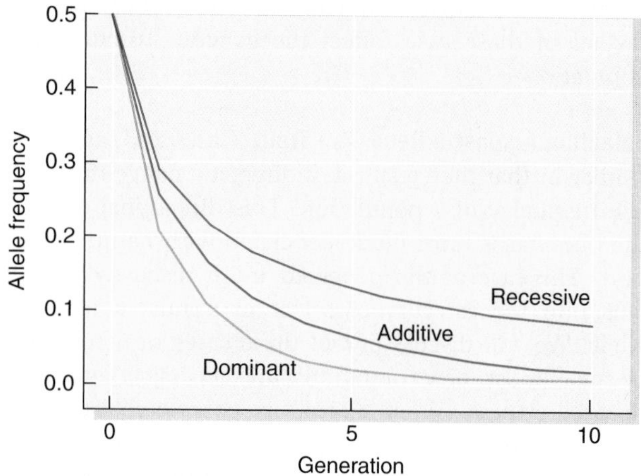


**Heterozygote Superiority.** Natural selection does not always result in a directional change in allele frequency and a decrease in genetic variation. Some forms of selection result in the maintenance of genetic variation and form the backbone of the balanced model of genetic variation discussed earlier. The simplest type of balancing selection is called **heterosis, overdominance**, or **heterozygote superiority.** Balancing selection is a form of natural selection that works to maintain genetic polymorphisms (multiple alleles of a gene) in a population. An equilibrium of allele frequencies arises when the heterozygote has higher fitness than either of the homozygotes. In this case (case 5), both alleles are

**Table 21.12 Formulas for Calculating Change in Allele Frequency After One Generation of Selection**

| Type of Selection | Fitnesses of Genotypes $A^1A^1$ | $A^1A^2$ | $A^2A^2$ | Calculation of Change in Allele Frequency |
|---|---|---|---|---|
| Selection against recessive homozygote | 1 | 1 | $1 - s$ | $\Delta q = \frac{-spq^2}{1 - sq^2}$ |
| Selection against a dominant allele | $1 - s$ | $1 - s$ | 1 | $\Delta p = \frac{-spq^2}{1 - s + sq^2}$ |
| Selection with no dominance | 1 | $(1 - s/2)$ | $1 - s$ | $\Delta q = \frac{-spq/2}{1 - sq}$ |
| Selection that favors the heterozygote (overdominance) | $1 - s$ | 1 | $1 - t$ | $\Delta q = \frac{pq(sp - tp)}{1 - sp^2 - tq^2}$ |
| Selection against the heterozygote | 1 | $1 - s$ | 1 | $\Delta q = \frac{spq(q - p)}{1 - 2spq}$ |
| General[a] | $w_{11}$ | $w_{12}$ | $w_{22}$ | $\Delta q = \frac{pq[p(w_{11} - w_{12}) - q(w_{22} - w_{12})]}{\overline{w}}$ |

[a]*Note:* For calculation of $\overline{w}$ see Table 21.10

maintained in the population because both are favored in the heterozygote genotype. Allele frequencies will change as a result of selection until the equilibrium point is reached and then will remain stable. The allele frequencies at which the population reaches equilibrium depend on the relative fitnesses of the two homozygotes. If the selection coefficient of *AA* is *s* and the selection coefficient of *aa* is *t*, it can be shown algebraically that at equilibrium

$$\hat{p} = f(A) = t/(s + t)$$

and

$$\hat{q} = f(a) = s/(s + t)$$

Notice that if selection against both homozygotes is the same (i.e., $s = t$), then the equilibrium allele frequency is 0.5. As selection against the homozygotes becomes less symmetrical, the equilibrium allele frequency shifts in the direction of the most fit homozygote.

The most famous example of heterozygote superiority operating in nature is provided by human sickle-cell anemia. Sickle-cell anemia results from a mutation in the gene coding for β-hemoglobin. In some populations, there are three hemoglobin genotypes: *Hb-A/Hb-A*, *Hb-A/Hb-S*, and *Hb-S/Hb-S*. Individuals with the *Hb-A/Hb-A* genotype have completely normal red blood cells, *Hb-S/Hb-S* individuals have sickle-cell anemia, and *Hb-A/Hb-S* individuals have sickle-cell trait, a mild form of sickle-cell anemia. In an environment in which malaria is common, the heterozygotes are at a selective advantage over the two homozygotes. The mild anemia suffered by heterozygotes is enough to inhibit growth and reproduction of the malarial parasite. The heterozygotes therefore have greater resistance to malaria and thus higher fitness than do *Hb-A/Hb-A* individuals. The *Hb-S/Hb-S* individuals are at a serious selective disadvantage because they have sickle-cell anemia. As a result, in malaria-infested areas in which the *Hb-S* gene is also found, an equilibrium state is established in which a significant number of *Hb-S* alleles are found in the heterozygotes as a result of the selective advantage of this genotype. The distributions of malaria and the *Hb-S* allele are illustrated in Figure 21.19. Thus, despite the problems faced by *Hb-S* homozygotes, natural selection cannot eliminate this allele from the population, because the allele has beneficial effects in the heterozygote state.

### Keynote

Natural selection involves differential reproduction of genotypes and is measured in terms of Darwinian fitness, the relative reproductive contribution of a genotype. The effects of selection depend on the relative fitnesses of the different genotypes. Directional selection results in the directional change in allele frequency, with the disfavored allele being eliminated from the population in the cases where it is dominant or codominant but persisting in the population at low frequencies if it is recessive and thus invisible in the heterozygote. In either case directional selection decreases the amount of genetic variation in a population. Balancing selection, exemplified here as heterozygote superiority, results in the maintenance of genetic variation in the population.

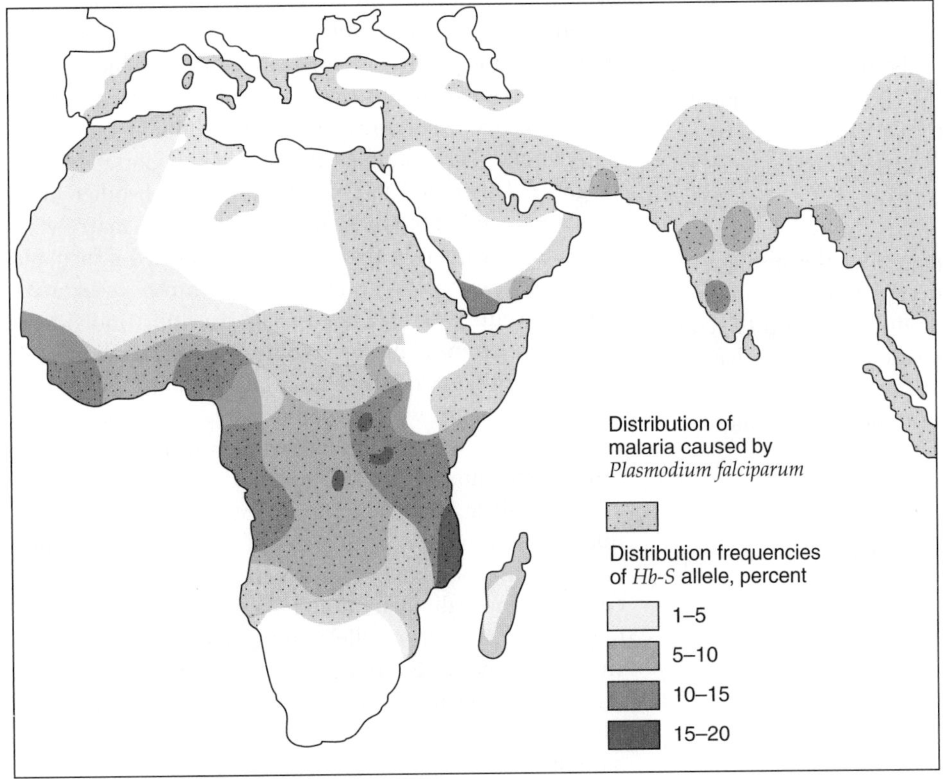


**Figure 21.19**

**The distribution of malaria caused by the parasite *Plasmodium falciparum* coincides with distribution of the *Hb-S* allele for sickle-cell anemia.** The frequency of *Hb-S* is high in areas where malaria is common because *Hb-A/Hb-S* heterozygotes are resistant to malarial infection.

## Balance Between Mutation and Selection

We have already seen two examples where we violated the Hardy–Weinberg law assumptions more than one at a time: the combination of drift and mutation and the combination of drift and migration. We saw that mutation, migration, and small population size interact to determine the genetic structure of a population when the alleles are selectively neutral. Population genetics theory has also been extended to accommodate other violations of several assumptions simultaneously; much has been learned by taking this approach. Here, as an example, we consider the simultaneous effects of mutation and natural selection.

As we have seen, natural selection can reduce the frequency of a deleterious recessive allele. As the frequency of the allele becomes low, the change in frequency diminishes with each generation. When the allele is rare, the change in frequency is very slight. Opposing this reduction in the allele's frequency due to selection is mutation pressure, which continually produces new alleles and tends to increase the frequency. Eventually a balance, or equilibrium, is reached, in which the input of new alleles by recurrent mutation is counterbalanced by the loss of alleles through natural selection. When equilibrium is obtained, the frequency of the allele remains stable, even though selection and mutation continue, unless the equilibrium is perturbed by some other process.

Consider a population in which selection occurs against a deleterious recessive allele, *a*. As we saw on pages 635–636, the amount *a* will change in one generation as a result of selection is

$$\Delta q = -spq^2/(1 - sq^2)$$

For a rare recessive allele, $q^2$ will be near 0, and the denominator in this equation, $1 - sq^2$, will be approximately 1, so that the decrease in frequency caused by selection is given by

$$\Delta q = -spq^2$$

At the same time, the frequency of the *a* allele increases as a result of mutation from *A* to *a*. Provided the frequency of *a* is low, the reverse mutation of *a* to *A* essentially can be ignored. Equilibrium between selection and mutation occurs when the decrease in allele frequency produced by selection is the same as the increase produced by mutation:

$$spq^2 = up$$

We can predict the frequency of *a* at equilibrium ($\hat{q}$) by rearranging this equation:

$$sq^2 = u$$
$$q^2 = u/s$$

and

$$\hat{q} = \sqrt{u/s}$$

If the recessive homozygote is lethal ($s = 1$), the equation becomes

$$\hat{q} = \sqrt{u}$$

As an example of the balance between mutation and selection, consider a recessive gene for which the mutation rate is $10^{-6}$ and *s* is 0.1. At equilibrium, the frequency of the gene will be $\hat{q} = \sqrt{10^{-6}/0.1} = 0.0032$ Most recessive deleterious traits remain within a population at low frequency because of equilibrium between mutation and selection.

For a dominant allele *A*, the frequency at equilibrium ($\hat{p}$) is

$$\hat{p} = u/s$$

If the mutation rate is $10^{-6}$ and *s* is 0.1, the frequency of the dominant gene at equilibrium is $10^{-6}/0.1 = 0.00001$, which is considerably less than the equilibrium frequency for a recessive allele with the same fitness and mutation rate. This is because selection cannot act on a recessive allele in the heterozygote state, whereas both the homozygote and the heterozygote for a dominant allele have reduced fitness. For this reason, detrimental dominant alleles are generally far less common than recessive ones.

## Assortative Mating

A fundamental assumption of the Hardy–Weinberg law is that members of the population mate randomly in regards to the gene in question. But many populations do not mate randomly for some traits, and when nonrandom mating occurs, the genotypes do not exist in the proportions predicted by the Hardy–Weinberg law. One form of nonrandom mating is **positive assortative mating,** which occurs when individuals with similar phenotypes mate preferentially. Positive assortative mating is common in natural populations. For example, humans mate assortatively for height; tall men and tall women marry each other more frequently, and short men and short women marry each other more frequently, than would be expected if men and women chose mates at random. **Negative assortative mating** occurs when phenotypically dissimilar individuals mate more often than do randomly chosen individuals. If humans exhibited negative assortative mating for height, tall men and short women would marry each other preferentially, and short men and tall women would marry each other preferentially. Positive assortative mating does not affect the allele frequencies of a population, but it will influence the genotype frequencies of the genes controlling the trait if the phenotypes for which assortative mating occurs are genetically determined. Negative assortative mating may affect both the allele and genotype frequencies of such genes, because types that are rare in the population may mate more frequently than those that are common. These effects may occur in small populations in which natural selection is an acting force, for example, sexual selection in which mating between individuals with particular phenotypes is not random.

## Inbreeding

Another important departure from random mating is **inbreeding**. Inbreeding involves preferential mating between relatives. In very small populations, even if individuals tossed a coin or used some other randomizing procedure to choose mates, their mates probably would end up being relatives. In this sense, small populations suffer the consequences of inbreeding even if there is no preferential tendency to select relatives as mates. Inbreeding often is measured in terms of the coefficient of inbreeding ($F_{IS}$), often simply referred to as $F$. The greater the value of $F_{IS}$, the greater the reduction in heterozygosity relative to that expected from the Hardy–Weinberg expectation. By definition,

$$F_{IS} = \frac{\overline{H}_e - \overline{H}_o}{\overline{H}_e}$$

where $\overline{H}_e$ is the mean expected heterozygosity in the subpopulation and $\overline{H}_o$ is the mean observed heterozygosity of individuals within the subpopulation.

If genotypes are in Hardy–Weinberg proportions, $F_{IS} = 0$ because observed heterozygosity and expected heterozygosity are equal. However, regular systems of inbreeding exist, such as self-fertilization, sib mating, and mating between first cousins. After one generation of mating in such systems, the value of $F_{IS}$ would be 0.5, 0.25, and 0.06, respectively.

The most extreme case of inbreeding is self-fertilization, which occurs in many plants and a few animals, such as some snails. The effects of self-fertilization are illustrated in Table 21.13. Assume that we begin with a population consisting entirely of *Aa* heterozygotes and that all individuals in this population reproduce by self-fertilization. After one generation of self-fertilization, the progeny will consist of $^1/_4$ *AA*, $^1/_2$ *Aa*, and $^1/_4$ *aa*. Now only half of the population consists of heterozygotes. When this generation undergoes self-fertilization, the *AA* homozygotes will produce only *AA* progeny, and the *aa* homozygotes will produce only *aa* progeny. When the heterozygotes reproduce, however, only half of their progeny will be heterozygous like the parents, and the other half will be homozygous ($^1/_4$ *AA* and $^1/_4$ *aa*). This means that in each generation of self-fertilization, the percentage of heterozygotes decreases by 50%. After a large number of generations, there will be no heterozygotes and the population will be divided equally between the two homozygous genotypes. Note that the population was in Hardy–Weinberg proportions in the first generation after self-fertilization, but after further rounds the proportion of homozygotes is greater than that predicted by the Hardy–Weinberg law.

Note that inbreeding has very similar effects to genetic drift in small populations. In both cases heterozygosity decreases and homozygosity increases. In the case of inbreeding in large populations, however, allele frequencies stay the same and homozygosity increases, whereas in the case of drift, allele frequency changes and homozygosity increases. Drift causes only small departures from Hardy–Weinberg; those caused by inbreeding can be extreme.

The result of continued self-fertilization is to increase homozygosity at the expense of heterozygosity. The frequencies of alleles *A* and *a* remain constant, while the frequencies of the three genotypes change significantly. When less intensive inbreeding occurs, similar but less pronounced effects occur.

**Keynote**

Inbreeding involves preferential mating between close relatives. Continued inbreeding increases homozygosity within a population and in most species results in reduced fitness.

**Table 21.13 Relative Genotype Distributions Resulting from Self-Fertilization over Several Generations Starting with an *Aa* Individual**

| | Genotype Frequencies | | |
|---|---|---|---|
| Generation | *AA* | *Aa* | *aa* |
| 0 | 0 | 1 | 0 |
| 1 | $^1/_4$ | $^1/_2$ | $^1/_4$ |
| 2 | $^1/_4 + {}^1/_8 = {}^3/_8$ | $^1/_4$ | $^1/_4 + {}^1/_8 = {}^3/_8$ |
| 3 | $^3/_8 + {}^1/_{16} = {}^7/_{16}$ | $^1/_8$ | $^3/_8 + {}^1/_{16} = {}^7/_{16}$ |
| 4 | $^7/_{16} + {}^1/_{32} = {}^{15}/_{32}$ | $^1/_{16}$ | $^7/_{16} + {}^1/_{32} = {}^{15}/_{32}$ |
| 5 | $^{15}/_{32} + {}^1/_{64} = {}^{31}/_{64}$ | $^1/_{32}$ | $^{15}/_{32} + {}^1/_{64} = {}^{31}/_{64}$ |
| $n$ | $[1 - (^1/_2)^n]/2$ | $(^1/_2)^n$ | $[1 - (^1/_2)^n]/2$ |
| $\infty$ | $^1/_2$ | 0 | $^1/_2$ |

## Summary of the Effects of Evolutionary Forces on the Genetic Structure of a Population

Let us now review the major effects of the different evolutionary processes on (1) changes in allele frequency within a population; (2) genetic divergence between populations; and (3) increases and decreases in genetic variation within populations.

### Changes in Allele Frequency Within a Population

Mutation, migration, genetic drift, and selection all have the potential to change the allele frequencies of a population over time. However, mutation usually occurs at such a low rate that the change resulting from mutation pressure alone is usually negligible. Genetic drift produces substantial changes in allele frequency when population size is small. Furthermore, mutation, migration, and selection may lead to equilibria where these processes continue to act, but the allele frequencies no longer change. While negative assortative mating can lead to changes in allele frequencies, other forms of nonrandom mating do not change allele frequencies. All forms of nonrandom mating affect the genotypic frequencies of a population. Inbreeding leads to increases in homozygosity, and if there are deleterious recessive alleles in the population (which there almost always are), then inbreeding leads to reduced fitness or inbreeding depression.

**Genetic Divergence Among Populations** Several evolutionary processes lead to genetic divergence between populations. Since genetic drift is a random process, allele frequencies in different populations may drift in different directions, sometimes increasing and sometimes decreasing. So genetic drift can produce genetic divergence among populations. Gene flow among populations has just the opposite effect, tending to equalize allele frequencies among populations. If the population size is small, different mutations may arise in different populations, so mutation may contribute to population differentiation. Natural selection can increase genetic differences among populations by favoring different alleles in different populations, or it can prevent divergence by keeping allele frequencies uniform among populations. Nonrandom mating, by itself, will not generate genetic differences between populations, although it may contribute to the effects of other processes.

### Increases and Decreases in Genetic Variation Within Populations

Migration and mutation tend to increase genetic variation within populations by introducing new alleles to the gene pool. Genetic drift produces the opposite effect, decreasing genetic variation within small populations through loss of alleles. Since inbreeding leads to increases in homozygosity, it also diminishes genetic variation within populations; outbreeding, on the other hand, increases genetic variation by increasing heterozygosity. Natural selection can increase or decrease genetic variation; if one particular allele is favored, other alleles decrease in frequency and can be eliminated from the population by selection. Alternatively, natural selection can increase genetic variation within populations through overdominance and other forms of balancing selection.

In natural populations, these evolutionary processes never act in isolation but combine and interact in complex ways. In most natural populations, the combined effects of these processes and their interaction determine the pattern of genetic variation observed in the gene pool over time.

### The Effects of Crossing-Over on Genetic Variation

In the 1980s, population geneticists became aware of the effects of crossing-over, which leads to recombination on patterns of genetic variation across the genome. Crossing-over may occur at any location across a chromosome and often leads to recombination between homologous chromosomes, as you learned in Chapters 12 and 14. This is the basic meiotic process used to create genetic maps. The number of recombination events during meiosis between two loci is a measure of the genetic distance between them, termed a centiMorgan, or map unit. If we examine genetic variation at two different loci in a natural population, we may find that both loci are in Hardy–Weinberg equilibrium, but that they do not segregate independently, thus violating Mendel's law of independent assortment. This can happen for two reasons. First, the loci may be very close together on the chromosome, and when one allele at locus *A* is passed on to offspring, the allele at locus *B* that is also on that same chromosome is also passed on. This is an example of linkage. The rate at which this linkage between locus *A* and locus *B* breaks down so they segregate independently depends on how often crossing-over occurs between the two loci. The second reason alleles at two loci may appear to segregate together is that hybridization, genetic drift, and migration can cause deviations from what is expected of loci that assort independently. This is called **gametic disequilibrium** and will decay with time after a demographic event such as a bottleneck, migration, or hybridization.

If we observe two loci that appear to be in Hardy–Weinberg equilibrium but do not segregate independently, we might first suspect that the loci are very close together on the chromosome. However, if we know that the loci are far apart, perhaps because they have been physically or genetically mapped, or we know that many pairs of loci in the population are not segregating independently, then we may suspect a recent population bottleneck, migration, or hybridization. Together, such deviations from the expectations of independent assortment and Hardy–Weinberg equilibrium caused either by physical linkage or population demography are lumped into a term called **linkage disequilibrium** that we can model to predict what factors may have caused the deviations we observe in natural populations. A simple measure of linkage disequilibrium is $D = g_{AB}g_{ab} - g_{Ab}g_{aB}$, where g is the frequency of a gamete at each locus. So, for example, $g_{Ab}$ is a gamete that has an *A* allele at locus *A* and a *b* allele at locus *B*. For this example, a high value of *D* from a sample of individuals in a population means that the *A* and *b* allele segregate together during meiosis more often than by chance.

Over time, *D* will always decay to zero, which can be described by the relationship $D_t = D_t\,(1 - r)\,D_0$, where $D_0$ is the parent generation; $D_t$ is generation $1(t = 1)$, generation $2(t = 2)$, and so on; and *r* is the rate of recombination from crossing-over that occurs between the two loci. If *r* is very high, then linkage disequilibrium will decay to 0 in very few generations. But if *r* is very low (say, because the two loci are very close together on the chromosome), then linkage disequilibrium will take many generations to reach 0. Recall, however, that at each individual locus, only one generation of random mating is required to produce Hardy–Weinberg genotype proportions. Thus, the decay in D is a special case of the relationship between two or more loci. Recent studies have shown that rates of recombination vary considerably across the genomes of most diploid organisms and that in some species, such as humans and yeast, the genome is punctuated with hot spots where recombination occurs very frequently. Furthermore, it appears that the recombination rate at very specific locations in the genome is itself heritable, making it subject to the action of natural selection. This dynamic interplay between genetic drift, natural selection, mutation, migration, and recombination is a hot area of research that promises to yield substantial insight into the evolutionary forces responsible for the levels and patterns of genetic variation across the genomes of eukaryotic organisms.

## The Role of Genetics in Conservation Biology

The current rate at which species are being driven to extinction is greater than it has ever been in recorded history. It is estimated that there are approximately 2 million known species and as many as 30 million that are yet to be described. As we alter the environment and reduce the amount of suitable habitat for species, their numbers frequently decline. It is important to consider the consequences on the gene pool of such species because the variability of the gene pool may affect the chances of long-term survival of the species. Evolution is a process of diversification and extinction, which maintains biodiversity. Many of the genetic principles and processes discussed in this chapter relate to this conservation problem. As we have seen, populations have genetic structure; conserving this structure may warrant special attention. For example, **population viability analysis** techniques are designed to estimate how large a population must be to keep from going extinct for a particular period of time with a degree of certainty. If one wants to ensure that a population has the potential to evolve over long periods of time, an adequate gene pool must be maintained. Clearly, determining the genetic structure of a population and how genetic variation within populations affects the probability of extinction requires a great deal of study. The problem is particularly acute for rare and already endangered species. The effects of unintentional inbreeding of species in zoos and game management programs are diminishing as population genetics principles are being used to manage genetic structures of populations more carefully. We have seen that inbreeding, genetic drift, and selection can all decrease genetic variation, and populations may need to be maintained at certain genetic effective sizes to ensure that ample amounts of variation remain. Fortunately, we now have tools to obtain quantitative assessments of the relationships among geographic populations and of the amount of genetic variability in each. These data provide essential information for management policies. However, there is some controversy regarding the utility of genetic information in conservation biology. The central problem in most cases is loss of habitat, and unless the rate of habitat destruction can be slowed, efforts to manage the genetic structure of populations may accomplish little. Nonetheless, insights from both population ecology and genetics are necessary to better understand the best course of action for maintaining the diversity of life in our ecosystems.

### Keynote

Rare and endangered species risk losing genetic variability and hence the ability to adapt to changing environmental conditions. Management practices are applied in an attempt to maintain genetic diversity by keeping adequate population size and avoiding inbreeding.

## Speciation

Our discussion of population genetics principles so far has considered only processes of changing allele frequencies within populations of interbreeding organisms. Population subdivision may be weak, or it may be extreme to the point that two populations never interbreed. If this occurs for a long period of time, one expects that eventually there will be fixation of different alleles in the subpopulations such that if they were to come together again, they would fail to mate or the hybrids would have low fitness. Recently, there has been great progress in identifying genes that play a role in the reproductive isolation of closely related species. Before examining how genetics provides the tools we needed to understand the basis for reproductive isolation, let us back up and consider some general principles about speciation.

## Barriers to Gene Flow

If a species is a group of reproductively compatible organisms, then the process of speciation is likely to involve the erection of barriers to gene flow. Eventually, geographically isolated populations, by a combination of drift and natural selection, will diverge until they can no longer reproduce with one another. The barriers to gene flow come in two major categories: those that result in poor fitness of hybrid offspring (postzygotic barriers) and those that keep the two species from mating in the first place (prezygotic barriers). In animals, it appears that mutations in many genes can result in infertility, especially male infertility. As a result, the earliest genetic change in our pair of isolated subpopulations results in hybrid offspring that are sterile. The adults from the two subpopulations still recognize one another and mate, but the offspring are a dead end. Thus, postzygotic isolation generally arises first. **Postzygotic isolation** may include *hybrid sterility*, *hybrid inviability* (the failure of the hybrids to perform well in future generations), or *hybrid breakdown*.

In the face of a postmating barrier, it seems that mating adults could increase their fitness if they could recognize the "wrong" species and avoid these unproductive matings. If the populations harbor genetic variation for mate recognition, then the alleles that allow the adults to discriminate successfully will increase in frequency. This simple model (called **reinforcement**) provides a mechanism whereby postzygotic isolation leads to **prezygotic isolation.** The genes that result in prezygotic isolation may keep the species apart in many different ways, including the following:

1. **Temporal isolation.** By changing the mating season or the activity periods such that they no longer overlap, the opportunity for mating is removed.
2. **Ecological isolation.** If the ecological niche of the two species is distinct, such that, for example, the two species' dietary preferences keep them geographically isolated, even on a small spatial scale, again the opportunity for mating is removed.

Consider now the cases in which the two species freely overlap and have plenty of opportunity for mating. There still are genetic means to prevent the formation of zygotes:

1. **Behavioral incompatibility.** If the two species recognize and avoid each other as mates, there is no opportunity for mating.
2. **Mechanical isolation.** The two species may not be able to discriminate one from another, but if their genitalia do not fit together, zygotes cannot be formed.
3. **Gametic isolation.** Even if they mate and gametes come in contact with one another, there still remains a highly complex process of gametic fusion that can fail. In the case of plants, pollen from the wrong species often lands on the stigma surface. If the chemical communication between the pollen (or pollen tube) and the stigma and style does not go correctly, either the pollen fails to germinate or the pollen tube fails to grow.

Once prezygotic isolation is partially achieved, there is a snowball effect in which the rate of divergence accelerates. Individuals who engage in interspecific matings suffer an increasing disadvantage until at last the barrier to gene flow is complete. There is good empirical support for speciation by geographical isolation, although support for other mechanisms remains controversial. The study of the genetic basis for species isolation and the mechanisms by which species differences arise remains an active field.

## Genetic Basis for Speciation

Because of the argument that younger species tend to rely more on postzygotic isolation, these species are sought after by evolutionary geneticists to understand the nature of the genes that partially isolate the species. Male hybrids often are sterile, but female hybrids are fertile. This is especially true when the males are heterogametic (they make two different kinds of gametes, namely, X- and Y-bearing sperm), and the females are homogametic (producing only X-bearing eggs). In birds and butterflies, where males are homogametic and females are heterogametic, the pattern is reversed. The pattern is so pervasive that J. B. S. Haldane noted it, and we call the phenomenon **Haldane's rule.**

The observation of Haldane's rule immediately begs the question, what is the genetic basis for hybrid male sterility? In many species of *Drosophila*, such as *D. simulans* and *D. mauritiana*, the $F_1$ females are viable and fertile, whereas the $F_1$ males are sterile. One can backcross the females to *D. simulans*, and a few of the backcrossed males (that are now roughly 3/4 *D. simulans*) are fertile. By doing tricks like this, or by crossing in markers and introgressing parts of the wrong species' genome, *Drosophila* population geneticists have learned that many genes are involved in the fertility of male hybrids.

Another intriguing example of prezygotic isolation occurs in species of abalone, a subtidal marine gastropod that sheds its gametes into the sea water. Sperm and eggs of more than one species may co-occur (although some species exhibit temporal and ecological isolation), so the only way to avoid interspecific matings is for the eggs to allow penetration only by conspecific sperm. This is accomplished by means of molecules in the sperm and the egg. The sperm protein lysin is able to disaggregate the egg glycoprotein VERL in a species-specific manner. Because the sperm and egg components must track one another, and because they must be able to respond quickly if they come in contact with new species, these molecules are undergoing rapid adaptive evolution. You will learn more about the inferences of adaptive change in proteins in Chapter 23.

## Summary

- Population genetics seeks to understand the genetic basis of evolutionary change by determining the mechanisms underlying the observed patterns of genetic variation within and among populations in nature. This field of study includes both empirical and theoretical approaches to testing hypotheses about the evolutionary processes that change gene frequencies.
- The genetic structure of a population is described by the number and frequency of alleles at each locus within and between populations. The gene pool of a population is the total of all alleles within the population, and it is described in terms of allele and genotype frequencies. It is the source of genetic information that is carried forward from one generation to the next.
- The Hardy–Weinberg law states that in a large, randomly mating population and assuming there is no mutation, no migration, and no natural selection, allele frequencies will not change and the genotype frequencies stabilize after one generation.
- When a population is in Hardy–Weinberg equilibrium, allele frequencies do not change from generation to generation, and genotype frequencies stabilize after one generation of random mating in the proportions $p^2$, $2pq$, $q^2$, where $p$ and $q$ equal the allele frequencies of the population.
- The classic, balanced, and neutral theory models of molecular evolution have generated testable hypotheses that help explain how much genetic variation should exist within natural populations and what mechanisms are responsible for generating and maintaining that genetic variation.
- Mutation is the source of all variation in a population. It changes allele frequencies at a very slow rate, so slow that almost any of the other forces swamp the effects of recurrent mutation on allele frequencies. Thus, although mutation is the initial source of all genetic variation, other forces predominate in determining its changes in frequency once it has been introduced.
- Genetic drift is a chance change in allele frequencies that arises from random sampling of gametes that occurs in each generation. Genetic drift leads to a loss of genetic variation within a population, genetic divergence among populations, and random change of allele frequency within a population. Because it is driven by chance, it has the largest effect on allele frequency changes in small populations.
- Migration, also called gene flow, is the movement of alleles among populations. It alters allele frequencies within populations and reduces genetic differences among populations.
- Natural selection is differential reproduction of genotypes. The relative reproductive contribution of genotypes is measured in terms of Darwinian fitness. The effects of natural selection depend on the fitnesses of the genotypes, the degree of dominance, and the frequencies of the alleles in the population.
- Nonrandom mating affects the effective population size and genotype frequencies of a population; the allele frequencies are unaffected, except in some cases of negative assortative mating. Positive assortative mating also decreases heterozygosity, although only for a single locus. One type of nonrandom mating, inbreeding, leads to an increase in homozygosity.
- New techniques of molecular genetics, including analysis of restriction fragment length polymorphisms and DNA sequences, have supported prior insights obtained from analyses of proteins and provided new insights onto the role of population size, mutation, migration, selection, and recombination on evolutionary processes in natural populations.
- Different parts of a gene are found to have different levels of polymorphism; the parts of the gene that have the least effect on fitness appear to evolve at the highest rates. This suggests that natural selection generally removes deleterious mutations. Exceptions occur in genes where it is advantageous to have high levels of variability.
- Principles of population genetics can inform us about our genetic heritage and can also be applied to the management of rare and endangered species. Genetic diversity is best maintained by establishing a population with adequate founders, expanding the population rapidly, avoiding inbreeding.



## Analytical Approaches to Solving Genetics Problems

**Q21.1** In a population of 2,000 gaboon vipers, a genetic difference with respect to venom exists at a single locus. The alleles are incompletely dominant. The population shows 100 individuals homozygous for the *t* allele (genotype *tt*, nonpoisonous), 800 heterozygous (genotype *Tt*, mildly poisonous), and 1,100 homozygous for the *T* allele (genotype *TT*, lethally poisonous).

**a.** What is the frequency of the *t* allele in the population?

**b.** Are the genotypes in Hardy–Weinberg equilibrium?

**A21.1.** This question addresses the basics of calculating allele frequencies and relating them to the genotype frequencies expected of a population in Hardy–Weinberg equilibrium.

**a.** The *t* frequency can be calculated from the information given because the trait is an incompletely dominant one. There are 2,000 individuals in the population under study, meaning a total of 4,000 alleles at the *T/t* locus. The number of *t* alleles is given by

$$(2 \times tt \text{ homozygotes}) + (1 \times Tt \text{ heterozygotes}) = (2 \times 100) + (1 \times 800) = 1{,}000$$

This calculation is straightforward because both alleles in the nonpoisonous snakes are *t*, whereas only one of the two alleles in the mildly poisonous snakes is *t*. Since the total number of alleles under study is 4,000, the frequency of *t* alleles is 1,000/4,000 = 0.25. This system is a two-allele system, so the frequency of *T* must be 0.75.

**b.** For the genotypes to be in Hardy–Weinberg equilibrium, the distribution must be $p^2\ TT + 2pq\ Tt + q^2\ tt$ genotypes, where $p$ is the frequency of the *T* allele and $q$ is the frequency of the *t* allele. In (a) we established that the frequency of *T* is 0.75 and the frequency of *t* is 0.25. Therefore, $p = 0.75$ and $q = 0.25$. Using these values, we can determine the expected genotype frequencies if this population is in Hardy–Weinberg equilibrium:

$$(0.75)^2\ TT + 2(0.75)(0.25)\ Tt + (0.25)^2\ tt$$

This expression gives 0.5625 *TT* + 0.3750 *Tt* + 0.0625 *tt*. Thus, with 2,000 individuals in the population we would expect 1,125 *TT*, 750 *Tt*, and 125 *tt*. These values are close to the values given in the question, suggesting that the population is indeed in genetic equilibrium.

To check this result, we should perform a chi-square analysis (see Chapter 11), using the given numbers (not frequencies) of the three genotypes as the observed numbers and the calculated numbers as the expected numbers. The chi-square analysis is as follows, where $d$ = (observed − expected):

| Genotype | Observed | Expected | $d$ | $d^2$ | $d^2/e$ |
|---|---|---|---|---|---|
| *TT* | 1,100 | 1,125 | −25 | 625 | 0.556 |
| *Tt* | 800 | 750 | +50 | 2,500 | 3.334 |
| *tt* | 100 | 125 | −25 | 625 | 5.000 |
| Totals | 2,000 | 2,000 | 0 | | 8.890 |

Thus, the chi-square value (i.e., the sum of all the $d^2/e$ values) is 8.89. For the reasons discussed in the text for a similar example, there is only one degree of freedom. Looking up the chi-square value in the chi-square table (Table 11.5, p. 313), we find a $P$ value of approximately 0.0025. Therefore, about 25 times out of 10,000 we would expect chance deviations of the magnitude observed. In other words, our hypothesis that the population is in Hardy–Weinberg equilibrium is not substantiated. In this case, our guess that it was in equilibrium was inaccurate. Nonetheless, the population is not greatly removed from an equilibrium state.

**Q21.2** Approximately one normal allele in 30,000 mutates to the X-linked recessive allele for hemophilia in each human generation. Assume for the purposes of this problem that one *h* allele in 300,000 mutates to the normal alternative in each generation. (Note that in reality it is difficult to measure the reverse mutation of a human recessive allele that is essentially lethal, such as the allele for hemophilia.) The mutation frequencies are indicated in the following equation:

$$h^+ \underset{v}{\overset{u}{\rightleftarrows}} h$$

where $u = 10v$. What allele frequencies would prevail at equilibrium under mutation pressures alone in these circumstances?

**A21.2.** This question seeks to test understanding of the effects of mutation on allele frequencies. In this chapter, we discussed the consequences of mutation pressure. The conclusion was that if *A* mutates to *a* at *n* times the frequency with which *a* mutates back to *A*, then at equilibrium the value of $q$ will be $\hat{q} = u/(u + v)$ or $\hat{q} = nv/(n + 1)v$. Applying this general derivation to this particular problem, we simply use the values given. We are told that the forward mutation rate is 10 times the reverse mutation rate, or $u = 10v$. At equilibrium the value of $q$ will be $\hat{q} = u/(u + v)$. Since $u = 10v$, this equation becomes $\hat{q} = 10v/11v$, so $q = 10/11$, or 0.909. Therefore, at equilibrium brought about by mutation pressures, the frequency of *h* (the hemophilia allele) is 0.909, and the frequency of $h^+$ (the normal allele) is $\hat{q}$, that is, $(1 - \hat{q}) = (1 - 0.909) = 0.091$.

# Questions and Problems

***21.1** In the European land snail *Cepaea nemoralis*, multiple alleles at a single locus determine shell color. The allele for brown ($C^B$) is dominant to the allele for pink ($C^P$) and to the allele for yellow ($C^Y$). The dominance hierarchy among these alleles is $C^B > C^P > C^Y$. In one population sample of *Cepaea*, the following color phenotypes were recorded:

| | |
|---|---|
| Brown | 236 |
| Pink | 231 |
| Yellow | 33 |
| Total | 500 |

Assuming that this population is in Hardy–Weinberg equilibrium (large, randomly mating, and free from evolutionary processes), calculate the frequencies of the $C^B$, $C^P$, and $C^Y$ alleles.

**21.2** Three alleles are found at a locus coding for malate dehydrogenase (MDH) in the spotted chorus frog. Chorus frogs were collected from a breeding pond, and each frog's genotype at the MDH locus was determined with electrophoresis. The following numbers of genotypes were found:

| | |
|---|---|
| $M^1M^1$ | 8 |
| $M^1M^2$ | 35 |
| $M^2M^2$ | 20 |
| $M^1M^3$ | 53 |
| $M^2M^3$ | 76 |
| $M^3M^3$ | 62 |
| Total | 254 |

**a.** Calculate the frequencies of the $M^1$, $M^2$, and $M^3$ alleles in this population.
**b.** Using a chi-square test, determine whether the MDH genotypes in this population are in Hardy–Weinberg proportions.

*__21.3__ As discussed in Chapter 4, cystic fibrosis (CF) is a recessive disorder caused by mutations in the gene for the CFTR protein. The prevalence of CF is estimated to be between 1/43,000 and 1/100,000 in the Indian subcontinent. One of the more frequent mutations in non-Hispanic Caucasians in the United States is the $\Delta$F508 mutation, a three-nucleotide deletion that removes a phenylalanine from the CFTR protein. When cord blood samples from 955 normal neonates in the Indian subcontinent were tested for the $\Delta$F508 mutation using PCR and gel electrophoresis, four were positive. Calculate the frequencies of carriers and homozygotes for the $\Delta$F508 mutation, carefully stating your assumptions. Explain why the frequency of homozygotes is not in the range of the estimates of the prevalence of CF in this population.

**21.4** In the Americas and Europe, multiple studies have assessed the frequency of CF mutations in individuals with no known family history of CF. The data from nine such studies are shown in the following table. Since these studies screened for only a subset of the known CF mutations, the number of carriers presented in this table has been corrected for the percentage of mutations that were assessed.

| Geographical Source of Persons Tested | Ethnicity | Number of Persons Tested | Carriers Detected |
|---|---|---|---|
| Copenhagen, Denmark | non-Hispanic Caucasians | 6,761 | 199 |
| Edinburgh, Scotland | non-Hispanic Caucasians | 3,275 | 135 |
| Maine, USA | non-Hispanic Caucasians | 4,413 | 193 |
| New York, NY, USA | Ashkenazi Jewish Caucasians | 3,792 | 159 |
| Northern California, USA | Hispanic Caucasians | 1,053 | 10 |
| Southern California, USA | Hispanic Caucasians | 1,040 | 8 |
| Rochester, NY, USA | Hispanic Caucasians | 78 | 2 |
| Rochester, NY, USA | African-Americans | 100 | 0 |
| Southern California, USA | African-Americans | 269 | 0 |

**a.** Use these data to predict the prevalence of CF in newborns in the different populations. Carefully state your assumptions.
**b.** The following table presents estimates of the prevalence of CF in neonates based on population surveys.

| Ethnicity | Prevalence |
|---|---|
| Ashkenazi Jewish Caucasians | 1 in 2,271 |
| Non-Hispanic Caucasians | 1 in 2,500 |
| Hispanic Caucasians | 1 in 13,535 |
| African-Americans | 1 in 15,100 |
| Asian Hawaiians | 1 in 90,000 |

Consider these data in light of your answers to part (a), and develop hypotheses to explain differences between the observed prevalence of CF and the frequency of carriers.

**21.5** In a large interbreeding population, 81% of the individuals are homozygous for a recessive character. In the absence of mutation or selection, what percentage of the next generation would be homozygous recessives? Homozygous dominants? Heterozygotes?

*__21.6__ Let *A* and *a* represent dominant and recessive alleles whose respective frequencies are *p* and *q* in a given interbreeding population at equilibrium (with $p + q = 1$).
**a.** If 16% of the individuals in the population have recessive phenotypes, what percentage of the total number of recessive genes exist in the heterozygous condition?
**b.** If 1.0% of the individuals were homozygous recessive, what percentage of the recessive genes would occur in heterozygotes?

*__21.7__ A population has eight times as many heterozygotes as homozygous recessives. What is the frequency of the recessive allele?

**21.8** In a large population of range cattle, the following ratios are observed: 49% red (*RR*), 42% roan (*Rr*), and 9% white (*rr*).
**a.** What percentage of the gametes that give rise to the next generation of cattle in this population will contain allele *R*?

**b.** In another cattle population, only 1% of the animals are white and 99% are either red or roan. What is the percentage of *r* alleles in this case?

**21.9** In a gene pool, the alleles *A* and *a* have initial frequencies of *p* and *q*, respectively. Show that the allelic frequencies and zygotic frequencies do not change from generation to generation as long as there is no selection, mutation, or migration, the population is large, and the individuals mate at random.

***21.10** One assumption of the Hardy–Weinberg law is that a population interbreeds randomly. Explain whether it is likely that a population interbreeds randomly for all traits, and if it is not, how the Hardy–Weinberg law can be valid.

**21.11** In some fish, sex is determined in response to environmental conditions. Some fish are born as males, and the dominant male in a group becomes a female. Breeding within this population is therefore nonrandom, as only the dominant male is guaranteed to pass on his germ line. Explain whether the Hardy–Weinberg law will hold for this population.

***21.12** The *S*–*s* antigen system in humans is controlled by two codominant alleles, *S* and *s*. In a group of 3,146 individuals, the following genotypic frequencies were found: 188 *SS*, 717 *Ss*, and 2,241 *ss*.

**a.** Calculate the frequencies of the *S* and *s* alleles.

**b.** Determine whether the genotype frequencies are in Hardy–Weinberg equilibrium by using the chi-square test.

**21.13** Refer to Problem 21.12. A third allele is sometimes found at the *S* locus. This allele $S^u$ is recessive to both the *S* and the *s* alleles and can be detected only in the homozygous state. If the frequencies of the alleles *S*, *s*, and $S^u$ are *p*, *q*, and *r*, respectively, what would be the expected frequencies of the phenotypes *S*–, *Ss*, *s*–, and $S^uS^u$?

***21.14** In a large interbreeding human population, 60% of individuals belong to blood group O (genotype *i/i*). Assuming negligible mutation and no selective advantage of one blood type over another, what percentage of the grandchildren of the present population will be type O?

***21.15** A selectively neutral, recessive character appears in 40% of the males and 16% of the females in a large, randomly interbreeding population. What is the frequency of the allele? What proportion of females are heterozygous for it? What proportion of males are heterozygous for it?

**21.16** Suppose you found two distinguishable types of individuals in wild populations of some organism in the following frequencies:

| | Type 1 | Type 2 |
|---|---|---|
| Females | 99% | 1% |
| Males | 90% | 10% |

The difference is known to be inherited. Are these data compatible with the trait being X-linked?

***21.17** Red-green color blindness is caused by an X-linked recessive gene. About 64 women out of 10,000 are color blind. What proportion of men would be expected to show the trait if mating is random?

**21.18** About 8% of the men in a population are red-green color blind (because of a sex-linked recessive allele). Answer the following questions, assuming random mating in the population, with respect to color blindness.

**a.** What percentage of women would be expected to be color blind?

**b.** What percentage of women would be expected to be heterozygous?

**c.** What percentage of men would be expected to have normal vision two generations later?

**21.19** What types of data prompted Kimura to develop the neutral theory to explain the amount of genetic variation in populations in the late 1960s? How does the neutral theory differ from classical models developed in the 1920s and 1930s? Summarize how our understanding of the origins of genetic variation in populations has changed since the neutral theory was developed.

***21.20** Two alleles of a locus, *A* and *a*, can be interconverted by mutation:

$$A \underset{v}{\overset{u}{\rightleftarrows}} a$$

where *u* is a mutation rate of $6.0 \times 10^{-7}$ and *v* is a mutation rate of $6.0 \times 10^{-8}$. What will be the frequencies of *A* and *a* at mutational equilibrium, assuming no selective difference, no migration, and no random fluctuation caused by genetic drift?

**21.21** Calculate the effective population size ($N_e$) for breeding populations having

**a.** 50 adult males and 50 adult females.

**b.** 60 adult males and 40 adult females.

**c.** 10 adult males and 90 adult females.

**d.** 2 adult males and 98 adult females.

**21.22** The land snail *Cepaea nemoralis* is native to Europe but has been accidentally introduced into North America at several localities. These introductions occurred when a few snails were inadvertently transported on plants, building supplies, soil, or other cargo. The snails subsequently multiplied and established large, viable populations in North America.

Assume that today the average size of *Cepaea* populations found in North America is equal to the average size of *Cepaea* populations in Europe. What predictions can you make about the amounts of genetic variation present in European and North American populations of *Cepaea*? Explain your reasoning.

*$\mathbf{21.23}$ A population of 80 adult squirrels resides on campus, and the frequency of the $Est^1$ allele among these squirrels is 0.70. Another population of squirrels is found in a nearby woods, and there, the frequency of the $Est^1$ allele is 0.5. During a severe winter, 20 of the squirrels from the woods population migrate to campus in search of food and join the campus population. What will be the allele frequency of $Est^1$ in the campus population after migration?

**21.24** Upon sampling three populations and determining genotypes, you find the following three genotype distributions.

| Population | *AA* | *Aa* | *aa* |
|---|---|---|---|
| 1 | 0.04 | 0.32 | 0.64 |
| 2 | 0.12 | 0.87 | 0.01 |
| 3 | 0.45 | 0.10 | 0.45 |

What does each of these distributions imply regarding selective advantages of population structure?

**21.25** The frequency of two adaptively neutral alleles in a large population is 70% *A* : 30% *a*. The population is wiped out by an epidemic, leaving only four individuals, who produce many offspring. What is the probability that the population several years later will be 100% *AA*? (Assume no new mutations occur.)

***21.26** A completely recessive allele, through changed environmental circumstances, becomes lethal when homozygous in a certain population. It was previously neutral, and its frequency was 0.5.

**a.** What was the genotype distribution when the recessive genotype was not selected against?
**b.** What will be the allele frequency after one generation in the altered environment?
**c.** What will be the allele frequency after two generations?

**21.27** Human individuals homozygous for a certain recessive autosomal gene die before reaching reproductive age. Despite this removal of all affected individuals, there is no indication that homozygotes occur less frequently in succeeding generations. To what might you attribute the continued appearance of these recessives?

***21.28** A completely recessive allele ($Q^1$) has a frequency of 0.7 in a large population, and the $Q^1Q^1$ homozygote has a relative fitness of 0.6.

**a.** What will be the frequency of $Q^1$ after one generation of selection?
**b.** If there is no dominance at this locus (the fitness of the heterozygote is intermediate to the fitnesses of the homozygotes), what will the allele frequency be after one generation of selection?
**c.** If $Q^1$ is dominant, what will the allele frequency be after one generation of selection?

**21.29** As discussed in this chapter, the gene for sickle-cell anemia exhibits heterozygote advantage. An individual who is an *Hb-A/Hb-S* heterozygote has increased resistance to malaria and therefore has greater fitness than the *Hb-A/Hb-A* homozygote, who is susceptible to malaria, and the *Hb-S/Hb-S* homozygote, who has sickle-cell anemia. Suppose that the fitness values of the genotypes in Africa are as presented here:

*Hb-A/Hb-A* = 0.88
*Hb-A/Hb-S* = 1.00
*Hb-S/Hb-S* = 0.14

Give the expected equilibrium frequencies of the sickle-cell allele (*Hb-S*).

***21.30** Achondroplasia, a type of dwarfism in humans, is caused by an autosomal dominant allele. The mutation rate for achondroplasia is about $5.0 \times 10^{-5}$ and the fitness of achondroplastic dwarfs has been estimated to be about 0.2, compared with unaffected individuals. What is the equilibrium frequency of the achondroplasia allele based on this mutation rate and fitness value?

**21.31** To answer the following questions, consider the spontaneous mutation frequencies given in Table 21.6 (p. 623).

**a.** In humans, why is the frequency of forward mutations to neurofibromatosis an order of magnitude larger than that for the other human diseases?
**b.** In *E. coli*, why is the frequency of mutations to arabinose dependence two to four orders of magnitude larger than the frequency of mutations to leucine, arginine, or tryptophan independence?
**c.** What factors influence the spontaneous mutation frequency for a specific trait?

**21.32** The frequencies of the $L^M$ and $L^N$ blood group alleles are the same in each of the populations I, II, and III; but the frequencies of the genotypes are not the same, as shown in the following table. Which of the populations is most likely to show each of the following characteristics: random mating, inbreeding, genetic drift? Explain your answers.

| | $L^ML^M$ | $L^ML^N$ | $L^NL^N$ |
|---|---|---|---|
| I | 0.50 | 0.40 | 0.10 |
| II | 0.49 | 0.42 | 0.09 |
| III | 0.45 | 0.50 | 0.05 |

***21.33** DNA was collected from 100 people randomly sampled from a given human population and was digested with the restriction enzyme *Bam*HI. The resulting fragments were separated by electrophoresis and then transferred to a membrane filter using the Southern blot technique. The blots were probed with a particular cloned sequence. Three different patterns of hybridization were seen on the blots. Some DNA samples (56 of them) showed a single band of 6.3 kb, others (6) showed a single band at 4.1 kb, and others (38) showed both the 6.3- and the 4.1-kb bands.

**a.** Interpret these results in terms of *Bam*HI sites.
**b.** What are the frequencies of the restriction site alleles?

**c.** Does this population appear to be in Hardy–Weinberg equilibrium for the relevant restriction sites?

***21.34** Fifty tiger salamanders from one pond in west Texas were examined for genetic variation by using the technique of protein electrophoresis. The genotype of each salamander was determined for five loci (AmPep, ADH, PGM, MDH, and LDH-1). No variation was found at AmPep, ADH, and LDH-1; in other words, all individuals were homozygous for the same allele at these loci. The following numbers of genotypes were observed at the MDH and PGM loci.

| *MDH* Genotypes | Number of Individuals | *PGM* Genotypes | Number of Individuals |
|---|---|---|---|
| *AA* | 11 | *DD* | 35 |
| *AB* | 35 | *DE* | 10 |
| *BB* | 4 | *EE* | 5 |

Calculate the proportion of polymorphic loci and the heterozygosity for this population.

**21.35** What is genetic hitchhiking, and what is its relationship to crossing-over during meiosis?

***21.36** Define *linkage disequilibrium* and describe its sources using two hypothetical examples. Explain how linkage disequilibrium differs from linkage.

***21.37** The success of a population depends in part on its reproductive rate, which may be affected by low genetic variability. Vyse and his colleagues have been interested in the conservation of small populations of grizzly bears, studying 304 members of 30 grizzly bear family groups in a population in northwestern Alaska. They have identified a set of polymorphic loci with these alleles, allele frequencies, and observed heterozygosities (obs. het.):

| Locus *G1A* | | Locus *G10X* | | Locus *G10C* | | Locus *G10L* | |
|---|---|---|---|---|---|---|---|
| Obs. het.: 0.776 | | Obs. het.: 0.783 | | Obs. het.: 0.770 | | Obs. het.: 0.651 | |
| Allele | Freq. | Allele | Freq. | Allele | Freq. | Allele | Freq. |
| *A194* | 0.398 | *X137* | 0.395 | *C105* | 0.355 | *L155* | 0.487 |
| *A184* | 0.240 | *X135* | 0.211 | *C103* | 0.257 | *L157* | 0.276 |
| *A192* | 0.211 | *X141* | 0.211 | *C111* | 0.240 | *L161* | 0.128 |
| *A180* | 0.086 | *X133* | 0.102 | *C113* | 0.092 | *L159* | 0.089 |
| *A190* | 0.036 | *X131* | 0.053 | *C107* | 0.043 | *L171* | 0.013 |
| *A200* | 0.016 | *X129* | 0.030 | *C101* | 0.010 | *L163* | 0.007 |
| *A186* | 0.007 | | | *C109* | 0.003 | | |
| *A188* | 0.006 | | | | | | |

The genotypes of a mother bear and her three cubs are shown here.

| Mother | Cub #1 | Cub #2 | Cub #3 |
|---|---|---|---|
| *A184, A192* | *A184, A194* | *A184, A192* | *A184, A194* |
| *X135, X137* | *X135, X137* | *X133, X135* | *X137, X141* |
| *C105, C113* | *C105, C111* | *C105, C105* | *C111, C113* |
| *L155, L159* | *L155, L157* | *L159, L161* | *L155, L155* |

**a.** How do the observed heterozygosities compare with the expected heterozygosities? Based on this information, can you tell whether this grizzly bear population is in Hardy–Weinberg equilibrium?

**b.** What can you infer about the paternity of the mother's three cubs? How might paternity information affect the genetic variability and the effective population size in this population of grizzly bears?

**21.38** What factors cause genetic drift?

***21.39** What are the primary effects of the following evolutionary processes on the gene and genotypic frequencies of a population?

**a.** mutation
**b.** migration
**c.** genetic drift
**d.** inbreeding

**21.40** When Sam was a child, his grandfather owned a small cabin on a few unfenced acres of land in a desert near a major metropolis. A population of about 30 desert tortoises, which can live a century and usually stay within a few miles of their natal nest, roamed freely on his grandfather's land and the neighbor's parcels. As the city expanded, Sam's neighbors gradually sold their plots to housing developers, and tortoises were increasingly found only on his grandfather's plot. By the time Sam turned 40 and inherited the land, multistory buildings, cement walkways, and traffic-filled roads surrounded his plot. Sam erected a low concrete border around it, and three remaining tortoises (one male and two females) now were restricted to its boundaries. When Sam finally agreed to develop the land 10 years later, he sought to save the tortoises. There were 30 of them now, though most were too young to reproduce, as these tortoises reproduce only after they are 7 to 15 years old. He gathered all of them and released half of them on a wild and remote desert island (that lacked tortoises) and the other half in a remote part of the desert where he had once seen a tortoise.

Describe what you would expect to find if you assessed the genetic variation in this tortoise population when Sam was a child, when he was 40, when he was 50, and, assuming that the tortoises survive and flourish after they are relocated, 10 and 200 years after their release. What population genetic processes could have contributed to differences in the amount of genetic variation during this population's history?

***21.41** Explain how overdominance leads to an increased frequency of sickle-cell anemia in areas where malaria is widespread.

*21.42 Since 1968, Pinter has studied the population dynamics of the montane vole, a small rodent in the Grand Teton mountains in Wyoming. For more than 25 years, severe periodic fluctuations in population density have been negatively correlated with precipitation levels: vole density sharply declines every few years when spring precipitation is extremely high.

**a.** Propose several hypotheses concerning the genetic structure of the population of montane voles in two separate sampling sites if
   **i.** there is negligible migration of voles between sampling sites.
   **ii.** there is substantial migration of voles between sampling sites.

**b.** How would you gather data to evaluate these hypotheses?

**21.43** What are some of the advantages of using DNA sequences to infer the strength of evolutionary processes?

*21.44 Explain how you would quantify genetic variation in geographically distinct human populations. What have we learned about the genetic structure and evolutionary history of human populations from such analyses?

*21.45 What percentage of the total genetic variance in humans do you expect to find between different populations, and how does this compare to the percentage you expect to find within a population? What conclusions can you draw from the results of this comparison?

**21.46** Reproductive isolation is important for speciation and can be prezygotic or postzygotic. Give several examples of each type of process to illustrate how they work, and how they differ in the ways that they lead to barriers to gene flow. Which type of reproductive isolation is generally thought to arise first during speciation, and how can this type of reproductive isolation lead to the other?

*21.47 Multiple, geographically isolated populations of tortoises exist on the Galápagos Islands off the coast of Ecuador. Several populations are endangered, partly because of hunting and partly because of illegal capture and trade. In principle, a significant number of tortoises in captivity (in zoos or private collections) could be used to repopulate some of the endangered populations.

**a.** Suppose you are interested in returning a particular captive tortoise to its native subpopulation, but you have no record of its capture. How could you determine its original subpopulation?

**b.** A researcher planned to characterize one subpopulation of tortoises. In her first field season, she tagged and collected blood samples from all animals in the subpopulation. When she returns a year later, she cannot locate two animals. She learns that two untagged, smuggled tortoises are being held by U.S. customs officials. How can she assess whether the smuggled animals are from her field site?

# 22 Quantitative Genetics

Various human eye colors.

## Key Questions

- What causes the continuous phenotypic variation characteristic of quantitative traits?
- How can statistical tools be used to describe and analyze quantitative traits?
- What is heritability, and how is it measured?
- How can the individual loci causing quantitative variation be identified?

### iActivity

JUST LIKE SNOWFLAKES, NO TWO FINGERPRINTS are alike, not even those of identical twins. Research shows, however, that the patterns of ridges on our fingers and palms are inherited. How can such a variety of phenotypes be produced? Is the trait encoded by more than one locus? Are environmental factors involved? Is there a relationship between a person's fingerprints and another trait, such as hair color or blood type? In this chapter, you will learn the answers to questions such as these. Then, in the iActivity, you can apply what you have learned as you investigate whether a relationship exists between fingerprint patterns and high blood pressure.

## The Nature of Continuous Traits

By isolating phenotypic mutants, then crossing and comparing the mutants with the wild type, Mendel was able to describe the basic laws of heredity. Later geneticists have extended our understanding down to the molecular basis of mutant phenotypes. The mutations used in these studies, and in fact most of the traits we have studied to this point, have been characterized by the presence of only a few distinct phenotypes. The seed coats of pea plants, for example, were either grey or white, the seedpods were green or yellow, and the plants were tall or short. In each trait, the phenotypes were markedly different, and each phenotype was easily separated from all other phenotypes. Traits such as these, with only a few distinct phenotypes, are called **discontinuous traits** (Figure 22.1).

For discontinuous traits, a simple relationship usually exists between the genotype and the phenotype. In most cases, the effects of variant alleles at the single locus are observable at the level of the organism, so the phenotype can be used as a quick assay for the genotype. When dominance occurs, the same phenotype may be produced by two different genotypes; but the relationship between the genes and the trait remains simple. Chapter 13 introduced situations where the relationship between genotype and phenotype is not so simple: variable **penetrance** and **expressivity,** as well as **pleiotropy** and **epistasis,** can lead to a complex relationship between genotype and phenotype. In addition, single genotypes can give rise to a range of phenotypes as the genotype interacts with variable environments during development to give rise to a **norm of reaction.** As a result of these and other factors, there are not many traits with phenotypes that fall into a few distinct categories. Many traits (probably most), such as human birth weight (illustrated in Figure 22.2) and adult height, protein content in corn, and the number of eggs laid by *Drosophila*, exhibit a wide range of possible phenotypes.

Figure 22.1

**Discontinuous distribution of shell color in the snail *Cepaea nemoralus* from a population in England.**

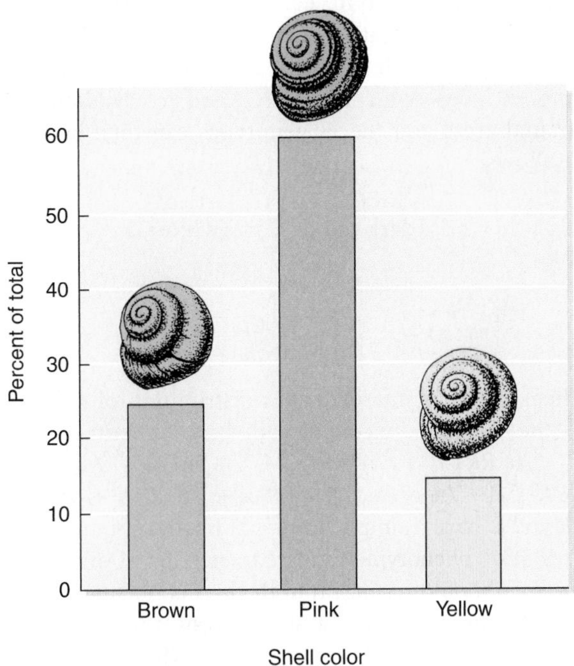


Traits such as these, with a continuous distribution of phenotypes, are called **continuous traits.** Since the phenotypes of continuous traits must be described by quantitative measures, such traits are also known as **quantitative traits,** and the field of **quantitative genetics** studies the inheritance of these traits.

Figure 22.2

**Distribution of birth weight of babies (males + females) born to teenagers in Portland, Oregon, in 1992.**

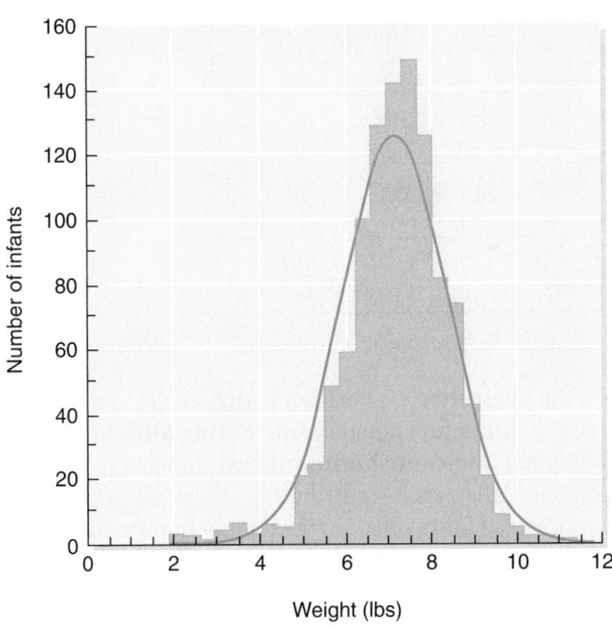


## Questions Studied in Quantitative Genetics

There is a great deal of genetic variation among individuals. The amount of variation and how it is distributed determines a population's genetic structure. In this chapter, we shift attention from a purely genetic perspective to consider the phenotypic structure of a population, as well as the relationship between the genetic structure and the phenotypic structure. As we will see, quantitative genetics plays an important role in our understanding of evolution, conservation, and complex human traits. Quantitative genetics is especially important in agricultural genetics, where traits such as crop yield, rate of weight gain, milk production, and fat content are all studied by quantitative genetics. In psychology, methods of quantitative genetics are used to study IQ, learning ability, and personality. Human geneticists also use these methods to study traits such as blood pressure, antibody titer, fingerprint pattern, and birth weight.

In transmission genetics, we frequently determine the probability of inheriting a particular phenotype. With quantitative traits, however, individuals differ in the quantity of a trait, so it makes no sense to ask about the probability of inheriting a continuous trait, as we did for simple discontinuous traits. Instead, the following are examples of questions frequently studied by quantitative geneticists:

1. To what degree does the observed variation in phenotype result from differences in genotype, and to what degree does this variation reflect the influence of different environments? In our study of discontinuous traits, this question assumed little importance because the differences in phenotype were assumed to reflect only genotypic differences.
2. How many genes determine the phenotype? When only a few loci are involved and the trait is discontinuous, the number of loci can often be determined by examining the phenotypic ratios in genetic crosses. With complex, continuous traits, however, determining the number of loci involved is more difficult.
3. Are the contributions of the determining genes equal? Or, do a few genes have major effects on the trait and other genes modify the phenotype only slightly?
4. Are the effects of alleles additive? To what degree do alleles at the different loci interact with one another?
5. When selection favors a particular phenotype, how rapidly can the trait change? Do other traits change at the same time?
6. What is the best method for selecting and mating individuals to produce desired phenotypes in the progeny?

## The Inheritance of Continuous Traits

Biologists began developing techniques for the study of continuous traits during the late nineteenth century, even before they were aware of Mendel's principles of heredity.

Francis Galton and his associate Karl Pearson demonstrated that for many traits in humans, such as height, weight, and cognitive traits, the phenotypes of parents and their offspring were statistically associated. From this result, they were able to infer that these traits were inherited, but they were not successful in determining how genetic transmission occurred. Even after the rediscovery of Mendel's work, considerable controversy arose over whether continuous traits also followed Mendel's principles or whether they were inherited in some different fashion.

## Polygene Hypothesis for Quantitative Inheritance

A trait may have a range of phenotypes because environmental factors affect the trait. When environmental factors exert an influence, the same genotype may produce a range of phenotypes (the norm of reaction), or multiple genotypes may produce the same phenotype. Which phenotype is expressed depends both on the genotype and on the specific environment in which the genotype is found. In 1903, Wilhelm Johansen published a study demonstrating that quantitative variation in seed weight in beans had both environmental and genetic determinants. His study was a crucial step in recognizing that both environment and genotype influence some quantitative traits; such traits are referred to as **multifactorial traits.** Because inheritance of quantitative traits cannot be explained by a single locus, the simplest alternative explanation is that they are controlled by many genes. This explanation, called the **polygene** or **multiple-gene hypothesis for quantitative inheritance**, is a landmark of genetic thought.

## Polygene Hypothesis for Wheat Kernel Color

The polygene hypothesis can be traced back to 1909 and the classic work of Hermann Nilsson-Ehle, who studied the color of wheat kernels. Like Mendel, Nilsson-Ehle started by crossing true-breeding lines of red kernel plants and white kernel lines. The $F_1$ had grains that were all the same shade of an intermediate color between red and white. When Nilsson-Ehle intercrossed the $F_1$s, the $F_2$ progeny displayed kernels that were white and many shades of red, in a ratio of approximately 15 red (all shades) : 1 white kernels. While this was clearly a deviation from a 3:1 ratio expected for a monohybrid cross, he could recognize four discrete shades of red among the progeny. When he counted the relative number of each class, he found a 1:4:6:4:1 phenotypic ratio of plants with dark red, medium red, intermediate red, light red, and white kernels.

*animation*

**Polygene Hypothesis for Wheat Kernel Color**

How can the data be interpreted in genetic terms? Recall from Chapter 13 (Analytical Approaches to Solving Genetics Problems, Answer to Question 13.3c, p. 392) that a 15:1 ratio of two alternative characteristics resulted from the interaction of the products of two genes that affect the same trait. Let us hypothesize that two independently segregating loci control the production of pigment: the *red* locus with alleles *R* and *r*, and the *crimson* locus with alleles *C* and *c*. Nilsson-Ehle's parental cross and the $F_1$ genotypes can then be shown as follows:

| | | | |
|---|---|---|---|
| P | *RR CC* (dark red) | × | *rr cc* (white) |
| | | ↓ | |
| $F_1$ | | *Rr Cc* (intermediate red) | |

When the $F_1$ is interbred, the distribution of genotypes in the $F_2$ is typical of dihybrid inheritance, that is, $^1/_{16}$ *RR CC* + $^2/_{16}$ *Rr CC* + $^1/_{16}$ *rr CC* + $^2/_{16}$ *RR Cc* + $^4/_{16}$ *Rr Cc* + $^2/_{16}$ *rr Cc* + $^1/_{16}$ *RR cc* + $^2/_{16}$ *Rr cc* + $^1/_{16}$ *rr cc*. If *R* and *C* have a simple dominant relationship to *r* and *c*, the 9:3:3:1 phenotypic ratio characteristic of dihybrid inheritance should result. From the kernel phenotypic ratio, then, dominance is not the simple answer because the observed phenotypes fell into five classes with a ratio that approximates 1:4:6:4:1.

Note that these numbers in the phenotypic ratio are the same as the coefficients in the binomial expansion of $(a + b)^4$. The following calculation demonstrates how we can arrive at the coefficients and their associated terms of this expansion. In essence, we multiply $(a + b)$ by $(a + b)$, then multiply the product by $(a + b)$, and so on:

$$
\begin{array}{rl}
 & a + b \\
 & \times\, a + b \\ \hline
= & a^2 + ab \\
 & \quad + ab + b^2 \\ \hline
= & a^2 + 2ab + b^2 \qquad \text{—this is } (a+b)^2 \\
 & \times\, a + b \\ \hline
= & a^3 + 2a^2b + ab^2 \\
 & \quad + a^2b + 2ab^2 + b^3 \\ \hline
= & a^3 + 3a^2b + 3ab^2 + b^3 \qquad \text{—this is } (a+b)^3 \\
 & \times\, a + b \\ \hline
= & a^4 + 3a^3b + 3a^2b^2 + ab^3 \\
 & \quad + a^3b + 3a^2b^2 + 3ab^3 + b^4 \\ \hline
= & a^4 + 4a^3b + 6a^2b^2 + 4ab^3 + b^4 \qquad \text{—this is } (a+b)^4
\end{array}
$$

An alternative explanation is that alleles can be classified as either functional (**contributing alleles**) or nonfunctional (**noncontributing alleles**) in pigment production, and that each contributing allele allows for the synthesis (addition) of a certain amount of pigment. Under this hypothesis, the intensity of kernel coloration is a function of the number of *R* and *C* alleles in the genotype: *RR CC* (term $a^4$ in the binomial expansion) would be dark red and $rr\ cc$ (term $b^4$ in the binomial expansion) would be white. Table 22.1 shows this would explain the five phenotypic classes observed. Nilsson-Ehle concluded that the inheritance of red kernel color in wheat is an example of a polygene series of two loci with as many as four contributing alleles.

We must be cautious in extending this explanation of the genetic basis of this trait to other situations. Some $F_2$ populations show only three phenotypic classes with a 3:1 ratio of red to white, whereas other $F_2$ populations show a 63:1 ratio of red to white, with other discrete classes of color falling between the dark red and the white. These results suggest that there is quite a bit of variability segregating among strains at the loci that contribute to kernel color, so that a locus does not always contribute to quantitative variation in a cross.

The multiple-gene hypothesis that fits the wheat kernel color example so well has been applied to other examples of quantitative inheritance. In its basic form, the multiple-gene hypothesis proposes that quantitative inheritance can be explained by the action and segregation of allele pairs at a number of loci, called **polygenes,** each with a small effect on the overall phenotype. For the most part, the multiple-gene hypothesis is satisfactory as a working hypothesis for interpreting many quantitative traits. The whole picture of quantitative traits is, of course, more complicated. For example, the proposal that a number of alleles each function to produce a particular amount of pigment is an attractive hypothesis, but what does that hypothesis mean at the molecular level? How is product output regulated? In a large polygenic series, how many biochemical pathways are controlled? Polygenic inheritance provides an explanation for the inheritance of continuous traits that is compatible with Mendel's laws and, as we shall see later in the chapter, with the application of molecular techniques, we are slowly developing a better understanding of the nature of the genes involved.

**Table 22.1 Genetic Explanation for the Number and Proportions of $F_2$ Phenotypes for the Quantitative Trait Red Kernel Color in Wheat**

| Genotype | Number of Contributing Alleles for Red | Phenotype | Fraction of $F_2$ |
|---|---|---|---|
| *RR CC* | 4 | Dark red | 1/16 |
| *RR Cc* or *Rr CC* | 3 | Medium red | 4/16 |
| *RR cc* or *rr CC* or *Rr Cc* | 2 | Intermediate red | 6/16 |
| *rr Cc* or *Rr cc* | 1 | Light red | 4/16 |
| *rr cc* | 0 | White | 1/16 |

**Keynote**

Discontinuous traits exhibit only a few distinct phenotypes and can be described in qualitative terms. Continuous traits, on the other hand, display a range of phenotypes and must be described in quantitative terms. The multiple-gene hypothesis assumes that multiple loci contribute to a quantitative phenotype, and as the number of contributing alleles increases, there is an additive (or occasionally multiplicative) effect on the phenotype. The relationship between the genotype and the phenotype for quantitative traits may be complex because multiple alleles at multiple loci allow many genotypes, and the genotypic response to environmental factors can modulate the range of phenotypes.



## Statistical Tools

When multiple genes and environmental factors influence a trait, the relationship between individual loci and their contribution to the phenotype may be obscured. The same rules of transmission genetics and gene function still apply, but defining the effect of a single locus requires the ability to determine the genotypes of many individuals at loci across the genome. Before the advent of modern genotyping methods, and even today in systems where such tools are lacking, it was impractical to try to understand the action and role of each individual gene, so quantitative geneticists applied statistical and analytical procedures to understand the overall influence of genes on continuous traits.

As indicated earlier, a fundamental question addressed in the study of quantitative traits is how much of the variation that exists among individuals in populations is genetically determined and how much is environmentally induced. This is an important question because in many situations an understanding of the contribution of both factors is needed to make informed decisions. For example, in agricultural yield trials, we may be interested in identifying superior lines of wheat (genotypes) from data gathered across many field sites (environments). Thus, at the heart of the field of quantitative genetics (the only field of science that addresses this question explicitly) is the perennial question of *nature versus nurture*, or the relative roles of genes versus environment in shaping patterns of phenotypic variation. Notice that the traditional phrasing of the problem pits one against the other, implying that they are mutually exclusive. This oversimplified phrasing has resulted in much bad science and needless debate. In quantitative genetic terms, we phrase the problem in terms of variation from these two sources: How much of the variation in some aspect of the phenotype ($V_P$) results from genetic variation ($V_G$) and how much from environmental variation ($V_E$)? This relationship is expressed as

$$V_P = V_G + V_E$$

To work this equation we must learn how to measure variation in phenotype and how to partition the variation into genetic and environmental components. To do this, we need to understand some statistical methodology, much of which was developed specifically to deal with quantitative genetics.

## Samples and Populations

Suppose we want to describe some aspect of a trait for a large group of individuals. For example, we might be interested in the average birth weight of infants born in New York City during 1987. One way to answer this question is to collect the weight of each of the thousands of babies born in New York City in 1987. An alternative method, which is less laborious, would be to collect these data from a subset of the group, say birth weights of 100 infants born in New York City during 1987, and then use the average obtained from this subset as an estimate of the average for all the infants from the entire city. Scientists commonly use this sampling procedure in data collection. The group of ultimate interest (in our example, all infants born in New York City during 1987) is called the **population,** and the subset (our set of 100 babies) used to give us an estimate of the population is called a **sample.** For a sample to give us confidence in our estimates for the population, it must be large enough that chance differences between the sample and the population are not misleading. If our sample consisted of only a few babies, and these infants were unusually large, then our estimate of the average birth weight of all babies would not be very accurate. The sample must also be a random subset of the population. If all the babies in our sample came from Hope Hospital for Premature Infants, then we would grossly underestimate the true average birth weight of the population. Although this might seem obvious, a great many errors are made because data are not collected randomly.

### Keynote

To describe and study a large group of individuals, scientists frequently examine a subset of the group. This subset is called a sample, and the sample provides estimates for the larger group, which is called the population. The sample must be of reasonable size, and it must be a random subset of the larger group for it to provide accurate information about the population.

## Distributions

When we studied discontinuous traits in Chapter 11 through 13, we defined alternate phenotypes found within a group of individuals and described the group by stating the proportion of individuals falling into each phenotypic class. Because continuous traits exhibit a range of phenotypes, describing a group of individuals is more complicated. One means of summarizing the phenotypes of a continuous trait is with a **frequency distribution,** which is a summary of a group in terms of the proportion of individuals that fall within a certain phenotypic range (see Figure 22.2).

To make a frequency distribution, first classes are constructed that consist of a specified range of the phenotypic measure, and the number of individuals in each class is counted. Table 22.2 presents the data from Johannsen's study of the inheritance of seed weight in the bean *Phaseolus vulgaris*. As shown in the table, Johannsen weighed 5,494 beans from the $F_2$ progeny of a cross and classified them into nine classes, each covering a 100-mg range of weight. Frequency data such as these can be displayed graphically in a frequency histogram, as shown in Figure 22.3. In the histogram, the phenotypic classes are indicated along the horizontal axis, and the number present in each class is plotted on the vertical axis.

There are certain shapes of frequency distributions that correspond to known, mathematically described probability distributions. For example, many continuous phenotypes exhibit a symmetrical, bell-shaped distribution similar to the curve overlaid on the data in Figure 22.3 (see also Figure 22.2). This type of distribution is called a **normal distribution.** Data that conform to a normal distribution can be described accurately by a few statistics, namely the mean and variance, which are described below. In addition, data that are normally distributed allow us to make simplifying assumptions that facilitate complicated analyses.

## The Mean

A frequency distribution of a normally distributed phenotypic trait can be summarized by two statistics, the mean and the variance. The sample **mean** ($\bar{x}$), also known as the average, tells us where the center of the distribution of the phenotypes from a sample is located. The mean of a sample is calculated by simply adding up all the individual measurements ($x_1, x_2, x_3, \ldots, x_n$) and dividing by the number of measurements we added ($n$).

The mean is one statistic used frequently in quantitative genetics to characterize the phenotypes of a group of

**Table 22.2 Weight of 5,494 $F_2$ Beans (Seeds of *Phaseolus vulgaris*) Observed by Johannsen in 1903**

| Weight (mg) | 50–150 | 150–250 | 250–350 | 350–450 | 450–550 | 550–650 | 650–750 | 750–850 | 850–950 |
|---|---|---|---|---|---|---|---|---|---|
| (Midpoint of range) | (100) | (200) | (300) | (400) | (500) | (600) | (700) | (800) | (900) |
| Number of beans | 5 | 38 | 370 | 1,676 | 2,255 | 928 | 187 | 33 | 2 |

**Figure 22.3**

**Frequency histogram for bean weight in *Phaseolus vulgaris* plotted from data in Table 22.2.** A normal curve has been fitted to the data and is superimposed on the frequency histogram.

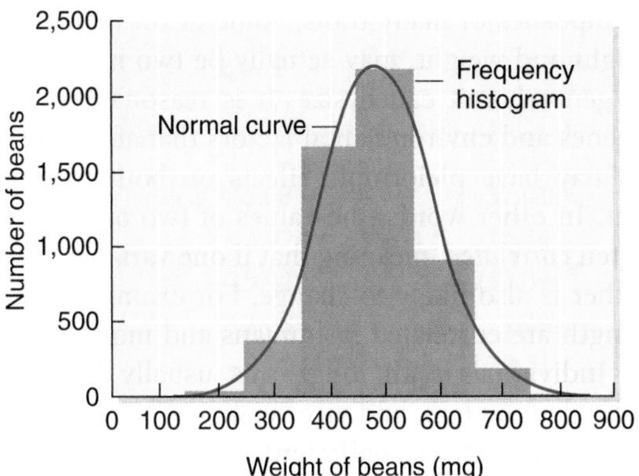


individuals. For example, in an early study of continuous variation, Edward M. East examined the inheritance of flower length using a cross between a long-flowered and a short-flowered strain of tobacco. Within each strain, flower length varied, so East reported that the mean phenotype of the short strain was 40.4 mm and the mean phenotype of the long strain was 93.1 mm. The $F_1$ progeny, which consisted of 173 plants, had a mean flower length of 63.5 mm. In this situation, the mean provides a convenient way for quickly characterizing and comparing the phenotypes of parents and offspring.

## The Variance and the Standard Deviation

A second statistic that provides key information about a distribution is the **variance.** The variance is a measure of how much the individual observations spread out around the mean—how variable the individuals and their measurements are. Two distributions may have the same mean, but when they have different variances, as shown in Figure 22.4, the distributions differ markedly. A broad curve implies high variability in the quantity measured and a correspondingly large variance. A narrow curve, in contrast, indicates little variability in the quantity measured and a correspondingly small variance. The sample variance, symbolized as $s^2$, is defined as the average squared deviation from the mean.

$$\text{Variance} = s^2 = \frac{\sum(x_i - \bar{x})^2}{n - 1}$$

The sample variance is calculated by first subtracting the sample mean from each individual measurement. This difference is then squared (so that the variance describes distance from the mean without regard to direction) and all the squared values are totaled. The sum of these squared values is then divided by the number of original measurements minus 1 (for mathematical reasons that we will not discuss here, the sample variance is obtained by dividing by $n - 1$ instead of by $n$).

**Figure 22.4**

**Graphs showing three distributions with the same mean but with different variances.**

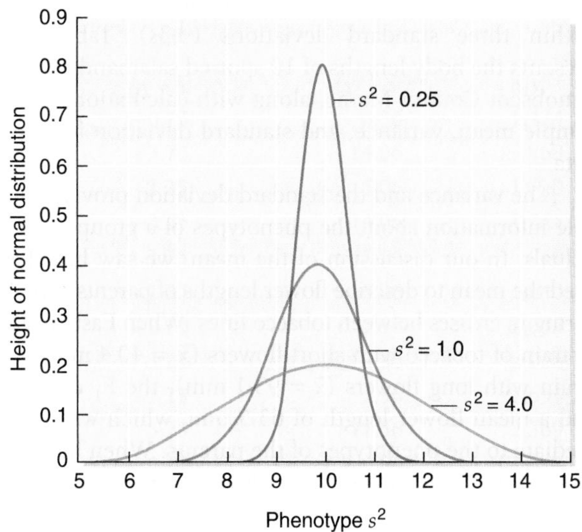


The **standard deviation** is often preferred over the variance because the standard deviation shares the same units as the original measurements (whereas the variance is in the units squared). The standard deviation for a sample is simply the square root of the sample variance:

$$\text{Standard deviation} = s = \sqrt{s^2}$$

A theoretical normal distribution is completely specified by the mean and standard deviation. It always has the shape indicated in Figure 22.5, where 66% of the

**Figure 22.5**

**Normal distribution curve showing the proportions of the data in the distribution that are included within certain multiples of the standard deviation.**

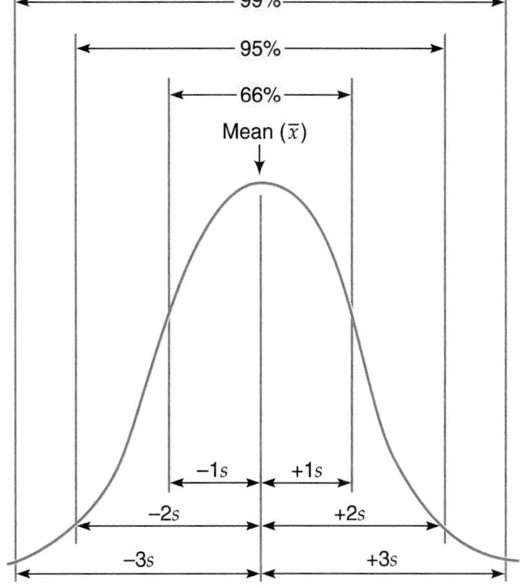

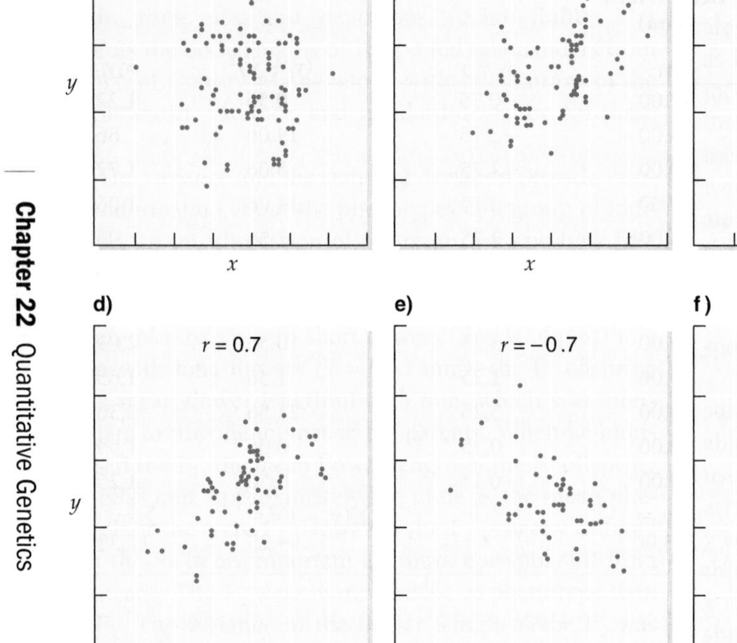


**Figure 22.6**

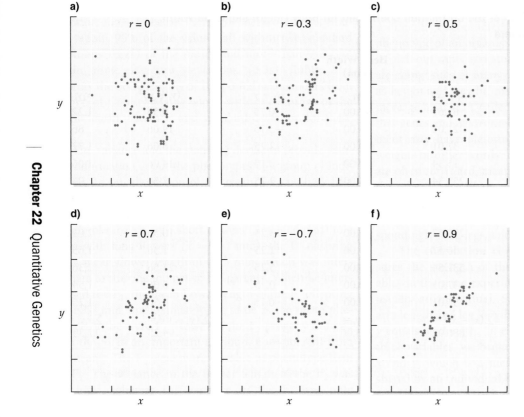

**Scatter diagrams showing the correlation of x and y variables.** Diagrams (b), (c), (d), and (f) show positive correlations, whereas diagram (e) shows a negative correlation. The absolute value of the correlation coefficient ($r$) indicates the strength of the association. For example, diagram (f) illustrates a strong correlation and diagram (b) illustrates a weak correlation. In diagram (a), the $x$ and $y$ variables are not correlated.

larger populations. Assuming that two factors are causally related because they are correlated often leads to erroneous conclusions.

Another important point is that because the correlation coefficient is unitless, correlation means only that a change in one variable is associated with a corresponding change in the other variable: *two variables can be highly correlated and yet have different values*. For example, the overall height and knee height of elderly Mexican females are highly correlated; however, the knee height is always much less than the overall height of a person. Thus, it is important to remember that correlation demonstrates only the trend between two variables.

## Keynote

The correlation coefficient is a measure of how strongly two variables are associated. A positive correlation coefficient indicates that the two variables change in the same direction: an increase in one variable usually is associated with a corresponding increase in the other variable. When the correlation coefficient is negative, an increase in one variable is most often associated with a decrease in the other. The absolute value of the correlation coefficient provides information about the strength of the association. Strong correlation does not imply that a cause–effect relationship exists between the two variables.

## Regression

The correlation coefficient tells us about the strength of association between variables and indicates whether the relationship is positive or negative, but it provides no information about the precise quantitative relationship between the variables. For example, if we know there is a correlation between heights of father and son, we might ask, "If a father is six feet tall, what is the most likely height of his son?" To answer this question, **regression** analysis is used.

The relationship between two variables can be expressed in the form of a **regression line,** as shown in Figure 22.7 for the relationship between the heights of fathers and sons. Each point on the graph represents the actual height of a father (value on the $x$ axis) and the height of his son (value on the $y$ axis). Regression finds the line that best fits the data, by minimizing the squared vertical distances from the points to the regression line. The regression line can be represented with the equation

$$y = a + bx$$

where $x$ and $y$ represent the values of the two variables (in Figure 22.7, the heights of father and son, respectively), $b$ represents the **slope of the line,** also called the **regression coefficient,** and $a$ is the $y$ intercept. The slope can be calculated from the covariance of $x$ and $y$ and the variance of $x$ in the following manner:

$$\text{slope} = b = \frac{\text{cov}_{xy}}{s_x^2}$$

**Figure 22.7**

**Regression of sons' height on fathers' height.** Each point represents a pair of data for the height of a father and his son. The regression equation is $y = 36.05 + 0.49x$.

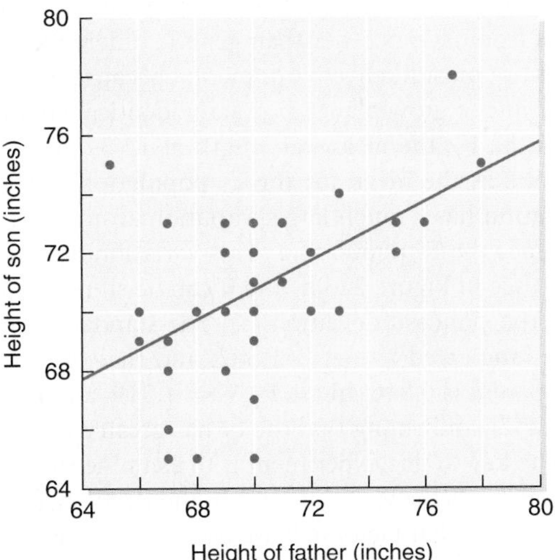


The slope indicates how much of an increase in the variable on the $y$ axis is associated with a unit increase in the variable on the $x$ axis. For example, a slope of 0.5 for the regression of father and son height would mean that for each 1-inch increase in height of a father, the expected height of the son would increase 0.5 inches. The $y$ intercept is the expected value of $y$ when $x$ is zero (the point at which the regression line crosses the $y$ axis). Examples of regression lines with different slopes are presented in Figure 22.8. Regression analysis is a commonly used method for measuring the extent to which variation in a trait is genetically determined, as will be described later in the section on heritability.

**Figure 22.8**

**Regression lines with different slopes.** The slope indicates how much of a change in the $y$ variable is associated with a change in the $x$ variable.

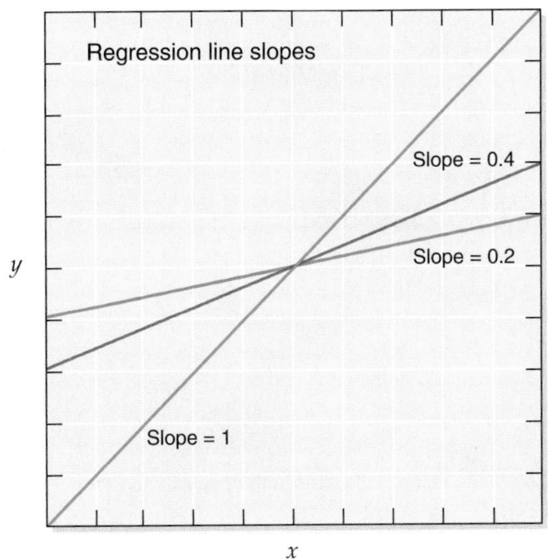


## Analysis of Variance

One last statistical technique that we will mention briefly is **analysis of variance (ANOVA).** Analysis of variance is a powerful statistical procedure for determining whether differences in means are significant (larger than we would expect from chance alone) and for dividing the variance into components. For example, we might be interested in knowing whether males with the XYY karyotype differ in height from males with a normal XY karyotype (see Chapter 12, p. 347). We would proceed by first calculating the mean height of a sample of XYY males and the mean height of a sample of XY males. Suppose we found that the mean height of our sample of XYY males was 74 inches, and the mean height of our sample of XY males was 70 inches. The means appear different, and ANOVA can provide us with the probability that the difference in means of the two samples results from chance. For example, our analysis might indicate that there is less than a 1% probability (often expressed as $p < 0.01$) that the difference we observed in the mean heights resulted from chance. We would probably conclude, then, that the difference in mean heights of XYY males and XY males is not caused by chance differences in our samples but results from some significant factor associated with the difference in chromosomes (i.e., karyotype).

More important to the understanding of genetic influence on traits such as height, ANOVA can also be used to determine how much of the variation in height is associated with the difference in karyotype. If we collected additional data on diet, exercise, other genetics differences, and so on, we might find that the difference in the karyotypes is associated with 40% of the overall variation in height among the individuals in our samples. Factors other than difference in the number of Y chromosomes (such as genes on other chromosomes, diet, and health care) would be responsible for the other 60% of the variation.

The calculations involved in ANOVA are beyond the scope of this book, but the concept of breaking down the variance into components—called *partitioning the variance*—is fundamental to quantitative genetics. For example, frequently we are interested in knowing how much of the variation in a trait is associated with genetic differences among individuals and how much of the variation is associated with environmental factors. Suppose we wanted to increase milk production in a herd of dairy cattle. We might use ANOVA to determine how much of the variation in milk production among the cows results from environmental differences and how much arises because of genetic differences. If much of the variation is genetic, we could increase milk production by selective breeding. On the other hand, if most of the variation is environmental, selective breeding will do little to increase milk production, and our efforts would be better directed toward providing the optimum environment for high production.

## Quantitative Genetic Analysis

Quantitative traits require a different kind of analysis from discontinuous traits. The genetic analysis of quantitative traits is illustrated in this section.

### Inheritance of Ear Length in Corn

With a statistical background, we can now see how quantitative geneticists apply these methods to study multifactorial traits. An organism that has been the subject of genetic and cytological studies for many years is corn (*Zea mays*). Ear length is one trait that was examined in a classic study using some of these techniques to demonstrate a genetic basis for a quantitative trait. In this study, reported in 1913, Rollins Emerson and Edward East started their experiments with two pure-breeding strains of corn, each of which displayed little variation in ear length. The two varieties were black Mexican sweet corn (which had short ears of mean length 6.63 cm) and Tom Thumb popcorn (which had long ears of mean length 16.80 cm).

Emerson and East crossed the two strains and then interbred the $F_1$ plants. The parental plants were inbred lines, so we can assume that each was homozygous for the genes controlling ear length. The $F_1$ plants were heterozygous for all genes, and all plants should have the same genotype. The top two panels of Figure 22.9 present the phenotypes of the parents and $F_1$ plants in photographs and histograms. The range of ear length phenotypes seen in the parents and $F_1$ plants presumably results from factors other than genetic differences. These factors are most likely environmental because it is almost impossible to grow plants in exactly identical conditions.

In the $F_2$, the mean ear length of 12.89 cm is about the same as the mean for the $F_1$ population, but the $F_2$ population has a much larger variation around the mean than does the $F_1$ population. This variation is intuitively easy to see in Figure 22.9b, and it can be shown by calculating the standard deviation ($s$). The standard deviation of the long-eared parent is 1.887, and that of the short-eared parent is 0.816. In the $F_1$, $s = 1.519$, and in the $F_2$, $s = 2.252$, confirming that the $F_2$ has greater variability.

The key to this experiment is to examine the patterns of variation. Is the variation seen in the $F_2$ solely the result of environmental factors? If we assume the environment was responsible for variation in the parental and the $F_1$ generations, then we have every reason to believe that it would have a similar effect on the $F_2$. Conversely, we have no reason to suppose that the environment would have a

**Figure 22.9**

**Inheritance of ear length in corn.** (a) Representative corn ears from the parental, $F_1$, and $F_2$ generations from an experiment in which two pure-breeding corn strains that differ in ear length were crossed and then the $F_1$s interbred. (b) Histograms of the distributions of ear length (in centimeters) of ears of corn from the experiment represented in (a); the vertical axes represent the percentages of the different populations found at each ear length.

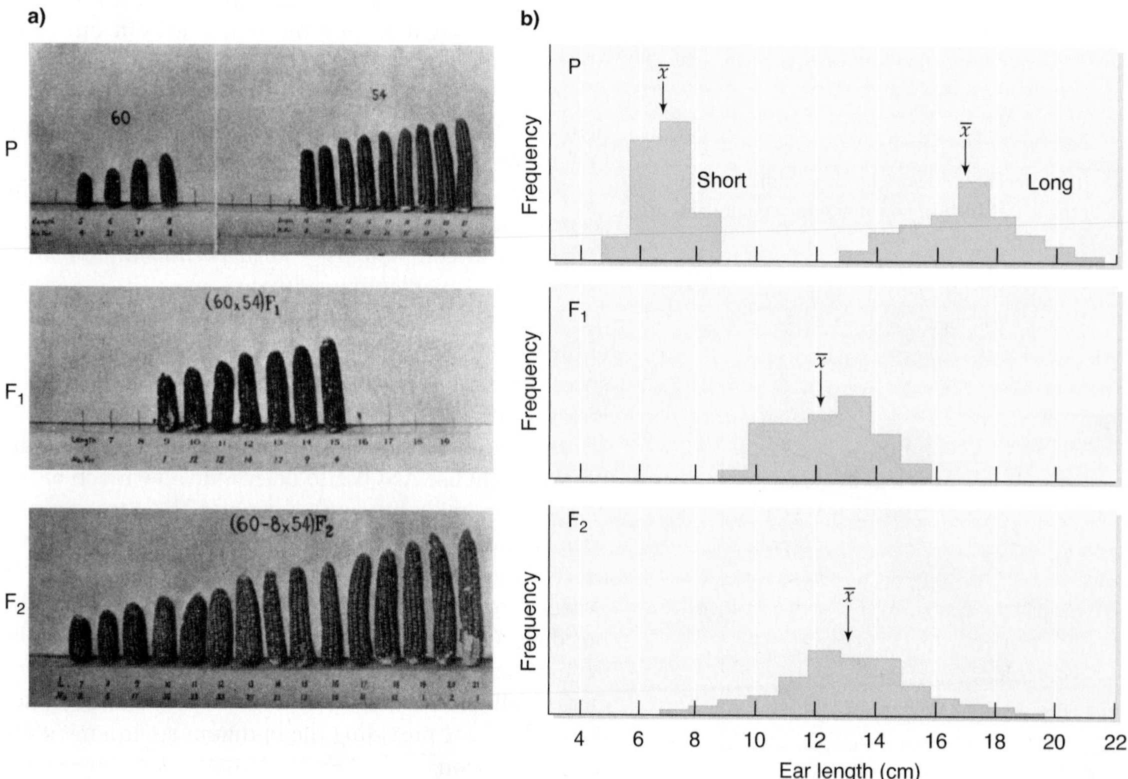

greater influence on the $F_2$ than on the other two generations, so there must be another explanation for the greater variation in ear length in the $F_2$ generation. A reasonable hypothesis is that the increased variability of the $F_2$ results from the presence of greater genetic variation in the $F_2$, which must have been inherited from the parents.

Setting aside the environmental influence for the moment, the data reveal four observations that apply generally to quantitative inheritance studies that incorporate crosses between genetically differentiated individuals or populations:

1. The mean value of the quantitative trait in the $F_1$ is usually intermediate between the means of the two true-breeding parental lines.
2. The mean value for the trait in the $F_2$ is usually approximately equal to the mean for the $F_1$ population.
3. The $F_2$ almost always shows more variability around the mean than the $F_1$ does.
4. The extreme values for the quantitative trait in the $F_2$ extend closer to the two parental values than do the extreme values of the $F_1$, and may sometimes even surpass the parental values.

### Keynote

For a quantitative trait, the $F_1$ progeny of a cross between two phenotypically distinct, pure-breeding parents usually has a phenotype intermediate between the parental phenotypes. The $F_2$ shows more variability than the $F_1$, with a mean phenotype close to that of the $F_1$. The extreme phenotypes of the $F_2$ extend well beyond the range of the $F_1$ and into the ranges of the two parental values. These patterns arise as a result of the heterozygosity of the $F_1$ and the different combinations of parental alleles in the $F_2$ created by independent assortment.

## Heritability

**Heritability** is the proportion of a population's phenotypic variation that is attributable to genetic factors. As we have seen, continuous traits are influenced by multiple genes and by environmental factors. To make informed management decisions, plant and animal breeders, for example, need to know the genetic contribution to traits such as weight gain in cattle, number of eggs laid by chickens, and amount of fleece produced by sheep. Moreover, many ecologically important traits, such as variation in body size, fecundity, and developmental rate, are also polygenic, and the genetic contribution to this variation is important for understanding how natural populations evolve. Heritability is nonetheless often misunderstood, and the term is frequently misused or tossed about without a firm scientific basis. For example, in humans, when individuals in a family resemble each other in some aspect of the phenotype, be it stature or intelligence, a genetic basis often is assumed to be responsible for the similarity. But all the resemblance among family members could just as easily result from their shared environment as opposed to their shared genes. Carefully planned quantitative genetic experiments are the only way to distinguish between these alternatives in any organism being studied. To assess heritability, we must first measure the variation in the trait, and then we must partition that variance into components attributable to different causes.



### Components of the Phenotypic Variance

As an analogy, we can consider the variance among individuals as a stick that can be divided into various pieces. The **phenotypic variance,** represented by $V_P$, is a measure of all variability for a trait (i.e., the whole stick). Recall that it is the sum of squared deviations from the mean for all individuals, as outlined in this chapter's section "Statistical Tools." Some of the differences from the mean for each individual may arise because of genetic differences between individuals (different genotypes within the group); alternatively, some of the differences from the mean may be due to different environments experienced by individuals. The genetic contribution to the phenotypic variation is called **genetic variance,** represented by the symbol $V_G$. Figure 22.10a shows a situation in which all the variation is due to genotype.

As noted, additional variation often results from environmental differences experienced by the individuals. The **environmental variance** is symbolized by $V_E$, and by definition includes any nongenetic source of variation. Temperature, nutrition, and parental care are examples of obvious environmental factors that may cause differences among individuals during development. Figure 22.10b shows a situation in which all the variation results from the environment, and Figure 22.10c shows a situation in which both genes and environment contribute to the variation. Thus, we have two pieces of our stick that correspond to the basic nature–nurture issue discussed earlier:

$$V_P = V_G + V_E$$

One hundred percent of the variation among individuals is accounted for by genetic and environmental influences; however, the partitioning of phenotypic variance can often be complicated. The sum of the genetically caused variance and environmentally caused variance may not add up to the total phenotypic variance, because the genetically caused variance and environmentally caused variance covary. For example, suppose that milk production in cows is influenced by genes, but it is also influenced by the amount of feed a farmer provides. The farmer knows his cows and provides the offspring of good milking cows more feed and those of poor milking cows less feed. Because individuals of above-average genetic quality therefore receive above-average resources,

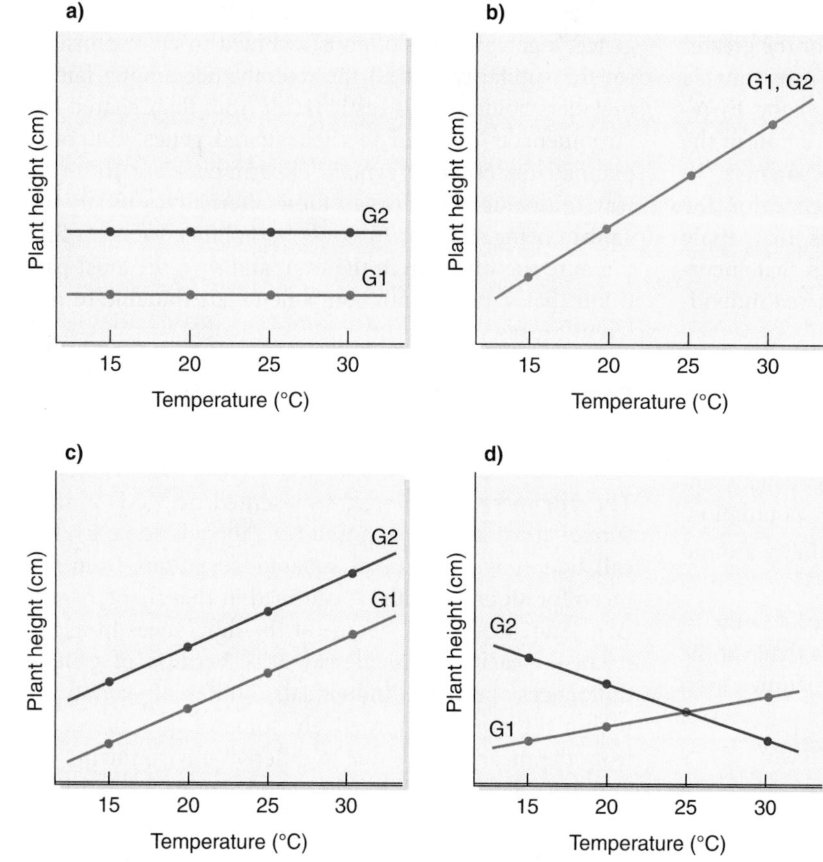


**Figure 22.10**

**Hypothetical example of genetic and environmental effects on plant height.** In each plot, the points represent the means for each genotype. The blue points represent genotype 1 (G1), red points represent genotype 2 (G2), and green points denote an overlap. **(a)** Plant height variation is influenced predominantly by genotype, with G1 individuals having on average a greater height than G2 individuals. Plant height is independent of the temperature in which the plants are raised. **(b)** In this case, variation is predominantly due to the environment, and the two genotypes are indistinguishable across the range of temperatures tested. **(c)** Both genotype and the environment exert an additive influence on height. **(d)** Both genotype and the environment exert an influence, but the response of each genotype is dependent on the environment. This is an example of genotype-by-environment interaction.

a covariance between genotype and environment is produced. In this way, the variance in milk production is increased beyond that expected on the basis of genes and environment operating independently. To account for situations such as these, another term ($COV_{G,E}$) is needed.

There is another source of phenotypic variance, genotype-by-environment interaction, or G×E. Variance caused by G×E exists when the relative effects of the genotypes differ among environments. For example, suppose that in a cold environment, *AA* plants are on average 40 cm tall and *aa* plants are on average 35 cm tall. However, when the genotypes are moved to a warm climate, *aa* plants are now on average 60 cm tall, and *AA* plants are on average 50 cm tall. In this example, both genotypes grow taller in the warm environment, so there is an environmental effect on variance. There is also a genetic effect, but the genetic effect depends on the environment. The relative performance of the genotypes switches in the two environments. Therefore, while both environmental differences (temperature) and genetic differences (genotypes) contribute to the phenotypic variance, the effects of genotype and environment cannot simply be added together. An additional component that accounts for how genotype and environment interact must be considered: $V_{G\times E}$. Figure 22.10d illustrates a significant G×E interaction.

The phenotypic variance, composed of differences arising from genetic variation, environmental variation, genetic–environmental covariation, and genetic–environmental interaction, can be represented by the following equation:

$$V_P = V_G + V_E + 2COV_{G,E} + V_{G\times E}$$

It is important to note that while this is the full equation, there may be situations in which individual components equal zero, depending on the genetic composition of the population, the specific environment, and how the genes interact with the environment.

## Keynote

Variation among individuals can be partitioned into genetic and environmental components, and the interaction between the two. Because genotypes might not be distributed randomly across environments, and genotypes may behave differently in different environments, care must be taken to identify the contributions of these factors correctly.

The pieces of the stick corresponding to genetic and environmental variation can be further divided to reveal more precisely the components of causal influence. Genetic variance, $V_G$, can be subdivided into components arising from different types of gene action and interactions between genes. Some of the genetic variance occurs as a result of the additive effects of the different alleles on the phenotype. For example, an allele g may, on average, contribute 2 cm in height to a plant, while the allele $G$ contributes on average 4 cm. In this case, the $gg$ homozygote would contribute $2 + 2 = 4$ cm in height, the $Gg$ heterozygote would contribute $2 + 4 = 6$ cm in height, and the $GG$ homozygote would contribute $4 + 4 = 8$ cm in height. To determine the genetic contribution to height, we would then add the effects of alleles at this locus to the effects of alleles at other loci that might influence the phenotype. Genes such as these are said to have additive effects, and variation resulting from this sort of gene action is called *additive genetic variance*, symbolized by $V_A$. You may recall that the genes studied by Nilsson-Ehle that determine kernel color in wheat are strictly additive in this way. Some alleles contribute to the pigment of the kernel and others do not; the added effects of all the individual contributing alleles determine the phenotype of the kernel. Thus, the genotypes *AA bb*, *aa BB*, and *Aa Bb* all produce the same phenotype because each genotype has two contributing alleles. The phenotypic variance arising from the additive effects of genes is the additive genetic variance.

Additivity among alleles at a locus is not always the case. Other genes may exhibit dominance, and this is the source of the *dominance variance* ($V_D$). When dominance is present, the individual effects of the alleles are not strictly additive and, as a result, so we must also factor in how genotypes contribute to the phenotypic variation. A locus exhibiting dominance will contribute to $V_P$ only when both the recessive homozygote and either the heterozygote or dominant homozygote are present in the population. Under these conditions, for example, an $F_2$ or a backcross to the recessive parent, loci showing dominance increase variability. In the case of a backcross to a homozygous dominant parent, loci with dominant alleles would not produce phenotypic variation. Finally, epistatic interactions may occur among alleles at different loci. Recall that when epistasis exists, alleles at different loci interact in determining the phenotype. Thus, we might have three genotypes at one locus, but their penetrance could be affected by variation at other loci. The presence of epistasis adds another source of genetic variation, called epistatic or **interaction variance ($V_I$)**. Thus, we can partition the genetic variance as follows:

$$V_G = V_A + V_D + V_I$$

and the total phenotypic variance to this point can then be summarized as

$$V_P = V_A + V_D + V_I + V_E + 2\text{COV}_{G,E} + V_{G\times E}$$

Just as we partitioned the genetic component of variance into three parts, the environmental component of variance can also be partitioned. Individuals in a population may be exposed to varying temperature or nutritional environments during development, resulting in somewhat irreversible differences called *general environmental effects* ($V_{Eg}$). For example, an individual that is raised in a nutritionally deprived region may have a smaller body size. Other environmental variation results in immediate, transient changes in the phenotype, such as skin pigment differences upon exposure to sun, called *special environmental effects* ($V_{Es}$). Finally, environmental effects may be shared by all members of a family. These common *family environmental effects* ($V_{Ecf}$) are especially important because they contribute to differences among families and can be confounded with genetic influences. For example, many insects deposit eggs on specific host plants, and their larvae develop by feeding on these plants. Individual plants vary with respect to nutritional quality and levels of toxins present. As a result, insects that sequester compounds from host plants for their own defense, such as monarch butterflies and the moth *Utetheisa*, can exhibit family differences simply because of the plant they feed on as larvae. Such differences often are misinterpreted as evidence of genetic variation.

A special category of common family environmental effects, **maternal effects ($V_{Em}$)**, are prevalent enough to deserve special mention. For example, variation in the size of mammals at birth has both genetic and environmental components. The genetic component results from the specific genotypic differences between offspring. Their environment up to birth is their mother's uterus, and because there is variation among mothers for variables such as litter size, gestation period, and so on, these constitute maternal effects. Maternal effects can continue after birth as well. In mammals, much of early growth is affected by the amount and nutritional makeup of milk available to offspring; therefore, milk quantity and quality are further maternal influences on offspring that can increase phenotypic variation.

At this point, our variation stick comprises many small segments, and the nature versus nurture equation is partitioned as follows:

$$V_P = V_A + V_D + V_I + V_{Eg} + V_{Es} + V_{Ecf} + V_{Em} + 2\text{COV}_{G,E} + V_{G\times E}$$

Partitioning the phenotypic variance into these components is useful for thinking about the contribution of different factors to the variation in phenotype. It is difficult to design experiments that can analyze all these components simultaneously, so assumptions about some components usually are made. For example, it is often assumed that there is no covariance between genotype and environment ($2\text{COV}_{G,E}$) or G×E variance, but the well-trained geneticist will always remember that the results of such an experiment must be presented with appropriate caution.

## Broad-Sense and Narrow-Sense Heritability

One of the most important questions for quantitative geneticists is the extent to which variation between individuals results from genetic differences. Thus they are interested in how much of the phenotypic variance, $V_P$, can be attributed to genetic variance, $V_G$. This quantity is called

the narrow-sense heritability by the following formula, known as the breeder's equation:

$$R = H^2_N S$$

With values for two of the three parameters in the preceding equation, the selection response (0.7 mg) and the selection differential (1.7 mg), we can solve for the narrow-sense heritability:

$$\text{Narrow-sense heritability} = H^2_N = \text{selection response/selection differential}$$

$$H^2_N = 0.7\text{ mg}/1.7\text{ mg} = 0.41$$

Selection experiments such as this provide another means for estimating the narrow-sense heritability.

A trait will continue to respond to selection, generation after generation, as long as heritable variation for the trait exists within the population. Recalling the example of dog evolution under domestication, Table 22.6 shows some of the heritabilities for a variety of morphological and behavioral traits in dogs. The results from an actual, long-term selection experiment on phototaxis in *Drosophila pseudoobscura* are presented in Figure 22.12. Phototaxis is a behavioral response to light. In this study, flies were scored for the number of times each moved toward light in a total of 15 light–dark choices. Two different experiments were carried out. In one, attraction to light was selected, and in the other, avoidance of light was selected. As can be seen in Figure 22.12, the fruit flies responded to selection for positive and negative phototactic behavior for a number of generations. Eventually, however, the response to selection tapered off, and finally no further directional change in phototactic behavior occurred. One possible reason for this lack of response in later generations is that no more genetic variation for phototactic behavior existed within the population. In other words, all flies at this point were homozygous for all the alleles affecting the behavior, and phototactic behavior could not undergo further evolution in this population unless input of additional genetic variation occurred. More often, some variation still exists for the trait, even after the selection response levels off, but the population fails to respond to selection because the genes for the selected trait have detrimental effects on other traits. These detrimental effects occur because of genetic correlations, which are discussed in the next section.

**Figure 22.12**

**Selection for phototaxis in *Drosophila pseudoobscura*.** The upper graph is the line selected for avoidance of light. The lower graph is the line selected for attraction to light. The phototactic score is the number of times the fly moved toward the light out of a total of 15 light–dark choices.

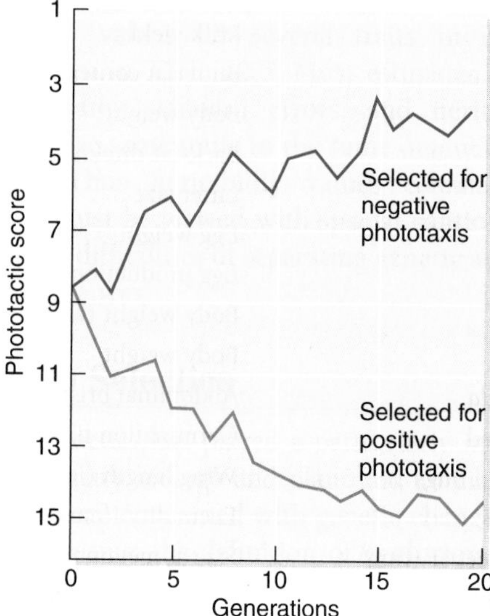


**Table 22.6 Approximate Heritabilities of Some Important Morphological and Behavioral Traits in Domestic Dogs**[a]

| Phenotype | $H^2$ |
|---|---|
| Litter size | 0.1–0.2 |
| Chest depth | 0.5 |
| Chest width | 0.8 |
| Muzzle length | 0.5 |
| Hip dysplasia | 0.2–0.5 |
| Nervousness | 0.5 |
| Hunting traits | 0.1–0.3 |
| Success as guide dog | 0.5 |

[a]Heritabilities of some traits depend on breed.

### Keynote

The amount that a trait changes in one generation as a result of selection on the trait is called the selection response. The magnitude of the selection response depends on both the intensity of selection, called the selection differential, and the narrow-sense heritability.

## Genetic Correlations

When two or more phenotypes are correlated, the traits do not vary independently. For example, fair skin, blond hair, and blue eyes often are found together in the same individual. The association is not perfect—sometimes we see individuals with dark hair, fair skin, and blue eyes—but the traits are found together with enough regularity for us to say that they are correlated. The **phenotypic correlation** between two quantitative traits can be computed by measuring the two phenotypes on a number of individuals and then calculating a correlation coefficient for the two traits.

One reason for a phenotypic correlation among traits is pleiotropy, where multiple phenotypic effects result from a single locus determining the traits (see Chapter 13). Genes rarely affect only a single trait, and this is particularly true for the polygenes that influence continuous traits. Indeed, this is the most likely reason for the association among hair color, eye color, and skin color in humans. For example, the genes that affect growth rates in humans also influence both weight and height, so these two phenotypes tend to be correlated. Pleiotropy is one of the main causes of **genetic correlations** for quantitative traits.

Another significant cause of genetic correlations is genetic linkage. Recall that genetic linkage is one violation of Mendel's law of independent assortment, and that the closer loci are on a chromosome, the greater the frequency that their alleles will be inherited together. When new alleles are first produced by mutation, they are associated with the other alleles that exist on that particular chromosome. These new alleles will be inherited with the other closely linked alleles, causing genetic correlations, and the persistence of these correlations over time depends in part on the amount of recombination between the loci. An important distinction between linkage and pleiotropy as causes of genetic correlations is that over evolutionary time, even the tightest linkages can be broken, allowing new associations between alleles. With pleiotropy, however, functional constraints of an individual protein might not be able to be dissociated, causing a correlation to persist. Table 22.7 presents some genetic correlations that have been detected in studies of quantitative genetics.

Care must be taken when calculating correlations, because beyond pleiotropy and linkage, environmental factors may also influence several traits simultaneously to cause nonrandom associations between phenotypes. For example, adding fertilizer to soil often causes plants both to grow taller and to produce more flowers. If we measured plant height and counted the number of flowers on a group of responsive plants, some of which received fertilizer and some of which did not, we would find that the two traits are correlated; plants receiving fertilizer would be tall and would have many flowers, and those without fertilizer would be short and have few flowers. This phenotype is due to gene action, but the correlation is a result of the common effect of an environmental factor, the fertilizer, on both traits.

Genetic correlations may be positive or negative. A positive correlation means that genes causing an increase in the magnitude of one trait bring about a simultaneous increase in the magnitude of the other. In chickens, body weight and egg weight have a positive genetic correlation. If breeders select for heavier chickens, both the size of the chickens and the mean weight of the eggs produced by these chickens will increase. This increase in egg weight occurs because the genes that produce heavier chickens presumably have a pleiotropic effect on egg weight. In the case of negative genetic correlations, genes that cause an increase in one trait tend to produce a corresponding decrease in another trait. For example, when breeders select for chickens that produce larger eggs, the average egg size increases, but the number of eggs laid by each chicken decreases.

Negative correlations between traits often represent trade-offs or genetic constraints that must be balanced when under selection pressures. For example, faster speed is important to garter snakes for both hunting and escaping predators. In the western United States, some garter snakes prey on toxic newts that produce the neurotoxin tetrodotoxin in their skin. The negative genetic correlation between speed and neurotoxin resistance in garter snake populations shown in Figure 22.13 is apparently an evolutionary constraint. Although tetrodotoxin resistance



**Table 22.7 Genetic Correlations between Traits in Humans, Domesticated Animals, and Natural Populations[a]**

| Organism | Traits | Genetic Correlation |
|---|---|---|
| Humans | IgG, IgM | 0.07 |
| Cattle | Butterfat content, milk yield | −0.38 |
| Pigs | Weight gain, back-fat thickness | 0.13 |
| | Weight gain, efficiency | 0.69 |
| Chickens | Egg weight, egg production | −0.31 |
| | Body weight, egg weight | 0.42 |
| | Body weight, egg production | −0.17 |
| Mice | Body weight, tail length | 0.29 |
| Jewelweed | Seed weight, germination time | −0.81 |
| Milkweed bugs | Wing length, fecundity | −0.57 |
| Wood frogs | Developmental rate, size at metamorphosis | −0.86 |
| *Drosophila* | Early life fecundity, resistance to starvation | −0.91 |

[a]The estimates given in this table apply to particular populations in particular environments; genetic correlations for other individuals may differ.

**Figure 22.13**

**Negative genetic correlation ($r = -0.45$) between speed and resistance to tetrodotoxin in garter snakes, illustrated by family means.** The correlation of family means approximates the genetic correlation when families are large.

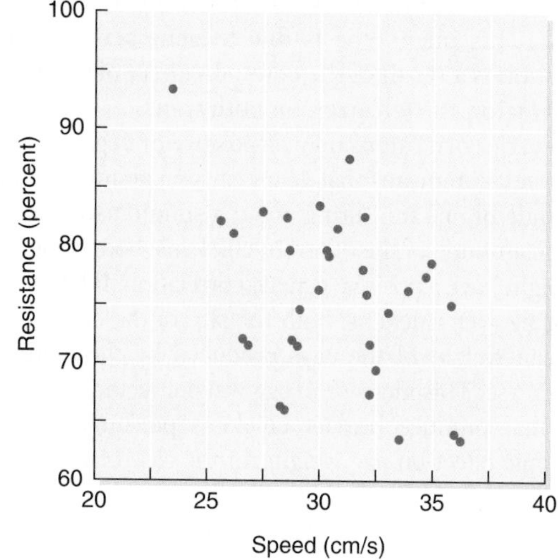


had evolved independently at least twice, it seems there have been no mutations that have increased resistance without decreasing speed.

Negative genetic correlations often place practical constraints on the ability of plant and animal breeders to make progress from selection. As an example, milk yield and butterfat content have a negative genetic correlation in cattle. The same genes that cause an increase in milk production bring about a decrease in butterfat content of the milk. Thus, when breeders select for increased milk yield, the amount of milk produced by the cows may go up, but the butterfat content decreases. Knowing the amount and type of genetic correlations before undertaking a breeding program is essential to ensure success.

The ability of an organism to adapt to a particular environment is strongly influenced by genetic correlations among traits, as we saw in the garter snake example; therefore, genetic correlations are of great interest to evolutionary biologists. As another illustration, consider two traits in tadpoles: developmental rate and size at metamorphosis. Most tadpoles are found in small ponds and pools, where fish (potential predators) are absent and food is abundant. A major liability in using this type of aquatic habitat is that ponds often dry up, frequently before the tadpoles have developed sufficiently to metamorphose into frogs and leave the water. One might expect, then, that natural selection would favor a maximum rate of development in tadpoles, so that the tadpoles could quickly metamorphose into frogs. However, many species of tadpoles fail to develop at maximum rates, contrary to this prediction. One reason for a slower rate of development is a negative genetic correlation between developmental rate and body size at metamorphosis. Genes that accelerate development also tend to cause metamorphosis at a smaller size, at least in some populations. Thus, selection for fast metamorphosis also produces smaller frogs, and size is extremely important in determining the survival of young frogs. Small frogs tend to lose water more rapidly in the terrestrial environment, are more likely to be eaten by predators, and have more difficulty finding sufficient food. The negative genetic correlation between developmental rate and body size at metamorphosis places constraints on the frogs' ability to develop rapidly and to attain a large body size at metamorphosis. Knowing about such genetic correlations is important for understanding how animals adapt or fail to adapt to a particular environment.

**Keynote**

Genetic correlations arise from pleiotropy or genetic linkage between different loci. When a trait is selected, any genetically correlated traits also exhibit a selection response. Thus, the evolution of a population from artificial breeding or in response to natural selection depends on the simultaneous integration of many aspects of the phenotype.

You are a researcher trying to determine whether fingerprint patterns are correlated with high blood pressure in the iActivity *Your Fate in Your Hands?* on the student website.

## Quantitative Trait Loci

The statistical approach to understanding quantitative inheritance has been useful in analyzing components of variation and response to selection. However, to understand quantitative traits fully, we need to characterize the individual loci that affect these traits. The individual loci that contribute to a quantitative trait are known as QTLs **(quantitative trait loci),** and the set of QTLs and their interactions that determine a trait is known as the genetic architecture of a trait. Recent advances combining molecular genotyping and statistical tools have enabled geneticists to begin to make the connection between quantitative phenotypes and the specific QTLs that control them.

QTL identification is an exercise in finding segments of the genome associated with phenotypic differences between individuals. As such, it is most powerful when analyzing a population with a detailed linkage map, substantial phenotypic variation, and a large number of individuals. Typically, inbred lines that have been selected for differing phenotypes are crossed and then either backcrossed, intercrossed to generate an $F_2$, or intercrossed and selfed to create a series of recombinant inbred strains. The population is then grown, measured, and genotyped. While the analytical methods used to determine which genomic regions are correlated with phenotypic variation are increasingly sophisticated, the essence of finding QTL is to split the individuals into groups on the basis of a marker genotype and then test to see whether the groups have similar or different means. In fact, the earliest QTL identification methods took the genotypic and phenotypic data from a population and conducted an ANOVA using each genetic marker as a factor and the phenotype as the response. If a marker locus is unlinked to a QTL, the average phenotype is the same for all genotype classes. If the marker locus is linked to a QTL, then genotypes should differ in their mean value of the trait examined. The difference in phenotypic means for the marker genotype classes depends on the size of the effect of a QTL and on how tightly linked the QTL is with the marker.

Because of the many QTL identification studies conducted to date, we can now begin to understand not only how many QTLs underlie these traits but also the magnitude of their effects and distribution in the genome. Some of the most significant applications of this work have led to the identification of QTLs responsible for important agronomic traits and QTLs responsible for adaptive differences between closely related species. For example, some of the most important differences between closely related plant species are the suites of floral traits that attract pollinators, including color, shape, and nectar rewards. Monkeyflowers have diverged into hummingbird-pollinated species, such as *Mimulus cardinalis*, exhibiting red coloration, deep tubular flowers with lots of nectar at the base, and flared back (reflexed) petals; and species such as *M. lewisii* with little nectar reward, broad petal landing pads, and pink flowers characteristic of bee-pollinated flowers (Figure 22.14). Crosses between these species have revealed that differences in each of the pollinator attraction and efficiency traits appear to be controlled by at least one QTL that influences 25% or more of the phenotypic variation (Figure 22.15). This



**Figure 22.14**

***Mimulus lewisii* (A, C) and *M. cardinalis* (B, D) flowers.** Flowers are shown from the front (A, B) as an approaching pollinator views them. In side views (C, D), the relative positions of the stigma and anthers are shown.

**Figure 22.15**

**QTL maps for 12 floral traits in monkeyflowers.** This figure shows the genetic maps constructed for this species and the locations of QTLs influencing floral traits. Because of the crossing design, there are two maps for each linkage group (the subscripts show whether the linkage group corresponds to the *M. lewisii* or *M. cardinalis* parent). Boxes show the position of markers correlated to phenotypic traits. Taller boxes indicate QTLs that explain 25% or more of the variance in a trait.

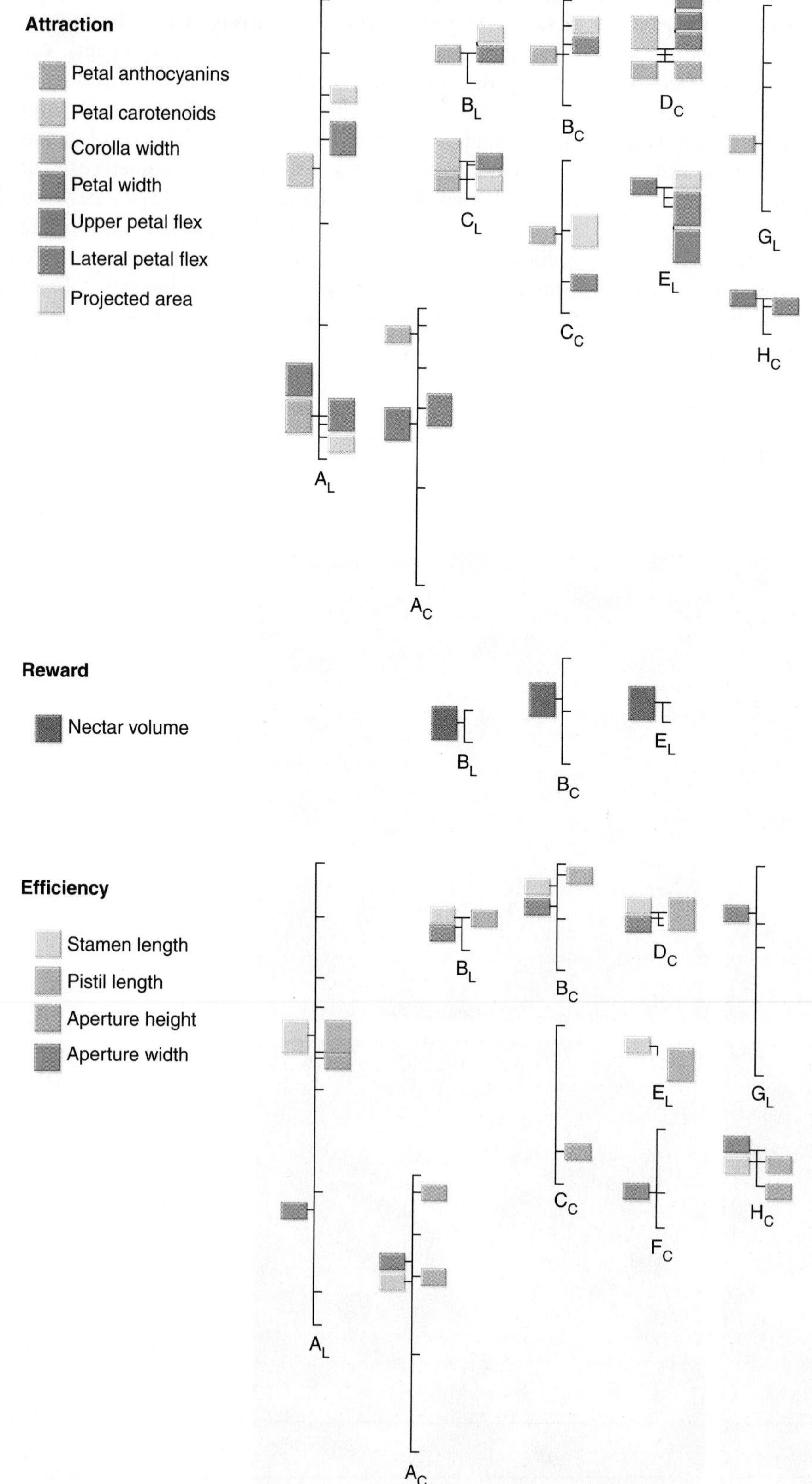

finding, along with data showing substantial adaptive differences between these QTL alleles, suggests that, in at least some cases, important phenotypic shifts between species may have occurred by mutations in single loci. Interestingly, many of the QTLs for different traits are physically close, suggesting that floral characters are genetically correlated.

Moving from associated genomic regions to cloning the actual QTL depends on the availability of a suite of molecular tools. Some of the better examples of QTL that have been characterized to date come from intensively studied model species, such as corn, tomato (*Solanum lycopersicum*), and *Drosophila melanogaster*. In corn, the *teosinte branched 1 (tb1)* QTL controlling the number of axillary branches was cloned in 1997. Branching is a key difference between cultivated corn and its wild relative teosinte, with both evolutionary and commercial importance. *tb1* has been shown to be a member of a DNA-binding transcriptional regulator gene family that acts in corn to suppress growth in specific tissues. The important allelic differences between corn and teosinte that appear to have been selected during domestication of maize are in a region of the gene that is not transcribed. The tomato QTL *fw2.2*, which controls up to 30% of the difference in fruit weight seen in crosses between cultivated and wild tomato species, was cloned in 2000. The FW2.2 protein is expressed early in fruit development, and variation between alleles has been found in the timing and levels of gene expression, but there are not major differences in the FW2.2 protein sequence. In addition, there appears to be a genetic correlation caused by a pleiotropic effect of the small fruit allele: lines that are homozygous for the small fruit allele also produce more fruits than plants that are homozygous for the large fruit allele.

Another approach to finding QTLs involves testing for associations between phenotypic differences and allelic variation at candidate loci. Candidate loci can be defined on the basis of known or suggested function, proximity to genomic regions implicated in QTL studies, or both. In *Drosophila*, for example, mapping experiments defined a QTL for sternopleural bristle number in a region of the genome that also contained a key developmental gene, *hairy (h)*, whose classically defined mutants have extra bristles. Surveys of natural variation in bristle number and molecular variation at the *h* locus have



## Focus on Genomics

### QTL Analysis of Aggression in *Drosophila melanogaster*

Genomic techniques, in combination with classical genetic techniques, were used by a group of scientists to estimate the number of genes involved in aggressive behavior by male *Drosophila melanogaster*. Behavior is clearly influenced both by the environment and the genetic makeup of the individual, but very few simple genetic behaviors have been found—most behaviors appear to be controlled by many genes, rather than just a few genes. The scientists wished to estimate how many genes influence aggressive behavior and to identify as many of these genes as possible. First, the scientists selected flies that were either more or less aggressive than normal flies, and used those flies to establish populations. Essentially, they established true-breeding, or nearly true-breeding, populations, although they assumed that the change in behavior was the result of many polymorphic alleles, each of which played a relatively minor role. They established two high-aggression populations, two low-aggression populations, and two control populations. They then used DNA microarrays (see Chapter 9, pp. 230–232) to compare how transcription differed between the different types of populations. They looked for transcripts that, for example, were expressed at higher levels in flies from both of the high-aggression populations and at the lower levels in flies from either the control or low-aggression populations. About 10% of the genes in the genome (roughly 1,500 genes) passed this test, and if we consider only the genes that differed between high-aggression and low-aggression populations, we are down to a group of about 1,000 genes. When the investigators looked at the functions of these genes, they saw some interesting trends. Many of the genes that were overexpressed in aggressive flies were involved in detecting and responding to chemical and biological stimuli, while the genes overexpressed in low-aggression flies tended to include genes involved in learning and memory and defense response. The investigators then decided to test individually some of the genes identified in this screen. They used classical genetics to find new alleles in each of 19 genes identified in their screens. They then took flies homozygous for one of these 19 new mutations and compared the aggression of the mutant flies to that of control flies. In 15 cases, the mutation caused a clear change in how aggressively the mutant flies behaved. This experiment shows that even very complex, multigenic traits can be dissected apart using the tools available to the modern geneticist.

shown that *h* is indeed a QTL for bristle number in laboratory populations. Interestingly, surveys of wild populations did not confirm the effect of this QTL, reinforcing the point that many such analyses are population dependent. Analysis of candidate loci does not, however, always yield similar results. An experiment to determine whether variation in the structural genes of the carotenoid biosynthetic pathway affected mature red fruit color in tomato showed that most of the QTL did not correspond to these genes. Continued refining of phenotypes and understanding of metabolic and genetic pathways, combined with new analytical techniques, will be crucial for identifying and analyzing QTLs for years to come.

The genes underlying quantitative traits in humans cannot be determined through traditional pedigree analysis as they are with discontinuous traits, because environmental differences through time and the action of other segregating genes tend to obscure the effects of single loci. However, important segments of the genome that play a part in phenotypic variation can still be identified through association studies. Association studies utilize widely distributed DNA markers (sequences at particular locations in the genome that vary in the population) in the human genome and populations showing a trait and random control populations to look for QTLs. The most useful DNA markers here are SNPs (see Chapter 8, pp. 192–193). The methodology for these studies is roughly similar to experimental populations, in that individuals are phenotyped and genotyped and then statistical methods are used to associate QTL and SNPs. The Focus on Genomics box on page 374 describes how genomics techniques were used to find genes that are expressed at different levels in high-aggression or low aggression fruit flies.

Applications of these methods to a variety of traits have begun to reveal the genes contributing to quantitative human variation, including traits ranging from the ability to taste PTC, to height, to susceptibility to diseases such as diabetes. The recent completion of the Human Genome Project, coupled with efforts to develop large panels of human SNP markers and the ability to characterize large numbers of individuals' genotypes, promises to reveal even more of these human QTLs in the near future.

**Keynote**

Marker-based mapping approaches can be used to correlate segments of the genome with phenotypic variation in quantitative traits in natural and experimental populations. Studies identifying quantitative trait loci (QTLs) provide estimates of the number of genes and size of effects influencing variation in continuous traits.

# Summary

- Discontinuous traits exhibit only a few distinct phenotypes. Continuous, or quantitative, traits display a range of phenotypes.
- Continuous traits have a range of phenotypes because many loci contribute to the phenotype (polygenic inheritance) and because environmental factors influence the phenotype produced by a genotype.
- Continuous traits can be described by using samples of populations and statistics such as the mean and variance.
- Continuous traits can be analyzed using statistical techniques such as correlation, regression analysis, and analysis of variance (ANOVA).
- Variation among individuals can be partitioned into genetic and environmental components. However, genotypes may respond differently in distinct environments, so caution must be exercised when designing and interpreting experiments that measure genetic and environmental contributions to phenotypic variation.
- The broad-sense heritability of a trait is the proportion of the phenotypic variance that results from genetic differences among individuals. The narrow-sense heritability is the proportion of the phenotypic variance due only to additive genetic variance. Both measures depend on a particular population in a particular environment.
- The amount that a trait changes in one generation as a result of selection on the trait is called the response to selection. The magnitude of the response to selection depends on the selection differential and the narrow-sense heritability.
- Genetic correlations arise when two traits are influenced by the same genes or linked genes. When a trait is selected, genetically correlated traits also exhibit a response to selection.
- Quantitative trait loci (QTLs) that determine continuous traits can be identified through marker-based mapping. QTL mapping provides an estimate of the number and relative importance of genes influencing quantitative genetic variation.

# Analytical Approaches to Solving Genetics Problems



**Q22.1** Assume that loci *A*, *B*, *C*, and *D* are members of a multiple-gene series that controls a quantitative trait, and that each gene assorts independently. Each *A*, *B*, *C*, and *D* allele has a cumulative effect that contributes 3 cm of height when present, and alleles *a*, *b*, *c*, and *d* do not contribute anything to the height of the organism. In addition, gene *L* is always present in the homozygous state, and the *LL* genotype contributes a constant 40 cm of height. If we ignore height variation caused by environmental factors, an organism with genotype *AA BB CC DD LL* would be 64 cm high, and one with genotype *aa bb cc dd LL* would be 40 cm. A cross is made of *AA bb CC DD LL* × *aa BB cc DD LL* and is carried into the $F_2$ by selfing of the $F_1$.

**a.** How does the size of the $F_1$ individuals compare with the size of each of the parents?

**b.** Compare the mean of the $F_1$ with the mean of the $F_2$, and comment on your findings.

**c.** What proportion of the $F_2$ population would show the same height as the *AA bb CC DD LL* parent?

**d.** What proportion of the $F_2$ population would show the same height as the *aa BB cc DD LL* parent?

**e.** What proportion of the $F_2$ population would breed true for the height shown by the *aa BB cc DD LL* parent?

**f.** What proportion of the $F_2$ population would breed true for the height characteristic of $F_1$ individuals?

**A22.1** This question explores our understanding of the basic genetics involved in a multiple-gene series that controls a quantitative trait. The approach we will take is essentially the same as the approach used with a series of independently assorting genes that controls different traits. That is, we make predictions on the basis of genotypes and relate the results to phenotypes, or we make predictions on the basis of phenotypes and relate the results to genotypes.

**a.** Each allele represented by a capital letter contributes 3 cm of height to the base height of 40 cm provided by *LL* homozygosity. Therefore, the *AA bb CC DD LL* parent, which has six capital-letter alleles from the *A–D* multiple-gene series, is 40 + (6 × 3) = 58 cm high. Similarly, the *aa BB cc DD LL* parent has four capital-letter alleles and therefore is 40 + 12 = 52 cm high. The $F_1$ from a cross between these two individuals would be heterozygous for the *A*, *B*, and *C* loci and homozygous for *D* and *L*, that is, *Aa Bb Cc DD LL*. This progeny has five capital-letter alleles apart from *LL* and therefore is 40 + 15 = 55 cm high.

**b.** The $F_2$ is derived from a self of the *Aa Bb Cc DD LL* $F_1$. All the $F_2$ individuals will be *DD LL*, making them at least 40 + 6 = 46 cm high. Now we must deal with the heterozygosity at the other three loci. What we need to calculate is the relative frequencies of all possible genotypes for the three loci and collect those with zero, one, two, three, four, five, and six capital-letter alleles. The probability of getting an individual with two capital-letter alleles for a particular locus is $^1/_4$, the probability of getting an individual with one capital-letter allele for the locus is $^1/_2$, and the probability of getting an individual with no capital-letter alleles for the locus is $^1/_4$. Therefore, the probability of getting an $F_2$ individual with six capital-letter alleles for the *A*, *B*, and *C* loci is $(^1/_4)^3 = {^1/_{64}}$, and the same probability is obtained for an individual with no capital-letter alleles. The simplest approach to find the expected numbers for genotypes with one to five capital-letter alleles is to compute the coefficients in the binomial expansion of $(a + b)^6$. Recall that in the wheat kernel color example, a phenotypic ratio of 1:4:6:4:1 was observed. These numbers are the coefficients in the binomial expansion of $(a + b)^4$ and correspond to the numbers of different genotypes with zero, one, two, three, and four contributing alleles. Similarly, the expansion of $(a + b)^6$ gives a 1:6:15:20:15:6:1 distribution of zero, one, two, three, four, five, and six capital-letter alleles, respectively. Since each capital-letter allele in the *A*, *B*, and *C* set contributes 3 cm of height over the 46-cm height given by the *DD LL* genotype common to all, the $F_2$ individuals would fall into the following distribution:

| Number of Capital-Letter Alleles | Height Added to 46 cm *DD LL* Genotype (cm) | Height of Individuals (cm) | Number Expected per 64 $F_2$ |
|---|---|---|---|
| 6 | 18 | 64 | 1 |
| 5 | 15 | 61 | 6 |
| 4 | 12 | 58 | 15 |
| 3 | 9 | 55 | 20 |
| 2 | 6 | 52 | 15 |
| 1 | 3 | 49 | 6 |
| 0 | 0 | 46 | 1 |

The distribution is clearly symmetrical, giving an average of 55 cm, the same height shown in $F_1$ individuals.

**c.** The *AA bb CC DD LL* parent was 58 cm, so we can read the proportion of $F_2$ individuals that show this same height directly from the table in part (**b**). The answer is $^{15}/_{64}$.

**d.** The *aa BB cc DD LL* parent was 52 cm, and from the table in part (**b**) the proportion of $F_2$ individuals that show this same height is $^{15}/_{64}$.

**e.** We are asked to determine the proportion of the $F_2$ population that would breed true for the height shown by the *aa BB cc DD LL* parent, which was 52 cm. To breed true, the organism must be homozygous. We have also established that *DD LL* is a constant genotype for the $F_2$ individuals, giving a basic height of 46 cm. Therefore, for a height of 52 cm, two additional, active, capital-letter alleles must be present apart from those at the *D* and *L* loci. With the requirement for homozygosity there are only three genotypes that give a 52-cm height; they are *AA bb cc DD LL*, *aa BB cc DD LL*, and *aa bb CC DD LL*. The probability of each combination occurring in the $F_2$ is $^1/_{64}$, so the answer to the problem is $^1/_{64} + {}^1/_{64} + {}^1/_{64} = {}^3/_{64}$. (Note that the individual probability for each genotype can be calculated. That is, probability of $AA = {}^1/_4$, probability of $bb = {}^1/_4$, probability of $cc = {}^1/_4$, and probability of $DD\ LL = 1$, giving an overall probability for *AA bb cc DD LL* of $^1/_{64}$.)

**f.** We are asked to determine the proportion of the $F_2$ population that would breed true for the height characteristic of $F_1$ individuals. Again, the basic height given by *DD LL* is 46 cm. The $F_1$ height is 55 cm, so three capital-letter alleles must be present in addition to *DD LL* to give that height because $(3 \times 3)$ cm = 9 cm, and 9 cm + 46 cm = 55 cm. However, because an individual must be homozygous to be true-breeding, the answer to this question is none, because 3 is an odd number, meaning that at least one locus must be heterozygous to get the 55-cm height.

**Q22.2** Five field mice collected in Texas had weights of 15.5 g, 10.3 g, 11.7 g, 17.9 g, and 14.1 g. Five mice collected in Michigan had weights of 20.2 g, 21.2 g, 20.4 g, 22.0 g, and 19.7 g. Calculate the mean weight and the variance in weight for mice from Texas and for mice from Michigan.

**A22.2** To answer this question, we use the formula given in the section "Statistical Tools." The formula for the mean is

$$\bar{x} = \frac{\Sigma x_i}{n}$$

The symbol $\Sigma$ means to add, and the $x_i$ represents all the individual values. We begin by summing up all the weights of the mice from Texas:

$$\Sigma x_i = 15.5 + 10.3 + 11.7 + 17.9 + 14.1 = 69.5$$

Next, we divide this summation by $n$, which represents the number of values added together. In this case, we added together five weights, so $n = 5$. The mean for the Texas mice is therefore

$$\frac{\Sigma x_i}{n} = \frac{69.5}{5} = 13.9$$

To calculate the variance in weight among the Texas mice, we use the formula

$$s^2 = \frac{\Sigma(x_i - \bar{x})^2}{n - 1}$$

We must take each individual weight and subtract it from the mean weight of the group. Each value obtained from this subtraction is then squared, and all squared values are added up, as shown below.

| | |
|---|---|
| 15.1 − 13.9 = 1.6 | $(1.6)^2 = 2.56$ |
| 10.3 − 13.9 = −3.6 | $(-3.6)^2 = 12.96$ |
| 11.7 − 13.9 = −2.2 | $(-2.2)^2 = 3.84$ |
| 17.9 − 13.9 = 4.0 | $(4.0)^2 = 16.00$ |
| 14.1 − 13.9 = 0.2 | $(10.2)^2 = 0.04$ |
| | 36.4 |

The sum of all the squared values is 36.4. All that remains for us to do is to divide this sum by $n - 1$, which is $5 - 1 = 4$.

$$s^2 = \frac{\Sigma(x_i - \bar{x})^2}{n - 1} = \frac{36.4}{4} = 9.1$$

The mean and the variance for the Texas mice are 13.9 and 9.1.

We now repeat these steps for the mice from Michigan.

$$\Sigma x_i = 20.2 + 21.2 + 20.4 + 22.0 + 19.7 = 103.5$$

$$\frac{\Sigma x_i}{n} = \frac{103.5}{5} = 20.7$$

$$s^2 = \frac{\Sigma(x_i - \bar{x})^2}{n - 1}$$

| | |
|---|---|
| 20.2 − 20.7 = −0.5 | $(-0.5)^2 = 0.25$ |
| 21.2 − 20.7 = 0.5 | $(0.5)^2 = 0.25$ |
| 20.4 − 20.7 = −3.0 | $(-0.3)^2 = 0.09$ |
| 22.0 − 20.7 = 1.3 | $(1.3)^2 = 1.69$ |
| 19.7 − 20.7 = −1.0 | $(-1.0)^2 = 1.0$ |
| | 3.28 |

$$s^2 = \frac{\Sigma(x_i - \bar{x})^2}{n - 1} = \frac{3.28}{4} = 0.82$$

The mean and the variance for the Michigan mice are 20.7 and 0.82.

We conclude that the Michigan mice are much heavier than the Texas mice, and the Michigan mice also exhibit less variance in weight.

# Questions and Problems

***22.1** Given the following sets of 30 phenotypic measurements for different traits, decide whether each trait is qualitative or quantitative and explain your answer.

**a.** Trait 1: 38.9, 47.0, 53.1, 39.1, 62.8, 46.8, 57.5, 54.9, 48.9, 56.3, 52.5, 60.8, 46.7, 48.0, 52.3, 40.7, 50.4, 51.0, 46.5, 47.9, 55.4, 53.1, 58.5, 51.1, 60.2, 50.6, 48.6, 52.5, 54.5, 51.4, 48.1, 49.5, 55.8, 52.9, 42.9, 44.4, 56.4, 38.9, 42.2, 42.2

**b.** Trait 2: 25.7, 8.8, 11.2, 5.7, 20.6, 34.3, 13.0, 28.8, 20.5, 24.1, 21.2, 14.3, 17.7, 18.7, 24.3, 30.2, 20.2,

25.1, 30.6, 21.2, 31.2, 23.0, 16.9, 10.5, 14.1, 10.2, 30.5, 22.5, 34.1, 10.6, 19.5, 21.0, 20.9, 27.7, 33.0, 7.7, 20.1, 16.9, 18.8, 15.7

**c.** Trait 3: 31.1, 22.0, 28.1, 14.1, 43.4, 52.8, 32.5, 39.0, 43.1, 52.2, 45.1, 35.8, 36.4, 38.7, 52.8, 42.6, 42.6, 54.8, 43.4, 45.1, 45.1, 49.5, 34.2, 26.1, 35.2, 25.6, 43.1, 48.3, 52.2, 26.4, 40.9, 44.5, 44.3, 36.4, 49.5, 19.4, 42.4, 34.2, 39.0, 31.1

***22.2** The following measurements of head width and wing length were made on a series of steamer ducks:

| Specimen | Head Width (cm) | Wing Length (cm) |
|---|---|---|
| 1 | 2.75 | 30.3 |
| 2 | 3.20 | 36.2 |
| 3 | 2.86 | 31.4 |
| 4 | 3.24 | 35.7 |
| 5 | 3.16 | 33.4 |
| 6 | 3.32 | 34.8 |
| 7 | 2.52 | 27.2 |
| 8 | 4.16 | 52.7 |

**a.** Calculate the mean and the standard deviation of head width and of wing length for these eight birds.

**b.** Calculate the correlation coefficient for the relationship between head width and wing length in this series of ducks.

**c.** What conclusions can you make about the association between head width and wing length in steamer ducks?

***22.3** Explain why the $F_1$ generation from a cross of two pure-breeding parents that differ in a size character usually is no more variable than the parents.

***22.4** Two pairs of independently segregating genes with two alleles each, *A/a* and *B/b*, determine plant height additively in a population. The homozygote *AA BB* is 50 cm tall, and the homozygote *aa bb* is 30 cm tall.

**a.** What is your prediction of the $F_1$ height in a cross between the two homozygous stocks?

**b.** What genotypes in the $F_2$ will show a height of 40 cm after an $F_1 \times F_1$ cross?

**c.** What will be the $F_2$ frequency of the 40-cm plants?

**d.** What assumptions have you made in answering this question?

**22.5** Assume that in squashes the difference in fruit weight between a 3-lb type and a 6-lb type results from three independently segregating allele pairs, *A/a, B/b*, and *C/c*. Each capital-letter allele contributes a half pound to the weight of the squash. From a cross of a 3-lb plant (*aa bb cc*) with a 6-lb plant (*AA BB CC*), what will be the phenotypes (weights) of the $F_1$ and the $F_2$? What will be their distribution?

**22.6** Refer to the assumptions stated in Problem 22.5. Determine the range in fruit weight of the offspring in the following squash crosses:

**a.** *AA Bb CC* × *aa Bb Cc*

**b.** *AA bb Cc* × *Aa BB cc*

**c.** *aa BB cc* × *AA BB cc*

**22.7** Three independently segregating genes (*A, B, C*), each with two alleles, determine height in a plant. Each capital-letter allele adds 2 cm to a base height of 2 cm.

**a.** What are the heights expected in the $F_1$ progeny of a cross between homozygous strains *AA BB CC* × *aa bb cc*?

**b.** What is the distribution of heights (frequency and phenotype) expected in an $F_1 \times F_1$ cross?

**c.** What proportion of $F_2$ plants will have heights equal to the heights of the original two parental strains?

**d.** What proportion of the $F_2$ will breed true for height?

**22.8** Repeat Problem 22.7, but assume that one of the loci shows dominance instead of additivity.

**22.9** Assume that three independently segregating, equally and additively contributing pairs of alleles control flower length in nasturtiums. A completely homozygous plant with 10-mm flowers is crossed to a completely homozygous plant with 30-mm flowers. The $F_1$ plants all have flowers about 20 mm long. The $F_2$ plants show a range of lengths from 10 to 30 mm, with about 1/64 of the $F_2$ having 10-mm flowers and 1/64 having 30-mm flowers. What distribution of flower length would you expect to see in the offspring of a cross between an $F_1$ plant and the 30-mm parent?

***22.10** An experiment found the mean internode length in spikes (the floral structures) of the barley variety *asplund* to be 2.12 mm. In the variety *abed binder*, the mean internode length was found to be 3.17 mm. The mean of the $F_1$ of a cross between the two varieties was approximately 2.7 mm. The $F_2$ population included individuals similar to both parents, as well as intermediate types. Analysis of the $F_3$ generation showed that 8 out of the total 125 $F_2$ individuals of the *asplund* type were true breeding, giving a mean of 2.19 mm. Nine other $F_2$ individuals were similar to *abed binder*, and they bred true to type, with a mean internode length of 3.24 mm. Is the internode length in spikes of barley a discontinuous or a quantitative trait? Why?

**22.11** Assume that the difference between a corn plant 10 dm (decimeters) high and one 26 dm high results from four pairs of equal and cumulative multiple alleles, with the 26-dm plants being *AA BB CC DD* and the 10-dm plants being *aa bb cc dd*. Make and detail your assumptions, then predict the following:

**a.** What will be the size and genotype of an $F_1$ from a cross between these two true-breeding types?

**b.** Determine the limits of height variation in the offspring from the following crosses:

**i.** *Aa BB cc dd* × *Aa bb Cc dd*

**ii.** *aa BB cc dd* × *Aa Bb Cc dd*

**iii.** *AA BB Cc DD* × *aa BB cc Dd*

**iv.** *Aa Bb Cc Dd* × *Aa bb Cc Dd*

**22.12** Refer to the assumptions given in Problem 22.11. For this problem, two 14-dm corn plants, when crossed, give nothing but 14-dm offspring (case A). Two other 14-dm plants give one 18-dm, four 16-dm, six 14-dm, four 12-dm, and one 10-dm offspring (case B). Two other 14-dm plants, when crossed, give one 16-dm, two 14-dm, and one 12-dm offspring (case C). What genotypes for each of these 14-dm parents (cases A, B, and C) would explain these results? Would it be possible to get a plant that is taller than 18 dm by selection in any of these families?

***22.13** Transgressive segregation is the phenomenon in which two pure-breeding strains, differing in a trait, are crossed and produce $F_2$ individuals with phenotypes that are more extreme than either grandparent (i.e., that are larger than the largest or smaller than the smallest in the original generation). Even if two pure-breeding strains are the same for a quantitative trait, it is possible to see transgressive segregation in an $F_2$. Propose scenarios with specific assumptions for each of these examples of transgressive segregation.

***22.14** Pigmentation in the imaginary river-bottom dweller *Mucus yuccas* is a quantitative character controlled by a set of five independently segregating polygenes with two alleles each: *A/a, B/b, C/c, D/d*, and *E/e*. Pigment is deposited at three different levels, depending on the threshold of gene products produced by the capital-letter alleles. Greyish brown pigmentation is seen if at least four capital-letter alleles are present, light tan pigmentation is seen if two or three capital-letter alleles are present, and whitish blue pigmentation is seen if these thresholds are not met. If an *AA BB CC DD EE* animal is crossed to an *aa bb cc dd ee* animal and the progeny are intercrossed, what kinds of phenotypes are expected in the $F_1$ and $F_2$?

***22.15** Alzheimer disease (AD) is the leading cause of dementia in older adults. Evidence that genetic alterations are involved in AD comes from three sources; the incidence of AD in first-degree relatives, the incidence in pairs of twins, and pedigree analysis. There is a 24–50% risk of AD by age 90 in first-degree relatives of individuals with AD, a 40–50% risk of AD in the identical (monozygotic) twin of an individual with AD, and a 10–50% risk of AD in the fraternal (dizygotic) twin of an individual with AD. Individuals with AD in a subset of families showing AD have an alteration in the *APP* (amyloid protein) gene on chromosome 21. Individuals with AD in another subset of AD families have a particular allele (*E4*) at the *APOE* (apolipoprotein E) gene on chromosome 19. Individuals homozygous for the *E4* allele have increased risk of AD and earlier disease onset than heterozygotes. Population studies have shown that 40–50% of AD cases are associated with alterations in the *APOE* gene, but less than 1% of AD cases are associated with mutations in the *APP* gene.

**a.** In what sense might AD be considered a polygenic trait?

**b.** If AD has a genetic basis, why are identical twins not equally affected?

**22.16** Since monozygotic twins share all their genetic material and dizygotic twins share, on average, half of their genetic material, twin studies sometimes can be useful for evaluating the genetic contribution to a trait. Consider the following two instances:

An intelligence quotient (IQ) assesses intellectual performance on a standardized test that involves reasoning, ability, memory, and knowledge of an individual's language and culture. IQ scores are transformed so that the population mean score is 100, and 95% of the individuals have scores in the range between 70 and 130. Observations in the United States and England found that monozygotic twins had an average difference of 6 IQ points, dizygotic twins had an average difference of 11 points, and random pairs of individuals had an average difference of 21 points.

In a large sample of pairs of twins in the United States where one twin was a smoker, 83% of monozygotic twins both smoked, whereas 62% of dizygotic twins both smoked.

From these data, can you infer the genetic determination of IQ or smoking?

**22.17** A quantitative geneticist determines the following variance components for leaf width in a population of wildflowers growing along a roadside in Kentucky:

Additive genetic variance ($V_A$) = 4.2
Dominance genetic variance ($V_D$) = 1.6
Interaction genetic variance ($V_I$) = 0.3
Environmental variance ($V_E$) = 2.7
Genetice–environmental variance ($V_{G \times E}$) = 0.0

**a.** Calculate the broad-sense heritability and the narrow-sense heritability for leaf width in this population of wildflowers.

**b.** What do the heritabilities obtained in (**a**) indicate about the genetic nature of leaf width variation in this plant?

***22.18** Domesticated house cats (*Felis silvestris catus*) are thought to have descended from a group of wildcats (*Felis silvestris*). Wildcats still can be found in Europe, the western part of Asia, and Africa, where they live solitary lives on territories of about 3 square kilometers. Which is more likely to have a higher heritability for size, a natural population of wildcats residing in a similar type of geographical environment or a domesticated population of house cats? Explain the reasoning behind your answer.

***22.19** Members of the inbred rat strain SHR are salt sensitive: they respond to a high-salt environment by developing hypertension. Members of a different inbred rat strain, TIS, are not salt sensitive. Imagine that you placed a population consisting only of SHR rats in an environment that was variable in regard to distribution of salt, so that some rats would be exposed to more salt than others. What would be the heritability of blood pressure in this population?

**22.20** In Kansas, a farmer is growing a variety of wheat called TK138. He calculates the narrow-sense heritability for yield (the amount of wheat produced per acre) and finds that the heritability of yield for TK138 is 0.95. The next year, he visits a farm in Poland and observes that another variety of wheat, UG334, growing there has only about 40% as much yield as the TK138 grown on his farm in Kansas. Since he found the heritability of yield in his wheat to be very high, he concludes that the TK138 wheat is genetically superior to the UG334 wheat, and he tells the Polish farmers that they can increase their yield by using TK138. Is his conclusion correct? Why or why not?

*__22.21__ Dermatoglyphics are the patterns of ridged skin found on the fingertips, toes, palms, and soles of the feet. (Fingerprints are dermatoglyphics.) Classification of dermatoglyphics frequently is based on the number of triradii: a triradius is a point from which three ridge systems separate at angles of 120°. The number of triradii on all 10 fingers was counted for each member of several families, and the results are tabulated here.

| Family | Mean Number of Triradii in the Parents | Mean Number of Triradii in the Offspring |
|---|---|---|
| I | 14.5 | 12.5 |
| II | 8.5 | 10.0 |
| III | 13.5 | 12.5 |
| IV | 9.0 | 7.0 |
| V | 10.0 | 9.0 |
| VI | 9.5 | 9.5 |
| VII | 11.5 | 11.0 |
| VIII | 9.5 | 9.5 |
| IX | 15.0 | 17.5 |
| X | 10.0 | 10.0 |

**a.** Calculate the narrow-sense heritability for the number of triradii by the regression of the mean phenotype of the parents against the mean phenotype of the offspring.

**b.** What does your calculated heritability value indicate about the relative contributions of genetic variation and environmental variation to the differences observed in number of triradii?

**22.22** The heights of nine college-age males and the heights of their fathers are presented here.

| Height of Son (inches) | Height of Father (inches) |
|---|---|
| 70 | 70 |
| 72 | 76 |
| 71 | 72 |
| 64 | 70 |
| 66 | 70 |
| 70 | 68 |
| 74 | 78 |
| 70 | 74 |
| 73 | 69 |

**a.** Calculate the mean and the variance of height for the sons and for the fathers.

**b.** Calculate the correlation coefficient for the relationship between the height of father and height of son.

**c.** Determine the narrow-sense heritability of height in this group by regression of the son's height on the height of father.

*__22.23__ A scientist wants to determine the narrow-sense heritability of tail length in mice. He measures tail length among the mice of a population and finds a mean tail length of 9.7 cm. He then selects the 10 mice in the population with the longest tails: mean tail length in these selected mice is 14.3 cm. He interbreeds the mice with the long tails and examines tail length in their progeny. The mean tail length in the $F_1$ progeny of the selected mice is 13 cm.

Calculate the selection differential, the response to selection, and the narrow-sense heritability for tail length in these mice.

**22.24** Assume that all phenotypic variance in seed weight in beans is genetically determined and is additive. From a population in which the mean seed weight was 0.88 g, a farmer selected two seeds, each weighing 1.02 g. He planted these and crossed the resulting plants to each other, then collected and weighed their seeds. The mean weight of their seeds was 0.96 g. What is the narrow-sense heritability of seed weight?

**22.25** The narrow-sense heritability of egg weight in a particular flock of chickens is 0.60. A farmer selects for increased egg weight in this flock. The difference in the mean egg weight of the unselected chickens and the selected chickens is 10 g. How much should egg weight increase in the offspring of the selected chickens?

**22.26** Members of a strain of white leghorn chickens are selectively crossed to produce two lines, A and B, that show improved egg production. The progeny from a cross of lines A and B are used for commercial egg production. The selection strategy is shown in Figure 22.A. The mean number of eggs produced in the first egg production year and the mean egg weight (in grams) from hens at an age of 240 days is given for animals at each step of the selection procedure.

assess whether any of the markers were linked to a locus contributing to the malting quality phenotype, a lod score—the $\log_{10}$ of the odds of linkage ratio (see Chapter 14, p. 416)—was calculated for each marker. Figure 22.B shows a plot of the lod score statistic (calculated at $\theta = 0$ using the malting quality data gathered for the 149 lines grown in Montana) against the relative chromosomal position of the markers.

**a.** There are four regions in the graph shown in Figure 22.B where the lod score rises above 4.0. How do you interpret a lod score of 4.0?

**b.** In general, what can you infer about the relative location of QTLs for malting extract quality on barley chromosome IV? More specifically, are there specific molecular markers that are associated with QTLs? Does each of the lod score regions above 4.0 necessarily identify a genome location that harbors a different QTL? Does each peak necessarily correspond to a single QTL?

**c.** When a lod score statistic for each marker is calculated using the malting quality data gathered for the 149 lines grown in Washington, Idaho, or Oregon, and lod scores are plotted against chromosomal position, three similarly shaped plots are obtained. However, the lod scores are lower, sometimes considerably. Why might this occur? Does it affect your general interpretation of where QTLs for malting quality lie on barley chromosome IV?

Figure 22.B

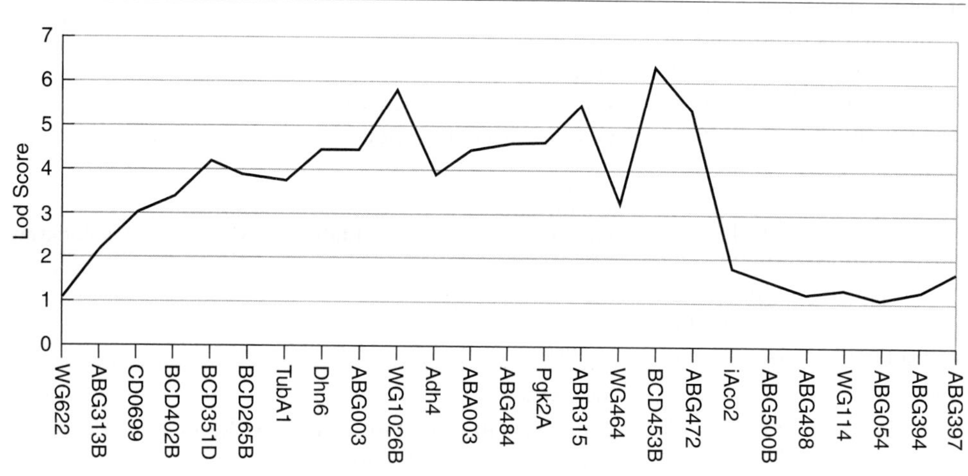

# 23 Molecular Evolution

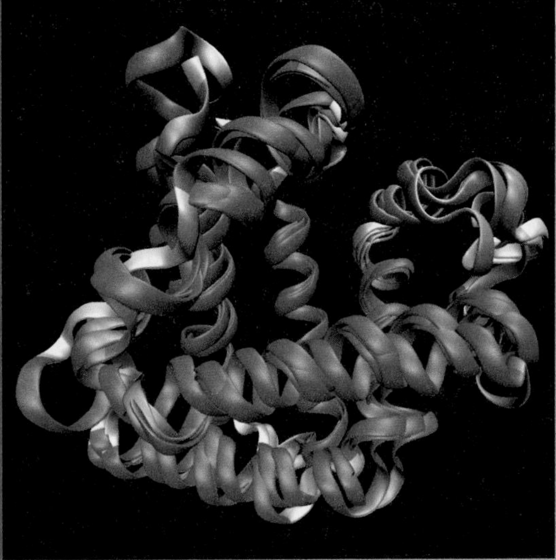

Superimposed globin chains from a human, a sperm whale, a chicken, and a soybean, showing conserved structures in purple and variations in other colors.

## Key Questions

- What effect does natural selection have upon the changes that are observed in molecules over time?
- Why do some genes accumulate changes at a faster rate than others?
- What portions of genes accumulate changes at the fastest rate?
- How can molecular data be used to infer phylogenetic relationships?
- How do proteins with new biological functions arise?

### iActivity

IN A HOT, HARSH REGION OF THE DESERT, YOU sift carefully through the sediment looking for fossils. Suddenly, you glimpse a fragment of bone. Eventually, you unearth a leg bone and part of a jaw, which lab analyses show to be those of an ancient hominid. Is this the distant predecessor of modern humans or merely an evolutionary cousin that became extinct hundreds of thousands of years ago? How could you find out? In this chapter, you will learn how population geneticists can apply molecular genetic techniques to answer questions about how species evolve. Then, in the iActivity, you will have the opportunity to use some of the same tools and techniques to determine whether we are the direct descendants of Neanderthals.

While individuals are the entities affected by natural selection, it is populations and genes that change over evolutionary time scales. **Molecular evolution** is evolution at the molecular level of DNA and protein sequences. The study of molecular evolution uses the theoretical foundation of population genetics to address two essentially different sets of questions: How do DNA and protein molecules evolve, and how are genes and organisms evolutionarily related?

Aside from the differences in the questions asked, population genetics and molecular evolution differ primarily in the time frame of their perspective. Population genetics (see Chapter 21) focuses on the changes in gene frequencies that occur from generation to generation, whereas molecular evolution typically considers the much longer time frames associated with speciation. Very small departures from the conditions needed to maintain Hardy–Weinberg equilibrium have small effects on gene frequencies in the short term but can take on great significance on evolutionary time scales. Moreover, random effects such as those associated with small amounts of sampling error along with exceptionally small differences in fitness tend to become the predominant process of genomic change when applied cumulatively over hundreds or thousands of generations.

The field of molecular evolution is multidisciplinary. It routinely invokes data and insights from genetics, ecology, evolutionary biology, statistics, and even computer science. However, before the widespread development of the tools of molecular biology in the 1970s and 1980s, researchers interested in the study of how biologically important molecules change over time had little data available to study. The ability to clone, sequence, and hybridize DNA removed the species barrier in population genetics studies. It also opened a window on a world that had been only dimly perceived where genes evolve by the accumulation of **mutations** (see Chapter 7), **transposition** (see Chapter 7), **duplication** (see Chapter 16), and **gene conversion** (a meiotic process of directed change in which one allele directs the conversion of a homologous allele on a sister chromosome to its own form). For the first time, studies of evolution had an abundance of parameters that could be measured and theories that could be tested. Molecular analyses made it clear that genomes are historical records that can be unraveled to identify the dynamics behind evolutionary processes and to reconstruct the chronology of change. The same approaches that allow these documents of evolutionary history to be put in order and deciphered also facilitate classification of the living world in true **phylogenetic relationships**—the hierarchical genealogical relationships between separated populations or species—across the vast distances of evolutionary time. Previously unimagined relationships between organisms became apparent, and even the kingdoms of life at the root of all systematics had to be rearranged.



In this chapter, we present the principles of molecular evolution studies and show how molecules with new functions arise and how the phylogenetic relationships of molecules and organisms are determined.

## Patterns and Modes of Substitutions

### Nucleotide Substitutions in DNA Sequences

**Substitutions in Protein and DNA Sequences.** An important question in the study of evolution is how the patterns and rates of substitution differ between different parts of the same gene. Studies addressing this question began in earnest in the 1970s and 1980s, when the best molecular data available came from the amino acid sequences in proteins. It quickly became apparent that some amino acid differences were more likely to be observed between two **homologous** proteins—proteins in different species that share a common ancestor—than were others. Specifically, amino acids were most likely to be replaced with amino acids that had similar chemical characteristics (see the groupings in Figure 6.2, p. 104) to those of the amino acid present in the ancestral protein. This replacement bias supported two evolutionarily important principles: (1) mutations are rare events; and (2) most dramatic alterations are removed from the gene pool by natural selection.

Chemically similar amino acids tend to have similar codons (see Figure 6.7, p. 108), and fewer changes at the level of DNA are required to change one into another. For example, a leucine codon (i.e., CUU) can be changed to an isoleucine codon (i.e., AUU) by a single base-pair change at the DNA level. In contrast, two base-pair changes must occur to convert that same leucine codon to a codon for the chemically dissimilar asparagine (i.e., AAU). DNA polymerase error rates are typically described in terms of errors per *millions* or *billions* of replicated nucleotides. Since nucleotide and amino acid changes are such rare events, scenarios that invoked the fewest numbers of changes were most likely to actually occur. At the same time, natural selection acting over many generations has caused most proteins to have amino acid sequences that make them very well-suited for their particular role and present environment. The more substantial an alteration to a protein's primary structure was, the more likely it was to have had a deleterious effect on its function that would not escape the scrutiny of natural selection.

**Sequence Alignments.** Analyzing changes at both the amino acid and nucleotide levels between two or more gene sequences begins with an alignment of all the sequences to be studied. Pairwise alignments between short, highly similar sequences can be done by hand, but computer programs are typically used to make more difficult alignments. Alignments represent specific hypotheses about the evolution of two or more sequences, and the best possible alignment reflects the true ancestral relationship at each position in the sequence of a protein or gene. For instance, the alignment of two sequences

```
G T A C C T
G - A T C T
```

can be interpreted to mean the following:

1. Four of the six nucleotide positions (first, third, fifth, and sixth from the left) have not undergone any change since the pair of sequences last shared a common ancestor.
2. A substitution has occurred at one site (the fourth from the left).
3. An insertion or deletion occurred in one of the sequences (second position).

Rapid divergence or long periods of evolution often leave little in common between sequences and can make generating alignments difficult. As a result, most studies are based on approximations of the true alignment, called **optimal alignments,** in which gaps are inserted to maximize the similarity between the sequences being aligned. Gaps in alignments are necessary because gene sequences are altered not just by changes to individual nucleotides (point mutations) but also by even less common insertion and deletion events. Since it is often difficult or impossible to distinguish an insertion in one sequence from a deletion in another, such gaps often are called **indels.** The number of possible alignments between two or more sequences of even modest length is enormous. Determining which alignment is optimal typically is left to computer algorithms that seek to maximize the number of matching amino acids or nucleotides between the sequences while invoking the smallest number of indel events possible.

**Substitutions and the Jukes–Cantor Model.** After two nucleotide sequences diverge from each other, they begin to accumulate nucleotide substitutions independently. The number of substitutions per site ($K$) observed in an alignment between two sequences is almost always the single most important variable in any molecular evolution analysis. If an alignment suggests that few substitutions have occurred between two sequences, then a simple count of the substitutions usually is sufficient to determine the number of substitutions that have occurred. However, in 1969, even before the nucleotide sequences of any genes were available for analysis, T. Jukes and C. Cantor realized that alignments between sequences with many differences might cause a significant underestimation of the actual number of substitutions since the sequences last shared a common ancestor. Where substitutions were common, there were no guarantees that a particular site had not undergone multiple changes such as those illustrated in Figure 23.1. To address that possibility, Jukes and Cantor assumed that each nucleotide was just as likely to change into any other nucleotide. Using that assumption, they created a mathematical model in which the rate of change to any one of the three alternative nucleotides was assumed to be $\alpha$, and the overall rate of substitution for any given nucleotide was $3\alpha$. In that model, if a site within a gene was occupied by a C at time 0 ($t = 0$ in Figure 23.1), then the probability ($P$) that that site would still be the same nucleotide at time 1 ($t = 1$) would be $P_{C(1)} = 1 - 3\alpha$. At subsequent points in time (e.g., $t = 2$) the possibility of reversions (back mutation) to C must also be considered. Specifically, if the original C changed to another nucleotide (say, an "A") in that first time span, at time 2 ($t = 2$) the probability, $P_{C(2)}$, would be equal to $(1 - 3\alpha) P_{C(1)} + \alpha[1 - P_{A(1)}]$. Further expansion suggested that at any given time ($t$) in the future, the probability that the site would contain a C was defined by the equation:

$$P_{C(t)} = 1/4 + (3/4)e^{-4\alpha t}$$

All that remained was for molecular biologists to determine the value for $\alpha$, the rate at which nucleotide substitutions occurred.

When those data arrived 10 years later, it became clear that the model used by Jukes and Cantor was an oversimplification. For instance, **transitions** (exchanging one purine for the other or one pyrimidine for the other) were seen to accumulate at a different, faster rate than **transversions** (exchanging a purine for a pyrimidine, or vice versa). Even so, the Jukes–Cantor model provided a useful framework for taking into account the actual number of substitutions per site ($K$) when multiple substitutions were possible. By manipulating the equation derived from the Jukes–Cantor model, it is possible to determine that $K$ can be calculated as

$$K = -3/4 \ln(1 - (4/3)p)$$

where $p$ is the fraction of nucleotides that a simple count reveals to be different between two sequences. This equation is consistent with the idea that when two sequences have few mismatches between them, $p$ is small and the chance of multiple substitutions at any given site is also small (e.g., when $p$ for a stretch of 100 nucleotides is 0.02, $K = 0.02$). It also suggests that when the observed number of mismatches is large, the actual number of substitutions per site can be substantially larger than what is counted directly (e.g., when $p$ for a stretch of 100 nucleotides is 0.50, $K = 0.82$).

**Figure 23.1**

**Two possible scenarios in which multiple substitutions at a single site would lead to underestimation of the number of substitutions that had occurred if a simple count was performed. T = time.**

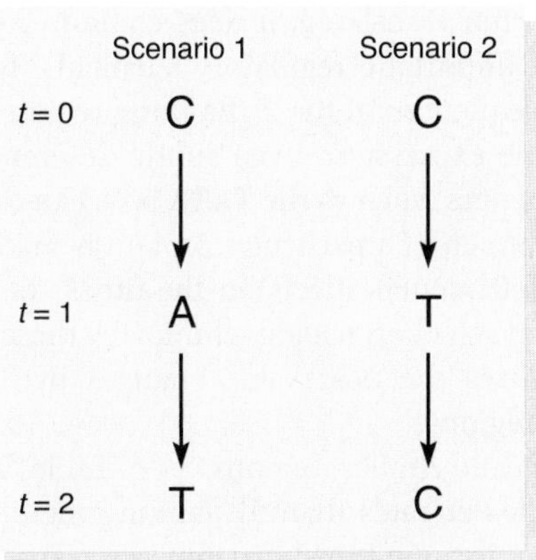


## Rates of Nucleotide Substitutions

The number of substitutions that two sequences have undergone since they last shared a common ancestor is a centrally important parameter to almost all molecular evolution analyses. When $K$ is expressed in terms of the number of substitutions per site and coupled with a divergence time ($T$), it is easily converted into a rate ($r$) of substitution. Because substitutions are assumed to accumulate simultaneously and independently in both sequences, the substitution rate is obtained by simply dividing the number of substitutions between two homologous sequences by $2T$, as shown in this equation:

$$r = K/(2T)$$

Note that to estimate substitution rates, those data must always be available from at least two species. Comparisons of substitution rates within and between genes can then give valuable insights into the mechanisms involved in molecular changes. And if evolutionary rates between several species are similar, substitution rates can give insights into the dates of evolutionary events for which no other physical evidence is available.

**Variation of Evolutionary Rates within Genes.** Studies of nucleotide sequences in numerous genes have revealed that different parts of genes evolve at widely differing rates that reflect the extent that they are subject to natural selection. Recall from our discussion of molecular genetics that a typical eukaryotic gene is made up of some nucleotides that specify the amino acid sequence of a protein (coding sequences) and other nucleotides that do not code for amino acids in a protein (noncoding sequences). Noncoding sequences include introns, leader regions, trailer regions (all of which are transcribed but not translated) and 5′ and 3′ flanking sequences that are not transcribed. Additional noncoding sequences include **pseudogenes,** which are nucleotide sequences that no longer produce functional gene products because they have accumulated inactivating mutations. Even within the coding regions of a functional gene, not all nucleotide substitutions produce a corresponding change in the amino acid sequence of a protein. In particular, many substitutions occurring at the third position of triplet codons have no effect on the amino acid sequence of the protein because such changes often produce synonymous codons, one that codes for the same amino acid (see Figure 6.7, p. 108).

**Synonymous and Nonsynonymous Sites.** Table 23.1 shows relative rates of change in different parts of mammalian genes. Within functional genes, notice that the highest rate of change involves synonymous changes in the coding sequences. The rate of synonymous nucleotide change is about five times greater than the observed rate of nonsynonymous changes. Synonymous changes do not alter the amino acid sequence of the protein. Thus, the high rate of evolutionary change seen there is not unexpected, because these changes do not affect a protein's functioning. Consider for a moment how variation in nucleotide sequences arise. All variation must arise either from errors in DNA replication or repair processes. The enzymes responsible for DNA replication and repair are in no way capable of distinguishing between synonymous and nonsynonymous changes to DNA sequences. As a result, synonymous and nonsynonymous mutations are likely to arise with equal frequency. However, nonsynonymous changes that arise within coding sequences often are detrimental to fitness and are eliminated by natural selection, whereas synonymous mutations usually are less detrimental, so they are tolerated. That raises an interesting and subtle distinction between the use of the words *mutation* and *substitution* in molecular evolution studies. **Mutations** are changes in nucleotide sequences that occur because of mistakes in DNA replication or repair processes (see Chapter 7). **Substitutions** are mutations that have passed through the filter of selection on at least some level. Synonymous substitution rates probably are fairly reflective of the actual mutation rate operating within a genome, whereas nonsynonymous substitutions rates are not.

**Table 23.1 Relative Rates of Evolutionary Change in DNA Sequences of Mammalian Genes**

| Sequence | Nucleotide Substitutions per Site per Year ($\times 10^{-9}$) |
|---|---|
| Functional genes | |
| 5′ flanking region | 2.36 |
| Leader | 1.74 |
| Coding sequence, synonymous | 4.65 |
| Coding sequence, nonsynonymous | 0.88 |
| Intron | 3.70 |
| Trailer | 1.88 |
| 3′ flanking region | 4.46 |
| Pseudogenes | 4.85 |

**Flanking Regions.** High rates of evolutionary change also occur in the 3′ flanking regions of functional genes (see Table 23.1). Like synonymous changes, sequences in the 3′ flanking regions have no effect on the amino acid sequence of a protein and usually have little effect on gene expression. Consequently, most substitutions that occur within a 3′ flanking region are tolerated by natural selection. Rates of change in introns are also high, but not as high as the synonymous changes and those in the 3′ flanking regions. Although the sequences in the introns usually are not used to spell out the sequence of amino acids in a protein, they must be properly spliced out for an mRNA to be translated into a functional protein. Some sequences within introns must be present for splicing to occur, including the 5′ and 3′ splice junctions and the branch point of the intron (see Chapter 5, pp. 93–94). And at least some introns occasionally code for protein coding regions in some tissues but not in others because of alternate splicing. As a result, not all changes in introns avoid detection by natural selection, but their overall rate of evolution is a bit lower than that seen in 3′ flanking regions and at synonymous sites within coding sequences.

Still lower rates of evolutionary change are seen in the 5′ flanking region. Although this region is neither transcribed nor translated, it does contain the promoter and other important regulatory elements for a gene; therefore, sequences in the 5′ flanking region are important for gene expression. Even subtle differences in promoter sequences, such as the TATA box, can dramatically affect how much of a particular protein is made and thus can have detrimental effects on the fitness of the organism. Natural selection usually eliminates these mutations and minimizes the observable changes in functionally important regions.

Leader and trailer regions (see Table 23.1) have somewhat lower rates than 5′ flanking regions. Leaders and trailers are not translated but are transcribed; and they provide important signals for processing and translation of the mRNA. Substitutions in these regions therefore are infrequent. Nonsynonymous substitutions in coding regions of genes occur least frequently, particularly those that significantly alter the amino acid sequence of a protein. As explained earlier, most proteins seem to have amino acid sequences that make them very well-suited for their particular role and environment, and most substitutions that cause departures from the current sequence are eliminated fairly quickly by natural selection.

**Pseudogenes.** The highest rate of evolution seen in Table 23.1 is that of nonfunctional pseudogenes. Among human globin pseudogenes, for example, the rate of nucleotide change is approximately five times that observed at the nonsynonymous sites within the coding sequence of functional globin genes. The high rate of evolution observed in these sequences occurs because pseudogenes no longer code for proteins. Since further changes in these genes do not affect an organism's fitness, changes are not eliminated through the process of natural selection. In summary, we usually observe what also makes intuitive sense: the stronger the functional constraints on a part (or the whole) of a macromolecule, the slower the rate of its evolution.

## Keynote

Rates of evolution vary between different portions of a gene. Sequences with the most functional importance to an organism, such as protein-binding sites within a promoter or nucleotides that would result in an amino acid substitution, evolve at the slowest rate. Pseudogenes are inactivated versions of genes that are no longer functionally constrained and tend to evolve at the fastest rate.

**Comparative Genomics.** The association between low evolutionary rate and high functional significance is particularly useful in **comparative genomics,** which entails comparisons of the sequences of entire genomes of different species. Any genome project like the Human Genome Project (see Chapter 8, p. 171) is an ambitious attempt to compile a kind of address list of where all the genes reside within an organism's chromosomes. Such a list or map is the foundation that one day may allow researchers to repair defective genes or add or replace a desired one that is missing or lost—thereby curing or preventing genetic-based diseases such as cystic fibrosis or even some kinds of cancer. However, having an address list for a large office building does not tell you what job each employee does; how (or whether) they work together or alone to accomplish tasks; and whether a worker's presence saves the company or sabotages it. Numerous genome projects have been generating large amounts of nucleotide sequence information, but more than 95% of the nucleotides within the genome of a complex organism do not appear to be functionally important. As a result, it is often difficult to tell which portions of a genome are associated with protein-coding regions and what role those proteins play. Comparison is proving to be one of the best ways to figure out the function of specific tracts of sequences within genomes. The discovery of a patch of significant sequence similarity between the genomes of two distantly related species is very suggestive of functional importance. For example, humans and mice last shared a common ancestor at least 80 to 100 million years ago. Given a pseudogene substitution rate of roughly $5 \times 10^{-9}$ changes per site per year, almost half of all nucleotides that are free of selective constraint should have undergone at least one change since humans and mice diverged. Regions under functional constraint, such as nonsynonymous sites within coding regions or protein-binding sites within the promoter of a gene, accumulate changes at one-fifth that rate or less and often are easy to recognize in pairwise comparisons. The functional consequences of an enormous number of changes are evaluated on evolutionary time scales through the process of natural selection. As a result, comparative genome analysis often takes the place of the experimentally challenging and time-consuming process of saturation mutagenesis, in which every position in a gene is mutated and the consequences are evaluated in a laboratory. Even better, narrowing down what an identified gene does in an organism as comparatively simple as yeast and then finding the same gene in human DNA greatly facilitates predicting the human version's function.

Comparative genomics is a powerful tool, but it does have limitations. Some human genes appear to have no counterparts in simpler organisms. And remember that many proteins have multiple functions—not all of which are shared by all their homologs.

**Codon Usage Bias.** The effect that even tiny differences in fitness can have when subject to natural selection over the course of thousands or millions of generations is evidenced by a phenomenon known as **codon usage bias.** Take note of the slightly lower rate of evolutionary change at synonymous sites relative to that of pseudogenes in Table 23.1. This observation suggests that synonymous substitutions are not completely neutral from the perspective of natural selection and that some triplet codons may be favored over others. This hypothesis is reinforced by the finding that synonymous codons are not used equally throughout the coding sequences of many organisms. For example, the redundancy of the genetic code allows six different codons (UUA, UUG, CUU, CUC, CUA, and CUG) to specify the amino acid leucine, but 60% of the leucine codons found in *Escherichia coli* are CUG and 80% in yeast are UUG. Since the alternative synonymous codons specify the very same amino acid, selection may be favoring some synonymous codons over others, or they would all be used equally. Remember that some synonymous codons pair with different tRNAs that carry the same amino acid. Therefore, a mutation to a synonymous codon does not change the amino acid, but it may change the tRNA used by a ribosome during translation. Studies of the different tRNAs reveal that within a cell, the amounts of the isoacceptor tRNAs (different tRNAs that accept the same amino acid) differ, and the most abundant tRNAs are those that pair with the most frequently used codons. Selection may favor one synonymous codon over another because the tRNA for the codon is more abundant, and translation of mRNAs containing that codon is more efficient (and even more accurate). Alternatively, the bonding energy between codon and anticodon of synonymous codons may differ slightly because different bases are paired. These extremely subtle differences in translation efficiency and bonding energy appear to be subject to natural selection. This is especially true in genes that are expressed at high levels and in organisms with short generation times and large population sizes, such as bacteria, yeast, and fruit flies, where codon usage is most apparent. The existence of codon usage bias is a profound testament to the sheer power of evolutionary processes. The difference in fitness between two bacteria that are identical in every way aside from a single synonymous codon (out of approximately 1 million within their genomes) must be infinitesimally small. Still, immeasurably small as it is, it is sufficient to result in only one of the two cells' ancestors being present for counting after evolutionary time scales have passed.

Again, the tendency to use preferred rather than less-used codons is most pronounced within the genes of bacterial organisms that are expressed at the highest levels. This makes sense because the genes that are expressed the most are also the most efficient and cost effective. A large company that overpays by 1% for something that it purchases thousands of times per year is likely to be worse off than a company that saves that 1% on the commonly purchased item but overpays by 20% on something that it purchases rarely. In fact, a gene's adherence to an organism's codon usage bias has proven to be one of the most reliable indicators of the relative amount of expression of the gene during the organism's lifetime. Given that this is the case, it should also be expected that highly expressed genes use energetically inexpensive amino acids in place of energetically expensive amino acids wherever possible. The energetic cost of synthesizing each of the 20 different amino acids actually varies greatly, with glycine requiring an investment of only 11.7 adenosine triphosphate (ATP) equivalents and tryptophan requiring an average of 78.3 ATP equivalents. There are some places within a protein's primary sequence where only a tryptophan will do, but wherever it is possible to substitute a glycine (or delete the residue all together), natural selection should lead to the change—especially in highly expressed genes where the cost savings would have the greatest effect. Analyses of the hundreds of completely sequenced prokaryotic genomes suggest that this is in fact the case (results for three prokaryotes are shown in Table 23.2).

**Keynote**

Even subtle differences in fitness, such as those associated with how efficiently one of several synonymous codons is read by ribosomes, can significantly affect the evolution of molecules after many generations of selection.

### Variation in Evolutionary Rates between Genes

Just as variation in evolutionary rates is readily apparent in comparisons of different regions within genes, striking differences in the rates of evolution between genes have been observed, even when genes from the same species are considered. As before, if stochastic factors (such as sampling distortions that arise from small population sizes) are ruled out, the difference must be attributable to one or some combination of two factors: (1) differences in mutation frequency; and (2) the extent to which natural selection affects the locus. Clever statistical analyses can aid in distinguishing between adaptive and random changes in nucleotide sequences. For example, the McDonald–Kreitman test, developed in 1991, compares the patterns of within-species polymorphism and between-species divergence at the synonymous and nonsynonymous sites in the coding region of a gene. If the ratio of nonsynonymous to synonymous substitutions within species is the same as that between species, then all the substitutions are likely to be neutral. If the ratios are not the same, then natural selection must be responsible—most likely by favoring nonsynonymous mutations that confer an adaptive advantage, although diversifying selection can also occur. Some regions of genomes seem to be more prone to random changes than are other regions, but synonymous substitution rates across a genome rarely differ by more than a factor of two. That difference is far from sufficient to account for the roughly thousandfold difference in nonsynonymous substitution rates observed between different classes of mammalian genes shown in Table 23.3. As was the case for variation in observed substitution rates within genes, variation of substitution rates between genes must result largely from differences in the intensity of natural selection at each locus.

Specific examples of two classes of genes, histones and apolipoproteins, illustrate the effects of different levels of functional constraint. Histones are positively charged, essential DNA binding proteins that are present in all eukaryotes. Almost every amino acid in a histone such as histone H4 interacts directly with specific

**Table 23.2 Amino Acid Utilization in the 10% of Genes with the Highest Expression Levels and the 10% with the Lowest Expression Levels in Three Different Prokaryotic Organisms[a]**

| | | | *Escherichia coli K12* | | | *Streptococcus pneumoniae R6* | | | *Bacillus subtilis subsp. subtilis str. 168* | | |
|---|---|---|---|---|---|---|---|---|---|---|---|
| **Amino Acid** | **Cost** | **Difference from Average** | **Low** | **High** | **Difference in Usage** | **Low** | **High** | **Difference in Usage** | **Low** | **High** | **Difference in Usage** |
| Gly | 11.7 | −15.66 | 6.40 | 8.16 | 1.76 | 6.01 | 8.30 | 2.29 | 6.12 | 8.11 | 1.99 |
| Ser | 11.7 | −15.66 | 6.65 | 4.99 | −1.67 | 6.47 | 4.96 | −1.51 | 6.24 | 5.84 | −0.40 |
| Ala | 11.7 | −15.66 | 8.48 | 9.76 | 1.27 | 5.85 | 9.80 | 3.95 | 7.40 | 8.70 | 1.29 |
| Asp | 12.7 | −14.66 | 4.81 | 5.95 | 1.15 | 4.85 | 6.04 | 1.20 | 5.06 | 5.42 | 0.36 |
| Asn | 14.7 | −12.66 | 4.45 | 3.92 | −0.52 | 3.64 | 4.60 | 0.96 | 3.32 | 4.24 | 0.92 |
| Glu | 15.3 | −12.06 | 5.23 | 7.10 | 1.87 | 6.54 | 7.91 | 1.37 | 6.71 | 8.11 | 1.40 |
| Gln | 16.3 | −11.06 | 4.48 | 3.83 | −0.66 | 4.43 | 3.33 | −1.10 | 3.98 | 3.46 | −0.52 |
| Thr | 18.7 | −8.66 | 5.41 | 5.22 | −0.19 | 4.78 | 5.97 | 1.19 | 4.92 | 5.82 | 0.89 |
| Pro | 20.3 | −7.06 | 4.22 | 4.08 | −0.15 | 3.17 | 3.72 | 0.55 | 3.87 | 3.62 | −0.25 |
| Val | 23.3 | −4.06 | 6.24 | 7.64 | 1.39 | 6.54 | 8.13 | 1.59 | 6.59 | 7.94 | 1.35 |
| Cys | 24.7 | −2.66 | 1.50 | 0.92 | −0.57 | 0.88 | 0.33 | −0.54 | 1.02 | 0.52 | −0.50 |
| Arg | 27.3 | −0.06 | 5.38 | 5.45 | 0.07 | 4.61 | 4.21 | −0.39 | 4.71 | 3.99 | −0.72 |
| Leu | 27.3 | −0.06 | 11.56 | 8.96 | −2.59 | 12.70 | 7.97 | −4.73 | 10.85 | 8.40 | −2.45 |
| Lys | 30.3 | 2.94 | 4.34 | 5.66 | 1.32 | 5.96 | 7.16 | 1.19 | 6.25 | 7.55 | 1.30 |
| Ile | 32.3 | 4.94 | 6.72 | 6.03 | −0.69 | 7.74 | 6.41 | −1.33 | 6.79 | 7.14 | 0.34 |
| Met | 34.3 | 6.94 | 2.39 | 2.70 | 0.31 | 2.29 | 2.07 | −0.22 | 2.65 | 2.28 | −0.37 |
| His | 38.3 | 10.94 | 2.50 | 2.05 | −0.45 | 2.21 | 1.61 | −0.60 | 2.74 | 1.80 | −0.93 |
| Tyr | 50 | 22.64 | 3.15 | 2.82 | −0.34 | 4.44 | 2.96 | −1.48 | 3.94 | 2.85 | −1.09 |
| Phe | 52 | 24.64 | 4.35 | 3.71 | −0.64 | 5.76 | 3.69 | −2.07 | 5.56 | 3.53 | −2.04 |
| Trp | 74.3 | 46.94 | 1.73 | 1.05 | −0.69 | 1.13 | 0.81 | −0.32 | 1.27 | 0.67 | −0.60 |

[a]Amino acids are listed in terms of ascending cost of biosynthesis (shown in terms of required average number of high-energy phosphate bonds from ATP to synthesize). Shown for each organism are the fraction of amino acids in the primary sequence of 10% of genes expressed at the lowest level and that of the 10% of genes expressed at the highest level.

Modeled after the work of H. Akashi and T. Gojobori, 2002. Metabolic efficiency and amino acid composition in the proteomes of *Escherichia coli* and *Bacillus subtilis*. *Proc. Natl. Acad. Sci. USA*, 99:3695–3700.

**Table 23.3 Relative Rates of Evolutionary Change in DNA Sequences of Different Mammalian Genes[a]**

| Gene | Nonsynonymous Rate | Synonymous Rate |
|---|---|---|
| Histone H4 | 0.004 | 1.43 |
| Insulin | 0.16 | 5.41 |
| Prolactin | 1.29 | 5.59 |
| α-Globin | 0.56 | 3.94 |
| β-Globin | 0.87 | 2.96 |
| Albumin | 0.92 | 6.72 |
| α-Fetoprotein | 1.21 | 4.90 |
| MHC | 5.10 | 2.40 |
| Apolipoprotein E | 0.98 | 4.04 |

[a]All rates are in nucleotide substitutions per site per year $\times 10^{-9}$.

chemical residues associated with negatively charged DNA. Thus any change to the amino acid sequence of histone H4 affects its ability to interact with DNA. As a result, histones are one of the slowest-evolving groups of proteins known, and it is possible to replace the yeast version of histone H4 with its human homolog with no effect despite hundreds of millions of years of independent evolution. Apolipoproteins, in contrast, are responsible for nonspecifically interacting with and carrying a wide variety of lipids in the blood of vertebrates. Their lipid-binding domains are made up predominantly of hydrophobic amino acids. Any similar amino acid (e.g., leucine, isoleucine, and valine) appears to function in those positions just as well as another as long as it too is hydrophobic. As a result, dozens of different versions (at the protein level) of human apolipoproteins have been characterized to date at the same time that only single versions of human histone H4 proteins have been identified.

### Keynote

The rate at which genes accumulate changes generally reflects the extent to which they are functionally constrained. Synonymous substitution rates are almost always much higher than nonsynonymous rates in genes that are subject to high levels of natural selection.

Although amino acid substitutions within many genes generally are deleterious, it should be pointed out that natural selection favors variability within populations for some genes. For instance, it is advantageous to mammals for there to be variation in the genes associated with the major histocompatability complex (MHC). As a result, the rate of nonsynonymous substitutions within the MHC is greater than that of synonymous substitutions (see Table 23.3). The MHC is a large multigene family whose protein products are involved with the immune system's ability to recognize foreign antigens. Within human populations, roughly 90% of individuals receive different sets of MHC genes from their parents, and a sample of 200 individuals can be expected to have 15 to 30 different alleles. Such high levels of diversity in this region are favored by natural selection because the number of individuals vulnerable to infection by any single virus (one that is not recognized by MHC proteins) is likely to be substantially less than it would have been if they all had similar immune systems. At the same time that host populations are driven to maintain diverse immune systems, viruses are driven to evolve rapidly. Error-prone replication, coupled with diversifying selection, causes the rate of nucleotide substitutions within the influenza *NS* genes to be $1.9 \times 10^{-3}$ per site per year, or roughly 1 million times greater than the synonymous substitution rate for mammalian genes in Table 23.1.

## Rates of Evolution in Mitochondrial DNA

Organelle genomes are transmitted in a manner distinct from that of nuclear genes, and the dynamics of their substitutions are substantially different as a result. In Chapter 13, we discussed the structure and function of the mitochondrial genome. The mammalian mitochondrial genome consists of a circular, double-stranded mitochondrial DNA (mtDNA) about 15,000 base pairs long. Human mtDNA is fairly typical and, at roughly 1/10,000 the size of the nuclear genome, encodes 2 rRNAs, 22 tRNAs, and 13 proteins. Changes in the proteins, tRNAs, and rRNAs encoded by the mitochondrial genome appear to be less detrimental to individual fitness than are changes in the proteins, tRNAs, and rRNAs encoded by nuclear genes. As a result, the average synonymous substitution rate in mammalian mitochondrial genes is approximately $5.7 \times 10^{-8}$ substitutions per site per year, about 10 times the average value for synonymous substitutions in nuclear genes. The small size of mtDNA and the discovery of its exceptionally high rate of substitution have stimulated substantial interest in its evolution.

Mammalian mtDNA also differs from nuclear DNA in that the overwhelming majority of mtDNA is inherited clonally from the mother (see Chapter 13, p. 386). Mitochondria are located in the cytoplasm, and only the mother's egg cell contributes cytoplasm to a zygote. Consequently, mtDNA does not undergo meiosis, and all offspring should be identical to the maternal genotype for mtDNA sequences (the offspring are clones for mtDNA genes). This pattern of inheritance allows matriarchal lineages (descendants from one female) to be traced and provides a means for examining family structure in some populations. This, in conjunction with the rapid and regular rate of accumulation of nucleotide sequence differences, has allowed mtDNA to become a valuable tool for comparing closely related lineages.

## Molecular Clocks

As described earlier, the differences in the nucleotide and amino acid replacement rates between nuclear genes can be striking but are likely to result primarily from differences in the selective constraint on each individual protein. However, rates of molecular evolution for loci with similar functional constraints can be relatively constant over long periods of evolutionary time.

In fact, the very first comparative studies of protein sequences performed by Emile Zuckerkandl and Linus Pauling in the 1960s suggested that substitution rates were essentially constant within homologous proteins over many tens of millions of years. Based on these observations, the accumulation of amino acid changes occurs in a way analogous to the steady ticking of a clock, they proposed the **molecular clock hypothesis** which states that the molecular clock may run at different rates in different proteins, but the number of differences between two homologous proteins appeared to be very well correlated with the amount of time since speciation caused them to diverge independently, as shown in Figure 23.2. This observation immediately stimulated intense interest in using biological molecules in evolutionary studies. Steady rates of change between homologous sequences would facilitate not only the determination of phylogenetic relationships between species but also the times of their divergence in much the same way that radioactive decay was used to date geological times.

Despite its great promise, however, a number of research results do not support Zuckerkandl and Pauling's molecular clock hypothesis. Classic evolutionists argued that the erratic tempo of morphological evolution was inconsistent with a steady rate of molecular change. Disagreements regarding divergence times have also placed

**Figure 23.2**

**The molecular clock runs at different rates in different proteins.** One reason is that the neutral substitution rate differs among proteins. Fibrinogen appears to be unconstrained and has a high neural substitution rate, whereas cytochrome *c* has a lower neutral substitution rate and may be more constrained. Data are from a wide variety of organisms.

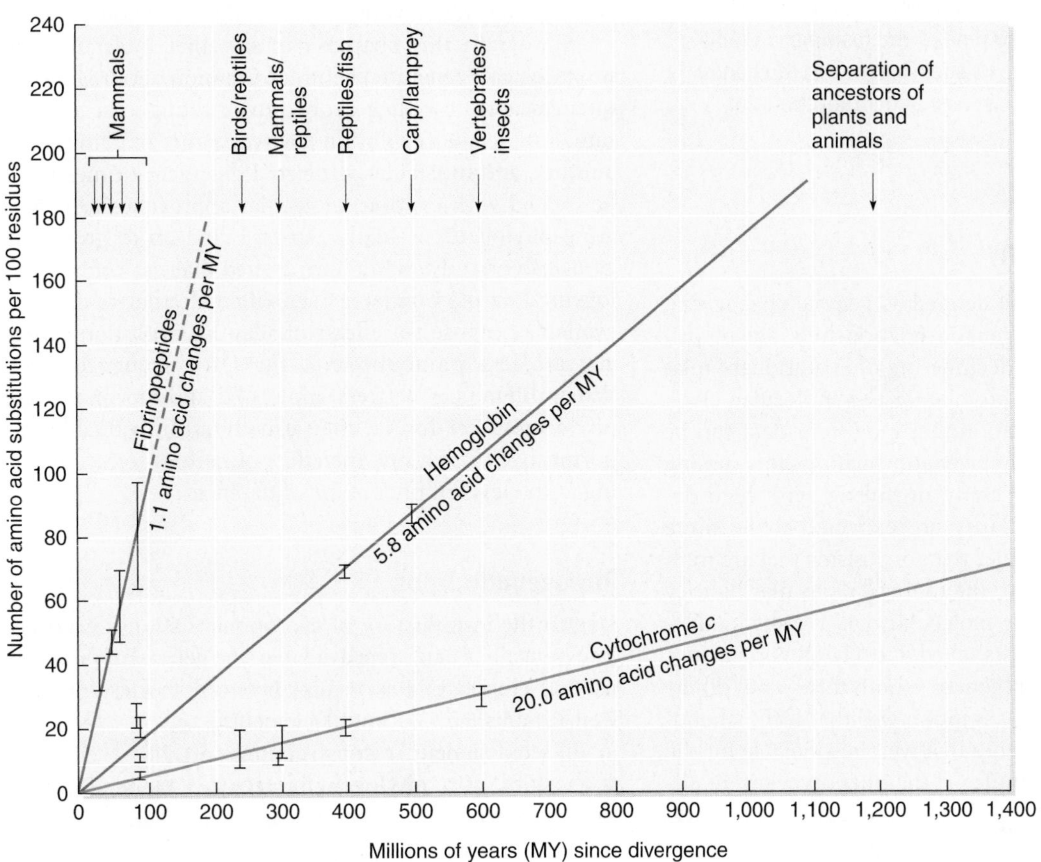


in question the uniformity of evolutionary rates at the heart of the idea.

**Variation in Rates.** Most divergence dates used in molecular evolution studies come from interpretations of the notoriously incomplete fossil record and are of questionable accuracy. Exponentially increasing amounts of DNA sequence data from a very wide variety of species are now available for testing the premise of the molecular clock, namely, that the rate of evolution for any given gene is constant over time in all evolutionary lineages. Substitution rates in rats and mice have been found to be largely the same. In contrast, molecular evolution in humans and apes appears to have been only half as rapid as that which has occurred in Old World monkeys since their divergence. Indeed, analyses of homologous genes in rats and humans suggest that rodents have accumulated substitutions at twice the rate of primates since they last shared a common ancestor during the time of the mammalian radiation 80 to 100 million years ago. The rate of the molecular clock clearly varies among taxonomic groups, and such departures from constancy of the clock rate pose a problem in using molecular divergence to date the times to recent common ancestors. Before such inferences can be made, it is necessary to demonstrate that the species being examined have a uniform clock such as the one observed within rodents.

Several possible explanations have been put forward to account for differences in evolutionary rates between lineages. For instance, generation times in monkeys are shorter than they are in humans, and the generation time of rodents is much shorter still. The number of germ-line DNA replications, occurring once per generation, should be more closely correlated with substitution rates than with simple divergence times. Differences may also result in part from a variety of other differences between two lineages since the time of their divergence, such as average repair efficiency, average exposure to mutagens, and the opportunity to adapt to new ecological niches and environments. The difficulty associated with estimating the extent to which these variables have affected evolutionary rates can sometimes be reduced by using independent dating information from the fossil record to calibrate molecular clocks for species within certain groups.

**Keynote**

Relative rate tests suggest that substitutions do not always accumulate at the same rate in different evolutionary lineages. Primates, humans in particular, appear to be evolving more slowly at a molecular level than are other mammals since the time of the mammalian radiation 80 to 100 million years ago. Faster rates of change may result from shorter average generation times or from a variety of other factors.

## Molecular Phylogeny

Because evolution can be defined as genetic change that takes place over time, genetic relationships are of primary importance in the deciphering of evolutionary relationships. The greatest promise of the molecular clock hypothesis is the implication that molecular data can be used to decipher the phylogenetic relationships among all living things. Quite simply, organisms with high degrees of molecular similarity are expected to be more closely related to each other than organisms that are molecularly dissimilar. Before the tools of molecular biology were available to provide molecular data for such analyses, evolutionary biologists relied entirely on comparison of phenotypes to infer genetic similarities and differences. The underlying assumption was that if the phenotypes were similar, the genes that were responsible for the phenotypes were also similar; if the phenotypes were different, the genes were different. Originally the phenotypes examined consisted largely of gross anatomical features. Later, behavioral, ultrastructural, and biochemical characteristics were also studied. Comparisons of such traits were used successfully to construct evolutionary trees for many groups of plants and animals and are still the basis of many evolutionary studies today.

However, relying on the study of such traits has limitations. Sometimes similar phenotypes can evolve in organisms that are distantly related, in a process called convergent evolution. For example, if a naive biologist tried to construct an evolutionary tree based on whether wings were present or absent in an organism, he might place birds, bats, and insects in the same evolutionary group because all have wings. In this particular case, it is fairly obvious that these three organisms are not closely related; they differ in many features other than the possession of wings, and the wings themselves are very different in their structure. But this extreme example shows that phenotypes can be misleading about evolutionary relationships, and phenotypic similarities do not necessarily reflect genetic similarities.

Another problem with relying on phenotypes to determine evolutionary relationships is that many organisms do not have easily studied phenotypic features suitable for comparison. For example, the study of relationships among bacteria has always been problematic because bacteria have few obvious traits that correlate with the degree of their genetic relatedness. A third problem arises when we try to compare distantly related organisms. What phenotypic features should be compared, for example, in an analysis of bacteria and mammals where so few characteristics are held in common?

Earlier in this chapter, we saw that molecular approaches can generate useful information about DNA sequences and how they evolve. Even though the relative rate of molecular evolution may vary from one lineage to another, and molecularly inferred divergence times must be treated with caution, molecular approaches to generating phylogenies usually can be relied on to group organisms correctly. Many have argued that molecular phylogenies are more reliable even when alternative data are available because the effects of adaptive evolution generally are less pronounced at the DNA sequence level. When differences between molecular and morphological phylogenies are found, they usually create valuable opportunities to examine the effect of natural selection acting at the level of phenotypic differences.

### Phylogenetic Trees

Due to the long history of evolutionary studies even before molecular data were available, the general approaches used to elucidate relationships between species are fairly well established. A central element in all phylogenetic reconstructions is the idea of a **phylogenetic tree** that graphically describes the relationship among different species. All living things on Earth, both in the present and in the past, share a single, common ancestor that lived roughly 4 billion years ago. Every phylogenetic tree portrays at least some portion of that ancestry with *branches* that connect two (occasionally more) adjacent *nodes*. Terminal nodes indicate the taxa for which molecular information has been obtained for analysis. Internal nodes represent common ancestors before the branching that gave rise to two separate groups of organisms. Branch lengths often are scaled to reflect the amount of divergence between the taxa they connect. Where it is possible to distinguish one internal node as representing a common ancestor to all the other nodes on a tree, it is possible to make a **rooted tree.** Unrooted trees specify only the relationship between nodes and say nothing about which nodes are more ancient. Roots for unrooted trees usually can be determined through the use of an outgroup. Outgroups are taxa that have unambiguously separated the earliest from the other taxa being studied. In the case of humans and gorillas, when baboons are used as an outgroup, the root of the tree can be placed somewhere along the branch connecting baboons to the common ancestor of humans and gorillas. When only three taxa are being considered, there are three possible rooted trees but only one unrooted tree (Figure 23.3).

@nimation
Phylogenetic Trees

Figure 23.3

**The relationship between three taxa (labeled 1, 2, and 3) can be described by three different rooted trees but only one unrooted tree.** The letter A denotes a common ancestor of all three species in the rooted tree.

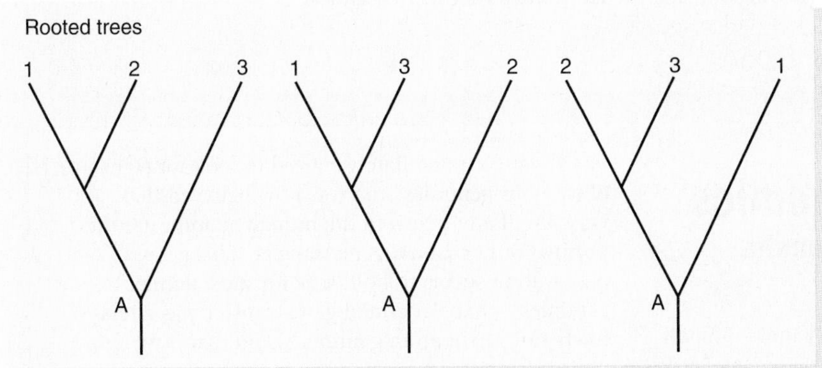


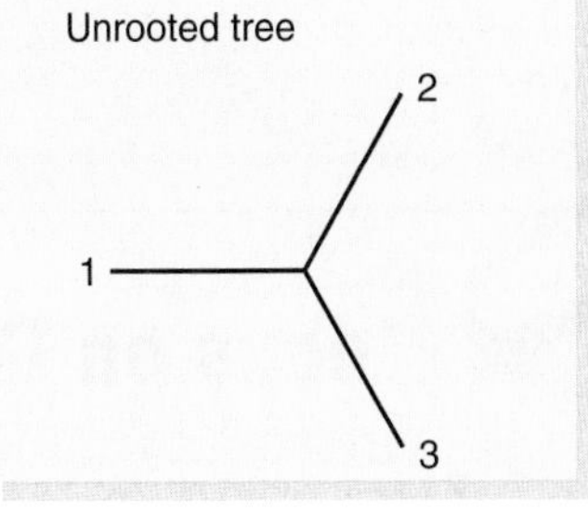


**Number of Possible Trees.** The number of possible rooted and unrooted trees quickly becomes staggering as more taxa are considered (Table 23.4). The actual number of possible rooted ($N_R$) and unrooted ($N_U$) trees for any number of taxa ($n$) can be determined with the following equations:

$$N_R = (2n - 3)!/[2^{n-2}(n - 2)!]$$
$$N_U = (2n - 5)!/[2^{n-3}(n - 3)!]$$

Values for $n$ must be greater than or equal to 2 for the first equation and greater than or equal to 3 for the second equation and can be extremely large. In practice, though, the value for $n$ is often described in terms of dozens or at most hundreds of taxa or individuals—where unimaginably large numbers of possible trees can describe the relationship between them.

**Gene versus Species Trees.** A phylogenetic tree based on the divergence observed within a single homologous gene is most appropriately called a **gene tree.** This type of tree may represent the evolutionary history of a gene but not necessarily that of the species in which it is found. **Species trees** usually are best obtained from analyses that use data from multiple genes. While this may sound counterintuitive, divergence within genes typically occurs before the splitting of populations that occurs when new species are created. For the locus being considered in Figure 23.4, some individuals in species 1 may actually be more similar to individuals in species 2 than they are to other members of their own population. The differences between gene and species trees tends to be particularly important when considering loci where diversity within populations is advantageous, as in the major histocompatability locus (MHC) described earlier. If MHC alleles alone were used to

**Table 23.4 Numbers of Rooted and Unrooted Trees That Describe the Possible Relationships between Different Numbers of Taxa**

| Number of Taxa | Number of Rooted Trees | Number of Unrooted Trees |
|---|---|---|
| 2 | 1 | 1 |
| 3 | 3 | 1 |
| 4 | 15 | 3 |
| 5 | 105 | 15 |
| 10 | 34,459,425 | 2,027,025 |
| 20 | $8.20 \times 10^{21}$ | $2.22 \times 10^{20}$ |
| 30 | $4.95 \times 10^{38}$ | $8.69 \times 10^{36}$ |

Figure 23.4

**Trans-species or shared polymorphism may occur if the ancestor was polymorphic for two or more alleles and if alleles persist to the present in both species.**

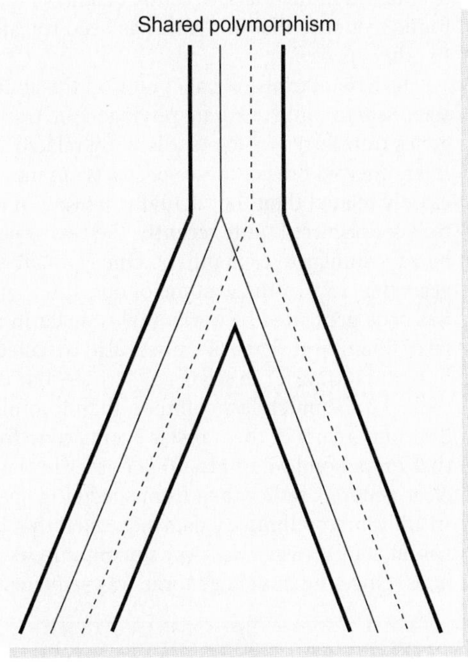

determine species trees, many humans would be grouped with gorillas rather than with other humans because the polymorphism they carry is older than the time of the split in the two lineages. A second advantage of species trees compared to gene trees is that species trees are less influenced by horizontal gene transfer (movement of a gene from a member of one species to a member of another). This movement of genes across species lines is discussed in more detail in the Focus on Genomics box in this chapter.

## Focus on Genomics

### Horizontal gene transfer

One of the events that can make it far more difficult to understand molecular evolution is *horizontal gene transfer*, or the transfer of genes across species lines. This appears to be quite common in bacteria, which is not too surprising since bacteria of different species have long been known to exchange genetic material by conjugation, transduction, and transformation. For instance, plasmids carrying genes that confer resistance to one or more antibiotics are transferred readily from one species to another. This was first described in bacteria in 1959. Horizontal transfer has been observed recently for one of the more troublesome current health threats. Some strains of *Staphylococcus aureus*, a common bacteria that lives on your skin, carry a gene that codes for a protein that breaks down more of the frequently used antibiotics. These strains are collectively called MRSA (Methicillin resistant *Staphylococcus aureus*), and cause infections that are very difficult to cure, and are sometimes fatal. Many MRSA infections occur in health care facilities, often because of improper sterilization of medical equipment or exposure of patients with compromised immune systems to infected individuals. Recently, scientists have found evidence that this gene has been transferred to other species.

Horizontal transfer can confound the ability of scientists to build accurate phylogenetic trees, since genes transferred from one distantly related species to another will make these species seem more closely related than they actually are (and if the transfer occurred quite recently, the two genes will be very similar to each other). One estimate suggests that 18% of the genome of one strain of *E. coli* has been generated by horizontal transfer in the last 100 million years. Scientists also detected true horizontal transfer in some eukaryotes, but it looked to be much less common than in prokaryotes (and many of the transfer events were from either mitochondrial or plastid genomes to the nuclear genome, rather than from species to species). In addition, preliminary data suggested that horizontal gene transfer was very uncommon in animals. Once the human genome was sequenced, investigators immediately started to look for signs of it in our genomes, and it is now believed that very few, if any, genes in the human genome are the product of horizontal gene transfer. This general observation seems to hold true for most animal genomes—that horizontal gene transfer was a relatively rare event among animals, and that it would not confuse phylogenetic trees for animals in the same way it confuses trees for prokaryotes.

Recently, a group of scientists have found one striking exception, and that is the Bdelloid rotifers. Rotifers are tiny aquatic animals that have one or more clumps of cilia near their mouth. These cilia are in constant motion, and the name *rotifer*, which means wheel bearer, refers to it. The Bdelloid rotifers are quite an oddity compared to other members of the animal kingdom, in that they appear to be unable to reproduce sexually. They also have genomes that are packed with transposable elements. One group of scientists looked at genomic DNA in the Bdelloid rotifer *Adineta vaga*, and found many genes that were almost certainly the result of horizontal gene transfer. For most of these genes, the best match (using BLASTp, or protein–protein comparisons, see Chapter 9, pp. 218–220) was to either a bacterial or fungal gene. Several of the genes had no matching animal genes. They also looked at another species of Bdelloid rotifer, and found similar results. Surprisingly, some of the genes that appear to have come from bacteria have gained classic eukaryotic introns since they entered the rotifer genome, and of the 22 genes identified as most likely the product of horizontal gene transfer, only 5 appeared to be clear pseudogenes, suggesting that at least some of the genes are expressed by the host organism. (Recall that a pseudogene is a sequence that resembles a gene but, for one reason or another, could not be expressed.) Most of the transferred genes seem to be clustered in the gene-poor, transposon-rich regions near the telomeres. The investigators estimate that about 6% of the genome in these telomeric regions is the result of horizontal gene transfer, while perhaps only 1% of the remainder of the genome (gene-rich, transposon-poor) comes from horizontal transfer. This kind of gene transfer may be limited only to these types of animals, since they live in ephemeral (short-lived) puddles and ponds, and are adapted to survive dessication (drying out), and this desiccation may alter the permeability of the cell, allowing foreign DNA to readily enter.

## Keynote

Sequence polymorphisms often predate speciation events. As a result, it is possible for phylogenetic trees made from a single gene not always to reflect the relationships between species. Species trees are best constructed by considering multiple genes.

Despite the staggering number of rooted and unrooted trees that can be generated even when using a small number of taxa (see Table 23.4), only one of the possible trees represents the true phylogenetic relationship between the taxa being considered. Since the true tree usually is known only when artificial data are used in computer simulations, most phylogenetic trees generated with molecular data from real organisms are called **inferred trees.** Distinguishing which of all the possible trees is most likely to be the true tree can be a daunting task and is typically left to high-speed computers. The computer algorithms used in these searches typically use one of a small number of different kinds of approaches: distance matrix, parsimony-based, maximum likelihood, and Bayesian methods. A basic understanding of the logic behind these approaches will help you understand exactly what information phylogenetic trees convey and what sort of molecular data are most useful for their generation.

### Reconstruction Methods

At least three fundamentally different approaches are commonly used to determine phylogenetic relationships using molecular data. Distance methods are based on statistical principles that group things based on their overall similarity to each other. This statistical approach is used for many kinds of data analysis other than just those of molecular evolution. In contrast, parsimony approaches group organisms in ways that minimize the number of substitutions that must have occurred since they last shared a common ancestor and are generally invoked only in molecular evolution studies. Maximum likelihood (or Bayesian) methods are intrinsically probabilistic/statistical and have only become feasible for typical data sets as the raw power of computers increased in the late 1990s.

**Distance Matrix Approaches to Phylogenetic Tree Reconstruction.** The oldest distance matrix method is also the simplest of all methods for tree reconstruction. Originally proposed in the early 1960s to help with the evolutionary analysis of morphological characters, the **unweighted pair group method with arithmetic averages (UPGMA)** is largely statistically based and requires data that can be condensed to a measure of genetic distance between all the pairs of taxa being considered. To illustrate the construction of a phylogenetic tree using the UPGMA method, consider a group of four taxa called A, B, C, and D. Assume that the pairwise distances between each of the taxa are given in the following matrix:

| Taxa | A | B | C |
|---|---|---|---|
| B | $d_{AB}$ | – | – |
| C | $d_{AC}$ | $d_{BC}$ | – |
| D | $d_{AD}$ | $d_{BD}$ | $d_{CD}$ |

In this matrix, $d_{AB}$ represents the distance (perhaps as calculated by the Jukes–Cantor model) between taxa A and B, $d_{AC}$ is the distance between taxa A and C, and so on. UPGMA begins by clustering the two taxa with the smallest distance separating them into a single, composite taxon. In this case, assume that the smallest value in the distance matrix corresponds to $d_{AB}$, in which case taxa A and B are the first to be grouped together (AB). After the first clustering, a new distance matrix is computed, with the distance between the new taxon (AB) and taxa C and D being calculated as $d_{(AB)C} = 1/2(d_{AC} + d_{BC})$ and $d_{(AB)D} = 1/2(d_{AD} + d_{BD})$. The taxa separated by the smallest distance in the new matrix are then clustered together to make another new composite taxon. The process is repeated until all taxa have been grouped together. If scaled branch lengths are to be used on the tree to represent the evolutionary distance between taxa, then branch points are positioned at a distance halfway between the taxa being grouped (i.e., at $d_{AB}/2$ for the first clustering).

A strength of distance matrix approaches in general is that they work equally well with morphological and molecular data as well as combinations of the two. They, like maximum likelihood analyses, also take into consideration all the data available for a particular analysis. In contrast, the alternative parsimony approaches discard many "noninformative" sites (described later). A weakness of the UPGMA approach in particular is that it presumes a constant rate of evolution across all lineages, something that is known to not always be the case. Several distance matrix-based alternatives to UPGMA such as the transformed distance method and the neighbor-joining method are more complex but capable of incorporating different rates of evolution within different lineages.

**Parsimony-Based Approaches to Phylogenetic Tree Reconstruction.** While the distance- and maximum likelihood-based methods of tree reconstruction are grounded in statistics, parsimony-based approaches rely more heavily on the biological principle that mutations are rare events. The word *parsimony* itself means "stinginess or cheapness" and refers to the fact that parsimony approaches attempt to minimize the number of mutations within a phylogenetic tree to account for the sequences of all taxa being considered. These parsimony approaches assume that the simplest tree (the one that invokes the fewest number of mutations) is considered to be the best and is deemed a tree of **maximum parsimony.**

As mentioned earlier, the parsimony-based approach does not use all sites when considering molecular data. Instead, it focuses only on positions within a multiple alignment that favors one tree over an alternative in terms of the number of substitutions they require. Not all positions within a multiple alignment favor one tree over an alternative from the perspective of parsimony. Consider the following alignment of four nucleotide sequences:

| | Site | | | | | |
|---|---|---|---|---|---|---|
| | 1 | 2 | 3 | 4 | 5* | 6* |
| **Sequence** | G | C | G | A | T | G |
| | G | T | G | T | T | G |
| | G | T | T | G | C | A |
| | G | T | C | C | C | A |

In such an alignment, only the fifth and sixth sites (marked with asterisks) qualify as *informative sites* from a parsimony perspective. As shown in Figure 23.5, only three possible unrooted trees can be drawn that describe the relationship between four taxa. The unrooted tree that groups sequences 1 and 2 separate from sequences 3 and 4 would require only one mutation to have occurred in the branch that connects both groupings. Either of the two alternative trees that group the taxa differently would require two mutations and therefore do not represent the most parsimonious arrangement of the sequences. In contrast, all three of the possible unrooted trees for site 1 are indistinguishable from the perspective of parsimony, because no mutations must be invoked for any of them. Similarly, site 2 is uninformative because one mutation occurs in all three of the possible trees. Likewise, site 3 is uninformative because all three trees require two mutations, and site 4 is uninformative because all three trees require three mutations. In general, for a site to be informative regardless of how many sequences are aligned, it has to have at least two different nucleotides, and each of these nucleotides has to be present at least twice.

Maximum parsimony trees are determined by first identifying all informative sites within an alignment and then determining which of all possible unrooted trees requires the fewest number of mutations for each of those sites. The tree or trees requiring the fewest number of mutations when all sites within an alignment are considered is the most parsimonious tree. A very useful

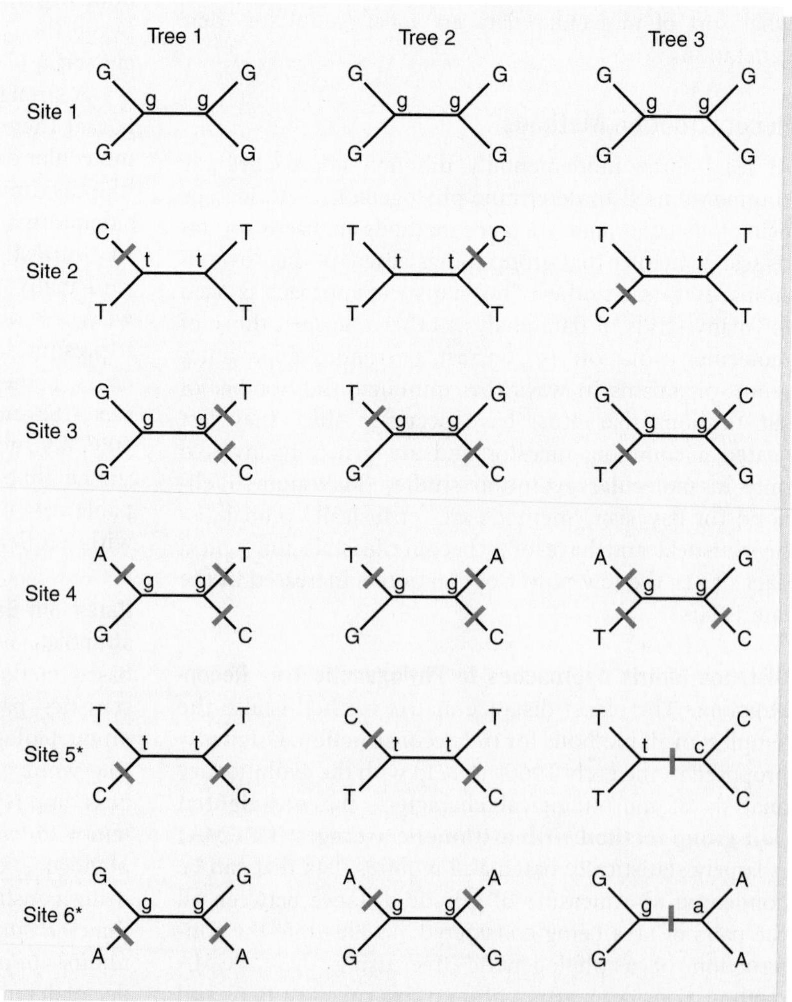

**Figure 23.5**

**Three different unrooted trees describe all possible relationships between four taxa.** Using the sequences (uppercase letters) and sites shown in the text, all three trees for each of the six sites are shown. Red lines are drawn on branches along which substitutions must have occurred, and inferred ancestral states are shown in lowercase letters. The sequence for site 1 requires no substitutions regardless of which tree is used, site 2 requires one for all trees, site 3 requires two for all trees, and site 4 requires three for all trees. Only sites 5 and 6 have one tree with fewer substitutions than the alternative tree; that makes them informative sites.

by-product of the parsimony approach is the generation of inferred ancestral sequences at each node of a tree (see Figure 23.5). These inferred ancestral sequences go a long way toward making a nonissue of the infamous "missing links" of the fossil record and, when analyzed carefully, can give remarkably clear insights into the nature of long-dead organisms and even the environment in which they lived. Of course, the parsimony approach described here assumes that all nucleotides are just as likely to mutate into any of the three alternative nucleotides. More complicated parsimony algorithms take the difference in transition and transversion frequencies into account, although none is particularly reliable when rates of substitutions between branches of a tree differ dramatically.

**Maximum Likelihood Approaches to Phylogenetic Tree Reconstruction.** Maximum likelihood approaches represent an alternative and purely statistically based method of phylogenetic reconstruction. With this approach, probabilities are considered for every individual nucleotide substitution in a set of sequence alignments. For instance, we know that transitions are observed roughly three times as often as transversions. In a three-way alignment where a single column is found to have a C, a T, and an A, it can be reasonably argued that a greater likelihood exists that the sequences with the C and the T are more closely related to each other than they are to the sequences with an A (because the C to T change represents a transition, while the C or T to A change represents a transversion). Calculation of probabilities is complicated because the sequence of the common ancestor to the sequences being considered is generally not known. Determining the most likely evolutionary history is further complicated by the fact that multiple substitutions may have occurred at one or more of the sites being considered, and all sites are not necessarily independent or equivalent. Still, objective criteria can be applied to calculating the probability for every site *and* for every possible tree that describes the relationship of the sequences in a multiple alignment. The number of possible trees for even a modest number of sequences (see Table 23.4) makes this a very computationally intensive proposition, yet the one tree with the single highest aggregate probability is, by definition, the most likely to reflect the true phylogenetic tree under the proposed model of nucleotide substitution.

The dramatic increase in the raw power of computers has made maximum likelihood approaches feasible, and trees inferred in this way are becoming increasingly common in the literature. Note, however, that no one substitution model is as yet close to general acceptance and, because different models can very easily lead to different conclusions, the model used must be carefully considered and described when using this approach.

### Keynote

The number of possible trees that describe the relationship between even a small number of taxa can be very large. Distance matrix and maximum likelihood methods rely on statistical relationships between taxa to group them. Parsimony approaches assume that the tree that invokes the fewest number of mutations is most likely to be the best. No method can guarantee that it will yield the true phylogenetic tree, but when multiple substitutions are not likely to have occurred and evolutionary rates within all lineages are fairly equal, all three methods have been demonstrated to work well.

**Bootstrapping and Tree Reliability.** Obviously, longer sequence alignments require a longer time to analyze than shorter ones when the parsimony approach is used. However, because of the relationship between the number of taxa and the corresponding number of unrooted trees illustrated in Table 23.4, the addition of more sequences has a much more dramatic effect on the time required to find a preferred tree. Once data sets involve 30 or more species, the number of possible trees is so large that it is simply not possible to examine all possible trees and assess the fit of the data to each, even when using the fastest computers. In addition, no tree reconstruction method is certain to yield the correct tree. Numerous variations on each approach have been suggested, and intensive simulation studies have been performed to compare the statistical reliability of almost all tree construction methods. The results of these simulations are easy to summarize: data sets that allow one method to infer the correct phylogenetic relationship generally work well with all the currently popular methods. However, if many changes have occurred in the simulated data sets or rates of change vary among branches, then none of the methods works very reliably. As a general rule, if a data set yields similar trees when analyzed by two or three of the fundamentally different tree reconstruction methods, that tree can be considered to be fairly reliable.

It is also possible for portions of inferred trees to be determined with varying degrees of confidence. **Bootstrap procedures** allow a rough quantification of those confidence levels by randomly changing the weighting of each site. The basic approach of the bootstrap procedure is straightforward: a subset of the original data is drawn (with replacement) from the original data set, and a tree is inferred from the new data set. In a physical sense, the process is equivalent to taking the print out of a multiple alignment; cutting it up into pieces, each one containing a different column from the alignment; placing all those pieces into a bag; randomly reaching into the bag and drawing out a piece; copying down the information from that piece before returning it to the bag; and then repeating the drawing step until an artificial data set has been created that is as long as the original alignment. This whole process is repeated to create hundreds or thousands of resampled data sets, and portions of the inferred tree that have the same groupings in many of the repetitions are those that are especially well supported by the entire original data set. Numbers that correspond to the fraction of bootstrapped trees yielding the same grouping are often placed next to the corresponding nodes in phylogenetic trees to convey the relative confidence in each part of the tree. Bootstrapping has become very popular in phylogenetic analyses even though some methods of tree inference can make it very time-consuming to perform.

Despite their often casual use in the scientific literature, bootstrap results need to be treated with some caution. First, bootstrap results based on fewer than several hundred iterations (rounds of resampling and tree generation) are not likely to be reliable, especially when large numbers of sequences are involved. Simulation studies have also shown that bootstrap tests tend to underestimate the confidence level at high values and overestimate it at low values. And, since many trees have very large numbers of branches, there is often a significant risk of succumbing to "the fallacy of multiple tests"—some results may appear to be statistically significant by chance simply because so many groupings are being considered. Still, some studies have suggested that commonly used solutions to these potential problems yield trees that are closer representations of the true tree than the single most parsimonious tree.

## Phylogenetic Trees on a Grand Scale

One of the most striking cases in which sequence data have provided new information about evolutionary relationships is in our understanding of the primary divisions of life. In the late 1800s, biologists divided all of life into three major groups: the plants, the animals, and the protists (a catchall category for everything that did not fit into the two eukaryotic categories). As more organisms were discovered and their features examined in more detail, this simple trichotomy became unworkable. It was later recognized that organisms could be divided into prokaryotes and eukaryotes on the basis of cell structure. Several additional primary divisions of life were subsequently recognized, such as the five *kingdoms* (prokaryotes, protista, plants, fungi, and animals) proposed by R. H. Whittaker in 1959.

**The Tree of Life.** In the mid-1980s, RNA and DNA sequences were used to uncover the primary lines of evolutionary history among all organisms. In one study, Carl Woese, Norm Pace, and colleagues constructed an evolutionary tree of life based on the nucleotide sequences of the 16S rRNA, which all organisms (as well as mitochondria and chloroplasts) possess. As illustrated in Figure 23.6, their evolutionary tree revealed three major evolutionary groups: the **Bacteria** (the traditional prokaryotes

**Figure 23.6**
**An evolutionary tree of life revealed by comparison of 16S rRNA sequences.**

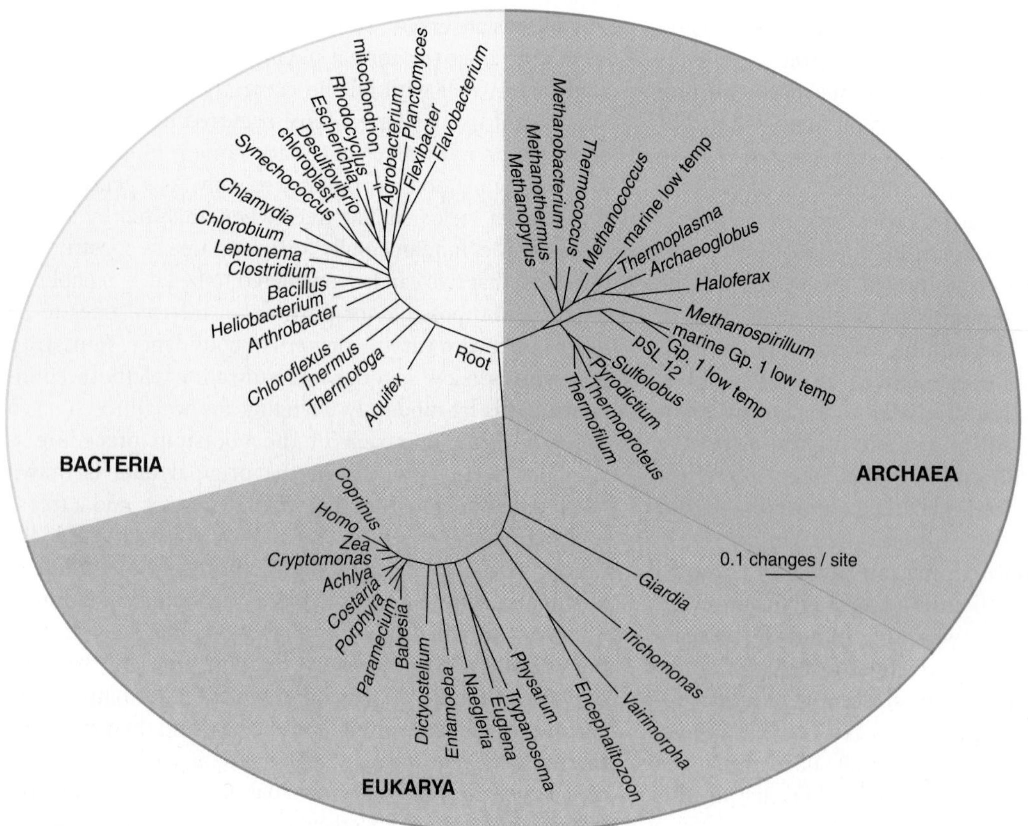

as well as mitochondria and chloroplasts), the **Archaea** (mostly extremophilic prokaryotes including many little-known organisms) and the **Eukarya.** Bacteria and archaeans, although both prokaryotic in that they had no internal membranes, were found to be as different genetically as bacteria and eukaryotes. The deep evolutionary differences that separate the bacteria and the archaeans were not obvious on the basis of phenotype, and the fossil record was silent on the issue. These molecular analyses support the idea that three major **evolutionary domains** (the Bacteria, the Archaea, and the Eukarya) exist among living organisms. Originally intended as a replacement for kingdoms, domains are now used as a higher-level rank with eukaryotes divided into four different kingdoms (protists, fungi, plants, and animals). These molecular phylogenies have led to many other surprising revelations, such as the observation that the genes of eukaryotic organelles like mitochondria and chloroplasts actually have separate, independent origins from their nuclear counterparts (Box 23.1).

**Human Origins.** Another field in which DNA sequences are being used is the study of modern human origins and modern human population diversification. In contrast to the extensive phenotypic variation observed in size, body shape, facial features, and skin color, genetic differences among human populations are surprisingly small. For example, analysis of mtDNA sequences shows that the mean difference in sequence between two human populations is about 0.33%. Other primates exhibit much larger differences. For example, the two subspecies of orangutan have mtDNA sequences that differ by as much as 5%. The high degree of genetic similarity among human groups indicates that all humans are closely related relative to groups of other primates. Another surprising observation emerges upon careful examination of those genetic differences that do exist between different human groups: the greatest differences are not found among populations located on different continents but rather are found between human populations residing in Africa. In fact, all human populations originating outside of Africa represent only a subset of the genetic diversity observed among African populations. Many experts interpret these findings to mean that humans originated and experienced their early evolutionary divergence in Africa. By this theory—called the **out-of-Africa theory**—small groups of humans migrated out of Africa and gave rise to all other human populations only after a number of genetically differentiated populations were present in Africa. Sequence data from both mitochondrial DNA and the nuclear Y chromosome (the male sex chromosome) are consistent with this theory in that they suggest that all people alive today have mitochondria that came from a "mitochondrial Eve," and that all men have Y chromosomes

### Box 23.1 The Endosymbiont Theory

The tree of life shown in Figure 23.6 suggests that the differences between Bacteria, Eukarya, and Archaea result from independent evolution that has been taking place for far longer than the time since plants and animals diverged (at least 1.5 billion years). Analyses such as those have also shed light on the long-standing question of how the compartmental organization of eukaryote cells could have evolved from the simpler condition still found in prokaryotes. The most important clue in providing a satisfying answer to that question came with the realization that the 16S ribosomal DNA of the nucleus, mitochondria, and chloroplasts were evolving independently even before the first eukaryotes appeared. In fact, the closest living relative of mitochondria today actually appears to be the bacteria *Rickettsia prowazekii*, the causative agent of epidemic typhus. A logical inference was that mitochondria and chloroplasts were free-living organisms that at some point in the past became engulfed by a prokaryote-like organism. The endosymbiotic (*endo* meaning "internal," and *symbiotic* meaning "cooperative relationship between two or more organisms") arrangement that resulted became the eukaryotes we see today. In other words, a merger of at least two or three evolutionary lineages gave rise to significantly different new forms of life.

The endosymbiont theory was originally suggested by a pioneering physiological ecologist, A. Schimper, in the early 1880s. It was championed by G. Mereschkovsky in the early 1900s based on microscopic examinations of plants and their plastids, which Mereschkovsky described as "little green slaves." More recent molecular analyses, especially those of Lynn Margulis, have led to general acceptance of this model for the origin of these eukaryotic organelles. Numerous additional similarities between Bacteria, mitochondria, and chloroplasts corroborate the 16S rRNA-based phylogenies. For instance, all organisms (including mitochondria and chloroplasts) in the Bacteria branch of the tree of life (see Figure 23.6) have circular chromosomes, similar genomic arrangements and replication processes, similar sizes, and similar drug sensitivities, all features that distinguish them from what is associated with the nucleus of eukaryotic cells. Mitochondria and chloroplasts share these properties.

In time, the endosymbionts in eukaryotic cells have become very specialized, with the nucleus (a likely endosymbiont of eukaryotic cells itself!) being the predominant site at which heritable information is stored, mitochondria being the primary site for oxidative phosphorylation, and chloroplasts being the site at which photosynthesis occurs. Many of the genes essential for organelle function have moved to the nucleus (which is also consistent with the observation that genes on mitochondrial chromosomes tend to have higher rates of substitution), and the relationship between organelles and their host cells has become an obligatory and elaborate one in which no unit (compartment) can live independently.

derived from a "Y-chromosome Adam" roughly 200,000 years ago. While the out-of-Africa theory is not universally accepted, DNA sequence data indicate that Africa has played a key role in the origin and migration of modern humans across the globe.

Join a team of molecular geneticists and anthropologists and use the technique they have developed to extract and analyze ancient DNA from Neanderthal fossils in the iActivity *Were Neanderthals Our Ancestors?* on the student website.



**Canine Origins.** The evolution of "man's best friend" has recently been studied in a similar way in a hybrid effort of phylogenetic reconstruction and comparative genomics. Over the past several centuries, artificial selection (selective breeding) has created hundreds of different breeds of dogs having a wide variety of physical features and temperaments. The inbreeding associated with the creation of these breeds has generated many breed-specific problems such as narcolepsy, arthritis, and the various forms of cancer that also afflict humans. Comparisons of closely related breeds that differ in their prevalence of diseases are allowing researchers to track down genes responsible for many illnesses in dogs as well as their counterparts in humans. As part of those disease-gene-finding efforts, a set of researchers working with Elaine Ostrander examined genetic variation associated with almost 100 different microsatellite markers in 85 different dog breeds in 2004. They found that dog breeds are a very real concept at a genetic level (e.g., each Saint Bernard is more closely related to other Saint Bernards than it is to any dog of a different breed). The phylogenetic trees the researchers constructed with their molecular data also revealed that all dog breeds clustered into only four different categories. The oldest (most genetically diverse) cluster (including Siberian husky, chow chow, and shar-pei) have the greatest similarity to wolf DNA and appear to trace their ancestry back to Asia and Africa. Subsequent European efforts seem to have been responsible for the creation of breeds specialized for guarding (including bulldog, rottweiler, and German shepherd), hunting (including golden retriever, bloodhound, and beagle), and herding (including collie, several sheepdogs, and Saint Bernard). Similar studies have also revealed interesting stories associated with the histories of wine-producing grapes, domestic pigs and cows, and important grain crops.

## Acquisition and Origins of New Functions

A long-standing question of deep interest to those who study molecular evolution is the issue of how genes with new functions arise. As early as 1932, British geneticist J. B. S. Haldane suggested that new genes arose from the process of mutating redundant copies of already existing genes. Although other means such as *transposition*, the movement of a segment of chromosome from one location to another in the genome (see Chapter 7), have since been described, Haldane's argument still does a good job of describing the origin of most new genes.

### Multigene Families

In eukaryotic organisms, we often find tandemly arrayed copies of genes, all having identical or very similar sequences. These **multigene families** are sets of related genes that have evolved from some ancestral gene through gene duplication. The globin gene family that encodes the proteins used to make up the oxygen-carrying hemoglobin molecule in our blood has become a classic example of such a multigene family. (The organization and expression pattern of this multigene family in humans was discussed in Chapter 19, pp. 552–553.) Briefly, the globin multigene family is composed of seven α-like genes found on chromosome 16 and six β-like genes found on chromosome 11.

Globin genes are also found in other animals, and globin-like genes are even found in plants, suggesting that this gene family is at least 1.5 *billion* years old. Almost all functional globin genes in animal species have the same general structure, consisting of three exons separated by two introns. However, the numbers of globin genes and their order vary among species, as is shown for the β-like genes in Figure 23.7. Since all globin genes have similarities in structure and sequence, it appears that an ancestral globin gene (perhaps most like the present-day myoglobin gene) duplicated and diverged to produce an ancestral α-like gene and an ancestral β-like gene. These two genes then underwent repeated, independent duplications, giving rise to the various α-like and β-like genes found in vertebrates today. Repeated gene duplication, such as that giving rise to the globin gene family, appears to be a frequent evolutionary occurrence. Indeed, the number of copies of globin gene varies even within some human populations. For example, most humans have two α-globin genes on chromosome 16 (as shown in Figure 21.8, p. 619). However, some individuals have a single α-globin gene; other individuals have three or even four copies of the α-globin gene on their chromosome 16. These observations suggest that duplication and deletion of genes in multigene families are part of a dynamic process that continues to operate today. Gene duplications and deletions in gene clusters often arise as a result of misalignment of sequences during crossing-over between homologous chromosomes, a process called **unequal crossing-over.** Duplications can also arise if matings in a population introduce a chromosome bearing a transposed segment from a second chromosome into a genome whose copies of that second chromosome are intact. Mobilization of transposons

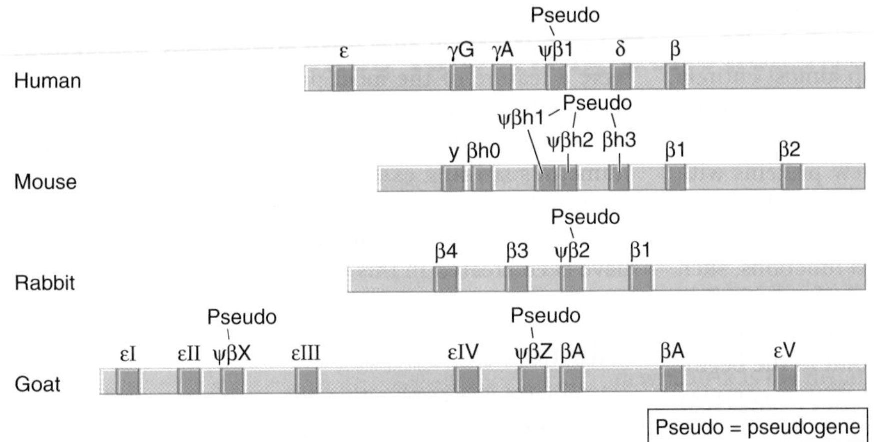


**Figure 23.7**
**Organization of the globin gene families in several mammalian species.**

(see Chapter 7) can result in a wide dispersal of copied sequences.

## Gene Duplication and Gene Conversion

Following gene duplication, one of the separate copies of a gene may undergo changes in sequence as if it were free from functional constraint as long as the other copy continues to function. As you might expect from the previous discussions in this chapter, most changes to the copy normally would have been selectively disadvantageous or even render it a nonfunctional pseudogene. On rare occasions, however, the changes may lead to subtle alterations of function or pattern of expression that are advantageous to the organism, and the change sweeps through a population. The pseudogenes found in mammalian globin gene families (see Figure 23.7) are thought to have occurred in just this way; that is, by mutation of a duplicated, active gene. This "tinkering" approach to evolution becomes even more of a "win/no-lose" scenario when misalignments between pseudogene copies and the functional copy occur during subsequent recombination events and the inactivating changes are corrected by *gene conversion*. Gene conversion is a process of genetic recombination in meiosis in which the DNA sequence of an allele on one homolog is copied and replaces the DNA sequence of the allele on the other homolog. In contrast to standard genetic recombination which involves a reciprocal exchange of genetic information, gene conversion is a nonreciprocal process. Thus, given an *A* allele on one homolog, gene conversion can lead to the replacement of an *a* allele on the homolog, resulting in both homologs now having an *A* allele. Gene conversion events can give an organism multiple chances to create a gene with a new function from the duplicate of an already functional one. Like gene duplication, gene conversion also continues to operate to this day, although it is usually most apparent when helpful substitutions to a gene copy are "corrected." For example, the two neighboring genes on the X chromosome that allow most humans to distinguish between red and green light are 98% identical at the nucleotide level, and most spontaneous occurrences of deficiencies in green-color vision occur as a result of gene conversions between the two. Approximately 8% of human males are color blind as a result of this kind of gene conversion event.

## *Arabidopsis* Genome

The extent to which organisms use gene duplication to generate proteins with new functions is becoming increasingly clear as more and more genome sequencing projects are concluding. For example, with only about 125 million nucleotides in its genome, *Arabidopsis thaliana* (thale cress) was the first plant genome to be completely sequenced (see Chapter 8, p. 204). Its short generation time and small size make it a favorite organism of plant geneticists, but it was a particularly appealing choice for genome sequencers because studies had indicated that its genome had undergone much less duplication than that seen in other, more commercially important plants. But when the sequencing was completed at the end of 2000, more than half of the 25,900 *Arabidopsis* genes were found to be duplicates. Phylogenetic analyses such as the distance matrix, parsimony, and maximum likelihood methods described earlier in this chapter revealed only about 11,600 distinct families of one or more genes. Even in this unusually nonredundant genome, the process of evolving through gene duplication followed by tinkering holds sway.

### Keynote

Gene duplication events appear to have occurred frequently in the evolutionary history of all organisms. Copies of genes provide the raw material for evolution in that they are free to accumulate substitutions that sometimes give rise to proteins with new, advantageous functions.

**Domain (Exon) Shuffling.** It should also be pointed out that an increase in the number of copies of a DNA sequence can also occur in segments of a genome that are smaller than complete genes. Numerous examples of

**inversion** A chromosomal mutation in which a segment of a chromosome is excised and then reintegrated in an orientation 180° from the original orientation.

**karyotype** A complete set of all the metaphase chromatid pairs in a cell.

**kinetochore** Specialized multiprotein complex that assembles at the centromere of a chromatid and is the site of attachment of spindle microtubules during mitosis.

**Klinefelter syndrome** A human clinical syndrome that results from disomy for the X chromosome in a male, which results in a 47,XXY male. Many of the affected males are mentally deficient, have underdeveloped testes, and are taller than average.

**knockout mouse** A mouse in which a nonfunctional allele of a particular gene has replaced the normal alleles, thereby knocking out the gene's function in an otherwise normal individual.

**lagging strand** In DNA replication, the DNA strand that is synthesized discontinuously from multiple RNA primers in the direction opposite to movement of the replication fork. *See also* **leading strand** and **Okazaki fragments.**

**leader sequence** *See* **5′ untranslated region (5′ UTR).**

**leading strand** In DNA replication, the DNA strand that is synthesized continuously from a single RNA primer in the same direction as movement of the replication fork. *See also* **lagging strand.**

**leptonema** The first stage in prophase I of meiosis during which the chromosomes begin to coil and become visible.

**lethal allele** An allele whose expression results in the death of an organism.

**light repair** *See* **photoreactivation.**

**LINEs (long interspersed elements)** One class of **dispersed repeated DNA** consisting of repetitive sequences that are several thousand base pairs in length. Some LINEs can move in the genome by **retrotransposition.**

**linkage** The association of genes located on the same chromosome such that they tend to be inherited together.

**linkage disequilibrium** Deviations from the expectations of **independent assortment** and Hardy–Weinberg equilibrium caused either by physical linkage or population demography.

**linkage map** *See* **genetic map.**

**linked genes** Genes that are located on the same chromosome and tend to be inherited together. A collection of such genes constitutes a *linkage group.*

**linker** *See* **restriction site linker.**

**locus** (*plural,* **loci**) The position of a gene on a genetic map; the specific place on a chromosome where a gene is located. More broadly, a locus is any chromosomal location that exhibits variation detectable by genetic or molecular analysis.

**lod score method** The lod (logarithm of odds) score method is a statistical analysis, usually performed by computer programs, based on data from pedigrees. It is used to test for linkage between two loci in humans.

**long interspersed elements** *See* **LINEs.**

**looped domains** Loops of supercoiled DNA that serve to compact the chromosomes.

**loss-of-function mutation** A mutation that leads to the absence or decreased biological activity of a particular protein.

**Lyon hypothesis** *See* **lyonization.**

**lyonization** A mechanism of dosage compensation, discovered by Mary Lyon, in which one of the X chromosomes in the cells of female mammals becomes highly condensed and genetically inactive.

**lysogenic** Referring to a bacterium that contains the genome of a temperate phage in the **prophage** state. On induction, the prophage leaves the bacterial chromosome, progeny phages are produced, and the bacterial cell lyses.

**lysogenic pathway** One of two pathways in the life cycle of temperate phages in which the phage genome is integrated into the host cell's chromosome and progeny phages are not formed.

**lysogeny** The phenomenon in which the genome of a temperate phage is inserted into a bacterial chromosome, where it replicates when the bacterial chromosome replicates. In this state, the phage genes are repressed and progeny phages are not formed.

**lytic cycle** Bacteriophage life cycle in which the phage takes over the bacterium and directs its growth and reproductive activities to express the phage genes and to produce progeny phages.

**mapping function** Mathematical formula used to correct the observed recombination frequencies for the incidence of multiple crossovers.

**map unit (mu)** A unit of measurement for the distance between two genes on a **genetic map.** A recombination (crossover) frequency of 1% between two genes equals 1 map unit. *See also* **centimorgan.**

**maternal effect** (a) The phenotype established by expression of **maternal effect genes** in the oocyte before fertilization. (b) An influence derived from the maternal environment (e.g., uterus size, quantity and quality of milk) that affects the phenotype of offspring, expressed as $V_{Em}$; one of the family environmental effects that influence the variation of **quantitative traits.**

**maternal effect gene** A nuclear gene, expressed by the mother during oogenesis, whose product helps direct early development in the embryo.

**maternal inheritance** A type of uniparental inheritance in which the mother's phenotype is expressed exclusively.

**mating types** In lower eukaryotes, two forms that are morphologically indistinguishable but carry different alleles and will mate; equivalent to the sexes in higher organisms. *See also* **genic sex determination.**

**maximum parsimony** Property of the **phylogenetic tree** (or trees) that invokes the fewest number of mutations and therefore is most likely to represent the true evolutionary relationship between species or their genes.

**MCS** *See* **multiple cloning site.**

**mean** The average of a set of numbers, calculated by adding all the values represented and dividing by the number of values.

**meiosis** Two successive nuclear divisions of a diploid nucleus, following one DNA replication, that result in the formation of haploid gametes or of spores having one-half the genetic material of the original cell.

**meiosis I** The first meiotic division, resulting in the reduction of the number of chromosomes from diploid to haploid.

**meiosis II** The second meiotic division, resulting in the separation of the chromatids and formation of four haploid cells.

**Mendelian factor** *See* **gene.**

**Mendelian population** A group of interbreeding individuals who share a common **gene pool;** the basic unit of study in population genetics.

**messenger RNA (mRNA)** Class of RNA molecules that contain coded information specifying the amino acid sequences of proteins.

**metabolomics** The study of all of the small chemicals that are intermediates or products of metabolic pathways.

**metacentric chromosome** A chromosome with the centromere near the center such that the chromosome arms are of about equal lengths.

**metagenomics** A branch of comparative genomics involving the analysis of genomes in entire communities of microbes isolated from the environment. Also called *environmental genomics*.

**metaphase** The stage in mitosis or meiosis during which chromosomes become aligned along the equatorial plane of the spindle.

**metaphase I** The stage in meiosis I when each homologous chromosome pair (bivalent) becomes aligned on the equatorial plate.

**metaphase II** The stage of meiosis II during which the chromosomes (each a sister chromatid pair) line up on the equatorial plate in each of the two daughter cells formed in meiosis I.

**metaphase plate** The plane in the cell where the chromosomes become aligned during metaphase.

**metastasis** The spread of malignant tumor cells throughout the body so that tumors develop at new sites.

**methyl-directed mismatch repair** An enzyme-catalyzed process for repairing mismatched base pairs in DNA *after* replication is completed; contrast to **proofreading,** a process for correcting mismatched base pairs *during* replication.

**methylome** The complete set of DNA methylation modifications in the cell.

**microbiome** The community of microorganisms in a particular environment.

**microRNA (miRNA)** Noncoding, single-stranded regulatory RNA molecule about 21–23 nt long derived from an RNA transcript. An miRNA regulates the expression of a target mRNA by binding to the 3′ UTR causing either inhibition of translation of the mRNA or degradation of that molecule, depending on the extent of complementary base-pairing between the two molecules.

**microsatellite** *See* **short tandem repeat.**

**minimal medium** For a microorganism, a medium that contains the simplest set of ingredients (e.g., a sugar, some salts, and trace elements) required for the growth and reproduction of wild-type cells.

**minisatellite** *See* **variable number tandem repeat.**

**missense mutation** A **point mutation** in a gene that changes one codon in the corresponding mRNA so that it specifies a different amino acid than the one specified by the wild-type codon.

**mitochondria** Organelles found in the cytoplasm of all aerobic animal and plant cells in which most of the cell's ATP is produced.

**mitosis** The process of nuclear division in haploid or diploid cells producing daughter nuclei that contain identical chromosome complements and are genetically identical to one another and to the parent nucleus from which they arose.

**moderately repetitive DNA** A class of DNA sequences, each of which is present from a few to about $10^5$ copies in the haploid chromosome set.

**modifier gene** A gene that interacts with another nonallelic gene causing a change in the phenotypic expression of the alleles of that gene.

**molecular clock hypothesis** The hypothesis that for any given gene, mutations accumulate at an essentially constant rate in all evolutionary lineages as long as the gene retains its original function.

**molecular cloning** *See* **cloning (a).**

**molecular evolution** Study of how genomes and macromolecules evolve at the molecular level and how genes and organisms are evolutionarily related.

**molecular genetics** Study of how genetic information is encoded within DNA and how biochemical processes of the cell translate the genetic information into the phenotype.

**monoecious** Referring to plant species in which individual plants possess *both* male and female sex organs and thus produce male and female gametes. Monoecious plants are capable of self-fertilization. *See also* **dioecious.**

**monohybrid cross** A cross between two individuals that are both heterozygous for the same pair of alleles (e.g., *Aa* × *Aa*). By extension, the term also refers to crosses involving the pure-breeding parents that differ with respect to the alleles of one locus (e.g., *AA* × *aa*).

**monoploidy** Condition in which a normally diploid cell or organism lacks one complete set of chromosomes.

**monosomy** A type of **aneuploidy** in which one chromosome of a homologous pair is missing from a normally diploid cell or organism. A monosomic cell is $2N - 1$.

**morphogen** A substance that helps determine the fate of cells in early development in proportion to its concentration.

**morphogenesis** Overall developmental process that generates the size, shape, and organization of cells, tissues, and organs.

**mRNA splicing** Process whereby an intron (intervening sequence) between two exons (coding sequences) in a **precursor mRNA (pre-mRNA)** molecule is excised and the exons ligated (spliced) together.

**multifactorial trait** A characteristic determined by multiple genes and environmental factors.

**multigene family** A set of genes encoding products with related functions that have evolved from a common ancestral gene through gene duplication.

**multiple alleles** Many alternative forms of a single gene. Although a population may carry multiple alleles of a particular gene, a single diploid individual can have a maximum of only two alleles at a locus.

**multiple cloning site (MCS)** A region within a cloning vector that contains many different restriction sites. Also called *polylinker.*

**multiple-gene hypothesis for quantitative inheritance** *See* **polygene hypothesis for quantitative inheritance.**

**mutagen** Any physical or chemical agent that significantly increases the frequency of mutational events above a spontaneous mutation rate.

**mutagenesis** The creation of mutations.

**mutant allele** Any form of a gene that differs from the wild-type allele. Mutant alleles may be **dominant** or **recessive** to wild-type alleles.

**mutation** Any detectable and heritable change in the genetic material not caused by genetic recombination; mutations may occur within or between genes and are the ultimate source of all new genetic variation.

**mutation frequency** The number of occurrences of a particular kind of mutation in a population of cells or individuals.

**mutation rate** The probability of a particular kind of mutation as a function of time.

**mutator gene** A gene that, when mutant, increases the spontaneous mutation frequencies of other genes.

**narrow-sense heritability** The proportion of the **phenotypic variance** that results from the additive effects of different alleles on the phenotype.

**natural selection** Differential reproduction of individuals in a population resulting from differences in their **genotypes.**

**negative assortative mating** Preferential mating between phenotypically dissimilar individuals that occurs more frequently than expected for **random mating.**

**neutral mutation** A **point mutation** in a gene that changes a codon in the corresponding mRNA to that for a different amino acid but results in no change in the function of the encoded protein.

**neutral theory** The hypothesis that much of the pattern of evolutionary changes in protein molecules can be explained by the opposing forces of mutation and random genetic drift.

**nitrogenous base** A nitrogen-containing **purine** or **pyrimidine** that, along with a pentose sugar and a phosphate, is one of the three parts of a **nucleotide.**

**noncontributing allele** An allele that has no effect on the phenotype of a **quantitative trait.**

**nondisjunction** A failure of homologous chromosomes or sister chromatids to separate at anaphase. *See also* **primary nondisjunction** and **secondary nondisjunction.**

**nonhistone** An acidic or neutral protein found in **chromatin.**

**nonhomologous chromosomes** Chromosomes that contain dissimilar genetic loci and that do not pair during meiosis.

**nonhomologous recombination** Recombination between DNA sequences that are not identical or highly similar. *See* **homologous recombination.**

**non-Mendelian inheritance** *See* **extranuclear inheritance.**

**nonsense codon** *See* **stop codon.**

**nonsense mutation** A **point mutation** in a gene that changes an amino-acid-coding codon in the corresponding mRNA to a stop codon.

**nonsynonymous** Referring to nucleotides in a gene that when mutated cause a change in the amino acid sequence of the encoded wild-type protein.

**normal distribution** Common probability distribution that exhibits a bell-shaped curve when plotted graphically.

**norm of reaction** Range of phenotypes produced by a particular genotype in different environments.

**northern blot analysis** A technique for detecting specific RNA molecules in which the RNAs are separated by gel electrophoresis, transferred to a nitrocellulose filter, and then hybridized with labeled complementary probes; also called *northern blotting. See also* **Southern blot analysis.**

**nuclease** An enzyme that catalyzes the degradation of a nucleic acid by breaking phosphodiester bonds.

**nucleic acid** High-molecular-weight polynucleotide. The main nucleic acids in cells are DNA and RNA.

**nucleoid** Central region in a bacterial cell in which the chromosome is compacted.

**nucleoside** A **purine** or **pyrimidine** covalently linked to a sugar.

**nucleoside phosphate** A nucleoside with an attached phosphate group. Also called *nucleotide.*

**nucleosome** The basic structural unit of eukaryotic **chromatin,** consisting of two molecules each of the four core histones (H2A, H2B, H3, and H4, the histone octamer), a single molecule of the linker histone H1, and about 180 bp of DNA.

**nucleosome remodeling complex** Large, multiprotein complex that uses the energy released by ATP hydrolysis to alter the position or structure of nucleosomes, thereby remodeling chromatin structure.

**nucleotide** The type of monomeric molecule found in RNA and DNA. Nucleotides consist of three distinct parts: a pentose (ribose in RNA, deoxyribose in DNA), a nitrogenous base (a purine or pyrimidine), and a phosphate group.

**nucleotide excision repair (NER)** *See* **excision repair.**

**nucleus** A discrete structure within eukaryotic cells that is bounded by a double membrane (the nuclear envelope) and contains most of the DNA of the cell.

**null hypothesis** A hypothesis that states there is no real difference between the observed data and the predicted data.

**nullisomy** A type of **aneuploidy** in which one pair of homologous chromosomes is missing from a normally diploid cell or organism. A nullisomic cell is 2N − 2.

**null mutation** A mutation that results in a protein with no function.

**nutritional mutant** *See* **auxotroph.**

**observed heterozygosity ($H_o$)** The number of individuals in the population that are heterozygous at that locus.

**Okazaki fragments** The short, single-stranded DNA fragments that are synthesized on the lagging-strand template during DNA replication and are subsequently covalently joined to make a continuous strand, the **lagging strand.**

**oligonucleotide** A short DNA molecule.

**oncogene** A gene whose protein product promotes cell proliferation. Oncogenes are altered forms of **proto-oncogenes.**

**oncogenesis** Formation of a tumor (cancer) in an organism.

**one-gene–one-enzyme hypothesis** The hypothesis that each gene controls the synthesis of one enzyme.

**one-gene–one-polypeptide hypothesis** The hypothesis that each gene controls the synthesis of a polypeptide chain.

**oogenesis** Development of female gametes (egg cells) in animals.

**open reading frame (ORF)** In a segment of DNA, a potential protein-coding sequence identified by a start codon in frame with a stop codon.

**operator** A short DNA region, adjacent to the promoter of a bacterial operon, that binds repressor proteins responsible for controlling the rate of transcription of the operon.

**operon** In bacteria, a cluster of adjacent genes that share a common operator and promoter and are transcribed into a single mRNA. All the genes in an operon are regulated coordinately; that is, all are transcribed or none are transcribed.

**optimal alignment** In the comparison of nucleotide or amino acid sequences from two or more organisms, an approximation of the true alignment of sequences where gaps are inserted to maximize the similarity among the sequences being aligned. *See also* **indels.**

**ORF** *See* **open reading frame.**

**origin** A specific site on a DNA molecule at which the double helix denatures into single strands and replication is initiated.

**origin recognition complex (ORC)** A multisubunit complex that functions as an **initiator protein** in eukaryotes.

**origin of replication** A specific region in DNA where the double helix unwinds and synthesis of new DNA strands begins.

**overdominance** *See* **heterosis.**

**ovum** (*plural,* **ova**) A mature female **gamete** (egg cell); the larger of the two cells that arise from a secondary oocyte by meiosis II in the ovary of female animals.

**pachynema** The stage in prophase I of meiosis during which the homologous pairs of chromosomes undergo **crossing-over.**

**paracentric inversion** A chromosomal mutation in which a segment on one chromosome arm that does not include the centromere is inverted.

**parental** *See* **parental genotype.**

**parental class** *See* **parental genotype.**

**parental genotype** The genetic makeup (allele composition) of individuals in the parental generation of genetic crosses. Progeny in succeeding generations may have combinations of linked alleles like one or the other of the parental genotypes or new (nonparental) combinations as the result of **crossing-over.**

**partial reversion** A **point mutation** in a mutant allele that restores all or part of the function of the encoded protein but not the wild-type amino acid sequence.

**particulate factors** The term Mendel used for the entities that carry hereditary information and are transmitted from parents to progeny through the gametes. These factors are now called *genes.*

**PCR** *See* **polymerase chain reaction.**

**pedigree analysis** Study of the inheritance of human traits by compilation of phenotypic records of a family over several generations.

**penetrance** The frequency with which a dominant or homozygous recessive gene is phenotypically expressed within a population.

**pentose sugar** A five-carbon sugar that, along with a nitrogenous base and a phosphate group, is one of the three parts of a **nucleotide.**

**peptide bond** A covalent bond in a polypeptide chain that joins the $\alpha$-carboxyl group of one amino acid to the $\alpha$-amino group of the adjacent amino acid.

**peptidyl transferase** Catalytic activity of an RNA component of the ribosome that forms the peptide bond between amino acids during translation.

**pericentric inversion** A chromosomal mutation in which a segment including the centromere and parts of both chromosome arms is inverted.

**P generation** The parental generation; the immediate parents of $F_1$ offspring.

**phage** Shortened form of **bacteriophage.**

**phage lysate** The progeny phages released after lysis of phage-infected bacteria.

**phage vector** A phage that carries pieces of bacterial DNA between bacterial strains in the process of transduction.

**pharmacogenomics** Study of how a person's unique genome affects the body's response to medicines.

**phenotype** The observable characteristics of an organism that are produced by the genotype and its interaction with the environment.

**phenotypic correlation** An association between two or more **quantitative traits** in the same individual.

**phenotypic variance ($V_P$)** A measure of all the variability for a **quantitative trait** in a population; mathematically is identical to the **variance.**

**phosphate group** An acidic chemical component that, along with a pentose sugar and a nitrogenous base, is one of the three parts of a **nucleotide.**

**phosphodiester bond** A covalent bond in RNA and DNA between a sugar of one nucleotide and a phosphate group of an adjacent nucleotide. Phosphodiester bonds form the repeating sugar–phosphate array of the backbone of DNA and RNA.

**photoreactivation** Repair of **thymine dimers** in DNA by exposure to visible light in the wavelength range 320–370 nm. Also called *light repair.*

**phylogenetic relationship** A reconstruction of the evolutionary history of groups of organisms (taxa) or genes.

**phylogenetic tree** A graphic representation of the evolutionary relationships among a group of species or genes. It consists of *branches* (lines) connecting *nodes*, which represent ancestral or extant organisms. *See also* **maximum parsimony.**

**physical map** A representation of the physical distances, measured in base pairs, between identifiable regions or markers on genomic DNA. A physical map is generated by analysis of DNA sequences rather than by genetic recombination analysis, which is used in constructing a **genetic map.**

**physical marker** Cytologically detectable visible (under the microscrope) changes in the chromosomes that make it possible to distinguish the chromosomes and, hence, the results of crossing-over.

**pistil** The female reproductive organ in flowering plants. It usually consists of a pollen-receiving stigma, stalklike style, and ovary.

**plaque** A round, clear area in a lawn of bacteria on solid medium that results from the lysis of cells by repeated cycles of phage lytic growth.

**plasmid** An extrachromosomal, double-stranded DNA molecule that replicates autonomously from the host chromosome. Plasmids occur naturally in many bacteria and can be engineered for use as **cloning vectors.**

**pleiotropic** Referring to genes or mutations that result in multiple phenotypic effects.

**point mutant** An organism whose mutant phenotype results from an alteration of a single nucleotide pair.

**point mutation** A heritable alteration of the genetic material in which one base pair is changed to another.

**poly(A)+ mRNA** An mRNA molecule in eukaryotes with a 3′ **poly(A) tail.**

**poly(A) polymerase (PAP)** The enzyme that catalyzes formation of the poly(A) tail at the 3′ end of eukaryotic mRNA molecules.

**poly(A) site** In eukaryotic precursor mRNAs (pre-mRNAs), the sequence that directs cleavage at the 3′ end and subsequent addition of adenine nucleotides to form the poly-A tail, during RNA processing.

**poly(A) tail** A sequence of 50 to 250 adenine nucleotides at the 3′ end of most eukaryotic mRNAs. The tail is added during processing of pre-mRNA.

**polycistronic mRNA** An mRNA molecule, transcribed from a bacterial or bacteriophage **operon,** that is translated into all the polypeptide encoded by the structural genes in the operon.

**polygene hypothesis for quantitative inheritance** The hypothesis that **quantitative traits** are controlled by many genes.

**polygenes** Two or more genes whose additive effects determine a particular **quantitative trait.**

**polylinker** *See* **multiple cloning site.**

**polymerase chain reaction (PCR)** A method for producing many copies of a specific DNA sequence from a DNA mixture without having to clone the sequence in a host organism.

**polynucleotide** A linear polymeric molecule composed of **nucleotides** joined by phosphodiester bonds. DNA and RNA are polynucleotides.

**polypeptide** A linear polymeric molecule consisting of **amino acids** joined by peptide bonds. *See also* **protein.**

**polyploidy** Condition in which a cell or organism has more than two sets of chromosomes.

**polyribosome (polysome)** The complex between an mRNA molecule and all the ribosomes that are translating it simultaneously.

**polytene chromosome** A special type of chromosome representing a bundle of numerous chromatids that have arisen by repeated cycles of replication of single chromatids without nuclear division. This type of chromosome is characteristic of various tissues of Diptera.

**population** A specific group of individuals of the same species.

**population genetics** Study of the consequences of Mendelian inheritance on the population level, including the mathematical description of a population's genetic composition and how it changes over time.

**population viability analysis** Analysis of the survival probabilities of different genotypes in the population.

**position effect** A change in the phenotypic effect of one or more genes as a result of a change in their position in the genome.

**positive assortative mating** Preferential mating between phenotypically similar individuals that occurs more frequently than expected for **random mating.**

**postzygotic isolation** Reduction in mating between closely related species by various mechanisms that act after fertilization, resulting in nonviable or sterile hybrids or hybrids of lowered fitness. *See also* **prezygotic isolation.**

**precursor mRNA (pre-mRNA)** The initial (primary) transcript of a protein-coding gene that is modified or processed to produce the mature, functional mRNA molecule.

**precursor rRNA (pre-rRNA)** The initial (primary) transcript produced from ribosomal DNA that is processed into three different rRNA molecules in prokaryotes and eukaryotes.

**precursor tRNA (pre-tRNA)** The initial (primary) transcript of a tRNA gene that is extensively modified and processed to produce the mature, functional tRNA molecule.

**prezygotic isolation** Reduction in mating between closely related species by various mechanisms that prevent courtship, mating, or fertilization. *See also* **postzygotic isolation.**

**Pribnow box** A part of the **promoter** sequence in bacterial genomes that is located at about 10 base pairs upstream from the transcription start site. Also called the *–10 box.*

**primary nondisjunction** A rare event in cells with a normal chromosome complement in which sister chromatids (in mitosis or meiosis II) or homologous chromosomes (in meiosis I) fail to separate and move to opposite poles. *See also* **nondisjunction** and **secondary nondisjunction.**

**primary oocytes** Diploid cells that arise by mitotic division of primordial germ cells (oogonia) and undergo meiosis in the ovaries of female animals.

**primase** *See* **DNA primase.**

**primer** *See* **RNA primer.**

**primosome** A complex of *E. coli* primase, helicase, and other proteins that functions in initiating DNA synthesis.

**principle of independent assortment** Mendel's second law stating that the factors (genes) for different traits assort independently of one another. In other words, genes on different chromosomes behave independently in the production of gametes.

**principle of segregation** Mendel's first law stating that two members of a gene pair (alleles) segregate (separate) from each other during the formation of gametes. As a result, one-half the gametes carry one allele and the other half carry the other allele.

**probability** The ratio of the number of times a particular event occurs to the number of trials during which the event could have happened.

**proband** In human genetics, an affected person with whom the study of a trait in a family begins. *See also* **proposita; propositus.**

**product rule** The rule that the probability of two independent events occurring simultaneously is the product of each of their probabilities.

**programmed cell death** *See* **apoptosis.**

**prokaryote** Any organism whose genetic material is not located within a membrane-bound nucleus. The prokaryotes are divided into two evolutionarily distinct groups, the **Bacteria** and the **Archaea.** *See also* **eukaryote.**

**prometaphase** Stage in **mitosis** in which the mitotic spindle that has been forming between the separating centriole pairs enters the former nuclear area, a kinetochore binds to each centromere, and kinetochore microtubules originating at one or other of the poles attach to each kinetochore.

**prometaphase I** Stage in **meiosis I** in which the nucleoli disappear, the nuclear envelope breaks down, the meiotic spindle that has been forming between the separating centriole pairs enters the former nuclear area, a kinetochore binds to each centromere, and kinetochore microtubules originating at one or other of the poles attach to each kinetochore.

**prometaphase II** Stage in **meiosis II** in which the nuclear envelopes (if formed in **telophase I**) break down, the spindle organizes across the cell, and kinetochore microtubules from the opposite poles attach to the kinetochores of each chromosome.

**promoter** A DNA region containing specific **gene regulatory elements** to which RNA polymerase binds for the initiation of transcription. *See also* **core promoter.**

**promoter-proximal elements** Gene regulatory elements in eukaryotic genomes that are located 50–200 base pairs from the transcription start site (upstream of the TATA **box**) and help determine the efficiency of transcription.

**proofreading** In DNA synthesis, the process of recognizing a base-pair error during the polymerization events and correcting it. Proofreading is carried out by some DNA polymerases in prokaryotic and eukaryotic cells.

**prophage** The genome of a temperate bacteriophage that has been integrated into the chromosome of a host bacterium in the **lysogenic pathway.** A prophage is replicated during replication of the host cell's chromosome.

**prophase** The first stage in mitosis or meiosis during which the replicated chromosomes condense and become visible under the microscope.

**prophase I** The first stage of meiosis, divided into several substages, during which the replicated chromosomes condense, homologues undergo **synapsis,** and **crossing-over** occurs.

**prophase II** The first stage of meiosis II during which the chromosomes condense.

**proportion of polymorphic loci (*P*)** A ratio calculated by determining the number of loci with more than one allele present and dividing by the total number of loci examined.

**proposita** In human genetics, an affected female person with whom the study of a trait in a family begins. *See also* **proband.**

**propositus** In human genetics, an affected male person with whom the study of a trait in a family begins. *See also* **proband.**

**protein** A macromolecule composed of one or more **polypeptides.** The functional activity of a protein depends on its complex folded shape and composition.

**protein array** A collection of different proteins, immobilized on a solid substrate, that serve as probes for detecting labeled target proteins that bind to those affixed to the substrate. Also called *protein microarray* and *protein chip.*

**proteome** The complete set of proteins in a cell.

**proteomics** The cataloging and analysis of the proteins in a cell to determine when they are expressed, how much is made, and which proteins interact.

**proto-oncogene** A gene that in normal cells functions to control the proliferation of cells and that when mutated can become an **oncogene.** *See also* **tumor suppressor gene.**

**prototroph** A strain of an organism that is wild type for all nutritional requirements and can grow on minimal medium. *See also* **auxotroph.**

**prototrophic strain** *See* **prototroph.**

**pseudodominance** The phenotypic expression of a single recessive allele resulting from deletion of a dominant allele on the homologous chromosome.

**pseudogene** A nonfunctional gene that has sequence homology to one or more functional genes elsewhere in the genome.

**Punnett square** A matrix that describes all the possible genotypes of progeny resulting from a genetic cross.

**pure-breeding strain** *See* **true-breeding strain.**

**purine** One of the two types of cyclic nitrogenous bases found in DNA and RNA. Adenine and guanine are purines.

**pyrimidine** One of the two types of cyclic nitrogenous bases found in DNA and RNA. Cytosine (in DNA and RNA), thymine (in DNA), and uracil (in RNA) are pyrimidines.

**pyrosequencing** A DNA sequencing technique using a single-stranded template DNA molecule attached to a bead in which the release of the pyrophosphate in DNA chain growth is detected enzymatically. Pyrosequencing does not involve chain termination.

QTL *See* **quantitative trait loci.**

**quantitative genetics** Study of the inheritance of complex characteristics that are determined by multiple genes.

**quantitative trait** A heritable characteristic that shows a continuous variation in phenotype over a range. Also called *continuous trait.*

**quantitative trait loci (QTL)** The individual loci that contribute to a **quantitative trait.**

**random mating** Matings between individuals of the same or different genotypes that occur in proportion to the frequencies of the genotypes in the population.

**rDNA repeat unit** Set of ribosomal RNA (rRNA) genes—encoding 18S, 5.8S, and 28S rRNAs—that are located adjacent to each other and repeated many times in tandem arrays in eukaryotic genomes.

**reading frame** Linear sequence of codons (groups of three nucleotides) in mRNA that specify amino acids during translation beginning at a particular start codon.

**real-time PCR** A PCR method for measuring the increase in the amount of DNA as it is amplified (which gives the technique its "real-time" name). Also called *real-time quantitative PCR.*

**recessive** Describing an allele or phenotype that is expressed only in the homozygous state.

**recessive lethal allele** An allele that results in the death of organisms homozygous for the allele.

**reciprocal cross** A pair of crosses in which the genotypes of the males and females for a particular trait is reversed. In the garden pea, for example, a reciprocal cross for smooth and wrinkled seeds is smooth female × wrinkled male and wrinkled female × smooth male.

**recombinant** A chromosome, cell, or individual that has nonparental combinations of **genetic markers** as a result of genetic recombination.

**recombinant chromosome** A daughter chromosome that emerges from meiosis with an allele composition that differs from that of either parental chromosome.

**recombinant DNA molecule** Any DNA molecule that has been constructed in the test tube and contains sequences from two or more distinct DNA molecules, often from different organisms.

**recombinant DNA technology** A collection of experimental procedures for inserting a DNA fragment from one organism into DNA from another organism and for cloning the new recombinant DNA.

**recombination** *See* **genetic recombination.**

**regression** A statistical analysis assessing how changes in one variable are quantitatively related to changes in another variable.

**regression coefficient** The slope of the **regression line** drawn to show the relationship between two variables.

**regression line** A mathematically computed line that represents the best fit of a line to the data values for two variables plotted against each other. The slope of the regression line indicates the change in one variable ($y$) associated with a unit increase in another variable ($x$).

**regulated gene** A gene whose expression is controlled in response to the needs of a cell or organism.

**reinforcement** A model which states that, if populations

harbor genetic variation for mate recognition, then the alleles that allow the adults to discriminate successfully will increase in frequency.

**release factor (RF)** One of several proteins that recognize stop codons in mRNA and then initiate a series of specific events to terminate translation.

**replica plating** Procedure for transferring the pattern of colonies from a master plate to a new plate. In this procedure, a velveteen pad on a cylinder is pressed lightly onto the surface of the master plate, thereby picking up a few cells from each colony to inoculate onto the new plate.



**replication bubble** A locally unwound (denatured) region of DNA bounded by replication forks at which DNA synthesis proceeds in opposite directions.

**replication fork** A Y-shaped structure formed when a double-stranded DNA molecule unwinds to expose the two single-stranded **template strands** for DNA replication.

**replicator** The entire set of DNA sequences, including the **origin of replication,** required to direct the initiation of DNA replication.

**replicon** A stretch of DNA in eukaryotic chromosomes extending from an origin of replication to the two termini of replication on each side of that origin. Also called *replication unit*.

**replisome** The complex of closely associated proteins that forms at the replication fork during DNA synthesis in bacteria.

**repressible operon** An **operon** whose transcription is reduced in the presence of a particular substance, often the end product of a biosynthetic pathway. The tryptophan (*trp*) operon is an example of a repressible operon. *See also* **inducible operon.**

**repressor** The major class of transcription regulatory proteins in prokaryotes. Bacterial repressors usually bind to the **operator** and prevent transcription by blocking binding of RNA polymerase. In eukaryotes, repressors act in various ways to control transcription of some genes. *See also* **activators.**

**repressor gene** A regulatory gene whose product is a protein that controls the transcriptional activity of a particular operon or gene.

**repulsion** In individuals heterozygous for two genetic loci, the arrangement in which each homologous chromosome carries the wild-type allele of one gene and the mutant allele of the other gene; also called *trans configuration*. *See also* **coupling.**

**restriction endonuclease** *See* **restriction enzyme.**

**restriction enzyme** Enzyme that cleaves double-stranded DNA molecules within or near a specific nucleotide sequence (restriction site), which often is present in multiple copies with a genome. These enzymes are used in analyzing DNA and constructing recombinant DNA. Also called *restriction endonuclease*.

**restriction fragment length polymorphism (RFLP)** Variation in the lengths of fragments generated by treatment of DNA with a particular restriction enzyme. RFLPs result from point mutations that create or destroy restriction enzyme cleavage sites.

**restriction mapping** Procedure for locating the relative positions of restriction enzyme cleavage sites in a cloned DNA fragment, yielding a restriction map of the fragment.

**restriction site** Sequence in DNA recognized by a **restriction enzyme.** Many restriction enzymes cut both strands of DNA within the restriction site. Some restriction enzymes cut both strands of DNA near the restriction site.

**restriction site linker** A double-stranded oligodeoxyribonucleotide about 8 to 12 base pairs long that contains the cleavage site for a specific restriction enzyme and is used in cloning cDNAs. Also called *linker*.

**retrotransposition** The movement of certain mobile genetic elements (retrotransposons) in the genome by a mechanism involving an RNA intermediate.

**retrotransposon** A type of mobile genetic element, found only in eukaryotes, that encodes **reverse transcriptase** and moves in the genome via an RNA intermediate.

**retrovirus** A virus with a single-stranded RNA genome that replicates via a double-stranded DNA intermediate produced by **reverse transcriptase,** an enzyme encoded in the viral genome. The DNA integrates into the host's chromosome where it can be transcribed.

**reverse genetics** An experimental approach in which investigators attempt to find what phenotype, if any, is associated with a cloned gene.

**reverse mutation** A point mutation in a mutant allele that changes it back to a wild-type allele. Also called *reversion*.

**reverse transcriptase** An enzyme (an RNA-dependent DNA polymerase) that makes a double-stranded DNA copy of an RNA strand.

**reverse transcriptase PCR (RT-PCR)** A two-step method for detecting and quantitating a particular RNA in an RNA mixture by first converting the RNAs to cDNAs and then performing the **polymerase chain reaction (PCR)** using primers specific for the RNA of interest.

**reversion** *See* **reverse mutation.**

**ribonuclease (RNase)** An enzyme that catalyzes degradation of RNA to nucleotides.

**ribonucleic acid (RNA)** A usually single-stranded polymeric molecule consisting of ribonucleotide building blocks. The major types of RNA in cells are **ribosomal RNA (rRNA), transfer RNA (tRNA), messenger RNA (mRNA), small nuclear RNA (snRNA),** and **microRNA (miRNA),** each of which performs an essential role in protein synthesis (translation). In some viruses, RNA is the genetic material.

**ribonucleotide** Any of the nucleotides that make up RNA, consisting of a sugar (ribose), a base, and a phosphate group.

**ribose** The pentose (five-carbon) sugar found in RNA.

**ribosomal DNA (rDNA)** The regions of the genome that contain the genes for rRNAs in prokaryotes and eukaryotes.

**ribosomal proteins** A group of proteins that along with rRNA molecules make up the ribosomes of prokaryotes and eukaryotes.

**ribosomal RNA (rRNA)** Class of RNA molecules of several different sizes that, along with ribosomal proteins, make up ribosomes of prokaryotes and eukaryotes.

**ribosome** A large, complex cellular particle composed of ribosomal protein and rRNA molecules that is the site of amino acid polymerization during protein synthesis (translation).

**ribosome-binding site (RBS)** The nucleotide sequence in an mRNA molecule on which the ribosome becomes oriented in the correct reading frame for the initiation of translation. More commonly called the **Shine–Dalgarno sequence.**

**ribosome recycling factor (RRF)** A protein shaped like a tRNA molecule that, after translation termination, participates with

EF-G in steps to release the uncharged tRNA and to cause the two ribosomal subunits to dissociate from the mRNA.

**ribozyme** An RNA molecule that has catalytic activity.

**RNA** *See* **ribonucleic acid.**

**RNA editing** Unusual type of RNA processing in which the nucleotide sequence of a pre-mRNA is changed by the posttranscriptional insertion or deletion of nucleotides or by conversion of one nucleotide to another.

**RNA enzyme** *See* **ribozyme.**

**RNA interference (RNAi)** Silencing of the expression of a specific gene by double-stranded RNA whose sequence matches a portion of the mature mRNA encoded by the gene. Also called *RNA silencing*.

**RNA polymerase** Any enzyme that catalyzes the synthesis of RNA molecules from a DNA template in a process called **transcription.**

**RNA polymerase I** An enzyme in eukaryotes that catalyzes transcription of 18S, 5.8S, and 28S rRNA genes.

**RNA polymerase II** An enzyme in eukaryotes that catalyzes transcription of mRNA-coding genes and some snRNA genes.

**RNA polymerase III** An enzyme in eukaryotes that catalyzes transcription of tRNA and 5S rRNA genes and of some snRNA genes.

**RNA primer** A short RNA chain, produced by DNA primase during DNA replication, to which DNA polymerase adds nucleotides, thereby extending the new DNA strand.

**RNA silencing** *See* **RNA interference (RNAi).**

**RNA splicing** *See* **mRNA splicing.**

**RNA synthesis** *See* **transcription.**

**RNA world hypothesis** Theory proposing that RNA-based life predates the present-day DNA-based life, with the RNA carrying out the necessary catalytic reactions required for life in the presumably primitive cells of the time.

**Robertsonian translocation** A type of nonreciprocal translocation in which the long arms of two nonhomologous **acrocentric chromosomes** become attached to a single centromere.

**rolling circle replication** Process that occurs when a circular, double-stranded DNA replicates to produce linear DNA.

**rooted tree** A **phylogenetic tree** in which one internal node is represented as a common ancestor to all the other nodes on the tree.

**rRNA transcription unit** *See* **ribosomal DNA.**

**RRF** *See* **ribosome recycling factor.**

**RT-PCR** *See* **reverse transcriptase PCR.**

**sample** Subset of individuals belonging to a population. Study of a sample can provide accurate information about the population if the sample is large enough and randomly selected.

**sampling error** Chance deviations from expected results that arise when the observed sample is small.

**secondary nondisjunction** Abnormal segregation of the X chromosomes during meiosis in the progeny of females with the XXY genotype produced by a primary nondisjunction. *See also* **nondisjunction,** and **primary nondisjunction.**

**secondary oocyte** The larger of the two daughter cells produced by unequal cytokinesis during meiosis I of a primary oocyte in the ovaries of female animals.

**second law** *See* **principle of independent assortment.**

**segmentation genes** Group of genes in *Drosophila* that determine the number and organization of segments in the embryo and adult.

**selection** The favoring of particular combinations of genes in a given environment.

**selection coefficient (*s*)** A measure of the relative intensity of selection against a genotype; equals $1 - w$ **(Darwinian fitness).**

**selection differential (*s*)** In natural and artificial selection, the difference between the mean phenotype of the selected parents and the mean phenotype of the unselected population.

**selection response (*R*)** The amount by which a phenotype changes in one generation when natural or artificial selection is applied to a group of individuals.

**self-fertilization (selfing)** The union of male and female gametes from the same individual.

**selfing** *See* **self-fertilization.**

**self-splicing** The excision of introns from some pre-RNA molecules that occurs by a protein-independent reaction in certain organisms.

**semiconservative model** A model for DNA replication in which each daughter molecule retains one of the parental strands. The results of the Meselson–Stahl experiment supported this model.

**semidiscontinuous** Concerning DNA replication, when one new strand (the **leading strand**) is synthesized continuously and the other strand (the **lagging strand**) is synthesized discontinuously.

**sex chromosome** A chromosome in eukaryotic organisms that differs morphologically or in number in the two sexes. In many organisms, one sex possesses a pair of visibly different chromosomes. One is an X chromosome, and the other is a Y chromosome. Commonly, the XX sex is female and the XY sex is male.

**sex-influenced trait** A characteristic controlled by autosomal genes that appears in both sexes, but either the frequency of its occurrence or the relationship between genotype and phenotype is different in males and females.

**sex-limited trait** A characteristic controlled by autosomal genes that is phenotypically exhibited in only one of the two sexes.

**sex-linked** *See* **X-linked.**

**sexual reproduction** Mode of reproduction involving the fusion of haploid gametes produced directly or indirectly by meiosis.

**Shine–Dalgarno sequence** A sequence in prokaryotic mRNAs upstream of the start codon that base-pairs with an RNA in the small ribosomal subunit, allowing the ribosome to locate the start codon for correct initiation of translation. Also called the **ribosome-binding site (RBS).**

**short interfering RNA (siRNA)** Short double-stranded RNAs that function in gene silencing by **RNA interference (RNAi).**

**short interspersed elements** *See* **SINEs.**

**short tandem repeat (STR)** A type of **DNA polymorphism** involving variation in the number of short identical sequences (2 to 6 bp in length) that are tandemly repeated at a particular locus in the genome. Also called *microsatellite* and *simple sequence repeat*.

**shuttle vector** A **cloning vector** that can be introduced into and replicate in two or more host organisms (e.g., *E. coli* and yeast).

**signal hypothesis** The hypothesis that secreted proteins are synthesized on ribosomes that are directed to the endoplasmic reticulum (ER) by an amino terminal **signal sequence** in the growing polypeptide chain.

**signal peptidase** An enzyme in the cisternal space of the endoplasmic reticulum that catalyzes removal of the **signal sequence** from growing polypeptide chains.

**signal recognition particle (SRP)** A cytoplasmic ribonucleoprotein complex that binds to the ER **signal sequence** of a growing polypeptide, blocking further translation of the mRNA in the cytosol.

**signal recognition particle (SRP) receptor** *See* **SRP receptor.**

**signal sequence** Hydrophobic sequence of 15–30 amino acids at the amino end of a growing polypeptide chain that directs the chain–mRNA–ribosome complex to the endoplasmic reticulum (ER) where translation is completed. The signal sequence is removed and degraded in the cisternal space of the endoplasmic reticulum.

**signal transduction** Process by which an external signal, such as a growth factor, leads to a particular cell response.

**silencer element** In eukaryotes, an **enhancer** that binds a repressor and acts to decrease RNA transcription rather than stimulating it, as most enhancers do.

**silent mutation** A **point mutation** in a gene that changes a codon in the mRNA to another codon for the same amino acid, resulting in no change in the amino acid sequence or function of the encoded protein.

**simple telomeric sequences** Short, tandemly repeated nucleotide sequences at or very close to the extreme ends of chromosomal DNA molecules. The same species-specific sequence is present at the ends of all chromosomes in an organism.

**SINEs (short interspersed elements)** One class of **dispersed repeated DNA** consisting of sequences that are 100 to 400 bp in length. SINEs can move in the genome by **retrotransposition.**

**single nucleotide polymorphism (SNP)** A difference in one base pair at a particular site (SNP locus) within coding or noncoding regions of the genome. SNPs that affect restriction sites cause **restriction fragment length polymorphisms (RFLPs).**

**single-strand DNA-binding (SSB) protein** A protein that binds to the unwound DNA strands at a **replication bubble** and prevents them from reannealing.

**sister chromatids** Two identical copies of a chromosome derived from replication of the chromosome during interphase of the cell cycle. Sister chromatids are held together by the replicated but unseparated centromeres.

**site-specific mutagenesis** Introduction of a mutation at a specific site in a particular gene by one of several in vitro techniques.

**slope of the line** *See* **regression coefficient.**

**small nuclear ribonucleoprotein particle (snRNP)** Large complex formed by small nuclear RNAs (snRNAs) and proteins in which the processing of pre-mRNA molecules occurs.

**small nuclear RNA (snRNA)** Class of RNA molecules, found only in eukaryotes, that associate with certain proteins to form small nuclear ribonucleoprotein particles (snRNPs).

**SNP (single nucleotide polymorphism) locus** Site of a simple, single base-pair alteration found between individuals that can be used as a **DNA marker.**

**somatic mutation** In multicellular organisms, a change in the genetic material of somatic (body) cells. It may affect the phenotype of the individual in which the mutation occurs but is not passed on to the succeeding generation.

**sonicate** The use of very high-frequency sound (well beyond what we can hear) to disrupt cells or molecules.

**Southern blot analysis** A technique for detecting specific DNA fragments in which the fragments are separated by gel electrophoresis, transferred from the gel to a nitrocellulose filter, and then hybridized with labeled complementary probes; also called *Southern blotting. See also* **northern blot analysis.**

**specialized transducing phage** A temperate bacteriophage that can transfer only a certain section of the bacterial chromosome from one bacterium to another.

**specialized transduction** A type of transduction in which only specific genes are transferred from one bacterium to another.

**species tree** A **phylogenetic tree** based on the divergence observed within multiple genes. A species tree is better than a **gene tree** for depicting the evolutionary history of a group of species.

**spermatogenesis** Development of male gametes (sperm cells) in animals.

**sperm cell** A mature male **gamete,** produced by the testes in male animals. Also called *spermatozoon* (plural: *spermatozoa*).

**spliceosome** Large complex in the nucleus of eukaryotic cells that carries out **mRNA splicing.** It consists of several small nuclear ribonucleoprotein particles (snRNPs) bound to a pre-mRNA molecule.

**spontaneous mutation** Any mutation that occurs without the use of a chemical or physical mutagenic agent.

**sporophyte** The haploid asexual generation in the life cycle of plants that produces haploid spores by meiosis.

**SRP receptor** The signal recognition particle (SRP) receptor is an integral protein in the membrane of the **endoplasmic reticulum (ER)** to which binds the complex of a growing **polypeptide, signal recognition particle (SRP),** and **ribosome.** This interaction facilitates binding of the ribosome to the outside surface of the ER and the insertion of the polypeptide into the lumen of the ER.

**stamen** The male reproductive organ in flowering plants. It usually consists of a stalklike filament bearing a pollen-producing anther.

**standard deviation** The square root of the **variance;** a common measure of the extent of variability in a population for **quantitative traits.**

**standard error of allele frequency** A statistical measure of the amount of variation in allele frequency among populations.

**steroid hormone response element (HRE)** DNA sequence to which a complex of a specific steroid hormone and its receptor binds, resulting in activation of genes regulated by that hormone.

**stop codon** One of three codons in mRNA for which no normal tRNA molecule exists and that signals the termination of polypeptide synthesis.

**STR** *See* **short tandem repeat.**

**submetacentric chromosome** A chromosome with the centromere nearer one end than the other such that one arm is longer than the other.

**substitution** A mutation that has passed through the filter of selection on at least some level.

**sum rule** The rule that the probability of either of two mutually exclusive events occurring is the sum of their individual probabilities.

**supercoiled** Referring to a double-stranded DNA molecule that is twisted in space about its own axis.

**suppressor gene** A gene that when mutated causes suppression of mutations in other genes.

**suppressor mutation** A mutation at a second site that totally or partially restores a function lost because of a primary mutation at another site.

**synapsis** The intimate association of replicated homologous chromosomes brought about by the formation of a zipper-like structure (the synaptonemal complex) between the homologues during prophase I of meiosis.

**synaptonemal complex** A complex structure that spans the region between meiotically paired (synapsed) chromosomes and facilitates **crossing-over.**

**synonymous** Referring to nucleotides in a gene that when mutated do not result in a change in the amino acid sequence of the encoded wild-type protein.

**tag SNP** One (or more) SNP locus used to test for and represent an entire haplotype.

**tandemly repeated DNA** Repetitive DNA sequences that are clustered together in the genome, so that each such sequence is repeated many times in a row within a particular chromosomal region.

**TATA box** A part of the **core promoter** in eukaryotic genomes; it is located about 30 base pairs upstream from the transcription start point. Also called the TATA element, or the **Goldberg–Hogness box.**

**tautomers** Alternate chemical forms in which DNA (or RNA) bases are able to exist.

**telocentric chromosome** A chromosome with the centromere more or less at one end such that only one arm is visible.

**telomerase** An enzyme that adds short, tandemly repeated DNA sequences **(simple telomeric sequences)** to the ends of eukaryotic chromosomes. It contains an RNA component complementary to the telomeric sequence and has **reverse transcriptase** activity.

**telomere** A specific set of sequences at the end of a linear chromosome that stabilizes the chromosome and is required for replication. *See also* **simple telomeric sequences** and **telomere-associated sequences.**

**telomere-associated sequence** Repeated, complex DNA sequence extending inward from the simple telomeric sequence at each end of a chromosomal DNA molecule.

**telophase** The stage in mitosis or meiosis during which the migration of the daughter chromosomes to the two poles is completed.

**telophase I** The stage in meiosis I, when chromosomes (each a sister chromatid pair) complete migration to the poles and new nuclear envelopes form around each set of replicated chromosomes.

**telophase II** The last stage of meiosis II, during which a nuclear membrane forms around each set of daughter chromosomes and **cytokinesis** takes place.

**temperate phage** A bacteriophage that is capable of following either the **lytic cycle** or **lysogenic pathway.** *See also* **virulent phage.**

**temperature-sensitive mutant** A strain that exhibits a wild-type phenotype in one temperature range but a defective (mutant) phenotype in another, usually higher, temperature range.

**template strand** DNA strand on which is synthesized a complementary DNA strand during replication or an RNA strand during transcription.

**terminator** A DNA sequence located at the distal (downstream) end of a gene that signals the termination of transcription.

**testcross** A cross of an individual of unknown genotype, usually expressing the dominant phenotype, with a homozygous recessive individual to determine the unknown genotype.

**testis-determining factor** Gene product in placental mammals that causes embryonic gonadal tissue to develop into testes; in the absence of this factor, the gonadal tissue develops as ovaries.

**tetrasomy** A type of **aneuploidy** in which a normally diploid cell or organism possesses four copies of a particular chromosome instead of two copies. A tetrasomic cell is 2N + 2.

**three-point testcross** A cross between an individual heterozygous at three loci with an individual homozygous for recessive alleles at the same three loci. Commonly used in mapping linked genes to determine their order in the chromosome and the distances between them.

**thymine (T)** A **pyrimidine** found in DNA but not in RNA. In double-stranded DNA, thymine pairs with adenine, a **purine,** by hydrogen bonding.

**thymine dimer** A common lesion in DNA, caused by ultraviolet radiation, in which adjacent thymines in the same strand are linked in an abnormal way that distorts the double helix at that site.

**topoisomerase** Any enzyme that catalyzes the supercoiling of DNA.

**totipotent** Describing a cell that has the potential to develop into any cell type of the organism.

**trailer sequence** *See* **3′ untranslated region (3′ UTR).**

**trait** *See* **hereditary trait.**

**transconjugant** A bacterial cell that incorporates donor DNA received during conjugation into its genome.

**transcription** The process for making a single-stranded RNA molecule complementary to one strand (the template strand) of a double-stranded DNA molecule, thereby transferring information from DNA to RNA. Also called *RNA synthesis*.

**transcriptome** The set of mRNA transcripts in a cell.

**transcriptomics** The study of gene expression at the level of the entire genome.

***trans*-dominant** Referring to a gene or DNA sequence that can control genes on different DNA molecules.

**transducing phage** Any bacteriophage that can mediate transfer of genetic material between bacteria by **transduction.**

**transducing retrovirus** Retrovirus that has picked up an **oncogene** from the genome of a host cell.

**transductant** In bacteria, a recombinant recipient cell generated by **transduction.**

**transduction** A process by which bacteriophages mediate the transfer of pieces of bacterial DNA from one bacterium (the donor) to another (the recipient).

**transfer RNA (tRNA)** Class of RNA molecules that bring amino acids to ribosomes, where they are transferred to growing polypeptide chains during translation.

**transformant** In bacteria, a recombinant recipient cell generated by **transformation.**

**transformation** (a) In bacteria, a process in which genetic information is transferred by means of extracellular pieces of DNA. (b) In eukaryotes, the conversion of a normal cell with regulated growth properties to a cancer-like cell that can give rise to tumors.

**transforming principle** Term coined by Frederick Griffith for the unknown agent responsible for the change in genotype via transformation in bacteria. DNA is now known to constitute the transforming principle.

**transgene** A gene introduced into the genome of an organism by genetic manipulation to alter its genotype.

**transgenic** Referring to a cell or organism whose genotype has been altered by the artificial introduction of a different allele or gene from the same or a different species.

**transition** *See* **transition mutation.**

**transition mutation** A type of **base-pair substitution mutation** that involves a change of one purine-pyrimidine base pair to the other purine–pyrimidine base pair (e.g., A-T to G-C) at a particular site in the DNA.

**translation** The process that converts the nucleotide sequence of an mRNA into the amino acid sequence of a polypeptide. Also called *protein synthesis*.

**translesion DNA synthesis** An inducible DNA repair process that allows the replication of DNA beyond a lesion that normally would interrupt DNA synthesis. In *E. coli*, this process is called the *SOS response*.

**translocation** (a) A chromosomal mutation involving a change in the position of a chromosome segment (or segments) and the gene sequences it contains. (b) In polypeptide synthesis, translocation is the movement of the ribosome, one codon at a time, along the mRNA toward the 3′ end.

**transmission genetics** Study of how genes are passed from one individual to another. Also called *classical genetics*.

**transposable element** A DNA segment that can move from one position in the genome to another (nonhomologous) position; also called *mobile genetic element*. Transposable elements are found in both prokaryotes and eukaryotes.

**transposase** An enzyme encoded by many types of mobile genetic elements that catalyzes the movement **(transposition)** of these elements in the genome.

**transposition** The movement of a transposable element within the genome. *See also* **retrotransposition.**

**transposon (Tn)** A mobile genetic element that contains a gene for transposase, which catalyzes transposition, and genes with other functions such as antibiotic resistance.

**transversion** *See* **transversion mutation.**

**transversion mutation** A type of **base-pair substitution mutation** that involves a change of a purine–pyrimidine base pair to a pyrimidine–purine base pair (e.g., A-T to T-A or G-C to T-A) at a particular site in the DNA.

**trihybrid cross** A cross between individuals of the same genotype that are heterozygous for three pairs of alleles at three different loci (e.g., *Ss Yy Cc* × *Ss Yy Cc*).

**trisomy** A type of **aneuploidy** in which a normally diploid cell or organism possesses three copies of a particular chromosome instead of two copies. A trisomic cell is 2N + 1.

**trisomy-13** The presence of an extra copy of chromosome 13, which causes Patau syndrome in humans.

**trisomy-18** The presence of an extra copy of chromosome 18, which causes Edwards syndrome in humans.

**trisomy-21** The presence of an extra copy of chromosome 21, which causes Down syndrome in humans.

**true-breeding strain** A strain in which mating of individuals yields progeny with the same genotype as the parents.

**true reversion** A **point mutation** in a mutant allele that restores it to the wild-type allele; as a result, the wild-type amino acid sequence and function of the encoded protein is restored.

**tumor** A tissue mass composed of transformed cells, which multiply in an uncontrolled fashion and differ from normal cells in other ways as well; also called *neoplasm*. Benign tumors do not invade the surrounding tissues, whereas malignant tumors invade tissue and often spread to other sites in the body.

**tumor suppressor gene** A gene in normal cells whose protein product suppresses uncontrolled cell proliferation. *See also* **proto-oncogene.**

**tumor virus** A virus that induces cells to dedifferentiate and to divide to produce a tumor.

**Turner syndrome** A human clinical syndrome that results from monosomy for the X chromosome in the female, which gives a 45,X female. Affected females fail to develop secondary sexual characteristics, tend to be short, have weblike necks, have poorly developed breasts, are usually infertile, and exhibit mental deficiencies.

**unequal crossing-over** The process of chromosomal interchange between misaligned chromosomes that may occur during meiosis.

**uniparental inheritance** A phenomenon, usually exhibited by mitochondrial and chloroplast genes, in which all progeny have the phenotype of only one parent.

**unique-sequence DNA** A class of DNA sequences, each of which is present in one to a few copies in the haploid chromosome set; includes most protein-coding genes. Also called *single-copy DNA*.

**3′ untranslated region (3′ UTR)** The untranslated part of an mRNA molecule beginning at the end of the amino acid-coding sequence and extending to the 3′ end of the mRNA.

**5′ untranslated region (5′ UTR)** In eukaryotes, the untranslated part of an mRNA molecule extending from the **5′** end to the first (start) codon. It contains coded information for directing initiation of protein synthesis at the translation start site.

**unweighted pair group method with arithmetic averages (UPGMA)** A statistically based approach used in constructing **phylogenetic trees** that groups taxa based on their overall pairwise similarities to each other. Also called *cluster analysis*.

**uracil (U)** A **pyrimidine** found in RNA but not in DNA.

**variable number tandem repeat (VNTR)** A type of **DNA polymorphism** involving variation in the number of identical sequences (7 bp to a few tens of base pairs in length) that are tandemly repeated at a particular locus in the genome. Also called a *minisatellite*.

**variance** A statistical measure of the extent to which values in a data set differ from the **mean.**

**virulent phage** A bacteriophage, such as T4, that always follows the **lytic cycle** when it infects bacteria. *See also* **temperate phage.**

**visible mutation** A mutation that affects the morphology or physical appearance of an organism.

**VNTR** *See* **variable number tandem repeat.**

**whole-genome shotgun approach for genome sequencing** An approach for sequencing an entire genome in which the whole genome is broken into partially overlapping fragments, each fragment is cloned and sequenced, and the genome sequence is assembled from the overlapping sequences by computer.

**wild type** Term describing an allele or phenotype that is designated as the standard ("normal") for an organism and is usually, but not always, the most prevalent in a "wild" population of the organism; also used in reference to a strain or individual.

**wild-type allele** *See* **wild type.**

**wobble hypothesis** A proposed mechanism that explains how one **anticodon** can pair with more than one **codon.**

**X chromosome** A sex chromosome present in two copies in the homogametic sex (the female in mammals) and in one copy in the heterogametic sex (the male in mammals).

**X chromosome–autosome balance system** A genotypic sex determination system in which the ratio between the numbers of X chromosomes and number of sets of autosomes is the primary determinant of sex.

**X chromosome nondisjunction** Failure of the two X chromosomes to separate in meiosis so that eggs are produced with two X chromosomes or with no X chromosomes instead of the usual one X chromosome.

**X-linked** Referring to genes located on the X chromosome.

**X-linked dominant trait** A characteristic caused by a dominant mutant allele carried on the X chromosome.

**X-linked recessive trait** A characteristic caused by a recessive mutant allele carried on the X chromosome.

**Y chromosome** A sex chromosome that when present is found in one copy in the heterogametic sex, along with an X chromosome, and is not present in the homogametic sex. Not all organisms with sex chromosomes have a Y chromosome.

**Y chromosome mechanism of sex determination** A genotypic system of sex determination in which the Y chromosome determines the sex of an individual. Individuals with a Y chromosome are genetically male, and individuals without a Y chromosome are genetically female.

**yeast artificial chromosome (YAC)** A vector for cloning large DNA fragments, several hundred kilobase pairs long, in yeast. A YAC is a linear molecular with a telomere at each end, a centromere, an autonomously replicating sequence (ARS), a selectable marker, and a **polylinker.**

**yeast two-hybrid system** Experimental procedure to find genes encoding proteins that interact with a known protein. Also called *interaction trap assay*.

**Y-linked trait** A characteristic controlled by a gene carried on the Y chromosome for which there is no corresponding gene locus on the X chromosome. Also called *holandric* or *"wholly male" trait.*

**zygonema** The stage in prophase I of meiosis during which homologous chromosomes begin to pair in a highly specific way along their lengths.

**zygote** The cell produced by the fusion of a male gamete (sperm cell) and a female gamete (egg cell).

# Suggested Readings

This section contains references to selected classic and relevant research papers and reviews, as well as selected websites for the topics presented in the chapters. To learn more about any topic, look for general information using keywords with the Google search engine (www.google.com). You may also search for specific research and review papers via keyword at the National Library of Medicine, PubMed website (www.pubmed.gov), and through Pearson's Research Navigator™ database, available at the *iGenetics* student website.


**Chapter 1:** Genetics: An Introduction

Genetics Review. http://www.ncbi.nlm.nih.gov/Class/MLACourse/Original8Hour/Genetics/

Sturtevant, A. H. 1965. *A history of genetics*. New York: Harper & Row.

**Chapter 2:** DNA: The Genetic Material

DNA Structure. http://www.johnkyrk.com/DNAanatomy.html

Arya, G., and Schlick, T. 2006. Role of histone tails in chromatin folding revealed by a mesoscopic oligonucleosome model. *Proc. Natl. Acad. Sci. USA* 103:16236–16241.

Avery, O. T., MacLeod, C. M., and McCarty, M. 1944. Studies on the chemical nature of the substance inducing transformation of pneumococcal types. Induction of transformation by a deoxyribonucleic acid fraction isolated from pneumococcus type III. *J. Exp. Med.* 79:137–158.

Blackburn, E. H. 1994. Telomeres: No end in sight. *Cell* 77:621–623.

Britten, R. J., and Kohne, D. E. 1968. Repeated sequences in DNA. *Science* 161:529–540.

Chargaff, E. 1951. Structure and function of nucleic acids as cell constituents. *Fed. Proc.* 10:654–659.

Clarke, L. 1990. Centromeres of budding and fission yeasts. *Trends Genet.* 6:150–154.

D'Ambrosio, E., Waitzikin, S. D., Whitney, F. R., Salemme, A., and Furano, A. V. 1985. Structure of the highly repeated, long interspersed DNA family (LINE or L1Rn) of the rat. *Mol. Cell. Biol.* 6:411–424.

Dickerson, R. E. 1983. The DNA helix and how it is read. *Sci. Am.* 249 (Dec):94–111.

Franklin, R. E., and Gosling, R. 1953. Molecular configuration of sodium thymonucleate. *Nature* 171:740–741.

Geis, I. 1983. Visualizing the anatomy of A, B, and Z-DNAs. *J. Biomol. Struct. Dyn.* 1:581–591.

Gierer, A., and Schramm, G. 1956. Infectivity of ribonucleic acid from tobacco mosaic virus. *Nature* 177:702–703.

Greider, C. W. 1999. Telomeres do D-loop–T-loop. *Cell* 97:419–422.

Griffith, F. 1928. The significance of pneumococcal types. *J. Hyg. (Lond.)* 27:113–159.

Griffith, J. D., Corneau, L., Rosenfield, S., Stansel, R. M., Bianchi, A., Moss, H., and de Lange, T. 1999. Mammalian telomeres end in a large duplex loop. *Cell* 97:503–514.

Grosschedl, R., Giese, K., and Pagel, J. 1994. HMG domain proteins: Architectural elements in the assembly of nucleoprotein structures. *Trends Genet.* 10:94–100.

Grunstein, M. 1998. Yeast heterochromatin: Regulation of its assembly and inheritance by histones. *Cell* 93:325–328.

Hershey, A. D., and Chase, M. 1952. Independent functions of viral protein and nucleic acid in growth of bacteriophage. *J. Gen. Physiol.* 36:39–56.

Jaworski, A., Hsieh, W. T., Blaho, J. A., Larson, J. E., and Wells, R. D. 1987. Left-handed DNA in vivo. *Science* 238:773–777.

Korenberg, J. R., and Rykowski, M. C. 1988. Human genome organization: Alu, LINES, and the molecular structure of metaphase chromosome bands. *Cell* 53:391–400.

Kornberg, R. D., and Klug, A. 1981. The nucleosome. *Sci. Am.* 244 (Feb):52–64.

Kornberg, R. D., and Lorch, Y. 1999. Twenty-five years of the nucleosome, fundamental particle of the eukaryote chromosome. *Cell* 98:285–294.

Krishna, P., Kennedy, B. P., van de Sande, J. H., and McGhee, J. D. 1988. Yolk proteins from nematodes, chickens, and frogs bind strongly and preferentially to left-handed Z-DNA. *J. Biol. Chem.* 263:19066–19070.

Mason, J. M., and Biessmann, A. 1995. The unusual telomeres of *Drosophila*. *Trends Genet.* 11:58–62.

Moyzis, R. K. 1991. The human telomere. *Sci. Am.* 265 (Aug):48–55.

Olins, A. L., Carlson, R. D., and Olins, D. E. 1975. Visualization of chromatin substructure: Nu-bodies. *J. Cell Biol.* 64:528–537.

Pauling, L., and Corey, R. B. 1956. Specific hydrogen-bond formation between pyrimidines and purines in deoxyribonucleic acids. *Arch. Biochem. Biophys.* 65:164–181.

Pluta, A. F., Mackay, A. M., Ainsztein, A. M., Goldberg, I. G., and Earnshaw, W. C. 1995. The centromere: Hub of chromosomal activities. *Science* 270:1591–1594.

Pruss, D., Bartholomew, B., Persinger, J., Hayes, J., Arents, G., Moudrianakis, E. N., and Wolfe, A. P. 1996. An asymmetric model for the nucleosome: A binding site for linker histones inside the DNA gyres. *Science* 274:614–617.

Singer, M. F. 1982. SINEs and LINEs: Highly repeated short and long interspersed sequences in mammalian genomes. *Cell* 28:133–134.

Sinsheimer, R. L. 1959. A single-stranded deoxyribonucleic acid from bacteriophage ΦX174. *J. Mol. Biol.* 1:43–53.

Wang, A. H. J., Quigley, G. J., Kolpak, F. J., Crawford, J. L., van Boom, J. H., van der Marel, G., and Rich, A. 1979. Molecular structure of a left-handed double helical DNA fragment at atomic resolution. *Nature* 282:680–686.

Wang, J. C. 1982. DNA topoisomerases. *Sci. Am.* 247 (Jul):94–109.

Watson, J. D. 1968. *The double helix*. New York: Atheneum.

Watson, J. D., and Crick, F. H. C. 1953. Genetical implications of the structure of deoxyribonucleic acid. *Nature* 171:964–969.

———. 1953. Molecular structure of nucleic acids. A structure for deoxyribose nucleic acid. *Nature* 171:737–738.

Wilkins, M. H. F., Stokes, A. R., and Wilson, H. R. 1953. Molecular structure of deoxypentose nucleic acids. *Nature* 171:738–740.

### **Chapter 3:** DNA Replication

DNA Replication. http://users.rcn.com/jkimball.ma.ultranet/BiologyPages/D/DNAReplication.html

Andrews, B., and Measday, V. 1998. The cyclin family of budding yeast: Abundant use of a good idea. *Trends Genet.* 14:66–72.

Armanios, M., and Greider, C. W. 2005. Telomerase and cancer stem cells. *Cold Spring Harbor Symp. Quant. Biol.* 70:205–208.

Blasco, M. A. 2005. Telomeres and human disease: Ageing, cancer, and beyond. *Nature Rev. Genet.* 6:611–622.

Chan, S. R. W. L., and Blackburn, E. H. 2004. Telomeres and telomerase. *Phil. Trans. R. Soc. Lond. B.* 359:109–121.

Cimbora, D. M., and Groudine, M. 2001. The control of mammalian DNA replication: A brief history of space and timing. *Cell* 104:643–646.

Cook, P. R. 1999. The organization of replication and transcription. *Science* 284:1790–1795.

DeLucia, P., and Cairns, J. 1969. Isolation of an *E. coli* strain with a mutation affecting DNA polymerase. *Nature* 224:1164–1166.

Diller, J. D., and Raghuraman, M. K. 1994. Eukaryotic replication origins: Control in space and time. *Trends Biochem.* 19:320–325.

Flores, I., Cayuela, M. L., and Blasco, M. A. 2005. Effects of telomerase and telomere length on epidermal stem cell behavior. *Science* 209:1253–1256.

Gilbert, W., and Dressler, D. 1968. DNA replication: The rolling circle model. *Cold Spring Harbor Symp. Quant. Biol.* 33:473–484.

Grabowski, B., and Kelman, Z. 2003. Archaeal DNA replication: Eukaryal proteins in a bacterial context. *Annu. Rev. Microbiol.* 57:487–516.

Greider, C. W., and Blackburn, E. H. 1996. Telomeres, telomerase and cancer. *Sci. Am.* 274 (Feb):92–97.

Huberman, J. A., and Riggs, A. D. 1968. On the mechanism of DNA replication in mammalian chromosomes. *J. Mol. Biol.* 32:327–341.

Kornberg, A. 1960. Biologic synthesis of deoxyribonucleic acid. *Science* 131:1503–1508.

Lendvay, T. S., Morris, D. K., Sah, J., Balasubramanian, B., and Lundblad, V. 1996. Senescence mutants of *Saccharomyces cerevisiae* with a defect in telomere replication identify three additional *EST* genes. *Genetics* 144:1399–1412.

Meselson, M., and Stahl, F. W. 1958. The replication of DNA in *Escherichia coli*. *Proc. Natl. Acad. Sci. USA* 44:671–682.

Nieduszynski, C. A., Knox, Y., and Donaldson, A. D. 2006. Genome-wide identification of replication origins in yeast by comparative genomics. *Genes Dev.* 20:1874–1879.

Ogawa, T., Baker, T. A., van der Ende, A., and Kornberg, A. 1985. Initiation of enzymatic replication at the origin of the *Escherichia coli* chromosome: Contributions of RNA polymerase and primase. *Proc. Natl. Acad. Sci. USA* 82:3562–3566.

Ogawa, T., and Okazaki, T. 1980. Discontinuous DNA replication. *Annu. Rev. Biochem.* 49:424–457.

Okazaki, R. T., Okazaki, K., Sakobe, K., Sugimoto, K., and Sugino, A. 1968. Mechanism of DNA chain growth. I. Possible discontinuity and unusual secondary structure of newly synthesized chains. *Proc. Natl. Acad. Sci. USA* 59:598–605.

Rossi, M. L., and Bambara, R. A. 2006. Reconstituted Okazaki fragment processing indicates two pathways of primer removal. *J. Biol. Chem.* 281:26051–26061.

Runge, K. W., and Zakian, V. A. 1996. *TEL2*, an essential gene required for telomere length regulation and telomere position effect in *Saccharomyces cerevisiae*. *Mol. Cell. Biol.* 16:3094–3105.

Shippen-Lentz, D., and Blackburn, E. H. 1990. Functional evidence for an RNA template in telomerase. *Science* 247:546–552.

Stillman, B. 1994. Smart machines at the DNA replication fork. *Cell* 78:725–728.

Taylor, J. H. 1970. The structure and duplication of chromosomes. In *Genetic organization*, E. Caspari and A. Ravin, eds., vol. 1 (pp. 163–221). New York: Academic Press.

Van der Ende, A., Baker, T. A., Ogawa, T., and Kornberg, A. 1985. Initiation of enzymatic replication at the origin of the *Escherichia coli* chromosome: Primase as the sole priming enzyme. *Proc. Natl. Acad. Sci. USA* 82:3954–3958.

Wright, W. E., Piatyszek, M. A., Rainey, W. E., Byrd, W., and Shay, J. W. 1996. Telomerase activity in human germline and embryonic tissues and cells. *Dev. Genet.* 18:173–179.

Zyskind, J. W., and Smith, D. W. 1986. The bacterial origin of replication, *oriC*. *Cell* 46:489–490.

### **Chapter 4:** Gene Function

Beadle, G. W., and Tatum, E. L. 1942. Genetic control of biochemical reactions in *Neurospora*. *Proc. Natl. Acad. Sci. USA* 27:499–506.

Bush, A., Chodhari, R., Collins, N., Copeland, F., Hall, P., Harcourt, J., Hariri, M., Hogg, C., Lucas, J., Mitchison, H. M., O'Callaghan, C., and Phillips, G. 2007. Primary ciliary dyskinesia: Current state of the art. *Arch. Dis. Child.* 92:1136–1140. [*Primary ciliary dyskinesia* is a pseudonym for *Kartagener syndrome*.]

Collins, F. 1992. Cystic fibrosis: Molecular biology and therapeutic implications. *Science* 256:774–779.

Garrod, A. E. 1909. *Inborn errors of metabolism*. New York: Oxford University Press.

Geremek, M., and Witt, M. 2004. Primary ciliary dyskinesia: Genes, candidate genes and chromosomal regions. *J. Appl. Genet.* 45:347–361. [*Primary ciliary dyskinesia* is a pseudonym for *Kartagener syndrome.*]

Gilbert, F., Kucherlapati, R., Creagan, R. P., Murnane, M. J., Darlington, G. J., and Ruddle, F. H. 1975. Tay–Sachs and Sandhoff's diseases: The assignment of genes for hexosaminidase A and B to individual human chromosomes. *Proc. Natl. Acad. Sci. USA* 72:263–267.

Gusella, J. F., Wexler, N. S., Conneally, P. M., Naylor, S. L., Anderson, M. A., Tanzi, R. E., Watkins, P. C., Ottina, K., Wallace, M. R., Sakaguchi, A. Y., Young, A. B., Shoulson, I., Bonilla, E., and Martin, J. B. 1993. A polymorphic DNA marker genetically linked to Huntington's disease. *Nature* 306:234–238.

Guttler, F., and Woo, S. L. C. 1986. Molecular genetics of PKU. *J. Inherit. Metab. Dis.* 9 (Suppl. 1):58–68.

Ha, M. N., Graham, F. L., D'Souza, C. K., Muller, W. J., Igdoura, S. A., and Schellhorn, H. E. 2004. Functional rescue of vitamin C synthesis deficiency in human cells using adenoviral-based expression of murine l-gulono-gamma-lactone oxidase. *Genomics* 83:482–492.

Inai, Y., Ohta, Y., and Nishikimi, M. 2003. The whole structure of the human nonfunctional L-gulono-gamma-lactone oxidase gene—the gene responsible for scurvy—and the evolution of repetitive sequences thereon. *J. Nutr. Sci. Vitaminol. (Tokyo)* 49:315–319.

Ingram, V. M. 1963. *The hemoglobins in genetics and evolution.* New York: Columbia University Press.

Kaput, J., and Rodriguez, R. L. 2004. Nutritional genomics: The next frontier in the postgenomic era. *Physiol. Genomics* 16:166–177.

McIntosh, I., and Cutting, G. R. 1992. Cystic fibrosis transmembrane conductance regulator and the etiology and pathogenesis of cystic fibrosis. *FASEB J.* 6:2775–2782.

Motulsky, A. G. 1973. Frequency of sickling disorders in U.S. blacks. *N. Engl. J. Med.* 288:31–33.

Neel, J. V. 1949. The inheritance of sickle-cell anemia. *Science* 110:64–66.

Pauling, L., Itano, H. A., Singer, S. J., and Wells, J. C. 1949. Sickle-cell anemia, a molecular disease. *Science* 110:543–548.

Riordan, J. R., Rommens, J. M., Kerem, B., Alon, N., Rozmahel, R., Grzelczak, Z., Zielenski, J., Lok, S., Plavsic, N., Chou, J. L., Drumm, M. L., Ianuzzi, M. C., Collins, F. S., and Tsui, L. C. 1989. Identification of the cystic fibrosis gene: Cloning and characterization of complementary DNA. *Science* 245:1066–1073.

Rommens, J. M., Ianuzzi, M. C., Kerem, B., Drumm, M. L., Melmer, G., Dean, M., Rozmahel, R., Cole, J. L., Kennedy, D., Hidaka, N., Zsiga, M., Buchwald, M., Riordan, J. R., Tsui, L. C., and Collins, F. S. 1989. Identification of the cystic fibrosis gene: Chromosome walking and jumping. *Science* 245:1059–1065.

Scriver, C. R., and Clow, C. L. 1980. Phenylketonuria and other phenylalanine hydroxylation mutants in man. *Annu. Rev. Genet.* 14:179–202.

Scriver, C. R., and Waters, P. J. 1999. Monogenic traits are not simple: Lessons from phenylketonuria. *Trends Genet.* 15:267–272.

Srb, A. M., and Horowitz, N. H. 1944. The ornithine cycle in *Neurospora* and its genetic control. *J. Biol. Chem.* 154:129–139.


### **Chapter 5:** Gene Expression: Transcription


Eukaryotic Transcription. http://www.mun.ca/biochem/courses/3107/Topics/euk_transcription.html

Gene Expression: Transcription. http://users.rcn.com/jkimball.ma.ultranet/BiologyPages/T/Transcription.html

Baker, T. A., and Bell, S. P. 1998. Polymerases and the replisome: Machines within machines. *Cell* 92:295–305.

Barabino, S. M. L., and Keller, W. 1999. Last but not least: Regulated poly(A) tail formation. *Cell* 99:9–11.

Bogenhagen, D. F., Sakonju, S., and Brown, D. D. 1980. A control region in the center of the 5S RNA gene directs specific initiation of transcription II: The 3′ border of the region. *Cell* 19:27–35.

Breathnach, R., and Chambon, P. 1981. Organization and expression of eucaryotic split genes coding for proteins. *Annu. Rev. Biochem.* 50:349–383.

Breathnach, R., Mandel, J. L., and Chambon, P. 1977. Ovalbumin gene is split in chicken DNA. *Nature* 270:314–318.

Brody, E., and Abelson, J. 1985. The "spliceosome": Yeast premessenger RNA associates with a 40S complex in a splicing-dependent reaction. *Science* 228:963–967.

Buratowski, S. 1994. The basics of basal transcription by RNA polymerase II. *Cell* 77:1–3.

Busby, S., and Ebright, R. H. 1994. Promoter structure, promoter recognition, and transcription activation in prokaryotes. *Cell* 79:743–746.

Cate, J. H., Yusupov, M. M., Yusupova, G. Z., Earnest, T. N., and Noller, H. F. 1999. X-ray crystal structures of 70S ribosome functional complexes. *Science* 285:2095–2104.

Cech, T. R. 1983. RNA splicing: Three themes with variations. *Cell* 34:713–716.

———. 1985. Self-splicing RNA: Implications for evolution. *Int. Rev. Cytol.* 93:3–22.

———. 1986. The generality of self-splicing RNA: Relationship to nuclear mRNA splicing. *Cell* 44:207–210.

Choi, Y. D., Grabowski, P. J., Sharp, P. A., and Dreyfuss, G. 1986. Heterogeneous nuclear ribonucleoproteins: Role in RNA splicing. *Science* 231:1534–1539.

Cook, P. R. 1999. The organization of replication and transcription. *Science* 284:1790–1795.

Cramer, P., Bushnell, D. A., Fu, J., Gnatt, A. L., Maier-Davis, B., Thompson, N. E., Burgess, R. R., Edwards, A. M., David, P. R., and Kornberg, R. D. 2000. Architecture of RNA polymerase II and implications for the transcription mechanism. *Science* 288:640–649.

Crick, F. H. C. 1979. Split genes and RNA splicing. *Science* 204:264–271.

Grabowski, P. J., Seiler, S. R., and Sharp, P. A. 1985. A multicomponent complex is involved in the splicing of messenger RNA precursors. *Cell* 42:355–367.

Green, M. R. 1986. Pre-mRNA splicing. *Annu. Rev. Genet.* 20:671–708.

———. 1991. Biochemical mechanisms of constitutive and regulated pre-mRNA splicing. *Annu. Rev. Cell Biol.* 7:559–599.

Guarente, L. 1988. UASs and enhancers: Common mechanism of transcriptional activation in yeast and mammals. *Cell* 52:303–305.

Guthrie, C. 1992. Messenger RNA splicing in yeast: Clues to why the spliceosome is a ribonucleoprotein. *Science* 253:157–163.

Guthrie, C., and Patterson, B. 1988. Spliceosomal snRNAs. *Annu. Rev. Genet.* 22:387–419.

Horowitz, D. S., and Krainer, A. R. 1994. Mechanisms for selecting 5′ splice sites in mammalian pre-mRNA splicing. *Trends Genet.* 10:100–105.

Jeffreys, A. J., and Flavell, R. A. 1977. The rabbit beta-globin gene contains a large insert in the coding sequence. *Cell* 12:1097–1108.

Kim, T. H., Barrera, L. O., Zheng, M., Qu, C., Singer, M. A., Richmond, T. A., Wu, Y., Green, R. D., and Ren, B. 2005. A high-resolution map of active promoters in the human genome. *Nature* 436:876–880.

Kim, M., Vasiljeva, L., Rando, O. J., Zhelkovsky, A., Moore, C., and Buratowski, S. 2006. Distinct pathways for snoRNA and mRNA termination. *Mol. Cell* 24:723–734.

Korzheva, N., Mustaev, A., Kozlov, M., Malhotra, A., Nikiforov, V., Goldfarb, A., and Darst, S. A. 2000. A structural model of transcription elongation. *Science* 289:619–625.

Marmur, J., Greenspan, C. M., Palecek, E., Kahan, F. M., Levine, J., and Mandel, M. 1963. Specificity of the complementary RNA formed by *Bacillus subtilis* infected with bacteriophage SP8. *Cold Spring Harbor Symp. Quant. Biol.* 28:191–199.

Narlikar, G. J., Fan, H. Y., and Kingston, R. E. 2002. Cooperation between complexes that regulate chromatin structure and transcription. *Cell* 108:475–487.

Nilsen, T. W. 1994. RNA–RNA interactions in the spliceosome: Unraveling the ties that bind. *Cell* 78:1–4.

Nomura, M. 1973. Assembly of bacterial ribosomes. *Science* 179:864–873.

O'Hare, K. 1995. mRNA 3' ends in focus. *Trends Genet.* 11:253–257.

Orphanides, G., and Reinberg, D. 2000. A unified theory of gene expression. *Cell* 108:439–451.

Padgett, R. A., Grabowski, P. J., Konarska, M. M., and Sharp, P. A. 1985. Splicing messenger RNA precursors: Branch sites and lariat RNAs. *Trends Biochem. Sci.* (April):154–157.

Proudfoot, N., Furger, A., and Dye, A. J. 2002. Integrating mRNA processing with transcription. *Cell* 108:501–512.

Reed, R. 2003. Coupling transcription, splicing and mRNA export. *Curr. Opin. Cell Biol.* 15:326–331.

Sharp, P. A. 1985. On the origin of RNA splicing and introns. *Cell* 42:397–400.

———. 1994. Split genes and RNA splicing. Nobel lecture. *Cell* 77:805–815.

Simpson, L., and Thiemann, O. H. 1995. Sense from nonsense: RNA editing in mitochondria of kinetoplastid protozoa and slime molds. *Cell* 81:837–840.

Sollner-Webb, B. 1988. Surprises in polymerase III transcription. *Cell* 52:153–154.

Thompson, C. C., and McKnight, S. L. 1992. Anatomy of an enhancer. *Trends Genet.* 8:232–236.

Tilghman, S. M., Curis, P. J., Tiemeier, D. C., Leder, P., and Weissman, C. 1978. The intervening sequence of a mouse β-globin gene is transcribed within the 15S β-globin mRNA precursor. *Proc. Natl. Acad. Sci. USA* 75:1309–1313.

Tilghman, S. M., Tiemeier, D. C., Seidman, J. G., Peterlin, B. M., Sullivan, M., Maizel, J. V., and Leder, P. 1978. Intervening sequence of DNA identified in the structural portion of a mouse beta-globin gene. *Proc. Natl. Acad. Sci. USA* 78:725–729.

Weinstock, R., Sweet, R., Weiss, M., Cedar, H., and Axel, R. 1978. Intragenic DNA spacers interrupt the ovalbumin gene. *Proc. Natl. Acad. Sci. USA* 75:1299–1303.

White, R. J., and Jackson, S. P. 1992. The TATA-binding protein: A central role in transcription by RNA polymerases I, II, and III. *Trends Genet.* 8:284–288.

Woychik, N. A., and Hampsey, M. 2002. The RNA polymerase II machinery: Structure illuminates function. *Cell* 108:453–463.

Zaug, A. J., and Cech, T. R. 1986. The intervening sequence RNA of *Tetrahymena* is an enzyme. *Science* 231:470–475.

**Chapter 6:** Gene Expression: Translation

Translation. http://users.rcn.com/jkimball.ma.ultranet/BiologyPages/T/Translation.html

Ban, N., Nissen, P., Hansen, J., Moore, P. B., and Steitz, T. A. 2000. The complete atomic structure of the large ribosomal subunit at 2.4Å resolution. *Science* 289:905–920.

Blobel, G., and Dobberstein, B. 1975. Transfer of proteins across membranes. I. Presence of proteolytically processed and unprocessed nascent immunoglobulin light chains on membrane-bound ribosomes of murine myeloma. *J. Cell Biol.* 67:835–851.

Brenner, S., Jacob, F., and Meselson, M. 1961. An unstable intermediate carrying information from genes to ribosomes for protein synthesis. *Nature* 190:576–581.

Carter, A. P., Clemons, W. M., Brodersen, D. E., Morgan-Warren, R. J., Wimberly, B. T., and Ramakrishnan, V. 2000. Functional insights from the structure of the 30S ribosomal sub-unit and its interactions with antibiotics. *Nature* 407:340–348.

Crick, F. H. C. 1966. Codon–anticodon pairing: The wobble hypothesis. *J. Mol. Biol.* 19:548–555.

Crick, F. H. C., Barnett, L., Brenner, S., and Watts-Tobin, R. J. 1961. General nature of the genetic code for proteins. *Nature* 192:1227–1232.

Garen, A. 1968. Sense and nonsense in the genetic code. *Science* 160:149–159.

Khorana, H. G. 1966–67. Polynucleotide synthesis and the genetic code. *Harvey Lect.* 62:79–105.

Kimchi-Sarfaty, C., Oh, J. M., Kim, I. W., Sauna, Z. E., Calcagno, A. M., Ambudkar, S. V., and Gottesman, M. M. 2007. A "silent" polymorphism in the *MDR*1 gene changes substrate specificity. *Science* 315:525–528.

Komar, A. A. 2007. SNPs, silent but not invisible. *Science* 315:466–467.

Kozak, M. 1983. Comparison of initiation of protein synthesis in procaryotes, eucaryotes, and organelles. *Microbiol. Rev.* 47:1–45.

———. 1989. Context effects and inefficient initiation at non-AUG codons in eukaryotic cell-free translation systems. *Mol. Cell. Biol.* 9:5073–5080.

McCarthy, J. E. G., and Brimacombe, R. 1994. Prokaryotic translation: The interactive pathway leading to initiation. *Trends Genet.* 10:402–407.

Meyer, D. I. 1982. The signal hypothesis: A working model. *Trends Biochem. Sci.* 7:320–321.

Morgan, A. R., Wells, R. D., and Khorana, H. G. 1966. Studies on polynucleotides. LIX. Further codon assignments from amino acid incorporation directed by ribopolynucleotides containing repeating trinucleotide sequences. *Proc. Natl. Acad. Sci. USA* 56:1899–1906.

Nierhaus, K. H. 1990. The allosteric three-site model for the ribosomal elongation cycle: Features and future. *Biochemistry* 29:4997–5008.

Nirenberg, M., and Leder, P. 1964. RNA code words and protein synthesis. *Science* 145:1399–1407.

Nirenberg, M., and Matthaei, J. H. 1961. The dependence of cell-free protein synthesis in *E. coli* upon naturally occurring or synthetic polyribonucleotides. *Proc. Natl. Acad. Sci. USA* 47:1588–1602.

Nissen, P., Hansen, J., Ban, N., Moore, P. B., and Steitz, T. A. 2000. The structural basis of ribosome activity in peptide bond formation. *Science* 289:920–930.

Noller, H. F., Hoffarth, V., and Zimniak, L. 1992. Unusual resistance of peptidyl transferase to protein extraction procedures. *Science* 256:1416–1419.

Ramakrishnan, V. 2002. Ribosome structure and the mechanism of translation. *Cell* 108:557–572.

Ryan, K. R., and Jensen, R. E. 1995. Protein translocation across mitochondrial membranes: What a long, strange trip it is. *Cell* 83:517–519.

Schnell, D. J. 1995. Shedding light on the chloroplast protein import machinery. *Cell* 83:521–524.

Shine, J., and Dalgarno, L. 1974. The 3′-terminal sequence of *Escherichia coli* 16S ribosomal RNA: Complementarity to nonsense triplet and ribosome binding sites. *Proc. Natl. Acad. Sci. USA* 71:1342–1346.

Watson, J. D. 1963. The involvement of RNA in the synthesis of proteins. *Science* 140:17–26.

Zheng, N., and Gierasch, L. M. 1996. Signal sequences: The same yet different. *Cell* 86:849–852.


## **Chapter 7:** DNA Mutation, DNA Repair, and Transposable Elements


Profiles in Science; The Barbara McClintock Papers. http://profiles.nlm.nih.gov/LL/

Transposable genetic elements. http://www.ndsu.nodak.edu/instruct/mcclean/plsc431/transelem/trans1.htm

Ames, B. N., Durston, W. E., Yamasaki, E., and Lee, F. 1973. Carcinogens are mutagens: A simple test system combining liver homogenates for activation and bacteria for detection. *Proc. Natl. Acad. Sci. USA* 70:2281–2285.

Boeke, J. D., Garfinkel, D. J., Styles, C. A., and Fink, G. R. 1985. Ty elements transpose through an RNA intermediate. *Cell* 40:491–500.

Boyce, R. P., and Howard-Flanders, P. 1964. Release of ultraviolet light-induced thymine dimers from DNA in *E. coli K12*. *Proc. Natl. Acad. Sci. USA* 51:293–300.

Cleaver, J. E. 1994. It was a very good year for DNA repair. *Cell* 76:1–4.

Cohen, S. N., and Shapiro, J. A. 1980. Transposable genetic elements. *Sci. Am.* 242 (Feb):40–49.

Devoret, R. 1979. Bacterial tests for potential carcinogens. *Sci. Am.* 241 (Aug):40–49.

Federoff, N. V. 1989. About maize transposable elements and development. *Cell* 56:181–191.

Fishel, R., Lescoe, M. K., Rao, M. R. S., Copeland, N. G., Jenkins, N. A., Garber, J., Kane, M., and Kolodner, R. 1993. The human mutator gene homolog *MSH2* and its association with hereditary nonpolyposis colon cancer. *Cell* 75:1027–1038.

Kingsman, A. J., and Kingsman, S. M. 1988. Ty: A retroelement moving forward. *Cell* 53:333–335.

Lederberg, J., and Lederberg, E. M. 1952. Replica plating and indirect selection of bacterial mutants. *J. Bacteriol.* 63:399–406.

Luria, S. E., and Delbrück, M. 1943. Mutations of bacteria from virus sensitivity to virus resistance. *Genetics* 28:491–511.

Makarova, K. S., Aravind, L., Wolf, Y. I., Tatusov, R. L., Minton, K. W., Koonin, E. V., and Daly, M. J. 2001. Genome of the extremely radiation-resistant bacterium *Deinococcus radiodurans* viewed from the perspective of comparative genomics. *Microbiol. Mol. Biol. Rev.* 65:44–79.

Makarova, K. S., Omelchenko, M. V., Gaidamakova, E. K., Matrosova, V. Y., Vasilenko, A., Zhai, M., Lapidus, A., Copeland, A., Kim, E., Land, M., Mavrommatis, K., Pitluck, S., Richardson, P. M., Detter, C., Brettin, T., Saunders, E., Lai, B., Ravel, B., Kemner, K. M., Wolf, Y. I., Sorokin, A., Gerasimova, A. V., Gelfand, M. S., Fredrickson, J. K., Koonin, E. V., and Daly, M. J. 2007. *Deinococcus geothermalis*: The pool of extreme radiation resistance genes shrinks. *PLoS ONE* 2:e955.

McClintock, B. 1939. The behavior in successive nuclear divisions of a chromosome broken at meiosis. *Proc. Natl. Acad. Sci. USA* 25:405–416.

———. 1950. The origin and behavior of mutable loci in maize. *Proc. Natl. Acad. Sci. USA* 36:344–355.

———. 1951. Chromosome organization and genic expression. *Cold Spring Harbor Symp. Quant. Biol.* 16:13–47.

———. 1953. Induction of instability at selected loci in maize. *Genetics* 38:579–599.

———. 1956. Controlling elements and the gene. *Cold Spring Harbor Symp. Quant. Biol.* 21:197–216.

———. 1961. Some parallels between gene control systems in maize and in bacteria. *Am. Naturalist* 95:265–277.

———. 1965. The control of gene action in maize. *Brookhaven Symp. Biol.* 18:162 ff.

———. 1984. The significance of responses of the genome to challenge. Nobel lecture. *Science* 226:792–801.

Morgan, A. R. 1993. Base mismatches and mutagenesis: How important is tautomerism? *Trends Biochem. Sci.* 18:160–163.

Pashin, Y. V., and Bakhitova, L. M. 1979. Mutagenic and carcinogenic properties of polycyclic aromatic hydrocarbons. *Env. Health Perspectives* 30:185–189.

Setlow, R. B., and Carrier, W. L. 1964. The disappearance of thymine dimers from DNA: An error-correcting mechanism. *Proc. Natl. Acad. Sci. USA* 51:226–231.

Tessman, I., Liu, S. K., and Kennedy, A. 1992. Mechanism of SOS mutagenesis of UV-irradiated DNA: Mostly error-free processing of deaminated cytosine. *Proc. Natl. Acad. Sci. USA* 89:1159–1163.


## **Chapter 8:** Genomics: The Mapping and Sequencing of Genomes


Cloning and Molecular Analysis of Genes. http://www.ndsu.nodak.edu/instruct/mcclean/plsc431/cloning/

Genome Research and Genetics News: *Nature* Genome Gateway. http://www.nature.com/genomics/

Recombinant DNA and Gene Cloning. http://users.rcn.com/jkimball.ma.ultranet/BiologyPages/R/RecombinantDNA.html

Adams, M. D., et al. 2000. The genome sequence of *Drosophila melanogaster. Science* 287:2185–2215.

*Arabidopsis* Genome Initiative. 2000. Analysis of the genome sequence of the flowering plant *Arabidopsis thaliana*. *Nature* 408:796–815.

Arber, W. 1965. Host-controlled modification of bacteriophage. *Annu. Rev. Microbiol.* 19:365–378.

Arber, W., and Dussoix, D. 1962. Host specificity of DNA produced by *Escherichia coli* I. Host controlled modification of bacteriophage lambda. *J. Mol. Biol.* 5:18–36.

Blattner, F. R., *et al.* 1997. The complete genome sequence of *Escherichia coli* K-12. *Science* 277:1453–1463.

Boyer, H. W. 1971. DNA restriction and modification mechanisms in bacteria. *Annu. Rev. Microbiol.* 25:153–176.

Bult, C. J., *et al.* 1996. Complete genome sequence of the methanogenic archaeon, *Methanococcus jannaschii. Science* 273:1058–1073.

The *C. elegans* Sequencing Consortium. 1998. Genome sequence of the nematode *C. elegans:* A platform for investigating biology. *Science* 282:2012–2018.

Danna, K., and Nathans, D. 1971. Specific cleavage of simian virus 40 DNA by restriction endonuclease of *Haemophilus influenzae. Proc. Natl. Acad. Sci. USA* 68:2913–2917.

Dujon, B. 1996. The yeast genome project: What did we learn? *Trends Genet.* 12:263–270.

Dunham, I., *et al.* 1999. The DNA chromosome of human chromosome 22. *Nature* 402:489–495.

Fleischmann, R. D., *et al.* Whole-genome random sequencing and assembly of *Haemophilus influenzae* Rd. *Science* 269:496–512.

Foote, S., Vollrath, D., Hilton, A., and Page, D. C. 1992. The human Y chromosome: Overlapping DNA clones spanning the euchromatic region. *Science* 258:60–66.

Fraser, C. M., *et al.* 1995. The minimal gene complement of *Mycoplasma genitalium. Science* 270:397–403.

Fraser, C. M., *et al.* 1998. Complete genome sequence of *Treponema pallidum,* the syphilis spirochete. *Science* 281:375–388.

Goffeau, A., *et al.* 1996. Life with 6000 genes. *Science* 274:546–567.

International Genome Sequencing Consortium. 2004. Finishing the euchromatic sequence of the human genome. *Nature* 431:931–945.

Klenk, H. P., *et al.* 1997. The complete genome sequence of the hyperthermophilic, sulphate-reducing archaeon *Archaeoglobus fulgidus. Nature* 390:364–370.

Kornberg, T. B., and Krasnow, M. A. 2000. The *Drosophila* genome sequence: Implications for biology and medicine. *Science* 287:2218–2220.

Luria, S. E. 1953. Host-induced modification of viruses. *Cold Spring Harbor Symp. Quant. Biol.* 18:237–244.

Mouse Genome Sequencing Consortium. 2002. Initial sequencing and comparative analysis of the mouse genome. *Nature* 420:520–562.

*Nature* 15 February 2001. An issue with a special section on "The Human Genome," an analysis of the draft sequence of the human genome.

Rat Genome Sequencing Project Consortium. 2004. Genome sequence of the Brown Norway rat yields insights into mammalian evolution. *Nature* 428:493–521

Rubin, G. M., and Lewis, E. B. 2000. A brief history of *Drosophila's* contributions to genome research. *Science* 287:2216–2218.

Sambrook, J., Fritsch, E. F., and Maniatis, T. 1989. *Molecular cloning: A laboratory manual*, 2nd ed. Cold Spring Harbor, NY: Cold Spring Harbor Laboratory.

Sanger, F., and Coulson, A. R. 1975. A rapid method for determining sequences in DNA by primed synthesis with DNA polymerase. *J. Mol. Biol.* 94:441–448.

*Science* 16 February 2001. An issue focused on "The Human Genome," an analysis of the draft sequence of the human genome.

Tang, C. M., Hood, D. W., and Moxon, E. R. 1997. *Haemophilus* influence: The impact of whole genome sequencing on microbiology. *Trends Genet.* 13:399–404.

Watson, J. D., Gilman, M., Witkowski, J., and Zoller, M. 1992. *Recombinant DNA*, 2nd ed. New York: Scientific American Books, Freeman.

Wayne, R. K., and Ostrander, E. O. 2007. Lessons learned for the dog genome. *Trends Genet.* 11:557–567.



**Chapter 9:** Functional and Comparative Genomics

Genome Research and Genetics News: *Nature* Genome Gateway. http://www.nature.com/genomics/

Mouse knockout project. http://www.nih.gov/science/models/mouse/knockout/

Polymerase Chain Reaction (PCR): Cloning DNA in the Test Tube. http://users.rcn.com/jkimball.ma.ultranet/BiologyPages/P/PCR.html

Allzadeh, A. A., *et al.* 2000. Distinct types of diffuse large B-cell lymphoma identified by gene expression profiling. *Nature* 403:503–511.

Bevan, M., and Murphy, G. 1999. The small, the large and the wild. The value of comparison in plant genomics. *Trends Genet.* 15:211–214.

Breitbart, M., Hewson, I., Felts, B., Mahaffy, J. M., Nulton, J., Salamon, P., and Rohwer, F. 2003. Metagenomic analyses of an uncultured viral community from human feces. *J. Bacteriol.* 185:6220–6223.

Cho, R. J., Campbell, M. J., Winzeler, E. A., Steinmetz, L., Conway, A., Wodicka, L., Wolfsberg, T. G., Gabriellan, A. E., Landsman, D., Lockhart, D. J., and Davis, R. W. 1998. A genome-wide transcriptional analysis of mitotic cell cycle. *Mol. Cell* 2:65–73.

Chu, S., DeRisi, J., Eisen, M., Mulholland, J., Botstein, D., Brown, P. O., and Herskowitz, I. 1998. The transcriptional program of sporulation in budding yeast. *Science* 282:699–705.

DeRisi, J. L., Iyer, V. R., and Brown, P. O. 1997. Exploring the metabolic and genetic control of gene expression on a genomic scale. *Science* 278:680–686.

Dib, C., Fauré, S., Fizames, C., Samson, D., Drouot, N., Vignal, A., Millasseau, P., Marc, S., Hazan, J., Seboun, E., Lathrop, M., Gyapay, G., Morissette, J., and Weissenbach, J. 1996. A comprehensive genetic map of the human genome based on 5,264 microsatellites. *Nature* 380:152–154.

Gill, S. R., Pop, M., Deboy, R. T., Eckburg, P. B., Turnbaugh, P. J., Samuel, B. S., Gordon, J. I., Relman, D. A., Fraser-Liggett, C. M., and Nelson, K. E. 2006. Metagenomic analysis of the human distal gut microbiome. *Science* 312:1355–1359.

Goetze, S., Mateos-Langerak, J., Gierman, H. J., De Leeuw, W., Giromus, O., Indemans, M. H. G., Koster, J., Ondrej, V., Versteeg, R., and van Driel, R. 2007. The three-dimensional structure of human interphase chromosomes is related to the transcriptome map. *Mol. Cell. Biol.* 27:4475–4487.

Green, R. E., Krause, J., Ptak, S. E., Briggs, A. W., Ronan, M. T., Simons, J. F., Du, L., Egholm, M., Rothberg, J. M., Paunovic, M., and Pääbo, S. 2006. Analysis of one million base pairs of Naeanderthal DNA. *Nature* 444:330–336.

Krings, M., Stone, A., Schmitz, R. W., Krainitzki, H., Stoneking, M., and Pääbo, S. 1997. Neanderthal DNA sequences and the origin of modern humans. *Cell* 90:19–30.

Mullis, K. B. 1990. The unusual origin of the polymerase chain reaction. *Sci. Am.* 262 (Apr):56–65.

Mullis, K. B., and Faloona, F. A. 1987. Specific synthesis of DNA *in vitro* via a polymerase-catalyzed chain reaction. *Methods Enzymol.* 155:335–350.

O'Brien, S. J., Wienberg, J., and Lyons, L. A. 1997. Comparative genomics: Lessons from cats. *Trends Genet.* 13:393–399.

Pääbo, S. 1993. Ancient DNA. *Sci. Am.* 269 (Nov):86–92.

Palacios, G., Druce, J., Du, L., Tran, T., Birch, C., Briese, T., Conlan, S., Quan, P. L., Hui, J., Marshall, J., Simons, J. F., Egholm, M., Paddock, C. D., Shieh, W. J., Goldsmith, C. S., Zaki, S. R., Catton, M., and Lipkin, W. I. 2008. A new arenavirus in a cluster of fatal transplant-associated diseases. *N. Engl. J. Med.* 358:991–998.

Pollard, K. S., Salamai, S. R., Lambert, N., Lambot, M. A., Coppens, S., Pedersen, J. S., Katzman, S., King, B., Onodera, C., Siepel, A., Kern, A. D., Dehay, C., Igel, H., Ares, M., Vanderhaegen, P., and Haussler, D. 2006. An RNA gene expressed during cortical development evolved rapidly in humans. *Science* 443:167–172.

Sambrook, J., Fritsch, E. F., and Maniatis, T. 1989. *Molecular cloning: A laboratory manual*, 2nd ed. Cold Spring Harbor, NY: Cold Spring Harbor Laboratory.

Schlabach, M. R., Luo, J., Solimini, N. L., Hu, G., Xu, O., Li, M. Z., Zhao, Z., Smogorzewska, A., Sowa, M. E., Ang, X. L., Westbrook, T. F., Liang, A. C., Chang, K., Hackett, J. A., Harper, J. W., Hannon, G. J., and Elledge S. J. 2008. Cancer proliferation gene discovery through functional genomics. *Science* 319:620–624.

Silva, J. M., Marran, K., Parker, J. S., Silva, J., Golding, M., Schlabach, M. R., Elledge, S. J., Hannon, G. J., and Chang K. 2008. Profiling essential genes in human mammary cells by multiplex RNAi screening. *Science* 319:617–620.

Various authors. 1999. The chipping forecast. *Nat. Genet.* 21(suppl.):1–60.

Versteeg, R., van Schaik, B. D. C., van Batenburg, M. F., Roos, M., Monajemi, R., Caron, H., Bussemaker, H. J., and van Kampen, A. H. C. 2003. The human transcriptome make revelas extremes in gene density, intron length, GC content, and repeat pattern for domains of highly and weakly expressed genes. *Genome Res.* 13:1998–2004.

Voight, B. F., Kudaravalli, S., Wen, X., and Pritchard, J. K. 2006. A map of recent positive selection in the human genome. *PLoS Biology* 4(3):e72.

Wang, D., Coscoy, L., Zylberberg, M., Avila, P. C., Boushey, H. A., Ganem, D., and DeRisi, J. L. 2002. Microarray-based detection and genotyping of viral pathogens. *Proc. Natl. Acad. Sci. USA* 99:15687–15692.

Watson, J. D., Gilman, M., Witkowski, J., and Zoller, M. 1992. *Recombinant DNA*, 2nd ed. New York: Scientific American Books, Freeman.

White, R., and Lalouel, J. M. 1988. Chromosome mapping with DNA markers. *Sci. Am.* 258 (February):40–48.

White, T. J., Arnheim, N., and Erlich, H. A. 1989. The polymerase chain reaction. *Trends Genet.* 5:185–188.

Young, R. A. 2000. Biomedical discovery with DNA arrays. *Cell* 102:9–15.


### **Chapter 10:** Recombinant DNA Technology


DNA Typing and Identification. http://faculty.ncwc.edu/toconnor/425/425lect15.htm

Gene Therapy. http://www.ornl.gov/sci/techresources/Human_Genome/medicine/genetherapy.shtml

What is Genetic Testing? http://www.lbl.gov/Education/ELSI/Frames/genetic-testing.html

Anderson, W. F. 1992. Human gene therapy. *Science* 256:808–813.

Antonarakis, S. E. 1989. Diagnosis of genetic disorders at the DNA level. *N. Engl. J. Med.* 320:153–163.

Cavazzana-Calvo, M., Havein-Bey, S., de Saint Basile, G., Gross, F., Yvon, E., Nusbaum, P., Selz, F., Hu, C., Certain, S., Casanova, J. L., Bousso, P., Le Deist, F., and Fischer, A. 2000. Gene therapy of human severe combined immunodeficiency (SCID)-X1 disease. *Science* 288:669–672.

Chien, C. T., Bartel, P. L., Sternglanz, R., and Fields, S. 1991. The two-hybrid system: A method to identify and clone genes for proteins that interact with a protein of interest. *Proc. Natl. Acad. Sci. USA* 88:9578–9582.

Collins, F. 1992. Cystic fibrosis: Molecular biology and therapeutic implications. *Science* 256:774–779.

Culver, K. V., and Blaese, R. M. 1994. Gene therapy for cancer. *Trends Genet.* 10:174–178.

Eisenstein, B. I. 1990. The polymerase chain reaction: A new method of using molecular genetics for medical diagnosis. *N. Engl. J. Med.* 322:178–183.

Feinberg, A. P., and Vogelstein, B. 1983. A technique for radiolabeling DNA restriction endonuclease fragments to high specific activity. *Anal. Biochem.* 132:6–13.

———. 1984. Addendum: A technique for radiolabeling DNA restriction endonuclease fragments to high specific activity. *Anal. Biochem.* 137:266–267.

Fields, S., and Sternglanz, R. 1994. The two-hybrid system: An assay for protein–protein interactions. *Trends Genet.* 10:286–292.

Geisbrecht, B. V., Collins, C. S., Reuber, B. E., and Gould, S. J. 1998. Disruption of a PEX1–PEX6 interaction is the most common cause of the neurological disorders Zellweger syndrome, neonatal adrenoleukodystrophy, and infantile Refsum disease. *Proc. Natl. Acad. Sci. USA* 95:8630–8635.

Harris, J. D., and Lemoine, N. R. 1996. Strategies for targeted gene therapy. *Trends Genet.* 12:400–405.

Huntington's Disease Collaborative Research Group. 1993. A novel gene containing a trinucleotide repeat that is expanded and unstable on Huntington's disease chromosomes. *Cell* 72:971–983.

Kay, M. A., and Woo, S. L. C. 1994. Gene therapy for metabolic disorders. *Trends Genet.* 10:253–257.

Knowlton, R. G., Cohen-Haguenauer, O., Van Cong, N., Frézal, J., Brown, V. A., Barker, D., Braman, J. C., Schumm, J. W., Tsui, L. C., Buchwald, M., and Donis-Keller, H. 1985. A polymorphic DNA marker linked to cystic fibrosis is located on chromosome 7. *Nature* 318:380–385.

Mulligan, R. C. 1993. The basic science of gene therapy. *Science* 260:926–932.

Mullis, K. B. 1990. The unusual origin of the polymerase chain reaction. *Sci. Am.* 262 (Apr):56–65.

Mullis, K. B., and Faloona, F. A. 1987. Specific synthesis of DNA *in vitro* via a polymerase-catalyzed chain reaction. *Methods Enzymol.* 155:335–350.

Murray, J. M., Davies, K. E., Harper, P. S., Meredith, L., Mueller, C. R., and Williamson, R. 1982. Linkage relationship of a cloned DNA sequence on the short arm of the X chromosome to Duchenne muscular dystrophy. *Nature* 300:69–71.

Rozsa, F. W., Shimizu, S., Lichter, P. R., Johnson, A. T., Othman, M. I., Scott, K., Downs, C. A., Nguyen, T. D., Polansky, J., and Richards, J. E. 1998. *GLC1A* mutations point to regions of potential functional importance on the TIGR/MYOC protein. *Mol. Vis.* 4:20.

Sambrook, J., Fritsch, E. F., and Maniatis, T. 1989. *Molecular cloning: A laboratory manual*, 2nd ed. Cold Spring Harbor, NY: Cold Spring Harbor Laboratory.

Southern, E. M. 1975. Detection of specific sequences among DNA fragments separated by gel electrophoresis. *J. Mol. Biol.* 98:503–517.

Stafford, H. A. 2000. Crown gall disease and *Agrobacterium tumefaciens:* A study of the history, present knowledge, missing information, and impact on molecular genetics. *Botanical Rev.* 66:99–118.

Watson, J. D., Gilman, M., Witkowski, J., and Zoller, M. 1992. *Recombinant DNA*, 2nd ed. New York: Scientific American Books, Freeman.

White, T. J., Arnheim, N., and Erlich, H. A. 1989. The polymerase chain reaction. *Trends Genet.* 5:185–188.

Wolfenbarger, L. L., and Phifer, P. R. 2000. The ecological risks and benefits of genetically engineered plants. *Science* 290:2088–2093.

### **Chapter 11:** Mendelian Genetics

Basic Principles of Genetics. http://anthro.palomar.edu/mendel/

Bateson, W. 1909. *Mendel's principles of heredity*. Cambridge, UK: Cambridge University Press.

Bhattacharyya, M. K., Smith, A. M., Ellis, T. H. N., Hedley, C., and Martin, C. 1990. The wrinkled-seed character of pea described by Mendel is caused by a transposon-like insertion in a gene encoding starch-branching enzyme. *Cell* 60:115–122.

Mendel, G. 1866. Experiments in plant hybridization (translation). In *Classic papers in genetics*, J. A. Peters, ed., 1959. Englewood Cliffs, NJ: Prentice Hall.

Peters, J. A., ed. 1959. *Classic papers in genetics*. Englewood Cliffs, NJ: Prentice Hall.

Sandler, I., and Sandler, L. 1985. A conceptual ambiguity that contributed to the neglect of Mendel's paper. *Hist. Phil. Life Sci.* 7:3–70.

Tschermak-Seysenegg, E. von. 1951. The rediscovery of Mendel's work. *J. Hered.* 42:163–171.

### **Chapter 12:** Chromosomal Basis of Inheritance

Cell Cycle and Mitosis Tutorial. http://www.biology.arizona.edu/cell_bio/tutorials/cell_cycle/cells3.html

Meiosis Tutorial. http://www.biology.arizona.edu/cell_bio/tutorials/meiosis/main.html

Barr, M. L. 1960. Sexual dimorphism in interphase nuclei. *Am. J. Hum. Genet.* 12:118–127.

Bridges, C. B. 1916. Nondisjunction as a proof of the chromosome theory of heredity. *Genetics* 1:1–52, 107–163.

———. 1925. Sex in relation to chromosomes and genes. *Am. Nat.* 59:127–137.

Egel, R. 1995. The synaptonemal complex and the distribution of meiotic recombination events. *Trends Genet.* 11:206–208.

Lyon, M. F. 1962. Sex chromatin and gene action in the mammalian X-chromosome. *Am. J. Hum. Genet.* 14:135–148.

McClung, C. E. 1902. The accessory chromosome: Sex determinant? *Biol. Bull.* 3:43–84.

McKusick, V. A. 1965. The royal hemophilia. *Sci. Am.* 213 (Aug):88–95.

Morgan, L. V. 1922. Non criss-cross inheritance in *Drosophila melanogaster. Biol. Bull.* 42:267–274.

Morgan, T. H. 1910. Sex-limited inheritance in *Drosophila. Science* 32:120–122.

———. 1911. An attempt to analyze the constitution of the chromosomes on the basis of sex-limited inheritance in *Drosophila. J. Exp. Zool.* 11:365–414.

Shonn, M. A., McCarroll, R., and Murray, A. W. 2000. Requirement of the spindle checkpoint for proper chromosome segregation in budding yeast meiosis. *Science* 289:300–303.

Stern, C., Centerwall, W. P., and Sarkar, Q. S. 1964. New data on the problem of Y-linkage of hairy pinnae. *Am. J. Hum. Genet.* 16:455–471.

Sutton, W. S. 1903. The chromosomes in heredity. *Biol. Bull.* 4:231–251.

Wilson, E. B. 1905. The chromosomes in relation to the determination of sex in insects. *Science* 22:500–502.

### **Chapter 13:** Extensions of and Deviations from Mendelian Genetic Principles

Gene Interactions. http://www.ndsu.nodak.edu/instruct/mcclean/plsc431/mendel/mendel6.htm

Birky, C. W. 1978. Transmission genetics of mitochondria and chloroplasts. *Annu. Rev. Genet.* 12:471–512.

Brown, M. D., Voljavec, A. S., Lott, M. T., MacDonald, I., and Wallace, D. C. 1992. Leber's hereditary optic neuropathy: A model for mitochondrial neurodegenerative diseases. *FASEB J.* 6:2791–2799.

Bultman, S. J., Michaud, E. J., and Woychik, R. P. 1992. Molecular characterization of the mouse *agouti* locus. *Cell* 71:1195–1204.

Chiu, W. L., and Sears, B. B. 1993. Plastome–genome interactions affect plastid transmission in Oenothera. *Genetics* 133:989–997.

Ephrussi, B. 1953. *Nucleo-cytoplasmic relations in microorganisms*. New York: Oxford University Press.

Freeman, G., and Lundelius, J. W. 1982. The developmental genetics of dextrality and sinistrality in the gastropod *Limnaea peregra. Wilhelm Roux Arch. Dev. Biol.* 191:69–83.

Ginsburg, V. 1972. Enzymatic basis for blood groups. *Methods Enzymol.* 36:131–149.

Grivell, L. 1983. Mitochondrial DNA. *Sci. Am.* 225 (Mar):78–89.

Gyllensten, U., Wharton, D., Josefsson, A., and Wilson, A. C. 1991. Paternal inheritance of mitochondrial DNA in mice. *Nature* 352:255–257.

Landauer, W. 1948. Hereditary abnormalities and their chemically induced phenocopies. *Growth Symp.* 12:171–200.

Lander, E. S., and Lodish, H. 1990. Mitochondrial diseases: Gene mapping and gene therapy. *Cell* 61:925–926.

Landsteiner, K., and Levine, P. 1927. Further observations on individual differences of human blood. *Proc. Soc. Exp. Biol. Med.* 24:941–942.

Siracusa, L. D. 1994. The *agouti* gene: Turned on to yellow. *Trends Genet.* 10:423–428.

Umesono, K., and Ozeki, H. 1987. Chloroplast gene organization in plants. *Trends Genet.* 3:281–287.

Van Winkle-Swift, K. P., and Birky, C. W. 1978. The nonreciprocality of organelle gene recombination in *Chlamydomonas reinhardi* and *Saccharomyces cerevisiae. Mol. Gen. Genet.* 166:193–209.

**Chapter 14:** Genetic Mapping in Eukaryotes

Gene Linkage and Genetic Maps. http://users.rcn.com/jkimball.ma.ultranet/BiologyPages/L/Linkage.html

Bateson, W., Saunders, E. R., and Punnett, R. G. 1905. Experimental studies in the physiology of heredity. *Rep. Evol. Committee R. Soc.* II:1–55, 80–99.

Blixt, S. 1975. Why didn't Mendel find linkage? *Nature* 256:206.

Creighton, H. S., and McClintock, B. 1931. A correlation of cytological and genetical crossing-over in *Zea mays. Proc. Natl. Acad. Sci. USA* 17:492–497.

Dib, C., Fauré, S., Fizames, C., Samson, D., Drouot, N., Vignal, A., Millasseu, P., Marc, S., Hazan, J., Seboun, E., Lathrop, M., Gyapay, G., Morissette, J., and Weissenbach, J. 1996. A comprehensive genetic map of the human genome based on 5,264 microsatellites. *Nature* 380:152–154.

Gusella, J. F. 1986. DNA polymorphism and human disease. *Annu. Rev. Biochem.* 55:831–854.

McKusick, V. A. 1971. The mapping of human chromosomes. *Sci. Am.* 224 (Apr):104–113.

Morgan, T. H. 1910. The method of inheritance of two sex-limited characters in the same animal. *Proc. Soc. Exp. Biol. Med.* 8:17.

———. 1910. Sex-limited inheritance in *Drosophila. Science* 32:120–122.

———. 1911. An attempt to analyze the constitution of the chromosomes on the basis of sex-limited inheritance in *Drosophila. J. Exp. Zool.* 11:365–414.

———. 1911. Random segregation versus coupling in Mendelian inheritance. *Science* 34:384.

Morgan, T. H., Sturtevant, A. H., Müller, H. J., and Bridges, C. B. 1915. *The mechanism of Mendelian heredity.* New York: Henry Holt.

Sturtevant, A. H. 1913. The linear arrangement of six sex-linked factors in *Drosophila* as shown by their mode of association. *J. Exp. Zool.* 14:43–59.

Sutton, W. S. 1903. The chromosomes in heredity. *Biol. Bull.* 4:231–251.

Szostak, J., Orr-Weaver, T., Rothstein, R., and Stahl, F. 1983. The double-strand break repair model for recombination. *Cell* 33:25–35.

Weissenbach, J., Gyapay, G., Dib, C., Vignal, A., Morissett, J., Millasseau, P., Vaysseix, G., and Lathrop, M. 1992. A second-generation linkage map of the human genome. *Nature* 359:794–801.

Yu, A., Zhao, C., Fan, Y., Jang, W., Mungall, A. J., Deloukas, P., Olsen, A., Doggett, N. A., Ghebranicus, N., Broman, K. W., and Weber, J. L. 2001. Comparison of human genetic and sequence-based physical maps. *Nature* 409:951–953.

**Chapter 15:** Genetics of Bacteria and Bacteriophages

Bacterial Conjugation (A History of its Discovery). http://www.mun.ca/biochem/courses/3107/Lectures/Topics/conjugation.html

Mapping within a Gene: The *rII* Locus. http://users.rcn.com/jkimball.ma.ultranet/BiologyPages/B/Benzer.html

Archer, L. J. 1973. *Bacterial transformation.* New York: Academic Press.

Benzer, S. 1959. On the topology of the genetic fine structure. *Proc. Natl. Acad. Sci. USA* 45:1607–1620.

———. 1961. On the topography of the genetic fine structure. *Proc. Natl. Acad. Sci. USA* 47:403–415.

———. 1962. The fine structure of the gene. *Sci. Am.* 206 (Jan):70–84.

Curtiss, R. 1969. Bacterial conjugation. *Annu. Rev. Microbiol.* 23:69–136.

Ellis, E. L., and Delbruck, M. 1939. The growth of bacteriophage. *J. Gen. Physiol.* 22:365–384.

Fincham, J. 1966. *Genetic complementation.* New York: W. A. Benjamin.

Hayes, W. 1968. *The genetics of bacteria and their viruses*, 2nd ed. New York: Wiley.

Hershey, A. D., and Rotman, R. 1949. Genetic recombination between host-range and plaque-type mutants of bacteriophage in single bacterial cells. *Genetics* 34:44–71.

Hotchkiss, R. D., and Gabor, M. 1970. Bacterial transformation with special reference to recombination processes. *Annu. Rev. Genet.* 4:193–224.

Jacob, F., and Wollman, E. L. 1951. *Sexuality and the genetics of bacteria.* New York: Academic Press.

Lederberg, J., and Tatum, E. L. 1946. Gene recombination in *Escherichia coli. Nature* 158:558.

Ravin, A. W. 1961. The genetics of transformation. *Adv. Genet.* 10:61–163.

Susman, M. 1970. General bacterial genetics. *Annu. Rev. Genet.* 4:135–176.

Vielmetter, W., Bonhoeffer, F., and Schutte, A. 1968. Genetic evidence for transfer of a single DNA strand during bacterial conjugation. *J. Mol. Biol.* 37:81–86.

Wollman, E. L., Jacob, F., and Hayes, W. 1962. Conjugation and genetic recombination in *E. coli K-12. Cold Spring Harbor Symp. Quant. Biol.* 21:141–162.

Zinder, N., and Lederberg, J. L. 1952. Genetic exchange in *Salmonella. J. Bacteriol.* 64:679–699.

**Chapter 16:** Variations in Chromosome Structure and Number

Barr, M. L., and Bertram, E. G. 1949. A morphological distinction between neurones of the male and female, and the behavior of the nucleolar satellite during accelerated nucleoprotein synthesis. *Nature* 163:676–677.

Borst, P., and Greaves, D. R. 1987. Programmed gene rearrangements altering gene expression. *Science* 235:658–667.

Caskey, C. T., Pizzuti, A., Fu, Y. H., Fenwick, R. G., and Nelson, D. L. 1992. Triplet repeat mutations in human disease. *Science* 256:784–789.

Dalla-Favera, R., Martinotti, S., Gallo, R., Erickson, J., and Croce, C. 1983. Translocation and rearrangements of the c-*myc* oncogene locus in human undifferentiated B-cell lymphomas. *Science* 219:963–997.

DeKlein, A., van Kessel, A. G., Grosveld, G., Bartram, C. R., Hagemeijer, A., Bootsma, D., Spurr, N. K., Heisterkamp, N., Groffen, J., and Stephenson, J. R. 1982. A cellular oncogene is translocated to the Philadelphia chromosome in chronic myelocytic leukemia. *Nature* 300:765–767.

Huntington's Disease Collaborative Research Group. 1993. A novel gene containing a trinucleotide repeat that is expanded and unstable in Huntington's disease chromosome. *Cell* 72:971–983.

Kremer, E., Pritchard, M., Lynch, M., Yu, S., Holman, K., Baker, E., Warren, S. T., Schlessinger, D., Sutherland, G. R., and Richards, R. I. 1991. Mapping of DNA instability at the fragile X to a trinucleotide repeat sequence p(CGG)*n*. *Science* 252:1711–1714.

Lyon, M. F. 1961. Gene action in the X-chromosomes of the mouse (*Mus musculus L*). *Nature* 190:372–373.

Penrose, L. S., and Smith, G. F. 1966. *Down's anomaly.* Boston: Little, Brown.

Richards, R. I., and Sutherland, G. R. 1992. Dynamic mutations: A new class of mutations causing human disease. *Cell* 70:709–712.

———. 1992. Fragile X syndrome: The molecular picture comes into focus. *Trends Genet.* 8:249–255.

Rowley, J. D. 1973. A new consistent chromosomal abnormality in chronic myelogenous leukemia identified by quinacrine fluorescence and Giemsa staining. *Nature* 243:290–293.

Shaw, M. W. 1962. Familial mongolism. *Cytogenetics* 1:141–179.

Tarleton, J. C., and Saul, R. A. 1993. Molecular genetic advances in fragile X syndrome. *J. Pediatr.* 122:169–185.

Verkerk, A. J. M. H., Piertti, M., Sutcliff, J. S., Fu, Y. H., Kuhl, D. P. A., Pizzuti, A., Reiner, O., Richards, S., Victoria, M. F., Zhang, F., Eussen, B. E., van Ommen, G. J. B., Blonden, L. A. J., Riggins, G. J., Chastain, J. L., Kunst, C. B., Galjaard, H., Caskey, C. T., Nelson, D. L., Oostra, B. A., and Warrent, S. T. 1991. Identification of a gene (*FMR-1*) containing a CGG repeat coincident with a breakpoint cluster region exhibiting length variation in fragile X syndrome. *Cell* 65:905–914.

## **Chapter 17:** Regulation of Gene Expression in Bacteria and Bacteriophages

The Operon. http://users.rcn.com/jkimball.ma.ultranet/BiologyPages/L/LacOperon.html

Bell, C. E., Frescura, P., Hochschild, A., and Lewis, M. 2000. Crystal structure of the λ repressor C-terminal domain provides a model for cooperative operator binding. *Cell* 101:801–811.

Bertrand, K., Korn, L., Lee, F., Platt, T., Squires, C. L., Squires, C., and Yanofsky, C. 1975. New features of the structure and regulation of the tryptophan operon of *Escherichia coli. Science* 189:22–26.

Bertrand, K., and Yanofsky, C. 1976. Regulation of transcription termination in the leader region of the tryptophan operon of *Escherichia coli* involves tryptophan as its metabolic product. *J. Mol. Biol.* 103:339–349.

Dickson, R. C., Abelson, J., Barnes, W. M., and Reznikoff, W. S. 1975. Genetic regulation: The *lac* control region. *Science* 187:27–35.

Fisher, R. F., Das, A., Kolter, R., Winkler, M. E., and Yanofsky, C. 1985. Analysis of the requirements for transcription pausing in the tryptophan operon. *J. Mol. Biol.* 182:397–409.

Gilbert, W., Maizels, N., and Maxam, A. 1974. Sequences of controlling regions of the lactose operon. *Cold Spring Harbor Symp. Quant. Biol.* 38:845–855.

Gilbert, W., and Muller-Hill, B. 1966. Isolation of the *lac* repressor. *Proc. Natl. Acad. Sci. USA* 56:1891–1898.

Jacob, F. 1965. Genetic mapping of the elements of the lactose region of *Escherichia coli. Biochem. Biophys. Res. Commun.* 18:693–701.

Jacob, F., and Monod, J. 1961. Genetic regulatory mechanisms in the synthesis of proteins. *J. Mol. Biol.* 3:318–356.

Lee, F., and Yanofsky, C. 1977. Transcription termination at the *trp* operon attenuators of *Escherichia coli* and *Salmonella typhimurium:* RNA secondary structure and regulation of termination. *Proc. Natl. Acad. Sci. USA* 74:4365–4369.

Lewis, M., Chang, G., Horton, N. C., Kercher, M. A., Pace, H. C., Schumacher, M. A., Brennan, R. G., and Lu, P. 1996. Crystal structure of the lactose operon repressor and its complexes with DNA and inducer. *Science* 271:1247–1254.

Maizels, N. 1974. *E. coli* lactose operon ribosome binding site. *Nature (New Biol.)* 249:647–649.

Pabo, C. O., Sauer, R. T., Sturtevant, J. M., and Ptashne, M. 1979. The λ repressor contains two domains. *Proc. Natl. Acad. Sci. USA* 76:1608–4612.

Ptashne, M. 1967. Isolation of the λ phage repressor. *Proc. Natl. Acad. Sci. USA* 57:306–313.

———. 1984. Repressors. *Trends Biochem. Sci.* 9:142–145.

———. 1992. *A genetic switch,* 2nd ed. Oxford: Cell Press and Blackwell Scientific Publications.

Ptashne, M., and Gilbert, W. 1970. Genetic repressors. *Sci. Am.* 222 (Jun):36–44.

Schlief, R. 2000. Regulation of the L-arabinose operon of *Escherichia coli. Trends Genet.* 16:559–566.

Schlief, R. 2003. AraC protein: A love-hate relationship. *BioEssays* 25:274–282.

Winkler, M. E., and Yanofsky, C. 1981. Pausing of RNA polymerase during in vitro transcription of the tryptophan operon leader region. *Biochemistry* 20:3738–3744.

Yanofsky, C. 1981. Attenuation in the control of expression of bacterial operons. *Nature* 289:751–758.

———. 1987. Operon-specific control by transcription attenuation. *Trends Genet.* 3:356–360.

## **Chapter 18:** Regulation of Gene Expression in Eukaryotes

Antisense RNA (*includes RNA interference*). http://users.rcn.com/jkimball.ma.ultranet/BiologyPages/A/AntisenseRNA.html

Control of Gene Expression (*includes prokaryotes*). http://themedicalbiochemistrypage.org/gene-regulation.html

Gene Regulation in Eukaryotes. http://users.rcn.com/jkimball.ma.ultranet/BiologyPages/P/Promoter.html

RNAi–Interference RNA. http://fig.cox.miami.edu/~cmallery/150/gene/siRNA.htm

RNA interference (*animation*). http://www.nature.com/focus/rnai/animations/index.html

Lifecyle of an miRNA (*video*). http://www.nature.com/ng/supplements/micrornas/video.html

Ambros, V., and Chen, X. 2007. The regulation of genes and genomes by small RNAs. *Development* 134:1635–1641.

Ashburner, M. 1990. Puffs, genes, and hormones revisited. *Cell* 61:1–3.

Bartel, D. P. 2004. MicroRNAs: Genomics, biogenesis, mechanism, and function. *Cell* 116:281–297.

Beermann, W., and Clever, U. 1964. Chromosome puffs. *Sci. Am.* 210 (Apr):50–58.

Blumenthal, T., and Gleason, K. S. 2003. *Caenorhabditis elegans* operons: Form and function. *Nature Rev. Genet.* 4:110–118.

Carpousis, A. J., Vanzo, N. F., and Raynal, L. C. 1999. mRNA degradation. A tale of poly(A) and multiprotein machines. *Trends Genet.* 15:24–28.

Chen, C. Y. A., and Shyu, A.-B. 1995. AU-rich elements: Characterization and importance of mRNA degradation. *Trends Biochem. Sci.* 20:465–470.

Claverle, J. M. 2005. Fewer genes, more noncoding RNA. *Science* 309:1529–1530.

Erkmann, J. A., and Kutay, U. 2004. Nuclear export of mRNA: From the site of transcription to the cytoplasm. *Exp. Cell Res.* 296:12–20.

Garneau, N. L., Wilusz, J., and Wilusz, C. J. 2007. The highways and byways of mRNA decay. *Nature Rev. Mol. Cell Biol.* 8:113–126.

Gasser, S. M. 2001. Positions of potential: Nuclear organization and gene expression. *Cell* 104:639–642.

Gellert, M. 1992. V(D)J recombination gets a break. *Trends Genet.* 8:408–412.

Green, M. R. 1989. Pre-mRNA processing and mRNA nuclear export. *Curr. Opin. Cell Biol.* 1:519–525.

Grewal, S. I. S., and Jia, S. 2008. Heterochromatin revisited. *Nature Rev. Genet.* 8:35–46.

Grunstein, M. 1992. Histones as regulators of genes. *Sci. Am.* 267 (Oct):68–74B.

Hochstrasser, M. 1996. Protein degradation or regulation: Ub the judge. *Cell* 84:813–815.

Horn, P. J., and Peterson, C. L. 2002. Chromatin higher order folding: Wrapping up transcription. *Science* 297:1824–1828.

Johnston, M., Flick, J. S, and Pexton, T. 1994. Multiple mechanisms provide rapid and stringent repression of *GAL* gene expression in *Saccharomyces cerevisiae*. *Mol. Cell. Biol.* 14:3834–3841.

Jones, P. A. 1999. The DNA methylation paradox. *Trends Genet.* 15:34–37.

Karlsson, S., and Nienhuis, A. W. 1985. Development regulation of human globin genes. *Annu. Rev. Biochem.* 54:1071–1078.

Kawasaki, H., Taira, K., and Morris, K. V. 2005. siRNA induced transcriptional gene silencing in mammalian cells. *Cell Cycle* 4:442–448.

Keyes, L. N., Cline, T. W., and Schedl, P. 1992. The primary sex determination signal of *Drosophila* acts at the level of transcription. *Cell* 68:933–943.

Kim, K., Lee, Y. S., and Carthew, R. W. 2007. Conversion of pre-RISC to holo-RISC by Ago2 during assembly of RNAi complexes. *RNA* 13:22–29.

Kornberg, R. D. 1999. Eukaryotic transcriptional control. *Trends Genet.* 15:M46–M49.

Lehner, B., and Sanderson, C. M. 2007. A protein degradation framework for human mRNA degradation. *Genome Res.* 14:1315–1323.

Mallory, A. C., and Vaucheret, H. 2006 Functions of microRNAs and related small RNAs in plants. *Nature Genet.* 38:S31–S37.

Mattick, J. S. 2005. The functional genomics of noncoding RNA. *Science* 309:1527–1528.

———. 2007. A new paradigm for developmental biology. *J. Exp. Biol.* 210:1526–1547.

Mattick, J. S., and Makunin, I. V. 2006. Non-coding RNA. *Hum. Mol. Gen.* 15:R17–R29.

Moore, M. J. 2005. From birth to death: The complex lives of eukaryotic mRNAs. *Science* 309:1514–1518.

Nilsen, T. W. 2007. Mechanisms of microRNA-mediated gene regulation in animal cells. *Trends Genet.* 23:243–249.

Oettinger, M. A., Schatz, D. G., Gorka, C., and Baltimore, D. 1990. *RAG-1* and *RAG-2*, adjacent genes that synergistically activate V(D)J recombination. *Science* 248:1517–1522.

O'Malley, B. W., and Schrader, W. T. 1976. The receptors of steroid hormones. *Sci. Am.* 234 (Feb):32–43.

Pankratz, M. J., and Jäckle, H. 1990. Making stripes in the *Drosophila* embryo. *Trends Genet.* 6:287–292.

Parker, R., and Sheth, U. 2007. P bodies and the control of mRNA translation and degradation. *Mol. Cell* 25:635–646.

Parthun, M. R., and Jaehning, J. A. 1992. A transcriptionally active form of GAL4 is phosphorylated and associated with GAL80. *Mol. Cell. Biol.* 12:4981–4987.

Postlethwait, J. H., and Schneiderman, H. A. 1973. Developmental genetics of *Drosophila* imaginal discs. *Annu. Rev. Genet.* 7:381–433.

Prasanth, K. V., and Spector, D. L. 2007. Eukaryotic regulatory RNAs: An answer to the 'genome complexity' conundrum. *Genes Dev.* 21:11–42.

Ptashne, M. 1989. How gene activators work. *Sci. Am.* 243 (Jan):41–47.

Rhodes, D., and Klug, A. 1993. Zinc fingers. *Sci. Am.* 259 (Feb):56–65.

Rivera-Pomar, R., and Jäckle, H. 1996. From gradients to stripes in *Drosophila* embryogenesis: Filling in the gaps. *Trends Genet.* 12:478–483.

Rogers, J. O., Early, H., Carter, C., Calame, K., Bond, M., Hood, L., and Wall, R. 1980. Two mRNAs with different 3′ ends encode membrane-bound and secreted forms of immunoglobin chain. *Cell* 20:303–312.

Rogers, S., Wells, R., and Rechsteiner, M. 1986. Amino acid sequences common to rapidly degraded proteins: The PEST hypothesis. *Science* 234:364–368.

Ross, J. 1996. Control of messenger RNA stability in higher eukaryotes. *Trends Genet.* 12:171–175.

Scott, M. P., Tamkun, J. W., and Hartzell, III, G. W. 1989. The structure and function of the homeodomain. *Biochim. Biophys. Acta* 989:25–48.

Segal, E., Fondufe-Mittendorf, Y., Chen, L., Thåström, A., Field, Y., Moore, I. K., Wang, P. Z., and Widom, J. 2006. A genomic code for nucleosome positioning. *Nature* 442:772–778.

Siomi, H., and Siomi, M. C. 2007. Expanding RNA physiology: MicroRNAs in a unicellular organism. *Genes Dev.* 21:1153–1156.

Struhl, K. 1999. Fundamentally different logic of gene regulation in eukaryotes and prokaryotes. *Cell* 98:104.

Varshavksy, A. 1996. The N-end rule: Functions, mysteries, uses. *Proc. Natl. Acad. Sci. USA* 93:12142–12149.

Vaucheret, H. 2007. Post-transcriptional small RNA pathways in plants: Mechanisms and regulations. *Genes Dev.* 20:759–771.

Verdel, A., Jia, S., Gerber, S., Sugiyama, T., Gygi, S., Grewal, S. I. S., and Moazed, D. 2004. RNAi-mediated targeting of heterochromatin by the RITS complex. *Science* 303:672–676.

Verdine, G. L. 1994. The flip side of DNA methylation. *Cell* 76:197–200.

Wilmut, I., Schnieke, A. E., McWhir, J., Kind, A. J., and Campbell, K. H. S. 1997. Viable offspring derived from fetal and adult mammalian cells. *Nature* 385:810–813.

Wolffe, A. P. 1994. Transcription: In tune with the histones. *Cell* 77:13–16.

Wolffe, A. P., and Pruss, D. 1996. Targeting chromatin disruption: Transcription regulators that acetylate histones. *Cell* 84:817–819.

Zamore, P. D., and Haley, B. 2005. Ribo-gnome: The big world of small RNAs. *Science* 309:1519–1524.

### **Chapter 19:** Genetic Analysis of Development

Zygote: A Developmental Biology Website (*by Scott Gilbert, Swarthmore College*). http://zygote.swarthmore.edu/

Albrecht, E. B., and Salz, H. K. 1993. The *Drosophila* sex determination gene *snf* is utilized for the establishment of the female-specific splicing pattern of *Sex-lethal*. *Genetics* 134:801–807.

Alvarez-Garcia, I., and Miska, E. A. 2005. MicroRNA functions in animal development and human disease. *Development* 132:4653–4662.

Beachy, P. A. 1990. A molecular view of the *Ultrabithorax* homeotic gene of *Drosophila*. *Trends Genet.* 6:46–51.

Boggs, R. T., Gregor, P., Idriss, S., Belote, J. M., and McKeown, M. 1987. Regulation of sexual differentiation in *Drosophila melanogaster* via alternative splicing of RNA from the transformer. *Cell* 50:739–747.

Capel, B. 1995. New bedfellows in the mammalian sex-determination affair. *Trends Genet.* 11:161–163.

De Robertis, E. M., and Gurdon, J. B. 1977. Gene activation in somatic nuclei after injection into amphibian oocytes. *Proc. Natl. Acad. Sci. USA* 74:2470–2474.

Efstratiadis, A., Posakony, J. W., Maniatis, T., Lawn, R. M., O'Connell, C., Spritz, R. A., DeRiel, J. K., Forget, B. G., Weissman, S. M., Slighton, J. L., Blechtl, A. E., Smithies, O., Baralle, F. E., Shoulders, C. C., and Proudfoot, N. J. 1980. The structure and evolution of the human β-globin gene family. *Cell* 21:653–668.

Eicher, E. M., and Washburn, L. L. 1986. Genetic control of primary sex determination in mice. *Annu. Rev. Genet.* 20:327–360.

Gurdon, J. B. 1968. Transplanted nuclei and cell differentiation. *Sci. Am.* 219 (Dec):24–35.

Gurdon, J. B., Laskey, R. A., and Reeves, R. 1975. The developmental capacity of nuclei transplanted from keratinized skin cells of adult frogs. *J. Embryol. Exp. Morph.* 34:93–112.

Haqq, C. M., King, C. Y., Ukiyama, E., Falsafi, S., Haqq, T. N., Donahoe, P. K., and Weiss, M. A. 1994. Molecular basis of mammalian sexual determination: Activation of Müllerian inhibiting substance gene expression by SRY. *Science* 266:1494–1500.

Hodgkin, J. 1987. Sex determination and dosage compensation in *Caenorhabditis elegans*. *Annu. Rev. Genet.* 21:133–154.

———. 1989. *Drosophila* sex determination: A cascade of regulated splicing. *Cell* 56:905–906.

———. 1993. Molecular cloning and duplication of the nematode sex-determining gene *tra-1*. *Genetics* 133:543–560.

Jiménez, R., Sánchez, A., Burgos, M., and Díaz de la Guardia, R. 1996. Puzzling out the genetics of mammalian sex determination. *Trends Genet.* 12:164–166.

Kay, G. F., Barton, S. C., Surani, M. A., and Rastan, S. 1994. Imprinting and X chromosome counting mechanisms determine *Xist* expression in early mouse development. *Cell* 77:639–650.

Kloosterman, W. P., and Plastork, R. H. A. 2006. The diverse functions of microRNAs in animal development and disease. *Dev. Cell* 11:441–450.

Koopman, P., Gubbay, J., Vivian, N., Goodfellow, P., and Lovell-Badge, R. 1991. Male development of chromosomally female mice transgenic for *Sry*. *Nature* 351:117–121.

Lee, J. T., Strauss, W. M., Dausman, J. A., and Jaenisch, R. 1996. A 450 kb transgene displays properties of the mammalian X-inactivation center. *Cell* 86:83–94.

Marahrens, Y., Loring, J., and Jaenisch, R. 1998. Role of the *Xist* gene in X chromosome choosing. *Cell* 92:657–664.

Meller, V. H. 2000. Dosage compensation: Making 1X equal 2X. *Trends Cell Biol.* 10:54–59.

Meyer, B. J. 2000. Sex in the worm. Counting and compensating X-chromosome dose. *Trends Genet.* 16:247–253.

Migeon, B. R. 1994. X-chromosome inactivation: Molecular mechanisms and genetic consequences. *Trends Genet.* 10:230–235.

Page, D. C. 1985. Sex-reversal: Deletion mapping of the male-determining function of the human Y chromosome. *Cold Spring Harbor Symp. Quant. Biol.* 51:229–235.

Page, D. C., de la Chapelle, A., and Weissenbach, J. 1985. Chromosome Y-specific DNA in related human XX males. *Nature* 315:224–226.

Page, D. C., Mosher, R., Simpson, E. M., Fisher, E. M. C., Mardon, G., Pollack, J., McGillivray, B., de la Chapelle, A., and Brown, L. G. 1987. The sex-determining region of the human Y chromosome encodes a finger protein. *Cell* 51:1091–1104.

Palmer, M. S., Sinclair, A. H., Berta, P., Ellis, N. A., Goodfellow, P. N., Abbas, N. E., and Fellous, M. 1990. Genetic evidence that ZFY is not the testis-determining factor. *Nature* 342:937–939.

Penny, G. D., Kay, G. F., Sheardown, S. A., Rastan, S., and Brockdorff, N. 1996. Requirement for *Xist* in X chromosome inactivation. *Nature* 379:131–137.

Peters, L., and Meister, G. 2007. Argonaute proteins: Mediators of RNA silencing. *Mol. Cell* 26:611–623.

Rivera-Pomar, R., and Jäckle, H. 1996. From gradients to stripes in *Drosophila* embryogenesis: Filling in the gaps. *Trends Genet.* 12:478–483.

Willard, H. F. 1996. X chromosome inactivation, *XIST*, and pursuit of the X-inactivation center. *Cell* 86:5–7.

Zhao, Y., and Srivastava, D. 2007. A developmental view of microRNA function. *Trends Biochem. Sci.* 32:189–197.

### **Chapter 20:** Genetics of Cancer

Oncogenes. http://users.rcn.com/jkimball.ma.ultranet/BiologyPages/O/Oncogenes.html

Tumor suppressor genes. http://users.rcn.com/jkimball.ma.ultranet/BiologyPages/T/TumorSuppressorGenes.html

Baltimore, D. 1985. Retroviruses and retrotransposons: The role of reverse transcription in shaping the eukaryotic genome. *Cell* 40:481–482.

Bishop, J. M. 1983. Cancer genes come of age. *Cell* 32:1018–1020.

———. 1983. Cellular oncogenes and retroviruses. *Annu. Rev. Biochem.* 52:301–354.

———. 1987. The molecular genetics of cancer. *Science* 235:305–311.

Brown, M. A., and Solomon, E. 1997. Studies on inherited cancers: Outcomes and challenges of 25 years. *Trends Genet.* 13:202–206.

Cavenee, W. K., and White, R. L. 1995. The genetic basis of cancer. *Sci. Am.* (Mar):72–79.

Fishel, R., Lescoe, M. K., Rao, M. R. S., Copeland, N. G., Jenkins, N. A., Garber, J., Kane, M., and Kolodner, R. 1994. The human mutator gene homolog MSH2 and its association with hereditary nonpolyposis colon cancer. *Cell* 75:1027–1038.

Jiricny, J. 1994. Colon cancer and DNA repair: Have mismatches met their match? *Trends Genet.* 10:164–168.

Kingston, R. E., Baldwin, A. S., and Sharp, P. A. 1985. Transcription control by oncogenes. *Cell* 41:3–5.

Levine, A. J. 1997. p53, the cellular gatekeeper for growth and division. *Cell* 88:323–331.

Negrini, M., Ferracin, M., Sabbioni, S., and Croce, C. M. 2007. MicroRNAs in human cancer: from research to therapy. *J. Cell Sci.* 120:1833–1840.

Rabbitts, T. H. 1994. Chromosomal translocations in human cancer. *Nature* 372:143–149.

Rebbeck, T. R., Couch, F. J., Kant, J., Calzone, K., DeShano, M., Peng, Y., Chen, K., Garber, J. E., and Weber, B. L. 1996. Genetic heterogeneity in hereditary breast cancer: Role of *BRCA1* and *BRCA2*. *Am. J. Hum. Genet.* 59:547–553.

Vousden, K. H. 2000. p53: Death star. *Cell* 103:691–694.

Weinberg, R. A. 1995. The retinoblastoma protein and cell cycle protein. *Cell* 81:323–330.

———. 1997. The cat and mouse games that genes, viruses, and cells play. *Cell* 88:573–575.

Welcsh, P. L., Owens, K. N., and King, M. C. 2000. Insights into the functions of *BRCA1* and *BRCA2*. *Trends Genet.* 16:69–74.

Wooster, R., and Stratton, M. R. 1995. Breast cancer susceptibility: A complex disease unravels. *Trends Genet.* 11:3–5.

Zakian, V. A. 1997. Life and cancer without telomerase. *Cell* 91:1–3.

Zhang, B., Pan, X., Cobb, G. P., and Anderson, T. A. 2007. microRNAs as oncogenes and tumor suppressors. *Dev. Biol.* 302:1–12.


**Chapter 21:** Population Genetics


Avise, J. C. 1986. Mitochondrial DNA and the evolutionary genetics of higher animals. *Phil. Trans. Roy. Soc. Lond., Ser. B* 321:325–342.

Buri, P. 1956. Gene frequency in small populations of mutant *Drosophila*. *Evolution* 10:367–402.

Clarke, C. A., and Sheppard, P. M. 1966. A local survey of the distribution of industrial melanic forms in the moth *Biston betularia* and estimates of the selective values of these in an industrial environment. *Proc. R. Soc. Lond. [Biol.]* 165:424–439.

Coop, G., and Przeworski, M. 2007. An evolutionary view of human recombination. *Nature Rev. Genet.* 8:23–24.

Crow, J. F. 1986. *Basic concepts in population, quantitative, and evolutionary genetics*. New York: Freeman.

Darwin, C. 1860. *On the origin of species by means of natural selection, or the preservation of favoured races in the struggle for life*. New York: Appleton.

Dobzhansky, T. 1951. *Genetics and the origin of species*, 3rd ed. New York: Columbia University Press.

Fisher, R. A. 1930. *The genetical theory of natural selection*. Oxford: Clarendon Press.

Ford, E. B. 1971. *Ecological genetics*, 3rd ed. London: Chapman & Hall.

Gillespie, J. H. 1991. *The causes of molecular evolution*. Oxford: Oxford University Press.

Glass, B., Sacks, M. S., Jahn, E. F., and Hess, C. 1952. Genetic drift in a religious isolate: An analysis of the causes of variation in blood group and other gene frequencies in a small population. *Am. Nat.* 86:145–159.

Hardy, G. H. 1908. Mendelian proportions in a mixed population. *Science* 28:49–50.

Hartl, D. L., and Clark, A. G. 1995. *Principles of population genetics*, 3rd ed. Sunderland, MA: Sinauer.

Hedrick, P. H. 2000. *Genetics of populations*. Boston: Science Books International.

Hillis, D. M., and Moritz, C. 1990. *Molecular systematics*. Sunderland, MA: Sinauer.

Hillis, D. M., Moritz, C., Porter, C. A., and Baker, R. J. 1991. Evidence for biased gene conversion in concerted evolution of ribosomal DNA. *Science* 251:308–309.

Kettlewell, H. B. D. 1961. The phenomenon of industrial melanism in the *Lepidoptera*. *Annu. Rev. Entomol.* 6:245–262.

Kreitman, M. 1983. Nucleotide polymorphism at the alcohol dehydrogenase locus of *Drosophila melanogaster*. *Nature* 304:412–417.

Lehman, N., Eisenhawer, A., Hansen, K., Mech, L. D., Peterson, R. O., Gogan, P. J., and Wayne, R. K. 1991. Introgression of coyote mitochondrial DNA into sympatric North American gray wolf populations. *Evolution* 45:104–119.

Lewontin, R. C. 1974. *The genetic basis of evolutionary change*. New York: Columbia University Press.

———. 1985. Population genetics. *Annu. Rev. Genet.* 19:81–102.

Lewontin, R. C., Moore, J. A., Provine, W. B., and Wallace, B. 1981. Dobzhansky's genetics of natural populations I–XLIII. New York: Columbia University Press.

Li, W. H. 1997. *Molecular evolution*. Sunderland, MA: Sinauer.

Li, W. H., Luo, C. C., and Wu, C. I. 1985. Evolution of DNA sequences. In *Molecular evolutionary genetics*, R. J. MacIntyre, ed. (pp. 1–94). New York: Plenum.

Maniatis, T., Fritsch, E. F., Lauer, L., and Lawn, R. M. 1980. The molecular genetics of human hemoglobin. *Annu. Rev. Genet.* 14:145–178.

Maynard Smith, J. 1989. *Evolutionary genetics*. Oxford: Oxford University Press.

Nei, M. 1987. *Molecular evolutionary genetics*. New York: Columbia University Press.

Nei, M., and Koehn, R. K. 1983. *Evolution of genes and proteins*. Sunderland, MA: Sinauer.

Powell, J. R. 1997. *Progress and prospects in evolutionary biology: The* Drosophila *model*. New York: Oxford University Press.

Selander, R. K., and Kaufman, D. W. 1975. Self-fertilization and genetic population structure in a colonizing land snail. *Proc. Natl. Acad. Sci. USA* 70:1186–1190.

Soulé, M. E., ed. 1986. *Conservation biology: The science of scarcity and diversity*. Sunderland, MA: Sinauer.

Stringer, C. B. 1990. The emergence of modern humans. *Sci. Am.* 263 (Dec):98–104.

Weir, B. S. 1996. *Genetic data analysis II.* Sunderland, MA: Sinauer.

Woese, C. R. 1981. Archaebacteria. *Sci. Am.* 244 (Jun):98–122.

### **Chapter 22:** Quantitative Genetics

Bradshaw, H. D., Otto, K. G., Frewen, B. E., McKay, J. K., and Schemske. D. W. 1998. Quantitative trait loci affecting differences in floral morphology between two species of monkeyflower (*Mimulus*). *Genetics* 149:367–382.

Dobzhansky, T., and Pavlovsky, O. 1969. Artificial and natural selection for two behavioral traits in *Drosophila pseudoobscura*. *Proc. Natl. Acad. Sci. USA* 62:75–80.

Doebley, J., Stec, A., and Hubbard, L. 1997. The evolution of apical dominance in maize. *Nature* 386:485–488.

East, E. M. 1910. A Mendelian interpretation of variation that is apparently continuous. *Am. Nat.* 44:65–82.

———. 1916. Studies on size inheritance in *Nicotiana*. *Genetics* 1:164–176.

Emerson, R. A., and East, E. M. 1913. The inheritance of quantitative characters in maize. *Bull. Nebr. Agric. Exper. Sta.* Bull. 2.

Frary, A. Nesbitt, T. C., Frary, A., Grandillo, S., van der Knaap, E., Cong, B., Liu, J., Meller, J., Elber, R., Alpert, K. B., and Tanksley, S. D. 2000. *fw2.2*: A quantitative trait locus key to the evolution of tomato fruit size. *Science* 289:85–88.

Kim, U-k., Jorgenson, E., Coon, H., Leppert, M., Risch, N., and Drayna, D. 2003. Positional cloning of the human quantitative trait locus underlying taste sensitivity to phenylthiocarbamide. *Science* 299:1221–1225.

Lynch, M., and Walsh, B. 1998. *Genetics and analysis of quantitative traits.* Sunderland, MA: Sinauer.

Nilsson-Ehle, H. 1909. Kreuzungsuntersuchungen an Hafer und Weizen. *Lunds Univ. Aarskr. N. F. Atd.*, Ser. 2, 5(2):1–122.

Robin, C., Lyman, R. F., Long, A. D., Langley, C. H., and Mackay, T. F. C. 2002. *hairy*: A quantitative trait locus for *Drosophila* sensory bristle number. *Genetics* 162:155–164.

Weedon, M. N., Lettre, G., Freathy, R. M., Lindgren, C. M., Voight, B. F., et al. 2007. A common variant of *HMGA2* is associated with adult and childhood height in the general population. *Nature Gen.* 39:1245–1250.

### **Chapter 23:** Molecular Evolution

Andersson, S. G. E., Zomorodipour, A., Andersson, J. O., Sicheritz-Ponten, T., Alsmark, U. C., Podowski, R. M., Naslund, A. K., Eriksson, A. S., Winkler, H. H., and Kurland, C. G. 1998. The genome sequence of *Rickettsia prowazekii* and the origin of mitochondria. *Nature* 396:133–140.

Cann, R. L., Stoneking, M., and Wilson, A. C. 1987. Mitochondrial DNA and human evolution. *Nature* 325:31–36.

Dobzhanksy, T. 1973. Nothing in biology makes sense except in the light of evolution. *Amer. Biol. Teacher* 35:125–129.

Fitch, W. M., and Ayala, F. J. Molecular clocks are not as bad as you think. In *Molecular Evolution of Physiological Processes*, D. M. Fambrough, ed. (pp. 3–12). New York: Rockefeller University Press, 1984.

Gillespie, J. H. 1997. Junk ain't what junk does: Neutral alleles in a selected context. *Gene* 205:291–299.

Gould, S. J., and Lewontin, R. C., 1979. The spandrels of San Marco and the Panglossian paradigm: A critique of the adaptationist programme. *Proc. Royal Soc. Lond., Series B*, 205:581–598.

Haldane, J. B. S. 1932. *The causes of evolution.* London: Longmans and Green.

Heizer, E. M., Raiford, D. W., Raymer, M. L., Doom, T. E., Miller, R. V., and Krane, D. E.. 2006. Amino acid cost and codon usage biases in six prokaryotic genomes: A whole genome analysis. *Mol. Biol. Evol.* 23:1670–1680.

Jacob, F. 1977. Evolution and tinkering. *Science* 196:1161–1166.

Jukes, T. H., and Cantor, C. R. 1969. Evolution of protein molecules. In *Mammalian protein metabolism*, H. N. Munro, ed. (pp. 21–123). New York: Academic Press.

Klein, J., and Figueroa, F. 1986. Evolution of the major histocompatability complex. *CRC Crit. Rev. Immunol.* 6:295–386.

Margulis, L. 1981. *Symbiosis in cell evolution: Life and its environment in the early earth.* San Francisco: W.H. Freeman.

Miklos, G. L. G. 1993. Emergence of organizational complexities during metazoan evolution: Perspectives from molecular biology, palaeontology and neo-Darwinism. *Memoirs of the Assoc. of Australasian Palaeontologists* 15:7–41.

Pace, N. 1997. A molecular view of microbial diversity in the biosphere. *Science* 276:735.

Papadopoulos, D., Schneider, D., Meier-Eiss, J., Arber, W., Lenski, R. E., and Blot, M. 1999. Genomic evolution during a 10,000-generation experiment with bacteria. *Proc. Natl. Acad. Sci. USA* 96:3807–3812.

Parker, H. G., Kim, L. V., Sutter, N. B., Carlson, S., Lorentzen, T. D., Malek, T. B., Johnson, G. S., DeFrance, H. B., Ostrander, E. A., and Kruglyak, L. 2004. Genetic structure of the purebred domestic dog. *Science* 304:1160–1164.

Perutz, M. F. 1983. Species adaptation in a protein molecule. *Mol. Biol. Evol.* 1:1–28.

Sarich, V. M., and Wilson, A. C. 1967. Immunological time scale for hominid evolution. *Science* 158:1200–1203.

Saxonov, S., and Gilbert, W. 2003. The universe of exons revisited. *Genetica* 118:267–278.

Vrba, E. S., and Gould, S. J. 1986. The hierarchical expansion of sorting and selection: Sorting and selection cannot be equated. *Paleobiology* 12:217–228.

Zuckerkandl, E., and Pauling, L. 1965. Molecules as documents of evolutionary history. *J. Theor. Biol.* 8:357–366.

**3.35** Telomerase synthesizes the simple-sequence telomeric repeats at the ends of chromosomes. The enzyme is made up of both protein and RNA, and the RNA component has a base sequence that is complementary to the telomere repeat unit. The RNA component is used as a template for the telomere repeat, so if the RNA component were altered, the telomere repeat would be as well. Thus, the mutant in this question is likely to have an altered RNA component.

### Chapter 4 Gene Function

**4.4** A double homozygote should have PKU, but not AKU. The PKU block should prevent most homogentisic acid from being formed, so it could not accumulate to high levels and cause AKU.

**4.6** Autosomes are chromosomes that are found in two copies in both males and females. That is, an autosome is any chromosome except the X and Y chromosomes. Since individuals have two of each type of autosome, they have two copies of each gene on an autosome. The alleles—alternative forms of a gene—of the two copies can be the same or different. Individuals have either two normal alleles (homozygous for the normal allele), one normal allele and one mutant allele (heterozygous for the normal and mutant alleles), or two mutant alleles (homozygous for the mutant allele). A recessive mutation is one that exhibits a phenotype only when it is homozygous. Therefore, an autosomal recessive mutation is a mutation that occurs on any chromosome except the X or Y and that causes a phenotype only when homozygous. Heterozygotes exhibit a normal phenotype.

Of the diseases discussed in this chapter, many are autosomal recessive. For example, phenylketonuria, albinism, Tay–Sachs disease, and cystic fibrosis are autosomal recessive diseases. Heterozygotes for the disease allele are normal, but homozygotes with the disease allele are affected. For phenylketonuria and albinism, homozygotes are affected because they lack a required enzymatic function. In these cases, heterozygotes have a normal phenotype because their single normal allele provides sufficient enzyme function.

Parents contribute one of their two autosomes to each of their gametes, so that each offspring of a couple receives an autosome from each parent. If in a particular conception each of two heterozygous parents contributes a chromosome with the normal allele, the offspring will be homozygous for the normal allele and be normal. If in a particular conception one of the two heterozygous parents contributes a chromosome with the normal allele and the other parent contributes a chromosome with the mutant allele, the offspring will be heterozygous but be normal. If in a particular conception each parent contributes the chromosome with the mutant allele, the offspring will be homozygous for the mutant allele and develop the disease. Therefore, heterozygous parents can have both normal and affected children. Since each conception is independent, two heterozygous parents can have all normal, all affected, or any mix of normal and affected children.

**4.7** A genetic disease such as sickle-cell anemia is caused by a change in DNA that alters levels or forms of one or more gene products. This leads to changes in cellular functions, which lead to a disease state. The examples given in this chapter demonstrate that genetic diseases can be associated with mutations in single genes that affect their protein products. For example, sickle-cell anemia is caused by mutations in the gene for β-globin. Mutations lead to amino acid substitutions that cause the β-globin polypeptide to fold incorrectly. This in turn leads to sickled red blood cells and anemia. Since the environment can affect disease severity significantly, many genetic diseases are treatable. For example, PKU can be treated by altering the diet. Unlike diseases caused by an invading microorganism or other external agent that are subject to the defenses of the human immune system and that generally have short-lived clinical symptoms and treatments, genetic diseases are caused by heritable changes in DNA that are associated with chronically altered levels or forms of one or more gene products.

**4.10** Wild-type T4 will produce progeny phages at all three temperatures. Consider what will happen under each model if *E. coli* is infected with a doubly mutant phage (one step is cold sensitive, one step is heat sensitive), and the growth temperature is shifted between 17°C and 42°C during phage growth. Suppose model 1 is correct, and cells infected with the double mutant are first incubated at 17°C and then shifted to 42°C. Progeny phages will be produced and the cells will lyse, as each step of the pathway can be completed in the correct order. In model 1, the first step, A to B, is controlled by a gene whose product is heat sensitive but not cold sensitive. At 17°C, the enzyme works, and A will be converted to B. While phage are at 17°C, the second, cold-sensitive step of the pathway prevents the production of mature phage. However, when the temperature is shifted to 42°C, the accumulated B product can be used to make mature phage so that lysis will occur. Under model 1, a temperature shift performed in the reverse direction does not allow for growth. When *E. coli* cells are infected with a doubly mutant phage and placed at 42°C, the heat-sensitive first step precludes the accumulation of *B*. When the culture is shifted to 17°C, B can accumulate; but now the second step cannot occur, so no progeny phage can be produced. Therefore, if model 1 is correct, lysis will be seen only in a temperature shift from 17°C to 42°C. If model 2 is correct, growth will be seen only in a temperature shift from 42°C to 17°C. Hence, the correct model can be deduced by performing a temperature shift experiment in each direction and observing which direction allows progeny phage to be produced.

**4.11** A strain blocked at a later step in the pathway accumulates a metabolic intermediate that can "feed" a strain blocked at an earlier step. It secretes the metabolic intermediate into the medium, thereby providing a nutrient to bypass the earlier block of another strain. Consequently, a strain that feeds all others (but itself) is blocked in the last step of the pathway, while a strain that feeds no others is blocked in the first step of the pathway. Mutant *a* is blocked in the earliest step in the pathway because it cannot feed any of the others. Mutant *c* is next because it can supply the substance *a* needs but cannot feed *b* or *d*. Mutant *d* is next, and mutant *b* is last in the pathway because it can feed all the others. The pathway is $a \rightarrow c \rightarrow d \rightarrow b$.

**4.12** Hypothesize that the normal alleles at the *rose-1* and *rose-2* genes produce enzymes lying in a linear biochemical pathway leading to the production of dark red eye color. Two alternative pathways are possible:

Pathway 1:

rose-colored precursor A $\xrightarrow{\textit{rose-1}}$ rose-colored precursor B $\xrightarrow{\textit{rose-2}}$ dark red pigment

Pathway 2:

rose-colored precursor A $\xrightarrow{\textit{rose-2}}$ rose-colored precursor B $\xrightarrow{\textit{rose-1}}$ dark red pigment

In pathway 1, *rose-1* mutants are blocked at the first step, so they accumulate precursor A. The *rose-2* mutants are blocked at the second step, so they accumulate precursor B. If extracts from *rose-2* mutants are fed to *rose-1* mutants, the *rose-1* mutants will obtain precursor B. This circumvents the block in their pathway: They can complete the pathway and produce dark red pigment. However, if extracts from *rose-1* mutants are fed to *rose-2* mutants, the *rose-2* mutants will obtain precursor A. They can convert this to precursor B, but they still cannot complete the pathway and are unable to produce dark-red pigment. In pathway 2, the steps of the pathway affected by the mutants are reversed. In this situation, if extracts from *rose-1* mutants are fed to *rose-2* mutants, the *rose-2* mutants will obtain precursor B. This circumvents the block in their pathway so they will be able to produce dark-red pigment. Feeding extracts from *rose-2* mutants to *rose-1* mutants will not allow *rose-1* mutants to complete pathway 2, so the *rose-2* mutants will still have rose-colored eyes. The data are consistent with the mutants affecting the steps shown in pathway 2.

**4.13** One approach to this problem is to try to fit the data to each pathway sequentially, as if each were correct. Check where each mutant *could* be blocked (remember, each mutant carries only one mutation), whether the mutant would be able to grow if supplemented with the single nutrient that is listed, and whether the mutant would not be able to grow if supplemented with the "no growth" intermediate. It will not be possible to fit the data for mutant 4 to pathway (a), the data for mutants 1 and 4 to pathway (b), or data for mutants 3 and 4 to pathway (c). The data for all mutants can be fit only to pathway (d). Thus, (d) must be the correct pathway.

A second approach to this problem is to realize that in any linear segment of a biochemical pathway (a segment without a branch), a block early in the segment can be circumvented by any metabolites that normally appear later in the same segment. Consequently, if two (or more) intermediates can support growth of a mutant, they normally are made after the blocked step in the same linear segment of a pathway. From the data given, compounds D and E both circumvent the single block in mutant 4. This means that compounds D and E lie after the block in mutant 4 on a linear segment of the metabolic pathway. The only pathway where D and E lie in an unbranched linear segment is pathway (d). Mutant 4 could be blocked between A and E in this pathway. Mutant 4 cannot be fit to a single block in any of the other pathways that are shown, so the correct pathway is pathway (d).

**4.14** If the enzyme that catalyzes the d $\rightarrow$ e reaction is missing, the mutant strain should accumulate d and be able to grow on minimal medium to which e is added. In addition, it should not be able to grow on minimal medium or on minimal medium to which X, c, or d is added but should grow if Y is added. Therefore, plate the strain on these media and test which intermediates allow for growth of the mutant strain and which intermediate is accumulated if the strain is plated on minimal medium.

**4.16 a.** In each of these diseases, the lack of an enzymatic step leads to the toxic accumulation of a precursor or its by-product. The proposed treatments are ineffective because they do not prevent the accumulation of the toxic precursor. For both diseases the symptoms would worsen as the precursor or by-product accumulated.

**b.** The loss of 25-hydroxycholecalciferal 1-hydroxylase should lead to increased serum levels of 25-hydroxycholecalciferol, the precursor it acts upon. Since administration of the end product of the reaction, 1,25-dihydroxycholecalciferol (vitamin D), is an effective treatment, this disease is unlike those in (a). It appears that this disease is caused by the loss of the reaction's end product and not the accumulation of its precursor.

**4.18 a.** Since normal parents have affected offspring, the disease appears to be recessive. However, since patients with 50% of GSS activity have a mild form of the disease, individuals may show mild symptoms if they are heterozygous (mutant/+) for a mutation that eliminates GSS activity. In a population, individuals having the disease may not all show identical symptoms, and some may have a more severe disease form than others. The severity of the disease in an individual will depend on the nature of the person's GSS mutation and, possibly, whether they are heterozygous or homozygous for the disease allele. The alleles discussed here appear to be recessive.

**b.** Patient 1, with 9% of normal GSS activity, has a more severe form of the disease; patient 2, with 50% of normal GSS activity, has a less severe form of the disease. Thus, increased disease severity is associated with less GSS enzyme activity.

**c.** The two different amino-acid substitutions may disrupt different regions of the structure of the enzyme (consider the effect of different amino acid substitutions on the function of hemoglobin, discussed in the text). As amino acids vary in their polarity and charge, different amino-acid substitutions within the same structural region could have different chemical effects on protein structure. This, too, could lead to different levels of enzymatic function. (For a discussion of the chemical differences between amino acids, see Chapter 6.)

**d.** By analogy with the disease PKU discussed in the text, 5-oxoproline is produced only when a precursor to glutathione accumulates in large amounts due to a block in a biosynthetic pathway. When GSS levels are 9% of normal, this occurs. When GSS levels are 50% of normal, there is sufficient GSS enzyme activity to partially complete the pathway and prevent high levels of 5-oxoproline.

**e.** The mutations are allelic (in the same gene), since both the severe and the mild forms of the disease are associated with alterations in the same polypeptide that is a component of the GSS enzyme. (Note that, although the data in this problem suggest that the GSS enzyme is composed of a single polypeptide, they do not exclude the possibility that GSS has multiple polypeptide subunits encoded by different genes.)

**f.** If GSS is normally found in fetal fibroblasts, one could, in principle, measure GSS activity in fibroblasts obtained via amniocentesis. The GSS enzyme level in cells from at-risk fetuses could be compared to that in normal control samples to predict disease due to inadequate GSS levels. Some variation in GSS level might be seen, depending on the allele(s) present. Since more than one mutation is present in the population, it is important to devise a functional test that assesses GSS activity, rather than a test that identifies a single mutant allele.

**4.22 a.** From Figure 4.11, p. 72, Hb-Norfolk affects the $\alpha$-chain whereas Hb-S affects the $\beta$-chain of hemoglobin. Since each chain is encoded by a separate gene, there remains one normal allele at the genes for each of $\alpha$- and $\beta$-chains in a double heterozygote. Thus, some normal hemoglobin molecules form, and double heterozygotes do not have severe anemia. However, unlike double heterozygotes for two different, completely recessive mutations that lie in one biochemical pathway, these heterozygotes exhibit an abnormal phenotype. This is because some mutations in the $\alpha$- and $\beta$-chains of hemoglobin show partial dominance. In particular, Hb-S/+ heterozygotes show symptoms of anemia if there is a sharp drop in oxygen tension, so these double heterozygotes exhibit mild anemia.

**b.** Both Hb-C and Hb-S affect the sixth amino acid of the $\beta$-chain. The Hb-C mutation alters the normal glutamate to lysine, while the Hb-S mutation alters it to valine. Since both mutations affect the β-chain, no normal hemoglobin molecules are present. According to the text, only one type of β-chain is found in any one hemoglobin molecule. Therefore, an Hb-C/Hb-S heterozygote has two types of hemoglobin: those with Hb-C β-chains and those with Hb-S β-chains.

**4.25 a.** Since the polypeptide in *Got-2^M^ Got-2^M^* homozygotes moves further toward the cathode (the negative pole), it is more positively charged and is therefore more basic.

**b.** The single bands that are seen in $Got\text{-}2^+\ Got\text{-}2^+$ and $Got\text{-}2^M\ Got\text{-}2^M$ homozygotes indicate that each has one type of homodimer, a protein composed of two identical polypeptides. The three bands seen in the $Got\text{-}2^+\ Got\text{-}2^M$ heterozygote are, in order from anode to cathode, a homodimer composed of $Got\text{-}2^+$ polypeptides, a heterodimer composed of $Got\text{-}2^M$ and $Got\text{-}2^+$ polypeptides, and a homodimer composed of $Got\text{-}2^M$ polypeptides. The different band intensities in the middle lane result from the random association of the two types of *Got-2* monomer to form dimers in the ratio of 1 $Got\text{-}2^+$ homodimer : 2 $Got\text{-}2^+\ Got\text{-}2^M$ heterodimer : 1 $Got\text{-}2^M$ homodimer.

**c.** A single cell produces only one type of β-globin polypeptide, so cells in $\beta^A\beta^S$ heterozygotes produce hemoglobin with either $\beta^A$ or $\beta^S$ globin. When hemoglobin is analyzed by gel electrophoresis, many cells are used, so heterozygotes have two bands. The gel electrophoresis result demonstrates that in contrast to what is seen for β-globin, a *Got-2* monomer is produced from both alleles in a cell of a $Got\text{-}2^+\ Got\text{-}2^M$ heterozygote. The monomers combine at random to produce three types of dimers in the 1:2:1 ratio described in part **b**.

**4.29 a.** In Caucasians, PKU occurs in about 1 in 12,000 births, while CF occurs in about 1 in 2,000 births. In African Americans and Asian-Americans, the CF frequency is 1 in 17,000 and 1 in 31,000, respectively. Given their relative frequencies in Caucasians, the decision to mandate testing for certain diseases is not based on disease frequency alone.

**b.** The Guthrie test is a simple clinical screen for phenylalanine in the blood. A drop of blood is placed on a filter paper disc, and the disc is then placed on a solid culture medium containing *B. subtilis* and β-2-thienylalanine. The β-2-thienylalanine normally inhibits the growth of *B. subtilis*, but the presence of phenylalanine prevents this inhibition. Therefore, the amount of growth of *B. subtilis* is a measure of the amount of phenylalanine in the blood. The test provides an easy, relatively inexpensive, and reliable means to quantify blood phenylalanine levels, making it an effective preliminary screen for PKU in newborn infants.

**c.** Mandated diagnostic testing requires a highly accurate test—one that has very low false-positive and false-negative rates—as misdiagnosis of a genetic disease in a genetically normal individual has significant potential for emotional distress in the family of the misdiagnosed child, and misdiagnosis of an affected individual as normal may delay necessary therapeutic treatment. A set of mutations with a range of different disease phenotypes may make it difficult to employ a single easy-to-use test. For example, different mutations may make it impossible to use just one DNA-based test, and non-DNA-based tests that are effective at diagnosing severe disease phenotypes may not be equally effective at diagnosing mild disease forms because they may give results that overlap with those from normal individuals.

**d.** Testing for PKU in newborns is essential for early intervention to prevent the toxic accumulation of phenylketones and the resulting neurological damage in early infancy. Unless it is documented that intervention in newborns is critical for CF disease management, testing for CF in newborns is less critical. Testing is warranted to confirm a diagnosis when severe CF symptoms are apparent in a newborn.

**4.31 a.** Tests can be DNA-based and determine the genotype of a parent or fetus, or they can be biochemically based and determine some aspect of the individual's physiology. For example, the Guthrie test determines the relative amount of phenylalanine in a drop of blood to assess whether an individual has PKU; enzyme assays can determine whether a person has a complete or partial enzyme deficiency; gel electrophoresis can determine whether a person has an altered α- or β-globin that might be associated with anemia. DNA-based tests assess the presence or absence of a specific mutation and are normally employed only when there is already suspicion that an individual may carry that mutation (e.g., the couple has already had an affected offspring). Biochemical tests typically focus on assessing gene function, so they are often used in screens. However, they may not provide detailed information about which gene or biochemical step is affected and require that the biochemical activity be present in the tested cell population, such as cells obtained from an amniocentesis.

**b.** Both PKU and Tay–Sachs disease are caused by autosomal recessive mutations through which each parent contributed one of their autosomes to the affected son, so you would use a DNA-based test to evaluate whether each parent is heterozygous for an allele present in the affected son. If either of the parents do not carry a mutation present in their affected son, that son has a new mutation.

**c.** There are multiple factors to weigh when making this decision. These include the chance that a child will be affected, the type of disease, and whether the disease can be treated effectively. Since each conception is independent, there is a one in four chance of having an affected offspring. There is no effective treatment for Tay–Sachs disease, and having witnessed your child suffer and die from this disease could strongly affect your decision. In contrast, there is an effective treatment for PKU, and knowing whether a fetus is affected could help you anticipate and prepare for your child's needs.

**4.32** Individuals with PKU who do not control phenylalanine intake accumulate phenylpyruvic acid, which is toxic to neurons. The accumulated phenylpyruvic acid can pass into the developing fetus and harm its developing nervous system. Therefore, even a fetus who is a heterozygote (having a normal allele contributed by the father and a mutant allele contributed by the mother) and is able to metabolize phenylalanine normally will be harmed by a maternal accumulation of phenylpyruvic acid. For this reason, women with PKU who are pregnant should limit their phenylalanine intake.

## Chapter 5 Gene Expression: Transcription

**5.1** While both DNA and RNA are composed of linear polymers of nucleotides, their bases and sugars differ. DNA contains deoxyribose and thymine, while RNA contains ribose and uracil. Their structures also differ. DNA is frequently double-stranded, while RNA is usually single-stranded. Single-stranded RNAs are capable of forming stable, functional, and complex stem–loop structures, such as those seen in tRNAs. Double-stranded DNA is wound in a double helix and packaged by proteins into chromosomes, either as a nucleoid body in prokaryotes or within the eukaryotic nucleus. After being transcribed from DNA, RNA can be exported into the cytoplasm. If it is mRNA, it can be bound by ribosomes and translated. Eukaryotic RNAs are highly processed before being transported out of the nucleus. DNA functions as a storage molecule, while RNA functions variously as a messenger (mRNA carries information to the ribosome), or in the processes of translation (rRNA functions as part of the ribosome; tRNA brings amino acids to the ribosome), and in eukaryotic RNA processing (snRNA functions within the spliceosome).

Both DNA polymerases and RNA polymerases catalyze the synthesis of nucleic acids in the 5′-to-3′ direction. Both use a DNA template and synthesize a nucleic acid polynucleotide that is complementary to the template. However, DNA polymerases require a 3′-OH to add onto, while RNA polymerases do not. That is, RNA polymerases can initiate chains without primers, while DNA polymerases cannot. Furthermore, RNA polymerases usually require specific base-pair sequences as signals to initiate transcription.

**5.3** Both eukaryotic and *E. coli* RNA polymerases transcribe RNA in a 5′-to-3′ direction, using a 3′-to-5′ DNA template strand. There are many differences between the enzymes, however. In *E. coli*, a single RNA polymerase core enzyme is used to transcribe genes. In eukaryotes, there are three types of RNA polymerase molecules: RNA polymerase I, II, and III. RNA polymerase I synthesizes 28S, 18S, and 5.8S rRNA and is found in the nucleolus. RNA polymerase II synthesizes hnRNA, mRNA, and some snRNAs and is nuclear. RNA polymerase III synthesizes tRNA, 5S rRNA, and some snRNAs and also is nuclear.

Each RNA polymerase uses a unique mechanism to identify those promoters at which it initiates transcription. In prokaryotes such as *E. coli*, a sigma factor provides specificity to the sites bound by the four-polypeptide core enzyme, so that it binds to promoter sequences. The holoenzyme loosely binds a sequence about 35 bp before transcription initiation (the −35 region), changes configuration, and then tightly binds a region about 10 bp before transcription initiation (the −10 region) and melts about 17 bp of DNA around this region. The two-step binding to the promoter orients the polymerase on the DNA and facilitates transcription initiation in the 5′-to-3′ direction. After about 8 or 9 bases are formed in a new transcript, sigma factor dissociates from the holoenzyme, and the core enzyme completes the transcription process. Although the principles by which eukaryotic RNA polymerases bind their promoters are similar in that they use a set of ancillary protein factors—transcription factors—the details are quite different. In eukaryotes, each of the three types of RNA polymerases recognizes a different set of promoters by using a polymerase-specific set of transcription factors, and the mechanisms of interaction are different.

**5.7** **a.** and **b.** There are multiple 5′-AG-3′ sequences in each strand, and transcription may proceed in either direction. Determine the correct initiation site by locating the −10 and −35 consensus sequences recognized by RNA polymerase and $\sigma^{70}$. Good −35 (TTGACA) and −10 (TATAAT) consensus sequences are found on the top strand, starting at the 8th and 32nd bases from the 5′ end, respectively, indicating that the initiation site is the 5′-AG-3′ starting at the 44th base from the 5′ end of that strand. The start codon, AUG, used to initiate translation, is downstream from the transcription start site and is not shown in this sequence.

**c.** Transcription proceeds from left to right in this example.

**d.** the bottom (3′-to-5′) strand

**e.** the top (5′-to-3′) strand

**5.9** **a.** *E. coli* promoters vary with the type of sigma factor that is used to recognize them. More than four types of promoters exist, each having different recognition sequences. Most promoters have −35 and −10 sequences that are recognized by $\sigma^{70}$. Other promoters have consensus sequences that are recognized by different sigma factors, which are used to transcribe genes needed under altered environmental conditions, such as heat shock and stress ($\sigma^{32}$), limited nitrogen ($\sigma^{54}$), or when cells are infected by phage T4 ($\sigma^{23}$).

**b.** Although there is one core RNA polymerase enzyme, different RNA polymerase holoenzymes are formed using different sigma factors. Promoter recognition is determined by the sigma factor.

**c.** Utilizing different sigma factors allows for a quick response to altered environmental conditions (for example, heat shock, low $N_2$, phage infection) by the coordinated production of a set of newly required gene products.

**5.11** RNA polymerase I transcribes the major rRNA genes that code for 18S, 5.8S, and 28S rRNAs; RNA polymerase II transcribes the protein-coding genes to produce mRNA molecules and some snRNAs; RNA polymerase III transcribes the 5S rRNA genes, the tRNA genes, and some small nuclear RNAs. All transcription occurs in the nucleus, and only some RNAs are transported into the cytoplasm.

In the cell, the 18S, 5.8S, 28S, and 5S rRNAs are structural and functional components of the ribosome, which functions during translation in the cytoplasm. After processing, mRNAs are transported into the cytoplasm, where they are translated to produce proteins. The tRNAs are also transported into the cytoplasm, where they bring amino acids to the ribosome to donate to the growing polypeptide chain during protein synthesis. Small nuclear RNAs function in nuclear processes such as RNA splicing and processing.

**5.15** **a.** This image shows that mature mRNAs are missing sequences present in DNA, and it provides evidence for the existence of introns that are spliced out of pre-mRNAs.

**b.** There are seven single-stranded, looped regions, so the pre-mRNA for ovalbumin has 7 introns and 8 exons.

**c.** The mRNA was purified from the cytoplasm. mRNAs purified from the nucleus would include pre-mRNAs that do not have all of their introns removed. As pre-mRNAs are processed, they are exported from the nucleus through nuclear pores.

**5.16** **a.** 1 (5′ $m^7G$ cap) + 40 (exon 1) + 60 (exon 2) + 200 (poly(A) tail) = 301 bases.

**b.** U1 snRNA will not recognize the 5′ splice site and so the intron will not be removed. The transcript would have an additional 135 intronic bases and be 436 bases long.

**c.** To pair with the G in the mutant U1 snRNA, the U at the asterisked site in the RNA would need to change to a C. The U in the RNA is encoded by an A in the DNA template strand, so an AT-to-GC DNA base-pair mutation would lead to a C at the asterisked position.

**5.21** A recessive lethal is a mutation that causes death when it is homozygous—that is, when only mutant alleles are present. Heterozygotes for such mutations can be viable. Recessive lethal mutations result in death because some essential function is lacking. Neither copy of the gene functions, so the organism dies.

**a.** Deletion of the U1 genes will be recessive lethal, since U1 snRNA is essential for the identification of the 5′ splice site in RNA splicing. Incorrect splicing would lead to nonfunctional gene products for many genes, a nonviable situation.

**b.** This mutation would prevent U1 from base pairing with 5′ splice sites and thus, by the same reasoning as in part (a), would be recessive lethal.

**c.** If a deletion within intron 2 did not affect a region important for its splicing (for example, the branch point or the regions near the 5′ or 3′ splice sites), it would have no effect on the mature mRNA produced. Consequently, such a mutation would lack a phenotype if it were homozygous. However, if the splicing of intron 2 were affected and the mRNA altered, such a mutation, if homozygous, could result in the production of only nonfunctional hemoglobin, leading to severe anemia and death.

**d.** The deletion described would affect the 3′ splice site of intron 2, leading to, at best, aberrant splicing of that intron. If the mutation were homozygous, only a nonfunctional protein would be produced, resulting in severe anemia and death.

**5.22 a.** The top (5′-to-3′) strand is the coding strand, and the bottom (3′-to-5′) strand is the template strand.

**b.** The 23rd base in the RNA has been posttranscriptionally edited from a U to a G.

**5.23** 1 (5′ $m^7G$ cap) + 100 (exon 1) + 50 (exon 2) + 25 (exon 3) 200 (poly(A) tail) = 376 bases.

**5.24** The first two bases of an intron are typically 5′-GU-3′ which are essential for base pairing with the U1 snRNA during spliceosome assembly. A GC-to-TA mutation at the initial base pair of the first intron impairs base pairing with the U1 snRNA, so that the 5′ splice site of the first intron is not identified. This causes the retention of the first intron in the *tub* mRNA and a longer mRNA transcript in *tub/tub* mutants. When the mutant *tub* mRNA is translated, retention of the first intron could result in the introduction of amino acids not present in the $tub^+$ protein—or, if the intron contained a chain termination (stop) codon, premature translation termination and the production of a truncated protein. In either case, a nonfunctional gene product is produced.

The *tub* mutation is recessive because the single $tub^+$ allele in a $tub/tub^+$ heterozygote produces mRNAs that are processed normally, and when these are translated, enough normal ($tub^+$) product is produced to obtain a wild-type phenotype. Only the *tub* allele produces abnormal transcripts. When both copies of the gene are mutated in *tub/tub* homozygotes, no functional product is made and a mutant, obese phenotype results.

## Chapter 6 Gene Expression: Translation

**6.2 a.** The hemoglobin will dissociate into its four component subunits, because the heat will destabilize the ionic bonds that stabilize the quaternary structure of the protein. An individual subunit's tertiary structure may also be altered, because the thermal energy of the heat may destabilize the folding of the polypeptide.

**b.** The protein will denature. Its secondary and tertiary structures are destabilized by heating, so it does not retain a pattern of folding that allows it to be soluble.

**c.** The protein will denature when its secondary and tertiary structures are destabilized by heating. Unlike albumin, RNase will renature if cooled slowly and will reestablish its normal, functional tertiary structure.

**d.** It is likely that the meat proteins will be denatured when their secondary, tertiary, and quaternary structures are destabilized by the acid conditions of the stomach. Then the primary structure of the polypeptides will be destroyed as they are degraded into their amino acid components by proteolytic enzymes in the digestive tract.

**e.** Valine is a neutral, nonpolar amino acid, unlike the acidic glutamic acid (see Figure 6.2 p. 104). A change in the chemical properties of the sixth amino acid may alter the function of the hemoglobin molecule by affecting multiple levels of protein structure. Since it is an amino acid substitution, it changes the primary structure of the β polypeptide. This change could affect local interactions between amino acids lying near it and, in doing so, alter the secondary structure of the β polypeptide. It could also affect the folding patterns of the protein and alter the tertiary structure of the β polypeptide. Finally, the sixth amino acid residue is known to be important for interactions between the subunits of hemoglobin molecules (see Figures 4.9–4.11, pp. 71–72), because some mutations which alter that amino acid result in sickle-cell anemia. Thus, this change could alter the quaternary structure of hemoglobin.

**6.3 a.** The primary structure, or amino acid sequence, of the prion protein would be unchanged because the disease is caused not by a mutation, but rather by misfolding of the prion protein. One misfolded protein can convert a normally folded protein to the misfolded state, so the misfolded proteins are infectious. The secondary structure is affected because α-helical regions are misfolded into β-pleated sheets. This is likely to lead to an altered tertiary structure that results in the formation of amyloid.

**b.** If a genetic mutation led to an amino acid substitution, it would affect the primary structure of the prion protein. A particular amino acid substitution in the prion protein could make it more susceptible to being misfolded and lead, as in (a), to changes in its secondary and tertiary structures.

**6.4** b

**6.6** The minimum word size must be able to uniquely designate 20 amino acids, so the number of combinations must be at least 20. The following table gives the number of combinations as a function of the word size.

| | Word Size | Number of Combinations |
|---|---|---|
| a. | 5 | $2^5 = 32$ |
| b. | 3 | $3^3 = 27$ |
| c. | 2 | $5^2 = 25$ |

**6.8 a.** Proflavin causes the addition or deletion of a single DNA base pair. If this occurs within a gene's protein-coding sequence, it causes a frameshift mutation that changes the reading frame after the mutant site.

**b.** Infect the mutagenized T4 phage into *E. coli B*. An *rII* mutant produces clear plaques, while the wild-type $r^+$ strain produces turbid plaques.

**c.** Wild-type $r^+$ phage, but not *rII* mutants, can grow on strain *E. coli K12*(λ). Select for revertants by infecting the *rII* mutants into *E. coli K12*(λ), plating the bacteria, and screening for plaques.

**d.** Mutation $rII^X$ is caused by a base-pair insertion (+ mutation) that disrupts the reading frame downstream of the insertion. Not all of the revertants must affect the same base pair because they need only to restore the reading frame. A deletion (– mutation) of the inserted base pair would precisely revert the mutation and restore the reading frame. A deletion (– mutation) of a nearby base pair, at a site just before or after the site mutated in $rII^X$, would restore the reading frame near the mutant site, and could lead to a functional protein. Figure 6.5, p. 107, depicts this type of situation for a deletion (–) mutation instead of an insertion mutation.

**e.** All of the revertants must result from a deletion (– mutation) of a base pair nearby the base pair inserted in the $rII^X$ mutations, so all are double mutants. The codons nearby the base pair inserted in $rII^X$ would be altered, so that the proteins produced in the revertants would have a short segment with one or more incorrect amino acids followed by the normal sequence of amino acids. Since the phage has a wild-type phenotype, the presence of the incorrect amino acids must not have eliminated the protein's function.

**f.** Recombination would separate the two *rII* mutations and give two products: the original one-base-pair insertion mutant ($rII^X$) and, since the cause of the reversion is a deletion, a one base-pair deletion (–) mutant. Select for revertants of the – mutant just as in part (c): Infect the mutant into *E. coli K12*(λ), plate the bacteria, and screen for plaques. Only $r^+$ phage can grow, so revertants will have a one-base-pair insertion nearby or at the deleted base. Revertants of – *rII* mutants will be + *rII* mutants.

**g.** The $rII^Y$ mutant is a – mutation, so combining it with another – mutant will give a double mutant having two nearby – mutations. The *rII* reading frame will not be restored, so the double mutant will be a *rII* mutant. Proflavin treatment causes a one-base-pair insertion or deletion. It will produce $r^+$ phage only if an additional – mutation occurs nearby, because only this event will restore the reading frame. Figure 6.6, p. 107, depicts this type of situation for three + mutations.

**h.** Obtaining an $r^+$ phage requires restoration of the reading frame. This can be accomplished only if the number of deleted bases is the same as the number of bases in a codon. The genetic code must be triplet since three nearby – mutations restore the reading frame and give an $r^+$ phenotype. Proflavin-induced revertants would not be recovered in part (g) unless the genetic code was triplet (see Figure 6.6).

**6.9** Determine the expected amino acids in each case by calculating the expected frequency of each kind of triplet codon that might be formed and inferring from these what types and frequencies of amino acids would be used during translation.

**a.** 4 A : 6 C gives $2^3 = 8$ codons—specifically, AAA, AAC, ACC, ACA, CCC, ACA, CAC, and CAA. Since there is 40% A and 60% C,

P(AAA) = 0.4 × 0.4 × 0.4 = 0.064, or 6.4% Lys
P(AAC) = 0.4 × 0.4 × 0.6 = 0.096, or 9.6% Asn
P(ACC) = 0.4 × 0.6 × 0.6 = 0.144, or 14.4% Thr
P(ACA) = 0.4 × 0.6 × 0.4 = 0.096, or 9.6% Thr
(24% Thr total)
P(CCC) = 0.6 × 0.6 × 0.6 = 0.216, or 21.6% Pro
P(CCA) = 0.6 × 0.6 × 0.4 = 0.144, or 14.4% Pro
(36% Pro total)
P(CAC) = 0.6 × 0.4 × 0.6 = 0.144, or 14.4% His
P(CAA) = 0.6 × 0.4 × 0.4 = 0.096, or 9.6% Gln

**b.** 4 G : 1 C gives $2^3 = 8$ codons—specifically, GGG, GGC, GCG, GCC, CGG, CGC, CCC, and CCG. Since there is 80% G and 20% C,

P(GGG) = 0.8 × 0.8 × 0.8 = 0.512, or 51.2% Gly
P(GGC) = 0.8 × 0.8 × 0.2 = 0.128, or 12.8% Gly
(64% Gly total)
P(GCG) = 0.8 × 0.2 × 0.8 = 0.128, or 12.8% Ala
P(GCC) = 0.8 × 0.2 × 0.2 = 0.032, or 3.2% Ala
(16% Ala total)
P(CGG) = 0.2 × 0.8 × 0.8 = 0.128, or 12.8% Arg
P(CGC) = 0.2 × 0.8 × 0.2 = 0.032, or 3.2% Arg
(16% Arg total)
P(CCC) = 0.2 × 0.2 × 0.2 = 0.008, or 0.8% Pro
P(CCG) = 0.2 × 0.2 × 0.8 = 0.032, or 3.2% Pro
(4% Pro total)

**c.** 1 A : 3 U : 1 C gives $3^3 = 27$ different possible codons. Of these, one will be UAA, a chain-terminating codon. Since there is 20% A, 60% U, and 20% C, the probability of finding this codon is 0.6 × 0.2 × 0.2 = 0.024, or 2.4%. All of the remaining 26 (97.6%) codons will be sense codons. Proceed in the same manner as in (a) and (b) to determine their frequency, and determine the kinds of amino acids expected. To take the frequency of nonsense codons into account, divide the frequency of obtaining a particular amino acid considering all 27 possible codons by the frequency of obtaining a sense codon. This gives

(0.8/0.976)% = 0.82% Lys
(3.2/0.976)% = 3.28% Asn
(12.0/0.976)% = 12.3% Ile
(9.6/0.976)% = 9.84% Tyr
(19.2/0.976)% = 19.67% Leu
(28.8/0.976)% = 29.5% Phe
(4.0/0.976)% = 4.1% Thr
(0.8/0.976)% = 0.82% Gln
(3.2/0.976)% = 3.28% His
(4.0/0.976)% = 4.1% Pro
(12.0/0.976)% = 12.3% Ser

It is likely that the chains produced would be relatively short, due to the chain-terminating codon.

**d.** 1 A : 1 U : 1 G : 1 C will produce $4^3 = 64$ different codons, all possible in the genetic code. The probability of each codon is 1/64, so there will be a 3/64 chance of a codon being chain terminating. With those exceptions, the relative proportion of amino acid incorporation is dependent directly on the codon degeneracy for each amino acid. Inspecting the table of the genetic code in Figure 6.7, p. 108, and taking the frequency of nonsense codons into account yields the following table:

| Amino Acid | Number of Codons | Frequency |
|---|---|---|
| Trp | 1 | 1/61 = 1.64% |
| Met | 1 | 1.64% |
| Phe | 2 | 2/61 = 3.28% |
| Try | 2 | 3.28% |
| His | 2 | 3.28% |
| Gln | 2 | 3.28% |
| Asn | 2 | 3.28% |
| Lys | 2 | 3.28% |
| Asp | 2 | 3.28% |
| Glu | 2 | 3.28% |
| Cys | 2 | 3.28% |
| Ile | 3 | 3/61 = 4.92% |
| Val | 4 | 4/61 = 6.56% |
| Pro | 4 | 6.56% |
| Thr | 4 | 6.56% |
| Ala | 4 | 6.56% |
| Gly | 4 | 6.56% |
| Leu | 6 | 6/61 = 9.84% |
| Arg | 6 | 9.84% |
| Ser | 6 | 9.84% |

**6.10** In population 1, the codons that can be produced encode Lys (AAA, AAG), Arg (AGG, AGA), Glu (GAG, GAA), and Gly (GGA, GGG). All of these are sense codons, so long polypeptide chains containing these amino acids will be synthesized. In population 2, the codons that can be produced encode Lys (AAA), Asn (AAU), Ile (AUA, AUU), Tyr (UAU), Leu (UUA), Phe (UUU), and stop (UAA). The frequency of the stop codon will be ($1/4 \times 3/4 \times 3/4$) = $9/64$ = 0.14, or 14%. Thus, the polypeptides formed in population 2 will, on average, be shorter than those formed in population 1. If a stop codon appears about 14% of the time, there will be, on average, 1/0.14 = 7.14 codons from one stop codon to the next. On average, six sense codons will lie in between the stop codons, so polypeptides will be synthesized that are about six amino acids long.

**6.13** **a.** 3

**b.** 1, 2, 3, 4

**c.** 3

**d.** 1, 2

**e.** 1 (note that some tRNAs and rRNAs have introns)

**f.** 4

**g.** 1

**6.15** **a.** 3′-TAC AAA ATA AAA ATA AAA ATA AAA ATA...-5′ (The first fMet or Met is removed following translation of the mRNA.)

**b.** 5′-ATG TTT TAT TTT TAT TTT TAT TTT TAT...-3′

**c.** 3′-AAA-5′ is the anticodon for Phe, and 3′-AUA-5′ is the anticodon for Tyr.

**6.16** Figure 6.7, p. 108, and Table 6.1, p. 109, aid in answering this question. The answer is given in the following table:

| Amino Acid | tRNAs Needed | Rationale |
|---|---|---|
| Ile | 1 | 3 codons can use 1 tRNA (wobble) |
| Phe | 1 | 2 codons can use 1 tRNA (wobble) |
| Tyr | 1 | " " " " " " " |
| His | 1 | " " " " " " " |
| Gln | 1 | " " " " " " " |
| Asn | 1 | " " " " " " " |
| Lys | 1 | " " " " " " " |
| Asp | 1 | " " " " " " " |
| Glu | 1 | " " " " " " " |
| Cys | 1 | " " " " " " " |
| Trp | 1 | 1 codon |
| Met | 2 | Single codon, but need 1 tRNA for initiation and 1 tRNA for elongation |
| Val | 2 | 4 codons: 2 can use 1 tRNA (wobble) |
| Pro | 2 | " " " " " " " |
| Thr | 2 | " " " " " " " |
| Ala | 2 | " " " " " " " |
| Gly | 2 | " " " " " " " |
| Leu | 3 | 6 codons: 2 can use 1 tRNA (wobble) |
| Arg | 3 | " " " " " " " |
| Ser | 3 | " " " " " " " |
| **Total** | 32 | 61 codons |

**6.18** Since a dipeptide is formed, translation initiation is not affected, nor is the first step of elongation—the binding of a charged tRNA in the A site and the formation of a peptide bond. However, since *only* a dipeptide is formed, it appears that translocation is inhibited.

**6.19** **a.** By saying that the genetic code is degenerate, we mean that more than one codon occurs for each amino acid.

**b.** Leu and Arg have codons where a mutation in the first nucleotide can result in a synonymous codon. Eight codons have this property: four of each of the six Leu and six Arg codons could have mutations in the first nucleotide that produce a synonymous codon. For Leu, synonymous codons would be produced by mutations causing a U-to-C change in the first nucleotide of codons UUA and UUG, and by mutations causing a C-to-U change in the first nucleotide of codons CUA and CUG. For Arg, synonymous codons would be produced by mutations causing a C-to-A change in the first nucleotide of codons CGA and CGG, and by mutations causing an A-to-C change in the first nucleotide of codons AGA or AGG.

**c.** No amino acids or codons show this property.

**d.** Met and Trp have codons where a mutation in the third nucleotide never generates a synonymous codon. Two codons, AUG (Met) and UGG (Trp), show this property.

**e.** $59/61$ = 96.7% of sense codons can be changed by a single nucleotide mutation to a synonymous codon. The code is highly degenerate. Since most of the degeneracy occurs at the third nucleotide position, mutations that affect this position often lead to synonymous codons.

**f.** Though silent mutations do not alter the amino acid sequence of a protein, they can affect the rate of translation. Not all aminoacyl–tRNA molecules are equally abundant, and a change from a wild-type to a synonymous codon may result in a codon being read by a rare aminoacyl-tRNA. This will result in a slower rate of translation. A slower rate of translation may affect how chaperones interact with the newly synthesized polypeptide and alter its folding. Two polypeptides with identical amino acid sequences that are folded differently may have different functional properties. As a result, silent mutations could affect progression of disease and response to drug treatments.

**6.22** The anticodon 5′-GAU-3′ recognizes the codon 5′-AUC-3′, which encodes Ile. The mutant tRNA anticodon 5′-CAU-3′ would recognize the codon 5′-AUG-3′, which normally encodes Met. The mutant tRNA would therefore compete with tRNA.Met for the recognition of the 5′-AUG-3′ codon, and if successful, insert Ile into a protein where Met should be. Since a special tRNA.Met is used for initiation, only AUG codons other than the initiation AUG will be affected. Thus, this protein will have four different N-terminal sequences, depending on which tRNA occupies the A site in the ribosome when the codon AUG is present there:

Met-Val-Ser-Ser-Pro-**Ile**-Gly-Ala-Ala-**Ile**-Ser
Met-Val-Ser-Ser-Pro-**Met**-Gly-Ala-Ala-**Ile**-Ser
Met-Val-Ser-Ser-Pro-**Ile**-Gly-Ala-Ala-**Met**-Ser
Met-Val-Ser-Ser-Pro-**Met**-Gly-Ala-Ala-**Met**-Ser

**6.28** Multiple lines of evidence support the view that the rRNA component of the ribosome serves more than a structural role. First, the 3′ end of the 16S rRNA is important for identifying where the small ribosomal subunit should bind the mRNA. It has a sequence that is complementary to the Shine–Dalgarno sequence, the ribosome-binding site (RBS) in the mRNA. Mutational analyses demonstrated that the 3′ end of the 16S rRNA must base-pair with this mRNA sequence for correct initiation of translation. Second, the 23S rRNA is required for peptidyl transferase activity. Evidence that the peptidyl transferase consists entirely of RNA comes from studies of the atomic structure of the large ribosomal subunit and is supported by experiments showing that peptidyl transferase activity remains following the depletion of the 50S subunit proteins, but not after the digestion of rRNA with ribonuclease T1.

**6.29** A eukaryotic mRNA is modified to contain a 5′ 7-methyl-G cap and a 3′ poly(A) tail. The 5′ cap is required early in translation initiation—it binds to the eIF-4F complex just before the binding of a complex of the 40S ribosomal subunit, the initiator Met–tRNA, and other eIF proteins. Transcription initiation is stimulated by the looping of the poly(A) tail close to the 5′ end. This occurs when the poly(A) binding protein (PAB) binds to eIF-4G, which is part of the eIF-4F complex.

**6.31** Rewriting the sequences to readily visualize the codons shows that mutants *a*, *b*, *c*, *d*, and *f* are point mutations in which one base has been substituted for another and that mutant *e* is a deletion of one base that causes a frameshift mutation. The following proteins are produced (alterations to the normal sequence are underlined):

Normal: Met-Phe-Ser-Asn-Tyr-. . . -Met-Gly-Trp-Val
Mutant *a*: Met-Phe-Ser-Asn
Mutant *b*: Starts at later AUG to give Met-Gly-Trp-Val
Mutant *c*: Met-Phe-Ser-Asn-Tyr-. . . -Met-Gly-Trp-Val
Mutant *d*: Met-Phe-Ser-Lys-Tyr-. . . -Met-Gly-Trp-Val
Mutant *e*: Met-Phe-Ser-Asn-Ser-. . .-Trp-Gly-Gly-Cys. . .
(no stop codon, protein continues)
Mutant *f*: Met-Phe-Ser-Asn-Tyr . . . -Met-Gly-Trp-Val-Trp . . .
(no stop codon, protein continues)

**6.35 a.** If the primary mRNA for this gene is 250 kb, it must be substantially processed by RNA splicing (removing introns) and polyadenylation to a smaller mature mRNA.

**b.** A 1,480-amino acid protein requires 1,480 × 3 = 4,440 bases of protein-coding sequence. This leaves 6,500 − 4,440 = 2,060 bases of 5′ untranslated leader and 3′ untranslated trailer sequence in the mature mRNA—about 32%.

**c.** The ΔF508 mutation could be caused by a DNA deletion for the three base pairs encoding the mRNA codon for phenylalanine. This codon is UUY (Y = U or C), and the DNA sequence of the nontemplate strand is 5′-TTY-3′. The segment of DNA containing these bases would be deleted in the appropriate region of the gene.

**d.** If positioned at random and solely within a gene's coding region (that is, not in 3′ or 5′ untranslated sequences or in intronic sequences), a deletion of three base pairs results either in an mRNA missing a single codon or an mRNA missing bases from two adjacent codons. If three of the six bases from two adjacent codons were deleted, the remaining three bases would form a single codon. In this case, an incorrect amino acid might be inserted into the polypeptide at the site of the left codon, and the amino acid encoded by the right codon would be deleted. If the 3′ base of the left codon were deleted, it would be replaced by the 3′ base of the right codon. Since the code is degenerate and wobble occurs in the 3′ base, this type of deletion might not alter the amino acid specified by the left codon. The adjacent amino acid would still be deleted, however.

**6.36 a.** Notice that the N-terminal sequence of the mRNA-encoded polypeptide contains many hydrophobic amino acids (see Figure 6.3, p. 105). It is a signal sequence. As the signal sequence is synthesized by the ribosome, it becomes bound by a signal recognition particle (SRP) that blocks further translation of the mRNA until the growing polypeptide–SRP–ribosome–mRNA complex becomes bound to the ER. When the SRP binds to an SRP receptor in the ER membrane, the ribosome becomes bound to the ER, the SRP is released, and translation resumes with the growing polypeptide extending through the ER membrane into its cisternal space. Once the signal sequence is fully within the cisternal space of the ER, it is cleaved from the polypeptide by a signal peptidase.

**b.** The mutation would disrupt the signal sequence, so the polypeptide would no longer be directed into the ER for further processing and targeting to the cell membrane. The ADAM12 protein would be synthesized, but not be positioned correctly in the cell membrane.

**6.38** Some genes can inhibit the activity of others. An increase in an enzyme's activity will be seen if actinomycin D blocks the transcription of a gene that codes for an inhibitor of the enzyme's activity.

## Chapter 7 DNA Mutation, DNA Repair, and Transposable Elements

**7.1** b

**7.2** False. Mutations occur spontaneously at more or less a constant frequency, regardless of selective pressure. Once they occur, however, they can be selected for or against, depending on the advantage or disadvantage they confer.

**7.4** c. The key to this answer is the word "usually." The other choices might apply rarely, but not usually.

**7.6 a.** If the normal codon is 5′-CUG-3′, the anticodon of the normal tRNA is 5′-CAG-3′. If a mutant tRNA recognizes 5′-GUG-3′, it must have an anticodon that is 5′-CAC-3′. The mutational event was a CG-to-GC transversion in the gene for the leucine tRNA. The mutant tRNA will carry leucine to a codon for valine.

**b.** Presumably Leu

**c.** Val

**d.** Leu

**7.8** Acridine is an intercalating agent that induces frameshift mutations. *lacZ*-1 probably is a frameshift mutation that results in a completely altered amino acid sequence after some point, although it might be truncated due to the introduction of an out-of-frame nonsense codon. In either case, the protein produced by *lacZ*-1 would most likely have a different molecular weight and charge. During gel electrophoresis (see Figure 4.8, p. 70), it would migrate differently than the wild-type protein. 5BU is incorporated into DNA in place of T. During DNA replication, it can be read as C by DNA polymerase because of a keto-to-enol shift. This results in point mutations, usually TA-to-CG transitions. *lacZ*-2 is likely to contain a single amino-acid difference, due to a missense mutation; although it, too, could contain a nonsense codon. A missense mutation might lead to the protein having a different charge, while a nonsense codon would lead to a truncated protein that would have a lower molecular weight. Both would migrate differently during gel electrophoresis.

**7.9 a.** Six

**b.** Three, since the UGG codon would be replaced by UAG, a nonsense (chain termination) codon, to give 5′-AUG ACC CAU UAG . . . -3′.

**7.11 a.** UAG: Gln (CAG), Lys (AAG), Glu (GAG), Leu (UUG), Ser (UCG), Trp (UGG), Tyr (UAC, UAU), chain terminating (UAA)

**b.** UAA: Lys (AAA), Gln (CAA), Glu (GAA), Leu (UUA), Ser (UCA), Tyr (UAC, UAU), chain terminating (UGA, UAG)

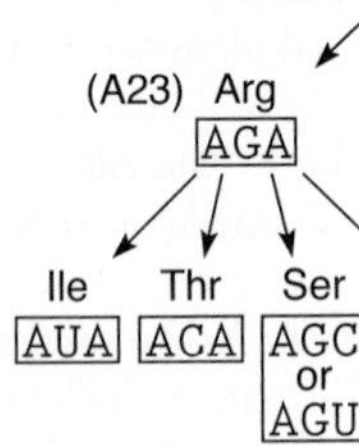


finding one specific base pair (A-T, T-A, G-C, or C-G) at a particular site is $^1/_4$. The chance of finding two specific base pairs at a site is $(^1/_4)^2$. In general, the chance of finding $n$ specific base pairs at a site is $(^1/_4)^n$. Here, $^1/_{4,096} = (^1/_4)^6$, so the enzyme recognizes a 6-bp site.

**8.5** **a.** Since 40% of the genome is composed of G-C base pairs, $P(\text{G}) = P(\text{C}) = 0.20$ and $P(\text{A}) = P(\text{T}) = 0.30$. Therefore, $P(\text{CCTAGG}) = (0.20)^4 \times (0.30)^2 = 0.000144$. A genome with $3 \times 10^9$ base pairs will have about $3 \times 10^9$ different groups of 6-bp sequences. Thus, the number of sites is $(0.000144) \times (3 \times 10^9) = 432{,}000$.

**b.** $3 \times 10^9$ bp/432,000 sites = 1/0.000144 = 6,944 bp between sites.

**c.** $P(\text{CCTAGG}) = (0.10)^4 \times (0.40)^2 = 0.000016$, so two *Avr*II sites are expected to be about 1/0.000016 = 62,500 bp apart.

**8.7** **a.** In a sequence that has a uniform distribution of A, G, C, and T, the chance of finding a 6-bp site is $(^1/_4)^6 = {^1/_{4,096}}$, and the chance of finding an 8-bp site is $(^1/_4)^8 = {^1/_{65,536}}$. In such a sequence, *Apa*I, *Hin*dIII, *Sac*I, and *Ssp*I should produce fragments that average 4,096 bp in size, and *Srf*I and *Not*I should produce fragments that average 65,536 bp in size.

**b. i.** The large variation in average fragment sizes when one restriction enzyme is used to cleave different genomes could reflect: (1) the nonrandom arrangements of base pairs in the different genomes (e.g., there is variation in the frequencies of certain sequences that are part of the restriction site in the different genomes); and/or (2) the different base compositions of the genomes (e.g., genomes that are rich in A-T base pairs should have fewer sites for enzymes recognizing sites containing only G-C base pairs).

**ii.** The large variation in fragment sizes when the same genome is cut with different enzymes that recognize sites having the same length could reflect: (1) the nonrandom arrangement of base pairs *in that genome*; and/or (2) the base composition *of that genome*.

**iii.** If the sequence of *Mycobacterium tuberculosis* was random and contained 25% each of A, G, T, and C, enzymes recognizing a 6-bp site should produce fragments that are about 16-fold smaller than enzymes recognizing an 8-bp site. This is not the case here, which suggests that at least one of these assumptions is incorrect. Two possibilities are that the genome of *Mycobacterium tuberculosis* is very rich in G-C base pairs and poor in A-T base pairs, and/or that there is a nonrandom arrangement of base pairs so that 5′-AA-3′, 5′-TT-3′, 5′-AT-3′, and/or 5′-TA-3′ sequences are rare. (The data given for *Sac*I suggest that the sites 5′-AG-3′ and 5′-CT-3′, which are parts of the *Hin*dIII site, are not rare.)

**8.8** Cloning vectors must have at least three features: the ability to replicate within a host cell conferred by an origin of replication (e.g., an *ori* in bacterial plasmids, an *ARS* in YACs), a dominant marker that allows for their selection in a host cell (e.g., antibiotic resistance in bacterial vectors, an auxotrophic marker in YACs), and one or more unique restriction sites for DNA insertion. Many different types of vectors have been developed. Three types are bacterial plasmids, bacterial artificial chromosomes (BACs), and yeast artificial chromosomes (YACs). In addition to the three required features mentioned previously, YACs also have *CEN* sequences to ensure their proper segregation during cell division. These vectors differ in the amount of DNA they hold and how they are used. Plasmids typically hold less than 10 kb of DNA, can replicate at a high copy number within bacterial cells, and are used for many different purposes (in addition to those described in this chapter, more are discussed in Chapters 9 and 10). When a genome is sequenced, they are used during the shotgun cloning of 2-kb and 10-kb inserts. BACs hold up to 300 kb of DNA and are present in a single copy in bacterial cells. They are the preferred vector for large clones in physical mapping studies of genomes because they do not undergo rearrangements, as do YACs. Two disadvantages to using *E. coli* cloning vectors are that very AT-rich sequences are difficult to clone in *E. coli*, and some sequences are poisonous to *E. coli* when cloned. YACs can hold between 0.2 and 2 Mb of DNA and are present in one copy per cell. Since they can hold large inserts, they have been useful for the construction of physical maps of the genome. However, their usefulness is limited because they can undergo rearrangements and are often chimeric (holding DNA from more than one site in the genome).

**8.10** **a.** Linearize the circular pBluescript II vector by digesting it with the enzyme *Pst*I. Then, treat the digested vector with alkaline phosphatase to remove its 5′ phosphates. This leaves only 5′-OH groups at its two ends, and so prevents its recircularization when mixed with DNA ligase. If the plasmid is not prevented from recircularizing, most of the colonies that are produced following transformation will not have inserts. After treating the vector with phosphatase, mix it with the 2-kb DNA fragment, and add DNA ligase. Since the insert DNA has not been treated with phosphatase, it retains 5′ phosphate groups and its 5′ ends can be ligated to the sticky ends of the digested vector. Then, transform *E. coli* with the ligation reaction, and plate the cells on medium containing ampicillin and X-gal. The presence of ampicillin in the medium ensures that only bacteria containing the pBluescript II plasmid will grow. The presence of X-gal allows colonies with inserts to be identified. If the fragment was not inserted into the *Pst*I site, the *lacZ* gene will function, β-galactosidase will be made, X-gal will be cleaved, and the colony will be blue. If the fragment was inserted into the *Pst*I site, it will have disrupted the *lacZ* gene, no β-galactosidase will be made, and the colony will be white.

**b.** Select white colonies and prepare plasmid DNA from each colony. Digest the prepared DNAs with *Pst*I, and separate the digestion products by size using agarose gel electrophoresis. A colony with the correct insert should produce two bands: a 2-kb band corresponding to the insert and a 3-kb band corresponding to the pBluescript II vector.

**8.11** If the enzyme is not inactivated, the restriction enzyme produced by the *hsdR* gene will cleave any DNA transformed into *E. coli* with the appropriate recognition sequence. This will make it impossible to clone DNA with the recognition sequence that is not already methylated at the A in this sequence.

**8.13** A genomic library made in a plasmid vector is a collection of plasmids that have different yeast genomic DNA sequences in them. Like two volumes of a book series, two plasmids in the library will have identical vector sequences but different yeast DNA inserts. Such a library is made as follows:

**i.** Isolate high-molecular-weight yeast genomic DNA by isolating nuclei, lysing them, and gently purifying their DNA.

**ii.** Cleave the DNA into fragments that are 5–10 kb, an appropriate size for insertion into a plasmid vector. This can be done by cleaving the DNA with *Sau*3A for a limited time (i.e., performing a *partial* digest) and then selecting fragments of an appropriate size by agarose gel electrophoresis.

**iii.** Digest a plasmid vector such as pBluescript II with *Bam*HI. This will leave sticky ends that can pair with those left by *Sau*3A.

**iv.** Mix the purified, *Sau*3A-digested yeast genomic DNA with the plasmid vector and DNA ligase.

**v.** Transform the recombinant DNA molecules into *E. coli*.

**vi.** Recover colonies with plasmids by plating on media with ampicillin (pBluescript II has a gene for resistance to this antibiotic) and with X-gal (to allow for blue-white colony screening to identify plasmids with inserts). Each colony will have a different yeast DNA insert, and all of the colonies comprise the yeast genomic library.

In a BAC vector, much larger DNA fragments—200 to 300 kb in size—would be used.

**8.15** Use a restriction site linker, a short segment of double-stranded DNA that contains a restriction site. The linker can be efficiently ligated onto blunt-ended DNA fragments. Digestion of the resulting DNA fragments with the restriction enzyme will then produce fragments with sticky ends. Their sticky ends allow for efficient ligation into plasmids digested with the same restriction enzyme (see Figure 8.16, p. 197). To clone DNA fragments that have the restriction site found in the linker, use an adapter—a short, double-stranded piece of DNA with one sticky end and one blunt end.

**8.16** From the text, $N = \ln(1 - p)/\ln(1 - f)$, where $N$ is the necessary number of recombinant DNA molecules, $p$ is the probability of including one particular sequence, and $f$ is the fractional proportion of the genome in a single recombinant DNA molecule. Here, $p = 0.90$ and $f = (2 \times 10^5)/(3 \times 10^9)$, so $N = 34{,}538$.

**8.18 a.** Approximately, $500 \times (13{,}543{,}099 + 10{,}894{,}467) = 1.22 \times 10^{10}$ nucleotides were sequenced, corresponding to ($1.22 \times 10^{10}$ nucleotides)/($3 \times 10^9$ bp/haploid genome) $\approx$ 4-fold coverage.

**b.** If a plasmid with a 2-kb insert has a unique sequence at one end but a repetitive sequence at the other end, it will not be possible to continue assembling the sequence past this plasmid because many clones in the library have the same repetitive sequence, and they come from all over the genome. Since many repetitive sequences are about 5 kb in length, sequencing plasmids with 10-kb inserts circumvents this problem. Some of the 10-kb inserts will have a sequence at one end that overlaps the unique sequence in the plasmid with the 2-kb insert as well as a unique sequence at their other end that lies past the repetitive element and can be assembled with unique sequence from other plasmids.

**c.** The sequence of the central region is obtained from the sequence of overlapping clones during sequence assembly.

**8.19 a.** In a DNA sequencing reaction, the annealing of a sequencing primer to one strand of a double-stranded DNA fragment defines the point from which DNA sequence can be obtained. If the sequence of an insert in pBluescript II is unknown, it is not possible to design and synthesize a sequencing primer targeted directly to it. To circumvent this issue, the pBluescript II vector has universal sequencing primer sites that flank the multiple cloning site. As shown in Figure 8.9, in pBluescript II, the T7 universal sequencing primer anneals near the *Kpn*I site, and the SP6 universal sequencing primer anneals on the other side of the multiple cloning site. These sequencing primers are positioned so that DNA polymerase can extend from the primer to obtain the sequence of the ends of the insert.

**b.** If dideoxy sequencing is used, only several hundred bases of sequence are obtained from one sequencing reaction. Let us consider the case where a sequencing reaction produces 500 bases of sequence. To obtain the sequence of the entire 7-kb insert, first obtain the sequence of the ends of the insert using the universal primers present in the pBluescript II vector. In a second step, use that sequence to design new primers that are about 450 bases from the ends of the insert, and use these to obtain an additional 500 bases of DNA sequence. Assemble this sequence with that obtained previously based on the overlap between the sequences—you will have about 950 bases of sequence at each end. Take a third step: design primers that are about 900 bases from the ends of the insert, use them in a third set of sequencing reactions to obtain an additional 500 bases of DNA sequence, and assemble this sequence into the one you already have. Continue to design primers based on the newly obtained sequence and use them to walk through the sequence in this manner until you have obtained the sequence of the entire insert. The sequence obtained from one end of the insert will be reversed and complementary to the sequence obtained from the other end of the insert.

**c.** While you could in principle use the "primer walking" method described in part (**b**) to obtain the sequence of a 200-kb insert in pBeloBAC11, it would be tedious and time-consuming. In addition, if there were repetitive sequences within the insert, you might run into problems—if you inadvertently designed a primer within a repetitive sequence, you would not obtain unambiguous sequence information from that primer. It is more efficient to obtain the insert's sequence by using a whole-genome shotgun cloning approach. Make a plasmid library with 2-kb and 10-kb inserts from the pBeloBAC11 clone, sequence the ends of the inserts from enough clones to obtain 7-fold coverage, and then assemble that sequence using computerized algorithms.

**8.21**

5′- T T C A G A T G C A T A T C C G G -3′

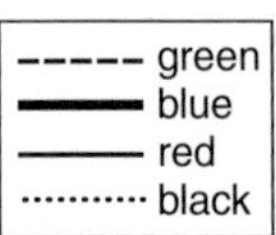


**8.25** D, G, I, M, N, R, and Y can serve as tag SNPs. The following table shows the haplotypes identified by alleles at these tag SNPs (the tag SNP allele is in boldface type):

| Tag SNP(s) | Haplotype |
|---|---|
| D | A1 B2 C1 **D1** E3 |
| | A3 B3 C2 **D2** E2 |
| | A2 B2 C1 **D3** D2 |
| | A1 B1 C3 **D4** E1 |
| G, I | F2 **G1** H2 **I2** J2 K2 L1 |
| | F1 **G2** H1 **I1** J1 K1 L1 |
| | F2 **G3** H1 **I3** J2 K1 L2 |
| M, N | **M1 N2** |
| | **M2 N1** |
| R | O1 P1 Q2 **R1** S2 T1 U1 V2 |
| | O2 P1 Q1 **R2** S1 T1 U2 V2 |
| | O1 P2 Q2 **R3** S1 T1 U2 V2 |
| Y | W3 X2 **Y1** Z1 |
| | W2 X1 **Y2** Z1 |
| | W1 X3 **Y3** Z2 |
| | W1 X1 **Y4** Z2 |

The 26 SNPs define 5 sets of haplotypes, so a minimum of 5 tag SNPs are needed to differentiate between them.

**8.27** Prepare a DNA microarray consisting of oligonucleotides that collectively represent the entirety of the normal dystrophin gene, including SNPs known to be present in normal individuals as well as known point mutations. It is important to consider that some sites in the gene will be polymorphic in normal individuals; multiple probes able to detect the various SNPs found in a particular region of the gene will need to be placed on the microarray. Isolate DNA from the blood of an individual affected with muscular dystrophy, label it with a fluorescent dye, and hybridize the chip with the labeled DNA under conditions that require a precise match between the labeled DNA and the oligonucleotide probes on the DNA microarray. The site of the mutation can be located by identifying the region of the gene where no hybridization signal is seen in any of the oligonucleotide probes that detect normal sequences. If the mutation corresponds to a previously known point mutation, the probe(s) able to detect that mutation should show a hybridization signal.

**8.30** Repetitive sequences pose at least two problems for sequencing eukaryotic genomes. Highly repetitive sequences associated with centromeric heterochromatin consist of short, simple repeated sequences. These are unclonable, making it impossible to obtain the complete genome sequence of organisms with them. More complex repetitive sequences such as those found within euchromatic regions can be cloned and sequenced. However, since they can originate from different genomic locations and a shotgun sequencing approach provides only short sequences, the assembly of overlapping sequences can be ambiguous. Some of the ambiguities can be resolved by comparing these sequences to overlapping sequences generated from sequencing clones with larger inserts.

**8.32 a.** Since prokaryotic ORFs should reside in transcribed regions, they should follow a bacterial promoter containing consensus sequences recognized by a sigma factor. For example, promoters recognized by $\sigma^{70}$ would contain −35 (TTGACA) and −10 (TATAAT) consensus sequences. Within the transcribed region, but before the ORF, there should be a Shine–Dalgarno sequence (UAAGGAGG) used for ribosome binding. Nearby should be an AUG (or GUG, in some systems) start codon. This should be followed by a set of in-frame sense codons. The ORF should terminate with a stop (UAG, UAA, UGA) codon.

**b.** Eukaryotic introns are transcribed but not translated sequences in the RNA-coding region of a gene. They will be spliced out of the primary mRNA transcript before it is translated. If not accounted for, they could introduce additional amino acids, frameshifts, and chain-termination signals.

**c.** The small average size of exons relative to the range of sizes for introns makes it challenging to predict whether a region with only a short set of in-frame codons is used as an exon. Such regions could have arisen by chance or be the remnants of exons that are no longer used due to mutation in splice site signals.

**d.** Eukaryotic introns typically contain a GU at their 5′ ends, an AG at their 3′ ends, and a YNCURAY branch-point sequence 18 to 38 nucleotides upstream of their 3′ ends. To identify eukaryotic ORFs in DNA sequences, scan sequences following a eukaryotic promoter for the presence of possible introns by searching for sets of these three consensus sequences. Then try to translate sequences obtained if potential introns are removed, testing whether a long ORF with good codon usage can be generated. Since alternative mRNA splicing exists at many genes, more than one possible ORF may be found in a given DNA sequence.

**8.34 a.** Comparison of cDNA and genomic DNA sequences can define the structure of transcription units by elucidating the location of the intron–exon boundaries, poly(A) sites, and the approximate locations of promoter regions. Comparison of different full-length cDNAs representing the same gene can identify the use of alternative splice sites, alternative poly(A) sites, and alternative promoters.

**b.** The analysis of full-length cDNAs provides information about an entire open reading frame, information about the site at which transcription starts and where the promoter lies, and the location of the poly(A) site. Partial-length cDNAs might provide some but not all of this information. While partial-length cDNAs could be compared and assembled to obtain more information, their assembly as challenging because alternative splice sites, alternative promoters, and/or alternative poly(A) sites can be used.

**c.** Genes are not uniformly distributed among different chromosomes, and some chromosomes have more genes than others. While consistent with the finding that chromosomes have gene-rich regions and gene deserts, more data is needed to infer the relationship between the density of genes on a chromosome and how gene-rich it is. For example, a chromosome with many small genes could still have regions classified as gene deserts.

**d.** Two possible explanations are that: (1) some regions of the genome sequence were not yet correctly assembled (e.g., due to the large numbers of repetitive sequences they contain), so the cDNAs are unable to be mapped to just one region; and (2) some of the genes are in regions that have not yet been assembled (e.g., because they are difficult to clone or sequence). As the genome sequence is revised, these issues should be resolved.

**8.35** Sequencing of genomes of the Archaea has shown that their genes are not uniformly similar to those of the Bacteria or the Eukarya. While most of the archaean genes involved in energy production, cell division, and metabolism are similar to their counterparts in Bacteria, the genes involved in DNA replication, transcription, and translation are similar to their counterparts in Eukarya.

## Chapter 9 Functional and Comparative Genomics

**9.2** Physically, a gene is a sequence of DNA that includes a transcribed DNA sequence and the regulatory sequences that direct its transcription (e.g., its promoter). Genes produce RNA (mRNA, rRNA, snRNA, tRNA, siRNA, and miRNA) and protein products. Functionally, genes can be identified by the phenotypes of mutations that alter or eliminate the functions of their products. In contrast, an ORF is a potential open reading frame, the segment of an mRNA (mature mRNA in eukaryotes) that directs the synthesis of a polypeptide by the ribosome. Therefore, genes have features that ORFs do not, including transcribed and untranscribed sequences and introns. We have experimental evidence for some ORFs (cDNA sequence, detected protein products) but the existence of other ORFs is predicted only from genomic sequence information.

Two general approaches can be used to determine the function of ORFs having unknown functions. First, computerized sequence similarity searches using programs such as BLAST can be performed to compare the ORF sequence and all sequences in a database. The extent of sequence similarity is used to infer whether the ORF encodes a protein with the same or similar function to that of a gene in a database. Since proteins can have multiple functional domains (e.g., a catalytic domain and a DNA-binding domain), sometimes this approach gives only partial insight into the function of the ORF's protein product. Second, an experimental approach can be used to investigate the function of the gene identified by the ORF. This may involve analyzing knockout or knockdown mutations and analyzing the resulting mutant phenotype. In organisms such as humans where this experimental approach would be unethical, a gene in a model organism that encodes its homolog can be investigated instead. This approach may include demonstrating that the ORF encodes a protein and characterizing its protein product.

**9.6** None of the inferences can be made without additional information and analyses. The question statement only indicates that the best match is to the *HprK* gene. The question statement does not describe the quality of the match or what region(s) have significant sequence similarity to *HprK*. It also leaves us without a critical piece of information we need to make inferences about the potential function of this DNA fragment—we do not know whether the DNA fragment contains a gene that encodes a protein product. To address this issue, the DNA fragment should be examined for the presence of an open reading frame (ORF), and the amino acid sequence of this ORF should be compared to that of *HprK* and other known kinases. To conclude that the DNA fragment encodes a gene encoding a kinase, a kinase domain should be found within the ORF. To infer that the DNA fragment encodes a gene homologous to *HprK*, it should have an ORF, and the ORF's amino acid sequence should have significant sequence similarity to the protein produced by *HprK* in regions that are important for its biological function. However, even if the DNA fragment contains a gene that appears to be homologous to *HprK*, that gene may not function to regulate carbohydrate metabolism. We cannot exclude that it has evolved to function in or regulate a different, though perhaps related, biochemical process. When a sequence alignment provides information about functional domains present in a new protein and its homology to known proteins, it suggests hypotheses about the functions of the new protein that still must be tested experimentally.

**9.8** To amplify a specific region, one needs to know the sequences flanking the target region so that primers able to amplify the target region can be designed. Once primers are synthesized, the polymerase chain reaction can be assembled. It contains a DNA template (genomic DNA, cDNA, or cloned DNA), the pair of primers that flank the DNA segment targeted for amplification, a heat-resistant DNA polymerase (such as *Taq*), the four dNTPs (dATP, dTTP, dGTP, and dCTP), and an appropriate buffer (see Figure 9.3, p. 222).

**9.10** PCR is a much more sensitive and rapid technique than cloning. Many millions of copies of a DNA segment can be produced from one DNA molecule in only a few hours using PCR. In contrast, cloning requires more DNA (ng to μg quantities) for restriction digestion and at least several days to proceed through all of the cloning steps.

**9.11** As shown in Figure 9.3, two unit-length, double-stranded DNA molecules (called amplimers) are produced after the third cycle of PCR from one double-stranded DNA template molecule. If each step of the PCR process is 100% efficient, the number of amplimers geometrically increases in each subsequent cycle: In the 4th cycle there will be 4 amplimers, in the 5th cycle there will be 8 amplimers, and more generally, in the $n$th cycle there will be $2^{n-2}$ amplimers. In the 30th cycle, there will be $2^{28} = 2.68 \times 10^8$ molecules. A larger number of initial template molecules will lead to a proportional increase in amplimer production.

**a.** $10 \times 2^{28} = 2.68 \times 10^9$ molecules

**b.** $1{,}000 \times 2^{28} = 2.68 \times 10^{11}$ molecules

**c.** $10{,}000 \times 2^{28} = 2.68 \times 10^{12}$ molecules

Consider these answers with respect to the experimental observation that about 5 ng of DNA (about $2.3 \times 2^9$ copies of a 200-bp DNA fragment) is detected readily on an ethidium bromide stained agarose gel.

**9.13 a.** ES cells are embryonic stem cells, cells derived from a very early embryo that retain the ability to differentiate into cell types characteristic of any part of the organism. They can be grown in culture and transformed with a gene-targeting vector. Then, cells that do not have a gene knockout are excluded by adding neomycin, which selects for cells with the gene-targeting vector, and ganciclovir, which selects for cells where the gene-targeting vector was incorporated via homologous recombination (see Figure 9.5, p. 226). The surviving cells are tested (using PCR) for the presence of the knockout mutation and then placed in a blastocyst mouse embryo that is implanted into a female for development. During development, the transformed ES cells provide progenitor cells for the germ line so that when the mouse is mated, the knockout mutation is passed on to the next generation.

**b.** A chimera is a genetic mosaic, an animal with two distinct tissue types. Chimeras arise because the ES cells containing the knockout mutation are genetically different from the embryos they are placed in for development—the ES cells are from a homozygous agouti mouse, while the blastocyst embryos they will be placed in have been harvested from a homozygous black mouse. This difference in coat color genes allows chimeric pups to be readily identified.

**c.** When the chimeric pups mature and are mated with non-transgenic black mice, they will pass the gene knockout to some of their progeny provided that some of their germ line consists of the transformed cells. Since the transformed cell had two copies of the agouti gene, these progeny will have one copy of the agouti gene (from the transformed cell) and one copy of the black gene (from the mate). The progeny will be agouti because agouti is dominant to black. To determine if an agouti offspring also is heterozygous for the knockout gene, isolate its DNA using a cheek scraping, a drop of blood, or a tail snip, and perform a PCR-based test to determine whether the *neo*$^R$ gene is present. Since the *neo*$^R$ gene was in the gene-targeting vector, animals with it are heterozygous for the gene knockout.

**9.16** Two approaches to knocking out or knocking down gene function without a gene-targeting vector are to generate mutants by transposon insertion and to use RNA interference methods to knock down gene function. A transposon inserted into the coding region of a gene should disrupt its function. However, since transposon insertion cannot be targeted to a specific gene, a collection of mutants with different transposon insertions must be screened to identify a mutation in a specific gene. In *Mycoplasma genitalium*, about 2,000 transposon insertions were characterized to identify how many protein-coding genes are required for the organism to survive. In a diploid organism, most gene knockouts due to a transposon insertion are likely to be viable as heterozygotes, so a collection of mutants generated by transposon insertion could be obtained and screened to identify transposon inserts in or near particular genes. This method requires detailed knowledge of the transposons in an organism, and how they can be mobilized to insert at different locations in the genome. Therefore, it is restricted to organisms where that information is available.

The RNA interference (RNAi) method introduces dsRNA molecules complementary to a specific mRNA into cells. Once in the cell, the dsRNAs trigger the cell's RNAi pathway to render the mRNA nonfunctional (see Figure 9.6, p. 228). This method requires information about an mRNA sequence so that a gene-specific dsRNA can be designed, and a means to introduce the dsRNA into cells. If these are available, RNAi can be used in a number of organisms without extensive modification. Indeed, it has been used in *Caenorhabditis elegans*, *Drosophila*, mice, and plants.

**9.18 a.** On chromosome V and the X chromosome, genes are distributed uniformly. However, especially on chromosome V, conserved genes are found more frequently in the central regions. In contrast, inverted and tandem-repeat sequences are found more frequently on the arms. It appears that at least on chromosome V, there is an inverse relationship between the frequency of inverted and tandem-repeats and the frequency of conserved genes.

**b.** Since there are fewer conserved genes on the arms, there appears to be a greater rate of change on chromosome arms than in the central regions.

**c.** Yes, since increased meiotic recombination provides for greater rates of exchange of genetic material on chromosomal arms.

**9.19** The transcriptome is the set of RNAs present in a cell at a particular stage and time, while the proteome is the set of proteins present in the cell at that stage and time.

**a.** It is likely that the proteome has both more total as well as unique members. It is likely that it has more total members since multiple copies of a protein can be translated from a single mRNA transcript. It is also likely that it has more unique members since many transcripts give rise to different protein isoforms. Once translated, proteins can be posttranslationally or cotranslationally modified in different ways: phosphorylation, glycosylation, methylation, proteolytic processing, etc. If proteins translated from a single transcript are modified in different ways, multiple protein isoforms are produced.

**b.** This analysis could be performed using DNA microarrays. RNA could be isolated from the nervous system of different developmental stages, and the transcriptional profile in each stage could be assessed using DNA microarrays. To do this type of analysis, RNA from one developmental stage would be reverse transcribed into cDNA in the presence of a fluorescently tagged nucleotide, so that the cDNA is labeled, say, to fluoresce green. RNA from a second developmental stage would be reverse transcribed in a similar manner, so that its cDNA is labeled, say, to fluoresce red. The labeled target DNAs would be hybridized to the DNA microarray, and the relative red : green fluorescence bound to a single site on the microarray would be used to infer the relative amounts of gene expression of the gene located at that site on the microarray. This could be done for many different developmental stages.

**c.** For the proteome, one would need to assess changes in the proteins produced over time. Thus, one would isolate proteins from the nervous system of different developmental stages and assess the relative abundance of individual proteins. For a specific set of proteins, this could be done by making measurements on the different samples in parallel using a protein array such as a capture array.

**9.21** Use microarray analysis to determine if patients who respond to therapy have a different pattern of gene expression in their blood cells than do patients who fail to respond to therapy. Prepare cDNA from the mRNA isolated from the blood cells of individual leukemia patients, label the cDNAs with fluorescent dyes, and use them in a DNA microarray analysis. For example, label cDNA from a patient who responds to the therapy with Cy3 and label cDNA from a patient who fails to respond to the therapy with Cy5. Mix the labeled cDNAs together and allow them to hybridize to a probe array containing oligonucleotides from many different genes as shown in Figure 9.7, p. 231. Then identify the set of genes whose pattern of expression differs in the two patients. Repeat the experiment using different pairs of patients to identify the set of genes that shows consistently greater (or lesser) expression in patients who respond to therapy. The hypothesis that two (or more) different types of leukemia are present in this patient population would be supported if there are consistent differences in the gene expression patterns between patients who respond to therapy and patients who fail to respond to therapy. The pattern of gene expression could be further evaluated as a clinical marker.

**9.22** One approach is to use model organisms (e.g., transgenic mice) that have been developed as models to study a specific human disease. Expose them and a control population to specific environmental conditions, and then simultaneously assess disease progression and alterations in patterns of gene expression using microarrays. This would provide a means to establish a link between environmental factors and patterns of gene expression that are associated with disease onset or progression.

**9.23** A DNA microarray has DNA probes (oligonucleotides, PCR-amplified cDNA products) bound to a solid substrate (a glass slide, membrane, microtiter well, or silicon chip), while a protein chip has proteins immobilized on solid substrates. Protein arrays are probed by labeling target proteins with fluorescent dyes, incubating the labeled target with the probe array, and measuring the bound fluorescence using automated laser detection. One type of protein chip is a capture array, where a set of antibodies is bound to a solid substrate and used to evaluate the level and presence of target molecules in cell or tissue extracts. A capture array can be used in disease diagnosis (to evaluate whether a specific protein associated with a disease state is present) and in protein expression profiling (evaluation of the proteome qualitatively and quantitatively).

**9.25 a., b.,** and **c.** Use representational oligonucleotide microarray analysis (ROMA) to identify which genes differ in copy number between a reference individual and sets of ASD and normal individuals. For each individual to be evaluated, isolate genomic DNA and digest it with a restriction enzyme that leaves a single-stranded overhang. Design and synthesize a single-stranded DNA adapter molecule so one of its ends is complementary to the overhang and its other end is complementary to a PCR primer. Anneal the adapter to the overhang of the restriction fragments and ligate it to the fragments using DNA ligase. Then use PCR to amplify all of the restriction fragments in the presence of nucleotides labeled with Cy3 (green) or Cy5 (red).

To determine which genes are deleted or duplicated in ASD individuals, prepare a reference target DNA by labeling DNA from a reference individual with Cy5 (red) and prepare experimental target DNAs by labeling DNA from different ASD individuals with Cy3 (green). Mix the reference target DNA with one experimental target DNA, and allow these to hybridize to a DNA microarray containing oligonucleotide probes representing genes in the 16p11.2 region (for part a) or, alternatively, probes representing genes throughout the genome (for part c). Red spots identify probes from genes that have a decreased copy number (deletion) in an ASD individual, green spots identify probes from genes that are present in increased copy number (duplication) in an ASD individual, and yellow spots identify probes from genes present in the same copy number in the reference and ASD individuals. To investigate if the same regions have altered copy number in normal individuals (for part b), compare the reference target DNA to target DNA samples prepared from normal individuals.

**9.28 a.** The Virochip is a DNA microarray with oligonucleotide probes for about 20,000 genes representing the very large number of viruses with sequenced genomes. When labeled target DNA is prepared from an unknown virus and mixed with the probes on the chip, it will hybridize to similar sequences. The pattern of sequence similarities that is observed can be used to identify the type of virus that the target DNA was derived from. In this way, it was determined that SARS patients all had a novel coronavirus.

**b.** Target DNA prepared from a new virus will hybridize to sequences on the Virochip that are similar to it. Thus, if the new virus is related to a known virus, the Virochip should be useful to detect and classify it.

### Chapter 10 Recombinant DNA Technology

**10.4** Use an expression vector. Expression vectors have the signals necessary for DNA inserts to be transcribed and for these transcripts to be translated. In prokaryotes, the vector should have a prokaryotic promoter sequence upstream of the site where the cDNA is inserted, and possibly, a terminator sequence downstream of this site. In eukaryotes, a eukaryotic promoter would be needed, and a poly(A) site should be provided downstream of the site where the cDNA is inserted. If the cDNA lacked a start codon, a start AUG codon embedded in a Kozak consensus sequence would be needed upstream of the site where the cDNA is inserted so that the transcript can be translated efficiently. In the event that the cDNA lacked a start codon, care must be taken during the design of the cloning steps to ensure that the open reading frame (ORF) of the cDNA is in the same reading frame with the start codon provided by the vector.

**10.5** It would be preferable to use cDNA. Human genomic DNA contains introns, while cDNA synthesized from cytoplasmic poly(A)+ mRNA does not. Prokaryotes do not process eukaryotic precursor mRNAs having intron sequences, so genomic clones will not give appropriate translation products. Since cDNA is a complementary copy of a functional mRNA molecule, the mRNA transcript will be functional, and when translated human (pro-)insulin will be synthesized.

**10.6** If genomic DNA had been used, there could be concerns that an intron in the genomic DNA was not removed, since *E. coli* does not process RNAs as eukaryotic cells do. However, the cDNA is a copy of a mature mRNA, so this should not be a concern. There are other potential concerns, however. First, depending on the nature of the sequence inserted, a fusion protein with β-galactosidase may have been produced, and not just human insulin. That is, in pBluescript II, the multiple cloning site (MCS) is within part of the *lacZ* (β-galactosidase) gene. Sequences inserted into the MCS, if inserted in the correct reading frame (the same one as used for β-galactosidase), will be translated into a β-galactosidase–fusion protein. If this was acceptable, it would be important to ensure that only the ORF (the open reading frame) of the insulin gene is inserted properly into the MCS of the pBluescript II vector. In order for the insulin ORF to be inserted properly, it must be inserted in the correct reading frame, so that premature termination of translation does not occur, and the correct polypeptide is produced. One could not use a complete copy of the human mRNA transcript for insulin. If transcribed, it would have features of eukaryotic transcripts but not features required for prokaryotic translation. Indeed, some of its 5′ UTR and 3′ UTR sequences may interfere with prokaryotic transcription and translation. For example, it will lack a Shine–Dalgarno sequence to specify where translation should initiate and identify the first AUG codon. In the pBluescript II vector, a Shine–Dalgarno sequence is supplied after the promoter for the *lacZ* gene, since without an insert in the MCS, β-galactosidase is produced. However, the cDNA may have 5′ UTR sequences which interfere with translation initiation in prokaryotes, or which contain stop codons, terminating translation of the β-galactosidase–fusion protein. Second, the cDNA may encode a protein that is processed post-translationally to become human insulin. The protein produced in *E. coli* may not be processed.

**10.11** Construct a map stepwise, considering the relationship between the fragments produced by double digestion and the fragments produced by single-enzyme digestion. Start with the larger fragments. The 1,900-bp fragment produced by digestion with both A and B is a part of the 2,100-bp fragment produced by digestion with A, and the 2,500-bp fragment produced by digestion with B. Thus, the 2,500-bp and 2,100-bp fragments overlap by 1,900 bp, leaving a 200-bp A–B fragment on one side and a 600-bp A–B fragment on the other. One has:

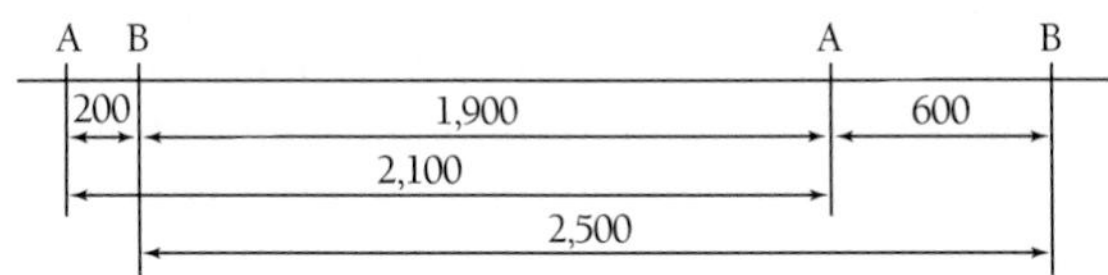


The map is extended in a stepwise fashion, until all fragments are incorporated into the map. The restriction map is:

A B A B A 5,000 bp
1,000 200 1,900 600 800 500
1,000 2,100 1,400 500 A fragments
1,200 2,500 1,300 B fragments

**10.12 a.** Table 8.1, p. 174, indicates that *Bgl*II enzyme leaves a 5′-GATC overhang, while the *Pst*I enzyme leaves an ACGT-3′ overhang. If the multiple cloning site (MCS) of the pBluescript II vector could be cleaved to leave these overhangs, the 4,500-bp fragment could be cloned directionally into the vector. Examination of the restriction enzyme sites available in the MCS of pBluescript II reveals a *Pst*I site, but no *Bgl*II site. One approach to obtain the required 5′-GATC overhang is to examine the MCS further to determine whether cleaving any of these sites leaves the same kind of overhang as *Bgl*II. A comparison of the sites in the MCS to the enzymes described in Table 8.1 identifies a *Bam*HI site that, if cut, would leave a 5′-GATC overhang, just like that of *Bgl*II. Thus, cleaving the vector with *Bgl*II would produce the appropriate sticky end. Therefore, to clone the insert directionally, cleave the pBluescript II vector with *Pst*I and *Bam*HI, allow the fragment to anneal to these sticky ends, and use DNA ligase to seal the gap in the phosphodiester backbones.

**b.** Transform the ligated DNA into a host bacterial cell, and plate the cells on bacterial medium containing ampicillin and a substrate (X-gal) for β-galactosidase that turns blue when cleaved by that enzyme. This selects for bacterial colonies that harbor pBluescript II plasmids and allows for blue-white screening to identify colonies that have plasmids with inserts. Pick white colonies (which have an interrupted *lacZ* gene, and so β-galactosidase is not produced and the substrate is not cleaved) and isolate plasmid DNA from them. Cleave the DNA with restriction enzymes and analyze the products using agarose gel electrophoresis to verify that the appropriate-sized fragments are recovered. Digestion with *Eco*RI should give two fragments, one that is 2,961 + 490 = 3,451 bp (vector plus the 490-bp *Eco*RI fragment of the insert) and one that is 4,500 − 490 = 4,010 bp (the insert minus the 490-bp fragment of the insert). A set of double digests (*Eco*RI + *Pst*I; *Eco*RI + *Bam*HI) will also be informative.

**10.16** She should clone the genes by complementation. Transform each mutant with a library containing wild-type sequences and then plate the transformants at an elevated, restrictive temperature. Colonies that grow have a plasmid that complements the cell division mutation—they are able to overcome the functional deficit of the mutation because the plasmid has provided a copy of the wild-type gene. Purify the plasmid from these colonies and characterize the cloned gene. The shuttle vector would also contain a selectable marker, such as *URA3*, for selection of transformants in yeast. If the cell division mutants were also made into *ura3* mutants, then transformants could be selected for using *URA3*. But, in this case, the temperature-sensitive phenotype of the cell-division mutants enables the direct selection for transformants receiving the wild-type gene.

**10.18** Compare the amino acid sequence to the genetic code, and design a "guessmer"—a set of oligonucleotides which could code for this sequence. Here, the guessmer would have the sequence 5′-ATG TT(T or C) TA(T or C) TGG ATG AT(T, C, or A) GG(A, G, T, or C) TA(T or C)-3′, and be composed of 96 different oligonucleotides. Synthesize and then label these oligonucleotides, and use them as a probe (in place of a radioactive antibody) to screen a cDNA library as described in Figure 10.5.

**10.19 a.** The lane with genomic DNA will have a smear: there are many *Eco*RI sites in a genome and the distances between these sites will vary. The smear reflects the large number and many different sizes of *Eco*RI fragments. Since *Eco*RI recognizes a 6-bp site, the average size will be about 4,096 bp (assume the genome is 25% A, G, C, and T, and the nucleotides are distributed uniformly), and more intense staining will be seen around this size. The pBluescript II plasmid has a single *Eco*RI restriction site into which the 10-kb insert has been cloned, so the lane with plasmid DNA will have two bands: the genomic DNA insert at 10 kb, and the plasmid DNA at 3 kb.

**b.** The probe will detect the 10-kb *Eco*RI fragment specifically, so a signal will be seen in each lane at 10 kb.

**10.20** The gel is soaked in an alkaline solution to denature the DNA to single-stranded form. It must be bound to the membrane in single-stranded form so that the probe can bind in a sequence-specific manner using complementary base pairing.

**10.21 a.** She should see a 2.0-kb band because the 2.0-kb probe is a single-copy genomic DNA sequence.

**b.** LINEs are moderately repetitive DNA sequences, which may be distributed throughout the genome. Since the LINE has an internal *Eco*RI site, each LINE in the genomic DNA will be cut by *Eco*RI during preparation of the Southern blot. When the blot is incubated with the probe, both fragments will hybridize to the probe. The size of the fragments produced from each LINE will vary according to where the element is inserted in the genome, and where the adjacent *Eco*RI sites are. Hence there will be many different-sized bands seen on the genomic Southern blot.

**c.** As in (b), there will be many different-sized bands on the genomic Southern blot. The sizes of the bands seen reflect the distances between *Eco*RI sites that flank a LINE. All of the bands will be larger in size than the element, as the element is not cleaved by *Eco*RI. Counting the number of bands can give an estimate of the number of copies of the element in the genome.

**d.** Since the heterozygote has one normal chromosome 14, the probe will bind to the 3.0-kb *Eco*RI fragment derived from the normal chromosome 14. If the translocation is a reciprocal translocation, the remaining chromosome 14 is broken in two, and attached to different segments of chromosome 21. Since chromosome 14 has a breakpoint in the 3.0-kb *Eco*RI fragment, the 3.0-kb fragment is now split into two parts, each attached to a different segment of chromosome 21. Consequently, the 3.0-kb probe spans the translation breakpoint and will bind to two different fragments, one from each of the translocation chromosomes. The sizes of the fragments are determined by where the adjacent *Eco*RI sites are on the translocated chromosomes. Thus, the blot will have three bands, one of which is 3.0 kb.

**e.** Since the *TDF* gene is on the Y chromosome, no signal should be seen in a Southern blot prepared with DNA from a female having only X chromosomes.

**10.22 a.** *Not*I recognizes an 8-bp site, while *Bam*HI recognizes a 6-bp site. 8-bp sites appear about 1/16 less frequently than 6-bp sites, hence the *Not*I fragments are larger, relatively speaking, than the *Bam*HI fragments.

**b.** There are many *Bam*HI fragments in the BAC DNA insert, while there are fewer fragments in each *Not*I fragment. Digesting first with *Not*I allows regions of the BAC to be evaluated in an orderly, systematic manner and allows for the *Bam*HI fragments containing the gene to be identified more precisely and then purified.

**c.** The 47-kb *Not*I fragment contains the gene, since it is the only *Not*I fragment that has sequences hybridizing to the cDNA.

**d.** The 10.5-, 8.2-, 6.1-, and 4.1-kb *Bam*HI fragments contain the gene, since they hybridize to the cDNA probe.

**e.** The RNA-coding region is about 28.9 kb. It is larger than the cDNA since genomic DNA contains intronic sequences.

**10.24** If the same gene functions in the brain, transcripts for the gene should be found in the brain. To evaluate this possibility, label the cloned DNA, and use it to probe a northern blot having either total RNA or purified poly(A)+ mRNA isolated from brain tissue. If the transcript is not abundant, preparing a northern blot with purified poly(A)+ mRNA should provide additional sensitivity. If the mRNA is particularly rare, it may be prudent to use mRNA isolated from a specific region of the brain, such as the hypothalamus.

**10.26** Isolate RNA from the livers of the alcohol-fed and control rats. Measure the levels of mRNA for alcohol dehydrogenase by either: (1) separating the RNA by size using gel electrophoresis, preparing a northern blot, and hybridizing it with a probe made from a cDNA for the alcohol dehydrogenase gene; or (2) using RT-PCR or real-time quantitative PCR.

**10.27 a.** Since *Taq* DNA polymerase lacks proofreading activity, base-pair mismatches that occur during replication go uncorrected. This means that some of the molecules produced in the PCR process will contain errors relative to the starting template. Enzymes with proofreading activity significantly reduce the introduction of errors.

**b.** If an error is introduced in the first few cycles of a PCR amplification, most of the derivative DNA molecules produced during subsequent cycles of PCR amplification will also contain the error. This happens since molecules produced in earlier cycles of PCR serve as templates for molecules synthesized in later cycles of PCR. Consequently, if an error is introduced in a later cycle in the PCR amplification process, fewer molecules will have the error.

**10.28** The insert Katrina sequenced was obtained from genomic DNA, while the inserts Marina sequenced were obtained from PCR. *Taq* DNA polymerase introduces errors during PCR, so that individual double-stranded molecules that are amplified during PCR may have small amounts of sequence variation. If PCR products are sequenced directly, the amount of variation is small enough that it may not be noticed—at a particular position in the sequence, only a very small number of molecules have an error. However, when PCR products are cloned, each independently isolated plasmid has an insert derived from a different double-stranded DNA PCR product, so that errors will be apparent.

**10.30** Design primers so that you can use PCR to amplify a segment of each orphan gene. Then prepare RNA from yeast at sequential stages of sporulation and use reverse transcriptase to reverse transcribe each RNA sample into cDNA. To measure the expression of each of the orphan genes in the different stages of sporulation, quantify the amount of each gene's cDNA in the different cDNA preparations using real-time PCR with SYBR® Green (see Figure 10.9).

**10.33 a.** The probe hybridizes to the same genomic region in each of the 10 individuals. Different patterns of hybridizing fragments are seen because of polymorphism of the *Eco*RI sites in the region. If a site is present in one individual but absent in another, different patterns of hybridizing fragments are seen. This provides evidence of restriction fragment length polymorphism. To distinguish between sites that are invariant and those that are polymorphic, analyze the pattern of bands that appear. Notice that the sizes of the hybridizing bands in individual 1 add up to 5 kb, the size of the band in individual 2 and the largest hybridizing band. This suggests that there is a polymorphic site within a 5.0-kb region. This is indicated in the diagram below, where the asterisk over site *b* depicts a polymorphic *Eco*RI site:

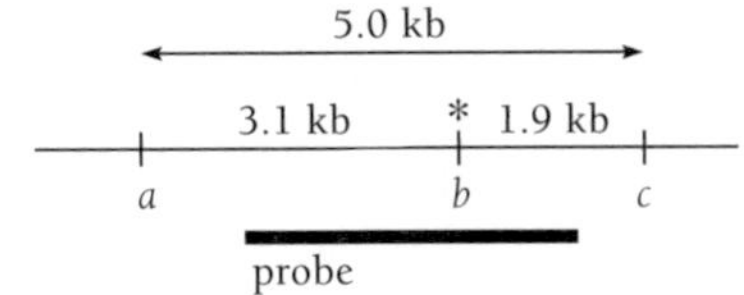


Notice also that the size of the band in individual 3 equals the sum of the sizes of the bands in individual 4. Thus, there is an additional polymorphic site in this 5.0-kb region. Since the 1.9-kb band is retained in individual 4, the additional site must lie within the 3.1-kb fragment. This site, denoted *x*, is incorporated into the diagram below. Notice that, because the 1.0-kb fragment flanked by sites *a* and *x* is not seen on the Southern blot, the probe does not extend into this region.

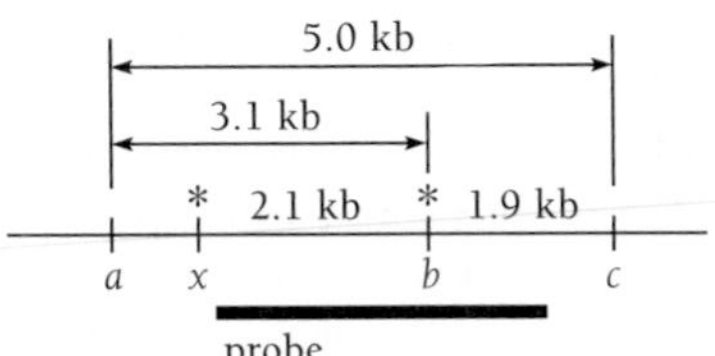


Depending on whether *x* and/or *b* are present, you will see either 5.0-, 3.1-, and 1.9-kb, 2.1- and 1.9-kb, or 4.0-kb bands. In addition, if an individual has chromosomes with different polymorphisms, you can see combinations of these bands. Thus, individual 5 has one chromosome that lacks sites *x* and *b* and one chromosome that has site *b*. The chromosomes in each individual can be tabulated as follows:

| Individual | Sites on Each Homolog | Homozygote or Heterozygote? |
|---|---|---|
| 1 | *a, b, c* | homozygote |
| 2 | *a, c* | homozygote |
| 3 | *x, c* | homozygote |
| 4 | *x, b, c* | homozygote |
| 5 | *a, c/a, b, c* | heterozygote |
| 6 | *x, c/a, b, c* | heterozygote |
| 7 | *a, b, c/x, b, c* | heterozygote |
| 8 | *a, c/x, c* | heterozygote |
| 9 | *a, c/x, b, c* | heterozygote |
| 10 | *x, c/x, b, c* | heterozygote |

**b.** Since individual 1 is homozygous, chromosomes with sites at *a*, *b*, and *c* will be present in all of the offspring, giving bands at 3.1 and 1.9 kb. Individual 6 will contribute chromosomes of two kinds, one with sites at *x* and *c* and one with sites at *a*, *b*, and *c*. Thus, if this analysis is performed on their offspring, two equally frequent patterns will be observed: a pattern of bands at 3.1 and 1.9 kb and a pattern of bands at 4.0, 3.1, and 1.9 kb. This is just like the patterns seen in the parents.

**10.34** Chromosomes bearing CF mutations have a shorter restriction fragment than chromosomes bearing wild-type alleles. Both parent lanes (M and P) have two bands, indicating that each parent has a normal and a mutant chromosome. The parents are therefore heterozygous for the CF trait. The fetus lane (F) shows only one (lower molecular weight) band. The size of the band indicates that the fetus has only mutant chromosomes. The intensity of the band is about twice that of the same-sized band in the parent lanes. This is because the diploid genome of the fetus has two copies of the fragment, while the diploid genome of each parent only has one. Since the fetus is homozygous for the CF trait, it will have CF.

**10.35 a.** Use the PCR-RFLP method: Isolate genomic DNA from the individual with Parkinson disease, and use PCR to amplify the 200-bp segment of exon 4; purify the PCR product, digest it with *Tsp*45I, and resolve the digestion products by size using gel electrophoresis. The normal allele will contain the *Tsp*45I site, and so produce 120- and 80-bp fragments. The mutant allele will not contain the *Tsp*45I site, and so produce only a 200-bp fragment.

**b.** Homozygotes for the normal allele will have 120- and 80-bp fragments; homozygotes for the mutant allele will have a 200-bp fragment; heterozygotes will have 200-, 120-, and 80-bp fragments.

**c.** Use RT-PCR to amplify a DNA copy of the mRNA, and digest the RT-PCR product with *Tsp*45I. First isolate RNA from the tissue. Then make a single-stranded cDNA copy using reverse transcriptase and an oligo(dT) primer. Then amplify exon 4 of the cDNA using PCR, digest the product with *Tsp*45I, and separate the digestion products by size using gel electrophoresis. If a 200-bp fragment is identified in a heterozygote, then the mutant allele is transcribed. If only 120- and 80-bp fragments are identified, then the mutant allele is not transcribed. Note that to accurately assess expression of either allele, it is essential that the RT-PCR reaction is performed on a purified RNA template without contaminating genomic DNA.

**10.36** Use the reverse ASO method described in the text.

**10.38** A SNP is a single nucleotide polymorphism. Since a single base change can alter the site recognized by a restriction endonuclease, a SNP can also be a RFLP, or restriction fragment length polymorphism. Since simple tandem repeats (STRs) and variable number of tandem repeats (VNTRs) are based on tandemly repeated sequences (2-to-6-bp repeats for STRs, 7 to tens of base pairs for VNTRs), they will not usually be SNPs.

**10.40** If an individual is homozygous for an allele at an STR, all of their gametes have the same STR allele. The STR cannot be used as a marker to distinguish the recombinant and parental gametes of the individual, and so will not be useful for mapping studies. In a population, individuals will be heterozygous more often for STRs with more alleles and higher levels of heterozygosity. The recombinant and parental offspring classes may be distinguished in individuals heterozygous for an STR, making crosses informative for mapping studies.

If an STR has few alleles and a low heterozygosity, many individuals in a population will share the same STR genotypes. Therefore, there will be many individuals in the population who, by chance alone, will share the same genotype as a test subject and the STR will not be very useful for DNA fingerprinting studies.

**10.42** James and Susan Scott are not the parents of "Ronald Scott." There are several bands in the fingerprint of the boy that are not present in either James or Susan Scott and thus could not have been inherited from either of them (e.g., bands a and b in the figure below). In contrast, whenever the boy's DNA exhibits a band that is missing from one member of the Larson couple, the other member of the Larson couple has that band (e.g., bands c and d). Thus, there is no band in the boy's DNA that he could not have inherited from one or the other of the Larsons. These data thus support an argument that the boy is, in fact, Bobby Larson. These data should be used together with other, non-DNA-based evidence to support the claim that the boy is Bobby Larson.

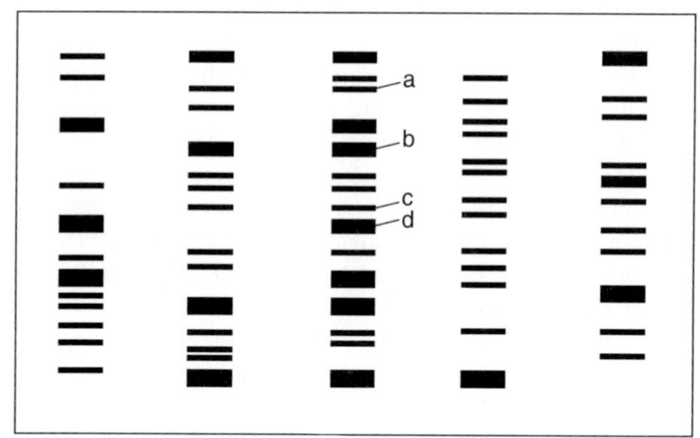


**10.43 a.** The PCR method requires very small (nanograms) amounts of template DNA, and if the primers are designed to amplify only small regions, the DNA can be degraded partially. In contrast, VNTR methods require larger amounts (micrograms) of intact DNA, as restriction digests are used to produce relatively large (kb-size) fragments that are then detected by Southern blotting. Some of the DNA samples used in forensic analysis are found in crime scenes and may be stored for years, so that they may often be degraded and not be present in large amounts. STR methods can still be used on such samples, while VNTR methods cannot.

**b.** Multiplexing PCR reactions ensures that: (1) the different STR results obtained in the reaction are all derived from a single DNA sample (laboratory labeling and pipetting errors are minimized); and (2) limited amounts of DNA samples are used efficiently.

**c.** $P(\text{random match}) = (0.112 \times 0.036 \times 0.081 \times 0.195) = 6.4 \times 10^{-5}$. About 1 person in 15,702 would be misidentified by chance alone using just these four markers.

**d.** $P(\text{random match}) = (0.112 \times 0.036 \times 0.081 \times 0.195 \times 0.062 \times 0.075 \times 0.158 \times 0.065 \times 0.067 \times 0.085 \times 0.089 \times 0.028 \times 0.039) = 1.7 \times 10^{-15}$. About 1 person in $594 \times 10^{14}$ would be misidentified by chance alone using all 13 markers.

**10.45** Use an interaction trap assay (the yeast two-hybrid system). Fuse the coding region of a protein produced by *fruitless* (obtained from an open reading frame within a cDNA) to the sequence of the Gal4p BD, and cotransform this plasmid into yeast with a plasmid library containing the Gal4p AD sequence, which is fused to protein sequences encoded by different cDNAs from the *Drosophila* brain. Purify colonies that express the reporter gene (see Figure 10.13, p. 269). In these colonies, the transcription of the reporter gene was activated when the AD and BD domains were brought together by the interactions of the *fruitless* protein with an unknown protein encoded by one of the brain cDNAs. Isolate and characterize the brain cDNA found in these yeast colonies.

## Chapter 11 Mendelian Genetics

**11.1 a.** Let $R$ represent red and $r$ represent yellow. The cross $RR \times rr$ gives all $Rr$. The $F_1$ are all red.

**b.** The $F_2$ is obtained from selfing the $F_1$. $Rr \times Rr$ gives $^3/_4\,R-$ : $^1/_4\,rr$. The $F_2$ are $^3/_4$ red and $^1/_4$ yellow.

**c.** $Rr \times RR$ gives all $R-$. The fruits all red.

**d.** $Rr \times rr$ gives $^1/_2\,Rr$ : $^1/_2\,rr$. The fruits are $^1/_2$ red and $^1/_2$ yellow.

**11.3** In Mendelian monohybrid crosses, $F_2$ plants display phenotypic ratios that are $^3/_4$ dominant : $^1/_4$ recessive. Since the $F_2$ ratio here is 3 colored : 1 colorless, we can infer that colored is dominant to colorless. Let $C$ represent colored and $c$ represent colorless. The $F_2$ has a 1 $CC$ : 2 $Cc$ : 1 $cc$ genotypic ratio, so there are two types of colored plants, $CC$ and $Cc$. If a $CC$ plant is picked and selfed, only colored plants will be seen in its offspring. In contrast, if a $Cc$ plant is picked and selfed, both colored and colorless plants will be seen in the offspring. To satisfy the conditions of the problem, a $Cc$ plant must be picked. Since the $F_2$ colored plants are present in a 1 $CC$ : 2 $Cc$ ratio, the chance of picking a $Cc$ plant is $^2/_3$.

**11.4 a.** Parents are $Rr$ (rough) and $rr$ (smooth); $F_1$ are $Rr$ (rough) and $rr$ (smooth).

**b.** $Rr \times Rr \rightarrow {}^3/_4\,R-$(rough) and $^1/_4\,rr$ (smooth).

**11.6** To obtain a 3 purple : 1 white ratio, the selfed plant must have been heterozygous, and purple ($P$) must be dominant to white ($p$). The purple-flowered progeny of a $Pp$ heterozygote have two genotypes, $PP$ and $Pp$, and they are present in a 2 $Pp$ : 1 $PP$ ratio. Since only $PP$ plants breed true and these are $^1/_3$ of the purple progeny, $^1/_3$ of the purple progeny will breed true.

**11.7** Black is dominant to brown. Let $B$ represent the black allele and $b$ represent the brown allele. Then female X is $Bb$, female Y is $BB$, and the male is $bb$.

**11.10**

| Parents | Progeny | | Female Parent |
|---|---|---|---|
| Female × Male | Grey | White | Genotype |
| grey × white | 81 | 82 | $Gg$ |
| grey × grey | 118 | 39 | $Gg$ |
| grey × white | 74 | 0 | $GG$ |
| grey × grey | 90 | 0 | $GG$ or $Gg$ ($G-$) |

**11.11** The farmer now has only black babbits, so he must breed animals that are either $BB$ or $Bb$. His initial pair gave both black and white progeny and is not true breeding. To obtain white offspring from them, each babbit must be heterozygous with a $Bb$ genotype. The unsold black babbit offspring should therefore have a 1 $BB$ : 2 $Bb$ ratio.

**a.** To obtain a white offspring from a cross of two black parents, both parents must be $Bb$ *and* a $bb$ offspring must be produced. The chance of picking a $Bb$ individual from among the $F_1$ offspring is $^2/_3$. The chance that a $bb$ offspring will be produced from a cross of two $Bb$ individuals is $^1/_4$.

$$\begin{aligned} P(\text{white offspring}) &= P(\text{both } F_1 \text{ babbits are } Bb \textit{ and } \text{a } bb \text{ offspring is produced}) \\ &= P(\text{both } F_1 \text{ babbits are } Bb) \times P(bb \text{ offspring}) \\ &= (^2/_3 \times {}^2/_3) \times (^1/_4) \\ &= {}^1/_9. \end{aligned}$$

**b.** If he crosses an $F_1$ male ($Bb$ or $BB$) to the parental female ($Bb$), two types of crosses are possible. The crosses and their probabilities are (1) $Bb$ ($F_1$ male) × $Bb$ (parental female), $P = {}^2/_3 \times 1 = {}^2/_3$ and (2) $Bb$ ($F_1$ male) × $Bb$ (parental female), $P = {}^1/_3 \times 1 = {}^1/_3$. Only the first cross can produce white progeny, $^1/_4$ of the time. Using the product rule, the chance that this strategy will yield white progeny is

$$P = {}^2/_3 \text{ (chance of } Bb \times Bb \text{ cross)} \times {}^1/_4 \text{ (chance of } bb \text{ offspring)} = {}^1/_6.$$

**c.** While it is more work initially, a productive long-term strategy is to remate the initial two black babbits (both are known to be $Bb$) to obtain a white male offspring [$P = {}^1/_4$ (white $bb$) × $^1/_2$ (male) = $^1/_8$]. Since only the fertility of white females and not that of white males is affected, retain this male and breed it back to its mother. This cross would be $Bb \times bb$ and give $^1/_2$ white ($bb$) and $^1/_2$ black ($Bb$) offspring. Use the progeny of this cross to develop a "breeding colony" consisting of black ($Bb$) females and white ($bb$) males. These would consistently produce half white and half black offspring.

**11.13 a.** Mutations 1, 3, 5, and 7 are loss-of-function mutations. Mutations 2, 4, 6, and 8 are gain-of-function mutations.

**b.** Mutations that cause sickness in heterozygotes will show dominant inheritance, while mutations that cause sickness only in homozygotes will show recessive inheritance. Mutation 1 will be recessive since homozygotes will have no enzyme activity and be sick, while heterozygotes will have 50% of reference activity and be normal. Mutation 2 will be dominant, since heterozygotes (and homozygotes) will have enzyme expression in the heart and be sick. Mutations that affect transcription initiation will affect the amount of mRNA available for translation and thus affect how much enzyme is produced. Whether mutations 3 and 4 lead to sickness and show an inheritance pattern depends on how much these mutations affect transcription initiation. Mutation 3 will not be dominant, since heterozygotes will have one normal allele and so have at least 50% of the reference activity. It will be recessive only if the decrease in transcription initiation at the two mutant alleles in homozygotes leads to less than 50% of reference activity. Mutation 4 will be dominant only if the increase in transcription initiation of the one mutant allele in a heterozygote, together with normal levels of transcription at the normal allele, leads to more than 150% of reference activity. If this is not the case, it will be recessive only if the increase in transcription initiation of the two mutant alleles in homozygotes leads to more than 150% of reference activity. Mutation 5 results in a truncated, nonfunctional protein, so it will be recessive just like mutation 1. Mutation 6 will be dominant since heterozygotes will be sick: they will have 250% of reference activity (200% from the mutant allele plus 50% from the normal allele). Mutation 7 will be recessive since homozygotes will be sick: they will have 20% of reference activity. Heterozygotes will be normal since they will have 60% of reference activity (10% from the mutant allele plus 50% from the normal allele). We cannot predict whether mutation 8 will have a phenotype, since there may or may not be a phenotypic consequence when the enzyme acts on additional substrates.

**11.17** Try fitting the data to a model in which catnip sensitivity/insensitivity is controlled by a pair of alleles at one gene. Since sensitivity is seen in all of the progeny of the initial mating between catnip-sensitive Cleopatra and catnip-insensitive Antony, hypothesize that sensitivity is dominant. Let $S$ represent the sensitive allele, and $s$ represent the insensitive allele. Then the initial cross is *S*– × *ss*, and the progeny are *Ss*. If two of the *Ss* kittens mate, a 3 sensitive (*S*–) : 1 insensitive (*ss*) progeny ratio is expected. In the mating with Augustus, the cross would be *Ss* × *ss*, and should give a 1 *Ss* (sensitive) : 1 *ss* (insensitive) progeny ratio. The observed progeny ratios are not far off from these expectations.

An alternative hypothesis is that sensitivity (*s*) is recessive and insensitivity (*S*) is dominant. For Antony and Cleopatra to have sensitive (*ss*) offspring, they would need to be *Ss* and *ss*, respectively. When two of their sensitive (*ss*) progeny mate, only sensitive (*ss*) offspring should be produced. Since this is not observed, this hypothesis does not explain the data.

**11.18 a.** *WW Dd* × *ww dd*

**b.** *Ww dd* × *Ww dd*

**c.** *ww DD* × *WW dd*

**d.** *Ww Dd* × *ww dd*

**e.** *Ww Dd* × *Ww dd*

**11.19** To determine the desired probabilities in the cross *Aa Bb Cc* × *Aa Bb Cc*, consider each gene separately and then use the product rule.

**a.** In the cross *Aa* × *Aa*, there is a 3/4 chance of obtaining an *A*– individual. Similarly, the chance of obtaining a *B*– individual from the cross *Bb* × *Bb* is 3/4 and the chance of obtaining a *C*– individual from the cross *Cc* × *Cc* is 3/4. Using the product rule, the probability of obtaining a phenotypically *A B C* (*A*– *B*– *C*–) offspring is 3/4 × 3/4 × 3/4 = 27/64.

**b.** There is a 1/4 chance of obtaining an *AA* offspring from the cross *Aa* × *Aa*. This is also the probability for obtaining a *BB* or *CC* offspring from a *Bb* × *Bb* or *Cc* × *Cc* cross, respectively. Using the product rule, the probability of obtaining an *AA BB CC* offspring is 1/4 × 1/4 × 1/4 = 1/64.

**11.22 a.** Let *Y* represent yellow and *y* green seeds, *P* represent purple and *p* white flowers, *A* represent axially positioned and *a* terminally positioned flowers, and *I* represent inflated and *i* pinched pods. The initial yellow seed produced a parent plant with purple, axially positioned flowers and inflated pods, so it had all four dominant alleles and was *Y*– *P*– *A*– *I*–. Determine whether the parent plant is homozygous or heterozygous at each gene by considering the types of offspring it produces when selfed. Selfing a homozygote never produces recessive offspring, while selfing a heterozygote produces 25% recessive offspring. Since selfing of the parent produces only yellow seeds, and since recessive traits for the flower color, flower position, and pod shape genes are seen when two $F_1$ seeds are sown, the parent is *YY Pp Aa Ii*. The $F_1$ plant with terminally positioned purple flowers, pinched pods, and yellow seeds is *YY P*– *aa ii*, and the $F_1$ plant with axially positioned white flowers, pinched pods, and yellow seeds is *YY pp A*– *ii*.

**b.** The seeds were produced by the cross *YY Pp Aa Ii* × *YY Pp Aa Ii*. A branch diagram will show that if the seeds are sown, they will produce yellow-seeded plants that are 27/64 purple, axially positioned flowers with inflated pods; 9/64 purple, axially positioned flowers with pinched pods; 9/64 purple, terminally positioned flowers with inflated pods; 9/64 white, axially positioned flowers with inflated pods; 3/64 purple, terminally positioned flowers with pinched pods; 3/64 white, terminally positioned flowers with inflated pods; 3/64 white, axially positioned flowers with pinched pods; and 1/64 white, terminally positioned flowers with pinched pods.

**11.24 a.** The cross is *aa BB CC* × *AA bb* $c^hc^h$. The $F_1$ trihybrids are all *Aa Bb* $Cc^h$ and are agouti and black. A branch diagram will show that the $F_2$ consists of 27/64 agouti and black; 9/64 agouti, black, Himalayan; 9/64 agouti, brown; 9/64 black; 3/64 agouti, brown, Himalayan; 3/64 black, Himalayan; 3/64 brown; and 1/64 brown, Himalayan.

**b.** $F_2$ animals that are non-Himalayan, black, and agouti are *A*– *B*– *C*–. Among the *A*– animals, 2/3 are *Aa*. Among the *B*– animals, 1/3 are *BB*. Among the *C*– animals, 2/3 are $Cc^h$, so the proportion of *Aa BB* $Cc^h$ animals is 2/3 × 1/3 × 2/3 = 4/27.

**c.** From the cross *Aa Bb* $Cc^h$ × *Aa Bb* $Cc^h$, 1/4 of the progeny will be *bb* and show brown pigment. This will be the case regardless of whether the animals are pigmented over their entire body or are Himalayan. Thus, 1/4 of the Himalayan mice will show brown pigment.

**d.** From the cross *Aa Bb* $Cc^h$ × *Aa Bb* $Cc^h$, 3/4 of the progeny will be *B*– and show black pigment. This will be the case regardless of whether the animals are agouti or nonagouti. Thus, 3/4 of the agouti mice will show black pigment.

**11.30** Mating type C is determined only by the genotype *aa bb*. Thus, C must be genotype *aa bb*. Crosses of the other strains to C, then, are testcrosses and the progeny ratios indicate the genotypes of the strains. Therefore, A is *Aa Bb*, B is *aa Bb*, and D is *Aa bb*.

**11.31 a.** The initial cross is *Ww Rr* × *W r*. The progeny females result from a queen's egg (1/4 *W R*, 1/4 *W r*, 1/4 *w R*, 1/4 *w r*) being fertilized by the drone's sperm (all *W r*). These will be workers and will be 1/2 *W*– *Rr* (black-eyed, wax sealers) and 1/2 *W*– *rr* (black-eyed, resin sealers).

**b.** Males arise solely from unfertilized eggs and receive chromosomes only from their mother. The progeny males will be 1/4 *W R* (black-eyed, wax sealers), 1/4 *W r* (black-eyed, resin sealers), 1/4 *w R* (white-eyed, wax sealers), 1/4 *w r* (white-eyed, resin sealers).

**c.** The egg fertilized by the mutation-bearing sperm results in a *Cc* female (Madonna). Since fertilization occurs in flight, males that fertilize a queen must be *C*. Hence, Madonna's first generation arises from the cross *Cc* × *C*. There is a 1/2 chance of obtaining daughters that are *Cc*. Since a *Cc* daughter can also be fertilized only by a *C* male, the chance of her having a *Cc* daughter is also 1/2. The chance of Madonna having a *Cc* granddaughter, who will produce 1/2 wingless males, is thus 1/2 × 1/2 = 1/4.

**d.** The chance that the $F_4$ generation great-great-granddaughter will be heterozygous is $(1/2)^4$ = 1/16.

**11.33 a.** The mother must be heterozygous *Aa* to have children that exhibit the recessive trait.

**b.** The father must be homozygous *aa*, since he expresses the trait.

**c.** Since the cross is *Aa* (mother) × *aa* (father), all offspring receive the recessive *a* allele from their father. If a child receives the recessive *a* allele from their mother, it will be affected and homozygous *aa* (II-2, II-5). If a child receives the normal *A* allele from their mother, it will be normal and heterozygous *Aa* (II-1, II-3, II-4).

**d.** In the cross *Aa* × *aa*, the prediction is that 1/2 of the progeny will be *Aa* (normal) and 1/2 will be *aa* (express the trait). There are five children, two affected and three normal. Thus, the ratio fits as well as it could for five children.

**11.36 a.** It is uncertain whether the brother of the man's wife's paternal grandmother had Gaucher disease. If this distant relative had the disease, a disease allele might have been passed on to the man's wife. Therefore, in a worst-case scenario, this distant relative would have had the disease. Under this scenario, the pedigree is as shown here:

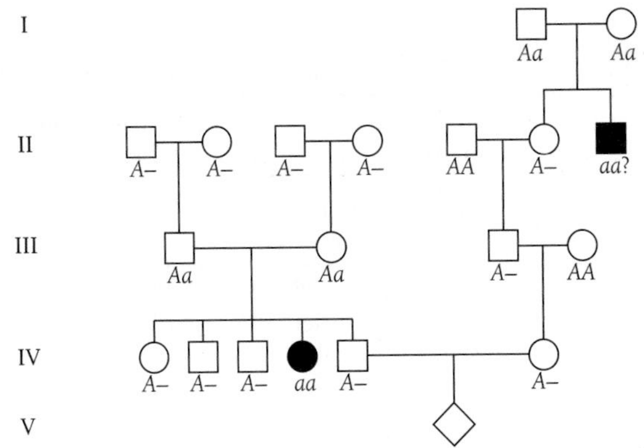


In this pedigree the man is IV-5, his affected sister is IV-4, his wife is IV-6, and the brother of his wife's paternal grandmother is II-7. Since II-7 is affected but his parents are not, the disease must be a recessive trait, and each of his parents must be heterozygous.

**b.** For the couple IV-5 and IV-6 to have an affected child (V-1), both IV-5 and IV-6 must give V-1 a recessive *a* allele. Since the trait is recessive and IV-5 and IV-6 are not affected, we know that IV-5 and IV-6 are *A*–. We must calculate the chance that they are *Aa* and that both pass on the *a* allele.

IV-5 has an affected sister, so his parents must both be *Aa*, and there is a $^2/_3$ chance that he is *Aa*. Therefore, there is a $^2/_3 \times {}^1/_2 = {}^1/_6$ chance that IV-5 will pass the *a* allele to V-1.

$$\begin{aligned} P(\text{IV-6 is } Aa) &= P(\text{III-3 is } Aa \textit{ and } \text{III-3 passed } a \text{ to IV-6}) \\ &= P[(\text{II-6 was } Aa \textit{ and } \text{II-6 passed } a \text{ to III-3}) \\ &\quad \textit{and } (\text{III-3 passed } a \text{ to IV-6})] \\ &= [({}^2/_3 \times {}^1/_2) \times {}^1/_2] = {}^1/_6. \end{aligned}$$

Therefore, there is a $^1/_6 \times {}^1/_2 = {}^1/_{12}$ chance that IV-6 will pass the *a* allele to V-1. In this worst-case scenario, the chance that both parents will pass on an *a* allele and have an affected child is $^1/_{12} \times {}^1/_3 = {}^1/_{36}$. If the brother of the wife's paternal grandmother did not have the disease, IV-6 would be *AA*, ensuring that V-1 will be *A*– and phenotypically normal.

**11.37** The $F_1$ cross is $a^+/a\ b^+/b\ c^+/c\ d^+/d \times a^+/a\ b^+/b\ c^+/c\ d^+/d$.

**a.** A colorless $F_2$ individual would result if an individual has an *a/a*, *b/b*, and/or *c/c* genotype. This would consist of many possible genotypes. Rather than identify all of these combinations, use the fact that the proportion of colorless individuals = 1 – the proportion of pigmented individuals. The proportion of pigmented individuals ($a^+/–\ b^+/–\ c^+/–$) is $^3/_4 \times {}^3/_4 \times {}^3/_4 = {}^{27}/_{64}$. The chance of not obtaining this genotype is $1 - {}^{27}/_{64} = {}^{37}/_{64}$.

**b.** A brown individual is ($a^+/–\ b^+/–\ c^+/–\ d/d$). The proportion of brown individuals is $^3/_4 \times {}^3/_4 \times {}^3/_4 \times {}^1/_4 = {}^{27}/_{256}$.

**11.39 a.** Since any of the normal alleles $a^+$, $b^+$, or $c^+$ is sufficient to catalyze the reaction leading to color, in order for color to fail to develop, all three normal alleles must be missing. That is, the colorless $F_2$ must be *a/a b/b c/c*. The chance of obtaining such an individual is $^1/_4 \times {}^1/_4 \times {}^1/_4 = {}^1/_{64}$.

**b.** Now, colorless $F_2$ are obtained if *either* one or both steps of the pathway are blocked. That is, colorless $F_2$ are obtained in either of the following genotypes: *d/d* –/– –/– –/– (the first or both steps blocked) or $d^+/–$ *a/a b/b c/c* (second step blocked). The chance of obtaining such individuals is $({}^1/_4 \times 1 \times 1 \times 1) + ({}^3/_4 \times {}^1/_4 \times {}^1/_4 \times {}^1/_4) = {}^{67}/_{256}$.

**11.40 a.** $^1/_2\ w/w^+\ bw/bw^+\ st/st^+$, fire-red-eyed daughters; $^1/_2\ w/Y\ bw/bw^+\ st/st^+$, white-eyed sons

**b.** $w/w^+\ se/se^+\ bw/bw^+$ and $w^+/Y\ se/se^+\ bw/bw^+$ all fire-red eyes

**c.** $w/w^+\ v/v^+\ bw/bw$ and $w^+/Y\ v/v^+\ bw/bw$, all brown eyes

**d.** $^1/_4\ w^+/w$ or $w^+/Y$, $bw/bw^+\ st/st^+$, fire-red eyes; $^1/_4\ w^+/w$ or $w^+/Y$, $bw/bw\ st/st^+$, brown eyes; $^1/_4\ w^+/w$ or $w^+/Y$, $bw/bw^+\ st/st$, scarlet eyes; $^1/_4\ w^+/w$ or $w^+/Y$, $bw/bw\ st/st$, (the color of 3-hydroxykynurenine plus the color of the precursor to biopterin, or colorless = white)

## Chapter 12 Chromosomal Basis of Inheritance

**12.1** c

**12.4** c

**12.7 a.** Yes, providing that the species has a sexual mating system in its life cycle. Meiosis can be initiated only in diploid cells. If a sexual mating system exists, two haploid cells can fuse to produce a diploid cell, which can then go through meiosis to produce haploid progeny. The fungi *Neurospora crassa* and *Saccharomyces cerevisiae* exemplify this positioning of meiosis in a life cycle.

**b.** No, because a diploid cell cannot be formed in a haploid individual and meiosis can be initiated only in a diploid cell.

**12.9** c. For example, in an organism with a haploid life cycle, gametes and somatic cells are both 1N.

**12.11 a.** Metaphase: Metaphase in mitosis, metaphase I and metaphase II in meiosis.

**b.** Anaphase: Anaphase in mitosis, anaphase I and anaphase II in meiosis.

**12.15 a.** The chance that a gamete would have a particular maternal chromosome is $^1/_2$. Applying the product rule, the chance of obtaining a gamete with all three maternal chromosomes is $({}^1/_2)^3 = {}^1/_8$.

**b.** The set of gametes with some maternal and paternal chromosomes is composed of all gametes *except* those that have only maternal or only paternal chromosomes. That is, *P*(gamete with both maternal and paternal chromosomes) = 1 – *P* (gamete with only maternal chromosomes or gamete with only paternal chromosomes). From (a), the chance of a gamete having chromosomes from only one parent is $^1/_8$. Using the sum rule, *P*(gamete with both maternal and paternal chromosomes) = $1 - ({}^1/_8 + {}^1/_8) = {}^3/_4$.

**12.16** Since the cells are normal and diploid, chromosomes should exist in pairs. There are pairs of medium and long chromosomes, leaving one short and one long chromosome. These could be members of a heteromorphic pair such as the X and Y chromosomes of a male mammal.

**12.18** False. Genetic diversity in the male's sperm is achieved during meiosis, when there is crossing-over between nonsister chromatids and independent assortment of the males' maternal and paternal chromosomes. These processes make it very unlikely that any two sperm cells are genetically identical.

**12.20 a.** 17 + 26 = 43 chromosomes.

**b.** Similar chromosomes pair in meiosis. The pairing pattern seen in the hybrid indicates that some of the chromosomes in these two species share evolutionary similarity, while others do not. Unpaired chromosomes will not segregate in an orderly manner, giving rise to unbalanced meiotic products with either extra or missing chromosomes. This can lead to sterility for two reasons. First, meiotic products that are missing chromosomes may not have genes necessary to form gametes. Second, even if gametes are able to form, a zygote generated from them will not have a complete chromosome set from the hybrid, the red, or the arctic fox. The zygote will be an aneuploid with missing or extra genes, causing it to be infertile.

**12.21** The chance of a particular paternal chromosome being present in a gamete is $^1/_2$. Using the product rule, the chance of all five paternal chromosomes being in one gamete is $(^1/_2)^5 = {}^1/_{32}$.

**12.24** Fathers always give their X chromosome to their daughters, so the woman must be heterozygous for the color-blindness trait and is $c^+c$. Her husband received his X chromosome from his mother and has normal vision, so he is $c^+$Y. The cross is therefore $c^+c \times c^+$Y. All daughters will receive the paternal X bearing the $c^+$ allele and have normal color vision. Sons will receive the maternal X, so half will be $c$Y and be color blind, and half will be $c^+$Y and have normal vision.

**12.26 a.** The parental cross is $ww\ vg^+vg^+ \times w^+$Y $vgvg$. This produces $F_1$ males that are $w$Y $vg^+vg$ (white, long wings) and $F_1$ females that are $w^+w\ vg^+vg$ (red, long wings).

**b.** In both males and females, the $F_2$ will be $^3/_8$ white, long; $^3/_8$ red, long; $^1/_8$ white, vestigial; $^1/_8$ white, vestigial.

**c.** If the $F_1$ males are crossed back to the female parent, the cross is $ww\ vg^+vg^+ \times w$Y $vg^+vg$. All the progeny are white, long. If the $F_1$ females are crossed back to the male parent, the cross is $w^+w\ vg^+vg \times w^+$Y $vgvg$. Male progeny: $^1/_4$ white, vestigial; $^1/_4$ white, long; $^1/_4$ red, vestigial; $^1/_4$ red, long. All female progeny are red, half are long, and half are vestigial.

**12.28 a.** Since the father of the calico cat is chocolate, he must be *o*Y *bb*. Deduce the genotype of his calico daughter by considering her phenotype and what paternal chromosomes she must have received. She has some black pigmentation, so she must also have a dominant *B* allele, and she received her father's X (with an *o* allele) and an autosome with a *b* allele, so she is *Oo Bb*. Since the parents of the chocolate male who mates with the calico cat were solid black, that cross was *Oo Bb* × *o*Y *Bb*, and their chocolate son is *o*Y *bb*. Thus, the cross between the calico female and chocolate male is *Oo Bb* × *o*Y *bb*. The progeny are $^1/_4$ orange females (*OO b*–), $^1/_8$ calico females with black and orange patches (*Oo Bb*), $^1/_8$ calico females with brown and orange patches (*Oo bb*), $^1/_4$ orange males (*O*Y *b*–), $^1/_8$ black males (*o*Y *Bb*) and $^1/_8$ chocolate males (*o*Y *bb*).

**b.** Sex-chromosome nondisjunction in meiosis I and II produces XXY, XO, XXX, and XYY animals. In Table 12.A, the parenthetical terms refer to the feline phenotypes corresponding to the Klinefelter, Turner, triplo-X, and XYY human phenotypes.

**12.30** The crisscross inheritance pattern (father to daughter) suggests an X-linked trait. Man A marries a normal woman and all his daughters have the trait, so the trait must be dominant. Let $X^B$ be the defective enamel allele and $X^b$ be the normal allele. Man A is $X^B$Y and his wife is $X^bX^b$, so all of their daughters are $X^BX^b$. As heterozygotes, they have defective enamel and 50% of their offspring receive the $X^B$ allele and are affected. The sons inherit the mother's $X^b$ allele, so they are normal and transmit only the normal allele.

**12.31** Since the inability to taste phenythiourea is recessive, the nontaster child must be homozygous for the recessive allele, and each of his parents must have given the child a recessive allele. Since both parents can taste, they must also bear a dominant allele. Let *T* represent the dominant (taster) allele, and *t* represent the recessive (nontaster) allele. Then the cross is

**Table 12.A**

| | Paternal Gametes | | | |
|---|---|---|---|---|
| | Nondisjunction in Meiosis I | | Nondisjunction in Meiosis II | |
| **Maternal Gametes** | *o*Y *b* | nullo-X *b* | *oo b* | YY *b* |
| *O B* | *Oo*Y *Bb* "Klinefelter" male, calico with black and orange patches | *O Bb* "Turner" female, orange | *Ooo Bb* "Triplo-X" female, calico with black and orange patches | *O*YY *Bb* "XYY" male, orange |
| *O b* | *Oo*Y *bb* "Klinefelter" male, calico with chocolate and orange patches | *O bb* "Turner" female, orange | *Ooo bb* "Triplo-X" female, calico with chocolate and orange patches | *O*YY *bb* "XYY" male, orange |
| *o B* | *oo*Y *Bb* "Klinefelter" male, black | *o Bb* "Turner" female, black | *ooo Bb* "Triplo-X" female, black | *o*YY *Bb* "XYY" male, black |
| *o b* | *oo*Y *bb* "Klinefelter" male, chocolate | *o bb* "Turner" female, chocolate | *ooo bb* "Triplo-X" female, chocolate | *o*YY *bb* "XYY" male, chocolate |

*Tt* × *Tt* and the chance that their next child will be a taster is the chance that the child will be *TT* or *Tt*, or $^3/_4$.

**12.32 a.** The unaffected parents have offspring affected with an autosomal recessive disorder, so both must be heterozygous. If $c^+$ is the normal allele and $c$ the affected allele, the cross is $c^+c \times c^+c$ and there is a $^1/_4$ chance of having a $cc$ offspring. Each conception is independent, so the probability that their next child will have cystic fibrosis is $^1/_4$.

**b.** Unaffected offspring are expected in a $1\ c^+c^+ : 2\ c^+c$ ratio, so there is a $^2/_3$ chance that an unaffected child is heterozygous.

**12.35 a.** In humans, sex type is determined by the presence or absence of a Y chromosome. The testis-determining factor gene present on the Y chromosome causes individuals with a Y to become males. In both *Drosophila melanogaster* and *Caenorhabditis elegans*, sex type is determined by the ratio of the number of X chromosomes to the sets of autosomes. In *Drosophila*, animals with an X:A ratio of 2:2 are female, while animals with an X:A ratio of 1:2 are male. In *Caenorhabditis elegans*, animals with an X:A ratio of 2:2 are hermaphrodites, while animals with an X:A ratio of 1:2 are males.

**b.** In humans, X-linked gene dosage is equalized by inactivating one X chromosome to form a Barr body. In flies, transcription of X-linked genes in males is higher than that in females so as to equal the sum of the expression levels of the two X chromosomes in females. In worms, genes on both of the X chromosomes of an XX hermaphrodite are transcribed at half the rate as the same gene on the single X chromosome in an XO male.

**12.37** Primary nondisjunction of sex chromosomes in a *ww* female produces two types of eggs: *ww* eggs having two

X chromosomes and eggs lacking an X chromosome. Red-eyed males have $w^+$-bearing and Y-bearing sperm. As shown in the Punnett square here, the only viable and fertile offspring produced from this cross are *ww*Y females:

| | | Sperm | |
|---|---|---|---|
| | | $w^+$ | Y |
| Eggs | *ww* | $www^+$ Usually dies | *ww*Y white ♀ |
| | O | $w^+$O Sterile red ♂ | YO Dies |

The *ww*Y females are the consequence of primary nondisjunction. They have XY (*w*Y) and X (*w*) gametes resulting from normal disjunction and, less frequently, XX (*ww*) and Y gametes resulting from secondary nondisjunction. The results of backcrossing a *ww*Y female to a $w^+$Y male are shown in the following Punnett square:

| | | | Sperm | |
|---|---|---|---|---|
| | | | $w^+$ | Y |
| Eggs | Normal X segregation | *w*Y | $w^+w$Y red ♀ | *w*YY white ♂ |
| | | *w* | $w^+w$ red ♀ | *w*Y white ♂ |
| | Secondary nondisjunction | *ww* | $w^+ww$ Triplo-X; usually dies | *ww*Y white ♀ |
| | | Y | $w^+$ Y red ♂ | YY dies |

**12.39** Turner females have just one X chromosome, so their X is not inactivated and Barr bodies are not produced.

**12.41 a.** Epigenetic. Patch color in calico cats depends on which X has been inactivated, rather than on a new DNA-based change. If the X bearing the *O* allele in an *Oo* individual has not been inactivated, the patch is orange; if it has been inactivated, the patch is black.

**b.** Epigenetic. X-linked gene transcription in a *Drosophila* male increases relative to that in a female to provide for dosage compensation.

**c.** Genetic. The first curly-winged male has a new mutation. The trait is heritable and autosomal dominant, since crossing the curly-winged male to a normal female produces a 1:1 ratio of curly-winged and normal males and females.

**d.** Epigenetic. Since diethylstilbestrol is not positive in the Ames test, it does not increase tumor frequency by inducing DNA mutations.

**e.** Epigenetic. The *in utero* hormonal environment, rather than any DNA-based change, activates a pattern of gene expression in the XX animal that leads to male sexual characteristics.

**f.** Genetic. The cinnamon-colored-stripe phenotype shows criss cross inheritance, indicating that it is inherited as a sex-linked trait. The first cinnamon female has a new Z-linked recessive mutation.

**12.42** This problem raises the issue that the precise mode of inheritance of a trait often cannot be determined when a pedigree is small and the trait's frequency in a population is unknown. For example, pedigree A could easily fit an autosomal dominant trait (*AA* and *Aa* = affected): The affected father would be heterozygous for the trait (*Aa*), the mother would be unaffected (*aa*), and half of their offspring would be affected (*A*–). However, pedigree A could also fit an autosomal recessive trait (*aa* = affected): The father would be homozygous for the trait (*aa*), the mother would be heterozygous (*Aa*), and half of their offspring would be affected (*aa*). Furthermore, pedigree A could fit an X-linked recessive trait: The mother would be heterozygous ($X^AX^a$), the father would be hemizygous ($X^aY$), and half of the progeny would be affected (either $X^aX^a$ or $X^aY$). An X-linked dominant trait would not fit the pedigree because it would require all the daughters of the affected father to be affected (because they all receive their father's X). Pedigrees B and C can be solved by similar analytical reasoning.

| | Pedigree A | Pedigree B | Pedigree C |
|---|---|---|---|
| Autosomal recessive | Yes | Yes | Yes |
| Autosomal dominant | Yes | Yes | No |
| X-linked recessive | Yes | Yes | No |
| X-linked dominant | No | No | No |

**12.44 a.** Since only males and no parents are affected, Duchenne muscular dystrophy most closely fits the profile of an X-linked recessive trait.

**b.** I-1, II-2, II-7

**c.** IV-1 and IV-2 will have an affected child only if III-2 is heterozygous ($P = 1/2$) *and* III-2 gives the X bearing the Duchenne muscular dystrophy mutation ($P = 1/2$) *and* the child is male (the child receives a Y, and not a normal X, from the father, $P = 1/2$). Using the product rule, $P = (1/2)^3 = 1/8$.

**d.** $P = 0$, since neither parent carries the disease allele (assume that IV-3 is homozygous for the normal allele).

**12.45 a.** Y-linked inheritance can be excluded because females are affected. X-linked recessive inheritance can also be excluded since an affected mother (I-2) has a normal son (II-5). Autosomal recessive inheritance can also be excluded since two affected parents, II-1 and II-2, have unaffected offspring.

**b.** The two remaining mechanisms of inheritance are X-linked dominant and autosomal dominant. Genotypes can be assigned to all members of the pedigree that satisfy either inheritance mechanism. Of these two, X-linked dominant inheritance may be more likely since II-6 and II-7 have only affected daughters, suggesting crisscross inheritance. If the trait were autosomal dominant, half of the daughters and half of the sons should be affected.

**12.48 a.** False. An affected father who is heterozygous should have only half affected children.

**b.** False. An affected mother who is heterozygous should have half affected offspring, regardless of sex type.

**c.** False. Two heterozygous parents should have 1/4 of their offspring be homozygous for the recessive, normal allele.

**d.** True. However, if the mutation were newly arisen in either child or his or her parents, his or her grandparent could have been unaffected.

**12.49 a.** True. Two affected individuals will always have affected children (*aa* × *aa* can give only *aa* offspring).

**b.** False. An autosomal trait is inherited independent of sex type.

**c.** May or may not be true. The trait could be masked by normal dominant alleles through many generations before two heterozygotes marry and produce affected, homozygous offspring.

**d.** May or may not be true. If the trait is rare, then it is likely that an unaffected individual marrying into the pedigree is homozygous for a normal allele. Since the trait is recessive, and the children receive the dominant, normal allele from the unaffected parent, the children will be normal. However, this statement would not be true if the unaffected individual was heterozygous. In this case, half of the children would be affected.

**12.51** Since hemophilia is an X-linked trait, the most likely explanation is that random inactivation of X chromosomes (lyonization) produces individuals with different proportions of cells with a functioning allele. This in turn leads to different amounts of clotting factor being made. Normal clotting times would be expected in females whose $X^h$ chromosome was very frequently inactivated. In these individuals, most cells have a functioning $h^+$ allele and near-normal amounts of clotting factor are made. In contrast, a clotting time consistent with clinical hemophilia would be seen in a woman having the $X^{h^+}$ chromosome inactivated, say, 90% of the time. In these individuals, only 10% of the cells have a functioning $h^+$ allele, and very little clotting factor would be made.

## Chapter 13 Extensions of and Deviations from Mendelian Genetic Principles

**13.2** Six genotypes are possible: $w^1/w^1$, $w^1/w^2$, $w^1/w^3$, $w^2/w^2$, $w^2/w^3$, $w^3/w^3$.

**13.5** The woman's genotype is $I^AI^B$ and the man's genotype is $I^Ai$. Their offspring have four equally likely genotypes ($I^AI^A$, $I^AI^B$, $I^Ai$, $I^Bi$) and three phenotypes: A ($P = 1/2$), AB ($P = 1/4$) and B ($P = 1/4$).

**a.** $P = 1/2 \times 1/2 = 1/4$.

**b.** $P = 0$, as there is no chance of producing a group O child.

**c.** *P* [(first child is male and AB) and (second child is male and B)] $= (1/2 \times 1/4) \times (1/2 \times 1/4) = 1/64$.

**13.7** The cross $C^R/C^W \times C^R/C^W$ gives a $1\ C^W/C^W : 2\ C^R/C^W : 1\ C^R/C^R$ progeny ratio, so half of the progeny resemble the parents in coat color.

**13.10 a.** *Y/Y R/R* (crimson) × *y/y r/r* (white) gives a *Y/y R/r* magenta-rose $F_1$. Selfing the $F_1$ gives an $F_2$ that is 1/16 crimson (*Y/Y R/R*), 1/8 orange-red (*Y/Y R/r*), 1/16 yellow (*Y/Y r/r*), 1/8 magenta (*Y/y R/R*), 1/4 magenta-rose (*Y/y R/r*), 1/8 pale yellow (*Y/y r/r*), and 1/4 white (*y/y –/–*). A backcross of the $F_1$ to the crimson parent will give 1/4 crimson (*Y/Y R/R*), 1/4 magenta-rose (*Y/y R/r*), 1/4 magenta (*Y/y R/R*), and 1/4 orange-red (*Y/Y R/r*).

**b.** *Y/Y R/r* (orange-red) × *Y/y r/r* (pale yellow) gives 1/4 orange-red (*Y/Y R/r*), 1/4 magenta-rose (*Y/y R/r*), 1/4 yellow (*Y/Y r/r*), and 1/4 pale yellow (*Y/y r/r*).

**c.** *Y/Y r/r* (yellow) × *y/y R/r* (white) gives 1/2 magenta-rose (*Y/y R/r*) and 1/2 pale yellow (*Y/y r/r*).

**13.13** Let *C/c* represent alleles at the locus controlling the pigment production and *Y/y* represent the alleles at the yellow/agouti locus. The 3 colored : 1 albino progeny ratio indicates that *C/–* individuals produce pigment, while *c/c* individuals do not and are albino. The 2 yellow : 1 agouti progeny ratio is a modified monohybrid cross ratio indicating recessive lethality: *Y/Y* individuals die, *Y/y* are yellow, and *y/y* are agouti. Since albino mice do not express alleles at the agouti locus, *c/c* is epistatic to alleles at the *Y/y* locus, and *c/c Y/y* and *c/c y/y* individuals are albino.

**a.** First, infer the partial genotypes: yellow mice are *C/– Y/y* and albino mice are *c/c –/y*. Then determine the complete genotypes from the progeny ratios for each trait. A 1 colored : 1 albino progeny ratio is expected from a *C/c* × *c/c* cross. A 2 yellow : 1 agouti progeny ratio is expected from a cross. Therefore, the parental genotypes were *C/c Y/y* × *c/c Y/y*.

**b.** The cross is *C/c Y/y* × *C/c Y/y*, and, since *Y/Y* progeny are inviable, will produce a phenotypic ratio of 1 albino : 2 yellow : 1 agouti. None of the yellow mice will be true breeding, as they are all *Y/y*.

**13.15 a.** The trait is not caused by an X-linked recessive allele, since affected females do not have all affected sons. It is also not caused by a maternally inherited mitochondrial mutation, since affected females do not have all affected offspring. It could be inherited as an autosomal recessive allele if both I-2 and II-1 are carriers and affected individuals are homozygotes. However, this is unlikely since the trait is not common. The pedigree is more consistent with either X-linked or autosomal dominant inheritance. If autosomal, affected members would be heterozygotes. Males are more severely affected, which could mean the trait is a sex-influenced trait, like pattern baldness or cleft lip and palate. If the trait were caused by an X-linked dominant allele, I-1, II-2, and III-2 would be heterozygotes while the spontaneously aborted males (II-3 and III-4) would be hemizygotes. In this case, males might be more severely affected because they lack a normal allele. Females might have a less severe phenotype because they are X-chromosome mosaics (due to inactivation of one X) and have some cells that express the normal allele.

**b.** If the trait results from an X-linked dominant allele, the death of two males during the fetal stage suggests that the allele shows some recessive lethality. However, the recessive lethal phenotype shows incomplete penetrance, as the problem statement indicates that some males survive. If the trait is caused by an autosomal dominant allele, then the two spontaneously aborted males are heterozygotes and there is no evidence of recessive lethality.

**c.** The dominant phenotype appears to be completely penetrant and exhibit variable, sex-influenced expressivity. The dominant phenotype does not appear to exhibit reduced penetrance, because half of the offspring of affected females are affected. This is the pattern expected for a dominant allele. Since males are affected more severely than females, the phenotype shows variable expressivity that is sex-influenced. As indicated in (b), the recessive lethal phenotype shows incomplete penetrance.

**13.16** A single $p^+$ allele provides 50% of the enzyme activity seen in a $p^+/p^+$ homozygote. Since $p^+$ is dominant (i.e., *P/– C/C* plants are purple), this appears to be enough activity to provide for a wild-type phenotype. If a plant with less than 50% of normal activity does not synthesize enough purple pigment for a wild-type phenotype (e.g., 25% of normal activity gives a light-purple flower) and a plant with more than 100% of normal activity produces noticeably darker purple pigmentation, the phenotypes in Table 13.A should be seen.

Table 13.A

| Genotype | Percent of +/+ Activity | Percent of + /+ Activity When Mixed 50:50 with +/+ Extract | (A) Homozygote Phenotype | (B) Heterozygote Phenotype | (C) Hemizygote Phenotype | (D) Allele Classification |
|---|---|---|---|---|---|---|
| $p^+/p^+$ | 100 | 100 | purple | purple | purple | wild type |
| $p^1/p^1$ | 20 | 60 | light purple | purple | very light purple | hypomorph |
| $p^2/p^2$ | 0 | 50 | white | purple | white | amorph |
| $p^3/p^3$ | 300 | 200 | very dark purple | dark purple | dark purple | hypermorph |
| $p^4/p^4$ | 0 | 5 | white | very light purple | white | antimorph |
| $p^5/p^5$ | 0 | 50 | red | reddish purple | red | neomorph |

**13.19** Hornless is a sex-influenced trait. In males, *H/H* and *H/h* are horned, and *h/h* is hornless. In females, *H/H* is hornless. The cross is *H/H W/W* ♂ × *h/h w/w* ♀. The $F_1$ is *H/h W/w*—horned white males and hornless white females. Interbreeding the $F_1$ gives the following $F_2$.

| | Male | Female |
|---|---|---|
| 3/16 *H/H W/–* | horned, white | horned, white |
| 6/16 *H/h W/–* | horned, white | hornless, white |
| 3/16 *h/h W/–* | hornless, white | hornless, white |
| 1/16 *H/H w/w* | horned, black | horned, black |
| 2/16 *H/h w/w* | horned, black | hornless, black |
| 1/16 *H/h w/w* | hornless, black | hornless, black |

In sum, the ratios for males are 9/16 horned white : 3/16 hornless white : 3/16 horned black : 1/16 horned black. The ratios for females are 3/16 horned white : 9/16 hornless white : 1/16 horned black : 3/16 hornless black.

**13.21** First, use the information in question 13.19 to infer partial genotypes from phenotypes:

| Individual | Phenotype | Inferred Partial Genotype |
|---|---|---|
| Male parent | horned white male | *H/– W/–* |
| Ewe A | hornless black female | *–/h w/w* |
| Ewe A offspring | horned white female | *H/H W/–* |
| Ewe B | hornless white female | *–/h W/–* |
| Ewe B offspring | hornless black female | *–/h w/w* |
| Ewe C | horned black female | *H/H w/w* |
| Ewe C offspring | horned white female | *H/H W/–* |
| Ewe D | hornless white female | *–/h W/–* |
| Ewe D offspring 1 | hornless black male | *h/h w/w* |
| Ewe D offspring 2 | horned white female | *H/H W/–* |

Then compare the offspring to their parents. Since ewe D's male offspring is *h/h w/w*, both ewe D and the male parent must have at least one recessive allele at each gene. Since ewe A and ewe D have *H/H* offspring, ewe A and ewe D must each have an *H* allele. Since ewe B has a *w/w* offspring, she must have a *w* allele. Therefore, the male parent and ewe D are *H/h W/w*, ewe A is *H/h w/w*, ewe B is either *H/h W/w* or *h/h W/w*, and ewe C is *H/H w/w*.

**13.22** c.

**13.26** The $F_1$ snail gives sinistral offspring when selfed, so it is *d/d*. Therefore, both parents had a *d* allele. Since the $F_1$ has a dextral pattern, its maternal parent was *D/d*. The paternal parent could have been either *D/d* or *d/d*.

**13.28 a.** The $F_1$ is normal, so *g* and *a* are mutants at different genes. The cross can be written as $g/g\ a^+/a^+ \times g^+/g^+\ a/a$, and the $F_1$ can be written as the dihybrid $g^+/g\ a^+/a$.

**b.** The cross produces a mutant $F_1$, so *g* and *b* are alleles at the same gene. The cross can be written as $g/g \times b/b$, and the $F_1$ can be written as the monohybrid *g/b*. Alternatively, if we assign the new symbol *x* to this gene, we can write its two alleles as $x^g$ and $x^b$. Then, the cross is $x^g/x^g \times x^b/x^b$, and the $F_1$ is $x^g/x^b$.

**c.** The three complementation groups identify three genes.

**d.** Mutants *a*, *f*, and *d* have defects in one gene; *b* and *g* are in a second gene; *c* and *e* are in a third gene.

**13.29** Complementation tests can be used only to determine whether two recessive mutations affect the same function. They cannot be used to determine whether two dominant alleles affect the same function, or whether a dominant allele affects the same function as a recessive allele. When two mutants are crossed in a complementation test, the phenotype of the heterozygous $F_1$ is used to infer whether they affect the same function. If it is normal, the two mutants affect different functions and complement each other. If it is abnormal, the two mutants affect the same function and do not complement each other. Since a dominant mutation always shows a phenotype in a heterozygote, a "complementation test" with a dominant mutant will always produce a mutant phenotype, whether or not the two mutants affect the same function. Therefore, complementation tests with dominant mutants are not interpretable.

**13.33 a.** *w/w C/– S/S*. Since the cat has a patch of pigmented hair over her left eye, she can produce some melanocytes and must be *w/w*. Since her eyes are not red, she can make pigment and must be *C/–*. Since she is white except for the patch over her left eye, she has one large spot and is most likely *S/S*. She has one brown eye and one blue eye because the brown eye is within the pigmented region where melanocytes are produced while the blue eye is not. She probably only acknowledges her human servant when he kneels in front of her because of a hearing deficit resulting from the absence of cochleal melanocytes.

**b.** The *S* and *W* alleles affect both pigmentation and hearing, so they are pleiotropic. Homozygotes for the *c* allele are white and probably have diminished vision since their retina lacks pigment, so this allele is also pleiotropic.

**c. i.** If the cross were *W/w C/C* (completely white with blue eyes) × *w/w c/c* (completely white with red eyes), half the offspring would be *w/w C/c* and have pigmented coats. However, the cross *W/W C/C* × *w/w c/c* would produce only *W/w C/c*, white-coated cats.

**ii.** If the cross were *W/w s/s* (completely white) × *w/w S/S* (white with a few grey hairs), half the offspring would be *w/w S/s* and be pigmented with one or more white spots.

**d.** If *W/– S/–* cats have no pigmentation and have the *W/–* phenotype, *W* is epistatic to *S*. For the *cc*, *S*, and *W* alleles, there are three phenotypes to consider: coat color, eye pigmentation, and hearing deficit. If *c/c W/–* and *c/c S/–* cats have red eyes, *cc* is epistatic to each of *W* and *S* for eye pigmentation. If they have a hearing deficit, *W* and *S* are each epistatic to *cc* for hearing. Since both *cc* and *W/–* produce completely white cats, these alleles are epistatic to each other for coat color. If *cc S/–* cats are white and not spotted, *cc* is epistatic to *S* for coat color.

**13.34** The 9:7 $F_2$ ratio is a modified 9:3:3:1 ratio obtained from the $F_1$ cross *A/a B/b* × *A/a B/b*. The $^9/_{16}$ colored plants are *A/– B/–* and will show the genotypic ratio 1 *A/A B/B* : 2 *A/a B/B* : 2 *A/A B/b* : 4 *A/a B/b*. Since both *A* and *B* are required for color, only if a true-breeding *A/A B/B* plant is selfed will there be "no segregation of the two phenotypes among its progeny." The *A/A B/B* plants are $^1/_9$ of the colored plants, so $P = {}^1/_9$.

**13.35** The 9:7 ratio in the $F_2$ is a modified 9:3:3:1 ratio, where the *A/– B/–* genotypes are "runner" and the *A/– b/b*, *a/a B/–*, and *a/a b/b* genotypes are "bunch." This is an example of duplicate recessive epistasis: Recessive alleles at either of the genes block (are epistatic to) the "runner" phenotype, resulting in the "bunch" phenotype.

**13.36 a.** The cross is *A/a B/b* × *A/a B/b*, which gives $^9/_{16}$ *A/– B–*, $^3/_{16}$ *A/– b/b*, $^3/_{16}$ *a/a B/–*, and $^1/_{16}$ *a/a b/b*. The *A/– b/b*, *a/a B/–*, and *a/a b/b* individuals are deaf because they are homozygous for either one or both recessive alleles. Only the *A/– B/–* individuals can hear. Therefore, the phenotypic ratio is 9 hearing rabbits : 7 deaf rabbits.

**b.** These alleles show duplicate recessive epistasis. Homozygous recessive alleles at either of two genes block hearing, and they are epistatic to the dominant alleles at the other gene.

**c.** The cross is *a/a B/b* × *A/a B/b*, which gives $^5/_8$ deaf progeny ($^1/_8$ *A/a b/b* + $^1/_2$ *a/a –/–*) and $^3/_8$ hearing progeny (*A/a B/–*).

**13.37 a.** In order for this pathway to produce black individuals, those individuals must have all three normal alleles: They must be *A/– B/– C/–*. The $F_1$ is the trihybrid *A/a B/b C/c* that, when selfed, gives black *A/– B/– C/–* individuals in ($^3/_4 \times {}^3/_4 \times {}^3/_4$) = $^{27}/_{64}$ of the progeny. The remaining $1 - {}^{27}/_{64} = {}^{37}/_{64}$ of the progeny are colorless, having *a/a*, *b/b*, and/or *c/c* genotypes. A 27 black : 37 colorless progeny ratio is expected in the $F_2$.

**b.** In order for this pathway to produce black individuals, those individuals must have the *A* and *B* functions, but not the inhibitor function provided by *C*: They must be *A/– B/– c/c*. The chance of obtaining this genotype in the $F_2$ progeny is ($^3/_4 \times {}^3/_4 \times {}^1/_4$) = $^9/_{64}$. The remaining $1 - {}^9/_{64} = {}^{55}/_{64}$ of the progeny will be colorless. A 9 black : 55 colorless progeny ratio is expected in the $F_2$.

**c.** Here, the ratio of black to colorless in the $F_2$ can be used to distinguish between hypotheses concerning the two pathways proposed in (a) and (b). A chi-square test can be used to evaluate whether the $F_2$ results fit either pathway.

**13.39 a.** Compare cytochrome oxidase activity in cybids made with platelets from diseased individuals and in cybids made with platelets from age-matched control individuals. It is important to assess several different enzyme activities associated with mitochondrial proteins to ensure that the deficits in cytochrome oxidase are specific.

**b.** The cells of individuals with diseases resulting from mitochondrial DNA defects have a mixture of mutant and normal mitochondria; that is, they are heteroplasmons (or cytohets). Thus, assays in cybids are measurements of the enzyme activity present in a population of mitochondria in a cell. It would be unlikely that each of the mitochondria of an affected individual has an identical defect.

**13.41 a.** The *tudor* mutation is a maternal effect mutation. Homozygous *tudor* mothers give rise to sterile progeny, regardless of their mate.

**b.** The grandchildless phenotype results from the absence of some maternally packaged component in the egg needed for the development of the $F_1$'s germ line.

**13.44** If the male lethality is caused by a sex-linked, male-specific lethal mutation (*l*), the cross can be written as *l/l* ♀ × +/Y ♂, giving *l*/+ (♀) and *l*/Y (dead ♂) progeny. A cross of the $F_1$ females to normal males (*l*/+ ♀ × +/Y ♂) will give a 2:1 ratio of females to males ($^1/_4$ *l*/+ ♀, $^1/_4$ +/+ ♀, $^1/_4$ +/Y ♂, and $^1/_4$ *l*/Y dead ♂). If the male lethality is caused by a maternally inherited cytoplasmic factor lethal to males, then the $F_1$ females will receive this factor in cytoplasm from their mother, so they (like their mothers) should have only female offspring when mated to wild-type males.

**13.46 a.** All of the progeny would inherit the *sigma* factor from the (sensitive) female parent. Consequently, all the progeny will be sensitive.

**b.** Since the resistant female parent lacks the *sigma* factor, all of the progeny will also lack the factor, and so be resistant.

**13.48 a.** I-1, II-2, II-7, III-2

**b.** IV-2, IV-3, IV-13

**c.** Disease severity is related to the relative amount of mutant mitochondria. The lack of disease penetrance in I-1, II-2, and III-2 probably results from their cells having mostly normal mitochondria. Since each has an affected offspring, each must have cells that are heteroplasmons with some normal and some mutant mitochondria.

**13.49 a.** If the normal cytoplasm is [*N*] and the male-sterile cytoplasm is [*Ms*], the cross is [*Ms*] *rf/rf* ♀ × [*N*] *Rf/Rf* ♂, and the $F_1$ would be [*Ms*] *Rf/rf* and is male-fertile.

**b.** The cross is [*Ms*] *Rf/rf* ♀ × [*N*] *rf/rf* ♂. Half of the progeny would be [*Ms*] *Rf/rf* and be male fertile, and half would be [*Ms*] *rf/rf* and be male sterile.

**13.50** Draw out two pedigrees to illustrate the lineage of Carlos. In one, include Mr. and Mrs. Escobar, Mr. and Mrs. Sanchez, their murdered children, and Carlos. In the other, include Mr. and Mrs. Mendoza and Carlos. Analyze each pedigree to determine who could have contributed mitochondrial DNA to Carlos.

**a.** Mitochondrial RFLP data can be helpful to trace the maternal line of descent. Carlos Mendoza will have inherited his mitochondrial DNA from his mother, and she will have inherited it from her mother. If Mrs. Escobar and Mrs. Mendoza have different mitochondrial RFLPs, it can be determined which of them contributed mitochondria to Carlos.

**b.** Only Carlos and individuals who might have maternally contributed his mitochondria (Mrs. Escobar and Mrs. Mendoza) need to be tested. The potential grandfathers need not be tested. Mrs. Sanchez also need not be tested; she may have given mitochondria to Carlos's father, but the father would not have passed them on to Carlos.

**c.** If Mrs. Mendoza and Mrs. Escobar do not differ in mitochondrial RFLPs, the data will not be helpful. If the mitochondrial RFLPs do differ, and Carlos matches Mrs. Mendoza, the case should be dismissed. If Carlos matches Mrs. Escobar, then the Escobar and Sanchez couples are indeed the grandparents, and the Mendozas have claimed a stolen child.

## Chapter 14 Genetic Mapping in Eukaryotes

**14.2** In a chi-square test of these data under the hypothesis of independent assortment, $\chi^2 = 16.1$ and $P < 0.01$: there is less than 1% likelihood of observing this much deviation from the expected values by chance alone. Linkage might seem reasonable initially, but it is inconsistent with the minority classes not being reciprocal classes (both carry the *a/a* phenotype). Consider the segregation at each locus: The *B/–* : *b/b* ratio is about 1:1 (203:197), while the *A/–* : *a/a* ratio is not (240:160). The large deviation, then, is due to a reduced number of *a/a* individuals. This reduction should be confirmed in other crosses that test segregation at the *A/a* locus. In corn, further evidence might show up as a class of ungerminated seeds or of seedlings that die early.

**14.5** If *mal* were X linked, the parental cross could be written as $X^{mal}/X^{mal}\ vg^+/vg^+$ ♀ × $X^{mal^+}/Y\ vg/vg$ ♂. The $F_1$ males would be $X^m/Y\ vg^+/vg$, and thus would have maroon eyes. Since the observed $F_1$ are all wild type, *m* cannot be X linked. If *vg* and *m* are linked, the crosses could be written as $vg^+\ mal/vg^+ mal$ ♀ × $vg\ mal^+/vg\ mal^+$ ♂. This would produce an $F_1$ that is $vg^+\ mal/vg\ mal^+$ and all wild type. The progeny of an $F_1 \times F_1$ cross are diagrammed below, recognizing that *Drosophila* females, but not males, exhibit recombination:

| | | **Gametes of $vg^+\ mal/vg\ mal^+$ ♂** | |
|---|---|---|---|
| | | $vg^+\ mal$ (parental) | $vg\ mal^+$ (parental) |
| **Gametes of $vg^+\ mal/vg\ mal^+$ ♀** | $vg^+\ mal$ (parental) | $vg^+\ mal/vg^+\ mal$ maroon | $vg^+\ mal/vg\ mal^+$ wild type |
| | $vg\ mal^+$ (parental) | $vg\ mal^+/vg^+\ mal$ wild type | $vg\ mal^+/vg\ mal^+$ vestigial |
| | $vg^+\ mal^+$ (recombinant) | $vg^+\ mal^+/vg^+\ mal$ wild type | $vg^+\ mal^+/vg\ mal^+$ wild type |
| | $vg\ mal$ (recombinant) | $vg\ mal/vg^+\ mal$ maroon | $vg\ mal/vg\ mal^+$ vestigial |

No recombinants are produced in the male parent, so it is impossible to obtain a *vg mal/vg mal* animal from this cross. Since a double mutant was found in the $F_2$ progeny, the genes cannot be linked. Finding a double mutant is enough evidence to conclude that the genes assort independently, as this is the only way to obtain *vg mal* gametes from *both* parents. Since *mal* is not on the X or chromosome 2, it must be on chromosome 3 or 4.

**14.6** The multiple crossovers that occur over distant intervals result in large recombination frequencies being less accurate measures of map distance than the small recombination frequencies observed between close neighbors. Therefore, construct a map starting with the smallest recombination frequencies, working upward.

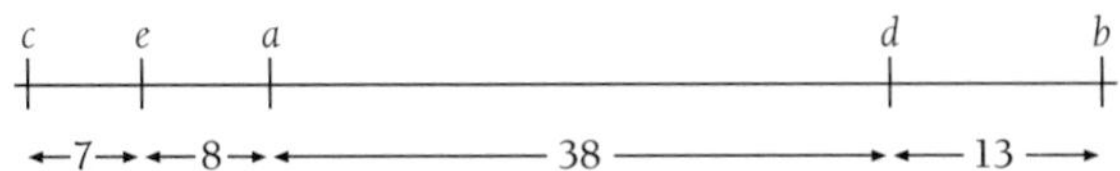


Due to the effects of multiple crossovers, recombination frequencies between loci are not strictly additive. Although recombination frequency will not exceed 50%, map distances can exceed 50 map units. The map distance between *c* and *b* is 66 map units (= 7 + 8 + 38 + 13), but *c* and *b* show 50% recombination.

**14.10** In this testcross, the genotypes of the dihybrid's gametes determine the progeny phenotypes. The $a\,b^+/a^+\,b$ parent has 90% parental type ($a\,b^+$, $a^+\,b$) and 10% recombinant ($a\,b$, $a^+\,b^+$) gametes, giving 45% $a\,b^+$, 45% $a^+\,b$, 5% $a\,b$, and 5% $a^+\,b^+$ offspring.

**14.11** The genes are 7 mu apart, so the female has 7% recombinant (3.5% each $a\,b$, $a^+\,b^+$) gametes and 93% parental (46.5% each $a^+\,b$, $a\,b^+$) gametes. The wild-type male has either $a^+\,b^+$ or Y gametes.

**a.** $P = 0.035(a^+\,b^+) + 0.465(a\,b^+) = 0.50$

**b.** $P = 1$, as all daughters receive their father's X, which is carrying $a^+\,b^+$.

**14.14 a.–d.** Inspection of the data shows that each phenotypic class has similar numbers of males and females and that none of the traits shows crisscross inheritance, suggesting that they are autosomal. For now, ignore sex type and combine males and females in each phenotypic class.

The cross of grey-bodied, light orange eyed Female 2 to a black-bodied, red-eyed male produces two equally frequent offspring phenotypes, so it is the simplest to analyze. All of the offspring have grey bodies, so grey body color is dominant to black body color, and each parent is homozygous. Let *b* represent black body color and $b^+$ represent grey body color. Female 2 is $b^+b^+$ and her mate is *bb*.

The offspring have a 1:1 ratio of light orange eyes to red eyes, consistent with this being a cross of the form *Aa* × *aa*. From just this cross, we cannot tell whether light orange eye color or red eye color is dominant. However, when the same male is crossed to females 1 and 3, four offspring phenotypes are produced with two frequencies. This result indicates that these crosses are dihybrid testcrosses involving two linked genes. Since the red-eyed, black-bodied male exhibits the recessive trait for body color, females 1 and 3 must be dihybrids exhibiting the dominant traits. Therefore, light orange eye color is dominant to red eye color. Let *O* represent light orange eye color and $O^+$ represent red eye color. With regards to both traits, females 1 and 3 are either $O^+\,b/O\,b^+$ or $O\,b/O^+\,b^+$ and their mate is $O\,b/O\,b$. (Note that if light orange (*o*) were recessive and red ($o^+$) dominant, females 1 and 3 would have to be *o/o*. For the parents to have the specified phenotypes and have had four types of offspring, each cross would have had to be $oo\,b^+b$ ♀ × $o^+o\,bb$ ♂. These crosses would have produced four equally frequent progeny phenotypes whether or not the genes were linked. This is not observed.) In a dihybrid testcross with two linked genes, different linkage arrangements produce different types of recombinant and nonrecombinant chromosomes, so the progeny in the crosses with females 1 and 3 have different phenotypic frequencies due to different linkage arrangements in the dihybrid females. The most frequent progeny phenotypes are nonrecombinants, while the less frequent progeny phenotypes are recombinants. Therefore, female 1 is $O\,b^+/O^+\,b$, and female 3 is $O\,b/O^+\,b^+$. Female 2 has only two types of offspring because she is $O\,b^+/O^+\,b^+$. Observing equally frequent classes of males and females in dihybrid testcrosses confirms that the genes are autosomal.

The original female must also have been a dihybrid: she expresses both dominant traits, so she must have dominant alleles at the eye-color and body-color loci; some of her offspring express both recessive traits, so she must have had a recessive allele to give to her offspring. Therefore, her linkage arrangement is either $O^+\,b/O\,b^+$ or $O\,b/O^+\,b^+$. We cannot determine with certainty which of these is correct, because her most frequent progeny phenotypes were not produced by a dihybrid testcross. In a dihybrid testcross, the most frequent phenotypic classes are nonrecombinants. Here, the two most frequent phenotypic classes have grey bodies, so her genotype would have had to be $b^+/b^+$. Since we know she was a dihybrid, this is not possible and her offspring do not arise from a testcross. Her offspring's phenotypes reflect the contribution of dominant and recessive alleles from multiple matings (recall that females store sperm). The observed frequencies of offspring phenotypes are possible if her mates included both $O^+\,b/O^+\,b^+$ and $O^+\,b^+/O^+\,b^+$ males. Unlike the original female, females 1, 2, and 3 were mated to a homozygous recessive male ($O^+\,b/O^+\,b$), so their linkage arrangements can be inferred from their progeny phenotypes and frequencies.

Use the data from the dihybrid crosses with females 1 and 3 to calculate a recombination frequency between the genes. Do not include the offspring of the original orange-eyed female, since you cannot tell which are recombinant and which are not (because the fathers were not homozygous recessive for each gene). The calculation of the recombination frequency is shown in the following equation:

$$\text{RF} = \frac{(4+4+6+4)}{(22+21+19+22+4+4+6+4)} + \frac{(4+5+5+7)}{(4+5+5+7+23+22+20+22)} = \frac{39}{210} = 18.6\%$$

This gives the following map:

$O/O^+$ ——— $b/b^+$

← 18.6 →

**14.15** The linkage of STR 1 cannot be evaluated, since the STR is homozygous and it is not possible to distinguish between recombinants and nonrecombinants. STR 3 does not show evidence of linkage, since there are four equally frequent progeny classes. STRs 2 and 4 are linked since there are not equally frequent progeny classes. Identify nonrecombinants as the more frequent progeny classes and recombinants as the less frequent progeny classes, and use this information to determine the map distance between the STR and the $O/O^+$ loci. STR 2 is $[(6+5)/(65+65+6+5)] \times 100\% = 12.7$mu, and STR 4 is $[(4+8+7+9)/(4+29+8+26+24+7+33+9)] \times 100\% =$ 20.0 mu from the gene for light orange eye color. We cannot determine the relative order of the three linked loci from the information in this problem. However, since we can know that the nonrecombinant progeny types are the most frequent progeny classes, we can infer that female 2 has the linkage arrangement $O^+\,STR4^6\,STR2^4/O\,STR4^3\,STR2^6$ (gene order unspecified).

**14.16** Infer the recombination frequency between two genes from the map distance between them. For example, *a* and *b* are 20 mu apart, so *A B*/*a b* will have 20% recombinant (10% each *A b*, *a B*) and 80% parental (40% each *A B*, *a b*) gametes. Then, since each chromosome pair segregates independently, use the product rule and multiply the probabilities of obtaining alleles from each chromosome.

**a.** $P(A\,B\,C\,D\,E\,F) = 0.40 \times 0.45 \times 0.35 = 0.063$, or 6.3%.
**b.** $P(A\,B\,C\,d\,e\,f) = 0.40 \times 0.05 \times 0.35 = 0.007$, or 0.7%.
**c.** $P(A\,b\,c\,D\,E\,f) = 0.10 \times 0.05 \times 0.15 = 0.00075$, or 0.075%.
**d.** $P(a\,B\,C\,d\,e\,f) = 0.10 \times 0.05 \times 0.35 = 0.00175$, or 0.175%.
**e.** $P(a\,b\,c\,D\,e\,F) = 0.40 \times 0.05 \times 0.15 = 0.003$, or 0.3%.

**14.17** With respect to the *D/d* and *P/p* loci, there are 95% parental type progeny and 5 percent recombinants. The *H/h* locus assorts independently.

**a.** 47.5% each *D P h* and *d p h*; 2.5% each *D p h* and *d P h*.
**b.** 23.75% each *d P H*, *d P h*, *D p H*, and *D p h*; 1.25% each *D P H*, *D P h*, *d p H*, and *d p h*.

**14.21** The double-crossover classes will always be the least frequent.

**a.** *F m W*, *f M w*
**b.** *M f W*, *m F w*
**c.** *F w M*, *f W m*

**14.23 a.** The crosses are (the correct order of the loci is not yet determined):

P: $b^+\,hk^+\,dp/b^+\,hk^+\,dp$ ♀ × $b\,hk\,dp^+/b\,hk\,dp^+$ ♂
$F_1$ testcross: $b^+\,hk^+\,dp/b^+\,hk^+\,dp$ ♀ × $b\,hk\,dp/b\,hk\,dp$ ♂

Tabulate the data to include parental, single-crossover (sco), and double-crossover (dco) classes (determined from their frequency):

| Phenotype | Gamete Genotype | Number | Class |
|---|---|---|---|
| dumpy | $b^+\,hk^+\,dp$ | 305 | parental |
| black, hooked | $b\,hk\,dp^+$ | 301 | parental |
| hooked, dumpy, black | $b\,hk\,dp$ | 171 | sco |
| wild type | $b^+\,hk^+\,dp^+$ | 169 | sco |
| dumpy, hooked | $b^+\,hk\,dp$ | 21 | sco |
| black | $b\,hk^+\,dp^+$ | 19 | sco |
| hooked | $b^+\,hk\,dp^+$ | 8 | dco |
| dumpy, black | $b\,hk^+\,dp$ | 6 | dco |

Compare the dco to parental classes to determine the correct gene order: *b* is in the middle. Rewrite the $F_1$ trihybrid using the correct gene order ($dp\,b^+\,hk^+/dp^+\,b\,hk$) and determine which scos belong to each interval: The $dp\,b\,hk$ and $dp^+\,b^+\,hk^+$ are scos between *dp* and *b*, while the $dp\,b^+\,hk$ and $dp^+\,b\,hk^+$ are scos between *b* and *hk*. Calculate the recombination frequencies (RFs):

$\text{RF}(dp\text{–}b) = [(171+169+6+8)/1{,}000] \times 100\% = 35.4\%$
$\text{RF}(b\text{–}hk) = [(21+19+6+8)/1{,}000] \times 100\% = 5.4\%$

The map distances are *dp–b*, 35.4 mu; *b–hk*, 5.4 mu; *dp–hk*, 40.8 mu.

**b.** Interference = 1 − coefficient of coincidence
= 1 − (observed dco frequency/ expected dco frequency)
= 1 − (14/1,000)/(0.354 × 0.054)
= 1 − 0.73 = 0.27

**14.27** Start by considering a simpler cross between a female heterozygous for an X-linked lethal (*l*) and a normal male: P: *l*/+♀× +/Y♂; $F_1$ : +/Y♂, *l*/Y♂ (dead), *l*/+♀, +/+♀. Half of the male progeny are not recovered due to the *l* allele, and the *l*-bearing chromosome is recovered only in the female offspring where it is masked by the dominant wild-type allele. Here, half of the male progeny are also not recovered due to the *l* allele, so each of the four classes seen represents one of the two reciprocal classes of progeny recovered in a three-point cross. The classes not seen bear the *l* allele. Tabulate the data from the male progeny into parental, single-crossover (sco) and double-crossover (dco) classes (determined from their frequency), including the third ($l/l^+$) locus:

| Gamete Genotype | Number | Class |
|---|---|---|
| $a\ b^+\ l^+$ | 405 | parental |
| $a^+\ b\ l^+$ | 44 | sco |
| $a^+\ b^+\ l^+$ | 48 | sco |
| $a\ b\ l^+$ | 2 | dco |

Comparison of the parental and dco classes indicates that the correct order is $a - b - l$. Since one of the parental-type chromosomes is $a\,b^+\,l^+$, the other is its reciprocal, $a^+\,b\,l$. This means the heterozygous female was $a\,b^+\,l^+/a^+\,b\,l$. The 44 $a^+\,b\,l^+$ progeny result from single crossovers between *b* and *l*, the 48 $a^+\,b^+\,l^+$ progeny result from single crossovers between *a* and *b*, and the 2 $a\,b\,l^+$ progeny result from crossovers in both intervals. For each class, the progeny are half of the total crossovers, as only one of the two reciprocal events in each class is viable. Since half of the progeny in each class are not recovered, use these numbers to estimate recombination frequencies (RF) and construct a map:

$$RF(a - b) = [(48 + 2)/499] \times 100\% = 10\%$$
$$RF(b - l) = [(44 + 2)/499] \times 100\% = 9.2\%$$

**14.28** Since the male parent is triply recessive, the phenotypes of both male and female progeny are determined by the female's gametes. The map distances between the loci give the frequency of recombinants (i.e., crossovers) in each gene interval. There are 14% recombinants in the *a–c* interval (7% each $a\,c$, $a^+\,c^+$), and 12% recombinants in the *c–b* interval (6% each $c^+\,b^+$, $c\,b$). These recombinants are distributed between both single- and double-crossover classes.

The coefficient of coincidence gives the percentage of expected double crossovers that are observed. The expected double crossover frequency is (0.12 × 0.14) × 100% = 1.68%. Since the coefficient of coincidence is 0.3, only 30% of expected double crossovers are observed, or 1.68% × 0.30 = 0.50% (0.25% each of $a\,c\,b$ and $a^+\,c^+\,b^+$).

The remaining recombinants will be single-crossover classes. In calculating this frequency, we must account for the double crossovers that result from crossovers in both gene intervals. The frequency of single crossovers in each gene interval equals the difference between the frequency of crossovers in that interval (inferred from a map distance) and the frequency of double crossovers. There will be 14% − 0.5% = 13.5% single crossovers between *a* and *c* (6.75% each $a\,c\,b^+$, $a^+\,c^+\,b$), and 12% − 0.5% = 11.5% single crossovers between *c* and *b* (5.75% each $a\,c^+\,b^+$, $a^+\,c\,b$).

The remaining progeny [100% − (13.5% + 11.5% + 0.5%) = 74.5%] will be parental types (37.25% each $a\,c^+\,b$, $a^+\,c\,b^+$). Therefore, the types of progeny are

| Genotype | Percent | Number |
|---|---|---|
| $a\ c^+\ b$ | 37.25 | 745 |
| $a^+\ c\ b^+$ | 37.25 | 745 |
| $a\ c\ b^+$ | 6.75 | 135 |
| $a^+\ c^+\ b$ | 6.75 | 135 |
| $a\ c^+\ b^+$ | 5.75 | 115 |
| $a^+\ c\ b$ | 5.75 | 115 |
| $a\ c\ b$ | 0.25 | 5 |
| $a^+\ c^+\ b^+$ | 0.25 | 5 |

**14.31 a.** Consider two genes at a time (see Table 14.A):

**Table 14.A**

| Gene Pair | # Parental-Type Progency | # Recombinant-Type Progeny | Recombination Frequency | Linked? |
|---|---|---|---|---|
| *a, b* | 902 | 98 | 9.8 | yes |
| *a, c* | 973 | 27 | 2.7 | yes |
| *a, d* | 957 | 43 | 4.3 | yes |
| *a, e* | 497 | 503 | 50.0 | no |
| *b, c* | 875 | 125 | 12.5 | yes |
| *b, d* | 945 | 55 | 5.5 | yes |
| *b, e* | 497 | 503 | 50.0 | no |
| *c, d* | 930 | 70 | 7.0 | yes |
| *c, e* | 498 | 502 | 50.0 | no |
| *d, e* | 496 | 504 | 50.0 | no |

The genes $a$, $b$, $c$, and $d$ are linked because the recombination frequencies between these genes are less than 50%. Gene $e$ is unlinked to the other four genes—it is either on a separate chromosome or far from the other loci. Develop a map starting with the smallest distances, as they are the most accurate:

**b.** Rewrite the cross using the correct gene order: $b^+ \, d \, a^+ \, c/b \, d^+ \, a \, c^+; e/e^+ \times b \, d \, a \, c/b \, d \, a \, c; e/e$. A $b^+ \, d^+ \, a^+ \, c^+$ fly is obtained from a triple crossover: There must be crossovers between $b$ and $d$, $d$ and $a$, and $a$ and $c$. The reciprocal products of the triple crossover are $b^+ \, d^+ \, a^+ \, c^+$ and $b \, d \, a \, c$. Among these, half will be $e^+$ and half will be $e$. Thus,

$$P(b^+ \, d^+ \, a^+ \, c^+ \, e^+) = P\text{ (receiving one of the two triple-crossover products and } e^+)$$
$$= \tfrac{1}{2} \times (0.055 \times 0.043 \times 0.027) \times \tfrac{1}{2}$$
$$= 1.6 \times 10^{-5}$$

**14.33** First, use the chi-square test to evaluate the hypothesis that there is no relationship between chestnut coat color and class. Further investigation into the potential linkage of this coat color gene and a hypothetical class gene is warranted if we can reject this hypothesis. The assumptions in this initial chi-square test involve the genotypes of the horses bred to Sharpen Up, as stated in the problem, and the hypothesis of the chi-square test, that chestnut coat color and class are unrelated.

From the initial hypothesis of *no relationship between class and chestnut coat color*, the likelihood of Sharpen Up siring a classy horse is uniform with regard to its coat color. There were 83 classy horses produced from a total of 367 + 260 = 627 progeny, so the chance of Sharpen Up siring a classy horse, independent of its coat color, is 83/627 = 13.24%. To perform the chi-square test, we need to compare the expected and observed number of classy chestnut and classy bay horses.

- *Assumption I: Sharpen Up is mated equally frequently to homozygous bay, heterozygous bay/chestnut, and homozygous chestnut mares.* Chestnut is recessive to bay, so let $c$ represent chestnut and $C$ represent bay. Sharpen Up is chestnut ($cc$), so the crosses and their progeny are (1) $cc \times CC \rightarrow$ all $C-$ (bay); (2) $cc \times Cc \rightarrow \tfrac{1}{2}\,Cc$ (bay), $\tfrac{1}{2}\,cc$ (chestnut); and (3) $cc \times cc \rightarrow$ all $cc$ (chestnut). If each cross is equally likely ($P = \tfrac{1}{3}$), the expected number of chestnut offspring, rounding up or down to the nearest whole horse, is $627 \times [(\tfrac{1}{3} \times 0) + (\tfrac{1}{3} \times \tfrac{1}{2}) + (\tfrac{1}{3} \times 1)] = 314$. The remaining 627 − 314 = 313 offspring are expected to be bay. Using assumption I and assuming that the frequency of classy offspring is uniform (13.24%) with respect to their coat color, the expected number of classy chestnut progeny is 314 × 0.1324 = 42, and the expected number of classy bay progeny is 313 × 0.1324 = 41. The observed numbers of classy horses were 45 chestnut and 38 bay. For these values, $\chi^2 = [(45 - 42)^2/42 + (38 - 41)^2/41] = 0.43$, df = 1, $0.50 < P < 0.70$. Under assumption I, then, the hypothesis that chestnut coat color and class are unrelated is accepted as possible.
- *Assumption II: Sharpen Up is mated equally frequently to heterozygous bay/chestnut and chestnut mares.* These crosses and their progeny are (1) $cc \times Cc \rightarrow \tfrac{1}{2}\,Cc$ (bay), $\tfrac{1}{2}\,cc$ (chestnut); and (2) $cc \times cc \rightarrow$ all $cc$ (chestnut). If each cross is equally likely ($P = \tfrac{1}{2}$), the expected number of chestnut offspring is $627 \times [(\tfrac{1}{2} \times \tfrac{1}{2}) + (\tfrac{1}{2} \times 1)] = 470$. The expected number of bay offspring is 627 − 470 = 157. Using assumption II and assuming that the frequency of classy offspring is uniform (13.24%) with respect to their coat color, the classy progeny are expected to be 470 × 0.1324 = 62 chestnut and 157 × 0.1324 = 21 bay. The observed numbers of classy horses were 45 chestnut and 38 bay. For these values, $\chi^2 = (45 - 62)^2/62 + (38 - 21)^2/21] = 18.4$, df = 1, $P < 0.001$. Under assumption II, then, the hypothesis that chestnut coat color and class are unrelated is rejected as being unlikely. It would be reasonable to consider the hypothesis that a gene closely linked to chestnut/bay coat color might contribute to class.

Notice that the evidence for a relationship between chestnut coat color and class hinges on knowing what alleles at the chestnut/bay gene were present in the mares bred to Sharpen Up. This information is available (although not in this problem). Additional assumptions required to test specifically for linkage to a class gene might include assumptions about the number of alleles in the population of horses, the dominance relationships between them, and which alleles reside on the same homolog with the chestnut allele.

**14.34** A physical exchange between two loci during meiosis will produce two recombinant and two nonrecombinant gametes. If 14% of meioses have physical exchanges, about seven percent of the gametes will be recombinants, $a^+ \, b$ or $a \, b^+$.

**14.35** A graph of θ versus lod score reveals a maximum lod score of 4.01 at a map distance of about 25 mu. Since the lod score is greater than 3, the marker is linked to the *waf* gene. However, the marker is not closely linked. Using the estimate from humans that 1 mu corresponds on average to 1 Mb, the marker is about 25 Mb away from the *waf* gene.

**14.36 a. i.** Males have just one X chromosome, so they have only one allele at each X-linked STR locus. Females with just one STR allele are homozygous for that STR allele. Females with two STR alleles are heterozygotes.

**ii.** Individual II-1 received an X from her father (I-2) with $A^+ \, B^9 \, C^2$. To account for her genotype, her mother (I-1) must have given her an X with $A^6 \, B^7 \, C^1$, so she is $A^+ \, B^9 \, C^2/A^6 \, B^7 \, C^1$.

**iii.** The paternal X chromosome with $A^+ \, B^9 \, C^2$ (from I-2) is a nonrecombinant chromosome since males have just one X and crossing over requires two homologous chromosomes. We cannot tell whether the maternal X chromosome with $A^6 \, B^7 \, C^1$ (from I-1) is recombinant or not, because we do not know what combinations of alleles were given to I-1 by each of her parents.

**iv, v.** The cross between II-1 and II-2 is $A^+ \, B^9 \, C^2/A^6 \, B^7 \, C^1 \times A^6 \, B^8 \, C^1/Y$. Two individuals received recombinant X chromosomes from II-1. The X chromosome of individual III-2 ($A^6 \, B^7 \, C^2$) was produced by a crossover between $B$ and $C$. Individual III-11 is $A^6 \, B^8 \, C^1/A^+ \, B^7 \, C^1$. A crossover between $A$ and $B$ produced her maternal X chromosome ($A^+ \, B^7 \, C^1$). There were no double crossovers.

**b.** The observed recombination frequency between $A$ and $B$, and between $B$ and $C$, is 1/11 = 9.1%. This gives the following map:

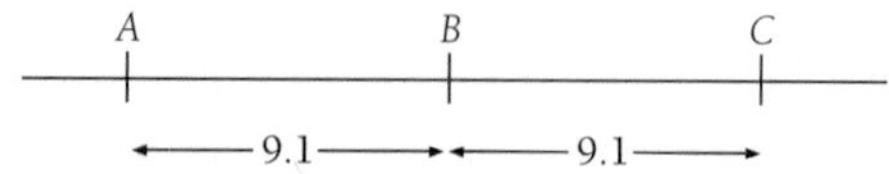


**c.** The sample size is quite small, so the map may not be very accurate. To increase the sample size and build a more accurate map, analyze these markers in additional three-generation pedigrees.

**d.** Average recombination rate = [(2/11) × 100] mu/1 Mb = 18.8 mu/Mb.

**14.38 a.** The lod score gives the odds of linkage at a particular recombination frequency, or distance, from the disease locus. The expected number of recombinants will vary depending on the map distance between the marker and disease locus, so lod score calculations to assess the likelihood of linkage must consider a range of potential distances between the marker and disease locus. If the marker locus shows linkage to a disease locus, the distance that gives a maximum lod score indicates whether the two loci are closely linked.

**b.** The increase in lod scores means that the likelihood of linkage increases. The disease locus is more likely to be close to those markers.

**c.** The lod score for some markers increases as $\theta$ decreases because the likelihood of linkage is increasing. If the lod score is sufficiently high, it suggests that the marker and disease locus are linked. If the lod score decreases as $\theta$ continues to decrease, the marker and the disease locus are not likely to be closely linked.

**d.** Markers AFMB041XB9, AFM296YG5, AFMB283XH5, AFM122XF6, and AFMB314YH5 all have lod scores above three and show evidence of linkage to this disease locus.

**e.** Marker AFM296YG5 shows the highest lod score of 6.97 at $\theta = 0.001$. This indicates that there is a $10^{6.97}$:1 = 9,332,542:1 chance that the disease locus is linked to this marker at a distance of 0.1% recombination. Put another way, it is very likely that the disease locus and this marker locus are closely linked.

**f.** The interval containing all of the markers noted in (d) corresponds to the cytological interval between 12p11.21 and 12q13.11. It spans the centromere, has a genetic map distance of 11.35 mu, and a physical distance of 15.8169 Mb. The markers AFM296YG5, AFMB283XH5, AFM122XF6 define a subinterval where maximal lod scores are seen at $\theta = 0.001$. Still substantial, this interval corresponds to cytological positions 12p11.21 to 12q13.11, includes the centromere, spans a genetic map distance of 8.25 mu, and has a physical distance of 13.4069 Mb.

## Chapter 15 Genetics of Bacteria and Bacteriophages

**15.1** For a recipient to be converted to a donor, a complete F factor must be transferred. In $F^+ \times F^-$ crosses, only the F factor is transferred, and this occurs relatively quickly. In *Hfr* × $F^-$ crosses, transfer starts at the origin within the F element and then must proceed through the bacterial chromosome before reaching the F factor. For transfer of the entire F factor, the whole chromosome would have to be transferred. This would take about 100 minutes, and usually the conjugal unions break apart before then.

**15.3**

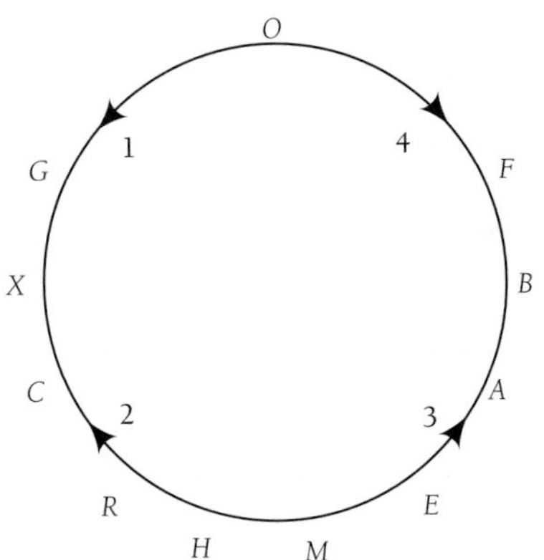


**15.6 a.** Initially select for $c^+$ $str^R$ recombinants by plating the progeny on minimal medium without compound C, but supplemented with streptomycin and compounds A, B, D, E, F, G, and H. To assess the complete genotype of the $c^+$ $str^R$ recombinants, replica plate them onto different minimal media supplemented with streptomycin and all but two of the compounds (compound C and one other). For example, to test if a $c^+$ $str^R$ colony was also $a^+$, replica plate it onto a medium that lacked compound A, but was supplemented with streptomycin and B, D, E, F, G, and H. If the colony were able to grow on this medium, it would be $a^+$ $c^+$ $str^R$. If it were unable to grow, it would be $a$ $c^+$ $str^R$.

**b.** Strain 1: Since no $c^+$ recombinants are ever obtained, strain 1 is unable to transfer $c^+$. This means it is either (1) $F^-$; (2) *Hfr* but with the F factor inserted either far from $c^+$ or close to it but in an orientation so that genes are transferred in a direction opposite to $c^+$; or (3) $F'$, with $c^+$ in the bacterial chromosome. It should not be $F^+$ because then, at a very low frequency, some $c^+$ recombinants would be obtained.

Strain 2: Since $c^+$ recombinants are obtained at 6 minutes, and $g^+$, $h^+$, $a^+$, and $b^+$ recombinants are obtained at subsequent time intervals, strain 2 is *Hfr*. The genes are transferred in the sequence $c^+$, $g^+$, $h^+$, $a^+$, and $b^+$. From the times of transfer of the genes, the map position of the genes is as follows: origin (0)–$c^+$(6)–$g^+$(8)–$h^+$(11)–$a^+$(14)–$b^+$(16). The location of genes $d^+$, $e^+$, and $f^+$ cannot be determined precisely; because they are not transferred in an *Hfr* × $F^-$ cross, they are either far away from the F factor insertion site, or close to it but near the fertility genes, which are rarely transferred by an *Hfr* strain. When the recombinants obtained from the strain 2 × $F^-$ mating at the 16-minute time period are crossed to an $amp^R$ $F^-$ strain, $c^+$ is not transferred. If these recombinants cannot conjugate with $F^-$, this indicates that although strain 2 is fertile, it did not transfer a complete F factor. Therefore, it must be *Hfr*.

Strain 3: Strain 3 transfers $c^+$ within 1 minute and $g^+$ by 3 minutes. From analysis of the strain 2 × $F^-$ cross, we knew that these genes are 2 minutes apart. These data support this conclusion. Since no other recombinants are obtained, no other genes are transferred. This suggests that strain 3 is $F'$, and that the segment of DNA containing $c^+$ and $g^+$ is in the $F'$ factor. If this is the case, the complete F factor will be transferred in a strain 3 × $F^-$ cross if the mating is allowed to proceed long enough. This is observed: $c^+$ recombinants from the strain 3 × $F^-$ cross obtained at 16 minutes are able to transfer $c^+$ to an $F^-$ $amp^R$ strain. Therefore, strain 3 is $F'$.

**c.** Information known with certainty is diagrammed here. The location of genes in strains 1 and 3 is inferred from crosses with strain 2, while that of genes $d^+$, $e^+$, and $f^+$ is unknown.

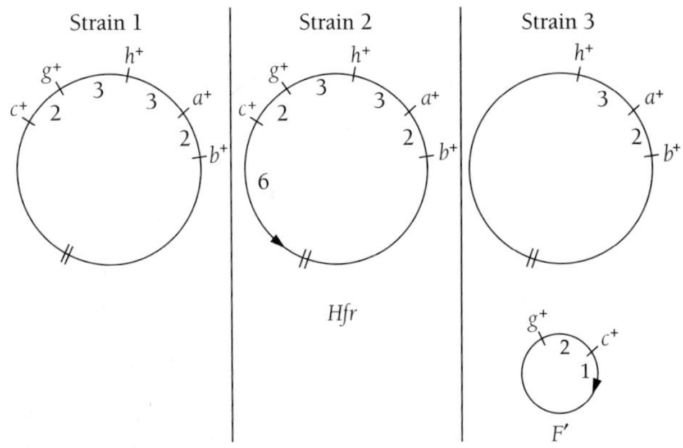

**15.7** Strain *A* is *thy*$^-$ *leu*$^+$ while strain *B* is *thy*$^+$ *leu*$^-$. DNA from *A* can transform *B* if DNA from *A* can transform the *leu*$^-$ allele of *B* to *leu*$^+$. Test this by adding DNA from *A* to a leucine-fortified culture of *B*. Incubate long enough for transformation to occur, and then plate the potentially transformed *B* cells on minimal medium or on medium supplemented only with thymine. This selects for *leu*$^+$ transformants.

**15.9 a.** Since the $F^-$ *leu arg str*$^R$ cells were treated with MMS, the colonies that are picked and plated into the grid-like pattern have randomly induced mutations. Plating a mixture of mutant $F^-$ *str*$^R$ and *Hfr leu*$^+$ *arg*$^+$ *str*$^S$ cells on minimal medium with streptomycin selects for *leu*$^+$ *arg*$^+$ *str*$^R$ exconjugants: growth will occur only if the *Hfr* strain donates DNA containing the *leu*$^+$ and *arg*$^+$ genes via conjugation and that DNA is incorporated into the genome of the mutant $F^-$ cell. Since nearly all of the 5,000 mixtures produced growth, the *leu*$^+$ and *arg*$^+$ genes are transferred relatively early from the *Hfr* strain. They must lie close to and behind the F-factor origin of replication.

**b.** After the *Hfr* cell transfers a single strand of DNA to the $F^-$ cell during conjugation, DNA polymerase synthesizes a complementary strand and a double crossover incorporates the linear donor DNA into the circular recipient chromosome. $F^-$ mutants unable to synthesize the complementary strand or recombine the double-stranded DNA into their chromosome would not produce exconjugants. Since the mutant $F^-$ cell can divide on medium supplemented with arginine and leucine, the mutation is unlikely to affect chromosomal DNA synthesis. Most mutants unable to generate *leu*$^+$ *arg*$^+$ exconjugants probably affect genes needed for recombination of the donor DNA into the recipient chromosome.

**15.12 a.** GT
**b.** ST
**c.** ST
**d.** GT
**e.** GT
**f.** B
**g.** B
**h.** B
**i.** N

**15.14** Closer genes have a higher cotransduction frequency. The *pyrD* and *cmlB* genes show the highest cotransduction frequency, so they are the closest together and (c) is eliminated. The genes *aroA* and *pyrD* show the lowest cotransduction frequency, so they are the farthest apart and (b) is eliminated. The *aroA* and *cmlB* genes show an intermediate cotransduction frequency, as would be expected if *cmlB* is between *aroA* and *pyrD*. Thus, the correct answer is (a), *aroA–cmlB–pyrD*.

**15.16** Since relatively closer loci show a relatively higher cotransduction frequency, pairs of loci can be ordered in terms of their proximity. The order (closest together to farthest apart) is *cheB–eda, cheA–eda, cheA–supD, cheB–supD, eda–supD*. A gene order consistent with these relationships is *eda–cheB–cheA–supD*.

**15.18** The genetic distance is 0.07 mu. The plaques produced on E. coli K12(λ) are $r^+$, while those on E. coli B may be either $r^+$ or $r^-$. Thus, the total number of progeny can be inferred from the number of plaques formed on *E. coli B*: Total number of progeny in 1 mL = (dilution × factor) × (progeny phage/mL) = 1,000 × (672/0.1) = $6.72 \times 10^6$/mL. Since E. coli B is coinfected with *rIIx* and *rIIy*, the only way to obtain an $r^+$ progeny phage is to have a crossover within the *rII* locus. The progeny resulting from a crossover would be ½ $r^+$ and ½ *rIIxy* recombinants. The number of recombinant phage is twice the number of $r^+$ phage, which can be assayed for by growth on *E. coli K12*(λ): Number of recombinant progeny in 1 mL = 2 × (number of $r^+$ phage/mL) = 2 × (470/0.2) = 4,700/mL. The map distance between *rIIx* and *rIIy* is [4,700/(6.72 × $10^6$)] × 100% = 0.07 mu.

**15.20** Analyze the data as you would a set of two-factor crosses. Two maps are compatible with the data:

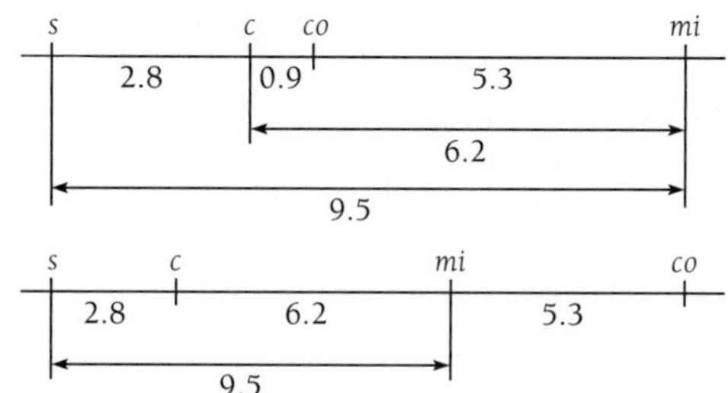


**15.22** Identify the region missing in each deletion mutant by applying two principles: (1) If no $r^+$ recombinants are obtained, the deletion removes the site of the point mutant. (2) If $r^+$ recombinants are obtained, the site of the point mutation is not within the boundaries of the deletion.

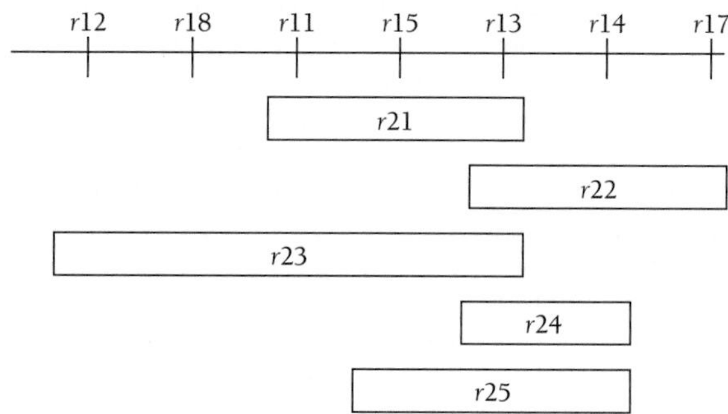


**15.23** Apply two principles to delineate the region where a point mutant lies: (1) If a point mutant can recombine with a deletion mutant, it must lie outside of the deleted region. (2) If a point mutant cannot recombine with a deletion mutant, it must lie within the deleted region. Use the complementation data to determine the positions of the A and B cistrons: If two mutants are unable to grow on *E. coli K12*(λ), they do not complement each other and cannot together provide the functions to complete the *rII* pathway. Both mutants must be defective in the same function, and either the *rIIA* or *rIIB* function is missing.

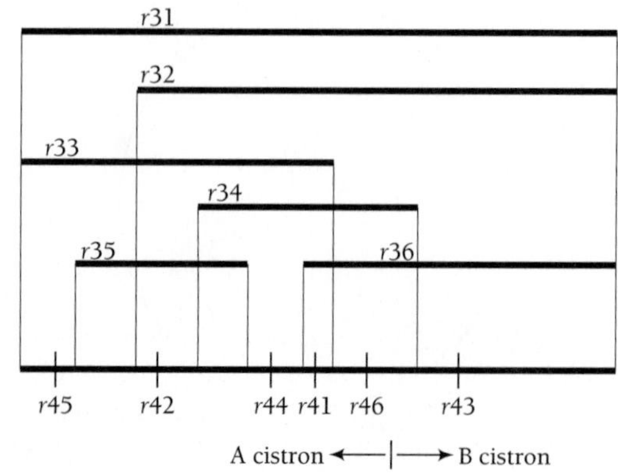

**15.26 a.** If DNA transduced into a particular *leu* mutant recombines with its chromosome to produce a *leu*$^+$ recombinant, the transduced DNA must contain a wild-type site that can replace the mutated *leu* site. Therefore, pairs of mutants that produce *leu*$^+$ recombinants affect different sites, while pairs of mutants unable to produce *leu*$^+$ recombinants affect one or more common sites. The sites may be single or multiple base-pair regions. Mutants that affect more than one site must be deletions. Deletions can be recognized by identifying mutants unable to produce *leu*$^+$ recombinants with pairs of mutants that lie in different sites. For example, mutants 3 and 8 can recombine to produce *leu*$^+$ recombinants, so they lie in different sites. Mutant 7 cannot recombine with either mutant 3 or mutant 8, so it must delete both sites. Using this logic, mutants 2, 4, and 7 must be deletions. In the map here, these deletions are shown by open boxes.

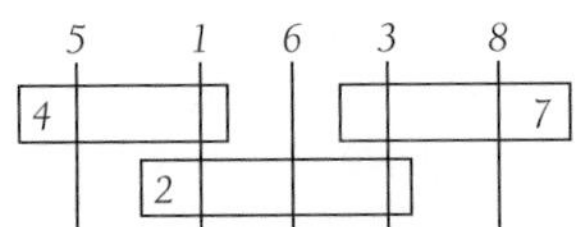


**b.** A site may be one or more base pairs. To address if a site is a point mutation, see if a mutant can be reverted. Point mutants, but not deletions, can be reverted.

**c.** This analysis does not address how many cistrons are present in this region. For this, complementation tests are needed.

**15.29** There are at least two options. First, the enzyme could be composed of multiple polypeptide subunits. The mutants affect different genes, each of which encodes a polypeptide that is part of the multimeric enzyme. Second, the polypeptide for the enzyme is modified before it becomes active as an enzyme. One mutant affects the gene for the polypeptide that will be modified to become the enzyme; the second mutant affects the gene for a modifier protein that is required to make the enzymatic polypeptide functional (e.g., it could affect a protease that cleaves a proenzyme to make an active enzyme; it could affect a kinase that phosphorylates a nonactive form of the enzyme to make it active.)

**15.31 a.** Since all nine very early mutants can be reverted, all are point mutants.

**b.** All nine very early mutants fail to complement each other (none produce enough virus in pairwise coinfections to be considered positive), so they affect one function.

**c.** Eight of the mutants (*B2, B21, B27, B28, B32, 901, LB2, D*) are able to recombine with each other, and so affect different sites. Mutant *c75* fails to recombine with mutant *D*, so these two mutants may affect the same site.

**d.** Mutant *c75* may be a mutant affecting multiple points. It reverts at a lower frequency than do the others and shows inconsistent recombination relative to the other mutants.

**e.** The mutants incompletely block viral growth, so some virus is produced by each mutant. *I* compares the amount of virus produced by coinfection of two mutants to the sum of the amounts produced by individual infections. If two viruses are blocked in the same function, a coinfection should produce low amounts of virus similar to the sum of two separate infections and *I* will be about 1.0. If two mutants are blocked in different functions, coinfection allows for complementation, as the function blocked in one mutant is provided by the other mutant. Substantial viral growth will occur, and *I* should be much larger than 1.

At the restrictive temperature of 39°C, neither parent nor doubly mutant recombinants can grow. Only wild-type recombinants can grow. These are half of the recombinants, so doubling the amount of virus produced at 39°C will estimate the number of recombinants between the two mutant sites. At the permissive temperature of 34°C, mutant and wild-type viruses can grow, so the amount of virus produced at 34°C measures the total amount of virus produced by coinfecting the two mutant strains. RF is calculated by doubling the amount of virus produced by coinfecting two mutants at 39°C and then dividing this number by the amount of virus produced by coinfecting the mutants at 34°C, so RF estimates recombination between two mutants.

**f.**

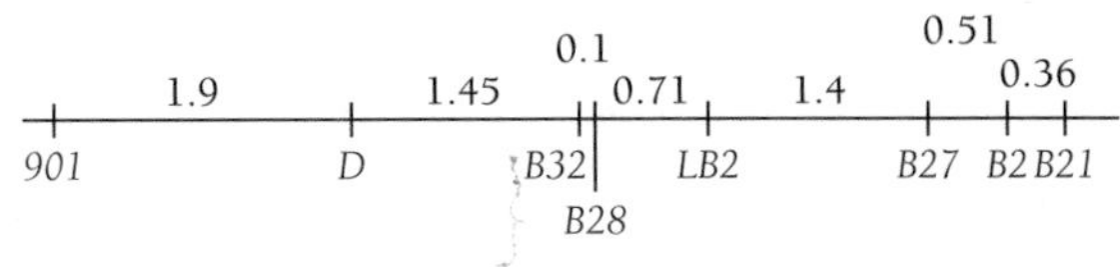


**g.** The reversion rate of mutant *c75* is less than the other mutants, suggesting that it is more complex than a simple point mutant. If it affected multiple sites, it would have recombination data inconsistent with the other mutants.

**15.32 a.** Cross each mutation to the wild-type, brick red strain and note the progeny phenotypes. A dominant mutation, by definition, appears in a heterozygote, so if the progeny have brownish eyes, the mutation is dominant. If they have brick red eyes, the mutation is recessive.

**b.** Set up pairs of crosses between the mutants to perform complementation tests. Mutations that affect the same gene function will produce brown-eyed progeny when crossed and belong to the same complementation group. Mutations that affect different functions will produce brick red progeny when crossed and belong to different complementation groups. Counting the number of different complementation groups will give the number of genes that are affected.

**c.** Allelic mutations are those that are members of the same complementation group, as determined in **(b)**.

**d.** One could determine if a particular mutant is allelic to a known eye color gene by performing complementation tests between it and mutants at all known eye color genes. This would involve crossing the mutant to mutants from a collection of strains with known eye color mutations and observing the progeny of each cross. If the progeny have a mutant eye color, one would infer that the mutations carried in the two strains are allelic. However, this would be a tremendous amount of work. There are many eye color mutations, and this would require a large number of crosses. It would be faster first to determine which of the six mutations are allelic, and then choose a representative allele from each complementation group and identify its chromosomal location using a set of two- and/or three-point mapping crosses. Once this is done, examine published genetic maps of the *Drosophila* genome (e.g., at the site http://www.flybase.org) to determine if any known eye color mutations lie in the same region. You could then obtain strains with these eye color mutations and perform complementation tests between these mutant strains and a representative mutant from each complementation group identified with the new eye color mutations.

## Chapter 16 Variations in Chromosome Structure and Number

**16.1 a.** pericentric inversion [*D* ○ *E F* inverted]

**b.** nonreciprocal translocation [*B C* moved from left to right arm]

**c.** tandem duplication [*E F* duplicated]

**d.** reverse tandem duplication [*E F* duplicated]

**e.** deletion [*C* deleted]

**16.2** A pericentric inversion includes the centromere, while a paracentric inversion lies wholly within one chromosomal arm (see Figure 16.7).

**16.4 a.** This is paracentric inversion, because the centromere is not included in the inverted DNA segment.

**b.**

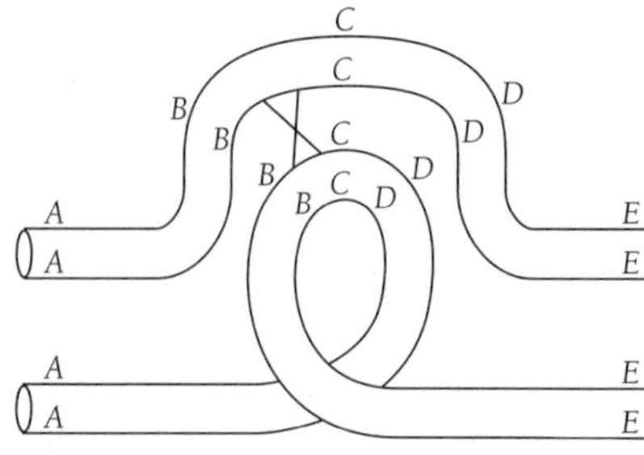


**c.** A crossover between B and C results in the following chromosomes:

*A B C D E* (normal order)

*A B C D A* (dicentric, duplication for *A*, deletion for *E*)

*E B C D E* (acentric, duplication for *E*, deletion for *A*)

*A D C B E* (inverted order)

**16.7** One series of sequential inversions is

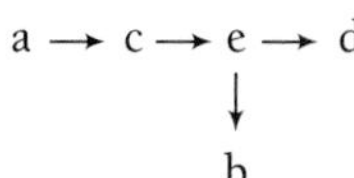


The regions inverted in each step are illustrated here.

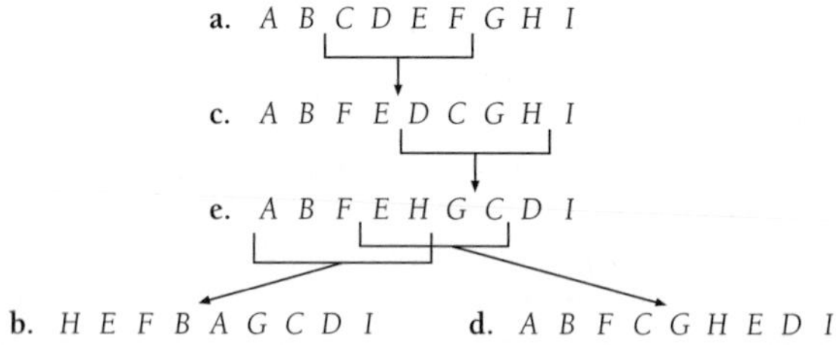


**16.9 a.** The following diagram shows a normal chromosome bearing $w^+$, the $w^{M4}$-associated inversion and a second inversion found in a $w^+$ revertant. The genes *a*, *b*, *c*, and *d* are inserted near the breakpoints of the different inversions to help visualize them. Euchromatin is represented by a thin line, centromeric heterochromatin by a thick line, and the centromere by an open circle. The brackets delineate inverted regions.

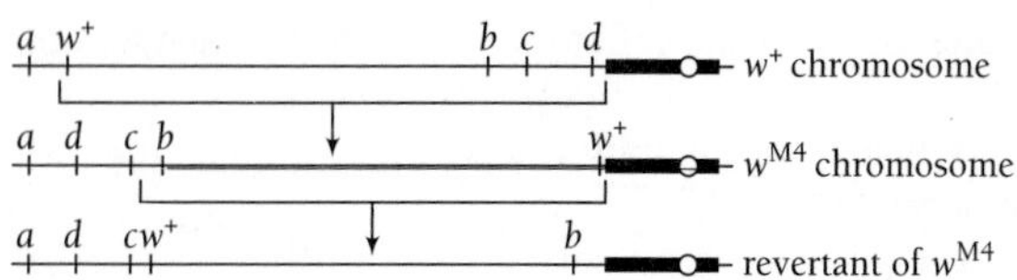


The mottled eye phenotype is associated with chromosomal rearrangements induced on a $w^+$-bearing chromosome that place the $w^+$ gene near heterochromatin. When a rearrangement is heterozygous with a *w* allele, only the $w^+$ gene on the rearranged chromosome can provide for normal eye pigmentation. The mottled appearance of the eye indicates that it functions in some, but not all cells. This suggests that the DNA sequence of the *white* gene is unaltered. It is more likely an epigenetic phenomenon caused by a position effect, a phenotypic change due to inactivation of the $w^+$ allele by neighboring heterochromatin.

**b.** The second inversion that occurs on the $w^{M4}$ chromosome repositions the $w^+$ gene to a euchromatic location. This supports the view that the mottled eye phenotype is caused by a position effect.

**16.10 a.** Parents of Rec(8) individuals are heterozygous for a pericentric inversion with breakpoints at 8p23.1 and 8q22.1. Rec(8) offspring with 8q duplication and 8p deletion probably arose from a single crossover within the pericentric inversion. Such an event is diagrammed in Figure 16.9, p. 471.

**b.** As shown in Figure 16.9, p. 471, a single crossover between two nonsister chromatids in an inversion heterozygote results in four products: two have the noncrossover chromosomes (one normal-ordered and one inverted) and two are duplication/deletion products. Here, the product with 8q duplication and 8p deletion contribute to a viable zygote with Rec(8) syndrome. The product with 8q deletion and 8p duplication is not discussed in the problem. It may be that zygotes with this product do not survive. In this case, $1/3$ of the surviving zygotes have Rec(8) syndrome. Of the $2/3$ normal zygotes, $1/2$ carry the chromosome 8 inversion.

**c.** The phenotypes of Rec(8) individuals could vary for one or a combination of reasons: (1) There could be several different chromosome 8 inversions in the population that vary slightly in their inversion breakpoints. The Rec(8) individuals resulting from single crossovers in inversion heterozygotes would differ symptomatically due to variation in genes that are duplicated and deleted or due to differences in gene activation or gene inactivation. (2) There may be a position effect. (3) The genetic background could vary. The phenotypic effects of gene deletion or duplication could depend on genetic interactions with other genes in the genome. In this case, alleles inherited from the father that are different from those inherited from the mother and grandmother could contribute to the phenotype. (4) Environmental effects could exacerbate the effects of the deleted and duplicated region. These effects may not be uniform and so could contribute to the observed phenotypic variability. Since many of the symptoms associated with Rec(8) syndrome are developmental abnormalities, variation in the environment during fetal development may contribute to phenotypic variability. (5) There may be other, cytologically invisible mutations associated with the Rec(8) individuals that could strongly affect their phenotype.

**d.** The child has the chromosome 8 inversion, but not the duplication/deletion chromosome that results from a single crossover in an inversion heterozygote; she is an atypical Rec(8) individual. There are several explanations for why some of her symptoms overlap with those of Rec(8) syndrome. She may have an additional mutation near one of the Rec(8) breakpoints, in a region that is duplicated or deleted in Rec(8) syndrome, or in a gene that interacts with genes in the duplicated or deleted regions. Alternatively, it is possible that the inversion disrupts the function of a gene or genes at one or both breakpoints, and that normally, the inversion is an asymptomatic condition. In this case, the inversion chromosome (in the child's mother and grandmother) would bear a recessive mutation. If she had a new allelic mutation, or her paternally contributed chromosome had an allelic mutation, she would be affected. This could also explain why she has only some of the symptoms of Rec(8) syndrome; she would have fewer genes affected than would most Rec(8) individuals.

Small deletions would be cytologically invisible, as would point mutations. Thus, the explanations given above could not be evaluated solely by karyotype analysis. DNA marker and/or DNA sequence analyses (see Chapters 8 and 9) could to be used to evaluate the integrity of the chromosomal regions near the breakpoints.

**16.11 a.** The irradiated chromosome has a paracentric inversion. Single crossovers produce dicentric chromosomes and fragments; a four-strand double crossover produces dicentric chromosomes with two bridges and two fragments.

**b.** The bridge chromosome would arise by a single crossover within the inversion loop (see Figure 16.8, p. 470).

**16.13 a.** Mr. Lambert is heterozygous for a pericentric inversion of chromosome 6. Relative to the centromere of the normal chromosome 6, one of the breakpoints is within the fourth light band up from the centromere, while the other is in the sixth dark band below the centromere. Mrs. Lambert's chromosomes are normal.

**b.** When Mr. Lambert's number 6 chromosomes paired during meiosis, they formed an inversion loop that included the centromere. Crossing-over occurred within the loop, and gave rise to the partially duplicated, partially deficient chromosome 6 that the child received.

**c.** The child is not phenotypically normal because the duplications and deletions for different parts of chromosome 6 lead to severe abnormalities. The child has three copies of some and only one copy of other chromosome 6 regions. The top part of the short arm is duplicated, and there is a deficiency of the distal part of the long arm.

**d.** Most future conceptions by this couple will produce an abnormal fetus. This is because the inversion covers more than half of the length of chromosome 6, and so the majority of meioses will have a crossover within the inverted region. A normal child will be produced in a minority of meioses where there is a two-strand double crossover inside the loop, where crossing-over occurs outside the loop, or where a crossover has occurred within the loop but the child receives a noncrossover chromosome.

**16.15**

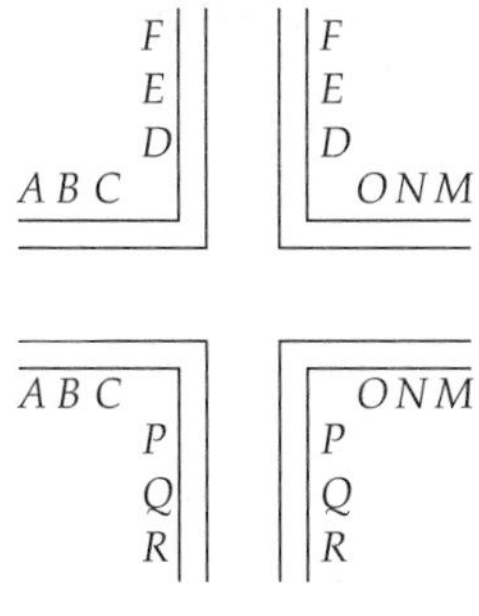

F F
E E
D D
ABC ONM

ABC ONM
P P
Q Q
R R

**16.17 a.** Mr. Denton has normal chromosomes. Mrs. Denton is heterozygous for a balanced reciprocal translocation between chromosomes 6 and 12. Most of the short arm of chromosome 6 has been reciprocally translocated onto the long arm of chromosome 12. The breakpoints appear to be in the first thick, dark band just above the centromere of 6 and in the third dark band below the centromere of 12.

**b.** The child received a normal chromosome 6 and a normal chromosome 12 from his father. In prophase I of meiosis in Mrs. Denton, chromosome 6 and 12 and the reciprocally translocated 6 and 12 paired to form a cruciform-like structure. Segregation of adjacent, nonhomologous centromeres to the same pole ensued, so that the child received a gamete containing a normal 6 and one of the translocation chromosomes. See Figure 16.11, p. 473, for an illustration of adjacent-1 segregation.

**c.** The child has a normal chromosome 6 and a normal chromosome 12 from Mr. Denton. The child also has a normal chromosome 6 from Mrs. Denton. However, the child also has one of the translocation chromosomes from Mrs. Denton. With this chromosome, the child is partially trisomic as well as partially monosomic. It has three copies of part of the short arm of chromosome 6 and only one copy of most of the long arm of chromosome 12. This abnormality in gene dosage is the cause of its physical abnormality.

**d.** Segregation of adjacent homologous centromeres to the same pole is relatively rare. The segregation pattern seen in this child (adjacent-1 segregation) and alternate segregation (see Figure 16.11, p. 473) are more common. About half the time, when alternate segregation occurs the gamete will have a complete haploid set of genes, and the embryo should be normal. However, half of the gametes resulting from alternate segregation will be translocation heterozygotes.

**e.** Prenatal monitoring of fetal chromosomes could be done, and given the severity of the abnormalities (high probability of miscarriage and multiple congenital abnormalities), therapeutic abortion of chromosomally unbalanced fetuses would be a consideration.

**16.22 a.** A reciprocal (interchromosomal) translocation results from the exchange of segments between two chromosomes. No genetic material is gained or lost. In a reciprocal translocation, part or all of one arm of one chromosome is exchanged for part or all of an arm of a second chromosome (see Figure 16.10c, p. 471). In contrast, a Robertsonian translocation occurs when two nonhomologous acrocentric chromosomes break near their centromeres and the long arms become attached to a single centromere. The short arms join to form the reciprocal product but are lost after a few cell divisions.

**b.** 45 (23 from the mother; a Y, 20 normal autosomes, and the Robertsonian translocation from the father).

**c.** Since chromosomes 13, 14, 15, 21, and 22 are acrocentric, they can be involved in Robertsonian translocations. Offspring inheriting a Robertsonian translocation involving two of these chromosomes from one parent and normal chromosomes from the other parent could be phenotypically normal if the short arms of the two chromosomes lack genes that are essential in two copies.

**d.** Different patterns of chromosome segregation during meiosis in the translocation-bearing male can give rise to offspring who are chromosomally normal, are translocation-bearing carriers, have Down syndrome, or are inviable. If the male donates normal chromosomes 14 and 21 or the translocation, he will produce phenotypically normal offspring. If he donates the translocation with his normal copy of chromosome 21, the offspring will have trisomy-21 and so have Down syndrome. If he donates the translocation with his normal copy of chromosome 14 (giving rise to trisomy-14), or if he donates only one copy of chromosome 14 or 21 (giving rise to monosomy-14 or monosomy-21), the offspring will be inviable.

Figure 16.A diagrams the pairing of the Robertsonian translocation during male meiosis, his gametes and zygotes that will be produced if he mates with a normal female. The thick grey lines represent chromosome 14 and the thick black lines represent chromosome 21.

Figure 16.A

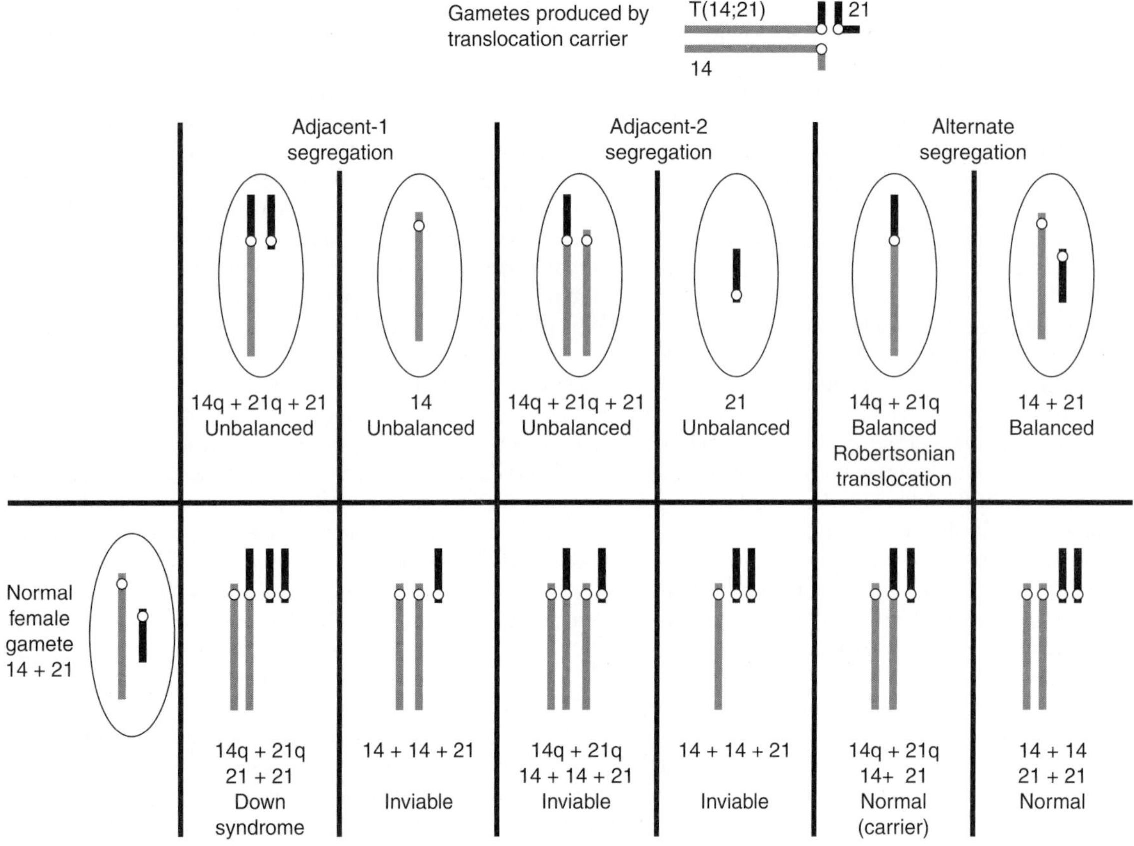


**16.23 a.** First, the pedigree is consistent with an X-linked recessive trait such as fragile X syndrome. Second, in fragile X syndrome, normal transmitting males carry a premutation that is passed to their daughters, and the sons of these daughters frequently show mental retardation. Here, the daughters of individual I-1 all have sons that have mental retardation, but neither he nor his son's children show mental retardation. Therefore, individual I-1 could have an X chromosome bearing a premutation that is passed to his daughters (but not his sons). During DNA replication in his daughters, the CGG triplet repeat in the *FMR-1* gene is amplified to generate a full mutation. Mental retardation is seen in their offspring when the X chromosome bearing the full mutation is transmitted.

**b.** Culture cells from the affected individuals, and examine chromosomes cytologically to determine whether a fragile site is present. Use PCR with primers that flank the CGG repeat in the *FMR-1* gene to evaluate the size of the repeat. Individuals with fragile X syndrome will exhibit a fragile site at Xp27.3, and have 200 to 1,300 copies of the CGG repeat.

**c. i.** I-1, II-3, II-8, II-11,

**ii.** I-1

**iii.** I-1, II-3, II-8, II-11, III-5, III-6, III-13, III-14, IIII-15, III-21

**iv.** III-7, III-8, III-20, III-26, III-27

**d.** In females, one X chromosome is inactivated. In some cells, the fragile X chromosome will be inactivated while the other X chromosome will have a normal *FMR-1* gene. This could underlie the less severe phenotype seen in females.

**16.26 a.** 45

**b.** 47

**c.** 23

**d.** 69

**e.** 48

**16.27 b.** trisomic

**16.28 a.** The cross can be written as $c^+/c^+ \times c/Y$. The $c/O$ child received its father's $c$. A chromosomally normal X-bearing sperm fertilized a nullo-X egg, so nondisjunction occurred in the mother.

**b.** The $c^+/O$ child received its mother's $c^+$. The chromosomally normal X-bearing egg was fertilized by a nullo-X, nullo-Y sperm, so nondisjunction occurred in the father.

**16.29** This problem considers what happens when a chromosome is lost at the very first mitotic division (and only at that division).

**a.** The cross is $y/y\ pal^+/pal^+$ (female) $\times$ $y^+/Y\ pal/pal$ (male), with progeny $y/y^+\ pal/pal^+$ (daughters) and $y/Y\ pal/pal^+$ (sons). The paternally contributed X is found only in the $y/y^+\ pal/pal^+$ daughter, so we need only consider the consequence of its loss in daughters. If a paternally contributed X chromosome ($y^+$) is lost during the first mitotic division in a $y/y^+\ pal/pal^+$ zygote, one daughter cell will lose an X chromosome and be $y\ pal/pal^+$. The other daughter cell will have two X chromosomes and be $y/y^+\ pal/pal^+$. The cell with two X chromosomes would be female (XX) and produce nonyellow cells ($y/y^+$), while the cell with one X chromosome would be male (XO) and produce yellow cells ($y$). The animal will be a mosaic with cells of two sex chromosome compositions that are marked by yellow (male) or grey (female) cuticle.

**b.** The cross is $pal^+/pal^+\ eye/eye$ ♀ $\times$ $pal/pal\ eye^+/eye^+$ ♂, with progeny $pal/pal^+\ eye/eye^+$ (daughters and sons). The paternally contributed fourth chromosome is $eye^+$. If it is lost during the first mitotic division in a *pal/pal*$^+$ *eye/eye*$^+$ zygote, one daughter cell will lose a fourth chromosome and be *pal/pal*$^+$ *eye*. The other daughter cell will have two fourth chromosomes and be normal, while the cell with one fourth chromosome will be *eye*. The animal will be a mosaic with some cells that are haploid for the fourth chromosome and some cells that are diploid for the fourth chromosome. If a patch of haplo-4 cells forms an eye during development, the eye will be reduced in size.

**c.** The cross is *pal*$^+$/*pal*$^+$ *e/e* (female) × *pal/pal* $e^+/e^+$ (male), with progeny *pal/pal*$^+$ $e/e^+$. The paternally contributed third chromosome is $e^+$. If it is lost during the first mitotic division in a *pal/pal*$^+$ $e/e^+$ zygote, one daughter cell will lose a third chromosome, and be *pal/pal*$^+$ *e*. This cell is inviable, and so will not be recovered in the organism, should the organism survive. Consequently, if the organism survives, it will be phenotypically normal (*pal/pal*$^+$ $e/e^+$).

**16.32** A general approach to answering this question is to model the appearance of males in the cleft species after a known sex-determination mechanism. Since there are multiple mechanisms for sex determination, this question has multiple solutions, each of which can be very instructive. The solution provided here models the appearance of males after mammalian sex-determination mechanisms. Recall that in organisms that use chromosomal sex-determination mechanisms, such as mammals and birds, sex chromosomes appear to have evolved from different autosome pairs, and that in humans, male sex determination results from the expression of the Y-linked testis-determining factor gene (see Chapter 12). Using this framework, hypothesize that the cleft females had pairs of autosomes, but no sex chromosomes, and developed as females due to a default sex-determination pathway until a chromosomal mutation occurred to generate a sex chromosome able to direct male sex determination. One scenario for this would be a nonreciprocal interchromosomal translocation between two different medium-sized chromosomes that produced a small Y-like chromosome and a larger X-like chromosome. If the translocation breakpoint affected the expression of a gene on the small Y-like chromosome that was already involved in sex determination, then some features of human sex determination would be in place. For example, we can hypothesize that altered expression of the gene would lead to an altered hormonal milieu and that this would initiate a male sexual determination pathway.

Explore whether this chromosomal mutation could explain the "increasingly frequent" birth of males by following the inheritance of the translocation products. For this purpose, hypothesize that cleft females reproduce parthenogenetically, that they are diploids who produce haploid eggs, and that the diploid number is restored in a developing egg when its first nuclear division proceeds without cytokinesis. During meiosis in a diploid organism, the products of a nonreciprocal interchromosomal translocation pair with the chromosomes they derived from, and they undergo alternate, adjacent-1 or adjacent-2 segregation (see Figure 16.11, p. 473). Figure 16.B illustrates the meiotic pairing of two acrocentric chromosomes A (black line) and B (grey line) with the products of a nonreciprocal interchromosomal translocation between them, T(A,B) and B′, and gametes that this produces.

Only two gametes would be viable: the gamete having the A + B chromosomes would have only autosomes and so develop as a female; the gamete having the T(A,B) + B′ chromosomes would develop as a male due to the new B′ chromosome. While this explains the appearance of one or a few males from a single cleft female, it does not explain the appearance of males with increasing frequency in a previously all-female population. Males would be seen with increasing frequency only if there were transfer of genetic information between cleft females (which does not happen, since they reproduce parthenogenetically), if the same translocation occurred nearly simultaneously in multiple cleft females (which is unlikely) or if the translocation-bearing cleft females mated with the newly produced males (which may not happen initially). To explain how multiple cleft females produce male offspring, we must modify our speculation about the male sex-determination gene on the B′ chromosome. Now suppose that all the T(A,B) + B′ individuals were initially females and that a male sex determination gene on the B′ chromosome evolved only later, after the T(A,B) + B′ individuals had become more common in the population. If at that point the expression of a sex determination gene on the B′ chromosome were altered, say, by a heritable genetic alteration such as the expansion of an unstable trinucleotide repeat, and this alteration in expression led

**Figure 16.B**

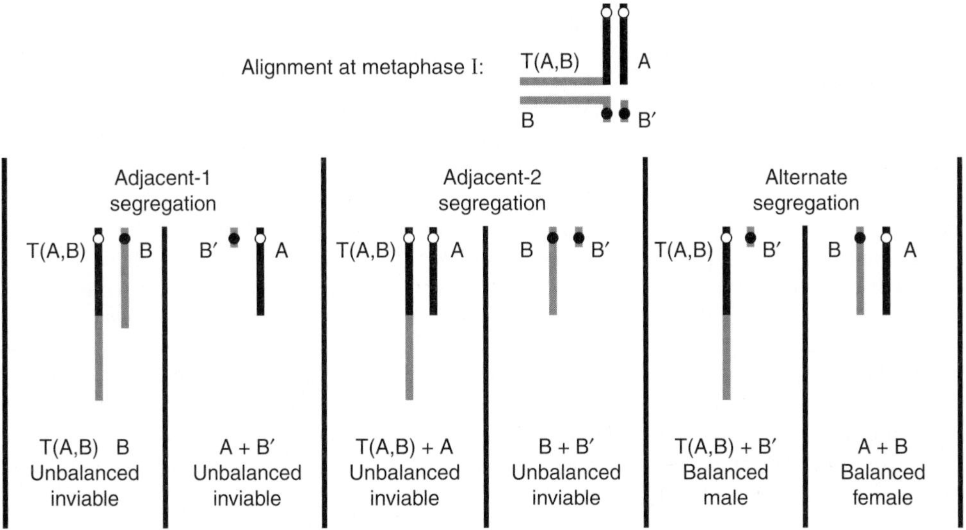

to male development, then males would begin to appear in the offspring of multiple females.

If males mated with translocation-bearing females, males and females would be produced in a 1:1 ratio. Balanced male gametes have T(A + B) + B′ and A + B, and translocation-bearing females produce A + B gametes. Zygotes would be half T(A + B) + B′ + A + B males and half A + A + B + B females. (To explore how the T(A + B) chromosome might eventually take on the role of the human X chromosome, consider the consequences to zygotes if the B′ element is lost during male meiosis.)

**16.34** **a.** The genotype of the $F_1$ peas will be *AAaa*.

**b.** If we label the four alleles in the $F_1$ $A^1, A^2, a^1$, and $a^2$, there are six possible gametes—$A^1A^2, A^1a^1, A^1a^2, A^2a^1, A^2a^2, a^1a^2$—giving 1/6 *AA*, 4/6 *Aa*, and 1/6 *aa*. As shown in the following Punnett square, selfing the $F_1$ gives a phenotypic ratio of 35 *A*– : 1 *aa*.

| | 1/6 *AA* | 4/6 *Aa* | 1/6 *aa* |
|---|---|---|---|
| 1/6 *AA* | 1/36 *AAAA* | 4/36 *AAAa* | 1/36 *AAaa* |
| 4/6 *Aa* | 4/36 *AAAa* | 16/36 *AAaa* | 4/36 *Aaaa* |
| 1/6 *aa* | 1/36 *AAaa* | 4/26 *Aaaa* | 1/36 *aaaa* |

**16.38** The initial allopolyploid will have 17 chromosomes. After doubling, the somatic cells will have 34 chromosomes.

## Chapter 17 Regulation of Gene Expression in Bacteria and Bacteriophages

**17.2** Allolactose and tryptophan are effector molecules that regulate the *lac* and *trp* operons, respectively. Effectors cause allosteric shifts in repressor proteins to alter their affinity for operator sites in DNA. When allolactose binds to the *lac* repressor, it loses its affinity for the *lac* operator, inducing transcription at the *lac* operon. When tryptophan interacts with the *trp* aporepressor, it is converted to an active repressor that can bind the *trp* operator, repressing transcription at the *trp* operon.

**17.5** A constitutive phenotype results from a $lacI^-$ or $lacO^c$ mutation.

**17.6** **a.** Since *lacA*, *lacY*, and *lacZ* are structural genes, loss-of-function mutations in them would prevent the bacterium from taking up and metabolizing lactose and show a *lac* phenotype. Since loss-of-function mutations in *lacI* would eliminate Lac repressor synthesis and allow synthesis of the structural genes even when lactose was not present (but in the absence of glucose), this mutation would not show a *lac* phenotype.

**b.** Loss-of-function mutations in the structural genes *trpA*, *trpB*, *trpC*, and *trpD* would show a *trp* phenotype while such mutations in *trpR*, which encodes the Trp aporepressor protein, would not. Without aporepressor sythesis, the *trp* operon would be transcribed even if tryptophan were present, though transcription would not be at maximal levels due to attenuation.

**c.** Loss-of-function mutations in the structural genes *araB*, *araA*, and *araD* would show an *ara* phenotype as the proteins needed to metabolize arabinose would not be synthesized. A loss-of-function mutation in *araC* would eliminate synthesis of the regulatory protein AraC. AraC plays two roles: it serves as a repressor of the operon when arabinose is absent and it serves as an activator when arabinose is present. In an *araC* loss-of-function mutation, the activator function would be absent, so the structural genes needed to metabolize arabinose would not be synthesized and the mutation would show an *ara* phenotype.

**17.8** The partial diploid genotype is $lacI^+\ lacO^c\ lacP^+\ lacZ^+\ lacY^-/lacI^+\ lacO^+\ lacP^+\ lacZ^-\ lacY^+$. (Only one $lacI^+$ gene is required, so one of the repressor genes may be $lacI^-$)

**17.10** The answer is given in Table 17.A.

**17.11** The *CAP*, in a complex with cAMP, is required to recruit RNA polymerase binding to the *lac* and *ara* promoters. RNA polymerase binds each promoter only in the absence of glucose. For the *lac* promoter, it occurs only if the operator is not occupied by repressor (i.e., lactose is present). For the *ara* promoter, it occurs only if AraC is not bound to the inducer site, $araI_1$ (i.e., arabinose is present). A loss-of-function mutation in the *CAP* gene, then, would render both the *lac* and *ara* operons incapable of expression because RNA polymerase would be unable to recognize the promoter. A constitutive mutation in the *CAP* gene would not affect the expression of these operons for two reasons. First, it is the binding of the CAP–cAMP complex, not CAP alone, that facilitates RNA polymerase binding to a promoter. In the presence of glucose, no new cAMP is produced, so the operons will still be subject to catabolite repression. Second, unless lactose (for the *lac* operon) or arabinose (for the *ara* operon) is present, the CAP–cAMP complex binding will be blocked by repressor molecules.

**17.13** **a.** A DNase protection experiment is an *in vitro* method to identify DNA sites that are bound by a protein. After a purified protein is allowed to bind a DNA segment, the complex is treated with DNase and sequences unprotected by protein binding are digested. Then, the sequence of the protected region is determined. DNase protection experiments defined the location of the operator, promoter, and CAP–cAMP binding site.

**b.** See Figure 17.14, p. 503.

**c.** **i.** This deletion disrupts the operator, so the operon would be expressed constitutively.

**ii.** A −12 transversion alters the −10 promoter consensus sequence, possibly decreasing the efficiency of transcription initiation. The operon may still be coordinately induced, but there will be diminished levels of β-galactosidase, permease, and transacetylase activity.

**iii.** A −69 transversion alters the consensus sequence for the CAP-binding site. If CAP–cAMP is unable to bind the CAP site, RNA polymerase will not be recruited to the promoter and the operon will not be coordinately induced.

**iv.** A +28 transition alters the Shine–Dalgarno sequence, and by affecting translation initiation, could result in diminished or absent β-galactosidase, and, due to polar effects, diminished or absent expression of permease and transacetylase.

**v.** A +9 transition alters the operator. It could either have no effect, cause the repressor to have more affinity for the operator (preventing coordinate induction in a *cis*-dominant manner), or cause the repressor to have less affinity for the operator (leading to constitutive expression).

**d.** None of the mutants will prevent catabolite repression.

**17.17** For a wild-type *trp* operon, the absence of tryptophan results in antitermination; that is, the structural genes are transcribed and the tryptophan biosynthetic enzymes are made. This occurs because a lack of tryptophan results in the absence of, or at least at a very low level of, Trp–tRNA.Trp. In turn, this causes the ribosome translating the leader sequence to stall at the Trp codons (see Figure 17.17a, p. 506). When the ribosome is stalled at the Trp codons, the RNA being synthesized just ahead of the ribosome by RNA polymerase assumes a particular secondary structure. This favors continued transcription of the structural genes by the polymerase. If the two Trp codons were mutated to stop codons,

Table 17.A

| | | Inducer Absent | | Inducer Present | |
|---|---|---|---|---|---|
| | Genotype | β-Galactosidase | Permease | β-Galactosidase | Permease |
| a. | $I^+\ P^+\ O^+\ Z^+\ Y^+$ | − | − | + | + |
| b. | $I^+\ P^+\ O^+\ Z^-\ Y^+$ | − | − | − | + |
| c. | $I^+\ P^+\ O^+\ Z^+\ Y^-$ | − | − | + | − |
| d. | $I^-\ P^+\ O^+\ Z^+\ Y^+$ | + | + | + | + |
| e. | $I^S\ P^+\ O^+\ Z^+\ Y^+$ | − | − | − | − |
| f. | $I^+\ P^+\ O^c\ Z^+\ Y^+$ | + | + | + | + |
| g. | $I^S\ P^+\ O^c\ Z^+\ Y^+$ | + | + | + | + |
| h. | $I^+\ P^+\ O^c\ Z^+\ Y^-$ | + | − | + | − |
| i. | $I^{-d}\ P^+\ O^+\ Z^+\ Y^+$ | + | + | + | + |
| j. | $I^-\ P^+\ O^+\ Z^+\ Y^+$ / $I^+\ P^+\ O^+\ Z^-\ Y^-$ | − | − | + | + |
| k. | $I^-\ P^+\ O^+\ Z^+\ Y^+$ / $I^+\ P^+\ O^+\ Z^-\ Y^-$ | − | − | + | + |
| l. | $I^S\ P^+\ O^+\ Z^+\ Y^-$ / $I^+\ P^+\ O^+\ Z^-\ Y^+$ | − | − | − | − |
| m. | $I^+\ P^+\ O^c\ Z^-\ Y^+$ / $I^+\ P^+\ O^+\ Z^+\ Y^-$ | − | + | + | + |
| n. | $I^-\ P^+\ O^c\ Z^+\ Y^-$ / $I^+\ P^+\ O^+\ Z^-\ Y^+$ | + | − | + | + |
| o. | $I^S\ P^+\ O^+\ Z^+\ Y^+$ / $I^+\ P^+\ O^c\ Z^+\ Y^+$ | + | + | + | + |
| p. | $I^{-d}\ P^+\ O^+\ Z^+\ Y^-$ / $I^+\ P^+\ O^+\ Z^-\ Y^+$ | + | + | + | + |
| q. | $I^+\ P^-\ O^c\ Z^+\ Y^-$ / $I^+\ P^+\ O^+\ Z^-\ Y^+$ | − | − | − | + |
| r. | $I^+\ P^-\ O^+\ Z^+\ Y^-$ / $I^+\ P^+\ O^c\ Z^-\ Y^+$ | − | + | − | + |
| s. | $I^-\ P^-\ O^+\ Z^+\ Y^+$ / $I^+\ P^+\ O^+\ Z^-\ Y^-$ | − | − | − | − |
| t. | $I^-\ P^+\ O^+\ Z^+\ Y^-$ / $I^+\ P^-\ O^+\ Z^-\ Y^+$ | − | − | + | − |

then the mutant operon would function constitutively in the same way as the wild-type operon in the absence of tryptophan. The ribosome would stall in the same place, and antitermination would result in transcription of the structural genes.

For a wild-type *trp* operon, the presence of tryptophan turns off transcription of the structural genes. This occurs because the presence of tryptophan leads to the accumulation of Trp–tRNA.Trp, which allows the ribosome to read the two Trp codons and stall at the normal stop codon for the leader sequence. When stalled in that position, the antitermination signal cannot form in the RNA being synthesized; instead, a termination signal is formed, resulting in the termination of transcription. In a mutant *trp* operon with two stop codons instead of the Trp codons, the stop codons cause the ribosome to stall, even though tryptophan and Trp–tRNA.Trp are present. This results in an antitermination signal and transcription of the structural genes.

In sum, in both the presence and the absence of tryptophan, the mutant *trp* operon will not show attenuation. The structural genes will be transcribed in both cases, and the tryptophan biosynthetic enzymes will be synthesized.

**17.18** *1*: If the aporepressor cannot bind to tryptophan, it will not be converted to an active repressor when tryptophan is present. This will lead to constitutive expression of tryptophan synthetase: In medium without tryptophan, expression will be the same as in the wild type; In medium with tryptophan, expression will be reduced only through attenuation, and so it will be about seventy-fold more than in the wild type.

*2*: The *trp* operon will exhibit constitutive expression, so mutant *2* will show the same expression patterns as mutant *1*.

*3*: The *trpE* gene is the first gene transcribed in the *trp* operon (see Figure 17.15, p. 504). A nonsense mutation would result in the translation termination while the ribosome is within the *trpE* coding region of *trp* polycistronic mRNA. Termination of translation in a region other than near the 3′ end of the *trpE* coding region could diminish the efficiency with which translation is reinitiated at downstream cistrons. This would result in diminished or absent translation of *trpB* and *trpA*, which encode tryptophan synthetase. Therefore, in medium without tryptophan where the operon is not repressed, mutant *3* would produce diminished levels of tryptophan synthetase compared to wild-type cells. In medium with tryptophan, the levels will be the same as in wild-type cells (very low).

*4*: Trp–tRNA.Trp molecules are needed to attenuate transcription at the *trp* operon. Therefore, if the levels of Trp–tRNA.Trp are always low, transcription of the *trp* operon will not be attenuated even when tryptophan levels are high. Therefore, in medium with tryptophan, mutant *4* will have about eight- to tenfold higher levels of tryptophan synthetase than will wild-type cells. In medium without tryptophan, attenuation does not occur, so mutant *4* will have levels of tryptophan synthetase that are similar to the wild type.

*5*: In mutant *5*, the 3:4 attenuator structure shown in Figure 17.17, p. 506 will not form, so attenuation will not occur when tryptophan levels are high. Tryptophan synthetase levels will be the same as in mutant *4*.

**17.19** If the products of the *ilvGMEDA* operon were required for the synthesis of the branched chain amino acids leucine, isoleucine, and valine, then it would make sense that attenuation of the operon is relieved by low levels of Leu–tRNA, Ile–tRNA, or Val–tRNA. Use attenuation at the *trp* operon as a model to hypothesize how this could occur. One hypothetical mechanism is that the mRNA leader sequence containing multiple codons for leucine, isoleucine, and valine forms a set of secondary structures that control the rate of transcription and whether transcripts are terminated in this region. During transcription of the leader sequence, regions within it pair to form a secondary structure—a pause signal—that causes RNA polymerase to pause. This pause allows the ribosome to load onto the mRNA and begin translating the leader peptide in close proximity to RNA polymerase. If the cells are starved for one or more of the three amino acids, levels of their aminoacyl–tRNAs will be low and the ribosome will pause at the codons for one or more of these amino acids. The pausing of the ribosome leads the mRNA to adopt a secondary structure that serves as an antitermination signal allowing RNA polymerase to transcribe past the attenuator and transcribe the operon's structural genes. If leucine, isoleucine, and valine are present in high enough amounts so that a paucity of Leu–tRNA, Ile–tRNA, or Val–tRNA does not cause the ribosome to pause in the region of the leader peptide, the mRNA adopts a different secondary structure that attenuates transcription by serving as a signal for transcription termination.

Evidence supporting this hypothesis would come from the analysis of mutants that failed to attenuate the expression of structural genes within the *ilvGMEDA* operon. If the hypothesis is correct, these mutations should destabilize the secondary structures predicted to form by base pairing within the leader transcript. Additional evidence would come from DNA manipulations in which the DNA sequences for each of the Leu, Ile, or Val codons within the leader peptide mRNA were changed to other amino acids. If the hypothesis is correct, the *ilvGMEDA* operon would now be attenuated by changing the levels of the amino acid(s) now specified by the altered set of codons and not by leucine, isoleucine, or valine.

**17.23** The *cI* gene product is a repressor protein that acts to keep the lytic functions of the phage repressed when λ is in the lysogenic state. A *cI* mutant strain would lack the repressor and be unable to repress lysis, so that the phage would always follow a lytic pathway.

**17.24** The *CAP* gene product functions as a positive regulator; the *lacI*, *cI*, and *trpR* gene products function as negative regulators; and the *araC* gene product can function as both. Only the CAP protein interacts directly with RNA polymerase. A CAP dimer bound to cAMP serves to positively regulate operons related to the catabolism of sugars other than glucose by binding to a CAP site in DNA and recruiting RNA polymerase to promoters. The *lacI*, *cI*, and *trpR* genes all produce repressor molecules that bind to operators near promoters. When bound to the operator sequences, they block the binding of RNA polymerase. The Lac repressor binds to the *lac* operator blocking RNA polymerase from binding the *lac* promoter. The product of the *cI* gene encodes a repressor that binds to two operator regions, $O_L$ and $O_R$, in phage λ. These overlap the $P_L$ and $P_R$ promoters and so repressor binding prevents the transcription of phage λ early operons from these promoters. The *trpR* gene produces an aporepressor protein. When tryptophan is present, it acts as an effector molecule to convert the aporepressor to an active Trp repressor. The active repressor binds to the *trp* operator and prevents transcription initiation. At the *ara* operon, AraC can serve as a negative regulator (repressor) or a positive regulator (activator) depending on whether arabinose is present. In the absence of arabinose, one subunit of AraC binds to the inducer site, $araI_1$, while the other subunit binds to the operator, $araO_2$. This causes the DNA to form a loop that blocks cAMP–CAP from binding to the CAP site so transcription cannot be initiated. AraC can also serve as a positive regulator because it undergoes an allosteric shift when bound by arabinose. In this form, the subunit of AraC bound to $araO_2$ is released and binds instead to $araI_2$ while the other subunit remains bound to $araI_1$. DNA no longer forms a loop, so cAMP–CAP can bind the CAP site to recruit RNA polymerase for transcription of the *ara* operon.

**17.25** **a.** Intact LexA binds to DNA sequences within promoters to block transcription, so functions as a negative regulatory protein.

**b.** It functions as an operator sequence.

**c.** Since the gene affected by mutant *A* lacks a consensus sequence able to bind LexA, it will be transcribed constitutively. Though it will be active and provide its "normal" function during an SOS response, it will not be repressed after the severe DNA damage is repaired by the SOS response. Since the SOS response is itself a mutagenic response, the constitutive expression of an SOS-response function in mutant *A* could cause it to show a mutator phenotype (see Chapter 7).

When the SOS response is triggered, RecA activation results in the LexA protein cleaving itself. Since cleaved LexA cannot bind DNA, and since the equilibrium between bound and unbound LexA is determined by the affinity of LexA for its binding sites, LexA self-cleavage leads to the additional release of DNA-bound LexA. This allows RNA polymerase access to the promoters of the genes repressed by LexA. In mutant *B*, LexA has a greater affinity to a LexA-binding site. It will be released from the gene more slowly during an SOS response and RNA polymerase will not gain access to the gene's promoter as readily as it normally would. This would delay the implementation of the SOS response. If this delay contributed to the cell's inability to recover from severe DNA damage, the cell would die.

**d.** Since the mutant LexA binds DNA more tightly than the normal LexA protein, it will replace normal LexA at its binding sites. Since the mutant LexA does not undergo self-cleavage, the genes normally activated in the SOS response would remain repressed following severe DNA damage. Therefore, the mutant would act as a dominant suppressor of the SOS response and show increased sensitivity to mutagenic agents such as UV light.

**17.28** See the following Table 17.B.

Table 17.B

| Mutant | Molecular Phenotype | Lytic Growth | Lysogenic Growth | Inducible by UV Light |
|---|---|---|---|---|
| *1* | The Cro protein is unable to bind DNA. | no | yes | no |
| *2* | The N protein does not function. | no | no | no |
| *3* | The cII protein does not function. | yes | no | yes |
| *4* | The Q protein does not function. | no | yes | no |
| *5* | $P_{RM}$ is unable to bind RNA polymerase. | yes | no | yes |

**Chapter 18** Regulation of Gene Expression in Eukaryotes

**18.3** See Table 18.A.

**18.5 a.** The disappearance of the 4-kb band indicates that the promoter region has a DNase I hypersensitive site—a less highly coiled site where DNA is more accessible to DNase I for digestion. Since DNase I digestion produces a 2-kb band, the site lies near the middle of the 4-kb *Eco*RI fragment. The 3-kb band does not diminish in intensity except at the highest DNase I concentration, so the region of the gene containing the 3-kb *Eco*RI fragment is more highly coiled by nucleosomes.

**b.** Following the ecdysone pulse, increased concentrations of DNase I lead to the disappearance of both the 4- and 3-kb bands, indicating that DNase I has increased access and the gene is less tightly coiled during transcription. The appearance of low molecular weight digestion products indicates that DNase I cannot access some regions of the 4- and 3-kb *Eco*RI fragments. These regions may be bound by proteins such as general transcription factors.

**18.6 a.** Histones repress gene expression, so if they are present on DNA, promoter-binding proteins cannot bind promoters and transcription cannot occur.

**b.** Histones will compete more strongly for promoters than will promoter-binding proteins, so transcription will not occur.

**c.** If promoter-binding proteins are already assembled on promoters, nucleosomes will be unable to assemble on these sites, so transcription will occur.

**d.** Enhancer-binding proteins will help promoter-binding proteins to bind promoters even in the presence of histones, so transcription will occur.

**18.8** The data indicate that the synthesis of ovalbumin is dependent upon the presence of the hormone estrogen. These data do not address the mechanism by which estrogen achieves its effects. Theoretically, it could act: (1) to increase transcription of the ovalbumin gene by binding to an intracellular receptor that, as an activated complex, stimulates transcription at the ovalbumin gene; (2) to stabilize the ovalbumin precursor mRNA; (3) to increase the processing of the precursor ovalbumin mRNA; (4) to increase the transport of the processed ovalbumin mRNA out of the nucleus; (5) to stabilize the mature ovalbumin mRNA once it has been transported into the cytoplasm; (6) to stimulate translation of the ovalbumin mRNA in the cytoplasm; or (7) to stabilize (or process) the newly synthesized ovalbumin protein. Experiments in which the levels of ovalbumin mRNA were measured have shown that the production of ovalbumin mRNA is primarily regulated at the level of transcription.

**18.10** Gene silencing occurs through changes in chromatin structure to produce heterochromatin, methylation of cytosines in the promoter upstream of a gene, and genomic imprinting associated with the inheritance of specific methylated sequences from one parent. In each case, chromatin remodeling leads to a promoter-inaccessible conformation so that a gene or zone of genes can no longer be transcribed. Heterochromatin formation is controlled by the acetylation and deacetylation of core histones and by ATP-dependent nucleosome remodeling complexes. Methylation of cytosines within CpG islands results in transcriptional repression via the recruitment of histone deacetylases to cause chromatin remodeling. Genomic imprinting is associated with the inheritance of specific methylated DNA sequences from one parent. Chromatin remodeling of the region containing the methylated DNA sequences blocks gene expression, so that only the gene inherited from the other parent is expressed.

**18.13 a.** In fragile X syndrome, the expanded CGG repeat results in hypermethylation and transcriptional silencing. A CAG codon specifies glutamine, so in Huntington disease, the expanded CAG repeat results in the inclusion of a polyglutamine stretch within the huntingtin protein. This causes it to have a novel, abnormal function.

**b.** A heterozygote with a CGG repeat expansion near one copy of the *FMR-1* gene will still have one normal copy of the *FMR-1* gene. The normal gene can produce a normal product, even if the other is silenced. (The actual situation is made somewhat more complex by the process of X inactivation in females, but in general one would expect that a mutation that caused

Table 18.A

| | Transcription Level Following Shift from Permissive to Restrictive Temperature for Gal80p in the Presence of: | | | |
|---|---|---|---|---|
| Gene | Glucose and Galactose | Glucose Only | Galactose Only | Neither Glucose Nor Galactose |
| *MIG1* | no change (on) | no change (on) | no change (presumably on) | no change (presumably on) |
| *GAL1* | no change (off) | no change (off) | no change (on) | increases (previously off) |
| *GAL4* | no change (off) | no change (off) | no change (on) | no change (on) increases |
| *GAL7* | no change (off) | no change (off) | no change (on) | (previously off) increases |
| *GAL10* | no change (off) | no change (off) | no change (on) | (previously off) |

transcriptional silencing of one allele would not affect a normal allele on a homolog.) In contrast, a novel, abnormal protein is produced by the CAG expansion in the disease allele in Huntington disease. Since the disease phenotype is due to the presence of the abnormal protein, the disease trait is dominant.

**c.** Transcriptional silencing may require significant amounts of hypermethylation, and so require more CGG repeats for an effect to be seen. In contrast, protein function may be altered by a stretch of more than 36 glutamines.

**18.15 a.** Four different protein isoforms differing in their C-terminus are produced.

**b.** The cDNA structures indicate that alternative mRNA splicing is used to generate the different protein isoforms. Specifically, alternative 5′ splice sites are used: the last exon of cDNA 4 contains a 5′ splice site that is used by cDNAs 1, 2, and 3; the last exon of cDNA 1 contains a 5′ splice site that is used by cDNAs 2 and 3; and the last exon of cDNA 2 contains a 5′ splice site that is used by cDNA 3.

**c.** Use RNA interference (RNAi) methods. Synthesize short, double-stranded RNAs targeted to regions specific to each of the cDNAs. Inject these individually into the (model) organism and examine the organism for phenotypic alterations. Phenotypes caused by the injection of a short double-stranded RNA targeted to a specific mRNA isoform most likely arise because of the posttranscriptional gene silencing of that mRNA and a decrease in the synthesis of the protein it encodes.

**18.18 a.** The *cortex* and *grauzone* mutants affect how much protein is produced by the maternally deposited *bicoid* and *toll* mRNAs, so they most likely affect how efficiently ribosomes select these mRNAs for translation. The translation of maternally deposited mRNAs increases significantly following fertilization. Prior to fertilization, proteins bind to the mRNAs to protect them and inhibit their translation. Subsequent to fertilization, translation is regulated by controlling the increasing the length of the poly(A) tail of maternally deposited mRNAs. Maternally deposited mRNAs generally have shorter poly(A) tails (15–90 As) than do active mRNAs (100–300 As). If the *cortex* and *grauzone* genes encoded proteins that functioned in protecting and limiting translation of the *bicoid* and *toll* mRNAs, deficits in *cortex* and *grauzone* would most likely result in either decreased amounts of these mRNAs or their increased translation. In contrast, if the *cortex* and *grauzone* genes function in the post-fertilization lengthening poly(A) tails, mutations in these genes would lead to decreased protein synthesis from mRNAs deposited with shorter poly(A) tails. Therefore, one hypothesis is that *cortex* and *grauzone* function to elongate poly(A) tails of maternally deposited mRNAs following fertilization.

**b.** The efficient translation of *nanos* mRNAs does not require lengthening of its poly(A) tail using the functions provided by *cortex* and *grauzone*. This could be because *nanos* mRNAs are deposited into the embryo with longer poly(A) tails, or because the post-fertilization poly(A)-tail elongation of *nanos* mRNAs is accomplished using functions provided by other genes.

**c.** In wild-type embryos, the *bicoid* and *toll* mRNAs would have short poly(A) tails (15–90 As) prior to fertilization and longer poly(A) tails (100–300 As) after fertilization. The mRNAs produced by *nanos* would have the same poly(A) tail length before and after fertilization (presumably long; 100–300 As). In *cortex* and *grauzone* embryos, the *bicoid*, *toll*, and *nanos* mRNAs would have the same poly(A) tail length before and after fertilization. The mRNAs of *bicoid* and *toll* would be short (15–90 As), while the mRNAs of *nanos* would (presumably) be long (100–300 As).

**d.** Similar amounts of Bicoid protein will be produced in i, ii, and iii; in ii, the poly(A) tail can be lengthened by wild-type *cortex* and *grauzone* functions; in iii, the poly(A) tail can be efficiently translated without further lengthening. Less Bicoid protein will be produced in iv, since the *bicoid* mRNA has a short poly(A) tail that cannot be lengthened due to a deficit in *cortex* function.

**18.20 a.** In bacterial operons, a common regulatory region controls the production of single mRNA from which multiple protein products are translated. These products function in a related biochemical pathway. Here, two proteins that are involved in the synthesis and packaging of acetylcholine are both produced from a common primary mRNA transcript.

**b.** Unlike the proteins translated from an mRNA synthesized from a bacterial operon, the protein products produced at the *VAChT/ChAT* locus are not translated sequentially from the same mRNA. Here, the primary mRNA appears to be alternatively processed to produce two distinct, mature mRNAs. These mRNAs are translated starting at different points, producing different proteins.

**c.** At least two mechanisms are involved in the production of the different ChAT and VAChT proteins: alternative mRNA processing and alternative translation initiation. After the first exon, an alternative 3′ splice site is used in the two different mRNAs. In addition, different AUG start codons are used.

### Chapter 19 Genetic Analysis of Development

**19.3** This experiment demonstrates the phenomenon of *determination* and the point at which it occurs during development. The tissue taken from the blastula or gastrula has not yet been committed to its final differentiated state in terms of its genetic programming; that is, it has not yet been *determined*. Thus, when the tissue is transplanted into the host, it adopts the fate of nearby tissues and differentiates in the same way that they do. Presumably, cues from the tissue surrounding the transplant determine its fate. In contrast, tissues in the neurula stage are stably determined. By the time the neurula developmental stage has been reached, a developmental program has been set. In other words, the fate of neurula tissue transplants is *determined*. Upon transplantation, they will differentiate according to their own set genetic program. Tissue transplanted from a neurula to an older embryo cannot be influenced by the surrounding tissues. It will develop into the tissue type for which it has been determined, in this case, an eye.

**19.6 a.** Based on the work of Wilmut and his colleagues, the nose cells would first be dissociated and grown in tissue culture. The cells would be induced into a quiescent state (the $G_0$ phase of the cell cycle) by reducing the concentration of growth serum in the medium. Then they would be fused with enucleated oocytes from a donor female and allowed to grow and divide by mitosis to produce embryos. The embryos would be implanted into a surrogate female. After the establishment of pregnancy, its progression would need to be maintained.

**b.** While the nuclear genome would generally be identical to that in the original nose cell, cytoplasmic organelles presumably would derive from those in the enucleated oocyte. Therefore, the mitochondrial DNA would not derive from the original leader. In addition, because telomeres in an older individual are shorter, one might expect the telomeres in the cloned leader to be those of an older individual.

**c.** In mature B cells, DNA rearrangements at the heavy and light chain immunoglobulin genes have occurred. One would expect the cloned leader to be immunocompromised, as he would be unable to make the wide spectrum of antibodies present in a normal individual.

**d.** It is likely that the cloning process will be very inefficient, with most clones dying before or soon after birth. The surviving clones are also likely to differ in body shape and personality, and they are unlikely to be normal since a nucleus donated from the differentiated nose cell is unlikely to be completely reprogrammed. One suggestion is for the government to hire a good plastic surgeon to alter the appearance of a good actor able to assume the role of the totalitarian leader.

**e.** There is no way to predict the psychological profile of the cloned leader based on his genetic identity. Even identical twins, who are genetically more identical than such a clone, do not always share behavioral traits.

**19.7 a.** Since Prometea's birth mother donated a nucleus and the donor egg was from a slaughtered horse, Prometea is identical to her birth mother with respect to her nuclear, but not mitochondrial genome.

**b.** Since the donor nucleus used to form Prometea is from the birth mother, Prometea and her birth mother will both be Haflinger horses. Since Prometea has the same phenotype as a sibling produced by a normal mating between her birth mother and a Haflinger stallion, the coat color phenotype cannot be used to demonstrate that Prometea is a clone of her birth mother. DNA testing would provide more compelling evidence. If Prometea is a clone having her mother's nuclear genome, she should show the same pattern of DNA sequence variation as her mother. Assess this by comparing the two horses using a panel of polymorphic STR markers (see Chapter 10, p. 272).

**c.** Nuclei taken from adult cells must be reprogrammed before the process of determination and differentiation can be started over. That only 22 of the 814 embryos survived 7 days to reach the blastocyst stage suggests the hypothesis that there was incomplete reprogramming of gene expression. If this were the case, the cloned embryos would be unable to recapitulate developmental processes and there would be poor cloning success. To test this hypothesis, establish cell lines from specific types of tissues biopsied from similarly staged normal and cloned embryos. Use DNA microarray analysis of gene expression in these lines to evaluate whether the gene expression profiles in cloned embryos are similar to those of normal embryos. Also, evaluate the patterns of DNA methylation, which influence patterns of transcription. Different patterns of gene expression and DNA methylation in cell lines established from cloned and normal embryos would provide support for the hypothesis that the nuclei underwent incomplete reprogramming of gene expression.

**19.11** There are several possibilities. One is that the $\gamma$-globin genes in bone marrow are under negative regulation by $\beta$ globin (or some metabolite of it). When $\beta$ globin is not formed, the $\gamma$-globin gene is derepressed.

**19.12** Polytene chromosomes occur in some terminally differentiated cells in dipteran insects such as *Drosophila*. Polytene chromosomes are formed by endoreduplication, in which repeated cycles of chromosome duplication occur without nuclear division or chromosome segregation. Since they can be 1,000 times as thick as the corresponding chromosomes in meiosis or in the nuclei of normal cells, they can be stained and viewed with a light microscope. Distinct bands, or chromomeres, are visible, and genes are located both in bands and in interband regions. A puff results when a gene in a band or interband region is transcribed at very high levels in a particular developmental stage. Puffs are accompanied by a loosening of the chromatin structure that allows for efficient transcription of a particular DNA region. When increased transcriptional activity of a gene ceases at a later developmental stage, the puff disappears, and the chromosome resumes its compact configuration. In this way, the appearance and disappearance of puffs provides a visual representation of differential gene activity during development.

**19.14** Experiment A results in all of the DNA becoming radioactively labeled. The distribution of radioactive label throughout the polytene chromosomes indicates that DNA is a fundamental and major component of these chromosomes. The even distribution of label suggests that each region of the chromosome has been replicated to the same extent. This provides support for the contention that band and interband regions are the result of different types of packaging, not of different amounts of DNA replication. Experiment B results in the radioactive labeling of RNA molecules. The finding that label is found first in puffs indicates that these are sites of transcriptional activity that arise from molecules that are in the process of being synthesized. The later appearance of label in the cytoplasm reflects the completed RNA molecules that have been processed and transported into the cytoplasm, where they will be translated. Experiment C provides additional support for the hypothesis that transcriptional activity is associated with puffs. The inhibition of RNA transcription by actinomycin D blocks the appearance of signal over puffs, indicating that it blocks the incorporation of $^3$H-uridine into RNA in puffed regions. The fact that the puffs are much smaller indicates that the puffing process itself is associated with the onset of transcriptional activity for the genes in a specific region of the chromosome.

**19.15** The ability to make $10^6$ to $10^8$ different antibodies arises from the combinatorial way in which antibody genes are generated in different antibody-producing cells during their development, and not from the existence of this many separate antibody genes in each and every mammalian cell. A template that exists in germ-line cells is processed differently during the development of different antibody-producing cells to generate antibody diversity.

Antibody molecules consist of two light (L) chains and two heavy (H) chains. The amino acid sequence of one domain of each type of chain is variable, and generates antibody diversity. In the germ-line DNA of mammals, coding regions for these immunoglobulin chains exist in tandem arrays of gene segments. For light chains, there are many variable (V)-region gene segments, a few joining (J) segments, and one constant (C) gene segment. Somatic recombination during development results in the production of a recombinant V–J–C DNA molecule that, when transcribed, produces a unique functional L chain. From a particular gene in one cell, only one L chain is produced. A large number of L chains are obtained by recombining the gene segments in many different ways. Diversity in these L chains results from variability in the sequences of the multiple V segments, variability in the sequences of the four J segments, and variability in the number of nucleotide pairs deleted at the V–J joints. H chains are similar, except that several D (diversity) segments can be used between the V and J segments, increasing the possible diversity of recombinant H chain genes. The type of C gene segment chosen for the constant domains of the H chain determines whether the antibody is IgM, IgD, IgG, IgE, or IgA.

**19.18 a.** Each cell should exhibit green fluorescence, since the gene is constitutively expressed.

**b.** About half the cells will exhibit green fluorescence. If more than one *Xic* is present, X inactivation will occur on one *Xic*-containing chromosome. Either the X chromosome or the *Xic*-bearing autosome will be inactivated, at random. If the X chromosome is inactivated, the *gfp* gene will not be expressed.

**c.** The *Xist* gene on the autosome is being expressed. Since the cell exhibits green fluorescence, the X chromosome with the *gfp* gene is not inactivated and the *Xic*-bearing autosome is inactivated. The *Xist* gene on the autosome is transcribed, and its RNA coats the autosome to trigger the methylation of histone H3. This initiates chromatin remodeling to silence genes on the *Xic*-bearing autosome.

**19.19 a.** The X:A ratio is detected by interactions of the protein products of three X-linked numerator genes (*sis-a*, *sis-b*, *sis-c*) and one autosomal denominator gene (*dpn*). The numerator gene products can form either homodimers or heterodimers with the denominator gene product. When the X:A ratio is 2:2, an excess of numerator gene products leads to the formation of many homodimers. These serve as transcription factors to activate *Sxl* transcription from $P_E$. When the X:A ratio is 1:2, most numerator subunits are found in heterodimers, so *Sxl* transcription is not activated. Therefore, activation of *Sxl* at $P_E$ by the homodimers serves to detect the X:A ratio and leads to the early sex-specific synthesis of SXL protein.

**b.** Transcription at $P_E$ is essential to generate a functional SXL protein in animals with an X:A ratio of 2:2. It is not used in individuals with an X:A ratio of 1:2, so these animals will be unaffected, and differentiate as males. In individuals with an X:A ratio of 2:2, SXL initiates a cascade of alternative mRNA splicing at *Sxl*, *tra*, and *dsx* that leads to the implementation of female differentiation. If there is no transcription from $P_E$, no functional SXL protein will be produced in these animals, and a default set of splice choices at *Sxl*, *tra*, and *dsx* will be used. In principle this would lead to male differentiation in individuals with an X:A ratio of 2:2. However, SXL also prevents the translation of *msl-2* transcripts so that dosage compensation does not normally occur in individuals with an X:A ratio of 2:2. Without SXL, *msl-2* transcripts will be translated so dosage compensation will occur, leading to four doses of X-linked gene products. The imbalance in X and autosomal gene product dosage is likely to be lethal to these animals.

**c.** This *tra* mutant will eliminate functional TRA protein. Since functional TRA is not normally present in animals with an X:A ratio of 1:2, this mutation will have no effect on these animals—they will differentiate normally into males. TRA is normally present in individuals with an X:A ratio of 2:2, where it functions to regulate alternative splicing at *dsx* and produce DSX-F, which implements female differentiation by repressing male-specific gene expression. Without TRA, default splicing will occur at *dsx* and produce DSX-M, which implements male differentiation by repressing female-specific gene expression. Thus, animals with an X:A ratio of 2:2 will be males.

**d.** Animals with knockout mutations at *dsx* will produce neither DSX-M, which represses female-specific gene expression, nor DSX-F, which represses male-specific gene expression. Therefore, neither male- nor female-specific gene expression will be repressed, and both male and female differentiation pathways will proceed.

**19.20** The early SXL protein binds to the *Sxl* pre-mRNA to cause alternative splicing: exons E1 and 3 are skipped, and exons L1, 2, 4, 5, 6, 7, and 8 are included. The resulting mRNA produces a functional late SXL protein. In the absence of the early SXL protein, default splicing occurs to produce a transcript that includes exon 3. This exon has a stop codon in frame with the start codon at the beginning of exon 2, so no functional SXL protein is produced. The SXL protein regulates *tra* in a similar manner: SXL binds to the *tra* pre-mRNA to produce an mRNA that encodes an active TRA protein. In the absence of SXL, a default stop-codon-containing exon is included and an active TRA protein is not produced.

In contrast to its role in alternative mRNA splicing at *Sxl* and *tra*, the late SXL protein serves to block translation of *msl-2* transcripts. In XX animals (females), the SXL late protein binds to the transcript of *msl-2*. This blocks its translation so that no MSL2 protein is produced. As a result, dosage compensation does not occur. In XY animals (males) where SXL protein is not produced, the *msl-2* transcript is translated and dosage compensation occurs.

**19.22** Preexisting, maternally packaged mRNAs that have been stored in the oocyte are recruited into polysomes as development begins following fertilization.

**19.24 a.** The *yobo* gene is a maternal effect gene since all offspring of homozygous, but not heterozygous, *yobo* females have delayed development and abnormal body plans, even if they have a *yobo*$^+$ allele contributed by their father. Offspring of *yobo/yobo* mothers are missing a maternally deposited gene product required for normal embryonic development.

**b.** Since it is a maternal effect gene, the *yobo*$^+$ gene must be transcribed during oogenesis and its mRNA or protein product must be deposited into the developing oocyte prior to fertilization.

**c.** The *yobo* gene product is used after fertilization during embryonic development. The *yobo/yobo* offspring of *yobo*$^+$*/yobo* females mated to *yobo/yobo* or *yobo*$^+$*/yobo* males are normal because *yobo*$^+$*/yobo* females have the capacity to deposit some *yobo*$^+$ gene product into the developing embryo. The *yobo* maternal effect phenotypes of slow development and abnormal head and tail morphology suggest that *yobo*$^+$ normally functions to regulate the rate of development and patterning in the head and tail regions. This might be achieved by the *yobo*$^+$ gene product functioning directly in the anterior and posterior regions of the developing embryo during a period critical for the development of head and tail structures. Alternatively, given its more general effects on the rate of development, *yobo* might function more widely and indirectly impact the development of these structures.

**19.26**

| Mutant | Class |
|---|---|
| *a* | Segmentation gene (segment polarity) |
| *b* | Maternal effect gene (anterior-posterior gradient) |
| *c* | Segmentation gene (gap) |
| *d* | Homeotic (transforms cell fate from eye to wing) |
| *e* | Segmentation gene (gap) |

**19.28** The primary signal that leads to the differential expression of the 534 genes is a pulse of the steroid hormone ecdysone during the late larval period. It does not act directly to control transcription at each of the 534 genes, but rather triggers a regulatory cascade that leads to their differential expression. Ecdysone binds to a receptor protein, and this complex binds to both early-puffing genes and late-puffing genes. The complex turns on the early genes, some of which encode DNA-binding proteins that could serve as regulatory genes, and represses the late genes. As one of the early gene's products accumulates, it displaces the ecdysone-receptor complex from both early and late genes. This turns off the early genes and derepresses the late genes. In this manner, the ecdysone signal triggers a cascade of gene activation and repression that leads to the widespread differential gene expression associated with the metamorphosis of the larval worm into an adult fly.

**19.29 a.** Regulatory noncoding RNAs have been found at the *bithorax* complex, whose genes function to determine the identity of the fly's posterior segments. The noncoding RNAs appear to silence *Ubx*, one of the genes within the *bithorax* complex, in early embryos as part of an RNA interference system to ensure correct developmental timing of its expression.

**b.** A complex of a miRNA and several proteins (including Ago1) silences gene expression by binding to the 3′ untranslated region (UTR) of one or more target mRNAs. Base pairing between the miRNA and the mRNA leads to either translation inhibition or mRNA degradation.

**c.** The primary evidence that miRNA-mediated gene silencing is essential for normal development in invertebrates and vertebrates comes from the analysis of loss-of-function mutations in individual miRNAs and genes for key proteins involved in miRNA-mediated gene silencing, such as Dicer and Argonaute. These analyses have revealed that miRNAs regulate many aspects of somatic cell and germ-line development in both invertebrates and vertebrates. Evidence that miRNA function plays an essential role in vertebrate development comes from the finding that Dicer knockouts display early developmental arrest and lethality in both mice and zebrafish.

**19.31** Preexisting mRNA that was made by the mother and packaged into the oocyte prior to fertilization is translated up to the gastrula stage. After gastrulation, new mRNA synthesis is necessary for the production of proteins needed for subsequent embryonic development.

## Chapter 20 Genetics of Cancer

**20.5** If FeSV contributed to the feline sarcoma, FeSV should be found in the neoplastic tissues (muscle and bone marrow). The Southern blot provides this evidence: A 1.2-kb DNA fragment hybridizes to the *fes* cDNA probe in the lanes with DNA from muscle and bone marrow, and in the control lane with FeSV cDNA. The size difference between the 1.2-kb hybridizing fragment and the 1.0-kb *fes* proto-oncogene *Hin*dIII-cut cDNA probe reflects their different origins. The *fes* proto-oncogene is found normally in a cat, while the FeSV *fes* oncogene is found in a retrovirus. The size of the fragment in the retrovirus may reflect a polymorphic *Hin*dIII site and/or a gene rearrangement. The *fes* proto-oncogene normally functions in the cat, so it is DNA that should be present in all tissues. The 3.4-kb DNA fragment is found in all of the cat tissues, so it is likely to be the genomic sequence. Since the *fes* proto-oncogene cDNA has a 1.0-kb *Hin*dIII fragment, the mRNA of this gene is very likely spliced to remove a 2.4-kb intron.

**20.6** The high degree of conservation of proto-oncogenes suggests that they function in normal, essential, conserved cellular processes. Given the relationship between oncogenes and proto-oncogenes, it also suggests that cancer occurs when these processes are not correctly regulated.

**20.7 a.** Proto-oncogenes encode a diverse set of gene products that includes growth factors, receptor and nonreceptor protein kinases, receptors lacking protein kinase activity, membrane-associated GTP-binding proteins, cytoplasmic regulators involved in intracellular signaling, and nuclear transcription factors. These gene products all function in intercellular and intracellular pathways that regulate cell division and differentiation.

**b.** In general, mutations that activate a proto-oncogene convert it into an oncogene. Since (i), (iii), and (viii) cause a decrease in gene expression, they are unlikely to result in an oncogene. Since (ii) and (vii) could activate gene expression, they could result in an oncogene. Mutations (iv), (v), and (vi) cannot be predicted with certainty. The deletion of a 3′ splice site acceptor would alter the mature mRNA and possibly the protein produced, and it may or may not affect the function and regulation of the protein. Similarly, it is difficult to predict the effect of a nonspecific point mutation or a premature stop codon. The text presents examples in which these types of mutations have caused the activation of a proto-oncogene and resulted in an oncogene.

**20.8 a.** Increased transcription of the mRNA will lead to increased levels of the growth factor, which in turn will stimulate fibroblast growth and division.

**b.** Constitutive activation of a nonreceptor tyrosine kinase could lead to abberant, unregulated phosphorylation and activation of many different proteins, including growth factor receptors, that are involved in signaling cascades used in regulating cellular growth and differentiation.

**c.** When the growth factor EGF binds its membrane-bound receptor, it stimulates its autophosphorylation, which allows for Grb2 to bind and recruit SOS. SOS displaces GDP from Ras, a membrane-associated G protein, so that it can bind GTP. GTP-bound Ras recruits and activates Raf-1 to initiate the cytoplasmic MAP kinase signaling cascade. In turn, this signaling cascade activates transcription factors such as Elk-1 to induce transcription of cell cycle-specific target genes. Therefore, if a membrane-associated G protein were unable to hydrolyze GTP, it would be constitutively active and lead to constant expression of genes needed for cell cycle progression.

**20.10** A proto-oncogene can be changed into an oncogene if there is an increase in the activity of its gene product or an increase in gene expression that leads to an increased amount of gene product. This can result from point mutation, deletion, or gene amplification.

**20.12** One hypothesis is that the proviral DNA has integrated near the proto-oncogene, and the expression of the proto-oncogene has come under the control of promoter and enhancer sequences in the retroviral long terminal repeat (LTR). This could be assessed by performing a whole-genome Southern blot analysis to determine whether the organization of the genomic DNA sequences near the proto-oncogene has been altered.

**20.13** Tumor growth induced by transforming retroviruses results either from the activity of a single viral oncogene or from the activation of a proto-oncogene caused by the nearby integration of the proviral DNA. The oncogene can cause abnormal cellular proliferation via the variety of mechanisms discussed in the text. The expression of a proto-oncogene, normally tightly regulated during cell growth and development, can be altered if it comes under the control of the promoter and enhancer sequences in the retroviral LTR.

DNA tumor viruses do not carry oncogenes. They transform cells through the action of one or more genes within their genomes. For example, in a rare event, the DNA virus can be integrated into the host genome, and the DNA replication of the host cell may be stimulated by a viral protein that activates viral DNA replication. This would cause the cell to move from the $G_0$ to the S phase of the cell cycle.

For both transducing retroviruses and DNA tumor viruses, abnormally expressed proteins lead to the activation of the cell from $G_0$ to S and abnormal cell growth.

**20.14** Experimentally fuse cells from the two cell lines and then test the resultant hybrids for their ability to form tumors. If the uncontrolled growth of the tumor cell line was caused by a mutated pair of tumor suppressor alleles, then the normal alleles present in the normal cell line would "rescue" the tumor cell line defect. The hybrid line would grow normally and be unable to form a tumor. If the uncontrolled growth of the tumor cell line was caused by an oncogene, the oncogene would also be present in the hybrid cell line. The hybrid line would grow uncontrollably and form a tumor.

**20.15** Hereditary cancer is associated with the inheritance of a germ-line mutation; sporadic cancer is not. Consequently, hereditary cancer runs in families. For some cancers, both hereditary and sporadic forms exist, with the hereditary form being much less frequent. For example, retinoblastoma occurs when both normal alleles of the tumor suppressor gene *RB* are inactivated. In hereditary retinoblastoma, a mutated, inactive allele is transmitted via the germ line. Retinoblastoma occurs in cells of an *RB/rb* heterozygote when an additional somatic mutation occurs. In the sporadic form of the disease, retinoblastoma occurs when both alleles are inactivated somatically.

**20.18** **a.** Studies of hereditary forms of cancer have led to insights into the fundamental cellular processes affected by cancer. For example, substantial insights into the important role of DNA repair and the relationship between the control of the cell cycle and DNA repair have come from analyses of the genes responsible for hereditary forms of human colorectal cancer. For breast cancer, studying the normal functions of the *BRCA1* and *BRCA2* genes promises to provide substantial insights into breast and ovarian cancers.

**b.** Genetic predisposition for cancer refers to the presence of an inherited mutation that, with additional somatic mutations during the individual's life span, can lead to cancer. For diseases such as retinoblastoma, a genetic predisposition has been associated with the inheritance of a recessive allele of the *RB* tumor suppressor gene. Retinoblastoma occurs in *RB/rb* individuals when the normal allele is mutated in somatic cells and the pRB protein no longer functions. Because somatic mutation is likely, the disease appears dominant in pedigrees.

Although there is a substantial understanding of the genetic basis for cancer and the genetic abnormalities present in somatic cancerous cells, there are also substantial environmental risk factors for specific cancers. Environmental risk factors must be investigated thoroughly when a pedigree is evaluated for a genetic predisposition for cancer.

**c.** Multiple genetic changes occur during the formation of a tumor cell. In an individual with an inherited predisposition to a particular cancer, a genetic difference that contributes to the formation of a tumor cell has been inherited. If that genetic difference was inherited from just one parent and, together with additional genetic changes that may be influenced by the environment, it contributes to the appearance of a cancer, the inheritance of that single genetic change has predisposed that individual to cancer. Since the genetic change that predisposes the individual to cancer has been inherited from just one parent, the trait is dominant. However, the additional genetic changes required to produce a cancer may not always occur, and so not every individual with the inherited genetic change will develop cancer. Consequently, the trait shows reduced penetrance. For the trait to be considered recessive, an individual must have two recessive alleles, one inherited from each parent.

**20.19** Tissues are produced during organismal development through regulated cell proliferation. Proto-oncogenes normally contribute to this process by positively controlling cell growth and division. Once cells become highly differentiated and are fully functional in a tissue, they typically no longer divide, and proto-oncogenes are no longer active. A mutation causing the constitutive activation of an oncogene is a gain-of-function mutation. It would lead to unregulated cell proliferation and the failure of cells to undergo terminal differentiation. It is likely that such a mutation in a heterozygous state would lead to lethality during development, so it would be a dominant embryonic lethal mutation and would not be inherited. This is in contrast to recessive loss-of-function mutations at tumor suppressor genes, where the one normal gene copy in a heterozygote can function to suppress uncontrolled cell proliferation and allow for a normal phenotype (until the normal copy is spontaneously mutated later in development; see Question 20.16, p. 600).

**20.25** **a.**

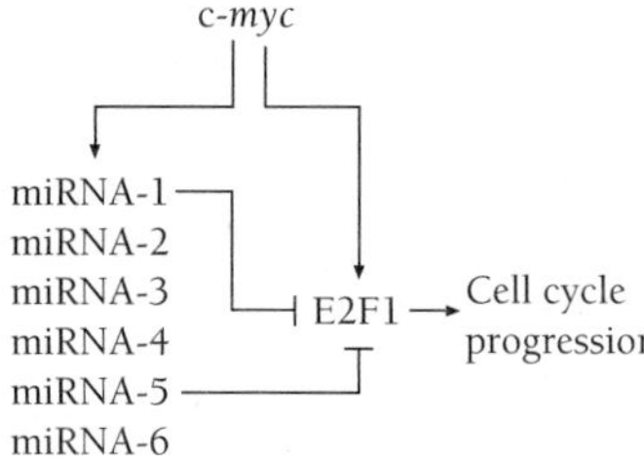


**b.** E2F1 and c-*myc* are proto-oncogenes, as they normally function to promote cell cycle progression, while the two miRNAs (arbitrarily chosen here as miRNA-1 and miRNA-5) that negatively regulate E2F1 are tumor suppressor genes, as they normally function to inhibit cell cycle progression.

**c.** The two miRNAs could act directly to silence E2F1 expression by base-pairing with sequences in the 3′ UTR of the E2F1 mRNA to inhibit its translation or target it for storage or degradation. Alternatively, they could act indirectly by silencing another gene whose product functions to positively regulate E2F1.

**d.** Different types of cancer, and different stages of a particular kind of cancer, are associated with different patterns of oncogene activation and tumor suppressor gene inactivation. Oncogenes and tumor suppressor genes act within regulatory gene networks. Sets of miRNAs act within these regulatory gene networks both as regulators of gene expression and as the regulatory targets of other genes. Therefore, one hypothesis why different cancer types show distinctive patterns of miRNA expression is that the pattern of miRNA expression is dependent on the pattern of oncogene activation and tumor suppressor gene inactivation in a particular type and stage of cancer. If this is correct, the pattern of miRNA expression could serve as a diagnostic indicator of a particular type and stage of cancer. Understanding how the miRNA expression pattern changes as a cancer progresses could provide insight into the specific changes in gene expression that occur during cancer progression and thereby define new therapeutic targets for halting or reversing the progression of a cancer.

**20.26** Apoptosis is programmed, or suicidal, cell death. Cells targeted for apoptosis are those that have high levels of DNA damage and so are at a greater risk for neoplastic transformation. (During the development of some tissues in multicellular organisms, cell death via apoptosis is a normal process.) Apoptosis is regulated by p53, among other proteins. In cells with large amounts of DNA damage, p53 accumulates and functions as a transcription factor to activate transcription of DNA repair genes and *WAF1*, whose product, p21, leads cells to arrest in $G_1$. If very high levels of DNA damage exist, p53 does not induce DNA repair genes and *WAF1*, but activates the *BAX* gene, whose product blocks the BCL-2 protein from repressing the apoptosis pathway. By blocking BCL-2 function, the apoptosis pathway is activated.

**20.27** The products of the normal alleles of mutator genes function in DNA repair processes. In the absence of a functional allele at such a gene, there is a dramatic increase in the accumulation of mutations. Some of these mutations will be at proto-oncogenes and lead to the formation of oncogenes, while others will be loss-of-function mutations at tumor suppressor genes. Therefore, error-prone DNA replication will lead to the formation of cancerous cells. Mutations in mutator genes give a phenotype of hereditary predisposition to cancer because just a single mutational event in a heterozygote will inactivate the remaining functional allele and result in error-prone DNA replication.

**20.31** Tumors result from multiple mutational events that typically involve both the activation of oncogenes and the inactivation of tumor suppressor genes. The analysis of hereditary adenomatous polyposis, an inherited form of colorectal cancer, has shown that the more differentiated cells found in benign, early stage tumors are associated with fewer mutational events, while the less differentiated cells found in malignant and metastatic tumors are associated with more mutational events. Although the path by which mutations accumulate varies between tumors, additional mutations that activate oncogenes and inactivate tumor suppressor genes generally lead to the breaking down of the multiple mechanisms that regulate growth and differentiation.

**20.32** **a.** The fact that some translocations are found as the only cytogenetic abnormality in certain cancers probably means they are a key event in tumor formation. It does not necessarily mean they are the primary cause of the tumor or the first of many mutational events.

**b.** A chimeric fusion protein may have different functional properties than do either of the two proteins from which it derives. If it results in the activation of a proto-oncogene product into a protein that has oncogenic properties, or if it results in the inactivation of a tumor suppressor gene product, it could play a key role in the genetic cascade of events leading to tumor formation.

**c.** Before drawing conclusions as to whether these chromosomal aberrations inactivate the function of tumor suppressor genes, or activate quiescent proto-oncogenes, it is necessary to have additional molecular information on the effects of the translocation breakpoints on specific transcripts. Finding that the translocation breakpoints result in a lack of gene transcription or in transcripts that encode nonfunctional products would support the hypothesis that the translocation inactivated a tumor suppressor gene. Finding that the translocation breakpoints result in activation of gene transcription or in the production of an active fusion protein would support the hypothesis that the translocation activated a previously quiescent proto-oncogene.

**d.** One hypothesis is that the various fusion proteins that result from different translocations involving the *EWS* gene somehow result in the transcription activation of different proto-oncogenes, and this leads to the different sarcomas that are seen. (Sarcomas are cancers found in tissues that include muscle, bone, fat, and blood vessels.)

**e.** If translocation breakpoints are conserved within a tumor type, molecular-based diagnostics can be developed to identify the breakpoints relatively quickly from a tissue biopsy. For example, if the genes at the breakpoints have been cloned, PCR methods can be used to address whether the gene is intact or disrupted, using the DNA from cells of a tumor biopsy. Primers can be designed to amplify different segments of the normal gene. Then, PCR reactions containing these primers and either normal, control DNA or tumor cell DNA can be set up to determine if each segment of a candidate gene is intact (a PCR product of the expected size is obtained) or disrupted (no PCR product will be obtained, because the gene has been rearranged).

Such molecular analyses would provide fast, accurate tumor diagnosis. If the different tumor types respond differentially to different regimens of therapeutic intervention, then a more rapid, unequivocal diagnosis of a particular tumor type should allow for the earlier prescription of a more optimized regime of therapeutic intervention. In addition, understanding the nature of the normal gene products of the affected genes may allow for the development of sarcoma-specific therapies.

**20.33** To proceed from $G_1$ into S, one or more $G_1$ cyclins bind to the cyclin-dependent kinase CDC28/cdc2 and activate it. The cyclin-dependent kinase then phosphorylates key proteins that are needed for progression into S. In the presence of heavy DNA damage, p53 is stabilized. The p53 protein acts as a transcription factor to activate *WAF1*, which produces p21. The p21 protein binds to cyclin–Cdk complexes and blocks the kinase activity required to activate the genes needed for the cell to make the transition through the cell cycle checkpoints, for example, from $G_1$ to S (see Question 20.22, p. 601). Thus, stabilization of p53 leads to cell cycle arrest at the $G_1$-to-S or other cell cycle checkpoints. This arrest allows the cell time to induce the necessary repair pathways and repair the DNA or, if damage is too severe, to undergo apoptosis.

**20.34** Direct-acting carcinogens are chemicals that bind to DNA and act as mutagens. Procarcinogens are chemicals that must be converted by normal cellular enzymes to become active carcinogens. These products, most of which also bind DNA and act as mutagens, are referred to as ultimate carcinogens.

## Chapter 21 Population Genetics

**21.1** Equate the frequency of each color with the frequency expected in Hardy–Weinberg equilibrium, letting $p = f(C^B)$, $q = f(C^P)$, and $r = f(C^Y)$.

Brown: $f(C^BC^B) + f(C^BC^P) + f(C^BC^Y) = p^2 + 2pq + 2pr = 236/500 = 0.473$
Pink: $f(C^PC^P) + f(C^PC^Y) = q^2 + 2qr = 231/500 = 0.462$
Yellow: $f(C^YC^Y) = r^2 = 33/500 = 0.066$

Now solve for $p$, $q$, and $r$, knowing that $p + q + r = 1$:

$$r^2 = 0.066, \text{ so } r = \sqrt{0.066} = 0.26$$

There are two approaches to solving for $q$. First, because $q^2 + 2qr = 0.462$, we can substitute $r = 0.26$, giving $q^2 + 2q(0.26) = 0.462$. Recognize this as a quadratic equation, set it equal to 0, and solve for $q$; that is, solve the equation $q^2 + 0.52q - 0.462 = 0$.

Solving the quadratic equation for $q$, we have

$$q = \frac{-0.52 \pm \sqrt{(0.52)^2 - 4(1)(-0.462)}}{2(1)} = 0.467$$

A second approach to solve for $q$ is to realize that

$$q^2 + 2qr = 0.462$$
$$r^2 = 0.066$$

Adding left and right sides of the equations together, we have

$$q^2 + 2qr - r^2 = 0.066 + 0.467$$
$$(q + r) = 0.528$$
$$q = 0.726 - r = 0.726 - 0.26 = 0.467$$

Since $p + q + r = 1$,

$$p = 1 - (q + r) = 1 - (0.26 + 0.467) = 0.273$$

**21.3** Let $c$ represent the ΔF508 mutant allele and $p$ its frequency, and let $C$ represent the normal allele and $q$ its frequency. In the sample of neonates tested, the frequency of $Cc$ carriers is $4/955 = 0.00402$, and the frequency of $CC$ homozygotes is $(995 - 4)/955 = 0.99598$. Assuming that the population is in Hardy–Weinberg equilibrium (individuals mate at random with respect to this allele, there is no selection, mutation or migration) these numbers also give the frequency of $Cc$ carriers ($= 2pq$) and $CC$ homozygotes ($= p^2$) in the general population. Since $p^2 = 0.99598$, $p = \sqrt{0.9959} = 0.99799$. Since $p + q = 1$, $q = 1 - 0.99799 = 0.00201$. The expected frequency of homozygotes is $f(cc) = p^2 = (0.00201) = 0.000004$. Therefore, the expected prevalence of homozygotes is one in $1/0.000004 =$ one in 250,000. This frequency is much lower than the prevalence of CF in the Indian subcontinent, for several possible reasons. One reasonable explanation is that additional mutations at the CFTR gene contribute to the incidence of CF. Indeed, the frequency of ΔF508 allele may be very different in this population than it is in non-Hispanic Caucasians. Alternatively, the sample assessed here is not representative of the population as a whole (genetic drift has occurred, and the population this sample is drawn from is not in Hardy–Weinberg equilibrium).

**21.6 a.** $\sqrt{0.16} = 0.40 = 40\% =$ frequency of recessive alleles: $1 - 0.4 = 0.6 = 60\% =$ frequency of dominant alleles; $2pq = (2)(0.4)(0.6) = 0.48 =$ probability of heterozygous diploids. Then, $(0.48)/[(2 \times 0.16) + 0.48] = 0.48/0.80 = 60\%$ of recessive alleles are heterozygotes.

**b.** if $q^2 = 1\% = 0.01$, then $q = 0.1$, $p = 0.9$, and $2pq = 0.18$ heterozygous diploids. Therefore, $(0.18)/[0.18 + 2(0.01)] = 0.18/0.20 = 0.90 = 90\%$ of recessive alleles would occur in heterozygotes.

**21.7** $2pq/q^2 = 8$, so $2p = 8q$; then $2(1 - q) = 8q$, and $2 = 10q$, or $q = 0.2$.

**21.10** Members of a population generally do not interbreed randomly for all traits. For example, humans mate preferentially for height, skin color, socioeconiomic status, and other traits. For many other traits however, mating is random. The Hardy–Weinberg law applies to any trait for which random mating occurs, even if mating is nonrandom for other traits.

**21.12 a.** Let $p$ equal the frequency of $S$ and $q$ equal the frequency of $s$. Then,

$$q = \frac{2(188)SS + 717Ss}{2(3{,}146)} = \frac{1{,}093}{6{,}292} = 0.1737$$

$$p = \frac{717Ss + 2(2{,}241)ss}{2(3{,}146)} = \frac{5{,}199}{6{,}292} = 0.8263$$

**b.**

| Class | Observed | Expected | $d$ | $d^2/e$ |
|---|---|---|---|---|
| $SS$ | 188 | 95 | +93 | 91.0 |
| $Ss$ | 717 | 903 | −186 | 38.3 |
| $ss$ | 2,241 | 2,148 | +93 | 4.0 |
| Total | 3,146 | 3,146 | 0 | 133.3 |

There is only one degree of freedom because the three genotypic classes are completely specified by two allele frequencies: $p$ and $q$ ($df$ = number of phenotypes − number of *alleles* = $3 - 2 = 1$). The $\chi^2$ value of 133.3, for one degree of freedom, gives $P < 0.0001$. Therefore, the distribution of gene types differs significantly from that expected if the population were in Hardy-Weinberg equilibrium.

**21.14** The population is in Hardy–Weinberg equilibrium, so neither the allele nor the zygote frequencies change from one generation to the next. Therefore, in two generations there should still be 60% type O individuals.

**21.15** Since the frequency of the trait is different in males than it is in females, the character might be caused by an X-linked recessive allele. If the frequency of this allele is $q$, females would occur with the character at a frequency of $q^2$, and males with the frequency of $q$. The frequency in males is 0.4, so we may predict that the frequency in females is $(0.4)^2 = 0.16$ if this is an X-linked gene. This result fits the data. Therefore, the frequency of heterozygous females is $2pq = 2(0.6)(0.4) = 0.48$. For X-linked genes, no heterozygous males exist.

**21.17** 64/10,000 women are color blind; that is, $0.0064 = q^2$, so $q = 0.08 =$ probability of a color-blind male.

**21.20**

$$q = \frac{u}{u + v} = \frac{6 \times 10^{-7}}{(6 \times 10^{-7}) + (6 \times 10^{-8})}$$

$$= \frac{6 \times 10^{-7}}{(6 \times 10^{-7}) + (0.6 \times 10^{-7})} = \frac{6}{6.6} = 0.91$$

Thus, the genotype frequencies are 0.0081 $AA$, 0.1638 $Aa$, and 0.8281 $aa$.

**21.23** $p'_x = mp + (1 + m)p$

Here, $p'_x = [20/(20 + 80)](0.50) + \{1 - [20/(20 + 80)]\}(0.70) = 0.66$

**21.26 a.** When selectively neutral, the alleles distribute themselves according to the Hardy–Weinberg law, so 0.25 are $AA$, 0.5 are $Aa$, and 0.25 are $aa$.

**b.** $q = 0.33$

**c.** $q = 0.66$

**21.28 a.** $q = 0.63$

**b.** $q = 0.64$

**c.** $q = 0.66$

**21.30** $q = u/s = (5 \times 10^{-5})/0.8 = 0.0000625$

**21.33 a.** The data fit the idea that a single *Bam*HI site varies. The probe is homologous to a region wholly within the 4.1-kb piece bounded on one end by the variable *Bam*HI site and on the other end by a constant site. When the variable site is present, the hybridized fragment is 4.1 kb. When the variable site is absent, the fragment extends to the next constant *Bam*HI site and is 6.7 kb long. People with only 4.1- or only 6.7-kb bands are homozygotes; people with both are heterozygotes.

**b.** The + allele of the variable site is present in $2(6) + 38 = 50$ chromosomes, and the − allele is present in $2(56) + 38 = 150$ chromosomes. Thus, $f(+)$ is 0.25 and $f(-)$ is 0.75.

**c.** If the population is in Hardy–Weinberg equilibrium, we would expect $(0.25)^2$ or 0.0625 of the sample to show only the 4.1-kb band. This would be 6.25 individuals. We observed 6. We expect $(0.75)^2$ or 0.5625 to be homozygous for the 6.7-kb band, which is 56.25 individuals. We saw 56. Finally, we would expect 2(0.25)(0.75) or 0.375 to be heterozygotes, or 37.5 individuals. We observed 38. The observed numbers are so close to the expected that a chi-square test is unnecessary.

**21.34** Only two of the five loci examined have more than one allele, so (2/5)(100%) = 40% of the loci are polymorphic. The heterozygosity is the average frequency of heterozygotes at each locus. At AmPep, ADH, and LDH-1, the frequency of heterozygotes is zero. At MDH, it is 35/50 = 0.70, and at PGM, it is 10/50 = 0.20. Therefore, the heterozygosity is 0 + 0 + 0 + 0.7 +0.2/5 = 0.18.

**21.36** Linkage disequilibrium is the nonrandom association of alleles at different loci. In linkage disqeuilibrium, some combinations of alleles occur more or less frequently than is expected from the random combination of alleles based on their frequencies. It can be caused by the physical linkage of the loci or can result from a recent population bottleneck, migration, or hybridization. In a first example, suppose two homologous chromosomes have three closely linked loci *A/a*, *B/b*, and *C/c*. One homolog has alleles *A b c* and the other has alleles *A B C*. Suppose further that a new mutation converts the *A* allele on the *A b c* homolog to an *a* allele. The newly induced *a* allele will be transmitted to gametes nonrandomly with the *b* and *c* alleles because it has been introduced onto the homolog that they are on. Thus, the *a* allele will show linkage disequilibrium with the *b* and *c* alleles. Over time, recombination between the homologs in different members of the population will separate *a* from the *b* and *c* alleles, and the amount of linkage disequilibrium will diminish.

For a second example, consider the consequences of a population bottleneck on alleles *A/a* and *B/b* at unlinked loci. Assume the population was in Hardy–Weinberg equilibrium before the bottleneck, so that alleles at these loci showed independent assortment and appeared in gametes at random based on their frequencies in the population. Now suppose that genetic drift during the bottleneck did not change the frequencies of the *A*, *a*, *B*, or *b* alleles, but did result in an increased number of *aa BB* and *AA bb* individuals. Then, immediately following the bottleneck, *a B* and *A b* gametes would be seen more frequently than expected solely based on the frequency of the *A*, *a*, *B*, or *b* alleles in the population, so the *A b* and the *a B* combinations of alleles would show linkage disequilibrium.

Linkage disequilibrium and linkage are not the same. Linkage disequilibrium involves the nonrandom associations of alleles at two or more loci that are not necessarily on the same chromosome. Linkage results from the association of two or more loci on a chromosome due to limited recombination and reflects loci that always do not assort independently.

**21.37 a.** The expected heterozygosity is 1 – (frequency of expected homozygotes). If the frequency of alleles in the population is $p_1, p_2, p_3, \ldots, p_n$, the expected frequency of homozygotes is $p_1^2 + p_2^2 + p_3^2 + \cdots p_n^2$. For locus *G1A*, the expected frequency of homozygotes is $(0.398)^2 + (0.240)^2 + (0.211)^2 + (0.086)^2 + (0.036)^2 + (0.016)^2 + (0.007)^2 + (0.006)^2 = 0.270$, and the expected heterozygosity is 1 – 0.270 = 0.730. The expected heterozygosities for the other loci are *G10X*, 0.741; *G10C*, 0.740; and *G10L*, 0.662. These are approximately the observed frequencies of heterozygosities. Since the numbers and types of different heterozygotes are given, it is not possible to use the chi-square test to directly evaluate whether the population is in Hardy–Weinberg equilibrium. The population appears to be close to Hardy–Weinberg equilibrium.

**b.** The three cubs of the mother show evidence of multiple paternity. For each of the loci *G10X* and *G10L*, three alleles present in the cub must have been contributed paternally (*G10X: X133* or *X137, X141; G10L: L155, L157, L161*). This could have happened only if the cubs were sired by at least two different fathers. Multiple paternity within one set of cubs would tend to increase the genetic variability in the population because it would allow a larger number of males to contribute gametes seen in the next generation. Since $N_e = (4 \times N_f \times N_m)/(N_f + N_m)$, a larger $N_m$ will tend to increase the effective population size.

**21.39 a.** Mutation leads to change in allele frequencies within a population if no other forces are acting and so introduces genetic variation. If population size is small, mutation may lead to genetic differentiation between populations.

**b.** Migration increases the population size and has the potential to disrupt Hardy–Weinberg equilibrium. It can increase genetic variation and may influence the evolution of allele frequencies within populations. Over many generations, migration reduces divergence between populations and equalizes allele frequencies between populations.

**c.** Genetic drift produces changes in allele frequencies within a population. It can reduce genetic variation and increase the homozygosity within a population. Over time, it leads to genetic change. When several populations are compared, genetic drift can lead to increased genetic differences between populations.

**d.** Inbreeding increases the homozygosity within a population and decreases genetic variation.

**21.41** Overdominance results when a heterozygote genotype has higher fitness than either of the homozygotes. In areas with malaria, heterozygotes for *Hb-A/Hb-S* are at a selective advantage because the hemoglobin mixture in these individuals provides an unfavorable growth environment for malarial parasites. Thus, heterozygotes have higher fitness than do *Hb-A/Hb-A* homozygotes who are susceptible to malaria, and they have a higher fitness than do *Hb-S/Hb-S* homozygotes who suffer from sickle-cell anemia. The favoring of the sickle-cell *Hb-S* allele in the heterozygote results in its relatively high frequency in areas with malaria.

**21.42 a.** The cyclical decline in population density could cause a cyclically repeated bottleneck effect at each of the sampling sites. This would lead to the loss of some alleles from the gene pool as a result of chance and founder effects. If there were negligible migration between sampling sites, the populations would most likely diverge in their allele frequencies through genetic drift. There would be more variance in allele frequency among the small populations at each sampling site. If there were substantial migration between sampling sites, there would be reciprocal gene flow among the populations. This would increase the amount of genetic variation in the populations at each sampling site, and reduce the divergence between the populations. If there were cyclical, sharp declines in the vole population, it is unlikely that the population would be in Hardy–Weinberg equilibrium.

**b.** First, randomly trap a subset of the voles and sample tissue or blood from them. Then, identify a set of polymorphic loci (using protein electrophoresis, RFLPs, STRs, DNA sequences, or SNPs) and estimate the heterozygosity and allele frequencies at these loci. Use this information to compare populations sampled at each of the study sites, and estimate the degree of homogeneity or divergence of the two populations.

**21.44** The most comprehensive analysis would involve obtaining DNA samples from a large number of individuals from each of several different geographical locations and then using high-throughput methods to assess the allele frequencies at highly polymorphic SNPs. Quantify the amount of genetic variation within and between populations from different geographical regions by analyzing the frequency of the different SNP alleles in the populations. These analyses can help sort out the effects of natural selection from migration, genetic drift, and mutation. A locus that shows similar patterns of genetic variation to many different loci across the genome may not have been influenced by natural selection, unlike a locus that shows a very different pattern of variation when compared to many different loci across the genome. These types of analyses have supported the view that human populations underwent serial population bottlenecks as they migrated out of Africa. They have also shown that Europeans harbor substantially more deleterious mutations than do ancestral African individuals, probably because the population genetic bottlenecks that followed migration out of Africa provided opportunities for deleterious mutations to rise in frequency.

**21.45** Human populations do differ in their genetic variation according to their geographical origin. However, only about 12–13% of the total genetic variance is found between different populations, whereas 87–88% is found within populations. One significant conclusion from this comparison is that different human populations, whether defined socially (e.g., race) or geographically, are far more similar to each other than they are different.

**21.47** In both instances, protein electrophoresis, RFLP analyses, and DNA sequence analysis of specific genes could be used to gather information on the genotype of the captured individuals and member of each island population. In (**a**), the captured individual should be returned to the subpopulation from which it shows the least genetic variation. In (**b**), evaluate the genotype of the two missing tortoises using DNA from the previously collected blood samples and compare these genotypes with the genotypes of the two captured tortoises. If a captured animal was taken from the field site, its genotype will exactly match a genotype obtained from one of the blood samples.

## Chapter 22 Quantitative Genetics

**22.1** When given a series of data, the best first step is to graph the data. We can determine the minimum and maximum values, and then create histograms for each series of data using different bin sizes (e.g., by 1, by 2, by 5, etc.) to get a feel for the distribution of the data. Some of these sample histograms can be found in the graphs presented here, along with notes on interpretation and the sources of the original data. One final note is that while many times data from a particular sample do not appear to have the bell-shaped distribution characteristic of a quantitative trait, if we know that we are dealing with a quantitative trait we often assume that the data are normally distributed so that we can apply certain statistical techniques in analyzing the data.

**a.**

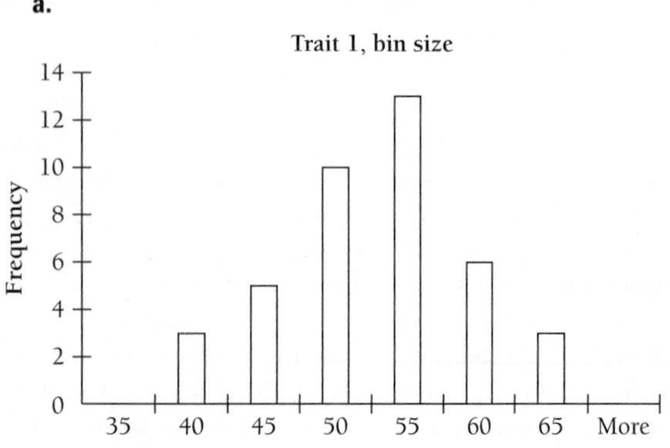


These data appear to be normally distributed, so we could assume that they are representative of the phenotypic data we would see for a quantitative trait. In fact, these are 40 sample values taken from a normal distribution with $\mu = 50$ and $s^2 = 5$.

**b.**

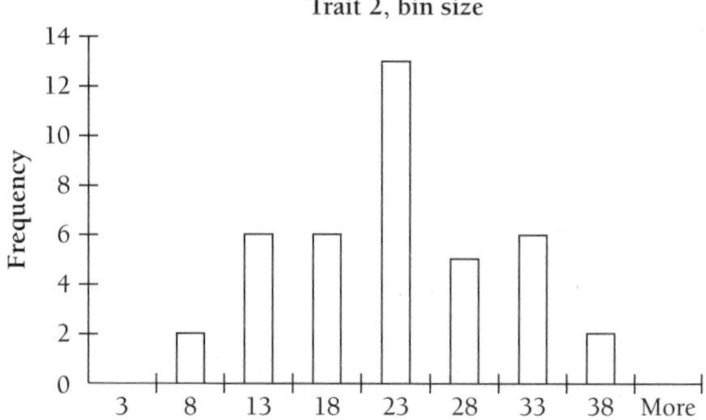


These data appear to have a more pronounced peak in the middle, without the "shoulders" next to the peak, such as those seen in (**a**). We could conclude that these data are not representative of a quantitative trait. In fact, these data are 10 sample values from a normal distribution, with $\mu = 10$ and $s^2 = 2$, 20 values from $\mu = 20$ and $s^2 = 2$, and 10 values from $\mu = 30$ and $s^2 = 2$. This is what we might expect to see from a simple additive Mendelian character with some environmental variance.

**c.**

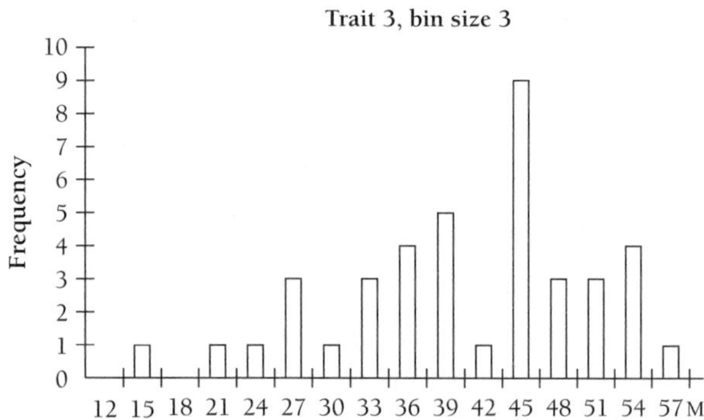


These data have a strong peak, but they do not have the characteristic shape of a normal distribution. You can see that there are no shoulders to the peak and that the peak trails off to the left, but not to the right. In fact, these data include 10 sample values from a normal distribution with $\mu = 25$ and $s^2 = 5$, 20 values from $\mu = 42$ and $s^2 = 5$, and 10 values from $\mu = 55$ and $s^2 = 5$, something we might expect to see from a trait showing simple Mendelian inheritance with a small degree of dominance and substantial environmental variance.

**22.2 a.** The mean of a sample is obtained by summing all the individual values and dividing by the total number of those values. The mean head width is 25.21/8 = 3.15 cm, and the mean wing length is 281.7/8 = 35.21 cm.

The standard deviation equals the square root of the variance ($s^2$). The variance is computed by summing the squares of the differences between each measurement and the mean value, and dividing this sum by the number of measurements minus one:

$$s_{\text{head width}} = \sqrt{s^2} = \sqrt{\frac{\Sigma(x_i - \bar{x})^2}{n - 1}}$$

$$= \sqrt{\frac{1.70}{7}} = \sqrt{0.24} = 0.49 \text{ cm}$$

$$s_{\text{wing length}} = \sqrt{s^2} = \sqrt{\frac{\Sigma(x_i - \bar{x})^2}{n - 1}}$$

$$= \sqrt{\frac{413.35}{7}} = \sqrt{59.05} = 7.68 \text{ cm}$$

**b.** The correlation coefficient, $r$, is calculated from the covariance, $cov$, of two sets of data. Let head width be represented by $x$, and wing length be represented by $y$. $r$ is defined as

$$r = \frac{cov_{xy}}{s_x s_y} = \frac{\dfrac{\Sigma x_i y_i - n\overline{xy}}{n - 1}}{s_x s_y}$$

The first factor ($\Sigma x_i y_i$) is obtained by taking the sum of the products of the individual measurements of head width and wing length for each duck. The next factor is the product of the number of individuals and the means of these two sets of measurements. The difference between these values is then divided by ($n - 1$), and then by the products of the standard deviations of each measurement. Thus,

$$r = \frac{\dfrac{913.50 - 8 \times 3.15 \times 35.21}{7}}{0.49 \times 7.68} = \frac{3.74}{3.76} = 0.99$$

**c.** Head width and wing length show a strong positive correlation, nearly 1.0. This means that ducks with larger heads will almost always have longer wings, and ducks with smaller heads will almost always have shorter wings.

**22.3** The degree of phenotypic variability is related to the degree of genetic variability. Since each pure-breeding parent is homozygous for the genes (however many there are) that control the size character, the variation seen within parental lines is due only to the environmental variation present. A cross of two pure-breeding strains will generate an $F_1$ heterozygous for those loci controlling the size trait, but genetically as homogeneous as each of the parents. Therefore, the only variation we would expect to see in the $F_1$ is that caused by the environment, and the $F_1$ should show no greater variability than do the parents.

**22.4** **a.** Since the cross is *AA BB* × *aa bb*, the $F_1$ genotype will be *Aa Bb*. Since capital-letter alleles additively determine height, and individuals with four capital-letter alleles have a height of 50 cm, while individuals with no capital-letter alleles have a height of 30 cm, each capital-letter allele appears to confer (50 − 30)/4 = 5 cm of height over the 30-cm base. *Aa Bb* individuals with two capital-letter alleles should have an intermediate height of 40 cm.

**b.** Any individuals with two capital-letter alleles will show a height of 40 cm. Thus, *Aa Bb*, *AA bb*, and *aa BB* individuals will be 40 cm high.

**c.** In the $F_2$, $^1/_{16}$ of the progeny are *AA bb*, $^4/_{16}$ are *Aa Bb*, and $^1/_{16}$ are *aa BB*. Thus $^6/_{16} = {}^3/_8$ of the progeny will be 40 cm high.

**d.** In answering this question, we have assumed that the *A* and *B* loci assort independently and that each locus and each allele contribute equally to the phenotype.

**22.10** Internode length shows the characteristics of a quantitative trait. These characteristics include $F_1$ progeny that show a phenotype intermediate between the two parental phenotypes, and an $F_2$ showing a range of phenotypes with extremes in the range of the two parents, some of which have all the original parental alleles.

**22.13** To see transgressive segregation, at least one of the parents must have some alleles that are "opposite" in effect of the expected direction. For example, imagine we are looking at height. If we assume that there are six loci that contribute to height, and that capital-letter alleles contribute a 5-cm increase over a base height of 1 meter, a cross between an *AA BB CC DD EE FF* individual (160 cm) and an *aa bb cc dd ee ff* individual (100 cm) will produce an $F_2$ with extreme individuals only as tall and as short as the original parents. If, however, the original parents have the genotypes *AA BB CC DD EE ff* (150 cm) and *aa bb cc dd ee FF* (110 cm), segregation in the $F_2$ can produce an *AA BB CC DD EE FF* genotype (160 cm) and an *aa bb cc dd ee ff* genotype (100 cm), which are taller and shorter than the original lines used. In this case, the taller parent has "shorter" alleles at one locus, and vice versa.

A more extreme case to consider is where the parents have the same phenotype, but produce segregating offspring. Imagine, for example, if an *AA BB CC dd ee ff* (130 cm) individual were crossed with an *aa bb cc DD EE FF* individual (130 cm). Their $F_1$ offspring would again be 130 cm (*Aa Bb Cc Dd Ee Ff*), but in the $F_2$, individuals from 160 cm (*AA BB CC DD EE FF*) to 100 cm (*aa bb cc dd ee ff*) could be seen!

**22.14** The $F_1$ is *A/a B/b C/c D/d E/e*, and so it is greyish brown. The $F_2$ phenotypes are determined by the number of capital-letter alleles contributed from each $F_1$ parent. The easiest way to proceed is to look for the proportions of individuals with the light tan pigmentation (3 or 4 capital-letter alleles) and the whitish blue pigmentation (0 or 1 capital-letter alleles), and the proportion of greyish brown offspring will be the rest. Start off by determining the chance of obtaining 0, 1, 2, or 3 capital-letter alleles in a gamete from each $F_1$ parent and then determining how these combinations make the desired genotypes.

Since the parent is heterozygous at all five loci, the chance of obtaining any specified set of five alleles in one gamete is $(^1/_2)^5$. The chance of obtaining a particular number of capital-letter alleles from an $F_1$ is the number of ways in which that number of alleles can be obtained, multiplied by $(^1/_2)^5$. There is one way to obtain 0 capital-letter alleles, 5 ways to obtain 1 capital-letter allele, 10 ways of obtaining two capital-letter alleles, and 10 ways of obtaining three capital-letter alleles. With this information, we can tabulate the ways in which the progeny classes we are interested in can be formed, as shown in Table 22.A. In the $F_2$, $11(^1/_2)^{10} = {}^{11}/_{1{,}024}$ will be bluish white and $165(^1/_2)^{10} = {}^{165}/_{1{,}024}$ will be light tan. The remaining $[1 - 176(^1/_2)^{10}] = {}^{848}/_{1{,}024}$ $F_2$ progeny will be greyish brown.

There is another method to solve this problem: Use the coefficients of the binomial expansion to determine the proportion of progeny with different numbers of capital-letter and lowercase alleles. Let $n$ = total number of alleles, $s$ = number of capital-letter alleles, $t$ = number of lowercase alleles, $a$ = chance of obtaining a capital-letter allele, $b$ = chance of obtaining a lowercase allele, and $x! = (x)(x - 1)(x - 2)\ldots(1)$, with $0! = 1$. Then the chance $p$ of obtaining progeny with a specified number of each type of allele is given by

Table 22.A

| F$_1$ Gamete #1 | | F$_1$ Gamete #2 | | F$_2$ Progeny | | |
|---|---|---|---|---|---|---|
| Capital Alleles | Gemete Fraction | Capital Alleles | Gamete Fraction | Capital Alleles | F$_2$ Fraction | Phenotype |
| 0 | $(1/2)^5$ | 0 | $(1/2)^5$ | 0 | $(1/2)^{10}$ | whitish blue |
| 0 | $(1/2)^5$ | 1 | $5(1/2)^5$ | 1 | $5(1/2)^{10}$ | whitish blue |
| 1 | $5(1/2)^5$ | 0 | $(1/2)^5$ | 1 | $5(1/2)^{10}$ | whitish blue |
| 0 | $(1/2)^5$ | 2 | $10(1/2)^5$ | 2 | $10(1/2)^{10}$ | light tan |
| 1 | $5(1/2)^5$ | 1 | $5(1/2)^5$ | 2 | $25(1/2)^{10}$ | light tan |
| 2 | $10(1/2)^5$ | 0 | $(1/2)^5$ | 2 | $10(1/2)^{10}$ | light tan |
| 0 | $(1/2)^5$ | 3 | $10(1/2)^5$ | 3 | $10(1/2)^{10}$ | light tan |
| 1 | $5(1/2)^5$ | 2 | $10(1/2)^5$ | 3 | $50(1/2)^{10}$ | light tan |
| 2 | $10(1/2)^5$ | 1 | $5(1/2)^5$ | 3 | $50(1/2)^{10}$ | light tan |
| 3 | $10(1/2)^5$ | 0 | $(1/2)^5$ | 3 | $10(1/2)^{10}$ | light tan |

$$p(s,t) = \frac{n!}{s!t!}a^s b^t$$

$$\left.\begin{aligned} p(0,10) &= \frac{10!}{0!10!}\left(\frac{1}{2}\right)^0\left(\frac{1}{2}\right)^{10} = \frac{1}{1{,}024} \\ p(1,9) &= \frac{10!}{1!9!}\left(\frac{1}{2}\right)^1\left(\frac{1}{2}\right)^{9} = \frac{10}{1{,}024} \end{aligned}\right\} \frac{11}{1{,}024} \text{ whitish blue}$$

$$\left.\begin{aligned} p(2,8) &= \frac{10!}{2!8!}\left(\frac{1}{2}\right)^2\left(\frac{1}{2}\right)^{8} = \frac{45}{1{,}024} \\ p(3,7) &= \frac{10!}{3!7!}\left(\frac{1}{2}\right)^3\left(\frac{1}{2}\right)^{7} = \frac{120}{1{,}024} \end{aligned}\right\} \frac{165}{1{,}024} \text{ light tan}$$

$$1 - \frac{11 + 165}{1{,}024} = \frac{848}{1{,}024} \text{ greyish brown}$$

**22.15 a.** From the data given, it appears that some proportion of cases of AD can be attributed to genetic factors. Multiple genes that increase the risk for AD have been identified, some of which appear to act in a dose-dependent manner. Thus, it could be that a number of different genes contribute to the onset of AD, with some having a greater contribution factor than others. This is somewhat similar to how polygenic traits control a phenotype, since there, alleles at multiple genes contribute in an additive, dose-dependent fashion to the phenotype.

**b.** Consider two explanations: First, if AD can be caused by environmental agents, mutation, and/or a combination of both environmental agents and mutation, the presence of AD in both twins could be due to the presence of one or more abnormal alleles in both and/or the exposure of both twins to adverse environmental conditions. The presence of AD in only one twin may be due to differences in the exposure of that twin to a contributing or causative environmental agent(s). Second, the presence of a particular allele or a specific mutation may only increase the risk of disease, and not determine its occurrence, since the penetrance of an allele may be strongly affected by the environment. In the case of AD, the environmental factors may not be clear cut or even small in number. There may be multiple environmental factors, some of which may be complex or subtle.

**22.18** For wildcats residing in similar environments, phenotypic variation in size would result primarily from variation in genetic differences between the cats and the differences resulting from genotype–environment interaction. For house cats residing in similar environments, one could make a similar argument. If domesticated cats descended from a group of wildcats, it is likely that they would have less genetic variation than the wildcats. Therefore, one might expect that wildcats would be more likely to exhibit a higher heritiability. However, both broad-sense and narrow-sense heritability are specific to a particular population in a particular environment, so any categorical response to this question is likely to have many exceptions.

**22.19** SHR rats will continue to respond to salt by developing hypertension. Since the strain is inbred, any variation in blood pressure will result from the amount of exposure to salt, and not from genetic variation. Therefore, heritability for this population will be zero. Similarly, the inbred TIS rats would also have a heritability of zero (and retain a low blood pressure).

**22.21 a.** The narrow-sense heritability of the number of triradii will equal the slope, $b$, of the regression line of the mean offspring phenotype on the mean parental phenotype.

$$b = \frac{cov_{xy}}{(s_x)^2} = \frac{\dfrac{\Sigma x_i y_i - n\overline{xy}}{n-1}}{\dfrac{\Sigma(x_i - \overline{x})^2}{n-1}} = \frac{\Sigma x_i y_i - n\overline{xy}}{\Sigma(x_i - \overline{x})^2}$$

For this data set, $x$ is the mean number of triradii in the parents and $y$ is the mean number of triradii in the offspring. Using either a calculator or a spreadsheet or statistics program, you can find the following:

$$\Sigma x_i = 111,\ \overline{x} = 11.1$$
$$\Sigma y_i = 108.5,\ \overline{y} = 10.85$$
$$\Sigma(x_i - \overline{x})^2 = 51.4$$
$$\Sigma x_i y_i = 1{,}257.5$$
$$b = \frac{1{,}257.5 - 10 \times 11.1 \times 10.85}{51.4} = \frac{53.15}{51.4} = 1.04$$

**b.** A slope of 1.04 indicates that additive genetic variation is responsible for essentially all of the observed variation in phenotype. Note that the estimate obtained for $h^2$ is greater than one, showing that methods for estimating narrow-sense heritability can overestimate the amount of additive genetic variation among individuals.

**22.23** The selection differential ($S$) equals $14.3 - 9.7 = 4.6$ cm. The response to selection ($R$) equals $13 - 9.7 = 3.3$ cm. The narrow-sense heritability $H^2_N$ equals $R/S = 3.3/4.6 = 0.72$.

**22.27** A response to selection depends on (a) variation on which selection can act and (b) a high narrow-sense heritability so that the selected individuals produce similar offspring. The narrow-sense heritability for each of the traits is $V_A/V_P$: 0.165 for body length, 0.061 for antenna bristle number, and 0.144 for egg production. The amount of raw variation is also greatest for body length. Thus, body length will respond most to selection, and antenna bristle number will respond least to selection.

**22.28** Assume that multiple loci contribute equally to fruit weight and days to first flower. To recover the cultivated phenotype most quickly from selection after crossing it with the wild genotype, we would like to find the cross where most of the variation is due to additive effects. A quick way to assess this is to look at the phenotype of the $F_1$: If most of the variation is due to additive effects, the phenotype of the $F_1$ will be intermediate to both parents. If the $F_1$ is closer in phenotype to one parent or the other, this can be taken as an indicator that that parent harbors some nonadditive variation. Using this criterion for both traits, crosses 2 and 4 appear to be the best initial crosses to work with.

**22.30 a.** As described in the text, analyses in model experimental organisms such as *Drosophila* have suggested that findings from QTL analyses can be population dependent. Genetic and environmental heterogeneity may contribute to the size of a QTL's effect, so a QTL that explains 15% of the risk for diabetes in a particular population may not explain a similar amount of risk in a different population.

**b.** Two complementary approaches are possible. In one, candidate loci are chosen based on known or suggested function, and then SNPs at these loci are tested for their association with disease. In the other, a genome-wide screen is performed: a panel of SNPs distributed throughout the genome is used as a set of DNA markers in association studies to identify segments of the genome where QTLs are located. Specific genes in these regions are then examined more closely for their association with disease.

**22.32 a.** Since the aim of QTL identification is to find segments of the genome associated with phenotypic differences between individuals, the first step in a typical QTL analysis is to develop inbred lines that have been selected for different phenotypes. These lines are crossed to generate an $F_1$, and then the $F_1$ is either backcrossed, intercrossed to generate an $F_2$, or intercrossed and selfed to generate a series of recombinant inbred strains. Each member of the set of recombinant inbred strains that is generated in this way received different parts of its genome from the two original inbred lines. After the phenotypes and genotypes of these strains are determined, QTLs are identified by correlating the genotypic and phenotypic differences among the different recombinant inbred strains.

A doubled haploid line is a diploid line generated from a single haploid gamete. In this case, the gametes used to form doubled haploid lines are produced by an $F_1$ hybrid between two highly inbred, phenotypically different lines. In the $F_1$ hybrid, random crossing-over and independent assortment led to the production of recombinant gametes, each having a unique combination of chromosomal segments drawn from the two original inbred lines. Therefore, each doubled haploid line is a different type of recombinant between the two original inbred lines. What is critically important here is that each line is homozygous when it is formed, and so additional crosses to develop inbred recombinant lines are unnecessary. When doubled haploid lines are not used, inbred recombinant lines must be developed through many generations of backcrosses or intercrosses, which requires considerable additional time.

**b.** See Figure 22.A below. When the barley lines are grown in four different states, malting quality values from each state are continuously distributed across a range of values. Therefore, malting quality is a quantitative and not a qualitative trait.

When the distributions of the same lines grown in four different states are compared, it is apparent that they have different means and variances (these data are quantified in Table 22.B, p. 681, and discussed further in part (e). Lines grown in Montana have a lower mean malting quality value than do lines grown in Washington, and lines grown in Idaho have a smaller variance than do lines grown in either Washington, Oregon or Montana. Therefore, we can infer that this trait also is affected by the environment.

**c.** Since the recombinant inbred lines were generated by doubling haploid gametes, each is homozygous for alleles at all loci. Thus, it is not possible to select for genetic differences in the offspring of line L87, and so it is not possible to select for blight resistance or further enhance its malting quality phenotype. To develop a fungal resistance in strain L87, it should be crossed to a resistant barley strain, and the $F_1$ should be backcrossed to L87. Recombination in the $F_1$ will produce hybrids, some of which are resistant. Repeated backcrossing to L87 under selection for fungal resistance will lead to a strain that is close to L87 in its malting quality phenotype.

**d. i.** $^1/_2$.

**ii.** 0 (doubled haploids are homozygous for alleles at all loci).

**iii.** $^1/_4$.

**iv.** $^1/_8$.

**Figure 22.A**

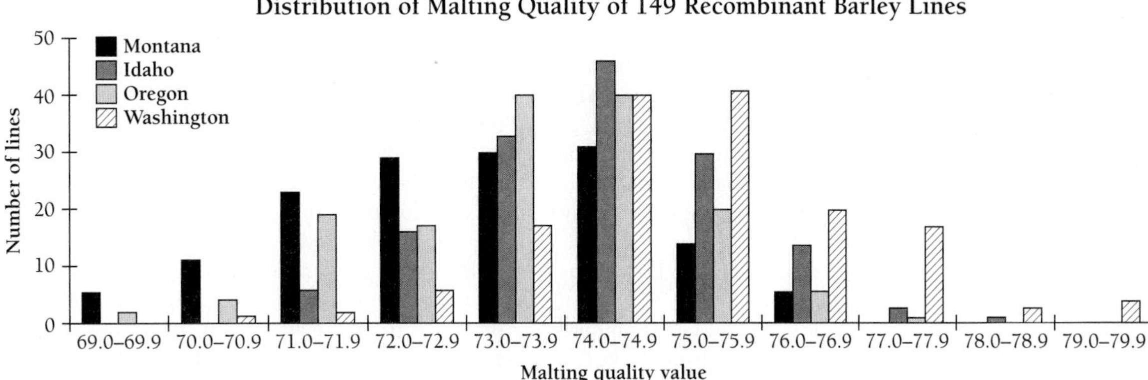

**v.** 0.

**vi.** 0.

**e.** The data show evidence of both genetic and environmental sources of phenotypic variance. The environment contributes to phenotypic variation since the mean, variance, and range of the malting quality values for the set of lines are similar in different environments. This is supported by an examination of the malting quality values of individual lines grown in different environments: no recombinant line gives identical malting quality values in all four environments. Support for a genetic contribution to phenotypic variance comes from the observation that most lines give values that are similar relative to the mean values seen in a particular environment. For example, most lines giving less than the mean malting quality value in one environment tend to give less than the mean malting quality value in all four environments. That some lines do not consistently follow this pattern (e.g., L51 is above the mean in Montana, but below the mean in the other three states; L126 is well above the mean in all states except Oregon, where it is well below the mean) suggests that there may also be covariance between genotype and environment (G × E variance).

**Chapter 23** Molecular Evolution

**23.2** Each of the three codon positions can change to three different nucleotides, so a total of 135 substitutions must be considered for each position. At the first position, only 9 of the 135 possible changes have no effect on the amino acid sequence of the polypeptide (0.07 are synonymous). Every change at the second codon position results in an amino acid substitution (0.00 are synonymous). A total of 98 changes at the third codon position have no effect on the amino acid sequence of the protein (0.73 are synonymous). Natural selection is much more likely to act on mutations that change amino acid sequences, such as the second and first codon positions, and those are the ones where the least change is likely to be seen as sequences diverge.

**23.3** $K = -3/4 \ln[1 - 4/3(p)]$

$K = -3/4 \ln[1 - 4/3(0.12)]$

$K = -3/4 \ln(1 - 0.16)$

$K = -3/4(-0.17)$

$K = 0.13$

**23.5** Mutation rates would be greater than or equal to the substitution rate for any locus. Mutations are any nucleotide changes that occur during DNA replication or repair, whereas substitutions are mutations that have passed through the filter of selection. Many mutations are eliminated through the process of natural selection.

**23.6** $1 \times 10^{-8}$ substitutions/nucleotide/year. $2 \times 10^{-8}$ substitution/nucleotide/year.

**23.9** The high substitution rate in mammalian mitochondrial genes is useful when determining the relationships between evolutionary closely related groups of organisms such as members of a single species. When longer divergence times are involved, such as those associated with the mammalian radiation, more slowly evolving nuclear loci are more convenient to study because multiple substitutions are less likely to have occurred.

**23.11** Sequence A is a pseudogene, and sequence B is a functioning gene.

**23.13** Regions involved in base pairing would not accumulate substitutions as quickly as those that are not. If the secondary structure of an RNA molecule is under selective constraint, then the only changes that would be found would be those for which a compensatory change also occurred in its complementary sequence.

**23.15** The appeal of analyzing ancient DNA samples is that they might allow ancestral sequences to be determined (and not just inferred). However, it is almost impossible to prove that an ancient organism is from the same lineage as an extant species. Increasing the number of taxa in any analysis increases the robustness of any phylogenetic inferences, but extant taxa are almost invariably easier to obtain.

**23.16** Divergence within genes typically occurs before the splitting of populations that occurs when new species are created. Preexisting polymorphisms such as those seen in the major histocompatibility locus could account for such a discordance.

**23.18** These chances are related to the total number of rooted $N_R$ and unrooted $N_U$ trees that can be generated using six taxa.

$N_R = (2n - 3)!/[2^{n-2}(n - 2)!]$

$N_R = 9!/[16 \times (4)!]$

$N_R = 362{,}880/384$

$N_R = 945$

$N_U = (2n - 5)!/[2^{n-3}(n - 3)!]$

$N_U = 7!/[8 \times (3)!]$

$N_U = 5{,}040/48$

$N_U = 105$

Since there are only 105 possible unrooted trees and 945 possible rooted trees, choosing a correct unrooted tree is more likely.

**23.22** The equations used to determine how many rooted and unrooted trees can describe the relationship between taxa have no parameter that considers the amount of data associated with each taxon but do consider the number of taxa.

**23.24** Informative sites are the only sites that are considered in parsimony analyses.

**23.26** Gene duplication allows most new genes to arise by mutating redundant copies of already existing genes. Copies of genes are free to accumulate substitutions, whereas the original version remains under selective constraint. When only part of a gene is duplicated, there is a potential for domain shuffling—the duplication and rearrangement of domains in proteins that provide specific functions. This can lead to the assemblage of proteins with more complex domain arrangements, which can result in proteins with novel functions. While not all changes to duplicated genes will be desirable or lead to new functions—many may result in a loss of function and result in a pseudogene—gene duplications are advantageous as a mechanism to generate genes with new functions because they provide a shortcut for modifying existing proteins via "tinkering."

Point mutations (or small deletion or insertion mutations) could alter the function of an existing gene or modify the way that existing RNA processing sites are used. The chance of a noncoding sequence randomly accumulating mutations that give it an open reading frame and appropriate promoter elements all at the same time is extremely small.

Chromosomal rearrangements can reposition DNA sequences that provide information necessary for gene transcription, translation, and function. They can alter the transcriptional structure of an existing gene, create a novel fusion protein, introduce new sites for mRNA processing, or place the gene under the control of another gene's regulatory elements and introduce new functions by altering where or when it is expressed during development.

Unequal crossing-over following misalignment between a duplicated gene or a pseudogene and the retained functional gene provide an opportunity for recombination. Gene conversion is a process that occurs during recombination and results in the replacement of an allele of one homolog with the allele of the other homolog. Consequently, if gene conversion occurs among misaligned, duplicated sequences, it can "repair" inactivated pseudogenes or restore altered functions.

These processes do not generally act independently of each other. Unequal crossing-over and the gene conversion events that occur during recombination involving misaligned sequences often involve prior gene duplication; point mutation following gene duplication can lead to new functions; and chromosomal rearrangements can accompany or result in gene duplication.

# Credits



## Text and Illustration Credits

**Figure 1.3:** From Robert H. Tamarin, *Principles of Genetics*, 5/e. Copyright © 1996 McGraw-Hill. Used by permission of McGraw-Hill Companies, Inc.

**Figure 1.5:** Peregrine Publishing.

**Figure 1.6:** Peregrine Publishing.

**Figure 2.18:** Reprinted from *Genes* IV by Benjamin Lewin. Copyright © 1990 with permission from Excerpta Medica, Inc.

**Figure 2.20a, b, and c:** Reprinted from James Watson et al., *Molecular Biology of the Gene*, Fifth Edition, fig. 7.21. Copyright © 2004 Benjamin Cummings. Reprinted by permission of Pearson Education, Inc.

**Figure 2.23a:** Figure 12.4 p. 421 from Hartwell, *Genetics: From Genes to Genomes*. © 2003 McGraw-Hill Companies, Inc.

**Figure 2.23b:** Reprinted from James Watson et al., *Molecular Biology of the Gene*, Fifth Edition, fig. 7.32b. Copyright © 2004 Benjamin Cummings. Reprinted by permission of Pearson Education, Inc.

**Figure 2.25:** Reprinted from *Cell*, Vol. 97, Greider, pp. 419–422, copyright © 1999 with permission from Excerpta Medica, Inc.

**Figure 3.4:** Reprinted from James Watson et al., *Molecular Biology of the Gene*, Fifth Edition, fig. 8.26. Copyright © 2004 Benjamin Cummings. Reprinted by permission of Pearson Education, Inc.

**Figure 3.16:** Reprinted from James Watson et al., *Molecular Biology of the Gene*, Fifth Edition, fig. 7.41. Copyright © 2004 Benjamin Cummings. Reprinted by permission of Pearson Education, Inc.

**Figure 5.7:** Reprinted from James Watson et al., *Molecular Biology of the Gene*, Fifth Edition, fig. 12.13. Copyright © 2004 Benjamin Cummings. Reprinted by permission of Pearson Education, Inc.

**Figure 5.11a, b:** Reprinted from *Cell*, Vol. 87, N. Proudfoot. "Ending the Message Is Not So Simple," pp. 779–781, copyright © 1996 with permission by Excerpta Medica, Inc.

**Figure 5.14:** From Maizels and Winer, "RNA Editing," *Nature*, Vol. 334, 1988, p. 469. Copyright © 1988 Macmillan Magazines Limited. Reprinted by permission.

**Figure 6.4:** Illustration, Irving Geis. Rights owned by Howard Hughes Medical Institute. Not to be used without permission.

**Figure 6.17:** Peregrine Publishing.

**Figure 7.16:** From Brooker, *Genetics: Analysis and Principles*, p. 474. Copyright © 1998. Reprinted by permission of Pearson Education, Inc.

**Figure 7.26:** "Ty-transposable element of yeast" adapted from Watson by permission of Gerald B. Fink. Reprinted by permission of Pearson Education, Inc.

**Box 7.1:** Excerpts from *Biographical Memoirs*, Vol. 68, 1996 by Nina Federoff. Copyright © 1996 National Academies Press. Used with permission.

**Figure 8.17:** Reprinted from James Watson et al., *Molecular Biology of the Gene*, Fifth Edition, fig. 7.2. Copyright © 2004 Benjamin Cummings. Reprinted by permission of Pearson Education, Inc.

**Figure 8.19:** Reprinted with permission from Fleischmann et al., *Science*, July 28, 1995, Vol. 269, p. 507. Copyright © 1995 American Association for the Advancement of Science.

**Figure 9.4:** Copyright © Stanford University. Used with permission.

**Figure 9.7a:** Reprinted with permission from Chu et al., *Science*, Vol. 282, No. 699, figure 1. Copyright © 1998 American Association for the Advancement of Science. Reprinted with permission from AAAS.

**Figure 9.7c:** Copyright © Patrick Brown. Used with permission.

**Figure 10.11:** From Johnston et al., *Molecular Cell Biology*, Vol. 14, pp. 3834–3841, 1994. Copyright © 1994 American Society for Microbiology. Used with permission.

**Figure 10.20a:** Roza et al., *Molecular Vision*, Vol. 4, No. 20, 1998, figure 3a. Used with permission.

**Figure 10.23:** From *Recombinant DNA* by J. D. Watson, M. Gillman, J. Witkowski, and M. Zoller. © 1983, 1992 by J. D. Watson, M. Gillman, J. Witkowski, and M. Zoller. Used with the permission of W. H. Freeman and Company.

**Figure 10.24:** From *Genetics* by Robert Weaver and Philip Hedrick, © 1989. Reprinted by permission of McGraw-Hill Companies, Inc.

**Table 11.5:** From Table IV in *Statistical Tables for Biological, Agricultural, and Medical Research*, 6/e by Fisher and Yates. © 1994 Pearson Education Ltd. Used with permission.

**Figure 12.10:** From *Biological Science*, Fourth Edition by William T. Keeton and James Gould with Carol Grant Gould. Copyright © 1986, 1980, 1979, 1978, 1972, 1967 by W. W. Norton & Company, Inc. Used by permission of W. W. Norton & Company, Inc.

**Figure 14.1:** From *Genetics*, Second Edition by Ursula W. Goodenough, copyright © 1978 Brooks/Cole, a part of Cengage Learning, Inc. Reproduced by permission of the publisher. This material may not be reproduced in any form or by any means without the prior written permission of the publisher.

**Figure 14.10:** From *General Genetics* by Srb. Owen, Edgar, © 1965 by W. H. Freeman and Company. Used with permission.

**Figure 16.12:** Reprinted from *Cancer and Cytogenetics*, Vol. 11, O. Prakash and J. J. Yunis, "High Resolution chromosomes of the +(922)Leukemias," pp. 361–368. Copyright © 1984 with permission from Elsevier Science, Inc.

**Figure 16.13b:** From Gerald Stine, *The New Human Genetics*. Copyright © 1989. Used with permission from McGraw-Hill Companies, Inc.

**Figure 17.9:** From Charles Yanofsky, "Attenuation in the Control of Expression of Bacterial Operons," *Nature*, Vol. 289, 1981. Copyright © 1981 by Macmillan Magazines Limited. Reprinted with permission.

**Figure 18.1:** From Peter J. Russell, *Genetics*, Fifth Edition, p. 538, fig. 17.1. Copyright © 1998. Reprinted by permission of Pearson Education, Inc.

**Figure 18.2a:** Reprinted from James Watson et al., *Molecular Biology of the Gene*, Fifth Edition, fig. 12.16. Copyright © 2004 Benjamin Cummings. Reproduced by permission of Pearson Education, Inc.

**Figure 18.2b:** Reprinted from James Watson et al., *Molecular Biology of the Gene*, Fifth Edition, fig. 17.1. Copyright © 2004 Benjamin Cummings. Reproduced by permission of Pearson Education, Inc.

**Figure 18.9b:** From Wolpert et al, *Principles of Development* 3/e. Copyright © 2007 Oxford University Press. Used with permission.

**Figure 18.10:** Reprinted from James Watson et al, *Molecular Biology of the Gene*, Fifth Edition. Copyright © 2004 Benjamin Cummings. Reproduced by permission of Pearson Education, Inc.

**Figure 18.13:** Reprinted from James Watson et al., *Molecular Biology of the Gene*, Fifth Edition. Copyright © 2004 Benjamin Cummings. Reproduced by permission of Pearson Education, Inc.

**Figure 20.2:** Adapted from Campbell et al., *Biology: Concepts and Connections*, Second Edition, fig. 8.10, p. 136. Copyright © 1997 Benjamin Cummings. Reprinted by permission of Pearson Education, Inc.

**Figure 20.6:** Reprinted from James Watson et al., *Molecular Biology of the Gene*, Fifth Edition. Copyright © 2004 Benjamin Cummings. Reproduced by permission of Pearson Education, Inc.

**Table 20.3:** Reprinted with permission from J. Marx, *Science*, Vol. 261, 1993, pp. 1385–1387. Copyright © 1993 American Association for the Advancement of Science.

**Table 21.3:** Data from R.K. Selander, "Behavior and Genetic Variation in Natural Populations (*Mus musculus*)" in *American Zoologist*, Vol. 10, 1970, pp. 53–66.

**Table 21.6:** From *Genetics*, 3/e, by Monroe W. Strickberger. Copyright © 1985. Adapted by permission of Pearson Education, Inc.

**Figure 21.2:** From *Ecological Genetics* by E. B. Ford. Copyright © 1975.

Reprinted by permission of The Natural History Museum Picture Library.

**Figure 21.3:** From P. Buri in *Evolution* 10 (1956), p. 367. Reprinted by permission of the Society for the Study of Evolution.

**Figure 21.6:** From R. K. Koehn et al., in *Evolution* 30 (1976), p. 6. Reprinted by permission of the Society for the Study of Evolution.

**Figure 21.9:** From P. Buri in *Evolution* 10 (1956), p. 367. Reprinted by permission of the Society for the Study of Evolution.

**Figure 21.10:** From Philip Hedrick, *Genetics of Populations*, 1983. Copyright © 1983 Jones and Bartlett Publishers, Sudbury, MA. www.jbpub.com. Reprinted with permission.

**Figure 21.11:** Courtesy of Andrew Clark.

**Figure 21.12:** Courtesy of Andrew Clark.

**Figure 21.18:** From A. C. Allison, "Abnormal Hemoglobin and Erythovute Enzyme-Deficiency Traits" in *Genetic Variation in Human Population*, by G. A. Harrison, ed.

**Figure 23.2:** From R. E. Dickerson, *Journal of Molecular Evolution*, Vol. 1, 1971, pp. 26–45. Copyright © 1971 Springer-Verlag. Used with permission.

**Figure 23.4:** From Hartl and Clark, *Principles of Population Genetics*, Third Edition, p. 373. Copyright Sinauer Associations. Reprinted by permission from the publisher.

**Figure 23.6:** From N. Pace, "A Molecular View of Microbial Diversity in the Biosphere," *Science*, Vol. 276, p. 735, 1007. Copyright © 1997. Reprinted with permission from AAAS.

**Table 23.1:** From W. Li, C. Luo, and C. Wu, "Evolution of DNA Sequences" in *Molecular Evolutionary Genetics*, Vol. 2, 1985, pp. 150–174 by R. J. MacIntyre, ed. Reprinted by permission of Kluwer Academic/Plenum Publishers.

## Photograph Credits

**Chapter 1** Opener: © Dorling Kindersley 1.1: © Steve Gschmeissner/Photo Researchers, Inc. 1.2: © Eli Lilly and Company. Used with permission. 1.4a: Photo Researchers, Inc. 1.4b: Max Westby 1.4c: Dr. Chin-Sang 1.4d: Photo Researchers, Inc. 1.4e: Alamy Images 1.4f: Shutterstock 1.4g: Prof. Katherine A. Borkovich, PhD 1.4h: Phototake/Carolina Biological Supply Company 1.4i: Pearson Science 1.4j: Visuals Unlimited 1.4k: © Detail Photography/Alamy 1.4l: istockphoto.com 1.4m: From Kimmel et al. "Stages of Embryonic Development of the Zebrafish." *Developmental Dynamics* 203:253–310 (1995)

**Chapter 2** Opener: © Pasieka/Photo Researchers, Inc. 2.1a: Peter Arnold, Inc. 2.1b-c: From "Pyruvate Oxidase Is a Determinant of Avery's Rough Morphology" Aimee E. Belanger, Melissa J. Clague, John I. Glass, and Donald J. LeBlanc J. Bacteriol. *American Society for Microbiology*, 186:8164–8171. Copyright © 2004, American Society for Microbiology 2.4: Courtesy of Dr. Harold W. Fisher, University of Rhode Island 2.10: National Cancer Institute 2.11a, left: Peter Arnold, Inc. 2.11a, right: Corbis/Bettmann 2.11b: Courtesy of Professor M.H.F. Wilkins, Biophysics Department, King's College, London. 2.15: Photo Researchers, Inc. 2.17a: Dr. Jack Griffith/University of North Carolina/School of Medicine 2.17b: Dr. Jack Griffith/University of North Carolina/School of Medicine 2.19: Dr. Jack Griffith/University of North Carolina/School of Medicine 2.21a: Barbara Hamkalo 2.22: Professor Ulrich K. Laemmli

**Chapter 3** Opener: Delft University of Technology Tremani TU Delft/Tremani 3.13: National Institutes of Health

**Chapter 4** Opener: © Ken Eward/Science Source/Photo Researchers, Inc. 4.5: National Geographic Image Collection 4.7: Photo Researchers, Inc. 4.12: Photo Researchers, Inc.

**Chapter 5** Opener: Courtesy of K. Kamada & S. K. Burley. From J. L. Kim, D. B. Nikolov, and S. K. Burley, "Co-crystal structure of TBP recognizing the minor groove of a TATA element," *Nature* 365, 520–527 (1993) 13.6. 5.6: Roger D. Kornberg

**Chapter 6** Opener: Venkitaraman Ramakrishnan 6.9c: Tripos, Inc. 6.12: Professor Harry Noller, University of California, Santa Cruz

**Chapter 7** Opener: The Protein Data Bank/RCSB 7.18: Visuals Unlimited Box 7.1: AP Wide World Photos 7.23: Virginia Walbot, Stanford University

**Chapter 8** Opener: National Human Genome Research Institute 8.8b: Dr. Fritz Thuemmler/Vertis Biotechnologie AG 8.14b: Alfred Pasieka/SPL/Photo Researchers, Inc. 8.18: Norbert Wu/Minden Pictures 8.20: J. Forsdyke/Gene Cox/Photo Researchers, Inc. 8.21: Dr. Chin-Sang

**Chapter 9** Opener: Richard Jenner, MA PhD 9.8: Radiological Society of North America, Figure 2c Zicherman JM, Weissman D, Griggin C, et al. "Best cases from the AFIP: primary diffuse large B-cell lymphoma of the epididymis and testis," *RadioGraphics* 2005; 25:243–248 9.9: Michael Wigler-ROMA lab/Cold Spring Harbor Laboratory

**Chapter 10** Opener: © Jean-Claude Revy/Phototake 10.22: Alec Jeffreys

**Chapter 11** Opener: © Nigel Cattlin/Holt Studios Int./Earth Scenes/Animals Animals 11.2: National Library of Medicine 11.15: Photo Researchers, Inc. 11.18a, left and right: Retna Ltd. 11.19a: © Itani/Alamy

**Chapter 12** Opener: © Biophoto Associates/Photo Researchers, Inc. 12.3a: Image courtesy of Applied Imaging a Genetix Company 12.3b: Image courtesy of Applied Imaging a Genetix Company 12.6a: © Michael Abbey/Photo Researchers, Inc. 12.6b: Photo Researchers, Inc. 12.6c: © Michael Abbey/Photo Researchers, Inc. 12.6d: Photo Researchers, Inc. 12.6e: © Michael Abbey/Photo Researchers, Inc. 12.6f: Photo Researchers, Inc. 12.7a: Armed Forces Institute of Pathology 12.7b: Visuals Unlimited 12.22a: Digamber S. Borgaonkar, Ph.D. 12.23a: © Dr. Glenn D. Braunstein, M.D.; Chairman, Dept. of Medicine, Cedars-Sinai Medical Center, Los Angeles, CA 12.24a: Visuals Unlimited 12.24b: Photo Researchers, Inc. 12.25: Peter J. Russell 12.26a: Courtesy of the Library of Congress 12.27a: National Library of Medicine

**Chapter 13** Opener: Animals Animals/Earth Scenes 13.8, top and bottom: Dr. James H. Tonsgard 13.10: istockphoto.com 13.15: Hans Reinhard/OKA PIA/Photo Researchers, Inc. 13.16: Photo Researchers, Inc.

**Chapter 14** Opener: Phototake NYC

**Chapter 15** Opener: © Dr. Dennis Kunkel/Phototake 15.1: Courtesy of Gunther S. Stent, University of California Berkeley 15.4a: © Dennis Kunkel/Phototake 15.4b: © Omikron/ Photo Researchers, Inc. 15.11: Bruce Iverson, Photomicrography 15.16: Courtesy of Gunther S. Stent, University of California Berkeley 15.17: Courtesy of Dr. D.P. Snustad, Department of Genetics and Cell Biology, College of Biological Sciences, University of Minnesota

**Chapter 16** Opener: © Addenbrookes Hospital/Photo Researchers, Inc. 16.4a: Dr. Laird Jackson, Thomas Jefferson University Hospital, Division of Medical Genetics. 16.4b: C. Weinkove and R. McDonald, *S Afr Med J* 43 (1969): 318 from *Syndromes of the Head and Neck*, 3rd ed., by Robert Golin, M. Michael Cohen, and L. Stefan Levin, Oxford University Press 16.13a: Courtesy of Christine J. Harrison, from the *American Journal of Medical Genetics*, Vol. 20, pp. 280–285, 1983. Reprinted by permission of Wiley-Liss, Inc., a division of John Wiley & Sons, Inc. 16.14: Used with permission from Warren & Nelson (*JAMA*, 2/16/94, 271; 536–542); Copyright 1994, American Medical Association. 16.17a: National Library of Medicine 16.17b: Getty Images-Stockbyte 16.20a-b: Dr. Laird Jackson, Thomas Jefferson University Hospital, Division of Medical Genetics. 16.21a-b: Dr. Laird Jackson, Thomas Jefferson University Hospital, Division of Medical Genetics.

**Chapter 17** Opener: © Ken Eward/ Science Source/Photo Researchers, Inc.

**Chapter 18** Opener: Medi-Mation/ Photo Researchers, Inc. 18.3: Riken BioResource Center

**Chapter 19** Opener: Courtesy of Stephen Paddock, James Langeland, Peter DeVries and Sean B. Carroll of the Howard Hughes Medical Institute at the University of Wisconsin (*Bio-Techniques*, January 1993) 19.1: Bill Stark, From Marcey, D. and Stark, W.S., "The morphology, physiology and neural projections of supernumerary compound eyes in Drosophila melanogaster." *Developmental Biology*, 1985, I07, I80–I97 19.2: Edward Kipreos 19.3, left and right: Yanofsky Martin 19.4, left and right: Gary C. Schoenwolf, From Kimmel et al., "Stages of Embryonic Development of the Zebrafish." *Developmental Dynamics* 203:253–310 (1995) 19.6a: Texas A & M University College of Veterinary Medicine 19.6b: Texas A & M University College of Veterinary Medicine 19.9: Photo Researchers 19.23: F. Rudolf Turner, Indiana University 19.24, top and bottom: Nipam H. Patel 19.26b-c: Edward B. Lewis, California Institute of Technology 19.27a-b: David Suzuki Foundation, From David Suzuki et al., *Introduction to Genetic Analysis*, p. 485

**Chapter 20** Opener: Courtesy of Y. Cho et al., kindly provided by N.P. Pavletich. Reprinted with permission from *Science* 265: 346–355, Fig. 6B. Copyright © 1994 American Association for the Advancement of Science. 20.1: GUSTOIMAGES/Photo Researchers, Inc. 20.7: Custom Medical Stock Photo, Inc.

**Chapter 21** Opener: Chip Clark 21.1a, c: Corbis/Bettmann 21.1b: Courtesy of James F. Crow 21.5: The Field Museum, Neg #CSA 118, Chicago 21.14: istockphoto.com 21.15a: Photo Researchers, Inc. 21.15b: The Granger Collection 21.16a-b: Breck P. Kent

**Chapter 22** Opener, top right: istockphoto.com; top left and bottom photos: Shutterstock 22.14a–d: Douglas W. Schemske

**Chapter 23** Opener: Michael L. Raymer, PhD

# Index

*Note*: Page numbers in *italics* indicate material in figures and tables; page numbers in **bold** indicate location of key term definition in text.

The *iGenetics* companion website contains 56 animations and 24 *iActivities* that help students grasp abstract concepts and dynamic processes, all described and referred to in the text in order to reinforce learning integration. In addition, hundreds of quiz questions that engage students in active problem-solving can either be completed as practice or submitted directly to the instructor online.

**Chapter 2**

***iActivity:*** Cracking a Viral Code

***Animations:*** DNA as Genetic Material: Avery's Transformation Experiment • DNA as Genetic Material: Hershey and Chase's Bacteriophage Experiment • DNA Supercoiling

**Chapter 3**

***iActivity:*** Unraveling DNA Replication

***Animations:*** The Meselson-Stahl Experiment • DNA Biosynthesis: How a New DNA Strand Is Made • Molecular Model of DNA Replication

**Chapter 4**

***iActivity:*** Pathways to Inherited Enzyme Deficiencies

***Animations:*** The One-Gene–One-Enzyme Hypothesis • Gene Control of Protein Structure and Function

**Chapter 5**

***iActivity:*** Investigating Transcription in Beta-Thalassemia Patients

***Animations:*** RNA Biosynthesis • mRNA Production in Eukaryotes • RNA Splicing

**Chapter 6**

***iActivity:*** Determining Causes of Cystic Fibrosis

***Animations:*** Initiation of Translation • Elongation of the Polypeptide Chain • Translation Termination

**Chapter 7**

***iActivities:*** A Toxic Town • The Genetic Shuffle

***Animations:*** Nonsense Mutations and Nonsense Suppressor Mutations • Mutagenic Effects of 5BU • Ames Test Protocol • Insertion Sequences in Bacteria • Transposable Elements in Plants

**Chapter 8**

***iActivity:*** Building a Better Beer

***Animations:*** DNA Cloning in a Plasmid Vector • The Whole-Genome Shotgun Approach to Sequencing

**Chapter 9**

***iActivity:*** Personalized Prescriptions for Cancer Patients

***Animations:*** Polymerase Chain Reaction (PCR) • Analysis of Gene Expression Using DNA Microarrays

**Chapter 10**

***iActivity:*** Combing Through "Fur"Ensic Evidence

***Animations:*** Restriction Mapping • The Yeast Two-Hybrid System • DNA Molecular Testing for Human Disease Gene Mutations • Plant Genetic Engineering

**Chapter 11**

***iActivity:*** Tribble Traits

***Animations:*** Mendel's Principle of Segregation • Mendel's Principle of Independent Assortment

**Chapter 12**

***iActivities:*** It Runs in the Family • Was She Charlie Chaplin's Child?

***Animations:*** Mitosis • Meiosis • X-Linked Inheritance • Nondisjunction • Gene and Chromosome Segregation in Meiosis

**Chapter 13**

***iActivity:*** Mitochondrial DNA and Human Disease

***Animations:*** Incomplete Dominance and Codominance • Maternal Effect

**Chapter 14**

***iActivity:*** Crossovers and Tomato Chromosomes

***Animations:*** Genetic Recombination and the Role of Chromosomal Exchange • The Chi-Square Test • Three-Point Mapping

**Chapter 15**

***iActivity:*** Conjugation in *E. coli*

***Animations:*** Mapping Bacterial Genes by Conjugation • Defining Genes by Complementation Tests

**Chapter 16**

***iActivity:*** Deciphering Karyotypes

***Animations:*** Crossing-over in an Inversion Heterozygote • Meiosis in a Translocation Heterozygote • Down Syndrome Caused by a Robertsonian Translocation

**Chapter 17**

***iActivity:*** Mutations and Lactose Metabolism

***Animations:*** Regulation of Expression of the *lac* Operon Genes • Positive Control of the *lac* Operon • Attenuation in the *trp* Operon of *E. coli*

**Chapter 18**

***iActivity:*** Sorting the Signals of Gene Regulation

***Animations:*** Regulation of Transcription in Animals by Steroid Hormones • RNA Processing Control

**Chapter 19**

***iActivity:*** The Great Divide

***Animations:*** Sex Determination and Dosage Compensation in *Drosophila* • Gene Regulation of the Development of the *Drosophila* Body Plan

**Chapter 20**

***iActivity:*** Tracking Down the Causes of Cancer

***Animations:*** Regulation of Cell Division in Normal Cells • The Tumor Suppressor Gene, *TP53*

**Chapter 21**

***iActivity:*** Measuring Genetic Variation

***Animation:*** Natural Selection

**Chapter 22**

***iActivity:*** Your Fate in Your Hands?

***Animation:*** Polygene Hypothesis for Wheat Kernel Color

**Chapter 23**

***iActivity:*** Were Neanderthals Our Ancestors?

***Animation:*** Phylogenetic Trees

---

**Web-Only Bonus: Tetrad Analysis**

***iActivity:*** Mapping Genes by Tetrad Analysis

***Animation:*** Mapping Linked Genes by Tetrad Analysis